ENCYCLOPEDIA *of* SEDIMENTS *and* SEDIMENTARY ROCKS

Kluwer Academic Encyclopedia of Earth Sciences Series

ENCYCLOPEDIA OF SEDIMENTS AND SEDIMENTARY ROCKS

Aim of the Series

The Kluwer Academic Encyclopedia of Earth Sciences Series provides comprehensive and authoritative coverage of all the main areas in the Earth Sciences. Each volume comprises a focused and carefully chosen collection of contributions from leading names in the subject, with copious illustrations and reference lists.

These books represent one of the world's leading resources for the Earth Sciences community. Previous volumes are being updated and new works published so that the volumes will continue to be essential reading for all professional earth scientists, geologists, geophysicists, climatologists, and oceanographers as well as for teachers and students.

See the back of this volume for a current list of titles in the Encyclopedia of Earth Sciences Series. Go to www.eseo.com to visit the "Earth Sciences Encyclopedia Online"—the online version of this Encyclopedia Series.

About the Editors

Professor Rhodes W. Fairbridge has edited more than 30 Encyclopedias in the Earth Sciences Series. During his career he has worked as a petroleum geologist in the Middle East, been a WW II intelligence officer in the SW Pacific and led expeditions to the Sahara, Arctic Canada, Arctic Scandinavia, Brazil and New Guinea. He is currently Emeritus Professor of Geology at Columbia University and is affiliated with the Goddard Institute for Space Studies.

Professor Michael Rampino has published more than 70 papers in professional journals including *Science*, *Nature*, and *Scientific American*. He has worked in such diverse fields as volcanology, planetary science, sedimentology, and climate studies, and has done field work on six continents. He is currently Associate Professor of Earth and Environmental Sciences at New York University and a consultant at NASA's Goddard Institute for Space Studies.

Volume Editor

Gerard V. Middleton is Professor Emeritus of Geology at McMaster University, Hamilton, Ontario, Canada.

He was co-author of "Origin of Sedimentary Rocks" (Prentice-Hall), "Mechanics in the Earth and Environmental Sciences" (Cambridge), and author of "Data Analysis in the Earth Sciences using MATLAB" (Prentice-Hall). He is a Fellow of the Royal Society of Canada, and in 2003 received the Twenhofel Prize of SEPM, the Society for Sedimentary Geology.

ENCYCLOPEDIA OF EARTH SCIENCES SERIES

ENCYCLOPEDIA of SEDIMENTS and SEDIMENTARY ROCKS

edited by

GERARD V. MIDDLETON
McMaster University

with Associate Editors

MICHAEL J. CHURCH
University of British Columbia

MARIO CONIGLIO
University of Waterloo

LAWRENCE A. HARDIE
Johns Hopkins University

FREDERICK J. LONGSTAFFE
University of Western Ontario

KLUWER ACADEMIC PUBLISHERS
DORDRECHT | BOSTON | LONDON

A C.I.P. Catalogue record for this book is available from the Library of Congress

ISBN 1-4020-0872-4

Published by Kluwer Academic Publishers
PO Box 17, 3300 AA Dordrecht, The Netherlands

Sold and distributed in North, Central and South America
by Kluwer Academic Publishers, 101 Philip Drive, Norwell, MA 02061, USA

In all other countries, sold and distributed
by Kluwer Academic Publishers
PO Box 322, 3300 AH Dordrecht, The Netherlands

Printed on acid-free paper

Every effort has been made to contact the copyright holders of the figures and tables which have been reproduced from other sources. Anyone who has not been properly credited is requested to contact the publishers, so that due acknowledgement may be made in subsequent editions.

All rights reserved
© 2003 Kluwer Academic Publishers
No part of this publication may be reproduced or utilized in any form or by any means, electronic, mechanical, including photocopying, recording or by any information storage and retrieval system, without written permission from the copyright owner.

Printed and bound in Great Britain by MPG Books, Bodmin, Cornwall.

Contents

List of Contributors	xiii	Authigenesis *James R. Boles*	27
Preface	xxvii		
Guide to the Reader	xxix	Autosuspension *Henry M. Pantin*	30
Algal and Bacterial Carbonate Sediments *Robert Riding*	1	Avalanche and Rock Fall *Michael J. Bovis*	31
Allophane and Imogolite *Roger L. Parfitt*	3	Avulsion *Norman D. Smith*	34
Alluvial Fans *Adrian M. Harvey*	5	Bacteria in Sediments *Nora Noffke*	37
Anabranching Rivers *Gerald C. Nanson and Martin R. Gibling*	9	Ball-and-Pillow (Pillow) Structure *Geraint Owen*	39
Ancient Karst *Brian Jones*	11	Bar, Littoral *Brian Greenwood*	40
Angle of Repose *Paul D. Komar*	15	Barrier Islands *Duncan M. FitzGerald and Ilya V. Buynevich*	43
Anhydrite and Gypsum *Lawrence A. Hardie*	16	Bauxite *Ray L. Frost*	48
Ankerite (in Sediments) *James P. Hendry*	19	Beachrock *Eberhard Gischler*	51
Armor *Rob Ferguson*	21	Bedding and Internal Structures *Franco Ricci-Lucchi and Alessandro Amorosi*	53
Atterberg Limits *Michael J. Bovis*	22	Bedset and Laminaset *John S. Bridge*	59
Attrition (Abrasion), Fluvial *Michael Church*	24	Bentonites and Tonsteins *D. Alan Spears*	61
Attrition (Abrasion), Marine *Hillert Ibbeken*	25		

Berthierine *Fred J. Longstaffe*	64	Clastic (Neptunian) Dykes and Sills *A. Demoulin*	136
Bioclasts *Paul Enos*	66	Clathrates *Miriam Kastner*	137
Bioerosion *Markus Bertling*	70	Clay Mineralogy *Stephen Hillier*	139
Biogenic Sedimentary Structures *S. George Pemberton*	71	Climatic Control of Sedimentation *Greg H. Mack*	142
Black Shales *Juergen Schieber*	83	Coal Balls *Andrew C. Scott*	146
Braided Channels *Peter Ashmore*	85	Coastal Sedimentary Facies *H. Edward Clifton*	149
Calcite Compensation Depth *Sherwood Wise*	88	Colloidal Properties of Sediments *Sandip and Devamita Chattopadhyay*	157
Caliche – Calcrete *V. Paul Wright*	89	Colors of Sedimentary Rocks *Paul Myrow*	159
Carbonate Diagenesis and Microfabrics *Robin G.C. Bathurst*	91	Compaction (Consolidation) of Sediments *Knut Bjørlykke*	161
Carbonate Mineralogy and Geochemistry *Fred T. Mackenzie*	93	Convolute Lamination *Gerard V. Middleton*	168
Carbonate Mud-Mounds *Pierre-André Bourque*	100	Coring Methods, Cores *Arnold H. Bouma*	168
Cathodoluminescence (applied to the study of sedimentary rocks) *Stuart D. Burley*	102	Cross-Stratification *David M. Rubin*	170
Cation Exchange *Balwant Singh*	106	Cyclic Sedimentation *Robert K. Goldhammer*	173
Cave Sediments *Brian Jones*	109	Debris Flow *Jon J. Major*	186
Cements and Cementation *Peter A. Scholle and Dana Ulmer-Scholle*	110	Dedolomitization *Mario Coniglio*	188
Chalk *Ida L. Fabricius*	119	Deformation of Sediments *John D. Collinson*	190
Charcoal in Sediments *Andrew C. Scott*	121	Deformation Structures and Growth Faults *John D. Collinson*	193
Chlorite in Sediments *Stephen Hillier*	123	Deltas and Estuaries *Janok Bhattacharya*	195
Classification of Sediments and Sedimentary Rocks *Gerald M. Friedman*	127	Depositional Fabric of Mudstones *Juergen Schieber*	203

Desert Sedimentary Environments *Joseph P. Smoot*	207	Features Indicating Impact and Shock Metamorphism *Wolf Uwe Reimold*	275
Desiccation Structures (Mud Cracks, etc.) *P.W. Geoff Tanner*	212	Feldspars in Sedimentary Rocks *Sadoon Morad*	278
Diagenesis *Kitty L. Milliken*	214	Flame Structure *Gerard V. Middleton*	281
Diagenetic Structures *Peter Mozley*	219	Flaser *Burghard W. Flemming*	282
Diffusion, Chemical *Bernard P. Boudreau*	225	Flocculation *Morten Pejrup*	284
Diffusion, Turbulent *Andrew J. Hogg*	226	Floodplain Sediments *Andres Aslan*	285
Dish Structure *Zoltán Sylvester and Donald R. Lowe*	230	Floods and Other Catastrophic Events *Victor R. Baker*	287
Dolomite Textures *Duncan Sibley*	231	Flow Resistance *Robert Millar*	293
Dolomites and Dolomitization *Hans G. Machel*	234	Fluid Escape Structures *Zoltán Sylvester and Donald R. Lowe*	294
Dunes, Eolian *Nicholas Lancaster*	243	Fluid Inclusions *Robert H. Goldstein*	297
Earth Flows *Rex L. Baum*	247	Flume *Basil Gomez*	300
Encrinites *William I. Ausich*	248	Forensic Sedimentology *Raymond C. Murray*	301
Eolian Transport and Deposition *Cheryl McKenna Neuman*	249	Gases in Sediments *Chris J. Clayton*	304
Erosion and Sediment Yield *Robert F. Stallard*	254	Geodes *Kitty L. Milliken*	306
Evaporites *Lawrence A. Hardie and Tim K. Lowenstein*	257	Geophysical Properties of Sediments (Acoustical, Electrical, Radioactive) *Anthony L. Endres*	308
Extraterrestrial Material in Sediments *Christian Koeberl*	263	Geothermic Characteristics of Sediments and Sedimentary Rocks *Daniel F. Merriam*	314
Fabric, Porosity, and Permeability *Gerard V. Middleton*	265	Glacial Sediments: Processes, Environments and Facies *Michael J. Hambrey and Neil F. Glasser*	316
Facies Models *Harold G. Reading*	268	Glaucony and Verdine *Alessandro Amorosi*	331
Fan Delta *George Postma*	272	Grading, Graded Bedding *Richard N. Hiscott*	333

Grain Flow *Charles S. Campbell*	335	Kaolin Group Minerals *Rossman F. Giese, Jr.*	398
Grain Settling *Paul D. Komar*	336	Kerogen *Raphael A.J. Wüst and R. Marc Bustin*	400
Grain Size and Shape *Michael J. Church*	338	Lacustrine Sedimentation *Robert Gilbert*	404
Grain Threshold *Paul D. Komar*	345	Laterites *Yves Tardy*	408
Gravity-Driven Mass Flows *Richard M. Iverson*	347	Lepispheres *Sherwood W. Wise, Jr.*	411
Gutters and Gutter Casts *Paul Myrow*	353	Liquefaction and Fluidization *Charles S. Campbell*	412
Heavy Mineral Shadows *Rick Cheel*	355	Load Structures *John R.L. Allen*	413
Heavy Minerals *Andrew C. Morton*	356	Lunar Sediments *Abhijit Basu*	415
Hindered Settling *Jon J. Major*	358	Magadiite *Robin W. Renaut*	417
Humic Substances in Sediments *Laura J. Crossey*	361	Magnetic Properties of Sediments *Mark J. Dekkers*	418
Hummocky and Swaley Cross-Stratification *Rick Cheel*	362	Mass Movement *Michael J. Bovis*	424
Hydrocarbons in Sediments *Martin Fowler*	364	Maturation, Organic *R. Marc Bustin and Raphael A.J. Wüst*	425
Hydroxides and Oxyhydroxide Minerals *Helge Stanjek*	366	Maturity: Textural and Compositional *Raymond V. Ingersoll*	429
Illite Group Clay Minerals *Jan Środoń*	369	Meandering Channels *Edward J. Hickin*	430
Imbrication and Flow-Oriented Clasts *Cyril Galvin*	371	Mélange; Melange *John W.F. Waldron*	434
Impregnation *Scott F. Lamoureux*	374	Micritization *Ian G. MacIntyre and R. Pamela Reid*	436
Iron–Manganese Nodules *Stephen E. Calvert*	376	Microbially Induced Sedimentary Structures *Nora Noffke*	439
Ironstones and Iron Formations *Bruce M. Simonson*	379	Milankovitch Cycles *Linda Hinnov*	441
		Mixed Siliciclastic and Carbonate Sedimentation *Robert K. Goldhammer*	443
Isotopic Methods in Sedimentology *Fred J. Longstaffe*	385	Mixed-Layer Clays *Jan Środoń*	447

Entry	Page
Mixing Models — *Bernard P. Boudreau*	450
Mudrocks — *Paul E. Potter*	451
Neomorphism and Recrystallization — *J. A. D. Dickson*	460
Nepheloid Layer, Sediment — *I. Nicholas McCave*	462
Neritic Carbonate Depositional Environments — *Noel P. James*	464
Numerical Models and Simulation of Sediment Transport and Deposition — *Rudy L. Slingerland*	475
Oceanic Sediments — *Robert G. Douglas*	481
Offshore Sands — *Serge Berné*	492
Oil Sands — *Daryl M. Wightman*	499
Oolite and Coated Grains — *Fredrick D. Siewers*	502
Ophicalcites — *Denis Lavoie*	506
Paleocurrent Analysis — *Andrew D. Miall*	509
Parting Lineations and Current Crescents — *Rick Cheel*	512
Peat — *Barry G. Warner*	514
Petrophysics of Sand and Sandstone — *Robert Ehrlich*	516
Phosphorites — *Craig R. Glenn and Robert E. Garrison*	519
Physics of Sediment Transport: The Contributions of R.A. Bagnold — *Colin R. Thorne*	527
Pillar Structure — *Zoltán Sylvester and Donald R. Lowe*	529
Placers, Fluvial — *W.K. Fletcher*	530
Placers, Marine — *David S. Cronan*	532
Planar and Parallel Lamination — *John S. Bridge*	534
Porewaters in Sediments — *Jeffrey S. Hanor*	537
Pressure Solution — *François Renard and Dag Dysthe*	542
Provenance — *Abhijit Basu*	544
Pseudonodules — *Geraint Owen*	549
Quartz, Detrital — *Harvey Blatt*	552
Red Beds — *Peter Turner*	555
Reefs — *Wolfgang Kiessling*	557
Relict and Palimpsest Sediments — *Donald J.P. Swift*	560
Relief Peels — *Robert W. Dalrymple*	561
Resin and Amber in Sediments — *Ken B. Anderson*	563
Ripple, Ripple Mark, Ripple Structure — *Jaco H. Baas*	565
Rivers and Alluvial Fans — *Gerald C. Nanson and Martin R. Gibling*	568
Sabkha, Salt Flat, Salina — *Lawrence A. Hardie*	584
Salt Marshes — *Daniel F. Belknap*	586
Sands, Gravels, and their Lithified Equivalents — *Andrew D. Miall*	588
Sapropel — *Stephen E. Calvert*	592
Scour, Scour Marks — *John R. L. Allen*	594
Seawater: Temporal Changes in the Major Solutes — *H. Wayne Nesbitt*	596

Sediment Fluxes and Rates of Sedimentation *James P.M. Syvitski*	600	Henry Clifton Sorby (1826–1908) *Gerard V. Middleton*	649
Sediment Transport by Tides *Robert W. Dalrymple and Kyungsik Choi*	606	William H. Twenhofel (1875–1957) *Robert H. Dott, Jr.*	650
		Johan August Udden (1859–1932) *Earle F. McBride*	651
Sediment Transport by Unidirectional Water Flows *John S. Bridge*	609	T. Wayland Vaughan (1870–1952) *Ellis L. Yochelson*	652
Sediment Transport by Waves *Brian Greenwood*	619	Johannes Walther (1860–1937) *Eugen and Ilse Seibold*	653
Sedimentary Geology *Gerard V. Middleton*	626	Sediments Produced by Impact *Christian Koeberl*	653
Sedimentary Structures as Way-up Indicators *Robert H. Dott, Jr.*	627	Septarian Concretions *Mark W. Hounslow*	657
Sedimentology, History *Gerard V. Middleton*	628	Silcrete *Mª Angeles Bustillo*	659
Sedimentology: History in Japan *Hakuyu Okada*	635	Siliceous Sediments *L. Paul Knauth*	660
Sedimentology—Organizations, Meetings, Publications *Gail M. Ashley*	637	Slide and Slump Structures *Ole J. Martinsen*	666
Sedimentologists:		Slope Sediments *Richard N. Hiscott*	668
Ralph Alger Bagnold (1896–1990) *Colin R. Thorne*	638		
Joseph Barrell (1869–1919) *Robert H. Dott, Jr.*	638	Slurry *Jon J. Major*	674
Robin G.C. Bathurst (1920–) *Gerard V. Middleton*	639	Smectite Group *Stephen P. Altaner*	675
Lucien Cayeux (1864–1944) *Albert V. Carozzi*	640	Solution Breccias *Derek Ford*	677
Carl Wilhelm Correns (1893–1980) *Karl Hans Wedepohl*	641		
Robert Louis Folk (1925–) *Earle F. McBride*	642	Speleothems *Derek Ford*	678
Grove Karl Gilbert (1843–1918) *Rudy Slingerland*	642	Spiculites and Spongolites *Paul R. Gammon*	681
Amadeus William Grabau (1870–1946) *Markes E. Johnson*	643	Stains for Carbonate Minerals *J.A.D. Dickson*	683
Amanz Gressly (1814–1865) *Gerard V. Middleton*	644		

William Christian Krumbein (1902–1979)
    Daniel F. Merriam
	644	Statistical Analysis of Sediments and
		    Sedimentary Rocks
		Daniel F. Merriam
			684

Paul Dimitri Krynine (1902–1964) *Gerard V. Middleton*	645	Storm Deposits *Paul Myrow*	686
Philip Henry Kuenen (1902–1976) *Gerard V. Middleton*	646		

John Murray (1841–1914) and the
    Challenger Expedition
    Gerard V. Middleton
	647	Stromatactis
		Pierre-André Bourque
			687

Francis J. Pettijohn (1904–1999) *Paul E. Potter*	648	Stromatolites *Brian R. Pratt*	688

Rudolf Richter (1881–1957) and the
    Senckenberg Laboratory
    S. George Pemberton
	648		

Stylolites 690
L. Bruce Railsback

Submarine Fans and Channels 692
Gerard V. Middleton

Substrate-Controlled Ichnofacies 698
S. George Pemberton

Sulfide Minerals in Sediments 701
Richard T. Wilkin

Surface Forms 703
John B. Southard

Surface Textures 712
W. Brian Whalley

Swash and Backwash, Swash Marks 717
Michael G. Hughes

Syneresis 718
P.W. Geoff Tanner

Talc 721
Richard H. April

Taphonomy: Sedimentological Implications of
   Fossil Preservation 723
Carlton E. Brett

Tectonic Control of Sedimentation 729
Greg H. Mack

Tidal Flats 734
Burghard W. Flemming

Tidal Inlets and Deltas 737
Duncan M. FitzGerald and Ilya V. Buynevich

Tides and Tidal Rhythmites 741
Erik P. Kvale

Tills and Tillites 744
John Menzies

Tool Marks 747
Gerard V. Middleton

Toxicity of Sediments 748
G. Allen Burton, Jr. and Peter F. Landrum

Tracers for Sediment Movement 752
Marwan A. Hassan

Tufas and Travertines 753
Martyn H. Pedley

Tsunami Deposits 755
Andrew Moore

Turbidites 757
Ben C. Kneller

Upwelling 761
Graham Shimmield

Varves 764
Robert Gilbert

Vermiculite 766
Prakash B. Malla

Weathering, Soils, and Paleosols 770
Gregory J. Retallack

X-ray Radiography 777
Arnold H. Bouma

Zeolites in Sedimentary Rocks 779
Richard L. Hay

Index of Authors Cited 781

Subject Index 805

Contributors

John R.L. Allen
Research Institute for Sedimentology
University of Reading
P.O. Box 227
Whiteknights, Reading Berkshire RG6 6AB
England, UK
e-mail: j.r.l.allen@reading.ac.uk
Load Structures
Scour, Scour Marks

Stephen P. Altaner
Department of Geology, 1301 W. Green St
University of Illinois
Urbana IL 61801
USA
e-mail: altaner@uiuc.edu
Smectite Group

Alessandro Amorosi
Department di Scienze della Terra
Universita di Bologna
Via Zamboni 67
Bologna 40127
Italy
e-mail: amorosi@geomin.unibo.it
Bedding and Internal Structures
Glaucony and Verdine

Ken B. Anderson
Chemistry Division
Argonne National Laboratory
Argonne IL 60439
USA
e-mail: kbanderson@anl.gov
Resin and Amber in Sediments

Richard April
Department of Geology
Colgate University
Hamilton NY 13346
USA
e-mail: Rapril@mail.colgate.edu
Talc

Gail M. Ashley
Department of Geological Sciences
Rutgers University
610 Taylor Road
Piscataway NJ 08854-8066
USA
e-mail: gmashley@rci.rutgers.edu
Sedimentology – Organizations, Meetings, Publications

Peter Z. Ashmore
Department of Geography
University of Victoria
PO Box 3050, Stn CSC
Victoria BC V8W 3P5
Canada
e-mail: pashmore@julian.uwo.ca
Braided Channels

Andres Aslan
Department of Physical and Env. Sciences
Mesa State College
Grand Junction CO 81501-7682
USA
e-mail: aaslan@mesastate.edu
Floodplain Sediments

William I. Ausich
Department of Geological Sciences
Ohio State University, 155 S. Oval
Columbus OH 43210-1398
USA
e-mail: ausich.1@osu.edu
Encrinites

Jaco H. Baas
Department of Earth Sci.
University of Leeds
Leeds LS2 9JT
England, UK
e-mail: j.baas@earth.leeds.ac.uk
Ripple, Ripple Mark, Ripple Structure

Victor R. Baker
Department Hydrology & Water Resources
University of Arizona
bldg 11, North Campus Drive
Tucson AZ 85721-0011
USA
e-mail: baker@hwr.arizona.edu
Floods and Other Catastrophic Events

Abhijit Basu
Department of Geological Sciences
Indiana University
1005 East 10th Street
Bloomington IN 47405
USA
e-mail: basu@indiana.edu
Provenance
Lunar Sediments

Robin G.C. Bathurst
Derwen Dêq Fawr
Llanfair D.C.
Ruthin
Denbighshire, LL15 2SN
North Wales, UK
Carbonate Diagenesis and Microfabrics

Rex L. Baum
Central Region Geol. Hazards
U.S. Geological Survey
Box 25046, M.S. 966
Denver CO 80225-0046
USA
e-mail: baum@usgs.gov
Earth Flows

Daniel F. Belknap
Department of Geological Sciences
117 Bryand Global Sciences Bldg.
University of Maine
Orono ME 04469-5790
USA
e-mail: belknap@maine.edu
Salt Marshes

Serge P. Berné
IFREMER, DRO-GM
B.P. 70
Plouzane 29280
France
e-mail: Serge.Berne@ifremer.fr
Offshore Sands

Markus Bertling
Geologisch-Palaeontologisches Institute u. Museum
Pferdegasse 3
Muenster D-48143
Germany
e-mail: bertlin@uni-muenster.de
Bioerosion

Janok Bhattacharya
Department of Geosciences
University of Texas at Dallas
P.O. Box 830688
Dallas TX 75083-0688
USA
e-mail: janokb@utdallas.edu
Deltas and Estuaries

Knut Bjorlykke
Department of Geology, Box 1047
University of Oslo
Oslo 0316
Norway
e-mail: knut.bjorlykke@geologi.uio.no
Compaction (Consolidation) of Sediments

Harvey Blatt
Institute of Earth Sci, Hebrew University
Givat Ram
Jerusalem 91904
Israel
e-mail: harvey@vms.huji.ac.il
Quartz, Detrital

James R. Boles
Department of Geological Sciences
University California Santa Barbara
Santa Barbara CA 93106-9630
USA
e-mail: boles@magic.ucsb.edu
Authigenesis

Bernard P. Boudreau
Department of Oceanography
Dalhousie University
Halifax
NS B3H 4JI
Canada
e-mail: bpboudre@is.dal.ca
Diffusion, Chemical
Mixing Models

Arnold H. Bouma
Department Geology/Geophysics
Louisiana State University
Baton Rouge LA 70803
USA
e-mail: bouma@geol.lsu.edu
Coring Methods, Cores
X-radiography

P. André Bourque
Department de Géologie et de Génie géologique
Université Laval
Québec PQ G1K 7P4
Canada
e-mail: bourque@ggl.ulaval.ca
Carbonate Mud-Mounds
Stromatactis

CONTRIBUTORS

Michael J. Bovis
Department of Geography
University of British Columbia 1984 West Mall
Vancouver BC V6T 1Z2
Canada
e-mail: mbovis@geog.ubc.ca
Atterberg Limits
Avalanch and Rockfall
Mass Movement

Carleton E. Brett
Department of Geology
University of Cincinnati
Cincinnati OH 45221-0013
USA
e-mail: brettce@email.uc.edu
Taphonomy: Sedimentological Implications of Fossil Preservation

John S. Bridge
Department Geol. Sci., SUNY
P.O. Box 6000
Binghamton NY 13902-6000
USA
e-mail: jbridge@binghamton.edu
Bedset and Laminaset
Planar and Parallel Lamination
Sediment Transport by Unidirectional Water flows

Stuart D. Burley
Subsurface Technology, BG Group
100 Thames Valley Park
Reading RG6 1PT
England, UK
e-mail: stuart.burley@bg-group.com
Cathodoluminescence

G. Allen Burton, Jr
Department Biological Sciences
Wright State University
Dayton OH 45435
USA
e-mail: allen.burton@wright.edu
Toxicity of Sediments

M. Angeles Bustillo
Museo Nacional de Ciencias Naturales
Dpto de Geologia
José Gutierrez Abascal no. 2
Madrid 28006
Spain
e-mail: abustillo@mncn.csic.es
Silcrete

Robert M. Bustin
Department Earth and Ocean Sciences, University British Columbia
6270 University Blvd
Vancouver BC V6T 1Z4
Canada
e-mail: bustin@unixg.ubc.ca
Kerogen
Maturation, Organic

Ilya V. Buynevich
Geology and Geophysics Department
Woods Hole Oceanographic Institution
Woods Hole, MA 02543
USA
e-mail: ibuynevich@whoi.edu
Barrier Islands
Tidal Inlets and Deltas

Stephen E. Calvert
Department of Oceanography, 1461-6270 University Blvd.
University of British Columbia
Vancouver BC V6T 1Z4
Canada
e-mail: calvert@eos.ubc.ca
Iron-Manganese Nodules
Sapropel

Charles S. Campbell
Department of Mechanical Engineering
University of Southern California
Los Angeles CA 90089-1453
USA
e-mail: campbell@rfc.usc.edu
Grain Flow
Liquefaction and Fluidization

Albert V. Carozzi
Department of Geology, University Illinois
245 Natural History Bldg. 1301 W Green
Urbana IL 61801-2999
USA
Present address:
7530 Lead Mine Road, #304
Raleigh, NC 27615-4897
USA
Sedimentologists: Lucien Cayeux

Devamitta Chattopadhyay
ARCADIS
6397 Emerald Parkway, Suite 150
Dublin, OH 43016
USA
e-mail: dchattopadhyay@arcadis-us.co
Colloidal Properties of Sediments

Sandip Chattopadhyay
Battelle Memorial Institute
Environmental Restoration Department
505 King Avenue
Columbus Ohio 43230
USA
e-mail: chattopadhyays@battelle.org
Colloidal Properties of Sediments

Richard J. Cheel
Dept of Geological Sciences
Brock University
St Catherines ON L2S 3A1
Canada
e-mail: rcheel@craton.geol.brocku.ca
Heavy Mineral Shadows
Hummocky and Swaley Cross-Stratification
Parting Lineation and Current Crescents

CONTRIBUTORS

Kyungsik Choi
Department of Geological Sciences and Geological
 Engineering
Queen's University
Kingston, Ontario K7L 3N6
Canada
Sediment Transport by Tides

Michael J. Church
Department of Geography
University of British Columbia
Vancouver BC V6T 1Z2
Canada
e-mail: mchurch@geog.ubc.ca
Grain Size and Shape
Attrition, Fluvial

Chris J. Clayton
16 Latimer Road
Teddington TW11 8QA
England, UK
e-mail: geochem@c-clayton.u-net.com
Gases in Sediments

H. Edward Clifton
1933 Fallen Leaf Lane
Los Altos CA 94024
USA
e-mail: eclifton@earthlink.net
Coastal Sedimentary Facies

John D. Collinson
1 Winchester Drive
Westlands
Newcastle-undor-Lyme Staffs ST5 3JH
United Kingdom
e-mail: johncollinson@c-j-c.demon.co.uk
Deformation of Sediments
Deformation Structures and Growth Faults

Mario Coniglio
Department of Earth Sciences
University of Waterloo
Waterloo ON N2L 3G1
Canada
e-mail: coniglio@sciborg.uwaterloo.ca
Dedolomitization

David S. Cronan
T.H. Huxley School of Environment
Imperial College
London SW7 2BP
England, UK
e-mail: d.cronan@ic.ac.uk
Placers, Marine

Laura J. Crossey
Department of Earth and Planetary Sciences
University of New Mexico
Albuquerque NM 87131
USA
e-mail: lcrossey@unm.edu
Humic Substances in Sediments

Robert W. Dalrymple
Department of Geological Sci.
Queens University
Kingston ON K7L 3N6
Canada
e-mail: dalrymple@geol.queensu.ca
Relief Peels
Sediment Movement by Tides

Mark J. Dekkers
Paleomagnetic Laboratory
Budapestlaan 17
Utrecht 3584 CD
The Netherlands
e-mail: dekkers@geo.uu.nl
Magnetic Properties of Sediments

A. Demoulin
Département de Géographie physique
Université de Liège
Allée du 6 Août, Sart Tilman
Liège B 4000
Belgium
e-mail: ademoulin@nlg.ac.be
Clastic (Neptunian) Dykes and Sills

J.A.D. Dickson
Department of Earth Sci., University of Cambridge
Downing St
Cambridge CB2 3EQ
England, UK
e-mail: jadd1@esc.cam.ac.uk
Neomorphism and Recrystallization
Stains for Carbonate Minerals

Robert H. Dott
Geol. & Geophysics, University Wisconsin
1215 W. Dayton St
Madison WI 53706-1692
USA
e-mail: rdott@geology.wisc.edu
Sedimentologists: Joseph Barrell
Sedimentologists: William H. Twenhofel
Sedimentary Structures as Way-up Indicators

Robert G. Douglas
Department of Earth Sciences
University of Southern California
Los Angeles CA 90089-0740
USA
e-mail: rdouglas@usc.edu
Oceanic Sediments

Dag Dysthe
Center for Advanced Study
78 Drammersveien
N-0271 Oslo
Norway
e-mail: d.k.dysthe@fys.uio.no
Pressure Solution

Robert Ehrlich
Energy and Geoscience Institute
University of Utah
423 Wakara Way Ste 300
Salt Lake City UT 84108-3537
USA
e-mail: bobehrlich@attbi.com
Petrophysics of Sand and Sandstones

Anthony L. Endres
Department of Earth Sciences
University of Waterloo
Waterloo ON N2L 3G1
Canada
e-mail: tendres@cgrnserc.uwaterloo.ca
Geophysical Properties of Sediments

Paul Enos
Department of Geology, 120 Lindley Hall
University of Kansas,
1475 Jayhawk Drive
Lawrence KS 66045
(785) 864-2744
e-mail: enos@falcon.cc.ukans.edu
Bioclasts

Ida L. Fabricius
IGG, DTU b204.
Technical University of Denmark
Lyngby DK 2800
Denmark
e-mail: iggill@pop.dtu.dk
Chalk

Rob Ferguson
Department of Geography, Winter St
Sheffield University
Sheffield S10 2TN
England, UK
e-mail: R.Ferguson@sheffield.ac.uk
Armor

Duncan M. Fitzgerald
Department Earth Sciences, Boston University
675 Commonwealth Ave
Boston MA 02215-2530
USA
e-mail: dunc@bu.edu
Barrier Islands
Tidal Inlets and Deltas

Burghard W. Flemming
Senckenberg Institute
Schleusenstrasse 39A
Wilhemshaven 26382
Germany
e-mail: bw.flemming@sam.terramare.de
Flaser
Tidal Flats

William K. Fletcher
Department of Earth & Ocean Sciences
6339 Stores Rd
University British Columbia
Vancouver BC V6T 1Z4
Canada
e-mail: fletcher@unixg.ubc.ca
Placers, Fluvial

Derek Ford
School of Geography & Geology
McMaster University, 1280 Main W
Hamilton ON L8S 4M1
Canada
e-mail: dford@mcmaster.ca
Solution Breccias
Speleothems

Martin Fowler
Geological Survey of Canada
3303-33rd St NW
Calgary Alberta T2L 2A7
Canada
e-mail: MFowler@NRCan.gc.ca
Hydrocarbons in Sediments

Gerald M. Friedman
Brooklyn College, Rensselaer Center
P.O. Box 746
Troy NY 12181
USA
e-mail: gmfriedman@juno.com
Classification of Sediments and Sedimentary Rocks

Ray L. Frost
Centre for Instrumental and Developmental Chemistry
School of Physical Sciences
P.O. Box 2434 GPO
Brisbane QLD 4001
Australia
e-mail: r.frost@qut.edu.au
Bauxite

Cyril Galvin
Coastal Engineer
P.O. Box 623
Springfield VA 22150
USA
e-mail: galvincoastal@juno.com
Imbrication and Flow-Oriented Clasts

Paul R. Gammon
Department of Geology and Geophysics
Adelaide University
Adelaide 5005
Australia
e-mail: paul.gammon@adelaide.edu.au
Spiculites and Spongolites

Robert E. Garrison
Department of Ocean Sciences
Earth and Marine Sciences Bldg
University of California
Santa Cruz CA 95064
USA
e-mail: regarris@cats.ucsc.edu
Phosphorites

Martin R. Gibling
Department of Earth Sciences
Dalhousie University
Halifax, Nova Scotia B3H 3J5
Canada
e-mail: mgibling@is.dal.ca
Anabranching Rivers
Rivers and Alluvial Fans

Rossman F. Giese
Department of Geology, SUNY
711 Natural Sci. Complex
Buffalo NY 14260
USA
e-mail: glgclay@acsu.buffalo.edu
Kaolin Group Minerals

Robert Gilbert
Department of Geography
Queens University
Kingston ON K7L 3N6
Canada
e-mail: gilbert@lake.geog.queensu.ca
Lacustrine Sedimentation
Varves

Eberhard Gischler
Geologisch-Palaeontologishes Institut
Johann Wolfgang Goethe-Universitaet
Senckenberglage 32–34
Frankfurt am Main D-60054
Germany
e-mail: gischler@em.uni-frankfurt.de
Beachrock

Neil F. Glasser
Centre for Glaciology
University of Wales
Aberystwyth Wales SY23 3DB
United Kingdom
e-mail: mjh@aber.ac.uk
Glacial Sediments: Processes, Environments and Facies

Craig R. Glenn
Department of Geology and Geophysics
University of Hawaii
Honolulu, HI 96822
USA
e-mail: glenn@soest.hawaii.edu
Phosphorites

Robert K. Goldhammer
Department of Geological Sciences
University of Texas
Austin TX 78712-1101
USA
e-mail: goldhrk@mail.utexas.edu
Cyclic Sedimentation
Mixed Siliciclastic and Carbonate Sedimentation

Robert H. Goldstein
Department of Geology, University of Kansas
120 Lindley Hall
Lawrence KS 66045-7613
USA
e-mail: gold@ukans.edu
Fluid Inclusions

Basil Gomez
Geomorphology Laboratory
Indiana State University
455 N 6th St
Terre Haute IN 47809-0001
USA
e-mail: bgomez@indstate.edu
Flume

Brian Greenwood
Environmental Sci., University Toronto
1265 Military Trail
Scarborough ON M1C 1A4
Canada
e-mail: greenw@scar.utoronto.ca
Bar, Littoral
Sediment Transport by Waves

Michael J. Hambrey
Centre for Glaciology
University of Wales
Aberystwyth Wales SY23 3DB
United Kingdom
e-mail: mjh@aber.ac.uk
Glacial Sediments: Processes, Environments and Facies

Jeffrey S. Hanor
Department of Geology and Geophysics
Louisiana State University
Baton Rouge LA 70803
USA
e-mail: hanor@unix1.sncc.lsu.edu
Porewaters in Sediments

Lawrence A. Hardie
Department of Earth and Planetary Sciences
The Johns Hopkins University
Baltimore MD 21218
USA
e-mail: hardie@ekman.eps.jhu.edu
Anhydrite and Gypsum
Evaporites
Sabkha, Salt Flat, Saline

Adrian M. Harvey
Department of Geography
University of Liverpool, P.O. Box 147
Liverpool L69 3BX
England, U.K.
e-mail: amharvey@liverpool.ac.uk
Alluvial Fan

Marwan Hassan
Department of Geography
University of British Columbia
Vancouver BC V6T 1Z2
Canada
e-mail: mhassan@geog.ubc.ca
Tracers for Sediment Movement

Richard L. Hay
4320 N Alvernon Way
Tucson AZ 85718-6180
USA
e-mail: rhay@dakotacom.net
Zeolites in Sedimentary Rocks

James P. Hendry
School of Earth and Environmental Sci.
University of Portsmouth, Burnaby Bldg.
Burnaby Rd., Portsmouth PO1 3QL
England (UK)
e-mail: jim.hendry@port.ac.uk
Ankerite

E.J. Hickin
Department of Earth Sciences
Simon Fraser University
8888 University Drive
Burnaby BC V5A 1S6
Canada
e-mail: hickin@sfu.ca
Meandering Channels

Stephen Hillier
Macauley Land Use Research Institute
Craigiebuckler
Aberdeen AB15 8QH
Scotland, UK
e-mail: S.Hillier@macauley.ac.uk
Chlorite in Sediments
Clay Mineralogy

Linda A. Hinnov
Department of Earth & Planetary Sci.
Johns Hopkins University
Baltimore MD 21218
USA
e-mail: hinnov@jhu.edu
Milankovitch Cycles

Richard N. Hiscott
Department of Earth Sciences
Memorial University of Newfoundland
St John's NF A1B 3X5
Canada
e-mail: rhiscott@sparky2.esd.mun.ca
Grading, Graded Bedding
Slope Sediments

Andrew J. Hogg
Centre for Environmental & Geophysical Flows
School of Mathematics, University Walk
Bristol BS8 1TW
England, UK
e-mail: A.J.Hogg@bristol.ac.uk
Diffusion, Turbulent

Mark W. Hounslow
Centre for Environmental Magnetism and Paleomagnetism
Lancaster Environmental Centre, Geography Department
Lancaster University
Bailrigg, Lancaster LA1 4YW
England, UK
e-mail: m.hounslow@lancaster.ac.uk
Septarian Concretions

Michael G. Hughes
Div. of Geology and Geophysics
Edgeworth David Bldg. (FO5)
University of Sydney
Sydney NSW 2006
Australia
e-mail: michaelh@mail.usyd.edu.au
Swash and Backwash, Swash Marks

Hillert Ibbeken
Instut für Geologie
Frei Universität Berlin
Berlin D 12249
Germany
e-mail: Hillert.Ibbeker@t-online.de
Attrition (Abrasion), Marine

Raymond V. Ingersoll
Department of Earth and Space Sciences
University of California
Los Angeles CA 90095-1567
USA
e-mail: ringer@ess.ucla.edu
Maturity: Textural and Compositional

Richard M. Iverson
US Geological Survey
5400 MacArthur Blvd.
Vancouver WA 98661
USA
e-mail: riverson@usgs.gov
Gravity-Driven Mass Flows

Noel P. James
Department of Geological Sciences
Queens University
Kingston ON K7L 3N6
Canada
e-mail: james@geol.queensu.ca
Neritic Carbonate Depositional Environments

Markes E. Johnson
Department of Geosciences, Williams College
947 Main St
Williamstown MA 01267-2606
USA
e-mail: markes.e.johnson@williams.edu
Sedimentologists: Amadeus William Grabau

Brian Jones
Department of Earth and Atmospheric Sciences
University of Alberta
Edmonton AB T6G 2E3
Canada
e-mail: brian.jones@ualberta.ca
Ancient Karst
Cave Sediments

Miriam Kastner
Geoscience Research Division
Scripps Institute of Oceanography
La Jolla CA 92093-0220
USA
e-mail: mkastner@ucsd.edu
Clathrates

Wolfgang Kiessling
Museum für Naturkunde
HU Berlin
Invalidenstrasse 43
10115 Berlin
Germany
e-mail: wolfgang.kiessling@museum.hu-berlin.de
Reefs

Paul L. Knauth
Department of Geology, Box 871404
Arizona State University
Tempe AZ 85287-1404
USA
e-mail: Knauth@asu.edu
Siliceous Sediments

Ben C. Kneller
Institute for Crustal Studies, Girvetz Hall
University of California
Sanata Barbara CA 93101
USA
e-mail: ben@crustal.ucsb.edu
Turbidites

Christian Koeberl
Institute of Geochemistry
University of Vienna, Althanstrasse 14
Vienna A-1090
Austria
e-mail: christian.koeberl@univie.ac.at
Extraterrestrial Material in Sediments
Sediments Produced by Impact

Paul D. Komar
College of Oceanography
Oregon State University
Corvallis OR 97331
USA
e-mail: pkomar@oce.orst.edu
Angle of Repose
Grain Settling
Grain Threshold

Erik P. Kvale
Indiana Geol. Survey/Department Geol. Sci.
Indiana University, 611 N. Walnut Grove
Bloomington IN 47405
USA
e-mail: kvalee@indiana.edu
Tides and Tidal Rhythmites

Peter F. Landrum
Great Lakes Environmental Research Laboratory NOAA
Ann Arbor, MI 48105
USA
Toxicity of Sediments

Scott F. Lamoureux
Department of Geography, EVEX Lab
Queens University
Kingston ON K7L 3N6
Canada
e-mail: lamoureux@lake.geog.queensu.ca
Impregnation

Nicholas Lancaster
Desert Research Institute
2215 Raggio Parkway
Reno NV 89512-1095
USA
e-mail: nick@dri.edu
Dunes, Eolian

Denis Lavoie
Centre Géoscientifique de Québec
Commission Géologique du Canada
880 Chemin Ste-Foy, P.O. Box 7500
Québec G1V 4C7
Canada
e-mail: delavoie@nrcan.gc.ca
Ophicalcites

Frederick J. Longstaffe
Department of Earth Sciences
University of Western Ontario
London ON N6A 5B7
Canada
e-mail: flongsta@julian.uwo.ca
Berthierine
Isotopic Methods in Sedimentology

Donald R. Lowe
Department of Geol./Environmental Sci.
Stanford University,
Bldg 320
Stanford CA 94305-2115
USA
e-mail: lowe@pangea.stanford.edu
Dish Structure
Fluid Escape Structures
Pillar Structure

Tim K. Lowenstein
Department of Geological Sciences
State University of New York
Binghamton NY 13901
USA
e-mail: lowenst@binghamton.edu
Evaporites

Hans G. Machel
Department of Earth and Atmospheric Sciences
University of Alberta
Edmonton AB T6G 2E3
Canada
e-mail: hans.machel@ualberta.ca
Dolomites and Dolomitization

Ian G. MacIntyre
Natl. Museum of Natural History MRC 125
Smithsonian Institution
Washington DC 20560
USA
e-mail: macintyre.ian@nmnh.si.edu
Micritization

Greg H. Mack
Geological Sciences
Department 3AB, P.O. Box 30001
Las Cruces NM 88003-8001
USA
e-mail: gmack@nmsu.edu
Climatic Control of Sedimentation
Tectonic Control of Sedimentation

Fred T. Mackenzie
Department of Oceanography
University of Hawaii
1000 Pope Road
Honolulu HI 96822
USA
e-mail: fredm@soest.hawaii.edu
Carbonate Mineralogy and Geochemistry

Jon J. Major
USGS Cascades Volcano Obs.
5400 Macarthur Blvd
Vancouver WA 98661-7049
USA
e-mail: jjmajor@usgs.gov
Debris Flow
Hindered Settling
Slurry

Prakash B. Malla
Manager, Paper research & Application Lab.
P.O. Box 1056, 520 Kaolin Road
Thiele Kaolin Company
Sandersville GA 31082
USA
e-mail: prakash.malla@thielekaolin.com
Vermiculite

Ole J. Martinsen
Norsk Hydro Research Centre
Bergen N-5020
Norway
e-mail: ole.martinsen@hydro.com
Slide and Slump Structures

Earle F. McBride
Department of Geological Sciences
University of Texas
Austin TX 78713-7909
USA
e-mail: efmcbride@mail.utexas.edu
Sedimentologists: Robert L Folk
Sedimentologists: Johan August Udden

Nicholas I. McCave
Department of Earth Sciences, Downing St
University of Cambridge
Cambridge CB2 3EQ
England, UK
e-mail: mccave@esc.cam.ac.uk
Nepheloid Layer, Sediment

Cheryl Z. McKenna-Neuman
Department of Geography
Trent University
1600 West Bank Drive
Peterborough ON K9J 7B8
Canada
e-mail: cmckneuman@trentu.ca
Eolian Transport and Deposition

Daniel F. Merriam
Kansas Geol. Survey
University of Kansas
Lawrence KS 66047
USA
e-mail: dmerriam@kgs.ukans.edu
Geothermic Characteristics of Sedimentary Rocks
Sedimentologists: William C. Krumbein
Statistics Analysis of Sediments and Sedimentary Rocks

John Merzies
Department of Earth Sciences
Brock University
St. Catherine's, ON L2S 3A1
Canada
e-mail: jmenzies@spartan.ac.brocku.ca
Tills and Tillites

Andrew D. Miall
Department of Geology
University of Toronto
Toronto ON M5S 3B1
Canada
e-mail: miall@quartz.geology.utoronto.ca
Sands, Gravels and their Lithified Equivalents
Paleocurrent Analysis

Gerard V. Middleton
School of Geography and Geology
McMaster University
1280 Main St West
Hamilton ON L8S 4K1
Canada
e-mail: middleto@mcmaster.ca
Convolute Lamination
Fabric, Porosity, Permeability
Flame Structure
Sedimentary Geology
Sedimentology: History
Sedimentologists: Robin G.C. Bathurst
Sedimentologists: Amanz Gressly
Sedimentologists: Paul D. Krynine
Sedimentologists: Philip H. Kuenen
Sedimentologists: John Murray and the Challenger Expedition
Sedimentologists: Henry Clifton Sorby
Submarine Fans and Channels
Tool Marks

Robert Millar
Department of Civil Engineering, 2324 Main Mall
University of British Columbia
Vancouver BC V6T 1Z4
Canada
e-mail: millar@civil.ubc.ca
Flow Resistance

Kitty L. Milliken
Department of Geological Sciences
University of Texas at Austin
Austin TX 78712
USA
e-mail: kittym@mail.utexas.edu
Diagenesis
Geodes

Andrew Moore
Disaster Control Research Center
Graduate School of Engineering
Tohoku University
Aramaki, Aoba 06, Sendai 980-8579
Japan
e-mail: moore@tsumani2.civil.tohoku.ac.jp
Tsunami Deposits

Sadoon Morad
Institute of Earth Sciences
Uppsala University
Uppsala S 752 36
Sweden
e-mail: sadoon.morad@geo.uu.se
Feldspars in Sedimentary Rocks

Andrew C. Morton
HM Research Associates
100 Main Street
Woodhouse Eaves Leics LE12 8RZ
England, UK
e-mail: a.c.morton@heavyminerals.fsnet.co.uk
Heavy Minerals

Peter Mozley
Department of Earth and Environmental Sciences
New Mexico Tech
Socorro NM 87801
USA
e-mail: mozley@nmt.edu
Diagenetic Structures

Raymond C. Murray
106 Ironwood Place
Missoula Montana 59803
USA
e-mail: rcm@selway.umt.edu
Forensic Sedimentology

Paul M. Myrow
Department of Geology
Colorado College
Colorado Springs CO 80903
USA
e-mail: pmyrow@cc.colorado.edu
Colors of Sedimentary Rocks
Gutter and Gutter Casts
Storm Deposits

Gerald C. Nanson
School of Geosciences
University of Wollongong
Wollongong NSW 2522
Australia
e-mail: gerald_nanson@uow.edu.au
Anabranching Rivers
Rivers and Alluvial Fans

H. Wayne Nesbitt
Department of Earth Sci.
University of Western Ontario
London ON N6A 5B7
Canada
e-mail: hwn@uwo.ca
Seawater: Temporal Changes in the Major Solutes

Nora Noffke
Department of Ocean, Earth, Atmospheric Sciences
Old Dominion University
4600 Elkhorn Ave
Norfolk VA 23529
USA
e-mail: nnoffke@odu.edu
Bacteria in Sediments
Microbially Induced Sedimentary Structures

Hakuyu Okada
Oyo Corporation Kyushu Office
2-21-36 Ijiri, Minami-ku
Fukuoka 811-1302
Japan
e-mail: hokada@bb.mbn.or.jp
Sedimentology: History in Japan

Geraint Owen
Department of Geography
University of Wales, Swansea
Singleton Park, Swansea SA2 8PP
Wales, UK
e-mail: g.owen@swansea.ac.uk
Ball-and-Pillow (Pillow) Structure
Pseudonodules

Henry M. Pantin
School of Earth Sciences
University of Leeds
Leeds LS5 9JT
England, UK
e-mail: h.m.pantin@earth.leeds.ac.uk
Autosuspension

Roger L. Parfitt
Landcare Research
PB 11052
Palmerston North
New Zealand
e-mail: ParfittR@landcare.cri.nz
Allophane and Imogolite

H. Martyn Pedley
Research Institute of Environmental Science
University of Hull
Cottingham Rd.
Hull East Yorkshire
England, UK
e-mail: H.M.Pedley@geo.hull.ac.uk
Tufas and Travertines

Morten Pejrup
Institute of Geography, University of Copenhagen
Oster Voldgade 10
Copenhagen K DK-1350
Denmark
e-mail: mp@geogr.ku.dk
Flocculation

George S. Pemberton
Department of Earth and Atmospheric Sciences
University of Alberta
Edmonton AB T6G 2E3
Canada
e-mail: gpembert@gpu.srv.ualberta.ca
Biogenic Sedimentary Structures
Sedimentologists: Rudolf Richter and the Senckenberg Laboratory
Substrate-Controlled Ichnofacies

George Postma
Department of Geology, Faculty of Earth Sci.
Utecht University, P.O. Box 80.021
Utrecht 3508 TA
The Netherlands
e-mail: gpostma@geo.uu.nl
Fan Delta

Paul E. Potter
Geosciencias/UFRGS
Campus Do Vale Sala 302B
Porto Allegre 91509-900
Brazil
Mudrocks
Sedimentologists: Francis J. Pettijohn

Brian R. Pratt
Department of Geol. Sci., 114 Science Place
University of Saskatchewan
Saskatoon SK S7N 5E2
Canada
e-mail: brian.pratt@usask.ca
Stromatolites

Bruce Railsback
Department of Geology
University of Georgia
Athens GA 30602-2501
USA
e-mail: rlsbk@gly.uga.edu
Stylolites

Harold G. Reading
Department of Earth Sciences
University of Oxford
Parks Road
Oxford OX1 3PR
United Kingdom
Facies Models

R. Pamela Reid
RSMAS-MGG
University of Miami
4600 Rickenbacker Causeway
Miami, FL 33149
USA
Micritization

Wolf Uwe Reimold
Department of Geology
University of Witwatersrand
Private Bag 3, P.O. Wits 2000
Johannesburg
South Africa
e-mail: 065wur@cosmos.wits.ac.za
Features indicating Impact and
Shock Metamorphism

François Renard
LGIT Université Joseph Fourier
BP 53
Grenoble 38041
France
e-mail: Francois.Renard@obs.ujf-grenoble.fr
Pressure Solution

Robin W. Renaut
Department of Geological Sci., 114 Science Place
University of Saskatchewan
Saskatoon SK S7N 5E2
Canada
e-mail: robin.renaut@sask.usask.ca
Magadiite

Gregory J. Retallack
Department of Geological Sciences
University of Oregon
Eugene OR 97403-1272
USA
e-mail: gregr@darkwing.uoregon.edu
Weathering, Soils, and Paleosols

Franco Ricci Lucchi
Instituto di Geologia, Universita di Bologna
Via Zamboni 67
40127 Bologna
Italy
e-mail: riccil@geomin.unibo.it
Bedding and Internal Structures

Robert Riding
Department of Earth Scences, Cardiff University
Park Place, P.O. Box 914
Cardiff CF10 3YE
Wales, UK
e-mail: riding@cardiff.ac.uk
Algal and Bacterial Carbonate Sediments

David M. Rubin
USGS Pacific Science Center
University of California
1156 High Street
Santa Cruz CA 95064
USA
e-mail: drubin@usgs.gov
Cross-Stratification

Juergen Schieber
Department of Geology
University of Texas
Arlington Box 19049
Arlington TX 76019-0049
USA
e-mail: schieber@uta.edu
Black Shales
Depositional Fabric of Mudstones

Peter A. Scholle
New Mexico Bureau of Mines and Mineral Resources
New Mexico Institute of Mining and Technology
801 Leroy Place
Socorro, NM 87801
USA
e-mail: pscholle@gis.nmt.edu
Cements and Cementation

Andrew C. Scott
Department of Geology
Royal Holloway College, University London
Egham Surrey TW20 OEX
England, UK
e-mail: scott@gl.rhul.ac.uk
Charcoal in Sediments
Coal Balls

E. Seibold
Richard Wagner Strasse 56
Freiburg D-7106
Germany
e-mail: Seibold-Freiburg@t-online.de
Sedimentologists: Johannes Walther

I. Seibold
Richard Wagner Strasse 56
Freiburg D-7106
Germany
e-mail: Seibold-Freiburg@t-online.de
Sedimentologists: Johannes Walther

Graham Shimmield
Dunstaffhage Marine Lab.
P.O. Box 3
Oban Argyll PA34 4AD
Scotland, UK
e-mail: gbs@dml.ac.uk
Upwelling

Duncan F. Sibley
Center for Integrative Studies
Michigan State University
100 North Kedzie Lab
East Lansing MI 48824
USA
e-mail: sibley@pilot.msu.edu
Dolomite Textures

Fredrick D. Siewers
Department of Geography and Geology
Western Kentucky University
Bowling Green, KY 42101
USA
e-mail: fred.siewers@wku.edu
Oolite and Coated Grains

Bruce M. Simonson
Department of Geology
Carnegie Bldg, Oberlin College
52 West Lorain St
Oberlin OH 44074-1044
USA
e-mail: bruce.simonson@oberlin.edu
Ironstones and Iron Formations

Balwant Singh
Department of Agricultural Chemistry & Soil Science
University of Sydney
Sydney NSW 2006
Australia
e-mail: b.singh@acss.usyd.edu.au
Cation Exchange

Rudy L. Slingerland
303 Deike Bldg
Pennsylvania State University
262 East Hamilton Ave
University Park PA 16801
USA
e-mail: sling@ustar.geosc.psu.edu
Numerical Models and Simulation
Sedimentologists: Grove Karl Gilbert

Norman D. Smith
Department of Geosciences, U. Nebraska
214 Bessey Hall
Lincoln NE 68588-0340
USA
e-mail: nsmith3@unl.edu
Avulsion

Joseph P. Smoot
U.S. Geological Survey
National Center MS 955
Reston VA 20192
USA
e-mail: jpsmoot@usgs.gov
Desert Sedimentary Environments

John B. Southard
EAPS, Bldg. 54-1026, Massachsetts Institute of Technology
77 Massachusetts Ave
Cambridge MA 02139
USA
e-mail: southard@mit.edu
Surface Forms

D.A. Spears
Environmental and Geological Sciences
Dainton Bldg., University of Sheffield
Sheffield S3 7HF
England, UK
e-mail: d.a.spears@sheffield.ac.uk
Bentonites and Tonsteins

Jan Środoń
Institute of Geol. Sciences, PAN
Senacka 1
Krakow 31-002
Poland
e-mail: ndsrodon@cyf-kr.edu.pl
Illite Group Clay Minerals
Mixed-Layer Clays

Robert F. Stallard
US Geological Survey, Campus Box 458
3215 Marine St, Room E146
Boulder CO 80303-1066
USA
e-mail: stallard@colorado.edu
Erosion and Sediment Yield

Helge Stanjek
Lehrstuhl für Bodenkunde
Technische Universität München
85350 Freising
Germany
e-mail: stanjek@wzm.tum.de
Hydroxides and Oxyhydroxide Minerals

Donald J.P. Swift
Department of Oceanography
Old Dominion University
Norfolk VA 23529-0276
USA
e-mail: dswift@odu.edu
Relict and Palimpset Sediments

Zoltán Sylvester
Shell International Exploration and Production, Inc.
P.O. Box 481
Houston, TX 77001-0481
USA
e-mail: zoltan.sylvester@shell.com
Dish Structure
Fluid Escape Structures
Pillar Structure

James P. Syvitski
Institute of Arctic and Alpine Research
University of Colorado, Campus Box 450
1560 30th St
Boulder CO 80309-0450
USA
e-mail: syvitski@stripe.colorado.edu
Sediment Fluxes and Rates of Sedimentation

P.W. Geoff Tanner
Div. of Earth Sci., Gregory Bldg.
University of Glasgow, Lilybank Gardens
Glasgow G12 8QQ
Scotland, UK
e-mail: G.Tanner@earthsci.gla.ac.uk
Structures (Mudcracks, etc.)
Syneresis

Yves Tardy
Institute National Polytechnique de Toulouse
Montegeard
Nailloux 31560
France
Laterites

Colin R. Thorne
Department of Geography
University of Nottingham
Nottingham NG7 2RD
England, UK
e-mail: Colin.Thorne@nottingham.ac.uk
Physics of Sediment Transport: The Contributions
of R.A. Bagnold
Sedimentologists: Reginald A. Bagnold

B.R. Turner
School of Earth Sciences
University of Birmingham,
Birmingham
England, UK
e-mail: p.turner@bham.ac.uk
Red Beds

Dana Ulmer-Scholle
Department of Earth and Environmental Sciences
New Mexico Institute of Mining and Technology
801 Leroy Place
Socorro, NM 87801
USA
e-mail: dulmer@nmt.edu
Cements and Cementation

John W. Waldron
Department of Earth & Atmospheric Sci.
University of Alberta
Edmonton AB T6G 2E3
Canada
e-mail: john.waldron@ualberta.ca
Mélange; Melange

Karl Hans Wedepohl
Geochemische Institut
Universität Göttingen
Goldschmidtstrasse 1
D-37077 Göttingen
Germany
e-mail: Hans.Wedepohl@geo.uni-goettingen.de
Sedimentologists: Carl Wilhelm Correns

Barry G. Warner
Wetland Research Centre
University of Waterloo
Waterloo ON N2L 3G1
Canada
e-mail: bwarner@watserv1.uwaterloo.ca
Peat

Brian W. Whalley
School of Geosciences
The Queen's College
Belfast BT7 1NN
Northern Ireland
e-mail: B.Whalley@Queens-Belfast.ac.uk
Surface Textures

Daryl M. Wightman
AEC East
3900 421-7th Avenue SW
Calgary AB T2P 4K9
Canada
e-mail: DarylWightman@aec.ca

Oil Sands

Rick T. Wilkin
U.S. EPA, Natl. Risk Management Res. Lab
P.O. Box 1198
Ada OK 74820
USA
e-mail: wilkin.rick@epa.gov

Sulfide minerals in Sediments

Sherwood W. Wise
Department of Geol. Sci. 4100, Florida State University
100 Antarctic Circle
Tallahassee FL 32306-4100
USA
e-mail: wise@gly.fsu.edu

Calcite Compensation Depth
Lepisphere

V. Paul Wright
Dept of Earth Sciences
Cardiff University
Cardiff Wales CF1 3YE
UK
e-mail: WrightVP@cardiff.ac.uk

Caliche – Calcrete

Raphael A.J. Wüst
Department Earth and Ocean Sciences
University of British Columbia
6270 University Blvd
Vancouver BC V6T 1Z4
Canada
e-mail: rwuest@interchange.ubc.ca

Kerogen
Maturation, Organic

Ellis L. Yochelson
Department of Paleobiology, National Museum of
  Nat. History
Smithsonian Institution, 10th St & Constitution Ave NW
Washington DC 20560-0121
USA
e-mail: yochelson.ellis@NMNH.SI.edu

Sedimentologists: T. Wayland Vaughan

Preface

The study of natural sediments and sedimentary rocks has been called *sedimentology*. This encyclopedia is a thorough revision of the original *Encyclopedia of Sedimentology*, published by Dowden, Hutchinson and Ross in 1978: The field has advanced so fast, however, that all the articles in the present volume are new, and this is recognized by a new title.

The present encyclopedia interprets *sedimentology*, both more narrowly and more broadly than is often the case (see *Sedimentary Geology*—an entry in this encyclopedia—for further discussion). More narrowly, because the encyclopedia contains relatively little information about *stratigraphy*, the science concerned with stratified rocks. Stratigraphy and sedimentology overlap, particularly in the area of facies analysis and sequence stratigraphy. In general, however, stratigraphic topics have been reserved for full treatment in a companion *Encyclopedia of Stratigraphy*, which is now in preparation. More broadly, this encyclopedia includes topics that some sedimentologists tend to exclude, for examples: the mineralogy of clays and other minerals common in sediments; geochemistry of sediments (sediments are included in the *Encyclopedia of Geochemistry)*; some features of sediments interesting to the general public but somewhat neglected by sedimentologists (e.g., clathrates, coal balls, geodes, resin and amber, speleothems, toxicity of sediments); contributions of engineering studies to sediment transport and soil mechanics; studies on the geophysical and petrophysical properties of sediments; and studies by physicists on granular matter.

Three other disciplines that are represented by separate encyclopedias also have overlapping interests in sediments: hydrology and hydrogeology (represented by *Encyclopedia of Hydrology and Water Resources*); geomorphology (represented by *Encyclopedia of Geomorphology and Lanforms*, and also by *Encyclopedia of Coastal Science*); and environmental studies (represented by *Encyclopedia of Environmental Science*).

A final word about the selection of topics: there is always a subjective element in the choice of topics, but if the reader does not see an entry for the particular topic that interests him, then he or she should look in the index. The topic may be covered (perhaps in more than one article) under a different name. The editors have tried to make the coverage comprehensive, but we are aware of some partial omissions: unfortunately, willing contributors cannot always be found for all the topics that might be suggested.

Encyclopedias are not generally places to look for extended acknowledgments. This is a tradition dating back to the days when contributors were anonymous, or only identified by their initials (and imagine the space that would be taken up by acknowledgments from every author: the Academy Awards would pale by comparison!). On behalf of all the editors and contributors, therefore, we extend thanks to those colleagues who have assisted us by providing data, figures, critical reviews, and sustaining personal and financial encouragement. Thank you all—we hope you realize that your generosity is not forgotten, even if your contribution remains anonymous.

Guide to the Reader

This encyclopedia is devoted to the science of sediments and sedimentary rocks, a science generally called *sedimentology*. It does not address those broader aspects of stratified rocks concerned with the naming of rocks units, their correlation from one place to another, and their dating in geological time. Those aspects belong to *stratigraphy*, the subject of another encyclopedia in this series.

Sediments and sedimentary rocks can be approached from three main points-of-view:

1. Like other rocks, they have a mineral and chemical composition, physical properties, and structures and textures, all of which need description and interpretation in order that we may understand their origin. These are the *geochemical, mineralogical, petrological and petrophysical (geophysical)* aspects of sediments and sedimentary rocks.
2. Sediments are first laid down in *sedimentary environments*. The (primary) aspects of sedimentary rocks that were formed at the time of deposition (particularly, but not exclusively, their structures), are generally called their *sedimentary facies*. Facies analysis is concerned with using primary aspects of sediments to determine the environment in which they were deposited: and, in a complementary way, with understanding how modern sedimentary environments control, or are determined by, the characteristics of the sediments deposited in them. Sediments interact with many other aspects of the environment, including their biology.
3. Many sedimentologists try to understand the basic *physical, chemical and biological processes* that form sediments, transport and deposit them, and later convert them into sedimentary rocks. Such studies may be carried out in the laboratory, in the field (particularly by studying processes active in modern environments), and by theoretical and numerical analysis and simulation.

For those readers not already familiar with sedimentology, Table I indicates the major introductory articles in each of these three categories (there is, of course, some overlap in the approaches used in most of the articles). Besides these there are also introductory articles on Sedimentary Geology; Sedimentology—Organizations, Meetings, Publications; Sedimentology—History; and Sedimentologists (brief biographic sketches).

Table 1 Major articles, classified by methodology: starred topics are general introductions

Geochemistry, Mineralogy, Petrology
*Bedding and Internal Structures
Biogenic Sedimentary Structures
Carbonate Mineralogy and Geochemistry
Cements and Cementation
*Classification of Sediments and Sedimentary Rocks
*Clay Mineralogy
Compaction (Consolidation) of Sediments
*Diagenesis
Diagenetic Structures
Dolomites and Dolomitization
Evaporites
*Fabric, Porosity, and Permeability
Geophysical Properties of Sediments
Grain Size and Shape
Ironstones and Iron Formations
Isotopic Methods in Sedimentology
Magnetic Properties of Sediments
*Mudrocks
Offshore Sands
*Paleocurrent Analysis
Phosphorites
Provenance
*Sands, Gravels and their Lithified Equivalents
Siliceous Sediments
Surface Forms
Surface Textures
Weathering, Soils, and Paleosols

Sedimentary Environments and Facies
*Climatic Control of Sedimentation
*Coastal Sedimentary Facies
Cyclic Sedimentation
Deltas and Estuaries
*Desert Sedimentary Environments
*Erosion and Sediment Yield
*Facies Models
Floods and Other Catastrophic Events
*Glacial Sediments: Processes, Environments and Facies
*Lacustrine Sedimentation
*Neritic Carbonate Depositional Environments
*Oceanic Sediments
*Rivers and Alluvial Fans
Slope Sediments
Submarine Fan and Channels

Table 1 Continued

*Taphonomy: Sedimentological Implications of Fossil Preservation
*Tectonic Controls of Sedimentation
Tidal Flats
Tidal Inlets and Deltas
Turbidites
Upwelling

Sedimentary Processes
Debris Flow
Eolian Transport and Deposition
Features Indicating Impact and Shock Metamorphism
Grain Settling
Grain Threshold
Gravity-Driven Mass Flows
Numerical Models and Simulation of Sediment Transport and Deposition
Sediment Fluxes and Rates of Sedimentation
Sediment Transport by Tides
Sediment Transport by Unidirectional Water Flows
Sediment Transport by Waves

A good approach for readers unfamiliar with the subject is to begin with a general article, then follow the *cross-references* listed at the end of the article to find related topics. For example, one might begin to learn something about Sands, Gravels and their Lithified Equivalents, go on to Bedding and Internal Structures, then Paleocurrent Analysis, then Cross-Stratification, or some other specific topic.

A reader with more knowledge, might begin searching for a specific topic, for example, concretions. As it happens, there is no article with that name, but reference to Diagenesis, or Diagenetic Structures (or to the Index) would soon lead to articles that describe concretions of various types. If the reader needs more than he can find in the encyclopedia, most articles give copious bibliographic references, to both general texts and research articles.

A

ALGAL AND BACTERIAL CARBONATE SEDIMENTS

Calcified algae and bacteria

Only a few algae and bacteria calcify (Figure A1), but their abundance and wide distribution make them important in limestones of many ages and environments (Figure A2). Microbial carbonates appeared in the Archaean and are significant in Proterozoic carbonate platforms. Calcified cyanobacteria became important in the Cambrian, and calcified green and red algae in the Ordovician. Additional extinct organisms have been regarded as calcified algae or bacteria, but are still of uncertain affinity. These problems of affinity hamper paleoecological and phylogenetic interpretations.

Calcification

Environmental range and variations in cellular site and mineralogy of calcification reflect the organism's control over calcification. With decreasing control, calcification site moves from intra- to extra-cellular, mineralogy shifts toward that of ambient abiotic carbonate precipitates, and environmental distribution becomes restricted to locations where inorganic precipitation is favored (e.g., warmer water in marine environments). Strong control (e.g., coralline red algae) allows wide environmental distribution of calcification and is linked to intracellular sites of $CaCO_3$ nucleation. Weak control (e.g., halimedaceans and cyanobacteria) limits the environmental distribution of calcification, and is linked to an extracellular site of $CaCO_3$ nucleation and a polymorph in equilibrium with the ambient environment. Consequently, calcified algae and bacteria have potential to reflect past fluctuations in environmental controls over carbonate precipitation. Cyanobacteria, for example, calcify only when environmental conditions are favorable (Arp et al., 2001). At present, this is only widespread in freshwater, but took place extensively in marine environments in the Paleozoic and Mesozoic.

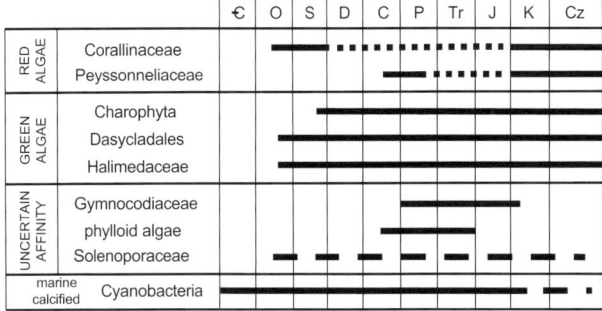

Figure A1 Principal groups of calcified benthic algae and cyanobacteria. Calcified red algae include corallines that are marine, calcitic, occur at all latitudes, and are important reef builders. In contrast, calcified marine green dasycladaleans and halimedaceans are aragonitic and mainly tropical. They mainly produce particulate sediment, although *Halimeda* creates reefs with its disarticulated segments. Charophyte green algae also produce bioclastic sediment, but are essentially freshwater, calcitic, and prefer temperate climates. Gymnocodiaceans and phylloids are certainly algae, but lack the distinctive features necessary to either subdivide or classify them. Solenoporaceans are a heterogeneous grouping.

Microbial carbonates

Microbes (bacteria, small algae, fungi) are widespread on wetted substrates. Carbonate precipitation, locally augmented by grain trapping, results in their accretion and preservation as microbial carbonates. Extracellular polymeric substances (EPS), produced by microbes for attachment and protection, provide nucleation sites and facilitate grain trapping. Precipitation is stimulated by photosynthetic uptake of CO_2 and/or HCO_3^- by algae and cyanobacteria, and by ammonification, denitrification, sulfate reduction and other metabolic processes in other bacteria.

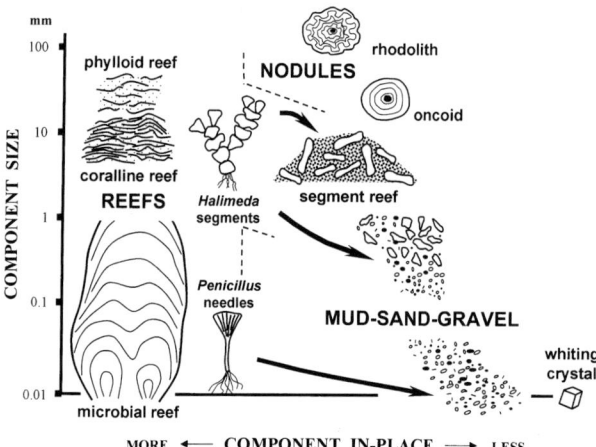

Figure A2 Algal-bacterial carbonate sediment. Variations in particle/component size and degree of movement. Sizes are for basic components in reefs (mud-grade to millimetric fabrics and calcified sheaths in microbial reefs, crustose thalli in coralline red algal reefs, and leaflike algal skeletons in phylloid reefs), and mud-sand-gravel. Internally, microbial domes may have stromatolitic, thrombolitic, dendrolitic, or leiolitic mesofabrics—or combinations of these. In nodules, size indicates overall rhodolith or oncoid size. Whereas components in microbial, coralline and phylloid reefs are essentially in place, segments of the green alga *Halimeda* in segment reefs are parauthochthonous. Centimetric nodules, built by red algae (rhodoliths) and calcified microbes (oncoids) are also commonly parautochthonous. *Halimeda* segment reefs typically accumulate at depths of 20–50 m on low angle shelves and atoll lagoon floors. In contrast, on slopes and in shallower water, carbonate mud-sand-gravel produced by disaggregation of *Halimeda* and *Penicillus* green algae is commonly transported. In addition, numerous other freshwater and marine algae, such as green charophytes and red articulated corallines, produce particulate sediment. Whiting crystals form in surface waters and settle out of suspension.

Biofilms are very thin layers, usually only a few hundreds of microns in thickness, of heterotrophic bacterial cells <2 μm in size in an EPS matrix, on solid substrates. Calcified biofilm, characterized by peloidal/clotted (grumous) nanofabric, internal microchannels and external architecture of elevated towers and plumes, is widespread as micritic veneers on grain and reef surfaces.

Microbial mats are turf-like structures dominated by intertwined, often filamentous, cyanobacteria and algae, on solid or grainy surfaces. They are believed to form the accreting surfaces of most stromatolites and other relatively large benthic microbial carbonates. Organic material produced by the photosynthetic surface community is recycled by chemoorganotrophic bacteria deeper in the mat, producing well-defined microbial stratification and steep chemical gradients, all in a depth of a few millimeters. Fossil mats preserve only a fraction of this complexity, but can nonetheless retain distinctive features (lamination, primary fenestrae, clotted and peloidal fabrics from calcification of degraded organic material, discrete microfossils such as calcified cyanobacteria) although these do not necessarily all co-occur.

Dome and column morphologies characterize thick microbial carbonates and display laminated (stromatolite), clotted (thrombolite), dendritic (dendrolite) or aphanitic (leiolite) macrofabrics. Nodules also form, usually with stromatolitic coats (oncoids). Stromatolites probably appeared at 3540 Ma and contributed significantly to Paleo-Mesoproterozoic (2500–1000 Ma) carbonate platforms. Their Neoproterozoic decline has been attributed to eukaryote competition and/or reduced lithification. However, thrombolites and dendrolites were major Cambrian and Late Devonian reef-builders. In addition to domes and columns, less conspicuous but volumetrically significant microbial masses and layers are widespread in Phanerozoic algal-invertebrate reefs.

Modern marine examples at Shark Bay, Western Australia, and Lee Stocking Island, Bahamas, are coarse-grained agglutinated columns with crudely layered macrofabrics built by cyanobacterial-algal mats on wave-swept hypersaline shorelines (Shark Bay) or in normal salinity tidal channels (Lee Stocking). Thick fine-grained microbial crusts also form on Neogene coral-coralline reefs, and mats and biofilms calcify heavily in present-day calcareous streams and lakes. Temporal variation in abundance of marine microbial carbonate has been attributed to dependence on supersaturation state of seawater facilitating synsedimentary calcification.

Cyanobacterial calcification, in which the protective mucopolysaccharide sheath is impregnated with $CaCO_3$, creates distinctive microfossils that contribute to dendrolites, some thrombolites, and skeletal stromatolites. Patterns of cyanobacterial calcification and microbial dome formation through time could reflect fluctuations in seawater chemistry (Riding 2000).

Reefs

Microbial. In the Paleo-Mesoproterozoic and Paleozoic, stromatolite and other microbial reefs are major components of carbonate platforms. Some of the largest examples, hundreds of meters in extent and with tens of meters of relief, formed in deepwater. Dendrolites and thrombolites built by millimetric calcified microbes, most likely cyanobacteria, such as *Angusticellularia*, *Epiphyton*, and *Renalcis*, are locally important as rigid microframes. In Cambrian reefs, (e.g., Siberia) they are often much more abundant than archaeocyath sponges, and can rival stromatoporoids in Late Devonian reefs (e.g., Canning Basin).

Phylloid algae. Carboniferous-Triassic phylloid algae are united by leaflike form more than affinity. Erect blades with internal medulla and cortex (e.g., *Anchicodium*, *Eugonophyllum*, *Ivanovia*) resemble halimedacean green algae. Prostrate crusts with internal cellular tissue and conceptacles (e.g., *Archaeolithophyllum*) resemble red algae, particularly peyssonneliaceans. Both forms build self-supporting skeletal frame reefs with substantial shelter cavities and abundant localized fine sediment that are common in the Late Carboniferous-Early Permian of the southwestern USA, Arctic Canada, and Russia.

Coralline algae. Cell-wall calcification and encrusting growth allow crustose coralline red algae (Cretaceous-Recent) to build reefs in wave-swept environments. Present-day coralline laminar frames are characteristic of Pacific algal ridges and Atlantic cup and boiler reefs. Corallines (e.g., *Lithoporella*, *Lithothamnion*, *Sporolithon*) have calcitic, often magnesium-rich, skeletons and range from tropical to cold and deep water. In mid-latitudes, for example, the Mediterranean, they form thick ledges (trottoir) close to sea-level, and reefs (coralligène) in deeper water. Branched forms create interlocking frameworks in high latitudes. Crustose corallines also form rhodolith nodules. Related articulated corallines disaggregate to sand.

Solenoporaceae, classically regarded as ancestral to corallines, are a heterogeneous group of algae and other organisms. Nonetheless, some Ordovician-Silurian fossils attributed to *Solenopora* closely resemble the extant coralline *Sporolithon*.

Peyssonneliaceae, (e.g., *Peyssonnelia, Polystrata*), also known as Squamaricaeae, resemble crustose corallines in morphology and also participate in rhodolith formation, but are aragonitic and only calcify in warm water. They certainly range Cretaceous-Recent, but also show similarities with some Carboniferous phylloid algae.

Halimeda **segment reefs.** The heavily calcified (aragonite) tropical marine green alga *Halimeda* (Halimedaceae) is a major Neogene sediment-producer to depths of 150 m. The large branched thallus consists of articulated segments that are shed during life and after death. Rapid growth, up to 35 segments in 13 days, produces copious coarse sediment. In the Florida Keys, *Halimeda* is the single most important component of carbonate sand and gravel, and rivals coral in abundance. At depths of 20–50 m *Halimeda* sediment accumulates as parauthochtonous matrix-supported Segment Reefs that extend for kilometers in the northern Great Barrier Reef, Indonesia and the Caribbean.

Cladophora. At calcareous lake margins, calcification of *Cladophora*-like green algae can form tufts of small ($\sim 100\,\mu m$) branching tubes that amalgamate into cones and ultimately build tufa bioherms several meters across, for example, Miocene of the Ries Crater, Germany.

Particulate sediment

Mud-silt

Post-mortem disintegration of the halimedacean *Penicillus* produces large quantities of aragonite mud-silt in modern tropical carbonate bays and back-reef lagoons. A similar role has been suggested for the calcified cyanobacterium *Girvanella* on Early Paleozoic carbonate platforms. Charophytes (Silurian-Recent, e.g., *Chara, Nitella*), large bushy algae that produce fine-grained calcite in freshwater calcareous oligotrophic lakes and streams, are important in the Cretaceous-Oligocene. Disintegration of their weakly calcified stems contributes to accretionary platforms around Holocene temperate 'marl' lakes. Sand-grade highly calcified female oosporangia (gyrogonites), 0.2–2 mm in size, remain intact.

Whitings. In calcareous lakes, photosynthetic uptake of CO_2 and HCO_3^- by seasonal blooms of picoplanktic cyanobacteria (e.g., *Synechococcus*), diatoms and the other planktic algae, stimulates water column precipitation of calcite crystals, mainly less than $\sim 20\,\mu m$ in size, that form milky suspensions in surface waters. Similar marine aragonite 'whitings' (e.g., Great Bahama Bank) may reflect the same biological stimulus, but inorganic precipitation has not been excluded. Whitings potentially account for abundant lime mud production on carbonate shelves. Stable isotope compositions may distinguish organic (e.g., photosynthetic) from inorganic (e.g., temperature induced) whiting precipitates in present-day sediments, but the origins of ancient lime mud are more problematic.

Sand-gravel

Dasycladaleans and halimedaceans, warm water aragonitic green algae that range Ordovician-Recent, readily disaggregate or fragment to coarse sediment. Dasycladaleans were important in tropical bays and lagoons during the Late Jurassic-Early Cretaceous and Paleogene. Halimedaceans extend into deeper water and were important in the Upper Triassic and Cenozoic. Gymnocodiaceans, possibly related to halimedaceans, have a similar role in the Permian. Articulated (geniculate) coralline red algae consist of segments, ~ 0.5–6 mm in size, separated by uncalcified nodes (genicula) and also dissociate after death.

Nodules

Rhodoliths are centrimetric-decimetric coralline red algal nodules. They occur in bays and reefal environments, but are most extensive on current-swept platforms, in both warm and cold water, to depths of $\sim 100\,m$. In northwestern France, rhodolith gravel is termed maërl. In the Neogene, rhodoliths form matrix-rich beds up to $\sim 15\,m$ thick. Locally, stabilized rhodoliths amalgamate into rigid coralline frame reefs ('Crustose Pavements').

Oncoids (oncolites, oncoliths) are millimetric-decimetric nodules with thick stromatolitic coats. They are common in shallow marine environments during much of the fossil record, and locally form marker beds in carbonate shelf sequences. In the Cenozoic they are more common in freshwater. Oncoid cortices are often complex associations of fine-grained and clotted microfabrics together with calcimicrobes. Oncoids in present day calcareous lakes and streams are usually dominated by calcified cyanobacteria.

Robert Riding

Bibliography

Arp, G., Reimer, A., and Reitner, J., 2001. Photosynthesis-induced biofilm calcification and calcium concentrations in Phanerozoic oceans. *Science*, **292**: 1701–1704.

Pratt, B.R., 2001. Calcification of cyanobacterial filaments: *Girvanella* and the origin of lower Paleozoic lime mud. *Geology*, **29**: 763–766.

Riding, R. (ed.), 1991. *Calcareous Algae and Stromatolites*. Berlin: Springer-Verlag, 571 pp.

Riding, R., 2000. Microbial carbonates: the geological record of calcified bacterial-algal mats and biofilms. *Sedimentology*, **47** (Supplement 1): 179–214.

Riding, R., and Awramik, S.M. (eds.), 2000. *Microbial Sediments*. Berlin: Springer-Verlag, 331 pp.

Cross-references

Bacteria in Sediments
Microbially Induced Sedimentary Structures
Reefs
Stromatolites

ALLOPHANE AND IMOGOLITE

Allophane and imogolite are clay-sized minerals commonly associated with tephra deposits. They are also found in some non-tephric soils and sediments, as well as in streambeds and drains.

Definitions

Ross and Kerr (1934) showed that allophane was an X-ray amorphous material commonly associated with the clay-mineral halloysite. They suggested that "the name allophane should be restricted to mutual solid solutions of silica, alumina, water and minor amounts of bases but should include all such materials, even though the proportions of these constituents may vary." Although this definition is still generally acceptable, some changes are required to (a) exclude imogolite, (b) allow for broad X-ray diffraction lines that are shown by allophane, and (c) allow for synthetic allophanes, that may not contain bases at low pH.

The definition given by Parfitt (1990a) is "Allophane is the name of a group of clay-size minerals with short-range-order which contain silica, alumina and water in chemical combination." There are three main types of allophane: these are the aluminum-rich allophanes; the silicon-rich allophanes; and the stream-deposit allophanes.

Imogolite, a mineral made up of bundles of fine tubes, is excluded from this definition because it has long-range-order in one dimension (Farmer and Russell, 1990).

Structures

Imogolite is a tubular mineral; the walls of individual tubes are 0.7 nm in thickness; the outer surface is a gibbsite-like curved sheet, and the inner surface consists of O_3SiOH, with the oxygen replacing the inner hydroxyls of the gibbsite sheet. Allophane is made up of hollow spherules with a diameter of 4–5 nm. Al-rich allophane has an Al/Si ratio of about 2 and an imogolite-like structure. Si-rich allophane contains polymerized silicate, and has an Al/Si ratio of about 1. Stream deposit allophanes have Al/Si ratios of 0.9–1.8, with Al substituting for some Si in the polymeric tetrahedra.

Properties

Allophane and imogolite have large specific surface areas (700–1500 m^2/g) and react strongly with anions, such as phosphate and arsenate. Organic matter is also strongly bound to allophane and imogolite, decomposing only slowly in allophanic deposits. Such deposits usually have large porosity and water contents; the pores being stabilized by the positive and negative charges on the surfaces of these minerals.

Identification and estimation

In the field, allophane deposits may be identified by their characteristic greasy feel. As little as 2% allophane can be detected in this way. Allophane and imogolite can best be estimated by dissolving in acid-oxalate, and measuring concentrations of Al and Si (Parfitt, 1990a). Imogolite can be estimated by electron microscopy and differential thermal analysis (Parfitt, 1990b). However, if the sediment contains more than about 0.5% carbon, the contribution of Al from Al-humus complexes that dissolve in oxalate must be accounted for and this can be achieved by using pyrophosphate reagent (Parfitt, 1990a).

Processes of formation

The rate of formation of allophane is chiefly controlled by macro- and micro-environmental factors, together with mineralogical and physicochemical composition of the parent deposits. The effect of time is subordinate to these factors (Lowe, 1986). The activity of silicic acid, the availability of Al species, and the opportunity for co-precipitation are very important. The Si and Al are controlled by leaching, organic matter and pH. Generally, allophane forms at pH values between 5 and 7, and a pH value of at least 4.8 is required for allophane to precipitate.

Allophane is commonly found in tephra layers under humid climates, where volcanic glass dissolves to produce allophane. On the face of open soil pits containing rhyolitic tephra, allophane has been observed to precipitate within a time frame of months (Parfitt, unpublished data). The mineral can also precipitate in drains. Allophane has been found in tuffs (Silber et al., 1994), lava (Jongmans et al., 1995), lacustrine sediments (Warren and Rudolph, 1997), and Silurian sediments (Moro et al., 2000). Further, allophane is involved in the formation of indurated layers or pans in soils and sediments (Thompson et al., 1996; Wilson et al., 1996; Jongmans et al., 2000).

Imogolite is usually found accompanying allophane, but the classical pure imogolite in Japan occurs as gel films over the surface of lapilli (Yoshinaga and Aomine, 1962).

Roger L. Parfitt

Bibliography

Farmer, V.C., and Russell, J.D., 1990. The structure and genesis of allophanes and imogolite: their distribution in non-volcanic soils. In *Soil Colloids and their Associations in Soil Aggregates*. Proc. NATO Advanced Studies Workshop, Ghent, 1985, NY: Plenum.

Jongmans, A.G., Verburg, P., Nieuwenhuyse, A., and Vanoort, F., 1995. Allophane, imogolite, and gibbsite in coatings in a Costa Rican andisol. *Geoderma*, 64: 327–342.

Jongmans, A.G., Denaix, L., Vanoort, F., and Nieuwenhuyse, A., 2000. Induration of C horizons by allophane and imogolite in Costa Rican volcanic soils. *Soil Science Society of America Journal*, 64: 254–262.

Lowe, D.J., 1986. Controls on the rate of weathering and clay mineral genesis in airfall tephras: a review and New Zealand case study. In Coleman, S.M., and Dethier, D.P. (eds.), *Rates of Chemical Weathering of Rocks and Minerals*. NY: Academic Press.

Moro, M.C., Cembranos, M.L., and Fernandes, A., 2000. Allophane-like materials in the weather zones of Silurian phosphate-rich veins from Santa Creu d'Olorda (Barcelona, Spain). *Clay Minerals*, 35: 411–421.

Parfitt, R.L., 1990a. Allophane in New Zealand—A Review. *Australian Journal of Soil Research*, 28: 343–360.

Parfitt, R.L., 1990b. Estimation of imogolite in soils and clays by DTA. *Communications in Soil Science and Plant Analysis*, 21: 623–628.

Ross, C.S., and Kerr, P.F., 1934. Halloysite and allophane. *U.S. Geological Survey Professional Paper*, 185–189.

Silber, A., Baryosef, B., Singer, A., and Chen, Y., 1994. Mineralogical and chemical composition of three tuffs from Northern Israel. *Geoderma*, 63: 123–144.

Thompson, C.H., Bridges, E.M., and Jenkins, D.A., 1996. Pans in humus podzols (Humods and Aquods) in coastal southern Queensland. *Australian Journal of Soil Research*, 34: 161–182.

Warren, C.J., and Rudolf, D.L., 1997. Clay minerals in basin of Mexico lacustrine sediments and their influence on ion mobility in groundwater. *Journal of Contaminant Hydrology*, 27: 177–198.

Wilson, M.A., Burt, R., Sobecki, T.M., Engel, R.J., and Hipple, K., 1996. Soil properties and genesis of pans in till-derived Andisols, Olympic Peninsula, Washington. *Soil Science Society of America Journal*, 60: 206–218.

Yoshinaga, N., and Aomine, S., 1962. Imogolite in some Ando soils. *Soil Science and Plant Nutrition*, 8: 22–29.

Cross-references

Clay Mineralogy
Weathering, Soils and Paleosols

ALLUVIAL FANS

Introduction

Alluvial fans are depositional features, formed of coarse gravel sediments, created where high-bed-load streams enter zones of reduced stream power, and deposit the coarser fraction of their loads. The resultant landforms are usually fan-shaped in plan and wedge-shaped in profile (Bull, 1977). They occur commonly in two topographic situations: at mountain fronts and at tributary junctions (Harvey, 1997). They are subaerial features, however if they extend into water they are known as fan-deltas.

Sediment transport and deposition processes on alluvial fans range from debris flows, to sheetfloods and channelized streamflows. Fans range in axial length up to 10s of km, though many fans described in the literature range in size between ca. 100 m and a few km. As scale increases there is a tendency for the dominant process to change, from small debris-flow dominated tributary-junction fans (debris cones), to mixed processes and sheetflood dominance at intermediate scales, and dominance by channelized fluvial flows at the largest scales.

Alluvial fans are important features within mountain fluvial systems, acting as sediment stores. They preserve a sedimentary record of environmental change, and act as major controls on the downstream fluvial system, often breaking the coupling between sediment source areas and distal fluvial environments (Harvey, 1997).

The published work deals with both the sedimentology and the geomorphology of alluvial fans, and to some extent the two are interdependent, however there are differences of emphasis. Much of the earlier work dealt with fans at faulted mountain fronts in the American southwestern deserts, hence there is an emphasis on fans as desert features and on tectonic controls. This emphasis occurs especially in the sedimentological literature, in relation to ancient fan sediments, perhaps because the preservation potential is high for fan sediments deposited at the margins of subsiding basins. However, fans do occur in many climatic environments, and in tectonically stable situations, in which case climate is often seen as a major control. This is perhaps the primary emphasis in the geomorphological literature, in relation to Quaternary and modern fans.

Historical context

Numerous studies in the American southwest provide the basis for our knowledge. Important papers by Blissenbach (1954), dealing with fan sedimens, by Denny (1965) dealing with the Death Valley fans, and by Hooke (1967) linking fan processes, sediments and morphology were followed by a series of papers by Bull (for details see Bull, 1977), examined the relations between tectonics, fan morphology and sediments. Research on modern fans has continued in the American southwest and has developed in other regions (see Rachocki and Church, 1990), in other dry regions (e.g., Wasson, 1979; Harvey, 1990; Nemec and Postma, 1993; Gerson et al., 1993), in arctic and alpine regions (e.g., Leggett et al., 1966; Ritter and Ten Brink, 1986), in humid temperate regions (e.g., Harvey and Renwick, 1987; Kochel, 1990) and in the humid tropics (e.g., Kesel and Lowe 1987).

Fan sedimentary sequences have been incorporated into the fluvial sediment sequence models of Miall (1977, 1978) and are also described in standard sedimentology texts (e.g., Reineck and Singh, 1973). Studies of alluvial fan sedimentary sequences, described from terrestrial environments throughout the geological timescale, are far too numerous to deal with in detail here. A comprehensive bibliography is given by Nilsen and Moore (1984).

Important studies have attempted to relate fan sedimentary sequences with fan processes (Bull, 1972; Blair and McPherson, 1994), especially in relation to individual flood events (Wells and Harvey, 1987; Blair and McPherson, 1998) and to infer catchment characteristics from past fan sediments (Hirst and Nichols, 1986; Mather et al., 2000). For reviews of the alluvial fan literature, with emphasis on sedimentology see Nilsen (1982), Nilsen and Moore (1984) and Blair and McPherson (1994), and with emphasis on geomorphology see Bull (1977) and Harvey (1997).

Essential concepts: fan processes and sediments

There is a close two-way link between alluvial fan morphology and sedimentology. Fans are products of deposition and therefore the morphology reflects the depositional style. For example, fan gradients are higher for deposition by debris-flows than by fluvial processes (Kostaschuk et al., 1986), and higher for sheetflood than for channelized fluvial deposition (Blair and McPherson, 1994). Similarly, deposition is conditioned by fan morphology, for example, by gradient and flow confinement. Because of this close relationship, fan morphological properties show relationships with the same catchment characteristics that control water and sediment delivery to the fan. Fan size generally shows a direct geometric relationship, and fan gradient an inverse geometric relationship to catchment area (Bull, 1977; Harvey, 1997).

In more detail, several fan styles can be identified, reflecting the relationships between erosion and deposition in fan environments. A non-trenched aggrading fan is one on which sedimentation takes place from the apex downfan. More common are telescopic or prograding fans, where the feeder channel is incised into the apex zone as a fanhead trench, but emerges onto the fan surface at a midfan intersection point, beyond which deposition occurs. Finally are various styles of dissecting fans, with dissection focused at the apex, in midfan or distally. The extreme would be a fan trenched throughout its length (Harvey, 1997).

Several groups of sedimentary processes are important: the primary debris-flow and fluvial processes delivering sediment to the fan, the secondary fluvial, and in arid areas eolian, processes reworking or eroding the sediment, and the tertiary pedogenic processes modifying the fan surface.

The primary processes depend on the water:sediment mix fed to the fan from the catchment during flood events (Wells and Harvey, 1987), and to a certain extent on the particle-size characteristics of the fine sediment. Debris-flow processes operate as sediment-rich flows, especially with a silt or clay matrix (Suwa and Okuda, 1983). Under greater dilution flows become transitional or hyperconcentrated and at higher dilution become fluvial flows. Debris flows are most common where sediment concentrations are high, for example, from small, steep catchments (Kostaschuk et al., 1986; Wells and Harvey, 1987). Fluvial flows may operate within channels or as sheetflows. Channelized flows would be more common in proximal fan environments especially where the feeder channel is in a fanhead trench, but may also occur throughout, especially on larger fans. On distal fan surfaces, channels may switch courses by avulsion, but sheetfloods are more common on the distal surfaces of desert fans (Blair and McPherson, 1994).

Fan sediments are coarse and texturally immature, having been deposited during floods either as debris flows or as fluvial deposits fed by steep mountain streams. Common sedimentary facies include matrix-supported gravels (Gms: after Miall, 1978), deposited by debris flows, and clast-supported gravels (Gm, Gt, Gp), deposited by both channelized fluvial flows and sheetfloods. Sandy facies and fines are of less importance except in distal environments. Debris flows will be represented by massive boulder- to cobble-conglomerates, with true debris-flow deposits exhibiting matrix-support and an internal fabric of crude oblique-to-vertical clast alignment resulting from compression and internal shear (Wells and Harvey, 1987; Blair and McPherson, 1998). Transitional deposits are usually massive, poorly sorted, clast-supported (fanglomerate) gravels, with less matrix than true debris-flow deposits. They have little or no internal structure or may exhibit weak bedding. Fluvial deposits within alluvial fans range from thin sheet gravels, often showing coarse and fine couplets, associated with maximum and waning flows (Blair and McPherson, 1994). They may show relatively little basal scour, in contrast with gravels deposited by channelized flows. These deposits will show characteristics of stream deposits, scoured bases, bedding (planar and locally cross-bedding), moderate sorting and often weak imbrication, though they are likely to show less order than perennial braided stream deposits, due to limited reworking and reorganization by moderate flows.

Because fans are generally flood-event related, receiving sediment only occasionally, and because sedimentation takes place on only parts of the fan, most of the fan surface undergoes weathering and soil formation for most of the time. This is accentuated on surfaces incised by a fanhead trench. Soils and paleosols are therefore important constituents of fans. They are especially useful for correlation and dating purposes, and have been used in this way on Quaternary fans in humid regions (e.g., Harvey, 1996a), in tropical areas (Kesel and Spicer, 1985), but above all in arid and semiarid regions (Wells et al., 1987; McFadden et al., 1989; Bull, 1991). Several aspects of desert soils on fan surfaces have been shown to change progressively with age, including desert pavement surfaces (Dan et al., 1982; McFadden et al., 1987; Al-Farraj and Harvey, 2000), soil B-horizon color, Fe-oxide and mineral magnetic properties (White and Walden, 1997; Harvey et al., 1999a), and pedogenic carbonate and calcrete (caliche) characteristics (Lattman, 1973; Machette, 1985; Alonso Zarza et al., 1998). There are implications for interpreting ancient as well as Quaternary fan sequences (Wright and Alonso Zarza, 1990).

The spatial and vertical variations of fan sediments depend on the relationships between fan morphology and processes. On many intermediate-sized fans debris flows may not travel much further than midfan. Sheetflood processes will occur only on non-dissected fan surfaces or on distal surfaces beyond the intersection point. Under both sheetflow and channelized fluvial flows transporting power may diminish downfan. These three trends lead to spatial patterns in sediment properties, with a tendency for proximal-to-distal variation in the relative importance of sediment types (Harvey, 1997) and an overall tendency for a downfan decrease in sediment size. Laterally there are also likely to be spatial variations in sediment types in relation to the position of the axial channel and the probability of sediment reworking.

Such trends are likely to be expressed in vertical sections with differences between sequences from small or intermediate-sized fans and those from larger fans, and differences between proximal and distal environments (Harvey, 1997). In the proximal zones of smaller fans, sequences are likely to show rapid vertical variations between debris-flow and fluvial channel deposits (the Trollheim type of Miall, 1978), but in distal zones debris flows will be rarer and thin sheetflood sediments more common. On fluvially dominant fans (with sequences of the Scott type of Miall, 1978) alternations between channel and sheetflood sediment are common. In both cases paleosols could be important horizons.

In fan distal zones sediments may exhibit simple stacked stratigraphy, but proximal zones prone to fanhead trenching, may exhibit inset stratigraphy, preserving deep incisions of buried fanhead trenches (Harvey, 1987). Indeed the presence of fanhead trenches produces two contrasting models of fan sedimentary sequences. On non-trenched aggrading fans an overall fining sequence would be expected, associated with the progressive burial of the mountain-front topography, whereas on fans undergoing fanhead trenching sediment sequences would be expected to show overall coarsening as formerly proximal deposits are reworked distally (Heyward, 1978; Steel et al., 1977).

Essential concepts: controls on fan development

Two main groups of controls influence the development of alluvial fans, those relating to the setting of the fan and those related to processes. Those related to the setting include the gross topography of the site, governed by tectonics and long-term geomorphic history (e.g., glaciation), and control the accommodation space. Those related to processes include the delivery of water and sediment from the mountain catchment, its transport to and deposition on the fan, and the potential for erosion of the fan surface. These processes are in turn controlled by the size, relief and bedrock geology of the catchment, by the climate, and by the fan morphology itself, including its relationship to local base levels.

These factors can be grouped into four partially related sets, any of which may be subject to change within the timescale of the existence of an alluvial fan. These sets are: (i) source area relief and geology, (ii) tectonic factors, (iii) climatic factors, and (iv) base level. Fan processes may respond to changes in these factors—the changes being recorded within the fan morphology and sediment sequences.

Source-area relief and geology influence the sediment yield of the catchment, but may change little over the short term. However, changes may occur in response to headwater stream capture (Mather et al., 2000) causing sudden changes in water and sediment supply, or to long-term changes in sediment availability related to progressive erosion of the source area.

Past tectonics may have a major influence on the fan setting. Ongoing tectonics may cause uplift of the source area and an increase in sediment production, or may cause a change in fan gradients by tilting. At the regional scale uplift-induced dissection may cause a change in base level, triggering fan-toe erosion (see below). However the most significant role of tectonics for basin-margin fans is the ongoing creation of accommodation space brought about by continued basin subsidence. Within the literature on Quaternary fans, tectonics has been seen as an important control on fan sedimentation (Calvache *et al.*, 1997) and on geomorphology (Bull, 1961, 1978; Silva *et al.*, 1992), but in the sedimentological literature as a whole is often seen as the primary control over fan sediment sequences in the ancient record (e.g., Sharp, 1948; Steel, 1974; Heyward, 1978).

Climate influences sediment availability within the catchment through weathering processes, and sediment delivery from hillslopes to channels through slope failure and slope erosion processes. Climate also influences both water and sediment supply to the fan through its influence on the flood hydrology of the stream system. Processes on the fan and the resultant morphology respond to climatically led sediment supply and flood hydrology regimes. The relationships between debris-flow and fluvial processes depend on the water-sediment mix during flood events (Wells and Harvey, 1987), and the erosional-depositional regime as a whole depends on the threshold of critical stream power (Bull, 1979), itself governed by the water-sediment regime. While it is recognized that alluvial fans may occur in any climatic environment, a climatic change may modify these relationships, resulting in changes in fan dynamics. Significant changes in erosion-sedimentation regimes resulting from Quaternary climatic changes have been identified in many dry regions including the American southwest (Bull 1991; Wells *et al.*, 1987; Dorn, 1994; Harvey *et al.*, 1999b), and the drier parts of the Mediterranean region (Roberts, 1995; Harvey, 1996a). Similarly in mountain areas glaciated during the Pleistocene, fan deposition has been identified as a paraglacial phenomenon, supplied with sediment from previously glaciated catchments (Ryder, 1971; Brazier *et al.*, 1988).

Most fans accumulate under relatively stable base-level conditions. However, a base-level change may cause a switch from deposition to erosion in fan-toe zones, which may progressively dissect the fan headwards. Base-level changes may be induced tectonically or as secondary effects of climatic change. These may be important on coastal fans following Quaternary eustatic sea-level changes, or on fans at the margins of pluvial lakes following changes in lake level. Similarly a base-level change may affect tributary-junction fans following climatically-induced incision of an axial drainage. However, the effects of base-level change are not necessarily straightforward. In the common case of a fall of base level, dissection will occur only if the newly exposed gradients are sufficient to trigger incision. On the other hand, a regional sea-level rise may cause fan dissection if it is accompanied by coastal erosion and fan-profile foreshortening (see Harvey *et al.*, 1999a). The effects are similar to erosional "toe-cutting" by a laterally migrating axial drainage (Leeder and Mack, 2001).

Interactions between tectonic, climatic and base-level factors form an ongoing discussion in the literature. The consensus is that, at least for Quaternary fans, climate appears to have the primary role (Bowman, 1978; Frostick and Reid, 1989; Ritter *et al.*, 1995).

Modern trends in alluvial fan research

In addition to considerations of interactions between the main groups of causal factors, several other trends in modern alluvial fan research can be identified. There has been detailed study of the relationships between the fundamental processes that affect alluvial fans and the properties of the sediments preserved. This is especially true of debris flows (see Costa, 1988; Blair and McPherson, 1998). Such an understanding provides the basis for a clearer interpretation of ancient fan sediments, related to knowledge of modern fan processes and sedimentology.

In relating Quaternary fan sequences to the climatic sequence, the interdisciplinary nature of such studies is becoming increasingly apparent (Harvey *et al.*, 1999b). Fundamental here are advances in dating and related techniques, for example the application of cosmogenic dating techniques (Harbor, 1999), and more sophisticated analyzes of soils (e.g., White and Walden, 1997) on alluvial fan surfaces will provide a much tighter framework for alluvial fan chronologies and environmental reconstruction.

Furthermore, the interdependence of geomorphology and sedimentology in alluvial fan research is becoming even more apparent. This is true not only for understanding the dynamics of alluvial fans themselves (e.g., Leeder and Mack, 2001), but also in relating alluvial fans to their source areas, where knowledge gained from the study of Quaternary and modern fans is being applied to interpreting older fan sequences (e.g., Mather *et al.*, 2000).

Summary

Alluvial fans are important sedimentary environments that are fundamental in a coupling/buffering role within mountain (especially but not exclusively dry-region) geomorphic systems. They preserve a sedimentary record, on modern fans of drainage-basin response to Quaternary climatic change, and in the longer term of the tectonic controls over sediment supply to sedimentary basins. There is a close two-way interaction between geomorphic and sedimentological processes on alluvial fans. Fan sedimentology and morphology respond to tectonic, climatic and base-level controls, and as such express the fundamental dynamics of geomorphic and sediment systems.

Adrian M. Harvey

Bibliography

Al Farraj, A., and Harvey, A.M., 2000. Desert pavement characteristics on wadi terrace and alluvial fan surfaces: Wadi Al-Bih, UAE and Oman. *Geomorphology*, **35**: 279–297.

Alonso-Zarza, A.M., Silva, P.G., Goy, J.L., and Zazo, C., 1998. Fan-surface dynamics and biogenic calcrete development: Interactions during ultimate phases of fan evolution in the semiarid SE Spain (Murcia). *Geomorphology*, **24**: 147–167.

Blair, T.C., and McPherson, J.G., 1994. Alluvial fan processes and forms. In Abrahams, A.D., and Parsons, A.J. (eds.), *Geomorphology of Desert Environments*. London: Chapman and Hall, pp. 354–402.

Blair, T.C., and McPherson, J.G., 1998. Recent debris-flow processes and resultant form and facies of the Dolomite alluvial fan, Owens Valley, California. *Journal of Sedimentary Research*, **68**: 800–818.

Blissenbach, E., 1954. Geology of alluvial fans in semi-arid regions. *Geological Society of America, Bulletin*, **65**: 175–190.

Bowman, D., 1988. The declining but non-rejuvenating base-level – the Lisan Lake, the Dead Sea, Israel. *Earth Surface Processes and Landforms*, **13**: 239–249.

Brazier, V., Whittington, G., and Ballantyne, C.K., 1988. Holocene debris cone evolution in Glen Etive, Western Grampian Highlands, Scotland. *Earth Surface Processes and Landforms*, **13**: 525–531.

Bull, W.B., 1961. Tectonic significance of radial profiles of alluvial fans in western Fresno County, California. *United States Geological Survey*. Professional Paper 424B, pp. 182–184.

Bull, W.B., 1972. Recognition of alluvial-fan deposits in the stratigraphic record. In Rigby, J.K., and Hamblin, W. (eds.),

Recognition of Ancient Sedimentary Environments. Society of Economic Palaeontologists and Mineralogists, Sp. Publ. 16: pp. 63–83.

Bull, W.B., 1977. The alluvial fan environment. *Progress in Physical Geography*, **1**: 222–270.

Bull, W.B., 1978. Geomorphic tectonic activity classes of the south front of the San Gabriel Mountains, California. *United States Geological Survey. Contract Report* 14-08-001-G-394, Menlo Park, California: Office of Earthquakes, Volcanoes and Engineering, 59pp.

Bull, W.B., 1979. Threshold of critical power in streams. *Geological Society of America, Bulletin*, **90**: 453–464.

Bull, W.B., 1991. *Geomorphic Responses to Climatic Change*. Oxford: Oxford University Press.

Calvache, M., Viseras, C., and Fernandez, J., 1997. Controls on alluvial fan development – evidence from fan morphometry and sedimentology; Sierra Nevada, SE Spain. *Geomorphology*, **21**: 69–84.

Costa, J.E., 1988. Rheologic, geomorphic, and sedimentologic differentiation of water floods, hyperconcentrated flows, and debris flows. In Baker, V.R., Kochel, R.C., and Patten, P.C. (eds.), *Flood Geomorphology*. New York: Wiley, pp. 113–122.

Dan, J., Yaalon, D.H., Moshe, R., and Nissim, S., 1982. Evolution of Reg soils in southern Israel and Sinai. *Geoderma*, **28**: 173–202.

Denny, C.S., 1965. Alluvial fans in Death Valley region, California and Nevada. *United States Geological Survey. Professional Paper* **466**: 59p.

Dorn, R.I., 1994. The role of climatic change in alluvial fan development. In Abrahams, A.D., and Parsons, A.J., (eds.), *Geomorphology of Desert Environments*. London: Chapman and Hall, pp. 593–615.

Frostick, L.E., and Reid, I., 1989. Climatic versus tectonic controls of fan sequences: lessons from the Dead Sea, Israel. *Journal of the Geological Society, London*, **146**: 527–538.

Gerson, R., Grossman, S., Amit, R., and Greenbaum, N., 1993. Indicators of faulting events and periods of quiescence in desert alluvial fans. *Earth Surface Processes and Landforms*, **18**: 181–202.

Harbour, J. (ed.), 1999. Cosmogenic isotopes in Geomorphology, Special Issue *Geomorphology*, **27**: 1–172.

Harvey, A.M., 1987. Alluvial fan dissection: relationships between morphology and sedimentation. In Frostick, L., and Reid, I. (eds.), *Desert sediments, ancient and modern*. Geological Society of London, Special Publication 35, Oxford: Blackwell, pp. 87–103.

Harvey, A.M., 1990. Factors influencing Quaternary alluvial fan development in southeast Spain. In Rachocki, A.H., and Church, M. (eds.), *Alluvial Fans: A Field Approach*. Chichester: Wiley, pp. 247–269.

Harvey, A.M., 1996a. Holocene hillslope gully systems in the Howgill Fells, Cumbria. In Anderson, M.G., and Brooks, S.M. (eds.), *Advances in Hillslope Processes*, Volume 2. Chichester: Wiley, pp. 731–752.

Harvey, A.M., 1996b. The role of alluvial fans in the mountain fluvial systems of southeast Spain: implications of climatic change. *Earth Surface Processes and Landforms*, **21**: 543–553.

Harvey, A.M., 1997. The role of alluvial fans in arid zone fluvial systems. In Thomas, D.S.G. (ed.), *Arid Zone Geomorphology: Process, Form and Change in Drylands*, 2nd edn., Chichester: Wiley, pp. 231–259.

Harvey, A.M., and Renwick, W.H., 1987. Holocene alluvial fan and terrace formation in the Bowland Fells, northwest England. *Earth Surface Processes and Landforms*, **12**: 249–257.

Harvey, A.M., Silva, P.G., Mather, A.E., Goy, J.L., Stokes, M., and Zazo, C., 1999a. The impact of Quaternary sea-level and climatic change on coastal alluvial fans in the Cabo de Gata ranges, southeast Spain. *Geomorphology*, **28**: 1–22.

Harvey, A.M., Wigand, P.E., and Wells, S.G., 1999b. Response of alluvial fan systems to the late Pleistocene to Holocene climatic transition: contrasts between the margins of pluvial Lakes Lahontan and Mojave, Nevada and California, USA. *Catena*, **36**: 255–281.

Heyward, A.P., 1978. Alluvial fan sequence and megasequence models, with examples for Westphalian D – Stephanian B coalfields, northern Spain. In Miall, A.D. (ed.), *Fluvial Sedimentology*. Memoir: Canadian Society of Petroleum Geologists, **5**: pp. 669–702.

Hirst, J.P.P., and Nichols, G.J., 1986. Thrust tectonic controls on Miocene alluvial distribution patterns, southern Pyrennees. *International Association of Sedimentologists*. Special Publication **8**: pp. 247–258.

Hooke, R. le B., 1967. Processes on arid region alluvial fans. *Journal of Geology*, **75**: 438–460.

Kesel, R.H., and Lowe, D.R., 1987. Geomorphology and sedimentology of Toro Amarillo alluvial fan in a humid tropical environment, Costa Rica. *Geografiska Annaler*, **69A**: 85–99.

Kesel, R.H., and Spicer, B.E., 1985. Geomorphic relationships and ages of soils on alluvial fans in the Rio General valley, Costa Rica. *Catena*, **12**: 149–166.

Kochel, R.C., 1990. Humid fans of the Appalachian Mountains. In Rachocki, A.H. and Church, M. (eds.), *Alluvial Fans: A Field Approach*. Chichester: Wiley, pp. 109–129.

Kostaschuk, R.A., MacDonald, G.M., and Putnam, P.E., 1986. Depositional processes and alluvial fan – drainage basin morphometric relationships near Banff, Alberta, Canada. *Earth Surface Processes and Landforms*, **11**: 471–484.

Lattman, L.H., 1973. Calcium carbonate cementation of alluvial fans in southern Nevada. *Geological Society of America, Bulletin*, **84**: 3013–3028.

Leeder, M.R., and Mack, G.H., 2001. Lateral erosion ("toe-cutting") of alluvial fans by axial rivers: implications for basin analysis and architecture. *Journal of the Geological Society, London*, **158**: 885–893.

Leggett, R.F., Brown, R.J.E., and Johnston, G.H., 1966. Alluvial fan formation near Aklavik, Northwest Territories, Canada. *Geological Society of America, Bulletin*, **77**: 15–30.

Machette, M.N., 1985. Calcic soils of the southwestern United States. In Weide, D.L. (ed.), *Soils and Quaternary Geology of the Southwestern United States*. Geological Society of America Special paper 203, pp. 1–21.

Mather, A.E., Harvey, A.M., and Stokes, M., 2000. Quantifying long-term catchment changes of alluvial fan systems. *Geological Society of America, Bulletin*, **112**: 1825–1833.

McFadden, L.D., Wells, S.G., and Jercinovich, M.J., 1987. Influences of eolian and pedogenic processes on the origin and evolution of desert pavements. *Geology*, **15**: 504–508.

McFadden, L.D., Ritter, J.B., and Wells, S.G., 1989. Use of multiparameter relative-age methods for age estimation and correlation of alluvial fan surfaces on a desert piedmont, eastern Mojave Desert, California. *Quaternary Research*, **32**: 276–290.

Miall, A.D., 1977. A review of the braided river depositional environment. *Earth Science Reviews*, **13**: 1–62.

Miall, A.D., 1978. Lithofacies types and vertical profile models in braided river deposits: a summary. In Miall, A.D. (ed.), *Fluvial Sedimentology*. Memoir: Canadian Society of Petroleum Geologists 5, pp. 597–604.

Nemec, W., and Postma, G., 1993. Quaternary alluvial fans in southwest Crete: sedimentation processes and geomorphic evolution. In Marzo, M. and Puigdefabrigas, C. (eds.), *Alluvial sedimentation*. International Association of Sedimentologists, Special Publication 17, pp. 235–276.

Nilsen, T.H., 1982. Alluvial fan deposits. In Scholle, P.A. and Spearing, P. (eds.), *Sandstone Depositional Environments*. Memoir: American Association of Petroleum Geologists 31: pp. 49–86.

Nilsen, T.H., and Moore, T.E., 1984. Bibliography of Alluvial-fan Deposits. Norwich: Geobooks, 96p.

Rachocki, A.H., and Church, M. (eds.), 1990. *Alluvial Fans: A Field Approach*. Chichester: Wiley, 391p.

Reinech, H.E., and Singh, I.B., 1973. *Depositional sedimentary environments with reference to terrigenous classics*. 3rd edn., Berlin: Springer Verlag, 1983. 549p.

Ritter, D.F., and Ten Brink, N.W., 1986. Alluvial fan development and the glacio-glaciofluvial cycle, Nemana Valley, Alaska. *Journal of Geology*, **94**: 613–625.

Ritter, J.B., Miller, J.R., Enzel, Y., and Wells, S.G., 1995. Reconciling the roles of tectonism and climate in Quaternary alluvial fan evolution. *Geology*, **23**: 245–248.

Roberts, N., 1995. Climatic forcing of alluvial fan regimes during the Late Quaternary in Konya basin, south central Turkey. In Lewin,

J., Macklin, M.G., and Woodward, J. (eds.), *Mediterranean Quaternary River Environments*. Rotterdam: Balkema, pp. 205–217.

Ryder, J.N., 1971. The stratigraphy and morphology of paraglacial alluvial fans in south central British Columbia. *Canadian Journal of Earth Sciences*, **8**: 279–298.

Sharp, R.P., 1948. Early Tertiary fanglomerate, Bighorn Mountains, Wyoming. *Journal of Geology*, **56**: 1–15.

Silva, P.G., Harvey, A.M., Zazo, C., and Goy, J.L., 1992. Geomorphology, depositional style and morphometric relationships of Quaternary alluvial fans in the Guadalentin depression (Murcia, southeast Spain). *Zeitschrift fur Geomorphologie*, N.F. **36**: 325–341.

Steele, R.J., 1974. New Red Sandstone floodplain and piedmont sedimentation in he Hebridean province, Scotland. *Journal of Sedimentary Petrology*, **44**: 336–357.

Steel, R.J., Moehle, S., Nilsen, H., Roe, S.L., and Spinnangr, A. 1977. Coarsening upwards cycles in the alluvium of Homelen Basin (Devonian), Norway: Sedimentary response to tectonic events. *Geological Society of America, Bulletin*, **88**: 1124–1134.

Suwa, H., and Okuda, S., 1983. Deposition of debris flows on a fan surface, Mt. Yakedale, Japan. *Zeitschrift fur Geomorphologie*. Supplementband **46**: 79–101.

Wasson, R.J., 1979. Sedimentation history of the Mundi Mundi alluvial fans, western New South Wales. *Sedimentary Geology*, **22**: 21–51.

Wells, S.G., and Harvey, A.M., 1987. Sedimentologic and geomorphic variations in storm generated alluvial fans, Howgill Fells, northwest England. *Geological Society of America, Bulletin*, **98**: 182–198.

Wells, S.G., McFadden, L.D., and Dohrenwend, J.C., 1987. Influence of late Quaternary climatic change on geomorphic and pedogenic processes on a desert piedmont, eastern Mojave Desert, California. *Quaternary Research*, **27**: 130–146.

White, K., and Walden, J., 1997. The rate of iron oxide enrichment in arid zone alluvial fan soils, Tunisian Southern Atlas, measured by mineral magnetic techniques. *Catena*, **30**: 215–227.

Wright, V.P., and Alonso Zarza, A.M., 1990. Pedostratigraphic models for alluvial fan deposits: a tool for interpreting ancient sequences. *Journal of the Geological Society, London*, **147**: 8–10.

Cross-references

Anabranching Rivers
Caliche–Calcrete
Climatic Control of Sedimentation
Debris Flow
Fan Delta
Grain Size and Shape
Gravity-Driven Mass Flows
Sediment Transport by Unidirectional Water Flows
Tectonic Controls of Sedimentation
Weathering, Soils, and Paleosols

ANABRANCHING RIVERS

Introduction

An *anabranching* river is defined as a system of multiple channels characterized by vegetated or otherwise stable alluvial islands that divide flows at discharges up to bankfull. These islands may be excised by channel avulsion from extant floodplain (see Avulsion), developed from within-channel deposition or formed by prograding distributary-channel accretion on splays or deltas. Common usage has confined the related term *anastomosing* to a specific subset of relatively distinctive low-energy anabranching systems associated with mostly fine-grained or organic deposition (Smith and Smith, 1980; Knighton and Nanson, 1993; Makaske, 2001). Neither of these terms now applies to braided rivers (see Rivers and Alluvial Fans) where the divided flow pattern is strongly stage-dependent around bars that are unconsolidated, ephemeral, poorly vegetated and overtopped at less than bankfull flow. In an anabranching system, the islands are about the same elevation as the adjacent floodplain, usually persist for decades or centuries, have relatively resistant banks, and support well-established vegetation. Anabranching always occurs concurrently with other patterns such that individual channels braid, meander or are straight. Such rivers occupy a wide range of environments, from low to high energy, and occur in arctic, alpine, temperate, humid tropical and arid climatic settings. They are more common than has been recognized previously; a total of 90% by length of the alluvial reaches of the world's ten largest rivers anabranch, and it is a particularly widespread river pattern for both large and small rivers in inland Australia. Numerous rivers in Europe used to anabranch but most of these have now been modified to more "convenient" single-thread forms in densely populated and heavily utilized valleys.

While research continues into the fundamental cause of anabranching, it has been argued that an advantage of anabranching is that islands concentrate stream flow and maximize bed-sediment transport per unit of stream power, particularly where there is little or no opportunity to increase channel gradient (Nanson and Huang, 1999).

Classification

On the basis of stream energy, sediment size and morphological characteristics, Nanson and Knighton (1996) recognize six types of anabranching river; Types 1–3 are lower energy and Types 4–6 are higher energy systems. Type 1 are *Cohesive Sediment* rivers (commonly termed anastomosing) with low w/d ratio channels that exhibit little or no lateral migration. They are divisible into three subtypes based on vegetative and sedimentary environment (Figure A3a). Type 2 are *Sand Dominated Island Forming* rivers and Type 3 are *Mixed Load Laterally Active* meandering rivers. Type 4 are *Sand Dominated Ridge Forming* rivers characterized by long, parallel channel-dividing ridges (Figure A3b). Type 5 are *Gravel Dominated Laterally Active* systems that interface between meandering and braiding in mountainous regions (Figure A3c). These have been described as wandering gravel-bed rivers (Church, 1983). Type 6 are *Gravel Dominated Stable* systems that occur as non-migrating channels in small, relatively steep basins.

Anastomosing rivers

Because of their fine-grained nature and tendency to accumulate substantial organic material, anastomosing rivers are a potentially economically important group of anabranching rivers. Modern examples were first described in detail in the alpine and humid environment of the Rocky Mountains of western Canada (e.g., Smith, 1973; Smith and Smith, 1980) but have subsequently been described in a wide variety of settings including arid environments (e.g., Gibling *et al.*, 1998; Makaske, 2001). In rapidly accreting humid settings, peats can accumulate in floodplain lakes and swamps to form coal, and sandy paleochannels may act as reservoirs for hydrocarbons. However, not all anabranching rivers are rapidly vertically accreting and in arid environments they do not accumulate organics. Makaske (2001) found no standard sedimentary

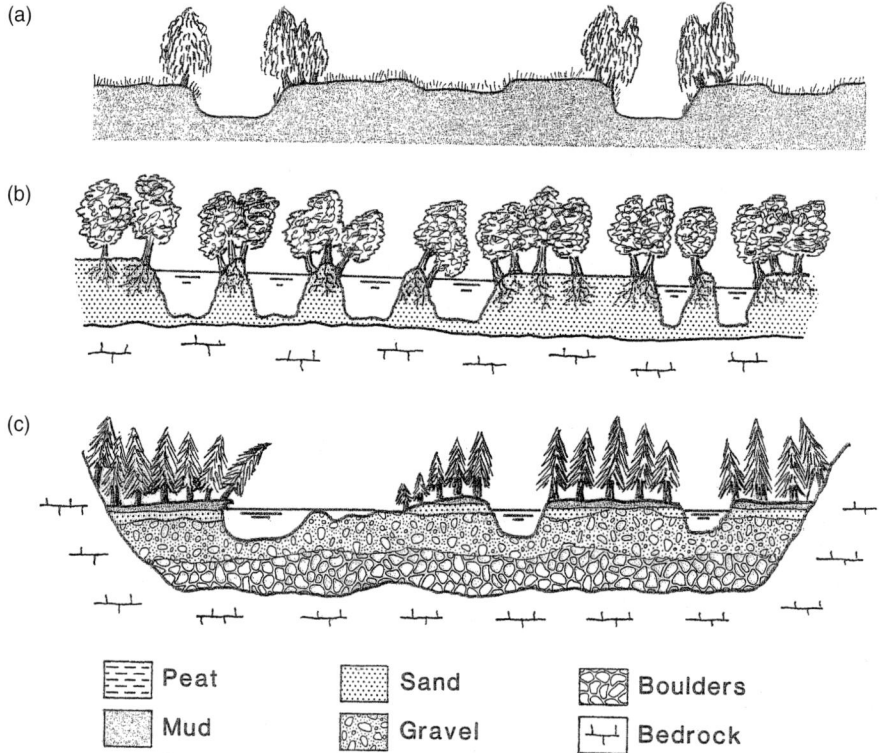

Figure A3 Examples from the range of anabranching river stratigraphy (a) Cohesive-sediment anabranching river; (b) Sand-dominated ridge-forming anabranching river; (c) Gravel-dominated laterally active anabranching river (after Nanson and Knighton, 1996).

succession for anastomosing rivers, however, he described them in three different settings and showed some common characteristics. His schematic diagram of the Columbia River is presented here as an example of the style of stratigraphy in a rapidly vertically-accreting humid montane setting with organic-clastic accumulation (Figure A4). Such anastomosing rivers (and delta distributary channels) tend to have fixed channels that aggrade with only limited lateral expansion, thus generating ribbons or narrow sheets (width : thickness ratio <15 and 15–100, respectively) encased in floodplain fines. Most fixed-channel bodies are less than 10 m thick, with width : thickness ratios typically less than 30 and commonly less than 15. Many examples in the rock record *probably* represent long-lived anastomosing rivers that formed trunk drainage systems or transverse alluvial cones. Many of these deposits lie in rapidly subsiding settings, especially foreland and extensional basins characterized by large sediment flux and low gradients. Deltaic distributary channel bodies, many probably formed from relatively straight channels, typically comprise fills of mixed sand and mud, with marine- or brackish-water fossils, slumped masses, and a common association with coal. The avulsion of a major channel into wetlands may generate a splay complex with suites of small, transient anastomosing channels, with the eventual establishment of a stable, single-channel course. In arid environments, alluvial and eolian deposits can be juxtaposed, whereas in vertically accreting humid environments channel fills are flanked by silty levee deposits, lacustrine clay and coal.

There is a recent tendency to attribute all ribbon bodies in the rock record to anastomosing river systems, however, some could also be assigned to the more general *fixed-channel model* (Friend, 1983) because it is difficult to show that the original channels formed a synchronous anastomosing network (Makaske, 2001). Indeed, in a review of fluvial facies models, Hickin (1993) suggested that the anastomosing model is a somewhat premature concept based on the study of relatively few rivers, the full development of which should await detailed investigation of a much wider range of types.

Conclusion

Anabranching rivers represent a diverse group from low energy, organic or fine sediment-textured, to relatively high energy gravel-transporting systems. They are commonly associated with flood-dominated flow regimes, banks that are resistant to erosion, and sometimes with mechanisms to block or constrict channels and induce channel avulsion. They can develop as erosional systems that scour channels into the floodplains or accretional systems that build islands within, or floodplains around (as in subaqueous deltas), existing channels. Anabranching is commonly associated with laterally stable channels but individual channels can meander, braid or be straight. Anabranching rivers represent a widespread and distinctive group that, because of particular sedimentary, energy-gradient, and other hydraulic conditions, operate most

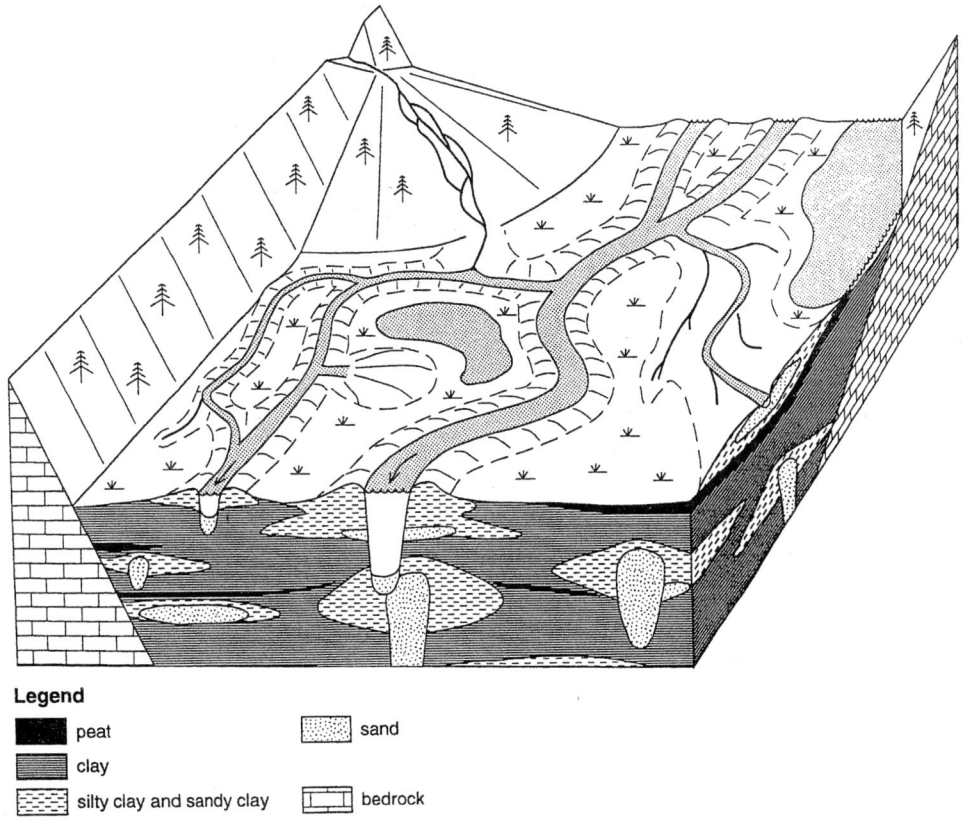

Figure A4 Textural facies model of the upper Columbia River (British Columbia, Canada), a rapidly aggrading anastomosing system in a temperate humid montane setting. Scale is approximately 2 km in width and alluvial thickness ~10 m (after Makaske, 2001).

effectively as a system of multiple channels separated by vegetated floodplain islands.

Gerald C. Nanson and Martin R. Gibling

Bibliography

Church, M., 1983. Anastomosed fluvial deposits: modern examples from Western Canada. In Collinson, J.D., and Lewin, J. (eds.), *Modern and Ancient Fluvial Systems*. International Association of Sedimentologists, Special Publication 6, pp. 155–168.

Friend, P.F., 1983. Towards the field classification of alluvial architecture or sequence. In Collinson, J.D., and Lewin, J. (eds.), *Modern and Ancient Fluvial Systems*. International Association of Sedimentologists, Special Publication, 6, pp. 345–354.

Gibling, M.R., Nanson, G.C., and Maroulis, J.C., 1998. Anastomosing river sedimentation in the Channel Country of central Australia. *Sedimentology*, **45**: 595–619.

Hickin, E.J., 1993. Fluvial facies models: a review of Canadian research. *Progress in Physical Geography*, **17**: 205–222.

Knighton, A.D., and Nanson, G.C., 1993. Anastomosis and the continuum of channel pattern. *Earth Surface Processes and Landforms*, **18**: 613–625.

Makaske, B., 2001. Anastomosing rivers: a review of their classification, origin and sedimentary products. *Earth-Science Reviews*, **53**: 149–196.

Nanson, G.C., and Huang, H.Q., 1999. Anabranching rivers: divided efficiency leading to fluvial diversity. In Miller, A.J., and Gupta, A. (eds.), *Varieties of Fluvial Form*. Chichester: Wiley, pp. 219–248.

Nanson, G.C., and Knighton, A.D., 1996. Anabranching rivers: their cause, character and classification. *Earth Surface Processes and Landforms*, **21**: 217–239.

Smith, D.G., 1973. Aggradation of the Alexandria-North Saskatchewan River, Banff Park, Alberta. In Morisawa, M. (ed.), *Fluvial Geomorphology*. Binghamton, NY: Publications in Geomorphology, New York State University, pp. 201–219.

Smith, D.G., and Smith, N.D., 1980. Sedimentation in anastomosed river systems: examples from alluvial valleys near Banff, Alberta. *Journal of Sedimentary Petrology*, **50**: 157–164.

Cross-references

Braided Channels
Meandering Channels
Rivers and Alluvial Fans

ANCIENT KARST

Introduction

Karst is characterized by spectacular surface topographies, caves, and subterranean drainage systems that have developed in soluble limestone, dolostone, or gypsum bedrocks (e.g., Jennings, 1971, 1985; Sweeting, 1973; Bögli, 1980; Trudgill, 1985). Karst is geologically complex because it represents the balance between the diametrically opposed processes of dissolution and precipitation which are mediated by groundwater that flows over and through the bedrock. Thus, from a geological perspective karst is a subaerial diagenetic terrain. Esteban and Klappa (1983) defined karst as "...a diagenetic facies, an overprint in subaerially exposed carbonate bodies, produced and controlled by dissolution and migration of calcium carbonate in meteoric waters, occurring in a wide variety of climatic and tectonic settings, and generating a recognizable landscape." Similarly, Choquette and James (1988) defined karst to "...include all of the diagenetic features—macroscopic and microscopic, surface and subterranean—that are produced during the chemical dissolution and associated modification of a carbonate sequence."

Ancient karst (or paleokarst) includes relict paleokarst (present landscapes formed in the past) and buried paleokarst (karst landscape buried under younger sediments) as defined by Jennings (1971), Sweeting (1973), and Choquette and James (1988). Ancient karst is characterized by an unconformity that represents the original erosional landform and the underlying "karsted" rocks, which are characterized by diagenetic fabrics formed by dissolution and precipitation. Recognition of ancient karst is important because the karst surface represents a break in deposition that commonly constitutes a sequence boundary, and the diagenetically modified rocks beneath the karst surface may be superb reservoir rocks for hydrocarbons or hosts for various economically important ores.

Features of ancient karst

An ancient karst surface may be characterized by (1) stratigraphic–geomorphic features, (2) macroscopic surface and subsurface karst features, and (3) microscopic features (Choquette and James, 1988).

Figure A5 General view of the Cayman Unconformity (U/C) that separates the Cayman Formation (CF) from the Pedro Castle Formation (PCF) in Pedro Castle Quarry, central-south coast of Grand Cayman. Although horizontal in this quarry, this karst surface has a relief of ~40 m when traced across the island (Jones and Hunter, 1994). Note paleocaves located beneath the unconformity.

Stratigraphic–geomorphic features

The recognition and reconstruction of the topography of large-scale, ancient karst landforms, such as towers and dolines, relies on the availability of good surface exposures, high-resolution seismic profiles, or numerous closely spaced wells. The profiles of such landforms will only become evident if they are mapped relative to a datum that represents a time surface and was flat at the time of formation (cf. Choquette and James, 1988). The lack of fluviatile sediments on the erosional surface indicates that it formed through karst rather than by fluvial erosion. An ancient karst surface may be difficult to recognize in small outcrops or core because the limited scale precludes recognition of the large-scale landforms. Indeed, on a small scale, an ancient karst surface may be identical to any other bedding plane in the succession. The Cayman Unconformity, for example, is a karsted surface that formed during the Messinian and now separates the Cayman Formation (Miocene) from the Pedro Castle Formation (Pliocene) on Grand Cayman (Jones and Hunter, 1994). In Pedro Castle Quarry it is apparent as a relatively flat surface that, at first glance, appears to be a bedding plane (Figure A5). Mapping across the island, however, shows that this unconformity has a rugged relief of at least 40 m (Jones and Hunter, 1994).

Potentially, a karst surface may separate sedimentary successions that formed in different depositional regimes. This, however, assumes that the depositional regime established following development of karst surface was significantly different from those that existed prior to subaerial exposure and karst development. This is not universally true. On Grand Cayman, for example, the facies in the Pedro Castle Formation are similar to those found in the underlying Cayman Formation despite the fact that the two formations are separated by a high-relief, karsted unconformity.

In the simplest sense, strata located beneath a karst surface should be characterized by fabrics that are indicative of meteoric diagenesis whereas the rocks above the unconformity should be devoid of such fabrics. In many situations, however, this is not the case because the entire carbonate succession evolved through a succession of depositional-karst cycles with each phase of deposition being followed by a phase of karst development. Accordingly, virtually all parts of a succession will exhibit the results of meteoric diagenesis and many parts of the succession may be subjected to more than one cycle of such diagenetic alteration.

Macroscopic surface and subsurface karst features

Macroscopic surface karst features include various types of karren, phytokarst, kamenitzas, soils, caliche (see caliche—this volume), nonsedimentary channels, lichen structures, brown- or red-dish fracture fillings, and mantling nonsedimentary breccias (Choquette and James, 1988, table 2). Macroscopic subsurface features include caves and other smaller cavities, *in situ* brecciated and fractured strata, collapse structures, dissolution-enlarged features, rubble-and-fissure fabrics, cave sediments, and irregular-shaped breccia bodies which may or may not be conformable with the host strata (Choquette and James, 1988, table 2).

Not all features will be found in every ancient karst terrain. Some may be absent simply because they were never developed. Other features may have been formed but then destroyed by erosion during the next transgression (e.g., Jones and Smith, 1988). Soils, including terrra rossa, are typically unconsolidated and therefore prone to removal as sea-level rises and drowns the karst terrain. Similarly, phytokarst is commonly formed of delicate structures that are prone to destruction as they are submerged by rising sea-level. In many cases it may be difficult to recognize surface karst features because they are buried beneath younger sediments. If surface features are present, such as those recognized on the unconformity at the top of the Lower Ordovician Romaine Formation in Quebec (Desrochers and James, 1988), they offer important evidence of karst development.

Microscopic karst features

Microscopic karst features, usually recognizable only in thin sections, include products of dissolution (e.g., formation of fabric-selective pores; etched carbonate cements), bioerosion (e.g., sparmicritization, micritized grains), precipitation (e.g., meniscus, pendent, and needle-fiber vadose cements), and/or sediment transportation (e.g., vadose silt, eluviated soil) (cf. Choquette and James, 1988, table 2). Typically, these fabrics have an irregular to random distribution on a scale of millimeters. Thus, dissolution fabrics may characterize one pore in a limestone whereas a neighboring pore may be partly filled or filled with precipitated cements and/or vadose silt. Similarly, individual pores may contain interlaminated sediments and precipitates that are a record of temporal changes in the operative processes.

Karst surface or hardground

As transgressive seas flood a karst surface, various opportunistic encrusting and/or boring organisms will establish themselves on the hard rock that forms the new seafloor. Such colonization typically takes place during the transgressive lag period when sediment production is minimal. The Cayman Unconformity on Grand Cayman, for example, is characterized by numerous sponge, worm, and bivalve borings which were formed when those animals colonized the seafloor as the Messinian karst landscape was flooded. Similarly, numerous borings characterize the unconformity at the top of the Romaine Formation (Desrochers and James, 1988). These surfaces, however, will have many features that are also found on submarine hardgrounds (see hardgrounds—this volume). A bored karst surface, however, will cap a succession of rocks that display a variety of subsurface karst features. By comparison, a hardground will be found in a succession of marine sediments that display no evidence of diagenesis in a karst setting.

Open and filled cavities

Open and filled cavities, which range in size from micropores to large caves, are common features in many ancient karst terrains. Such cavities may have formed through the selective dissolution of metastable mineral components in heterogeneous carbonates whereas others are independent of fabric. Preferential leaching of aragonitic fossils, for example, produces fossilmoldic cavities with morphologies that reflect the shape of the original fossils. Fabric-independent cavities, which cross-cut textures in the host rocks, tend to be much larger than the fabric-dependent cavities. The aggressiveness of the waters and the length of time over which they interacted with the bedrock controlled development of these cavities. In many cases, they are solution-widened features that were formed by aggressive waters widening preexisting features, such as joints, fractures, and bedding planes.

Cavities in ancient karst terrains may be open or contain speleothems and/or sediments. Although speleothems (e.g., stalactites, stalagmites, flowstone) are common in modern caves, there are relatively few examples of speleothems known from ancient karst terrains. Exceptions to this include the speleothems found in paleocaves in the Cayman Formation (U. Miocene) and Pedro Castle Formation (L. Pliocene) on Cayman Brac (Jones, 1992, figures 12–14). On a small scale, pores may be partly or completely filled with aragonite, calcite, and dolomite cements. Microstalactitic and/or menicus cements may characterize the vadose cements whereas coarse, cavity-filling spar calcite may form in the phreatic zone.

Many cavities in ancient karst terrains contain sediments that range from mudstones to breccias (see cave sediments—this volume). Their composition is ultimately related to their point of origin because they may have been derived from sources external to the bedrock (exogenic) or from the bedrock itself (endogenic). Cave sediments are commonly characterized by complex arrays of sedimentary structures, including graded bedding, cross-bedding, and/or desiccation cracks. Cavities in the dolostones of the Bluff Group of the Cayman Islands, for example, are commonly filled or partly filled with a wide range of sediments that reflect complex depositional histories (Jones, 1992).

Complex intercalations of speleothems and cave sediments characterize cavities in some ancient karst terrains. Interpretation of these precipitates and deposits may offer important information regarding the conditions that existed following development of the original cavities. Speleothems, for example, typically formed during periods of high rainfall that engendered high groundwater flow and hence, speleothem precipitation. Unconformities in the speleothems typically mark periods of relatively dry climates when speleothem corrosion rather than speleothem precipitation was operative. Similarly, sediments found in a cavity are protected from erosion that may have removed their surface counterparts (cf. Smart et al., 1988; Jones and Smith 1988). Accordingly, analysis and interpretation of the cave sediments may provide information on the geological evolution of an area that cannot be obtained in any other way. In many cases, the sediments found in a cavity can be divided into discrete, unconformity-bounded packages that reflect different phases of deposition. In other settings, unconformity-bounded packages of sediment alternate with speleothems (e.g., Jones, 1992).

Karst breccia

Brecciated masses that formed as a result of bedrock collapse or by angular lithoclasts being transported and deposited in open cavities (e.g., sinkholes) are found in many ancient karst terrains. These concordant and discordant karst breccia bodies form a substantial part of the succession in some karst terrains.

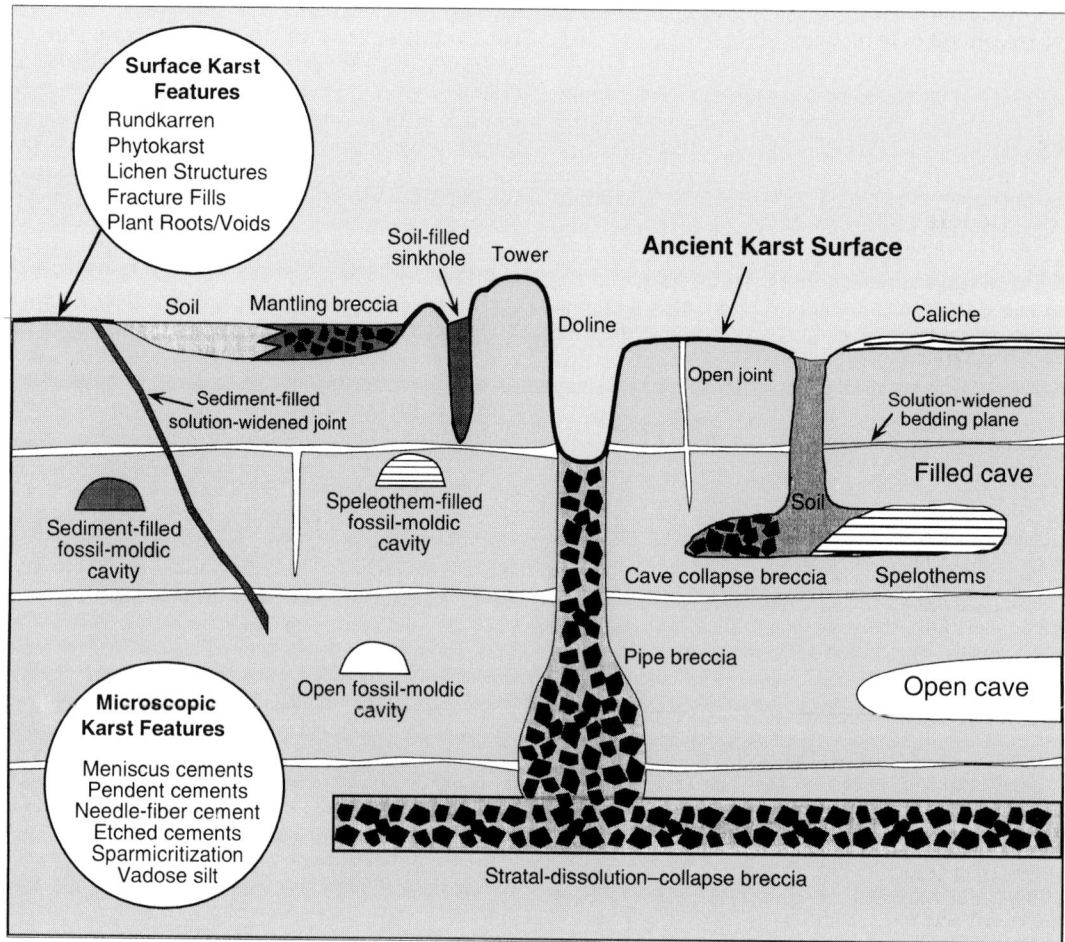

Figure A6 Schematic diagram showing features that may be associated with an ancient krast terrain.

Such breccias are economically important because they may host economically important deposits such as "Mississippi Valley type" lead-zinc-sulfide deposits (e.g., Sangster, 1988).

Karst breccia is virtually impossible to classify because it is so highly variable in terms of its composition, fabric, and external form. Choquette and James (1988, figure 5) suggested that karst breccias can only be classified by virtue of their location and inferred genesis. Mantling breccia, which lies on a karst surface, is formed of collapsed, surface or near-surface country rock that is intermixed with soil and/or sediment (Figure A6). Breccia pipes, which are typically formed by the collapse of strata beneath sinkholes, are usually formed of clasts derived from the bedrock, extraneous clasts that may have been washed in the sinkhole, soil, and sediment. The external shape of the resultant breccia body distinguishes it from mantle breccia (Figure A6). Cave collapse breccia, which generally has a tabular shape, is formed of poorly sorted clasts derived from the bedrock that once formed the cave roof and from external sources (Figure A6). These clasts are commonly held in matrix formed of soil and/or cave sediments that may be exogenic or endogenic in origin. Stratal-dissolution collapse breccia, which has a tabular morphology, commonly appears to be part of the bedded succession (Figure A6). They usually form when brittle bedrock collapsed following removal of highly soluble rocks, such as salt and evaporates, by subterranean dissolution.

Summary comments

1. An ancient karst surface is an unconformity that formed during a period of subaerial exposure. As such it denotes an important stage in the evolution of ancient carbonate successions and commonly constitutes a sequence boundary.
2. Ancient karst is a diagenetically complex terrain that encompasses features formed by dissolution (e.g., caves), precipitation (e.g., speleothems), and deposition (e.g., cave sediments).
3. "Karsted rocks" are commonly characterized by complex diagenetic fabrics that are highly variable at all scales of observation.
4. Ancient karst terrains can be recognized by their stratigraphic—geomorphic features, their macroscopic surface and subsurface features, and their microscopic features.

Many of these features are subtle and can only be recognized through careful examination of the rocks.

Brian Jones

Bibliography

Bögli, J., 1980. *Karst Hydrology and Physical Speleology*. Berlin/Heidelberg: Springer-Verlag, 285 pp.

Choquette, P.W., and James, N.P., 1988. Introduction, in *Paleokarst*. In James, N.P. and Choquette, P.W. (eds.), New York: Springer-Verlag, pp. 1–21.

Desrochers, A., and James, J.P., 1988. Early Paleozoic surface and subsurface paleokarst: Middle Ordovician carbonates, Mingan Island, *Québec, Paleokarst*. In James, N.P. and Choquette, P.W. (eds.), New York: Springer-Verlag, pp. 183–210.

Esteban, M., and Klappa, C.F., 1983. Subaerial exposure environments, in *Carbonate Depositional Environments*. In Scholle, P.A., Bebout, D.G., and Moore, C.H. (eds.), American Association of Petroleum Geologists Memoir **33**: pp. 1–54.

James, N.P., and Choquette, P.W., 1988. *Paleokarst*. New York: Springer-Verlag.

Jennings, J.N., 1971. *Karst*. Cambridge, Massachusetts: MIT-Press.

Jennings, J.N., 1985. *Karst Geomorphology*. Oxford: Basil Blackwell.

Jones, B., 1992. Void-filling deposits in karst terrains of isolated carbonate oceanic islands: a case study from Tertiary carbonates of the Cayman Islands. *Sedimentology*, **39**: 857–876.

Jones, B., and Hunter, I.G., 1994. Messinian (Late Miocene) karst on Grand Cayman, British West Indies: an example of an erosional sequence boundary. *Journal of Sedimentary Research*, **64**: 531–541.

Jones, B., and Smith, D.S., 1988. Open and filled karst features on the Cayman Islands: implications for the recognition of paleokarst. *Canadian Journal of Earth Sciences*, **25**: 1277–1291.

Sangster, D.F., 1988. Breccia-hosted lead-zinc deposits in carbonate rocks, in *Paleokarst*. In James, N.P., and Choquette, P.W. (eds.), New York: Springer-Verlag, pp. 102–116.

Smart, P.L., Palmer, R.J., Whitaker, F., and Wright, V.P., 1988. Neptunian dikes and fissure fills: an overview and account of some modern examples, in *Paleokarst*. In James, N.P., and Choquette, P.W. (eds.), New York: Springer-Verlag, pp. 149–163.

Sweeting, M.M., 1973. *Karst Landforms*. London: MacMillan Publ Co.

Trudgill, S., 1985. *Limestone Geomorphology*. New York: Longman.

Cross-references

Bioerosion
Caliche–Calcrete
Cave Sediments
Diagenesis
Diagenetic Structures
Solution Breccias
Speleothems

ANGLE OF REPOSE

When sand cascades down the face of a dune, it comes to rest at a slope that has nearly the same angle as a cone-shaped pile of sand formed by a colony of ants. A pile of wheat develops a nearly identical cone having essentially the same slope. This slope angle must therefore be an intrinsic property of loose grains, a property that is generally termed the *angle of repose*, the acute angle that the slope makes with the horizontal. On average this angle is approximately 33°, but varies by a few degrees depending on the sizes and shapes of the grains, as well as on other factors such as the moisture content.

Investigations of the angle of repose have produced the slope in the laboratory by: (a) carefully pouring loose particles into a pile, (b) removing one side of a container to release the particles, or (c) rotating a trough containing grains until movement forms a slope (Van Burkalow, 1945; Allen, 1969; Carrigy, 1970). The rotating trough in particular reveals the existence of two distinct critical angles rather than one. The slope can be slowly increased by rotation until it reaches the critical angle for avalanching, Φ_a, at which point the grains begin to flow, reducing the overall slope to the true angle of repose, Φ_r, the slope after avalanching has ceased. In a similar fashion, when piled to form a cone, the grains accumulate to the angle Φ_a, the mass then slumps to form a new surface at the repose angle Φ_r. The difference between these two angles is typically only 2° to 5°. However, in some cases the difference can be large, especially when moisture introduces surface tension that holds the grains together, acting to increase the angle needed for avalanching, while having little effect on the post-avalanche repose angle.

The studies agree that of the many factors that determine the angle of repose, particle shape has the greatest effect. For example, experiments by Carrigy (1970) comparing spherical glass beads and crushed angular quartz having nearly the same grain sizes revealed that Φ_r is about 10° greater for the angular quartz grains than for the smooth spheres, and when there is a mixture of the two, Φ_r is directly dependent on the proportions of rounded versus angular grains; the commonly observed repose angle of 33° in natural sands occurred when the mixture was about 80% crushed quartz. These results are what one might expect in that the angular grains provide more interlocking surfaces, thereby resisting slippage until a larger avalanche angle is reached, and then coming to rest at a higher angle of repose. Experiments such as these illustrate the fact that the angle of repose represents a measure of the internal friction of the granular material, the degree of friction or interlocking between grains.

The experiments of Carrigy (1970) and others also indicate that slightly different values are obtained for the avalanche angle when the grains are immersed in water rather than in air, while the repose angles are about the same. It had been suggested that the angle of repose should be less in water than in air, and therefore one could use measurements of maximum dips angles of cross-bedding strata to establish whether a sandstone had been deposited subaerially or under water. However, cross-bedding should mainly reflect the angle of repose, and the uniformity of Φ_r in water and air does not support such an interpretation. In addition, Carrigy made measurements on thirty different sands from various natural environments, and found that Φ_r was the same (within measurement error), but that the Φ_a avalanche angles were without exception greater in air than in water. This was due largely to the effects of surface moisture which tends to bind the grains together in air, a binding effect that is lost when immersed in water. Carrigy demonstrated for one sand sample that the avalanche angle increased from 35.6° when oven dried to 43.4° when 0.95% moisture by weight is present, although the angles of repose following avalanching were nearly the same. When moisture contents are greater than 1%, the sample ceased to behave as a loose granular material, even being able to maintain a vertical escarpment. Static-electricity charges on grain surfaces can have a similar effect in binding the grains.

Paul D. Komar

Bibliography

Allen, J.R.L., 1969. The maximum slope angle attainable by surfaces underlain by bulked equal spheroids with variable dimensional ordering. *Bulletin Geological Society of America*, **80**: 1923–1930.

Carrigy, M.A., 1970. Experiments on the angles of repose of granular materials. *Sedimentology*, **14**: 147–158.

Van Burkalow, A., 1945. Angle of repose and angle of sliding friction: An experimental study. *Bulletin Geological Society of America*, **56**: 669–708.

Cross-references

Fabric, Porosity and Permeability
Grain Size and Shape

ANHYDRITE AND GYPSUM

Anhydrite ($CaSO_4$) and gypsum ($CaSO_4 \cdot 2H_2O$) are the two most abundant minerals of ancient marine evaporite deposits and are also common in non-marine evaporite deposits. Sedimentary gypsum forms by direct precipitation out of evaporating seawater under arid climatic conditions in hydrologically restricted marine and marginal marine environments (e.g., tidal flats, coastal lagoons, "inland seas", etc.). In non-marine, arid closed basin systems, gypsum precipitates from evaporating meteoric waters with chemical compositions dependent on the bedrock types and their proportions in the drainage areas. In contrast, at temperatures and pressures typical of sedimentary environments, anhydrite does not precipitate directly from evaporating waters but, as discussed in the next section, forms by dehydration of precursor gypsum precipitates. Sedimentary anhydrite is, in essence, a diagenetic mineral.

Stability fields of anhydrite and gypsum

In modern arid sedimentary environments gypsum readily precipitates directly from evaporating waters that carry calcium and sulfate ions, a phenomenon that is easily reproduced in the laboratory. In contrast, at typical Earth surface and shallow burial temperatures anhydrite can only be formed experimentally by dehydration of gypsum, a mechanism of formation of anhydrite that appears to operate in modern evaporite environments (e.g., Hardie and Shinn, 1986, p. 49). At higher temperatures, anhydrite nucleates directly from hydrothermal solutions such as the hot brines venting on the modern seafloor beneath mid-ocean ridges. Primary anhydrite precipitates are typical components of ancient hydrothermal Pb-Zn deposits.

Since anhydrite will not precipitate directly from aqueous solutions at sedimentary temperatures, the stability fields of gypsum and anhydrite cannot be accurately determined by classic solubility measurements (Hardie, 1967, pp. 185–188). However, the gypsum-anhydrite transition has been determined via gypsum-anhydrite conversion experiments (Hardie, 1967). For the reversible reaction $CaSO_4 \cdot 2H_2O_{(s)} = CaSO_{4(s)} + 2H_2O_{(l)}$ at any given temperature and pressure, the following thermodynamic relationship must be true:

$$K_{P,T} = a_{CaSO_4} \cdot a^2_{H_2O} / a_{CaSO_4 \cdot 2H_2O}$$

where K is the equilibrium constant at a given P and T, and a is the activity of each component in the reaction. Since the standard states for gypsum and anhydrite are the pure solids, which by definition have activities equal to unity, then the above relationship simplifies to:

$$K_{P,T} = a^2_{H_2O}$$

It follows that the gypsum-anhydrite transition at any given P and T is a function only of the activity of H_2O in the solution phase. Reversible experiments in which gypsum was converted to anhydrite and anhydrite converted to gypsum in solutions with fixed activities of H_2O allowed the construction of a T-a_{H_2O} diagram that predicts the gypsum-anhydrite transition temperature in any aqueous solution of known activity of H_2O (Figure A7).

This experimental study underscores the role of kinetics in the behavior of gypsum and anhydrite. *At surface temperatures and pressures, gypsum will nucleate directly from supersaturated aqueous solutions but anhydrite will not*. In the anhydrite stability field (Figure A7), gypsum will be the initial (but metastable) precipitate. With time this metastable gypsum will convert to anhydrite (Hardie, 1967, pp. 183–184).

Depositional features of sedimentary gypsum and anhydrite

In modern and ancient sedimentary deposits, both gypsum and anhydrite are found in the form of (1) beds, (2) as isolated crystals or clusters of crystals embedded in carbonate or siliciclastic sediments, and (3) as nodules or clusters of nodules embedded in carbonate or siliciclastic sediments.

Bedded gypsum

There are three main environments in which bedded gypsum can accumulate: ephemeral, shallow perennial, and deep perennial. In these settings bedded gypsum may be found as (1) *chemically precipitated layers*, as settle-out of gypsum crystals nucleated at the brine surface and/or as vertically-oriented crystals grown on the bottom by syntaxial overgrowth, and as (2) *detrital layers* made of sand and gravel size crystals and crystal fragments eroded from primary gypsum sources and redeposited in other parts of the depositional basin.

The distinction between ephemeral and shallow perennial marine environments in which bedded gypsum can accumulate is a narrow, but critical, one. In marginal marine systems, it is the distinction between a supratidal and a shallow subtidal setting. In both settings, bedded gypsum is deposited under subaqueous conditions, conditions that produce identical bottom syntaxial growth fabrics. However, in arid supratidal settings gypsum layers are deposited subaqueously in shallow, ephemeral seawater "lakes" (gypsum pans, see *Sabkha, Salt Flat, Salina*) formed by storm-driven flooding of the tidal flat. As such a "lake" shrinks with progressive evaporative concentration, the newly formed bed of crystalline gypsum is exposed to the air. This leads to evaporative pumping of brine from the shallow brine table, up through the vadose zone to the now dry pan surface. The subaqueously deposited gypsum bed becomes overprinted by penecontemporaneous vadose diagenetic processes, such as vadose cementation, that leads to the formation of enterolithic folds, tepees and diapiric structures (see under *Sabkha, Salt Flat, Salina*). However, there is an

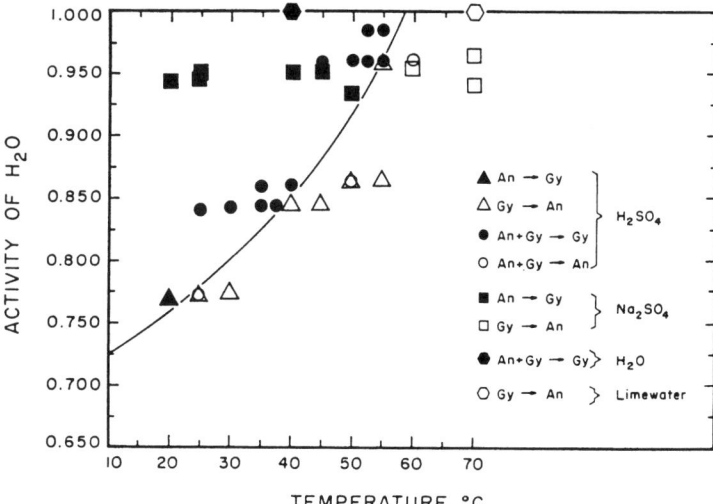

Figure A7 The stability of gypsum and anhydrite determined experimentally as a function of temperature and activity of H_2O at atmospheric pressure (from Hardie, 1967, Figure A8).

additional set of penecontemporaneous diagenetic modifications that are produced during the next flooding stage when gypsum-undersaturated seawater partly dissolves and erodes the exposed gypsum crystals. But during the subsequent evaporative concentration stage syntaxial overgrowth can "repair" the partly dissolved crystals (e.g., Demicco and Hardie, 1994, Figure 154C) and gypsum cements can fill or partly fill the intercrystalline cavities.

The absence (or rarity) of dissolution and syntaxial repair features in beds of vertically oriented gypsum crystals points unerringly to a *shallow perennial* subaqueous depositional environment, such as a coastal, back-barrier lagoon, as exemplified by the bedded gypsum of the Upper Miocene Solfifera Series of Sicily (Hardie and Eugster, 1971, Figures 4–7; Demicco and Hardie, 1994, Figure 154B). This Solfifera Series also carries bedded gypsum in the form of mm-scale lamination composed of couplets of calcite and finely crystalline gypsum laminae (so-called "balatino", Hardie and Eugster, 1971, Figures 8, 10), planar and cross-laminated gypsum-calcite sandstones, and gypsum crystal conglomerates (Hardie and Eugster, 1971, Figures 12, 13, 14, 16).

Figure A8 shows a hypothetical comparison of the position and characteristics of bedded gypsum in (1) a marginal marine sabkha setting in which the cap of layered gypsum was deposited in an ephemeral supratidal pan, and (2) shallow restricted subtidal lagoon which developed into a closed ephemeral salt pan.

Bedded gypsum can form and accumulate in *deep water evaporite environments*. The clearest modern example is provided by the modern Dead Sea (Warren, 1989, pp. 140–145). Before 1979 the Dead Sea (maximum depth 400 m) was stratified and only tiny needles of aragonite and polygonal plates of gypsum were precipitated from the Upper Water Mass (UWM, 40 m thick), which was undersaturated with respect to halite (Neev and Emery, 1967). These mud-sized aragonite and gypsum crystals slowly sank to the bottom of the Dead Sea where they collected as well-defined laminae. The laminae are not annual deposits; as of 1967 only 15–20 laminae had been deposited in the previous 70 years. The gypsiferous laminae were particularly well preserved where the Dead Sea floor was shallower than about 40 m. In 1979 the salinity of the UWM had increased to the point of overturn, and mixing with the underlying Lower Water Mass ensued. Since then halite, nucleated at the water–air interface, has been added to the saline mineral accumulation on the floor of the Dead Sea (Garber *et al.*, 1987).

A well-documented example of an ancient deep water gypsum deposit is the Miocene of the Periadriatic trough. For this gypsum deposit Parea and Ricchi Lucchi (1972) made a convincing case for transportation ("resedimentation") of shallow water gypsum sediment into a deep water basin by turbidity currents and mass flow slope processes (for other examples, see Warren, 1989, pp. 150–163).

Bedded anhydrite

As discussed above, the evidence from experimental work as well from observations of modern active evaporite settings points overwelmingly to the inability of anhydrite to nucleate as a primary mineral at normal Earth surface temperatures and pressures. Therefore, we must conclude that *beds of anhydrite, both modern and ancient, are formed by diagenetic alteration (dehydration) of primary gypsum*. The conversion of gypsum to anhydrite can be penecontemporaneous or it can take place at any stage during burial. The depositional environment recorded by bedded anhydrite, then, must be that of its bedded gypsum precursor. The criteria used to interpret the depositional setting of bedded gypsum, whether ephemeral, shallow perennial or deep perennial, should be applied to the bedded anhydrite.

Nodular and "chicken wire" gypsum and anhydrite: origin and significance

Individual ovoid to irregular shaped nodules (mm to cm scale) made of a fine felted mass of anhydrite laths (~100 microns long) embedded in carbonate or siliciclastic host sedimentary

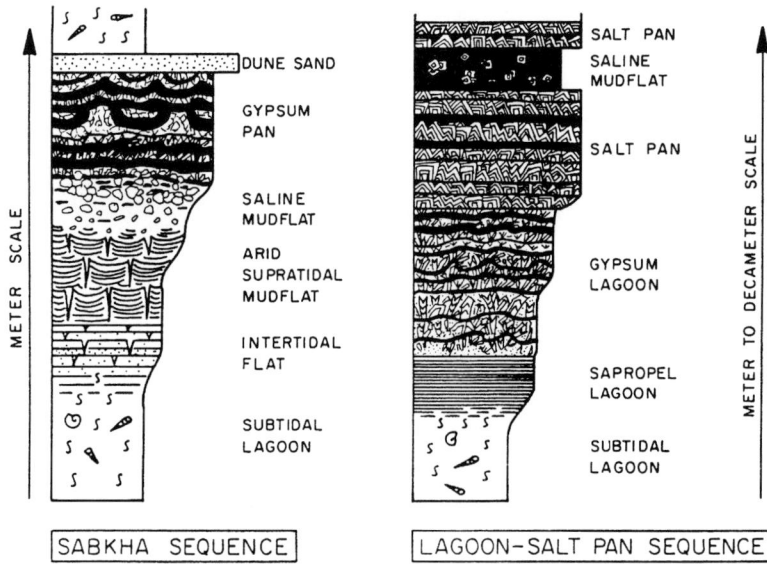

Figure A8 Hypothetical shallowing-upward successions comparing a lagoon-sabkha cycle to a lagoon-salt pan cycle (from Hardie and Shinn, 1986, Figure 69).

rocks have been described from many Phanerozoic evaporite-bearing deposits. Other nodular fabrics observed in anhydrite have been called "chicken wire lattice structure" (Imlay, 1940), "flaser" anhydrite (Jung, 1958) and "mosaic" anhydrite (Riley and Burne, 1961) (see also Bebout and Maiklem, 1973). Of these latter fabrics, the most commonly encountered is "chicken wire" anhydrite, which is a mosaic of irregular nodules separated by very thin, dark, wavy remnants of the sediment in which the nodules grew. The wavy boundaries in many cases have a distinct stylolitic character suggestive of modification by pressure solution. Both "chicken wire" anhydrite (Butler, 1970) and "chicken wire" gypsum (Ali and West, 1983) have been found in modern sabkha environments.

As discussed in the preceeding sections, at typical Earth surface temperatures and pressures anhydrite nodules form by the conversion of a precursor gypsum crystal or cluster of crystals. Based on the discovery of nodular anhydrite in the modern tidal flat sediments of the Persian Gulf sabkhas (e.g., Butler, 1970), it is common practice to assign a sabkha environment to any ancient sedimentary rock that carries any type of nodular anhydrite or gypsum. Some workers have suggested that nodules with morphologies similar to those of modern anhydrite but composed of chert, mega-quartz, calcite, dolomite or celestite found embedded in ancient carbonate rocks are pseudomorphs after sabkha anhydrite. However, *the presence of gypsum or anhydrite with nodular or chicken wire fabrics is not, on its own, sufficient evidence to assign a sabkha origin to the host sedimentary rocks*. This follows because anhydrite nodules can form under a variety of conditions, the most unusual being the nodules of anhydrite found in cores of seafloor sediments under 2 km of water in the Atlantis II Deep of the Red Sea, a site of upwelling hydrothermal brines (Degens and Ross, 1969, pp. 366–367). Clearly, modern nodular anhydrite is not unique to sabkha settings. Instead, the common factor is the process of replacement of gypsum by anhydrite, that is, single crystals of gypsum are altered to a mass of tiny anhydrite laths plus liquid water, a rather cohesionless mixture easily deformed by the weight of the enclosing sediment. The observed tendency is for the anhydrite mass to change toward a spherical or ovoid shape. Dehydration of clusters of gypsum crystals or layers of crystalline (or detrital) gypsum would yield "chicken wire" anhydrite. The *timing* of the formation of the nodular anhydrite would depend on the P, T and activity of H_2O of the pore fluids. Anhydritization could have taken place at the surface contemporaneously with sedimentation or much later during burial. Nodular and "chicken wire" gypsum is likely to have had a more complex history, one in which primary gypsum crystals were first replaced by a felted aggregate of anhydrite laths which in turn were later rehydrated to a mass of small gypsum crystals. In this regard, Murray (1964) has pointed out that gypsum formed at the surface will inevitably dehydrate to anhydrite on burial (depth of conversion dependent on the prevailing T, P and groundwater salinity). In turn, on uplift and contact with dilute groundwaters, anhydrite will rehydrate to gypsum.

In conclusion, gypsum and anhydrite nodular fabrics are at best ambiguous criteria for determining depositional environment. Nodules composed of other minerals such as calcite, chert or other non-sulfate mineral presumed to be pseudomorphous replacements of anhydrite or gypsum must carry even less weight. Instead, reliance can only be placed on the vertical succession of subfacies and their internal primary sedimentary structures and fabrics, fossils assemblages, etc., in diagnosing the depositional settings of ancient gypsum/anhydrite deposits (e.g., Figure A8)

Lawrence A. Hardie

Bibliography

Ali, Y.A., and West, I.M., 1983. Relationships of modern gypsum nodules in sabkhas of loess to compositions of brines and sediments in northern Egypt. *Journal of Sedimentary Petrology*, 53: 1151–1168.

Bebout, D.G., and Maiklem, W.R., 1973. Ancient anhydrite facies and environments, Middle Devonian Elk Point Basin, Alberta. *Bulletin of Canadian Petroleum Geology*, **21**: 287–343.
Butler, G.P., 1970. Holocene gypsum and anhydrite of the Abu Dhabi sabkha: an alternative explanation of origin. In Rau, J.L., and Dellwig, L.F. (eds.), *Proceedings of the Third Salt Symposium*. Cleveland: Northern Ohio Geological Society, pp. 120–152.
Degens, E.T., and Ross, D.A. (eds.), 1969. *Hot Brines and Recent Metal Deposits in the Red Sea*. Springer-Verlag.
Demicco, R.V., and Hardie, L.A., 1994. *Sedimentary Structures and Early Diagenetic Features of Shallow Marine Carbonate Deposits*. SEPM Atlas Series Number 1, 265pp.
Garber, R.A., Levy, Y., and Friedman, G.M., 1987. The sedimentology of the Dead Sea. *Carbonates and Evaporites*, **2**: 43–57.
Hardie, L.A., 1967. The gypsum-anhydrite equilibrium at one atmosphere pressure. *American Mineralogist*, **52**: 171–200.
Hardie, L.A., and Eugster, H.P., 1971. The depositional environments of marine evaporites: a case for shallow clastic accumulation. *Sedimentology*, **16**: 187–220.
Hardie, L.A., and Shinn, E.A., 1986. Carbonate depositional environments modern and ancient, part 3: tidal flats. *Colorado School of Mines Quarterly*, **81**: 1–74.
Imlay, R.W., 1940. Lower Cretaceous and Jurassic formations of southern Arkansas and their oil and gas possibilites. Information *Circular 12, Arkansas Resources and Development Commission*, Little Rock, 64 pp.
Jung, W., 1958. Zur Feinstratigrafie de Werraanhydrite (Zechstein 1) im Bereich der Sangerhauser und Mansfelder Mulde. *Geologie*, Beihefte, **24**: 312–325.
Murray, R.C., 1964. Origin and diagenesis of gypsum and anhydrite. *Journal of Sedimentary Petrology*, **34**: 512–523.
Neev, D., and Emery, K.O., 1967, The Dead Sea: depositional processes and environments of evaporites. *Geological Survey of Israel Bulletin*, **41**: 147 pp.
Parea, G.C., and Ricchi Lucchi, F., 1972. Resedimented evaporites in the Periadriatic trough. *Israel Journal of Earth Science*, **21**: 125–141.
Riley, C.M., and Byrne, J.V., 1961. Genesis of primary structures in anhydrite. *Journal of Sedimentary Petrology*, **31**: 553–559.
Warren, J.K., 1989. *Evaporite Sedimentology*, Prentice Hall.

Cross-references

Evaporites
Sabkha, Salt Flat, Salina

ANKERITE (IN SEDIMENTS)

Ankerite is a frequent but usually minor burial diagenetic phase in sandstones. It can also be present in early diagenetic mudrock-hosted concretions, and rarely as a replacive or void filling burial precipitate in limestones. Ankerite precipitation may result from local (bed-scale) to regional mass transfer, particularly during interaction of compositionally and thermally distinct subsurface fluids.

Mineralogy and chemistry

Ankerite has a general formula $Ca(Mg, Fe^{2+}, Mn)(CO_3)_2$ and is part of a solid solution series resulting from substitution of Fe^{2+} ($\pm$ subordinate Mn^{2+}) for Mg^{2+} in the dolomite lattice. It is usually defined as having a $Mg^{2+} : Fe^{2+}$ ratio of $\leq 4 : 1$, and diagenetic ankerite typically contains 10–25 percent $FeCO_3$. Ankerite in sedimentary rocks commonly contains up to 10 mol percent excess $CaCO_3$, particularly where it has replaced calcite. Ankerite is more stable than both calcite and dolomite in many iron-rich fluids, even if they have relatively high Ca^{2+}/Mg^{2+} ratios.

Chemical analysis is usually required to distinguish ankerite from ferroan dolomite, although ankerite may be discriminated from pure dolomite by X-ray diffractometry. It is identical to dolomite in thin section, and gives a turquoise color with alizarin red-S and potassium ferricyanide mixed stain (see "Stains"). Ankerite is black in cathodoluminescence, but may display complex compositional zonation when imaged with backscatter scanning electron microscopy. It sometimes has curved crystal faces comparable the high-temperature "baroque" form of dolomite (e.g., Spötl et al., 1996).

Ankerite in sedimentary rocks is commonly investigated using stable isotope geochemistry. However, O isotope fractionation between water and ankerite at diagenetic temperatures is poorly constrained. Many studies extrapolate experimental high temperature fractionation factors, or assume an "average" +3 percent difference between cogenetic calcite and dolomite/ankerite (e.g., Dutton and Land, 1988). Theoretical considerations based on crystal structure, chemical composition and cation-oxygen bond strengths suggest that the ankerite-calcite fractionation may actually be less than 3 percent in low-temperature situations (Zheng, 1999). Valid interpretations of ankerite geochemistry critically require evaluation of its replacive or pore-filling nature, which can be difficult in sandstones.

Occurrence of ankerite in sedimentary rocks

Sandstones

Disseminated ankerite and ferroan dolomite are very common diagenetic precipitates in both marine and continental sandstones, albeit rarely amounting to >10 percent of the mineral assemblage. Ankerite commonly overgrows earlier dolomite and/or replaces bioclastic or authigenic calcite. Quartz grains enveloped by ankerite cements are corroded, and partial to complete replacement of feldspars is also common. Many examples are described from the Mesozoic North Sea (e.g., Kantorowicz, 1985), Gulf Coast and Eastern Texas (e.g., Land and Fisher, 1987) and Alberta Basin. Ankerite is also recorded from Permo-Triassic continental sandstones in a variety of locations, and from a number of Paleozoic sandstones, notably in the southern USA. Pervasive cementation by ankerite is rare, although m-dm sized concretions have been described from deeply buried (>5 km) Jurassic sandstones of the Central North Sea (Hendry et al., 2000).

Ankerite in sandstones tends to be paragenetically late, strongly depleted in ^{18}O and variably depleted in ^{13}C. This suggests formation at elevated temperatures (deep burial) with a proportion of the carbon derived by thermal decarboxylation of organic matter and the remainder remobilized from bioclasts and early cements or from adjacent limestones. Radiogenic $^{87}Sr/^{86}Sr$ values commonly indicate ankerite formation after a substantial amount of silicate diagenesis. Mudrocks are commonly believed to be principal sources of Mg^{2+} and Fe^{2+}, and ankerite formation has been causally related to mobilization of these cations following illitization of smectite and thermal reduction of ferric oxides (e.g., Boles, 1978). This interpretation has been questioned because Cenozoic strata in the Gulf Coast contain very little ankerite despite large volumes of illitized mudrock, and because much of the Fe^{2+}

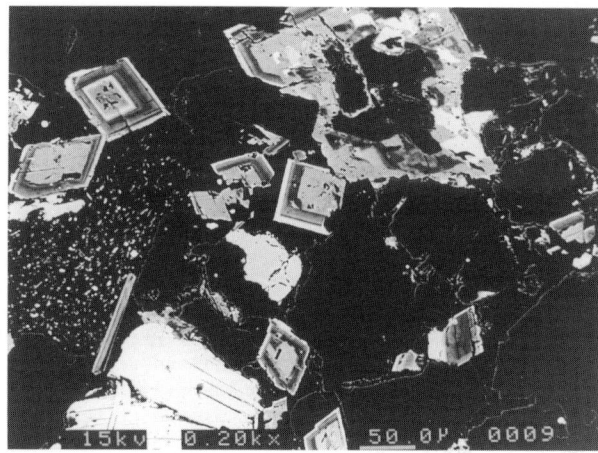

Figure A9 High contrast backscatter electron photomicrograph of compositionally zoned ankerite—ferroan dolomite in Upper Jurassic sandstones, Central Graben, North Sea. Quartz grains and porosity appear black, K-feldspars are bright.

and Mg^{2+} produced in mudrocks can be consumed by *in situ* mineral authigenesis. An alternative source of Fe^{2+} and Mg^{2+} may be volcaniclastic materials within sandstones (Morad *et al.*, 1996). However, mudrocks appear to be the only viable Fe^{2+} and Mg^{2+} source for ankerite formation in many sandstones.

A variety of models have been proposed for fluid and mass transfer related to ankerite formation in sandstones. Variable ankerite compositions and/or concentration in close proximity to intercalated mudrocks suggest local solute transfer in stagnant pore fluids (e.g., Macaulay *et al.*, 1993). Other ankerites yield isotopic and fluid inclusion data that require long distance and/or cross-formational fluid flow. Ankerite associated with sulfate-sulfide mineralization in sandstones has been linked to mixing of hypersaline fluids derived from juxtaposed evaporites and metaliferous hot fluids transmitted up faults from source rocks deeper in the basin (Burley *et al.*, 1989).

Limestones

Ankerite and ferroan dolomite do not constitute major replacive phases in limestones except where stratigraphically associated with mudrocks that released Fe^{2+} ($\pm Mg^{2+}$) during burial diagenesis (e.g., Hendry, 2002; Taylor and Sibley, 1986). Minor ^{18}O-depleted ankerite may form late stage vug and fracture cements in some limestone hydrocarbon reservoirs (e.g., Woronick and Land, 1985). Such cements have variably negative ^{13}C values depending on mass balance of carbon derived from organic matter oxidation in source rocks versus local remobilization of $CaCO_3$. Ankerite may form localized replacive and void filling precipitates in some limestones influenced by volcanic or tectonically driven hydrothermal fluid circulation (e.g., Searl, 1992), as well as being present as gangue cements associated with epigenetic Fe-Pb-Zn ore deposits.

Mudrocks and coals

Ankerite has been recorded from a variety of marine mudrocks, both as disseminated crystals, grain replacements and as concretionary cement. Stable isotopic data suggest that it formed at shallow to intermediate burial depths (10's–100's m) with carbon sourced by bacterial fermentation or thermal decarboxylation of organic matter and Fe^{2+} from reduction of detrital ferric oxides (e.g., Irwin, 1980). A limited supply of Mg^{2+} from residual seawater might explain why such ankerite is usually calcian. Ankerite does not typically form in the shallower sulfate reduction zone because pyrite is a more effective sink for Fe^{2+}.

Ankerite is less common than siderite and calcite in non-marine mudrocks because of a paucity of Mg^{2+}. Nevertheless, it can be a significant diagenetic precipitate in siltstones and mudrocks of coal measures as well as replacing authigenic siderite in coal seams. High concentrations of ferric oxides and hydroxides were eroded from lateritic soils and deposited in anaerobic coastal plain lagoons and swamps. Pore fluid mixing during marine inundation and/or between intercalated brackish and marine facies during shallow burial introduced Mg^{2+} and Fe^{2+} into the non-marine strata allowing ankerite to form (Matsumoto and Iijima, 1981).

Summary

Ankerite exists in sedimentary rocks as a late diagenetic cement or replacement of pre-existing carbonate. It can be formed in a number of ways, provided an adequate supply of Fe^{2+} and Mg^{2+} relative to Ca^{2+}. The iron is mostly derived from ferric oxides deposited in mudrocks, but sources of Mg^{2+} are more varied and can be difficult to constrain in individual cases. Ankerite can be formed as a response to basin-scale fluid flow or by localized mass transfer between juxtaposed porous sediments and mudrocks. Ankerite cements may also precipitate during early diagenesis given sufficient Mg^{2+} in pore fluids and provided that pyrite formation did not exhaust the local supply of labile iron.

James P. Hendry

Bibliography

Boles, J.R., 1978. Active ankerite cementation in the subsurface Eocene of Southwest Texas. *Contributions to Mineralogy and Petrology*, **68**: 13–22.

Burley, S.D., Mullis, J., and Matter, A., 1989. Timing diagenesis in the Tartan Reservoir (U. K. North Sea): constraints from combined cathodoluminescence microscopy and fluid inclusion studies. *Marine and Petroleum Geology*, **6**: 98–120.

Dutton, S.P., and Land, L.S., 1988. Cementation and burial history of a low permeability quartzarenite, Lower Cretaceous Travis Peak Formation, East Texas. *Geological Society of America Bulletin*, **100**: 1271–1282.

Hendry, J.P., 2002. Geochemical trends and palaeohydrological significance of shallow burial calcite and ankerite cements in Middle Jurassic strata on the East Midlands Shelf (onshore UK): *Sedimentary Geology*, **151**: 149–176.

Hendry, J.P., Wilkinson, M., Fallick, A.E., and Haszeldine, R.S., 2000. Ankerite cementation in deeply buried Jurassic sandstone reservoirs of the Central North Sea. *Journal of Sedimentary Research*, **70**: 227–239.

Irwin, H., 1980. Early diagenetic carbonate precipitation and pore fluid migration in the Kimmeridge Clay of Dorset, England. *Sedimentology*, **27**: 577–591.

Kantorowicz, J.D., 1985. The origin of authigenic ankerite from the Ninian Field, UK North Sea. *Nature*, **315**: 214–216.

Land, L.S., and Fisher, R.S., 1987. Wilcox sandstone diagenesis, Texas Gulf Coast: a regional isotopic comparison with the Frio

Formation. In Marshall, J.D. (ed.), *Diagenesis of Sedimentary Sequences*. Geological Society of London, Special Publication, 36, pp. 219–235.

Macaulay, C.I., Haszeldine, R.S., and Fallick, A.E., 1993. Distribution, chemistry, isotopic composition and origin of diagenetic carbonates: Magnus Sandstone, North Sea. *Journal of Sedimentary Petrology*, **63**: 33–43.

Matsumoto, R., and Iijima, A., 1981. Origin and diagenetic evolution of Ca-Mg-Fe carbonates in some coalfields of Japan. *Sedimentology*, **28**: 239–259.

Searl, A., 1992. Dolomite-carbonate replacement textures in veins cutting Carboniferous rocks in East Fife. *Sedimentary Geology*, **77**: 1–14.

Spötl, C., Houseknecht, D.W., and Burns, S.J., 1996. Diagenesis of an 'overmature' gas reservoir: the Spiro sand of the Arkoma Basin, USA. *Marine and Petroleum Geology*, **13**: 25–40.

Taylor, T.R., and Sibley, D.F., 1986. Petrographic and geochemical characteristics of dolomite types and the origin of ferroan dolomite in the Trenton Formation, Ordovician, Michigan Basin, U.S.A. *Sedimentology*, **33**: 61–86.

Woronick, R.E., and Land, L.S., 1985. Late burial diagenesis, Lower Cretaceous Pearsall and Lower Glen Rose Formations, South Texas. In Schneidermann, N., and Harris, P.M. (eds.), *Carbonate Cements*, SEPM (Society for Sedimentary Geology), Special Publication, 36, pp. 265–276.

Zheng, Y.-F., 1999. Oxygen isotope fractionation in carbonate and sulfate minerals. *Geochemical Journal*, **33**: 109–126.

Cross-references

Carbonate Mineralogy and Geochemistry
Cements and Cementation
Diagenesis
Diagenetic Structures
Dolomites and Dolomitization
Fluid Inclusions
Ironstones and Iron Formations
Isotopic Methods in Sedimentology
Stains for Carbonate Minerals

ARMOR

The term 'armor' (or 'armour' in English usage outside the US) refers to clastic deposits in which the surface layer is coarser than the substrate. The phenomenon is widespread in gravel-bed rivers, also occurs in stony deserts (where it is usually referred to as 'desert pavement'), and has occasionally been reported from pebbly beaches. The existence of armor is generally obvious if surface clasts are removed from a small area, and the degree of armoring can be quantified by comparing surface and subsurface grain-size distributions.

The commonest explanation for armoring is that it develops when a poorly-sorted noncohesive sediment is eroded by water or wind. The smaller clasts are more easily removed so the larger ones become concentrated on the surface as it is lowered. Thus many desert pavements are regarded as deflation lag deposits, and fluvial armor was originally recognized in degrading rivers below dams. However, as discussed below, fluvial armor can form without degradation and several alternative explanations have been proposed for desert pavement.

No matter how an armor develops, the word itself hints at a key consequence of the phenomenon: a greatly reduced rate of entrainment from the bed surface, compared to an unarmored arrangement of the same bulk mixture of grain sizes. Armoring is therefore associated with a low rate of sediment transport, and provides a link between local sedimentology and larger-scale sediment flux.

Fluvial armor

The wide range of grain sizes present in gravel-bed rivers allows the possibility of size sorting and segregation through selective transport and deposition. From this perspective the development of a coarse surface layer is a kind of vertical sorting, and is almost as prevalent as the downstream- or downbar-fining trends that reflect spatial sorting (see Powell, 1998, for a useful review of armoring in this wider context). Fine material (sand and/or granules) is usually visible only in the sheltered lee of large clasts, but fills a substantial proportion of the subsurface voids in the gravel framework. The maximum surface and subsurface grain sizes may be similar, but the respective median diameters typically differ by one to two phi units.

Gravel armor can form in more than one way. The traditional explanation is preferential winnowing of finer sediment from the surface during degradation, for example, below dams which cut off the gravel flux from upstream so that the river erodes its bed to regain a capacity load. This degradation is self-limiting (Gessler, 1970) because, as the surface coarsens, the transport capacity of the flow declines (to zero immediately below the dam) and the coarser grains (if not the whole bed) become immobile. This sequence of events has been investigated in flume experiments with no sediment feed and has been modeled mathematically (see Sutherland, 1987, for a good review of both approaches). It can occur in sand-bed rivers that contain a little gravel, which becomes concentrated on the surface as a stable lag deposit.

In the 1980s fluvial scientists recognized that coarse surface layers also exist in many unregulated rivers with an ongoing sediment supply and peak flows which can transport all sizes of bed material. These surfaces were termed "pavements" at first but "mobile armor" has come to be preferred, as distinct from the "static" armor just described. This terminology is not yet standard, though, and some workers restrict "armor" to what is here termed static armor. Mobile armor allows a gravel-bed river reach to be in equilibrium (neither degrading nor aggrading, neither coarsening nor fining) despite the size-selective nature of bed-load transport. Without armor the annual bed-load flux would be finer-grained than the bed surface, since shear stresses in gravel-bed rivers are very rarely high enough for all sizes to be equally mobile. Armoring makes the intrinsically less mobile coarse fractions preferentially available for transport, whereas the more mobile fine fractions are mainly hidden in the subsurface (Parker and Klingeman, 1982). The armor forms by vertical winnowing during active bed-load transport. Entrainment of coarse clasts during floods creates gaps which are filled mainly by finer grains, in much the same way that the smallest bits of breakfast cereal end up at the bottom of the packet. Extreme floods may wash out the armor, but it re-forms during intermediate flows in most environments. However, in ephemeral streams there are no such flows and armoring is generally absent (Laronne *et al.*, 1994).

The two types of fluvial armor were initially treated separately but, following Dietrich *et al.* (1989) and Parker and Sutherland (1990), they have come to be seen as members of a continuum. In this view some degree of armoring is the response to any reduction in sediment supply rate below the capacity transport rate for an unarmored bed. Transport capacity depends on excess shear stress over a threshold for movement which, in turn, depends on the median diameter of the surface. Armoring raises the threshold and reduces the transport rate. In this view static armor is simply the limiting case as the bed-load flux vanishes, and different degrees of armouring represent channel self-adjustment to equalize bed-load capacity with supply (Dietrich *et al.*, 1989). Parker and Sutherland (1990) developed the concept by inverting transport equations for size mixtures and successfully retrodicted armor composition from measured bed-load flux and substrate size distribution.

Armoring can involve changes in grain packing as well as grain size distribution, and these may be at least as important in increasing resistance to entrainment and thus regulating sediment flux (Church *et al.*, 1998). The development of static armor is accompanied by progressively stronger imbrication and interlocking of the coarser clasts, as illustrated in Sutherland (1987), and this further increases the threshold stress. Additional structures form in streams with a mobile armor: pebble clusters, irregular reticulate stone cells, and transverse boulder steps in steep headwater streams. Surface structures associated with armoring are therefore important in understanding fluvial processes and the role of the bed surface in responding to, and regulating, sediment flux.

Desert pavement

Stony deserts are sufficiently extensive to have acquired local names (e.g., reg in Africa, gibber in Australia). One widely-held view about their origin is essentially the same as for static armor in a river: winnowing of finer material by wind, leaving a coarse lag. Sediment flux tends to zero over a fully-developed deflation lag surface, though in the early stages it may increase slightly because of enhanced elastic rebound of saltating grains (e.g., Nickling and Neuman, 1995). Pavement formation by deflation is inhibited by vegetation which reduces near-ground windspeed, and by surface crusting, so it is probably most likely in hyperarid conditions.

In semi-arid environments pavements can form through dislodgment of finer material by rainsplash followed by removal in sheetwash (e.g., Cooke, 1970), or stones can be forced upward through a clayey soil horizon by wetting and drying cycles (Jessup, 1960). More radically, McFadden *et al.* (1987) proposed that desert pavement can form during depositional episodes through a combination of eolian dust deposition and pedogenic processes beneath an evolving armor initiated by rock weathering.

Beach armor

Cobble beaches and spits are common on coasts with high wave energy, and coastal sedimentologists regard armoring as a fairly widespread phenomenon. Most interpretations emphasize clast sorting into shore-parallel bands under the action of wave swash and backwash. Shape as well as size sorting is involved (Bluck, 1967; Isla, 1993), with the most spherical clasts overpassing to higher up the beach. In certain instances gravel berms high on a beach may become armored by eolian deflation, with the sand winnowed out to accumulate in coastal dunes (Bascom, 1951).

Rob Ferguson

Bibliography

Bascom, W.N., 1951. The relationship between sand size and beach-face slope. *Transactions of the American Geophysical Union*, **32**: 866–874.
Bluck, B.J., 1967. Sedimentation of beach gravels: examples from South Wales. *Journal of Sedimentary Petrology*, **37**: 128–156.
Church, M., Hassan, M.A., and Wolcott, J.F., 1998. Stabilizing self-organized structures in gravel-bed stream channels: Field and experimental observations. *Water Resources Research*, **34**: 3169–3179.
Cooke, R.U., 1970, Stone pavements in deserts. *Annals of the Association of American Geographers*, **60**: 560–577.
Dietrich, W.E., Kirchner, J.F., Ikeda, H., and Iseya, F., 1989. Sediment supply and the development of the coarse surface layer in gravel-bedded rivers. *Nature*, **340**: 215–217.
Gessler, J., 1970. Self-stabilizing tendencies of alluvial channels. *Journal of the Waterways and Harbors Division, American Society of Civil Engineers*, **96**: 235–249.
Isla, F.I., 1993. Overpassing and armoring phenomena on gravel beaches. *Marine Geology*, **110**: 369–376.
Jessup, R.W., 1960. The stony tableland soils of the south-eastern portion of the Australian arid zone and their evolutionary history. *Journal of Soil Science*, **11**: 188–197.
Laronne, J.B., Reid, I., Frostick, L.C., and Yitshak, Y., 1994. The non-layering of gravel streambeds under ephemeral flood regimes. *Journal of Hydrology*, **159**: 353–363.
McFadden, L.D., Wells, S.G., and Jercinovich, M.J., 1987. Influences of eolian and pedogenic processes on the origin and evolution of desert pavements. *Geology*, **15**: 504–508.
Nickling, W.G., and Neuman, C.M., 1995. Development of deflation lag surfaces. *Sedimentology*, **42**: 403–414.
Parker, G., and Klingeman, P.C., 1982. On why gravel bed streams are paved. *Water Resources Research*, **18**: 1409–1423.
Parker, G., and Sutherland, A.J., 1990. Fluvial armor. *Journal of Hydraulics Research*, **28**: 529–544.
Powell, D.M., 1998. Patterns and processes of sediment sorting in gravel-bed rivers. *Progress in Physical Geography*, **22**: 1–32.
Sutherland, A.J., 1987. Static armour layers by selective erosion. In Thorne, C.R., Bathurst, J.C., and Hey, R.D. (eds.), *Sediment Transport in Gravel-bed Rivers*. Chichester: Wiley, pp.141–169.

Cross-references

Grain Size and Shape
Grain Threshold
Rivers and Alluvial Fans
Sediment Transport by Unidirectional Water Flows

ATTERBERG LIMITS

Atterberg Limits are the water contents which define transitions between the solid, plastic, and liquid states of a given soil material. The tests are restricted to cohesive soils with appreciable silt or clay fraction, and cannot be conducted readily on either sands or silts with a high sand fraction. Detailed mineralogical studies of cohesive soils have shown that Atterberg Limits are strongly related to both clay content and clay mineral species, and hence ultimately to the climatic and geologic conditions prevailing in the environment of deposition and during the post-depositional weathering cycle.

The modern test procedures for Atterberg Limits were formalized by engineers in the first half of the 20th century, and are described in most modern soil mechanics texts (Craig, 1997). Although the tests yield simple index numbers of soil behavior at various water contents, an important focus of

Table A1 Atterberg Limits of selected soils and associated engineering properties

Material type	Liquid limit, w_L	Plastic limit, w_P	Plasticity index, I_P	Liquidity index, I_L	Activity index, I_A	Φ'_r
Mexico City clay	400	100	300	0.9	3	—
Pierre Shale, USA	200	40	160	−0.2	2.5	8°
London Clay	80	30	50	0	0.9	10°
Leda Clay, Canada	55	25	30	1.4	0.5	27°
Glaciolacustrine silt	25	20	5	0	0.5	30°
Loess, Nebraska	35	15	20	—	1.3	—
Montmorillonite clay	650	70	580	—	9	5°
Illite clay	110	50	60	—	0.5	10°
Kaolinite clay	50	40	10	—	0.4	15°

Tabled values are averages, based in part on Lambe and Whitman (1969), Selby (1993), Kenney (1984), Higgins and Modeer (1996), and Lefebvre (1996).
Φ'_r = residual angle of shearing resistance.

modern soils engineering research has been to establish correlations between Atterberg Limits and more complex soil properties, such as shear strength.

Atterberg Limits tests require the preparation of an air-dried sample through a 425 µm screen, followed by mixing with distilled water to form a soil paste. In each test, the water content, w, is calculated as the mass percentage of water per mass of oven dry soil. The Liquid Limit, w_L, defines the boundary between the plastic solid and liquid states, and is traditionally measured in the Casagrande device. A 5.4 cm radius brass cup is partially filled with wetted soil paste to a depth of about 1 cm. A grooving tool is then used to scribe the center of the soil pat. An eccentric cam attached to a crank handle is used to raise and drop the cup twice per second through a 1 cm vertical distance onto a hard rubber base plate. The number of blows required to close the base of the groove over a distance of one-half inch (12 mm) is noted, and this failing soil is sampled, weighed and oven dried. The test is repeated at least three times at different water contents to obtain a semi-logarithmic plot of number of blows (log axis) versus water content. The water content interpolated at 25 blows is then the Liquid Limit.

The procedure described above requires some technical experience to obtain consistent results, and an operationally simpler drop-cone test is now used in many soils laboratories. The drop cone, with a mass 80 g and cone angle of 30°, is lowered until its tip just contacts the sample surface. The cone is then released for 5 s and the depth of penetration is noted. The penetration depths at various water contents are noted, with the Liquid Limit defined as the water content interpolated at a cone penetration of 20 mm.

The Plastic Limit, w_P, is the water content below which the soil becomes brittle, and cracks when remolded. Traditionally, the Plastic Limit is determined by rolling of a soil worm on a ground glass plate, using the heel of the hand, to a thickness of 0.125 inches (3 mm). At the Plastic Limit, the soil worm can no longer be thinned by rolling, and cracking and breakage occurs. The sample is then weighed and oven dried. Three more tests are conducted to ensure consistent results. By the drop-cone method, w_P is the water content at which the cone penetrates the sample approximately 3 mm.

The Shrinkage Limit, w_S, is determined by filling a brass, semi-circular conduit, 12.5 mm in radius and 140 mm in length, with remolded soil paste, then allowing it to air dry. Linear shrinkage of the sample is monitored, with w_S defined as the water content at which no further volume decrease occurs.

Various indices of soil behavior can be derived from basic Atterberg Limits data. The Plasticity Index, $I_P = (w_L - w_P)$, represents the range of water contents between the solid and liquid states. An engineering convention is to construct a plasticity chart by plotting I_P as ordinate versus Liquid Limit, w_L. The plasticity chart provides a valuable summary of the geotechnical properties of a large number of soil samples, and allows a rapid separation of low plasticity silty materials of relatively high frictional strength from highly plastic clays, the latter usually associated with high swelling potential and lower shear strength.

The Liquidity Index, $I_L = (w - w_P)/(w_L - w_P)$, where w is the natural field water content. I_L is a measure of field moisture content normalized by plasticity, since I_P is denominator. I_L values above 1.0 indicate a field water content above the Liquid Limit, a condition common in soft unconsolidated clays (Lefebvre, 1996). These soils tend to liquefy when disturbed. Conversely, negative I_L values indicate moisture values below the Plastic Limit, which is common in stiff clays and shales in drier climates. Such soils exhibit much higher shear strength. I_L is therefore a useful hydrotechnical index of the relative stability of various earth materials.

The Activity Index, $I_A = I_P/\%$ clay, and measures soil plasticity normalized by clay content. Activity values are strongly correlated with clay mineral type. Values above 2 indicate expanding montmorillonite clays; values closer to 0.5 are typical of kaolinite clay soils. High I_A values generally indicate soils of low strength, high plasticity and high swell-shrink potential, thus indicating potential slope stability and foundation engineering problems.

Table A1 summarizes typical values of Atterberg Limits and associated indices for a range of soil materials. The very large ranges in values emphasize the importance of both clay mineral species and clay content as controls. Of particular note is the contrast between very high plasticity, montmorillonite clay soils (e.g., Mexico City volcanogenic lacustrine clay and Pierre Shale marine clay-shale), in comparison with very low plasticity silts and silty-clay soils (e.g., glaciolacustrine silts and glaciomarine Leda Clay). In the latter two deposits, much of the clay-sized fraction typically is not true phyllosilicate but

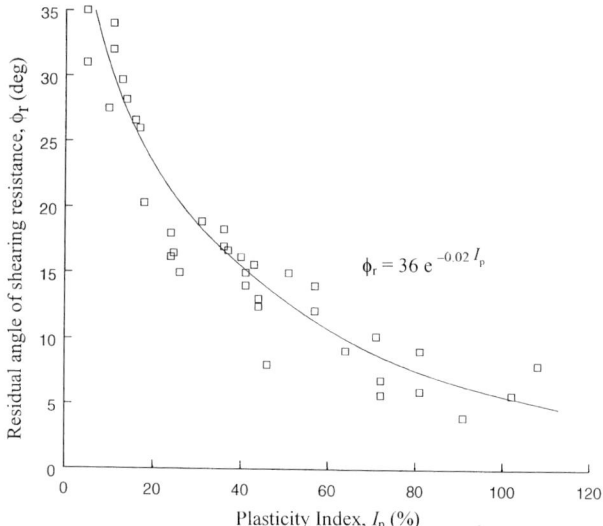

Figure A10 Relation between plasticity index and residual angle of shearing resistance. Data points are derived from Voight (1973), Kanji (1974) and the author's files.

instead quartz feldspar rock flour derived from glacial abrasion. Average values of residual friction angle are also given for each soil type. Residual strength refers to the lowest value obtainable from large-displacement, drained shear tests. Strong inverse relations exist between Φ'_r and I_P (Figure A10) and I_A, as previously noted by Voight (1973), Kanji (1974), and Wroth and Wood (1979).

Michael J. Bovis

Bibliography

Craig, R.F., 1997. *Soil Mechanics*, 6th edn. London: Spon Press.
Higgins, J.D., and Modeer, V.A., 1996. Loess. In Turner, A.K., and Schuster, R.L. (eds.), *Landslides: Investigation and Mitigation*. Washington, DC: National Academy Press, pp. 585–606.
Kanji, M.A., 1974. The relationship between drained friction angles and Atterberg limits of natural soils. *Géotechnique*, **24**: 671–674.
Kenney, C., 1984. Properties and behaviours of soils relevant to slope stability. In Brunsden, D., and Prior, D.B. (eds.), *Slope Instability*. New York: J. Wiley & Sons, pp. 27–66.
Lambe, T.W., and Whitman, R.V., 1969. *Soil Mechanics*. New York: J. Wiley & Sons.
Lefebvre, G., 1996. Soft sensitive clays. In Turner, A.K., and Schuster, R.L. (eds.), *Landslides: Investigation and Mitigation*. Washington, DC: National Academy Press, pp. 607–619.
Selby, M.J., 1993. *Hillslope Materials and Processes*, 2nd edn. Oxford: Oxford University Press.
Voight, B., 1973. Correlation between Atterberg plasticity limits and residual strength of natural soils. *Géotechnique*, **23**: 265–267.
Wroth, C.P., and Wood, D.M., 1979. The correlation of index properties with some basic engineering properties of soils. *Canadian Geotechnical Journal*, **15**: 137–145.

Cross-references

Bentonites and Tonsteins
Colloidal Properties of Sediments
Grain Size and Shape
Gravity-Driven Mass Flows
Liquefaction and Fluidization
Mass Movement
Mudrocks
Slurry
Smectite Group
Weathering, Soils, and Paleosols

ATTRITION (ABRASION), FLUVIAL

Fluvial attrition is generally understood to refer to any process of mechanical grain wear leading to grain size reduction in the course of sediment transport through rivers. Fluvial abrasion is generally accepted as an equivalent term, though it might also be interpreted to mean modification of the stream bed or banks. Mikos (1993) has presented a modern treatise on the subject of fluvial attrition of clastic grains.

"Abrasion" in the present context specifically implies surface wear caused by the relative motion of two surfaces in frictional contact, especially when one surface is harder than the other ("grinding", when one of the grains is at rest, is sometimes distinguished). Additional processes that have been invoked as contributing to fluvial "abrasion" include impact or percussive effects (e.g., chipping, splitting and sandblasting), and crushing, when the impact of a larger rock causes a small grain to disintegrate. Weathering and solution have also been invoked, but this appears to confuse preparatory processes with mechanical destructive mechanisms. In rivers, abrasion appears mainly to affect pebble and larger sizes.

The characteristic downstream reduction in size, hence weight, of fluvial gravels was formulated by Sternberg (1875) into a "law" of weight loss—"the wear of a grain is proportional to its submerged weight (W) and the distance (L) it has travelled". Mathematically, $dW/dL = -\alpha W$, in which α is the weight loss rate coefficient. The statement integrates to the familiar result

$$W = W_0 e^{-\alpha L}$$

in which W_0 is the initial grain weight before travel. Since $W \propto D^3$, grain diameter, the rule can be interpreted as a size reduction rule, in which $\alpha_D = \alpha_W/3$. Sternberg's result has been abundantly confirmed empirically and, since river engineers supposed for a long time that abrasion was the chief source of the effect, it became known as the law of fluvial abrasion.

However, experimental measurements of abrasion nearly always return rate coefficients much smaller than ones inferred from field data (Adams, 1978 provides a list of experimentally derived abrasion coefficients for a wide range of lithologies). It has consequently been recognized that size-selective transport (e.g., Plumley, 1948; Bradley et al., 1972) and weathering of grains during storage in bars and floodplains (Bradley, 1970) represent additional mechanisms for gravel size reduction along rivers.

Adams (1979) investigated the wear of weathered grains recruited into river headwaters. He found that the size reduction coefficient (α) itself changes with distance downstream, at a declining rate, so that the rate of change of grain size with distance follows a power law, rather than the exponential rule of Sternberg. He also showed convincingly that the result was due to wear and interpreted it to reflect the rapid destruction of weathered material. Farther downstream, the Sternberg rule applies to the remaining, competent material. However, Jones and Humphrey (1997) demonstrated rapid changes in the wearing coefficient of material liberated from long-term storage in fluvial deposits, where substantial weathering again occurs. At the other extreme, Rice and Church (1998) demonstrated dramatic changes in grain size over reach-scale distances far too short for appreciable particle wear to occur, indicating the efficacy of size-selective transport. Both processes occur simultaneously, particularly in the presence of multiple lithologies, when less resistant lithologies may be consumed by abrasion whilst the size of the more resistant ones varies mainly by selective transport. Recent work has sought to examine the relative significance of and interaction between abrasion and selective transport of grains (e.g., Parker, 1991a,b).

There remain significant complexities in the problem of determining wear and its relation to grain size variation downstream. Mikos (1993) criticized tumbling mill experiments, in which many experimental abrasion coefficients have been determined, and showed (also in Mikos, 1995) that, when grain size is established on a weight basis (e.g., by customary sieve analysis), the relation between size loss and weight loss of measured aggregations of grains depends upon whether or not abrasion products are preserved in the analysis and also upon the measurement principle used in the analysis.

Michael Church

Bibliography

Adams, J., 1978. Data for New Zealand pebble abrasion studies. *New Zealand Journal of Science*, **21**: 607–610.
Adams, J., 1979. Wear of unsound pebbles in river headwaters. *Science*, **203**: 171–172.
Bradley, W.C., 1970. Effect of weathering on abrasion of granitic gravel, Colorado River (Texas). *Geological Society of America Bulletin*, **81**: 61–80.
Bradley, W.C., Fahnestock, R.K., and Rowekamp, E.T., 1972. Coarse sediment transport by flood flows on Knik River, Alaska. *Geological Society of America Bulletin*, **83**: 1261–1284.
Jones, L.S., and Humphrey, N.F., 1997. Weathering-controlled abrasion in a coarse-grained, meandering reach of the Rio Grande: implications for the rock record. *Geological Society of America Bulletin*, **109**: 1080–1088.
Mikos, M., 1993. Fluvial abrasion of gravel sediments. ETH, Zurich, Versuchsanstalt für Wasserbau, Hydrologie und Glaziologie. *Mitteilungen*, **123**: 322pp.
Mikos, M., 1995. Fluvial abrasion: converting size reduction coefficients into weight reduction rates. *Journal of Sedimentary Research*, **A65**: 472–476.
Parker, G., 1991a. Selective sorting and abrasion of river gravel. I: Theory. *Journal of Hydraulic Engineering*, **117**: 131–149.
Parker, G., 1991b. Selective sorting and abrasion of river gravel. II: Applications. *Journal of Hydraulic Engineering*, **117**: 150–171.
Plumley, W.J., 1948. Black Hills terrace gravels: a study in sediment transport. *Journal of Geology*, **56**: 526–577.
Rice, S.P., and Church, M., 1998. Grain size along two gravel-bed rivers: statistical variation, spatial pattern and sedimentary links. *Earth Surface Processes and Landforms*, **23**: 345–363.
Sternberg, H., 1875. *Untersuchungen über längen-und Querprofil geschiebeführende Flüsse. Zeitschrift für Bauwesen*, **25**: 483–506.

Cross-references

Armor
Attrition (Abrasion), Marine
Grain Size and Shape
Grain Threshold
Sediment Transport by Unidirectional Water Flows

ATTRITION (ABRASION), MARINE

Marine attrition is the mechanical wearing down of rock particles and minerals by wave action both longshore and shore-normal, leading to increasing textural and compositional maturity. The term textural maturity describes the prevalence of finer and better rounded grains, compositional maturity comprises the extinction of less resistant rock particles or minerals. Well-rounded fine-grained quartz sand theoretically represents the final stage of maturity.

The role of sphericity as a measure of marine attrition is debated, as is the effect of sorting or selective transportation. Because of their differing settling velocities, particles of different size or weight, of different sphericity and different roundness behave differently during marine transportation both longshore and shore-normal, leading possibly to similar results as attrition. The same is true for longshore fining of the particles.

Marine attrition of sand

Literature dealing with marine attrition of sand is extremely scarce. Experiments demonstrate that rounding and thus attrition of quartz sand is extremely low or practically negligible (Thiel, 1940; Kuenen, 1960; Schubert, 1964). The relative smoothness of foreshore grains due to marine attrition is discussed by Lee and Osborne (1995), the creation of a more mature heavy mineral suite of beach placers in southeastern Australia by Roy (1999). Most modifications of quartz sand or heavy mineral suites along beaches are assumed to be the result of sorting or selective transportation (Bascom, 1951; McBride *et al.*, 1996; Trenhaile *et al.*, 1996).

Marine attrition of pebbles

Weight loss and the modification of petrographic composition

Matthews (1983) put up to 76 tonnes of limestone pebbles into the surf of Palliser Bay, New Zealand, at three sites. A weight loss of 41 percent per year due to rounding of the pebbles was found at the most exposed site and 15 percent or 7 percent at the two other sites. Bartholomä *et al.* (1998) investigated the 8 km Bianco Beach in Calabria, southern Italy, a tide-less, wave dominated high energy beach, sampling the beach-parallel profile and the beach-normal profiles (back-, fore- and offshore). This beach is fed by a fluvial delta at its southern end. Thirty nine samples comprise a total of 2.6×10^3 kg with grain sizes between 16 mm and 63 mm. The beach sediment contains, among others, two reference groups of rocks, a granitic group (more or less isotropic granites, pegmatites and vein quartz) and a gneiss group (more or less anisotropic strongly foliated gneisses, schists and phyllites). The granites are obviously more resistant to abrasion than the gneisses. In all fractions and beach zones the proportion of gneiss clasts decreases in the direction of net transport from 56 percent to 31 percent, corresponding to a mean reduction of 3 percent per kilometer (Figure A11(1), the zero value corresponds to the fluvial input), indicating strong marine attrition.

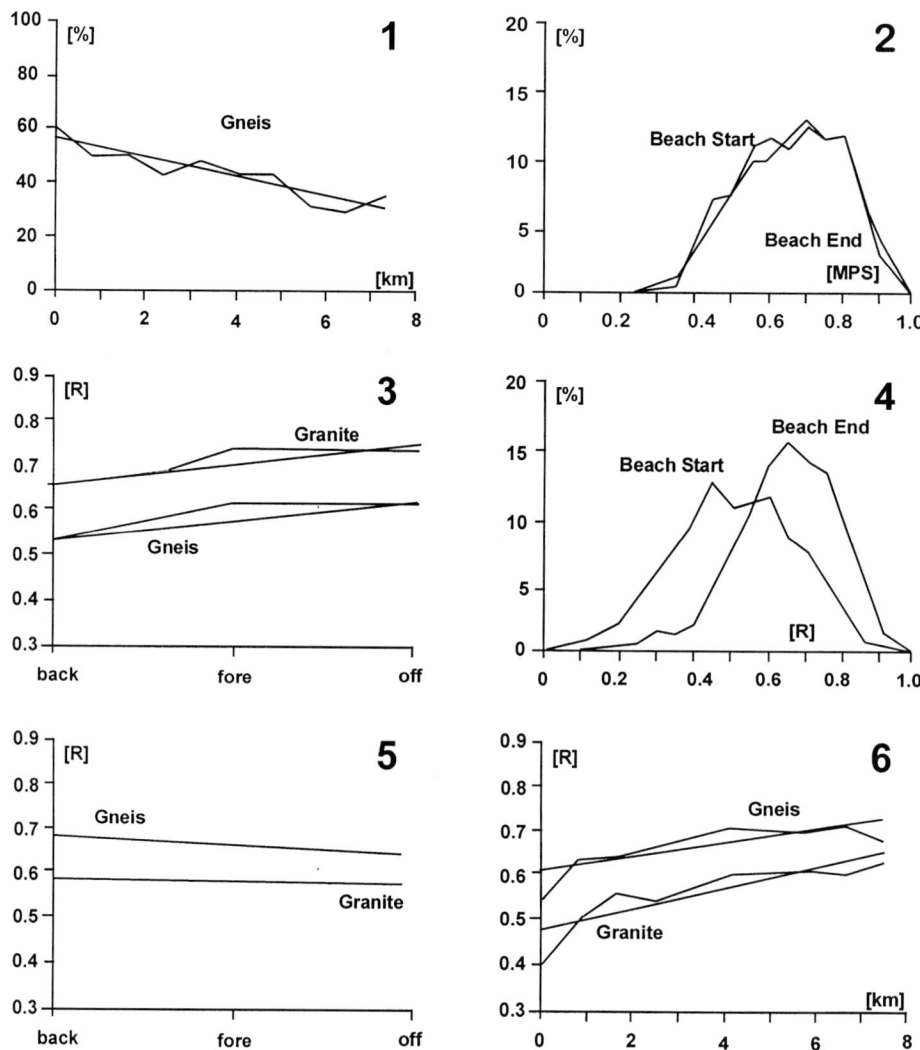

Figure A11 (1) Reduction of less resistant gneiss clasts longshore, measurements and trend; (2) Sphericity frequency distributions of the gravel populations at beach start and beach end. MPS: Maximum projection sphericity, Sneed and Folk, 1958; (3) Sphericity (MPS) values normal to the beach, granite and gneiss clasts, measurements and trends, mean values of the backshore, foreshore, and offshore zone; (4) Roundness frequency distributions of the gravel populations at beach start and beach end. R: Krumbein Roundness, Krumbein 1941; (5) Roundness (R) values normal to the beach, gneiss and granite clasts, measurements and trends, mean values of the backshore, foreshore, and offshore zone; (6) Longshore rounding (R) of gneiss and granite clasts, measurements and trends. (From Bartholomä et al., 1998, printed with permission of SEPM (Society for Sedimentary Research).

Sphericity

Dobkins and Folk (1970) grouped authors into "sorters" who believe that disc-shaped gravels remain high on the beach where geologists can easily find them, whereas rods and equants selectively roll back to the foot of the beach and are lost from sight, and "abraders", who believe that surf mechanics develops discs by attrition, presumably by sliding the gravel back and forth. Dobkins and Folk (1970) defined themselves as abraders: the production of discs on the beaches is predominantly a result of attrition caused by pebbles sliding over sand or small pebbles in the surf zone. In contrast, Kuenen (1964) states that pebbles were flattened very little by sliding. The flatness of beach pebbles as compared to fluvial ones must be a matter of selection. Also Bluck (1967) demonstrated that particle shapes are not so much made as used on these beaches, and that discs are not produced by a special feature of marine attrition. The most oblate discs are found in areas least worked on by the sea. Even according to Bartholomä et al. (1998) sphericity fails as an appropriate indicator of marine attrition. The sphericity frequency distribution of more than 20,000 clasts is almost identical comparing beach start and beach end (Figure A11 (2)). There is a slight increase of sphericity of both granites and gneisses in the foreshore zone (Figure A11 (3)), but roundness is equal when the three beach environments are compared (Figure A11 (5)), indicating that this increase in sphericity is caused by sorting rather than attrition.

Roundness

Roundness values of older articles are hard to compare, because only printed roundness images of the Krumbein tables (1941) were used. Bartholomä et al. (1998) determined the roundness values of the entire data set via Fourier analysis (Diepenbroek et al., 1992) and found a remarkable difference between beach start and beach end (Figure A11 (4)). Because the roundness-profile crossing the beach does not show any trend (Figure A11 (5)), the foreshore zone is ruled out as a locus of forced marine abrasion. The authors show, that the uniform roundness along the cross-shore profiles and the increasing alongshore roundness in the direction of net transport (Figure A11 (6)) can only be interpreted as the result of sporadic high energy events, which cause the entire set of pebbles to migrate offshore. The cross-shore profile of the beach then is uniform and very flat, being composed entirely of coarse sand. All pebble grain-sizes and all rock types are abraded and rounded simultaneously. This means that during rounding due to attrition the proportions of the axes of both anisotropic gneiss clasts and isotropic granite clasts remain preserved, and no change of sphericity takes place.

Longshore fining

The loss of weight due to rounding and the extinction of less resistant rock types certainly go hand in hand with a longshore fining of the clasts. It is not possible, however, to discriminate this effect from longshore sorting and selective transportation.

Summary

Increasing roundness and decreasing percentages of less resistant rock types or minerals are the main indicators of marine attrition. The meaning of sphericity is uncertain, but the majority of the authors tend to negate sphericity as an useful tool in examining marine attrition. Longshore fining can be the product of both marine attrition and selective transportation or sorting, and as a result it fails as an indicator of marine attrition alone.

Hillert Ibbeken

Bibliography

Bartholomä, A., Ibbeken, H., and Schleyer, R., 1998. Modification of gravel during longshore transport (Bianco Beach, Calabria, Southern Italy). *Journal of Sedimentary Research*, **68**: 138–147.
Bascom, W.N., 1951. The relationship between sand-size and beach-face slope. *Transactions of the American Geophysical Union*, **32**: 866–874.
Bluck, B.J., 1967. Sedimentation of beach gravel: examples from south Wales. *Journal of Sedimentary Petrology*, **37**: 128–156.
Diepenbroek, M., Bartholomä, A., and Ibbeken, H., 1992. How round is round? A new approach to the topic "roundness" by Fourier grain shape analysis. *Sedimentology*, **39**: 411–422.
Dobkins, J.E., and Folk, R.L., 1970. Shape development on Tahiti-Nui. *Journal of Sedimentary Petrology*, **40**: 1167–1203.
Krumbein, W.C., 1941. Measurement and geological significance of shape and roundness of sedimentary particles. *Journal of Sedimentary Petrology*, **11**: 61–72.
Kuenen, Ph.H., 1960. Experimental abrasion of sand grains. *Report of the International Geological Congress*, **10**: 50–53.
Kuenen, Ph.H., 1964. Experimental abrasion, 6: surf action. *Sedimentology*, **3**: 29–43.
Lee, A.C., and Osborne, R.H., 1995. Quarz grain-shape of Southern California beaches. *Journal of Coastal Research*, **11**: 1336–1345.
Matthews, E.R., 1983. Measurements of beach pebble attrition in Palliser Bay, southern North Island, New Zealand. *Sedimentology*, **30**: 787–799.
McBride, E.F., Abel, W.A., and McGivery, Th.A., 1996. Loss of sand-size feldspar and rock fragments along the South Texas barrier island, USA. *Sedimentary Geology*, **107**: 37–44.
Roy, P.S., 1999. Heavy mineral beach placers in southeastern Australia; their nature and genesis. *Economic Geology and the Bulletin of the Society of Economic Geologists*, **94**: 567–588.
Schubert, C., 1964. Size-frequency distributions of sand-sized grains in an abrasion mill. *Sedimentology*, **3**: 288–295.
Sneed, E.D., and Folk, R.L., 1958. Pebbles in lower Colorado River, Texas. *Journal of Geology*, **27**: 140–150.
Thiel, G.A., 1940. The relative resistance to abrasion of mineral grains of sand size. *Journal of Sedimentary Petrology*, **10**: 103–104.
Trenhaile, A.S., van der Nol, L.V., and LaValle, P.D., 1996. Sand grain roundness and transport in the swash zone. *Journal of Coastal Research*, **12**: 1017–1023.

Cross-references

Attrition (Abrasion), Fluvial
Grain Size and Shape
Maturity: Textural and Compositional
Sands and Sandstones

AUTHIGENESIS

Overview

Authigenesis refers to processes by which minerals form, *in place*, within sediments and sedimentary rocks. Such minerals are referred to as *authigenic minerals*. In contrast, *detrital* minerals are erosion products deposited in the depositional environment. Diagenesis refers to the physical and chemical changes that occur in sediments after they are deposited (or precipitated). All authigenic minerals form within this context. The formation of authigenic minerals is a result of the presence of unstable mineral grains, biological modification of the sediment, grain deformation and fracturing, flux of various fluids through the sediment, and changing burial conditions. The intrinsic value of most sediments is their ability to hold fluid such as water and hydrocarbons. From an economic viewpoint, authigenesis is important because it influences the *porosity* (void volume in the rock) and *permeability* of the rock (ability of the rock to transmit fluid).

Authigenic minerals can reveal clues to the conditions the sediment has been subjected to after deposition. For example, petrographic thin sections reveal sequences of authigenic minerals, which allows a reconstruction of the relative timing or *paragenesis* of events. Early formed authigenic minerals are characterized by euhedral terminating crystal faces indicating growth into open pore space. Later pore-filling cements or grain replacements are characteristically anhedral crystals. Trace elements, isotopic, and fluid inclusion data can reveal the source of authigenic mineral components, temperatures of crystallization, and the type of water with which the sediment has interacted (e.g., Schultz et al., 1989; Stallard and Boles, 1989; Thyne and Boles, 1989). In some cases, quantitative crystallization ages can be obtained from authigenic minerals using isotopic dating methods (e.g., Lee et al., 1985).

Cause of authigenesis

Authigenic minerals form because sedimentary particles are not in thermodynamic equilibrium with the fluid composition, pressure, and temperature conditions at the time of deposition. Biological activity is common near the sediment-water interface. The organisms have the ability to control oxidation-reduction reactions and the generation of methane or carbon dioxide. Initially, the sediment is deposited with high porosity and permeability, which implies that large fluid volumes can easily move through the sediment. While the sediment resides near the surface, salinity and water composition may change drastically by changes in sea level, changes in hydrologic drainage, and evaporation. As a result, early authigenesis may be controlled by biological activity and changing fluid composition in an open system. Over time the sediment loses pore space and permeability is reduced (due to compaction and cementation); thus fluid composition becomes a less important control on the formation of authigenic minerals.

As the sediment becomes buried, pressure and temperature increases, causing further mineralogic (and physical changes) to the sediment. Typically, temperature increases 25–35°C per kilometer of burial. This causes mineral-solution reaction rates to increase by a factor of 10 for every kilometer of burial (Krauskopf and Bird, 1995). As a result of increased reaction rates, authigenic mineral replacement of pre-existing minerals becomes more common at higher temperature. From a thermodynamic viewpoint, reactions sensitive to temperature involve net changes in entropy (or degree of disorder). Thus, dehydration reactions occur at elevated temperatures due to the higher entropy associated with water released from hydrated minerals.

Pressure typically increases between 11.3 MPa/km (hydrostatic gradient) and 22.6 MPa/km (lithostatic gradient) during burial. Pressure has a less marked effect on the formation of authigenic minerals. From a thermodynamic perspective, mineral changes (reactions) which are pressure dependent involve an overall net volume reduction.

Eventually, conditions of pressure and temperature at deep burial become so extreme that we refer to the processes as *metamorphic* and the rocks as metamorphic rocks. This transition is not recognized by nature and many authigenic minerals span the boundary between diagenesis and metamorphism. In general though, most petrologist would agree that 200–250°C would be the upper temperature limit of what would be called authigenic minerals.

Examples of authigenesis

Authigenic minerals forming at shallow depths

Shallow sediment can be subjected to marked chemical changes and/or high flux of pore water. As a result, large volumes of early cements can form such as carbonate concretions of calcite, siderite, dolomite, or pyrite (e.g., Curtis *et al.*, 1986; Hayes and Boles, 1993). High fluid flux can result in leaching of detrital aluminosilicate grains from the rock, leaving a residue of authigenic kaolinite (e.g., Hayes and Boles, 1992). One of the earliest consequences of seawater being present within the sediment is biogenic reduction of the sulfate to sulfide and precipitation of authigenic pyrite. Further evidence of biological activity in shallow buried sediment is the production of ammonium, which can substitute for potassium in authigenic feldspar (Ramseyer *et al.*, 1992). In many cases, grains in clastic sediments show early coatings of authigenic iron oxides or clay minerals, including hematite, smectite, illite or chlorite (e.g., Heald and Larese, 1974).

Volcanic glass is an unstable component in volcaniclastic sediments and it is rarely preserved. During shallow burial, the main reaction products are authigenic clay minerals (smectite) and zeolites (Mumpton, 1981). Zeolite producing reactions are widely dispersed in deep sea pelagic sediments including formation of phillipsite, usually found in young sediments and clinoptilolite, found in older sediment. Saline alkaline lakes have abundant altered ash beds including the zeolites phillipsite, clinoptilolite, mordenite and analcime. Diagenetic K-feldspar is found in the most concentrated pore waters of these lacustrine environments.

Authigenic minerals forming during burial

Calcite, dolomite, and ankerite cements can form in clastic sediments at burial depths as deep as 4–5 km (e.g., Boles and Franks, 1979). The source of calcium for these cements may include clay minerals and feldspars, and carbon derived from organic matter (e.g., Boles, 1998).

Clays and silts undergo mineral transformations, which are largely a response to temperature increase during burial. One of the most studied reactions is the dehydration of detrital smectite to illite (Hower *et al.*, 1976). Most smectite has some interlayering of illite and observations show that the proportion of illite layers increases with increasing burial or temperature. This transition occurs at about 100°C in many sedimentary basins. The source of the potassium is typically detrital K-feldspar or in some cases potassium micas. The smectite-illite transformation releases Si, Mg, and Fe, which forms cements of quartz, carbonates, and clays (Boles and Franks, 1979). Other clay mineral changes that occur are the transition of kaolinite to chlorite or illite (Boles and Franks, 1979; Bjorlykke and Agaard, 1992).

Authigenic quartz cements commonly form euhedral overgrowths on detrital quartz grains. Most recent work has shown that quartz cements are controlled largely by the kinetics of quartz precipitation. Thus, at temperatures of 90–100°C quartz cement becomes significant in many sedimentary basins (Walderhaug, 1996).

Biogenic siliceous sediment is deposited as opal or amorphous silica. Opal is an unstable component of sediments and it becomes progressively altered from cristobalite to chert composed of quartz (Murata and Larson, 1975; Pisciotto, 1980). The opal-cristobalite transition occurs at about 35–55°C and the cristobalite-quartz transition occurs from about 55°C to 110°C.

Albitization of plagioclase is widely recognized in deeply buried sediments (e.g., Coombs, 1954; Land and Milliken, 1981). To a lesser extent K-feldspar becomes albitized during burial. The albitization of plagoclase usually starts at about 100°C in sedimentary sequences and most calcium plagioclase has been altered to albite by about 150°C (Boles, 1982). Albitization consumes sodium and releases calcium to the pore water. Thus, if the rock has very low permeability the reaction is inhibited due to the inability of material to be transported (Boles and Ramseyer, 1988). Authigenic products of feldspar dissolution and albitization are clay minerals (kaolinite or chlorite) and calcite or laumontite (Boles and Coombs, 1975).

Early formed zeolites are eventually replaced in a series of dehydration reactions by laumontite, albite, K-feldspar, and chlorite (Boles and Coombs, 1977) or other minerals such as prehnite or pumpellyite with increasing age and burial depth. Laumontite requires a minimum temperature of about 70°C but is stable beyond 200°C and is considered by many as a transition to conventional metamorphism.

Summary

In summary, authigenic mineral reactions are initially controlled by fluid dominated open systems and later by rock-dominated relatively closed system. Although we have a good descriptive understanding of the type of authigenic mineral reactions that occur, there remains much to be done to quantify these processes. Present and future research trends are interdisciplinary studies to understand the mineral/fluid interface, the role of organisms in mineralization, the rates of authigenesis, the role of faulting as pathways and barriers to fluid flow, and interpreting early authigenic minerals in terms of sea level and climate/atmospheric changes.

James R. Boles

Bibliography

Bjorlykke, K., and Aagaard, P., 1992. Clay minerals in North Sea sandstones In *Origin, Diagenesis and Petrophysics of Clay Minerals in Sandstones*, SEPM, Special Publication, 37, 65–80.

Boles, J.R., 1982. Active albitization of plagioclase, Gulf Coast Tertiary. *American Journal of Science*, **282**: 165–180.

Boles, J.R., 1998. Carbonate cementation in Tertiary sandstones of the San Joaquin Basin. In Morad, S. (ed.), *Carbonate Cementation in Sandstones*. International Association of Sedimentology Special Publication, 26, pp. 261–284.

Boles, J.R., and Coombs, D.S., 1975. Mineral reactions in zeolitic Triassic tuff, Hokonui Hills, New Zealand. *Geological Society of America Bulletin*, **86**: 163–173.

Boles, J.R., and Coombs, D.S., 1977. Zeolite facies alteration of sandstone in the Southland syncline, New Zealand. *American Journal of Science*, **77**: 982–1012.

Boles, J.R., and Franks, S.G., 1979. Clay diagenesis in Wilcox sandstones of southwest Texas: Implications of smectite diagenesis on sandstone cementation. *Journal of Sedimentary Petrology* **49**: 55–70.

Boles, J.R., and Ramseyer, K., 1988. Albitization of plagioclase and vitrinite reflectance as paleothermal indicators, San Joaquin Basin. In Graham, S.A. (ed.), *Studies of the Geology of the San Joaquin Basin*, Pacific Section Society of Economic Paleontologists and Mineralogists, 60, pp. 129–139.

Coombs, D.S., 1954. The nature and alteration of some Triassic sediments from Southland, New Zealand. *Transaction of the Royal Society of New Zealand*, **82**: 65–109.

Curtis, C.D., Coleman, M.L., and Love, L.G., 1986. Pore water evolution during sediment burial from isotopic and mineral chemistry of calcite, dolomite, and siderite. *Geochimica et Cosmochimica Acta*, **50**: 2321–2334.

Hayes, M.J., and Boles, J.R., 1992. Volumetric relations between dissolved plagioclase and kaolinite in sandstones: implications for aluminum mass transfer in the San Joaquin basin, California. *Society of Economic Paleontologists and Mineralogists*, Special Publication, 47, pp. 110–123.

Hayes, M.J., and Boles, J.R., 1993. Evidence for meteoric recharge in the San Joaquin basin, California provided by isotope and trace element geochemistry. *Marine and Petroleum Geology*, **10**: 135–144.

Heald, M.T., and Larese, R.E., 1974. Influence of coatings on quartz cementation. *Journal of Sedimentary Petrology* **44**: 1269–1274.

Hower, J., Eslinger, E.V., Hower, M.E., and Perry, E.A., 1976. Mechanisms of burial metamorphism of argillaceous sediments: 1. Mineralogic and chemical evidence. *Bulletin of the Geological Society of America*, **87**: 725–737.

Krauskopf, K.B., and Bird, D.K., 1995. *Introduction to Geochemistry*. 3rd edn., New York, NY: McGraw-Hill Publ. Co., p. 647.

Land, L.S., and Milliken, K.L., 1981. Feldspar diagenesis in the Frio Formation, Brazoria County, Texas Gulf Coast. *Geology*, **9**: 314–318.

Lee, M., Aronson, J.L., and Savin, S.M., 1985. K/Ar dating of gas emplacement in Rotliegendes Sandstone, Netherlands. *American Association of Petroleum Geologist Bulletin*, **69**: 1381–1385.

Mumpton, F.A. (ed.), 1981. Mineralogy and geology of natural zeolites. In *Reviews in Mineralogy*, 4, Mineralogic Society America, Special Publication, p. 225.

Murata, K.J., and Larson, R.R., 1975. Diagenesis of Miocene siliceous shales, Temblor range, California. *U.S. Geological Survey Journal of Research*, **3**: 553–566.

Pisciotto, K.A., 1980. Diagenetic trends in the siliceous facies Monterey Shale in the Santa Maria region, California. *Sedimentology*, **28**: 547–571.

Ramseyer, K., Diamond, L.W., and Boles, J.R., 1992. Authigenic K-NH_4-feldspar in sandstones: a fingerprint of the diagenesis of organic matter. *Journal of Sedimentary Petrology*, **63**: 1092–1099.

Schultz, J.L., Boles, J.R., and Tilton, G.R., 1989. Tracking calcium in the San Joaquin basin, California: a strontium isotopic study of carbonate cements at North Coles Levee. *Geochimical et Cosmochimica Acta*, **53**: 1991–1999.

Stallard, M.L., and Boles, J.R., 1989. Oxygen isotope measurements of albite-quartz-zeolite mineral assemblages, Hokonui Hills, Southland, New Zealand. *Clays and Clay Minerals*, **37**: 409–418.

Thyne, G., and Boles, J.R., 1989. Isotopic evidence for origin of the Moeraki septarian concretions, New Zealand. *Journal of Sedimentary Petrology*, **59**: 272–279.

Walderhaug, O., 1996. Kinetic modeling of quartz cementation and porosity loss in deeply buried sandstone reservoirs. *American Association of Petroleum Geologist Bulletin*, **80**: 731–745.

Cross-references

Anhydrite and Gypsum
Ankerite
Bacteria in Sediments
Bentonites and Tonsteins
Cathodoluminescence
Cements and Cementation
Chlorite in Sediments
Compaction (Consolidation) of Sediments
Diagenesis
Diagenetic Structures
Dolomite Textures
Fabric, Porosity, and Permeability
Fluid Inclusions
Gases in Sediments
Hydrocarbons in Sediments
Hydroxides and Oxyhydroxide Minerals
Illite Group Clay Minerals
Isotopic Methods in Sedimentology
Kaolin Group Minerals
Magadiite
Mixed-Layer Clays
Oceanic Sediments
Porewaters in Sediments
Pressure Solution
Sabka, Salt Flats, Salina
Seawater
Septarian Concretions
Siliceous Sediments
Smectite Group
Sulfide Minerals in Sediments
Zeolites in Sedimentary Rocks

AUTOSUSPENSION

Definition

An autosuspension current may be defined as a particle-driven gravity flow that can persist indefinitely, without any external supply of energy. The ultimate criterion for autosuspension must, therefore, be lack of net deposition. This criterion is reinforced if the current is also capable of erosion.

Theoretical development

Daly (1936) and Kuenen (1938) both realized the possibility of the process now called "autosuspension", although the first analysis of this phenomenon was made by Knapp (1938). He realized that the gravitational energy expended by a particle-driven gravity flow must be at least equal to that needed to maintain the suspended load, and derived a simple criterion for this situation: $US/v_s > 1$, where U is the velocity, S is the bed slope (small), and v_s is the settling velocity of the sediment particles.

The same concept was discovered independently, and considerably extended, by Bagnold (1962). He was apparently the first to use the term "auto-suspension". Bagnold recognized that although $US/v_s > 1$ was a necessary condition for this phenomenon, it was not sufficient, and he placed additional restrictions on the occurrence of autosuspension.

More extensive analyzes were carried out independently by Pantin (1979) and Parker (1982), both of whom described time-developing currents that could either erode or deposit, depending on the initial velocity and sediment content (both depth-averaged). The bed slope and other parameters were assumed constant. Eventually, a given current would either collapse and deposit all of its load, or enter a state in which erosion would occur (i.e., autosuspension), accompanied by a progressive increase in both velocity and sediment content. Both of the foregoing models used coupled differential equations to express changes in volume, sediment content, and momentum (Pantin) or mean-flow energy (Parker), but neither explicitly accounted for the turbulent energy balance. Parker (1982) named the erosive currents "ignitive", by analogy with the ignition and continued burning of combustible material, and predicted "catastrophic equilibrium", a state in which increasing sediment concentration eventually imposes a limit to further sediment entrainment, and thus to further acceleration of an ignitive current. These two models showed that autosuspension is promoted by large flow thickness, high bottom gradient, and a low sediment-settling velocity, as well as high initial current velocity and sediment concentration.

Further investigation (Fukushima et al., 1985; Parker et al., 1986) showed that the above "three-equation" models made physically-impossible demands on the available energy, and that an equation for turbulent energy balance would have to be included. This resulted in a "four-equation" model, which kept the energy requirement within appropriate limits, and predicted autosuspension under physically reasonable conditions. The process of ignitive autosuspension may be regarded as a positive-feedback loop, and is shown diagrammatically in Figure A12.

Other models that predicted autosuspension soon followed. Stacey and Bowen (1988) and Eidsvik and Brørs (1989)

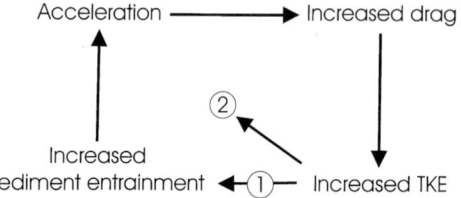

Figure A12 The positive-feedback loop in ignitive autosuspension. TKE = turbulent kinetic energy. (1) = portion of TKE expended on the entrainment of new sediment; (2) = portion of TKE expended on viscous dissipation, maintaining the existing sediment load, and thickening of the flow by water entrainment (see Parker et al., 1986, Eq. 8). Continuation of the loop may lead to "catastrophic equilibrium" (see text). Failure of the loop will result in deceleration, decrease in drag and TKE, and deposition of sediment.

considered the problems of stratified flows; Pantin (1986) studied the generation of autosuspension by a periodic forcing process, and later (1991) developed an autosuspension model for underflows with a light interstitial fluid. These models, like all of those described earlier, were two-dimensional Cartesian, and did not allow for lateral changes.

Emms (1999) advanced the autosuspension concept further, by developing models that included the Coriolis parameter. In addition, his "four-equation" and "five-equation" models allowed for lateral spread, and could thus be applied to the open slope.

Autosuspension in natural systems

There is widespread indirect evidence that autosuspension occurs in natural systems, including storms in submarine canyons (Inman et al., 1976, interpreted by Fukushima et al., 1985 and by Pantin, 1986); flooded rivers in fiords (Prior et al., 1987, interpreted by Pantin, 1991); and earthquake-induced failure on the continental slope (Hughes Clarke et al., 1990). Prior et al. (1986) have described channeling due to erosion by high-density flows entering bodies of lower density fluid (hyperpycnal flows), a phenomenon entirely consistent with autosuspension. There is also good evidence of autosuspension in powder-snow avalanches (Fukushima and Parker, 1990), while the severe erosion reported from some pyroclastic flows (Sparks et al., 1997) may well involve autosuspension.

Autosuspension can also help to explain some of the main features of the turbidite beds that occur widely in stratigraphic sequences. These beds evidently represent major, time-discontinuous events associated with high sediment transport rates and strong currents; they can be interpreted as the product of turbidity currents, initially in autosuspension, generated periodically under exceptionally severe conditions (e.g., storms, floods, or submarine landslides). These currents would deposit their load when, inevitably, they reached the base of the slope or a closed basin, and encountered bottom gradients too low to sustain the autosuspension process (Komar, 1977; Pantin, 1979).

Experimental work

Until very recently, there have been no successful attempts to produce unequivocal evidence of autosuspension in the laboratory. Neither Southard and Mackintosh (1981) nor

Parker et al. (1987) succeeded in producing autosuspension. However, Pantin (2001), using specially designed equipment, was able to obtain clear evidence of autosuspension in a series of experimental flows. The method consisted of running a suspension of 20 μm silica through a sloping 5 cm Perspex tube with special inserts, over a test bed of the same material (stained blue), and measuring the increase in flow velocity, together with the proportion of test bed entrained.

Deliberate attempts to generate turbidity currents in submarine canyons (Buffington, 1961; Dill, 1964) have met with little success; fortunately for the operators, perhaps, they did not meet the conditions necessary for triggering an autosuspension current.

Henry M. Pantin

Bibliography

Bagnold, R.A., 1962. Auto-suspension of transported sediment; turbidity currents. *Proceedings of the Royal Society of London*, A **265**: 315–319.
Buffington, E.C., 1961. Experimental turbidity currents on the sea floor. *Bulletin of the American Association of Petroleum Geologists*, **45**: 1392–1400.
Daly, R.A., 1936. Origin of submarine "canyons". *American Journal of Science*, **31**: 401–420.
Dill, R.F., 1964. Sedimentation and erosion in Scripps Submarine Canyon head. In Miller, R.L. (ed.), *Papers in Marine Geology: Shepard Commemorative Volume*. New York: Macmillan, pp. 23–41.
Eidsvik, K.J., and Brørs, B., 1989. Self-accelerated turbidity current prediction based upon (k-ε) turbulence. *Continental Shelf Research*, **9**: 617–627.
Emms, P.W., 1999. On the ignition of geostrophically rotating turbidity currents. *Sedimentology*, **46**: 1049–1063.
Fukushima, Y., and Parker, G., 1990. Numerical simulation of powder-snow avalanches. *Journal of Glaciology*, **36**: 229–237.
Fukushima, Y., Parker, G., and Pantin, H.M., 1985. Prediction of ignitive turbidity currents in Scripps Submarine Canyon. *Marine Geology*, **67**: 55–81.
Hughes Clarke, J.E., Shor, A.N., Piper, D.J.W., and Mayer, L.A., 1990. Large-scale current-induced erosion and deposition in the path of the 1929 Grand Banks turbidity current. *Sedimentology*, **37**: 613–629.
Inman, D.L., Nordstrom, C.E., and Flick, R.E., 1976. Currents in submarine canyons: an air-sea-land interaction. *Annual Review of Fluid Mechanics*, **8**: 275–310.
Knapp, R.T., 1938. Energy-balance in stream-flows carrying suspended load. *Transactions of the American Geophysical Union*, **19**: 501–505.
Komar, P.D., 1977. Computer simulation of turbidity current flow and the study of deep-sea channels and fan sedimentation. In Goldberg, E.D., McCave, I.N., O'Brien, J.J., and Steele, J.H. (eds.), *The Sea, Volume 6*. New York: John Wiley and Sons, pp. 603–621.
Kuenen, Ph.H., 1938. Density currents in connection with the problem of submarine canyons. *Geological Magazine*, **75**: 241–249.
Pantin, H.M., 1979. Interaction between velocity and effective density in turbidity flow: phase-plane analysis, with criteria for autosuspension. *Marine Geology*, **31**: 59–99.
Pantin, H.M., 1986. Triggering of autosuspension by a periodic forcing function. In *River Sedimentation, Volume III*. Proceedings of the Third International Symposium on River Sedimentation, pp. 1765–1770.
Pantin, H.M., 1991. A model for ignitive autosuspension in brackish underflows. In *Sand Transport in Rivers, Estuaries, and the Sea*. Proceedings of the EUROMECH 262 Colloquium, pp. 283–290.
Pantin, H.M., 2001. Experimental evidence for autosuspension. In *Particulate Gravity Currents*. International Association of Sedimentologists, Special Publication 31, pp.189–205.
Parker, G., 1982. Conditions for the ignition of catastrophically erosive turbidity currents. *Marine Geology*, **46**: 307–327.
Parker, G., Fukushima, Y., and Pantin, H.M., 1986. Self-accelerating turbidity currents. *Journal of Fluid Mechanics*, **171**: 145–181.
Parker, G., García, M., Fukushima, Y., and Yu, W., 1987. Experiments on turbidity currents over an erodible bed. *Journal of Hydraulic Research*, **25**: 123–147.
Prior, D.B., Yang, Z.-S., Bornhold, B.D., Keller, G.H., Lin, Z.H., Wiseman, W.J. Jr., Wright, L.D., and Lin, T.C., 1986. The subaqueous delta of the modern Huanghe (Yellow River). *Geo-Marine Letters*, **6**: 67–75.
Prior, D.B., Bornhold, B.D., Wiseman, W.J. Jr., and Lowe, D.R., 1987. Turbidity current activity in a British Columbia fjord. *Science*, **237**: 1330–1333.
Southard, J.B., and Mackintosh, M.E., 1981. Experimental test of autosuspension. *Earth Surface Processes and Landforms*, **6**: 103–111.
Sparks, R.S.J., Gardeweg, M.C., Calder, E.S., and Matthews, S.J., 1997. Erosion by pyroclastic flows on Lascar Volcano, Chile. *Bulletin of Volcanology*, **58**: 557–565.
Stacey, M.W., and Bowen, A.J., 1988. The vertical structure of turbidity currents and a necessary condition for self-maintenance. *Journal of Geophysical Research*, **93** (C4): 3543–3553.

Cross-references

Gravity-Driven Mass Flows
Numerical Models and Simulation of Sediment Transport and Deposition
Oceanic Sediments
Slope Sediments
Submarine Fans and Channels
Turbidites

AVALANCHE AND ROCK FALL

Avalanche and rock fall are slope movement processes by which masses are detached from jointed rock outcrops and accelerate by self weight to lower elevations by a combination of sliding, rolling, and flow. Since the sedimentological record imprinted by snow avalanches is very localized and transient, this article is restricted to rock-slope instability processes.

Rock avalanche

General attributes

Rock avalanche is a large-scale mass movement process involving the disintegration of a rock slide failure to form a rapidly flowing, granular mass. Many of the largest rock avalanches are known to have been triggered by either volcano collapse or seismic shock. Rock avalanches attain their largest size and destructive potential in high mountain areas and have been recorded on virtually every rock type. The avalanche attributes of high volume (10^6–10^{10} m^3), high velocity (20–50 ms^{-1}), long running distance (3–20 km), and extensive deposits (hundreds of km^2 for the largest events) distinguish them as the most hazardous and destructive geomorphic events on Earth. Some deposits are extensive enough to warrant the status of stratigraphic formations. Very large events, estimated to lie in the range 10^{10}–10^{14} m^3, have also been identified on Mars and the Moon.

Slope failures derived from unconsolidated colluvial, glacial, and volcanic materials, termed *debris avalanches*, and are mechanically similar to rock avalanches. Debris avalanches derived from the failure of colluvial and glacial materials are generally small by comparison (typically 10^3–10^5 m^3); however, debris avalanches derived from stratovolcanic collapse are among the largest slope failures on Earth. The world record is presently held by the 300–360 ka event from Mt. Shasta, California, having an estimated volume of $2.6 \times 10^{10}\,\mathrm{m}^3$ and a deposit surface area of 450 km².

Sedimentology and morphology of rock avalanche deposits

Rock avalanche deposits are diamictons comprising mainly angular, poorly sorted rock debris of local provenance, ranging in size from very large boulders, tens of meters in diameter, to fine silt and clay-sized fractions. The wide spectrum of sizes is produced in part by fracture and kinetic comminution of rock material as it moves downslope from the avalanche detachment zone, and partly by entrainment of water-saturated fine materials from the valley floor overridden by the avalanche. The vertical grading of rock avalanche deposits is highly variable, though some show slight inverse grading, caused by upward migration of large boulders by dispersive pressure (Middleton and Wilcock, 1994). Inverse grading may also develop from kinetic sieving, in which a preferential downward movement of finer material occurs through voids between larger clasts.

Rock avalanches discharging into lowland areas produce extensive, spatulate, hummocky sheets of rubbly diamicton, tens of meters thick and kilometers in extent. Longitudinal grooves and ridges commonly occur sub-parallel to the flow direction, with pressure ridges developed transverse to the flow near its terminus (Shreve, 1968). Avalanches exhibit a range of morphologies depending on their mobility. Relatively dry avalanches, moving orthogonal to the valley axis, have lower mobility through expending much energy by collision with the adverse valley wall. Major landslide dams impounding lakes are common in these cases (Costa and Schuster, 1988). Avalanches discharging along a valley axis are more likely to entrain water and saturated valley floor materials. This in turn causes high basal water pressure and a reduction of internal friction, leading to a high velocity, streaming type of motion. Avalanche running distances exceeding 5 km are common. Rock avalanches confined by valley walls often exhibit evidence of run-up and superelevation against obstacles and valley side slopes, allowing estimates to be made of avalanche flow velocity (Hungr, 1995).

Rock avalanche mechanics

A long-standing problem in rock avalanche mechanics is their high mobility on relatively low slope angles. Figure A13 shows a plot of avalanche volume, V, against angle of reach, defined by the quotient of fall height, H, and avalanche travel distance, L. H/L defines the running slope, f, of the avalanche, which approximates the average friction angle mobilized by the moving debris (Middleton and Wilcock, 1994, p.107). The inverse relation between V and f is the well documented 'size effect' (Hsü, 1975). A range of mechanisms has been proposed to account for this relation. Basal water pressure is a possibility, given that many avalanches traverse low-lying areas where saturated, fine grained materials are readily entrained. Very long-running volcanogenic debris avalanches or *lahars* (Reid et al., 2001) result from the incorporation of melting snow and ice. Fluid-escape craters and fine debris cones on the surfaces of some rock and debris avalanches demonstrate that high fluid pressures prevailed during the motion. In the fully geostatic case, where total overburden pressure equals fluid pressure, effective stress would be close to zero, and frictional resistance in the Mohr-Coulomb sense would be lost. The avalanche size effect due to water pressure could then be accounted for by the longer fluid drainage paths required in larger, thicker avalanches, which would favor a longer retention of excess fluid pressures.

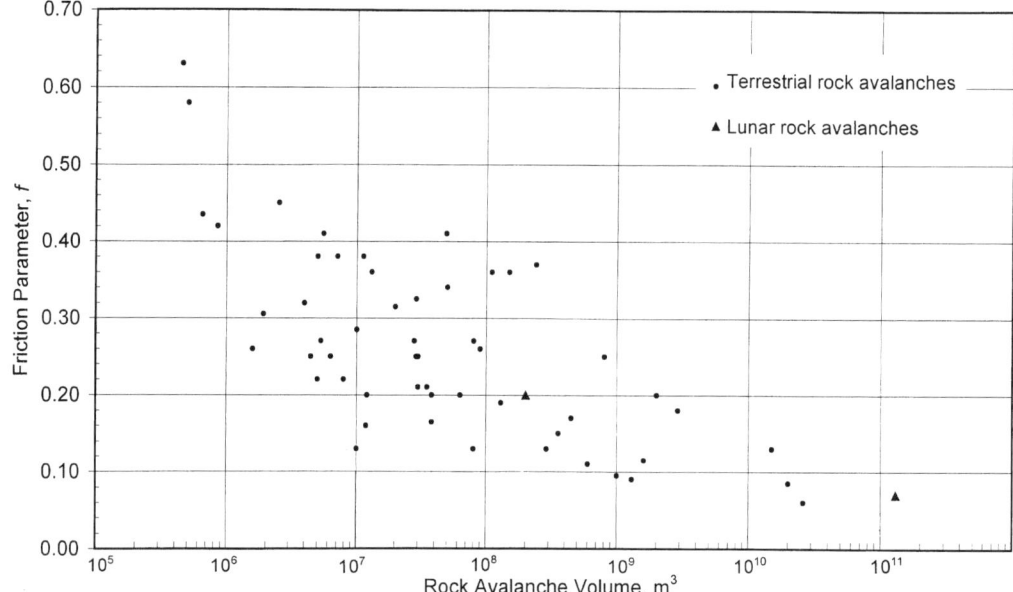

Figure A13 Rock avalanche volume versus avalanche friction parameter, *f*. Data are derived from Hsü (1975) and Evans et al. (1989).

The 'air-layer lubrication' theory (Shreve, 1968) requires that avalanches be air launched at some point along their path, followed by the avalanche running on a cushion of compressed air. This is a less appealing mechanism given the high compressibility of air. Acoustic fluidization (Melosh, 1987) requires propagation through the moving mass of acoustical energy with wavelengths longer than the average particle diameter. Wave propagation creates grain zones in compression, which move with little internal deformation, and zones of rarefaction where slip is possible. The result is a lesser dissipation of energy than would occur if all grain zones experienced shear strain. Finally, the 'self-lubrication' mechanism (Campbell, 1989) is based on the development of a relatively low-density, rapidly sheared zone at the base of an avalanche, upon which the bulk of the mass is efficiently transported with significantly less internal deformation. The size effect is then explained by the greater length of time required to dissipate the greater kinetic energy of very large avalanches.

Rock fall

General attributes

Rock fall is a generally small-scale mass movement process involving the detachment of individual rock particles, or relatively small groups of particles, from cliffs and other bedrock outcrops. The total volume of material detached by a rockfall event is generally $10-10^4 \, m^3$, which is several orders of magnitude smaller than volumes delivered by rock avalanches. Detachment of individual rockfall blocks from a rock face typically occurs along pre-existing planes of weakness such as joints, bedding, and foliation, with most failures occurring by either sliding or toppling. Downslope movement typically occurs by a combination of sliding, free-fall, saltation, and rolling. Rock fall is especially active during episodes of multiple freeze-thaw cycles in autumn and spring periods, or whenever high water pressure exists along joints. Rocks with closely spaced joints are most prone to production of rockfall debris, yielding a high frequency of blocks in the range $0.01-10 \, m^3$. However, in sparsely jointed massive rocks, rock fall is a less frequent though sometime catastrophic process. Huge monoliths totaling several thousands of cubic meters may topple or fall (Schumm and Chorley, 1964).

Rock fall deposits

The prolonged operation of rock fall at a site leads to cone-shaped accumulations of angular rubble known as talus. Most talus slopes produced by rock fall develop characteristic angles of rest in the range $34°-36°$, the so-called angle of repose of the rubble material (Carson, 1977). Somewhat steeper angles can develop when talus is built slowly by small-scale rock fall. Angles significantly flatter than $35°$ indicate that other processes, notably snow avalanche, rock avalanche, or debris flow, have reworked talus material (Luckman, 1988), implying that some talus deposits are composite features. It is common to find finer material interbedded with coarse-grained rockfall deposits, so the surface rubble layer is often not indicative of the texture of the entire deposit.

Rock fall mechanics

The loss of potential energy by a falling particle must be balanced by the sum of its kinetic energy gain and energy losses caused by friction with, and deformation of, the talus surface underlying the moving particle. The total travel distance of a particle across a talus slope is a function of its initial angle of impact, its initial velocity, and its size relative to that of material comprising the talus slope. Particle impact angle on the talus slope is a critical factor, since more than 80 percent of kinetic energy can be dissipated by cratering work at the first impact. A transition from bouncing to rolling eventually occurs, since progressively less of the particle's remaining momentum is directed normal to the slope during successive impacts. Once rolling has commenced, larger particles tend to run further by virtue of their lower rolling-friction angles. Over time, the talus slope develops a basal concave section with larger blocks deposited in the "shadow" zone, well beyond the base of the main talus deposit (Evans and Hungr, 1993). A distinct fall-sorted texture thus evolves comprising a progressive downslope increase in mean particle size.

Because of the sporadic nature of the rock fall process, and the difficulty of measuring its effects in remote areas, some of the most detailed process studies have been conducted along highways and railways where rock fall poses a hazard (Hungr et al., 1999). Computer simulations, corroborated by field studies, are now routinely used in determining rockfall risk assessments wherever public safety is a concern (Evans and Hungr, 1993).

Michael J. Bovis

Bibliography

Campbell, C.S., 1989. Self-lubrication for long runout landslides. *Journal of Geology*, **97**: 653–665.

Carson, M.A., 1977. Angles of repose, angles of shearing resistance and angles of talus slopes. *Earth Surface Processes and Landforms*, **2**: 363–380.

Costa, J.E., and Schuster, R.L., 1988. The formation and failure of natural dams. *Geological Society of America Bulletin*, **100**: 1054–1068.

Cruden, D.M., and Hungr, O., 1985. The debris of the Frank Slide and theories of rockslide-avalanche mobility. *Canadian Journal of Earth Sciences*, **23**: 425–432.

Evans, S.G., Clague, J.J., Woodsworth, G.J., and Hungr, O., 1989. The Pandemonium Creek rock avalanche, British Columbia. *Canadian Geotechnical Journal*, **26**: 427–446.

Evans, S.G., and Hungr, O., 1993. The assessment of rockfall hazard at the base of talus slopes. *Canadian Geotechnical Journal*, **30**: 620–636.

Hungr, O., 1995. A model for the runout analysis of rapid flow slides, debris flows, and avalanches. *Canadian Geotechnical Journal*, **32**: 610–623.

Hungr, O., Evans, S.G., and Hazzard, J., 1999. Magnitude and frequency of rock falls and rock slides along the main transportation corridors of southwestern British Columbia. *Canadian Geotechnical Journal*, **36**: 224–238.

Hsü, K., 1975. Catastrophic debris streams (sturzstroms) generated by rockfalls. *Geological Society of America Bulletin*, **86**: 129–140.

Luckman, B.H., 1988. Debris accumulation patterns on talus slopes in Surprise Valley, Alberta. *Géographie Physique et Quaternaire*, **42**: 247, 278.

Melosh, J.H., 1987. The mechanics of large rock avalanches. In Costa, J.E., and Wieczorek, G.F. (eds.), *Debris Flows/Avalanches: Process, Recognition, and Mitigation*. Boulder, CO: Geological Society of America, pp. 41–49.

Middleton, G.V., and Wilcock, P.R., 1994. *Mechanics in the Earth and Environmental Sciences*. Cambridge, UK: Cambridge University Press.

Reid, M.E., Sisson, T.W., and Brien, D.L., 2001. Volcano collapse promoted by hydrothermal alteration and edifice shape, Mount Rainier, Washington. *Geology*, **29**: 779–782.

Schumm, S.A., and Chorley, R.J., 1964. The fall of Threatening Rock. *American Journal of Science*, **262**: 1041–1054.

Shreve, R.L., 1968. *The Blackhawk Landslide*. Geological Society of America, Special Paper 108.

Cross-references

Angle of Repose
Floods and Other Catastrophic Events
Fluid Escape Structures
Grain Flow
Grain Size and Shape
Gravity-Driven Mass Flows
Liquefaction and Fluidization
Mass Movement
Slope Sediments
Surface Textures

AVULSION

Avulsion is the process in which river (or other) flows are diverted out of an established channel into a new course at a lower elevation on the adjacent surface. Although the term usually refers to major discharge diversions that result in new channels, it is sometime used in reference to short-term flow switching within braided channels. Avulsions may be *full*, in which all flow is transferred out of the parent channel, or *partial*, in which only a portion of the flow is transferred. Full avulsions result in abandonment of the parent channel downstream of the diversion site, whereas partial avulsions lead to new channels that coexist with the parent channel. Partial avulsion is a principal mechanism that forms anastomosing channels (if the channels rejoin) and distributary channels (if the channels do not rejoin), the latter a common feature of alluvial fans and deltas. Avulsions may be abrupt or gradual. For example, in 1855 the Yellow River (China) breached one of its levees during a large flood, and within a single day, all flow had diverted through the breach into the floodplain (Qian, 1990). In contrast, full avulsions in the Meuse-Rhine delta often require several centuries to complete, and durations of up to 1250 years have been measured (Stouthamer and Berendsen, 2001).

Avulsions are features of aggrading river systems and should not be expected in actively incising stream valleys. They are not restricted to any particular channel form or size and will recur in any river system for as long as aggradation continues. Frequency of avulsion recurrence varies widely among the few modern rivers for which such data exists, ranging from as low as 28 years for the Kosi River (India) up to 1400 years for the Mississippi (see Table 4.2 in Stouthamer and Berendsen, 2001). Data from the geologic record suggest that much larger intervals between avulsions have existed (e.g., Kraus and Aslan, 1993). By forcing large-scale repositioning of stream channels, avulsion is a dominant mechanism for constructing river floodplains, deltas, and alluvial fans and for generating laterally extensive sedimentary deposits in the stratigraphic record.

Causes of avulsion

Two basic conditions are required for avulsion: (1) a long-term "setup" in which the channel gradually increases its susceptibility to avulsion, and (2) a short-term "trigger" event which initiates the flow diversion. The setup requirement is attained by the tendency for a channel belt to rise above the level of the adjacent floodplain under conditions of net aggradation. Such *superelevation* of the channel belt develops because overbank deposition rates tend to be greatest near the channel (forming levees) and to decrease exponentially away from the channel across the floodplain. Such elevated channel belts are known as *alluvial ridges*. With continued aggradation and elevation differentiation, this topology becomes increasingly unstable and is eventually relieved by avulsion when a sufficiently large trigger event occurs. Such events are typically floods, but also include such processes as abrupt neotectonic movements, ice jams, log jams, vegetative blockages, debris dams, or bank failures (see Jones and Schumm, 1999, for review).

Implicit with the setup requirement is the presumption that some critical value of channel-to-floodplain slope must be exceeded before avulsion can take place. Mackey and Bridge (1995), for example, consider the probability of avulsion occurring at a particular site as a product of the ratio of cross-valley to down-channel slope and the return period of flood discharge in excess of some threshold value. Note that the slope ratio can be increased by either increasing cross-valley slope or decreasing down-channel slope. The former is readily attained by increasing the elevation of the channel belt relative to the floodplain, whereas down-channel slope may be reduced by increasing sinuosity or, in the case of prograding deltas, by channel lengthening. Cross-valley or down-channel slope can also be altered by tectonic tilting. As an alternative to channel-to-floodplain slope, Mohrig *et al*. (2000) argue that elevation difference between the channel water surface and the lowest areas of the floodplain exerts the primary control on avulsion setup. In either case, continuing elevation of the alluvial ridge relative to the rest of the floodplain eventually makes avulsion inevitable.

When general setup conditions are sufficient to permit avulsion, the site at which the avulsion takes place is likely to be opportunistic and determined by such local factors as channel geometry, bank stability, or topography. For example, avulsion sites commonly occur at the outer banks of meander bends where flow velocities are high, confining levees are narrow, and flood flows impinge the banks at high angles. Floods continually test weaknesses in the channel banks, and local areas of relative instability, for example, intersected sand bodies of former channels, are potential locations for avulsion (Chrzastowski *et al*., 1994; Smith *et al*., 1998). Avulsion sites commonly begin as crevasses that enlarge until flow is permanently diverted from the parent channel. Many crevasse channels, however, become plugged with sediment and do not result in permanent flow transfer. Modeling results (Slingerland and Smith, 1998) suggest that whether a crevasse-channel plugs, runs to full avulsion, or reaches a steady state of partial avulsion depends on the ratio of crevasse to main-channel bed slope, ratio of crevasse to main channel depth, and bed grain size. For fine to medium sand sizes, crevasse-channel slopes greater than about 5 to 8 times main-channel slope are predicted to develop into full avulsions regardless of initial crevasse size. Lower crevasse-channel slopes will more likely become plugged or sustain partially diverted flows for long periods.

Styles of avulsion

It is generally presumed that the initial pathways of newly diverted flow will follow the direction of greatest slope or maximum flow efficiency away from the parent channel. The extent of avulsive flooding and nature of channel patterns that subsequently develop are then governed in large part by the character of the floodplain surface, for example, its gradient, relief, vegetative cover, presence of existing channels (active or abandoned), elevation of water table, and floodplain width. Abandoned alluvial ridges and narrow valley walls, for example, limit the lateral extent of avulsive flow normal to the parent channel and redirect it downvalley. Dense vegetation covers, low floodplain gradients, and high water tables (associated with shallow floodplain lakes and marshy wetlands) promote slow runoff, leading to ponding and deposition of fine sediment. Rapid runoff is enhanced by high gradients, low water tables, and sparse vegetative covers.

Three broadly different styles of floodplain response to avulsion have been recognized: (1) appropriation or reoccupation of preexisting channels, (2) creation of new channels by incision, and (3) deposition of multichaneled progradational sediment wedges. (1) Appropriation of active floodplain channels or reoccupation of abandoned channels is a common avulsion style that results from the tendency for the diverted flow to seek pathways of maximum transport efficiency. If the annexed channel is large enough to accommodate the diverted flow, little floodplain modification takes place. Commonly, however, the annexed channel is too small for the newly captured flow; deepening and/or widening of the channel usually follows, sometimes accompanied by extensive crevassing and deposition of crevasse splays (Pérez-Arlucea and Smith, 1999). (2) Incision of new channels is most likely to occur in well-drained floodplains where avulsive floodwaters run off quickly and little sediment deposition takes place (Mohrig et al., 2000). Such incisional channels may form by downstream extension from the initial avulsion site or upstream migration of a knickpoint from the site where the diverted flows reenter the parent channel or tributary (Schumm et al., 1996). (3) Deposition and downfloodplain progradation of multichaneled sediment wedges is a dominant avulsion style in poorly drained marshy floodplains (Smith et al., 1989; Tye and Coleman, 1989). Scales of progradation range from small crevasse splays to regional-scale complexes of coalesced splays and lacustrine deltas. The active margins of the progradational wedge are supplied by many small, often interconnected, channels which later are succeeded by a single large channel as the avulsion evolves to completion. In cases of regional-scale full avulsions, this final channel becomes the new parent channel for the next avulsion.

Avulsion deposits

Because avulsions are expectable consequences of aggradation and the principal mechanisms by which laterally extensive alluvial deposits are formed, we should expect abundant evidence of avulsion in alluvial stratigraphic successions. As yet, however, only a few studies of ancient river sediments have drawn specific attention to avulsion processes or deposits. Evidence for avulsion is often straightforward but indirect. For example, the mere presence of a well-developed paleosol within an alluvial succession suggests that avulsion resituated the contemporaneous river channel to a location too distant to flood the paleosol site with fine sediment.

Direct evidence of avulsions can be found in the character of their deposits. Avulsion involving channel reoccupation can be inferred by well-defined, multistoried, often stepped, sand bodies within a single channel-fill succession, or different stratigraphic levels of levee deposits associated with the same channel-fill. Avulsion dominated by incisions of new channels may be manifested by abundant ribbon (low width/thickness ratio) sand bodies encased in fine floodplain sediments and defined by sharp erosional contacts (Mohrig et al., 2000). Deposits of progradational avulsions tend to be the most heterogeneous. They include ribbon sands formed by anastomosing distributary networks, sheet sands formed by crevasse splays and channel-mouth bars, and a variety of fine-grained sediments deposited in interchannel lows and in ponded or otherwise slow-moving water bodies formed by the avulsion (Smith et al., 1989). Because the avulsive sediment wedge is predominantly progradational in character, upward-coarsening successions, with thicknesses scaled to water depth, are typical (Pérez-Arlucea and Smith, 1999). Sedimentary successions formed by progradational avulsions are likely to be capped by paleosols or fine-grained flood deposits of succeeding channel belts. Ancient examples are described by Kraus (1996) and Kraus and Wells (1999).

Norman D. Smith

Bibliography

Chrzastowski, M.J., Killey, M.M., Bauer, R.A., DuMontelle, P.B., Erdmann, A.L., Herzog, B.L., Masters, J.M., and Smith, L.R., 1994. The Great Flood of 1993. *Illinois State Geological Survey Special Report 2*, pp. 1–45.

Jones, L.S., and Schumm, S.A., 1999. Causes of avulsion: an overview. In *Fluvial Sedimentology VI*, SEPM (Society for Sedimentary Geology), Special Publication, 28, pp. 171–178.

Kraus, M.J., 1996. Avulsion deposits in Lower Eocene alluvial rocks, Bighorn Basin, Wyoming. *Journal of Sedimentary Research*, **66**: 354–363.

Kraus, M.J., and Aslan, A., 1993. Eocene hydromorphic paleosols: Significance for interpreting ancient floodplain processes. *Journal of Sedimentary Petrology*, **63**: 453–463.

Kraus, M.J., and Wells, T.M., 1999. Recognizing avulsion deposits in the ancient stratigraphical record. In *Fluvial Sedimentology VI*, SEPM (Society for Sedimentary Geology), Special Publication, 28, pp. 251–268.

Mackey, S.D., and Bridge, J.S., 1995. Three-dimensional model of alluvial stratigraphy: Theory and application. *Journal of Sedimentary Research*, **B65**: 7–31.

Mohrig, D., Heller, P.L., Paola, C., and Lyons, W.J., 2000. Interpreting avulsion process from ancient alluvial sequences: Guadalope-Matarranya system (Northern Spain) and Wasatch Formation (Western Colorado). *Geological Society of America Bulletin*, **112**: 1787–1803.

Pérez-Arlucea, M., and Smith, N.D., 1999. Depositional patterns following the 1870s avulsion of the Saskatchewan River (Cumberland Marshes, Saskatchewan, Canada). *Journal of Sedimentary Research*, **69**: 62–73.

Qian Ning, 1990. Fluvial processes in the Lower Yellow River after levee breaching at Tongwaxiang in 1855. *International Journal of Sediment Research*, **5**: 1–13.

Schumm, S.A., Erskine, W.D., and Tilleard, J.W., 1996. Morphology, hydrology, and evolution of the anastomosing Ovens and King Rivers, Victoria, Australia. *Geological Society of America Bulletin*, **108**: 1212–1224.

Slingerland, R., and Smith, N.D., 1998. Necessary conditions for a meandering-river avulsion. *Geology*, **26**: 435–438.

Smith, N.D., Cross, T.A., Dufficy, J.P., and Clough, S.R., 1989. Anatomy of an avulsion. *Sedimentology*, **36**: 1–23.

Smith, N.D., Slingerland, R.L., Pérez-Arlucea, M., and Morozova, G.S., 1998. The 1870s avulsion of the Saskatchewan River. *Canadian Journal of Earth Sciences*, **35**: 453–466.

Stouthamer, E., and Berendsen, H.J.A., 2001. Avulsion frequency, avulsion duration, and interavulsion period of Holocene channel belts in the Rhine-Meuse Delta, the Netherlands. *Journal of Sedimentary Research*, **71**: 589–598.

Tye, R.S., and Coleman, J.M., 1989. Depositional processes and stratigraphy of fluvially dominated lacustrine deltas: Mississippi Delta Plain. *Journal of Sedimentary Petrology*, **59**: 973–996.

Cross-references

Alluvial Fans
Anabranching Rivers
Braided Channels
Floodplain Sediments
Meandering Channels
Rivers and Alluvial Fans
Submarine Fans and Channels

B

BACTERIA IN SEDIMENTS

Epibenthic bacteria forming biofilms and microbial mats

Bacteria live in an extremely wide range of habitats, and their occurrence is only restricted by the requirement for water and the physicochemical stability limits of biomolecules (Knoll and Bauld, 1989). In sediments, epibenthic bacteria attach firmly to the surfaces of mineral particles by their adhesive and mucous, 'extracellular polymeric substances (EPS)' often more abundant than the cell material itself (Decho, 1990 for overview and introduction; Decho, 2000). The mucilaginous substances aid the microbes to sequester nutrients, to protect themselves against osmotic pressure caused by changing salinities, and to maintain an optimal chemical microenvironment for activities of extracellular enzymes (Decho, 1990). These coatings, composed of single cells and their mucilages enveloping mineral particles are known as biofilms (Marshall, 1984; Charaklis and Wilderer, 1989; compare also Stolz, 2000). Further biomass enrichment leads to the formation of discrete, tissue-like organic layers, microbial mats, that cover extensive areas of the seafloor (definition of term see Krumbein, 1983; Cohen and Rosenberg, 1989; Stal and Caumette, 1994; Stolz, 2000; brief historic outline included in Stal, 2000). Microbial mats are biologically stratified communities. Different bacterial populations organize in horizontal layers in dependence to intrasedimentary, chemical gradients (Stolz, 2000, and references therein). Their highly active metabolic cycles are in steady interaction with chemical parameters of their surroundings, and thus the microbes powerfully influence their environment (Stal, 2000). This bacterially mediated element transfer causes diverse intrasedimentary mineral deposits and structures of high preservation potential (Krumbein, 1986). In consolidated rocks, occurrences of the minerals permit conclusions on the biology of ancient microbiota, and on the environmental conditions of the past.

Epibenthic bacterial populations as 'bioreactors'

Epibenthic microorganisms and their slimy secretions contribute to a high porosity of the surface sediments, and their intensive metabolic and catalysatory activities raise the amount of chemical reactions substantially. The transfer capacity of a multi-layered microbial mat is even greater, because the different metabolic cycles interlock with each other. Therefore a microbial mat was compared with an intrasedimentary 'bioreactor' (Hanselmann, 1989; compare also model of 'bioid' by Krumbein, 1986). Within the stratified community of a microbial mat (Figure B1), each population depends on the metabolic products of the other below or above (Cohen and Rosenberg, 1989; Hanselmann, 1989; Stal and Caumette, 1994; Ehrlich, 1996; Nealson, 1997; Stal, 2000; Stolz, 2000; Nisbet and Sleep, 2001). In the photic zone, primary producers, mainly cyanobacteria (and diatoms), build up biomass by oxygenic or anoxygenic photosynthesis. Below the cyanobacteria, a layer of purple sulfur bacteria may establish (depending on the relative depth gradients of light and oxygen). Their purely anaerobic photosynthesis makes use of lower light intensities than cyanobacteria. Sulfur bacteria require sulfide, which is provided by sulfate-reducing bacteria colonizing deeper parts of the sediments. Here, a great variety of heterotrophic bacteria decompose the primary organic material. With the aid of their highly effective enzymes, they degrade polysaccharides, lipids, or proteins, of the primary biomass, and release monomers. These are further decomposed by sulfate reducing and methanogenic bacteria to carbon dioxide, hydrogen sulfide, methane, and inorganic nutrients.

Chemical processes as a consequence of decomposition and mineralization of organic material lead to precipitation of sulfides, carbonates, silicates, and many other minerals (Krumbein, 1986). This will be outlined in the following.

Decomposition of organic material causing precipitation of minerals

In microscopic scale, the cell walls of bacteria are composed of macromolecules that give an overall electro-negative charge to

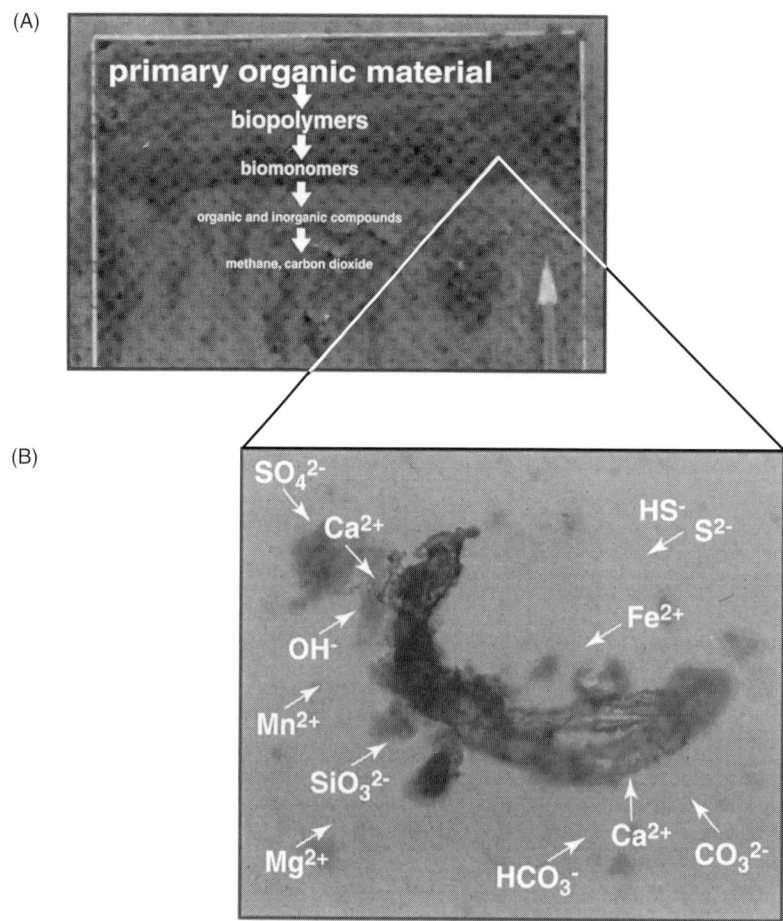

Figure B1 Bacterial decomposition of organic material leads to precipitation of sulfides, carbonates, silicates, and other minerals. (A) Within microbial mats, primary organic material is decomposed in several steps until inorganic nutrients are formed. (B) In microscale, ions dissolved in the surroundings react with reactive chemical groups along the outer cell wall. This might induce mineral precipitation. Photo shows initial precipitates (? pyrite) along cyanobacterial filament.

the outer cell walls (Beveridge, 1989; Beveridge and Doyle, 1989; Schulze-Lam et al., 1996). The reactive chemical groups of the cell wall project into the surrounding environment, but protons released through the cell membrane in course of ATP-generation of the living cell occupy the negatively charged sites. Additionally, the heterogeneously charged microsites of the cell wall adsorb water molecules (that have bipolar chemical structure), and as a consequence a film of water permanently envelopes and buffers the cells (Schulze-Lam et al., 1996). It is assumed that after bacterial cells die, the increasingly electro-negative charge of the cell surfaces react with electro-positive metal ions dissolved in the seawater. In laboratory experiments, the first metal ions serve as nucleus for further metal enrichment, and counter-ions are attracted (Beveridge, 1989; Beveridge and Doyle, 1989; Schulze-Lam et al., 1996; Douglas and Beveridge, 1998; Ferris, 2000). Because of the water molecules enveloping bacterial cells, initial precipitates are hydrated and amorphous (Beveridge, 1989; early steps of aragonite formation on cell envelopes are documented by Krumbein, 1979a; see also Schulze-Lam et al., 1996). Over time, dehydratation of the precipitates leads to a crystalline mineral phase, and during this diagenetic alteration for example, aragonite changes to calcite (Beveridge, 1989). Depending on microenvironmental geochemistry induced by decomposing heterotroph bacteria, counter ions like sulfate, carbonate, or sulfide ions determine formation of specific minerals, (Figure B1). For example, pyrite is formed in presence of Fe^{2+} and HS^- or S^{2-} ions, and often can be found intimately related to carbonate layers in subrecent, decaying microbial mats (Gerdes et al., 2000).

Microbial ecosystems are ancient and nearly as old as the earliest geological record, so knowledge of bacterially induced chemical processes provides important clues on the biological and environmental evolution of Earth and other planets.

Nora Noffke

Bibliography

Beveridge, T.J., 1989. Role of cellular design in bacterial metal accumulation and mineralization. *Annual Reviews Microbiology*, **43**: 147–171.

Beveridge, T.J., and Doyle, R.J. (eds.), 1989. *Metal Ions and Bacteria*. New York: Wiley.

Charaklis, W.G., and Wilderer, P.A., 1989. *Structure and Function of Biofilms*. New York: Wiley.
Cohen, Y., and Rosenberg, E., 1989. *Microbial Mats. Physiological Ecology of Benthic Microbial Communities*. Washington, DC: American Society of Microbiologists.
Decho, A.W., 1990. Microbial exopolymer secretions in ocean environments: their role(s) in food webs and marine processes. *Oceanographic Marine Biology Annual Review*, **28**: 73–153.
Decho, A.W., 2000. Exopolymer microdomains as a structuring agent for heterogeneity within microbial biofilms. In Riding, R., and Awramik, S.M. (eds.), *Microbial Sediments*. Berlin: Springer Verlag, pp. 9–15.
Douglas, S., and Beveridge, T.J., 1998. Mineral formation by bacteria in natural microbial communities. *FEMS Microbial Ecology*, **26**: 79–88.
Ehrlich, H.L., 1996. *Geomicrobiology*. New York, Basel: Marcel Dekker.
Ferris, F.G., 2000. Microbe-metal interactions in sediments. In Riding, R., and Awramik, S.M. (eds.), *Microbial Sediments*. Berlin: Springer Verlag, pp. 121–126.
Gerdes, G., Klenke, Th., and Noffke, N., 2000. Microbial signatures in peritidal siliciclastic sediments: a catalogue. *Sedimentology*, **47**: 279–308.
Hanselmann, K., 1989. Rezente Seesedimente. Lebensraeume fuer Mikroorganismen. *Die Geowissenschaften*, **4**: 98–113.
Knoll, A.H., and Bauld, J., 1989. The evolution of ecological tolerance in prokaryotes. *Transactions of the Royal Society Edinburgh: Earth Science*, **80**: 209–223.
Krumbein, W.E., 1979a. Photolithotrophic and chemoorganotrophic activity of bacteria and algae as related to beachrock formation and degradation (Gulf of Aqaba, Sinai). *Geomicrobiology Journal*, **1**: 156–202.
Krumbein, W.E., 1983. Stromatolites – The challenge of a term in time and space. *Precambrian Research*, **20**: 493–531.
Krumbein, W.E., 1986. Biotransfer of minerals by microbes and microbial mats. In Leadbeater, B.S., and Riding, R. (eds.), *Biomineralization in Lower Plants and Animals*. Oxford: Clarendon, pp. 55–72.
Marshall, K., 1984. *Microbial Adhesion and Aggregation*. Berlin: Springer Verlag.
Nealson, K., 1997. Sediment bacteria: whos there, what are they doing, and whats new? *Annual Review Earth Planet Science*, **25**: 403–434.
Nisbet, E.G., and Sleep, N.H., 2001. The habitat and nature of early life. *Nature*, **409**: 1083–1091.
Schulze-Lam, S., Fortin, D., Davis, B.S., and Beveridge, T.J., 1996. Mineralization of bacterial surfaces. *Chemical Geology*, **132**: 171–181.
Stal, L.J., and Caumette, P., 1994. *Microbial mats. Structure, Development and Environmental Significance*. NATO ASI Series Ecological Sciences, Volume 35: Berlin: Springer Verlag.
Stal, L.J., 2000. Cyanobacterial mats and stromatolites. In Whitton, B.A., and Potts, M., (eds.), *The Ecology of Cyanobacteria*. Dordrecht: Kluwer Academic Publishers, pp. 62–120.
Stolz, J.F., 2000. Structure of microbial mats and biofilms, In Riding, R., and Awramik, S.M. (eds.), *Microbial Sediments*. Berlin: Springer Verlag.

Cross-references

Algal and Bacterial Carbonate Sediments
Oceanic Sediments
Sulfide Minerals in Sediments

BALL-AND-PILLOW (PILLOW) STRUCTURE

Definition and description

Ball-and-pillow structure is a soft-sediment deformation structure comprising rounded masses of clastic sediment (pseudonodules, q.v.) set in similar or finer grained matrix, that are vertically stacked within one horizon (Figure B2(A), (B)). Much of the discussion in the entry for pseudonodules, relating to their description and their origin through loading processes related to variations in gravitational potential energy, is relevant to ball-and-pillow structure. Distinctive features of ball-and-pillow structure include: the pseudonodules are commonly contorted into spiral or zig zag shapes; there may be very little matrix between the pseudonodules; and some ball-and-pillow horizons are consistently underlain by convolute stratification and overlain by dish structure.

Historical development

Ball-and-pillow structure is usually traced back to Smith (1916), in a paper entitled 'Ball or pillow-form structures in sandstones'. The term 'ball-and-pillow structure' was introduced by Potter and Pettijohn (1963), who mistakenly cited Smith's (1916) title as 'Ball- or pillow-form structure', although Smith (1916) clearly did not intend 'ball' and 'pillow' to be part of a single name, referring instead to 'pillow-form

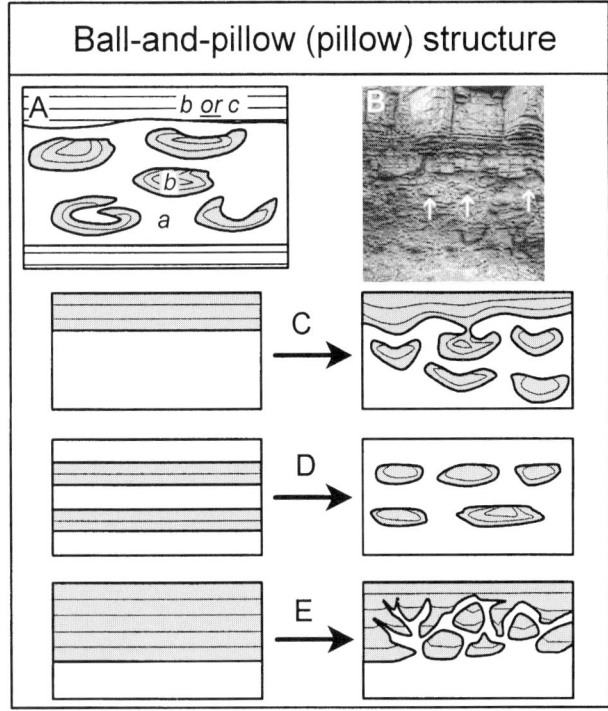

Figure B2 Definition and interpretation of ball-and-pillow structure. For simplicity, stratification is not shown in the lower, less dense layer. (A) Definition diagram for ball-and-pillow structure, where a, b and c are distinct sedimentation units. (B) Ball-and-pillow structure from an Upper Carboniferous delta front sequence at Amroth, west Wales, UK. Arrows point to individual pseudonodules. The section illustrated is 3 m thick. (C) Formation by repeated detachment from the base of a layer undergoing load deformation. (D) Formation by the simultaneous loading of several couplets of denser and less dense sediment. (E) Formation by a process of sedimentary stoping in which mobile sediment rises through an essentially passive layer.

or ball structures'. The structures described by Smith (1916) and Potter and Pettijohn (1963) are very similar to those that have subsequently been described as both ball-and-pillow structure and pseudonodules (or pseudo-nodules). The history and use of these terms are inextricably entwined, and are discussed fully in the entry for pseudonodules.

Interpretation

Examples of deformation pathways that lead to ball-and-pillow structure are shown in Figure B2. Some ball-and-pillow horizons represent isolated fold noses formed by downslope slumping, but most result from loading processes related to variations in gravitational potential energy and driven by variations in bulk density. Factors that favor the development of pseudonodules, in contrast to load casts that affect the base of a laterally continuous bed, include a driving force of large magnitude, a combination of driving forces, a very weak lower layer, or a long duration of deformation. Loading processes that can result in ball-and-pillow structure include: lateral drift of pseudonodules as they sink (e.g., Kuenen, 1958: Plate I); repeated detachment of masses from the base of a 'source layer' (Figure B2(C)); simultaneous liquidization and loading of several couplets of denser and less dense sediment (Figure B2(D)); and a process of 'sedimentary stoping', in which liquidized sediment (usually fluidized sand) rises through an essentially passive upper layer (Figure B2(E)). Criteria need to be developed to characterize the products of each of these pathways. For example, in ball-and-pillow horizons formed by sedimentary stoping, the pseudonodules are unlikely to have flat upper surfaces, they are likely to be relatively undeformed internally, and the upper surface of the ball-and-pillow horizon may be highly irregular.

Summary

Ball-and-pillow structure represents an extreme form of load structure. It can form as the end product of several soft-sediment deformation pathways. Further studies are needed to develop criteria that differentiate these various pathways.

Geraint Owen

Bibliography

Allen, J.R.L., 1982. *Sedimentary Structures: their Character and Physical Basis.* Amsterdam: Elsevier.
Kuenen, Ph.H., 1958. Experiments in geology. *Transactions of the Geological Society of Glasgow*, **23**: 1–28.
Potter, P.E., and Pettijohn, F.J., 1963. *Paleocurrents and Basin Analysis.* Berlin: Springer-Verlag.
Smith, B., 1916. Ball or pillow-form structures in sandstones. *Geological Magazine*, **53**: 146–156.

Cross-references

Bedding and Internal Structures
Convolute Lamination
Deformation of Sediments
Fluid Escape Structures
Liquefaction and Fluidization
Load Structures
Pseudonodules
Slide and Slump Structures

BAR, LITTORAL

Introduction

Ridges of unconsolidated sediment, which characterize the upper shoreface of littoral zones dominated by waves, and are generally larger than bedforms are termed *wave-formed bars* (e.g., see Figure B3). They were recognized as early as 1845 on the marine coasts of Europe, by 1851 in the Great Lakes of North America and subsequently on marine and lacustrine coasts worldwide (see Zenkovich, 1967). Confusion surrounds the term *bar* because: (a) it is used for similar forms in other environments (see Bar, Fluvial); (b) bars occur within the littoral zone with a wide range of size, shape, location and orientation relative to the shoreline, and within a wide range of littoral environments; (c) no single classification system has proved satisfactory (e.g., Greenwood and Davidson-Arnott, 1975; Wright and Short, 1984; Lippmann and Holman, 1990); (d) explanation of the origin(s) and dynamics of *wave-formed bars* is far from complete. Detailed reviews of the origin, form and dynamics of bars are given in van Rijn (1998) and Greenwood (2002, in press).

Origin

The boundary conditions for bar formation depend upon the bed material (mineralogy, size, etc.), the bathymetric setting, and the geographic location, which controls the wave climate, tidal regime, etc. In general, barred profiles are associated with large values of wave steepness and wave height-to-grain size ratios; however, the degree of wave energy reflection or dissipation is an important control and thus shoreface slope is critical. A number of hypotheses have been proposed for bar formation: (i) *break point hypotheses* relate bars to a seaward transport of sediment entrained by roller or helical vortices under plunging or spilling breakers, or a convergence of sediment at the breakpoint through onshore transport associated with wave asymmetry and skewness, and offshore transport through set up induced undertow; (ii) *infragravity wave hypotheses* propose that low frequency waves generated within the surf zone or offshore and reflected, produce a convergent pattern of drift velocities, which interact with large incident short waves to induce a range of forms from simple 2-dimensional to 3-dimensional crescentic forms; (iii) *self-organization hypotheses* suggest that processes associated with the complex, non-linear feedback between the hydrodynamics and the sand bed give rise to a range of topographic forms. Alongshore and offshore sediment transport was proposed for bar formation: (a) under meandering or cellular nearshore circulations produced by the instability of longshore flows; (b) a coupling between morphodynamic instability and mean flows; and (c) a non-linear action between shoaling waves and the bed.

Morphology and morphodynamics

Modern *wave-formed bars* are near symmetrical, to strongly asymmetrical undulations in the upper shoreface profile (Figure B3). They occur both intertidally and subtidally, as well as in nontidal areas, and may range in number from 1 to >30, covering a belt up to 1.5 km wide. In plan view, they form

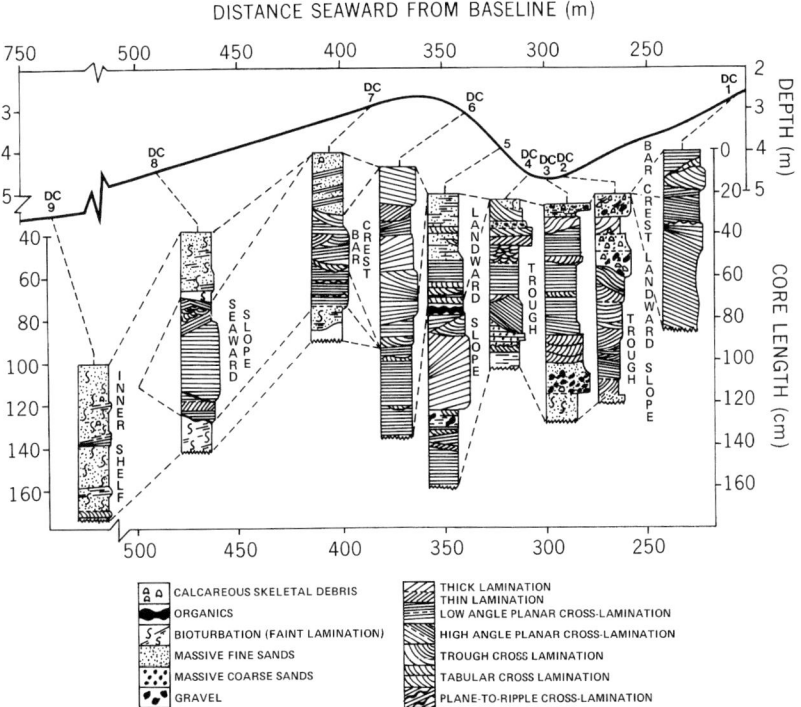

Figure B3 Typical barred profile from a marine environment in the southern Gulf of St. Lawrence, Canada. Note: the structures shown are from a series of nine tube cores taken across the bar; the lateral correlation of facies is shown by the dotted lines (modified after Greenwood and Mittler, 1985).

continuous or compartmentalized, linear, sinuous, or crescentic ridges, and range in orientation from shore-parallel to shore-normal, often producing periodic or rhythmic forms, both alongshore and cross-shore. Bars increase in size and spacing in the offshore direction, range from a few decimeters to more than 4.75 m in height (Greenwood and Mittler, 1979), and may stretch alongshore for tens of kilometers. Shepard (1950), Hands (1976) and others present detailed morphometry based on topographic surveys. Recently, analysis of video imagery has given new insights on bar morphodynamics during storm events and over long (decadal) time periods (e.g., Lippmann and Holman, 1990; Reussink et al., 2000). Van Rijn (1998) summarizes much of the vast literature on bar dynamics.

Texture

The vast majority of *wave-formed bars* consist of sand-size materials, although they have been recorded in sands mixed with small gravels; the associated troughs often contain gravels plus silt and clay-sized sediments. A distinct size fractionation is associated with the bar form, which is superimposed on the expected coarsening of sediments shoreward across the shoreface. Crest sediments are better sorted than those of the associated trough, but may be finer or coarser depending upon whether the latter are erosional lags or suspension deposits from quiescent periods (Mothersill, 1969; Greenwood and Davidson-Arnott, 1972; Keranen, 1985). In general sediments are negatively (coarse) skewed, although this is not universal (see Fraser et al., 1991). Sizes vary significantly with wave energy levels (Mulrennan, 1992; Stauble and Cialone, 1996).

Bedforms and internal stratification

Bed shear over nearshore bars is controlled by the complex, non-linear interactions within combined wave-current boundary layers. Sediment transport (Osborne and Greenwood, 1992), bedform development (Greenwood and Sherman, 1986) and internal stratification (Greenwood and Mittler, 1985) depend on the relative importance of a spectrum of oscillatory flows and any superimposed quasi-steady flow (Table B1). A number of studies document the bedforms that occur over bar topographies (Hunter et al., 1979; Dabrio and Polo, 1981; Shipp, 1984; Short, 1984; Ollerhead and Greenwood, 1992) and Clifton (1976) and Clifton and Dingler (1984) provide conceptual models for analyzing modern and ancient littoral systems. The internal structures of bars have been described by several authors (Howard and Reineck, 1972; Reineck and Singh, 1973; Davidson-Arnott and Greenwood, 1974, 1976; Hunter et al., 1979; Greenwood and Mittler 1979, 1985; Shipp, 1984), and summarized in Allen (1982). Stratification depends on position (see Figure B3): (a) *offshore slope*: couplets of small-scale trough (occasionally chevron) cross-lamination and low-angle parallel lamination; (b) *bar crest*: sub-horizontal, parallel lamination is most common, with medium-to-large scale trough cross-lamination dipping onshore (most common) and offshore; (c) *landward slope*: similar structures to those on the seaward slope except the parallel lamination dips onshore; (d) *trough*: parallel bedding and small-to-medium-scale trough cross-lamination; dips are highly variable reflecting both waves and bidirectional longshore currents; (e) *rip channel*: offshore-dipping, small-to-medium-scale units of tabular and trough cross-bedding occur in association with

Table B1 Flow regimes, bedform sequences and structural assemblages in littoral bars

Symmetrical oscillatory flow	
Bedform	Symmetric (vortex) ripples ⇒ Flat bed
Structure	Small, trough/chevron X-lamination ⇒ Parallel lamination
Asymmetrical oscillatory flow	
Bedform	Asymmetric ripples ⇒ Lunate megaripples ⇒⇒ Flat bed
	Asymmetric ripples ⇒ Post-vortex ripples ⇒⇒ Flat bed
	Asymmetric ripples ⇒ Cross Ripples ⇒⇒⇒ Flat Bed
Structure	Unimodal, trough X-lamination ⇒⇒⇒⇒⇒⇒⇒⇒⇒⇒⇒⇒⇒ Parallel lamination
	Unimodal, trough X-lamination ⇒ Medium, trough X-lamination ⇒ Parallel lamination
	Unimodal, trough X-lamination ⇒ Bimodal, trough X-lamination ⇒ Parallel lamination
Combined wave-current flow (oscillatory dominant)	
Bedform	Ripples ⇒ Lunate megaripples ⇒⇒⇒ Flat bed
	Ripples ⇒ Hummocky megaripples ⇒ Flat bed
Structure	Small, X-lamination ⇒ Trough X-lamination ⇒⇒⇒ Parallel lamination
	Small, X-lamination ⇒ Hummocky X-lamination ⇒ Parallel lamination
Unidirectional flow	
Bedform	Ripples ⇒ Dunes (Megaripples) ⇒ Flat bed
Structure	Small, trough X-lamination ⇒ Medium, trough X-lamination ⇒ Parallel lamination

quasi-steady rip currents. Rapid bar migration in shallow water produces medium-to-large-scale sets of onshore dipping tabular cross-lamination by avalanching on the bar front (Davidson-Arnott and Greenwood, 1974; Fraser *et al.*, 1991). Bioturbation is controlled by the degree of sediment reworking by the hydraulic regime and therefore is greatest on the lower offshore slope.

Facies models

The vertical and lateral sequences of stratification which occur with bar migration/accretion are described by Howard and Reineck (1972), Shipp (1984) and Greenwood and Mittler (1985) and models have been proposed by several authors (Davidson-Arnott and Greenwood, 1976; Hunter *et al.*, 1979; Shipp, 1984). Preservation potential depends upon the rate and direction of bar migration, the rate of sediment supply and water level change (Hunter *et al.*, 1979). For subaqueous bars, the most likely facies to be preserved are those of the trough, lower seaward slope and inner shelf. In bars that weld to the beach face, seaward progradation may preserve the large-scale tabular sets associated with the migrating slip face (van den Berg, 1977). However, the likelihood of preservation is generally small and conclusive evidence of wave-formed bars is lacking (see Ly, 1982, for one possible example).

Brian Greenwood

Bibliography

Allen, J.R.L., 1982. *Sedimentary Structures: Their Character and Physical Basis*, Volume 2. Amsterdam: Elsevier Scientific Publishing Company.

Clifton, H.E., 1976. Wave-generated structures—a conceptual model. In Davis, R.A. Jr., and Ethington, R.L. (eds.), *Beach and Nearshore Sedimentation*. Tulsa, Oklahoma: Society of Economic Paleontologists and Mineralogists, Special Publication 24, pp. 126–148.

Clifton, H.E., and Dingler, J.R., 1984. Wave-formed structures and paleoenvironmental reconstruction. *Marine Geology*, **60**: 165–198.

Dabrio, C.J., and Polo, M.D., 1981. Flow regime and bedforms in a ridge and runnel system. *Sedimentary Geology*, **18**: 97–110.

Davidson-Arnott, R.G.D., and Greenwood, B., 1974. Bedforms and structures associated with bar topography in the shallow-water wave environment. *Journal of Sedimentary Petrology*, **44**: 698–704.

Davidson-Arnott, R.G.D., and Greenwood, B., 1976. Facies relationships on a barred coast, Kouchibouguac Bay, New Brunswick, Canada. In Davis, R.A. Jr., and Ethington, R.L. (eds.), *Beach and Nearshore Sedimentation*. Tulsa, Oklahoma: Society of Economic Paleontologists and Mineralogists, Special Publication 24, pp. 140–168.

Fraser, G.S., Thompson, T.A., Kvale, E.P., Carlson, C.P., Fishbaugh, D.A., Gruver, B.L., Holbrook, J., Kairo, S., Kohler, C.S., Malone, A.E., Moore, C.H., Rachmanto, B., and Rhoades, L., 1991. Sediments and sedimentary structures of a barred non-tidal coastline, southern shore of Lake Michigan. *Journal of Coastal Research*, **7**: 1113–1124.

Greenwood, B., 2002. Wave-formed bars. In Schwartz, M. (ed.), *Encyclopedia of Coastal Science*. Amsterdam: Kluwer Academic Publishers (in press).

Greenwood, B., and Davidson-Arnott, R.G.D., 1972. Textural variation in the subenvironments of the shallow-water wave zone, Kouchibouguac Bay, New Brunswick Canada. *Canadian Journal of Earth Science*, **9**: 679–688.

Greenwood, B., and Davidson-Arnott, R.G.D., 1975. Marine bars and nearshore sedimentary processes, Kouchibouguac Bay, New Brunswick, Canada. In Hails J., and Carr, A. (eds.), *Nearshore Sediment Dynamics and Sedimentation*. London: John Wiley, pp. 123–150.

Greenwood, B., and Mittler, P.R., 1979. Structural indices of sediment transport in a straight, wave-formed, nearshore bar. *Marine Geology*, **32**: 191–203.

Greenwood, B., and Davidson-Arnott, R.G.D., 1979. Sedimentation and equilibrium in wave-formed bars: a review and case study. *Canadian Journal of Earth Sciences*, **16**: 312–332.

Greenwood, B., and Mittler, P.R., 1985. Vertical sequence and lateral transitions in the facies of a barred nearshore environment. *Journal of Sedimentary Petrology*, **55**: 366–375.

Greenwood, B., and Sherman, D.J., 1986. Hummocky cross-stratification in the surf zone: flow parameters and bedding genesis. *Sedimentology*, **33**: 33–45.

Hands, E.H., 1976. Observations of barred coastal profiles under the influence of rising water levels, Eastern Lake Michigan, 1967–71. *U.S. Army Corps of Engineers, CERC Technical Report* 76-1, 113 pp.

Howard, J.D., and Reinick, H.-E., 1972. Georgia coastal region, Sapelo Island, U.S.A.: sedimentology and biology, IV. Physical and biogenic sedimentary structures of the nearshore shelf. *Senckenbergiana Maritima*, **4**: 81–123.

Hunter, R.E., Clifton, H.E., and Phillips, R.L., 1979. Depositional processes, sedimentary structures and predicted vertical sequences in barred nearshore systems, southern Oregon coast. *Journal of Sedimentary Petrology*, **49**: 711–726.

Keranen, R., 1985. Wave-induced sandy shore formations and processes in Lek Oulujarvi, Finland. *Nordia*, **19**: 1–58.

Lippmann, T.L., and Holman, R.A., 1990. The spatial and temporal variability of sand bar morphology. *Journal of Geophysical Research*, **95**: 11575–11590.
Ly, C.K., 1982. Sedimentology of nearshore marine bar sequences from a Paleozoic depositional regressive shoreline deposit of the central coast of Ghana, West Africa. *Journal of Sedimentary Petrology*, **52**: 199–208.
Mothersill, J.S., 1969. A grain size analysis of longshore bars and troughs, Lake Superior, Ontario. *Journal of Sedimentary Petrology*, **39**: 1317–1324.
Mulrennan, M.E., 1992. Ridge and runnel beach morphodynamics: an example from the central east coast of Ireland. *Journal of Coastal Research*, **8**: 906–918.
Ollerhead, J.W., and Greenwood, B., 1990. Bedform geometry and dynamics in the upper shoreface, Bluewater Beach, Ontario, Canada. *Proceedings, Canadian Coastal Conference 1990*, Kingston, Ont., NRCC Associate Comm. on Shorelines, Ottawa, pp. 337–348.
Osborne, P.D., and Greenwood, B., 1992. Frequency dependent cross-shore suspended sediment transport 2: a barred shoreface, Bluewater Beach, Ontario, Canada. *Marine Geology*, **106**: 25–51.
Reineck, H.-E., and Singh, I.B., 1973. *Depositional Sedimentary Environments*. New York: Springer Verlag.
Ruessink, B.G., van Enckevort, I.M.J., Kingston, K.S., and Davidson, M.A., 2000. Analysis of observed two- and three-dimensional nearshore bar behaviour. *Marine Geology*, **64**: 237–257.
Shepard, F.P., 1950. Longshore bars and longshore troughs. *Beach Erosion Board, Technical Memorandum*, **15**: 32pp.
Shipp, R.C., 1984. Bedforms and depositional sedimentary structures of a barred nearshore system Eastern Long Island, New York. *Marine Geology*, **60**: 235–259.
Short, A.D., 1984. Beach and nearshore facies: southeast Australia. *Marine Geology*, **60**: 261–282.
Stauble, D.K., and Cialone, M.A., 1996. Sediment dynamics and profile interactions: Duck94. *25th International Conference on Coastal Engineering*, Orlando, pp. 3921–3934.
Van den Berg, J.H., 1977. Morphodynamic development and preservation of physical sedimentary structures in two prograding recent ridge and runnel beaches along the Dutch coast. *Gelogie en Mijnbouw*, **56**: 182–202.
Van Rijn, L.C., 1998. *Principles of Coastal Morphology*. Amsterdam: Aqua Publications, pp. 4.30–4.279.
Wright, L.D., and Short, A.D., 1984. Morphodynamic variability of surf zones and beaches: a synthesis. *Marine Geology*, **56**: 93–118.
Zenkovich, V.P., 1967. *Processes of Coastal Development*. London: Oliver and Boyd.

Cross-references

Coastal Sedimentary Facies
Cross-Stratification
Facies Models
Planar and Parallel Lamination
Ripple, Ripple Mark, Ripple Structures
Sediment Transport by Waves
Surface Forms

BARRIER ISLANDS

Introduction

Barrier islands are wave-built accumulations of sediment that accrete vertically due to wave action and wind processes and are separated from adjacent barriers or headlands by tidal inlets. Most are linear features that tend to parallel the coast, generally occurring in groups or chains. Barriers are separated from the mainland by a region termed the *backbarrier* consisting of tidal flats, shallow bays, lagoons and/or marsh systems. Barrier islands may be less than 100 m wide or more than several kilometers in width. Likewise, they range in length from several hundred meters to certain barriers along open coasts that extend for more than 100 km. Generally, barrier islands are wide where the supply of sediment has been abundant and relatively narrow where erosion rates are high or where the sediment was scarce during their formation. Their length is partly a function of sediment supply but is also strongly influenced by wave versus tidal energy of the region.

Coastal barriers consist of many different types of sediment depending on their geological setting. Sand, which is the most common constituent, comes from a variety of sources including rivers, deltaic and glacial deposits, eroding cliffs, and biogenic material. The major components of land-derived sand are the minerals quartz and feldspar. In northern latitudes where glaciers have shaped the landscape, gravel is a common constituent of barriers, whereas, in southern latitudes carbonate material, including shells and coral debris, may comprise a major portion of the barrier sands. Along the southeast coast of Iceland, as well as parts of Alaska and Hawaii, the barriers are composed of black volcanic sands derived from upland volcanic rocks.

Global distribution

Barriers comprise approximately 15 percent of the worlds coastline. They are found along every continent except Antarctica, in every type of geological setting, and in every kind of climate. Tectonically, they are most common along Amero-trailing edge coasts where low gradient continental margins provide ideal settings for barrier formation (Inman and Nordstrom, 1971). They are also best developed in microtidal to mesotidal regimes and in mid- to low-latitudes (Hayes, 1979). Climatic conditions control the vegetation on the barriers and in backbarrier regions, the type of sediment on beaches, and in some regions such as the Arctic, the formation and modification of barriers themselves. The disappearance of barriers toward the very low energy northwest ("Big Bend") coast of Florida in the Gulf of Mexico attests to the requirement of wave energy in the formation of barriers.

Low-relief coastal plains and continental shelves provide a platform upon which barriers can form and migrate landward during periods of eustatic sea-level rise. The longest barrier chains in the world coincide with Amero-trailing edges and include the east coast of the United States (3,100 km) and the Gulf of Mexico coast (1,600 km). There are also sizable barrier chains along the east coast of South America (960 km), east coast of Indian coast (680 km), North Sea coast of Europe (560 km), eastern Siberia (300 km), and the North Slope of Alaska (900 km).

Barrier island sub-environments

Many sub-environments make up a barrier and their arrangement differs from location to location reflecting the type of barrier and the physical setting of the region. Generally, most barriers can be divided in three zones: the beach, the barrier interior, and the landward margin (Figure B4). Some barrier systems, such as those along the Mississippi coast, have active beaches on their landward sides due to expansive adjacent bays. The continual sediment reworking by wind, waves, and tides, establishes the beach as the most dynamic part of the barrier. Beaches exhibit a wide range of morphologies depending upon a number of factors including the sediment grain size and abundance, and the influence of storms.

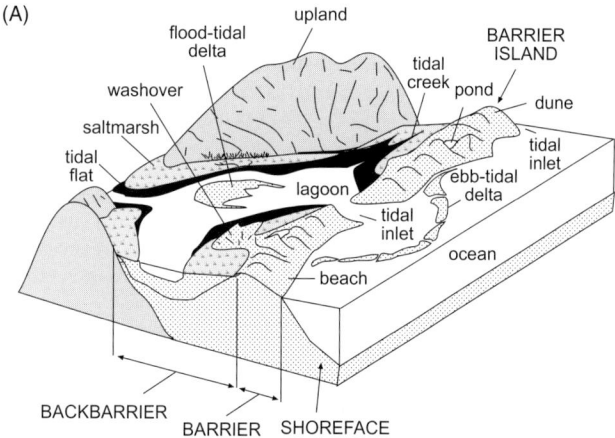

Figure B4 (A) Barrier island sub-environments, (B) Oblique aerial photograph of highly retrogradational barrier island, showing extensive washovers, Masonboro Island, North Carolina.

Sediment is removed from the beach during storms and returned during more tranquil wave conditions. During the storm and post-storm period, the form of the beach evolves in a predictable fashion. Along sandy barriers, the beach is backed by a frontal dune ridge (*foredune*) which may extend almost uninterruptedly along the length of the barrier provided the supply of sediment is adequate and storms have not incised the dune ridge. The frontal dune ridge is the first line of defense in protecting the interior of the barrier from the effects of storms. Landward of this region are the secondary dunes which have a variety of forms depending on the historical development of the barrier and subsequent modification by wind processes. The low areas between individual beach ridges, called *swales*, commonly extend below the water table and are the sites of fresh and brackish water ponds or salt marshes.

Along the backside of many barriers the dunes diminish in stature and the low relief of the barrier gradually changes to an intertidal sand or mud flat or a salt marsh. In other instances, the landward margin of the barrier abuts an open water area of a lagoon, bay, tidal creek, or coastal lake. Along coasts where the barrier is migrating onshore, the landward margin may be dominated by aprons of sand that have overwashed the barrier during periods of storms (Leatherman, 1979; Figure B4). In time, these sandy deposits may be colonized by salt grasses producing arcuate-shaped marshes. In still other regions, the backside of a barrier may be fronted by a sandy beach bordering a lagoon or bay. In this setting, wind-blown sand from the beach may form a rear-dune ridge outlining the landward margin of the barrier. The ends of the barrier island abutting tidal inlets are usually the most dynamic regions due to the effects of inlet migration and associated sediment transport patterns.

Origin of barrier islands

The widespread distribution of barrier islands along the worlds coastlines and their occurrence in many different environmental settings has led numerous scientists to speculate on their origin for more than 150 years (Schwartz, 1973). The different explanations of barrier island formation can be grouped into three major theories: (1) *Offshore bar theory* (de Beaumont–Johnson), (2) *Spit accretion theory* (Gilbert–Fisher), and (3) *Submergence theory* (McGee–Hoyt) (Figure B5).

Elie de Beaumont believed that waves moving into shallow water churned up sand which was deposited in the form of a submarine bar when the waves broke and lost much of their energy. As the bars accreted vertically, they gradually built above sea level forming barrier islands (Figure B5). de Beaumonts idea that barriers were formed from offshore sand sources was countered by G.K. Gilbert, who argued that the barrier sediments came from alongshore sources. Gilbert proposed that sediment moving in the breaker zone through agitation by waves would construct spits extending from headlands parallel to the coast. The subsequent breaching of spits by storm waves would form barrier islands (Figure B5). Before the end of the nineteenth century, a third barrier island theory was published by W.D. McGee in 1890. He believed that during submergence, coastal ridges were separated from the mainland forming lagoons behind the ridges. In 1919, D.W. Johnson re-investigated the various theories and studied the shore-normal profile of barrier coasts. He reasoned that if barriers had formed from spits, then the offshore profile should intersect the mainland at the edge of the lagoon.

Hoyt (1967) correctly argued that if barrier islands had developed from offshore bars or through the breaching of spits, then open-ocean conditions would have existed along the mainland prior to barrier formation. Before becoming sheltered, breaking waves and onshore winds would have formed beaches and dune systems at these locations. Hoyt was able to show that for most barrier systems an open-ocean coast never existed along the present mainland coast. In situations where rising sea level and marsh development has led to an encroachment of lagoonal deposits onto the mainland, sediment cores from these regions have showed no evidence

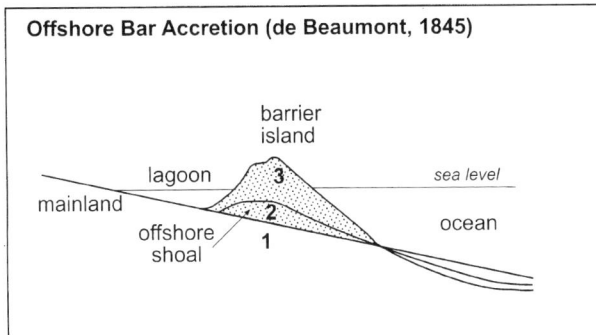
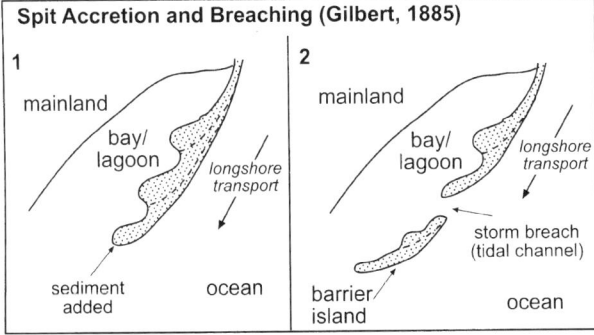
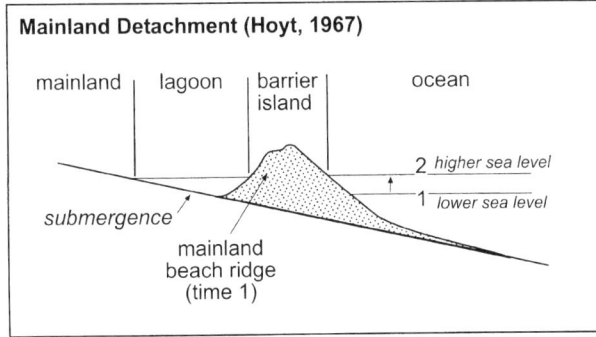

Figure B5 Models of barrier island formation (modified after Penland and Boyd, 1985).

of beach and nearshore sediments or the shells of organisms that are commonly inhabit the nearshore region. Armed with sedimentologic data, Hoyt (1967) had made a compelling case for barrier islands forming by submergence of coastal dunes or beach ridges due to rising sea level (Figure B5). However, in 1968, John Fisher produced a critical review of the submergence theory pointing out that long, straight, and continuous dune ridges would not occur along a coast being inundated by rising sea level. In light of these observations, Fisher became a strong proponent of Gilberts spit accretion theory.

Recent barrier island research

Since the late-1960s there have been many studies of barrier islands aimed at determining the sedimentary layers making up the barrier and deciphering the manner in which these layers were deposited. This research has been aided by the radiometric dating of organic material contained within the barriers sediments, such as shells, peat, and wood, thereby providing a chronology of barrier construction (Davis, 1994). In addition to traditional stratigraphic techniques, ground-penetrating radar (GPR) is also employed to study barriers (Jol *et al.*, 1996; van Heteren *et al.*, 1998). Sediment cores are taken in conjunction with the GPR surveys to determine the composition of the sediment layers and the environment in which they were deposited. Additional information concerning former barriers and their associated tidal inlets and lagoons has been gathered from the inner continental shelf using high-resolution shallow seismic surveys, a technology similar in principle to GPR. These advancements have provided new insights about the formation of coastal barrier systems.

Scientists are beginning to accept the idea that barriers can form by a number of different mechanisms. In considering the formation of barrier islands, it is important to recognize that almost all the worlds barriers are less than 6,500 years old and most are younger than 4,000 years old. Most barriers formed in a regime of rising sea level but during a time when the rate of rise began to slow. Sedimentological data from the inner continental shelves off the East Coast of the United States, in the North Sea, and in southeast Australia suggest that barriers once existed offshore and have migrated to their present positions.

Morphosedimentary barrier types

The overall form of barriers, their stability, and future erosional or depositional trends are related to the supply of sediment, the rate of sea-level rise, storm cycles, and the topography of the mainland. When a barrier builds in a seaward direction it is said to prograde and is called a *prograding* barrier. Prograding barriers form in a regime of abundant sand supply during a period of stable or slow rise in sea level. The sand to build these barriers may come from along the shore or from offshore sources. These conditions were met along much of the East Coast of the United States and many other regions of the world about 4,000 to 5,000 years ago when the rate of sea-level rise decreased and sediment was contributed to the shore from *eroding headlands* (examples: Provinceland Spit in northern Cape Cod, Massachusetts (FitzGerald *et al.*, 1994); Lawrencetown barrier along the Northeast Shore of Nova Scotia), from the *inner continental shelf* (examples: barrier system along Bogue Banks, North Carolina; Algarve barrier chain, in southern Portugal; Tuncurry barrier in southeast Australia (Roy *et al.*, 1994); East Friesian Islands along the German North Sea coast), and *directly from rivers* (examples: most barrier systems in northern New England; barrier chain situated north of the Columbia River along southern coast of Washington).

A *retrograding* barrier forms when the supply of sand is inadequate to keep pace with relative sea-level rise and/or with sediment losses (Moslow and Colquhoun, 1981). Stated in terms of a sediment budget, a barrier becomes retrogradational when the amount of sand contributed to the barrier is less than the volume transported away from the barrier. The sand may be lost offshore during storms, moved along shore in the littoral system, or transported across the barrier by overwash. *Overwash* is a cannibalistic process whereby storm waves transport sand from the beach through the dunes, depositing it along the landward margin of the barrier (Figure B4). In this way the barrier is preserved by retreating landward. Because these barriers have been migrating over various types of backbarrier settings, including lagoons and marshes, the

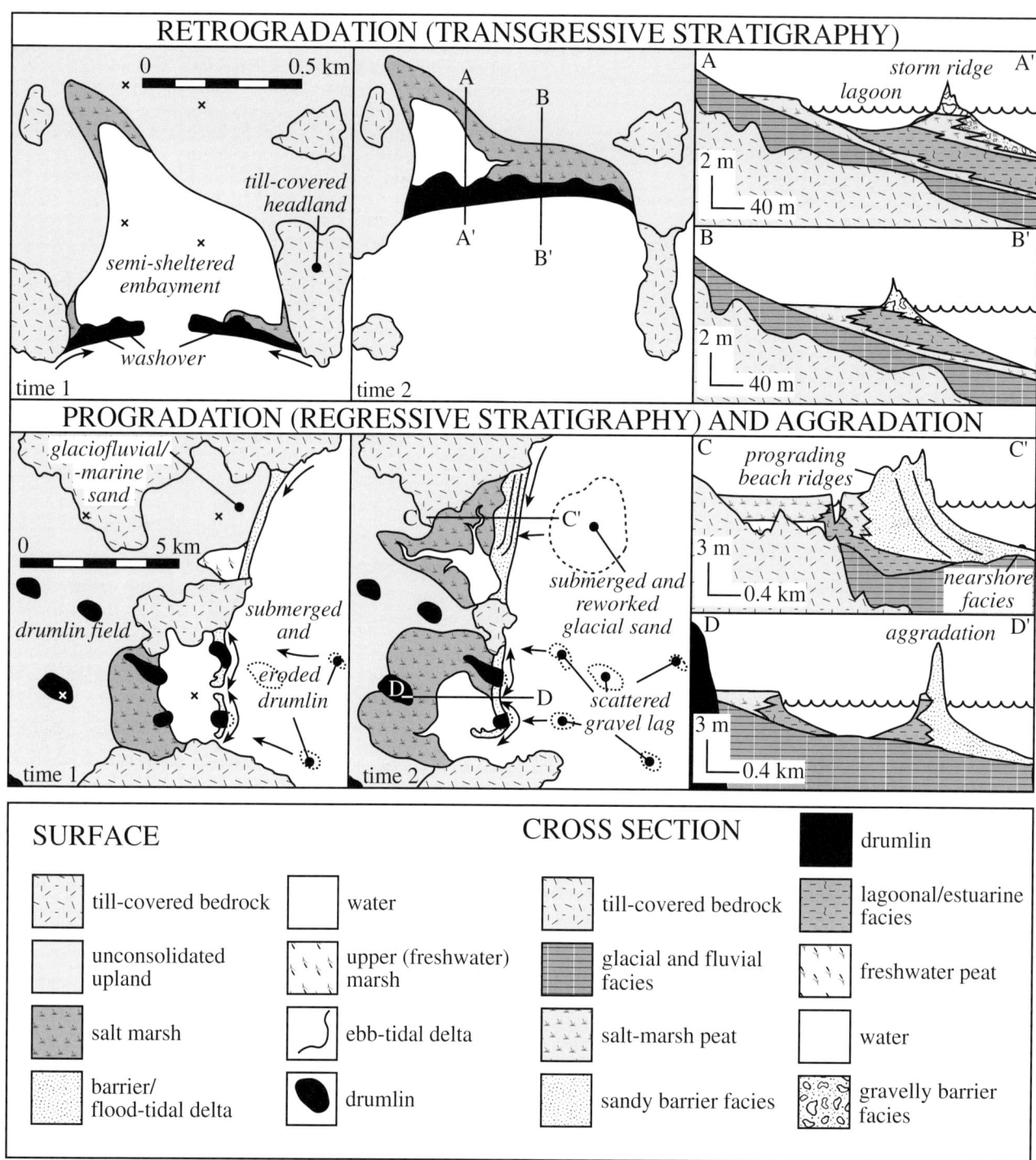

Figure B6 Barrier island stratigraphies as a function of different evolutionary styles (modified from FitzGerald and van Heteren, 1999).

sedimentary components comprising these environments (lagoonal muds, marsh peats, upland tree stumps) are commonly exposed along the front side of the barrier, usually in the intertidal zone.

If a barrier has built vertically through time in a regime of rising sea level and occupies approximately the same footprint as it did when it first formed or stabilized, it is termed an *aggrading* barrier. These barriers are rare because it requires that the rate of sediment supply to the barrier exactly compensates for rising sea level. Too little sand and the barrier migrates onshore (retrograding) whereas too large a supply and the barrier builds seaward. Without subsurface information they are difficult to recognize, because morphologically they may appear similar to non-beach ridge, prograding barrier or even a retrograding barrier that has stopped moving onshore.

Barrier island stratigraphy

Barriers exhibit a variety of architectures consisting of many different types of sedimentary sequences depending upon their evolutionary development. The stratigraphy of a barrier is defined by a set of grain size, mineralogical, and other characteristics of the layers comprising the barrier deposit. Factors such as sediment supply, rate of sea-level rise, wave and tidal energy, climate, and topography of the land dictate how a barrier develops and its resulting stratigraphy. Another important factor affecting barrier thickness is accommodation space which defines how much room is available for the accumulation of barrier sands.

Barrier sequences often contain tidal inlet deposits, especially along barrier coasts where tidal inlets open and close and/or where tidal inlet migration is an active process (Reinson, 1984, 1992). A tidal inlet migrates by eroding the downdrift side of its channel while at the same time sand is added to the updrift side of its channel. In this way, the updrift barrier elongates, the downdrift barrier retreats, and the migrating inlet leaves behind channel fill deposits underlying the updrift barrier. Independent studies along New Jersey and the Delmarva Peninsula, North Carolina, and South Carolina indicate that 20 percent to 40 percent of these barrier coasts are underlain by tidal inlet fill deposits (Moslow and Heron, 1978).

Each morphosedimentary barrier type has a diagnostic stratigraphy that reflects the manner in which it developed (Figure B6). Prograding barriers typically exhibit *regressive* stratigraphy. Because this type of barrier builds in a seaward direction, the barrier sequence is commonly thick (10–20 m) and overlies offshore deposits, usually composed of fine-grained sands and silts (Bernard et al., 1970). The barrier sequence consists of nearshore sands, overlain by beach deposits, and topped by dune sands. The contacts between the units are gradational and for the most part the sedimentary sequence coarsens upward except for the uppermost fine-grained dune sands. The retrograding barrier type migrates in a landward direction over the marsh and lagoon by overwash processes, resulting in a *transgressive* stratigraphic sequence (Kraft and John, 1979). The Holocene sequence typically bottoms in lagoonal muds, however, if the barrier has retreated far enough landward, mainland deposits may be preserved forming the base of the sequence. In these instances, we may find tree stumps, soils, and other deposits. The mainland units are overlain by backbarrier sediments including a variety of units such as lagoonal silts and clays and marsh peats which had formed in intertidal areas. In the vicinity of tidal inlets, backbarrier deposits consist of channel and flood-tidal delta sands. Overlying the backbarrier deposits is the thin barrier sequence (<3 m to 4 m) consisting of washovers, beach deposits, and dune sediments if they are present. Aggrading barriers build upward in a regime of rising sea level and in an ideal case, the deposits from the same environmental setting are stacked vertically. In most cases, however, the barrier has shifted slightly landward and seaward through time due to changes in sediment supply and rates of sea-level rise. Therefore, most aggrading barriers exhibit some interstacking of various units. For example, in the rear of the barrier the sequence may consist of washover and dune units inter-layered with marsh and lagoonal deposits. Aggrading barriers tend to be thick (10 m to 20 m) and for reasons stated previously, are uncommon.

Duncan M. FitzGerald and Ilya V. Buynevich

Bibliography

Bernard, H.A., Major, C.F., Parrott, B.S., and LeBlanc, R.J., 1970. Recent sediments of Southeast Texas: a field guide to the Brazos alluvial and deltaic plains and the Galveston barrier island complex. Univ. of Texas—Austin, *Bureau of Economic Geology, Guidebook No. 11*.

Davis, R.A. Jr. (ed.), 1994. *Geology of Holocene Barrier Island Systems*. Springer-Verlag.

FitzGerald, D.M., Rosen, P.S., and van Heteren, S., 1994. New England barriers. In Davis, R.A. Jr. (ed.), *Geology of Holocene Barrier Island Systems*. Springer-Verlag, pp. 305–394.

FitzGerald, D.M., and van Heteren, S., 1999. Classification of paraglacial barrier systems: coastal New England, USA. *Sedimentology*, **46**: 1083–1108.

Hayes, M.O., 1979. Barrier island morphology as a function of tidal and wave regime. In Leatherman, S.P. (ed.), *Barrier islands: from the Gulf of St. Lawrence to the Gulf of Mexico*. New York: Academic Press, pp. 1–28.

Hoyt, J.H., 1967. Barrier island formation. *Geological Society of America Bulletin*, **78**: 1123–1136.

Inman, D.L., and Nordstrom, C.E., 1971. On the tectonic and morphologic classification of coasts. *Journal of Geology*, **79**: 1–21.

Jol, H.M., Smith, D.G., and Meyers, R.A., 1996. Digital ground penetrating radar (GPR): an improved and very effective geophysical tool for studying modern coastal barriers (examples for the Atlantic, Gulf and Pacific coasts, U.S.A.). *Journal of Coastal Research*, **12**: 960–968.

Kraft, J.C., and John, C.J., 1979. Lateral and vertical facies relations of transgressive barriers. *Bulletin of American Association of Petroleum Geologists*, **63**: 2145–2163.

Leatherman, S.P., 1979. *Barrier Island Handbook*. Amherst, Massachusetts.

Moslow, T.F., and Heron, S.D., 1978. Relict inlets: preservation and occurrence in the Holocene stratigraphy of southern Core Banks, North Carolina. *Journal of Sedimentary Petrology*, **48**: 1275–1286.

Moslow, T.F., and Colquhoun, D.J., 1981. Influence of sea-level change on barrier island evolution. *Oceanis*, **7**: 439–454.

Penland, S., and Boyd, R., 1985. Transgressive depositional environments of the Mississippi River Delta plain: a guide to the barrier islands, beaches, and shoals of Louisiana. Baton Rouge, LA, *Louisiana Geological Survey, Guidebook Series #3*.

Reinson, G.E., 1984. Barrier island and associated strand-plain systems. In Walker, R.G. (ed.), *Facies Models*. Geoscience Canada Reprint Series **1**: pp. 119–140.

Reinson, G.E., 1992. Transgressive barrier island and estuarine systems. In Walker, R.G., and James, N.P. (eds.), *Facies Models: Response to Sea Level Change*. Geological Association of Canada, pp. 179–194.

Roy, P.S., Cowell, P.J., Ferland, M.A., and Thom, B.G., 1994. Wave-dominated coasts. In Carter, R.W.G., and Woodroffe, C.D. (eds.), *Coastal Evolution: Late Quaternary shoreline morphodynamics*. Cambridge University Press, pp. 121–186.

Schwartz, M.L., 1973. *Barrier islands*. Stroudsburg, PA: Dowden, Hutchinson and Ross.

van Heteren, S., FitzGerald, D.M., McKinlay, P.A., and Buynevich, I.V., 1998. Radar facies of paraglacial barrier systems: coastal New England, USA. *Sedimentology*, **45**: 181–200.

Cross-references

Bar, Littoral
Tidal Inlets and Deltas

BAUXITE

Introduction

Bauxite is found in many parts of the world, but more particularly in tropical areas. Bauxite is of supergene origin commonly produced by weathering and leaching of silica from aluminum bearing rocks. Bauxite may occur *in situ* as a direct result of weathering or it may be transported and deposited in sedimentary formation. Gibbsite [Al(OH)$_3$], boehmite and diaspore [AlO(OH)] are the three principal hydrates of aluminum and form the main constituents of bauxite and laterites with gibbsite often being the predominant mineral. Gibbsite or hydrargillite, bayerite and nordstrandite are all polytypes of aluminum trihydroxide. Diaspore is a dimorph of boehmite. Australian soils often contain gibbsite, particularly in the soils of hot humid climates where the topography is suitable for its accumulation as occurs in Northern Queensland. Gibbsite often occurs in association with kaolinite as exemplified by the Weipa deposits of North Queensland (Wilke and Schwertmann, 1977). Gibbsite is the end product of granitic weathering and is formed from the diagenetic sequence: plagioclase ⇒ amorphous or allophanic minerals ⇒ kaolinite ⇒ gibbsite. Such a sequence shows why the impurity in gibbsites is often kaolinite. Australian bauxites are predominantly composed of mixtures of gibbsite and boehmite with no diaspore. Impurities include hematite, kaolinite, quartz, and other derived minerals. The Weipa bauxite composition varies from 70 percent to 95 percent gibbsite and from 25 percent to 5 percent boehmite.

Thermal transformations

Bauxites are a complex admixture of minerals with a number of alumina phases present in any mix. Figure B7 summarizes the thermal transformations of the components of bauxite. These transformations are dependent upon the particle size, the rate of heating and the vapor pressure above the bauxite. In principle if gibbsite is heated it alters structure to form Chi-alumina, then kappa-alumina and at temperatures above 1100 °C alpha-alumina. Boehmite on the other hand transforms to gamma-, delta-, theta- and then alpha-alumina. Bayerite, which is not commonly found in bauxites either, transforms to boehmite or to eta-alumina. Diaspore transforms directly to alpha-alumina. Such complex phase relations affect the way in which bauxite is processed for aluminum production.

Thermal analysis

The thermal transformations of the bauxitic minerals may be studied by a number of techniques including thermal analysis. The use of thermal techniques to study the dehydration and dehydroxylation of bauxite has been widely documented (Lodding, 1969). It is clear that many variables must be taken into account when using techniques such as DTA, DSC, TGA, CRTA and quasi-isothermal TGA and isobaric TGA (Frost *et al.*, 1999a,b,c; Ruan *et al.*, 2001a,b). Such variables include heating rate, external pressure, water vapor pressure, sample particle size and even thickness of sample size in the DTA crucible (Paulik *et al.*, 1983; Naumann *et al.*, 1983). Figure B8 illustrates the thermal analysis patterns of gibbsite bauxites. It has been shown that boehmite and gamma-alumina were formed under the dehydration of gibbsite. The initial step in the thermal decomposition of gibbsite is the diffusion of protons and the reaction with hydroxyl ions to form water (Frost *et al.*, 1999a). This process removes the binding forces between the layers of the gibbsite structure and causes changes in the chemical composition and density within the layers. Published DTA patterns of a coarse-grained gibbsite show an endotherm centerd on 230°C followed by a second at 280°C (Frost *et al.*, 1999a,b,c). This latter endotherm is attributed to

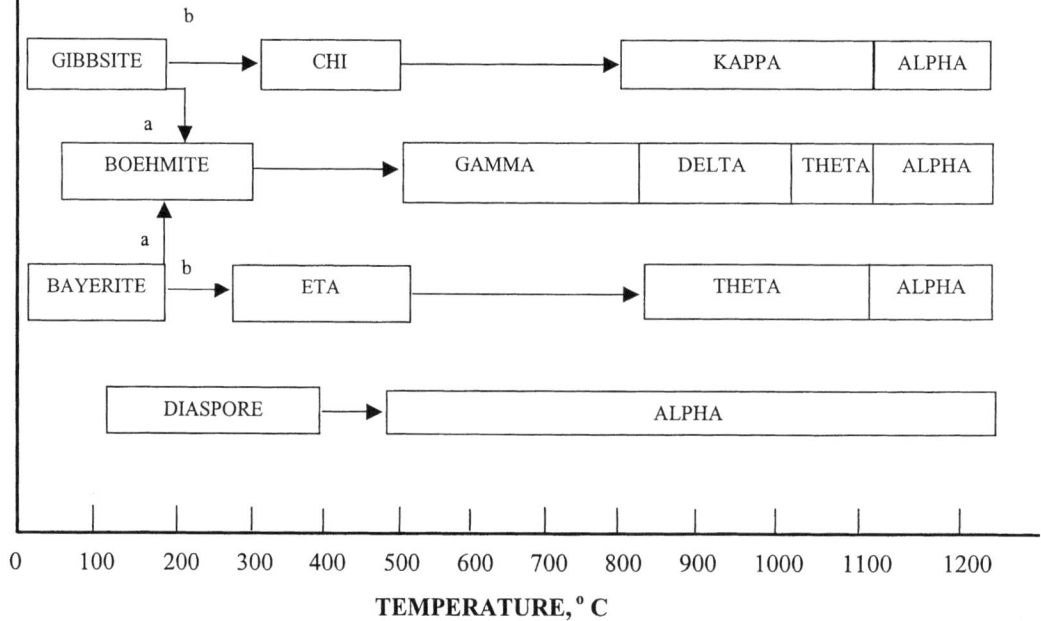

Figure B7 Relationship between the oxo(hydroxy) phases of aluminium.

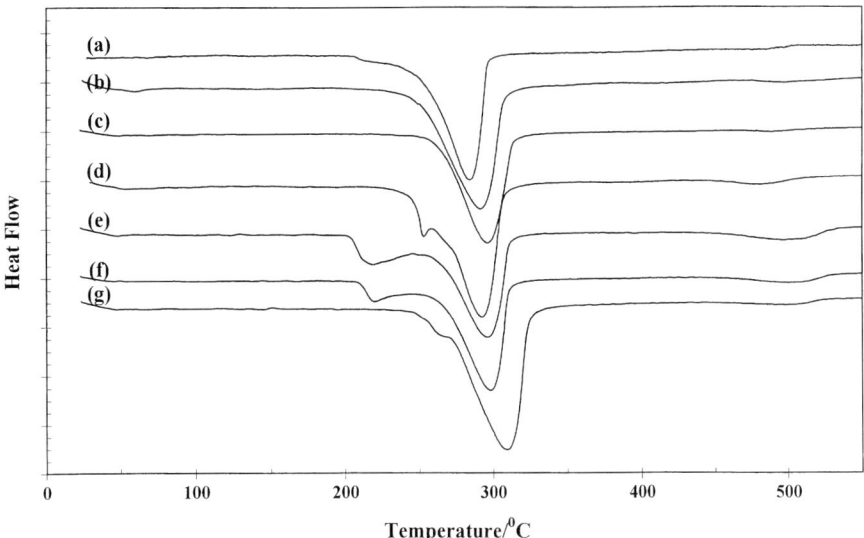

Figure B8 Differential thermal analysis of gibbsitic bauxites from (a) Weipa (Australia) (b) Slovakia (c) Surinam (d) California (e and f) synthetic gibbsite (g) Brazil.

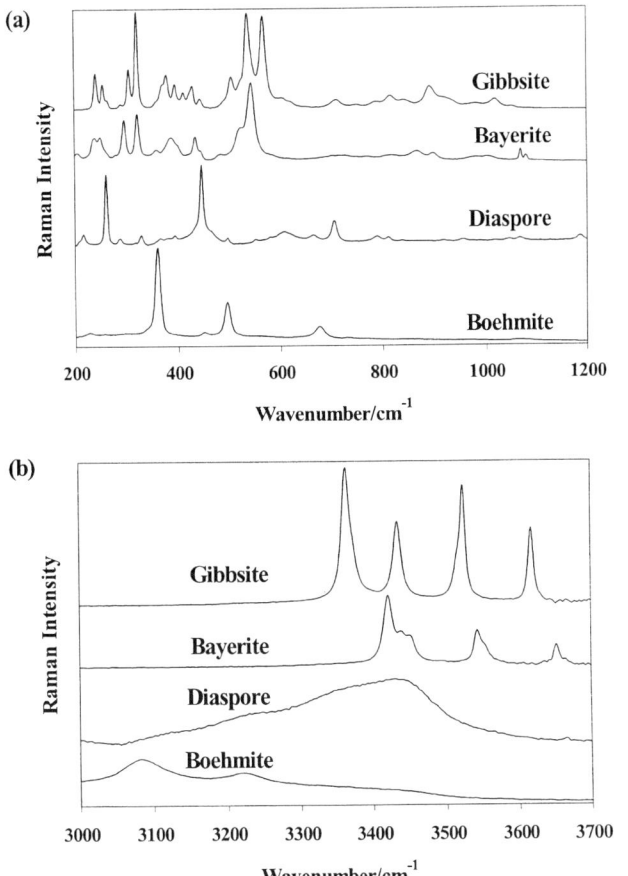

Figure B9 Raman spectra of (a) the low frequency region (b) hydroxyl-stretching region of the bauxite oxo(hydroxy) minerals.

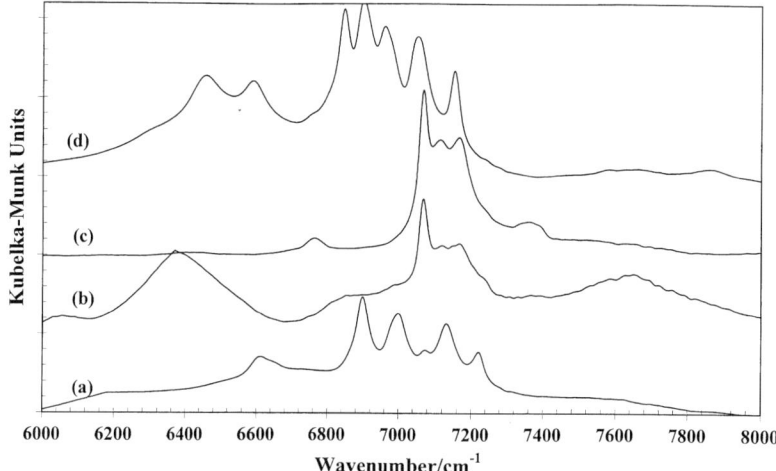

Figure B10 Near-IR spectra of the first hydroxyl stretching fundamental of the bauxitic minerals.

the formation of boehmite by hydrothermal conditions due to the retardation diffusion of water out of the larger grains. This exothermic reaction does not occur in the DTA patterns of finely grained gibbsite. A shallow endotherm may be observed between 500°C and 550°C and is attributed to the formation of boehmite. There is general agreement that boehmite and a disordered transition alumina are formed upon the thermal treatment of coarse gibbsite up to 400°C. When fine-grained gibbsite is heated rapidly, an X-ray amorphous product labeled rho-alumina is obtained.

Spectroscopic characterization of bauxites

One valuable suite of techniques that are suited for the characterization of bauxites are based upon vibrational spectroscopic techniques. These techniques include those of infrared including near-IR (NIR), far-IR, DRIFT spectroscopic techniques. Raman spectroscopy is also useful in the characterization of the bauxitic phases.

Raman spectroscopy

Raman spectroscopy has the advantage that it is a scattering technique, does not require sample preparation and is not sensitive to the presence of water. The Raman spectrum of gibbsite shows four strong, sharp bands at 3617, 3522, 3433 and 3364 cm^{-1}. The spectrum of bayerite shows seven bands at 3664, 3652, 3552, 3542, 3450, 3438, and 3420 cm^{-1} in the hydroxyl-stretching region. Five broad bands at 3445, 3363, 3226, 3119, and 2936 cm^{-1} and four broad and weak bands at 3371, 3220, 3085, and 2989 cm^{-1} are present in the Raman spectrum of the hydroxyl-stretching region of diaspore and boehmite, respectively. The hydroxyl-stretching bands are related to the surface structure of the minerals. The Raman spectra of bayerite, gibbsite, and diaspore are complex while the Raman spectrum of boehmite only shows four bands in the low wavenumber region. These bands are assigned to deformation and translational modes of the alumina phases. A comparison of the Raman spectrum of bauxite with those of boehmite and gibbsite shows the possibility of using Raman spectroscopy for online processing of bauxites.

Near infrared spectroscopy

Another most useful technique for studying bauxites and their phase components is the reflectance technique of NIR spectroscopy. NIR is known as the spectroscopy of protons as so as a consequence all phases containing water or hydroxyl units can be measured by NIR. This means that NIR is also most useful in the online processing of bauxites. Figure B9 displays the first overtone of the fundamental vibrations of the hydroxyl stretching vibrations.

NIR spectroscopy distinguishes between alumina oxo and hydroxy phases. Two NIR spectral regions are identified for this function: (a) the high frequency region between 6400 cm^{-1} and 7400 cm^{-1}, attributed to the first overtone of the hydroxyl stretching mode, and (b) the 4000–4800 cm^{-1} region attributed to the combination of the stretching and deformation modes of the AlOH units. NIR spectroscopy allows the study and differentiation of the hydroxy and oxo(hydroxy) alumina phases, since each phase has its own characteristic spectrum. The spectrum of bayerite resembles that of gibbsite whereas the spectrum of boehmite is similar to that of diaspore. Bayerite has four characteristic NIR bands at 7218, 7128, 6996, and 6895 cm^{-1}. Gibbsite shows five major bands at 7151, 7052, 6958, 6898, and 6845 cm^{-1}. Boehmite displays three NIR bands at 7152, 7065, and 6960 cm^{-1}. Diaspore shows a prominent band at around 7176 cm^{-1}. The use of NIR reflectance spectroscopy to study alumina surfaces has wide application, particularly with thin films and surfaces. The technique is rapid and accurate. NIR, because of its sensitivity can be used in reflectance mode for the on-line processing of bauxitic minerals.

Infrared emission spectroscopy

Details of the experimental part of infrared emission spectroscopy have been detailed in a number of publications (Frost et al., 1999a,b,c). The emission spectra were collected at intervals of 50°C over the range 200–750°C. The time between scans (while the temperature was raised to the next hold point) was approximately 100 seconds. It was considered that this was sufficient time for the heating block and the powdered

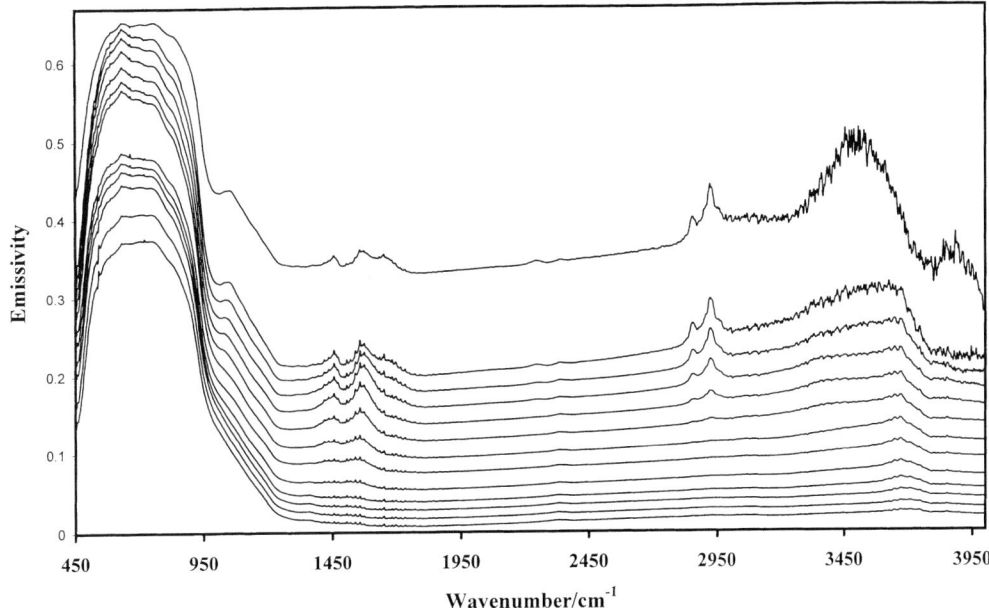

Figure B11 Infrared emission spectra of Weipa bauxite.

sample to reach temperature equilibrium. The spectra were acquired by coaddition of 64 scans for the whole temperature range (approximate scanning time 45 seconds), with a nominal resolution of $4\,cm^{-1}$. Good quality spectra can be obtained providing the sample thickness is not too large. If too large a sample is used then the spectra become difficult to interpret because of the presence of combination and overtone bands. Figure B11 displays a series of infrared emission spectra for a Weipa bauxite. The spectra clearly show the temperature at which the bauxite dehydrates and dehydroxylates. Phase changes may be observed through the loss of intensity in the OH deformation band at around $1020\,cm^{-1}$.

Summary

Modern techniques of analysis including NIR, infrared emission and Raman spectroscopy enable on-line and *in situ* studies of bauxites and its components to be studied. When these spectroscopic techniques are combined with modern thermal analysis, new and essential information to the understanding of bauxites is obtained. Such information is critical to aluminum processing. Normally the bauxite is dissolved in hot caustic at 100°C. In such processing the gibbsite component of the bauxite dissolves but not the boehmite and other components. To dissolve the boehmite, higher temperatures and pressures are required. It is useful to modify the bauxite surfaces to enhance this dissolution.

Ray L. Frost

Bibliography

Frost, R.L., Kloprogge, J.T., Russell, S.C., and Szetu, J., 1999a. The dehydroxylation of aluminium (oxo)hydroxides using infrared emission spectroscopy: gibbsite. *Applied Spectroscopy*, **53**: 423–434.
Frost, R.L., Kloprogge, J.T., Russell, S.C., and Szetu, J., 1999b. The dehydroxylation of aluminium (oxo)hydroxides using infrared emission spectroscopy: boehmite. *Applied Spectroscopy*, **53**: 570–582.
Frost, R.L., Kloprogge, J.T., Russell, S.C., and Szetu, J., 1999c. The dehydroxylation of aluminium (oxo)hydroxides using infrared emission spectroscopy, Part 3 diaspore. *Applied Spectroscopy*, **53**: 829–835.
Lodding, E., 1969. The gibbsite dehydroxylation fork. In Schwenker, R.F., and Gran, P.D. (eds.), *Thermal Analysis*, Volume 2. Inorganic Materials and Physical Chemistry, New York: Academic Press, pp. 1239–1250.
Ruan, H.D., Frost, R.L., and Kloprogge, J.T., 2001a. Mid-infrared and near-infrared spectroscopic study of aluminas. *Applied Spectroscopy*, **55**: 190–196.
Ruan, H.D., Frost, R.L., and Kloprogge, J.T., 2001b. Comparison of Raman spectra in characterizing gibbsite, bayerite, diaspore and boehmite. *Journal of Raman Spectroscopy*, **32**: 745–750.
Paulik, F., Paulik, J., Naumann, R., Kohnke, K., and Petzold, D., 1983. Mechanism and kinetics of the dehydration of hydrargillites, Part I. *Thermochimica Acta*, **64**: 1–5.
Naumann, R., Kohnke, K., Paulik, J., and Paulik, F., 1983. Kinetics and mechanism of the dehydration of hydrargillites, Part II. *Thermochimica Acta*. **64**: 6–12.
Wilke, B.M., and Schertmann, U., 1977. Gibbsite and halloysite decompostition in strongly acid podzolic solids developed from granite saprolite of the Bayerischer Wald. *Geoderma*, **19**: 51–63.

Cross-references

Clay Mineralogy
Weathering, Soils and Paleosols

BEACHROCK

Beachrock is a sedimentary rock that results from rapid lithification of sand and/or gravel by calcium carbonate cements in the intertidal zone. It occurs predominantly on tropical coasts, but is also found as far north and south as 60° latitude. In contrast to the implications of the name, beachrock outcrops are not restricted to beaches but some are found on tidal flats, in tidal channels, and on reef ridges. To date the origin of beachrock is not fully understood having been variously attributed to physicochemical precipitation, biologically induced cementation, or a combination of both.

Occurrence, composition, texture, and cements

Beachrock exposures are typically patchy with bedded layers that dip gently (<10°) toward the sea. Hopley (1986) suggested that the largely intertidal occurrence of beachrock makes its fossil occurrence a potential indicator of sea level. Problems with this approach include possible failure to recognize other deposits that might get cemented close to sea level such as dune calcarenite, eolianite, or cay sandstone (carbonate sand cemented by calcium carbonate from fresh water above high tide level on reef islands), especially when diagenetic environments frequently changed, and the fact that radiometric dating of beachrock can only produce an average age of the constituent particles and cements.

The composition of beachrock constituent particles is more or less similar to that of the adjacent consolidated sediment. Grain-size ranges from sand to gravel: sediments are moderate to very well sorted, with sorting usually better than that of the adjacent subtidal sediments.

Beachrock cements are dominated by aragonite and high-magnesium calcite. The aragonite cement includes isopachous fringes of needles that are up to 100 μm long, and often overlie a dark layer at their base that consists of aragonite platelets rich in iron and sulfur (Strasser et al., 1989). Small aragonite (<10 μm long) needles may form "micritic" cement (e.g., Webb et al., 1999). High-magnesium calcite cements often include microcrystalline ("micritic") cement with crystals less than 5 μm in diameter. Less common are cement fringes of high-magnesium calcite blades or scalenohedral crystals (up to 70 μm long). Peloidal cements with approximately 40 μm diameter peloids, and equant crystal crusts of high-magnesium calcite also occasionally occur. Cements indicative of the vadose diagenetic environment are encountered rarely and these result from cementation in the wave spray zone or during low tide exposure. In this case, typical meniscus or gravitational ("dripstone") cement fabrics are common. Detailed descriptions of beachrock cements are provided in Bricker (1971: 1–43), Moore (1973), Meyers (1987), Strasser et al. (1989), and Gischler and Lomando (1997) among others. Scoffin and Stoddart (1983) provided a comprehensive review on beachrock and intertidal cementation up to that time.

Origin

Early 19th century reports on beachrock confirm its rapid formation in the intertidal zone (e.g., Moresby, 1835). For example, natives to Indo-Pacific islands were known to harvest beachrock for building stone where new occurrences formed on the same beach within less than a year. High resolution radiometric dating of cements in coral reef slopes have shown that marine aragonite cements, comparable to those observed in beachrock, reach growth rates of 80–100 μm per year (Grammer et al., 1993).

The mechanisms used to explain beachrock cementation include both physicochemical and biologically induced precipitation of calcium carbonate. Physicochemical models explain precipitation of carbonate cement by evaporation of seawater during low tide (e.g., Ginsburg, 1953; Hanor, 1978), and by degassing of CO_2 during falling tides (Meyers, 1987) or during higher tides (Pigott and Trumbly, 1985). Apart from CO_2 pore water saturation states, water agitation is probably another important factor in beachrock formation. A large scale study on Belize beachrock showed that the vast majority of beachrock exposures occured on windward beaches of reef islands, suggesting that beachrock cementation only occurred where beaches experienced intensive and persistent flushing by seawater (Gischler and Lomando, 1997). Variations in pore water pressure during pumping of seawater through the beach may also lead to calcium carbonate precipitation. Beachrock formation in Grand Cayman island was interpreted by Moore (1973) to be largely a product of cementation under mixed meteoric-marine conditions. In contrast, Hanor (1978) used thermodynamic calculations to show that precipitation of beachrock cement cannot be induced by mixing of marine and meteoric water. Rather, these calculations favor CO_2 degassing as a means of supersaturating pore-water with calcium carbonate to the point of inducing cement precipitation.

There is also evidence suggesting the importance of biological processes in beachrock formation. Algal coatings may form on the beach, and stabilize grains so that they can be preferentially cemented (Davies and Kinsey, 1973). Withdrawal of CO_2 by photosynthesis may furthermore induce calcium carbonate precipitation. Krumbein (1979) noted the occurrence of high concentrations of organic matter during initial stages of beachrock formation in the Gulf of Aqaba and suggested that anaerobic and later aerobic decay processes were responsible for cement precipitation. Decay processes may include ammonification leading to higher pH, or sulfate reduction that elevates alkalinity, or hydrolysis of urea forming ammonium carbonate and eventually calcium carbonate. Further suggestions of a biological influence on beachrock formation are provided by Webb et al. (1999). These workers found microbial filaments in both beachrock cement and microbialites within beachrock cavities on Heron Island, Great Barrier Reef. They stressed the importance of acid organic macromolecules with Ca^{2+} binding carboxyl groups in biofilms in forming nucleation zones for cements. The common dark zones at the base of isopachous fringes of acicular aragonite beachrock cement may reflect the occurrence of organic mucus (Davies and Kinsey, 1973). According to the SEM study of Chafetz (1986), nuclei of marine peloids, which may also form beachrock cement, are composed of bacterial clumps.

Open questions remain with regard to the formation of beachrock. If physicochemical (inorganic) processes are sufficient to induce calcium carbonate cementation, why are beachrock outcrops so patchy in distribution, even on the windward beaches of reef islands? The same patchy distribution argues against organic processes as a pre-requisite for precipitation of $CaCO_3$ cement, unless and until it can be shown that specific organic processes are peculiar to beachrock formation and therefore are not ubiquitous in occurrence as is the presence of microbes. It would seem that aspects of both inorganic and organic processes have to be taken into account to explain the origin of beachrock. To date, however, exclusively inorganic or exclusively organic models do not suffice to explain the distribution of intertidal marine cementation.

There are only very few reports on fossil, pre-Holocene occurrences of beachrock, which probably is to a large extent a consequence of its poor preservation potential. As soon as beachrock is exposed it becomes subject of intensive intertidal erosion, so only in cases of rapid burial would beachrock outcrops be preserved.

Eberhard Gischler

Bibliography

Bricker, O.P. (ed.), 1971. *Carbonate Cements*. Baltimore: Johns Hopkins Press.
Chafetz, H.S., 1986. Marine peloids: a product of bacterially induced precipitation of calcite. *Journal of Sedimentary Petrology*, **56**: 812–817.
Davies, P.J., and Kinsey, D.W., 1973. Organic and inorganic factors in recent beachrock formation, Heron Island, Great Barrier Reef. *Journal of Sedimentary Petrology*, **43**: 59–81.
Ginsburg, R.N., 1953. Beachrock in south Florida. *Journal of Sedimentary Petrology*, **23**: 85–92.
Gischler, E., and Lomando, A.J., 1997. Holocene cemented beach deposits in Belize. *Sedimentary Geology*, **110**: 277–297.
Grammer, G.M., Ginsburg, R.N., Swart, P.K., McNeill, D.F., Jull, A.J., and Prezbindowsky, D.R., 1993. Rapid growth rates of syndepositional marine aragonite cements in steep marginal slope deposits, Bahamas and Belize. *Journal of Sedimentary Petrology*, **63**: 983–989.
Hanor, J.S., 1978. Precipitation of beachrock cements: mixing of marine and meteoric waters vs. CO_2-degassing. *Journal of Sedimentary Petrology*, **48**: 489–501.
Hopley, D., 1986. Beachrock as sea-level indicator. In van de Plassche, O. (ed.), *Sea-level Research*. Great Yarmouth: Galliard Printers, pp. 157–173.
Krumbein, W.E., 1979. Photolithotropic and chemoorganotrophic activity of bacteria and algae as related to beachrock formation and degradation (Gulf of Aqaba, Sinai). *Journal of Geomicrobiology*, **1**: 139–203.
Meyers, J.H., 1987. Marine vadose beachrock cementation by cryptocrystalline magnesian calcite–Maui, Hawaii. *Journal of Sedimentary Petrology*, **57**: 755–761.
Moore, C.H. Jr., 1973. Intertidal carbonate cementation Grand Cayman, West Indies. *Journal of Sedimentary Petrology*, **43**: 591–602.
Moresby, R., 1835. Extracts from Commander Moresbys report on the northern atolls of the Maldives. *Journal of the Royal Geographical Society*, **5**: 398–404.
Pigott, J.D., and Trumbly, N.I., 1985. Distribution and origin of beachrock cements, Discovery Bay (Jamaica). *Proceedings of the 5th International Coral Reef Symposium*, Tahiti, **3**: 241–247.
Scoffin, T.P., and Stoddart, D.R., 1983. Beachrock and intertidal cements. In Goudie, A.S., and Pye, K. (eds.), *Chemical Sediments and Geomorphology*. London: Academic Press, pp. 401–425.
Strasser, A., Davaud, E., and Jedoui, Y., 1989. Carbonate cements in Holocene beachrock: example from Baihret el Biban, southeastern Tunisia. *Sedimentary Geology*, **62**: 89–100.
Webb, G.E., Jell, J.S., and Baker, J.C., 1999. Cryptic intertidal microbialites in beachrock, Heron Island, Great Barrier Reef: implications for the origin of microcrystalline beachrock cement. *Sedimentary Geology*, **126**: 317–334.

Cross-references

Carbonate Mineralogy and Geochemistry
Cements and Cementation
Coastal Sedimentary Facies

BEDDING AND INTERNAL STRUCTURES

Bedding

Bedding is the arrangement of sedimentary rocks in beds or layers of varying thickness and character. A bed (or stratum) is a relatively conformable succession of genetically related laminae or lamina-sets (Campbell, 1967) bounded by surfaces (called bedding surfaces) of erosion, nondeposition, or their correlative conformities.

The *bed* is the smallest formal lithostratigraphic unit of formal *stratigraphy*, and as such is defined solely on the basis of its lithology, including color, grain size, and thickness. A bed may range in thickness from a few millimeters to over 10 m. A key bed (or key horizon) is a well-defined, easily identifiable stratum or body of strata that can be easily distinguished from the overlying and underlying strata by sufficiently distinctive characteristics, such as lithology, thickness, texture or fossil content and their lateral gradients. These allow physical tracing and stratigraphic correlation of the key bed over long distances, up to several tens of km.

In *sedimentology*, the bed has a different, genetic connotation: in other words, it represents a distinctive depositional event, related to some processes or mechanisms of deposition. In this meaning, a bed may show a change of lithology from bottom to top, for example, from sand(stone) to mud(stone); this is the case for most turbidites, storm layers (tempestites), crevasse deposits, etc.

There exists a hierarchy of bedding in terms of scale, geometry and spatial relationships of physical surfaces. This hierarchy defines units of different rank, ranging from the individual lamina to the basin-fill complex, and temporally spans over a wide range of timescales (from 10^{-6} years to 10^7 years). Not always is the hierarchical arrangement recognizable, especially where the main bedding surfaces (*master bedding*) are obscured by superposition of lower-rank surfaces. In this case, individual beds cannot be clearly identified and counted, and only the bedding style, or pattern, is described (e.g., cross-bedding, sand/mud interbedding, clinoforms, etc.), with a rough specification of scale (small, medium, large).

The *lamina* is the smallest sedimentation unit and is characterized more by being a part of a bed than by its thickness, although most laminae are in the submillimeter to centimeter range. A single set of conformable laminae (laminaset) may form a bed (solitary set and form-set of J.R.L. Allen) or be a part of it: it can be associated either with other laminasets in a wholly *laminated bed* or with structureless (massive, graded) portions in a partly laminated bed (see below, Bouma sequence). There exist also entirely structureless beds, which can be either emplaced by mass flows or have their original structures destroyed by bioturbation or liquefaction after deposition.

Bed packets or bundles distinct from those above and below in a stratigraphic succession are called bedsets by Campbell (1967); within a bedset, there can be a dominant facies (i.e., all beds are similar in terms of lithology, thickness, texture, structures) or a systematic vertical change corresponding to a *facies sequence* (coarsening- or fining-up, thickening- or thinning-up, etc.). Bedsets are delimited by set boundaries and form *sedimentary bodies* of simple or composite character.

Repetitive bedsets can be grouped in cosets separated by coset boundaries (base of the lowest set and top of the youngest).

The spatial arrangement of individual component bodies in a complex body is called *stacking pattern* or *architecture* (Miall, 1985). The term architecture is also applied to the *internal organization* of an individual body showing a hierarchy of volumes and surfaces: in this case, an *architectural element* is defined at each level of organization or rank. All such elements have a *facies* connotation, but the term facies is too often used quite loosely because there is a lack of agreed-upon rules; for the sake of clarity and communication, it is advisable to specify at which rank the facies is defined, this being left to the choice of the operator. After that, all smaller scale elements will be *sub-facies*, all larger scale elements will be *facies associations*. An example of facies can be a trough cross-bedded sand(stone), representing bar crest deposits in a deltaic depositional system. An example of facies association is a mouth bar, for example, including bar back, bar crest, bar front and distal bar facies.

When a facies association shows some kind of vertical order, it is usually called a *facies sequence*. Lateral facies changes are much less documented due to lack of sufficiently long exposures; lateral transitions are rather inferred than observed, with some notable exceptions (see *Facies Tracts* by Mutti, 1992).

Beds, bedsets, architectural elements, facies, facies associations, and stacking patterns are all three-dimensional units, but for most practical purposes they can be recognized (or at least inferred) and described in two (cross-sections), and even in one (vertical profiles, well logs). The hierarchical classification of depositional units and architectural scale concepts are thus useful for basin analysis and the petroleum industry, since reservoir geometry and internal heterogeneity may involve different ranks and physical scales.

Internal structures

These are sedimentary structures located within beds or layers and observable on surfaces cut at high angles to bedding planes, that is, in normal stratigraphic sections. In clastic deposits, most internal structures are physical, or purely mechanical in origin, with the exception of trace fossils and bioturbation (bio-mechanical). In carbonates, evaporites, cherts and other sedimentary rocks of chemical and biochemical origin, a color-compositional differentiation (e.g., color banding) or peculiar structures (e.g., stromatolites, birds eyes, structures related to crystal growth or dissolution) can be observed, but a detailed description of them is not given here (see *Diagenetic Structures*). Attention will be focused here on mechanical structures. They are subdivided into two main groups: those produced by transportation and deposition, and those resulting from deformation soon after deposition (soft-sediment deformation as contrasted with tectonic or structural deformation).

Structures related to transportation processes

Most of these structures consist in geometrical arrangements of laminae or kinds of *lamination*, and occur in sands and sandstones; they are also called *tractive structures*, traction referring to bed-load movement of sedimentary particles. Lamination can also be found in silty sediments, in which case resulting from differential settling from suspensions. Laminated or structureless mud(stone) can be associated in alternating patterns with sand(stone); *draping* indicates a thin muddy cover, laterally continuous or discontinuous, on sand beds or an irregular topography. When traction is simultaneous with settling of silt and sand (which is put in suspension by highly turbulent flows, such as turbidity currents), *traction-plus-fallout* lamination forms (Jopling and Walker, 1968).

Local erosion is involved in the formation of tractive structures, producing truncation surfaces and discordances of various geometry and scale; some of them can be used for the hierarchical differentiation of bedding. Tractive structures may form when a current is in equilibrium (no net erosion or deposition) with its lower boundary (sedimentary interface), or in conditions of net deposition due to loss of carrying capacity. In the former case, the result is a diastem or a single, thin bed (solitary set, form-set with bedforms preserved on top: Allen) of tractive laminae due to either a fresh sand supply or *in situ* reworking of preexisting sand. Net deposition implies a higher preservation potential of laminated intervals, even if the flow is oscillating and some erosion is produced; continuos feeding of sand to the bottom not only buries previously deposited laminae, but can build up decimeters or even meters of laminated sediment (see the "Contessa" bed and other similar "megaturbidites" in the Apennines of Italy).

Unidirectional flows

Tractive currents produce bedforms of various scales: small (ripples), medium (subaqueous dunes), large (eolian dunes, subaqueous bars and sandwaves). A set thickness of 4–5 cm separates ripple from dune scale but no agreement exists for a boundary between medium and large structures. Bedforms can be extremely shallow or the bottom even and flat when the Froude number is in excess of or near unity. One can thus distinguish subcritical (lower flow regime) from supercritical tractive structures, and the distinction is applicable even at a small scale (e.g., core examination), because subcritical forms have a separation zone at their lee side, where gravity avalanching takes place developing *foreset laminae* inclined at the angle of repose of the sand (up to 30°–35°). Strata therefore exhibit, in sections cut at a small angle with the flow direction, *high angle lamination* with respect to their bedding planes. No flow separation, and hence *low angle lamination* ($<10°$) characterizes supercritical bedforms. Low angle laminae are not reliable indicators of current direction, whereas foreset laminae are.

In case of a flat bottom, horizontal or *plane-parallel lamination* occurs. A bed can part along weaker laminae showing *parting lineation* (or primary current lineation), that is, slightly terraced, shallow ridges formed by strips of sand grains aligned parallel to the flow. They are present also on low amplitude bedforms of upper flow regime, such as antidunes and humpback dunes.

At outcrop scale, the terms cross-bedding, cross-stratification and cross-lamination are used when laminasets are bound by nonparallel, intersecting surfaces; cross-bedding is the net result of local deposition and erosion related to the migration of bedforms (those of small and medium size are more commonly represented in subaqueous deposits).

Cross-bedding should be qualified by specifications, for example: (a) laminaset thickness as a scale indicator; (b) presence or absence of foreset (high-angle) laminae, which means lower or upper flow regime; (c) relation with master bedding or other evidence of an original flat bottom;

(d) morphology of set boundaries (erosional vs non-erosional; tabular vs concave or trough). Such characters can help, for instance, to distinguish subaqueous from eolian deposits (the latters show thick to very thick beds and lack ripple scale lamination) or bedforms grown on flat bottoms from those topping larger forms, such as ripples on dunes or dunes on sandwaves (typical of bars and sandwaves is the hierarchical organization in cross-bedded sets).

Backset laminae (dipping opposite to the flow) can form on the stoss side of so-called regressive antidunes (Allen, 1982), but are not usually preserved; the few examples reported in the literature for ancient deposits, especially turbidites, have been reinterpreted as regular foreset laminae left by inverted or reflected flows. Only in pyroclastic strata, in particular those deposited by turbulent surge flows, are backset laminae commonly found (Allen, 1982, figures 10–15), probably owing to the peculiar conditions (temperature/moisture, density contrast between solid particles and fluid). In general, tractive lamination is beautifully preserved in these deposits, and foreset laminae inclined at 20° or less can be observed in dune-like or dunoid (Ricci Lucchi, 1995a) bedforms. *Counter-flow ripples* are sometimes seen at the toe of large subaqueous structures, for example, tidal sandwaves; they are the product of recirculation in separation bubbles.

Sets and cosets of cross-laminae can show a *climbing* attitude of the corresponding bedforms (more commonly ripples, but also subaqueous dunes), with a variable angle of climb indicated by the stoss sides. This is a clear evidence of traction combined with fallout, and the cross-bedding style has been called *ripple-drift cross-lamination* (Jopling and Walker, 1968); three varieties (types A, B, and C) have been distinguished on the basis of relative amounts of erosion and deposition on the two sides of the bedforms. Climbing cross-bedding normally derives from decaying currents and shows bedform profiles preserved by a mud cover.

Low angle laminae occur in sets usually separated by low angle erosional surfaces, both planar and curved; resulting laminasets are wedge-shaped to sigmoidal. It is not easy, especially in limited outcrops, to decide whether they are related to a current, an oscillatory motion or a combination of the two (see below). Low angle lamination has been described in "proximal" sandy turbidites (Mutti and Ricci-Lucchi, 1972, facies B) and fluvial deposits (couplets of planar laminated sand and gravel are a common feature of sheetflood deposits within alluvial plain/alluvial fan depositional systems).

Oscillatory and combined flows

Two kinds of "short term" (non-geological) periodicity are recorded in sediments, that of wind waves and that of tidal waves.

Wind-generated waves may form within any body of water and exert traction below a critical depth referred to as the wave base. Wave orbits increase their ellipticity toward the bottom, where they culminate as straight lines representing to-and-fro motion. This motion results in grain movement by rolling and saltation on an initially plane sand bottom. Wave-formed ripples vary greatly in size, with h (height) commonly ranging between 0.003 m and 0.25 m, and λ (wavelength) between 0.009 m and 2 m. Their ripple index varies between 4 and 13.

Wave ripples are characterized by a symmetrical profile near wave base and tend to become more asymmetrical approaching the shoreline (steeper side facing landward). In 2D-sections, they display diagnostic internal features, such as inconsistency between ripple morphology and internal structure, structural dissimilarity between adjacent sets, presence of chevron and bundled upbuildings of foreset laminae, and irregular and undulating lower set boundaries (wave-truncated ripples of Campbell, 1967; de Raaf *et al.*, 1977). In plan view, wave ripples display straight to slightly sinuous crests, with characteristic bifurcations, that are uncommon in current ripples. Wave ripples are typically found within lower shoreface sands, but have also high preservation potential in shallow-water environments affected by oscillatory waves, such as lagoons and large lakes.

In intertidal environments, partial erosion of wave ripples at low tide conditions results in characteristic truncation of their tops. Actually, in very shallow water depths (foreshore deposits, swash/backwash zone) upper flow regime conditions prevail and planar surfaces, both erosional and depositional, are the dominant feature. Foreshore sands are thus characterized by the stacking of gently seaward-dipping sets of parallel laminated sand, formed during fair-weather conditions, separated by erosional surfaces related to storm events (wedge-shaped, low-angle cross-stratification).

While wave ripples are generated by fair-weather waves, storm waves tend to produce a structure of similar or greater size, called *hummocky cross-stratification* (HCS) by Harms *et al.*, (1975). More precisely, HCS is interpreted as the result of combined, waning unidirectional *and* oscillatory flows, but anyway a diagnostic feature of storm-dominated processes. Recently, however, sedimentary structures very similar to HCS have also been observed to characterize flood-dominated fan delta and river delta systems (Mutti *et al.*, 1996).

HCS consists of convex-up, large amplitude (1–5 m) and low-relief (10–50 cm), irregular domal features (hummocks), separated by broad troughs (swales). Gently undulating (5–15°) lamination is the most common internal feature, displaying a general concordance with the basal erosion surfaces and systematic lateral variations of laminae thickness (pinching and swelling of individual laminae). Where concave-up sets are well preserved, swaley cross-stratification (SCS) is developed (Leckie and Walker, 1982). HCS is generally restricted to coarse silt to fine sand, and is particularly abundant within lower shoreface deposits, in association with symmetrical wave ripples. SCS may be abundant in shallower (upper shoreface) and coarser grained deposits. Storm events in the offshore-transition area, above storm wave base, characteristically develop idealized sequences (Dott and Bourgeois, 1982) that rest on scoured surfaces, passing upward into HCS, and capped by flat lamination and symmetrical ripples at top of the sandy layers. The sequences are capped by bioturbated mudstone intervals.

In flood-generated delta-front sandstone lobes, large-scale HCS is developed within 3–15 m thick sandstone packets, separated by highly fossiliferous and bioturbated muddier deposits.

Tidal structures and bedding

Tidal currents regularly change direction from the flood tide current, which flows landward between the low and high tide, and the ebb tide current, which flows in the opposite direction as the water level returns to low tide. Bipolar cross-bedding, forming the characteristic herringbone cross-stratification, is the typical, expression of alternating tidal currents in high-energy, subtidal settings. Herringbone cross-stratification, however, is not very common in the rock record, because of the fact that one half of the current generally is much stronger than the other, and ebb and flood currents may follow different paths. Bedforms migrating in one direction, under the influence of the dominant current, can be only slightly modified by the subordinate current. Diurnal changes in bedform migration controlled by asymmetrical tidal currents may thus result in discrete sets of unidirectional cross-strata (sand foresets form on the lee side of large bedforms), separated by gently sloping erosion surfaces, which are termed *reactivation surfaces*. This pattern of alternating undirectional cross beds and reactivation surfaces is characteristic of large, linear bedforms (sand waves), with wavelengths of tens to hundreds of meters and heights exceeding 10 m, which are commonly observed in open-shelf environments.

Regular changes in magnitude and direction of tidal currents over the tidal cycle allow transport and deposition of sand at some stages and fallout of mud from suspension during periods of slack waters. Tidal structures resulting from this rhythmic alternation of traction and fallout processes are termed tidal rhythmites (Tessier, 1993) and generally characterize intertidal flats. Tidal rhythmites include sand layers, showing small-scale current ripples, alternating with mud drapes, reflecting deposition during stillstand phases.

Distinctive sedimentary structures of intertidal flat environments include flaser bedding, wavy bedding and lenticular bedding (Reineck and Wunderlich, 1968). *Flaser bedding* (see Flaser) is characterized by isolated thin drapes of mud amongst a cross-laminated sandy unit. In contrast, lenticular bedding is made up of isolated sand ripples within a mud-dominated deposit. Wavy bedding is intermediate between the two, consisting of approximately equal proportion of sand and mud.

In higher-energy, subtidal settings, a characteristic feature of tidal sedimentation at the scale of diurnal, ebb-flood tidal cycles is double mud drapes, formed during the two slack water stages of the tidal cycle. Double mud drapes, commonly a few mm thick, typically separate unidirectional, bipolar cross beds showing current reversals. These diagnostic tidal structures, forming a characteristic sigmoidal stratification, are referred to as *tidal bundles* (Visser, 1980), and are commonly developed in subtidal (estuarine) channels. In intertidal settings, the *tidal bundles* include just one mud drape. Cyclic variations in strength of tidal currents at the scale of neap-spring cycles result in cyclic variations in thickness of tidal bundles, forming characteristic tidal bundle sequences (Yang and Nio, 1985).

Internal structures in episodic (catastrophic) deposits

Episodic events include tsunami waves, seismic shocks, violent hurricanes, catastrophic flash floods, and sediment gravity-flows. Resulting beds are lenticular to tabular and internally disorganized (random arrangement of particles) to organized. Organization means size grading (graded bedding), grading in the lower part plus lamination in the upper portion of sandstone, or thoroughly laminated sandstone. Beds are commonly capped by mud, but can be separated by erosion surfaces ("amalgamated"). The different parts of organized beds are called *intervals* or *divisions*.

Gravity flows are moving dispersions (mixtures of solid particles and water) classified on the basis of the different proportions of the two phases and the grain-support mechanisms. Following Middleton, Lowe, Postma and Mutti (see Mutti, 1992, for references), different flow types can be identified. In cohesive debris flows, the fluid has pseudoplastic characteristics and grain transport is a result of matrix strength, which is in many instances adequate to support large boulders. Coarser grains move to regions of lower shear, and are rafted on the top or edge of flow, resulting in inverse grading. In grain flows, shear stress is transmitted through the flow by a dispersive pressure resulting from intergranular collisions. Frictional freezing, which occurs when the applied shear can no longer overcome the flow strength, is the dominant depositional mechanism from both cohesive debris flows and grain flows.

Debris flow deposits are characteristically very poorly sorted and lack almost any internal structure (the preponderance of plug flow hampers the development of shear fabric or sorting). When large clasts are present, they are randomly distributed (the bed has a chaotic aspect) or concentrated near the top, locally protruding. A rough development of inverse grading can thus be observed throughout the bed or at the base, combined with lack of significant basal scours.

Debris flows show marked downslope transformations, as they decelerate and mix with ambient waters; such transformations are reflected in internal fabric and structures. A high-density, inertia-driven mixture of sediment and water that moves downslope under conditions of high pore-fluid pressure is termed *hyperconcentrated flow*. The typical structureless debris flow deposits may thus be replaced in a downcurrent direction by deposits from hyperconcentrated flows; these are characterized by deep basal scours, large rip-up mudstone clasts, a coarser grained matrix, and tendency of larger clasts to concentrate in the lower part of the bed. Deposits from hyperconcentrated flows may also include evidence for tractional bedforms.

Traction carpets, resulting from a combination of fluid turbulence, hindered settling, and dispersive pressure produced by grain collision, may develop in the basal portion of high-density flows, and are a relatively common feature of grain flows. Traction carpets, which are commonly followed by *en-masse* deposition, display characteristic, inversely graded stratification bands and show upward transition to ungraded, crudely graded or well-graded beds.

At modest levels of sediment concentration, the flow is cohesionless and may be internally turbulent. In fluidized and liquefied flows, escaping pore fluid provides a full or partial supporting mechanism, respectively. In low-density turbidity currents, particles are supported by the flow turbulence, and sediment deposition occurs through grain settling and traction-plus-fallout processes. The typical pattern of grain size change within a bed which is formed from a decelerating flow is referred to as *normal grading*. Larger particles achieve a higher terminal velocity and settle out of suspension faster than smaller grains. Two types of grading have been distinguished in turbidites (Middleton, 1967): distribution and coarse-tail. The first type is shown by all size classes, the second one by coarser particles only. Grading may pass upward into, and/or be superimposed by, tractive lamination, as observed since the thirties of last century by geologists in the Apennines looking for way-up criteria in zone of vertical and overturned beds. Later Bouma (1962) formalized a vertical sequence of internal structures, the now familiar "Bouma sequence".

The *Bouma sequence* includes, from base to top: normally graded, structureless medium to fine sand, with frequent dewatering pipes and dish-and-pillar structures (division a); planar laminated to undulating fine sand, with parting lineations and inversely graded stratification (division b); small-scale cross-laminated very fine sand, with abundant climbing ripples (division c); current-laminated coarse silts and interlaminated silts and mud (division d). The succession is capped by a homogeneous and massive mudstone (division e), actually a pelitic bed or parting. The transition from thoroughly current-laminated sand to structureless mud is more or less gradual and marks the passage from traction-plus-fallout processes to pure grain settling from suspension. In more proximal turbidites, richer in sand, Lowe (1982) defined different varieties of structure sequences, but they seem to have a less universal application.

Sharp-based layers with distinctive internal fining-upward tendencies, being the expression of decelerating, waning turbulent flows, are not exclusive of submarine gravity flows, but may be commonly encountered in different depositional environments, such as lakes, river deltas (hyperpycnal plumes related to river floods) and shallow-marine realms (storm-generated currents).

Intraformational mud clasts, referred to usually as clay chips and a variety of other terms, and deriving from intrabasinal sources, are a common feature of debris flow, hyperconcentrated flow and structureless to graded portions of turbidity flow deposits. They are typically embedded in sand and occur: a) concentrated at or near the base of the bed; b) scattered throughout; c) aligned (and relatively sorted) in the uppermost structureless or graded sandstone, often at the transition with tractive laminae. These situations reflect different distances of travel within the flow and a marked tendence to buoyancy. This should be considered a fabric more than a structure; the same applies to *clast imbrication* (see *Imbrication*), showing in both massive and selective (e.g., beach) deposits. Flattened clasts are inclined to the horizontal, so that the plane formed by the long and intermediate axis of the clasts dips up-current, according to the most stable position. Imbrication represents a very useful paleocurrent indicator and is a characteristic feature of gravel-bed rivers and gravel beaches (where dip is seaward). It indicates a transport of pebbles in temporary, or intermittent, suspension.

Internal scours may be found in gravity flow and other episodic deposits, owing to flow pulsations or bottom irregularities during deposition; care should be taken, however, not to mistake such local features with parts of larger erosion or amalgamation surfaces separating distinct beds (events).

Internal structures and facies sequences

The "Bouma sequence" just described represents an example of a *genetic package*, or association, of structures, that is, a systematic vertical succession of structures consistent with a specific mechanism evolving in time (decaying suspension flow). In this case, the time is short (hours, days) being the duration of a single depositional event of episodic nature. Similar genetic packages can be found in multilayer sedimentary bodies, built up by several events, and in normal or selective deposits. The best known case is that of the fining-upward sequence displayed by fluvial and tidal point bars formed by lateral accretion (Allen, 1982, his figures 2–23). Notice that Allen uses the term structure not only for the cross-bedding, parallel lamination, etc., inside the body, but also for its overall geometry ("lateral accretion structure"). Alternatively the same object can be described as a facies sequence, a facies association or an architectural element.

The following description summarizes that of Allen (1982): in tidal flats, mud-sand interbedded points bars form the inner bank of small meandering gullies. They are rarely thicker than 1.5 m and consist of curved, sigmoidal layers, from parallel-laminated to current rippled, both ebb and flood directions often being preserved. The base of the bar is an irregular erosional surface, generally overlain by gravel of bivalve and other shells mixed with pebbles of muddy sediment derived from cut banks.

River point bars, in intermediate to large sand-bedded rivers, where the secondary [transversal] flow is fully developed, have also a basal, diachronous erosion surface, with flutings, pot-holes and other scours. Sediments in the lateral accretion deposit fine upward in general but interfinger in detail, and Internal sedimentary structures appear upward in order of declining flow power, an arrangement which reflects the distribution of bedforms over the bar surface, in turn revealing the inward decline of current strength. Curved bedding surfaces, generally sigmoidal in radial profile, define the *lateral accretion structure* [= master bedding].

Structures related to deformational processes

Primary sedimentary structures are subjected to disturbances by fluid movement, compaction and gravitational instability if the sediment remains soft and deformable. *Soft sediment deformation* is the general term for changes to the fabric and layering of a bed of unconsolidated sediment. It affects most commonly multilayer deposits, especially sand-mud alternations, and can be diffused in the whole packet or concentrated along bedding planes or within individual beds. No systematic classification exists, apart from a broad subdivision into structures characterized by vertical movement and structures showing a lateral component of motion (gravitational or current shearing). For nomenclature, see Table B2.

Concerning the first group, trapping of water induced by rapidly deposited sediment may lead to disruption of previous structures within a bed (more commonly sand) or a bedset. This may occur due to shocks by earthquakes or oscillations in water pressure, which may temporarily liquefy the sediment. Beside liquefaction, deformation can result from sediment contraction (e.g., desiccation) or dilatation, unequal loading, upward injection of fluid or paste-like material, downward injection in fractures, compaction, downslope movements, current or ice drag or combined effects.

Sedimentary structures formed by liquefaction, injection, etc. are often associated with deposits from high-density turbidity currents and subaqueous slumps; those that can be related to earthquakes may be called *seismites*, but this attribution is often conjectural.

Sand volcanoes are formed by the violent expulsion from the subsurface of sand and water mixtures. Dewatering (fluid-escape) structures result from forceful expulsion of water from a consolidating fluidized bed, due to loading or a shock. The water moves upward, disrupting overlying layers and forming *dish-and-pillar* structures. Dishes are subhorizontal, gently concave-upward clay or organic-rich laminae, deformed at their margins by upward flow. Pillars are near-vertical

Table B2 Soft-sediment deformation structures reported in modern and ancient sediments. For references, see Ricci Lucchi (1995b)

Modern	Ancient
Liquefaction structures	Ball-and-pillow structure (Smith, 1961; Pettijohn and Potter, 1964)
Sand boil	Load structures
Sand volcano, sand-vented volcano (Obermeier et al., 1986)	Flame structure (Cooper, 1943)
Sand blow (Sieh, 1978; Doig, 1990)	Injection structures, diapiric structures
Sand slough (Fuller, 1914; Rascoe, 1975)	Flow rolls (Pepper et al., 1954)
Sand slough (Fuller, 1914; Rascoe, 1975)	Pseudonodule (Macar, 1948)
Patterned ground (Washburn, 1950)	Convolute lamination (Kuenen, 1953)
Mima mounds (Fuller, 1914; Arkley and Brown, 1954; Berg, 1990)*	Cycloid (Hempton and Dewey, 1983)
Patterned ground (Washburn, 1950)	Dish structure (Wentworth, 1967; Lowe and Lo Piccolo, 1974)
	Pillar structure, or dish-and-pillar structure (Lowe and Lo Piccolo, 1974)
	Water escape structures
	Sand volcano
	Sandstone dike
	Sandstone pipe
	Clastic (sedimentary) sill
	Clastic (sedimentary) dike
	Slump ball (Kuenen, 1948)
	Slump structures
	Boudinage
	Pull-apart (Natland and Kuenen, 1950)
	Ice-shear (ice-push) structures
	Glaciotectonic structures
	Drag fold
	Hydroplastic deformation
	Wrinkle marks (Häntzschel and Reineck, 1968; Allen, 1985)

* Fuller (1912) regarded this structure as aseismic, Berg (1990) reinterpreted it as seismic

dewatering pipes, interrupting the dishes and made up of structureless sand. Dishes are found in both massive and crudely laminated sandstones and are related to horizontal movements of water or liquidized sand. In the dark laminae, preexisting (semipermeable membranes?) or newly formed, fine particles are concentrated by elutriation.

Convolute laminations are deformational structures developed in rapidly deposited fine-grained sands and silts, and commonly associated to water-escape structures and climbing ripple cross-lamination in turbidites and river flood deposits. They are either symmetrical about a vertical plane or leaning and asymmetrical, and usually exhibit narrow vertical upturned laminae, often truncated at the top, separated by broader synclinal downfolds, with wavelengths of a few centimeters or decimeters. Some convolutions are associated with water-escape dishes, pillars, etc.

Many explanations have been offered for the origin of this structure; the asymmetrical varieties involve shearing, probably caused by an overflowing dense current. Downslope movement (slumping) seems to be excluded by the common occurrence in basinal sediments, especially turbidites, and the dying out of the folds both upward and downward into flat surfaces, without any evidence of *décollement*. Liquefaction or hydroplasticity of some or all laminae involved (which contain variable amounts of organic material or clay) and density instability are the most invoked mechanisms. Timing of formation can be during bed deposition (syndepositional type of Allen, 1982), just before or immediately after deposition ceases (metadepositional), or definitely after (postdepositional).

Load casts are the commonest type of deformation of sand-mud interfaces and "represent the instability of discontinuous layered systems" (Allen, 1982) due to inverted density gradients (Dzulynski, 1966). The structure consists of lobes of sagging sand alternating laterally with sharp-crested diapers ("flames"), projecting from a mud that was quite fluid when it was loaded. Load casts can be found at various scales ranging from millimeters to decameters (in outcrops).

Similar to load casts, but more deformed and complex in shape and internal structure are *pseudonodules* or *ball-and-pillow* structures. In this case, sand is pierced or completely disrupted and detached from the "mother bed" and embedded in the underlying mud (something analogous to stoping in magmas). Internal laminae follow the curvature of the outer surfaces, indicating that they are formed by viscous shearing within the ball or pillow during its descent. The most plausible explanation is again the physical state of the materials (Rayleigh-Taylor instability of sand-mud interbeds) combined with an external perturbation, that is, seismic shocks. Downslope movements are not necessarily implied, even though the pillows may end up in slumped masses.

Deformations in slump sheets is best observed where sliding materials are heterolithic and/or in different state of compaction and cohesion. A variety of geometries is thus displayed, from disharmonic folds, cascade folds, box folds, concentric folds, isolated hinges, shear planes, slabs and blocks, brecciation and even, in most conspicuous slump sheets (tens of meters thick), cleavage in mudstones (Ricci Lucchi, 1995b). Sand volcanoes, other fluid-escape structures and burrows may occur at the top. Axial surfaces vary from steeply inclined to flat-lying, and axes tend to be parallel to depositional strike, but with a wide dispersion of orientation.

Structures related to advancing ice or sand bodies are shown by layered muds in the form of folding, crumpling and contortion, at places extremely "baroque" (complicated). Antiform folds can be arranged *en echelon*, more or less parallel to the edge of the encroaching body, and may be associated with shear planes. These structures are the combined effect of differential loading and drag.

Sedimentary dykes can be produced at various scales by both forceful injection of fluid or hydroplastic sediment from below and infilling of open cracks from above. Ptygmatic folds can develop in dykes owing to subsequent compaction or to increasing resistance encountered by the injection.

Smaller injected dykes are indistinguishable from the pillars associated to dishes in sandstones; larger structures are called pseudodiapirs, mudlumps, etc., and originate in multilayered successions rich in mud. Infilled dykes range from desiccation cracks to large fissures and fractures formed in lithified sediments, especially limestone, subjected to extensional stresses ("neptunian dikes").

Deformed cross-bedding has various aspects. It can affect either the topmost part of foreset laminae, overturning or disharmonically folding them, or the mid-lower portions, where small folds, discordant slabs, small normal (gravity) faults or patches of vaguely laminated or structureless sediment appear (Allen, 1982). Fluid drag on subaqueous or subaerial, water-saturated sediment is generally held responsible for the first case, whereas in the other, some form of gravity induced downslope slip is more plausible. The lee side of eolian dunes, for example, is prone to sliding when wet. Detachment and sliding of slightly cohesive sand laminasets can occur without internal deformation.

Franco Ricci-Lucchi and Alessandro Amorosi

Bibliography

Allen, J.R.L., 1982. *Sedimentary Structures: Their Character and Physical Basis*. Elsevier, Developments in Sedimentology, 2 volumes: 593p. and 693p.

Bouma, A.H., 1962. Sedimentology of Some Flysch Deposits. Elsevier.

Campbell, C.V., 1967. Lamina, laminaset, bed and bedset. *Sedimentology*, 8: 7–26.

de Raaf, J.F.M., Boersma, J.R., and Van Gelder, A., 1977. Wave-generated structures and sequences from a shallow marine succession, Lower carboniferous, County Cork, Ireland. *Sedimentology*, 24: 451–484.

Dott, R.H. Jr., and Bourgeois, J., 1982. Hummocky stratification: Significance of its variable bedding sequences. *Bulletin of Geological Society of America*, 93: 663–680.

Dzulynski, S., 1966. Sedimentary structures resulting from convection-like pattern of motion. *Annales de la Société géologique de Pologne*, 361: 3–21.

Harms, J.C., Southard, J.B., Spearing, D.R., and Walker, R.G., 1975. *Depositional Environments as Interpreted from Primary Sedimentary Structures and Stratification Sequences*. Society of Economic Paleontologists and Mineralogists Short Course No.2, Lecture Notes, Dallas, 162p.

Jopling, A.V., and Walker, R.G., 1968. Morphology and origin of ripple-drift cross-lamination, with examples from Pleistocene of Massachusetts. *Journal of Sedimentary Petrology*, 38: 971–984.

Leckie, D.A., and Walker, R.G., 1982. Storm and tide-dominated shorelines in Cretaceous Moosebar—Lower Gates interval: outcrop equivalents of Deep Basin gas trap, western Canada. *American Association of Petroleum Geologists Bulletin*, 66: 138–157.

Lowe, D.R., 1982. Sediment gravity flows, II. Depositional models with special reference to the deposits of high-density turbidity currents. *Journal of Sedimentary Petrology*, 52: 279–297.

Miall, A.D., 1985. Architectural-element analysis: a new method of facies analysis applied to fluvial deposits. *Earth Science Reviews*, 22: 261–308.

Middleton, G.V., 1967. Experiments on density and turbidity currents, III. Deposition of sediment. *Canadian Journal of Earth Sciences*, 4: 475–505.

Mutti, E., and Ricci-Lucchi, F., 1972. Le torbiditi dellAppennino Settentrionale: introduzione allanalisi di facies. *Memorie Società Geologica Italiana*, 11 (2): 161–199.

Mutti, E., 1992. *Turbidite Sandstones*. San Donato Milanese: AGIP, 275 pp.

Mutti, E., Davoli, G., Tinterri, R., and Zavala, C., 1996. The importance of ancient fluvio-deltaic systems dominated by catastrophic flooding in tectonically active basins. *Memorie di Scienze Geologiche* (Padova, Italy), 48: 233–291.

Reineck, H.E., and Wunderlich, F., 1968. Zur Unterscheidung von asymmetrischen Oszillationsrippeln und Schömungsripplen. *Senckenbergiana Lethaea*, 49: 321–345.

Ricci Lucchi, F., 1995a. *Sedimentographica. A Photographic Atlas of Sedimentary Structures*. Columbia University Press.

Ricci Lucchi, F., 1995b. *Sedimentological Indicators of Paleoseismicity*. Association of Engineering Geolgists, Special Publication, 6, 7–18.

Tessier, B., 1993. Upper intertidal rhythmites in the Mont-Saint-Michel Bay (NW France); perspectives for paleoreconstruction. *Marine Geology*, 110: 355–367.

Visser, M.J., 1980. Neap-spring cycles reflected in Holocene subtidal large-scale bedform deposits: a preliminary note. *Geology*, 8: 543–546.

Yang, C.-S., and Nio, S.-D., 1985. The estimation of paleohydrodynamic processes from subtidal deposits using time series analysis methods. *Sedimentology*, 32: 41–58.

Cross-references

Accretion, Lateral and Vertical
Bedset and Laminaset
Clastic (Neptunian) Dykes and Sills
Convolute Lamination
Cross-Stratification
Deformation of Sediments
Dish Structure
Flaser
Fluid Escape Structures
Grading, Graded Bedding
Grain Flow
Grain Size and Shape
Gravity-Driven Mass Flows
Hummocky and Swaley Cross-Stratification
Imbrication and Flow-Oriented Clasts
Liquefaction and Fluidization
Pillar Structure
Planar and Parallel Lamination
Sediment Transport by Tides
Sediment Transport by Unidirectional Water Flows
Sediment Transport by Waves
Storm Deposits
Surface Forms
Tides and Tidal Rhythmites
Turbidites

BEDSET AND LAMINASET

According to McKee and Weir (1953), a *bed* is a sedimentary stratum greater than 10 mm thick, whereas a *lamina* is less than 10 mm thick. The divide at 10 mm is arbitrary, and not

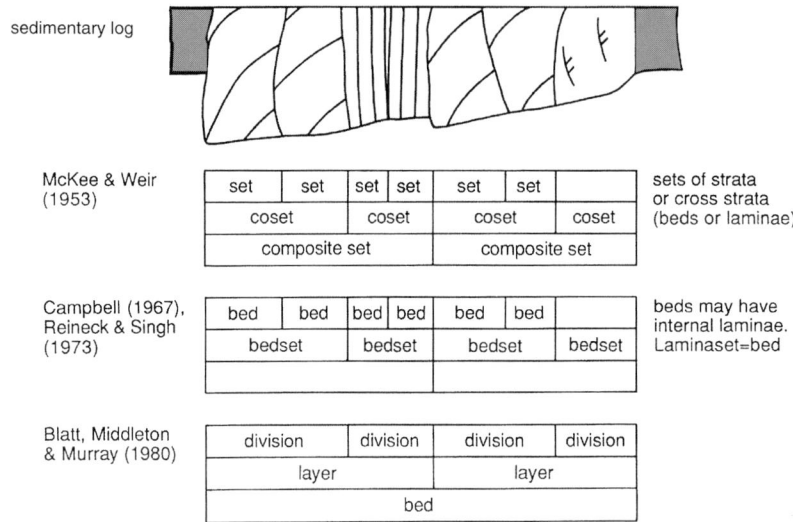

Figure B12 Comparison of terminology for describing stratification (from Bridge, 1993).

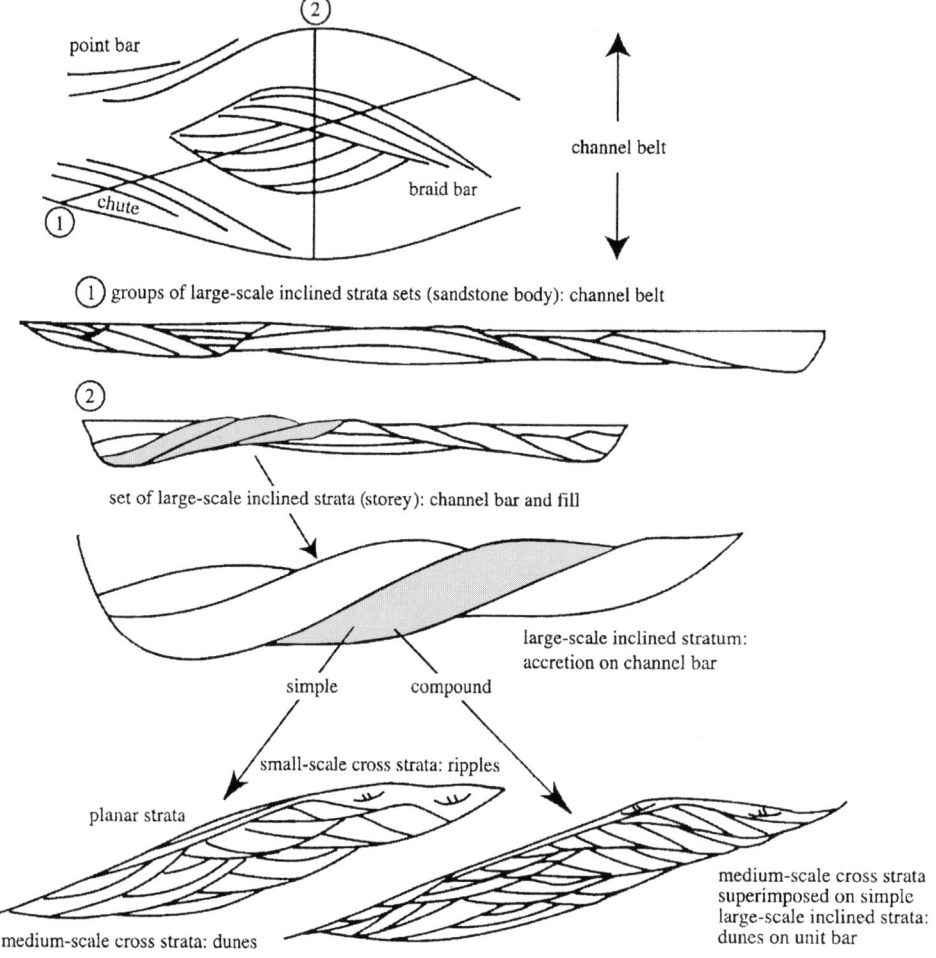

Figure B13 Proposed terminology for the case of superimposed scales of strata, using river channel deposits as an example (modified from Bridge, 1993).

based on objective statistical analysis of stratal thickness, nor on any genetic implication. Beds and laminae are recognized on the basis of changes in sediment texture and/or composition within or between them. Beds and laminae generally occur in sets (*bedsets, laminasets*) that are distinctive in terms of the orientation, texture and composition of beds or laminae within the set (Figure B12). If the beds or laminae within a set are inclined at more than a few degrees relative to the set boundaries, the beds or laminae are given the prefix *cross* or *inclined* (e.g., set of cross laminae, or cross-lamina set). Bedsets or laminasets also occur in sets. For example, a set of similar cross-lamina sets or cross-bed sets is referred to as a simple set or *coset* (Figure B12), according to McKee and Weir (1953). A set or different types of bedsets or laminasets is called a *composite set* (Figure B12).

This terminology for defining beds, laminae, bedsets, and laminasets is widely used but is not without its problems and detractors. Modifications to the terminology shown in Figure B12 suggest that there was a desire to use the term bed for a larger scale of stratum than that defined by McKee and Weir (1953). The term bed may therefore refer to a set of strata rather than a single stratum with no internal subdivision. Another problem with McKee and Weirs terminology is that both laminae and beds may occur in the same set of strata. This problem arises because of the arbitrary division of strata into beds and laminae. A single strataset containing beds and laminae could logically be called a composite set, but this term has already been used for a set of different sets rather than a set of different strata. Perhaps the greatest difficulty with all of the different methods for defining strata and sets of strata is that there is no consistent use of terms for referring to the many different superimposed scales of strata and sets of strata in sedimentary basins. The terms used for the smaller stratasets (bedset, laminaset) have not been applied to the larger-scale stratasets. Instead, poorly defined, inexplicit terms such as *storey*, *architectural element* and *parasequence* have been used.

Bridge (1993) argued that, in order to describe hierarchies of different-scale strata, it is desirable to use a reasonably small number of explicit terms consistently irrespective of scale, and to use qualifying terms to describe relative scale of strata. Figure B13 shows an example of this approach applied to river deposits, and includes interpretation as well as descriptive terminology. An alternative method of describing hierarchies of strata of different scale is by numerically ordering their bounding surfaces (e.g., Brookfield, 1977; Allen, 1983; Miall, 1996). As discussed by Bridge (1993), this approach is difficult to use in practice.

John S. Bridge

Bibliography

Allen, J.R.L., 1983. Studies in fluviatile sedimentation: bars, bar-complexes and sandstone sheets (low-sinuosity braided streams) in the Brownstones (Lower Devonian), Welsh Borders. *Sedimentary Geology*, **33**: 237–293.
Blatt, H., Middleton, G.V., and Murray, R., 1980. *Origin of Sedimentary Rocks*, 2nd edn., Prentice-Hall.
Bridge, J.S., 1993. Description and interpretation of fluvial deposits: a critical perspective. *Sedimentology*, **40**: 801–810.
Brookfield, M.E., 1977. The origin of bounding surfaces in ancient eolian sandstones. *Sedimentology*, **24**: 303–332.
Campbell, C.V., 1967. Laminae, lamina set, bed and bedset. *Sedimentology*, **8**: 7–26.
McKee, E.D., and Weir, G.W., 1953. Terminology for stratification and cross-stratification in sedimentary rocks. *Bulletin of the Geological Society of America*, **64**: 381–390.
Miall, A.D., 1996. *The Geology of Fluvial Deposits*. New York: Springer-Verlag.
Reineck, H.E., and Singh, I.B., 1973. *Depositional Sedimentary Environments*. Berlin: Springer.

Cross-references

Bedding and Internal Structures
Cross-stratification
Grading, Graded Bedding

BENTONITES AND TONSTEINS

Definition

Bentonites are thin, widespread clay-altered ash layers that are commonly dominated by smectite (Fisher and Schminke, 1984). This is the definition used by sedimentologists. Knight (1898) first used the name bentonite for an unusual clay in the Benton Shale of Cretaceous age in Wyoming. The absorbent properties and swelling nature of this and similar clays were later shown to be due to the presence of the smectite group of minerals (*q.v.*) of which montmorillonite is the most common. In the definition of bentonite the emphasis on the origin is generally attributed to Ross and Shannon (1926). They noted that the clay was a product of the "alteration of a glassy igneous material, usually a tuff or volcanic ash" and the resultant clay "is usually the mineral montmorillonite, but less often beidellite" (a member of the smectite group). Smectite-rich bentonites are valuable raw materials and some authors would restrict the bentonite definition to smectite-rich deposits irrespective of origin (Grim and Guven, 1978; Roen and Hosterman, 1982). This definition cannot be reconciled with that used by sedimentologists and it has to be accepted that both definitions will continue to be used, albeit for different purposes.

Mesozoic bentonites usually consist of smectite. As temperature rises during burial diagenesis smectite layers are progressively converted into illite, via a mixed-layer illite-smectite. These beds are known as K-bentonites because of the introduction of potassium in the illitization reaction. Most Paleozoic clay-altered volcanic ashes are of this type. In coal-bearing strata of all ages the volcanic ash usually alters to kaolinite forming thin beds known as *tonsteins*. Only since the 1980s has the volcanic origin been generally accepted.

Fisher and Schmincke (1984) suggest that the nomenclature of bentonites should reflect the dominant clay mineral species with smectite (s)-bentonite, kaolinite (k)-bentonite and illite (i)-bentonite. K-bentonite is a well-established term and Huff (1983) has made a good case for its retention. Likewise the term tonstein is also well-established and will continue to be used by coal geologists. Possibly smectite-bentonite should be adopted as this would eliminate potential confusion between bentonite as a general term for the whole group and a specific member within the group.

Recognition of bentonites

Field characteristics

Bentonites typically occur as thin beds with a wide lateral extent of up to hundreds of kilometers. The contacts with the enclosing sediment are nearly always sharp. If the material is unweathered internal lamination and graded bedding may be visible. The expansion of smectite during weathering causes preferential loss of bentonites in surface exposures. Tonsteins on the other hand may be more durable than the host sediments because kaolinite does not expand and there is an interlocking texture. Many tonsteins have the appearance of flint clays, but others are more granular. Bentonites are mainly associated with shales and coals because a low energy depositional environment is required for their preservation. Ashfalls may incorporate minor amounts of organic matter, but the content is significantly less than in the enclosing sediments and hence there is a usually a color contrast.

Mineralogy

The clay mineralogy of bentonites differs significantly from that of the associated sediments. The normal detrital clay assemblage of kaolinite, illite and chlorite is absent and in tonsteins the kaolinite is well-ordered in contrast to most detrital kaolinites. Bentonites are dominated by one clay mineral to the extent that many are monomineralic and in this they are exceptional.

The clay minerals form from unstable volcanic glass and to a lesser extent from some of the pyrogenic minerals. Many of the latter are relatively stable and survive the ash alteration. These minerals are diagnostic of the volcanic origin and also of the original ash composition. Quartz and biotite are common, sanidine and high temperature plagioclase feldspars less so and olivine, pyroxene and amphibole are rare. The relative abundances largely reflect the dominance of acid volcanics, but with weathering stability superimposed. In bentonites there is a restricted heavy mineral suite consisting mainly of zircon, apatite and ilmenite/magnetite. Weaver (1963) emphasized the interpretative value of this restricted suite, which is unlike that normally encountered in detrital sediments.

As the thickness of the volcanic ash decreases away from the eruptive source so the grain size of the particles decreases and their proportions change. Minerals are generally most abundant between 2 mm and 1/16 mm and are practically absent below 10 microns, whereas glass shards can be much smaller (Fisher and Schmincke, 1984). These features are inherited by bentonites. There is also mineral fractionation in air-fall ashes and in bentonites a noteworthy result of this eolian fractionation is the enrichment of biotite at the base of some units (Diessel, 1985; Huff and Morgan, 1990).

Textures

In the magma the pyrogenic minerals develop relatively freely resulting in euhedral crystals. These are found in bentonites and also high temperature forms such as beta quartz crystals. During explosive volcanism crystals are shattered and quartz splinters are very common. As would be anticipated minerals with well-developed cleavages also fragment. There is therefore a contrast with the shape of grains in normal detrital sediments.

Vitric particles are the other main component of volcanic ashes. Cuspate and platy shards result from the shattering of bubbles in silicic magmas and the characteristic shape may survive the alteration to clay. Clay-altered pumice shards may retain the shape but lose the internal structure and consequently are more difficult to identify positively.

Composition

The major element composition of bentonites essentially reflects the diagenetic conditions and the resulting clay mineral proportions rather than the magmatic origin. There are rare instances where vitric particles have been preserved, usually by an early diagenetic cement. Glass inclusions in pyrogenic minerals are more common and are sufficiently large to be analyzed by an electron microprobe, as in the work of Lyons *et al.* (1993).

Not all elements are mobile in the alteration process and the immobile elements are concentrated. The radioactivity of many bentonites in borehole logs drew attention many years ago to the high concentrations of Th and U. More recently it has been appreciated that classification schemes for altered volcanic ashes using immobile elements (Winchester and Floyd, 1977) also apply to bentonites (Lyons *et al.*, 1993). The use of these elements confirms the volcanic origin, but more importantly provides information on the original ash composition and the tectonic setting of the vulcanicity.

Classification of tuffs and bentonites

Tuffs are classified on the glass-mineral proportions and range from vitric tuffs to crystal tuffs (Pettijohn, 1975). As some minerals and textures survive the alteration process such a classification should be applicable to bentonites. The petrography of tonsteins has been studied in the greatest detail because of their essential role in coalfield correlations and also tonsteins represent the worst case scenario in terms of the extent of alteration. Most authors use Schullers classification (see Bohor and Triplehorn, 1993), which is kristal tonstein (crystals of kaolinite), graupen tonstein (grains of kaolinite), psudomorphosen tonstein (kaolinite pseudomorphs after silicates) and dichte tonstein (fine-grained kaolinite). Bouroz *et al.* (1983) and Bohor and Triplehorn (1993) have made some progress in relating the Schuller classification to the tuff nomenclature. Dichte tonstein is interpreted as a clay-altered vitric tuff. A graupen tonstein could be equivalent to a vitric tuff containing pumice shards. There is also the possibility raised by Diessel (1985) that highly altered biotites which lose the original lath-shape of the grains and become barrel-shaped could be responsible for a texture characteristic of graupen tonsteins. The pseudomorphosen tonsteins are clearly altered crystal tuffs and so are some of the kristal tonsteins. The difficulty with other kristal tonsteins is that the crystals of kaolinite are vermicular and may have formed from the matrix rather than being a pyrogenic mineral. If this was resolved then bentonites could be related to the equivalent tuff category with greater certainty.

Rate of ash alteration and clay stability

The alteration of volcanic glass is reviewed in detail by Fisher and Schmincke (1984). The rate of alteration and formation of intermediate products are further discussed in Bohor and Triplehorn (1993) with greater emphasis on the formation of the clay minerals. There is textural evidence from bentonites that clay minerals sometimes formed relatively rapidly. The rate does depend on a number of factors including ash composition and the pore solution chemistry, which is also influenced by the depositional environment and the permeability of the enclosing sediments.

In the earlier literature there is an emphasis on low pH for the formation of kaolinite. The work of Garrels and Christ (1965) and others shows low ionic activities of component elements are also important. In seawater, activities are higher and smectite is stable. For pore solutions to remain within the kaolinite stability field during diagenesis the system must remain open.

Use of bentonites

Those bentonites consisting of smectite are of economic value and have many commercial uses (Grim and Guven, 1978). Tonsteins are generally not of such value with the exception of some deposits in China.

Bentonites are comparable to tephra layers in that they are chronostratigraphic markers and are very important in many parts of the stratigraphical column. Tonsteins came to prominence over a hundred years ago because of their stratigraphic values, and correlation of coal seams in the paralic coalfields of mainland Europe would have been inconceivable without them. The survival of pyrogenic minerals in bentonites creates excellent opportunities for absolute age dating, often from parts of the column where other suitable materials are absent. The time consuming task of separating a sufficient mass of minerals for analysis is now redundant with current instrumentation (SHRIMP) capable of producing age profiles within grains.

Summary

Bentonites are clay-altered volcanic ash layers. They are formed during explosive volcanism and as a consequence have a wide distribution. Bentonites are characteristically preserved as thin beds in low energy depositional environments, such as shales and coals, and have considerable stratigraphic value. Many bentonites in Mesozoic sediments consist of smectite. Some authors would restrict the term bentonite to smectite rich deposits irrespective of origin. As the diagenetic level increases the smectite is converted to illite via a mixed-layer clay. These are the K-bentonites. Another group were subjected to greater alteration, which produced kaolinite and these are the tonsteins. The terms K-bentonite and tonstein are well established in the literature and should be retained. If smectite-bentonite were generally adopted potential confusion could be avoided.

Bentonites are recognized as altered volcanic ashes on bed form, structures, textures, mineralogy and trace element geochemistry. The latter two also enable the original ash composition and the tectonic setting of the vulcanism to be determined. Although tonsteins result from the greatest level of alteration their classification and that of tuffs have many features in common.

D. Alan Spears

Bibliography

Bohor, B.F., and Triplehorn, D.M., 1993. Tonstein: altered volcanic ash layers in coal-bearing sequences. *Geological Society of America Special Paper*, 285, 40p.

Bouroz, A., Spears, D.A., and Arbey, F. 1983. *Essai de synthèse des données acquises sur la genèse et l'évolution des marqueurs petrographiques dans les bassins houillers*. Memoire XVI, Société Géologique du Nord.

Diessel, C.F.K., 1985. Tuffs and tonsteins in the Coal Measures of New South Wales, Australia. *10th Congres International de Stratigraphie et de Géologie du Carbonifere, Madrid, 1983*, **4**: 197–210.

Fisher, R.V., and Schmincke, H.U., 1984. *Pyroclastic Rocks*. Springer Verlag.

Garrels, R.M., and Christ, C.L., 1965. *Solutions, Minerals and Equilibria*. Freeman, Cooper and Co.

Grim, R.E., and Guven, N., 1978. *Bentonites: Geology, Mineralogy, Properties and Use*. Developments in Sedimentology, 24: Amsterdam: Elsevier.

Huff, W.D., 1983. Misuse of the term "bentonite" for ash beds of Devonian age in the Appalachian basin. Discussion and reply. *Geological Society of America Bulletin*, **94**: 681–683.

Huff, W.D., and Morgan, D.J., 1990. Stratigraphy, mineralogy and tectonic setting of Silurian K-bentonites in Southern England and Wales. In Farmer, V.C., and Tardy, Y. (eds.), *Proceedings, 9th International Clay Conference, Strasbourg 1989*. Sci. Geol. Mem., **88**: 33–42.

Knight, W.C., 1898. Mineral soap. *Engineering Mining Journal*, **66**: 481.

Lyons, P.C., Spears, D.A., Outerbridge, W.F., Congdon, R.D., and Evans, H.T. Jr., 1993. Euramerican tonsteins: overview, magmatic origin and depositional-tectonic implications. *Palaeogeography, Palaeoclimatology, Palaeoecology*, **106**: 113–139.

Pettijohn, F.J., 1975. *Sedimentary Rocks*. 3rd edn. Harper and Row.

Roen, J.B., and Hosterman, J.W., 1982. Misuse of the term "bentonite" for ash beds of Devonian age in the Appalachian basin. *Geological Society of America Bulletin*, **93**: 921–925.

Ross, C.S., and Shannon, E.V., 1926. The minerals of bentonite and related clays and their physical properties. *Journal of the American Ceramic Society*, **9**: 77–96.

Weaver, C.E., 1963. Interpretative value of heavy minerals from bentonites. *Journal of Sedimentary Petrology*, **33**: 343–349.

Winchester, J.A., and Floyd, P.A., 1977. Geochemical discrimination of different magma series and their differentiation products using immobile elements. *Chemical Geology*, **20**: 325–343.

Cross-references

Diagenesis
Kaolin Group Minerals
Mixed-Layer Clays
Mudrocks
Smectite Group

BERTHIERINE

Introduction

Berthierine is an iron-rich, aluminous, 1 : 1-type layer silicate belonging to the serpentine group (Bailey, 1980; Brindley, 1980, 1981). The name honors Berthier, who reported the first studies of this mineral in 1827. In the past, this clay mineral was also known as septechlorite, septechamosite, chamosite, 7Å-chamosite, and 7Å-chlorite, but berthierine is now the accepted name (Brindley et al., 1968). Berthierine was considered rare until recently. The failure to recognize berthierine arose mostly from its chemical similarity to Fe-rich chlorite (chamosite); these phases share many significant X-ray powder diffraction (XRD) lines. The ease with which berthierine can be mistaken for kaolinite in XRD patterns of oriented crystal aggregates has also contributed to its misidentification (Moore and Reynolds, 1997, p. 139).

Berthierine is chemically and structurally similar to odinite. However, odinite is now formally recognized as a distinct 0.7 nm phase (Bailey, 1988a), and exhibits some key differences from berthierine, the most significant being the abundance of ferric iron. Odinite is a member of the verdine facies, a group of green clay minerals present on many shallow, marine shelves and reef lagoonal areas in tropical locations (Odin, 1988). It has been previously identified as berthierine or 7Å-chamosite in some localities (e.g., Porrenga, 1967; Rohrlich et al., 1969).

Berthierine is best known from shallow marine environments, particularly ironstones (Van Houten and Purucker, 1984). However, it is increasingly recognized in brackish to nonmarine environments. These include laterites, flint clays and other pedogenic settings, and estuarine/deltaic sandstones, where it may have precipitated directly or formed by alteration of a precursor (kaolinite, aluminous goethite, odinite). Berthierine and odinite themselves may be precursors to diagenetic chamosite (Bailey, 1988a, b; Hornibrook and Longstaffe, 1996).

Structure and chemistry

Berthierine occurs in two structural forms, orthohexagonal and monoclinic, the former being most abundant. Both can be present in the same sample. The cell parameters of berthierine are $a \cong 0.540$ nm, $b \cong 0.936$ nm, $c \sin\beta \cong 0.170$ nm, and $\beta \cong 104.5°$ and $90°$ (Bailey, 1980). Disorder within the structure is common, arising from variations in layer stacking and chemistry (Brindley, 1980).

Berthierine is characterized in XRD by a (001) diffraction at 0.7 nm, which defines its 1 : 1 layer structure, and a (060) diffraction at 0.1555 nm, which indicates its dominantly trioctahedral character. The XRD identification of berthierine is described in Bailey (1980, 1988b) and Brindley (1980). Identification by XRD alone is difficult, however, because of the presence of two berthierine polytypes and the similarity to chamosite. Chlorite ought to be distinguishable from berthierine in XRD by its characteristic 1.4 nm (001) diffraction. However, this diffraction is very poorly developed or absent in chamosite, while its (002) diffraction at 0.7 nm overlaps that of the berthierine (001). Berthierine also can occur with chamosite (as a physical mixture and/or an interstratification), which further complicates identification. Unambiguous recognition of berthierine commonly requires direct measurement of layer thickness using TEM (transmission electron microscope) lattice-fringe images (e.g., Jiang et al., 1992).

The generalized formula of berthierine is $(R^{2+}_a R^{3+}_b \square_c)(Si_{2-x}Al_x)O_5(OH)_4$, where R^{2+}(Fe, Mg, Mn) and R^{3+}(Fe, Al) are cations in octahedral positions, $\square$ describes possible vacant octahedral sites, and $a + b + c = 3$ (Brindley, 1982). Berthierine is dominantly trioctahedral, with a typical composition of $(Fe^{2+}_{1.5}Mg_{0.375}Al_{1.0}\square_{0.125})(Si_{1.25}Al_{0.75})O_5(OH)_4$ (after Moore and Reynolds, 1997). However, a wide range in composition is known (Brindley, 1980, 1982). For example, end-member Fe-berthierine $[(Fe^{2+}_2Al)(SiAl)O_5(OH_4)]$ occurs in Cretaceous laterite deposits of Minnesota (Toth and Fritz, 1997). By comparison, di-trioctahedral berthierine is present in oil-sands from Alberta, $(Fe^{2+}_{1.01}Al_{0.82}Mg_{0.46}Fe^{3+}_{0.28}Mn_{<0.01}\square_{0.43})(Si_{1.74}Al_{0.26})O_5(OH)_4$ (Hornibrook and Longstaffe, 1996). Its composition suggests the possibility of a series between berthierine and odinite. Odinite is similar to berthierine in that it is Al- and Fe-rich, has a 1 : 1 layer structure, and shares many of the same XRD lines. However, it is richer in Fe^{3+}, Mg and Si, and contains less Al than most berthierines. It has a representative formula of $(Fe^{3+}_{0.8}Mg_{0.8}Al_{0.53}Fe^{2+}_{0.3}\square_{0.57})(Si_{1.8}Al_{0.2})O_5(OH)_4$ (after Moore and Reynolds, 1997).

Occurrence and formation

Marine

Berthierine is most commonly associated with marine-oolitic ironstone formations, which are generally considered to have formed in low energy, marginal marine shelf environments in tropical to subtropical regions. There, berthierine is typically a component of pisoids and ooids, consisting of concentric sheathes around a nucleus of clay, ferric oxide, other ooids, fossil fragments or quartz grains (Bhattacharyya, 1983; Van Houten and Purucker, 1984; Maynard, 1986; Siehl and Thein, 1989).

Van Houten and Purucker (1984) summarized possible origins for such berthierine and other 'chamositic' minerals. Neoformation of berthierine is possible where grain-coatings, pore-linings and void-fillings are observed. However, the general consensus favors early diagenetic, anoxic modification of a precursor, which had accumulated as granules on an oxygenated seafloor. Numerous identities for this early substrate have been proposed. Bhattacharyya (1983) showed that remobilization of iron from hydroxides could react with detrital kaolinite to form berthierine in seawater. Odin (1988) and his coauthors suggested that odinite could evolve to berthierine during early diagenesis, producing a better organized 0.7 nm structure with only moderate chemical modification. Morphological relationships are usually cited as the main difficulty in accepting odinite as a precursor to berthierine. Odinite in Recent sediments almost always occurs as peloids of presumed fecal origin, whereas there are no reports of grain-coating or pore-lining odinite. Reworked, transported and transformed lateritic minerals are another possible precursor for berthierine in shallow marine sediments.

Nonmarine

Nonmarine occurrences of berthierine are reported with increasing frequency, including laterites (Toth and Fritz,

1997), Arctic desert soils (Kodama and Foscolos, 1981), coal swamps (Iijama and Matsumoto, 1982), fresh to brackish floodplain and estuarine/deltaic sediments (Taylor, 1990; Hornibrook and Longstaffe, 1996), and flint clays (Moore and Hughes, 2000).

Berthierine from laterites and estuarine sandstones is likely of most significance. For example, the Lower Cretaceous Weald Clay of southeast England contains transported berthierine-bearing pisoids and ooids derived from Fe-rich soils. These grains were deposited in a scour within a fresh to brackish water mudplain (Taylor, 1990). *In situ* formation of berthierine has been reported in a laterite buried beneath a lignitic horizon in Minnesota (Toth and Fritz, 1997; Fritz and Toth, 1997). There, berthierine formed by alteration of kaolinite and by neoformation. Berthierine formation is attributed to percolation of fresh water under reducing conditions through a pisolitic laterite (gibbsite, kaolinite, and goethite), which had developed on a low-relief peneplain. Reduction of Fe^{3+} was accomplished via oxidation of organic carbon. The absence of sulfate-inhibited pyrite formation, thus facilitating berthierine and siderite formation. Radial bladed and radial blocky coats of berthierine on pisoids formed first. As the soil became saturated by groundwater, precipitation of macroscopic crystals of Fe-berthierine followed in the voids between pisoids. The raised water table was associated with a regional transgression of the Western Interior Seaway during Late Cretaceous time. Such laterites represent possible sources of detritus for some varieties of marine ironstones. Likewise, the presence of such berthierine and siderite in nonmarine sediments may be correlatable with sequence or parasequence boundaries (Toth and Fritz, 1997).

Well-crystallized laths of berthierine, together with Fe-rich saponite and traces of chamosite, occur as early diagenetic pore-linings and grain-coatings on sand grains in the estuarine/deltaic Lower Cretaceous Clearwater Formation oil-sand deposits of Alberta (Hornibrook and Longstaffe, 1996). The stable isotope compositions of this berthierine indicate formation in the presence of meteoric water at 25–45°C. Berthierine formation was followed by calcite precipitation, and then emplacement of hydrocarbons.

Volcanic rock fragments in these sands provided a leachable source of Fe, which accreted about grains in the form of hydroxides or odinite in the freshwater estuary. High sedimentation rates helped to establish reducing conditions in the estuarine sediments, which favored transformation of Fe-minerals to berthierine. The presence of calcite rather than siderite suggests that most available iron had been consumed by Fe-clays. Calcite $\delta^{13}C$ values of up to $+23^a$ indicate that extreme reducing conditions (microbial CO_2-reduction) had been established by the time of carbonate precipitation. In the absence of sulfate and prior to CO_2 reduction, microbes may have acted first to reduce Fe^{3+}, and triggered berthierine generation. Once Fe was reduced and more or less fixed into clays, calcite crystallization and microbial CO_2 reduction ensued. The possibility remains that odinite was the precursor to berthierine, and that ferric iron was reduced *in situ*. Subtropical conditions existed in this area during the Early Cretaceous, and the estuarine setting would have favored odinite formation. While the blades and laths of Clearwater berthierine are unlike the morphology known for Recent odinite, it almost certainly becomes unstable in most ancient sediments, transforming to better crystallized phases, including berthierine.

Reactions during burial diagenesis

Reactions involving berthierine during burial diagenesis are of growing interest. For example, reaction between siderite and kaolinite at 65–150°C under reducing conditions produced aluminous, low-Mg berthierine in shales associated with coal swamps (Iijima and Matsumoto, 1982). Replacement of kaolinite by chlorite during burial diagenesis of mudstones is also suspected to proceed through a berthierine intermediary (Burton *et al.*, 1987). Berthierine is increasingly considered to be a diagenetic precursor to chamosite, based on the numerous observations of berthierine-chamosite intercalation (e.g., Ahn and Peacor, 1985; Hillier and Velde, 1992). Temperatures as low as 70°C have been suggested for initiation of this reaction (Jahren and Aagaard, 1989; Hornibrook and Longstaffe, 1996).

Chamosite is a common grain-coating and early pore-lining in many sandstones. These rims, where they are not too thick, preserve porosity and permeability by inhibiting further diagenetic mineral growth. In contrast, development of thick rims reduces the potential for hydrocarbon saturation and increases the likelihood of formation damage during hydrocarbon recovery. Berthierine (and possibly odinite) may be low-temperature precursors of this chloritic rim. Hence, depositional and early diagenetic enviroments like those in which the Clearwater berthierine formed may illustrate one set of conditions under which formation of Fe-clay rims is most favorably initiated.

Fred J. Longstaffe

Bibliography

Ahn, J.H., and Peacor, D.R., 1985. Transmission electron microscopic study of diagenetic chlorite in Gulf Coast argillaceous sediments. *Clays and Clay Minerals*, **33**: 228–236.

Bailey, S.W., 1980. Structures of layer silicates. In Brindley, G.W., and Brown, G. (eds.), *Crystal Structures of Clay Minerals and their X-ray Identification*. London: Mineralogical Society, Monograph No. 5, pp. 1–124.

Bailey, S.W., 1988a. Odinite, a new dioctahedral-trioctahedral Fe^{3+}-rich 1:1 clay mineral. *Clay Minerals*, **23**: 237–247.

Bailey, S.W., 1988b. Structures and compositions of other trioctahedral 1:1 phyllosilicates. In Bailey, S.W. (ed.), *Hydrous Phyllosilicates (Exclusive of Micas)*. Mineralogical Society of America, Reviews in Mineralogy, Volume 19, pp. 169–188.

Bhattacharyya, D.P., 1983. Origin of berthierine in ironstones. *Clays and Clay Minerals*, **31**: 173–182.

Brindley, G.W., 1980. Order-disorder in clay mineral structures. In Brindley, G.W., and Brown, G. (eds.), *Crystal Structures of Clay Minerals and their X-ray Identification*. London: Mineralogical Society, Monograph No. 5, pp. 125–195.

Brindley, G.W., 1981. Structures and chemical compositions of clay minerals. In Longstaffe, F.J. (ed.), *Clays and the Resource Geologist*. Mineralogical Association of Canada, Short Course No. 7, pp. 1–21.

Brindley, G.W., 1982. Chemical compositions of berthierines—a Review. *Clays and Clay Minerals*, **30**: 153–155.

Brindley, G.W., Bailey, S.W., Faust, G.T., Forman, S.A., and Rich, C.I., 1968. Report of the Nomenclature Committee (1966–67) of the Clay Minerals Society. *Clays and Clay Minerals*, **16**: 322–324.

Burton, J.H., Krinsley, D.H., and Pye, K., 1987. Authigenesis of kaolinite and chlorite in Texas Gulf Coast sediments. *Clays and Clay Minerals*, **35**: 291–296.

Fritz, S.J., and Toth, T.A., 1997. An Fe-berthierine from a Cretaceous laterite: Part II. Estimation of Eh, pH and pCO_2 conditions of formation. *Clays and Clay Minerals*, **45**: 580–586.

Hillier, S., and Velde, B., 1992. Chlorite interstratified with a 7Å mineral: an example from offshore Norway and possible

implications for the interpretation of the composition of diagenetic chlorites. *Clay Minerals*, **27**: 475–486.
Hornibrook, E.R.C., and Longstaffe, F.J., 1996. Berthierine from the Lower Cretaceous Clearwater Formation. *Clays and Clay Minerals*, **44**: 1–21.
Iijima, A., and Matsumoto, R., 1982. Berthierine and chamosite in coal measures of Japan. *Clays and Clay Minerals*, **30**: 264–274.
Jahrens, J.S., and Aagaard, P., 1989. Compositional variations in diagenetic chlorites and illites, and relationships with formation water chemistry. *Clay Minerals*, **24**: 157–170.
Jiang, W.-T., Peacor, D.R., and Slack, J.F., 1992. Microstructures, mixed layering, and polymorphism of chlorite and retrograde berthierine in the Kidd Creek massive sulfide deposit, Ontario. *Clays and Clay Minerals*, **40**: 501–514.
Kodama, H., and Foscolos, A.E., 1981. Occurrence of berthierine in Canadian arctic desert soils. *Canadian Mineralogist*, **19**: 279–283.
Maynard, J.B., 1986. Geochemistry of oolitic iron ores, an electron microprobe study. *Economic Geology*, **81**: 1473–1483.
Moore, D.M., and Hughes, R.E., 2000. Ordovician and Pennsylvanian berthierine-bearing flint clays. *Clays and Clay Minerals*, **48**: 145–149.
Moore, D.M., and Reynolds, R.C. Jr., 1997. *X-ray Diffraction and the Identification and Analysis of Clay Minerals*. Oxford, New York: Oxford University Press.
Odin, G.S. (ed.), 1988. *Green Marine Clays*. Developments in Sedimentology, No. 45, Amsterdam: Elsevier.
Porrenga, D.H., 1967. Glauconite and chamosite as depth indicators in the marine environment. *Marine Geology*, **5**: 495–501.
Rohrlich, V., Price, N.B., and Calvert, S.E., 1969. Chamosite in the Recent sediments of Loch Etive, Scotland. *Journal of Sedimentary Petrology*, **39**: 624–631.
Siehl, A., and Thein, J., 1989. Minette-type ironstones. In Young, T.P., and Taylor, W.E.G. (eds.), *Phanerozoic Ironstones*. London: Geological Society, Special Publication No. 46, pp. 175–193.
Taylor, K.G., 1990. Berthierine from the non-marine Wealden (Early Cretaceous) sediments of south-east England. *Clay Minerals*, **25**: 391–399.
Toth, T.A., and Fritz, S.J., 1997. An Fe-berthierine from a Cretaceous laterite: Part I. Characterization. *Clays and Clay Minerals*, **45**: 564–579.
Van Houten, F.B., and Purucker, M.E., 1984. Glauconitic peloids and chamositic ooids—favorable factors, constraints, and problems. *Earth Science Reviews*, **20**: 211–243.

Cross-references

Chlorite in Sediments
Clay Mineralogy
Diagenesis
Glaucony and Verdine
Ironstones and Iron Formations
Mixed-Layer Clays
Oil Sands

BIOCLASTS

Bioclasts, a.k.a. fossils, shells, skeletal particles, biotics, etc., are an important component of many limestones, shales, and sandstones and are volumetrically the dominant to exclusive particulate building block of many limestones. They are typically the major tool for age determination and paleoecological information. Most shells are constructed of calcium carbonate (aragonite, calcite, high-magnesium calcite), silica (opal), or phosphate (apatite or collophane) with varying proportions of organic compounds. In the rock record, the original mineralogy is commonly altered to a more stable calcite, dolomite, or microcrystalline quartz. The original mineralogy is rarely taxonomically diagnostic, but strongly influences the type of preservation. Aragonite shells can be neomorphosed to calcite with varying degrees of fidelity to the original structure, but are commonly dissolved wholesale and preserved only as filled molds or lithified casts of the body cavity (steinkerns). In contrast, calcite or Mg-calcite skeletons typically preserve fine structural details and rarely form molds. The proportions of minerals within some shells and of minor elements, trace elements, and stable isotopes within the minerals can vary with the taxa and with environmental parameters, notably temperature and salinity. Thus they are sources of paleoenvironmental information, if the signal is preserved and can be calibrated.

Identification of bioclasts is easiest, most detailed, and most reliable where well-preserved specimens can be observed on bedding planes or etched surfaces or can be isolated from the rock by disaggregation or dissolution of the matrix. This is possible in limestones only where the mineralogy of the bioclasts differs from that of the matrix, typically through selective silicification of the fossils, as in the famous Permian biotas from the Glass Mountains, Texas. In the more common case, where both bioclasts and matrix or cement are calcite, identification must proceed through polished sections, thin sections, or acetate peels. Recognition depends on distinctive cross-sections, diagnostic structures, or skeletal wall structure.

Outstanding, well-illustrated general references on bioclast description and identification include Majewske (1969), Horowitz and Potter (1971), Milliman (1974, pp. 52–137; identification key pp. 315–318), Bathurst (1975, pp. 1–76), Scholle (1978), Flügel (1982), Adams and Mackenzie (1998). In the highlights of bioclast identification that follow, geologic range and original mineralogy of taxa and typical dimensions of specific features are indicated parenthetically.

Algae

Skeletons of various algae are major rock formers in the Phanerozoic. In addition algae and other microbes bound sediment (stromatolites, now cyanobacteria or microbialites: see *Microbially Induced Sedimentary Structures*) and produced carbonate mud through much of geologic time. Coralline red algae ¿Ordovician-Silvrian? E. Cretaceous–Recent; see *Algal and Bacterial Carbonate Sediments*; Mg calcite) are typified by a layer of rectangular chambers ($<10-20\,\mu m$). Nodules formed by red algae (rhodoliths), in consortium with encrusting bryozoans, worms, corals, foraminifers, etc., are conspicuous markers of slow deposition, for example, ushering in transgressive sequences. Solenoporids (Cambrian–Paleocene; Mg calcite?) also form nodules or frameworks of larger, polygonal cells ($10s-100s\,\mu m$). Halimedacean green algae (Ordovician–Recent; aragonite) are prodigious producers of needle mud ($1-10\,\mu m$) and, in the Cenozoic, producers of segments (1–5 mm) calcified around filaments to give the appearance of Swiss cheese in section. Phylloid algae is a form designation for skeletal fragments of both halimedacean-like and coralline-like algae that resemble cabbage leaves (cm-scale; <1mm thick) and accumulated in broad mounds in the Late Paleozoic. Dasyclad green algae (¿Cambrian?–Ordovician–Recent; aragonite) are perforated cylinders, kegs, or spheres ($100s\,\mu m$) with a hollow central axis. Calcispheres (Devonian?–Recent), in their most common and simplest form are a hollow spheres ($50-225\,\mu m$) with a thin ($3-30\,\mu m$)

calcareous wall. Some have multi-layered walls or protruding hollow spines. The central cavity is invariably filled with cement, suggesting exclusion of sediment, but the nearest living analogs, aragonitic reproductive cysts of dasyclad algae, have a covered hole. The modern forms are benthic as were the associations of Paleozoic forms, but Cretaceous associations were pelagic. Charophytes (Silurian–Recent; calcite) are fresh- or brackish-water green algae with calcified subspherical sex organs (100s μm) typified by spiral ridges on the exterior. Coccoliths (Late Jurassic–Recent) are calcareous discs (2–20 μm) produced by planktonic yellow-green algae. In very thin sections, typical forms appear as dual concentric rings with a swastika cross in cross-polarized light.

Foraminifers (Cambrian–Recent)

Foraminifers are important guide fossils from the Late Paleozoic to Recent and are locally rock formers. They are 'microfossils' that exceptionally reach 10 cm in size with chambers in a multitude of shapes, arrangements, and wall structures. Planktonic forms, numerous since the Jurassic, typically have globular chambers (100s μm) with perforate walls of calcite. Benthic and encrusting forms have walls of Mg calcite (microgranular, hyaline, porcellaneous textures; Fig. B14), calcite, agglutinated sediment, chitin, or, rarely, aragonite.

Radiolarians (Proterozoic–Recent)

Radiolarians are marine plankton whose porous skeletons (siliceous, 10s μm) have superficially radial symmetry commonly adorned with spines. They are useful guide fossils and local rock formers.

Sponges (Proterozoic–Recent)

Porous walls that may be convoluted or chambered surround a central cavity (mm–m) with a single large opening in a typical sponge. Hard parts, where present, are calcareous skeletons, isolated calcareous or siliceous spicules (10s–100s μm), or frameworks of fused spicules. Stromatoporoids (Cambrian to Recent), now recognized as sponges, are important rock formers Silurian through Devonian and locally in the Mesozoic. Typical skeletons (aragonite in Recent) consist of planar, wavy, or even spherical laminae separated by pillars. Megascopic clusters of radial furrows on the surface (astrorhizae) are characteristic, but scarcely discernible in thin section. Archeocyathids are possible sponges that formed 'reefs' in the Early and Middle Cambrian. Typical skeletons are shaped like saucers, bowls, or inverted cones with porous calcareous walls, in most cases double walls, joined by radial partitions.

Cnidarians (Ordovician–Recent)

Corals, octocorals or alcyonarians, and milleporids comprise the calcified cnidarians. Only the corals are important rock formers, although the smooth to roughly warted Mg-calcite spicules (0.1–1 mm) of octocorals can form appreciable carbonate sand. The basic structure of the outer walls and septa of corals is bundles of acicular crystals that radiate from a few up to 90° from a dark central axis (Fig. B14). The needles are aragonite in the scleractinians (Triassic–Recent), but were calcite in the rugosans and tabulates (both Ordovician–Permian). In the other elements that partitioned the living quarters of the Paleozoic corals, the tabulae and dissepiments, the fibrous crystals are perpendicular to the surface. Paleozoic corals include solitary forms, typically conical or cylindrical with radiating septa, and massive colonies of nearly parallel-growing corallites that expand by budding new individuals. Scleractinians are colonial with hemispherical, spherical, branching, or encrusting forms.

Bryozoans (Ordovician–recent)

Bryozoans colonies are typically dense branching cylinders (ramose) or lacy reticulate sheaths (fenestrate), although massive, box-like, and encrusting habits are not uncommon. Wall structure is generally a thin inner layer of granular carbonate overlain by laminated layers that locally pucker toward the exterior into dense rods. The mineralogy in modern forms is calcite, Mg calcite, aragonite, or mixed. The living chambers are sub-spherical in fenestrates and partitioned tubes that bend to an opening in the outer wall in branching forms. Bimodal size of living chambers and hooded openings are characteristic of some bryozoans. Fenestrate colonies in longitudinal cuts appear as a line of isolated beads (30–50 μm) spaced about 1 mm apart. Modern marine bryozoans are most abundant on temperate shelf settings; Paleozoic bryozoans appear to have more tropical associations.

Brachiopods (Early Cambrian–Recent)

Taxonomically useful variety is not wanting in brachiopods, but of all the phyla, they have perhaps the most consistent form, two shells, each bilaterally symmetrical, and wall structure, a very thin outer layer and a thicker inner layer, both commonly of calcite. The inner layer typically consists of fibers oriented about 10° to the shell surface that produce a sweeping extinction band nearly perpendicular to the surface under crossed polars (Fig. B14). The internal structures of the shell, such as teeth and the distinctive loops or spirals that support the aeration apparatus have the same structure. The large order of strophomenids (Ordovician–Triassic) has a laminar inner layer that gives a similar extinction pattern. The calcite composition makes the shells very stable; they are rarely dissolved and have been widely used to provide pristine material for stable isotopic analyzes. The valves typically remain articulated after death, in contrast to the bivalves. Other distinctive features are series of hollow tubes called punctae (10s μm diameter) that extend across the shells in certain suborders. Other suborders have superficially similar rods of calcite, pseudopunctae, that warp the shell laminae toward the interior surface. The occurrence of long, slender spines along the hinge line or in rows along the shell or simply as fragments associated with the shells helps distinguish brachiopods from bivalves. In section, the hollow calcite spines show concentric lamination and a pseudouniaxial cross under crossed polars. Brachiopods were major bioclast contributors throughout the Paleozoic; a few survived the end-Permian extinction and staged a moderate resurgence in the Jurassic, but they have since declined to minor importance.

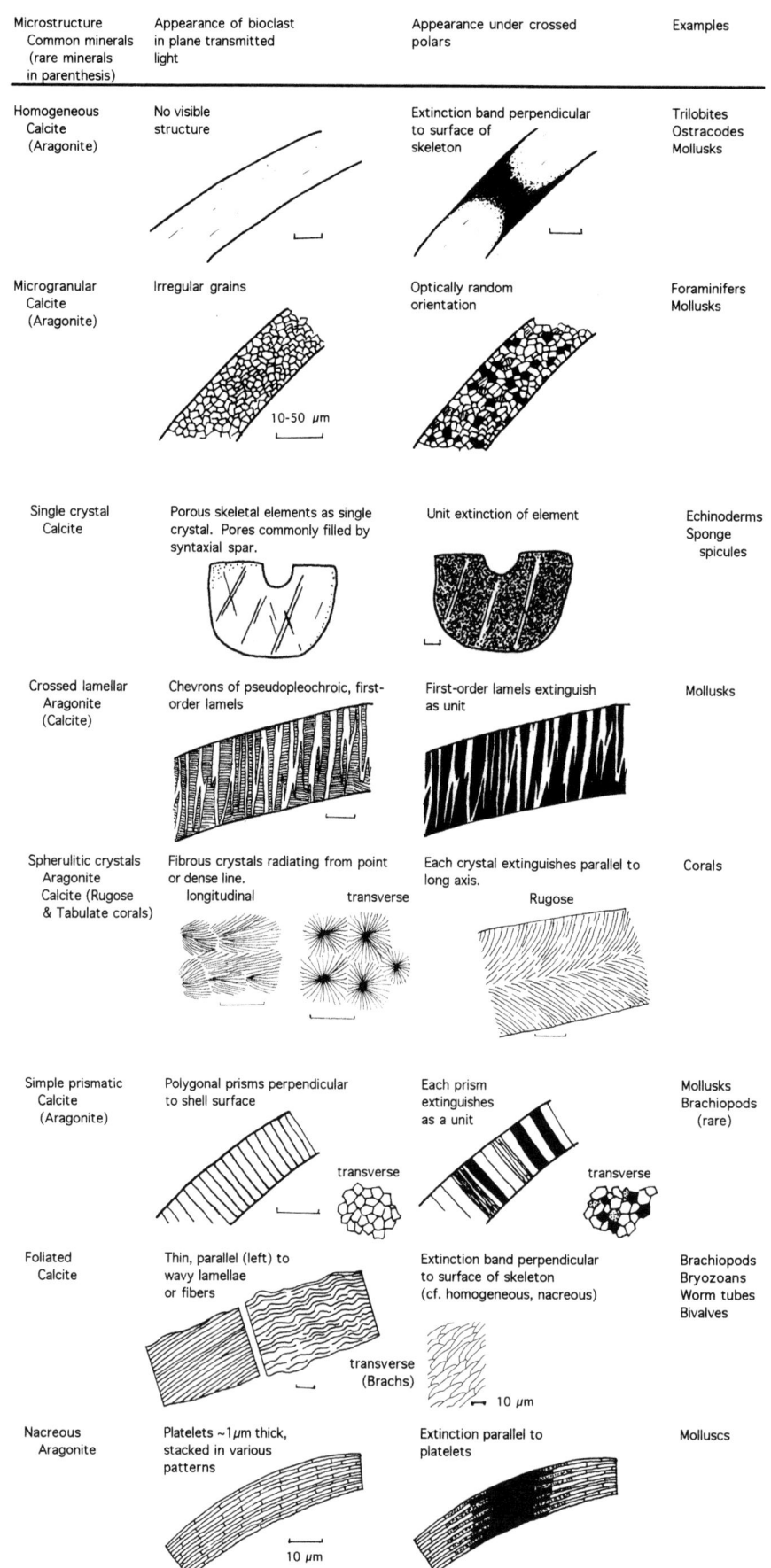

Figure B14 Common types of skeletal microstructure as seen in thin section. Scale bars are 200 μm, except as noted. Modified from Scoffin (1987) and Majewske (1969).

Mollusks (Cambrian–Recent)

Three classes of mollusks are geologically important; gastropods and bivalves as bioclasts of steadily increasing volume and diversity from the Cambrian to the present and the cephalopods as excellent guide fossils, primarily in the Mesozoic. The wall structures of mollusks are diverse; six types (nacreous, prismatic, homogeneous, foliated, cross-lamellar, complex cross-lamellar; (Fig. B14)) were identified in the pioneering work of Bøggild (1930) and some subsequent work has subdivided them extensively. Foliated and some simple prismatic layers are calcite; the others are aragonite. The combinations of layers have some taxonomic value, but preservation of the aragonite layers is commonly poor to nil.

Gastropods are most readily recognized by their spiral shells with large, generally smooth and unsegmented, internal cavities. They have two to six wall layers, commonly three, with most or all composed of aragonite.

Bivalves can be confused with other mollusks in small fragments or with brachiopods if a foliated layer is prominent. However, they generally have two or three shell layers, of which at least one is originally aragonite, so it has a structure not found in brachiopods or is extensively dissolved. Bivalve shells are not bilaterally symmetrical and lack spines or internal structures, except teeth. Exceptional bivalves are the oysters and some pectens, which have shells entirely of calcite, and the rudists, which resembled giant solitary corals and formed reefs in the Tethyan Cretaceous.

Cephalopods were represented primarily by nautiloids (Cambrian–Recent) in the Paleozoic and ammonoids (Devonian-Cretaceous) in the Mesozoic. Both had coiled and straight conical forms, partitioned internally by regularly spaced septa that were perforated by a tube, the siphuncle, running the length of the shell. The septa of nautiloids are simple, slightly curved, and turned back along the siphuncle. In ammonoids the septa are fluted; intensely fluted and turned forward along the siphuncle in advanced forms. The pattern of fluting where it intersects the shell wall and the external ornamentation are primary taxonomic features of ammonoids. The shells are entirely aragonite, except for the small, unarticulated cover of the aperture, the aptychus, which is preferentially preserved in some strata because it is calcite. Belemnite cephalopods (Mississippian–Eocene) are represented by dense, black, torpedo-shaped fragments of an internal 'counterweight' composed of radial calcite crystals.

Annelids

Among the annelid worms, calcified, encrusting serpulids have a fossil record extending back perhaps into the Precambrian. They secrete tight planispiral coils (mm) and long, straight to irregular tubes (mm–cm). Wall structure, which helps distinguish them from vermetid gastropods and scaphopods, is subparallel laminae of microcrystalline carbonate that appear as concentric circles in cross-section, typically giving a pseudouniaxial cross under crossed polars. Mineralogy is aragonite, Mg calcite, or a mixture. Both aragonite and Mg contents increase dramatically with temperature. Serpulids occur in abundance only locally, but as encrusters on soft as well as hard substrates, they can indicate the existence of forms, for example, aquatic grasses, that have scant fossil records.

Arthropods (Cambrian–Recent)

Arthropods, the most abundant and diverse of animals, are represented in the fossil record mainly by trilobites and ostracodes. Trilobites (Early Cambrian–Permian) are the most useful guide fossils in the Early Paleozoic but were rarely major sediment producers. The segmented exoskeleton bends under around the margins to produce the characteristic "shepherds crook" in cross-section. Microcrystalline calcite of the skeleton resembles the homogeneous structure of mollusks; the optic axis is oriented perpendicular to the surface, producing a broad extinction band under crossed polars (Fig. B14). Ostracodes (Cambrian?–Recent) have two valves (0.5–20 mm), typically ovoid in outline, with smooth to highly ornamented surfaces. The valves are essentially mirror images, but one is slightly larger and overlaps the other. The adult valve has a narrow, acute inward projection from the margin. The shell is calcite with 1–10 mole percent $MgCO_3$; the structure appears similar to that of trilobites. Ostracodes can be useful paleoenvironmental indicators as they inhabit marine, brackish, and fresh waters.

Echinoderms (Cambrian–Recent)

Echinoderms are readily recognized by their apparent pentagonal symmetry and by uniform extinction under crossed polars (Fig. B14). Syntaxial overgrowths of cement commonly fill the orderly network of pores (10s µm) in the original Mg calcite skeleton; occasionally staining by organic or iron compounds preserves the outline of the network within the resulting single crystal. Recognition of subgroups within the echinoderms depends on characteristic shapes of segments, such as the short cylinders (mm) with axial tubes of crinoid stems (Ordovician-Recent), prodigious sediment producers in the mid-Paleozoic.

Conclusions

Identification of bioclasts relies on a number of features such as distinctive cross sections, diagnostic structures, or skeletal wall structure. The taxonomic level of identifications varies greatly with the taxa and the preservation, as well as the exigencies of the study. Foraminifer and bryozoan species are routinely determined in thin section, but the most common purpose of study is to draw paleoenvironmental conclusions from the overall biotic assemblage.

Paul Enos

Bibliography

Adams, A.E., and Mackenzie, W.S., 1998. *A Colour Atlas of Carbonate Sediments and Rocks under the Microscope*. London: Manson.

Bathurst, R.G.C., 1975. Carbonate Sediments and their Diagenesis. *Developments in Sedimentology, no. 12, 2nd edn.* Amsterdam: Elsevier.

Bøggild, O.B., 1930. The shell structure of the molluscs. *Kongelige Danske Videnskabernes Selskab, Matematisk-Fysiske Meddelelser*. 9: 231–325.

Flügel, E., 1982. *Microfacies Analysis of Limestones*. Berlin: Springer.

Horowitz, A.S., and Potter, P.E., 1971. *Introductory Petrography of Fossils*. New York: Springer.

Majewske, O.P., 1969. *Recognition of Invertebrate Fossil Fragments in Rocks and Thin Sections*. Leiden: E. J. Brill.

Milliman, J.D., 1974. *Marine Carbonates*. Berlin: Springer.

Scholle, P.A., 1978. *Carbonate Rock Constituents, Textures, Cements, and Porosities*. Tulsa: American Association of Petroleum Geologists. Mem. 27.

Cross-references

Algal and Bacterial Carbonate Sediments
Carbonate Mineralogy and Geochemistry
Foramol-Chlorozoan-Heterozoan-Brymol Assemblages
Neomorphism and Recrystallization
Reefs
Stromatolites

BIOEROSION

The term bioerosion was coined by Neumann (1966) for hard substrate destruction by organisms (Greek, "bios" life; Latin "erodere" erode). It was originally restricted to calcareous rocks but is now being used for biogenic removal of all types of rocks, engineering works, and mineral and organic skeletons. The term should only apply to activities removing grains or fibers, for example, an organism which just pushes its substrate aside does not qualify as a bioeroder.

Classification

Bioeroders leave characteristic traces of their action which may remain visible in the sedimentary record. These traces are named like other trace fossils; their taxonomic treatment is independent of their producers (e.g., a sponge boring is not named "*Cliona*" after its presumed producer but is cited as the trace fossil taxon "*Entobia*"). Bioerosion may be classified according to several overlapping schemes.

- origin of substrate attack: surface (rasping or etching) or interior (boring);
- type of substrate: rock (calcareous or non-calcareous) and carbonate hardparts (lithic); wood (xylic) or bone (osteic);
- size of traces: micro- or macroscale.

The ecology of bioeroders is highly variable: they use their substrate as food (e.g., boring fungi, shipworms, termites or hyenas), as their domicile (e.g., boring sponges or bivalves, some ants), as a brooding place (e.g., woodpeckers, printer beetles), or they remove it in search of food (e.g., sea urchins, parrotfish, moon snails).

Importance

Rasping or boring organisms rapidly recycle substrates and produce fine-grained detritus. Their impact on the global carbon cycle and sediment production largely depends on the nature of the substrate: just a small percentage of bones is scraped off by carnivores, and only a limited array of coastal borers (e.g., some bivalves and sea urchins) can tackle all type of material. Insect wood boring is a prominent factor in forest ecology, and this book probably would appear in Spanish if shipworms had not devastated the Armada sailing against England in 1588. Most significant (in ecological and sedimentological terms) is the action of marine calcareous bioeroders. They provide cavities for secondary dwellers as well as bare space for settlement of larvae, thus enhancing diversity; they produce enormous amounts of lime mud, and they shape the morphology of limestone coasts.

Substrate types

Wood

Wood bioerosion basically is a terrestrial phenomenon. Apart from minute tunnels produced by oribatid mites, the resulting traces are macroborings. The substrate is almost exclusively removed from the inside, independent of the animal group involved (structures in wood made by fungi or microorganisms are not bioerosive — see first paragraph). Several groups of wood borers are able to digest lignin thanks to their intestinal symbionts, others cultivate fungi in their tunnels, yet others excavate holes as nests or as permanent domiciles. Insects are the most prominent and most diverse wood borers (ecologically and taxonomically) but given the paucity of terrestrial sediments their traces rarely enter the fossil record. Driftwood in fluviatile and brackish realms is bored by only a few highly specialized groups, in contrast to the importance of marine borers. Here, rather few genera of the bivalve families, Pholadidae and Teredinidae, account for most of the fossil wood bioerosion. Borings in driftwood or coal seams do not necessarily indicate marine conditions, however, instead they may signal sea-level lowstand (e.g., Bertling and Hermanns, 1996).

Non-calcareous rocks

Bioerosion of crystalline rocks and siliciclastic sediments mainly occurs in the coastal zone. Fungi, snails, sea urchins, shrimps and pholadid bivalves may attack even the hardest substrates. In their search for shelter and/or food, they bore exclusively by mechanical means. Fungi are primary colonizers, mainly found above the low-tide level. Together with snails which rasp off green algae, they have limited bioerosive effect. The other groups require permanent water cover. Sea urchins are most effective, annually removing 1–10 mm of substrate, or up to $2\,kg/m^2 a$ (per square meter per year) (Allouc et al., 1996).

Carbonates

Carbonate bioerosion is characterized by an intricate interaction of rasping and micro- and macroboring, each with different organisms involved (Kiene, 1989). The process is further complicated by penecontemporaneous biogenic encrustation and physical erosion: biogenic loss of substrate may be compensated or aggravated within a few months time. Bioerosion is crucial in carbonate sedimentology, especially in reef environments, reducing the grain size of components (skeletons) to mud and exporting large masses of carbonate from reefs: 30 percent to 40 percent of reef fine sediment is made up of chips produced by boring sponges (see Hutchings, 1986).

Experimentally measured bioerosion rates vary with substrate, type and agents of bioerosion (Trudgill, 1983; Hutchings, 1986). From 0.5 mm to 10 mm of substrate may be eroded annually, corresponding to values of 1 to $20\,kg/m^2 a$. Most efficient are clionid sponges and parrotfish which remove 2–3 t/ha of reef coral each year. This silt-sized material is deposited either in cavities of the reef itself or is exported to the fore-reef and lagoon, respectively.

Carbonate macroborers largely use chemicals such as calcium-complexing proteins (acids have not been identified) in addition to mechanical abrasion. The mostly suspension-feeding animals (bivalves of families Mytilidae and

Gastrochaenidae, various polychaetes and sipunculans, clionid and other sponges, thoracic barnacles, among others) use the substrate for their dwellings. They tend to resemble the producers body outline closely (e.g., Bromley, 1994). The fossil borings hence are highly characteristic, sometimes down to the species level, allowing substantial inferences about the paleoenvironment. Modern macroborers exhibit a depth zonation: bivalves, sipunculan and polychaete worms are frequent and abundant down to the normal-wave base whereas sponges may still occur in bathyal depths (Bromley and Allouc, 1992). This distribution reflects the ecological demands of the borers: they require low sedimentation rates and at least fair water movement and nutrient conditions; in addition, a denser substrate and longer exposure will result in more intense macroboring. Rasping bioeroders as well as encrusters are antagonistic to borers of all types.

Microborings in carbonates are produced by fungi, cyanobacteria, chlorophytes, and rhodophytes. Diameters between 5 and 2000 μ necessitate (electron) microscopic examination. A strong apparative effort is rewarded by excellent depth indicator values of microborings: maximum diversity is above the storm-wave base with chlorophytes and cyanobacteria being the dominant producers. Between 30 m and 100 m depth, chlorophyte borings govern the spectrum, followed by fungi in depths below 150 m (Vogel *et al.*, 2000). The bioerosive action of microborers is strongest in the supratidal zone; honeycombed limestone coasts ("biokarst") and notches are morphological features exclusively due to microborers. Without continuous physical removal of substrate, microborers are most efficient in the first year after substrate availability; later they cannot compete with macroborers.

Rasping bioeroders are less coherent ethologically. Sea urchins, amphineurans and snails graze on filamentous chlorophytes on limestone just as much as on other substrates (see above), removing their substrate accidentally. Various families of reef fish have the same effect on coral skeletons while feeding either on boring green algae (e.g., parrotfish and surgeonfish) or the coral polyps (butterflyfish and puffers). A third group, octopuses and the gastropod families Muricidae and Naticidae, only affects the calcareous shells of molluscs as these predators drill through their victims tests in order to access the soft parts.

Taphonomy

The preservation of bioerosion traces is at least to some extent hindered by their self-destructive nature. As long as no significant sedimentation occurs, successive generations of bioeroders obliterate the structures of their predecessors. This process may be modified by rapid infill and cementation of vacated borings, followed by renewed bioerosion. The original substrate may thus be replaced by micrite filling bored boreholes with the outer shape still intact. Surface raspings and very shallow microborings have the least preservation potential, whereas the deeply penetrating borings of larger animals may still be recognized even after several centimeters of substrate have been eroded. For this reason, the fossil record of bioerosion is strongly skewed toward the borings of bivalves, worms and sponges. Preservation is optimal when the producers are killed by overgrowth or sedimentological events such as sudden burial (obrution) or emersion.

Markus Bertling

Bibliography

Allouc, J., le-Campion-Alsumard, T., and Leung Tack, D., 1996. La bioérosion des substrates magmatiques en milieu littoral: lexemple de la presqîle du Cap Vert (Sénégal occidental). *Geobios*, **29**: 485–502.
Bertling, M., and Hermanns, K., 1996. Autochthone Muschelbohrungen im Neogen des Rheinischen Braunkohlenreviers und ihre sedimentologische Bedeutung. *Zentralblatt für Geologie und Paläontologie, Teil I*, 1995: 33–44.
Bromley, R.G., 1994. The palaeoecology of bioerosion. In Donovan, S.K. (ed.), *The Palaeobiology of Trace Fossils*. Chichester, Wiley: pp. 134–154.
Bromley, R.G., and Allouc, J., 1992. Trace fossils in bathyal hardgrounds. *Ichnos*, **2**: 43–54.
Hutchings, P.A., 1986. Biological destruction of coral reefs. *Coral Reefs*, **4**: 239–252.
Kiene, W.E., 1989. A model of bioerosion on the Great Barrier Reef. *Proceedings of the 6th International Coral Reef Symposium*, **3**: 449–454.
Neumann, A.C., 1966. Observations on coastal erosion in Bermuda and measurements of the boring rate of the sponge Cliona lampa. *Limnology and Oceanography* **11**: 92–108.
Trudgill, S.T., 1983. Measurements of rates of erosion of reefs and reef limestones. In Barnes, D.J. (ed.), *Perspectives on Coral Reefs*. Manuka: Clouston, pp. 256–262.
Vogel, K., Gektidis, M., Golubic, S., Kiene, W.E., and Radtke, G., 2000. Experimental studies on microbial bioerosion at Lee Stocking Island, Bahamas and One Tree Island, Great Barrier Reef, Australia: implications for paleoecological reconstructions. *Lethaia*, **33**: 190–204.

Cross-references

Attrition (Abrasion), Marine
Erosion and Sediment Yield
Neritic Carbonate Depositional Environments
Reefs
Taphonomy: Sedimentological Implications of Fossil Preservation

BIOGENIC SEDIMENTARY STRUCTURES

Biogenic sedimentary structures (trace fossils or ichnofossils) are biologically produced structures that include tracks, trails, burrows, borings, fecal pellets and other traces made by organisms. Excluded are markings that do not reflect a behavioral function. Owing to their nature, trace fossils can be considered as both paleontological and sedimentological entities, thereby bridging the gap between two of the main subdivisions in sedimentary geology. Four major categories are recognized: (1) *bioturbation structures*—they reflect the disruption of biogenic and physical stratification features or sediment fabrics by the activity of an organism: includes tracks, trails, burrows, and similar structures; (2) *biostratification structures*—they consist of stratification imparted by the activity of an organism: biogenic graded bedding, byssal mats, certain stromatolites, and similar structures; (3) *biodepositional structures*—they reflect the production or concentration of sediments by the activity of an organism: includes fecal pellets, pseudofeces, products of bioerosion, and similar structures; and (4) *bioerosion structures*—a biogenic structure excavated mechanically or biochemically by an organism into a ridged substrate: includes borings, gnawings, scrapings, bitings, and related traces.

The contributions of ichnology to sedimentary geology are considerable, including: (1) the production of sediment by boring organisms; (2) the consolidation of sediment by suspension feeders; (3) the alteration of grains by sediment-ingesting organisms; (4) the destruction of sedimentary fabrics and sedimentary structures; (5) the construction of new fabrics and sedimentary structures; (6) the initial history of lithification; (7) the interpretation of depositional environments; (8) the delineation of facies and facies successions; (9) rates of deposition; (10) substrate coherence and stability; (11) aeration of water and sediments; and (12) the amounts of sediment deposited or eroded. Recent summaries dealing with general ichnological principles can be found in Pemberton et al. (1992), Bromley (1996) and Pemberton et al. (2002).

The conceptual framework of ichnology

The importance of ichnology to the fields of stratigraphy, paleontology, and sedimentology stems from the fact that trace fossils display the following characteristics: (1) long temporal range—although a disadvantage in refined biostratigraphy, it greatly facilitates paleoecological comparisons of rocks differing in age; (2) narrow facies range—reflects similar responses by tracemaking organisms to given sets of paleoecological parameters; (3) no secondary displacement—biogenic sedimentary structures, where preserved intact, are closely related to the environment in which they were produced; (4) occurrence in otherwise unfossiliferous rocks—trace fossils are commonly enhanced by the very diagenetic processes that can obliterate body fossils, especially in siliciclastic regimes; (5) creation by soft-bodied biota—many trace fossils are formed by the activities of organisms that generally are not preserved because they lack hard parts; such organisms, in many environments, represent the greatest biomass; (6) a particular structure may be produced by the work of two or more different organisms living together, or in succession, within the structure; (7) the same individual or species of organism may produce different structures corresponding to different behavior patterns; (8) the same individual may produce different structures corresponding to identical behavior but in different substrates (e.g., in sand, in clay, or at sand–clay interfaces); and (9) identical structures may be produced by the activity of systematically different trace-making organisms, where behavior is similar (Ekdale et al., 1984).

Such characteristics make trace fossils very useful in facies analyzes, including reconstruction of individual paleoecological factors, sedimentary dynamics, and the documentation of local and regional facies changes.

Classification of trace fossils

Unique classification schemes have been developed in order to interpret and decipher trace fossils. Historically, the more important classifications have included the preservational, behavioral, and phylogenetic aspects of the tracemaker.

Preservational classifications

Classifications of the stratigraphic arrangements and modes of preservation of trace fossils are both descriptive and interpretive. These preservational concepts may be reduced to two basic facets: (1) toponomy, including modes of occurrence and the mechanical-sedimentological processes of alteration and preservation, and (2) physiochemical (pre- through post-diagenetic) processes of preservation and alteration. Toponomic (or stratinomic) classification schemes have been devised (see Pemberton et al., 1992 for details). Most of these schemes attempt to relate the position of the trace fossil with respect to the main casting medium, which is commonly sandstone or siltstone. Diagenetic preservations have been stressed by numerous authors (e.g., Ekdale et al., 1984). The burrowing activities of benthic organisms can result in significant changes in porosity-permeability patterns that will have a significant effect on later diagenesis (Gingras et al., 1999). Likewise, burrow linings that may be simple secretions of mucus or particulate walls, agglutinated by organic compounds are preferred sites for subsequent mineralization. Therefore, diagenetic processes that obliterate other fossils may even enhance the preservability of trace fossils.

Behavioral classification

Perhaps the single most important facet of ichnology is the behavioral interpretation of trace fossils. Fundamental behavioral (or ethological) patterns are dictated and modified not only by genetic preadaptations, but also by prevailing environmental parameters. Ekdale et al. (1984) recognized seven basic categories of behavior (Figure B15); resting traces (*cubichnia*), locomotion traces (*repichnia*), dwelling traces (*domichnia*), grazing traces (*pascichnia*), feeding burrows (*fodinichnia*), farming systems (*agrichnia*), and escape traces (*fugichnia*). Ekdale (1985) added predation traces (*praedichnia*), and Frey et al. (1987) emphasized the importance of equilibria (*fugichnia*) to all other behavioral patterns. The basic ethological categories evolved early and have generally persisted throughout the Phanerozoic. Although individual tracemakers have evolved, basic benthic behavior has not.

Phylogenetic classification

One of the most frustrating—albeit most fascinating—aspects of ichnology is the attempt to establish the zoological affinities of specific ichnofossils (Figure B16). This results because ichnofossils reflect the behavior of animals, and only to a small extent reflect their anatomy or morphology. The result is that more than one genus or species of ichnofossil may have been constructed by a single species of animal, or conversely different species of animals may have made identical species or genera of trace fossils (Frey and Seilacher, 1980). For example, specimens of *Skolithos linearis* at one locality may show affinities to the phoronids whereas at another locality they may show affinities to onuphid polychaetes. Finally, a nestler or commensal organism in some instances may be better suited for preservation within a burrow than the animal that constructed the burrow. Therefore, each occurrence of a given ichnofossil most be treated on an individual basis; sweeping generalizations on their zoological affinities should be avoided. Notable exceptions include certain types of distinctive, hard substrate borings and those rare instances where the tracemaking organism is preserved within the burrow. In most cases, however, attributing a particular trace to a particular soft-bodied organism depends on a uniformitarian approach and other indirect lines of evidence.

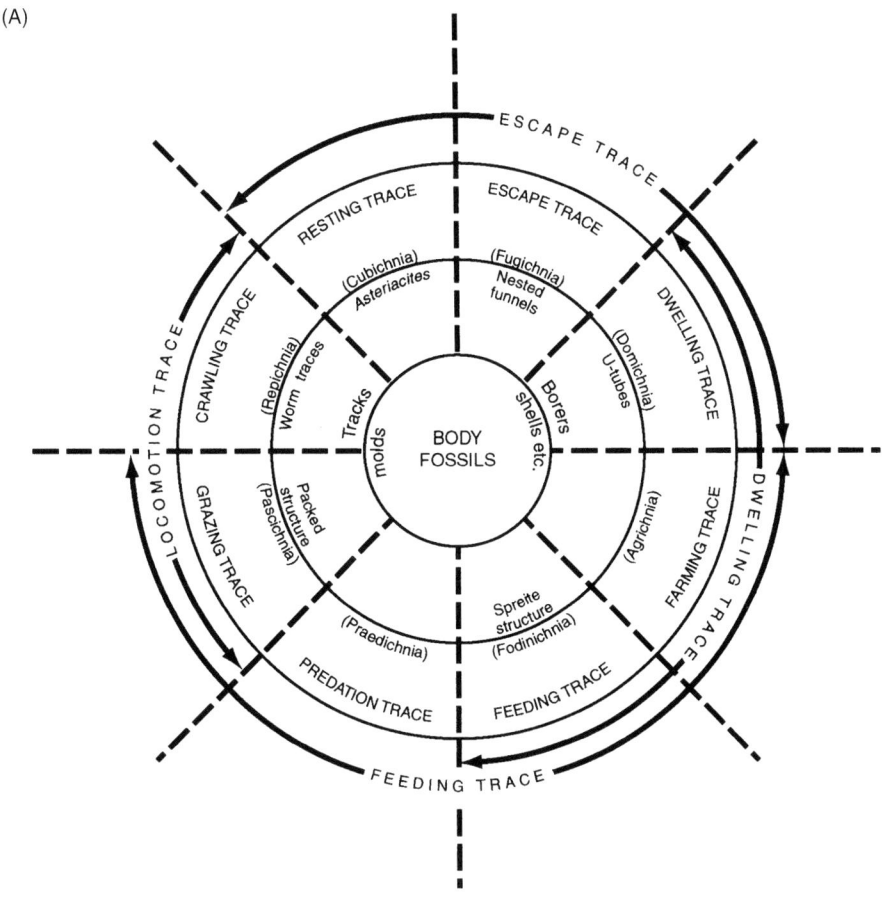
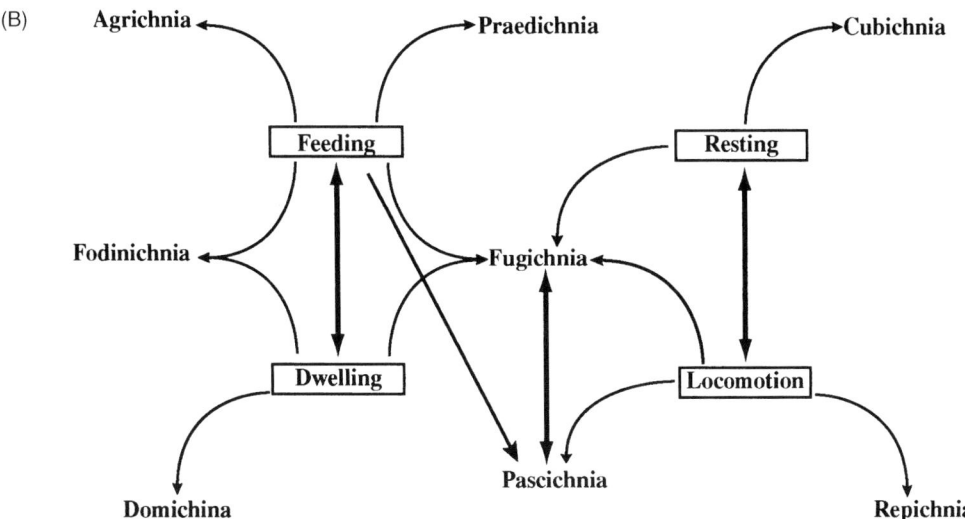

Figure B15 Classification of trace fossils. (A) Ethological classification of trace fossils; (B) Interactions between behavioral categories. (Modified after Pemberton *et al.*, 1992).

Phylogenetic Classification

Although very difficult sometimes it is possible to match the trace making organism to the individual trace fossil. One must approach this with caution but it can assist in paleoecological

Lockeia is thought to represent marks left by the foot of a bivalve

Schaubcylindrichnus is thought to represent the dwelling burrow of a Sabellarid polychaete

Figure B16 Phylogenetic classification of trace fossils is based on the investigation of modern trace-making organisms. (Modified from Pemberton *et al.*, 2002).

The ichnofacies concept

Perhaps the essence of trace fossil research involves the grouping of characteristic ichnofossils into recurring ichnofacies. This concept developed by Adolf Seilacher in the 1950s and 1960s, was based originally on the fact that many of the parameters that control the distribution of tracemakers tend to change progressively with increased water depth (Figure B17). Because of the potential geological value of this bathymetric relationship, the Seilacherian ichnofacies sequence soon came to be regarded almost exclusively (albeit erroneously) as a relative paleobathometer. Today the ichnofacies concept remains valuable in paleoenvironmental reconstruction, but paleobathymetry is only one aspect of the modern ichnofacies concept (Frey *et al.*, 1990).

Ichnofacies are part of the total aspect of the rock and therefore, like lithofacies, are subject to Walthers Law (see *Sedimentologists: Johannes Walther*). For example, isolated bored shells or clasts do not in themselves constitute the *Trypanites* ichnofacies and a bored log is not an example of the *Teredolites* ichnofacies. Rather, there should be some semblance of stratification, lateral continuity, and vertical succession.

Archetypal ichnofacies

Nine recurring ichnofacies have been recognized, each named for a representative ichnogenus: *Scoyenia, Trypanites, Teredolites, Glossifungites, Psilonichnus, Skolithos, Cruziana, Zoophycos,* and *Nereites*. These trace fossil associations (Figure B17) reflect adaptations of tracemaking organisms to numerous environmental factors such as substrate consistency, food supply, hydrodynamic energy, and salinity and oxygen levels (Pemberton *et al.*, 1992). Traces in nonmarine and brackish marine settings are in need of further study; the marine softground ichnofacies (*Psilonichnus, Skolithos, Cruziana, Zoophycos,* and *Nereites*) are distributed according to numerous environmental parameters; traces in the firmground (*Glossifungites*), woodground (*Teredolites*), and hardground (*Trypanites*) ichnofacies are distributed on the basis of substrate type and consistency. Representative occurrences of the various ichnofacies are summarized below. Each may appear in other settings, however, as dictated by characteristic sets of recurrent environmental parameters.

Nonmarine ichnofacies

The use of trace fossils in the interpretation of freshwater deposits is becoming increasingly important. Recent work by Buatois and Mángano (1995), among others, have stressed the abundance and diversity of tracemaking organisms in freshwater environments and emphasized their potential importance in paleoenvironmental reconstruction. Distinct differences in trace fossil types and abundance have been

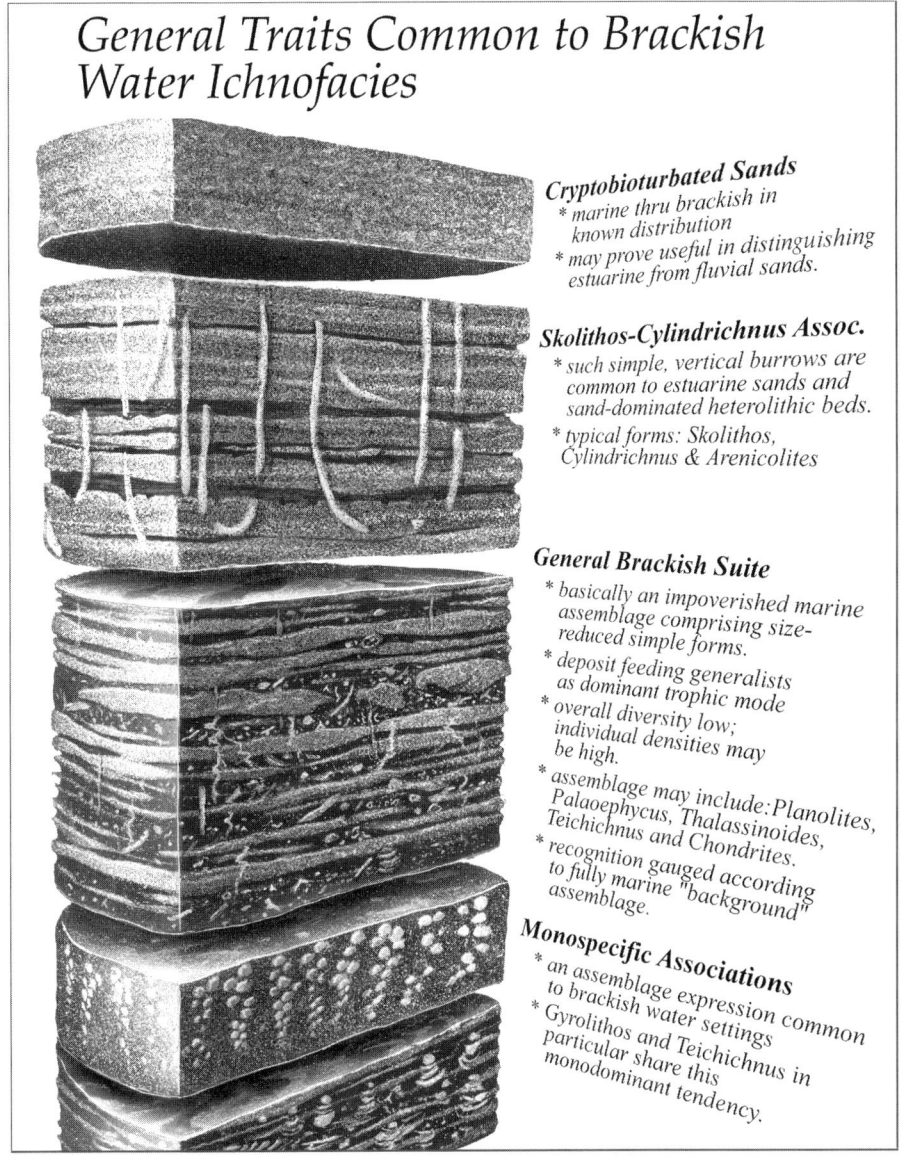

Figure B17 General brackish water trace fossil assemblage. (Modified from Pemberton et al., 2002).

reported from a wide range of freshwater-terrestrial environments, in both ancient and recent settings (Ekdale *et al.*, 1984). At present, nonmarine environments have been characterized by three main ichnofacies: the *Scoyenia*, the *Mermia*, and the *Termitichnus* ichnofacies. Frey *et al.* (1984) concluded that the *Scoyenia* ichnofacies remains a valid concept within appropriate limits and suggests deposition on the shore of ephemeral lakes and the overbank of sluggish rivers. The *Scoyenia* assemblage is characterized by: (1) small horizontal, lined, backfilled feeding burrows; (2) curved to tortuous feeding burrows; (3) sinuous crawling traces; (4) vertical, unlined, cylindrical to irregular shafts; and (5) vertebrate and arthropod tracks and trails. Buatois and Mángano (1995) concluded that the *Mermia* ichnofacies typifies unconsolidated, fine-grained, permanent subaqueous substrates, and well-oxygenated, low-energy lake bottoms. The *Mermia* assemblage is characterized by: (1) dominance of horizontal to sub-horizontal grazing and feeding traces produced by mobile deposit feeders; (2) subordinate occurrence of locomotion traces; (3) generally high to moderate diversity and abundance; and (4) low abundance of specialized grazing structures. Smith *et al.* (1993) named the *Termitichnus* ichnofacies as a distinct terrestrial assemblage characterizing paleosols. The *Termitichnus* ichnofacies is characterized by: (1) vertebrate tracks and trails; (2) low diversity and abundance; (3) dominated by bee, beetle, and termite nests; (4) few feeding structure; and (5) rhizoliths, coprolites, and leaf miners. Prospects for the recognition of additional archetypal nonmarine ichnofacies remain encouraging. For example, Ekdale *et al.* (1984) and Frey and Pemberton (1987) noted that distinct suites of trace fossils characterize eolian dunes, fluvial, paleosol, and lake environments, among others.

Brackish water ichnofacies

Brackish marginal marine environments are widespread in the modern, and are being documented with more frequency from the rock record. The general brackish water trace fossil assemblage (Figure B18) reflects inherently fluctuating environmental parameters (salinity, exposure, sedimentation rates, temperature, and shifting substrates). These factors suggest that the benthic biota of brackish water environments is indeed unique and in contrast to Bromley and Asgaard (1991) the influence of salinity gradients can hardly be divorced from the ichnocoenose concept. For the same reasons that distinct non-marine associations are now being recognized, unique brackish water assemblages are also widespread. In summary, brackish water environments tend to be characterized by: (1) an impoverished marine suite of benthic organisms; (2) trace-making infaunal animals are more abundant than epifaunal animals; (3) soft bodied organisms are more prevalent than shelled organisms; (4) some animals show a distinct size reduction; (5) some animals are bathymetrically displaced; (6) a higher percentage of the assemblage of benthic organisms are trophic generalists whose activities result in morphologically simple burrows; and (7) although diversity is reduced, high individual densities may be attained (Pemberton *et al.*, 1992).

Marine soft ground ichnofacies

The *Psilonichnus* ichnofacies is associated with supralittoral/upper littoral, moderate to low-energy marine and/or eolian conditions typically found in beach to backshore to dune environments. The comments by Bromley and Asgaard (1991) notwithstanding, the *Psilonichnus* ichnofacies was founded on fossil examples (Frey and Pemberton, 1987) and is no more theoretical than any other recurrent ichnofacies; rather, the modern ichnocoenoses were emphasized to show the richness that one might reasonably expect to have existed for various ancient ichnofaunas. Furthermore, one of the major tenets of ichnofacies reconstruction is that the namebearer need not be present in every occurrence of the ichnofacies; thus, just as *Cruziana* is rare in post-Paleozoic occurrences of the *Cruziana* ichnofacies, *Psilonichnus* may well be absent in pre-Mesozoic occurrences of the *Psilonichnus* ichnofacies.

The *Skolithos* ichnofacies (Figure B19) is generally associated with high-energy, sandy, shallow-marine environments. The trace fossils are characterized by: (1) predominantly vertical, cylindrical or U-shaped burrows; (2) few horizontal structures; (3) few structures produced by mobile organisms; (4) low diversity, although individual forms may be abundant; and (5) mostly dwelling burrows constructed by suspension feeders or passive carnivores.

The *Cruziana* ichnofacies (Figure B20) usually is associated with infralittoral/shallow circalittoral marine substrates below minimum wave base and above maximum wave base. The trace fossils are characterized by: (1) high diversity; (2) low individual densities; (3) a mixed association of vertical, inclined, and horizontal structures; (4) presence of structures produced by mobile organisms; (5) they are dominated by feeding and grazing structures constructed by deposit feeders some of which show quite complex behavior patterns resulting in complex trace fossils (Pemberton *et al.*, 1992).

The *Zoophycos* ichnofacies ideally is found in circalittoral to bathyal, quiet-water marine muds or muddy sands, below maximum wave base to fairly deep water in areas free of turbidity flows and subject to oxygen deficiencies. The trace fossils are characterized by: (1) low diversity, though individual traces may be abundant; (2) grazing and feeding structures produced by deposit feeders; and (3) horizontal to gently inclined spreiten structures (i.e., reptitious parellel or concentric burrows or traces). However, the ichnofacies also may occur in restricted intracoastal settings, particularly in Paleozoic sequences; the ichnogenus *Zoophycos* possibly represents greater depths of burrowing in Mesozoic and Cenozoic deposits than in Paleozoic deposits (Pemberton *et al.*, 1992), hence the character of the *Zoophycos* ichnofacies may vary from one part of the stratigraphic column to the next.

The *Nereites* ichnofacies typically is associated with bathyal/abyssal, low-energy, oxygenated marine environments subject to periodic turbidity flows. The trace fossils are characterized by: (1) high diversity but low abundance; (2) complex horizontal grazing traces and patterned feeding/dwelling structures; (3) numerous crawling/grazing traces and sinuous fecal castings; and (4) structures produced by deposit feeders, scavengers, or possibly harvesters (Ekdale *et al.*, 1984). As presently understood, the *Nereites* ichnofacies is restricted primarily to flysch or turbidite sequences; sediments in the great expanses of seafloor beyond influence of turbidity flows consist chiefly of bioturbate textures rather than discrete traces hence there is no well-preserved record of specific ichnocoenoses.

Substrate controlled ichnofacies

The remaining three ichnofacies represent specialized, substrate-controlled tracemakers and, environmentally, are very general in scope. The *Glossifungites* ichnofacies develops in firm but unlithified substrates (i.e., dewatered muds) the *Trypanites* develops in fully cemented substrates, and the *Teredolites* ichnofacies forms in woody substrates. Such horizons may be critical markers in the evolving concept of sequence stratigraphy (see *Substrate-Controlled Ichnofacies*).

Evaluation of the models

These archetypical models, particularly the marine ones, have proven to be valuable indicators of general environmental conditions. For instance, except for a capping layer of planar bedforms, climbing ripples, or antidunes, primary physical sedimentary structures of fluvial point bars may be strikingly like those of estuarine point bars. However, biogenic sedimentary structures are very different in the two settings.

Perhaps the most misunderstood aspect of these recurrent ichnofacies is their use in paleobathymetry. Although some workers have been complacent in this aspect of environmental reconstruction (cf. Frey *et al.*, 1990), various ichnologists have long and persistently emphasized that local sets of environmental factors are most important in controlling the distribution of tracemakers, whether or not these parameters occur at specific water depths

Nevertheless, many kinds of environmental parameters do tend to change progressively with water depth and distance from shore (grain size, energy levels, food resources, etc.), and these gradients affect corresponding changes in the distribution of physical and biogenic sedimentary structures. To

Seilacher's Concept of Recurring Ichnofacies

TRACE FOSSILS ▶ **BEHAVIOR** ▶ **ENVIRONMENT**

Trace fossils are a manifestation of behavior which can be modified by the environment.

ECOLOGICAL CONTROLS

The distribution and behavior of benthic organisms is limited by a number of interrelated ecological controls, including:

1. Sedimentation Rate
2. Substrate Coherence
3. Salinity
4. Oxygen Level
5. Turbidity
6. Light
7. Temperature
8. Water Energy

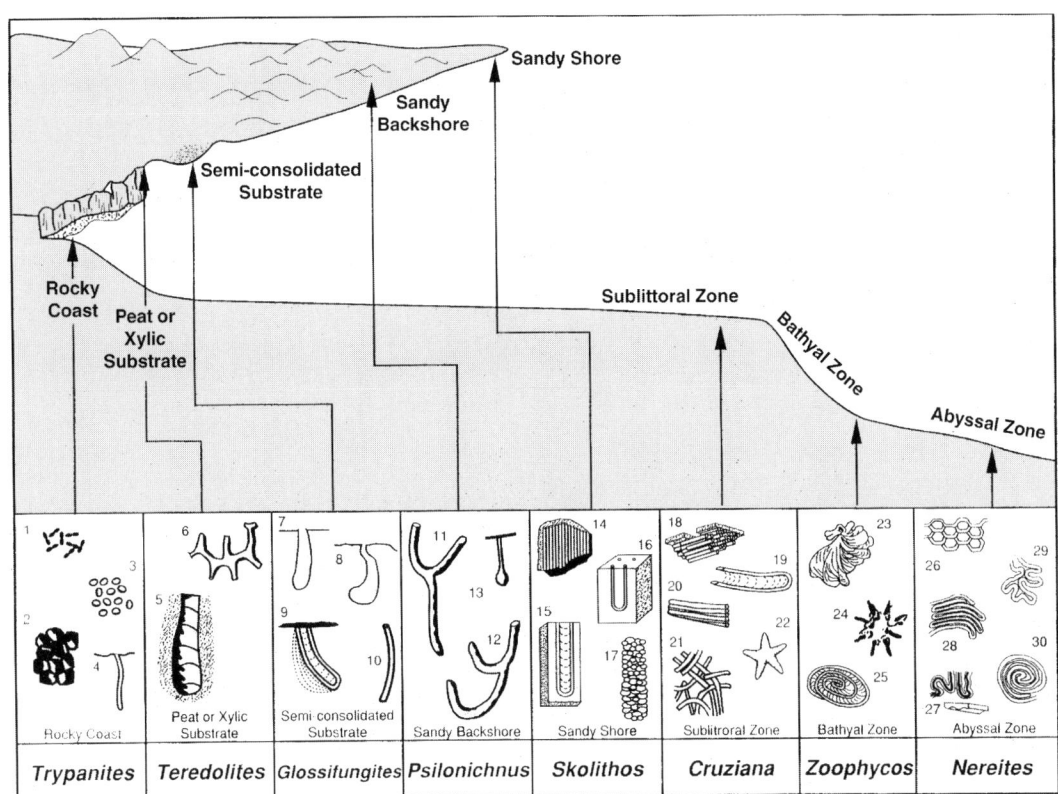

Distribution of Common Marine Ichnofacies

Typical trace fossils include: *1) Caulostrepsis; 2) Entobia; 3) echinoid borings; 4) Trypanites; 5) Teredolites; 6) Thalassinoides; 7, 8) Gastrochaenolites or related genera; 9) Diplocraterion (Glossifungites); 10) Skolithos; 11,12) Psilonichnus; 13) Macanopsis; 14) Skolithos; 15) Diplocraterion; 16) Arenicolites; 17) Ophiomorpha; 18) Phycodes; 19) Rhizocorallium; 20) Teichichnus; 21) Planolites; 22) Asteriacites; 23) Zoophycos; 24) Lorenzinia; 25) Zoophycos; 26) Paleodictyon; 27) Taphrhelminthopsis; 28) Helminthoida; 29) Cosmorhaphe; 30) Spirorhaphe.*

Figure B18 The concept of recurring ichnofacies (Modified after Pemberton *et al.*, 1992).

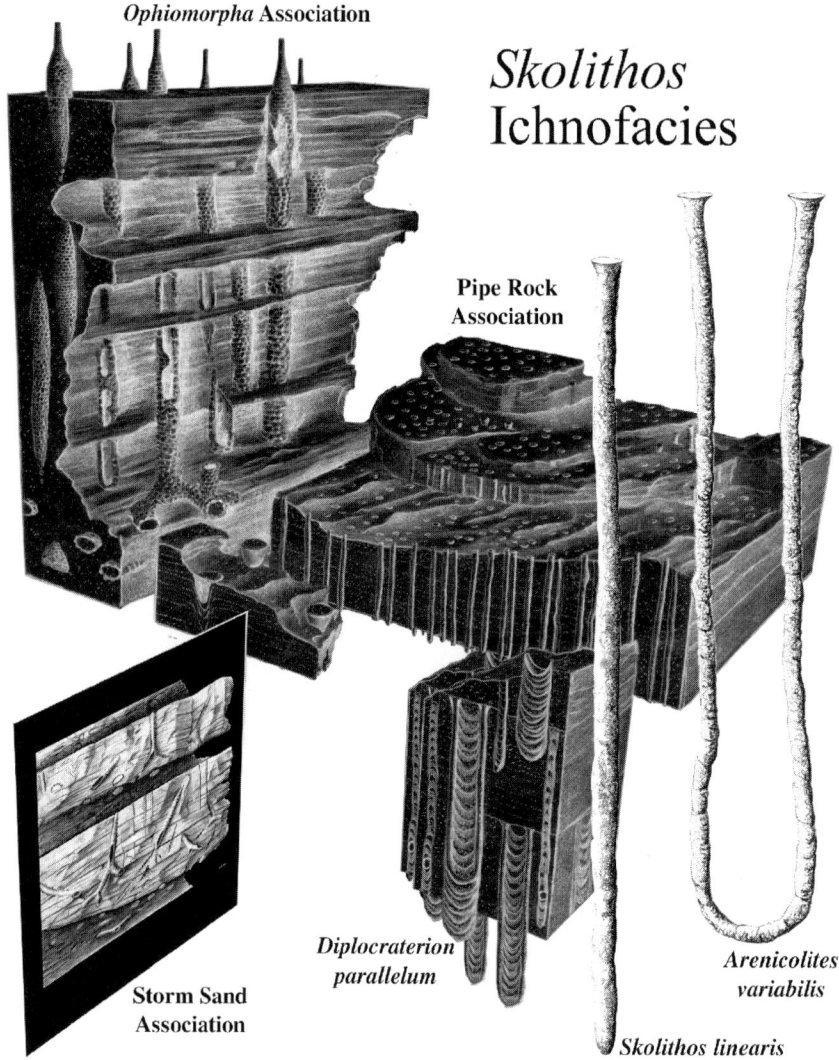

Figure B19 The different expressions of the *Skolithos* ichnofacies (Modified from Pemberton et al., 2002).

that extent, trace fossil associations are indeed useful in paleobathymetry.

Equally important is the long temporal duration of most kinds of trace fossils. These basic benthic behavioral patterns are more nearly like stable ecologic niches than individualistic records of particular animal species (Frey and Seilacher, 1980). As long as the functional niche remains advantageous under given environmental conditions, many different animal species, over long intervals of geologic time, may be expected to exploit it: their preserved traces are strikingly similar and have equivalent significance. Hence, although we conveniently speak of the "*Skolithos* animal" as the architect for a particular kind of dwelling structure, numerous different animal species actually were involved. The longevity of recurrent ichnofacies thereby exceeds the longevity of recurrent biofacies by a considerable margin, and are correspondingly more useful as archetypical models not only for environmental interpretation but also for comparisons of depositional environments of widely differing ages.

The purpose for recognizing these recurrent ichnofacies must not be overlooked. Interpreted in terms of the original trace fossil assemblage, these ichnofacies are merely archetypal facies models with which the local ichnofacies may be compared. The archetypes are intended to supplement, not supplant, local ichnofacies designations, some of which are quite distinctive.

The idealized ichnofacies succession works well in most "normal" situations (Frey and Pemberton, 1987), including distributions according to salinity gradients (cf. Bromley and Asgaard, 1991); yet one should not be surprised to find nearshore assemblages in offshore sediments, and vice versa, for example, if these accumulated under conditions otherwise like those preferred by the tracemaking organisms. The basic consideration rests not with such inanimate backdrops as water depth or distance from shore, or some particular tectonic or physiographic setting, but rather with such innate, dynamic controlling factors as substrate consistency, hydraulic energy, rates of deposition, turbidity, oxygen and salinity levels, toxic

CRUZIANA ICHNOFACIES

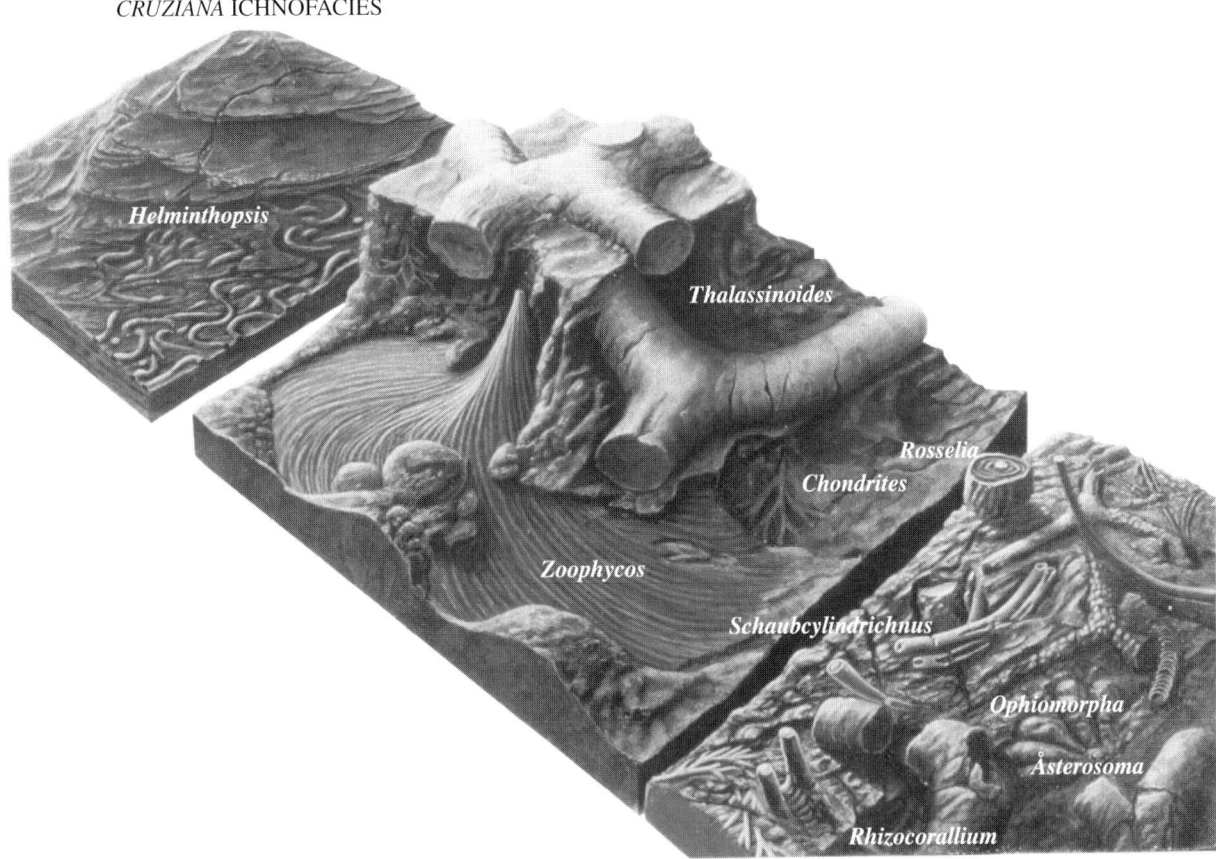

Figure B20 Schematic diagram of a typical Mesozoic example of the *Cruziana* ichnofacies (Modified from Pemberton et al., 2002).

substances, the quality and quantity of available food, and the ecologic or ichnologic prowess of tracemakers themselves.

Finally, the models should not be divorced from associated patterns of bioturbation. Numerous local ichnofacies, particularly those representing low-energy conditions and slow rates of deposition, are set in a complex biogenic fabric. Several generations of burrows may be discernible via their crosscutting relationships, showing that the same volume of sediment passed repeatedly through various styles of reworking. In environmental reconstruction, such ichnologic fabrics may be equally as important as the individual, named trace fossils (Frey and Pemberton, 1984).

Paleoenvironmental significance of trace fossils

The concept of functional morphology, a basic premise employed by ecologists and paleoecologists in environmental reconstruction, is equally applicable to ichnology. In fact, trace fossils are unique in that they represent not only the morphology and ethology of the tracemaking organism but also the physical characteristics of the substrate; they are closely linked to the environmental conditions prevailing at the time of their construction. Frey and Seilacher (1980) re-emphasized that such variables as bathymetry, temperature, volumes of sediment deposited or eroded, aeration of water and sediment, and substrate coherence and stability, have a profound effect on resultant trace fossil distributions and morphology, and hence can be used in the determination of original biological, ethnological, and sedimentological conditions.

The application of ichnology to paleoenvironmental analysis goes far beyond the mere establishment of gross or archetypal ichnofacies. For instance, shallow-water, coastal marine environments comprise a multitude of sedimentological regimes, which are subject to fluctuations in many physical and ecological parameters (see *Coastal Sedimentary Facies*). In order fully to comprehend the depositional history of such zones in the rock record, it is imperative to have some reliable means of differentiating subtle changes in these parameters. Detailed investigations of many of these coastal marine zones in Georgia have shown the value of utilizing biogenic sedimentary structures (in concert with physical sedimentary structures) in delineating them (Frey and Pemberton, 1987). The application of these studies in deciphering paleoenvironments has also proven invaluable. For example, the shoreface consists of a seaward-sloping sediment wedge extending from the low tide mark, generally to the fairweather (minimum) wave base, corresponding to approximately 10–20 m of water depth. The shoreface setting is dominated by wave energy and, as a result of decreasing wave interaction with the substrate in a seaward direction, shows a pronounced basinward fining. The shoreface is classically divided into three subenvironments (from seaward to landward): the lower, the middle and the upper shoreface. The boundaries between these are not always clearly defined (Reinson, 1984). Although the large-scale

BIOGENIC SEDIMENTARY STRUCTURES

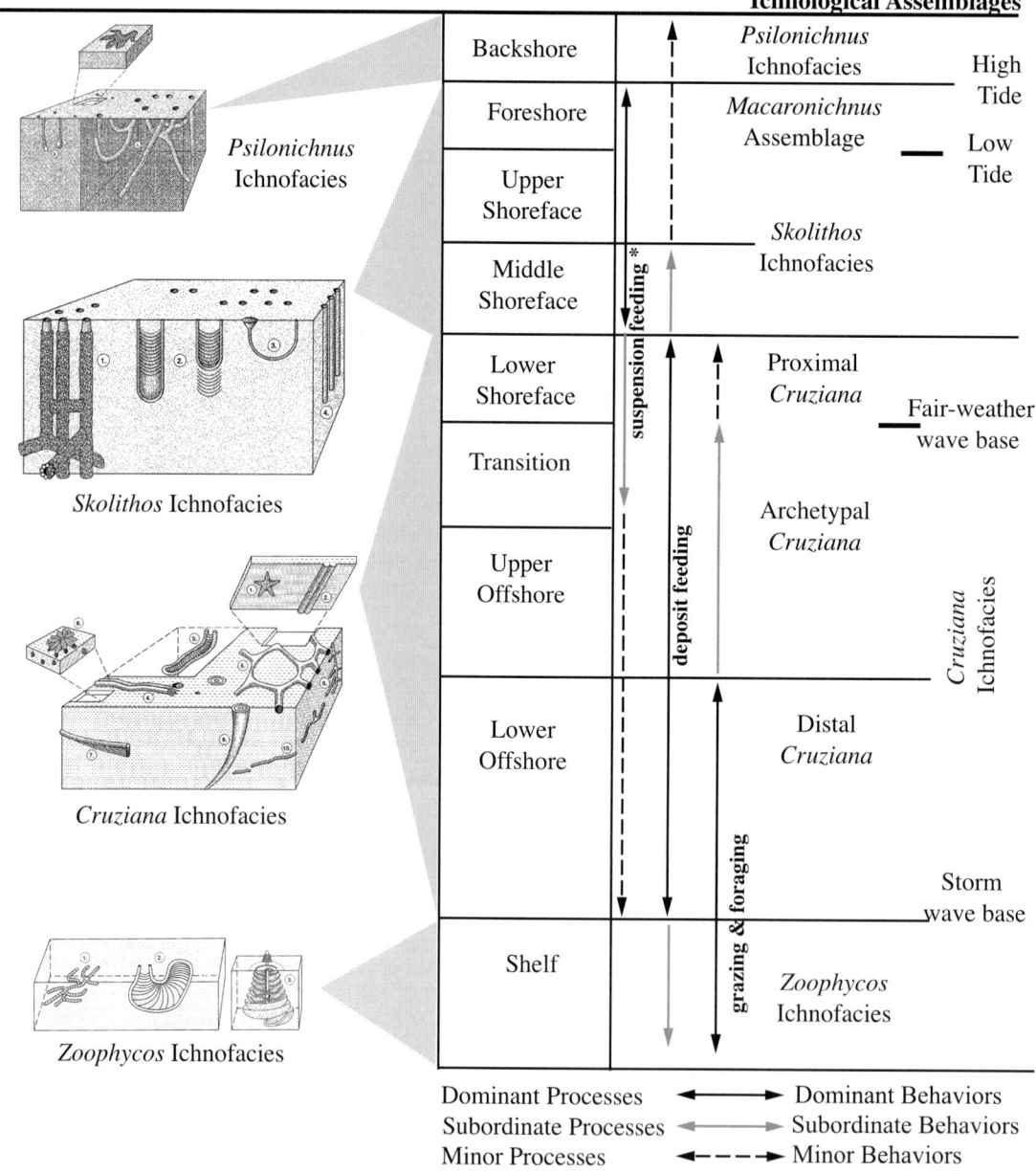

Figure B21 Shoreface model showing the distribution of Ichnological assemblages based on examples from the Cretaceous of the Western Interior of North America (Modified after Pemberton et al., 1992).

regional context may vary, the specific subenvironments of the shoreface are not significantly different whether they occur as part of a strandplain, barrier island or wave/storm-dominated delta (free from interference from active distributary channels). The shoreface grades distally into offshore units and landward into foreshore deposits. A complete shoreface progradational succession reflects offshore to foreshore environments; consequently, these adjacent environments bear inclusion in any discussion of shoreface deposits *per se*. There are several excellent outcrop examples of Cretaceous shoreface deposits in the Western Interior Seaway. The integration of the sedimentology and ichnology within

Figure B22 Schematic diagram of typical recurring trace fossil assemblages in brackish water bay-like environments (Modified from Pemberton et al., 2002).

these deposits affords the opportunity to characterize the facies and facies successions (Figure B21), and to explain the observed facies variability.

Hudson *et al.* (1995) recently stressed that Jurassic benthic molluscan assemblages in the Great Estuarine Group of Great Britain are controlled largely by salinity with different genera or groups of organisms characterizing a spectrum of mainly brackish water environments. As pointed out by Pemberton and Wightman (1992) salinity gradients also influence the nature of the ichnofauna. Although some behavioral patterns may occur on each side of the salinity transition, distinct ichnocoenoses are present in freshwater, brackish water and fully marine environments. Recently, marginal marine environments (including tidal channels, estuaries, bays, shallow lagoons, delta plains, etc.) have been recognized with more frequency in the rock record. Such environments characteristically display steep salinity gradients, which, when combined with corresponding changes in temperature, turbulence, exposure, and oxygen levels, result in a physiologically stressful environment for numerous groups of organisms. Brackish water bay-like environments include bays, sounds, lagoons, inter-distributary bays, and the central basins of wave-dominated estuaries to name just a few (Figure B22).

One of ichnologys greatest strengths, the bridging of sedimentology and paleontology, in some respects can be its greatest liability. Sedimentologists tend to use a strict uniformitarian approach to paleoenvironmental interpretation and rely heavily on modern analogues. Paleontologists, on the other hand, must temper their observations in the light of organic evolution. Although trace fossils can be considered as biogenic sedimentary structures and are difficult to classify phylogenetically, they are constructed by biological entities and are thus subjected at least to some degree to evolutionary trends. For example, occurrences of well-developed terrestrial trace fossil assemblages are much more prevalent in post-Cretaceous rocks. This development corresponds to the evolutionary explosion of the insects, brought on by the diversification of the angiosperms in the Late Cretaceous. Prior to this time terrestrial substrates may not have been as extensively bioturbated due to a paucity of tracemakers. Likewise, patterned grazing traces, which characterize deep-sea environments, show a trend toward more complex organization through most of the Phanerozoic. This trend may be related to the evolution of more efficient foraging strategies (Seilacher, 1986). For these reasons, paleoenvironmental interpretations based on trace fossils must be considered not in strict uniformitarian terms, but rather, in actualistic ones.

Equally important, unique quantitative environmental indicators are indeed rare in the geological record, and ichnology is no exception (Frey and Seilacher, 1980). However, trace fossils can supply a wealth of environmental information that cannot be obtained in any other way and which should not be ignored. Their potential usefulness is accentuated when fully integrated with other (chemical, physical, and biological) lines of evidence. Combined studies of physical and biogenic sedimentary structures constitute a powerful approach to facies analysis.

Summary

Ichnology and its significance to sedimentary geology is based on the following concepts:

(a) Biogenic structures represent the activity of soft-bodied organisms that are not generally preserved. Such organisms (including many entire phyla) are commonly the dominant component of the biomass of many environments.

(b) Biogenic structures are commonly enhanced by diagenesis and can be used in horizons where physical sedimentary structures have been masked. For example, in oil sands deposits bitumen staining obliterates most physical structures, but due to the concentration of clay minerals, it enhances the visibility of biogenic structures.

(c) Biogenic structures can be associated with facies that do not contain any other fossils. In many siliciclastic regimes, diagenesis dissolves most of the shelly fauna and trace fossils represent the only clue as to the original biogenic component of the unit.

(d) Biogenic structures can be used for the paleoecological reconstruction of depositional environments.

(e) Biogenic structures are sensitive to fluctuations in sedimentary dynamics and are important in recognizing event beds and distinct sedimentation patterns.

(f) Biogenic structures are sensitive indicators of substrate coherence and substrate-controlled ichnofacies are emerging as important elements in genetic stratigraphic paradigms.

(g) Biogenic structures ordinarily cannot be transported and therefore represent the original environmental position of the trace making animal.

(h) Biogenic structures are sensitive to changes in certain ecological parameters that are otherwise difficult to ascertain such as salinity and oxygen levels.

(i) An integrated approach utilizing physical, chemical, biological, and ichnological lines of evidence constitutes a powerful tool for facies interpretation.

S. George Pemberton

Bibliography

Bromley, R.G., 1996. *Trace Fossils, Biology and Taphonomy*, 2nd edn. London: Unwin Hyman.

Bromley, R.G., and Asgaard, U., 1991. Ichnofacies: a mixture of taphofacies and biofacies. *Lethaia*, **24**: 153–163.

Buatois, L.A., and Mángano, M.G., 1995. The palaeoenvironmental and palaeoecological significance of the *Mermia* ichnofacies: an archetypal subaqueous non-marine trace fossil assemblage. *Ichnos*, **4**: 151–161.

Ekdale, A.A., 1985. Paleoecology of the marine endobenthos. *Palaeogeography, Palaeoclimatology, Palaeoecology*, **50**: 63–81.

Ekdale, A.A., Bromley, R.G., and Pemberton, S.G., 1984. *Ichnology: The Use of Trace Fossils in Sedimentology and Stratigraphy*. Society of Economic Paleontologists and Mineralogists, Short Course Notes 15.

Frey, R.W., and Pemberton, S.G., 1987. The *Psilonichnus* ichnocoenose and its relationship to adjacent marine and nonmarine ichnocoenoses along the Georgia coast. *Bulletin of Canadian Petroleum Geology*, **35**: 333–357.

Frey, R.W., and Seilacher, A., 1980. Uniformity in marine invertebrate ichnology. *Lethaia*, **13**: 183–207.

Frey, R.W., Pemberton, S.G., and Fagerstrom, J.A., 1984. Morphological, ethological, and environmental significance of the ichnogenera *Scoyenia* and *Ancorichnus*. *Journal of Paleontology*, **58**: 511–528.

Frey, R.W., Pemberton, S.G., and Saunders, T.D.A., 1990. Ichnofacies and bathymetry: a passive relationship. *Journal of Paleontology*, **64**: 155–158.

Gingras, M.K., Pemberton, S.G., Mendoza, C., and Henk, B., 1999. Modeling fluid flow in trace fossils; assessing the anisotropic permeability of *Glossifungites* surfaces. *Petroleum Geosciences*, **5**: 349–357.

Hudson, J.D., Clements, R.G., Riding, J.B., Wakefield, M.I., and Walter, W., 1995. Jurassic paleosalinities and brackish-water communities—a case study. *Palaios*, **10**: 392–407.

Pemberton, S.G., MacEachern, J.A., and Frey, R.W., 1992. Trace fossil facies models: environmental and allostratigraphic significance. In Walker, R.G., and James, N. (eds.), *Facies Models: Response to Sea Level Change*. St. Johns, Newfoundland: Geological Association of Canada, pp. 47–72.

Pemberton, S.G., Spila, M.V., Pulham, A.J., Saunders, T., MacEachern, J.A., Robbins, D., and Sinclair, I., 2001. *Ichnology and Sedimentology of Shallow and Marginal Marine Systems: Ben Nevis and Avalon Reservoirs, Jeanne DArc Basin*. St. Johns, Newfoundland: Geological Association of Canada, Short Course Notes 15: 353 p.

Pemberton, S.G., and Wightman, D.M., 1992. Ichnological characteristics of brackish water deposits. In Pemberton, S.G. (ed.), *Applications of Ichnology to Petroleum Exploration*. Tulsa: Society of Economic Paleontologists and Mineralogists, Core Workshop 17, pp. 141–167.

Reinson, G.E. 1984. Barrier-island and associated strand-plain systems. In Walker, R.G. (ed.), *Facies Models, second edition*. St. Johns, Newfoundland: Geological Association of Canada, pp. 119–140.

Seilacher, A. 1986. Evolution of behavior as expressed in marine trace fossils. In Nitecki, N.W., and Kitchell, J.A. (eds.), *Evolution of Animal Behavior*. New York: Oxford University Press, pp. 62–87.

Smith, R.M.H., Mason, T.R., and Ward, L.F., 1993. Flash flood sediments and ichnofacies of the Late Pleistocene Homeb Silts, Kuiseb River, Namibia. *Sedimentary Geology*, **85**: 579–599.

Cross-references

Amanz Gressly (1814–1865)
Facies Models
Johannes Walther (1860–1937)
Rudolf Richter (1881–1957) and the Senckenberg Laboratory
Substrate-Controlled Ichnofacies
Taphonomy: Sedimentological Implications of Fossil Preservation

BLACK SHALES

Black Shales are fine grained, generally organic carbon-rich sedimentary rocks that primarily consist of a mixture of clay minerals, quartz silt, organic particles (mostly planktonic algae and plant debris), and kerogen. They may also contain variable amounts of disseminated finely crystalline calcite and dolomite, as well as phosphate (commonly as concretions). Most black shales are found in marine sediments (Potter *et al.*, 1980), but they can also form prominent deposits in lacustrine successions (Bohacs *et al.*, 2000). Their black color is due to two constituents: (1) the contained organic matter, and (2) finely disseminated pyrite. The reducing conditions indicated by the latter have long led geologists to believe that ancient black shales required anoxic bottom waters for their formation, and were a typical deposit of the distal, deepest portions of sedimentary basins (via comparisons with the abyssal Black Sea where carbonaceous muds currently accumulate).

The Black Sea is a stratified, silled basin, in which a lower marine water body is overlain by a layer of brackish water due large input of freshwater from rivers. Because of density contrasts a halocline is developed at a depth of approximately 200 m, severely restricting vertical advection of oxygen-rich surface waters. Oxygen demand by decaying and descending organic matter exceeds oxygen replenishment by vertical advection, rendering the water column beneath the halocline anoxic and facilitating the accumulation of organic-rich muds. These observations have suggested to many geologists that organic carbon enrichment in the bottom sediments was basically a result of enhanced preservation (Wignall, 1994).

Although the Black Sea model dominated the study of black shales for many years, it is of limited use when studying ancient black shales. A major difference between the Black Sea and practically all black shales in the rock record is that the substantial depth ($\sim$2000 m) of the former, in conjunction with a stable halocline, greatly restricts circulation, whereas ancient black shales accumulated in much shallower epicontinental seas with a much larger width/depth ratio. Thus, on a geologic timescale at least, storms should have caused frequent mixing of the water column. With continued research into the origin of black shales, many occurrences that were initially interpreted according to the Black Sea model, are now in the process of re-evaluation.

Evolving views of the origin of the Late Devonian Chattanooga Shale of the eastern US may serve as an example for these changing perspectives in black shale deposition. On the basis of well developed fine lamination and apparently little evidence of benthic life, it was initially thought of as a very good representative of a basin with a stratified water column, and enhanced preservation as the primary cause for black shale accumulation. Recent research, however, has uncovered abundant evidence for wave reworking and intermittent erosion (Figure B23) of black shale (Schieber *et al.*, 1998). The observation that erosion surfaces of the type illustrated in Figure B23 can be traced for hundreds of kilometers actually forms the basis for a sequence stratigraphic interpretation of this succession (Schieber, 1998). Careful examination of shale fabrics has also revealed subtle but nonetheless widespread evidence (bioturbation features mostly) for benthic colonization of the seafloor. The high organic carbon contents in the Chattanooga Shale (up to 20 per cent) initially suggest that preservation due to oxygen restriction may have been mainly what controlled its accumulation. Yet, sedimentary and bioturbation features suggest that anoxia, if they indeed occurred, were intermittent, due to frequent water column mixing by storms or seasonal temperature variations. In the case of the Chattanooga Shale, therefore, surface productivity probably was also an important factor for organic carbon enrichment (Schieber *et al.*, 1998).

Although the "deep anoxic basin" model has enjoyed prominence for many years, when we look at black shales more closely we find increasingly that they can actually form in a wide range of depositional settings. Black shales do not constitute a large proportion of the sedimentary rock column, but they are of great economic interest as the main source rock for hydrocarbon production, as well as containing unusually high metal concentrations (Potter *et al.*, 1980). Estimates suggest that more than 90 percent of the world's recoverable oil, and gas reserves were generated from black shales (Klemme and Ulmishek, 1991). Equally significant is the fact that the latter are restricted to six stratigraphic intervals that represent a mere one third of Phanerozoic time (Silurian, Upper Devonian-Tournaisian, Pennsylvanian-Lower Permian, Upper Jurassic, Middle Cretaceous, Oligocene-Miocene).

Because organic carbon results from photosynthesis by plants and algae, each atom of carbon buried implies a molecule of oxygen added to the atmosphere. Carbon burial is

Figure B23 Erosion surfaces in the Chattanooga Shale (entire exposure consists of black shale). Picture shows truncation of black shale packages by erosion surfaces (dashed white lines marked with large white arrows), and illustrates how succeeding black shale packages are draped conformably over these erosion surfaces. Small white arrows point out truncation of shale beds by erosion surface. In places erosion has removed in excess of 1 m of black shale between successive black shale packages (at a total Chattanooga thickness of less than 10 m).

linked to other global biogeochemical cycles and is a critical variable in our attempts to understand the history and evolution of the oceans and atmosphere, as well as global climate change. Over geologic time, carbon burial probably was responsible for a gradual rise in atmospheric oxygen levels. Also, by reducing the greenhouse effect it can lead to lower global temperatures and even ice ages (Berner, 1997). In that context, considering that the geological record is punctuated by global episodes of widespread black shale formation (Klemme and Ulmishek, 1991), understanding what combinations of variables are required to produce a "black shale world" is of considerable significance.

Three factors control organic matter accumulation in sediments: (1) organic matter input; (2) organic matter decay; and (3) dilution of organic matter by other ingredients. Factor 1, organic matter input, has two components, detrital organic matter washed in by rivers and organic matter from primary production in the water column. The latter is strongly dependent on nutrient availability. Factor 2 relates to breakdown of organic matter, mostly by microbes. Its efficiency and the types of bacteria involved depend on oxygen availability in the water column and the sediment. Factor 3, dilution, has a terrigenous component (fluvial and eolian contributions), and an intrabasinal component of skeletal remains (e.g., coccoliths, radiolaria, foraminifera, diatoms, etc.). The fluvial component is linked to climate and relief in the source region, and to sea level fluctuations. Because the magnitude of this component rises and falls with continental runoff, it is directly linked to the nutrient input that impacts primary production. Relief and sediment supply are linked to tectonism, which in addition can affect subsidence rates, basin geometry, and water depth (Schieber et al., 1998). Because each factor is a function of several other variables that are in part cross-linked to other factors, determining the magnitude of each of these factors in the rock record is far from trivial. Obviously, black shale formation is a multifaceted problem and depending on the interaction of the variables, there are multiple scenarios that are conducive to black shale formation (Tyson and Pearson, 1991). Microbial mats can also promote accumulation, and preservation of organic-rich sediments in a wide range of environments (Schieber, 1999), a currently under appreciated fact (see *Microbially Induced Sedimentary Structures*).

Black shales are not only common in thick, basinal, successions, but also occur as thin tongues within basin-margin successions. The latter are thought to owe their existence to high stands of sea level when terrigenous input was at a minimum, and as such may mark maximum flooding surfaces (Creaney and Passey, 1993). Whether these shallow water black shale tongues reflect deposition beneath a nearshore zone of oxygen restriction, or simply are the outer edge of an expanding anoxic water body that usually only occupies the basin center (Wignall, 1994), is still being debated. Nearshore oxygen restriction could, for example, be produced by salinity stratification due to freshwater plumes, or by high nearshore productivity fueled by nutrients in continental runoff. Alternatively, rapid sea-level rise could have caused expansion of central anoxic water bodies into shallow nearshore waters. Recent efforts to integrate black shales into a sequence stratigraphic context suggest that at the onset of transgression, black shale deposition may commence in the basin interior, that nearshore black shale deposition is possible early in transgressions, and that black shales related to maximum flooding can form during high stands of sea level (Wignall, 1994; Schieber et al., 1998). Sea level variations may also have lead to conditions with recurring water column mixing events and recycling of key nutrients (N, P) to surface waters, making possible a "productivity-anoxia feedback" mechanism to maintain high rates of carbon burial (Ingall et al., 1993).

Black shales are the end product of the complex interplay among a range of geologic variables and processes, and there are no easy answers and no "one size fits all" models. The seemingly drab and uniform nature of many of these rocks poses a considerable challenge and requires us to extract information by all means possible. Sedimentological study includes recognition and tracing of facies changes, sedimentary features, shale fabrics, erosion surfaces and internal stratigraphy, as well as information that can be derived from interbedded non-shale lithologies. Paleontological studies may provide data on paleo-oxygenation, primary production, substrate conditions, paleocurrents, bathymetry, and paleosalinity (Schieber et al., 1998). Petrographic investigations provide the basic inventory of shale constituents, and may yield clues about provenance, compaction history, sedimentary processes (via small scale sedimentary structures), the origin and maturation of organic matter (Taylor et al., 1998), and even paleo-oxygenation (via analysis of pyrite framboid size distribution; Wilkin et al., 1997). Chemical analysis of major and trace elements can provide information on provenance (Roser and Korsch, 1986), and a range of proxies for paleo-oxygenation (Jones and Manning, 1994). Organic geochemistry can furnish information on the source of organic matter, its dispersal, and maturation (Engel and Macko, 1993). Certain organic molecules, such as photosynthetic pigments, may survive as "carbon skeletons" (biomarkers), and may identify the original source of diagenetically altered organic matter and potential indicators of water column anoxia (Brassell, 1992).

Summary

Black shales do not give up their secrets easily. The tools are at hand, however, to extract a wide range of data from them and allow for multiple avenues of inquiry. Past experience shows that interpretation from just one perspective (e.g., petrography, fabric study, trace element geochemistry, or organic geochemistry) leads to conclusions that often are in conflict with those coming from a different line of inquiry. Black shales need to be investigated in a multidisciplinary way and at multiple scales because of the complex interplay of variables that produces them. Conclusions derived from microscopic features must be in agreement with insights coming out of basin scale studies, as well as with findings from all scales in between. Once this is accomplished, patterns are likely to emerge that will not only tell us the critical factors for the formation of a specific example of a black shale, but also reveal the fundamental conditions that make a "black shale world" tick.

Juergen Schieber

Bibliography

Berner, R.A., 1997. The rise of plants and their effect on weathering and atmospheric CO_2. *Science*, **276**: 544–546.
Bohacs, K.M., Carroll, A.R., Neal, J.E., and Mankiewicz, P.J., 2000. Lake-basin type, source potential, and hydrocarbon character: an integrated sequence-stratigraphic-geochemical framework. In Gierlowski-Kordesch, E.H., and Kelts, K.R. (eds.), *Lake Basins through Space and Time*. AAPG Studies in Geology, Volume 46, pp. 3–34.
Brassell, S.C., 1992. Biomarkers in sediments, sedimentary rocks and petroleums. In Pratt, L.M., Brassell, S.C., and Comer, J.B. (eds.), *Geochemistry of Organic Matter in Sediments and Sedimentary Rocks*. SEPM Short Course Notes 27, pp. 29–72.
Creaney, S., and Passey, Q.R., 1993. Recurring patterns of total organic carbon and source rock quality within a sequence stratigraphic framework. *AAPG Bulletin*, **77**: 386–401.
Engel, M.H., and Macko, S.A., 1993. *Organic Geochemistry: Principles and Applications*. New York: Plenum Press.
Ingall, E.D., Bustin, R.M., and Van Capellen, P., 1993. Influence of water column anoxia on the burial and preservation of carbon and phosphorous in marine shales. *Geochimica et Cosmochimica Acta*, **57**: 303–316.
Jones, B., and Manning, D.A.C., 1994. Comparisons of geochemical indices used for the interpretation of paleoredox conditions in ancient mudstones. *Chemical Geology*, **111**: 111–129.
Klemme, H.D., and Ulmishek, G.F., 1991. Effective petroleum source rocks of the world: stratigraphic distribution and controlling depositional factors. *AAPG Bulletin*, **75**: 1809–1851.
Potter, P.E., Maynard, J.B. and Pryor, W.A., 1980. *Sedimentology of Shale*. New York: Springer Verlag.
Roser, B.P., and Korsch, R.J., 1986. Determination of tectonic setting of sandstone-mudstone suites using SiO_2 content and K_2O/Na_2O ratio. *Journal of Geology*, **94**: 635–650.
Schieber, J., 1998. Developing a Sequence Stratigraphic Framework for the Late Devonian Chattanooga Shale of the southeastern US: Relevance for the Bakken Shale. In Christopher, J.E., Gilboy, C.F., Paterson, D.F., and Bend, S.L. (eds.), *Eighth International Williston Basin Symposium*. Saskatchewan Geological Society, Special Publication No. 13, pp. 58–68.
Schieber, J., 1999. Microbial mats in terrigenous clastics: the challenge of identification in the rock record. *Palaios*, **14**: 3–12.
Schieber, J, Zimmerle, W., and Sethi, P. (eds.), 1998. *Shales and Mudstones*. (Volume 1 and 2): Stuttgart: Schweizerbartsche Verlagsbuchhandlung.
Taylor, G.H., Teichmüller, M., Davis, A., Diessel, C.F.K., Littke,R., and Robert, P., 1998. *Organic Petrology*. Stuttgart: Borntraeger.
Tyson, R.V., and Pearson, T.H. (eds.), 1991. *Modern and Ancient Continental Shelf Anoxia*. Geological Society of London, Special Publication 58.
Wignall, P.B., 1994. *Black Shales*. Oxford: Oxford University Press.
Wilkin, R.T., Arthur, M.A., and Dean, W.E., 1997. History of water-column anoxia in the Black Sea indicated by pyrite framboid size distributions. *Earth and Planetary Science Letters*, **148**: 517–525.

Cross-references

Depositional Fabric of Mudstones
Hydrocarbons in Sediments
Mudrocks
Oceanic Sediments
Sulfide Minerals in Sediments

BRAIDED CHANNELS

Characteristics and definitions

Braided channels are a distinctive alluvial river morphology characterized by multiple, inter-woven branches separated by ephemeral braid bars (Bridge, 1993, Figure B24). Braided channels are also distinguished by the bewildering rapidity and complexity of the plan-form changes, channel migration and bar development processes during periods of high flow and intense bed load transport. There are several different descriptive criteria for defining braided channels (Bridge, 1993) but no strict quantitative criteria. Braided and anastamosed were originally synonymous but they are now regarded as distinctly different types of river morphology. Braided channels form in non-cohesive sand and gravel and are often associated with glaciated mountain regions but also occur on alluvial fans, interior and coastal plains, and in a variety of climatic regions.

The braided appearance varies with flow stage but even at high flow, when many bars are partially submerged, flow and bed-load transport in a braided channels remains divided

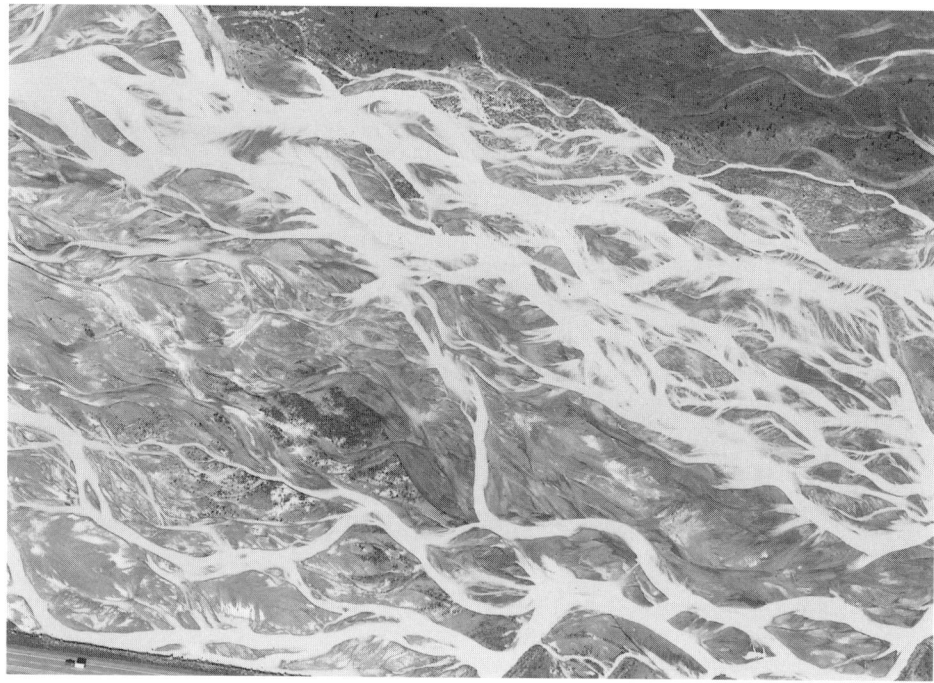

Figure B24 A reach of the braided gravel-bed Sunwapta River, Canada. Flow is left to right and length of channel shown is about 500 m. National Air Photo Library (Canada) A31610-32.

among several branches. Thus, the appearance of braiding caused only by local erosion of minor channels on bar surfaces during falling flow stage is excluded from the definition. The degree of braiding (braiding intensity or braid index) is measured either by the average number of branches across the river, or the total length of all branches relative to the length of the river segment (Bridge, 1993).

Causes and development of braiding

Braiding occurs spontaneously in unconfined flow over mobile, non-cohesive sediment (Paola, 2001). It occurs in rivers in which stream power is high relative to the size of sediment in the river (van den Berg, 1995). The transition to braiding occurs at much lower stream power in fine-grained sediment (sand) than in gravel or cobbles. High stream power (steep gradients and/or high discharge) contributes to development of a large width/depth ratio which is identified, theoretically, as a fundamental control on the occurrence of braiding and this, in turn, may be a response to a large sediment supply rate relative to the discharge (causing steeper gradient) and/or low bank resistance. Bank resistance due to cohesive bank material or vegetation cover inhibits braiding (Ferguson, 1987). Braiding can occur at a constant discharge so that variability of flow is not a primary factor, except that occasional large floods may cause a transient braided morphology, especially if the river is already close to a braided state. Aggradation is not necessary for braiding but short-term increases in sediment input may cause braiding or an increase in degree of braiding.

Beginning from a straight channel, braiding develops from the formation of single (alternating) bars, single mid-channel bars or multiple (laterally), mid-channel bars. These are low-amplitude bedforms, with a linguoid or diagonal crest, sometimes having downstream avalanche faces. Bars occupy most of the channel width, have a length several times the channel width and are connected to upstream scour pools. The bars cause development of channel sinuosity and braiding follows by either cut-offs of single-row bars or by bifurcation around mid-channel bars (Bridge, 1993, Figure B24). The braided pattern is maintained by repetition of these processes in the individual channels of the braided network because of the inherent instability of individual channels, the large stream-wise and temporal variations in transport rate and the frequent re-distribution of the river discharge among the individual channels. New channels may also form by avulsion of flow between branches and re-occupation of abandoned branches.

Morphology and bed material transport processes

Braided channel morphology follows similar functional relations to that of single channel streams. The braided channel is a network of branches, with links of varying length and nodes formed by channel confluences and bifurcations. The average distance between nodes (and, therefore, the length of braid bars), scales with the discharge of the whole channel, in much the same way as the wavelength of meanders in single channels (Ashmore, 2001). This length scale is related to the bars and bends formed within the individual branches, which often closely resemble low-sinuosity meanders. The cross-section dimensions of branches follow hydraulic geometry relations very similar to those of single channel streams. In addition, the total width of braided channels (the sum of individual branch widths) varies in proportion to the square root of the discharge of the whole channel, as it does for single channel streams. Finally, the braiding intensity tends to increase with increasing total stream power and with finer bed-material grain size.

After initiation of braiding, braid bars grow and develop by episodic deposition associated with migration of the adjacent channels and the downstream migration or progradation of submerged bars or gravel sheets. In simple cases the process resembles the development of back-to-back point bars in a single channel, although the two sides of the bar need not develop simultaneously. However, in most cases, bars are built in complex sequences involving both progressive deposition and erosional modification. Systematic gradual accretion related to steady migration in one direction is largely absent because of the rapid and erratic shifts in channel configuration that characterize braided channels. Bed scour occurs in bends of individual branches but is greatest at confluence zones, which often have bars deposited immediately downstream. The orientation of confluence zones may change rapidly in response to migration of the upstream channels and changes in their relative discharge and sediment load, with the result that the confluences act as control points for the downstream distribution of sediment, sending it in different directions as they change orientation.

The rapid, unstable changes in channel morphology cause large spatial and temporal variations (at least 10 fold) in the total bed-load transport rate in braided rivers at a given (constant) total discharge (Shvidchenko and Kopaliani, 1998). Average total bed-load transport rate for braided rivers correlates well with variation in total stream power or discharge, but the local and short-term rates are extremely variable so that transport rate appears to be controlled more by local triggering of abrupt channel changes than by hydraulic conditions. Although braided channels have multiple branches, it is likely that bed-load transport at any particular time is restricted to only two or three branches, even at high flow stages.

Prospect

While the main elements of the morphology and dynamics of braided channels have been described, quantitative understanding of the morpho-dynamics is still rudimentary (Ashmore, 2001), in part because of the difficulties of obtaining field data from these vigorous rivers. However, the combination of improved field technology, continued use of physical models, recent theoretical advances and the development of numerical models that can reproduce basic elements of the known characteristics of these rivers (Paola, 2001), promises to stimulate rapid advances in understanding braided river morpho-dynamics and therefore of the formation and characteristics of braided river sediments.

Peter Ashmore

Bibliography

Ashmore, P., 2001. Braiding phenomena: statics and kinetics. In Mosley, M.P. (ed.), *Gravel-Bed Rivers V*. Christchurch: New Zealand Hydrological Society, pp. 95–114.
Bridge, J.S., 1993. The interaction between channel geometry, water flow, sediment transport and deposition in braided rivers. In Best, J.L., and Bristow, C. (eds.), *Braided Rivers: Form, Process and Economic Applications*. London: Geological Society, Special Publication 75, pp.13–71.
Ferguson, R.I., 1987. Hydraulic and sedimentary controls of channel pattern. In Richards, K. (ed.), *River Channels: Environment and Process*. Oxford: Blackwell, Institute of British Geographers, Special Publication, 17, pp. 129–158.
Paola, C., 2001. Modelling stream braiding over a range of scales. In Mosley, M.P. (ed.), *Gravel-Bed Rivers V*. Christchurch: New Zealand Hydrological Society, pp. 11–38.
Shvidchenko, A.B., and Kopaliani, Z.D., 1998. Hydraulic modeling of bed load transport in gravel-bed Laba River. *Journal of Hydraulic Engineering*, **124**: 778–785.
Van den Berg, J.H., 1995. Prediction of alluvial channel pattern of perennial rivers. *Geomorphology*, **12**: 259–279.

Cross-references

Anabranching Rivers
Avulsion
Facies Models
Floodplain Sediments
Meandering Channels
Numerical Models and Simulation of Sediment Transport and Deposition
Rivers and Alluvial Fans
Sediment Transport by Unidirectional Water Flows
Surface Forms

C

CALCITE COMPENSATION DEPTH

The calcite compensation depth (calcite compensation depth), a term coined by Bramlette (1961), is the depth in the oceans at which the rate of calcium carbonate accumulation equals its rate of dissolution. Dissolution of carbonate being supplied from the surface waters increases progressively with depth, thus the calcite compensation depth is the level at which the net accumulation of calcite is zero. The calcite compensation depth is often compared to the snow line along a mountain front where the supply of snow is exactly balanced by melting, leaving the elevations above draped with snow. By analogy, the seafloor and bathymetric highs in the oceans above the calcite compensation depth are draped in carbonate sediment.

The discovery that carbonate is largely absent in the oceans below about 4,500 m as a result of selective dissolution of carbonate goes back to the *HMS Challenger* expedition of the 1870s. Today, it is well established that the depth of the calcite compensation depth varies throughout the modern oceans, being in general deeper in the Atlantic than the Pacific. Similarly, the North Atlantic is more carbonate-rich than the South Atlantic, and there is variation among the sub-basins of the latter, this being a function of the circulation patterns within the oceans (Berger, 1970).

The ultimate sources of calcium carbonate to the oceans are riverine input from the weathering of terrestrial rocks and contributions from hydrothermal vents along the mid-ocean ridges. Carbonate is extracted from seawater biochemically by calcite- or aragonite-secreting benthic, nektonic, and planktonic organisms that form tests or skeletons, mainly within the photic zone. Beyond the shallow-water carbonate banks and atolls, which are dominated by megafaunal organisms, the tests of micro- and nannoplankton account for most of the carbonate "snow" supplied to the seafloor. Much of this material is packaged and transported to the bottom as fecal pellets produced by grazing organisms such as copepods (Honjo, 1976), where it accumulates above the calcite compensation depth as nannofossil, foraminiferal, and more rarely, aragonitic pteropod ooze. Geographically, the rate of supply of pelagic carbonate may vary across the oceans by a factor of 10, depending on the biological productivity. Productivity is generally highest along the margins of oceanic gyres where upwelling is induced, as along the equatorial divergence. Conversely it is lowest in the gyre centers. Hence the calcite compensation depth is deepest along the equatorial Pacific and along some coastal upwelling zones.

Dissolution of carbonate at depth is a function of the corrosiveness of the water, which increases with depth along with decreasing carbonate-ion content, decreasing temperature, increasing hydrostatic pressure and increasing partial pressure of CO_2. Carbon dioxide is produced by the respiration of organisms and the combustion of organic matter, and tends to increase toward the ocean floor. Field experiments have verified laboratory and theoretical predictions that the oceans should be undersaturated with respect to calcium carbonate beneath the upper few hundred meters (Figure C1). This relationship was quantified during experiments where calcite spheres and later foraminiferal and nannofossil assemblages were suspended at intervals along deep-moored buoys in the oceans. Calcite loss indicated that a marked increase in dissolution occurred at around 3,700–4,000 m, a level named the "lysocline" (Berger, 1967), below which rapid dissolution ensues until the calcite compensation depth is reached. The lysocline may coincide with water mass boundaries, such as the top of the Antarctic Bottom Water in the South Atlantic.

Further experiments and observations, particularly those from the first four cruises of the Deep Sea Drilling Project, showed that planktonic foraminifers are more susceptible to dissolution than calcareous nannofossils, probably because of a lower magnesium content in the calcite lattice of the latter. More dissolution susceptible still are the aragonitic pteropod shells. Hence progressively deeper compensation depths and lysoclines could be measured for these various groups. Furthermore, it was conclusively demonstrated that the levels of these calcite compensation depths have varied throughout Mesozoic and Cenozoic time. Studies of these variations (e.g., Ramsay, 1974) have laid the groundwork for calculations of global carbonate sediment budgets over geologic time. Such

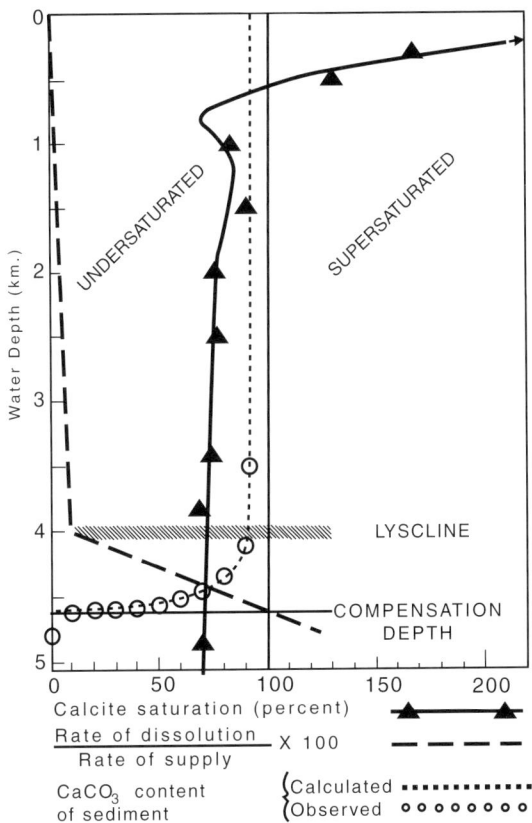

Figure C1 Characteristics of the oceanic water column that affect the dissolution of calcium and the level of the CCD (from van Andel et al., 1975, figure 24).

studies must take into account the partitioning of carbonate deposition between epiric seas and continental margins, carbonate platforms and atolls, and the deep sea, all of which affect the level of the calcite compensation depth. The refinement and extension of these kinds of studies is a fruitful area for future research.

Sherwood Wise

Bibliography

van Andel, T.H., Heath, G.R., and Moore, T.C., 1975. *Cenozoic History and Paleoceanography of the Central Equatorial Pacific Ocean*. Geological Society of America Memoir, 143.
Berger, W.H., 1967. Foraminiferal ooze: solution at depths. *Science*, **156**: 383–385.
Berger, W.H., 1970. Biogenous deep-sea sediments: fractionation by deep-sea circulation. *Bulletins of the Geological Society of America*, **81**: 1385–1402.
Bramlette, M.N., 1961. Pelagic sediments. In Sears, M. (ed.), *Oceanography*. Publications of the American Association for the Advancement of Science, 67, pp. 345–366.
Hay, W.W., 1970. Calcium carbonate compensation. *Initial Reports of the Deep Sea Drilling Project 4*. Washington: U.S. Government Printing Office, pp. 669, 672–673.
Honjo, S., 1976. Coccoliths: production, transportation and sedimentation, *Marine Micropaleontology*, **1**: 65–79.
Ramsay, A.T.S., 1974. Sedimentological clues to palaeo-oceanography. In Ramsay, A.T.S. (ed.), *Oceanic Micropaleontology*. Academic Press, Volume 2, pp. 1371–1453.

Cross-references

Carbonate Mineralogy and Geochemistry
Neritic Carbonate Depositional Environments
Oceanic Sediments
Seawater: Temporal Changes in the Major Solutes
Sedimentologists

CALICHE – CALCRETE

Calcrete, a term effectively now synonymous with caliche, refers to near surface, terrestrial accumulations of predominantly calcium carbonate within soil profiles, the vadose zone or associated with shallow groundwaters, where waters are saturated with respect to calcium carbonate (Wright and Tucker, 1991). In the past these terms have been used in a much more general way to describe any surface or near surface type of carbonate occurrence. There are problems with the currently used definition and the terms are not used to describe tufas, travertines, beachrocks or cemented dunes even though all these could be regarded as falling into the basic definition. Suggestions have been made to limit the usage of the terms to carbonate accumulations in soil or paleosol profiles, however the similarities between many pedogenic (soil formed) calcretes and ones developed in the phreatic zone indicate that the mechanisms of formation are so similar that to restrict the term to one type of setting would be wholly artificial. Goudie (1973) provides a detailed review of terminology and Milnes (1992) gives an historical review of the ideas on calcrete development. Dolcretes are calcretes in which the dominant mineral is dolomite, typically as a primary precipitate, although it can be replacive after calcite.

The terms are most commonly used for indurated materials but many forms are only weakly cemented, and powdery and granular types are known. During the development of calcrete horizons or profiles there is a tendency toward induration and increase in the bulk content of carbonate as a result of progressive cementation, coupled and displacive and replacive introduction or redistribution of calcium carbonate into a host material. That host could be soil, sediment (including carbonate sediments), or bed rock.

Morphology

A very wide range of morphologies are known from laminar, nodular, powdery, mottled, massive, rhizocretionary (composed wholly or partly of the calcareous remains or coatings around roots), pisolitic, prismatic, platy, and brecciated (Netterberg, 1967, 1980; Goudie, 1983). The term 'hardpan' is commonly used to describe highly indurated horizons. Laminar forms are particulary distinctive, and can form in a variety of settings (Wright and Tucker, 1991).

Classification and development

The simplest way to classify calcretes is to use their morphology and many calcretes exhibit only one or two basic morphologies. However, it is common to see regular relationships between the morphology and the amount of secondary carbonate in the calcrete, such that it is possible to define profiles with two or more horizons which exhibit different morphologies (Gile *et al.*, 1965). It has long been known that calcretes exhibit a progression in development from small, dispersed concentrations to massive horizons, especially in pedogenic forms, and this has been used as the basis for widely used classifications by Gile *et al.* (1966) and Machette (1985). These classifications are particularly useful for describing calcretes developed in siliciclastic sediments where the growth of carbonate nodules is a distinctive feature. However, calcretes formed in carbonate hosts develop differently and lack distinct nodules but commonly display sub-horizontal to sub-vertical sheets of secondary carbonate known as stringers (Wright, 1994). The growth of discrete carbonate nodules requires some element of displacive growth, which is triggered in non-carbonate hosts by the inability of calcite to form adhesive bonds with non-calcitic grains such as silica. However, in carbonate-rich hosts calcite forms cohesive bonds with other carbonate crystals (Chadwick and Nettleton, 1990). A slightly modified version of the Machette classification is used to describe calcretes produced largely or wholly by roots (Wright *et al.*, 1995).

The progessive, time dependent development of the features in these calcretes can be used to define chronosequences. By dating the geomorphic surfaces associated with the different stages of calcrete development, it has been possible to identify time ranges required for their development. These approximate time intervals can even be used, with care, to identify relative (not absolute) time relationships in ancient successions (Leeder, 1975; Wright and Marriott, 1996).

A critical distinction must be made between pedogenic and groundwater calcretes. The former, more easily studied in soil pits and excavations, and widespread in the stratigraphic record in red bed successions, are what most sedimentologists will encounter and for which we have well documented modern analogues. What is less well appreciated is that in many regions, especially Australia and Oman, there are extensive areas of carbonate cemented alluvium, petrographically similar to some forms of pedogenic calcrete, produced by cementation and replacive growth of carbonate from shallow groundwaters (Arakel, 1986). Few ancient examples have been documented but more remain to be discovered and Pimentel *et al.* (1996) provide criteria for their recognition.

Most calcretes do form in soils yet there are no soils called calcretes. Accumulations of calcium carbonate are prominent features of several major soil orders including the aridisols, alfisols and vertisols. Pedogenic calcretes form horizons (ca horizons) within soil profiles, but in many cases they are sufficiently prominent to develop several horizons constituting a sub-profile within the main profile. These can be refered to a K horizons with various sub-horizons designated K2, etc. (Gile *et al.*, 1965). Thus it is technically incorrect to refer to calcrete soils or calcrete paleosols, when what is meant are calcrete-bearing soils (e.g., aridisols or alfisols).

Distribution

Calcretes occur in almost all climatic regions. A net moisture deficit is required whereby soil solutions become concentrated enough for carbonate precipitation to be triggered (commonly biogenically). If the moisture regime is arid or hyperarid there may be too little moisture to generate and transport the carbonate in solution. Climates with a strong seasonally moisture regime, with a prolonged dry season, are the ones most likely to produce soil carbonates. However, if the through put of soil water in the wet season is too great, the precipitates of the dry season may be leached away. There is a relationship between the depth of carbonate accumulation in a soil and the mean annual rainfall (Retallack, 1994). Clear proof that calcretes are unreliable climate indicators comes from their occurrence in cold or cool temperate climates, and even free draining limestone gravels in northern England can develop meter-scale horizons with fabrics identical to calcretes formed in Mediterranean climates such as north-east Spain and South Australia (Strong *et al.*, 1992). Groundwater calcretes appear to also develop in areas where there is a strong moisture deficit such that shallow, alkaline groundwaters become highly concentrated.

Sources of carbonate and mechanisms of formation

Pedogenic carbonates are illuvial in origin, meaning that the carbonate is leached from the upper soil horizons, transported in solution and reprecipitated at depth. There are several likely sources for the carbonate; wind blown dust is the most likely source (Machette, 1985), especially in carbonate-poor hosts. Evaporation, evapotranspiration, and degassing are regarded as major mechanisms causing carbonate precipitation in soils. In the case of groundwater calcretes precipitation by the common ion effect is a contributory process. Precipitation triggered directly or indirectly by biological processes is widespread, with fungi playing a key role in the indirect precipitation of carbonate. Plant roots act in a variety of ways to fix carbonate in soils. Surprisingly, despite bacteria having the potential to precipitate calcite in soils, there are very few records of such a process having contributed to calcretes. There is a growing body of evidence that specific vegetation types can produce different calcretes, for example, the distinctive calcretes of north-east Spain appear to be produced by fungal activity associated with pine tree root mats (Wright *et al.*, 1998).

Uses

Calcretes have been used in many countries for road construction. Groundwater forms are the hosts for extensive vanadium and uranium deposits (Arakel, 1982). They have been used extensively in paleoclimatic reconstruction, and studies of their O and C stable isotopes have proved invaluable for such diverse purposes as determining paleovegetation, paleotemperatures and paleoatmospheric composition (Cerling, 1999). Calcretes can present problems for soil management as hardpans can prevent infiltration by water and promote soil erosion.

Controversies and future research

There are many unresolved issues relating to calcretes. We understand little about the actual causes of precipitation in many cases. The origin of the ubiquitous laminar calcretes is still debated (Verrecchia *et al.*, 1995). The exact origin of calcrete structure *Microcodium*, which constitutes the bulk of many Cainozoic paleosols in Europe, is still unresolved and hotly disputed (e.g., see Freytet *et al.*, 1997). We need a much better understanding of the rates of formation of calcretes before we can use them for time resolution in the fossil record. A problem is that even where we have age ranges for individual stages of calcrete development from Quaternary deposits we cannot simply extrapolate because even though a calcrete occurs beneath a surface that is 150 K years old it does not mean that calcrete formation was continuous and uniform over that interval of time. Considering the complexities of climate and vegetation change during the Quaternary calcrete growth may have only taken place during a small part of that time interval. Whereas progress has been made in understanding the diversity of biogenic fabrics in many calcretes, we understand little about the crystalline textures typical of calcretes found in modern deserts and in ancient red bed successions. Of critical importance is a better appreciation of the link between the characteristics of calcrete profiles and specific vegetation types, as this could provide a new approach to using calcretes for paleoenvironmental reconstruction.

V. Paul Wright

Bibliography

Arakel, A.V., 1986. Evolution of calcrete in palaeodrainages of the Lake Napperby area, central Australia. *Paleogeography, Palaeoclimatology, Palaeoecology*, **54**: 282–303.
Cerling, T.E., 1999. Stable carbon isotopes in paleosol carbonates. In *Palaeoweathering, Palaeosurfaces and Continental Deposits*. International Association of Sedimentologists, Special Publication, 27, pp. 43–60.
Chadwick, O.A., and Nettleton, W.D., 1990. Micromorphologicval evidence of adhesive and cohesive forces in soil cementation. *Developments in Soil Science*, **19**: 207–212.
Freytet, P., Plaziat, J.C., and Verrecchia, E.P., 1997. A classification of rhizogenic (root-formed) calcretes, with examples from the Upper Jurassic—Lower Cretaceous of Spain and Upper Cretaceous of France—Discussion. *Sedimentary Geology*, **110**: 299–303.
Gile, L.H., Peterson, F.F., and Grossman, R.B., 1965. The K horizon: a master soil horizon of carbonate accumulation. *Soil Science*, **99**: 74–82.
Gile, L.H., Peterson, F.F., and Grossman, R.B., 1966. Morphological and genetic sequences of carbonate accumulation in desert soils. *Soil Science*, **100**: 347–360.
Strong, G.E., Giles, J.R.A., and Wright, V.P., 1992. A Holocene calcrete from North Yorkshire, England: implications for interpreting palaeoclimates using calcretes. *Sedimentology*, **39**: 333–347.
Goudie, A.S., 1973. *Duricrusts in Tropical and Subtropical Landscapes*. Oxford: Clarendon Press.
Goudie, A.S., 1983. Calcrete. In Goudie, A.S., and Pye, K. (eds.), *Chemical Sediments and Geomorphology*. Academic Press, pp. 93–131
Leeder, M.R., 1975. Pedogenic carbonates and floodplain sediment accretion rates: a quantitative model for alluvial arid-zone lithofacies. *Geological Magazine*, **112**: 257–270.
Machette, M.N., 1985. Calcic Soils of the South-Western United States. Geological Society of America Special Paper, 203, pp. 1–21
Milnes, A.R., 1992. Calcrete. In Martini, I.P., and Chesworth, W. (eds.), *Weathering, Soils and Paleosols*. Amsterdam: Elsevier, pp. 309–347.
Netterberg, F., 1967. Some road making properties of South African calcretes. *Proceedings of the 4th Regional Conference of African Soil Mechanics and Foundation Engineers, Cape Town*, 1, pp. 77–81.
Netterberg, F., 1980. Geology of South African calcretes I: terminology, description macrofeatures and classification. *Transactions of the Geological Society of South Africa*, **83**: 255–283.
Pimentel, N.L., Wright, V.P., and Azevedo, T.M., 1996. Distinguishing early groundwater alteration effects from pedogenesis in ancient alluvial basins: examples from the Palaeogene of southern Portugal. *Sedimentary Geology*, **105**: 1–10.
Retallack, G.J., 1994. The Environmental Approach to the Interpretation of Paleosols. Soil Science Society of America Special Publication, 33, pp. 31–64.
Verrecchia, E.P., Freytet, P., Verrecchia, K.E., and Dumont, J.L. 1995. Spherulites in calcrete laminar crusts: biogenic $CaCO_3$ precipitation as a major contributor to crust formation. *Journal of Sedimentary Research*, **A65**: 690–700.
Wright, V.P., 1994. Paleosols in shallow marine carbonate sequences. *Earth Science Reviews*, **35**: 367–439.
Wright, V.P., and Marriott, S.B., 1996. A quantitative approach to soil occurrence in alluvial deposits and its application to the Old Red Sandstone of Britian. *Journal of the Geological Society of London*, **153**: 907–913.
Wright, V.P., Platt, N.H., Marriott, S.B., and Beck, V.H., 1995. A classification of rhizogenic (root-formed) calcretes, with examples from the Upper Jurassic–Lower Cretaceous of Spain and Upeer Cretaceous of southern France. *Sedimentary Geology*, **100**: 143–158.
Wright, V.P., Sanz, E.M., and Beck, V.H., 1998. Rhizogenic origin for laminar-platy calcretes, Plio-Quaternary of Spain. In Canavaras, J.C., Garcia del Cura, M.A., and Soria, J. (eds.), *15th International Sedimentology Congress, Alicante, Abstracts*, 827p.
Wright, V.P., and Tucker, M.E., 1991. Calcretes: an introduction. *International Association of Sedimentologists Reprint Series*, Volume 2, pp. 1–22.

Cross-references

Carbonate Mineralogy and Geochemistry
Cements and Cementation
Desert Sedimentary Environments
Rivers and Alluvial Fans
Weathering, Soils, and Paleosols

CARBONATE DIAGENESIS AND MICROFABRICS

The four decades from about 1950 are of interest because they encompassed a flowering of our understanding of the carbonate fabrics we see with the microscope. The history of this period is dealt with in greater detail in Bathurst (1993: errors no fault of author) with extensive references. This growth from a topic that was almost unheard of in universities (my "sedimentary petrology" lessons in the postwar years were limited to the optical identification of sand grains) was much helped by funding from the oil industry.

The use of the microscope is only a natural extension of field work. What insight had William Blake when he wrote "To see a world in a grain of sand". Until the 1950s, the study of sedimentary petrology (as it was then known) was rare, and that of diagenesis virtually unknown. However, by then, the economic importance of underground sources of water and hydrocarbon was clearly apparent and the study of sedimentology took wing. The foundations upon which post-World War II researchers had to build were noble but scant. Sorby (see *Sedimentologists*), in his paper to the Geological Society of London in 1897, had grasped the skeletal composition of limestones, also the dissolution of aragonite with simultaneous precipitation of calcite, as space filler or replacement. Cullis, in his communication to the Royal Society in 1904, had described the cores of the Pacific Funafuti Atoll, much helped by the two stains, Meigen and Lemberg, to distinguish aragonite, calcite, and dolomite (see *Stains for Carbonate Minerals*). Vaughan (see *Sedimentologists*), in 1910, publishing from the Laboratory on Dry Tortugas (all that remains now is a memorial plaque on the beach) had analyzed a variety of marine sediments, and Black (who taught sedimentary petrology in the University of Cambridge) had, in 1933, revealed the importance of cyanobacteria in the Bahamas. Cayeux (see *Sedimentologists*), in 1935, had provided a superb range of black and white photomicrographs. Sander in 1936 (English version 1951) in a book remarkable for its scientific discipline, had given us the terms "geopetal", "internal sediment", and "fabric" (from "gefüge"). Hadding (1941–1959) had written superb analyzes of Swedish Paleozic limestones.

The first sign of a new start came from work in the universities. From Cambridge, in 1954, Illing made a major lithofacies study of the sediments on the Great Bahama Bank. From Liverpool, Bathurst, in 1958 and 1959, offered criteria, based on metalurgical fabrics, for distinguishing space-filling from replacement calcite. At much the same time, in the University of Texas, Austin, Folk (see *Sedimentologists*) gave us in 1959, the extremely successful system for recognizing limestone types, such as biomicrite and biosparite. An extraordinary group of workers in Shell Development Company were producing innovative ideas: for example Ginsburg, in 1956, continuing his earlier research done in the University of Miami, published an extensive study of the Florida seafloor. Murray, in 1960, examined a range of porosity-forming processes and Dunham, in 1962, 1969, and 1971, compiled a classification of allochems involving the valuable concepts of "grain-supported" and "mud-supported", also "vadose silt" and "meniscus cement".

Now a search could be made for diagenetic environments. Where had lithification taken place? Ginsburg in 1957, showed that the Pleistocene Miami Oolite, while friable in seawater, was cemented with calcite in freshwater. The supply of carbonate for the calcite came from dissolved aragonite. Schlanger in 1963, confirmed that freshwater lithification had taken place just below unconformities. Research on the Pleistocene of Bermuda in the 1960s, by Friedman, Gross, Land, Mackenzie, and Gould, led to a concept of freshwater lithification, the new calcite being regarded as either sparry cement or neomorphic replacement. All sparry calcite was assumed then to be of freshwater origin. The idea of meteoric lithification had great appeal—no heat, no pressure. The interpretation of process was aided by the new alliance between fabric and analysis of elements and isotope ratios. The concept of changing composition of pore water was clarified by the development of stains in the 1960s by Evamy, Shearman and Dickson. Cathodoluminescence (CL) was applied by Sippel and Glover following the pioneer work of Amieux.

Marine diagenesis was examined as well. Illing in his 1954 paper recorded beach rock and Bathurst in 1966 described micritization of shells by boring cyanobacteria followed by cementation. Yet, far from today's seas, in New Mexico, Pray was mapping Mississippian bioherms with clastic dikes of synsedimentary origin which were clearly in a marine cemented substrate.

Indeed, in the 1970s a new world of submarine cementation was opening up with the work of Schroeder and Ginsburg on Bermuda reefs and on other reefs off Jamaica by Land and Goreau. In the Persian Gulf cemented crusts in shallow water had been located by Shinn, Taylor, and Illing. Evidence of cementation in deep sea sediments in dredged blocks was revealed by Fischer and Garrison in 1967.

Older marine cemented limestones were recorded too by Purser in Jurassic hardgrounds in the Paris Basin and by Zankl in the Jurassic and Trias of Germany and Austria. Void-filling cements were found in Devonian reefs in Germany by Krebs.

Further support came from Bromley's Chalk hardgrounds in Denmark.

The study of the finer fabrics was greatly helped by use of the transmission electron microscope, as in the book by Fischer, Honjo, and Garrison, and even finer detail was revealed by the scanning electron microscope pioneered especially by Alexandersson and Loreau.

A most valuable international conference on carbonate cements took place on Bermuda in 1969 when four diagenetic environments were distinguished: the intertidal marine, the submarine, the vadose freshwater, and the phreatic freshwater. Many of these were later described from Belize by James and Ginsburg in 1979.

An understanding of aquifer hydrology became increasingly pressing and was helped by work such as that by Back and Hanshaw published in 1970. There was thus a successful Penrose Conference in Vail, Colorado, combining hydrologists with sedimentologists. At that meeting a significant leap forward was taken by Meyers who introduced the use of CL zoning to reveal cement stratigraphy.

The study of freshwater aquifers led to the recognition of caliches by Esteban, Read, and Klappa and a range of soil textures.

Now some consolidation of ideas was necessary and five important books appeared: Bathurst on carbonate sediments and their diagenesis, Wilson's great summary of carbonate lithofacies in the Phanerozoic, Flügel's acute study of microfacies, with visual identification of allochems supported by the books of Majewske, and of Horowitz and Potter.

The deformation of carbonate sediments during burial was being revealed by Mossop's research on pressure-dissolution in Devonian reefs, by Mimran in steep limbs of folded Cretaceous Chalk, and by Dickson and Coleman in 1980 who showed with isotopic analysis of calcite cement zones a long history of growth during burial. The later cement zones are commonly ferroan.

It was becoming clear that the amount of calcium carbonate required to fill the pores with cement must have been considerable, even exceeding the original mass of the sediments as proposed by Choquette and Pray in 1970. The importance of a local source through pressure-dissolution was suggested by Hudson in 1975. Calcite cementation in some Jurassic limestones is late burial because it post-dates fracture of micrite envelopes, as demonstrated by Emery, Dickson, and Smalley in 1987. A useful terminology for pressure-dissolution fabrics was presented by Buxton and Sibley in 1981, i.e., "fitted fabric", "dissolution seam", and "stylolite". With the author's permission Bathurst in 1991 tightened up their definition of "fitted fabric". Lithification of coccolithic chalk by pressure-dissolution over tens of millions of years was demonstrated by Matter, and by Scholle, in 1974. Bathurst in 1983 dealt with burial calcitization in terms of compaction fabrics.

Dolomite, too, could have a deep crustal origin, both as replacement and cement. Choquette found it could be a fracture fill, as did Grover and Read, Moldovanyi and Lohmann, and Dorobek. Mattes and Mountjoy found dolomite related to pressure-dissolution. Wardlaw revealed fine detail by using resins to make pore casts.

The role of early cementation in preserving sediment from compaction was emphasized for Jurassic hardgrounds by Purser in 1969, for the Jurassic Smackover by Swirydczuk in 1988 and for the Carboniferous Limestone of England by Hurd and Tucker in 1988. James and Bone in 1989 showed how, in the Cenozoic of South Australia, two grain-stones with identical burial histories were pressure-welded or cemented depending on the content of aragonitic molluscan debris.

Interest in the chemistry of the new microfabrics grew also, it being clear that whole rock analysis was of little use. The road to success lay in selective analysis of different components identified with the microscope, as by Hudson in 1977 and Lohmann in 1988. Marshall and Ashton exhibited the dual use of trace elements and stable isotope ratios, a study elaborated by Moldvanyi and Lohmann. Dickson and Coleman, in 1980, as already mentioned, had identified their zonal sequence by staining.

Another interesting diagenetic environment received attention in the 1970s. This was the early subsurface sediment, isolated from the overlying seawater. Chemical reactions continue in a closed system where sediments are buried with their original seawater, bearing a mixture of solid carbonate debris with organic material containing microorganisms. In 1980 Berner published a book on the early diagenesis of sediments within the first 100 m or so of burial. This exposed a new world of sulfate reduction, fermentation, redox potential and reactions of methane. Here was a framework for the growth of concretions investigated by Berner, Raiswall, Hudson, Curtis, Irwin and Marshall. This work provided an essential background for understanding how concretions coalesce to form lithified beds. It has become clear that many marine sediments, especially fine-grained, were buried in this way. Their development was treated in a remarkable book by Einsele and Seilacher in 1982 and its successor by Einsele, Ricken and Seilacher (1991).

The importance of a late diagenetic overprint, giving rise to new bedding, though always in response to a generally unknown but fluctuating syndeposional signal, has been treated in both books and by Simpson in 1985, by Ricken in 1986, and Bathurst in 1987 and 1990.

Important advances were made in understanding the calcitization of aragonite in the 1970s and 1980s. Relics of aragonite are commonly preserved inside the neomorphic calcite spar. These were revealed using SEM by Sandberg, Schneidermann and Wunder, and by Hudson. Using this information Lasemi and Sandberg were able to characterize micrites and microspars as ADP (aragonite-dominated precursor) or CDP (calcite-dominated precursor). The ability to distinguish ADP and CDP made it possible for Sandberg to recognize more surely periods of aragonite or calcite precipitation in seawater and to relate these to Fischer's secular variations of icehouse and greenhouse—or to a mixture of these as shown by Wilkinson, Buczynski, and Owen.

In conclusion, the principles followed by Sorby remain the same. We still need to look, describe, identify and find geometric relations. The complexity has been explored by Schroeder and Purser. We can now make out at least four major diagenetic environments: (1) the near subsea (first 40 cm or so), (2) the first 100 m or thereabouts which is a closed system, (3) the freshwater vadose and phreatic, and (4) the deep burial affected by high pressure and temperature. In addition there are the various lacustrine situations.

We still need to grasp more clearly just how pores in so many limestones are almost totally filled with cement. Vast volumes of water must pass through a pore in order that enough ions arrive to fill it with cement—some 10,000 to 100,000 pore volumes. What stimulates all this water to be supersaturated throughout such large volumes of sediment? We need more help from sympathetic geochemists such as Morse and Mackenzie and their "Journey of a rain drop".

Robin G.C. Bathurst

Bibliography

Bathurst, R.G.C., 1993. Microfabrics in carbonate diagenesis: a critical look at forty years in research. In Rezak, R., and Lavoie, D.L. (eds.), *Carbonate Microfabrics*. Berlin: Springer-Verlag, pp. 3–14.

Einsele, G., Ricken, W., and Seilacher, A. (eds.), 1991. *Cycles and Events in Stratigraphy*. Berlin: Springer-Verlag.

Jeans, C.V., and Rawson, P.F. (eds.), 1980. *Andros Island, Chalk and Oceanic Oozes: Unpublished Work of Maurice Black*. Yorkshire Geological Society, Occasional Publication 5.

Cross-references

Beachrock
Bioclasts
Caliche–Calcrete
Carbonate Mineralogy and Geochemistry
Cathodoluminescence
Cements and Cementation
Diagenesis
Diagenetic Structures
Dolomite Textures
Micritization
Neomorphism and Recrystallization
Sedimentologists
Sedimentology, History
Stains for Carbonate Minerals
Stylolites

CARBONATE MINERALOGY AND GEOCHEMISTRY

Introduction

Sediments and sedimentary rocks contain a variety of carbonate minerals. Calcite (trigonal $CaCO_3$) and dolomite [trigonal $CaMg(CO_3)_2$] are by far the most important carbonate minerals in ancient sedimentary rocks. Aragonite (orthorhombic $CaCO_3$) is rare and other carbonate minerals are largely confined to special deposits such as evaporites or iron formations or present as minor minerals. Calcitic and dolomitic sedimentary rocks constitute 10–15 percent of the mass of sedimentary rocks and approximately 20 percent of the sedimentary rock mass younger than 600 Ma (the Phanerozoic). Carbonate rocks are important reservoirs of oil and gas and contain commercial ore bodies and in the form of biological, chemical, mineralogical, and isotopic signatures, important information concerning the evolution of the Earth. Thus, the mineralogy and geochemistry of the carbonate minerals constituting carbonate sediments and rocks have been investigated in great detail.

There are three polymorphs (similar composition, different crystal structure) of $CaCO_3$ that are found in sediments and in the structures of organisms. The rhombohedral mineral *calcite* is the most abundant and is thermodynamically stable. *Aragonite*, the orthorhombic form, is also abundant but found primarily in young sediments and the skeletal structures of marine organisms. *Aragonite* is the more dense form and hence

Table C1 Some properties of the carbonate minerals. ΔG_{f298} is the standard Gibbs free energy of formation, and K_{sp} is the solubility product of the phase. Full references for the data cited are found in Morse and Mackenzie (1990). (after Morse and Mackenzie, 1990)

Mineral	Formula	Formula wt (g)	Density (g cm^{-3})	Crystal System	ΔG_{f298} (J mole^{-1})	$-\log K_{sp}$ (calc)	$-\log K_{sp}$ (ref 2)	$-\log K_{sp}$ (other)
Calcite	$CaCO_3$	100.09	2.71	Trig.	-1128842^3	8.30	8.35	$8.48^{6,8}$, 8.46^9
Aragonite	$CaCO_3$	100.09	2.93	Ortho.	-1127793^3	8.12	8.22	8.34^6, 8.30^8
Vaterite	$CaCO_3$	100.09	2.54	Hex.	-1125540^3	7.73	—	7.91^6
Monohydrocalcite	$CaCO_3 \cdot H_2O$	118.10	2.43	Hex.	-1361600^3	7.54	—	7.60^{17}
Ikaite	$CaCO_3 \cdot 6H_2O$	208.18	1.77	Mono.	—	—	—	7.12^5
Magnesite	$MgCO_3$	84.32	2.96	Trig.	-1723746^3	8.20	7.46	5.10^7, 8.10^{10}
Nesquehonite	$MgCO_3 \cdot 3H_2O$	138.36	1.83	Trig.	-1723746^3	5.19	4.67	
Artinite	$Mg_2CO_3(OH)_2 \cdot 3H_2O$	196.68	2.04	Mono.	-2568346^3	18.36	—	
Hydromagnesite	$Mg_4(CO_3)_3(OH)_2 \cdot 3H_2O$	359.27	—	Mono.	-4637127^4	36.47	—	30.6^7
Dolomite	$CaMg(CO_3)_2$	184.40	2.87	Trig.	-2161672^3	17.09	—	
Huntite	$CaMg_3(CO_3)_4$	353.03	2.88	Trig.	-4203425_3	30.46	—	
Strontianite	$SrCO_3$	147.63	3.70	Ortho.	-1137645_3	8.81	9.03	9.13^9, 9.27^{11}
Witherite	$BaCO_3$	197.35	4.43	Ortho.	-1132210^3	7.63	8.30	$8.56^{9,18}$
Barytocalcite	$CaBa(CO_3)_2$	297.44	—	Trig.	-2271494^4	17.68	—	
Rhodochrosite	$MnCO_3$	114.95	3.13	Trig.	-816047^3	10.54	9.30	10.59^{14}
Kutnahorite	$CaMn(CO_3)_2$	215.04	—	Trig.	-195058^1	55.79	—	
Siderite	$FeCO_3$	115.85	3.80	Trig.	-666698^3	10.50	10.68	10.91^{12}
Cobaltocalcite	$CoCO_3$	118.94	4.13	Trig.	-650026^1	11.87	9.68	
—	$CuCO_3$	123.56	—	Trig.	—	—	9.63	11.51^{12}
Gaspeite	$NiCO_3$	118.72	—	Trig.	-613793^1	7.06	6.87	
Smithsonite	$ZnCO_3$	125.39	4.40	Trig.	-731480^3	9.87	10.00	
Otavite	$CdCO_3$	172.41	4.26	Trig.	-669440^3	11.21	13.74	
Cerussite	$PbCO_3$	267.20	6.60	Ortho.	-625337^3	12.80	13.13	12.15^{15}
Malachite	$Cu_2CO_3(OH)_2$	221.11	4.00	Mono.	—	—	33.78	33.46^{13}
Azurite	$Cu_3(CO_3)_2(OH)_2$	344.65	3.88	Mono.	—	—	45.96	
Ankerite	$CaFe(CO_3)_2$	215.95	—	Trig.	-1815200	19.92^{19}	—	
Natronite	$Na_2CO_3 \cdot 10H_2O$	285.99	—	Ortho.	-3428997^4	1.03	—	
Thermonatrite	$Na_2CO_3 \cdot H_2O$	124.00	—	Ortho.	-1286538^4	-0.403	—	-0.54^{16}
Trona	$NaHCO_3 \cdot Na_2CO_3 \cdot 2H_2O$	229.00	2.25	Mono.	-2386554^4	2.07	—	1.00^{16}
Nahcolite	$NaHCO_3$	84.01	2.16	Mono.	-851862^4	0.545	—	0.39^{16}
Natron	$NaHCO_3 \cdot H_2O$	102.03	—	—	—	—	—	0.80^{16}

[1] Weast, 1985;
[2] Smith and Martel, 1976
[3] Robie et al., 1979
[4] Garrels and Christ, 1965
[5] Suess, 1982 at 1.6°C
[6] Plummer and Busenberg, 1982
[7] Langmuir, 1973
[8] Sass et al., 1983
[9] Millero et al., 1984
[10] Christ and Hostetler, 1970
[11] Busenberg et al., 1984
[12] Reiterer et al., 1981
[13] Symes and Kester, 1984
[14] Johnson, 1982
[15] Bilinski and Schindler, 1982
[16] Monnin and Schott, 1984
[17] Hull and Turnbull, 1973
[18] Busenberg and Plummer, 1986a
[19] Woods, 1988

is the $CaCO_3$ phase stable at high temperature but metastable relative to calcite at low temperature. It is about 1.5 times more soluble than calcite. *Vaterite* is the third anhydrous $CaCO_3$ phase, has a hexagonal structure, and is metastable relative to aragonite and calcite under the environmental conditions found in sediments and sedimentary rocks. It is approximately 3.7 times more soluble than calcite and 2.5 times more soluble than aragonite. In addition to the anhydrous $CaCO_3$ minerals found in sediments, there are scarce occurrences of hydrated $CaCO_3$ minerals, such as ikaite ($CaCO_3 \cdot (6H_2O)$).

Dolomite is one of the most abundant carbonate phases found in carbonate rocks. However, it does not occur as a skeletal component of organisms as do calcite and aragonite. Even after years of study, the mode of formation of dolomite remains controversial. Its properties under the environmental conditions at and near the surface of the Earth are less well-known than those of calcite and aragonite. This is partly a reflection of the fact that the synthesis of dolomite in low temperature laboratory experiments has proven to be difficult.

Table C1 lists the variety of carbonate minerals found in nature and some of their properties. In this article because of the abundance of calcite and dolomite in the sedimentary rock record, their mineralogy and geochemistry are discussed in some detail.

Calcite

Calcites are very important biogenic and inorganic constituents of modern marine sediments and Pleistocene rocks. They are complex minerals containing up to 30 mole percent $MgCO_3$. Calcites containing more than a few percent $MgCO_3$ are usually referred to as magnesian calcites. Generally, for the same $MgCO_3$ content, the biogenic calcites have greater concentrations of sodium, sulfate, water, hydroxide, and bicarbonate and tend to have larger cell volumes and greater carbonate positional disorder in their structure than natural or synthetic inorganic phases. Chemical and microstructural heterogeneities characterize biogenic magnesian calcites, and dislocations and plane defects are common in magnesian calcites. All of these characteristics may affect the thermodynamic and kinetic properties and reactivity of these phases in aqueous solution. In contrast to younger sediments, calcites found in Paleogene and older rock units commonly contain only a few percent $MgCO_3$. This contrast in composition and its interpretation have been the principal driving force behind studies of the basic crystal chemistry, thermodynamic, and kinetic properties of calcites. The literature concerning these properties is substantial and is summarized in Mackenzie *et al.* (1983), Morse and Mackenzie (1990), Tribble *et al.* (1995), and Arvidson and Mackenzie (1999).

Crystal structure and chemistry

During recent years, a number of new observations concerning the crystal structure and chemistry of calcites has been made. These observations have proved useful in interpreting the experimental data on the solubility of calcites in aqueous solution, and in turn, the behavior of calcites during diagenesis and their usefulness in environmental interpretation. The following is a summary of the more important recent conclusions:

(1) Unit cell parameters of synthetic calcites vary smoothly as a function of $MgCO_3$ content. These phases have larger c/a axial ratios than those obtained from a straight line connecting the c/a axial ratios of calcite and magnesite or disordered dolomite (Figure C2(A)). Furthermore, the synthetic phases exhibit negative excess cell volumes in the composition range 0–20 mole percent $MgCO_3$ and positive excess volumes for $MgCO_3$ contents greater than 20 mole percent (Figure C2 (B)). In contrast, the unit cell parameters of biogenic calcites do not vary smoothly as a function of $MgCO_3$ content, and their axial ratios and cell volumes are larger than those of synthetic solids of the same $MgCO_3$ content (Figure C3).

(2) The excess c/a axial ratios in synthetic calcite phases are the result of the positional disorder of the CO_3-anion group in the vicinity of Mg^{2+} ions. In essence, the anion group becomes progressively more inclined to the c crystal axis with increasing magnesium concentration in the calcite phase. This phenomenon has been documented in Raman spectroscopic studies of calcites that show an increase in the halfwidths of Raman spectral bands with increasing magnesium concentration (Figure C4). Single crystal X-ray refinements of two biogenic magnesian calcites have confirmed the hypothesis of positional disorder in these phases with increasing Mg content (Paquette

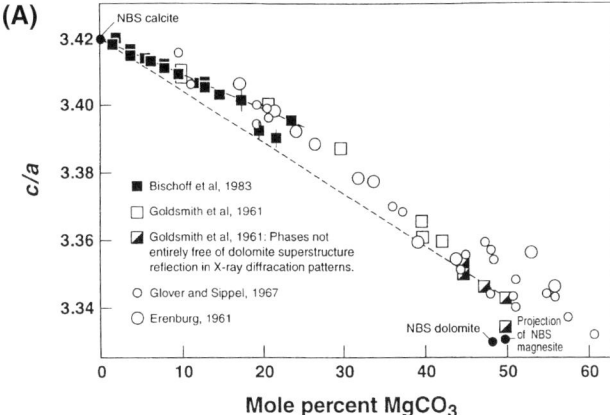

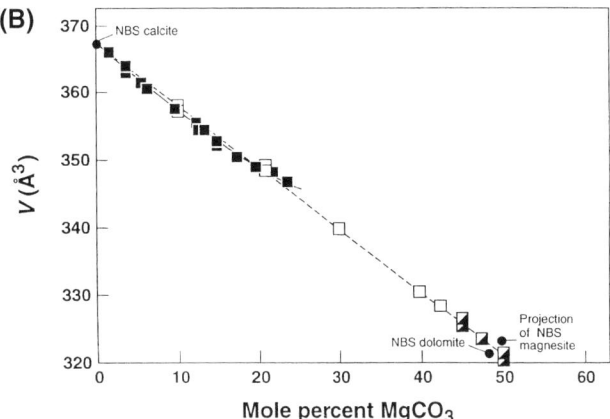

Figure C2 Unit cell axial ratios (A) and volumes (B) versus composition for synthetic magnesian calcites. Solid line is quadratic, least-squares fit for synthetic phases. The dashed line connects pure calcite and least-ordered dolomite (after Mackenzie *et al.*, 1983).

and Reeder, 1990). Furthermore, Paquette and Reeder have shown that this disorder is not simply limited to the CO_3 group but also affects the cation positions.

(3) In general, biogenic calcites exhibit a greater degree of heterogeneity with respect to the distribution of magnesium in their structure than synthetic phases. Furthermore, some biogenic calcites and synthetic calcites precipitated at high supersaturations from multicomponent aqueous solutions contain trace amounts of Na^+ and SO_4^{2-} owing to point defects and dislocations. It is likely that some substitutions of Na^+ for Ca^{2+} in the calcite lattice are balanced by incorporation of HCO_3^- into the structure.

(4) Busenberg and Plummer (1989) modeled the solid solution behavior of calcites based on experimentally obtained dissolution data. Compositionally pure binary solid solutions of $CaCO_3$ and $MgCO_3$ that include metamorphic and hydrothermal phases, synthetic phases prepared at high temperatures and pressures, and synthetic phases prepared from aqueous solution at low temperatures and very low calcite super-saturations were modeled as sub-regular solid solutions between calcite and dolomite or disordered dolomite. For the biogenic calcites and synthetic calcites synthesized at high calcite supersaturations from aqueous solution, a

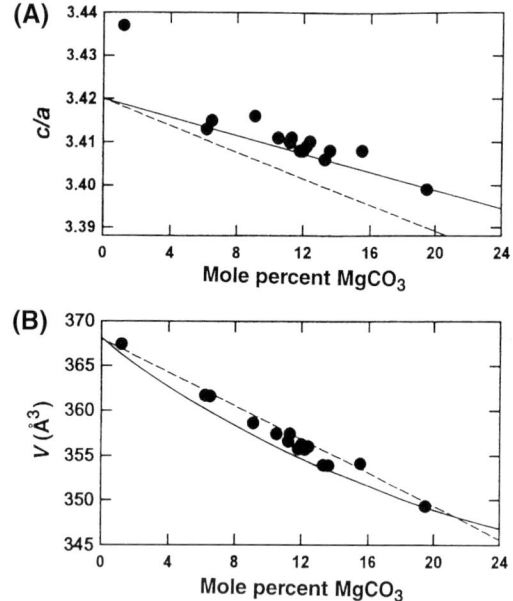

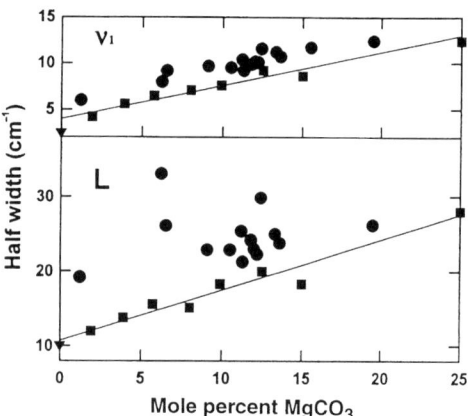

Figure C3 Unit cell axial ratios (A) and volumes (B) versus composition for biogenic magnesian calcites. Solid line is quadratic, least-squares fit for synthetic phases. The dashed line connects pure calcite and least-ordered dolomite (after Bischoff et al., 1983).

Figure C4 Halfwidths of the ν_1 and L Raman modes versus composition for synthetic (squares) and biogenic (dots) magnesian calcites. The triangle represents pure calcite. The straight line is the least-squares fit to the synthetic phase data (after Bischoff et al., 1985).

sub-regular solid solution between defective calcite and disordered dolomite appeared to fit best the experimental data.

The solubility of a calcite in aqueous solution depends significantly on its physical and chemical characteristics. These characteristics must be defined and quantified before one can interpret the laboratory solubility or kinetic data necessary to studies of the reactivity of calcites in natural environments.

Solubility and solid solution behavior

The solubilities of magnesian calcites as a function of their $MgCO_3$ content have been a subject of considerable debate.

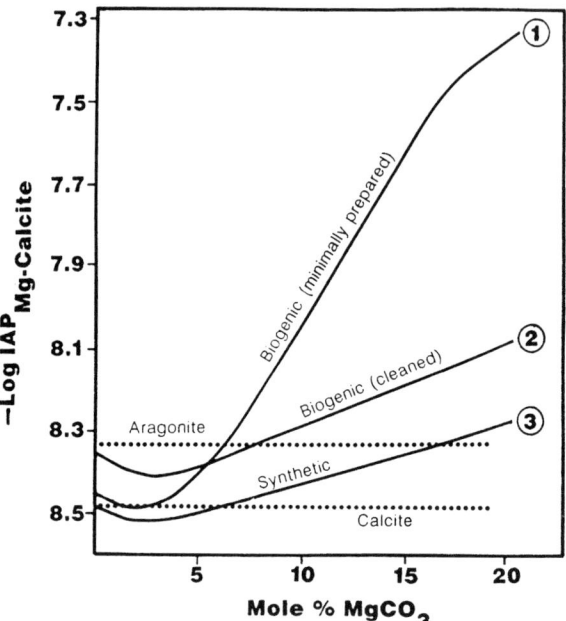

Figure C5 Solubility of magnesian calcites as a function of the $MgCO_3$ in the phase. See text for further explanation (after Bischoff et al., 1993).

However, with the work of Bischoff et al. (1987, 1993), Busenberg and Plummer (1989), and Bertram et al. (1991), the solubility trend with composition of these phases at 25°C and 1 atmosphere total pressure appears reasonably well defined (Figure C5). The solubilities of these phases are expressed as a function of their $MgCO_3$ content and the ion activity product (IAP) of a solution in equilibrium with the solid:

$$IAP_{calcite} = K_{sp} = [Ca^{2+}]^{1-x}[Mg^{2+}]^{x}[CO_3^{2-}]$$

where brackets denote the activity of the ion in aqueous solution, x is the mole fraction of $MgCO_3$ in the solid, and K_{sp} is the solubility product.

The solubilities of magnesian calcites exhibit a minimum around two mole percent $MgCO_3$, beyond which they increase nearly linearly. There appear to be two trends in the solubility-composition data; one for inorganic solids that are well crystallized, compositionally homogeneous, and chemically pure (Figure C5, curve 3) and the other for biogenic phases and inorganic solids that exhibit substantial lattice defects, carbonate positional disorder, and substitution of foreign ions like sulfate and sodium (Figure C5, curve 2). A third solubility-composition curve is shown in Figure C5, that of Plummer and Mackenzie (1974) for biogenic materials subjected to minimal cleaning procedures in the laboratory before use in dissolution experiments designed to determine the solubilities of these phases (Figure C5, curve 1). This curve probably reflects primarily kinetic rather than thermodynamic factors, a result of the lack of annealing of the biogenic materials and the non-removal of fines from the reactant solids used in these experiments.

As can be seen in Figure C5, biogenic and some inorganic calcites (Figure C5, curve 2) are relatively more unstable in aqueous solutions than synthetic phases of similar $MgCO_3$

content (Figure C5, curve 3). Their solubilities differ by roughly 0.15 pIAP units (pIAP = −log IAP) or about 4.6 kJ/mole. This difference arises because of physical and chemical differences between the two types of solids. The complicated microarchitecture, greater positional disorder of the CO_3 anion group, and chemical impurity of the biogenic phases and some inorganic calcites result in higher solubilities of these solids. It is likely that the solubility-composition curve 3 (Figure C5) for pure, compositionally homogeneous and structurally well-ordered calcite phases represents the true metastable equilibrium solubility of magnesian calcites in aqueous solution at 25°C and 1 atmosphere. This difference between the equilibrium and biogenic solubilities of calcite phases of the same $MgCO_3$ content has important implications for diagenesis of the metastable assemblage of carbonate minerals deposited on the seafloor today and in the geologic past.

Dolomite

The mineral dolomite, $CaMg(CO_3)_2$, is a major constituent of carbonate rocks. For geologists dolomite has been the center of an enduring debate regarding its mode of formation in surface environments, its past abundance relative to other sedimentary carbonates, and its overall significance in the geologic record (Machel and Mountjoy, 1986; Hardie, 1987; Mackenzie and Morse, 1992). This debate has motivated the search for a modern analogue to the surface or near-surface environment that is capable of producing large quantities of dolomite, whether as a primary precipitate, or (as is more commonly seen in ancient carbonate rocks) an early diagenetic phase replacing primary $CaCO_3$. Although this labor has yielded possible scenarios in which certain sedimentary dolomites no doubt formed, it has failed to offer a reasonable model for the formation of *massive* dolomite in the geologic record. This lack of a modern analogue for such an important mineral has challenged experimentalists to observe and measure directly the physicochemical conditions and rates under which dolomite precipitation and replacement reactions occur.

Mineralogy and phase relations

Dolomite is distinguished from calcite and the other rhombohedral carbonates by its stoichiometry, ideal dolomite having equal numbers of Ca and Mg atoms, and by the segregation of Ca and Mg in distinct lattice planes. These planes are oriented normal to the *c*-axis, each alternating with *c*-normal trigonal CO_3 groups, that are also in essentially planar orientation. This cation ordering results in a coupled rotation of the carbonate groups and a reduction of symmetry relative to calcite from R3c to R3. Each cation is octahedrally coordinated by six oxygen atoms, which are themselves threefold coordinated with a calcium, magnesium, and carbon atom.

Determination of the univarient temperature-CO_2 curve for the calcite-dolomite-magnesite system was made by Harker and Tuttle (1955a) at high to moderate temperatures and pressures. The order of decomposition with increasing temperature was shown to be magnesite, dolomite, and calcite. The location of calcite-dolomite and dolomite-magnesite solvi along the $CaCO_3$-$MgCO_3$ binary join (Figure C6) was established through the collective work of Harker and Tuttle (1955b), Graf and Goldsmith (1955, 1958), and Goldsmith and Heard (1961). Lattice constants for Ca-Mg carbonates have

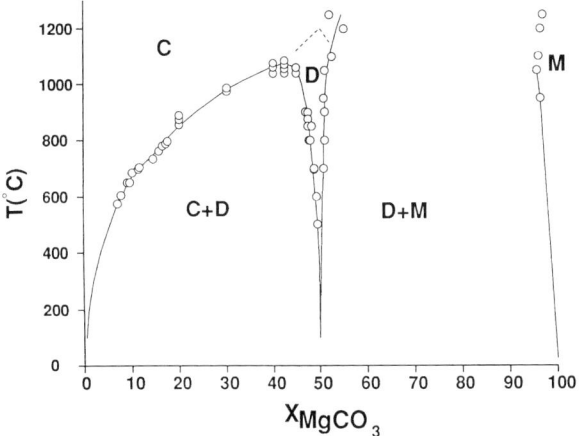

Figure C6 Subsolvus relationships for the $CaCO_3$-$MgCO_3$ join. The top of the calcite-dolomite immiscibility gap occurs at 1075°C, at this temperature, the phase composition is $Ca_{57.5}Mg_{42.5}(CO_3)_2$. Dotted line represents limit of detectable ordering in the dolomite structure. C = calcite; M = magnesite; and D = dolomite [after Tribble *et al.*, 1995; data from Goldsmith and Graf (1958) and Goldsmith and Heard (1961)].

also been determined by Goldsmith and Graf (1958), Goldsmith *et al.* (1961), and Graf (1961). Those of Goldsmith and Graf have been abandoned.

The top of the calcite-dolomite solvus was located by Goldsmith and Heard (1961) at 1075°C, giving a composition of $Ca_{57.5}Mg_{42.5}(CO_3)_2$ (Figure C6). Above this temperature, complete solid solubility is seen between calcite and a carbonate of dolomite composition. However, a phase change must still separate the disordered Mg calcite (R3c space group) from ordered dolomite (R3). To distinguish this phase change as discontinuous (first order) or continuous (order ≥2) is problematic: a first-order transition would demand coexistence of ordered and disordered phases at the temperature of transition. Goldsmith and Heard (1961) observed a continuous increase in disorder from ~1000°C through 1150°C, based on the decreasing intensity of cation ordering reflections. The intensity of ordering reflections depends on the difference between *average* scattering factors for the atoms occupying distinct sites. Thus, Goldsmith and Heard concluded that the phase change is high order, with the caveat that a two-phase field, if present, may be so narrow as to have eluded detection.

With decreasing subsolvus temperatures, the amount of excess $CaCO_3$ that can be accommodated in stable dolomite is greatly reduced (Figure C6). Dolomites formed at temperatures < ~500°C with more than one mole percent excess $CaCO_3$ must thus be regarded as metastable. Goldsmith and Graf (1958) recognized the difficulty in accommodating this excess Ca in Mg sites, and showed that natural dolomites whose compositions deviate from stoichiometric proportions ($0.49 \leq X_{CaCO_3, dol} \leq 0.57$) also showed evidence of some cation disorder.

Early work on synthesis and mechanism

The work of Graf and Goldschmidt (1956) was one of the first experimental attempts to obtain information on the rate and mechanism of dolomite formation as a function of temperature and reactant composition. Important conclusions of this work are:

(1) Ordered, stoichiometric dolomite similar to natural samples can be readily synthesized under dry conditions during brief reaction times at temperatures of 500°C to 800°C using a variety of reactants. However, the reaction below 400°C is so slow that no dolomite product is identifiable after reaction times of thousands of hours.

(2) Dolomite can be synthesized in hydrothermal (wet) conditions over a range in temperatures. The phases formed at a given temperature depend on the partial pressure of CO_2 and H_2O. At temperatures down to 200°C, ordered, stoichiometric dolomite can be synthesized. At lower temperatures or for shorter reaction times, the product dolomites may exhibit 1:1 Ca:Mg ratios, but cation ordering is usually incomplete.

(3) At temperatures less than 100°C, regardless of reaction time, dolomite-like materials are formed having compositions that deviate significantly from ideal and are typically calcium rich. These Ca-rich dolomites completely lack evidence of ordering. The term *protodolomite* was introduced by Graf and Goldschmidt (1956) to describe Ca-Mg carbonates that deviate from stable, ordered, stoichiometric dolomite. Protodolomite can transform to dolomite with sufficient time or temperature.

Since Graf and Goldschmidt's pioneering work, subsequent experimental work focused on the stability relations in the $CaCO_3$-$MgCO_3$ system under hydrothermal conditions in chloride solutions (Rosenberg and Holland, 1964; Rosenberg et al., 1967; Gaines, 1974, 1980; Katz and Matthews, 1977). Gaines (1974) was able to produce ordered, stoichiometric dolomite from calcite and aragonite at 100°C reacted in Ca-Mg chloride solutions for only ~200 hours. He demonstrated qualitatively that the dolomite reaction rate depended on temperature, ionic strength, Mg:Ca ratio in solution, and the mineralogy of the reactant carbonate solid (calcite, magnesian calcite, and aragonite increasing in respective reactivity). The activation energy of the formation reaction of an ordered dolomite recalculated from Gaines' experiments by Arvidson and Mackenzie (1997) was found to be 42.1 kcal mol^{-1}.

Recent and current work in mineralogy and synthesis

Sibley (1990) produced stoichiometric, well-ordered dolomite from calcite at 218°C. Intermediate products included various fractions of magnesian calcite and non-stoichiometric dolomite whose compositions and rate of formation were apparent functions of the Mg/Ca ratio in solution. Sibley concluded that the reaction sequence of magnesian calcite → Ca-rich dolomite → dolomite is controlled by the relative (decreasing) nucleation and growth rates of these phases.

Lumsden et al. (1989) synthesized ordered, stoichiometric dolomite from calcite in Mn-doped $MgCl_2$ solutions at 192°C and 224°C, with magnesite ($MgCO_3$) and brucite [$Mg(OH)_2$] appearing as reactive intermediates. At a given temperature, Mn addition apparently decreased the reaction rate. Cation ordering in product dolomites appeared to decrease with increased residence time (coincident with an increase in $X_{CaCO_3,dol}$); however, these data are not well constrained, and the problems inherent in intensity ratio measurements are well-known. It is thus difficult to ascertain whether this trend in order/disorder reflects Mn substitution (Peacor et al., 1987).

Schultz-Guttler (1986) explored the relationship between disordering in dolomite and the $X_{MgCO_3,cal}$ in coexisting calcite at relatively high temperatures and pressures, using pure dolomite and calcite reacted in water with Ag oxalate over long reaction times (~100 hours). Schultz-Guttler documented a positive relationship between isothermal $MgCO_3$ solubility in calcite and the disorder of coexisting dolomite, and suggested that discrepancies among previous data in the location of the calcite limb of the solvus may reflect the time required to achieve a completely ordered structure, even at high temperatures and pressures. One would expect this effect to be exacerbated at lower temperatures.

Although cation disorder in dolomite has typically been estimated from intensity ratios of ordering versus nonordering reflections, structural disorder resulting from other defects may play a critical role in reactivity. Wenk et al. (1983) summarized the occurrence of a diverse set of features known collectively as *modulated structures*. These are seen as diffuse intensity contrasts in TEM images of a variety of carbonates, including dolomite (particularly Ca-rich varieties, although often absent in stoichiometric dolomite). The role of these microstructures and their relationship to cation ordering is not yet understood, but this area holds potential in terms of documenting the relationship between dolomite of replacement origin and the carbonate precursor, and thus yielding insight into the reaction mechanism. The most common microstructure is a modulation parallel to (1014) having a wavelength of 200 Å.

Wenk et al. (1993) recently reported possible evidence of direct precipitation of ordered dolomite in a modern sabkha facies at Abu Dhabi. TEM images and electron diffraction data were used to differentiate coexisting aragonite, disordered magnesian calcite (grading to partially ordered calcian dolomite), and ordered dolomite. Despite the coexistence of these phases, Wenk et al. argued that there is no evidence to suggest that the ordered dolomite formed as a result of ordering of a calcite precursor, and thus must have formed by direct precipitation, possibly with concurrent aragonite dissolution. If this observation is correct, it suggests that dolomite growth can occur at low to moderate temperatures on relatively short timescales, given an appropriate solution chemistry.

Finally, Arvidson and Mackenzie (1997, 1999, 2000) showed from theoretical considerations and continuous flow, dolomite-seeded, reactor experiments that the dolomite precipitation reaction rate is strongly dependent on temperature and moderately dependent on the saturation state of the solution from which the dolomite is forming. The dolomite produced in their experiments was variable in composition but typically was a calcium-rich protodolomite that formed syntaxial overgrowths on the seed material. The activation energy for this protodolomite precipitation reaction was found to be 31.9 kcal mol^{-1}, lower than that for reactions involving ordered dolomite. In addition, Arvidson and Mackenzie (2000) found that the energy required to convert a calcium-rich protodolomite to an ordered dolomite is about 5.5 kcal mol^{-1}.

Some geological considerations

Questions concerning the diagenesis of calcites in carbonate sediments and rocks and the occurrence and mode of formation of dolomite have plagued geologists for nearly a century. Our knowledge of the mineralogy, chemistry, and phase relations of these minerals acquired from experimental work has formed the necessary foundation to obtain answers to some of the questions. For example, the finding that calcites rich in magnesium are not stable relative to pure calcite and dolomite is the principal reason that the former phases convert to the latter with increasing age of the sedimentary rock in which the minerals are found. Modern calcareous sediments of shallow-water origin are characterized by a metastable assemblage of calcites and aragonite, primarily of biogenic origin, and sparse inorganic dolomite. With time and burial, this metastable assemblage generally will be converted to the stable assemblage of calcite and dolomite; aragonite will alter to calcite, the stable dimorph, and magnesium-rich calcite will lose magnesium and be converted to nearly pure calcite and contemporaneously, or later in the history of the sedimentary deposit, dolomite. Evidence for the diagenetic stabilization of aragonite and magnesian calcite is ample, particularly from limestones subjected to vadose and phreatic meteoric diagenesis.

In general, the alteration of magnesian calcite to calcite is petrographically described as a pseudomorphic replacement, in which the original skeletal or non-skeletal materials retain their original microarchitecture. The original fabrics are retained because the chemical reactions that transform magnesian calcites to calcite probably occur along a microscopic diagenetic front. Incongruent reaction is the reaction mechanism most commonly invoked to explain the retention of microarchitecture in stabilized magnesian calcite skeletal materials or other substrates. The reaction involves dissolution of the magnesian calcite accompanied by precipitation of calcite with a lower concentration of magnesium than that of the initial solid. Based mainly on experimental and theoretical arguments, Bischoff *et al.* (1993) showed that there are several potential stabilization pathways by which a magnesian calcite is converted to calcite. However, these pathways are poorly documented in the rock record because of lack of adequate field studies.

The mineral dolomite and the uncertainties surrounding its origin have attracted the attention of earth scientists for more than a century. The core of the dolomite problem is the apparent paradox posed by the paucity of dolomite in modern marine depositional environments and its relative abundance in the sedimentary rock record. There are two major contrasting paradigms that have been promoted to explain this observation. The first is that most dolomite is not primary in nature but forms over time from the alteration of the other principal carbonate phases of calcite and aragonite. The second is that the atmosphere/ocean environmental conditions of today are not favorable to dolomite precipitation but were at times during the geologic past. This latter paradigm requires changes in seawater chemistry during geologic time (see *Sea water: Temporal changes to the major solutes*). To obtain further insight into this latter paradigm, Arvidson and Mackenzie (1999) studied the rate of dolomite precipitation in a steady-state, dolomite-seeded, reflux reactor over a range of temperatures at fixed bulk composition of the reacting solution and at variable temperature and solution composition. They found that the rate was a strong function of temperature and the degree of saturation of the solution with respect to dolomite, as has been shown with calcite and aragonite. The strong dependence on temperature and saturation state observed in the experiments led the authors to conclude that it is the overall rate of dolomite precipitation relative to competing carbonate phases of calcite and aragonite at surface temperatures that determines dolomite's abundance in the sedimentary regime. They also concluded that small changes in temperature can substantially affect the dolomite precipitation rate. Thus, it is possible that the abundance of dolomite in the sedimentary rock record reflects its absolute precipitation rate in the environment, the rate being controlled by the environmental variables of temperature and solution composition, i.e., the saturation state of seawater. There is little doubt that the global mean temperature of the planet has varied over geologic time as has the chemistry of seawater. Thus, the abundance of dolomite might reflect to some degree these changes. This assertion is currently being debated (Holland and Zimmermann, 2000) and further investigated, as is the role of bacteria in the dolomite precipitation reaction (Warthmann *et al.*, 2000).

Fred T. Mackenzie

Bibliography

Arvidson, R.S., and Mackenzie, F.T., 1997. Tentative kinetic model for dolomite precipitation rate and its application to dolomite distribution. *Aquatic Geochemistry*, **2**: 273–298.

Arvidson, R.S., and Mackenzie, F.T., 1999. The dolomite problem: control of precipitation kinetics by temperature and saturation state. *American Journal of Science*, **299**: 257–288.

Arvidson, R.S., and Mackenzie, F.T., 2000. Temperature dependence of mineral precipitation rates along the $CaCO_3$-$MgCO_3$ join. *Aquatic Geochemistry*, **6**: 249–256.

Bertram, M.A., Mackenzie, F.T., Bishop, F.C., and Bischoff, W.D., 1991. Influence of temperature on the stability of magnesian calcite. *American Mineralogist*, **76**: 1889–1896.

Bischoff, W.D., Bishop, F.C., and Mackenzie, F.T., 1983. Biogenically produced magnesian calcite: inhomogeneities in chemical and physical properties: comparison with synthetic phases. *American Mineralogist*, **68**: 1183–1188.

Bischoff, W.D., Sharma, S.K., and Mackenzie, F.T., 1985. Carbonate ion disorder in synthetic and biogenic magnesian calcites: a Raman spectral study. *American Mineralogist*, **70**: 581–589.

Bischoff, W.D., Mackenzie, F.T., and Bishop, F.C., 1987. Stabilities of synthetic magnesian calcites in aqueous solution: comparison with biogenic materials. *Geochimica et Cosmochimica Acta*, **51**: 1413–1423.

Bischoff, W.D., Bertram, M.A., Mackenzie, F.T., and Bishop, F.C., 1993. Diagenetic stabilization pathways of magnesian calcites. *Carbonates and Evaporites*, **8**: 82–89.

Busenberg, E., and Plummer, N.J., 1989. Thermodynamics of magnesian calcite solid-solutions at 25°C and 1 atm total pressure. *Geochimica et Cosmochimica Acta*, **53**: 1189–1208.

Gaines, A.M., 1974. Protodolomite synthesis at 100°C and atmospheric pressure. *Science*, **183**: 178–182.

Gaines, A.M., 1980. Dolomitization kinetics, recent experimental studies. In Zenger, D.H., Dunham, J.B., and Ethington, R.L. (eds.), *Concepts and Models of Dolomitization*. Tulsa: Society of Economic Paleontologists and Mineralogists, Special Publication, 28, pp. 81–86.

Goldsmith, J.R., and Graf, D.L., 1958. Structural and compositional variations in some natural dolomite. *Journal of Geology*, **66**: 678–793.

Goldsmith, J.R., and Heard, H.C., 1961. Subsolidus phase relations in the system $CaCO_3$-$MgCO_3$. *Journal of Geology*, **69**: 45–74.

Goldsmith, J.R., Graf, D.L., and Heard, H.C., 1961. Lattice constants of calcium-magnesium carbonates. *American Mineralogist*, **46**: 456–457.

Graf, D.L., 1961. Crystallographic tables for the rhombohedral carbonates. *American Mineralogist*, **46**: 1283–1316.

Graf, D.L., and Goldsmith, J.R., 1955. Dolomite-magnesian relations at elevated temperatures and CO_2 pressures. *Geochimica et Cosmochimica Acta*, **7**: 109–128.

Graf, D.L., and Goldsmith, J.R., 1956. Some hydrothermal syntheses of dolomite and protodolomite. *Journal of Geology*, **64**: 173–186.

Graf, D.L., and Goldsmith, J.R., 1958. The solid solubility of MgCO$_3$ in CaCO$_3$: a revision. *Geochimica et Cosmochimica Acta*, **13**: 218–219.

Hardie, L.A., 1987. Dolomitization: a critical review of some current views. *Journal of Sedimentary Petrology*, **57**: 166–183.

Harker, R.I., and Tuttle, O.F., 1955a. Studies in the system CaO-MgO-CO$_2$, Part 1. The thermal dissociation of calcite, dolomite, and magnesite. *American Journal of Science*, **255**: 209–224.

Harker, R.I., and Tuttle, O.F., 1955b. Studies in the system CaO-MgO-CO$_2$, Part 2. Limits of solid solution along the join CaCO$_3$-MgCO$_3$. *American Journal of Science*, **253**: 274–282.

Holland, H.D., and Zimmermann, H., 2000. The dolomite problem revisited. *International Geology Review*, **42**: 481–490.

Katz, A., and Matthews, A., 1977. The dolomitization of CaCO$_3$, an experimental study at 252–295 degrees C. *Geochimica et Cosmochimica Acta*, **41**: 297–308.

Lumsden, D.N., Snipe, L.G., and Lloyd, R.V., 1989. Mineralogy and Mn geochemistry of laboratory synthesized dolomite. *Geochimica et Cosmochimica Acta*, **53**: 2325–2329.

Machel, H.G., and Mountjoy, E.W., 1986. Chemistry and environments of dolomitization-a reappraisal. *Earth Science Review*, **23**: 175–222.

Mackenzie, F.T., and Morse, J.W., 1992. Sedimentary carbonates through Phanerozoic time. *Geochimica et Cosmochimica Acta*, **56**: 3281–3295.

Mackenzie, F.T., Bischoff, W.D., Bishop, F.C., Loijens, M., Schoonmaker, J., and Wollast, R., 1983. Magnesian calcites: low-temperrature occurrence, solubility, and solid-solution behavior. In Reeder, R.J. (ed.), *Carbonates: Mineralogy and Chemistry, Mineralogical Society of America Reviews in Mineralogy No 11*. (2nd edn), pp. 97–144.

Morse, J.W., and Mackenzie, F.T., 1990. *Geochemistry of Sedimentary Carbonates*. Amsterdam: Elsevier.

Paquette, J., and Reeder, R.J., 1990. Single crystal X-ray structure refinements of two biogenic magnesian calcite crystals. *American Mineralogist*, **75**: 1151–1158.

Peacor, D.R., Essene, E.J., and Gaines, A.M., 1987. Petrologic and crystal-chemical implications of cation-disorder in kutnahorite [CaMn(CO$_3$)$_2$]. *American Mineralogist*, **72**: 319–328.

Plummer, L.N., and Mackenzie, F.T., 1974. Predicting mineral solubility from rate data: application to the dissolution of magnesian calcites. *American Journal of Science*, **274**: 61–83.

Rosenberg, P.E., and Holland, H.D., 1964. Calcite-dolomite-magnesite stability relations in solutions at elevated temperatures. *Science*, **145**: 700–701.

Rosenberg, P.E., Burt, D.M., and Holland, H.D., 1967. Calcite-dolomite-magnesite stability relations in solutions: the effect of ionic strength. *Geochimica et Cosmochimica Acta*, **31**: 391–396.

Schultz-Guttler, R., 1986. The influence of disordered, non-equilibrium dolomites on the Mg solubility in calcite in the system CaCO$_3$-MgCO$_3$. *Contributions to Mineralogy and Petrology*, **93**: 395–398.

Sibley, D.F., 1990. Unstable to stable transformations during dolomitization. *Journal of Geology*, **98**: 739–748.

Tribble, J.S., Arvidson, R.S., Lane III, M., and Mackenzie, F.T., 1995. Crystal chemistry, and thermodynamic and kinetic properties of calcite, dolomite, apatite, and biogenic silica: applications to petrologic problems. *Sedimentary Geology*, **95**: 11–37.

Warthmann, R., van Lith, Y., Vasconcelos, C., McKenzie, J.A., and Karpoff, A.M., 2000. Bacterially induced dolomite precipitation in anoxic culture experiments. *Geology*, **28**: 1091–1094.

Wenk, H.R., Barber, D.J., and Reeder, R.J., 1983. Microstructures in carbonates. In Reeder, R.J. (ed.), *Carbonates: Mineralogy and Chemistry, Mineralogical Society of America Reviews in Mineralogy No. 11*. pp. 301–367.

Wenk, H.R., Meisheng, H., and Frisia, S., 1993. Partially disordered dolomite: microstructural characteristics of Abu Dhabi carbonates. *American Mineralogist*, **78**: 769–774.

Cross-references

Ankerite
Authigenesis
Bioclasts
Calcite Compensation Depth
Cathodoluminescence
Cements and Cementation
Chalk
Dedolomitization
Diagenesis
Dolomites and Dolomitization
Isotopic Methods in Sedimentology
Neomorphism and Recrystallization
Neritic Carbonate Depositional Environments
Oceanic Sediments
Seawater: Temporal Changes in the Major Solutes
Speleothems
Stains for Carbonate Minerals
Tufas and Travertines

CARBONATE MUD-MOUNDS

Mud-mounds (also mud mounds, and mudmounds) are important geological bodies in the carbonate system. Wilson (1975) is the one who used the term most extensively. His conception of mud mounds had a profound influence on subsequent perception of this type of carbonate body, particularly through the last chapter of his book where he refers to "foreslope mud mounds" on type I carbonate shelf margin profile (Wilson, 1975, 361, and Figure XII-3, a widely cited figure in subsequent geological literature). For most workers then, mud-mounds became these geological structures found associated with basin-type facies.

Definition

There is no widely accepted definition of mud-mound. Several have been proposed (e.g., several papers in Monty *et al.*, 1995), based on various parameters such as percentage of mud versus organisms, presence or absence of specific elements such as stromatactis, relief above seafloor at time of deposition, or type of organisms controlling surface accretion. On a strictly descriptive base, mud-mound can be broadly defined as a rock or sediment body, having a mound or lens shape, massive to crudely bedded, within which finely crystalline carbonate predominates (without connotation on the origin of the last), and more commonly associated with deep water facies (modified from Bourque, 2001). This roughly corresponds to the objects described by Wilson (1975) as mud-mounds. It excludes mudbanks that are shallow water structures (AGI Glossary of Geology), like for instance those of the Florida Bay and Shelf.

A key point of this definition is that mud-mound is a geological object that resulted from a more or less modified primary depositional structure which did not necessarily have an important relief above the seafloor at time of deposition (even if the object exhibits today a mound or lens shape), nor a mud content equal to the present finely crystalline carbonate content.

Facies spectrum and origin of mud-mounds

The origin of mud-mounds is controversial. First of all, there is an inherent ambiguity related to the term itself: the "mud" part of it, which is the fundamental component of mud-mounds, evokes a primary composition of loose lime mud. However, most studies during the last two decades stressed the necessary role of soft-bodied and/or poorly fossilizable organisms in mound accretion, in particular that of the microbes and the sponges, and showed that a significant portion of the finely crystalline material may correspond to an early lithification of these organisms or parts of them. An important character of mud mounds is the multigeneration nature of the finely crystalline material (polymuds of Lees and Miller, 1985): the first generation corresponds to the primary, early cemented, often pelleted mudstone, whereas the subsequent generations correspond to geopetal muds originating from collapse of uncemented material and/or internal sedimentation, all types of generation forming a highly structured mosaic.

On the one hand, several workers put forward an all-microbial origin of mud-mounds following Monty (1976, and in Monty et al., 1995). In particular, the pelleted nature of the finely crystalline carbonate is viewed as the product of in situ calcifying microbes and therefore considered as diagnostic for a microbial origin of the mound (several papers in Monty et al., 1995). This view is emphasized by the mud-mound spectrum proposed by Bosence and Bridges (in Monty et al., 1995) who recognized only two types of mud-mounds: the microbial mud-mounds, and the biodetrital mud-mounds, leaving no room for other builders than the microbial community.

On the other hand, several studies have shown that most Paleozoic mud-mounds are not made up of a single component, but rather of a mosaic of facies (e.g., James and Bourque, 1992). For one, the red stromatactis facies commonly found in the basal part in Paleozoic mud-mounds has been shown to be very rich in sponge bodies and/or in sponge spicules everywhere it has been studied (references in Bourque and Boulvain, 1993). The facies has been interpreted to have originated from the calcification of a primary sponge network through microbial decay during early diagenesis, destroying the original shape of the sponges and much of the spicules, and ending with a finely crystalline limestone having a pelleted texture (Bourque and Boulvain, 1993).

There is no doubt that microbial communities may have acted as primary builders in mud-mound accretion, for instance in trapping and binding allochthonous mud, producing in situ loose lime mud or mediating pelleted micrite-microspar precipitation via their vital activity (e.g., by photosynthesis). They may also have acted as the main agent of transformation of a primary organic-rich substrate into micrite-microspar via organic matrices. For instance, decaying former microbial community (Défarge et al., 1996) or decaying sponges (Reitner et al., 1995) may be used as substrate for calcite nucleation. In both cases, the end product is a pelleted finely crystalline limestone (micrite-microspar), a situation that renders difficult the recognition of the primary community.

Figure C7 presents a genetic classification of mud-mounds that account for the primary facies spectrum found in mud-mounds. Microbial and skeletal mud-mounds are those which were primarily accreted by microbial communities or skeletal organisms (sponges, bryozoans, delicate corals, etc.), respectively. Detrital mud-mounds are those which resulted from the piling up of biodetrital material and/or loose lime mud.

The above discussed types of mud-mounds were built mainly by autotrophic and/or heterotrophic communities. These last years, ancient mud-mounds related to chemosynthesis have been described, and we should expect that more will be recognized in the future. In this case, deciphering which group of organisms is involved in chemosynthesis is not readily achieved.

Gaps in current knowledge

Beyond the controversies on the origin of mud-mounds, two main gaps in the current knowledge of mud-mounds are particularly teasing. Firstly, the origin of the enormous amount of marine cement and the driving mechanism for marine fluid circulation remain poorly understood. For instance, the stromatactis facies into which early marine cement may constitute more than half of the rock volume testifies to rapid, massive cementation. An enormous volume of water supersaturated with respect to $CaCO_3$ is needed. The mechanisms that drive the necessary fluids to the sites of early cementation in relatively deep-water mounds are presently poorly circumscribed.

Secondly, mud-mound workers can rely on a very limited number of modern analogs for mud-mounds. One of the strengths that lead to satisfactorily understand the ancient reef system through geological time is the actualistic approach favored by the accessibility of the modern reef system. This is far from being the case for the mud-mound system. The only modern objects that may represent analogs to mud-mounds are the deep-water mud-rich ahermatypic coral mounds. These can be classified as skeletal and/or detrital mud-mounds. Obviously, they represent only a very small portion of the mud-mound spectrum in the geological record, and therefore carbonate mud-mounds are geological bodies important for our understanding of the past carbonate system and life.

Pierre-André Bourque

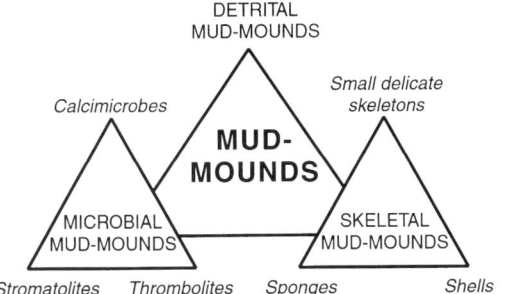

Figure C7 Primary facies spectrum of mud-mounds., from Bourque (2001) (modified from James and Bourque, 1992).

Bibliography

Bourque, P.-A., 2001. Mud-mounds: do they still constitute an enigma? Géologie Méditerranéenne, 28: 27–32.

Bourque, P.-A., and Boulvain, F., 1993. A model for the origin and petrogenesis of the red stromatactis limestone of Paleozoic carbonate mounds. Journal of Sedimentary Petrology, 63: 607–619.

Défarge, C., Trichet, J., Jaunet, A.-M., Robert, M., Tribble, J., and Sansone, F.J., 1996. Texture of microbial sediments revealed by cryo-scanning electron microscopy. Journal of Sedimentary Research, 66: 935–947.

James, N.P., and Bourque, P.-A., 1992. Reefs and Mounds. In Walker, R.G., and James, N.P. (eds.), Facies Models—Response to Sea-Level Change. Geological Association of Canada, pp. 323–347.

Lees, A., and Miller, J., 1985. Facies variation in Waulsortian buildups, Part 2; Mid-Dinantian buildups from Europe and North America. *Geological Journal*, **20**: 159–180.

Monty, C.L.V., 1976. The origin and development of cryptalgal fabrics. In Walter, M.R. (ed.), Stromatolites. *Developments in Sedimentology, Elsevier*, 20, pp. 193–249.

Monty, C.L.V., Bosence, D.W.J., Bridges, P.H., and Pratt, B.R. (eds.), 1995. Carbonate mud-mounds, their origin and evolution. International Association of Sedimentologists, Special Publication, 23.

Reitner, J., Neuweiler, F., and Gautret, P., 1995. Modern and fossil automicrites: implications for mud mound genesis. In Reitner, J. and Neuweiler, F. (coord.), *Mud Mounds: A Polygenetic Spectrum of Fine-Grained Carbonate Buildups*. *Facies*, **32**: pp. 4–17.

Wilson, J.L., 1975. *Carbonate Facies in Geological History*. Springer-Verlag

Cross-references

Algal and Bacterial Carbonate Sediments
Bacteria in Sediments
Cements and Cementation
Diagenesis
Neritic Carbonate Depositional Environments
Oceanic Sediments
Spiculites and Spongolites
Stromatactis

CATHODOLUMINESCENCE (applied to the study of sedimentary rocks)

Introduction

Cathodoluminescence (CL) is the term used to describe the emission of light from crystalline materials caused by irradiation with high energy electrons emitted from a cathode source. The detailed physical principles of CL are explained in Walker (1985). When a beam of electrons impinge on a solid material a variety of interactions take place. Some energy is absorbed, some transmitted (but only through very thin sheets of matter) and some is reflected in both particulate and wave forms. It is the light emission that constitutes CL. This light is not always emitted in the visible region of the spectrum, although there are few studies in sedimentology of CL in the UV or IR portions of the electromagnetic spectrum.

Sedimentologists use CL extensively in clastic and carbonate petrography to reveal fabric and paragenetic information not apparent with other imaging techniques. CL is now an established routine tool that provides qualitative information on grain provenance, sediment fabric, compaction processes, cement growth fabric, and by inference, diagenetic history. It also enables more precise quantification of constituents and fabrics. CL petrography is an important prerequisite to reconstructing paleo-fluid evolution of carbonate and sandstone aquifers in order to define the relationship between growth zones or healed fractures and multiple generations of fluid inclusions.

Analytical techniques

The basic requirements for recording CL in sediments are detailed in Marshall (1988). In essence, a source of high energy electrons generated under near vacuum is focused onto a polished thin section causing excitation of atoms in the surface layers of the mineral lattice. CL is emitted as the excited electrons fall back to their ground state. The emission can be recorded photographically, digitally or spectroscopically. Levels of CL emission are commonly very low so efficient light transfer instrumentation has been developed to optimize capture of CL emissions. Several complicating factors arise. Firstly, electron irradiation damages some minerals (especially quartz) resulting in changes to or destruction of the CL emission. To record the initial state of luminescence either low energy electron beams or short exposure times are required (Ramseyer *et al.*, 1989). Instruments with scanning beams and low beam energies that operate at high vacuum (and thus produce fewer damaging negative particles) provide the best results with quartz. Secondly, the character of the CL emission changes with temperature so instruments with cold stages enabling cooling of the sample to liquid N or He temperatures have been developed. Moreover, color as observed with the eye is a subjective description of overlapping spectral bands so accurate determination of CL emission can only be conducted if spectra are measured. Commercial 'cold cathode' instruments are available and are adequate for the study of CL in carbonate minerals, whilst either so-called 'hot cathode' research instruments or CL attachments for the SEM are required if accurate information is to be collected on CL in quartz (Walker and Burley, 1991).

The nature of CL emission

The characteristics of CL in many common minerals encountered in sediments are well documented, although some important questions remain. Luminescence emanates from emission centers. These are an expression of changes from an ideal, perfect lattice structure (Walker, 1985), usually grouped as either: (i) intrinsic centers, effectively structural imperfections of the lattice, alternatively referred to as 'defect centers'; and (ii) extrinsic centers, the result of trace elements being incorporated into the crystal reflecting the composition of the solution or melt from which the mineral crystallized, commonly referred to as 'impurity centers'. Defects can be growth distortions, 'holes' in the lattice or particular vibrational states of atoms, whilst impurities can behave as activators, sensitizors or quenchers of CL. However, the distinction between these two categories is artificial as many defect centers occur as the result of an impurity. Most natural crystalline materials exhibit some CL (see the compilation in Marshall, 1988), unless they contain significant amounts of Fe^{2+}, an effective quencher of CL when incorporated into the mineral lattice. Carbonate cements (aragonite, calcite, dolomite, magnesite, but not siderite or ankerite), quartz, detrital and authigenic quartz and feldspars, authigenic clay minerals (kaolinite but not illite or chlorite), sulfates (barite and anhydrite) and phosphates (apatite) all exhibit significant CL emissions.

The nature of the emission depends on the interaction between the CL center and the vibrational energy of the lattice, which is strongly temperature dependent. The number and type of both emission and quenching centers present determine the nature of the emission spectrum and color observed. An emission spectrum reveals how many spectral bands contribute to the overall color but also resolves the nature and number of emission centers present (Walker and Burley, 1991). This is particularly important in determining the cause of CL in a mineral. Without spectroscopy, CL remains a poorly constrained tool for describing fabrics.

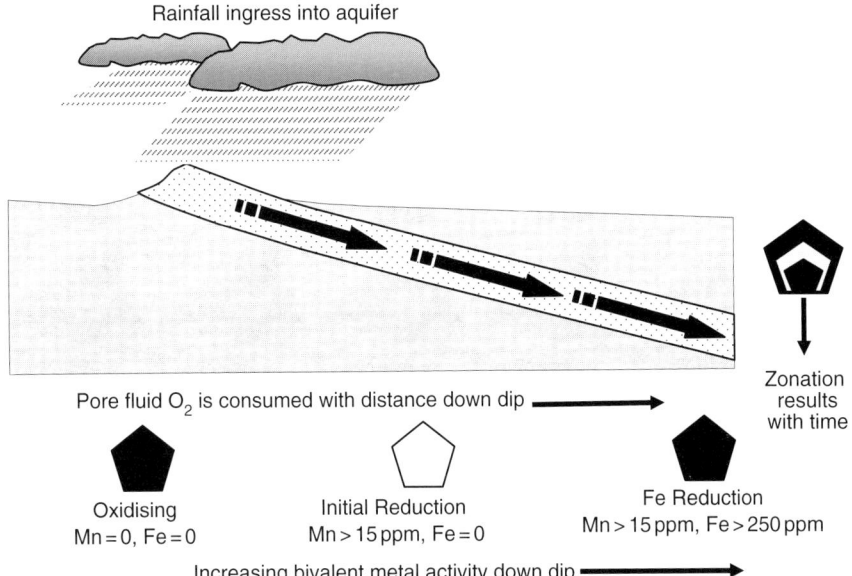

Figure C8 The evolution of carbonate CL fabrics during gradual burial as Mn and Fe become available for incorporation into cement. Nonluminescent cement precipitated near a source of oxygenated fluid recharge may be the lateral correlative of bright cement some distance down-aquifer, and low intensity CL cement further into the subsurface (based on Hendry 1993).

CL of carbonates

Sedimentary carbonates of different major and trace element composition exhibit distinct CL emission spectra and resulting colors, revealing a variety of growth fabrics. Activators in sedimentary carbonate minerals are mostly transition metals (especially Mn^{2+}) and rare earth elements. Intrinsic CL in carbonates is a low intensity dark blue emission, that is swamped if more than a few ppm of Mn^{2+} are present. Mn^{2+} produces a yellow CL emission in calcites when substituted in Ca^{2+} sites. Substitution of Mn^{2+} into Mg^{2+} lattice sites shifts the emission to longer wavelengths. Thus Mn^{2+} activated Mg-calcites are characterized by more orange CL than yellow and dolomite is bright red, whilst Mn^{2+} substitution into a dolomite Ca^{2+} lattice site produces yellow CL emission. Some dolomites exhibit a luminescence that is green to the eye but when resolved spectroscopically is shown to be a combination of the intrinsic blue and a low intensity yellow emission. Fe^{2+} is the dominant quencher in carbonates, and at concentrations of ca >250 ppm shifts the yellow-orange red luminescence emission toward longer wavelength resulting in brown-appearing emissions that become less intense as the Fe^{2+} content increases.

CL color in carbonates depends on both absolute concentrations of Mn^{2+} and Fe^{2+} and the ratio between them (ten Have and Heijen, 1985). The maximum CL intensity is controlled by the Fe/Mn ratio, whilst the CL intensity below the maximum is a function of the absolute amounts of Mn^{2+} for a given Fe/Mn ratio (Hemming et al., 1989). Incorporation of Mn^{2+} and Fe^{2+} into carbonates is thus largely controlled by the redox state of the diagenetic environment (Barnaby and Rimstidt, 1989). Oxidized Mn^{3+} and Fe^{3+} are both insoluble in aqueous fluids, but as the oxygen content decreases, first Mn^{3+} and then Fe^{3+} are reduced to their soluble bivalent form, successively becoming available for incorporation into the carbonate lattice. If calcite is precipitated during burial its CL emission character tends to evolve from nonluminescent (although with intrinsic blue emission being much more common than reported because of its low emission intensity) through bright yellow to dull brown (Figure C8).

CL of quartz

CL emissions in quartz are much more variable, and are ascribed to a variety of emission centers (Table C2). Additionally, the spectrum and resulting CL color of some types of quartz changes drastically during irradiation, indicating that an emission center is created during the excitation process, although this center remains poorly understood. The CL characteristics of quartz crystallized from a melt are different to that precipitated from an aqueous fluid, such as authigenic quartz in sandstones and some veins. Anhydrous, magmatic quartz is characterized by two broad emission bands, one in the blue, and one in the red range of the visible spectrum. In fact the broad blue emission is considered by solid state physicists to represent two, and possibly three, separate but overlapping intrinsic center emissions (Gorton et al., 1997). The visible CL colors of quartz, ranging from bright blue, though purple to red, reflect variation in the relative intensity of these emission bands. The red peak also has a long tail that extends into the infrared, and this peak increases in intensity at the expense of the blue peak with electron irradiation, producing a range of reddish 'brown' colors.

Work on authigenic quartz precipitated from aqueous solutions has demonstrated that hydrous quartz exhibits a wide variety of other CL emission bands that include relatively low intensity, short lived blue, green, yellow and red colors as well as combinations of these (Ramseyer and Mullis, 1988). The blue emissions also extend into the UV range (Demars et al., 1996). Microprobe analysis and spectral studies suggest that these transient emissions are impurity-related and

Table C2 Summary of suggested CL emission centers in quartz

CL emission center	CL color	Wavelength	Quartz Type
Electron irradiation	red	650	Magmatic and α-quartz
Lattice defects	Broad red	650–690	Magmatic and α-quartz
Defect center	blue	430	Anhydrous quartz
O_2 vacancy	UV	260	Anhydrous quartz
SiO_4 tetrahedra	blue	455	Anhydrous quartz
Self trapped excitons	blue	430–450	Anhydrous quartz
Structural water	Green-blue	510	Diagenetic α-quartz
Li-compensated Al^{3+}	UV	340	Diagenetic α-quartz
Interstitial Li^+	Blue, green, yellow	440–580	Diagenetic α-quartz
Interstitial Na^{2+}	UV	290	Diagenetic α-quartz
Fe	red	640	Magmatic quartz

Figure C9 Complex CL zoning patterns in calcite. Early non-luminescent cements are followed by cyclical banding whilst final pore fills are brightly luminescing suggestive of initial burial followed by unconformity related meteoric recharge.

Figure C10 'Hot-CL' photomicrograph of plutonic detrital quartz grains enclosed in zoned syntaxial quartz overgrowths.

correlate with abundance of lithium-compensated aluminum substitution in the quartz lattice, and with the abundance of structurally bound water in the lattice. This raises the possibility that CL in quartz may record pore fluid changes or specific pore fluid conditions, much as CL fabrics do in carbonates.

Applications of CL in sedimentology

CL is routinely used as a qualitative tool to reveal details of sediment texture. In carbonates depositional grains and bioclasts are easily distinguished from cements, in some cases even when recystallization has taken place. In sandstones, the contrast in CL character between detrital quartz of various origins and authigenic quartz overgrowths enables accurate quantification of cement abundance and determination of porosity-loss processes (Houseknecht, 1988), including mechanical compaction and pressure dissolution. Zinkernagel (1978) was the first to demonstrate that the various CL colors of quartz grains can be related to their geological history, making quartz CL a very powerful provenance tool. Magmatic α-quartz with dominant blue emission is typical of plutonic environments, volcanic quartz is dominated by red CL whilst reddish 'brown' colors are typical of metamorphic terrains.

Figure C11 SEM-CL showing compactional microfracturing of a detrial quartz grain in a quartz cemented sandstone. Note that overgrowths postdate fracturing.

CL of both carbonate and quartz cements reveals details of growth fabric and growth processes (Figures C9 and C10). These fabrics include zonation, dissolution events and microfracturing. Concentric zoning consists of parallel bands of different CL emission colors indicative of growth through time. Discontinuities and irregular crosscutting surfaces in these

Table C3 Compilation of CL characteristics of growth zone patterns in carbonate cements and their CL hydrological interpretation. Fluctuations in redox state that produce cyclical non-bright luminescent zoning are a strong indicator of active but near surface fluid flow conditions prior to fluid-rock stabilization. They most commonly form in shallow groundwater aquifers that respond to recharge. In contrast, poorly zoned, low-luminescent sparry cements indicate stagnant anoxic conditions that typify deeper burial

Zoning pattern	Environment	Hydrology	Examples
Intrinsic blue or nonluminescent, almost always unzoned	Hardgrounds, shallow subsurface, marine or meteoric	Oxygenated, open to seawater circulation, active to stagnant aquifers	Syn-depositional calcite marine cements; tufa; unconformity cements
Concentric non (or intrinsic blue)—bright—low intensity luminescent zones	Shallow burial, marine or meteoric aquifer	Falling oxygenation, active or stagnant aquifers	Gradual burial in stagnant or slow moving aquifer
Nonluminescent with multiple bright concentric subzones	Shallow burial meteoric aquifer	Fluctuating oxygenated to suboxic, active recharge	Transient fresh water lenses beneath emergent shoals or cyclothems
Concentric non, bright and low intensity luminescent zones	Deeper burial or distal aquifer	Low but fluctuating oxygenation levels, active aquifer	Major unconformity sourced meteoric phreatic cements
Dominantly low intensity luminescence, with sector zoning, cements tend be coarsely crystalline; recrystallization may obliterate early fabrics	Deep burial beneath influence of active surface hydrology	Variably anoxic, typically stagnant, with slow cementation	Typical of burial cements formed in chemical compaction regime or from dewatering of mudstones and organic matter-rich source rocks

zones document intervening episodes of dissolution or replacement. Sector zoning comprises polygonal to irregular zones representing adjacent crystal faces that result from differential trace element up-take (Reeder, 1991). Microfracturing may record compactional processes or tectonic events (Boiron et al., 1992; Figure C11). CL of cements is thus most powerful as a means of resolving fine scale detail of textural fabrics and determining paragenetic relationships between successive cement generations, and gave rise to the term 'cement stratigraphy', the study of spatial relationships of coeval cement generations (Meyers, 1991). As a consequence of changing redox relationships, growth fabrics of carbonate cements revealed by CL are often related to paleohydrology, burial and uplift history (Table C3). However, although CL emission is qualitatively related to geochemistry, there are many other factors that contribute to CL color and intensity (Machel and Burton, 1991). In practice, therefore, erroneous inferences can be made unless other geochemical techniques, such as fluid inclusions and stable isotope analysis, are used to fully characterize cements. Equivalent zoning patterns, for example, may be produced by different processes or be diachronous, even in closely spaced samples. Cement stratigraphy of quartz overgrowths is less well documented, largely because the relationship between pore fluid conditions and resulting CL fabrics is less well understood (but see Demars et al., 1996). CL fabrics in quartz overgrowths were used by Burley et al. (1989) to establish a detailed cement and fluid inclusion paragenesis, enabling burial diagenetic histories to be reconstructed.

Future research directions

CL petrography is now an established routine tool in sedimentology because of its value in distinguishing depositional from authigenic components, revealing cement fabrics, and in reconstructing diagenetic histories. However, the use of CL in providing quantitative constraints on paleo-pore fluid geochemistry requires a more complete understanding of the nature of CL centers, especially in silicates. Quantitative spectroscopic studies are required, linked with geochemical analysis of cements, if CL emissions and cement growth fabrics are to provide more useful information on lattice defects, trace element compositions and pore fluid evolution.

Stuart D. Burley

Bibliography

Barnaby, R.J., and Rimstidt, J.D., 1989. Redox conditions of calcite cementation interpreted from Mn and Fe contents of authigenic calcites. *Geological Society of America Bulletin*, **101**: 795–804.

Boiron, M.C., Essarraj, E., Sellier, M., Cathelineau, M., Lespinasse, M., and Poty, B., 1992. Identification of fluid inclusions in relation to their host microstructural domains in quartz by cathodoluminescence. *Geochemica et Cosmochimica Acta*, **56**: 175–185.

Burley, S.D., Matter, A., and Mullis, J., 1989. Timing diagenesis in the Tartan reservoir (UK North Sea): constraints from combined cathodoluminescence microscopy and fluid inclusion studies. *Marine and Petroleum Geology*, **6**: 98–120.

Demars, C., Pagel, M., Deloule, E., and Blanc, P., 1996. Cathodoluminescence of quartz from sandstones: interpretation of the UV range by determination of trace element distributions and fluid inclusion P-T-X properties in authigenic quartz. *American Mineralogist*, **81**: 891–901.

Gorton, N.T., Walker, G., and Burley, S.D., 1997. Experimental analysis of the composite blue cathodoluminescence emission in quartz. *Journal of Luminescence*, **72**: 669–671.

Hemming, N.G., Meyers, W.J., and Grams, J.C., 1989. Cathodoluminescence in diagenetic calcites: the roles of Fe and Mn as deduced from electron microprobe analysis and spectroscopic measurements. *Journal of Sedimentary Petrology*, **59**: 404–411.

Hendry, J.P., 1993. Calcite cementation during bacterial manganese, iron and sulphate reduction in Jurassic shallow marine carbonates. *Sedimentology*, 40: 87–106.

Houseknecht, D., 1988. Intergranular pressure solution in four quartzose sandstones. *Journal of Sedimentary Petrology*, 58: 228–246.

Machel, H.G., and Burton, E.A., 1991. Factors governing cathodoluminescence in calcite and dolomite, and their implications for studies of carbonate diagenesis. In Barker, C.E., and Kopp, O.C. (eds.), *Luminescence Microscopy: Quantitative and Qualitative Analysis*. SEPM Short Course, 25, pp. 37–57.

Marshall, D.J., 1988. *Cathodoluminescence of Geological Materials: An Introduction*. Winchester, MA: Allen and Unwin.

Meyers, W.J., 1991. Calcite cement stratigraphy: an overview. In Barker, C.E., and Kopp, O.C. (eds.) *Luminescence Microscopy: Quantitative and Qualitative Analysis*. SEPM Short Course, 25, pp. 133–148.

Ramseyer, K., and Mullis, J., 1988. Factors influencing the short-lived blue cathodoluminescence of α-quartz. *American Mineralogist*, 75: 791–800.

Ramseyer, K., Fischer, J., Matter, A., Eberhardt, P., and Geiss, J., 1989. A cathodoluminescence microscope for low intensity luminescence. *Journal of Sedimentary Petrology*, 59: 619–622.

Reeder, R.J., 1991. An overview of zoning in carbonate minerals. In Barker, C.E., and Kopp, O.C. (eds.), *Luminescence Microscopy: Quantitative and Qualitative Analysis*. SEPM Short Course, 25, pp. 77–81.

ten Have, T., and Heijen, W., 1985. Cathodoluminescence activation and zonation in carbonate rocks: an experiemental approach. *Geologie en Mijnbouw*, 64: 297–310.

Walker, G., 1985. Mineralogical applications of luminescence techniques. In Berry, F.J., and Vaughan, D.J. (eds.), *Chemical Bonding and Spectroscopy in Mineral Chemistry*, Chapman and Hall, pp. 103–140.

Walker, G., and Burley, S.D., 1991. Luminescence petrography and spectroscopic studies of diagenetic minerals. In Barker, C.E., and Kopp, O.C. (eds.), *Luminescence Microscopy: Quantitative and Qualitative Analysis*. SEPM Short Course 25, pp. 83–96.

Zinkernagel, U., 1978. *Cathodoluminescence of Quartz and its Application in Sandstone Petrology*. Contributions to Sedimentology, 8, 69p.

Cross-references

Authigenesis
Carbonate Mineralogy and Geochemistry
Cements and Cementation
Diagenesis
Fluid Inclusions
Provenance

CATION EXCHANGE

Cation exchange is a process by which cations are reversibly adsorbed on charged surfaces of sediments from solution. Isomorphous substitution and broken edges in the phyllosilicates, and deprotonation of acid groups in the organic matter provide net negative charge. The negative charge is balanced by the electrostatic attraction of cations. Cation exchange is a rapid and reversible process, and cations are exchanged on an equivalent charge basis. All components of sediments contribute to some extent to cation exchange. However, minerals in the clay fraction ($<2\,\mu m$) and organic matter are the most important components involved in this process.

Because of their charge and size, the alkaline earth and alkali metal cations (Ca^{2+}, Mg^{2+}, Na^+, and K^+) do not form stable secondary minerals and are left in the solution phase to neutralize surface charge of sediments. Under acidic conditions Al^{3+} (and monomers of Al) and H^+ are also present on the exchange sites. Since the cations are held by a weak electrostatic force by sediments and are easily exchangeable, they are called exchangeable cations. The total amount of exchangeable cations that can be held by sediments is called the cation exchange capacity (CEC). The CEC and exchangeable cations are expressed in mmoles of charge kg^{-1} ($mmol_c\ kg^{-1}$). The CEC values of some common clay minerals in sediments are given in Table C4.

Historical development

Following Thompson's (1850) discovery of the cation exchange phenomenon, Way (1850) conducted the first comprehensive studies of cation exchange in Rothamsted, England. He concluded that cation exchange is instantaneous, cations are exchanged on equivalent basis, adsorption of cations increases with increasing concentration of added salt and clays are responsible for cation exchange reactions. Way also made some wrong conclusions including that organic matter takes no part in cation exchange and cation adsorption is irreversible. Several scientists reviewed the work by Way and the two wrong conclusions were corrected by Johnson (1859). Van Bemmelen (1888) presented a more complete picture of the cation exchange process and demonstrated that cation exchange is not restricted to Ca^{2+} in the soil, but other cations also participate in the process. The controversy about the nature of exchanging materials was resolved after the advent of X-ray diffraction when Hendricks and Fry (1930), and Kelley et al. (1931) independently established the crystalline nature of the clay particles. Within next 20 years, structures of phyllosilicates, which are the major components of the clay fraction of sediments, and their cation exchange were well established.

In the cation exchange process, there is selectivity or preference for smallest hydrated radius among cations of the same valence and for cations of different valencies generally the higher valence cations are preferred. The nature of sediments and the amount of CEC also influence the relative replacing power of cations on sediments. There is no single universal order of replacing power of cations on sediments. For high charge sediments the relative order of cation replaceability (lyotropic series) is $Li^+ \approx Na^+ > K^+ \approx NH_4^+ > Rb^+ > Cs^+ \approx Mg^{2+} > Ca^{2+} > Sr^{2+} \approx Ba^{2+} > La^{3+} \approx "H" (Al^{3+}) > Th^{4+}$ (Bohn et al., 1985).

Table C4 Cation exchange capacities of some common clay minerals found in sediments

Clay mineral	Cation exchange capacity ($mmol_c\ kg^{-1}$)
Kaolinite	30–150
Mica	100–400
Vermiculite	1200–2070
Smectite	800–1350
Chlorite	100–400
Palygorskite-sepiolite	50–450
Allophane-imogolite	100–400
Zeolites	2200–4600

Table C5 Cation exchange selectivity coefficients for Na–K and Na–Ca exchange on sediments (after White and Zelazny, 1986)

Selectivity coefficient	Na–K exchange (K-sediment + Na$^+$-solution → Na-sediment + K$^+$-solution)	Na–Ca exchange (Ca-sediment + 2Na$^+$-solution → 2Na-sediment + Ca^{2+}-solution)
Kerr[1]	$K_K = \dfrac{\{Na^+\text{-sediment}\}[K^+]}{\{K^+\text{-sediment}\}[Na^+]}$	$K_K = \dfrac{\{Na^+\text{-sediment}\}^2[Ca^{2+}]}{\{Ca^{2+}\text{-sediment}\}[Na^+]^2}$
Vanselow[2]	$K_V = \dfrac{\{Na^+\text{-sediment}\}[K^+]}{\{K^+\text{-sediment}\}[Na^+]}$	$K_V = \left[\dfrac{\{Na^+\text{-sediment}\}^2[Ca^{2+}]}{\{Ca^{2+}\text{-sediment}\}[Na^+]^2}\right] \times \left[\dfrac{1}{\{Ca^{2+}\text{-sediment}\} + \{Na^+\text{-sediment}\}}\right]$
Krishnamoorty and Overstreet[2]	$K_{KO} = \dfrac{\{Na^+\text{-sediment}\}[K^+]}{\{K^+\text{-sediment}\}[Na^+]}$	$K_{KO} = \left[\dfrac{\{Na^+\text{-sediment}\}^2[Ca^{2+}]}{\{Ca^{2+}\text{-sediment}\}[Na^+]^2}\right] \times \left[\dfrac{1}{1.5\{Ca^{2+}\text{-sediment}\} + \{Na^+\text{-sediment}\}}\right]$
Gaines and Thomas[2]	$K_{GT} = \dfrac{\{Na^+\text{-sediment}\}[K^+]}{\{K^+\text{-sediment}\}[Na^+]}$	$K_{GT} = \left[\dfrac{\{Na^+\text{-sediment}\}^2[Ca^{2+}]}{\{Ca^{2+}\text{-sediment}\}[Na^+]^2}\right] \times \left[\dfrac{1}{2[2\{Ca^{2+}\text{-sediment}\} + \{Na^+\text{-sediment}\}]}\right]$
Gapon[3]	$K_G = \dfrac{\{Na^+\text{-sediment}\}[K^+]}{\{K^+\text{-sediment}\}[Na^+]}$	$K_G = \dfrac{\{Na^+\text{-sediment}\}\sqrt{[Ca^{2+}]}}{\{Ca^{2+}_{1/2}\text{-sediment}\}[Na^+]}$

[1] Concentration in solution in mol l^{-1} and on the sediment in mol kg^{-1}.
[2] Concentration in solution as activities and on the sediment in mol kg^{-1}.
[3] Concentration in solution in mol l^{-1} and on the sediment in mol$_c$ kg^{-1}.

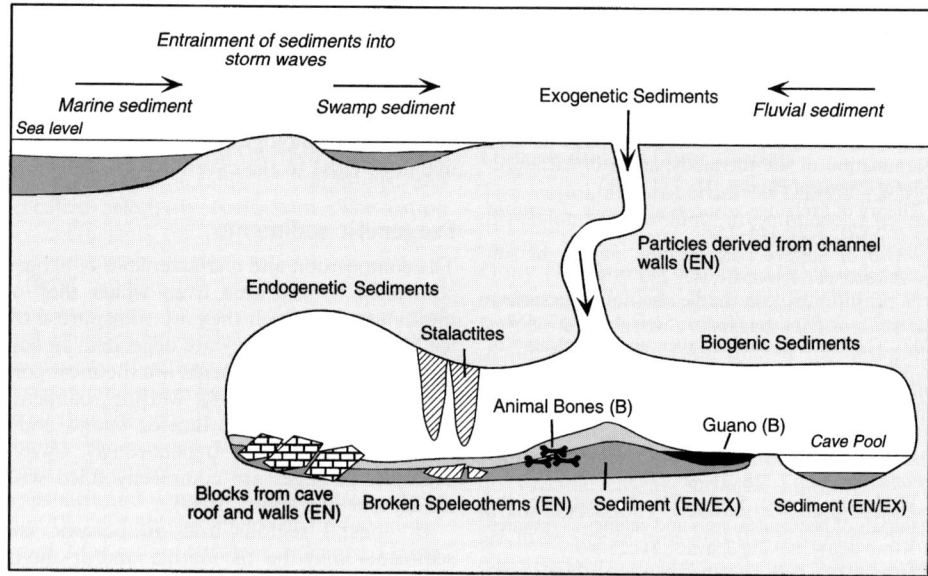

Figure C12 Schematic diagram showing potential source areas for cave sediments. EN = endogenic; EX = exogenic; B = biological.

invertebrates and vertebrates that are adapted to life in these dark environs. Bones or waste products from these animals may become an integral and important component of cave sediments. The bones may also serve as nuclei around which speleothems are precipitated. Guano, which is common in some caves, is formed by bats, birds, and even cave crickets (Jefferson, 1978).

Summary comments

Cave sediments are formed of sediments derived from many different sources (Figure C12). Determining the origin of the sediment may be difficult because grains of the same composition, size, and composition may be exogenic or endogenic in origin. Nevertheless, study of these sediments is important because they may provide evidence of deposits that are not preserved elsewhere in the succession.

Brian Jones

Bibliography

Dunham, R.J., 1969. Vadose pisolite in the Capitan Reef, Permian—New Mexico and Texas: *Society of Economic Paleontologists and Mineralogists, Special Publication*, 14, pp. 182–191.

Ford, D., 1988. Characteristics of dissolution cave systems in carbonate rocks. In James, N.P., and Choquette, P.W. (eds.), *Paleokarst*. New York: Springer-Verlag, pp. 25–57.

Jefferson, G.T., 1978. Cave faunas, In Ford, T.D., and Cullingford, C.H.D. (eds.), *The Science of Speleology*. New York: Academic Press, pp. 359–421.

Jones, B., 1992a. Void-filling deposits in karst terrains of isolated oceanic islands: a case study from Tertiary carbonates of the Cayman Islands. *Sedimentology*, **39**: 857–876.

Jones, B., 1992b. Caymanite, a cavity-filling deposit in the Oligocene-Miocene Bluff Formation of the Cayman Islands. *Canadian Journal of Earth Sciences*, **29**: 720–736.

Jones, B., and Kahle, C.F., 1995. Origin of endogenetic micrite in karst terrains: a case study from the Cayman Islands. *Journal of Sedimentary Research*, **65**: 283–293.

Smart, P.L., Palmer, R.J., Whitaker, F., and Wright, V.P., 1988. Neptunian dikes and fissure fills: an overview and account of some modern examples, In James, N.P., and Choquette, P.W. (eds.), *Paleokarst*. New York: Springer-Verlag, pp. 149–163.

Cross-references

Ancient Karst
Speleothems

CEMENTS AND CEMENTATION

Introduction

Cementation is the process of precipitation of mineral matter (cements) in pores within sediments or rocks. It is one of several processes, including mechanical and chemical compaction and mineral replacement, that constitute diagenesis and, taken collectively, produce progressive porosity reduction and lithification of sedimentary strata with increasing age and/or depth of burial. Cementation occurs in open intergranular or intragranular pores (i.e., between or within grains), and also takes place in larger openings such as vugs, caves or fractures. Cements even form crusts on surfaces at sediment-water or sediment-air interfaces. Precipitation of cements can occur at any stage from deposition, through burial, to uplift and re-exposure.

Cements occur in all types of siliciclastic, carbonate and evaporite strata and include an enormous variety of minerals. The most common cements are carbonates (especially calcite, aragonite, dolomite, and siderite), silicates (primarily quartz, opal, clay minerals, and zeolites), sulfates (especially gypsum and anhydrite) and chlorides (mainly halite). Cement morphologies vary widely, and morphological description, coupled with detailed geochemistry, may yield information on timing of precipitation as well as the physical and geochemical conditions under which cements formed.

CEMENTS AND CEMENTATION

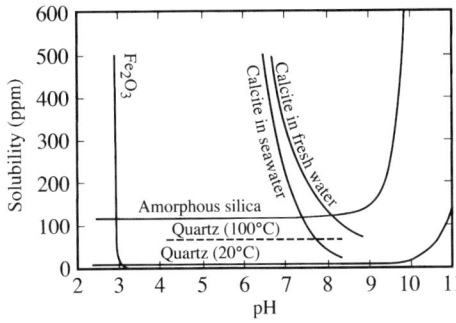

Figure C13 Solubility curves at approximately 25°C for hematite (Fe_2O_3), amorphous silica, quartz and calcite in marine and fresh waters as a function of pH. Dashed line shows quartz solubility at 100°C. (Adapted from Correns, 1950; Blatt *et al.*, 1972; Rosler and Lange, 1972).

Precipitation of cements is controlled by a variety of physical, chemical and biological factors. Crystallization kinetics, Eh, pH, P_{CO_2}, temperature-pressure conditions, and ionic concentrations and interactions of dissolved chemical constituents are common physiochemical controls. For example, Figure C13 illustrates the effects of pH on the solubility of a variety of common cementing agents. For calcite and dolomite, as temperature and pressure increase, aqueous solubility decreases, favoring precipitation at higher temperatures (Mackenzie and Bricker, 1971; Holser, 1979). Calcite and dolomite solubilities also increase with increased P_{CO_2} and decreased pH. Ions that appear to inhibit calcite precipitation include magnesium, sulfate, chloride and phosphate.

For quartz (and its polymorphs), solubility increases slightly with temperature, therefore decreasing the likelihood of significant quartz precipitation at very high burial temperatures (Mackenzie and Bricker, 1971). The pH of diagenetic fluids exerts a major influence on the formation of quartz and amorphous silica (Figure C13), with pH levels below 9 favoring precipitation.

Anhydrite solubility decreases with increasing temperatures, and gypsum solubility increases slightly to a maximum at approximately 40°C, but decreases at higher temperatures (Holser, 1979). Halite solubility increases with temperature. Pressure affects the solubility of halite slightly, but affects anhydrite more substantially. Elevated levels of sodium, potassium, and magnesium appear to decrease solubilities of gypsum/anhydrite and thus enhance its precipitation (Holser, 1979).

Iron oxides, sulfides and some carbonates (siderite and ankerite) are strongly controlled by Eh and pH conditions. Under oxidizing conditions, iron oxides form whereas under reducing conditions, pyrite or siderite are formed, depending on the relative sulfide and carbonate concentrations in the solution. Acidic pore fluids increase the solubility of any iron-bearing minerals present.

Biological metabolism and decay affect physiochemical conditions and mineral precipitation. Furthermore, because cementation involves precipitation from aqueous solutions, the presence of hydrophobic organic coatings (such as hydrocarbons) on grains can inhibit dissolution and retard or prevent cement formation.

Rates of cement precipitation vary greatly. The presence of suitable substrates for mineral precipitation plays a major role in determining rates, because crystal nucleation is kinetically more difficult than continued growth on an already-nucleated crystal. Where suitable substrates exist, cementation by minerals that are compositionally similar or identical to the substrate may proceed rapidly (Wollast, 1971), commonly producing overgrowths that are in optical continuity with substrate crystals.

Some strata are completely cemented in as little as a few years after deposition; others retain as much as 40–50 percent porosity after tens to hundreds of millions of years and hundreds to thousands of meters of burial. An enormous body of research has focused on the factors that promote or inhibit cementation as well as controls on rates of precipitation (e.g., Moore, 1989; Bjørkum *et al.*, 1998). Such extensive studies were conducted not only to gain a general understanding of the post-depositional history of sedimentary strata, but also because such knowledge is essential for proper evaluation of hydrocarbon and metallic mineral prospects.

Cements in carbonate rocks

The dominant cements in most carbonate rocks are, not surprisingly, carbonate minerals. Aragonite and high-Mg calcite are the dominant precipitates in modern marine carbonate sediments. Both minerals are unstable under most non-marine conditions, however, and so undergo rapid alteration when removed from a marine setting. Aragonite normally is dissolved in meteoric fluids; high-Mg calcite most commonly converts to low-Mg calcite during meteoric or burial diagenesis. Low-Mg calcite is the most common cement in surficial, non-marine carbonate settings and is also the dominant cement in subsurface settings along with dolomite and, far less commonly, siderite. Carbonate cements (see Folk, 1965, for shape classification) are precipitated in a variety of diagnostic fabrics and crystal morphologies that may or may not survive later diagenesis—the most common of these are illustrated in Figure C14 (see also Scholle, 1978).

Sulfate or chloride cements, especially gypsum-anhydrite and halite, are additional important cements, especially in carbonate deposits from arid to extremely arid climates. Silica forms cements in carbonate rocks under some limited conditions, as do phosphate and glauconite.

Syndepositional (eogenetic) cements in marine carbonate settings

Synsedimentary carbonate cements are widespread in modern oceans and are volumetrically important in a number of settings ranging from coastal areas to oceanic depths. The main controls on marine carbonate cementation are sedimentation rates, degree of carbonate saturation, and rates of water exchange through the sediment. Where water throughput is high (because of a strong pumping mechanism coupled with permeable sediment) and saturation levels are high, cementation is extensive. Cementation also occurs, however, where water throughput is low, provided that sedimentation rates also are very low. Areas of extensive water throughput occur mainly where wave energy or tidal pumping force seawater through the sediment—such areas include windward reef fronts and beaches. In such settings, even though they are environments with high sedimentation rates, enough new carbonate is added from seawater that cement precipitation and lithification occur quite rapidly. Many tropical carbonate

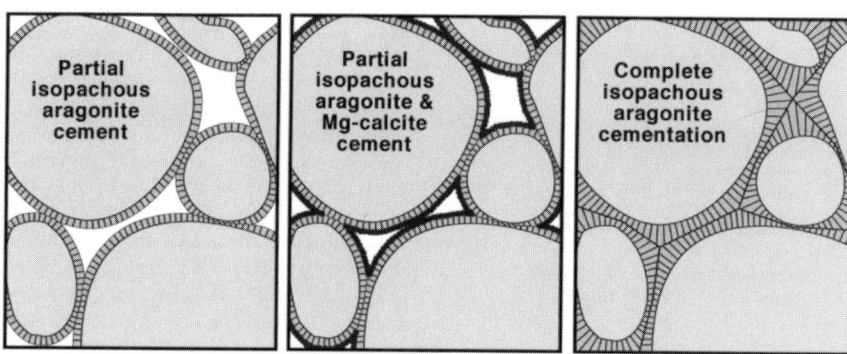

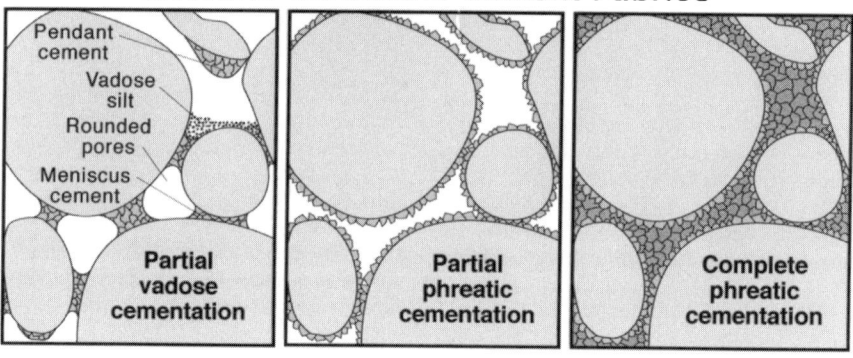

Figure C14 Sketches of typical shallow marine and meteoric (eogenetic) carbonate cement fabrics found in modern sediments. The marine cements are aragonite and high-Mg calcite; the meteoric cements are all low-Mg calcite.

beaches, for example, are armored by lithified carbonate sand, termed *beachrock* (*q.v.*), that forms in years to tens of years, based on the presence of incorporated modern artifacts such as coins (Scoffin and Stoddart, 1983). Although primarily a marine process, beachrock formation may involve mixing with coastal groundwaters in some areas. Subtropical to tropical platform or shallow shelf settings also are sites in which extensive syndepositional carbonate cements are widely distributed and volumetrically abundant, typically forming grapestone deposits or hardground layers.

Several time periods in the geologic record (e.g., the late Precambrian and late Permian) appear to have had widespread conditions that were especially favorable for syndepositional shallow-marine and coastal cementation (Grotzinger and Knoll, 1995). Large botryoids of aragonite and/or high-Mg calcite formed as seafloor crusts by direct precipitation from seawater during those time intervals (Mazzullo and Cys, 1979).

Modern shallow marine carbonate cements are mainly aragonitic with subordinate high-Mg calcite (James and Choquette, 1983). This probably results from inhibition of low-Mg calcite formation by one or more ions in seawater and pore waters including Sr, Mg, and PO_4 (e.g., Folk, 1974; Burton and Walter, 1990). Aragonite cement is found mainly as isopachous (uniform thickness) crusts or palisades of fibrous crystals (Figures C14 and C15*). Coarser botryoids of bladed aragonite are known from large voids in reef-front settings (James and Ginsburg, 1979). High-Mg calcite cements typically occur as peloidal micritic precipitates with crystal sizes of less than 1–2 µm. Because aragonite and high-Mg calcite cements are unstable during later burial, recognition of synsedimentary cementation may depend on recognition of remnants of primary fabrics, the occurrence of intraclasts (synsedimentary rock fragments), borings that crosscut grains and cements, or attached/encrusting organisms. Several studies indicate that the mineralogy of inorganic marine precipitates has varied through geologic time as a function of a number of large-scale climatic and oceanographic variables (e.g., Sandberg, 1983).

Synsedimentary marine cementation also occurs in finer grained and deeper water carbonates, especially nannofossil oozes or chalks. Although deeper ocean waters generally are undersaturated with respect to calcium carbonate, precipitation still can occur interstitially below the sediment-water interface where biological processes, decomposition of organic matter, and diffusive exchange of seawater and pore waters can lead to carbonate supersaturation. Enhanced water exchange and bacterial activity in burrows, leads to preferential cementation of these structures. In areas or times of essentially zero sediment accumulation (hiatus zones), such cementation may extend beyond burrows and become pervasive within a sediment layer, leading to the formation of firmgrounds or, with more extensive cementation, hardgrounds (Kennedy and Garrison, 1975) in both shallow and deep-water marine settings. True hardground creation can take

* Figures C15, C17, C18A and C18B appear on Plate I, facing page 114.

as little as a few years to hundreds of years in lagoonal or open shelf settings, but may require hiatal periods of tens to hundreds of thousands of years in outer shelf to deep ocean floor settings.

The most common cement in such deeper-water settings is extremely finely crystalline high-Mg calcite that is very difficult to differentiate optically from the sediment matrix. Non-carbonate precipitates such as calcium phosphates and glauconite (glaucony) are also common, and many hardgrounds are typified by a yellowish-brown or greenish coloration where such non-carbonate minerals are present. Other factors aiding hardground recognition are the presence of synsedimentary clasts, borings, and encrusting or attached fauna.

Early diagenetic (eogenetic) cements in non-marine carbonate settings

Nonmarine carbonate cements are produced in terrestrial settings both above and below the water table (that is in undersaturated or vadose zones as well as saturated or phreatic zones). Such cements typically form as a result of: (1) degassing of CO_2-rich groundwaters emerging at the surface or in caves, (2) by evapotranspiration of carbonate-bearing surface waters and near-surface groundwaters, or (3) by mixing of near-surface waters of different compositions. Travertine deposits forming at hot- and cold-springs exemplify the first process; these deposits include substantial amounts of interstitial carbonate cement as well as surficial cement crusts. Likewise, cave *speleothems* (*q.v.*) (stalactites, stalagmites, flowstone) represent cements formed in caves and solution-enlarged fractures from meteoric waters that lose CO_2 as they emerge into large, partially air-filled cavities. Most speleothems are composed of low-Mg calcite; although a few examples include aragonite, dolomite, or gypsum. Most speleothems have distinctive palisades of coarse crystals oriented perpendicular to the sequential growth banding that reflects the frequently changing conditions that characterize vadose to shallow phreatic settings.

Nonmarine cements resulting from evapotranspiration of near-surface waters are common in soils formed on carbonate or siliciclastic sediments and extend from the soil zone down to the water table and below. Such exposure-related carbonate precipitates include root coatings (rhizoliths), soil nodules, and massive calcrete or caliche horizons. Cements produced under freshwater vadose conditions consist of low-Mg calcite with distinctive structures (Figure C14), including meniscus, needle-fiber, and pendant fabrics, and may be associated with vadose silt (James and Choquette, 1984).

Non-carbonate cements and crusts can also be formed at nonmarine exposure surfaces, including precipitates of iron oxides (ferricretes: see *Laterites*) and of silica (silcretes). In most cases, the cementing agents in such crusts are derived by local dissolution and reprecipitation of primary constituent grains and thus form mainly on siliciclastic substrates.

Exposure surfaces associated with hypersaline waters have very different cementation patterns. In particular, areas with intense evaporation of marine-derived waters (e.g., areas of coastal sabkhas or salinas such as the Persian Gulf or southern and western Australia) have a wide variety of carbonate and evaporite cements (see *Sabkha, Salar, Salt Flat*). Aragonite and high-Mg cements are common in such settings, as are evaporite minerals such as gypsum, anhydrite, and halite. The removal of calcium from pore waters by carbonate and sulfate precipitation, in turn, leaves a dense residual brine depleted in calcium and enriched in magnesium. Such brines can replace primary aragonite crystals with dolomite and perhaps also form dolomite cements (McKenzie, 1981). Typically, such penecontemporaneous dolomites are non-stoichiometric, aphanocrystalline, and contain numerous solid inclusions.

Pore fluids that form as a result of evaporative concentration of seawater or waters of mixed continental/seawater origin can be extremely saline—commonly as much as 8 times normal seawater concentration. In sabkha and playa environments, these waters precipitate euhedral, tabular to lenticular gypsum crystals that displace or engulf the carbonate and siliciclastic sediments in which they form. Anhydrite typically occurs as displacive nodules or contorted layers composed of white, felted masses of lath-shaped crystal fragments (Kinsman, 1966). Kinsman (1974) described primary anhydrite precipitates as well as anhydrite formed by syndepositional dehydration of gypsum. Climate cycles now are known to cause calcium sulfates to undergo multiple episodes of dehydration and rehydration prior to significant burial, making evaluation of primary fabrics and mineralogies difficult.

At the landward edges of coastal sabkhas, pore water salinity dilution through influx of continental groundwater commonly leads to dissolution and reprecipitation of early-formed anhydrite and gypsum as clear, poikilotopic gypsum that can completely cement sabkha and playa sands (carbonate or siliciclastic). However, as sabkha and playa sediments undergo deeper burial, increasing temperatures and pressures result in reconversion of gypsum to anhydrite.

The third process leading to cementation in nonmarine to marginal marine settings, the mixing of waters of dissimilar compositions is widespread, but not yet well understood. This process can involve mixing of marine with fresh or saline meteoric waters as well as mixing of purely terrestrial waters of different compositions (e.g., vadose and phreatic waters or groundwaters from two different sources). Mixing can lead either to dissolution or precipitation, even when two under-saturated fluids are mixed. The main cements attributed to this process are low-Mg calcite and dolomite. Low-Mg calcite forms in the mixing zone in fabrics similar to normal phreatic precipitates—i.e., as isolated euhedral crystals that coat grains and that generally are randomly oriented unless substrate microstructure influences orientation. Modern mixing-zone dolomite commonly consists of bands of euhedral, limpid, pore-lining dolomite crystals, interlayered with bands of syntaxial calcite (Humphrey, 2000).

In situations in which mixing of marine with fresh or brackish water takes place under locally reducing conditions, siderite ($FeCO_3$) cements can form (e.g., Mozley and Burns, 1993). Siderite cements, however, also form in deep marine settings lacking water mixing, but where bacterial processes provide reducing conditions.

Most early diagenetic carbonate cements, other than those formed in such unusually reducing conditions, have a common set of trace-element geochemical characteristics. In particular, virtually all have low iron and manganese concentrations (and thus are non- or weakly-luminescent under cathodoluminescence (*q.v.*); they are pale pink to red when stained with Alizarin Red-potassium ferricyanide solutions—see Stains, and Dickson, 1966). The low iron and manganese concentrations result from the fact that these elements are substantially incorporated into calcium carbonates only under reducing conditions.

Distinguishing carbonate cements formed in marine environments from those formed in meteoric settings is complex. Cement fabrics and morphologies, discussed above, are most effectively used when combined with relevant isotopic information. Marine cements typically are relatively enriched in ^{18}O and ^{13}C, which helps to distinguish them from isotopically much "lighter" cements formed from low salinity meteoric fluids or under deeper burial conditions.

Burial-stage (mesogenetic) cements

Most porosity loss in carbonate strata occurs during progressive burial at depths well below the environment of near-surface water circulation (e.g., Schmoker and Halley, 1982). Although physical and chemical compaction account for some of the porosity reduction, much of the loss (especially in grainstones) results from carbonate cementation. The subsurface precipitation of up to 30–40 percent cement in carbonate rocks creates a substantial mass transport problem if the material is imported from external sources. In most instances, however, burial-stage cements are derived primarily by dissolution of primary carbonate grains or unstable eogenetic carbonate cements, either locally or in nearby strata. Dissolution can occur at high-pressure grain contacts, along solution seams and stylolites, or in associated shales, marls or other organic-rich strata. Elevated temperatures and pressures, forced fluid movements due to subsidence and compaction, a variety of diagenetic changes in associated siliciclastic rocks or organic deposits, changes in Eh (generally to progressively more reducing conditions), and many other factors affect burial cementation (Scholle and Halley, 1985; Choquette and James, 1987).

Low-Mg calcite is the main mesogenetic cement in most carbonate rocks; dolomite and quartz cements (in several varieties) also are common. Evaporite cements (especially anhydrite) or hydrothermal precipitates (e.g., fluorite and barite) are important in some special settings.

Most mesogenetic calcite cements, like many meteoric phreatic cements, have equant to bladed crystals that increase in size toward the centers of large pores. The crystal size and orientation of substrate grains and eogenetic cements can affect mesogenetic cement morphologies. Echinodermal bioclasts exert especially strong substrate control because each grain is a single crystal of calcite. This encourages the formation of very large cement overgrowths, which may obliterate all surrounding pore space, a common situation in pelmatozoan limestones. In some cases, single-crystal overgrowths surround several framework grains, producing a "poikilotopic" fabric. Overgrowth cements also are found at much smaller scales. In chalks and other micritic carbonate deposits, calcite overgrowths are documented on sub-micron-sized material, including the individual elements of coccolith "shields". Such overgrowths are only discernable using scanning electron microscopy (Figure C16).

Many, but not all, burial-stage calcites are compositionally zoned (Figure C17). Recognition of zonation typically requires observations using staining techniques (Dickson, 1966), cathodoluminescence microscopy (Miller, 1988; Reeder, 1991), or backscattered electron or microprobe mapping (Reed, 1989). Zonation provides evidence of fluctuating geochemical conditions during an extended period of crystal growth. Corrosion zones preserved between stages of crystal

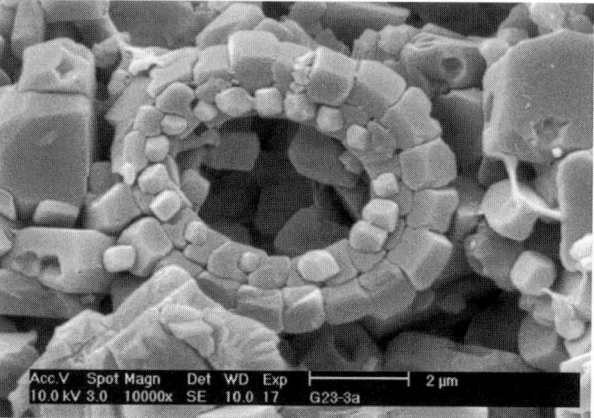

Figure C16 Scanning electron photomicrograph of a Maastrichtian pelagic chalk from the Danish North Sea. The large coccolith shield (probably *Cribrosphaerella ehrenbergii*) has multiple calcite overgrowths formed in optical continuity with individual crystal elements (note irregular sizes of individual elements indicating variable overgrowths). Scale bar = 2 μm.

growth may reveal episodes of crystal growth alternating with dissolution and therefore even greater geochemical variations.

Burial-stage dolomite also is common, resulting from either deep reflux of surface-derived brines or remobilization of Mg^{2+} from precursor high-Mg calcite or earlier-formed non-stoichiometric dolomite at the higher temperatures and pressures encountered at depth. In general, dolomite is far more common as a replacement than as cement, but dolomite cements do occur, especially in strata where the rock framework has already been largely or completely dolomitized. Burial-stage dolomites commonly are stoichiometric, zoned, relatively inclusion free, and medium to coarsely crystalline. They typically show substantial amounts of iron substitution and may be termed ferroan dolomite or ankerite, depending on iron content (see *Ankerite*).

Dolomites formed at temperatures of about 60–150°C and higher, especially from hydrothermal fluids, hydrocarbon-bearing brines, or waters associated with sulfate reduction, have distinctive morphologies. They are coarsely crystalline, rich in fluid inclusions, and have curved crystal faces, curved cleavage planes, and undulose extinction due to lattice deformation as a result of calcium enrichment. These are termed "saddle", "baroque", or "hydrothermal" dolomites (Radke and Mathis, 1980; and see *Dolomite Textures*). Unlike other dolomites, saddle dolomites occur with roughly equal frequency as cements and replacements, and commonly are accompanied by a suite of associated precipitates, including fluorite, metallic sulfides (e.g., sphalerite, galena), and barite.

Evaporite minerals, especially gypsum and anhydrite, may be remobilized during burial (mesogenetic) and uplift (telogenetic) phases of basin history. Where primary or eogenetic evaporite minerals are present in arid-region deposits, they can be dissolved and reprecipitated. In some cases, substantial subsurface transport occurs, allowing mesogenetic evaporite cements to precipitate in strata that were not directly associated with arid climates, and therefore, carbonate rocks, irrespective of their original depositional environments, may undergo filling of primary or secondary porosity by remobilized evaporite minerals. Anhydrite commonly precipitates

PLATE I

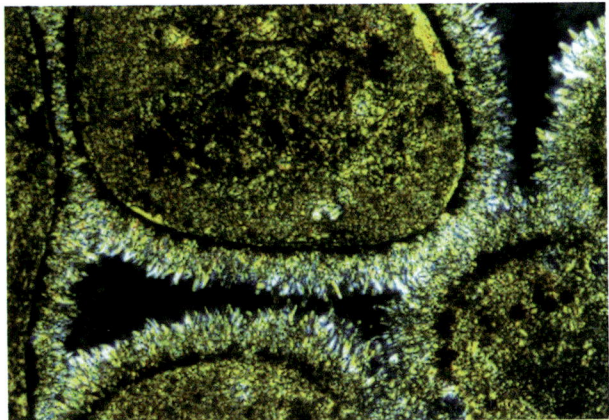

Figure C15 Thin-section photomicrograph (cross-polarized light) of modern beachrock at Salt Cay, Bahamas Islands. Dark, lumpy material surrounding grains is peloidal high-Mg calcite cement; it is followed by isopachous rim of fibrous, aragonitic, marine cement. Long axis of photo = 0.52 mm.

Figure C18A Thin-section photomicrograph (plane-polarized light) of a quartz arenite from the Devonian Hoing Sandstone in Illinois. Note the apparent interlocking, closely-packed fabric of sand grains with sutured grain boundaries possibly indicating substantial grain-to-grain pressure solution (but see Figure C18B for clarification). A few quartz overgrowths are weakly visible (e.g., lower left). Long axis of photo = 0.9 mm.

Figure C17 Thin-section photomicrograph (plane-polarized light; Alizarin-Fe stained) showing the value of staining in recognizing cement generations. This fracture in a Permian limestone from East Greenland is filled with successive bands of ferroan (blue stained) and non-ferroan (pink stained) calcite indicative of fluctuations in redox conditions during precipitation. Long axis of photo = 4.5 mm.

Figure C18B Photograph of same area as Figure C18A but under cathodoluminescence and using very long time-exposure. Detrital quartz cores appear blue-white or orange, and well-defined quartz overgrowths (transparent) surround virtually every grain. Boundaries between overgrowths are not sutured, and sandstone is clearly not pressure dissolved. From Scholle (1979, p. 114, photographs by Robert Sippel).

PLATE II

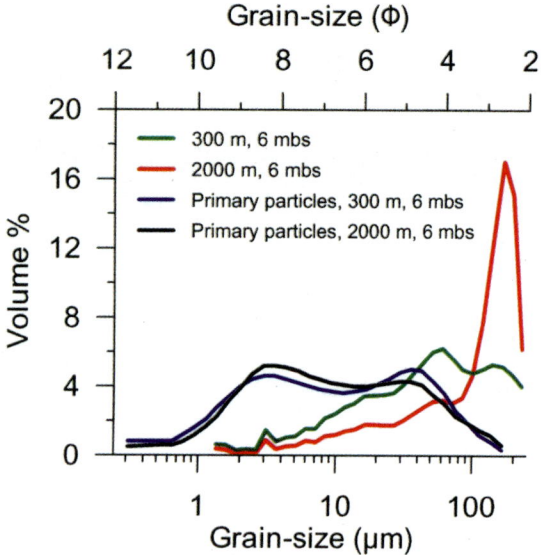

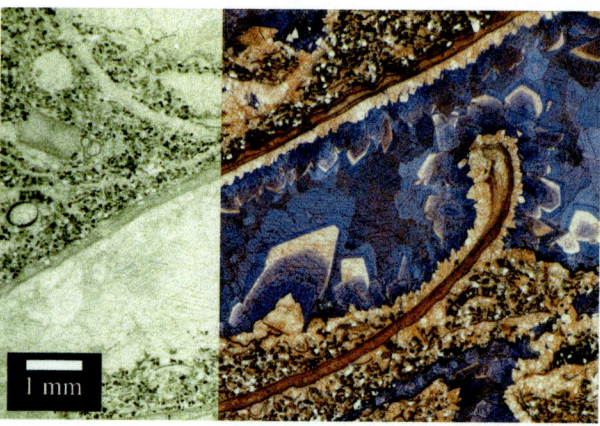

Figure F15 *In situ* grain size spectra of inorganic lime particles from a dredging plume measured with a LISST-100 *in situ* laser 300 m and 2000 m from the dredging site respectively. The grain size spectra of the primary particles have been analysed in the laboratory by use of a Malvern MasterSizer/E laser. The spectra indicate the significance of pure salt flocculation because no organic material was present in the sediment plume (Mikkelsen and Pejrup, 2000).

Figure S37 Photomicrograph of grainstone from the Derbyhaven Formation, Carboniferous Limestone, Isle of Man. Stain highlights the distribution of detrital quartz (unstained) in peloidal sediment caught between calcite strophomenid brachiopods that have pseudopuntate shell structure stained orange/red. The differentiation of early non-ferroan calcite cement (pink) from later zoned ferroan calcite cement (blue and mauve) is only visible in the stained segment of this photomicrograph.

from hot, geochemically evolved, basin-derived brines. Unlike the felted masses of crystals that form under near-surface conditions, mesogenetic hydrothermal anhydrite forms as large, blocky crystals that may also replace the carbonate host rock.

Silica cements can form at relatively early stages of burial where primary opal (opal-A) is present in the form of siliceous fossils (diatoms, radiolarians, siliceous sponges) or volcanic ash. Small particle sizes, high surface areas, high solubility of opal relative to quartz, and increasing temperatures with depth all drive opal alteration (Williams *et al.*, 1985). Opal-A may undergo diagenetic alteration to opal-CT, a transformation that can yield cement *lepispheres* (*q.v.*). However, the ultimate transformation product normally consists of one or more forms of quartz: megaquartz, microquartz (chert), or chalcedony that form either as replacement of carbonate grains or as cement. Fibrous quartz (chalcedony) occurs in both length-fast (chalcedonite) and length-slow (quartzine and/or lutecite) forms; the presence of length-slow chalcedony can indicate a possible association with evaporites or former evaporites.

Understanding the formation of burial-stage cements depends largely on observations of the sequence of cement formation (paragenesis) coupled with geochemical information. Paragenetic determinations show the relative temporal relationships between any given cement and a variety of other geological events, the absolute timing of which may or may not be known. Such events can include earlier cementation episodes, compaction effects (including stylolitization), uplift and possible exposure, development of fractures or other deformation features, hydrocarbon emplacement, and hydrothermal mineralization.

Several types of geochemical information are used to assist in the recognition of burial-stage cements. Trace-element data is useful mainly because most near-surface cements are produced in oxic environments and many burial-stage cements are formed at lower Eh levels. Thus, Fe, Mn and other metals concentrations typically are elevated in burial cements relative to earlier-formed precipitates. Stable isotope geochemistry (primarily $^{16}O/^{18}O$) is widely applied to the study of timing of carbonate cementation because $^{16}O/^{18}O$ ratios depend, in part, on temperature of precipitation. Thus, burial-stage cements typically are enriched in ^{16}O (and, largely as a result of organic matter alteration, ^{12}C) relative to most marine precipitates. Unfortunately, the isotopic composition of waters from which precipitation occurs also strongly affects the isotopic ratios of the cement, making it impossible to determine uniquely the temperature of formation from isotopic measurements alone on a single precipitate. The same "light" O and C ratios that characterize burial cements are also found in many near-surface meteoric precipitates. Analyzes of two or more minerals (e.g., calcite and quartz) precipitated at the same time and from the same waters can resolve the problem of dealing with two "unknowns", but proving simultaneous precipitation is difficult.

Homogenization measurements on *fluid inclusions* (*q.v.*) incorporated within cements can be used to directly determine precipitation temperatures, and thus, their time of formation if the geothermal history is well constrained (Goldstein and Reynolds, 1994). In addition, freezing-point determinations on the fluid inclusions can provide data on the chemical composition of precipitating fluids. Such studies have been applied very successfully to a variety of minerals, but should be approached with care for several reasons. Inclusions can be incorporated into crystals when they are precipitated but also at later stages through the healing of fractures, cleavages and other dislocations. Only "primary" inclusions provide accurate temperatures and fluid chemistries of initial crystal formation; thus, their proper recognition is critical, but often difficult. In addition, some minerals, especially calcite, are unlikely to maintain intact primary fluid inclusions during deeper burial and elevated temperatures. Several studies have shown that as little as 15°C of "overheating" can cause primary inclusions in calcite to re-equilibrate, thus altering their apparent formation temperatures (Goldstein and Reynolds, 1994).

In recent years, $^{87}Sr/^{86}Sr$ determinations have been applied to carbonate cement studies. Compilation of a seawater $^{87}Sr/^{86}Sr$ secular variation curve for the Phanerozoic (Burke *et al.*, 1982) can aid in the identification of unaltered marine precipitates. Many burial cements, by contrast, have $^{87}Sr/^{86}Sr$ ratios far more radiogenic than marine cements in the same rocks, largely through incorporation of strontium derived from subsurface alteration of feldspars, clays and other clastic terrigenous minerals.

Absolute age determinations using a variety of radiometric methods (^{14}C, Th/Th, $^{230}Th/^{234}U/^{238}U$, $^{238}U/^{206}Pb$, and others) have been used to constrain the timing of carbonate cement formation and continue to be perfected. Additional dating techniques (K/Ar, Ar/Ar, U-Th, and others) are applicable to non-carbonate precipitates, such as authigenic clays or feldspars.

Uplift-stage (telogenetic) cements in carbonate rocks

Uplift typically leads to reduced temperatures and pressures and may be accompanied by substantial changes in pore fluid composition due to deep penetration of meteoric fluids from elevated recharge areas. This generally leads to substantial dissolution, with generation of solution-enlarged fractures, vugs and caves. Carbonate cements precipitated during this stage are volumetrically subordinate to earlier precipitates and are difficult to distinguish from either early meteoric or burial cements except by their paragenetic relationships. Thus, they are discussed only very briefly here.

Carbonate cements normally remain unaltered or may undergo partial dissolution during telogenesis, but evaporite cements may undergo dramatic transformations. Anhydrite cements, because of their high solubility in meteoric waters, may be extensively dissolved and reprecipitated as gypsum or may be leached entirely during uplift. Calcite cements commonly fill some of the porosity that is newly generated by such dissolution (Scholle *et al.*, 1992).

Cements in sandstones and other siliciclastic strata

Cementation of siliciclastic sandstones and conglomerates can be quite complex—where siliciclastic strata have a diverse primary mineralogical composition, they also undergo complex subsurface alteration processes. Many of the varied detrital grains in siliciclastic strata act as substrates for the precipitation of overgrowth cements. Other grains are unstable in surficial or shallow- to moderate-burial conditions and dissolve to yield a variety of complex aqueous solutions capable of precipitating a diverse range of cements.

Three precipitates account for the vast majority of cementation in clastic terrigenous strata: various forms of silica (quartz and opal), carbonate minerals (mainly calcite), and authigenic clay minerals. Many other minerals, including carbonates (dolomite and siderite), evaporites, zeolites, iron oxides, iron sulfides, and feldspars occur as cements in lesser volumes or under specialized conditions; many of these cements are illustrated in Scholle (1979).

Syndepositional to early diagenetic (eogenetic stage) cements in siliciclastic deposits

It is uncommon for siliciclastic sediments to be extensively cemented during eogenetic diagenesis, especially as compared to their carbonate counterparts. Some hardgrounds, especially phosphatic ones, are known from modern siliciclastic shelf sands in areas of very low sediment accumulation rates. Beachrock cementation occurs on siliciclastic shorelines in areas of marine carbonate formation (e.g., the Red Sea and Gulf of Elat/Aqaba) and carbonate cementation of shelf sands occurs in extensive areas of methane seepage. Aragonite and high-Mg calcite are the predominant marine carbonate cements in modern marine siliciclastic settings, as they are in carbonate environments.

The most common sites of eogenetic cementation of siliciclastic deposits are surficial to shallow subsurface meteoric environments. In surficial settings, calcium carbonate-bearing waters are drawn to the surface by capillary forces and evaporatively concentrated. This leads to precipitation of micritic low-Mg calcite cements forming highly lithified calcretes and caliche crusts. This process occurs primarily in areas of shallow water tables in semiarid to arid settings and, in some instances, is substantially influence by transpiration (sometimes yielding cemented root structures, termed "rhizocretions"). Dissolution and reprecipitation of silica- and/or iron-bearing minerals by surface evaporation creates siliceous or ferruginous surface crusts, termed silcretes and ferricretes, respectively. In arid regions, it is even more common to observe near-surface dissolution, evaporative concentration, and reprecipitation of sulfate evaporites, producing surficial layers of siliciclastic sediment that are completely cemented by coarsely crystalline, poikilotopic gypsum (gypcretes).

Influx of calcium carbonate-rich groundwaters in tectonically active basins occurs by subsurface flow from adjacent highlands that may contain carbonate strata undergoing surficial weathering or subsurface dissolution. Surface re-emergence and evaporation of those waters produces spring-fed low-Mg calcite travertines as well as interstitial cementation and surface coating of fluvial sediments downstream from groundwater emergence sites.

Early diagenetic infiltration and lateral flow of groundwaters can also lead to localized concretionary cementation of siliciclastic sediments. Carbonate concretions are most common, but iron oxides and hydroxides, iron sulfides, silica, barite, and gypsum also form concretions. Concretions are spherical or elongate masses of sediment cemented by finely crystalline precipitates. Septarian concretions (see *Septarian Concretions*) are precipitated carbonate nodules that also have fractures filled with coarse, multigeneration carbonate. Individual concretions can range from millimeters to several meters in length, and these sometimes coalesce to form larger sheets or irregular masses. Most concretions are syngenetic to very early diagenetic precipitates, although some have long precipitation histories or form during late diagenesis.

Carbonate concretions normally are composed of multiple generations of non-ferroan low-Mg calcite, reflecting the consistently high Eh and compositional variability of near-surface waters. However, where eogenetic reducing conditions exist, siderite and/or ferroan calcite form concretionary precipitates. Root traces or other permeability corridors, localized concentrations of organic matter, distribution of carbonate fossil fragments, and other factors play roles in localizing sites of concretion formation, and the directions of groundwater flow may control the elongation of concretions.

Siliceous cements also are formed in early diagenetic settings, especially in volcaniclastic rocks. Amorphous silica (opal) is a common precipitate in sandstones that have volcanic glass as a primary constituent. Near-surface devitrification of glass shards proceeds rapidly in meteoric waters and the increase in aqueous silica concentrations is followed by opal precipitation. The opal cements, however, normally are transformed to one or more forms of quartz (chert, chalcedony, or megaquartz) at the increased temperatures encountered during deeper burial. Zeolites (especially analcime, clinoptilolite, laumontite, and phillipsite), clays (especially smectite and kaolinite), and calcite are other common cementing agents in early- to intermediate-stage diagenesis of volcanogenic sediments.

Iron oxide grain coatings, once considered entirely detrital, in many cases now are recognized to be early diagenetic precipitates (Walker, 1976). Amorphous or extremely finely crystalline hematite and hematite precursors (mainly goethite) form as grain coatings in oxidizing near-surface settings, producing reddened sediments (redbeds, *q.v.*). The iron is derived from dissolution of unstable iron-bearing minerals, including amphiboles, olivene, and magnetite.

Burial (mesogenetic stage) cements in siliciclastic deposits

Much more extensive cementation of siliciclastic strata takes place during moderate burial and commonly is dominated by quartz precipitation, especially in strongly quartzose sediments (quartz arenites, orthoquartzites). Mesogenetic sources of silica for cementation are varied, but pressure solution of detrital quartz grains was once considered to be the primary source. That importance is now questioned, and other sources of silica are considered to be more significant by many authors (e.g., Land, 1984). Alternative silica sources include dissolution and reprecipitation of less stable forms of silica (especially opal from radiolarians, diatoms and volcanic glass, and extremely fine-grained quartzose eolian dust), subsurface alteration or dissolution of amorphous aluminosilicates, feldspars, and other silicate minerals, and diagenetic transformations of clays (especially conversion of smectite to illite) within sandstones or associated shales (Füchtbauer, 1978). Hot waters from deep basinal sources that cool as they move upward can also precipitate quartz, because silica solubility decreases with decreasing temperature. Although such basin-scale fluid movements may be slow, the tens to hundreds of millions of years over which burial diagenesis can act allows extensive quartz cementation of siliciclastic strata (Sibley and Blatt, 1976).

Quartz cements form primarily on detrital quartz grains and almost always grow as syntaxial overgrowths (i.e., in optical continuity with the substrate). Monocrystalline quartz grains have single-crystal overgrowths; polycrystalline detrital grains can have more complex overgrowths. A variety of sedimentary or early diagenetic coatings on detrital grains (including iron oxides or hydroxides, organic films, and clays) can inhibit formation of quartz overgrowths and help to preserve porosity in quartz-rich sandstones, but such coatings rarely completely cover the surface of grains. Overgrowths thus may be initiated at breaks in the coatings, initially forming as small protrusions from the detrital grain surface. Multiple initial growths may eventually coalesce into a complete overgrowth with euhedral crystal outlines or may completely fill pore spaces, forming compromise boundaries with overgrowths on adjacent grains. Extensive overgrowth precipitation yields a completely cemented fabric of interlocked quartz grains that can erroneously appear to be the result of extensive compaction and grain-boundary suturing (Figure C18A). The presence of visible grain coatings surrounding the detrital quartz grains, where they have not inhibited cementation, are a great aid in the proper identification of overgrowths as the cause of porosity destruction. They can clearly delineate the contacts between detrital cores and overgrowths, even though both act as a single crystal. Where such "dust rims" are absent, cathodoluminescence microscopy may aid in the recognition of overgrowths (Figure C18B), through accentuation of trace-element variations between detrital and authigenic constituents (Sippel, 1968; Miller, 1988).

Other forms of silica—mainly chalcedony and chert (microquartz)—can also cement siliciclastic strata in mesogenetic settings, especially where detrital quartz substrates are relatively scarce or where silica precipitation rates are very high. Chalcedony is a fibrous growth form that typically forms palisade rims and radiating, spherulitic growths in interstitial pores. Microquartz form as tiny, equant, often inclusion-rich crystals, with anhedral to subhedral morphologies, commonly increasing in crystal size from pore wall to pore center.

As with carbonate cements, fluid inclusion geothermometry and stable isotope geochemical studies are very useful in determining the temperature and geochemical conditions of quartz cement formation. Because of the physical strength and lack of cleavage in quartz, fluid inclusions studies of this mineral are far more reliable than those done on calcite, dolomite or most evaporite minerals.

Burial-stage calcite is an important mesogenetic cementing agent in many siliciclastic strata and can form throughout the history of mesogenetic alteration. Early-stage calcites form predominantly where a significant volume of unstable carbonate (commonly aragonitic shell fragments) was present in the original sediment. These grains are readily dissolved during early diagenesis in the presence of nonmarine or low-salinity (mixed) pore fluids. The presence of calcitic shells or other carbonate grains can provide isolated nucleation sites for the growth of calcite cements from carbonate-saturated pore waters. These isolated sites of cement formation allow nucleated crystals to grow unimpeded by competition from neighboring precipitates. This commonly produces coarsely crystalline cements and even poikilotopic fabrics in which single calcite crystals engulf several detrital grains. The force of crystallization of such early mesogenetic calcite can lead to displacement or fracturing of detrital grains and produce apparently undercompacted or "floating" fabrics.

Late-stage calcite cements typically postdate quartz overgrowths and authigenic clays. They result from the fact that calcite solubility decreases at increased temperatures and basinal brines may have elevated pH and high Ca and CO_3 concentrations. In addition, under most deep burial conditions, waters are strongly reducing, and late-stage calcites therefore are characterized by elevated iron and manganese concentrations.

Mesogenetic dolomite and siderite cements can be locally important in siliciclastic deposits. Dolomite cements can be anhedral to euhedral, generally are zoned and tend to be ferroan, reflecting reducing conditions during formation. Siderite cements tend to be finely crystalline with yellow-brown flattened rhombs; they too require reducing conditions for formation.

Clay minerals in siliciclastic rocks were once considered to be entirely detrital. In recent decades, however, numerous studies have demonstrated that authigenic clays are extremely common and precipitate under a variety of conditions (Blatt, 1979), in some cases substantially affecting the petrophysical properties of sandstones (Bjørlykke, 1988). Illite, smectite, and mixed-layer smectite-illite and smectite-chlorite and kaolinite all occur as pore-filling and grain-coating cements and can be difficult to distinguish from detrital counterparts. Some authigenic clays, especially smectite-illite grain coatings, clearly predate all other cements and may be of eogenetic origin. As such, these clays can inhibit later quartz or calcite cementation and thus help to preserve sandstone porosity. Other clays form later in the cementation history. Kaolinite (or its polymorph, dickite) occurs as authigenic vermicular stacks ("books" of mica-like plates) that generally postdate eogenetic cements. Because kaolinite precipitation is favored by low pH conditions, its formation sometimes is associated with decomposition of organic matter or meteoric water influx. Alteration and recrystallization of both detrital and authigenic clays during progressive burial can further complicate recognition of primary versus secondary origins or timing of precipitation.

A number of other cements, including authigenic feldspar overgrowths, pyrite and other iron sulfides, barite, fluorite, sphalerite, and other hydrothermal minerals, can form during various stages of siliciclastic diagenesis. However, they normally have only local importance in controlling rock properties and so will not be detailed here.

Summary

Many cements and cement fabrics are known from carbonate and siliciclastic strata. In combination, they occlude 20–40 percent porosity in many sedimentary rocks and contribute substantially to sediment lithification. The fabrics and paragenetic relationships of cements provide important clues to the time and conditions under which they were formed. Those interpretations are strengthened when done in combination with trace-element analyzes, isotopic studies, and fluid-inclusion geothermometry. Absolute age determinations on cements are becoming more precise and as techniques are perfected for analyzing small, *in situ* samples, such determinations will substantially improve our understanding of cementation. Future studies may clarify the poorly understood role of microbial activity in mediating cementation, especially in the subsurface. Likewise, refining our understanding of the factors that prevent or retard cementation remains of fundamental importance, because it aids in predicting the location and quality of possible unconventional oil and gas reservoirs and economic mineral deposits.

Peter A. Scholle and Dana Ulmer-Scholle

Bibliography

Bjørkum, P.A., Oelkers, E.H., Nadeau, P.H., Walderhaug, O., and Murphy, W.M., 1998. Porosity prediction in quartzose sandstones as a function of time, temperature, depth, stylolite frequency, and hydrocarbon saturation. *AAPG Bulletin*, **82**(4): 637–648.

Bjørlykke, K., 1988. Sandstone diagenesis in relation to preservation, destruction and creation of porosity. In Chilingarian, G.V., and Wolf, K.H. (eds.), *Diagenesis*. New York: Elsevier, pp. 555–588.

Blatt, H., 1979. Diagenetic processes in sandstones. In Scholle, P.A., and Schluger, P.R. (eds.), *Aspects of Diagenesis*. Tulsa, OK: SEPM Special Publication, 26, pp. 141–157.

Blatt, H., Middleton, G.V., and Murray, R.C., 1972. *Origin of Sedimentary Rocks*. Englewood Cliffs, NJ: Prentice Hall.

Burke, W.H., Denison, R.E., Hetherington, E.A., Koepnick, R.B., Nelson, H.F., and Otto, J.B., 1982. Variation of seawater $^{87}Sr/^{86}Sr$ throughout Phanerozoic time. *Geology*, **10**: 516–519.

Burton, E.A., and Walter, L.M., 1990. The role of pH in phosphate inhibition of calcite and aragonite precipitation rates in seawater. *Geochimica et Cosmochimica Acta*, **54**: 797–808.

Choquette, P.W., and James, N.P., 1987. Diagenesis #12. Diagenesis in limestones–3. The deep burial environment. *Geoscience Canada*, **14**: 3–35.

Correns, C.W., 1950. Zur Geochimie der Diagenese. I. Das Verhalten von $CaCO_3$ und SiO_2. *Geochimica et Cosmochimica Acta*, **1**: 49–54.

Dickson, J.A.D., 1966. Carbonate identification and genesis as revealed by staining. *Journal of Sedimentary Petrology*, **36**: 491–505.

Folk, R.L., 1965. Some aspects of recrystallization in ancient limestones. In Pray, L.C., and Murray, R.S. (eds.), *Dolomitization and Limestone Diagenesis*. Tulsa, OK: Society of Economic Paleontologists and Mineralogists, Special Publication, 13, pp. 14–48.

Folk, R.L., 1974. The natural history of crystalline calcium carbonate: effect of magnesium content and salinity. *Journal of Sedimentary Petrology*, **44**: 40–53.

Füchtbauer, H., 1978. Zur Herkunft des Quarzzements: Abschätzung der Quarzauflösung in Silt- und Sandsteinen. *Geologische Rundschau*, **67**(3): 991–1008.

Goldstein, R.H., and Reynolds, T.J., 1994. *Systematics of Fluid Inclusions in Diagenetic Minerals*. Tulsa, OK: SEPM Short Course 31.

Grotzinger, J.P., and Knoll, A.H., 1995. Anomalous carbonate precipitates: is the Precambrian the key to the Permian? *Palaios*, **10**: 578–596.

Holser, W.T., 1979. Mineralogy of evaporites. In Burns, R.G. (ed.), *Marine Minerals* Washington, DC: Mineralogical Society of America Short Course Notes, 6, pp. 211–235.

Humphrey, J.D., 2000. New geochemical support for mixing-zone dolomitization at Golden Grove, Barbados. *Journal of Sedimentary Research*, **70**(5): 1160–1170.

James, N.P., and Choquette, P.W., 1983. Diagenesis 6. Limestones—the sea floor diagenetic environment. *Geoscience Canada* **10**: 162–179.

James, N.P., and Choquette, P.W., 1984. Diagenesis 9. Limestones–the meteoric diagenetic environment. *Geoscience Canada*, **11**: 161–194.

James, N.P., and Ginsburg, R.N., 1979. *The Seaward Margin of Belize Barrier and Atoll Reefs*. Oxford: International Association of Sedimentologists, Special Publication No. 3.

Kennedy, W.J., and Garrison, R.E., 1975. Morphology and genesis of nodular chalks and hardgrounds in the Upper Cretaceous of southern England. *Sedimentology* **22**: 311–386.

Kinsman, D.J.J., 1966. Gypsum and anhydrite of Recent age, Trucial Coast, Persian Gulf. In Rau, J.L. (ed.), *Second Symposium on Salt*. Volume 1, Cleveland, OH: Northern Ohio Geological Society, pp. 302–326.

Kinsman, D.J.J., 1974. Calcium sulfate minerals of evaporite deposits: their primary mineralogy. In Coogan, A.J. (ed.), *Fourth Symposium on Salt*, Volume 1, Cleveland, OH: Northern Ohio Geol. Soc., pp. 343–348.

Land, L.S., 1984. Frio Sandstone diagenesis, Texas Gulf Coast: a regional isotopic study. In Mcdonald, D.A., and Surdam, R.C. (eds.), *Clastic Diagenesis*. Tulsa, OK: American Association of Petroleum Geologists Memoir, 37, pp. 47–62.

MacKenzie, F.T., and Bricker, O.P., 1971. Cementation of sediments by carbonate minerals. In Bricker, O.P. (ed.), *Carbonate Cements*. Baltimore, MD: Johns Hopkins Press, pp. 239–246.

Mazzullo, S.J., and Cys, J.M., 1979. Marine aragonite sea-floor growths and cements in Permian phylloid algal mounds, Sacramento Mountains, New Mexico. *Journal of Sedimentary Petrology*, **49**: 917–936.

McKenzie, J.A., 1981. Holocene dolomitization of calcium carbonate sediments from the coastal sabkhas of Abu Dhabi, U.A.E.: a stable isotope study. *Journal of Geology*, **89**: 185–198.

Miller, J., 1988. Cathodoluminescence microscopy. In Tucker, M. (ed.), *Techniques in Sedimentology*. Oxford: Blackwell Scientific Publications, pp. 174–190.

Moore, C.H., 1989. *Carbonate Diagenesis and Porosity*. New York: Elsevier.

Mozley, P.S., and Burns, S.J., 1993. Oxygen and carbon isotopic composition of marine carbonate concretions: an overview. *Journal of Sedimentary Petrology*, **63**: 73–83.

Radke, B.M., and Mathis, R.L., 1980. On the formation and occurrence of saddle dolomite. *Journal of Sedimentary Petrology*, **50**: 1149–1168.

Reed, S.J.B., 1989. Ion microprobe analysis—a review of geological applications. *Mineralogical Magazine*, **53**: 3–24.

Reeder, R.J., 1991. An overview of zoning in carbonate minerals. In Barker, C.E., and Kopp, O.C. (eds.), *Luminescence Microscopy and Spectroscopy: Qualitative and Quantitative Applications*. Tulsa, OK: SEPM Short Course 25, pp. 77–82.

Rosler, H.J., and Lange, H., 1972. *Geochemical Tables*. Amsterdam: Elsevier.

Sandberg, P.A., 1983. An oscillating trend in Phanerozoic non-skeletal carbonate mineralogy. *Nature*, **305**: 19–22.

Schmoker, J.W., and Halley, R.B., 1982. Carbonate porosity versus depth: a predictable relation for south Florida. *Bulletin of the American Association of Petroleum Geologists*, **66**: 2561–2570.

Scholle, P.A., 1978. *A Color Illustrated Guide to Carbonate Rock Constituents, Textures, Cements, and Porosities*. Tulsa, OK: American Association of Petroleum Geologists Memoir 27.

Scholle, P.A., 1979. *A Color Illustrated Guide to Constituents, Textures, Cements, and Porosities of Sandstones and Associated Rocks*. Tulsa, OK: American Association of Petroleum Geologists Memoir 28.

Scholle, P.A., and Halley, R.B., 1985. Burial diagenesis: out of sight, out of mind!. In Schneidermann, N., and Harris, P.M. (eds.), *Carbonate Cements*. Tulsa, OK: SEPM Special Publication No. 36, pp. 309–334.

Scholle, P.A., Ulmer, D.S., and Melim, L.A., 1992. Late-stage calcites in the Permian Capitan Formation and its equivalents, Delaware Basin margin, west Texas and New Mexico: evidence for replacement of precursor evaporites. *Sedimentology*, **39**: 207–234.

Scoffin, T.P., and Stoddart, D.R., 1983. Beachrock and intertidal cements. In Goodie, A.S., and Pye, K. (eds.), *Chemical Sediment and Geomorphology*. London: Academic Press, pp. 401–425.

Sibley, D.F., and Blatt, H., 1976. Intergranular pressure solution and cementation of the Tuscarora Quartzite. *Journal of Sedimentary Petrology*, **46**: 881–896.

Sippel, R.F., 1968. Sandstone petrology, evidence from luminescence petrography. *Journal of Sedimentary Petrology*, **38**: 530–554.

Walker, T.R., 1976. Diagenetic origin of continental red beds. In Falke, H. (ed.), *The Continental Permian in Central West and South Europe*. Dordrecht (Netherlands): D. Reidel Publishing Co., pp. 240–282.

Williams, L.A., Parks, G.A., and Crerar, D.A., 1985. Silica diagenesis, I. Solubility controls. *Journal of Sedimentary Petrology*, **55**: 301–311.

Wollast, R., 1971. Kinetic aspects of the nucleation and growth of calcite from aqueous solutions. In Bricker, O.P. (ed.), *Carbonate Cements*. Baltimore, MD: Johns Hopkins Press, pp. 264–273.

Cross-references

Ankerite
Beachrock
Caliche–Calcrete
Cathodoluminescence

Chalk
Chert, Chalcedony and Flint
Compaction (Consolidation) of Sediments
Diagenetic Structures
Diffusion, Chemical
Dolomites and Dolomitization
Evaporites
Glaucony and Verdine
Gypsum
Hardground (and Firmground)
Hydrocarbons in Sediments
Kaolin Group Minerals
Laterites
Lepisphere
Micritization
Porewaters in Sediments
Pressure Solution
Quartz, Detrital
Red Beds
Sabkha, Salar, Salt Flat
Sand and Sandstones (Siliciclastics)
Seawater
Septarian Concretions
Silcrete
Siliceous Sediments
Smectite Group
Speleothems
Stylolites
Travertine
Zeolites in Sedimentary Rocks

CHALK

The Ocean Drilling Program defines chalk as a firm pelagic sediment composed predominantly of calcareous pelagic grains (e.g., Kroenke et al., 1991, see *Oceanic Sediments*). Calcareous pelagic grains comprise calcareous nannofossils of silt size and calcareous microfossils of mm size, as well as fine fossil debris. The term chalk can also refer to a calcareous nannofossil-rich facies irrespective of induration. The grains are mixed into sediments of mudstone, wackestone, or packstone texture (see *Classification of Sediments and Sedimentary Rocks*). Calcareous nannofossils include coccoliths which originate from spores of golden brown algae, and are found in sediments ranging from Late Triassic to present (Siesser and Winter, 1994). Microfossils are typically foraminifers.

Chalk facies sediments contain varying amounts of siliceous micro- and nannofossils wich upon burial may form chert (see *Siliceous Sediments*). Similarly, chalk also contains varying amounts of clay which upon burial may form stylolites or flaser structures (Garrison and Kennedy, 1977, see *Diagenetic Structures*).

Oceanic chalk deposits form in areas where terrestrial input is low, and where the seafloor is above the carbonate compensation depth, typically on deep sea plateaus (see *Oceanic Sediments*). These sediments are almost exclusively of Cretaceous age or younger. Upon the large transgression in Upper Cretaceous, chalk facies sediments accumulated on the continental shelf (Håkansson et al., 1974). The Chalk Group of the North Sea Basin is primarily composed of chalk sediments of Upper Cretaceous and Paleocene age.

Freshly deposited chalk facies sediments have porosities around 70 percent; and porosities of around 35 percent may be maintained to an effective burial of c.1 km (Scholle, 1977, Figure C19). This preservation of porosity probably reflects that the fossils constituting the chalk are mainly composed of chemically stable low Mg calcite (see *Carbonate Mineralogy and Geochemistry*).

Chalk is utilized as a source of lime and for improvement of acid soils. In North West Europe it is a reservoir for groundwater (Downing et al., 1993) and in the Central North Sea for hydrocarbons (Andersen, 1995). Chalk is used for rinsing smoke after coal combustion in power plants and as a raw material for production of pure calcite or aragonite for industrial purposes: pigment in paper, plastic and paint. Indurated chalk has been used as building stone (a beautiful example is the cathedral of York, England).

Until the 1960s chalk was seen as a uniform white powder, possibly formed by precipitation of calcite in the seawater, and when studied in exposures, the purpose was mainly extracting macrofossils. With the introduction of electron

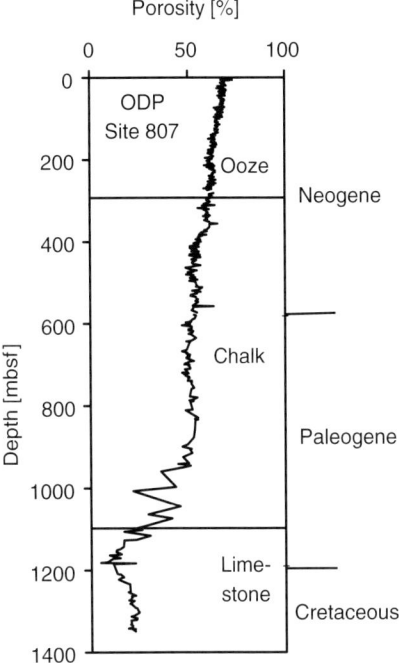

Figure C19 Chalk facies sediments on the Ontong Java Plateau at a water depth of 2,805 m, investigated by the Ocean Drilling Program (Kroenke et al., 1991). Laboratory porosity versus depth in meters below sea floor (mbsf). With burial, the soft calcareous ooze grades into firmer chalk and this again to hard limestone. Compaction experiments indicate that the gradual porosity loss through ooze and chalk is a consequence of physical compaction (Lind, 1993, see *Compaction of Sediments*), whereas geochemical data indicate that the limestone is formed via porosity reducing precipitation of calcite caused by diffusion of ions from where flint or stylolites form or from the basement (Delaney et al., 1991; see *Cements and Cementation*). The small scale porosity variation in the chalk section primarily reflects varying proportion of carbonate mud and hollow microfossils (Masters et al., 1993), whereas the more dramatic porosity shift within the limestone reflects an upper wackestone interval rich in cemented microfossils overlying an interval of re-crystallized carbonate mudstone (Borre and Fabricius 1998; see *Fabric, Porosity and Permeability*). The interval from 970 mbsf to 1100 mbsf is rich in flint beds (see *Siliceous Sediments*).

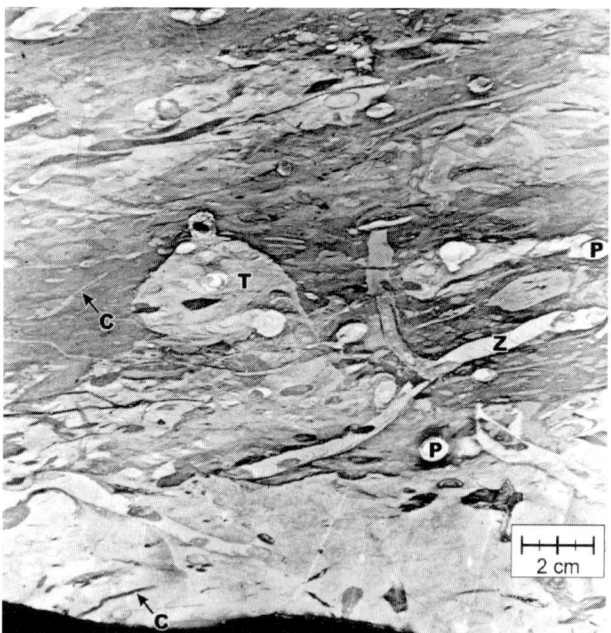

Figure C20 White chalk containing about 5–10% clay minerals, Maastrichtian, form the quarry "Dania" in northern Jutland, Denmark. The chalk has been sectioned vertically and smoothed. The surface has been lightly oiled with common lubricating oil (see Bromley, 1981). The treatment greatly increases the visibility of the sediment structures. The sediment has been fully bioturbated by burrowing animals. The trace fossils represented are: (T) *Thalassinoides*, (Z) *Zoophycos*, (P) *Planolites* and (C) *Chondrites* (see *Biogenic Structures*). R. G. Bromley.

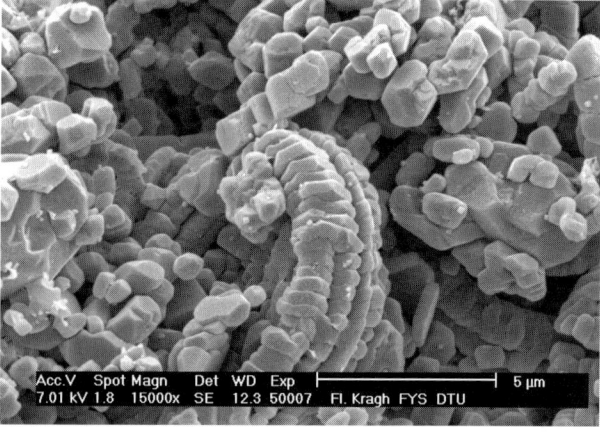

Figure C21 Electron micrograph of Maastrictian chalk from the central North Sea. The sample has a mudstone texture, a calcite content of 98.8 wt.% and a porosity, $\emptyset$, of 37%. The mud is composed of coccoliths and fossil debris. All calcite particles are apparently thoroughly re-crystallized and because the clay content is low, the specific surface, s, as measured by BET is relatively low ($1.6\,m^2g^1$), and the resulting gas permeability 4.3 mD. Chalk is petrophysically an exceptionally simple sediment which obeys Kozeny's equation relating permeability, specific surface, and porosity: $k = C^* \emptyset^3/S^2$, where S is the specific surface relative to bulk volume, k is the Klinkenberg corrected gas permeability (in m^2) and C is near constant, varying from 0.21 for a porosity of 20% to 0.24 for a porosity of 40% (Mortensen et al., 1998).

microscopy it became clear that the white powder is composed of nannofossils and fossil debris; and the expansion of drilling activity in the oceans and in the North Sea directed the scientific interest to the study of nanno- and microfossils with the purpose of establishing a biostratigraphy usable on drilling cores and cuttings. In the sixties and seventies scientists intensified the study of sedimentary structures and it appeared that chalk as a rule is bioturbated (Figure C20). It was also found that chalk may contain re-sedimented chalk-clast bearing intervals (e.g., in the central North Sea, Nygaard et al., 1983). Chalk may exhibit cyclicity in silica content (frequently seen as series of chert beds) or clay content (reflected as series of marl beds). By some researchers this was seen as a signature of geochemical segregation from an originally more homogeneous sediment, but from studies of the pattern of bed thickness variations and oxygen isotopic variations (see *Isotopic Methods in Sedimentology*) it became more and more probable that the cyclicity reflects an original sedimentary variation which from the eighties have been related to paleo-climatic fluctuations (e.g., Gale, 1989; and see *Cyclic Sedimentation, Milankovitch Cycles*).

Major constructions in chalk (e.g., the Channel tunnel between England and France) and the search for and production of hydrocarbons from chalk reservoirs has in the nineties directed research attention to the mechanical, hydraulic and petrophysical properties of chalk (e.g., Burland et al., 1990; Maliva and Dickson, 1992; Megson and Hardmann 2001; Figure C21). Specific current research questions are how the porosity loss proceeds in geologic time in the sedimentary basin, and the rock mechanical behavior during hydrocarbon production and by the change in pore fluid from oil to water.

Ida L. Fabricius

Bibliography

Andersen, M.A., 1995. *Petroleum Research in North Sea Chalk*. Stavanger, Norway: RF-Rogaland Research.

Borre, M., and Fabricius, I.L., 1998. Chemical and mechanical processes during burial diagenesis of chalk: an interpretation based on specific surface data of deep-sea sediments. *Sedimentology*, **45**: 755–769.

Bromley, R.G., 1981. Enhancement of visibility of structures in marly chalk: modification of the Bushinsky oil technique. *Bulletin of the Geological Society of Denmark*, **29**: 111–118.

Burland, J.B. et al., 1990. *Chalk*. London: Thomas Telford.

Delaney, M.L., and Shipboard Scientific Party, 1991. Inorganic Geocheistry Summary. In Kroenke, L.W., Berger, W.H., and Janecek, T.R. (eds.), *Proceedings of the Ocean Drilling Program Initial Reports 130*. College Station TX: Ocean Drilling Program, pp. 549–551.

Downing, R.A., Price, M., and Jones, G.P., 1993. *The Hydrogeology of the Chalk of North-West Europe*. Oxford: Claredon Press.

Gale, A.S., 1989. A Milankovitch scale for Cenomanian time. *Terra Nova*, **1**: 420–425.

Garrison, R.E., and Kennedy, W.J., 1977. Origin of solution seasms and flaser structures in Upper Cretaceous chalks of Southern England. *Sedimentary Geology*, **19**: 107–137.

Håkansson, E., Bromley, R., and Perch-Nielsen, K., 1974. Maastrichtian chalk of north-west Europe—a pelagic shelf sediment. In Hsü, K., and Jenkyns, H.C. (eds.), *Pelagic Sediments: on Land and Under the Sea, Special Publication IAS*, **1**, pp. 211–233.

Kroenke, L.W., Berger, W.H., Janecek, T.R. et al., 1991. **Proc. ODP, Initial Repts. 130**. College Station TX: Ocean Drilling Program.

Lind, I.L., 1993. Loading experiments on carbonate ooze and chalk from Leg 130, Ontong Java Plateau. In Berger, W.H., Kroenke,

L.W., and Mayer, L.A. (eds.), **Proceedings of the Ocean Drilling Program, Scientific Results 130**. College Station TX: Ocean Drilling Program, pp. 673–686.

Maliva, R.G., and Dickson, J.A.D., 1992. Microfacies and Diagenetic Control of Porosity in Cretaceous/Tertiary Chalks, Eldfisk Field, Norwegian North Sea. *AAPG Bulletin*, **76**: 1825–1838.

Masters, J.C., Resig, J.M., and Wilcoxon, J.A., 1993. Relationship between physical properties and microfossil content and preservation in calcareous sediments of the Ontong Java Plateau. In Berger, W.H., Kroenke, L.W., and Mayer, L.A. (eds.), **Proc. ODP Sci. Results 130**. College Station TX: Ocean Drilling Program, pp. 641–652.

Megson, J., and Hardman, R., 2001. Exploration for and development of hydrocarbons in the Chalk of the North Sea: a low permeability system. *Petroleum Geoscience*, **7**: 3–12.

Mortensen, J., Engstrøm, F., and Lind, I. 1998. The relation among porosity, permeability, and specific surface of chalk from the Gorm field, Danish North Sea. *SPE 31062, SPE Reservoir Evaluation & Engineering*, **1**, pp. 245–251.

Nygaard, R., Lieberkind, K., and Frykman, P. 1983. Sedimentology and reservoir parameters of the Chalk Group in the Danish Central Graben. *Geologie en Mijnbouw*, **62**: 177–190.

Scholle, P.A., 1977. Chalk diagenesis and its relationship to petroleum exploration: oil from chalks, a modern miracle? *AAPG Bulletin*, **61**: 982–1009.

Sieser, W.G., and Winter, A., 1994. Composition and morphology of coccolithophore skeletons. In Winter, A., and Sieser, W.G. (eds.), *Coccolithophores*. Cambridge University Press, pp. 51–62.

Cross-references

Biogenic Sedimentary Structures
Carbonate Mineralogy and Geochemistry
Cements and Cementation
Classification of Sediments and Sedimentary Rocks
Compaction (Consolidation) of Sediments
Cyclic Sedimentation
Diagenesis
Diagenetic Structures
Fabric, Porosity and Permeability
Isotopic Methods in Sedimentology
Milankovitch Cycles
Oceanic Sediments
Siliceous Sediments

CHARCOAL IN SEDIMENTS

Definition

When plant material is burnt by fire there is often an incomplete combusted residue known as charcoal. All plant organs, from leaves to flowers to wood, may be preserved as charcoal. During the charring process wood may break up into characteristic sizes and shapes. It has been shown that wood fragments may typically be about 1 cm cubed. The charcoal is brittle (getting hands dirty upon touch) and relatively inert with the cell walls being converted to almost pure carbon (Scott, 1989). Charcoal usually shows exceptional anatomical detail when seen, for example, under the scanning electron microscope (Scott and Jones, 1991) (Figures C27, C28, C29). Even small flowers may be preserved as charcoal (Scott et al., 2000).

Transport

During wild fires small particulate fragments (around 1–10 µm, including charcoal and soot particles) may be blown away from the fire sites. These small particles may enter lakes and other wetland environments or into the sea and be incorporated into accumulating sediments (Clark and Patterson, 1997). In contrast, larger charcoal fragments may either, in the case of fires within mires, be incorporated into peat or else be transported by water (Scott, 2000). Severe fires may destroy plants which help bind soil and it is possible that rainstorms following major fires may result in significant erosion of the fire site. Sediment and charcoal may be washed into a depositional site (Nichols and Jones, 1992). It has been shown, for example, that major alluvial fans may be triggered by fires, such as in Yellowstone National Park (Meyer et al., 1992).

Macroscopic charcoal fragments usually float in water. The length of time for a charcoal fragment to be waterlogged depends on a series of variables from size, temperature of charring, organ and species (Nichols et al., 2000).

Deposition

It has been argued that larger charcoal fragments in a sediment implies a nearby fire (Clark and Patterson, 1997). Whilst this may be the case for classes of windblown microscopic charcoal, however, it is not necessarily true for larger macroscopic charcoal fragments. Charcoal may be incorporated into bed load sediments. Flume experiments have shown that charcoal may be entrained in sand ripples of velocities around $35\,\text{cm s}^{-1}$ (Nichols et al., 2000). Marine sediments may also yield charcoal fragments (Bird, 1997; Scott, 2000).

Sedimentary distribution

Macroscopic charcoal when encountered in fossil deposits is often called fusain (Scott, 1989). This is one of the four fundamental lithotypes of coal described by Marie Stopes. In the Quaternary there has been a number of significant studies of microscopic charcoal from lake and peat sediments to help unravel more recent fire histories (Patterson et al., 1987). In marine clastic sediments also most research has centered upon microscopic charcoal and soot particles which are likely to record larger and more frequent regional fires (Bird, 1997). In contrast, studies of Pre-Quaternary charcoal has centered upon larger charcoal fragments that are visible in rocks (Scott, 2000). There is a good fossil record of charcoal from the latest Devonian and evidence of extensive wildfire from the Early Carboniferous (Scott, 2000).

Most often a geologist will encounter charcoal (fusain) fragments in terrestrial clastic rocks and also commonly they may be abundant in estuarine, deltaic and shallow marine deposits. These may be recognized as small black cubes with a lustrous sheen which marks the finger upon touch (Figures C22, C23). This is in contrast to coalified plant material, which is often black, shiny, having a concoidal fracture and is clean to touch. Charred leaf material may also be recognized (Jones and Chaloner, 1991) (Figure C24). Smaller charred plant fragments, which may include flowers, may be difficult to see on a bedding surface. In these cases only bulk maceration, using HF, or sieving of unconsolidated sediment, will yield the smaller charcoal fragments (Figures C25, C26).

Charcoal is often encountered within coal seams (Figure C22) where it may represent 10–50 percent of the

Figure C22 Wood charcoal (fusain) in coal.

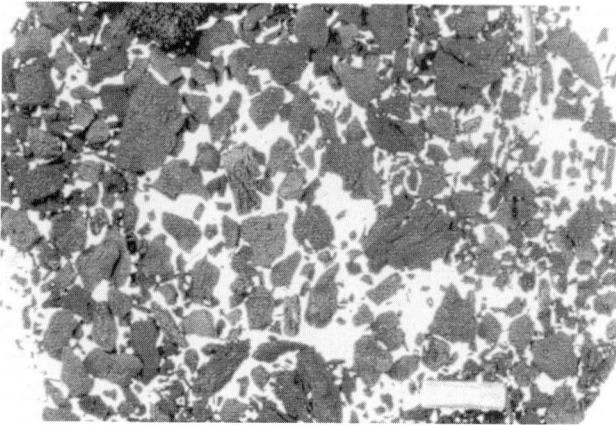

Figure C25 Charcoal (fusain), macerated from lacustrine clays.

Figure C23 Wood charcoal (fusain) in fluvial sandstone.

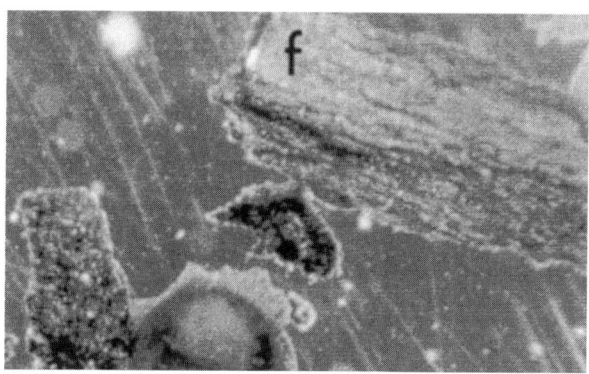

Figure C26 Charcoal (fusain) macerated from coal.

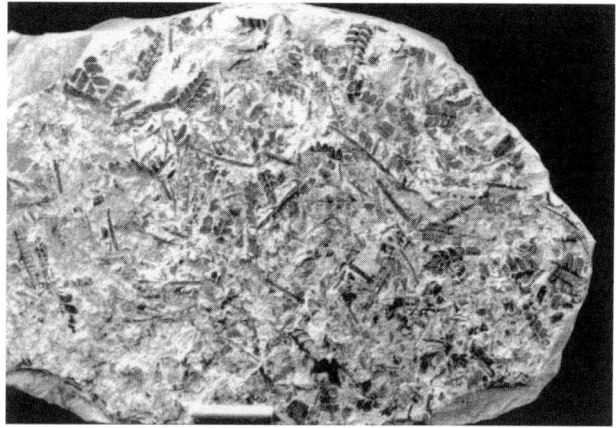

Figure C24 Fern charcoal (fusain) in fluvial siltstone.

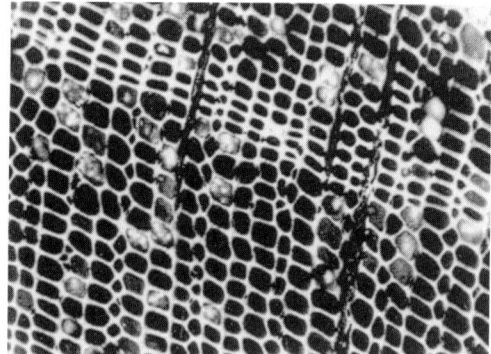

Figure C27 Reflectance micrograph of Recent charcoal.

seam. Charcoal may also be encountered in volcanic rocks, especially pyroclastic deposits where charcoal has been entombed in hot lava or ash (Scott, 2001). Charcoal is often not recorded by sedimentologists in the field.

Implications

The recording of charcoal (fusain), as opposed to other forms of organic matter, has considerable implications for the interpretation of paleoenvironments and diagenesis. In addition, microscopic and isotopic studies of charcoal may yield data on

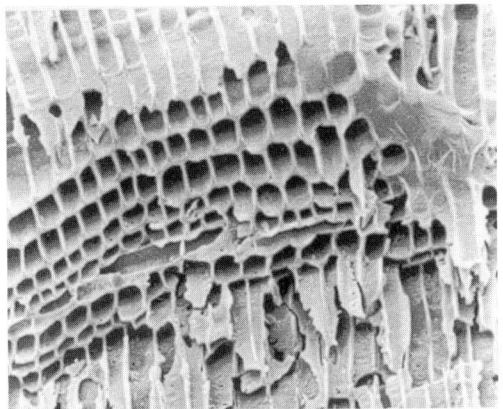

Figure C28 SEM of Cretaceous conifer wood charcoal.

Figure C29 SEM of Carboniferous lycophyte wood charcoal.

fire type, frequency, climate, vegetation, terrestrial ecosystem dynamics as well as an atmospheric composition (both O_2 and CO_2) (Chaloner, 1989; Scott, 2000).

Andrew C. Scott

Bibliography

Bird, M.I., 1997. Fire in the earth sciences. *Episodes*, **20**: 223–226.
Chaloner, W.G., 1989. Fossil charcoal as an indicator of palaeo-atmospheric oxygen. *Journal of the Geological Society of London*, **146**: 171–174.
Clark, J.S., and Patterson III, W.A., 1997. Background and local charcoal in sediments: scales of fire evidence in the fossil record. In Clark, J.S., Cachier, H., Goldammer, J.G., and Stocks, B. (eds.), *Sediment Records of Biomass Burning and Global Change*. NATO ASI Series, Volume 151, Berlin: Springer, pp. 23–48.
Jones, T.P., and Chaloner, W.G., 1991.Fossil charcoal, its recognition and palaeoatmospheric significance. *Palaeogeography, Palaeoclimatology, Palaeoecology (Global and Planetary Change Section)*, **97**: 39–50.
Meyer, G.A., Wells, S.G., Balling, R.C., and Jull, A.J.T., 1992. Response of alluvial systems to fire and climate change in Yellowstone National Park. *Nature*, **357**: 147–150.
Patterson III, W.A., Edwards, K.J., and Maguire, D.J., 1987. Microscopic charcoal as an indicator of fire. *Quaternary Science Reviews*, **6**: 3–23.
Nichols, G., Cripps, J., Collinson, M.E., and Scott, A.C., 2000. Experimental taphonomy of charcoal and implications for the palaeoenvironmental interpretation of fossil charcoal deposits. *Palaeogeography, Palaeoclimatology, Palaeoecology*, **164**: 43–56.
Nichols, G.J., and Jones, T.P., 1992. Fusain in Carboniferous shallow marine sediments, Donegal, Ireland: the sedimentologicl effects of wildfire. *Sedimentology*, **39**: 487–502.
Scott, A.C., 1989. Observations on the nature and origin of fusain. *International Journal of Coal Geology*, **12**: 443–475.
Scott, A.C., 2000. The pre-Quaternary history of fire. *Palaeogeography, Palaeoclimatology, Palaeoecology*, **164**: 281–329.
Scott, A.C., 2001. Preservation by fire. In Briggs, D.E.G., and Crowther, P.R. (eds.), *Palaeobiology II*. Oxford: Blackwells, pp. 277–280.
Scott, A.C., and Jones, T.P., 1991. Microscopical observations of Recent and fossil charcoal. *Microscopy and Analysis*, **24**: 13–15.
Scott, A.C., Cripps, J., Nichols, G., and Collinson, M.E., 2000. The taphonomy of charcoal following a recent heathland fire: and some implications for the interpretation of fossil charcoal deposits. *Palaeogeography, Palaeoclimatology, Palaeoecology*, **164**: 1–31.

CHLORITE IN SEDIMENTS

Chlorite is a common phyllosilicate mineral found in all kinds of sediments and sedimentary rocks. In fact the name chlorite, derived from the green color of most specimens, encompasses a group of minerals characterized by a wide range of chemical and structural variation (Bailey, 1988a). In sediments, much chlorite falls by definition into the category of minerals known as "clay minerals" (see *Clay Mineralogy*). Small amounts of chlorite are common in many sediments and it may be both detrital or formed diagenetically. Overall, diagenetic chlorite is probably rarer than other diagenetic clay minerals but it is common in certain kinds of sedimentary rocks. To understand some types of diagenetic chlorite its relationship to serpentine minerals and to mixed-layer chlorite minerals must be considered (Reynolds, 1988). Indeed, the petrology of chlorite in the sedimentary realm is diverse. This contribution attempts to summarize what is known about the characteristics, occurrence, and origins of various kinds of chlorite in sediments.

Structure and composition

All varieties of chlorite have a common structure consisting of a negatively charged 2:1 layer and a positively charged 'interlayer' octahedrally coordinated hydroxide sheet (Bailey, 1988a). Most chlorites are trioctahedral and as such they are normally rich in Fe and Mg. Less commonly, chlorites may be dioctahedral in which case they tend to be enriched in aluminum. For many years the nomenclature and classification of trioctahedral chlorites was plagued by the existence of numerous species names. This has been greatly simplified with Fe-rich chlorites classified as chamosites, and Mg-rich chlorites as clinochlores, with suitable adjective modifiers used if necessary (Bayliss, 1975). The chemical compositions of chlorites are complex, but a generalized structural formula distinguishing octahedral and tetrahedral cations and encompassing both trioctahedral and dioctahedral species can be written as:

$$(R^{2+}_{6-x-3y}R^{3+}_{x+2y}\square_y)(Si_{4-x}R^{3+}_x)O_{10}(OH)_8$$

In this form, R^{2+} represents divalent cations, mainly Fe and Mg, R^{3+} represents trivalent cations mainly Al, and $\square$

represents a vacant octahedral site. Fully trioctahedral chlorites do not have octahedral vacancies so that the octahedral cation total in the above formula is six. Di-trioctahedral chlorites, such as the species sudoite, have a dioctahedral 2:1 layer, and a trioctahedral interlayer hydroxide sheet. Thus in the general formula given above they have one vacant site and an octahedral cation total of five. Similarly a chlorite in which both the 2:1 layer and the interlayer octahedral sheets were both dioctahedral would have two vacant sites and thus only four octahedral cations.

In terms of crystal structure the 2:1 layer and the interlayer hydroxide sheet in chlorites may be stacked together in different ways so defining different chlorite polytypes. For some time it has been recognized that type I polytypes are common in sedimentary rocks formed at diagenetic temperatures whereas the IIb polytype is the common polytype for high temperature metamorphic chlorites. Recent summaries are provided by de Caritat et al. (1993) and Walker (1993).

In contrast to metamorphic chlorites formed at higher temperatures, recalculation of chemical analyzes to structural formulae for trioctahedral chlorites from sedimentary rocks often indicates the presence of octahedral vacancies, i.e., octahedral totals fall short of six cations. It is not yet clear, however, if such vacancies are present in the structure of sedimentary chlorites or are simply apparent due the assumptions made about the (usually unanalyzed) anions and/or analytical problems inherent in the analysis of clay minerals, e.g., avoiding contamination (Hillier and Velde, 1991; Jiang et al., 1994).

Occurrence and origin

Much of what is known about chlorite in sediments can be rationalized into a number of broad categories or associations, each probably with some degree of genetic significance (Figure C30). These categories include detrital and weathered chlorite; the odinite-berthierine-chamosite shallow-marine association; the Mg-rich chlorite evaporite/dolomite association; the aluminous dioctahedral chlorite red-bed association; the lithic/volcaniclastic sediment association, and the shale diagenesis/low grade-metamorphism association. However, the nature of chlorite in sediments is diverse, and undoubtedly many occurrences will not fall neatly into one of these associations; others, such as chlorites in limestones, have not yet received the study they deserve. Nonetheless, the use of these associations provides a conceptual framework to help understand the occurrence of chlorite in sediments in many instances.

Detrital and weathered chlorite

Detrital chlorite in sediments reflects provenance. In fact, chlorite is very easily weathered being one of the first minerals to disappear from soil profiles, even in temperate climates (Bain, 1977; Barnhisel and Bertsch, 1989). As a result it is quite unusual to find abundant detrital chlorite in sediments, unless they were deposited in close proximity to chlorite-rich parent materials. Nonetheless, small amounts of detrital chlorite can be found in many sediments. In modern oceanic sediments, chlorite is largely confined to, and concentrated in, sediments from high latitudes (Chamley, 1989; Weaver, 1989). This is undoubtedly a direct reflection of the fact that chlorite is such an easily weathered mineral. Indeed, estimates of its abundance in the clay size fraction of high latitudes oceanic sediment are only rarely more than 20–30 percent (Windom, 1976). The effect of weathering on the distribution of detrital chlorite in marine sediments is also evident from studies of oceanic sediments from dated cores of the western Atlantic region, which record a general increase in chlorite content from the Eocene to the Quaternary (Chamley, 1989). This appears to reflect the general pattern of global cooling over this time and consequent decrease in the intensity of chemical weathering in many source areas. To some extent it may also reflect a general increase in relief resulting from Alpine tectonism with the associated stripping of more mature soils and more widespread exposure of low-grade (chlorite-rich)

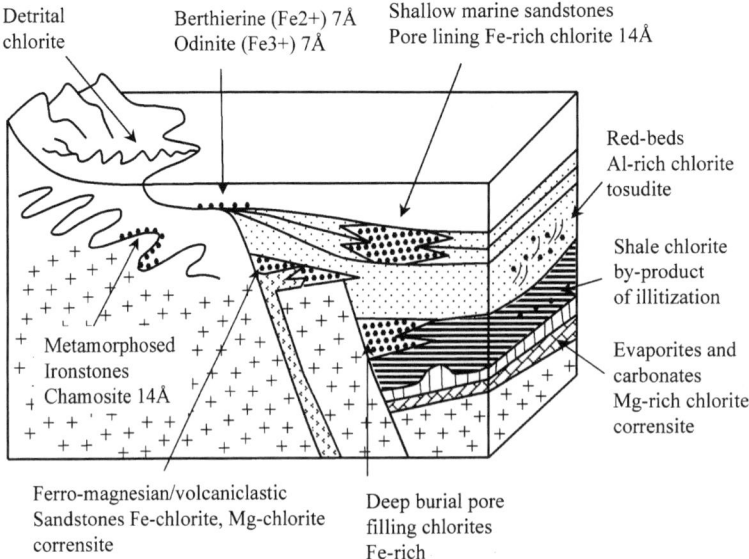

Figure C30 Common occurrences of chlorite and related minerals in sediments.

metamorphic rocks. In summary, the conditions necessary to supply detrital chlorite to sediments in great quantity are probably only rarely established. Therefore, the simple presence of abundant chlorite in sediments is more often than not an indication of its diagenetic formation, or at least that such an origin should be seriously investigated. Often it is assumed, quite correctly, that the presence of the IIb polytype in un-metamorphosed sediments indicates a detrital origin for chlorite. Caution must be applied, however, since the polytypic forms of chlorite are not entirely independent of compositional controls, for example Mg-rich chlorites formed at diagenetic temperatures, such as those in the evaporite/dolomite association described below, often crystallize as the IIb polytype (Hillier, 1993; Hillier et al., 1996).

Odinite-berthierine-chamosite shallow marine association

The Fe-rich variety of chlorite known as chamosite is named after its occurrence in an oolitic ironstone, near the town of Chamoson in the French speaking part of the Swiss Alps. The ironstone at Chamoson has been affected by low-grade metamorphism and similar ironstones from other localities not so affected are characterized instead by the presence of the 0.7 nm (7Å) serpentine mineral berthierine (e.g., Hughes, 1989). Oolitic ironstones are deposited in shallow marine environments and increasingly the presence of Fe-rich chlorite (chamosite), often as a clearly authigenic isopachous pore-lining (Figure C31), has been recognized in sandstones deposited in similar shallow marine settings. Indeed, iron ooids are frequently present in small quantities in such sandstones (Ehrenberg, 1993). This evidence is enough to suggest that the geochemical conditions that lead to chlorite formation must be similar in both cases. In modern environments examples of berthierine formation in shallow marine tropical deltaic environments have been known for some time (Porrenga, 1967), although they have more recently been reinterpreted and the 0.7 nm (7Å) mineral renamed odinite (Odin, 1988; Bailey, 1988b). Reinterpretation rests largely on

Figure C31 Diagenetic pore-lining chlorite from a shallow marine sandstone (Dogger beta Haupsandstein, see Hillier, 1994).

the basis that much of iron present is ferric rather than ferrous as in berthierine per se. Pore-lining chamosites from sandstones often show evidence for interstratification with a 0.7 nm (7Å) mineral, probably berthierine. Thus it has been suggested that odinite and/or berthierine may represent the precursors of diagenetic chamosite, since all appear to share a common depositional/geochemical environment characterized by a high supply of colloidal iron from a tropically weathered landscape. Taken together one may thus define an odinite-berthierine-chamosite association. Although many questions have yet to be answered concerning the precise genetic links amongst these minerals and the precise conditions that promote their formation, the alteration of syn-sedimentary Fe-rich 0.7 nm (7Å) minerals to Fe-rich chlorite as a result of diagenetic processes is a common hypothesis.

Diagenetic alteration of chlorites from shallow marine sandstones has recently been studied in some detail for (Hillier, 1994) including the well-known Tuscaloosa sandstone of the US Gulf Coast (Ryan and Reynolds, 1996). Thus polytypes are known to change from 1b to 11b with increasing temperature of alteration, with the 1a polytype an intermediate step. Additionally, 0.7 nm (7Å) layers are eventually, if erratically, lost and crystal size may increase with temperature.

Mg-rich chlorites and the evaporite/dolomite association

It has long been known that Mg-rich chlorites and corrensite, (R1 trioctahedral mixed-layer chlorite(50)-smectite, see *mixed-layer clays*) are common in many dolomite and/or evaporite bearing rocks. Early work in both Europe and North America was summarized by Millot (1964) and later by Hauff (1981) who focused on occurrences of corrensite. Examples from Permo-Triassic rocks are well-known almost worldwide and it is probably fair to say that some of them may represent the most chlorite-rich sediments known. Generally, there is a consensus that occurrences of chlorite and corrensite in this category represent authigenic minerals, but there are still two schools of thought concerning the timing of the authigenesis. One hypothesis is that these minerals are syn-sedimentary, the other that they form as a result of burial diagenesis. Hillier (1993) summarized the arguments in favor of the role of burial diagenesis, such as the absence of any modern syn-sedimentary analogues. He also proposed that diagenetic reactions amongst dioctahedral clay minerals (the most common detrital clay mineral assemblage) and dolomite, may be an important but hitherto overlooked pathway for the formation of Mg-rich chlorites and corrensite in rocks that fall into this category. A further pathway is by the diagenetic alteration of precursor Mg-rich smectites that may be altered to corrensite and subsequently to chlorite with increasing diagenesis. This is analogous to the reaction pathways that have now been well documented from andesitic and basaltic volcanic rocks affected by low grade metamorphism (Frey and Robinson, 1999). The Mg-rich chlorites of the evaporite/dolomite association are rarely restricted to one lithology and may be as common in sandstones (Hillier et al., 1996) as they are in mudstones.

Aluminous dioctahedral chlorites red beds association

In most types of sediments and sedimentary rocks dioctahedral chlorites are rarer than trioctahedral types. In some red beds, however, this trend may be reversed. In fact the consistent and pervasive occurrence of dioctahedral chlorites in some red beds, as well as the allied mixed-layer dioctahedral chlorite-smectite mineral called tosudite, indicates a clear association which hitherto has not generally been recognized. Examples include the Triassic of the Colorado Plateau (Schultz, 1963), some Permo-Triassic sandstones of Germany e.g., Kulke (1969), and parts of the Old Red Sandstone of the United Kingdom (Garvie, 1992). One way to recognize dioctahedral chlorite is by an intense 003 peak in an X-ray powder diffraction pattern; this feature is directly is related to the aluminous composition. Indeed, other features such as abundant kaolinite or dickite in association with dioctahedral chlorites indicates quite clearly that the type of reds beds in which they occur have aluminous diagenetic mineral assemblages. In red beds this might be related to widespread leaching and flushing by meteoric water during early diagenesis, with consequent alteration of feldspars and micas. Dioctahedral chlorites are also common in hydrothermal alterations of aluminous rocks and in sedimentary basins they can occur as a result of localized hydrothermal activity associated with fault systems. However, the pervasive nature of their distribution in some red beds argues against a localized hydrothermal origin. Pathways for their diagenetic formation have not yet been adequately established but might involve either a kaolin precursor, or perhaps a sequence beginning with aluminous (beidellitic) smectite that evolves to dioctahedral chlorite via tosudite as an intermediate step. Such a sequence would be entirely analogous to the well-documented saponite to corrensite to trioctahedral chlorite sequence in Mg-rich environments. Indeed, tosudite may be easily misidentified as corrensite when it occurs in red beds simply because the association of corrensite with red beds is more widely known. Any clay mineral identified as "corrensite" but associated with abundant kaolinite (aluminous system) in a red bed is certainly worthy of close scrutiny.

Lithic and volcaniclastic sediment association

One of the well-known pathways by which chlorite may form diagenetically is by the alteration of detrital ferromagnesian minerals, for example biotite is commonly altered to chlorite during diagenesis (Jiang and Peacor, 1994). Clearly, this process may operate in any sediment where detrital ferromagnesium minerals are present, but is particularly common in sediments that are mineralogically immature. Volcaniclastic sediments also are prone to the diagenetic formation of chlorite, often with associated formation of corrensite (Almon et al., 1976; Stalder, 1979). In volcaniclastic sandstones there are undoubtedly many parallels with the low temperature formation of corrensite and chlorite in altered volcanic rocks (Frey and Robinson, 1999).

Shale diagenesis/low grade-metamorphism association

The smectite to illite reaction, or more generally illitization, is one of the most important reactions that occur during diagenesis. Although there are different ways to write the overall reaction, it is generally agreed that magnesium and especially iron are liberated and become available for chlorite formation. Thus in the classic study of Hower et al. (1976) of shales from the Gulf Coast chlorite was first observed at 2500m, increased in abundance to 3700m and thereafter remained constant to the greatest depth examined at 5500m. Hower et al. (1976) interpreted all chlorite they observed as diagenetic and later TEM work by Ahn and Peacor (1985) supported the hypothesis that chlorite was formed diagenetically in Gulf Coast shales as a result of the illitization of smectite. Recalculation of the data of Hower et al. (1976) indicates the formation of between 5–7 percent chlorite by weight in the shales they studied. Formation of chlorite as a consequence of the processes of illitization is probably common place in mudrocks. Furthermore it has been suggested that this process may account for the observation that older rocks (e.g., Paleozoic shales) tend to have higher chlorite contents than younger rocks (e.g., Weaver 1989), simply because they are more likely (longer thermal histories) to have been affected by advanced diagenesis. This seems a logical explanation since sedimentary rocks that have entered the realm of low-grade metamorphism are typified by a phyllosilicate assemblage of white mica and chlorite, culminating in the greenschist facies. It is unlikely, however, that late diagenetic chlorite formation in mudrocks is entirely a consequence of illitization processes. More likely there is a spectrum of reactants that may be involved in chlorite formation depending on mudrock composition (Burton et al., 1987). This spectrum may encompass many of the processes that characterize any of the different associations outlined above thereby resulting in Fe-rich, Mg-rich or Al-rich chlorite.

Stephen Hillier

Bibliography

Ahn, J.H., and Peacor, D.R., 1985. Transmission electron-microscopic study of diagenetic chlorite in Gulf-Coast argillaceous sediments. *Clays and Clay Minerals*, **33**: 228–236.

Almon, W.R., Fullerton, L.B., and Davies, D.K., 1976. Porespace reduction in Cretaceous sandstones through chemical precipitation of clay minerals. *Journal of Sedimentary Petrology*, **46**: 89–96.

Bailey, S.W., 1988a. Chlorites: Structures and crystal chemistry: In Bailey, S.W. (ed.), *Hydrous Phyllosilicates (exclusive of micas)*. Reviews in Mineralogy, Volume 19, Mineralogical Society of America, pp. 347–403.

Bailey, S.W., 1988b. Odinite: a new dioctahedral-trioctahedral Fe^{3+}-rich 1:1 clay mineral. *Clay Minerals*, **23**: 237–247.

Bain, D.C., 1977. Weathering of chlorite minerals in some Scottish soils. *Journal of Soil Science*, **28**: 144–164.

Barnhisel, R.I., and Bertsch, P.M., 1989. Chlorites and hydroxy-interlayered vermiculite and smectite. In Dixon, J.B., and Weed, S.B. (eds.) *Minerals in Soil Environments*. Soil Science Society of America, 2nd edn, pp. 729–788.

Bayliss, P., 1975. Nomenclature for the trioctahedral chlorites. *Canadian Mineralogist*, **13**: 178–180.

Burton, J.H., Krinsley, D.H., and Pye, K., 1987. Authigenesis of kaolinite and chlorite in Texas Gulf-Coast sediments. *Clays and Clay Minerals*, **35**: 291–296.

Chamley, H., 1989. *Clay Sedimentology*. Springer Verlag.

DeCaritat, P., Hutcheon, I., and Walshe, J.L., 1993. Chlorite geothermometry—a review. *Clays and Clay Minerals*, **41**: 219–239.

Ehrenberg, S.N., 1993. Preservation of anomalously high porosity in deeply buried sandstones by grain-coating chlorite: examples form the Norwegian continental shelf. *American Association of Petroleum Geologists Bulletin*, **77**: 120–1286.

Frey, M., and Robinson, D., 1999. *Low Grade Metamorphism*. Oxford: Blackwell Science.

Garvie, L.A.J., 1992. Diagenetic tosudite from the lowermost St-Maughans-group, Lydney harbor, Forest of Dean, UK. *Clay Minerals*, **27**: 507–513.

Hauff, P.L., 1981. Corrensite: mineralogical ambiguities and geologic significance. *United States Geological Survey, Open File Report 81-850*. 1–45.

Hillier, S., 1993. Origin, diagenesis, and mineralogy of chlorite minerals in Devonian lacustrine mudrocks, Orcadian basin, Scotland, *Clays and Clay Minerals*, **41**: 240–259.

Hillier, S., 1994. Pore-lining chlorites in siliciclastic reservoir sandstones - electron-microprobe, SEM and XRD data, and implications for their origin. *Clay Minerals*, **29**: 665–679.

Hillier, S., Fallick, A.E., and Matter, A., 1996. Origin of pore-lining chlorite in the aeolian Rotliegend of northern Germany. *Clay Minerals*, **31**: 153–171.

Hillier, S., and Velde, B., 1991. Octahedral occupancy and the chemical-composition of diagenetic (low-temperature) chlorites. *Clay Minerals*, **26**: 149–168.

Hower, J., Eslinger, E.V., Hower, M.E., and Perry, E.A., 1976. Mechanism of burial metamorphism of argillaceous sediment: 1. Mineralogical and chemical evidence. *Geological Society of America Bulletin*, **87**: 725–737.

Hughes, C.R., 1989. The application of analytical transmission electron microscopy to the study of oolitic ironstones: a preliminary study. In Young, T.P., and Taylor, W.E.G. (eds.), *Phanerozoic Ironstones*. Geological society, Special Publication, 46, pp. 121–131.

Jiang, W.T., and Peacor, D.R., 1994. Formation of corrensite, chlorite and chlorite-mica stacks by replacement of detrital biotite in low-grade pelitic rocks. *Journal of Metamorphic Geology*, **12**: 867–884.

Jiang, W.T., Peacor, D.R., and Buseck, P.R., 1994. Chlorite geothermometry—contamination and apparent octahedral vacancies. *Clays and Clay Minerals*, **42**: 593–605.

Kulke, H., 1969. Petrographie und Diagenese des Stubensandsteins (mittlerer Keuper) aus Tiefbohrungen im Raum. Memmingen (Bayern). *Contributions to Mineralogy and Petrology*, **20**: 135–163.

Millot, G., 1964. *Géology des Argiles*. Paris: Masson et Cie.

Odin, G.S., 1988. *Green Marine Clays*. Elsevier.

Porrenga, D.H., 1967. Clay mineralogy and geochemistry of recent marine sediments in tropical areas. *Fysisch-Geographisch Lab. Univ. Amsterdam* Dordt-Stolk, Amsterdam, pp. 1–145.

Reynolds, R.C.J., 1988. Mixed layer chlorite minerals. In Bailey, S.W. (ed.), *Hydrous Phyllosilicates (exclusive of micas)*. Reviews in Mineralogy 19, Mineralogical Society of America, pp. 601–629.

Ryan, P.C., and Reynolds, R.C., 1996. The origin and diagenesis of grain-coating serpentine-chlorite in Tuscaloosa formation sandstone, US Gulf Coast. *American Mineralogist*, **81**: 213–225.

Schultz, L.G., 1963. Clay minerals in Triassic rocks of the Colorado plateau. *Geological Survey Bulletin*, **1147-C**: C1–C71.

Stalder, P.J., 1979. Organic and inorganic meatmorphism in the Taveyannaz Sandstone of the Swiss Alps and equivalent sandstones in France and Italy. *Journal of Sedimentary Petrology*, **49**: 463–482.

Walker, J.R., 1993. Chlorite polytype geothermometry. *Clays and Clay Minerals*, **41**: 260–267.

Weaver, C.E., 1989. *Clays, Muds, and Shales*. Elsevier.

Windom, H.L., 1976. Lithogenous material in marine sediments. In Riley, J.P., and Chester, R. (eds.), *Chemical Oceanography*. Academic Press, pp. 103–135.

Cross-references

Berthierine
Clay Mineralogy
Glaucony and Verdine
Illite Group Clay Minerals
Kaolin Group Minerals
Mixed-Layer Clays
Smectite (Montmorillonite) Group
Talc
Vermiculite

CLASSIFICATION OF SEDIMENTS AND SEDIMENTARY ROCKS

Sediments and sedimentary rocks may be divided into two kinds, intrabasinal or autochthonous and extrabasinal or allochthonous. Intrabasinal sediments and sedimentary rocks or autochthonous deposits are those whose particles were derived from within the basin of deposition. Most carbonate sediments and rocks (including limestones and dolomites or dolostones) were precipitated within a basin of deposition. Terrigenous particles or sedimentary rocks belong to the extrabasinal or allochthonous group and were derived from outside the basin of deposition; examples are sandstones and shales (Lajoie, 1979; Friedman et al., 1992).

Intrabasinal or autochthonous deposits

Intrabasinal or autochthonous particles include various solids that grew biochemically or chemically in the waters of the depositional basin. These include carbonate biocrystals and other carbonate particles, silica biocrystals, particles composed of evaporite minerals, and certain authigenic minerals, such as glaucony (glauconite), that grew at the water/sediment interface.

Carbonate particles, sediments and rocks

Carbonate sediments and rocks are composed of carbonate particles and cement. They grew as solids in the depositional basin and include skeletal and non-skeletal calcium-carbonate materials. Carbonate skeletal debris includes whole skeletons of calcium carbonate-secreting organisms, as well as broken pieces of the hard parts secreted by organisms. Ordinarily, organisms must die to contribute their skeletal material to sediments. However, several exceptions are known. Ostracodes and trilobites discard their shells during molting and coccoliths shed plates during life.

The clay- and silt-size particles of carbonate sediments are collectively designated as lime mud. Lime mud for the most part consists of tiny needles and platelets of carbonate crystals. Organisms secrete the tiny solids that compose lime mud. Lime mud may form by the accumulation of tiny skeletal components secreted by algae. During the Cretaceous pelagic algae, known as coccoliths, were abundant enough to form an ooze that, when lithified, became chalk (*q.v.*).

The minerals secreted by organisms form biocrystals, that is, solids having the lattice properties of minerals but distinctive shapes that are not crystal faces. In the modern marine environment, widely varying groups of organisms produce skeletal debris. In the past, some groups of organisms that are now important sediment producers were insignificant, and some once-important groups are now minor or no longer exist at all.

In modern carbonate sediments the two principal minerals are calcite and aragonite. This fact was established in 1859 by Henry Clifton Sorby (1826–1908; see *Sedimentologists*), who also showed that calcite tends to predominate in sand-size skeletal debris, whereas aragonite predominates in lime mud.

Although the chemical formulas of both aragonite and calcite are the same, $CaCO_3$, these minerals differ in crystallographic arrangment, in density, in hardness along certain crystallographic directions, in their contents of trace elements, and in their solubilities in various fluids (see *Carbonate Mineralogy and Geochemisty*). One variety of calcite has been termed low-magnesian calcite and the other high-magnesian calcite. Modern skeletal carbonate particles consist mostly of high-magnesian calcite and aragonite, whereas ancient limestone consists of low-magnesion calcite. Low-magnesian calcite has less than 8 mol percent $MgCO_3$. A calcite containing 8 mol percent to 10 mol percent $MgCO_3$ may be considered either high or low-magnesian calcite. The composition of the mineral calcite, which represents both kinds, can be given by the formula $(Ca_{1-x}Mg_x)CO_3$, where $0<x<0.30$ (Bathurst, 1975; Sanders and Friedman, 1967; Tucker and Wright, 1990).

Lime mud: silt- and clay-size

The silty-and clay-size components of carbonate sediments are collectively designated as lime mud. Most lime mud consists of tiny needles and platelets of carbonate crystals that are too small to resolve with binocular- or petrographic microscopes. Accordingly, interpretations of the particles of lime mud have been based on the notion that they shared a common origin, namely inorganic precipitation from saline water. Studies with the scanning-electron microscope have shown that, among the supposedly uniform constituents of lime muds, great variety exists. Three mechanisms for forming lime mud include: (1) inorganic precipitation, (2) biochemical secretion, and (3) breakup of sand-size (or larger) aragonitic skeletal debris.

Inorganic precipitation may be an important source of lime mud. According to an estimate of Shinn *et al*. (1989), the amount of lime mud precipitated inorganically on the Great Bahama Bank may be much greater than that secreted by algae. The tiny needles of aragonite precipitated inorganically are dispersed in the water and form drifting "clouds" of milky-white water called whitings. Sedimentation rates from whitings are high, yet the whitings appear to persist for months. This implies that, despite the losses from the fallout to the bottom, the components of the whitings are being renewed. A plausible mechanism for such renewal is rapid precipitation in the water column of aragonite needles (and possibly small particles of high-magnesian calcite). The tiny individual aragonite crystals precipitated by whitings are indistinguishable from aragonite biocrystal needles secreted by codiacean green algae.

Within their cells, certain algae secrete solid skeletal components that consist of tiny biocrystals of aragonite. Much lime mud forms by the accumulation of such tiny biocrystals, which are released onto the bottom when the algae die.

Codiacean or green algae, especially the alga *Penicillus* sp. which resembles a shaving brush and hence is popularly known as the shaving-bruch alga, produce fine biocrystals (needles) of aragonite ($<15\,\mu m$) within the sheaths of their filaments. When the organism dies, the filaments disintegrate and release the needles to the sea, where they accumulate as lime mud. The modern-day rate of production of aragonite needles by *Penicillus* sp. alone can account for all the mud in the inner Florida Reef tract and one third of the mud in northeastern Florida Bay. Two more-or-less-equally abundant green algae (*Udotea* and *Rhipocephalus*) are likewise active contributors of lime mud in this area. Codiacean algae are known in rocks as old as Ordovician, and they may have contributed aragonite needles to ancient lime muds.

Coralline red algae secrete skeletons composed of high-magnesian calcite. High-magnesian calcite, probably derived from the breakdown of red-algal skeletons, is abundant in carbonate muds in Florida Bay and on the Great Bahama Bank west of Andros Island. Red algae and serpulid worms living on leaves of the marine grass *Thalassia* may produce as much as or more lime mud than does *Penicillus*.

Some lime mud results from the physical breakage of sand-size-and larger aragonitic skeletal material. Such mud can originate by (1) decomposition of organic matter, such as conchiolin, which binds the small hard parts together; (2) weakening or comminution of the skeletal material by boring organisms such as fungi, algae or sponges, (3) feeding activities of predatory organisms, and (4) physical breakage by abrasion in agitated waters.

Intraclasts

The term intraclast refers to sand-size or larger particles, texturally analogous to rock fragments, broken from consolidated-or hardened materials in one locality and redeposited at another locality within the basin of deposition. "Intra" means within, in this context within the basin of deposition, and "clast" denotes broken. Many intraclasts are recycled fragments of coherent sediment. Intraclasts are of various sizes and shapes. Many are angular and their diameters may exceed 2 mm.

Micrite

Lithfied lime mud that was deposited mechanically, and which may form a matrix among sand-size particles in limestone or be the only particle in a fine-textured limestone is known as a **micrite**. This term is a shorthand expression for microcrystalline calcium carbonate.

Pellets

Spherical- or ellipsoidal sand-size particles of calcium carbonate are called pellets. Internally, pellets are commonly homogeneous. Although they are of sand size, most pellets may be compared to tiny mudballs, and commonly consist of aragonite. Most pellets are formed by deposit-feeding organisms that eat the mud. These organisms digest organic matter from the mud and excrete the undigested lime mud in the form of fecal pellets.

Peloids

Peloids are particles that resemble pellets but for which no particular origin is implied. Not all pelletlike particles are of fecal origin; some are lime-mud aggregates that originated when lime mud dried out on exposure to the atmosphere. When lime mud is so exposed, desiccation cracks (the so-called mud cracks) and small chips of dried-out mud spall off. These particles may be larger than sand-size pellets, but they are very fragile, and are quickly comminuted to rounded, sand-size, pellet like particles. Many peloids are sand-size, rounded intraclasts. Fecal pellets and pellet-shaped mud aggregates commonly cannot be distinguished.

Ooids

The sand-size coated particles known as ooids derive their name from a Greek word which means egg or egglike because under the microscope these particles resemble the roe of fish. Ooids usually are spherical or elliptical.

The rims of most modern marine ooids consist of aragonite; a few are composed of high-magnesian calcite. In most modern aragonitic ooids the long axes of the individual aragonite crystals are tangential to the rims. This is known as concentric fabric. In some ooids the fabric is radial; the long axes of the aragonite crystals are normal to the rims and therefore diverge away from the center of the ooid. Modern ooids having radial fabric (mostly composed of high-magnesian calcite) occur in hypersaline waters, such as in the Great Salt Lake of Utah; in Baffin Bay, off Laguna Madre, Texas; or in thick accumulations of algal mats as found in sea-marginal ponds of the modern Red Sea and along the west coast of Australia. In addition to higher salinity, locations of formation of radial ooids are characterized by less-energetic water movement. It appears, therefore, that low-energy conditions favor growth of radial ooid fabric whereas high-energy conditions favor growth of tangential (concentric) ooid fabric. However, this applies only to aragonitic ooids; modern marine high-magnesian calcite ooids are exclusively radial (see *Oolites and other coated grains*).

On the basis of their internal fabric, most ancient marine ooids may be classified into two groups; (1) Group one consists of ooids having either (a) concentric aragonitic fabric or (b) coatings consisting of crystalline mosaics of calcite (or dolomite), commonly lacking any vestige of tangential or radial structure. These mosaics may consist of minute euhedral rhombs only a few micrometers across or of anhedral spar crystals nearly as large as the ooids themselves. The structures of these ooids have been substantially altered by diagenesis.

The ooids in algal mats may be precipitated by the algae themselves, yet the origin of most ooids is still problematic. Most modern ooids occur in or close to the intertidal position in a zone in which the waves break and pound. Many geologists consider that in this turbulent, shallow-water environment, ooids may form inorganically as the cooler water from adjoining deep zones spreads across the shoals, gives up some of its CO_2, and becomes warmer.

Pisolites

Spherical- or elliptical coated particles that exceed 2 mm in diameter are known as pisolites. The division between pisolites and ooids is one of size; ooids are smaller than 2 mm. Despite this seemingly arbitrary size differentiation, the origins of ooids and pisolites are not the same. Pisolites that are merely oversized ooids are not common, either in modern carbonate sedimentary environments or in ancient limestones. Three common kinds of pisolites are known: oncolites, caliche or vadose pisolites, and cave pearls or cave pisolites.

Oncolites (also known as algal pisolites) very closely resemble vadose pisolites. Their origins are, however, quite different. Oncolites consist of encrustations on various particles. When particles, commonly skeletal, roll about intermittently on the sedimentation surface, microorganisms, especially cyanobacteria, but also green algae, bacteria, and diatoms, attach themselves to and repeatedly coat these particles with concentric laminae by trapping and binding sediment. The microorganisms coat the exposed portions of the oncolites during periods of quiescence, and may be partially abraded from the surfaces of the oncolites during the intervening periods of rolling. Thus oncolites commonly are composed of irregular-and incomplete laminae.

The most common pisolites are known as caliche- or vadose pisolites. Vadose pisolites form in the weathering zone as caliche (*q.v.*). Cave pearls, or cave pisolites, are pisolites that are true particles; they form in a manner analogous to ooids, but in pools of water in caves (Bathurst, 1975; Friedman, 1985; Friedman *et al.*, 1992; Shinn *et al.*, 1989; Tucker and Wright, 1990).

Limestones

Except for reefs which framework builders construct, for consolidated lime mud (micrite) and for limestones formed by the bacterial breakdown of calcium sulfate, the basic constituents of most limestones are recognizable sand-size particles and the spaces between these particles. The interparticles spaces may be occupied by (1) micrite, (2) optically clear spar (calcite), and (3) void space, which may be filled with fluids, such as water or petroleum (oil or gas) (Friedman *et al.*, 1992; Scoffin, 1987; Jones and Goodbody, 1985; Tucker and Wright, 1990).

Dunham classification

Dunham (1962) recognized two first categories among limestones: (1) those in which the original particle components were bound together (named collectively as boundstones) and (2) those in which the original components were not bound together (no collective name) (Figure C32).

Boundstones are *limestones showing evidence that particles being deposited were bound by organisms or that they consist of frameworks constructed by organisms*. This group includes reefs, stromatolites, and travertine. Embry and Klovan (1971) have divided Dunham's boundstone group into subgroups according to the nature of the binding. Thus *limestones bound together with microbial laminae* are bindstones. *Limestones that originated in place* (autochthonous limestones) *as frame-built reefs* are framestones. *Limestones that consist predominantly of sediment trapped by baffling organisms* are bafflestones. This expansion of the Dunham classification is congruent with ecological classifications of dominant reef organisms into guilds (constructors, binders, bafflers).

Particle-supported rocks are subdivided into two further classes: (1) *particle-supported limestones devoid of lithified lime mud*, known as grainstones, and (2) *particle-supported limestones containing some lithified lime-mud matrix, but not enough to keep the sand-size particles from touching one another*; known as packstones.

Matrix-supported limestones are also subdivided into two classes; (1) *mud-supported limestones containing at most 10 percent sand-size- or larger particles*, known as mudstones, and (2) *mud-supported limestones containing at least 10 percent sand-size- or larger particles*, known as wackestones. Notice that in the Dunham classification the term micrite is not used. Instead, the fine- size carbonate rock is called mudstone (better, lime mudstone) (Figure C32).

The rationale of this classification is that it allows one to map gradients in rate of production of sand-size particles relative to rate of accumulation of lime mud. In calm waters,

CLASSIFICATION ACCORDING TO DEPOSITIONAL TEXTURE

DEPOSITIONAL				TEXTURE
Original components not bound together during deposition				Original components were bound together during deposition... as shown by intergrown skeletal matter, lamination contrary to gravity, or sediment-floored cavities that are roofed over by organic- or questionably organic matter and are too large to be interstices.
Contains mud (particles of clay- and fine silt size)		Lacks mud and is particle supported		
Mud supported		Particle supported		
Less than 10% particles	More than 10% particles			
Mudstone	Wackestone	Packstone	Grainstone	Boundstone

Figure C32 Classifying and naming limestones according to depositional texture; the scheme of Dunham (1962).

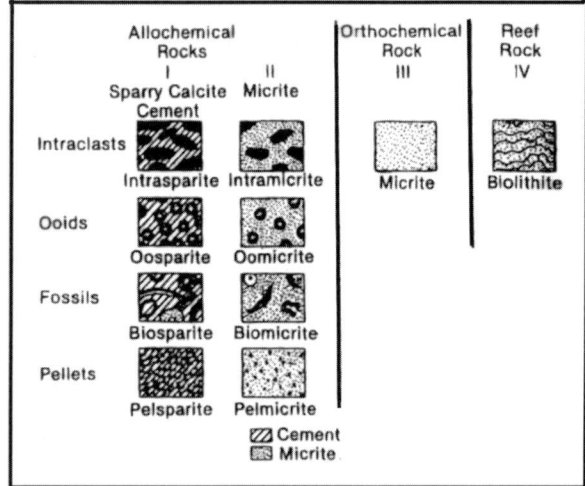

Figure C33 Classifying and naming limestones according to the scheme of Folk (1959) (modified by Friedman et al., 1992).

lime mud, if present, settles on the bottom and remains there. Hence limestones derived by lithification of an original lime mud deserve to be contrasted with those derived by lithification of carbonate sediment devoid of lime mud. This relationship between sand-size particles and lime mud generally distinguishes an original sediment deposited in calm water from a sediment deposited in agitated water. This distinction is fundamental (Dunham, 1962).

Folk classification

R.L. Folk (1959) subdivided the constituents of limestones into two categories: (1) allochemical constituents or allochems (particles, meaning sand-size or larger), and (2) orthochemical constituents (collectively designated by the word orthochems and consisting of micrite, presumably lithified original lime mud; and cement, sparry calcite; Figure C33). Folk's classification lists four kinds of allochems (particles): (1) intraclasts, (2) ooids, (3) skeletal particles and (4) pellets (including peloids).

The interstices among the allochems are filled with the orthochemical constituents (micrite or sparry calcite cement). Combinations of allochems and orthochems provide a basis for recognizing eight kinds of limestones. In addition, micrite may lack sand-size particles and hence stand alone as a ninth kind of limestone. Finally, a tenth kind is a reef rock (biolithite) (Figure C33).

Folk assigned names to eight of these ten kinds of limestone by using composite words consisting of two parts: (a) an initial abbreviated expression for the allochems, and (b) a word designating the orthochemical constituents, based on one of the two groups listed above. These words for the orthochems are (1) micrite, and (2) sparite (*limestones having a cement consisting of sparry calcite*). The prefixes for the allochems are abbreviated as follows: intraclasts = intra; ooids = oo; fossils = bio; pellets = pel.

Thus the names of these eight kinds of limestone are intrasparite, intramicrite, oosparite, oomicrite, biosparite, biomicrite, pelsparite, and pelmicrite. If several kinds of particles are important constituents of a limestone, their abbreviations are strung together in order of increasing abundance. For example, a limestone composed of micrite and having allochems composed of 10 percent intraclasts, 20 percent skeletal debris, and 70 percent peloids would be an intrabiopelmicrite.

The Folk classification allows for particles larger than sand size, which are however, not particularly common in limestones. For particles of pebble size, such as the intraclasts of flat-pebble conglomerates, the shells of lag concentrates, or the pellets of organisms having large anal diameter, Folk continued the tradition of one of Grabau's terms, rudite. A limestone composed of large pellets, if accumulated in a micrite, is known as pelmicrudite; flat pebbles in a micrite would be intramicrudite. If sparry cement occurs between these same pebble-size particles, the limestones would be designated pelsparrudite and intrasparudite.

The distinctive aspect of Folk's classification (Folk, 1959) is its use of names for eight major kinds of limestone based on what lies between the particles (spar or micrite).

Dolostones

Dolostones are carbonate rocks consisting of the mineral dolomite CaMg $(CO_3)_2$. Two kinds of dolostone occurs: (1) Calcareous dolostones, carbonate rocks containing 50 to 10 percent of the mineral dolomite, and (2) Dolostones, carbonate rocks containing 90 percent or more of the mineral dolomite. Dolostones may or may not preserve the original texture of the antecedent limestone, such as skeletal particles or ooids, which they have replaced

A dolomite fabric in which the predominant constituents are euhedral crystals is known as idiotopic. Idiotopic dolostone is one of the most-important carbonate reservoirs for oil and gas because it is abundant, porous, and permeable. Dolomite crystals partly bounded by crystal faces are subhedal. A dolomite fabric in which most of the constituent crystals are subhedral is hypidiotopic. Dolomite crystals bounded by anhedral crystal facies constitute a fabric known as xenotopic (Friedman, 1965).

Other crystals may be added to the name as modifiers. Such modifiers for crystal size include: micron-sized: 0 μm to 10 μm (0.001 to 0.01 mm); decimicron-sized: 10 μm to 100 μm (0.010 to 0.100 mm); centimicron-sized: 100 μm to 1000 μm (0.100 mm to 1.000 mm); millimeter-sized: 1 mm to 10 mm; centimeter-sized: 10 mm to 100 mm.

A further breakdown of the centimicron-size range may be made by using the terms fine crystalline (100 μm to 250 μm), medium crystalline (250 μm to 500 μm), and coarse crystalline (500 to 1000 μm). This further breakdown into three additional size classes is based on empirical observations of the distribution of crystal sizes in dolostones and recrystallized limestones and is therefore useful; it does not, however, follow the metric framework of the above-given size classification.

The naming and classifying of these various fabrics and size classes can also be extended to limestones that have undergone neomorphism or to precipitates, such as evaporites.

Dolostones may also be classified genetically into three major groups: (1) syngenetic dolostone, (2) diagenetic dolostone, and (3) epigenetic dolostone. Syngenetic dolostone is here defined as *dolostone that has formed penecontemporaneously in its environment of deposition*. This kind of dolostone contrasts with diagenetic dolostone, which is *dolostone formed by replacement of calcium-carbonate sediments or of limestones during or following consolidation, such as within beds of carbonate sediments or limestones*. In borderline cases, the distinction between syngenetic and diagenetic becomes difficult or impossible to make. Epigenetic dolostone is here defined as *dolostone that has formed by replacement of limestone along post-depositional structural elements, such as faults and fractures*. Many, but not all, epigenetic dolostones are genetically associated with metallic ore deposits, notably of lead and zinc minerals. These dolostones have recently become known as hydrothermal dolostones (Friedman, 1965; Friedman and Sanders 1967; Bathurst, 1975; Baker and Burns, 1985; Land, 1985; Hardie 1987; Tucker and Wright, 1990; Friedman *et al.*, 1992).

Authigenic rocks

Authigenic is a term designating rocks that have grown in place subsequent to the formation of the sediment or rock of which they constitute a part. The most-important kinds of authigenic rocks are (1) chert, (2) phosphate rock, and (3) sedimentary iron ores. Evaporites formed by precipitation in the interstitial waters of a preexisting sediment likewise qualify as being authigenic.

Chert

Chert is a tough, brittle siliceous rock exhibiting a splintery-to conchoidal fracture and a vitreous luster. The silica to form most cherts is thought to come from the tests of siliceous organisms.

The tests of siliceous organisms, such as those of radiolarians and diatoms, as well as the spicules of sponges consist of opal. Likewise, siliceous shells in cherts of Tertiary age are mostly opaline. In contrast, nearly all silica in Paleozoic cherts occurs as quartz and chalcedony.

Silica minerals present in chert include quartzine and lutecine, which are fibrous varieties of quartz, and cristobalite (opal-CT). Quartzine and lutecine may form at the expense of sulfate minerals in evaporite deposits. These silica minerals may be the only testimony to the former presence of such sulfates, now replaced and vanished. These silica minerals may therefore be useful indicators of ancient environments.

The silica of chert may be (1) an alteration product of volcanic rock, such as smectite and volcanic glass, or (2) a precipitate derived from the dissolution of tests of siliceous organisms. The most-probable source of silica in the cherts of the central Pacific Ocean is dissolution of radiolarian tests. The silica in most oceanic cherts probably is of biogenic origin. Biogenic opal is dissolved and reprecipitated as finely crystalline cristobalite or opal-CT, which inverts to quartz. The end product is a classic dense, vitreous chert.

Chert occurs (1) in nodules, and (2) in strata (bedded or ribbon cherts). Common synonyms for chert or varieties of chert include jasper, flint, novaculite, and porcellanite.

Although the source of silica for many cherts evidently is the dissolved siliceous tests of marine organisms, many chert nodules are inferred to have grown by the simultaneous action of (a) dissolution of carbonate, and (b) precipitation of silica in thin films of water at the boundaries of the carbonate being dissolved (Jones and Segnit, 1971; Laschet, 1984; Tada and Iijima, 1983; Williams and Crear, 1985).

Phosphate rock

Sedimentary phosphate deposits usually consist of a carbonate fluorapatite [$Ca_5(PO_4)_3(F, CO_3)$]; a variety of apatite known as francolite. When the apatite-like phase cannot be identified, the name collophane is commonly applied. Such phosphates, or phosphorites (the two terms are synonymous) are here referred to as phosphate rock.

Phosphate rock occurs on the modern sea bottom and in ancient deposits. Ancient deposits of phosphate rock form extensive phosphogenic provinces (Bentor, 1980; Germann *et al.*, 1984; Kolodny, 1981; Riggs, 1986; Sheldon, 1987; Notholt *et al.*, 1989; Notholt and Jarvis, 1990;).

Sedimentary iron ores

Valuable iron-rich sedimentary rocks, true ores in the fullest sense of the term, have influenced modern history. England achieved greatness through exploitation of her sedimentary iron ores, while at the same time mining her rich coal deposits. During the nineteenth- and early part of the twentieth centuries Germany and France were at each other's throats over possession of the vital Lorraine deposits, the famous oolitic minette ores of Jurassic age. Discovery of the vast Lake Superior iron ore deposits in 1844 ushered in the industrial age for the United States. In Europe and in North America, where sedimentary iron ores and coal met, large industrial centers arose, particularly in England, France, Germany, and the United States.

The term sedimentary iron ores designates those consisting predominantly of iron minerals, notably oxides, hydroxides, carbonates, silicates, and sulfides. The oxides are hematite and magnetite; the hydroxide mineral is goethite (including limonite); the carbonate mineral is siderite; the silicate minerals are chamosite, greenalite, and glauconite; and the sulfide minerals are pyrite and pyrrhotite.

Sedimentary iron ores are classified into three broad groups: (1) *bog-iron deposits*, (2) *ironstones*, and (3) *iron formations*. The basis of the classification combines mineral composition, occurrence, and geologic age. In addition, sedimentary iron ores may be named according to their dominant iron mineral; e.g., siderite rock, hematite rock, or magnetite rock (James, 1954; Maynard, 1983; Guilbert and Park, 1986; Young and Taylor, 1989; Van Houton, 1990).

Evaporites

Evaporites form by precipitation from brines whose salinity values have been greatly increased by evaporation. Evaporation may take place in closed or in open systems in numerous environmental settings.

Although almost forty different precipitate-type minerals have been recorded from evaporite deposits, only about twenty are present in more than trace amounts. Of these, only two kinds are common, sulfates and halides, both of which form extensive deposits of sedimentary rock. Two kinds of sulfate minerals are common in sedimentary rocks: gypsum ($CaSO_4 \cdot 2H_2O$) and anhydrite ($CaSO_4$).

Among the halides, halite rock (rock salt) is the most common. It forms successions up to 1,000 meters (m) thick (Schreiber and Friedman, 1976; Warren, 1989; Tucker and Wright, 1990).

Carbonaceous sediments and rocks

Carbonaceous sediments and rocks, primarily coal, have been classified in two ways: by rank (content of carbon and caloric value) and by petrographic characteristics. In order of increasing rank, the classes of coal are (1) lignite; (2) subbituminous coal; (3) bituminous coal; and (4) anthracite. The progression of carbonaceous material through the continuous series of lignite (or its precursor, peat) through bituminous coal to anthracite is known as coalification.

Coal has been recorded in strata ranging in age from Precambrian to Holocene. However, only after the Late Silurian Period, when land plants had become established, was it possible for plant material to accumulate on a scale to form large deposits of coal. In the history of the earth, two particularly rich coal-forming periods are known, the Carboniferous (principally Pennsylvanian) and Early Tertiary Periods. Most coals form from accumulated plant debris in sea-marginal swamps or in closed fluvial basins (Sackett *et al.*, 1974; Hunt, 1979).

Extrabasinal or allochthonous deposits

Extrabasinal or allochthous rocks consist mainly of one group. These are the terrigenous rocks.

A fundamental basis for classification is particle size. On the basis of increasing particle size, we recognize *three main kinds of terrigenous rocks:* (1) shales, (2) sandstones, and (3) conglomerates and sedimentary breccias. According to the Grabau system of classification, we use the terms lutite for shale, arenite for sandstone, and rudite for conglomerates. Some geologists recognize a fourth kind of terrigenous rock, siltstone. For it the term siltite has been introduced.

Shales

Fine-textured terrigenous rocks compose about two thirds of the sedimentary rock record, yet because their particles are so small, they are incompletely understood. Various bases for naming these rocks have been proposed, including (1) particle size, (2) proportion of clay minerals, and (3) fissility, *the property of splitting easily into thin layers parallel to the bedding*.

Probably the most-satisfactory definition of shale is according to particle size. In terms of size, the name shale is the lithified equivalent of mud. Mud is a sediment consisting of clay- and silt-size particles. Likewise, shales are *terrigenous rocks composed of clay- and silt-size particles*.

The proportion of clay minerals in shales varies widely. In most shales, quartz may compose between one quarter and one half of the total; in some shales, quartz together with feldspar generally contain greater proportions of quartz and feldspar, whereas fine shales contain less of these and a greater proportion of clay minerals. Fine-textured terrigeneous sedimentary rocks having abundant quartz (and commonly also feldspar) particles of silt size and that as a result, closely resemble sandstones.

On surface exposures, where weathering has taken place, it is easy to distinguish fissile from nonfissle fine-textured rocks. The terms mudrock, mudstone, claystone, and argillite have been employed for shales lacking fissility; argillites are more-indurated shales than mudstones and claystones. In the Dunham classification of carbonates, the term mudstone refers to fine-textured limestone (in which case the rock should be called lime mudstone to avoid confusion with fine-textured siliciclastic rocks).

A dark, thinly laminated carbonaceous shale, exceptionally rich in organic matter (5 percent or more of total organic carbon) and sulfides (FeS_2), is known as black shale. It forms by partial anaerobic decay of organic matter in a reducing setting in which water circulation is restricted and deposition is slow (Chamley, 1989; Dunbar and Rodgers, 1957; Füchtbauer and Leggewie, 1984; Heling, 1988; Pettijohn, 1975; Potter *et al.*, 1980; Weaver, 1989).

Sandstones

Sandstones are terrigenous sedimentary rocks in which sand-size particles predominate. As with limestones, sandstones can be classified on the basis of (1) particles, (2) matrix, and (3) cement. Particles are mostly derived from the lands, hence are terrigenous, but pyroclastic and even clastic-carbonate particles may be abundant. The essential particles of sandstones are (1) quartz, (2) feldspar, and (3) rock fragments. Many other kinds of particles occur in sands and sandstones,

End-member sand-size particles	Idealized end-member rock names	
	< 15% matrix	> 15% matrix
Quartz	Quartz sandstone	Argillaceous Quartz sandstone
Feldspar	Feldspar sandstone	Argillaceous feldspar sandstone
Rock fragments	Rock-fragment sandstone	Argillaceous rock-fragment sandstone

Figure C34 Classifying and naming sandstones.

but those other are not common and are not considered essential in naming and classifying sandstones. The matrix is the mud: physically deposited material that consists mostly of clay minerals and quartz. Cement is the chemically precipitated filling of original void spaces. Some sandstones lack any matrix; only mineral cement occupies the spaces among the framework particles. The argillaceous matrix of sandstones implies (1) availability of mud, and (2) a low-energy regime in the depositional environment.

Depending on the presence or absence of argillaceous matter, we recognize two end-member groups of sandstones: (1) argillaceous sandstones (sandstones containing about 15 percent or more of clay-size material), and (2) ordinary sandstones, which contain less than about 15 percent clay-size material. The value of 15 percent is arbitrary; in a qualitative study, the obvious presence of a fair amount of clay suffices to classify a sample as argillaceous.

Based on the proportions of quartz, feldspar, and rock fragments, we can divide each of these two first-order groups (argillaceous sandstones and ordinary sandstones) into three second-order groups. Thus, as indicated in Figure C34, we can recognize six sandstone end members: (1) quartz sandstone, (2) feldspar sandstone, (3) rock-fragment sandstone, (4) argillaceous quartz sandstone, (5) argillaceous feldspar sandstone, and (6) argillaceous rock-fragment sandstone.

The sand-size debris of most sandstones includes chiefly quartz, but with the quartz may be variable amounts of feldspar or rock fragments or feldspar and rock fragments. These mixtures are named by incorporating the names of all kinds of sand-size debris present, listed in increasing order of abundance. Thus if a sandstone contains 70 percent quartz, 20 percent feldspar, and 10 percent rock fragments, we would name it a rock fragment-feldspar-quartz sandstone.

The kinds of feldspars or rock fragments reflect (1) the parent terrain or provenance from which the particles in a sample of sandstone have been derived, and (2) their preservation during weathering, transportation, and diagenesis.

The pioneer sedimentary petrographer Paul D. Krynine (1902-1964; see *Sedimentologists*) provided the modern impetus in naming and classifying sandstones. Krynine's (1948) original article is a classic in geology; it combines astute observation in the field and of samples with a deep insight and appreciation for the fundamental principles of geology. In his classification of sandstones, Krynine introduced triangular diagrams for plotting his three fundamental end members: orthoquartzite, graywacke, and arkose (Figure C35).

The earliest description of graywacke dates from 1789, and the term arkose was introduced in 1823. According to Krynine, a graywacke is *a sandstone composed of angular quartz (and chert) particles and abundant metamorphic rock fragments, with little or no cement and feldspar, and containing more than 12 percent to 17 percent micas and chlorite (either in the clay matrix or as metamorphic rock framents)*. Krynine felt that the mica and chlorite between the particles were a mechanically deposited matrix and represented poor sorting; we now know that this can no longer be assumed.

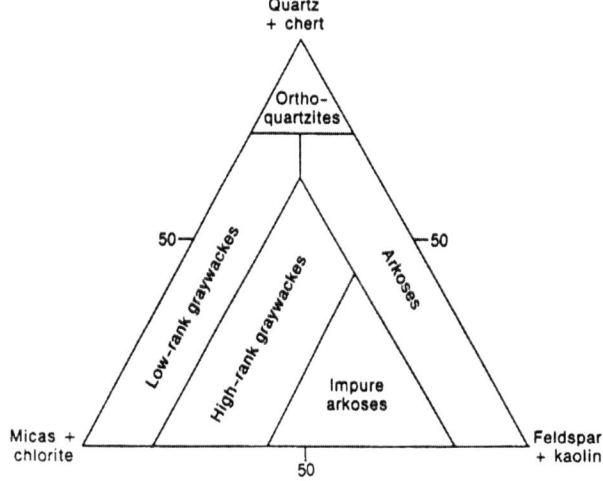

Figure C35 Triangular diagram; Krynine's classification of sandstones. (after P.D. Krynine, 1948, Figure C35. p.137).

When deeply buried and subjected to appropriate geothermal gradients and migrating solutions, framework particles may be dissolved and new clay minerals, such as chlorite and illite, may form. This diagenetic neoformation of clay-mineral cements within sandstones may generate a rock composition fitting the term graywacke. Yet Krynine and many others use this term to refer to a rock having a mechanically deposited matrix and poor sorting. To avoid the term graywacke, some authors substituted the term wacke. In his classification, Krynine used the term arkose for *a sandstone with more than 30 percent feldspar*.

A rock defined by Krynine as a graywacke would be named *argillaceous rock fragment-quartz sandstone* if it contained no feldspar, and an argillaceous rock fragment-feldspar-quartz sandstone if if contained feldspar in greater abundance than rock fragments, or an argillaceous feldspar-rock fragment-quartz sandstone if the rock fragments were more abundant than the feldspar. The arkose of Krynine would be named a feldspar-quartz sandstone if the quartz is more abundant than the feldspar, but a quartz-feldspar sandstone if feldspar abundance exceeds quartz abundance (Folk, 1956; McBride, 1962; Klein, 1963; Sanders, 1978; Füchtbauer *et al.*, 1988).

Conglomerates and sedimentary breccias

Coarse terrigenous rocks are formed by the lithification of gravel. The degree of rounding of the particles defines two categories. Conglomerate is defined *as a coarse terrigenous sedimentary rock formed by the lithification of rounded gravel.* Sedimentary breccia is *a coarse terrigenous sedimentary rock formed by the lithification of angular gravel-size or larger particles.* In conglomerates and sedimentary breccias more than 30 percent of the large particles exceed 2 mm in diameter. The particles may consist of pebbles, cobbles, boulders, or mixtures of these sizes.

Conglomerates and sedimentary breccias may be named and classified by (1) the proportion of gravel size particles, (2) the kind of matrix, and (3) the kinds of gravel-size particles.

According to the proportion of gravel-size particles and the kinds of matrix, (1) a sample containing 80 percent pebbles, cobbles, and/or boulders is termed a conglomerate proper, and (2) one having between 30 percent and 80 percent of such coarse particles is (a) a sandy or arenaceous conglomerate or (b) a shaly or argillaceous conglomerate. The matrix between the coarse particles in a conglomerate may also be calcareous or sideritic.

Based on the variety of pebbles, cobbles, and/or boulders composing them, we can classify conglomerates into two kinds: (1) *conglomerates consisting only of pebbles, cobbles, and/or boulders of a single kind of rock* (such as one or various varieties of chert and quartzite or of other rock), or oligomictic conglomerates; and (2) conglomerates containing pebbles, cobbles, and/or boulders of many kinds of rock, known as polymictic conglomerates.

The term diamictite has been proposed for a *nonsorted noncalcareous, terrigenous sedimentary rock composed of sandsize and/or larger particles dispersed through a fine-textured matrix.* Many, if not most, diamictites are conglomerates. Diamictites originate by landslides, earth flows, mudflows, solifluction, ice rafting, subaqueous slumping and sliding, and/or by glaciers. A nonlithified diamictite is a diamicton (Flint *et al.*, 1960; Pettijohn, 1975; Friedman and Sanders, 1978; Füchtbauer *et al.*, 1988).

Pyroclastic rocks

Pyroclastic rocks are lithified tephra. Their particles originate as explosive igneous material but are deposited as sediments.

Volcanoes and their products, both on land and in the sea, are the major source of particles for pyroclastic rocks, especially in areas of tectonic activity and in island arcs. Particles originating by explosive ejection from volcanic vents spread out over great areas. On the continents, pyroclastic rocks far exceed the volume of extrusive igneous rocks.

Tephras are *chemically and texturally distinctive*, so ancient tephras may be used for correlation. Mapping of particle size and thickness of tephra deposits assists in locating the parent volcano.

Figure C36 shows how sizes of tephra particles can be used to classify pyroclastic rocks.

Pyroclastic rocks composed mostly of blocks are volcanic breccia; pyroclastic rocks composed mostly of lapilli are lapilli tuffs; and *pyroclastic rocks composed of ash* are tuffs in the strict sense of the word.

The three end members of pyroclastic rocks are (1) volcanic rock fragments, (2) crystals, and (3) glass. If by size a rock is a

Limiting Particle Diameter mm / φ units	Standard Size Classes of Sediments	Size Classes of Tephra Particles	Size Classes of Pyroclastic Rocks
64 / −6	Cobbles	Blocks / Bombs	Agglomerate / Volcanic breccia
2 / −1	Pebbles	Lapilli	Lapilli tuffs
1/16 / +4	Sand	Coarse ash	Tuffs
	Silt	Fine ash	

Figure C36 Size classes and terms for tephra particles and pyroclastic rocks (modified from R.V. Fisher, 1966).

tuff and rock fragments exceed glass, this tuff may be described as a glass, rock-fragment tuff. End members are listed in order of increasing abundance. By this approach one can recognize crystal-, rock-fragment-, and glass tuffs. Adjectival modifiers may be added.

Because of the chemical reactivity and instability of their constituents, pyroclastic rocks are highly susceptible to diagenetic alteration. Glass alters to clay minerals, especially to smectite, zeolites, chalcedony, opal, quartz, or to a microcrystalline material, which in a thin slice under the petrographic microscope resembles chert. An alteration product of tephra is bentonite, *a plastic clay or shale composed for the most part of the clay mineral montmorillonite (a variety of smectite) that formed by the alteration of tephra.*

One kind of tuff that carries a special name is ignimbrite or welded tuff, *a nonsorted pyroclastic rock deposited from a* nuée ardente *while the particles were still in a plastic condition.* Ignimbrites may occupy large areas and attain considerable thicknesses (Fisher, 1961, 1966; Lajoie, 1979; Schmincke, 1988; USDOE, 1988).

Gerald M. Friedman

Bibliography

Baker, P.A., and Burns, S.J., 1985. The occurrence and formation of dolomite in organic-rich continental margin sediments: *Bulletin of the American Association of Petroleum Geologists*, **69**: 1917–1930.

Bathurst, R.G.C., 1975. *Carbonate Sediments and Their Diagenesis*. 2nd edn, Elsevier Scientific Publishing Company.

Bentor, Y.K. (ed.), 1980. *Marine Phosphoritez*. Society of Economic Paleonotologists and Mineralogists, Special Publication, 29.

Chamley, H., 1989. *Clay Sedimentology*. Springer-Verlag.

Dunbar, C.O., and Rodgers, J., 1957. *Principles of Stratigraphy*. John Wiley & Sons.

Dunham, R.J., 1962. Classification of carbonate rocks according to depositional texture. In Ham, W.E. (ed.), *Classification of Carbonate Rocks*. Tulsa, OK: American Association of Petroleum Geologists Memoir 1, pp. 108–121.

Embry, A.F., and Klovan, J.E., 1971. A late Devonian reef tract on Northeastern Banks Island, N.W.T. *Canadian Petroleum Geology Bulletin*, **19**: 730–781.

Fisher, R.V., 1961. Proposed classification of volcaniclastic sediments and rocks. *Geological Society of America Bulletin*, **72**: 1409–1414.

Fisher, R.V., 1966. Mechanism for deposition from pyroclastic flows. *American Journal of Science*, **264**: 350–363.
Flint, R.F., Sanders, J.E., and Rodgers, J., 1960. Diamictite, a substitute term for symmictite. *Geological Society of America Bulletin*, **71**: 1809–1810.
Folk, R.L., 1956. The role of texture and composition in sandstone classification. *Journal of Sedimentary Petrology*, **26**: 166–171.
Folk, R.L., 1959. Practical petrographic classification of limestones. *American Association of Petroleum Geologists Bulletin*, **43**: 1–38.
Friedman, G.M., 1965. Terminology of crystallization textures and fabrics in sedimentary rocks. *Journal of Sedimentary Petrology*, **35**: 643–655.
Friedman, G.M., 1985. The problem of submarine cement in classifying reefrock: an experience in frustration. In Schneidermann, N., and Harris, P.M. (eds.), *Carbonate Cements*. Society of Economic Paleontologists and Mineralogists, Special Publication, 36 pp. 117–121.
Friedman, G.M., and Sanders, J.E., 1967. Origin and occurrence of dolostones. In Chilingar, G.V., Bissell, H.J., and Fairbridge, R.W. (eds.), Carbonate rocks, origin, occurrence, and classification. Elsevier Scientific Publishing Company, pp. 267–348.
Friedman, G.M., and Sanders, J.E., 1978. Principles of sedimentology. New York: John Wiley & Sons, 792 p.
Friedman, G.M., Sanders, J.E., and Kopaska-Merkel, D.C., 1992. Principles of sedimentary deposits. New York: Macmillan Publishing Co., 717p.
Füchtbauer, H., and Leggewie, R., 1984. Korngrössenbeziehungen zwischen Silt- und Sandsteinen: *Neues Jahrbuch für Geologische Paläontologische Abhandlungen*, **167**: 133–161.
Füchtbauer, H., 1988. *Sediment und Sedimentgesteine*. Sediment-Petrologie, Part II, Stuttgart, E. Schweizerbart'sche Verlagsbuchhandlung.
Germann, K., Bock, W.D., and Schroter, T., 1984. Facies development of Upper Cretaceous phosphorites in Egypt: sedimentological and geochemical aspects. In Klitzsch, E., Said, R., and Schrank, E. (eds.), SFB 69: Results of the special research project Arid Areas, Period 1981–1984: *Berliner Geowisseuscheft*, Abh. A., Volume 50, pp. 354–361.
Guilbert, J.M., and Park, C.F. Jr., 1986. *The Geology of Ore Deposits*. W.H. Freeman and Company.
Hardie, L.A., 1987. Dolomitization: a critical view of some current views. *Journal of Sedimentary Petrology*, **57**: 166–183.
Heling, D., 1988. Ton und Siltsteine. In Füchtbauer, H. (ed.), *Sedimente und Sedimentgesteine*. Stuttgart: E. Schweitzerbart, pp. 185–232.
van Houton, F.B., 1990. Paleozoic oolitic ironstones on North American craton (abstract): *Geological Society of America, Northeastern Section, Abstracts with Programs*, **22**(2): 76 (only).
Hunt, J.M., 1979. *Petroleum Geochemistry and Geology*. W.H. Freeman.
James, H.L., 1954. Sedimentary facies of iron formation. *Economic Geology*, **49**: 235–293.
Jones, B., and Goodbody, Q.H., 1985. Oncolites from a shallow lagoon, Grand Cayman Island. *Bulletin of Canadian Petroleum Geology*, **32**: 254–260.
Jones, J.B., and Segnit, E.R., 1971. The nature of opal 1. Nomenclature and constituent phases. *Journal of the Geological Society of Australia*, **18**: 57–68.
Klein, G.deV., 1963. Analysis and review of sandstone classifications in the North American geological literature 1940–1960, *Geological Society of America Bulletin*, **74**: 555–576.
Kolodny, Y., 1981. Phosphorites. In Emiliani, C. (ed.), *The Sea*. **7**: Wiley-Interscience, pp. 981–1023.
Krynine, P.D., 1948. The megascopic study and field classification of sedimentary rocks. *Journal of Geology*, **56**: 130–165.
Lajoie, J., 1979. Facies models 15. Volcanoclastic rocks: *Geoscience Canada*, **6**: 129–139.
Land, L.S., 1985. The origin of massive dolomite: *Journal of Geological Education*, **33**: 112–125.
Laschet, C., 1984. On the origin of cherts, *Facies*, **10**: 257–290.
Maynard, J.B., 1983. *Geochemistry of Sedimentary Ore Deposits*. Springer-Verlag.
Mcbride, E.F., 1962. Flysch and associated beds of the Martinburg Formation (Ordovician) central Appalachians. *Journal of Sedimentary Petrology*, **32**: 32–91.

Notholt, A.J., Sheldon, R.P., and Davidson, D.F., 1989. *Phosphate Deposits of the World, volume 2, Phosphate Rock Resources*. International Geological Correlation Programme Project 156: Phosphorites, Cambridge University Press.
Notholt, A.J.G., and Jarvis, I. (eds.), 1990. *Phosphate Research and Development*. Geological Society of London, Special Publication, 52.
Pettijohn, F.J., 1975. *Sedimentary Rocks*, 3rd edn, Harper & Row.
Potter, P.E., Maynard, J.B., and Pryor, W.A., 1980. *Sedimentology of Shale*. Springer-Verlag.
Riggs, S.R., 1986. Proterozoic and Cambrian phosphorites—specialist studies: phosphogenesis and its relationship to exploration for Proterozoic and Cambrian phosphorites. In Cook, P.J., and Shergold, J.H. (eds.), *Proterozoic and Cambrian Phosphorites*. Cambridge University Press.
Sackett, W.M., Poag, C.W., and EDIE, B.J., 1974. Kerogen recycling in the Ross Sea, Antarctica, *Science*, **185**: 1045–1047.
Sanders, J.E., 1978. Graywacke. In Fairbridge, R.W., and Bourgeois, J. (eds.), *The Encyclopedia of Sedimentology*. Encyclopedia of earth sciences, Volume VI: Stroudsburg, PA: Dowden, Hutchinson, and Ross, pp. 389–391.
Sanders, J.E., and Friedman, G.M., 1967. Origin and occurrence of limestones. In Chilingar, G.V., Bissell, H.J. and Fairbridge, R.W. (eds.), *Carbonate Rocks*. Elsevier Scientific Publishing Company, pp. 169–365.
Schmincke, H.U., 1988. Pyroklastische Gesteine. In Füchtbauer, H., (ed.), *Sediment und Sedimentgesteine*. Sediment-Petrologie Part II: Stuttgart, E. Schweizerbart, pp. 731–778.
Schreiber, B.C., and Friedman, G.M., 1976. Depositional environments of Upper Miocene (Messinian) evaporites of Sicily as determined from analysis of intercalated carbonates. *Sedimentology*, **23**: 255–270.
Scoffin, T.P., 1987. *An Introduction to Carbonate Sediments and Rocks*. Blackie, Glasgow.
Sheldon, R.P., 1987. Association of phosphorites, organic-rich shales, chert, and carbonate rocks. *Carbonates and Evaporites*, **2**: 7–14.
Shinn, E.A., Steinen, R.P., Lidz, B.H., and Swart, P.K., 1989. Whitings, a sedimentologic dilemma. *Journal of Sedimentary Petrology*, **59**: 147–161.
Tada, R., and Iijima, A., 1983. Petrology and diagenetic changes of Neogene siliceous rocks in northern Japan. *Journal of Sedimentary Petrology*, **53**: 911–930.
Tucker, M.E., and Wright, V.P., 1990. *Carbonate Sedimentology*. Blackwell Scientific Publications.
USDOE, 1988. What is Tuff? Office of Civilian Radioactive Waste Management Nevada Nuclear Waste Storage Investigations, 2p.
Warren, J.K., 1989. *Evaporite sedimentology*. Prentice Hall.
Weaver, C.E., 1989. *Clays, Muds, and Shales*. Elsevier.
Williams, L.A., and CREAR, D.A., 1985. Silica diagenesis II, general mechanisms. *Journal of Sedimentary Petrology*, **55**: 312–321.
Young, T.P., and Taylor, W.E.G. (eds.), 1989. *Phanerozoic Ironstones*. London: Geological Society of London.

Cross-references

Algal and Bacterial Carbonates Sediments
Anhydrite and Gypsum
Bioclasts
Caliche–Calcrete
Carbonate Mineralogy and Geochemistry
Cave Sediments
Dolomite Textures
Dolomites and Dolomitization
Encrinites
Evaporites
Glaucony and Verdine
Grain Size and Shape
Ironstones and Iron Formations
Laterites
Oolite and Coated Grains
Phosphorites
Sands and Sandstones
Sedimentologists

Silcrete
Siliceous Sediments
Speleothems
Spiculites and Spongolites
Tills and Tillites
Varves

CLASTIC (NEPTUNIAN) DYKES AND SILLS

Clastic dykes are tabular bodies of clastic material, mostly fine sand, cutting across sedimentary formations or, more rarely, volcanic and granitic rocks (Allen, 1984; Peterson, 1968). They rank among the earliest described deformational sedimentary structures. Already in 1791, Werner published his theory of vein genesis based on the observation of marl veins in the middle Trias of central Germany. Since then, innumerable reports of sedimentary dykes throughout the world have accumulated (see Diller (1890), Williams (1927), Strauch (1966), Allen (1984), and Maltman (1994) for successive reviews). Clastic sills are also encountered, with the frequent difficulty of distinguishing them from normal beds (Archer, 1984). Features described as clastic dykes range from filled centimetric crevices to several-m-wide, several-km-long structures. A main genetic distinction is to be made between true clastic dykes and sills, forcefully injected from below, and passive fissure-fills, sometimes called neptunian dykes.

Intrusive clastic dykes are discordant—and sills concordant—planar to irregular and often branching sheet-like bodies which commonly occur in swarms. Their walls are generally smooth though locally disturbed and sometimes slickensided or erosionally grooved (Allen, 1984). The filling material is generally fine sand or clay and includes pieces of the host sediments. It is mostly structureless but in some cases a wall-parallel orientation of the sediment grains or even a wall-parallel layering is observed. Clastic dykes are often associated with other soft-sediment deformation structures or otherwise mobilized material (slumps). When the intrusion breaks through to the contemporaneous sediment surface, sand volcanoes or extruded sand sheets (often difficult to distinguish from clastic sills) can be formed. Unless they can be traced down to the source layer, and in spite of suggestive upward branching, wall-parallel grain fabric and irregular or striated walls, the intrusive origin of the dykes is not easily demonstrated.

Clastic dykes are formed as the result of forceful injection, commonly from below and more rarely sideways or even from above, after unconsolidated sediments have been liquidized (see *Liquefaction and Fluidization*). Therefore, such soft-sediment deformation features are penecontemporaneous with deposition or at least occur before complete lithification. Liquefaction is achieved through either thixotropic behavior of some cohesive materials (e.g., certain clays) or liquefaction of cohesionless saturated sands, whose primary strength results from grain interlocking and friction at particle contacts. Liquefaction is caused by the pore fluid pressure increase to the level of the lithostatic stress (thus making the effective stress equal to zero) in a system more or less closed by overlying less permeable, cohesive materials (Lowe, 1975). The consequence is a dramatic loss of strength of the sand whose particles are dispersed in the pore fluid. This will eventually result in closer grain packing and forcing up of pore water, leading to fluidization when escaping excess pore water drags sediment. Fluidization occurs in an open system after overpressure has caused hydraulic fracturing in the impermeable overlying strata. In the case of true clastic dykes, fissuring and filling therefore occur simultaneously. In horizontal strata, the formation of dykes or sills respectively depends on the horizontal or vertical orientation of the σ_3 principal stress in the fracturing layers. Increase of pore fluid pressure and liquefaction may occur in different ways, either static, dynamic (impulsive) or cyclic. Main triggers for static liquefaction are artesian groundwater movements and escape of pore water from underlying compacting sediment. Rapid sediment deposition, slope failure, breaking waves and flood surges may be responsible for impulsive liquefaction. Finally, seismic shaking, pressure variations associated with waves or with flow separation will cause cyclic stresses and liquefaction (Owen, 1987). Unfortunately, since clastic dykes depend primarily on overpressuring which may occur in so many different environments, they are generally no clear diagnostic feature of the trigger agent, and the latter has rather to be deduced from the sedimentological context (Owen, 1987). Indeed, clastic dykes occur in a wide variety of environments, ranging from deep-water turbidite associations (Truswell, 1972) to shallow-water marine and nonmarine sediments to subaerial mass-flow deposits (Collinson et al., 1989). They are also known from seismic (Obermeier, 1996) and glacial (Amårk, 1986) environments.

Though in many cases difficult to distinguish from injected sedimentary dykes, neptunian dykes fed from above sometimes exhibit characteristic features. Often, they will show a clear downward tapering, while their filling may more frequently include coarse material and is in some cases horizontally layered. Genetically, the fissuring and filling stages are not necessarily simultaneous and the cracks may widen progressively over years, then allowing for vertical layering of the fill. Many different causes can produce open cracks. Earthquake-related ground cracks are common feature, while other tectonic cracks like enlarged joints and tension gashes can also trap sediments and develop in neptunian dykes. Subaqueous slumps and subaerial landslides as well are another frequent cause of ground cracking. Karstification of carbonate rocks leads to extensive fissuring which again can and often do catch sediments from above. Finally, desiccation and syneresis cracks are also mentioned as features possibly evolving into neptunian dykes. Especially when they are synsedimentary subaqueous features, filling of the fissures may be slow with horizontal stratification (Strauch, 1966). However, subaerial progressive filling has also been shown to result in vertical layering and fining upward of the material brought in the fissure by rainwash and clay infiltration (Demoulin, 1996). Owing to weathering and collapse of the fissure walls, slow fissure filling frequently displays a brecciated structure. In other cases, rapid filling can occur, provided overlying sediment is available. It can occasionally involve injection from above, and yields a mostly structureless or faintly vertically laminated infill. Neptunian dykes have been described from as varying environments as those of true clastic dykes. Moreover, complex structures coupling forceful injection from below and collapse of surface sediments within the cracks can also form, namely as the result of seismic shaking (Thorson et al., 1986). Note however that although passive fissure-fills of different origin have sometimes been mistakenly interpreted as periglacial ice-wedge casts, the latter are generally not referred to as neptunian dykes.

As synsedimentary features, clastic and many neptunian dykes can experience later compaction-induced deformation and folding. Displacement across the dykes or even across their infill as well as slickensided walls have in some cases been interpreted as the imprint of subsequent tectonic use of the weakened cracked zones. They should anyway be carefully distinguished from cataclastic shear zones which sometimes mimic them.

A. Demoulin

Bibliography

Allen, J. R. L., 1984. *Sedimentary structures*. Amsterdam: Elsevier.
Amårk, M., 1986. Clastic dykes formed beneath an active glacier. *Geologiska Föreningens i Stockholm Förenhandlingar*, **108**: 13–20.
Archer, J., 1984. Clastic intrusions in deep-sea fan deposits of the Rosroe Formation, lower Ordovician, western Ireland. *Journal of Sedimentary Petrology*, **54**: 1197–1205.
Collinson, J., Bevins, R., and Clemmensen, L., 1989. Post-glacial mass flow and associated deposits preserved in palaeovalleys: the Late Precambrian Moræneso Formation, North Greenland. *Meddelelser om Grønland. Geoscience*, **21**: 3–26.
Demoulin, A., 1996. Clastic dykes in east Belgium: evidence for upper Pleistocene strong earthquakes west of the Lower Rhine rift segment. *Journal of the Geological Society*, **153**: 803–810.
Diller, G., 1890. Sandstone dykes. *Geological Society of America Bulletin*, **1**: 411–442.
Lowe, D., 1975. Water escape structures in coarse-grained sediments. *Sedimentology*, **22**: 157–204.
Maltman, A., 1994. Introduction and overview. In Maltman, A. (ed.), *The Geological Deformation of Sediments*. London: Chapman & Hall, pp. 1–35.
Obermeier, S., 1996. Using liquefaction-induced features for paleoseismic analysis. In McCalpin, J. (ed.), *Paleoseismology*. San Diego: Academic Press, pp. 331–396.
Owen, G., 1987. Deformation processes in unconsolidated sands. In Jones, M., and Preston, R. (eds.), *Deformation of Sediments and Sedimentary Rocks*. London: Geological Society, Special Publication, 29, pp. 11–24.
Peterson, G., 1968. Flow structures in sandstone dikes. *Sedimentary Geology*, **2**: 177–190.
Strauch, F., 1966. Sedimentgänge von Tjörnes (Nord-Island) und ihre geologische Bedeutung. *Neues Jahrbuch für Geologischen und Paläontologischen Abhandlungen*, **124**: 259–288.
Thorson, R., Clayton, W., and Seeber, L., 1986. Geologic evidence for a large prehistoric earthquake in eastern Connecticut. *Geology*, **14**: 463–467.
Truswell, J., 1972. Sandstone sheets and related intrusions from Coffee Bay, Transkei, South Africa. *Journal of Sedimentary Petrology*, **42**: 578–583.
Werner, A., 1791. *Neue Theorie von Entstehung der Gänge*. Freiberg/Sa.
Williams, W., 1927. Sandstone dikes in southeastern Alberta. *Transactions of the Royal Society of Canada*, **21**: 153–174.

Cross-references

Compaction (Consolidation) of Sediments
Convolute Lamination
Deformation of Sediments
Desiccation Structures (Mud Cracks, etc.)
Dish Structure
Fabric, Porosity and Permeability
Flame Structure
Fluid Escape Structures
Liquefaction and Fluidization
Pillar Structure
Porewaters in Sediments
Slope Sediments
Storm Deposits
Syneresis

CLATHRATES

Clathrate hydrates

Clathrate hydrates (or gas hydrates) are solid ice-like crystalline compounds of hydrogen bonded water molecules that form a rigid lattice of cages that host small guest gas molecules. They are stable only when the cages contain gas molecules. They are stable at moderate to high pressures and low temperatures (Sloan, 1998). The stability depends on the composition of the gas molecules and of the pore fluid (Englezos and Bishnoi, 1988). Methane and other clathrates most likely formed in the solar nebula and thus played a role in the accretion of planets and satellites and probably comets (Lunine and Stevenson, 1985). On Earth, methane hydrate is the principal clathrate hydrate.

Methane hydrate has a body centered cubic structure I, which is one of the three hydrate structures known to occur in the natural environment. Structure I forms with gas molecules smaller than propane, hence with CH_4, CO_2, H_2S. Structure II forms with molecules greater than ethane but smaller than propane. The two most important clathrate hydrate structures I and II consist of different combinations of three types of cages; each has two cage sizes. A third hexagonal structure H was described from the Gulf of Mexico having three types of cages.

The predominant natural gas hydrate on Earth, methane hydrate, has a density of 0.93 g/cm^3. Clathrate hydrates have large storage capacity for gas molecules; for example, the ratio of gas to water molecules in structure I methane hydrates is $1CH_4 : 5.75 \text{ } H_2O$, when decomposed, the volumetric ratio is $164 \text{ } CH_4 : 0.8 \text{ } H_2O$ (Kvenvolden, 1993). When hydrates form they exclude ions and fractionate O and H isotopes; they become enriched in ^{18}O and D isotopes, like sea ice.

Most of the modern ocean seafloor is within the temperature and pressure stability field for methane hydrate. Methane concentration constraints restrict its occurrence to two main environments: (1) in the oceans to the uppermost few hundred meters of sediments on submerged continental margins slope and rise sediments, where water depths exceed $\sim$500 m, and (2) in polar regions associated with permafrost. A conservative estimate of the oceanic gas hydrate reservoir is enormous: $\sim 10^{19}$ g of methane C is stored in them (Kvenvolden, 1988). This C reservoir is larger than the C in all other known fossil fuel deposits. The permafrost reservoir is considerably smaller than the oceanic one by about two orders of magnitude ($\sim 4 \times 10^{17}$ g).

The methane in hydrates originates either from bacterial anaerobic degradation of organic matter or from thermal breakdown of organic matter at elevated temperatures (>80 to 90° C). The conventional bacterial methane is characterized by $\delta^{13}C$ values that range from about −55‰ to 85‰ (PDB) and $C_1/(C_2+C_3) > 200$, whereas conventional thermogenic methane typically has $\delta^{13}C$ values of −20‰ to −50‰, and $C_1/(C_2+C_3) < 100$. The source of methane in the hydrates is from either *in situ* production of bacterial methane or upward transport of methane-rich fluids into the hydrate stability zone. The latter source may have a bacterial, thermogenic, or mixed origin (Hyndman and Davis, 1992). *In situ* production and/or transport of methane by fluids into the hydrate stability zone is the limiting factor for methane hydrate formation in most regions of the modern ocean. Geochemical studies, especially of C isotope values of the methane, indicate that in methane hydrates, the methane is largely bacterial in origin, but in some regions, such as the Gulf of Mexico or Caspian Sea, it is thermogenic.

Only a fraction of the marine sediment organic C is available for methane production bacterially or thermogenically. Reasonable estimates of the efficiency for methane generation from organic matter vary from 2–3 percent to 20 percent. Accordingly, the maximum methane hydrate that can be generated *in situ* in continental margin slope and rise sediments, assuming an average porosity of 50 percent and 1 percent organic C content is 4–6 percent of the void space. The higher concentrations of 20–35 percent in convergent margins based on geochemical indicators, such as pore fluid Cl concentrations, or on geophysical interpretations, indicate that transport of methane into the stability field is widespread. Because advection of methane-rich fluids is more pervasive and aggressive in convergent than in rifted and sheared margins, Kastner (2001) suggested that convergent margins host 60–65 percent of the enormous oceanic methane hydrate reservoir.

Much of our knowledge about methane hydrate occurrence, distribution, and geochemistry in marine sediments is from recovered samples and geochemical data obtained through the Deep Sea Drilling Project (DSDP) and Ocean Drilling Project (ODP). All the successful recoveries were from about 100–400 mbsf, primarily from samples in the two-phase region of the stability field. The direct observations and inferences obtained from geophysical and pore fluid geochemical data indicate that methane hydrates are distributed inhomogeneously in ocean sediments, irregularly disseminated in turbidite sediments and in silty mudstones with increasing abundance near and in the sediment at the base of the three-phase equilibrium field of methane hydrate. A Bottom Simulating Reflector (BSR) of reverse polarity that parallels the seafloor often characterizes this boundary (Shipley *et al.*, 1979). In clay-rich sediments methane hydrate occurs in microfractures, or in thin platelets parallel to fissility or scaly fabric; it cements coarser sediment horizons of higher permeability, such as sandy silts or ash layers; and concentrates in fractures, fault zones, or along lithologic boundaries where the occurrence is modular to massive. Gas hydrates thus affect the sediment physical properties, particularly the permeability, seismic velocity, thermal conductivity, and electric resistivity. Sedimentological processes such as compaction or cementation by silicates or carbonates are inhibited where gas hydrate fills void space.

Because of continuous sedimentation, methane hydrate in the sediment at the BSR dissociates. The methane released is recycled, it migrates upward into the stability field, thus stimulates the formation of the methane hydrate near the base of the 3-phase stability field, and produces the generally larger accumulations near the BSR. This overall vertical distribution is amplified in convergent margins where upward transport of methane-rich fluids is widespread. The recently observed moderate to large accumulations of methane hydrate near the sea-floor at the Cascadia margin, Eel River basin in northern California, Gulf of Mexico, Okhotsk Sea, Black and Caspian Seas thus require active focused transport of methane-rich fluids or methane gas to the sea-floor.

Recent interest in natural clathrate hydrates has resulted from recognition that global warming may destabilize some of the vast quantities of methane hydrate in shallow marine slope sediments (and permafrost). The potential environmental consequences of rapid release of large quantities of methane for both ocean and atmosphere are important questions surrounding the very large amounts of gas hydrates in the shallow geosphere. In the ocean, intense microbial oxidation of methane would reduce the amount released into the atmosphere, but this would result in extensive local oxygen consumption and CO_2 production, therefore in some reduction of the capacity of the ocean to incorporate fossil fuel CO_2. Because methane gas is an important contributor to the atmospheric radiation balance as it is a significantly more effective greenhouse gas than CO_2, any additional flux of methane into the atmosphere beyond the present annual growth rate of ~ 0.8 percent per year would accelerate global warming.

Significant environmental stresses or geologic perturbations, such as global warming, rapid deglaciation or glaciation, earthquakes, or tectonic uplift, may cause clathrate hydrate dissociation. If at the BSR it may trigger geologic hazards such as giant landslides that could cause tsunamis and perhaps rapidly release large quantities of methane to the ocean and atmosphere with complex climatic feedbacks. New evidence exists for massive methane releases from gas hydrate and their possible association with global warming in the geologic past, for example in the late Paleocene, ~ 55.6 Ma (Dickens *et al.*, 1997); this event was accompanied by a thermal maximum.

It also has been suggested that clathrate hydrates may have played a role in past climate change. For example, during deglaciation, an increase in atmospheric methane from hydrate dissociation may have accelerated the retreat of continental ice sheets (Nisbet, 1990); or during glaciation, the decomposition of gas hydrate resulting from lowering sea level may have moderated the extent of glaciation (Paull *et al.*, 1991).

A clathrate structure of silica

Melanophlogite is a rare cubic, low-density (2.06 g/cm^3) polymorph of SiO_2 having a clathrate structure. The guest material in the cavities consists of small organic molecules containing S that are essential structural elements stabilizing the Melanophlogite structure. Melanophlogite occurs as authigenic single crystals and interlocking intergrowths on S crystals in Sicilian sulfur deposits (Skinner and Appleman, 1963; Kamb, 1965).

Miriam Kastner

Bibliography

Dickens, G.R., and Castillo, M.M., and Walker, G., 1997. A blast of gas in the latest paleocene: simulating first-order effects of massive dissociation of oceanic methane hydrate. *Geology*, **25**: 259–262.

Englezos, P., and Bishnoi, P.R., 1988. Prediction of gas hydrate formation conditions in aqueous electrolyte solutions. *AIChE Journal*, **34**: 1718–1721.

Hyndman, R.D., and Davis, E.E., 1992. A mechanism for the formation of methane hydrate and seafloor bottom-simulating reflectors by vertical fluid expulsion. *Journal of Geophysical Research*, **97**: 7025–7041.

Kamb, B., 1965. A clathrate crystalline form of silica. *Science*, **148**: 232–234.

Kastner, M., 2001. Gas hydrates in convergent margins: formation, occurrence, geochemistry, and global significance. In *Natural Gas Hydrates: Occurrence, Distribution, and Detection*, AGU (American Geophysical Union), Geophysical Monograph, 124, pp. 67–86.

Kvenvolden, K.A., 1988. Methane hydrate—a major reservoir of carbon in the shallow geosphere. *Chemical Geology*, **71**: 41–51.

Kvenvolden, K.A., 1993. Gas hydrates—geological perspectives and global change. *Reviews of Geophysics*, **31**: 173–187.
Lunine, J.I., and Stevenson, D.J., 1985. Thermodynamics of clathrate hydrate at low and high pressure with application to the outer solar system. *Astrophysical Journal Supplementary Series*, **58**: 493–531.
Nisbet, E.G., 1990. The end of the ice age. *Canadian Journal of Earth Sciences*, **27**: 148–157.
Paull, C.K., Ussler, W., and Dillon, W., 1991. Is the extent of glaciation limited by marine gas hydrates? *Geophysical Research Letters*, **18**: 432–434.
Shipley, T.H., Houston, M., Buffler, R. *et al.*, 1979. Seismic reflection evidence for the widespread occurrence of possible gas-hydrate horizons in continental slopes and rises. *American Association of Petroleum Geologists Bulletin*, **63**: 2204–2213.
Skinner, B.J., and Appleman, D.E., 1963. Melanophlogite, a cubic polymorph of silica. *American Mineralogist*, **48**: 854–867.
Sloan, D.D. Jr., 1998. *Clathrate Hydrates of Natural Gases*, 2nd edn, New York: Marcel Dekker.

CLAY MINERALOGY

Definitions of clay and clay mineral

The term "clay" is common to many disciplines and frequently used in two quite different ways. On the one hand, it is a term for the finest division of many particle size schemes; on the other hand it is a term applied to many naturally occurring materials that may otherwise be classified as soil, sediment or rock. Not surprisingly then, there is no universally accepted definition of the word "clay", although in context its meaning is generally understood. Most recently Guggenheim and Martin (1995) offered the following definition of clay as a material: "The term clay refers to a naturally occurring material composed primarily of fine-grained minerals, which is generally plastic at appropriate water contents and will harden when dried or fired. Although clay usually contains phyllosilicates, it may contain other materials that impart plasticity and harden when dried or fired. Associated phases in clay may include materials that do not impart plasticity and organic matter". In its other context, i.e., particle size, clay is variously defined depending on discipline or country. The fraction less than 2 micrometers (equivalent spherical diameter) is a common definition, but the boundary with silt can be as high as 6 microns in some schemes. When dealing with natural materials, such as sediments, there is a conceptual link between both uses of the word clay, since clay materials inevitably contain a large proportion of particles that are clay size. Additionally, the clay size fraction, even if only a minor fraction of any given soil, sediment or rock, will in most cases be composed predominantly of minerals that are known as "clay minerals". Again, just like clay, there is no universally accepted definition of the term "clay mineral". Guggenheim and Martin (1995) define it as follows: 'The term "clay mineral" refers to phyllosilicate minerals and to minerals which impart plasticity to clay and which harden upon drying or firing'. Most "clay minerals" are phyllosilicates (synonym: layer silicates) and a correspondingly narrower use of the term is common place, but it is important to realize that minerals that are not phyllosilicates are also included in the definition of clay mineral if they contribute to the properties associated with clay. This encompasses many oxides of iron and aluminum, that like the phyllosilicate clay minerals are similarly predisposed toward a natural occurrence in the clay size fraction; additionally minerals like allophane and imogolite are included. In contrast, minerals such as quartz and feldspars, although frequently found in clay size fractions, would never be considered "clay minerals". A further difficulty with a more precise definition of "clay mineral" arises in part because there are no meaningful boundaries to continuous properties such as particle size. Hence many hydrous layer silicates that occur as clay minerals for example, micas and chlorites have larger relatives whose size flouts any such description. For other types, for example, kaolinite and smectite, it is rare indeed to find specimens that have outgrown the description "clay mineral" probably because their crystal structures place severe constraints on the size to which they can grow. Clay minerals thus occur as fine-grained particles with consequent high surface to volume ratios, further augmented by the platy shapes of many, such that surface properties are accentuated. The historical development of the modern concept of "clay minerals" is described in more detail by Grim (1968); there have long been points of contention in relation to definitions and some will no doubt remain (Moore, 1996; Guggenheim and Martin, 1996).

Structure and classification of clay minerals

The crystal structures of phyllosilicate clay minerals have been elucidated mainly from the study of larger analogous specimens (Brindley and Brown, 1980). They are hydrous layer silicates consisting of planes of atoms arranged in layers, their crystal habits and morphologies reflecting this arrangement, so that most are platy and have perfect cleavage. There are two basic modular components of phyllosilicate clay minerals; sheets of tetrahedrally coordinated atoms and sheets of octahedrally coordinated atoms, know as tetrahedral and octahedral *sheets* (Figure C37).

Tetrahedra are formed of a cation, usually silicon, surrounded by four oxygens. By sharing three of these oxygens between adjacent tetrahedra they may be linked together to form a continuous tetrahedral sheet, in which the unshared oxygens all point in the same direction. Thus one side of a tetrahedral sheet consists of an hexagonal mesh of shared oxygens, whilst the other side is formed by the remaining so-called "apical" oxygens.

An octahedral sheet is formed from two planes of close packed oxygens and/or hydroxyls. In the center of such a sheet, and adjacent to every anion, there are three octahedral sites which may be occupied by cations such as aluminum, iron and magnesium, each cation being surrounded by six anions. There are two kinds of common octahedral sheets distinguished by different cation to anion ratios as required for electrical neutrality. If divalent cations fill all three sites the octahedral sheet is know as trioctahedral. Trivalent cations need only occupy two out of every three sites to maintain neutrality and the sheet is then called dioctahedral. Octahedral sites are linked together by sharing of edges. The vacancies in the dioctahedral sheet lead to a more distorted structure than the completely occupied trioctahedral sheet.

By joining tetrahedral and octahedral sheets together, two basic clay mineral units known as *layers* can be formed. The unit formed by linking one tetrahedral sheet and one octahedral sheet together is called a 1:1 layer, sometimes also designated T-O. The linkage is achieved by replacing two out of every three anions of an octahedral sheet by the apical oxygens of a tetrahedral sheet, so that at the 'junction plane' the apical oxygens are shared jointly by tetrahedral and

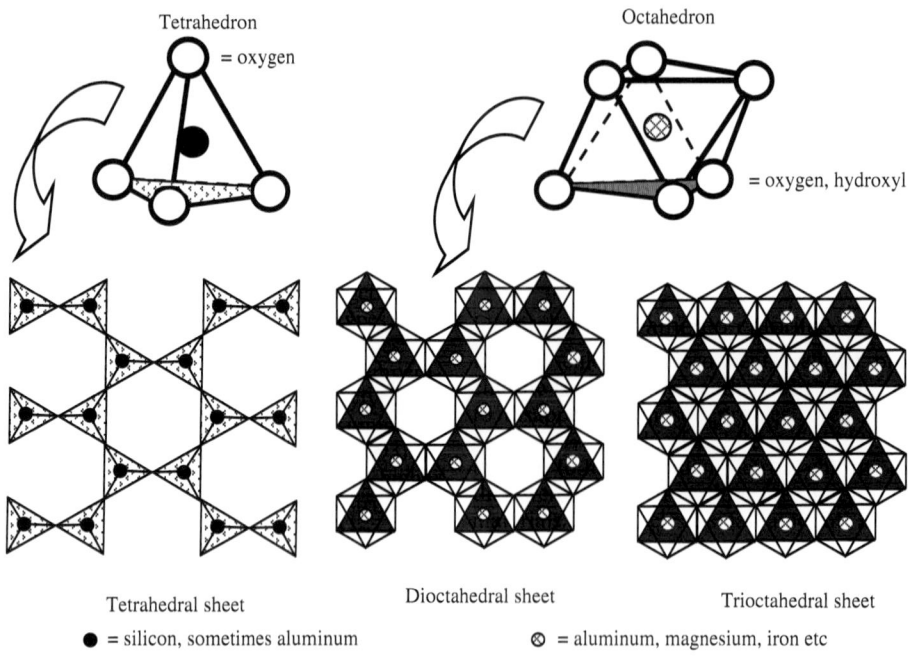

Figure C37 Schematic diagram of tetrahedron, octahedron, and fragments of tetrahedral, dioctahedral and trioctahedral sheets.

octahedral sheets. The remaining unshared anion of the octahedral sheet in this junction plane is an hydroxyl that lies in projection at the center of the hexagonal ring of apical oxygens. A tetrahedral sheet can be similarly linked to the other side of an octahedral sheet and the unit formed is known as a 2:1 or T-O-T layer. In reality, the dimensions of tetrahedral and octahedral sheets are different and a variety of structural adjustments, such as tetrahedral rotation, and atomic substitutions are necessary to enable the sheets to fit together. Furthermore, substitutions of cations by others of lower valence such as aluminum for silicon in tetrahedral sheets, and magnesium or ferrous iron for aluminum in octahedral sheets are common place. Any substitutions in 1:1 layers are usually fully compensated by others, such that there is no net layer charge. Substitutions in 2:1 layers, however, frequently give rise to layers that carry a net negative charge. This charge may be neutralized by cations, hydrated cations, or by octahedrally coordinated hydroxyl groups or sheets, which occupy a position between the layers known as the interlayer. Additionally, because of the similarity of many clay mineral structures it is common to find minerals that consist of two or more layer types, so-called interstratified or mixed-layer clay minerals (Środoń, 1999). The different layers in mixed-layer clay minerals can be visualized as stacked in a sequence normal to the plane of the layers, and the stacking sequence may be random or more or less organized into in a regular ordered pattern of layer types. Most mixed-layer clay minerals involve two kinds of layers, such as mixed-layer illite-smectite, or chlorite-smectite, more rarely interstratifications of three kinds of layers are reported. Some mixed-layer clays in which the layers are organized into a regular sequence are given specific mineral names (Brindley and Brown, 1980); rectorite and corrensite are examples.

The classification of the phyllosilicate clay minerals (Figure C38) is based collectively, on the features of layer type (1:1 or 2:1), the dioctahedral or trioctahedral character of the octahedral sheets, the magnitude of any net negative layer charge resulting from atomic substitutions, and the nature of the interlayer material. The most important members of the 1:1 kaolin-serpentine group, are kaolinite, dickite, halloysite, and berthierine. Of the 2:1 clay minerals, pyrophyllite and talc represent the uncharged dioctahedral and trioctahedral species, respectively. The distinction between smectites and vermiculites is arbitrarily placed at a layer charge of 0.6 per formula unit. Cations required to compensate these net negative layer charges may be hydrated such that both smectites and vermiculites exhibit intracrystalline swelling. Such cations may also be exchanged with others. This property when measured is known as the cation exchange capacity, abbreviated CEC, although exchangable cations may also reside at other sites, so that most clay minerals exhibit some CEC. The amount of swelling can be related to the kind and number of interlayer cations, humidity or the nature of the solution in which the clay occurs and to the magnitude of the layer charge. In micas and illites the net negative charge is stronger and the cations that compensate it are unhydrated and more strongly held such that they are not exchangable. In chlorites the 2:1 layer charge is compensated by an interlayer hydroxide sheet. The clay minerals sepiolite and palygorskite are classified with the other 2:1 minerals but they differ from them because the 2:1 layers are not laterally continuous. Instead, tetrahedra are inverted periodically dividing the structure into ribbons of three hexagonal chains width in sepiolite and two in palygorskite so that only the tetrahedral basal oxygens form continuous planes throughout the structure. Tetrahedra in the structure with apical oxygens facing up and down are linked by octahedral sheets to form a 2:1 layer structure that is limited in extent to the direction perpendicular to the ribbons.

Layer type	Layer charge (q)	Group	Subgroup	Species
1:1	q ~ 0	Kaolin-serpentine	Kaolin Serpentine	Kaolinite Berthierine
2:1	q ~ 0	Pyrophyllite-talc (q ~ 0)	Pyrophylite Talc	Pyrophylite Talc
	Increasing layer charge ↓	Smectite (q ~ 0.2-0.6)	Di. smectite Tri. smectite	Montmorillonite, Beidellite Saponite
		Vermiculite (q ~ 0.6-0.9)	Di. vermiculite Tri. vermiculite	Di. vermiculite Tri. vermiculite
	q ~ 1	Mica (q ~ 1)	Di. mica Tri. mica	Illite, Muscovite Biotite
	q Variable	Chlorite	Di. chlorite Tri. chlorite	Sudoite Chamosite, clinochlore
		Sepiolite-palygorskite	Sepiolite Palygorskite	Sepiolite Palygorskite
Variable		Mixed-layer	Di. mica-di.smectite Tri. chlorite-tri.smectite	Rectorite Corrensite

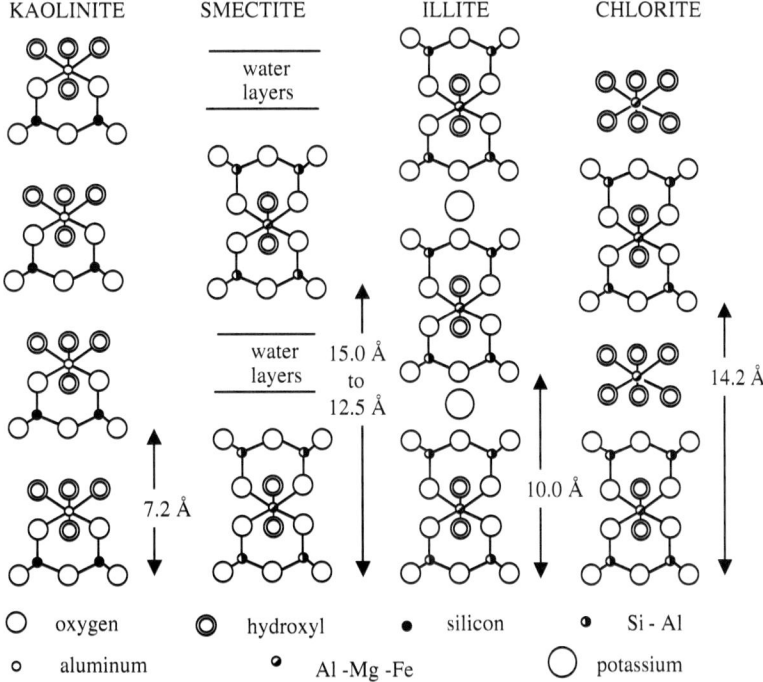

Figure C38 Classification of some common phyllosilicate clay minerals based on layer type, layer charge, and type of octahedral sheet (di = dioctahedral; tri = trioctahedral).

Figure C39 Diagrammatic representation of some phyllosilicate clay minerals and characteristic 'basal spacings' representing the distance between repeating layers.

Identification of clay minerals

The precise identification of clay minerals often requires analytical data from several techniques. X-ray powder diffraction (XRPD) is one of the most import (Brindley and Brown, 1980; Moore and Reynolds, 1997), but it is often essential to obtain supplementary data from chemical, spectroscopic, thermal or other techniques (Wilson, 1987). The importance of XRPD lies in the ability to measure the so-called basal spacings, and changes in these spacings after diagnostic treatments; these measurements can be easily related to the layer structure and hence to the type of clay minerals present. A diagrammatic representation of some clay minerals and the corresponding layer spacings is shown in Figure C39.

Origins, occurrence and distribution of clay minerals

Clay minerals are the characteristic minerals of the Earth's near surface hydrous environments including those of weathering, sedimentation, diagenesis, and hydrothermal alteration. Within all of these environments clay minerals may have one of three different origins; they may be neoformed, transformed, or inherited (Millot, 1970; Eberl, 1984). Neoformation describes the direct formation by crystallization of a clay mineral from solution or by ageing of amorphous gels. Neoformation is limited by the constraints of time and temperature, such that longer times and higher temperatures promote neoformation. The sluggishness of neoformation and the apparent stability of clay minerals at low temperatures means that clay minerals formed in one environment are often recycled into another. This is an origin known as inheritance. Most clay minerals found in modern sedimentary environments represent the redistributed products of weathering and as such they are inherited (Chamley, 1989; Weaver, 1989). The origin known as transformed lies between these two poles since it describes the formation of clay minerals where some structural elements of a precursor mineral (usually another clay mineral) are retained but new ones are added.

As far as sediments and sedimentary rock are concerned, the importance of weathering is as a primary and often the ultimate source for many clay minerals. Weathering and the formation of various kinds of clay minerals in soils is controlled by many variables operating at various scales including, climate, drainage, topography, parent rock type, and time. At a global scale, a broadly latitudinal climatically controlled pattern of clay mineral distribution is recognized in zonal soils. Thus in polar, tundra, temperate and desert climatic zones the processes of inheritance and transformation are dominant and the clay minerals assemblages are, very generally speaking, characterized by assemblages containing illite and chlorite inherited from parent materials and mixed-layer clay minerals and vermiculite formed by transformation from illite and chlorite. In tropical climates neoformation processes are much more evident: kaolin is formed where soil solutions are thoroughly depleted of bases cations, usually as a result of constant leaching, and smectite is formed where conditions allow base cations to accumulate.

The broadly latitudinal pattern of clay mineral distribution in zonal soils is mirrored in the recent sediments of the ocean basins and attests to the largely detrital origin of clay minerals in the modern marine environment (Chamley, 1989; Weaver, 1989). However, neoformation of clay minerals in certain sedimentary environments is commonplace and in most cases the clay minerals involved are rich in iron or magnesium and poor in aluminum. In the marine environment examples include the iron-rich clay minerals glauconite, berthierine, odinite, and nontronite. Magnesium-rich clay minerals, such as sepiolite and palygorskite, and some species of trioctahedral smectites, tend to be most commonly encountered in peri-marine evaporitic settings or in alkaline lakes.

Clay minerals are found in most sediments and sedimentary rocks, but they are most abundant in mudrocks and shales, where they typically account for about 50 percent of the mineral components. When sediments are buried by yet more sediment then diagenetic processes begin to act and their action is commonly recorded by the changes they induce in the clay minerals and clay mineral assemblages present. Probably the best known and most widespread is the smectite-to-illite reaction, whereby originally deposited smectitic clays are progressively transformed to illite via mixed-layer illite-smectite, although there is much debate about the nature of the reaction(s) involved. The article by Środoń (1999) provides a recent review, and summarizes those aspects that are still contentious.

Stephen Hillier

Bibliography

Brindley, G.W., and Brown, G., 1980. Crystal structures of clay minerals and their X-ray identification. *Mineralogical Society Monograph No.5.*
Chamley, H., 1989. *Clay Sedimentology*. Springer Verlag.
Eberl, D.D., 1984. Clay mineral formation and transformation in rocks and soils. *Philosophical Transactions of the Royal Society of London*, A 311, pp. 241–257.
Grim, R.E., 1968. *Clay Mineralogy*, 2nd edn, McGraw-Hill.
Guggenheim, S., and Martin, R.T., 1995. Definition of clay and clay mineral: joint report of the AIPEA nomenclature and CMS nomenclature committees. *Clays and Clay Minerals* 43: 255–256.
Guggenheim, S., and Martin, R.T., 1996. Reply to the comment by D.M. Moore on "Definition of clay and clay mineral: joint report of the AIPEA nomenclature and CMS nomenclature committees". *Clays and Clay Minerals*, 44, pp. 713–715.
Millot, G., 1970. *The Geology of Clays*. Paris: Masson.
Moore, D.M., 1996. Comment on: "Definition of clay and clay mineral: joint report of the AIPEA nomenclature and CMS nomenclature committees". *Clays and Clay Minerals*, 44, pp. 710–712.
Moore, D.M., and Reynolds, R.C. Jr., 1997. *X-ray Diffraction and the Identification and Analysis of Clay Minerals*, 2nd edn, Oxford University Press.
Środoń, J., 1999. Nature of mixed-layer clays and mechanisms of their formation and alteration. *Annual Review of Earth and Planetary Sciences*, 27: 19–53.
Weaver, C.E., 1989. *Clays, muds and shales*. Elsevier.
Wilson, M.J., 1987. *A Handbook of Determinative Methods in Clay Mineralogy*. New York: Blackie & Son.

Cross-references

Bentonites and Tonsteins
Berthierine
Black Shales
Cation Exchange
Chlorite in Sediments
Classification of Sediments and Sedimentary Rocks
Colloidal Properties of Sediments
Diagenesis
Glaucony and Verdine
Grain Size and Shape
Hydroxides and Oxyhydroxide Minerals
Illite Group Clay Minerals
Kaolin Group Minerals
Mixed-Layer Clays
Mudrocks
Smectite Group
Talc
Vermiculite
Weathering, Soils and Paleosols

CLIMATIC CONTROL OF SEDIMENTATION

Introduction

Sedimentation is influenced by three extrinsic variables, tectonics, sea level, and climate, with climate potentially dependent upon the other two variables. Global climate has undergone substantial changes on a variety of temporal and geographic scales throughout earth history. Interpreting climates of the past (paleoclimate) is an important aspect of understanding earth history (Crowley and North, 1991; Parrish, 1998).

Global climate

Climate may be defined as the condition of the atmosphere at a specific location near the earth's surface averaged over several years or tens of years. The climatic conditions that most affect sedimentation are mean annual temperature, mean annual precipitation, seasonal variations in temperature and precipitation, the rate of evaporation compared to precipitation, and the direction, velocity, and seasonal variation of winds and ocean currents.

The first-order variables affecting global climate are the angle at which solar radiation strikes the earth's surface and the Coriolis effect. The Coriolis effect results from the earth's rotation and deflects currents to the right in the northern hemisphere and to the left in the southern hemisphere. These variables produce climatic zones that are roughly parallel to latitude (Figure C40). Solar radiation striking the equator at nearly right angles effectively heats the atmosphere, causing the air to rise and move toward the poles (Hadley cell). The air descends at about 30°N and 30°S and flows back toward the equator, where it is deflected by the Coriolis effect, producing the easterly trade winds. The equatorial climatic zone where the trade winds meet, called the intertropical convergence zone, is characterized by low atmospheric pressure and high precipitation associated with convective cooling of the rising air mass. In contrast, the descending part of the Hadley cell results in high atmospheric pressure and dry conditions (subtropical high). Solar heating is at a minimum near the poles, because the sun's rays strike the earth at a very low angle. Cold air at the poles flows toward the equator and is deflected by the Coriolis effect to become the polar easterlies. At about 60°N and 60°S the polar air begins to rise and flow back toward the poles, completing the polar cell and producing a zone of low pressure (subpolar low) characterized by abundant precipitation. Air moving poleward from the subtropical high creates upper-level, westerly winds at mid-latitudes. Collision of warm tropical and cold polar air along the polar front results in eastward-moving, low-pressure troughs that are responsible for much of the precipitation in the mid-latitudes. A narrow zone of upper-level, high-velocity, east-flowing wind (jet stream) is also created along the polar front and influences movement of the low-pressure troughs.

The simple global pattern of climatic zones shown in Figure C40 is disrupted by other variables (Figure C41). The tilt of the earth on its axis of rotation results in seasonal shifts of the climatic zones by five to ten degrees of latitude. Different heat capacities of land and sea tend to intensify the subpolar low and subtropical high over the oceans, especially during winter. Disruption of zonal circulation may also result from the development of a high pressure cell in winter and low pressure cell in summer over a large land mass at mid to high latitudes, such as occurs in the Asian monsoons. Other effects more localized in time and space include increased precipitation on the windward side of mountain ranges (orographic effect), variations in reflectance of solar radiation among different surfaces (albedo), variations in the amount of solar radiation that strikes the earth due to cloud cover or volcanic ash or eolian dust in the atmosphere, and the redistribution of heat by ocean currents.

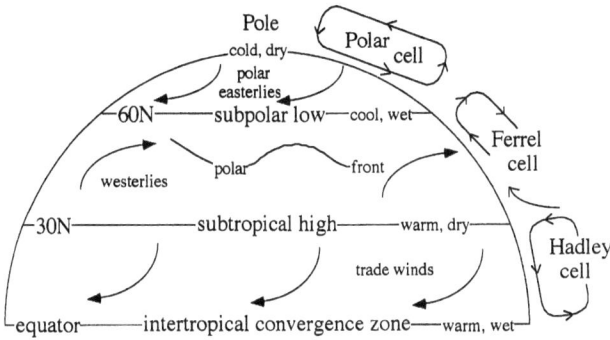

Figure C40 Climatic zones of the northern hemisphere. Arrows indicate wind directions. Climatic zones in the southern hemisphere are a mirror image of those in the northern hemisphere.

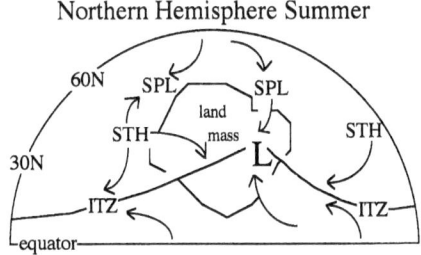

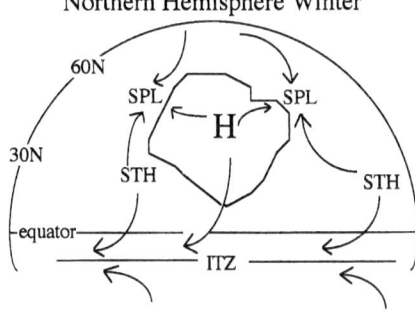

Figure C41 Example of the disruption of the general climatic zones of Figure C40. ITZ = intertropical convergence zone; STH = subtropical high; SPL = subpolar low; arrows indicate wind directions. The ITZ, STH, and SPL shift north during northern hemisphere summer and south during southern hemisphere summer, and the STH and SPL are intensified over the oceans. During summer, a large low pressure cell develops over the large land mass, deflecting the ITZ into it. In winter, a high pressure cell develops over the cold land mass.

Effects of climate on sedimentation

Climate plays a major role in determining sediment yield, which is the amount of siliciclastic sediment produced by weathering and transported out of a drainage basin to a sedimentary basin. Empirical data suggest that sediment yield in modern drainage basins, when normalized to temperature, is primarily related to annual precipitation and the competing effects of surface runoff and vegetative stabilization of soil (Figure C42; Langbein and Schumm, 1958; Leeder et al., 1998). Under arid conditions, sediment yield is low, because weathering rates are low and there is little surface runoff.

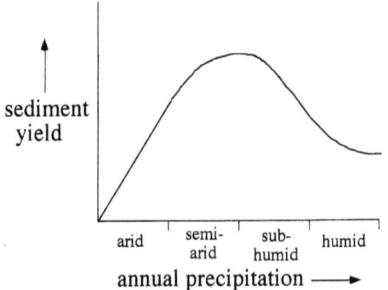

Figure C42 Schematic bivariate plot showing the relationship between sediment yield and annual precipitation.

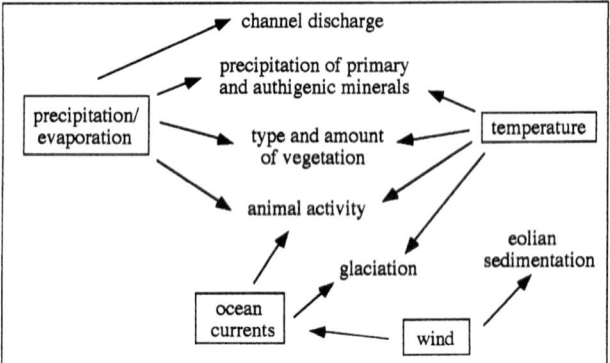

Figure C43 Climatic influence on some of the processes and variables that affect depositional environments.

Sediment yield reaches a maximum in semiarid and sub-humid climates, because there is sufficient precipitation for weathering and surface runoff but vegetative cover remains low to moderate. Included in the semiarid and sub-humid climatic groups are climates characterized by strongly seasonal precipitation, which may have large annual values of precipitation but relatively sparse vegetative cover due to a protracted dry season. Under humid conditions, lush vegetation tends to hold the soil in place, reducing sediment yield. Variations in sediment yield may affect deposition or incision of alluvial fans and rivers (Bull, 1991), the type of river and its bars and bedforms (Schumm, 1977), and whether shorelines prograde into the sea or undergo retreat in the face of rising sea level. In regions of extremely low sediment yield, carbonate and/or evaporite sedimentation may occur along the shoreline and in the shallow oceans.

In addition to sediment yield, climate also affects most depositional environments in terms of depositional processes, the type of sediment deposited, and syndepositional modification of the sediment. Some of the most important relationships between climatic variables and sedimentation are shown in Figure C43.

Interpretation of paleoclimate

Interpretation of the climates that existed in the geologic past is fundamental to the study of earth history and provides a template for interpreting future climatic change. There are two basic approaches to paleoclimate interpretation, modeling and paleoclimatic indicators. Whenever possible these two approaches should be used in concert to provide the most reliable interpretation.

Paleoclimate modeling involves applying the physics of modern climate to paleogeography, and includes energy balance (e.g., North and Crowley, 1985; Schmidt and Mysak, 1996) and general circulation models (e.g., Rees et al., 1999; Poulsen et al., 1999). A third type of modeling is concerned with global geochemical cycles, such as variations through time in the concentration of atmospheric carbon dioxide (Berner, 1994). Numerical models provide testable hypotheses regarding global paleoclimate and are especially useful in predicting the effects of specific variables on paleoclimate.

Paleoclimatic indicators are types of rock or features within rocks that only develop under specific climatic conditions. Perhaps the best terrestrial paleoclimatic indicators are plant fossils, because plant species and their relative abundance on earth are sensitive to precipitation, temperature, and evaporation rate (e.g., Wolfe, 1979, 1993; DiMichele et al., 2001). Vertebrate fossils may also be used as paleoclimatic indicators, particularly the distribution of large ectothermic animals (Markwick, 1994). Paleosols are effective terrestrial paleoclimatic indicators, especially the mature, well-drained types (Mack and James, 1994). Particularly diagnostic of paleoclimate are calcic, gypsic, and natric paleosols, which indicate relatively dry paleoclimate (Watson, 1990), extremely chemically weathered paleosols (oxisols, laterites, bauxites), which indicate humid paleoclimate (Bardossy and Aleva, 1990), cryoturbated paleosols indicative of permafrost (Retallack, 1999), and vertic paleosols that contain features produced by shrinking and swelling of expandable clays and that are best developed in regions with seasonal precipitation (Gustavson, 1991). The formation of peat, which may later convert to coal, is favored in regions with humid climate, but may also be related to locally high water table regardless of climate. Other terrestrial paleoclimatic indicators include widespread eolian deposits, which provide evidence about wind directions and are best but not exclusively developed in regions of dry climate (Peterson, 1988), tillites deposited by glaciers (Eyles et al., 1983), and the composition, sedimentary structures, and depositional cycles of lacustrine sediment (Hardie et al., 1978; Olsen, 1990).

Some marine sedimentary rocks also provide important evidence about paleoclimate. Many marine organisms are sensitive to water temperature, salinity, and ocean currents and the distribution of their fossils provides information about global paleoclimate (Huber, 1992). Moreover, the ratio of ^{18}O to ^{16}O in calcite or aragonite shells of some benthic and planktonic organisms has been used to estimate the temperature of the seawater from which the shells precipitated (Savin, 1977; Wilson and Opdyke, 1996). The oxygen isotopic ratio of marine shells is also influenced by preferential storage of the lighter isotope (^{16}O) in continental glaciers and its subsequent release during melting, providing a high-resolution record of glacial onset and fluctuations (Shackleton, 1988). Marine bedded chert, phosphorite, and organic-rich sediment tend to be associated with zones of oceanic upwelling, which are related to global wind patterns (Parrish, 1995). Finally, marine gypsum/anhydrite and halite precipitate in areas where high evaporation rates create hypersaline water.

Using modern climate as a guide is a useful approach to understanding ancient climate and is the foundation of paleoclimatic modeling and use of paleoclimatic indicators.

Some climatic variables or climate-related processes may have been quite different in the geologic past, however, affecting the boundary conditions of models and the applicability of certain paleoclimatic indicators. Several examples illustrate the problem: (1) lower solar luminosity in the Archean compared to today; (2) greater ratio of sea to land in the Archean compared to today; (3) higher global sea level in the Cambrian and Cretaceous, which may have ameliorated terrestrial climate and affected ocean circulation; (4) prior to the appearance of vascular land plants in the Devonian, sediment yield was not affected by vegetation; and (5) the global distribution of extinct plants and animals may not have been the same as that of their closest surviving relatives.

Temporal scales of climate change

Climatic cycles refer to repetitive changes in earth's climate related to global processes and may occur on the scale of a few years to tens of millions of years. At the largest timescale, the earth may have experienced as many as nine changes from warm to cool conditions over the past 700 million years (Frakes et al., 1992). These climatic intervals are commonly referred to as "icehouse" and "greenhouse" periods, because most of the cool intervals have evidence for continental glaciation and because the cycles appear to be strongly influenced by changes in the atmospheric concentration of carbon dioxide, a greenhouse gas.

At much smaller temporal scales are Milankovitch cycles, which result from orbital perturbations that significantly affect the distribution of solar energy on the earth, especially at high latitudes. One of the perturbations involves changes in the shape of the earth's orbit around the sun, which occurs at periods of 100 k years and 400 k years (thousand years). In addition, a complete cycle of change in the angle of tilt of the earth's axis of rotation takes place every 41 k years, and the precession effect, which is primarily related to the earth's wobble as it spins, produces cycles of 19 k years and 23 k years. The waxing and waning of Pleistocene glaciers have been correlated with Milankovitch cycles (Hays et al., 1976), as have other sedimentologic patterns in the stratigraphic record (Olsen and Kent, 1996).

Not all climatic changes in the geologic past were necessarily driven by global cycles, however. Climate changes may occur on a continent as it drifts across latitudinal climate zones, and local climate change may result from the rise and fall of mountains or from changes in ocean and wind currents. Given the multitude of variables that affect climate, it is unwise to assume that all climatic changes in earth history will occur at the same time scale and for the same reason.

Summary

1. Climate is the average condition of the atmosphere near the earth's surface and includes temperature, precipitation, evaporation, winds, and ocean currents. Climatic zones on the earth are generally distributed parallel to latitude, although local conditions may disrupt this pattern.
2. Climate affects the amount of siliciclastic sediment made available to sedimentary basins, as well as sediment composition, depositional processes, and syndepositional modification by organisms and precipitation of authigenic minerals.
3. The interpretation of ancient climate (paleoclimate) is based on numeral modeling and paleoclimatic indicators in the rocks, such as plant fossils, evaporite minerals, and paleosols. Uniformitarian considerations are necessary when interpreting paleoclimate, especially for the Precambrian.
4. Cyclic changes in climate may occur on the scale of years to tens of millions of years, although not all the climate changes in the geologic past resulted from global cycles.

Greg H. Mack

Bibliography

Bardossy, G., and Aleva, G.J.J., 1990. *Lateritic Bauxites*. Developments in Economic Geology 27. Amsterdam: Elsevier.

Berner, R.A., 1994. GEOCARB II: a revised model of atmospheric CO_2 over Phanerozoic time. *American Journal of Science*, **294**: 56–91.

Bull, W.B., 1991. *Geomorphic Responses to Climate Change*. Oxford: Oxford University Press.

Crowley, T.J., and North, G.R., 1991. *Paleoclimatology*. Oxford: Oxford University Press.

DiMichele, W.A., Pfefferkorn, H.W., and Gastaldo, R.A., 2001. Response of Late Carboniferous and Early Permian plant communities to climate change. *Annual Review of Earth and Planetary Sciences*, **29**: 461–487.

Eyles, N., Eyles, C.H., and Miall, A.D., 1983. Lithofacies types and vertical profiles; an alternative approach to the description and environmental interpretation of glacial diamict and diamictite sequences. *Sedimentology*, **30**: 393–410.

Frakes, L.A., Francis, J.E., and Syktus, J.I., 1992. *Climate Modes of the Phanerozoic*. Cambridge: Cambridge University Press.

Gustavson, T.C., 1991. Buried Vertisols in lacustrine facies of the Pliocene Fort Hancock Formation, Hueco bolson, west Texas and Chihuahua, Mexico. *Geological Society of America Bulletin*, **103**: 448–460.

Hardie, L.A., Smoot, J.P., and Eugster, H.P., 1978. Saline lakes and their deposits: a sedimentological approach. In *Modern and Ancient Lake Sediments*. International Association of Sedimentologists, Special Publication, 2, pp. 7–42.

Hays, J.D., Imbrie, J., and Shackleton, N.J., 1976. Variations in the earth's orbit: pacemaker of the ice ages. *Science*, **194**: 1121–1132.

Huber, B.T., 1992. Paleobiogeography of Campanian-Maastrichtian foraminifera in the southern high latitudes. *Palaeogeography, Palaeoclimatology, Palaeoecology*, **92**: 325–360.

Langbein, W.B., and Schumm, S.A., 1958. Yield of sediment in relation to mean annual precipitation. *American Geophysical Union Transactions*, **39**: 1076–1084.

Leeder, M.R., Harris, T., and Kirkby, M.J., 1998. Sediment supply and climate change: implications for basin stratigraphy. *Basin Research*, **10**: 7–18.

Mack, G.H., and James, W.C., 1994. Paleoclimate and the global distribution of paleosols. *Journal of Geology*, **102**: 360–366.

Marwick, P.J., 1994. "Equability", continentality, and Tertiary "climate": the crocodilian perspective. *Geology*, **22**: 613–616.

North, G.R., and Crowley, T.J., 1985. Application of a seasonal climate model to Cenozoic glaciation. *Journal of the Geological Society of London*, **142**: 475–482.

Olsen, P.E., 1990. Tectonic, climatic, and biotic modulation of lacustrine ecosystems-examples from Newark Supergroup of eastern North America. In *Lacustrine Basin Exploration*. American Association of Petroleum Geologists Memoir 50, pp. 209–224.

Olsen, P.E., and Kent, D.V., 1996. Milankovitch forcing in the tropics of Pangaea during the Late Triassic. *Palaeogeography, Palaeoclimatology, Palaeoecology*, **122**: 1–26.

Parrish, J.T., 1995. Paleogeography of organic-rich rocks and the preservation versus production controversy. In *Paleogeography, Paleoclimate, and Source Rocks*. American Association of Petroleum Geologists, Studies in Geology 40, pp. 1–20.

Parrish, J.T., 1998. *Interpreting Pre-Quaternary Climate from the Geologic Record*. New York: Columbia University Press.

Peterson, F., 1988. Pennsylvanian to Jurassic eolian transportation systems in the western United States. *Sedimentary Geology*, **56**: 207–260.

Poulsen, C.J., Barron, E.J., Johnson, C.C., and Fawcett, P., 1999. Links between major climatic factors and regional oceanic circulation in the mid-Cretaceous. In *Evolution of the Cretaceous Ocean-Climate System*. Geological Society of America, Special Paper, 332, pp. 73–89.

Rees, P.McA., Gibbs, M.T., Ziegler, A.M., Kutzbach, J.E., and Behling, P.J., 1999. Permian climates: evaluating model predictions using global paleobotanical data. *Geology*, **27**: 891–894.

Retallack, G.J., 1999. Carboniferous fossil plants and soils of an early tundra ecosystem. *Palaios*, **14**: 324–336.

Savin, S.M., 1977. The history of the earth's surface temperature during the past 100 million years. *Annual Review of Earth and Planetary Sciences*, **5**: 319–355.

Schmidt, G.A., and Mysak, L.A., 1996. Can increased poleward oceanic heat flux explain the warm Cretaceous climate? *Paleoceanography*, **11**: 579–593.

Schumm, S.A., 1977. *The Fluvial System*. New York: John Wiley and Sons.

Shackleton, N.J., 1988. Oxygen isotopes, ice volume, and sea level. *Quaternary Science Review*, **6**: 183–190.

Watson, A., 1990. Desert soils. In *Weathering, Soils and Paleosols*. Developments in Earth Surface Processes 2. Amsterdam: Elsevier, pp. 225–260.

Wilson, P.A., and Opdyke, B.N., 1996. Equatorial sea-surface temperatures for the Maastrichtian revealed through remarkable preservation of metastable carbonate. *Geology*, **24**: 555–558.

Wolfe, J.A., 1979. Temperature parameters of humid to mesic forests of eastern Asia and relation to forests of other regions of the northern hemisphere and Australasia. *U.S. Geological Survey Professional Paper 1106*.

Wolfe, J.A., 1993. A method of obtaining climatic parameters from leaf assemblages. *U.S. Geological Survey Bulletin 2040*.

Cross-references

Alluvial Fan
Anhydrite and Gypsum
Bauxite
Caliche–Calcrete
Coal Balls
Coastal Sedimentary Facies
Cyclic Sedimentation
Debris Flow
Deltas and Estuaries
Desert Sedimentary Environments
Dunes, Eolian
Erosion and Sediment Yield
Evaporites
Glacial Sediments
Lacustrine Sedimentation
Loess
Maturation, Organic
Milankovitch Cycles
Neritic Carbonate Depositional Environments
Peat
Phosphorites
Red Beds
Rivers and Alluvial Fans
Sabkha, Salt Flat, Salina
Sapropel
Smectite Group
Tills and Tillites
Upwelling
Varves
Weathering, Soils, and Paleosols
Zeolites in Sedimentary Rocks

COAL BALLS

Definition

Coal balls are permineralized peat, mainly found in Upper Carboniferous coal seams of Europe and North America but also in some Chinese Permian coals. Coal balls are predominantly calcium carbonate which has precipitated in the cell lumina and spaces between the plants within a peat formed in a mire (Scott and Rex, 1985).

Formation

Peat accumulates in wetland environments (mires such as swamps and bogs) where the accumulation of organic matter is faster than its decay. Studies on modern peat-forming environments indicate that the two main sources of water are from throughflow (rheotrophic) and rainfall (ombrotrophic) (Moore, 1987). In general the breakdown of plant material (cellulose and lignin) predominantly through the action of bacteria causes a decrease in pH. However, during the Carboniferous in Tropical Euramerica, in particular, mineral charged waters flowing through rheotrophic peats precipitated calcium carbonate, usually in the form of calcite. The calcite precipitated not only within cell lumina and spaces with the plants but also between plant parts and fragments (Scott and Rex, 1985) (Figure C44). This gave rise to a limestone 'nodule' within the peat which entombed the plants but did not replace them (Winston, 1986). These nodules are known as coal balls (Figures C45, C46). Discovered first in England in the 1850s these coal balls ranged in size from a few centimeters to tens of centimeters across. They appeared to be statigraphically restricted to the base of the British Coal Measures.

However, subsequently coal balls were found in continental Europe from Spain in the South to the Ukraine in the east. By far the most diverse occurrences of coal balls have been discovered in North America (mainly in the USA, but also some finds in Nova Scotia, Canada). In the USA coal balls may replace entire coal seams in part and coal ball masses may be several meters in diameter (DiMichele and Phillips, 1994; Phillips, 1980; Greb *et al.*, 1999).

When cut coal balls reveal excellent anatomical preservation of the plants which formed the peat (Figure C47). The original organic cell wall material (albeit slightly chemically altered) is preserved (Collinson *et al.*, 1994). The calcite is precipitiated as fibrous calcite which grows on organic walls into voids (Figure C44). There may be subsequent alterations, as in the case of some Belgium coal balls, to dolomite. In many cases, the coal seam containing coal balls is overlain by marine rocks. In England, roof nodules within marine shales contained anatomically preserved plants. This led to the suggestion by Stopes and Watson (1908) that the source of the carbonate was marine. Research in North America by Mamay and Yochelson (1962) showed that some coal balls had a marine core thus strengthening the idea of a marine source for the carbonate. Isotopic studies, however, suggest that not all of the carbonate is marine and that there is likely to be a complex origin for the carbonate, some of which appears to be derived from meteoric fresh water (Scott *et al.*, 1996; DeMaris, 2000). At least five models of formation for coal balls have been proposed, each with supporting data.

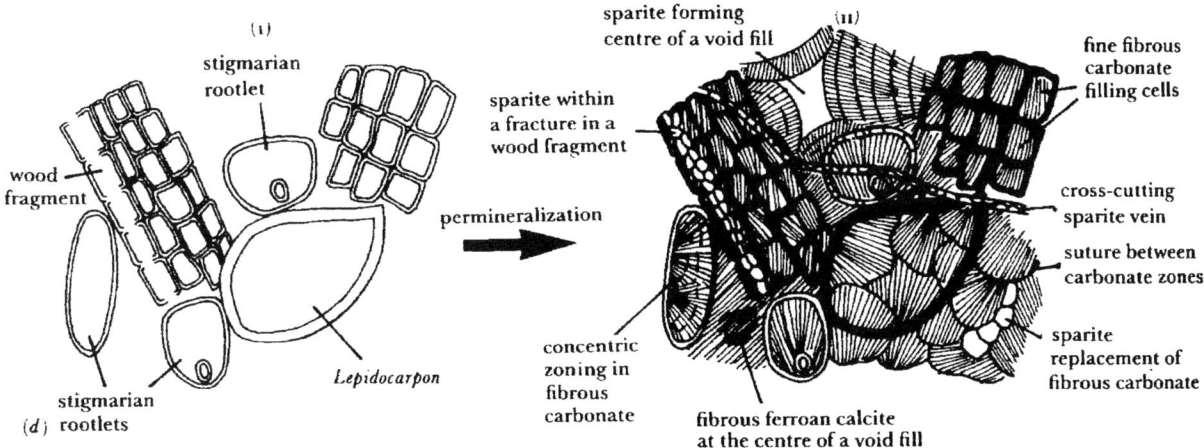

Figure C44 Petrography of a Carboniferous coal ball (from Scott and Rex, 1985).

Figure C45 Coal ball, *in situ*, Springfield Coal, Pennsylvania, Indiana, USA.

Figure C46 Section cut through coal ball shown in Figure C45.

Whilst most coal balls are found in Upper Carboniferous coal seams from tropical Euramerica some coal balls have been described from the Permian of China (Tian 1985). We do not understand why carbonate coal balls are restricted to Late Paleozoic coal seams.

Content

A wide variety of plant species and plant parts may be preserved in coal balls (DiMichelle and Phillips, 1994). Permineralization may happen at any time during the formation of the peat from immediately the plant material has been incorporated into the peat to when there has been some considerable decay within the peat (Scott *et al.*, 1996). The speed of permineralization in some cases may be illustrated by the preservation of ephemeral structures such as pollen drops. In many cases, however, plants may have started to rot before permineralization so that strips of bark or cortex of larger trunks may be preserved (DiMichele and Phillips, 1994). A feature of coal balls is that the original organic cells walls are preserved. Because they are entombed within limestone they are not compacted, as normal peat would be, by burial. This ensures anatomical preservation of the plants. The organic cell walls are subject to temperature rise during burial and the organic material undergoes several changes whilst still retaining features of their original cell wall chemistry.

The detailed plant anatomy has allowed the identity and the life history of the preserved plants to be elucidated. Serial peels allow the reconstruction of the morphology of the plants. The reconstruction of a number of coal ball plants from small scramblers to large trees (Phillips and DiMichele, 1992; Bateman *et al.*, 1992) has been made possible from the connection of plant organs, recurrent association and specific anatomical or morphological features of the plants. As coal balls represent peat which may have acted as a substrate for other plants, rhizomes, roots and rootlets are common, often growing through other plants and plant organs (DiMichele and Phillips, 1994).

Figure C47 Thin section of Carboniferous coal ball showing anatomical preservation of *Lepidodendron*, Lancashire, England (×2).

In Namurian and Westphalian coal balls of Europe and North America arborescent lycopsids are the dominant plant (Galtier, 1997). However ferns, sphenopsods and pteridosperms may also be locally abundant. In some coal balls of the North American mid-continent basin some coal balls are dominated by Cordaites, a close relative of conifers. Some Chinese coal balls of Permian age are also dominated by Cordaites. In Stephanian coals of USA, coal balls are dominated by tree ferns (Phillips, 1980).

Spores and pollen may be extracted from fertile structures allowing comparison of the vegetation in coal balls with that from coals where spores and pollen are the only way of identifying the coal-forming vegetation (Balme, 1995).

Significance

Detailed anatomical and studies of reproductive biology have allowed the life habits and ecology of the plants to be understood and in addition allowed phylogenetic reconstruction of several important plant groups (Bateman *et al.*, 1992).

Stratigraphic studies of coal ball plants and spores have indicated major vegetation changes of coal-ball floras through the Upper Carboniferous. Extinctions of plants have been linked to distinctive climatic changes, in particular decreasing rainfall which affected the more wet-loving plants (DiMichele and Phillips, 1994). Coal balls also provide evidence of plant-arthropod interaction (Scott and Taylor, 1983). Recent studies have indicated an extensive arthropod decomposer community as recorded by coprolitic debris (Labendeira, 1998).

Andrew C. Scott

Bibliography

Balme, B.E., 1995. Fossil *in situ* spores and pollen grains: an annotated catalogue. *Review of Palaeobotany and Palynology*, **87**: 81–323.

Bateman, R.M., DiMichele, W.A., and Willard, D.A., 1992. Experimental cladistic analysis of anatomically preserved arborescent lycopsids from the Carboniferous of Euramerica: an essay on palaeobotanical phylogenetics. *Annals of the Missouri Botanical Garden*, **79**: 500–559.

Collinson, M.E., van Bergen, P.F., Scott, A.C., and De Leeuw, J.W., 1994. The oil generating potential of plants from coal and coal-bearing strata through time: a review. In Scott, A.C. and Fleet, A.J. (eds.), *Coal and Coal-Bearing Strata as Oil-Prone Source Rocks?*. Geological Society of London, Special Publication, **77**: pp. 31–70.

DeMaris, P.J., 2000. Formation and distribution of coal balls in the Herrin Coal (Pennsylvanian), Franklin County, Illinois Basin, USA. *Journal of the Geological Society of London*, **157**: 221–228.

DiMichele, W.A., and Phillips, T.L., 1994. Palaeobotanical and Palaeoecological constraints on models of peat formation in the Late Carboniferous of Euamerica. *Palaeogeography, Palaeoclimatology, Palaeoecology*, **106**: 39–90.

Galtier, J., 1997. Coal ball floras of the Namurian-Westphalian of Europe. *Review of Palaeobotany and Palynology*, **95**: 51–72.

Greb, S.F., Eble, C.F., Chesnut, D.R., Phillips, T.L., and Hower, J.C., 1999. An *in situ* occurrence of coal balls in the Amburgy Coal Bed, Pikeville Formation (Duckmantian), Central Appalachian Basin, U.S.A. *Palaios*, **14**: 432–450.

Labendeira, C., 1998. Plant-insect associations from the fossil record. *Geotimes*, **43**(9): 18–24.

Mamay, S.H., and Yochelson, S.H., 1962. Occurrence and significance of marine animal remains in American Coal Balls. *US Geological Survey Professional Paper*, 354, pp. 193–224.

Moore, P.D., 1987. Ecological and hydrological aspects of peat formation. In Scott, A.C. (ed.) *Coal and Coal-Bearing Strata: Recent Advances*. Geological Society of London, Special Publication, 32, pp. 7–15.

Phillips, T.L., 1980. Stratigraphic and geographic occurrence of permineralized coal-ball plants—Upper Carboniferous of America and Europe. In Dilcher, D.L., and Taylor, T.N. (eds.), *Biostratigraphy of Fossil Plants*. Dowden: Hutchison and Ross, Stroudsburg, pp. 25–92.

Phillips, T.L., and DiMichele, W.A., 1992. Comparative ecology and life-history biology of arborescent lycopsids in Late Carboniferous swamps of Euramerica. *Annals of the Missouri Botanical Garden*, **79**: 560–588.

Phillips, T.L., and DiMichele, W.A., 1999. Coal ball sampling and quantification. In Jones, T.P., and Rowe, N.P. (eds.), *Fossil Plants and Spores: Modern Techniques*. Geological Society of London, pp. 206–209.

Scott, A.C., Mattey, D., and Howard, R., 1996. New data on the formation of Carboniferous Coal Balls. *Review of Palaeobotany and Palynology*, **93**: 317–331.

Scott, A.C., and Rex, G., 1985. The formation and significance of Carboniferous coal balls. *Philosophical Transactions of the Royal Society of London B*, **311**: 123–137.

Scott, A.C., and Taylor, T.N., 1983. Plant-animal interactions during the Upper Carboniferous. *Botanical Review*, **49**: 259–307.
Stopes, M.C., and Watson, D.M., 1908. On the present distribution and origin of the calcareous concretions known as 'coal balls'. *Philosophical Transactions of the Royal Society of London B*, **200**: 167–218.
Winston, R.B., 1986. Characteristic features and compaction of plant tissues traced from permineralised peat to coal in Pennsylvanian coals (Desmoinsian) from the Illinois Basin. *International Journal of Coal Geology*, **6**: 21–41.
Tian Baolin, 1985. Coal balls in coal seams of China. *Compte Rendu Neuvième Congrès International de Stratigraphie et de Géologie du Carbonifère. Urbana, Illinois 1979*, **5**: 102–108.

Cross-references

Diagenetic Structures
Peat

COASTAL SEDIMENTARY FACIES

Introduction

This section describes some of the more common clastic sedimentary facies associated with open coasts. For a review of carbonate coastal facies, the reader is referred to Demicco and Hardy (1994). The discussion here also does not address facies of high-latitude coasts or low-latitude coasts dominated by mangrove swamps (see Hill *et al.*, 1995, and Cobb and Cecil, 1993, respectively). Many of the facies described here originate in both marine and lacustrine settings, although tidally influenced facies are restricted to the marine environment.

For more detailed information on coastal sedimentary facies, the reader is referred to the excellent summaries provided by Reineck and Singh (1973), Walker and James (1992), Galloway and Hobday (1996), and Reading (1996). In addition, the following complementary sections are available in this volume: Barrier Islands, Beach, Deltas and Estuaries, Hummocky and Swaley Cross-Stratification, Off-Shore Sands, Salt Marsh, Storm Deposits, Swash and Backwash, and Tidal Delta and Inlet.

Coastal sedimentary processes

A complex array of processes influence coastal sedimentary facies. Of these, the most important are waves, tides, and biogenic processes. Sediment input is also critical to the facies character (in its influence on grain size) and to the nature of the preserved deposits (the effects of sedimentation rate).

Waves may exist as "seas", driven by local winds, or as "swell", generated by distant storms. Swell tends to have longer period and to influence the seabed to greater depths than do local sea waves. "High-energy" coasts are likely to be dominated by swell. "Low-energy" coasts receive smaller everyday waves, but can experience very large waves during storms.

Waves move sediment by two mechanisms. The passage of waves induces a back and forth orbital motion at the bottom. In water that is deep relative to the size of the wave, this motion is symmetrical in velocity and duration of flow. In shallower water, the motion becomes asymmetric, with short relatively strong landward flow under the crest of a wave and a more prolonged weaker seaward flow under the trough

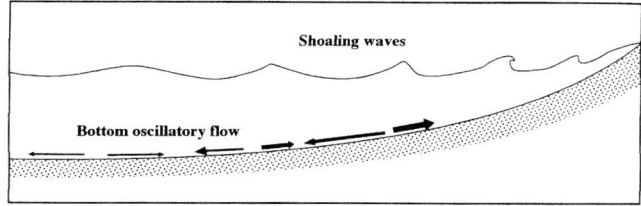

Figure C48 Nature of bottom oscillatory currents under shoaling waves. Length of arrow indicates duration of flow. Thickness of arrow indicates flow velocity.

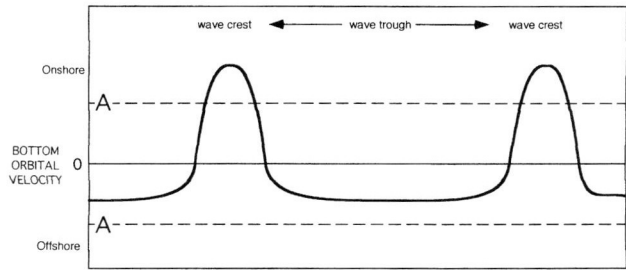

Figure C49 Schematic showing direction and velocity of bottom oscillatory currents at a point on the nearshore profile under shoaling waves. Line "A" represents a threshold velocity that is exceeded by landward flow as wave crests pass, but not by seaward flow as troughs pass (see Komar, 1976).

(Figure C48). The orbital velocity asymmetry is important for promoting the onshore transport of coarser material (Figure C49) and bed load in general. For the conditions under which this asymmetric flow occurs, see the "Stokes wave" field in Komar (1976, Figure C50–17).

Waves also, as they encounter the shoreline, generate sustained flow in the form of shore-parallel longshore currents and seaward-directed rip currents (Figure C50). These currents, which form nearshore circulation cells, not only transport sediment but also shape the nearshore morphology into bars and troughs.

The intensity and nature of wave-generated processes can differ significantly as conditions change from fair-weather to storm. Storms, which on most coasts obtain for only a small percentage of time, mobilize the seabed at greater depths and intensify nearshore circulation. Rip currents are stronger and extend into deeper water. In addition, wind-driven currents can augment the offshore flow (Swift *et al.*, 1985). As a consequence, relatively large volumes of sediment are transported alongshore and offshore in a short period of time. Much of this material, however, can be reworked in the long intervals between storms and returned to its pre-storm setting.

Tides influence sedimentary facies in two ways (see *Sediment Movement by Tides*). The rhythmic rise and fall of the sea changes water depth and locally exposes extensive intertidal flats. It also generates tidal currents that flow alternately landward (flood tides) and seaward (ebb tides). Tides in an open coast setting mostly only raise and lower the water level, exposing the seabed to different wave-energy regimes. Tidal currents here are likely to be important only near inlets, or, in epeiric seas, further offshore.

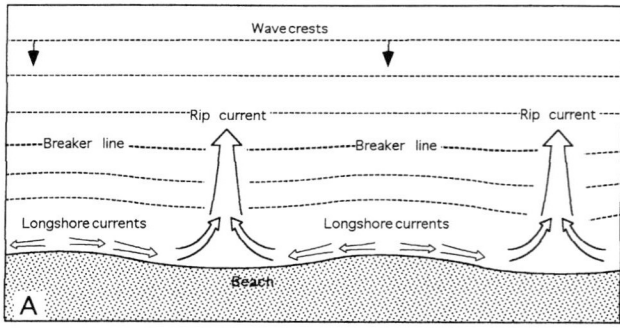

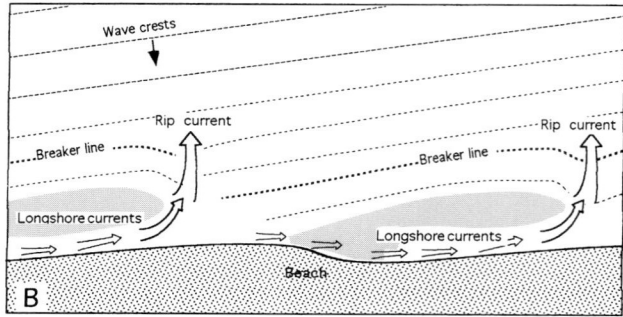

Figure C50 Nearshore circulation cells where wave incidence is parallel to coast (A) or oblique (B). Circulation consists of unidirectional rip and longshore currents generated by the hydraulic head caused by set-up/setdown of the sea surface in and near the breaker zone.

Biogenic processes are important in the development of coastal facies. Biogenic effects include bioturbation, the generation of distinctive traces, contribution of shell detritus, the fixing of suspended fine sediment by filter feeding and vegetative baffling, and the binding and bioerosion of the substrate. These processes depend on many factors, but faunal activities are particularly sensitive to variations in salinity. Faunal traces in brackish environments tend to be simpler, less diverse, and more likely to be dominated by a single ichnogenus (Pemberton and Wightman, 1992) than those made in oceanic salinities.

The response of sediment to fluid motions depends in large part on its textural nature, which is determined partly by local dynamic processes and partly by what is available to the system. Grain size is also a major influence on the rate and style of bioturbation, which combined with the frequency and intensity of physical processes determine the degree of preservation of physical structures. On most open coasts the sediment lies somewhere in the spectrum of very fine-grained sand to gravel, although coastline adjacent to or influenced a large delta may be dominated by mud.

The shoreface

A fundamental aspect of most open clastic coasts is the *shoreface* (Figure C51), a relatively steeply-inclined (1:200) ramp that extends offshore to a nearly flat (1:2000) inner shelf or basin floor (Johnson, 1919). This feature is probably

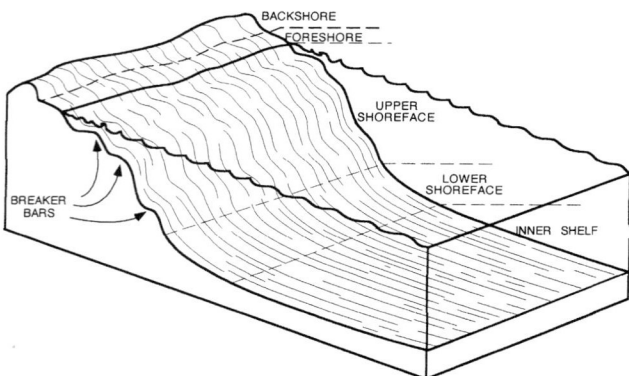

Figure C51 Typical morphologic zonation of a wave-dominated open coast. Many shorefaces have 1 to 3 breaker bars.

maintained by a combination of sediment transport to shoreward under shoaling waves (as described above) and transport to seaward by rip currents or geostrophic flow during storms (Niedoroda and Swift, 1981).

As noted by Niedoroda *et al.* (1985), the boundaries between the shoreface and its adjacent coastal environments can be difficult to define, and as a result, subsequent work on the shoreface has led to conflicting variations on the original definitions. Many workers and texts follow Johnson's (1919) lead in placing the landward boundary of the shoreface at the low-tide line, but others, however, place the boundary just seaward of the surf zone, and a few place it at the top of the swash zone.

The seaward boundary of the shoreface is even more problematic, because it involves a subtle change in slope. On many modern coasts, the inner shelf is characterized by a gently inclined, nearly planar surface, whereas the adjacent shoreface is not only more steeply inclined, but the angle of slope increases to landward. Many workers place the base of the shoreface at this increase in slope, but a few do not. Some extend the shoreface to a location where the seabed texture changes (typically from sand to mud). Others place the base of the shoreface at the mean fair-weather wave base, with or without morphologic connotation.

Most workers refer to the area seaward from the shoreface as "shelf" or "offshore". Howard and Reineck (1981) identify a "transition zone" separating the shoreface and the offshore in which the sediment is siltier and/or more bioturbated than that of the shoreface.

The depth to which a shoreface extends obviously depends on its definition, but the break in slope used by most workers can occur in water depths ranging from a few meters to 25–30 m. The depth of the slope break is in part a function of wave energy, but much of the variation results from the seafloor profile in a transgressive or erosional setting. In progradational settings, where sufficient sediment exists to provide for an equilibrium profile with the shoaling waves, the break in slope occurs in relatively shallow depths. Even where wave energy is high (as on the Pacific Northwest coast of the United States), progradational shorefaces do not extend to depths much greater than 10 m.

In the study of ancient coastal sediment, shoreface deposits are identified on the basis of a variety of criteria. General agreement exists that the preservation of shoreface successions requires shoreline progradation, and that, accordingly, the deposits display an upward-shallowing facies trend and upward textural coarsening. Otherwise there is little consistency of usage. Some workers attribute the entire sandy component of a progradational succession to the shoreface. Others restrict the shoreface to thick-bedded sandstone as opposed to interbedded sand and shale. Some associate shoreface deposits with specific physical structures, such as trough cross-bedding and swaley cross-stratification. Others use the degree of bioturbation to separate the shoreface from underlying shelf facies. Some workers propose evidence of deposition above fair-weather wave base as a criterion of shoreface deposits.

An assumption that shoreface deposits are sandy and shelf deposits muddy is unsupportable. In some areas the base of the shoreface marks the transition between sand and mud, but in many places sand continues well out onto the inner shelf. On much of the Pacific coast of California, Oregon, and Washington, for example, fine to very-fine sand extends to water depths of 60 m or more. In contrast, the shoreface of the Suriname coast of South America is underlain by mud and the adjacent shelf by sand (Rine and Ginsburg, 1985). The proportions of sand, silt and mud, depend on the sediment available and the wave-energy regime (Galloway and Hobday, 1996) and are not specifically related to the shoreface morphology.

An assumption that fair-weather wave base defines the base of the shoreface is similarly unfounded. Again, although this can be the case in some settings, it is demonstrably not in many others. Wave base is commonly regarded as a water depth equal to half the wave-length. Long-period waves (≤ 8 s) characterize many coasts during fair weather and such waves have deep-water wave lengths of 100 m or more. Ten-second waves, which prevail during summer fair-weather conditions on the western coast of the US, have a wave base of nearly 80 m, well beyond any shoreface.

The thickness of inferred individual ancient shoreface successions also is highly variable. Some workers place them in the range of several tens of meters, whereas others place the thickness at only a few meters. Given the water depth to which modern prograding shorefaces extend, most individual ancient shoreface successions will be less than 10–15 m thick, unless influenced by conditions of exceptional accommodation.

Shoreface subdivisions

Galloway and Hobday (1996) describe the active surf zone as "upper shoreface", a zone of breaker bars as "middle shoreface", and the area from the outermost bar to the break in slope as "lower shoreface". Many interpreters of ancient coastal deposits use texture, physical structures or trace fossils to divide the shoreface into upper, middle and lower parts, and a few subdivide these further into "proximal" and "distal" portions. These divisions may be valid lithologic entities, but they also may be unrelated to the morphologic shoreface.

Sedimentary facies of the shoreface and adjacent environments

The *upper shoreface* is equivalent to the nearshore zone, where wave-generated currents predominate (see Clifton *et al.*, 1971). Upper shoreface facies are generally the coarsest part of the coastal system and commonly are characterized by high-angle trough cross-bedding that is directed parallel to the shoreline (in response to longshore currents) or offshore (in response to rip currents). Modern medium- to coarse-grained nearshore zones are dominated during fair-weather by medium-scale lunate megaripples (dunes) that face and migrate in an onshore direction (Figure C52). Cross-bedding produced by these structures seems rarely preserved, however, in ancient upper shoreface deposits, which seem dominated by storm effects. If gravel is present, it typically is well-sorted into discrete beds and layers, and gravel and shells commonly form erosional lags within the deposit. Small, shore-normal gravel-filled gutter casts (*q.v.*) exist at the base of some gravel beds. Fine-grained sandy shorefaces are likely to lack larger bedforms and to be dominated by planar beds and rippled surfaces (Figure C53), unless breaker bars provide a mechanism for focusing longshore or offshore flow (Figure C54) or wave energy is so low that bioturbation predominates in the troughs (Figure C55).

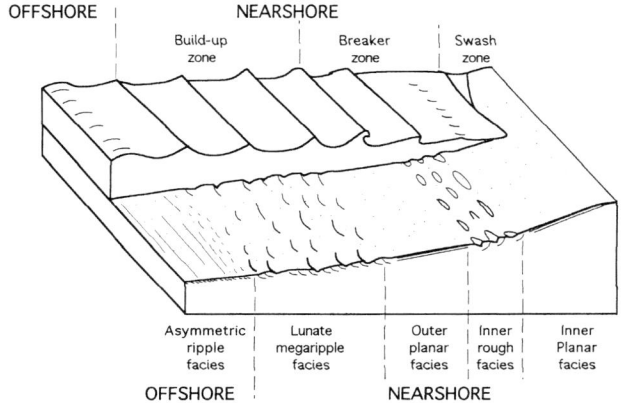

Figure C52 Sedimentary structural facies in the unbarred nearshore (upper shoreface) in coarse sandy sediment on the high-energy coast of southern Oregon under fair-weather conditions.

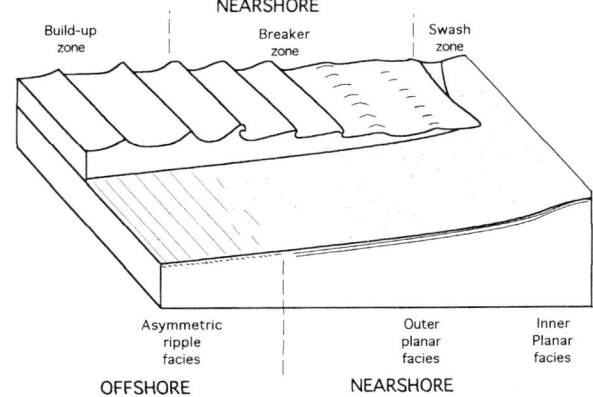

Figure C53 Sedimentary structural facies in the unbarred nearshore (upper shoreface) in fine sandy sediment on a high-energy coast under fair-weather conditions. Flatter beach/nearshore profile expands the surf zone relative to coarser shorelines. No medium-large scale bedforms. Surf and swash zones underlain by planar parallel lamination.

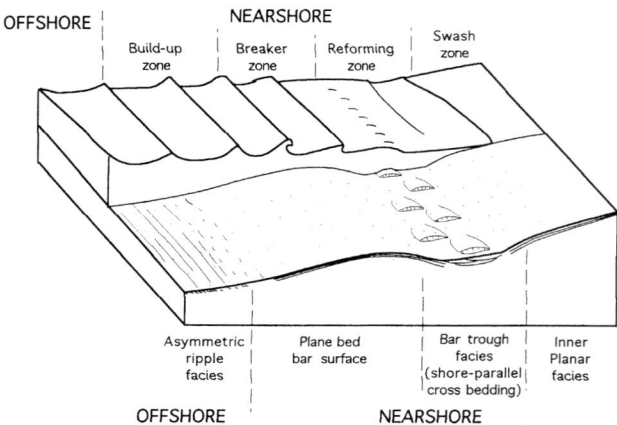

Figure C54 Sedimentary structural facies in a barred high-to-moderate-energy fine-grained nearshore under fair-weather conditions. Longshore flow in the trough of the shore-parallel bar is strong enough to create small dunes that migrate alongshore in the trough.

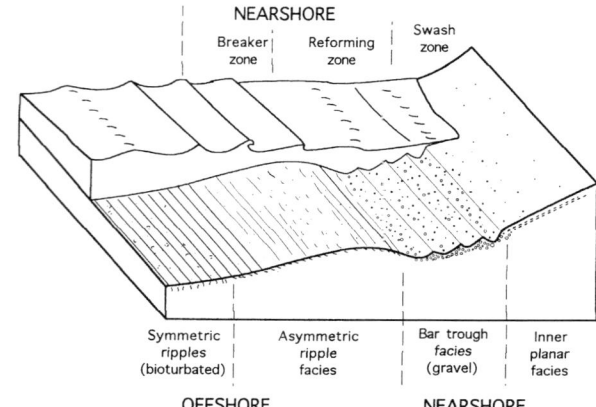

Figure C56 Sedimentary structural facies in a barred low energy coarse-grained nearshore under fair-weather conditions. Coarse sediment is concentrated in the trough of the shore-parallel bar.

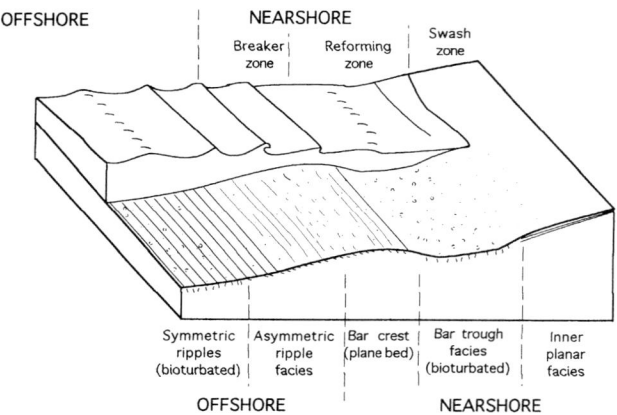

Figure C55 Sedimentary structural facies in a barred low-energy fine-grained nearshore under fair-weather conditions (example, Padre Island, Texas, after Hill and Hunter, 1976). Sediment in the longshore trough and seaward from the bar is intensely bioturbated.

Coarse sediment, if present, typically is concentrated in troughs behind breaker bars (Figure C56).

Burrowing in upper shoreface deposits is dominated by the *Skolithos* ichnofacies, characterized by mostly vertical cylindrical and U-shaped burrows, the dwelling structures of suspension feeders or passive carnivores (Pemberton *et al.*, 1992). The trace *Macaronichnus* is common in fine sand of high-energy coastal deposits and feeding pits of rays or other vertebrates may occur. The thickness of upper shoreface deposits generally ranges from 1 to 8 m, depending on the intensity of wave energy.

Many interpreters of the stratigraphic record postulate a *middle shoreface* facies in their deposits based on the presence of bar-trough facies or other evidence, but this identification is questionable. On many coasts the innermost breaker bar is part of the active surf zone, and during storms, the largest waves commonly break on the outermost breaker bar. As a consequence, preserved bar-trough facies may be indistinguishable from that of the inner surf zone.

A *lower shoreface* facies is identifiable in most pebbly shoreface deposits. This facies, 1–3 m thick, consists of small pebbles of relatively uniform size in a matrix of well-sorted fine-to very fine-grained sand. The pebbles occur as scattered clasts, in small clusters, or as thin discontinuous stringers 1–2 pebbles thick. The associated sand may be laminated or structureless and is indistinguishable from inner shelf sand. The facies results from a combination of offshore transport from the upper shoreface during storms and onshore transport of most of this material by the shoaling waves in the aftermath of a storm. Winnowed by the waves, the coarser material moves landward more slowly and some is trapped in burrows, swales, or other depressions on the seafloor. In the absence of pebbles, the upper shoreface facies may be the only recognizable shoreface component.

On transgressive coasts, *shoreface-attached ridges* extend obliquely into the sea. These ridges, composed of fine to coarse sand, are tens of km long, several km wide, and 5–15 m high. Their origin has been attributed to longshore migration of tidal inlets during transgression, leaving behind a ridge formed by one side of the tidal delta. (McBride and Moslow, 1991). Cores through shoreface-attached ridges show that they consist of an upward-coarsening accumulation of storm-event beds (Rine *et al.*, 1991). Some shoreface-attached ridges show evidence of both storm and tidal influences in their internal structures (Antia *et al.*, 1994).

The *beach foreshore* facies lies within an intertidal zone subject to the swash and backwash of the waves. Foreshore sands are typically very well-sorted and characterized by planar lamination. Foreshore deposits are generally 1–2 m thick, and if the grain size permits, show an upward-fining. Individual lamina may show inverse grading. Placers of heavy minerals define the upper part of some foreshore deposits. Small vertical burrows and the trace *Macaronichnus* typify the ichnologic signature of the foreshore.

Nonmarine coastal facies of the open coast include eolian dunes, characterized by large-scale landward dipping foresets and backshore facies of wind/tidal flats. The character of the *backshore* facies depends on the specific setting and the amount of vegetative growth. Backshore sand is commonly fine and well sorted; layers of pebbles and coarser sand are introduced during floods, and layers of muddy and/or peaty sediment can develop under wetter conditions. Structures such as climbing adhesion ripples or vertebrate footprints document subaerial exposure. Roots are common, and substantial coals may directly overlie the foreshore facies in some ancient deposits.

Mudflats dominate some open coasts proximal to, or downcurrent from, the mouths of large mud-bearing streams. In settings such as the coast of Surinam, inshore fluid mud maintains its presence by frictionally reducing the energy of incoming waves. Episodically, mud deposition ceases, either during storms or because the mud moves along the coast in discrete waves (Rine and Ginsburg, 1985), and sandy beaches develop. A return to muddy deposition isolates the sandy sediment in shore-parallel ridges or *cheniers*. Cheniers on the coast of Surinam are about 2 m high and contain a combination of steeply and gently landward dipping cross-beds or, where fine-grained, gently landward and seaward-dipping cross-beds (Augustinius, 1989). Cheniers adjacent to the Mississippi River contain shell aggregates concentrated by storm waves. Mud in the associated flats may be laminated to structureless; mud flats covered by salt marshes will contain root or rhizome structures.

Sedimentary facies of inlet areas

Tidal inlets, fluid interfaces between bays and the open sea, are common on many contemporary open coasts. Inlets may range from simple channels a few meters deep and tens of meters wide to complexes of channels and shoals more than 10 km or more across wide at the mouths of large embayments. Individual channels in this setting may be tens of meters deep and hundreds of meters wide. Inlet sediment is mostly sand, and occurs in a setting where tidal flow is likely to be at its highest. As a consequence, inlet channels are typically floored by 2- and 3-dimensional dunes. Individual channels may be dominated by either ebb or flood tide. Lag deposits of shells and or pebbles (where the adjacent shoreface is gravelly) are generally present, and may contain species representative of both open coast and bay. Trough and tabular cross-bedding and abundant burrows characterizes the resulting deposit.

Small inlets tend to migrate laterally along the coast, particularly where wave-driven longshore sediment transport is significant. Larger inlets may be more stationary, but interior inlet channels may migrate rapidly. In Willapa Bay, Washington, for example, a primary inlet channel, 25 m deep and 600–700 m wide, has moved laterally more than a kilometer in the past 50 years. Inlet facies in such a setting are likely to consist of a lower part composed of cross-bedded sand produced in the primary channel overlain by the deposits of smaller tidal channels and wave- and tide-influenced shoals. Channel migration can produce clinoforms, the thickness of which depends on the depth of the channel. The lateral extent of inlet deposits depends on the degree of inlet migration. Extensive migration will generate laterally extensive inlet facies within a coastal succession.

Lobate bodies of sand, *flood tidal-deltas*, form sandy ramps on the bayward side of many inlets. Where the tidal inlet migrates, flood tidal deltas may compose laterally extensive sand bodies along the seaward side of a lagoon enclosed by a barrier island. Initially these deltas form as a set of lobes building into the bay, covered by straight to sinuous 2-dimensional landward-facing dunes (Hine, 1975). As they mature, tidal flow is concentrated into channels, ultimately producing a ramp dissected by flood-dominated tidal channels. The resulting deposit is dominated by landward-directed trough and tabular cross-bedding and contains internal erosional surfaces capped by shell lags.

Ebb tidal deltas form on the seaward side of inlets, where they are subject to the influence of both waves and tides. Off large embayments, these deltas may be quite large, exceeding 10 km in width and extending 5 km or more into the adjacent sea. Off smaller inlets on a transgressive coast, the deltas may form the nucleus of shoreface-attached ridges. Flood- and ebb-dominated tidal channels cross the shoal constituting the delta: typically the ebb flow is concentrated in the deeper channels. Upward-fining storm beds with erosional bases are common, particularly in the shallower parts of an ebb tidal delta.

Preservation of coastal sedimentary facies and resulting facies successions

Thick accumulations of coastal facies typically reflect a record of fluctuating sea level superimposed on a setting of overall tectonic subsidence. The shoreline response to sea-level change depends on the rate of sedimentation. In general, the influx of sediment at the shoreline promotes progradation, whereby the additional sediment forces the shoreline to shift basin-ward. Most prograding open coasts today are characterized by a series of each ridges, each marking an earlier position of the shoreline. The resulting accumulation conforms to Walther's Law, which holds that a vertical arrangement of facies conforms to their lateral distribution at the time of deposition. Progradational successions show an upward-shallowing, produced as shallower water facies encroach over their deeper water counterparts (Figure C57). On most open coasts, the result typically generates an upward-coarsening interval. Figures C58–C62 show examples of progradational successions that reflect differing combinations of grain size and wave energy.

Although progradation is the most common circumstance of preservation of coastal sedimentary facies, not all facies are preserved. Those associated with features that stand in relief above other, more landward, features are likely to be destroyed as progradation progresses. Although breaker bars, for example, are common on many upper shorefaces, their potential for preservation is limited. During progradation, the seaward migration of the bar troughs that lie to landward will erase much, if not all, of the bar.

Where relative sea level falls during progradation a "forced regression" ensues (Figure C63). Wave erosion associated with the readjustment to a new, lower base level can create a "regressive surface of marine erosion". The amount of sediment removed by this erosion depends on the rate and degree of base level fall and prevailing wave energy. Commonly the surface marks an abrupt change from shelf to shoreface deposits.

Progradation may occur during relative sea-level rise, if the rate of sedimentation is sufficiently high. But commonly a rise in base level results in deposition within the feeding river

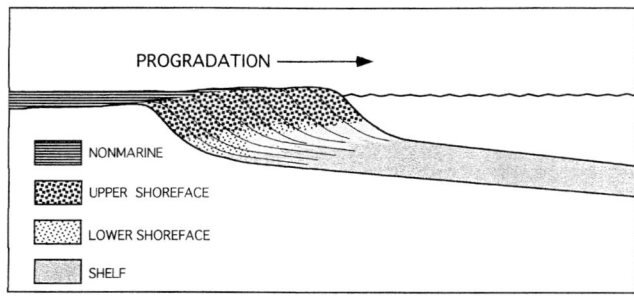

Figure C57 Characteristic mode of accumulation of shoreline deposits through progradation whereby shallow water facies build laterally over deeper water counterparts, generating an upward-shallowing succession or parasequence.

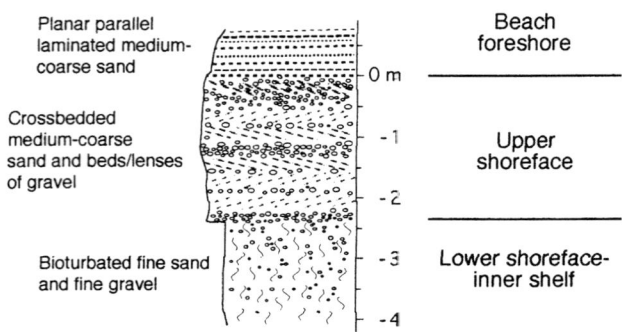

Figure C59 Shoreface succession on a prograding coarse, pebbly, low-energy coast (Pleistocene, southeastern Spain).

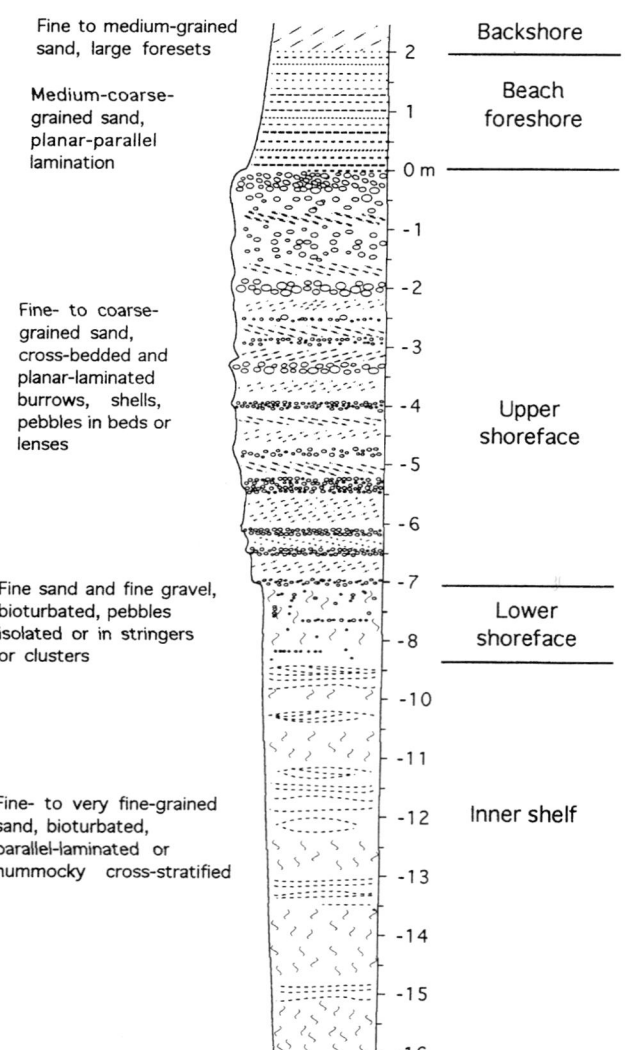

Figure C58 Shoreface succession on a prograding pebbly, high-energy coast (Pleistocene, Central California).

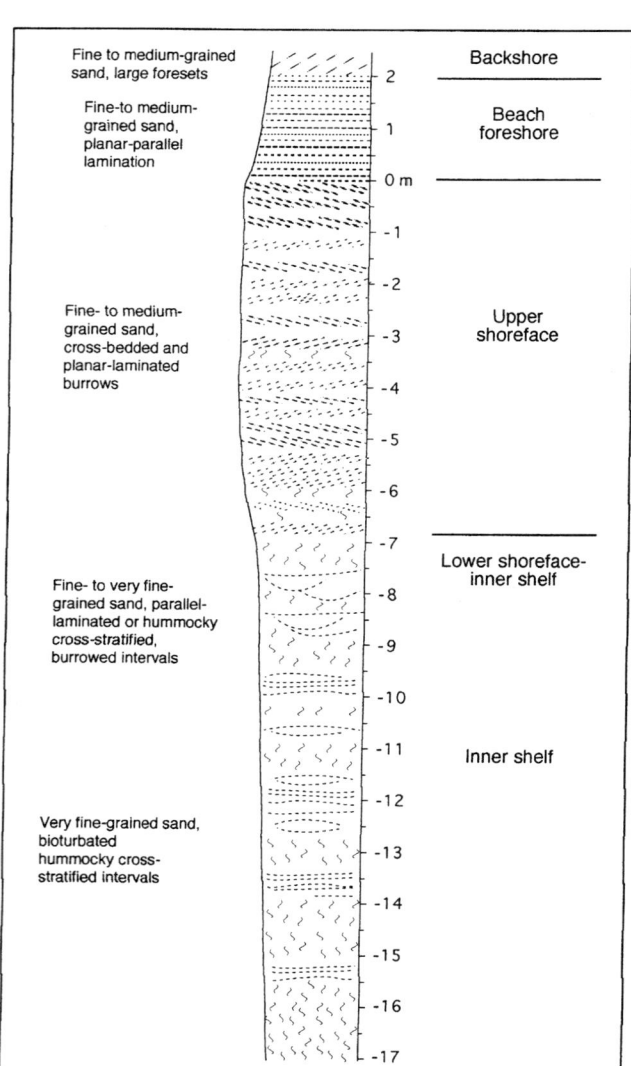

Figure C60 Typical shoreface succession on a prograding finer-grained, non-pebbly, unbarred, high-energy coast (hypothetical).

COASTAL SEDIMENTARY FACIES 155

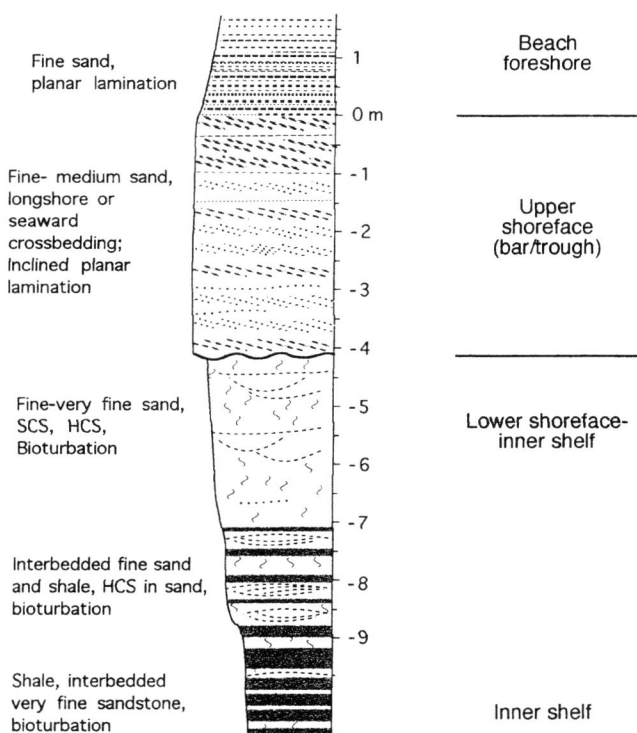

Figure C61 Shoreface succession on a prograding fine-grained, non-pebbly, moderate-low energy coast (Cretaceous, Book Cliffs, Utah).

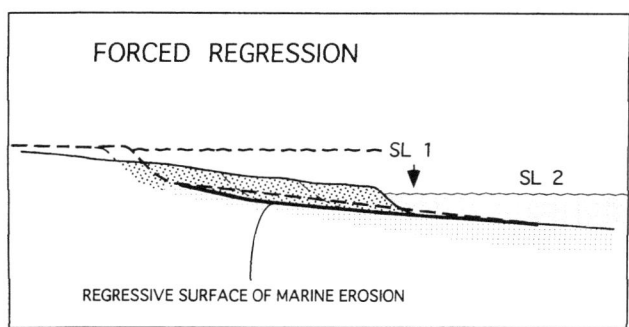

Figure C63 Progradation accompanied by a fall in sea level (forced regression) occurs over a regressive surface of marine erosion, cut as wave base is lowered.

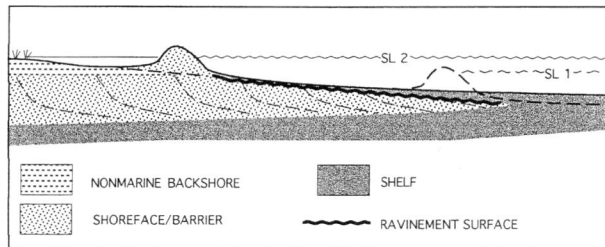

Figure C64 Transgressive erosion caused by the lateral translation of the shoreface (see Brunn, 1962). None of the transgressive coastal facies are preserved, and the transgression is recorded by an erosional ("ravinement") surface overlain by deeper water (typically shelf) deposition.

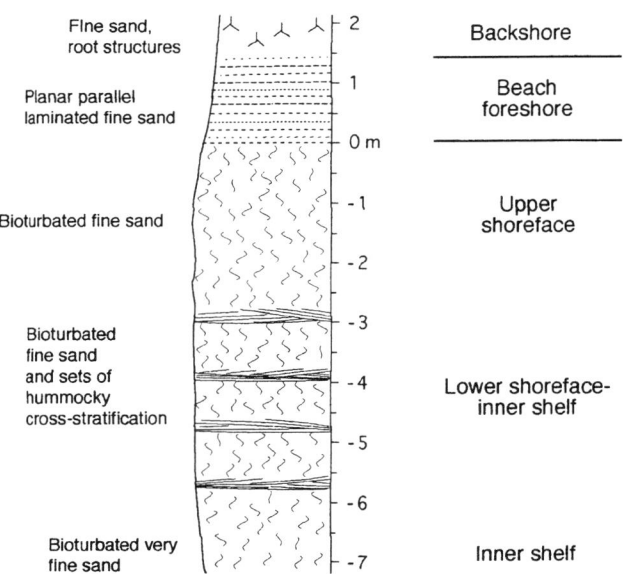

Figure C62 Shoreface succession on a prograding very fine-grained, low-energy coast (Eocene, West Texas).

systems. With less sediment contributed to the coast, marine transgression occurs, whereby the shoreline shifts landward. Where the sand supply is limited, landward transport by shoaling waves and coastal winds concentrates it in a shoreface-foredune ridge system. In most wave-dominated settings, enough sand exists that a small rise in relative sea level is met by an upbuilding of the shoreline to a topographically higher position. If the sediment input is sufficiently reduced, a basin develops behind the topographically high beach ridge and a barrier island is created. Most present-day coasts fronted by barrier islands are in a state of transgression.

Barrier islands formed during transgression have a limited potential for preservation. The landward translation of a shoreface erases the deposits that lie landward from its base, leaving an erosional (*ravinement*) surface overlain by shelf facies (Figure C64). Commonly this surface is the only preserved expression of a transgression.

Under conditions of rapid accommodation and sufficient sedimentation, however, lagoonal and inlet facies are preserved (Figure C65). These may occur in inverted succession whereby the landward-most facies (lagoonal marshes and back-bay muds) are overlain by sandier deposits associated with the approach of the transgressing barrier and capped by tidal inlet facies just below the ravinement surface (Figure C66). The thickness and organization of these

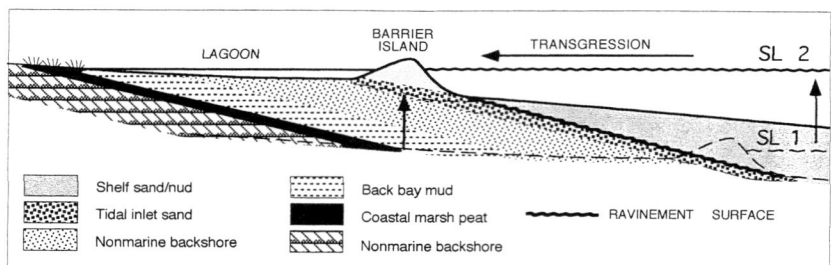

Figure C65 Transgressive succession produced in settings with rapid accommodation and adequate sedimentation. A lagoonal succession develops as the lagoonal facies are buried by the landward-migrating barrier. It is capped by a "ravinement" surface overlain by shelf facies.

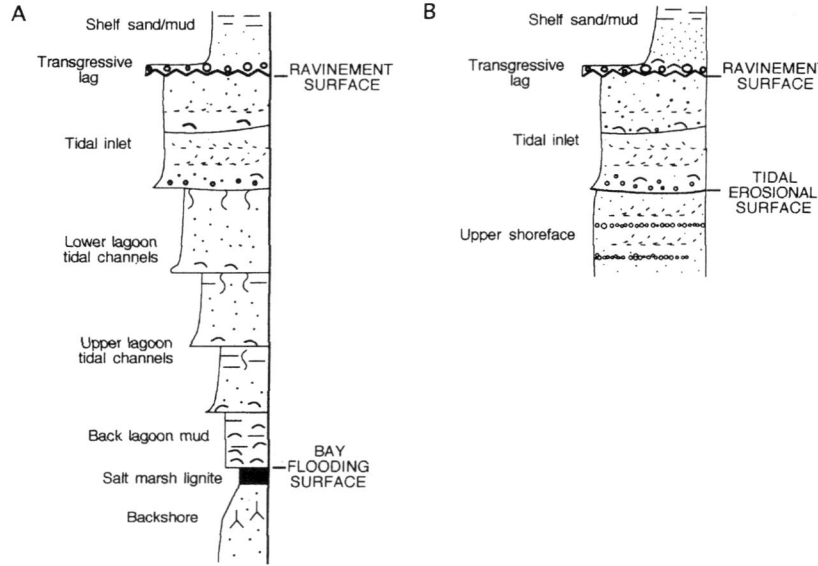

Figure C66 Vertical successions produced by preservation of transgressing lagoons. A. High rate of accommodation. B. Modest rate of accommodation.

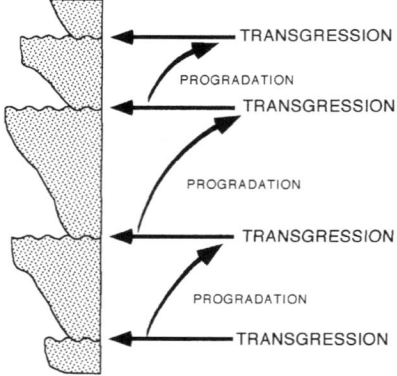

Figure C67 Typical pattern of stacked shoreline successions. Shallowing-upward progradational parasequences are separated by transgressive surfaces of erosion (ravinement surfaces). Falls in sea level prior to transgressions can produce erosional sequence boundaries and incised-valley fill deposits at the top of the parasequences.

transgressive lagoonal deposits differs depending on the balance between accommodation and sedimentation. Although most are but a few meters thick, some can exceed 20 m. Most display a basal "bay-flooding surface" and a capping ravinement surface, although vigorous tidal currents and low rates of accommodation may in some places lead to the erosion of the lower part of the succession (Figure C66).

Most successions of coastal deposits consist of stacked sets of progradational deposits separated by erosional (ravinement) surfaces (Figure C67). The degree of preservation of individual progradational intervals depends on the amount of accommodation provided during the interval between progradation and transgression (Figure C68).

H. Edward Clifton

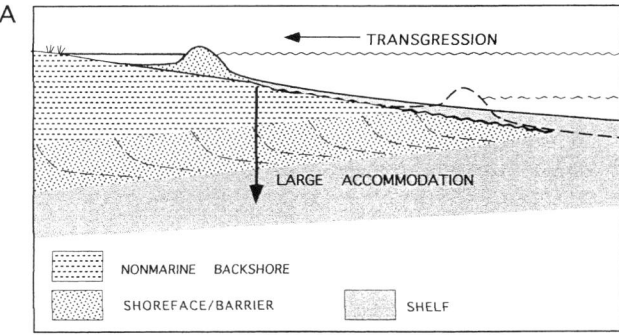

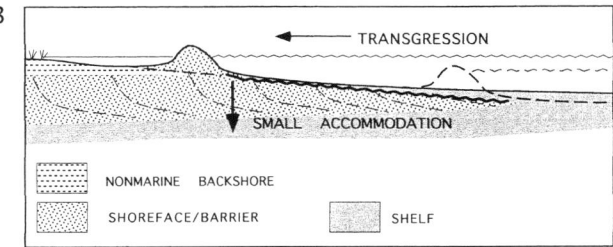

Figure C68 Preservation of progradational sequences as a function of accommodation in an area of alternating regression and transgression. Where the amount of accommodation is large (subsidence is rapid relative to the rate of sea-level change), complete parasequences, including nonmarine facies at their tops are preserved. Where the amount of accommodation is small, erosion associated with the marine transgression is likely to remove a significant part of the parasequence deposited as part of the previous regression. Such "beheaded" successions could prove difficult to interpret. The location relative to the point of maximum regression is also a factor. Near the turnaround to transgression, much of regressive shoreline deposit is likely to be lost.

Bibliography

Antia, E.E., Flemming, B.W., and Wefer, G., 1994. Sedimentary facies associations of the shoreface-connected ridge systems in the German Bight (southern North Sea). *Deutches Hydrographiche Zeitschrift*, **46**: 229–244.

Augustinius, P.G.E.F., 1989. Cheniers and chenier plains: a general introduction. *Marine Geology*, **90**: 219–229.

Brunn, P., 1962. Sea-level rise as cause of shore erosion. *Journal of Waterways, Harbor Division, American Society of Civil Engineers*, **88**: 117–130.

Clifton, H.E., Hunter, R.E., and Phillips, R.L., 1971. Depositional structures and processes in the nonbarred high-energy nearshore. *Journal of Sedimentary Petrology*, **41**: 651–670.

Cobb, J.C., and Cecil, C.B., 1993. *Modern and Ancient Coal-forming Environments*. Geological Society of America, Special Publication, 286, 198p.

Demicco, R.V., and Hardy, L.A., 1994. *Sedimentary Structures and Early Diagenetic Features of Shallow Marine Carbonate Deposits*. SEPM (Society for Sedimentary Geology), Atlas Series No. 1, 265 p.

Galloway, W.E., and Hobday, D.K., 1996. *Terrigenous Clastic Depositional Systems: Applications to Fossil Fuel and Groundwater Resources*, 2nd edn, New York: Springer.

Hill, G.W., and Hunter, R.E., 1976. Interaction of biological and geological processes in the beach and wearshore environment, northern Padre Island. In Davis, R.A., and Ethington, R.L. (eds.), *Beach and Nearshore Processes*. SEPM Special Publication, 24, pp. 169–187.

Hill, P.R., Barnes, P.W., Héquette, A., and Ruz, M.-H., 1995. Arctic coastal plain shorelines. In Carter, R.W.G., and Woodroffe, C.D. (eds.), *Coastal Evolution: Late Quaternary Shoreline Morphodynamics*. Cambridge University Press, chapter 9, pp. 341–372.

Hine, A.C., 1975. Bedform distribution and migration patterns on tidal deltas in the Chatham Harbor estuary, Cape Cod, Massachusetts. In Cronin, L.E. (ed.), *Estuarine Research, v. II, Geology and Engineering*. London: Academic Press, pp. 235–252.

Howard, J.D., and Reineck, H.-E., 1981. Depositional facies of high-energy beach to offshore sequence: comparison with the low-energy sequence. *American Association of Petroleum Geologists Bulletin*, **65**: 807–830.

Johnson, D.W., 1919. *Shore Processes and Shoreline Development*. John Wiley and Sons.

Komar, P.D., 1976. *Beach Processes and Sedimentation*. Englewood Cliffs, New Jersey: Prentice-Hall.

McBride, R.A., and Moslow, T.F., 1991. Origin, evolution, and distribution of shoreface sand ridge, Atlantic inner shelf, USA. *Marine Geology*, **21**: 1063–1066.

Niedoroda, A.W., and Swift, D.J.P., 1981. Maintenance of the shoreface by wave orbital currents and mean flow. *Geophysical Research Letters*, **8**: 337–348.

Niedoroda, A.W., Swift, D.J.P., Hopkins, 1985. The shoreface. In Davis, R.A. Jr. (ed.), *Coastal Sedimentary Environments*, 2nd edn, New York: Springer-Verlag, pp. 533–624.

Pemberton, S.G., Van Waggoner, J.C., and Wach, G.D., 1992. Ichnofacies of a wave-dominated shoreline. In Pemberton S.G. (ed.), *Applications of Ichnology to Petroleum Exploration*. SEPM Core Workshop No. 17, pp. 339–382.

Pemberton, S.G., and Wightman, D.M., 1992. Ichnologic characteristics of brackish water deposits. In Pemberton, S.G. (ed.), *Applications of Ichnology to Petroleum Exploration*, SEPM Core Workshop No. 17, pp. 141–167.

Reading, H.G., 1996. *Sedimentary Environments: Processes, Facies, and Stratigraphy*. 3rd edition, Blackwell Scientific.

Reineck, H.-E., and Singh, I.B., 1973. *Depositional Sedimentary Environments – with Reference to Terrigenous Clastics*. Berlin: Springer-Verlag.

Rine, J.M., and Ginsburg, R.N., 1985. Depositional facies of a mud shoreface in Surinam, South America—a mud analogue to sandy shallow-marine deposits. *Journal of Sedimentary Petrology*, **55**: 633–652.

Rine, J.M., Tillman, R.W., Stubblefeld, W.L., and Swift, D.J.P., 1991. Lithostratigraphy of Holocene sand ridges from the nearshore and middle continental shelf of New Jersey, USA. In Moslow, T.F., and Rhodes, E.G. (eds.), *Modern and Ancient Shelf Clastics: A Core Workshop*. SEPM Core Workshop No. 9, pp. 1–72.

Swift, D.J.D.P., Niedoroda, A.W., Vincent, C.E., and Hopkins, T.S., 1985. Barrier island evolution, Middle Atlantic Shelf, USA, Part 1: shoreface dynamics. *Marine Geology*, **63**: 331–361.

Walker, R.G., and James, N.P., 1992. *Facies Models Response to Sea Level Change*. Geological Association of Canada.

COLLOIDAL PROPERTIES OF SEDIMENTS

Sediment is a mass of organic or inorganic solid fragmented material that comes from the weathering of rock and is carried by, suspended in, or dropped by air, water, or ice, and that forms layers on the earth's surface as gravel, sand, silts, or mud. In sediments, the clay-sized and the fine silt-sized particles are classified as colloids, which are made up of particles having dimensions of 10–10,000Å. The term "colloids" indicates microparticles or macromolecules that are small enough to move primarily by Brownian motion, rather than by gravitational settling, and are large enough to provide a microscopic "environment" into or onto which molecules of interest can escape the aqueous solution. (The theoretical gravitational settling velocity of colloid particles is less than $0.01\,\mathrm{cm\,s^{-1}}$.) Colloids found in soil and sediment are aggregates of highly disperse or loosely cohering organic, mineral and organo-mineral particles, which are capable of acting as sorbents and as ion exchangers.

Colloids are ubiquitous, they are found everywhere in concentrations of above 10^7 to 10^8 particles per liter of water. Colloidal dispersions bridge the gap between the true dissolved solutes and true bulk phases. The different varieties of colloidal materials found in natural system are: (i) soil colloids (aluminosilicates; humus and colloidal humic acids; iron hydrous oxides; polymeric coating of soil particles by humus, by hydrous Fe(III) oxides and hydroxy-Al(III) compounds); (ii) river-borne particles (weathering products and soil colloids, along with colloidal Fe(III) oxides stabilized by humic and fulvic acids; phytoplankton, biological debris, and colloidal humic acid); and (iii) sulfide and polysulfide colloids in anoxic sediments.

The fine size of the colloidal particles is responsible for their very high surface area/volume ratio. Chemical and physical reactions are promoted in colloidal suspension due to the large area of solid–liquid interface characteristic of systems containing many small particles. The high surface area of the colloids also means that there are many broken bonds, and both positively and negatively charged ions and molecules will be present on the solid surface. According to Gouy–Chapman model, the excess ions are non-uniformly distributed in the vicinity of the colloid surface, their concentration is the largest at the surface, and decreasing non-linearly till they reach bulk concentration (Stumm, 1992). The net charge on the solid colloidal particles will be the sum of all these surface charges together with the contributions from the adsorbed ions. The charge will depend on the chemical and crystallographic structure of the solid and on the pH of the solution surrounding the solid (Yariv and Cross, 1979). For each type of colloidal solid, there is a pH value at which the surface concentrations of H^+ and OH^- are equal and there is no net charge on the particle, called the point of zero charge (PZC); in the absence of ions other than H^+ and OH^-, the PZC will be at a pH referred to as the isoelectric point (IEP). At pH values above the IEP, the colloidal particles will carry a net negative charge, and the particles will have a net positive charge at pH values below the IEP.

Colloidal sediments can be grouped into two distinct classes as lypophilic (i.e., liquid loving) and lyophobic (liquid-hating). Lyophilic colloids display strong affinity between the dispersed particles and the liquid. The liquid is strongly adsorbed to the surface of the particle, so the interfacial tension between the particle and the liquid is very similar to that between the liquid molecules, leading to an intrinsically stable system. In case of lyophobic colloids, the liquid does not show much affinity for the particles. If attractive forces exist between the particles, there will be a strong tendency for them to stick together when they come in contact, rendering the system generally unstable, and resulting in flocculation (q.v.). A system containing colloidal particles is said to be stable if during the period of observation it is slow in changing its state of dispersion. Colloidal stability can be affected by the presence of electrolytes, adsorbates that affect the surface charge of the colloids, polymers that affect particle interaction by forming bridges between them or sterically stabilizing them. Colloidal flocculation, induced by a rise in electrolyte concentration, is an important sedimentation process, and has significant implications regarding dispersion of potential pollutants that may be bound to the charged particle surface.

Interactions of different interfacial forces, such as electrostatic repulsion, van der Waals attraction, hydration phenomenon, hydrodynamic interactions on particle collision, and forces due to adsorbed substances dictate colloidal stability, and these factors can be categorized into two groups as (i) the frequency of collisions between particles and (ii) the probability of particles adhering if they collide. Particles in suspension collide with each other as a consequence of at least three mechanisms of particle transport: (1) particles move because of their thermal energy (Brownian motion); (2) if colloids are sufficiently large or the fluid shear rate is high, the relative motion from velocity gradients exceeds that caused by Brownian effects; (3) in settling, particles of different gravitational settling velocities may collide and agglomerate (Ryan and Elimelech, 1996).

Colloids are stable when they are electrically charged. The stability is usually explained by Derjaguin–Landau–Verwey–Overbeek (DLVO) theory. The attractive forces and repulsive forces between two particles are calculated as a function of particle–particle separation distance. The curves of potential energy versus particle–particle separation distance typically show an energy barrier that must be overcome by colliding particles to adhere. The primary attractive forces are dispersion forces, and the primary repulsive forces are long-range electrical forces and short-range Born repulsion. The double-layer potential energy is affected by the changes in solution chemistry, though the attractive van der Waals forces are independent of it. Short-range repulsive forces (up to a few nanometers separation distance) may also be affected by changes in solution chemistry (Ryan and Elimelech, 1996).

Figure C69 shows the plot of normalized potential energy as a function of separation distance between a colloid and collector. A typical total potential energy curve is characterized by a deep well at a very small separation distance, which is the primary minimum ($\phi_{\min 1}$), a repulsive energy barrier, which is the primary maximum ($\phi_{\max}$), and a shallow well at a larger separation distance ($\phi_{\min 2}$).

The attractive energy is primarily the result of London–van der Waals dispersion forces and is related to the polarizability of the particles and the distance of separation between the particles. For colloidal particles, the attractive energy declines with the separation according to

$$\phi^{\mathrm{vdW}} = -A_H/d^n \qquad \text{(Eq. 1)}$$

where A_H is the Hamaker constant, which depends on the difference in polarizability between dispersed particles and the dispersing medium, d is the separation, and n is an exponent, the value of which is $1<n<3$, depending on the geometry of the system (flat–flat, flat–sphere, sphere–sphere, etc.). If a colloidal particle and the dispersing solvent are of similar polarizability, then the tendency for the particles to seek out one another and to separate from the solvent is small and there is little tendency for the particles to adhere. The electrical repulsion energy is due to the repulsion of like-charged surfaces. There are three components of surface charge

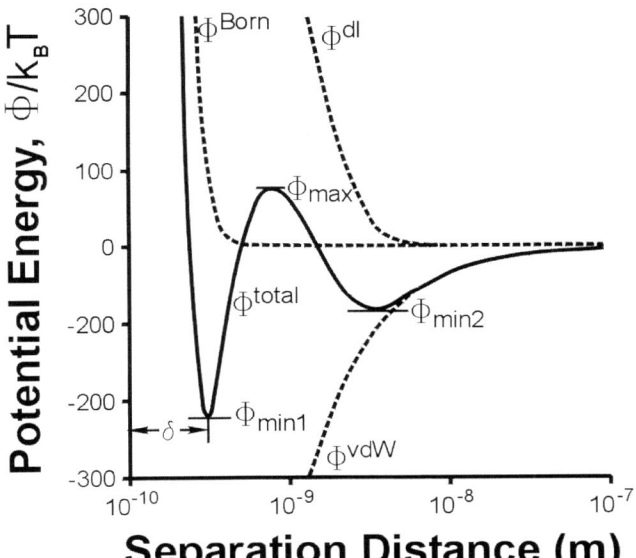

Figure C69 The resultant curve for the total potential energy, ϕ^{total} (solid curve), is the sum of the van der Waals potential energy, ϕ^{vdW}, the double layer potential energy, ϕ^{dl}, which is due to the repulsion between the particles of similar charge, and the Born potential energy, ϕ^{Born}, the short range repulsion energy. k_B is the Boltzmann's constant, and T is the absolute temperature.

resulting from specific chemical interactions or specific adsorption: (1) constant charge surfaces (e.g., clays); (2) constant potential surfaces (e.g., AgI); and (3) constant chemical affinity surfaces (e.g., oxides).

Another component of surface charge is counter charge that is held electrostatically to balance the specific-adsorption charge. The counter charge is usually treated through an electric double layer based on the Gouy–Chapman–Stern theory. According to this theory, the repulsive energy drops off exponentially with distance from the surface according to

$$\phi^{dl} \propto \exp(-\kappa d) \quad \text{(Eq. 2)}$$

where κ is the reciprocal thickness of double layer, which is proportional to the square root of the ionic strength, I, for a z : z electrolyte, $\kappa \propto I^{1/2}$.

Another repulsive interaction that has not been widely documented is steric repulsion. This repulsion occurs when particles that are coated with large polymers collide and the polymeric coatings overlap. Factors affecting the extent of the steric repulsion are the similarity of the coating to the solvent and to the particle, and the expansion or contraction of the polymer according to the properties of the solvent.

Finally, it can be said that colloids represent the most valuable soil fraction, as colloids can retain mineral nutrients (including trace elements) in the mobile state, and retain moisture and support formation of soil structure favorable to aeration. It is the colloidal component that dominates the leachable chemistry component of the sediments.

Sandip and Devamita Chattopadhyay

Bibliography

Ryan, J.N., and Elimelech, M., 1996. Colloid mobilization and transport in groundwater. *Colloids and Surfaces A: Physicochemical and Engineering Aspects*, **107**: 1–56.

Stumm, W., 1992. *Chemistry of the Solid–Water Interface: Processes at the Mineral–Water and Particle–Water Interface in Natural Systems*. New York: John Wiley & Sons, Inc.

Yariv, S., and Cross, H., 1979. *Geochemistry of Colloid Systems for Earth Scientists*. Berlin: Springer-Verlag.

Cross-references

Clay Mineralogy
Flocculation

COLORS OF SEDIMENTARY ROCKS

The colors of sedimentary rocks can have complex origins and in cases are secondary. However, colors are commonly primary and reflect important aspects of depositional environments including redox conditions and rates of deposition of organic matter. These parameters in turn affect fauna and thus a strong correlation exists between color and biofacies patterns, including macrofaunal distributions and burrowing type and depth (Leszcynski, 1993), particularly in deposits of oxygen-stratified basins (Myrow and Landing, 1992). Color may be useful for the interpretation of variations in such factors as relative sea level, oceanic circulation, sedimentation rate and primary productivity.

Colors are generally controlled by accessory minerals and compounds of iron and organic carbon (see reviews by Pettijohn, 1975; Potter *et al.*, 1980; and Myrow, 1990). Compounds of other transition metals (e.g., Ti, Mn, Co, Cu, and Zn) also impart a pigment in some cases. For the most part the colors of sediment and sedimentary rock fall within two spectra: green-gray to red and olive-gray to black (Figure C70). The first color spectrum is controlled by the oxidation state of iron, specifically the Fe^{+3}/Fe^{+2} ratio in various minerals, and not the total iron content (Tomlinson, 1916; McBride, 1974). In particular, the color is a function of the mole fraction representing the proportion of iron in the +2 state ($Fe^{+2}/Fe^{+2} + Fe^{+3}$) per gram of rock. In relatively unoxidized strata, green colors result from the presence of iron-bearing phyllosilicates such as chlorite, illite, and in cases, glauconite. Kaolinite and smectite provide for white to light neutral colors. Red coloration is due to the presence of hematite, whereas less common yellows and browns generally result from limonite and goethite, respectively. Red hematitic staining is generally an early diagenetic phenomenon resulting from: (1) dehydration reactions in which limonite stains on detrital particles are altered to hematite; (2) dissolution of iron silicates and precipitation of the released iron; and (3) direct oxidation of magnetite and ilmenite grains (Hubert and Reed, 1978). Very low levels of hematite can impart a deep red coloration to a rock. The conversion from red to green colors occurs by reduction of the iron, which is either carried away in solution or reprecipitated as iron-rich clay minerals such as chlorite (Thompson, 1970). The strength of red or green color is a function of grain size, with fine-grained rocks having higher iron content and thus more intense color. This

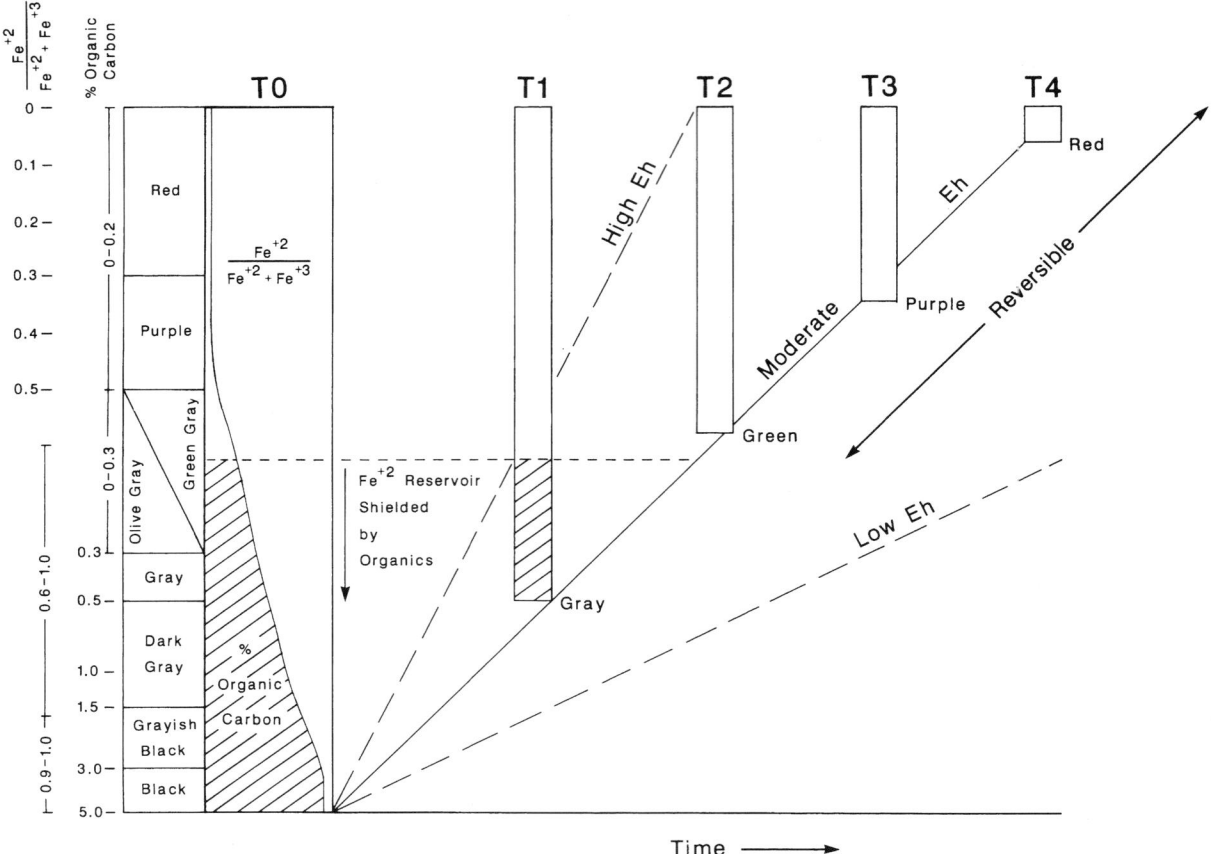

Figure C70 Graph relating color of rock, organic content and oxidation state of iron to time. "Time" refers to the length of time that pore fluids interact with sediment prior to lithification. Dark colors, olive-gray to black, are controlled by organic content. Green-gray to red colors, controlled by the oxidation state of iron, are reversible. The color history of sediment is dependent upon the Eh conditions. Three representative paths (high, moderate and low Eh) are shown for sediment with initially high organic content. Figure from Myrow (1990).

relationship is due to the presence of amorphous or poorly crystalline iron oxides attached to clays (McPherson, 1980). Changes in the oxidation state of iron, and hence the color, result from interactions with altering fluids at any time in the postdepositional history of a unit, and are thus controlled in part by permeability. Oxidation or reduction spots are due to incomplete diagenetic alteration of a layer.

The green-gray to black color spectrum in rocks is a function of total organic carbon (TOC), with darker colors corresponding to higher carbon content. This relationship is empirically confirmed in studies of modern sediment (e.g., Sheu and Presley, 1986). Main controls on organic carbon content include accumulation rate of organic matter, sediment accumulation rate, rate of decay of organics, and oxygen levels (Potter *et al.*, 1980). Diagenetic loss of carbon, which leads to lighter colors, is an irreversible process. The carbon reservoir acts to shield the iron-bearing minerals from oxidation until most of the carbon is oxidized, at which point the green to red color spectrum is found. Oxidative loss of carbon, and red coloration, occurs when bottom water and/or pore-fluids are highly oxygenated. Green colors result from deposition of sediment with low organic content and weakly reducing to oxidizing conditions. Gray to black colors are associated with high TOC and dysaerobic to anaerobic bottom waters.

Gray colors also occur as a result of the presence of disseminated pyrite. During burial diagenesis metastable minerals such as mackinawite (FeS) and greigite (FeS_4) are transformed into submicron sized framboids of pyrite (Berner, 1971). Thermal maturity may also be an important factor in some rock. For instance, Lyons (1988) demonstrated that the black color of some limestone is due to carbonization of very small quantities of organic matter (<0.06 percent TOC). Darker colors may also be imparted by Mn oxides or Fe-Mn oxides such as pyrolusite. Shale with siderite is gray to bluish on fresh surfaces but altered to brown tones by weathering (Pettijohn, 1975).

Paul Myrow

Bibliography

Berner, R.A., 1971. *Principles of Chemical Sedimentology*. McGraw-Hill Book Company.

Hubert, J.F., and Reed, A.A., 1978. Red-bed diagenesis in the East Berlin Formation, Newark Group, Connecticut Valley. *Journal of Sedimentary Petrology*, **48**: 175–184.

Leszcynski, S., 1993. Ichnocoenosis versus colour in Upper Albian to Lower Eocene turbidites, Guipuzcoa province, northern Spain. *Paleogeography, Paleoclimatology, and Paleoecology*, **100**: 251–265.

Lyons, T.W., 1988. Color and fetidness in fine-grained carbonate rock. *Geological Society of America Abstracts with Programs*, **20**: A211.

McBride, E.F., 1974. Significance of color in red, green, purple, olive, brown and gray beds of Difunta Group, northeastern Mexico. *Journal of Sedimentary Petrology*, **44**: 760–773.

McPherson, J.G., 1980. Genesis of variegated redbeds in fluvial Aztec Siltstone (Late Devonian), Southern Victoria Land, Antarctica. *Sedimentary Geology*, **27**: 119–142.

Myrow, P.M., 1990. A new graph for understanding colors of mudrocks and shales. *Journal of Geological Education*, **38**: 16–20.

Myrow, P.M., and Landing, E., 1992. Mixed siliciclastic-carbonate deposition in an Early Cambrian oxygen-stratified basin, Chapel Island Formation, southeastern Newfouldland. *Journal of Sedimentary Petrology*, **62**: 455–473.

Pettijohn, F.J., 1975. *Sedimentary Rocks*. Harper and Row.

Potter, P.E., Maynard, J.B., and Pryor, W.A., 1980. *Sedimentology of Shale*. Springer-Verlag.

Sheu, D.-D., and Presley, B.J., 1986. Variations of calcium carbonate, organic carbon and iron sulfides in anoxic sediment from the Orca Basin, Gulf of Mexico. *Marine Geology*, **70**: 103–118.

Thompson, A.M., 1970. Geochemistry of color genesis in red-bed sequence, Juniata and Bald Eagle formations, Pennsylvania. *Journal of Sedimentary Petrology*, **40**: 599–615.

Tomlinson, C.W., 1916. The origin of red beds. *Journal of Geology*, **24**: 153–179.

Cross-references

Black Shales
Chlorite in Sediments
Red Beds
Sulfide Minerals in Sediments

COMPACTION (CONSOLIDATION) OF SEDIMENTS

Compaction may be defined as the process by which the sediment volume is reduced and the sediment density increased. Sediments change their physical properties after deposition due to the stress from the overburden (gravitational compaction) and as a result of biological or chemical reactions involving dissolution and precipitation of minerals. In some sediments compaction of amorphous materials like organic material (kerogen) and biogenic silica (Opal A) also play an important role. All natural processes modifying sediments after deposition are in geological literature referred to as diagenesis (*q.v.*) and this term also includes compaction processes. Soil engineers use the term compaction to describe densifications on the laboratory and in the field by vibrations of mechanical equipment to reduce the pore volume and consolidation for the expulsion and flow of water in soils. Geologist tends to use the term consolidation to describe a strengthening of the rocks, often by precipitation of cements (lithification).

Compaction of sediments and rocks (soil and rock mechanics) has been developed as a quantitative engineering subject since Terzaghi (1925) from the need to predict the response of sediments due to changes in stress caused by different types of constructions. Laboratory tests determining the mechanical properties of different types of sediments (soils) have formed the basis for such quantitative prediction (Lambe and Whitman, 1979). Such tests measure the friction coefficients between grains and the mechanical strength of grains. Later Terzaghi's principles were applied to petroleum engineering and structural geology (Hubbert and Rubey, 1959). The terminology used in the engineering literature is rather different from that used by most geologists and that may cause confusion.

Sedimentologists and petroleum geologists have taken an increasing interest in sediment compaction. This is stimulated by the need to predict the porosity of reservoir rocks as a function of burial history in sedimentary basins (Wilson, 1994). The densities of the different lithologies are required to compute subsidence and backstripping in basin modeling (Giles, 1997; Sclater and Christie, 1980). There is also a need to understand the distribution of rock strength, seismic velocities and attributes in sedimentary basins.

Until recently, sediment compaction has been mostly treated as mechanical compaction as a function of effective stress. Compaction processes involving chemical dissolution and precipitation of grains (chemical compaction) are principally different since chemical processes then control the rate of compaction and the strength of the rocks.

It is much more difficult to study chemical compaction in the laboratory, particularly in silicate rocks (i.e., sandstones) due to the slow kinetic reaction rates involved in the compaction processes at low temperatures.

Clastic sediments are granular materials and the grains may be rock fragments, minerals or amorphous materials. The fluids between the grains may be water, oil or gasses or a mixture of these fluids.

Porosity (φ) is the volume fraction of the fluids in the sediment. The volume fraction of the solids then becomes $1-\varphi$. The ratio between the volume of the pore space (void) filled with fluids and the solid grains and matrix (mainly clay minerals) is also referred to as the void ratio (v_r):

$$v_r = \varphi/(1 - \varphi) \qquad \text{(Eq. 1)}$$

Stresses in sedimentary basins

The effective stress of the overburden is an average of the stress transmitted through the solid phases through a network of load-bearing grains. The stress at grain contacts may be very high because of small contact areas and focused on some grains, which may then bend or fracture. With increasing compaction the rearrangements or breakage of grains will increase both the contact area and the number of grain contacts. Precipitation of cement in the grain surfaces increases the area of contact further so that the stress per contact is reduced enough to prevent further grain crushing or mechanical compaction all together.

The total overburden stress (σ_v) at a certain depth in sedimentary basins is equal to the weight of the overlying rock and fluids per unit area and is often referred to as the lithostatic stress.

At a certain depth (Z_s) below the seafloor in a sedimentary basin the total vertical stress is:

$$\sigma_v = Z_s \cdot \rho_b + Z_w \cdot \rho_w \qquad \text{(Eq. 2)}$$

Here the average bulk density of the sedimentary rocks over a depth range Z_w is ρ_b. The water depth above the seafloor is Z_w and the density of water (ρ_w).

The bulk density (ρ_b) of the rocks varies as a function of the porosity (ϕ), the density of the fluid (ρ_f) in the pore space, and the density of the solid phase (ρ_m), which are mainly minerals:

$$\rho_b = \rho_r \phi + \rho_m (1 - \phi) \quad \text{(Eq. 3)}$$

Usually the density of the mineral matrix in sandstones and shales is close to 2.65–2.70 g/cm^3 because this is the density of the most common minerals. If there are significant contents of denser minerals such as siderite or pyrite the bulk density will be higher. Smectite and mixed layer minerals have variable but generally lower densities. The fluid density also varies with the composition of water and petroleum. The weight per unit area of the rock column including fluids is called the lithostatic pressure (stress).

The total overburden stress (load) (σ_v) can be obtained by integrating the sediment density (ρ) over the entire overburden sequence (Z) or by estimating an average density (ρ_s). This stress is partly carried by the grain framework (solid phase) which is called the effective stress (σ_e) and by the stress (pressure) carried by the fluid phase which is pore pressure (P_p)

$$\sigma_v = \sigma_v e + P_p \quad \text{(Eq. 4)}$$

The effective stress is thus

$$\sigma_v e = \sigma_v - P_p \quad \text{(Eq. 5)}$$

This was first formulated by Terzaghi in 1925 and it formed the basis for modern soil and rock mechanics.

If the average sediment density (ρ_s) in the upper 3 km in a sedimentary basin is 2.0 g/cm^3, the total stress at 3 km depth is 60 MPa. At hydrostatic conditions (no overpressure) the pore pressure (P_p) is equal to the average density of the pore fluid multiplied by the height (Z) of a water column in the sediments and any water column above (water depth). Saline water may have a density of 1.033 g/cm^3 and the pore pressure at 3 km depth is then 31 MPa and the effective stress (σ_e) is then 29 MPa at 3 km depth. For very fine-grained sediments and rocks with low porosity and poorly connected pores, the fluid phase will carry a smaller fraction of the overburden load following poro-elastic theory. A constant α (Biot constant) was therefore introduced in the Terzaghi's law:

$$\sigma_e = \sigma - \alpha P_p \quad \text{(Eq. 6)}$$

For porous and compressible sediments α may be taken to be 1, but for stiff sedimentary rocks α may have significantly lower values meaning that more of the total load is carried by the solid phase.

Geomechanical properties

The reduction in porosity with burial depth may be described as an exponential function following Athy (1930):

$$\phi = \phi_0 e^{-cZ} \quad \text{(Eq. 7)}$$

Here the porosity (ϕ) is an exponential function of an initial porosity ϕ_0, a compressibility parameter c and the burial depth (Z). This may be a useful approximate expression for compaction trends in sedimentary basins and it is useful in mathematical modeling. However there is nothing in the mechanical and chemical compaction processes that should cause the porosity to be a negative exponential function of depth or the effective stress. Sedimentary basins consist of sedimentary layers with different compressibility, and the compaction curves showing observed porosity loss with depth may be very complex.

In the laboratory we can measure the induced deformation (strain) in a sample of sediment or rock for a given effective stress and construct stress–strain curves. Experimental data show degree of time dependent compaction (creep) at constant stress. In sedimentary basins the rates of increase in effective stress due to loading are very slow compared to the experimental loading rates in the laboratory and it is somewhat uncertain how significant mechanical creep is in a sedimentary basin.

If one assumes that the grain structure may be considered linearly elastic and isotropic for very small deformations, one may define the following deformational characteristics: Young's modulus (E) is the ratio between increase in normal effective stress ($\Delta\sigma$) and the resulting strain ($\Delta\varepsilon_z$) in the stress direction (z), when there is no change in the orthogonal normal stresses.

$$E = \Delta\sigma_z / \Delta\varepsilon_z \quad \text{(Eq. 8)}$$

The bulk modulus (K) is the ratio between the increase in equal all-round stress ($\Delta\sigma$) and the resulting volumetric compression, $\Delta\varepsilon_{vol} = \Delta V/V$:

$$K = \Delta\sigma / \Delta\varepsilon_{vol} \quad \text{(Eq. 9)}$$

The bulk compressibility (c) is the inverse of the bulk modulus K and is a parameter describing the rate of compaction as a function of stress. In sedimentary basins with only minor tectonic stress, the vertical stress (σ_v) is normally the principle stress as in uniaxial compression tests. The compressibility of a sedimentary layer may the be expressed as the shortening (ΔL) over a bed thickness L and the strain (ε) is then ($\Delta L/L$).

The ratio between the lateral expansion (lateral strain ε_l) and vertical shortening (axial strain ε_a) is expressed as the Poissons ratio (v):

$$v = (\varepsilon_l / \varepsilon_a) \quad \text{(Eq. 10)}$$

If the lateral expansion in both directions is equal to half the shortening, there is no change in volume. This corresponds to $v = 0.5$.

For uniaxial compaction (one dimensional compaction) which may be assumed for wide sedimentary basins, the lateral strain is zero and a lateral effective stress will be built up. For linearly elastic and isotropic grain structure, the lateral effective stress increase is:

$$\Delta\sigma_{h_e} = \Delta\sigma_{ve} \cdot v/(1-v)$$

The compaction processes determine rock properties such as rock strength and the mechanical responses to changes in stress during drilling and production of petroleum.

Compaction processes in sedimentary rocks also determine well log responses and the velocity of seismic signals and their attributes.

Types of compaction

Modeling of sediment compaction requires that we understand the processes involved. A clear distinction must be made between purely mechanical compaction and chemical compaction:

(1) Mechanical compaction
 (a) Rearrangement (sliding and reorientation) of grains.
 (b) Bending and ductile deformation of grains.
 (c) Breakage of grains (grain crushing).
(2) Chemical compaction
 (a) Dissolution at contacts between grains or along stylolites (pressure solution) and a corresponding precipitation on minerals (cement) in the pore space.
 (b) Dissolution of thermodynamically unstable grains that are load-bearing.

Mechanical compaction

The vertical stress is normally the maximum stress (major principle stress) in sedimentary basins that are not subjected to strong tectonic compression, extension or uplift. During progressive burial at hydrostatic pore pressures (drained conditions), the sedimentary rocks always experience increasing effective stress, and such sediments are referred to as normally consolidated. Sediments that previously have experienced higher effective stresses than at present are referred to as over-consolidated.

The term under-consolidation is sometimes used to characterize poorly compacted sediments, particularly mud with high pore pressure. This term should however be avoided since the degree of consolidation always refers to the maximum effective stress sediments have experienced, and they can therefore not be under-consolidated. The term under-compaction may also cause confusion.

The degree of compaction (porosity loss) varies greatly with the lithology at the same effective stress (Figure C71). During uplift and erosion or a build up of over-pressure (higher than hydrostatic pore pressures) the effective stress may be reduced so that the sediments become over-consolidated and brittle. Sediments are likely to respond in a ductile manner also in

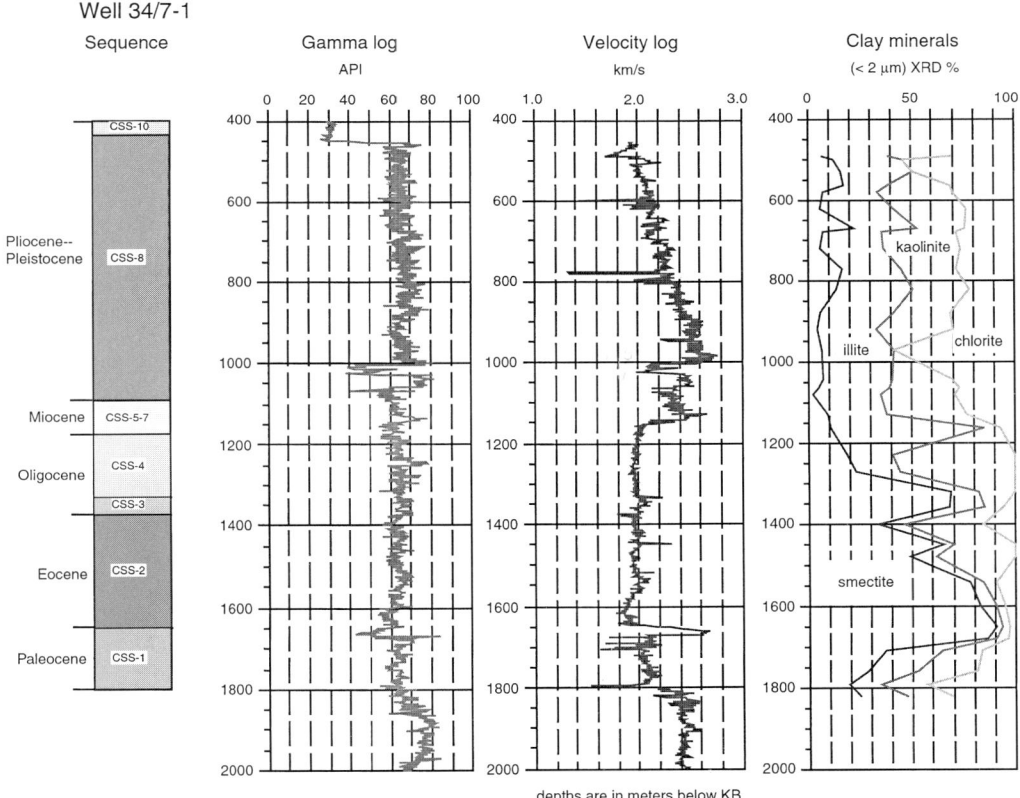

Figure C71 The velocity from well logs (or seismic data) may be a good indicator of the degree of compaction. The velocity in a well penetrating tertiary sediments from the Northern North Sea show that Pliocene and Pleistocene muds show a normal compaction trend (higher velocities) with depth. Smectite rich Eocene and Oligocene muds have very much lower velocities and is less compacted despite greater burial depth. This may be partly due to overpressure but is mostly due to the poor compressibility of smectitic clays. Modified from Thyberg et al. (2000).

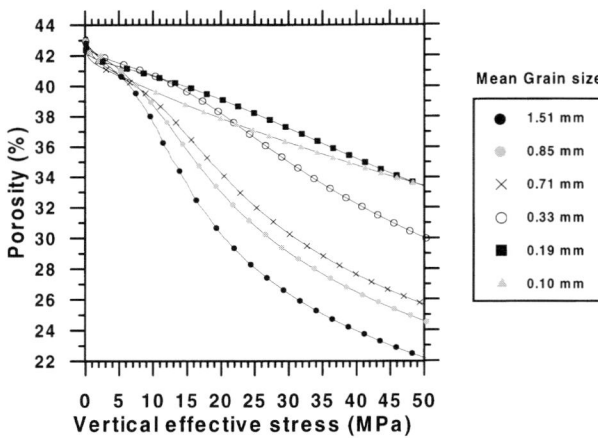

Figure C72 Experimental uniaxial compaction of well sorted quartz rich and showing that coarse sands have a much higher porosity loss than fine grained sand. From Chuhan et al. (2002).

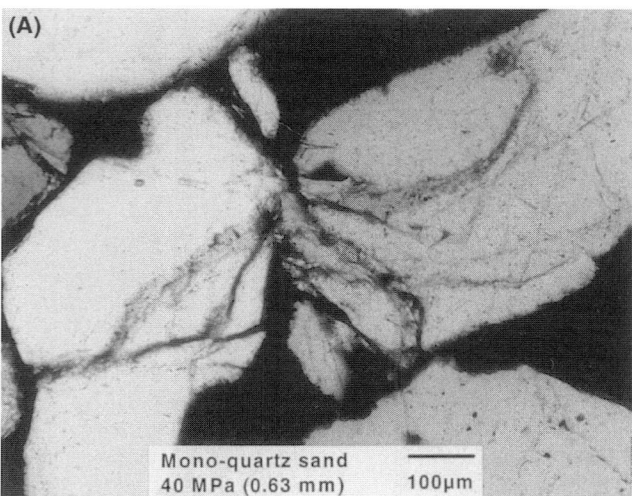

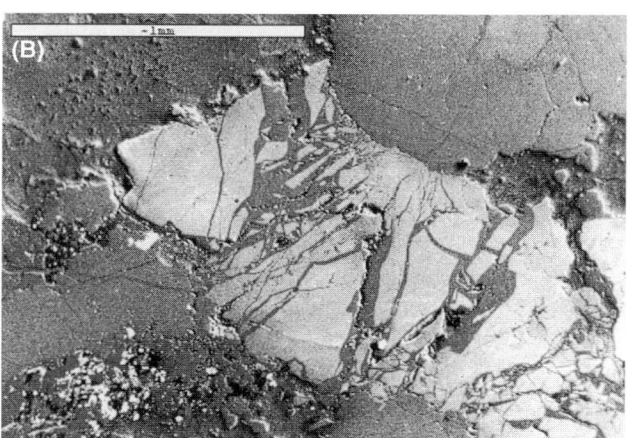

Figure C73 Fracturing of sand grains: (A) Experimental fracturing of sand at 40 MPa stress. This is equivalent to the effective stress below 3.5–4 km of overburden; (B) Natural fracturing of sand grains in Jurassic reservoir rocks at 4,800 m depth, Lavrans Field, Haltenbanken, offshore Norway. The fractures were produced by mechanical compaction prior to quartz cementation, and subsequently healed by authigenic quartz, which is darker on the cathodoluminescence microphotgraph. The present burial depth (4,800 m) is probably the maximum, and the reservoir pressure is hydrostatic. Chlorite coating the grains delayed quartz cementation until fractures produced a suitable substratum.

response to tectonic stress if the strain rates are very low so that the rocks can deform by chemical compaction (Bjørlykke and Høeg, 1997).

Experimental compaction of sand

Compaction of sand depends on the compostion and primary sorting of grains. Sands with lithic fragments show more compaction at low to moderate stresses than well-sorted sand (Pittman and Larese, 1991).

Experimental uniaxial compression at high stress (50 MPa) on well-sorted mono-quartz rich sands shows that stress–strain curves are non-linear over the entire stress range. Initial compaction is due to grain sliding rotation and re-orientation. A stable grain framework is then established and further compaction requires grain breakage. Recent experimental work (Chuhan et al., 2002) show that coarse, well-sorted sands have a higher porosity loss than fine-grained sand subjected to the same stress (Figure C72). Coarse sands also show evidence of more frequent grain fracturing than fine sands in sandstone reservoirs, probably because the stress per contact increases with increasing grain size (Figure C73). Fracturing reduces the grain size and gradually distributes the load onto more grain contacts. At higher stresses porosity/stress curves for different sands approach similar slopes indicating similar compressibility. The porosity at higher stresses is therefore mostly a result of the differences in compaction up to 20–25 MPa or 2.0–2.5 km burial depth.

The grain crushing can thus be considered a part of an adjusting (self-organizing) process where the sediment is adapting itself to higher stress. Grain crushing is particularly intense in sandstones where chlorite or micro-quartz coatings retard the precipitation of quartz overgrowth (Fisher et al., 1999). This is also true in cold basins (<20°C/km) where the onset of quartz cementation may start at 5–6 km. Fracturing of grains may destroy grain coatings, thus opening fresh surfaces for quartz cementation (Figure C73(B)).

Compaction of mudstones

Mudstones compact mechanically as function of the effective stress and the compressibility of the mud (Aplin et al., 1999).

Laboratory tests show that pure clay minerals may retain a relatively high porosity even at high stresses. Smectite which is the most fine grained clay retains porosities of about 45 percent at stresses equivalent to about 5 km of overburden (Rieke and Chillingarian, 1974) while illite and kaolinite which form larger crystals compact more readily.

Smectitic clays consist of very small particles, down to 10 Å thick, and have very large specific surface areas (Nadeau and Bain, 1986). This implies that the stress is distributed on a very large number of grain contacts so that the stress per grain and unit area is very low. This may explain the low compressibility of smectitic shales down to 2–3 km burial depth (Bjørlykke,

1998). Some of the water constituting the high porosity is actually bound water due to the large surface area of smectitic clays.

Fine-grained clay and mudstones tend to compact slowly when loaded in sedimentary basins as well as when loaded by constructions. The rate of compaction is a function of the gradual increase in effective stress as the pore fluid escapes with a corresponding reduction in pore pressure.

The composition of the pore water has very significant effects on the rate of compaction in fine-grained mudstones. The cation concentration (salinity) in the pore water plays an important role because the cations adsorbed on the mineral surfaces. Marine clays have a "house of card structure" from the flocculation process, and may become unstable (quick) and collapse if the saline pore water has been diluted by meteoric ground water (Rosenquist, 1958). Such clays can gain strength and stiffnes by adding salt (NaCl or KCl). The compressibility of mud is thus a complex function of mineralogy, grain size and pore water composition.

Chemical compaction

Compaction processes involving dissolution and precipitation of minerals or other solids may be referred to as chemical compaction. We then have to consider the chemical processes involved and the thermodynamics and kinetics of the critical reactions. Chemical compaction is therefore different in principal from mechanical compaction. We distinguish between two types of dissolution:

(a) Dissolution of grain or matrix because minerals or associations of minerals are chemically unstable, for example, aragonite, gypsum, and smectite. Kaolinite becomes unstable in the presence of K-feldspar.
  Dissolution of amorphic materials like Opal A may cause compaction when replaced by Opal CT and quartz. Transformation of solids like kerogen into fluids (petroleum) changes the fluid/solid ratio (void ratio). This type of dissolution will only result in compaction if the dissolving grains are load-bearing. Dissolution of pore-filling kaolinite and smectite and precipitation of illite in sandstones ($>120–140°C$) will normally not cause compaction in sandstones. There will however be a porosity increase corresponding to the dehydration involved in mineral reactions. In mudstones, however, kaolinite crystals may be a load-bearing part of the grain framework and therefore cause a closer packing of grains as they dissolve.
(b) Dissolution at grain contacts and precipitation of the same mineral in the pore space (pressure solution).

For reactions with very low kinetic reaction rates at low temperature like in most silicate reactions, temperature rather than stress at the grain contacts may be rate limiting for the reactions and thereby for the rate of compaction.

Chemical compaction of sandstones

The stress at the contact between sand grains is very high prior to precipitation of mineral cement because of the small contact area. This causes increased solubility of minerals like quartz, and there is thus a thermodynamic drive for dissolving at grain contact and precipitation on mineral surfaces of quartz that are not under stress (Renard et al., 2000). The increase in solubility has been estimated to be 1–3 percent for an increase in effective stress of 10 MPa (Jahren and Ramm, 2000). Due to very high activation energies precipitation of quartz is however very slow and at normal pH values (5–6) quartz does not precipitate in sedimentary basins at temperatures below 70–80°C (Bjørlykke and Egeberg, 1993).

Chemical compaction of quartz sandstones consists of four steps:

(1) Dissolution of quartz at grain contacts or along stylolites;
(2) Transport of silica in solution along a thin water film at the contact;
(3) Diffusion of silica in the pore water;
(4) Precipitation of quartz on grain surfaces.

In the case of highly soluble minerals with a low kinetic activation energy like halite steps 1–3 may be rate limiting (Renard et al., 2001) and the system is then dissolution- or transport-controlled. In the case of dissolution at contacts between quartz grains the transport of silica along thin water film is very complex and is not a clear function of the stress (Renard et al., 2001) and the rate of pressure solution can then be ignored. The rate of quartz cementation should also be a function of the degree of supersaturation with respect to quartz in the pore water, and therefore influenced by dissolution and transport of silica. There is considerable evidence that the precipitation of cement is rate limiting (Oelkers et al., 1996) for short diffusion distances (a few cm). The kinetics of the precipitation process will therefore be controlled by temperature. The rate of quartz cementation at constant temperature is then a function of the surface area available for quartz cementation (Walderhaug, 1996; Bjørkum and Nadeau, 1998). Based on these assumptions, compaction and porosity loss of sandstones can be modeled as a function of subsidence rates, geothermal gradients, and grain textures determining the surface area available for quartz cementation. As a result of the larger surface area, finegrained sandstones will compact faster than coarse grained sandstones under the same temperature condition. This is in contrast to mechanical compaction where well-sorted coarse grained sand is most compressible. Compaction of sand at grain contact or along stylolites requires some stress even if the magnitude of stress is not rate limiting. Petrographic observations suggest that pressure solution of quartz does not require very high stress and that dissolution preferentially occurs along contacts between mica and quartz (Bjørkum, 1996).

Surface controlled quartz cementation occurs as a function of temperature and time and sandstones compact even when there is no burial or temperature increase (Figure C74) (Walderhaug et al., 2001).

Relationships between compaction and porosity loss

During mechanical compaction the volume of the solid phase remains practically constant and the change in rock volume (shortening) is therefore a direct function of the porosity (fluid) loss. Chemical compaction implies that the volume of the solid phase may change due to dehydration of minerals and other mineral reaction. Generation of oil and gas from solid load-bearing kerogen is another phase change that has consequences for compaction because it causes changes in the fluid/solid ratio (void ratio) due to temperature rather than stress.

Changes in the volume of solids in a volume of rock may be due to net export (leaching) or import (precipitation) of solids by diffusion or fluid flow (advection). If all the pore space between grains is filled by mineral cement where the solids constituents are imported from outside, the porosity may be reduced to zero with little or no compaction. This can occur by diffusion on a local scale and is typical of concretion, which can form at shallow depths and provide an indication of porosity at shallow burial. Dissolution of chemically unstable grains may increase the porosity when there is a net transport of solids from the rock volume. This may be the case when feldspar and carbonate minerals are dissolved during meteoric water flushing.

Flow of compaction-driven pore water normally has fluxes, which are very low and insufficient to transport and precipitate large volumes of cement. When we consider larger volumes of rocks (e.g., $>1,000\,m^3$), the import and export of solids in solution can be ignored except in very special settings.

Compaction of carbonate sand

Normal marine carbonate sediments consist of grains composed of fossils fragments, ooids, pellets or lithic fragments which may primarily be composed of aragonite, Mg-calcite or calcite. In terms of mechanical compaction they are in principal like siliceous sand and mud. Carbonate grains are mechanically weaker than quartz grains, but the difference as measured by experimental compaction of carbonate sand is not very great. Chemical compaction of carbonate sediments carbonate is however very different from that of siliceous sediments because of the faster reaction kinetics also at low temperatures. Chemical compaction may therefore start at shallow depths (low temperatures) shortly after deposition.

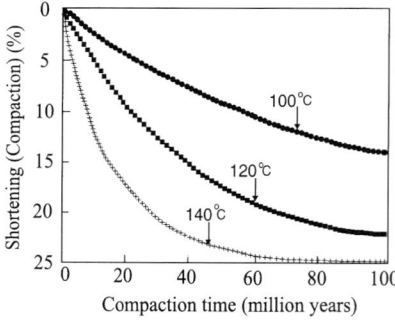

Figure C74 Modeling of compaction as a function of temperature and time even if there is no further subsidence. The rate of quartz cementation and thereby also the rate of compaction is a function of the activation energy E ($J\,mol^{-1}$) and the surface area A ($mol\,cm^{-2}\,s^{-1}$) available for quartz cementation. From Walderhaug et al. (2001).

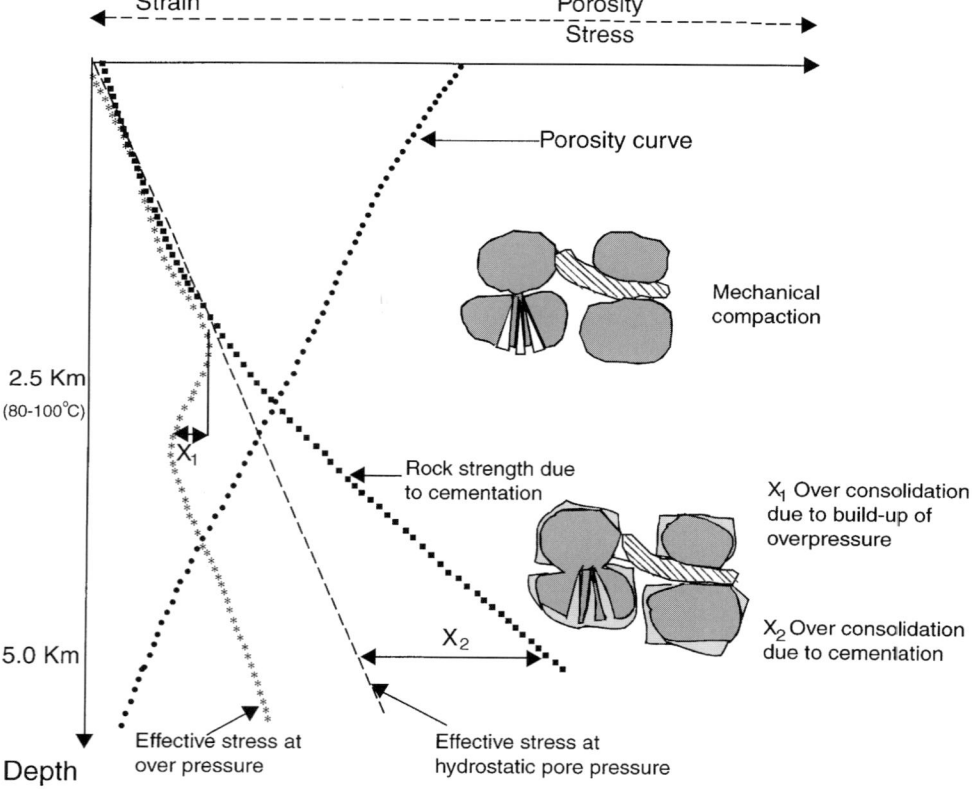

Figure C75 Schematic illustration of the effects of mechanical and chemical compaction. During progressive burial sediments may become over-consolidated by a build up of over-pressure (X_1), and by cementation (X_2). Further compaction then requires a stress exceeding the over-consolidation stress due to mechanical compaction, or the "pseudo over-consolidation" stress due to cementation.

Chemical compaction in carbonate sediments includes two different processes:

(1) Dissolution of thermodynamically unstable mineral. Dissolution of load-bearing grains that is, of aragonite may the cause compaction at shallow depth (low stresses). Dissolution of aragonite may be accompanied by precipitation of early calcite cement, which will stabilize grain structure and thus preserve porosity during burial.

(2) Pressure solution. The driving force here is the increased solubility of minerals like calcite due to stress at grain contacts, and the temperature plays a minor role. The rate of precipitation may also depend on grains surface properties. Large calcite crystals, for example, from crinoids, may provide very suitable sites (low solubilities) for precipitation, while very small crystals and organic coatings may retard precipitation of cement, thus influencing the rate of compaction.

Chemical compaction and rock strength

Sediments become over-consolidated when the effective stress is reduced by uplift and erosion or overpressure and can then be loaded up to the previous maximum stress before any significant compaction occurs (Figure C75). Precipitation of cement in the pores between grains causes an increase in the strength of the grain framework and produces a pseudo overconsolidation effect, that is not due to high mechanical stresses. This "over-consolidation" may allow a rather high increase in overburden stress without any significant mechanical compaction. In the case of sandstones, quartz cement will normally become significant at 2.5 km to 3 km depth (80–100°C). Only 2–3 percent quartz cement may be enough to strength the grain framework and virtually stop further mechanical compaction. Fractures in quartz grains formed by mechanical compaction usually do not penetrate the quartz overgrowths, indicating that mechanical compaction stops after significant quartz cementation (Figure C75). Well-cemented sandstones have a high rock strength. Consequently, reducing fluid pressure during production of water or petroleum will normally not cause mechanical compaction and surface subsidence.

Summary

Mechanical compaction tests of sediments form the basis for predicting subsidence (settlement) due to the loading of constructions like buildings, pipelines or oil platforms. Compaction tests at high stresses can also simulate mechanical compaction in the upper 2–3 km (at temperatures <80–120°C) in clastic sedimentary basins, while chemical compaction is dominant in the deeper and hotter parts. In carbonate sediments compaction is more a function of stress and is less dependent on temperature.

Modeling of chemical compaction may now provide a basis for quantitative prediction of porosity in deep reservoir rocks. Compaction processes also control the rock mechanical and petrophysical properties, seismic velocities and attributes but this is still poorly understood.

Knut Bjørlykke

Bibliography

Aplin, A.C., Fleet, A.J., and Macquaker, J.-H.-S. (eds.), 1999. *Mud and Mudstones: Physical and Fluid Flow Properties*. Geological Society of London, Special Publication, 158.

Athy, L.F., 1930. Density, porosity and compaction of sedimentary rocks. *Bulletin of the American Association of Petroleum Geologists*, **14**: 1–24.

Bjørkum, P.-A., 1996. How important is pressure solution in causing dissolution of quartz in sandstones? *Journal Sedimantary Research*, **66**: 147–154.

Bjørkum, P.A., and Nadeau, P.H., 1998. Temperature controlled porosity/permeability reduction; fluid migration and petroleum exploration in sedimentary basins. *APPEA Journal*, **38**: 453–464.

Bjørlykke, K., 1998. Clay Mineral diagenesis in sedimentary basins—a key to the prediction of rock properties. *Clay Minerals*, **33**: 15–34.

Bjørlykke, K., and Egeberg, P.K., 1993. Quartz cementation in sedimentary basins. *Bulletin of the American Association of Petroleum Geologists*, **77**: 1538–1548.

Bjørlykke, K., and Høeg, K., 1997. Effects of burial diagenesis on stresses, compaction and fluid flow in sedimentary basins. *Marine and Petroleum Geology*, **14**: 267–276.

Chuhan, F.A., Kjeldstad, A., Bjørlykke, K., and Høeg, K., 2002. Porosity loss in sand by grain crushing experimental evidence and relevance to reservoir quality. *Marine and Petroleum Geology*, **19**: 39–53.

Fisher, Q.J., Casey, M., Clenell, M.B., and Knipe, R.J., 1999. Mechanical compaction of deeply buried sandstones of the North Sea. *Marine and Petroleum Geology*, **16**: 605–618.

Giles, M.R., 1997. *Diagenesis: A Quantitative Perspective. Implications for Basin Modelling and Rock Property Prediction*. Kluwer Academic Publishers.

Hubbert, M.K., and Rubey, W.W., 1959. Role of fluid pressure in mechanics of overthrust faulting I. Mechanics of fluid filled porous solids and its application to overthrust faulting. *Bulletin of the Geological Society of America*, **70**: 115–166.

Jahren, J., and Ramm, M., 2000. The porosity preserving effects of microcrystalline quartz coatings in arenitic sandstones; examples from the Norwegian continental shelf. In *Quartz Cementation in Sandstones*. Association of Sedimentologists, Special Publication, 29, pp. 271–280.

Lambe, T.W., and Whitman, R.V., 1979. *Soil Mechanics, SI Version*. John Wiley.

Nadeau, P.H., and Bain, D.C., 1986. Composition of some smectite and diagentic illitic clays and implications for their origin. *Clays and Clay Minerals*, **34**: 455–464.

Oelkers, E.H., Bjørkum, P.A., and Murphy, W.M., 1996. A petrographic and computational investigations of quartz cement and porosity reduction in North Sea sandstones. *American Journal of Science*, **296**: 420–452.

Pittman, E.D., and Larese, R.E., 1991. Compaction of lithic sand: experimental results and applications. *American Association of Petroleum Geologists Bulletin*, **75**: 1279–1299.

Renard, F., Brosse, E., and Gratier, J.P., 2000. The different processes involved in the mechanism of pressure solution in quartz-rich rocks and their interactions. In Worden, R.H., and Morad, S. (eds.), *Quartz Cementation in Sandstones*. International Association of Sedimentologists, Special Publication, 29, pp. 67–78.

Renard, F., Dysthe, D., Feder, J., Bjørlykke, K., and Jamtveit, B., 2001. Enhanced pressure solution creep rated induced by clay particles: experimental evidence from salt aggregate. *Geophysical Research Letters*, **28**: 1295–1298.

Rieke, H.H., and Chillingarian, G.V., 1974. *Compaction of Argillaceous Sediments*. Developments in Sedimentology, 16, Elsevier.

Rosenquist, I.Th., 1958. Remarks to the mechanical properties of soils—water systems. *Geologiske Foreningens Stockholm*, **80**: 435–457.

Sclater, J.G., and Christie, P.A.F., 1980. Continental stretching. An explanation of the post Mid-Cretaceous subsidence of the Central North Sea Basin. *Journal of Geophysical Research*, **85**(B7): 3711–3739.

Terzaghi, K., 1925. *Erdbaumeckanik Auf Bodenphysikalischer Grunglage*. Leipzig: Deuticke.

Thyberg, B.I., Jordt, H., Bjørlykke, K., and Faleide, J.I., 2000. Relationships between sequence stratigraphy, mineralogy and geochemistry in Cenozoic sediments of the northern North Sea. In Nøttvedt, A. *et al.* (ed.), *Dynamics of the Norwegian Margin*. Geological Society of London, Special Publication, 167, pp. 245–272.

Walderhaug, O., 1996. Kinetic modeling of quartz cementation and porosity loss in deeply buried sandstones reservoirs. *Bulletin of the American Association of Petroleum Geologists*, **80**: 731–745.

Walderhaug, O., Bjørkum, P.A., Nadeau, P.H., and Langnes, O., 2001. Quantitative modelling of basin subsidence caused by temperature-driven silica dissolution and reprecipitation. *Petroleum Geoscience*, **7**: 107–114.

Wilson, M.D., 1994. *Reservoir Quality Assessment and Prediction in Clastic Rocks*. SEPM Short Course 30, 432p.

Cross-references

Atterberg Limits
Cements and Cementation
Diagenesis
Diagenetic Structures
Diffusion, Chemical
Fabric, Porosity, and Permeability
Geophysical Properties of Sediments
Porewaters in Sediments
Pressure Solution
Stylolites

CONVOLUTE LAMINATION

A term introduced by Kuenen (1953) for irregularly wavy laminae confined within a single sediment layer. The deformation dies out both downward, and upward, and may be confined to a part of a bed: it does not affect the geometry of the lower or upper boundaries of the bed. It is commonly found in turbidites, but is also present in other facies (Figure C76). The convolutions appear to develop gradually above some level within the bed, frequently distorting otherwise plane or ripple cross-lamination, and may show one or more surfaces of erosional truncation before dying out upward, generally at or just below the top of the bed. Thus they are generally interpreted as having developed during or immediately after deposition. Good figures are provided by Allen (1982) and Ricci Lucchi (1995).

The actual structure was described earlier by several authors using several different names (see Potter and Pettijohn, 1963, p. 152; Allen, 1982, p. 349), and its abundance was noted particularly by Rich (1950) who thought the structure had been formed postdepositionally by intra-stratal sliding, and had therefore developed on submarine slopes. Bouma (1962) thought that the structure developed in, or as a replacement of, the rippled (C) division of his ideal turbidite "sequence," but it is now clear that it may affect either the plane laminated (B) or rippled (C) divisions. The structure has much in common with Pillow Structure (*q.v.*), which has also been misinterpreted as a slope indicator.

Most convolute lamination is found in fine grained sand or coarse silts, and has been reported (rarely) from limestones as well as from siliciclastics. Allen (1982) distinguished three types: postdepositional, syndepositional, and metadepositional, depending on the geometry (the metadepositional variety shows erosional truncation at the top of the bed). The mechanism of formation is deposition that is rapid enough to produce pore pressure exceeding hydrostatic, and consequent loss of strength (partial liquefaction) of the sediment either during or immediately following deposition. Gravitational instability then produces "diapir-like" structures, and shear by the depositing flow commonly results in a predominant orientation and direction of overturning of the (generally very irregular) folds. Thus, some convolute lamination can be used as a paleocurrent indicator.

Gerard V. Middleton

Bibliography

Allen, J.R.L., 1982. *Sedimentary Structures: Their Character and Physical Basis*. Elsevier, 2, pp. 349–354.
Bouma, A.H., 1962. *Sedimentology of Some Flysch Deposits*. Elsevier.
Kuenen, Ph.H., 1953. Significant features of graded bedding. *American Assciation of Petroleum Geologists Bulletin*, **37**: 1044–1066.
Potter, P.E., and Pettijohn, F.J., 1963. *Paleocurrents and Basin Analysis*. Academic Press.
Ricci Lucchi, F., 1995. *Sedimentographica: Photographic Atlas of Sedimentary Structures*. 2nd Edn, Columbia University Press.
Rich, J.L., 1950. Flow markings, groovings, and intrastratal crumplings as criteria for recognition of slope deposits, with illustrations from Silurian rocks of Wales. *American Association of Petroleum Geologists Bulletin*, **34**: 717–741.

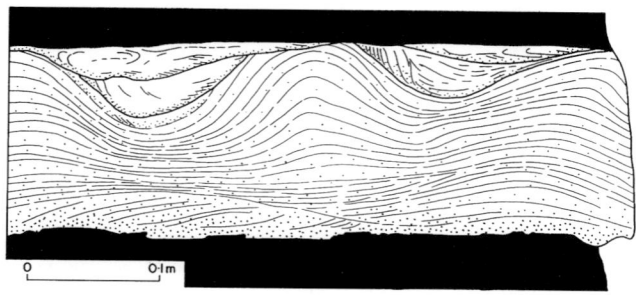

Figure C76 Syndepositional convolute lamination. Redrawn by Allen (1982, p.349, his Figure 9.4) from observations made on the Pliocene turbidites of the Ventura Basin, California. See Allen's book for original source.

Cross-references

Liquefaction and Fluidization
Pillar Structure

CORING METHODS, CORES

Cores are a very important source for the study of sediments. The oil and gas industry prefers to eliminate or strongly reduce coring by collecting a suite of well logs and thereby reducing well costs. Exploration and production people nevertheless need some cores for grain size, grain shape and surface characteristics, mineralogy and diagenesis, and for the calibration of seismic and well log data. Most sedimentary geologists, working at universities, governments and the mining industry prefer to deal with cores. Because of the variety in interests and the many types and hardnesses of sediment, an extensive array of coring methods has been developed. They range from simple hand-operated samplers on land and shallow water, to large and expensive drilling operations.

The specific need and the available budget dictate the size and type of a selected coring device. It is also possible to categorize coring equipment by the desired length and diameter of needed core, the hardness of the sediment, and if the coring occurs onshore or offshore. The small hand-operated tools have the advantage that they are light and can be transported everywhere. The large variety of coring devices fills more than a book and therefore only a few major types will be discussed. Typically a core is a long, cylindrical object. Nevertheless the term "core" refers also to other shapes, such as box core, box sample, square long box core, and some others.

Sampling unconsolidated sediments on land or in wading depth is easy and inexpensive as long as a short sample is sufficient. Metal pipes, plastic or PVC tubes, or even metal cans with a hole in the bottom to let air or water out, have provided great results. For work off a small boat one can use metal, plastic or PVC tubes with extension rods (Bouma, 1969; Figures C77-A, C78-A). In most cases a core catcher is required. A second pair of hands facilitates the operation and prevents hitting the boat several times, which often results in core loss.

Small boxes, such as the Senckenberg box (Reineck, 1957) do well on land and at wading depth, even in sands. Petri dishes are being used successfully because the sample can be radiographed while still in the dish, or photographed, dried, impregnated, etc.

Moving into deeper water requires a ship, the size depending on water depth, and the size and weight of both the corer and the winch. The most common corers are gravity corers and piston corers. Both categories collect the core in a pipe. In most cases one uses a metal pipe and a plastic liner inside, sometimes both can be replaced by one PVC tube. A beveled coring nose, with a core catcher inside, is secured to the base of the pipe. The top of the pipe fits inside a slightly wider pipe, which has a hook (bail) at the top with stabilizing fins underneath and weights below the fins. A valve on top of the upper pipe closes directly after the corer terminates penetration.

Rather than letting the corer free-fall over the last 3–5 m, a tripping device is used that gives a pre-measured free-fall (Bouma, 1969). It consists of an arm connected to the main cable via a cable-clamping device (Figure C77-B1, B2). At the end of the arm is a cable (corer length + free-fall long) with a weight. As soon as the weight hits the sediment it unhooks the cable at the clamping device and free-fall starts. Due to friction between the core pipe and the sediment, one seldom collects a core over 2–3 m in length. The friction along the inside of the liner results in a core shortening, ranging from 10–40%.

The piston corer is very similar to the gravity corer, except that a piston is inserted at the base of the liner. The main cable is connected to the piston. A loop of cable (free-fall length) is rolled up between the fins (Figure C77-B1). The trigger cable has the length of the corer plus free-fall. If everything is

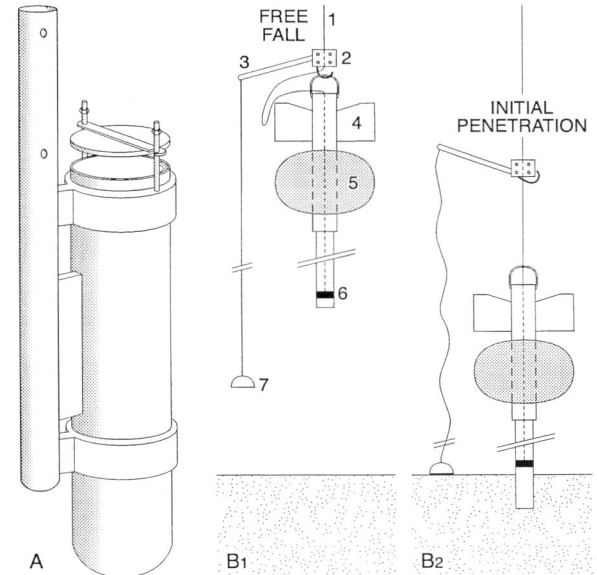

Figure C77 Two types of coring devices. (A) van Straaten tube used offshore in water up to 10 m deep. The valve can be swung to the side and the core pushed out, unless a liner is used. Three meter long extension rods make it possible to work from a small boat. (B) Piston corer operation. 1, main cable; 2, trigger device wire-clamp; 3, trigger arm; 4, stabilizing fans; 5, weights; 6, piston; 7, trigger weight. (B1) lowering, (B2) initial penetration; piston stays just above the bottom.

connected properly, the piston should stop as soon as it is near the water–sediment interface, while the core pipe slides over the piston, penetrating the sediment (Figure C77-B2). Because of the piston a slight vacuum is created underneath it which reduces the friction of sediment within the liner. This type of piston corer, weighing about one ton, and having a pipe length of 12–13 m, may collect a core up to 12 m long. Nine meters is more common. If the length setting was wrong, or the clamping device opens too early, the piston can suck in sediment, characterized by flow-in structures (Figure C78-B).

A Giant Piston Corer was developed at the Woods Hole Oceanographic Institution with the goal to collect 50 m long cores. The lighter Jumbo Piston Corer (2,300 kg) has been used successfully in the Gulf of Mexico. Recovered cores were up to 19.2 m in length, retrieved from 800–2,600 m water depth (Silva and Bryant, 2000). The extended length is necessary for geological and geotechnical studies to better understand the processes of slumping, sliding, debris flows and turbidity currents.

Anything longer than the devices mentioned above has to be done with rotary drilling from a special vessel such as operated by the Deep Sea Drilling Project, its successor the Ocean Drilling Project, or the drilling vessels working for the oil industry. Deep drilling onshore is simpler because the rig is stable rather than heaving and other motions. Coring at sea or land can produce cores 8 to $3\frac{1}{2}$ "(20.5 to 9 cm) in diameter with a common diameter of 4" (10 cm). The mining industry, and most other shallow drilling requirements, uses the slimhole version, producing cores that are $2^{1/2}$ to $1^{1/2}$ "(6.3 to 3.8 cm) in diameter with a common size of $1^{7/8}$" (4.7 cm). Those rigs are

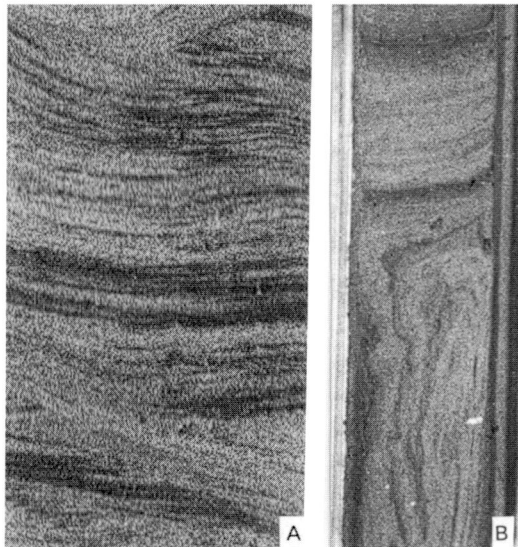

Figure C78 (A) Part of a core (2.5 cm wide) collected by hand with a short tube. Point bar, Mississippi River. Large current ripples. Courtesy J.M. Coleman. (B) Piston core in plastic liner. Upper 1/3 normal core, lower 2/3 flow-in. Core 5 cm wide. Surinam shelf, water depth 72 m.

for land use and typically are mounted on a medium-sized truck.

Cochran (1959) and Bouma (1969) have described several varieties of coring devices. Amateurs with some experience can handle small devices. Professionals should handle the larger tools to avoid serious accidents and poor core recovery.

Arnold H. Bouma

Bibliography

Bouma, A.H., 1969. *Methods For The Study Of Sedimentary Structures*. New York: Wiley-Interscience. (Can be obtained from Krieger Publishing Co., Huntington, New York, 1979.)
Cochran, J.B., 1959. *Sampling Techniques*. 2nd edn, Wiley and Sons.
Reineck, H.E., 1957. Stechkasten und Deckweisz, Hilfsmittel des Meeresgeologen. *Natur Volk*, **87**: 132–134.
Silva, A.J., and Bryant, W.R., 2000. Jumbo piston coring in deep water Gulf of Mexico for seabed geohazard and geotechnical investigations. In *Tenth International Offshore and Polar Engineering Conference*, pp. 424–433.

Cross-references

Bedding and Internal Structures
X-ray Radiography

CROSS-STRATIFICATION

Introduction

Cross-strata are layers of sediment that are inclined relative to the base and top of the set in which the inclined layers are grouped. Each group is called a set of cross-strata or a

Figure C79 Example of eolian cross-stratification. Bed "A" (between heavy white lines) is a cross-stratified bed or set of cross-strata. Because the cross-strata (such as bed "B", between thin white lines) are themselves cross-stratified (bed "C", for example), this structure is called compound cross-stratification.

cross-stratified bed (Figure C79). Individual cross-strata can be classified (McKee and Weir, 1953) as cross-laminae (<1 cm thick) or cross-beds (>1 cm thick). In general, each set of cross-strata is deposited by a migrating bedform. Thin sets are deposited by small migrating bedforms such as ripples, small dunes, or small antidunes, and thick sets are deposited by larger dunes, antidunes, bars, or other large bedforms (Allen, 1962; Harms and Fahnestock, 1965). Cross-strata are a natural record of transported sediment and are therefore useful for understanding the behavior of modern bedforms and for interpreting environments in ancient deposits.

Because most cross-strata are deposited by migrating bedforms, study of cross-strata is inextricably entwined with studies of bedforms, sediment transport, and lee-side processes including granular flow. Advances in the understanding of cross-stratification include recognition of cross-laminae deposited by current ripples (Sorby, 1859), recognition of climbing eolian dune deposits (Barrell, 1917; Shotton, 1937), experimental studies of grainflows (Bagnold, 1941), discovery of large subaqueous dunes using sonar, recognition that subaqueous dunes deposit cross-strata (Allen, 1962, 1963a; Potter and Pettijohn, 1963; Harms and Fahnestock, 1965), recognition of stratification types in eolian dunes (Hunter, 1977), and the use of 3-D computer graphics to visualize the migration of complex bedforms (Rubin, 1987).

Origin of cross-stratified beds

Deposition of cross-stratified beds requires two conditions: migration of geomorphic surfaces that are inclined relative to the generalized depositional surface and deposition during migration. The inclined surfaces can either be individual landforms (such as delta-like bodies) or can occur within a train (bedforms); most sets of cross-strata are produced by trains of migrating bedforms.

Bedforms can only leave deposits where deposition occurs while the bedforms are migrating. This combination is defined as bedform climbing, because deposition causes the bedforms to move upward (climb) while they migrate. Where the rate of deposition is rapid enough (relative to the rate of migration), both the stoss and lee slopes of the migrating bedforms are preserved, resulting in complete preservation of the original bedforms (supercritical climb of Hunter, 1977). In other cases, the stoss sides of bedforms can undergo erosion while the lower lee sides are preferentially preserved (subcritical climb).

Origin of individual cross-strata

Sediment grains within cross-strata are deposited by a variety of processes: settling out from the water or air, avalanching (grainflow), or deposition by superimposed bedforms. On steep lee slopes, initial deposition is primarily by grains settling out of the flow. In both air and water, sediment can be sorted by grain size or density as it settles out on the lee slopes. On wind ripples, segregation by size can mask or obscure cross-lamination. Instead of producing visible cross-lamination, wind ripples typically deposit inversely graded laminae, because the grains on the lee sides of the wind ripples are sorted from coarse at the crest to fine at the trough (Hunter, 1977).

Cross-stratification deposited by grainflow is common on steep lee slopes of bedforms. Grainflow occurs because sediment that is transported across a bedform crest tends to be deposited at higher rate on the upper lee side than the lower lee side. Consequently, the lee slope can steepen until the slope exceeds the angle of repose, reaches the higher angle of initial yield, and causes the slope to fail by grainflow. In air, grainflow is discontinuous in time and space, producing grainflow deposits that are lenticular in horizontal cross-section. In contrast, subaqueous grainflow can occur more continuously in both time and space (Hunter and Kocurek, 1986). In both air and water, grainflow sorts grains by size, with the finest grains tending to sink to the base of the flow (Bagnold, 1941). On low-angle lee slopes, cross-strata are commonly deposited by superimposed bedforms.

Classifications of cross-stratification

McKee and Weir (1953) classified cross-stratification on the basis of descriptive features such as the shape of the set and shape of individual cross-strata. This is a useful approach, because some of these attributes are related to the kinds of bedforms that deposited the cross-strata (Allen, 1962). For example, sets of cross-strata with a planar base and uniform dip azimuth are deposited by two-dimensional bedforms (straight-crested bedforms without scour pits in their troughs), whereas trough-shaped (festoon) sets of cross-strata are deposited by more sinuous bedforms with scour pits in their troughs (Figure C80). Rubin (1987) also followed a geometric approach and classified cross-stratification based on properties that reflect whether the strata are constant along-strike (i.e., deposited by two-dimensional bedforms), constant down-dip (i.e., deposited by bedforms that maintained the same shape while migrating), and properties of the cross-stratification that indicate whether the original bedforms were transverse, oblique, or parallel to the resultant sediment-transport vector. Other classifications have been proposed for cross-stratification and bounding surfaces deposited by specific subaqueous or eolian bedforms.

Kinds of cross-stratification

Compound cross-stratification

Some cross-stratification is complicated by containing cross-strata that are internally cross-stratified (Figure C79), a structure defined as compound cross-stratification (Harms et al., 1975), double cross-bedding, or furious cross-bedding (Reiche, 1938). Compound cross-stratification is formed where

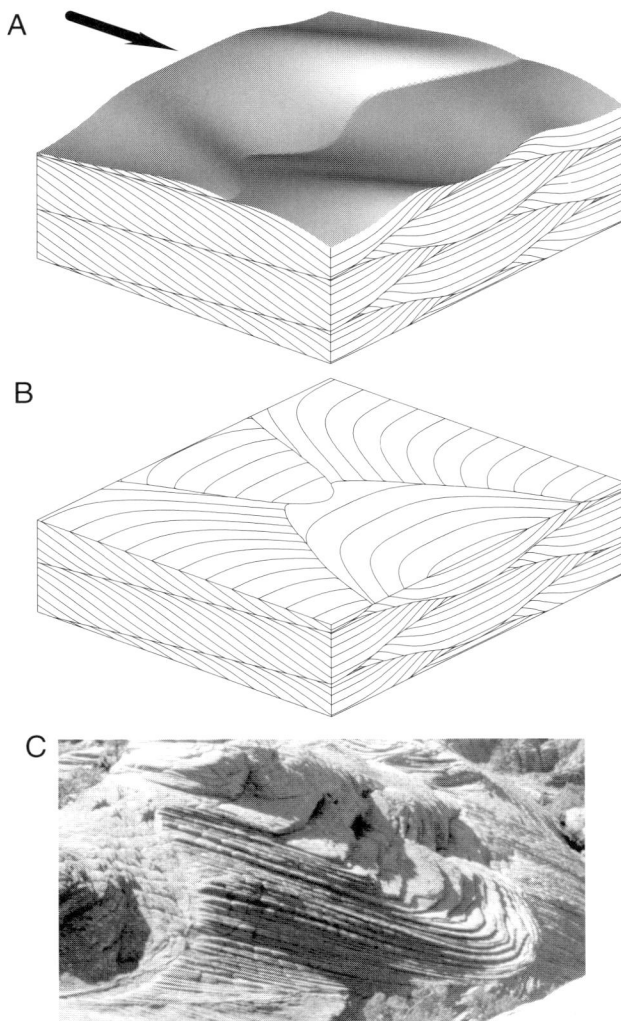

Figure C80 Trough cross-stratification. (A) Simulated bedform surface and stratification. Each trough-shaped set of strata is deposited by a migrating scour pit in the bedform trough. Arrow shows migration direction. (B) Horizontal section through structure in (A). (C) Eolian example of structure shown in (A) and (B). Dunes and scour pit migrated left-to-right out of the outcrop.

the lee side of a bedform undergoes changes in shape (often accompanied by erosion). Shape can change in response to changes in flow (such as ebb-flood reversals in tidal currents or annual changes in wind direction), or by migration of bedforms superimposed on a larger bedform as described by Allen (1963b, epsilon cross-stratification) or McCabe and Jones (1977). Some workers refer to bounding surfaces within sets of cross-strata as reactivation surfaces, but if "reactivation" is taken literally the term is more appropriate to surfaces produced by fluctuating-flow, because superimposed bedforms can produce bounding surfaces under steady conditions (no reactivation required).

Compound cross-strata deposited by bedform assemblages are particularly useful for determining the flow direction relative to the bedform crest. For example, the best method of identifying an oblique bedform is to demonstrate that bedforms superimposed on the lee side of the main bedform systematically migrated along the crest of the main bedform in a preferred direction (Rubin and Hunter, 1983).

Cyclic cross-stratification

Some cross-stratification exhibits cyclic changes in grain size (such as mud drapes in tidal bundles), dip azimuth or inclination, stratification type (e.g., alternations of grainflow and grainfall), or spacing of bounding surfaces. In herringbone cross-stratification (attributed to reversing tidal flows) successive sets dip in opposing directions. Scalloped cross-bedding contains cyclically truncated trough-shaped subsets. Some cyclic cross-strata arise from flow cycles, and others arise from the passage of superimposed bedforms.

Deformed cross-stratification

Because cross-strata are deposited on slopes, they are particularly susceptible to slumping. Slumping can disturb individual beds or a sequence of beds near the free surface. Bedforms can also fail by the crestal region of a bedforms subsiding and sediment flowing into trough regions. This kind of failure may include the lower set boundary as well as sediment in the subsurface.

Relation of set thickness to bedform size

Where bedforms in a train are similar to each other and have a relatively uniform trough elevation, the thickness of the preserved sets depends on the bedform wavelength and the angle of climb (Rubin and Hunter, 1982). In such situations, the thickness of cross-stratified beds can be a very small fraction of the original bedform height. For example, eolian dunes in large deserts are commonly 100 m in height, whereas eolian cross-stratified beds in the rock record are typically a full order of magnitude smaller.

On the other extreme, where bedforms have irregular trough elevations, the thickness of cross-stratified beds depends on the particular sequence of troughs passing a point on the bed (Paola and Borgman, 1991). In this case, the thickness of cross-stratified beds can be a substantial proportion of the original bedform height.

Relation of dip direction to flow direction

Dip directions of cross-stratification are routinely used to infer paleocurrent directions. Although this approach probably gives reasonable results when sufficient measurements are made, the results can be biased in individual cases, particularly where bedforms are oblique to flow (Rubin and Hunter, 1987). Trough axes may be a more representative measure of the flow direction, but even trough axes can diverge from the actual transport direction. For example, in oblique bedforms, both the cross-strata dips and trough axis trends can diverge substantially from the transport direction. A more general rule can be demonstrated for compound cross-stratification deposited by bedform assemblages: the transport direction lies between the dip direction of the bounding surfaces scoured by the superimposed bedforms and the dip direction of foresets deposited by superimposed bedforms (Rubin and Hunter, 1983; Rubin, 1987).

Techniques

Several new techniques are particularly useful for studying the geometry of cross-stratification. Ground-penetrating radar has proven successful at mapping internal stratification within eolian dunes, and borehole imagery can determine vertical profiles of dip directions. Three-dimensional graphics modeling is essential for reconstructing the morphology and behavior of bedforms that are too complicated to be visualized otherwise.

Future studies

Perhaps the greatest advance in understanding cross-stratification will come with the development of coupled flow and sediment-transport models that can accurately predict the morphology, migration, and evolution of bedforms, including interaction between successive bedforms in a train. Such a model would allow computational experiments to recreate specific bedding geometries from simulated flow sequences.

David M. Rubin

Bibliography

Allen, J.R.L., 1962. Asymmetrical ripple marks and the origin of cross-stratification: *Nature*, **194**: 167–169.

Allen, J.R.L., 1963a. Asymmetrical ripple marks and the origin of water-laid cosets of cross-strata. *Liverpool Manchester Geological Journal*, **3**: 187–236.

Allen, J.R.L., 1963. The classification of cross-stratified units with notes on their origin. *Sedimentology*, **2**: 93–114.

Allen, J.R.L., 1981. Paleotidal speeds and ranges estimated from cross-bedding sets with mud drapes. *Nature*, **293**: 394–396.

Bagnold, R.A., 1941. *The Physics of Blown Sand and Desert Dunes*. London: Metheun.

Barrell, J., 1917. Rhythms and the measurement of geological time. *Bulletin of the Geological Society of America*, **28**: 45–904.

Harms, J.C., and Fahnestock, R.K., 1965. Stratification, bed forms, and flow phenomena, with an example from the Rio Grande. In Middleton, G.V. (ed.), *Primary Sedimentary Structures and Their Hydrodynamic Interpretation*. SEPM Special Publication, 12, pp. 84–115.

Harms, J.C., Southard, J.B., Spearing, D.R., and Walker, R.G., 1975. *Depositional Environments as Interpreted from Primary Sedimentary Structures and Stratification Sequences*. SEPM Short Course 2, Dallas: 161 pp.

Hunter, R.E., 1977. Basic types of stratification in small eolian dunes. *Sedimentology*, **24**: 362–387.

Hunter, R.E., 1985. A kinematic model for the structure of lee-side deposits. *Sedimentology*, **32**: 409–422.

Hunter, R.E., and Kocurek, G., 1986. An experimental study of subaqueous slipface deposition. *Journal of Sedimentary Petrology*, **56**: 387–394.

Hunter, R.E., and Rubin, D.M., 1983. Interpreting cyclic cross-bedding, with an example from the Navajo Sandstone. In Brookfield, M.E., and Ahlbrandt, T.S., (eds.), *Eolian Sediments and Processes*. Amsterdam: Elsevier, , pp. 429–454.

Jopling, A.V., 1965. Laboratory study of the distribution of grain sizes in cross-bedded deposits. In Middleton, G.V., (ed.), *Primary Sedimentary Structures and their Hydrodynamic Interpretation*. SEPM Special Publication, 12, pp. 53–65.

McCabe, P.J., and Jones, C.M., 1977. Formation of reactivation surfaces within superimposed deltas and bedforms. *Journal of Sedimentary Petrology*, **47**: 707–715.

McKee, E.D., and Weir, G.W., 1953. Terminology for stratification and cross-stratification in sedimentary rocks. *Bulletin of the Geological Society of America*, **64**: 381–389.

Paola, C., and Borgman, L., 1991. Reconstructing random topography from preserved stratification. *Sedimentology*, **38**: 553–565.
Potter, P.E., and Pettijohn, F.J., 1963. *Paleocurrents and Basin Analysis*, 2nd edn, 1977, Springer-Verlag.
Reiche, P., 1938. An analysis of cross-lamination: the Coconino Sandstone. *Journal of Geology*, **46**: 905–932.
Rubin, D.M., 1987. *Cross-Bedding, Bedforms, and Paleocurrents*. SEPM Concepts in Sedimentology and Paleontology, 1, 187p.
Rubin, D.M., and Hunter, R.E., 1982. Bedform climbing in theory and nature. *Sedimentology*, **29**: 121–138.
Rubin, D.M. and Hunter, R.E., 1983. Reconstructing bedform assemblages from compound crossbedding. In Brookfield, M.E., and Ahlbrandt, T.S., (eds.), *Eolian Sediments and Processes*. Elsevier, Amsterdam, pp. 407–427.
Rubin, D.M., and Hunter, R.E., 1987. Bedform alignment in directionally varying flows. *Science*, **237**: 276–278.
Sorby, H.C., 1859. On the structures produced by the currents present during the deposition of stratified rocks. *The Geologist*, **2**: 137–147.
Shotton, F.W., 1937. The lower Bunter Sandstones of North Worchestershire and East Shropshire. *Geological Magazine*, **74**: 534–553.

Cross-references

Angle of Repose
Bedding and Internal Structures
Bedset and Laminaset
Convolute Lamination
Cyclic Sedimentation
Deformation of Sediments
Dunes, Eolian
Eolian Transport and Deposition
Grain Flow
Hummocky and Swaley Cross-Stratification
Paleocurrent Analysis
Ripple, Ripple Mark, and Ripple Structure
Sediment Transport
Sedimentologists
Turbidites

CYCLIC SEDIMENTATION

Introduction and background

Cyclic or rhythmic sedimentation is an old stratigraphic problem that has received considerable attention in the past (Suess, 1888; Grabau, 1913; Wanless and Shepard, 1936; Wheeler and Murray, 1957; Wells, 1960; Merriam, 1964; Vella, 1965; Duff *et al.*, 1967; Elam and Chuber, 1972; Schwarzacher, 1975), and remains today an unresolved controversial topic (Wilkinson, 1982; James, 1984; Algeo and Wilkinson, 1988; Einsele *et al.*, 1991; Goldhammer *et al.*, 1993; Drummond and Wilkinson, 1993a,b,c; de Boer and Smith, 1994; Read, 1995; Grammer *et al.*, 1996; Wilkinson *et al.*, 1996; Miall, 1997). Most, if not all, stratigraphic successions display repetitions of strata, at different scales, that reflect a succession of related depositional processes and environmental conditions that are repeated in the same order. The repetition of such events is termed cyclic or rhythmic sedimentation, and this leads to the formation of stratigraphic successions that display repetitive, orderly arrangement of different kinds of sediments or rock types. Studies of high-frequency cyclicity span all of geologic time from the Proterozoic (e.g., Grotzinger, 1986; Eriksson and Simpson, 1990; Clough and Goldhammer, 2000) to the Pleistocene (e.g., Mesolella *et al.*, 1969; Bloom *et al.*, 1974).

The essential concept of cyclic sedimentation is that of repetition of a series of events rather than their mere recurrence or random distribution. Thus, the term "cycle" refers to a series of connected events which return to a starting point. When applied to a sedimentary sequence, "cycle" refers to a distinctive series of lithologies that are arranged vertically in a predictable pattern in which at least one rock type, which is regarded as the starting point, is repeated (Schwarzacher, 1975). Accompanying the concept of "lithologic cyclicity" is the notion of "time cyclicity" (Schwarzacher, 1975). The term "cycle" should refer to regularities in rock pattern, whereas the term "rhythm" should be reserved for identical rock types that have been formed at equal time intervals. However, often this distinction is not made (Wells, 1960), or is impossible to judge. In this review I will address both lithologic cycles and temporal rhythms, at the macro-scale (i.e., outcrop-scale through seismic-scale), and exclude varve-scale laminations (mm-scale; see Einsele *et al.*, 1991 for a summary).

Types of cycles: allocycles versus autocycles

Vertical facies successions and their lateral distribution observed in the stratigraphic record are the results of changes in the environment of deposition through time, and these changes may be induced by autocyclic or allocyclic phenomena (Figure C81; Beerbower, in Merriam, 1964). Autocyclic processes are internal to a sedimentary system, totally independent of external influences. To develop cyclic deposits, these processes have to be self-regulating and include a built-in feedback mechanism. Excellent examples of autocyclic processes which lead to clastic cyclic deposits are: (i) delta lobe switching due to repeated progradation and retreat perhaps due to local compaction of pro-delta deposits or changes in rates of sediment supply; and (ii) migration and superposition of channel and lobe systems in fluvial environments. Other examples include non-periodic storm bed deposition and deep-water turbidite deposition. As these processes are limited areally to the basin depocenter itself, resultant "cycles" will be of limited lateral extent. Additionally, the time period of autocyclic deposits is often less than that of allocyclic deposits (Einsele *et al.*, 1991).

Allocyclic processes are external to the depositional system and they impart a net effect on stratigraphic accumulation. Fundamentally, external variations are driven by changes in climate and/or regional (or global) tectonics. Climate can affect sea-level changes through the advance and retreat of continental glaciers, and obviously influences deposition of evaporites in closed (lacustrine) or semi-enclosed marginal marine settings. Tectonic movements influence sea-level changes which in turn effect *accommodation space*, that is the vertical space available for sediment accumulation, as well as water depths which influence facies. In contrast to autocyclic processes, allocyclic packages may extend over great distances within a basin, as well as from one basin to another, and their time periodicity is often greater than that of autocycles (Einsele *et al.*, 1991).

Terminology, cyclic hierarchies, and orders of cyclicity

Perhaps the largest impediment to furthering our understanding of cyclicity is the development and utilization of a

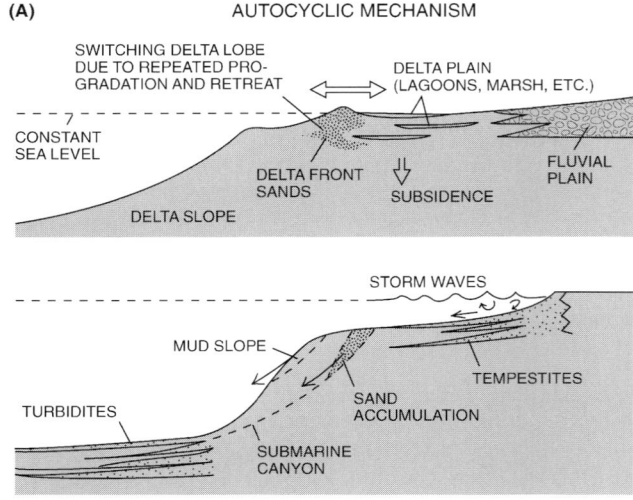

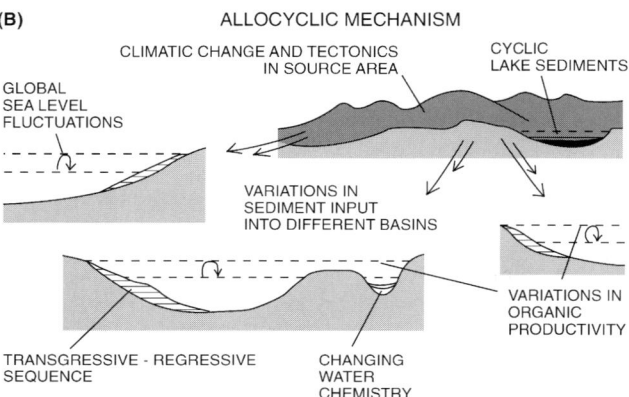

Figure C81 Autocyclic (A) and allocyclic (B) mechanisms (after Einsele et al., 1991).

consistent terminology which satisfies a person's perception of the nature of cyclicity in the stratigraphic record. As advances are made in the accuracy of age-dating, cycle duration will become better constrained and this will certainly assist in a more widely adopted terminology and classification scheme. At present, however, absolute age dates fall short of the required precision, and thus geologists continue to wrestle with a consistent terminology. Perhaps a more severe problem is the fact that workers have different criteria regarding the recognition of cycles and particularly the surfaces that separate them.

For example, the high-frequency building block of shallow-marine clastic deposits is termed the *parasequence* defined by Van Wagoner et al. (1990) as "a relatively conformable succession of genetically related beds or bedsets bounded by marine-flooding surfaces or their correlative surfaces". A marine-flooding surface is defined as "a surface separating younger from older strata across which there is evidence of an abrupt increase in water depth. This deepening commonly is accompanied by minor submarine erosion or nondeposition (but not by subaerial erosion), with a minor hiatus indicated".

This definition of the highest frequency building block in shallow marine clastic systems, which many workers would equate with the term "cycle", emphasizes the physical bounding surface to the point of excluding the internal succession of rock types or facies from the definition.

By contrast, in shallow marine carbonate systems, Goldhammer *et al.* (in Franseen *et al.*, 1991; 1993), following Wilson (1975), James (1984) and Hardie and Shinn (1986), define the high-frequency building block as a *cycle* which they define as:

a relatively conformable succession of genetically related subtidal subfacies bounded by peritidal subfacies, subaerial exposure surfaces, and/or marine flooding surfaces (very thin intervals characterized by slow rates of deposition). Cycles are the thinnest recognizable allocyclic or autocyclic depositional unit and they may be progradational and/or aggradational and thus subfacies within cycles shoal upward. Individual cycles typically contain both a "transgressive" and "regressive" component, and the relative proportion of these components within any given cycle is a function of the cycle's position within a lower frequency cycle or sequence.

Cycles by this definition are highly accurate depictions of laterally coeval facies or subfacies associations in accordance with Walther's Law.

Goldhammer *et al.* (in Franseen *et al.*, 1991, 1993) recognize two types of shallow marine platform carbonate cycles, based on physical bounding surfaces : cycles bounded by features indicative of subaerial exposure (e.g., mudcracked peritidal laminites, caliches, etc), termed "exposure cycles" or "peritidal cycles", and those that do not shoal to fill level which are bounded by marine flooding surfaces across which subfacies deepen, termed "subtidal cycles" (see also Osleger and Read, 1991). Below, I briefly review the approaches to a consistent terminology and classification, attempting to provide some historical context.

Hierarchical subdivision based on duration of the "cycle"

The sedimentary record throughout all of geologic history is characterized by stratigraphic cycles of different orders of magnitude, both in terms of thickness, duration and regional extent. Suess (1888) early on appreciated the notion that several orders of cyclicity existed within the stratigraphic record and Barrell introduced the term "composite cycle" to successions in which cycles of different wavelengths are superimposed upon each other (Schwarzacher, 1975). This concept is now well-established as indicated by Exxon's global cycle charts (Vail *et al.*, 1977; Haq *et al.*, 1987) constructed for the Phanerozoic, which consist of three superimposed orders of cyclicity. Extending Sloss' original idea of regional "unconformity-bounded sequences" Vail and his Exxon coworkers proposed the existence in the stratigraphic record of "global cycles of relative sea level change" and they presented a hierarchy of global cycles based on their inferred temporal duration (Figure C82): (1) first-order cycles with durations of 200–300 m year; (2) second-order cycles with durations of 10–80 m year; (3) third-order cycles with durations of 1–10 m year (Vail *et al.*, 1977). At that time, in the Exxon scheme, cycles referred strictly to relative sea-level changes and the term "sequence" was reserved for the stratigraphic record

Tectonic-Eustatic/Eustatic Cycle Order	Sequence Stratigraphic Unit	Duration (m year)	Relative sea-level amplitude (m)	Relative sea-level rise/fall rate (cm/1,000 years)
First		>100		<1
Second	Supersequence	10–100	50–100	1–3
Third	Depositional Sequence Composite Sequence	1–10	50–100	1–10
Fourth	High Frequency Sequence Parasequence and Cycle Set	0.1–1	1–150	40–500
Fifth	Parasequence High-Frequency Cycle	0.01–0.1	1–150	60–700

Figure C82 Orders of stratigraphic cyclicity (modified from Goldhammer *et al.*, in Franseen *et al.*, 1991, and Kerans and Tinker, 1997)

of such a global cycle of relative sea-level change, such that a depositional sequence was defined as "a conformable succession of genetically related strata bounded by interregional unconformities or their correlative conformities (i.e., sequence boundaries)". Extending their 1977 concepts, Vail *et al.* (1984), Vail (1987), and Van Wagoner *et al.* (1987) expanded on the Exxon viewpoint such that a depositional sequence was interpreted to be deposited during a cycle of eustatic change in sea level starting and ending in the vicinity of the inflection points on the falling limbs of a eustatic sea level curve. Hence the term sequence was originally linked to "third-order" changes in accommodation driven by relative sea-level changes interpreted as eustatic in origin.

Making the logical connection between low-frequency seismic-scale "sequences" and outcrop-scale "high-frequency" cycles, Miall (1984) proposed a logical extension of the original Vail *et al.* (1977) terminology for stratigraphic cycles and introduced the term fourth-order cycles for cycles with durations between 0.2 m year and 0.5 m year. Soon the hierarchy was expanded further for both clastics (Van Wagoner *et al.*, 1990) and platform carbonate cycles (Goldhammer and Harris, 1989; Koerschner and Read, 1989; Goldhammer *et al.*, 1990) and the term fifth-order cycle was proposed for high-frequency cycles with durations between 0.01–0.1 m year. Vail *et al.* (in Einsele *et al.*, 1991) introduced yet another subdivision with the term sixth-order cycle for cycles with durations between 0.01 and 0.03 m year, a term also recognized and used by Grammer *et al.* (1996).

Currently, many stratigraphers use the "order" subdivision realizing that although absolute age durations are nearly impossible to obtain and thus define the duration of a cycle with precision, the idea of a "building-block" approach with a hierarchy of cycles based on some inference of duration is a valid and useful first approach, regardless of the type of deposit (e.g., Mutti, 1992; Kerans and Tinker, 1997; Miall, 1997). A limitation of this approach is that each of these "orders" implies a temporal order-of-magnitude for a given rock-stratigraphic unit, independent of origin, whether relative sea-level, climate, or sediment supply, thus potentially suggesting some link between duration and process. Although these terms are useful order-of-magnitude indicators of cycle duration, as pointed out by Sonnenfeld (1996) predictions with regard to associated rate or amplitude of sea-level change, vertical thickness, lateral extent, correlation range, or even depositional style for a given sequence cannot be confidently made simply based on "order."

Subdivision for clastic successions based on a hierarchy of stratal units

This approach, set forth by Van Wagoner *et al.* (1990), recognizes a hierarchy of stratal units from the mm-scale lamination to the km-scale seismic sequence. Although divorced from the "order" approach of his Exxon colleagues and the inferred eustatic interpretation of cycles, the Van Wagoner scheme recognizes that the basic building-block, the parasequence (defined above) is equivalent to the "upward-shoaling cycle" (Van Wagoner *et al.*, 1990, p. 3), and thus it is reviewed in this overview of cyclicity. Principally, an objective approach based on stratal attributes, each stratal unit in the hierarchy is defined and identified only by the physical relationships of the strata, including lateral continuity and geometry of the surface bounding the units, vertical stacking patterns of the units, and lateral geometry of the strata within the units. Significantly, thickness, time for formation, and the interpretation of regional or global origin are not used to define stratal units or place them within the hierarchy.

As related to this discussion of cyclicity, the following terms were proposed by Van Wagoner *et al.* (1990):

(1) the *parasequence* (defined above) would be the highest-frequency "fundamental building block" of clastic sequences, which could be either autocyclic or allocyclic in origin. For example, within a parasequence from a beach deposit, Van Wagoner *et al.* (1990) recognize predictable vertical successions in lower-shoreface, upper-shoreface, foreshore and backshore subenvironments and associated variations in bed and bedset thickness, sand-shale ratios, and ichnofossil assemblages.

(2) the *parasequence set* defined as "a succession of genetically related parasequences forming a distinctive stacking pattern bounded by major marine-flooding surfaces and their correlative surfaces" (Figure C83). A "stacking pattern", as defined by Sonnenfeld (1996) represents the progressive lateral translation in space of a set of cycles forming a geometric arrangement describable by the terms "retrogradational" or "landward-stepping," "progradational" or "seaward-stepping," and "aggradational" or "vertically stacked". Cycle stacking patterns are interpreted as the product of systematic changes in the ratio of accommodation to sediment supply (Van Wagoner et al., 1990; Sonnenfeld and Cross, 1993; Sonnenfeld, 1996), and such patterns form the basis for recognizing

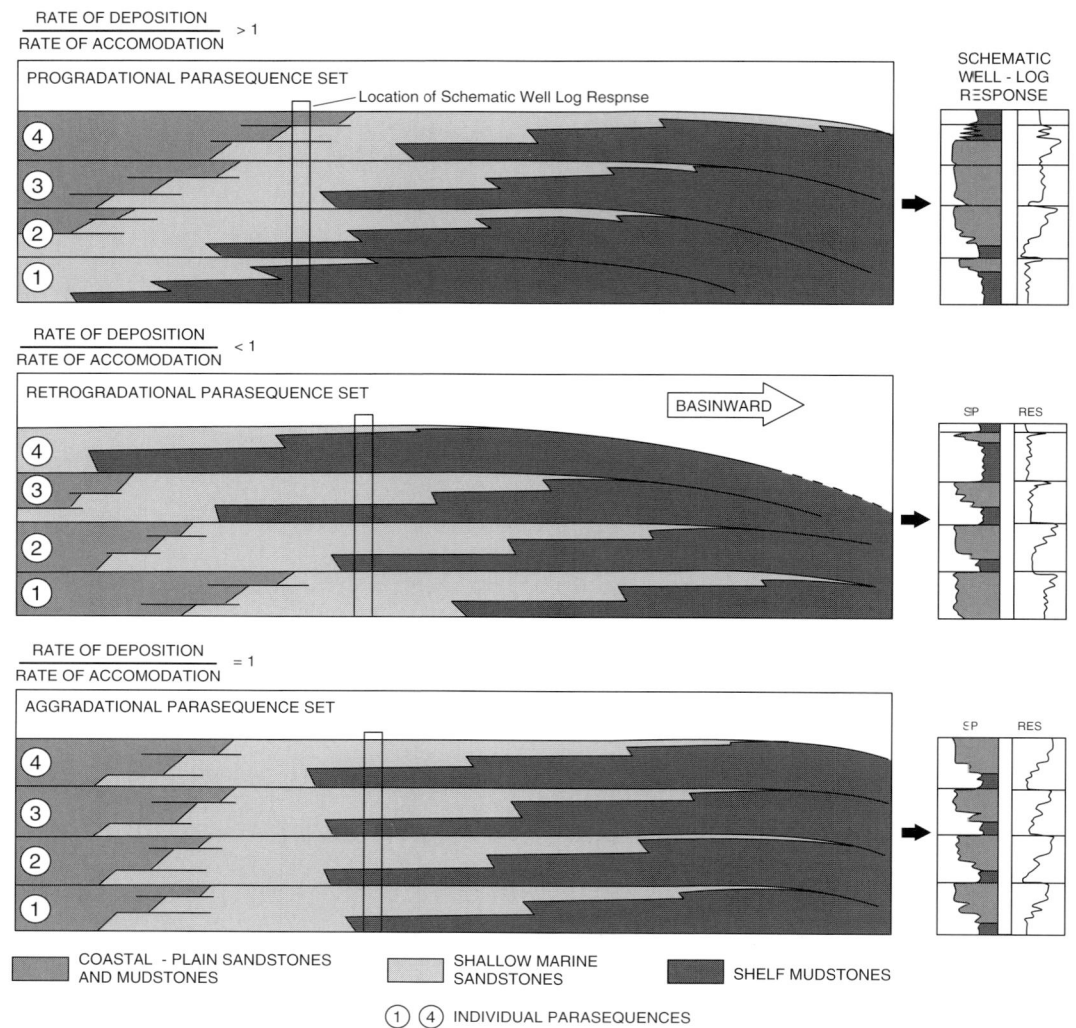

Figure C83 Parasequence set stacking patterns (modified form Van Wagoner et al., 1990).

progressively larger-scale sequences within a hierarchy (e.g., Goldhammer et al., in Franseen et al., 1991, 1993). Accommodation is defined as the cumulative sum of space available for potential sediment deposition; it is the product of eustasy, subsidence, and compaction and is equivalent in shallow marine settings to relative sea-level. Although two-dimensional in implication (Van Wagoner et al., 1990; Fitchen, 1997), where lateral continuity of cycles is lacking owing to limited outcrop or in the case of subsurface cores, stacking patterns may be inferred from one-dimensional analysis of vertical cycle trends (Kerans and Tinker, 1997; Read and Goldhammer, 1988; Lehrmann and Goldhammer, 1999).

(3) A *sequence* is defined as "a relatively conformable succession of genetically related strata bounded by unconformities or their correlative conformities" (Van Wagoner et al., 1990). Parasequence sets are the building blocks of sequences. The "transgressive" portions of sequences are marked by retrogradational parasequence sets, and the "regressive" portions of sequences are marked by progradational parasequence sets.

Hierarchical subdivision based on accommodation/sediment supply ratios: expansion of the Van Wagoner et al. (1990) scheme

In this scheme proposed by Sonnenfeld and Cross (1993) and Sonnenfeld (1996), a cycle or sequence of any physical or temporal scale is defined as the stratigraphic record of a full cycle of increasing followed by decreasing accommodation/sediment supply ratios (A/S)—the base-level transit cycle of Wheeler (1964). Cycles, of any scale, that are bounded by conformable surfaces cannot be defined solely based upon the erosive or altered nature of the bounding surfaces; in such cases, the cycle boundary is defined as the "turnaround" from decreasing to increasing accommodation/sediment supply ratios as inferred from stacking pattern analysis (e.g., Van Wagoner et al., 1987; Lehrmann and Goldhammer, 1999).

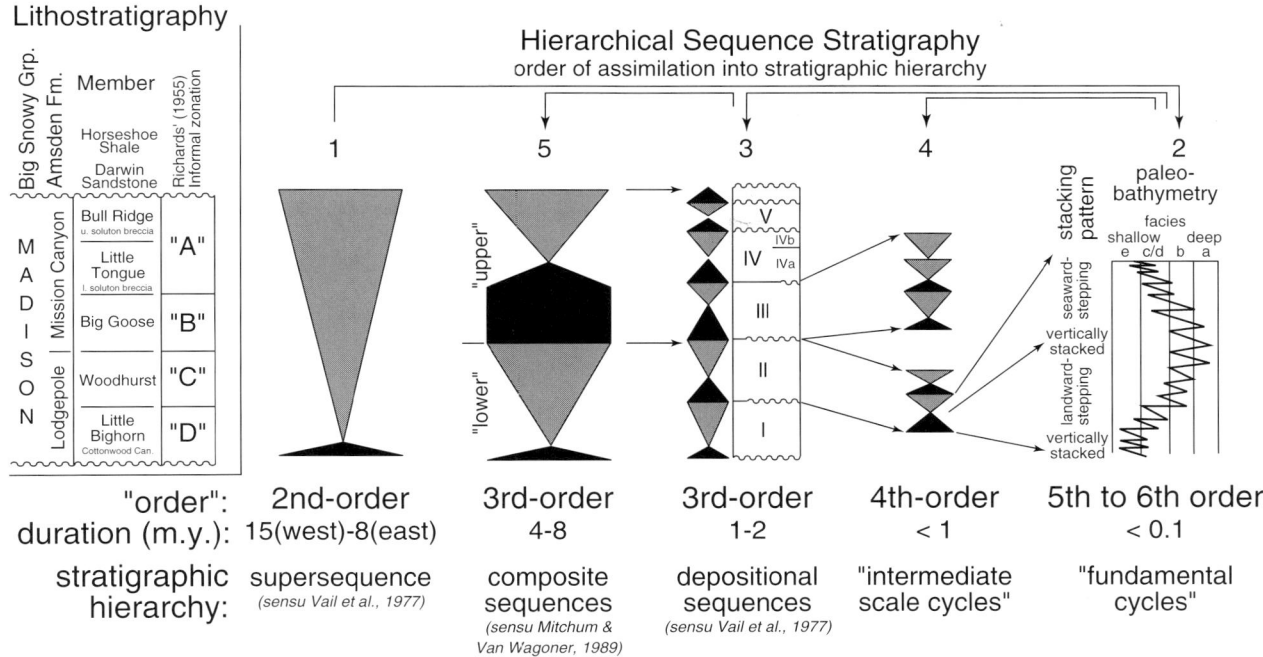

Figure C84 Hierarchical subdivision of cyclicity based on accommodation/sediment supply ratios (modified from Sonnenfeld, 1996).

In the Sonnenfeld (1996) scheme (Figure C84) triangles pointing down are used to graphically depict portions of sequences at any scale characterized by decreasing accommodation/sediment supply ("A/S") ratios as inferred from progradational stacking patterns. Conversely, triangles pointing up denote inferred increasing "A/S" as inferred from retrogradational stacking patterns. Squares separating the triangular symbols represent a condition of balanced "A/S" (= 1) as inferred from aggradational stacking patterns. The triangles described above may be drawn for cycles of any physical and temporal scale, and provide an effective graphical shorthand for portraying the hierarchical embedding of sequences within sequences.

Sonnenfeld (1996) introduces the term "fundamental cycle" as the "smallest resolvable vertical facies succession which forms a "small-scale" cycle (<1–20 ft), and define cycle boundaries at their accommodation minima. Fundamental cycles are in turn grouped into "intermediate-scale cycles" which are the highest-order cycle that can be correlated across large areas (e.g., >100 km). These in turn are nested into "third-order depositional sequences" (in the sense of Vail et al., 1977), which make up third-order "composite sequences" (defined below), in turn grouped into "second-order supersequences". Cycle boundaries of lower-frequency sequences occur at the "turnaround" from decreasing to increasing "A/S". Note that these critical turnarounds may either be represented by a unique surface or an interval of rock.

Integrated hierarchical subdivision utilizing sequence stratigraphic terminology

In a practical effort aimed at subsurface geologists, a scheme has evolved which integrates much of the previous approaches, balancing both the objective approach (emphasizing physical attributes and facies relationships observed in core and outcrop) with the somewhat subjective approach of assigning temporally-constrained orders (Figure C85). This "sequence stratigraphic" approach proposed by Kerans (1995), and utilized by Kerans and Fitchen (1995), Kerans and Tinker (1997), Fitchen (1997), and Tinker (1998), recognizes an ordered hierarchy of cycle types based on a number of objective, rock-based criteria used to define the terms and orders of the cycles. These include: vertical lithofacies successions, cycle stacking patterns, lithofacies proportions, cycle symmetry, stratal preservation, lithofacies tract offset, and stratal geometry.

The technique first defines the "high-frequency cycle" (the basic building block, equivalent to the parasequence for clastic deposits) as "the smallest set of genetically related lithofacies deposited during a single base-level cycle, which can be mapped across multiple lithofacies tracts, which include multiple vertical lithofacies, and thus are inferred to be chronostratigraphic units" (Kerans and Tinker, 1997). High-frequency cycles are grouped into "cycle sets" which are "bundles of cycles that show a consistent trend, either progradational, aggradational, or retrogradational (transgressive)" (Kerans and Tinker, 1997). The next level of the hierarchy introduces the term "high-frequency sequence" which are intermediate-order cycles bounded locally by unconformities. High-frequency sequences may have all the attributes of the original Vail et al. (1977) sequence, with the complete suite of systems tracts defined by parasequences and parasequence sets, and they are bounded by surfaces or thin intervals (zones) which demonstrate "base-level fall to base-level rise turn-arounds" (Kerans and Tinker, 1997). High-frequency sequences *can equate to or can be part of* "composite

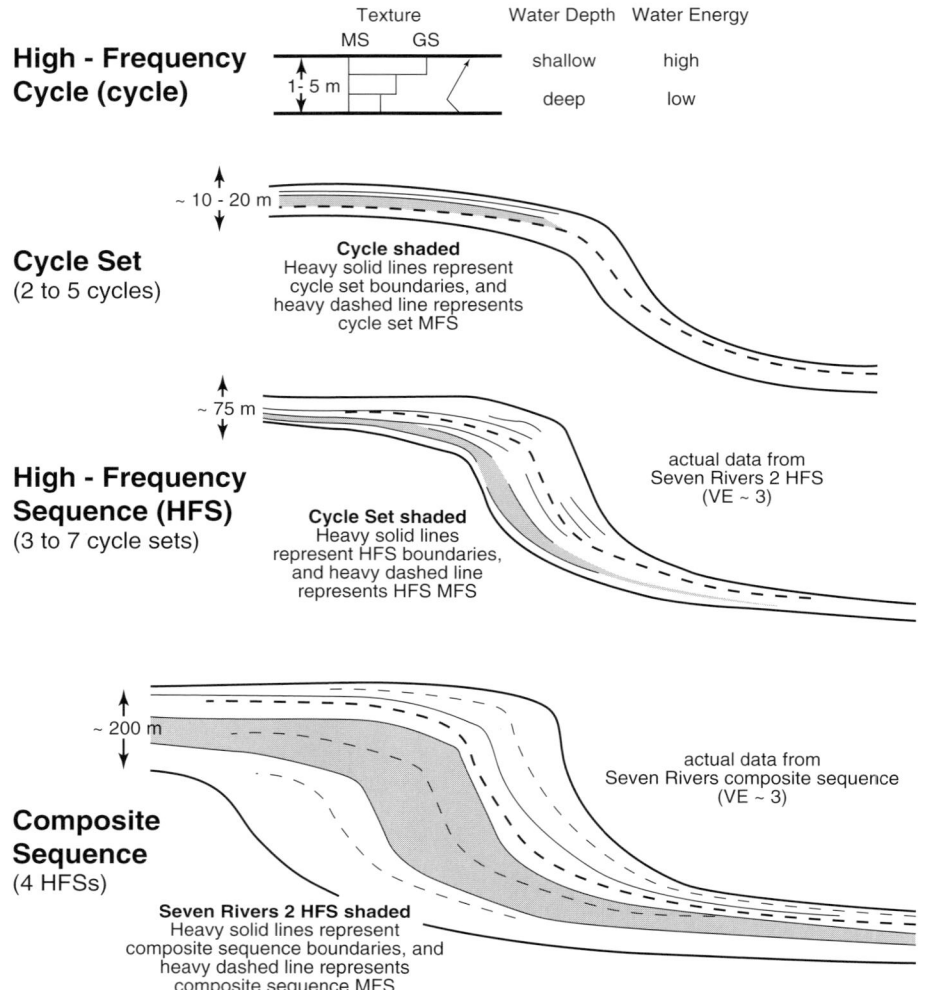

Figure C85 An integrated hierarchical subdivision utilizing sequence stratigraphic terminology (modified from Tinker, 1998).

sequences" which is the next level in the hierarchy, often, but not necessarily, reserved for the Vail et al. (1977) original "third-order" sequence.

The term composite sequence, proposed by Mitchum and Van Wagoner (1991), resulted from the realization by these workers that in clastic depositional systems the original Vail et al. (1977) "sequence" itself could be composed of several unconformity-bounded packages or "sequences". On a parallel course in the carbonate arena, Goldhammer et al. (in Franseen et al., 1991) grouped sets of "fourth-order sequences" (30–50 m thick) bounded by subaerial exposure surfaces into "third-order" sequences for mixed shallow marine clastic and carbonates.

In summary, the hierarchical approach universal in the recognition, classification and interpretation of cycles recognizes that cycles of progressively shorter duration, or "higher-order," are physically contained within longer-term cycles (Goldhammer et al., 1990; Mitchum and Van Wagoner, 1991). Unraveling the hierarchy of cycles should be a task which is independent of assigning an origin to the cycles or quantifying forcing functions responsible for cycle formation (e.g., calculating amplitudes of relative sea-level rise/fall). The cyclo-stratigrapher's goal is to create a physical stratigraphic hierarchy of cycles contained within cycles contained within cycles, whereby one can then evaluate the manner by which facies arrangements within cycles change in a progressive, organized manner according to position within the next larger scale (e.g., "lower-order") cycle (Goldhammer et al., 1993; Sonnenfeld, 1996). Goldhammer et al. (1990) termed this concept "the hierarchy of stratigraphic forcing" in their study of cyclic Alpine Triassic carbonates, an objective hierarchy which they state is independent of the forcing function (allocyclic or autocyclic).

Postulated causes of cycles—first, second, and third-order

Once one accepts the plausibility of an ordered, yet imperfect, existence of stratigraphic cycles, the next challenge is unravelling the causes of cycles, at all scales. For ideas, concepts and discussions of this topic, the reader is referred to Allen and Allen (1990), Einsele *et al.* (1991), Walker and James (1992), Miall (1990, 1997), Reading (1996), and Emery and Myers (1996). Stratigraphic cycles at all orders result from the interplay of tectonics (typically in the form of basin subsidence), eustatic sea-level changes (of differing magnitude and duration), and sediment supply. Other more localized factors, such as antecedent topography, compaction, windward-leeward effects, etc, while clearly significant and worth evaluating, are generally not regarded as the prime suspects influencing the development of cycles. Together, tectonics and eustasy basically create accommodation space available for sedimentation, and stratigraphic cycles result from the intricate interplay of accommodation and sediment supply.

Allen and Allen (1990), Miall (1990, 1997), Walker and James (1992), and Reading (1996) summarize both tectonic and eustatic mechanisms that dictate the creation of accommodation within different types of basins. These sources and many others provide numerous examples of how such space is filled, and what influences sediment supply which together generate the cyclicity observed in the stratigraphic record. Reduced to the essential (Allen and Allen (1990), tectonic mechanisms that effect accommodation include: (1) effects of lithospheric flexure in (divergent margin) basins due to stretching (either brittle crustal extension or thermal subsidence); (2) the effects of flexural rigidity on peripheral bulge formation within (convergent margin) foreland basins; (3) changes in regional stress fields of the lithosphere inducing vertical movements ("intra-plate" stress).

Mechanisms for inducing eustatic changes in sea-level (Allen and Allen, 1990) which in turn influence accommodation include: (1) continuing differentiation of lithospheric material as a result of plate tectonic processes, whereby the volume of water in the oceans may be added by contributions from mid-oceanic ridge and island arc volcanism; (2) changes in the volumetric capacity of the ocean basins caused by sediment accumulation or by sediment abstraction, such that inequalities might exist between rates of erosion and sediment input into oceanic basins versus rates of sediment removal at subduction zones; (3) changes in the volumetric capacity of the ocean basins induced by volume changes in the mid-ocean ridge system; as the bathymetry of the ocean is dependent on age, variations in rates of spreading causes major ridge-volume changes with the faster spreading ridges being hotter and thus more voluminous; (4) reduction of available water by locking up in polar ice caps and glaciers (and the reverse), termed "glacio-eustasy"; (5) desiccation events due to evaporation of inland basins and marginal seas, for example the Miocene Mediterranean salinity crisis; such events could cause global sea-level rise in other oceanic basins.

As these factors control accommodation space over potentially large regions, it is the interaction of these factors, at different magnitudes and over varying timescales that yields the cyclic rock record. Given the fact that these variables operate independently and their effects (in terms of magnitudes and rates of space creation or destruction) are not accurately constrained, it is little wonder why the origin of cycles has long been and continues to be a major issue of heated controversy in the geologic community. Having stated the obvious there is general agreement over the postulated origins of the different orders of cyclicity. First-order cycles (200–400 m year duration) are ascribed to supercontinent assembly (i.e., Pangaea) by seafloor spreading and their subsequent breakup and dispersal. Eustatic sea level falls during times when supercontinents are assembled and rises during times of rapid seafloor spreading when continents have rifted apart and are drifting. Two, first-order, or supercontinent, cycles appear to have occurred during the Phanerozoic (Figure C86), and global first-order global climate fluctuations, from icehouse to greenhouse conditions, have accompanied these cycles.

Second-order cycles are attributed to volume changes in oceanic spreading centers—with more rapid spreading and hotter ridges, ridge volume increases and eustatic sea level rises; with a decrease in spreading rate and cooler ridges, ridge volume decreases and sea level falls. As discussed above, broad, regional cycles of changes in basement elevation due to lithospheric flexure which occur in response to divergent, convergent and transcurrent plate motions are also candidates. For example the Sloss (1963) second-order supersequences, major unconformity-bounded packages that correlate not only across North America, but also with other regional cycles on other continents demand a global origin.

Of the various possible causes of third-order eustatic changes only two are plausible candidates for third-order cyclicity (see Fairbridge, 1961; Donovan and Jones, 1979; Allen and Allen, 1990; Einsele *et al.*, 1991; Walker and James, 1992; Miall, 1990, 1997; Emery and Myers, 1996; Reading, 1996): (1) waxing and waning of polar ice caps (glacio-eustasy), and (2) changes in the volume of ocean basins (tectono-eustasy). Glacio-eustatic rates of change (up to 1 cm/year) and total amount of change (150 m) can readily account for third-order rates and magnitude (rates = 1–10 cm/1000 years; magnitude = 50–300 m; Kendall and Schlager, 1981; Hancock and Kauffman, 1979; Hine and Steinmetz, 1984) but for geologic periods lacking evidence for major continental glaciation, tectono-eustasy must be invoked as the cause. Hays and Pitman (1973) and Pitman (1978) have illustrated that changes in rates of mid-oceanic spreading centers can induce sea-level changes of up to 300 m at a maximum rate of 1 cm/1000 years. However, tectono-eustasy is unable to account for all of the third-order cycles in the stratigraphic record, because most third-order cycles demand faster rates. Thus the origin and control of third-order cycles remains problematic, especially for periods lacking evidence for major glaciation.

Postulated causes of cycles—fourth and fifth-order

Essentially four mechanisms have been proposed to explain the successive stacks of fifth/fourth-order cycles that characterize shallow marine platform carbonate, shallow shelf clastic, mixed clastic-carbonate, and mixed carbonate evaporite deposits of all ages: (1) high-frequency glacio-eustatic oscillations forced by deterministic Milankovitch climatic rhythms; (2) high-frequency aperiodic eustatic oscillations forced by stochastic processes; (3) autocyclic mechanisms specific to either carbonate or clastic systems; (4) tectonic pulsing of marine shelves related to successive down-dropping of platform tops or tectonic "yo-yoing".

(1) Glacio-eustatic (Milankovitch) cycles

Recently, many workers have sought to explain the origin of high-frequency cycles through a glacio-eustatic, allocyclic mechanism driven by climatic cycles under orbital control (Hays *et al.*, 1976; Berger, 1984; Imbrie, 1985; Hinnov, 2000),

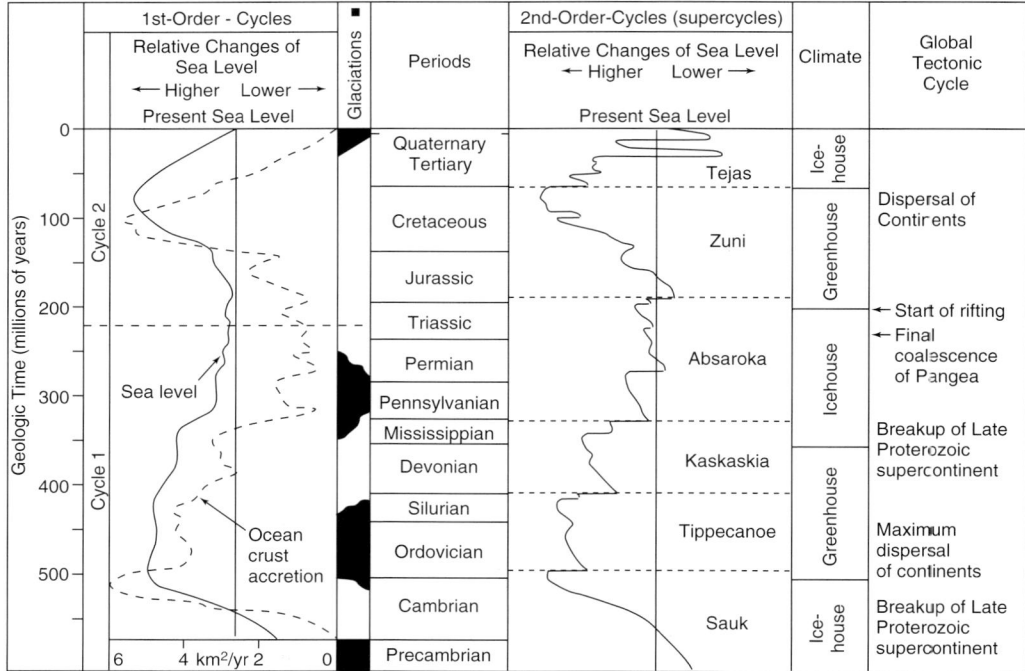

Figure C86 Illustration of first-order and second-order global sea-level cycles (after Boggs, 2001).

which is a rather deterministic mechanism. Deterministic processes are in essence entirely predictable and involve order with periodicity. Deterministic sytems are appealing because of their predictive capability. The general acceptance of the Milankovitch theory for Pleistocene cycles has been provided by the correlation of the Pleistocene deep sea and coral reef record with Milankovitch climatic rhythms, for example, as suggested for the Pleistocene carbonate stratigraphy of south Florida (Broecker and Thurber, 1965; Osmond et al., 1965; Perkins, 1977). An outgrowth of the Milankovitch revival is the question of whether orbital variations have produced a legible stratigraphic signal in pre-Pleistocene cyclic stratigraphy. Pursuit of the feasability of pre-Pleistocene Milankovitch forcing in carbonate systems is exemplified by studies published over the last several years (Grotzinger, 1986; Goldhammer et al., 1987; Strasser, 1988; Koerschner and Read, 1989; Jimenez De Cisneros and Vera, 1993; D'Argenio et al., 1997; Hinnov, 2000).

Milankovitch astronomical rhythms might influence clastic, carbonate and evaporite sedimentation in three ways: (1) by causing variations in continental ice cap volumes, which in turn would cause eustatic sea-level changes; (2) by causing ocean temperature to change sufficiently that ocean water volume would change; (3) by causing climatic changes in the tropics so that shallow water carbonate production would change.

Waxing and waning of continental ice sheets would yield rhythms of submergence and emergence across shallow marine shelves (Hardie and Shinn, 1986). Such submergence-emergence glacio-eustatic oscillations would produce repetitive shallowing-upward cycles, given proper conditions of sedimentation and subsidence. More specifically, the earth experiences three climatic cycles: (1) approximately 20,000 year precessional cycle; (2) approximately 40,000 year obliquity cycle; (3) approximately 100,000 year eccentricity cycle. Each results from the gravitational interaction of the Earth with the moon and other planetary bodies (see *Milankovitch Cycles*, and Hinnov, 2000). These climatic cycles occur with different, yet somewhat predictable frequencies, and cause the Earth's average temperature in and around the polar regions to increase and decrease in a cyclical manner. Alternate warming and cooling of the polar regions triggers the melting and growth of land-locked glaciers, which in turn modulates sea-level change.

To establish the existence of Milankovitch forcing, some workers have simply compared calculated cycle periods or a range of periods with Milankovitch values (e.g., Heckel, 1986; Koerschner and Read, 1989). Others have noted the bundling of small-scale, fifth-order cycles into larger-scale fourth-order cycles and compared this stratigraphic hierarchy to the Milankovitch hierarchy of superimposed orders of climatic cyclicity (Goodwin and Anderson, 1985; Goldhammer et al., 1987; Strasser, 1988; Crevello, in Franseen et al., 1991). In light of age-date errors, the calculation of cycle periods and a comparison with predicted Milankovitch frequencies is fraught with hazards (Hardie et al., 1986; Algeo and Wilkinson, 1988). Analysis of cycle stacking patterns and comparison of cycle bundles with Milankovitch ratios alone provides ambiguous results.

Time series analysis provides the most reliable means of demonstrating Milankovitch frequencies within platform carbonates (e.g., de Boer and Smith, 1991; Einsele et al., 1991; Hinnov, 2000). This approach basically scans stratigraphic data sets for cyclic signatures and compares the extracted frequencies with known Milankovitch frequency ratios. However a serious problem regarding time series analyzes of ancient platform cyclic sequences pertains to the completeness of the cyclic succession. Inherent in traditional time series analysis is the assumption that the cyclic record is complete and that each recorded stratigraphic rhythm equates directly to one rhythmic pulse of the cyclic-producing mechanism (e.g., glacio-eustasy; Duff et al., 1967, p.13). In examining the relationship between cyclic events in time (i.e., rhythms) and the resulting stratigraphic record, it must be pointed out that shallow marine shelves (clastic or carbonate) may depict at best a rather "faulty recording mechanism" (Scwarzacher, 1975). Despite a regular, rhythmic time pattern of cyclic-producing events (e.g., Milankovitch-driven glacio-eustasy), the corresponding stratigraphic record may not preserve each and every potential cyclic-producing pulse due to complexities of sedimentation.

Glacio-eustatic control on cyclicity has received considerable attention for mixed clastic-carbonate cyclothems of the Mid-continent USA and Europe, which have been inferred to result from the waxing and waning of large Gondwanan ice caps (Wanless and Shepard, 1936; Wheeler and Murray, 1957; Wanless and Cannon, 1966; Crowell, 1978; Ramsbottom, 1979). More recently, Heckel (1986, 1989) and Ross and Ross (1987) have advocated a direct Milankovitch control on the cycles characteristic of the interior and southwestern regions of the USA. Original glacial proponents (Wanless and Shepard, 1936) drew upon the widespread distribution of cyclothems, extreme lateral continuity of lithofacies, and the occurrence of age-equivalent cyclothems on different continents to support their viewpoint.

(2) High-frequency stochastic cycles

The "Milankovitch revival" is currently under attack largely due to the difficulties associated with "proving" a connection between cyclic sections and deterministic astronomical forcing (Drummond and Wilkinson, 1993a,b,c; Satterly, 1996; Wilkinson et al., 1997). This has opened the door for evaluating the influence/control of stochastic mechanisms. Stochastic processes are entirely random and depict disorder without periodicity. Stochastic phenomena are unpredictable. High-frequency eustasy may be in-part or entirely stochastic. Either the periodicity of high-frequency oscillations, or the amplitudes, or both, may be forced by stochastic mechanisms. Likewise, carbonate, clastic and evaporite sedimentation processes, or the manner whereby such processes act to fill accommodation space may be stochastic. Either way, it is feasible that high-frequency cycles in the geologic record may be the result of stochastic high-frequency eustasy.

For example in shallow marine carbonate platform systems, an emerging "random" camp has attacked the fundamental concept of a hierarchical grouping of cyclicity in carbonates as part of an effort to demonstrate that cycle stacking patterns are entirely random (Drummond and Wilkinson, 1993a,b,c; Satterly, 1996; Wilkinson et al., 1997). They maintain that cycle populations are erratic, bounced about by random unpredictable factors wiping out whatever deterministic signal may exist. These authors are thus shifting attention away from the predictive capacity of cyclic stratigraphy and the utility of using carbonate cycle stacking patterns as a tool for evaluating the origin of depositional sequences. This camp has applied statistical analyzes to cyclic carbonate successions and conclude that the vertical arrangement of high-frequency cycles and subfacies is random. Furthermore, as the distribution is random, they claim that the process is more than likely random and appeal to autocyclic phenomena in the spirit of complex Holocene facies models. In contrast, Sadler et al. (1993), using similar statistical tests, concluded that the vertical organization of long strings of high-frequency cycles are decidedly non-random.

Time series analyzes of stratigraphic sections typically do not reveal resonant periodicity and thus it is tempting to conclude that stochastic processes dominate the formation of cyclic carbonates. But recalling that the stratigraphic machine is at best a faulty recording mechanism, a deterministic driver may yield a stochastic output via the "stratigraphic filter". The burden of proof is just as challenging as with deterministic systems. If empirical observations of vertical stacking patterns, and the vertical/lateral distribution of facies are rendered entirely unpredictable, then perhaps the system is random and stochastic stratigraphic models are appropriate. It is this author's opinion however that there is always some element of order, and this is both age-dependent reflecting secular geologic changes, and scale-dependent (e.g., Lehrmann and Goldhammer, 1999).

(3) Autocycles

Based on Holocene insight, Ginsburg (1971) formulated a progradation-driven autocyclic mechanism that readily accounts for repetitious stacks of meter-scale fifth- and fourth-order platform carbonate cycles. This model couples continuous carbonate sedimentation and progradation of tidal flat complexes across gently sloping carbonate platforms under conditions of constant subsidence and stationary sea-level (e.g., Hardie and Shinn, 1986). The subtidal carbonate factory provides the sediment source which is transported shoreward, accreting against exposed land masses or islands. The tidal flat complex progrades out and over the subtidal deposits yielding a shallowing-upward cycle capped by progradational tidal flat facies. The key aspect of the model is that carbonate production rates will be unable to keep pace with subsidence as tidal flat progradation diminishes the size of the "subtidal factory" and eventually the progradational cycle will halt. With passive subsidence, a minimum "lag depth" (the depth required to resume carbonate production) of a meter or two is achieved initiating carbonate sedimentation, and hence a new cycle.

Utilizing typical rates of passive margin subsidence (1–20 cm/1000 years) and known Holocene tidal flat progradation rates (0.5–20 km/1000 years), Ginsburg autocycles can easily prograde across platforms up to 100 km in width within the Milankovitch time band (Hardie and Shinn, 1986). During geologic periods when no continental land masses covered the polar regions, for example the Cambrian and Lower Ordovician, the influence of Milankovitch-forced glacio-eustasy would be minimized, and the feasibilty of autocycles would be enhanced.

In clastic systems, autocyclic processes are well-known. For example, in Pennsylvanian mixed clastic and carbonate cycles, autocyclic influences were suggested by Van der Heide (1950) who speculated that compaction of peat layers in clastic portions of cyclothems in the Netherlands could have induced sudden marine transgressions contributing to cyclic deposition. More recently, autocyclic models of delta-lobe shifting have been applied to Upper Pennsylvanian mixed clastic-carbonate cyclothems (Galloway and Brown, 1973; Elliot, 1976).

Recognizing that most of the deltaic deposits occur within the regressive portions of cylothems, Heckel (1989) viewed autocyclic delta-lobe switching as a subordinate process operating within an overall eustatic framework. Other examples of clastic autocyclicity include fluvial avulsion in nonmarine deposits, channel avulsion in deep marine submarine fan deposits, and so on. Other examples of proposed autocyclic control on cyclic sequences includes Bosellini and Hardie (1973), Matti and McKee (1976), Mossop (1979), and Wilkinson (1982).

(4) Tectonic cycles

Another viable cycle-producing mechanism is that of tectonically dictated cycles. Repeated tectonic pulses of down-dropping and uplift (reversal or "yo-yo" tectonics) could induce meter-scale cyclicity by alternately submerging and exposing (subaerially) shallow shelf areas (Hardie and Shinn, 1986). Episodic, meter-scale down-dropping driven by faulting could also create accommodation space leading to shallowing-upward cycles (Hardie et al., 1991). With both models, the accommodation space required for each cycle would be controlled by meter-scale pulses that might be somewhat rhythmic if a threshold stress must be achieved to induce crustal rupturing (Hardie and Shinn, 1986).

Modern data concerning reversal tectonics is scarce and information regarding recurrence frequency, amount of vertical displacement, and the size of areas involved is lacking. Cisne (1986) has outlined a model of stick-slip faulting at the edges of carbonate platforms that demands a fault-bounded shelf-edge. In this model, the platform edge and the updip area located within a hundred kms of the shelf edge would experience high-frequency crustal oscillations due to reversal tectonics (Cisne, 1986). While this model is intriguing, there is no way of estimating recurrence frequencies and it necessitates block faulting coincident with a platform shelf edge.

Presently, there is sufficient data regarding tectonic forcing through step-wise subsidence to render episodic subsidence a plausible mechanism for the generation of meter-scale sedimentary cycles and larger sequences (Hardie et al., 1991). For example, modern manifestations of coseismic subsidence over large areas (equal to or greater than the size of many carbonate platforms) associated with very large earthquakes are reported along convergent margins and in intraplate regions (Plafker, 1965; Atwater, 1987). Although large areas can be downdropped a few to several meters, data regarding recurrence frequencies is sparse, suggesting only that recurrence frequencies span the 10–100 ka band (Hardie et al., 1991).

Early proponents of tectonically-driven high-frequency cyclcity include Stout (1931), who interpreted Mid-Continent cyclothems in Ohio as records of periodic downdropping of the basin, inducing transgression, followed by steady sediment infill (i.e., a form of episodic subsidence). Weller's (1930, 1956) "diastrophic control theory" centered upon alternate pulsating uplift and depression (reversal tectonics) of large continental areas with the magnitude being greatest in the uplands, declining basinward. Periodic episodes of uplift resulted in termination of marine deposition (inferred to occur at the top of a cyclothem), erosion and influx of clastic (nonmarine) material.

Quantitative analyzes of cycles

Our understanding of cyclic sedimentation has been greatly advanced through the use of quantitative analyzes and computer-aided modeling of cycles and sequences. There are basically two approaches: (1) inverse modeling, whereby stacks of cyclic strata are analyzed by some statistical method, rigorous time series or spectral analysis, and graphical techniques for evaluating long strings of cyclic successions; (2) forward modeling, which normally entails computer-assisted simulations of synthetic stratigraphic sections by user-defined manipulation of the stratigraphic governing factors such as eustasy (periodicities, amplitudes, curve shapes), subsidence (rates, curve shape, compaction corrected, etc.), and sedimentation (carbonates, clastics and evaporites, etc.).

Statistical evaluation of vertical facies successions which define cycles has often been invoked (e.g., Merriam, 1964; Duff et al., 1967; Krumbein and Dacey, 1969; Carr, 1982; Davis, 1986) either to assist in defining cycles, or to verify their inferred existence. More recently, (Drummond and Wilkinson, 1993a,b,c; Satterly, 1996; Wilkinson et al., 1997; Lehrmann and Goldhammer, 1999) have employed statistical routines, including runs tests, Durbin-Watson test, and entropy analysis (Hattori, 1976) to evaluate the long-term order of thick stacks of cyclic platform carbonates. The point of such analyzes is to somehow quantify the degree of order in vertical successions of high-frequency cycles. Application of time series and spectral analysis (e.g., Schwarzacher, 1975; Schawarzacher and Fischer, 1982; Weedon, 1989; Hinnov and Park, 1998; Bazykin and Hinnov, 2001) to a succession of cycles enables one to detect superimposed orders of relative sea-level changes expressed as a hierarchy of stratigraphic cycles. The main assumption is that of equal periodicity per cycle and essentially no gaps in the data set.

Perhaps the most popular graphical technique applied to the analysis of cyclic strata is the method of Fischer (1964), coined "Fischer Plots" by Goldhammer et al. (1987) and Read and Goldhammer (1988). These plots graphically portray deviations in a successive string of cycles from the mean thickness for the string as a whole (Sadler et al., 1993). Such plots are intended as a graphically-appealing manner by which one can visualize long-term stacking patterns of one-dimensional stacks of cycles, particularly when used in conjunction with cycle-specific facies proportions (e.g., Goldhammer et al., 1993; Kerans and Tinker, 1997). Several workers have used Fischer plots to attempt to correlate third-order cyclic sequences from one part of a basin to another or from one basin to another (Read and Goldhammer, 1988; Osleger and Read, 1991; Montanez and Osleger, 1993).

The development of computer simulation models that simulate synthetic stratigraphy dramatically evolved from rather simple one-dimensional models of the early 1980s (Turcotte and Willeman, 1983; Read et al., 1986; Goldhammer et al., 1987) to sophisticated two- and three-dimensional models through the mid-1990s (Dunn, 1991; Franseen et al., 1991; Miall, 1997). The essential features of these and many other simulation programs are similar in as far as they all allow the user to define a variety of parameters (such as eustasy, subsidence, etc.) according to a variety of rules (e.g., depth-dependent sedimentation, sedimentation driven by diffusion gradients, etc.) and generate synthetic (and often very realistic) output. In this author's opinion, their value lies not in the precise simulation of a particular section; rather they are invaluable as an educational tool which can aid one in evaluating the effects of certain parameters and the manner in which they interact.

Conclusions

Sedimentary cycles have several important implications for furthering our understanding of the stratigraphic machine, including: (a) as shallow-marine siliciclastic and carbonate (and evaporite) sedimentation normally exceeds background subsidence, such deposits often accrete to sea level, and hence repetitive shoaling sequences (i.e., cycles) record information about relative changes in sea level; (b) high-frequency cycles commonly show remarkable lateral continuity and the cycles, as well as associated unconformities, are normally regarded as synchronous, not diachronous stratigraphic deposits. Thus their potential for stratigraphic correlation should not be overlooked; (c) marine high-frequency cycles often correspond to one relative rise and fall of sea level and cut across lateral facies changes and thus they are high-resolution time-stratigraphic units; (d) if cycles are indeed rhythmic, then cycles can be used to calibrate geologic time, provided the cycle periodicity is known; (e) high-frequency evaporite cycles in marginal-marine and nonmarine settings, and lacustrine cycles record perhaps the clearest record of stratigraphic accumulation driven by ancient climate variations controlled by orbital forcing (Milankovitch theory of climate).

Robert K. Goldhammer

Bibliography

Algeo, T.J., and Wilkinson, B.H., 1988. Periodicity of mesoscale Phanerozoic sedimentary cycles and the role of Milankovitch orbital modulation. *Journal of Geology*, **96**: 313–322.

Allen, P.A., and Allen, J.R., 1990. *Basin Analysis, Principles and Applications*. Oxford: Blackwell Scientific.

Astin, T.R., 1990. The Devonian lacustrine sediments of Orkney, Scotland; implications for climate cyclicity, basin structure and maturation history. *Journal of the Geological Society of London*, **147**: 141–151.

Atwater, B.F., 1987. Evidence for great Holocene earthquakes along the outer coast of Washington state. *Science*, **236**: 942–944.

Bazykin, D.A., and Hinnov, L.A., 2001. Orbitally-driven depositional cyclicity of the Lower Paleozoic Aisha-Bibi seamount (Malyi Karatau, Kazakstan): integrated sedimentological and time series study. In Zempolich, W.G., and Cook, H.E. (eds.), *Paleozoic Carbonates of the Commonwealth of Independent States (CIS): Subsurface Reservoirs and Outcrop Analogs*. SEPM (Society for Sedimentary Geology), Special Publication, in press.

Berger, A.L., 1984. Accuracy and frequency stability of the Earth's orbital elements during the Quaternary. In Berger, A., Imbrie, J., Hays, J., Kukla, G., and Saltzman, B. (eds.), *Milankovitch and Climate*. Boston: Reidel Publishing Co., pp. 3–39.

Bloom, A.L., Broecker, W.S., Chappell, J.M.A., Matthews, R.K., and Mesolella, K.J., 1974. Quaternary sea level fluctuations on a tectonic coast. *Quaternary Research*, **4**: 185–205.

Boggs, S. Jr., 2001. *Principles of Sedimentology and Stratigraphy*. New Jersey: Prentice Hall.

Bosellini, A., and Hardie, L.A., 1973. Depositional theme of a marginal marine evaporite. *Sedimentology*, **20**: 5–17.

Broecker, W.S., and Thurber, D.L., 1965. Uranium-series dating of corals and oolites from Bahaman and Florida Key limestones. *Science*, **149**: 58–60.

Carr, T.R., 1982. Log-linear models, Markov chains and cyclic sedimentation. *Journal of Sedimentary Petrology*, **52**: 905–912.

Cisne, J.L., 1986. Earthquakes recorded stratigraphically on carbonate platforms. *Nature*, **323**: 320–322.

Clough, J.G., and Goldhammer, R.K., 2000. Evolution of the Neoproterozoic Katakturuk Dolomite ramp complex, northeastern Brooks Range, Alaska. In Grotzinger, J.P., and James, N.P. (eds.), *Carbonate Sedimentation and Diagenesis in the Evolving Precambrian World*. SEPM (Society for Sedimentary Geology). 67, pp. 209–241.

Crowell, J.C., 1978. Gondwana glaciation, cyclothems, continental positioning, and climate change. *American Journal of Science*, **278**: 1345–1372.

D'Argenio, B.D., Ferreri, V., Amodio, S., and Pelosi, N., 1997. Hierarchy of high-frequency orbital cycles in Cretaceous carbonate platform strata. *Sedimentary Geology*, **113**: 169–193.

Davis, J.C., 1986. *Statistics and Data Analysis in Geology*, 2nd edn. New York: Wiley.

de Boer, P.L., and Smith, D.G. (eds.), 1994. *Orbital Forcing and Cyclic Sequences*. International Association of Sedimentologists, Special Publication, 19. Oxford: Blackwell Scientific.

Donovan, D.T., and Jones, E., 1979. Causes of world-wide changes in sea level. *Journal of the Geological Society of London*, **136**: 187–192.

Drummond, C.N., and Wilkinson, B.H., 1993a. On the use of cycle thickness diagrams as records of long-term sealevel change during accumulation of carbonate sequences. *The Journal of Geology*, **101**: 687–702.

Drummond, C.N., and Wilkinson, B.H., 1993b. Aperiodic accumulation of cyclic peritidal carbonates. *Geology*, **21**: 1023–1026.

Drummond, C.N., and Wilkinson, B.H., 1993c. Carbonate cycle stacking patterns and hierarchies of orbitally forced eustatic sealevel change. *Journal of Sedimentary Petrology*, **63**: 369–377.

Duff, P.McL.D., Hallam, A., and Walton, E.K., 1967. *Cyclic Sedimentation*. Developments in Sedimentology 10. New York: Elsevier.

Dunn, P.A., 1991. *Diagenesis and cyclostratigraphy: an example from the Middle Triassic Latemar platform, Dolomites Mountains, northern Italy*. Baltimore, Maryland: unpublished Ph.D. dissertation, The Johns Hopkins University.

Einsele, G., Ricken, W., and Seilacher, A. (eds.), 1991. *Cycles and Events in Stratigraphy*. New York: Springer-Verlag.

Elam, J.C., and Chuber, S. (eds.), 1972. *Cyclic Sedimentation in the Permian Basin*, 2nd edn, Midland, Texas: West Texas Geologic Society.

Elliot, T., 1976. Upper Carboniferous sedimentary cycles produced by river-dominated, elongate deltas. *Journal of the Geological Society of London*, **132**: 199–208.

Emery, D., and Myers, K.J. (eds.), 1996. *Sequence Stratigraphy*. London: Blackwell Science Ltd.

Eriksson, K.A., and Simpson, E.L., 1990. Recognition of high-frequency sea-level fluctuations in Proterozoic siliciclastic tidal deposits, Mount Isa, Australia. *Geology*, **18**: 474–477.

Fairbridge, R.W., 1961. Eustatic changes in sea level. In Ahrens, L.A. (ed.), *Physics and Chemistry of the Earth*. London: Pergamon Press, pp. 99–185.

Fischer, A.G., 1986. Climatic rhythms recorded in strata. *Annual Reviews of Earth and Planetary Science*, **14**: 351–376.

Fitchen, W.M., 1997. Carbonate sequence stratigraphy and its application to hydrocarbon exploration and reservoir development. In Ibrahim, P., and Marfurt, K.J. (eds.), *Carbonate Seismology*. Society of Exploration Geophysics, Geophysical Development Series no. 6, pp. 121–178.

Franseen, E.K., Watney, W.L., Kendall, G.C.St.C., and Ross, W. (eds.), 1991. *Sedimentary modeling: Computer simulations and methods for improved parameter definition*. Kansas Geological Survey, Special Publication, 233.

Galloway, W.E., and Brown, L.F., Jr., 1973. Depositional systems and shelf-slope relations on cratonic basin margin, uppermost Pennsylvanian of north-central Texas. *American Association of Petroleum Geologists Bulletin*, **57**: 1185–1218.

Ginsburg, R.N., 1971. Landward movement of carbonate mud: new model for regressive cycles in carbonates, (abstract). *American Association of Petroleum Geologists Bulletin*, **55**: 340.

Goldhammer, R.K., Dunn, P.A., and Hardie, L.A., 1987. High frequency glacio-eustatic sea level oscillations with Milankovitch characteristics recorded in Middle Triassic platform carbonates in northern Italy. *American Journal of Science*, **287**: 853–892.

Goldhammer, R.K., Dunn, P.A., and Hardie, L.A., 1990. Depositional cycles, composite sea level changes, cycle stacking patterns, and the hierarchy of stratigraphic forcing: examples from platform carbonates of the Alpine Triassic. *Geological Society of America Bulletin*, **102**: 535–562.

Goldhammer, R.K., and Harris, M.T., 1989. Eustatic controls on the stratigraphy and geometry of the Latemar buildup (Middle Triassic), the Dolomites of northern Italy. In Crevello, P., Sarg, J.F., Read, J.F., and Wilson, J.L. (eds.), *Controls on Carbonate Platform to Basin Development*. Society of Economic Paleontololologists and Mineralogists Special Publication no. 44, pp. 323–338.

Goldhammer, R.K., Lehmann, P.J., and Dunn, P.A., 1993. The origin of high frequency platform carbonate cycles and third-order sequences (Lower Ordovician El Paso Gp, West Texas); constraints from outcrop data and stratigraphic modeling. *Journal of Sedimentary Research*, **63**: 318–359.

Goodwin, P.W., and Anderson, E.J., 1985. Punctuated aggradational cycles: a general hypothesis of episodic stratigraphic accumulation. *Journal of Geology*, **93**: 515–533.

Grabau, A.W., 1913. *Principles of Stratigraphy*. New York: Seiler and Co.

Grammer, G.M., Eberli, G.P., Van Buchem, F.S., Stevenson, G.M., and Homewood, P., 1996. Application of high-resolution sequence stratigraphy to evaluate lateral variability in outcrop and subsurface—Desert Creek and Ismay intervals, Paradox Basin. In Longman, M.W., and Sonnenfeld, M.D. (eds.), *Paleozoic Systems of the Rocky Mountain Region*. Rocky Mountain Section SEPM (Society for Sedimentary Geology), pp. 235–266.

Grotzinger, J.P., 1986. Cyclicity and paleoenvironmental dynamics, Rocknest platform, northwest Canada. *Geological Society of America Bulletin*, **97**: 1208–1231.

Hancock, J.M., and Kauffman, E.G., 1979. The great transgressions of the Late Cretaceous. *Journal of the Geological Society of London*, **136**: 175–186.

Haq, B.U., Hardenbol, J., and Vail, P.R., 1987. Chronology of fluctuating sea levels since the Triassic. *Science*, **235**: 1156–1166.

Hardie, L.A., and Eugster, H.P., 1971. The depositional environment of marine evaporites: a case for shallow, clastic accumulation. *Sedimentology*, **16**: 187–220.

Hardie, L.A., and Shinn, E.A., 1986. Carbonate depositional environments, modern and ancient, 3, Tidal Flats. *Colorado School of Mines Quarterly*, **81**: 1–74.

Hardie, L.A., Bosellini, A., and Goldhammer, R.K., 1986. Repeated subaerial exposure of subtidal carbonate platforms, Triassic, northern Italy: evidence for high frequency sea level oscillations on a 10,000 year scale. *Paleoceanography*, **1**: 447–457.

Hardie, L.A., Dunn, P.A., and Goldhammer, R.K., 1991. Field and modelling studies of Cambrian carbonate cycles, Virginia Appalachians: a discussion. *Journal of Sedimentary Petrology*, **61**: 636–646.

Hattori, I., 1976. Entropy in Markov chains and discrimination of cyclical patterns in lithologic successions. *Journal of Mathematical Geology*, **8**: 477–497.

Hays, J.D., and Pitman, W.C., 1973. Lithospheric plate motion, sea level changes, and climatic and ecological consequences. *Nature*, **246**: 18–22.

Hays, J.D., Imbrie, J., and Shackleton, N.J., 1976. Variation in the earth's orbit: pacemaker of the ice ages. *Science*, **194**: 1121–1132.

Heckel, P.H., 1986. Sea-level curve for Pennsylvanian eustatic marine transgressive-regressive depositional cycles along midcontinent outcrop belt, North America. *Geology*, **14**: 330–334.

Heckel, P.H., 1989. Current view of Midcontinent Pennsylvanian cyclothems. In Boardman, D.R. (ed.), *Middle and Late Pennsylvanian Chronostratigraphic Boundaries in North-Central Texas: Glacial-eustatic Events, Biostratigraphy, and Paleoecology*. Texas Tech Univ. Studies in Geology 2, pp. 17–34.

Hine, A.C., and Steinmetz, J.C., 1984. Cay Sal Bank, Bahamas—a partially drowned carbonate platform. *Marine Geology*, **59**: 135–164.

Hinnov, L.A., 2000. New perspectives on orbitally forced stratigraphy. *Annual Review of Earth and Planetary Sciences*, **28**: 419–475.

Hinnov, L.A., and Park, J., 1998. Detection of astronomical cycles in the sedimentary record by frequency modulation analysis. *Journal of Sedimentary Research*, **68**: 524–539.

House, M.R., 1985. A new approach to an absolute time scale from measurements of orbital cycles and sedimentary microrhythms. *Nature*, **315**: 721–725.

Imbrie, J., 1985. A theoretical framework for the Pleistocene ice ages. *Journal of the Geological Society of London*, **142**: 417–432.

James, N.P., 1984. Shallowing-upward sequences in carbonates. In Walker, R.G. (ed.), *Facies Models*. Geoscience Canada Reprint Series 1, pp. 213–228.

Jimenez De Cisneros, C., and Vera, J.A., 1993. Milankovitch cyclicity in Purbeck peritidal limestones of the Prebetic (Berriasian, southern Spain). *Sedimentology*, **40**: 513–537.

Kendall, G.St.C., and Schlager, W., 1981. Carbonates and relative changes in sea level. *Marine Geology*, **44**: 181–212.

Kerans, C., 1995. Use of one- and two-dimensional cycle analysis in establishing high-frequency sequence frameworks. In Read, J.F., Kerans, C., and Weber, L.J. (eds.), *Milankovitch Sea Level Changes, Cycles and Reservoirs on Carbonate Platforms in Greenhouse and Ice-House Worlds*. SEPM (Society for Sedimentary Geology) Short Course Notes no. 35, pp. 1–20.

Kerans, C., and Fitchen, W.M., 1995. Sequence stratigraphy and facies architecture of a carbonate ramp system: San Andres Formation of Algerita Escarpment and western Guadalupe Mountains, West Texas and New Mexico. *Bureau of Economic Geology, University of Texas at Austin, Report of Investigations*, No. 235, pp. 1–86.

Kerans, C., and Tinker, S.W., 1997. *Sequence Stratigraphy and Characterization of Carbonate Reservoirs*. SEPM (Society for Sedimentary Geology) Short Course Notes no. 35, pp. 1–130.

Koerschner III, W.F., and Read, J.F., 1989. Field and modelling studies of Cambrian carbonate cycles, Virginia Appalachians. *Journal of Sedimentary Petrology*, **59**: 654–687.

Krumbein, W.C., and Dacey, M.F., 1969. Markov chains and embedded Markov chains in geology. *International Association of Mathematical Geology Journal*, **1**: 79–96.

Lehrmann, D.J., and Goldhammer, R.K., 1999. Secular variation in parasequence and facies stacking patterns of platform carbonates: a guide to applications of stacking patterns analysis in strata of diverse ages and settings. In Harris, P.M., and Simo, J.A.T. (eds.), *Advances in Carbonate Sequence Stratigraphy: Application to Reservoirs, Outcrops and Models*. Society of Economic Paleontologists and Mineralogists, Special Publication, 63, pp. 187–227.

Lowenstein, T.K., 1988. Origin of depositional cycles in a Permian "saline giant": The Salado (McNutt zone) evaporites of New Mexico and Texas. *Geological Society of America Bulletin*, **100**: 592–608.

Matti, J.C., and McKee, E.D., 1976. Stable eustacy, regional subsidence and a carbonate factory: self-regulating model for onlap-offlap cycles in shallow water carbonate sequences (abstract). *Geological Society of America Abstracts with Programs*, **8**: 1000–1001.

Merriam, D.F. (ed.), 1964. *Symposium on Cyclic Sedimentation*. Kansas Geological Survey Bulletin, 169.

Mesolella, K.L., Matthews, R.K., Broecker, W.S., and Thurber, D.L., 1969. The astronomical theory of climatic change: Barbados data. *Journal of Geology*, **77**: 250–274.

Miall, A.D., 1984. *Principles of Sedimentary Basin Analysis*, 2nd edn, New York: Springer Verlag, 1990.

Miall, A.D., 1997. *The Geology of Stratigraphic Sequences*. New York: Springer-Verlag.

Mitchum, R.M., and Van Wagoner, J.C., 1991. High-frequency sequences and their stacking patterns: sequence-stratigraphic evidence of high-frequency eustatic cycles. *Sedimentary Geology*, **70**: 131–160.

Montanez, I.P., and Osleger, D.A., 1993. Parasequence stacking patterns, third order accomodation events, and sequence stratigraphy of Middle to Upper Cambrian platform carbonates, Bonanza King Formation, southern Great Basin. In Loucks, R.G., and Sarg, J.F. (eds.), *Carbonate Sequence Stratigraphy: Recent Developments and Applications*. American Association of Petroleum Geologists Memoir 57, pp. 305–326.

Mossop, G.D., 1979. The evaporites of the Ordovician Baumann Fiord Formation, Ellesmere Island, Arctic Canada. *Geological Survey of Canada Bulletin*, 298, pp. 1–52.

Mutti, E., 1992. *Turbidite Sandstones*. Special Publication of the AGIP Petroleum Corporation. Milan: San Donato Milanese.

Olsen, P.E., 1986. A 40-million-year lake record of early Mesozoic orbital climatic forcing. *Science*, **234**: 842–848.

Osleger, D., and Read, J.F., 1991. Relation of eustasy to stacking patterns of meter-scale carbonate cycles, Late Cambrian, U.S.A. *Journal of Sedimentary Petrology*, **61**: 1225–1252.

Osmond, J.K., Carpenter, J.R., and Windom, H.L., 1965. 230Th/234U age of the Pleistocene corals and oolites of Florida. *Journal of Geophysical Research*, **70**: 1843–1847.

Perkins, R.D., 1977. Depositional framework of Pleistocene rocks in south Florida. *Geological Society of America Memoir* 147, pp. 131–198.

Pitman, W.C., 1978. Relationship between eustasy and stratigraphic sequences of passive margins. *Geological Society of America Bulletin*, **89**: 1389–1403.

Plafker, G., 1965. Tectonic deformation associated with the 1964 Alaskan earthquake. *Science*, **148**, 1675–1687.

Ramsbottom, W.H.C., 1979. Rates of transgression and regression in the Carboniferous of NW Europe. *Geological Society of London Journal*, **136**: 147–153.

Read, J.F., 1995. Overview of carbonate platform sequences, cycle stratigraphy and reservoirs in greenhouse and ice-house worlds. In Read, J.F., Kerans, C., and Weber, L.J. (eds.), *Milankovitch Sea Level Changes, Cycles and Reservoirs on Carbonate Platforms in Greenhouse and Ice-House Worlds*. SEPM (Society for Sedimentary Geology) Short Course Notes no. 35, pp. 1–102.

Read, J.F., Grotzinger, J.P., Bova, J.A., and Koerschner, W.F., 1986. Models for generation of carbonate cycles. *Geology*, **14**: 107–110.

Read, J.F., and Goldhammer, R.K., 1988. Use of Fischer plots to define third-order sea-level curves in Ordovician peritidal cyclic carbonates, Appalachians. *Geology*, **16**: 895–899.

Reading, H.G. (ed.), 1996. *Sedimentary Environments: Processes, Facies and Stratigraphy*, 3rd edn, New York: Blackwell Scientific.

Ross, C.A., and Ross, J.R.P., 1987. Late Paleozoic sea levels and depositional sequences. Cushman Foundation for Foraminiferal Research, Special Publication 24. pp. 137–149.

Sadler, P.M., Osleger, D.A., and Montanez, I.P., 1993. On the labeling, length, and objective basis of Fischer plots. *Journal of Sedimentary Petrology*, **63**: 360–368.

Satterley, A.K., 1996. Cyclic carbonate sedimentation in the Upper Triassic Dachstein Limestone, Austria—the role of patterns of sediment supply and tectonics in a platform-reef-basin system. *Journal of Sedimentary Research*, **66**: 307–323.

Schwarzacher, W., and Fischer, A.G., 1982. Limestone-shale bedding and perturbations of the earth's orbit. In Einsele, G., and Seilacher, A. (eds.), *Cyclic and Event Stratification*. New York: Spinger Verlag, pp. 72–95.

Schwarzacher, W., 1975. *Sedimentation Models and Quantitative Stratigraphy*. Developments in Sedimentology 19. New York: Elsevier.

Sloss, L.L., 1963. Sequences in the cratonic interior of North America. *Geological Society of America Bulletin*, **74**: 93–113.

Sonnenfeld, M.D., 1996. Sequence evolution and hierarchy within the Lower Mississippian Madison Limestone of Wyoming. In Longman, M.W., and Sonnenfeld, M.D. (eds.), *Paleozoic Systems of the Rocky Mountain Region*. Rocky Mountain Section SEPM (Society for Sedimentary Geology), pp. 165–192.

Sonnenfeld, M.D., and Cross, T.A., 1993. Volumetric partitioning and facies differentiation within the Permian Upper San Andres Formation of Last Chance Canyon, Guadalupe Mountains, New Mexico. In Loucks, R.G., and Sarg, J.F. (eds.), *Carbonate Sequence Stratigraphy: Recent Developments and Applications*. American Association of Petroleum Geologists Memoir, 57, pp. 435–474.

Sprenger, A., and Ten Kate, W.G., 1993. Orbital forcing of calcilutite-marl cycles in southeast Spain and an estimate for the duration of the Berriasian stage. *Geological Society of America Bulletin*, **105**: 807–818.

Stout, W.E., 1931. Pennsylvanian cycles in Ohio. *Illinois State Geological Survey Bulletin*, **60**: 195–216.

Strasser, A., 1988. Shallowing-upward sequences in Purbeckian peritidal carbonates, lowermost Cretaceous, Swiss and French Jura Mountains. *Sedimentology*, **35**: 369–383.

Suess, E., 1888. *Das Antlitz der Erder, Volume 2* (English Translation by Sollas, 1906). Oxford: Clarendon Press, pp. 1–254.

Tinker, S.W., 1998. Shelf-to-basin facies distribution and sequence stratigraphy of a steep-rimmed carbonate margin: Capitan depositional system. McKittrick Canyon, New Mexico and Texas. *Journal of Sedimentary Research*, **68**, 1146–1175.

Turcotte, D.L. and Willemann, J.H., 1983. Synthetic cyclic stratigraphy. *Earth and Planetary Science Letters*, **63**: 89–96.

Vail, P.R., 1987. Seismic stratigraphy interpretation procedure. In Bally, A.W., ed., *Atlas of Seismic Stratigraphy, Volume 1*. American Association of Petroleum Geologists Studies in Geology 27, Tulsa: Oklahoma, pp. 1–11.

Vail, P.R., Mitchum, R.M., Jr., and Thompson, S., III, 1977. Seismic stratigraphy and global changes of sea level. In Payton, C.E., (ed.), *Seismic Stratigraphy—Applications to Hydrocarbon Exploration*. American Association of Petroleum Geologists Memoir 26, pp. 83–97.

Vail, P.R., Hardenbol, J., and Todd, R.G., 1984. Jurassic unconformities, chronostratigraphy, and sea-level changes from seismic stratigraphy and biostratigraphy. In Schlee, J.S. (ed.), *Interregional Unconformities and Hydrocarbon Accumulation*. American Association of Petroleum Geologists Memoir 36, pp. 129–144.

Van der Heide, S., 1950. Compaction as a possible factor in upper Carboniferous rhythmic sedimentation. *Report of the 18th International Geologic Congress, part 4*. pp. 38–45.

Van Wagoner, J.C., Mitchum, R.M. Jr., Posamentier, H.W., and Vail, P.R., 1987. The key definitions of stratigraphy. In Bally, A.W. (ed.), *Atlas of Seismic Stratigraphy, Volume 1*. American Association of Petroleum Geologists Studies in Geology 27, Tulsa: Oklahoma, pp. 11–14.

Van Wagoner, J.C., Mitchum, R.M., Campion, K.M., and Rahmanian, V.D., 1990. Siliciclastic sequence stratigraphy in well-logs, cores, and outcrops. *American Association of Petroleum Geologists Methods in Exploration Series*, No. 7, 55p.

Vella, P., 1965. Sedimentary cycles, correlation, and stratigraphic classification. *Transactions of the Royal Society of New Zealand*, **3**: 1–9.

Walker, R.G., and James, N.P. (eds.), 1992. *Facies Models: Response to Sea Level Change*. Geological Association of Canada.

Wanless, H.R., and Shepard, F.P., 1936. Sea level and climate changes related to late Paleozoic cyclothems. *Geological Society of America Bulletin*, **47**: 1177–1206.

Wanless, H.R., and Cannon, J.R., 1966. Late Paleozoic glaciation. *Earth Science Reviews*, **1**: 247–286.

Weedon, G.P., 1989. The detection and illustration of regular sedimentary cycles using Walsh power spectra and filtering, with examples from the Lias of Switzerland. *Journal of the Geological Society of London*, **146**: 133–144.

Weller, J.M., 1930. Cyclical sedimentation of the Pennsylvanian period and its significance. *Journal of Geology*, **38**: 97–135.

Weller, J.M., 1956. Argument for diastrophic control of late Paleozoic cyclothems. *American Association of Petroleum Geologists Bulletin*, **40**: 17–50.

Wells, A.J., 1960. Cyclic sedimentation: a review. *Geological Magazine*, **97**: 389–403.

Wheeler, H.E., 1964. Baselevel, lithosphere surface, and time-stratigraphy. *Geological Society of America Bulletin*, **75**: 599–610.

Wheeler, H.E., and Murray, H.H., 1957. Base level control patterns in cyclothemic sedimentation. *American Association of Petroleum Geologists Bulletin*, **41**: 1985–2011.

Wilkinson, B.H., Diedrich, N.W., and Drummond, C.N., 1996. Facies successions in peritidal carbonate sequences. *Journal of Sedimentary Research*, **66**: 1065–1078.

Wilkinson, B.H., Drummond, C.N., Rothman, E.D., and Diedrich, N.W., 1997. Stratal order in peritidal carbonate sequences. *Journal of Sedimentary Research*, **67**: 1068–1082.

Wilkinson, B.H., 1982. Cyclic cratonic carbonates and Phanerozoic calcite seas. *Journal of Geological Education*, **30**: 189–203.

Wilson, J.L., 1975. *Carbonate Facies in Geologic History*. New York: Springer-Verlag.

D

DEBRIS FLOW

A debris flow is a type of sediment gravity flow. It is a rapid mass movement of a concentrated mixture of sediment, organic matter, and water that can flow like a liquid yet can stop on sloping surfaces and form a nearly rigid deposit. Observed debris flows commonly have been likened to flowing masses of wet concrete. Generically, debris flow includes mudflow (debris containing mostly sand, silt, and clay), lahar (volcanic debris flows), and till flow (debris flow from active or stagnant ice).

Debris flows are a significant geomorphological and sedimentological agent in a wide range of geologic and climatic settings. Subaerial debris flows pose a natural hazard than can cause fatalities, damage structures, and diminish land productivity. Subaqueous debris flows can produce turbidity currents, and they can transport sand from continental slopes to abyssal plains, threatening underwater infrastructures and creating important hydrocarbon reservoirs. Modern debris-flow deposits that line numerous mountain channels and blanket many subaerial and subaqueous fans, and ancient debris-flow deposits that form distinctive strata in many sedimentary sequences attest to their widespread sedimentological importance. Although the discussion below describes recent developments regarding subaqueous debris flows, it focuses primarily on subaerial debris flows.

Debris flows have distinctive physical characteristics. They have bulk densities approximately twice that of clear water, and they can easily transport clasts larger than 1 m in diameter. Debris flows have volumes that range from a few tens of cubic centimeters to more than 1 billion cubic meters; they have maximum velocities that range from a few to a few tens of meters per second; they can affect lowlands more than 100 km from their initial source; and they can flow tens to hundreds of km across nearly flat subaqueous plains. A single subaerial debris flow can transport about as much sediment as that transported annually by small mountainous rivers, and it can introduce abundant organic matter (trees) to river channels, modifying channel dynamics and affecting aquatic habitat.

Debris-flow origins

A debris flow forms when a mass of sediment is mobilized and transformed into a flowing slurry. Most commonly, a debris flow results from a landslide triggered by torrential rainfall, rapid snowmelt, or an earthquake, and in subaqueous environments by additional combinations of rapid sedimentation, gas release, and storm-wave loading. A debris flow can form during a volcanic eruption when hot volcanic debris melts and mixes with snow and ice, or when crater lakes or near-surface groundwater are explosively expelled. A debris flow may also form when pore water is expelled during consolidation of a landslide deposit. A rapidly released flood surge, resulting from a glacier outburst or from breaching of a natural or engineered dam, may entrain a sufficient amount of sediment to become a debris flow.

Behavior of flowing debris

A debris flow moves in an unsteady and nonuniform manner. It commonly flows in a series of surges having periods that range from a few seconds to several minutes or longer. Debris surges commonly develop bulbous heads that contain a concentration of the coarsest particles and precede gradually waning, finer-grained bodies, but surges can also have agitated watery flow fronts that precede more sediment-rich waning flow. Despite its commonly coarse grain size, a debris flow can move across very gentle slopes. When a debris flow stops, it commonly leaves a lobate or digitate deposit having a steep terminal snout and coarse-grained, steep-sided lateral levees. A more watery debris flow forms a deposit that drapes valley slopes and terraces and leaves a thin veneer near its inundation limit.

Scientific advances in understanding debris flows

Conceptual mechanics

Concepts and analyzes of debris-flow motion and deposition have evolved over several decades. From the 1920s through 1970s, debris-flow studies progressed from descriptive accounts of flows and deposits toward efforts to model flow behavior and interpret deposit character in terms of material rheology (e.g., a Newtonian fluid, Sharp and Nobles, 1953; a Bingham viscoplastic material, Johnson, 1970, 1984) or particle collisions (e.g., Takahashi, 1978, 1991). Although many studies proposed various constitutive equations that relate shear stress to shear strain rate, support for these models remains largely anecdotal or based primarily upon small-scale laboratory experiments. Small-scale experiments, however, do not necessarily scale appropriately with natural processes. Many laboratory-scale experiments lack gravel and use amounts of silt and clay that are disproportionate for most debris flows. Physico-chemical effects related to cohesion, important at the small scale, are negligible at the field scale, and inertial effects important at the field scale are poorly simulated at the small laboratory scale.

Debris flows undergo significant changes of state from initiation through deposition (Iverson and Vallance, 2001). As a result, others view debris flows as granular materials having flaw behavior modified by interstitial fluid (Iverson, 1997). Iverson's research has challenged efforts aimed at describing and modeling a debris flow in terms of fixed rheology. Most debris flows are fundamentally granular materials mixed with water; therefore, Iverson (1997) hypothesized that a debris flow can be modeled most parsimoniously as a simple grain-fluid mixture having properties that depend on position and time. In his model, the Coulomb criterion for frictional material relates shear stress to normal stress, and intergranular pore-fluid pressure that varies with position and time mediates the frictional resistance of the granular debris. Rather than treating a debris flow as a material having a fixed rheology, Iverson's model focuses on conserving mass and momentum in a deformable body and allows debris-flow "rheology" to evolve as a flow forms, moves, and deposits sediment. Iverson and colleagues (e.g., Iverson, 1997; Major and Iverson, 1999; Iverson and Denlinger, 2001) tested Iverson's hypothesis in a series of large-scale flume experiments using up to $10\,m^3$ of poorly sorted sandy muddy gravel. The experiments demonstrate that fluid pressure, solids concentration, and mixture agitation evolve as a debris flow proceeds from initiation to deposition and that the evolution of these variables strongly influences flow dynamics, depositional process, and the characteristics of the deposit.

Depositional process and deposit characteristics

At any given location, debris-flow deposition can occur abruptly or it can be prolonged over a significant time interval. Recent field and experimental studies show that common characteristics of debris-flow deposits, such as poor sorting, massive to stratified texture, and graded to ungraded bedding can result from progressive, incremental accretion rather than exclusively by abrupt *en masse* deposition. Debris flows that are longitudinally segregated—those having coarse flow fronts followed by finer-grained flow bodies, and those that coarsen from head to tail—can produce massively textured deposits exhibiting normally graded or inversely graded clastic fragments (Vallance, 2000). Deposits having pervasive particle alignment, strong vertical changes in particle composition or sorting, and clear evidence of emplacement by recurrent debris surges also result from incremental accretion (Vallance and Scott, 1997; Major, 1997, 1998). Unfortunately, single outcrops of a debris-flow deposit commonly do not display clear evidence of depositional process. Owing to possible accumulation by prolonged accretion, deposit thickness cannot be used to infer rheological properties of a debris flow, contrary to a common field procedure that assumes deposition by a Bingham viscoplastic material (Johnson, 1984).

Advances in theoretical and experimental studies indicate that flow-margin dynamics play a pivotal role in debris-flow deposition, contrary to previous suggestions that deposition is controlled by intrinsic rheological properties of flow, such as uniform yield strength, or pervasive dissipation of pore-fluid pressure. Iverson (1997) demonstrated that leading edges of experimental debris flows of $10\,m^3$ exhibited little or no positive pore-fluid pressure, whereas trailing flow was nearly liquefied, and he successfully modeled the experimental debris flows by incorporating high-resistance flow fronts dominated by solid forces. Major and Iverson (1999) further demonstrated that high pore-fluid pressure developed within debris-flow interiors during flow initiation and acceleration persisted during flow deceleration and deposition. Deposition of the experimental debris flows was therefore controlled by grain-contact friction concentrated at flow margins that acted as dynamic debris dams rather that by an intrinsic uniform material strength or by pervasive dissipation of high pore-fluid pressure. The dynamic dams that developed in these flows resulted from segregation and concentration of the coarsest clastic material at surge fronts.

Flow-front dynamics affect deposition by submarine debris flows as well as subaerial debris flows. For example, Mohrig *et al.* (1998) observed that the fronts of subaqueous experimental debris flows, composed of nearly equal mixtures of sand and silt, accelerated rapidly and sometimes detached from flow bodies if the flow fronts hydroplaned over the substrate. Hydroplaning occurs when a debris flow cannot displace leading ambient fluid rapidly enough and as a result high basal fluid pressure develops at the flow front. Thus, when high fluid pressure is developed at the flow front, flow resistance markedly declines, a finding compatible with results from subaerial debris-flow experiments.

Sustained high pore-fluid pressure within bodies of debris flows can influence deposit characteristics as well as depositional process. This can be demonstrated by monitoring post-depositional pore-fluid pressure in debris-flow deposits. Major and Iverson (1999) demonstrated that significant dissipation of high pore-fluid pressure in a debris flow occurs only after deposition, except in narrow regions around flow margins, and Major (2000) demonstrated that high pore-fluid pressure in a debris-flow deposit dissipates over minutes to months depending on deposit thickness and material diffusivity. By comparing typical periods of debris surges with characteristic times of dissipation of deposit fluid pressure, Major (2000) demonstrated that freshly deposited debris could be remobilized easily. Deposit remobilization may be preserved as soft-sediment deformation (e.g., Mohrig *et al.*, 1999). However, deposit remobilization may simply mute or obliterate stratigraphic distinction among recurrent debris surges and lead to a homogeneous, incrementally accreted deposit that appears to be the product of a single debris surge.

Regardless of depositional process or depositional environment, all debris-flow deposits share common sedimentological characteristics. In general, debris-flow deposits typically are unstratified, poorly sorted mixtures of sediment ranging in size from microns to meters. Deposits may show little particle-size segregation or exhibit normal or inverse grading of their coarsest fragments. Deposit thicknesses typically range from several centimeters to several meters. Elongate clasts in deposits may show preferential alignment or imbricate structure, particularly along deposit margins. Deposit termini, where preserved, typically are lobate and have blunt margins that commonly are studded with the coarsest debris. Although deposit interiors may appear to represent deposition from a single debris surge, deposit surfaces can exhibit arcuate ridges, clusters and bands of gravel or sand, and particles that align along debris-surge perimeters, all sedimentologic features that indicate that deposition resulted from emplacement by multiple debris surges.

Debris-flow transformation

Although the composition and size distribution of sediment in a debris-flow deposit is commonly similar to that of the source material, debris-flow character and volume can evolve during transit and result in a deposit having characteristics that are different than those of the initiating flow. After mobilization, a debris flow can entrain channel-bed and bank sediment and increase in volume several fold. Such "bulking", which usually occurs in steep, proximal channel reaches (e.g., Vallance and Scott, 1997), can modify the initial particle-size distribution and introduce "exotic" clasts into the mixture. In more distal reaches a debris flow can interact with streamflow and undergo volume loss and a gradual transformation into so-called *hyperconcentrated streamflow* (Pierson and Scott, 1985; Vallance, 2000) and eventually into sediment-laden floodflow. A hyperconcentrated-streamflow deposit typically is composed of massive to weakly stratified, moderately sorted sand, and has a sedimentologic texture that is intermediate between those of debris-flow and alluvial deposits (Scott, 1988).

Summary

Researchers have made significant strides developing greater insights regarding debris-flow initiation, transport, and deposition since the 1980s. Studies have moved from descriptions of flow behavior and sedimentologic properties of deposits toward analyzes of event frequency and magnitude, development of more realistic mechanical models, and quantification of the subtleties that influence mobilization, transport, and deposition of debris. New insights show that a debris flow can be modeled as a simple grain-fluid mixture having flow behavior that is influenced by distributions of fluid pressure and solids concentration, which evolve with position and time. Characteristics of deposits formerly thought to form only by sudden *en masse* deposition have been shown to develop through prolonged, incremental deposition as well.

Jon J. Major

Bibliography

Iverson, R.M., 1997. The physics of debris flows. *Reviews of Geophysics*, **35**: 245–296.
Iverson, R.M., and Denlinger, R.P., 2001. Flow of variably fluidized granular masses across three-dimensional terrain: 1. Coulomb mixture theory. *Journal of Geophysical Research*, **106**(B1): 537–552.
Iverson, R.M., and Vallance, J.W., 2001. New views of granular mass flows. *Geology*, **29**: 115–118.
Johnson, A.M., 1970. *Physical Processes in Geology*. San Francisco: Freeman, Cooper, and Co.
Johnson, A.M., 1984. Debris flow. In Brunsden, D., and Prior, D.B. (eds.), *Slope Instability*. New York: Wiley and Sons, pp. 257–361.
Major, J.J., 1997. Depositional processes in large-scale debris-flow experiments. *Journal of Geology*, **105**: 345–366.
Major, J.J., 1998. Pebble orientation on large experimental debris-flow deposits. *Sedimentary Geology*, **117**: 151–164.
Major, J.J., 2000. Gravity-driven consolidation of granular slurries: implications for debris-flow deposition and deposit characteristics. *Journal of Sedimentary Research*, **70**: 64–83.
Major, J.J., and Iverson, R.M., 1999. Debris-flow deposition: effects of pore-fluid pressure and friction concentrated at flow margins. *Geological Society of America Bulletin*, **111**: 1424–1434.
Mohrig, D., Whipple, K.X., Hondzo, M., Ellis, C., and Parker, G., 1998. Hydroplaning of subaqueous debris flows. *Geological Society of America Bulletin*, **110**: 387–394.
Mohrig, D., Elverhøi, A., and Parker, G., 1999. Experiments on the relative mobility of muddy subaqueous and subaerial debris flows, and their capacity to remobilize antecedent deposits. *Marine Geology*, **154**: 117–129.
Pierson, T.C., and Scott, K.M., 1985. Downstream dilution of a lahar: transition from debris flow to hyperconcentrated streamflow. *Water Resources Research*, **21**: 1511–1524.
Scott, K.M., 1988. *Origins, Behavior, and Sedimentology of Lahars and Lahar-runout Flows in the Toutle-Cowlitz River system*. U.S. Geological Survey Professional Paper, 1447-A.
Sharp, R.P., and Nobles, L.H., 1953. Mudflow of 1941 at Wrightwood, southern California. *Geological Society of America Bulletin*, **64**: 547–560.
Takahashi, T., 1978. Mechanical aspects of debris flow. *American Society of Civil Engineers, Journal of the Hydraulics Division*, **104**: 1153–1169.
Takahashi, T., 1991. *Debris Flow*. Rotterdam: A.A. Balkema.
Vallance, J.W., 2000. Lahars. In Sigurdsson, H., Houghton, B., McNutt, S.R., Rymer, H., and Stix, J. (eds.), *Encyclopedia of Volcanoes*. San Diego: Academic Press, pp. 601–616.
Vallance, J.W., and Scott, K.M., 1997. The Osceola mudflow from Mount Rainier: sedimentology and hazards implications of a huge clay-rich debris flow. *Geological Society of America Bulletin*, **109**: 143–163.

Cross-references

Alluvial Fans
Avalanche and Rock Fall
Earth Flows
Fabric, Porosity, and Permeability
Flume
Grading, Graded Bedding
Grain Flow
Grain Size and Shape
Gravity-Driven Mass Flows
Hindered Settling
Imbrication and Flow-Oriented Clasts
Liquefaction and Fluidization
Mass Movement
Slope Sediments
Slurry
Submarine Fans and Channels
Tills and Tillites

DEDOLOMITIZATION

Introduction

The term "*dedolomitization*" refers to the partial to wholesale transformation of former dolomite rocks (dolostones—see *Dolomites and Dolomitization*) to limestones, or partial to complete transformation of dolomite to calcite on the scale of individual crystals. The transformation process includes both neomorphic replacement (recrystallization—see *Neomorphism and Recrystallization*) of the dolomite by calcite, and dolomite dissolution followed by calcite cementation.

Dedolomitization has been recognized as an important process in the diagenetic histories of a wide range of Precambrian to Phanerozoic marine, and to a lesser extent, lacustrine carbonates. As with dolomitization, dedolomitization histories can be multistaged and protracted. Dedolomitization has also been shown by geochemical modeling to be an ongoing and important reaction in modern-day karstic aquifers (Plummer *et al.*, 1990; Raines and Dewers, 1997).

Dedolomitization process

Several authors have described the dedolomitization process as the reverse but analogous situation to replacement dolomitization (e.g., Fairbridge, 1978; Dockal, 1988). As with dolomitization, dedolomitization involves consideration of thermodynamics, kinetics, and fluid flux. But unlike replacement dolomitization, dedolomitization can also refer to a pore-filling process. Porosity changes associated with dedolomitization are inconsistent and difficult to predict.

Dedolomitization can be represented simply by the following reaction:

$$Ca^{2+}_{(aq)} + CaMg(CO_3)_{2(s)} \rightarrow 2CaCO_{3(s)} + Mg^{2+}_{(aq)}$$

The above reaction should proceed to the right if $aCa^{2+}/aMg^{2+} > 1$ (Abbott, 1974). Ca^{2+} is usually supplied by dissolution of a mineral source, for example gypsum or anhydrite. Mg^{2+} liberated into solution may accumulate, causing the solution to reach saturation with respect to both calcite and dolomite, and thus halting the reaction. Alternatively, dedolomitization may continue until one of the reactants is no longer present if Mg^{2+} is incorporated into another diagenetic phase, or if the fluid flux is sufficiently high that dolomite saturation is not attained. The above reaction reverses, theoretically, when $aCa^{2+}/aMg^{2+} < 1$ and dolomite is precipitated. The dedolomitization reaction is generally considered to be slow under natural geological conditions, but reactive transport modeling by hydrogeologists has suggested that, in some instances, the process is geologically quite rapid, occurring in hundreds of years.

von Morlot (1847) considered the dissolution of evaporites (gypsum or anhdrite) to be essential in the dedolomitization process, as shown in the following reaction:

$$CaSO_{4(s)} + CaMg(CO_3)_{2(s)} \rightarrow 2CaCO_{3(s)} + Mg^{2+}_{(aq)} + (SO_4)^{2-}_{(aq)}$$

Numerous others have also suggested that evaporite dissolution was essential in driving dedolomitization, where the continual addition of Ca^{2+} from gypsum or anhydrite results in the precipitation of calcite due to the common ion effect (e.g., Back *et al.*, 1983; Plummer *et al.*, 1990). Decreases in pH, calcium and carbonate from solution cause further dissolution of dolomite. Back *et al.* (1983) and others, have also suggested that dedolomitization will not occur in low-temperature environments unless gypsum, or some other source of additional Ca^{2+} ions, is present to maintain a high Ca^{2+}/Mg^{2+} ratio. Oxidation of pyrite has also been proposed as a source of aqueous $(SO_4)^{2-}$ in some situations (e.g., Zenger, 1973). Others have indicated that the presence of $(SO_4)^{2-}$ is not a prerequisite for dedolomitization to occur (e.g., Stoessel *et al.*, 1987).

Numerous field, petrographic and geochemical studies have supported the conclusion that dedolomitization is a surficial phenomenon, related to post-burial surface weathering and other near-surface processes associated with subaerial exposure, including the development of unconformities and karstic aquifers. Dedolomite often occurs within the most readily weathered parts of the rock and grades into "fresh" dolomite away from weathered surfaces. Dedolomites from near-surface diagenetic settings may show negative carbon and oxygen isotopic values reflecting soil-gas influences in meteorically-derived water, including fluids derived from buried paleoaquifers.

Dedolomitization can occur during deeper burial diagenesis and may be associated with stylolites, hydrocarbons and other late diagenetic mineral phases, including fracture or vug-filling calcites (e.g., Budai *et al.*, 1984). The influence of hydrocarbons may be recognized in light carbon isotopic values of the dedolomite. Light oxygen isotopic values should reflect precipitation at elevated temperatures characteristic of the burial environment. Other authors have noted dedolomitization associated with mixing zones, contact metamorphism and hot Ca-rich brines derived from albitization of plagioclase in nearby sandstones.

Recognition of dedolomite

Limestone formed by dedolomitization may have very little in common with the parent dolostone, or its precursor limestone. In the field, the occurrence of limestone in otherwise dolomitic strata in the vicinity of faults, or along weathered surfaces and below unconformities, may indicate dedolomitization.

A range of microscopic fabrics can be used to identify dedolomitization (Figure D1), although not all are reliable indicators. Empty rhomb-shaped pores indicate dissolution of dolomite and perhaps dedolomitization if it can be shown that calcite once occupied these pores. Calcite partially or entirely filling rhomb-shaped pore space, or replacing dolomite crystals, constitutes the most straightforward evidence of dedolomitization. Differentiation of calcite precipitated as cement versus replacement is made on the same type of textural evidence used to interpret aragonite to calcite transformations. Isolated calcite rhombs may represent very selective dedolomitization because calcite does not commonly develop a rhombohedral habit in limestones. However, exceptions have been reported.

Mosaics of calcite crystals showing rhombic patterns of ferric oxides or other inclusions could represent the palimpsest remnants or "ghost" outlines of former dolomite crystals or internal zoning in these crystals. In some dedolomite rocks, the dedolomite matrix is characterized by a clotted "grumeleuse" texture resembling peloidal grainstone. This fabric results from the retention of impurities from the cloudy, inclusion-rich cores of the original dolomite crystals in the newly formed dedolomite crystals, which otherwise obscure the former dolomite crystal boundaries (Evamy, 1967; Kenny, 1992).

Ferroan dolomites appear to be more susceptible to dedolomitization due to their metastability in the surface environment, resulting in the oxidation and hydration of the ferrous iron previously bound in the dolomite lattice, and a

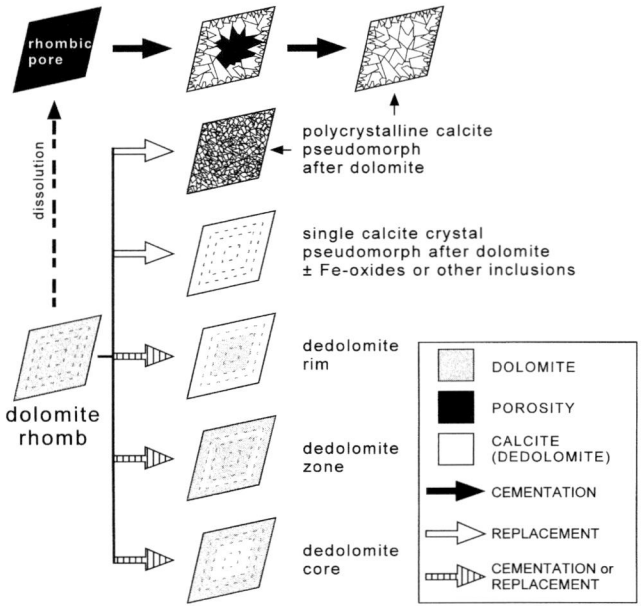

Figure D1 Schematic summary of common dedolomite microfabrics.

close association of hematite and goethite with the newly formed dedolomite. These iron-bearing minerals are usually responsible for the rusty weathering color of many dedolomitized rocks, and they may also retain a secondary chemical remanent magnetization useful for dating diagenetic events. Dedolomitization can also be zone-selective, typically after ferroan or calcian internal zones, rims or cores of dolomite crystals, leaving more stoichiometric dolomite relatively unaltered (Figure D1).

Terminology

Modern usage of the term *dedolomite* typically infers calcite as the mineral occupying the rock volume formerly taken by dolomite, but other minerals, such as quartz, celestite, and iron oxides, are also known to replace dolomite. As terms, *dedolomite* and *dedolomitization* have been criticized as being ambiguous and inconsistent with the nomenclature for other types of mineral replacements, where names of newly formed minerals were based on mineralogy of the replacement, rather than the replaced phase (Smit and Swett, 1969). *Dedolomitization* has also been used as an all-encompassing definition to refer to any alteration affecting dolomite (Fairbridge, 1978). However, the terms *dedolomite* and *dedolomitization*, as used in this article, are well-entrenched in the modern literature and their current meanings are clear.

Mario Coniglio

Bibliography

Abbott, P.L., 1974. Calcitization of Edwards Group dolomites in the Balcones fault zone aquifer, south-central Texas. *Geology*, 1: 359–362.
Back, W., Hanshaw, B.B., Plummer, L.N., Rahn, P.H., Rightmire, C.T., and Rubin, M., 1983. Process and rate of dedolomitization: mass transfer and ^{14}C dating in a regional carbonate aquifer. *Geological Society of America Bulletin*, **94**: 1415–1429.
Budai, J.M., Lohmann, K.C., and Owen, R.M., 1984. Burial dedolomite in the Mississippian Madison Limestone, Wyoming and Utah thrust belt. *Journal of Sedimentary Petrology*, **54**: 276–288.
Dockal, J.A., 1988. Thermodynamic and kinetic description of dolomitization of calcite and calcitization of dolomite (dedolomitization). *Carbonates and Evaporates*, **3**: 125–141.
Evamy, B.D., 1967. Dedolomitization and the development of rhombohedral pores in limestones. *Journal of Sedimentary Petrology*, **37**: 1204–1215.
Fairbridge, R.W., 1978. Dedolomitization. In Fairbridge, R.W., and Bourgeois, J. (eds.), *The Encylopedia of Sedimentology*. Stroudsburg, Pennsylvania: Dowden, Hutchison and Ross Inc., pp. 233–235.
Kenny, R., 1992. Origin of disconformity dedolomite in the Martin Formation (Late Devonian, northern Arizona). *Sedimentary Geology*, **78**: 137–146.
von Morlot, A., 1847. Ueber Dolomit und seine künstliche Darstellung aus Kalkstein. *Naturwissenschaftliche Abhandlungen Gesammelt, und durch Subscription Lursg, von Wilhelm Haidinger*, **1**: 305–315.
Plummer, L.N., Busby, J.F., Lee, R.W., and Hanshaw, B.B., 1990. Geochemical modeling of the Madison Aquifer in parts of Montana, Wyoming, and South Dakota. *Water Resources Research*, **26**: 1981–2014.
Raines, M.A., and Dewers, T.A., 1997. Dedolomitization as a driving mechanism for karst generation in Permian Blaine Formation, southwestern Oklahoma, USA. *Carbonates and Evaporites*, **12**: 24–31.
Smit, D.E., and Swett, K., 1969. Devaluation of "dedolomitization". *Journal of Sedimentary Petrology*, **39**: 379–380.
Stoessell, R.K., Klimentidis, R.E., and Prezbindowski, D.R., 1987. Dedolomitization in Na–Ca–Cl brines from 100° to 200° at 300 bars. *Geochimica et Cosmochimica Acta*, **51**: 847–855.
Zenger, D.H., 1973. Syntaxial calcite borders on dolomite crystals, Little Falls Formation (Upper Cambrian), New York. *Journal of Sedimentary Petrology*, **43**: 118–124.

Cross-references

Ancient Karst
Anhydrite and Gypsum
Dolomites and Dolomitization
Magnetic Properties of Sediments
Neomorphism and Recrystallization
Porewaters in Sediments

DEFORMATION OF SEDIMENTS

Introduction

Most sediments undergo some degree of deformation after deposition. Burial under later sediment leads to expulsion of pore fluids and to adjustments in grain packing, involving both physical and chemical processes. The main result is a reduction in thickness of sedimentary units and, in some cases, the development of horizontal fabrics of platy and elongate grains. In the case of fine-grained sediments, containing clay minerals and deposited with high initial porosities, such compactional reduction in thickness can be large so that depositional lamination and bedding are compressed beyond recognition. For coarser sediments, the depositional grain framework usually involves lower initial porosities and a greater resistance to compaction. In the extreme, organic-rich sediments such as peat may undergo compaction to only 10 percent of initial thickness whilst, for well-sorted sands, thickness changes are negligible. Where sediments of contrasting grain size are interbedded in a lenticular fashion, differential compaction can distort depositional geometries (see *Compaction*). Sediments that have undergone compaction and other diagenetic changes are typically lithified. In this more rigid state, they may be deformed by tectonic processes, most typically compressional folding and thrusting or extensional faulting, often at great depths where temperatures are high and rates of deformation slow. These processes fall within the realm of tectonics and are not dealt with here.

Prior to significant compaction, sediments may be deformed, in appropriate circumstances, soon after deposition and this early, soft-sediment deformation is the subject of this article. Some early deformation is virtually contemporaneous with deposition whilst some occurs because of shallow burial and early compaction. In all cases, the sediment experiences a loss of strength and behaves, for a short period, as a plastic or a fluid. In some cases, loss of strength may result in total remobilization of the sediment and transformation into a gravitational mass movement, leading to a phase of resedimentation. This section concentrates on situations where the loss of strength is sufficiently short-lived for sediment to regain its strength and preserve structures that record more or less *in situ* deformation. However, these distinctions are not always clear-cut. Tectonic structures are often geometrically identical with syndepositional structures and boundaries between tectonic stress regimes and more local, sediment-induced regimes may be blurred.

Deformation within a body of sediment occurs when intergranular forces are unable to resist applied stresses, which are usually, but not exclusively, of gravitational origin. The shear strength of sediment is normally expressed by the equation:

$$\tau = C + (\sigma - p)\tan\varphi$$

where τ is shear strength, C is grain cohesion, σ is pressure normal to shear, p is excess pore fluid pressure, and φ is the angle of internal friction. This means that sediment will fail when the applied stress exceeds τ and such a condition will be favored by lower cohesive strength C, by changes in grain packing to reduce $\tan\varphi$ or by pore fluid pressure p increasing beyond some critical value. Such conditions can occur for several reasons, which are discussed below. Failure, when it does occur, may be isotropic, extending throughout the body of sediment or it may be concentrated on discrete surfaces.

Loss of strength

Loss of strength and sediment deformation are most likely in sands and finer-grained sediments. They are rare in conglomerates unless these are very poorly sorted with a high proportion of fine-grained matrix. The presence of fine-grained, especially muddy sediment tends to increase the role of cohesive strength, which is largely a function of grain size; in better sorted, mud-free sediment, frictional forces dominate. In both cohesive and non-cohesive sediments, elevated pore-water pressure is the main means by which strength is lost but that condition can have different causes and durations in contrasting sediments.

In fine-grained, muddy sediments, excess pore fluid pressures are created by relatively rapid deposition. Low permeabilities inhibit escape of pore water and hence sediment compaction, giving over-pressured conditions. In addition, the decay of organic matter within a body of sediment can generate gases, which, if unable to escape, also contribute to pore pressure. Over-pressured conditions can occur close to the sediment surface where highly mobile mud may be susceptible to deformation or flowage. They can also develop at a larger scale and on a longer timescale within a thick pile of sediments such as that produced by a prograding delta or deltaic margin (e.g., Mississippi, Niger). In such cases, increasing pore-water pressure with burial may eventually exceed lithostatic pressure, leading to failure. This may result in plastic flowage (Fig. D2) of a thick, over-pressured layer or it may result in the development of discrete slip surfaces, which cut up section as listric faults or the bounding surfaces of sediment gravity slides. Such conditions, and the resultant movements, develop and persist over long periods of time, in the case of the largest deltaic margins, over millions of years.

Rapidly deposited sands, lacking significant fine-grained sediment, commonly have rather loose grain packing. They are particularly susceptible to shock, such as might be generated by an earthquake, by heavy wave action, by further rapid deposition or by a sudden rise in water level. Shock breaks grain contacts and induces tighter packing and, as a result, excess pore water will be present, raising the pore-water pressure and causing temporary liquefaction (Fig. D2). The duration of liquefaction will be determined by the rate at which pore water escapes. In the case of sands, high permeability usually means that liquefaction is short-lived. Whilst liquefied, the sediment-water mixture deforms readily as a Newtonian fluid, in response to gravitational and other applied stresses. As pore fluid escapes, usually upward to the sediment surface, grain contacts are re-established as a rising front of reconsolidation "freezes" the deformed sediment. Deformation is usually visualized as the distortion of original depositional lamination but, in the case of extreme or sustained liquefaction and deformation, depositional lamination is destroyed and structureless sand or slurried textures are produced. Upward movement of escaping pore water may also create deformation, both by distorting depositional laminae and by creating new vertical structures such as dishes, pipes and sheets within which sediment-water mixtures move as fluidized flows (Fig. D2).

Deforming forces

Virtually all the forces that act upon sediment weakened by the above processes are gravitational in origin. These can be divided into two major classes, a down-slope component of gravity and the action of gravity on inverted density gradients. In addition, deforming forces can also be produced by shear at the sediment surface caused by flowing water, ice and sediment gravity flows.

Where muddy sediment occurs on a slope, the down-slope gravitational component may be sufficient to overcome its cohesive strength. This can lead to both detachment of a slab of sediment on a basal shear surface and to internal deformation within the moving layer. The style of deformation will vary depending upon the position within the moving sheet. Stretching, particularly in the up-slope end, may lead to tensile failure and to the development of brittle extensional structures whilst compression, commonly in a down-slope setting, causes plastic folds. These styles of deformation are associated with a family of mass movements (see *Mass Movement; Slide and Slump Structures*). Also partly driven by down-slope gravitational components are sedimentary growth faults. These are related to failures in deeply buried over-pressured muds. Extension on the faults is also driven by lateral flowage

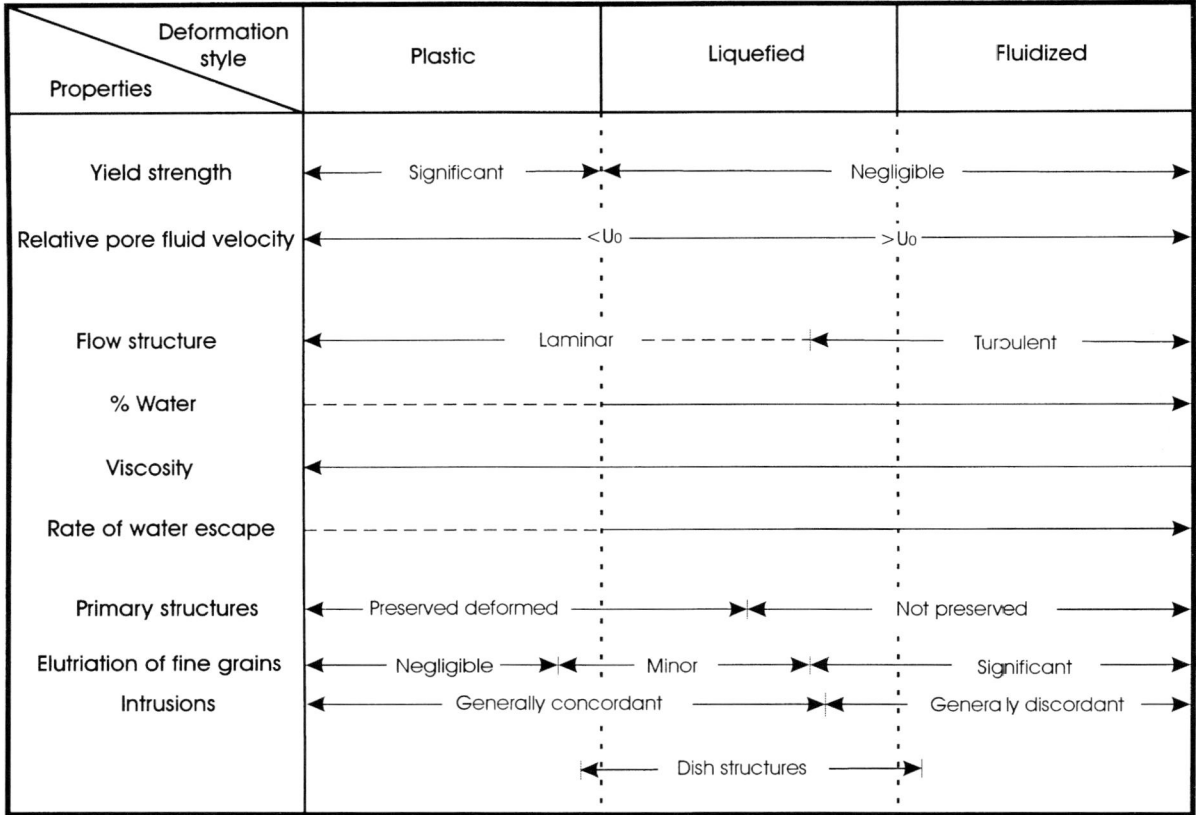

Figure D2 Main processes of sediment deformation, the associated characteristics and the resultant effects and products. Based on Lowe (1975).

of those muds due to vertical loading by the overlying sediment pile.

Density inversions occur commonly, as a result both of depositional processes and of early loss of pore water. A sand layer, deposited on mud, will commonly have a higher density than the mud and, if the mud is sufficiently weak, the instability may lead to loading of sand into the mud (see *Load Structures*). Where the sand is of uneven thickness, as with ripple lenses or erosional sole marks, loading may be concentrated where the sand is thickest. Extreme cases of loading show detached balls of sand, usually encased in slurried mud or silt.

Within liquefied sand layers, which commonly result from shock, patterns of density inversion may be more subtle. Internal depositional lamination may be contorted in convolute lamination (*q.v.*) some of which reflects the upward escape of water or low-density sediment-water mixtures. Very rapidly deposited sands may lack depositional lamination and undergo liquefaction as part of the depositional process. De-watering of such sands and the associated elutriation of fines can lead to the development of subtle structures such as dish structures and water-escape pillars and pipes.

Deeper burial and associated differential compaction due to gravitational loading may lead to deformation of larger-scale depositional geometries. The cross-sections of channel sand-bodies, in particular, are susceptible to change from planar-convex forms to concavo-convex patterns where encased in muds. Where channel sands are encased in mud, as in some deep-water depositional settings, de-watering of the muds during burial can lead to high pore-water pressures in the sands. This can eventually lead to rupture of the mud seal and to injection of liquefied sands as sills and dykes (see *Clastic (Neptunian) Dykes and Sills*). When injection occurs close to the sediment surface, liquefied sand may be extruded as slurry flows or as sand volcanoes.

Where sands are deposited by powerful currents as dunes or sand waves, they may be susceptible to liquefaction as a result mainly of rapid deposition on the lee slope of the bed form. If this happens whilst the current is still active, the cross-bedded sands, developed by lee-side deposition, may be sheared and the foresets dragged into folds, giving overturned cross-bedding.

The movement of ice over unconsolidated sediment also generates large-scale and intense deformation of those sediments. This occurs near the front of a glacier, where glacial advance operates as a bulldozer to fold and thrust proglacial deposits. In addition, basal layers of a mobile glacier may be rich in sediment and shearing across the basal layers can lead to intense deformation though the rheology has more to do with the plastic deformation of ice than with processes of liquefaction or sediment cohesion.

John D. Collinson

Bibliography

Boulton, G.S., and Hindmarsh, R.C.A., 1987. Sediment deformation beneath glaciers: rheology and geological consequences. *Journal of Geophysical Research*, **92**: 9059–9082.
Maltman, A. (ed.), 1994. *The Geological Deformation of Sediments*. London: Chapman & Hall.
Jones, M.E., and Preston, R.M.F. (eds.), 1987. *Deformation of Sediments and Sedimentary Rocks*. London: Geological Society of London, Special Publication, 29.
Lowe, D.R., 1975. Water escape structures in coarse grained sediments. *Sedimentology*, **23**: 285–308.

Cross-references

Bedding and Internal Structures
Clastic (Neptunian) Dykes and Sills
Compaction (Consolidation) of Sediments
Convolute Lamination
Debris Flow
Deformation Structures and Growth Faults
Dish Structure
Flame Structure
Fluid Escape Structures
Gases in Sediments
Liquefaction and Fluidization
Load Structures
Pillar Structure
Ball-and-Pillow (Pillow) Structure
Pseudonodules
Slide and Slump Structures
Syneresis

DEFORMATION STRUCTURES AND GROWTH FAULTS

This section deals with so-called soft-sediment deformation. It documents how different types of deforming stress, acting on sediment that has been temporarily weakened, lead to a range of deformational structures. The physical processes are described in more detail in the section on *Deformation of Sediments*. The structures themselves do not fall naturally into well-defined classes but Figure D3 shows a scheme, which helps to relate structures to processes. This diagram illustrates interactions between different causes of loss of strength and deforming forces and shows that processes are gradational with tectonic deformation and with re-sedimentation. It shows the wide range of scales across which deformation can occur, from micro-loading to growth faults at the scale of continental slopes and highlights differences in timing and duration of deformation. Structures that form during deposition, such as overturned cross-bedding, contrast with structures dependant upon significant later burial, such as growth faults. The most common deformation structures are described below, roughly in order of increasing scale but moderated by considerations of process. Several are described in more detail in other sections.

Load casts and pseudo-nodules

These distortions occur at the interface between a mud layer and an overlying sand layer. Downwards-facing bulbous load casts of sand are separated by upward-pointing flame structures of mud. Some occur on otherwise flat bedding surfaces whilst others accentuate erosional or depositional irregularities. In extreme cases, sand may be detached from its source bed and occur as isolated load balls or pseudonodules (*q.v.*) within highly disturbed muds or silts. Internal lamination within both sand and mud is highly contorted. Loading is caused by gravity acting on an unstable density stratification where the sand layer is denser than the muds and where both layers are weakened by excess pore fluid.

Convolute lamination

This commonly occurs within sand or silt units and involves distortion of depositional lamination by folding of variable intensity. In some cases, the folding is chaotic whilst in other cases upward movement of escaping water has dragged lamination into upright folds. Cuspate folds with sharp anticlines and more rounded synclines are typical. In many cases, depositional lamination can still be identified, commonly cross-bedding or cross-lamination. The good sorting and high porosities of eolian sands makes them particularly susceptible to liquefaction (*q.v.*) when water table rises rapidly through them.

Convolute lamination (*q.v.*) is a common feature of turbidite sandstones, often associated with ripple laminated (Bouma C) interval of internal lamination. Some of the largest examples occur in eolian sands where a rising water table has caused instability. All examples reflect liquefaction of the sand, which can happen for a variety of reasons, including seismic shock, rapid sedimentation, breaking waves or a shift of water table. Where seismic shock was the cause of liquefaction, deformed layers may have very widespread distribution. Upward escape of excess pore water, internal density inversions and downslope components of gravity may all contribute to deformation. De-watering leads to reconsolidation of the sediment from the bottom up, "freezing" the deforming laminae. Protracted liquefaction may lead to total homogenization.

Overturned cross-bedding

Overturned cross-bedding is a special case of convolute lamination where liquefaction occurred in an actively migrating bed form. Current shear on the sediment–water interface dragged the liquefied sand in a down-current direction in an essentially laminar style. An upward migrating front of reconsolidation allowed higher parts of the sediment to be sheared longer, giving down-current facing recumbent folds.

Dish and pillar structures

These structures, which are most common in thick, otherwise massive sands, result from rapid de-watering following rapid deposition. Dish and pillar structures (*q.v.*) are direct products of de-watering and are not distortions of pre-existing lamination. Dish structures are thin concave-upward, sub-horizontal zones of slight clay enrichment produced by local filtering out of the elutriated fines. Their upturned edges merge with pillar structures, which are vertical conduits of water escape and record fluidized or near-fluidized conditions.

Applied Force \ Loss of Strength	Exceed Strength			Liquidize	
	Internal Tensile (Brittle)	Internal Cohesive (Plastic)	External Surface Cohesive (Plastic)	Liquefied	Fluidized
Gravitational body force on slope	Slides	Slumps	Slumps and slides	Debris flows	
Unequal confining load	Growth faults	Loaded ripples Shale ridges and diapirs		Loaded ripples and sole marks	
					Clastic dykes Sand volcanoes
Gravitationally unstable density gradient — Continuous	Soft sediment faults			Convolute lamination	
Gravitationally unstable density gradient — Within a single layer				Dish structures	Water escape pipes and pillars
Gravitationally unstable density gradient — Multiple layer, not pierced				Bedding surface load casts	
Gravitationally unstable density gradient — Multiple layer, pierced		Shale ridges and mud diapirs		Ball and pillow/pseudonodules Isolated load balls	
Shear stress — Current drag				Overturned cross-bedding	
Shear stress — Vertical					Water escape pipes and pillars
Physical — Physical	Liquidize				

Figure D3 Styles of sedimentary deformation classified in terms of the interaction between the cause of loss of strength and the applied deforming force. Adapted from Owen (1987).

Sand injection structures

Where a sandbody is encased in mud and progressively buried, compaction and de-watering of the muds can lead to overpressure in the sands. If the fluid pressure exceeds the tensile strength of the overlying mud, rupturing occurs and complex networks of dykes and sills of liquefied sand are intruded into the muds. This is quite common in channel sandbodies in deep-water and slope settings though outcrop examples are rare.

Small-scale injection dykes occur in mudstones of interbedded sandstone and mudstone successions. These may be folded ptygmatically because of differential compaction. Sand within dykes is typically structureless though some examples show foliation parallel with dyke margins.

Sand volcanoes and extruded sheets

Some upward-intruding liquefied sand, both dykes and pipes, penetrates to the free sediment surface where it is extruded. Where the escape conduits are pipes, sand volcanoes form, at scales from a few centimeters to a few meters. Where the intrusions are dykes, fissure extrusion occurs and liquefied sand may flow as sheets before de-watering and "freezing". Where sand dykes penetrate muddy debris flows, extruded sand may become inter-folded with the upper levels of the debris flow.

Desiccation and other cracks

When surface layers of mud or silt dry out they contract and create isotropic horizontal tension. If this exceeds the tensile strength of the mud, polygonal (commonly hexagonal) cracks develop which may be filled with wind-blown or water-lain sand. Dimensions of polygons are a function of mud layer thickness, and several scales may coexist. Sub-aqueous shrinkage of surface mud layers also leads to cracks though these often form less regular patterns. They are thought to result from volume changes of clays within the surface layer. Large-scale polygonal cracks also develop through protracted freeze-thaw processes in periglacial settings.

Slumps and slides

Gravitational body forces acting on sediments lying on a slope lead to slumps and slides where cohesive (plastic) and tensile (brittle) strengths are exceeded. In both types of movement, most shearing is concentrated on discrete basal surfaces. They differ principally in their internal deformation. Slides are essentially undeformed but may have internal faults whilst slumps show plastic folding. The two styles commonly co-exist in the same mass movement. Internal deformation falls short of the total mixing and homogenization that characterizes debris flows but both slumps and slides may develop into such flows.

Mud diapirs

Overpressure develops where muds are buried quite rapidly, as in a prograding delta. The resultant loss of strength and the significant overburden lead to both vertical and horizontal flowage of the muds. Where the overlying sediment is denser than the mud, as in an upward-coarsening deltaic succession, mud may rise vertically as a diapir or mudlump, pushing aside or penetrating mouth-bar sands. In the Mississippi delta, mudlumps rise to sea level and create short-lived islands offshore the distributary mouths. Mud diapirs lead to very variable thicknesses in the mouth bar sands, which thicken into withdrawal synclines between diapirs.

Horizontal flowage of overpressured muds leads to extrusion of muds at the base of the prograding slope as imbricate thrusts and folds. The same motion is partly responsible for extensional stresses that drive sedimentary growth faults in deltaic successions.

Growth faults

Sedimentary growth faults occur in many deltaic successions. They are large-scale features and commonly require exceptional exposures to be readily identified at outcrop. In small exposures, they may be confused with later tectonic faults though a lack of mineralization and sediment smearing on the fault plane may suggest a syndepositional origin. Growth faults are more commonly imaged in seismic data where upward truncation by overlying undisturbed sediments can be a criterion for their recognition. They are caused by shear failure at discrete surfaces within the sedimentary pile, and they commonly occur as listric surfaces that sole out on a basal decollement. In smaller examples, confined within a single progradational cycle, the basal surface may be a thin bed of particularly fine-grained mudstone. In larger examples, which penetrate several progradational units within a delta complex at a continental margin, the decollement may be within a thick interval of overpressured muds. Lateral flowage of those muds toward the free surface slope and gravitational forces on the hanging wall block both help to drive the movement of that block down slope. Growth faults are characterized by slow and continual movement and are aided by progressive sediment loading. Their hanging wall areas act as local depocenters in which thickened delta-front sediments are deposited and preserved. With continued progradation of individual deltas or of the continental slope the position of active faulting shifts progressively basinwards.

Large-scale examples imaged on seismic at continental margins show complex fault geometries with both synthetic and antithetic fault elements. In plan, the major faults have cuspate traces, concave toward the basin. Protracted fault movement and the listric geometry of the faults leads to roll-over anticlines in the hanging walls which may be important hydrocarbon traps as they also coincide with enhanced reservoir thickness. Clay smearing on the fault surfaces may enhance the hydrocarbon seal.

John D. Collinson

Bibliography

Allen, J.R.L., and Banks, N.L., 1972. An interpretation and analysis of recumbent-folded deformed cross-bedding. *Sedimentology*, **19**: 257–283.

Collinson, J.D., 1994. Sedimentary deformational structures. In Maltman, A. (ed.), *The Geological Deformation of Sediments*. London: Chapman & Hall, pp. 95–125.

Edwards, M.B., 1976. Growth faults in Upper Triassic deltaic sediments, Svalbard. *Bulletin of the American Association of Petroleum Geologists*, **60**: 341–355.

Jones, M.E., and Preston, R.M.F. (eds.), 1987. *Deformation of Sediments and Sedimentary Rocks*. London: Geological Society of London, Special Publication, 29.

Lowe, D.R., and Lopiccolo, R.D., 1974. The characteristics and origins of dish and pillar structures. *Journal of Sedimentary Petrology*, **44**: 484–501.

Owen, G., 1987. Deformation processes in unconsolidated sands. In Jones, M.E., and Preston, R.M.F. (eds.), *Deformation of Sediments and Sedimentary Rocks*. Geological Society of London, Special Publication, 29, pp. 11–24.

Winker, C.D., and Edwards, M.B., 1983. Unstable progradational clastic shelf margins. In Stanley, D.J., and Moore, G.T. (eds.), *The Shelfbreak; Critical Interface on Continental Margins*. Society of Economic of Paleontologists and Mineralogists, Special Publication, 33, pp. 139–157.

Cross-references

Bedding and Internal Structures
Convolute Lamination
Debris Flow
Deformation of Sediments
Dish Structure
Flame Structure
Fluid Escape Structures
Gases in Sediments
Liquefaction and Fluidization
Load Structures
Pillar Structure
Ball-and-Pillow (Pillow) Structure
Pseudonodules
Slide and Slump Structures
Syneresis

DELTAS AND ESTUARIES

Definition of a delta

A delta is a discrete bulge of the shoreline formed at the point where a river enters an ocean, sea, lake, lagoon or other standing body of water (Figures D4 and D5). The bulge is formed because sediment is supplied more rapidly than it can be redistributed by basinal processes, such as waves and tides. Deltas are thus fundamentally regressive in nature, which means that their deposits record a seaward migration or progradation of the shoreline. The term delta has also been used to describe any regressive deposit built by any terrestrial feeder system into any standing body of water (Nemec, 1990 in Colella and Prior, 1990). In this scheme, terrestrial feeder systems can be alluvial (rivers, alluvial fans, braidplains, scree-cones) or non-alluvial (volcanic lavas or pyroclastic flows). The terms ebb- and flood-tidal delta have also been generally applied to sediment accumulations that form around tidal inlet channels in barrier-lagoon depositional systems (see *Tidal Inlets and Deltas*), but these do not fit the above definition of a delta because they are not linked to a river or terrestrial feeder. This entry focuses on river deltas as a discrete depositional system in which both environments and their deposits will be described.

Definition of an estuary

Sedimentologists define an estuary (Figure D5) as the seaward portion of a drowned river valley which receives sediment from both fluvial and marine sources (modified after Dalrymple *et al.*, 1992). An estuary may be affected by tide, wave, and river processes and is defined as extending from the landward limit of tidal influence to the seaward limit of coastal influence.

The term estuary is also defined on an oceanographic basis as a semi-enclosed body of marine water that is measurably

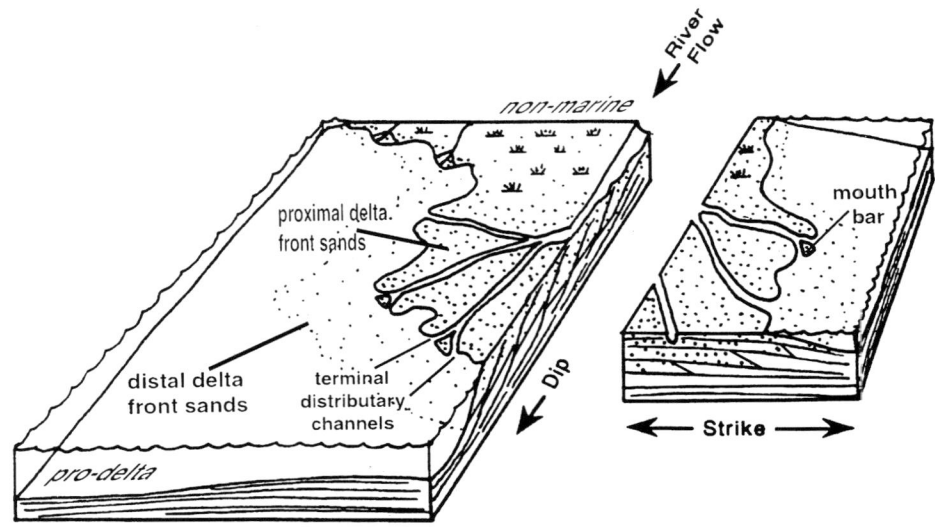

Figure D4 Block diagram of a lobate, river-dominated delta showing numerous channel bifurcations. Terminal distributary channels fed triangular-shaped mouth bar deposits. Seaward progradation forms a series of offlapping inclined strata in dip-view. The strike view shows a lens-shaped sediment body with beds dipping away from the depositional axis of the delta lobe.

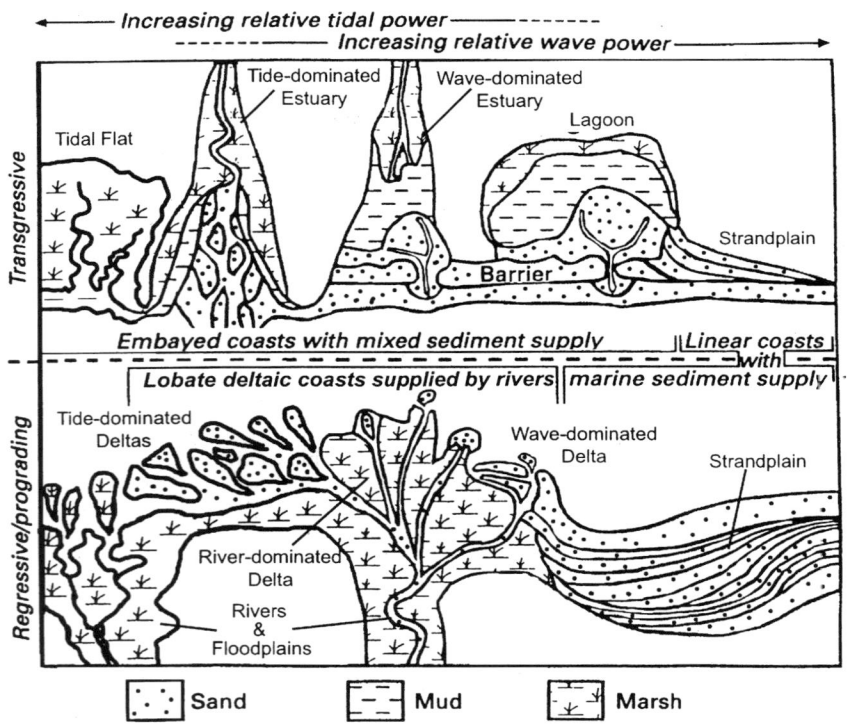

Figure D5 Comparison of regressive versus transgressive coastal depositional systems. Deltas form lobate sediment bodies. The character and distribution of sediment in a delta is highly sensitive to the relative proportion of wave, tide, and fluvial processes. Differences in these processes result in different shaped sand bodies. Also, note river-dominated delta at the head of the wave-influenced estuary. Deltas can thus form smaller components of other types of coastal depositional systems. Changes in sediment supply and sea level change can cause one type of depositional system to evolve into another type (modified from Reading and Collinson, 1996 based on Heward, 1981, and Boyd, Dalrymple, and Zaitlin, 1992).

diluted by land-derived fresh water (Pritchard, 1967; Nichols and Biggs, 1985). Sedimentologists tend to reject this usage as being too broad since it includes all brackish water environments including lagoons and many marine-influenced delta fronts which are not generally thought of as estuarine by sedimentologists. The etymology of the word estuary also means tides, so by definition all estuaries are formed adjacent to a marine body of water. There are no entirely fresh-water estuaries because measurable tides do not occur in lakes. Consequently, drowned river valleys filled during rising lake levels are not generally considered as estuarine.

In an estuary, the seaward portion of the valley is filled with marine sediments and estuaries are fundamentally transgressive in nature, unlike deltas, which are regressive. Deltas and estuaries are not mutually exclusive, however, because regressive bayhead delta deposits can readily form within the up-valley reaches of an estuary (Figure D5). In this case, the delta deposits would form a minor component of a larger, overally transgressive estuarine valley-fill. If fluvial discharge became high enough to "flush out the valley" and begin to form a broadly regressive deposit then the estuary would evolve into a delta.

Estuaries form by the interaction of waves, tides and fluvial processes. Two end members have been described (Figure D5). In tide-dominated estuaries (such as the Bay of Fundy in Nova Scotia, Canada) the mouth of the estuary is kept open by strong tidal currents. The center of the estuary tends to be dominated by sandy bedforms and tidal bars whereas the margins tend to be muddy tidal flats. In wave-dominated estuaries, the mouth of the estuary is partly closed by a wave-formed barrier island. A brackish lagoon or bay lies behind the barrier and is commonly filled with fine-grained mud. At the head of the bay, fluvial or tide-influenced bayhead delta deposits form, which may be sandy or muddy. Mud accumulates primarily in the so-called "central basin" bounded by the barrier and bay head delta.

Distinguishing deltas, barriers, estuaries, and strandplains

Most of the sediment in a delta is derived directly from the river that feeds it; in contrast with estuaries in which much of the sediment is derived from the marine realm and in which deposits are fundamentally transgressive. In barrier-island systems (Figure D5), sediment is supplied by alongshore transport. Alongshore transport occurs because waves approaching the shoreline are deflected and migrate parallel to the shoreline causing longshore drift. The longshore drift system may carry sand and gravel, which can be deposited as linear shoreface deposits, or they may carry large quantities of fluidized mud that can be deposited to form muddy coastlines. In mud-dominated coastlines, silt, sand, or shelly material may be winnowed out in the intertidal zone, forming thin and narrow beach deposits termed cheniers. Barrier-islands may also form components of wave-influenced deltas. Rivers can act as a baffle or groin that causes sediment carried in the longshore drift system to be deposited on the updrift side of the distributary channel (see wave-dominated delta in Figure D5).

Where basinal processes redistribute sediment to the point that the fluvial source and delta morphology can no longer be recognized, more general terms such as paralic, strandplain or coastal plain may be preferable.

Scale and importance of deltas

Deltas occur at a wide variety of scales ranging from basin-scale depositional systems, such as the modern Mississippi delta with an area of about 28,500 km^2, to smaller components of other depositional systems, such as bayhead deltas within estuarine or lagoonal systems. Many large deltas, such as the Danube in Romania and the Mississippi in the Gulf of Mexico, contain several scales of delta lobes associated with the fact that as a channel splits into several smaller channels, each new channel can feed its own lobe.

A large number of the earth's peoples live on or near deltaic coastlines and they are thus important from an environmental perspective. Ancient deltas are also economically important because they are commonly associated with major coal, oil and gas reserves. As a consequence, much has been written about deltas, although there is still much interest in new delta research. There has recently been an increase in documented examples of delta types that have been historically lacking, such as tide-influenced systems (e.g., Maguregui and Tyler, 1991; Reading and Collinson, 1996; Willis et al., 1999). Readers are referred to several summaries for general information on deltas (Reading and Collinson, 1996; Bhattacharya and Walker, 1992; Colella and Prior, 1990; Whateley and Pickering, 1989; Coleman and Prior, 1982; Broussard, 1975) and estuaries (Nichols and Biggs, 1985; Dalrymple et al., 1992; Dalrymple et al., 1994).

History of delta research

Historical overviews of delta research are provided by Bhattacharya and Walker (1992) and Reading and Collinson (1996) and summarized here. The concept of a delta dates back to the time of Herodotus (c. 400 BC) who recognized that the alluvial plain at the mouth of the Nile had the form of the Greek letter Δ. The first study of ancient deltas was that of Gilbert, 1885 (see Bhattacharya and Walker, 1995), who described Pleistocene fresh-water gravelly deltas in Lake Bonneville, Utah. Gilbert recognized a threefold cross-sectional division of a delta into topset, foreset and bottomset deposits (note sigmoidal stratal geometry in Figures D4 and D7). Barrell (1912 ibid.) extended these subdivisions to the much larger scale of the Devonian Catskill wedge in the Appalachians, and also provided the first explicit definition of the essential features of an ancient delta deposit. Barrell considered the recognition of overlying nonmarine facies crucial in recognizing ancient deltas, although recent research demonstrates that in a significant number of deltas, non-marine topset facies may not be preserved (e.g., Bhattacharya and Willis, 2001). Barrell actually wrote his 1912 paper in order to address what he perceived as an over-application of an estuarine interpretation to many ancient sedimentary successions that contained a marine to non-marine transition and that he believed should be better interpreted as deltas.

The Mississippi river and its deltas have long been a focus for understanding continental-scale delta systems (see summary of early papers by LeBlanc, 1975). Scruton (1960, see LeBlanc, 1975) recognized that deltas are cyclic in nature and consist of a progradational constructive phase usually followed by a thinner retrogradational destructive phase coinciding with delta abandonment. He also illustrated a vertical facies succession of coarsening- and sandier-upward facies related to the progradation of bottomset, foreset and topset strata

(Figure D7). Kolb and Van Lopik (1966, see LeBlanc, 1975) summarized much of the work that described the way in which the Mississippi river constructed its delta plain over the past 9000 years, suggesting that the development was autocyclic. This work was critical in establishing the idea that apparently cyclic vertical successions of sedimentary strata can result from the intrinsic way in which river channels naturally avulse and cause switching of delta lobes.

Coleman and Wright (in Broussard, 1975) compiled data on 34 modern deltas and developed a sixfold classification based on sand distribution patterns. One of the most widely used classification schemes still used today is that of Galloway (in Broussard, 1975), who subdivided deltas according to the dominant processes controlling their morphology; rivers, waves and tides (Figure D5). This scheme has since been expanded to include grain calibre (Orton and Reading, 1993; Reading and Collinson, 1996). This scheme has also been extended to show how deltas, estuaries and barrier-lagoons may evolve into one another in the context of changing sea-level and sediment supply (Dalrymple et al., 1992).

Improvements in seafloor-imaging and seismic data acquisition led to the recognition of the abundance and importance of synsedimentary deformation in the subaqueous parts of modern deltas (Coleman, Prior and Lindsay, 1983; Winker and Edwards, 1983; Bhattacharya and Walker, 1992). Similar features have now been recognized in ancient deltas (e.g., Bhattacharya and Davies, 2001).

More recently the evolution of modern deltas has been interpreted in the context of sea-level changes and plate tectonics, rather than just sediment supply (e.g., Dominguez et al., 1987; and Boyd, et al., 1989; see Bhattacharya and Walker, 1992; Hart and Long, 1996). This resulted in widespread abandonment of the idea that apparently cyclic sedimentary successions are autocyclic, but rather are controlled by allocyclic processes such as subsidence, sea level change, and climate change (see *Cyclic Sedimentation*). Integration of basin-scale seismic stratigraphic concepts with ideas about cyclic sea-level change were applied to many ancient deltas in the development of *sequence stratigraphy* (e.g., Galloway, 1989; Van Wagoner et al., 1990; Bhattacharya, 1993). More recent work is beginning to elucidate the nature of tide-influenced deltas that have heretofore been under-recognized in ancient sedimentary rocks because of lack of well-studied modern or ancient examples (e.g., Maguregui and Tyler, 1991; see Reading and Collinson, 1996; Jenette and Jones, 1995; Willis et al., 1999).

River mouth processes

A delta forms when a river of sediment-laden freshwater enters a standing body of water, loses its competence to carry sediment, and deposits it. Much of the active sand deposition occurs in distinctive *distributary mouth bars*, formed directly at the mouth of the channel (Figure D6). The mouth bars form naturally as a consequence of the decrease in discharge and bed shear stresses associated with the loss of flow competence, although very little work has been done that quantifies the mechanics of how mouth bars scale to the flow behavior at the river mouth. River mouths may be deflected downdrift by waves and can be scoured and modified by tidal currents (Figure D5).

The general form of the deltaic deposit depends upon (1) whether the river outflow is more dense (*hyperpycnal flow*),

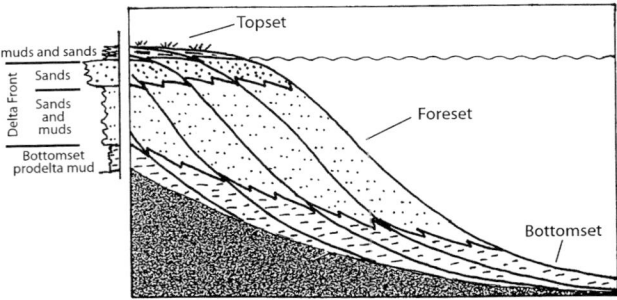

Figure D6 Seaward migration (termed progradation or regression) of a delta builds an upward coarsening facies succession, that shows a transition from marine into nonmarine topset deposits (modified after Scruton, 1960). Topset facies may be eroded by waves as the sea floods back over the land during transgression.

equally dense (*homopycnal*) or less dense (*hypopycnal*) than the standing body of water (Figure D6), and (2) the extent to which the deposits are reworked by wave and tidal processes, although tidal processes are insignificant in lakes. At any given river mouth, inertial, frictional and buoyant forces may be operative in varying proportions (Figure D6). In hyperpycnal deltas that have exceedingly high sediment concentrations at the river mouth, the slurry-like mixture of sediment and water just keeps on moving downslope as a sediment gravity flow (see sediment gravity flow). These sediments can end up being deposited in deepwater systems. Hyperpycnal rivers tend to be small and/or "dirty" such as the modern Sepik River, in Papua New Guinea, the Eel River in Northern California, and the rather larger Yellow River (Yangste) in China (Mulder and Syvitsky, 1995). In hypopycnal settings buoyant, less dense fresh water rides over a marine salt-wedge and carries suspended mud far offshore. Sand tends to be segregated and deposited at the mouth of the channel as a distinct sandy mouth bar with muddier sediments deposited farther offshore in the prodelta area (Figure D6). In homopycnal deltas a great degree of mixing occurs and frictional forces cause rapid deposition of all of the mud, sand, and/or gravel carried by the river (Figure D6). This results in deposition of a large poorly sorted "Gilbert-delta"-type mouth bar commonly with angle-of-repose foresets (Figure D6).

Rapid sedimentation in all delta types can cause sediment instability at the river mouth (see *Synsedimentary Structures and Growth Faults*). Slumping commonly occurs after the mouth bar builds up to a slope that exceeds the failure threshold. Mouth bars are especially susceptible to failure because of the high amount of water trapped within the sediment that causes high pore pressure. Slumped sediment at the river mouth can evolve into various types of sediment gravity flows and are an important way of providing sediment to the deep oceans.

Many rivers will experience dramatic changes in discharge as a function of seasonal climate change or as a result of major storms. Sediment discharge and sediment caliber of the river plays an important role in determining the sediment concentration discharged at the river mouth (Figure D6). These changes in sediment concentration can cause a river to change from hypopycnal to hyperpycnal, even in fully marine settings (Mulder and Syvitsky, 1995). Recent work has attempted to distinguish the relative importance of these processes in

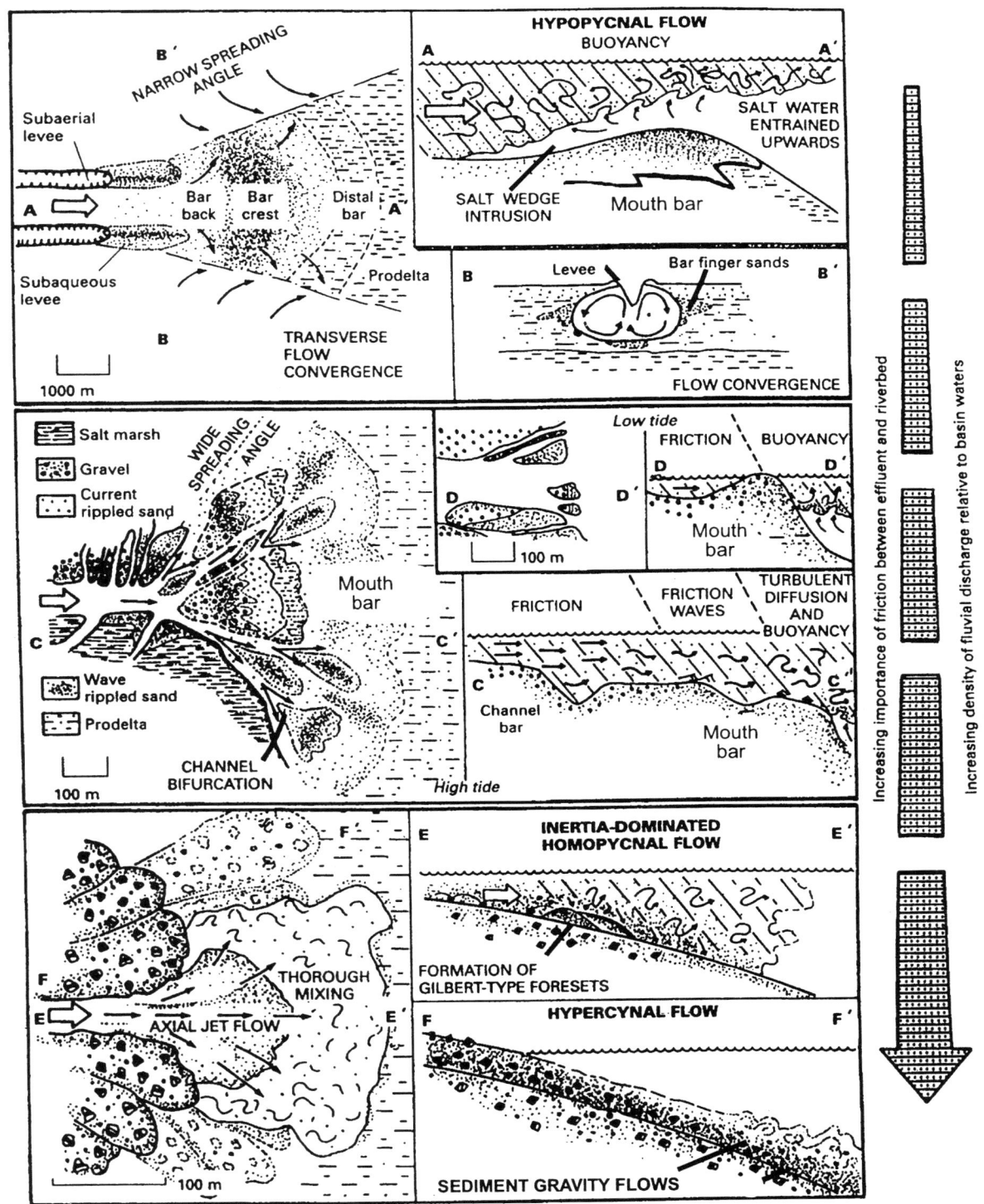

Figure D7 River Mouth processes and their deposits. Hypopycnal flow (top panel) is defined where inflowing water (typically fresh) is less dense than water in the receiving basin (typically saline). This favors the formation of a salt wedge and elongate bar-finger type mouth bars. The less dense fresh water is buoyed-up by the denser seawater and mud is carried far offshore. These delta fronts are typically referred to as buoyancy dominated. As the sediment becomes coarser-grained (middle panel), more friction is generated. This is common where the density of inflowing and receiving waters are the same (homopycnal) such as in many lacustrine deltas and in rivers with high sediment loads. These delta fronts are sometimes referred to as friction-dominated and favor development of triangular (radial) mouth bars. In rivers with exceptionally high sediment concentrations (lower panel), especially small muddy rivers, hyperpycnal conditions may result. In these cases, sediment-laden water is more dense than water in the receiving basin. The inflowing mixture of water and sediment may have enough inertia to continue flowing down the sea bed as a sediment gravity flow. These types of delta fronts are said to be inertia-dominated. These types of deltas may be linked to deep water depositional systems and provide a means of supplying sediment to submarine fan deposits. These mouth bar types are not mutually exclusive. Rivers can change from hypopycnal to hyperpycnal as discharge increases during exceptional flooding events and the degree of buoyant, frictional and inertial forces can vary across a bar (middle and lower panel). Modified after Reading and Collinson, 1996, based on Orton and Reading, 1993.

classifying ancient river mouth sediments (e.g., Martinsen, 1990; see Reading and Collinson, 1996).

River mouth processes can also be important in the autocyclic process of river avulsion that results in delta lobe switching. In this process, the river builds a large deltaic edifice that creates enormous form friction at the mouth of the river. This friction causes a loss of discharge at the river mouth. In addition, the river extends seaward, which decreases the effective slope. The loss of discharge and decreasing slope propagates upstream and eventually causes the river to avulse.

At a more local scale, friction at the river mouth may cause upstream deposition of sand. In a study of Cretaceous-age growth faulted deltaic strata in Utah, growth faults initiate both upstream and downstream as the loci of rapidly deposited sands shift (Bhattacharya and Davies, 2001).

Delta environments

No one environment is characteristic of a delta. In plan view, deltas comprise three main environments: the delta plain (where river processes dominate), the delta front (where river and basinal processes are both important) and the prodelta (where basinal processes dominate). In cross-sectional views, these three environments roughly coincide with the topset, foreset and bottomset strata of early workers (Figure D7).

Delta plains usually contain distributary channels and a wide variety of nonmarine to brackish, paralic environments including swamps, marshes, tidal flats, and interdistributary bays (Figure D4). The line that demarcates the landward limit of tidal incursion of seawater is referred to as the bay line and marks the boundary between the upper and lower delta plain. In ancient settings, the bay line may be indicated by the landward limit of marine or brackish-tolerant fossils or trace fossils. Lacustrine deltas have no bayline as such, since there are no significant tides in lakes.

Many deltas contain distributary channels that show different sizes and shapes in different positions on the delta. There is therefore no such thing as "a distributary channel" in many deltas. Typically, a trunk fluvial system will first avulse at the point where the river becomes unconfined forming a nodal avulsion point (Mackey and Bridge, 1995). Delta plain channels tend to be few in number and are separated by wide areas of interdistributary bays, swamps, marshes, or lakes in the upper delta plain, although these interdistributary areas can be replaced with channel deposits depending on the avulsion frequency (Mackey and Bridge, 1995). Upper delta plain channels may be exceedingly difficult to distinguish from fluvial channels, especially in lacustrine deltas. In fact, these types of distributaries are, strictly speaking, non-marine in nature and have been described in the context of fluvial depositional systems (Mackey and Bridge, 1995). Distributaries can show several orders of branching. The smallest scale channels are referred to as terminal distributary channels and are associated with the proximal delta front. Commonly these are only a few meters deep and will tend to lie on top of associated distal delta front deposits.

In many ancient examples, thick channelized deposits overlying marine prodelta shales have been interpreted as distributary channels (e.g., Rasmussen *et al.*, 1985) eroding into their associated delta fronts. Some of these sandstones are over 30 m thick and cut into delta front sandstones that are only 10 m or so thick. Terminal distributaries should actually be difficult to recognize in subsurface because they tend to be shallow and completely contained within the sandy delta front facies. Some of these deeply incised channels are now interpreted to be fluvially-incised valleys (e.g., Dalrymple *et al.*, 1994; Willis, 1997).

The delta front is the site of much of the active deposition, particularly at the mouths of distributary channels, where the coarsest sediment is deposited in distinct bars. Much of the deposition occurs in distributary mouth bars (also referred to as stream mouth or middle-ground bars) separated by shallow "terminal" distributary channels (Figures D4 and D6). Mouth bars can be on the order of several hundred meters wide and several kilometers long, such as in the Atchafalaya delta in the Gulf of Mexico (e.g., Van Heerden and Roberts, 1988). These mouth bars typically coalesce to form broader depositional lobes. Sediments deposited in mouth bars can also be profoundly reworked by waves, to form shore-parallel beach and strandplain deposits, or they can be reworked by tides to form shore-normal tidal bars (Figure D5).

Some distributary channels are fixed in the same position for long time periods, especially in deep-water mud-dominated deltas that build into relatively quiet marine basins (low tidal range, low wave action, Figure D6). The seaward building forms elongate bar fingers, as in the modern Mississippi "birdfoot" delta and in many bayhead lagoonal deltas. By contrast in siltier or sandier systems, deposited in shallower water, distributary channels switch more rapidly and coalesce to form more lobate sediment bodies, as in the Lafourche and Atchafalaya deltas in the Gulf of Mexico. Some researchers have used the term "braid" delta or "braidplain" delta to refer to a sandy or gravelly delta front fed by a braided river and characterized by a fringe of active mouth bars such as in the Canterbury Plains of New Zealand or in many glacial outwash plains (e.g., McPherson, Shanmugam and Moiola, 1987; see Reading and Collinson, 1996).

The seaward-dipping slope associated with distributary mouth bars is referred to as the distal delta front and can form a relatively continuous sandy fringe in front of the active mouth bar (Figure D4).

The prodelta is the area where fine material settles out of suspension. It may be burrowed or largely unburrowed, depending on sedimentation rates. Prodelta muds tend to merge seaward with fine-grained sediment of the basin floor. The preservation of silty or sandy lamination is commonly taken to mark the influence of the river, as opposed to total bioturbation of the basin floor sediments in areas away from the active river. Where the sediments are rhythmically laminated, there may be a tidal influence. Also, because of the abundant suspended sediment, certain types of vertical filter feeders and other organisms that produce open vertical burrows of the *Skolithos* ichnofacies tend to be suppressed (e.g., Gingras *et al.*, 1998). Because the delta front and prodelta areas are characterized by high levels of suspended sediment and the mixing of fresh and marine water, they tend to form very stressful environments for organisms resulting in low biological diversity (see *Bioturbation and Trace Fossils*). In New Orleans oyster bars, the best oysters (filter feeders) come from areas away from active river mouths whereas deposit-feeding crawfish favor more muddy prodelta and brackish lake environments.

Vertical facies successions

Progradation of a delta lobe will tend to produce a single, relatively thick coarsening-upward facies succession

(Figure D7) showing a transition from muddier facies of the prodelta into the sandier facies of the delta front and mouth bar environments (see numerous examples in Bhattacharya and Walker, 1992; and Reading and Collinson, 1996). Thicknesses may range from a few meters to a hundred meters depending on the scale of the delta and the water depth. Continued progradation may result in delta plain facies overlying the delta front sands in a continuous succession. However, delta front sands may be partially eroded by progradation of distributary channels or fluvial channels over its own mouth bar. Commonly, progradational delta lobe successions are truncated by thin transgressive abandonment facies.

In deltas deposited at the edge of the continental shelf, thick coarsening-upward delta front successions are commonly completely preserved within thicker deposits of the hanging wall of growth faults (e.g., Winker and Edwards, 1983; see Bhattacharya and Walker, 1992).

The specific nature of the facies in prograding prodelta and delta front successions will depend on the processes influencing sediment transport, deposition, and reworking. In addition, coarsening-upward facies successions can be produced by the progradation of other types of shoreline depositional systems.

Facies successions through distributary channels are erosionally based and typically fine upward. Filling commonly takes place after channel switching and lobe abandonment. At this time, the distributary channel may develop into an estuary, and the fill may be transgressive. The facies succession will tend to fine upward, with some preserved fluvially derived facies at the base, and a greater proportion of marine facies in the upper part of the channel fill. The extent of marine facies development will depend on the degree of fluvial dominance. Numerous examples of these different types are presented in Bhattacharya and Walker (1992) and Reading and Collinson (1996).

The overall proportion of distributary channel facies is a function of the type of delta. In general, the more wave-dominated the delta, the greater will be the proportion of lobe sediment with more limited amounts of interlobe and distributary channel facies.

Interdistributary and interlobe areas tend to be less sandy, and commonly contain a series of relatively thin, stacked coarsening- and fining-upward facies successions. These are usually less than ten meters thick, and much more irregular than the successions found in prograding deltaic lobes (see examples in Reading and Collinson, 1996). The proportion of lobe versus interlobe successions will depend on the nature and type of delta system and will tend to be greater in more river- or tide-influenced systems and less in wave-dominated deltas, although wave-influenced systems, like the Danube Delta in the Black Sea, can contain significant lagoonal and bay mudstones in regions downdrift of the river mouth (Bhattacharya and Giosan, *in press*).

Lateral facies variability

The lateral facies variability of many depositional systems requires a detailed understanding of bedding geometry. Bedding geometry and lateral facies variability can be addressed by the use of seismic data, Ground Penetrating Radar, continuous outcrop data (e.g., Willis *et al.*, 1999), and interpolation of well data. Analysis of lateral facies variability by tracing facies and bedding is termed *Facies Architectural Analysis*, in which discrete sediment bodies are analyzed on the basis of bounding surfaces and bedding geometry (Miall, 1985). Facies variations associated with bounding surfaces can have significant impact on fluid flow in subsurface reservoirs. Considerable research in the petroleum industry has thus been dedicated to investigating the importance of bounding surfaces at a variety of scales. Much of this research emphasis has been on fluvial systems (see examples in Miall and Tyler, 1991; Flint and Bryant, 1993) but more recent studies of deltaic systems are becoming available (e.g., Ainsworth *et al.*, 1999; Willis *et al.*, 1999; Tye *et al.*, 1999).

Dip-variability

In cross-sectional view, deltas show the distinct topset, foreset and bottomset geometry, with strata typically organized into an offlapping pattern, formed as the delta progrades (Figures D4 and D7). Strata associated with progradation of the entire continental shelf-slope and basins also form larger-scale "clinoforms". Delta-front foreset-dips range from a few degrees up to the angle of repose in coarse-grained Gilbert-type deltas. Prodelta bottomsets typically dip at less that $1°$ whereas nonmarine topset facies are typically flat to undulating. The seaward migration of these geomorphic areas of deposition builds the vertical facies successions as shown in Figure D7.

Berg (1982, see Bhattacharya and Walker, 1992) discussed typical seismic facies in deltaic depositional systems and suggested that sandy wave-dominated systems tend to be characterized by a more shingled pattern. The muddier river-dominated delta types tend to show more of an oblique-sigmoidal pattern. The steeply dipping, sigmoid-shaped portions are probably characteristic of the mud-dominated prodelta facies whereas the more flat-lying upper reflectors represent the sandier delta front and delta plain facies. Frazier (1974, see Bhattacharya and Walker, 1992) showed a similar geometry based on geological studies of the Mississippi delta plain. Offlapping clinoform geometries have also been recognized in the Late Quaternary wave-dominated sediments of the Rhone shelf (Tesson *et al.*, 1990; see Bhattacharya and Walker, 1992) and in many other studies (e.g., Hart and Long, 1996).

Strike variability

Along strike, facies relationships may be less predictable and depositional surfaces may dip in different directions (Figure D4). This is particularly so in more river-dominated deltas where along-strike reworking is not significant and abrupt facies transitions may occur between distributaries and interdistributary areas. Overlapping delta lobes result in lens-shaped stratigraphic units that exhibit a mounded appearance on seismic lines (Figure D4). 3D bedding geometries have been simulated in computer-based models developed by Tetzlaff and Harbaugh (1989).

Conclusions

A good case can be made for regarding deltas as the single most important clastic sedimentary environment (Leeder, 1999, p. 383) but because of their complexity deltas cannot be classified according to any one simplistic model. Parameters now considered to be essential in understanding facies distribution in deltas include: feeder type, river discharge, sediment caliber, water depth, basin physiography, storms, waves and tides, sea level, physical position in the basin, and degree of soft-sediment deformation. Clearly, combination of these parameters results in a nearly infinite number of possible delta types reflecting a chaotic, nonlinear, dynamic sedimentary continuum.

Many depositional systems (e.g., barrier-islands, deltas, estuaries) are not mutually exclusive and components of one can be found in another. Also, because of changes in parameters through time, one type of depositional environment can evolve into another. There are now too many end-member delta models. Future facies models must take a more quantitative, dynamic, predictive, parametric approach, such as used by Tetzlaff and Harbaugh (1989) in simulating deltaic deposition. These approaches will necessarily focus on the mechanics of delta formation and resulting facies distribution, informed by focused field studies, rather than merely classifying the type of delta observed in an outcrop or core.

Janok Bhattacharya

Bibliography

Ainsworth, R.B., Sanlung, M., Theo, S., and Duivenvoorden, C., 1999. Correlation techniques, perforation strategies, and recovery factors: An integrated 3-D reservoir modeling approach Sirkit Field, Thailand. *AAPG Bulletin*, **83**: 535–1551.

Bhattacharya, J.P., 1993. The expression and interpretation of marine flooding surfaces and erosional surfaces in core; examples from the Upper Cretaceous Dunvegan Formation in the Alberta foreland basin. In Summerhayes, C.P., and Posamentier, H.W. (eds.), *Sequence Stratigraphy and Facies Associations*. IAS, Special Publication, No. 18, pp. 125–160.

Bhattacharya, J.P., and Davies, R.K., 2001. Growth faults at the prodelta to delta-front transition, Cretaceous Ferron Sandstone, Utah. *Marine and Petroleum Geology*, **18**: 525–534.

Bhattacharya, J.P., and Giosan, L., *in press*, Wave influenced deltas. *Sedimentology*.

Bhattacharya, J.P., and Walker, R.G., 1992. Deltas. In Walker, R.G., and James, N.P. (eds.), *Facies Models: Response to Sea-level Change*. Geological Association of Canada, p. 157–177.

Bhattacharya, J.P., and Willis, B.J., 2001. Lowstand Deltas in the Frontier Formation, Wyoming Powder River Basin, Wyoming: Implications for sequence stratigraphic models, U.S.A., *AAPG Bulletin*, **85**: 261–294.

Boyd, R., Dalrymple, R.W., and Zaitlin, B.A., 1992. Classification of clastic coastal depositional environments. *Sedimentary Geology*, **80**: 139–150.

Boyd, R., Suter, J., and Penland, S., 1989. Sequence stratigraphy of the Mississippi delta: Gulf Coast Association of Geological Societies. *Transactions*, **39**: 331–340.

Broussard, M.L. (ed.), 1975. *Deltas, Models for Exploration*. Houston: Houston Geological Society, 555pp.

Busch, D.A., 1971. Genetic units in delta prospecting. *American Association of Petroleum Geologists Bulletin*, **55**: 1137–1154.

Colella, A., and Prior, D.B., 1990. *Coarse-grained Deltas*. International Association of Sedimentologists, Special Publication, 10, 357 p.

Coleman, J.M., and Prior, D.B., 1982. Deltaic environments. In Scholle, P.A., and Spearing, D.R. (eds.), *Sandstone Depositional Environments*. American Association of Petroleum Geologists, Memoir, 31, pp. 139–178.

Dalrymple, R.W., Zaitlin, B.A., and Boyd, R., 1992. Estuarine facies models: Conceptual basis and stratigraphic implications. *Journal of Sedimentary Petrology*, **62**: 1130–1146.

Dalrymple, R.W., Boyd, R., and Zaitlin, B.A., 1994. *Incised-Valley Systems: Origin and Sedimentary Sequences*. SEPM, Special Publication No. 51, 391pp.

Dominguez, J.M.L., 1996. The São Francisco strandplain: a paradigm for wave-dominated deltas? In De Baptist, M., and Jacobs, P. (eds.), *Geology of Siliciclastic Shelf Seas*. Geological Society Special Publication, 117, pp. 217–231.

Dominguez, J.M.L., Martin, L., and Bittencourt, A.C.S.P., 1987. Sea-level history and Quaternary evolution of river mouth-associated beach-ridge plains along the east- southeast Brazilian Coast: a summary. In Nummedal, D., Pilkey, O.H., and Howard, J.D. (eds.), *Sea-level fluctuations and coastal evolution: Society of Economic Paleontologists and Mineralogists*, Special Publication, **41**, pp. 115–127.

Flint, S.S., and Bryant, I.D., 1993. *The Geological Modelling of Hydrocarbon Reservoirs and Outcrop Analogues*. International Association of Sedimentologists Special Publication, 15, 269pp.

Galloway, W.E., 1989. Genetic stratigraphic sequences in basin analysis I: architecture and genesis of flooding-surface bounded depositional units. *American Association of Petroleum Geologist Bulletin*, **73**: 125–142.

Gingras, M.K., MacEachern, J.A., and Pemberton, S.G., 1998. A comparative analysis of the ichnology of wave- and river-dominated allomembers of the Upper Cretaceous Dunvegan Formation. *Bulletin of Canadian Petroleum Geology*, **46**: 51–73.

Hart, B.S., and Long, B.F., 1996. Forced regressions and lowstand deltas: Holocene Canadian examples. *Journal of Sedimentary Research*, **66**: 820–829.

van Heerden, I.L., and Roberts, H.H., 1988. Facies development of Atchafalaya delta, Louisiana: a modern bayhead delta, American Association of Petroleum Geologists. *Bulletin*, **72**: 439–453.

Heward, A.P., 1981. A review of wave-dominated clastic shoreline deposits. *Earth-Science Reviews*, **17**: 223–276.

Jenette, D.C., and Jones, C.R., 1995. Sequence stratigraphy of the Upper Cretaceous Tocito Sandstone: a model for tidally influenced incised valleys, San Juan basin, Mexico. In Van Wagoner, J.C., and Bertram, G.T. (eds.), *Sequence Stratigraphy of Foreland Basin Deposits: Outcrop and Subsurface Examples from the Cretaceous of North America*. American Association of Petroleum Geologists Memoir, 64, 311–347.

LeBlanc, R.J., 1975. Significant studies of modern and ancient deltaic sediments. In Broussard, M.L. (ed.), *Deltas, Models for Exploration*. Houston: Houston Geological Society, pp. 13–84.

Leeder, M., 1999. *Sedimentology and Sedimentary Basins, from Turbulence to Tectonics*. Blackwell Science.

Mackey, S.D., and Bridge, J.S., 1995. Three dimensional model of alluvial stratigraphy: theory and application. *Journal of Sedimentary Research*, **65**: 7–31.

Maguregui, J., and Tyler, N., 1991. Evolution of Middle Eocene tide-dominated deltaic sandstones, Lagunillas Field, Maracaibo Basin, western Venezuala. In Miall, A.D., and Tyler, N. (eds.) *The three-dimensional facies architecture of terrigenous clastic sediments, and its implications for hydrocarbon discovery and recovery*. SEPM Concepts and Models in Sedimentology and Paleontology, **3**: 233–244.

Martinsen, O.J., 1990. Fluvial, inertia-dominated deltaic deposition in the Namurian (Carboniferous) of Northern England. *Sedimentology*, **37**: 1099–1113.

Miall, A.E., 1985. Architectural-element analysis: a new method of facies analysis applied to fluvial deposits. *Earth Science Reviews*, **22**: 261–308.

Miall, A.D., and Tyler, N. (eds.), 1991. *The Three-Dimensional Facies Architecture of Terrigenous Clastic Sediments, and its Implications for Hydrocarbon Discovery and Recovery*. SEPM Concepts and Models in Sedimentology and Paleontology, Volume 3: 309pp.

Mulder, T., and Syvitsky, J.P.M., 1995. Turbidity currents generated at river mouths during exceptional discharge to the world's oceans. *Journal of Geology*, **103**: 285–298.

Nichols, M.M., and Biggs, R.B., 1985. Estuaries. In Davis, R.A. (ed.), *Coastal Sedimentary Environments*. Springer-Verlag, pp. 77–186.

Orton, G.J., and Reading, H.G., 1993. Variability of deltaic processes in terms of sediment supply, with particular emphasis on grain size. *Sedimentology*, **40**: 475–512.

Rasmussen, D.L., Jump, C.J., and Wallace, K.A., 1985. Deltaic systems in the Early Cretaceous Fall River Formation, southern Powder River Basin, Wyoming. *Wyoming Geological Association, 36th Annual Field Conference Guidebook*, pp. 91–111.

Reading, H.G., and Collinson, J.D., 1996. Clastic coasts. In Reading, H.G., (ed.), *Sedimentary Environments: processes, facies and stratigraphy*, 3rd edn. Blackwell Science, pp. 154–231.

Scruton, P.C., 1960. Delta building and the deltaic sequence. In Shepard, F.P., Phleger, F.B., and van Andel, T.H. (eds.), *Recent Sediments Northwest Gulf of Mexico*. American Association of Petroleum Geologists, pp. 82–102.

Tetzlaff, D.M., and Harbaugh, J.W., 1989. Simulating Clastic Sedimentation. Van Nostrand Reinhold.

Tye, R.S., Bhattacharya, J.P., Lorsong, J.A., Sindelar, S.T., Knock, D.G., Puls, D.D., and Levinson, R.A., 1999. Geology and stratigraphy of fluvio-deltaic deposits in the Ivishak Formation: applications for development of Prudhoe Bay Field, Alaska. *AAPG Bulletin*, **83**: 1588–1623.

Whateley, M.K.G., and Pickering, K.T., 1989. *Deltas: Sites and Traps for Fossil Fuels*. Geological Society of London, Special Publication, 41, 360pp.

Willis, B.J., 1997. Architecture of fluvial-dominated valley-fill deposits in the Cretaceous Fall River Formation. *Sedimentology*, **44**: 735–757.

Willis, B.J., Bhattacharya, J.B., Gabel, S.L., and White, C.D., 1999. Architecture of a tide-influenced delta in the Frontier Formation of Central Wyoming, USA. *Sedimentology*, **46**: 667–688.

Van Wagoner, J.C., Mitchum, R.M., Campion, K.M., and Rahmanian, V.D., 1990. *Siliciclastic sequence stratigraphy in well logs, cores, and outcrops*. American Association of Petroleum Geologists, Methods in exploration series 7, 55pp.

Cross-references

Angle of Repose
Barrier Islands
Biogenic Sedimentary Structures
Coastal Sedimentary Facies
Cyclic Sedimentation
Deformation Structures and Growth Faults
Gravity-Driven Mass Flows
Lacustrine Sedimentation
Neritic Carbonate Depositional Environments
Ripple, Ripple Mark, Ripple Structure
Rivers and Alluvial Fans
Sediment Transport by Tides
Sediment Transport by Waves
Tidal Flats
Tidal Inlets and Deltas

DEPOSITIONAL FABRIC OF MUDSTONES

Definition

The fabric of a rock is the total of all textural and structural features (Whitten and Brooks, 1972). Texture is defined as the relationship between the particles that form a rock, and structure has to do with discontinuities and major inhomogeneities. Both are intrinsic properties of the rock that serve as guides to its origins (Pettijohn and Potter, 1964), as well as providing a basic rationale for the study of mudstone fabrics.

Studying mudstone fabrics

The structural dimension of mudstone fabrics (discontinuities and major inhomogeneities) is best examined in hand specimens (preferably cut and polished surfaces) and in petrographic thin sections (Figure D8). For hand specimens, one starts with visual inspection, followed by closer examination under a binocular microscope. This approach is suitable to

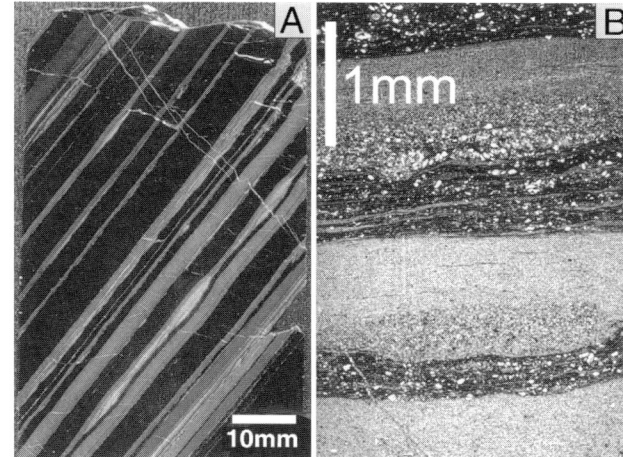

Figure D8 (A) example of hand specimen with alternating carbonaceous layers (black) and silt/clay couplets (light). Note lenticular and cross-laminated silt layer. (B) thin section of same shale as in A. Note graded silt/clay couplets (storm layers), and wavy crinkly laminae in carbonaceous layers (microbial mat deposit; Schieber, 1986).

study features in the millimeter range, such as small-scale sedimentary structures. Petrographic thin sections should initially be examined at low magnification (binocular microscope). Further study by petrographic microscope at higher magnification can reveal additional details (e.g., nature of lamina contacts), as well as providing information on mineral composition.

For the investigation of the textural dimension of mudstone fabrics (relations between particles) high magnifications are required to see individual grains and their contact relations. The scanning electron microscope is the tool of choice here (Trewin, 1988), operating in either secondary (SEM) or backscattered (BSE) mode (Figure D9). Observation in SEM mode is popular because sample preparation is easy (a freshly broken surface will suffice), but BSE imaging has gained popularity because it provides compositional information in addition to much clearer contact relations. The main drawback of the BSE mode is the need for a polished surface.

Fabric elements

From a sedimentologist's perspective, small-scale sedimentary structures tend to be the most informative fabric elements. Inconspicuous or even invisible in outcrop (they range in size from a few centimeters to less than a millimeter), they can nonetheless reveal a wealth of information about depositional conditions and history of a mudstone. When studied in polished slabs and petrographic thin sections, even seemingly drab shales can show a large range and variability of sedimentary features (Figure D10). If density contrasts are sufficient, small-scale sedimentary features may also be enhanced through X-radiography.

Laminae are the most typically observed sedimentary feature in shales, and exhibit a large range in thickness and lamination styles (even, discontinuous, lenticular, wrinkled, etc.), which can represent quiet settling, sculpting of the sediment surface by bottom currents, and growth of microbial mats respectively (Schieber, 1986). Internal lamina features include: (a) grading; (b) random clay orientation; (c) preferred clay orientation; (d) sharp basal contacts; and (e) sharp top contacts. They may be interpreted as indicative of: (a) event-sedimentation (e.g., floods, storms, turbidites); (b) flocculation or sediment trapping by microbial mats; (c) settling from dilute suspension; (d) current flow and erosion prior to deposition; and (e) current flow and erosion/reworking after deposition (Schieber, 1990). Silt laminae are usually the most easily observed lamina type in shales because of their somewhat larger grain size. They also imply somewhat more energetic conditions, and may, for example, indicate deposition by density currents (grading, fading ripples), storm reworking and transport (graded rhythmites), wave winnowing (fine even laminae with scoured bases), and bottom currents (silt layers with sharp bottom and top). Gradual compositional changes between clay and silt dominated laminae are another commonly observed feature, and are suggestive of continuous (although slow) deposition, possibly from deltaic sediment plumes and shifting nepheloid flows.

Other small scale sedimentary features that can be observed in shales are mudcracks, load casts, flame structures, bioturbation, graded rhythmites (Reineck and Singh, 1980), fossil concentrations and lags, cross-lamination, and loop structures (Cole and Picard, 1975), all of which carry information about conditions of sedimentation. Shales may also reveal clay-filled mud cracks, brecciation due to desiccation, and sands or conglomerates that consist entirely of shale particles (Schieber *et al.*, 1998). Because in the latter features the fills in cracks and between shale particles consist of the same components as the cracked substrate or the shale particles, there is little compositional contrast to reveal them on broken surfaces in outcrop. Polished slabs or petrographic thin sections are typically required to detect shale-filled cracks and shale sands and conglomerates. The latter, for example, can form as a result of soil erosion (pedogenic particles; Rust and Nanson, 1989), erosion of cracked mud crusts, and submarine scouring of mud substrates by strong currents.

Biologic agents may produce microbial laminae and protection of mud surfaces from erosion (Schieber, 1986; O'Brien and Slatt, 1990), or may manifest themselves as bioturbation and destruction of primary fabrics. In many instances, however, sufficient primary features survive bioturbation, and observation of bioturbation features can thus provide additional information about substrate firmness, event deposits (escape traces), and rates of deposition. Fecal pellets and pelletal fabrics are another by-product of organic activity,

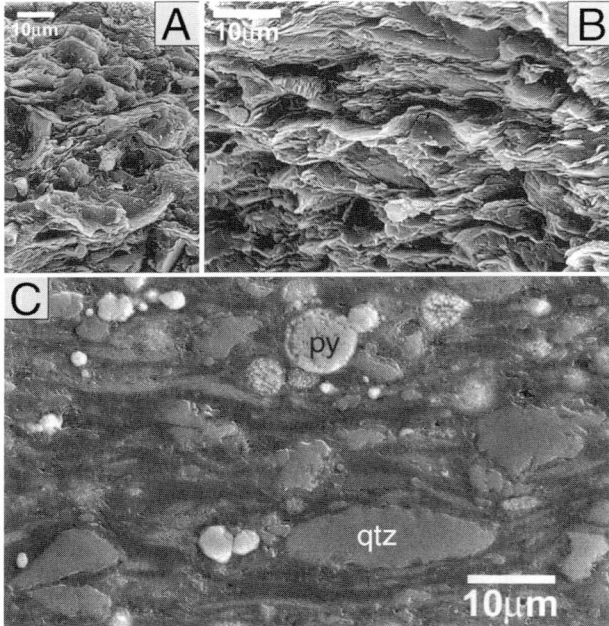

Figure D9 (A) SEM picture of bioturbated mudstone fabric; (B) SEM picture of laminated mudstone; (C) BSE picture of carbonaceous shale. Minerals with high molecular weight (py = pyrite) appear brighter than those of lower molecular weight (qtz = quartz). Gray streaks consist of clay minerals, dark-black streaks are kerogen. Note horizontal alignment of sediment grains, clay streaks, and kerogen stringers.

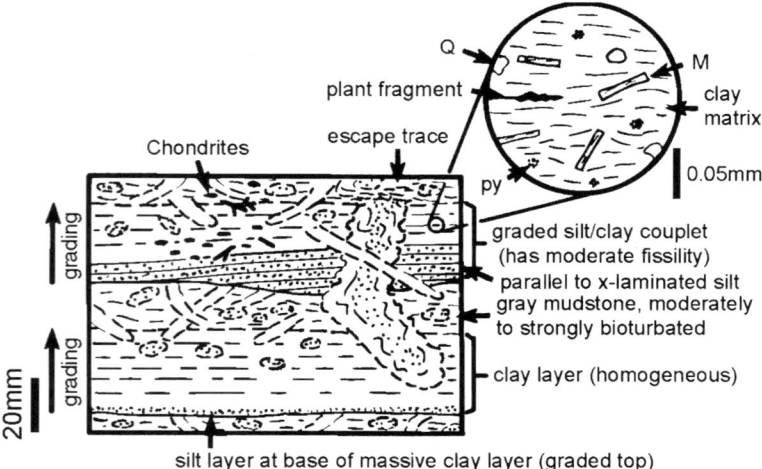

Figure D10 Line drawing that summarizes features observed in a mudstone from the Late Devonian Sonyea Group in New York (Schieber, 1999b). M = mica; Q = quartz; py = pyrite.

Figure D11 The "self-similar" nature of sediments. Magnification jumps by a factor of 10–20 between successive photos. (A) shale packages alternate with sandstone-rich units (Blackhawk Fm., Cretaceous, Utah); (B) sandstone-rich units from A consist of alternating sandy ledges and shaley intervals; (C) shaley intervals from B show softer shale alternate with centimeter-thick sandstone/siltstone beds; (D) thin section of soft interval from C shows graded siltstone layers and variably bioturbated mudstone. The resistant beds are storm deposits throughout (from HCS beds in B, to distal tempestites in D).

and they are probably more abundant in mudstones and shales than commonly recognized (Potter *et al.*, 1980; Cuomo and Rhoads, 1987). They are indicators of organic activity, and are best seen in thin section, where they typically differ from the shale matrix with respect to texture, color, and organic content. They may range in size from 0.2 mm to 2 mm and are generally of elliptical outline. Compaction tends to flatten the pellets to varying degrees, and if matrix and pellets are similar in composition and composed of particles of similar grain size, their identification can be challenging. The composition of fecal pellets can give clues on whether they were produced by benthic or planktonic organisms. Investigation of modern environments suggests that potential microbial colonization of mud surfaces has to be taken into account when studying ancient mudstones. Although difficult, recognition of microbial laminae is possible, and can lead to substantially different interpretations of mudstone environments (Schieber, 1999a).

Notwithstanding the usefulness of small-scale sedimentary structures, the study of particle relationships (texture) also constitutes an important avenue of inquiry. This aspect of shale fabrics, studied by electron microscope, is frequently referred to in the literature as *microfabric* (O'Brien and Slatt, 1990). Microfabric features of mudstones record the combined impact of processes active during transport, deposition, and burial at the scale of individual grains and grain aggregations. The *Argillaceous Rock Atlas* by O'Brien and Slatt (1990) contains case examples from a variety of sedimentary environments (as well as many references) and illustrates mudstone fabrics with SEM photos and photomicrographs.

Experimental work has shown that microfabrics can for example be related to depositional processes, such as flocculation and single-grain sedimentation (Mattiat, 1969; O'Brien and Slatt, 1990). The book *Microstructure of Fine-Grained Sediments* (Bennett *et al.*, 1991) contains a series of papers that employ microfabric analysis in the study of modern muds. A major drawback with the application of modern microfabrics to the interpretation of depositional conditions is the fact that muds undergo serious compaction and diagenetic changes on the way to becoming rocks. To determine from an SEM study which fabrics are secondary and which features are reflective of original depositional fabrics is difficult and requires considerable experience. In many instances the successful interpretation of microfabrics hinges on the availability of other information, such as small-scale sedimentary features, paleogeographic setting, and paleoecological information (e.g., O'Brien *et al.*, 1994). Nonetheless, a comparison of modern and fossil examples suggests that under favorable circumstances, fabrics produced by flocculation, settling of dispersed clays, bioturbation, agglutination on microbial mats, deposition of biosediment aggregates ("marine snow"), compaction, and fecal pellets can still be recognized and differentiated (Bennett *et al.*, 1991).

Fabric and environment

The broader significance of mudstone fabrics relates to the observation that sedimentary rocks and sedimentary successions have an inherent fractal quality (Mandelbrot, 1983). For example, large-scale stratification features can be repeated at different scales within individual stratigraphic packages, in the details of bedding, and even in the minutiae of lamination (Figure D11). Figure D11 illustrates that at increasingly higher levels of magnification, the basic theme of alternating soft and resistant intervals remains unchanged. Likewise, and regardless of scale, scoured bases, cross-stratification, and grading characterize resistant intervals, and bioturbation is more intense in soft intervals. This "self-similarity" from one magnification level to the next is at the heart of fractal geometry (Mandelbrot, 1983), and in our example (Figure D11) it reflects the overall conditions under which the succession accumulated. In that context we may even be tempted to speculate whether the information that can be extracted from fabric elements of a thin section or hand specimen (Figure D11(D)), will lead us to conclusions that are comparable to those derived from the study of an entire outcrop as shown in Figure D11(B). Studies of shale fabrics from rocks ranging in age from Proterozoic to the Tertiary have positively shown that shale fabrics do reflect the conditions of the broader sedimentary setting (Schieber, 1986, 1989, 1990; O'Brien and Slatt, 1990; Schieber *et al.*, 1998). Comparing what can be learned about depositional conditions in a mudstone-rich sedimentary succession from paleoecological observations and the study of interbedded sandstones, with the information derived from a detailed study of shale fabrics, shows that shale fabrics can provide a more concise account of depositional conditions than the other two approaches (Schieber, 1999b).

Summary

Mudstone fabrics are the cumulative record of the history of transport, deposition, and burial of muddy sediments. Their final "look" depends upon the interplay between a whole range of physical, biological, and chemical processes. As such, successful studies of mudstone fabrics require the integration of all available paleontologic (body and trace fossils), sedimentologic, and petrographic observations. If done thoroughly, the information derived from a sample of mudstone (fabric study of hand specimen and thin section) can accurately portray depositional conditions and environments of the interval from which it was collected. If there is a choice between studying an entire outcrop of more resistant interbedded lithologies (e.g., sandstone), and the careful study of a mudstone sample from the same locality, the information derived from mudstone fabrics may well be superior.

Juergen Schieber

Bibliography

Bennett, R.H., Bryant, W.R., and Hulbert, M.H. (eds.), 1991. *Microstructure of Fine-Grained Sediments*. New York: Springer-Verlag.
Cole, R.D., and Picard, M.D., 1975. Primary and secondary sedimentary structures in oil shale and other fine-grained rocks, Green River Formation (Eocene), Utah and Colorado. *Utah Geology*, **2**: 49–67.
Cuomo, M.C., and Rhoads, D.C., 1987. Biogenic sedimentary fabrics associated with pioneering polychaete assemblages: modern and ancient. *Journal of Sedimentary Petrology*, **57**: 537–543.
Mandelbrot, B.B., 1983. *The Fractal Geometry of Nature*. New York: W.H. Freeman and Company.
Mattiat, B., 1969. Eine Methode zur elektronenmikroskopischen Untersuchung des Mikrogefüges in tonigen Sedimenten. *Geologisches Jahrbuch*, **88**: 87–111.
O'Brien, N.R., and Slatt, R.M., 1990. *Argillaceous Rock Atlas*. New York: Springer-Verlag.
O'Brien, N.R., Brett, C.E., and Taylor, W.L., 1994. Microfabric and taphonomic analysis in determining sedimentary processes in marine mudstones: example from Silurian of New York. *Journal of Sedimentary Research*, **A64**: 847–852.
Potter, P.E., Maynard, J.B., and Pryor, W.A., 1980. *Sedimentology of Shale*. New York: Springer-Verlag.
Pettijohn, F.J., and Potter, P.E., 1964. *Atlas and Glossary of Primary Sedimentary Structures*. New York: Springer-Verlag.
Reineck, H.-E., and Singh, I.B., 1980. *Depositional Sedimentary Environments*, 2nd edn. Berlin: Springer-Verlag.
Rust, B.R., and Nanson, G.C., 1989. Bedload transport of mud as pedogenic aggregates in modern and ancient rivers. *Sedimentology*, **36**: 291–306.
Schieber, J., 1986. The possible role of benthic microbial mats during the formation of carbonaceous shales in shallow Proterozoic basins. *Sedimentology*, **33**: 521–536.
Schieber, J., 1989. Facies and origin of shales from the Mid-Proterozoic Newland Formation, Belt basin, Montana, U.S.A. *Sedimentology*, **36**: 203–219.
Schieber, J., 1990. Significance of styles of epicontinental shale sedimentation in the Belt basin, Mid-Proterozoic of Montana, U.S.A. *Sedimentary Geology*, **69**: 297–312.
Schieber, J., 1999a. Microbial mats in terrigenous clastics: the challenge of identification in the rock record. *Palaios*, **14**: 3–12.
Schieber, J., 1999b. Distribution and deposition of mudstone facies in the Upper Devonian Sonyea Group of New York. *Journal of Sedimentary Research*, **69**: 909–925.
Schieber, J., Zimmerle, W., and Sethi, P. (eds.), 1998. *Shales and Mudstones*, 2 vols. Stuttgart: Schweizerbart'sche Verlagsbuchhandlung.
Trewin, N., 1988. Use of the scanning electron microscope in sedimentology. In Tucker, M. (ed.), *Techniques in Sedimentology*. Oxford: Blackwell Scientific Publications, pp. 229–273.
Whitten, D.G.A., and Brooks, J.R.V., 1972. *The Penguin Dictionary of Geology*. New York: Penguin Books.

Cross-references

Algal and Bacterial Carbonate Sediments
Bedding and Internal Structures

Black Shales
Desiccation Structures (Mud Cracks, etc.)
Flocculation
Mudrocks
X-ray Radiography

DESERT SEDIMENTARY ENVIRONMENTS

Introduction

Deserts are defined as areas of sparse vegetation due to aridity (cf. Glennie, 1970; Cooke and Warren, 1973; Cooke and others, 1993; Thomas, 1997). The United Nations study of desertification (UNEP, 1992) measured aridity by the ratio of precipitation to potential evapotranspiration and defines hyperarid areas with ratios of less than 0.05 and arid areas with ratios of 0.05–0.20. These areas comprise nearly 20 percent of the earth's land surface, primarily concentrated along the high pressure belts of both the subtropical horse latitudes and the poles. Many large deserts are found along these belts as well as the midlatitude intracontinental steppes of Asia and North America in areas classified as semiarid (0.20–0.50). Smaller desert areas less dependent on latitude occur in orographic rain shadows and in areas associated with cold oceanic upwelling. The distribution of deserts in the geologic past was strongly influenced by the configuration of continental land masses (for instance, the Pangean supercontinent), tectonic activity, and the cycles of variation in solar influx.

The range of sedimentary environments encountered in deserts is summarized in Strakhov (1970), Glennie (1970, 1987), Cooke and Warren (1973), Hardie and others (1978), Warren (1989), Fryberger and others (1990), Smoot and Lowenstein (1991), Cooke and others (1993), Lancaster (1995), and Thomas (1997). Desert sedimentary environments are characterized by: (1) poor vegetation, resulting in greater erosion by wind and water, (2) low rainfall, leading to greater desiccation and evaporation and highly sporadic sedimentation, and (3) water input from outside the desert, producing local facies associations. Desert environments are strongly influenced by the interplay of eolian processes, intermittent wetting and drying of sediment, and evaporation of standing water and groundwater.

The characteristically low vegetation cover of desert environments allows wind to erode surfaces and transport large quantities of sediment. Wind erosion may sculpt bedrock into fantastic forms or polish the surfaces of pebbles and boulders into characteristic forms called ventifacts. Wind erosion (deflation) may form small catchment basins a few tens of feet in diameter or larger basins that are hundreds of meters deep and thousands of square kilometers in area. Precipitation of saline minerals on exposed bedrock enhances its breakdown and makes it more susceptible to wind erosion.

Soils in deserts reflect low vegetation, intermittent wetting, and frequent complete desiccation. Many soils are regoliths with little horizon development. Microphytes (cyanobacteria, lichen, moss, etc.) may form a significant cover to the soil surface, protecting it from erosion or acting as a trap for windblown dust. Microphyte trapping of windblown dust is also thought to contribute to the formation of desert varnish on rock outcrops and boulders. Wetting and drying of desert soils produce complex desiccation crack patterns and may build up compressive forces that churn the sediment and force pebbles to the surface as pavements. These compressive forces may also produce slickensided shear planes. Vesicular fabrics are produced by air trapped below wet sediment or rock pavements. Windblown dust may introduce soluble minerals that are reworked by rainfall in the soil, forming caliche, silcrete, or evaporite minerals that precipitate as isolated nodules, crystals, or cemented crusts.

Evaporative concentration of solutes in groundwater and surface water makes these waters less hospitable to organisms and may lead to precipitation of saline minerals. Groundwater evaporation occurs throughout the capillary zone and may be aided by evapotranspiration by plants. A process called evaporative pumping has been considered an important source of hydrostatic head in the capillary zone, but Macumber (1995) argues that evaporation actually depresses the surface of regional groundwater discharge. The path of solute concentration is controlled by the initial groundwater composition, which reflects the drainage area lithology, and the early precipitation of less soluble phases (Eugster and Hardie, 1978). Similarly, the initial inflow waters control the composition of lakes or isolated marine embayments. At higher concentrations, the brine composition is complicated by a variety of other parameters, such as temperature, biological activity, mixing of different waters, solute recycling, dehydration, and reactions with minerals in the sediment. Where heavy saline brines sink into the water table, producing a lens-shaped groundwater body, the fresher groundwater may be refracted upward along the contact to produce artesian springs.

The geologically most important desert environments are summarized under the broad headings of eolian, fluvial, lacustrine, playa, and marine coastal environments. The discussions of these environments will focus on the features unique to desert settings. Additional details about these environments can be found under other headings.

Eolian environments

Vast deposits of wind-blown sand, called ergs or sand seas, may dominate the desert landscape. The primary components of these eolian sand deposits are dunes and sand sheets. Eolian dunes have a variety of morphologies and sizes that reflect sediment supply and the variability of the sand-transporting winds. Dune heights may reach hundreds of meters and lengths may extend kilometers. Eolian dunes are primarily classified by their crest configuration. Dune ridges that are transverse to wind direction include barchan, crescentic ridges, and parabolic dunes; ridges parallel to wind direction are called linear dunes; and dunes with complex multiple ridges are called star dunes. All of these dunes internally produce tabular cross-bedding from slip-face migration and low-angle tangential cross-bedding primarily from wind ripple migration. Eolian deposits dominated by transverse dune ridges have largely unidirectional cross-bedding parallel to the net wind direction, whereas those dominated by linear dunes have bimodal cross-bedding at right angles to the net wind direction. Cross-bedding sets are divided by erosional bounding surfaces that represent the stoss-side erosion of climbing dunes, or larger scale bounding surfaces that represent the stoss-side erosion of large compound dunes (alsocalled draa) or deflation of a dune field to the local groundwater table. Interdune areas may be represented by pebble lags from

Figure D12 Eolian dune environments. (A) Crescentic ridge dunes in foreground and barchan dunes in background. Note muddy interdune area in foreground. (B) Deflation flat with small coppice dunes. Note alluvial fan in background. (C) Linear dunes cut by ephemeral lake margin.

deflation, thin mudcracked mud layers or evaporites from ponding of water, or flat bedding from the migration of wind ripples or adhesion ripples.

Sand sheets are broad areas of eolian sand that have low relief. These form around the margins of sand seas or in areas that are unfavorable for dune formation due to coarse grain size, vegetative cover, frequent flooding, cementation, or a high groundwater table. Most sedimentation in eolian sand sheets is from the migration of wind ripples producing relatively flat lamination. Zibars are low-relief (1m to 10m) and long-wavelength (100–500m) dune forms with rounded crests and no slip face. These produce low-angle inclined lamination. Coppice dunes (or nebkhas) are mounds of wind-blown sand that are trapped by vegetation. Eolian sand sheets that form near evaporite-producing areas may have their surfaces cemented and deformed by efflorescent salt crusts. Sedimentary layers may be deformed into upturned polygons or blister like knobs. Wind-blown dust trapped on the efflorescent salt crust surface produces irregular thin mud partings that accentuate these features.

Large areas of some deserts are dominated by wind deflation leaving exposed bedrock eroded into characteristic shapes and lags of wind-blasted cobbles. Deflation surfaces covered by lags of wind-smoothed pebbles and cobbles are also called gibber plains or regs. Windblown dust (loess) may form thick deposits in areas marginal to deserts that have some vegetative cover, and it is often incorporated into the deposits of other desert environments. Blowouts may produce small depressions that host lakes or saline pans. Mud pellets and sand eroded by wind may accumulate on the downwind edges of these basins, producing lunettes or parna dunes. Salts eroded from the flats, particularly gypsum, may contribute significant portions of sand to adjacent dune fields.

Fluvial environments

Fluvial systems in desert regions can be divided into those whose drainage extends out of an arid region (allogenic) or those whose drainage area is entirely within the arid environment (endogenic). Allogenic fluvial systems typically have flowing water all the time as perennial streams and rivers. Endogenic fluvial systems frequently are dry, flowing only during sporadic rainfall events. These fluvial systems include ephemeral streams and alluvial fans. Endogenic streams terminate within the arid region (endoreic), whereas allogenic streams and rivers may be endoreic or they may flow through the desert region (exoreic).

Perennial rivers and streams in deserts may have meandering, braided, or anastomosing channel morphologies similar to rivers of more humid regions. The size and distribution of cross-bedding from bedforms and bars in a channel deposit and the lateral distribution of channel deposits reflect the channel morphologies similar to rivers in humid regions. Since perennial rivers in deserts are a rare source of fresh water, vegetation tends to cluster around channel margins. Roots may preferentially obscure channel bedding features rather than those of the adjacent floodplain unlike the relationship for humid regions.

Ephemeral streams have channels that are dry except during sporadic floods. Most ephemeral streams have small braided channels that are dominated by horizontal lamination and low-angle tabular foresets. Ephemeral streams dominated by pebbles and cobbles display gravel bar lenses, pebble clusters, and shadow fabrics. Larger ephemeral streams include abundant trough cross-bedding and other features similar to perennial braided streams. Channel deposits of ephemeral streams may be capped by mudcracked mud drapes or they may have eolian overprints. Mud intraclasts derived by reworking of mud drapes or desiccated flood-plain deposits may be a major component of ephemeral stream deposits.

Fluvial floodplain deposits in deserts lack the vegetative cover typical of humid floodplains. Thick muddy floodplains are disrupted by complex desiccation cracks and may have desert soil features such as evaporites, caliche, or silcrete. Windblown sand may fill desiccation cracks. Sandy floodplains

DESERT SEDIMENTARY ENVIRONMENTS

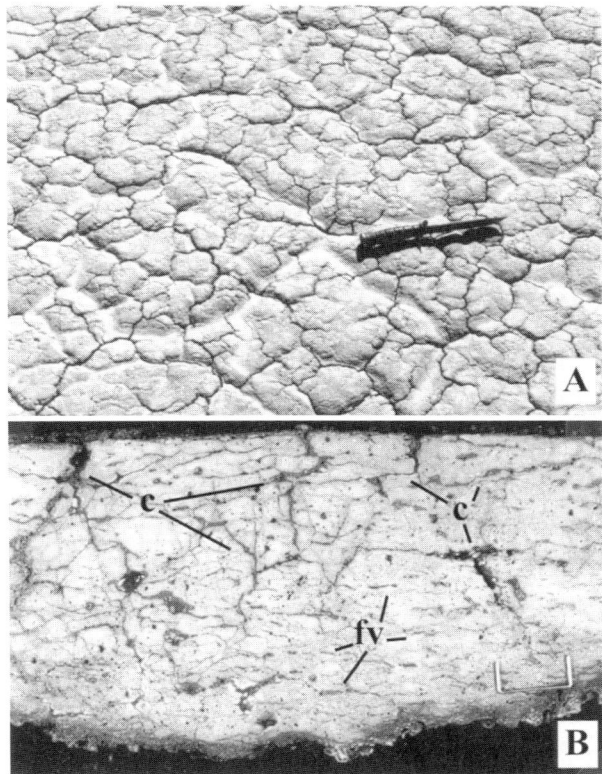

Figure D13 Mudcracked dry mudflat. (A) Surface view of complex mudcracks including partially filled older cracks cut by newer open cracks. Knife is 35 cm long. (B) Cross-section of above showing narrow irregular cracks (c), ovate vesicles (black), and flattened vesicles (fv). Scale is 7 mm wide.

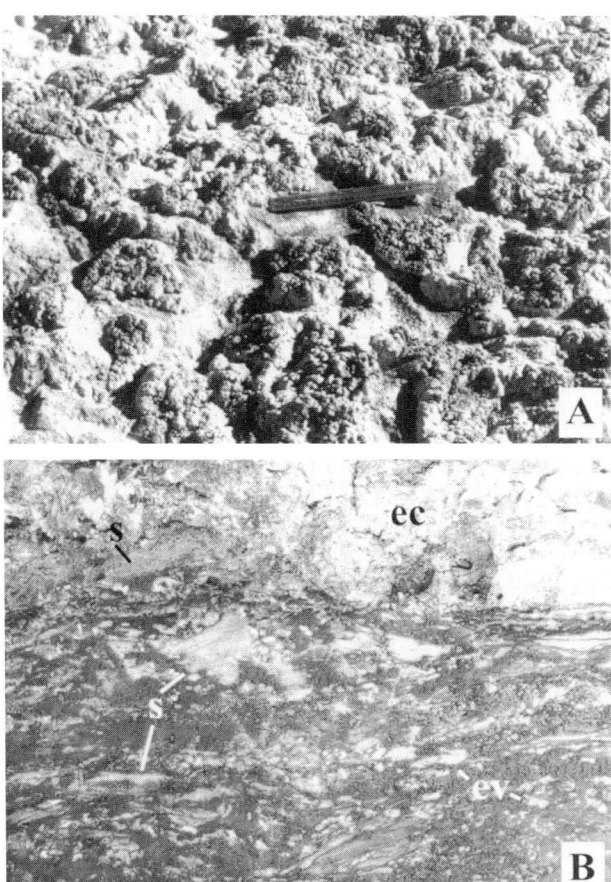

Figure D14 Efflorescent salt crust on a saline mudflat. (A) Surface view of aeolian sand partially filling depressions within a thick, mud-rich, efflorescent crust. Scale is 35 cm long. (B) Cross-section of crust above showing mud-rich, porous efflorescent crust (ec) with pockets of aeolian sand (s) that are also visible as patches in poorly sorted mud with scattered microcrystalline evaporites (ev). View is about 30 cm high.

may be reworked into eolian deposits including small dunes and sandsheets. Thin mud drapes break up into polygonal cracks that have upcurled edges that may be buried by windblown sand or incorporated as intraclasts in subsequent floods.

Alluvial fans are cones of sediment deposited where stream drainages debouch from mountain fronts. Alluvial fans are not restricted to deserts, but are commonly the dominant fluvial deposits in mountainous desert areas. Sedimentation on alluvial fans occurs during infrequent floods producing stream and/or debris-flow deposits. Streams are typically shallow and high-gradient with deposits of thin, poorly sorted gravel bars and planar bedding. Larger stream channels may develop tabular cross-bedding from bars and low-angle antidune backsets. Debris flows form levees, lobes, and sheets of poorly sorted muddy conglomerates with steep tightly packed boulder-cobble margins. During the long periods between debris-flow events, the deposits are commonly modified by rainfall, stream erosion of fines, and wind deflation. Debris-flow deposits tend to be more sheet-like and thinner at the toes of alluvial fans. Some stream-dominated fans have an apron of unchaneled sand at the toe called a sandflat. These deposits consist of sandy thin beds with thin, mudcracked mud partings. Broad, low-angle alluvial fans may develop where perennial rivers enter a desert basin over a mountain front.

Lacustrine environments

Lakes in desert conditions are commonly temporary features formed by sporadic flooding events. Perennial lakes form at the terminus of allogenic rivers or in basins with high groundwater flow. Like lakes in more humid settings, desert lakes are sites for accumulation of fine-grained sediment with coarser sediment accumulating near the margins. Lake sediments may be massively bioturbated or they may have flat lamination from settling of different grain sizes or types, or they may have layering defined by sandy storm layers. The surface area and depth of the lake controls the amount of wave-sorting that occurs within the lake and the development of shoreline deposits. Wedges of deltaic sediment form where river or streams intersect the lakes. Unlike most lakes in humid regions where lake depth is controlled by the spill level, desert lakes are typically hydrologically closed with their depth controlled by the balance between inflow and evaporation. These so-called terminal lakes change depth and surface area radically over short periods of time, even completely desiccating,

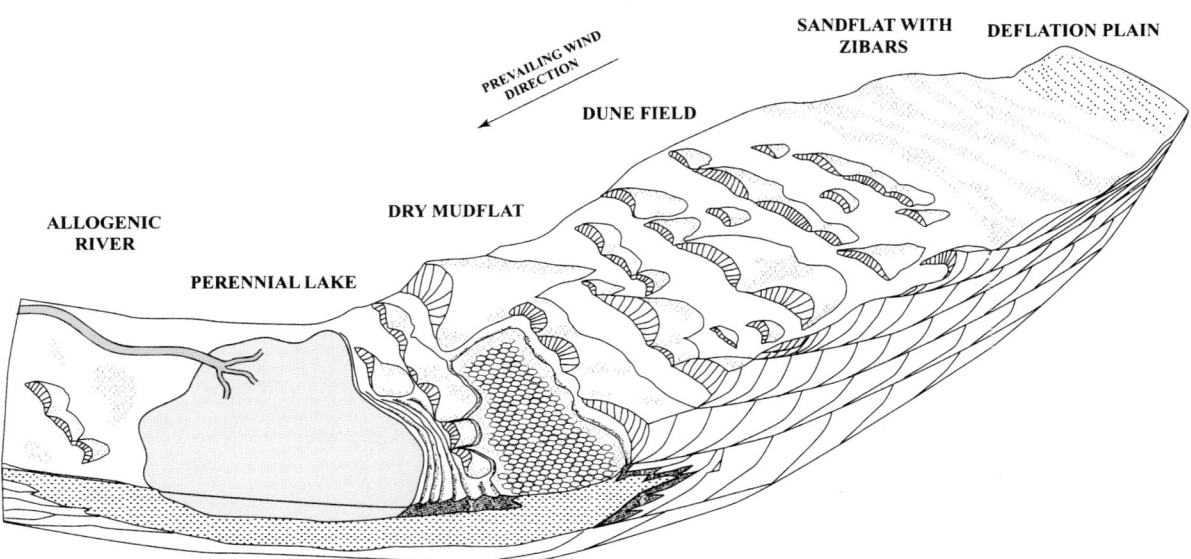

Figure D15 Distribution of sedimentary environments in a sand sea setting. Perennial lake has beach ridges and a birdfoot delta. Dry mudflat remains from a former lake embayment within the dunes.

and they accumulate solutes the longer they remain as standing water bodies.

Shoreline deposits range from well-developed spits and bars to thin sheets of rippled sand or imbricated pebbles. Shifting shorelines often leave the deposits stranded away from the lake to be modified by wind, flooding, and desiccation. Deltaic deposits include steep forset "Gilbert-type" deltas and stream-dominated birdfoot deltas. The frequent changes in lake level make these delta deposits more sheet-like and may result in stacks of delta fronts separated by subaerial surfaces. Where alluvial fans impinge on lake margins, the boulders and cobbles may be sorted by waves forming tabular sets that dip lakeward. Algal tufa deposits may coat cobbles or form mounds and towers along the shorelines, particularly where spring water mixes with carbonate-saturated lake water.

Desert lakes may accumulate enough solutes over time to supersaturate with saline minerals, such as gypsum, halite, trona, or borates. The minerals precipitated primarily depends upon the composition of the inflow water, but they are also controlled by variables such as temperature, depth, mixing, and organic productivity. Evaporite minerals may precipitate at the water surface producing cumulate crystal layers as they settle, or the crystals may precipitate on the lake floor producing crusts or vertically oriented surface crystals. Dense saline brines may sink through the underlying sediment precipitating crystals or altering matrix composition as they interact with previously deposited sediment or mix with interstitial water. Deep saline lakes are often strongly stratified with dense brine at the lake bottom inhibiting mixing and inducing bottom anoxia.

Playa environments

Playas include a variety of geomorphic features such as salars, salinas, chotts, inland sebkhas, and pans, among others, that comprise the "dry lakes". These deposits occur in depressions within the desert and are mostly dry with infrequent flooding. Some playas are superimposed on the remnants of former desert lakes, whereas others are aggrading. Playas may be dry muddy surfaces or they may be dominated by saline minerals. Wind erosion of mud and saline minerals may be very important on playa surfaces, particularly where powdery efflorescent crusts are developed. The major varieties of playa environments are dry mudflats, saline mudflats, and saline pans.

Dry mudflats are muddy surfaces dominated by desiccation cracks. They form in areas of intermittent surface flooding where the groundwater table is well below the surface. Bedding, when present, consists of graded sand-mud layers representing individual flood deposits. More commonly, the only sediment accumulation is within mudcracks producing a massive mud with complex sediment-filled cracks and open vesicles. Dry mudflat surfaces may develop powdery "puffy ground" efflorescent crusts where windblown soluble minerals are dissolved by rainfall or flooding and reprecipitated within the sediment surface.

Saline mudflats are muddy surfaces that contain abundant saline minerals precipitated within the sediment and/or on the surface as efflorescent crusts. These form in areas with shallow groundwater tables or groundwater discharge near the sediment surface and where there are abundant windblown soluble minerals available. Intrasediment crystals form in equilibrium with groundwater along its flowpath as it is evaporated producing large euhedral crystals in the phreatic zone and smaller crystal clusters or dehydrated forms in the vadose zone. Windblown dust may contribute to saline mineral formation in vadose zones when soluble materials are moved down by rainfall or flooding. This produces vertical profiles of upward coarsening crystals, opposite of the groundwater situation. Efflorescent crusts form by the complete evaporation of porewater at the sediment–air interface and are comprised of a mixture of tiny crystals including the most soluble phases. These crusts are rarely preserved as they are dissolved on the bottom by the undersaturated groundwater. Hard efflorescent crusts made of halite trap windblown dust on the crystal surfaces and coarser material in surface pits and irregularities thereby producing a poorly sorted mud with irregular sand patches. Powdery efflorescent crusts disrupt the upper sediment surface and form pelleted textures in mud that may be redeposited into cracks and surface irregularities.

Saline pans are salt-encrusted surfaces within saline mudflats that are intermittently occupied by shallow saline lakes. Saline pan lakes form during flooding events and the solutes are mostly derived by the dissolution of efflorescent crusts. This causes the saline pans to be dominated by the most soluble minerals. As the temporary lake waters evaporate, crystals are precipitated first as cumulus crystals then a bottom growth crystal crust. During prolonged periods between flooding, the crust may be modified by polygonal expansion cracks with formation of efflorescent crusts along their edges. The sinking brine will precipitate crystals in pore spaces and within sediments. The next flood will dissolve the efflorescent crust and part of the bottom crust before the cycle of deposition begins again. This mechanism may produce thick monomineralic deposits that are thin bedded.

Marine coastal environments

Where deserts border marine environments, the influence of seawater and marine sediment produces distinctive deposits. Many of the environments found within inland deserts may also occur in a coastal setting. Gypsum and halite characteristically dominate saline soils, saline mudflats, and saline pans along desert coastlines due to the influence of seawater chemistry. Eolian sands may contain a significant component of marine shelly material.

Sabkhas are broad, salt-encrusted flats that bound a marine coast. Clastic sedimentation on sabkhas is largely by storm surge tides, producing graded thin beds or lamination, or by wind. Thick algal mats commonly cover the sediment surface in nearshore areas modifying or controlling the sediment layering. Gypsum crystals form within the sediments that are saturated with seawater. Halite efflorescent crusts cover most of the inland part of the sabkha. Saline pans precipitating halite or gypsum crusts are common. Dehydration of gypsum crystals in distal areas produces anhydrite nodules and layers. Marine carbonate sediment is commonly altered to dolomite in this setting. Refraction of groundwater by the hypersaline seawater commonly produces artesian springs or spring-fed ponds along the coast within which carbonate crusts and micritic mud precipitate.

Isolated embayments of seawater in deserts become more saline by evaporation and less hospitable to normal marine organisms. If circulation with normal seawater is strongly curtailed, these embayments may become supersaturated with gypsum or halite producing cumulus crystal layers or bottom crusts similar to those of saline perennial lakes. Shallow, intermittently desiccated marine pools are called salterns. These produce layered evaporites similar to those of saline pans. In the geologic past, large embayments such as the Mediterranean Sea have become hypersaline producing extensive bedded evaporites.

Desert environment associations

Sedimentary deposits of desert environments have been recognized in rocks throughout the geologic record to the Precambrian. The distribution of these desert deposits through time reflects changes in the Earth's tectonic regime and global changes in climate. Some desert environments appear to have left little record in the pre-Quaternary despite their areal abundance in the modern world. Polar deserts and high-latitude saline lakes have not been noted in the geologic record. Deflation plains have only been recognized within eolian sand deposits and never as a regional surface in the pre-Quaternary. The dearth of these deposits in the geologic record probably reflects the fact that they do not aggrade making them less likely to be preserved.

Although the combinations of desert environments are as varied as the potential combination of climatic and tectonic histories, there are certain associations that are commonly repeated. Sand sea deposits are commonly developed on stable cratons or broad structural sags where low relief is ideal for erosion and transport by wind. Perennial lakes associated with these sand seas are broad and shallow with superimposed dry playa mudflat deposits common. Foreland basins of thrust belts in desert regions produce extensive ephemeral and perennial river plains that terminate into broad basins with perennial lakes and playas. Isolated marine basins may produce thick evaporite deposits ranging from deep water to subaerial. Rift basins produce thick sediment accumulations with alluvial fans, perennial lakes, and playas as the common associations. Thick sabkha and marine embayment deposits may also be associated with alluvial fans in this setting.

Climatic instability can shift a region from humid conditions to arid and back again repeatedly in the span of thousands of years. Desert deposits may be interbedded with non-desert deposits at the meter scale, or desert fabrics may be superimposed on non-desert deposits or vice versa. This is particularly true for orographic desert situations that exist closer to the equator or at higher latitudes than the high-pressure belts. Sea-level shifts in response to global climate change may result in marine inundation of coastal desert deposits or isolation of marine bays to form saline embayments or salterns.

Joseph P. Smoot

Bibliography

Cooke, R.U., and Warren, A., 1973. *Geomorphology in Deserts*. Los Angeles: University of California Press.
Cooke, R.U., Warren, A., and Goudie, A.S., 1993. *Desert Geomorphology*. London: University College London Press.
Eugster, H.P., and Hardie, L.A., 1978. Saline lakes. In Lerman, A. (ed.), *Lakes: Geochemistry, Geology, and Physics*. Springer-Verlag, pp. 237–293.
Fryberger, S.G., Krystinik, L.F., and Schenk, C.J., 1990. *Modern and Ancient Eolian Deposits: Petroleum Exploration and Production*. Denver: Rocky Mountain Section, SEPM.
Glennie, K.W., 1970. *Desert Sedimentary Environments*. Elsevier.
Glennie, K.W., 1987. Desert sedimentary environments, past and present—a summary. *Sedimentary Geology*, 50: 135–165.
Hardie, L.A., Smoot, J.P., and Eugster, H.P., 1978. Saline lakes and their deposits: a sedimentological approach. In Matter, A., and Tucker, M.E. (eds.), *Modern and Ancient Lake Sediments*. International Association of Sedimentology, Special Publication, 2, pp. 7–41.
Lancaster, N., 1995. *Geomorphology of Desert Dunes*. New York: Routledge.
Macumber, P.G., 1991. *Interaction between Groundwater and Surface Systems in Northern Victoria*, Melbourne, Department of Conservation and Environment—Victoria.
Rosen, M.R., 1994. The importance of groundwater in playas: A review of playa classifications and the sedimentology and hydrology of playas. In Rosen, M.R. (ed.), *Paleoclimate and Basin*

Evolution of Playa Systems. Geological Society of America, Special Paper, 289, pp. 1–18.

Smoot, J.P., and Lowenstein, T.K., 1991. Depositional environments of non-marine evaporites. In Melvin, J.L. (ed.), *Evaporites, Petroleum and Mineral Resources.* Elsevier, pp. 189–347.

Thomas, D.S.G., 1997. *Arid Zone Geomorphology.* John Wiley and Sons.

UNEP, 1992. *World Atlas of Desertification.* Seven Oaks, UK: Edward Arnold.

Warren, J.K., 1989. *Evaporite Sedimentology: Its Importance in Hydrocarbon Accumulations.* Prentice-Hall.

DESICCATION STRUCTURES (MUD CRACKS, etc.)

Desiccation structures originate as shrinkage cracks formed by the evaporation of water from the surface of clay-rich sediment. Previously called *mud cracks*, they are of subaerial origin, and are caused by the slow drying-out of muddy sediments which have been exposed to the action of sun and wind. The volume decrease that results from this loss of fluid gives rise to tensile stresses distributed equally in all directions within the bedding plane, that are relieved by the formation of a characteristic pattern of open polygonal cracks on the surface of the sediment.

Arrays of polygonal cracks resulting from this process are commonly seen on dried-out surfaces in puddles and slurry pits, around the margins of lakes, reservoirs and playas, and on tidal mudflats. Where well-developed, the desiccation structures form striking prismatic columns a few centimeters to several meters across, separated by deep fissures, and divided internally into secondary sets of cracks. Following a change in local climatic conditions and a rise in water level, the cracked surfaces become submerged, and the cracks and fissures may be filled with sediment and preserved. Once these sediments have been lithified, and later exposed at the Earth's surface, the host mudrock is more easily eroded than the sandy material infilling the cracks, and a cast of the sand-filled cracks is generally preserved on the *base* of the overlying sandstone bed.

Ancient desiccation cracks preserved in this way are not only useful as paleoenvironmental indicators, but provide vital evidence of way-up in rock sequences affected by folding and faulting. The crack patterns have geometric features in common with many other types of shrinkage cracks, such as cooling joints in basaltic dykes and sills, ice-wedge cracks in frozen ground (see *Tills and Tillites*), and with microcracks formed during the drying of thin layers of ceramic glaze, polymer, corn starch, paint and other materials. This feature can be used to advantage in experimental and theoretical studies (Weinberger, 1999, 2001), but their similarity to other sedimentary structures, and especially interstratal dewatering structures (Tanner, 1998), and syneresis cracks (see *Syneresis*) means that all sand-filled cracks have to be examined with great care, to determine their origin.

Morphology

The terminology used to describe desiccation structures was reviewed and illustrated by Allen (1984), following Lachenbruch (1962). The morphology of intertidal desiccation cracks was analyzed in a seminal study by Allen (1987), and that of filled desiccation cracks in vertic paleosols is described by Paik and Lee (1998).

In plan view, desiccation cracks vary from straight to curved. They may form random, unconnected, even radial, patterns (*incomplete cracks*) but are most commonly seen as closed shapes bounded by 3, 4, or 5 cracks, with Y- or T-shaped intersections (*complete cracks*). Although the linked hexagonal pattern (*nonorthogonal type*) is that most commonly portrayed in textbooks for mud cracks, the polygons in naturally occurring desiccation cracks are most commonly 4–5 sided, with the majority of individual cracks meeting at right angles (*orthogonal type*) (Figure D16). Polygon size is a function of the thickness of the desiccated layer, with thick layers producing the larger polygons.

In cross-section, individual desiccation cracks form at right angles to bedding and generally taper downward (V-shaped profile) from the surface of the cracked layer. In three dimensions, the sets of polygonal cracks divide the mud layer into separate columns or pillars which may be cut by horizontal cracks which connect with the bounding fractures. At an advanced stage of desiccation, and especially in thinly bedded mudrocks, bedding laminae at the top of the pillars become dish-shaped and may become distorted to form tricorn shapes or "mud curls" (Figure D17).

The infilling of the cracks with sand and silt grains may occur in a single event, or be multi-stage as is common in intertidal situations. Repeated cycles of desiccation and infilling also occur, leading to a complex internal crack morphology. Vertically stratified crack-infills have been reported from vertic paleosols. Alternatively, cracks formed by drying-out of the surface of a very wet substrate may subsequently be intruded by mud or fluidized sediment from below, to form polygonal patterns of clastic dykes.

History

Geologists working in the Appalachian Mountains had already recognized the environmental significance of mud cracks by the mid-19th century; the first use of mud cracks, or "suncracks", as a way-up indicator is not known, but according to Schrock (1948) the technique was already established by 1908. This author provides an excellent account of the morphology of desiccation cracks, together with a comprehensive review of earlier work. Fundamental advances in understanding the mechanics of the desiccation process were made by investigators studying the formation of closely related structures: ice-wedge cracks in frozen ground (Lachenbruch, 1962; Corte and Higashi, 1964). This phase was followed by further careful documentation of the morphology of mud cracks, foremost amongst these being the studies by Allen (1984, 1987). It was also a period in which there was considerable debate over the criteria that could be used to distinguish desiccation cracks from syneresis, or subaqueously-formed, cracks (see *Syneresis*). In recent years there has been a revival of interest in these superficially simple desiccation structures and further progress has been made in understanding their mechanics of formation (see below).

Mode of formation

The desicccation process requires subaerial exposure of a muddy layer of sediment. As moisture is lost from the upper

Figure D16 Orthogonal desiccation cracks in lime-rich mud from a tailings pit in SW England (see text for description).

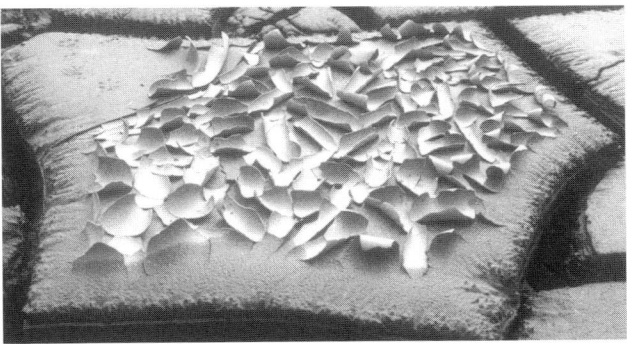

Figure D17 Mud curls surmounting a desiccation polygon at the same locality as Figure D16.

surface of the layer, the traditional model based on field relationships postulates that shrinkage cracks initiate at nucleation points on the surface, that in some cases can be identified as imperfections on the surface caused by bird's footprints, raindrops, bivalve shells, vegetation, etc. Aided by high local stress concentrations at their tips, the early cracks propagate both laterally and downward, and in most cases link up to form closed networks of polygonal cells. As in the case of cooling joints in igneous rocks, the pseudo-hexagonal form of these cells is controlled by the principle of least work: a detailed mechanical analysis of the process was given by Lachenbruch (1962).

Crack formation takes place in stages, with a network of primary or first-order cracks dividing the drying layer into a series of polygonal shapes, each of which is then subdivided by second-order cracks (Figure D16). Once a crack has formed it represents a principal or "free" surface, and new cracks propagating toward it turn to meet with it at right angles, so explaining the preponderance of orthogonal junctions over "ideal" 120° contacts (Lachenbruch, 1962). The more fine-grained, smecite-rich, and homogeneous the starting material, the more perfectly formed and nearly hexagonal the pattern of cracks which results. A striking feature of some sets of mud cracks is the development of secondary and even tertiary sets of fractures sub-dividing the polygons formed by their predecessors (Figure D16).

Current investigations

Recent work has been aimed at analyzing the desiccation process in detail, both in the field and by studying the formation of desiccation cracks in the laboratory using analog materials such as corn starch–water mixtures.

In both natural and experimental situations, the surfaces of the cracks are commonly decorated with plumose structures, which consist of a series of faint ridges radiating out from a central axis. Plumose markings enable the actual nucleation points for each crack to be identified, as well as showing the direction of propagation of the rupture surface, so enabling a detailed kinematic picture of the growth of the array to be assembled (Corte and Higashi, 1964; Weinberger, 1999, 2001). In contradiction to the accepted model, new field evidence shows that, in some cases at least, desiccation cracks nucleate at the base of the bed and propagate upward (Weinberger, 1999, 2001). In experimental situations the speed of crack propagation can be monitored using a video camera, and was found to vary from 100 mm per second at initiation to <10 mm/min during amplification (Müller, 2001).

Summary

Desiccation structures result from the uniform layer-parallel shrinkage of homogeneous layers of mud that are rich in swelling clays. This process occurs as a result of subaerial exposure of recently deposited sediment and gives rise to familiar patterns of polygonal cracks that are analogous to the cooling joints found in igneous dykes and sills. Individual cracks are V-shaped in cross-section and when subsequently filled with silt or sand, and preserved in the geological record, provide a reliable way-up indicator. A feature diagnostic of sub-aerial exposure is the formation, in thin, dried-out layers of mud at the top of the polygonal columns, of concave-upward plates or "mud curls".

Desiccation structures are indicative of subaerial exposure of muddy sediment whether in an intertidal, lacustrine, or playa-lake environment. However the simple patterns shown by these structures should not be confused with those shown by interstratal dewatering structures (see *Fluid Escape Structures*) or syneresis cracks (see *Syneresis*). Polygonal patterns of sand-filled cracks seen on bedding planes are not necessarily proof of subaerial exposure and their misinterpretation can, and has, led to serious errors in identifying the depositional environment.

P.W. Geoff Tanner

Bibliography

Allen, J.R.L., 1984. *Sedimentary Structures, their Character and Physical basis*. Developments in Sedimentology 30. Amsterdam: Elsevier.

Allen, J.R.L., 1987. Desiccation of mud in the temperate intertidal zone: studies from the Severn Estuary and eastern England. *Philosophical Transactions of the Royal Society*, **B315**: 127–156.
Corte, A.E., and Higashi, A., 1964. Experimental research on desiccation cracks in soil. *U.S. Army Snow, Ice and Permafrost Establishment*, 66.
Lachenbruch, A.H., 1962. Mechanics of thermal contraction cracks and ice wedge polygons in permafrost. *Geological Society of America Special Paper*, **70**: 1–69.
Müller, G., 2001. Experimental simulation of joint morphology. *Journal of Structural Geology*, **23**: 45–49.
Paik, I.S., and Lee, Y.I., 1998. Desiccation cracks in vertic palaeosols of the Cretaceous Hasandong Formation, Korea: genesis and palaeoenvironmental implications. *Sedimentary Geology*, **119**: 161–179.
Shrock, R.R., 1948. *Sequence in Layered Rocks*. New York: McGraw-Hill.
Tanner, P.W.G., 1998. Interstratal dewatering origin for polygonal patterns of sand-filled cracks: a case study from Late Proterozoic metasediments of Islay, Scotland. *Sedimentology*, **45**: 71–89.
Weinberger, R., 1999. Initiation and growth of cracks during desiccation of stratified muddy sediments. *Journal of Structural Geology*, **21**: 379–386.
Weinberger, R., 2001. Evolution of polygonal patterns in stratified mud during desiccation: the role of flaw distribution and layer boundaries. *Geological Society of America Bulletin*, **113**: 20–31.

Cross-references

Clastic (Neptunian) Dykes and Sills
Fluid Escape Structures
Septarian Concretions
Syneresis

DIAGENESIS

Introduction

Diagenesis encompasses all the physical and chemical processes that affect sediments and sedimentary rocks in the portion of the rock cycle that includes the period of burial between deposition and weathering, and also between deposition and the onset of metamorphism. Processes that induce the lithification of sediments constitute a major aspect of diagenesis. Determination of historical sequences of physical and chemical processes is an important element of diagenetic studies. Diagenesis is key to understanding the evolution of rock composition and texture with depth, time, and temperature, and to deciphering the mechanisms by which elements are cycled between the atmosphere, ocean, and crust.

Diagenetic processes are responsible for the generation of coal, oil, gas, and "hydrothermal" ore deposits, and exert a major control on the composition of subsurface water and the distribution of porosity in the subsurface. The effects of diagenesis may erect barriers to fluid flow or establish selective passageways for the migration of fluids. Thus, there are compelling practical aspects to diagenetic studies.

Recognition of the substantial magnitude of chemical and physical reorganization that takes place during transformation of sediments into sedimentary rocks has been a major conclusion of diagenetic studies over the past quarter century. Many processes formerly viewed in terms of a narrow affiliation with either weathering, metamorphism, or tectonic deformation are now seen as integral and essential elements of diagenesis. Study of the widespread role of microbial activity in diagenetic processes, both physical and chemical, is an area of currently active research.

Conditions of diagenesis

Although it begins at the sediment/water or sediment/air interface, diagenesis takes place largely in the subsurface and primarily under conditions in which the rock pore spaces are filled with water. No particular depth is implied, the scale ranging, potentially, from millimeter to kilometers. Important diagenetic fluids include seawater, meteoric water, hypersaline evaporative fluids, brackish mixed fluids, and pore fluid of any of these types that has been altered in varying degrees by fluid/rock interactions. Extreme degrees of diagenesis can yield pore fluids that are so substantially modified that their original composition (e.g., marine versus meteoric) is undecipherable. The temperature range for diagenesis extends from subzero (in sublimation brines) to 250°C and perhaps as high as 300°C. Rocks heated above this maximum temperature range may still have an identifiable sedimentary protolith, but tend to have lost the fabrics (grain-cement-pore space) and pervasive solid-phase disequilibrium that characterize sedimentary rocks. The pressure regime of diagenesis ranges from the conditions at the Earth's surface up to and including conditions of fully lithostatic pressure ($P_{fluid} = P_{load}$). Few subsurface samples are available below 5 km, but samples obtained from as deep as 7 km are still clearly sedimentary as long as their temperature has not exceeded 250°C for very long.

Rock components in diagenesis

Recognition of the initial versus the modified chemical and textural states of sedimentary rock components is essential to deciphering diagenetic processes. Rock components that exist at the time of deposition are known as *primary*; diagenetic modifications of primary components yield features described as *secondary*. Primary mineral components in sediments and sedimentary rocks are of two types. Primary chemical precipitates are characteristic of evaporative environments and, by definition, must precipitate from fluids concentrated by solar evaporation (Warren, 1999). A larger group of primary minerals are called *detrital* in recognition of their existence as transportable particles or *grains*, either formed within the depositional environment (*allochems*) or derived from older rocks outside the basin of deposition (*terrigenous debris*). Pore space constitutes another important primary component of sediments and may be either *intergranular*, occurring between the detrital grains, or primary *intragranular* pore space, occurring within the grain interior. Primary intragranular pore space is observed in all types of grains, but occurs most commonly and most extensively within allochems.

Some rock types are completely authigenic and diagenetic in the sense that they contain little or no strictly detrital material and instead are constituted almost entirely of minerals precipitated from aqueous solution either at the depositional surface or in the shallow subsurface. Such "chemical sediments" include caliche, chert, and nodular anhydrite, and halite. However, most sediments, at least initially, are constituted primarily of detrital grains and pore space.

All of the primary rock components described above are subject to profound modifications by both chemical and physical processes during diagenesis. Primary pore space can be substantially reduced by *compaction* and also through partial or complete filling by authigenic minerals (see *Authigenesis*), a process known as *cementation*. With the possible exception of detrital quartz and certain other minerals of high chemical stability, primary detrital minerals are subject to dissolution that can lead to the formation of *secondary intragranular pores*.

Secondary components, both authigenic minerals and pore space, are also subject to modifications during diagenesis. Early-formed authigenic phases may dissolve to leave secondary pores as ambient temperature and fluid composition evolve to conditions unfavorable for the mineral's continued stability. Authigenic phases precipitated during deep burial may be prone to dissolve when exposed to fluid flow at low temperatures following uplift.

Secondary pore spaces, within both primary and secondary minerals, may become filled with authigenic minerals, in a process known as *replacement*. Some of the most notable compositional and textural changes in diagenesis, for example dolomitization and albitization, take place largely through replacement. Replacement of primary or secondary minerals by new crystals of the same gross mineralogy, *recrystallization*, is notably difficult to document with complete certitude. This difficulty arises because a single mineral may be present in both primary and secondary forms or as co-existing authigenic crystals representing multiple episodes of precipitation, yet display only subtle heterogeneity in texture and composition. Demonstration of a cementation versus a replacement (or, recrystallization) origin for microcrystalline components (e.g., clay minerals, micrite) is particularly challenging because heterogeneities in texture and composition that can prove a primary versus an authigenic origin cannot be readily imaged at submicron scales.

Physical processes in diagenesis

Physical processes are important across the entire range of diagenetic conditions, but are prominently observed in the early stages of burial, prior to substantial lithification, when grain rearrangement by bioturbation and compaction are most readily accomplished. Major categories of physical processes in diagenesis include bioturbation, elutriation, compaction, and tectonic deformation.

Physical grain rearrangement in very early diagenesis of unconsolidated sediment is an important control on the subsequent path of chemical diagenesis because of the effects on permeability and grain surface characteristics. *Bioturbation* (see *Biogenic Sedimentary Structures*) is responsible for substantial grain rearrangement in certain depositional environments. Primary depositional fabrics (e.g., sorting, layering, etc.) may be greatly rearranged and new fabrics established by bioturbation, profoundly reducing the permeability of the sediment and the rocks that develop from them if fine and coarse detrital components, for example, clay and sand, become mixed. *Elutriation* is a process by which fine-grained particles are infiltrated through pore spaces following deposition, leading in some cases to rocks with reduced permeability. Thick grain coating clays introduced by elutriation and adhered tightly to grain surfaces by repeated wetting and drying can greatly alter the path of diagenesis in the sediment both by reducing overall permeability and also by creating a partial barrier to reaction between pore fluids and grain surfaces. This is one illustration of how physical and chemical processes in diagenesis are linked.

Compaction refers to the reduction of primary intergranular space during burial and deformation. Distinct from cementation, compaction arises solely from the rearrangement of detrital grains. In well-sorted sand-size sediments the intergranular volume (IGV), also known as the "minus-cement porosity", is the sum of primary intergranular pore space and intergranular cements. Volume loss by compaction is calculated as follows (e.g., Lundegard, 1992):

Compactional porosity loss =

$$P_i - (((100 - P_i) * IGV)/(100 - IGV))$$

At deposition, the initial porosity (P_i) ranges from around 40 percent of sediment volume for sand-size materials to much larger values (ranging to 80 percent?) for sediments with abundant clay-size particles and sediments containing grains with abundant primary intragranular pores. IGV reduced solely by grain rearrangement is believed to reach a minimum value of around 26 percent. Larger IGV reduction (to a maximum of IGV = 0) is observed in both sandstones and limestones, requiring the operation of other mechanisms of compaction such as ductile deformation, grain crushing by brittle deformation (see Pressure Dissolution). Near-depositional values for IGV can be preserved by early cementation, a phenomenon that is more widespread in limestones than in sandstones.

Deformation at a variety of scales is increasingly recognized as an integral element of diagenesis. In addition to prominent grain crushing during burial compaction, grains may also succumb to crushing by stresses applied tectonically, especially if the tectonism occurs while the rock is still relatively porous. Planar zones of crushed and cemented grains known as deformation bands are prominent in some porous sandstones. In rocks that are highly lithified, deformation may be throughgoing, cutting across detrital and authigenic components. Substantial modifications of porosity, both increases and decreases, can result from deformation. Fracture porosity is one of the main categories of secondary porosity.

Chemical processes in diagenesis

Chemical processes in diagenesis involve interactions between solid phases and fluids contained within the rock's pore spaces. Such aqueous reactions, both dissolution and precipitation, commence in unconsolidated sediment within the depositional environment and continue throughout the range of diagenetic conditions into metamorphism. All reaction in diagenesis manifests a relative slowness of rate compared to reactions at higher temperatures. This tendency for reactions to be sluggish has profound implications for the kinds of mineral assemblages that can exist and also for the kinds of textures that can arise in diagenesis.

It is reasonable to assert that overall chemical equilibrium between solid phases in rocks is not achieved in diagenesis. Even if a phase is in gross equilibrium (e.g., detrital albite in contact with albite-saturated water), the actual crystal lattice may still have trace element, isotopic, and structural disequilibrium with ambient conditions. Such disequilibrium represents a sort of driving force that can ultimately lead to dissolution, even in the presence of fluids that are near equilibrium with the bulk mineral. Transient local, microscale equilibrium between an individual authigenic phase and pore fluid occurs in diagenesis, at least theoretically, but does not apply more generally to the bulk rock which contains a diversity of phases that are not in equilibrium either with pore fluids or one another. The equilibrium that does exist is important, in the sense that it is the basis for using isotopic and elemental analysis of authigenic phases to interpret the compositions of ancient fluids responsible for precipitation.

Dissolution

A key factor in chemical diagenesis is *metastability*, which refers to the tendency of mineral crystals to persist for geologically long periods under conditions far outside the range of their thermodynamic stability because the rate of their reequilibration is slow. A metastable mineral is said to have a "kinetic barrier" to reaction. Amazingly, some feldspars, originally formed by crystallization from magma, can persist for billions of years in aqueous environments at surface temperatures. Some detrital feldspar persists even in Archean sandstones, having survived erosion, transport, deposition, and a prolonged history of diagenesis.

The degree of metastability differs greatly for different sedimentary materials (both crystalline and amorphous) and can vary greatly even among individual components of similar bulk composition, for example among different grains of K-feldspar in a sandstone or mudrock. Hence, there is a tremendous range in the conditions required to initiate dissolution of primary and secondary mineral components. Most evaporite and many carbonate minerals manifest extensive recrystallization (dissolution followed by reprecipitation of a more stable crystal lattice) over short time periods under essentially syndepositional conditions. Highly unstable materials may recrystallize repeatedly through a range of diagenetic conditions and environments, leading to extensive and complex textural and chemical modifications. For example, dolomites typically retain almost no discernible primary detrital textures at any scale, being so highly prone to recrystallization. In a less extreme case, albitized detrital feldspars may retain a gross sand grain morphology as seen at low magnifications, but as observed on the scale of a few micrometers are revealed to be highly microporous aggregates of the authigenic feldspar albite. In stark contrast with dolomite and feldspars, detrital quartz fails to react with pore fluids and maintains its primary isotopic and trace element composition up through the incipient stages of metamorphism.

Crystal defects are an important control on the microscale localization of dissolution in diagenesis. In scanning electron microscope observation, dissolution of detrital minerals is typically observed to be localized at pits that correspond to the intersection of defects with the crystal surface. Important textural implications arise from the defect control on dissolution. At the scale of a few 10s or 100s of micrometers, dissolution of both primary and secondary minerals is typically observed to be highly irregular. Crystals are not generally observed to dissolve progressively from the outer surface, but rather in patches, at sites anywhere within the crystal that correspond to energetically favorable defects. Secondary pores formed in this manner show a wide range of size, varying from sub-micron micropores to pores of 10s to 100s of microns surrounded only by skeletal remnants of the original grain. Complete dissolution of detrital grains can lead ultimately to the formation of oversized pores. Compactive collapse of oversized pores can completely erase evidence for the former existence of some detrital grains. Dissolution can ultimately induce a considerable loss of information on the nature of primary rock components. In limestones for example, aragonitic allochems that may have been abundant in the depositional environment are rare in rocks of all ages and have never been observed in rocks older than Pennsylvanian. Such allochems are typically replaced by low-Mg calcite early in their diagenetic history. In the sandstones known as *diagenetic quartzarenites*, detrital quartz is the only remaining unreacted primary mineral, all feldspars having been lost to dissolution.

Precipitation

The sluggishness of precipitation reactions in diagenesis is manifested as a difficulty with nucleation. As a result, in both cementation and replacement, authigenic minerals typically show a highly localized spatial distribution, an effect that is apparent at a variety of scales. For example, some minerals show a marked tendency to nucleate only on the surfaces of pre-existing (i.e., mostly primary) crystals, a phenomenon that is apparent on the scale of a few microns to hundreds of microns. In essence, precipitation of such authigenic crystals enlarges the earlier-formed crystal, forming an outer zone known as an *overgrowth*. Authigenic quartz, feldspar, calcite, and ankerite are all notable examples of this mode of nucleation and growth. In the absence of appropriate substrates, these minerals may fail to grow even in the presence of fluids of appropriate composition. Substrate control of cement localization is observed in all types of porosity. Coatings on the surfaces of potential nuclei, formed through elutriation or early cementation, can markedly inhibit the growth of minerals in later diagenesis.

Replacement can be thought of as cementation occurring within secondary pores (however tiny) and, thus, replacement minerals, too, tend have highly localized distribution. Replacements tend to occur only at the sites of defect-controlled dissolution, as described above, and thus, typically have a highly patchy distribution within the crystal that is being replaced. As with dissolution, replacement that is localized near the outermost crystal surfaces (halo texture) generally is not observed in diagenesis.

An example of authigenic mineral localization on a somewhat larger scale (typically 100s of microns to meters) is the formation diagenetic structures known as nodules (mineral aggregates distinct from the host rock and typically formed very early in diagenesis) and concretions (regions of preferential cementation, formed at any time in diagenesis). As with overgrowths, this localization reflects, at least in part, the difficulty of nucleation at low temperatures. Nucleation of one crystal appears to have a dampening effect on further nucleation in its vicinity, leading to subsequent precipitation that is preferred (as overgrowths?) on the early-formed nucleus. The detailed mechanics of how concretions and nodules are localized is one of the enduring enigmas of diagenesis.

Interrelated physical and chemical processes

Some diagenetic processes take place through intimately linked chemical and physical processes. For example, a weak connection between physical and chemical processes is seen in localization of cementation in fracture porosity (vein formation) which takes place partly as a result of generation of fresh, uncoated surfaces favorable to nucleation and partly as a result of enhanced fluid flow. A stronger link between chemical and physical processes is seen in the case of pressure solution (see *Pressure Dissolution*) including both intergranular pressure solution and the formation of stylolites. In this instance, dissolution is localized along certain crystal contacts (in evaporites, carbonates, and quartz-rich sandstones), leading to prominent porosity loss and development of sutured boundaries. A variety of mechanisms have been proposed for the dissolution step in the pressure solution phenomenon, but the stress field is, in every case, the fundamental determinant of form and orientation of the dissolution surface. In simple burial compaction sutures are oriented parallel to bedding (perpendicular to the principal, compactional stress), whereas in tectonic deformation, stylolite orientation may be determined by the orientation of tectonic stresses. Pervasive tectonically driven stylolite development in late diagenesis, grades, ultimately, into metamorphic foliation. Another intriguing example of closely linked physical and chemical processes is dissolution driven by force of crystallization. In certain circumstances it appears that nucleation of one crystalline phase can actually induce the dissolution of surrounding crystals, an alternative mechanism for overcoming the kinetic barrier to mineral dissolution, leading to replacement (Sorby, 1863; Weyl, 1959; Maliva and Siever, 1988).

Fluid flow and diagenetic heterogeneity

An on-going controversy in diagenetic studies concerns the degree to which diagenetic processes of dissolution, cementation, replacement, and recrystallization are accomplished in so-called "open" versus "closed" chemical systems. The opening volley of this debate was the paper by Land and Dutton (1978) which demonstrated a disparity of tremendous magnitude between the amount of quartz cement in a Pennsylvanian sandstone and the amount of silica available from pore fluids, given the fluid flow limitations of a simple burial compaction model. Similar conundrums immediately became apparent in connection with many instances of cementation, dissolution, and replacement in all types of sedimentary rocks, especially with regard to the transport of components such as Al which are relatively insoluble in pore fluids and also components of volumetrically significant reactions in the deep subsurface. In general, studies that focus on petrographic documentation of authigenic reactions find evidence for a greater degree of "openness" in terms of the amounts of material transport required to accomplish the observed amounts of cementation and replacement. Studies that focus on rock permeability, fluid flow, and fluid compositions generally find potential for lesser amounts of material transport.

One of the obstacles to resolving this controversy concerns the tendency of diagenetic processes to manifest a high degree of spatial heterogeneity, making it difficult to obtain statistically meaningful data on sedimentary rock compositions. Within a given rock unit, the degree of cementation or the degree of replacement can vary due to the controls on reaction localization as discussed above as well as from historical vagaries in primary composition and fluid flow. As a result of compositional heterogeneity it is difficult to ascertain quantitatively whether materials apparently lost from one volume of rock have been gained by another. Likewise, some workers have called into question whether our understanding of permeability in the subsurface is sufficiently complete. Studies to determine the appropriate scale for rock property observations are continuing as are efforts to better understand the role of endolithic microbes in element mobility in the subsurface.

Microbial influences in diagenesis

Microbial activity is recognized as an important element of diagenetic processes in a wide range of subsurface environments (e.g., Banfield and Nealson, 1997). Mineral surfaces within endolithic pore spaces provide attachment sites and nutrient sources to bacteria, archaea, and fungi across environments extending from near-syndepositional to several kilometers depth. Microbes affiliated with mineral surfaces can, either actively or passively, be a significant factor in overcoming kinetic barriers to both dissolution and precipitation in diagenesis. The mere presence of microbial communities within diagenetic environments may exert controls on the microscale morphology and distribution of authigenic precipitates because biofilms exuded by living cells change mineral surfaces in ways that can promote, inhibit, or modify nucleation.

There is currently vigorous research activity focused on the microscopy and geochemistry of microbe/mineral interactions seen in laboratory and field experiments, in modern surface and subsurface environments, and in ancient rocks. These studies are beginning to contribute a view in which the diagenetic realm is also a vital region of the biosphere on the Earth, and perhaps, other planets.

Boundaries between diagenesis and depositional processes, weathering, and metamorphism

No sharp division between diagenesis and metamorphism, nor between diagenesis and weathering, can be readily specified. Historically, various temperature/pressure conditions, textural, and mineral assemblage criteria have been applied in making distinctions between these different realms of rock alteration. All of these approaches have been rendered somewhat obsolete by the growing recognition of the role of initial composition on textural modification and the fact that many reactions and mineral assemblages once placed within the zone of metamorphism (e.g., zeolite precipitation, albitization) are now known to occur within rocks that are clearly sedimentary. Rocks might be considered to be "metamorphic" once their sedimentary textural components (grains, cements, and pore spaces) have been "erased". Development of through-going metamorphic fabrics such as pressure shadows or foliation is clearly a marker for the transition out of diagenesis. Of course such degrees of textural alteration are achieved under very different conditions for materials of different starting compositions. For example, limestones and especially dolomites may manifest a high degree of textural and chemical modification by one kilometer of burial ($<30°C$) whereas an interbedded quartz-rich sandstone may survive to nearly $300°C$ with clearly sedimentary textures intact. The method of observation is another control on such a determination (e.g., rocks that appear as metamorphic "quartzites" in light microscope observation, may still appear as "quartz-cemented sandstones" in cathodoluminescence). Many processes identified as "diagenetic" (e.g., albitization, quartz and feldspar precipitation in fractures, clay mineral formation) also take place within crystalline rocks at relatively low temperatures.

Both weathering and metamorphism encompass physical and chemical processes, involve aqueous reactions, and assign great importance to metastability. In metamorphism there is a role for solid state diffusion that is not observed in diagenesis (halo textures do occur prominently in metamorphic rocks) and also a degree of solid phase equilibrium. In weathering there is a physical decomposition of rocks that is not affiliated with diagenesis.

Given that the boundaries between diagenesis and metamorphism and between diagenesis and weathering are so obscure, it is especially difficult to subdivide diagenesis in a meaningful and consistent way (Fairbridge, 1967). Diagenetic processes are often divided loosely on the basis of their timing relative to compaction, those processes occurring nearest the depositional interface, prior to substantial compaction, and at temperatures <50°C being designated "early", whereas processes occurring in deeper burial, following substantial compaction, and at temperatures around 80°C and higher, are designated "late".

In general, chemical processes are assigned to diagenesis more broadly than physical processes. For example, dissolution of unstable minerals during transport or through bioerosion (chemical boring) would be widely considered as elements of diagenesis whereas bioturbation affecting unconsolidated sediments within the depositional environment is typically studied by experts familiar with deposition rather than diagenesis. Similarly, growth of clay minerals through replacement of feldspars in weathering would be considered diagenetic whereas physical disaggregation of rock by wind or ice would be most commonly thought of as weathering.

Methods for the study of diagenesis

Documentation of the primary versus secondary character of sedimentary rock components is accomplished through the integrated use of microscopical and geochemical techniques. The pervasive chemical disequilibrium that characterizes rocks throughout diagenesis lessens the importance of bulk methods of chemical analysis in diagenetic studies in comparison to studies of igneous and metamorphic systems. Diagenetic studies over the past several decades have benefited greatly from advances in imaging and spatially resolved chemical analysis. Such methods allow characterization of the microscale chemical heterogeneities that persist throughout the realm of diagenesis and permit a correlation of such heterogeneities with information on the sequential development of rock components (e.g., cement stratigraphy).

A technology of particular importance in diagenetic studies is the use of colored impregnation media, first applied in the 1970s, that allows reliable discrimination of natural versus artificially induced porosity in thin sections. Widespread use of this technique in limestones and sandstones led to a revolution in understanding of the importance of grain dissolution in the subsurface and also inspired a renaissance of data collection that greatly improved understanding of compaction. Imaging with scanning electron microscopes and electron microprobes created a vastly greater appreciation of the heterogeneity that exists for any given mineral phase across a range of components (grains, cements, and replacements) within sedimentary rocks. For example, scanned cathodoluminescence imaging and back-scattered electron imaging may permit discrimination of grains, cements, and grain replacements all composed of the same mineral phase, but having small differences in defect structure or trace element composition. Likewise, different generations of a single authigenic phase can be identified on images that are sensitive to subtle differences in trace element composition. Such distinctions may be highly obscure in other modes of imaging such as transmitted light microscopy that only allows discrimination between phases or different crystal orientations of a single phase. Diagenetic studies typically employ a combination of imaging (for historical aspects) and analysis (in order to quantify T and fluid compositions). Important petrographic methods in diagenetic studies include transmitted plane- and cross-polarized microscopy, secondary electron imaging, back-scattered electron imaging, X-ray mapping, and cathodoluminescence. Important analytical approaches include X-ray diffraction, major and trace element analysis by electron microprobe, and analysis of the stable isotopes of O, C, and Sr. Sampling of detrital and authigenic phases by microdrilling techniques has been widely employed in combination with isotopic analyzes.

Diagenetic studies

Published work on diagenesis is extensive and is summarized in numerous textbooks, review papers, and reprint collections (e.g., Pettijohn, Potter *et al.*, 1973; Bathurst, 1975; Pettijohn, 1975; Berner, 1980; Folk, 1980; Land, 1985; Horbury and Robinson, 1993; Giles, 1997; Land, 1997). Diagenetic studies are found primarily in mainstream geological journals as well as in publications in the areas of petroleum and environmental engineering, chemistry, and physics.

Summary

Diagenesis encompasses all physical and chemical modifications that affect sediments and sedimentary rocks following deposition and prior to the advent of weathering or metamorphism. The tendency of minerals to manifest sluggish reequilibration reactions under diagenetic conditions is responsible for the preservation in sedimentary rocks of considerable historical information concerning both the origins of detrital grains and the progress of chemical reactions involving dissolution, cementation, and replacement. Even though reactions in diagenesis are comparatively slow, modern diagenetic studies have concluded that most primary components in sediments are ultimately subject to profound modifications during burial as metastable components react with the surrounding pore fluids. The microbial role in diagenetic reactions is an exciting area of research that is contributing important new perspectives on the nature of diagenesis.

No absolute boundaries are assignable between diagenesis and either weathering or metamorphism, but rocks in the diagenetic realm are characterized by pervasive chemical nonequilibrium between the various solid phases that comprise the rock and, to a high degree, nonequilibrium between the solid phases in the rock and the contained pore fluids. With the exception of so-called chemical sediments that are constituted almost entirely of authigenic precipitates, sedimentary rocks in the diagenetic realm manifest a clearly sedimentary fabric with distinct grains, pores, and cements. Erasure of such fabrics is a hallmark of the transition into metamorphism.

Kitty L. Milliken

Bibliography

Banfield, J.F., and Nealson, K.H. (eds.), 1997. *Geomicrobiology: Interactions between Minerals and Microbes*. Reviews in Mineralogy. Washington, DC: Mineralogical Society of America.
Bathurst, R.G. C., 1975. *Carbonate Sediments and their Diagenesis*. Amsterdam: Elsevier Scientific.
Berner, R.A., 1980. *Early Diagenesis: A Theoretical Approach*. Princeton University Press.
Fairbridge, R.W., 1967. Phases of diagenesis and authigenesis. In Larsen, G. and Chillingar, G.V. (eds.), *Diagenesis in Sediments*. Amsterdam: Elsevier, 8, pp. 2–89.
Folk, R.L., 1980. *Petrology of Sedimentary Rocks*. Austin, Texas: Hemphill Publishing.
Giles, M.R., 1997. *Diagenesis: A Quantitative Perspective, Implications for Basin Modelling and Rock Property Prediction*. Dordrecht: Kluwer Academic.
Horbury, A.D., and Robinson, A.G. (eds.), 1993. *Diagenesis and Basin Development*. AAPG Studies in Geology. Tulsa: American Association of Petroleum Geologists.
Land, L.S., 1985. The origin of massive dolomite. *Journal of Geological Education*, **33**: 112–125.
Land, L.S., 1997. Mass-transfer during burial diagenesis in the Gulf of Mexico sedimentary basin. In Montañez, I., Gregg, J.M., and Shelton, K.L. (eds.), *Basin-wide Diagenetic Patterns: Integrated Petrologic, Geochemical, and Hydrologic Considerations*. Tulsa: Oklahoma, Society for Sedimentary Geology (SEPM), 57, pp. 29–39.
Land, L.S., and Dutton, S.P., 1978. Cementation of a Pennsylvanian deltaic sandstone: isotopic data. *Journal of Sedimentary Petrology*, **48**: 1167–1176.
Land, L.S., and Dutton, S.P., 1979. Cementation of sandstones, reply. *Journal of Sedimentary Petrology*, **49**: 1359–1361.
Lundegard, P.D., 1992. Sandstone porosity loss—a 'big picture' view of the importance of compaction. *Journal of Sedimentary Petrology*, **62**: 250–260.
Maliva, R.G., and Siever, R., 1988. Diagenetic replacement controlled by force of crystallization. *Geology*, **16**: 688–691.
Pettijohn, F.J., 1975. *Sedimentary Rocks*. New York: Harper and Row.
Pettijohn, F.P., Potter, P.E., and Siever, R., 1973. *Sand and Sandstone*. Berlin: Springer-Verlag.
Sorby, H.C., 1863. On the direct correlation of mechanical and chemical forces. *Royal Society of London Proceedings*, 12, 538–550.
Warren, J., 1999. *Evaporites. Their Evolution and Economics*. Oxford: Blackwell Science.
Weyl, P.K., 1959. Pressure solution and force of crystallization—a phenomenological theory. *Journal of Geophysical Research*, **64**: 2001–2025.

Cross-references

Authigenesis
Biogenic Sedimentary Structures
Cathodoluminescence
Cements and Cementation
Compaction (Consolidation) of Sediments
Dedolomitization
Deformation of Sediments
Diagenetic Structures
Dolomites and Dolomitization
Feldspars in Sedimentary Rocks
Geodes
Isotopic Methods in Sedimentology
Maturation, Organic
Micritization
Neomorphism and Recrystallization
Porewaters in Sediments
Septarian Concretions
Stylolites

DIAGENETIC STRUCTURES

The results of diagenetic alterations are visible whenever we examine sedimentary rocks in the field. Indeed, the very hardness of the rocks is the result of the diagenetic process of lithification, in which loose sediment is transformed into solid rock by cementation and compaction. However, there are numerous structures that can be readily observed in outcrop that result wholly from diagenetic alterations. These diagenetic structures include features such as concretions, cone-in-cone structure, geodes, liesegang bands, reduction features, sand crystals, and stylolites. Of these, concretions and stylolites are by far the most common and have been intensively studied, whereas the other features, although locally abundant, are relatively rare and have received less attention in the literature.

Concretions

Concretions are hard masses of sedimentary and, more rarely, fragmental volcanic rock, that form by the preferential precipitation of minerals (cementation) in localized portions of the rock. They are commonly subspherical (Figure D18), but frequently form a variety of other shapes, including disks, grape-like aggregates, and complex shapes that defy description. Concretions often form around a nucleus, such as a fossil fragment or piece of organic matter. This occurs because, for a variety of reasons, the nucleus created a more favorable site for cement precipitation than other sites in the rock. Concretions are usually very noticeable features, because they have a strikingly different color and/or hardness than the rest of the rock. The term nodule is often used as a synonym for concretion, but some workers prefer to reserve the term nodule for concentrations of authigenic minerals that do not enclose detrital grains (Selles-Martinez, 1996) or are relatively small and largely of replacement origin (Bates and Jackson, 1987).

Concretions often preserve features of the original sediment—such as burrows, fossils, and sedimentary layering—that cannot be seen in the rest of the rock (e.g., Kidder, 1985). This occurs because concretions usually begin to form relatively early, before sediment compaction and other processes disrupt the original sediment. Thus, cementation of

Figure D18 Concretion in shale. Lens cap for scale.

the concretions "freezes" the early sediment structure, forming a rigid mass that resists later chemical and mechanical alterations.

Concretions most commonly result from carbonate precipitation, although a variety of minerals also can form concretions, including pyrite, apatite, iron oxides, and silica (as opal, chert, or chalcedony). Of the carbonate minerals, calcite most commonly forms concretions, followed by dolomite, and siderite. Carbonate and iron oxide concretions are most commonly found in shales and sandstones, whereas apatite and pyrite concretions are mainly found in shales, and silica concretions normally form in limestones and shales. Carbonate concretions, given their abundance, and the ease of obtaining trace-element and isotopic data from them, have been studied in much greater detail than any other type. These studies have provided a wealth of information concerning the nature of early diagenetic reactions, and have conclusively demonstrated the importance of the microbial oxidation of organic matter in the early diagnetic environment (e.g., Raiswell and Fisher, 2000).

Concretions grow either by precipitation between grains in the host rock (usually the case for carbonates and apatite) or by replacement of the host rock (usually the case for chert). In either case, the conventional model for concretion growth is concentric, in which the authigenic mineral is progressively added (by replacement or direct precipitation) to the outer edge of the growing concretion. Consequently, early zones are at the center and late zones near the outer edge (e.g., Raiswell, 1971). Most studies of concretions have assumed this mode of growth. However, recent studies indicate that concretion structure and growth often does not follow the concentric model. Small-scale zonation incompatible with such a mode of cementation has been observed in calcite, siderite, dolomite, and chert concretions (Mozley, 1996; McBride *et al.*, 1999; Raiswell and Fisher, 2000). In some cases, it is clear that late-stage and early-stage cements precipitated throughout the concretion, indicating that early cementation was incomplete, and that significant porosity and permeability remained to allow precipitation of late-stage cements in the concretion interior.

Some concretions are cut by a radiating network of fractures (septaria) filled with coarsely crystalline calcite and other minerals. Such concretions are known as septarian concretions, and a variety of hypotheses have been proposed for the formation of the fracture systems, including: (1) Shrinkage of the interior of the concretion, a result of the dehydration or "chemical desiccation" of either a clay-mineral-rich center (Richardson, 1919; Raiswell, 1971), or dehydration of a gel-like precursor (Lippmann, 1955). (2) Expansion of the concretion interior and expansive growth of the fractures (Davies, 1913; Todd, 1913), presumably accomplished by the force of crystallization. (3) Development as tensile fractures in response to stresses induced during burial (Astin, 1986). (4) Fracturing induced by propagation of seismic waves through poorly lithified sediment (Pratt, 2001). Of these various hypotheses, the shrinkage hypothesis remains the most favored. However, the geochemical conditions necessary for significant dehydration of clay mineral centers have not been quantitatively explored. The gel-like precursor model has the problem that the nature of the gel is poorly-defined (some workers have proposed an organic soap), and such gels have never been observed in modern concretions. Thus, the origin of septarian fractures is currently a topic of debate and ongoing research.

Recently hydrologists have become interested in elongate concretions. These concretions range in size from pencil- and cigar-like bodies to those that resemble large fallen logs. They are thought to form from flowing groundwater, with the long axis of the concretion oriented parallel to the groundwater flow direction (McBride *et al.*, 1994; Mozley and Davis, 1996). Elongate concretions have even been found in faults, where they record the past flow orientation of groundwater in the fault zone (Mozley and Goodwin, 1995).

Cone-in-cone structure

Cone-in-cone structure consists of nests of cones of authigenic minerals, one inside another, which form either in thin beds, as veins, or at the margins of carbonate concretions (Figure D19). Concretions that consist almost entirely of cone-in-cone carbonate are also observed. In such concretions, the apices of the cones invariably point toward a shale core that bisects the concretions (e.g., Franks, 1969). Individual cones are often separated from one another by a thin layer of clay. The structure mainly forms in shales, but has been reported for argillaceous sandstones as well. Most cone-in-cone is composed of calcite, but it has also been observed in, siderite, gypsum and pyrite (Carstens, 1985).

The carbonate material making up the cones has variable characteristics. It is often fibrous calcite (6 microns or less in diameter, but up to 3 cm long; Franks, 1969), however, microcrystalline calcite has also been reported. This microcrystalline calcite is generally interpreted to represent recrystallized fibrous calcite, based upon the presence of relict textures (Franks, 1969). When present in sandstone, the cone-in-cone structure is often thinner and less distinct than in shales.

A variety of mechanism have been proposed for the origin of cone-in-cone structure. These include: (1) Volume increase accompanying inversion from aragonite to calcite (Tarr, 1922;

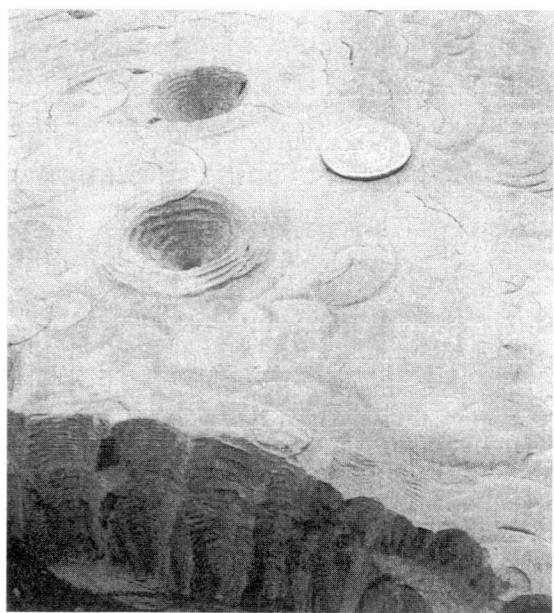

Figure D19 Cone-in-cone structure. Coin for scale.

Gilman and Metzger, 1967), in which expansion of conical aragonite pushed cones apart and allowed clay intrusion; (2) burial-induced pressure solution, with clay films originating as insoluble residues (Tarr, 1932); (3) fracturing of crystalline aggregates that grow in overpressured chambers, with the fractures induced by a decrease in pore pressure (Selles-Martinez, 1994); and (4) formation during early diagenesis by expansive mineral growth (i.e., force of crystallization), in which the cones are produced by the growth of cone-shaped aggregates of fibrous calcite (Franks, 1969), and the clay films originate as the crystals displace and disturb the original clay-rich sediment. By the latter mechanism, the orientation of the growing crystal is assumed to be controlled by the local stress configuration (direction of easiest crystal growth). Of these hypotheses, the displacive crystal growth mechanism has proven to be the most popular, thought the debate is by no means settled.

The inferred displacive nature of the crystal growth, and common deflection of host rock laminae around cone-in-cone concretions has led most authors to conclude that much of the precipitation occurs very early, during shallow burial. However, some authors have concluded, largely on the basis of ^{18}O-depleted values of some of the cone-in-cone material, that they can form later, at perhaps hundreds of meters of burial depth (overpressuring would be necessary to allow displacive growth at such depths; Marshall, 1982; Pirrie et al., 1994).

Geodes

Geodes are subspherical accumulations of authigenic minerals (mainly silica minerals and to a lesser degree calcite and celestite) which occur in a variety of rock types, particularly limestone and shale (Figure D20). They are superficially similar to concretions. However, unlike concretions, which contain considerable detrital material, they consist almost entirely of authigenic minerals. Geodes typically have a thin outermost layer of dense chalcedony and a hollow interior, in which well formed crystals project into a central cavity. They can also be completely filled with authigenic minerals (some authors use the term "filled geode" for such features), including a variety of exotic minerals, such as millerite,

Figure D20 Silica geode. Fibrous mineral in interior is millerite. Coin for scale.

anatase, chalcopyrite, and galena (Goldstein and McKenzie, 1997). Some geodes even contain oil in the central cavity (Anonymous, 1938).

The term geode is sometimes applied to features in volcanic rock, including basalt (precipitated in gas cavities; Garlick and Jones, 1990), though some authors indicate that such features are more properly termed amygdules and that the term should be restricted to features observed in sedimentary rocks (Chowns and Elkins, 1974). Thunder eggs are related structures of spherical or ellipsoidal shape, which contain cores of chalcedony and sometimes quartz. They are completely filled with minerals (i.e., contain no internal cavity) and are found exclusively in welded tuffs and rhyolites (Staples, 1965).

Numerous theories have been proposed to explain the origin of geodes, including: (1) Precipitation within dissolved fossil material, particularly sponges (Van Tuyl, 1916); (2) a complex process of replacement and solution/precipitation alteration of calcite concretions (Hayes, 1964); and (3) dissolution and replacement of evaporite nodules. Of these, hypothesis-3, alteration of evaporite nodules, is the most commonly cited mechanism. Indeed, many workers use the term "silicified evaporite nodule" rather than geode to describe these structures (e.g., Milliken, 1979; Ulmer-Scholle et al., 1993; Ulmer-Scholle and Scholle, 1994). Geodes have been reported in rocks that originated in sabkha environments (Chowns and Elkins, 1974), however, they can also form in "normal marine" sediments that have experienced circulation of concentrated brines (Maliva, 1987). Geode formation (i.e., silicification) can apparently proceed during early diagenesis (e.g., prior to compaction; Chowns and Elkins, 1974), as well as during late-stage diagenesis in association with meteoric influx and hydrocarbon migration (Ulmer-Scholle et al., 1993).

Liesegang banding

Liesegang bands were first described by Liesegang (1896) in experimental precipitates. His original experiment involved placing a drop of silver nitrate solution on a surface of potassium dichromate gel. This resulted in the formation of concentric rings of silver dichromate. Liesegang and subsequent workers noted similar features in rock, which are now referred to as Liesegang bands or rings. These features consist of bands of authigenic minerals arranged in a regular repeating pattern. They typically cut across-stratification and occur in a variety of rock types including sandstone and chert. Despite the frequent occurrence of Liesegang bands in rocks (Merino, 1984), only a few authors have examined their mineralogy and texture in any detail (Fu et al., 1994). Liesegang bands are frequently cited as prime examples of geochemical self-organization, in that their distribution in the rock often does not seem to be directly related to preexisting features (Chen et al., 1990).

The exact mechanism of band formation is poorly understood and remains an area of active research (e.g., Krug et al., 1996). A commonly cited mechanism is precipitation in a Ostwald-Liesegang supersaturation–nucleation–depletion cycle. By this mechanism diffusion of reactants leads to supersaturation and nucleation; this precipitation results in localized band formation and depletion of reactants in adjacent zones (DeCelles and Gutschick, 1983; Sultan et al., 1990).

Reduction features

Gray-green reduction features are common in many redbeds. They occur as isolated, subspherical to irregular reduction spots, and as reduction zones associated with fractures, faults, and permeable sandstone beds (Miller, 1957; Shawe, 1976; Chan et al., 2000). These features result from localized reducing conditions in the pore water, most commonly resulting from the presence of organic matter (in the case of reductions spots), or migration of reducing groundwater along zones of higher permeability (in the case of reduction zones). The exact origin of the reduced areas is poorly understood. Some authors ascribe the phenomenon mainly to reduction of ferric to ferrous iron, without net transport of iron from the reduced area (e.g., McBride, 1974; Mykura and Hampton, 1984). Others note either no change in ferric/ferrous ratio across reduced areas and contend that they are the result of Fe transport (e.g., Manning, 1975), or they describe changes in both ferric/ferrous ratio and total iron content (Keller, 1929; Picard, 1965; Durrance et al., 1978). Significant mineralogical differences occur in association with these features, including dissolution of iron oxides and ilmenite in the reduced portion and precipitation of authigenic feldspars and biotite (Miller, 1957; van de Poll and Sutherland, 1976). Some trace-element results suggest that Fe, Ca, Cr, and Nb are leached, and Co, Ni, V, and Y may be added during production of reduced areas (Shawe, 1976).

Sand crystals

Sand crystals are large (1 to >50 cm long) euhedral to subhedral poikilotopic crystals of authigenic minerals (commonly barite, gypsum or calcite) that formed in sandstone (Figure D21). Such crystals commonly have well-developed crystal faces despite the fact that they can contain greater than 50 percent included sand grains. The percentage of sand within the crystals is variable, and reflects the mode of growth of the crystal. Passive precipitation of poikilotopic crystals results in intergranular volumes (IGVs) that are comparable to that of uncompacted or slightly compacted sand (i.e., 40–50 percent). Expansive crystal growth results in much higher IGVs and disrupted fabrics (Dapples, 1971; McBride et al., 1992). In extreme cases, expansive growth can completely exclude sand grains from the crystal's interior. Some crystals contain alternating zones of greater and lesser sand percentages, which are interpreted to reflect alternating passive and displacive growth. This is thought to result from alternating fast and slow crystal growth, with exclusion of sand grains occurring during periods of slow growth (McBride et al., 1992; Shearman, 1981).

Stylolites

Stylolites are irregular serrated surfaces in sedimentary rock. They frequently consist of interlocking finger- and column-like extensions that are subparallel to one another and perpendicular to the trend of the surface. This can give stylolites an irregular zig-zag appearance when viewed in cross section (Figure D22). They form by a process of pressure dissolution, in which portions of minerals under relatively high stress, such as at grain contacts, undergo preferential dissolution due to a pressure-induced solubility increase. Stylolites form perpendicular to the maximum compressive stress, which is commonly vertically oriented, due to overburden pressure, resulting in horizontal, bedding parallel stylolites. In areas that have experienced tectonism, however, the principal compressive stress is commonly horizontally oriented, resulting in subvertical stylolites that cut across-bedding. Sutured grain contacts are similar to stylolites, but unlike sutured contacts, stylolites are large-scale features, which traverse numerous grains and often entire outcrops. They are most commonly observed in carbonate rocks, which are particularly susceptible to pressure solution, but are found in sandstones and other rock types as well. Undulatory surfaces that show evidence for pressure solution, but are not serrated in cross section, are referred to as solution seams (or wispy pressure solution seams).

Lithologic heterogeneity appears to be the main variable controlling stylolite morphology (Andrews and Railsback, 1997). If the rock is relatively homogeneous, dissolution proceeds equally along the stylolite and a linear or undulatory seam develops. High heterogeneity results in unequal dissolution, with variable amounts of dissolution occurring along the stylolite due to differences in solubility. This results in a "zig-zag" columnar appearance. As dissolution proceeds, the nature of the heterogeneity may too, resulting in changing stylolite morphology through time.

Stylolites commonly contain a concentration of clay-rich material, which is assumed at least in part to be a build-up of

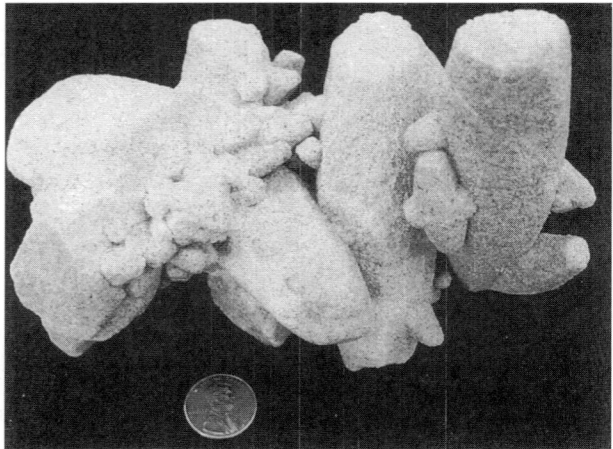

Figure D21 Calcite sand crystals. Coin for scale.

Figure D22 Cross-sectional view of several stylolites. Pencil for scale.

insoluble minerals from the host rock that underwent dissolution. Thus, as dissolution proceeds along the stylolite, this insoluble residue becomes thicker. One cannot simply assume that all the insoluble material along a given stylolite originated in this way, because in some cases much of the clay may have originated as a depositional clay seam, which subsequently became a stylolite. Some authors have proposed that clays promote pressure solution by providing a pathway for ion diffusion (Weyl, 1959). Andrews and Railsback (1997), however, demonstrated that stylolites can form in limestones with as little as 2.1 wt percent insoluble residue and concluded that abundant clay is not a necessary prerequisite for stylolite development. Meike and Wenk (1988) provided TEM evidence suggesting that some of the clay in these seams is authigenic.

There are two competing hypotheses for how pressure solution proceeds along the stylolite: the undercutting hypothesis and the solution film hypothesis. The undercutting hypothesis (Bathurst, 1975) proposes that dissolution occurs initially at point contacts resulting in preferential dissolution at the contact, which causes the grains to become closer to one another and transforms the point contact to a surface contact. This is followed by undercutting of the surface contact at the edges (water being absent along the surface contact), which results in undercutting and eventually in the formation of another point contact. The solution film hypothesis proposes that a contiguous film of liquid separates the two grains in a manner similar to that operative during displacive crystal growth (Weyl, 1959). This water is assumed to be under higher pressure at grain contacts and selective dissolution proceeds along the contact as a result. The second hypothesis (thin film) is currently in favor with most workers (Morse and Mackenzie, 1990).

As dissolution proceeds, ions migrate away from the site of dissolution either along the stylolite or away from the stylolite surface. Bathurst (1975) notes that much of the material must leave the stylolite, migrating into overlying or underlying host rock in order for dissolution to proceed. This material commonly reprecipitates near the stylolite (e.g., Wong and Oldershaw, 1981). The combination of cementation adjacent to stylolites, and the build-up of low-permeability clay-rich insoluble residue (discussed above) can produce a very low-permeability zone, that can act as a barrier to petroleum migration and groundwater flow.

Loss of material by dissolution along stylolites can have a significant impact on the sedimentary record. Bathurst (1975) indicates that reductions of vertical thickness due to stylolite growth of 20–35 percent is common, and that crustal shortening accomplished by tectonically generated stylolites can exceed that caused by folding and other deformation mechanisms.

Summary

Diagenetic structures are macroscopic features that are the result of diagenetic alteration of the host rock. The principal diagenetic structures are concretions, cone-in-cone structure, geodes, liesegang bands, reduction features, sand crystals, and stylolites.

Concretions are concentrations of authigenic minerals (mainly carbonate, silica, apatite, and iron oxides) that most commonly form in shales, sandstones, and limestones. They originate either by filling pores between framework grains or by replacement of the host rock. Concretions form either concentrically, in which concretion growth initiates at the concretion core and material is progressively added to the exterior of the growing concretion; or in a complex manner, involving precipitation at multiple nucleation sites throughout the concretion body. Septarian concretions contain internal radial fracture networks that may result from internal shrinkage. Elongate concretions have been used to map out paleogroundwater flow orientations.

Cone-in-cone structures consist of nested cones of authigenic minerals, most commonly calcite, that are separated from one another by thin films of clay. They occur mainly in shales, but have also been reported in argillaceous sandstones. Numerous explanations have been put forth to explain these structures, with the most popular being formation through expansive crystal growth of the cones.

Geodes are subspherical concentrations of authigenic minerals (most commonly silica) that occur mainly in shales and limestones. They typically have a fine-grained external layer of chalcedony and an interior of coarser crystals that project into a cavity. Most appear to have formed by dissolution and silicification of evaporite nodules, though other origins are possible.

Liesegang bands are nested zones of authigenic minerals arranged in a regular repeating pattern. Their origin remains poorly understood, however, a commonly cited mechanism is precipitation in a Ostwald-Liesegang supersaturation–nucleation–depletion cycle.

Reduction features occur in redbeds as gray-green subspherical "spots" and elongate "zones" that are commonly associated with fractures. They form due to the presence of locally reducing pore waters.

Sand crystals are large, well-formed poikilotopic crystals in sandstone. They are most commonly composed of gypsum, barite, or calcite. Some grow passively, simply filling available pore space in the sand, whereas others grow displacively, excluding sand grains and disturbing the detrital fabric.

Stylolites are irregular, serrated surfaces that originate through pressure-induced dissolution of the host rock, most commonly a limestone. The surfaces typically contain a buildup of clay-rich material that is commonly assumed to represent an insoluble residue. The morphology of the surface is largely governed by the degree of heterogeneity of the host rock, with more heterogeneous lithologies having more serrated and irregular stylolites. Stylolites form perpendicular to the principal compressive stress axis, resulting in subhorizontal stylolite formation during normal burial and subvertical stylolites of tectonic origin. Stylolite formation accounts for the loss of considerable thicknesses of host rock by dissolution. The material removed by dissolution is precipitated adjacent to the stylolite in some cases, but can also migrate great distances before reprecipitating.

Peter Mozley

Bibliography

Andrews, L.M., and Railsback, L.B., 1997. Controls on stylolite development: morphologic, lithologic, and temporal evidence from bedding-parallel and transverse stylolites from the U.S. Appalachians. *The Journal of Geology*, **105**: 59–73.

Anonymous, 1938. Petroleum filled quartz geodes. *The Mineralogist*, **6**: 11 and 33.

Astin, T.R., 1986. Septarian crack formation in carbonate concretions from shales and mudstones. *Clay Minerals*, **21**: 617–632.

Bates, R.L., and Jackson, J.A., 1987. *Glossary of Geology*. Alexandria: American Geological Institute.

Bathurst, R.G.C., 1975. *Carbonate Sediments and their Diagenesis*. Amsterdam: Elsevier.

Carstens, H., 1985. Early diagenetic cone-in-cone structures in pyrite concretions. *Journal of Sedimentary Petrology*, **55**: 105–108.

Chan, M.A., Parry, W.T., and Bowman, J.R., 2000. Diagenetic hematite and manganese oxides and fault-related fluid flow in Jurassic sandstones, southeastern Utah. *American Association of Petroleum Geologists Bulletin*, **84**: 1281–1310.

Chen, W., Park, A., and Ortoleva, P., 1990. Diagenesis through coupled processes: modeling approach, self-organization, and implications for exploration. American Association of Petroleum Geologists Memoir 49, *Prediction of Reservoir Quality Through Chemical Modeling*. pp. 103–130.

Chowns, T.M., and Elkins, J.E., 1974. The origin of quartz geodes and cauliflower cherts through the silicification of anhydrite nodules. *Journal of Sedimentary Petrology*, **44**: 885–903.

Dapples, E.C., 1971. Physical classification of carbonate cement in quartzose sandstones. *Journal of Sedimentary Petrology*, **411**: 196–204.

Davies, A.M., 1913. The origin of septarian structure. *Geological Magazine*, **10**: 99–101.

DeCelles, P.G., and Gutschick, R.C., 1983. Mississippian wood-grained chert and its significance in the western interior United States. *Journal of Sedimentary Petrology*, **53**: 1175–1191.

Durrance, E.M., Meads, R.E., Ballard, R.R.B., and Walsh, J.N., 1978. Oxidation state of iron in the Littleham mudstone formation of the new red sandstone series (Permian-Triassic) of southeast Devon, England. *Geological Society of America Bulletin*, **89**: 1231–1240.

Franks, P.C., 1969. Nature, origin, and significance of cone-in-cone structures in the Kiowa formation (Early Cretaceous), north-central Kansas. *Journal of Sedimentary Petrology*, **39**: 1438–1454.

Fu, L., Milliken, K.L., and Sharp, J.M. Jr., 1994. Porosity and permeability variations in fractured and liesegang-banded Breathitt sandstones (Middle Pennsylvanian), eastern Kentucky: diagenetic controls and implications for modeling dual-porosity systems. *Journal of Hydrology*, **154**: 351–381.

Garlick, G.D., and Jones, F.T., 1990. Deciphering the origin of plume-textured geodes. *Journal of Geological Education*, **38**: 298–305.

Gillman, R.A., and Metzger, W.J., 1967. Cone-in-cone concretions from western New York. *Journal of Sedimentary Petrology*, **37**: 87–95.

Goldstein, A., and McKenzie, B., 1997. Famous mineral localities: Halls Gap, Lincoln County, Kentucky. *The Mineralogical Record*, **28**: 369–384.

Hayes, J.B., 1964. Geodes and concretions from the Mississippian Warsaw Formation, Keokuk region, Iowa, Illinois, Missouri. *Journal of Sedimentary Petrology*, **34**: 123–133.

Keller, W.D., 1929. Experimental work on red bed bleaching. *American Journal of Science*, **18**: 65–70.

Kidder, D.L., 1985. Petrology and origin of phosphate nodules from the Midcontinent Pennsylvanian epicontinental sea. *Journal of Sedimentary Petrology*, **55**: 809–816.

Krug, H.-J., Brandtstädter, H., and Jacob, K.H., 1996. Morphological instabilities in pattern formation by precipitation and crystallization processes. *Geologische Rundschau*, **85**: 19–28.

Liesegang, R.E., 1896. A-linien. Liesegangs photographisches Archiv, **21**: 321–326.

Lippmann, F., 1955. Ton, Geoden und Minerale des Barreme von Hoheneggelsen. *Geologische Rundschau*, **43**: 475–503.

Maliva, R.G., 1987. Quartz geodes: early diagenetic silicified anhydrite nodules related to dolomitization. *Journal of Sedimentary Petrology*, **57**: 1054–1059.

Manning, P.G., 1975. Mössbauer studies of the reduction spots in welsh purple roofing slates. *Canadian Mineralogist*, **13**: 358–360.

Marshall, J.D., 1982. Isotopic composition of displacive fibrous calcite veins: reversals in pore-water composition trends during burial diagenesis. *Journal of Sedimentary Petrology*, **52**: 615–630.

McBride, E.F., 1974. Significance of color in red, green, purple, olive, brown, and gray beds of Difunta Group, northeastern Mexico. *Journal of Sedimentary Petrology*, **44**: 760–773.

McBride, E.F., Honda, H., and Abdel-Wahab, A.A., 1992. Fabric and origin of gypsum sand crystals, Laguna Madre, Texas. *Gulf Coast Association of Geological Societies*, **42**: 543–551.

McBride, E.F., Picard, M.D., and Folk, R.L., 1994. Oriented concretions, Ionian Coast, Italy: evidence of groundwater flow direction. *Journal of Sedimentary Research*, **64**: 535–540.

McBride, E.F., Abdel-Wahab, A., and El-Younsy, A.R.M., 1999. Origin of spheroidal chert nodules, Drunka Formation (Lower Eocene), Egypt. *Sedimentology*, **46**: 733–755.

Meike, A., and Wenk, H.R., 1988. A TEM study of microstructures associated with solution cleavage in limestone. *Tectonophysics*, **154**: 137–148.

Merino, E., 1984. Survey of geochemical self-patterning phenomena. In Nicolis, G., and Baras, F. (eds.), *Chemical Instabilities*. Dordrecht: D. Reidel Publishing Company, pp. 305–328.

Miller, D.N., 1957. Authigenic biotite in spheroidal reduction spots, Pierce Canyon Redbeds, Texas and New Mexico. *Journal of Sedimentary Petrology*, **27**: 177–180.

Milliken, K.L., 1979. The silicified evaporite syndrome—two aspects of silicification history of former evaporite nodules from southern Kentucky and northern Tennessee. *Journal of Sedimentary Petrology*, **49**: 245–256.

Morse, J.W., and Mackenzie, F.T., 1990. *Geochemistry of Sedimentary Carbonates*. Amsterdam: Elsevier.

Mozley, P.S., 1996. The internal structure of carbonate concretions in mudrocks: a critical evaluation of the conventional concentric model of concretion growth. *Sedimentary Geology*, **103**: 85–91.

Mozley, P.S., and Davis, J.M., 1996. Relationship between oriented calcite concretions and permeability correlation structure in an alluvial aquifer, Sierra Ladrones Formation, New Mexico. *Journal of Sedimentary Research*, **66**: 11–16.

Mozley, P.S., and Goodwin, L., 1995. Patterns of cementation along a Cenozoic normal fault: a record of paleoflow orientations. *Geology*, **23**: 539–542.

Mykura, H., and Hampton, B.P., 1984. On the mechanism of formation of reduction spots in the Carboniferous/Permian red beds of Warwickshire. *Geology Magazine*, **121**: 71–74.

Picard, M.D., 1965. Iron oxides and fine-grained rocks of red peak and crow mountain sandstone members, Chugwater (Triassic) Formation, Wyoming. *Journal of Sedimentary Petrology*, **35**: 464–479.

Pirrie, D., Ditchfield, P.W., and Marshall, J.D., 1994. Burial diagenesis and pore-fluid evolution in a Mesozoic back-arc basin: the Marambio Group, Vega Island, Antarctica. *Journal of Sedimentary Research*, **64**: 541–552.

van de Poll, H.W., and Sutherland, J.K., 1976. Cupriferous reduction spheres in Upper Mississippian redbeds of the Hopewell Group at Dorchester Cape, New Brunswick. *Canadian Journal of Earth Science*, **13**: 781–789.

Pratt, B.R., 2001. Septarian concretions: internal cracking caused by synsedimentary earthquakes. *Sedimentology*, **48**: 189–213.

Raiswell, R., 1971. The growth of Cambrian and Liassic concretions. *Sedimentology*, **17**: 147–171.

Raiswell, R., and Fisher, Q.J., 2000. Mudrock-hosted carbonate concretions: a review of growth mechanisms and their influence on chemical and isotopic composition. *Journal of the Geological Society of London*, **157**: 239–251.

Richardson, W.A., 1919. On the origin of septarian structure. *Mineralogical Magazine*, **18**: 327–338.

Selles-Martinez, J., 1994. New insights in the origin of cone-in-cone structures. *Carbonates and Evaporites*, **9**: 172–186.

Selles-Martinez, J., 1996. Concretion morphology, classification, and genesis. *Earth Science Reviews*, **41**: 177–210

Shawe, D.R., 1976. Sedimentary rock alteration in the Slick Rock District, San Miguel and Dolores Counties, Colorado. *US Geological Survey Professional Paper* 576-D, D1–D51.

Shearman, D.J., 1981. Displacement of sand grains in sandy gypsum crystals. *Geology Magazine*, **118**: 303–306.

Staples, L.W., 1965. Origin and history of the thunder egg. *The Ore Bin*, **27**: 195–204.

Sultan, R., Ortoleva, P., DePasquale, F., and Tartaglia, P., 1990. Bifurcation of the Ostwald-Liesegang supersaturation-nucleation-depletion cycle. *Earth Science Reviews*, **29**: 163–173.

Tarr, W.A., 1922. Cone-in-cone. *American Journal of Science*, **4**: 199–213.

Tarr, W.A., 1932. Cone-in-cone. In Twenhofel, W.H. (ed.), *Treatise on Sedimentation*. Baltimore: Williams and Wilkins, pp. 716–733.

Todd, J.E., 1913. More about septarian structure. *Geological Magazine*, **10**: 361–364.

Ulmer-Scholle, D.S., and Scholle, P.A., 1994. Replacement of evaporites within the Permian Park City Formation, Bighorn Basin, Wyoming, USA. *Sedimentology*, **41**: 1203–1222.

Ulmer-Scholle, D.S., Scholle, P.A., and Brady, P.V., 1993. Silicification of evaporites in Permian (Guadalupian) back-reef carbonates of the Delaware basin, west Texas and New Mexico. *Journal of Sedimentary Petrology*, **63**: 955–965.

Van Tuyl, F.M., 1916. The geodes of the Keokuk beds. *American Journal of Science*, **42**: 34–42.

Weyl, P.K., 1959. Pressure-solution and the force of crystallization—a phenomenological theory. *Journal of Geophysical Research*, **64**: 2001–2025.

Wong, P.K., and Oldershaw, A., 1981. Burial cementation in the Devonian, Kaybob reef complex, Alberta, Canada. *Journal of Sedimentary Petrology*, **51**: 507–520.

Cross-references

Carbonate Mineralogy and Geochemistry
Cements and Cementation
Diagenesis
Geodes
Pressure Solution
Septarian Concretions
Stylolites

DIFFUSION, CHEMICAL

Diffusion of chemicals is one of the three main processes of solute movement in sedimentary porewaters, that is, diffusion, advection (flow) and biologically induced exchange near the sediment water interface (mainly irrigation). Chemical diffusion is the net transport that is caused by thermally generated random motions of particles in "solution", may they be ions, molecules or colloids. The net result of this phenomenon is the transfer of a solute, that is, a flux, from regions of high concentration to those of low concentration. In a closed system, this flux leads to an eventual homogenization of the solute concentration, if no other processes act to maintain the gradients (Figure D23).

The diffusive flux is governed by Fick's First Law of Diffusion, which in 1-dimension has the form

$$(F_D)_i = -D_i \frac{\partial C_i}{\partial x} \quad \text{(Eq. 1)}$$

where $(F_D)_i$ is the diffusive flux of solute i, D_i is its diffusion coefficient, C_i is its concentration and $\partial C_i/\partial x$ is its vertical gradient. Values of D_i for various solutes of interest to diagenesis can be found in Boudreau (1997, chapter 4), and these typically fall in the range of 0.3–$1.25 \times 10^{-5}\,\text{cm}^2\,\text{s}^{-1}$.

Diffusion in aqueous sediments is slower than in an equivalent volume of water because of the convoluted path(s) the diffusing particles must follow to circumvent the solid sediment particles (Figure D24); therefore, the effective solute diffusivity, D_i', in sediments is smaller than D_i and given by

$$D_i' = \frac{D_i}{\theta^2} \quad \text{(Eq. 2)}$$

where θ is the *tortuosity* (dimensionless). In diagenetic studies, θ is defined as the ratio of the average incremental distance, dL, that a diffusing particle must travel to cover a direct (vertical) distance, dx, that is, dL/dx (Berner, 1980).

The tortuosity can be related to a measured quantity called the "formation factor", ff, which is obtained from electrical resistivity measurements,

$$f = \frac{\text{bulk sediment specific electrical resistivity}}{\text{resistivity of porewater alone}} \quad \text{(Eq. 3)}$$

The relationship between tortuosity and resistivity is assumed to be of the form (Berner, 1980)

$$\theta^2 = \varphi f \quad \text{(Eq. 4)}$$

where φ is the sediment porosity, that is, porewater volume fraction. When measurements of f are lacking, θ can be estimated by Archie's law in sands, that is,

$$\theta^2 \approx \varphi^{-2} \quad \text{(Eq. 5)}$$

and, in fine-grained sediments (Boudreau, 1997, chapter 4),

$$\theta^2 \approx 1 - ln(\varphi^2) \quad \text{(Eq. 6)}$$

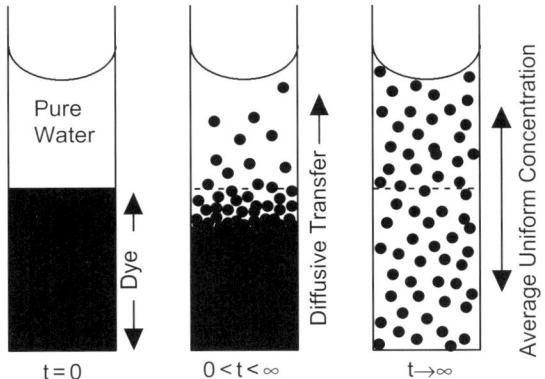

Figure D23 Illustration of diffusion is a tube containing pure water and water with dissolved dye particles (black circles).

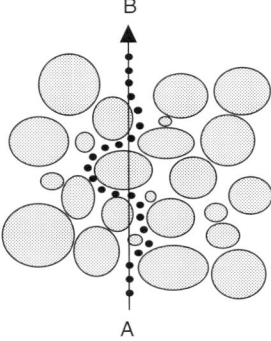

Figure D24 A particle trying to diffuse from point A to point B cannot pass through solid sediment particles (filled shapes); it must travel around these particles and so traverses the *tortuous* dotted path, rather than the direct solid-line path.

Molecular/ionic diffusion can be an effective transport process over "small" distances in sediments. Small in this context is defined by the so-called Peclet number

$$Pe = \frac{\varphi v L}{D_i'} \quad \text{(Eq. 7)}$$

where L is the distance of interest and v is the porewater flow speed. If $Pe \gg 1$, flow is the dominant transport process; conversely, if $Pe \ll 1$, diffusion is dominant, while both contribute significantly if $Pe \approx 1$.

An example of a situation where solute transport is strongly dominated by chemical diffusion is the penetration of oxygen into muddy organic-rich sediments. The scale of penetration is only a few tens of centimeters, at most, for example, Figure D25 where the Peclet number is of the order of 10^{-3} for $v \leq 0.1$ cm per year, $D_i = 10^2$ cm^2 per year, and $L = 1$ cm. In such situations, the oxygen mass balance is between the reaction that consumes it (decay) and the supply by diffusion. Because such oxygen profiles are essentially time invariant when observed from the sediment–water interface, that is, steady state, the mathematical statement of mass balance is given by the diagenetic equation (Berner, 1980; Boudreau, 1997)

$$\underbrace{\frac{\partial}{\partial x}\left[D'\frac{\partial C_i}{\partial x_i}\right]}_{\text{Change of the diffusive flux with depth}} - \underbrace{R_o}_{\text{Rate of oxygen consumption}} = 0 \quad \text{(Eq. 8)}$$

where R_o is generally assumed to be a constant. The solution to equation 8 with constant D_i' is of the form

$$C_i = C_0 + C_1 x + C_2 x^2 \quad \text{(Eq. 9)}$$

where C_0, C_1, C_2 are constants. The excellent fit of equation 9 to data is also illustrated in Figure D25.

Outside of areas affected by groundwater flow, porewater velocities in recently deposited sediments are of the same order of magnitude as the accumulation velocity of the sediment, that is, from about 10^{-1} cm per year in the nearshore to 10^{-3} cm per year in abyssal clays. With solute diffusion coefficients typically of the order of 10^2 cm per year, the Peclet number for these porewaters will be less than unity for L ranging from tens to hundreds of meters. Thus, diffusion continues to act as a significant transport mechanism during latter diagenesis and even over fairly large distances, as illustrated by the modeling of Deep Sea Drilling Project data (Lerman, 1979; Berner, 1980).

Other chemical transport processes that can occur in sediments are often modeled in analogy to chemical diffusion. Thus, dispersion of solutes that accompanies strong hydrological flow of porewater ($Pe > 1$) is almost universally represented in models as equation 1 with a new empirical diffusion coefficient, known as the dispersion coefficient. Likewise, mixing of porewaters by wave-induced water motions or by infaunal movement (not irrigation) is also conveniently represented as diffusion. An important point to remember in this respect is that molecular diffusion is always present, while these other "diffusions" may or may not occur in a given situation.

Bernard P. Boudreau

Bibliography

Berner, R.A., 1980. *Early Diagenesis: A Theoretical Approach*. Princeton University Press.
Boudreau, B.P., 1997. *Diagenetic Models and Their Implementation*. Springer-Verlag.
Lerman, A., 1979. *Geochemical Processes: Water and Sediment Environments*. John Wiley & Sons.

DIFFUSION, TURBULENT

A key feature of turbulent flows is that they are diffusive, so that the mean separation between initially adjacent fluid particles progressively increases in time. Sedimentary particles, suspended in a turbulent flow, move in response to fluid motions in addition to settling under gravity. The ratio of the settling velocity of the particles, w_s, to the magnitude of the turbulent fluid fluctuations, u_*, is therefore an important ratio for characterizing the motion and distribution of sediment. (This ratio, w_s/u_*, sometimes known as the *Rouse parameter*, is further discussed below.) In this article we present models for the diffusive motion of particles. We begin by reviewing the forces acting on individual particles and developing an equation of motion. We then consider how to model the evolution of a suspension of relatively heavy particles within a fluid by introducing a volumetric concentration, which varies both spatially and temporally. Even in a nonturbulent flow, suspended particles are subject to random fluid motion due to molecular movements (Brownian motion). This leads to a molecular diffusivity and we demonstrate how to develop a mathematical model of this process. In a turbulent flow, there are random fluid motions associated with unsteady eddies. We show how to form a mathematical description for the evolution of the average concentration and how, in analogy with molecular diffusive processes, it is possible to describe

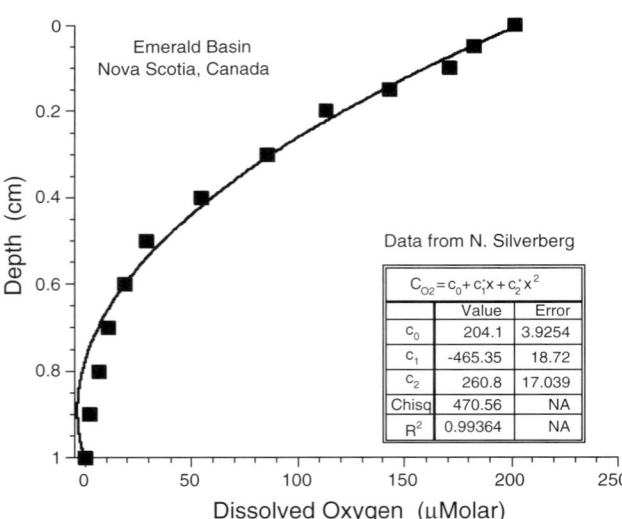

Figure D25 Dissolved oxygen profile (squares) from a muddy sediment, taken off the coast of Nova Scotia, Canada. The line through the data is equation 9.

these turbulent processes using a diffusion model. This model is empirical and we indicate some of the limits of its validity. We show how it can be used to model the vertical distribution of relatively heavy sediment suspended in a horizontal turbulent flow.

Particles suspended in fluids are subject to a number of forces, which means that they are not passively advected by the fluid motions. If the concentration of particles is high then these forces may arise from interactions between the suspended material (e.g., Bagnold, 1954), but even when the suspension is dilute, there are a range of possible effects. Maxey and Riley (1983) have formulated an expression for the forces on a single spherical particle in an unbounded fluid. These include the buoyancy force, which arises due to the density difference between the solid and fluid phases; the drag force, which is the force exerted by the fluid on the surface of the particle and is a function of the relative velocity between the phases; the added mass term, which is due to the displacement and acceleration of fluid surrounding the particle; the pressure gradient term which arises due to a nonuniform distribution of pressure within the fluid; and the Basset history term, which is associated with the acceleration of fluid surrounding the particle in response to the unsteady particle motion. In addition forces may arise due to the rotation of the particle, vorticity in the background fluid motion and the interaction with boundaries (see Clift et al., 1978; Auton et al., 1988). Thus, the calculation of the trajectory of a single particle is complicated even in an idealized, simple flow. However, the buoyancy and drag forces dominate the motion of most natural particles, with particle inertia playing a small role in some circumstances. Denoting the particle and fluid velocity by **v** and **u**, respectively, we find that when the drag force is linearly proportional to the relative velocity between the two phases, the equation of motion is then given by

$$\frac{d\mathbf{v}}{dt} = \frac{1}{\tau_p}(\mathbf{u} - \mathbf{v} - w_s\hat{\mathbf{Z}}), \quad \text{(Eq. 1)}$$

where w_s is the Stokes settling velocity, $\tau_p = w_s/g$ is the particle relaxation time and $\hat{\mathbf{Z}}$ is a unit vector aligned with the vertical axis.

Given the complexity of this equation of motion, it is infeasible to calculate the trajectory of each constituent particle within a suspension. Rather, it is usual to introduce a volumetric concentration, $C(\mathbf{x},t)$, which measures the volume of particles relative to the total volume of the suspension and which is a scalar function of time and space. (Some investigators employ a mass concentration which is the mass of particulate relative to the total mass of the suspension.) The dispersed particles are treated as a continuum and the evolution of the concentration field is calculated.

Whenever a quantity, such as temperature, salinity or the concentration of suspended particles, is nonuniformly distributed, random motions give rise to diffusion, whereby it is gradually spread throughout space. At a microscopic level, diffusive spreading is caused by the progressive increase of the mean separation between adjacent molecules, which is a consequence of Brownian motion. The botanist Robert Brown observed the motion of small pollen grains suspended in a fluid. Even though the system had come close to equilibrium and the pollen grains were uniformly distributed, he observed individual grains undergoing seemingly irregular motion which we now term Brownian motion. For a thorough discussion of mathematical models of molecular interactions and the kinetic theory of gases see Chapman and Cowling (1970). The effects of diffusive spreading may be illustrated by considering a concentration distribution, $C(z,t)$, within a randomly moving fluid. Here, for simplicity, only one spatial dimension is considered although the analysis can be generalized to three dimensions. The random motion of the fluid may be modeled probabilistically so that $P(y,\delta t)$ denotes the probability that concentration field at position z is displaced to a new position $z+y$ after a time δt. The probability distribution has the following features: first, the probability of making a displacement is unity; and displacements are in either direction with equal probability, so that the first moment of the distribution vanishes. These properties may be written as

$$\int_{-\infty}^{\infty} P(y,\delta t)dy = 1 \quad \text{and} \quad \int_{-\infty}^{\infty} yP(y,\delta t)dy = 0. \quad \text{(Eq. 2)}$$

Also the second moment is given by

$$\int_{-\infty}^{\infty} y^2 P(y,\delta t)dy = 2D\,\delta t. \quad \text{(Eq. 3)}$$

Thus the mean square displacement grows linearly in time with rate $2D$. The concentration $C(z,t+\delta t)$, can be expressed as follows

$$C(z,t+\delta t) = \int_{-\infty}^{\infty} C(z-y,t)P(y,\delta t)dy. \quad \text{(Eq. 4)}$$

Forming a Taylor series expansion for the concentration field, which is valid if the displacements are not large, yields

$$C(z-y,t) = C(z,t) - y\frac{\partial C}{\partial z} + \frac{y^2}{2}\frac{\partial^2 C}{\partial z^2} + \cdots \quad \text{(Eq. 5)}$$

Finally multiplying by the probability of a displacement and integrating gives an expression for the change in the concentration field during a time δt

$$C(z,t+\delta t) - C(z,t) = D\delta t \frac{\partial^2 C}{\partial z^2}. \quad \text{(Eq. 6)}$$

In this expression D is termed the diffusivity and is the rate at which the random fluid motion leads to the increase of the mean-square separation of initially adjacent fluid particles. Diffusion leads to particle movement from regions of high to low concentration ("transport down the concentration gradient") and the diffusive flux is given by

$$-D\frac{\partial C}{\partial z}. \quad \text{(Eq. 7)}$$

Note that individual particles are moving randomly and independently of the concentration gradient. However collectively their motion is such that there is net migration.

For small particles it is possible to estimate the diffusion coefficient due to Brownian motion, using arguments first presented by Langevin (1908) and Einstein (1905). Both of these yield

$$D = \frac{kT}{6\pi\mu a}, \quad \text{(Eq. 8)}$$

where T is the temperature, k the Boltzman coefficient ($k = 1.38 \times 10^{-23}$ JK^{-1}), such that kT is a measure of the energy of the molecular vibrations, μ is the dynamic viscosity and a the particle radius.

We are now able to formulate the concentration transport equation on the assumption that the particles are advected with the fluid flow, have settling velocity w_s, and are diffused by random processes associated with Brownian motion. For a concentration field that is dependent on z and t, and a fluid flow denoted w, this gives

$$\frac{\partial C}{\partial t} = \frac{\partial}{\partial z}\left(D\frac{\partial C}{\partial z} - wC + w_s C\right). \quad \text{(Eq. 9)}$$

The relative magnitude of the diffusion to the advection terms is UL/D, where U and L are the velocity and length scales of the flow, respectively, and this ratio is termed the *Peclet number*. For most naturally occurring flows with sediment, the Peclet number is much larger than unity, indicating that the diffusive transport of particles by Brownian motion is much smaller than advection with the fluid. Thus we may approximate the sediment motion to be advection with the fluid flow and settling under the action of gravity.

Many natural flows are turbulent and an essential characteristic of them is that the velocity fields are irregular and fluctuating. These apparently random fluid motions are able to induce diffusive migration, in an analogous manner to that generated by the random molecular motion described above (Hinze, 1959). A mathematical model of this turbulent diffusion is most simply developed by making the Reynolds decomposition in which the velocity and concentration fields are decomposed into mean and fluctuating parts. In what follows we treat only one-dimensional fields, which depend only on z and t. Thus we write

$$C(z,t) = \overline{C}(z) + c'(z,t) \quad \text{and} \quad w(z,t) = \overline{w}(z) + w'(z,t). \quad \text{(Eq. 10)}$$

where an overbar denotes the mean quantity and the means of the fluctuating components vanish. In passing we note that this decomposition requires some care if the state around which the fluctuations are occurring is itself evolving in time. In such a situation it is still possible to make a Reynolds decomposition provided a typical timescale of the fluctuations is much shorter than the timescale over which the flow is evolving. In a similar vane, the Reynolds decomposition of flows which have a periodic variation requires the period of the oscillation to be much longer than the timescale of the fluctuations (Dyer and Soulsby, 1988). The mean evolution equation in the absence of molecular diffusion, is given by

$$\frac{\partial \overline{C}}{\partial t} = \frac{\partial}{\partial z}\left(w_s\overline{C} - \overline{w}\,\overline{C} - \overline{w'c'}\right). \quad \text{(Eq. 11)}$$

The terms on the RHS of this equation are the mean settling flux, $w_s\overline{C}$, the mean advective flux, $\overline{w}\,\overline{C}$, and the turbulent flux, $\overline{w'c'}$ (which is known as the *Reynolds flux*). If the fluctuations in the velocity and concentration fields are completely decorrelated then the Reynolds flux vanishes. However if the distribution of concentration is nonuniform, the correlation between these fluctuating fields provides a mechanism for the transport of particles. A model of the Reynolds flux as a function of the flow field and the property of the particles awaits a complete description of two-phase turbulent flows. However a simple closure assumption is to draw an analogy between molecular diffusion and these turbulent motions and write the diffusive migration in terms of the gradient of the mean field,

$$\overline{w'c'} = -K_s\frac{\partial \overline{C}}{\partial z}, \quad \text{(Eq. 12)}$$

where K_s is the *eddy diffusivity* (sometimes known as the sediment diffusivity). This relationship between the Reynolds flux and the eddy diffusivity is empirical and is justified by the apparently diffusive nature of turbulent flows in many situations. Here we reiterate that this diffusion is not associated with molecular fluctuations but is entirely due to the seemingly random turbulent fluid motion. One of the key weaknesses in the analogy between molecular and turbulent processes in some situations the scale of the turbulent eddies excursions, which comprise the "random" flow, can be quite large. In some boundary layer flows it is possible for the scale of these eddies to be comparable with the length scale over which there is significant variation within the concentration field. Thus the Taylor series approximation used to derive the dependence of the diffusive flux on the gradient of the concentration field may not be valid. However in many circumstances this eddy diffusion model has led to a good model of the distribution of sediment within a naturally occurring flow.

In addition to transporting particles, turbulent eddies also transport fluid and thus provide a means for interchanging fluid elements with different momenta. Their action, therefore, leads to an effective stress. By forming a Reynolds decomposition of the momentum equations it is possible to identify *Reynolds stresses* which are correlations between fluctuating velocity fields. These can then be related empirically to gradients of the mean velocity through the use of an eddy viscosity. For example for the two-dimensional, horizontally-aligned, shear flow $\overline{\mathbf{u}} = (U(z), 0)$ with fluctuating velocity field $\mathbf{u}' = (u', w')$, the Reynolds stress can be written

$$\overline{u'w'} = -K\frac{\partial \overline{U}}{\partial z}, \quad \text{(Eq. 13)}$$

where K is the eddy viscosity (Hinze, 1959).

In this context the *Schmidt number*, $\beta = K_s/K$, is the ratio between the sediment diffusivity and the eddy viscosity. Since both processes depend on the same turbulent eddies, it is natural to investigate whether the diffusion of sediment is more rapid than the diffusion of momentum or vice versa. Since a complete model of particle motion within a turbulent flow remains elusive, it is difficult to estimate the magnitude of β. This led Dyer and Soulsby (1988) to suggest that in most practical problems it is sufficient to set β equal to unity. There are, however, some physical processes in these flows which mean that the motion of the particles and the fluid are not identical and so β should be different from unity. This discussion is aided by identifying three timescales within the flow. The first is the particle relaxation timescale, τ_p, which is the time over which the relative velocity between the fluid and particle diminishes. This timescale is associated with the inertia of the particle. The other two timescales are associated with the turbulent flow and are the integral timescale of the turbulence, T_L and a time scale L/u_0, where L and u_0 are length and velocity scales of the flow. Note that the ratio $T_L u_0 / L$ is not necessarily of order unity (Hunt *et al.*, 1994).

We noted above that the gravitational force and the drag govern the motion of individual particles, which is a function of the relative velocity between the two phases. If the timescale over which the particle accelerates, τ_p, is much smaller than the timescale of the flow (either of the turbulent timescales) then the particle moves with the fluid motion and sediments under the action of gravity. However within a turbulent flow there are irregular motions with very rapid variations. Thus there will be periods in which the particle acceleration is non-negligible and thus the particle lags behind the fluid motion. Hence the particulate phase is not dispersed as widely as the fluid phase and the sediment diffusivity should be expected to be smaller than the eddy viscosity. Csanady (1963) discusses this phenomenon and proposes that β is a decreasing function of τ_p/T_L. However, for suspensions of dust in the atmosphere, this reduction is very small.

A second mechanism is the effect of crossing trajectories. Heavy particles, suspended in fluid, continuously change their fluid particle neighbors. Thus the velocity history of the particle is different from the surrounding fluid. For example, a heavy particle will sediment out of the eddy in which it is initially located. Thus the spatial decay of the correlation of the velocity field around it is more rapid than the decay of the coherent eddies. Hence the diffusivity of the particles should be expected to be smaller than the diffusivity of the fluid (Csanady, 1963; Hunt *et al.*, 1994).

One of the most fundamental problems in sediment transport is to establish the vertical distribution of suspended sediment within a turbulent flow over a horizontal, rigid boundary, such as a turbulent pipe flow or a free-surface flow such as a river. In these flows there can exist steady vertical distributions of sediment in which the weight of the suspended sediment is supported by the turbulent motions. (See the experimental data reported by Coleman, 1970; Vanoni, 1946). In terms of the diffusion equation, this balance is expressed by equating the settling flux and the vertical Reynolds flux. This balance is given by

$$\overline{w'c'} \equiv -K_s(z)\frac{d\overline{C}}{dz} = w_s\overline{C}. \tag{Eq. 14}$$

In this expression the eddy diffusivity varies spatially which reflects the non-isotropic nature of the turbulent eddies. Close to the boundaries the size of the largest eddy is limited. Since these dominate the eddy-induced transport of sediment, it is to be expected that the eddy diffusiivity should increase with distance from the boundary. Vanoni (1946) suggests that

$$K_s(z) = \beta\kappa u_* z(1 - z/H) \tag{Eq. 15}$$

where β is the Schmidt number (assumed constant), $\kappa = 0.4$ is Von Karman's constant, H is the flow depth and the boundary shear stress $\tau_0 = \rho u_*^2$. Thus the vertical profile of sediment is given by

$$\frac{\overline{C}(z)}{\overline{C}(\delta)} = \left[\left(\frac{H-z}{z}\right)\left(\frac{\delta}{H-\delta}\right)\right]^{w_s/(\beta\kappa u_*)}, \tag{Eq. 16}$$

where the concentration is known at a height δ. The key parameter in this distribution is the ratio of the settling velocity to a measure of magnitude of the turbulent velocity fluctuations (w_s/u_*) and this determines the shape of the profile.

Some relativity recent observations of turbulent, boundary-layer flow have indicated that the notion of purely random fluid movement is inappropriate; rather there are coherent fluid motions which may dominate the dynamics close to the boundary (Kline *et al.*, 1967; Grass, 1971). Typically there is an inrush of relatively fast moving fluid toward the boundary and an ejection of slowly moving fluid away from the boundary. Collectively these are termed the turbulent bursting phenomenon. These types of fluid motion may have important consequences for the suspension of sediment, since the ejection events are a coherent means for lifting relatively heavy sediment upward from the boundary (Grass *et al.*, 1991). Furthermore since these events are coherent over relatively long length scales, it is doubtful whether a gradient-diffusion model is an appropriate framework for analyzing the effect of the turbulent motion. Further research is required to understand the dynamics of these coherent structures and to analyze their effects on the movement of particles

Andrew J. Hogg

Bibliography

Auton, T.R., Hunt, J.C.R., and Prud'homme, M., 1988. The force exerted on a body in inviscid rotational flow. *Journal of Fluid Mechanics*, **197**: 241–257.

Bagnold, R.A., 1954. Experiments on a gravity-free dispersion of large solid spheres in Newtonian fluid under shear. *Proceedings of the Royal Society of London*, **225**: 49–63.

Chapman, S., and Cowling, T.G., 1970. *The Mathematical Theory of Non-uniform Gases*. Cambridge University Press.

Clift, R., Grace, J.R., and Weber, M.E., 1978. *Bubbles, Drops and Particles*. Academic Press

Coleman, N.L., 1970. Flume studies of the sediment transfer coefficient. *Water Resources Research*, **3**: 801–809.

Csanady, G.T., 1963. Turbulent diffusion of heavy particles in the atmosphere. *Journal of Atmospheric Science*, 201–208.

Dyer, K.R., and Soulsby, R.L., 1988. Sand transport on the continental shelf. *Annual Review of Fluid Mechanics*, **20**: 295–324.

Einstein, A., 1905. On the motion of small particles suspended in liquids at rest required by the molecular-kinetic theory of heat. *Annales de Physique*, **17**: 549–560.

Grass, A.J., 1971. Structural features of turbulent flow over smooth and rough boundaries. *Journal of Fluid Mechanics*, **50**: 233–255.

Grass, A.J., Stuart, R.J., and Mansour-Tehrani, M., 1991. Vortical structures and coherent motions in turbulent flow over smooth and rough boundaries. *Philosophical Transactions of the Royal Society of London*, **336**: 36–65.

Hinze, J.O., 1959. *Turbulence*. McGraw-Hill.

Hunt, J.C.R., Perkins, R.J., and Fung, J.C.H., 1994. Problems in modelling disperse two-phase flows. *Applied Mechanics Reviews*, **47**: S49–S60.

Kline, S.J., Reynolds, W.C., Schraub, F.A., and Runstadler, P.W., 1967. The structure of turbulent boundary layers. *Journal of Fluid Mechanics*, **30**: 741–773.

Langevin, P., 1908. Sur la théorie du mouvement browien. *Comptes Rendus de l'Académie des Sciences, Paris*, **146**: 530–533.

Maxey, M.R., and Riley, J.J., 1983. Equations of motion of a small rigid sphere in a non-uniform flow. *Physics of Fluids A*, **26** (4): 883–889.

Vanoni, V.A., 1946. Transportation of suspended sediment by water. *Transactions of the American Society of Civil Engineers*, **111**: 67–133.

Cross-references

Autosuspension
Diffusion, Chemical

Grain Flow
Gravity-Driven Mass Flows
Hindered Settling
Numerical Models and Simulation of Sediment Transport and Deposition
Physics of Sediment Transport
Sediment Transport by Tides
Sediment Transport by Unidirectional Water Flows
Settling

DISH STRUCTURE

Dish structures are thin, roughly horizontal, usually dark, flat to concave-upward laminations in sandstone and coarse siltstone beds (Figure D26). The dark color usually results from a local enrichment in mud, organic matter, micas, and other hydraulically fine or low-density grains. Dish structures were first described and named by Wentworth (1967) and Stauffer (1967) from thick-bedded deep-water sandstones of western California. Initially they were interpreted to be primary sedimentary structures related to the growth and breaking of antidunes during deposition from turbidity currents (Wentworth, 1967) or to inhomogeneous shearing within grain flows (Stauffer, 1967). Lowe and LoPiccolo (1974) provided evidence that they are post-depositional structures formed during liquefaction and dewatering of rapidly deposited sediments. Experiments designed to reproduce dish structures reinforce this interpretation (Tsuji and Miyata, 1987; Nichols *et al.*, 1994).

Dish structures are most common in thick sandstone beds that are usually interpreted as deposits of high-density turbidity currents, although they have also been reported from alluvial sandstones. In general, each concave-up lamination or dish is underlain by a zone up to a few millimeters thick of lighter-colored sand representing the primary water flow path from which clays, organic particles, micas, and other hydraulically fine grains have been elutriated (Figure D27). Many dishes have upturned ends that define vertical water escape conduits called pillars or pillar structures (Figure D27). The width of individual dish structures ranges from about 1 cm to over 50 cm and their vertical spacing from 0.5 cm to over 30 cm. The bases of dish structures are sharp, whereas the tops tend to be gradational. The dishes range from relatively flat, continuous laminae with widely spaced interruptions (Figure D26) to more strongly curved dishes with frequent breaks (Figure D27). Rare bedding-plane views of dish-structures show a polygonal pattern of ridges separating shallow concave-upward depressions (Lowe and LoPiccolo, 1974).

Figure D26 Concave-upward dish structures cross-cut primary cross-laminations. The upturned ends of the dishes form peaks or pillar structures. Scale at top in millimeters. Turbidite from the Pennsylvanian Jackfork Group, Oklahoma.

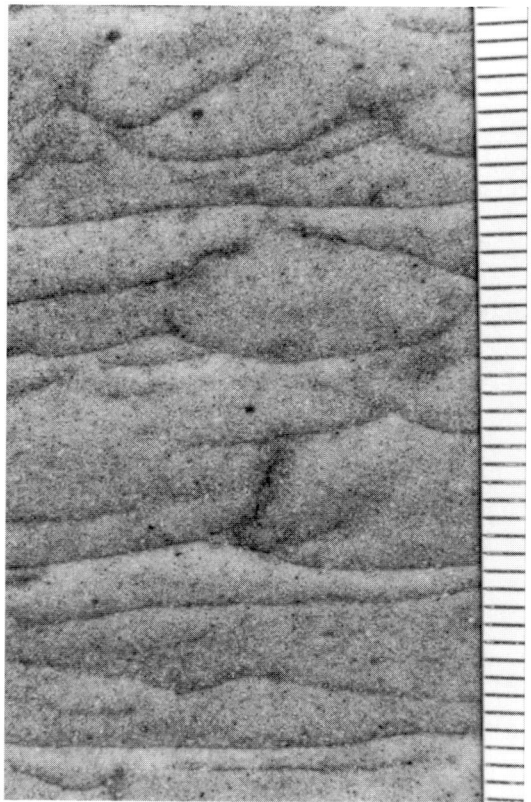

Figure D27 Dish structures showing well-developed underlying light-colored water-flow paths from which dark clays, micas, and organic particles have been removed. Scale on right in millimeters. Turbidite from the Pennsylvanian Jackfork Group, Oklahoma.

Dish structures are interpreted to form through dewatering of rapidly deposited, cohesionless sand and silt. Rapid deposition results in loose packing of grains. Such sediments often undergo partial liquefaction in response to shocks associated with tectonic or sedimentological processes. The escaping water entrains hydraulically mobile grains, such as clays, organic materials, and micas. Semi-permeable laminations act as partial barriers to the upward flow of water within the liquefied sediment, forcing it to move laterally toward points where continued vertical escape is possible. Slow seepage through the bounding semi-permeable laminations results in filtering and concentration of the dark, hydraulically mobile grains to form the dark dish structures. Subsidence of the central parts of the dishes as the underlying water and sediment escape increases dish curvature. Primary semi-permeable laminae are not always present in dish-structured beds and might not be necessary for dish formation. Minor inhomogeneities in permeability, which are likely to be present even in apparently structureless beds, as well as the inhomogeneous development of flow paths due to hydrodynamic instabilities, can probably initiate dish-structure formation.

Allen (1982: 373–374) proposed that some or most dish structures can form through a process of stoping at the tops of water-filled cavities during dewatering of a loosely-packed, slightly cohesive water-saturated sediments. Tsuji and Miyata (1987) observed formation of dish structures as fine-grained sediment in water-filled cavities settled to the bottom, followed by the collapse of the roof. Some cavities had slowly subsiding roofs, and the water was filtered through the subsiding, relatively coherent sand mass. It is probable that dishes can form through both filtration along permeability barriers in liquefied sediments and stoping within water-filled cavities. The latter process is likely to result in dishes with more irregular shapes.

Zoltán Sylvester and Donald R. Lowe

Bibliography

Allen, J.R.L., 1982. *Sedimentary Structures; Their Character and Physical Basis*, Volume II. Amsterdam: Elsevier.
Lowe, D.R., and LoPiccolo, R.D., 1974. The characteristics and origins of dish and pillar structures. *Journal of Sedimentary Petrology*, **44**: 484–501.
Nichols, R.J., Sparks, R.S.J., and Wilson, C.J.N., 1994. Experimental studies of the fluidization of layered sediments and the formation of fluid escape structures. *Sedimentology*, **41**: 233–253.
Stauffer, P.H., 1967. Grain-flow deposits and their implications, Santa Ynez Mountains, California. *Journal of Sedimentary Petrology*, **37**: 487–508.
Tsuji, T., and Miyata, Y., 1987. Fluidization and liquefaction of sand beds—experimental study and examples from Nichinan Group. *Journal of the Geological Society of Japan*, **93**: 791–808.
Wentworth, C.M. Jr., 1967. Dish structure, a primary sedimentary structure in coarse turbidites. *American Association of Petroleum Geologists Bulletin*, **51**: 485.

Cross-references

Fluid Escape Structures
Gravity-Driven Mass Flows
Liquefaction and Fluidization
Pillar Structure
Turbidites

DOLOMITE TEXTURES

Rock texture is a composite characteristic determined by the size, shape, distribution and orientation of crystals, grains and pores. Texture carries genetic implications about kinetics of nucleation and growth of crystals and directly relates to porosity and permeability. Therefore, classification based on textural characteristics is appropriate when researchers are interested in the origin of dolomite or its petrophysical properties. The most useful textural characteristics for classifying dolomites are crystal shape (planar and nonplanar) and the degree to which textures of replaced allochems and cements are preserved (mimetic versus nonmimetic replacement).

Planar and nonplanar dolomite

Crystal shape is the most commonly used textural feature for classifying dolomite. Crystal shape may be categorized as planar or nonplanar (Sibley and Gregg, 1987). Planar dolomite crystals (Figure D28(A)) have straight dolomite-dolomite crystal boundaries. Nonplanar dolomites (Figure D28(B))

Figure D28 (A) Planar dolomite. Arrows point to straight enfacial junctions at a compromise growth boundary between adjacent crystals. Seroe Domi Fm., Pliocene, Bonaire, N.A. Cenozoic, Scale bar = 0.1 mm. (B) Nonplanar dolomites have both straight and irregular compromise boundaries (arrows). Trenton Fm., Ordovician, MI basin. Also notice the irregular extinction pattern in the dark crystal in the center and left-center of the field of view. Scale bar = 0.5 mm.

have curved, lobate or otherwise irregular dolomite-dolomite boundaries. Most petrographers use the terms planar and nonplanar to describe dolomite crystal shape, but the terms anhedral, euhedral, xenotopic and idiotopic are still commonly used (see *Classification of Sediments and Sedimentary Rocks*). Euhedral crystals are bounded by their faces, anhedral crystals are not. Idiotopic refers to a texture in which the majority of crystals are euhedral and xenotopic refers to a texture in which the majority of crystals are anhedral. Anhedral, euhedral, xenotopic and idiotopic are not appropriate descriptors of interlocking arrays of crystals observed in thin section because one cannot determine whether or not crystal-crystal boundaries represent crystal faces without using a universal stage. The terms planar and nonplanar are also flawed. These terms refer to three dimensions when, of course, thin section observations are two-dimensional. However, most dolomite crystal shapes are isotropic and, therefore, planar and nonplanar shapes are adequately represented in two dimensions. More importantly, segments of nonplanar dolomite-dolomite boundaries may be straight and straight boundaries may appear curved due to overlapping grains. As a result, qualitative distinction between planar and nonplanar dolomites may be very difficult even for experienced petrographers. Arbitrary criteria allow a quantitative distinction between planar and nonplanar (Gregg and Sibley, 1984) but the vast majority of published descriptions are based on qualitative analysis. Fortunately, it is usually easy to distinguish planar from nonplanar dolomite based on extinction characteristics. Planar dolomite has straight extinction and nonplanar dolomite usually has sweeping or segmented extinction (Figure D28).

Nonplanar dolomite cement, commonly referred to as saddle dolomite (Radke and Mathis, 1980) or baroque dolomite (Folk and Asserto, 1974), is volumetrically very minor but common. The distinguishing features are curved, saddle-shaped crystal faces (Figure D29) in open pore space and sweeping extinction. Dolomite-dolomite boundaries between saddle dolomite crystals are nonplanar. Saddle dolomite is a textural term that is commonly applied to dolomite $CaMg(CO_3)_2$, ferroan dolomite (<10 mole percent $FeCO_3$) and ankerite $Ca(Fe, Mg)(CO_3)_2$.

Mimetic and nonmimetic replacement

Most dolomites are composed of equant rhombs tens to hundreds of micrometers in diameter. In these, it is generally impossible to determine whether the dolomite replaced a precursor calcium carbonate phase or precipitated directly from solution. When evidence of a precursor allochem (composite carbonate grain) or carbonate cement is present, the replacement dolomite is referred to as mimetic or nonmimetic (Kaldi and Gidman, 1982). Mimetic replacement of carbonate precursors occurs when micrometer to submicrometer-size dolomite crystals replace the carbonate in such a manner as to preserve the shape and internal texture of cements or allochems (Figure D30(A)). Poor preservation of allochems or cement texture is referred to as nonmimetic replacement. Nonmimetic textures often include "ghosts" which are inclusions of non-carbonate material which outline or show internal structure of pre-dolomitization allochems or cements (Figure D30 (B)).

Dolomite tends to nucleate epitaxially on calcite. Therefore, mimetic replacement may preserve the crystallographic orientation of calcite cements and allochems. When the host material is aragonite, dolomite tends to nucleate non-epitaxially. Therefore, mimetic replacement of aragonite will

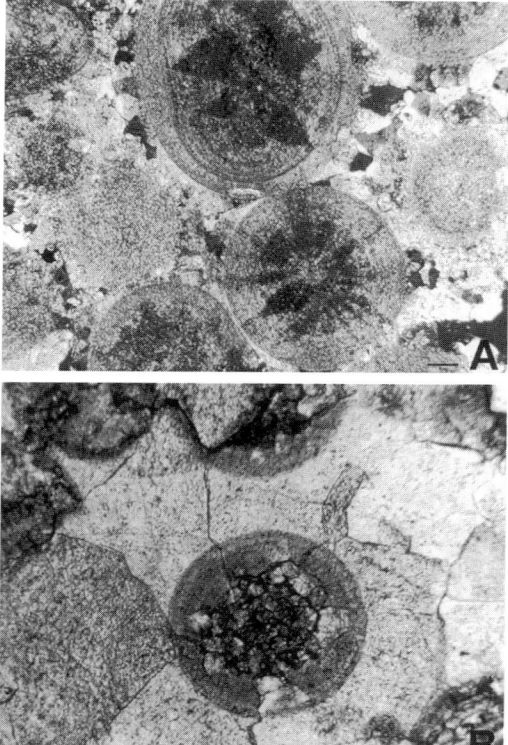

Figure D30 Mimetic (A) and nonmimetic (B) replacement of ooids. The mimetically replaced dolomite preserves the radial structure common in ooids. Nonmimetically replaced ooids are "ghosts" made visible by inclusions of organic material. Both samples from the Praire du Chien Gp, Ordovician, MI Basin. Scale bars = 0.1mm.

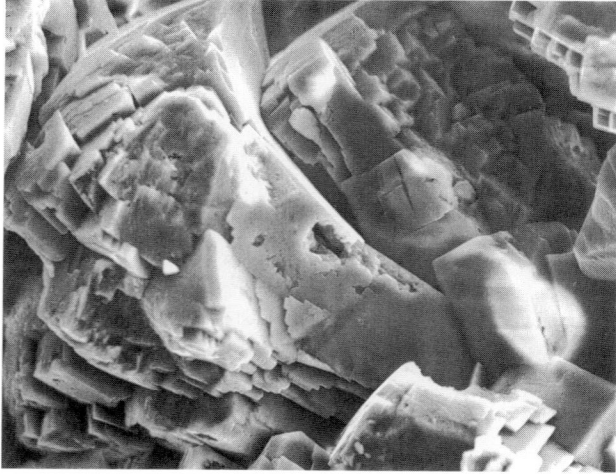

Figure D29 Synthetic saddle dolomite with characteristic curved faces produced in a hydrothermal bomb at 250°C (see Gregg, 1983). Field of view is 0.1 mm across.

preserve the structure of allochems or cements but not the original crystallographic orientation of aragonite crystals.

Crystal habit

Dolomites are not generally classified according to crystal habit because the only common form of dolomite crystals is the {10 1̄} rhombohedron, although complex polyhedra occur (Naiman *et al.*, 1983; Gregg *et al.*, 1992). Spheroidal dolomites have been found associated with submarine hydrothermal vents (Pichler and Humphrey, 2001), hydrocarbons seeps (Gunatilaka, 1989) and possibly bacteria (Nielsen *et al.*, 1997).

Crystal size distribution

Dolomites crystal size distributions are lognormal to normal coarse skewed regardless of age or mean size (Gregg *et al.*, 1992; Sibley *et al.*, 1993). Similar crystal size distributions are found in other diagenetic minerals, as well as minerals in igneous and metamorphic rocks (Eberl *et al.*, 1998). Nonplanar dolomite tends to be coarser crystalline and more lognormal than planar dolomite (Sibley *et al.*, 1993; Woody *et al.*, 1996). Trends in crystal size and characterization of crystal size distributions have been of limited use in understanding dolomite petrogenesis because size distributions have not been directly linked to nucleation and growth theory.

Dolomite rock textures

Dolomite selectivity

Dolomite selectively replaces the fine crystalline portions of limestone. Therefore, the distribution of dolomite in a rock may reflect the original limestone texture rather than the chemistry or flow paths of dolomitizing solutions.

Fossil moldic porosity

Because dolomite selectively replaces fine crystalline components of a rock, coarse crystalline allochems such as brachiopod fragments often resist dolomitization. These resistant allochems may be subsequently leached. Fossil moldic porosity is also common in limestones but the fossils which make up the molds in limestones are typically not brachiopods or other coarse crystalline low Mg-calcite allochems.

Sucrosic dolomite

Porous, planar dolomites are generally equigranular giving the rock a sugary texture. The sugary texture is noticeable in hand specimen as light reflects off dolomite crystal faces. These rocks have attracted attention because they have relatively high permeability making them valuable hydrocarbon reservoir rocks.

Interpretations of dolomite textures

Interpretation of planar and nonplanar dolomite texture (Gregg and Sibley, 1984) is based on theoretical relationships between growth rate, supersaturation and/or temperature and crystal morphology (Sunagawa, 1984). Planar dolomite forms a smooth interface due to dislocation or 2-dimensional nucleation growth (Figure D31(A)). Nonplanar dolomite forms a rough interface by 2-dimensional nucleation and/or

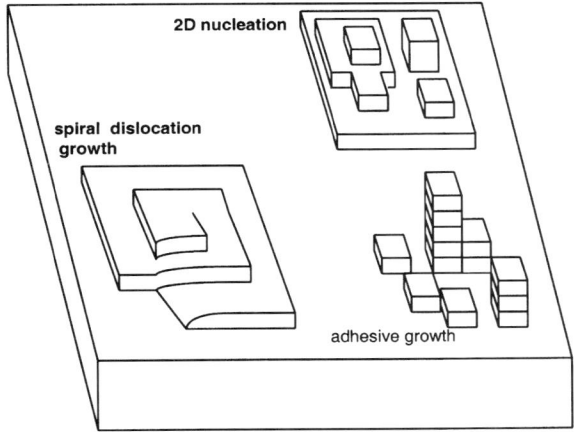

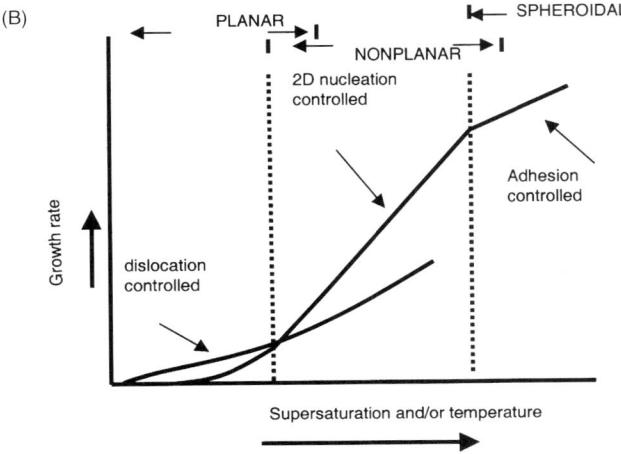

Figure D31 Models for crystal growth (A) and the associated geometry of crystal surface (B). Figure D31(B) is modified after Sunagawa (1984).

adhesive growth. Boundaries between dolomite types and crystal growth mechanism overlap because 2D nucleation can forms plane crystal surfaces at relatively low temperature and/or degrees of supersaturation and irregular surfaces at higher temperature and/or supersaturation (Figure D31(B)). Spheroidal crystals are direct evidence that low temperature dolomites may form by adhesive growth but the majority of nonplanar dolomites have been attributed to growth at >50°C based on stable isotope and fluid inclusions (Spötl and Pitman, 1998).

Mimetic and nonmimetic textures reflect the density of nucleation sites present on the substrate being dolomitized. Mimetic texture requires a relatively high density of nucleation sites whereas nonmimetic texture reflects a low density of nucleation sites. The density of nucleation sites on a substrate is determined by the crystal size and mineralogy of the substrate as well as the degree of supersaturation and temperature of the solution. Mimetic replacement is favored in cryptocrystalline allochems. Coarse crystalline, low Mg-calcite allochems such as brachiopods are not mimetically replaced (Bullen and Sibley, 1984). Because the texture and mineralogy of allochems can change during diagenesis, pre-dolomitization diagenesis can determine whether or not allochems are mimetically replaced.

In Paleozoic rocks, echinoderm fragments are commonly mimetically replaced whereas brachiopods are never mimetically replaced. In Cenozoic rocks, coralline algae and echinoderms are commonly mimetically replaced and generally the first allochems to be replaced (Sibley, 1982). Allochems that are normally mimetically replaced may be nonmimetically replaced if pre-dolomitization neomorphism results in significant crystal coarsening.

Lucia (1995) has shown correlations between porosity and permeability of dolomites, particularly when samples are grouped according to crystal size. Woody *et al.* (1996) have shown a correlation between porosity and permeability in planar and a lack of correlation in nonplanar dolomites. Therefore, improved correlations between porosity and permeability should be found if both crystal size and shape were considered.

Duncan Sibley

Bibliography

Bullen, S.B., and Sibley, D.F., 1984. Dolomite selectivity and mimic replacement. *Geology*, **12**: 655–658.
Eberl, D.D., Drits, V.A., and Srodon, L., 1998. Deducing growth mechanisms for minerals from the shapes of crystal size distribution. *American Journal of Science*, **298**: 499–533.
Folk, R.L., and Asserto, R., 1974. Giant aragonite rays and baroques dolomite in tepee-fillings, Triassic of Lombardy, Italy (abstract). *American Association of Petroleum Geologists*, Abstracts With Program, Annual Meeting, San Antonio, pp. 34–35.
Gregg, J.M., 1983. On the formation and occurrence of saddle dolomite-discussion. *Journal of Sedimentary Petrology*, **53**: 1025–1026.
Gregg, J.M., and Sibley, D.F., 1984. Epigenetic dolomitization and the origin of xenotopic dolomite texture. *Journal of Sedimentary Petrology*, **54**: 908–931.
Gregg, J.M., Scott, H., and Mazzullo, S.J., 1992. Early diagenetic recrystallization of Holocene (<300 years old) pertidal dolomites, Ambergis Cay, Belize. *Sedimentology*, **39**: 143–160.
Gunatilaka, A., 1989. Spheroidal dolomites—origin by hydrocarbon seepage? *Sedimentology*, **36**: 701–710.
Kaldi, J., and Gidman, J., 1982. Early diagenetic dolomite cements: examples from the Permian lower magnesian Limestone of England and the Pleistocene carbonates of the Bahamas. *Journal of Sedimentary Petrology*, **52**: 1073–1085.
Lucia, F.J., 1995. Rock-fabric/petrophysical classification of carbonate pore space for reservoir characterization. *American Association of Petroleum Geologists Bulletin*, **79**: 1275–1300.
Naiman, E.R., Bein, A., and Folk, R.L., 1983. Complex polyhedral crystals of limpid dolomite associated with halite, Permian upper Clear Fork and Glorieta formations, Texas. *Journal of Sedimentary Petrology*, **53**: 549–555.
Nielsen, P., Swennen, R., Dickson, J.A.D., Fallick, A.E., and Keppens, E., 1997. Spheroidal dolomites in a Visean karst system—bacterial induced origin. *Sedimentology* **44**: 177–195.
Pichler, T., and Humphrey, J.D., 2001. Formation of dolomite in recent island-arc sediments due to gas-seawater-sediment interaction. *Journal of Sedimentary Research*, **71**: 394–399.
Radke, B.M., and Mathis, R.L., 1980. On the formation and occurrence of saddle dolomite. *Journal of Sedimentary Petrology*, **50**: 1149–1168.
Sibley, D.F., 1982. The origin of common dolomite fabrics. *Journal of Sedimentary Petrology*, **52**: 1087–1100.
Sibley, D.F., and Gregg, J.M., 1987. Classification of dolomite rock texture. *Journal of Sedimentary Petrology*, **57**: 967–975.
Sibley, D.F., Gregg, J.M., Brown, R.G., and Laudon, P.R., 1993. Dolomite crystal size distribution. In Rezak, R., and Lavoie, D. (eds.), *Carbonate Microfabrics*. New York: Springer-Verlag, pp. 195–204.
Spötl, C., and Pitman, J.K., 1998. Saddle (baroque) dolomite in carbonates and sandstones: a reappraisal of a burial-diagenetic concept. In Morad, S. (ed.), *Carbonate Cementation in Sandstones*. Special Publications of Association of Sedimentologists **26**: pp. 437–460.
Sunagawa, I., 1984. Growth of crystals in nature. In Sunagawa, I. (ed.), *Material Science in the Earth's Interior*. Tokyo: Terra Scientific Publishing Co., pp. 63–1205.
Woody, R.E., Gregg, J.M., and Koederitz, L.F., 1996. Effect of texture on petrophysical properties of dolomite: Evidence from the Cambro-Ordovician of Southeastern Missouri. *American Association of Petroleum Geologists Bulletin*, **80**: 119–132.

Cross-references

Carbonate Mineralogy and Geochemistry
Cements and Cementation
Classification of Sediments and Sedimentary Rocks
Diagenesis
Dolomites and Dolomitization
Fabric, Porosity, and Permeability
Grain Size and Shape
Surface Textures

DOLOMITES AND DOLOMITIZATION

The mineral *dolomite* is widely distributed in the Earth's crust, especially in sedimentary/diagenetic settings, in rocks that range in age from the Precambrian to the Recent. Ideal, ordered dolomite has a formula of $CaMg(CO_3)_2$ and consists of alternating layers of Ca^{2+}-CO_3^{2-}-Mg^{2+}-CO_3^{2-}-Ca^{2+} etc. perpendicular to the crystallographic c-axis. Most natural dolomite has up to a few percent Ca-surplus (and corresponding Mg-deficit), as well as less than ideal ordering. *Protodolomite* has about 55 percent to 60 percent Ca and is partly to completely disordered, that is, the alternating layer structure is poorly developed to non-existing. Protodolomite or partially ordered dolomite with a few percent Ca-surplus are common as metastable precursors of well-ordered, nearly stoichiometric dolomite. *Dolostone* is a rock that consist of >75 percent dolomite. Dolostones are also called *dolomites*, especially in the literature prior to about 1990, but the term dolomites should be used only for different types of dolomite.

Two types of dolomite formation are common, that is, *dolomitization*, which is the replacement of $CaCO_3$ by $CaMg(CO_3)_2$, and *dolomite cementation*, which is the precipitation of dolomite from aqueous solution in primary or secondary pore spaces. The term dolomitization is also, yet incorrectly, applied to dolomite cementation. A third type of dolomite formation, where dolomite precipitates from aqueous solution to form sedimentary deposits ("primary dolomite"), appears to be rare and is restricted to some evaporitic lagoonal and/or lacustrine settings.

Despite intensive research for several decades, the origin of dolomites and dolostones is subject to considerable controversy. This is because some of the chemical and/or the hydrological conditions of dolomite formation are poorly understood, and because the available data permit more than one viable genetic interpretation in many case studies. One key problem is that well-ordered, stoichiometric dolomite has never been successfully grown inorganically in laboratory experiments at near-surface conditions of 20–30°C and 1 atm pressure (to date, the lowest temperature/pressure at which (proto-) dolomite has been grown under well-controlled laboratory conditions is 60°C/1 atm: Usdowski 1994). Hence, geochemical parameters that are necessary for back-calculating the composition(s) of the dolomitizing fluid(s), such as the equilibrium oxygen isotope fractionation and trace element partitioning coefficients, have to be extrapolated from high-temperature experiments, which renders them notoriously inaccurate and often leads to ambiguous genetic interpretations in case studies of dolomitization. Furthermore, dolomite is fairly rare in Holocene environments and sediments, yet very abundant in older rocks. Collectively, these aspects make up the so-called "dolomite problem", and they have led to a number of models for dolomitization.

Chemical constraints

Dolomite formation is chemically, that is, thermodynamically and/or kinetically, favored by (a) low Ca^{2+}/Mg^{2+}-ratios; (b) low Ca^{2+}/CO_3^{2-}-ratios ($=$ high carbonate alkalinity); (c) high temperatures; and (d) salinities substantially lower or higher than that of seawater (e.g., Machel and Mountjoy, 1986; Arvidson and MacKenzie, 1999; Figures D32, D33). These constraints translate into three essential and common conditions and natural settings to form dolostones.

(1) A sufficient supply of Mg^{2+} and CO_3^{2-}; this condition favors marine settings and burial-diagenetic settings with pore fluids of marine parentage because seawater is the only common Mg-rich natural fluid in sedimentary/diagenetic settings.

(2) A long-lasting and efficient delivery system for Mg^{2+} and/or CO_3^{2-} (also exporting Ca^{2+} in the case of calcite replacement), which favors settings with an active and long-lasting hydrologic drive.

(3) Carbonate depositional settings and/or limestones.

In addition, dolomite may form when rapidly rising formation fluids suddenly release CO_2 (Leach et al. 1991).

Considering that the above chemical constraints allow dolomite formation in almost the entire range of surface and subsurface diagenetic settings, the question arises why there are so many undolomitized limestones. The essential conditions for the common lack of dolomitization probably are: (d) ion pair formation, inactivating much of the Mg^{2+} and CO_3^{2-} in solution; (e) insufficient flow because of the lack of a persistent hydraulic head, too small a hydraulic head, or insufficient diffusion, resulting in insufficient magnesium and/or carbonate ion supply; (f) the limestones are cemented and not permeable enough; and (g) the diagenetic fluids are incapable of forming dolomite because of kinetic inhibition, for example, because the environment is too cold (most kinetic inhibitors of dolomite nucleation and growth are rather potent at temperatures below about 50°C, and the Ca^{2+}/Mg^{2+}-ratio of many cold diagenetic fluids is not low enough for dolomitization.

Mass balance constraints

Dolomitization is a diagenetic dissolution-replacement process for which many mass balance equations can be written. Dolomitization in its most simplistic form is represented by

$$2CaCO_3(s) + Mg^{2+}(aq) - - > CaMg(CO_3)_2(s) + Ca^{2+}(aq)$$

(reaction 1)

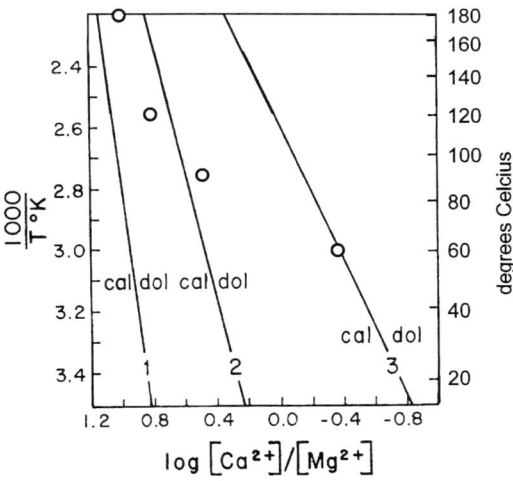

Figure D32 Bivariate thermodynamic stability diagram for the system calcite-dolomite-water. Square brackets denote activities. Lines are calculated from experimental data: line 1 = calcite + ideal, fully ordered dolomite; line 2 = calcite + ordered dolomite with slight Ca-surplus; line 3 = calcite + fully disordered protodolomite. The four open circles denote the results of Usdowski (1994), whose up to 7 year-long runs represent the lowest-temperature experimental dolomite formation performed to date. Usdownski's (1994) data for 90, 120 and 180°C plot close to line 2, but his data for 60°C plots on line 3, which probably reflects that protodolomite rather than dolomite formed at 60°C. Data from natural aquifers cluster close to line 2 (not shown), which can be considered representative of most natural dolomite. Diagram is modified from Carpenter (1980).

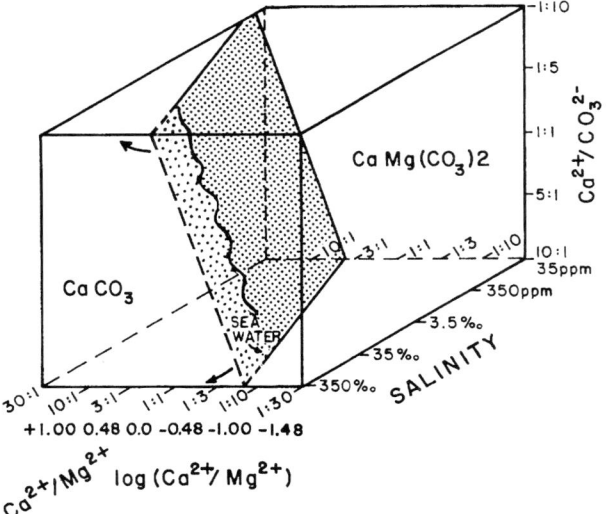

Figure D33 Trivariate kinetic stability diagram for the system calcite-dolomite-water. The ionic ratios are molar ratios. Seawater plots just in the calcite field. The stippled field boundary is bent towards higher Ca/Mg ratios at salinities greater than 35 permil. From Machel and Mountjoy (1986).

or by

$$CaCO_3(s) + Mg^{2+}(aq) + CO_3^{2-}(aq) \rightarrow$$
$$CaMg(CO_3)_2(s) \quad \text{(reaction 2)}$$

Reactions 1 and 2 are end members of a range of possible stoichiometries, that is,

$$(2-x)\,CaCO_3(s) + Mg^{2+}(aq) + xCO_3^{2-}(aq) \rightarrow$$
$$CaMg(CO_3)_2(s) + (1-x)Ca^{2+}(aq) \quad \text{(reaction 3)}$$

For $x = 0$, reaction 3 becomes reaction 1, and for $x = 1$ reaction 3 becomes reaction 2. Magnesium has to be imported to and calcium has to be exported from the reaction site in the case of reaction 1, whereas there is no export of calcium in the case of reaction 2. Intermediate cases are represented by values of x between zero and one. Dolomite cementation is most simplistically represented by

$$Ca^{2+}(aq) + Mg^{2+}(aq) + 2CO_3^{2-}(aq) \rightarrow$$
$$CaMg(CO_3)_2(s) \quad \text{(reaction 4)}$$

In any given case, the amount of the diagenetic fluid needed for dolomite formation depends on the stoichiometry of the reaction, temperature, and composition of the fluid. For example, if dolomitization proceeds via reaction 1, about 650 m^3 of normal seawater are needed to dolomitize 1 m^3 of limestone with 40 percent initial porosity at 25°C (Land, 1985). Dolomitization may not take place with 100 percent-efficiency, and larger water/rock ratios are required for complete dolomitization by seawater in a near-surface setting. If seawater is diluted to 10 percent of its original concentration, ten times as much water is needed. On the other hand, about 30 m^3 of brine are needed per m^3 of limestone at 100 percent-dolomitization efficiency in the case of a halite-saturated brine. The role of increasing temperature in the underlying thermodynamic calculations is to reduce the amount of magnesium necessary for dolomitization because the equilibrium constant (and hence the equilibrium Ca/Mg-ratio) is temperature-dependent (Figure D32). For example, at 50°C only about 450 m^3 of seawater are needed for complete dolomitization of 1 m^3 of limestone with 40 percent initial porosity at 100 percent-efficiency. The amounts of dilute and hypersaline waters change accordingly.

These calculations imply that, in all cases, large water/rock ratios are required for complete dolomitization. This necessitates the presence of active hydrologic flow systems for extensive, pervasive dolomitization. The exception are natural environments where carbonate muds or limestones can be dolomitized via diffusion of magnesium from seawater rather than by hydrologic flow.

Porosity and permeability

Dolomitization is important for petroleum and economic geology because most dolostones are more porous and permeable than limestones. However, it is not justified to presume that dolostones are always better hydrocarbon reservoir rocks than limestones. Schmoker and Halley (1982) have demonstrated with porosity/depth profiles of southern Florida that many dolostones that have not been buried too deeply (less than about 1 km) have porosities equal to or less than those of adjacent limestones. A similar pattern has been found in Devonian dolostones of western Canada that were buried to more than 4 km (Amthor et al., 1994). Furthermore, the shape and extent of dolostone reservoirs are determined by the process of dolomitization, that is, the direction and duration of the fluid flow regime(s). Therefore, and nevertheless, most dolomitized carbonates are better reservoirs than limestones, and dolostone reservoirs are diagenetic traps or combination traps with a diagenetic component. A related aspect is that dolostones often form aquifers and/or preferential migration pathways for hydrocarbons. For example, the dolomitized margin of the Devonian Cooking Lake platform is known as a migration pathway over a length of several hundred km (e.g., Amthor et al., 1994).

Porosity may be retained, lost, or generated during dolomite formation. In the case of dolomite cementation, porosity (and permeability) loss is obvious and occurs in proportion to the amounts of dolomite cement formed. In contrast, about 13 percent of porosity is generated, such that a limestone with 40 percent porosity is replaced by a dolostone with about 45 percent porosity, in the so-called "mole-per-mole" replacement of calcite by dolomite according to reaction 1 (whereby two moles of calcite are replaced by one mole of dolomite). In such a case, the fabrics of the original limestone must be at least partially obliterated in order to account for the volume change during replacement.

More generally, porosity gains or losses can be represented by the values of x in reaction 3. In the special cases of $x = 0.11$ and $x = 0.25$ (for aragonite and calcite, respectively), there is no volume loss or gain (Morrow 1982). Hence, limestones dolomitized in a volume-per-volume replacement should not contain secondary pores or dolomite cements, and the primary textures may be partially or largely preserved. However, partial or complete obliteration of primary textures can occur even in a volume-per-volume replacement because of a marked change in crystal size (usually an increase, due to Ostwaldt ripening) and/or porosity redistribution.

All the above cases of dolomitization may, and many invariably do, involve reorganization of permeability pathways. Permeability modifications are a function of the textures and permeabilities of the precursor carbonates, as well as of reaction kinetics during dolomitization, that is, crystal nucleation and growth rates. While porosity changes can be estimated from the amounts of dolomite cement and/or intercrystalline secondary porosity, permeability changes during dolomitization are almost impossible to quantify.

Dolomite textures

The most widely used classification of dolomite/dolostone textures is the one by Sibley and Gregg (1987), based on Gregg and Sibley (1984). This classification is popular because it is simple and largely descriptive, although it carries genetic implications. Crystal size distributions are classified as "unimodal" and "polymodal", whereas crystal boundary shapes are classified as "planar-e" (euhedral), "planar-s" (subhedral), and "nonplanar" (anhedral). Where possible, a textural description includes recognizable allochems or biochems, matrix, and void fillings. Particles and cements may be unreplaced, partially replaced, or completely replaced. Replacement may be mimetic or nonmimetic. Unimodal size distribution generally results from a single nucleation event and/or a unimodal primary (pre-dolomite) size distribution of the substrate. Polymodal size distributions indicate multiple nucleation events and/or a differential nucleation on an originally polymodal substrate. Planar crystal boundaries tend to develop at growth below about 50°C (the so-called "critical roughening temperature"), whereas nonplanar boundaries tend to develop at T>50°C and/or high degrees of supersaturation. An exception are the planar yet curved crystal faces of saddle dolomite, which is known to form only at elevated temperatures of more than about 80°C (Radtke and Mathis, 1980; Machel, 1987). Saddle dolomite is different also in other respects, for example, it almost invariably is sparry and commonly milky-white, it has distinctively warped crystal lattices, undulose extinction, pearly luster, and abundant fluid inclusions.

Textures and reservoir characteristics of natural dolostones are highly variable. Where dolomitization is matrix-selective, significant porosity tends to be remnant primary porosity with relatively poor interconnection. Undolomitized calcite fossils commonly are dissolved during or after dolomitization, leaving a rock resembling Swiss cheese, that is, a relatively tight matrix with large holes. Various types of cement, often anhydrite, partially or completely occlude such secondary porosity. Where dolomitization is not matrix-selective, or where the matrix is relatively coarse-grained, sucrosic dolostones with intercrystalline porosity and excellent permeability tend to form.

Dolomite geochemistry

In addition to textures and their distribution relative to stratigraphy and structure, a wide range of geochemical methods may be used to characterize dolomites and dolostones, and to decipher their origin. The most widely used methods are stable isotopes (O, C), Sr-isotopes, trace elements, fluid inclusions, and paleomagnetics (e.g., various case studies in Purser *et al.*, 1994). Care must be taken to assess the degree of recrystallization, which may result in dolomite compositions that are characteristic of the environment(s) of recrystallization rather than the original environment(s) of dolomitization (e.g., Machel, 1997).

Natural environments and models of dolomitization

Dolomite occurs in many diagenetic environments that range from the surface to deep subsurface settings of several km burial depth. Most dolostones originate by the replacement of limestones that form(ed) in shallow–marine environments with normal seawater salinity, which is shown by relics of primary sedimentary features, such as ripple marks, reef fossils, burrows, etc., that have survived the replacement process. In the Recent, however, dolomite is almost absent from such carbonate depositional environments, despite the fact that seawater is many times supersaturated with respect to dolomite. Thus, various *models of dolomitization* have been devised to explain the origin of these types of replacement dolostones, and almost every time a new model has been proposed, it became a bandwagon until the next popular model arose. There also are dolostones that are devoid of primary sedimentary features, and where an origin without a limestone precursor appears possible. Dolomite is also common, albeit generally not abundant, as a cement in limestones and clastic rocks.

The potential of natural environments to form dolomite and dolostone can be assessed on the basis of the above chemical considerations, and using hundreds of case studies that have been published over the last five decades (see compilations in Zenger *et al.*, 1980; Shukla and Baker, 1988; Purser *et al.*, 1994). Genetically, all natural dolomites can be grouped into two families (Budd, 1997). *Penecontemporaneous dolomites* form while the host sediments are in their original depositional setting. Most known penecontemporaneous dolomites are of Holocene age. However, there probably are many older examples in the geologic record, yet they are much more difficult to prove. *Postdepositional dolomites* form after deposition has ceased and the host carbonates have been removed from the zone of active sedimentation by progradation of the depositional interface, burial, uplift, eustatic sea-level change, or any combination of these factors. Almost all dolostones are postdepositional. Additionally, small amounts of dolomite may be formed syndepositionally or postdepositionally by the magnesium derived from within the dolomitized sediments, that is, originally contained in or adsorbed by Mg-calcites, organic matter, chlorophyll, biogenic silica, certain clay minerals, or older detrital dolomites, if present (e.g., Lyons *et al.*, 1984; Baker and Burns, 1985).

Penecontemporaneous dolomites and the microbial/organogenic model

There are two preferred settings of penecontemporaneous dolomite formation, that is, shallow-marine to supratidal, and hemipelagic to pelagic. In either case, and with few exceptions, the amounts of dolomite formed are small, and these dolomites are texturally as well as geochemically distinct.

Penecontemporaneous dolomites in shallow-marine to supratidal settings usually form in quantities of <5 vol-percent, mostly as Ca-rich and poorly ordered, fine-crystalline cements and/or directly from aqueous solution. One especially important type of penecontemporaneous (proto)dolomite forms lenses and layers of up to 100 percent in sabkhas, as discussed further below. The other occurrences are as lithified supratidal crusts (e.g., Andros Island; Sugarloaf Key; Ambergris Cay); as thin layers in salinas (e.g., Bonaire; West Caicos Island) and evaporative lagoons/lakes (e.g., Coorong); as fine-crystalline cements and replacements in peritidal sediments (e.g., Florida Bay; Andros Island); and as fine-crystalline supratidal weathering product of basic rocks (Capo *et al.*, 2000; this is the only known case to date where penecontemporaneous dolomite is ordered and stoichiometric). The dolomite-forming fluid is normal seawater and/or evaporated seawater, rarely with admixtures of evaporated groundwater.

Penecontemporaneous dolomite formation in hemipelagic to pelagic settings also form very small quantities of microcrystalline protodolomite, if any, that is, generally less than 1 weight-percent (Lumsden, 1988). However, under favorable circumstances the amount of dolomite can reach up to 100 percent locally. For example, Miocene hemipelagic carbonate sediments from the margin of the Great Bahama Bank are partially to completely dolomitized over a depth of range of about 50–500 m subsea. Dolomite occurs as a primary void-filling cement and by replacing micritic sediments, red algae, and echinoderm grains (Swart and Melim, 2000).

Both settings of penecontemporaneous dolomite formation appear to be linked to the "microbial model" or "organogenic model" of dolomitization (Vasconcelos and McKenzie, 1997; Burns et al., 2000; Mazzullo, 2000). According to this model, dolomite may be formed syndepositionally or early-postdepositionally at depths of a few cm to a few hundred meters under the influence of or promoted by bacterial sulfate reduction and/or methanogenesis, as shown by $\delta^{13}C$-values. The specific role of microbial activity in reducing the kinetic notorious barriers to dolomitization is presently uncertain, although reduction of Mg- and Ca-hydration barriers, an increase in alkalinity and changes in pH, seem to be involved. Most microbial/organogenic dolomites are cements, but some are replacive. They typically are fine-crystalline (less than 10 μm), calcic and poorly ordered protodolomites. The chief modes of Mg-supply are diffusion from the overlying seawater and/or release from Mg-calcites and clay minerals. Microbial/organogenic dolomites may act as nuclei for later, more pervasive dolomitization during burial.

Hyposaline environments and the mixing zone model

Hyposaline environments are those with salinities below that of normal seawater (35 mg/l). This includes coastal and inland freshwater/seawater mixing zones, marshes, rivers, lakes, and caves. Postdepositional dolomite has been found to form in all of these environments, but only in small amounts and commonly as cements.

One hyposaline environment, the coastal freshwater/seawater mixing zone (often simply called mixing zone) has given rise to one of the oldest and most popular models, that is, the "mixing zone model". However, the mixing zone model has been consistently overrated since it was first proposed (Badiozamani, 1973; Land, 1973). For example, Humphrey and Quinn (1989) suggested that coastal mixing zones may form thick sections of dolomite in a platform-margin setting, and that such dolostones may be common in the geologic record. Other authors have proposed mixing zone dolomitization for entire carbonate platforms many hundreds of km^2 in size (e.g., Xun and Fairchild, 1987).

Dolomite may form in freshwater/seawater mixing zones. This is due to the fact that many freshwater/seawater mixing zones are thermodynamically undersaturated for calcium carbonate yet supersaturated with respect to dolomite in at least a part of the mixing spectrum (commonly in mixtures of about 10–50 percent seawater). However, not a single location in the world has been shown to be extensively dolomitized in a freshwater/seawater mixing zone, in recent or in ancient carbonates. Rather, dolomite formation appears to be kinetically inhibited in the vast majority of coastal mixing zones (Smart et al., 1988; Machel and Mountjoy, 1990). Furthermore, calcite and aragonite dissolve at rates many times faster than dolomite is formed, thus, the most common result of mixing zone diagenesis is rocks riddled with secondary porosity up to the dimensions of caves, with or without small amounts of dolomite (e.g., Florida, Yucatan: Back et al., 1986; confirmed by modeling: Sanford and Konikow, 1989). In addition, most coastal mixing zones are only a few hundreds of meters wide and pass relatively quickly through the rocks in response to eustatic sea level fluctuations and/or subsidence. Hence, even where mixing zones are capable of forming dolomite, the dolomitized rock volume is relatively small and restricted to the platform margin(s) (Figure D34(A)).

Most mixing zone dolomite is petrologically and geochemically distinct. Most crystals are relatively clear, planar-e or planar-s, stoichiometric, well-ordered rhombs. However, some mixing zone dolomite is protodolomite. Crystal sizes commonly range from 1 micron to 100 microns but reach several millimeters in some cases. Most mixing zone dolomite occurs as cements in microscopic interstices and macroscopic voids, molds, vugs, and caverns. Alternating generations or growth zones of calcite/dolomite are common in coastal mixing zones with rapid and cyclical changes of salinity.

The main role of coastal mixing zones in dolomitization might be that of a hydrologic pump for seawater dolomitization, rather than that of a geochemical environment favorable for dolomitization (Machel and Mountjoy, 1990). At the seaward margin of a coastal mixing zone seawater is pumped

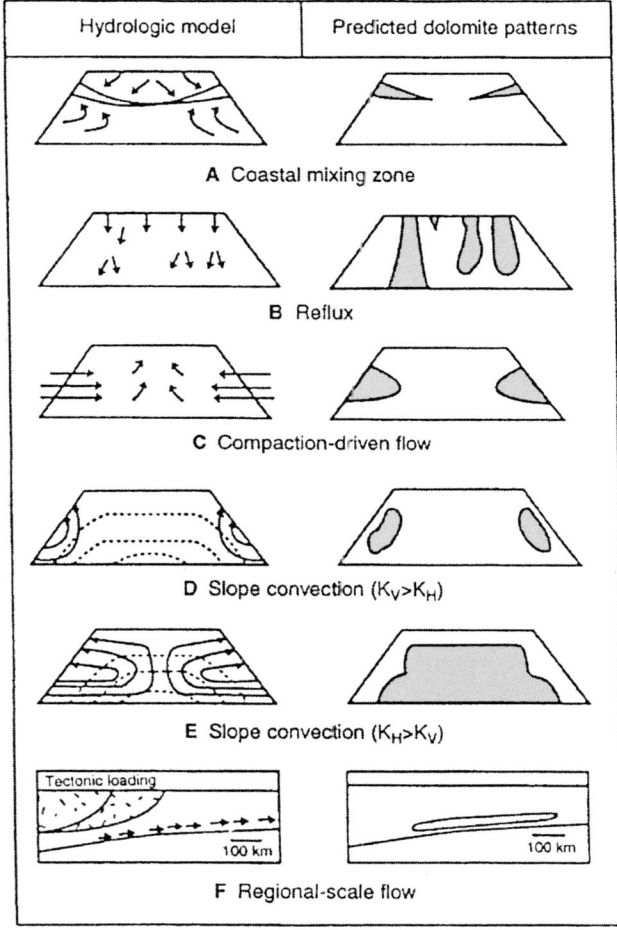

Figure D34 Selected models of dolomitization, illustrated as groundwater flow systems (left) and predicted dolomitization patterns (right). Examples are of incomplete dolomitization of carbonate platforms or reefs. Arrows denote flow directions; dashed lines show isotherms. Predicted dolomitization patterns are shaded. Models A–E are km scale; model F is basin-scale. Reproduced from Amthor et al. (1993).

through the sediments by the hydrologic action of freshwater–seawater mixing, whereby the water movement is influenced also by tidal pumping. Hence, substantial amounts of seawater may be pumped through the affected sediments by the hydrologic forces of freshwater–seawater mixing and the tides.

Hypersaline environments and the reflux model

Intertidal to supratidal environments with salinities greater than that of normal seawater occur throughout the world at latitudes of less than about 30°. Extensive, pervasive dolomitization of shallow water carbonates, up to the dimensions of huge and thick carbonate platforms, from density-driven flow (reflux) of the associated evaporitic brines has been suggested by many authors on the basis of case studies and modeling (e.g., Shields and Brady, 1995). This "reflux model" (Figure D34(B)) has been another very popular model of dolomitization throughout the last 30 years.

The sabkha of the Trucial Coast of Abu Dhabi, the probably best researched recent hypersaline intertidal to supratidal flat (Patterson and Kinsman, 1982; Baltzer et al., 1994), is representative of prolific reflux dolomite formation. Fine-crystalline protodolomite is formed penecontemporaneously by evaporating floodwaters that are propelled periodically onto the lower supratidal zone and along remnant tidal channels by strong onshore winds. Dolomite formation takes place in a narrow (1–1.5 km) fringe next to the strandline and within the channels within the upper 1 m to 2 m of the sediments, and preferentially where the environment is chemically reducing (sulfate reduction, enhanced carbonate alkalinity). The distribution of protodolomite is uneven across the sabkha, and the best dolomitized parts contain 5 to about 65 weight percent protodolomite. Dolomite forms as a cement and aragonite is replaced, but lithification does not occur, or only partially. A variety of evaporite minerals, most notably multigenetic calcium sulfates, form within the dolomitized part of the sabkha.

In most respects, the Abu Dhabi sabkha appears to be a good recent analog for dolomitization in many ancient intertidal to supratidal flats, such as landward of the famous Permian carbonates of Texas and New Mexico. Rather than forming reservoir rocks, such dolostones—including the associated evaporites—generally form tight seals for underlying hydrocarbon reservoirs (e.g., Major et al., 1988).

In general, sabkhas and 'similar' intertidal to supratidal environments in more humid climates typically form small quantities of fine crystalline protodolomite in thin beds, crusts, or nodules, either within the upper 1 m to 2 m of sediment, or at the sediment surface. Repeated transgressions and regressions may stack such sequences upon one another. Furthermore, it appears that the efficiency of reflux is different during transgression and regression. Mutti and Simo (1994) found in Permian carbonates that dolomitization occurred during transgressive cycles only in intertidal and supratidal facies, whereas dolomitization affected supra-, inter-, and subtidal facies during regression. They also speculated that tidal pumping and evaporative pumping may aid reflux in supplying Mg. In contrast, Montañez (1997) found in Ordovician carbonates that transgressive cycles were not dolomitized by reflux, whereas the facies within regressive cycles were almost completely replaced by tight fine-crystalline reflux dolomite. Be this as it may, deep penetration by refluxing evaporative brines and/or platform-wide dolomitization has not been demonstrated anywhere in the world.

On the other hand, some the latest sophisticated modeling suggests that refluxing brines originating on 'permanently' flooded carbonate platforms may penetrate the platforms rather deeply, that is, 1–2 km (Jones et al., 2002a,b). This situation differs from sabkhas and similar intertidal to supratidal flats, as the latter are flooded only intermittently. Even in flooded carbonate platforms, however, dolomitization is predominant and/or restricted to the upper few meters or tens of meters. Below these depths, the fluxes are too low and the brines, although still refluxing, have exhausted their dolomitization potential. Modeling the Devonian Grosmont platform in western Canada, based on extensive stratigraphic, lithologic, and geochemical data, Jones et al. (2002b) found that after 100 k years, during which brines concentrated up to 150 per mil (slightly beyond gypsum saturation) overlay the platform, reflux develops and shuts off geothermal circulation. The onset of reflux results in a reversal in groundwater flow direction as brines are transported from the platform interior basinwards. Refluxing brines enter the platform in a zone that extends approximately 150 km from the coast. As the platform evolves the brine plume, which is preferentially orientated in a horizontal direction, sinks into the platform subsurface and migrates basinwards. Computed distributions of fluid flux in conjunction with Mg mass-balance calculations permit the tracking of dolomite fronts as the Grosmont platform evolves. Within the range of parameter uncertainty, reflux could have delivered sufficient Mg in 1.6 m years to account for the pervasive dolomitization of the Grosmont Formation. However, even in these modeling results the amount of dolomite formed is not 100 percent, and dolomitization is restricted to layers and beds of dolomite that are interrupted by limestones, which results from repeated transgressions and regressions.

A previously unrecognized variant of reflux, proposed on the basis of numerical modeling by Jones et al. (2002a), is "latent reflux". This type of reflux describes the phenomenon whereby brines of reflux origin continue to sink into a platform after the cessation of brine generation on the platform top. Hence, latent reflux could deliver magnesium to the platform carbonates after the cessation of evaporation, thereby forming additional dolomite. However, latent reflux, as active reflux during evaporation, would exit along or near the platform margin, and the brines would not lead to any dolomitization at greater depths. The efficacy of this process with respect to dolomitization remains to be proven.

Seawater dolomitization model

Postdepositional formation of massive dolostones has been attributed to the "seawater dolomitization model" (Purser et al., 1994), with specific reference to the Cenozoic dolostones of the Bahama platform, which are often used as an analog for older dolomitized carbonate platforms elsewhere. Various hydrologic systems been invoked to drive large amounts of seawater through the Bahama platform, that is, thermal convection (Sanford et al., 1998); a combination of thermal seawater convection and reflux of slightly evaporated seawater derived from above (Whitaker et al., 1994); or seawater driven by an overlying freshwater/seawater mixing zone during partial platform exposure (Vahrenkamp and Swart, 1994), possibly layer-by-layer in several episodes (Vahrenkamp et al., 1991). All these studies agree that the dolomitizing fluid was essentially seawater that was, at best, only slightly altered by water–rock interaction.

It is noteworthy that the Bahama dolostones represent a hybrid regarding classification. This is because the dolomitizing solution probably was nearly normal seawater, yet thermal convection, as a hydrologic system and drive for dolomitization, is better classified under burial (subsurface) models. Similarly, Devonian dolostones in western Canada, which probably formed at depths of 300–1,500 m from slightly altered seawater, have been classified as burial dolostones (Amthor et al., 1994), yet they could equally well be classified as seawater dolostones. This classification dilemma arose from the historical evolution of our understanding of these dolostones.

Burial (subsurface) environments and related models

Burial (subsurface) environments are those removed from active sedimentation by burial, and where the pore fluid chemistry is no longer entirely governed by surface processes, that is, where water–rock interaction has modified the original pore waters to a significantly degree (whereby 'significant' is ill defined). All burial (subsurface) models for dolomitization essentially are hydrologic models that differ mainly in the hydrologic drives and direction(s) of fluid flow (e.g., Morrow, 1999). Four main types of fluid flow take place in subsurface diagenetic settings: (1) compaction flow; (2) topography-driven flow; (3) thermal convection; and (4) tectonically-driven flow. Mixtures of these flow regimes and fluids are possible under certain circumstances.

The oldest burial model of dolomitization is the "compaction model" (Illing, 1959; Jodry, 1969). In this model (Figure D34(C)), seawater and/or its subsurface derivative is pumped through the rocks at several tens to several hundreds of meters. The model was never especially popular because it became clear that burial compaction can only generate fairly limited amounts of dolostone due to the limited amounts of compaction water.

Convection through carbonate platforms may be driven by heat from below or by density differences from above. One type of thermal convection originates from the horizontal density gradient caused by the regional temperature difference between the interior of uplifted platforms and the surrounding seawater. In this case, convection may take place in half-cells that are open to the adjacent and overlying ocean, with marked differences in the fluxes and heat transport as a function of the ratios of horizontal to vertical permeability (Figures D34(D), (E); Sanford et al., 1998). The amounts of dolomite that can be formed in such convection half-cells are essentially unlimited, that is, dolomite can be formed as long as convection takes place. In contrast, the amounts of dolomite are much smaller where convection takes place in full-cells that are not open to the ocean. This may happen where convection is driven by a relatively weak regional heat gradient or by a local heat source, such as a magmatic intrusion. Thermal convection has been suggested for extensive, pervasive dolomitization in several carbonate platforms and/or regional carbonate aquifers (e.g., Morrow, 1999). In particular, thermal convection has been advocated as the chief mode of dolomitization for the Cenozoic carbonates of the Bahamas platform (here discussed under "Seawater dolomitization model").

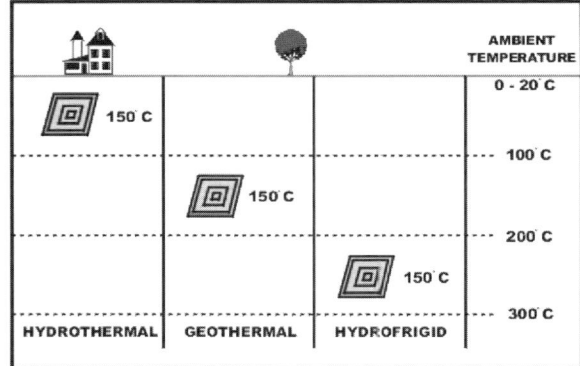

Figure D35 Hydrothermal, geothermal (formed in thermal equilibrium with the surrounding rocks), and hydrofrigid mineral formation. From Machel and Lonne (2002).

Convection cells invariably have rising limbs that penetrate the overlying and cooler strata, linking thermal convection to hydrothermal dolomitization. There are well-documented examples of hydrothermal dolomite, in local as well as in regional flow systems (e.g., Spencer-Cervato and Mullis, 1992; Qing and Mountjoy 1992, 1994; Duggan et al., 2001). However, many dolomites that formed at elevated temperatures, especially but not exclusively saddle dolomites that formed at T>80°C, have mistakenly been called hydrothermal (e.g., Davies, 1997). The term "hydrothermal" is and should be used according to White's (1957) time-honored definition as "aqueous solutions that are warm or hot relative to the surrounding environment", with no genetic implications regarding the fluid source, including Stearns et al.'s (1935) rationale for a "significant" temperature difference of at least 5–10°C (Figure D35). Hence, any mineral, including dolomite, should be called "*hydrothermal*" only if it can be demonstrated to have formed at a higher (by >5–10°C) than ambient temperature, regardless of fluid source or drive (Figure D35, left). This definition does not carry a lower or upper temperature limit. A dolomite formed at only 40°C could be hydrothermal, if the surrounding rocks were significantly colder than that at the time of dolomite formation. If a mineral was formed at or near the same temperature as the surrounding rocks (within 5–10°C), it should be called "*geothermal*" (Figure D35, center), whatever the geothermal gradient. In keeping with common usage, the qualifier "geothermal" may simply be omitted, unless special emphasis needs to be placed on the geothermal nature of a particular mineralization event. Minerals formed at temperatures significantly lower than ambient (by >5–10°C) may be called "*hydrofrigid*", even if they formed at a rather high temperature (Figure D35, right). For example, seawater or freshwater may penetrate a rock sequence through highly permeable pathways, such that the water is heated to 150°C at a depth where the surrounding rock has a temperature of 250°C. If a mineral were formed at that time from the incompletely heated water, the mineral would be hydrofrigid.

Topography-driven flow takes place in uplifted sedimentary basins that are exposed to meteoric recharge. With time enormous quantities of meteoric water may be pumped through a basin, often concentrated by water–rock interaction (especially salt dissolution), and preferentially funnelled through aquifers. Significant dolomitization can only take place, however, where the meteoric water dissolves enough Mg en route before encountering limestones. Gregg (1985) advocated topography-driven flow for regionally extensive dolomitization in Cambrian carbonates of Missouri.

Another type of flow that has been suggested to result in pervasive dolomitization is tectonically-driven, that is, metamorphic fluids expelled from crustal sections affected by tectonic loading that also drive basinal fluids toward the basin margin (Figure D34(F)). Such fluid flow could be injected into compactional and/or topography-driven flow, with attendant fluid mixing. Dorobek (1989), Drivet and Mountjoy (1997), and others have invoked tectonically-driven flow for dolomitization of the Siluro-Devonian Helderberg Group, USA, and for Devonian reefs in Canada, respectively. However, it is hypothetical at the present time whether tectonically induced flow, or the related fluid mixing, can form massive dolostones, because the fluxes of tectonically induced flow appear to be low (e.g., Machel *et al.*, 2000).

Textures and reservoir characteristics of dolostones formed in burial (subsurface) environments are variable. Where dolomitization is matrix-selective, significant porosity tends to be remnant primary porosity with relatively poor interconnection. Undolomitized calcite fossils may be dissolved during or after dolomitization, leaving molds and vugs. Where dolomitization is not matrix-selective, or where the matrix is relatively coarse-grained, sucrosic dolostones with intercrystalline porosity and excellent permeability tend to be formed. Unfortunately, these textures are not indicative of any of the genetic possibilities discussed above, with the exception of saddle dolomite, which appears to form only at T>80°C. This dolomite is texturally and geochemically distinct. Saddle dolomite occurs as cements in voids, as replacive masses (often as part of MVT-sulfide mineralization), or as so-called zebra dolomite in alternating layers with much finer-crystalline matrix dolomites. Zebra dolomite is much debated and may originate from repeated fracturing or some type of geochemical self-organization (Krug *et al.*, 1996).

Secular distribution of dolostones

The relative abundance of dolostones that originated from the replacement of marine limestones appears to have varied cyclically through time, with two discrete maxima of "significant early" dolomite formation (massive early diagenetic replacement of marine limestones) during the Phanerozoic, that is, the Early Ordovician/Middle Silurian and the Early Cretaceous (Given and Wilkinson, 1987). Various explanations have been proposed for this phenomenon, that is, that periods of enhanced early dolomite formation were related to or controlled by plate tectonics and related changes in the compositions of the atmosphere and seawater, such as an increased atmospheric CO_2-level, high eustatic sea level, low saturation state of seawater with respect to calcite, changes in the marine Mg/Ca ratio, or low atmospheric O_2-levels that coincided with enhanced rates of bacterial sulfate reduction (e.g., Stanley and Hardie, 1999; Burns *et al.*, 2000). At present, the reason(s) for the alleged or observed secular variations in dolostone abundance remain(s) uncertain.

Concluding remarks

There are many environments and possibilities to form dolomites and dolostones. However, the (paleo-) hydrologic characteristics are rarely known well enough for an unambiguous interpretation of the dolomitizing fluid flow regime. Dolostones often consist of different generations of dolomite formed by different flow systems and mechanisms. Metastable (proto-) dolomites may be formed, especially in near-surface and shallow burial settings, which later and deeper recrystallize to more stable dolomites, either in one or in several recrystallization steps.

Hans G. Machel

Bibliography

Amthor, J.E., Mountjoy, E.W., and Machel, H.G., 1994. Regional-scale porosity and permeability variations in Upper Devonian Leduc buildups: implications for reservoir development and prediction in carbonates. *American Association Petroleum Geologists Bulletin*, **78**: 1541–1559.

Arvidson, R.S., and MacKenzie, F.T., 1999. The dolomite problem: control of precipitation kinetics by temperature and saturation state. *American Journal of Science*, **299**: 257–288.

Badiozamani, K., 1973. The Dorag dolomitization model—application to the Middle Ordovician of Wisconsin. *Journal of Sedimentary Petrology*, **43**: 965–984.

Back, W., Hanshaw, D.B., Herman, J.S., and van Driel, J.N., 1986. Differential dissolution of a Pleistocene reef in the groundwater-mixing zone of coastal Yucatan, Mexico. *Geology*, **14**: 137–140.

Baker, P.A., and Burns, S.J., 1985. Occurrence and formation of dolomite in organic-rich continental margin sediments. *American Association Petroleum Geologists Bulletin*, **69**: 1917–1930.

Baltzer, F., Kenig, F., Boichard, R., Plaziat, J.-C., and Purser, B.H., 1994. Organic matter distribution, water circulation and dolomitization beneath the Abu Dhabi sabhka (United Arab Emirates). In Purser, B., Tucker, M., and Zenger, D. (eds.), *Dolomites—A Volume in Honor of Dolomieu*. International Association of Sedimentologists, Special Publication, 21, pp. 409–428.

Budd, D.A., 1997. Cenozoic dolomites of carbonate islands: their attributes and origin. *Earth-Science Reviews*, **42**: 1–47.

Burns, S.J., McKenzie, J.A., and Vasconcelos, C., 2000. Dolomite formation and biochemical cycles in the Phanerozoic. *Sedimentology*, **47**: 49–61.

Capo, R.C., Whipkey, C.E., Blachère, J.R., and Chadwick, O.A., 2000. Pedogenic origin of dolomite in a basaltic weathering profile, Kohala Peninsula, Hawaii. *Geology*, **28**: 271–274.

Carpenter, A.B., 1980. The chemistry of dolomite formation I: the stability of dolomite. In Zenger, D.H., Dunham, J.B., and Ethington, J.B. (eds.), *Concepts and Models of Dolomitization*. Society of Economic Paleontologists and Mineralogists Special Publication, 28, pp. 111–121.

Davies, G.R., 1997. Hydrothermal dolomite (HTD) reservoir facies: global perspectives on tectonic-structural and temporal linkage between MVT and Sedex Pb-Zn ore bodies, and subsurface HTD reservoir facies. *Canadian Society of Petroleum Geologists Short Course Notes*, 167 p.

Drivet, E., and Mountjoy, E.W., 1997. Dolomitization of the Leduc Formation (Upper Devonian), southern Rimbey-Meadowbrook reef trend, Alberta. *Journal of Sedimentary Research*, **67**: 411–423.

Duggan, J.P., Mountjoy, E.W., and Stasiuk, L.D., 2001. Fault-controlled dolomitization at Swan Hills Simonette oil field (Devonian), deep basin west-central Alberta, Canada. *Sedimentology*, **48**: 301–323.

Dorobek, S., 1989. Migration of orogenic fluids through the Siluro-Devonian Helderberg Group during late Paleozoic deformation: constraints on fluid sources and implications for thermal histories of sedimentary basins. *Tectonophysics*, **159**: 25–45.

Given, R.K., and Wilkinson, B.H., 1987. Dolomite abundance and stratigraphic age: constrains on rates and mechanisms of Phanerozoic dolostone formation. *Journal of Sedimentary Petrology*, **57**: 1068–1078.

Gregg, J.M., 1985. Regional epigenetic dolomitization in the Bonneterre dolomite (Cambrian), southeastern Missouri. *Geology*, **13**: 503–506.

Gregg, J.M., and Sibley, D.F., 1984. Epigenetic dolomitization and the origin of xenotopic dolomite texture. *Journal of Sedimentary Petrology*, 54: 908–931.

Humphrey, J.D., and Quinn, T.M., 1989. Coastal mixing zone dolomite, forward modeling, and massive dolomitization of platform-margin carbonates. *Journal of Sedimentary Petrology*, 59: 438–454.

Illing, L.V., 1959. Deposition and diagenesis of some upper Paleozoic carbonate sediments in western Canada. *Fifth World Petroleum Congress New York, Proceedings Section 1*, pp. 23–52.

Jodry, R.L., 1969. Growth and dolomitization of Silurian Reefs, St. Clair County, Michigan. *American Association Petroleum Geologists Bulletin*, 53: 957–981.

Jones, G.D., Whitaker, F.F., Smart, P.L., and Sanford, W.E., 2002a. Fate of reflux brines in carbonate platforms. *Geology* 30: 371–374.

Jones, G.D., Smart, P.L., Whitaker, F.F., Rostron, B.J., and Machel, H.G., 2002b. Numerical modeling of reflux dolomitization in the Grosmont Platform complex (Upper Devonian), Western Canada Sedimentary Basin. *American Association of Petroleum Geologists Bulletin*, submitted.

Krug, H.-J., Brandstädter, H., and Jacob, K.H., 1996. Morphological instabilities in pattern formation by precipitation and crystallization processes. *Geologische Rundschau*, 85: 19–28.

Land, L.S., 1973. Contemporaneous dolomitization of middle Pleistocene reefs by meteoric water, North Jamaica. *Bulletin Marine Science*, 23: 64–92.

Land, L.S., 1985. The origin of massive dolomite. *Journal of Geological Education*, 33: 112–125.

Lasemi, Z., Boardman, M.R., and Sandberg, P.A., 1989. Cement origin of supratidal dolomite, Andros Island, Bahamas. *Journal of Sedimentary Petrology*, 59: 249–257.

Leach, D.L., Plumlee, G.S., Hofstra, A.H., Landis, G.P., Rowan, E.L., and Viets, J.G., 1991. Origin of late dolomite cement by CO_2-saturated deep basin brines: evidence from the Ozark region, central United States. *Geology*, 19: 348–351.

Lumsden, D.N., 1988. Characteristics of deep-marine dolomites. *Journal of Sedimentary Petrology*, 58: 1023–1031.

Lyons, W.B., Hines, M.E., and Gaudette, H.E., 1984. Major and minor element pore water geochemistry of modern marine sabkhas: the influence of cyanobacterial mats. In Cohen, Y., Castenholz, R.W., and Halvorson, H.O. (eds.), *Microbial Mats*. New York: A. R. Liss, pp. 411–423.

Machel, H.G., 1987. Saddle dolomite as a by-product of chemical compaction and thermochemical sulfate reduction. *Geology*, 15: 936–940.

Machel, H.G., 1997. Recrystallization versus neomorphism, and the concept of 'significant recrystallization' in dolomite research. *Sedimentary Geology*, 113: 161–168.

Machel, H.G., and Lonne, J., 2002. Hydrothermal dolomite—a product of poor definition and imagination. *Sedimentary Geology*, 152: 163–171.

Machel, H.G., and Mountjoy, E.W., 1986. Chemistry and environments of dolomitization—a reappraisal. *Earth Science Reviews*, 23: 175–222.

Machel, H.G., and Mountjoy, E.W., 1990. Coastal mixing zone dolomite, forward modeling, and massive dolomitization of platform-margin carbonates—Discussion. *Journal of Sedimentary Petrology*, 60: 1008–1012.

Machel, H.G., Mountjoy, E.W., and Amthor, J.E., 1996. Mass balance and fluid flow constraints on regional-scale dolomitization, Late Devonian, Western Canada Sedimentary Basin. *Bulletin of Canadian Petroleum Geology*, 44: 566–571.

Machel, H.G., Cavell, P.A., Buschkuehle, B.E., and Michael, K., 2000. Tectonically induced fluid flow in Devonian carbonate aquifers of the Western Canada Sedimentary Basin. *Journal of Geochemical Exploration*, 69–70: 213–217.

Major, R.P., Bebout, D.G., and Lucia, F.J., 1988. Depositional facies and porosity distribution, Permian (Guadalupian) San Andres and Grayburg Formations, PJWDM Filed Complex, Central Basin Platform, West Texas. In Lomando, A.J., and Harris, P.M. (eds.), *Giant Oil and Gas Fields—A Core Workshop*. SEPM Core Workshop 12, pp. 615–648.

Mazzullo, S.J., 2000. Organogenic dolomitization in peritidal to deepsea sediments. *Journal of Sedimentary Research*, 70: 10–23.

Montañez, I.P., 1997. Secondary porosity and late diagenetic cements in the Upper Knox Group, central Tennessee region: a temporal and spatial history of fluid flow conduit development within the Knox regional aquifer. In Montañez, I.P., Gregg, J.M., and Shelton, K.L., (eds.), *Basin-Wide Diagenetic Patterns: Integrated Petrologic, Geochemical, and Hydrologic Considerations*. SEPM Special Publication, 57, pp. 101–117.

Morrow, D.W., 1982. Diagenesis II. Dolomite—part II: Dolomitization models and ancient dolostones. *Geoscience Canada*, 9: 95–107.

Morrow, D.W., 1999. Regional subsurface dolomitization: models and constraints. *Geoscience Canada*, 25: 57–70.

Mutti, M., and Simo, J.A., 1994. Distribution, petrography and geochemistry of early dolomite in cyclic shelf facies, Yates Formation (Guadalupian), Capitan Reef Complex, USA. In Purser, B., Tucker, M., and Zenger, D. (eds.), *Dolomites – A Volume in Honor of Dolomieu*. International Association of Sedimentologists, Special Publication, 21, pp. 91–107.

Patterson, R.J., and Kinsman, D.J.J., 1982. Formation of diagenetic dolomite in coastal sabkha along the Arabian (Persian) Gulf. *American Association Petroleum Geologists Bulletin*, 66: 28–43.

Purser, B., Tucker, M., and Zenger, D. (eds.), 1994. *Dolomites—A Volume in Honor of Dolomieu*. International Association of Sedimentologists, Special Publication, 21, 451pp.

Qing, H., and Mountjoy, E.W., 1992. Large-scale fluid flow in the Middle Devonian Presqu'ile barrier, Western Canada Sedimentary Basin. *Geology*, 20: 903–906.

Qing, H., and Mountjoy, E.W., 1994. Formation of coarsely crystalline, hydrothermal dolomite reservoirs in the Presqu'ile Barrier, Western Canada Sedimentary Basin. *American Association Petroleum Geologists Bulletin*, 78: 55–77.

Radke, B.M., and Mathis, R.L., 1980. On the formation and occurrence of saddle dolomite. *Journal of Sedimentary Petrology*, 50: 1149–1168.

Schmoker, J.W., and Halley, R.B., 1982. Carbonate porosity versus depth: a predictable relation for south Florida. *American Association Petroleum Geologists Bulletin*, 66: 2561–2570.

Sanford, W.E., and Konikov, L.F., 1989. Porosity development in coastal carbonate aquifers. *Geology*, 17: 249–252.

Sanford, W.E., Whitaker, F.A., Smart, P.L., and Jones, G., 1998. Numerical analysis of seawater circulation in carbonate platforms: I. Geothermal convection. *American Journal of Science*, 298: 801–828.

Shields, M.J., and Brady, P.V., 1995. Mass balance and fluid flow constraints on regional-scale dolomitization, Late Devonian, Western Canada Sedimentary Basin. *Bulletin of Canadian Petroleum Geology*, 43: 371–392.

Shukla, V., and Baker, P.A. (eds.), 1988. *Sedimentology and Geochemistry of Dolostones*. Society of Economic Paleontologists and Mineralogists, Special Publication, 43, 266 pp.

Sibley, D.F., and Gregg, J.M., 1987. Classification of dolomite rock textures. *Journal of Sedimentary Petrology*, 57: 967–975.

Spencer-Cervato, C., and Mullis, J., 1992. Chemical study of tectonically controlled hydrothermal dolomitization: an example from the Lessini Mountains, Italy. *Geologische Rundschau*, 81/2: 347–370.

Stanley, S.M., and Hardie, L.A., 1999. Hypercalcification: paleontology links plate tectonics and geochemistry to sedimentology. *GSA Today*, 9(2): 2–7.

Stearns, N.D., Stearns, H.T., and Waring, G.A., 1935. *Thermal Springs in the United States*. United States Geological Survey Water Supply Paper, 679-B, pp. 59–191.

Smart, P.L., Dawans, J.M., and Whitaker, F., 1988. Carbonate dissolution in a modern mixing zone. *Nature*, 335: 811–813.

Swart, P.K., and Melim, L.A., 2000. The origin of dolomites in Tertiary sediments from the margin of Great Bahama Bank. *Journal Sedimentary Research*, 70: 738–748.

Usdowski, E., 1994. Synthesis of dolomite—and geochemical implications. In Purser, B., Tucker, M., and Zenger, D. (eds.), *Dolomites - A Volume in Honor of Dolomieu*. International Association of Sedimentologists Special Publication 21, pp. 345–360.

Vahrenkamp, V.C., Swart, P.K., and Ruiz, J., 1991. Episodic dolomitization of Late Cenozoic carbonates in the Bahamas: evidence from strontium isotopes. *Journal of Sedimentary Petrology*, 61: 1002–1014.

Vahrenkamp, V.C., and Swart, P.K., 1994. Late Cenozoic dolomites of the Bahamas: metastable analogues for the genesis of ancient platform dolomites. In Purser, B., Tucker, M., and Zenger, D. (eds.), *Dolomites—A Volume in Honor of Dolomieu*. International Association of Sedimentologists, Special Publication, 21, pp. 133–153.

Vasconcelos, and McKenzie, J.A., 1997. Microbial mediation of modern dolomite precipitation and diagenesis under anoxic conditions (Lagoa Vermelha, Rio de Janeiro, Brazil). *Journal of Sedimentary Research*, **67**: 378–390.

Wanless, H.R., 1979. Limestone response to stress: pressure solution and dolomitization. *Journal of Sedimentary Petrology*, **49**: 437–462.

Ward, W.C., and Halley, R.B., 1985. Dolomitization in a mixing zone of near-seawater composition, Late Pleistocene, Northeastern Yucatan Peninsula. *Journal of Sedimentary Petrology*, **55**: 407–420.

Whitaker, F.F., Smart, P.L., Vahrenkamp, V.C., Nicholson, H., and Wogelius, R.A., 1994. Dolomitization by near-normal seawater? Field evidence from the Bahamas. In Purser, B., Tucker, M., and Zenger, D. (eds.), *Dolomites—a Volume in Honor of Dolomieu*. International Association of Sedimentologists, Special Publication, 21, pp. 111–132.

White, D.E., 1957. Thermal waters of volcanic origin. *Geological Society of America Bulletin*, **68**: 1637–1658.

Xun, Z., and Fairchild, I.J., 1987. Mixing zone dolomitization of Devonian carbonates, Guangxi, South China. In Marshall, J.D. (ed.), *Diagenesis in Sedimentary Sequences*. Geological Society of London, Special Publication, 36, pp. 157–170.

Zenger, D.H., Dunham, J.B., and Ethington, R.L. (eds.), 1980. *Concepts and Models of Dolomitization*. Society Economic Paleontologists Mineralogists, Special Publication, 28, 320 pp.

Cross-references

Carbonate Mineralogy and Geochemistry
Cements and Cementation
Dedolomitization
Dolomite Textures
Evaporites
Fluid Inclusions
Isotopic Methods in Sedimentology
Porewaters in Sediments
Sabkha, Salt Flat, Salina
Seawater: Temporal Changes in the Major Solutes
Stains for Carbonate Minerals

DUNES, EOLIAN

Introduction

Eolian dunes form part of a hierarchical system of bedforms developed in wind-transported sand which comprises: (i) wind ripples (spacing 0.1–1 m); (ii) individual simple dunes or superimposed dunes on compound and complex dunes (spacing 50–500 m); and (iii) compound and complex dunes (spacing >500 m). Dunes occur wherever there are sufficient supplies of sand-sized sediment, winds to transport that sediment, and conditions that promote deposition of the transported sediment. These requirements are satisfied in two main environments: (1) coastal areas with sandy beaches and onshore winds; and (2) desert areas. Most dunes occur in contiguous areas of eolian deposits called sand seas (>100 km^2) or dune fields. Sand seas and dune fields are dynamic sedimentary bodies that form part of well-defined regional- and local-scale sediment transport systems in which sand is moved by wind from source areas (e.g., distal fluvial deposits, sandy beaches) via transport pathways to depositional sinks (Kocurek and Lancaster, 1999).

Dune types

Eolian dunes occur in five main morphologic types (Table D1). Although some authors have argued that coastal dunes have a distinct morphology, many recent workers argue that the only characteristic coastal dune form is the foredune because it is an integral part of the complex of nearshore processes forming the beach-dune system (Bauer and Sherman, 1999). Three varieties of each dune type can occur: simple (the basic form); compound (superimposition of small dunes of the same type on larger dunes); and complex (superimposition of different dune types on the primary form (e.g., crescentic dunes on linear dunes).

Eolian dunes occur in a self-organized pattern (Figure D36) that develops as the response of sand surfaces to the wind regime (especially its directional variability) and the supply of sand (Werner and Kocurek, 1997). Relations between dune types and wind regimes indicate that the main control of dune type is the directional variability of the wind regime (Figure D37). Grain size, vegetation cover, and sediment supply play subordinate roles.

Migration of crescentic dunes occurs as a result of erosion of the stoss slope and deposition on the lee. Rates of dune migration are inversely proportional to dune height. Linear

Table D1 A morphological classification of eolian sand dunes (after Lancaster 1995).

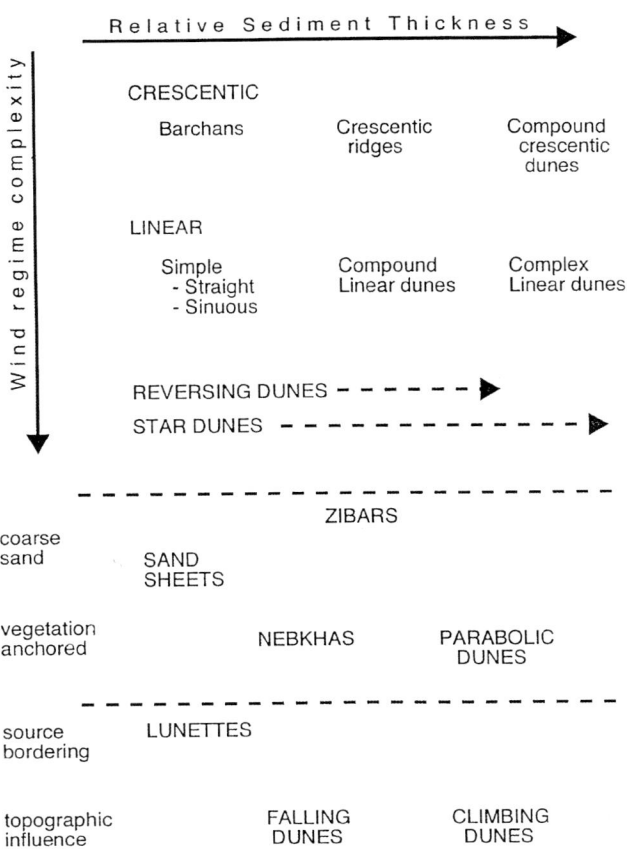

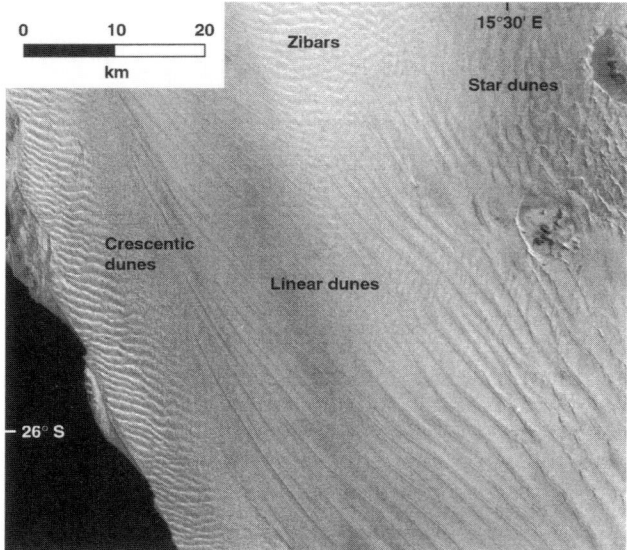

Figure D36 LANDSAT image of the southern part of the Namib Sand Sea, showing occurrence of different dune types as wind regime changes from coast to inland areas.

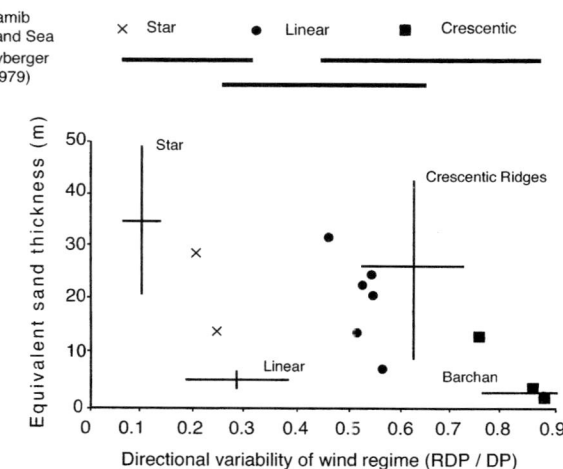

Figure D37 Relations between dune types and wind regimes. From Wasson and Hyde (1983) with Namib Sand Sea data superimposed (symbols) and range of wind regimes for three major dune types from Fryberger (1979). Equivalent sand thickness is a measure of the sand available for dune building and represents the thickness of sand if the dunes were levelled. The directional variability of the wind regime is characterized by the ratio between resultant (RDP) and total sand drift potential (DP).

dunes do not migrate laterally, but extend downwind. Star dunes are essentially stationary forms.

The simplest dune types and patterns form in areas characterized by a narrow range of wind directions (unidirectional wind regime). In the absence of vegetation, crescentic dunes will be the dominant form. Parabolic dunes are the equivalent in partially vegetated areas. Isolated crescentic dunes or barchans occur in areas of limited sand supply, and coalesce laterally to form crescentic or barchanoid ridges that consist of a series of connected crescents in plan view as sand supply increases. Larger forms with superimposed dunes are termed compound crescentic dunes.

Linear dunes are characterized by their length (often more than 20 km), sinuous crestline, parallelism, and regular spacing, and high ratio of dune to interdune areas. They form in areas of bimodal or wide unimodal wind regimes. The essential mechanism for linear dune formation is the deflection of winds that approach at an oblique angle to the crest to flow parallel to the lee side and transport sand along the dune. Thus any winds from a 180° sector centered on the dune will be diverted in this manner and cause the dune to elongate downwind (Tsoar, 1983).

Star dunes have a pyramidal shape, with three or four sinuous sharp-crested arms radiating from a central peak and multiple avalanche faces. Star dunes occur in multi-directional or complex wind regimes and are the largest dunes in many sand seas, reaching heights of more than 300 m. The development of star dunes is strongly influenced by the high degree of form—flow interaction which occurs as a result of seasonal changes in wind direction, and the existence of a major lee-side secondary circulation. Most of the erosion and deposition involves the reworking of deposits deposited in the previous wind season. Sand, once transported to the dune, tends to stay there and add to its bulk (Lancaster, 1989).

Parabolic dunes are characterized by a U or V shape with a "nose" of active sand and two partly vegetated arms that trail up wind. They are common in many coastal dunefields and semi arid inland areas and often develop from localized blowouts in vegetated sand surfaces (Wolfe and David, 1997).

Other important dune types include nebkhas or hummock dunes (common in may coastal dunefields) anchored by vegetation; lunettes (often comprised of sand-sized clay pellets) that form downwind of small playas; and a variety of topographically controlled dunes (climbing and falling dunes, echo dunes).

Dune processes and dynamics

The development and morphology of all eolian dunes are determined by a complex series of interactions between dune morphology, airflow, vegetation cover, and sediment transport rates (Lancaster, 1995). As dunes grow, they project into the atmospheric boundary layer so that streamlines are compressed and winds accelerate up the stoss or windward slope. The degree of flow acceleration (the speed-up factor) is determined by the aspect ratio of the dune. Wind speed at the crest of the dune is typically 1.1 times to 2.5 times that measured immediately upwind of the dune (Figure D38(A)). Flow acceleration coupled with effects of stream line curvature on the windward slopes of dunes give rise to an exponential increase in sediment transport rates (Figure D38(B)) toward the dune crest (McKenna Neuman et al., 1997), resulting in erosion of the stoss slope. In the lee of dunes, there is a complex pattern of flow separation, diversion, and re-attachment determined by the angle between the wind and the dune crest (angle of attack) and the dune aspect ratio (Sweet and Kocurek, 1990). On high aspect ratio (steep) dunes, high angles of attack result in flow separation, and lower angles of attack result in flow diversion. Low aspect ratio dunes are characterized by flow expansion. Flow separation (Figure D38(C)) results in the development of a back-flow

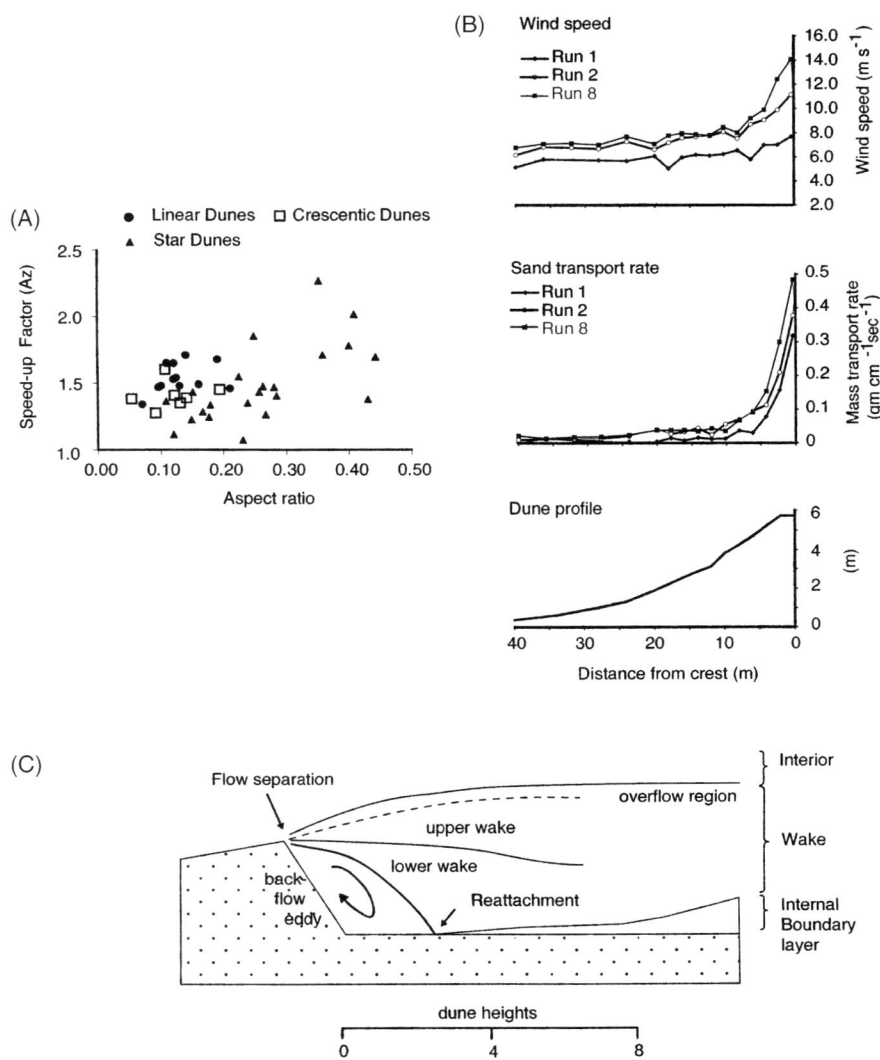

Figure D38 Elements of dune dynamics: (A): velocity speed-up (Lancaster, 1995); (B) winds and sediment transport rates on the stoss slope (McKenna Neuman et. al. 1997) (C) lee side flow separation and wake mixing (Frank and Kocurek, 1996).

eddy in the lee of the dune and a series of wakes that gradually diffuse and mix downwind (Frank and Kocurek, 1996). Flow separation also causes fallout of previously saltating sand grains and development of avalanche (slip) faces (Nickling et al., 2002). Secondary flows, including lee-side flow diversion, are especially important where winds approach the dune obliquely, and are an important process on linear and many star dunes.

Dune sediments

Dunes are comprised of moderately- to well-sorted sands (1000–63 μm), with a mean grain size in the range 160 μm to 300 μm. Most dunes are composed of quartz, but may include significant quantities of feldspar (e.g., Kelso Dunes, California). Dunes composed of volcaniclastic materials occur in some areas (e.g., Great Sand Dunes, Colorado). Carbonate-rich dune sands are found adjacent to many sub-tropical coastlines (e.g., Arabian Gulf) and gypsum dunes occur close to playas in Tunisia and at White Sands, New Mexico.

Three primary modes of deposition occur on dunes: (1) migration of wind ripples; (2) fallout from temporary suspension of grains in the flow separation zone (grainfall); and (3) avalanching of grains initially deposited by grainfall (grainflow). These create three basic types of eolian sedimentary structures: (1) wind ripple laminae; (2) grainfall laminae; and (3) grainflow cross-strata (Hunter, 1977). The primary sedimentary structures may be separated by a hierarchy of bounding surfaces representing episodes of reactivation, dune growth, and migration of dunes in conditions of bedform climb.

Crescentic dunes are dominated by grainfall and grainflow cross strata (Figure D39(A)) (Hunter, 1977; McKee, 1979).

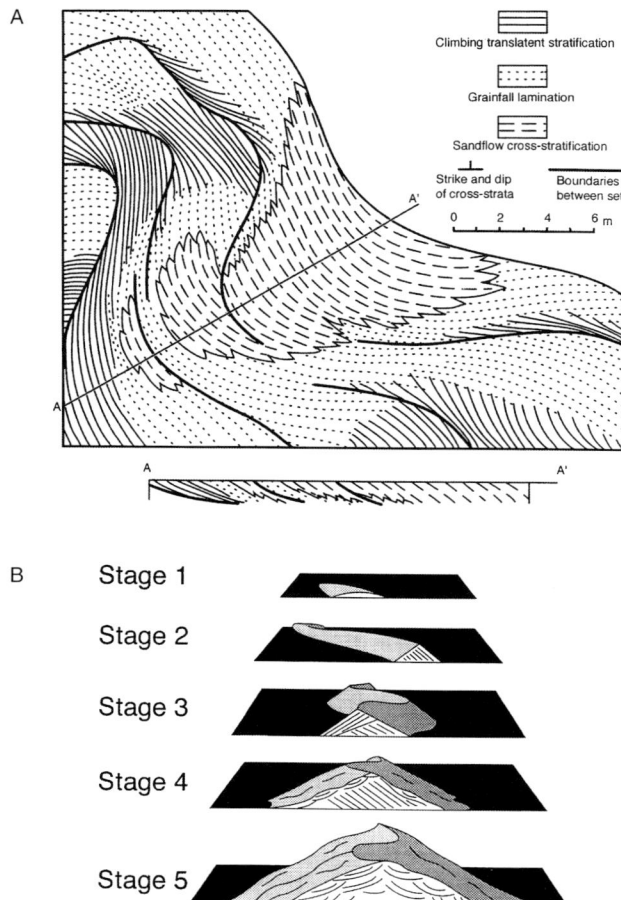

Figure D39 Sedimentary structures of dunes (A) crescentic dunes (Hunter, 1977); (B) linear dunes (from Bristow et al., 2000).

Linear dune structures recently imaged using ground penetrating radar (Figure D39(B)) show a change in the dominant type of sedimentary structures as the dune grows (Bristow et al., 2000). Small dunes are dominated by stacked sets of cross strata that dip in opposite directions and are formed by the migration of the elements of a sinuous crestline up the dune. The structures of larger dunes are dominated by trough cross-strata created by the migration of superimposed bedforms. Sedimentary structures of star dunes are not well-known, but likely complex.

Conclusions

Traditional field research, use of satellite image data, and some innovative modeling have provided a comprehensive picture of the basic elements of eolian dune forms and processes (Bauer and Sherman, 1999; Lancaster, 1995; Nickling and McKenna Neuman, 1999; Pye and Tsoar, 1990). Significant areas for future research include the dynamics of lee face avalanching and numerical modeling of dunes and dunefields.

Nicholas Lancaster

Bibliography

Bauer, B.O., and Sherman, D.J., 1999. Coastal dune dynamics: problems and prospects. In Goudie, A.S., Livingstone, I., and Stokes, S. (eds.), *Eolian Environments, Sediments, and Landforms*. New York: Wiley, Chichester, pp. 71–104.

Bristow, C.S., Bailey, S.D., and Lancaster, N., 2000. Sedimentary structure of linear sand dunes. *Nature*, **406**: 56–59.

Frank, A., and Kocurek, G., 1996. Toward a model for airflow on the lee side of eolian dunes. *Sedimentology*, **43**(3): 451–458.

Fryberger, S.G., 1979. Dune forms and wind regimes. In McKee, E.D. (ed.), *A Study of Global Sand Seas*. United States Geological Survey, Professional Paper. U.S.G.S. Professional Paper, pp. 137–140.

Hunter, R.E., 1977. Basic types of stratification in small eolian dunes. *Sedimentology*, **24**: 361–388.

Kocurek, G., and Lancaster, N., 1999. Eolian sediment states: theory and Mojave Desert Kelso dunefield example. *Sedimentology*, **46** (3): 505–516.

Lancaster, N., 1989. The dynamics of star dunes: an example from the Gran Desierto, Mexico. *Sedimentology*, **36**: 273–289.

Lancaster, N., 1995. *Geomorphology of Desert Dunes*. London: Routledge.

McKee, E.D., 1979. Sedimentary structures in dunes. In McKee, E.D. (ed.), *A Study of Global Sand Seas*. United States Geological Survey, pp. 83–136.

McKenna Neuman, C., Lancaster, N., and Nickling, W.G., 1997. Relations between dune morphology, air flow, and sediment flux on reversing dunes, Silver Peak, Nevada. *Sedimentology*, **44** (6): 1103–1114.

Nickling, W.G., and McKenna Neuman, C., 1999. Recent investigations of airflow and sediment transport over desert dunes. In Goudie, A.S., Livingstone, I., and Stokes, S. (eds.), *Eolian Environments, Sediments and Landforms*. Chichester: John Wiley and Sons.

Nickling, W.G., McKenna Neuman, C., and Lancaster, N., 2002. Grainfall processes in the lee of transverse dunes, Silver Peak, Nevada. *Sedimentology*, **49** (1): 191–209.

Pye, K., and Tsoar, H., 1990. *Eolian Sand and Sand Dunes*. London: Unwin Hyman.

Sweet, M.L., and Kocurek, G., 1990. An empirical model of eolian dune lee-face airflow. *Sedimentology*, **37** (6): 1023–1038.

Tsoar, H., 1983. Dynamic processes acting on a longitudinal (seif) dune. *Sedimentology*, **30**: 567–578.

Wasson, R.J., and Hyde, R., 1983. Factors determining desert dune type. *Nature*, **304**: 337–339.

Werner, B.T., and Kocurek, G., 1997. Bed-form dynamics: does the tail wag the dog? *Geology*, **25** (9): 771–774.

Wolfe, S.A., and David, P.P., 1997. Parabolic dunes: examples from the Great Sand Hills, southwestern Saskatchewan. *The Canadian Geographer*, **41** (2): 207–213.

Cross-references

Coastal Sedimentary Facies
Cross-Stratification
Desert Sedimentary Environments
Eolian Transport and Deposition
Sediment Transport by Waves
Surface Forms

E

EARTH FLOWS

Earth flows are mass movements of fine-grained soils that range from rapid earth flows formed in highly sensitive clay deposits to relatively slower and drier earth flows common in plastic, fine-grained soils (Sharpe, 1938; Varnes, 1978; Cruden and Varnes, 1996). Common and widespread, earth flows can cause extensive damage to property and infrastructure; rapid earth flows have taken the lives of scores of people. Earth flows are a primary agent of erosion in many areas and contribute large amounts of sediment to streams and rivers. Scientists have recognized earth flows as a distinct type of mass movement from at least the early twentieth century; since then, many investigators have studied individual earth flows, their geology and underlying physical processes (e.g., Keefer and Johnson, 1983; Lefebvre, 1996). Earth flows as defined in older classifications, such as Sharpe's (1938), cross the divisions of modern landslide classification schemes to include earth flows, earth slides, composite earth slide-earth flows, as well as liquefaction spreads, and certain debris slides (Varnes, 1978; Cruden and Varnes, 1996).

Rapid earth flows

Most rapid earth flows occur in areas of highly sensitive clay deposits (quick clays). The largest known deposits of sensitive clay occur in Scandinavia and eastern Canada, but deposits also exist in Alaska, Japan, the former Soviet Union, and New Zealand (Lefebvre, 1996). Most sensitive clay deposits are young marine clays deposited in bodies of salt water that existed during retreat of the Wisconsin Ice Sheet. As a result of isostatic rebound, the formerly submerged deposits have since been uplifted, subjected to fresh-water leaching, and eroded. The marine clays typically contain 30–70 percent silt and their clay-sized fractions ($<2\mu$) typically contain more rock flour than clay minerals. The clay minerals are predominantly illite. Most workers have attributed the sensitivity to loss of salt from the pore water, an effect known to reduce the liquid limits and remolded shear strength of clays (Lefebvre, 1996). Others have suggested that the high proportion of rock flour and resultant low plasticity of quick clays is a significant cause of their high sensitivity (Cabrerra and Smalley, 1973).

A rapid earth flow typically begins as a small landslide on a steep bank where a stream or river has eroded a valley into a sensitive clay deposit. Excess precipitation, elevated groundwater levels, earthquakes, pile driving and long-term erosion have triggered such earth flows (Sharpe, 1938; Lefebvre, 1996). Once the initial slide begins moving, it liquefies spontaneously and begins to flow. Consequently, the scarp is left unsupported and failure rapidly extends away from the bank as a series of rotational or translational slides. The overall movement is translational and retrogressive; as the basal quick clay liquefies and flows out of the scar, the relatively dry upper soil separates into blocks that may rotate or sink into the flowing quick clay. The entire failure process is usually completed within several minutes to several hours. Rates of movement as high as several meters per second have been observed (Cruden and Varnes, 1996). The retrogressive failure may result in either an arcuate scar that is elongate parallel to the bank or more commonly, a so-called "bottle-neck" landslide, in which the quick clay has flowed through a narrow notch in the bank from an irregular landslide area that may extend as much as 1 km from the original bank. The earth-flow deposit is hummocky, several meters to a few tens of meters lower than the original ground surface, and slopes gently toward the stream channel.

Slow earth flows

Slow earth flows are found in temperate and tropical climates across the globe, wherever clay or weathered, clay-bearing rocks occur in combination with moderate slopes and adequate moisture. Thus, earth flows are common in areas of clay-bearing sedimentary rocks and weathered volcanic rocks, as well as some areas of weathered metamorphic and plutonic rocks. Slow earth flows occur in fine-grained soils that consist dominantly of plastic silt or clay as well as rocky soils that are supported by a plastic silt-clay matrix. Atterberg limits of earth-flow material range widely between individual earth flows, but generally are consistent with moderate to high plasticity. The materials also have low to moderate shear strength and low sensitivity.

Earth flows commonly have a teardrop or bulbous shape in map view, a sinusoidal profile, are elongate in the direction of downslope movement and several times wider than thick (Keefer and Johnson, 1983). Discrete lateral boundaries, characterized by shear zones that resemble strike-slip faults, are typical of slow earth flows. Depending on the rate of movement and amount of displacement, the lateral boundaries may be manifest as en echelon cracks or as continuous shear zones. Locally, flank ridges form inboard of the lateral boundaries (Fleming and Johnson, 1989). Such ridges form as a result of internal deformation as the earth-flow mass translates downslope.

The majority of slow earth flows move primarily by sliding on discrete basal and lateral slip surfaces. Distributed internal deformation accounts for a small fraction of the movement and contributes to the appearance of flowing movement. Numerous measurements and observations have confirmed however, that sliding is the dominant mode of movement and that internal deformation is consistent with the observed mechanism of sliding (Keefer and Johnson, 1983). In unusual cases, earth flow deposits absorb enough water and become sufficiently remolded that they actually flow; however, rapid earth flows of plastic, low-sensitivity clays are uncommon. Earth flows typically move at slow to moderate rates ranging from 10^{-10} m/s to 10^{-3} m/s; in a few instances, earth flows have surged briefly to rates ranging from 10^{-2} m/s to about 10^{-1} m/s (Keefer and Johnson, 1983).

The translational movement of earth flows makes them prone to persistent, long-term instability, thus earth flows commonly reactivate during periods of above-average precipitation or other disturbance, such as earthquakes, grading or irrigation (Keefer and Johnson, 1983; Skempton et al., 1989). Long-term persistence of slip surfaces at residual strength contributes to earth-flow reactivation (Skempton, 1985). Mechanical and hydrologic isolation of earth flows by clay-rich layers at their boundaries also contribute to the persistent instability of earth flows (Baum and Reid, 2000).

Water is a significant factor in the movement and activity of slow earth flows. Increasing water content increases the unit weight, thereby increasing the driving stresses that tend to cause movement. Also, increasing pore-water pressure decreases the effective stress acting between soil particles, and thus the shear strength of the soil. Pore-pressure fluctuations, in combination with soil properties and shear-surface roughness, directly influence the rate of earth-flow movement (Keefer and Johnson, 1983; Iverson, 2000).

Summary

Earth flows are mass movements of fine-grained soils that result from composite modes of failure and movement. Most rapid earth flows form in highly sensitive clay deposits and progress by a rapid sequence of smaller failures that spontaneously liquefy and flow out of the source area. Slow earth flows form in deposits of plastic clay or silt and progress gradually by brief episodic movements or periods of sustained, relatively steady movement. Most slow earth flows move primarily by sliding on a distinct basal shear surface, accompanied by internal deformation of the earth-flow material.

Rex L. Baum

Bibliography

Baum, R.L., and Reid, M.E., 2000. Groundwater isolation by low-permeability clays in landslide shear zones. In Bromhead, E., Dixon, N., and Ibsen, M. (eds.), *Landslides in Research, Theory and Practice*. London: Thomas Telford, Proceedings of the 8th International Symposium on Landslides, pp. 139–144.
Cabrerra, J.G., and Smalley, I.J., 1973. Quickclays as products of glacial action—A new approach to their nature, geology, distribution, and geotechnical properties. *Engineering Geology*, **7**: 115–133.
Cruden, D.M., and Varnes, D.J., 1996. Landslide types and processes. In Turner, A.K., and Schuster, R.L. (eds.), *Landslides—Investigation and Mitigation*. Washington D.C.: National Academy Press, Transportation Research Board Special Report 247, pp. 36–75.
Fleming, R.W., and Johnson, A.M., 1989. Structures associated with strike-slip faults that bound landslide elements. *Engineering Geology*, **27**: 39–114.
Iverson, R.M., 2000. Landslide triggering by rain infiltration. *Water Resources Research*, **36**: 1897–1910.
Keefer, D.K., and Johnson, A.M., 1983. Earth flows—morphology, mobilization and movement. US Geological Survey Professional Paper 1264.
LeFebvre, Guy, 1996. Soft sensitive clays. In Turner, A.K., and Schuster, R.L. (eds.), *Landslides—investigation and Mitigation*. Washington D.C.: National Academy Press, Transportation Research Board Special Report 247, pp. 607–619.
Sharpe, C.F.S., 1938. *Landslides and Related Phenomena—A study of Mass-movements of Soil and Rock*. New York: Columbia University Press.
Skempton, A.W., Leadbeater, A.D., and Chandler, R.J., 1989. The Mam Tor landslide, north Derbyshire. *Philosophical Transactions of the Royal Society of London*, **A329**: 503–547.
Skempton, A.W., 1985. Residual strength of clays in landslides, folded strata and the laboratory. *Géotechnique*, **35**: 3–18.
Varnes, D.J., 1978. Slope movement types and processes. In Schuster, R.L., and Krizek, R.J. (eds.), *Landslides—Analysis and Control*. Washington, D.C.: National Academy of Sciences, Transportation Research Board Special Report 176, pp. 12–33.

Cross-references

Atterberg Limits
Avalanche and Rock Fall
Debris Flow
Illite Group Clay Minerals
Liquefaction and Fluidization
Mass Movement
Sediment Gravity Flows
Slope Sediments

ENCRINITES

Definition

Modern usage of encrinite was established by Bissell and Chilingar (1967: 156) as a crinoidal limestone with crinoidal ossicles in excess of 50 percent of the bulk of the rock. It could be either a crinoidal packstone or crinoidal grainstone (Dunham, 1962) or a crinoidal biomicrite, crinoidal biosparite, crinoidal biomicrudite, or crinoidal biosparudite (Folk, 1959).

In the first known reference to crinoids, Agricola (1546) used *Encrinos* to distinguish isolated pentagonal crinoid columnals. *Encrinos* or *Encrinus* was widely used, but interestingly, when this word was validly proposed in zoological nomenclature, it was used for a crinoid that has circular columnals, i.e., *Encrinus liliiformis* Lamarck, 1801 (Lane, 1978). *Encrinus* is a common Triassic crinoid in western Europe (Hagdorn, 1993). The word encrinite is a derivative of Agricola's *Encrinos,* and through the centuries it has been used to generally refer a crinoidal limestone.

Abundance and diversity of encrinites

Encrinites are common fossiliferous limestones from the Ordovician through the Jurassic. The most common occurrence is thin 1s to 10s of cm thick tabular to lenticular beds where crinoidal debris was concentrated either in a tempestite or a turbidite or in a lag due to winnowing of the matrix. Thicker encrinite deposits are common as flanking beds of reefs, mudmounds, and other carbonate buildups. The most extensive encrinite deposits are regional encrinites (Ausich, 1997). Regional encrinites are encrinite deposits that average more than 5 m to 10 m in stratigraphic thickness and have an areal extent in excess of 500 km^2. Although not exclusively, regional encrinites are most common on carbonate ramps during times when reefs did not form rimmed shelves. Examples include the Lower Silurian Brassfield Formation of Ohio; the Lower Mississippian Burlington Limestone of Missouri, Iowa, and Missouri; and the Triassic Muschelkalk of Germany. Regional encrinites are an "extinct" lithofacies.

William I. Ausich

Bibliography

Agricola, G., 1546. *De ortu det causis subterraneorum lib. V. De natura eorum quae effluunt ex terra lib. III. De natura fossilium lib. X. De veteribus et novis metallis lib. II. Bermannus, sive De re metallica dialogus. Interpretatio germanica vocum rei metallicae, addito indice foecundissimo.* Basel: Forben Press.
Ausich, W.I., 1997. Regional encrinites: a vanished lithofacies. In Brett, C.E., and Baird, G.C. (eds.), *Paleontological Events Stratigraphic, Ecological, and Evolutionary Implications.* New York: Columbia University Press, pp. 509–519.
Bissell, H.J., and Chilingar, G.V., 1967. Classification of sedimentary carbonate rocks. In Chilingar, G.V., Bissell, H.J., and Fairbridge, R.W. (eds.), *Carbonate Rocks.* Amsterdam: Elsevier, pp. 87–168.
Dunham, R.J., 1962. Classification of carbonate rocks according to depositional texture. In Ham, W.E. (ed.), *Classification of Carbonate Rocks,* American Association of Petroleum Geologists Memoir, Volume 1, pp. 108–121.
Folk, R.L., 1959. Practical petrographic classification of limestone. *American Association of Petroleum Geologists Bulletin,* **43**: 1–38.
Hagdorn, H., 1993. Encrinus liliiformis im Trochitenkalk Süddeutschlands. In Hagdorn, H., and Seilacher, A. (eds.), *Muschelkalk, Schöntaler Symposium 1991, Sonderbände der Gesellschaft für Naturkunde in Württemberg 2.* Stuttgart, Korb: Goldschneck, pp. 245–260.
Lamarck, J.B.P.A. de M.de, 1801. *Système des Animaux sans Vertèbres.* Paris: (published by the author).
Lane, N.G., 1978. Historical review of classification of Crinoidea. In Moore, R.C., and Teichert, C. (eds.), *Treatise on Invertebrate Paleontology, Part T, Echinodermata, 2,* Volume 2, Lawrence and Boulder, Geological Society of America and University of Kansas Press, pp. T348–T359.

Cross-references

Bioclasts
Neritic Carbonate Depositional Environments

EOLIAN TRANSPORT AND DEPOSITION

Eolian processes operate at a range of spatial and temporal scales, from the microscopic to the global and from microseconds to millennia. Sand and dust "storms", surface deflation and abrasion, as well as sand dune formation and migration are all part of the eolian process spectrum.

Landforms associated with eolian processes are conservatively estimated to cover 20 percent to 25 percent of the terrestrial land surface. Arid and semi-arid regions are primarily affected, but on a local scale, coastlines, proglacial and periglacial settings, and surfaces disturbed by cultivation, mining and construction are also recognized for eolian transport. Annual global dust emissions are in the order of 2×10^9 tonnes (D'Almeida, 1989), with the Sahara alone producing up to 35 percent of this amount. Windblown silt and clay deposits (loess) cover as much as 10 percent of Earth's terrestrial surface in depths between 1 m and 100 m. These deposits form some of Earth's richest farmlands. Deep ocean sediments also are dominated by deposition of wind borne dust particles. At present, eolian transport is estimated to be roughly comparable on a global scale to the amount of sediment carried by rivers (Livingstone and Warren, 1996). Ancient sand seas, the massive Mesozoic and Permian (Colorado Plateau) sandstones, and the extensive Pleistocene eolian deposits throughout North America and Europe bear witness to an even greater role during varied periods of Earth's history.

It now has been 60 years since Bagnold produced his seminal work *The Physics of Blown Sand and Desert Dunes* in 1941. Since its first printing this text has been the benchmark reference for studies on eolian processes, and it remains on the reading list of many undergraduate and graduate courses even today. In the last two decades, the publication of papers addressing particle transport by wind has escalated. Five International Conferences on Eolian Research (at Aarhus, 1985 and 1991; at Zzyzx, 1994; and at Oxford, 1998 at Lubbock, 2002) have been organized and no fewer than ten comprehensive textbooks have been prepared, addressing all aspects of eolian transport and deposition. They include Brookfield and Ahlbrandt (1983), Greeley and Iversen (1985), Nickling (1986), Pye (1987), Thomas (1989), Pye and Tsoar (1990), Cooke *et al.* (1993), Lancaster (1995), Livingstone and Warren (1996), and Goudie *et al.* (1999).

The following sections attempt to introduce some of the central concepts and research questions surrounding sediment erosion and deposition by wind.

Initiation of motion

Surface shearing stress

One of the central concepts in eolian process geomorphology concerns the minimum force required to move particles of a given size. In practice, this force is very difficult to measure directly, and so, the threshold wind speed (u_t) is commonly determined. Where detailed wind speed profile data do exist for the inner constant stress region of a deep, fully adjusted boundary layer, it is possible to estimate the surface shear stress (τ_o) at the threshold of motion from

$$\tau_o = \rho u_*^2 \quad \text{(Eq. 1)}$$

where u_* is the friction velocity as defined in the Karman-Prandtl log model:

$$\frac{u_z}{u_*} = \frac{1}{\kappa}\ln\left(\frac{z}{z_0}\right) \quad \text{(Eq. 2)}$$

The fluid density is represented as ρ, von Karman's constant as κ, the reference height as z, and the roughness length as z_o. Employment of this model has been the mainstay of shear stress estimation in a very great number of field and wind tunnel studies that depend upon robust and relatively inexpensive rotating cup and pitot anemometers for wind speed profiling. Extension of this model to strongly accelerated and decelerated flows (i.e., over dunes and surfaces of greatly varied roughness), especially those carrying sediment, has prompted much recent discussion concerning an acceptable range of application. Turbulent shear stresses, i.e., *Reynolds stresses* can also be measured via eddy correlation: this technique is witnessing increased usage in eolian process research, but constant temperature and sonic anemometers are expensive and not without their own particular set of limitations.

Grain entrainment models

The most widely employed relation describing the initiation of grain motion remains that of Bagnold (1941). In this simple model, the force moments associated with the particle weight and the drag of the airstream are balanced at the threshold of movement to give

$$u_{*_t} = A\sqrt{\left(\frac{\sigma-\rho}{\rho}\right)gd} \quad \text{(Eq. 3)}$$

where u_{*_t} is the friction velocity at threshold, σ is the grain density, g is gravitational acceleration, d is the mean grain diameter, and A is a dimensionless coefficient (see also *Grain Threshold*). Since the grain and fluid densities are approximately constant at any given time and place, it is clear that the wind speed required for grain entrainment varies principally as the square root of the mean grain diameter. The most easily entrained sand particle has a diameter of approximately 80 μm. Coarser grains require a higher friction velocity for entrainment because of their greater weight ($A \sim 0.1$). Grains finer than 80 μm lie either partially or fully immersed within the viscous sublayer, so that the value of the entrainment coefficient A is found to increase with decreasing grain diameter. Small cohesive forces, associated with weak van der Waal effects and thin films of adsorbed water, further add to the resistance to motion (Iversen *et al.*, 1976).

Bagnold's model of the *static* or *fluid* threshold is based upon a number of simplifying assumptions that are important to identify. The force moments are derived for idealized spherical particles of uniform diameter. Lift and cohesive forces are omitted, as are turbulent fluctuations of shear stress on the bed.

A number of modellers have attempted to improve on Bagnold's entrainment function through developing a more explicit, physically-based specification for the entrainment coefficient A (e.g., Iversen *et al.*, 1976). Wind tunnel simulation has played an important role in the development, validation and calibration of these model revisions. Even so, the direct measurement of threshold carries its own error of estimation. Specifically, there is no agreement as to the precise definition or identification of "threshold" for a natural sand containing a wide range of particle sizes. The uncertainty associated with the measurement of u_* is in the order of 10 percent in wind tunnel testing (Gillette and Stockton, 1989).

Once fluid entrainment has commenced, impact maintains the system through the grain-borne transfer of kinetic energy from the air to the bed. Consequently, it is estimated that the wind speed can drop to about 80 percent of the *static* threshold before saltation bombardment is no longer sufficient to initiate motion. This point is termed the *dynamic* or *impact* threshold. The partitioning of momentum among rebounding and ejected grains following collision is represented in numeric models of eolian saltation as the "splash function" (Anderson and Hallet, 1986; Anderson and Haff, 1988, 1991; Anderson and Willetts, 1991). It is measured directly in wind tunnel simulation using particle tracking velocimetry or PTV (Willetts and Rice, 1985, 1988, 1989; Rice *et al.*, 1995).

Eolian transport

Transport modes

In order of increasing velocity, wind transported grains travel in four principal modes: *creep, reptation, saltation* and *suspension*. The central process is *saltation* from the Latin *saltare*, to leap. Following their ejection into the air stream, saltating grains travel along smooth trajectories unaffected by turbulence. They gain momentum from the air stream during their ascent in association with the greater streamwise wind speeds above the surface, and eventually descend to collide with the surface several centimeters (on average) downwind of the point from which they emerged. The remaining transport modes all depend to varying degrees upon the momentum transfer between these saltators and other loose grains residing within the bed. *Creep* refers to grains that are rolled along the surface. These grains are generally coarser than the saltators. The creep:saltation ratio was found (by Willetts and Rice, 1985) to vary between 1:1 and 1:3. When a saltator strikes a surface, it also dislodges about 10 other grains which move at low velocity in short, low hops. Unlike saltating grains, these *reptating grains* have insufficient energy to eject other surface grains upon impact. Reptating grains constitute the majority of all grains in transport. Similar to saltators, their numbers are strongly dependent upon the speed of incoming grains and therefore, the friction velocity. *Suspended grains* follow the turbulent motion within the airstream. Over a relatively large range of friction velocity, the transition from saltation to suspension coincides roughly with a drop in grain diameter below 100 μm (Nalpanis, 1985). Pye (1987) also reports that sand sized grains rarely go into suspension. Since the entrainment of silt and clay sized particles requires relatively high wind speeds, the entrainment of these grains usually is dependent upon the impact of saltating grains. An intermediate transport mode, modified saltation, is sometimes identified wherein irregularities in the trajectories of saltating grains are introduced by turbulent eddies. The vertical velocity in turbulence is strongly affected by bed and saltation (Owen, 1964) roughness and by topography. As a result, some sand grains do become suspended in the lee of transverse dunes, and dust particles may be found within the saltation layer where turbulence appears to be dampened.

Mass transport rate measurement

The transport rate (q) is defined by the mass of sediment moving through a unit width of surface per unit time (SI units: $kg\,m^{-1}\,s^{-1}$, or alternatively as the volume rate of transport: $m^3\,m^{-1}\,s^{-1}$).

Accurate measurement of q in field settings is especially difficult because the sediment motion is: (1) omni-directional (unlike stream and wind tunnel flows); and (2) deflected by the presence of any given measurement device in the airstream. Mechanical trap designs fall broadly into 3 groups: horizontal, vertical passive, and vertical active (i.e., airflow is maintained by a pumping system). Sand traps designed to rotate into the wind require constant maintenance to keep them moving freely. The response time in gusty winds of highly variable direction is largely unknown, and is likely poor at best. No mechanical trap design has been successful as yet in monitoring omni-directional transport on sloping surfaces. Though flow acceleration around fixed traps does lead to scour at the base and under-collection, many workers have been successful in stabilizing the surface in the immediate area of the trap inlet with cementing agents and flat plates. Flow deflection within the saltation curtain is never completely avoidable, but samplers which are close to iso-kinetic can reduce under-collection considerably (Nickling and McKenna Neuman, 1997). Surface disturbance during trap installation and maintenance, as well as selectivity in particle size collection, further affect collection efficiency.

The last decade also has been witness to the development of several electronic devices for measurement of the mass transport rate and kinetic energy flux. The Sensit™ outputs small electronic pulses at a rate dependent upon the kinetic energy of saltators striking a piezoelectric crystal while the Saltiphone™ records the sound of saltating grains striking the device. Butterfield (1999) reports on a device that measures the extinction of light across a narrow width of surface as it is scattered by mobile grains. All of these electronic instruments are considerably more expensive than any given mechanical trap, and they require regular calibration.

Mass transport rate models

Given the many challenges and uncertainties associated with direct measurement of the mass transport rate, especially over an extended period of time, a very large effort has been expended in the development of predictive models. These models emanate from theoretical investigation, wind tunnel simulation and model calibration, and empirical field studies. The general form of the relation between Q and u_* is given as follows:

$$q \propto C \frac{\rho}{g} u_*^a (u_* - u_{*_t})^b \qquad (Eq.\ 4)$$

where the exponents a and b sum to 3 (Livingstone and Warren, 1996). A detailed listing of these models is contained in table 3.5 in Greeley and Iversen (1985). A further review of their derivation and performance is provided in Sarre (1988).

As illustrated in Figure E1, very substantive differences exist in the predictions from these models, particularly at high

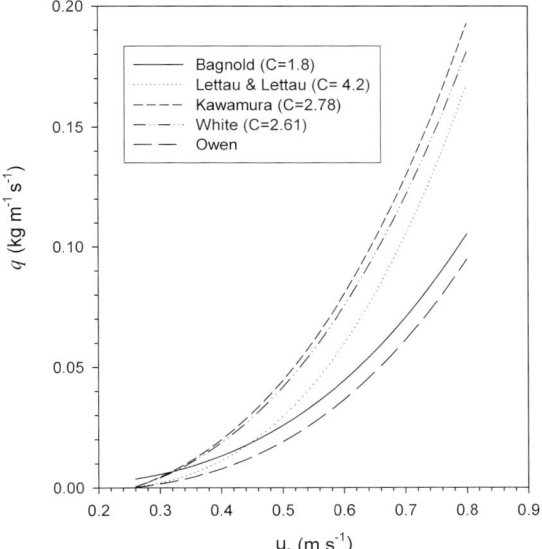

Figure E1 Comparison of mass transport rate model performance: Bagnold (1941), Kawamura (1951), Lettau and Lettau (1978), Owen (1964), White (1979).

friction velocities. Clearly no universal relationship between sediment transport and wind velocity has been established as yet. Even if it were to exist, its application would be limited to the simplest of field settings as represented by an extensive, horizontal surface of loose, well sorted sand. Herein, no restrictions on the sediment supply to the wind are assumed to exist, the boundary layer is deep and fully adjusted to the surface roughness, the wind is steady, the lower atmosphere is stable, and the velocity profile is semi-logarithmic.

Supply limited transport

Unfortunately for those wishing to fully model eolian transport, the idealized conditions described above rarely occur in nature. The presence of moisture, nonerodible roughness elements, crusting agents such as salts and microphytic organisms, organic matter, and macrophytic vegetation (i.e., grasses and shrubs) all play a role in limiting the supply of particles to the airstream. A good deal of current research on eolian processes is aimed at understanding, defining, and quantifying this role. Basically, all of these supply-limiting factors operate through one of more of the following effects: (1) introduction of interparticle force; (2) redistribution of momentum partitioning at the surface; and (3) alteration of the turbulence structure and intensity.

Moisture and crusting are most obviously associated with the introduction of interparticle force which opposes drag and lift to raise the (fluid) threshold friction velocity, often to levels exceeding those representative of most natural winds. However, moisture effects are characterized by strong temporal variability as the evaporation rate also depends on turbulent transfer, and therefore, wind speed. Little is known about this relationship at present. Particle impact also complicates modeling efforts considerably. It is well demonstrated that loose grains not only can saltate across damp and crusted surfaces, but also, the bombardment will eject some exposed grains. This ejection contributes to further impacts, selective particle removal, and increased surface roughness and turbulence intensity (McKenna Neuman and Maljaars Scott, 1998; McKenna Neuman and Maxwell, 1999). Through this strong positive feedback, the ultimate destruction of the surface can take place relatively quickly, especially in the context of crusts.

Vegetation and other nonerodible roughness elements (NERE) in the form of gravel sized particles alter the partitioning of momentum to the sand bed, and the turbulence structure of the airstream. Grasses, shrubs, and protruding gravels shelter the surrounding sand surface from the drag of the wind, as described in the Raupach (1992) model:

$$\frac{\tau_s}{\tau} = \frac{1}{(1 - m\sigma\lambda)(1 + m\beta\lambda)} \quad \text{(Eq. 5)}$$

where τ_s/τ represents the proportion of the total fluid stress which acts on the intervening fine grains, λ is the frontal area of the NERE per unit ground area (roughness density), β is the ratio of the surface drag coefficient to that of the NERE, σ is the ratio of the basal area of the NERE to its frontal area and m is an empirical coefficient. Grains in transport along intervening sand corridors can become trapped in the vicinity of vegetative elements, forming low sand shadow deposits and on a larger scale, coppice dunes. In the case of a developing gravel lag, the high momentum retention associated with grains ricocheting from the fixed elements can give rise to a temporary increase in the mass transport rate over a limited range of roughness density (McKenna Neuman, 1998).

Eolian deposition

Modes of deposition

There are three basic processes, and similarly three primary sedimentary structures, associated with the accumulation of wind blown sand (Hunter, 1977): (1) tractional or accretional deposition; (2) grainfall deposition; and (3) grainflow deposition or avalanching.

In tractional deposition (also termed accretion), a decline in the transport capacity of the wind causes grains moving by saltation, reptation and creep to come to rest in sheltered positions, typically on the lee slopes of sand ripples. This accumulation of excess sediment takes the form of wind ripple laminae. Pin-stripe bedding (Fryberger and Schenk, 1988) is also described for this mode of deposition with the concentration of relatively coarse grains on the upper part of ripple lee slopes forming the "stripes". Accretion deposits are tightly packed and firm to walk across. They are very widespread, occurring on dune stoss slopes, dune plinths or aprons, interdune areas and most sand sheets.

Grainfall deposition refers to particles that settle out of the air in the lee of dune crests as a result of a very rapid decrease in the transport capacity of the wind within the zone of flow separation. When saltating grains overshoot the dune brink, they can become temporarily suspended in strong updrafts within the wake flow so that they travel further than predicted by purely ballistic models (McDonald and Anderson, 1995). Measurement of the grainfall flux requires collection of the descending grains in small open cups spaced at regular intervals along a supporting horizontal boom. This flux (SI units: $kg\,m^{-2}\,s^{-1}$) typically follows an exponential decline with horizontal distance, the value of the decay coefficient scaling with the dune height (Nickling et al., 2002). Grainfall deposits are characterized by intermediate packing. Grainfall lamination is preserved only if no subsequent avalanching takes place, and it is most commonly observed on small dunes (Hunter, 1977).

Grainflow or avalanching occurs along the dune slip face (upper lee slope) as it is built to an angle greater than the angle of repose of the sediment. Very little is known about the frequency and point of origin of these localized mass failures. The grain flow cross-strata which result appear as a series of overlapping tongues of sediment with coarse grains concentrated on their upper surfaces and near the toe of the deposit. Most grainflow strata narrow toward to crest of the dune. In damp sand, slump structures and blocks may be preserved.

The sediment continuity equation

The kinematics of grain transport require that the mass or volume of sediment is preserved (Middleton and Southard, 1984) so that patterns of erosion and deposition can be modeled over a wide range in scale using the following continuity equation:

$$\frac{dz}{dt} + \frac{1}{\gamma}\left(\frac{dq}{dx}\right) = 0 \quad \text{(Eq. 6)}$$

where z is the surface level, x is the horizontal distance, γ is the bulk density of the sediment, and t is the time. A local decrease in the mass transport rate with distance, for example, is interpreted to mean that sediment storage or accumulation is occurring with the influx of sediment to the area exceeding the outflux. The reverse also is true. Spatial changes in the mass transport rate are fundamental to understanding dune development and maintenance.

Eolian morphodynamic processes

There is a huge literature on eolian bedforms that includes reviews by Pye and Tsoar (1990) and Lancaster (1995; and see Dunes, Eolian).

Dune morphodynamics encompass the following processes: (1) initiation; (2) replication; (3) stoss side adjustments; and (4) grainfall and grainflow in the dune lee (addressed above).

Theories surrounding the origin of dunes focus on the inception of sediment accumulation around small obstacles such as plants, though more subtle features such as slight surface indentations or changes in roughness also may serve as nodes of accumulation (Kocurek et al., 1992). Some field workers report calving of small dunes from larger ones, while others implicate the convergence of sand laden winds in association with secondary flow patterns. Very few sand patches formed by any of these means actually survive and grow into free moving dunes. The controls on protodune survival are not well understood. Very few direct observations have been made of dune initiation.

Eolian bedforms are dynamically similar to those formed in subaqueous shearing flows, and so, theories designed to explain their regular form and spacing bear many of the same features as those proposed for water (see Surface Forms). All require an initial disturbance on the bed, which may include development of the first protodune. This disturbance is believed by some to fix in place underlying structured patterns in the turbulent flow, which ultimately control dune spacing. Evidence for this theory, though sparse, can be found in dune patterns that appear to be related to wave-like atmospheric motion in the lee of hills and mountain ranges. Other workers suggest that it is the initial, fully formed dune that produces the regular flow disturbance downstream. In this explanation, the sequential dune is thought to form beyond the point of flow reattachment associated with the upwind dune, as a new internal boundary layer develops. The reattachment point itself, which lies beyond the flow separation bubble or lee eddy, is noted for intense turbulence that prevents sand from accumulating (Frank and Kocurek, 1996).

Growth of a dune upward into the airflow causes a pressure drop in the outer flow over the dune crest, and therefore, flow acceleration along the direction of increasing elevation. High Reynolds stresses near the dune toe, as measured by Wiggs et al. (1996) in wind tunnel simulation, are believed to prevent sand accumulation within this zone of flow stagnation. The generalized pattern of increasing velocity and shear stress toward the dune crest should lead to an overall flattening out of the bedform through time. Clearly though, dunes do grow and so, this explanation must represent an oversimplification of stoss side transport processes. Detailed field measurements of stoss side transport and deposition on a reversing dune by McKenna Neuman et al. (2000) do suggest that in unsteady winds: (1) the proportion of the stoss slope length which is active is highly variable; (2) the local sand transport is rarely in full equilibrium with the air flow; and (3) spatial variations in stoss slope transport are influenced by outer wake flow shed from the upwind dune. The shape of the dune crest is also implicated in a number of studies. Rounded forms with a clear crest-brink separation are usually associated with shear stress reduction and deposition at the upper elevations, while erosion and dune lowering is observed to be more typical of sharp crested features. Wiggs et al. (1996) suggest that streamline curvature in the area of the dune crest is responsible for these patterns.

Summary

The last two decades have witnessed major advances in the identification and understanding of eolian transport mechanics. In parallel, technological developments have allowed eolian geomorphologists to explore in detail small scale, short term processes involved sediment entrainment and transport. Wind tunnel and numerical simulation has played a key role in these efforts. Grain scale processes are now relatively well understood and modeled. Still, great challenges lie ahead for the advancement of eolian process geomorphology, particularly in the content of fieldwork. Surprisingly few studies have attempted direct measurement of eolian dune morphodynamics. A high level of uncertainty still surrounds measurement of the mass transport rate and the surface shearing stress. Foremost, the bulk of model construction and experimental work has been predicated on the assumption that the airflow is uniform and steady, and that the sand transport is fully adjusted to this condition. In nature, this is perhaps rarely so. Future efforts also must work toward improved integration of the temporal and spatial scales over which eolian processes operate.

Cheryl McKenna Neuman

Bibliography

Anderson, R.S., and Haff, P.K., 1988. Simulation of eolian saltation. *Science*, **241**: 820–823.

Anderson, R.S., and Haff, P.K., 1991. Wind modification and bed response during saltation of sand in air. *Acta Mechanica Supplementum*, **1**: 21–52.

Anderson, R.S., and Hallet, B., 1986. Sediment transport by wind: toward a general model. *Geological Society of America Bulletin*, **97**: 523–535.

Anderson, R.S., and Willetts, B.B., 1991. A review of recent progress in our understanding of aeolian sediment transport. *Acta Mechanica Supplementum*, **1**: 1–19.

Bagnold, R.A., 1941. *The Physics of Blown Sand and Desert Dunes*. London: Methuen.

Brookfield, M.E., and Ahlbrandt, T.S., 1983. *Eolian Sediments and Processes*. Amsterdam: Elsevier.

Butterfield, G.R., 1999. Near-bed mass flux profiles in aeolian sand transport: high-resolution measurements in a wind tunnel. *Earth Surface Processes and Landforms*, **24**: 393–412.

Cooke, R.U., Goudie, A.S., and Warren, A., 1993. *Desert Geomorphology*. London: UCL Press.

D'Almeida, G.A., 1989. Desert aerosol: characteristics and effects on climate. In Leinen, M., and Sarnthein, M. (eds.), *Paleoclimatology and Palaeometeorology: Modern and Past Patterns of Global Atmospheric Transport*. Dordrecht: Kluwer Academic.

Frank, A., and Kocurek, G., 1996. Toward a model of airflow on the lee side of aeolian dunes. *Sedimentology*, **43**: 451–458.

Fryberger, S.G., and Schenk, C.J., 1988. Pin stripe lamination: a distinctive feature of modern and ancient eolian sediments. *Sedimentary Geology*, **55**: 1–15.

Gillette, D.A., and Stockton, P.H., 1989. The effect of nonerodible particles on wind erosion of erodible surfaces. *Journal of Geophysical Research*, **94** (12): 885–912.

Goudie, A.S., Livingstone, I., and Stokes, S., 1999. *Aeolian Environments, Sediments and Landforms*. Chichester: Wiley.

Greeley, R., and Iversen, J.D., 1985. *Wind as a Geological Process on Earth, Mars, Venus and Titan*. Cambridge University Press.

Hunter, R.E., 1977. Basic types of stratification in small eolian dunes. *Sedimentology*, **24**: 361–388.

Iversen, J.D., Pollack, J.B., Greeley, R., and White, B.R., 1976. Saltation threshold on Mars: the effect of interparticle force, surface roughness, and low atmospheric density. *Icarus*, **29**: 381–393.

Kawamura, R., 1951. Study on sand movement by wind. In *Institute of Science and Technology, Report 5(3–4)*. University of Tokyo.

Kocurek, G., Townsley, M., Yeh, E., Havholm, K., and Sweet, M.L., 1992. Dune and dunefield development on Padre Island, Texas, with implications for interdune deposition and water-table-controlled accumualtion. *Journal of Sedimentary Petrology*, **62** (4): 622–635.

Lancaster, N., 1995. *Geomorphology of Desert Dunes*. London: Routledge.

Lettau, K., and Lettau, H.H., 1978. Experimental and micrometeorological field studies on dune migration. In Lettau, K., and Lettau, H.H. (eds.), *Exploring the World's Driest Climate*. University of Wisconsin-Madison, Institute for Environmental Studies, pp. 110–147.

Livingstone, I., and Warren, A., 1996. *Aeolian Geomorphology: An Introduction*. Harlow: Addison Wesley Longman.

McDonald, R.R., and Anderson, R.S., 1995. Experimental verification of aeolian saltation and lee side deposition models. *Sedimentology*, **42** (1): 39–56.

McKenna Neuman, C., 1998. Sediment flux and boundary layer adjustments over rough surfaces with an unrestricted, upwind sediment supply. *Geomorphology*, **25**: 1–17.

McKenna Neuman, C., and Maljaars Scott, M., 1998. A wind tunnel study of the influence of pore water on aeolian sediment transport. *Journal of Arid Environments*, **39** (3): 403–420.

McKenna Neuman, C., and Maxwell, C., 1999. A wind tunnel study of the resilience of three fungal crusts to particle abrasion during aeolian sediment transport. *Catena*, **38** (2): 151–173.

McKenna Neuman, C., Lancaster, N., and Nickling, W.G., 2000. The effect of unsteady winds on sediment transport on the stoss slope of a transverse dune, Silver Peak, NV, USA. *Sedimentology*, 47: 211–226.

Middleton, G.V., and Southard, J.B., 1984. *Mechanics of Sediment Movement*. Tulsa: Oklahoma, SEPM.

Nalpanis, P., 1985. Saltating and suspended particles over flat and sloping surfaces, II. Experiments and numerical simulations. In Barndorff-Nielsen, O.E., Møller, J.T., Rasmussen, K.R., and Willetts, B.B. (eds.), *Proceedings of International Workshop on the Physics of Blown Sand*. Aarhus: University of Aarhus, pp. 37–66.

Nickling, W.G., 1986. *Aeolian Geomorphology*. Binghamton Symposia in Geomorphology, London: Allen and Unwin.

Nickling, W.G., and McKenna Neuman, C., 1997. Wind tunnel evaluation of a wedge-shaped aeolian sediment trap. *Geomorphology*, 18: 333–345.

Nickling, W.G., McKenna Neuman, C., and Lancaster, N., 2002. Grainfall processes in the lee of transverse dunes, Silver Peak, Nevada. *Sedimentology*, 49 (1): 191–209.

Owen, P.R., 1964. Saltation of uniform grains in air. *Journal of Fluid Mechanics*, 20 (2): 225–242.

Pye, K., 1987. *Aeolian Dust and Dust Deposits*. London: Academic Press.

Pye, K., and Tsoar, H., 1990. Aeolian sand and sand dunes. London: Unwin Hyman.

Raupach, M.R., 1992. Drag and drag partition on rough surfaces. *Boundary Layer Meteorology*, 60: 375–395.

Rice, M.A., Willetts, B.B., and McEwan, I.K., 1995. An experimental study of multiple grain size ejecta produced by collisions of saltating grains with a flat bed. *Sedimentology*, 42 (4): 695.

Sarre, R.D., 1988. Evaluation of aeolian sand transport equations using intertidal zone measurements, Saunton Sands, England. *Sedimentology*, 35: 671–679.

Thomas, D.S.G., 1989. *Arid Zone Geomorphology*. London: Bellhaven/Halsted Press.

White, B.R., 1979. Soil transport by winds on Mars. *Journal of Geophysical Research*, 84: 4643–4651.

Wiggs, G.F.S., Livingstone, I., and Warren, A., 1996. The role of streamline curvature in sand dune dynamics: evidence from field and wind tunnel measurements. *Geomorphology*, 17: 29–46.

Willetts, B.B., and Rice, M., 1985. Wind tunnel tracer experiments using dyed sand. In Barndorff-Nielsen, O.E., Møller, J.T., Rasmussen, K.R., and Willetts, B.B. (eds.), *Proceedings of International Workshop on the Physics of Blown Sand*. Aarhus: University of Aarhus, pp. 271–300.

Willetts, B.B., and Rice, M.A., 1988. Particle dislodgement from a flat sand surface by wind. *Earth Surface Processes and Landforms*, 13: 717–728.

Willetts, B.B., and Rice, M.A., 1989. Collisions of quartz grains with a sand bed: the influence of incident angle. *Earth Surface Processes and Landforms*, 14: 719–730.

Cross-references

Cross-Stratification
Dunes, Eolian
Grain Flow
Grain Size and Shape
Grain Threshold
Bedding and Internal Structures
Ripple, Ripple Mark, and Ripple Structure
Sediment Fluxes and Rates of Sedimentation
Sedimentologists
Surface Forms

EROSION AND SEDIMENT YIELD

Erosion is the suite of processes that lower topography through the breakdown and transport of the substrate that forms the land surface. The substrate may either be bedrock that must first be degraded chemically or physically to produce mobile solutes and loose particles that can be transported by water or wind, or the substrate may already be loose material, such as sand, loess, or volcanic ash. From a research perspective, the characterization of erosion depends on both spatial or temporal scales. Processes that are important on short timescales and small spatial scales average out as scales increase. Moreover, upon progressing to ever larger spatial scales depositional processes become more important, such that at the largest scale, the whole Earth over geologic time, uplift, erosion, and deposition appear to roughly balance.

Hillslope scale considerations

At the scale of a hillslope or a small plot of land, erosion appears relatively simple, a combination of chemical reactions, referred to as chemical weathering, and physical processes, referred to as physical weathering. The latter include phenomena such as freezing and thawing, salt crystallization, raindrop impaction, and wind abrasion, that degrade the substrate, loosening it and making it susceptible to transport processes. Dissolved weathering products are typically removed from the site of weathering, entering rivers and transported down to the ocean, unless these are incorporated into plants or reprecipitated within the soil profile or downslope because of drying or changes in oxidation state. The suite of dominant weathering processes changes with climate. Chemical weathering tends to dominate in warmer and wetter settings, while physical erosion tends to dominate in colder and drier settings (Garrels and Mackenzie, 1971; Stallard, 1995a,b).

The susceptibility of bedrock to erosion reflects both the chemical and bulk physical properties of the bedrock. Accordingly, susceptibility to erosion is a bulk property of bedrock. Rocks of identical geologic classification may erode at different rates. For example, very old granodiorites of Precambrian age appear to weather at about half the rate of much younger granodiorites. This may be due to the loss of or translocation volatiles—water, halides, and sulfides—over billions of years, making the bedrock more resistant to weathering. Likewise, densely jointed or sheared granodiorites are more susceptible than massive granodiorites. Certain classes of rocks, such as massive evaporites, carbonates, and quartzites are eroded almost exclusively by chemical weathering. The interaction between bedrock, tectonic regime, and erosion produces characteristic landscapes (Stallard, 1988, 2000).

The mobilization and transport of solid erosion products off of a hillslope is decidedly more complex than for solutes. The energy available for transport processes depends on the steepness and length of the slope. Loose material tends to migrate down the slope gradient, and water flows down gradient as well. The balance between the supply of loose material and the ability of physical transport processes to move that material defines two types of erosion regimes (Carson and Kirkby, 1972; Stallard, 1985). One is a weathering- or supply-limited regime, where the capacity of transport processes to move loose material, averaged over time, exceeds the rate at which loose material is supplied through weathering. Weathering limited settings typically have steep slopes. The other regime is transport limited, where the weathering processes generates more loose material than can be moved by transport processes. The thickening layer of loose material develops into a soil. Transport limited settings typically have flat to gentle topography. If soils are deep and permeable such that even intense rains generate little surface runoff, even steeper slopes may be transport limited. If erosion is weathering limited and if the rate limiting step is chemical weathering, then the rates and chemistry of chemical and physical erosion products can be linked in models (Stallard 1995a,b).

Several rules of thumb can be applied to sediments produced during this more classical erosion. Where physical erosion and transport predominate, sediments are rich in primary minerals with only a small contribution from weathering products. The greater the role of chemical weathering the more depleted the sediments are in alkalis (sodium, potassium, etc.) and alkali-earth (magnesium, calcium, etc.). The most intense chemical weathering leaves quartz, kaolinite, and aluminum and iron sesquioxides. Chemical weathering is greatest under a warm, wet climate and transport-limited erosion.

Episodic erosion and the role of organisms

Many aspects of the breakdown of parent material, its processing within soil and its protection from immediate erosion are influenced by a huge suite of organisms (Stallard, 2000). The supply of loose material to physical transport is often strongly mediated by biological processes. The top of the soil profile is often reworked and churned by an ensemble of biological processes. Included are the actions of soil fauna, the growth and decay of roots, the toppling of trees, and in agricultural lands, plowing. Depths of this zone range from a few centimeters to a few meters. Moreover, animal burrowing and the formation and decay of roots increases the permeability of this zone, increasing infiltration of intense rain and reducing surficial runoff and physical transport. Surficial erosion processes, such as, raindrop impact, sheet wash, and rilling detach material not anchored by vegetation (roots and fungal mycorrhizae) and biotic crusts (lichens, mosses, liverworts) and coverings (litter, duff), moving it downslope. Roots and fallen trunks can dam surficial runoff promoting its infiltration and reducing its erosiveness.

Major, often catastrophic erosional events are often associated with the development and collapse of plant and animal communities, often defining what is often referred to as a "disturbance regime." The growth and failure of vegetation as a soil anchor is related to dramatic episodic erosion when the anchoring fails for a variety of reasons. Two types of episodic erosion stand out: post-fire erosion (fire/flood events) and soil avalanches (landslides). Of these, post-fire erosion appears to have the shortest repeat times. Post-fire erosion requires an intense forest fire that is capable of burning away the organic duff or litter and the root network that protects the soil. The intense heat of the fire may also drive hydrophobic organic compounds into the soil as these materials burn. The heat also drys the soil making it hydrophobic. Whatever the immediate cause of this hydrophobicity, the frequent observation if that if an intense rain follows a major fire by a year or two, typically the soil fails to soak up the rain, which instead flows overland producing dramatic sheet erosion, rilling, and channel excavation (McNabb and Swanson, 1990). So much erosion occurs that major debris flows form. Much of the eroded material is temporarily stored in alluvial fans and within low gradient streams (Meyer *et al.*, 1995). Regrowing vegetation eventually anchors material on the hillslope, but it often takes several hundred years before enough burnable plant material (fuels) accumulates for a fire to be sufficiently hot to repeat the process. During the time between fires, chemical and physical breakdown of bedrock produces the soils and other loose material that will erode in a subsequent fire.

Soil avalanching is remarkably similar in its requirement that vegetation anchor a developing soil. In contrast to fire erosion, soil avalanching requires a prolonged and continuous protection of soil such that surficial erosion does not balance the production of loose material, and the soil profile thickens. During this time, the erosion regime appears to be transport limited. Rooting depths are limited, and eventually roots are predominately in less cohesive parts of the soil profile. If the slope is sufficiently steep, and if there is a triggering event that cause the soil to shear internally, such as a major rainfall or an earthquake, then a soil avalanche forms, and a mass of hillslope is hurtles toward the valley bottom, where it is in turn, eroded by fluvial processes. Subsequently, for a number of years, surficial erosion removes more loose material from the scar. The slide scar eventually revegetates and soil profiles develop and thicken once more (Scott, 1975; Stallard, 1985). Because a critical accumulation of soil is require, typically a meter or more, the repeat time for landslides is much longer than for major fires.

In setting where episodic erosion operates, it contributes much of the sediment delivered to rivers, often well in excess of surficial physical erosion by sheet wash and rilling. In protected forests of eastern Puerto Rico, landslides may contribute 80 percent to 90 percent of the sediment in rivers. In parts of Yellowstone National Park, fire-related erosion may contribute half of the sediment. Sediments generated by episodic erosion are derived from the entire loose soil profile. Typically, this material has lost a larger proportion of sodium and calcium-bearing minerals than potassium and magnesium bearing minerals.

Human activities

Human activities are responsible for mobilizing vast quantities of material for physical erasion. Humans accelerate erosion by removing plant cover and altering hillslope profiles. A feature that distinguishes this erosion from fire-related erosion or soil avalanching is the human erosion is almost never in equilibrium with the processes that generate loose material, and the erosion proceeds by degrading soils or exposing to transport processes vulnerable substrates that had long been protected by vegetation. Many rural land-use practices promote or accelerate erosional processes that already operate on a landscape. In the case of landslides and debris flow, this phenomena has cause great loss of life and property. In 1998 more than 10,000 people died during Hurricane Mitch and in 1999, 15,000 to 30,000 people died following rains along the north coast of Venezuela. Many have argued that fire related erosion has accelerated in the American West because a century of fire suppression has allowed for sufficient buildup of fuels that eventual fires are so intense that all biomass burns leaving especially unprotected and vulnerable hydrophobic soils. Stratigraphic data from deposits in parts of the American West indicate that fire-related erosion was operating at similar rates long before European settlement. In terms of eroded mass, the most intense human-related erosion appears to result from the destruction of natural ground cover on vulnerable substrates such as loess terrains (Yellow River, central Mississippi Valley) and deeply weathered tropical soils.

The accelerated erosion of uplands by human activities has not translated to a major export of sediment to the ocean (Milliman and Mead, 1983). Instead much of the sediment appears to be trapped as colluvium, alluvium, and lacustrine deposits. The construction of dams worldwide, about 70,000 in the United States alone has promoted lacustrine deposition. The storage of carbo in these deposits may be a significant component in today's carbon cycles (Stallard, 1998).

Role of storage and remobilization

The highest rates of physical erosion typically are measured on hillslopes in steep upland catchments. Frequently, because flows are less energetic on less steep slopes near valley bottoms, loose material is deposited as colluvium. Then, as streams leave uplands and gradients become less steep and flows less flashy, sediment is deposited in channel systems as alluvium. Colluvial, alluvial, and lacustrine sedimentation can intercept and store sediment coming from upland regions over long, even geologic, timescales. For example, vast quantities of sediment is stored in alluvial fans and foreland basins. A large fraction of the sediment generated by human-induced physical erosion is stored as colluvium, alluvium, and lacustrine sediment in lakes and reservoirs.

Chemical weathering in alluvium can completely change the composition of sediment migrating downriver. For example, sands crossing the Llanos alluvial plane in Venezuela change from lithic arenites to quartz arenites (Stallard, *et al.*, 1991).

Glacial erosion

During the past approximately two million years, the Earth has experienced episodic glaciations with repeat times of a bit more than 100,000 years. Associated with these ice ages has been the development of large continental glaciers, the largest being the Laurentide Ice Sheet, along with global cooling and drying. The flow of glacial ice is often associated with profound physical erosion, particularly where glaciers have warm bases, where liquid water exists at the glacial bed. While exact mechanisms are uncertain, freezing and thawing near the glacial bed and perhaps more narrow shear zones appear to enhance the breakdown of bedrock. Cold-based glaciers, where the base is frozen, can be remarkably nonerosive, sometimes preserving pre-glacial soils. Marine cores indicate that the continental ice sheets were quite erosive, lowing the Canadian Shield, for example, more than would chemical and physical weathering operating over the same period of time. The chemical weathering associated with glaciation is quite distinctive, reflecting a high ratio of fresh rock to water. Of note is the large contribution from easily weathered minerals that are disseminated through most siliceous rocks. These include carbonates (dissolved calcium), sulfides (dissolved sulfate), and micas (dissolved potassium). The exposure of large amounts of biotite to water and oxidative breakdown produces and releases large amounts of vermiculites to soils and sediments (Stallard, 2000).

Ice ages are times of exceptional wind erosion and deposition. The wind-borne dust forms great deposits of loess around the peripheries of the glacierized landscapes. Ice Age samples from deep ice cores from Greenland show far more dust than was deposited after the Ice Ages ended.

The study of erosion and sediment production is one of the most interdisciplinary fields of investigation in the Earth sciences. As such it still presents many barely-explored frontiers for researchers. Of particular note is: (1) the role of organisms, especially plants, in the erosion process; (2) the chemical and physical changes of sediments deposited on land; and (3) the details of erosion under continental ice sheets. An understanding of these issues is critical for the reconstructing sedimentary record of the history of the Earth.

Robert F. Stallard

Bibliography

Carson, M.A., and Kirkby, M.J., 1972. *Hillslope, Form and Process.* Cambridge University Press.
Garrels, R.M., and Mackenzie, F.T., 1971. *Evolution of Sedimentary Rocks*. New York: W.W. Norton.
McNabb, D.H., and Swanson, F.J., 1990. Effects of fire on soil erosion. In Walstad, J.D., Radosevich, S.L., and Sandberg, D.V. (eds.), *Natural and Prescribed Fire in the Pacific Northwest*. Corvallis, Oregon: Oregon State University Press, pp. 159–176.
Meyer, G.A., Wells, S.G., and Jull, A.J.T., 1995. Fire and alluvial chronology in Yellowstone National Park: climatic and intrinsic controls on Holocene geomorphic processes. *Geological Society of America Bulletin*, **107**: 1211–1230.
Milliman, J.D., and Meade, R.H., 1983. World-wide delivery of river sediment to the oceans. *Journal of Geology*, **91**: 1–21.
Moody, J.A., and Martin, D.A., 2001. Initial hydrologic geomorphic response following a wildfire in the Colorado front range. *Earth Surface Processes and Landforms*, **26**: 1049–1070.
Scott, G.A.J., 1975. Relationships between vegetation cover and soil avalanching in Hawaii. *Proceedings of the Association of American Geographers*, **7**: 208–212.
Stallard, R.F., 1985. River chemistry, geology, geomorphology, and soils in the Amazon and Orinoco basins. In Drever, J.I. (ed.), *The Chemistry of Weathering*. Dordrecht, Holland: D. Reidel, pp. 293–316.
Stallard, R.F., 1988. Weathering and erosion in the humid tropics. In Lerman, A., and Meybeck, M., (ed.), *Physical and Chemical Weathering in Geochemical Cycles*. Dordrecht, Holland: Kluwer, pp. 225–246.
Stallard, R.F., 1995a. Relating chemical and physical erosion. In White, A.F., and Brantley, S.L. (eds.), *Chemical Weathering Rates of Silicate Minerals*. Washington, DC: Mineralogical Society of America, pp. 543–564.
Stallard, R.F., 1995b. Tectonic, environmental, and human aspects of weathering and erosion: a global review using a steady-state perspective. *Annual Review of Earth and Planetary Sciences*, **12**: 11–39.
Stallard, R.F., 1998. Terrestrial sedimentation and the carbon cycle: coupling weathering and erosion to carbon burial. *Global Biogeochemical Cycles*, **12**: 231–252.
Stallard, R.F., 2000. Tectonic processes and erosion. In Jacobson, M.C. *et al.* (eds.), *Earth System Science: From Biogeochemical Cycles to Global Change*. San Diego: Academic Press, pp. 195–229.
Stallard, R.F., Koehnken, L., and Johnsson, M.J., 1991. Weathering processes and the composition of inorganic material transported through the Orinoco River System, Venezuela and Colombia. *Geoderma*, **51**: 133–165.

Cross-references

Climatic Control of Sedimentation
Cyclic Sedimentation
Debris Flow
Earth Flows
Mass Movement
Sands, Gravels and their Lithified Equivalents
Tectonic Controls of Sedimentation
Weathering, Soils, and Paleosols

EVAPORITES

Introduction

Evaporites are sedimentary deposits chemically precipitated from standing bodies of seawater (marine evaporites) or lakewaters (nonmarine evaporites) as a result of evaporative concentration. The major minerals of evaporites are salts of sodium, potassium, calcium, magnesium, chloride, sulfate, bicarbonate and carbonate ions (Table E1). Evaporite deposits are found in stratigraphic units of all ages back to the Late Precambrian. Indirect evidence of evaporite deposition is found in the form of molds and casts of saline minerals such as halite and gypsum in rocks as old as Archean. The scientific importance of evaporites lies mainly in the information they carry about: (1) the chemistry of ancient seawaters and lakewaters; (2) the distribution in space and time of hydrologically-restricted sedimentary basins with arid climates; and (3) the early stages of continental breakup and the formation of "proto-oceans" when thick and extensive saline deposits accumulated (e.g., the huge thicknesses of Jurassic and Cretaceous salt deposits preserved in initial rift valleys beneath the continental margins of both the North and South Atlantic Oceans).

Evaporite deposits are of major economic importance, mined on all continents as basic materials for the manufacture of inorganic chemicals used extensively in the chemical, agricultural and construction industries. Principal mining targets are halite (NaCl, common table salt), gypsum ($CaSO_4 \cdot 2H_2O$, used to make "drywall"), potash salts (such as sylvite, KCl, known as "muriate of potash" in industry), sodium carbonate salts (such as trona, $NaHCO_3 \cdot Na_2CO_3 \cdot 2H_2O$ from which "soda ash" used in the glass, paper and pulp industries is made), and sodium sulfate salts (such as mirabilite, $Na_2SO_4 \cdot 10H_2O$ from which "salt cake" or "Glauber's salt" is made). Modern surface and subsurface saline brines (the latter are known as basinal brines or oilfield brines), both primary and secondary (the latter formed by dissolution of buried evaporites), are major sources of important minor chemical elements such as bromine, iodine, lithium, etc. Evaporite deposits also play a significant role as impermeable seals of petroleum reservoirs. Despite the fact that evaporites constitute less than 5 percent of the volume of sedimentary rocks in the geologic record, they cap about half the world's known oil and gas resources.

Common table salt has been a critical trading commodity for millenia. Roman soldiers were given money to buy salt, a payment that became known as a *salarium*, from which the English word "salary" was derived. As early as 500 B.C. the Roman government took over control of salt works near the River Tiber. Three hundred years later the Roman government imposed taxes on salt to raise money to finance the Second Punic War. Medieval France copied this old Roman practice of salt monopolies and taxes. The leaders of the French Revolution made the repeal of the salt tax a major objective, which they achieved in 1790 (Multhauf, R.P., 1978).

Environmental conditions required for the formation of evaporite deposits

First and foremost, evaporites require for their formation: (1) an *arid climate*, that is, annual rate of evaporation >annual rate of inflow of water, where inflow = surface sources (seawater and/or river water + rainfall + surface spring inflow) + subsurface input of groundwater; and (2) a *hydrologically closed or restricted basin* (outflow<inflow). These two basic conditions will ensure that any standing water on the floor of the depositional basin will undergo the evaporative concentration necessary to induce evaporite minerals to precipitate. However, for accumulation of any measurable thicknesses of evaporites there are two additional requirements; (3) the closed arid basin must be either a deep receptacle before evaporite accumulation or must actively subside during evaporite accumulation; and (4) the inflow of solutes into the basin must be substantial over a long period of time.

If seawater is the primary inflow water then the evaporating basin must be a coastal depression (shallow or deep) with a constricted, very shallow inlet channel or a "leaky" sill of some kind. Such a hydrologic system must be fundamentally dependent on the position of sea level. The absence of any major marine evaporite depocenters today is testimony to this particular constraint. The rising Holocene sea level has overrun all the preexisting restricted marine basins in arid settings, such as the Red Sea and the Mediterranean Sea. During the Miocene these two basins were essentially isolated from direct surface inflow of seawater as sea level dropped to or below the shallow sills at their "mouths". During this period of basin restriction, very thick (km-scale) deposits of marine evaporites accumulated on the floors of the Red and Mediterranean Seas, as has been revealed by drilling.

Table E1 Chemical composition of some of the common saline minerals of evaporites

Anhydrite	$CaSO_4$	kainite	$MgSO_4 \cdot KCl \cdot 11/4H_2O$
Antarcticite	$CaCl_2 \cdot 6H_2O$	kieserite	$MgSO_4 \cdot H_2O$
Aragonite	$CaCO_3$	langbeinite	$2MgSO_4 \cdot K_2SO_4 \cdot$
Bassanite	$CaSO_4 \cdot 1/2H_2O$	mirabilite	$Na_2SO_4 \cdot 10H_2O$
Bischofite	$MgCl_2 \cdot 6H_2O$	nahcolite	$NaHCO_3$
Bloedite	$Na_2SO_4 \cdot MgSO_4 \cdot 6H_2O$	natron	$Na_2CO_3 \cdot 10H_2O$
Calcite	$CaCO_3$	polyhalite	$2CaSO_4 \cdot MgSO_4 \cdot K_2SO_4 \cdot 2H_2O$
Carnallite	$MgCl_2 \cdot KCl \cdot 6H_2O$	rinneite	$FeCl_2 \cdot NaCl \cdot 3KCl \cdot$
Dolomite	$MgCO_3 \cdot CaCO_3$	shortite	$2CaCO_3 \cdot Na_2CO_3$
Epsomite	$MgSO_4 \cdot 7H_2O$	sylvite	KCl
Glauberite	$CaSO_4 \cdot Na_2SO_4$	tachyhydrite	$CaCl_2 \cdot 2MgCl_2 \cdot 12H_2O$
Gypsum	$CaSO_4 \cdot 2H_2O$	thenardite	Na_2SO_4
Halite	$NaCl$	thermonatrite	$Na_2CO_3 \cdot H_2O$
Hexahydrite	$MgSO_4 \cdot 6H_2O$	trona	$NaHCO_3 \cdot Na_2CO_3 \cdot 2H_2O$

For nonmarine basins, the inflow waters may be meteoric waters (dilute river waters or groundwaters), upwelling basinal brines, upwelling hydrothermal brines, upwelling acid volcanogenic waters, or any combination of these source waters. Of particular significance is that many modern nonmarine evaporite deposits are accumulating in active continental rift and strike-slip basins. The floors of these tectonic basins are fed by groundwaters convected upward along fault and fracture zones by thermally-induced density instabilities (e.g., generated by magmatic heat sources beneath the rifts) or forced to the surface by large topographic gradients resulting from the high relief between the basin floor and the surrounding uplifted fault-block mountains.

Types of evaporite deposits

Marine evaporites (formed by evaporation of seawater). There are two major groups, related to their tectonic setting: (1) *Shallow shelf marine evaporites* deposited along the coasts of continental shelves (e.g., modern Lake MacLeod, Western Australia) or foreland basin ramps (e.g., modern Persian Gulf) in arid settings. Stratigraphically, these evaporites are typically made up of stacks of thin (meter scale), sheet-like bodies deposited in back-barrier tidal flats and shallow lagoons.

The tidal flat (or sabkha, an Arabic term for salt flat) marine evaporites are supratidal deposits that cap subtidal-intertidal marine detrital deposits (siliciclastic or carbonate composition), producing a classic "shallowing-upward cycle". Each cycle is formed by seaward progradation of the shoreline, with the vertical thickness dictated by the tidal range and the depth of the subtidal lagoon (typically a few meters).

The lagoonal evaporites are subtidal deposits that consist principally of layers of subaqueous gypsum and halite crystal accumulations of: (1) "pelagic" crystals which nucleated at the brine-air interface then settled gravitationally to the lagoon floor; and (2) arrays of vertically-oriented crystals grown on the lagoon floor. These gypsum-halite units may be punctuated by thin interbeds of marine detrital sediment that record "freshening" events when new seawater entered the lagoon.
(2) *Rift basin marine evaporites* deposited in active tectonic pull-apart basins formed during the early stages of continental breakup and development of ocean basins by seafloor spreading (e.g., Cretaceous of Brazil and Gabon, proto-Atlantic Ocean; Miocene beneath the Red Sea). These evaporites are commonly very thick (10^2 m to 10^3 m) deposits, fringed by alluvial fan deposits. The type of evaporite depositional environment can vary from tidal flat to lagoonal to deep perennial basin. This latter setting is characterized by the deposition of "pelagic" gypsum and halite precipitated from the evaporating surface brines and by incursions of gypsiferous turbidity flows derived from the "gypsum factories" on the shallow fringing shelves.

Nonmarine evaporites (formed by evaporation of continental waters)

Ancient nonmarine evaporite deposits are uncommon. The Eocene Green River Formation of Wyoming, a major source of sodium carbonate and oil shale, is one of the best known of the nonmarine type of evaporite deposit, renowned for its huge inventory of unusual minerals and spectacular specimens of fossil plants, fish, insects, birds, turtles, crocodilians, lizards, snakes, bats, and a host of invertebrates.

Much of what we know about deposition of nonmarine evaporites comes from the study of active modern systems. The vast majority of such depositional settings are hydrologically closed basins formed by continental rifting (e.g., Lake Magadi, East African rift) or strike-slip faulting (e.g., Dead Sea). Importantly, the high relief of the upfaulted mountains enclosing these two types of continental tectonic basins can act as rainfall–snowfall traps and barriers, making the basin floors local rain–shadow deserts independent of latitude ("orographic deserts", e.g., the modern Basque Lakes evaporites are located in western Canada at latitude 50°N). For those tectonic basins outside of the dry "horse-latitudes" the atmospheric precipitation trapped by the enclosing high mountains provides a significant source of inflow waters (e.g., modern Death Valley, California; modern Great Salt Lake, Utah). Such basins are capable of accumulating very thick nonmarine evaporite deposits.

Chemically and mineralogically, non-marine evaporites are far more varied than marine evaporites because of the very wide range of chemical compositions displayed by continental waters (Table E2). Nonetheless, non-marine evaporites can be separated into two major chemical groups, an alkaline group and a neutral group.

The *alkaline group* of composition Na-K-CO_3-SO_4-Cl are characterized by sodium carbonate minerals such as trona,

Table E2 Representative chemical compositions of inflow waters to modern evaporite basins (concentrations in parts per million)

	1	2	3	4	5	6	7	8	9
SiO_2	35.0	49.0	7.7	10.4	369.0	<50	400	—	—
Ca	13.0	13.0	43.0	13.4	6.5	32,800	28,000	805	413
Mg	4.3	9.0	1.2	3.4	0.0	6,440	54	0	1,296
Na	8.4	6.6	1.5	5.2	243.0	51,500	50,400	11,725	10,800
K	3.5	2.8	0.7	1.3	61.0	2,460	17,500	1,009	407
HCO_3	72.0	88.0	133.0	52.0	0.0	0	500	0	137
SO_4	6.9	4.9	3.2	8.3	454.0	140	5	0	2,717
Cl	3.8	6.9	2.2	5.8	408.0	160,000	155,000	20,530	19,010
TDS	148.0	189.0	199.0	99.6	1,570.0	256,000	258,000	34,069	35,000
pH	7.0	7.7	7.3	—	2.5	—	5.2	—	8.1

1. Groundwater (granite, White *et al.*, Table 1, #7); 2. Groundwater (basalt, White *et al.*, Table 2, #8); 3. Groundwater (limestone, White *et al.*, Table 6, #2); 4. "Average" modern river water (Spencer and Hardie, 1990, Table 1, #9); 5. Acid volcanic hot spring (White *et al.*, Table 19, #1); 6. Basinal brine (White *et al.*, Table 13, #7); 7. Continental hydrothermal brine (Salton Sea, Calif., Hardie, 1991, Table 6); 8. Mid-ocean ridge hydrothermal brine (Spencer and Hardie, 1990, Table 1, #4); 9 = Modern seawater (Spencer and Hardie, 1990, Table 1, #1).

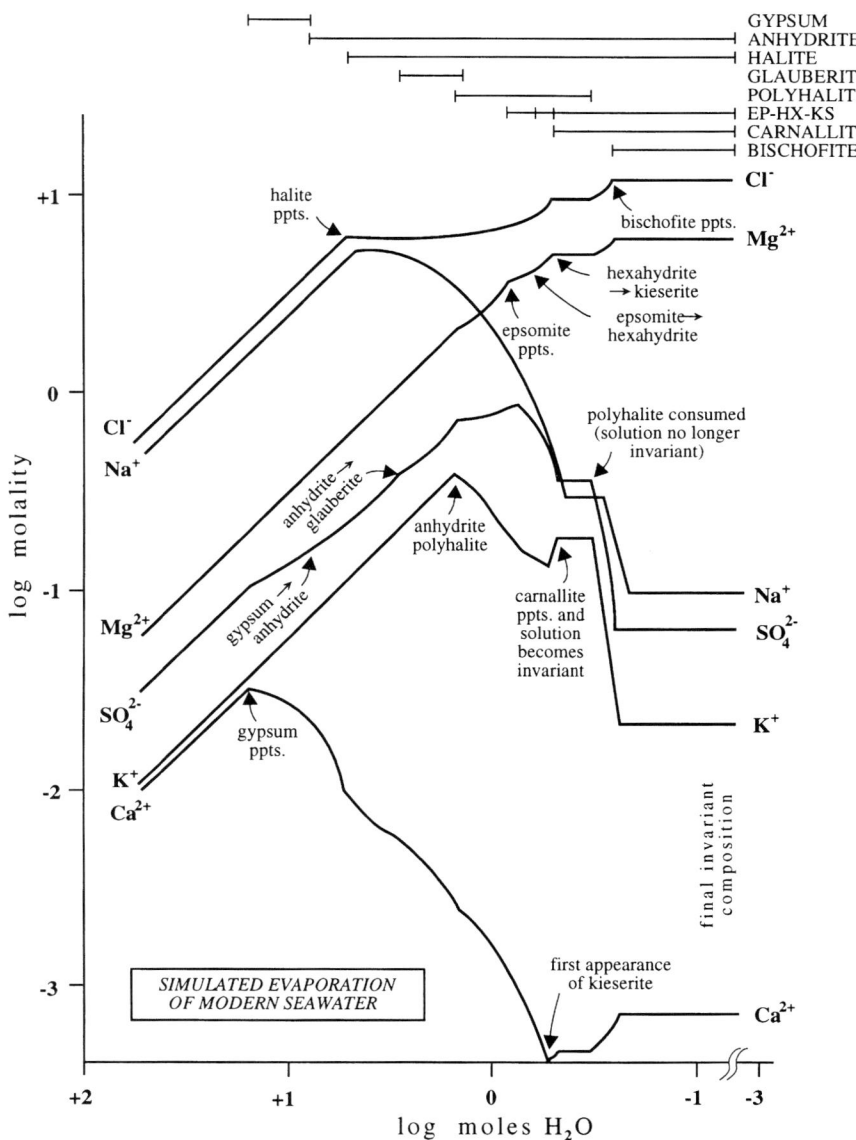

Figure E2 Evaporative concentration of modern seawater at 25°C and 1 atm. pressure simulated using the computer program of Harvie et al. (1980). The curves show the progressive changes in major ion concentrations in the seawater brine as evaporation and mineral precipitation and resorption proceed (initial mass of seawater contained 1 kg of H_2O) (from Hardie, 1990).

nahcolite and natron (Table E1) precipitated from high pH alkaline brines. The parent waters of this alkaline group are river waters and shallow continental groundwaters, which are classic bicarbonate-rich meteoric waters formed by chemical weathering of continental bedrocks by CO_2-charged rainwater. Evaporative concentration of these bicarbonate-rich meteoric waters produces alkaline brines rich in sodium and carbonate ions and poor in calcium and magnesium ions (see next section on brine evolution).

The *neutral group* of composition Na-K-Ca-Mg-SO_4-Cl are characterized by gypsum, halite, and a variety of sodium and magnesium sulfate minerals, among others (Table E1), precipitated from near-neutral pH sulfate-chloride brines.

The chemical evolution of these bicarbonate/carbonate-poor brines during evaporative concentration pivots around the early precipitation of gypsum, which determines whether the brines will become sulfate-poor or calcium-poor (the so-called gypsum "chemical divide", see next section on brine evolution). This leads to two main chemical subgroups, an Na-K-Mg-Ca-Cl subgroup ("$CaCl_2$ evaporites", characterized by potash minerals such as sylvite and rare $CaCl_2$ minerals such as tachyhydrite, Table E1) and an Na-K-Mg-SO_4-Cl subgroup ("$MgSO_4$-$NaSO_4$ evaporites", characterized by minerals such as epsomite, thenardite and glauberite, Table E1). These subgroups owe their chemical signatures to the influence of unusual bicarbonate-poor inflow waters, such as SO_4-rich acid volcanogenic waters, SO_4-rich meteoric waters derived by weathering of sulfide ore-bearing bedrocks, or $CaCl_2$-rich basinal brines and hydrothermal brines (Table E2).

Brine evolution and evaporite mineral sequences: geochemical principles

The reluctance of sparsely soluble minerals to precipitate from aqueous solutions at Earth surface temperatures is well-known. Marine carbonates, so common in the sedimentary record, are classic examples of non-equilibrium precipitation of such sparsely soluble minerals. Modern marine cements are typically aragonite and high-magnesian calcite (HMC), minerals that are theoretically metastable at Earth surface temperatures and pressures with respect to calcite and mixtures of dolomite + calcite respectively. Experimental work has shown that the Mg/Ca ratio, the temperature and the salinity of the parent water determines whether calcite, HMC + aragonite, or aragonite will precipitate, in agreement with the observation that metastable aragonite + HMC are the phases that preferentially nucleate from modern seawater with its high Mg/Ca mole ratio of 5.1. In contrast to all this, thermodynamics predicts that the only stable carbonate phase in equilibrium with modern seawater should be ordered stoichiometric dolomite. However, experimental work has failed to nucleate ordered stoichiometric dolomite in typical surface waters such as modern seawater at Earth surface temperatures, supporting the long recognized absence of ordered dolomite as the primary sedimentary precipitate in both modern and ancient carbonate deposits (the "dolomite problem"). Clearly, kinetics rather than equilibrium thermodynamics controls the important alkaline earth carbonate system in modern and ancient sedimentary systems, including evaporites.

While the kinetically-controlled alkaline earth carbonates are the typical minerals that precipitate during the early stages of evaporative concentration of natural waters, evaporite deposits are characterized by the more soluble salts such as gypsum and halite that for the most part behave in accordance with thermodynamic predictions. Using a combination of equilibrium thermodynamics and experimentally determined solubilities of evaporite minerals, it is possible to predict quantitatively: (1) the chemical evolution of waters undergoing evaporative concentration at a given temperature and 1 atm. pressure; and (2) the sequence of minerals (and their masses) precipitated during this evolution (Harvie and Weare, 1980; Harvie et al., 1980). Figure E2 illustrates this approach as applied to evaporative concentration of modern seawater (see Table E2 for composition of modern seawater). Figure E2 clearly highlights the critical steps in the chemical evolution of modern seawater as it undergoes progressive evaporative concentration. These critical steps occur at those points where new minerals first precipitate. The progressive crystallization of a new mineral causes drastic changes in the concentrations in the brine of those ionic species making up that mineral. Halite provides a simple but powerful example (Figure E2). As the evaporating brine first reaches saturation with halite, the equilibrium between halite and brine is defined by the chemical reaction:

$$Na^+_{(aq)} + Cl^-_{(aq)} = NaCl_{(s)} \qquad (Eq. 1)$$

At equilibrium between halite and the brine, the activities ("thermodynamic concentrations" = molality × activity coefficient) of the components are related as follows:

$$K_a = a_{NaCl}/a_{Na} \cdot a_{Cl}. \qquad (Eq. 2)$$

where K_a is the equilibrium constant for the T and P of interest. Since halite is a pure crystalline phase the activity of the component NaCl in halite is unity by definition. Thus, as long as equilibrium is maintained between halite and the brine during evaporative concentration, the product of the activities of Na and Cl ions in the brine is fixed even though these ions are being extracted from the brine to produce halite crystals. It follows, then, that during halite precipitation, as Na and Cl ions are *removed in equal molar proportions* while the ion activity product ($a_{Na} \cdot a_{Cl}$) remains constant, the concentrations of Na and Cl ions *must change antipathetically*. Typically, as evaporative concentration proceeds the ion with the lower concentration at the initial point of mineral precipitation will progressively decrease in concentration while that of the other ion will progessively increase (but at a much reduced rate). This type of critical step in brine evolution driven by evaporative concentration (see Hardie and Eugster, 1970) has been dubbed a "*chemical divide*" by Drever (1982, pp. 201–209). This behavior is clearly illustrated in Figure E2 for both halite and gypsum. Experimental work and observations of natural saline lakes and commercial salt pans have verified this predicted behavior.

The significance of the principle of "chemical divides" is that it explains the origin of the following three major compositional types of brines (Hardie and Eugster, 1970):

(1) Na-K-Ca-Mg-Cl brines ("calcium chloride brines", e.g., hydrothermal brines, basinal brines)
(2) Na-K-Mg-SO_4-Cl brines ("magnesium sulfate brines", e.g., modern seawater bitterns)
(3) Na-K-CO_3-SO_4-Cl brines ("alkaline brines", e.g., soda lake brines).

As illustrated in Figure E3, the initial precipitation of $CaCO_3$ (as either calcite or aragonite) during the very early stages of evaporative concentration of either seawater or lakewaters determines whether the resulting brine will become

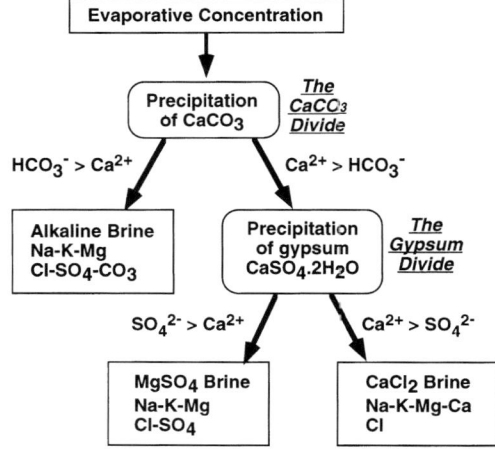

Figure E3 "Chemical divides" and the chemical evolution of the three major types of brines during evaporative concentration of surface waters (after Hardie and Eugster, 1970).

an "alkaline brine" (rich in $HCO_3 + CO_3$ ions, poor in Ca ions) or a "neutral brine" (poor in $HCO_3 + CO_3$ ions, rich in Ca ions). The outcome is determined by the relative concentrations of Ca and HCO_3 ions at the initial point of precipitation of $CaCO_3$: if $Ca > HCO_3$ then the resulting brine will be a "neutral brine", if $HCO_3 > Ca$ then the brine will be an "alkaline brine" (Step 1 in Figure E3). Further evaporation of the alkaline brine will lead to removal of essentially all the Ca ions while the CO_3 ion concentration and the pH of the brine both increase, ultimately producing an Na-K-CO_3-SO_4-Cl type brine. For those waters initially richer in Ca than HCO_3 ions (Table E2), evaporation beyond the "$CaCO_3$ divide" leads to a second evolutionary step, the "gypsum divide" (Step 2 in Figure E3). The outcome at this "gypsum divide" is determined by the relative concentrations of Ca and SO_4 ions: if $Ca > SO_4$ then the resulting brine will be a "$CaCl_2$ brine" (Na-K-Ca-Mg-Cl type), if $SO_4 > Ca$ then the brine will be an "$MgSO_4$ brine" (Na-K-Mg-SO_4-Cl type) (Figure E3).

The exact pathways followed on evaporative concentration of a given water belonging to any one of the three main types depends mainly on the particular ion ratios at each chemical divide but other factors that can alter the chemical evolution of the evaporating brine may come into play. In this latter regard, two of the more important factors are: (1) nucleation of metastable instead of stable minerals, e.g., precipitation of magnesian calcite instead of calcite will influence the ultimate Mg/Ca ratio of the brine; (2) equilibrium brine evolution may depend on certain early formed minerals that *back-react* with the chemically evolving brine, i.e., the reaction

$$2CaSO_4 \cdot 2H_2O_{(gypsum)} + 2K^+_{(aq)} + 2Mg^{2+}_{(aq)} + 2SO_4^{2-}(aq) \rightarrow$$
$$2CaSO_4 \cdot MgSO_4 \cdot K_2SO_4 \cdot 2H_2O_{(polyhalite)} + 2H_2O$$

predicted for modern seawater. In this example, if the early formed gypsum becomes separated from the mother-liquor by settle-out or by brine migration then both the chemical evolution of the brine and the mineral succession will be significantly different from those involving back-reaction (Hardie, 1984, figure E4).

Evaporites and the chemistry of phanerozoic seawater

Ancient "potash evaporites", rich in economically important potassium minerals, have for more than a century been assumed to be of marine origin, formed by evaporation of seawater with the chemistry of modern ocean water. Indeed, J.H. van't Hoff, the first Noble Prize winner for Chemistry spent the later years of his career (1896–1908) experimentally investigating the phase relations among the saline minerals in the system Na-K-Ca-Mg-SO_4-Cl-H_2O at temperatures from 0 to 110°C at atmospheric pressure. His objective was to better understand the origin of the economically important Stassfurt potash evaporites of Germany. His extensive series of studies (carried out with the aid of 33 collaborators) constitutes the first systematic application of chemical experiments to the solution of a geological problem, and gave birth to modern experimental geochemistry (Eugster, 1971).

Unfortunately, van't Hoff's pioneering experimental efforts were unable to resolve the fundamental dilemma presented by the stratigraphic record of evaporite deposits, namely, that the majority of ancient potash evaporites thought to be of marine

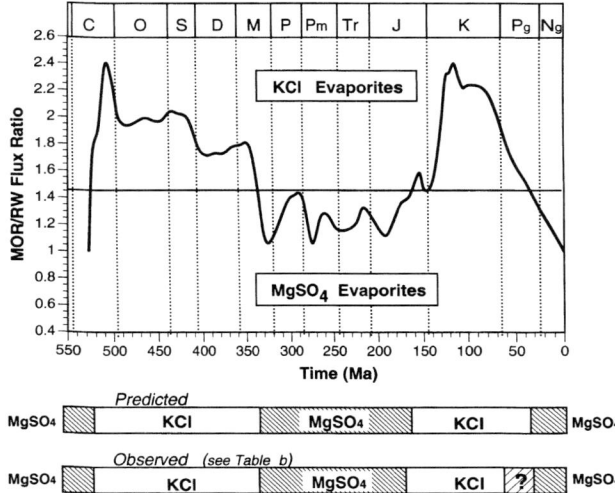

Figure E4 Plot of MOR/RW flux ratio versus geologic time over the past 550 m.y. using the new calculation method of Hardie (2000). The horizontal line at MOR/RW = 1.45 is the chemical divide that separates paleoseawaters from which either KCl or $MgSO_4$ evaporites will form on evaporative concentration. Comparison of the predicted and observed potash evaporite types is shown in the bars below the graph.

origin are of the "KCl-type" (formed from $CaCl_2$ brines of the system Na-K-Ca-Mg-Cl, see above) rather than the "$MgSO_4$-type" (formed from brines of the system Na-K-Mg-SO_4-Cl, to which modern seawater belongs) (Hardie, 1990, Tables 1 and 2). If the major ion chemistry of seawater has remained close to its present composition for the last 1–2 billion years, as has been assumed by almost all 20th Century geologists and paleontologists as well as aqueous geochemists (e.g., Rubey, 1951; Mackenzie and Garrels, 1966; Sillen, 1967; Holland, 1972), then how are we to explain the preponderance of KCl over $MgSO_4$ marine potash evaporites in the Phanerozoic record? A variety of explanations have been proposed, mainly involving penecontemporaneous or later burial diagenesis of primary $MgSO_4$ potash evaporites, contemporaneous alteration of seawater entering a marginal marine basin by dolomitization of early formed $CaCO_3$ deposits, by the activity of sulfate-reducing bacteria, or by mixing of seawater with Ca-HCO_3-rich river waters. A radical solution to this dilemma has been suggested by Spencer and Hardie (1990). They proposed that the chemistry of global seawater over much of geologic time has been controlled principally by simple mixing of the two principal sources of major ions to the ocean, namely, river water (RW) and midocean ridge (MOR) hydrothermal brine. They presented quantitative predictions of secular variations in the major ion chemistry of seawater as a function of hypothetical secular variations in the ratio of the input fluxes of MOR brine and river water. Their calculations indicated that relatively small changes in the MOR/RW flux ratio would result in seawater chemistry switching from the Na-K-Mg-SO_4-Cl type (the modern seawater type, from which $MgSO_4$ potash evaporites would precipitate) to the Na-K-Ca-Mg-Cl type (from which KCl potash evaporites would precipitate). Hardie (1996, 2000),

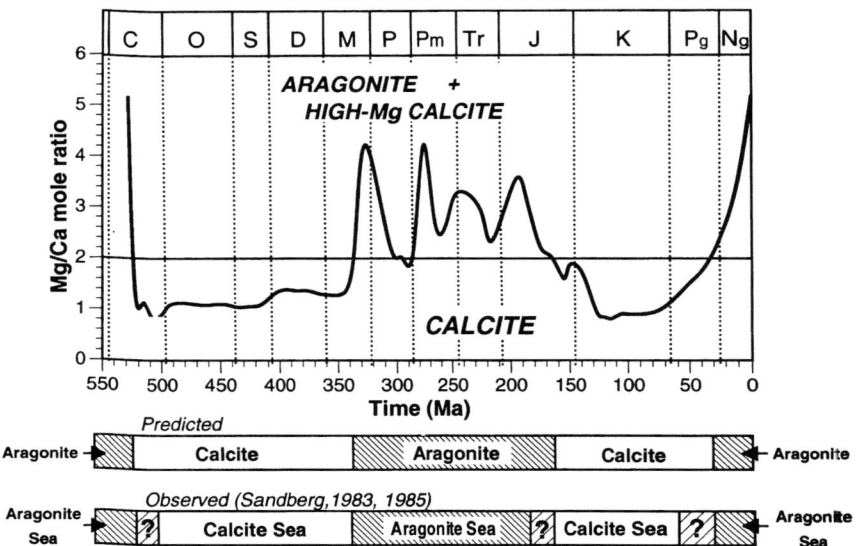

Figure E5 Plot of predicted oscillations in Mg/Ca mole ratio in seawater over the past 550 m.y. using the new calculation method of Hardie (2000). The horizontal line at Mg/Ca = 2.00 is the divide that separates waters from which either calcite or aragonite (±high-Mg) are predicted to precipitate (Stanley and Hardie, 1998, 1999). Comparison of the predicted and observed mineralogies of nonskeletal limestones is shown in the bars below the graph.

using the Spencer–Hardie seawater approach and the assumption that secular variations in the MOR brine flux were proportional to secular variations in ocean floor production, showed that the temporal records of oscillations during the Phanerozoic in KCl versus $MgSO_4$ marine potash evaporites, and in "aragonite seas" versus "calcite seas" determined by Sandberg (1983, 1985) are in very good agreement with those predicted by this simple "mixing model" (see Figures E4 and E5, which are based on the improved mixing model of Hardie, 2000). Furthermore, the predictions are in full accord with the observed synchroneity of KCl potash evaporites with calcite seas and $MgSO_4$ potash evaporites with aragonite seas. Chemical analyzes of fluid inclusions in ancient halite crystals from evaporite deposits ranging in age from Late Precambrian to Late Miocene carried out by a number of workers using several different techniques are also in very good agreement with the predicted oscillations in global seawater chemistry based on the MOR brine–river water mixing model (e.g., Lowenstein et al., 2001). Finally, Stanley and Hardie (1998, 1999) have drawn attention to the possible consequences of oscillations in the Mg/Ca ratio and the Ca concentrations of Phanerozoic seawaters on the carbonate mineralogies of invertebrate skeletons. For example, the very low Mg/Ca ratio and very high Ca concentration predicted for Cretaceous seawater (compared to modern seawater) provide a plausible explanation for the spectacular production of coccolith chalk during this period. The progessive increase in Mg/Ca ratio and decrease in Ca concentration predicted for the period Paleocene to Quaternary can explain the observed progressive loss of coccolith mass and the final demise of the *Discoaster* coccolithoporids during this period (Stanley and Hardie, 1998, 1999). The combined weight of all these results provides strong support for the hypothesis that seawater chemistry has oscillated significantly during the Phanerozoic Eon as a result of oscillations in the MOR/RW flux ratio driven mainly by secular variations in rates of seafloor spreading.

Lawrence A. Hardie and Tim K. Lowenstein

Bibliography

Drever, J.I., 1982. The Geochemistry of Natural Waters, 3rd edn. Prentice-Hall (see also later editions).
Eugster, H.P., 1971. The beginnings of experimental petrology. *Science*, **173**: 481–489.
Eugster, H.P., and Hardie, L.A., 1978. Saline lakes. In Lerman, A. (ed.), *Lakes: Chemistry, Geology, Physics*. New York: Springer-Verlag, pp. 237–293.
Hardie, L.A., 1984. Evaporites: marine or nonmarine? *American Journal of Science*, **284**: 193–240.
Hardie, L.A., 1990. The roles of rifting and hydrothermal $CaCl_2$ brines in the origin of potash evaporites: an hypothesis. *American Journal of Science*, **290**: 43–106.
Hardie, L.A., 1991. On the significance of evaporites. *Annual Reviews of Earth and Planetary Sciences*, **19**: 131–168.
Hardie, L.A., 1996. Secular variation in seawater chemistry: an explanation for the coupled secular variation in the mineralogies of marine limestones and potash evaporites over the past 600 my. *Geology*, **24**: 279–283.
Hardie, L.A., 2000. A new method of predicting paleoseawater compositions. *Geologial Society of America Abstracts with Programs*, 2000, Reno, Nevada, p.A-68.
Hardie, L.A., and Eugster, H.P., 1970. *The Evolution of Closed basin Brines*. Mineralogical Society of America, Special Paper, 3, pp. 273–290.
Hardie, L.A., Lowenstein, T.K., and Spencer, R.J., 1985. The problem of distinguishing between primary and secondary features in evaporites. *Proceedings of the Sixth International Symposium on Salt* 1983, Alexandria, Virginia, The Salt Institute, pp. 11–39.
Hardie, L.A., Smoot, J.P., and Eugster, H.P., 1978. *Saline Lakes and Their Deposits: A Sedimentological Approach*. International Association of Sedimentologists, Special Publication, 2, pp. 7–42.

Harvie, C.E., Moller, N., and Weare, J.H., 1984. The prediction of mineral solubilities in natural waters: the Na-K-Mg-Ca-H-Cl-SO$_4$-OH-HCO$_3$-CO$_3$-H$_2$O system to high ionic strengths at 25 °C. *Geochimica et Cosmochimica Acta*, **48**: 734–751.

Harvie, C.E., Weare, J.H., Hardie, L.A., and Eugster, H.P., 1980. Evaporation of sea water: calculated mineral sequences. *Science*, **208**: 498–500.

Holland, H.D., 1972. The geologic history of seawater—an attempt to solve the problem. *Geochimica et Cosmochimica Acta*, **36**: 637–651.

Lowenstein, T.K., and Hardie, L.A., 1985. Criteria for the recognition of salt pan evaporites. *Sedimentology*, **32**: 627–644.

Lowenstein, T.K., Timofeeff, M.N., Brennan, S.T., Hardie, L.A., and Demicco, R.V., 2001. Oscillations in Phanerozoic seawater chemistry: evidence from fluid inclusions. *Science*, **294**: 1086–1088.

Mackenzie, F.T., and Garrels, R.M., 1966. Chemical mass balance between rivers and oceans. *American Journal of Science*, **264**: 507–525.

Multhauf, R.P., 1978. *Neptune's Gift: A History of Common Salt*. Johns Hopkins University Press.

Rubey, W.W., 1951. Geologic history of seawater. *Geological Society of America Bulletin*, **62**: 1111–1148.

Sandberg, P.A., 1983. An oscillating trend in Phanerozoic non-skeletal carbonate mineralogy. *Nature*, **305**: 19–22.

Sandberg, P.A., 1985. Non-skeletal aragonite and pCO$_2$ in the Phanerozoic and Proterozoic. In Sundquist, E.T., and Broecker, W.S. (eds.), *The Carbon Cycle and Atmospheric CO$_2$: Natural Variations Archean to Present*. American Geophysical Union Monograph, 32, pp. 585–594.

Sillen, L.G., 1967. The ocean as a chemical system. *Science*, **156**: 1187–1197.

Spencer, R.J., and Hardie, L.A., 1990. Control of seawater composition by mixing of river waters and mid-ocean ridge hydrothermal brines. In Spencer, R.J., and Chou, I-M. (eds.), *Fluid-Mineral Interactions: A Tribute to H.P. Eugster*. San Antonio, Geochemical Society, Special Publication, 2, pp. 409–419.

Stanley, S.M., and Hardie, L.A., 1998. Secular oscillations in the carbonate mineralogy of reef-building and sediment-producing organisms driven by tectonically forced shifts in seawater chemistry. *Palaeogeography, Palaeoclimatology, Palaeoecology*, **144**: 3–19.

Stanley, S.M., and Hardie, L.A., 1999. Hypercalcification: paleontology links plate tectonics and geochemistry to sedimentology. *GSA Today*, **9** (2): 1–7.

White, D.E., Hem, J.D., and Waring, G.A., 1963. *Chemical Composition of Subsurface Waters*. US Geological Survey Professional Paper, 440-F, 67pp.

Some Textbooks and Monographs on Evaporites

Braitsch, O., 1971. *Salt Deposits: Their Origin and Composition*. NY: Springer-Verlag.

Borchert, H., and Muir, R.O., 1964. *Salt Deposits: The Origin, Metamorphism and Deformation of Evaporites*. Van Nostrand.

Handford, C.R., Loucks, R.G., and Davies, G.R. (eds.), 1982. *Depositional and Diagenetic Spectra of Evaporites—A Core Workshop*. Calgary, Canada: SEPM Core Workshop, No. 3, 395pp.

Schreiber, B.C. (ed.), 1988. *Evaporites and Hydrocarbons*. Columbia University Press.

Sonnenfeld, P., 1984. *Brines and Evaporites*. Academic Press.

Stewart, F.H., 1963. *Marine evaporites*. US Geological Survey Professional Paper, 440-Y, 53pp.

Smoot, J.P., and Lowenstein, T.K., 1991. Depositional environments of non-marine evaporites. In Melvin, J.L. (ed.), *Evaporites, Petroleum and Mineral Resources*, Elsevier, Developments in Sedimentology 50, pp. 189–347.

Warren, J.K., 1989. *Evaporite Sedimentology*. Prentice Hall.

Warren, J., 1999. *Evaporites: Their Evolution and Economics*. Blackwell Science.

Zharkov, M.A., 1981. *History of Paleozoic Salt Accumulation*. Springer-Verlag.

Zharkov, M.A., 1984. *Paleozoic Salt Bearing Formations of the World*. Springer-Verlag.

Cross-references

Anhydrite and Gypsum
Seawater: Temporal Changes to the Major Solutes

EXTRATERRESTRIAL MATERIAL IN SEDIMENTS

Ultimately, the entire Earth is composed of "extraterrestrial" material, which accreted (mostly in the form of kilometer-sized and larger bodies) about 4.56 billion years ago to form our planet. But even today, the Earth continues to accrete large amounts of extraterrestrial matter, about 40,000 tons per year of mostly fine dust (within the mass range 10^{-16} kg to 10^{-4} kg), which is derived from asteroids, comets, and even from the Moon, Mars (as rare meteorites), and the interstellar medium, although proportions vary widely. This extraterrestrial matter settles through the atmosphere and accumulates, much diluted, in terrestrial and marine sediments, comprising one class of extraterrestrial material in sediments. On occasion, larger bodies (cm- to m-size) fall as meteorites, which mostly weather away in geologically short time periods, although a few fossil meteorites have been found in Ordovician rocks (Peucker-Ehrenbrink and Schmitz, 2001). Even less frequent, but still often in geological terms, asteroid- and comet-nucleus-sized bodies impact the Earth with cosmic velocity, leading to the formation of an impact crater, and, depending on the size of the impactor, a regional to global distribution of impact ejecta, being the other class of extraterrestrial material in sediments.

Extraterrestrial dust is mainly derived from the collision of asteroids in the asteroid belt and the subsequent migration of the dust due to the Poynting-Robertson effect toward the Earth and Sun. The other main source is the disintegration of comets ("dusty snowballs"), which release dust into the inner solar system when they get close to the sun and the ice, which makes up the bulk of their mass, evaporates. The flux from both sources varies considerably over time (Peucker-Ehrenbrink and Schmitz, 2001). An excellent tracer for extraterrestrial material in marine sediments is the presence of ^{3}He, which is carried to the seafloor in the finest fraction of interplanetary dust and is retained in some sediments for up to several hundred million years (Farley, 2001). Such isotopic studies of oceanic sediments provided evidence for a shower of long-period comets at around 35 Ma.

Ejecta from large impact events on Earth comprise a mixture of uniquely shocked and melted terrestrial material and remnants of the extraterrestrial impactor (Montanari and Koeberl, 2000; Koeberl, 2001). The discovery of shock-metamorphosed mineral grains (e.g., quartz, feldspar) in sediments of Cretaceous-Tertiary (K-T) boundary age (65 Ma) from locations around the world (Bohor *et al.*, 1987) provided confirming evidence for the hypothesis that a large impact event marked the end of the Cretaceous and was responsible for the mass extinction at that time (Koeberl and MacLeod, 2002). Impact craters with diameters of about $\geq$80 km may be associated with globally distributed impact ejecta; the Chicxulub impact structure, which is buried beneath Tertiary sediments in Yucatan, Mexico, and which is the source of the K-T boundary ejecta, has a diameter of almost 200 km.

Geochemical methods can be employed to detect the chemical signature of the remnant of the impactor in sediments (Koeberl, 1998). This is mainly done by searching for anomalously high concentrations of siderophile elements, especially the platinum-group elements (PGEs—e.g., Ir, Os), because these elements typically show low concentrations in crustal rocks and sediments, and much higher abundances (by several orders of magnitude) in most meteorites. Most often contents of the element Ir are reported, because it is the easiest of all PGEs to measure. Crustal Ir contents are about 0.02 ppb, whereas impact ejecta commonly have 0.2 ppb to 10 ppb. Also, the isotopic compositions of the elements Os (Koeberl, 1998) and Cr (Shukolyukov *et al.*, 2000) show significant differences between terrestrial crustal rocks and meteorites, providing very sensitive and selective tracers for meteoritic matter. Impact ejecta are found from the beginning of the geological history of the Earth as recorded in sedimentary rocks, although the evidence for traces of the so-called late heavy bombardment at 3.85 Ga is still contradictory (Ryder *et al.*, 2000). Rather unusual impact ejecta are found in 3.4 Ga rocks from Barberton, South Africa (Shukolyukov *et al.*, 2000). Geochemical evidence for an extraterrestrial component was found in 2 Ga spherule layers from the Transvaal Supergroup, South Africa (Simonson *et al.*, 2000), which are similar to approximately coeval spherule layers in Australia.

Christian Koeberl

Bibliography

Bohor, B.F., Modreski, P.J., and Foord, E.E., 1987. Shocked quartz in the Cretaceous/Tertiary boundary clays: evidence for global distribution. *Science*, **236**: 705–708.

Farley, K.A., 2001. Extraterrestrial helium in seafloor sediments: identification, characteristics, and accretion rate over geologic time. In Peucker-Ehrenbrink, B., and Schmitz, B. (eds.), *Accretion of Extraterrestrial Matter throughout Earth's History*. New York: Kluwer Academic/Plenum Publishers, pp. 179–204.

Koeberl, C., 1998. Identification of meteoritical components in impactites. In Grady, M.M., Hutchison, R., McCall, G.J.H., and Rothery, D.E. (eds.), *Meteorites: Flux with Time and Impact Effects*. Geological Society of London, Special Publication, 140, pp. 133–152.

Koeberl, C., 2001. The sedimentary record of impact events. In Peucker-Ehrenbrink, B., and Schmitz, B. (eds.), *Accretion of Extraterrestrial Matter Throughout Earth's History*. New York: Kluwer Academic/Plenum Publishers, pp. 333–378.

Koeberl, C., and MacLeod, K. (eds.), 2002. *Catastrophic Events and Mass Extinctions: Impacts and Beyond*. Geological Society of America, Special Paper 356, 746 pp.

Montanari, A., and Koeberl, C., 2000. *Impact Stratigraphy: The Italian Record*. Lecture notes in earth sciences, Volume 93, Heidelberg: Springer Verlag, 364 pp.

Peucker-Ehrenbrink, B., and Schmitz, B. (eds.), 2001. *Accretion of Extraterrestrial Matter throughout Earth's History*. New York: Kluwer Academic/Plenum Publishers. 466 pp.

Ryder, G., Koeberl, C., and Mojzsis, S.J., 2000. Heavy bombardment on the earth ~3.85 Ga: the search for petrographic and geochemical evidence. In Canup, R., and Righter, K. (eds.), *Origin of the Earth and Moon Tucson*. University of Arizona Press, pp. 475–492.

Shukolyukov, A., Kyte, F.T., Lugmair, G.W., Lowe, D.R., and Byerly, G.R., 2000. The oldest impact deposits on earth—first confirmation of an extraterrestrial component. In Gilmour, I., and Koeberl, C. (eds.), *Impacts and the Early Earth*. Lecture Notes in Earth Sciences, Volume 91. Heidelberg-Berlin: Springer, pp. 99–115.

Simonson, B.M., Koeberl, C., McDonald, I., and Reimold, W.U., 2000. Geochemical evidence for an impact origin for a late Archean spherule layer, Transvaal Supergroup, South Africa. *Geology*, **28**: 1103–1106.

Cross-reference

Sediments Produced by Impact

F

FABRIC, POROSITY, AND PERMEABILITY

Introduction

Rocks are generally classified by their *texture* and *composition*. Texture includes the size and shape of the various solid particles, and their arrangement in space: it therefore also includes the geometry of the pore spaces included within the rock (or sediment). There is no generally accepted term to include all the types of solid particles, but in this article we follow Lucia (1995) in using the term *particle* to include both chemically precipitated *crystals*, and clastic *grains*, which may themselves be composed of one or more crystals. Unlithified sediments are generally composed exclusively of grains which originated either outside the basin of deposition (so are *allogenic*) or within the basin of deposition (so are *autogenic*), but a few sediments or sedimentary rocks may be composed of crystals precipitated directly on the sediment surface. The size and shape of sediment particles is an important part of texture (see *Grain Size and Shape*) but it is not the whole story: it is very important how the particles are arranged in space, because this determines some of the bulk (geophysical) properties, including the size, shape, and orientation of the pore spaces. These in turn determine two of the most economically important bulk properties of a sediment: (1) The percentage volume of pore space, or *porosity*; and (2) The ease with which fluids can flow through a rock, which is generally quantified as its *permeability*.

Fabric

The arrangement of particles in space is described as the *fabric*. Early workers (for a review, see Martini, 1978) described fabric in terms of two main properties: (1) The *orientation* of the particles: either the actual orientation of the grains themselves (*dimensional orientation*), or the orientation of the crystals composing the particles (*crystallographic orientation*); and (2) The *packing* of the particles. Packing can be described using theoretical models for regular packings of sphere, or (somewhat more realistically) for random packings. For real particles, the *packing density* (see Kahn, 1956; or Martini, 1978, for quantitative definitions) is a measure of the porosity, and the *packing proximity* describes the number of contacts between particles. The types of contacts are also very important: for grains, Taylor (1950) introduced the terms tangential, long (essentially planar), concavo-convex, and sutured to describe increasing close contact between grain in sandstones, and showed that only tangential and a few long contacts are present in uncompacted sands, but concavo-convex and sutured contacts become more significant as sands are compacted by deep burial. Later studies have stressed the importance of using techniques such as cathodoluminescence (*q.v.*) to distinguish original clastic grains from chemically deposited (diagenetic) cements in assessing the nature of grain contacts. Many apparently concavo-convex and sutured contacts are not produced by pressure solution (*q.v.*) but by deposition of cement (see *Cements and Cementation*). For more recent approaches to packing see Allen (1982, Chapter 4), Bideau and Dodds (1991) and Rogers *et al.* (1994).

Grain orientation

The way that grains become oriented is an interesting part of the depositional process. It has also been studied as an indicator of the depositional process itself, particularly for deposits that do not show obvious internal stratification, such as tills, debris flow deposits, and some turbidites (for a review see Allen, 1982, Chapter 5). Most types of deposits show elongate grains oriented in the direction of the flow, but transverse and bimodal orientations are also known, particularly for roller-shaped pebbles (see *Size and Shape of Grains*). Flat pebbles in fluvial deposits commonly dip in the up-flow direction, but dip directions on beaches are more variable (see *Imbrication and Flow-Oriented Clasts*).

Most investigations, of grain orientation have been motivated by the need to obtain paleocurrents from such deposits (see *Paleocurrent Analysis*). The statistical techniques required to establish the fabric type and the preferred orientation (if any) have been reviewed by Fisher (1993) and Fisher *et al.* (1987): see Middleton (2000, Chapter 9) for references. Particle orientation may be produced or modified by compaction (*q.v.*)

or diagenesis (*q.v.*), particularly in fine grained, clay rich sediments (see *Depositional Fabric of Mudrocks*). Most sedimentary rocks have anisotropic fabrics, i.e., the grains and pore spaces show a preferred orientation, with the result that bulk properties, especially permeability, are also markedly anisotropic. Permeability is almost always much larger in the plane of stratification than normal to that plane. Fabrics generated by physical depositional processes are frequently modified by biological processes (see *Biogenic Structures*, and the symposium organized by Kemp, 1995).

Porosity

Techniques to measure porosity are reviewed by Monicard (1980). It is generally useful to distinguish total porosity from that part that consists of interconnected pores (*effective porosity*): e.g., pumice may have a large porosity, but fluids cannot move through the pores because they are not interconnected. Geophysical techniques may measure the total water (or hydrogen) content of a rock, some of which may be too tightly bound to mineral surfaces to move freely. Porosities of unconsolidated sands are generally 35 to 40 percent but may be higher (particularly for irregularly shaped bioclasts): for recently deposited muds they are generally 50 to 80 percent, with considerable consolidation taking place in the first meter or two of burial.

Geologists are generally interested in classifying pore spaces according to their genesis. In sands part of the porosity may be depositional *intergranular porosity*, due to the original depositional packing (or what is left of it, after compaction): another part may be *secondary porosity* produced by dissolution of grains during diagenesis. Some may even be *intragranular*, due to pore spaces within the grains, either present already at the time of deposition, or produced later during diagenesis. Diagenetic processes tend to decrease porosity, especially during deep burial (see *Compaction*) but there are exceptions. For further details, see *Petrophysics of Sand and Sandstone*.

Carbonates, even original carbonate sands, tend not to retain their original intergranular porosity for long, due to extensive early diagenesis (see *Neomorphism and Recrystallization; Carbonate Diagenesis and Microfacies*). The result is that extensive classifications of pore types have been developed for carbonate rocks. Choquette and Pray (1970) divided pores into two major categories, according to whether their presence was determined by a preexisting fabric (*fabric selective*) or not. Much secondary porosity is fabric selective, because it is determined by dissolution or recrystallization or replacement of fossil fragments. Large vugs and fracture systems, however, are generally not fabric selective, and cut indiscriminately across the original depositional fabric. Lucia (1995) has proposed a revised system, based more closely on petrophysical rock properties (see below). It recognizes two major categories of pore space: (i) *interparticle*; and (ii) *vuggy*, subdivided into *separate*, and *touching* (i.e., interconnected vugs).

Permeability

Permeability expresses the ease with which fluids flow through rocks. Henry Darcy first established the law (now called *Darcy's law*) that the discharge of fluid per unit cross-sectional area of the porous medium, q, is proportional to the hydraulic gradient (dh/dx). The coefficient of proportionality is called the *hydraulic conductivity*. Hydraulic conductivity is commonly used by hydrogeologists (e.g., Freeze and Cherry, 1979), but has the disadvantage that it depends not only on the porous medium but also on the properties of the fluid. A form of Darcy's law which separates the fluid and rock properties is:

$$q = -k \, (\gamma/\mu)(dh/dx) \qquad \text{(Eq. 1)}$$

Here γ is the specific weight of the fluid, and μ is the fluid viscosity. The coefficient of proportionality, k, in equation 1 is called the permeability, or *intrinsic permeability*, and for rocks saturated with a single fluid (e.g., water or air) it is determined only by the rock properties. Its units, therefore, are length squared. Permeability is related to some "average size" of the pore system, though the "average cross-section" of the pores, is not given by the units of intrinsic permeability. For intergranular porosity, permeability is related to the size and sorting of the grains (see *Petrophysics of Sands and Sandstone*; Chapman, 1981; Middleton and Wilcock, 1994, Chapter 6).

Unfortunately the same law does not apply (without major modification) for rocks or sediments that contain more than one fluid, e.g., air and water in the vadose zone; oil or gas and water in petroleum reservoirs. It is possible, however, to define a "relative permeability" which expresses the permeability to a single fluid, at a given fluid content, as a proportion of the intrinsic permeability: but once again, relative permeability is not a property only of the porous medium, but also of the other fluids present in the pores.

Permeability of relatively porous, high permeability rocks is generally determined by forcing gas through dry rock samples, or by pumping tests on wells. The older unit of permeability was the *darcy*. Very permeable coarse sediments may have a permeability of the order of one darcy, but the permeability of most reservoir rocks is best measured in *millidarcies*. Many good reservoirs have permeabilities of only tens of millidarcies. The SI unit is meters squared, a very large value (for comparison, a darcy is *roughly* one micrometer squared!). A portable "micropermeameter" that can be used on outcrops in the field has been developed (Davis *et al.*, 1994), though the results do not always correlate well with measurements made on cores (Tidwell *et al.*, 1998). Almost all sedimentary rocks have some permeability, and although it is not easy to determine the *in situ* permeability of rocks with very low permeability, such as shales, it can be done (see Neuman and Neretnieks, 1990). Such rocks have measured permeabilities as low as 10^{-9} darcies.

Petrophysics

The term petrophysics was introduced by Archie (1942, 1952) as a discipline that is concerned with relating the physical properties of porous rocks, as measured, for example, by electrical conductivity, to the texture and composition of the rock. Most petrophysical studies are concerned with the detailed geometry of the pore space, because this determines not only the average bulk properties of the rock, such as porosity and permeability, and electrical resistance when saturated with water, but also with the way the rock transmits two fluids. In the petroleum industry, the two fluids are water (because essentially all reservoir rocks contain water) and oil or natural gas. For introductions see Greenkorn (1983) and Bideau and Dodds (1991).

To understand two-phase flow, consider the simplest case: movement of two fluid phases in a straight uniform capillary, with circular cross-section of diameter d. If the two fluids are water and air, then the water tends to be drawing into the tube, displacing air. The interface or *meniscus*, makes a *contact angle* θ (theta) with the solid surface. The *displacement pressure*, P, tending to drive the water along the capillary is given by

$$P = -4\gamma \cos\theta/d \qquad (Eq.\ 2)$$

where γ (gamma) is the *interfacial tension* between water and air (it is about 70×10^{-3} N/m for water, and 20–40×10^{-3} N/m for crude oil, decreasing with temperature). For water, air and glass (or most mineral surfaces) the contact angle is about $30°$, for oil and mercury it is greater than $90°$. What this means is that water *wets* the mineral surface, but oil and mercury do not. Experience shows that almost all rocks are water-wet below the vadose zone: even when they are "saturated" with oil or natural gas, there is always a residual film of water between the mineral surface and the nonwetting fluid.

The geometry of pore space in sedimentary rocks is complex (see *Petrophysics of Sand and Sandstone*; Pitman and Duschatko, 1970; Enos and Sawatsky, 1981) and differs significantly from a bundle of capillaries. Nevertheless, the simple relationship linking displacement pressure with capillary size makes it possible to determine some important geometrical properties of interconnected pores by monitoring the displacement pressure needed to force a known volume of a nonwetting fluid, such as mercury into a dry, air-saturated sample of the rock, up to pressures high enough to (almost) saturate the rock with mercury, and then monitor the pressure again as the mercury is allowed to flow out. A typical graph (Figure F1) shows the displacement pressure (on a logarithmic scale, pressure increasing downward on the ordinate) plotted again the percentage of mercury saturation. It is also a way of displaying the size distribution of the pore system. As can be seen in the figure there is a hysteresis loop: No mercury enters the pore space until a certain critical pressure is exceeded (determined by the size of the largest pores), and progressively more mercury enters as mercury is forced into smaller pores (or the interconnecting *throats* between pores) up to 100 percent saturation, but when the pressure is released some mercury remains in the pores, because the internal pressure is no longer sufficient to force the mercury through the small throats connecting the larger pores. This factor limits not only the *recovery* of mercury from test samples, but also the recovery of oil from petroleum reservoirs. Note that extremely high displacement pressures are required to force fluids such as petroleum through very small pores: this is the reason why rocks of very low permeability, such as shales, will slowly "leak" water from underlying aquifers, but will serve as effective seals to form traps for petroleum.

Displacement pressure on such mercury injection plots serves as a proxy for pore or throat size (the smallest throats require the highest displacement pressure). For further discussion see *Petrophysics of Sands and Sandstone*. Diagenetic processes, particularly compaction and cementation (*q.v.*), frequently decrease the size of the throats (and the coordination) more than they do the absolute porosity. Burial generally produces a roughly linear decrease in porosity, but an exponential decrease in permeability, with depth. In sandstones, clay cements (illite and mixed layer clays) are

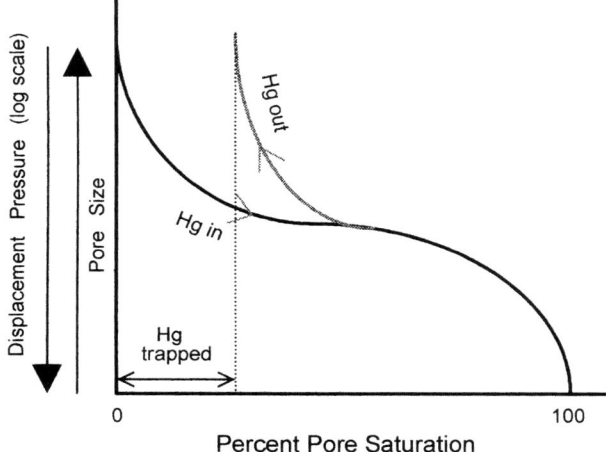

Figure F1 Schematic graph showing result of a mercury injection test. After the displacement pressure reaches a threshold value, mercury is driven into the (gas filled) pore space: as the pressure increases, more mercury enters smaller pores, until saturation is approached asymptotically. On decreasing the pressure, mercury flow out but not all it is can be recovered. The displacement pressure can be considered to be a proxy for pore (and throat) size: the largest pores correspond to the least displacement pressure, and the smallest pores the most pressure, so the "in" curve corresponds roughly to the size distribution of the pores.

particularly effective in clogging throats. There are, however, exceptions, where diagenetic dissolution at depth produces secondary porosity with large interconnecting throats (see *Diagenesis*).

Mercury injection is not the only technique available to provide a more detailed measure of pore structure than bulk measurements of porosity and permeability or their geophysical proxies (see *Geophysical Properties of Sediments*; Prensky, 1994).

Gerard V. Middleton

Bibliography

Allen, J.R.L., 1982. *Sedimentary Structures: Their Character and Physical Basis*. Volume I. Elsevier.
Archie, G.E., 1942. The electrical resistivity log as an aid in determining some reservoir characteristics. *Petroleum Transactions of the AIME*, **146**: 54–62.
Archie, G.E., 1952. Classification of carbonate reservoirs and petrophysical considerations. *American Association of Petroleum Geologists Bulletin*, **36** (2): 145–152.
Bideau, D., and Dodds, J. (eds.), 1991. *Physics of Granular Media*. Commack, NY: Nova Science Publishers.
Chapman, R.E., 1981. *Geology and Water: An Introduction to Fluid Mechanics for Geologists*. Martinus Nijhoff.
Choquette, P.W., and Pray, L.C., 1970. Geological nomenclature and classification of porosity in sedimentary carbonates. *American Association of Petroleum Geologists Bulletin*, **54**: 207–250.
Davis, J.M., Wilson, J.L., and Phillips, F.M., 1994. A portable air-minipermeameter for *in situ* field measurement. *Ground Water*, **32**: 258–266.
Enos, P., and Sawatsky, L.H., 1981. Porenetworks in Holocene carbonate sediments. *Journal of Sedimentary Petrology*, **51**: 961–985.

Freeze, R.A., and Cherry, J.A., 1979. *Groundwater*. Prentice-Hall.
Kahn, J.S., 1956. The analysis and distribution of packing in sand-size sediments. *Journal of Geology*, **64**: 385–395.
Kemp, A.E.S., 1995. Early sediment fabrics. *Journal of the Geological Society of London*, **152**: 117–118.
Lucia, F.J., 1995. Rock-fabric/petrophysical classification of carbonate pore space for reservoir characterization. *AAPG Bulletin*, **79**: 1275–1300.
Greenkorn, R.A., 1983. *Flow Phenomena in Porous Media: Fundamentals and Applications in Petroleum, Water, and Food Production*. Marcel Dekker.
Martini, I.P., 1978. Fabric, sedimentary. In Fairbridge, R.W., and Bourgeois, J. (eds.), *The Encyclopedia of Sedimentology*. Dowden: Hutchinson and Ross, pp. 320–322.
Middleton, G.V., 2000. *Data Analysis in the Earth Sciences using MATLAB*. Prentice Hall.
Middleton, G.V., and Wilcock, P.R., 1994. *Mechanics in the Earth and Environmental Sciences*. Cambridge University Press.
Monicard, R.P., 1980. *Properties of Reservoir Rocks: Core Analysis*. Paris: Editions Technip.
Neuman, S.P., and Neretnieks, I. (eds.), 1990. *Hydrogeology of Low Permeability Environments*. Verlag Heinz Heise.
Pitman, E.D., and Duschatko, R.W., 1970. Use of pore casts and scanning electron microscope to study pore geometry. *Journal of Sedimentary Petrology*, **40**: 1153–1157.
Prensky, S.E., 1994. A survey of recent developments and emerging technology in well logging and rock characterization. *The Log Analyst*, **35** (2): 15–45; **35** (5): 78–84.
Rogers, C.D.F., Dijkstra, T.A., and Smalley, I.J., 1994. Particle packing. *Earth Science Reviews*, **36**: 59–82.
Taylor, J.M., 1950. Pore-space reduction in sandstones. *Bulletin of the American Association of Petroleum Geologists*, **34**, 701–716.
Tidwell, V.C., Gutjahr, A.L., and Wilson, J.L., 1998. What does an instrument measure? Empirical spatial weighting functions calculated for permeability data sets measured on multiple sample supports. *Water Resources Research*, **35** (1): 43–54.

Cross-references

Cements and Cementation
Compaction (Consolidation) of Sediments
Diagenesis
Petrophysics of Sand and Sandstone

FACIES MODELS

Models are idealized simplifications set up to aid our understanding of complex natural phenomena and processes. They have been extensively used in the interpretation of sedimentary rock facies.

The concept of facies, that is a body of rock with specified characteristics, has been used ever since geologists, engineers and miners recognized that features found in particular rock units were useful in correlation and in predicting the occurrence of coal, oil, and mineral ores. The term was introduced by Gressly (1838) who used it to embrace the sum total of lithological and paleontological aspects of a stratigraphic unit. Since that time the term has been the subject of considerable debate, well summarized by Middleton (1973) and discussed extensively by Anderton (1985), Walker and James (1992), Reading (1987, 1996) and Miall (1999).

Ideally a rock facies should be a distinctive body of rock that formed under certain conditions of sedimentation, reflecting a particular process, set of conditions, or environment. It should differ from those bodies of rock above, below and laterally adjacent. It may be a single bed or group of similar beds. A facies may be subdivided into subfacies or grouped into associations of facies (facies associations). Where sedimentary rocks can be handled at outcrop or in cores, a facies can be defined on the basis of color, bedding, composition, texture, fossils, and sedimentary structures. If the biological content of the rock is its dominant aspect, then it should be called a biofacies, an ichnofacies being distinguished by its suite of trace fossils. If fossils are absent or of little consequence and emphasis is to be placed on the physical and chemical characteristics of the rock, then it should be called lithofacies. Where definition depends on features seen in thin section, as is often the case with carbonates, the term microfacies is used.

Facies models developed out of facies classifications that had been based on observable and measurable features, but, as time went by and our understanding of process and of environment increased, the word facies came to be used in more genetic senses, that is for the product of a process by which the rock was thought to have formed or the environment in which it was deposited. Although all these senses overlap, it is necessary to be aware of the sense in which the word facies or the phrase facies model is being used, in an objectively defined descriptive sense, or as an interpretation of the generating process or the environment in which it was deposited.

Facies classifications

The earliest facies classification in clastic sediments was that of Mutti and Ricci Lucchi (1972) for deep water facies. It has continued, with slight modifications, to the present day. Deep water sediments are divided into 7 classes, 16 groups, and 59 facies based on grain size, internal organization, composition, bed thickness, and sedimentary structures. In alluvial sediments, a scheme initiated by Miall (1977) is also widely used with extensive modifications to suit particular sequences. It recognizes three major grain size classes, gravel, sand and fines, and several modes of bedding (e.g., massive, trough cross-bedded, planar/tabular cross-bedded, rippled, horizontal laminated, shallow scours, etc.). These facies schemes are invaluable in giving a rigorous objective record of the rocks and enabling different geologists to produce consistent results; and they also force the observer to look at every aspect of the lithology. However they can become unwieldy and time consuming. The observer should therefore be prepared to modify such schemes according to the time available for the study and the objective of the exercise.

Classification schemes for limestones began, as with clastic sediments, with ones based on grain size. Limestones and dolomites were divided into calcilutite/dololutite (grains <62 μm), calcarenite/dolarenite (62 μm to 2 mm) and calcirudite/dolorudite (>2 mm). From this facies scheme, built on grain size, there developed the two classical, and still used, classifications of Folk (1959) and Dunham (1962). These depended more on texture than on grain size, in particular on the proportion of grains to matrix, and on the manner in which the grains were supported. Whilst utterly objective and descriptive, the divisions reflected, as with any grain size subdivision of facies, the energy level of the depositional environment. For example, a wackestone, where the grains are supported by the surrounding mud and fine-grained carbonate (micrite) indicate deposition in quiet water; a grainstone formed of detrital grains, supporting each other and lacking mud indicates agitated water.

Process facies models

Facies models based on the inferred process by which a sediment is considered to have formed generally have the suffix "-ite". The term *evaporite* (*q.v.*) has long been used for precipitates formed by the evaporation of sea or lake waters in climates where losses by evaporation exceed additions by precipitation and influx of less saline waters either fresh or from the sea. In the 1950s, as it began to be realized that many "graywackes", that is sandstones with a high matrix content, had been deposited by turbidity currents it became common to use the term "graywacke facies" for the deposits of a turbidity current. However, not all graywackes were formed by turbidity currents; nor does every deposit of a turbidity current have the high matrix content of a graywacke. The problem was solved by the introduction of the word *turbidite* (*q.v.*) for all rocks thought to have been deposited by a turbidity current. The term "graywacke" could then be restricted to a sandstone of a well defined composition regardless of whether the high matrix content was the result of deposition from a turbid current or diagenetic alteration of unstable Fe–Mg minerals as some had postulated.

The turbidite model, the classic facies model, was first developed by Kuenen (1950) and then more widely publicized in the paper by Kuenen and Migliorini (1950). They combined knowledge of graded beds as seen in outcrops of ancient rocks, flume experiments, and the increasing knowledge of present day deep oceanic processes into a genetic turbidite model. Their pioneering work was further developed by Bouma (1962) whose field measurements quantified the range and variety of turbidite facies. Bouma's study was essentially descriptive. Harms and Fahnestock (1965) and Walker (1965), on the other hand, used our expanding understanding of the physics of sedimentation to explain the Bouma sequence of sedimentary structures in terms of flow regimes.

The turbidite model is only one of an increasing number of models for deep water mass-gravity flows. Soon after its arrival it was realized: (1) that turbidity currents could be of low density as well as of high density, as Kuenen had thought; (2) that grain flows (with grain-to-grain interactions), liquefied flows (where grains settle downward, displacing fluid upward) and fluidized flows (where fluid moves upward) may each produce their own suite of sedimentary structures. In addition, coarser material than sand can be transported into deep water by sudden falls of rock, by slumping, by sliding, and by slurry-like debris flows (*q.v.*) that yield a facies known as a debrite.

Since the processes of transport and deposition can never be known with surety, and since post-depositional modifications can add to or obliterate depositional structures, debate continues fiercely to this day as to how these deep-water facies should be categorized. Massive sandstones, for example, though generally thought to have been formed from high density turbidity currents, are argued by Shanmugam (2000) to have formed as sandy debris flows and therefore should be *debrites*, not turbidites.

Other deep water facies are *pelagites*, pelagic deposits consisting of various proportions of calcareous and siliceous microfossils and clay formed by slow settling in the oceans. Rocks formed in this way include chalk, chert, Fe-Mn nodules, phosphates and red clays. *Hemipelagites* are fine-grained sediments that have not only the same biogenic and clay material as pelagites but also contain a significant terrigenous component (25 to 90 percent) derived from an adjacent land area.

Contourites are somewhat ambiguously termed. They form from relatively continuous semi-permanent currents in all depths of oceanic waters, and in lakes. Until the 1960s it was generally thought that the oceans were rather stagnant bodies of water where deposition below wave-base (approximately 200 m) was confined to continuous pelagic settling interspersed with rare catastrophic slumps and turbidity current flows that could bring sands and other coarser material into deep water. As it became possible to measure mass water flows in the ocean it was realized (e.g., Heezen *et al.*, 1966) that the oceans contained an enormous range of semi-permanent currents which are the deep-water expression of oceanic thermohaline circulation. Deep-ocean bottom water is formed by the cooling and sinking of surface water in polar regions that flows toward more equatorial regions with surface waters, such as the Gulf Stream, returning waters from the tropics to the poles. The distribution of land and ocean-bottom topography, and the Earth's spin, all cause complications in any general pattern. Of particular importance is the Coriolis Force, due to the Earth's rotation, that causes an anticlockwise rotation in the Northern Hemisphere and thus produces currents such as the Western Boundary Undercurrent that flows southwestward, paralleling the contours on the continental slope and rise. Such currents are known as contour currents and the facies is termed contourites.

In shallow marine and coastal waters, the two principal physical processes of sedimentation are tidal currents and storms. On some confined continental shelves, particularly in estuaries, tidal currents may be very powerful, giving rise to a facies known as *tidalites*, that can record the daily or, more commonly twice daily (semidiurnal) tidal rise and fall that give rise to flood and ebb tides. The very strong currents produced by these tides can give a facies dominated by tidal currents with their own distinctive pattern of cross beds, sometimes even bipolar, formed during the two ebb and flood current stages and mud drapes deposited during the intervening slack periods.

Where tidal currents are weak or absent the facies on continental shelves are documented by storm deposits (*tempestites*), coarse-grained sandy or bioclastic sediments, often graded, interbedded with finer, quiet water sediments. These graded, sharp-based sandstones have often been confused with graded turbidites because the sequence of sedimentary structures is very similar. They can be distinguished by the presence of wave ripples indicating that they formed above the level of wave base (see *Storm Deposits*).

Facies relationships, associations, and sequences

The interpretation of a facies can seldom be made in isolation. A rootlet bed can indicate deposition very close to or just above water level. But it does not tell us whether it formed in a backswamp, on an alluvial fan or river levee. Graded sandstone beds may be formed by a river flood, a storm in the sea or by the collapse of continental slope sediments into deeper water. Only the context in which we find the graded bed tells us the environment or the process by which it formed. We have to recognize the limitations of individual facies taken in isolation.

Facies have to be interpreted by reference to their neighbors and are consequently grouped into *facies associations* that are thought to be genetically or environmentally related (Collinson, 1969). The association provides additional evidence which makes interpretation of the environment easier than treating each facies in isolation, particularly in the elimination of alternative interpretations.

Facies associations are therefore the essential building blocks of facies analysis and model groups of facies associations have been constructed for most environments. For example, Miall (1985) postulates eight architectural elements in fluvial environments and Mutti and Normark (1991) suggest there are five primary elements in deep-sea fans, including large-scale erosional features as one of their elements.

In some successions, the facies within a succession are interbedded randomly. In others, the facies may lie one above another in a preferred order with predictable upward and downward changes. The importance of facies sequences has long been recognized, at least since Walther's Law of Facies (quoted in Middleton, 1973) which states that "only those facies and facies areas can be superimposed, without a break, that can be observed beside each other at the present time". This concept has been taken to indicate that facies occurring in a conformable vertical sequence were formed in laterally adjacent environments and that facies in vertical contact must be the product of geographically neighboring environments.

However, Walther's Law only applies to successions without major breaks. A break in the succession marked either by an erosive contact, or simply by a hiatus in deposition may represent the passage of any number of environments whose products, if they were deposited, were subsequently removed. Some breaks in the succession are essentially autocyclic, that is they reflect natural sedimentary processes such as the switching of a delta, the lateral migration and avulsion of a river channel or the build up of a carbonate platform. Other breaks, generally more laterally extensive, may be the result of allocyclic controls, external to the local environment due to tectonic movements or changes in climate, sediment supply or sea level.

Environmental facies models

The ultimate object of facies modeling is to interpret not only the depositional environment of the sedimentary body being examined, but the total context within which it formed, the climate, oceanic circulation patterns, rates of sedimentary supply, rates and nature of sea-level changes both local and eustatic, and local, regional and global tectonics. The stratigraphic age is also important; facies models change as organisms have evolved, and Precambrian oceans were very different in their chemistry to those of Mesozoic or Cenozoic times.

To this end we have not only to describe the facies and determine the physical, chemical, and biological processes involved, but to understand the depositional environment.

Environmental facies models were developed initially from studies of modern environments. Classic studies of these environments include: (1) Glennie (1970) on desert sedimentary environments: he was sent to study deserts by Shell after the discovery of huge amounts of gas in Permian desert deposits in the Netherlands; (2) Oomkens and Terwindt (1960) on estuaries, especially deposition in tidal channels; (3) Fisk (1944) and Bernard and Major (1963) on fluvial, especially meandering river deposits; (4) Fisk *et al.* (1954) on delta deposits; (5) Illing (1954) on the carbonate platform of the Bahamas; and (6) Shearman (1966) on evaporites forming in sabkhas (*q.v.*).

Environmental facies models were then further developed by measurement of outcrops, particularly analysis of sequences and cycles, research into which had been considerable in previous decades (see *Cyclic Sedimentation*), the interpretation of sedimentary structures in terms of physical processes, and of composition by analogy with chemical theory and experimentation, and by comparing structures and composition with comparable features in modern sediments.

The best known and most influential studies were those of Allen 1964, 1965, 1970 who published a series of papers that interpreted fining upward cycles of the alluvial Old Red Sandstone of England and Wales as a consequence of lateral migration of point bars in meandering rivers (see *Meandering Channels*). The full cycle consisted of a sharp, channelled base overlain by an intraformational conglomerate of caliche fragments that passed gradually upward through parallel-bedded, cross-bedded, cross-laminated sandstones to siltstones capped by a calcrete soil bed. This fining upward meandering river model was used extensively by workers all over the world, sometimes to the detriment of alternative explanations for such fining upward cycles. One alternative model, particularly suitable for sediments deposited in semi-arid climates such as that of the Old Red Sandstone, is that the cycles are the result of ephemeral rivers that flow for limited periods of time, the channelling and deposition of sandstones occurring during the wet (pluvial) periods and with flows diminishing as rainfall is reduced. Another explanation for many of the smaller cycles is that they were caused by flash floods of very limited duration that can give rise to graded beds of some thickness.

Models for a wider spectrum of alluvial deposits were gradually developed over many years by Schumm (1972) who separated alluvial systems into streams dominated by bedload, by mixed load and by suspended load sediments, by Miall (1985) who emphasized successions made up of building blocks termed architectural elements and by Orton and Reading (1993), who showed that models needed to be developed for a spectrum that went from alluvial fans (*q.v.*) through braidplains to the finer grained, low gradient normal river systems.

Delta models were developed very early on by the convergence of studies of coarsening upward cyclothems, especially in the Upper Carboniferous sediments of the USA and Europe, and studies of modern deltas, initially the Mississippi. The abundance of petroleum reservoirs in such deltaic deposits led to petroleum financed studies of other deltas such as the Rhône, Rhine, Niger, and Ganges/Brahmaputra and it was soon realized that delta shape and sediment accumulation patterns, especially reservoir sand bodies varied substantially according to whether the dominant processes of sedimentation were by river discharge or by marine reworking by waves, storms or by tidal currents and in particular the calibre of sediment supplied by the alluvial system (Orton and Reading, 1993).

Environmental facies models for carbonate sediments were initiated by Illing (1954) working on the Bahama platform and by Ginsburg from 1956 onward studying the Florida shelf. Their work on modern environments was then extended in the 1970s by Wilson (1975) who combined knowledge of ancient carbonate facies with the increasing research on modern carbonate environments into a model of nine lateral facies belts that extended from a deep water basinal, euxinic environment through open, oxygenated marine, slope, organic build up to open or restricted platform. Each environment could be recognized by its facies, especially biotas. This very general facies model was then supplemented by a range of lateral facies models for isolated platforms, shelves attached to the land and carbonate ramps with gently sloping surfaces of less than 1°. In addition, models had to be created for platforms that do not exist today such as the very extensive epeiric platforms that were widespread in the late Precambrian and Paleozoic. They were some hundreds or thousands of kilometers across, developing on stable cratonic interiors and are characterized by distinctive sags or intraplatform basins.

At the platform margins, reefs (*q.v.*) may develop. Here, biological sediment production is realized to its maximum extent. Over geologic time a range of different organisms have contributed to the building of reefs so the biofacies is stratigraphically controlled. Corals and algae may predominate today, but in Cretaceous times rudist bivalves built reefs; at other times building may have been by bryozoa, sponges or crinoids.

Beyond the platform margin, a range of different slopes occur. Some margins, especially below reefs, are steep-sided; others are more gentle. Some are erosional; others are accretionary. Windward and leeward margins also differ. At the base of slope carbonate fans and aprons may form.

The three principal environments of deep water deposition, basin floor, submarine fan, and slope apron were first delineated in modern Californian basins by Gorsline and Emery (1959). However, facies models for submarine fans were developed primarily from the study of ancient rocks in the English Pennines (Walker, 1966) and Italian Apennines (Mutti and Ricci Lucchi, 1972) with the single "all-purpose" fan model of Walker (1978) widely used by everyone in the next decade. The reason for models being driven by studies of ancient rocks, rather than modern sediments was because, whilst the shape of modern submarine fans could be very roughly outlined (e.g., Normark, 1970) it was not possible to determine the nature of the sediments other than in the top few decimeters. On land, however, whilst submarine fan shapes and channels were only rarely exposed, vertical sections, kilometers in length could be measured. By analogy with the fining upward fluvial model and the coarsening upward delta model great emphasis was placed on sequences of bed thickness in submarine fans, thinning-upward sequences being thought to indicate channels and thickening-upward sequences to indicate prograding lobes. As DSDP, ODP and petroleum company cores have been recovered, vertical sequence models have now been developed for a much wider range of deep sea environments. For example, Stow (1985) differentiates 18 vertical sequences for turbidites, contourites, debrites, pelagites, and hemipelagites in the deep seas.

However, in the last 20 years, these earlier simplistic facies models have been shown to be inadequate. Ironically, our understanding of deep-sea sediments has been increasingly based on studies of modern environments rather than ancient rocks. This is not only due to the increasing number of cores in oceanic sediments, but is mainly the result of seismic profiling. Because seismic profiling is much easier in deep water than in shallow water or on land, models for deep water sediments are now derived more than any other facies models from studies of modern seas. Since the paper of Mitchum *et al.* (1977) in the classic memoir edited by Payton (1977), seismic facies, based on reflection configuration, continuity, amplitude, frequency and interval velocity, together with the external form of the unit has become the standard tool for 3-dimensional facies patterns which are now well established for deep-sea sediments. The architectural elements of Mutti and Normark (1991) and the 12 distinctive facies models of Reading and Richards (1994) are largely based on data obtained from seismic profiles of base-of-slope depositional systems. The latter showed that separate models needed to be created, distinguished by the volume and calibre (grain-size) of the sediment supply and the nature of the supplying system whether it is a single point source, a multiple source or a linear slope apron.

Summary

Facies classifications were initially mainly descriptive, based on observable, measurable features such as the composition and texture of sedimentary rocks. As time went by, facies models were developed based on the inferred process of formation. Such genetic facies models generally have the suffix "-ite", e.g., "turbidite". Simultaneously, it was appreciated that individual facies could not be interpreted in isolation but had to be studied with reference to their neighbors. Such models emphasized the association of facies, sequences of facies, in particular those that coarsened or fined upward, and the presence or absence of breaks in the succession.

The establishment of environmental facies models is the ultimate objective of facies modeling. Only in this way is it possible to establish the nature of past climates, oceans, sea levels and global and local tectonic patterns. Environmental models are developed by the interaction of studies on modern and ancient rock sequences. Initially relatively simple, all embracing models were developed. These have become increasingly complex as our knowledge of the variability of nature has increased.

Complex though these models are, they are only simplifications of reality. In nature there are no models and the majority of past environments differed in some respect from modern environments. It must be remembered therefore that each environment and rock sequence is unique.

Harold G. Reading

Bibliography

Allen, J.R.L., 1964. Studies in fluviatile sedimentation: six cyclothems from the Lower Old Red Sandstone, Anglo-Welsh basin. *Sedimentology*, **3**: 163–198.

Allen, J.R.L., 1965. Fining upwards cycles in alluvial successions. *Geological Journal*, **4**: 229–246.

Allen, J.R.L., 1970. Studies in fluviatile sedimentation: a comparison of fining-upwards cyclothems with special reference to coarse-member composition and interpretation. *Journal of Sedimentary Petrology*, **40**: 298–323.

Anderton, R., 1985. Clastic facies models and facies analysis. In *Geological Society of London Special Publication* 18, pp. 31–47.

Bernard, H.A., and Major, C.F. Jr., 1963. Recent meander belt deposits of the Brazos River: an alluvial 'sand' model. *Bulletin American Association of Petroleum Geologists*, **47**: 350.

Bouma, A.H., 1962. *Sedimentology of Some Flysch Deposits: A Graphic Approach to Facies Interpretation*. Amsterdam: Elsevier.

Collinson, J.D., 1969. The sedimentology of the Grindslow Shales and the Kinderscout Grit: a deltaic complex in the Namurian of northern England. *Journal of Sedimentary Petrology*, **39**: 194–221.

Dunham, R.J., 1962. Classification of carbonate rocks according to depositional texture. In *American Association of Petroleum Geologists Memoir* 1, pp. 108–121.

Fisk, H.N., 1944. *Geological Investigations of the Alluvial Valley of the Lower Mississippi River*. Vicksburg: Mississippi River Commission.

Fisk, H.N., McFarlan, E. Jr., Kolb, C.R., and Wilbert, L.J. Jr., 1954. Sedimentary framework of the modern Mississippi delta. *Journal of Sedimentary Petrology*, **24**: 76–99.

Folk, R.L., 1959. Practical petrographic classification of limestones. *Bulletin American Association of Petroleum Geologists*, **43**: 1–38.

Ginsburg, R.N., 1956. Environmental relationships of grain size and constituent particles in some south Florida carbonate sediments. *Bulletin American Association of Petroleum Geologists*, **40**: 2384–2427.

Glennie, K.W., 1970. *Desert Sedimentary Environments*. Amsterdam: Elsevier.

Gorsline, D.S., and Emery, K.O., 1959. Turbidity-current deposits in San Pedro and Santa Monica basins off Southern California. *Bulletin Geological Society of America*, **70**: 279–290.

Gressly, A., 1838. Observations géologiques sur le Jura Soleurois. *Neue Denkschriften Der Allgemeinen Shweizerischen Gesellschaft fur die gesammten Naturwissenschaften*, **2**: 1–112.

Harms, J.C., and Fahnestock, R.K., 1965. Stratification, bed forms, and flow phenomena (with an example from the Rio Grande). In *Society of Economic Paleontologists and Mineralogists Special Publication* 12, pp. 84–115.

Heezen, B.C., Hollister, C.D., and Ruddiman, W.F., 1966. Shaping of the continental rise by deep geostrophic contour currents. *Science*, **152**: 502–508.

Illing, L.V., 1954. Bahaman calcareous sands. *Bulletin American Association of Petroleum Geologists*, **38**: 1–95.

Kuenen, Ph.H., 1950. Turbidity currents of high density. *18th International Geological Congress, London*, 1948; Report part 8, pp. 44–52.

Kuenen, Ph.H., and Migliorini, C.I., 1950. Turbidity currents as a cause of graded bedding. *Journal of Geology*, **58**: 91–127.

Miall, A.D., 1977. A review of the braided-river depositional environment. *Earth Science Reviews*, **13**: 1–62.

Miall, A.D., 1985. Architectural-element analysis: a new method of facies analysis applied to fluvial deposits. *Earth Science Reviews*, **22**: 261–308.

Miall, A.D., 1999. Perspectives: in defense of facies classifications and models. *Journal of Sedimentary Research*, **69**: 2–5.

Middleton, G.V., 1973. Johannes Walther's law of the correlation of facies. *Bulletin Geological Society of America*, **84**: 979–988.

Mitchum, R.M. Jr., Vail, P.R., and Sangree, J.B., 1977. Seismic stratigraphy and global changes of sea level, Part 6: stratigraphic interpretation of seismic reflection patterns in depositional sequences. In *American Association of Petroleum Geologists Memoir* 26, pp. 117–133.

Mutti, E., and Ricci Lucchi, F., 1972. Le torbiditi dell'Appennino Settentrionale: introduzione all'analisi di facies. *Memorie della Societa geologica Italiana* 11, pp. 161–199.

Mutti, E., and Normark, W.R., 1991. An integrated approach to the study of turbidite systems. In Weimer, P., and Link, M.H. (eds.), *Seismic Facies and Sedimentary Processes of Submarine Fans and Turbidite Systems*. New York: Springer-Verlag, pp. 75–106.

Normark, W.R., 1970. Growth patterns of deep-sea fans. *Bulletin American Association of Petroleum Geologists*, **54**: 2170–2195.

Oomkens, E., and Terwindt, J.H.J., 1960. Inshore estuarine sediments in the Haringvliet, Netherlands. *Geologie Mijnbouw*, **39**: 701–710.

Orton, G.J., and Reading, H.G., 1993. Variability of deltaic processes in terms of sediment supply, with particular emphasis on grain size. *Sedimentology*, **40**: 475–512.

Payton, C.E. (ed.), 1977. Seismic stratigraphy—applications to hydrocarbon exploration. *American Association of Petroleum Geologists Memoir* 26.

Reading, H.G., 1987. Fashions and models in sedimentology: a personal perspective. *Sedimentology*, **34**: 3–9.

Reading, H.G. (ed.), 1996. *Sedimentary Environments: Processes, Facies and Stratigraphy*. Oxford: Blackwell Science.

Reading, H.G., and Richards, M., 1994. Turbidite systems in deep-water basin margins classified by grain size and feeder system. *Bulletin American Association of Petroleum Geologists*, **78**: 792–822.

Schumm, S.A., 1972. Fluvial paleochannels. In *Society of Economic Paleontologists and Mineralogists Special Publication* 16, pp. 98–107.

Shanmugam, G., 2000. 50 years of the turbidite paradigm (1950s–1990s): deep-water processes and facies models—a critical perspective. *Marine and Petroleum Geology*, **17**: 285–342.

Shearman, D.J., 1966. Origin of marine evaporites by diagenesis. *Transactions Institution of Mining and Metallurgy, Series B*, **75**: 208–215.

Stow, D.A.V., 1985. Deep-sea clastics: where are we and where are we going? In *Geological Society of London Special Publication* 18, pp. 67–93.

Walker, R.G., 1965. The origin and significance of the internal sedimentary structures of turbidites. *Proceedings Yorkshire Geological Society*, **35**: 1–32.

Walker, R.G., 1966. Shale Grit and Grindslow Shales: transition from turbidite to shallow water sediments in the Upper Carboniferous of northern England. *Journal of Sedimentary Petrology*, **36**: 90–114.

Walker, R.G., 1978. Deep-water sandstone facies and ancient submarine fans: models for exploration for stratigraphic traps. *Bulletin American Association of Petroleum Geologists*, **62**: 932–966.

Walker, R.G., and James, N.P. (eds.), 1992. *Facies Models: Response to Sea Level Change*. Waterloo: Geological Association of Canada.

Wilson, J.L., 1975. *Carbonate Facies in Geologic History*. Berlin, Heidelberg, New York: Springer-Verlag.

Cross-references

Alluvial Fan
Bioturbation and Trace Fossils
Caliche–Calcrete
Classification of Sediments and Sedimentary Rocks
Cyclic Sedimentation
Debris Flow
Evaporites
Gravity-Driven Mass Flows
Meandering Channel
Reefs
Sabkha, Salt Flat, Salina
Sedimentologists
Submarine Fans and Channels
Tills and Tillites
Turbidites

FAN DELTA

The most commonly adapted definition of fan delta is a coastal prism of sediments derived from an *alluvial-fan* feeder system and deposited mainly or entirely subaqueously at the interface between the active fan and a standing body of water (Nemec and Steel, 1988). The importance of the feeder system as a criterion for fan delta recognition roused significant discussion among students of fan deltas on what is an alluvial fan and how to recognize it in the fossil record, an issue that is still not resolved. Most fan delta researchers, however, refer to alluvial fan systems as steep gradient, often gravelly, cone-shaped fluvial systems, which can be dominated by either *sediment gravity-flow* processes or by shallow streams feeding the delta front essentially as a line source.

Fan deltas are taken to be sensitive recorders of climate change and tectonics. The response time of fan deltas is short compared to river deltas, because their drainage basins are generally small and the gradient of their feeder system is relatively steep permitting high rates of sediment transport.

Hence, supply pulses generated by climate and tectonics are likely to be preserved in the delta body in the form of bed thickness variation, grain size and sediment composition, while changes in relative sea-level would leave its trace in the form of stacking patterns of delta components.

Seismic studies of the subaqueous part of modern fan deltas that started in the late seventies (Prior *et al.*, 1981) added significantly to researchers' insight into the sedimentary processes that governed the morphology of the delta slopes. Due to the combination of extensive marine and outcrop studies of fan deltas in various tectonic and climate settings, progress in fan-delta research was significant in the eighties and early nineties (Kosters and Steel, 1984; Nemec and Steel, 1988b; Colella and Prior, 1990; Dabrio *et al.*, 1991; Chough and Orton, 1995).

Fan delta concept

Fan delta morphology and architecture is the result of the interaction of the alluvial fan feeder system with the receiving basin. Figure F2 shows how delta morphology is defined by alluvial processes (fluvial regime) on the one hand and basinal processes (basinal regime) on the other. The fluvial regime is controlled by hinterland characteristics such as drainage area, geology (e.g., lithological variation) and topographical relief of the catchment, which are all subject to continuous change. The rates at which these changes take place depend on the rate of variation in climate, tectonic and sea-level change. For instance, periods with increased run-off at times of climate deterioration, or periods of tectonic uplift of the hinterland will lead to an increase in sediment supply to the delta due to increased erosion rates in the hinterland. But also other factors, like vegetation and not to forget civilization, add to the total of the fluvial regime. The basinal regime is controlled by characteristics of the basin. Geographical setting, shape, depth and size of the basin define, by and large, its dynamics in terms of available tide and wave energy, but also other variables like the density of the basin water relative to the river water plays a role in the morphology of the fan-delta.

The control of the fluvial regime on fan delta architecture is best shown by fan deltas of low energy basins, i.e., in basins where wave and tidal influence is small. Figure F3 shows 6 prototype deltas based on the type of feeder system and the basin depth. Feeder system type-A represents steep gradient (more than a few degrees), often gravelly, cone-shaped systems, dominated by mass-flow processes and type-B represents a steep-gradient ($\sim 0.4°$) feeder system that is characterized by highly mobile (unstable) bed-load channels feeding the delta front essentially as a line source (cf. Postma, in Colella and Prior, 1990). The steepness of the fan-delta front depends most on the initial basin relief (e.g., absence or presence of a shelf) and the fluvial regime at the river outlet. In settings where the depth ratio (channel depth over basin depth) is small (e.g., fault scarps) and where bed-load transport is high, deltas with steep gradient delta slope (Gilbert-type deltas called after G.K. Gilbert: see *Sedimentologists*, types 3 to 6, in Figure F3) will form. In settings of high depth ratio and with high-competence streams of high suspension-bed-load ratio (occuring for instance during flooding), deltas will be characterized by shoal water profiles, possibly with mouth bars in the stream outlets (Hjulström-type deltas, types 1 and 2 in Figure F3).

The role of the basinal regime in creating delta morphology can be pictured as the relative contribution of wave and tide processes to river processes. Wave-dominated fan deltas have been described frequently from the modern and ancient record. They are often recognized by straightened, wave-worked delta fronts and gravelly beach facies overlying a flat erosion surface paved with a gravel lag and truncating the delta foreset. Tidal influence in steep gradient fan deltas is commonly assumed to be minimal.

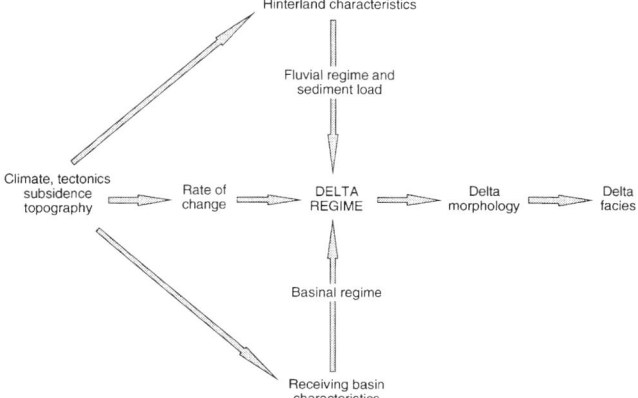

Figure F2 Conceptual framework for fan delta morphology, architecture and facies.

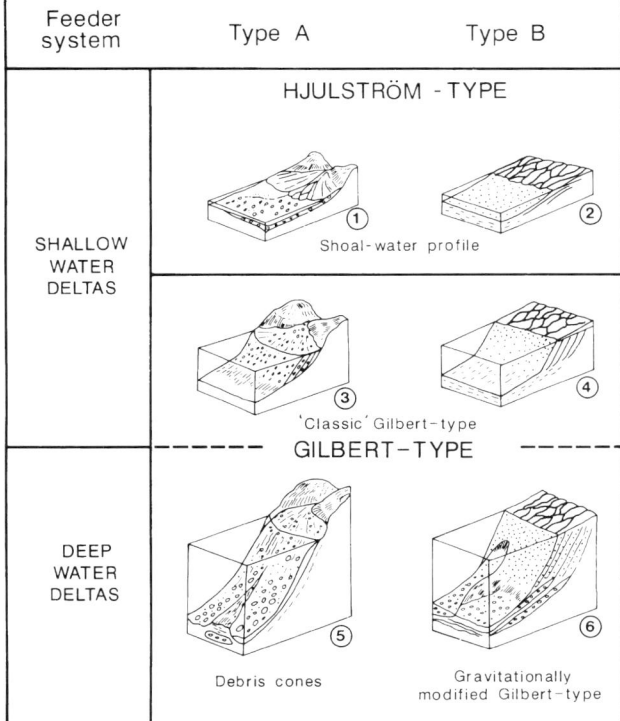

Figure F3 Morphology and architecture of various types of fan delta. Deltas built by the high gradient alluvial streams are generally small in size (few km's radius).

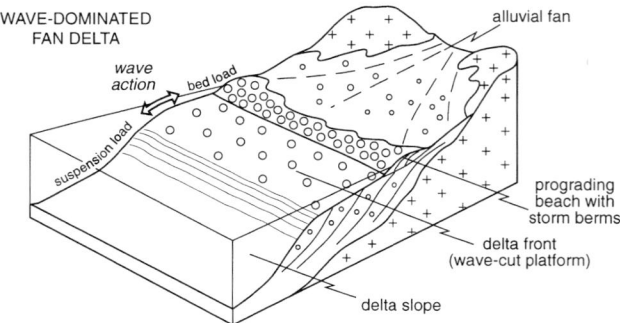

Figure F4 Wave-dominated fan delta with gravelly beach and storm-berms prograding onto the wave-cut platform (modified from Dabrio et al., 1991b).

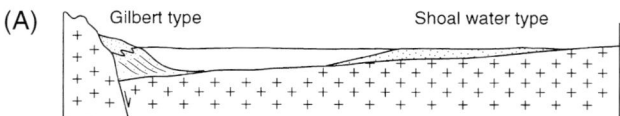

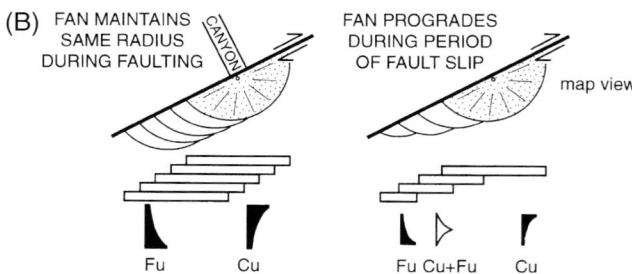

Figure F5 (A) Tectonically controlled delta architecture (Gulf of Corinth, redrawn from Leeder et al., 1988); (B) Stacking of fan deltas in strike slip settings resulting in fining and coarsening upward cycles (Hornelen Basin, redrawn from Steel, 1988).

Physical signatures of allocyclic change

Physical signatures of climate change in delta successions occur at various scales and frequencies. Very high frequency changes occur in the form of catastrophic flood deposits. On a time scale of *Milankovitch cycles* (*q.v.*), orbital-forced climate change have been found to be recorded as distinct turbidite sequences in the prodelta of deep-water delta systems, where turbidite deposition occurred mainly during the wet periods of the precession cycle (Postma, 2001).

Sea-level changes are preserved in the form of distinct stratigraphical styles of delta development: Progradation, aggradation and retrogradation define the large-scale architecture of a delta and are principally governed by sediment yield and relative sea-level change (Postma, in Chough and Orton, 1995; Steel and Marzo, 2001).

Physical signatures of tectonics comprise distinct patterns of stacking, rotation, supply and erosion of delta systems. The development of grabens can significantly influence the supply to delta systems by closure and opening of sediment transfer zones. Gilbert-type deep water delta systems are likely to form against the footwall, while shoal-water type deltas develop on the hanging wall crest. Strike-slip related subsidence may lead to a stack of rotated and offset stacked fan bodies, whose architecture is controlled by the relative rates of fan progradation and fault slip (Figure F5).

Current investigations and concluding remarks

Current research is focusing on the impact of allocyclic controls on the architectural style and distribution. Recently documented studies of the Eocene Montserrat and Sant Llorenç del Munt clastic wedges in northern Spain in Marzo and Steel (2000) illustrate nicely the progress that has been made in linking fan delta architecture and facies with tectonically controlled subsidence and relative sea-level changes. The importance of the work lies in careful 3-dimensional high-resolution reconstruction of the fan delta complexes, which allows detailed estimates of sediment volume and distribution in relation to relative sea-level change and climate. However, the values of the allocyclic variables cannot be determined from the architecture of deposits alone and need to be verified against other independent evidence.

George Postma

Bibliography

Chough, S.K., and Orton, G.J. (ed.), 1995. *Fan Deltas: Depositional Styles and Controls*. Sedimentary Geology, Special Volume 98, 292pp.

Colella, A., and Prior, D.B. (ed.), 1990. *Coarse Grained Deltas*. International Association of Sedimentologists, Special Publication, 10, 357pp.

Dabrio, C.J., Zazo, C., and Goy, J.L. (eds.), 1991a. *The Dynamics of Coarse-grained Deltas*. Madrid: Cuadernos de Geologia Iberica, Universidad Complutense de Madrid, 405pp.

Dabrio, C.J., Bardaji, T., Zazo, C., and Goy, J.L., 1991b. Effects of sea-level changes on a wave-worked Gilbert-type delta (Late Pliocene, Aguilas Basin, SE Spain). In Dabrio, C.J., Zazo, C., and Goy, J.L. (eds.), *The Dynamics of Coarse-grained Deltas*. Madrid: Cuadernos de Geologia Iberica, Universidad Complutense de Madrid, pp. 103–137.

Kosters, E.H., and Steel, R.J. (eds.), 1984. *Sedimentology of Gravels and Conglomerates*. Canadian Society of Petroleum Geologists Memoir, 10, 441pp.

Leeder, M.R., Ord, D.M., and Collier, R., 1988. Development of alluvial fans and fan deltas in neotectonic extensional settings: implications for the interpretation of basin fills. In Nemec, W., and Steel, R.J. (eds.), *Fan Deltas: Sedimentology and Tectonic Settings*. Glasgow: Blackie and Son, pp. 173–185.

Nemec, W., and Steel, R.J. (eds.), 1988. *Fan Deltas: Sedimentology and Tectonic Settings*. Glasgow: Blackie and Son, 444pp.

Postma, G., 2001. Physical climate signatures in shallow- and deep-water deltas. *Global and Planetary Change*, **28**: 1–4, 93–106.

Prior, D.B., Wiseman, W.J. JR., and Bryant, W.R., 1981. Submarine chutes on the slopes of fjord deltas. *Nature*, **290**: 326–328.

Steel, R.J., 1988. Coarsening upward and skewed fan bodies: symptoms of strike-slip and transfer fault movement in sedimentary basins. In Nemec, W., and Steel, R.J. (eds.), *Fan Deltas: Sedimentology and Tectonic Settings*. Glasgow: Blackie and Son, pp. 75–83.

Steel, R.J., and Marzo, M. (eds.), 2001. High resolution sequence stratigraphy and sedimentology of tectonic clastic wedges (SE Ebro Basin, NE Spain). *Sedimentary Geology*, Special Issue, 138(1–4).

Cross-references
Alluvial Fan
Gravity-Driven Mass Flows
Milankovitch Cycles
Sedimentologists

FEATURES INDICATING IMPACT AND SHOCK METAMORPHISM

Asteroid/comet impact is now recognized as a fundamental process on the surfaces of all solid bodies in the Solar System. Some 170 impact structures have been identified on Earth—only a fraction of the predicted terrestrial impact record (Grieve, 1998). A few impactogenic horizons and two large impact structures have been recognized in the Archean-early Proterozoic record. One global mass extinction horizon, the Cretaceous–Tertiary boundary, has been positively linked with a giant impact event at Chicxulub (Mexico). Several other mass extinction horizons, including the largest one at the Permian–Triassic boundary, are still being investigated for conclusive evidence of impact (e.g., Dressler and Sharpton, 1999; Montanari and Koeberl, 2000; Koeberl and McLeod, 2001). Some impact structures have sedimentary fills that have remained virtually undisturbed since deposition and could provide outstanding sedimentological and paleoenvironmental records (e.g., Partridge, 1998). Detailed accounts of the impact process, the history of its investigation, and its environmental effects are given by, for example, Melosh (1989, 1992), Chapman and Morrison (1994), and French (1998).

Much remains to be learnt about Earth's impact record and the role that this process has played throughout Earth's evolution. Likely, much impact evidence is still hidden in the sedimentological record. This is reason enough to review the phenomena used for recognition of impact facies—particularly the mineralogical phenomena known collectively as *shock metamorphism*.

Impact cratering process and impactites

Upon impact of an extraterrestrial projectile traveling at hypervelocity (11–72 km per second), an intense shock wave is generated, with shock pressures in excess of several 100 GPa and shock temperatures of 10,000s°C near the point of impact. The rapidly expanding shock wave imparts a particle velocity onto the target material, thus excavating a crater, but quickly loses energy along its outward path, with concomitant deformation decrease. In the immediate impact area, target rock and much of the projectile are completely vaporized. The next zones are characterized by wholesale impact melting, followed by successive zones of mineral melting, plastic deformation, and, eventually, elastic deformation (cataclasis). In some structures it may be possible to identify this zonation long after the impact event, but crater collapse, erosion, and tectonic overprint may obscure this pattern (e.g., Melosh, 1989; French, 1998).

The deformation caused during propagation of the shock wave through the target and later structural modification lead to a distinct crater geology and impact rock types in specific settings in and around the crater. The crater rim zone comprises up- and overturned target strata, is somewhat elevated above the pre-impact surface, and may display injections of *allogenic*, or formations of *authigenic*, impact breccia. Before the onset of erosion, the rim and wider environs are covered with a blanket of *fall-out* impact breccia that quickly decreases in thickness away from the crater. The crater interior is filled with *fall-back* impact breccia. Three impact breccia types are distinguished: *impact melt breccia*, formed in relatively close proximity to the impact site, resembles clast-laden, glassy or crystalline igneous rocks; *suevite* is composed of a matrix of clastic material containing clasts of impact glass or melt; and *fragmental (lithic) impact breccia* comprises clastic constituents only. Due to the turbulent processes in the crater, each breccia type may contain clasts of other types. All breccia types may also occur in the form of injections into the crater rim and floor. The floor zone is strongly brecciated, containing authigenic, polymict and monomict, fragmental breccias, as well as injections of allogenic breccia of different types. In addition, friction melting can occur and form pseudotachylitic breccia (Reimold, 1998). Fragmental impact breccia may macroscopically resemble sedimentary breccias, particularly those classified as *diamictites*. Indeed, some diamictites have been regarded in the past as possible impact ejecta, and vice versa (Reimold *et al.*, 1998).

Macroscopic recognition criteria of impact structures

Remnants of a meteoritic projectile provide unambiguous proof for the presence of an impact structure (cf. Koeberl, this volume). Where this is absent, circular crater morphology, occurrences of breccias, and other macroscopic indicators, e.g., massive occurrences of pseudotachylitic breccias, presence of shatter cones (Nicolaysen and Reimold, 1999), or geophysical anomalies (especially circular gravity and magnetic anomalies that contrast with regional potential fields) may provide useful indications for the presence of an impact structure. However, none of these criteria can be taken as unambiguous evidence. Only the detection of the uniquely impact-produced, ultra-high pressure mineral deformation effects known as *impact* or *shock metamorphism* represent acceptable proof for impact. They may be observed within *in situ* impact formations, in allogenic impact breccias in the environs of impact structures, or—in widely dispersed form—in sedimentary strata deposited after erosion of impact structures and impact breccia formations. Generally, these effects are studied by optical microscopy, but electron microscopic techniques may be required to confirm the true nature of some deformation features.

Shock metamorphism

Conventional metamorphism of upper crustal rocks is limited to the regime of $<1,000°C$ and $<5\,GPa$. In contrast, hypervelocity impact is capable of attaining shock pressures of $>100\,GPa$ and temperatures in excess of thousands of degree centigrade (Figure F6). In addition, the strain rates associated with impact processes (10^4–10^6) are many orders of magnitude higher than those associated with endogenous geological processes (10^{-3}–10^{-6}). Under these conditions, rock-forming minerals will be subjected to *elastic deformation* (irregular and planar, crystallographically controlled

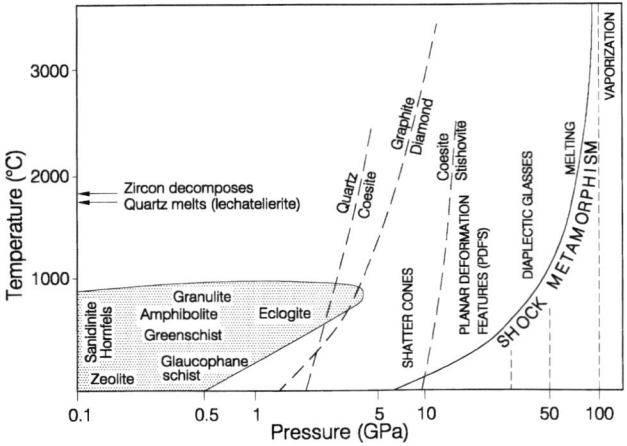

Figure F6 Conditions of shock metamorphism, in comparison to the range of pressures and temperatures normally attained in regional/contact metamorphism of crustal rocks (modified after French, 1998).

fracturing, cataclasis, twinning, kinkbanding in micas) below the Hugoniot elastic limit of a mineral (generally 0.6–1 GPa), plastic deformation (*planar deformation features, diaplectic mineral glasses, mosaicism*), and transformation to *high-pressure polymorphs* (Stöffler, 1972; Stöffler and Langenhorst, 1994; Grieve et al., 1996; French, 1998; Deutsch and Langenhorst, 1998; Bischoff and Stöffler, 1999). All these shock effects form at pressures below 30 GPa to 35 GPa. Above these pressures, shock energy is completely used for *isotropization* (melting) of minerals, and above ca. 50 GPa for *bulk melting* of rocks. The progressive stages of shock metamorphism have been extensively studied in natural impact-deformed rocks and through shock experiments. Pressure ranges in which certain shock metamorphic effects occur have been well calibrated for many minerals (Figure F7), especially in quartz and feldspars that represent the most abundant rock-forming minerals in crustal rocks. Many of these data have, however, been determined through shock experimentation with minerals at room temperature. Target materials at higher temperatures (especially in deeper levels of very large impact structures that excavate mid-crustal levels) will experience the same shock metamorphism as "cold" targets, but certain

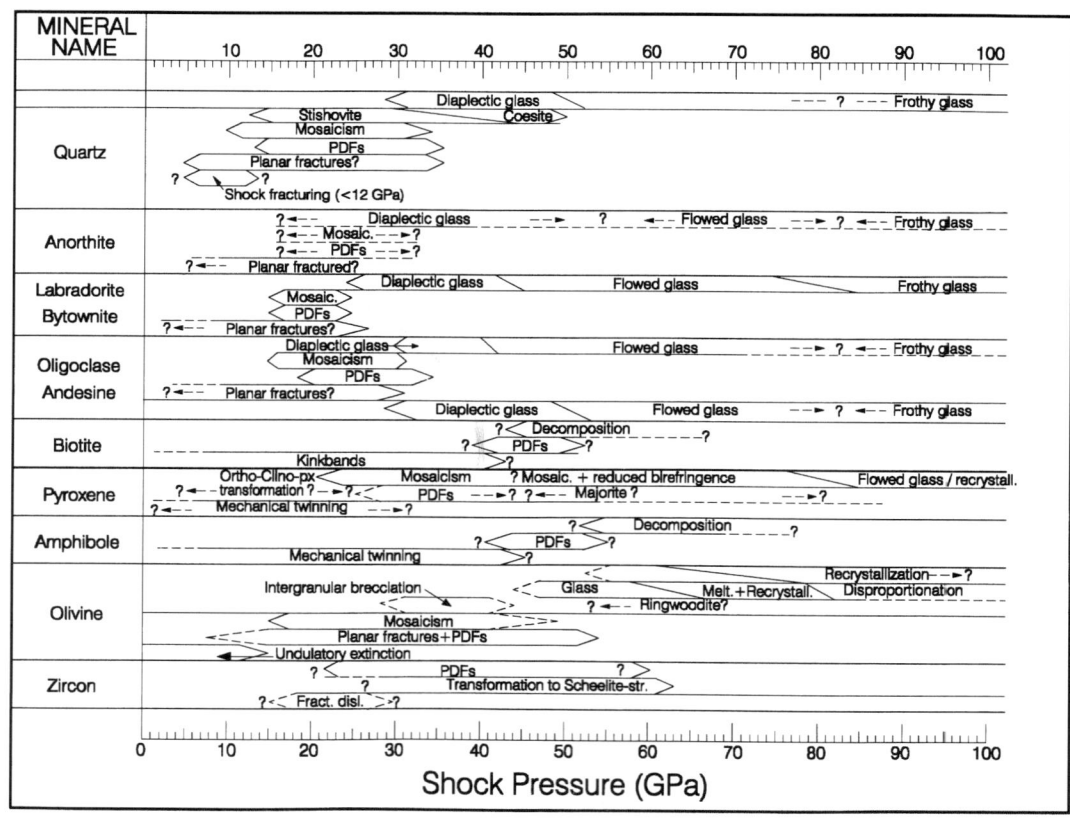

Figure F7 A summary of pressure ranges of specific shock metamorphic effects, as determined for major rock-forming minerals through comparative study of natural and experimentally induced mineral deformation. Modified after Bischoff and Stöffler (1992) and Deutsch and Langenhorst (1998).

shock effects may already occur at comparatively lower shock pressures (Huffman and Reimold, 1996).

Planar deformation features (PDFs) occur as single or, at higher shock pressures, multiple sets of extremely narrow (<2 μm) and closely spaced (2–10 μm) isotropic features (Figures F8(A) and (B)). Optical and TEM investigations show that PDFs are either composed of amorphous, or amorphous and extremely fine-grained crystalline material. They occur in distinct crystallographic orientations, and the relative abundance of PDF orientations in quartz may be used to constrain the shock pressure range in which they formed (predominance of (0001) orientations of Brazil twins is characteristic of shock pressures <10 GPa, {10–13} orientations form abundantly between 10 GPa and 20 GPa, and {10–12} orientations between 20 GPa and 30 GPa). Where PDFs have been subjected to annealing, they may only be represented by so-called *decorated PDFs*, whereby planar fluid inclusion or solid inclusion trails may still be visible in lieu of the original PDFs. Orientation measurements of such decorated features may provide the proof that these features are of impact origin. Recently, much attention has been paid to sets of narrow-spaced, planar features in the refractory and weathering-resistant mineral zircon. In this case, it has not been possible yet to identify the true nature—planar fractures, PDFs, or high-density dislocation bands—of such features (Figure F8(C)), but they can be considered as diagnostic of impact deformation (Leroux et al., 1999).

Diaplectic (or thetomorphic) mineral glass is different from conventional glasses formed by melting of a mineral to a liquid phase. Diaplectic glass forms through solid-state transformation and, thus, preserves the original textures of a mineral. Although optically isotropic, features such as original fractures, PDFs, or inclusions/fluid inclusions can still be recognized. The refractive indices of various SiO_2 shock phases have been calibrated against shock pressures (Stöffler and Langenhorst, 1994).

Transformations of minerals to their high-pressure polymorphs typically occur at moderate to high shock pressures: α-quartz is converted at 12 GPa to 15 GPa into stishovite and at >30 GPa to coesite. About 10 GPa to 15 GPa are required to convert graphite to diamond, and the highly refractory mineral zircon, which is especially important in rocks that may have been subjected to post-impact thermal metamorphism capable of erasing shock effects in less refractory minerals, converts between 40 GPa and 60 GPa into a scheelite-structure phase. Planar microdeformation features in zircon form at shock pressures between 20 GPa and 40 GPa, and a granular recrystallization texture (strawberry texture) is thought to represent a high-temperature/high-shock pressure stage of shock metamorphism.

Whilst the recognition of shock deformation in a single mineral may suffice to prove an impact effect and presence of an impact formation, shock pressure determination studies should, if possible, consider shock effects in as many minerals as possible.

Summary

Large meteorite impact has been (and still is) a very important geological process on Earth. As only a fraction of the terrestrial impact cratering record—especially of the Archean-Proterozoic record—has been deciphered to date, and impact has played a major role in the evolution of life on Earth,

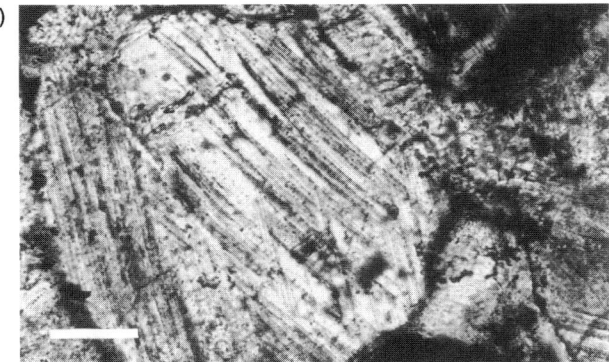

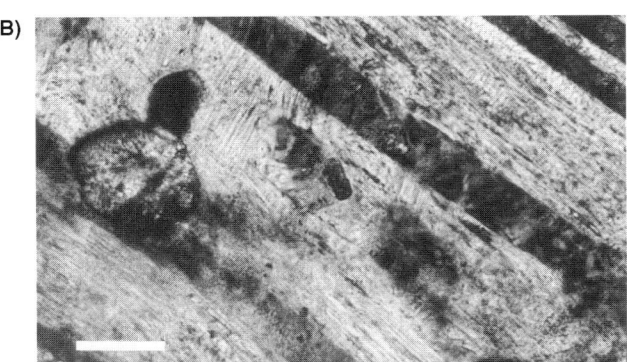

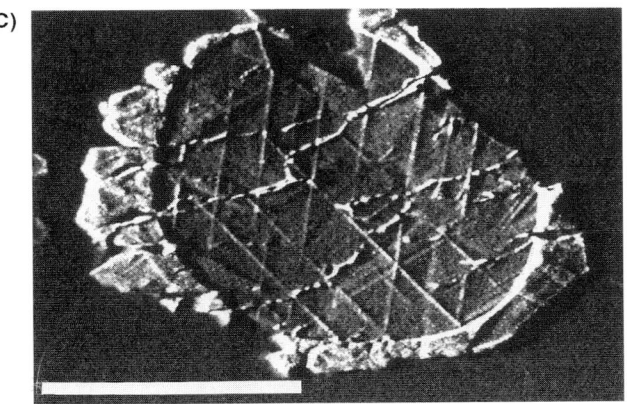

Figure F8 Typical examples of planar deformation features (PDFs): (A) Quartz from a granite in the Manson impact structure (Iowa, USA) displaying multiple sets of PDFs, many of which are decorated with fluid and solid inclusions. Scale bar: 50 μm, parallel nicols.
(B) Plagioclase (also from Manson) showing alternating twin lamellae that are either melted (locally devitrified) or display numerous PDFs, mostly parallel to or at high angles to the lamellae orientation. Scale bar: 50 μm, crossed polarizers. (C) Planar features in a zircon crystal from the Vredefort Central Granite. The core of this grain, inherited from target crust, shows three orientations of planar features. It is overgrown by an undeformed generation of zircon that yielded a U-Pb age corresponding to that of the impact event. Scale bar: 200 μm, back-scattered electron-cathodoluminescence image (courtesy Dr. R.A. Armstrong, ANU).

impact cratering studies are a significant component of modern geoscience.

Shock metamorphic effects related to shock pressures between 10 GPa and 35 GPa provide unambiguous evidence for the identification of impact formations. In the absence of other definite evidence for impact (remnants of the meteoritic projectile), shock effects such as PDFs and diaplectic glasses are required as proof of impact.

Wolf Uwe Reimold

Bibliography

Bischoff, A., and Stöffler, D., 1992. Shock metamorphism as a fundamental process in the evolution of planetary bodies: information from meteorites. *European Journal of Mineralogy*, **4**: 707–755.
Chapman, C.R., and Morrison, D., 1994. Impacts on the earth by asteroids and comets: assessing the hazard. *Nature*, **367**: 33–40.
Deutsch, A., and Langenhorst, F., 1998. Mineralogy of astroblemes—terrestrial impact craters (sections 1.10.1–1.10.5). In Marfunin, A.S. (ed.), *Advanced Mineralogy: Mineral Matter in Space, Mantle, Ocean Floor, Biosphere*. Springer-Verlag, pp. 76–119.
Dressler, B.O., and Sharpton, V.L. (eds.), 1999. *Large Meteorite Impacts and Planetary Evolution II*. Geological Society of America Special Paper, 339, 464pp.
French, B.M., 1998. *Traces of Catastrophe. A Handbook of Shock-Metamorphic Effects in Terrestrial Meteorite Impact Strutures*. LPI Contribution No. 954, Houston: Lunar and Planetary Institute, 120pp.
Grieve, R.A.F., 1998. Extraterrestrial impacts on earth: the evidence and the consequences. In Grady, M.M., *et al.* (eds.), *Meteorites: Flux with Time and Impact Effects*. Geological Society of London Special Publication, 140, pp. 105–131.
Grieve, R.A.F., Langenhorst, F., and Stöffler, D., 1996. Shock metamorphism of quartz in nature and experiment: II. significance in geosciences. *Meteoritics and Planetary Science*, **31**: 6–35.
Huffman, A.R., and Reimold, W.U., 1996. Experimental constraints on shock-induced microstructures in naturally deformed silicates. *Tectonophysics*, **256**: 165–217.
Koeberl, C., and McLeod, K.G., 2001. *Catastrophic Events and Mass Extinctions: Impacts and Beyond*. Geological Society of America Special Paper, 356, 746 pp.
Leroux, H., Reimold, W.U., Koeberl, C., Hornemann, U., and Doukhan, J.-C., 1999. Experimental shock metamorphism in zircon: a transmission electron microscopic study. *Earth and Planetary Science Letters*, **169**: 291–301.
Melosh, H.J., 1989. *Impact Cratering: A Geologic Process*. Oxford University Press, 245 pp.
Melosh, H.J., 1992. Impact crater geology. In Nierenberg, W.A. (ed.), *Encyclopedia of Earth System Science*, Volume 2. Academic Press, pp. 591–605.
Montanari, A., and Koeberl, C., 2000. *Impact Stratigraphy: The Italian Record*. Lecture Notes in Earth Sciences, 93, Springer-Verlag, 364 pp.
Nicolaysen, L.O., and Reimold, W.U., 1999. Vredefort shatter cones revisited. *Journal of Geophysical Research*, **104**: 4911–4930.
Partridge, T.C. (ed.), 1998. Investigations into the origin, age and palaeoenvironments of the Pretoria Saltpan. *Council for Geoscience Memoir*, 85, 198pp.
Reimold, W.U., 1998. Exogenic and endogenic breccias: a discussion of major problematics. *Earth Science Reviews*, **43**: 25–47.
Reimold, W.U., von Brunn, V., and Koeberl, C., 1998. Are diamictites impact ejecta?—no evidence from South African occurrences. *Journal of Geology*, **105**: 517–530.
Stöffler, D., 1972. Deformation and transformation of rock-forming minerals by natural and experimental shock processes: I. Behavior of minerals under shock compression. *Fortschritte der Mineralogie*, **49**: 50–113.
Stöffler, D., and Langenhorst, F., 1994. Shock metamorphism of quartz in nature and experiment: I. Basic observation and theory. *Meteoritics*, **29**: 155–181.

Cross-references

Extraterrestrial Material in Sediments
Lunar Sediments

FELDSPARS IN SEDIMENTARY ROCKS

Feldspars are the most abundant mineral group in the Earth's crust, mainly because they form at a wide range of temperature and pressure encountered in igneous, metamorphic and sedimentary environments. Feldspars in sedimentary rocks can be of detrital and diagenetic origins. Feldspars are a group of alumino-silicate minerals whose structures are composed of corner-sharing AlO_4 and SiO_4 tetrahedra linked in an infinite, three-dimensional array. Charge balancing, M cations with radius greater than 1.0Å occupy large, irregular cavities in the tetrahedral framework. The general chemical formula of feldspars is MT_4O_8, where T is Al and Si, M is Na and/or K for alkali feldspars [$(K,Na)AlSi_3O_8$] and Ca for plagioclase ($CaAl_2Si_2O_8$). There are several polymorphs of K-feldspar minerals, including microcline (triclinic), orthoclase (monoclinic) and sanidine (monoclinic). Si/Al is ordered in orthoclase but disordered in sanidine. The Na-plagioclase ($NaAlSi_3O_8$) is albite, whereas Ca-plagioclase is anorthite ($CaAl_2Si_2O_8$). There are several types of plagioclase minerals in igneous rocks according to Ca/Na ratios, which increase with increase in temperature and increase in Ca/Na ratio of the magma, as shown in page 279.

Detrital feldspars

Detrital feldspars, which are derived mainly from the erosion of igneous and metamorphic rocks, are second most abundant (average 10–15 volume percent), detrital mineral in sandstones after quartz. Sandstones that are considerably enriched in feldspars (>25 percent) are called arkoses. Due to their cleavage and twinning planes, feldspars have lower mechanical stability than quartz, and are thus more abundant in fine- than in coarse-grained sandstones. Feldspars are also common components in mudstones and siltstones, forming on average about 4.5 bulk volume percent. In carbonate rocks, detrital feldspars occur as a few scattered grains. Detrital feldspar content in clastic rocks decreases subsequent to deposition due to various diagenetic reactions, such as dissolution, alteration into clay minerals and replacement by calcite.

Detrital feldspars are dominated by K-feldspar, albite, and Na-Ca plagioclase. Albite and K-feldspars are derived mainly from felsic igneous rocks (such as granite in uplifted basement rocks), whereas Na-Ca plagioclase is derived from intermediate (e.g., diorite) and mafic (e.g., gabbro) igneous and companion metamorphic rocks, which occur in dissected island arcs. Ca-rich plagioclase is rare in clastic rocks due to their sensitivity to chemical weathering and diagenesis. Chemical weathering of feldspars is most extensive under warm, humid climate being enhanced by carbonic and organic acids.

Detrital feldspars (particularly plagioclase) are often riddled with inclusions of sericite and epidote due to hydrothermal/deuteric alteration which takes place due to the interaction of

Magma Composition	Ultramafic		Mafic	Intermediate		Felsic
Plagioclase Composition			──── increasing Ca/Na ratio ────→			
	anorthite An_{90-100}	bytownite An_{70-90}	labrodorite An_{50-70}	andesine An_{30-50}	oligoclase An_{10-30}	albite An_{0-10}
(Volcanic/ plutonic)	komatiite/ peridotite	basalt/ gabbro		andesite/ diorite		rhyolite/ granite

igneous rocks with hot fluids released during and subsequent to the final stages of magma solidification. This kind of alteration is referred to as saussuritization.

Sediments derived from tectonically stable cratons and from recycled, fold-thrust sedimentary and low-grade metamorphic rocks have consistently low contents of feldspar. Elimination of feldspars from basement rocks in cratons occurs during the prolonged weathering subsequent to exposure, long sediment transportation distance, and reworking prior to burial. Conversely, sediments enriched in feldspar and/or rock fragments are derived from the erosion of uplifted basement, such as along rifts and transform ruptures within continental crust. Such settings provide little time for chemical weathering of subaerially exposed bedrock and shorter transportation distances and reworking of the sand particles prior to final settlement and burial. Rapid sediment burial is caused by high rates of sediment supply to basins in the vicinity of uplifted areas (i.e., high rates of mechanical weathering and erosion). It is thus apparent that the types and amounts of detrital feldspars in clastic rocks provide important clues to provenance, tectonic setting of the area and climatic conditions prevailed during weathering.

Provenance of detrital feldspars

The source rock of detrital feldspar grains can be deciphered based on chemical and mineralogical zonation, chemical composition and structural state (Helmold, 1985). The chemical composition of feldspar grains is obtained by means of electron microprobe analyzes performed on thin sections. Feldspars are defined chemically based on the solid solution between three end members: albite, orthoclase, and anorthite. However, some feldspars may contain significant amounts of barium and strontium. Trevena and Nash (1981) have used the potassium content to decipher the origin of plagioclase and alkali feldspars. The potassium content of plagioclase decreases gradually from plutonic to metamorphic rocks. However, the fields of high-K volcanic and plutonic plagioclase partially overlap, and those of low-K metamorphic and plutonic plagioclase are similar. Alkali feldspars from volcanic rocks vary widely in potassium content and their molar solid-solution compositions range between $Ab_{74}An_{13}Or_{13}$ and $Ab_{12}An_1Or_{87}$. Conversely, alkali feldspars from plutonic and metamorphic rocks are K-rich, and have composition ranging from $Ab_{43}An_1Or_{56}$ to $Ab_2An_0Or_{98}$ and from $Ab_{47}An_2Or_{51}$ to $Ab_2An_0Or_{98}$, respectively. Trevena and Nash (1981) presented a ternary diagram of the molar proportion of albite, anorthite and orthoclase solid solutions in which they delineated eight provenance groups of compositional ranges of feldspars from various crystalline sources. Chemical zonation occurs in both plagioclase and alkali feldspars of volcanic and plutonic origins, whereas those of metamorphic origin are unzoned.

Parameters controlling the structural state (i.e., Si-Al disorder) of alkali feldspars in igneous rocks include the equilibration temperatures, rate of subsolidus cooling and the activity of H_2O. K-feldspars derived from volcanic sources are highly disordered structures (sanidine), whereas those derived from plutonic sources are moderately- (orthoclase) to well-ordered (microcline) structures.

Diagenetic feldspars

Diagenetic feldspars include K-feldspar and albite that are characterized by nearly stoichiometric, end-member composition (i.e., $KAlSi_3O_8$ and $NaAlSi_3O_8$, respectively) and lack of cathodoluminescence. Diagenetic K-feldspar crystals are monoclinic and, less commonly, triclinic with various degrees of Al-Si ordering, whereas diagenetic albite is well-ordered, triclinic. Authigenic buddingtonite, which is a hydrous, ammonium feldspar ($NH_4AlSi_3O_8 \cdot nH_2O$) with considerable amounts of orthoclase solid solution occurs in organic-matter rich mudstones and sandstones.

Diagenetic feldspars have three main habits: (i) overgrowths around detrital feldspars, (ii) discrete crystals that are devoid of detrital feldspar cores, and (iii) replacement of detrital feldspars by albite (albitization). Types (i) and (iii) dominate in sandstones whereas type (ii) is important both in sandstones and carbonate rocks. Types (i) and (ii) form by direct precipitation from pore waters, whereas type (iii) forms by dissolution of detrital feldspar and concomitant precipitation of diagenetic feldspar.

Feldspar overgrowths

Feldspar overgrowths are formed by precipitation of K-feldspars around detrital orthoclase and microcline (Figure F9), and of albite around detrital plagioclase. The overgrowths, particularly K-feldspar, are usually not in full optical continuity with their detrital cores. K-feldspar overgrowths do not show twinning, but albite overgrowths inherit the twinning pattern of the core.

Although K-feldspar overgrowths usually have a pure, end member composition, they may in some cases display small-scale zonation due to considerable variations in barium contents. Conversely, albite overgrowths display a pure, end-member composition and a high degree of crystal structure ordering.

K-feldspar overgrowths form mainly in continental and coastal sandstones that are enriched in detrital K-feldspars. In such sandstones meteoric waters can induce early diagenetic dissolution of detrital K-feldspars, and hence increase in the activities of K^+, Al^{3+}, and Si^{4+} in the pore waters. Albite overgrowths dominate in sandstones rich in volcanic

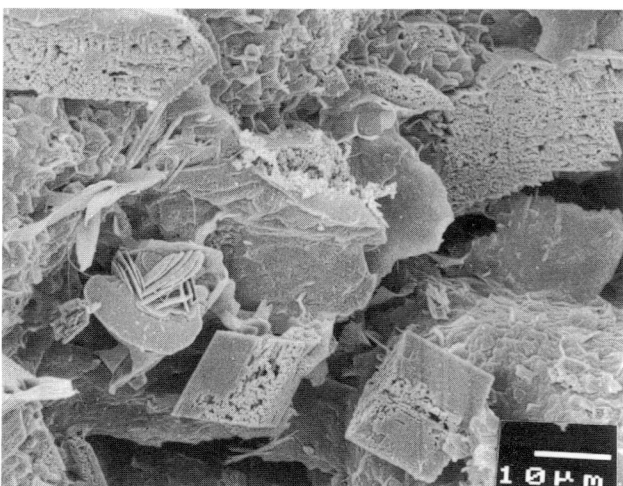

Figure F9 Scanning electron micrograph showing discrete K-feldspar crystals and K-feldspar overgrowths.

Figure F10 Scanning electron micrograph showing an albitized detrital feldspar comprised of numerous, euhedral authigenic albite crystals.

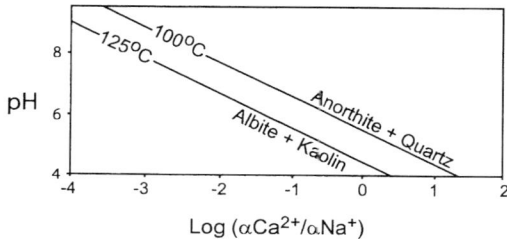

Figure F11 Equilibrium diagram showing the temperature-related plagioclase albitization (see equation 1).

and metamorphic grains derived from magmatic and orogenic arcs.

Discrete authigenic feldspar crystals

Small, discrete (usually $<100\,\mu(m)$), untwined, authigenic K-feldspar (Figure F9) and twinned albite occur in sandstones as crystals scattered in pore space or associated with other diagenetic cements (e.g., clay minerals). Authigenic albite dominates in carbonate rocks, whereas K-feldspar dominates in sandstones (Kastner and Siever, 1979). In marine and non-marine, volcanic and volcaniclastic sediments, which are enriched in highly reactive grains (e.g., volcanic glass), discrete K-feldspar crystals may extensively fill the pore space. In saline alkaline, lacustrine environments, authigenic K-feldspar may form almost entire beds by alteration of tuffs and ash deposits.

Albitization of feldspars

During burial diagenesis and increase in temperature, detrital feldspars are transformed into chemically pure albite. The original outline of the feldspar grain is well preserved, making immediate identification of albitized feldspars difficult. Each detrital feldspar grain is replaced by numerous tiny, euhedral crystals of albite that are arranged parallel to cleavage and/or twinning planes (Figure F10), causing destruction of twinning pattern in the detrital feldspar. Hence, they can be easily mistaken for untwined feldspar, such as orthoclase. Albitization of detrital feldspar does not occur through solid-state replacement but by small-scale dissolution-reprecipitation.

The albitization of plagioclase commences at shallower burial depths (2 km) and temperatures (65°C) compared to potassium feldspars (Morad *et al.*, 1990). The higher the molar anorthite solid solution content, the more sensitive is plagioclase to albitization. Albitization consumes Si^{4+} and liberates Al^{3+} and Ca^{2+}, which often re-precipitate as kaolin and calcite, respectively, as follows:

$$CaAl_2Si_2O_8 + 2SiO_{2aq.} + 0.5H_2O + Na^+ + H^+$$
anorthite

$$= NaAlSi_3O_8 + 0.5Si_2Al_2O_5(OH)_4 + Ca^{2+} \quad \text{(Eq. 1)}$$
albite   kaolin

Albitization of plagioclase is thus controlled by temperature, pH and Ca^{2+}/Na^+ activity ratio (Figure F11). Albitization of K-feldspar grains is strongly controlled by temperature and K^+/Na^+ activity ratio:

$$KAlSi_3O_8 + Na^+ = NaAlSi_3O_8 + K^+ \quad \text{(Eq. 2)}$$
K-feldspar   albite

Importance of feldspars for reservoir quality of sandstones

Detrital feldspars exert important control on reservoir quality of sandstones in the various following ways:

(1) Compaction: feldspar grains contain abundant planes of weakness (cleavage and twinning planes), and hence have lower mechanical stability than quartz. Therefore, feldspar-rich sandstones undergo a more rapid and a greater degree of loss of primary porosity due to mechanical compaction than quartz-rich sandstones.

(2) Secondary porosity: the dissolution of detrital feldspars occurs in all diagenetic regimes. Detrital feldspars are in thermodynamic disequilibrium with meteoric waters (weak carbonic acids), and hence undergo dissolution and the formation of secondary pores. However, as Si and, particularly, Al liberated from dissolution of feldspars have very low solubility in meteoric waters, feldspar dissolution is accompanied by crystallization of kaolinite in continental and transitional sandstones. The dissolution of feldspar and formation of kaolinite is most pervasive under warm, humid climatic conditions, and occur both during early, shallow burial diagenesis and below unconformities.

Feldspar dissolution and the formation of secondary porosity may also occur during burial diagenesis by means of acidic formation waters charged with CO_2 or organic acids derived from thermal maturation of organic matter or biodegradation of oil. By-products of feldspar dissolution include dickite, illite, chlorite, and albite. Progressive feldspar dissolution with increasing burial depth is common in mudstones too, but it is usually associated with reprecipitation of secondary minerals such as illite and albite rather the formation of secondary pores.

(3) Illitization of kaolinite: unlike kaolinite, illite induces considerable deterioration to permeability, and hence to reservoir quality of sandstones because illite occurs commonly as elongated filaments that may efficiently block the pore throats. Kaolinite formed at shallow burial diagenesis is illitized when subjected to elevated burial temperatures (>100°C; >3 km), as follows:

$$3Al_2Si_2O_5(OH)_4 + 2K^+ = 2KAl_3Si_3O_{10}(OH)_2 + 2H^+ + 3H_2O$$
$$\text{kaolinite} \qquad\qquad\qquad \text{illite}$$
(Eq. 3)

Potassium needed to accomplish the above reaction is derived from the dissolution of detrital K-feldspars in both sandstones and interbedded mudstones. Hence the overall illitization reaction can be envisaged as follows:

$$Al_2Si_2O_5(OH)_4 + KAlSi_3O_8$$
$$\text{kaolinite} \qquad \text{K-feldspar}$$

$$= KAl_3Si_3O_{10}(OH)_2 + 2SiO_2 + H_2O \quad \text{(Eq. 4)}$$
$$\text{illite} \qquad\qquad \text{quartz}$$

The albitization of K-feldspar is a viable source of potassium for illitization reactions at the expense of kaolinite, as both processes occur over a similar range of burial temperatures of 100°C to 150°C (Figure F12). In the absence of kaolinite, illitization of detrital K-feldspar in sandstones may occur directly as follows:

$$3KAlSi_3O_8 + 2H^+ = KAl_3Si_3O_{10}(OH)_2 + 6SiO_2 + 2K^+$$
$$\text{K-feldspar} \qquad\qquad \text{illite} \qquad\qquad \text{quartz}$$
(Eq. 5)

Oxygen isotopic composition and dating of diagenetic feldspars

The $\delta^{18}O$ values of feldspar decrease with increase in temperature. Hence, diagenetic feldspars have higher $\delta^{18}O_{SMOW}$ values (+18‰ to +28‰) than igneous (mostly +6‰ to +12‰) and metamorphic (mostly +12‰ to +16‰) feldspars.

Reliable dating of diagenetic K-feldspars by K/Ar or Rb/Sr methods is performed *in situ* by using the laser ablation or ion probe techniques, which encounter little risk for contamination with detrital feldspars. Dating of diagenetic albite is not possible yet due to their very low of K, Rb, and Ar contents.

Sadoon Morad

Bibliography

Helmold, K.P., 1985. Provenance of feldspathic sandstones—the effect of diagenesis on provenance interpretations: a review. In Zuffa, G.G. (ed.), *Provenance of Arenites*. Dordrecht, D: Reidel Publishing Company, pp. 139–163.

Kastner, M., and Siever, R., 1979. Low temperature feldspars in sedimentary rocks. *American Journal of Science*, **279**: 435–479.

Morad, S., Bergan, M., Knarud, R., and Nystuen, J.P., 1990. Albitization of detrital plagioclase in Triassic reservoir sandstones from the Snorre Field, Norwegian North Sea. *Journal of Sedimentary Petrology*, **60**: 411–425.

Trevena, A.S., and Nash, W.P., 1981. An electron microprobe study of detrital feldspar. *Journal of Sedimentary Petrology*, **51**: 137–150.

Cross-references

Authigenesis
Clay Minerals
Diagenesis
Provenance

FLAME STRUCTURE

A term introduced by Walton (1956, p. 267) for a sedimentary structure consisting of sharp-crested wave- or flame-shaped plumes of mud that have risen irregularly upward into an overlying layer, generally a rapidly deposited sand. The flames, though irregular in shape, are generally overturned predominantly in one direction, which is the paleocurrent direction of the overlying sand.

Flame structures are closely allied to load structures (*q.v.*): they are probably formed by sinking of rapidly-deposited, therefore easily liquefied, sand into an underlying

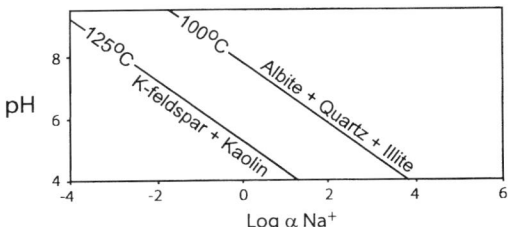

Figure F12 Equilibrium diagram showing the temperature-dependent reaction: 2K-feldspar + 2.5kaolin + Na^+ = albite + 2illite + 2quartz + 2.5H_2O + H^+.

unconsolidated mud, while continued flow of the current depositing the sand produces a shear stress that results in down-flow vergence of the flames. They were formed experimentally by Kuenen and Menard (1952). Flames may also be roughly parallel to the current, forming part of the structure called "longitudinal ridges and furrows" by Dzulynski and Walton (1965, p. 66).

Gerard V. Middleton

Bibliography

Dzulynski, S., and Walton, E.K., 1965. *Sedimentary Features of Flysch and Greywackes*. Elsevier.
Kuenen, Ph.H., and Menard, H.W., 1952. Turbidity currents, graded and non-graded deposits. *Journal of Sedimentary Petrology*, 22: 83–96.
Walton, E.K., 1956. Limitations of graded bedding: and alternative criteria of upward sequence in the rocks of the southern Uplands. *Edinburgh Geological Society Transactions*, 16: 262–271.

Cross-references

Load Structures
Turbidites

FLASER

Flaser bedding commonly occurs in association with lenticular and wavy bedding. All three sedimentary structures and their interrelationships are hence discussed in this section.

Definitions

Quoting Reineck and Wunderlich (1968), the Glossary of Geology (Bates and Jackson, 1987, 3rd edn.) defines the term "flaser structure" as "ripple cross-lamination in which mud streaks are preserved in the troughs but incompletely or not at all on the crests" (p. 246). The term "lenticular bedding" is defined as "a form of interbedded mud and ripple cross-laminated sand, in which the ripples or lenses are discontinuous not only in the vertical but also more or less in the horizontal direction" (p. 376), and the term "wavy bedding" as "a form of interbedded mud and ripple-cross-laminated sand, in which the mud layers overlie ripple crests and more or less fill the ripple troughs, so that the surface of the mud layer only slightly follows the concave or convex curvature of the underlying ripples (p. 734).

Origin of the terms

The term "flaser" (German word for streak) was probably borrowed from petrology where it was introduced earlier to describe streaky layers of parallel, scaly aggregates surrounding lenticular bodies of granular minerals in metamorphic rocks, giving the appearance of a crude flow structure (Bates and Jackson, 1987). In a sedimentological context the terms "flaser" and "lenticular bedding", or rather their German equivalents, appear to have been introduced by Richter (1931). He described sand lenses as depositional features involving sand transport by moving water (p. 314) and provided photographic evidence of fossil examples, calling the internal structure "Linsenschichtung" (lenticular bedding), adding "Flaserung" (flaser structure) in brackets (p. 317). The genesis of flaser bedding was subsequently described by Schwarz (1933, p. 102) who correctly interpreted the structure as resulting from the deposition of mud in ripple troughs which, in cross-section, produces a thin mud layer pinching out in both directions. Excellent documents of lenticular and flaser bedding structures were published by Häntzschel (1936) who systematically applied the impregnation technique invented by Schwarz (1929) to box cores from modern intertidal flats to visualize and preserve internal sedimentary structures in unconsolidated sediments (Figure F13). The term "wavy laminae" was already used by van Straaten (1951), whereas "wavy bedding" in the present context appears to have been introduced by Reineck (1960).

Genesis and classification

Flaser, lenticular, and wavy bedding structures form under conditions where periods of strong flow capable of transporting sand in bed-load alternate with periods of quiescence with little or no water movement in the presence of high concentrations of suspended matter in the water column, and

Figure F13 The first photographic document of a box-core resin cast displaying flaser structures and sand lenses draped in mud (from Häntzschel, 1936; scale added).

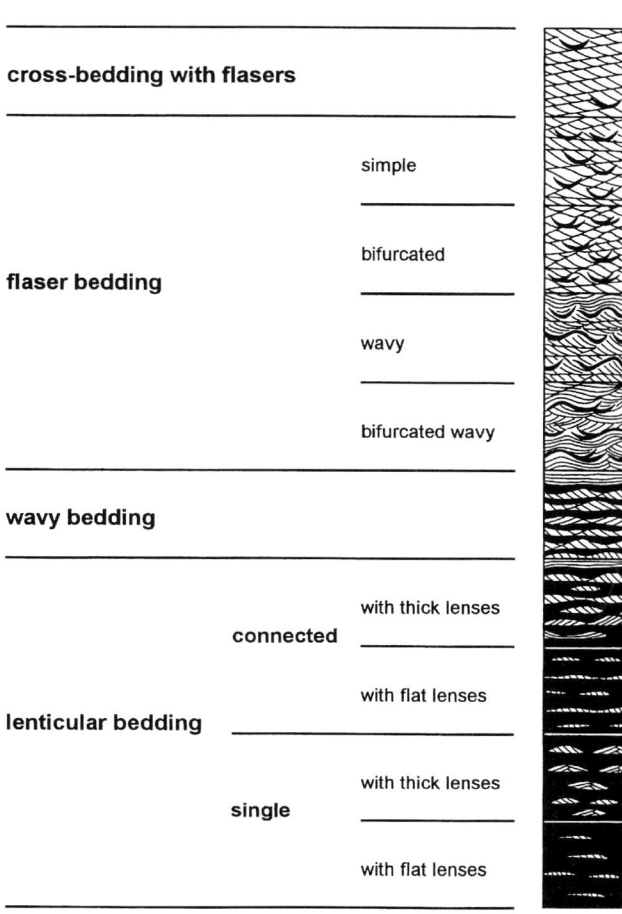

Figure F14 Composite succession illustrating the association of lenticular, wavy, and flaser bedding structures in a continuous, upward-coarsening tidal deposit (modified after Reineck and Wunderlich, 1968).

where the slack water periods are sufficiently long to allow some of the suspended matter to settle out. Although most examples documented in the literature are associated with tidal environments, the conditions are also met in a number of non-tidal environments, e.g., fluvial and deltaic (Coleman and Gagliano, 1965; Conybeare and Crook, 1968), marginal sea and continental shelf (Werner, 1968), deep sea fan (Mutti, 1977; Ricci Lucchi, 1995).

A systematic description, classification, and illustration of the variability of sedimentary structures associated with lenticular, flaser, and wavy bedding is given in Reineck and Wunderlich (1968). Four basic bedding types are distinguished, two of which are further subdivided. These are:

(a) *lenticular bedding*: this group is subdivided into *connected* or *disconnected* (single) *sand lenses* which, in turn, may either be *thick* or *flat*.
(b) *wavy bedding*: this group comprises thin, continuous mud layers which drape ripple crests while more or less filling ripple troughs, thereby assuming a wavy shape.
(c) *flaser bedding*: this group is subdivided into (i) *simple flaser bedding*, in which the individual structures are neither in vertical nor horizontal contact with each other; (ii) *bifurcated flaser bedding*, in which individual flasers frequently appear to split up or bifurcate because they are in contact with the remnants of an older generation of flasers which have not been completely eroded; (iii) *wavy flaser bedding*, in which individual flaser structures are frequently connected with their neighbors; (iv) *bifurcated wavy flaser bedding*, in which horizontally connected flasers show similar signs of splitting or bifurcation as described above.
(d) *cross-bedding with flasers*: in this group, massive cross-bedded sands occasionally display isolated flaser structures.

The classification has been summarized in a composite vertical bedding succession showing the association of the various forms in a continuous, upward-coarsening tidal depositional sequence (Figure F14). Additional comments together with some excellent photographic documents are contained in Reineck and Singh (1980).

Burghard W. Flemming

Bibliography

Bates, R.L., and Jackson, J.A. (eds.), 1987. *Glossary of Geology*, 3rd edn.. Alexandria (Virginia, USA): American Geological Institute.
Coleman, J.M., and Gagliano, S.M., 1965. Sedimentary structures: Mississippi river deltaic plain. In Middleton, G.V. (ed.), *Primary Sedimentary Structures and their Hydrodynamic Interpretation*. SEPM (Society for Sedimentary Research), Special Publication, 12, pp. 133–148.
Conybeare, C.E.B., and Crook, K.A.W., 1968. Manual of sedimentary structures. *Bureau of Mineral Resources, Geology and Geophysics (Commonwealth of Australia) Bulletin*, **102**: 1–327.
Häntzschel, W., 1936. Die Schichtungs-Formen rezenter Flachmeer-Ablagerungen im Jade-Gebiet. *Senckenbergiana*, **18**: 316–356.
Mutti, E., 1977. Distinctive thin-bedded turbidite facies and related depositional environments in the Eocene Hecho Group (south-central Pyrenees, Spain). *Sedimentology*, **24**: 107–131.
Reineck, H.-E., 1960. Über die Entstehung von Linsen- und Flaserschichten. *Abhandlungen der deutschen Akademie der Wissenschaften zu Berlin*, **1960**/1, 369–374.
Reineck, H.-E., and Singh, I.B., 1980. *Depositional Sedimentary Environments*, 2nd edn. Berlin: Springer-Verlag.
Reineck, H.-E., and Wunderlich, F., 1968. Classification and origin of flaser and lenticular bedding. *Sedimentology*, **11**: 99–104.
Ricci Lucchi, F., 1995. *Sedimentographica. Photographic Atlas of Sedimentary Structures*, 2nd edn. New York: Columbia University Press.
Richter, R., 1931. Tierwelt und Umwelt im Hunsrückschiefer; zur Entstehung eines schwarzen Schlammsteins. *Senckenbergiana*, **13**: 299–342.
Schwarz, A., 1929. Ein Verfahren zum Härten nichtverfestigter Sedimente. *Natur und Museum*, **59**: 204–208.
Schwarz, A., 1933. Meerische Gesteinsbildung. I. *Senckenbergiana*, **15**: 149–160.
van Straaten, L.M.J.U., 1951. Texture and genesis of Dutch Wadden Sea sediments. In *Proceedings of the Third International Congress of Sedimentology*. Groningen, Netherlands, 5–12 July 1951, pp. 225–244.
Werner, F., 1968. Gefügeanalyse feingeschichteter Schlicksedimente der Eckernförder Bucht (westliche Ostsee). *Meyniana*, **18**: 79–105.

Cross-references

Bedding and Internal Structures
Cross-stratification
Ripple, Ripple Mark, Ripple Structure

FLOCCULATION

Definition

Flocculation is the process where small particles suspended in water lump and form larger aggregates or flocs. A sediment floc consists of many small primary particles and the settling velocity of the sediment flocs normally are orders of magnitude higher than the single particles contained within the floc. There are three different processes causing aggregation of particles suspended in water. *Salt Flocculation*: clay and small silt particles are generally negatively charged because of broken bonds and isomorph substitution of the Si^{4+} ion with the Al^{3+} ion in the crystal lattice. Therefore they are surrounded by a cloud of positive ions when suspended in water, the so-called diffuse double layer. The cloud of positive ions surrounding the small particles repel each other and prevent the particles from aggregating. However, when the ionic strength of the water increases (typically where salt and fresh water mixes) an apparent compression of the double layer occurs and the particles can come so close together that the van der Waals intermolecular attractive forces dominate the electrostatic Coulombs expelling forces (van Olphen, 1963). *Bioflocculation*: in both fresh and salt water organic polymers may adsorb on the surfaces of sediment particles and bind particles together in large aggregates with high organic content (van Leussen, 1994). *Fecal pellet formation*: filter- and deposit feeders such as mussels and snails feed on fine-grained sediment particles and excrete these as fecal pellets or pseudo pellets (Edelvang and Austen, 1997). This process binds the sediment tightly together by organic glue and create very strong aggregates with characteristic shapes for different species.

Controlling processes

The formation of a sediment floc in a turbulent flow of natural water is mainly controlled by two factors. The number of sediment grains in the water and the probability that two particles collide and lump together. Number of grains are generally expressed as mass concentration and the probability of collision is mainly controlled by water turbulence, mostly expressed as the root mean square velocity gradient, (G):

$G = \sqrt{\varepsilon/\nu}$, which can also be expressed as

$$G = \sqrt{\frac{u_f^3[1-z/H]}{\nu \kappa z}} \quad \text{(Eq. 1)}$$

where: ε is turbulent dissipation rate per unit mass ($Nm\,kg^{-1}$ per second), ν is the kinematic water viscosity (m^2 per second), u_f is friction velocity (m per second), z is height above the bottom (m), H is total depth (m), and κ is von Karman's constant. Collision caused by Brownian motions is unimportant in the natural environment whereas collision caused by differential settling may be important in quiescent water. Different clay minerals have different ability for flocculation, however, such specific abilities are often overshadowed by coating of the sediment surfaces by iron oxides and/or organic coatings.

Floc size, composition, strength and density

The smallest flocs are of the same size as the smallest turbulent eddies in the turbulent flowing water, the Kolmogorov micro scale. Krone (1986) defined such small flocs as zero order flocs. When zero order flocs lump together first order flocs are formed and so on. The higher the order of the floc the more water it contains and the lower the floc density becomes. But along with the growth in size floc strength decreases and eventually the floc will be broken up by turbulence and the flocculation process will start again. Therefore, floc size in any given turbulent flow is a dynamic equilibrium controlled by both sediment concentration and water turbulence. Another way of describing the floc structure is by describing its fractal dimension. It has been suggested that aggregates and flocs in suspension tends to arrange in a self-similar geometrical system. Kranenburg (1994) describes the process as follows: if the basic element of a fixed structure (e.g., a microfloc) is made up of m_1 primary particles and m_1 of these basic elements are connected in such a way that the geometric arrangement of the new elements is the same as the arrangement of the primary particles in the basic aggregate then the total number N of particles in such a fractal aggregate will be:

$$N \approx \left[\frac{R_a}{R_p}\right]^D \quad \text{(Eq. 2)}$$

where R_a is the diameter of the aggregate and R_p is the diameter of the primary particle and D is the fractal dimension given by the expression:

$$D = \left[\frac{\ln m_1}{\ln m_2}\right] \quad \text{(Eq. 3)}$$

where m_2 is a factor by which the aggregate size increases independent of R_a during each flocculation step similar to moving from one floc order to one higher order. Kranenburg (1994) states that the fractal dimension D can vary within the interval 1.0–2.7, where values of $D<2$ indicates loose aggregates and values of $D>2$ indicates more compact aggregates. Dyer and Manning (1999) found a mean value of D of 2.1 for *in situ* analyzes in the Humber estuary using the video device INSSEV, whereas Mikkelsen and Pejrup (2001) found values for coastal waters between 2.1 and 2.7 based on measurements of *in situ* floc size spectra.

Flocculation time

Because floc size is the product of a dynamic equilibrium process it must be anticipated that a characteristic time for this equilibrium to be reached exists in any given stationary turbulent flow. Few studies have been conducted on this parameter but Mikkelsen and Pejrup (2000) found *in situ* flocculation times in the order of 50 min whereas Milligan (1995) determined the time scale of flocculation to 10–15 min using a mesocosm (i.e., a part of the natural environment that has been closed off so that specific parameters can be controlled).

In situ settling velocity of sediment flocs

The settling velocity of a floc is controlled by its size, shape, and density. It is difficult to estimate *in situ* settling velocity theoretically. Therefore it must be determined by *in situ* measurements. Such measurements were first carried out by Owen (1971) who used a 1 m long perspex tube deployed in the water to collect a water sample. The sample was taken with the tube in a horizontal position and subsequently used as a settling column raised to a vertical position. Owens experiments from the Thames for the first time suggested, that settling velocities of flocs measured *in situ* were much higher than measured in the laboratory. The so-called Owen tube has been used by numerous investigators in a slightly modified version, the BrayStoke SK 110 settling tube, as described by Pejrup (1988). Generally there is consensus that the following simple equation describe the relationship between median settling velocity and mass concentration:

$$W_{50} = a \cdot C^b \qquad \text{(Eq. 4)}$$

(Burt, 1986) but no generic equation has been established probably because of the omission of the parameter G and biological and mineralogical parameters.

Current investigations and state of the art

When modeling cohesive sediment transport in estuaries and harbors knowledge of the field settling velocity of the suspended flocs is extremely important. Because of the difficulties in measuring the field settling velocity of sediment flocs numerous devices have been constructed to serve this purpose. Besides many different kinds of settling tubes, video techniques have been applied to directly measure both size and settling velocity of flocs (e.g., Fennessey et al., 1994; van Leussen and Cornelisse, 1996). Different techniques for measuring floc settling velocity and size are summarized and compared by Dyer et al. (1996) and Eisma et al. (1996). A new technique that has proved useful is laser diffraction and laser backscatter devices. Bale and Morris (1987) were the first to use a laser diffraction instrument *in situ*. Later devices have been improved and today *in situ* lasers are manufactured commercially. During the last few years results with especially the LISST-100 *in situ* laser have been published (e.g., Mikkelsen and Pejrup, 2000, 2001; and Agrawal and Pottsmith, 2000). An example of such *in situ* measurements of aggregate spectra are seen in Figure F15*. Mikkelsen and Pejrup (2001) furthermore suggested a method to compute mean settling velocity of a flocculated suspension from measurement of mass concentration, volume concentration and mean floc size.

Summary or conclusion

It is well-known that estuaries concentrate fine-grained sediment because of the estuarine processes settling-scour lag, tidal asymmetry, and estuarine circulation (see *Sediment Transport by Tides*). However, most of the fine sediment grains concentrated would never settle to the bottom during the short period of slack water if the field settling velocity was not increased dramatically by flocculation of the primary particles and aggregates. Therefore, the flocculation process is essential for the formation of muddy sediments in estuaries. All three forms of flocculation plays a role here but salt flocculation and pelletization are considered to be the two most important processes.

Morten Pejrup

*Figure F15 appears in Plate II, facing p.115.

Bibliography

Agrawal, Y.C., and Pottsmith, H.C., 2000. Instruments for particle size and settling velocity observations in sediment transport. *Marine Geology*, **168**: 89–114.

Bale, A.J., and Morris, A.W., 1987. In situ measurements of particle size in estuarine waters. Estuarine, *Coastal and Shelf Science*, **24**: 253–263.

Burt, N.T., 1986. Field settling velocities of estuary flocs. In Mehta, A.J. (ed.), *Estuarine Cohesive Sediment Dynamics*. Lecture notes on Coastal and Estuarine Studies 14, pp. 126–150.

Dyer, K.R., Cornelisse, J., Jago, C., McCave, I.N., Pejrup, M., van Leussen, W., Wolfstein, K., Puls, W., Kappenburg, J., and Dearnaley, M., 1996. A comparison of *in-situ* techniques for estuarine floc settling velocity measurement. *Journal of Sea Research*, **36**: 15–29.

Dyer, K.R., and Manning, A.J., 1999. Observation of the size, settling velocity and effective density of flocs, and their fractal dimensions. *Journal of Sea Research*, **41**: 87–95.

Edelvang, K., and Austen, I., 1997. The temporal variation of flocs and fecal pellets in a tidal channel. *Estuarine, Coastal and Shelf Science*, **44**: 361–367.

Eisma, D., Bale, A.J., Dearnaley, M.J., Fennessy, M.J., van Leussen, W., Maldiney, M.A., Pfeiffer, A., and Wells, J.T., 1996. Intercomparison of *in situ* suspended matter floc size measurements. *Journal of Sea Research*, **36**: 3–14.

Fennessey, M.J., Dyer, K.R., and Huntley, D.A., 1994. INSSEV: an instrument to measure the size and settling velocity of flocs *in situ*. *Marine Geology*, **117**: 107–117.

Kranenburg, C., 1994. The fractal structure of cohesive sediment aggregates. *Estuarine, Coastal and Shelf Science*, **39**: 451–460

Krone, R.B., 1986. Aggregation of suspended particles in estuaries. In Kjerfve, B. (ed.), *Estuarine Transport Processes*. University of South Carolina Press, pp. 177–190.

van Leussen, W., 1994. *Estuarine Macroflocs and their Role in Fine-grained Sediment Transport*. University of Utrecht, 487pp.

van Leuessen, W., and Cornelisse, J., 1996. The underwater video system VIS. *Journal of Sea Research*, **35**: 83–86

Mikkelsen, O.A., and Pejrup, M., 2000. *In situ* particle size spectra and density of particle aggregates in a dredging plume. *Marine Geology*, **170**: 443–459.

Mikkelsen, O.A., and Pejrup, M., 2001. The use of a LISST-100 laser particle sizer for *in-situ* estimates of floc size. *Geo-Marine Letters*, **20**(4): 187–195.

Milligan, T., 1995. An examination of the settling behaviour of a flocculated suspension. *Netherlands Journal of Sea Research*, **33**: 163–171.

Owen, M.W., 1971. The effect of turbulence on the settling velocities of silt flocs. In: sedimentation in estuaries and rivermouths. *Proceedings of the 14th IAHR Congress Paris*, **4**: 27–32.

van Olphen, H., 1963. *An Introduction to Clay Colloid Chemistry*. Interscience Publishers (John Wiley & Sons), 301pp.

Pejrup, M., 1988. Flocculated suspended sediment in a micro-tidal environment. *Sedimentary Geology*, **57**: 249–256.

Cross-references

Clay Minerals
Grain Settling
Grain Size and Shape
Sediment Transport by Tides

FLOODPLAIN SEDIMENTS

Introduction

Floodplain sediments are represented by a continuum of sediment types that range from clay- to gravel-size particles and include both terrigenous and organic deposits. Prior investigations of modern floodplain deposits (e.g., Wolman and Leopold, 1957; Kesel et al., 1974; Farrell, 1987; Brakenridge, 1989; Nanson and Croke, 1992; Zwolinski, 1993) provide a basis for interpreting ancient floodplains (Flores, 1981; Bridge, 1984; Platt and Keller, 1992; Kraus and Aslan, 1993; Willis and Behrensmeyer, 1994). Syntheses of modern and ancient floodplain sediments are presented in Allen (1965) and Miall (1996). The importance of floodplain sediments is two fold. First, these deposits represent economically important reservoirs of oil, natural gas, and water, and include significant coal reserves. Second, floodplain sediments provide detailed records of past and present geologic processes and continental environments.

Types of floodplain sediment

Floodplain sediments are represented by channel and overbank deposits. *Channel deposits* are dominated by coarse-grained sediments such as sand and gravel. The major exception to this generalization are abandoned channel fills, which are often fine grained (Fisk, 1947). *Overbank deposits* are fine grained and consist primarily of clay, silt, and lesser amounts of sand. Organic sediments are also important constituents in some settings (Flores, 1981). Previous workers have referred to channel and overbank sediments as substratum and topstratum deposits, respectively (Allen, 1965). Although overbank sediments have received considerably less attention than channel deposits (Farrell, 1987), studies of paleosols have significantly increased interest in overbank environments (Kraus, 1999).

Channel deposits

Channel deposits consist of channel-bar and channel-fill sediments. Channel-bar sediments are typically represented by stratified sands and gravels and accumulate in response to river migration and changes in stream hydrology. Channel fills range from coarse-grained (active) to fine-grained (passive) deposits. Active fills are represented by interbedded sand and silt that accumulates in response to gradual channel abandonment by chute cutoff or similar processes. Passive fills consist of silt and clay or organic sediments that are deposited within lakes or oxbows formed by neck cutoff (Fisk, 1947). More detailed treatments of channel deposits are found elsewhere (see Cross-references).

Overbank deposits

Overbank sediments accumulate during floods in natural-levee, crevasse-splay, and flood-basin environments. Modern and ancient examples of overbank deposits are numerous (cf. Gersib and McCabe, 1981; Bridge, 1984; Smith et al., 1989; Aslan and Autin, 1999). Classifications of these deposits are presented in Platt and Keller (1992) and Miall (1996). Although overbank deposits are sometimes viewed as monotonous sequences of fine-grained alluvium, the recognition of paleosols in ancient overbank sediments has proved useful for subdividing and interpreting these deposits (Kraus, 1999). Analyzes of modern Rhine-Meuse overbank sediment show that these deposits can be subdivided statistically, and that floodplain heterogeneity is closely linked to channel pattern (Weerts and Bierkens, 1993).

Natural-levee sediments

Natural levees are represented by wedges of sand, silt, and clay that typically thin and fine away from channel margins. Natural levees have widths of 2 to 3 km and are up to 5 m thick. These features are best developed along the outer bends of meanders in low-gradient sinuous rivers with large suspended-sediment loads such as the Mississippi. Natural-levee sediments are represented by rhythmically bedded sand and silt with lesser amounts of clay (Farrell, 1987). Sedimentary structures include ripple lamination. In proximal levee settings, beds are up to 1 m thick whereas distal positions contain thinner beds, and beds are typically disrupted by bioturbation. In ancient floodplain settings, natural-levee deposits consist of packages of interbedded fine sandstone, siltstone, and mudrock that commonly overlie channel-belt sandstones (Kraus, 1999).

Crevasse-channel and crevasse-splay sediments

Crevasse splays are tabular to lenticular accumulations of sand and silt that accumulate in floodplain depressions such as wetlands. In plan view, crevasse splays are lobate to sinuous and contain distributary channel networks that extend up to 10 km into flood basins (Smith, 1986; Smith et al., 1989). Crevasse-channel deposits are typically lenticular, sandy, and coarser-grained than other types of overbank sediments. However, fine-grained crevasse-channel fills also occur (Smith and Perez-Arlucea, 1994). Stratification types include small-scale trough-cross-bedding and ripple lamination.

Crevasse-splay deposits are tabular sheets of interbedded sand and silt with lesser amounts of clay that are <10 m thick. Deposits thin or thicken away from channel margins depending on local floodplain topography (Willis and Behrensmeyer, 1994). Strata are commonly bioturbated and ripple laminated (Farrell, 1987). Individual splays show either upward-coarsening or upward-fining sequences (Smith and Perez-Arlucea, 1994). Upward-coarsening sequences reflect crevasse-splay progradation whereas deposition of overbank fines during waning flood stages produces upward-fining deposits. Crevasse-splay deposition is particularly important within avulsion belts (*sensu* Smith et al., 1989) because this process causes rapid fine-grained floodplain aggradation.

Flood-basin sediments

Tabular deposits of clay, silt, and organic sediment fill flood-basin depressions. In large river systems such as the Mississippi, flood-basin sediments are up to 40 m thick, encase channel sand bodies, and interfinger with crevasse-splay and natural-levee deposits (Aslan and Autin, 1999). These deposits can extend along valley axes for hundreds of kilometers. Slow sediment accumulation rates in flood basins favor bioturbation and soil development. Shallow fluctuating water levels and large amounts of organic matter produce gray sediment colors and brown or red iron-oxide mottles. Authigenic minerals such as iron and manganese oxides, calcite, gypsum, pyrite, vivianite, and jarosite are also common. In settings where water levels fluctuate seasonally and smectitic clays are present, pedogenic slickensides are abundant. In contrast, perennial saturation and low amounts of clastic input leads to widespread peat and coal formation (Flores, 1981).

Summary

Floodplain sediments provide records of terrestrial environments that are essential for deciphering continental sedimentary rocks. These sediments are also important for evaluating petroleum, coal, and water resources. Future lines of potentially fruitful research include: (1) further evaluation of the influence of avulsion on floodplain construction; (2) quantification of relationships between flood hydrology and floodplain stratigraphy; and (3) continued study of overbank deposits and paleosols. This latter effort may be particularly important because fine-grained floodplain deposits are less studied than sands and gravels, yet overbank sediments volumetrically dominate many ancient alluvial sequences.

Andres Aslan

Bibliography

Allen, J.R.L., 1965. A review of the origin and characteristics of recent alluvial sediments. *Sedimentology*, **5**: 91–191.
Aslan, A., and Autin, W.J., 1999. Evolution of the Holocene Mississippi river floodplain, ferriday, louisiana: insights on the origin of fine-grained floodplains. *Journal of Sedimentary Research*, **69**: 800–815.
Brakenridge, G.R., 1989. River flood regime and floodplain stratigraphy. In Baker, V.R., Kochel, R.C., and Patton, P.C. (eds.), *Flood Geomorphology*. New York: John Wiley and Sons, pp. 139–156.
Bridge, J.S., 1984. Large-scale facies sequences in alluvial overbank environments. *Journal of Sedimentary Petrology*, **54**: 583–588.
Farrell, K.M., 1987. Sedimentology and facies architecture of overbank deposits of the Mississippi River, False River region, Louisiana. In *Recent Developments in Fluvial Sedimentology*. SEPM (Society of Sedimentary Geology) Special Publication 39, pp. 111–120.
Fisk, H.N., 1947. *Fine-grained Alluvial Deposits and their Effect on Mississippi River Activity*. Vicksburg: Mississippi River Commission.
Flores, R.M., 1981. Coal deposition in fluvial paleoenvironments of the Paleocene Tongue River Member of the Fort Union Formation, Powder River Area, Powder River Basin, Wyoming and Montana. In *Recent and Ancient Nonmarine Depositional Environments: Models for Exploration*. SEPM (Society of Sedimentary Geology) Special Publication 31, pp. 169–190.
Gersib, G.A., and McCabe, P.J., 1981. Continental coal-bearing sediments of the Port Hood Formation (Carboniferous), Cape Linzee, Nova Scotia, Canada. In *Recent and Ancient Nonmarine Depositional Environments: Models for Exploration*. SEPM (Society of Sedimentary Geology) Special Publication 31, pp. 95–108.
Kesel, R.H., Dunne, K.C., McDonald, R.C., Allison, K.R., and Spicer, B.E., 1974. Lateral erosion and overbank deposition on the Mississippi River in Louisiana caused by 1973 flooding. *Geology*, **2**: 461–464.
Kraus, M.J., 1999. Paleosols in clastic sedimentary rocks: their geologic applications. *Earth Science Reviews*, **47**: 41–70.
Kraus, M.J., and Aslan, A., 1993. Eocene hydromorphic paleosols: significance for interpreting ancient floodplain processes. *Journal of Sedimentary Petrology*, **63**: 453–463.
Miall, A.D., 1996. *The Geology of Fluvial Deposits*. Berlin: Springer-Verlag.
Nanson, G.C., and Croke, J.C., 1992. A genetic classification of floodplains. *Geomorphology*, **4**: 459–486.
Platt, N.H., and Keller, B., 1992. Distal alluvial deposits in a foreland basin setting—the Lower Freshwater Molasse (Lower Miocene), Switzerland: sedimentology, architecture and palaeosols. *Sedimentology*, **39**: 545–565.
Smith, D.G., 1986. Anastomosing river deposits, sedimentation rates and basin subsidence, Magdalena River, northwestern Colombia, south America. *Sedimentary Geology*, **46**: 177–196.
Smith, N.D., and Perez-Arlucea, M., 1994. Fine-grained splay deposition in the avulsion belt of the lower Saskatchewan River, Canada. *Journal of Sedimentary Research*, **B64**, 159–168.
Smith, N.D., Cross, T.A., Dufficy, J.P., and Clough, S.R., 1989. Anatomy of an avulsion. *Sedimentology*, **36**: 1–23.
Weerts, H.J.T., and Bierkens, M.F.P., 1993. Geostatistical analysis of overbank deposits in anastomosing and meandering fluvial systems: Rhine-Meuse Delta, the Netherlands. *Sedimentary Geology*, **85**: 221–232.
Willis, B.J., and Behrensmeyer, A.K., 1994. Architecture of Miocene overbank deposits in northern Pakistan. *Journal of Sedimentary Research*, **B64**: 60–67.
Wolman, M.G., and Leopold, L.B., 1957. River Flood Plains: some observations on their formation. *US Geological Survey Professional Paper*, **282-C**: 87–109.
Zwolinski, Z., 1993. Sedimentology and geomorphology of overbank flows on meandering river floodplains. *Geomorphology*, **4**: 367–379.

Cross-references

Avulsion
Mudrocks
Rivers and Alluvial Fans
Weathering, Soils, and Paleosols

FLOODS AND OTHER CATASTROPHIC EVENTS

Floods occur when the stage of water flow exceeds some level, commonly marked by the banks of the stream or river that usually confine the level of flow. What one considers to be a flood is a notion very much limited by human experience, in which observations occur over days, months, or some rather short period of years. In contrast, the timescales of the natural world operate over centuries, millennia, and much longer time periods. When the time scale is enlarged, then the rare occurrence of an extreme flood event can yield magnitudes that seem bizarre relative to the observations made on shorter timescales. The concern of this article will be with those floods that are both rare and of great magnitude. Smaller, common floods are usually considered to be a part of the average process regime of rivers, though lying on the upper tail of the frequency distribution for flow events.

Extreme flood phenomena defy direct measurement in the field. Their recurrence intervals can greatly exceed the life spans of potential human observers. Even when very rare, extreme floods do occur, measurements of flow phenomena and sediment transport are not made because of: (1) difficulties in accessing measurement sites or inserting measurement equipment during conditions of extreme danger; and (2) destruction of any in-place measurement devices because of the intensity of the flow processes. The resulting lack of quantitative data, so essential for the testing of models for physical processes, may explain why catastrophic flood phenomena have received relatively little attention in fluvial studies.

What is a catastrophic flood?

To be considered catastrophic a flood must possess a combination of rare occurrence and very great intensity of operation. Flood processes have probably been most intensively studied for low-energy intensity, alluvial streams. The latter have channel beds and banks that are composed of the same types of sediment that they convey in transport.

Moreover, fluvial geomorphological studies in the 1950s and 1960s firmly established that the intensity of process operation, usually measured as stream power per unit area of bed, is actually adjusted to moderate, optimal values by the morphological response of channels and floodplains. Downstream channel evolution is directly related to an upstream drainage basin in which a network of contributory channels is optimally adjusted to rainfall-runoff conditions and to prevailing rock types. Such rivers are the self-adjusted authors of their own geometries (paraphrasing a quip by Luna Leopold and Walter Langbein).

These considerations do not apply to the less-studied, high-energy processes that can occur during extreme floods in bedrock channel situations. Moreover, there are causes that are much more rare and extreme that the usual rainfall-runoff processes of contributory drainage basins. These causes include the failure of landslide and glacial dams, the cutting of huge spillways by the overflowing of immense ice-marginal lakes, volcano-ice processes, and even the spilling of huge waves across the land. Especially intense, rare rainfall-runoff process from tropical storms or local thunderstorms may also yield catastrophic responses. Thus, catastrophic flooding can be characterized first by rarity, both in the temporal distribution of occurrence and/or in the unusual or observationally unprecedented nature of occurrence or causation. Second, the catastrophic flood should display an intensity of operative processes. This may be quantified in terms of velocity, bed shear stress, or stream power, though the real measure of catastrophe is the impact of the processes on landscape. When the flood generates spectacular effects of erosion and sedimentation that are similarly rare or unusual, then it is readily designated catastrophic. Finally, the human context can designate as catastrophic any flood that produces especially great loss of life or damage to property. That the rare or unusual floods of immense intensity are also less readily anticipated means that there is some overlap of the natural and human contexts of catastrophe. Of course, this overlap leads to some practical applications for the science.

Historical and philosophical perspective

When the terms were first introduced in 1832 by the polymath William Whewell, there was a clear distinction between "catastrophists" and "uniformitarians". The former maintained or accepted that certain geological or biological phenomena are caused by sudden and violent disturbances of nature. The latter maintained that phenomena have always been and still are due to causes or forces operating continuously or uniformly in time. (The obvious corollary being that the causes or forces operative in the past are in continuous operation today and therefore accessible to current observation and study.) Thus, both scientific camps, as originally defined, made substantive claims about the world that could presumably be scientifically tested. Unfortunately, the original distinction became confused through advocacy by Sir Charles Lyell, who promoted "uniformitarianism" as a presumed basis for sound methodological reasoning in geology (Baker, 1998). One does not test a metaphysical claim as to how science is to be done; one either follows its precepts or one does not.

The substantive notion of past catastrophic events of causal character not readily observable in operation today has been thoroughly confirmed by scientific study subsequent to Whewell's original definition of the problem. That it took so long to achieve this recognition is illustrated by the great scablands debate of the 1920s. The debate began when Bretz (1923) described landforms in the Chaneled Scabland region of Washington state that required cataclysmic flooding for their origin. Blindly applying Lyell's methodological uniformitarianism, a great many geologists rose in sharp criticizm of Bretz's hypothesis (Baker, 1978). The resolution of the debate did not come until the 1960s and 1970s, when distinctive bedforms (giant current ripples) were documented, and quantitative methods showed the physical consistency of the cataclysmic flooding hypothesis (Baker, 1973). Of course, numerous other examples of catastrophic geological processes have also been documented, most notably the effects of meteor impacts and extremely large, explosive volcanic eruptions. We can now look upon the catastrophist versus uniformitarianist debates with hindsight: "...the paleocatastrophists of the early 19th century seem not to have been so much the bad chaps in the black hats, at war with the white knights of the uniformitarian camp, as they were precursors of today's neocatastrophists" (Albritton, 1989, p. 177).

While the great scablands debate of the 1920s and 1930s opened awareness to catastrophic flooding as a geological agent, its resolution in the 1970s coincided with another important stimulus that was also a remarkable surprise. The discovery of immense flood channels on Mars in the 1970s (Baker, 1978) spread awareness of catastrophic flooding as a process of significance at the planetary scale. The Martian outflow channels were produced by the largest known flood discharges, and their erosional and depositional effects have been preserved for billions of years. In addition, the past 30 years has seen the recognition of evidence for great cataclysmic flows from spillways and ice-dammed lake failures in the basins of many of the world's great river systems, including the Columbia, Yukon, Mackenzie, Mississippi, St. Lawrence, Irtysh, Ob, Yenesei, Lena, and Amur (Baker, 2002).

Flood dynamics and processes

Paleohydraulic calculations

Because of difficulties in the direct measurement of on-going processes, physical understanding of high-energy megaflooding has only been possible through the use of quantitative models applied inversely to the calculation of paleodischarges and other parameters, given preserved or inferred flow geometries. Computer flow models compute energy-balanced water-surface profiles for steady, gradually-varied flow on the basis of a form of the Bernoulli Equation in conjunction with the step-backwater method of profile computation. For application to catastrophic flood channel geometries, various water-surface profiles are computed with different combinations of discharge and energy loss coefficients until a profile is achieved that closely matches the water-surface profile suggested by field evidence of the past flow (O'Connor and Baker, 1992).

Based on paleohydraulic studies, the properties of past cataclysmic floods can be compared to the flow processes in modern rivers. Table F1 compares the hydraulic parameters for various catastrophic paleofloods with flooding on the largest alluvial river on Earth, the Amazon, and with flooding in a narrow, deep bedrock gorge, the Katherine Gorge of northern Australia. Note that slopes and flow depths for the ancient catastrophic floods, both on Earth and Mars, are much

Table F1 Paleohydraulic parameters for various cataclysmic flood channelways in comparison to a large alluvial river (Amazon) and a small bedrock gorge (Katherine River, Australia)

Location	Width (km)	Depth (m)	Velocity (m/s)	Discharge ($m^3 s^{-1}$)	Slope	Bed shear stress (Nm^{-2})	Power per unit area (Wm^{-2})
Amazon	2	60	2	3×10^5	1×10^{-5}	6	12
Katherine Gorge	0.05	45	7.5	6×10	3×10^{-3}	1.5×10^3	1×10^4
Missoula Flood	6	150	25	2×10^7	0.01	1×10^4	2×10^5
Altai Flood (Chuja)	2.5	400	25	2×10^7	0.01	4×10^4	1×10^6
Maja Vallis (Mars)	80	100	38	3×10^8	0.02	2×10^4	8×10^5
Kasei Vallis (Mars)	80	400–1,300	30–75	$1–2 \times 10^9$	0.01	1×10^5	3×10^6

larger than for the modern Amazon River. Despite its rather large discharge, the Amazon has flow velocities, bed shear stress values, and stream power per unit area values that are in a similar range to all alluvial rivers. The cataclysmic flood features of very deep, high-gradient flows are generated by power values many orders of magnitude larger than those characteristic of alluvial rivers. Nevertheless, local bedrock gorges, such as that of the Katherine River, may experience modern high-energy conditions, somewhat comparable to those of the ancient catastrophic floods.

Flow dynamics

The enormous velocities achieved in high-energy floods (Table F1) leads to reduced absolute pressure in the flowing water, reaching the vapor pressure. Shock waves produced by the collapse of the vapor bubbles that form in such situations can produce intense erosion of rock, a process called *cavitation*. Another fundamental characteristic of such flows is the development of secondary circulation, flow separation, and the birth and decay of vorticity around obstacles and along irregular flow boundaries. Collectively known as *macroturbulence*, this three-dimensional flow phenomenon is poorly understood in theoretical terms. Immense pressure and velocity gradients in the turbulent vortices can produce phenomenal hydraulic lift forces. The most intense development of macroturbulence occurs during the following conditions: (1) a steep energy gradient and great flow depth; (2) a low ratio of sediment transported to potential sediment transport; and (3) irregular, rough flow boundaries optimal for generating flow separation. Macroturbulent processes are best evidenced in the erosional forms of bedrock rivers that experience extreme floods. The forms include longitudinal grooves, potholes, and inner channels that reflect the evolution interaction between the flow dynamics and the bedrock flow boundary.

Sediment transport

Boulder transport by large floods is well-documented (Costa, 1983). Many observations show that in very high-energy floods the boulders move in suspension, probably buoyed by the macroturbulent forces described above. At the energy levels of catastrophic floods the usual distinctions of bed-load, suspended load, and washload occur at much coarser grain sizes than in common river flows. Komar (1980) analyzed criteria for these distinctions, and O'Connor (1993) extended the relationships further for the analysis of catastrophic floods. At sustained bed shear stresses of $1,000\,Nm^{-2}$ particles as large as 10 cm to 20 cm move in suspension, and coarse sand moves as washload. From the values in Table F1 it can be seen that even more extremes can occur in catastrophic floods, resulting in larger sizes moving in the various modes of transport.

Depositional forms and structures

Much of the evidence for the past occurrence of catastrophic floods is erosional, particularly for the bedrock-controlled settings of the floods. However, local depositional environments occur even within otherwise high-energy settings. This is possible because of the flow-separation effects noted earlier. While such deposits have little potential for long-term geological preservation in the sedimentary record, their study can be of great importance for understanding the flood dynamics and for hazard studies related to more recent flood regimes. The impact of catastrophic floods beyond the bedrock reaches may be best appreciated by studies of sedimentation in nearby continental or marine basins.

Large-scale patterns

The megafloods associated with the last glaciation show a pattern of spilling between tectonic basins. High-energy flows and erosional environments are generally confined to areas of cross-axial drainage across mountain or other uplands. Depositional environments, including very large fan complexes, occur in the tectonic basins that separate the mountain areas. Examples of this pattern can be found in the tectonic basins of the central Columbia Plateau region, impacted by Missoula Flooding (Baker, 1981), and in the upper Yenesei Valley of central Asia.

The role of catastrophic flooding has been documented for several large-scale submarine fans. In the Iceland Basin of the northern Atlantic the Maury Fan is dominated by turbidites that were emplaced by large Icelandic jokulhlaups (Lacasse *et al.*, 1998). The turbidites all consist of monolithologic glass volcanic glass shards produced by the subglacial volcanic eruptions of the source area. Long-distance submarine effects of catastrophic flooding are also documented for the Missoula Floods, which generated hyperpycnal gravity flows that induced turbidity currents flowing across 1,000 km of the Pacific Ocean floor from sources at the head of the Astoria Fan

(Zuffa et al., 2000). It is likely that many fans, particularly those fed from tectonically active basins, show the influence of catastrophic flooding (Mutti et al., 1996).

Bedforms

A classification of bedforms for catastrophic flood channels is shown in Table F2. This classification was developed for features studied in the Chaneled Scabland (Baker, 1981) and in the Altai of central Asia (Rudoy and Baker, 1993). The classification relates macroforms to flow width such that local flow conditions are not the primary factors in their origin. These forms persist through multiple flood events, and may show complex internal evidence of such a history. In contrast, mesoforms respond to the dynamics of an individual flow, and are scaled to the channel depth. The classification includes erosional forms, both those scoured in rock and those scoured in sediment, and depositional forms. The discussion here will emphasize the latter.

Longitudinal bars are elongated parallel to the flow direction. They may form at local flow expansions or they may develop because of flow separations near the channel walls. Often such separations alternate in a downstream direction. They may also form as pendant bars, downstream of an obstruction in the bed of the flood channel. At the mouths of tributaries and other zones of dead water relative to the high-velocity flood channel, eddy bars may develop (Figure F16). In narrow, deep flood channels these various bars tend to be high and mounded. The great flood bars of the Chuya River valley have a sedimentary thickness of nearly 200 m. They consist of thick foreset beds of coarse sand and gravel that was carried in suspension in the main flood channel.

Large-scale transverse gravel ripples were recognized in the Chaneled Scabland (Figure F17), and their study contributed to the general acceptance of the catastrophic flood origin of that landscape (Baker, 1973, 1981). Although they were first named "giant current ripples" by J Harlen Bretz in his now-classic studies, their size and hydraulic relationships indicate that they are best classified as fluvial dunes (Baker, 1973). Nevertheless, the original name is often retained for historical reasons. In both the Chaneled Scabland and in central Asia,

Table F2 Cataclysmic flood landform assemblage (Earth and Mars)

Scale	Eroded in rock	Eroded in sediment	Depositional
Macroforms (Scaled to width)	Anastomosing pattern	Streamlined hills	Alternate bars Expansion bars Fans Pendant bars Eddy bars
Mesoforms (Scaled to depth)	Longitudinal grooves Butte-and-basin topography Inner channels Cataracts Potholes	Scour marks	Giant current Ripples (Dunes, Gravel waves)

Figure F16 Large eddy bar a the mouth of a tributary canyon along the Flathead River, Perma, Montana. The bar formed during the rapid late Pleistocene drainage of glacial Lake Missoula (Baker, 1973) as high-energy flood water moved from right to left in the main channel. Gravel in suspension in the channel was rapidly deposited in the slack-water zone of the tributary mouth to create the bar. Very similar bars were produced by late Pleistocene glacial outburst floods in the Altai Mountains of Siberia (Rudoy and Baker, 1993).

Figure F17 Giant current ripples (dunes) near Marlin, Washington. These gravel bedforms have average heights of 6 m and average spacings of 60 m (Baker, 1973). They occur on the divide between two preflood creek valleys, thereby illustrating the immense scale of the responsible late Pleistocene cataclysmic flooding.

the giant current ripples range in spacing from about 30 m to 200 m and are composed of pebble and cobble gravel. Internally they consist of foreset-bedded gravel dipping at angles of about 20° to 30°. They are generally superimposed on longitudinal, alternate bars or on expansion bars. Trains of giant current ripples as much as 10 km or 15 km long are found in parts of the Chaneled Scabland, and also in the Altai and Tuva regions of central Asia. Bedform dimensions have been related to hydraulic factors for both these areas (Baker, 1973; Carling, 1996).

Slack-water deposits

Slack-water sediments accumulate in areas marginal to main channelways, such as in the mouths of backflooded tributary valleys or in alcoves along channel margins. The sediments derive from the rapid fall of suspended coarse particles, usually sand and gravel, as these materials move from the high-velocity main thread of flow into the marginal zones of slack water. Because such deposits can be emplaced at high elevation, up to the level of the peak flood stage, they can serve as indicators of that stage. Moreover, their sequence of emplacement can reveal the history of flood events, especially when they are preserved in the low-energy flow environments marginal to main catastrophic floodways. This situation was recognized in the Chaneled Scabland region (Baker, 1973). There, the slack-water deposits display a great variety of sedimentary structures and stratigraphic relationships that can be used to infer aspects of the flood dynamics and history (Smith, 1993).

Paleoflood hydrology

Although its name was not introduced until 1982 (Kochel and Baker, 1982), the science of paleoflood hydrology has its origin in a century of geological studies of past floods. Using the experience from studies of the Chaneled Scabland flooding (Baker, 1973), it was found that similar methods of paleoflood reconstruction could be applied to late Holocene paleofloods, thereby extending discharge records used in flood hazard evaluation (Baker, 1987). In essence, paleoflood hydrology relies upon studies of those rivers that are, "...chroniclers of their own cataclysms" (Baker and Pickup, 1987). A wide variety of paleostage indicators are used to determine past flood magnitudes, but the study of flood slack-water deposits are especially important, given their preservation of the time sequence of past floods. The sequence of paleoflood magnitudes and their temporal distribution can be used to evaluate the flood-frequency estimates that assign risk to various hazardous sites or structures (House et al., 2002).

Floods on Mars

The ancient floods of Mars are inferred from features recognized in terrestrial catastrophic flood channels, such as the Chaneled Scabland (Baker, 1982). The flows emanated from collapse zones or fractures at the heads of the flood-scoured troughs. Many of the largest such "outflow channels" surround the Tharsis volcanic province of Mars, an immense plateau of volcanic rocks, cut by a huge rift-related canyon system, the Valles Marineris. The flooding was likely genetically related to the volcanism, and its occurrence is related to the formation of extensive areas of ponded water and associated climate change on a planet that is otherwise extremely cold and dry (Baker, 2001).

Continuing controversies

Giant wave deposits

Recent work on the possible action of giant waves along coastlines has documented a sequence of erosional and depositional landforms with considerable similarity to those described for catastrophic flooding in areas like the Chaneled Scabland (Bryant, 2001). The giant waves are presumed to be tsunami or direct splash effects of meteor impacts into the ocean. Such impacts on geological timescales that would lead to preserved effects are highly likely, given knowledge of impact frequencies on Earth. The best examples of the boulder deposits, streamlined deposits, and various erosional forms are documented along tectonically stable coastlines, like those of Australia (Bryant, 2001). However, the mechanics of the processes for generating these phenomena are not well understood, and the regional importance of such features is still a matter of on-going investigation.

Subglacial flooding

The landforms that develop beneath thick continental glaciers have long been controversial for their genesis. As in the catastrophic flooding problem, the modern analogs (modern glaciers) do not provide satisfactory information on the ancient ice sheets. During the glacial maxima of the Quaternary, the great ice sheets of North America, Eurasia, and Antarctica were warm based and associated with very large amounts of meltwater along their margins. Shaw (1996) developed a model that envisions a whole sequence of erosional/depositional processes beneath these ice sheets, particularly cataclysmic flooding of magnitudes comparable to those illustrated in Table F1 (10^7 m^3 per second). The flooding produces all the distinctive subglacial landforms, including drumlins, tunnel valleys, bedrock erosional marks, and Rogen moraines. The drumlins and Rogen moraines are generated by deposition into cavities formed on the sole of ice sheet by flood erosion. There is considerable controversy over this model (Benn and Evans, 1998), but its continuing testing against field evidence is leading to discovery of overall coherence in erosional and depositional features produced under the ancient ice sheets. The consistency of some of Shaw's documented patterns with those displayed by recognized catastrophic flooding is one of the major arguments in favor of this hypothesis.

Central Asia

During the Pleistocene glaciations the large, north-flowing Siberian rivers were dammed by ice sheets situated on the shelf areas of northern Eurasia. The resulting innundation produced spillways that carried flows southward to the present-day basins of the Aral, Caspian, and Black Seas. Additionally, these flows were augmented by the huge floods of ice-dammed lake bursts from the mountain regions of central Asia (Baker, 1997). A system of huge spillways connected the various elements of this system, which may temporarily have comprised the largest drainage basin on the planet. In a study of linear erosional and depositional features, visible on satellite imagery and extending across divides through central Asia, Grosswald (1999) proposed the provocative hypothesis that immense volumes of water were conveyed from the margins of the great northern ice sheets. The flows were so great that they not only inundated the spillways, but also flowed across divides in a flow width of hundreds of kilometers. The mechanism for this water burst is the pressurized release of water from underneath the northern ice sheets, which are thought to have formed a circumpolar dam of arctic ocean water and subglacial water storage.

Noah's flood?

Based mainly on evidence for the transition from freshwater to marine conditions in the Black Sea basin, Ryan et al. (1997) proposed a catastrophic inflow of Mediterranean water through the Bosporus about 7.5 ka. In subsequent popular accounts this innundation, estimated to have occurred in less than 2 years, was linked to the biblical flood of Noah. However, a climbing delta, dated at 10 ka to 9 ka, indicates that freshwater outflow from the Bosporus was occurring prior to the presumed catastrophic flooding episode (Aksu et al., 2002). If these observations are correct then the Bosporus could not have passed later Holocene cataclysmic flood flows that would have removed these sediments. This issue is not completely resolved, but it illustrates the difficulty (and the controversies) that commonly surround catastrophic flood investigations. Moreover, it shows the importance of marshalling many diverse lines of evidence in support and any catastrophic flooding hypothesis.

Conclusions

Despite considerable controversy in their early study, the distinctive erosional and sedimentary features of catastrophic superfloods are now well-known. They include scabland erosion, streamlining of residual uplands, large-scale scour around obstacles, depositional bars, giant current ripples, and huge sediment fans. Moreover, the sequence of paleoflood events can be inferred from study of the deposition in bars and in slack-water areas marginal to the high-energy flood channelways. Further opportunites exist for study of catastrophic flood sequences in the deposits of both continental and marine sedimentary basins. For paleofloods of less intense magnitude, occuring in time frames closer to those of probable human hazard, the science of paleoflood hydrology has arisen to generate flood risk information from the geological evidence of past floods.

Victor R. Baker

Bibliography

Aksu, A.E., Hiscott, R.N., Mudie, P.J., Rochon, A., Kaminski, M.A., Abrajano, T., and Yasar, D., 2002. Persistent holocene outflow from the Black Sea to the Eastern Mediterranean contradicts Noah's flood hypothesis. *GSA Today*, **12** (5): 4–10.
Albritton, C.C., 1989. *Catastrophic Episodes in Earth History*. New York: Chapman and Hall.
Baker, V.R., 1973. *Paleohydrology and Sedimentology of Lake Missoula Flooding in Eastern Washington*. Geological Society of America Special Paper, 144.
Baker, V.R., 1978. The spokane flood controversy and the Martian outflow channels. *Science*, **202**: 1249–1256.
Baker, V.R. (ed.), 1981. *Catastrophic Flooding: The Origin of the Channeled Scabland*. Stroudsburg, PA: Dowden, Hitchinson and Ross.
Baker, V.R., 1982. *The Channels of Mars*. Austin: University of Texas Press.
Baker, V.R., 1987. Paleoflood hydrology and extraordinary flood events. *Journal of Hydrology*, **96**: 79–99.
Baker, V.R., 1997. Megafloods and glaciation. In Martini, I.P. (ed.), *Late Glacial and Postglacial Environmental Changes*. Oxford University Press, pp. 98–108.
Baker, V.R., 1998. Catastrophism and uniformitarianism: logical roots and current relevance in geology. In Blundell, D.J., and Scott, A.C. (eds.), *Lyell: The Past is the Key to the Present*. London: Geological Society, Special Publication, 143, pp. 171–182.
Baker, V.R., 2001. Water and the martian landscape. *Nature*, **412**: 228–336.
Baker, V.R., 2002. High-energy megafloods: planetary settings and sedimentary dynamics. In Martini, I.P., Baker, V.R., and Garzon, G. (eds.), *Flood and Megaflood Processes and Deposits: Recent and Ancient Examples*. International Association of Sedimentologists Special Publication, 32, pp. 3–15.
Baker, V.R., and Pickup, G., 1987. Flood geomorphology of the Katherine Gorge, Northern Territory, Australia. *Geological Society of America Bulletin*, **98**: 635–646.
Benn, D.I., and Evans, D.J.A., 1998. *Glaciers and Glaciation*. London: Arnold.
Bretz, J.H., 1923. The channeled Scabland of the Columbia Plateau. *Journal of Geology*, **31**: 617–649.
Bryant, E., 2001. *Tsunami: The Underrated Hazard*. Cambridge: Cambridge University Press.
Carling, P.A., 1996. Morphology, sedimentology, and paleohydraulic significance of large gravel dunes: Altai Mountains, Siberia. *Sedimentology*, **43**: 647–664.
Costa, J.E., 1983. Paleohydraulic reconstruction of flash flood peaks from boulder deposits in the Colorado front range. *Geological Society of America Bulletin*, **94**: 986–1004.
Grosswald, M., 1999. *Cataclysmic Megafloods in Eurasia and the Polar Ice Sheets*. Moscow: Scientific World (in Russian).
House, P.K., Webb, R.H., Baker, V.R., and Levish, D.R. (eds.), 2002. *Ancient Floods, Modern Hazards: Principles and Applications of Paleoflood Hydrology*. American Geophysical Union Water Science and Application, No. 5.
Kochel, R.C., and Baker, V.R., 1982. Paleoflood hydrology. *Science*, **215**: 353–361.
Komar, P.D., 1980. Modes of sediment transport in channelized water flows with ramifications for the erosion of the Martian outflow channels. *Icarus*, **42**: 317–329.
Lacasse, C., Carey, S., and Sigurdsson, H., 1998. Volcanogenic sedimentation in the Iceland Basin: influence of subaerial and subglacial eruptions. *Journal of Volcanology and Geothermal Research*, **83**: 47–73.
Mutti, E., Davoli, G., Tinterri, R., and Zavala, C., 1996. The importance of ancient fluvio-deltaic systems dominated by catastrophic flooding in tectonically active basins. **Mem. Sci. Geol. 48**: 233–291.
O'Connor, J.E., 1993. *Hydrology, Hydraulics and Sediment Transport of Pleistocene Lake Bonneville Flooding on the Snake River, Idaho*. Geological Society of America Special Paper, 274.
O'Connor, J.E., and Baker, V.R., 1992. Magnitudes and implications of peak discharges from Glacial Lake Missoula. *Geological Society of America Bulletin*, **104**: 267–279.
Rudoy, A.N., and Baker, V.R., 1993. Sedimentary effects of cataclysmic late Pleistocene glacial outburst flooding, Altay Mountains, Russia. *Sedimentary Geology*, **85**: 53–62.
Ryan, W.B.F., Pittman, III W.C., Major, C.O., Shimkus, K., Maskalenko, V., Jones, G.A., Dimitrov, P., Gorur, N., Sakinc, M., and Yuce, H., 1997. Abrupt drowning of the Black Sea shelf. *Marine Geology*, **138**: 119–126.
Shaw, J., 1996. A meltwater model for Laurentide subglacial landscapes. In McCann, S.B., and Ford, D.C. (eds.), *Geomorphology sans Frontiere*. Wiley, pp. 182–226.
Smith, G.A., 1993. Missoula flood dynamics and magnitudes inferred from sedimentology of slack-water deposits on the Columbia Plateau, Washington. *Geological Society of America Bulletin*, **105**: 77–100.

Zuffa, G.G., Normack, W.R., Serra, F., and Brumer, C.A., 2000. Turbidite megabeds in an oceanic rift valley recording jokulhlaups of Late Pleistocene Glacial Lakes of the Western United States. *Journal of Geology*, **108**: 253–274.

Cross-references

Gravity-driven Mass Flows
Rivers and Alluvial Fans
Sediment Fluxes and Rates of Sedimentation
Sedimentology, History
Storm Deposits
Submarine Fans and Channels
Tsunami Deposits
Turbidites

FLOW RESISTANCE

Introduction

Flow of any real viscous fluid (such as water or air) results in energy dissipation because the fluid must do work to overcome resistance due to fluid viscosity. This resistance produces an "energy loss" that must often be determined when computing water levels and velocity. (Note that the term energy loss is a convenient way to refer to the energy that is dissipated as turbulence or heat.)

Resistance develops in the presence of a velocity gradient, and can be likened to a shear stress acting between adjacent layers sliding overtop one another (Figure F18). Shear stress, τ, develops at the interface between adjacent layers, which are moving at different relative velocities. In open channel flow, the velocity gradient (Figure F18(A)) and the shear stress, τ (Figure F18(B)) usually originate at a solid boundary, such as the bed or banks of a river or stream. The shear stress within the fluid, τ, typically varies linearly from a maximum at the boundary, where τ equals the boundary shear stress τ_0, to zero at the free surface. Flow resistance and shear stress can also develop at the interface between bodies of fluid moving at different velocities such as the interface between overbank and in-channel flow, at river confluences, and along the interface of ocean and air currents.

Viscosity that gives rise to flow resistance can be either laminar or turbulent depending upon the value of the flow Reynolds Number, $\boldsymbol{R} = VD/v$, in which V = mean velocity, D = depth of flow, and v = kinematic viscosity. For a wide channel or sheet flow, the fluid will remain laminar for values of $\boldsymbol{R}$ up to about 500. For Newtonian fluids such as water, laminar viscosity is strictly a property of the fluid. For values of $\boldsymbol{R} > 2,000$ flow becomes turbulent. For values of $\boldsymbol{R}$ between 500 and 2,000, the flow is transitional between laminar and turbulent. Turbulent or eddy viscosity is not only a fluid property, but varies with velocity gradient, distance from the boundary, and other flow conditions. Because the turbulent viscosity is related in part to flow properties, and flow properties are influenced by fluid viscosity, it remains exceedingly difficult to fully describe turbulent fluid flow and the resulting flow resistance.

General theory

A common requirement in fluvial geomorphology or river hydraulics is estimation of flow resistance based on the mean velocity V. Several equations are available, of which the Darcy equation is most common:

$$f = \frac{8gDS}{V^2} \quad \text{(Eq. 1)}$$

in which f = friction factor, which is a coefficient of flow resistance; g = gravitational acceleration; D = flow depth; and S = slope of the energy grade line. Equation 1 applies to wide channels where the contribution of the banks can be neglected. For narrow channels, closed conduits or pipes, the hydraulic radius R should be used in place of D.

For laminar flow it has been determined experimentally that $f = 64/\boldsymbol{R}$, which when substituted into equation 1 yields a form of the Hagan-Poiseuille law:

$$V = \frac{gD^2 S}{64v} \quad \text{(Eq. 2)}$$

In practice, equation 2 is generally limited to very shallow and/or slow moving flows, such as overland sheet flow, or where the fluid is highly viscous, as in the case of some mud and debris flows (see *Debris Flow*).

Typical flow in rivers and streams and other open channels is usually described as fully-rough turbulent flow in which $R > 2000$, and f is independent of R. In fully-rough turbulent flow, the flow resistance is dependent only on the relative boundary roughness D/k_s, as in the well-known Keulegan equation (Keulegan, 1938):

$$\frac{1}{\sqrt{f}} = 2.03 \log\left(\frac{12.2D}{k_s}\right) \quad \text{(Eq. 3)}$$

where k_s = equivalent sand roughness, and was originally defined as the diameter of the sand grains used to roughen pipe walls during early flow resistance experiments (Nikuradse, 1933). The Manning roughness coefficient, n, is also used as a measure of boundary roughness.

In flow over alluvial boundaries hydraulic roughness can occur at a range of scales from individual grains, to clusters of grains, bedforms such as ripples and dunes, to bars, meanders and other planform variables. Vegetation and large woody debris can also contribute significantly to flow resistance. It is commonly assumed that resistance can be divided or partitioned into individual components (e.g., Millar, 1999):

$$f = f' + f'' + f''' + \cdots + f^n \quad \text{(Eq. 4)}$$

where $f', f'', \ldots, f^n$ represent contributions of the different components of the total resistance. In practice the term f' is

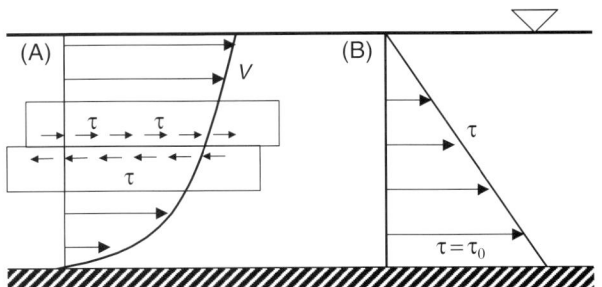

Figure F18 (A) Velocity gradient and (B) shear stress in open-channel flow.

often referred to as the "grain" component due to the flow resistance that develops from the roughness of the grains on the bed. The additional components are collectively referred to as the "form" component that is due to bedforms, bars, vegetation, etc. (Einstein and Barbarossa, 1952).

Where these additional roughness elements exist, it may be difficult or even questionable to express the hydraulic roughness in terms of k_s or n. In some instances, k_s may actually exceed D. Nonetheless, the value of k_s is often assumed to be approximately equal to some multiplier times a characteristic grain diameter, such as $6.8D_{50}$ or $3.5D_{84}$ for gravel-bed rivers (Bray, 1982).

Flow resistance is directly related to the boundary shear stress by:

$$\tau_0 = \frac{f}{8}\rho V^2 \quad \text{(Eq. 5)}$$

The boundary shear stress can also be partitioned into components (e.g., Buffington and Montgomery, 1999):

$$\tau = \tau' + \tau'' + \tau''' + \cdots + \tau^n \quad \text{(Eq. 6)}$$

where the terms $\tau', \tau'', \ldots, \tau^n$ represent the component of the total shear stress acting on different roughness elements. As with the resistance, the τ' term usually denotes the grain component. The value of τ' can be calculated by substituting f' into equation 6. Several bed material transport theories are formulated on the assumption that it is only the grain shear τ' that contributes to transport, a concept first proposed by Einstein (1950).

The preceding discussion considers the most common application of flow resistance, where a fully developed boundary layer (Schlichting, 1979) extends from the solid boundary to the surface. This is sufficient to describe steady flow in rivers, streams and other open channels, for currents in relatively shallow water ($D < 25$ m), sheet flow across surfaces, and flow in closed conduits or pipes. It is also adequate to describe moderately unsteady flows such as passage of a flood wave along a river, or tidally-forced currents.

For oscillatory flows due to ocean gravity waves, the flow resistance, f_w, can be expressed as:

$$f_w = \frac{2\tau_0}{\rho u_b |u_b|} \quad \text{(Eq. 7)}$$

where $u_b = $ bottom orbital velocity (Raudkivi, 1998, 327).

Measurement and prediction of flow resistance

There is no simple direct measurement of flow resistance or hydraulic roughness. Values of f, k_s, or n must in general be back calculated using measured values of V, D, and S, or obtained through model calibration. Different values of the coefficients will be associated with different model formulations. For instance, one-dimensional (1D) and 2D or 3D models will yield different values of f or n, even when calibrated using the same data. This is because in a 1D formulation, f or n are used to account for all resistance components. 2D and 3D models typically yield lower values for the resistance coefficients because eddy losses and other form components are better simulated by the physics of the model.

There remains considerable uncertainty associated with predictions of flow resistance in ungauged rivers and streams (Millar, 1999). In particular, the contributions of bank vegetation (Darby, 1999), nonemergent vegetation (Lopez and Garcia, 2001), gravel bars, riffles and cluster bedforms (Lawless and Roberts, 2001) have yet to be quantified adequately. An ongoing research effort is directed toward understanding and predicting resistance due to interactions between overbank and in-channel flow in compound channels (Lyness et al., 2001).

Robert Millar

Bibliography

Bray, D.I., 1982. Flow resistance in gravel-bed rivers. In Hey, R.D., Bathurst, J.C., and Thorne, C.R. (eds.), *Gravel-Bed Rivers*. Chichester, England: John Wiley and Sons., pp. 109–133.
Buffington, J.M., and Montgomery, D.R., 1999. Effects of hydraulic roughness on surface textures of gravel-bed rivers. *Water Resources Research*, **35** (11): 3507–3521.
Darby, S.E., 1999. Effect of riparian vegetation on flow resistance and flood potential. *Journal of Hydraulic Engineering-ASCE*, **125**(5): 443–454.
Dudley, S.J., Fishenich, J.C., and Abt, S.R., 1998. Effect of woody debris entrapment on flow resistance. *Journal of the American Water Resources Association*, **34** (5): 1189–1197.
Einstein, H.A., 1950. The bedload function for sediment transport in open channel flows. *Technical Bulletin* 1026, US Department of Agriculture, Washington DC, 71 p.
Einstein, H.A., and Barbarossa, N., 1952. River channel roughness. *Transactions of the American Society of Civil Engineers*, **117**: 1121–1146.
Keulegan, G.H., 1938. Laws of turbulent flow in open channels. *Journal of the National Bureau of Standards*, **21**: 707–741.
Lawless, M., and Robert, A., 2001. Scales of boundary resistance in coarse-grained channels: turbulent velocity profiles and implications. *Geomorphology*, **39**(3–4): 221–238.
Lopez, F., and Garcia, M.H., 2001. Mean flow and turbulence structure of open-channel flow through non-emergent vegetation. *Journal of Hydraulic Engineering-ASCE*, **127**(5): 392–402.
Lyness, J.F., Myers, W.R.C., Cassells, J.B.C., and O'Sullivan, J.J., 2001. The influence of planform on flow resistance in mobile bed compound channels. *Proceedings of the Institution of Civil Engineers—Water and Maritime Energy*, 5–14.
Millar, R.G., 1999. Grain and form resistance in gravel-bed rivers. *Journal of Hydraulic Research*, **37** (3): 303–312.
Nikuradse, J., 1933. Strömungsgesetze in rauhen Rohren. English Translation: laws of flow in rough pipes. *Technical Memorandum* 1292, National Advisory Commission for Aeronautics, Washington, DC.
Raudkivi, A.J., 1998. *Loose Boundary Hydraulics*. In Balkema, A.A. (ed.), Rotterdam.
Schlichting, H., 1979. *Boundary Layer Theory*, 7th edn. New York: McGraw-Hill.

Cross-references

Armor
Debris Flow
Grain Size and Shape
Imbrication and Flow-Oriented Clasts
Ripple, Ripple Mark, and Ripple Structure
Rivers and Alluvial Fans
Sediment Transport by Unidirectional Water Flows
Surface Forms

FLUID ESCAPE STRUCTURES

Fluid escape structures are sedimentary structures that form during the escape of pore fluids from loose, unconsolidated sediments. Most are postdepositional in origin, but some, including convolute lamination, can form during sedimentation. Fluid escape structures may represent modified primary sedimentary structures, modifications of previously formed postdepositional structures, or entirely new structures. Deformation may result directly from: (1) pore-fluid movement; (2) current-induced shear; or (3) gravity forces acting on density contrasts within low-strength liquefied sediments. In most cases, the pore fluid is water, and the resulting structures are water-escape structures. However, gas escape structures can form within sediments deposited by dry pyroclastic flows (Cas and Wright, 1987). In general, fluid escape structures are most common in rapidly deposited, poorly sorted, fine- to medium-grained sands and less common in more slowly deposited, coarser- and finer-grained, and texturally mature deposits.

Processes of fluid escape

The development of denser packing due to rearrangement of sediment grains accompanied by fluid expulsion from a loosely-packed sediment is called consolidation. Fluid escaping from sediments during consolidation can interact with the sedimentary grains in four ways: seepage, liquefaction, fluidization, and elutriation.

Seepage

When the escaping pore fluids have low velocities, the rearrangement of grains takes place slowly, most grain-to-grain contacts are preserved, and even the hydraulically fine grains remain largely unaffected. This fluid escape process is called seepage (Lowe, 1975; Owen, 1987).

Liquefaction

During liquefaction, most of the grain-to-grain contacts are suddenly lost, the unconsolidated sediment collapses, and pore fluid pressure increases rapidly but temporarily. Grains settle briefly through their own pore fluid, fluid escape takes place in short time, and the result is a more tightly packed grain framework. Grains of different sizes and shapes will behave differently during settling and under the influence of the escaping pore fluid. This differential behavior commonly results in particle segregation, the formation of focused fluid flow paths separated by more quiescent sediments, and gravity instabilities associated with the presence of masses of sediment of differing densities within the liquefied material.

Fluidization

Fluidization occurs when escaping pore fluid is able to fully support a significant proportion of the grains; that is, it exerts an upward drag force on the grains equal to their immersed weight. The fluid flow rates are so high that the sediment ceases to be grain-supported and becomes fluid-supported, often accompanied by dilation and density reduction. Fluidized sediment masses are highly mobile and may flow from their original location into overlying or adjacent strata, forming clastic dykes and sills, respectively. In many cases however, fluidization occurs along discrete vertical fluid-flow paths with little bulk movement of the fluidized sediment. These paths can be preserved in the sediment as vertical fluid escape structures. Fluid escape can be further complicated by the development of ephemeral localized pockets or layers of fluid in sediment undergoing consolidation, especially if slight cohesiveness and/or permeability heterogeneities are present (Allen, 1982: 366; Nichols et al., 1994).

Elutriation

The fluid force exerted on grains by escaping pore fluids can exceed the fall velocities of individual grains, especially the smaller and less dense grain components, and these particles, where loose, will be carried upward and separated from the surrounding larger and/or denser grains, a process termed elutriation. Clays, organic particles, and micas are especially likely to be elutriated from associated quartz and feldspar sand during water escape to form dish and other water-escape structures.

Fine-grained silt and clay are cohesive and not readily liquefied; hydroplastic intrusions and soft-sediment folds can form, but they are usually load structures formed through density instabilities unrelated to water escape. Gravels are difficult to fluidize and tend to resediment rapidly when liquefied. Both fine-grained cohesive and coarse-grained sediments tend resist grain-by-grain reorganization by escaping pore fluids and dewater by seepage. As a result, fluid escape structures are most common in fine- to coarse-grained sands, which are easily liquefied and fluidized and from which more mobile grains can be easily elutriated.

Classification

Consolidation laminations and dish structures

The presence of slight cohesiveness and/or permeability heterogeneities in a sediment leads to the development of consolidation laminations and dish structures during seepage and liquefaction. Consolidation laminations are subhorizontal laminations that develop as a result of: (1) gravitational segregation of grains of different densities during settling within liquefied sediments; and (2) hydraulic segregation of particles during flow associated with fluid escape (Lowe, 1975). Primary semi-permeable laminations act as partial barriers to vertical fluid flow and cause escaping pore fluids to move laterally to points where the laminations disappear or have been breached. With time, flow beneath and slow seepage across such semi-permeable laminations result in the concentration of hydraulically finer grains carried within the escaping fluids along these laminations. Differential subsidence associated with fluid withdrawal may cause the laminations to become concave upward. The resulting structures are called dish structures (Lowe and LoPiccolo, 1974). Dish structures can also form through stoping, when ephemeral pockets of pore fluid develop in the sediment (Allen, 1982: 373–374; Tsuji and Miyata, 1987). Larger, more irregular zones of partially or totally homogenized sediment that underwent liquefaction or fluidization (Figure F19), with the possible involvement of ephemeral cavities, can be referred to as liquefaction or fluidization pockets and layers (Lowe, 1975).

Vertical fluid escape channels

Once discrete zones of vertical fluid escape are established within a dewatering sediment, the focused fluid flow is commonly strong enough locally to fluidize the sediment and elutriate hydraulically finer grains. Thus the permeability is increased and the fluid escape rate correspondingly rises. The resulting subvertical to vertical zones of mud-poor and usually

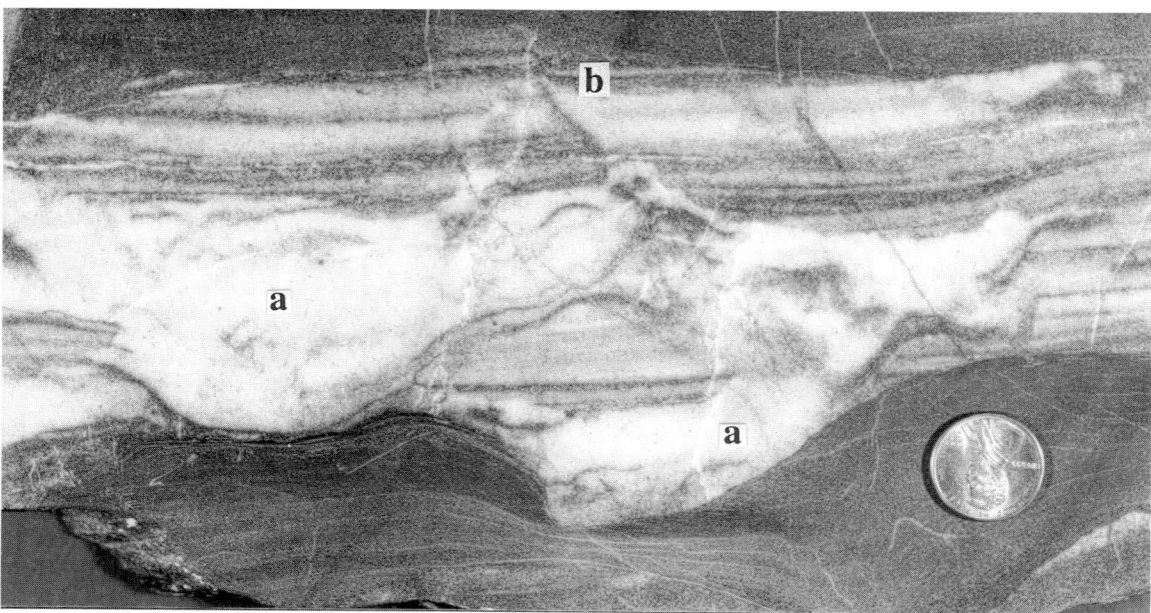

Figure F19 Irregular pockets (a) formed by sediment liquefaction accompanying water escape in a thin layer of flat- to cross-laminated sandstone. Within the liquefaction pockets, primary current lamination has been largely erased by mixing. A small subvertical water-escape conduit leads upward from the larger liquefaction pockets (b). Coin is 1.8 cm in diameter.

coarser-grained sediment are preserved as water-escape channels, sheets, or conduits, which have been termed pillar structures.

Soft-sediment folds

Soft-sediment folds can form due to processes other than fluid escape (Owen, 1987). However, two types of soft-sediment folds commonly form during and as a result of sediment liquefaction and water-escape: convolute lamination (convolute stratification) and recumbent or oversteepened cross-bedding. Convolute lamination formed during water escape is most common in the cross-laminated divisions of fine sandy turbidites and forms during concurrent deposition of the upper finer parts of the beds and dewatering of the underlying coarser sediments (Lowe, 1975). In many cases, the usually narrow anticlines represent subvertical zones of water escape, with sediment locally homogenized by fluidization. Upward decreasing permeability in individual turbidites and Rayleigh–Taylor instabilities may play a role in the development of convolute lamination. Oversteepened or recumbent cross-bedding occurs mainly in cross-bedded fluvial and cross-laminated turbiditic sands. Deposition through avalanching on the lee face of dunes or ripples results in loose packing of sediment that is easily liquefied. The liquefied foreset laminations are sheared in a downcurrent direction by the overlying current.

Soft-sediment intrusions

Sediment that is liquefied or fluidized during fluid escape long after burial can become mobilized and intruded into overlying and adjacent strata. Where tabular, such intrusions are commonly termed clastic dykes or sills, but more equant masses are locally developed. The driving forces for these processes include gravity forces resulting from low density of liquefied and fluidized sediment or sudden pressure release through joint formation. Gravity-driven soft-sediment intrusions result from density instabilities in soft sediment, even in the absence of escaping pore fluids, in which case they represent load structures but not, strictly speaking, fluid-escape structures. The latter can generate sufficient fluid velocities that large volumes of sediment are not just supported but entrained by the escaping fluid.

Zoltán Sylvester and Donald R. Lowe

Bibliography

Allen, J.R.L., 1982. *Sedimentary Structures; Their Character and Physical Basis*, Volume II: Amsterdam: Elsevier.

Cas, R.A.F., and Wright, J.V., 1987. *Volcanic Successions: Modern and Ancient*. London: Allen and Unwin.

Lowe, D.R., and LoPiccolo, R.D., 1974. The characteristics and origins of dish and pillar structures. *Journal of Sedimentary Petrology*, **44**: 484–501.

Lowe, D.R., 1975. Water-escape structures in coarse-grained sediments. *Sedimentology*, **22**: 157–204.

Nichols, R.J., Sparks, R.S.J., and Wilson, C.J.N., 1994. Experimental studies of the fluidization of layered sediments and the formation of fluid escape structures. *Sedimentology*, **41**: 233–253.

Owen, G., 1987. Deformation processes in unconsolidated sands. In *Deformation of Sediments and Sedimentary Rocks. Geological Society Special Publication*, **29**: 11–24.

Tsuji, T., and Miyata, Y., 1987. Fluidization and liquefaction of sand beds—experimental study and examples form Nichinan group. *Journal of the Geological Society of Japan*, **93**: 791–808.

Cross-references

Clastic (Neptunian) Dykes and Sills
Convolute Lamination
Dish Structure

Liquefaction and Fluidization
Pillar Structure
Turbidites

FLUID INCLUSIONS

Fluid inclusions are fluid-filled vacuoles trapped within minerals. They may preserve the composition and density of any or all fluids present throughout the history of a sedimentary rock. Fluid inclusions typically are very small, micrometers in size (Figure F20), so they must be studied with a good microscope in relatively coarse minerals. An inclusion could consist of a single fluid phase, such as liquid water or natural gas, could contain solid mineral phases, organics or bacteria, and could contain multiple fluid phases such as oil, aqueous liquid, natural gas, or water vapor. Commonly, fluid inclusions contain only one or two fluid phases when viewed at room temperature, one of liquid and another of gas (Figure F20).

In addition to fluid inclusions being interesting curiosities within crystals, they provide one of the most powerful tools for reconstructing temperature, pressure and fluid-composition history in sedimentary systems. Unlike many other geochemical tools, the fluid inclusion tool is normally petrographically based and provides a direct sample of conditions within ancient sedimentary systems. Useful compilations helpful in delving into the details of fluid inclusions in sedimentary rocks are Hollister and Crawford (1981), Roedder (1984), Goldstein and Reynolds (1994), and Andersen et al. (2001).

Entrapment

In sedimentary systems, sediments and rocks spend their entire history bathed in fluids. As a mineral precipitates from an aqueous solution, it is well-known that crystal growth is seldom perfect. Irregularities in the face of a growing crystal are commonly engulfed by further crystal growth, sealing inside a bit of the fluid. In samples where there is a petrographic relationship indicating that fluid inclusions were trapped during growth of the crystal, the fluid inclusions are considered "primary" in origin. Similarly, many diagenetic minerals tend to recrystallize through a fine-scale dissolution-reprecipitation process (see *Neomorphism and Recrystallization*). If there is petrographic evidence that inclusions were trapped during recrystallization, they are also considered as primary in origin. Minerals present in sedimentary systems are commonly fractured. As tiny micrometer-scale cracks form, they are filled with whatever fluid happens to be present in pore systems. These cracks are thermodynamically unstable, so the minerals on walls of cracks tend to dissolve and reprecipitate to "neck down" the crack to a planar array of fluid inclusions sealed within the mineral. Where petrographic evidence exists that inclusions were trapped in this way, after growth of the mineral phase, they are called secondary fluid inclusions. Where petrographic evidence exists that cracks have healed during and not after precipitation of the sequence of precipitated minerals, inclusions are termed pseudosecondary in origin. Determining the timing of entrapment of fluid inclusions is normally the most important part of any fluid inclusion study.

Where there is only a single fluid phase present during the entrapment of fluid inclusions, the composition and density of the entrapped fluid generally is representative of the bulk fluid that was present. However, if immiscible fluid phases are present at the time of entrapment of inclusions, the bulk composition of inclusions typically is not the same as the bulk composition of the pore fluids, because one phase is normally preferentially entrapped over another.

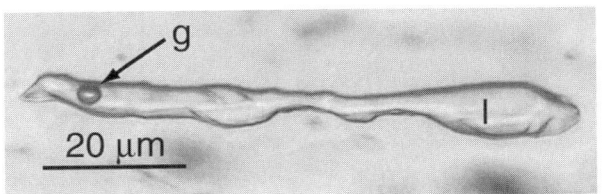

Figure F20 Transmitted light photomicrograph of large elongate fluid inclusion containing a gas bubble (g) within an aqueous liquid (l).

Preservation

It is important to evaluate the degree to which fluid inclusions are preserved in their original state of entrapment. Any fracturing or recrystallization of a mineral containing fluid inclusions has potential for resetting the inclusions. Thus, recrystallization and fracturing of minerals must be evaluated in all fluid inclusion studies.

Under normal conditions, it appears that many minerals containing fluid inclusions are good "bottles", sealing in the fluid at its original density and composition over geologic time. However, aqueous liquid fluid inclusions develop internal overpressure during burial or hydrothermal heating. Because mineral volume changes little during burial, the volume of an inclusion cavity, in theory, would change little. Heating a liquid confined to such a fixed volume would induce high internal pressure. Thus, during burial heating, inclusions develop high internal pressures and follow steep, constant volume (isochoric) P–T paths. At a given temperature, external confining pressure normally is constrained by a thermobaric P–T path (Figure F21), and is lower than inclusion internal pressure. In many cases, the mineral contains the internal overpressure, but for some, pressure is relieved by expanding the inclusion's volume through permanent plastic deformation of the crystal around the inclusion wall. This decreases the inclusion fluid's density, but preserves its composition. Alternatively, the inclusion may burst, and the contents may leak and refill with new pore fluids. To place fluid inclusion data into a paragenetic context, such thermal reequilibration must be evaluated in all studies of fluid inclusions in sedimentary systems.

Fluid inclusions can change their shape over time through a process of dissolution and reprecipitation to yield more stable, negative crystal or globular shapes. A single large elongate inclusion may "neck down" into multiple inclusions during this process. If this happens while more than one phase is present in the inclusion, the new inclusions will have densities and compositions that differ from the original one. Necking down in the presence of multiple phases also must be evaluated to interpret fluid inclusion data.

Philosophy and practice

Normally, a fluid inclusion study begins just as any other scientific study, by defining a specific geologic problem to be solved. Then, samples are taken within a paragenetic, tectonic and stratigraphic context. Rock samples cannot have been heated in the lab before or during preparation. Polished thick sections are prepared without heating or fracturing. The overall paragenesis is worked out and the timing of entrapment of fluid inclusions is integrated using transmitted light, cathodoluminescence, UV epi-illumination and back-scattered electron imaging. If the fluid inclusions needed to answer the question posed are present, then the study is continued, but if they are not, the study is aborted. The most finely petrographically discriminated events of fluid inclusion entrapment (fluid inclusion assemblages; FIAs) are then defined. Composition and phase distribution of inclusions in FIAs are evaluated, providing preliminary interpretations about temperature and pressure of entrapment, extent of thermal reequilibration, and fluid composition. Problems with necking down in the presence of multiple phases and some thermal reequilibration can be recognized at this stage. Then, inclusions can be heated, frozen and melted on a microscope heating and freezing stage; the temperatures and types of phase changes observed with this microthermometry yield compositional information on fluid inclusions and information on pressure and temperature of their entrapment. Problems with thermal reequilibration normally are recognized at this stage. Later, more sophisticated microanalytical techniques might be employed to further explore the composition of the inclusions.

Composition

Initially, simple petrographic approaches may be the best means for evaluating the composition of inclusions. Gas phases are typically dark in transmitted light whereas liquid phases are typically bright (Figure F20). Oil inclusions normally fluoresce with UV-epi-illumination and aqueous liquids do not.

The most commonly applied technique for evaluating composition is microthermometry. The composition of gas phases can be determined by their phase behavior at low temperature. Aqueous inclusions may be frozen, and as long as a bubble is present during warming, the temperatures of phase changes can be used to evaluate the major salts in solution and their concentration. For example, the initial melting temperature (Te) can be used to identify major salts. The higher the salt concentration, the more is the melting temperature of ice depressed. If major salts are known or assumed, the temperature of final melting of ice (Tm ice) can be used to measure salinity (Figure F22). Normally, the exact composition of the salts cannot be known, so a model of the salts must be assumed, such as NaCl equivalent or seawater salt equivalent. The assumption of the model composition makes a small difference in calculating the overall salinity from a measurement of Tm ice.

Temperature

If an aqueous fluid inclusion is trapped as a single homogeneous liquid phase at high temperature (Figure F21(A)), and then is cooled, the P–T path followed is prescribed by the fact that the fluid inclusion vacuole is largely a constant density (isochoric) system. Thus, as the fluid inclusion cools, its pressure drops along an isochoric path (Figure F21(B)). After further cooling, the inclusion eventually intersects a vapor/liquid phase boundary (Figure F21(C)), where theoretically, a tiny bubble of vapor would be stable. With further cooling, the inclusion would follow the liquid/vapor phase boundary and the tiny bubble would grow slightly (Figure F21(D)) along its path toward earth-surface temperature (Figure F21(E)). In the laboratory,

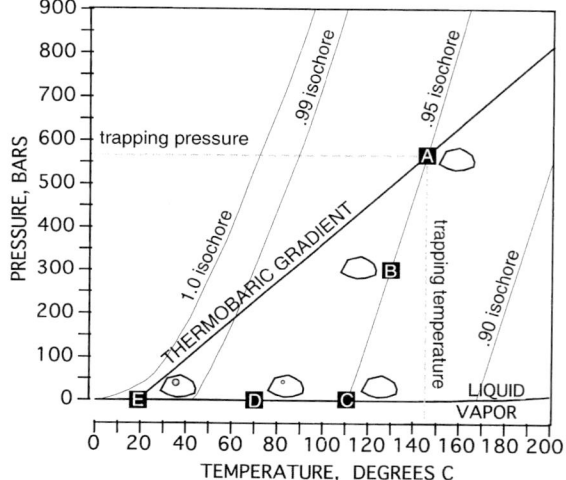

Figure F21 Pressure–temperature behavior of low-salinity aqueous fluid inclusions, with illustrations of fluid inclusion appearance.

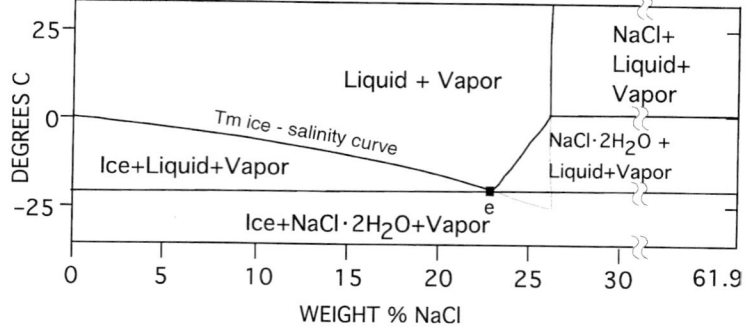

Figure F22 Low-temperature phase behavior of the H_2O–NaCl system.

the inclusion can be reheated until the bubble disappears (Figure F21(C)). The temperature of disappearance is called the homogenization temperature (Th) and it is a measure of the minimum temperature of entrapment of the fluid inclusion. Validity of Th data can be evaluated by examining the variability of Th within each FIA. To correct unaltered Th to actual entrapment temperature, the composition of the inclusion must be known, and either there must be evidence for immiscibility at the time of entrapment, or the pressure or thermobaric gradient must be known at the time of entrapment (Figure F21(A)). In aqueous systems that have not cooled significantly below the original entrapment temperature, the inclusion may remain all-liquid at room temperature, and depending on the degree of metastability, these all-liquid inclusions may provide evidence for low temperature of entrapment.

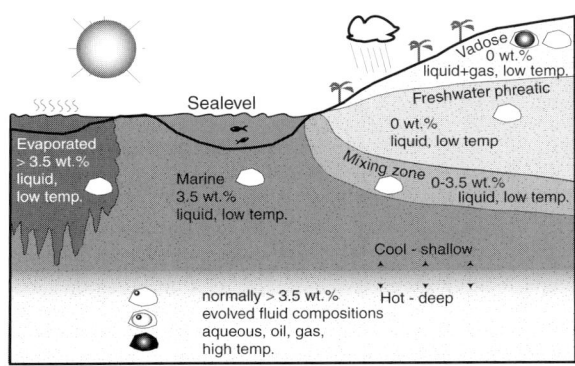

Figure F23 Linkage between temperature and fluid composition in diagenetic environments, with illustrations of fluid inclusion appearance at room temperature.

Pressure

Fluid inclusions provide a geobarometer for sedimentary rocks. With low-temperature entrapment of immiscible gas and aqueous liquid fluid inclusions, the pressure in inclusions, at surface temperature, is close to entrapment pressure. For higher temperature immiscible entrapment, where compositions of inclusions are known, the intersection of the isochore of the gas-rich end member and the homogenization temperature of the aqueous end member yields pressure of entrapment. For immiscible systems in which the composition of the aqueous end member is known, Th can be used to determine the position on a bubble point curve to yield pressure. For nonimmiscible systems, in which compositions of the inclusions are known, the homogenization temperature can be used to determine an isochore; the intersection of an assumed thermobaric gradient with the isochore yields the pressure of entrapment (Figure F21(A)).

Applications

Fluid inclusions are developing as tools in global-change research. They provide actual samples of the composition of the ancient ocean-atmosphere system (Johnson and Goldstein, 1993; Lowenstein et al., 1994). Fluid inclusions from non-marine settings provide paleoclimate information from composition of lake and groundwaters and their temperatures (Wenbo et al., 1995; Roberts and Spencer, 1995). Inclusions may provide some of the best-preserved samples of ancient microbes (Vreeland et al., 2000).

In petroleum exploration and development, fluid inclusions are an important tool, providing records of the history and pathways of hydrocarbon migration and generation. Petroleum fluoresces with UV epi-illumination, and its color of fluorescence is tied to maturity and API gravity (Tsui, 1990) providing a record of composition of petroleum fluid inclusions. Extraction and analysis of hydrocarbons from inclusions is useful for studying hydrocarbon systems (Burruss, 1987). Integrating hydrocarbon fluid inclusions, and their homogenization temperatures, into an overall basin and reservoir history is important in developing better understanding of reservoirs in general. Geothermometry using fluid inclusion Th has proven helpful in constraining thermal history in sedimentary basins (Tobin and Claxton, 2000).

Most diagenetic systems are defined by the temperature, pressure and salinity of the fluids active in them (Figure F23). For that reason, fluid inclusions are one of the best techniques for constraining the diagenetic history of sedimentary rocks. Fluid inclusions have been successfully applied to all diagenetic systems in sandstones, carbonates and evaporites (Goldstein and Reynolds, 1994).

Summary

The study of fluid inclusions is an important discipline in sedimentary geology. It is the only technique that can directly provide temperature, pressure and compositional information for sedimentary systems. It is a discipline that is rapidly expanding and finding new applications.

Robert H. Goldstein

Bibliography

Andersen, T., Frexxotti, M.-L., and Burke, E.A.J. (eds.), 2001. *Fluid Inclusions: Phase Relationships-Methods-Applications Special Volume in Honor of Jacques Touret*. Lithos, **55**(1–4), pp. 1–320.
Burruss, R.C., 1987. Crushing-cell, capillary column gas chromatography of petroleum fluid inclusions: method and application to petroleum source rocks, reservoirs, and low temperature hydrothermal ores. *American Current Research on Fluid Inclusions*. Socorro, NM, Abstracts (unpaginated).
Goldstein, R.H., and Reynolds, T.J., 1994. *Systematics of Fluid Inclusions in Diagenetic Minerals*. SEPM (Society for Sedimentary Geology) Short Course 31.
Hollister, L.S., and Crawford, M.L., 1981. *Short Course in Fluid Inclusions: Applications to Petrology*. Mineralogical Association of Canada Short Course Handbook 6.
Johnson, W.J., and Goldstein, R.H., 1993. Cambrian seawater preserved as inclusions in marine low-magnesium calcite cement. *Nature*, **362**: 335–337.
Lowenstein, T.K., Spencer, R.J., Wenbo, Y., Casas, E., Pengxi, Z., Baozhen, Z., Haibo, F., and Krouse, H.R., 1994. Major-element and stable-isotope geochemistry of fluid inclusions in halite, Qaidam Basin, western China: implications for late Pleistocene/holocene brine evolution and paleoclimates. In *Paleoclimate and Basin Evolution of Playa Systems*. Geological Society of America, Special Paper 289, pp. 19–32.
Roberts, S.M., and Spencer, R.J., 1995. Paleotemperatures preserved in fluid inclusions in halite. *Geochimica et Cosmochimica Acta*, **59**: 3929–3942.
Roedder, E., 1984. *Fluid Inclusions*. Mineralogical Society of America, Reviews in Mineralogy 12.

Tobin, R.C., and Claxton, B.L., 2000. Multidisciplinary thermal maturity studies using vitrinite reflectance and fluid inclusion microthermometry; a new calibration of old techniques. *American Association of Petroleum Geologists Bulletin*, **84**: 1647–1665.

Tsui, T.F., 1990. Characterizing fluid inclusion oils via UV fluorescence microspectrophotometry—a method for projecting oil quality and constraining oil migration history. *American Association of Petroleum Geologists Bulletin*, **74**: 781.

Vreeland, R.H., Rosenzweig, W.D., and Powers, D.W., 2000. Isolation of a 250 million-year-old halotolerant bacterium from a primary salt crystal. *Nature*, **407**: 897–900.

Wenbo, Y., Spencer, R.J., Krouse, H.R., Lowenstein, T.K., and Casas, E., 1995. Stable isotopes of lake and fluid inclusion brines, Dabusun Lake, Qaidam Basin, western China: Hydrology and paleoclimatology in arid environments. *Paleogeography, Paleoclimatology, Paleoecology*, **117**: 279–290.

Cross-references

Amber and Resin
Anhydrite and Gypsum
Ankerite
Bacteria in Sediments
Cathodoluminescence
Cements and Cementation
Climatic Control on Sedimentation
Diagenesis
Dolomites and Dolomitization
Evaporites
Neomorphism and Recrystallization
Porewaters in Sediments
Seawater: Temporal Changes to the Major Solutes
Speleothems

FLUME

Most flumes comprise a straight, open-ended, inclined, rectangular trough with either a fixed or mobile bed, down which water flows under the influence of gravity. Water is either supplied continuously to the upstream end of the flume by an external feed, or recirculated. In an annular flume the flow is propelled through a closed, semi-circular conduit. Flumes have been used since the middle of the nineteenth century to study sediment transport. They range in size from channels one or two particle diameters wide and a meter or so long, in which the flow is a few centimeters deep, to channels several meters wide and many tens of meters long, in which the flow may be more than a meter deep. Compared to rivers, they have a significant advantage for direct investigations of sediment transport because clear water and transparent sidewalls make the near-bed region more accessible. The most common objectives of flume experiments have been to evaluate competence and the force required to move and transport different bed materials, determine the conditions under which primary sedimentary structures form in unidirectional flows, and investigate the dynamics of the transport process.

Perhaps the most well-known flume experiments are those of Gilbert (1914) and Shields (1936). These and the plethora of other experiments conducted in the first half of the twentieth century sought to determine empirically the laws governing the initiation of motion and transport of bed load in natural channels. In Gilbert's (1914) experiments graded sediments were supplied to the upstream end of the flume and transported over a bed with a slope (and flow depth) in adjustment with a given discharge. The sediment discharge was computed on the basis of the weight of material retained in a stilling basin located at the downstream end of the flume. This is still a standard experimental procedure, though a larger range of initial bed slope is obtainable in a tilting flume and, as in Shields' (1936) experiments, sediment may be placed in the flume and the bed leveled prior to starting a run. Shields (1936) measured bed load transport under conditions in excess of those required to initiate motion and determined the critical shear stress for the initiation of sediment transport by extrapolation to the point of zero transport. His innovative analytical approach relied on dimensional analysis.

Flume studies are usually experiments designed to investigate distinctive aspects of a natural phenomenon's development, and reproduce significant aspects of its form and function. They are not strictly scale models whose properties are defined, using dimensional analysis, through the principles of similitude (Herbertson, 1968). Dimensional analysis simplifies the scaling procedure by generalizing the conditions that must exist for similarity between the real world and a model and thereby produce results that are independent of the scale of the system (Shields, 1936). Two systems behave similarly if they are geometrically (all the details of the system, ratios of critical dimensions, are in direct proportion); kinematically (corresponding velocities and velocity gradients are in the same ratios at corresponding locations); and dynamically similar (ratios of forces acting within the two systems are equal). Models of fluvial systems rely on the criterion of dynamic similarity. For complete dynamic similarity the Reynolds number (which determines the behavior and characteristics of viscous flows) and Froude number (which represents free-surface effects in systems where the effect of gravity influences the flow) must be the same in both the model and the real world. This implies that all length measures (e.g., width, depth, and representative particle size) in the real world and the model vary in the same ratio. But it is difficult to model sediment mixtures on the basis of strict similarity between the flow depth and grain size (Kramer, 1935); and because most liquids have densities close to that of water Froude and Reynolds scaling are, for practical reasons, impossible to achieve simultaneously. Parsimonious solutions are to adjust viscosity by heating the water (Boguchwal and Southard, 1990), and to relax the Reynolds similarity criterion for hydraulically rough and turbulent flows (Young and Davies, 1991).

Gilbert (1914) and Shields (1936) found that the characteristic bed form changed with the flow conditions. The initiation of bed forms remains a topic for investigation (Coleman and Melville, 1996), but data from subsequent experiments helped clarify the geometry and hydraulic relationships of bed configurations (Vanoni and Brooks, 1957; Guy et al., 1966). These and many other experiments were conducted with bed material in the sand range, in flumes capable of recirculating both water and sediment. For sediment mixtures, transport relations derived from experiments undertaken in recirculating flumes may differ from those obtained in a sediment feed flume because the final equilibrium state achieved (of steady, uniform flow over an erodible bed) is dependent on the initial conditions; the rate and size distribution of the transport depend on the flow and size distribution of the bed surface (Parker and Wilcock, 1993). In the sediment feed configuration the final state is independent of initial conditions, and the transport rate of every size range equals the feed rate. Neither configuration can simulate natural transport exactly, because in rivers the pattern of transport reflects the influence both of the bed state at an earlier time (as in a recirculating flume), and the sediment supply from upstream (as in the sediment feed configuration).

The dominant transport condition in many gravel-bed rivers may be one of partial transport, where only some of the particles in a given size range exposed on the bed surface move over the duration of a transport event. It affects the pattern of vertical sediment exchange between the bed surface and substrate, and therefore size sorting in both the vertical and downstream directions (Harrison, 1950; Paola *et al.*, 1992). Color-coding each size fraction permits the bed surface size distribution to be monitored directly by noninvasive means (Wilcock and McArdell, 1993). A variety of other innovative techniques, including high-speed photography and laser Doppler anemometry (Francis, 1973; Bennett and Best, 1995; Nelson *et al.*, 1995), have been used to investigate the dynamics of particle motion and the turbulent motions of the fluid and sediment phases. Many such experiments are conducted under simplified conditions, using a fixed rather than a mobile bed and solitary particles.

Basil Gomez

Bibliography

Bennett, S.J., and Best, J.L., 1995. Mean flow and turbulence structure over two-dimensional dunes: implications for sediment transport and bedform stability. *Sedimentology*, **42**: 491–513.

Boguchwal, L.A., and Southard, J.B., 1990. Bed configurations in steady unidirectional water flows. Part 1. Scale model study using fine sands. *Journal of Sedimentary Petrology*, **60**: 649–657.

Coleman, S.E., and Melville, B.W., 1996. Initiation of bed forms on a flat sand bed. *Journal of Hydraulic Engineering*, **122**: 301–310.

Francis, J.R.D., 1973. Experiments on the motion of solitary grains along the bed of a water-stream. *Proceedings of the Royal Society of London*, **A332**: 443–471.

Gilbert, G.K., 1914. The transportation of débris by running water. *US Geological Survey Professional Paper*, 86, 263 pp.

Guy, H.P., Simons, D.B., and Richardson, E.V., 1966. Summary of alluvial channel data from flume experiments, 1956–61. *US Geological Survey Professional Paper* 462-I, 96pp.

Harrison, A.S., 1950. Report on special investigation of bed sediment segregation in a degrading bed. Berkeley: University of California. *Institute of Engineering Research Series*, **33** (1), 205pp.

Herbertson, J.G., 1968. Scaling procedures for mobile bed hydraulic models in terms of similitude theory. *Journal of Hydraulic Research*, **7**: 315–353.

Kramer, H., 1935. Sand mixtures and sand movement in fluvial models. *Transactions of the American Society of Civil Engineers*, **100**: 798–838.

Nelson, J.M., Shreve, R.L., McLean, S.R., and Drake, T.G., 1995. Role of near-bed turbulence structure in bed load transport and bed form mechanics. *Water Resources Research*, **31**: 2071–2086.

Paola, C., Parker, G., Seal, R., Sinha, S.K., Southard, J.B., and Wilcock, P.R., 1992. Downstream fining by selective deposition in a laboratory flume. *Science*, **258**: 1757–1760.

Parker, G., and Wilcock, P.R., 1993. Sediment feed and sediment recirculating flumes: fundamental difference. *Journal of Hydraulic Engineering*, **119**: 1192–1204.

Shields, A., 1936. Anwendung der Aehnlichkeitsmechanik und der turbulenzforschung auf die Geschiebebewegung. Mitteilungen Preussischen Versuchsanstalt für Wasserbau und Schiffbau, Berlin, 26. English translation: Application of Similarity Principles and Turbulence Research to Bed-load Movement. W.M. Keck Laboratory of Hydraulics and Water Resources, California Institute of Technology, Report 167, 43pp.

Vanoni, V.A., and Brooks, N.H., 1957. Laboratory studies of the roughness and suspended load of alluvial streams, *US Army Corps of Engineers Missouri River Division, Sediment Series*, 11, 121pp.

Wilcock, P.R., and McArdell, B.W., 1993. Surface-based fractional transport rates: mobilization thresholds and partial transport of a sand-gravel sediment. *Water Resources Research*, **29**: 1297–1312.

Young, W.J., and Davies, T.R.H., 1991. Bedload transport processes in a braided gravel-bed river model. *Earth Surface Processes and Landforms*, **16**: 499–511.

Cross-references

Bedding and Internal Structures
Sediment Transport by Unidirectional Water Flows
Surface Forms

FORENSIC SEDIMENTOLOGY

The use of geologic materials as trace evidence in criminal cases has existed for approximately one hundred years. Palenick (1993) provides an overview and reminds us that it began as so many of the other types of evidence with the writings of Conan Doyle. Doyle wrote the Sherlock Holmes series between 1887 and 1893. He was a physician who apparently had two motives: writing salable literature and using his scientific expertise to encourage the use of science as evidence. In 1893 Hans Gross wrote his book *Handbook for Examining Magistrates* in which he suggested that perhaps you could tell more about where someone had last been from the dirt on their shoes than you could from toilsome inquiries. A German chemist, Georg Popp, in 1908 examined the evidence in the Margarethe Filbert case. In this homicide a suspect had been identified by many of his neighbors and friends because he was known to be a poacher. The suspect's wife testified that she had dutifully cleaned his dress, shoes the day before the crime. Those shoes had three layers of soil adhering to the leather in front of the heel. Popp, using the methods available at that time, said that the uppermost layer, thus the oldest, contained goose droppings and other earth materials that compared with samples in the walk outside the suspect's home. The second layer contained red sandstone fragments and other particles that compared with samples from the scene where the body had been found. The lowest layer, thus the youngest, contained brick, coal dust, cement and a whole series of other materials that compared with samples from a location outside a castle where the suspect's gun and clothing had been found. The suspect said that he had walked only in his fields on the day of the crime. Those fields were underlain by porphyry with milky quartz. Popp found no such material on the shoe although the soil had been wet on that day. In this case, Popp had developed most of the elements involved in present day forensic soil examination. He had compared two sets of samples and identified them with two of the scenes associated with the crime. He had confirmed a sequence of events consistent with the theory of the crime and he had found no evidence supporting the alibi.

Sedimentary and related materials have evidential value. The value lies in the almost unlimited number of kinds of materials and the large number of measurements that we can make on these materials. For example, the number of sizes and size distributions of grain combined with colors, shapes and mineralogy is almost unlimited. There are an almost unlimited number of kinds of rocks and fossils. These are identifiable and recognizable. It is this diversity in earth materials, combined with the ability to measure and observe the different kinds, which provides the forensic discriminating power.

There have been many contributions to the discipline over the last 100 years. Many have been made by the Laboratory of the Federal Bureau of Investigation, the Central Research Establishment, McCrone Associates, The Centre for Forensic Sciences, Microtrace, The Japanese National Research Institute of Police Science, government, private and academic researchers.

Because much of the evidential value of earth materials lies in the diversity and the differences in the minerals and particles, microscopic examination at all levels of instrumentation is the most powerful tool. In addition such examination provides an opportunity to search for man made artifact grains and other kinds of physical evidence.

Individualization, that is, uniquely associating samples from the crime scene with those of the suspect to the exclusion of all other samples is not possible in most of the cases. In this sense earth material evidence is not similar to DNA, fingerprints and some forms of firearms and tool mark evidence. However, in a South Dakota homicide case, soil from the scene where the body was found and from the suspect's vehicle both contained similar material including grains of the zinc spinel gahnite. This mineral had never before been reported from South Dakota. Such evidence provides a very high level of confidence and reliability.

One of the most interesting types of studies is the *aid to an investigation*. There are many examples of cases where a valuable cargo in transit is removed and rocks or bags of sand of the same weight are substituted. If the original source of the rocks or sand can be determined, then the investigation can be focused at that place. In a high visibility case, DEA agent Enrique Camarena was murdered in Mexico. His body was exhumed as part of a cover-up. When his body was found, it contained rock fragments that were different from the country rock at that place and represented the rocks from the original burial site. Petrographic examination of those rocks combined with a detailed literature search of Mexican volcanic rock descriptions enabled the finding of the original burial location.

Most examinations involve *comparison*. Comparison is establishing a high probability that two samples have a common source. In comparison studies of soils, it is difficult to overestimate the value of findings in the soil artifacts or some other type of evidence. In an Upper Michigan rape case, three flowerpots had been tipped over and spilled on the floor during the struggle. It was shown that potting soil on the suspect's shoe compared with a sample collected from the floor and represented soil from one of the pots. In addition, small clippings of blue thread existed both in that flowerpot sample and on the shoe of the suspect. The thread provided additional trace evidence. In a New Jersey rape case, the suspect had soil samples in the cuffs of his pants. In addition to the glacial sands that compared with soil samples collected from the crime scene, the soil contained fragments of clean Pennsylvania anthracite coal. Such coal fragments are not uncommon in the soils of most of the older cities in eastern North America. In this sample there was too much coal when compared with samples in the surrounding area. Further investigation showed that some 60 years before the crime scene was the location of a coal pile for a coal burning laundry. Again, the tying in of soil-evidence with an artifact and history increased the evidential value.

The alertness of those who collect samples, and the quality of collection, is critical to the success of any examination. If appropriate samples are not collected during the initial evidence gathering, they will never by studied and never provide assistance to the court. There is the case in which an alert police officer happened to look at an individual arrested for a minor crime. He observed, "that is the worst case of dandruff I have ever seen." It was not dandruff but diatomaceous earth, which was essentially identical with the insulating material of a safe that had been ripped the previous day.

The future of Forensic Geology holds much promise. However that future will see many changes and new opportunities. New methods are being developed that take advantage of the discriminating power inherent in earth materials. Quantitative X-ray diffraction may possibly revolutionize forensic soil examination. McVicar and Graves (1997) have reported important progress in this area. When developed to the point that this or similar methods become ordinary lab techniques, it will be possible to do a quantitative mineralogical analysis that is easily reproducible. However, the microscope will remain an important tool in the search for the unusual grain or artifact. Sampling methods should be improved and those people who collect samples for forensic purposes should be thoroughly and completely trained. Soils are extremely sensitive to change over short distances, both horizontally and vertically. Soil sampling in many cases is the search for a sample that compares. The collecting of all the other samples serves only the purpose of demonstrating the range of local differences. In collecting soil samples for comparison, we are searching for one that would have the possibility of comparison. Screening techniques during sampling that eliminate samples that are totally different is appropriate. For example, a surface sample offers little possibility for comparison with material collected at a depth of four feet in a grave.

Studies that demonstrate the diversity of soils are important. One approach is to take an area that one would normally assume was fairly homogenous in its soil character and collect a hundred samples on a grid. Each pair of samples would then be compared until all the pairs are shown to be different. Starting with color and moving on to size distribution and mineralogy, different methods are used to eliminate all of these pairs that appear similar. Junger (1996) performed several such studies and suggested methods for soil examination.

The qualifications of examiners are a very major problem. How do you learn to do forensic soil examinations? This requires a thorough knowledge of mineralogy and the ability effectively to use a microscope and the other techniques used in earth material examination. It is also important that examiners be familiar with the other kinds of trace evidence, and also with the law and practice of forensic examination.

Raymond C. Murray

Bibliography

Junger, E.P., 1996. Assessing the unique characteristics of close-proximity soil samples: just how useful is soil evidence. *Journal of Forensic Sciences, JFSCA*, **41** (1): 27–34.

McVicar, M.J., and Graves, W.J., 1997. The forensic comparison of soils by automated scanning electron microscopy. *Canadian Society Forensic Science Journal*, **30** (4): 241–261.

Murray, R.C., and Tedrow, J.C.F., 1975. *Forensic Geology: Earth Science and Criminal Investigation*. New Brunswick, N.J.: Rutgers University Press.

Murray, R.C., and Tedrow, J., 1992. *Forensic Geology*. Englewood Cliffs, N.J.: Prentice Hall.

Palenik, S.J., 1993. The analysis of dust traces. In *Proceedings of the International Symposium on the Forensic Aspects of Trace Evidence*. Government Printing Office, Washington, DC.

G

GASES IN SEDIMENTS

Gases dissolved in pore waters of sediments are ubiquitous although free gases are confined to specific facies and sedimentary environments. Most dissolved gases are derived by equilibration with the atmosphere and their concentrations subsequently modified by diagenetic (principally biological) processes during shallow burial. Free gases, where present, derive from either diagenetic bacterial reactions (indigenous gases) or migration of thermally derived gases from greater depth.

Free gases of indigenous origin include hydrogen sulfide (H_2S), methane (CH_4), sometimes carbon dioxide (CO_2) and more rarely traces of nitrogen (N_2) or ammonia (NH_3). All of these are formed by microbial communities (see Bacteria in Sediments) operating on organic substrates and are therefore limited to organic-rich sediments, (see Organic Sediments). The relevant diagenetic zones for gas generation are summarized in Figure G1, although local microenvironments may result in disruption of this general vertical stratification. With progressive burial, bacteria utilize the most energy efficient oxidizing agent available until consumed, at which point a different bacterial ecosystem becomes dominant to utilize the next available oxidant (strictly speaking, the next ultimate electron acceptor since all the reactions represent consortia of bacterial groups working in concert). If oxygen is available in bottom waters, an aerobic bacterial community dominates shallowest burial. The end product of this is carbon dioxide although this is generally able to diffuse upward into the water column and rarely reaches concentrations high enough for a free gas phase to develop. Below this, in the suboxic zone, dissolved nitrate and iron and manganese oxides act as the oxidizing agents. In extreme cases this may give rise to free nitrogen and carbon dioxide although generally the carbon dioxide is either lost by diffusion or precipitated as solid carbonates and insufficient nitrogen is generated to form a free gas. Below this are the principal zones of gas generation—the sulfate reduction zone when hydrogen sulfide is formed and

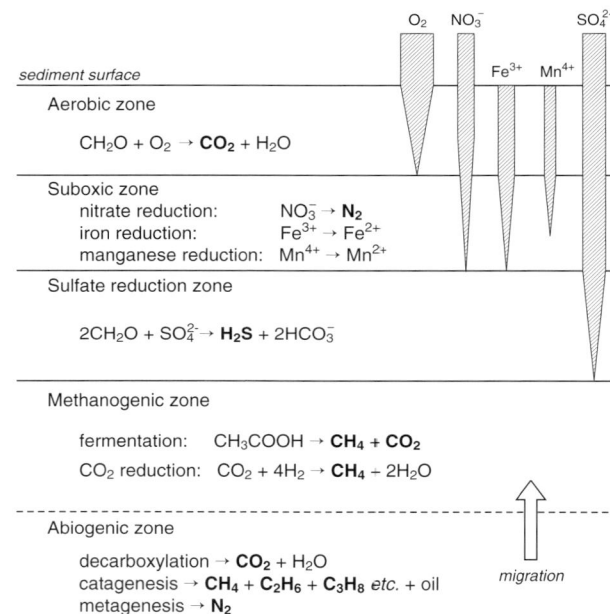

Figure G1 Principal diagenetic zones responsible for the formation of free gas within sediments. The lower limit of the aerobic, suboxic and sulfate reduction zones are dictated by availability of the oxidizing agent concerned (diffusion from the overlying water column in the case O_2, NO_3^- and SO_4^{2-} or indigenous to the detrital minerals for Fe and Mn). The methanogenic zone is limited by the presence of sulfate above it and approximately the 65°C isotherm below.

the methanogenic zone where methane and carbon dioxide are formed. Below this, further carbon dioxide may be formed by thermal "decarboxylation" of the residual organic matter.

Hydrogen sulfide

When sulfate is available from the overlying water column (generally in marine environments), hydrogen sulfide is the principal end product within the sulfate-reduction zone. Many bacterial groups operate in this zone to oxidize organic matter but the key reaction is that of the sulfate-reducing bacteria (SRB, notably *Desulfovibrio sp.*). Taking carbohydrate to represent the bulk organic matter, the net reaction can be represented by the reaction:

$$2CH_2O + SO_4^{2-} \rightarrow H_2S + 2HCO_3^- \quad \text{(Eq. 1)}$$

If iron is freely available within this zone, the H_2S will be buffered to form pyrite (see Sulfides in Sediments). Where insufficient iron is present, free H_2S may build up although the ultimate concentration is limited by bacterial oxidation of the sulfide as it diffuses upward into the suboxic and oxic zones.

Methane

The deepest bacterial zone in sediments is represented by the methanogenic zone. Here, two competing bacterial reactions generate free methane gas:

acetate (or similar) fermentation:

$$CH_3COOH \rightarrow CH_4 + CO_2 \quad \text{(Eq. 2)}$$

carbon dioxide reduction:

$$CO_2 + 4H_2 \rightarrow CH_4 + 2H_2O \quad \text{(Eq. 3)}$$

The acetate, CO_2 and H_2 substrates in both cases are provided by parallel, largely fermentation type, bacterial reactions. In general, CO_2 reduction is dominant in marine sediments while fermentation reactions are dominant, at least in the early stages, in freshwater environments although in all cases both reactions operate to some extent (Clayton, 1995). The so-called "biogenic" or "bacterial" gas formed in this way is economically important and may account for as much as 20% of the world's natural gas resources (Rice and Claypool, 1981).

Since no external oxidizing agents are required for methane production, the ultimate methane concentration is limited largely by the availability of organic matter although some methane is lost by anaerobic oxidation at the base of the sulfate-reduction zone (Iversen and Jørgensen, 1985) and methane generation ceases when the sediment is buried to temperatures above approximately 60–65°C due to metabolic limitations of the bacteria involved. Laboratory experiments show that the optimum temperature for methanogenesis using natural assemblages is around 35–45°C (Zeikus, J.G. and Winfrey, M.R., 1976). Under optimum conditions, up to about 10% of the organic carbon may be converted to methane (Clayton, 1992).

Carbon dioxide

Carbon dioxide is the ultimate fate of organic matter oxidation in all bacterial zones although it is not usually as abundant as a free gas as is CH_4. There are a number of reasons for this. Firstly, during very shallow diagenesis, the majority of the CO_2 is able to diffuse upward back into the overlying water column (bottom waters are frequently enriched in CO_2 relative to surface or intermediate depth waters). Secondly, the collective bacterial reactions within each zone generally result in CO_2 released in the form of dissolved bicarbonate (HCO_3^-) or carbonate (CO_3^{2-}) which can be precipitated as diagenetic concretions (Irwin *et al.*, 1977). Finally, CO_2 is itself consumed by the CO_2 reduction pathway within the methanogenic zone which generally results in lower CO_2 concentrations in marine environments (where CO_2 reduction is dominant over acetate fermentation) relative to freshwater environments (Clayton, 1995, 146).

Deeper within the sediment, thermal decarboxylation (i.e., abiogenic removal of carboxyl and similar oxygen bearing functional groups from the organic matter) contributes further CO_2 which may reach appreciable concentrations. This process occurs principally between 50 and 100°C (Sweeney and Burnham, 1990) and to some extent overlaps bacterial methanogenesis and deeper thermogenic production of hydrocarbons.

Nitrogen

Nitrogen is also released by bacterial processes during early diagenesis but is generally not as abundant as H_2S and CH_4 as a free gas. It is sourced by both dissolved nitrate (NO_3^-) and organic bound nitrogen (particularly in proteins) although the latter is overwhelmingly dominant. Nitrate reduction to form N_2 is generally considered to be a suboxic process but in practice there is considerable nitrate reduction within the aerobic zone also. The decomposition of proteins (deamination) occurs both within the aerobic zone and deeper and produces ammonia (NH_3) as an end product, usually in the form of dissolved ammonium (NH_4^+). In the presence of oxygen, the ammonia may then be oxidized to nitrate (by *Nitrosomonas* sp.) or nitrite (NO_2^-, by *Nitrobacter* sp.) and can then be bacterially reduced to molecular nitrogen deeper in the sediment.

Migrated gases

Gases migrated from greater depth are locally abundant in surface sediments within petroliferous sedimentary basins (i.e., those containing economic oil and natural gas accumulations). These gases result from thermal breakdown of the residual organic matter which has survived early diagenesis (see Kerogen), typically at temperatures above 100°C (Pepper and Corvi, 1995). They are characteristically dominated by methane with varying concentrations of higher hydrocarbons (ethane, propane, etc.), CO_2 and N_2. Where migration from depth is highly active, such gases may escape to the water column as active "seeps", often associated with a number of associated sedimentary features such as diapirism, pockmarks, chemosynthetic communities, clathrates (q.v.) and carbonate cementation (see Hovland and Judd, 1988, for a full discussion of these features).

Summary

Where present, free gases in sediments are dominated by methane, hydrogen sulfide, carbon dioxide and sometimes traces of nitrogen or ammonia. Although occasional migrated deep-sourced thermogenic gases containing higher hydrocarbons are locally important, the gases are dominantly the result of *in situ* bacterial reactions during early burial diagenesis of organic rich sediments. Carbon dioxide is generated at all depths, N_2 is formed mainly in the suboxic zone, H_2S is formed below this under sulfate reducing conditions and CH_4 is formed deeper still.

Chris J. Clayton

Bibliography

Clayton, C.J., 1992. Source volumetrics of biogenic gas generation. In Vially, R. (ed.), *Bacterial Gas*. Paris: Édition Technip, pp. 191–204.
Clayton, C.J., 1995. Microbial and Organic Processes. In Parker, A. and Sellwood, B.W. (eds.), *Quantitative Diagenesis: Recent Developments and Applications to Reservoir Geology*. The Netherlands: Kluwer Academic Publishers, pp. 125–160.
Hovland, M., and Judd, A.G., 1988. *Seabed Pockmarks and Seepages. Impact on Geology, Biology and the Marine Environment*. London: Graham & Trotman.
Irwin, H., Curtis, C., and Coleman, M.L., 1977. Isotopic evidence for the source of diagenetic carbonates formed during burial of organic-rich sediments. *Nature*, **269**: 209–213.
Iversen, N., and Jørgensen, B.B., 1985. Anaerobic methane oxidation rates at the sulfate-methane transition in marine sediments from the Kattegat and Skagerrak (Denmark). *Limnology and Oceanography*, **30**: 944–955.
Pepper, A.S., and Corvi, P.J., 1995. Simple kinetic models of petroleum formation. Part 1: oil and gas generation from kerogen. *Marine & Petroleum Geology*, **12**: 291–319.
Rice, D.D., and Claypool, G.E., 1981. Generation, accumulation and resource potential of biogenic gas. *American Association of Petroleum Geologists Bulletin*, **65**: 5–25.
Sweeney, J.J., and Burnham, A.K., 1990. Evaluation of a simple kinetic model of vitrinite reflectance based on chemical kinetics. *American Association of Petroleum Geologists Bulletin*, **74**: 1559–1570.
Zeikus, J.G., and Winfrey, M.R., 1976. Temperature limitation of methanogenesis in aquatic sediments. *Applied Environmental Microbiology*, **31**: 99–107.

Cross-references

Bacteria in Sediments
Black Shales
Clathrates
Hydrocarbons in Sediments
Kerogen
Maturation, Organic
Peat
Sulfide Minerals in Sediments

GEODES

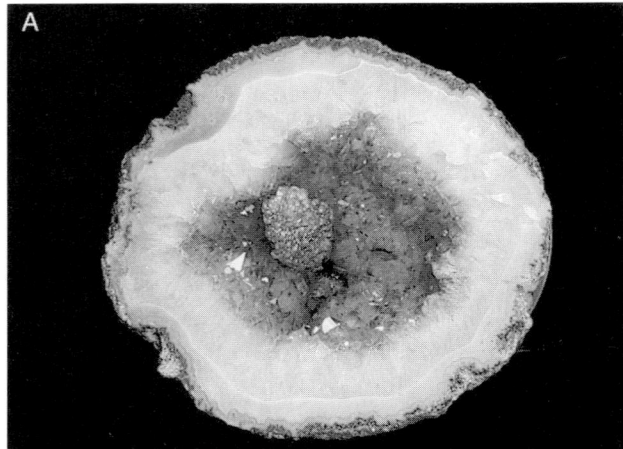

Figure G2 (A) Geode from a volcanic host rock. A chalcedony rim followed by coarse quartz crystals is a widely observed sequence in both volcanic and sedimentary geodes. A late precipitate of calcite is observed in the geode interior. Diameter 15 cm. Barron Collection, Texas Memorial Museum, University of Texas at Austin. (B) Calcite geode in a dolomitic limestone. Many geodes of this type form by replacement of earlier-formed nodular anhydrite. Diameter approximately 12 cm. Collection of the author. Photos by Joe Jaworski.

Introduction

Geodes are nodules (i.e., mineral aggregates having a composition that contrasts strongly with that of the surrounding rock) containing an interior cavity lined with macroscopic crystals (i.e., crystals sufficiently large to be distinguishable without magnification). Geodes occur primarily within limestones, dolomites, and mudrocks, but some are found in volcanic rocks. The term geode is largely synonymous with "vug" (a crystal-lined cavity), but has the additional connotation that the object has, or could, weather free by virtue of its compositional and textural distinctiveness from the host rock. The sizes and shapes of geodes vary widely, but most are in the range of 5 to 10 cm across and roughly spherical or oblate. Geodes are recognized as a group partly on the basis of their appearance and distinctiveness, in particular their attractiveness to mineral collectors. (Figure G2).

Geode formation

All geodes share certain genetic aspects. Foremost, crystals in geodes are emplaced by precipitation from aqueous solution (authigenesis) within cavities that are anomalously large in comparison to typical pore spaces in the surrounding rock. These essential characteristics and the attendant necessity for substantial volumes of mineral-bearing fluids to flow through low-permeability rocks were recognized in some of the earliest reports on geode formation (e.g., Dana and Shepard, 1845; Shaler, 1899). Abnormally large pore spaces form in a variety of ways. Cavities created by dissolution of earlier-formed nodules, gas vesicles in volcanic rocks, cavities within fossils, desiccation cracks within early-formed concretions (see Septarian Nodules), and pore spaces associated with tectonic fractures can all become the locus of geode formation.

Categories of geodes

Most geodes belong to one of two major categories. Quartz replacement of former gypsum and anhydrite is responsible for the *cauliflower nodules* found in Carboniferous shallow-marine limestones and dolomites of the upper Mississippi Valley region of the central United States. The classic locality for such geodes is at Keokuk, Iowa (e.g., Van Tuyl, 1916; Hayes, 1964; Sinotte, 1969). Similar silicified evaporites are known worldwide in shallow marine rocks ranging in age from Precambrian to Cenozoic (e.g., Chowns and Elkins, 1974; Siedlecka, 1976; Fisher, 1977; Milliken, 1979; Radke and Nicoll, 1981; Matzko and Naqvi, 1983; Elorza and Rodriguez-Lazaro, 1984; Maliva, 1987; Ulmer-Scholle, Scholle *et al.* 1993). Many calcite geodes in sedimentary rocks also have features consistent with formation by evaporite replacement.

An intriguing aspect of silicified evaporite nodules is expansive growth during their early development. Formation of evaporite nodules is a syn-sedimentary or very early diagenetic process, occurring prior to lithification of the enclosing sediment. The exact nature of microenvironments responsible for evaporite localization is uncertain, but it is clear that in some cases precipitation was localized within the skeletal remains of marine invertebrates leading to dramatic expansion or "explosion" of the fossils (Figure G3) (Maliva, 1987; Foerster, 1991). Fossils expanded in this manner attracted great interest from early students of geodes (e.g., Shaler, 1899; Bassler, 1908), but a convincing mechanism for expansion eluded these workers. Growth within unconsolidated sediment, either through direct displacive precipitation of anhydrite, or by expansion (approximately 30%) during hydration of anhydrite to gypsum provides a plausible mechanism for the development of exploded fossils. Replacement of the early-formed nodules during later diagenesis is accomplished by dissolution of the evaporite minerals followed by precipitation of authigenic quartz into the resulting pore space. In *klappersteins* some of the later-formed quartz crystals, perhaps once encased in the now-vanished evaporite mineral, are loose within the geode interior and rattle audibly against the inner walls when the geode is shaken.

Figure G3 Cauliflower chert nodules (exploded fossils) formed by silicification of anhydrite. Specimen on the left (2cm tall) is a silicified blastoid of the genus *Pentremites*. The three specimens on the right show progressive disruption of similar blastoids by displacive growth of the anhydrite, now completely replaced by quartz. Specimen on the far right is a klapperstein (see text). Collection of the author. Photo by Joe Jaworski.

A second important category of geodes corresponds to silicified vesicles in tuffaceous volcanic rocks ranging in composition from rhyolite to basalt. Weathering of unstable minerals and volcanic glass provides high concentrations of dissolved silica to pore fluids in such rocks. In the northern Mexico state of Chihuahua, quartz-lined vesicles weathered from rhyolitic and andesitic tuffs supply an abundance of geodes to mineral collectors (Finkleman, 1974; Keller, 1979; Cross, 1996).

Precipitation of cavity-filling minerals

Although geode development must postdate the formation of the enclosing rock, the timing of mineral precipitation can vary from near-syndepositional to much later in the rock's history. Temperatures in the range of 25° to 100°C have been estimated for quartz in silicified evaporite nodules, based on two-phase fluid inclusions and stable isotopic evidence (Milliken, 1979; Roedder, 1979); similar temperatures are estimated for quartz in volcanic geodes (Keller, 1979). The necessity for a large, open pore space through which significant volumes of mineral-bearing fluids can flow most likely restricts geode formation to relatively shallow depths in the crust (<6 km?).

Quartz in a wide variety of textural and color subtypes is the most common geode-filling mineral. In both sedimentary and volcanic rocks, there is a common sequence of mineral formation from the outer rim to the center of the geode. Microcrystalline quartz, either equant or fibrous (chalcedony) microquartz, forms the initial precipitate on the interior surface of the cavity, followed by megaquartz which may develope inward-projecting euhedra around the central cavity.

Other common geode-forming minerals are calcite, dolomite, and, in volcanic rocks, zeolites. Most minerals that occur in geodes postdate the early-formed quartz and belong to a mineral assemblage that also occurs in associated veins and fracture fills. Aragonite, ankerite, celestite, magnetite, hematite, marcasite, pyrite, millerite, chalcopyrite, galena, sphalerite, fluorite, and kaolinite are widely reported in silicified-evaporite-type geodes (e.g., Goldstein and McKenzie, 1997). Many of the same minerals are reported to occur in volcanic-associated geodes along with a wide variety of zeolites and authigenic feldspars (e.g., Finkleman, 1974; Ernst and Hall, 1975).

Particularly intriguing to collectors are geodes that retain fluids within the interior cavity. Water-bearing geodes, *enhydros*, are obtained from volcanic deposits in Brazil (e.g., Boutakoff and Whitehead, 1952; Prashnowsky, 1990). Petroleum-filled geodes are reported in limestones near Niota, Illinois (e.g., Robertson, 1942; Spitznas, 1949).

Research on geodes

Much information on geode occurrence is found within the mineral collecting literature. In the academic arena, geodes have received surprisingly little attention despite the fact that the large pore spaces within geodes permit the growth of relatively large crystals that represent accessible samples of authigenic minerals that may otherwise be pervasive through the host rock but also rare or cryptic. The sequence of mineral precipitation observable within geodes and the elemental and isotopic zoning within the geode's large crystals can record an eminently readable history of fluid/rock interaction across a substantial range of diagenetic conditions (e.g., Milliken, 1979).

Summary

Authigenic mineral precipitation on the walls of large cavities in sedimentary and volcanic rocks produces the nodules known as geodes. The large crystals of authigenic minerals found within geodes have interest from the dual standpoints of their esthetic qualities and the intriguing record of rock chemical history that they represent.

Kitty L. Milliken

Bibliography

Bassler, R.S., 1908. The formation of geodes, with remarks on the silicification of fossils. *Proceedings of the United States National Museum*, 133–154.
Boutakoff, N., and Whitehead, S. 1952. Enhydros or water-stones. *Mining and Geological Journal*, **4**(5): 14–18.
Chowns, T.M., and Elkins, J.E. 1974. The origin of quartz geodes and cauliflower cherts through the silicification of anhydrite nodules. *Journal of Sedimentary Petrology*, **44**: 885–903.
Cross, B.L., 1996. *The Agates of Northern Mexico*. Edina, Minnesota, Burgess International Publishing Group.
Dana, J.D., and Shepard, C.U., 1845. Origin of the constituent and adventitious minerals of trap and allied rocks. *American Journal of Science*, **49**: 49–64.
Elorza, J.J., and Rodriguez-Lazaro, J., 1984. Late Cretaceous quartz geodes after anhydrite from Burgos, Spain. *Geological Magazine*, **121**: 107–113.
Ernst, W.G., and Hall, C.A. Jr., 1975. Feldspathic geodes near Black Mountain, western San Luis Obispo County, California. *American Mineralogist*, **60**: 1105–1112.
Finkleman, R., 1974. A guide to the identification of minerals in geodes from Chihuahua, Mexico. *Lapidary Journal*, **27**: 1742–1744.
Fisher, I.S., 1977. Distribution of Mississippian geodes and geodal minerals in Kentucky. *Economic Geology*, **72**: 864–869.
Foerster, R., 1991. Fossil geodes. *Lapidary Journal*, **45**(3): 102–105.
Goldstein, A., and McKenzie, B. 1997. Halls Gap, Lincoln County, Kentucky. *The Mineralogical Record*, **28**: 369–384.
Hayes, J.B., 1964. Geodes and concretions from the Mississippian Warsaw Formation, Keokuk region, Iowa, Illinois, Missouri. *Journal of Sedimentary Petrology*, **34**: 123–133.
Keller, P.C., 1979. Quartz geodes from near the Sierra Gallego area, Chihuahua, Mexico. *The Mineralogical Record*, **10**(4): 207–214.
Maliva, R.G., 1987. Quartz geodes: early diagenetic silicified anhydrite nodules related to dolomitization. *Journal of Sedimentary Petrology*, **57**: 1054–1059.
Matzko, J.J., and Naqvi, J.M. 1983. Geodes from Saudi Arabia. *Lapidary Journal*, **37**(8): 1140–1156.
Milliken, K.L., 1979. The silicified evaporite syndrome: two aspects of silicification history of former evaporite nodules from southern Kentucky and northern Tennessee. *Journal of Sedimentary Petrology*, **49**: 245–256.
Prashnowsky, A.A., 1990. Biogeochemical study of enhydros (Brazil). *Mitteilungen aus dem Geologisch-Palaeontologischen Institut der Universitaet Hamburg*, **69**: 35–43.
Radke, B., and Nicoll, R.S. 1981. Evidence for former evaporites in the Carboniferous Moogooree Limestone, Carnarvon Basin, Western Australia. *BMR Journal of Australian Geology and Geophysics*, **6**(1): 106–108.
Robertson, P., 1942. Bituminous matter in Warsaw geodes [Illinois]. *Transactions of the Illinois State Academy of Science*, **35**: 138–140.
Roedder, E., 1979. Fluid inclusion evidence on the environments of sedimentary diagenesis, a review. In Scholle, P.A. and Schluger, P.R. (eds.), *Aspects of Diagenesis*. Tulsa, Oklahoma, SEPM, **26**: pp. 89–107.
Shaler, N.S., 1899. Formation of dikes and veins. *Geological Society of America Bulletin*, **10**: 253–262.
Siedlecka, A., 1976. Silicified Precambrian evaporite nodules from northern Norway; a preliminary report. *Sedimentary Geology*, **16**: 161–175.
Sinotte, S.R., 1969. *The Fabulous Keokuk Geodes-M Volume 1. their origin, formation, and development in the Mississippian lower Warsaw beds of southeast Iowa and adjacent states*. Des Moines, Iowa, Wallace-Homestead Co.
Spitznas, R.L., 1949. Petroliferous geodes, their occurrence and origin [Tyson Creek area, Illinois]. *Earth Science Digest*, **3**(11): 15–18.
Ulmer-Scholle, D.S., Scholle, P.A. *et al.*, 1993. Silicification of evaporites in Permian (Guadalupian) back-reef carbonates of the Delaware Basin, West Texas and New Mexico. *Journal of Sedimentary Petrology*, **63**: 955–965.
Van Tuyl, F.M., 1916. The geodes of the Keokuk beds. *American Journal of Science*, **42**(192): 34–42.

Cross-references

Authigenesis
Cements and Cementation
Diagenesis
Diagenetic Structures
Septarian Concretions

GEOPHYSICAL PROPERTIES OF SEDIMENTS (ACOUSTICAL, ELECTRICAL, RADIOACTIVE)

Introduction

Geophysics is an important tool for investigating sediments in the Earth's subsurface. The conversion of geophysical data into geological information requires an understanding of the connection between the nature of the sediments and the physical properties that govern their geophysical response, a discipline called petrophysics or rock physics.

Sediments are heterogeneous systems; their effective physical properties are a function of the properties and volume fraction of the individual components (i.e., the minerals that compose the grains and the fluid species filling the pores), as well as the geometric configuration of these components. This heterogeneity occurs on a wide range of size scales, and these scaling effects on the macroscopic properties of rocks are an active field of investigation. This article will assume for simplicity that size scale of the heterogeneities are on the order of the individual rock grains and pores, forming a single lithologic unit for the larger size scale at which the effective properties are measured.

Even with this assumption, sediments are very complex systems from the petrophysical point of view. A specific physical property can have a broad range of values for a given lithology, making the analysis and interpretation of the geophysical properties of sediments in terms of geological factors a difficult task. With this in mind, the aim of this article to introduce the reader to petrophysical principles involved in the acoustical, electrical and radioactive properties of sediments; in-depth information on this subject can be found in a number of recent texts (e.g., Guéguen and Palciauskas, 1994; Mavko *et al.*, 1998; Schön, 1996).

Acoustical properties

The propagation of elastic waves through geological materials is the basis for seismic methods and acoustic (or sonic) logging.

Table G1 A list of compressional (V_P) and shear (V_S) wave velocities for a variety of sedimentary rocks

Lithology	V_P (in m/s)	V_S (in m/s)
Chalks (water saturated)[1]	1530–4300	1590–2510
Dolomites (water saturated)[1]	3410–7020	2010–3640
Limestones (water saturated)[1]	3390–5790	1670–3040
Sandstones (water saturated)[1]	3130–5520	1730–3600
High porosity sandstones (water saturated)[1]	3460–4790	1950–2660
Poorly consolidated sandstones (water saturated)[1]	2430–3140	1210–1660
Tight-gas sandstones (dry)[1]	3810–5570	2590–3500
Shales[2]	1830–5180	
Anhydrite[2]	6090	3470
Gypsum[2]	5800	
Halite[3]	4590	2660

Sources: [1]Mavko et al., 1998; [2]Schön, 1996; [3]Guéguen and Palciauskas, 1994.

With minor exceptions, these methods use body waves (i.e., waves that travel through a body's interior) to investigate the Earth's subsurface. There are two types of body waves that differ in terms of their particle motion relative to the propagation direction of the wave. P waves (or compressional waves) have particle displacement in the direction of wave propagation; S waves (or shear waves) have particle displacement perpendicular to the propagation direction. The commonly measured and analyzed acoustic properties of geological materials are the velocities of these two body waves.

For a homogeneous and isotropic medium, the P and S wave velocities are given by

$$V_P = \sqrt{(k + 4\mu/3)/\rho} \quad \text{and} \quad V_S = \sqrt{\mu/\rho}, \quad \text{(Eq. 1)}$$

respectively, where k and μ are the effective bulk and shear moduli of the medium and ρ is its density. The elastic moduli quantify the deformation a material undergoes when subjected to different types of applied stress. The bulk modulus k relates volumetric changes to variations in applied pressure. The shear modulus μ describes the shape distortion due to a pure shear stress; there is no volumetric change accompanying the pure shear stress. The dependence on density is due to inertial effects, such as those described by Newton's laws of motion. The overwhelming majority of seismic and acoustic well logging data involves the propagation of P waves; however, information about S wave propagation is becoming more readily available. A list of P and S wave velocities for a variety of sedimentary rocks is given in Table G1.

The velocity ratio V_P/V_S is also used in the analysis of velocity data; Table G2 give the ranges of theis ratio for different sedimentary lithologies. It can be seen from Equation (1) that this ratio only depends on the elastic moduli. The composition and microstructural elements of sediments affect k and μ differently; these differences are reflected in the V_P/V_S ratio.

Acoustic velocities of carbonates and sandstones

Due to their importance as potential hydrocarbon reservoirs, the acoustic velocities of carbonates and sandstones have been more extensively studied than other sedimentary lithologies. An obvious factor affecting these velocities is the composition of the solid grains. Typically, carbonates have higher V_P than sandstones having the same porosity (Table G1). This is related to the differences in V_P for the rock forming minerals calcite (6650 m/s), dolomite (7370 m/s) and quartz (6060 m/s).

Table G2 Typical range of the velocity ratio V_P/V_S for common lithologies (Domenico, 1984)

Lithology	V_P/V_S
Dolomites[1]	1.78–1.84
Limestones[1]	1.84–1.99
Sandstones[1]	1.59–1.76
Shales[2]	1.70–3.00

Sources: [1]Domerico, 1984; [2]Schön, 1996.

However, variations in porosity and microstructure produce a range of acoustic velocities for each lithology. These factors produce significant overlapping of the velocity ranges that makes it difficult to differentiate between carbonates and sandstones using only V_P.

Ambiguity is reduced in the lithology-velocity relationship by analyzing the velocity ratio V_P/V_S. It can be seen from Table G2 that this ratio is commonly larger for carbonates than for sandstones. Further, the velocity ratio V_P/V_S has been used to infer composition in mixed siliceous carbonate systems. Lithological discrimination using the velocity ratio is based on the large difference in V_P/V_S between quartz ($V_P/V_S = 1.47$) and carbonate minerals ($V_P/V_S = 1.93$ for calcite and $V_P/V_S = 1.85$ for dolomite).

Another lithological element affecting acoustic velocities is clay content in siliciclastic rocks. Small amounts of clay in sandstones can reduce its elastic moduli, producing lower velocities. The effects of clay content are greater for V_S due to the lessened rigidity or stiffness (i.e., lower μ) associated with clay; this produces higher V_P/V_S ratio with increasing clay content. Further, it has been found that clay effects are determined by its position among the sand grains. Clay in load-bearing positions, such as contact clay cements, affects sandstone stiffness; non-load-bearing clay that occupies pores has little impact of sandstone stiffness.

A significant factor affecting acoustic velocities of carbonates and sandstones is the porosity and pore structure. Increasing porosity lowers V_P and V_S; the magnitude of this effect depends on the shapes of the added pore space. Experimental and modeling results show that the cracks and fractures lowers the acoustic velocities of rocks much more than the same amount of more spherical pores such as vugs and main intergranular voids. This effect is due to the increasing ease with which flatter pores can be deformed.

Further, pore shape affects μ more than k; hence, V_S is more sensitive to the presence of crack-like pores than V_P. Increasing V_P/V_S ratio is used as indicator of fracturing.

The grain composition and pore structure are dependent on many geological factors such as the source material, depositional environment and diagenesis. Attempts have been made to connect these geological factors with elastic wave velocities, starting with Faust (1951) relating V_P of siliciclastic rocks to age of burial. More recently, a number of systematic studies have been published that analyze the relationship between acoustic velocities and geological parameters for carbonate rocks (e.g., Anselmetti and Eberli, 1997) and siliciclastic rocks (e.g., Vernik and Nur, 1992).

Acoustic properties of shales

While shales constitute a majority of the sedimentary rock mass, their acoustic properties have not been well documented in the past due to their lack of importance as reservoir rocks and the exceptional difficulty of performing laboratory measurements on them. However, their importance as source rocks and their effects on seismic imaging is leading to greater interest in their properties. From Tables G1 and G2, it can be seen that shales have a wide range of acoustic properties. Their acoustic velocities rise with an increasing degree of induration due to growing stiffness; hence, the V_P/V_S decreases with induration.

The preferred alignment of clay platelets in shales produces significant velocity anisotropy (i.e., directionally dependent acoustic velocities). Both V_P and V_S are consistently lower in the direction perpendicular to bedding than parallel to bedding. Velocity differences up to 30% for V_P and 35% for V_S have been measured.

Both the presence of kerogen and its maturity are factors that influence the acoustic properties of shales. Acoustic velocities decrease with rising organic content due the lower elastic moduli of the kerogen. Acoustic velocities also decrease with kerogen maturity; several factors have been suggested for this behavior, such as the formation of free phase hydrocarbons and fracturing associated with the development of overpressuring. The orientation of these fractures will affect the velocity anisotropy of the shale.

Unconsolidated sediments acoustic velocities

Unconsolidated sediments have distinctly lower velocities than their equivalent consolidated lithologies. For instance, a typical range of V_P is 200–500 m/s for dry sediments and 1600–2000 m/s when water-saturated. These lower velocities are due to the low rigidity possessed by unconsolidated sediment, especially near the transition to a fluid supported suspension. This condition produces the extreme low V_S valves observed in situ for some unconsolidated sediments (e.g., ~100 m/s in water saturated marine sediments). Hence, very large values of V_P/V_S (i.e., 3 and greater) are possible for unconsolidated media.

These sediments can be divided into two groups: granular media, such as sands and gravels, and muds principally composed of clay and silt. Experimental and theoretical investigations have been chiefly directed toward the behavior of granular sediments. In granular materials, acoustic wave propagation through the solid framework is dependent on the number of grain contacts per particle and the contact micromechanics.

The number of contacts per grain determines how elastic energy is transmitted throughout a granular medium. The number of contacts per grain is determined by the nature of grain packaging, particle shape and grain-size distribution. For example, the compact hexagonal packing of identical spheres gives 12 contacts per sphere, whereas the simple cubic packing has 6 contacts. A dense random packing is a more realistic representation of unconsolidated sediments, these systems have an average of 6.8 contacts per sphere with 2.1 near contacts that can close to form additional contacts as pressure conditions change.

The contact micromechanics describe how elastic energy is transmitted between adjacent grains at an individual contact. Different types of contact micromechanics have been used to explain the velocity data for granular sediments. These contact phenomena are dependent on a number of parameters such as bulk grain properties, grain shapes, surface energy, surface roughness and the macroscopic sense of the applied stresses. In unlithified sediments, these interactions occur at the relatively small contact areas. As lithification occurs, pressure solution and cementation modify the nature of the contact micromechanics by strengthening the bond between spheres and increasing the contact area. The lithification process enhances the rigidity of the sediments and is reflected in increased acoustic velocities, especially V_S. Hence, the lithification process is accompanied by decreasing V_P/V_S.

Acoustic velocities of evaporates and coal

Evaporates (e.g., halite, gypsum) are essentially zero porosity rocks; the velocities of these lithologies strongly reflect their mineral content (Table G1). However, these velocities are not significantly different from porous sedimentary rocks. Coal is distinguished from other sedimentary rocks by its very low V_P. P-wave velocities in coals are a function of grade, ranging from a low of 1790 m/s for lignite to 3050 m/s for anthracite.

Dependence on pore fluids

Pore fluid properties determine the elastic behavior of the pore space; they are strongly dependent on pore fluid composition, temperature and fluid pressure. At seismic frequencies (i.e., less than 2 kHz), V_P is affected by variations in pore fluid properties. V_S is relatively insensitive to these changes, responding only to fluid density variations. This condition results from lack of rigidity in fluids (i.e., $\mu_{\text{fluids}} = 0$) and complete pressure equilibration throughout the pore structure.

Pressure equilibration between the pore spaces means its contents act as a single effective pore filling. The effective bulk modulus k_{eff} for a pore space containing water (k_w) and gas (k_g) phases is given by

$$k_{\text{eff}}^{-1} = S_w k_w^{-1} + (1 - S_w) k_g^{-1} \qquad \text{(Eq. 2)}$$

where S_w is water saturation (i.e., volume fraction of the porosity filled with water). Since the bulk moduli of liquids is much greater than those of gases, Equation (2) predicts that the presence of a gas phase, even in small amounts, will significantly lowers the bulk modulus of the pore filling. For example, consider a pore space containing 98% water ($k_w = 2.2 \times 10^9$ Pa (or N/m^2)) and 2% air ($k_g = 1.55 \times 10^5$ Pa); it acts as if it is filled with an effective fluid with $k_{\text{eff}} = 7.7 \times 10^6$ Pa.

Hence, the introduction of gas can produce large decreases in V_P relative to its value when the medium is liquid saturated. Given the insensitivity of V_S to changing pore fluid properties, a decreased V_P/V_S ratio is used an indicator for natural gas. Further, the presence of a dissolved gas phase, such as occurs in live oils, can reduce the pore fluid bulk modulus and lower V_P.

Effects of pressure and temperature conditions on velocities

Pressure effects are primarily dependent on effective (or differential) pressure Δp which is the difference between the external confining (or overburden) pressure p_c and the internal fluid pressure p_f. Increasing Δp causes pore closures in consolidated reservoir rocks, resulting in velocity increases. The velocity changes are more rapid at lower Δp values and slacken as Δp grows. This behavior is due to the swift closure of the more compliant intergranular cracks and fractures, leaving the more spherical pores that are much harder to collapse. In unconsolidated materials, increasing Δp narrows the distance between grains in the medium. This phenomenon produces stronger bonding at existing grain contact and creates new contact as smaller intergranular gaps close. Conversely, the existence of geopressure will reverse this closure process and produce lower velocities. Variations in p_f can also affect pore fluid properties, particularly the gas phase compressibility and density, as well as gas solubility in liquids.

Temperature change generally has a relatively small affect on the elastic wave velocities of reservoir rocks with increasing temperature causing minor velocity decreases. Major variations in velocities that are observed are primarily due to changes in pore fluid properties. Significant temperature induced change can results if it associated with fluid phase transformations (e.g., freezing) or large bulk modulus and viscosity changes in heavy oils.

Velocity dispersion

Dispersion is the variation of a physical property with measurement frequency. Most laboratory measurements of elastic wave velocity are performed at ultrasonic frequencies (i.e., above 100 kHz) while acoustic well logging and seismic exploration are done at lower frequency ranges (i.e., 10–20 kHz and below 1 kHz, respectively). Since there is a significant difference in measurement frequencies, an understanding of dispersion is necessary for relating laboratory results to the interpretation of geophysical field measurements.

Both V_P and V_S increase with increasing measurement frequency for porous rocks containing pore liquids. While there are very few direct measurements of velocity dispersion; it can be estimated using ultrasonic measurements on dry samples to predict low frequency velocities. This method predicts that the magnitude of velocity dispersion can range up to 20% or more; however, it is frequently 10% or less for water-saturated sedmentary rocks. Dispersion is enhanced by the presence of intergranular cracks; hence, dispersion decreases as cracks close with increasing effective pressure. In addition, dispersion can be significantly large for sedimentary rocks containing gas-liquid mixtures and viscous oils.

Several mechanisms have been proposed as the cause of velocity dispersion in porous rocks. Current evidence indicates that the local flow mechanism (Murphy et al., 1986) is the dominant mechanism in most porous rocks. At low frequencies, fluid pressure equilibration is maintained by fluid flow between pores in response to varying degrees of deformation within sections of the pore space. Dispersion due to this mechanism results from the breakdown of interpore fluid flow as frequency increases due to viscous forces. In the absence of this flow, individual pores act as isolated units within the porous rock at high frequencies. Under these conditions, pore spaces, in particular the intergranular cracks, become harder to deform and result in larger k and μ for the porous rock.

Another dispersion mechanism that is proposed for sedimentary materials is the Biot (1956) dynamic poroelasticity. Dispersion due to this mechanism is due to the progressive decoupling of the average motion of the solid component from that of the pore fluid as frequency increases. At low frequencies, the motions of the solid and pore fluid are perfectly coupled. At high frequencies, the decoupled motions lead to inertial effects that change the effective elastic moduli and density of the rock. Laboratory results and theoretical modelling indicate that this mechanism is important in materials with open pore structures where local flow effects are small.

Petrophysical relationships for acoustic velocities

A wide variety of relationships are available for analyzing and predicting the elastic wave velocities of reservoir rocks. Some are empirical relationships between velocity and parameters (e.g., lithology, porosity, clay content, pressure) obtained from the statistical analysis of laboratory and field measurements. However, an empirical relationship could have limited accuracy when applied to sediments different from those used in its development.

One of the most commonly used relationships is the time-average equation of Wyllie et al. (1956); this relationship has been extensively used in acoustic well logging analysis. It assumes that the traveltime for a P-wave through a unit thickness of porous medium is the sum of the traveltimes through the volumetric fractional thicknesses of the solid matrix and pore fluid. The resulting relationship between the velocities of the porous medium, solid matrix and fluids (V, and V_m, V_f, respectively) is

$$\frac{1}{V} = \frac{(1-\phi)}{V_m} + \frac{\phi}{V_f} \qquad \text{(Eq. 3)}$$

where ϕ is porosity. Table G3 gives a list of solid matrix and pore fluid velocities for use in this relationship. While this mixing formula can give reasonable estimates for V_P, its neglects the effects due to the geometrical configuration of the constituents, in particular the pore structure. This element of the time-averaging approach is partially reflected in the range of V_m values in Table G3 used for each lithology.

Relationships given by Gassmann (1951) are frequently used to analyze the effects of pore fluids on acoustic velocities. These relationships are valid for any porous medium when complete fluid pressure communication occurs throughout the pore space (i.e., perfect interpore fluid flow); this condition is approximated during seismic wave propagation. However, the Gassmann relationships use effective moduli to implicitly incorporate pore structure effects; this feature limits its predictive capacity.

Table G3 Range of solid matrix and pore fluid velocities used in Wyllie's time average relationship (Matrix values from Schön, 1996)

Material	V_m or V_f (in m/s)
Sandstone (general)	3049–5977
Sandstone (consolidated)	5797–5882
Sandstone (semi-consolidated)	5291–5482
Sandstone (unconsolidated)	2500–5184
Limestone	6402–7008
Dolomite	6969–7008
Fresh water	1524
Brine (200 kppm NaCl)	1737
Oil	1311
Air	332

Another approach used to analyze the elastic wave velocities of sedimentary rocks is the inclusion-based formulation. This formulation treats a porous rock as a solid matrix and embedded inclusions representing individual pore spaces. Hence, the pore structure of the rock is defined in terms of the inclusion shapes and concentration. These formulations commonly assume that no fluid pressure communication between pores (i.e., complete interpore fluid flow breakdown); this condition replicates the high frequency condition for the local flow dispersion mechanism. Therefore, inclusion-based formulations have been found to be more accurate than the Gassmann's relationships when predicting ultrasonic velocity measurements. The connection between Gassmann's relationships and inclusion-based models has now been established, permitting a framework to analyze the effects of pore geometry and pore fluid properties on velocity dispersion.

Electrical properties

Electrical measurements of various types are made along the Earth's surface or within boreholes. These geophysical methods involve either the flow of an electrical current between electrodes or the propagation of an electromagnetic (EM) field from a transmitter to a receiver. The behavior of the current or field in a medium is dependent its electrical resistivity, or its reciprocal electrical conductivity, which quantifies the transport of electric current. It is one of the most variable physical properties for geological materials, extending over a range of values greater than 10^5 for typical sediments.

Resistivity of porous sediments

The extensive use of electrical and induction logging in reservoir evaluation has lead to wide-ranging research on the electrical resistivity of sandstones and carbonates. Since the sediment grains are essentially insulators, the electrical resistivity of clean (i.e., clay-free) sediments is controlled by electrolytic (or ionic) conduction through the pore water. The magnitude of the electrolytic conduction is a function of the pore water resistivity ρ_w and the pore structure.

The value of ρ_w is directly related to the concentration of the dissolved ions, and both can vary widely. For instance, at 75°C water resistivity will decrease from 10 to 0.04 ohm-meters (Ω-m) as the dissolved NaCl concentration increases from 500 to 250000 ppm (i.e., from freshwater to brine). Further, ionic species differ in terms of valence and mobility; hence, the ionic composition of the pore water is also important. While the porosity establishes the number of charge carriers available in the porous medium, it is the pore space connectivity that determines the paths followed by the ionic charge carriers. Materials with open pore structure, such as those occurring in well-sorted sands and gravels, permit more efficient current flow than in media with narrow, tortuous pore spaces.

Electrical resistivity is also affected by surface conduction along silica-fluid interfaces; this conduction is the combined result of different surface phenomena. The magnitude of the surface conduction for a material is proportional to its specific surface area (i.e., surface area per unit pore volume); hence, it is important conduction mechanism in clayey materials such as shaly sandstones. It has been found that the style of clay distribution influences this mechanism. Dispersed clay that occurs between grains and along grain surfaces has more surface area than clay laminations or framework grains. Clay mineralogy is also a factor as surface area varies widely with clay type. The strength of the surface conduction has been observed to be nearly independent of ρ_w; it becomes the dominant conduction mechanism in clayey materials when fresh or low salinity water (i.e., large ρ_w) is present.

Archie's equation for resistivity

The petrophysical relationship most commonly used for the electrical resistivity of clean sediments was given by Archie (1942). This relationship states that the electrical resistivity of a water saturated porous medium ρ is given by

$$\rho = a \phi^{-m} \rho_w \quad \text{(Eq. 4)}$$

where a and m are empirical parameters. The combined terms $a\phi^{-m}$ are also referred to as the formation factor F ($=\rho/\rho_w$). The proportionality coefficient a takes on values between 0.5 and 3.5.

The parameter m implicitly incorporates the effects of pore structure on resistivity. The observed value of m systematically varies with the degree of consolidation in sandstones (Table G4a); hence, it is often called the cementation exponent. A correlation between the value of m and porosity type also has been noted for carbonates (Table G4b). For unconsolidated samples, the exponent m has been related to particle shape, varying from 1.3 for spherical grains to 1.9 for thin disc-like grains.

The presence of immiscible pore fluid species (e.g., air, hydrocarbons, chlorinated solvents) is another factor affecting electrolytic conduction porous sediments. Since immiscible pore fluids are commonly electrical insulators, their presence impedes electrolytic conduction through the pore space, leading to higher resistivity values. Immiscible fluid effects on the resistivity of porous media are described by the second form of Archie (1942) equation:

$$\rho = a \phi^{-m} S_w^{-n} \rho_w \quad \text{(Eq. 5)}$$

where S_w is the water saturation and n is an empirical constant called the saturation exponent. The observed values of n for sandstones and limestones range between 1.1 and 2.6; it is commonly assumed to be 2 when laboratory measurements are not available. Further, it has been found that n for a given sample is dependent on wetting conditions, immiscible species present and saturation history. These factors influence the pore scale fluid distribution, affecting the geometry of the electrolytic conduction path through the water phase.

Table G4 The typical values of the cementation exponent m for (a) sands with varying degrees of consolidation and (b) carbonates with different porosity types (Doveton, 1986)

(a) Unconsolidated sands	$m = 1.3$
Very slightly cemented sands	$m = 1.4–1.5$
Slightly cemented sands	$m = 1.5–1.7$
Moderately cemented sands	$m = 1.8–1.9$
Highly cemented sands	$m = 2.0–2.2$
(b) Fractured carbonates	$m = 1.4$
Chalky limestones	$m = 1.7–1.8$
Crystalline and granular carbonates	$m = 1.8–2.0$
Carbonates with vugs	$m = 2.1–2.6$

Petrophysical relationship for shaley sediments

The combined effects of electrolytic and surface conduction in clay bearing sediments are generally treated as an equivalent circuit consisting of parallel conductors. A commonly used petrophysical relationship based on this model for water-saturated shaly sands given by Waxman and Smits (1968) is

$$\frac{1}{\rho} = \frac{1}{a\phi^{-m}\rho_w} + \frac{BQ_v}{a\phi^{-m}} \qquad \text{(Eq. 6)}$$

The electrolytic conduction term $1/a\phi^{-m}\rho_w$ is obtained from the first Archie equation (Equation 4). In the surface conduction term, Q_v is the total charge in the electrical double layer per unit pore volume and is related to the cation exchange capacity of the material; hence, it can vary widely with clay mineralogy. The empirical parameter B is the counterion mobility. A similar relationship is available for partially saturated shaly sands that accounts for the pore space occupied by the insulating nonaqueous fluid phases.

Resistivity of shales, coal and evaporites

Shales generally have lower resistivity than sandstones and carbonates due to their water content and the surface conduction associated with clay minerals. An important factor that can affect shale resistivity is the degree of thermal maturity of their organic matter. Hydrocarbons are released and displace pore water as kerogen matures, increasing shale resistivity. Coals are often considered to have high resistivity. However, both anthracite and lignite can have low resistivity. In the case of lignite, resistivity can be used as a measure of water content. Due to their lack of porosity, evaporites have very high resistivity; this is used as a diagnostic property in well log analysis.

Temperature and pressure effects on resistivity

Temperature changes affect aqueous solutions resistivity with declining ρ_w as temperature rises. This response is due to lowering of fluid viscosity with warmer temperatures that leads to greater ion mobility. The dependence of ρ_w on temperature is significant (e.g., a 2% change in ρ_w for a 1°C variation at 20°C); hence, temperature effects need to be considered in the analysis of resistivity data.

Fluid pressure changes have a minimal influence on ρ_w, and its effect is generally neglected. However, applied pressure variations alter the pore structure and affect pore connectivity. Pore closure with higher effective pressure increases tortuosity that results in higher ρ for porous rocks.

Table G5 The abundance of potassium, uranium and thorium content in sediments and rock forming minerals. Brackett indicates average value (Bateman, 1985)

	K (%)	U (ppm)	Th (ppm)
Lithologies			
Carbonates	0.0–2.0 [0.3]	0.1–9.0 [2.2]	0.1–7.0 [1.7]
Sandstones	0.7–3.8 [1.1]	0.2–0.6 [0.5]	0.7–2.0 [1.7]
Shales	1.6–4.2 [2.7]	1.5–5.5 [3.7]	8–18 [12]
Colorado Oil Shale	<4.0	up to 500	1–30
Minerals			
Quartz, Calcite, Dolomite	0	0	0
Plagioclase	0.54	–	<0.01
Orthoclase	11.8–14.0	–	<0.01
Glauconite	5.08–5.30	–	–
Montmorillonite	0.16	2–5	14–24
Kaolinite	0.42	1.5–3	6–19
Illite	4.5	1.5	–
Micas	6.7–9.8	–	<0.01
Phosphates	–	100–350	1–5
Zircon	–	300–3000	100–2500

Radioactive properties

Radiometric methods, such as natural γ-ray well logging, are used to measure the radioactivity of sediments due to the spontaneous decay of radionuclides present in these materials. There are three series of natural radionuclides that are common in the Earth's crust: the uranium (^{235}U and ^{238}U) series, thorium (^{232}Th) series and potassium (^{40}K). Both the thorium and uranium series involve a decay chains with a number of daughter products. The isotope ^{40}K decays to ^{40}Ca and emits a γ-ray. The relative abundances of these radionuclides are variable in sedimentary rocks (Table G5). While total γ-ray tools count all γ-ray without differentiation, spectral γ-ray tools determine the contribution due to each series.

Shales and clays produce a higher γ-ray count due to clay minerals that contain K and Th; hence, total γ-ray count is used as measure of shale content. Since clay minerals differ widely in terms of their Th/K ratios (Table G5), it has been suggested that spectral γ-ray data be used for clay mineral identification. Organic matter in black shales preferentially adsorb U ions; the U concentration is a function of U ion species, amount and degree of organic matter maturity, and pH/Eh conditions. Therefore, the Th/U ratio is useful in source rock characterization.

Pure quartz sandstones contain very small concentrations of radioactive isotopes and exhibit low γ-ray count. The presence of K-rich minerals (i.e., feldspars and micas) produces significantly higher γ-ray count and can be used as a sediment maturity indicator. Some heavy minerals (e.g., zircon) are Th and U bearing, and sandstones with these minerals also produce significantly higher γ-ray counts. These two cases can be identified using the Th/K ratio. Both calcite and dolomite have very low K, Th and U contents that explain the very low γ-ray activity commonly measured in carbonates. Elevated γ-ray count in carbonates can be attributed to the presence of clay mineral or U associated with organic material and phosphate minerals.

Coals commonly have little, if any, K or Th. Their γ-ray activity is very low; coals may have readings even lower than clean sandstones on γ-ray logs. Natural γ-ray logging is an excellent method for distinguishing coal from interbedded shales. Potassium bearing evaporites (e.g., sylvite) are very radioactive which combined with their very high resistivity can be used for lithologic identification. Other evaporites are non-radioactive, and their low γ-ray readings separate them from most shales.

Summary

The physical properties of sediments are dependent on the properties and abundance of the individual components as well as their geometrical configuration. Since sediments are complex heterogeneous systems, their physical properties can vary widely for a given lithology. However, there are general petrophysical principles that govern the physical properties of sediments.

The acoustic wave velocities for sediments are a function of the elastic moduli and densities of the rock forming minerals composing the solid grains and the fluid species that fill the pore space. For seismic waves, the pore fluids act as a single effective fluid that is very sensitive to the presence of gases. The acoustic velocities decrease with increasing porosity; this effect is greater for crack and fractures.

Since solid grains are insulators, the electrical resistivity of sediments is primarily controlled by electrolytic conduction through the pore water. The transport of electrical current by this mechanism is dependent on the water resistivity and the pore structure. In addition, electrical current can flow along solid-fluid interfaces due to surface conduction. This mechanism can significantly contribute to transport of electrical current in materials with large surface area such as clayey lithologies.

The natural radioactivity of sediments is a function of their potassium, uranium and thorium content. Total γ-ray count is used as measure of shale content due to clay minerals containing potassium and thorium. Source rocks can have elevated levels of uranium due to preferential absorption by organic matter. The presence of potassium bearing minerals can significantly increase the radioactivity of sandstones and evaporites that otherwise have low γ-ray count.

Anthony L. Endres

Bibliography

Anselmetti, F.S., and Eberli, G.P., 1997. Sonic velocities in carbonate sediments and rocks. In Palaz, I., and Marfurt, K.J. (eds.), *Carbonate Seismology*. Tulsa: Society of Exploration Geophysicists, pp. 53–74.
Archie, G.E., 1942. The electrical resistivity log as an aid in determining some reservoir characteristics. *Transactions AIME*, **146**: 54–62.
Bateman, R.M., 1985. *Open-Hole Log Analysis and Formation Evaluation*. Boston: International Human Resources Development Corporation.
Biot, M.A., 1956. Theory of propagation of elastic waves in fluid-saturated porous solids. *Journal of the Acoustical Society of America*, **28**: 168–191.
Domenico, S.N., 1984. Rock lithology and porosity determination from shear and compressional wave velocity. *Geophysics*, **49**: 1188–1195.
Doveton, J.H., 1986. *Log Analysis of Subsurface Geology*. New York: Wiley-Interscience.
Faust, L.Y., 1951. Seismic velocity as a function of depth and geologic time. *Geophysics*, **16**: 192–206.
Gassmann, F., 1951, Über die Elastizität poröser Medien. *Viertelijahresschr. Der Naturforsh. Gesellschaft Zurich*, **96**: 1–23.
Guéguen, Y., and Palciauskas, V., 1994. *Introduction to the Physics of Rocks*. Princeton University Press.
Mavko, G., Mukerji, T., and Dvorkin, J., 1998. *The Rock Physics Handbook: Tools for Seismic Analysis of Porous Media*. Cambridge University Press.
Murphy, W.F., Winkler, K.W., and Kleinberg, R.L., 1986. Acoustic relaxation in sedimentary rocks,: Dependence on grain contacts and fluid saturation. *Geophysics*, **51**: 757–766.
Schön, J.H., 1996. *Physical Properties of Rocks: Fundamentals and Principles of Petrophysics*. Tarrytown: Pergamon.
Vernik, L., and Nur, A., 1992. Petrophysical classification of siliciclastics for lithology and porosity prediction from seismic velocities. *AAPG Bulletin*, **76**: 1295–1309.
Waxman, M.H., and Smits, L.J.M., 1968. Electrical conductivity in oil-bearing shaly sands. *Society of Petroleum Engineers Journal, Transactions AIME*, **243**: 107–122.
Wyllie, M.R.J., Gregory, A.R., and Gardner, L.W., 1956. Elastic wave velocities in heterogenous and porous media. *Geophysics*, **21**: 41–70.

Cross-references

Classification of Sediments and Sedimentary Rocks
Compaction (Consolidation) of Sediments
Geothermal Properties of Sediments
Heavy Minerals
Kerogen
Magnetic Properties of Sediments
Maturation, Organic
Mudrocks
Porewaters in Sediments

GEOTHERMIC CHARACTERISTICS OF SEDIMENTS AND SEDIMENTARY ROCKS

It has been known for centuries that temperature in the Earth increases with depth. This increase towards the interior is the natural process of general cooling of the lithosphere and where a portion of the heat flow is generated by the decay of radioactive constituents contained in the crust. The escape of the heat (the heat flow), is measured by determining differences in temperature with depth (the gradient), at various places usually in mines, tunnels, or well bores and the type of rock through which the heat is transmitted (rock conductivity). Heat flow is expressed as:

$$Q = -K \, (dT/dx) \qquad \text{(Eq. 1)}$$

where Q is the heat flow, K is the thermal conductivity, T is the temperature, and x is the distance (depth).

Heat flow is computed either by the: (1) interval method; or (2) Bullard method (thermal depth method) (Jessop, 1990). Both methods utilize the temperature gradient (measured in a well bore, mine, or tunnel) and rock conductivity value(s) for each geological unit through which the gradient is determined.

Where the crystalline basement (crust) is covered by younger sediments or sedimentary rocks, the heat flow, may be influenced by the blanket effect of the cover. Physical properties of the sedimentary cover, including the rock type and contained radioactive elements, thermal conductivity, and their contained fluids or gas, affect the thermal regime. In addition, topography and past climatic conditions may be important factors.

Table G6 Average heat-production values of some igneous rocks (from Haenel, Ryback, and Stegena, 1988, p. 136)

Rock type	Heat production μWm^{-3}
Granite	2.45
Granodiorite/dacite	1.48
Diorite, quartzdiorite/andesite	1.08
Gabbro/basalt	0.309
Peridotite	0.012
Dunite	0.002

Table G7 Representative heat generation units (HGUs) for sediments/sedimentary rocks (Haenel, Rybach, and Stegena, 1988, p. 136)

Rock type	Heat production (μWm^{-3})
Limestone	0.62
Dolomite	0.36
Quartzitic sandstone	0.32
Arkose	0.84
Graywacke	0.99
Shale and siltstone	1.80
Black shale	5.50
Salt	0.012
Anhydrite	0.090

Table G8 Thermal conductivities of some sedimentary rocks (from Blackwell and Steele, 1992; Jessop, 1990)

Rock type	Conductivity range (W/mK)
Limestone	2.50–3.10
Dolomite	3.75–6.30
Sandstone	2.50–4.20
Clay and siltstone	0.80–1.25
Shale	1.05–1.45
Anhydrite	4.80–5.80
Salt	4.80–6.05
Basalt	1.69–1.96
Granite	307–3.50

The heat (H), partly generated by the radioactive isotopes of Uranium (U), Thorium (Th), or Potassium (K), is dissipated through the crust and overlying sediments or sedimentary rocks and is measured in μWm^{-3}. The heat is transmitted by conduction, convection, or radiation. The amount of heat, timing, and cause of heating is of prime importance to many geological applications such as understanding cratonic sedimentary basin development, generation and migration of petroleum, emplacement of ore deposits, and sedimentary diagenesis.

Igneous and metamorphic rocks which constitute the Earth's crust are naturally radioactive and have different heat-production rates. For example, granites usually contain more radioactive elements than gabbros and basalts. Some representative values are given in Table G6.

Some heat may be generated by radioactive elements within the sediments themselves (Table G7). For example, black shales may be highly radioactive, but other rock types such as limestones and sandstone contain few radioactive elements, except in special conditions, and, therefore, their heat generation unit (HGU) values are small. Evaporites essentially are devoid of any radioactive elements and therefore the HGUs are extremely small or nil. The radioactive sediments, of course, contribute to the overall heat flow, but generally this contribution is small (because of the thinness of the sedimentary package) relative to the basement rocks (= crust), which constitute infinitely more mass.

The heat flow in sedimentary basins, for the most part, is computed from temperatures measured in boreholes, usually drilled for purposes other than measuring temperature. A vast amount of these data are available as bottomhole temperatures (BHTs) or temperatures taken in drillstem tests (DSTs) during logging of commercial wells drilled in search of petroleum, ore deposits, or water. Temperatures also may be recorded in well bores by conventional continuous temperature logging techniques or by the new Distributed Optical-Fibre Temperature Sensing technique (DTS). Temperatures measured during well logging may be of questionable value depending on the circumstances and conditions during their recording and collection. If certain conditions are known and controlled, then the values can be corrected to obtain a 'true' temperature.

Topography

In lowland areas or areas with little topographic relief, a correction for topographic irregularities is not needed. In other areas with considerable topographic relief, however, a correction factor needs to be applied.

Rock type and conductivity

Different rock types exhibit different thermal conductivities and these values have been measured carefully in laboratories under controlled conditions. A summary of values is presented in Table G8.

Porosity and contained fluid or gas

The type of fluid or gas contained in pores of a rock affect their conductivity and thus the computation of any value dependent on the thermal conductivity (Table G8). Water, oil, or natural gas all have different conductivities and alter the overall rock conductivity, but usually the interporosity substance is known and can be considered in the computations.

Past climatic conditions

This factor affects only the upper hundred meters or so of the subsurface and the effect can be determined from any reliable temperature log taken in a borehole. The correction for the climatic change is relatively small, but could be important, especially in areas of permafrost.

The computed heat-flow values, then, can be compared from one location to another, plotted on a map, and contoured for further interpretation (see for example Blackwell and Steele, 1992; Čermák and Hurtig, 1979; Hamza and Munoz, 1996).

For reference, the average heat flow on continents is approximately 62 mWm^2 and for oceanic areas, it is slightly higher at about 80 mWm^2 (Jessop, 1990, p. 205).

Daniel F. Merriam

Bibliography

Blackwell, D.D., and Steele, J.L. (eds.), 1992. Geothermal map of North America. *Geological Society of America, Map CSM007*; scale: 1:5,000,000.

Čermák, V., and Hurtig, E., 1979. Heat flow map of Europe. In Čermák, V., and Rybach, L., (eds.), *Terrestrial Heat Flow in Europe*. Springer-Verlag, pp. 3–40.

Haenel, R., Rybach, L., and Stegena, L., 1988. *Handbook of Terrestrial Heat-flow Density Determination*. Kluwer Academic Publishers.

Hamza, V., and Munoz, M., 1996, Heat flow map of South America. *Geothermics*, **25**(6): 599–646.

Jessop, A.M., 1990. *Thermal Geophysics*. Elsevier.

GLACIAL SEDIMENTS: PROCESSES, ENVIRONMENTS AND FACIES

Introduction

Significance of glacial deposits

The importance of glacial sediments can be gauged from the fact that 10 percent of the Earth's land surface currently is covered by glacier ice, a figure that exceeded 30 percent during the Quaternary glaciations of the last 2 Ma. Glacier ice has left a complex, often patchy, record of deposition on land, and offshore has contributed substantially to the build up of continental shelves. In earlier geological history, the Earth experienced several continental-scale glaciations, some of them even more extensive than those of the Quaternary Period. Glacial deposition is intimately associated with a wide range of other processes, including fluvial, mass flowage, eolian, lacustrine, and marine. The resulting facies associations are highly variable, and without detailed investigation can be subject to a wide range of interpretations. It is only within the last three decades that studies of glacial processes in modern settings have made it possible to develop plausible models of past glacial depositional environments.

Understanding the nature of Quaternary glacial sediments and their associated landforms is vital in glaciated areas of North America and Europe, where sand and gravel extraction is essential for construction purposes, and for sensible management of water resources and waste disposal. Ancient glacial deposits and associated facies are also economically important; in some regions, since they control the presence of petroleum resources, as for example in the Permo-Carboniferous and Neoproterozoic sequences of the Middle East and South America.

Historical background

Today, it is common knowledge that the Earth experienced a series of ice ages but, when the concept was first mooted in the early 19th century, it met with fierce opposition, as most of the unconsolidated deposits (*drift*) that are familiar to geologists were attributed to Noah's flood of the Old Testament, often with the proviso that larger boulders ("erratics") were deposited from icebergs (Hambrey 1994, Chapter 1 for review). The Swiss natural historian, Agassiz, became the chief protagonist of the "Ice Age Theory", and when he delivered his ideas in 1837, they had a Europe-wide impact. In the following decades, through Agassiz's influence, geologists in the UK and North America gradually accepted the theory as being applicable to their areas. In the second half of the 19th century, "ancient" glacial deposits (*tillites*) were recognized in many parts of the world. However, even in the second half of the 20th century, the glacial origin of supposed tillites was challenged, notably those of Neoproterozoic age. It has taken systematic sedimentological investigations, coupled with an appreciation of modern glacial processes, to settle these debates.

Extent of glacier ice today

The areal extent of glacier ice today has been documented by the World Glacier Monitoring Service (1989) (Table G9). By far the greatest expanses of ice are the ice sheets of Antarctica (85.7 percent of area) and Greenland (10.9 percent). However, it is the remaining 3.4 percent that impinges directly on human civilization, and most detailed sedimentological studies have focused on these smaller ice masses. The potential contribution of the Antarctic ice sheet to sea level rise is 56 m, with the Greenland ice sheet adding a further 7 m. The remaining glaciers account for a mere fraction of this. Fluctuations of the world's ice masses continue to affect global sea levels, providing on-going eustatic controls on the sedimentary processes on continental shelves.

Characteristics of glaciers

Mass balance

The state of health of a glacier is a reflection of the balance between accumulation and ablation over periods of decades to hundreds of years, or thousands of years in the case of the polar ice sheets. The difference between accumulation and ablation in any one year is referred to as *mass balance*, which is positive if accumulation exceeds ablation and negative if the reverse. Accumulation mainly takes the form of snow, which is transformed or metamorphosed by burial, through *firn*, to glacier ice. Ablation is largely accomplished by melting on temperate glaciers, while in some regions calving from ice masses into the sea accounts for the bulk of ice losses. Glaciers are typically subdivided into an *accumulation zone* and an *ablation zone*, separated by an *equilibrium line* where there is no net gain or loss of mass (Figure G4).

Table G9 Distribution of glacierized areas of the world (World Glacier Monitoring Service, 1989)

Region	Area (km^2)
Africa	10
Antarctica	13,593,310
Asia and Eastern Europe	185,211
Australasia (i.e., New Zealand)	860
Europe (Western)	53,967
Greenland	1,726,400
North America excluding Greenland	276,100
South America	25,908
World total	15,861,766

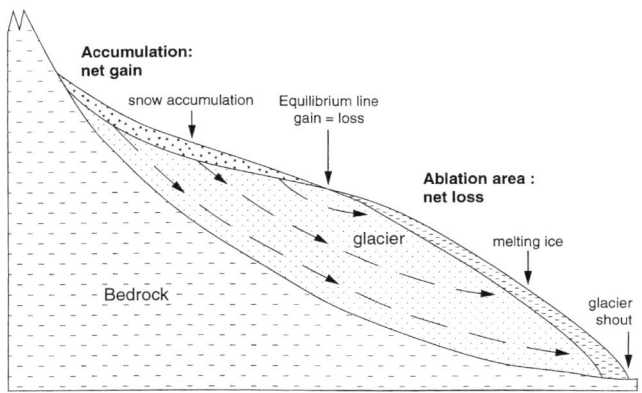

Figure G4 Longitudinal profile through a valley glacier, illustrating flow lines (particle paths) in relation to the equilibrium line.

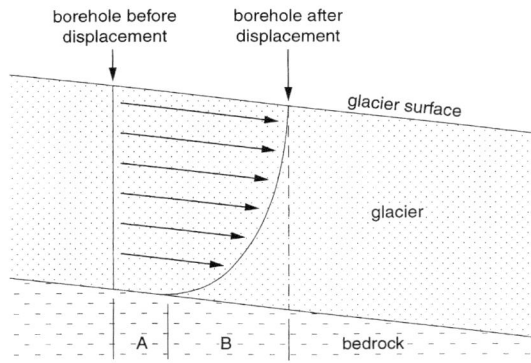

A – basal sliding component
B – internal deformation component

Figure G5 Vertical longitudinal profile through a glacier, illustrating the two main components of glacier flow, internal deformation and basal sliding. Arrows indicate relative displacements, as determined from borehole studies.

Glacier dynamics

In order to interpret the origin of glacial sediments and landforms, it is necessary to understand the mechanisms of ice deformation and glacier flow. Glaciers flow by one or more of three main mechanisms: internal deformation, basal sliding, and movement over a soft, deformable bed (Paterson, 1994). *Internal deformation* is best explained by Glen's flow law for polycrystalline ice; this law relates the effective shear-strain rate ($\dot{\varepsilon}$) to shear stress (τ) in the following equation:

$$\dot{\varepsilon} = A\tau^n$$

where n is a constant, typically 3, and A depends on ice temperature, crystal size and orientation, and impurities. Internal deformation results in the slowest flow occurring at the surface and at the base of a glacier (Figure G5). *Basal sliding* is important where rain or melt-water is able to lubricate the bed (Figure G5), and varies according to the season and time of day or night. Frictional and geothermal heating may add to the availability of meltwater. Many glaciers flow over a bed of unconsolidated sediment which, when saturated, is readily deformed. Deformable beds exist beneath modern fast-flowing ice streams in Antarctica. Much of the movement of the Quaternary ice sheets in North America, has also been linked to subsole deformation. The relative importance of these mechanisms is highly variable. In moist temperate regions as much as 80 percent of glacier flow is from sliding. Cold polar glaciers, which are frozen to their beds, flow almost entirely by internal deformation. Where a deformable bed exists, the bulk of movement may be within the sediment layer beneath the ice. In *surge-type glaciers*, flow is unstable, with long periods (often decades) of quiescence punctuated by short bursts (several months to a few years) of high velocity (*surges*). At its peak, the velocity may reach several orders magnitude above normal, and the glacier may advance rapidly, and redistribute large volumes of sediment.

Glacier structure

Glacier structures are principally the product of internal deformation, and are intimately associated with the transport of debris. Glacier ice is similar to any other type of geological material in that it comprises strata that progressively deform to produce a wide range of structures. Primary structures include sedimentary stratification derived from snow and superimposed ice, unconformities, and regelation layering resulting from pressure melting and refreezing at the base of a glacier. Secondary structures are the result of deformation, and include both brittle features (crevasses, crevasse traces, faults and thrusts) and ductile features (foliation, folds, boudinage). Typically, a glacier reveals a sequential development of structures as in deformed rocks, so that by the time the glacier snout is reached, ice may record several phases of deformation.

Thermal regime

The temperature distribution or *thermal regime* of a glacier is fundamental to glacier flow, meltwater production and routing, and to styles of glacial erosion and deposition. The most active glaciers are so-called *temperate* or *warm*, in which the ice is at the pressure-melting point throughout. These glaciers slide rapidly on their beds and induce much erosion; they are typical of alpine regions. At the opposite end of the spectrum are *cold* glaciers, which are below the pressure melting point throughout. Since they are frozen to the bed, cold glaciers are generally regarded as having little ability to erode, although this assumption has been challenged by observations in the Dry Valleys of Antarctica recently. An intermediate type of glacier, found in the High-Arctic, is referred to as *polythermal*. In such glaciers, it is typical for the snout, margins and surface layer of the glacier to be below the pressure-melting point, whereas thicker, higher level ice is warm-based.

Glacier hydrology

Meltwater within, beneath and beyond glaciers plays a vital role in the processes of erosion and deposition. Water, derived from melting snow and ice, flows in channels on the glacier surface, until plunging *via* moulins into the interior or bed of the glacier, finally emerging at the snout. Drainage routes differ according to the thermal regime of the glacier. In cold and polythermal glaciers, meltwater is forced toward the glacier margins, while in temperate glaciers water flows in discrete channels at the bed, often emerging from the glacier at a single portal. Glaciers act as natural storage reservoirs, retaining water in winter and releasing it in summer. Thus discharge is markedly seasonal. Furthermore, summer discharge is strongly diurnal, peaking in the early afternoon. On the braid-plain, beyond the glacier, marked fluctuations in discharge result in rapid continuous channel shifts. Meltwater also accumulates in ice-constrained lakes, including ice-dammed, proglacial, supraglacial, and subglacial types. Many of these are ephemeral, and those dammed by ice are particularly prone to catastrophic failure, resulting in outburst floods called *jökulhlaups*. Glacial meltwater is not only a powerful erosive agent, but is also responsible for the most important sedimentary facies in glacierized regions.

Morphological classification of glaciers

Glaciers range in size from ice masses only a few hundred meters across to the huge ice sheets that cover Antarctica and Greenland, and there are thus many different types (Hambrey, 1994; Bennett and Glasser, 1996; Benn and Evans, 1999). The largest features are *ice sheets*, arbitrarily defined as exceeding 50,000 km^2 in area. *Ice caps* form dome-like masses, burying the topography. Ice sheets and ice caps commonly have zones of faster flow called *ice streams*, which is where erosion is most strongly focused, and where debris transfer is most marked. *Highland icefields* also bury much of the topography, but are punctuated by upstanding peaks or *nunataks*. *Valley* and *cirque (corrie) glaciers* are constrained by topography in alpine terrain. Where coalescing ice streams in Antarctica meet the sea, floating slabs called *ice shelves* are formed. The extension of an individual ice stream into the sea results in an *ice tongue*. One of the key roles of the sedimentologist and geomorphologist is to reconstruct the morphological characteristics of former ice masses, to constrain the interpretation of past climates and sea-level response.

Glacier fluctuations

Glaciers, together with their sediments and landforms, are important sources of information about environmental change. Reconstructions of pre-Pleistocene ice masses are based on correlation of strata (especially those of marine origin) and erosion surfaces over wide areas. Reconstruction of Late Quaternary glacier variations are usually based on the distribution of glacigenic landforms and surficial sediments, including moraines and trimlines, coupled with radiometric, lichenometric, or dendrochrological dating. Reconstructions of more recent historical fluctuations rely on repeated measurements of glacier frontal positions, aerial photograph interpretation, or historical records such as maps and reports of early expeditions to glacierized areas. Most of the world's mountain glaciers are currently in recession from their historical maxima, attained during the Little Ice Age between AD 1600 and AD 1850. Glaciers do not generally respond instantaneously to climatic change. The lag-time varies from just several years for dynamic alpine glaciers to hundreds or thousands of years for the polar ice sheets.

Glacial processes

Erosion

Glaciers and ice sheets are agents of net erosion because they flow outwards from a source area toward their margins. Glacial erosion occurs by three main processes: *glacial abrasion*, *glacial plucking* (or *quarrying*), and *glacial meltwater erosion* (Hambrey, 1994; Bennett and Glasser, 1996; Benn and Evans, 1998), and gives rise to a wide range of landforms (e.g., Figure G6). Since glacial erosion is the starting point for debris entrainment and transport by glaciers and ice sheets, an understanding of this topic is essential if we are to understand the sedimentary products of ice masses.

Glacial abrasion is the process by which particles entrained in the basal layer of a glacier are dragged across the subglacial surface. The scratching and polishing associated with glacial abrasion tends to create smoothed rock surfaces, often with striations or grooves.

Glacial plucking (or *quarrying*) is the process whereby a glacier removes and entrains large fragments of its bed. The processes of rock fracture are controlled by the density, spacing and depth of pre-existing joints in bedrock, together with the stresses applied by the glacier. Fracturing of bedrock is also be aided by the presence of meltwater beneath the glacier, where bedrock is loosened by subglacial water pressure fluctuations. Plucked blocks and fine material are then entrained in the basal ice by freezing-on (regelation), when surrounded by flowing ice, or when incorporated along thrusts.

Glacial meltwater erosion involves both mechanical and chemical processes. Mechanical erosion occurs primarily through fluvial abrasion by the transport of suspended sediment and gravel in traction within the meltwater. Locally, fluvial cavitation (the sudden collapse of bubbles within turbulent meltwater under high pressure) may also be important. Chemical erosion involves the removal in solution of rock and rock debris, especially in areas of carbonate lithology. Rates of chemical erosion beneath ice masses are high because of high flushing rates, the availability of large

Figure G6 The pyramidal peak or 'horn' is the product of glacial plucking on three or four sides of a mountain, here epitomized the Matterhorn (4,476 m) and Dent d'Hérens (4,171 m), Switzerland in this photograph. Rock faces provide an intermittent supply of angular supraglacial debris to the glaciers on the flanks of the mountain.

amounts of chemically-reactive rock flour, and the enhanced solubility of CO_2 at low temperatures.

To these three main processes we may add a fourth: the relatively poorly understood process of *subglacial sediment deformation*. Sediment deformation beneath ice sheets contributes to glacial erosion wherever there is a net removal of material from the bed during sediment deformation (Boulton, 1996).

Debris entrainment and transport

It is widely recognized that debris is incorporated mainly at the bed and on the surface of a glacier, and that the transport paths and textural character of glacial sediments are related to the dynamic and thermal characteristics of the glacier. In general terms, polythermal glaciers tend to carry a high basal debris load, and their surfaces rarely have a substantial cover of debris. In contrast, temperate glaciers, especially those in alpine terrain, normally carry little debris at the bed, but their surfaces commonly have extensive areas of supraglacial debris. The resulting sedimentary products can thus be used to infer the thermal and topographic regimes of the glacier.

The entrainment of debris at the bed, by a combination of pressure melting and refreezing (regulation), to create a basal debris layer is well-known from studies of both temperate and polythermal glacier margins (e.g., Knight, 1997) (Figure G7). Basally derived debris is subject to comminution at the ice/bedrock interface, and typically is dominated by clasts up to boulder-size, with subangular and subrounded shapes, faceted surfaces, and striations (if the lithologies are fine-grained) (Figure G8). Much clay- or silt-grade sediment is produced by abrasion. In contrast, debris that falls on the glacier surface as a result of frost shattering of the overlooking cliffs is generally very angular to angular, and the proportion of fines is small.

Ice deformation results in reorganization of the debris that is incorporated at the base and surface. In polythermal glaciers in Svalbard, three main modes of entrainment have been recorded in addition to regulation:

(i) Incorporation of angular rockfall material within the stratified sequence of snow, firn, and superimposed ice (Figure G9(A)). This debris takes an englacial path through the glacier, and becomes folded. Near the snout the debris emerges at the surface on the hinges and upper limbs of the folds, producing medial moraines that merge toward the snout. The resulting lines of debris are deposited on the proglacial area in the form of regular trains of angular debris, as the glacier recedes.

(ii) Incorporation of debris of both supraglacial and basal character within longitudinal foliation. This is particularly evident at the margins and at flow unit boundaries where the folding is commonly isoclinal. The folding has an axial planar relationship with the foliation. As these features melt out, "foliation-parallel ridges" form.

(iii) Thrusting, whereby debris-rich basal ice (including regulation ice) and rafts of subglacial sediments are uplifted into an englacial position, sometimes emerging at the ice surface (Figures G9(B), G10). This material is more varied than rockfall debris, and reflects the substrate lithologies: typically poorly sorted sediment (diamicton) with striated clasts, and sandy gravel. As thrusts melt out, groups of roughly aligned hummocks are formed.

Figure G7 Deformed debris-rich basal ice zone (strongly layered), overlain by clean glacier ice, and underlain by basal debris released from the ice. Taylor Glacier, Dry Valleys, Antarctica. A frozen lateral melt-stream is at the bottom of the picture; glacier flow is toward the right.

Figure G8 Striated and faceted clast from Neoproterozoic diamictite (originally deposited as a basal till), Nordaustlandet, Svalbard.

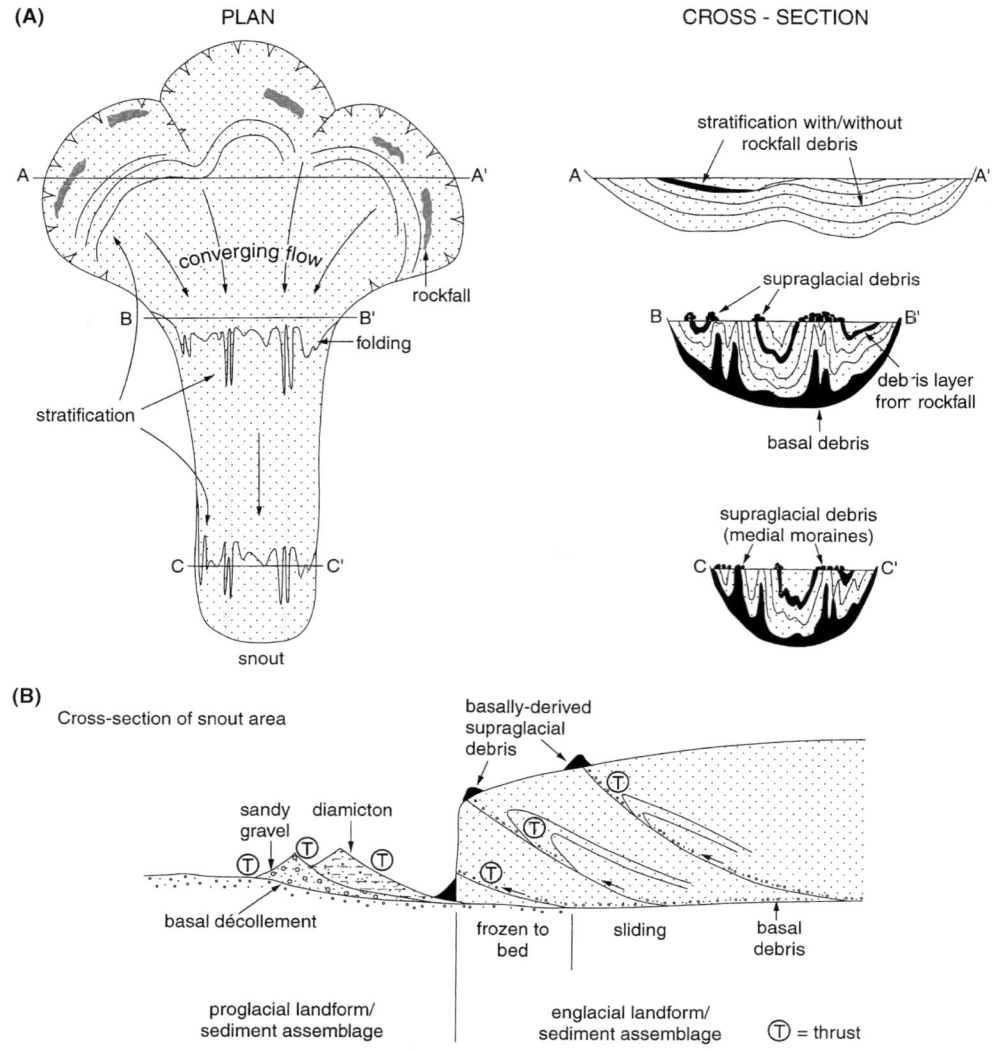

Figure G9 Conceptual models of debris entrainment. (A) Folding of supraglacial debris with stratified ice and of basal debris as a result of converging flow from multiple accumulation basins (simplified from Hambrey et al., 1999). (B) Thrusting of basal, subglacial, and proglacial debris in a polythermal glacier as found in the maritime High-Arctic (after Hambrey et al., 1997).

Of these mechanisms, only thrusting is well-known from temperate glaciers, but usually the amount of debris involved is insufficient to leave a landform imprint. The other mechanisms are likely to exist in temperate glaciers, but have not yet been documented, although it is well-known that shearing at the margins of alpine valley glaciers, and at the confluences of two flow units, incorporate debris.

Deposition

Glacial deposition involves the release of debris that has been transported on or within glacier ice. Debris is modified during transport primarily by basal processes (e.g., abrasion and quarrying during intra-clast collision, subglacial sediment deformation), and by water in subglacial, englacial, and supraglacial stream channels. Debris that follows a passive transport path (supraglacially or englacially) tends to retain its primary characteristics.

Sediment may be deposited directly beneath the glacier or at its margins, or it can be transported significant distances from the glacier itself by other agents such as rivers or by iceberg calving. During release from the ice, numerous glacier-related processes, including reworking in marginal streams and lakes, debris flows and eolian activity, may modify sediment. Many glacigenic sediments may be related to specific glacial environments. For example, the temperate terrestrial glacier system is commonly regarded as being dominated by a mixture of basal (actively transported) and supraglacial (passively transported) sediment, with a strong element of glaciofluvial modification upon release (Figure G11(A)). Glaciers terminating in fiords produce a facies association that is also dependent on thermal regime. Temperate and polythermal

Figure G10 Thirty meter-high terminal cliff of advancing Thompson glacier, Axel Heiberg Island, Canada, showing debris-laden thrusts, and ice-debris apron in front. Note person for scale, slightly left of base of waterfall.

glaciers, such as those in Alaska and Greenland, respectively (Figure G11(B)), not only provide basal and supraglacial debris inputs, but also sediments released from subglacial streams emanating at or below water level close to the ice margin, sediment released from suspension over the whole depositional basin, and iceberg-rafted debris. The resulting facies associations reflect thermal regime, which primarily controls the balance between direct glacial deposition and fluvial inputs. For the coldest glaciers, terminating as ice shelves on the continental shelf, as in Antarctica today, direct glacial deposition is restricted to the grounding-line, the volume of meltwater sediments is limited, whereas biogenic sedimentation in the form of diatom ooze may become dominant (Figure G11(C)). Indeed, rather than releasing sediments, some ice shelves accrete saline ice at their base, trapping sediment, which is released only when the tabular icebergs, calved from the ice shelf disintegrate.

Glaciotectonism

Glaciotectonic deformation is now recognized as a widespread phenomena. Not only is deformation associated with internal processes, such as folding and thrusting, as noted above, but it is also transmitted subglacially and proglacially (e.g., Maltman et al., 2000). Glaciotectonic deformation operates in any topographic setting, both during advancing and recessional phases, and involves all types of material, including frozen, saturated and dry unconsolidated sediments, as well as bedrock. Deformation may detach blocks of rock and sediment, occasionally hundreds of meters across, incorporating them into the ice by thrusting, or pushing them in front of the glacier. Faults and brecciated zones are common in such materials. Sediments may also be deformed in a ductile fashion, especially if wet and fine-grained. Beneath ice sheets, deformation may affect sediment and bedrock to depths of several hundred meters.

Reworking of glacigenic sediments

Glacigenic sediments are typically subject to syndepositional and post-depositional modification by fluvial, mass-movement, and eolian processes. In terrestrial settings, fluvial modification by proglacial streams is particularly important in temperate climates, and many temperate glaciers terminate at the head of large *outwash* or *sandur* plains composed almost entirely of reworked glacial sediments (Figure G12). Resedimentation by mass-movement processes is common in ice-cored terrain where water, released by the melting of buried glacier ice or permafrost, mixes with sediment to create *glacigenic sediment flows*. Eolian modification involves the redistribution by wind of smaller, readily entrained particles, particularly in more arid areas, creating *deflation surfaces* and *ventifacts*. The extent to which each of these processes operates is controlled to a great extent by the local topographic, meteorological and climatological conditions. Resedimentation by subaquatic gravity flows is also important in glaciomarine and glaciolacustrine environments, where large amounts of sediment may accumulate on relatively steep ice-contact slopes, which become unstable during recession.

Chemical processes

As noted above, glacial meltwater is often enriched in solutes. Mineral-rich waters may leave thin deposits of calcite, silica or iron oxides, especially on the lee side of bedrock obstacles.

Depositional landforms

Glaciers and ice sheets produce a huge variety of depositional landforms. These are commonly grouped according to their origin into ice-marginal and subglacial landforms (Hambrey, 1994; Bennett and Glasser, 1996; Benn and Evans, 1998).

Ice-marginal landforms. Ice-marginal landforms can be produced by advancing, static or receding ice margins, as well as during seasonal fluctuations of an ice front. They may be deposited directly from glacier ice or be composed of facies previously deposited by other processes (Figure G13). Ice-marginal landforms are commonly used to reconstruct changes in glacier size, morphology and extent over time.

Glaciotectonic moraines encompass a broad range of different types of moraine formed by deformation of ice, sediment and rock. *Push moraines* (both seasonal and annual) are formed when a glacier flows into sediment and bulldozes material into a ridge. Other types of glaciotectonism include: thrust-block moraines, where large slabs of material are entrained and partially overridden; englacial thrust moraines, where material is elevated along thrusts as a result of longitudinal compression near the ice margin; and proglacial thrust moraines where compressive stresses from the glacier propagate into the foreland.

Dump moraines are formed at stationary or near-stationary ice fronts where debris accumulates along the margin or front of a glacier to form a ridge of sediment. The size of these features is controlled by ice velocity, debris concentration and the rate (if any) of marginal recession.

Ablation moraines (sometimes referred to as *ice-cored moraines*) form wherever ice melt is retarded beneath a cover of supraglacial debris (Figure G13). This supraglacial debris may be derived from rockfalls and avalanches in the accumulation area, often arranged in flow-parallel medial moraines, or it

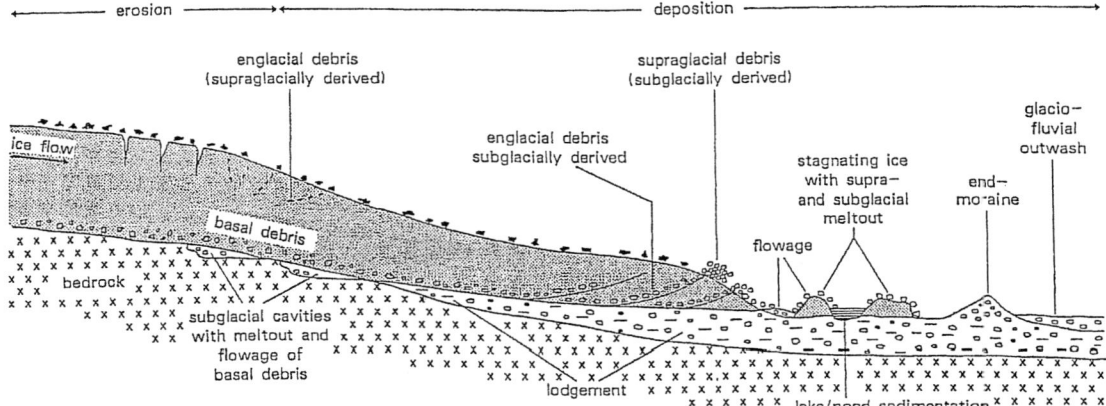
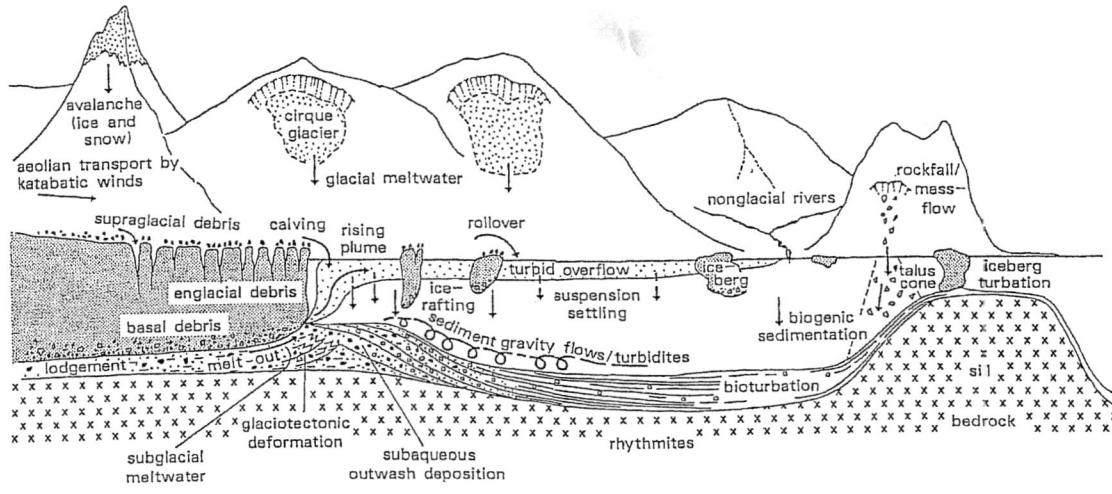
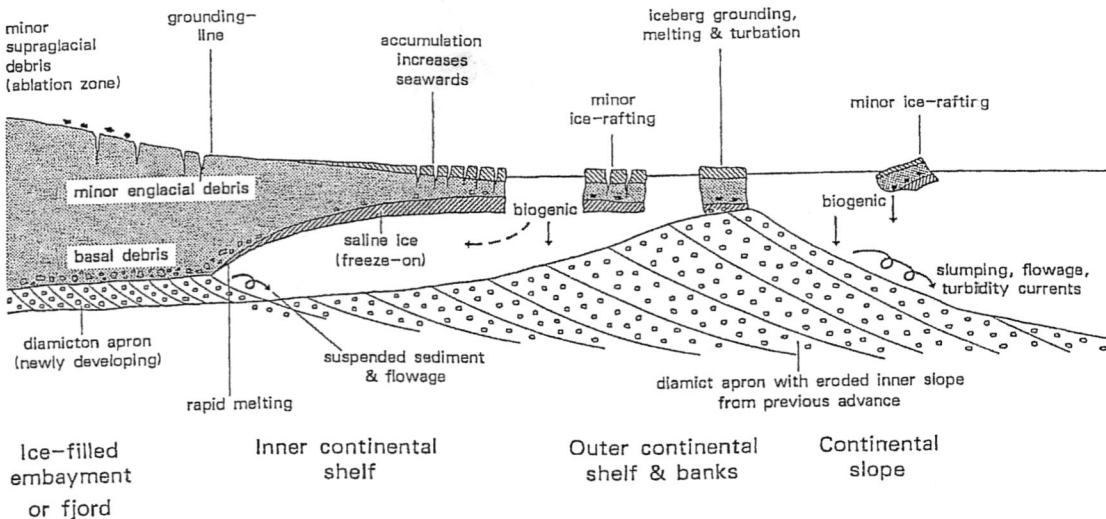

Figure G11 Depositional processes and products in a selection of glacial environments. (A) Terrestrial. (B) Fjord with temperate tidewater glacier. (C) Ice shelf with freeze-on of marine ice, typical of the Antarctic continental margin (from Hambrey, 1999; published with permission of *Terra Antartica Publications*, Siena).

Figure G12 Braided river system emanating from debris-covered Casement Glacier (left), southern Alaska. Small pits on the outwash plain are kettle holes arising from slow melting of buried ice blocks.

Figure G13 Ice-marginal landforms, especially ablation moraine, comprising a mixture of supraglacially and basally derived debris, Vadrec da Morteratsch, Switzerland.

may be composed of subglacial or englacial debris elevated to the ice surface by folding, thrusting or upward-directed flow lines in the ablation area. If the cover of insulating debris is irregular, as is commonly the case, irregular ridges and mounds of debris, often known as *hummocky moraines*, will develop as ablation proceeds. Although initially large and with a pronounced morphology, ablation moraines reduce dramatically in size as the ice-core melts and material is subject to debris flowage.

Outwash fans and *outwash plains* (or *sandar*, plur; *sandur*, sing.) are formed as glacial meltwater emerges from the glacier and sediment is deposited at or beyond the ice margin. Outwash fans form at stationary ice margins where meltwater is concentrated at a particular point for a length of time. Outwash plains are much larger features, formed where individual fans merge away from the glacier to create a braided river facies association (Figure G12). Characteristics of the glaciofluvial environment are braided river channels with rapidly migrating bars, terraces, frequent channel avulsions and the formation of *kettle holes* where sediment is deposited over buried ice.

Kame terraces are formed when sediment is deposited by meltwater flowing laterally along an ice margin. *Kames* are more fragmentary features, formed in a similar manner, but often in ice-walled tunnels and against steep valley sides. *Kame-and-kettle topography* is the term used to describe the landform-sediment assemblage often found on glacier forefields where there was formerly a high proportion of buried ice.

Subglacial landforms (sometimes referred to as *subglacial bedforms*) are produced beneath actively flowing ice. They provide information about former subglacial conditions, including ice-flow directions, thermal regime and paleohydrology. The most common of these is a family of ice-molded landforms, all of which are parallel to ice-flow. *Flutes* are low (typically <3 m), narrow (<3 m), regularly spaced ridges that are rarely continuous for less than 100 m. *Megaflutes* are taller (>5 m), broader and longer (>100 m) than flutes. *Mega-scale lineations* are much larger (tens of kilometers in length and hundreds of meters in width) that are often only visible when viewed on satellite imagery. *Drumlins* are typically smooth, oval-shaped or elliptical hills composed of a variety of glacigenic sediments. They are generally between 5 m and 50 m high and 10 m to 3,000 m in length. Drumlins normally have length-to-width ratios of less than 50. Their steeper, blunt end often faces up-ice and they are often found in large groups known as *drumlin swarms*. The origin of drumlins is unclear, but they have been ascribed to subglacial deformation, to lodgement, to the melt out of debris-rich basal ice and to subglacial sheet floods.

Ribbed moraines (also known as *Rogen moraines*) are large, regularly and closely spaced moraine ridges consisting of glacigenic sediment. They are often curved or anastomosing, but their general orientation is transverse to ice flow. They often show drumlinoid elements or superimposed fluting. Ribbed moraines may represent the fracturing of a pre-existing subglacial till sheet at the transition from cold- to warm-based, presumably during deglaciation.

Geometric ridge networks and *crevasse-fill ridges* are subglacial landforms that are not generally ice-molded. These features, composed of subglacial material, are low (1–3 m high) ridges that, when viewed in plan, show a distinct geometric pattern. The traditional explanation for these features is that they form by the squeezing of subglacial material into basal crevasses or former subglacial tunnels, commonly during surges. Alternatively, geometric-ridge networks also form beneath glaciers as a result of the intersection of foliation-parallel ridges and englacial thrusts.

Eskers are glaciofluvial landforms created by the flow of meltwater in subglacial, englacial or supraglacial channels. They are usually sinuous in plan and composed of sand and gravel. Some eskers are single-crested, whilst others are braided in plan. *Concertina eskers* are deformed eskers, created by compression beneath overriding ice.

Bathymetric forms resulting from glacial processes

Erosional forms

Various erosional phenomena, mainly associated with grounded ice or subglacial meltwater are found in marine settings. The larger scale forms are filled by sediment and may be recognizable in seismic profiles (e.g., Anderson, 1999). *Submarine troughs* are found on continental shelves, and are genetically equivalent to fiords and other glacial troughs, but are generally much broader. The largest occur in Antarctica where they attain dimensions of over 400 km in length, 200 km in width and 1,100 m in depth. They are formed by ice streams and, where two streams merge, an *ice-stream boundary ridge* is formed. Steep-sided channels a few kilometers wide, carved out by subglacial meltwater and subsequently filled by sediment, are known as *tunnel valleys*. These are well-known from the NW European continental shelf around Britain, the Scotian Shelf off Canada, and in Antarctica. Icebergs can also cause considerable erosion as they become grounded on the seafloor. Large tabular bergs can scour the bed of the sea for several tens of kilometers, leaving impressions up to 100 m wide and several meters deep. *Slope valleys* are groups of gullies forming a dendritic pattern, and develop just beyond the ice margin, on the continental slope, as a result of erosion by sediment gravity flows emanating from sediment that accumulated at the ice margin. On continental shelf areas, where the ice repeatedly becomes grounded, and then releases a large amount of rain-out sediment, alternations of diamicton and *boulder pavements* may be observed. The pavements build up by accretion of boulders around an obstacle, by subglacial erosion, or as a lag deposit from winnowing by bottom currents.

Depositional forms

The morphology and sediment composition of subaquatic features, particularly in fjords and on continental shelves are less well-known than their terrestrial counterparts, but major strides have been made in identifying such features in the last two decades. As on land, depositional assemblages reflect the interaction of a wide range of processes.

Ice-contact features form when a glacier terminus remains quasi-stationary in water, particularly in fiords (Powell and Alley, 1997) (Figure G14). *Morainal banks* form by a combination of lodgement, meltout, dumping, push and squeeze processes, combined with glaciofluvial discharge; poorly sorted deposits are typical of such features. *Grounding-line fans* extend from a subglacial tunnel that discharges meltwater and sediment into the sea, and are typically composed of sand and gravel. Developing out of grounding-line fans are *ice-contact deltas* that form when the terminus

Figure G14 Tidewater glaciers in Harriman Fiord, southern Alaska. Glaciers such as this deliver large amounts of sediment, especially via subglacial streams to the marine environment, and sediment accumulation rates may be tens of meters per year.

remains stable long enough for sediment to build up to the surface of the fiord. Where a glacier becomes disconnected from the water body, alluvial sediments may prograde to form *fluviodeltaic complexes*, again dominated by sand and gravel. In addition to these large-scale forms (measured in 100s of meters), there are small-scale features (measured in meters or less), found particularly on beaches, such as iceberg tool-marks, ridges, depressions from melting of buried ice, bounce-marks, chattermarks and roll-marks. Icebergs can also churn up submarine sediment, particularly on shoals, producing iceberg turbates.

Depositional forms on continental shelves are best known from a combination of deep drilling and seismic profiling. Some, as in Alaska, are simply larger scale analogues of fiordal features, such as delta-fan complexes, but others form under floating tongues that are typical of the colder ice of the polar regions. These include subglacial deltas, till tongues, diamicton aprons and the immense (up to 400 km wide) trough-mouth fans. Other features are similar to those on land, including shelf moraines, flutes and transverse ridges.

Terminology and classification of glacigenic sediments

The terminology used to describe glacigenic sediment has always been in a state of flux, and the evolution of terms may be linked to progressive improvement of understanding glacial processes. For an objective study of sedimentary facies, a non-genetic classification of poorly sorted sediments is required, before process-related terms such as "till" are used. The terms *diamicton* (unlithified), *diamictite* (lithified) and *diamict* (both) are now well established for "a non-sorted or poorly sorted terrigenous sediment that contains a wide range of particle sizes" (Flint *et al.*, 1960). However, this definition hides a vast range of textures, and the literature abounds with conflicting ideas about what constitutes a "diamicton". A textural classification, based on the relative proportions of sand and gravel, has subsequently been developed (Hambrey, 1994) (Table G10).

The genetic terminology is perhaps even more confused. First, we need to define some widely used general terms:

Glacigenic sediment: "of glacial origin"; the term is used in the broad sense to embrace sediments that have been influenced by glacier ice.

Glacial debris: material being transported by a glacier and thus in contact with glacier ice.

Glacial drift: all rock material in transport by glacier ice, all deposits released by glaciers, and all deposits predominantly of glacial origin deposited in the sea by icebergs, or from glacial meltwater.

For a more detailed process-based classification of glacigenic sediments, the recommendations of a commission of the International Quaternary Association are followed (Dreimanis, 1989; Hambrey 1994). Although somewhat out-of-date, this is the only scheme that has achieved a wide measure of consensus (Table G11). In this genetic classification, the key term is *till* (*tillite* where lithified). *Primary till* is defined as 'an unsorted sediment with a wide range of grain sizes deposited directly from glacier ice, without subsequent disaggregation and flow". Secondary tills are the products of resedimentation of glacial debris that has already been deposited by the glacier, with little or no sorting by water. Within these categories several varieties of till have been documented, although these represent end members in a continuous spectrum of depositional types (Dreimanis, 1989). Benn and Evans (1998: 386–399) have provided a concise review of the characteristics of tills, and the main types (with their textural characteristics) are:

Meltout till: deposited by slow release of glacial debris from ice that is not moving. Typically, meltout tills inherit the

Table G10 Textural classification of poorly sorted sediments (from Hambrey, 1994; modified from Moncrieff, 1989). In order to derive sediment name, percent of gravel is first estimated (horizontal axis) and then proportion or percent of sand in the matrix (vertical axis)

	Percent gravel (<2 mm) in whole sediment						
Mud <0.06 mm	Trace (<0.01)	<1%	1–5%	5–50%	50–95%	>95%	0
0.11	MUD(STONE)	MUD(STONE) with dispersed clasts	clast-poor muddy DIAMICT(ON/ITE)	clast-rich muddy DIAMICT(ON/ITE)	Muddy GRAVEL/BRECCIA/ CONGLOMERATE		
	Sandy MUD(STONE)	Sandy MUD(STONE) with dispersed clasts					20
Sand/mud ratio of matrix 1			clast-poor intermediate DIAMICT(ON/ITE)	clast-rich intermediate DIAMICT(ON/ITE)	GRAVEL/ BRECCIA/ CONGLOMERATE		40 Percent sand in matrix
	Muddy SAND(STONE)	Muddy SAND(STONE) with dispersed clasts					60
			clast-poor sandy DIAMICT(ON/ITE)	clast-rich sandy DIAMICT(ON/ITE)			80
9	SAND(STONE)	SAND(STONE) with dispersed clasts	Gravelly SAND(STONE)		Sandy GRAVEL/BRECCIA/ CONGLOMERATE		
Sand 2.0–0.06 mm				50% gravel			100

Table G11 Genetic classification of terrestrial glacigenic sediments (adapted from Dreimanis, 1989). The vertical columns are independent of each other and no correlation horizontally is implied. Not all combinations are feasible

Release of glacial debris and its deposition or redeposition			Depositional genetic varieties of till		
I Environment	II Position	III Process	IV By environment	V By position	VI By process
Glacioterrestrial	Ice-marginal Frontal Lateral	A. Primary Melting out Lodgement Sublimation Squeeze-flow Subsole drag	Terrestrial non-aquatic till	Ice-marginal till Supraglacial till Subglacial till	A. Primary till Meltout till Lodgement till Sublimation till Deformation till Squeez-flow till
	Supraglacial				
	Subglacial	B. Secondary Gravity flow Slumping Sliding and rolling			B. Secondary till Flow till
	Substratum				

granulometry and particle-size distribution of the parent ice, so may be massive or layered, commonly with a strong clast orientation parallel to flow.

Lodgement till: deposited by "plastering" of glacial debris from the sliding base of a moving glacier by pressure melting or other mechanical processes. This process also normally produces a massive diamicton, although sometimes with anastomosing planar discontinuities resulting from shear (Figure G15(A)). A consistent flow-parallel clast orientation is evident in lodgement tills.

Basal till: embraces both the above.

Sublimation till: is released by sublimation direct from debris-rich ice in sub-zero temperatures. Texturally, this is a diamicton, but is a rare deposit, only found in a few arid parts of Antarctica.

Deformation till: comprises weak rock or unconsolidated sediment that has been detached by the glacier at its source and subsequently deformed or disaggregated, and some foreign material admixed. Texturally, this is variable, often breccia-like; rafts of rock or sediment may be many meters in length.

Flow till: till which has been remobilized by gravity flowage, although better known as a *glacigenic sediment flow*. A degree of sorting into gravel-rich and gravel-poor components accompanies this process, and the clast fabric is weak and highly variable.

The above discussion refers primarily to terrestrial glacial sediments. There is no agreed genetic classification for glaciomarine sediments. The following terms, however, are commonly used:

Ice-proximal glaciomarine sediment: sediment that is released either from floating basal glacier ice or by continuous rain-out from icebergs, without subsequent winnowing by currents and waves; this embraces the term waterlain till which is no longer favored. Typically, this is a diamicton, either massive where

Figure G15 Some typical lithofacies from glacial environments: (A) Clast-rich massive diamictite interpreted as lodgement till, overlain successively by stratified diamictite of glaciolacustrine origin, and more massive diamictite, Sirius Group (Neogene), Shackleton Glacier, Transantarctic Mountains. (B) Clast-poor weakly stratified diamictite, interpreted as proximal glaciomarine sediment, Wilsonbreen Formation (Neoproterozoic), Backlundtoppen, Spitsbergen. (C) Dropstone-laminite, interbedded with diamictite from Paleoproterozoic Gowganda Formation, Whitefish Falls, Ontario; interpreted as cyclopsams and cyclopels (tidal), and subaquatic debris flows. (D) Thin section of laminated clay (dark) and sand/silt (light) with dropstone, Neoproterozoic Petrovbreen Member, NE Spitsbergen; these couplets are typical of glaciolacustrine sediments, each pair formed in one year (a varvite).

iceberg rainout is dominant and continuous, or has a faint wispy stratification (Figure G15(B)). Clast fabrics are weak or random.

Ice-distal glaciomarine sediment: sediment that is released from icebergs, but subject to winnowing, and admixed with other marine sediment, including biogenic components. Texturally, this facies is sandy mud or muddy sand with dispersed stones, and in modern environments often rich in diatoms.

Iceberg turbate: sediment deposited on the seafloor that is subsequently reworked by grounded icebergs. The end product is commonly a massive diamicton with a heavily grooved upper surface.

Cyclopels (silt/mud couplets) and *cyclopsams* (graded sand/mud couplets): rhythmically laminated sediments derived from turbid overflow plumes from subglacial discharge, especially in fiords, typically producing two couplets a day according to the tidal cycle. If deposition is accompanied by ice-rafting, dropstone structures may be evident (Figure G15(C)).

Glaciolacustrine facies are similar to those in glaciomarine environments. In addition, lakes commonly have rhythmically laminated sands and muds called *varves* (or *varvites* if lithified), similar to cyclopsams and cyclopels, but with each couplet forming over one year. Sand laminae represent input to the lake during summer, and silt/clay represent settling out from suspension in winter. Iceberg debris, in the form of dropstone structures, may disrupt the laminae (Figure G15(D)). Other laminated sediments in lakes or the sea may be deposited from turbidity currents and also include dropstone structures. Differentiation of the processes responsible for the formation of laminated facies is challenging.

Facies analysis of glacigenic sediments

The facies approach

Compared with other sedimentologists, glacial geologists have been rather slow in adopting the *facies* approach. However, it is now recognized that for objective treatment this approach must be used.

The first step is to describe the lithofacies found in a vertical section or core, using a wide range of criteria (texture, sedimentary structures, bed geometry and boundary relations, clast characteristics), and avoiding genetic terms such as "till". Some authors use formal lithofacies codes (Eyles *et al.*, 1983), but this is thought by others (e.g., Hambrey, 1994; Miller, 1996) to be too inflexible and include interpretative elements. Here, we advocate that workers devise their own scheme to suit the distinctiveness of their sections. The next step is to group the lithofacies into facies associations that contain a set of attributes that will allow the depositional setting to be determined, such as glacioterrestrial or glaciomarine. On a regional scale, given sufficient three-dimensional exposure on land, or extensive seismic and borehole stratigraphy offshore, derivation of the *facies architecture* is a desirable aim, allowing one to address larger issues, such as basin evolution, ice sheet-scale fluctuations and sea-level changes. Representative facies associations from terrestrial, lacustrine, and glaciomarine settings are illustrated in Figure G16.

Glacial sediment/landform associations

The concept of *glacial sediment/landform associations* (the "*landsystems*" approach) is relatively new to glacial geology. The approach rests on the assumption that at the large-scale it is possible to identify areas of land with common attributes, distinct from the surrounding areas, that can be related to the processes involved in their development (Benn and Evans, 1998). The landsystems approach was pioneered for glaciated terrain by Boulton and Paul (1976) in an attempt to identify tracts of land created by similar till-forming processes and therefore of similar geotechnical properties. The *glacial landsystem* concept was then applied to former ice-sheet beds. Originally, three distinct landsystems were defined:

The subglacial landsystem, in which the dominant glacial sediment/landform associations are formed at the glacier bed.

The supraglacial landsystem, in which the dominant glacial sediment/landform associations are largely composed of a drape of supraglacial debris.

The glaciated valley landsystem, in which the dominant glacial sediment/landform association is that formed by mountain glaciation.

As the concept has developed, other landsystems have been added to this simple list. These include landsystems formed in proglacial and glaciomarine environments, as well as those found in settings where surge-type glaciers are common. It is also possible to relate landsytems to glacier thermal regime. Thus we can identify landsystems related to warm-based ("temperate"), polythermal and cold-based ("polar") glacier margins. A criticizm of the landsystems approach is that it takes no account of spatial patterns of ice-sheet erosion and deposition, and how these vary during ice-sheet growth and decay. Elucidating these links remains a challenge for glacial geologists.

Glacial sediments in the geological record

Identification of glaciation in the geological record

The recognition of glacigenic sediments in the rock record is of fundamental importance to paleoclimatological and paleoenvironmental reconstruction. Even where exposure is good, evidence for glaciation may be equivocal as, for example, in alpine regions where mass-movement and fluvial processes may overprint direct evidence for glacial deposition, or in those areas where tectonic deformation has overprinted depositional features. However, detailed analysis may yield sufficient criteria that, taken together, could form the basis of a glacial interpretation. Table G12 lists the principal criteria for establishing the glacial origin of a diamict-bearing succession.

Preservation potential of glacigenic successions

Tectonic setting is the principal control on whether a glacial succession is preserved. Eyles (1993) has provided a detailed review of the wide range of tectonic settings suitable for preservation. Intracratonic basins, rifts, forearc and backarc basins, and subsiding continental shelves are particularly suitable receptacles for preservation of a glacial record, especially if those sites are first buried and later uplifted and exposed. Except for deposits of Quaternary age, land areas are less likely to preserve glacial sediment, as continued erosion tends to remove traces of glaciation. There are, however, notable exceptions in the geological record.

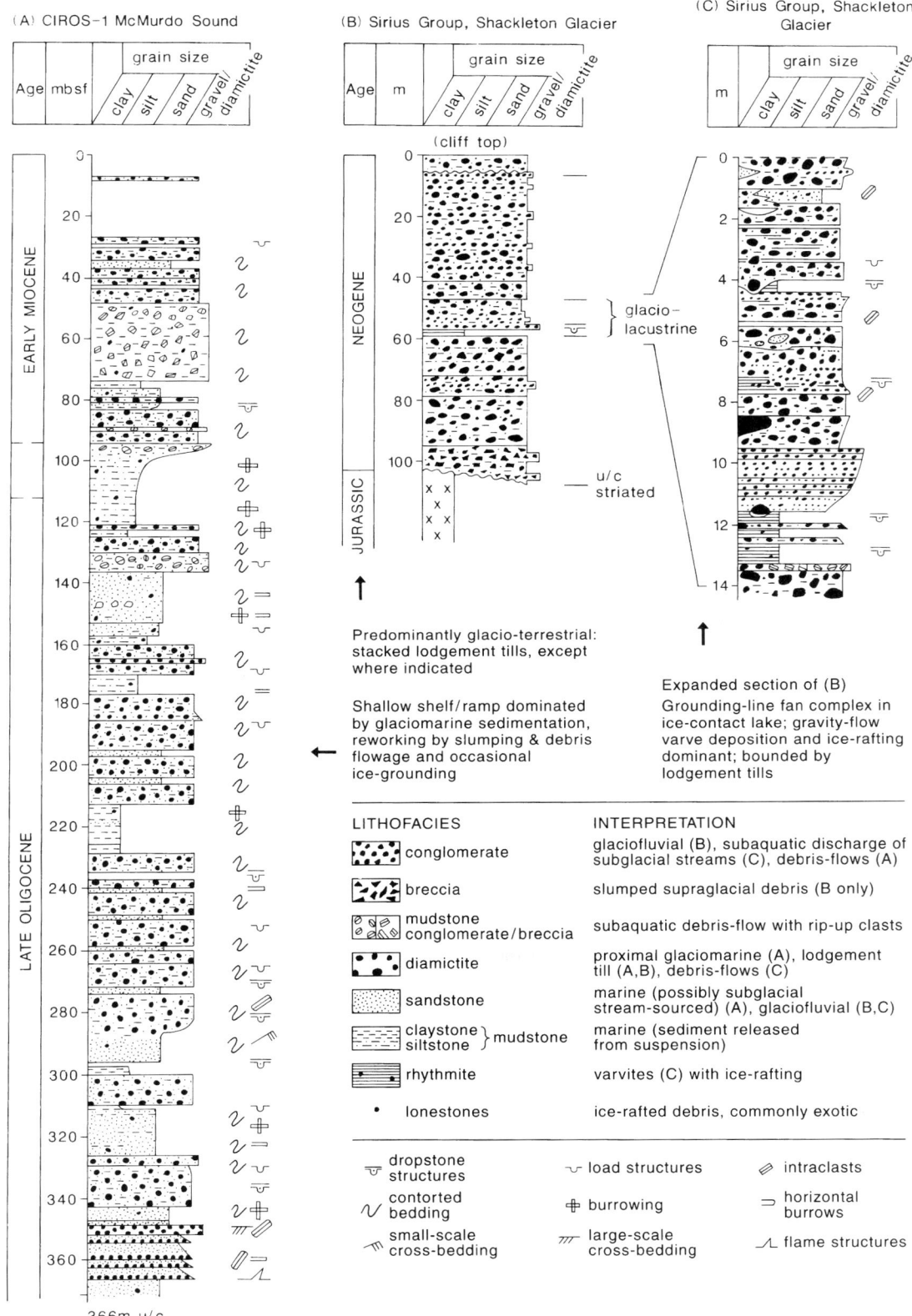

Figure G16 Characteristic facies associations from: (A) Late Oligocene-Miocene proximal continental shelf facies association, upper part of CIROS-1 drill-core, McMurdo Sound, Antarctica, illustrating deposition in a subsiding rift-basin, with occasional ice-grounding events (data from Barrett, 1989). (B) A terrestrial setting dominated by repeated lodgement till deposition, represented by stacked diamictites; Sirius Group (Neogene), Shackleton Glacier, Transantarctic Mountains (cf. Figure G15(A)). (C) Glaciolacustrine association in middle of section shown in B, illustrating ice-proximal deposition dominated by slumping and debris-flowage of rain-out diamicts and varve-sedimentation.

Table G12 Criteria for establishing a glacial origin of diamict-bearing successions for different environments (from Hambrey, 1999)

(a) Evidence for terrestrial glaciation

Abraded surfaces
  Grooved striated and/or polished surfaces
  Crescentic gouges, friction cracks, chattermarks
  Striated boulder pavements
  Roches moutonnées
  Nye channels, "p-forms"

Diamict and muddy boulder gravel beds with:
  Little or no stratification
  Irregular thickness, typically several m (but may be up to 10s of m)
  Preferred clast orientation
  lenses of sand/gravel (glaciofluvial)
  Shear structures parallel to depositional surface

Depositional landforms
For example, moraines, eskers, but rarely recognisable in pre-Quaternary successions

Association with extensive sheets of sand and gravel—inferred glaciofluvial origin

(b) Evidence of glaciomarine/glaciolacustrine deposition
Massive or stratified diamicts, often tens to hundreds of m thick, with gradational boundaries
Dropstones in stratified units
Random clast fabric
Slight sorting or winnowing at top of beds, to give lag deposits
Association with *in situ* fossils
Association with rhythmites (varves in lakes; tidal rhythmites in fjords; turbidites in both)
Association with resedimented deposits (subaquatic gravity flows)

(c) Evidence common to both environments
Clasts
Range in size up to a few meters
Range in lithologies, reflecting terrain over which ice flowed
Constant mix of clasts over wide area common
Range in shape from very angular to rounded (e.g., subangular to subrounded predominant in subglacial debris), sometimes influenced by prior transport history
Surface features include facets and striations from transport within the basal ice zone, and flat-iron or bullet-nosed shapes from persistent glacial transport
Fragile clasts survive well
Calcareous crusts in some deposits

Grains
Surface of quartz grains show typical fracture characteristics
Surface of quartz grains show chattermarks
Unstable (ferromagnesian) minerals remain relatively unweathered

(d) Other evidence of cold climate
Ice wedge casts
Sorted stone circles, polygons and stripes
Solifluction lobes
Association with lithified loess (loessite)

Earth's glacial record

Far more is known about the Quaternary glaciations than all previous glaciations put together, even though some of those earlier events were equally dramatic in terms of scale. Since the International Geological Correlation Programme compiled an inventory of all known pre-Quaternary glacigenic sequences (Hambrey and Harland, 1981), several detailed syntheses have been undertaken (e.g., Eyles, 1993; Deynoux *et al.*, 1994; Crowell, 1999). For the Quaternary Period there is a vast literature to chose from.

The oldest known glacigenic sediments are of late Archean age (2,600–3,100 Ma) from South Africa. Extensive Paleoproterozoic (2,500–2,000 Ma) tillites are known from South Africa, Australia and Finland. The most prolonged and globally extensive glacial era took place in Neoproterozoic time (1,000–550 Ma). Evidence for glaciation occurs on all continents, leading to the "snowball earth" hypothesis that envisages the Earth being totally ice-covered. Sporadic Early Proterozoic glacial events are best represented by the late Ordovician to early Silurian deposits and erosional features of Gondwana, notably in Africa. The most prolonged and extensive phase of glaciation during the Phanerozoic Eon spanned about 90 Ma of the Carboniferous and Permian Periods, affecting all Gondwana continents. Finally, after a phase of global warmth, with little evidence of ice on Earth, the Cenozoic glaciations began in Antarctica at the Eocene/Oligocene transition (35 Ma). Northern Hemisphere glaciation followed, with minor ice-rafting events recorded in the North Atlantic from late Miocene time, until full-scale glaciations began in late Pliocene time (2.4 Ma).

State-of-the-art and future prospects

Considerable strides have been made in the last three decades in understanding glacial processes, which, when combined with rigorous facies analysis, have led to well-founded reconstructions of past glacial environments (see, e.g. compilations and reviews by Dowdeswell and Scourse, 1990; Hambrey, 1994; Menzies, 1995, 1996; Bennett and Glasser, 1996; Benn and Evans, 1998). Major advances in recent years have included: (i) elucidation of oceanographic and sedimentary processes in the glaciomarine environment; (ii) the recognition of the importance for ice dynamics of soft, deformable beds (particularly beneath large ice streams in the Antarctic); (iii) the role played by glacier surges in terrestrial glacier sedimentation; (iv) linking deformation in glacier ice, for example, folding and thrusting, to facies and landforms in terrestrial settings; and (v) establishing the relationship between glacier fluctuations and climatic change. These advances have provided the means for enhanced evaluation of Quaternary and ancient glacial sequences, but there remain considerable opportunities to undertake such work. Several important questions and controversies remain: what determines the nature of the glacier bed; how do conditions at the bed influence long-term dynamics and stability of ice sheets; what is the role of glaciotectonics in the generation of landforms; how may proximal glaciomarine sequences, that may not provide clear indications of water-depth, be reconciled with sequence stratigraphy; and how do sediment facies associations vary according to climatic and topographic regimes?

Michael J. Hambrey and Neil F. Glasser

Bibliography

Anderson, J.B., 1999. *Antarctic Marine Geology*. Cambridge: Cambridge University Press.
Barrett, P.J. (ed.), 1989. *Antarctic Cenozoic History from the CIROS-1 Drillhole, McMurdo Sound*. Wellington: DSIR Publishing, DSIR Bulletin, 245.

Benn, D.I., and Evans, D.J.A., 1998., *Glaciers and Glaciation*. London: Arnold.
Bennett, M.R., and Glasser, N.F., 1996. *Glacial Geology: Ice Sheets and Landforms*. Chichester: John Wiley & Sons.
Boulton, G.S., 1996. Theory of glacial erosion, transport and deposition as a consequence of subglacial sediment deformation. *Journal of Glaciology* **42**: 43–62.
Boulton, G.S., and Paul, M.A., 1976. The influence of genetic processes on some geotechnical properties of tills. *Journal of Engineering Geology*, **9**: 159–194.
Crowell, J.C., 1999. *Pre-Mesozoic Ice Ages: Their Bearing on Understanding the Climate System*. Geological Society of America, Memoir 192.
Deynoux, M., Miller, J.M.G., Domack, E.W., Eyles, N., Fairchild, I.J., and Young, G.M., 1994. *Earth's Glacial Record*. Cambridge: Cambridge University Press.
Dowdeswell, J.A., and Scourse, J.D. (eds.), 1990. *Glaciomarine Environments: Processes and Sediments*. London: Geological Society of London, Special Publication, 53.
Dreimanis, A., 1989. Tills, their genetic terminology and classification. In Goldthwait, R.P., and Matsch, C.L. (eds.), *Genetic Classification of Glacienic Deposits*. Rotterdam: Balkema, pp. 17–84.
Eyles, N., 1983. Glacial geology: a landsystems approach. In Eyles, N. (ed.), *Glacial Geology*. Oxford: Pergamon, pp. 1–18.
Eyles, N., 1993. Earth's glacial record and its tectonic setting. *Earth-Science Reviews*, **35**: 1–248.
Eyles, N., Eyles, C.H., and Miall, A.D., 1983. Lithofacies types and vertical profile models: an alternative approach to the description and environmental interpretation of glacial diamict and diamictites sequences. *Sedimentology*, **30**: 393–410.
Fairchild, I.J., Hambrey, M.J., Spiro, B., and Jefferson, T.H., 1989. Late proterozoic glacial carbonates in northeast Spitsbergen: new insights into the carbonate-tillite association. *Geological Magazine*, **126**: 469–490.
Flint, R.F., Sanders, J.E., and Rodgers, J., 1960. Diamictite: a substitute term for symmictite. *Geological Society of America Bulletin*, **71**: 1809–1810.
Hambrey, M.J., 1994. *Glacial Environments*. Vancouver: University of British Columbia Press, and London: UCL Press.
Hambrey, M.J., 1999. The record of earth's glacial history during the last 3000 Ma. In Barrett, P.J., and Orombelli, G. (eds.), *Geological Records of Global and Planetary Changes*. Terra Antartica Reports, 3, pp. 73–107.
Hambrey, M.J., and Harland, W.B., 1981. *Earth's Pre-Pleistocene Glacial Record*. Cambridge: Cambridge University Press.
Hambrey, M.J., Huddart, D., Bennett, M.R., and Glasser, N.F., 1997. Dynamic and climatic significance of "hummocky moraines": evidence from Svalbard and Britain. *Journal of the Geological Society of London*, **154**: 623–632.
Hambrey, M.J., Bennett, M.R., Dowdeswell, J.A., Glasser, N.F., and Huddart, D., 1999. Debris entrainment and transfer in polythermal valley glaciers. *Journal of Glaciology*, **45**: 69–86.
Knight, P.G., 1997. The basal ice layer of glaciers and ice sheets. *Quaternary Science Reviews*, **16**: 975–993.
Maltman, A.J., Hubbard, B., and Hambrey, M.J. (eds.), 2000. *Deformation of Glacial Materials*. Geological Society of London, Special Publication, 176.
Menzies, J. (ed.), 1995. *Modern Glacial Environments: Processes, Dynamics and Sediments*. Oxford: Butterworth-Heinemann.
Menzies, J. (ed.), 1996. *Past Glacial Environments: Sediments, Forms and Techniques*. Oxford: Butterworth-Heinemann.
Miller, J.M.G., 1996. Glacial sediments. In Reading, H.G. (ed.), *Sedimentary Environments: Processes, Facies and Stratigraphy*, 3rd edn., Oxford: Blackwell Science, Chapter 11, pp. 454–484.
Moncreiff, A.C.M., 1989. Classification of poorly sorted sediments. *Sedimentary Geology*, **65**: 191–194.
Paterson, W.S.B., 1994. *Physics of Glaciers*. Oxford: Pergamon.
Powell, R.D., and Alley, R.B., 1997. Grounding-line systems: processes, glaciological inference and the stratigraphic record. In Barker, P.F., and Cooper, A.K. (eds.), *Geology and Seismic Stratigraphy of the Antarctic Margin, 2*. Antarctic Research Series, Volume 71, pp. 169–187. Washington DC: American Geophysical Union.
World Glacier Monitoring Service, 1989. *World Glacier Inventory*. IAHS (ICSI)-UNEP-UNESCO.

Cross-references

Climatic Control of Sedimentation
Debris Flow
Floods and Other Catastrophic Events
Lacustrine Sedimentation
Sediments Produced by Impact
Tills and Tillites
Varves

GLAUCONY AND VERDINE

Glaucony

Glaucony, a general term introduced by Odin and Létolle (1980) to represent a depositional facies, corresponding to sand-sized green glauconitic grains regardless of their mineralogical structure, consists of 2 : 1 layered, potassium and iron-rich, dioctahedral minerals with a high Fe^{3+}/Fe^{2+} ratio. Glauconitic grains generally occur as light to dark green pellets.

Glauconitization is characterized by the formation of a glauconitic precursor, corresponding to a K-poor and Fe-rich smectite. Maturation includes progressive incorporation of potassium and iron and change of mineralogical structure toward an end member constituted by a non-expandable, K-rich glauconitic mica (Odin and Matter, 1981; Odin and Fullagar, 1988), namely glauconite (Burst, 1958; Hower, 1961). Common substrates of glauconitization include microfossil tests, carbonate grains, fecal pellets, and rock fragments.

Several types of glaucony can be identified based upon mineralogical structure and chemical composition. X-ray diffraction patterns from well-ordered, highly-evolved glaucony reveal sharp and narrow (001) reflections at 10Å (illite-type clay mineral). A predominantly ordered structure is also indicated by well shaped (111-) and (021) reflections, and high-intensity (112) and (112-) reflections of comparable size to the (003) peak. Poorly evolved glaucony is structurally disordered, with broad peaks between 12Å and 14Å, a small number of diffractions and low-intensity (112) and (112-) peaks.

Four stages of evolution of glaucony (nascent, slightly evolved, evolved and highly evolved), each reflecting specific morphological habits, can be differentiated on the basis of potassium content of the green grains (Odin and Matter, 1981). K_2O generally shows a direct correlation with other attributes, such as the color of grains. K-poor, nascent ($K_2O = 2$ percent to 4 percent) and slightly evolved ($K_2O = 4$ percent to 6 percent) glaucony generally has a pale to light green color, with a few exceptions, and generally shows obvious traces of the primary substrate of glauconitization. In contrast, evolved ($K_2O = 6$ percent to 8 percent) and highly evolved ($K_2O > 8$ percent) glaucony, where substrate has been completely dissolved, is green to dark green and display characteristic fractures (*craquelures* of French geologists) at grain surfaces. Incorporation of iron mostly predates any potassium uptake into the structure of glauconitic minerals, as shown by high (>15 percent) Fe_2O_3 values also in the least evolved, K-poor samples. However, the direct correlation between K_2O and Fe_2O_3 observed within evolved and highly evolved glaucony, combined with the characteristic higher paramagnetic susceptibility of the most evolved samples, indicate that additional

Fe_2O_3 can be incorporated as potassium begins to enter the smectite structure of glaucony.

Glauconitization typically occurs in submarine, low-energy conditions, under slow sedimentation rates, within confined microenvironments at the interface between oxidizing seawater and slightly reducing interstitial water. Typical depths for glauconitization are comprised between 50 m and 500 m, with temperatures below 15°C. In modern environments, authigenic glaucony typically develops on the outer margins of continental shelves and adjacent slope areas. Rapid burial limits sediment residence time in the appropriate sub-oxic redox regime, preventing glaucony evolution. The maturity of glaucony thus reflects mostly the duration of nondeposition before burial. Present-day forming glaucony is a poorly evolved (nascent to slightly evolved) glauconitic smectite. Glaucony can undergo more or less pronounced modifications due to changes in local subsidence, terrigenous supply, and other factors, such as availability and abundance of iron, pH/Eh, and size and composition of substrates.

Despite its importance, the origin of glaucony is not fully understood and considerable debate exists about both the nature of the chemical reactions involved and the parameters that control its development. The major models of glauconitization include the layer lattice theory of Burst (1958) and Hower (1961), the *verdissement* theory of Odin and Matter (1981) and the two-stage evolutionary model of Clauer *et al.* (1992). All these models involve derivation of the constituents ions from both parent sediment and seawater.

In common usage glaucony is depicted as an autochthonous constituent of marine sediments. Since the first half of this century, glaucony has been regarded as one of the most reliable indicators of low sedimentation rate in marine settings, and glaucony-bearing horizons have been considered as diagnostic of transgressions, because of their presence in lower parts of transgressive/regressive cycles. Glaucony has been commonly reported from condensed sections (Loutit *et al.*, 1988), in association with phosphate grains and abundant fossils. Glaucony may also occur as a coating and incrusting film facies, associated with hardgrounds or burrowed omission surfaces.

Despite the widely held concept that the accumulation of glaucony generally takes place in open-marine environments and far from zones of active sedimentation, preferably during long periods of sediment starvation due to relative sea-level rise (Odin and Fullagar, 1988), glaucony is commonly encountered within a variety of deposits in the rock record. Penecontemporaneous remobilization of the green grains by storms, tidal currents and waves, or reworking caused by subaerial shelf exposure during relative sea-level fall are likely to lead to important concentrations of allochthonous glaucony in a variety of environments poorly suited to glauconitization, such as nearshore, lagoonal, estuarine, incised-valley and turbidite systems, and even alluvial plains.

A reliable interpretation of glaucony-bearing deposits thus requires that spatial distribution, maturity, and genetic attributes of glaucony be determined. This implies the evaluation of autochthonous versus allochthonous glaucony, with a further division of allochthonous glaucony to parautochthonous (intrasequential) versus detrital (extrasequential). The integration of sedimentological, petrographic and mineralogical studies provides a comprehensive framework to differentiate autochthonous from allochthonous glaucony (Amorosi, 1997). Criteria that might be useful in recognizing an allochthonous origin of glaucony include: (i) association with nonmarine deposits; (ii) selective spatial distribution of grains; (iii) high degree of sorting and roundness; (iv) absence of fractures in the most evolved samples, since such features represent zones of weakness that are likely to facilitate mechanical breakdown of grains in the case of prolonged transport or reworking.

The distinction between parautochthonous and detrital glaucony can be sometimes performed by radiometric dating. More commonly, it requires that compositional attributes of glaucony be matched against those of putative sources. The combination of factors that primarily control glaucony development is unique during each transgressive-regressive cycle, and glauconies from distinct parent rocks generally have different maturity and carry unique geochemical and mineralogical fingerprints.

Glaucony is a valid tool for the understanding of depositional cyclicity in the sedimentary record (Amorosi, 1995). Stratigraphically condensed intervals, interpreted to have formed during prolonged breaks in sediment accumulation at sequence scale, may include up to 90 percent autochthonous glaucony. Within these condensed sections, glaucony abundance may show rhythmic fluctuations, with maximum concentration near the base of each cycle and a systematic upward decrease. In these instances, the burrowed, glaucony-rich cycle boundaries correspond to marine flooding surfaces, and the vertically stacked cycles reflect short term sea-level fluctuations, at the scale of parasequences or parasequence sets (Amorosi and Centineo, 2000). Although autochthonous glaucony is most commonly associated with condensed sections and marks the major marine flooding surfaces within the transgressive systems tract, the green grains may be present at any site of the third-order depositional sequence. Parautochthonous glaucony can be widespread throughout the sequence, generally showing lower concentration and maturity than its autochthonous counterpart. Detrital glaucony is present mainly within falling-stage and lowstand deposits, its concentration and composition depending on the characteristics of the source horizon.

Radiometric dates of K-rich glauconies have been widely used to determine the depositional age of sedimentary rocks, and have provided numerous age constraints for the calibration of the relative time scale in strata lacking reliable high-temperature chronometers (Odin, 1982). Although glaucony supplies 40 percent of the absolute-age database for the geological time scale of the last 250 million years, radiometric dating of glaucony often presents large practical limitations, because glaucony can be affected by tectonics, alteration and diagenetic effects, and K-poor glauconies and (to some extent) evolved glaucony are likely to provide apparent ages that are significantly different from the actual age of glaucony.

Verdine

Verdine is a 1 : 1 layered, trioctahedral, 7Å iron-rich silicate, known for long under the name berthierine, but also referred to as phyllite V or odinite. In terms of color and morphology, the verdine facies is similar to and often visually indistinguishable from the glaucony facies. Verdine minerals, however, differ from glauconitic minerals in their mineralogical structure (major peaks at 7.2Å and 14Å) and chemical composition (contain significantly less K_2O than glaucony).

As with glauconitic minerals, minerals of the verdine facies form in marine settings near the sediment–water interface, under conditions of low sedimentation rates and iron availability (they contain iron in both its reduced and oxidized states). The environment of verdine formation, however, is different than that for glaucony, corresponding to warmer, tropical (low-latitude) settings, with water temperature near or greater than 25°. Verdine is abundant at shallower depths than glaucony (15 m to 60 m), and has been observed to replace glaucony on present continental shelves at about 60 m depth (Thamban and Rao, 2000). Verdine generally occurs in proximity of river mouths, which largely contribute the iron needed for verdine formation (Odin and Sen Gupta, 1988).

Verdine is a newly distinguished facies that has been reported from less than 20 locations in the the world (Rao et al., 1995). However, most of the occurrences are restricted to Holocene deposits from tropical oceans, and verdine is very rarely present in the rock record. Similarly to glaucony, verdine is abundant in coincidence with condensation intervals. To date, the only sequence stratigraphic interpretation for verdine deposits is the one by Kronen and Glenn (2000). More work thus is needed to constrain the range of occurrence of verdine in the stratigraphic record.

Summary

Glaucony is one of the most reliable indicators of low sedimentation rate in marine settings and is a potentially powerful tool for basin analysis and stratigraphic correlations, because of its occurrence within stratigraphically condensed intervals formed during major transgressive events. However, the presence of glaucony in the record is not diagnostic *per se* of sediment starvation and transgression.

Glaucony can provide useful information only if the variability of its physicochemical properties and spatial/temporal characteristics are taken into account. This approach requires that the following types of grains be differentiated: (i) low-maturity (K-poor) versus mature (K-rich) glaucony; (ii) autochthonous versus allocthonous glaucony; (iii) intrasequential versus extrasequential glaucony. Glaucony characterization should be performed by integrated mineralogical and geochemical investigations. Caution should be taken when interpreting glaucony-rich deposits based upon visual estimates of the green grains.

Verdine is visually indistinguishable from glaucony, but can be easily identified by its peculiar mineralogical structure and chemical composition. As with glaucony, verdine is concentrated in relatively condensed sections, although it appears to form at shallower depths (<50 m) than glaucony (50 m to 500 m), under strongly river-influenced conditions. The geological significance of the verdine facies is not fully understood, due to its extremely low abundance in the rock record.

Alessandro Amorosi

Bibliography

Amorosi, A., 1995. Glaucony and sequence stratigraphy: a conceptual framework of distribution in siliciclastic sequences. *Journal of Sedimentary Research*, **B65**: 419–425.
Amorosi, A., 1997. Detecting compositional, spatial, and temporal attributes of glaucony: a tool for provenance research. *Sedimentary Geology*, **109**: 135–153.
Amorosi, A., and Centineo, M.C., 2000. Anatomy of a condensed section: the lower cenomanian glaucony-rich deposits of Cap Blanc-Nez (Boulonnais, Northern France). In *Marine Authigenesis: From Globial to Microbial*. SEPM (Society for Sedimentary Geology), Special Publication, 66, pp. 405–413.
Burst, J.F., 1958. "Glauconite" pellets: their mineral nature and applications to stratigraphic interpretations. *American Association of Petroleum Geologists Bulletin*, **42**: 310–327.
Clauer, N., Keppens, E., and Stille, P., 1992. Sr isotopic constraints on the process of glauconitization. *Geology*, **20**: 133–136.
Hower, J., 1961. Some factors concerning the nature and origin of glauconite. *American Mineralogist*, **46**: 313–334.
Kronen, J.D. Jr., and Glenn, C.R., 2000. Pristine to reworked verdine: keys to sequence stratigraphy in mixed carbonate-siliciclastic forereef sediments (Great Barrier Reef). In *Marine Authigenesis: From Globial to Microbial*. SEPM (Society for Sedimentary Geology), Special Publication, 66, pp. 387–403.
Loutit, T.S., Hardenbol, J., Vail, P.R., and Baum, G.R., 1988. Condensed sections: the key to age determination and correlation of continental margin sequences. In *Sea Level Changes: An Integrated Approach*. SEPM (Society for Sedimentary Geology), 42, pp. 183–213.
Odin, G.S., 1982. *Numerical Dating in Stratigraphy*. Chichester: J. Wiley.
Odin, G.S., and Létolle, R., 1980. Glauconitization and phosphatization environments: a tentative comparison. In *Marine Phosphorites*. SEPM (Society for Sedimentary Geology), Special Publication, 29, pp. 227–237.
Odin, G.S., and Matter, A., 1981. De glauconiarum origine. *Sedimentology*, **28**: 611–641.
Odin, G.S., and Fullagar, P.D., 1988. Geological significance of the glaucony facies. In Odin, G.S. (ed.), *Green Marine Clays*. Amsterdam: Elsevier, pp. 295–332.
Odin, G.S., and Sen Gupta, B.K., 1988. Geological significance of the verdine facies. In Odin, G.S. (ed.), *Green Marine Clays*. Amsterdam: Elsevier, pp. 205–219.
Rao, V.P., Thamban, M., and Lamboy, M., 1995. Verdine and glaucony facies from surficial sediments of the eastern continental margin of India. *Marine Geology*, **127**: 105–113.
Thamban, N., and Rao, V.P., 2000. Distribution and composition of verdine and glaucony facies from the sediments of the western continental margin of India. In *Marine Authigenesis: From Global to Microbial*. SEPM (Society for Sedimentary Geology), Special Publication, 66, pp. 233–244.

Cross-references

Authigenesis
Berthierine
Clay Mineralogy
Cyclic Sedimentation
Illite Group Clay Minerals
Phosphorites
Provenance
Sediment Fluxes and Rates of Sedimentation
Smectite Group
Substrate-Controlled Ichnofacies

GRADING, GRADED BEDDING

Graded bedding (Bailey, 1930) characterizes a clastic sedimentary deposit if there is a progressive upward change in the mean, maximum, or modal grain size. If the particle-size variation is repetitive rather than progressive, then the deposit is *stratified* (>1 cm scale) or *laminated* (<1 cm scale), but not necessarily graded. Stratified or laminated beds may also be graded if the average particle size changes progressively upward at a scale exceeding that of single laminae; this is the case in graded turbidites containing *Bouma sequences*. Graded beds range in thickness from ~1 cm to many meters. The thickest graded beds are so-called "megaturbidites" which may be thicker than 100 m (Labaume *et al.*, 1985). Thick graded beds may be composed of gravel and sand or predominantly mud (Weaver and Rothwell, 1987). The term "grading" should not be used to describe gradual changes in rock type across-bedding planes. Instead, it is better in such cases to refer to *gradational bed contacts*.

Graded bedding may be *normal*, with an upward decline in particle size, or *reverse* (sometimes called *inverse*) if there is an upward increase in particle size. In some conglomerates, reverse grading at the base of a bed passes upward into normal grading (Walker, 1975). Graded bedding must be distinguished from so-called *upward fining* or *upward coarsening* successions, which involve many beds and generally more than one facies (Nichols, 1999, pp. 42–43). Because normal grading is much more common than reverse grading, its presence is often used for *way-up determination*.

Graded beds are commonly the result of deposition from short-lived *events* that introduce sediment by downslope transport, offshore transport, or flooding. Examples are *turbidity currents*, storms on shallow-marine shelves, and overbank flooding of rivers. If the energy available for transport increases during the depositional event, then reverse grading may result. Alternatively, reverse grading may result from *grain interaction* (i.e., collisions and near-collisions) in a basal grain-rich layer sheared along by the overriding current (Bagnold, 1956), or by downward percolation of finer particles through coarser fractions during near-bed shearing (Allen, 1984, p. 157).

The normal grading that characterizes most turbidites (Allen, 1984, p. 404) has been interpreted to result from a progressive decline in the carrying capacity of the current for all size fractions initially suspended in the flow (Hiscott, 1994). This decline in capacity occurs at each point along the transport path, and is ascribed to the temporal deceleration of a waning flow (Kneller, 1995). Alternatively, depositing turbidity currents can produce normally graded beds as they pass over a fixed point if there is a lateral size grading from the front to the rear of the flow (Walker, 1965). The lateral grading of the suspended load develops because the body of turbidity currents moves faster than the head of the flow on basin slopes. The head does not grow unchecked as suspension from the body of the current overtakes it. Instead, the rate of body flow into the head region is balanced by loss of suspension from the wake at the back of the head. This ejected material settles back into the flow top according to fall velocity, with the coarsest grains settling into the body nearest the head and the finest grains returning far behind the head, leading to lateral size grading in the flow.

Kneller (1995) modeled the development of grading in turbidites depending on whether flow velocity at a single site decreases, remains steady, or increases with time (temporal deceleration and acceleration); and whether velocity decreases, remains uniform, or increases along the transport path (spatial deceleration and acceleration). Five combinations of spatial and temporal acceleration and deceleration are predicted to generate deposits with distinctive grading profiles (Figure G17). This theoretical treatment applies whenever transport velocity changes systematically through time or along the transport path, and is not limited to turbidites.

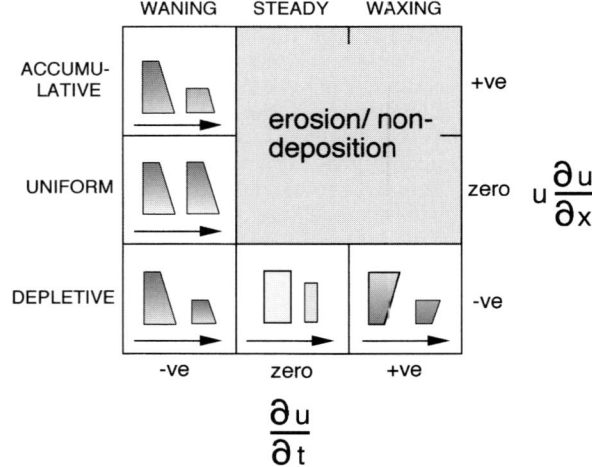

Figure G17 Acceleration matrix where u/t is temporal acceleration at a fixed point, and $u \cdot \partial u / \partial x$ is spatial acceleration in the downflow direction. Three combinations of these variables produce normal grading, and one produces reverse grading. Arrows point downflow. Reproduced by permission of Ben Kneller (Kneller, 1995).

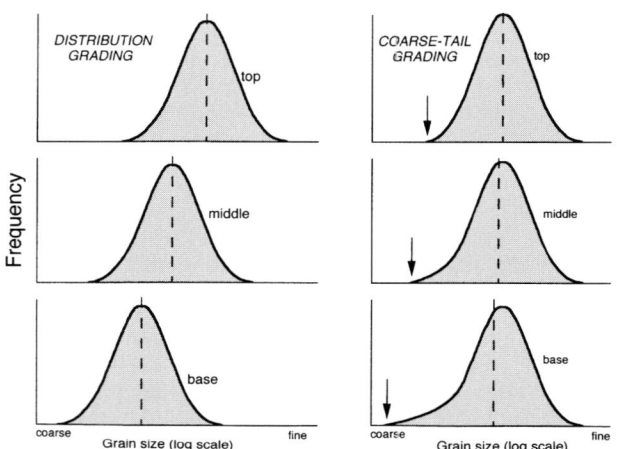

Figure G18 Vertical changes in grain-size distributions for beds showing *distribution grading* (left) and *coarse-tail grading* (right). The dashed lines indicate the mean size. An alternative type of coarse-tail grading would have a symmetrical (bell-shaped) basal sample with no tail of coarse material, and overlying size distributions truncated at their coarse end.

Normal grading can involve the entire size population (*distribution grading*) or only the coarsest 1 percent to 5 percent of the population (*coarse-tail grading*; Middleton, 1967). In distribution grading, there is a progressive shift to finer sizes, so that central measures like mean and modal size clearly reflect the grading (Figure G18). In coarse-tail grading, the percentage of the coarsest grains, and the maximum size, both decrease upward. However, the mean size decreases only very slightly and the mode remains unchanged. Coarse-tail grading is best developed in the deposits of high-concentration turbidity currents.

Although graded bedding is generally used to describe upward changes in grain size, it is permissable to extend the concept to compositional grading, for example in pyroclastic deposits (Cas and Wright, 1987, p. 194). White *et al.* (2001) describe reverse grading in marine pyroclastic deposits, formed because larger pumice takes longer to become saturated with water (and sink) than smaller particles.

The term "grading" is generally used to describe vertical changes in average grain size. However, it is acceptable to refer to *lateral grading* if the grain size changes systematically along or across the transport path.

Richard N. Hiscott

Bibliography

Allen, J.R.L., 1984. *Sedimentary Structures: Their Character and Physical Basis*, Volume II. Amsterdam: Elsevier.
Bailey, E.B., 1930. New light on sedimentation and tectonics. *Geological Magazine*, 67: 77–92.
Bagnold, R.A., 1956. The flow of cohesionless grains in fluids. *Philosophical Transactions of the Royal Society of London (A)*, 249: 235–97.
Cas, R.A.F., and Wright, J.V., 1987. *Volcanic Successions: Modern and Ancient: A Geological Approach to Processes, Products and Successions*. London: Unwin-Hyman.
Hiscott, R.N., 1994. Loss of capacity, not competence, as the fundamental process governing deposition from turbidity currents. *Journal of Sedimentary Research*, A64: 209–214.
Kneller, B., 1995. Beyond the turbidite paradigm: physical models for deposition of turbidites and their implications for reservoir prediction. In Hartley, A.J., and Prosser, D.J. (eds.), *Characterization of Deep Marine Clastic Systems*. Oxford: Geological Society (London), Special Publication, 94, pp. 31–50.
Labaume, P., Mutti, E., and Seguret, M., 1985. Megaturbidites: a depositional model from the eocene of the SW-Pyrenean foreland basin, Spain. *Geo-Marine Letters*, 7: 91–101.
Middleton, G.V., 1967. Experiments on density and turbidity currents: III. deposition of sediment. *Canadian Journal of Earth Sciences*, 4: 475–505.
Nichols, G., 1999. *Sedimentology and Stratigraphy*. Oxford: Blackwell Scientific.
Walker, R.G., 1965. The origin and significance of the internal sedimentary structures of turbidites. *Yorkshire Geological Society, Proceedings*, 35: 1–32.
Walker, R.G., 1975. Generalized facies model for resedimented conglomerates of turbidite association. *Geological Society of America, Bulletin*, 86: 737–748.
Weaver, P.P.E., and Rothwell, R.G., 1987. Sedimentation on the madeira abyssal plain over the last 300 000 years. In Weaver, P.P.E., and Thomson, J. (eds), *Geology and Geochemistry of Abyssal Plains*. Oxford: Geological Society (London), Special Publication, 31, pp. 71–86.
White, J.D.L., Manville, V., Wilson, C.J.N., Houghton, B.F., Riggs, N., and Ort, M., 2001. Settling and deposition of 181 A.D. taupo pumice in lacustrine and associated environments. In White, J.D.L., and Riggs, N.R. (eds.), *Volcaniclastic Sedimentation in Lacustrine Settings*. Oxford: International Association of Sedimentologists, Special Publication, 30, pp. 141–150.

Cross-references

Bedset and Laminaset
Gravity-Driven Mass Flows
Physics of Sediment Transport
Planar and Parallel Lamination
Sedimentary Structures as Way-up Indicators
Sedimentologists
Turbidites

GRAIN FLOW

Perhaps the most familiar example of a granular material is the sand on a beach, which by itself clearly illustrates the dual solid/fluid nature of a granular material. When walking on the beach the sand easily supports your weight like a solid material. Yet, a handful of sand will run out your fingers as if it were a liquid. As it runs out, the same sand will form a pile with sloping sides that shows that material can support weight (at least up to a point) and thus has once again, assumed a solid behavior. Granular flow concerns this intermediate fluid-like movement of a material consisting of solid particles that in other circumstances will behave as a solid. This fluid-like behavior is completely a result of the particulate nature of the material. While the interstitial space between the particles is filled with some sort of fluid such as air or water, there are many situations for which the fluid has no significant effect on the flow behavior.

The particles interact by the elastic deformation of frictional contacts between each particle and its neighbors. The area of these contacts is a very small fraction of the surface area of the particles and as a result the bulk granular material will be much softer than the material that makes up its constituent particles. Any force applied to the bulk material is distributed between the particles and supported across the contacts. However, the force will seldom be distributed evenly. Instead, networks of highly loaded particles known as *force chains* will support the majority of the force, while neighboring particles will experience little or none of the force. Force chains were first observed by Drescher and de Josselin de Jong (1972).

When will the material behave like a solid and when will it behave as a fluid? First of all, a fluid is classically defined as a material that cannot withstand a shear force, hence a classical fluid-like movement is a shear motion (one with velocity gradients) that comes about in response to a shear force. Imagine now that a force is applied to the material, such as the weight of your foot or gravity which makes the material run out between your fingers. Basically, if the particle contacts can support the force, the granular material will behave as an elastic solid and if not, the contacts will break and the material will flow. But it is not simply a function of the magnitude of the applied force as clearly a much larger force is applied by walking on a material than is applied by gravity as the same material flows from your hand. The difference between the two cases lies in their different configurations. When resting on the beach the sand distributes your weight to the particles beneath it. Between your fingers nothing supports the weight and it flows freely.

To induce flow, it is usually necessary to first decrease the particle concentration or to move the particle centers apart increasing the interstitial void space. This process is known as *dilatation* and can be observed while walking on a wet beach; even though the sand still behaves as a solid, it will dilatate under the pressure of the footfall which will suck water into the increased pore space, causing an easily seen local drying of the surface in the area around the foot. The first step in inducing flow is to force this dilatation which generally requires doing work against the applied forces. Thus, unlike a normal fluid, one must first force a granular material to yield before flow can occur. This yield determines the slope of the sides of a granular pile which is known as the *angle of repose* (*q.v.*); steepening of the slope beyond the angle of repose will cause yield and the surface material will flow until the slope is equal to or slightly gentler than the angle of repose.

When static, the majority of the force is carried by force chains and when the material begins to flow, the particles in the force chain will move as a unit. In this way, flow is possible while particles remain locked at the same relative positions to their neighbors in the chain. This flow state is particularly likely at large concentrations and for particles with angular shapes whose protrusions may lock together giving extra strength to the chain. But even spherical particles can be held together by frictional forces and flow in the same manner. The life cycle of a force chain in a shear flow may be likened to that of a pole-vaulter's pole. The chain is formed as the shear flow forces particles together. It will start nearly horizontal, and be compressed by the shear motion initially generating horizontal forces. However, the shear motion will cause the chain to rotate until it is oriented vertically and exerts a nearly vertical force. Rotating beyond the vertical eventually causes the chain to collapse. A new chain then forms and the cycle repeats. As the direction of force rotates with the chain, both shear and normal forces are generated. Now, the average force generated depends on the degree to which the chain is compressed which is largely a function of the particle concentration and in particular, does not depend on the shear rate. The rate at which chains are formed is proportional to the shear rate but the duration of the chain is inversely proportional to the shear rate so that the product of the two is shear-rate independent. Thus, the resultant stresses are shear rate independent or *quasistatic*.

Quasistatic flows have historically been modeled using techniques borrowed from metal plasticity by incorporating a Mohr-Coulomb failure criterion. Such techniques were originally used in the Civil Engineering science of soil mechanics, which describes how granular soils support buildings. However, in that field the concern stops at the initiation of flow, after which time the foundation fails and the building falls down. Plasticity models are less successful in predicting flow behavior, partially because they generally assume that the ratio of maximum shear to normal stress is a constant material property known as the "internal angle of friction." However, computer simulations have shown that the ratio is far from a constant.

At lower concentrations (just how low depends on particle shape and frictional properties) it is possible to have flows in which the individual particles are not locked into chains but can move independently of their neighbors. In an extreme case, the flow shears so rapidly that, except during collisions, the particles lose contact with their neighbors and move freely with a random motion reminiscent of the thermal motion of molecules in a gas. This is the *rapid-flow* regime and is typically modeled using techniques borrowed from the kinetic theory of gases (see Campbell, 1990). Forces in this regime must be small as they are carried by the inertia of the particles and imparted by impact; large forces would require such large thermal velocities and such strong impacts that particles would shatter. Furthermore, very large shear rates (of the order of 100 inverse seconds) are required to generate such flows under Earth gravity. As a result there are essentially no rapid granular flows of geological importance as geological forces are large and the corresponding shear rates small. The more common regime where particles remain in nearly perpetual contact with their neighbors but still move freely, is not well understood. However in this case, the stresses are generated inertially and dimensional analysis dictates that the stresses must vary as the square of the shear rate. But as the stresses are inertial, it is again unlikely the shear rates will ever be large enough to support the large forces encountered in geological applications. Consequently, most geological flows will fall into the quasistatic flow regime. The various flow regimes, the transitions between them, including the many orders of magnitude changes in the stresses that accompany transition are discussed in Campbell (2002).

But even when flowing as a liquid, a granular material retains some of its solid character. One example already mentioned is that a granular material must yield before flow can begin. But also, when confined in a vertical tube, the friction contacts between the granules and the walls will help support the weight of the particles. Beyond a certain height, all of the remaining particle weight will be supported by wall friction so that the force on the material on the bottom of the column is independent of the height of material above it. (This result is due to Janssen, 1895, and is one of the earliest engineering treatments of granular materials.) For this reason, sand is used to fill an hourglass. Were the hourglass filled with a fluid, the pressure at the opening and thus the flowrate depend on the height of liquid above it, making it difficult to predict the time required to empty the glass. But this calibration is simple for a granular material, as the pressure at the opening and thus the flowrate are height independent so that doubling the amount of material in the glass simply doubles the emptying time.

Charles S. Campbell

Bibliography

Campbell, C.S., 1990. Rapid granular flows. *Annual Review of Fluid Mechanics*, **22**: 57
Campbell, C.S., 2002. Granular shear flows at the elastic limit. *Journal of Fluid Mechanics*, **465**: 261–291.
Drescher, A., and De Josselin de Jong, G., 1972. Photoelastic verification of a mechanical model for the flow of a granular material. *Journal of the Mechanics of Physics and Solids*, **20**: 337.
Janssen, H.A., 1895. Versuche uber getreidedruck in silozellen. *Z. Ver. Dt. Ing.*, **39**: 1045.

Cross-references

Angle of Repose
Debris Flow
Grain Size and Shape
Gravity-Driven Mass Flows

GRAIN SETTLING

Pebbles settle faster in water than do grains of sand, and any particle settles more slowly in a viscous fluid such as oil. An understanding of grain settling is important in analyzes sediment transport and deposition. The turbulent eddies of flowing water in a river or ocean current lift sediment particles above the bottom and transport them in suspension. This

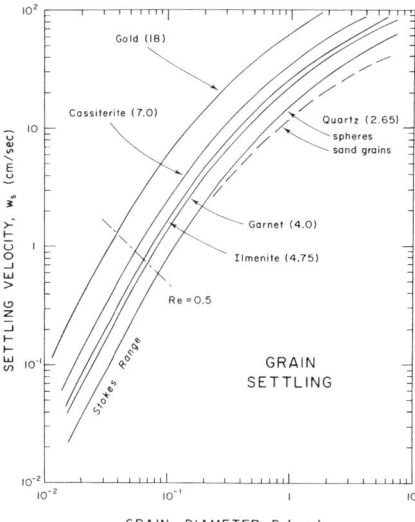

Figure G19 Curves for the settling velocities of spherical grains having densities that range from quartz (2.65 g/cm³) to gold (18 g/cm³). The dashed curve is for the settling of natural sand grains as measured by Baba and Komar (1981), giving some indication of the importance of the grain's nonsphericity on its settling velocity.

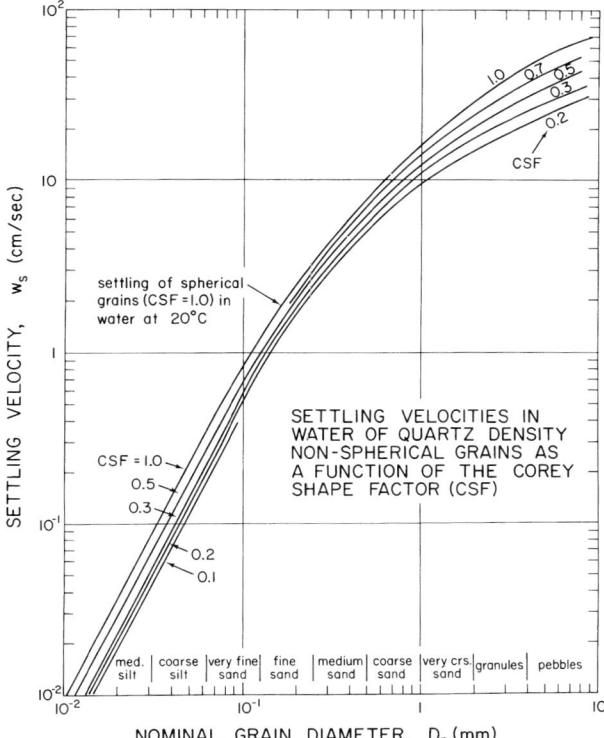

Figure G20 Curves for the settling of particles having the shape of a triaxial ellipsoid, with the nominal diameter being that of an equivalent sphere having the same volume and weight. The settling velocity depends on the Corey Shape Factor (CSF), a measure of the grain's sphericity. After Komar and Reimers (1978).

process depends on the settling velocities of the grains, the lower their settling rate the higher the grains are lifted above the bottom and the greater their concentration. As the strength of the current wanes, the grains are deposited on the bed at rates that depend on their settling velocities.

A number of studies have been conducted to measure settling velocities of sediment grains, with the analyzes relating the settling to the sizes and densities of the grains. Such measurements are usually made in a cylindrical tube filled with water. When the grain is released in the water, it initially accelerates, increasing in velocity until a balance is reached between the immersed weight of the grain in water, which is the net downward force causing it to settle, versus the drag force produced by the friction of the water that resists the grain's movement. When a balance is achieved between these opposing forces, the grain settles at a constant velocity, its *terminal velocity*. It is this terminal velocity that is measured in settling-tube experiments, and used in most applications.

Equations can be derived for the prediction of this terminal velocity by balancing relationships for the immersed weight of the grain and fluid drag. If one considers a spherical grain with a diameter D and density ρ_s, its immersed weight is equal to the difference between its density and that of the fluid, ρ, times the acceleration of gravity g, and the volume of the sphere, giving $(\rho_s-\rho)g(\pi D^3/6)$. For small particles settling at a slow rate, Sir George Stokes derived a relationship for the fluid drag, equal to $3\pi\mu D w_s$ where μ is the fluid's viscosity and w_s is the settling velocity. If the force of drag is equated to the immersed weight and then solved for w_s, one obtains

$$w_s = \frac{1}{18}\frac{1}{\mu}(\rho_s - \rho)gD^2$$

the Stokes settling equation. It is seen that in the Stokes range the settling velocity of a sphere depends on the square of its diameter, the density difference between the grain and fluid, and is inversely proportional to the viscosity of the fluid. Application of this equation is limited to values of the *Reynolds number*, $\mathrm{Re} = \rho D w_s/\mu$, that are less than about 0.5, that is to small combinations of D and w_s. For quartz-density spheres settling in water, this limits its use to particles of diameter less than about 0.08 mm, that is primarily to silt and clay. For large particles that settle at high velocities (pebbles), the drag becomes proportional to the square of the velocity. The derived relationship for the terminal velocity is then proportional to $\sqrt{D}$. The relationship for the settling of large particles is also complicated by the inclusion of a drag coefficient that must be evaluated from empirical curves that are based on actual measurements of settling spheres.

Curves of settling velocities in water versus grain diameters are given in Figure G24 for a range of sediment grain densities, from quartz ($\rho_s = 2.65$ g/cm³) through the range of common heavy minerals, and including native gold with its very high density of 18 g/cm³. As expected, for a given grain diameter the denser the mineral the higher its settling velocity. The cut-off Reynolds number $\mathrm{Re} = 0.5$ for application of the Stokes equation is shown in the diagram. In the Stokes range of small particles, the increase in w_s is proportional to D^2; the proportion to $\sqrt{D}$ does not occur for quartz-density grains until the diameter is on the order of 10 mm. There is a zone of transition, which unfortunately in the case of quartz-density grains corresponds to diameters in the sand-size range, making it more difficult to evaluate their settling velocities.

The curves in Figure G19 are for the settling of perfect spheres. Most sediment grains are not smooth spheres, so the effects of particle shape (sphericity and angularity) must be considered. The dashed curve in Figure G19 for natural sand grains is based on the measurements by Baba and Komar (1981) relating the terminal settling velocities to the intermediate axial diameters of the natural grains whose overall shapes are roughly that of a triaxial ellipsoid so the intermediate diameter is representative of the grain's overall size. The resulting curve is seen to be at lower settling velocities than the curve for quartz spheres, since the nonsphericity and angularity of the natural grains increases their drag and thereby reduces their settling velocities.

Experiments have been conducted on the settling of regularly shaped but nonspherical grains, such as perfect triaxial ellipsoids or cylinders. Curves based on experiments by Komar and Reimers (1978) with triaxial ellipsoids are given in Figure G25, with each curve depending on a value of the Corey Shape Factor

$$CSF = \frac{D_c}{\sqrt{D_a D_b}}$$

where D_a, D_b, and D_c are respectively the longest, intermediate and shortest axial diameters of the triaxial ellipsoid. This shape factor evaluates the degree of departure from a sphere, which has a value $CSF = 1$, the greater the departure the smaller the value of CSF. The series of curves in Figure G20 is seen to depend on CSF, progressively shifting to reduced settling velocities compared with the curve for perfect spheres as the value of CSF decreases and the grains become less spherical. Although the natural sand grains employed by Baba and Komar (1981) in their experiments, yielding the dashed curve in Figure G19, where not perfect triaxial ellipsoids, the degree of departure of the measured settling velocities from the curve for perfect spheres was still found to depend in large part on CSF.

The extreme case of particle shape affecting settling occurs for flat mica plates. Experiments have been undertaken with some success by Komar *et al.* (1984) to measure settling velocities of mica and to relate them to the grain's degree of flatness (the ratio of the thickness of the plate to its average two-dimensional diameter). However, the analysis of mica settling can be complex due to the oscillations they sometimes undergo, particularly for larger mica plates (Berthois, 1962).

The degree of angularity or roundness of the grain can also have some effect on its settling velocity. However, grain roundness has been found to be less important than sphericity (Dietrich, 1982).

While most applications involve assessments of the settling velocities of silt and sand grains, there is also interest in the settling of clay, which is complicated by its tendency to form low-density flocs, the density of which often decreases as the sizes of the flocs increase (Kajihara, 1971). Another application is to the settling of fecal pellets and shells of organisms such as foraminifera, important to deep-sea sedimentation. Fecal pellets are commonly elliptical or cylindrical in shape, and measurements of their settling velocities have been shown to agree with equations for the settling of such particle shapes (Komar *et al.*, 1981).

Paul D. Komar

Bibliography

Baba, J., and Komar, P.D., 1981. Measurements and analysis of settling velocities of natural quartz sand grains. *Journal of Sedimentary Petrology*, **51**: 631–640.
Berthois, L., 1962. Etude du comportement hydraulique du mica. *Sedimentology*, **1**: 40–49.
Dietrich, W.E., 1982. Settling velocity of natural particles. *Water Resources Research*, **18**: 1615–1626.
Kajihara, M., 1971. Settling velocity and porosity of large suspended particles. *Journal of the Oceanographical Society of Japan*, **27**: 158–162.
Komar, P.D., and Reimers, C.E., 1978. Grain shape effects on settling rates. *Journal of Geology*, **86**: 193–209.
Komar, P.D., Morse, A.P., and Small, L.F., 1981. An analysis of sinking rates of natural copepod and euphausiid fecal pellets. *Limnology and Oceanography*, **26**: 172–180.
Komar, P.D., Baba, J., and Cui, B., 1984. Grain-size analyses of mica within sediments and the hydraulic equivalence of mica and quartz. *Journal of Sedimentary Petrology*, **54**: 1379–1391.

Cross-references

Flocculation
Grain Size and Shape
Hindered Settling

GRAIN SIZE AND SHAPE

Introduction

Size and shape are conceptually distinct properties of sedimentary grains. The former derives from reference to an external metric to determine the absolute magnitude of a grain, while the latter represents relative measures of the grain to establish its geometry. In practice, the two are related, first by the means that are customarily employed to establish the size of a grain, and second by the fact that grain shape in a sample might vary systematically with grain size. Size and shape of constituent grains are component properties of the texture of granular aggregates. Most grains originate by disintegration of bedrock, so size and shape are set initially by characteristics of the rock and by the disintegration process. Grains may also be formed by chemical precipitation or by cementation of previously existing grains, whence the initial properties are set by the endpoint of those processes. Size and shape are subsequently modified in the surface environment by weathering and by the damage imposed by processes of sediment transport.

Size and shape are fundamental properties of clastic sediments. The distributions of size and shape in sediment deposits influence and index other important physical properties of the sediment, such as porosity, permeability, and surface roughness, they carry important information about the origin of a deposit, they affect the stability of the deposit, and they influence habitat quality for small organisms. In this article we first examine conceptual and operational measures of grain size and shape. We then consider sampling procedures for determining representative size and shape in sedimentary deposits, since sampling places some constraints on the information that practically can be had. Finally, the distributions of size and shape properties of sedimentary aggregates are considered. A more extended review is given by Pye (1994), while the classic work on sampling and analysis of sediment properties remains that of Griffiths (1967).

Grain size

The size of a compact, three-dimensional object such as a sedimentary grain might be indexed by some measure of its volume, or by some linear measure of its geometry. For geometrically regular objects, either carries equivalent information. For irregular objects, including sediment grains, they do not. Both approaches have been used for characterizing grain size. Conceptually, the most thorough measure is provided by grain volume, or by the equivalent linear measure $\sqrt[3]{V}$. The most common explicit measure, however, is the notional length of the second principal axis (b-axis) of the grain (see Figure G21). This measure gains its prominence from the operational method for sizing grains by washing or dry passage through a sequence of sieves with defined openings (DIN 18123, 1983; BS812-103, 1985; ASTM D422-63, 1990; D75-87, 1992) It is always possible for the a-axis to pass through on end, so grain size is determined by the b-axis length.

Sieves with round openings gauge the b-axis accurately, but square-mesh screen sieves are customarily used. Passage through a square opening depends jointly on b- and c-axis dimensions of the grain (Figure G21), so the measure is biased in comparison with a caliper measurement. The sieve opening, D, is customarily accepted as the practical measure of grain size. This technique is commonly applied down to 0.063 mm (63 μm), the lower limit of sand sizes, but the dry technique is inefficient below 125 μm. Size determination in this manner is in fact a classification procedure: sizes are determined only to within the limits established by successive sieves. The conventional classification scale is a geometric progression with ratio 2, known as the Udden-Wentworth scale (Table G13, which includes the equivalent ϕ-scale employed by sedimentologists). In research practice, clasts larger than about 64 mm commonly are measured individually using a ruler or calipers (usually without bias correction for D), although grains up to boulder size may be classified through screens in industrial operations.

Very small grains defy efficient size separation by screens. Two methods dominate size determination of fine materials. The traditional method entails particle settlement in water. The density of an aqueous suspension of particles is periodically measured, traditionally, by pipette withdrawal or hydrometer (BS3406-2, 1984; ASTM D422-63, 1990), but in modern methods by optical or X-ray beam transmission. The limit of sizes that have cleared the measured suspension or the beam path at the time of measurement is determined from Stokes' Law,

$$v_t = g(\rho_s - \rho)D^2/18\eta \qquad \text{(Eq. 1)}$$

in which v_t is the terminal fall velocity of a spherical grain in the laminar regime, g is the acceleration of gravity, ρ_s and ρ are respectively the solid and fluid densities, and η is the dynamic viscosity of the fluid. The notional grain size assigned by this technique is the diameter of a sphere settling in still water in the laminar regime at the same rate as the observed particle (the *nominal* or *settling diameter*). Stokes Law holds (i.e., the settling regime is satisfactorily laminar) for grains finer than 50 μm in suspensions more dilute than about

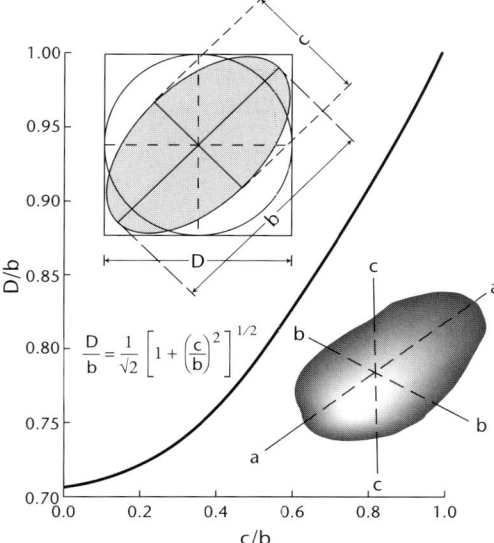

Figure G21 The ratio of square sieve opening (D) to b-axis length as a function of clast flatness, c/b. D/b gives the bias between commonly employed measures of grain size, and is the correction factor required to convert the caliper-determined b-axis to the equivalent square sieve opening. c must be measured to calculate it. Insets: 1. definition of principal axes of a grain; 2. Geometry of a grain in a square mesh opening.

1 percent. The practical lower limit of size determination by settling is around 0.5 μm, below which Brownian effects dominate the settling behavior. A similar technique can be used to determine the settling diameter of sands (USFIASP, 1957) but, in this case, terminal velocities fall in a transitional region between laminar and fully turbulent settling

In settling techniques, the actual grain size of the irregular particles is nearly always different than the settling diameter. The assigned grain size is no longer simply related to particle geometry at all (Figure G22). The measure is nonetheless convenient since, in sedimentology, the settling behavior of the grain in water is often a critical element of interpretation. Hallermeier (1981) presented empirical functions for the settlement of natural quartz grains and Dietrich (1982) presented generalized functions for settling velocity that subsume variations in size, density, shape and roundness, while Komar and Cui (1984) have systematically studied the variation between geometrical diameter (b-axis) and the measures realized by sieving and settling methods (see *Grain Settling*).

A modern technique for determining the size of fine grains entails the measurement of laser or optical beam interruption intervals induced by a suspension pumped through the beam (BS3406-7, 1998). This technique measures an arbitrary diameter or "cut" across the grain. The relation between the distribution of random cuts and the true diameter for grains of fixed shape is known for dilute particulate phases (Nicholson, 1970), and is used to derive a size distribution.

The methods for sizing grains are almost exclusively applied to aggregations of grains to return a frequency distribution of sizes in the aggregation. The size of individual particles is not distinguished. If individual sizes are of interest, they can be

Sampling grains

Sampling enters into the definition of grain size because, in most instances, grains are collected and sized in aggregate (such as by sieving), so that a distribution of sizes is obtained. Grains are not measured individually. To the extent that the sampling procedure biases the selection of grains, it influences the observed results. For example, point (grid), linear, areal and volumetric methods for selecting material predetermine 0, 1, 2, and all 3 dimensions, respectively, of the sample volume, whilst number and weight as bases for assigning frequency to separated size classes also vary by a factor L^3. Kellerhals and Bray (1971) presented a formal conversion procedure based upon the dimensions of a sample that are predetermined by the sampling and analysis techniques so that observed results can be represented in terms of any preferred measurement conventions. The general conversion formula is

$$f_{ci} = f_{oi} D_i^x \bigg/ \left[\sum f_{oi} D_i^x \right] \qquad \text{(Eq. 3)}$$

in which f_{oi} is the observed frequency, D_i is mean grain size for size class i, x is the integer dimension required for the conversion, n is the number of classes, and the sum effects a renormalization of the distribution to yield the converted frequencies f_{ci}. Table G14 gives the dimensions for conversions amongst any usual combination of sampling and frequency assignment methods. For the conversion grid-by-number (the usual means to select and analyze individual grains from outcrop or surface) to volume-by-weight (classified bulk samples), $x = 1$; these two standard procedures are equivalent in principle. Photographic samples are area-by-number samples, for which the conversion factor to equate the result with standard analyzes is $x = 2$. A number of subsequent analysts have criticized the Kellerhals–Bray procedure on the grounds that the simple packed cubes model employed by Kellerhals and Bray to illustrate their concept is defective since it does not realistically model the porosity or grain arrangements in real earth materials. It should be emphasized that the formal model does not depend upon the illustration at all. Properly conducted tests (Church et al., 1987) confirm the efficacy of the formal conversions.

Counts are used to establish frequency by size of grains when relatively small numbers or comparatively large grains are selected from outcrop or a sedimentary surface, and when grains are selected on a microscope slide. They would normally be selected on a grid or by application of randomly drawn coordinates (see, for example, Wolman, 1954). Weights are recovered when large numbers of grains are sorted in sieves, and when settling methods are used. An important distinction in the quality of the information in these two cases is that number counts constitute legitimate quantities for statistical manipulation; frequencies derived from weights do not. Knowledge of grain size, packing and mineral density make the derivation of approximate grain number equivalents feasible but, in practice, the numbers become so large for fine materials that no useful statistical resolution may be gained anyway.

The question arises how large must a sample be in order to obtain an unbiased representation of sizes, or an estimate of some summary measure—such as the mean size—to within some desired level of precision. For samples in which frequency is assigned by enumeration (typically, grid-by-number samples—commonly known as "Wolman samples"), these are statistical questions that have been considered by many analysts. Table G15 summarizes some results.

Frequency is assigned in a bulk sample by relative weight, so the numbers of grains present in one size class influences the observed frequency in all other classes. The critical class is that which contains the largest grains, since these will be the least numerous in comparison with their aggregate weight. One requires a sample large enough to ensure that this class is adequately represented. The requisite weight for a given maximum grain size is the subject of several standards (ISO 4364, 1977; ISO 9195, 1992; ASTM D75-87, 1987; BS3680-10C, 1996; BS3680-1-E, 1998). In the sedimentary literature, Church et al. (1987) and Milan et al. (1999) have pursued the question of adequate sample size for gravels. Church et al. determined that the largest grain should not constitute more than about 0.1 percent of the sample weight in order to obtain a reliable analysis. The criterion is not conservative (compare ISO criteria, based on the D_{84} size: Figure G25), but it nevertheless leads to sample sizes that are impractical for manual sampling of material coarser than gravel. Ferguson and Paola (1997) have estimated by numerical simulations the bias and precision of percentile estimates in bulk size distributions. Their results show that, if only information about relatively central measures is required, then substantially smaller samples are tolerable than those suggested for characterizing the full distribution, and that results depend upon the gradation (range of sizes) of the material.

There remains the additional sampling question raised by the probability that the sampled materials are inhomogeneous over the sampling surface or volume. If one is interested only in average measures of the material, it is far more convenient to sample over the variability than to attempt to identify and sample each homogeneous subunit (which may be very difficult to identify in some circumstances). Wolcott and Church (1991) and Crowder and Diplas (1997) have addressed this question for bulk and surface sampling, respectively. A special problem presented by fluvial gravels and by some other deposits is that the surface is characteristically different than the subsurface because of the loss of fines from the surface by settling or by winnowing (armoring). Consequently, surface (by point, linear or areal means) and subsurface sampling (by volumetric means) are customarily conducted separately in coarse materials with evident segregation.

Because the distribution of grain sizes in a deposit may be very wide (possibly surpassing five orders of magnitude), so that it may not be feasible to use a single technique to sample or analyze the entire range of sizes that occur, and because

Table G14 Weighting factors for conversion of grain size frequency distributions[a]

Conversion from	Conversion to sieve by weight	Grid by number	Grid by weight	Area by number	Area by weight
Sieve by weight	1	1	D^3	D^{-2}	D
Grid by number	1	1	D^3	D^{-2}	D
Grid by weight	D^{-3}	D^{-3}	1	D^{-5}	D^{-2}
Area by number	D^2	D^2	D^5	1	D^3
Area by weight	D^{-1}	D^{-1}	D^2	D^{-3}	1

[a] After Kellerhals and Bray (1971, their table 2).

Table G15 Criteria for grid-by-number sample size[a]

Authority	Basis of recommendation	Parameter	Recommendation
Wolman (1954)	9 replicate samples	median	100 grains
Hey and Thorne (1983)	Operator variance	mean	100
Mosley and Tindale (1985)	Cumulative histogram	mean	70
Church et al. (1987)	Statistical theory	Precision of the mean	Variable
Fripp and Diplas (1993)	Binomial theory	Percentiles	200–400
Rice and Church (1996)	Bootstrap analysis	Percentiles and size distribution	400

[a] Mainly after Rice and Church (1996).

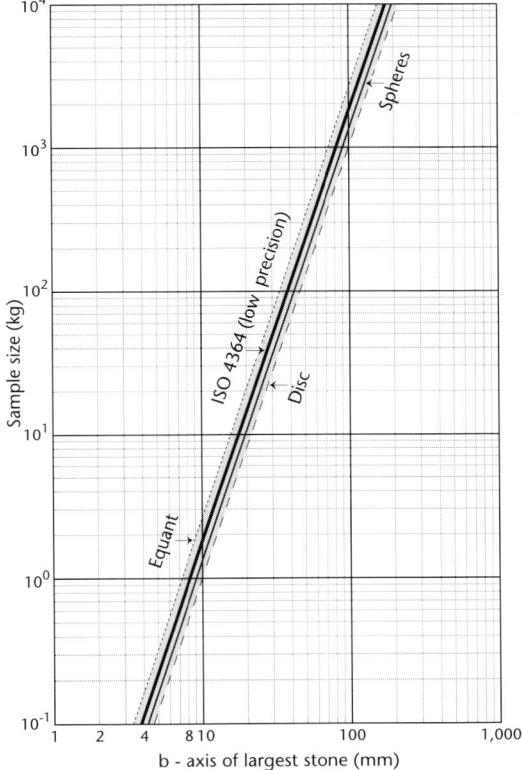

Figure G25 Bulk sample size standards based on the largest grain constituting 0.1 percent of the total sample weight (after Church et al., 1987; Milan et al., 1999). ISO 4364-1977(E), based on the D_{84}, is shown for comparison.

sampling requirements may become unrealizable for deposits that include very large materials, samples often represent only a portion of the entire size distribution. The sample may still contain significant information (as, e.g., when it is of interest to sample the matrix materials in a diamict sediment). Because frequencies are assigned on a relative basis, it is important to constrain comparisons between samples to the common range of analyzed sizes. Such samples may be regarded as truncated or censored samples of the entire size distribution. Combining data from different sampling techniques may require adjustment of the assigned sizes in one or the other portion of the range, as the consequence of the nonequivalence of size assignments amongst different techniques (Komar and Cui, 1984). The most frequent example of this problem arises in the joining of sieve data of sands with the data of silts derived by settling methods.

The distribution of grain sizes

The distribution of grain sizes arrived at by enumeration is a statistically valid distribution, but the distribution of sizes arrived at by weights is not. Since the relation between the two bases goes as D^3, frequency by weight represents a nonlinear transform of frequency by number. Despite the ill-conditioned nature of distribution by weight, sedimentologists have long used mathematical functions to approximate grain size distributions (GSD). The procedure is essentially descriptive, and no rigorous goodness of fit tests are available for comparing distributions with data or with each other. Analysis is characteristically conducted in ϕ-space, since the logarithmic transform is necessary to homogenize the variance of grain size data.

Two distributions dominated classical thinking about GSD: the so-called *Rosin distribution* (see Kittleman, 1964) which is, in fact, the two-parameter Weibull distribution, and the *log-normal distribution* (Krumbein, 1938), both customarily applied to GSD by weight. The former is justified on the basis of the size distribution of crushed material (Bennett, 1936; see also Turcotte, 1992), whilst Inman (1949) and Middleton (1970) have found justifications for log-normal distributions on the basis of sediment sorting in transport. The Rosin–Weibull distribution has been successfully applied to a range of naturally broken materials, but mostly with $D_{max} < 32$ mm ($\phi > -5$). Exceptions exist, but a plausible reason for an upper size limit is that the size of larger "broken" grains will almost always reflect primary jointing and bedding planes at outcrop, which are often not random. Ibbeken (1983) gave a review of empirical results. The log-normal distribution appears to describe tolerably well a wide range of sedimentary materials, but it often has been adopted without critical analysis as a matter of descriptive expediency.

In recent years, additional distributions have been proposed to describe transported sediments, including the log-hyperbolic distribution and the log-skew-Laplace distribution. Both are related to the normal distribution (the hyperbolic distribution being the most general). Fieller *et al.* (1992) give a brief but comprehensive description of the application of distribution functions to describe GSD.

Many GSD are clearly not simple. Bimodal or multimodal distributions have been described using mixture distributions. However, insofar as multiple modes usually signify subpopulations that have been influenced by distinctive processes, it seems most sensible to partition the distribution into unimodal components and to study each one. The partition procedure, formerly tedious, is easy today using algorithms designed for this purpose (which usually require that the characteristic distribution be specified).

In the past, GSD have customarily been summarized by statistics, usually the first three moment statistics or graphical approximations of them. Approximate moments based on ϕ-percentiles have been advocated (Inman, 1952; Folk and Ward, 1957) and have the advantages that their consistent interpretation does not presuppose a particular distribution function, nor a complete expression of the distribution function (the extreme tails often remaining poorly sampled or unknown in sediment analysis), or a statistically valid description of the distribution. Strenuous efforts have been expended to associate moment domains with particular depositional environments, which implicitly supposes that some characteristic GSD is associated with an environment. Whilst some general associations were recognized the project did not identify clean discriminations and is not now much pursued.

Distributions of grain shape

The distribution of shape characteristics has received much less attention. A significant issue is the definition of a homogeneous population of grains for consideration, since shape variations may systematically be affected by grain size and by grain lithology. Shape measures have mainly been studied in terms of changes in median or modal values along sedimentation gradients. No general distributions have been proposed and there appears little reason to expect any, in homogeneous samples, beyond normal or log-normally distributed random variations.

Michael J. Church

Bibliography

ASTM (American Society for Testing and Materials), 1992. D75-87: standard practices for sampling aggregates.
ASTM (American Society for Testing and Materials), 1990. D422-63: Standard test method for particle size analysis of soils.
Bennett, J.G., 1936. Broken coal. *Journal of the Institute of Fuel*, **10**: 22–39.
BSI (British Standards Institution), 1985. BS812-103. Methods for determination of particle size distribution.
BSI (British Standards Institution), 2000. BS3406-2(1984). Recommendations for gravitational liquid sedimentation methods for powders and suspensions.
BSI (British Standards Institution), 1998. BS3406-4(1993). Guide to microscope and image analysis methods (1 μm to 1 mm).
BSI (British Standards Institution), 1998. BS3406-7(1995). Recommendations for single particle light interaction methods.
BSI (British Standards Institution), 1996. BS3680-10C. Guide to methods of sampling sand bed and cohesive bed materials.
BSI (British Standards Institution), 1998. BS3680-10E. Sampling and analysis of gravel bed material.
Butler, J.B., Lane, S.N., and Chandler, J.H., 2001. Automated extraction of grain-size data from gravel surfaces using digital image processing. *Journal of Hydraulic Research*, **39**: 519–529.
Church, M.A., McLean, D.G., and Wolcott, J.F., 1987. River bed gravels: sampling and analysis. In Thorne, C.R., Bathurst, J.C., and Hey, R.D. (eds.), *Sediment Transport in Gravel-bed Rivers*. Chichester: John Wiley & Sons. pp. 43–88.
Crowder, D.W., and Diplas, P., 1997 Sampling heterogeneous deposits in gravel-bed streams. *Journal of Hydraulic Engineering*, **123**: 1106–1117.
Dietrich, W.E., 1982. Settling velocity of natural particles. *Water Resources Research*, **18**: 1615–1626.
DIN (Deutsche Ingenieure Normung), 1983. DIN18123: Subsoil: testing of soil samples: determination of the particle size distribution.
Ferguson, R., and Paola, C., 1997. Bias and precision of percentiles of bulk grain size distributions. *Earth Surface Processes and Landforms*, **22**: 1061–1077.
Fieller, N.R.J., Flenley, E.C., and Olbricht, W., 1992. Statistics of particle size data. *Applied Statistics*, **41**: 127–146.
Folk, R.L., and Ward, W.C., 1957. Brazos river bar: a study in the significance of grain size parameters. *Journal of Sedimentary Petrology*, **27**: 3–26.
Fripp, J.B., and Diplas, P., 1993. Surface sampling in gravel streams. *Journal of Hydraulic Engineering*, **119**: 473–490.
Griffiths, J.C., 1967. *Scientific Method in Analysis of Sediments*. McGraw-Hill.
Hallermeier, R.J., 1981. Terminal settling velocity of commonly occurring sand grains. *Sedimentology*, **28**: 859–865.
Hey, R.D., and Thorne, C.R., 1983. Accuracy of surface samples from gravel bed material. American Society of Civil Engineers. *Proceedings, Journal of the Hydraulics Division*, **109**: 842–851.
Ibbeken, H., 1983. Jointed source rock and fluvial gravels controlled by Rosin's Law: a grain-size study in Calabria, south Italy. *Journal of Sedimentary Petrology*, **53**: 1213–1231.
Inman, D.L., 1949. Sorting of sediments in the light of fluid mechanics. *Journal of Sedimentary Petrology*, **19**: 51–70.
Inman, D.L., 1952. Measures for describing the size distribution of sediments. *Journal of Sedimentary Petrology*, **22**: 125–145.
ISO (International Standards Organization), 1977. ISO4364-1977. Liquid flow measurement in open channels—bed material sampling.
ISO (International Standards Organization), 9195-1992. Sampling and analysis of gravel bed material.
Kellerhals, R., and Bray, D.I., 1971. Sampling procedures for coarse fluvial sediments. American Society of Civil Engineers, *Journal of the Hydraulics Division* **97**: 1165–1179.
Kellerhals, R., Shaw, J., and Arora, V.K., 1975. On grain size from thin sections. *Journal of Geology*, **83**: 79–96.
Kittleman, L.R. Jr., 1964. Application of Rosin's distribution in size-frequency analysis of clastic rocks. *Journal of Sedimentary Petrology*, **34**: 483–502.2
Komar, P.D., and Cui, B., 1984. The analysis of grain size measurements by settling tube techniques. *Journal of Sedimentary Petrology*, **54**: 603–614.
Krumbein, W.C., 1938. Size-frequency distributions and the normal phi-curve. *Journal of Sedimentary Petrology*, **8**: 84–90.
Krumbein, W.C., 1941. Measurement and geological significance of shape and roundness of sedimentary particles. *Journal of Sedimentary Petrology*, **11**: 64–72.
Logan, B.E., and Kilps, J.R., 1995. Fractal geometry of particle aggregates formed in different fluid mechanical environments. *Water Research*, **29**: 443–453.
Middleton, G.V., 1970. Generation of the log-normal frequency distribution in sediments. In Romanova, M.A., and Sarmonova, O.V. (eds.), *Topics in Mathematical Geology*. N.Y.: Consultants Bureau: pp.34–42. [Originally published in Russian, 1968 as the Andrew Vistelius Anniversary Volume]
Milan, D.J., Heritage, G.L., Large, A.R.G., and Brunsdon, C.F., 1999. Influence of particle shape and sorting upon sample size estimates for a coarse-grained upland stream. *Sedimentary Geology*, **129**: 85–100.

Mosley, M.P., and Tindale, D.S., 1985. Sediment variability and bed material sampling in gravel-bed rivers. *Earth Surface Processes and Landforms*, **10**: 465–482.
Nicholson, W.L., 1970. Estimation of linear properties of particle size distributions. *Biometrika*, **57**: 273–297.
Pye, K., 1994. Properties of sediment particles. Chapter 1 In Pye, K. (ed.), *Sediment Transport and Depositional Processes*. Oxford: Blackwell Scientific, pp. 1–24.
Rice, S.P., and Church, M., 1996. Sampling surficial fluvial gravels: the precision of size distribution percentile estimates. *Journal of Sedimentary Research*, **66**: 654–665.
Sneed, E.D., and Folk, R.L., 1958. Pebbles in the lower Colorado River, Texas: a study in particle morphogenesis. *Journal of Geology*, **66**: 114–150.
Truesdell, P.E., and Varnes, D.J., 1950. *Chart Correlating Various Grain-size Definitions of Sedimentary Material*. Washington: United States Geological Survey.
Turcotte, D.L., 1992. *Fractals and Chaos in Geology and Geophysics*. Cambridge University Press.
USFIASP (United States Federal Interagency Sedimentation Project.), 1957. *The Development and Calibration of the Visual-Accumulation Tube*. Minneapolis: St.Anthony Falls Hydraulic Laboratory.
Wadell, H., 1933. Sphericity and roundness of rock particles. *Journal of Geology*, **41**: 310–331.
Wentworth, C.K., 1922. A scale of grade and class terms for clastic sediments. *Journal of Geology*, **30**: 377–392.
Wolcott, J., and Church, M., 1991. Strategies for sampling spatially heterogeneous phenomena: the example of river gravels. *Journal of Sedimentary Petrology*, **61**: 534–543.
Wolman, M.G., 1954. A method of sampling coarse river bed material. *American Geophysical Union, Transactions*, **35**: 951–956.
Zingg, T., 1935. Beitrag zur Schotteranalyse. *Schweizerische Mineralogische Und Petrologische Mitteilungen*, **15**: 39–140.

Cross-references

Angle of Repose
Armor
Fabric, Porosity and Permeability
Flocculation
Grading, Graded Bedding
Grain Settling
Imbrication and Flow-oriented Clasts
Sedimentologists
Statistics Applied to Sediments

GRAIN THRESHOLD

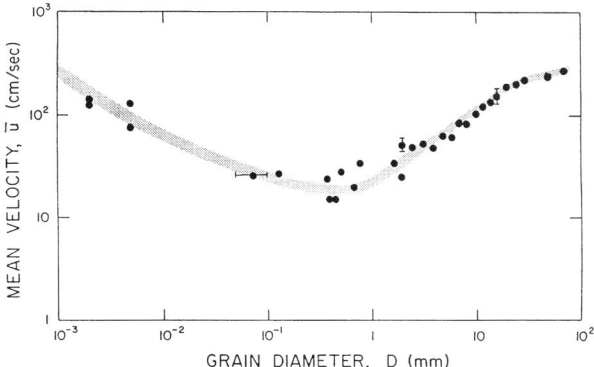

Figure G26 The threshold curve developed by Hjulstrøm (1939), but with the data he used included.

As the velocity of flowing water progressively increases over a bed of sediments, a condition is eventually reached when individual grains are dislodged and begin to be transported. This condition of first grain movement is called the *threshold of sediment motion*, or more concisely, *grain threshold*. It is the first stage in producing a transport of sediment, and in that context equations have been developed to correlate the quantity of sediment being transported to the difference between the measured flow strength (velocity, discharge, or mean stress) and that required to initiate sediment motion, thereby predicting zero transport prior to the threshold condition. In another application geologists have employed evaluations of grain threshold in interpretations of modern and ancient sedimentary deposits, for example to assess discharges of extreme floods from the maximum sizes of cobbles or boulders that were moved.

It might appear that the analysis of sediment threshold is a simple problem that can be solved through a series of laboratory flume experiments, each using a different grain size. A number of such experiments have been performed, usually with sediments having nearly uniform grain sizes, each experiment being conducted with a single sieve fraction. However, natural deposits of sediments contain ranges of particle sizes, and often include minerals that have different densities. Selective grain entrainment can then be important, for example involving the movement of various sizes of cobbles in a stream by floods having different discharges, or the formation of heavy-mineral placers in sands when the light minerals are selectively entrained and transported away, leaving behind the heavy minerals.

The earliest and still best known attempt to quantify the grain-threshold condition was that of Hjulstrøm (1939), yielding the graph in Figure G26. Hjulstrøm compiled the threshold measurements in the sand-size range, obtained by previous investigators in flume experiments, and data on flow velocities required to erode cohesive clays and silty soils. An interesting result seen in his grain-threshold curve is the presence of a minimum at a grain diameter of about 0.5 mm to 1 mm, a minimum threshold velocity of about 20 cm per second. Accordingly, it would be easiest to erode fine sand, whereas silt and clay, as well as coarser sand and gravel, require greater velocities for grain movement. The increased velocity needed to transport coarser grains is what one would expect. That it takes higher velocities to erode the fine-grained clays and silts results from cohesion binding the particles.

Although the Hjulstrøm curve is reproduced in many textbooks, generally it should not be used in applications. This is because the evaluated threshold velocity is the average velocity as might be measured in a river, related to the river's discharge. It has no reasonable counterpart to a current in the ocean, which is characterized by a bottom boundary layer of increasing velocities with distance above the seafloor. In such a case it is necessary to relate the grain threshold to the velocity measured at some specified level above the bottom. In their review of the available threshold data, Miller *et al.* (1977) present such a threshold graph where the mean-flow threshold velocity is measured at 1 m above the bottom.

Rather than using a measure of the current's velocity, generally it is preferable to express the grain-threshold condition in terms of the mean stress exerted by the current on the sediment, the force per unit area. This stress is related to the gradient of increasing flow velocities above the bottom, and thus to the frictional drag coupling the water flow and seafloor, representing in part the forces that are exerted on the individual sediment grains. The curve derived by Miller *et al.* (1977) for the threshold flow stress, τ_t, is shown in Figure G27. Of interest is the absence of a pronounced minimum like that seen in the Hjulstrom curve. This is because Miller *et al.* (1977) utilized only laboratory flume data extending down through silt grain sizes, where the cohesive effects of clay and organic matter were excluded. Even without the presence of cohesion to produce a distinct minimum in the curve, there still is an inflection with different slopes to either side. This is attributed to the changing character of the water flow immediately above the bed of sediments. With small grain sizes and generally reduced flow strengths required for their movement, there may be a viscous sublayer at the bed which has a reduced level of turbulence that acts to move the grains. In contrast, with larger grains and the absence of a viscous sublayer, individual particles shed eddies and absorb more of the drag. Such relationships between the nature of the flowing fluid in close proximity to the sediment bed and its importance to the initiation of grain movement were initially investigated by the German engineer Albert Shields in 1936, and more recently by Yalin and Karahan (1979). Of importance, such considerations led to the development of dimensionless graphs for the presentation of grain-threshold data, where the sediment grains might be composed of magnetite or some other mineral, and the experiments included viscous fluids such as glycerin and oils, as well as water. The result is a "universal curve" that is applicable to various combinations of grain densities and fluids, contrasting with the curves of Figures G26 and G27 that are limited in applications to the threshold of quartz-density grains in water.

Figure G27 reveals a considerable degree of data scatter above and below the threshold curve, perhaps surprisingly so in that the measurements were obtained in the controlled conditions of laboratory flumes. Furthermore, systematic differences are found between the data derived from different studies, evident in Figure G27 by the plotting position of the gravel threshold measurements obtained by Neill (1968). This scatter is due largely to the rather subjective nature of sediment threshold determinations, it being the experimenter's judgement as to just how much grain movement constitutes "threshold". The condition has been variously taken as "weak movement", "general bed movement", and for a single stone to be "first displaced" or to experience "scattered movement". This subjectivity in the determination of the grain-threshold condition needs to be recognized in applications where Figure G27, or any other threshold curve, is used. In addition, the idealized conditions of the flume experiments must be recognized, which excluded grain cohesion, represented experiments with only a narrow range of grain sizes (a sieve fraction), and where the bed of sediments was carefully smoothed in order to avoid irregularities that might affect the results. Smoothing the sediment of course eliminated the effects of bedforms such as previously formed ripple marks, which can affect the grain-threshold condition in natural environments.

The use of a narrow range of grain sizes in the experiments also departs from most natural sediments where there can be large ranges of sizes available for transport, and also different grain densities. This leads to the occurrence of *selective grain entrainment*, where with increasing flow velocities and stresses, progressively larger and more dense grains can be moved. Laboratory and field measurements have been undertaken to document this process, and example results are given in Figure G28, compiled by Komar (1987). Shown is a series of selective-entrainment curves that obliquely cross the threshold curve (dashed) for uniform grain sizes, the same curve given in Figure G27. The data of Milhous and Carling are for gravel entrainment in streams, while that of Hammond is from the observed movement of gravel on the continental shelf. The experiments of Day were undertaken in laboratory flumes, using various gravel-size mixtures. Each curve represents the largest gravel particle that is moved at a particular flow stress. Such an assessment is referred to as *flow competence*, with one important application being to evaluate the magnitudes (stresses, velocities and discharges) of extreme floods from the maximum sizes of the cobbles and boulders that were

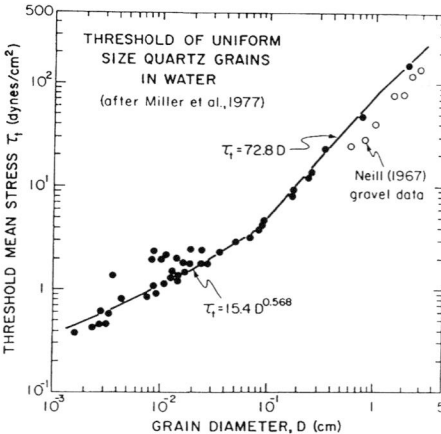

Figure G27 The flow stress τ_t required for the threshold of quartz-density grains of diameter D in water, based on the data compilation of Miller *et al.* (1977).

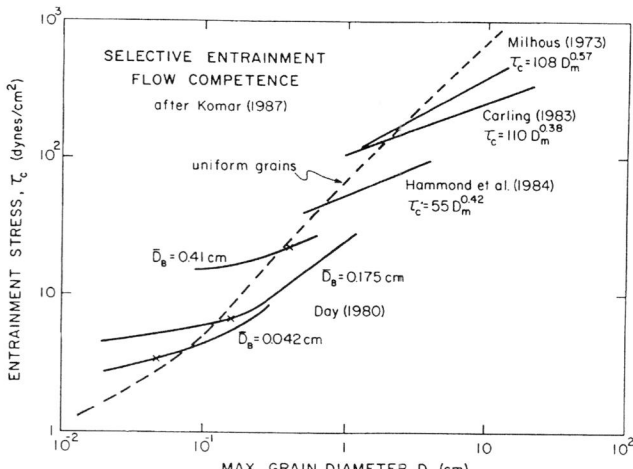

Figure G28 A series of selective entrainment or flow-competence threshold curves for the flow stress that moves the maximum grain diameter D_m within a deposit of mixed grain sizes. The dashed curve is for the threshold of uniform sediments, the same as given in Figure G27 [after Komar (1987)].

transported. The curves shown in Figure G28 all demonstrate that with a mixture of grain sizes available for transport, the stronger the flow (the higher the value of the flow stress), the larger the maximum sizes of gravel particles moved. Each selective entrainment curve crosses the curve based on uniform grain sizes at approximately its median grain size, so in general the use of the curve in Figure G27 represents the threshold of that median size within the deposit of mixed sizes. A more detailed review of selective grain entrainment is provided by Komar (1996), including the presentation of curves and equations for flow competence evaluations, together with a review of grain sorting by particle density in the sand-size range, important to the formation of placers.

Paul D. Komar

Bibliography

Hjulstrøm, F., 1939. Transport of detritus by moving water. In Trask, P.O. (ed.), *Recent Marine Sediments*. American Association of Petroleum Geologists, pp. 5–31.

Komar, P.D., 1987. Selective grain entrainment by a current from a bed of mixed sizes: a reanalysis. *Journal of Sedimentary Petrology*, **57**: 203–211.

Komar, P.D., 1996. Entrainment of sediments from deposits of mixed grain sizes and densities. In Carling, P.A., and Dawson, M.R. (eds.), *Advances in Fluvial Dynamics and Stratigraphy*. John Wiley & Sons Ltd., pp. 127–181.

Miller, M.C., McCave, I.N., and Komar, P.D., 1977. Threshold of sediment motion under unidirectional currents. *Sedimentology*, **24**: 507–527.

Neill, C.R., 1968. Note on initial movement of coarse, uniform bed-material. *Journal of Hydraulic Research*, **6**: 173–176.

Yalin, M.S., and Karahan, E., 1979. Inception of sediment transport. *American Society of Civil Engineers, Journal of the Hydraulics Division*, **105**(HY11) 1433–1443.

Cross-references

Armor
Eolian Transport and Deposition
Flow Resistance
Flume
Grain Size and Shape
Placers
Sediment Transport by Unidirectional Water Flows
Surface Forms

GRAVITY-DRIVEN MASS FLOWS

Introduction

Gravity-driven mass flows, also known as sediment gravity flows, include a spectrum of phenomena in which more-or-less coherent mixtures of grains and intergranular fluid flow down slopes. At one end of this spectrum are dilute flows in which momentum is transferred mostly by fluid forces and sediment is largely a passive cargo that increases the effective fluid density. These dilute mass flows are part of a larger class of fluid dynamical phenomena known as gravity currents (Simpson, 1987; Bonnecaze *et al.*, 1993). At the other end of the spectrum are concentrated mass flows such as granular avalanches and debris flows. Forces exerted by interacting solid grains dominate momentum transfer in such flows, although fluid forces can mediate grain interactions (Savage and Hutter, 1989; Iverson, 1997).

A complete gradation of mass flows probably exists between the dilute (fluid-dominated) and concentrated (grain-dominated) end members of the spectrum. Most research emphasizes one end member or the other, but this article considers the full spectrum of mass flows and emphasizes unifying physical principles that help explain similarities and differences among flows with diverse origins, compositions, and modes of deposition. A fully unified physical theory applicable to all gravity-driven mass flows does not yet exist, however.

A variety of nomenclature is commonly used to describe and categorize gravity-driven mass flows that occur in subaerial, subaqueous, and volcanic environments (Table G16). This article does not emphasize nomenclature, because nomenclature is less important than is understanding the phenomena that govern flow behavior and sedimentation.

Macroscopic dynamics: unsteady flow with a free surface

All gravity-driven mass flows share some macroscopic traits that are largely independent of sediment concentration and the detailed behavior of solid and fluid constituents. One universal trait is unsteady, nonuniform motion—that is, motion that changes with time and position. Unlike ocean currents and river floods, gravity-driven mass flows typically have distinct starting and ending points in space and time. Most flows originate from static rock or sediment masses in steeply sloping source areas, accelerate and elongate as they move downslope, and then decelerate on flatter slopes, where they eventually form deposits. The character and composition of mass flows can evolve as changes in mass and momentum distributions occur during various stages of motion.

Another universal feature of gravity-driven mass flows is an unbounded upper surface that is free to change its shape. The presence of this deformable free surface fundamentally distinguishes gravity-driven mass flows from flows in pipes or other containers and also from rigid landslides that translate down slopes. The free upper surfaces of mass flows evolve rapidly after motion commences; most flows quickly develop blunt heads or snouts and tabular bodies with horizontal dimensions much greater than their thicknesses (Figure G29).

The tabular geometry of gravity-driven mass flows allows their macroscopic motion to be analyzed using shallow-flow theory. This well-known theory results from applying the principles of mass conservation and momentum conservation (Newton's laws of motion) to a deformable body flowing down a slope, and from using approximations that are generally suitable if the thickness of the body is much less than its areal extent. Vreugdenhil (1994) and Gray *et al.* (1999) provide detailed derivations of shallow-flow theory for water floods and granular avalanches, respectively, and Parker *et al.* (1986) and Iverson and Denlinger (2001) provide derivations for dilute and concentrated grain-fluid mixtures, respectively.

The equations of shallow-flow theory express the effects of macroscopic mass and momentum conservation on mass-flow behavior. A simple but instructive form of the equations that

Table G16 Names and attributes of some common gravity-driven mass flows. "Ambient fluid" is the fluid surrounding the flow. "Intergranular fluid" is the fluid within the flow. "Sediment concentration" is high or low depending on whether grains or fluid dominate momentum transfer

	Attributes of flow			
Name of flow	Ambient fluid	Intergranular fluid	Sediment concentration	Predominant grain size
Rock avalanche	Air	Air	High	Sand and larger
Debris flow	Air or water	Water	High	Sand and larger
Mud flow	Air or water	Water	High	Sand and smaller
Lahar	Air (at volcano)	Water	High	Undifferentiated
Pyroclastic flow	Air (at volcano)	Air	High	Sand and larger
Ash flow	Air (at volcano)	Air	Low	Sand and smaller
Turbidity current	Water	Water	Low	Sand and smaller

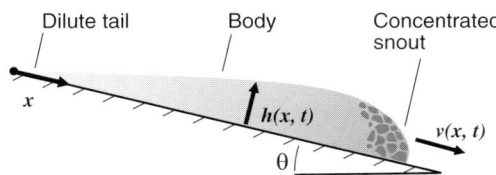

Figure G29 Schematic vertical cross section of a generic gravity-driven mass flow, with identification of morphological features and variables defined in the text. Vertical exaggeration is about 10–100×.

describe flow in one direction is

$$\rho\left[\frac{\partial h}{\partial t} + \frac{\partial (hv)}{\partial x}\right] = 0 \qquad (\text{Eq. 1})$$

$$\rho\left[\frac{\partial (hv)}{\partial t} + \frac{\partial (hv^2)}{\partial x}\right] = (\rho - \rho_a)gh\left(\sin\theta - k\cos\theta\frac{\partial h}{\partial x}\right) - \tau \qquad (\text{Eq. 2})$$

in which the independent variables are time, t, and downslope distance, x, and the dependent variables are the flow thickness $h(x,t)$ and velocity $v(x,t)$ (Figure G29). Parameters in the equations include the mass density of the flowing mixture, ρ, the mass density of the ambient fluid through which the mixture flows, ρ_a, the magnitude of gravitational acceleration, g, the slope angle, θ, and a lateral stress coefficient, k (equal to 1 if the mass flow behaves as a fluid but varying from about 1/5 to 5 if grain-contact stresses are important (Iverson, 1997)). All of these parameters can be measured or estimated with fair accuracy, and all are assumed to be constants in equations 1 and 2. A more enigmatic quantity in equation 2 is the resisting shear stress, τ, which can vary greatly depending on whether solid or fluid forces are dominant. Before considering the differing effects of solid and fluid forces, however, it is important to assess the terms in equations 1 and 2 that are universal and responsible for strong similarities among gravity-driven mass flows.

Each term in the mass-conservation equation 1 involves the product of ρ (mass per unit volume) and a velocity that describes the rate at which mass is redistributed within a deforming flow. The velocity $\partial h/\partial t$ expresses the change in the free-surface height due to movement of mass normal to the bed, and the velocity $\partial(hv)/\partial x$ expresses mass redistribution parallel to the bed. A term can be added to equation 1 to express the velocity at which erosion or deposition normal to the bed adds or subtracts mass from the moving flow, but this term is omitted for the present because it significantly complicates the accompanying momentum equation 2.

The momentum-conservation equation 2 includes terms in brackets that are like the terms in brackets in the mass-conservation equation 1, except the terms in equation 2 contain an additional v. This relationship between mass-conservation and momentum-conservation terms is no accident, as the additional v's in equation 2 reflect the definition of momentum: mass times velocity. Overall, the terms in brackets on the left side of equation 2 express the rate at which a segment of the deforming flow accelerates in response to the evolving balance between driving and resisting stresses. These inertial terms represent the "ma" term in Newton's famous equation, "F = ma." Inertial terms play a crucial role in mass-flow dynamics, and they cannot be neglected in efforts to understand mass-flow deposition.

The terms on the right side of equation 2 represent the "F" term in "F = ma," and they express the balance of stresses that drive and resist mass-flow motion. The first term, $(\rho-\rho_a)gh\sin\theta$, is the driving stress due to the component of gravity that pulls the mass downslope. The gravitational pull is effectively reduced by the presence of ambient fluid with density ρ_a, because the ambient fluid exerts an upward buoyancy force on the flowing mass. However, this buoyancy force is negligible in concentrated subaerial flows, which typically have $\rho/\rho_a > 1,000$. The second term on the right side of equation 2 is the longitudinal stress due to the buoyant weight of flowing material that pushes against adjacent material upslope and downslope, $-(\rho-\rho_a)ghk\cos\theta(\partial h/\partial x)$. This longitudinal stress resists motion if the flowing mass thickens in the downslope direction ($\partial h/\partial x < 0$), but it helps drive motion if the mass tapers downslope ($\partial h/\partial x > 0$). The third term on the right side of equation 2 is the resisting stress due to internal and basal shearing, $-\tau$. If the terms on the right side of equation 2 sum to zero, the mass is static or has constant velocity; otherwise the mass accelerates or decelerates.

Equation 2 reveals that evolution of gravity-driven mass flows into tabular forms is attributable to gravity itself. When a mass flow commences, the leading edge of the flow experiences more driving force than resistance as gravity pulls it downslope. (Otherwise it would not move!) Moreover, the leading edge is pushed forward by the weight of the mass behind it, because $\partial h/\partial x < 0$ necessarily applies in the tapered leading edge. Behind the tapered leading edge there necessarily exists a region in which $\partial h/\partial x > 0$, and in this region the distribution of weight exerts a backward (upslope) force on the trailing mass, which causes it to lag behind. The flowing mass thereby flattens into a tabular body, and continued tapering of the leading edge persists until the mass begins to decelerate or additional forces locally increase resistance and produce a blunt front or snout.

Two distinct mechanisms produce blunt fronts or snouts at the heads of gravity-driven mass flows. In dilute flows the mechanism involves the ambient fluid (water or air) that is displaced by an advancing flow front (Simpson, 1987). Acceleration of the ambient fluid produces an inertial reaction stress approximately equal to $\frac{1}{2}\rho_a v_{head}^2$, where v_{head} is the velocity of the advancing flow front. The reaction stress pushes against the flow front and produces resistance in addition to that included in the momentum equation 2. However, this additional resistance plays an important role only if $\frac{1}{2}\rho_a v_{head}^2$ has a magnitude comparable to that of the internal stresses affecting the flow. As indicated on the right side of equation 2, internal stresses have magnitudes proportional to $(\rho - \rho_a)gh$. A useful criterion for evaluating the importance of the flow-front reaction stress compares the values of $\frac{1}{2}\rho_a v_{head}^2$ and $(\rho - \rho_a)gh$ by employing their ratio $\mathrm{Fr}_{head}^2 = [\rho_a/(\rho - \rho_a)](v_{head}^2/gh)$, where Fr is known as a *densiometric Froude number*.

Experiments have shown that $\mathrm{Fr}_{head}^2 \approx 0.5$ is typical of advancing fronts of dilute mass flows (Middleton, 1993). For example, for a dilute subaqueous flow with a thickness of 1 m and a density 10 percent greater than that of the ambient water through which it flows, $\mathrm{Fr}_{head}^2 \approx 0.5$ occurs when flow speeds reach about 0.7 m per second. Such speeds are readily attainable, and dilute flows can therefore develop blunt snouts solely as a consequence of the inertia of the ambient fluid.

Concentrated subaerial mass flows develop blunt snouts through an entirely different mechanism, which entails local thickening due to locally enhanced internal friction. Ambient fluid has negligible effect on concentrated subaerial flows, in which $\rho_a \ll \rho$ and $\mathrm{Fr}_{head}^2 \ll 0.1$ are typical. However, intense grain interactions in such flows cause efficient grain-size segregation that almost invariably leads to a cumulation of coarse clasts at flow fronts (Iverson, 1997; Vallance and Savage, 2000). Commonly these coarse clasts are more angular than the smaller clasts in the flow, and they produce more internal friction (Pouliquen *et al.*, 1997). Coarse-grained fronts also dissipate pore-fluid pressures in liquid-saturated mass flows, thereby enhancing local resistance and growth of blunt snouts (Major and Iverson, 1999). Feedbacks involving grain-size segregation, pore-fluid pressure, and flow-front resistance indicate that micromechanics on the scale of individual grains can play a key role in macroscopic flow dynamics.

Concentrated subaqueous mass flows (such as submarine debris flows) constitute a hybrid case in which blunt fronts may develop through a combination of grain-size segregation and the inertia of ambient fluid. Such flows can begin to hydroplane on a thin film of water when flow is fast enough that $\mathrm{Fr}_{head}^2 \approx 0.2$; in such instances the snout may detach from the mass-flow body and run out in advance of it (Mohrig *et al.*, 1998).

Micromechanics: steady uniform shear flows

Momentum transfer on the scale of individual grains and similarly small elements of fluid occurs differently in dilute and concentrated mass flows. In dilute flows viscosity and turbulence of the fluid phase cause most micromechanical momentum transfer. In concentrated flows enduring grain contacts and brief grain collisions are more important, although fluid may mediate grain interactions. These diverse modes of micromechanical momentum transfer are not represented explicitly in macroscopic models of mass flow, but are instead represented by stresses. Thus, micromechanics provides a basis for assessing the factors that determine the resisting shear stress τ and the lateral stress coefficient k that appear in equation 2.

Micromechanics can be quantified most simply by focusing on ideal, steady, uniform, shear flows (Figure G30), which are called "rheometric" flows because exact knowledge of the macroscopic state of stress aids determination of rheological relationships between stress and deformation. Many investigations of mass-flow mechanics have emphasized ideal, rheometric flows (e.g., Bagnold, 1954; Lowe, 1976; Chen, 1987), but it is important to recognize that this idealization balances the driving stress $(\rho - \rho_a)gh \sin \theta$ against just one term (τ) in a general mass-flow momentum equation such as equation 2. Therefore, the following characterization of steady, uniform, shear flows provides limited insight about unsteady flows in which all the terms in equation 2 are important.

Sediment concentration and the strength of the effective gravitational force per unit volume $(\rho - \rho_a)g$, are the most obvious causes of differing styles of micromechanical momentum transfer in dilute and concentrated mass flows. Effects of volumetric sediment concentration (v_s) and gravitational forces can be quantified in a rudimentary way by evaluating the ratio of the buoyant unit weight of the flowing mixture to its total unit weight, which yields the normalized bulk density

$$\rho' = \frac{\rho - \rho_a}{\rho} = \frac{\rho_s v_s + \rho_f(1 - v_s) - \rho_a}{\rho_s v_s + \rho_f(1 - v_s)} \quad \text{(Eq. 3)}$$

where ρ_s and ρ_f are the densities of the solid grains and intergranular fluid, respectively, and $\rho = \rho_s v_s + \rho_f(1 - v_s)$ is the bulk density of the grain-fluid mixture. The fundamental importance of ρ' is evident from the fact that this quantity appears as a governing parameter on the right-hand side of the macroscopic momentum equation 2 if the equation is normalized by dividing all terms by ρgh.

Unfortunately, no universal criterion exists for distinguishing concentrated, grain-dominated flows from dilute,

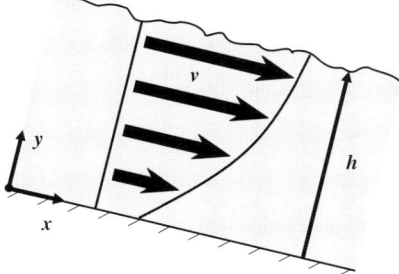

Figure G30 Schematic vertical cross section of a steady, uniform, gravity-driven shear flow with a hypothetical velocity profile. The mixture shear rate is related to the velocity profile slope by $\dot{\gamma} = dv/dy$ and is approximated by $\dot{\gamma} = \bar{v}/h$, where $\bar{v}$ is the average velocity over the interval $y = 0$ to $y = h$.

fluid-dominated flows solely on the basis of values of ρ'. For steady shear flows in which the ambient fluid is water and the solid phase consists of lithic grains with density $\sim 2{,}700\,\mathrm{kg/m^3}$, flows are probably in the concentrated regime if $\rho' > 0.4$, and are likely in the dilute regime if $\rho' < 0.2$. Intermediate values of ρ' indicate a transitional regime that is neither concentrated nor dilute. If the ambient fluid is air, $\rho_a \approx 0$ and $\rho' \approx 1$ apply in most cases, and $\rho_s v_s / \rho_f (1 - v_s)$ is a more useful parameter than ρ' for distinguishing concentrated and dilute flows (Iverson and Vallance, 2001).

Concentrated flows

Large values (~ 1) of the dimensionless parameters ρ' and $\rho_s v_s / \rho_f (1 - v_s)$ indicate that solid grains are likely to dominate momentum transfer, but these parameters provide no information about the effects of flow size, the rate of deformation, or the vigor of grain interactions. Additional dimensionless parameters include these effects by including flow thickness, h, grain diameter, δ, fluid viscosity, μ, and mixture shear rate, $\dot{\gamma}$ in their definitions. These dimensionless parameters may be termed the *Savage*, *Stokes*, and *Bagnold numbers*, defined here as (cf. Iverson, 1997; Iverson and Vallance, 2001)

$$N_{Sav} = \frac{\rho_s \dot{\gamma}^2 \delta^2}{(\rho_s - \rho_f) g h} \quad N_{Sto} = \frac{\mu \dot{\gamma}}{(\rho_s - \rho_f) g h} \quad N_{Bag} = \frac{\rho_s \dot{\gamma} \delta^2}{\mu} \lambda^{1/2}$$

(Eq. 4a,b,c)

where λ is the "linear concentration" of grains defined by Bagnold (1954). For spherical grains, the linear concentration is related to the volumetric concentration by $\lambda = v_s^{1/3} (v_*^{1/3} - v_s^{1/3})$, where v_* is the maximum concentration that is geometrically possible. The numbers defined in equation 4a,b,c take no account of fluid turbulence, which is assumed negligible owing to the high concentration of solid grains. (Although concentrated mass flows can be highly agitated when they move rapidly down rough slopes, agitation of the granular mass is not the same as fluid turbulence.)

The dimensionless parameters defined in equations 4a,b,c have the following physical significance: N_{Sav} estimates the ratio of grain-collision stresses to grain-contact stresses induced by gravity; N_{Sto} estimates the ratio of viscous fluid stresses to grain-contact stresses induced by gravity; and N_{Bag} estimates the ratio of grain-collision stresses to viscous fluid stresses if gravity-induced grain-contact stresses are negligible ($N_{Sav} \to \infty$; $N_{Sto} \to \infty$), as in the experiments of Bagnold (1954). Importantly, the parameters defined in equation 4 are interrelated by $N_{Bag} = (N_{Sav}/N_{Sto}) \lambda^{1/2}$.

Figure G31(A) illustrates a scheme for distinguishing modes of micromechanical momentum transfer in concentrated, steady shear flows on the basis of values of N_{Sav}, N_{Sto} and N_{Bag}. The figure indicates that flows with sufficiently small values of N_{Sav} and N_{Sto} generate stresses primarily through enduring, gravitational grain contacts that produce intergranular rubbing (i.e., Coulomb friction). The Coulomb friction regime is typical of relatively thick flows ($h > 1$ m) with shear rates less than 10 per second, such as most rock avalanches and debris flows (Iverson and Denlinger, 2001). The Bagnold number of such flows is largely irrelevant, because gravity stresses dominate. On the other hand, if either N_{Sav} or N_{Sto} has a value that is of order 1 or larger—as may be true in thin, rapidly shearing flows—Coulomb friction is relatively unimportant, and the Bagnold number distinguishes flow regimes in which grain collisions and fluid viscosity dominate momentum transfer. Boundaries between various concentrated-flow regimes are indistinct, and a combination of Coulomb friction, grain collisions, and fluid viscosity may be important in some circumstances. Moreover, transitions may occur between concentrated flow regimes and dilute flow regimes, particularly if the ambient fluid density ρ_a is similar to the intergranular fluid density ρ_f.

Dilute flows

Dilute mass flows occur most commonly where the ambient fluid and intergranular fluid are the same, a condition that facilitates fluid entrainment and sediment dispersal. Examples of such flows include subaqueous turbidity currents and subaerial volcanic ash flows. Buoyancy forces due to the ambient fluid are more important in dilute flows than in concentrated flows, and for dilute flows the parameter ρ' has correspondingly smaller values. [Equation 3 defines ρ' slightly differently than do some works on dilute, buoyancy-dominated mass flows, which use ρ_a rather than ρ in the denominator (e.g., Bonnecaze et al., 1993; Simpson, 1987). However, the two definitions differ negligibly if the ambient fluid and intergranular fluid are the same and the sediment concentration is small.]

Commonly, ρ' is multiplied by g to define a "reduced gravity," $g' = \rho' g$, which appears in a densiometric Froude number applicable to steady, uniform flows,

$$\mathrm{Fr}^2 = \frac{v^2}{g' h} = \frac{\dot{\gamma}^2 h}{g'} = \frac{\rho}{\rho - \rho_a} \frac{\dot{\gamma}^2 h}{g}$$

(Eq. 5)

The last two forms of equation 5 employ the relation, $\dot{\gamma} = v h$, which is a suitable approximation for uniform shear flows (Figure G3). For cases with $\rho \gg \rho_a$, equation 5 reduces to the conventional definition of the Froude number for open-channel flow of dense liquid (e.g., water) beneath ambient air. Furthermore, Fr^2 defined in equation 5 bears a strong resemblance to N_{Sav} defined in equation 4a. Both definitions involve a ratio of inertial forces to gravity forces, but in N_{Sav} the forces are those of discrete solid grains, whereas in Fr^2 the forces are those of continuous fluid.

Micromechanical momentum transfer in dilute, gravity-driven mass flows depends not only on fluid inertia but also on

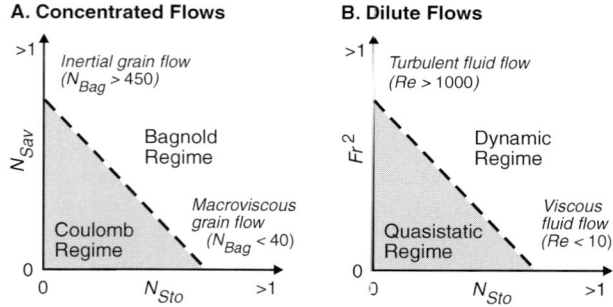

Figure G31 A scheme that uses values of dimensionless parameters to distinguish dominant modes of micromechanical momentum transfer in steady, uniform, gravity-driven shear flows of grain-fluid mixtures (See Figure G30).

the effective fluid viscosity, which is enhanced by the presence of suspended sediment. No generally applicable formula describes this enhancement, but for dilute ($v_s < 0.1$) aqueous mixtures of silt and clay-sized sediment, Einstein's (1906) equation provides satisfactory estimates. The Einstein equation, $\mu_{effective} = \mu(1 + 2.5v_s)$, indicates that effective viscosity is increased only moderately by suspended sediment.

Reynolds numbers are generally used to assess the relative importance of fluid inertia and viscosity, which together determine the propensity for fluid turbulence. An applicable Reynolds number (Re) for dilute mass flows is defined by the ratio of Fr^2 and N_{Sto} (just as a Bagnold number is defined by the ratio of N_{Sav} and N_{Sto} for concentrated flows). The form of this Reynolds number is simplest if it is assumed that $\rho_a = \rho_f$, so that $\rho' = v_s(\rho_s - \rho_f)/\rho$. In this case

$$\text{Re} = \frac{Fr^2}{N_{Sto}} = \frac{1}{v_s} \frac{\rho \dot{\gamma} h^2}{\mu} \quad \text{(Eq. 6)}$$

If the factor $1/v_s$ is neglected, equation 6 is identical to the conventional Reynolds number describing steady, uniform shear flow of pure fluid with $\dot{\gamma} = v/h$.

Figure G31(B) illustrates a scheme that employs Fr^2, Re, and N_{Sto} for distinguishing regimes of micromechanical momentum transfer in dilute, steady shear flows. Two basic regimes are identified: a quasistatic regime in which small values of Fr^2 and N_{Sto} indicate that gravity forces dominate, and a dynamic regime in which either viscous or inertial forces dominate, depending on the value of Re. Most dilute mass flows have relatively large values of Fr^2 and Re, indicating that inertial forces and turbulent flow are prevalent (Parker et al., 1986; Bonnecaze et al., 1993; Middleton, 1993).

The idealized steady flow regimes depicted in Figure G31 help identify parallels between dilute and concentrated shear flows. For example, Figure G31(A) and (B) illustrate an analogy between the Froude number of dilute flows and Savage number of concentrated flows. They also illustrate an analogy between the Reynolds number of dilute flows and Bagnold number of concentrated flows. Large values of Re and N_{Bag} indicate agitated flow regimes in which inertial forces dominate, and small numbers indicate more quiescent flows. However, owing to strong buoyancy effects that reduce the effect of gravity on most dilute flows, the Reynolds regime in dilute flows is more prevalent than is the Bagnold regime in concentrated flows.

Micromechanics: unsteady flow, changes of state, and sedimentation

Gravity-driven mass flows seldom resemble ideal, steady, uniform, rheometric flows. As mass-flow motion ceases, for example, $\dot{\gamma} \to 0$ and behavior necessarily shifts toward the origins of Figure G31(A) and/or (B). Thus, the apparent rheology of mass flows can change as flow speeds diminish and quasistatic gravitational forces surpass dynamic forces.

It is instructive to view departures from steady, uniform, rheometric flow in terms of changes in "state variables," which characterize partitioning of internal mechanical energy as a function of position and time (Iverson and Vallance, 2001). The central idea of the state-variable concept is that grain-fluid mixtures can exist in differing states, somewhat analogous to conventional solid and fluid states (Jaeger et al., 1996). For example, a rapidly flowing granular mixture can behave much like a fluid, whereas a static deposit of the same mixture can behave much like a rigid solid. Values of state variables depend on the composition of a mass flow but also evolve as flow proceeds from initiation to deposition. Therefore, state variables differ fundamentally from rheological constants.

Two readily identifiable state variables are the nonequilibrium intergranular fluid pressure, p, which is the observed fluid pressure minus the local hydrostatic equilibrium pressure, and the granular temperature, T, which is a measure of the intensity of grain velocity fluctuations around a mean value. Both p and T are zero in static equilibrium states. Both p and T can increase during a mass flow through conversion of bulk potential and kinetic energy to internal mechanical energy, and both can decrease through degradation of internal mechanical energy to thermodynamic heat. Increases in either p and T decrease the rigidity of a grain-fluid mixture and enhance its tendency to flow.

Large values of p exist when fluid pressures approximately balance the total stress due to the weight of the solid-fluid mixture. In dilute flows, large values of p exist continuously because the solid grains contribute little of the mixture weight. In concentrated flows, large values of p can result from subtle increases in v_s that occur if the sediment mixture settles gravitationally and transfers stresses to the adjacent fluid—a process known as *liquefaction*. Liquefaction transiently mimics the effects of enhanced buoyancy because it reduces Coulomb friction at grain contacts and inhibits further settlement. Therefore, a liquefied, concentrated mass flow can to some degree mimic the behavior of a dilute flow and exhibit great flow efficiency, but this behavior continues only as long as liquefaction persists (see *Liquefaction and Fluidization*).

Interdependence of changes in p, T, and v_s in unsteady mass flows has important consequences for sedimentation. A rudimentary understanding of this interdependence can be gained by considering the relative timescales of vertical grain settling and downslope mass-flow motion. Because downslope mass-flow motion is driven by the potential for gravity-driven freefall, the timescale for motion is $\sqrt{L/g'}$, where L is the flow length, typically similar to the slope length (Savage and Hutter, 1989). The timescale for gravity-driven settling of grains is the same as the timescale of dissipation of p in the absence of energy conversions that provide new inputs of p and T (Gidaspow, 1994, chapter 12). This settling timescale is h^2/D, where h is the flow thickness and D is the pore-pressure diffusivity or "consolidation coefficient." Values of D for mass-flow mixtures depend on sediment concentrations and grain-size distributions and can range over many orders of magnitude; typical values range from about 10^{-7}m^2 per second in concentrated, clay-rich slurries to about 0.1m^2 per second in dilute suspensions of sand in water. The wide variation in D values produces wide variation in the ratio of the pressure-diffusion time scale to the mass-flow motion time scale, $(h^2/D)/\sqrt{L/g'}$.

If $(h^2/D)/\sqrt{L/g'} \gg 1$, settlement of grains from a moving mixture occurs much more slowly than does motion of the mixture itself. In such cases little tendency exists for grain-by-grain sedimentation, and deposition occurs by relatively abrupt deceleration and stoppage of the mass flow as a unit or series of rapidly accreted units. Such deposition generally produces massive bedding with few sedimentary structures (Major, 1997). On the other hand, if $(h^2/D)/\sqrt{L/g'} \ll 1$, grains can readily settle from a mass flow while it is still moving and thereby produce bedding and structures reminiscent of fluvial deposits (Middleton, 1993). From a sedimentological perspective, differences between gradual, grain-by-grain deposition and abrupt deposition may constitute the central distinction between dilute and concentrated mass flows. Importantly, this distinction depends not only on flow composition but also on flow size (through dependence of $(h^2/D)/\sqrt{L/g'}$ on h and L). Thus, large-size geological flows may produce deposits that differ from those produced by small-scale laboratory flows of the same composition.

Effects of sorting and mass change

The preceding sections emphasize idealized mass flows with more-or-less constant compositions, but in natural environments, mass flows can be highly heterogenous and can alter their compositions through gain or loss of sediment and fluid in transit. This section considers some of these complications.

Grain-size sorting affects both the dynamics and deposits of mass flows. Generally, concentrated mass flows selectively sort large grains upward and toward flow margins, and thereby produce inversely graded deposits with coarse-grained perimeters. The sorting mechanism may be multifaceted (Vallance and Savage, 2000), but the dominant mechanism appears to be *kinetic sieving* (Middleton, 1970). Kinetic sieving occurs only when v_s is sufficiently large that spaces between grains are typically smaller than the grains themselves. Then agitation of the mixture allows unusually small grains but not unusually large grains to fall downward though intergranular voids, leaving a residue of large grains at the flow surface. Once this surface residue forms, a cumulation of coarse clasts tends to form at mass-flow margins because surficial sediment generally moves downslope more rapidly than does sediment nearer the bed. Coarse clasts that accumulate at flow margins produce blunt snouts as noted above, and can control the geometry of lobate deposits (Pouliquen *et al.*, 1997; Major and Iverson, 1999).

Sorting in dilute mass flows is influenced more by fluid forces than grain-to-grain interactions. In dilute flows sorting has much in common with that in typical fluvial systems, and it generally produces normally graded beds and a variety of sedimentary structures that may occur in characteristic sequences (Middleton, 1993). Dilute flows are unlikely to produce inversely graded beds or concentrations of coarse clasts at the perimeters of deposits.

Mass flows can not only sort sediment but can also gain or lose sediment (and fluid) in transit, with fundamental implications for flow dynamics and deposits (e.g., Imran *et al.*, 1998). Mass acquisition or loss is easily represented in the mass-conservation equation 1 by adding a term $\rho E(x, t)$ to the right-hand side, where E is the rate of erosion (or $-E$ is the rate of deposition). A corresponding term $-\rho v E(x, t)$ must then be added to the right-hand side of the momentum-conservation equation 2 to account for the force (expressed as a stress) required to accelerate the eroded sediment from zero velocity to velocity v. However, this added momentum-expenditure term provides no information about the forces that determine the magnitude of E. The term does not account of the additional force required to detach sediment eroded from a bed at rest, nor for the additional force exerted by the bed to decelerate sediment that is deposited.

The difficulty of accounting completely for the forces associated with nonzero E has hindered rigorous inclusion of mass-change terms in models of concentrated mass flows. For dilute mass flows this difficulty has been circumvented by omitting terms such as $-\rho v E(x, t)$ (Imran *et al.*, 1998). The omission is justifiable only if sediment entrainment (or deposition) has negligible effect on momentum conservation in the mass flow as a whole.

Conclusion

Gravity-driven mass flows occur in a variety of subaerial and subaqueous environments. The principles of mass and momentum conservation provide a framework for understanding the behavior of all gravity-driven mass flows, but dilute and concentrated flows have some distinct differences.

Dilute flows generally occur where the ambient fluid and intergranular fluid are very similar and the sediment is sufficiently fine that it is readily dispersed by fluid turbulence. The ambient fluid affects the dynamics of dilute flows by exerting quasi-static buoyancy forces that reduce the effect of gravity and by exerting inertial reaction forces that retard motion of advancing flow fronts. Deposition by dilute flows can occur by settling of individual grains, and can produce sedimentary structures reminiscent of fluvial deposits.

Concentrated mass flows involve significant momentum exchange by solid grains, which is absent in dilute flows. The density of concentrated flows substantially exceeds that of the surrounding ambient fluid, particularly in subaerial environments. Therefore, effects of buoyancy are relatively small in concentrated flows, and effects of grain-size sorting are amplified by strong interactions between sediment clasts. Sorting commonly produces inversely graded vertical profiles and coarse-grained perimeters that enhance friction and growth of blunt snouts at flow fronts. The timescale of grain settling in concentrated flows commonly exceeds the timescale of downslope mass-flow motion. Thus, concentrated flows can produce massive beds with few features reminiscent of fluvial deposits.

Richard M. Iverson

Bibliography

Bagnold, R.A., 1954. Experiments on a gravity-free dispersion of large solid spheres in a Newtonian fluid under shear. *Proceedings of the Royal Society of London*, Ser. A, **225**: 49–63.

Bonnecaze, R.T., Huppert, H.E., and Lister, J.R., 1993. Particle-driven gravity currents. *Journal of Fluid Mechanics*, **230**: 339–369.

Chen, C.L., 1987. Comprehensive review of debris-flow modeling concepts in Japan. In Costa, J.E., and Wieczorek, G.F. (eds.), *Debris Flows/Avalanches: Process, Recognition, and Mitigation*. Geological Society of America Reviews in Engineering Geology, 7, pp. 13–29.

Einstein, A., 1906. A new determination of molecular dimensions. *Annalen der Physik*, **19**: 289–306.

Gidaspow, D., 1994. *Multiphase Flow and Fluidization*. Academic Press.

Gray, J.M.N.T., Wieland, M., and Hutter, K., 1999. Gravity driven free surface flow of granular avalanches over complex basal topography. *Proceedings of the Royal Society of London*, Ser. A, **455**: 1841–1874.

Imran, J., Parker, G., and Katopodes, N., 1998. A numerical model of channel inception on submarine fans. *Journal of Geophysical Research*, **C103**: 1219–1238.

Iverson, R.M., 1997. The physics of debris flows. *Reviews of Geophysics*, **35**: 245–296.

Iverson, R.M., and Denlinger, R.P., 2001. Flow of variably fluidized granular masses across three-dimensional terrain: 1. Coulomb mixture theory. *Journal of Geophysical Research*, **B106**: 537–552.

Iverson, R.M., and Vallance, J.W., 2001. New views of granular mass flows. *Geology*, **29**: 115–118.
Jaeger, H.M., Nagel, S.R., and Behringer, R.P., 1996. Granular solids, liquids, and gases. *Reviews of Modern Physics*, **68**: 1259–1272.
Lowe, D.R., 1976. Grain flow and grain flow deposits. *Journal of Sedimentary Petrology*, **46**: 188–199.
Major, J.J., 1997. Depositional processes in large-scale debris-flow experiments. *Journal of Geology*, **105**: 345–366.
Major, J.J., and Iverson, R.M., 1999. Debris-flow deposition—Effects of pore-fluid pressure and friction concentrated at flow margins. *Geological Society of America Bulletin*, **111**: 1424–1434.
Middleton, G.V., 1970. Experimental studies related to problems of flysch sedimentation. *Geological Society of Canada Special Paper*, **7**: 253–272.
Middleton, G.V., 1993. Sediment deposition from turbidity currents. *Annual Review of Earth and Planetary Sciences*, **21**: 89–114.
Mohrig, D., Whipple, K.X., Hondzo, M., Ellis, C., and Parker, G., 1998. Hydroplaning of subaqueous debris flows. *Geological Society of America Bulletin*, **110**: 387–394.
Parker, G., Fukushima, Y., and Pantin, H.M., 1986. Self-accelerating turbidity currents. *Journal of Fluid Mechanics*, **171**: 145–181.
Pouliquen, O., Delour, J., and Savage, S.B., 1997. Fingering in granular flows. *Nature*, **386**: 816–817.
Savage, S.B., and Hutter, K., 1989. The motion of a finite mass of granular material down a rough incline. *Journal of Fluid Mechanics*, **199**: 177–215.
Simpson, J.E., 1987. Gravity currents in the environment and the laboratory. Chichester: Ellis Horwood Limited.
Vallance, J.W., and Savage, S.B., 2000. Particle segregation in granular flows down chutes. In Rosato, A., and Blackmore, D. (eds.), *Segregation in Granular Flows*. Dordrecht: International Union of Theoretical and Applied Mechanics, pp. 31–51.
Vreugdenhil, C.B., 1994. *Numerical Methods for Shallow-water Flow*. Dordrecht: Kluwer Academic.

Cross-references

Debris Flow
Grading, Graded Bedding
Grain Flow
Liquefaction and Fluidization
Physics of Sediment Transport
Turbidites

GUTTERS AND GUTTER CASTS

The term "gutter cast" was coined by Whitaker (1973) for elongate downward-bulging, deep, narrow erosional structures on the base of sandstone beds. The overlying bed may be a small fraction of the thickness of the gutter cast to several times its thickness or, in cases, the gutter casts are isolated scour-fills surrounded by mudstone or shale. These represent the casts of small-scale erosional channels cut into consolidated mud. A wide variety of sizes and shapes have been described for these features using many different terms, including priels, furrows, rinnen, erosionrinnen, large groove casts, rills, etceteras (see Myrow, 1994). The term gutter cast is preferred and is in general use. The wide range of size, shape, lithology and internal structures of these erosional structures suggest that their origin is polygenetic. The sizes and geometries of gutter casts are likely a function of many variables including the intensity and nature of the eroding flows, length of time that erosion takes place, and the grain size and diagenetic history (e.g., degree of compaction or lithification) of the underlying eroded substrate.

Gutter casts occur in association with pot casts, which are cup-shaped to cylindrical pillars of sandstone that formed from the infilling of potholes, or rounded nonlinear erosional depressions. They range in shape from discs to rounded loaflike forms to tall pillars, and in size from 1 cm to 20 cm in diameter or more. The bottoms of pot casts are commonly deepest around the outside, with a central erosional high. Pot casts have geometries and markings that indicates vertically oriented vortex flow (Myrow, 1992a) similar to that which forms potholes in bedrock in glaciated regions (Alexander, 1932).

Gutter casts are generally composed of sandstone but in some cases they contain conglomeratic lags. Internal structures include normal grading, parallel lamination, and wave-generated stratification including small-scale hummocky cross-stratification and 2d wave ripple lamination. The cross-sectional shapes of these gutter casts range from symmetrical to strongly asymmetrical, and include u-shaped, bilobate, v-shaped, semicircular, flat-based and wide, shallow forms (see Myrow, 1992a). The plan-view shapes of gutter casts are highly variable in width, and range from straight to sinuous to highly irregular. Some gutter casts bifurcate and pinch out along strike. Sinuous gutter casts may have geometries comparable to modern meandering rivers with steep to overhanging walls on the outside of meander bends. Sole markings are common features along the sides and bases of these gutter casts, and include groove marks, prods, and flute marks, and post-depositional trace fossils. The presence of delicate groove marks on the sides of gutter casts indicates that sand-sized sediment may have aided in their erosion as an abrasive agent.

Gutter casts are described from ancient deposits with paleoenvironments that range from tidal flat to submarine fan (Whitaker, 1973). Most gutter casts are described from shallow-marine rocks, and are considered to be storm-generated features (e.g., Aigner and Futterer, 1978; Kreisa, 1981; Aigner, 1985; Myrow, 1992a,b). Currents responsible for their erosion were likely highly variable. Unoriented (Allen, 1962) to bidirectional (Bloos, 1976; Aigner, 1985) prod marks on gutter casts indicating erosion by multidirectional and bidirectional currents/waves or combined flows (Aigner, 1985; Duke, 1990). The occurrence of spiraling or ropelike pattern of grooves on the soles of gutter casts has led to the suggestion that some formed by horizontal helical flows (Whitaker, 1973; Myrow, 1992a). In many cases, gutters were likely formed by powerful unidirectional flows that characterize the initial stages of deposition of tempestites (Myrow, 1992a). Aigner and Futterer (1978) suggested that gutters are produced by currents interacting with obstacles forming horseshoe hollows that were later developed into channels in a downstream direction. This is not readily apparent from most ancient deposits, although evidence for such and origin would likely be lost during continued erosion of the gutter. A series of laboratory experiments on fine-grained cohesive sediment are needed to constrain the nature of flows that could potentially be responsible for gutter erosion, and thus help constrain their conditions of formation within storm depositional models.

The long axes of gutter casts are generally oriented perpendicular to shoreline (e.g., Daley, 1968; Myrow, 1992a,b) although some studies indicate shore-parallel orientations (Aigner, 1985; Aigner and Futterer, 1978). Gutter cast orientation can vary within depositional sequences. McKie (1994) shows a shift from shore-parallel orientations in transgressive systems tracts that were dominated by geostrophic flow to shore-normal orientations in highstand deposits that reflect downwelling flows in friction-dominated conditions. Gutter casts have also been shown to develop at the base of highstand systems as a result of forced regression (Graham and Ethridge, 1995). They are also known from nearshore areas of sediment bypass under thick sediment-laden flows (Myrow, 1992b). In these cases, isolated gutter casts (or those connected by very thin overlying beds) are up to orders of magnitude thicker than intercalated tempestites and represent sediment that accumulated almost exclusively in the gutters during bypass to deeper shelf environments.

Paul Myrow

Bibliography

Aigner, T., 1985. Storm depositional systems. *Lecture Notes in Earth Sciences*, 3. New York: Springer-Verlag, 174pp.

Aigner, T., and Futterer, E., 1978. Kolk-topfe und -rinnen (pot and gutter casts) im Muschelkalk – anzeger für Wattermeer? *Neues Jahrbuch Für Geologie Und Paläontologie Abhandlungen*, **156**: 285–304.

Alexander, H.S., 1932. Pothole erosion. *Journal of Geology*, **40**: 305–337.

Allen, P., 1962. The Hastings Beds Deltas: recent progress and Easter field meeting report. *Proceedings of the Geologists Association*, **73**: 219–243.

Bloos, G., 1976. Untersuchungen uber Bau und Entstehung der feinkornigen Sandsteine des schwarzen Jura alpha (Hettangian und tiefstes Sinemurian) im schwabischen Sedimentationsbereich. *Arbeiten Institut Für Geologie Paleontologie University Stuttgart*, N.F., **71**: 1–296.

Daley, B., 1968. Sedimentary structures from a non-marine horizon in the Bembridge Marls (Oligocene) of the Isle of Wight, Hampshire, England. *Journal of Sedimentary Petrology*, **38**: 114–127.

Duke, W.L., 1990. Geostrophic circulation or shallow marine turbidity currents? the dilemma of paleoflow patterns in storm-influenced prograding shoreline systems. *Journal of Sedimentary Petrology*, **60**: 870–883.

Graham, J., and Ethridge, F.G., 1995. Sequence stratigraphic implications of gutter casts in the Skull Creek Shale, Lower Cretaceous, northern Colorado. *The Mountain Geologist*, **32**: 95–106.

Kreisa, R.D., 1981. Storm-generated sedimentary structures in subtidal marine facies with examples from Middle and Upper Ordovician of southwestern Virginia. *Journal of Sedimentary Petrology*, **51**: 823–848.

McKie, T., 1994. Geostrophic versus friction-dominated storm flow: paleocurrent evidence from the Late Permian Brotherton Formation, England. *Sedimentary Geology*, **93**: 73–84.

Myrow, P.M., 1992a. Pot and gutter casts from the Chapel Island Formation, southeast Newfoundland. *Journal of Sedimentary Petrology*, **62**: 992–1007.

Myrow, P.M., 1992b. Bypass-zone tempestite facies model and proximality trends for an ancient muddy shoreline and shelf. *Journal of Sedimentary Petrology*, **62**: 99–115.

Myrow, P.M., 1994. Pot and gutter casts from the Chapel Island Formation, southeast Newfoundland—reply. *Journal of Sedimentary Research*, **A64**: 706–709.

Whitaker, J.H. McD., 1973. 'Gutter casts', a new name for scour-and-fill structures: with examples from the Llandoverian of Ringerike and Malmoya, Southern Norway. *Norsk Geologisk Tidsskrift*. **53**: 403–417.

Cross-references

Bedding and Internal Structures
Coastal Sedimentary Facies
Paleocurrent Analysis
Parting Lineation and Current Crescents
Scour, Scour Marks
Storm Deposits
Tool Marks

H

HEAVY MINERAL SHADOWS

First described by Cheel (1984) from the fluvial facies of the Silurian Whirlpool Sandstone of southern Ontario, heavy mineral shadows are accumulations of dark heavy mineral grains (e.g., magnetite, leucoxene) that form on bedding surfaces of predominantly quartz sand. Shadows have been observed on flat, planar surfaces associated with upper flow regime plane beds (Cheel, 1984) and on the stoss sides of some large dunes (Brady and Jobson, 1973). The accumulations display an asymmetric gradient in the concentration (Figures H1 and H2) with the highest concentration on the upstream side of the shadow, decreasing in the downstream direction. Observations in a flume indicated that shadows migrate downstream by the periodic erosion at their upstream edges with subsequent deposition a short distance downstream; the amount of heavy minerals that are deposited decreases downstream, resulting in the downstream decrease in concentration of heavy minerals that characterize shadows. Active heavy mineral shadows were observed to be overlain by much faster moving quartz-rich sand. A series of flume experiments (Cheel, 1984) suggested that heavy mineral shadows were stable at the lower end (e.g., lower velocity) of the upper plane bed stability field and that they "washed out" as flow strength increased.

The main value of heavy mineral shadows is that they can be used as indicators of paleocurrent direction. On bedding plane exposures of upper plane bed deposits (internally displaying horizontal lamination), current (Allen, 1964) and

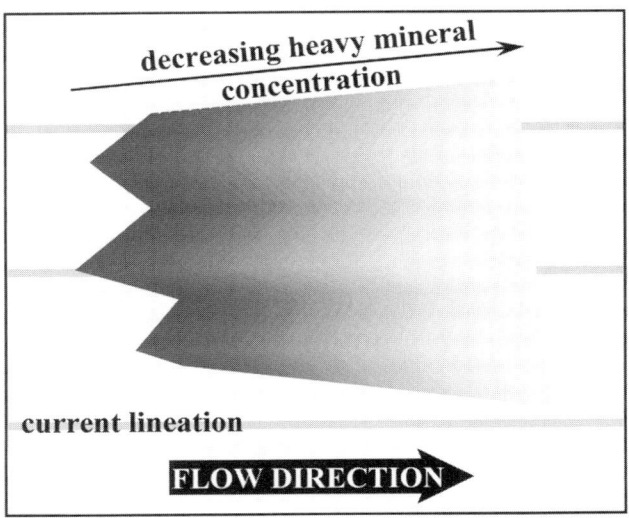

Figure H1 Schematic illustration of a heavy mineral shadow in relation to flow direction and current lineations on an upper flow regime plane bed.

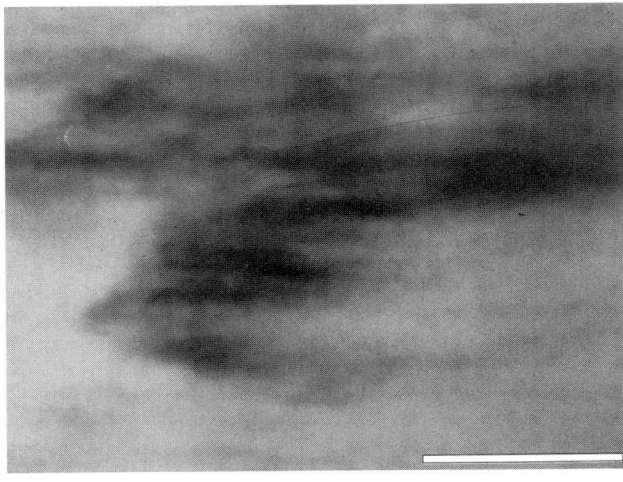

Figure H2 Plan view of a heavy mineral shadow on an active upper flow regime plane bed. The heavy mineral is chromite and the underlying bed is medium grained quartz sand. Flow is from left to right. White bar for scale is 5 cm long.

parting (McBride and Yeakle, 1963) lineations are commonly present (see *Parting Lineations and Current Crescents*), giving the sense of the flow direction but not the absolute direction. When heavy mineral shadows are present, the asymmetry of the heavy mineral concentration gradient indicates the absolute direction, along the lineations (Figure H1).

Rick Cheel

Bibliography

Allen, J.R.L., 1964. Primary current lineations in the Lower Old Red Sandstone (Devonian), Anglo-Welsh Basin. *Sedimentology*, **3**: 89–108.

Brady, L.L., and Jobson, H.E., 1973. An experimental study of heavy-mineral segregation under alluvial-flow conditions. *US Geological Survey Professional Paper*, 562-k, 38pp.

Cheel, R.J., 1984. Heavy mineral shadows: a new sedimentary structure formed under upper-flow-regime conditions: its directional and hydraulic significance. *Journal of Sedimentary Petrology*, **54**: 1175–1182.

McBride, E.F., and Yeakel, L.S., 1963. Relationship between parting lineation and rock fabric. *Journal of Sedimentary Petrology*, **33**: 779–782.

Cross-reference

Parting Lineations and Current Crescents

HEAVY MINERALS

Heavy minerals are high-density components of siliciclastic sediments. They comprise minerals that have specific gravities greater than the two main framework components of sands and sandstones, quartz (s.g. 2.65) and feldspar (s.g. 2.54–2.76). In practice, heavy minerals are usually considered to be those with specific gravities greater than 2.8 to 2.9, the limit being dependent on the density of the liquid used to separate them from the volumetrically more abundant light minerals. Density separation is required because heavy minerals rarely comprise more than 1 percent of sandstones, making it necessary to concentrate them before analysis can take place. Concentration is achieved by disaggregation of the sandstone, followed by mineral separation using dense liquids such as bromoform, tetrabromoethane or the more recently developed nontoxic polytungstate liquids. Heavy minerals are generally identified on the basis of their optical properties under the polarizing microscope. This is usually achieved using grain mounts but some workers prefer to use thin sections. Increasingly, identification is being made on the basis of grain composition, using microbeam techniques. See Carver (1971) and Mange and Maurer (1992) for more details on laboratory methods for the separation and preparation of heavy mineral residues.

Most heavy mineral studies are undertaken to determine sediment provenance, because heavy mineral suites provide important information on the mineralogical composition of source areas (Table H1). Over 50 different non-opaque heavy mineral species have been recognized in sandstones, many of which have specific parageneses that enable a direct match between sediment and source lithology. See Mange and Maurer (1992) for comprehensive information on the optical properties and provenance of a large number of non-opaque heavy minerals.

Geographic and stratigraphic variations in heavy mineral suites within a sedimentary basin can be used to infer differences in sediment provenance. Such differences result either from the interplay between a number of sediment transport systems draining different source regions, or from erosional unroofing within a single source area. Heavy mineral data therefore play an important role in the understanding of depositional history and paleogeography. In some cases, sophisticated mathematical and statistical treatment of heavy mineral data may be required to elucidate the interplay between multiple sediment transport systems.

Although many heavy mineral species are provenance-diagnostic, it is rarely possible to directly match heavy mineral suites in sandstones with those of their source area. This is because a number of processes have the potential to alter the original provenance signal at various points in the sedimentation cycle. The identification of provenance on the basis of heavy mineral data therefore requires careful consideration of all the factors that may have influenced the composition of the assemblages. The factors identified as having significant potential to overprint the original provenance signal (Morton and Hallsworth, 1999) are:

- *Weathering in the source region*: This has the potential to affect the mineralogy prior to incorporation of sediment in the transport system, by increasing the relative abundance of stable to unstable minerals (see Table H1). The effects of weathering in the source region are unlikely to be extensive except under transport-limited denudation regimes (Johnsson et al., 1991).
- *Abrasion during transport*: It may cause destruction of mechanically unstable minerals. Experimental studies have shown that there are variations in the mechanical stability of heavy minerals subjected to abrasion. However, the effects of abrasion on heavy mineral assemblages appear to be negligible. For example, monazite, determined experimentally as the least mechanically stable mineral of all, is commonly found as a placer mineral in beach deposits, even though beach sediments probably suffer the greatest actual degree of mechanical abrasion of all environments, with the possible exception of eolian dune sands.
- *Weathering during periods of alluvial storage on the floodplain*: It may cause depletion of chemically unstable components. There is evidence to suggest this process plays an important role in modifying sand composition, both in terms of framework components (Johnsson et al., 1988) and heavy mineral assemblages (Morton and Johnsson, 1993).
- *Hydrodynamics during transport and at the final depositional site:* Since heavy minerals are denser than quartz and feldspar, their hydrodynamic behavior is different. Furthermore, there is a range of hydrodynamic behavior within the heavy mineral group owing to the wide range of densities (Table H1). This is manifested by variations in both settling velocity and entrainment potential (Rubey, 1933; Slingerland, 1977). The variation in hydrodynamic behavior of heavy minerals is one of the major controls on the relative abundance of heavy minerals in sands and sandstones.
- *Diagenesis*: It causes depletion of minerals that are unstable in the subsurface. Case studies from widely separated

Table H1 Provenance of the most commonly recorded non-opaque detrital heavy mineral species. Note that high-stability minerals (such as zircon, rutile and tourmaline) are commonly recycled

Mineral	Density	Hardness	Stability in acidic weathering	Stability in burial diagenesis	Most common provenance
Anatase	3.82–3.97	$5\frac{1}{2}$–6	High	High	Various igneous and metamorphic rocks; commonly authigenic
Andalusite	3.13–3.16	$6\frac{1}{2}$–$7\frac{1}{2}$	High	Low	Metapelites
Amphibole	3.02–3.50	5–6	Low	Low	Various igneous and metamorphic rocks
Apatite	3.10–3.35	5	Low	High	Various igneous and metamorphic rocks
Cassiterite	6.98–7.07	6–7	High	Uncertain	Acid igneous rocks
Chloritoid	3.51–3.80	$6\frac{1}{2}$	Moderate	Moderate	Metapelites
Chrome spinel	4.43–5.09	$7\frac{1}{2}$–8	High	High	Ultramafic igneous rocks
Clinopyroxene	2.96–3.52	5–$6\frac{1}{2}$	Low	Low	Basic igneous and metamorphic rocks
Epidote group	3.12–3.52	6–$6\frac{1}{2}$	Low	Low	Low-grade metamorphic rocks
Garnet	3.59–4.32	6–$7\frac{1}{2}$	Moderate	Moderate-high	Metasediments
Kyanite	3.53–3.65	$5\frac{1}{2}$–7	High	Low-moderate	Metapelites
Monazite	5.00–5.30	5	High	High	Metamorphic rocks; granites
Orthopyroxene	3.21–3.96	5–6	Low	Low	Ultramafic rocks: high-grade metamorphic rocks
Rutile	4.23–5.50	6–$6\frac{1}{2}$	High	High	High-grade metamorphic rocks
Sillimanite	3.23–3.27	$6\frac{1}{2}$–$7\frac{1}{2}$	High	Low	Metapelites
Staurolite	3.74–3.83	$7\frac{1}{2}$	High	Moderate	Metapelites
Titanite	3.45–3.55	5	Moderate	Low-moderate	Various igneous and metamorphic rocks
Tourmaline	3.03–3.10	7	High	High	Metasediments; granites
Zircon	4.60–4.70	$7\frac{1}{2}$	High	High	Granites and other acidic igneous rocks

sedimentary basins around the world, including the North Sea (Morton, 1984) and the US Gulf Coast (Milliken, 1988) have shown that mineral suites become progressively less diverse as burial depth (and, in consequence, pore fluid temperature) increases. This is the result of dissolution of unstable minerals by circulating pore waters, a process known as *intrastratal solution*. The effects of burial diagenesis on heavy mineral suites can be considerable, since the process has the ability to alter an originally diverse, provenance-diagnostic suite to one solely comprising the stable group zircon-rutile-tourmaline-apatite. The relative stability of detrital heavy minerals during burial diagenesis is shown in Table H1.

Although the combined effect of the processes that operate during the sedimentation cycle obscures the original provenance signal, in many cases to a profound degree, all heavy mineral suites retain important provenance information. Morton and Hallsworth (1994) recommended a combined approach to isolate the provenance-sensitive component of heavy mineral assemblages. The approach concentrates on the stable components of the assemblages, and involves the determination of ratios of minerals with similar hydraulic behavior in conjunction with single-grain geochemical analysis of one or more mineral components (e.g., by electron microprobe). In order to minimize hydrodynamic sorting, ratio determinations should be made on minerals with similar grain sizes and densities (the two main controls on hydraulic behavior).

Provenance studies that concentrate on heavy mineral assemblages have become considerably more sophisticated with the advent of single-grain geochemical analysis equipment. Most of the components of heavy mineral suites have variable compositions that can be readily determined using the electron microprobe. For instance, constraints on plate-tectonic settings of sedimentary basins have been inferred from microprobe analysis of detrital augite (Cawood, 1983), and exhumation of high-pressure metamorphic belts has been evaluated through analysis of amphibole populations (Mange-Rajetzky and Oberhänsli, 1982). Tourmaline compositions can be used to distinguish a variety of granitic and metasedimentary sources (Henry and Guidotti, 1985). Garnet compositions are particularly useful in provenance studies, in view of their response to differences in metamorphic grade and protolith composition. Some heavy minerals, notably zircon, can be dated radiometrically (Pell et al., 1997), and this, combined with conventional heavy mineral analysis, is a very powerful tool for provenance studies.

Heavy minerals have important economic applications, both in the hydrocarbons and minerals sectors. Their use in paleogeographic reconstructions, especially in elucidating sediment transport pathways, is of particular value in hydrocarbon exploration. The recognition of changes in provenance within a sedimentary sequence provides a basis for correlation of strata that is independent of more traditional biostratigraphic methods. Heavy mineral assemblages have proven to be especially useful for correlating sandstones that lack age-diagnostic fossils, especially nonmarine sequences (e.g., Allen and Mange-Rajetzky, 1992). The use of heavy minerals in correlation has important applications in hydrocarbon reservoir evaluation and production. Recent advances have made it possible to utilise the technique on a 'real-time' basis at the well site, where it is used to help steer high-angle wells within the most productive reservoir horizons (Morton *et al.*, in press).

Heavy minerals may become concentrated naturally by hydrodynamic sorting, usually in shallow marine or fluvial depositional settings. Naturally occurring concentrates of economically valuable minerals are known as *placers* (*q.v.*), and such deposits have considerable commercial significance. Cassiterite, gold, diamonds, chromite, monazite and rutile are among the minerals that are widely exploited from placer deposits.

Andrew C. Morton

Bibliography

Allen, P.A., and Mange-Rajetzky, M.A., 1992. Devonian-Carboniferous sedimentary evolution of the Clair Area, offshore northwestern UK: impact of changing provenance. *Marine and Petroleum Geology*, **9**: 29–52.

Carver, R.E., 1971. Heavy mineral separation. In Carver, R.E. (ed.), *Procedures in Sedimentary Petrology*. New York: Wiley, pp. 427–452.

Cawood, P.A., 1983. Modal composition and detrital clinopyroxene geochemistry of lithic sandstones from the New England fold belt (east Australia): a Paleozoic forearc terrain. *Bulletin of the Geological Society of America*, **94**: 1199–1214.

Henry, D.J., and Guidotti, C.V., 1985. Tourmaline as a petrogenetic indicator mineral: an example from the staurolite-grade metapelites of NW Maine. *American Mineralogist*, **70**: 1–15.

Johnsson, M.J., Stallard, R.F., and Lundberg, N., 1991. Controls on the composition of fluvial sands from a tropical weathering environment: sands of the Orinoco drainage basin, Venezuela and Colombia. *Bulletin of the Geological Society of America*, **103**: 1622–1647.

Johnsson, M.J., Stallard, R.F., and Meade, R.H., 1988. First-cycle quartz arenites in the Orinoco River Basin, Venezuela and Colombia. *Journal of Geology*, **96**: 103–277.

Mange, M.A., and Maurer, H.F.W., 1992. *Heavy Minerals in Colour*. London: Chapman and Hall.

Mange-Rajetzky, M.A., and Oberhänsli, R., 1982. Detrital lawsonite and blue sodic amphibole in the Molasse of Savoy, France and their significance in assessing Alpine evolution. *Schweizerische Mineralogische und Petrographische Mitteilungen*, **62**: 415–436.

Milliken, K.L., 1988. Loss of provenance information through subsurface diagenesis in Plio-Pleistocene sediments, northern Gulf of Mexico. *Journal of Sedimentary Petrology*, **58**: 992–1002.

Morton, A.C., 1984. Stability of detrital heavy minerals in Tertiary sandstones of the North Sea Basin. *Clay Minerals*, **19**: 287–308.

Morton, A.C., and Hallsworth, C.R., 1994. Identifying provenance-specific features of detrital heavy mineral assemblages in sandstones. *Sedimentary Geology*, **90**: 241–256.

Morton, A.C., and Hallsworth, C.R., 1999. Processes controlling the composition of heavy mineral assemblages in sandstones. *Sedimentary Geology*, **124**: 3–29.

Morton, A.C., and Johnsson, M.J., 1993. Factors influencing the composition of detrital heavy mineral suites in Holocene sands of the Apure river drainage basin, Venezuela. In Johnsson, M.J., and Basu, A. (eds.), *Processes Controlling the Composition of Clastic Sediments*. Geological Society of America, Special Paper, 284, 171–185.

Morton, A.C., Spicer, P.J., and Ewen, D.F., in press. Geosteering of high-angle wells using heavy mineral analysis: the Clair Field, West of Shetland. In Carr, T., Feazel, C., Mason, E., and Sorensen, R. (eds.), *Horizontal Wells, Focus on the Reservoir*. Memoir of the American Association of Petroleum Geologists.

Pell, S.D., Williams, I.S., and Chivas, A.R., 1997. The use of protolith zircon-age fingerprints in determining the protosource areas for some Australian dune sands. *Sedimentary Geology*, **109**: 233–260.

Rubey, W.W., 1933. The size distribution of heavy minerals within a water-lain sandstone. *Journal of Sedimentary Petrology*, **3**: 3–29.

Slingerland, R.L., 1977. The effect of entrainment on the hydraulic equivalence relationships of light and heavy minerals in sands. *Journal of Sedimentary Petrology*, **47**: 753–770.

Cross-references

Attrition (Abrasion), Fluvial
Grain Settling
Grain Threshold
Placers
Provenance

HINDERED SETTLING

Gravitational segregation and settling of particles in an aqueous suspension are strongly affected by suspension concentration. Lone particles or particles in very low-concentration suspensions settle freely through a fluid unencumbered by hydrodynamic influences of other particles (see *Grain Settling*). However, at a sufficient volume concentration of solids, changes in fluid density and viscosity, particle interactions, and upwelling of fluid caused by downward particle motion (Hawksley, 1951; Davies, 1968) hinder particle settling. As a result, enhanced particle segregation occurs and overall particle-settling velocity is less than that of a single particle of the same size settling in an infinite fluid.

For nearly a century scientists have recognized that particle concentration affects the physical properties of fluids and particle settling behavior. Sedimentation of any given particle in a suspension is affected by both the mere presence of other particles as well as by motion of neighboring particles. The presence of rigid particles in a fluid distorts fluid motion as the fluid is at rest at particle surfaces. Such distortion leads to an increase in effective fluid viscosity, which is reasonably predictable for dilute (≤ 10 percent solids volume) suspensions (Einstein, 1906). Particle motion also distorts fluid motion by dragging fluid along as particles settle. However, owing to mass conservation, downward particle motion induces fluid upwelling which can enhance particle buoyancy and help hinder particle settling.

Over the past several decades, various models have been proposed to predict sedimentation velocities of different particle species in suspensions. Each of these models defines a so-called hindered settling function that describes the reduction of particle settling velocity as a function of particle concentration (e.g., Richardson and Zaki, 1954; Davis and Gecol, 1994). Efforts have evolved from modeling suspensions of monosized spheres at low Reynolds number (Re<0.5) to multicomponent (i.e., multiple grain size and density) suspensions at finite (1–100) Reynolds numbers (Davis and Gecol, 1994; Manasseh et al., 1999), and toward development of numerical models that simulate deposition from multicomponent suspensions containing large numbers of particles (e.g., Zeng and Lowe, 1992; Höfler et al., 2000).

The following discussion outlines the effects of solids concentration on particle settling and the effects of particle settling behavior on the character of resulting sedimentary deposits. The discussion focuses on the behavior of non-flocculating suspensions in static fluids. Flocculent suspensions

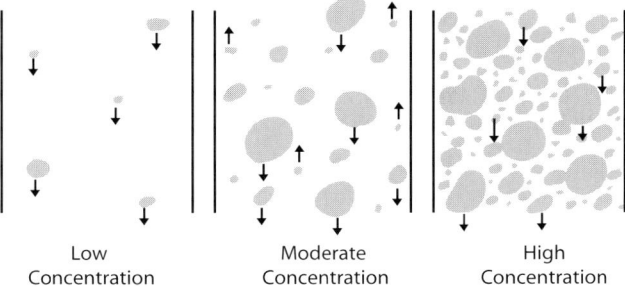

Figure H3 Particle settling in suspensions of various concentrations. At low concentration, particles settle according to Stokes' Law on the basis of size, density, and shape. At moderate concentrations hydrodynamic interactions among particles and concentration-induced changes in the physical properties of the suspending fluid affect particle settling and segregation. Settling of larger and denser particles displaces fluid and causes counterflow that carries smaller and lighter particles upward through the suspension, enhancing particle segregation and altering settling velocities. At high concentrations particle contacts suppress segregation, and all particles settle at the same velocity. Interstitial counterflow consists of nearly pure fluid. Arrows indicate direction of particle movement. (Reprinted from Journal of Volcanology and Geothermal Research, Volume 65, T.H. Druitt, Settling behavior of concentrated dispersions and some volcanological applications, pp. 27–39, Copyright 1995, with permission from Elsevier Science.)

(i.e., cohesive particles) form aggregates of various size and shape that can trap fluid, leading to greater effective particle volume than that of the primary particles (see *Flocculation*). The effective increase in solids volume increases suspension viscosity. In general, aggregate suspensions behave as multi-component suspensions that settle differentially at low and moderate concentration; however, predictions of settling behavior are complicated by floc formation. In turbulent and oscillatory flows, influences other than solids concentration and particle morphology strongly affect particle settling behavior. Such effects, however, are beyond the scope of this discussion.

Effect of concentration on particle settling

Low concentration

Significant interactions among nonflocculating particles generally are negligible at particle concentrations of a few percent or less. In dilute suspensions at low Reynolds number particles settle according to the principles of Stokes' Law, in which buoyant particle weight is balanced by drag forces owing to slow fluid motion around the particle. Settling velocity is affected by particle size, density, and shape as well as by fluid properties, but particles that are hydraulically equivalent settle at the same rate (Figure H3). In a homogeneous, uniform single-component suspension, all particles settle at the same rate, suspension concentration remains constant, and the suspension develops a sharp interface between the settling particles and supernatant fluid. In an initially homogenous, uniform, multicomponent suspension the particles settle independently at different rates and are not hydrodynamically influenced by the presence of other particles (Cheng, 1980; Davis and Acrivos, 1985). As a result, the suspension becomes stratified into settling zones. The lowest zone contains all sediment components and maintains the initial suspension concentration. The overlying region contains all but the fastest settling component, and thus has a lower concentration. Successively higher regions contain one less component than underlying regions, and the uppermost region contains only the slowest settling sediment. The regions are separated by discontinuities in sediment concentration. In a continuous sediment dispersion, the concentration gradient varies gradually with height.

Moderate concentration

At volume concentrations of greater than a few percent particles no longer settle freely and Stokes' Law breaks down. At volume concentrations up to about 50 percent, segregation occurs not only in response to differences in particle size and density, but also to concentration-induced changes in fluid density and viscosity and to fluid displacement by settling of larger and denser particles (Davies, 1968). At these moderate concentrations, sedimentation from multicomponent suspensions commonly is unstable and develops convective fingers (Weiland et al., 1984). Fluid counterflow elutriates finer and lighter particles from lower depths in the suspension, carries them upward through the suspension, and concentrates them near the suspension surface. Settling behavior depends in detail upon solids concentration and on particle size, density, and distribution. In general, the effect of fluid counterflow is to enhance particle segregation. Greatest enhancement occurs in suspensions having large contrasts in particle size and/or density, and in suspensions having relatively high concentrations (Phillips and Smith, 1971).

High concentration

Particle contacts suppress segregation when solids concentration exceeds about 50 percent by volume. At these concentrations, sediment fractionation is negligible, all component particles settle at the same rate, interstitial counterflow consists of nearly pure fluid, and the initial concentration and species distribution are maintained (Figure H3). Strong reduction of vertical size segregation of the suspension results in *en masse* settling or zone sedimentation (Cheng, 1980). The concentration at which this mode of sedimentation dominates suspension behavior varies with particle sorting, size, and shape. In general, the more irregularly shaped or better sorted the sediment' components, the lower the concentrations at which this mode of sedimentation occurs (Davies, 1968).

Effects of suspension concentration and settling behavior on deposit character

Experiments by Druitt (1995) illustrate several aspects of settling behaviors of variably concentrated, multicomponent suspensions on deposit character. The experiments consisted of poorly sorted mixtures of variable size and density plastic and silicon carbide particles having total solids concentrations ranging from 10 percent to 60 percent by volume.

Low-concentration suspension

Despite violating the general conditions for free settling, a 10 percent concentration mixture generally settled as a typical low-concentration suspension. The densest particles segregated rapidly according to particle size and produced a normally graded deposit devoid of sedimentary structures. The lighter, but larger, plastic particles settled with their hydraulically equivalent silicon carbide counterparts, and were dispersed through the deposit according to their hydraulic equivalence. If settling was at all hindered by hydrodynamic particle interactions at this concentration, such hinderance was not readily apparent in the resulting deposit.

Moderate-concentration suspension

Hindered settling affected the sedimentologic characteristics of deposits from suspensions having concentrations ranging from 25 percent to 55 percent. Initially well dispersed suspensions segregated rapidly according to particle size and density, and formed suspensions having vertical concentration gradients. Settling of the dense silicon carbide particles generated vigorous upwelling that swept the finest and lightest particles upward through the suspensions. The light plastic particles segregated and rose upward to their levels of neutral buoyancy (near the surface in the more concentrated suspensions), and at the tops of the suspensions zones of fine particles developed that thickened with time.

Sedimentation by these suspensions produced normally graded deposits having *elutriation pipes* (finger-like zones of well sorted sediment resulting from upward fluid flow), "floating" rafts of light particles, and capping layers of fines elutriated by escaping pore fluid.

High-concentration suspension

A 60 percent concentration mixture settled as a typical high-concentration suspension. There was little contrast between the sedimentologic character of the suspension and its resulting deposit. Minor segregation of particles of different sizes or densities occurred, escaping fluid focused along discrete, high-permeability zones, and the suspension dewatered and consolidated upward from the deposit base (e.g., Major, 2000). The resulting deposit was poorly sorted and ungraded, and was capped by a very thin layer of elutriated fines.

Summary

The settling behavior of suspensions of nonflocculating solids is strongly affected by solids concentration in addition to particle size, shape, and density. Particles segregate and settle according to size and density at low concentrations. At higher concentrations, particle segregation and settling are modified by changes in fluid properties wrought by increased particle volume, by hydrodynamic interactions among particles, and by fluid displacement and counterflow, which can greatly enhance size and density segregation. At sufficiently high concentrations, particle segregation is suppressed resulting in *en masse* settling of the entire suspension.

Jon J. Major

Bibliography

Cheng, D.C.-H., 1980. Sedimentation of suspensions and storage stability. *Chemistry and Industry*, **10**: 407–414.
Davies, R., 1968. The experimental study of the differential settling of particles in suspensions at high concentrations. *Powder Technology*, **2**: 43–51.
Davis, R.H., and Acrivos, A., 1985. Sedimentation of noncolloidal particles at low Reynolds numbers. *Annual Reviews of Fluid Mechanics*, **17**: 91–118.
Davis, R.H., and Gecol, H., 1994. Hindered settling function with no empirical parameters for polydisperse suspensions. *American Institute of Chemical Engineers Journal*, **40**: 570–575.
Druitt, T.H., 1995. Settling behaviour of concentrated dispersions and some volcanological applications. *Journal of Volcanology and Geothermal Research*, **65**: 27–39.
Einstein, A., 1906. Eine neue Bestimmung der Molekuldimensionen. *Annalen der Physik*, **19**: 289–306.
Hawksley, P.G., 1951. The effect of concentration on the settling of suspensions and flow through porous media. In *Some Aspects of Fluid Flow*. London: Edward Arnold and Co., pp. 114–135.
Höfler, K., Müller, M., and Schwarzer, S., 2000. Design and application of object oriented parallel data structures in particle and continuous systems. In Krause, E., and Jager, W. (eds.), *High Performance Computing in Science and Engineering '99*. Berlin: Springer, pp. 403–412.
Major, J.J., 2000. Gravity-driven consolidation of granular slurries: implications for debris-flow deposition and deposit characteristics. *Journal of Sedimentary Research*, **70**: 64–83.
Manasseh, R., Metcalfe, G., and Wu, J., 1999. Polydisperse sedimentation processes: macro- and micro-scale experiments. In Celata, G.P., DiMarco, P., and Shah, R.K. (eds.), *Two-phase Flow Modelling and Experimentation 1999*. Pisa: Edizioni ETS, pp. 881–888.
Phillips, C.R., and Smith, T.N., 1971. Modes of settling and relative settling velocities in two-species dispersions. *Industrial and Engineering Chemistry Fundamentals*, **10**: 581–587.
Richardson, J.F., and Zaki, W.N., 1954. Sedimentation and fluidisation: part 1. *Transactions of the Institution of Chemical Engineers*, **32**: 35–53.
Weiland, R.H., Fessas, Y.P., and Ramarao, B.V., 1984. On instabilities arising during sedimentation of two-component mixtures of solids. *Journal of Fluid Mechanics*, **142**: 383–389.
Zeng, J., and Lowe, D.R., 1992. A numerical model for sedimentation from highly-concentrated multi-sized suspensions. *Mathematical Geology*, **24**: 393–415.

Cross-references

Fabric, Porosity, and Permeability
Flocculation
Grading, Graded Bedding
Grain Settling
Grain Size and Shape
Gravity-Driven Mass Flows
Liquefaction and Fluidization
Slurry

HUMIC SUBSTANCES IN SEDIMENTS

Introduction

Organic carbon compounds are a ubiquitous and essential constituent in sedimentary systems. They fuel virtually all biogeochemical cycles, and are important reservoirs for the major elements of living systems: carbon, hydrogen, nitrogen, oxygen, phosphorous, and sulfur. The physiology of living organisms combined with physicochemical transformations occurring during sediment burial and lithification imparts compositional, structural and isotopic characteristics to organic matter that serve as an important facet of the sedimentary record. Humic substances are high-molecular weight organic compounds defined operationally on the basis of solubility properties and high molecular weight (hundreds to 300,000 Da). They comprise a significant fraction (up to 70–80 percent) of organic matter in soils and recent sediments, and represent a complex organic mixture of primary biochemical compounds and their by-products formed by microbial degradation, oxidation, and incipient chemical polymerization/condensation. Continued thermal exposure and time result in more extensive chemical modification including the formation of kerogen and coal, as well as the release of organic acids, hydrocarbons, and carbon dioxide.

History and importance

Humic substances (HS) have had a long history of examination (compilations include Schnitzer and Khan, 1972; Gjessing, 1976; Christman and Gjessing, 1983; Aiken et al., 1985; and Thurman, 1985). Since the late 1700s, workers have focused on the important role of humic substances in soils and agriculture. Stevenson (1985) provides a detailed history of the study of humic substances. Since 1970, advances in analytical techniques in organic chemistry and petroleum-related research resulted in great advances in understanding sedimentary organic matter as the precursor to kerogen (*q.v.*, and see Durand, 1980). In the 1980s and beyond, interest in HS expanded within ecology and environmental sciences. Many of these advances were related to efforts to better characterize HS (a major fraction of dissolved organic carbon) in a variety of aquatic environments, including groundwater systems (Aiken et al., 1985). Now, at the turn of the millennium, the importance of humic substances in global earth systems and biogeochemical cycling continues to be a growing field. Biomedical technology has led to powerful new methods of analyzing important biomolecules such as DNA, which may be applied to HS. Yet even as analytical capabilities continue to better resolve the individual components of HS, the most fundamental questions are currently being revisited: what are humic substances? Are the traditional concepts of humic substances appropriate (Tate, 2001; Burdon, 2001)? And important new questions may be answerable: how does organic carbon sequestration in soils affect atmospheric CO_2? In what ways can/do humic substances attenuate the migration and bioavailability of anthropogenic organic chemicals and metals?

Composition

Humic substances are complex mixtures of organic compounds. They are operationally defined on the basis of solubility. Humic substances are separated by soil or sediment extraction into humic acid (soluble in aqueous media at pH > 2), fulvic acid (soluble in aqueous media across the range of pH) and humin (insoluble in aqueous media). In the traditional view, HS are mixtures of plant and microbial carbohydrates, proteins, and lipids, with degraded lignins and tannins combined in high molecular weight "polymers" through hydrophobic (e.g., van der Waals) and hydrogen bonds. Although each HS "molecule" is essentially unique, generalized formulas are based on average composition (C, H, N, S, and O content and proportions of various organic functional groups). Table H2 presents average elemental compositional ranges for humic and fulvic acids. Variations in initial organic matter composition, the availability of water and presence of oxygen all play a role in the transformation of labile biogenic compounds into HS.

Structure

Structural determinations of HS have historically been performed on HS after separation by a variety of isolation and fractionation procedures. Isolation generally involves extraction by use of sodium pyrophosphate or other solvents buffered over several pH ranges. Fractionation has been performed by use of resin columns (singly or in tandem), and more recently, through use of gel filtration techniques. Hayes and Malcolm (2001) summarize these procedures. Hatcher and others (2001) provide a thorough overview of structural determinations based on a variety of methods, including pyrolysis coupled with gas chromatography/mass spectrometry (GC–MS). The basic structure for a "typical" humic acid molecule consists of aromatic rings bridged by a variety of CH_2, N, S, and O groups, with peripherally attached protein and carbohydrate residues and simple carboxyl and phenolic moieties. Reconstructions of high-molecular weight substances consistent with observed products and considering the degradation associated with extraction and fractionation procedures has led to the conventional mega-structures thought to exist. Piccolo (2001) takes issue with these constructs and argues for an interpretation of simpler (and smaller) naturally occurring molecules. Advanced techniques in solid-state nuclear magnetic resonance (NMR; Hatcher et al., 2001) which look directly at untreated organic structures, hold great promise for unraveling of these difficulties. Additional tools developed for biomolecular analysis with applications to HS include electrospray ionization and matrix-assisted laser desorption ionization (Hatcher et al., 2001).

Table H2 Average Values of Elemental Compositions of Humic Substances

	Humic acid	Fulvic acid
Carbon	53.8–58.7%	40.7–50.6%
Hydrogen	3.2–6.2	3.8–7.0
Oxygen	32.8–38.3	39.7–49.8
Nitrogen	0.8–4.3	0.9–3.3
Sulfur	0.1–1.5	0.1–3.6

from Steelink (1985).

Despite advances in analytical methodology, often the characterized component of humic substances remains at only 20 percent.

Diagenesis

Humic substances are a transitional form of organic compounds between biomolecules and important fossil fuels (kerogen, coal, oil, and gas). Systematic changes in both structure and composition result as sediments undergo diagenesis and lithification during burial.

Terrestrial sediments

Terrestrial environments with significant components of humic substances include soils, peatlands, lakes, streams, and groundwater. Humic substances differ across these environments based on initial type of organic matter, subsequent exposure to organic matter recycling, and the extent of subsurface oxidation. Lakes sediments dominated by algal organic matter have higher than average H/C ratios do to the high proportion of sugars, carbohydrates and lipids as compared to greater oxygen content of cellulose and lignin-derived organic matter. Continued burial of terrestrial humic substances rich in oxygen (cellulose and lignin derivatives) leads to the formation of peat and eventually coal. The sapropel (algal-dominated) deposits of saline lakes eventually yield hydrocarbons due to the high H/C mentioned above.

Marine sediments

Humic substances in marine sediments derive from both degraded terrestrial organic matter and marine sources. Fulvic acid contents decrease with increasing burial depth in marine sediments. In the humic acid fraction, O/C and N/C ratios also decrease with burial. It is probable that fulvic acid is a product of oxidation of organic matter, and that its production decreases with depth due to a decrease in oxygen availability. Fulvic acid also decreases due to solubilization in the aqueous media and consumption by microbial populations. Humic acid becomes progressively less soluble with depth due to condensation reactions forming humin and eventually kerogen.

Summary

Humic substances in soils and sediment constitute a significant and dynamic reservoir of organic carbon, essential for considerations of carbon cycling. They are also critical in consideration of metal mobility and bioavailability due to metal binding capacities of immobile humic substances and the complexation stability of labile aqueous humic acids. Despite their study for more than two centuries, major advances in understanding HS with applications to agriculture, environmental change, and the energy industry await further investigation.

Laura J. Crossey

Bibliography

Aiken, G.R., McKnight, D.M., Wershaw, R.L., and MacCarthy, P. (eds.), 1985. *Humic Substances in Soil, Sediment and Water*. New York: Wiley Interscience.
Burden, J., 2001. Are the traditional concepts of the structures of humic substances realistic? *Soil Science*, **166**: 752–769.
Christman, R.F., and Gjessing, E.T. (eds.), 1983. *Aquatic and Terrestrial Humic Materials*. Ann Arbor: Ann Arbor Science Publications.
Durand, B., 1980. *Kerogen—Insoluble Organic Matter from Sedimentary Rocks*. Paris: Editions Technip.
Engel, M.H., and Macko, S.A. (eds.), 1993. *Organic Geochemistry: Principles and Applications*. New York: Plenum Press.
Gjessing, 1976. *Physical and Chemical Characteristics of Aquatic Humus*. Ann Arbor, Michigan: Ann Arbor Science.
Hatcher, P., Dria, K., Kim, S., and Frazier, S., 2001. Modern analytical studies of humic substances. *Soil Science*, **166**: 770–794.
Hayes, M., and Malcolm, R., 2001. Structures of humic substances. In Clapp, C., Hayes, M., Senesi, N., Bloom, P., and Jardine, P. (eds.), *Humic Substances and Chemical Contaminants*. WI: Soil Science Society of America, pp. 3–40.
Piccolo, A., 2001. The supramolecular structure of humic substances. *Soil Science*, **166**: 810–832.
Schnitzer, M., and Khan, S.U. (eds.), 1972. *Humic Substances in the Environment*. New York: Marcel Dekker.
Steelink, C., 1985. Implications of elemental characteristics of humic substances. In Aiken, G., McKnight, D., Wershaw, R., and MacCarthy, P. (eds.), *Humic Substances in Soil, Sediment and Water*. New York: Wiley Interscience, pp. 457–476.
Stevenson, F.J., 1985. Geochemistry of soil humic substances. In Aiken, G.R., McKnight, D.M., Wershaw, R.L., and MacCarthy, P. (eds.), *Humic Substances in Soil, Sediment and Water*. New York: Wiley Interscience.
Tate, R., 2001. Soil organic matter: evolving concepts. *Soil Science*, **166**: 721–722.
Thurman, E.M., 1985. *Organic Geochemistry of Natural Waters*. Dordrecht, Netherlands: Martinus Nijhoff/Dr. W. Junk Publishers.

HUMMOCKY AND SWALEY CROSS-STRATIFICATION

Hummocky and swaley cross-stratification are two closely related forms of stratification that are generally attributed to the action of oscillating (wave-generated) currents or combined (oscillating and unidirectional) flows. While these structures were once thought to be ubiquitous to shallow marine storm deposits, similar forms of stratification have been recognized in both clastic and carbonate sediments of a variety of depositional environments.

Hummocky cross-stratification (HCS) was first described by that name by Harms *et al.* (1975) although the structure had been described in earlier literature by other names (e.g., "truncated wave ripple laminae", Campbell, 1966; "crazy bedding", Howard, 1971; "truncated megaripples", Howard, 1972). This structure is characterized by internal laminae that locally dome upward (hummocks) passing laterally into laminae that are concave upward (swales) as shown in Figure H4. HCS is generally thought to be limited to coarse silt and fine sand (Dott and Bourgeois, 1982; Brenchley, 1985; Swift *et al.*, 1987) although it is rarely reported in coarse sand (e.g., Brenchley and Newall, 1982). The fundamental characteristics of HCS were listed by Harms *et al.* (1982, pp. 3–30) and remain important for the recognition of this structure: "(1) lower bounding surfaces of sets are erosional and commonly slope at angles less than 10°, though dips can reach 15°; (2) laminae above these erosional set boundaries are parallel to that surface, or nearly so; (3) laminae can systematically thicken laterally in a set, so that their traces on a vertical sequence are fan-like and dip diminishes regularly; and (4) dip directions of erosional set boundaries and of the overlying laminae are scattered." Harms *et al.* (1982) further stated that the scale of the stratification varies from medium to large scale (decimeters to meters from hummock crest to hummock crest). They attributed the form of the stratification to deposition on bedforms that consisted of, more or less, circular hummocks (positive topographic features) and swales (negative topographic features); thus, the form of the stratification mimics the morphology of the bedform.

Figure H4(A) shows the hypothetical morphology of the bedforms that result in HCS, approximately circular hummocks surrounded by circular swales, along with the internal laminae that make up the structure. Figure H4(B) shows the relationship between the internal laminae and the hummocky and swaley surfaces in terms of a hierarchy of surfaces (Campbell, 1966; Dott and Bourgeois, 1982). In Figure H4(B) the first-order surfaces bound one or more sets of HCS. Second-order surfaces are internal surfaces of scour that bound sets of hummocky laminae; these surfaces display the steepest angles of dip, ranging from 10° to 20°. Presumably second-order surfaces are produced by erosion that makes up a hummocky/swaley surface on which internal laminae (bounded by third-order surfaces) are draped as sediment is deposited on the underlying hummocky surface. This mode of formation is supported by the common observation that internal laminae thicken from hummocks into the swales and become flatter upward within an HCS set (Figures H4 and H5).

Swaley cross-stratification (Figure H6) was first described by Leckie and Walker (1982). SCS is generally similar to HCS but lacks the positive relief associated with hummocks. In addition, Harms *et al.* (1982) stated that the maximum dip of second-order sets in SCS is somewhat shallower than HCS, not exceeding 10°.

While HCS-like stratification (HCS "mimics", for example, Prave and Duke, 1990; Rust and Gibling, 1990) has been

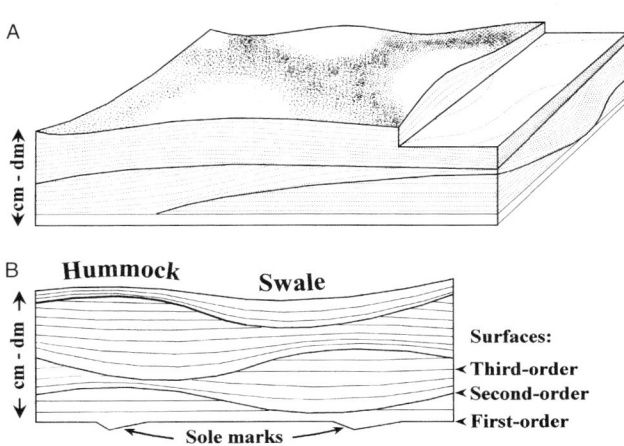

Figure H4 (A) Block diagram illustrating the characteristics of HCS (Harms *et al.*, 1975). (B) The form of stratification and ordered bounding surfaces commonly found in HCS sandstone beds (Cheel and Leckie, 1993).

Figure H6 SCS in out crop. Note the large, low-angle internal surfaces that define that structure. Bar scale is approximately one meter long. Photo courtesy of Dale Leckie.

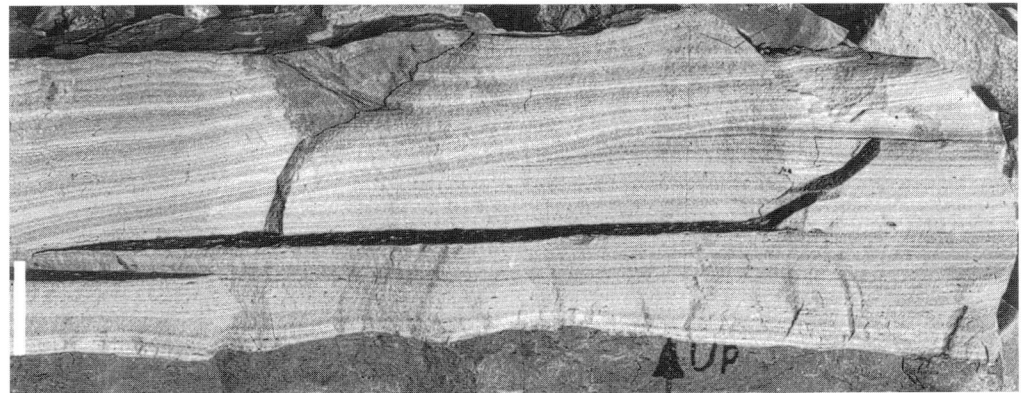

Figure H5 HCS in outcrop showing a hummocky surface cutting into horizontally laminated sandstone, draped by laminae that thin from left to right from the swale to the hummock. Bar for scale is approximately 10 cm. Photo courtesy of Dale Leckie.

described from the deposits of various depositional environments, HCS and SCS are most commonly reported in the deposits of shallow marine shelf settings that were influenced by periodic intense storms (see *Storms Deposits*). In a progradational (shallowing upward) sequence HCS first appears in basal, discrete sandstone beds, bounded by shales, which are commonly overlain by thick sandstones of amalgamated beds displaying HCS. In some such sequences SCS may be present stratigraphically above the amalgamated HCS beds.

HCS and its origin has been the focus of several studies but there remains no consensus as to the mechanism(s) of its formation. Harms *et al.* (1975, 1982) attributed HCS to deposition under powerful oscillatory currents generated by storm waves on the water surface. Others (e.g., Allen, 1985) have suggested that purely oscillating currents could not produce the bedforms that form HCS and that at least a unidirectional component (i.e., a combined flow) was necessary to produce the necessary bed topography. However, experiments by Arnott and Southard (1990) indicated that even a very weak unidirectional component of a combined flow led to the development of asymmetric bedforms that would generate cross-bedding which dipped preferentially in one direction (contrary to the definition of HCS). Similarly, a detailed analysis of grain orientation in within beds displaying HCS identified fabrics with a strong signature of an oscillating current and no recognizable component of the fabric due to unidirectional currents (Cheel, 1991). Cheel and Leckie (1993) suggested that under storm conditions on the open shelf, rogue waves may scour the bed into a hummocky/swaley topography that make up the second-order internal surfaces and, as the waves diminished, previously suspended sand would drape the bedform under the action of weaker waves which resulted in the internal laminae which drape second-order surfaces. SCS would develop under similar conditions but in shallower water where the hummocks have a low preservation potential due to their positive relief. Given the range of settings in which HCS-like stratification has been reported, it is likely that the same style of internal lamination and associated surfaces may be produced by a variety of mechanisms.

Rick Cheel

Bibliography

Allen, P.A., 1985. Hummocky cross-stratification in not produced purely under progressive gravity waves. *Nature*, **313**: 562–564.
Arnott, R.W., and Southard, J.B., 1990. Exploratory flow-duct experiments on combined flow bed configurations and some implications for interpreting storm-event stratification. *Journal of Sedimentary Petrology*, **60**: 211–219.
Brenchley, P.J., 1985. Storm influenced sandstone beds. *Modern Geology*, **9**: 369–396.
Brenchley, P.J., and Newall, G., 1982. Storm influenced inner-shelf sand lobes in the Caradoc (Ordovician) of Shropshire, England. *Journal of Sedimentary Petrology*, **52**: 1257–1269.
Brenchley, P.J., Newall, G., and Stanistreet, I.G., 1979. A storm surge origin for sandstone beds in an epicontinental platform sequence, Ordovician, Norway. *Sedimentary Geology*, **22**: 185–217.
Campbell, C.V., 1966. Truncated wave-ripple laminae. *Journal of Sedimentary Petrology*, **36**: 825–828.
Cheel, R.J., 1991. Grain fabric in hummocky cross-stratified storm beds: genetic implications. *Journal of Sedimentary Petrology*, **61**: 102–110.
Cheel, R.J., and Leckie, D.L., 1993. Hummocky cross-stratification. *Sedimentology Reviews*, **1**: 103–122.
Dott, R.H., and Bourgeois, J., 1982. Hummocky stratification: significance of its variable bedding sequences. *Geological Society of America Bulletin*, **93**: 663–680.
Duke, W.L., 1990. Geostrophic circulation or shallow marine turbidity currents? The dilemma of paleoflow patterns in storm-influenced prograding shoreline systems. *Journal of Sedimentary Petrology*, **60**: 870–883.
Harms, J.C., Southard, J.B., Spearing, D.R., and Walker, R.G., 1975. *Depositional Environments as Interpreted from Primary Sedimentary Structures and Stratification Sequences*. Society of Economic Paleontologists and Mineralogists, Short Course 2, 2nd edn, 1982.
Howard, J.D., 1971. Comparison of the beach-to-offshore sequence in modern and ancient sediments. In Howard, J.D., Valantine, J.W., and Warme, J.E. (eds.), *Recent Advances in Paleoecology and Ichnology: Short Course Lecture Notes*. Washington, DC: American Geological Institute, pp. 148–183.
Howard, J.D., 1972. Trace fossils as criteria for recognizing shorelines in the stratgraphic record. In Rigby, J.K., and Hamblin, W.K. (eds.), *Recognition of Ancient Sedimentary Environments*. Society of Economic Paleontologists and Mineralogists, Special Publication, 16, pp. 215–255.
Leckie, D.L., and Walker, R.G., 1982. Storm- and tide-dominated shorelines in Cretaceous Moosebar-Lower Gates interval—outcrop equivalents of Deep Basin gas traps in Western Canada. *American Association of Petroleum Geologists*, **66**: 138–157.
Prave, A.R., and Duke, W.L., 1990. Small-scale hummocky cross-stratification in turbidites: a form of antidune stratification? *Sedimentology*, **37**: 771–783.
Rust, B.R., and Gibling, D.A., 1990. Three-dimensional antidunes as HCS mimics in a fluvial sandstone: the Pennsylvanian South Bar Formation near Sydney, Nova Scotia. *Journal of Sedimentary Petrology*, **60**: 540–548.
Swift, D.J.P., Hudelson, P.M., Brenner, R.L., and Thompson, P., 1987. Shelf construction in a foreland basin: storm beds, shelf sandstones, and shelf-slope depositional sequences in the Upper Cretaceous Mesaverde Group, Book Cliffs, Utah. *Sedimentology*, **34**: 423–457.

Cross-references

Bedding and Internal Structures
Coastal Sedimentary Facies
Cross-Stratification
Storm Deposits

HYDROCARBONS IN SEDIMENTS

A hydrocarbon is strictly a compound composed of only the elements hydrogen and carbon. However, the term "hydrocarbons" is often extended within the Earth Sciences to include all the organic compounds that comprise petroleum and bitumen (the portion of organic matter that can be extracted from a rock using organic solvents) even though these include compounds containing other elements, especially nitrogen, sulfur and oxygen (NSO compounds). The elements carbon and hydrogen make up about 97.5 percent of an average oil (Hunt, 1996, p.24). Hence it is not surprising that hydrocarbons make up 80 percent to 90 percent of a normal crude oil with the remainder being NSO compounds and asphaltenes. The proportion of hydrocarbons in bitumen is more variable than in oils, depending on the thermal maturity and lithology of the rock, the type of organic matter and the solvent used for the extraction. Generally less than 50 percent of the bitumen, and commonly much less, are hydrocarbons. The difference between the composition of bitumen and crude oils is due to functionalized (NSO) compounds not being able to migrate as easy as hydrocarbons, and thus being selectively retained within the source rock and to a lesser extent on mineral surfaces along migration pathways.

Occurrences

Although hydrocarbons have been reported almost everywhere in the earth's crust, they are far more abundant in sediments than in igneous or metamorphic rocks. This reflects the biogenic origin of most hydrocarbons in sedimentary rocks. Hydrocarbons make up a very low proportion of the biomass and hence their abundance in recent sediments is generally low, in the 20 ppm to 100 ppm range (Tissot and Welte, 1984, p.95). In ancient non-reservoir sediments, especially organic-rich, fine-grained rocks, their abundance is much higher, up to several thousand ppm. The large scale accumulations of hydrocarbons within commercial oil fields are due to their migration out of large volumes of source rocks and into more permeable and porous beds where they have been trapped.

A diverse range of alkanes and aromatic hydrocarbons have been reported in petroleum and sediments. The average hydrocarbon composition of a crude oil has been stated to be 33 percent acyclic (normal and branched alkanes), 32 per cent cycloalkanes and 35 percent aromatic hydrocarbons (Killops and Killops, 1993, p.137). These proportions will vary depending on the nature of the organic matter in the oil source rock and the level of thermal maturity, with saturated hydrocarbons becoming more predominant at higher maturity. Organic geochemists usually analyze for compounds with less than forty carbon atoms, but the recent development of high temperature gas chromatography has shown higher molecular weight hydrocarbons, up to C_{120} may be ubiquitous in oils (Hsieh and Philp, 2001). The simplest and usually the most abundant hydrocarbons are the straight-chain n-alkanes, which can range from C_1 (methane) to C_{80+}. Particularly in recent sediments, the C_{25}–C_{33} n-alkanes often show a pronounced odd carbon number preference. This is due to the predominance of odd carbon number n-alkanes in plant waxes. Simple branched alkanes such as the iso-alkanes (2-methylalkanes) and anteiso-alkanes (3-methylalkanes) and cyclic alkanes such as the cyclohexanes and cyclopentanes with a similar carbon number range to the n-alkanes are also ubiquitous in crude oils and ancient sediments. Many hydrocarbons have carbon skeletons that can be easily related to a precursor originally synthesized by an organism (Figure H7). Such compounds are called *biomarkers* (short for biological markers) or *chemical fossils* (Peters and Moldowan, 1993). Many of the most commonly found and abundant biomarkers are those that are based on the isoprene unit. These include acyclic isoprenoids such as pristane (C_{19}i) and phytane (C_{20}i) and cyclic compounds such as steranes, hopanes, carotenoids (e.g., β-carotane) and other terpanes (Figure H8). Sterane and terpane distributions (usually hopane and tricyclic terpanes) are those most commonly used by petroleum geochemists to "fingerprint" oils so that they can be compared to other oils or potential source rocks for oil–oil and oil–source correlation. Aromatic compounds are also abundant in petroleum and extracts from ancient sediments. Benzene and toluene are the major light aromatics while heavier fractions of petroleum contain alkylated benzenes, naphthalenes, phenanthrenes, triaromatic steroids and triterpenoids and other polycyclic aromatic hydrocarbons (PAHs) (Figure H8). Alkylated compounds predominate in oils as opposed to anthropogenic combustion sources where parent PAHs are dominant. The distributions of several saturated and aromatic hydrocarbons (e.g., steranes and methylphenanthrenes) can be used as an indicator of the thermal stress that organic matter has been subjected to. Although a range of 0.02 percent to 5–10 percent have been reported for olefins in oils (Curiale and Frolov, 1998), their concentration is usually toward the lower part of this range. Olefins are generally much more abundant in recent sediments where they are directly inherited from compounds originating from living organisms. Some of the olefins found in petroleum are thought to be biogenic compounds directly derived from the source rock or picked up during migration when petroleum came into contact with lower maturity sediments. Other olefins have been attributed an abiogenic origin, either from radiolytic dehydrogenation of saturated hydrocarbons due to exposure to radioactive elements or to enhanced thermal activity such as when an igneous intrusion comes into close contact with a reservoir (Curiale and Frolov, 1998).

Origin

Although a biogenic origin of hydrocarbons is easily the most currently favored theory, this has not always been the case and nor is it accepted by everybody today. There are three general classes of ideas that have been put forward to explain the

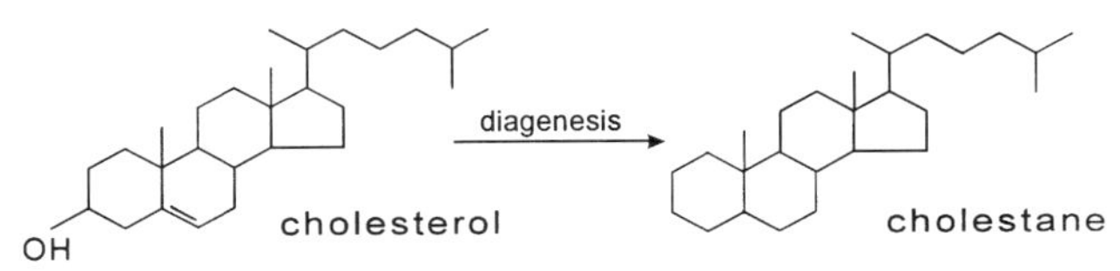

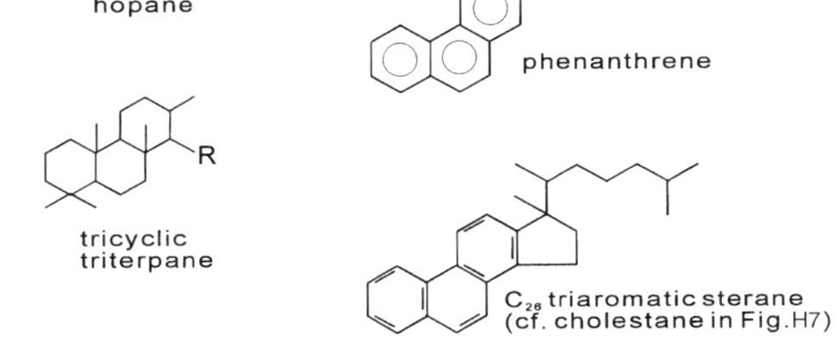

Figure H7 Example of biological precursor—biomarker relationship. Cholesterol is an example of a sterol, compounds found in all higher (eukaryotic) organisms including humans. During diagenesis, dehydration and hydrogenation of cholesterol occurs to give cholestane, a sterane that is almost ubiquitous in petroleum and sediments.

Figure H8 Examples of hydrocarbons commonly found in sediments. R indicates alkyl chain.

origin of petroleum, and by implication, hydrocarbons in sedimentary rocks: inorganic, biogenic, and duplex origins.

In the nineteenth century, many people did not think that the formation of petroleum involved the biosphere. Reactions using inorganic substrates such as the Fischer–Tropsch reaction, which commercially is used to synthesize hydrocarbons from carbon monoxide and hydrogen on a metallic catalyst (Olah and Molnár, 1995, p. 66–75), were suggested as likely mechanisms. During the twentieth century, most people came to accept the biogenic theory outlined below although some workers still support an inorganic origin (e.g., Porfir'ev, 1974; Gold and Soter, 1982). A vast body of evidence indicates that this is very unlikely for most sedimentary hydrocarbons. Small amounts of abiogenic hydrocarbons, mostly methane, are found in Precambrian Shield and mantle-derived rocks but not enough to constitute economic deposits (Sherwood-Lollar et al., 1993; Jenden et al., 1993; Sugisaki and Mimura, 1994). The duplex theory holds that petroleum is a mixture of compounds having inorganic and biogenic origins. It was proposed (Robinson, 1963) at a time when it was more difficult to relate many of the constituents of petroleum to a biogenic origin. Many of today's proponents of an inorganic origin for petroleum tend toward a duplex origin in order to explain some of the more obvious biogenic compositional characteristics of oils, such as the occurrence of biomarkers.

Most geologists and geochemists believe that petroleum is the product of the thermal degradation of insoluble organic matter derived from dead organisms buried in sedimentary rocks. This is the only theory that satisfactorily explains the composition and geographical and geological distribution of petroleum. In the past, it was suggested that petroleum was simply an accumulation of the hydrocarbons derived from living flora and fauna that occur in recent sediments. Hence there was no need for any chemical conversion of organic matter into petroleum with time or depth of burial. This is no longer thought to be credible for a number of reasons, most notably because hydrocarbons are found in very low abundance in living organisms and recent sediments (these contain mostly non-hydrocarbon compounds), and petroleum contains abundant hydrocarbons not found in living organisms, especially in the lower molecular weight C_4–C_8 range.

There are three stages in the transformation of biogenic organic matter into petroleum (see also *Maturation, Organic*). *Diagenesis* is the low temperature (up to 50–60°), largely microbial mediated, irreversible alteration of organic debris in the water column, at the water/sediment interface and in near surface sediments. The biogenically derived organic matter that survives bacterial attack at this stage is preserved as mostly insoluble organic matter called kerogen. During diagenesis, hydrocarbons are in low abundance. Most of the compounds that can be extracted from sediments still contain functional groups and are directly inherited from living organisms. Biogenic methane due to methanogenic bacteria activity is the only hydrocarbon generated in significant amounts during diagenesis. *Catagenesis* is the second stage of organic maturation and takes place between about 50°C and 150°C. This is a thermally mediated process in which kerogen cracks to form oil and gas. The catagenesis stage thus includes the principle zone of oil formation and later, the initial stages of gas formation when C_2–C_4 hydrocarbon gases with an increasing amount of methane are generated in large amounts from the cracking of previously generated larger molecules and kerogen. Biogenically derived molecules are diluted by these neo-formed hydrocarbons which have a less obvious structural relationship to their biogenic precursors. *Metagenesis* is the final thermal degradation of both residual kerogen and the bitumen formed during catagenesis to the most thermodynamically stable products, carbon (pyrobitumen or graphite) and methane. It occurs at temperatures between 175°C and 250°C.

Martin Fowler

Bibliography

Curiale, J.A., and Frolov, E.B., 1998. Occurrence and origin of olefins in crude oils. A critical review. *Organic Geochemistry*, **29**: 397–408.
Gold, T., and Soter, S., 1982. Abiogenic methane and the origin of petroleum. *Energy Exploration and Exploitation*, **1**: 89–104.
Hsieh, M., and Philp, R.P., 2001. Ubiquitous occurrence of high molecular weight hydrocarbons in crude oil. *Organic Geochemistry*, **32**: 955–956.
Hunt, J.M., 1996. *Petroleum Geochemistry and Geology*, 2nd edn. New York: W.H. Freeman.
Jenden, P.D., Hilton, D.R., Kaplan, I.R., and Craig, H., 1993. Abiogenic hydrocarbons and mantle helium in oil and gas fields. In Howell, D.G. (ed.), *The Future of Energy Gases*. US Geological Survey Professional Paper 1570, pp. 31–56.
Killops, S.D., and Killops, V.J., 1993. *An Introduction to Organic Geochemistry*. Longman Scientific & Technical.
Olah, G.A., and Molnár, Á., 1995. *Hydrocarbon Chemistry*. John Wiley & Sons.
Peters, K.E., and Moldowan, J.M., 1993. *The Biomarker Guide. Interpreting Molecular Fossils in Petroleum and Ancient Sediments*. Prentice Hall.
Porfir'ev, V.B., 1974. Inorganic origin of petroleum. *American Association of Petroleum Geologists Bulletin*, **58**: 3–33.
Robinson, R., 1963. Duplex origin of petroleum. *Nature*, **199**: 113–114.
Sherwood Lollar, B., Frape, S.K., Weise, S.M., Fritz, P., Macko, S.A., and Welhan, J.A., 1993. Abiogenic methanogenesis in crystalline rocks. *Geochimica et Cosmochimica Acta*, **57**: 5087–5097.
Sugisaki, R., and Mimura, K., 1994. Mantle hydrocarbons: abiotic or biotic? *Geochimica et Cosmochimica Acta*, **58**: 2527–2542.
Tissot, B.P., and Welte, D.H., 1984. *Petroleum Formation and Occurrence*. 2nd edn., Springer-Verlag.

Cross-references

Diagenesis
Gases in Sediments
Kerogen
Maturation, Organic
Oil Sands
Mutation, Organic

HYDROXIDES AND OXYHYDROXIDE MINERALS

The group of oxides, hydroxides and oxyhydroxides comprises several cations such as Al, Fe, Mn or Ti, where the anionic part is either an oxygen (–O), a hydroxyl group (–OH) or an oxygen and a hydroxyl (–OOH). This article deals essentially with the iron minerals (collectively called iron oxides), because aluminum minerals such as gibbsite (α-Al(OH)$_3$), diaspore (α-AlOOH) and boehmite (γ-AlOOH) are handled in the article on bauxites (*q.v.*), whereas manganese oxides and phyllomanganates are essential constituents of iron–manganese nodules (*q.v.*). The TiO_2 polymorphs rutile, anatase and brookite are of minor importance in sediments and sedimentary rocks.

Minerals

In sediments and sedimentary rocks, oxides (hematite $\alpha\text{-}Fe_2O_3$, magnetite $Fe^{2+}Fe_2^{3+}O_4$, maghemite $\gamma\text{-}Fe_2O_3$, wuestite FeO, ilmenite $FeTiO_3$) and oxyhydroxides (goethite α-FeOOH, lepidocrocite γ-FeOOH, ferrihydrite $\sim$FeOOH) do occur. Schwertmannite, $Fe_8O_8(OH)_6SO_4$, is transitional between oxyhydroxides and sulfates (for details see Cornell and Schwertmann, 1996).

Occurrences

The iron contents of sediments are highly variable. On average, carbonate rocks contain *ca.* $6\,\text{mg}\,\text{g}^{-1}$ Fe_2O_3, sandstones *ca.* $14\,\text{mg}\,\text{g}^{-1}$ and claystones *ca.* $71\,\text{mg}\,\text{g}^{-1}$. The iron oxides are either of detrital origin or neoformed in the sediments. Titanomagnetites and ilmenite belong to those detrital minerals that survive weathering due to high stability in surface environments, and, hence, may concentrate, for example, in placers (*q.v.*), up to mining quality. Hematite, which is the major pigmenting mineral in red beds (*q.v.*) or any other red-colored rock, may be either of detrital origin or it formed diagenetically. The *banded iron formations* (BIF) or *itabirites* are Precambrian, thin-bedded or laminated chemical sediments, where hematite and magnetite are interbedded with chert layers. In other kinds of sedimentary iron ores, hematite may be associated with goethite and/or magnetite (see *Ironstones and Iron Formations*, and Füchtbauer, 1988). Yellowish to brown colors indicate the presence of goethite, which is present in small concentrations ubiquitously but in Paleozoic and older rocks to a lesser extent than is hematite. Laterites (*q.v.*), ferricretes and bauxites contain goethite, hematite and also maghemite. Quaternary bog ores, which are the younger analogs to ferricretes and bauxites, are dominated by goethite and ferrihydrite, whereas hematite is absent. Tertiary sediments may contain lepidocrocite in addition to goethite. Wuestite has been reported as rims around magnetite in sandstones.

Properties

The basic structural unit in most of the iron oxides are Fe^{3+} cations coordinated by six anions (O^{2-}, and/or OH^-). These units are connected either by sharing one oxygen across a corner, two oxygens across an edge or three along a face. With the addition of fourfold coordinated Fe^{2+} and/or Fe^{3+} cations in the spinel phases, distinct structures with densities between $4\,\text{g}\,\text{mL}^{-1}$ and $5.2\,\text{g}\,\text{mL}^{-1}$ can arise by mixing several kinds of bonds. In cases where one iron oxide is present, the color of sediments (*q.v.*) is an indicative property. Variations in crystal size, however, or mixtures of two minerals (e.g., goethite and hematite) not only change the hues, but also might mask other iron oxides. Identification and characterization is therefore better done by X-ray diffraction using Co radiation. Depending on "crystallinity", detection limits may range from a few $\text{mg}\,\text{g}^{-1}$ for micron-sized hematites in sandstones to $\sim 0.2\,\text{g}\,\text{g}^{-1}$ for ferrihydrite. Mössbauer spectroscopy offers even lower detection limits because of being sensitive to solely ^{57}Fe.

Crystal size and deviation from the ideal chemical composition reflect the physico–chemical environment in which authigenic iron oxides have grown. In laterites, ferricretes, and bauxites, crystal sizes of goethite and hematite range from tens to hundreds of nanometers, whereas in sandstones hematites may have grown to several microns in size. The first mentioned environments with their intensive chemical weathering yield also goethites and hematites, in which Al^{3+} substitutes for Fe^{3+}. Maximum substitutions of $0.33\,\text{mol}\,\text{mol}^{-1}$ Al/(Al + Fe) have been found for goethites; in hematites the substitution seems to be limited to $0.16\,\text{mol}\,\text{mol}^{-1}$. Smaller crystal sizes and Al substitution distinguishes iron oxides derived from supergene weathering processes from those found in sandstones, carbonates and other rocks, where crystal sizes are usually much larger and substitution of foreign cations is almost absent.

Besides Al, other heavy metals such as V, Cr, Co, or Ni may substitute in iron oxides, although with much lower absolute contents. In ferricretes so much phosphate is associated with iron oxides that it spoils their use as iron ores. Phosphate sorbs strongly on iron oxide surfaces by forming a bidentate mononuclear complex, in which two oxygens of the PO_4^{3-} anion are shared with two surface oxygens (which were OH groups before adsorption) bound to two Fe^{3+} beneath the surface. Selenite and sulfate adsorb similarly, but less strongly. This kind of inner-sphere complexation has also been observed for arsenate and chromate bound to schwertmannite in sediments from acid mine drainages.

Unpaired electrons of structural Fe(III) and Fe(II) result in magnetic moments, which interact with each other and depending on thermal motion produce magnetic ordering in all iron oxides. The ordering temperatures of micronsized crystals of magnetite, hematite and to some extent also goethite are high enough (850, 956, and 400 K, respectively) to preserve magnetizations even on a geological timescale. When such particles settle freely during sediment formation, they orient along the Earth's magnetic field and build up a *detrital remanent magnetization*. Alternatively, a very stable *chemical remanent magnetization* is acquired, when these iron oxides grow authigenically in a sediment (see Dunlop and Özdemir, 1997).

Formation

Chemical weathering of primary, Fe(II)-containing silicates liberates Fe^{2+}, which oxidizes and hydrolyses to form any of the above-mentioned phases. Which iron oxide forms is governed by pH, the rate of oxidation, the presence of inorganic and organic anions, temperature, activity of Fe^{2+} and other foreign compounds. Near-neutral pH values and higher temperatures favor formation of hematite over goethite. With decreasing oxidation rates of Fe^{2+}, ferrihydrite, lepidocrocite or goethite are formed. In agreement with the occurrences of less crystalline and less stable iron oxides in younger sediments, kinetic factors seem to outweigh thermodynamic stabilities, because at conditions prevailing in surficial environments (i.e., low temperature, activity of water = 1, presence of oxygen), large crystals of goethite (α-FeOOH) are the only thermodynamically stable phase. The ubiquitous presence of hematite ($\alpha\text{-}Fe_2O_3$) can be explained by lower activities of water due to stronger binding of water in small pores and/or due to higher temperatures, where hematite becomes stable despite of $a_{H_2O} = 1$ (*cf.* the formation of hematitic red beds (*q.v.*) in a warmer climate). A further factor, which influences thermodynamic stability, is crystal size because of increasing contributions of surface energy.

Transformation

By heating in dry conditions, all OH-containing oxides transform finally to hematite via solid state reactions, but depending on the nature of the compound, its crystallinity, the extent of isomorphous substitution and other chemical impurities, temperatures between 140°C and 500°C are required. In the presence of water, dehydroxylation of goethite to hematite requires hydrothermal conditions, whereas the reverse reaction (hydroxylation of hematite) has not been observed *in situ* because of kinetic hinderance. ^{18}O isotope measurements of such transformations have shown that a solution step with transport *via* water molecules is always involved (Bao and Koch, 1999).

The solubilities of Fe(III) oxides at naturally occurring pH values are so low that the concentration of Fe^{3+} in solution is orders of magnitudes lower than the solubility as Fe^{2+}. Although complexing organic compounds increase the total solubility, the most efficient way to mobilize iron is by reductive processes mediated by bacteria or reducing compounds (e.g., HS^-). The high mobility of Fe^{2+} then allow redistribution of iron from the nanometer scale to landscapes. An indicator for Fe^{2+} as the feeding source is substitution of V^{3+} in lateritic goethites and hematites. Under natural pH and Eh values, where Fe^{2+} is stable, tetravalent VO^{2+} coexists, but not the unstable V^{3+}. A reduction VO^{2+} to V^{3+} followed by incorporation is feasible, however, by a surface-mediated reduction after sorption of Fe^{2+} onto an oxide surface, which decreases the redox potential of Fe^{3+}/Fe^{2+} in solution from 0.77 Volt to a value of ~ 0.35 Volt low enough to reduce sorbed V^{4+} to V^{3+} (Schwertmann and Pfab, 1996).

Helge Stanjek

Bibliography

Bao, H., and Koch, P.L., 1999. Oxygen isotope fractionation in ferric oxide-water systems: low temperature synthesis. *Geochimica et Cosmochimica Acta*, **63**: 599–613.

Cornell, R.M., and Schwertmann, U., 1996. *The Iron Oxides*. Weinheim: VCH.

Dunlop, D.J., and Özdemir, Ö., 1997. *Rock Magnetism*. Cambridge University Press.

Füchtbauer, H., 1988. *Sedimente und Sedimentgesteine*. Stuttgart: Schweizerbart.

Schwertmann, U., and Pfab, G., 1996. Structural vanadium and chromium in lateritic iron oxides: genetic implications. *Geochimica et Cosmochimica Acta*, **60**: 4279–4283.

Cross-references

Bauxite
Colors of Sedimentary Rocks
Diagenesis
Iron-Manganese Nodules
Ironstones and Iron Formations
Laterites
Magnetic Properties of Sediments
Red Beds

ILLITE GROUP CLAY MINERALS

The term *'illite'* usually is used in its petrographic sense as a name for the K-rich, argillaceous component of sedimentary rocks, identified by ca. 1 nm spacing for its 001 X-ray diffraction (XRD) peak. At 30 percent, illite is the second most abundant mineral component of sedimentary rocks (after quartz, 34 percent). Defined that way, it is not a mineral, but is the "illitic fraction" of a rock, which, in addition to the mineral illite, may contain admixtures of similar minerals, most often fine-grained micas, glauconite and mixed-layer illite–smectite. Used in the mineralogical sense, "illite" identifies a clay structure, which is of 2:1 type, dioctahedral, non-expanding, aluminous, and contains nonexchangeable K as the major interlayer cation. The potassium balances a negative layer charge, generated by Al for Si substitution in the tetrahedral sheet and (Mg, Fe^{2+}) for Al in the octahedral sheet (Figure I1). This definition refers both to illite as a separate mineral (discrete illite), and as a non-expanding component of mixed-layer illite–smectite (I–S), known as the *illite fundamental particle* (Figure I1). Due to its complicated chemical composition and broad range of possible isomorphous substitutions, illite becomes an important sink for most major elements liberated at the Earth surface by interaction between rocks and the hydrosphere. Thin illite crystals contribute considerably to a rock's cation exchange capacity and they reduce its permeability. Illite group minerals are covered in detail in recent reviews by Środoń (1999a,b).

Origin and evolution of illite in the rock cycle

Common surface environments are not favorable for the formation of the mineral illite. Reports of illite formation in weathering or marine environments often are misleading because of eolian contamination (Šucha *et al.*, 2001). Illite may be partially stripped of its potassium by weathering and may fix it back in sedimentary environment without changing its silicate layers. Experiments indicate that smectites also may fix potassium at surface temperatures without changing its silicate layers, thus forming mixed-layer I–S during the process of alternate wetting and drying, however, the geological significance of this process has not been assessed. Illite can be crystallized at surface temperatures at pH > 10 (Bauer *et al.*, 2000). In natural high pH environments (lacustrine, playa), ferruginous illite, having a composition intermediate between illite and glauconite, is a common product of low-temperature diagenesis.

Illitization becomes a major chemical reaction in sedimentary rocks above approximately 70°C. Three types of processes have been recognized: (1) alteration of smectite into illite *via* mixed-layer I–S; (2) direct alteration of kaolinite and feldspar into discrete illite; (3) crystalization of filamentous illite in the pore space. The first process, volumetrically dominant, is known in the most detail. Illitization of smectite is a continuous process, producing pure non-expandable illite at the bottom of the *anchizone* (ca. 275°C—this zone is the highest grade of burial diagenesis). If not uplifted and eroded, it gradually recrystallizes into coarse-grained mica up to 325°C (biotite isograd, McDowell and Elders, 1980).

In the weathering environment, eroded illite is very stable and commonly becomes recycled into new sediments either chemically intact or partially stripped of its potassium (soil vermiculite).

Mixed-layering, polytypes, and structural defects

Illite fundamental particles, which are only a few layers thick, are very flexible. They can join each other or smectite monolayers face-to-face to form multi-particle crystals (Figure I1), called *mixed-layer crystals* because the particle interfaces display not illite but smectite characteristics (e.g., exchangeable cations, swelling properties). In the course of diagenesis, smectite particles dissolve and illite particles grow in thickness (Środoń *et al.*, 2000), thus mixed-layering evolves along with increasing percent of illite layers from so-called R0 (smectite monolayers plus illite bilayers) into R1 (no smectite monolayers, bilayers dominant) and then R > 1 (3 nm or thicker particles dominant).

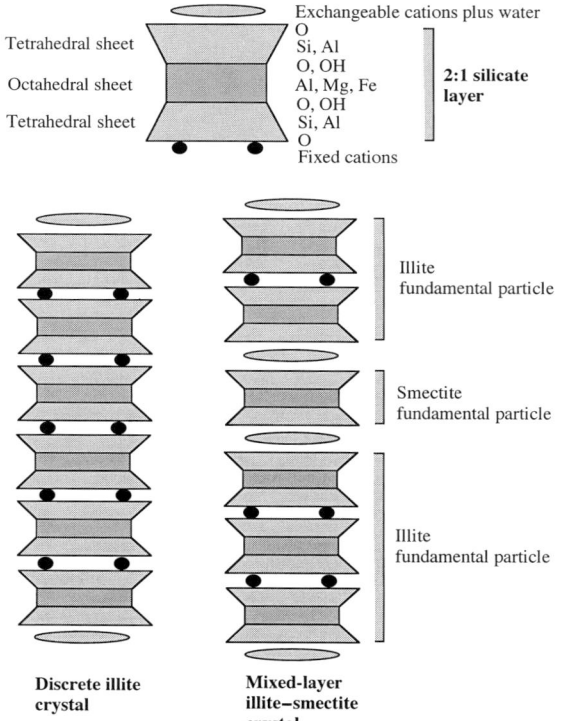

Figure I1 Schematic structure of the 2:1 silicate layer. The layer has two tetrahedral and one octahedral sheet and is dioctahedral, i.e., 2 out of 3 octahedral positions are occupied by cations. The lower part of the figure shows two types of occurrence of illite layers: as discrete crystals and as fundamental particles in mixed-layer illite-smectite crystals.

Mixed-layer crystals produce 00l XRD reflections at positions intermediate between the positions of end-members. The illite:smectite ratio is measured from these XRD positions (Moore and Reynolds, 1997) and used for tracing the increasing grade of diagenesis. Mixed-layering becomes undetectable at approximately 20 layers mean thickness.

Because of the pseudohexagonal symmetry of layer surfaces, subsequent layers in illite particles can be superimposed not only in the same direction, but also rotated by 60° or 120°. If such pattern repeats systematically a *polytype* is created. If the pattern is random, the structure contains stacking faults (structural defects). A second type of structural defect recognized in illites arises from the location of cation vacancies in the octahedral sheet. Only two of the three sites are occupied in dioctahedral minerals and the sites are not equivalent (two "*cis*" and one "*trans*" site). Illite layers can be either trans or cis vacant and both types are commonly found in one mineral (Moore and Reynolds, 1997).

Among common illite polytypes, 1M is characteristic of diagenesis and $2M_1$ appears in the anchizone (intermediate between diagenesis and metamorphism). While $2M_1$ is typically devoid of structural defects, 1M most often contains defects, which broaden and weaken its XRD diffraction peaks (so-called $1M_d$ illite). Polytypes and defects can be quantified by computer modeling of XRD patterns (Moore and Reynolds, 1997) and used for tracing geological processes. For example, the $1M_d/2M_1$ ratio is measured to quantify the ratio of diagenetic to detrital illite in the diagenetic zone, and the progress toward metamorphism in the anchizone. Generally, the density of defects decreases during diagenesis, with trans-vacant layers becoming dominant, but exceptions to this rule are known.

Crystal size, shape, and "crystallinity" (Kubler index)

Illite crystals and fundamental particles are plates terminated by 001 planes, either equidimensional (the shape common for shales) or highly elongated (the shape typical of "hairy illite" that clogs sandstone porosity). These shapes are controlled by crystal growth rates (Bauer *et al.*, 2000).

The thickness of illite fundamental particles, first measured by Nadeau *et al.* (1984), was studied in detail for bentonites (Środoń *et al.*, 2000). Illite nucleates as particles 2 layer (2 nm) thick, which grow during the course of diagenesis, producing lognormal distributions of increasing mean thickness, broadening and skewness. The growth of fundamental particles is the major factor responsible for a decrease in the XRD 001 peak breadth ("Kubler index") during anchimetamorphism. Particle thickness distributions can be now measured directly by XRD (Eberl *et al.*, 1998).

Layer charge and fixed cations of illite

The typical structural formula for pure illite from bentonites:

$$(K, Na, NH_4)_{0.90}(Al_{1.85}Fe_{0.05}Mg_{0.10})(Si_{3.20}Al_{0.80})O_{10}(OH)_2$$

corresponds to infinitely thick illite crystals. The thinner the crystals, the lower their bulk fixed charge, because the number of interlayers is always smaller by one than the silicate layers (Figure I1). In the extreme case of 2 nm illite crystals, the fixed charge equals 0.45 and there is a substantial number of exchangeable cations at crystal surfaces.

The charge of illite layers is satisfied mostly by fixed potassium. Minor Na substitution is well-known but there is no complete solid solution, and if sodium is abundant, Na-mica (paragonite) is formed in the anchizone (Frey, 1987). Solid solution between K and NH_4 is possible and NH_4-rich illites (tobelites) have been reported from deep diagenesis (Schroeder and McLain, 1998) and anchizone (Lindgreen *et al.*, 2000).

Illite as paleogeothermometer

Illite crystal growth has been traced in the diagenetic zone by measuring the illite:smectite ratio of mixed-layer I-S and in the anchizone by measuring the Kubler index. Students of the anchizone believed that the Kubler index indicates paleotemperatures; and anchizone limits were defined using its values (Frey, 1987).

In the diagenetic zone, the effect of bulk rock chemistry on the illite:smectite ratio has been demonstrated and explained by a deficiency of K in bentonites. Because of abundant K and relatively closed system, shales are considered the most suitable lithology for studying illite response to basin temperature. The approach, introduced by Hoffman and Hower (1979), assumes that the illitization reaction in shales is so fast that the time factor can be neglected and that maximum paleotemperatures can be evaluated from measured illite:smectite ratios. Other investigators take into account the time factor and chemistry and use various kinetic equations to model the experimental illitization profiles.

K–Ar dating of illite

The interpretation of K–Ar dates depends on the type of host rock. In pure beach or eolian sandstones all illite can be authigenic (hairy illites) and the radiometric date indicates the time of hot fluid flow and perhaps oil emplacement (Hamilton *et al.*, 1992). Shales, even in the finest fractions, usually contain a mixture of authigenic and detrital illite. Techniques for obtaining both ages have been proposed. If extracted properly, the K–Ar date of diagenetic illite from shale represents the mean time that has passed between the onset of illitization and the time of maximum paleotemperature. This mean strongly depends on burial history (Środoń *et al.*, 2002). Many bentonites are free of detrital contamination, but the K-Ar date represents an even longer period of time than I-S from surrounding shales because of slow diffusion of potassium into the bentonite. This period of time can be evaluated by dating bentonites that are zoned, that is, less highly illitized in the center of the bed, and/or fractions of fundamental particles separated from a bentonite.

Jan Środoń

Bibliography

Bauer, A., Velde, B., and Gaupp, R., 2000. Experimental constraints on illite crystal morphology. *Clay Minerals*, **35**: 587–597.
Eberl, D.D., Nuesch, R., Sucha, V., and Tsipursky, S., 1998. Measurement of fundamental particle thicknesses by X-ray diffraction using PVP-10 intercalation. *Clays and Clay Minerals*, **46**: 89–97.
Frey, M., 1987. *Low Temperature Metamorphism*. Blackie.
Hamilton, P.J., Giles, M.R., and Ainsworth, P., 1992. K-Ar dating of illites in Brent Group reservoirs: a regional perspective. Geological Society Special Publication, 61, pp. 377–340.
Hoffman, J., and Hower, J., 1979. Clay mineral assemblages as low grade metamorphic geothermometers: application to the thrust faulted disturbed belt of Montana, USA. SEPM, Special Publication, 26, pp. 55–79.
Lindgreen, H., Drits, V.A., Sakharov, B.A., Salyn, A.L., Wrang, P., and Dainyak, L.G., 2000. Illite–smectite structural changes during metamorphism in black Cambrian Alum shales from the Baltic area. *American Mineralogist*, **85**: 1223–1238.
McDowell, S.D., and Elders, W.A., 1980. Authigenic layer silicate minerals in borehole Elmore 1, Salton Sea geothermal field, California, USA. *Contributions to Mineralogy and Petrology*, **74**: 293–310.
Moore, D.M., and Reynolds, R.C., 1997. *X-Ray Diffraction and the Identification and Analysis of Clay Minerals*. Oxford University Press.
Nadeau, P.H., Wilson, M.J., McHardy, W.J., and Tait, J., 1984. Interstratified clay as fundamental particles. *Science*, **225**: 923–925.
Schroeder, P.A., and McLain, A.A., 1998. Illite-smectites and the influence of burial diagenesis on the geochemical cycling of nitrogen. *Clay Minerals*, **33**: 539–546.
Środoń, J., 1999a. Use of clay minerals in reconstructing geological processes: current advances and some perspectives. *Clay Minerals*, **34**: 27–37.
Środoń, J., 1999b. Nature of mixed-layer clays and mechanisms of their formation and alteration. *Annual Review of Earth and Planetary Sciences*, **27**: 19–53.
Środoń, J., Eberl, D.D., and Drits, V., 2000. Evolution of fundamental particle-size during illitization of smectite and implications reaction mechanism. *Clays and Clay Minerals*, **48**: 446–458.
Środoń, J., Clauer, N., and Eberl, D.D., 2002. Interpretation of K–Ar dates of illitic clays from sedimentary rocks aided by modelling. *American Mineralogist*, **87**: 1528–1535.
Sucha, V., Środoń, J., Clauer, N., Elsass, F., Eberl, D.D., Kraus, I., and Madejova, J., 2001. Weathering of smectite and illite-smectite in Central-European temperate climatic conditions. *Clay Minerals*, **36**: 403–419.

Cross-references

Bentonites and Tonsteins
Clay Mineralogy
Diagenesis
Fabric, Porosity, and Permeability
Glaucony and Verdine
Mixed-Layer Clays
Mudrocks

IMBRICATION AND FLOW-ORIENTED CLASTS

Imbrication is the overlapping arrangement of similar parts, as of roof tiles or fish scales. The earliest use of "imbricate" or related words by geologists quoted in the *Oxford English Dictionary* are by Dana (1852, 1862) and by A. Geike (1858) to describe biologic or paleontologic objects. An illustration of sedimentary imbrication is given as early as Jamieson (1860), reproduced here in Figure I2. A modified version of this figure appears in the 11th and 12th editions of Lyell's (1872, 1875, p. 342) *Principles*, but neither Jamieson nor Lyell use any form of the word "imbricate" in their discussions of the figures.

Imbrication, which is a special case of flow-oriented clasts, is a common phenomenon, but observations and theory concerning it are published rarely. The flow-orientation of a single clast develops in many ways. To understand the flow orientation of clasts in different environments can improve interpretation of rocks, and lead to a better understanding of sediment transport.

Imbricated clasts must have unequal lengths for their a, b, c axes (a longest, c shortest). Usually, imbricated clasts are approximately tabular in shape, where the a and b axes significantly exceed the c axis. Roof tiles, or fish scales, or tabular rock fragments can be imbricated; basketballs, or grapes, or spherical sand grains cannot. As suggested by Figure I2, imbricated clasts must be approximately similar in absolute size: flat pebbles can form an imbricated group, but a flat pebble cannot imbricate with a flat boulder.

Jamieson (1860, p. 350) suggests that flow orientation, such as shown in Figure I2, "is best exemplified when the stones are pretty large, say, from six to twelve inches in length." However, flow orientation and imbrication occurs in finer sediments, including sand (Gibbons, 1972), although compared to pebbles or boulders, sand grains have more random orientations because their size is smaller relative to the turbulent eddies, and their submerged weight is relatively less than the drag exerted by such eddies. On the larger scale, boulders have been imbricated in debris flows and floods (Blair, 1987, his figure 7C).

Sediment sorting determines the likelihood of imbrication. A well-sorted collection of flat pebbles commonly contains an imbricated array such as in Figure I2, but a poorly sorted sand

Figure I2 An early illustration of imbricated and flow-oriented clasts (Jamieson, 1860) and photos of the same phenomena from the Colorado Plateau. These illustrations and photos show regularly oriented clasts, where the a–b planes of the clasts dip down toward the upstream direction. Full scale is 30 cm.

with a gravel component will rarely contain imbricated pebble arrays.

Regularly-oriented clasts

Isolated tabular clasts are easily oriented by flowing water. Analysis of the fluid mechanics involved in overturning a tabular clast suggests that there is a critical overturning *Froude number* ($\mathbb{F}_i$) above which the clasts overturns,

$$\mathbb{F}_i = V/\sqrt{gc} \qquad \text{(Eq. 1)}$$

where V is the mean flow velocity affecting the clasts, g is the acceleration due to gravity, and c is the thickness (short axis) of the tabular clasts (Galvin, 1987). The critical overturning value of $\mathbb{F}_i$, above which the tabular clast is oriented by the flow, is about 3.0 to 3.5. This critical value decreases as the suspended mud content of the water increases. The analysis assumes that flow depth exceeds the a axis of the clast, a condition where the immersed weight of the clast resisting the overturning is significantly less than its weight in air.

Usually, the acceleration, g, in a Froude number can be treated as just another constant necessary to make the units balance, but Pathfinder photos of clasts on the surface of Mars suggest that these clasts may be imbricated or flow-oriented. The lower Martian value of g to be used in equation 1 for explaining this imbrication or clast orientation implies that a given size clast overturns in flowing water on Mars when the velocity is only about 62 percent that needed to overturn it on Earth.

The expected orientation of an imbricated group of tabular clasts or an isolated tabular clast is one in which the planes containing the longer (a, b) axes dip down toward the source of the flow (Figure I2). For a clast in a stream, this means that the $a–b$ plane usually dips toward the upstream direction. Clasts having this expected dip toward the upstream direction are *regularly oriented clasts*.

Reversely-oriented clasts

On active foreset beds, there may be exceptions to this upstream dip (Figure I3, top). On active foresets, isolated clasts may dip in the down-slope direction parallel to, or more steeply than, the slope itself, if turbulence shed by the clast scours a hole at the downstream edge of the clast, and (possibly) if grain flow exerts an overturning moment on the upslope edge of the clast. Clasts whose $a–b$ planes dip in the downstream direction are *reversely oriented clasts*, so-called because the dip is opposite that expected from intuition developed by observations such as shown in Figure I2.

Figure I3 Reversely oriented clasts. Above: downstream dip of larger pebbles in a sandy member of the Cretaceous Potomac Formation, Lorton Station, Virginia. The horizontal is indicated by the central bubble of the level. Inferred flow is from right to left. Notice the collection of smaller pebbles in the wake of the large reversely oriented pebble. At a higher level in the same outcrop where the bedding is more nearly horizontal, flow-oriented pebbles have regularly inclined orientation. Below: in Brushy Basin Member of Morison Formation, reversely oriented blocky sandstone on "popcorn" montmorillorite shale, apparently the result of scour on the downslope edge of the clasts during sheet flow runoff.

In arid regions it is common to find clasts on eroding slopes whose a–b planes dip downslope more steeply than the slope on which they rest. This commonly occurs when volcanic rocks capping the uplands break up and slide down slopes of fine-grained volcanic ash. Such an orientation of clasts (flints) in surficial gravels were reported toward the end of his life by Charles Darwin from the south of England. Probably, the cause is runoff by overland sheet flow during rare, heavy thunderstorms. For sheet flow, the water depth is typically less than the c-axis, a condition where the effective weight of the clast on the bottom is nearly that of its weight in air. The sheet flow scours on the downslope side of the clasts, and the clast settles into the scoured depression with the result that the a–b plane of the clast dips more steeply than the slope (Figure I3, bottom). With respect to the presentation shown in Figure I2, the clasts in Figure I3 are reversely oriented because they dip toward the downstream direction of the flow. Such orientations of the a–b planes steeper than the underlying slope may superficially resemble the condition produced by waves on beaches.

Finally, a form of reversely oriented clasts develops when wedge-shaped or pear-shaped clasts are subject to sheet flow. The thick end of the clast exerts more weight on the underlying

Figure I4 Imbricated boulders at Monument Cove, Acadia National Park, Maine. Boulders are locally derived and abraded (Waag and Ogren, 1984). The long axes of boulders dip into the waves, usually more steeply than the foreshore slope.

Table I1 Classification of imbricated and flow-oriented clasts

Class	Flow	Condition
Regularly inclined	Stream flow (clear water)	$\mathbb{F}_i > 3.0$ to 3.5
	Debris flow	$\mathbb{F}_i < 3.0$
Reversely inclined	Delta flow (foreset beds)	Scour at downstream end of clasts with tilt due to fluid flow, gravity, and grain flow.
	Sheet flow (erodible bed)	Scour at downstream end of clast with tilt due to gravity.
	Sheet flow (pavement)	Pivot about thick end of clasts. Thick end fixed by friction
Wave-oriented	Waves	Onshore–offshore flow dominates. Clasts dip toward waves, more steeply than foreshore.
	Longshore currents	Significant longshore flow accompanies waves. Clasts dip into longshore current.
Vertically oriented	Steady flows in streams or surf	High concentrations of plate-like clasts.

surface. If the weight produces sufficient friction along the underlying surface, the thin edge of the clast will be moved by the current, pivoting like a rudder about its post, to point downstream. Such orientations might be expected on large clasts after sudden extreme floods such as the floods that produced the chaneled scablands on the Columbia Plateau or jökulhlaup deposits in Iceland.

Wave-oriented clasts

For clasts on beaches, the orientations of a–b planes have aspects of both regularly and reversely oriented clasts. For a clast subject to wave action on a beach foreshore, the a–b plane dips toward the incoming waves (Figure I4). The incoming wave is the dominant flow, so this seaward dip may be considered the expected regular orientation. Relative to the foreshore on which the clast rests, the clast dips more steeply so that the lower end of the clast is embedded in the foreshore. It appears that this embedding of the clast at its lower end is at least in part due to the currents draining seaward from the swash, in which perspective, the clast is reversely oriented. When the waves produce a significant longshore current, especially at the distal ends of spits, the a–b plane of the clast will reorient to dip toward (into) the longshore currents, producing a regular orientation from the perspective of the longshore current.

On low spits during times when waves actively overtop the spit, the overtopping due to the waves acts as a uniform stream which orients flat pebbles with their a–b planes dipping into that current. This results in clasts on the land-facing shore of spits having a seaward dip to the a–b plane.

Vertically-oriented clasts

Finally, there is the vertical orientation of a–b planes introduced when exceptionally thin clasts, such as relatively flat shells or slate, are subject to strong flows, such as surging waves (Sanderson and Donovan, 1974). This phenomenon is relatively rare on a worldwide basis but locally common, particularly where clams are shucked commercially. The origin of beds of shells with vertically oriented a–b plane appears to be similar to the origin of the more common phenomenon of leaf jams in streams flowing through deciduous woodlands. There, a heavy concentration of large leaves, such as oak leaves, in flowing water produces a stable orientation with the leaf planes vertical.

Summary

The relationship of the various factors to flow direction and slope are summarized in Table I1.

Cyril Galvin

Bibliography

Blair, T.C., 1987. Sedimentary processes, vertical stratification sequences, and geomorphology of the Roaring River Alluvial Fan, Rocky Mountain National Park, Colorado. *Journal of Sedimentary Research*, **57**: 1–18.
Galvin, C., 1987. Imbrication of tabular boulders. In *Proceedings of the National Conference on Hydraulic Engineering*. American Society of Civil Engineering, pp. 279–284.
Gibbons, G.S., 1972. Sandstone imbrication study in planar sections: dispersion, biasses, and measuring methods. *Journal of Sedimentary Petrology*, **42**: 966–972.
Jamieson, T.F., 1860. On the drift and rolled gravel of the north of Scotland. *Quarterly Journal of the Geological Society of London*, **16**: 347–371.
Sanderson, D.J., and Donovan, R.N., 1974. The vertical packing of shells and stones on some recent beaches. *Journal of Sedimentary Petrology*, **44**: 680–688.
Waag, C.J., and Ogren, D.E., 1984. Shape evolution and fabric in a boulder beach, Monument Cove, Maine. *Journal of Sedimentary Petrology*, **54**: 98–102.

IMPREGNATION

Impregnation is a technique used to obtain high quality sections of soft or poorly consolidated sediments for preservation or petrographic analysis. The hardened sediment samples are also valuable for elemental and microfabric analyzes. It is also frequently beneficial to impregnate partially consolidated samples (e.g., diamicts) prior to facing and thin section preparation. The primary goal of any sediment impregnation method is to stabilize the material of interest in order to obtain undisturbed samples. Due to the range of sample characteristics and the varying needs of researchers, a large number of impregnation methods have been developed. Workers in the fields of sedimentology and soil science have made parallel developments in impregnation techniques and both fields, often in isolation, have adopted many common techniques. However, despite the wide use of impregnation techniques in sedimentology, geologists have not adopted one or more standard methods for sample preparation.

The main developments in sediment impregnation have resulted from the availability of new embedding materials. In particular, the advent of low viscosity epoxy resins has made impregnating clay-rich and compacted samples possible. Key criteria for selecting a particular method and embedding media depend on the sample cohesiveness, texture, and cementation. In most cases, the most appropriate method will produce an imbedded sample that is fully stabilized and has undergone minimal disturbance, particularly shrinkage. Thorough discussions of various materials and the comparative advantages of a large number of different techniques can be found in Bouma (1969) and Murphy (1986).

Main approaches

One of the earliest approaches used to obtain an impregnated sample from unconsolidated sediment samples was the production of a tape or glue peel (see *Peels*). This method is most appropriate for unconsolidated sediments typically found in sections and from split cores. The common approach is to saturate the sediment surface with diluted, water-soluble glue that is allowed to dry over several days. In coarse-grained samples, the glue typically penetrates to a depth of 0.5 cm to 2 cm. When dry, the glue surface can be separated from the remaining sediment, yielding a permanent sample for storage, mounting, and analysis. Although suitable for large exposures and field use, the chief disadvantage of water-soluble glues is that some degree of shrinkage occurs. The use of a tacky acetate sheet applied to the sediment surface is an alternate approach that avoids the subsequent shrinkage problem. However, the acetate peel can only bond to surface grains and may not provide a complete, representative sample.

A more common approach to imbedding involves filling the entire sample with a cement to preserve the sedimentary structures. In practice, a large number of possible embedding materials exist, and availability varies in different areas. In general, embedding media can be characterized as either water-soluble or insoluble. Soluble materials are useful when the sample is wet, such as recent lacustrine or marine sediments. The two most common soluble materials include diluted ordinary glue and polyethylene glycol (PEG). The sample is immersed in the liquid embedding material and allowed to soak for several days or longer to allow the embedding fluid to fully infiltrate the sample. Generally, the PEG used for embedding is a high molecular weight (>800) variety and is a waxy solid at room temperature. The PEG flakes must be melted by the application of heat during the soaking process. Although the water-soluble media are nontoxic and require no specialized equipment to use, they often result in some shrinkage, are generally too soft for thin sectioning, require non-aqueous grinding fluids, and they have poor optical properties.

Water insoluble media are generally either epoxy or acrylic resins that require the sample to be dehydrated prior to embedding to obtain full polymerization of the resin compounds. While air or oven drying may be suitable for sandy or gravelly samples, finer textured samples will crack and shrink during drying, and the dry permeability of the sample will often be too low to allow full infiltration of the embedding material, especially into the center of larger samples.

Two common methods for dehydrating samples to minimize shrinkage are to sublimate the sample water under vacuum (freeze drying) or by using repeated liquid-liquid replacements. Freeze drying effectively dehydrates samples of all textures, although many commercial units are limited to small samples (less than a few centimeters). Where suitable equipment is available, freeze-drying produces minimal disturbance of the sample if the sample is first frozen with liquid nitrogen (Bouma, 1969). Pre-freezing is particularly important for clay-rich samples, as the rapid freezing in nitrogen prevents the formation of large ice crystals in the sample that result in voids in the embedded sample. However, some ice crystal formation tends to occur in very fine clay units, particularly if the sediment sample is more than 1 cm thick. Depending on the rate of sample drying and the sediment composition, some minor shrinkage may also occur and cracks may form at sedimentary contacts. Moreover, the freeze-dried samples are extremely fragile and coarse units may require support to prevent collapse during handling.

The second approach to dehydrating samples involves replacing the pore water with a water and resin miscible solution. Most epoxy resins are compatible with acetone and various alcohols. By replacing the pore water with such a fluid, there is no shrinkage or disturbance to the sedimentary structures. This method is suitable for a variety of sediments including those from clay-rich lacustrine, marine, and glaciogenic environments. The wet samples are submerged in the replacement fluid that is left to fully infiltrate into the sample, often for one or more days. The fluid is then drawn off and replaced with fresh solution until the specific gravity of the supernatant indicates that the water content is below 1 percent (or less). Liquid-liquid replacement dehydration of samples results in the least disturbance of samples, as there is no shrinkage or danger of ice crystal casts in the final embedded sample. However, the process is time consuming and uses a large amount of the exchange fluid and resin. In some cases, application of a light vacuum during the resin infiltration can avoid extra resin waste.

Selection of the embedding resin varies considerably among users, ranging from slow curing acrylics (months) to rapid curing epoxies (hours). Some epoxy resins that cure rapidly are unsuitable because they are strongly exothermic and can disturb the sample. Depending on circumstances, a variety of resins can be used (Tippkötter and Ritz, 1996). Low viscosity resins used for embedding histological and material samples are readily available and suitable for fine-grained lacustrine and marine sediments. Less specialized, higher viscosity resins are suitable for coarser samples. Most require curing in an oven to fully polymerize. The advantages of impregnating samples with resin include standard handling procedures for thin section preparation, excellent optical characteristics, and permanent preservation of the sample. Compared to water-soluble media, impregnating samples with resin is often time consuming and relatively expensive. Moreover, more specialized equipment is required, including facilities to handle what are typically toxic chemicals.

Partial curing of an impregnated block, especially with epoxy resins, usually indicates the presence of water in the sample or poor resin infiltration. Increased infiltration times using resins with long working times can avoid this problem, and light vacuum may also accelerate infiltration. In some situations, samples with incomplete infiltration can still be cut and the exposed face treated with resin to complete the stabilization prior to mounting (Carr and Lee, 1998).

Impregnation of sedimentary samples has proven to be a valuable technique for geologists working in a wide range of environments where sample preservation and microscopic analysis has been limited by poorly consolidated materials. A number of methods are available for impregnation, depending on sample texture and moisture content. Samples that are dehydrated and embedded with epoxy resins are suitable for most applications, although simpler methods are available as well.

Scott F. Lamoureux

Bibliography

Bouma, A.H., 1969. *Methods for the Study of Sedimentary Structures*. John Wiley & Sons.
Carr, S.J., and Lee, J.A., 1998. Thin-section production of diamicts: problems and solutions. *Journal of Sedimentary Research*, **68**: 217–220.
Murphy, C.P., 1986. *Thin Section Preparation of Soils and Sediments*. Berkhansted: A B Academic.
Tippkötter, R., and Ritz, K., 1996. Evaluation of polyester, epoxy and acrylic resins for suitability in preparation of soil thin-sections for *in situ* biological studies. *Geoderma*, **69**: 31–57.

Cross-references

Bedding and Internal Structures
Fabric, Porosity, and Permeability
Relief Peels

IRON–MANGANESE NODULES

Black to dark brown spheroidal to discoidal concretions of iron and manganese oxyhydroxides, generally a few cm in diameter, that cover extensive areas of the ocean floor in all water depths as well as the bottoms of some temperate latitude lakes (Figures I5 and I6). The ferromanganese oxyhydroxides are intimately mixed with mineral and rock fragments and fossil debris, and occur as concentric layers arranged around a central nucleus or core consisting of rock fragments (generally volcanic), fossils (e.g., fish teeth, whale ear bones, etc.) or broken nodules. Surfaces are either smooth or granular. Nodules are highly porous and have dry bulk densities ranging from $1.2\,\mathrm{g\,cm^{-3}}$ to $1.6\,\mathrm{g\,cm^{-3}}$ depending on the chemical composition.

Distribution and abundance

Nodules are most abundant on the sediment surface and occur sporadically within the seafloor sediment. Sampling and bottom photography has shown that average surface concentrations of $10\,\mathrm{kg\,m^{-2}}$, reaching a maximum of about $35\,\mathrm{kg\,m^{-2}}$, are typical of vast areas of the Indian and Pacific seafloors between 4 km and 5 km water depth. The highest concentrations are found where the sediment accumulates very slowly, so that the nodules are not covered and buried. Hence, dense pavements of nodules are characteristic of the central ocean basins most remote from continental sediment supply. Similar Mn and Fe deposits are found on northern European shelves, the Baltic Sea and some fjords. Ferromanganese oxyhydroxides also occur as crusts and coatings on rock and mineral surfaces, especially on seamounts and similar topographic elevations throughout the ocean basins.

Ferromanganese nodules are locally abundant on the floors of temperate zone lakes, such as those of Scandinavia and eastern North America (Cronan and Thomas, 1970). They occur as spherical and discoidal nodules and as coatings on rock surfaces.

Growth rate

The rate of accretion of ferromanganese nodules and crusts has been determined by dating rock nuclei by the K/Ar method (Barnes and Dymond, 1967), and by measuring the decay of natural uranium series (^{230}Th, ^{231}Pa) or cosmogenic (^{10}Be) nuclides in the oxyhydroxide layers (Ku, in Glasby, 1977). The median rate obtained with these different methods is around 5 mm/Ma, with a range of <1 mm/Ma to >50 mm/Ma. In contrast, the rate of accumulation of associated sediment is of order 1 mm/Ka to 5 mm/Ka. This discordance, and the puzzling fact that most deep-sea nodules are found at the sediment surface, implies that they must be maintained at the sediment-water interface for very long time periods. Occasional disturbance of the seafloor, by currents, slumping or the activities of benthic fauna, is probably sufficient to remove sediment from the tops of the nodules and to turn them over from time to time (Piper and Fowler, 1980).

Chemical composition

The chemical composition of ferromanganese nodules and crusts reflects their constituent mixtures of mineral phases, comprising oxyhydroxides, aluminosilicates, carbonates and phosphates. They are significantly enriched in Mn, Fe and other transition metals compared with crustal rocks. Mn/Fe ratios are generally greater than unity and Cu and Ni concentrations are highest in the Pacific Ocean, whereas Mn/Fe ratios are less than unity and Zn and Pb concentrations are highest in the Atlantic Ocean (Table I2). On an ocean-wide

Figure I5 Photographs of Pacific seafloor ferromanganese nodules. The upper pair of nodules have smooth surfaces and were collected from the northern central region where sedimentation rates are very low. The lower pair of nodules have granular, botryoidal surfaces and were collected from the northern equatorial region where sedimentation rates are higher.

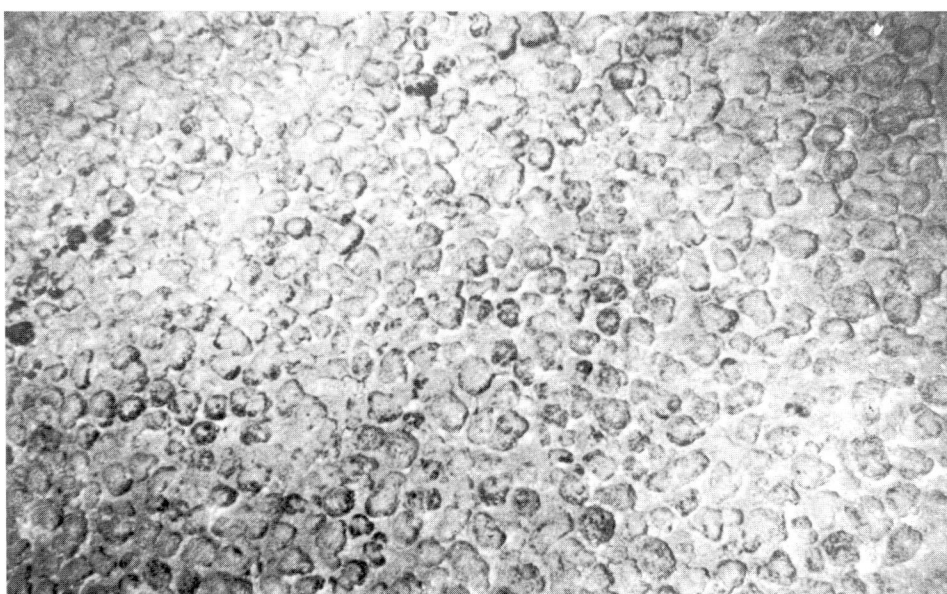

Figure 16 Seafloor photograph from the Southern Ocean (61° 41S; 170° 40'E; water depth 5160 m) showing dense pavement of nodules 2 cm to 8 cm in diameter with a light sediment coating. Surface concentration is approximately 25 kg m^{-2}. Photograph taken by S.E. Calvert using equipment owned by the US Navy.

Table 12 Average chemical composition of oceanic ferromanganese deposits (wt. per cent). From Cronan (1980)

Element	Pacific Ocean	Atlantic Ocean	Crustal abundance
Mn	19.8	15.8	0.1
Fe	12.0	20.8	5.6
Co	0.33	0.32	0.002
Ni	0.63	0.33	0.007
Cu	0.39	0.17	0.005
Zn	0.07	0.08	0.007
Pb	0.08	0.13	0.001

scale, minor element concentrations appear to be related to the Mn or the Fe contents; Ba, Cu, Mo, Ni, and Zn vary proportionally with the Mn content, whereas Co, Pb, Ti, and Ce concentrations correlate with the Fe content. These relationships reflect the substitution of the minor metals for Mn or Fe in the major oxyhydroxide mineral phases. *Buserite* is a fibrous manganate in which a number of divalent cations, including Mn^{2+}, Cu, Ni, Zn, etc., substitute for Mn^{4+} in an hexagonal array of edge-shared $Mn(IV)O_6$ octahedra thereby stabilizing the structure (Giovanoli and Arrhenius, in Halbach et al., 1988); vernadite is a highly disordered manganate with the approximate formula MnO_2 which is intimately associated with ferrihydrite (FeOOH), the principal Fe phase (Burns and Burns, in Glasby, 1977). MnO_2 and FeOOH host minor metals that can exist in their higher oxidation states in the ocean (Co^{3+}, Pb^{4+}, Ti^{4+}, Ce^{4+}), but lack sites of the appropriate size for the substitution of divalent cations. Thus, buserite is only found in nodules with high Mn, Cu, and Ni concentrations, whereas vernadite occurs in nodules with high Fe and Co contents. Nodules from shallow water environments consist of a mineral phase similar to buserite, which contains very low minor element contents and is unstable when dried in air (Calvert and Price, in Glasby, 1977).

Regional variations

The chemical composition of ferromanganese nodules varies markedly within a given ocean basin. This is best illustrated in the Pacific (Figure 17), where nodules that form in the marginal areas and in east–west zones north and south of the equator, have the highest Mn/Fe ratios. The abyssal Mn-rich nodules also have the highest Cu and Ni contents and they contain stable buserite as the principal Mn mineral phase (Cronan, 1976; Piper and Williamson, 1977; Calvert, 1978). The marginal Mn-rich nodules are impoverished in minor elements, and consequently they contain metastable buserite. Nodules and crusts that form in areas of the seafloor with very low sedimentation rates and on sediment-starved elevations above the abyssal seafloor (seamounts and banks) have the lowest Mn/Fe ratios, they lack buserite and have the highest

Cu and Ni contents, and the highest Co, Pb, Ti, and Ce contents. *Diagenetic* nodules form by the more rapid precipitation of Mn and divalent metals supplied to the seafloor in association with settling planktonic organic matter and recycled from the underlying sediment. Such nodules generally have granular surfaces and buserite is the principal Mn phase. They have much higher Mn/Fe ratios, and the highest Cu and Ni contents (Calvert and Price, 1977). These two types of precipitates can occur in the same nodule; in this case a distinctly discoidal nodule from the equatorial region of the Pacific Ocean has a smooth, vernadite-bearing upper surface with a low Mn/Fe ratio and high Co content, and a granular, buserite-bearing lower surface with a much higher Mn/Fe ratio and high Cu and Ni contents growing within the surficial sediment (Raab, 1972). Distinct compositional regions in the Pacific therefore identify environments where metals are either supplied mainly from bottom waters or additionally from the bottom sediment (Halbach *et al.*, 1988). Marginal nodules with high Mn/Fe ratios have low minor metal contents because diagenetic reactions in the subsurface anoxic sediment, fueled by high burial rates of organic matter, trap them as sulfides or cause them to be adsorbed on particle surfaces, thereby preventing their recycling to the sediment surface where they would otherwise be incorporated into surficial oxyhydroxides (Calvert and Price, in Glasby 1977). Lacustrine nodules form by very similar processes and have compositions that reflect the relative importance of metal precipitation from lake waters and the bottom sediments (Harriss and Troup, 1970).

Summary

Ferromanganese nodules and crusts represent a common type of authigenic deposition in all oxygenated aquatic environments in all water depths. They grow very slowly by the accretion of very fine-grained and colloidal oxyhydroxides onto solid substrates, and often form dense pavements over large areas of the seafloor. They can be greatly enriched in transition metals, which are derived from normal ocean waters and from sediment pore waters. Their mineralogies and bulk compositions vary from ocean to ocean, within an ocean basin and on the scale of a single nodule. There are two end-member nodule and crust types, with all gradations between these forms. Hydrogenetic nodules and crusts are formed from bottom seawater, whereas diagenetic nodules also have a metal source from the associated sediment via diagenetic metal recycling. These processes are also responsible for the formation of similar deposits on shallow areas of the seafloor and in lakes.

Stephen E. Calvert

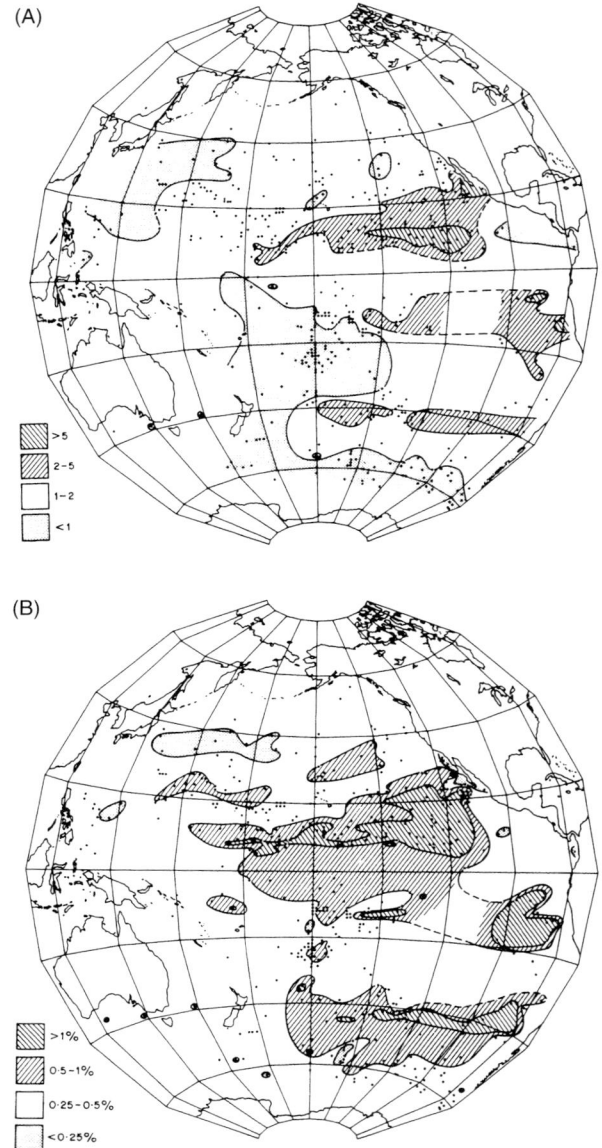

Figure 17 Ocean-wide variation in (A) the Mn/Fe ratio and (B) the Ni concentration in seafloor nodules from the Pacific Ocean. From (Cronan, 1980).

Co contents. This compositional variability reflects the two processes by which ferromanganese deposits are formed.

Origin

Hydrogenetic nodules and crusts form by the very slow precipitation of seawater Mn and Fe, supplied to ocean waters by rivers and seafloor hydrothermal sources, as very fine-grained or colloidal oxyhydroxides onto solid substrates (Halbach *et al.*, 1981). The nodules and crusts formed in this manner generally have smooth surfaces, and vernadite is the principal Mn phase. They have the lowest Mn/Fe ratios and

Bibliography

Barnes, S.S., and Dymond, J.R., 1967. Rates of accumulation of ferromanganese nodules. *Nature*, **213**: 1218–1219.

Burns, R.G., and Burns, V.M., 1977. Mineralogy. In Glasby, G.P. (ed.), *Marine Manganese Deposits*. Elsevier, pp. 185–248.

Calvert, S.E., 1978. Geochemistry of oceanic ferromanganese deposits. Philosophical Transcations of the Royal Society of London, 290: 43–73.

Calvert, S.E., and Price, N.B., 1977a. Geochemical variation in ferromanganese nodules and associated sediments from the Pacific Ocean. *Marine Chemistry*, **5**: 43–74.

Calvert, S.E., and Price, N.B., 1977b. Shallow water, continental margin deposits: distribution and geochemistry. In Glasby, G.P. (ed.), *Marine Manganese Deposits*. Amsterdam: Elsevier, pp. 45–86.
Cronan, D.S., 1976. Manganese nodules and other ferromanganese oxide deposits. In Riley, J.P., and Chester, R. (eds.), *Chemical Oceanography*. San Diego, California: Academic, pp. 217–263.
Cronan, D.S., 1980. *Underwater Minerals*. London: Academic Press, 362p.
Cronan, D.S., and Thomas, R.L., 1970. Ferromanganese concretions in Lake Ontario. Canadian Journal of Earth Sciences, 1346–1349.
Giovanoli, R., and Arrhenius, G., 1988. Structural chemistry of marine manganese and iron minerals and synthetic model compounds. In Halbach, P., Friedrich, G., and von Stackelberg, U. (eds.), *The Manganese Nodule Belt of the Pacific Ocean*. Stuttgart, Enke, pp. 20–37.
Glasby, G.P., (ed.), 1977. *Marine Manganese Deposits*. Elsevier.
Halbach, P., Friedrich, G., and von Stackelberg, U., 1988. *The Manganese Nodule Belt of the Pacific Ocean*. Stuttgart, Enke, 254 p.
Halbach, P., Scherhag, C., Hebisch, V., and Marchig, V., 1981. Geochemical and mineralogical control of different genetic types of deep-sea nodules from the Pacific Ocean: *Mineralia Deposita*, **16**: 59–84.
Harriss, R.C., and Troup, A.G., 1970. Chemistry and origin of freshwater ferromanganese concretions: *Limnology and Ocenaography*, **15**: 702–712.
Ku, T.L., 1976. Rates of accretion, In Glasby, G.P. (ed.), *Marine Manganese Deposits*. New York: Elsevier, pp. 249–267.
Piper, D.Z., and Fowler, B. 1980.. New constraint on the maintenance of Mn nodules at the sediment surface: *Nature*, **286**: 880–883.
Piper, D.Z., and Williamson, M.F., 1977. Regional variations in the elemental and mineralogical composition of ferromanganese nodules from the pelagic environment of the Pacific Ocean. *Marine Geology*, **23**: 285–303.
Raab, W., 1972. Physical and chemical features of Pacific deep-sea manganese nodules and their implications to the genesis of nodules. In Horn, D.R. (ed.), *Ferromanganese Deposits on the Ocean Floor*. Washington, DC: National Science Foundation, pp. 31–49.

Cross-references

Hydroxides and Oxyhydroxide Minerals
Oceanic Sediments

IRONSTONES AND IRON FORMATIONS

Iron-rich sedimentary rocks contain ≥15 percent metallic iron by weight (James, 1966) and form a class of chemical sediments comparable to evaporites and phosphorites. Most workers recognize two main categories. *Iron Formations* are generally cherty, thinly laminated, and Precambrian in age. *Ironstones* are generally less siliceous, more aluminous, not laminated, smaller, and Phanerozoic in age. This distinction highlights time-related changes which constitute one of the most interesting aspects of iron-rich sedimentary rocks and has important implications for the evolution of Earth's atmosphere and hydrosphere. Both types have long been sources of iron for human use. Ironstones were mined in Roman times and were a key resource in the industrial revolution (Young, 1993). Mineralogical investigations of ironstone began in the mid-1800s. Studies of iron formations were begun in the late 1800s by geologists from the fledgling U.S. Geological Survey in the Lake Superior region. This work resulted in classic monographs focused on specific "iron ranges" as well as Van Hise and Leith (1911). Fresh studies triggered by the demand for iron during World War II yielded a new round of U.S.G.S. publications and the seminal article by James (1954). Large iron formations and associated deposits of iron ore were still being discovered. The subsequent discovery and study of large iron formations in Western Australia have been particularly influential in shaping current perceptions of iron formations, even though they are exceptional in many ways (Trendall, 2002).

The vast majority of iron that is ever mined will be extracted from Precambrian iron formations or derivative deposits. Large iron formations are currently mined all over the world. In 2000, Australia produced over 160 million metric tons of iron ore worth in excess of US$2.5 billion, and China and Brazil produced even more. The largest ore deposits occur where silica was leached and iron was oxidized in iron formations during the Precambrian (Morris, 1987). Ores produced by post-Precambrian weathering generally have lower iron contents and more impurities. Iron can also be extracted from unaltered iron formations by crushing them and separating the magnetite.

Iron formations have also yielded economic quantities of other materials. Iron formations in Archean greenstone belts host world-class gold deposits, and manganese is extracted in large quantities from iron formations in Brazil, India, and especially South Africa. Blue asbestos (crocidolite) was extracted from iron formations during the mid-1900s, but production dropped off precipitously as the highly carcinogenic nature of blue asbestos became clear.

Iron formations

The main iron minerals fall into four groups which James (1954) used to define four "facies" of iron formation: oxide, silicate, carbonate, and sulfide. In low-grade iron formations, the dominant minerals are magnetite and hematite in the oxide facies; greenalite and minnesotaite in the silicate facies; siderite and ankerite in the carbonate facies; and pyrite in the sulfide facies. Where detrital textures are not obscured by metamorphism, iron formations can be sub-divided into *banded* versus *granular* varieties. Banded iron formations or BIFs, by far the more abundant of the two, were originally chemical muds. Most BIFs exhibit a quasi- to highly rhythmic alternation of layers rich in iron and chert on a scale of millimeters to centimeters (Figure I8(A)). Granular iron formations or GIFs originated as well-sorted chemical sands (Figure I8(B)) in which the clasts were largely derived via intrabasinal erosion and redeposition of pre-existing BIF. GIFs lack the even, continuous banding of BIFs and generally belong to the oxide or silicate mineral facies, whereas BIFs show a much broader spectrum of iron minerals, including a greater abundance of reduced facies (James, 1954; Simonson, 1985). Layers of pure GIF thicker than a few meters are rare, whereas BIF can continue uninterrupted by GIF for tens to hundreds of meters stratigraphically. Iron formations with a mixture of BIF and GIF are actually more abundant than pure GIFs, and they show bedding that is more irregular than pure BIF, but less massive than pure GIF. In mixed iron formations, GIF takes the form of discontinuous lenses that probably represent bedforms generated by storm waves and currents (Simonson, 1985). Chert is not used to classify iron formations because it is a ubiquitous component, but its abundance helps distinguish them from ironstones. Another

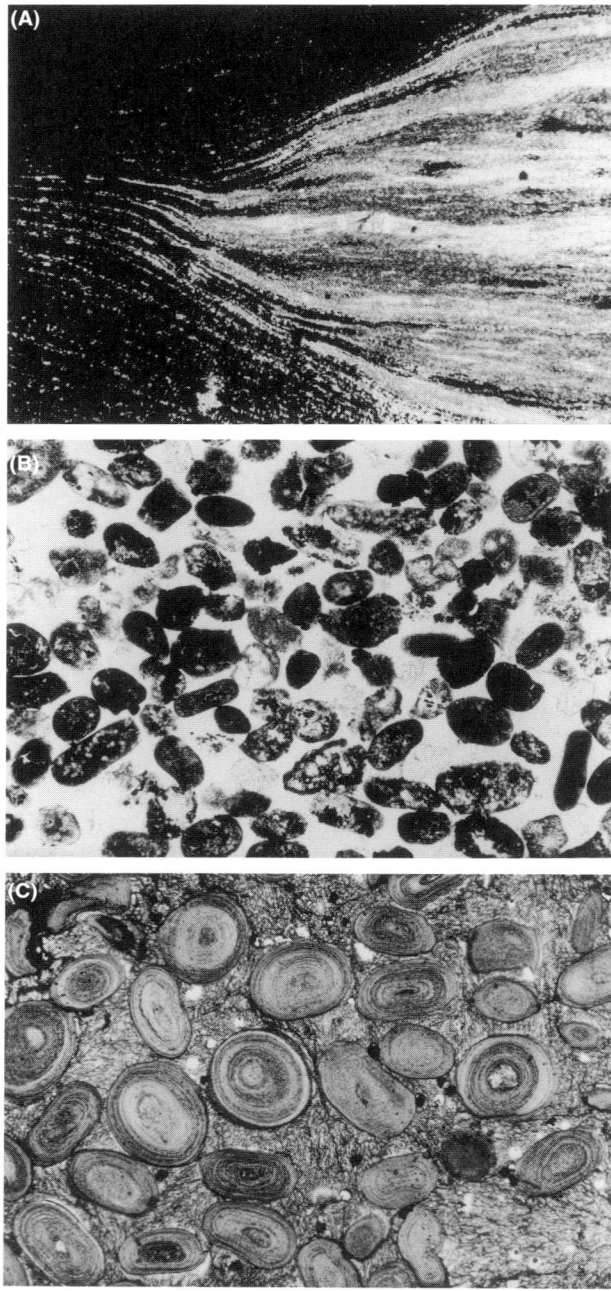

Figure 18 Photomicrographs of iron-rich sedimentary rocks in plane polarized light. (A) Banded iron formation from the early Paleoproterozoic Dales Gorge Member of the Brockman Iron Formation, Hamersley basin of Western Australia. Clear part is chert pod with opaque hematite impurities, dark part is dominantly magnetite; note the contrast in thickness and sharpness of laminae as they exit the chert pod into the surrounding BIF. Field of view is ∼5 mm wide. (B) Granular iron formation from the Paleoproterozoic Sokoman Iron Formation, Labrador trough. Peloid-like granules consist of chert and hematite with lesser amounts of magnetite; the interstices are filled with slightly recrystallized chalcedonic cement. Field of view is ∼2.5 mm wide. (C) Oolitic ironstone from the late Ordovician to early Silurian Don Braulio Formation, Sierra de Villicúm of the Argentine Pre-Cordillera. The ooids consist of silicate, show nice pseudouniaxial extinction crosses between crossed polarizers, and are locally truncated along dark stylolitic seams; the composition of the fine-grained matrix is undetermined, but it is probably a mixture of fine quartzose silt and chert. Field of view is ∼5 mm wide.

scheme subdivides iron formations into Superior-type versus Algoma-type, as described below. Given their great age, most iron formations are tectonically deformed or metamorphosed to some degree, yet iron formations retain their distinctive chemical composition, even at high grades. The metamorphism of iron formations is a complex topic in its own right (Trendall and Morris, 1983, pp. 417–469).

Intrinsic characteristics

Despite their great antiquity, many iron formations have well-preserved sedimentary features. Like all arenites, GIFs consist of a combination of: (1) sand-size clasts (long referred to as "granules," for example, Van Hise and Leith, 1911); (2) finer grained material (matrix), which is relatively rare; and (3) cements (interstitial, void-filling authigenic minerals) (Dimroth, 1976). In addition, diagenetic phases replace these primary constituents in part or whole in all iron formations. The granules, which are analogous to carbonate peloids and intraclasts, are mostly fine to coarse sand size (Mengel, 1973) and range from well-rounded to angular (Figure I8(B)). Ooids with concentrically laminated cortices are abundant locally (Dimroth, 1976; Simonson, 1987), but much rarer than granules. GIFs had depositional porosities of 40 percent to 50 percent, much of which was occluded by siliceous cement early in diagenesis (Figure I8(B); Mengel, 1973; Simonson, 1987). Other GIFs are heavily compacted and show tight frameworks and distorted clasts. Many GIFs have intergranular cavities, cracks, and vugs that are largely filled with chalcedonic and drusy quartz cements prior to compaction. Similar cracks and vugs are widely developed in the stromatolitic cherts found in some GIFs. These cracks and vugs probably originated via post-depositional shrinkage due to the dewatering of gelatinous precursors, and some contain internal sediment that is distributed geopetally (Simonson, 1987).

Depositional structures are often obscured by diagenetically redistributed minerals, but dune-scale cross-stratification is widespread in GIFs (Simonson, 1985). The few paleocurrents which have been measured show complex polymodal patterns with hints of herringbone typical of shallow marine sands (Ojakangas, 1983). Flat pebble conglomerates are a minor but widespread component of GIFs with the pebbles generally derived from more siliceous layers. Stromatolites that vary in width from a centimeter to over a meter and range in morphology from columnar to domal are another minor but distinctive sedimentary structure found in GIFs but not BIFs. Although they are usually interpreted as products of sediment trapping and/or precipitation by microbial mats, some stromatolites in GIFs have characteristics like those of siliceous sinters deposited in and around hot springs. Thanks to the early silica cementation, iron formations contain some of the best-preserved early Precambrian biotas in the world (Trendall and Morris, 1983, pp. 373–400).

In contrast to GIFs, BIFs generally lack any obvious detrital textures. Even though they are diagenetically reorganized, many BIFS are still remarkably fine-grained and uniform (Figure I8(A)). As noted above, BIFs show more diversity of iron minerals than GIFs, including substantial thicknesses of all four of James' facies. Most of these minerals are thought to be derived from phases precipitated from the original basin waters, but some BIFs contain aluminous silicate phases, most notably stilpnomelane, due to the presence of volcaniclastic contaminants. Exquisite volcanic shards replaced by stilpnomelane occur in some iron formations (LaBerge, 1966a,b).

As the name implies, most BIFs have well-developed thin lamination to thin bedding (Figure I8(A)). This is no surprise, given the lack of burrowers in the Precambrian, but the layers in BIFs (particularly those rich in iron oxides) can be quite striking, and in some cases they can be correlated for hundreds of kilometers (Trendall and Blockley, 1970; Ewers and Morris, 1981). The bedding in BIFs can also be highly cyclic via alternations of iron-rich and iron-poor precipitates or alternations of BIF and fine shaly or volcaniclastic strata (Trendall and Blockley, 1970; Ewers and Morris, 1981; Beukes, 1984). Trendall (1972) attempted to relate these cycles to orbital parameters, but they have yet to be tested for the periodicities typical of Milankovitch forcing. Sedimentary structures other than banding are rare in BIFs, except for chert pods (Figure I8(A)), which are best interpreted as localized pockets of early cementation. Chert-poor BIF surrounding the pods offers textbook examples of differential compaction following localized early cementation (Dimroth, 1976; Beukes, 1984; Simonson, 1987) which indicate chert-poor BIFs lost up to 90 percent or more of their original thickness during compaction; this is typical of muds in general. The minerals shielded within chert pods also indicate the original sediment had a range of compositions similar to the ones shown by present-day BIFs.

Stratigraphic context and secular variation

Although they are associated with a wide variety of lithologies, the nature of the rocks associated with iron formations changes progressively through time. Gross (1965, 1983) used this and other criteria to subdivide iron formations into Algoma-type and Superior-type. His original definitions have evolved through time, but the modifier Superior-type is still used for uniquely large iron formations of the late Archean and Paleoproterozoic that are associated with non-volcanic sedimentary rocks. The modifier Algoma-type is used for iron formations that are generally smaller and occur in Archean greenstone belts where they are associated with volcanic rocks (Fralick and Barrett, 1995). Virtually all of the Algoma-type iron formations are BIFs, and they run the full gamut from oxide to sulfide facies mineralogies. In contrast, the thicker, more laterally extensive Superior-type iron formations contain GIFs as well as BIFs; the BIFs again show a full range of iron mineralogies, but the GIFs are dominated by oxide and silicate facies rocks. Some of the Superior-type iron formations are associated with shallow-water deposits, but many of them as well as most of the Algoma-type iron formations are associated with deeper water deposits.

Iron formations range from early Archean to Paleoproterozoic in age and are found among the oldest sedimentary strata on earth in the Isua supracrustal belt in western Greenland. Iron accumulation climaxed during an interval of ≤ 800 million years in the Late Archean to Paleoproterozoic and ended rather abruptly on or before 1.8 Ga (Gole and Klein, 1981; Trendall and Morris, 1983, pp. 471–490). Algoma-type iron formations were forming in North America as the oldest Superior-type iron-formations formed in Western Australia and South Africa, indicating a gradual transition between the two.

A noteworthy episode of global sedimentary iron accumulation also took place in the Neoproterozoic, at which time iron-rich rocks often referred to as iron formations were deposited on several continents. These late Precambrian iron formations tend to be richer in iron and have a simple iron mineralogy dominated by hematite (Beukes and Klein, 1992). Many of these iron formations are also closely associated with glaciogenic sediments (Young, 1988), so they may differ in origin.

Origins and implications for earth history

The slow rise and more rapid collapse of iron formation deposition early in Earth's history must be linked to co-evolution of the atmosphere, hydrosphere, biosphere, and lithosphere, but controversy continues over the details. Many hypotheses championed by early workers have been abandoned. For example, suggestions that iron formations originated via the replacement of carbonates, represent evaporites unique to the early Precambrian, or were deposited in lacustrine basins are no longer tenable. Sedimentological studies have made it clear that iron formations are primary precipitates deposited mostly in open marine settings (Ojakangas, 1983; Beukes, 1984; Simonson, 1985). The genetic model currently presented in most textbooks is likewise at odds with our current knowledge of iron formations. Cloud (1973) first suggested that dissolved ferrous iron was ubiquitous in the early ocean and iron formations formed wherever photosynthesizing microbes produced free oxygen, causing iron oxides to precipitate. It was an elegant hypothesis at the time, but we now know the iron minerals in iron formations are much more than just oxides and the iron concentrations in carbonate sediments contemporaneous with iron formations are not high enough to fit this scenario (Veizer et al., 1990, 1992; Simonson and Hassler, 1996)

It is now widely believed that hydrothermal sources in the deep ocean supplied the iron to make iron formations. BIF deposition has been linked to hydrothermal activity with reasonably high confidence via stratigraphic context and facies relationships for some Algoma-type iron formations, and all types of iron formations show various geochemical signatures of hydrothermal sources (Klein and Beukes, 1992). It is also widely believed that iron formations precipitated along chemoclines in a stratified ocean (Beukes and Klein, 1992; Simonson and Hassler, 1996), which would explain why iron formations occur in many different tectonic settings and in association with many different rock types (Gross, 1983; Fralick and Barrett, 1995). Isley (1995) outlined a plausible scenario whereby iron-rich plumes released from deep-sea hydrothermal systems could travel to distant depocenters. The main point of contention is the mechanism(s) responsible for precipitating such great quantities of iron undiluted by other types of clastic or chemical sediment. Recent isotopic work indicates microbes played a role in the precipitation and/or diagenetic reorganization of the iron (Beard et al., 1999).

Consensus has also been reached on the cause of the shift from Algoma- to Superior-type iron formations through time. The appearance of large iron-formations probably reflects the first development of extensive shelf environments in the Late Archean (Eriksson, 1995). This is consistent with the simultaneous appearance of laterally extensive examples of other stable-shelf deposits such the first large carbonate platforms (Grotzinger, 1994). Age differences among the largest iron formations on different continents could reflect the fact that the "cratonization" of shields was a highly diachronous process (Eriksson and Donaldson, 1986). The smaller size of Algoma-type iron formations is generally believed to signal smaller depositional environments, but Gole and Klein (1981) point out that they are more highly deformed and some "may have been quite extensive prior to deformation and disruption." The growth in the size of iron formations through time could also involve changes in the bathymetry and/or chemistry of the world ocean that permitted the dispersal of iron over longer distances. However, this would require an increase in the mobility of dissolved iron, which runs counter to the prevailing view that the concentration of free oxygen in the atmosphere was increasing at that time (Eriksson, 1995). Even within the Superior-type category of iron formations, secular changes are evident. The oldest Superior-type iron formations are almost pure BIF, whereas GIFs reached their acme around 2.0 Ga. This suggests the locus of iron formation deposition shifted into progressively shallower environments, which may signal a progressive shallowing of a global chemocline through time.

The abrupt end to the deposition of iron formations in the Paleoproterozoic is presumably related more to atmospheric and hydrospheric than lithospheric evolution. Many lines of evidence signal a dramatic rise in atmospheric oxygen around 2.2 Ga to 1.9 Ga (Holland, 1994), which at some point ventilated the world ocean sufficiently well to prevent the storage and long distance transport of large quantities of dissolved iron. However, the disappearance of iron formations around 1.8 Ga may have actually been a result of the deep oceans becoming sulfidic rather than oxic (Anbar and Knoll, 2002). In either case a brief reappearance by iron formations late in the Precambrian may signal a return to these earlier conditions. The Neoproterozoic glaciations were so extreme that Hoffman et al. (1998) have argued they gave rise to a "snowball earth." Perhaps the ice capped the world ocean in such a way that it could become highly stratified for the first time in over an eon, thereby reactivating some of the mechanisms at work in the Paleoproterozoic. Like the older iron formations, the source of the iron for the Neoproterozoic iron formations appears to have been hydrothermal (Young, 1988).

Iron formations also differ from younger iron-rich sediments in their high silica content. Various lines of evidence including the cement-filled cracks and vugs in GIFs and the apparent syneresis of the original sediment indicate the silica in iron formation is a primary component. Given the absence of silica-secreting organisms, higher silica concentrations should have been present in Precambrian seawater to co-precipitate with the iron, but again the precise mechanism for this is unclear. On average, GIFs are probably more siliceous than BIFs, and oolitic and stromatolitic layers probably have the lowest iron contents of all. This is consistent with a deep-ocean source of iron, but it could also reflect a greater abundance of silica cement in GIFs.

Ironstones

The term ironstones encompasses a diverse group of rocks ranging in age from Proterozoic to Pliocene (Van Houton, 2000) which consist of materials ranging in size from sand to mud. Ironstones are one of several types of iron-rich sedimentary rocks formed during the Phanerozoic. The latter includes exhalative sediments formed directly by hydrothermal activity and "green sands," which are clastic sediments rich in glauconitic silicates. Such deposits have traditionally been differentiated from ironstones and that practice is followed here, even though glauconitic sediments and ironstones have much in common (Van Houton and Purucker, 1984; Odin, 1988). Given their young geologic age, ironstones are much less likely to be deformed or metamorphosed than iron formations and commonly contain fossil debris. This has permitted detailed analyzes of the depositional setting of ironstones (e.g., Hunter, 1970; Van Houton and Bhattacharyya, 1982), but consensus on the specific mechanisms and timing of iron precipitation has remained elusive.

Intrinsic characteristics

Ironstones contain representatives of all four of James' (1954) mineral facies, but differ from iron formations in some ways. Most ironstones consist of oxides and/or silicates, although siderite is also common in some (Odin, 1988). This is similar to the assemblages found in GIFs, but more oxidized than iron formations as a whole, given the widespread occurrence of reduced facies in BIFs. The oxides and silicates found in ironstones versus iron formations also differ systematically. Hematite is found in both, but goethite is only common in ironstones whereas magnetite is only common in iron formations. Likewise, chamosite and berthierine are the chief iron-rich silicates in ironstones but rare in iron formations. In addition to iron-rich phases, ironstones typically contain quartz sand, calcite, dolomite, and authigenic phosphorite and chert. All of these phases are rare in iron formations with the exception of chert. These mineralogical differences are reflected in lower iron and silica and higher alumina contents in ironstone versus iron formation, which is partly, but not entirely, a reflection of the age disparity between the two (see below).

Ooids are a distinctive feature of many ironstones (Figure I8(C)), but they are rarely its main constituent. Like those in iron formations, the ooids in ironstones consist of a nucleus plus a cortex with thin concentric laminations; no radial textures have been reported from ooids in either iron formations or ironstones. Ironstone ooids consist of iron silicates and/or iron oxides; these two minerals occur in alternating laminae in some ooids. The nulei of the ooids are diverse in composition, ranging from calcareous fossil debris to detrital quartz sand grains to fragments of oolitic cortices. Replacement textures are commonly observed in the ooids, involving both iron-rich and iron-poor materials (including fossils) as both replacer and replacee. Despite the attention lavished on the ooids, other constituents probably form the bulk of ironstones. One common constituent is interstitial calcite or dolomite cement, but the most abundant are probably finely crystalline iron oxides, iron-rich silicates, and/or siderite. Much of this finer material is authigenic rather than detrital in origin (Odin, 1988). Many oolitic ironstones are cross-bedded, a sure sign they originated as well-sorted sands, but other ironstones are highly bioturbated, and show evidence of hardground-like induration in the form of local sediment reworked as intraclasts (Odin, 1988).

Stratigraphic context and secular variation

Most oolitic ironstones take the form of thin interbeds amidst successions of clastic and carbonate strata deposited in shallow marine environments. These are typically sediments deposited close to paleocoastlines in water on the order of a few tens of meters deep. In general, ironstones are formed closer to shore and in shallower water than time-equivalent glauconitic deposits (e.g., Hunter, 1970), but the abundant remains of normal marine organisms and the evidence of repeated reworking indicate ironstones accumulated on the open seafloor rather than in restricted settings. Nevertheless, most workers believe ironstones did not accumulate in the highest energy environments, which is consistent with their common association with clastic mudrocks. Ironstones also tend to occur at stratigraphic breaks which coincide with transgressions (Van Houton and Bhattacharyya, 1982). Essentially, ironstones are now believed to constitute condensed sequences formed largely during times of maximum transgression and reduced clastic influx. This is also true on a grander scale; the two main pulses of ironstone accumulation coincide with Fischer's two "greenhouse" phases of sea-level highstand from the Ordovician to the Devonian and from the Jurassic to the Paleogene (Van Houton and Arthur, 1989).

Ironstones also vary in mineralogical composition as a function of age. Goethite and berthierine are both relatively common in the Mesozoic to Tertiary ironstones, whereas hematite and chamosite dominate the Paleozoic ironstones. This contrast probably reflects a higher degree of diagenetic equilibration by the older ironstones.

Origins and implications for earth history

The fundamental problem in the origin of ironstones is reconciling the evidence for deposition under open-marine, oxidizing conditions with the fact that dissolved iron is only mobile under reducing conditions (Odin, 1988). This leaves three options: (1) ironstones were originally deposited as something entirely different and replaced by iron-rich phases after burial; (2) conditions fluctuated between oxidizing and reducing during deposition; or (3) high concentrations of iron somehow managed to accumulate under oxidizing conditions. Some of the earliest investigators took the presence of ooids in ironstones to mean that they were diagenetically replaced carbonates. Even though carbonate fossils are locally replaced by iron-rich minerals, this interpretation has fallen out of favor due to the evidence of synsedimentary reworking of various types of ironstone and the apparent lack of any units of carbonate oolite that are half-converted to oolitic ironstone. Faced with a lack of sedimentological evidence for bottom water anoxia during deposition, most workers now envision ironstones as forming under a fully oxidized water column. This is clearly happening today in the case of glauconitic sediments (Van Houton and Purucker, 1984), and thin oolitic coatings of chamosite have been reported from pellets on the modern seafloor at one site (Rohrlich et al., 1969). The latter are well below wave base, which suggests the coatings may be able to grow *in situ* without benefit of regular physical movement. Some researchers have even proposed the ooids originated entirely *in situ*, but the widespread breakage and regrowth of the ooids argue against this. Other workers have suggested they are pedogenic ooids that were transported then redeposited on the seafloor, but the highly concentrated nature of ironstones and their stratigraphic context makes this seem highly unlikely also. The current thinking is that ironstones originally accumulated mainly as ferric materials, then were modified considerably via early diagenetic authigenesis of the iron-rich phases very close to the sediment–water interface (Odin, 1988).

In summary, ironstones can probably originate in different ways, but most formed more or less where they were ultimately buried via a process of slow iron precipitation during times of highly reduced sedimentation rates. Proximity to a river mouth is also thought to be advantageous because the source of the iron is generally attributed to deep weathering on adjacent landmasses. However, this is somewhat at odds with the temporal correlation of ironstone and black shale deposition (Van Houton and Arthur, 1989). Whatever the source of the iron and the mechanism of deposition, both the iron-rich and iron-poor phases in ironstones were clearly affected by a host of early diagenetic transformations; many took place so close to the sediment/water interface that new phases were detritally reworked. In conclusion, the separation of iron formations and ironstones into two separate categories seems to be quite justifiable, but a closer comparison between these two might be very instructive in terms of revealing what processes were taking place on or near the early Precambrian seafloor.

Bruce M. Simonson

Bibliography

Anbar, A.D., and Knoll, A.H., 2002. Proterozoic ocean chemistry and evolution: A bioinorganic bridge? *Science*, **297**: 1137–1192.

Beard, B.L., Johnson, C.M., Cox, L., Sun, H., Nealson, K.H., and Aguilar, C., 1999. Iron isotope biosignatures. *Science*, **285**: 1889–1892.

Beukes, N.J., 1984. Sedimentology of the Kuruman and Griquatown Iron-formations, Transvaal Supergroup, Griqualand West, South Africa. *Precambrian Research*, **24**: 47–84.

Beukes, N.J., and Klein, C., 1992. Models for iron-formation deposition. In Schopf, J.W., and Klein, C. (eds.), *The Proterozoic Biosphere: A Multidisciplinary Study*. Cambridge University Press, pp. 147–151.

Cloud, P., 1973. Paleoecological significance of the banded iron-formation. *Economic Geology*, **68**: 1135–1143.

Dimroth, E., 1976. Aspects of the sedimentary petrology of cherty iron formation. In Wolf, K.H. (ed.), *Handbook of Strata-Bound and Stratiform Ore Deposits*, 7. Elsevier, pp. 203–254.

Eriksson, K.A., 1995. Crustal growth, surface processes and atmospheric evolution of the early Earth. In Coward, M.P., and Ries, A.C. (eds.), *Early Precambrian Processes*. Geological Society of London, Special Publication, 95, pp. 11–25.

Eriksson, K.A., and Donaldson, J.A. 1986. Basinal and shelf sedimentation in relation to the Archaean-Proterozoic boundary. *Precambrian Research*, **33**: 103–121.

Ewers, W.E., and Morris, R.C., 1981. Studies of the Dales Gorge Member of the Brockman Iron Formation, Western Australia. *Economic Geology*, **76**: 1929–1953.

Fralick, P.W., and Barrett, T.J., 1995. Depositional controls on iron formation associations in Canada. In Plint, A.G. (ed.), *Sedimentary Facies Analysis*. International Association of Sedimentologist, Special Publication, 22, pp. 137–156.

Gole, M.J., and Klein, C., 1981. Banded iron-formations through much of Precambrian time. *Journal of Geology*, **89**: 169–183.

Gross, G.A., 1965. *Geology of Iron Deposits of Canada, Volume I, General Geology and Evaluation of Iron Deposits*. Ottawa: Geological Survey of Canada, Economic Geology, Report 22.

Gross, G.A., 1983. Tectonic systems and the deposition of iron-formation. *Precambrian Research*, **20**: 171–187.

Grotzinger, J.P., 1994. Trends in Precambrian carbonate sediments and their implications for understanding evolution. In Bengston, S. (ed.), *Early Life on Earth. Nobel Symposium, 84*. Columbia University Press, pp. 245–258.

Hoffman, P.F., Kaufman, A.J., Halverson, G.P., and Schrag, D.P., 1998. A Neoproterozoic snowball earth. *Science*, **281**: 1342–1346.

Holland, H.D., 1994. Early Proterozoic atmospheric change. In Bengston, S. (ed.), *Early Life on Earth. Nobel Symposium No. 84*. Columbia University Press, pp. 237–244.

Hunter, R.E., 1970. Facies of iron sedimentation in the Clinton Group. In Fisher, G.W., Pettijohn, F.J., Reed, J.C. Jr., and Weaver, K.N. (eds.), *Studies of Appalachian Geology: Central and Southern*. Wiley, pp. 101–124.

Isley, A.E., 1995. Hydrothermal plumes and the delivery of iron to banded iron formation. *Journal of Geology*, **103**: 169–185.

James, H.L., 1954. Sedimentary facies of iron-formation. *Economic Geology*, **49**: 235–293.

James, H.L., 1966. Chemistry of the Iron-Rich Sedimentary Rocks. *United States Geological Survey Professional Paper 440-W*.

LaBerge, G.L., 1966a. Altered pyroclastic rocks in iron-formation in the Hamersley Range, Western Australia. *Economic Geology*, **61**: 147–161.

LaBerge, G.L., 1966b. Pyroclastic rocks in South African iron-formations. *Economic Geology*, **61**: 572–581.

Klein, C., and Beukes, N.J., 1992. Proterozoic iron formations. In Condie, K.C. (ed.), *Proterozoic Crustal Evolution*, Elsevier, pp. 383–418.

Mengel, J.T., 1973. Physical sedimentation in Precambrian cherty iron formations of the Lake Superior type. In Amstutz, G.C., and Bernard, A.J. (eds.), *Ores in Sediments*, Springer-Verlag, pp. 179–193.

Morris, R.C., 1987. Iron ores derived by enrichment of banded iron-formation. In Hein, J.R. (ed.), *Siliceous Sedimentary Rock-hosted Ores and Petroleum*. Van Nostrand Reinhold Co., pp. 231–267.

Odin, G.S. (ed.), 1988. *Green Marine Clays*. Developments in Sedimentology 45, Elsevier.

Ojakangas, R.W., 1983. Tidal deposits in the early Proterozoic basin of the Lake Superior region—the Palms and Pokegama Formations: evidence for subtidal shelf deposition of Superior-type banded iron-formation. In Medaris, L.G. Jr. (ed.), *Early Proterozoic Geology of the Great Lakes Region*. Geological Society of America Memoir, 60, pp. 49–66.

Rohrlich, V., Price, N.B., and Calvert. S.E., 1969. Chamosite in the Recent sediments of Loch Etive, Scotland. *Journal of Sedimentary Petrology*, **39**: 624–631.

Simonson, B.M., 1985. Sedimentological constraints on the origins of Precambrian iron-formations. *Geological Society of Amererica Bulletin*, **96**: 244–252.

Simonson, B.M., 1987. Early silica cementation and subsequent diagenesis in arenites from four early Proterozoic iron formations of North America. *Journal of Sedimentary Petrology*, **57**: 494–511.

Simonson, B.M., and Hassler, S.W., 1996. Was the deposition of large Precambrian iron formations linked to major marine transgression? *Journal of Geology*, **104**: 665–676.

Trendall, A.F., 1972. Revolution in earth history. *Journal of the Geological Society of Australia*, **19**: 287–311.

Trendall, A.F., 2002. The significance of iron-formation in the Precambrian stratigraphic record. In Altermann, W., and Corcoran, P.L. (eds.), *Precambrian Sedimentary Environments: A Modern Approach to Ancient Depositional Systems*. International Association of Sedimentologists, Special Publication, 33, pp. 33–66.

Trendall, A.F., and Blockley, J.G., 1970. The Iron Formations of the Precambrian Hamersley Group, Western Australia. Perth: *Geological Survey of Western Australia Bulletin*, 119.

Trendall, A.F., and Morris, R.C. (eds.), 1983. *Iron-Formation: Facts and Problems*. Elsevier.

Van Hise, C.R., and Leith, C.K., 1911. Geology of the Lake Superior Region. *United States Geological Survey Monograph 52*.

Van Houton, F.B., 2000. Ooidal ironstones and phosphorites—a comparison from a stratigrapher's view. In Glenn, C.R., Prévôt-Lucas, L., and Lucas, J. (eds.), *Marine Authigenesis: From Global to Microbial*. Society of Economic Paleontologists and Mineralogists, Special Publication, 44, pp. 79–106.

Van Houton, F.B., and Arthur, M.A., 1989. Temporal patterns among Phanerozoic oolitic ironstones and oceanic anoxia. In Young, T.P., and Taylor, W.E.G. (eds.), *Phanerozoic Ironstones*. Geological Society of London, Special Publication, 46, pp. 33–49.

Van Houton, F.B., and Bhattacharyya, D.P., 1982. Phanerozoic ooltitic ironstones—geologic record and facies model. *Annual Reviews of Earth and Planetary Sciences*, **10**: 441–457.

Van Houton, F.B., and Purucker, M.E., 1984. Glauconitic peloids and chamositic ooids—favorable factors, constraints, and problems. *Earth-Science Reviews*, **20**: 211–243.

Veizer, J., Clayton, R.H., Hinton, R.W., von Brunn, V., Mason, T.R., Buck, S.G., and Hoefs, J., 1990. Geochemistry of Precambrian carbonates: 3—shelf seas and non-marine environments of the Archean. *Geochimica et Cosmochimica Acta*, **54**: 2717–2729.

Veizer, J., Clayton, R.H., and Hinton, R.W., 1992. Geochemistry of Precambrian carbonates: IV—Early Paleoproterozoic (2.25 0.25 Ga) seawater. *Geochimica et Cosmochimica Acta*, **56**: 875–885.

Young, G.M., 1988. Proterozoic plate tectonics, glaciation and iron-formations. *Sedimentary Geology*, **58**: 127–144.

Young, T., 1993. Sedimentary iron ores. In Patrick, R.A.D., and Polya, D.A. (eds.), *Mineralization in the British Isles*. London: Chapman and Hall, pp. 446–489.

Young, T.P., 1989. Phanerozoic ironstones: an introduction and review. In Young, T.P., and Taylor, W.E.G. (eds.), *Phanerozoic Ironstones*. Geological Society of London, Special Publication, 46, pp. ix–xxv.

Cross-references

Berthierine
Chlorite in Sediments
Glaucony and Verdine
Oolite and Coated Grains
Seawater: Temporal Changes in the Major Solutes
Siliceous Sediments
Stromatolites
Syneresis

ISOTOPIC METHODS IN SEDIMENTOLOGY

Introduction

Isotopes are atoms whose nuclei contain the same number of protons but a different number of neutrons. The nuclei of unstable (i.e., radioactive) isotopes decay spontaneously toward the stable condition through the emission of particles and energy. Radioactive decay of a particular unstable nuclide is generally accomplished by one or more of five decay schemes, broadly characterized by beta (β^-), positron (β^+) and alpha (α) particle emission, electron capture (e.c.) and fission. The decay schemes are accompanied by other nuclear emissions, including neutrinos, anti-neutrinos, gamma radiation, and X-radiation. The element undergoing radioactive decay (the parent) is transformed into another element (the daughter). The end-point of a radioactive decay scheme is the production of a stable nuclide. The daughter(s) of radioactive decay are referred to as radiogenic nuclides.

The nuclei of stable isotopes do not undergo radioactive decay to form other isotopes. Nevertheless, the relative abundances of an element's stable isotopes vary within and among phases bearing that element. These differences, commonly described as "fractionation" between two phases, are controlled by a variety of equilibrium and kinetic processes, including temperature and reaction rate.

Isotopic methods are utilized in sedimentology in two main ways. First, they provide a means to date sedimentary materials (commonly referred to as geochronology or radiometric dating). This is done almost exclusively using radioactive isotopes. Geochronometers include the Rb–Sr, K–Ar, Sm–Nd, and U–Th–Pb systems (including U-series disequilibrium methods), and various cosmogenic and related radionuclides, including ^{14}C. The choice of geochronometer depends on the composition and nature of the material to be studied, and its estimated age. Radiometric methods can be used to provide many types of sedimentological information, including depositional ages, ages of detrital grains, crustal residence times for sediments, residence times of oceans, deposition rates, a temporal chronology for Pleistocene and Holocene paleoclimate proxies (minerals, organic matter, etc.), ages of erosion surfaces and landscapes, dates for diagenetic mineral formation, and temporal constraints on porewater evolution and its movement in sedimentary basins.

Second, isotopes can be used to trace the origin of inorganic and organic solids, liquids and gases within the sedimentary system, and their interactions within and among the hydrosphere, biosphere, atmosphere and deeper lithosphere. The stable isotopes of hydrogen, oxygen, carbon, nitrogen and sulfur are of special interest because they can be fractionated on a large scale by low-temperature processes. Commonly utilized radiogenic isotope tracers include strontium-87 (expressed as ^{87}Sr/^{86}Sr ratios) and neodymium-143 (expressed as ^{143}Nd/^{144}Nd ratios). These isotopic tracers can be used in identification of sediment provenance, characterization of physicochemical conditions during deposition and weathering/soil formation, determination of the temperature and fluids involved in diagenesis, characterization of organic-inorganic interactions within pore systems, description of fluid evolution and migration in sedimentary basins, and reconstruction of paleoclimate.

Dating methods and radiogenic isotopic ratios

The principles of radiometric dating are described by Faure (1986) and Bowen (1988). You are directed to these books for detailed discussion and references concerning much of what follows.

Basic equations

The decay rate of a radioactive parent nuclide (N) with time (t) is described by:

$$-dN/dt = \lambda N \qquad \text{(Eq. 1)}$$

where λ is the decay constant. Long-lived radionuclides (e.g., ^{147}Sm, ^{87}Rb, ^{232}Th, ^{238}U, ^{235}U, and ^{40}K have very small values of λ (e.g., 6.54×10^{-12} yr^{-1} for ^{147}Sm) and hence long half-lives (the time it takes for 50 percent of the parent to decay). They can be used to date systems that are millions to billions of years old. Other radionuclides have much larger decay constants (e.g., ^{14}C, $\lambda = 1.21 \times 10^{-4}$ yr^{-1}) and hence much shorter half-lives. This restricts their application to much younger systems (e.g., $\sim 10^5$ years for ^{14}C).

Equation 1 can be integrated to produce the basic equation of radioactive decay:

$$N/N_0 = e^{-\lambda t} \qquad \text{(Eq. 2)}$$

where N_o is the number of parent atoms at $t = 0$. Equation 2 can be expressed in terms of the number of daughter atoms (D) as

$$D = D_0 + N(e^{\lambda t} - 1), \qquad \text{(Eq. 3)}$$

where D_0 is the number of daughter atoms initially present in the system ($t = 0$). For radiometric dating to work, the system (rock, mineral, etc.) cannot have lost or gained parent or daughter atoms except through radioactive decay (a "closed" system), and the material analyzed must be representative of the system. If daughter atoms were present in the system prior to closure, their abundance must be determined accurately.

K-Ar methods

For some geochronometers (e.g., decay of ^{40}K to ^{40}Ar), there are few daughter atoms in the system at the time of closure ($D_0 \approx 0$), thus allowing the dating equation to take the form:

$$^{40}\text{Ar} = (\lambda_{\text{Ar}}/\lambda_{\text{total}})(^{40}\text{K})(e^{\lambda t} - 1) \quad \text{(Eq. 4)}$$

In the K–Ar system, a factor expressing a ratio of decay constants ($\lambda_{\text{Ar}}/\lambda_{\text{total}}$) must be introduced, since the parent (^{40}K) undergoes decay not only to ^{40}Ar but also to ^{40}Ca. Additional complications arise because the daughter is a gas, and is easily lost from the system upon heating or over time. Correction for diffusive argon loss is possible in some minerals (e.g., K-feldspar) if stepwise heating and the ^{40}Ar/^{39}Ar method are employed (^{39}Ar is produced by neutron bombardment of ^{39}K in a reactor).

The ^{40}K/^{40}Ar and ^{40}Ar/^{39}Ar methods can be used to obtain depositional ages for some sediments. Sanidine and unaltered biotite are preferred for such measurements in bentonites, but clay minerals derived from the ash can also be used, provided that they formed very early in diagenesis and that their crystal growth is fully understood (Clauer and Chaudhuri, 1995; Clauer et al., 1997). Syndepositional, K-rich glauconite (glaucony) formed close to the sediment-water interface in marine environments is also useful for obtaining depositional ages.

Dating of clay minerals presents a challenge. First, their large surface area and general reactivity encourages sorption of ^{40}Ar and K on one hand, and their potential loss during post-crystallization processes (both natural and during analysis) on the other hand. Minimum depositional ages are commonly obtained from poorly crystallized, low-K glauconite for these and other reasons (Clauer and Chaudhuri, 1995). Second, the same clay mineral in sediment can be of multiple origins; for example, detrital illite from several sources may be mixed with illite formed during several stages of diagenesis. During burial diagenesis of shales, K–Ar dates for illitic clays tend to decrease with increasing temperature and decreasing particle size, in tandem with the smectite to illite reaction. Diagenetic illite is concentrated in the finer clay fractions. However, even when the finest clay size-fractions (e.g., <0.1 μm) are isolated, contamination by detrital illite can still occur. This can induce a substantial error in the calculated age. Likewise, diagenetic clays formed by transformation rather than direct precipitation may inherit radiogenic isotopes (including ^{40}Ar) from their precursor. Such problems of contamination and inheritance are much less severe for authigenic clays that have precipitated directly from solution in the pores of sandstones.

A novel ^{40}Ar/K$_2$O "pseudoisochron" method to identify sources of detrital clay assemblages has been described by Jeans et al. (2001). Pevear (1992) described a method of ^{40}K–^{40}Ar analysis that allows both the detrital and diagenetic end-member ages to be obtained over a range of illitic clay particle sizes. Such information contributes to our understanding of both the depositional and thermal histories of many potential petroleum source rocks. Authigenic illitic clays and K-feldspar, which are common in sandstone pores, can also be dated. Ziegler and Longstaffe (2000a,b), for example, used K–Ar dates for early K-feldspar and late illite to recognize multiple stages of fluid flow in sandstones of the Appalachian Basin, and related this behavior to regional and local tectonic activity. Girard and Onstott (1991) used laser-probe sampling and ^{40}Ar/^{39}Ar step-heating methods to obtain separate dates for detrital K-feldspars and their secondary overgrowths at a precision and on a microscopic scale not possible using whole crystals. Likewise, ^{40}Ar/^{39}Ar dating of authigenic K-feldspar in carbonate rocks has been used to discern episodes of fluid flow and to relate their timing to regional tectonic events (Spötl et al., 1998).

Rb–Sr and Sm–Nd isochron methods

Radiometric dating using many geochronometers, including the decay of ^{87}Rb to ^{87}Sr and decay of ^{147}Sm to ^{143}Nd, is complicated by an *initial* quantity of stable daughter in the system at $t = 0$. Estimation of D_0 is normally accomplished using the isochron method. In this approach, isotopic analyzes are obtained for several cogenetic samples containing varying abundances of the parent and daughter elements (i.e., a range of Rb/Sr or Sm/Nd ratios). The isotopic abundances are expressed as a ratio using a common stable isotope of the daughter as the denominator. For the Rb–Sr and Sm–Nd geochronometers, these expressions are:

$$(^{87}\text{Sr}/^{86}\text{Sr})_{\text{total}} = (^{87}\text{Sr}/^{86}\text{Sr})_{\text{initial}} + (^{87}\text{Rb}/^{86}\text{Sr})(e^{\lambda t} - 1) \quad \text{(Eq. 5)}$$

and

$$(^{143}\text{Nd}/^{144}\text{Nd})_{\text{total}} = (^{143}\text{Nd}/^{144}\text{Nd})_{\text{initial}} + (^{147}\text{Sm}/^{144}\text{Nd})(e^{\lambda t} - 1) \quad \text{(Eq. 6)}$$

On an isochron diagram for the Rb–Sr system, for example, ^{87}Sr/^{86}Sr is plotted versus ^{87}Rb/^{86}Sr for each sample. The slope of the straight line ($e^{\lambda t} - 1$) formed by the data from a closed system provides the date, and the intercept of this line with the y-axis at ^{87}Rb/^{86}Sr = 0 provides the *initial* ratio for strontium (^{87}Sr/^{86}Sr$_{\text{initial}}$). In sedimentary rocks, straight lines on isochron diagrams can also arise from mixing between source materials of different ages ("mixing" lines). If the system has been disturbed following closure (e.g., by weathering, interaction with formation waters, hydrothermal alteration), a straight-line fit to the isotopic data is not possible on the isochron diagram, and a date cannot be obtained.

Feldspars, micas and illitic clays are typical targets of the Rb–Sr isochron dating method in sediments. Ages of deposition that are consistent with K–Ar dates and other stratigraphic information are commonly preserved in well-crystallized glauconite (K$_2$O > 6.5 percent) (Clauer and Chaudhuri, 1995). Rb–Sr dates for secondary K-feldspar (adularia) in porous and permeable units can be used to define episodic brine migration in sedimentary basins (Harper et al., 1995). Rb–Sr isochron dates have also been used to study the smectite to illite reaction during burial diagenesis of shales, and in particular to suggest that this reaction can occur episodically rather than continuously (Morton, 1985; Ohr et al., 1991).

Use of the Rb–Sr method to study clay mineral formation during weathering is more complex, because of the open nature of these systems, and the potential for gain and loss of parent and daughter isotopes. Progress is being made in this area by using the "internal" isochron method described by Clauer and Chaudhuri (1995). In this approach, each of acid-leached clay, untreated clay and leachate solutions are analyzed to determine whether the radiogenic ^{87}Sr in these fractions evolved together or was derived from different sources.

Sm and Nd are generally not fractionated during sedimentary processes. Hence, Sm–Nd isochron dating is rarely attempted in sediments because the spread in Sm/Nd ratios of cogenetic phases is normally insufficient. However, such dating of diagenesis in argillaceous sediments has been performed using authigenic apatite, Fe-oxides/hydroxides and fine clays ($<0.2\,\mu m$) (Awwiller and Mack, 1991; Ohr et al., 1991). These Sm–Nd dates can survive isotopic rehomogenization during deeper burial, hydrothermal activity or metamorphism, unlike the Rb–Sr isotopic system, which is typically reset.

As for the K–Ar system, detailed mineralogical characterization (microscopy, crystal structure, chemical composition) that permits detrital and diagenetic minerals to be discerned and separated is required when applying Rb–Sr and Sm–Nd geochronometers to sedimentary rocks. "Best-practices" have been elaborated by Clauer and Chaudhuri (1995).

Model ages and epsilon (ϵ) values

In many sedimentary systems, it is not possible to obtain the *initial* ratio using the isochron method. In these cases, model ages can be calculated by assuming an *initial* ratio, although such results must be used with care. One approach is to derive the *initial* ratio from some generalized reservoir by assuming its linear evolution over Earth history. A common choice for this reservoir in the Nd-isotope system is CHUR (Chondritic Uniform Reservoir; DePaolo and Wasserberg, 1976). Differences between the $^{143}Nd/^{144}Nd$ ratio (or $^{87}Sr/^{86}Sr$) of a sample and the values of CHUR (or some other baseline) at a given time are commonly expressed as the epsilon (ϵ) parameter. The $^{143}Nd/^{144}Nd$ value of ϵ^{o} (i.e., at present time) is described by:

$$\epsilon^{o}_{CHUR} = \left[(^{143}Nd/^{144}Nd)_{measured} / (^{143}Nd/^{144}Nd_{CHUR*}) - 1\right] \times 10^4$$

(Eq. 7)

where CHUR* is the ratio at the present time. This approach allows small but significant differences between these ratios to be more easily expressed (Clauer and Chaudhuri, 1995).

The sediment load of modern rivers, as well as fine-grained clastic sedimentary rocks in general, acquire Sm/Nd ratios and Sm–Nd isotopic compositions similar to their igneous and metamorphic source rocks. These have $^{143}Nd/^{144}Nd$ ratios and hence ϵ_{Nd} values characteristic of their age (e.g., <-20 for Archean crust versus >2 for recent island arc volcanic rocks; Patchett et al., 1999). The ϵ_{Nd} values of clastic sedimentary rocks can therefore provide useful information about their provenance, plus insight into the crustal residence time of the detritus (i.e., time elapsed since Nd was separated from the mantle into the continental crust). Patchett et al. (1999) used ϵ_{Nd} values for sedimentary rocks from cratonic and synorogenic sequences to infer temporal patterns in the dominant sources of sediment in North America (600–450 Ma, continental shield; 450–150 Ma, Appalachian orogen; <150 Ma, Cordilleran orogen). Unterschutz et al. (2002) use a similar approach to demonstrate a mixed volcanic arc and cratonic provenance for metasedimentary rocks comprising the Triassic Quesnel terrane of the southern Canadian Cordillera. These data imply a pericontinental rather than exotic origin for the Quesnel terrane during the assembly of this portion of the Canadian Cordillera.

$^{87}Sr/^{86}Sr$ ratios in seawater and sediments

The $^{87}Sr/^{86}Sr$ ratios of stratigraphically well-identified marine carbonates (brachiopod shells, abiotic marine carbonates), phosphates (e.g., conodonts) and sulfates are used for regional and global chemostratigraphic correlation. They also provide a means to understand the interactions among global tectonic and climatic processes (mountain building, seafloor spreading, glaciation, erosion and weathering rates, sea level change, etc.) over geological time (Veizer, 1989; Edmond, 1992). Because Rb is not incorporated into most carbonates, phosphates and sulfates, they acquire and retain the $^{87}Sr/^{86}Sr$ ratio of the water from which they precipitated, unless modified by interaction with later fluids. Such modification of strontium- (and oxygen-) isotope compositions is a significant concern, given the relative ease with which carbonate minerals recrystallize. Alteration is normally detected using stable isotopic, major and trace element, and cathodoluminescence methods, or avoided through the use of more resistant phases (e.g., euconodonts) (Derry et al., 1992; Asmerom et al., 1991; Ebneth et al., 2001).

Water acquires its $^{87}Sr/^{86}Sr$ ratio through rock–water interaction (hydrothermal alteration, weathering, exchange). Interaction with mantle-related volcanic rocks produces water with abundant Sr but generally low $^{87}Sr/^{86}Sr$ ratios (~ 0.7035), which characterizes deep-sea hydrothermal input to the oceans. The submarine hot-spring contribution to the oceans may have varied with changes in the rate of seafloor spreading, particularly during Paleozoic time, but likely represents a more or less constant flux of constant Sr-isotope composition during the Phanerozoic (Edmond, 1992).

Water associated with erosion and weathering of continental rocks and their soils normally acquires low Sr contents but relatively high $^{87}Sr/^{86}Sr$ ratios (~ 0.712), mostly as a consequence of interaction with feldspars and micas. The $^{87}Sr/^{86}Sr$ ratio of the leachate tends not to be as high as the rocks because of preferential dissolution of plagioclase relative to more radiogenic K-feldspar and biotite (Clauer and Chaudhuri, 1995). However, much higher contents of highly radiogenic Sr are released during erosion of Himalayan-style metamorphic core complexes (Edmond, 1992).

The flux of continental Sr is carried to the ocean by rivers and groundwater (Edmond, 1992; Basu et al., 2001). The $^{87}Sr/^{86}Sr$ ratio and Sr content of the meteoric water reflect the lithologies intersected, and can also vary with rainfall amounts in a region (Galy et al., 1999). Continental hot springs associated with collisional orogens also contribute highly radiogenic Sr ($^{87}Sr/^{86}Sr \approx 0.77$) to the Sr-budget of some globally important fluvial systems, such as Himalayan rivers (Evans et al., 2001). Variations in continental glaciation, atmospheric CO_2 levels, climate, vegetation, orogeny and eustatic sea level all help to determine the rock types exposed at the Earth's surface, and the rates at which weathering and erosion occur.

Marine carbonates and evaporites are the sink for Sr from seawater, removing it from solution with no isotopic fractionation. These rocks typically release high Sr contents and intermediate $^{87}Sr/^{86}Sr$ ratios (~ 0.7075–0.7080) to river water. These ratios correspond to the balance between crustal weathering and mantle-related processes at the time of chemical sedimentation. The carbonates and evaporites can be exposed by tectonic processes and eustatic changes in sea level. When this happens, they tend to dominate the fluvial flux of Sr to the ocean. This feedback attenuates the variations in Sr-isotope composition of seawater that otherwise would ensue if it was determined only by the balance between mantle-related activity and the weathering of sialic rocks (Edmond, 1992).

The oceans presently have a uniform $^{87}Sr/^{86}Sr$ ratio of 0.70906 ± 0.00003, which reflects the long residence time of Sr in seawater. This value corresponds well with its proxy in modern marine carbonate (0.70910 ± 0.00004) (Burke et al., 1982). Its composition has fluctuated widely over geological time. Values were high in the Late Cambrian (> 0.7092), but gradually declined through wide fluctuations to as low as 0.70676 in the Jurassic; such low values also occurred in the Permian. Since the Jurassic, the Sr-isotopic composition of seawater has increased irregularly, with a particularly rapid climb toward the present value beginning in the mid-Tertiary (Burke et al., 1982; Edmond, 1992; Denison et al., 1998; Ebneth et al., 2001).

The strongest variations in the strontium isotopic composition of seawater reflect changes in continental input. Edmond (1992) proposed that only Himalayan-style continental collisions can produce a sufficiently high flux of radiogenic Sr to account for the positive extremes in the seawater $^{87}Sr/^{86}Sr$ record. During such major orogenic events, the Rb–Sr systems of ancient continental rocks are disturbed, leading to the redistribution of highly radiogenic Sr ($^{87}Sr/^{86}Sr \sim 0.769$ in Himalayan core complexes) into minerals such as plagioclase, which weather easily following their exposure. During the late Proterozoic, $^{87}Sr/^{86}Sr$ ratios also rose from very low (~ 0.707) to high (~ 0.709) values. This may suggest a fundamental change from rifting and production of juvenile oceanic crust to continent–continent collisions, production of recycled continental crust, and its erosion (Asmerom et al., 1991). Many research efforts continue to be directed toward finer-scale definition of the Sr-isotope seawater curve, and the specific variations arising from second-order processes, such as glaciation (Blum and Erel, 1995).

$^{143}Nd/^{144}Nd$ ratios in seawater and sediments

The $^{143}Nd/^{144}Nd$ ratios of seawater (commonly recast in terms of ε_{Nd}) are controlled principally by erosion of the land masses adjacent to the oceans. Nd, unlike Sr, has a very short residence time in seawater relative to the mixing time of the Earth's oceans. The Atlantic Ocean, which is surrounded mainly by older sialic rocks, has a lower ε_{Nd} (-15 to -10) than the Pacific Ocean (-5 to 0), which is surrounded mostly by younger volcanic rocks related to subduction of oceanic crust (Piepgras et al., 1979). As a result, the spatial variation in Nd-isotope composition of modern seawater can be used to trace circulation within and among oceans.

Variations in seawater $^{143}Nd/^{144}Nd$ ratios can be tracked through geological time through analysis of phosphates (e.g., conodonts) and carbonates (e.g., brachiopods). They acquire Nd-isotope compositions characteristic of ambient seawater shortly after deposition at the sediment–water interface, and are largely unaffected by later alteration, including interaction with meteoric water (Banner et al., 1988). This makes $^{143}Nd/^{144}Nd$ ratios more robust proxies of oceanic conditions than strontium- or oxygen-isotope ratios. The Nd-isotope compositions can be used to reconstruct the variation in the exposed crust of the Earth, and to deduce plate tectonic activity (i.e., the opening and closing of ocean basins) that determined the distribution of the Earth's oceans (Shaw and Wasserburg, 1985).

U–Th–Pb methods

The U–Th–Pb system provides a wide range of geochronometers useful in the study of sediments. It comprises three separate and complex decay series, ^{238}U to ^{206}Pb, ^{235}U to ^{207}Pb and ^{232}Th to ^{208}Pb. In each case, the decay scheme contains numerous intermediate daughters before the stable daughter is achieved. The intermediate daughters have much shorter half-lives than the parent, and under the condition described as secular equilibrium, the rate of the decay of these daughters is controlled by the rate of decay of the parent (Faure, 1986). For each of the U–Th–Pb decay schemes, model ages can be obtained by assuming reasonable *initial* Pb-isotope ratios (given relative to ^{204}Pb). Combination of any two of the U–Th–Pb geochronometers permits the concordia approach to dating to be used. The latter is particularly useful in studies of sediment provenance, which focus on detrital zircon and other U- or Th-rich phases.

The U–Th–Pb geochronometers that are perhaps most valuable in sedimentology are provided by certain intermediate radioactive nuclides (commonly with high values of λ and hence short half-lives). These intermediate daughters are used in a wide variety of ways, collectively described as U-series disequilibrium dating. Some allow dating of intervals between those accessible using the ^{14}C ($< 10^5$ years) and ^{40}K–^{40}Ar ($> 10^6$ years) methods; others (e.g., ^{210}Pb) permit dating of modern materials (i.e., < 100 years old). U-series disequilibrium dating is possible because many intermediate daughters in the decay series have geochemical properties different from their parent. This allows parent and daughter to be separated during processes such as chemical weathering, (bio)chemical sedimentation, and adsorption onto clay minerals. For example, uranium (as the uranyl ion) tends to remains in solution in seawater, whereas Th is preferentially removed by precipitation of zeolites or by adsorption onto particulate surfaces (e.g., clays, organic matter, etc.). Radiometric dating is thus accomplished using an isolated intermediate daughter (e.g., ^{230}Th, ^{231}Pa, ^{210}Pb) that, free from replenishment by decay of the absent parent, disintegrates at a rate determined solely by its own (shorter) half-life. Alternatively, a daughter may be excluded from a geologic material at the time of its formation, leaving only the radioactive parent. The new daughter production from that parent can be used to obtain the time since the system was last open, at least until reestablishment of secular equilibrium (Faure, 1986).

U-series disequilibrium dating methods have proven most useful in the study of Pleistocene sediments, commonly through examination of $^{230}Th/^{232}Th$, $^{234}U/^{238}U$, $^{230}Th/^{238}U$,

^{230}Th/^{234}U, and ^{230}Th/^{231}Pa ratios. For example, oceanic sedimentation rates can be determined by obtaining these ratios with increasing distance from the sediment-water interface. Likewise, dates can commonly be obtained for young marine and nonmarine carbonates, given that their crystal structure incorporates uranium upon precipitation, but almost always excludes thorium. The activity of ^{230}Th in such carbonate derives from the decay of ^{234}U, which in turn is derived from ^{238}U. The ability to date growth bands in speleothem and other terrestrial carbonate cements allows temporal calibration of the paleoclimate information obtained from oxygen, hydrogen, carbon and strontium isotopic compositions of the same material (Schwarcz, H.P., 1986; Dorale et al., 1998; Kaufman et al., 1998). Corals can be used in a similar fashion (Edwards et al., 1993), and provide a chronology for sea-level fluctuations.

Because of its short half-life (22.26 years), the abundance of ^{210}Pb can be used to obtain sedimentation rates in lacustrine and marine sediments that are <100 years old. It is a radioactive daughter of gaseous ^{222}Rn within the ^{238}U decay series. Upon production, ^{222}Rn is released to the atmosphere, where it decays rapidly to ^{210}Pb. The ^{210}Pb is swept quickly from the atmosphere by precipitation, and accumulates in snow, ice, lakes, vegetation and sediments. The declining level of ^{210}Pb activity can then be used to calculate the time elapsed since separation from its parent (Faure, 1986).

Cosmogenic and anthropogenic radionuclides

Another useful dating tool in Pleistocene and Recent sedimentary environments is provided by cosmogenic nuclides. These isotopes are produced in the Earth's atmosphere by interactions between cosmic radiation and oxygen, nitrogen and argon. The most familiar of these reactions is the neutron-induced formation of ^{14}C from ^{14}N, and its subsequent return to ^{14}N via β^- decay with a half-life of 5730 years. Carbon-14 is incorporated into CO_2, which mixes rapidly within the atmosphere and hydrosphere, and is fixed by photosynthesis into plants and the animals that eat them. Once an organism dies, replenishment of $^{14}CO_2$ ends, and the time elapsed since death can be determined from the remaining activity of ^{14}C. This model is complicated by many factors, including variations in the radiocarbon content of the atmosphere (e.g., Edwards et al., 1993; Beck et al., 2001). These have resulted from changes in the flux of cosmic radiation to the Earth, release of "dead" carbon by combustion of fossil fuels, addition of ^{14}C via nuclear devices, and mass-dependent fractionation among ^{12}C, ^{13}C, and ^{14}C.

In addition to organic matter from sediments and soils, considerable effort has been devoted to ^{14}C-dating of chemically or biologically precipitated calcium carbonate. For accurate dating of fresh-water shells, their carbon must have been derived wholly from the atmosphere, and the ^{14}C content must have not been modified by postmortem alteration. Complications also arise in marine environments in environments where older, deep ocean waters may have mixed with modern waters, both of which can supply carbon to the shells of organisms (Faure, 1986). Carbon-14 measurements have also been used to date groundwater and describe fluid mixing in the subsurface (Clark and Fritz, 1997).

Tandem accelerator mass spectrometry permits much smaller quantities of ^{14}C to be analyzed than possible using standard radiochemical methods. It has also made feasible measurement of other cosmogenic (e.g., ^{3}H, ^{3}He, ^{7}Be, ^{10}Be, ^{14}C, ^{26}Al, ^{32}Si, ^{36}Cl, ^{39}Ar, ^{81}Kr) and anthropogenic radionuclides (e.g., ^{3}H, ^{90}Sr, ^{137}Cs) (Faure, 1986; Clark and Fritz, 1997). Cosmogenic radionuclides with long residence times in the atmosphere (e.g., ^{39}Ar, ^{81}Kr) provide the potential to date air bubbles trapped in ice cores (Faure, 1986). Others are removed quickly from the atmosphere by precipitation (e.g., ^{10}Be, ^{26}Al, ^{3}H, ^{36}Cl, ^{129}I). They can be used to determine sediment transport and deposition rates in marine and terrestrial environments, to measure accumulation rates of ice sheets, and to date groundwater and brines (Lao et al., 1993; Wang et al., 1996; Yechielli et al., 1996).

The abundance of cosmogenic nuclides has also been used to determine the age of erosion surfaces (Phillips et al., 1996). The ^{10}Be and ^{26}Al contents of terrestrial surfaces, especially quartz, provide estimates of surface exposure time, and from this, erosion rates and the ages of exposed terrestrial surfaces can be deduced (Nishiizumi et al., 1993; Bierman and Turner, 1995). Cosmogenic ^{36}Cl production in limestones has been used for similar purposes (Stone et al., 1998). Anthropogenic radionuclides (e.g., ^{3}H, ^{137}Cs) have also been used to date and track mixing in modern groundwater, and to trace modern geomorphic processes in sediments and soils (Clark and Fritz, 1997; Vanden Bygaart and Protz, 2001).

Stable isotopes

The stable isotopes of hydrogen (^{2}H (or D), ^{1}H), carbon (^{13}C, ^{12}C), nitrogen (^{15}N, ^{14}N), oxygen (^{18}O, ^{17}O, ^{16}O) and sulfur (^{36}S, ^{34}S, ^{33}S, ^{32}S) are of most interest in the study of sediments. However, boron (^{11}B, ^{10}B; Vengosh et al., 1991; Williams et al., 2001), silicon (^{30}Si, ^{29}Si, ^{28}Si; Ding et al., 1996; De La Rocha et al., 1998), chlorine (^{37}Cl, ^{35}Cl; Kaufmann et al., 1993; Eggenkamp et al., 1995) and iron (^{58}Fe, ^{57}Fe, ^{56}Fe, ^{54}Fe; Zhu et al., 2000; Bullen et al., 2001) are attracting increasing interest. Reviews of stable isotope systematics pertinent to sedimentology include Fritz and Fontes (1980, 1986, 1989), Arthur et al. (1983), Faure (1986), Kyser (1987), Longstaffe (1987, 1989, 1994, 2000), Savin and Lee (1988), Bowen (1988, 1991), Swart et al. (1993), Clauer and Chaudhuri (1995), Clark and Fritz (1997), Hoefs (1997) and Kendall and McDonnell (1998).

Stable isotope results are reported using the delta (δ) notation in parts per thousand (‰ or permill). For phase A,

$$\delta_A = [(R_A - R_{standard})/R_{standard}] \times 1000, \quad \text{(Eq. 8)}$$

where R is normally D/H, ^{13}C/^{12}C, ^{15}N/^{14}N, ^{18}O/^{16}O or ^{34}S/^{32}S. The δ-value for the standard is defined as 0‰ exactly. For hydrogen, this standard is VSMOW (Vienna Standard Mean Ocean Water). Two standards are in use for oxygen, VSMOW and VPDB (Vienna Peedee Belemnite). While the VSMOW standard is preferred amongst isotopists, the VPDB standard is in common use among sedimentologists; VPDB has a δ^{18}O value of +30.91‰ relative to VSMOW. Stable carbon-isotope results are reported relative to VPDB, sulfur isotopes to CDT (Canyon Diablo Troilite), and nitrogen isotopes to atmospheric nitrogen (air). Most δ-values can be measured with a precision of at least ±0.2‰, except for hydrogen (±2‰).

For oxygen and sulfur, which comprise three and four stable isotopes respectively, it is normally sufficient to measure only ^{18}O/^{16}O and ^{34}S/^{32}S ratios. However, non-mass dependent fractionation in the Earth's stratosphere makes ^{17}O/^{16}O,

^{36}S/^{32}S and ^{33}S/^{32}S ratios of interest in sedimentary systems where feedback from the atmosphere might be measurable (e.g., the Earth's early sulfur cycle, Farquhar *et al.*, 2000; studies of oceanic productivity, Luz and Barkan, 2000).

Recent advances in analytical methods

Most stable isotope measurements are made using the gas-source, triple-collecting, dual-inlet, stable isotope ratio mass-spectrometer. In the traditional approach, which dominated until the late 1990s, samples of solids or liquids were converted quantitatively into the appropriate gas (hydrogen, carbon dioxide, nitrogen, sulfur dioxide, etc.) using complex "off-line" preparation methods. These gases were then brought separately to the mass spectrometer for analysis. However, recent advances in stable isotope analysis of hydrogen, carbon, oxygen, nitrogen and sulfur now make collection of these data much more accessible to sedimentologists. In the new approach, known as continuous-flow, stable isotope ratio mass-spectrometry, "on-line" preparation devices (peripherals) are used to produce and concentrate the appropriate pure gas from a wide variety of solids, liquids and gases. This pure gas is swept by a stream of helium from the peripheral directly into the attached mass spectrometer for analysis.

This conceptually simple, but fundamental change in analytical approach has enormous implications for stable isotope research in sedimentology. Very small quantities of gas (routinely to ~ 200 nmols) can now be analyzed. Sample size largely ceases to be a limitation on the ability to measure stable isotope ratios. Lasers, for example, allow precision sampling and appropriate gas production from small (~ 30–$100\,\mu m$) regions of many solids (e.g., carbonate, sulfides, some silicates). Other peripherals (gas chromatographs, elemental analyzers, thermochemical convertors, thermogravimetric balances, etc.) allow similarly small quantities of water, organic matter and other materials to be analyzed in continuous-flow mode. Continuous-flow coupling of a gas chromatograph to a stable isotope ratio mass-spectrometer, for example, allows compound-specific stable isotopic fingerprints to be obtained for complex organic materials. Likewise, the coupling of a thermogravimetric balance and a stable isotope ratio mass-spectrometer makes possible separation and isotopic analysis of waters of hydration and hydroxyl groups from clays and other hydrous minerals. Such data can be used to determine temperatures and fluid evolution in many sedimentary environments.

The mass spectrometers and their on-line peripherals are automated, which allows large numbers of samples to be analyzed in short periods of time. This approach largely replaces labor-intensive, arcane and complex conventional procedures with a methodology that is accessible to most researchers. Accordingly, research problems requiring a large number of analyzes (typical of sedimentology) can be undertaken routinely.

Development of the ion-probe is also of interest to sedimentologists. Direct stable isotopic measurement of solids in thin section is now possible for some materials, with a precision that begins to approach that of conventional dual-inlet or continuous-flow methods. The ion-probe, for example, has been used to obtain the oxygen-isotope compositions of quartz overgrowths in sandstones, and from that, to infer temperature and fluid compositions during secondary quartz cementation (Hervig *et al.*, 1995). Like the laser methods described earlier, the ion-probe has the potential to replace tedious and error-prone physical and chemical methods previously used to isolate such samples for stable isotopic analysis (Ayalon and Longstaffe, 1992).

Stable isotopic variations in water

Seawater and fresh (meteoric) water have characteristic hydrogen- and oxygen-isotope compositions, as do formation waters whose isotopic compositions evolved during rock-water interaction and mixing in sedimentary basins. Direct isotopic measurements of oceans, lakes, rivers, shallow porewaters, and formation waters from deep in sedimentary basins reveal much about the origin and evolution of these fluids. The stable carbon and hydrogen isotopic compositions of hydrocarbons in sedimentary systems provide analogous information about their sources, as well as hydrocarbon generation and degradation (Hoefs, 1997).

Seawater. Unmodified ocean water has $\delta^{18}O$ and δD values of about 0‰, which generally vary only slightly because of dilution by fresh water near major rivers. High latitude, shallow epicontinental seas that receive a large influx of surface runoff also can acquire isotopic compositions substantially less than 0‰, especially in surface layers (e.g., the Cretaceous Western Interior Seaway). Evaporation can also cause variation in the stable isotopic composition of seawater (and fresh water). Because ^{16}O and ^{1}H are preferentially partitioned into the vapor phase during evaporation, surface waters within a restricted basin are progressively enriched in ^{18}O and D, at least until extreme levels of solute concentration are reached. The isotopic composition of evaporated bodies of water is described by:

$$\delta D = m\delta^{18}O. \quad \text{(Eq. 9)}$$

The value of m decreases as relative humidity declines. It is characteristic of a given region, and the resulting distribution of δD and $\delta^{18}O$ water values along this slope defines a "local evaporation line".

Continental glaciation can affect the oxygen and hydrogen isotopic composition of seawater. Continental ice caps are major reservoirs of ^{16}O- and ^{1}H-rich water; melting of the present Greenland and Antarctic ice caps would lower the $\delta^{18}O$ value of the ocean by ~ 0.7‰. Studies of Pleistocene ice-cover fluctuations suggest that the variation then was no larger than ~ 1‰ (Savin and Yeh, 1981).

Whether the oxygen isotopic composition of the ocean has changed with time on a larger scale remains unresolved. "Least altered" shell materials (mostly brachiopods) and marine carbonate cements suggest a trend toward ^{18}O depletion of seawater with increasing age (to -7‰ for the Cambrian) (Wadleigh and Veizer, 1992). Data from ophiolites and mudrocks support the concept of a 0‰ ocean throughout much of geologic time (Muehlenbachs and Clayton, 1976; Land and Lynch, 1996). This behavior may reflect greater susceptibility of carbonates versus silicates to post-formational isotopic alteration. It may also arise from different water compositions in the deep, open ocean versus shallow, epicontinental seas.

Seawater sulfate. Ocean water is the Earth's major reservoir of dissolved sulfate. Modern seawater sulfate has an average $\delta^{34}S$ value of $+21.0$‰, and $\delta^{18}O$ value of $\sim +9.6$‰. The $\delta^{34}S$ values of sulfate minerals are very similar to the seawater from which they evaporated, whereas their $\delta^{18}O$ value are enriched by ~3.5‰ (Hoefs, 1997). The stable isotope composition of seawater sulfate has varied throughout geological time, especially its $\delta^{34}S$ values, which have ranged from ~+10 to +30‰ (Claypool et al., 1980). The variations in $\delta^{18}O$ values are smaller and do not necessarily track those of sulfur, for reasons yet poorly understood (Hoefs, 1997).

Stable sulfur-isotope behavior in sedimentary environments is complex. The isotopic composition of seawater sulfate at any time is the baseline from which the δ-values of diagenetic marine sulfide and residual sulfate are derived. However, their δ-values also depend on whether open communication with seawater was maintained in the sediment during sulfide formation (i.e., an open versus closed system), and the nature, rate and temperature of the processes that drive reduction of sulfate to sulfide. The isotopic fractionations can be considerable; on average, diagenetic pyrite in sediments is depleted of ^{34}S relative to the dissolved sulfate by ~50‰ (Longstaffe, 1989; Hoefs, 1997).

The changes in sulfur and oxygen isotopic compositions of dissolved sulfate in seawater arise from variations in the input and output of sulfate to and from the oceans. The volumes and isotopic composition of these fluxes are controlled by microbial reduction of sulfate to sulfide and evaporite formation in seawater, contributions of sulfate and sulfide to the oceans from continental weathering, the cycling of seawater through mid-oceanic ridges, and the volume of seawater itself. Variations in seawater-sulfate isotopic composition through geological time can be linked to the same kinds of global tectonic and eustatic processes that drive variations in the $^{87}Sr/^{86}Sr$ ratio of seawater. Variations in the S- and Sr-isotope seawater curves show many similarities through geological time. By comparison, the $\delta^{13}C$ values of marine carbonates, which capture the fluctuations in inorganic versus organic contributions to marine bicarbonate, tend to be antipathetic to marine sulfate $\delta^{34}S$ values (Hoefs, 1997).

Porewater in shallow marine sediments. The $\delta^{18}O$ and δD values of porewater in the first ~800 m of Cenozoic oceanic sediments commonly decrease to values less than 0‰ with increasing depth from the sediment–water interface (Savin and Yeh, 1981). Numerous explanations have been offered for this behavior, including primary variations in porewater composition, and production of secondary clays, carbonates and zeolites. Low-temperature crystallization of ^{18}O-rich secondary minerals has the potential to deplete the porewater reservoir of ^{18}O in a quasi-closed system with a high rock/water ratio (Longstaffe, 2000). However, neoformation of secondary minerals cannot easily account for the lowering of δD values in oceanic porewater, which can exceed 10‰. Low-D water (<200‰) formed during oxidation of organic matter, methane and hydrogen may contribute to the composition of porewater in oceanic sediments, as may exchange with methane or hydrogen from deeper sources (Lawrence and Taviani, 1988).

Meteoric (fresh) water. Meteoric water has a large but systematic variation in isotopic composition ($\delta^{18}O \approx -55$ to 0‰; $\delta D \approx -400$ to +10‰) described by the Global Meteoric Water Line (GMWL) (Craig, 1961):

$$\delta D = 8\delta^{18}O + 10(‰). \qquad (Eq. 10)$$

This regularity is largely the product of Rayleigh distillation within the hydrological cycle. In the simplest situation, water evaporated from the ocean is progressively depleted of ^{18}O and D through successive evaporation and condensation while travelling across the continents. This distillation produces a geographically controlled distribution of oxygen- and hydrogen-isotope compositions, with fresh water becoming progressively depleted of the heavier isotope at higher altitudes and latitudes. The depletion is further accentuated by lower average air-mass temperatures and larger values of $\alpha_{liquid-vapor}$ toward higher latitudes and altitudes, and by the re-evaporation of meteoric waters from the continental surface (Yurtsever and Gat, 1981).

Formation water. The $\delta^{18}O$ and δD values of formation water from sedimentary basins commonly (but not always) define a trend ("formation-water line") with a lower slope than the GMWL. Along this trend, hotter and more saline formation waters are generally more ^{18}O-rich than cooler, less saline samples. The formation-water line for a specific sedimentary basin commonly intersects the GMWL at an isotopic composition typical of local surface and shallow groundwater. This suggests that meteoric water comprises an important fraction of many formation waters (Clayton et al., 1966). However, the deviation of the formation-water line from the GMWL (equation 10) indicates that other processes are also at play. These include evaporation (see equation 9), isotopic exchange between pore-lining minerals and porewater, release of water by the smectite to illite reaction, mixing between seawater and fresh water or their evolved equivalents, and perhaps membrane filtration (see Longstaffe, 2000 and references therein).

The extent of formation-water ^{18}O-enrichment *via* exchange depends on the isotopic composition of the exchanging solid phases, temperature, mineral/water ratio, and the exchangeability of the minerals in the pore system. Extensive exchange may be plausible in carbonate-rich rocks, but is unlikely in siliciclastic units, which exchange oxygen isotopes much less easily than carbonates. Isotopic exchange involving clays or organic matter may also contribute to the δD values of formation waters. Fluid movement from shales to sandstones in settings where the smectite to illite reaction controls shale porewater chemistry can explain the ^{18}O-rich nature of some formation waters, but this requires very high (originally smectite-rich) shale to sand ratios within the basin. Mixing between evaporitic brines derived from seawater and (later) meteoric water can also explain the stable isotopic composition of many formation waters.

It seems unlikely that there is a single explanation for the δD and $\delta^{18}O$ values of formation waters in sedimentary basins. Porewater evolution and movement follow different pathways, depending on the geological circumstances of each sedimentary basin. By understanding the range of stable isotopic compositions that can be expected for seawater, meteoric water and formation water, one can track the evolution of fluids in a sedimentary basin, using the proxies provided by secondary minerals formed in the pore system.

Stable isotopic variations in sediments and sedimentary rocks

Chemical sediments. Chemical sediments and sedimentary rocks generally have very high $\delta^{18}O$ values (up to +45‰), because they acquire their isotopic compositions at low temperatures in oxygen isotopic equilibrium with water, during which enrichment in ^{18}O of the solid phase is large. Modern marine limestones, for example, commonly have $\delta^{18}O$

values near 30‰ (VSMOW); fresh-water varieties have variably lower values that reflect the composition of the meteoric water involved. Dolomitization and other diagenetic modifications have a substantial impact on the $\delta^{18}O$ values of carbonates; for example, post-crystallization interaction with meteoric water invariably leads to a lowering of their $\delta^{18}O$ values (Arthur et al., 1983). Care is needed in the study of ancient marine carbonates to obtain samples that have preserved their original oxygen isotopic compositions (Carpenter et al., 1991).

Stable carbon-isotope variations. Marine carbonate rocks normally have $\delta^{13}C$ values of ~ 0‰, which reflect the balance between inorganic (^{13}C-rich) and organic (^{12}C-rich) carbon in the total dissolved marine carbonate reservoir at the time of formation (Land, 1980; Hoefs, 1997). Variations in the stable carbon-isotope composition of the oceanic reservoir through geologic time are normally related to changes in the amount of organic (^{13}C-poor) versus inorganic (^{13}C-rich) carbon added to or removed from the dissolved carbon reservoir.

The shifts in stable carbon-isotope composition of marine carbonates have normally been no more than ± 2‰ throughout most of geological time, but relatively sudden shifts of larger magnitude and global distribution are known. Such decreases at the Precambrian/Cambrian, Permian/Triassic, and Cretaceous/Tertiary boundaries are commonly correlated to upwelling of ^{13}C-poor, anoxic, oceanic bottom-waters and substantially reduced organic productivity (extinction events?) in shallow marine environments (Veizer et al., 1980; Hoefs, 1997; Kimura and Watanabe, 2001). Very high $\delta^{13}C$ values for marine carbonates ($> +10$‰) at ~ 2.2 Ga to 2.0 Ga, and at the end of the Proterozoic, have been linked to periods of vastly increased burial of organic carbon. This behavior has been attributed by some to the complete freezing of the Earth's oceans (the "Snowball Earth" hypothesis) (Knoll et al., 1986; Baker and Fallick, 1989; Derry et al., 1992; Hoffman et al., 1998).

Most carbonate-bearing sediments existed in zones of microbial activity during sedimentation and early diagenesis. In sediments with open access to the marine bicarbonate reservoir (presently $\delta^{13}C \approx 0$‰), the volume of carbon dioxide added to the system from microbial activity is too small to detect. However, when organic carbon comprises a significant fraction of the bicarbonate supply to the porewater, distinctive carbonate $\delta^{13}C$ values result (see Longstaffe, 1989 and references therein). For example, carbonate rocks invaded by meteoric water typically acquire lower $\delta^{13}C$ values because of the flux of low-^{13}C, soil-derived CO_2 ($\delta^{13}C \approx -25$‰). Large variations in carbonate $\delta^{13}C$ values ($\delta^{13}C \approx -30$ to $+30$‰) are also characteristic of carbonate rocks intercalated with coals and other organic matter, such as organic-rich shales, oil-shales and oil-sands. Calcite with $\delta^{13}C$ values as low as -70‰ is associated with methane oxidation at seeps.

Growth of carbonate concretions in organic-rich mudstones and shales represents a special case where the CO_2 contribution from marine bicarbonate trapped with the porewater is small compared to that produced during diagenetic reactions. In the simplest case, each carbonate growth band acquires a $\delta^{13}C$ value characteristic of the processes dominant at that depth from the sediment-water interface (Irwin et al., 1977). Consumption of organic matter in the shallowest zones is dominated by microbial oxidation. A zone of microbial sulfate reduction then follows in marine and brackish water environments. Both zones are characterized by CO_2 with $\delta^{13}C$ values of ~ -25‰. Still deeper zones are characterized by microbial fermentation, which leads to production of ^{12}C-rich CH_4; in some extreme case, CO_2 reduction follows. Residual CO_2 available for fixation in carbonate minerals is therefore variably enriched in ^{13}C (avg. $\delta^{13}C \approx +15$‰). At still greater depths, residual organic matter is gradually destroyed by a variety of abiotic, thermally driven reactions, which release CO_2 with $\delta^{13}C$ values of ~ -25‰ to -10‰ (see Longstaffe, 1989 and references therein).

Clastic sedimentary rocks. The bulk oxygen-isotope composition of clastic sedimentary rocks is highly variable ($\delta^{18}O = +8$‰ to $+25$‰) because it reflects varying contributions of (i) detrital minerals and rock fragments that normally retain oxygen-isotope compositions characteristic of their original high-temperature source, (ii) ^{18}O-rich weathering products ($\sim +10$‰ to $+30$‰; clays, altered feldspars and rock fragments, etc., and (iii) ^{18}O-rich cements. To summarize, clastic sedimentary rocks are typically richer in ^{18}O than igneous rocks ($+5$‰ to $+10$‰) because they contain some minerals that acquired their isotopic compositions at low temperatures in oxyen isotopic equilibrium with water (Savin and Epstein, 1970a,b; Longstaffe and Schwarcz, 1977). Shales and mudstones generally are enriched in ^{18}O relative to sandstones because of the former's much higher content of clay minerals and secondary silica.

Diagenetic minerals such as carbonate, silica and clays are of particular interest because they are the product of a complex mineralogical, fluid and thermal history. They form by precipitation from solution, and by transformation or alteration of pre-existing phases. Their isotopic composition is controlled by several parameters, including composition of the fluid, the degree of isotopic equilibrium during crystallization (including microbial effects), the mass balance for the element of interest between the mineral and the pore fluid, and the extent of isotopic exchange following crystallization.

Clay minerals are the most important mineral reservoirs of hydrogen in sedimentary rocks, with δD values that normally range from -100‰ to -20‰, typical of phases formed during weathering (Lawrence and Taylor, 1971, 1972; Savin and Hsieh, 1998; Longstaffe, 2000). However, lower values (~ -150‰) are known for clay minerals formed or exchanged at high latitudes or altitudes or during diagenesis in the presence of low-D porewater (Longstaffe, 1989; Longstaffe and Ayalon, 1990). Soil clays typically display a systematic variation in δD and $\delta^{18}O$ values that describe a trend parallel to, but offset from the GMWL. The position of these "clay weathering lines" relative to the GMWL is determined by the hydrogen- and oxygen-isotope mineral-water fractionations for the temperature of clay formation (Sheppard et al., 1969; Lawrence and Taylor, 1971, 1972; Mizota and Longstaffe, 1996).

Equilibrium stable isotope fractionation and geothermometry

The difference between two phases in the ratios (R) of an element's "heavy" to "light" isotopes (e.g., oxygen in $CaCO_{3(solid)}$ and $H_2O_{(liquid)}$) is known as fractionation. The stable isotope fractionation factor between two phases A and B is defined as:

$$\alpha_{A-B} = R_A/R_B. \quad \text{(Eq. 11)}$$

Values of α are normally very close to unity. For example, $CO_{2(gas)}$ in equilibrium with $H_2O_{(liquid)}$ is enriched in ^{18}O relative to the water, hence the value of α is positive. At 25°C, $\alpha^{18}O_{CO_2\text{-}H_2O} = 1.0412$ (Friedman and O'Neil, 1977). As for all temperature-dependent fractionations, this value increases at lower temperatures and decreases at higher temperatures. Equilibrium fractionations derive from small differences in the thermodynamic properties between compounds containing an isotope with a higher (e.g., ^{18}O) versus a lower mass number (e.g., ^{16}O). This relative mass difference exerts important control on the size of equilibrium fractionations. As a result, the largest fractionations in nature are typically observed for stable hydrogen isotopes.

The ability to calculate values of α from δ_A and δ_B

$$\alpha_{A-B} = (\delta_A + 1000)/(\delta_B + 1000), \quad \text{(Eq. 12)}$$

is of special significance in equilibrium systems, because α is temperature (T) dependent. Experimentally determined α-values for mineral-mineral or mineral-water pairs (α_{A-B}) normally fall on smooth curves when $10^3\ln\alpha_{A-B}$ (known as permill fractionation) is plotted versus $10^6/T^2$ or some variation of that function (T in Kelvin). The smectite and water system provided a typical example of an equilibrium oxygen-isotope geothermometer (Savin and Lee, 1988):

$$10^3\ln\alpha_{\text{smectite-water}} = 2.58(10^6)T^{-2} - 4.19. \quad \text{(Eq. 13)}$$

Such relationships are commonly used to deduce the crystallization temperature of an authigenic mineral, provided that the composition of the mineralizing fluid is known. Stable isotope geothermometers of use in sedimentary systems have been summarized by Kyser (1987), Savin and Lee (1988), Longstaffe (1989) and Sheppard and Gilg (1996). New or revised equations appear regularly (e.g., Bird et al., 1994; Vitali et al., 2001). Stable isotope geothermometers are useful in low temperature systems provided that (i) equilibrium was attained between phases A and B, and (ii) subsequent isotopic exchange was negligible. Knowledge of the rock/water ratio for the element of interest is also important when considering the stable isotopic behavior of mineral-water systems (Longstaffe, 1989, 1994).

Isotopic Exchange. Temperature, grain-size, surface area, crystal chemistry, mineral/water ratio and, in some cases, microbial activity are important factors controlling isotopic exchange between minerals and water (see Longstaffe, 2000 and references therein). Isotopic exchange between minerals and water requires dissolution and recrystallization of the solid phase. Consequently, post-crystallization isotopic exchange between many minerals and water is insignificant at surface temperatures. For example, post-crystallization oxygen-isotope exchange between quartz and water is not detected at <200°C, even for the finest particles. By comparison, such exchange can affect calcite and, to a lesser extent, dolomite, because these minerals are much more easily recrystallized at low temperatures in the presence of porewater. The $\delta^{13}C$ value of carbonates is generally less affected by such processes, as the mineral/water ratio for carbon is very high in most cases.

Stable isotopic exchange in clay-water systems is complex, and manifested differently in clay-dominated shales and mudstones (where porewater is most affected) versus fluid-dominated systems such as porous and permeable sandstones (where clay is most affected). Exchange of oxygen isotopes between water and structural sites in clay minerals is extremely slow in low-temperature environments, including modern ocean sediments (Savin and Epstein, 1970a,b; Yeh and Eslinger, 1986). Without neoformation, oxygen-isotope exchange is insignificant in marine environments, even for very fine (<0.1 μm), smectite-rich clay assemblages. The rate of oxygen-isotope exchange between clay minerals and water increases as temperatures rise, with significant exchange occurring by 300°C. Mineralogical reactions, such as the smectite to illite conversion, facilitate this behavior (Yeh and Savin, 1977).

Hydrogen-isotope exchange between clay minerals and water also occurs slowly at low temperatures, but the rate is faster than for oxygen. Measurable hydrogen-isotope exchange can take place at <100°C; by 200°C, extensive hydrogen-isotope exchange can occur unaccompanied by oxygen isotopic effects. Some studies suggest that hydrogen-isotope exchange between seawater and detrital clay minerals from 2 million to 3 million year old ocean sediments was restricted to the <0.1 μm size-fraction (Yeh and Epstein, 1978); others suggest that exchange with expandable clay minerals occurs readily at surface temperatures. Significant post-formational hydrogen-isotope exchange between non-expandable clays and porewater can also occur over millions of years at 30°C to 40°C without a corresponding effect on the oxygen-isotope composition of the clay (Longstaffe and Ayalon, 1990).

Applications of equilibrium stable isotope fractionation

Stable isotope geothermometry underlies the vast body of research in which oxygen-isotope compositions of ice cores, foraminifera, brachiopods, diatoms and biogenic phosphate in the oceans, and speleothem carbonates on land are used to reconstruct the Earth's past climate (Shackleton, 1968; Emiliani and Shackleton, 1974; Schwarcz, 1986; Bowen, 1991; Petit et al., 1999; McDermott et al., 2001; Norris et al., 2002; Stanton et al., 2002). These stable isotope proxies for temperature are coupled with radiometric, biostratigraphic and paleomagnetic information, which provide an absolute chronology for the scale and rate of climate change.

Stable isotope geothermometers for oxygen and hydrogen are also used to identify the type(s) of water (seawater, fresh water, brine) involved in formation of chemical sediments, provided that independent temperature data are available (e.g., see equation 13). Likewise, this approach has been combined with stable carbon-isotope, $^{87}Sr/^{86}Sr$ and $^{143}Nd/^{144}Nd$ ratios to explain dolomitization, and to identify multiple fluid regimes responsible for diagenesis of carbonate rocks (e.g., Land, 1980, 1991; Banner et al., 1988; Coniglio et al., 1988; Chaftez and Rush, 1995; Benito et al., 2001).

The stable isotopic compositions of carbonate and silicate cements from siliciclastic rocks have been used in a similar fashion. A common application is to deduce the mixing and movement of seawater, meteoric water and their evolved brines on basinwide scales over time, and to relate this behavior to the generation, migration and degradation of hydrocarbons (e.g., Longstaffe and Ayalon, 1987, 1991; Ayalon and Longstaffe, 1988; Tilley and Longstaffe, 1989; Longstaffe, 1993, 1994, 2000; Ziegler and Longstaffe, 2000a,b). Neoformed minerals (pedogenic clays, hydroxides, carbonates) from paleosols and regoliths can also be used to deduce the $\delta^{18}O$ and δD values of ancient meteoric water. These results provide critical insight into continental paleoclimate because the oxygen- and hydrogen-isotope compositions of meteoric water are strongly determined by temperature and the precipitation–evaporation regime at the time of soil formation (Lawrence and Taylor, 1971, 1972; Bird et al., 1992, 1994; Mizota and Longstaffe,

1996; Stern *et al.*, 1997; Savin and Hsieh, 1998; Yapp, 1998; Girard *et al.*, 2000).

Kinetic stable isotope fractionation and its applications

Kinetic isotope effects, which occur during incomplete and unidirectional processes, are a second important source of stable isotopic fractionation in sedimentary systems. These occur during rapid evaporation of water, and are also especially important in biologically mediated systems. For example, $^{12}CO_2$ is selected preferentially over $^{13}CO_2$ during photosynthesis (Hoefs, 1997). The magnitude of the fractionation depends mostly on reaction rates and pathways. For example, significantly larger stable carbon-isotope fractionations characterize the C_3 versus the C_4 photosynthetic pathways. Hence, stable carbon-isotope analysis of ancient organic matter and pedogenic carbonates in sediments and soils can provide significant paleoecological and paleoclimatic information (Cerling, 1984; Cerling *et al.*, 1989, 1997).

Kinetic isotope fractionation associated with microbial activity in sedimentary systems is of special importance. The stable carbon- and hydrogen-isotope compositions of methane and carbon dioxide emanating from wetlands, for example, can be used to identify the reaction pathways that produce these greenhouse gases (Hornibrook *et al.*, 1997). As discussed earlier, many early diagenetic processes are also driven by microbially mediated reactions. These produce characteristic stable carbon- and sulfur-isotope compositions in secondary carbonates and pyrite (Irwin *et al.*, 1977; Curtis *et al.*, 1986; Longstaffe, 1989; Habicht and Canfield, 1997, 2001; Canfield, 2001). A full understanding of these reactions allows stable carbon and sulfur isotopic data for these phases, particularly from concretions, to be used to reconstruct ancient depositional environments. Likewise, one can deduce the geochemical conditions at and beneath the sediment-water interface during early diagenesis. Depositional and early diagenetic interplay between fresh and marine waters as sea level fluctuated in shallow epeiric seas can also be elucidated, using the stable carbon-, sulfur-, and oxygen-isotope results for secondary minerals (Irwin *et al.*, 1977; Curtis *et al.*, 1986; Longstaffe, 1987, 1989, 1994; Mozley and Burns, 1993; McKay *et al.*, 1995; McKay and Longstaffe, 2002).

Equilibrium oxygen-isotope mineral-water geothermometers are generally used to identify the water involved during formation of early diagenetic minerals in sedimentary systems. However, microbial activity is known to mediate the precipitation of some oxygen-bearing phases, such as siderite. The oxygen-isotope fractionation associated with microbially produced siderite at low temperatures is different than the equilibrium (inorganic) fractionation for this system. Use of the latter will result in calculated $\delta^{18}O$ values for porewater that are lower than produced by the former; this can lead to incorrect identification of paleoenvironments (Mortimer and Coleman, 1997). The potential consequences of such microbial interventions on mineral-water oxygen isotopic fractionation in sedimentary and early diagenetic environments remain to be explored fully.

Conclusion

The use of radioactive, radiogenic and stable isotopes is now common in sedimentology. These methods are used to study rates of deposition and erosion, characterize depositional environments, identify the provenance and dispersal of sediment, make regional and global chemostratigraphic correlations, understand diagenetic processes and geochemical conditions at and near the sediment-water interface, define the timing and nature of regional fluid flow in sedimentary basins, deduce global paleoclimate both in terrestrial and marine environments, infer variations in global biogeochemical cycles and their significance, and make global paleogeographic and tectonic reconstructions. The most important advances over the last decade are the ease with which isotopic measurements can now be made, and the fact that a comprehensive approach to the use of isotopic methods is now the norm. Multiple isotopic chronometers and tracers are employed together to test and constrain hypotheses much more rigorously than possible from use of a single isotopic tracer or geochronometer. The hallmark of effective utilization of isotopic methods is their integration with sedimentological, stratigraphic and other geological data.

Fred J. Longstaffe

Bibliography

Arthur, M.A., Anderson, T.F., Kaplan, I.R., Veizer, J., and Land, L.S., 1983. *Stable Isotopes in Sedimentary Geology*. Tulsa: SEPM Short Course, 10.

Asmerom, Y., Jacobsen, S.B., Knoll, A.H., Butterfield, N.J., and Swett, K., 1991. Strontium isotopic variations of Neoproterozoic seawater: implications for crustal evolution. *Geochimica et Cosmochimica Acta*, **55**: 2883–2894.

Ayalon, A., and Longstaffe, F.J., 1988. Oxygen-isotope studies of diagenesis and pore water evolution in the Western Canada Sedimentary Basin: evidence from the Upper Cretaceous basal Belly River sandstone, Alberta. *Journal of Sedimentary Petrology*, **58**: 489–505.

Ayalon, A., and Longstaffe, F.J., 1992. Isolation of diagenetic silicate minerals in clastic sedimentary rocks for oxygen isotope analysis: a summary of methods. *Israel Journal of Earth Sciences*, **39**: 139–148.

Awwiller, D.N., and Mack, L.E., 1991. Diagenetic modification of Sm–Nd model ages in Tertiary sandstones and shales, Texas Gulf Coast. *Geology*, **19**: 311–314.

Baker, A.J., and Fallick, A.E., 1989. Heavy carbon in two-billion-year-old marbles from Lofoten-Vesteralen, Norway: implications for the Precambrian carbon cycle. *Geochimica et Cosmochimica Acta*, **53**: 1111–1115.

Banner, J.L., Hanson, G.N., and Meyers, W.J., 1988. Rare earth element and Nd isotopic variations in regionally extensive dolomites from the Burlington-Keokuk formation (Mississippian): implications for REE mobility during carbonate diagenesis. *Journal of Sedimentary Petrology*, **58**: 415–432.

Basu, A.R., Jacobsen, S.B., Poreda, R.J., Dowling, C.B., and Aggarwal, P.K., 2001. Large groundwater strontium flux to the oceans from the Bengal Basin and the marine strontium isotope record. *Science*, **293**: 1470–1473.

Beck, J.W. *et al.*, 2001. Extremely large variations of atmospheric ^{14}C concentration during the last glacial period. *Science*, **292**: 2453–2458.

Benito, M.I., Lohmann, K.C., and Mas. R., 2001. Discrimination of multiple episodes of meteoric diagenesis in a Kimmeridgian reefal complex, North Iberian Range, Spain. *Journal of Sedimentary Research*, **71**: 380–393.

Bierman, P.R., and Turner, J., 1995. ^{10}Be and ^{26}Al evidence for exceptionally low rates of Australian bedrock erosion and the likely existence of pre-Pleistocene landscapes. *Quaternary Research*, **44**: 378–382.

Bird, M.I., Longstaffe, F.J., Fyfe, W.S., and Bildgen, P., 1992. Oxygen-isotope systematics in a multiphase weathering system in Haiti. *Geochimica et Cosmochimica Acta*, **56**: 2831–2838.

Bird, M.I., Longstaffe, F.J., Fyfe, W.S., Tazaki, K., and Chivas, A.R., 1994. Oxygen-isotope fractionation in gibbsite: synthesis experiments versus natural samples. *Geochimica et Cosmochimica Acta*, 58: 5267–5277.

Blum, J.D., and Erel, Y., 1995. A silicate weathering mechanism linking increases in marine $^{87}Sr/^{86}Sr$ with global glaciation. *Nature*, 373: 415–418,

Bowen, R., 1988. *Isotopes in the Earth Sciences*. Barking: Elsevier Applied Science Publishers Ltd.

Bowen, R., 1991. *Isotopes and Climates*. Barking: Elsevier Science Publishers Ltd.

Bullen, T.D., White, A.F., Childs, C.W., Vivit, D.V., and Schulz, M.S., 2001. Demonstration of significant abiotic iron isotope fractionation in nature. *Geology*, 29: 699–702.

Burke, W.H., Denison, R.E., Hetherington, E.A., Koepnick, R.B., Nelson, H.F., and Otto, J.B., 1982. Variation of seawataer $^{87}Sr/^{86}Sr$ throughout Phanerozoic time. *Geology*, 10: 516–519.

Canfield, D.E., 2001. Isotopic fractionation by natural populations of sulfate-reducing bacteria. *Geochimica et Cosmochimica Acta*, 65: 1117–1124.

Carpenter, S.J., Lohmann, K.C., Holden, P., Walter, L.M., Huston, T.J., and Halliday, A.N., 1991. $\delta^{18}O$ values, $^{87}Sr/^{86}Sr$ and Sr/Mg ratios of Late Devonian abiotic marine calcite: implications for the composition of ancient seawater. *Geochimica et Cosmochimica Acta*, 55: 1991–2010.

Cerling, T.E., 1984. The stable isotopic composition of modern soil carbonates and its relation to climate. *Earth and Planetary Science Letters*, 71: 229–240.

Cerling, T.E., Quade, J., Wang, Y., and Bowman, J., 1989. Carbon isotopes in soils and paleosols as ecology and paleoecology indicators. *Nature*, 341: 138–139.

Cerling, T.E. *et al.*, 1997. Global vegetation change through the Miocene/Pliocene boundary. *Nature*, 389: 153–158.

Chafetz, H.S., and Rush, P.F., 1995. Two-phase diagenesis of Quaternary carbonates, Arabian Gulf: insights from $\delta^{13}C$ and $\delta^{18}O$ data. *Journal of Sedimentary Research*, A65: 294–305.

Clark, I.D., and Fritz, P., 1997. *Environmental Isotopes in Hydrogeology*. Boca Raton: CRC Press LLC.

Clauer, N., and Chaudhuri, S., 1995. *Clays in Crustal Environments. Isotope Dating and Tracing*. Berlin-Heidelberg: Springer-Verlag.

Clauer, N., Srodon, J., Francu, J., and Sucha, V., 1997. K–Ar dating of illite fundamental particles separated from illite-smectite. *Clay Minerals*, 32: 181–196.

Claypool, G.E., Holser, W.T., Kaplan, I.R., Sakai, H., and Zak, I., 1980. The age curves of sulfur and oxygen isotopes in marine sulfate and their mutual interpretation. *Chemical Geology*, 28: 199–260.

Clayton, R.N., Friedman, I., Graf, D.L., Mayeda, T.K., Meents, W.F., and Shimp, N.F., 1966. The origin of saline formation waters. 1. isotopic composition. *Journal of Geophysical Research*, 71: 3869–3882.

Coniglio, M., James, N.P., and Aissaoui, D.M., 1988. Dolomitization of Miocene carbonates, Gulf of Suez, Egypt. *Journal of Sedimentary Petrology*, 58: 100–119.

Craig, H., 1961. Isotopic variations in meteoric waters. *Science*, 133: 1702–1703.

Curtis, C.D., Coleman, M.L., and Love, L.G., 1986. Pore water evolution during sediment burial from isotopic and mineral chemistry of calcite, dolomite and siderite concretions. *Geochimica et Cosmochimica Acta*, 50: 2321–2334.

De La Rocha, C.L., Brzezinski, M.A., DeNiro, M.J., and Shemesh, A., 1998. Silicon-isotope composition of diatoms as an indicator of past oceanic change. *Nature*, 395: 680–683.

Denison, R.E., Koepnick, R.B., Burke, W.H., and Hetherington, E.A., 1998. Construction of the Cambrian and Ordovician seawater $^{87}Sr/^{86}Sr$ curve. *Chemical Geology*, 152: 325–340.

DePaolo, D.J., and Wasserburg, G.J., 1976. Nd isotopic variations and petrogenetic models. *Geophysical Research Letters*, 3: 249–252.

Derry, L.A., Kaufman, A.J., and Jacobsen, S.B., 1992. Sedimentary cycling and environmental change in the Late Proterozoic: evidence from stable and radiogenic isotopes. *Geochimica et Cosmochimica Acta*, 56: 1317–1329.

Ding, T., Jiang, S, Wan, D., Li, Y., Li, J., Song, H., Liu, Z., and Yao, X., 1996. *Silicon Isotope Geochemistry*. Beijing: Geological Publishing House.

Dorale, J.A., Edwards, R.L., Ito, E., and Gonzalez, L.A., 1998. Climate and vegetation history of the midcontinent from 75 to 25 ka: a speleothem record from Crevice Cave, Missouri, USA. *Science*, 1871–1874.

Ebneth, S., Shields, G.A., Veizer, J., Miller, J.F., and Shergold, J.H., 2001. High-resolution strontium isotope stratigraphy across the Cambrian-Ordovician transition. *Geochimica et Cosmochimica Acta*, 65: 2273–2292.

Edmond, J.M., 1992. Himalayan tectonics, weathering processes, and the strontium isotope record in marine limestones. *Science*, 258: 1594–1597.

Edwards, R.L., Beck, J.W., Burr, G.S., Donahue, D.J., Chappell, J.M.A., Bloom, A.L., Druffel, E.R.M., and Taylor, F.W., 1993. A large drop in atmospheric $^{14}C/^{12}C$ and reduced melting in the Younger Dryas, documented with ^{230}Th ages of corals. *Science*, 260: 962–968.

Eggenkamp, H.G.M., Kreulen, R., and Koster van Groos, A.F., 1995. Chlorine stable isotope fractionation in evaporites. *Geochimica et Cosmochimica Acta*, 59: 5169–5175.

Emiliani, C., and Shackleton, N.J., 1974. The Brunhes Epoch: isotopic paleotemperatures and geochronology. *Science*, 183: 511–514.

Evans, M.J., Derry, L.A., Anderson, S.P., and France-Lanord, C., 2001. Hydrothermal source of radiogenic Sr to Himalayan rivers. *Geology*, 29: 803–806.

Farquhar, J., Bao, H., and Thiemens, M., 2000. Atmospheric influence of Earth's earliest sulfur cycle. *Science*, 289: 756–758.

Faure, G., 1986. *Principles of Isotope Geology*, 2nd edn. New York: John Wiley & Sons.

Friedman, I., and O'Neil, J.R., 1977. Compilation of stable isotope fractionation factors of geochemical interest. In Fleischer, M. (ed.), *Data of Geochemistry, 6th edn.*, United States Geological Survey Professional Paper, 440-KK, Washington: US Government Printing Office.

Fritz, P., and Fontes, J.Ch. (eds.), 1980. *Handbook of Environmental Isotope Geochemistry, Volume 1, The Terrestrial Environment, A*. Amsterdam: Elsevier Science Publishing Company.

Fritz, P., and Fontes, J.Ch. (eds.), 1986. *Handbook of Environmental Isotope Geochemistry, Volume 2, The Terrestrial Environment, B*. Amsterdam: Elsevier Science Publishing Company.

Fritz, P., and Fontes, J.Ch. (eds.), 1989. *Handbook of Environmental Isotope Geochemistry, Volume 3, The Marine Environment, A*. Amsterdam: Elsevier Science Publishers B.V.

Galy, A., France-Lanord, C., and Derry, L.A., 1999. The strontium isotopic budget of Himalayan rivers in Nepal and Bangladesh. *Geochimica et Cosmochimica Acta*, 63: 1905–1925.

Girard, J.-P., and Onstott, T.C., 1991. Application of $^{40}Ar/^{39}Ar$ laser-probe and step-heating techniques to the dating of diagenetic K-feldspar overgrowths. *Geochimica et Cosmochimica Acta*, 55: 3777–3793.

Girard, J.-P., Freyssinet, P., and Chazot, G., 2000. Unraveling climatic changes from intraprofile variation in oxygen and hydrogen isotopic composition of goethite and kaolinite in laterites: an integrated study from Yaou, French Guiana. *Geochimica et Cosmochimica Acta*, 64: 409–426.

Habicht, K.S., and Canfield, D.E., 1997. Sulfur isotope fractionation during bacterial sulfate reduction in organic-rich sediments. *Geochimica et Cosmochimica Acta*, 61: 5351–5361.

Habicht, K.S., and Canfield, D.E., 2001. Isotope fractionation by sulfate-reducing natural populations and the isotopic composition of sulfide in marine sediments. *Geology*, 29: 555–558.

Harper, D.A., Longstaffe, F.J., Wadleigh, M.A., and McNutt, R.H., 1995. Secondary K-feldspar at the Precambrian-Paleozoic unconformity, southwestern Ontario. *Canadian Journal of Earth Sciences*, 32: 1432–1450.

Hervig, R.L., Williams, L.B., Kirkland, I.K., and Longstaffe, F.J., 1995. Oxygen isotope microanalyses of diagenetic quartz: possible low temperature occlusion of pores. *Geochimica et Cosmochimica Acta*, 59: 2537–2543.

Hoefs, J., 1997. *Stable Isotope Geochemistry*, 4th edn, Berlin-Heidelberg: Springer-Verlag.

Hoffman, P.F., Kaufmann, A.J., Halverson, G.P., and Schrag, D.P., 1998. A Neoproterozoic snowball Earth. *Science*, 281: 1342–1346.

Hornibrook, E.R.C., Longstaffe, F.J., and Fyfe, W.S., 1997. Spatial distribution of microbial methane production pathways in temperate zone wetland soils: stable carbon and hydrogen isotope evidence. *Geochimica et Cosmochimica Acta*, **61**: 747–753.

Irwin, H., Curtis, C., and Coleman, M., 1977. Isotopic evidence for source of diagenetic carbonates formed during burial of organic-rich sediments. *Nature*, **269**: 209–213.

Jeans, C.V., Mitchell, J.G., Fisher, M.J., Wray, D.S., and Hall, I.R., 2001. Age, origin and climatic signal of English Mesozoic clays based on K/Ar signatures. *Clay Minerals*, **36**: 515–539.

Kaufman, A., Wasserburg, G.J., Porcelli, D., Bar-Matthews, M., Ayalon, A., and Halicz, L., 1998. U-Th isotope systematics from the Soreq Cave, Israel and climatic correlations. *Earth and Planetary Science Letters*, **156**: 141–155.

Kaufmann, R.S., Frape, S.K., McNutt, R., and Eastoe, C., 1993. Chlorine stable isotope distribution of Michigan Basin formation waters. *Applied Geochemistry*, **8**: 403–407.

Kendall, C., and McDonnell, J.J. (eds.), 1998. *Isotope Tracers in Catchment Hydrology*. Amsterdam: Elsevier Science B.V.

Kimura, H., and Watanabe, Y., 2001. Oceanic anoxia at the Precambrian-Cambrian boundary. *Geology*, **29**: 995–998.

Knoll, A.H., Hayes, J.M., Kaufman, A.J., Swett, K., and Lambert, I.B., 1986. Secular variation in carbon isotope ratios from Upper Proterozoic successions of Svalbard and East Greenland. *Nature*, **321**: 832–838.

Kyser, T.K., 1987. Equilibrium fractionation factors for stable isotopes. In Kyser, T.K. (ed.), *Stable Isotope Geochemistry of Low Temperature Fluids*. Mineralogical Association of Canada, Short Course, 13, pp. 1–84.

Land, L.S., 1980. The isotopic and trace element geochemistry of dolomite: the state of the art. In Zenger, D.H., Dunham, J.B., and Ethington, R.L. (eds.), *Concepts and Models of Dolomitization*. Society of Economic Paleontologists and Mineralogists, Special Publication, 28, pp. 87–110.

Land, L.S., 1991. Dolomitization of the Hope Gate Formation (North Jamaica) by seawater: reassessment of mixing-zone dolomite. In Taylor, H.P., O'Neil, J.R., and Kaplan, I.R. (eds.), *Stable Isotope Geochemistry: A Tribute to Samuel Epstein*. The Geochemical Society, Special Publication, 3, pp. 121–133.

Land, L.S., and Lynch, F.L., 1996. $\delta^{18}O$ values of mudrocks: more evidence for an ^{18}O-buffered ocean. *Geochimica et Cosmochimica Acta*, **60**: 3347–3352.

Lao, Y., Anderson, R.F., Broecker, W.S., Hofmann, H.J., and Wolfli, W., 1993. Particulate fluxes of ^{230}Th, ^{231}Pa, and ^{10}Be in the northeastern Pacific Ocean. *Geochimica et Cosmochimica Acta*, **57**: 205–217.

Lawrence, J.R., and Taylor, H.P. Jr., 1971. Deuterium and oxygen-18 correlation: clay minerals and hydroxides in Quaternary soils compared to meteoric waters. *Geochimica et Cosmochimica Acta*, **35**: 993–1003.

Lawrence, J.R., and Taylor, H.P. Jr., 1972. Hydrogen and oxygen isotope systematics in weathering profiles. *Geochimica et Cosmochimica Acta*, **36**: 1377–1393.

Lawrence, J.R., and Taviani, M., 1988. Extreme hydrogen, oxygen and carbon isotope anomalies in the pore waters and carbonates of the sediments and basalts from the Norwegian Sea: Methane and hydrogen from the mantle? *Geochimica et Cosmochimica Acta*, **52**: 2077–2083.

Longstaffe, F.J., 1987. Stable isotope studies of diagenetic processes. In Kyser, T.K. (ed.), *Stable Isotope Geochemistry of Low Temperature Fluids*. Mineralogical Association of Canada, Short Course, 13, pp. 187–257.

Longstaffe, F.J., 1989. Stable isotopes as tracers in clastic diagenesis. In Hutcheon, I.E. (ed.), *Burial Diagenesis*. Mineralogical Association of Canada, Short Course, , pp. 201–277.

Longstaffe, F.J., 1993. Meteoric water and sandstone diagenesis in the Western Canada Sedimentary Basin. In Horbury, A.D., and Robinson, A.G. (eds.), *Diagenesis and Basin Development*. American Association of Petroleum Geologists, Studies in Geology, 36, pp. 49–68.

Longstaffe, F.J., 1994. Stable isotopic constraints on sandstone diagenesis in the Western Canada Sedimentary Basin. In Parker, A., and Sellwood, B.W. (eds.), *Quantitative Diagenesis: Recent Developments and Applications to Reservoir Geology*. Dordrecht: Kluwer Academic Publishers, pp. 223–274.

Longstaffe, F.J., 2000. An introduction to stable oxygen and hydrogen isotopes and their use as fluid tracers in sedimentary systems. In Kyser, T.K. (ed.), *Fluids and Basin Evolution*. Mineralogical Association of Canada, Short Course, 28, pp. 115–163.

Longstaffe, F.J., and Ayalon, A., 1987. Oxygen-isotope studies of clastic diagenesis in the Lower Cretaceous Viking Formation, Alberta: implications for the role of meteoric water. In Marshall, J.D. (ed.), *The Diagenesis of Sedimentary Sequences*. Geological Society, Special Publication, 36, pp. 277–296.

Longstaffe, F.J., and Ayalon, A., 1990. Hydrogen-isotope geochemistry of diagenetic clay minerals from Cretaceous sandstones, Alberta, Canada: evidence for exchange. *Applied Geochemistry*, **5**: 657–668.

Longstaffe, F.J., and Ayalon, A., 1991. Mineralogical and O-isotope studies of diagenesis and pore water evolution in continental sandstones, Cretaceous Belly River Group, Alberta, Canada. *Applied Geochemistry*, **6**: 291–303.

Longstaffe, F.J., and Schwarcz, H.P., 1977. $^{18}O/^{16}O$ of Archean clastic metasedimentary rocks: a petrogenetic indicator for Archean gneisses? *Geochimica et Cosmochimica Acta*, **41**: 1303–1312.

Luz, B., and Barkan, E., 2000. Assessment of oceanic productivity with the triple-isotope composition of dissolved oxygen. *Science*, **288**: 2028–2031.

McDermott, F., Mattey, D.P., and Hawkesworth, C., 2001. Centennial-scale Holocene climate variability revealed by a high-resolution speleothem $\delta^{18}O$ record from SW Ireland. *Science*, **294**: 1328–1331.

McKay, J.L., and Longstaffe, F.J., 2002. Sulphur isotope geochemistry of pyrite from the Upper Cretaceous Marshybank Formation, Western Interior Basin. *Sedimentary Geology*. (in press).

McKay, J.L., Longstaffe, F.J., and Plint, A.G., 1995. Early diagenesis and its relationship to depositional environment and relative sea-level fluctuations (Upper Cretaceous Marshybank Formation, Alberta and British Columbia). *Sedimentology*, **42**: 161–190.

Mizota, C., and Longstaffe, F.J., 1996. Origin of Cretaceous and Oligocene kaolinites from the Iwaizumi clay deposit, Iwate, Northeastern Japan. *Clays and Clay Minerals*, **44**: 408–416.

Mortimer, R.J.G., and Coleman, M.L., 1997. Microbial influence on the oxygen isotopic composition of diagenetic siderite. *Geochimica et Cosmochimica Acta*, **61**: 1705–1711.

Morton, J.P., 1985. Rb/Sr evidence for punctuated illite/smectite diagenesis in the Oligocene Frio Formation, Texas, Gulf Coast. *Geological Society of America Bulletin*, **96**: 1043–1049.

Mozley, P.S., and Burns, S.J., 1993. Oxygen and carbon isotopic composition of marine carbonate concretions: an overview. *Journal of Sedimentary Petrology*, **63**: 73–83.

Muehlenbachs, K., and Clayton, R.N., 1976. Oxygen isotope composition of the oceanic crust and its bearing on seawater. *Journal of Geophysical Research*, **81**: 4365–4369.

Nishiizumi, K. *et al.*, 1993. Role of *in situ* cosmogenic nuclides ^{10}Be and ^{26}Al in the study of diverse geomorphic processes. *Earth Surface Processes and Landforms*, **18**: 407–425.

Norris, R.D., Bice, K.L., Magno, E.A., and Wilson, P.A., 2002. Jiggling the tropical thermostat in the Cretaceous hothouse. *Geology*, **30**: 299–302.

Ohr, M., Halliday, A.N., and Peacor, D.R., 1991. Sr and Nd isotopic evidence for punctuated clay diagenesis. Texas Gulf Coast. *Earth and Planetary Science Letters*, **105**: 110–126.

Patchett, P.J., Ross, G.M., and Gleason, J.D., 1999. Continental drainage in North America during the Phanerozoic from Nd isotopes. *Science*, **283**: 671–673.

Petit, J.R. *et al.*, 1999. Climate and atmospheric history of the past 420,000 years from the Vostok ice core, Antarctica. *Nature*, **1999**: 429–436.

Pevear, D.R., 1992. Illite age analysis, a new tool for basin thermal history analysis. In Kharaka, Y.K., and Maest, A.S. (eds.), *Water-Rock Interaction, Volume 2*. Rotterdam: A.A. Balkema, pp. 1251–1254.

Phillips, F.M., Zreda, M.G., Flinsch, M.R., Elmore, D., and Sharma, P., 1996. A reevaluation of cosmogenic ^{36}Cl production rates in terrestrial rocks. *Geophysical Research Letters*, **23**: 949–951.

Piepgras, D.J., Wasserburg, G.J., and Dasch, E.J., 1979. The isotopic composition of Nd in different ocean masses. *Earth and Planetary Science Letters*, **45**: 223–236.

Savin, S.M., and Epstein, S.M., 1970a. The oxygen and hydrogen isotope geochemistry of clay minerals. *Geochimica et Cosmochimica Acta*, **34**: 25–42.

Savin, S.M., and Epstein, S.M., 1970b. The oxygen and hydrogen isotope geochemistry of ocean sediments and shales. *Geochimica et Cosmochimica Acta*, **34**: 43–63.

Savin, S.M., and Hsieh, J.C.C., 1998. The hydrogen and oxygen isotope geochemistry of pedogenic clay minerals: principles and theoretical background. *Geoderma*, **82**: 227–253.

Savin, S.M., and Lee, M., 1988. Isotopic studies of phyllosilicates. In Bailey, S.W. (ed.), *Hydrous Phyllosilicates (Exclusive of Micas)*. Mineralogical Society of America, Reviews in Mineralogy, 19, pp. 189–223.

Savin, S.M., and Yeh, H.-W., 1981. Stable isotopes in ocean sediments. In Emiliani, C. (ed.), *The Sea, Volume 7, The Oceanic Lithosphere*. New York: John Wiley and Sons, pp. 1521–1554.

Schwarcz, H.P., 1986. Geochronology and isotopic geochemistry of speleothems. In Fritz, P., and Fontes, J.Ch. (eds.), *Handbook of Environmental Isotope Geochemistry, Volume 2, The Terrestrial Environment, B*. Amsterdam: Elsevier Science Publishing Company, pp. 271–303.

Shackleton, N.J., 1968. Depth of pelagic foraminifera and isotopic changes in Pleistocene oceans. *Nature*, **218**: 79–80.

Shaw, H.F., and Wasserberg, G.J., 1985. Sm–Nd in marine carbonates and phosphates: implications for Nd isotopes in seawater and crustal ages. *Geochimica et Cosmochimica Acta*, **49**: 503–518.

Sheppard, S.M.F., and Gilg, H.A., 1996. Stable isotope geochemistry of clay minerals. *Clay Minerals*, **31**: 1–24.

Sheppard, S.M.F., Nielson, R.L., and Taylor, H.P. Jr., 1969. Oxygen and hydrogen isotope ratios of clay minerals from porphyry copper deposits. *Economic Geology*, **64**: 755–777.

Spötl, C., Kunk, M.J., Ramseyer, K., and Longstaffe, F.J., 1998. Authigenic potassium feldspar: a tracer for the timing of palaeofluid flow in carbonate rocks, Northern Calcareous Alps, Austria. In Parnell, J. (ed.), *Dating and Duration of Fluid Flow and Fluid-Rock Interaction*. Geological Society, Special Publications, 144, 107–128.

Stanton, R.J., Jeffery, D.L., and Ahr, W.M., 2002. Early Mississippian climate based on oxygen isotope compositions of brachiopods, Alamogordo Member of the Lake Valley Formation, South-central New Mexico. *GSA Bulletin*, **114**: 4–11.

Stern, L.A., Chamberlain, C.P., Reynolds, R.C., and Johnson, G.D., 1997. Oxygen isotope evidence of climate change from pedogenic clay minerals in the Himalayan molasse. *Geochimica et Cosmochimica Acta*, **61**: 731–744.

Stone, J.O.H., Evans, J.M., Fifield, L.K., Allan, G.L., and Cresswell, R.G., 1998. Cosmogenic chlorine-36 production in calcite by muons. *Geochimica et Cosmochimica Acta*, **62**: 433–454.

Swart, P.K., Lohmann, K.C., McKenzie, J., and Savin, S. (eds.), 1993. *Climate Change in Continental Isotopic Records*. American Geophysical Union, Geophysical Monograph 78.

Tilley, B.J., and Longstaffe, F.J., 1989. Diagenesis and isotopic evolution of pore waters in the Alberta Deep Basin: the Falher member and Cadomin formation. *Geochimica et Cosmochimica Acta*, **53**: 2529–2546.

Unterschutz, J.L.E., Creaser, R.A., Erdmer, P., Thompson, R.I., and Daughtry, K.L., 2002. North American margin origin of Quesnel terrane strata in the southern Canadian Cordillera: inferences from geochemical and Nd isotopic characteristics of Triassic metasedimentary rocks. *GSA Bulletin*, **114**: 462–475.

Vanden Bygaart, A.J., and Protz, R., 2001. Bomb-fallout ^{137}Cs as a marker of geomorphic stability in dune sands and soils, Pinery Provincial Park, Ontario, Canada. *Earth Surface Processes and Landforms*, **26**: 689–700.

Veizer, J., 1989. Strontium isotopes in seawater through time. *Annual Reviews in Earth and Planetary Science*, **17**: 141–167.

Veizer, J., Holser, W.T., and Wilgus, C.K., 1980. Correlation of $^{13}C/^{12}C$ and $^{34}S/^{32}S$ secular variations. *Geochimica et Cosmochimica Acta*, **44**: 579–587.

Vengosh, A., Kolodny, Y., Starinsky, A., Chivas, A.R., and McCulloch, M.T., 1991. Coprecipitation and isotopic fractionation of boron in modern biogenic carbonates. *Geochimica et Cosmochimica Acta*, **55**: 2901–2910.

Vitali, F., Longstaffe, F.J., Bird, M.I., Gage, K.L., and Caldwell, W.B.E., 2001. Hydrogen-isotope fractionation in aluminum hydroxides: synthesis products versus natural samples from bauxites. *Geochimica et Cosmochimica Acta*, **65**: 1391–1398.

Wadleigh, M.A., and Veizer, J., 1992. $^{18}O/^{16}O$ and $^{13}C/^{12}C$ in Lower Paleozoic articulate brachiopods: implications for the isotopic composition of seawater. *Geochimica et Cosmochimica Acta*, **56**: 431–443.

Wang, L., Ku, T.L., Lui, S., Southon, J.R., and Kusakabe, M., 1996. $^{26}Al-^{10}Be$ systematics in deep-sea sediments. *Geochimica et Cosmochimica Acta*, **60**: 109–119.

Williams, L.B., Hervig, R.L., and Hutcheon, I., 2001. Boron isotope geochemistry during diagenesis. Part II. applications to organic-rich sediments. *Geochimica et Cosmochimica Acta*, Volume **65**: 1783–1794.

Yapp, C.J., 1998. Paleoenvironmental interpretations of oxygen isotope ratios in oolitic ironstones. *Geochimica et Cosmochimica Acta*, **62**: 2409–2420.

Yechielli, Y., Ronen, D., and Kaufman, A., 1996. The source and age of groundwater brines in the Dead Sea Area, as deduced from ^{36}C and ^{14}C. *Geochimica et Cosmochimica Acta*, **60**: 1909–1916.

Yeh, H.-W., and Epstein, S., 1978. Hydrogen isotope exchange between clay minerals and sea water. *Geochimica et Cosmochimica Acta*, **42**: 140–143.

Yeh, H.-W., and Eslinger, E.V., 1986. Oxygen isotopes and the extent of diagenesis of clay minerals during sedimentation and burial in the sea. *Clays and Clay Minerals*, **34**: 403–406.

Yeh, H.-W., and Savin, S.M., 1977. Mechanism of burial metamorphism of argillaceous sediments: 3. O-isotope evidence. *Geological Society of America Bulletin*, **88**: 1321–1330.

Yurtsever, Y., and Gat, J.R., 1981. Atmospheric waters. In Gat, J.R., and Gonfiantini, R. (eds.), *Stable Isotope Hydrology: Deuterium and Oxygen-18 in the Water Cycle*. Vienna: IAEA, Technical Report Series, 210, pp. 103–142.

Zhu, X.-K., O'Nions, K., Guo, Y., and Reynolds, B.C., 2000. Secular variation of iron isotopes in North Atlantic deep water. *Science*, **287**: 2000–2002.

Ziegler, K., and Longstaffe, F.J., 2000a. Clay mineral authigenesis along a mid-continental scale fluid conduit in Paleozoic sedimentary rocks from southern Ontario, Canada. *Clay Minerals*, **35**: 239–260.

Ziegler, K., and Longstaffe, F.J., 2000b. Multiple episodes of clay alteration at the Precambrian/Paleozoic unconformity, Appalachian basin: isotopic evidence for long-distance and local fluid migrations. *Clays and Clay Minerals*, **48**: 474–493.

Cross-references

Cements and Cementation
Climatic Control of Sedimentation
Diagenesis
Mutation, Organic
Porewaters in Sediments
Provenance
Seawater
Tectonic Controls of Sedimentation

K

KAOLIN GROUP MINERALS

Kaolin is a name given to a group of phyllosilicate minerals whose layers have a 1:1 structure with a composition of $Al_2Si_2O_5(OH)_4$. The 1:1 designation indicates the individual layers are composed of sheets of corner sharing tetrahedra in a distorted hexagonal arrangement and a sheet of edge sharing octahedra. The tetrahedra are occupied by silicon atoms and the octahedra are occupied by aluminum atoms in a dioctahedral manner. The join between the two sheets is effected by sharing of what would be the apical oxygen atoms of the tetrahedra with atoms of the octahedra. This creates an asymmetrical layer since one of the surfaces of the layer consists of basal oxygen atoms of the tetrahedral sheet while the opposite surface is populated by hydroxyl groups. The stacking of the layers and the nature of the material, if any, between the layers generate the individual minerals in the group.

The bonding between the layers is due to hydrogen bonding involving hydrogen atoms of one layer and the basal oxygen atoms of the adjacent layer (Hendricks, 1939). To this should be added the interfacial interactions of the layer surfaces. The geometric requirement necessary to create a number of different stable stackings is to enable the hydroxyls (the hydrogen bond donors) of one surface to be close to the basal oxygen atoms (the hydrogen bond acceptors) of the adjacent layer. Newnham (1961) analyzed this and incorporated other factors such as the superposition of cations of the two layers that would be a destabilizing contribution. He determined that there were theoretically 36 possible stackings of the layers taken two at a time. When considering 1-layer structures, two of the stackings were judged to be most stable. These correspond to a right and left-handed kaolinite structure. For two layer structures only 2 of the 36 were likely because the other stackings were destabilized by other factors such as tetrahedral rotation and tilting and the distortions of the individual layers. These two correspond to dickite and nacrite.

Thus, of the large number of possible stackings of the 1:1 kaolin layers, only three are observed. These are kaolinite, dickite and nacrite. There is normally a fourth mineral added to the kaolin group; halloysite, a hydrated (interlayer water) version. Of these four minerals, kaolinite is by far the most common and, along with halloysite, the only mineral that is commercially exploited.

Structures

Kaolin minerals, in common with many clay minerals, occur only as small particles so that traditional single crystal diffraction studies are not possible. There are occasions when one finds what appear to be macroscopic crystals, but these are always either not single crystals or they are highly disordered and not suitable.

Kaolinite derives its name from a locality in China where it was first mined (Grim, 1968). Early attempts to understand the structure of these minerals were based on X-ray powder diffraction experiments. This is difficult because of the low symmetry. Hendricks (1936) observed the characteristic 7.1Å basal spacing. Knowing the thickness of the layer structure allowed Pauling (1930) to outline the architecture of the kaolinite structure. There were a number of attempts to refine the structure suggested by Pauling. Initially, the structure was thought to be monoclinic (Gruner, 1932; Hendricks, 1936), but later Brindley and Robinson (1945) demonstrated that the unit cell is really triclinic and this allowed them to determine the positions of the non-hydrogen atoms. The structure is based on a simple shift of $\sim -a/3$ where a is the unit cell axis. The vacant octahedral site is the same in each of the layers.

The hydrogen atoms clearly are extremely important in the interlayer bonding of these minerals so there has been great interest in locating their positions so as to understand the interlayer hydrogen bonding. Various attempts were made to infer the hydroxyl orientations using infrared absorption spectroscopy (Farmer, 1964). There were great problems in these efforts. Later, potential energy calculations were used to locate the OH orientations that corresponded to the minimum energy (Giese and Datta, 1973; Giese, 1982).

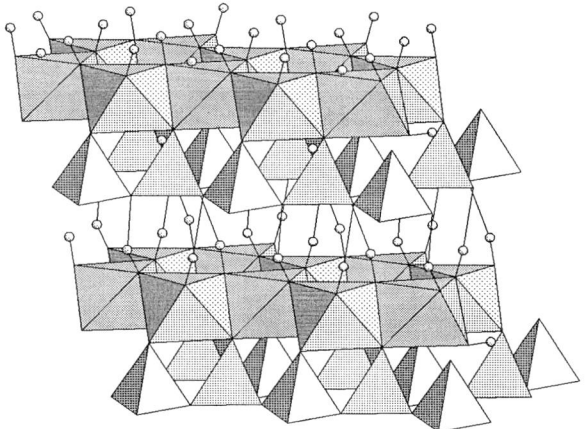

Figure K1 Clinographic view of the kaolinite structure showing two 1:1 layers. The hydrogen atoms are shown as small circles and the hydrogen bonding is indicated by the longer solid lines to oxygen atoms in the adjacent layer (drawing produced with the program ATOMS, by Shape Software, www.shapesoftware.com).

The kaolinite structure has been refined several times since then with increasing accuracy aided by improvements in the refinement of structures based largely on powder diffraction data (see Giese, 1988, for a discussion). Oddly enough, refinements still seem to be in progress even though there seems little justification for such work.

Dickite is named after the mineralogist (Dick, 1888: see Newnham and Brindley, 1956). The major features of the dickite structure were determined by Newnham and Brindley (1956) and there are a number of modern refinements. The stacking is $\sim -a/3$ as for kaolinite. The dickite structure differs from that of kaolinite in that the vacant octahedral site alternates its position in adjacent layers giving a two-layer structure. It is possible to find small but useful single crystals of dickite and these have eased the search for the hydrogen atoms.

Nacrite was named for its nacreous appearance. The structure is different from those of kaolinite and dickite (Blount et al., 1969). The stacking between layers follows the 8.9 Å a axis and there is a rotation of the 1:1 layer by 180° between successive layers. The unit cells and space groups can be found in recent refinements of the structures; viz. Bish and von Dreele (1989) for kaolinite, Joswig and Drits (1986) for dickite and Zheng and Bailey (1994) for nacrite.

Defect structure

It was noticed early on that there were wide variations in the X-ray diffraction patterns of kaolinite samples from different localities. These varied from patterns with narrow Bragg reflections to those in which many of the typical kaolinite diffraction bands were diffuse and there was typically a high background between peaks. This was correctly interpreted as some sort of stacking disorder. The peaks most affected had indices with $k \neq 3$. This suggested that the disorder was due to random shifts of $\pm b/3$ (Brindley and Robinson, 1945). It was also pointed out that rotations of $\pm 2\pi/3$ would result in a shift in the vacant octahedral site from layer to layer while maintaining the interlayer hydrogen bonding (Murray, 1954). The variations in the diffraction patterns were of some use in determining the physical properties of kaolinites, so there were several attempts to formulate an empirical estimate of the degree of disorder. The most widely used index was that of Hinckley (1963). It should be noted that Hinckley used the term "crystallinity" in the title of the paper placed in quotes, presumably to indicate that he was not using the term in strict sense of the word. Unfortunately, subsequent usage has emphasized the normal meaning of crystallinity. Typical values of the Hinckley index range from roughly 0.2 to 1.7 with the larger values corresponding to a low level of structural defects. Recent work has shown that for values of the index greater than 0.43, the samples are mixtures of two kinds of kaolinite, one having few defects and the other having a high concentration of defects (Plancon and Giese, 1988, Bookin et al., 1989, Plancon et al., 1989).

Chemistry

The kaolin minerals in general are reasonably pure with small amounts of iron and titanium. The chemistry is somewhat clouded because kaolinite is now known to occur with interstratified smectite. The admixed smectite introduces alkali and alkaline earth elements, increases the surface area and accounts for the small amount of water often found in samples. Pure kaolinites have small surface areas (approximately $<20\,m^2/g$ and very modest cation and anion exchange capacities. Typical chemical analyzes are shown below (Murray and Keller, 1993).

Component	Georgia kaolinite	England kaolinite	Theoretical
SiO_2	45.30	46.77	46.55
Al_2O_3	38.38	37.79	39.50
Fe_2O_3	0.30	0.36	—
TiO_2	1.44	0.02	—
MgO	0.25	0.24	—
CaO	0.05	0.13	—
Na_2O	0.27	0.05	—
K_2O	0.04	1.49	—
LOI (water)	13.97	12.97	13.96

Common mineral impurities in kaolinite deposits are anatase, smectite, and ilmenite.

Recent measurements of the enthalpy of formation of the kaolin minerals has allowed more accurate determinations of the standard free energy of formation (de Ligney and Navrotsky, 1999). The values show that kaolinite is the stable mineral while the other forms are metastable. Further, the variation in the standard free energy of formation does not vary measurably as a function of the defect structure.

Origin and occurrence

There are two types of kaolin deposit; primary and secondary. The primary kaolins are derived from feldspar and muscovite in granite or rhyolite under conditions of high rainfall, moderate to high temperatures above the water table. The large amounts of water are necessary to remove the potassium and prevent the formation of illite rather than kaolin. A higher temperature regime can result in hydrothermal kaolins. If the kaolin remains in place, the deposit is termed residual. On the other hand if the kaolin is transported by running water and deposited elsewhere, one has a secondary deposit. The formation of kaolinite is favored by low pH and low concentrations of alkali ions as shown in Faure (1998).

Hydrothermal deposits are known in China, Japan, and Mexico. These deposits are exploited primarily for ceramic clay. Residual deposits are found in Australia, Czechoslovakia, Germany and Indonesia. These are used in paper filling and ceramic manufacturing. Deposits which were formed from a combination of hydrothermal and chemical weathering are found in Cornwall, England and New Zealand. The Cornwall kaolinite is particularly well-ordered and pure, and is extensively used in paper coating. There are large deposits of sedimentary kaolin in Australia, Brazil, Germany, Spain, and the United States, principally in Georgia and South Carolina (Murray and Keller, 1993). There are many occurrences of kaolinite "books" in sandstones where they have formed under diagenetic conditions.

Rossman F. Giese, Jr.

Bibliography

Bish, D.L., and von Dreele, R., 1989. Rietveld refinement of the crystal structure of kaolinite. *Clays and Clay Minerals*, **37**: 289–296.
Blount, A.M., Threadgold, I.M., and Bailey, S.W., 1969. Refinement of the crystal structure of nacrite. *Clays and Clay Minerals*, **17**: 185–194.
Bookin, A.S., Drits, V.A., Plancon, A., and Tchoubar, C., 1989. Stacking faults in kaolin-group minerals in the light of real structure features. *Clays and Clay Minerals*, **37**: 297–307.
Brindley, G.W., and Robinson, K., 1945. Structure of kaolinite. *Nature*, **156**: 661–663.
Farmer, V.C., 1964. Infrared absorption of hydroxyl groups in kaolinite. *Science*, **145**: 1189–1190.
Faure, G., 1998. *Principles and Applications of Geochemistry*, 2nd edn. Prentice Hall.
Giese, R.F., 1982. Theoretical studies of the kaolin minerals: electrostatic calculations. *Bulletin Mineralogy*, **105**: 417–424.
Giese, R.F., 1982. Kaolin minerals: structures and stabilities. Chapter 3. In Bailey, S.W. (ed.), *Hydrous Phyllosilicates*. Reviews in Mineralogy, **19**: 29–66.
Giese, R.F., 1988. Kaolin minerals: structures and stabilities. In Bailey, S.W. (ed.), *Hydrous Phyllosilicates (exclusive of micas)*. Mineralogical Society of America, pp. 29–66.
Giese, R.F., and Datta, P., 1973. Hydroxyl orientations in kaolinite, dickite, and nacrite. *American Mineralogist*, **58**: 471–479.
Grim, R.E., 1968. *Clay Mineralogy*. McGraw-Hill.
Gruner, J.W., 1932. The crystal structure of kaolinite. *Zeitschrift für Kristallographie*, **83**: 75–88.
Hendricks, S.B., 1936. Concerning the crystal structure of kaolinite, $Al_2O_3 \cdot 2SiO_2 \cdot 2H_2O$ and the composition of anauxite. *Zeitschrift für Kristallographie*, **95**: 247–252.
Hendricks, S.B., 1939. The crystal structure of nacrite $Al_2O_3 \cdot 2SiO_2 \cdot 2H_2O$ and the polymorphism of the kaolin minerals. *Zeitschrift für Kristallographie*, **100**: 509–518.
Hinckley, D.N., 1963. Variability in "crystallinity" values among the kaolin deposits of the coastal plain of Georgia and South Carolina. *Clays and Clay Minerals*, **11**: 229–235.
Joswig, W., and Drits, V.A., 1986. The orientation of the hydroxyl groups in dickite by X-ray diffraction. *Neues Jahrbuch für Mineralogie Mh.*, **H1**: 19–22.
de Ligny, D., and Navrotsky, A., 1999. Energetics of kaolin polymorphs. *American Mineralogist*, **84**: 506–516.
Murray, H.H., 1954. Structural variations of some kaolinites in relation to dehydrated halloysite. *American Mineralogist*, **39**: 97–108.
Murray, H.H., and Keller, W.D., 1993. Kaolins, kaolins and kaolins. In Murray, H.H., Bundy, W.M., and Harvey, C.C. (eds.), *Kaolin Genesis and Utilization*. The Clay Minerals Society, pp. 1–24.
Newnham, R.E., 1961. A refinement of the dickite structure and some remarks on polymorphism in kaolin minerals. *Mineralogical Magazine*, **32**: 683–704.
Newnham, R.E., and Brindley, G.W., 1956. The crystal structure of dickite. *Acta Crystallographica*, **9**: 759–764.
Pauling, L., 1930. The structure of chlorites. *Proceedings of the National Academy of Sciences U.S.A.*, **16**: 578–582.
Plancon, A., and Giese, R.F., 1988. The hinckley index for kaolinites. *Clay Minerals*, **23**: 249–260.
Plancon, A., Giese, R.F., Snyder, R., Drits, V.A., and Bookin, A.S., 1989. Stacking faults in the kaolin-group minerals defect structures of kaolinite. *Clays and Clay Minerals*, **37**: 203–210.
Zheng, H., and Bailey, S.W., 1994. Refinement of the nacrite structure. *Clays and Clay Minerals*, **42**: 46–52.

Cross-references

Clay Minerals
Diagenesis
Weathering, Soils and Paleosols

KEROGEN

Kerogen is the solid, high molecular-weight fraction of sedimentary organic matter (OM) that is insoluble in organic solvents (e.g., chloroform) as opposed to the soluble fraction called bitumen. Kerogen is a macromolecule of condensed cyclic nuclei linked by heteroatomic bonds or aliphatic chains. The importance of differentiating the soluble and insoluble fractions of OM in standard organic solvents is based on the origin and formation of petroleum in sediments. The soluble fraction contains "free" hydrocarbons (HC) formed during geological processes, whereas the insoluble fraction has large proportions that are liable to form HCs (Durand, 1980). The insoluble nature is linked to condensation of OM and thus represents polycondensed or polymerized states of OM in contrast to the biogenic polymers of the living matter.

Introduction

Kerogen has been defined in several studies, which are summarized in Durand (1980). Durand (1980) proposed to call kerogen sensu stricto the fraction of sedimentary OM that is insoluble in standard organic solvents. The distribution of kerogen in the earth's crust is estimated to be 10^{16}t, with most of the kerogen occurring in the fine silt or clay fraction deposited on continent shelf and slope (Figure K2). Kerogen makes up the largest part of the total organic matter and can represent more than 95 percent of the weight of recent OM. Kerogen content is generally estimated on the basis of organic carbon content of the sediment, which yields often an underestimated value that has to be corrected by a factor of 1.5 or 1.6 for slightly evolved sediments. The proportion of kerogen progressively changes during burial and the formation of soluble and/or volatile products, such as HCs, which makes it often difficult to estimate the average proportion of kerogen of the total sedimentary OM (Durand, 1980).

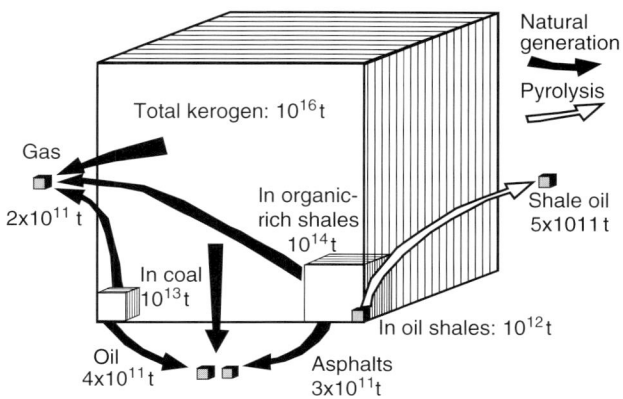

Figure K2 Comparison of kerogen abundance in the earth's crust and the ultimate resources of fossil fuels. Modified after Durand (1980).

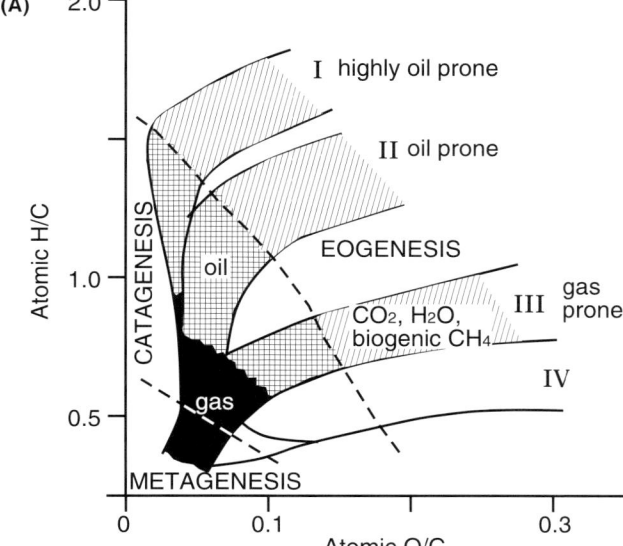

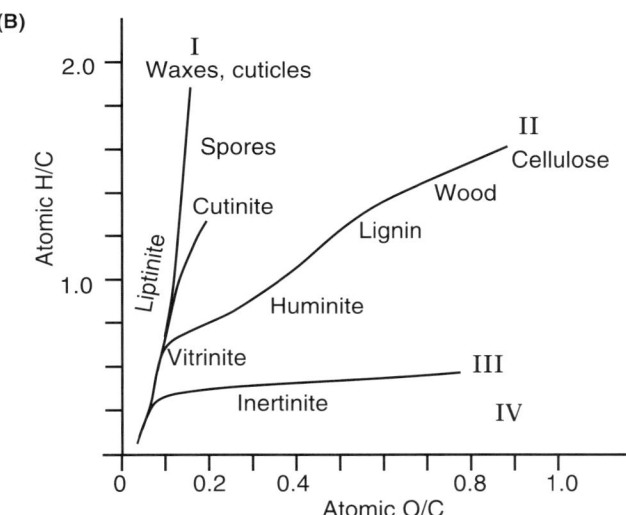

Figure K3 (A) Kerogen evolution illustrated on a van Krevelen diagram. Several evolution stages occur during progressive diagenesis, which have different principal HC production depending on original type of kerogen. During progressive diagenesis, kerogens become enriched in carbon. Note that type IV organic matter does not have an evolution path. Modified after Tissot and Welte (1978). (B) van Krevelen diagram of the diagenetic pathway of coal macerals (kerogen types determined on morphology). During progressive diagenesis, the three maceral groups (liptinite, inertinite, huminite) become enriched in carbon. The diagram shows the changes in terms of H/C and O/C ratios. Modified after Bustin et al. (1983).

Historical development

Scientists studying the oil-shales of the Lothians of Scotland end of the 19th century named the oil-yielding, waxy part of OM "kerogen" (from Greek keros = wax) or "pyrobitumen" (Bailey, 1927, p.158). Kerogen and kerogen maturation received much attention throughout the 20th century, because of its importance for fossil fuels. Numerous methods identifying kerogen type and kerogen maturity have been developed and are summarized in Barnes et al. (1990), Bustin et al. (1983), Bustin et al. (1990), Durand (1980), Hunt (1979), Peters (1986), Taylor et al. (1998), and Tissot and Welte (1984).

Characterization of kerogen

Accumulated organic matter in sedimentary deposits provides the potential source for kerogen and hence HC formation during diagenesis. The kerogen concentration, composition, and state of maturation are important parameters for the formation of oil and gas. The OM in sedimentary deposits derives from either living or dead particulate OM or dissolved OM. Amount and type of kerogen will determine the amount of oil or gas that has been or will be generated from OM in sedimentary deposits.

Kerogen type is commonly determined either by optical (petrography) or chemical methods. Petrographic analyzes allow morphological classification of OM of various depositional environments, such as marine or lacustrine deposits, and the level of maturation, while geochemical analysis allow the classification based on chemical properties. In coal geology, three main groups of *macerals* (kerogen constituents of organic matter) are distinguished: liptinite, huminite/vitrinite, and inertinite (Figure K3). During progressive diagenesis, these maceral groups become enriched in carbon but each component follows a predetermined and distinct coalification path (Figure K3). Vitrinite reflectance (Ro percent) is widely used as a parameter of the digenetic level of kerogen (see *Maturation, Organic*).

In petroleum geochemistry, four kerogen types (I to IV) are commonly differentiated (Figure K3). All kerogen types are determined solely on H-, O-, and C-content. Type I, consisting of mainly algal and amorphous kerogen that is highly likely to generate light oil; Type II, mixed terrestrial and marine material (algae, spores, cuticles) that can generate waxy oil and some gas; Type III, woody terrestrial source material that typically generates gas; and Type IV, alsocalled residual or refractory kerogen, composed of resedimented, oxidized OM, that has no potential for oil and gas generation (Tissot and Welte, 1984).

Diagenesis of kerogen

Upon deposition and burial of sedimentary deposits, OM (e.g., biogenic polymers, such as proteins, carbohydrates) is transformed (biochemical degradation) and polycondensed structures (geopolymers) are formed which are the precursors of kerogen. The initial stage of diagenesis, called *eogenesis* (Barnes *et al.*, 1990), includes processes of bacterial oxidation, sulfate reduction and fermentation. The most important HC formed during this stage is CH_4 (Figures K3, K4). Oxygen elimination of kerogen takes place and results in the formation of CO_2, H_2O, and some heteroatomic (N, S, O)-petroleum compounds of high-molecular weight (e.g., asphaltenes and resins) are liberated. Most carboxyl groups are removed (Barnes *et al.*, 1990; Bustin *et al.*, 1990). Type III kerogen may produce some gas (Figure K3).

Increasing burial and progressive diagenesis (increase in temperature and pressure) of the sediments results in the production of liquid petroleum from kerogen. During this stage of thermal maturation, called *catagenesis*, both liquid oil and condensate, including methane, are formed (wet gas phase) (Figure K4). HC molecules, such as aliphatic chains, are produced from kerogen and cracking results in the production of light HCs. Because of the chemical structures present, immature kerogens that are hydrogen-rich will generate crude oil and hydrogen-poor kerogens will yield only gas. As burial depth and temperature increase, the residual kerogen becomes depleted in H (labile compounds) and has an atomic H/C ratio of about 0.5 by the end of the catagenesis (Tissot and Welte, 1984).

Further increase in pressure and temperature leads to the stage of *metagenesis*. During this stage, kerogen is structurally reorganized and a higher degree of ordering is obtained. Only minor amounts of methane are formed from the residual kerogen during the last stage of diagenesis.

Determining level of kerogen maturity

With increasing levels of organic diagenesis, the color of kerogen in transmitted light changes progressively and several methods have utilized such characteristics for kerogen classification, for example the *thermal alteration index*, a numeric scale used to classify the color changes of palynomorphs. Because of the widespread abundance of palynomorphs in sedimentary deposits, optical determination of the level of kerogen maturity is seldom a problem. Another optical method is fluorescence microscopy, which determines the irradiation level of liptinite components (algae, spores, resin, cutin). In coals, similar changes are recorded by analyzing the *vitrinite reflectance* (R_o). The vitrinite reflectance is commonly used to classify coal rank and OM maturity level (see *Maturation, Organic*).

Classification of kerogen by chemical means is limited to the pre and early oil generation maturities because generation and expulsion of hydrogen-rich oil and gas alters the hydrogen content of the original kerogen (Figure K3). Numerous methods are used for determining the geochemistry of kerogen. The most important methods are: pyrolysis, ultimate analysis, infrared spectroscopy, electron paramagnetic resonance, carbon isotope studies, proximate analysis, caloric value, etc., (Bustin *et al.*, 1990; Garcette-Lepecq *et al.*, 2000). *Pyrolysis* or progressive heating of the sample up to 550°C, utilizing Rock-Eval® instruments (see Peters, 1986), measures three types of gases: (1) the HCs residual in the sample (S_1); (2) HCs generated during thermal cracking between 300°C and 550°C (S_2, T_{max}), and (3) CO_2 (S_3). This allows determination of the level of organic diagenesis and type of HCs the kerogen will yield. This method is widely used despite the fact that some surface samples tend to be prone to alterations and thus erroneous T_{max} (see Copard *et al.*, 2002).

Ultimate analysis determines elemental concentrations of C, H, and O and H/C and O/C ratios are plotted against each other in a *van Krevelen* diagram. van Krevelen diagrams allow characterization of kerogen type and maturity level (Figure K3).

Infrared spectroscopy quantifies the main functional groups. During progressive diagenesis of OM, carboxyl, carbonyl, and saturated HCs decrease and aromatic C-H groups are formed and removed (Bustin *et al.*, 1990). Kerogen type, the precursor of the functional groups as well as diagenetic stage can be determined.

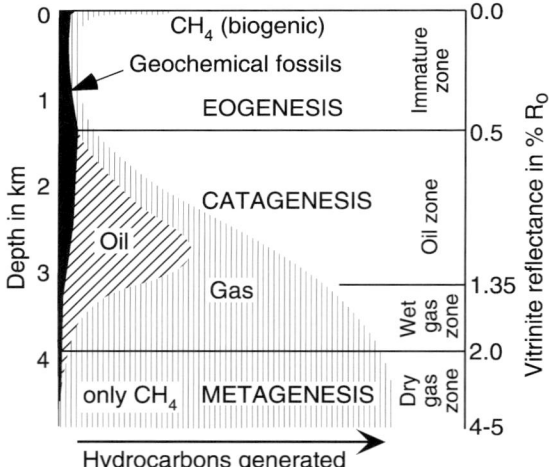

Figure K4 Hydrocarbon generation changes in sediments during their evolution of progressive diagenesis. Vitrinite reflectance (R_o %) corresponds to the degree of organic matter maturation and is dependent on type of kerogen, burial history and geothermal gradient. Modified after Bustin *et al.* (1990).

Summary

Kerogen is the fraction of OM in sedimentary deposits that is insoluble in standard organic solvents. It is a macromolecule of condensed cyclic nuclei linked by heteroatomic bonds or aliphatic chains. Different types of kerogen (I to IV) occur depending on the composition of the organic matter, most of which generate oil and gas upon diagenesis. Burial of the sediments, including increasing temperature and pressure, results in the rearrangement of kerogen. During eogenesis, elimination of heteroatomic bonds and functional groups occurs and CO_2, H_2O and some heavy N, S, O compounds are released, while CH_4 is produced. During catagenesis, HC chains and cycles are eliminated and crude oil and gas are formed. During metagenesis, aromatic sheets of kerogen are rearranged and form clusters while some gas is generated. Progressive changes of HCs occur with progressive diagenesis. Whilst biomolecules are dominant during eogenesis, medium to low molecular weight HCs become predominant during catagenesis and CH_4 is practically the remainder during metagenesis.

Raphael A.J. Wüst and R. Marc Bustin

Bibliography

Bailey, E.M., 1927. The chemistry of the oil-shales. In Carruthers, R.G., (ed.), *The Oil-shales of the Lothians*. Geological Survey of Great Britain, Memoirs, pp. 158–239.

Barnes, M.A., Barnes, W.C., and Bustin, R.M., 1990. Chemistry and diagenesis of organic matter in sediments and fossil fuels. In McIlreath, I.A., and Morrow, D.W. (eds.), *Diagenesis*. Geological Association of Canada, pp. 189–204.

Bustin, R.M., Barnes, M.A., and Barnes, W.C., 1990. Determining levels of organic diagenesis in sediments and fossil fuels. In McIlreath, I.A., and Morrow, D.W. (eds.), *Diagenesis*. Geological Association of Canada, pp. 205–226.

Bustin, R.M., Cameron, A.R., Grieve, D.A., and Kalkreuth, W.D., 1983. *Coal Petrology: Its Principles, Methods, and Applications* Canada, Short Course Notes 3. Geological Association of Canada, 230p.

Copard, Y., Disnar, J.-R., and Becq-Giraudon, J.F., 2002. Erroneous maturity assessment given by Tmax and HI Rock-Eval parameters on highly mature weathered coals. *International Journal of Coal Geology*, **49** (1): 57–65.

Durand, B., 1980. *Kerogen. Insoluble Organic Matter from Sedimentary Rocks*. Paris: Éditions Technip.

Garcette-Lepecq, A., Derenne, S., Largeau, C., Bouloubassi, I., and Saliot, A., 2000. Origin and formation pathways of kerogen-like organic matter in recent sediments off the Danube delta (Northwestern Black Sea). *Organic Geochemistry*, **31**: 1663–1683.

Hunt, J.M., 1979. *Petroleum Geochemistry and Geology*. W.H. Freeman.

Peters, K.E., 1986. Guidelines for evaluating petroleum source rock using programmed pyrolysis. *AAPG Bulletin*, **70** (3): 318–329.

Taylor, G.H., Teichmüller, M., Davis, A., Diessel, C.F.K., Littke, R., and Robert, P., 1998. *Organic Petrology*. Berlin: Gebrüder Bornträger.

Tissot, B.P., and Welte, D.H., 1978. Petroleum formation and occurrence. *A new approach to oil and gas exploration*, 2nd edn. Springer-Verlag. 1984.

Cross-references

Diagenesis
Maturation, Organic

L

LACUSTRINE SEDIMENTATION

The lake is a system, that is, a bounded entity possessing a set of unique, interconnected elements, especially of process and response, that can be recognized as such regardless of how they are dealt with by scientific enquiry (Strahler, 1980). Further, the lake is an open system; processes beyond the boundaries relate to those within the system to produce mass and energy exchange across the boundaries. For a lake in equilibrium with its environment, the components of these exchanges lead, over the long term, to a zero-sum result; what goes in must come out. While that is generally true of energy and of water, sediment flux is normally not balanced because input and storage much exceed loss from the system.

Sedimentary processes and deposits in lakes are determined by the physical, chemical and biological environment of the lake and of its watershed and airshed. The processes and products of this environment are controlled in complex interactions among climate, hydrology, biology, geomorphology, geology, and, increasingly, human actions. Figure L1 sets out a conceptual framework that associates these factors and indicates qualitatively the way in which the sedimentology of the lake is controlled.

Limnologic processes affecting sedimentation

The processes in a lake are driven by the energy of the lacustrine system. The most important input of energy is through solar heating as part of the radiation balance which reaches an average of about $200\,W/m^2$ in summer in mid latitudes. The major ways in which heat energy is lost from the lake are by sensible heat transfer and evaporation. Solar heating is particularly effective because the mean albedo (reflectivity) of the water surface is less than 10 percent, one of the lowest of any natural surface. Solar radiation penetrates water up to several tens of meters before being absorbed and heating the upper part of the lake, thus reducing the density of the water and leading to a tripartite structure of the water column: an upper epilimnion separated from a colder, higher density hypolimnion by a zone of thermal change, the metalimnion. In mid-latitude lakes the isolation of these zones breaks down twice each year in spring and autumn as water in the epilimnion warms or cools, respectively, through the temperature of the hypolimnion. This semi-annual overturn and mixing of the lake water, referred to as *dimixis*, is important in mass (including sediment) and energy transfer throughout the lake. In high latitudes and tropical to subtropical regions heating or cooling to the near-bottom temperature occurs only in summer or winter, respectively, giving rise to one period of overturn, or *monomixis*. In lakes less than about 2 m to 10 m deep depending on fetch (Gorham and Boyce, 1989) wind-generated forces frequently overcome thermal stabilization of the water column, leading to repeated or near continuous overturn of the lake water referred to as *polymixis*.

Mass exchange of water (the water balance) is important to sedimentary processes because normally a significant amount of sediment is delivered to lakes by rivers. The annual pattern of precipitation is especially important, leading to flushes of water and its sediment load in seasonal events, for example related to snowmelt in the mid to high latitudes or monsoons in the sub-tropics, and to irregular, catastrophic, major storm events. Where evaporation from the lake exceeds precipitation and runoff into the lake, outflow does not occur and dissolved salts accumulate in the lake water. This commonly leads to a persistent structure based on lower density (fresher) water near the surface, and high-density, in some lakes hypersaline, water at depth which does not permit overturn and is referred to as *meromixis*. In extremely arid environments the entire lake may be hypersaline. Salts may also be present in lakes lifted from the sea by tectonic processes. A review of the water and energy balances of lakes is provided by Lerman *et al.* (1995).

The components of the water balance determine the *residence time* of water in the lake which is an important factor influencing sedimentary processes. This is the time that water spends in the lake on average between when it enters the lake as runoff or precipitation and when it leaves the lake as outflow or through evaporation. It is a function of the relative volume of the lake with respect to the volume of input, and can

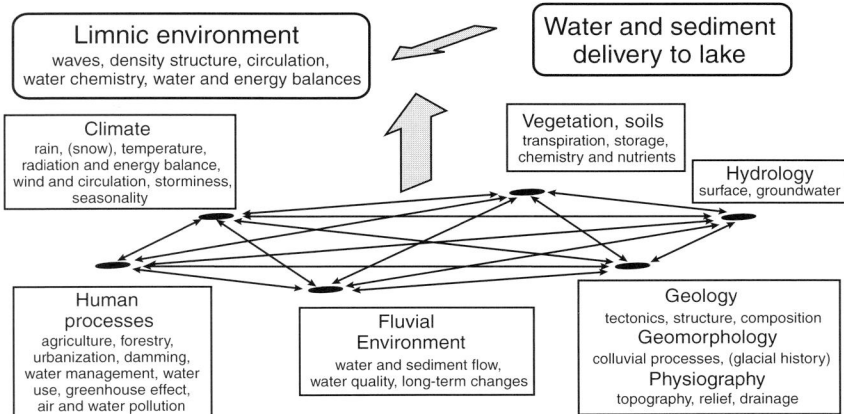

Figure L1 Conceptual model of the relation between the characteristics of the drainage basin and the nature of lacustrine sedimentation.

vary from a few days to many centuries, the latter for large lakes with small drainage basins. The longer the residence time, the more likely that a given sediment particle will have time to settle to the lake floor rather than being swept from the lake. Thus, depending on the formation of precipitates, most of the sediment in solution passes from the lake, while greater amounts of progressively coarser-grained clastic and biogenic sediment are trapped. As a general rule, lakes having residence time of 0.1 years trap 80 percent to 95 percent of the sediment input, while those with residences times of 1 year or more trap 95 percent or more (Brune, 1953).

Of the many other processes that occur in lakes, the action of waves generated by wind is the most important to the sedimentology of the lake, even though as a source of energy to the lake as a whole, waves represent a relatively minor component. The size and energy of waves is governed by the velocity and duration of the wind, and the effective fetch (the distance the wind blows across open water, accounting for the irregularities of the shores: Håkanson and Jansson, 1983). There are a number of graphical and numerical methods by which calculations of wave size may be made (e.g., Sinclair and Smith, 1972). Because of the role of waves in shaping coastal environments, wave energy plays a very important role in the sedimentology of the entire water body. Fundamental is the concept of *wave base*, the water depth to which waves erode sediments on the lake floor. This depth is about one quarter of the length (crest to crest distance) of the wave in deep water (Sly, 1978), and so is a function of wind velocity and fetch (Figure L2). In large lakes having an effective fetch of 100 km or more, wave base may reach 40 m in the largest storms. Fine-grained sediments are winnowed from deposits on the lake floor above wave base, and are transported to deeper water for redeposition. In addition, waves generate circulation of the lake water, moving mass and energy through the lake and affecting sedimentary processes.

Types of lacustrine sediment

Classifications of lacustrine sediments include those based on their place of origin, the processes of their transportation and deposition, their location in the lake, and their composition (Håkanson and Jansson, 1983). Sediments which are transported from the watershed or the airshed to the lake, and

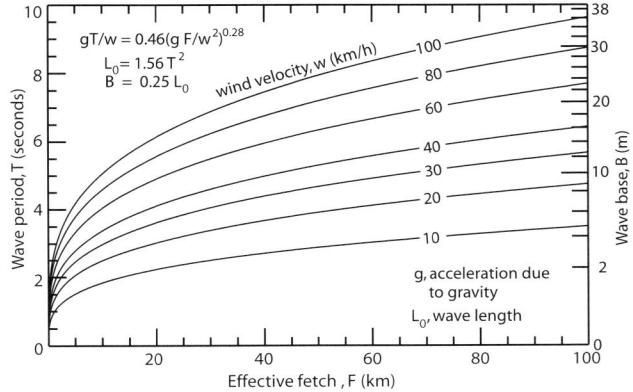

Figure L2 Relation of effective fetch to wave period according to Sinclair and Smith (1972), and wave base. Results correspond with observations of sediment erosion on the lake floor (Håkanson and Jansson, 1983).

are deposited in the lake with little physical or chemical alteration are referred to as *allochthonous*. Those that originate within the lake itself are referred to as *autochthonous*. Most commonly these are created as a result of precipitation of salts dissolved in the lake water or from the remains of aquatic plants and animals. At least some of the components of autochthonous sediments come from outside the lake (as the dissolved or nutrient loads of incoming streams, for example), but the form in which they are deposited is very much altered from the form in which the constituents arrived at the lake.

Primary deposits are those which have undergone only minor changes since they were deposited, while *secondary* or *reworked* deposits have been significantly altered since deposition. Gravitational processes, erosion by waves and currents, and bioturbation are the principal ways in which sediments are reworked and redistributed in the lake. The distinction between primary and secondary deposits becomes important, for example, where ancient sediments are eroded from the lake floor and redeposited elsewhere along with sediments newly arrived in the lake, mixing an ancient paleoenvironmental signal (that might be interpreted from the physical, chemical or biological nature of the sediment) with the signal from the modern environment.

Classification based on the composition of the sediment, divides deposits into *lithogenic*, *biogenic*, and *hydrogenic* sediments. Lithogenic sediments are the clastic detrital products of the physical and chemical weathering of rock or unconsolidated sediment. These sediments may be further classified on the basis of mineral composition, grain size (from boulders to clay size), sedimentary structures and stratigraphy, color, magnetic characteristics, and so on. Biogenic sediment is the remains of aquatic organisms with normally lesser amounts of allochthonous organic material. It includes detritus dominated by carbon-based molecules, including those in association with calcium (as calcite and aragonite), phosphorous (as phosphate), and silica. This black organic ooze referred to as *gyttja* dominates many mid-latitude lakes. Hydrogenic sediment results from precipitation from solution of salts in the lake water or the interstitial water of previously deposited sediments. Included in this class are the *evaporites* (q.v.) of saline lakes and the *marl* deposits of hard-water lakes.

A geochemical classification considers the presence of oxygen in the sedimentary environment. *Oxic* sediments are those in which there is oxygen in sufficient quantity (more than about 10^{-6} moles per litre) to determine processes dominated by oxidization. *Anoxic* sediments where oxygen is insufficient to allow oxidation are divided into *sulfidic* and *non-sulfidic*, depending on the presence of dissolved sulfide. Non-sulfidic environments are divided again into *post-oxic*, where oxygen removal has occurred without sulfate reduction and *methanic* where sulfate reduction and methane gas formation have occurred. These environments commonly change as the redox boundary migrates upward as sediment accumulates, and so remains just below the sediment surface. Oxic deposition is followed by post-oxic, sulfidic and methanic. Deposition of iron and manganese in the sediments or as nodules at the surface is closely associated with this chemistry.

Lacustrine sedimentary processes

Figure L3 shows a generalized, conceptual model of sedimentation in a lake. Not included are components that are unimportant in many lakes (such as the action of seasonal ice cover). Three sources of sediment to lakes are indicated in Figure L3. Fluvial sediment, including coarse-grained bedload and fine-grained suspended load (wash load), is the most important source of sediment for most lakes. Fine-grained airborne sediment (dust), including mineral and organic material of natural and human origin is important in many lakes, especially where pollutants such as the sulfur and nitrogen compounds that create acid precipitation, and phosphorous in various forms, for example, are delivered in significant quantity. Autochthonous sediments are commonly created from the dissolved load (especially calcium carbonate and similar minerals) in inflowing streams and groundwater, as well as from biological activity, especially photosynthesis.

A delta commonly forms at the mouths of rivers bearing large clastic loads. Flow decelerates in hydraulic backwater, allowing bed-load deposition in a wedge of foreset beds created near the subaqueous angle of repose by quasi-continuous avalanching, overlain by topset beds deposited in response to changing fluvial gradient as the delta advances. Variations occur in lakes having fluctuating water levels and

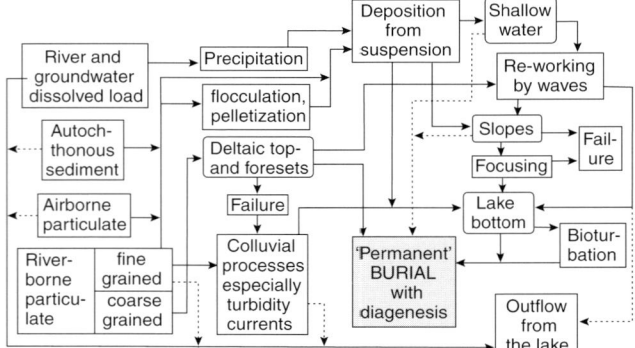

Figure L3 Generalized components of the lacustrine sedimentary environment. This model does not consider coastal processes and sediments.

where alluvial fans extend from land to water. Finer-grained material in suspension is transported into the lake in a decelerating jet to be deposited distally from suspension as bottomset beds. Deltas are absent or poorly developed at many river mouths where clastic sediment input is low or where vigorous waves and currents redistribute sediment as quickly as it is delivered.

Fine-grained sediment in suspension settles to the lake floor through calm water according to Stokes Law:

$$v = (\rho_s - \rho_w) g \, d^2 / 18\eta$$

where ρ is the density of the sediment ($_s$) and water ($_w$), g is the acceleration due to gravity, d is the particle diameter, and η is the dynamic viscosity of the water. Accordingly, a 10 μm diameter particle requires about one to two weeks to settle 100 m, while a 1 μm particle requires several years. An analysis that includes sand size particles where viscous resistance is small in comparison to the weight of the particle (i.e., beyond the range of Stokes Law) and the effect of particle shape is presented by Dietrich (1982). Actual rates of transfer to the lake floor are greater than settling predicted by these models for two principal reasons. First, flocculation even in fresh water increases the effective particle diameter especially of clay minerals such as kaolinite and illite and settling time is reduced by an order of magnitude or more (Droppo et al., 1997). Complimenting this is the incorporation of fine-grained sediments into fecal pellets produced by zooplankton. Second, general circulation carries parcels of water and suspended sediment to depth, although the presence of thermal or chemical stratification prevents mass transport throughout depth, except during overturn.

Colluvial (gravitationally driven) processes commonly redistribute sediment from shallow to deeper basins in a lake. Slides move large quantities of sediment along a well-defined glide plane with little internal deformation, so the deposit remains as a coherent mass retaining clearly recognizable sedimentary characteristics of the original deposit. In slumps the sediment is deformed but the final deposit is recognizable as a single event (Figure L4). In *gravity flows*, which are significant where deposition of clastic sediment predominates, the sediment is completely reworked during transport (see

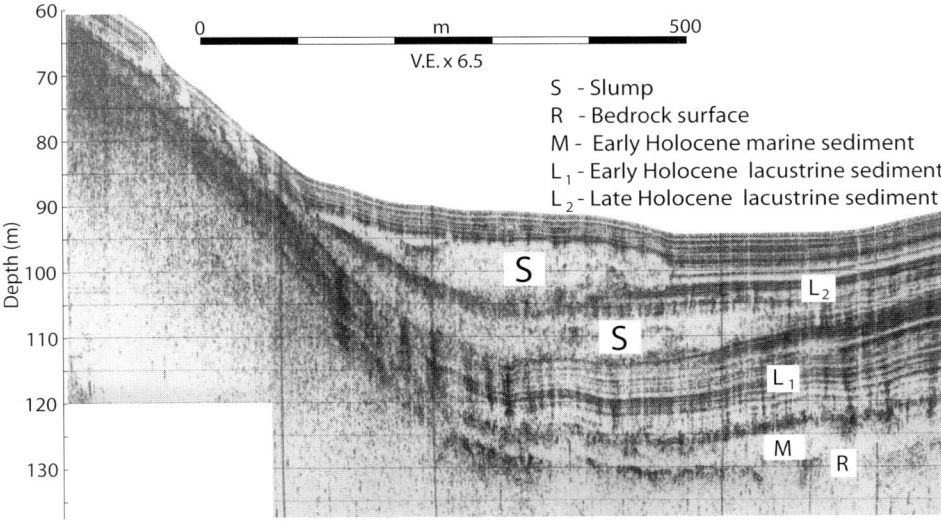

Figure L4 3.5 kHz subbottom acoustic profile from Stave Lake, British Columbia, Canada showing marine sediments deposited over bedrock immediately following deglaciation, then lacustrine sediments deposited and focused to the lake basin, including by slumping, through the Holocene.

Gravity Driven Mass Flows). Middleton and Hampton (1976) distinguish *fluidized flow* in which the sediment is supported by escaping pore water, *grain flow* where particles are supported by grain-to-grain interaction, *debris flow* as a slurry where coarse particles are supported by the matrix of fine particles, and *turbidity currents* where sediment is supported by turbulent flow. Turbidity currents are generated directly from inflowing streams or as a result of colluvial processes on the lake floor; in either case, the current is driven down slope by its greater density due to suspended particulate matter. Behaving as a river under the water of the lake (considering the buoyant effect of the ambient water and the interfacial shear stress with the flow—Middleton, 1966), turbidity currents intermittently distribute relatively coarse-grained sediment widely over the floors of some lakes, even creating leveed channels in a few cases.

An important consequence of the reworking of sediments by waves from the coast to the depth of wave base, and by rapid, episodic, and slow, continuous colluvial processes throughout the lake is that sediment is focused to the deepest basins of the lake (Figure L4). The result is that rates of accumulation vary significantly throughout lakes (Blais and Kalff, 1995). This must be assessed when deciding the location of point samples (cores) used to reconstruct depositional rates and sedimentary history of a lake.

Sediments deposited in deep water are normally subject only to reworking by bioturbation and to diagenesis as they are sequentially buried. Diagenesis refers to the changes that the sediment undergoes after it has been deposited on the lake floor, normally as it is buried in the sediment column (Berner, 1980). The common forms of diagenesis occurring in lacustrine sediment include: (1) compaction and dewatering caused by loading with associated increase of shear strength of the sediment; (2) diffusion or mass transport in pore water of dissolved salts, including metals through the sediment; (3) dissolution or precipitation of, for example of marl (calcium carbonate); and (4) decomposition of organic material by aerobic or anaerobic bacteria. Thus, after the sediment is buried in the lake floor, it may change significantly, and assessment of environmental change from the characteristics of the sediment must take into account these possible changes. For example, a pollutant that migrates upward from older to more recently deposited sediment as compaction and dewatering occur may appear to indicate more recent contamination of the aquatic system than has actually occurred.

Lacustrine sediment as paleoenvironmental proxy

Figure L1 indicates a strong, if complex relation between terrestrial systems and the nature of sedimentation in lakes. If these pathways are understood, the sedimentary record has potential to provide a wealth of knowledge about the lake and its contributing region, both in the present and throughout the history of the lake. Facilitating this are the characteristics of limnic processes and lacustrine sediments:

1. Lacustrine sediments are normally deposited quasi-continuously without subsequent erosion and only minor diagenesis, so according to the Law of Superposition, a long, uninterrupted, decipherable record from present to the beginning of the lake can be assessed.
2. Most of the sediment entering the lake is trapped there so that whatever has occurred in the lake or its basin is registered in the sedimentary record.
3. Commonly, the finest grained sediments are deposited in the lake, while gravel and even sand are left in the rivers and deltas, and on the land. It is this fine organic and inorganic material that is most diagnostic of the environment of the watershed and airshed contributing to the lake.
4. Sediments deposited in a lake represent the entire drainage basin and so integrate and smooth catastrophic events. Thus, while a major storm or localized soil erosion event

may be represented in the lacustrine record, they do not overwhelm a sedimentary sequence as may occur in the terrestrial record.

5. Commonly, annual cycles (varves, *q.v.*) are preserved in lacustrine sediments. These are powerful as dating tools and in high-resolution paleoenvironmental assessment.

There is an extensive and rapidly growing literature interpreting changes in climate, hydrology, geomorphology, and human actions on scales from intra-annual to hundreds of thousands of years. Proxies include the physical characteristics of clastic sediments (Lamoureux, 1999), the chemistry of precipitates (Last and Slezak, 1988) and biologic material including diatoms (Stoermer and Smol, 1999) and pollen.

Investigating the lacustrine sedimentary record

The nature of sediment in a lake is commonly investigated at two scales. First, at a synoptic scale, acoustic techniques provide remotely sensed imagery of the entire lacustrine sedimentary package (Figure L4). Even conventional high frequency (50 kHz and above) echo sounding penetrates soft, organic sediments such as gyttja but lower frequency (commonly centered on 3.5 kHz) provides penetration of tens of meters into most types of lacustrine sediment. In a few cases, low frequency seismic surveys have provided very deep penetration through extremely thick lacustrine sequences. From these records the distribution and thickness of sediment in the lake is mapped, sedimentary processes and rates of accumulation are inferred, and changes in sedimentary processes and deposits through time are documented.

Second, samples of the sediment are recovered for detailed analyzes. Many samplers have been devised, each designed to particular needs. Dredges, including trawls and samplers with closing jaws, recover surficial material. Samplers having a chamber to retain sediment include box corers, gravity and piston corers, and side-opening samplers. Adhering samplers cause sediment to accumulate on their outer surfaces, normally by freezing due to a mixture of dry ice and alcohol inside the sampler. All are designed to recover material from the surface or near-surface of the sediment to depth of up to tens of meters with as little disturbance as practicable. Unlike oceanographic surveys where a large ship with heavy equipment is normally available, a major concern in sampler design and deployment in many lake studies is light weight, portability and ease of deployment from small vessels, or where conditions permit, from an ice cover. Details of lacustrine sedimentary analyzes are documented in a number of sources including Håkanson and Jansson (1983).

Robert Gilbert

Bibliography

Berner, R.A., 1980. *Early Diagenesis. A Theoretical Approach*. Princeton University Press.
Blais, J.M., and Kalff, J., 1995. The influence of lake morphology on sediment focusing. *Limnology and Oceanography*, **40**: 582–588.
Brune, G.M., 1953. Trap efficiency of reservoirs. *Transactions of the American Geophysical Union*, **34**: 407–418.
Dietrich, W.E., 1982. Settling velocity of natural particles. *Water Resources Research*, **18**: 1615–1626.
Droppo, I.G., Leppard, G.G., Flannigan, D.T., and Liss, S.N., 1997. The freshwater floc: a functional relationship of water and organic and inorganic floc constituents affecting suspended sediment properties. *Water Air and Soil Pollution*, **99**: 43–54.
Gorham, E., and Boyce, F.M., 1989. Influence of lake surface area and depth upon thermal stratification and the depth of the summer thermocline. *Journal of Great Lakes Research*, **15**: 233–245.
Håkanson, L., and Jansson, M., 1983. *Principles of Lake Sedimentology*. Springer-Verlag.
Lamoureux, S., 1999. Spatial and interannual variations in sedimentation patterns recorded in nonglacial varved sediments from the canadian high arctic. *Journal of Palæolimnology*, **21**: 73–84.
Last, W.M., and Slezak, L.A., 1988. The salt lakes of western canada: a paleolimnological overview. *Hydrobiologia*, **105**: 245–326
Lerman, A., Imboden, D., and Gat, J. (eds.), 1995. *Physics and Chemistry of Lakes, Second Edition*. Springer-Verlag.
Middleton, G.V., 1966. Experiments in density and turbidity currents II. Uniform flow of density currents. *Canadian Journal of Earth Sciences*. **3**: 627–637.
Middleton, G.V., and Hampton, M.A., 1976. Subaqueous sediment transport and deposition by sediment gravity flows. In Stanley, D.J., and Swift, D.J.P. (eds.), *Marine Sediment Transport and Environmental Management*. John Wiley & Sons, pp. 197–218.
Sinclair, I.R., and Smith, I.J., 1972. Deep water waves in lakes. *Freshwater Biology*, **2**: 387–399.
Sly, P.G., 1978. Sedimentary processes in lakes. In Lerman, A. (ed.), *Lakes Chemistry Geology Physics*. Springer-Verlag, pp. 65–89.
Stoermer, E.F., and Smol, J.P., 1999. *The Diatoms: Applications for the Environmental Earth Sciences*. Cambridge University Press.
Strahler, A.N., 1980. Systems theory in physical geography. *Physical Geography*, **1**: 1–27.

Cross-references

Climatic Control of Sedimentation
Coring Methods, Cores
Deformation of Sediments
Diagenesis
Deltas and Estuaries
Flocculation
Geophysical Properties of Sediments
Grain Settling
Gravity-Driven Mass Flows
Iron-Manganese Nodules
Sediment Fluxes and Rates of Sedimentation
Sediment Transport by Waves
Slide and Slump Structures
Varves

LATERITES

Laterites, products of tropical weathering, cover about one third of the world land surfaces. Inventories and interpretations concerning the worldwide distribution of laterites have considerably progressed during the last 50 years: new concepts have been emerging in geomorphology, soil science, petrography and geochemistry, while the nomenclature has been progressively revised (Millot, 1964; King, 1967; Michel, 1973; Nahon, 1976, 1991; Bardossy and Aleva, 1990; Thomas, 1994; Segalen, 1994, 1995; Tardy, 1969, 1993, 1994, 1997; Tardy and Roquin, 1998).

Definition

The terms laterite and lateritic materials no longer refer to hardening. Even if several types are effectively indurated (when concretionary and nodular), they may be also soft (if massive) or loose (if pisolitic and granular). They do not refer to color even if most of them are dominantly red. They may be yellow, white, pink or even black in color. The major constituent minerals are dominantly kaolinite, goethite or hematite (secondary minerals) and very often quartz (primary mineral). Frequently the aluminum hydroxide gibbsite, and rarely other oxides or oxihydroxides such as boehmite, pyrolusite, cryptomelane, etc., are sometimes formed in large proportions, depending on the nature of the parent rock and climate. Laterites are products of humid tropical or subtropical weathering. Profiles developed under arid or semiarid climates, in which carbonates, salts, smectites or amorphous silica could be formed are excluded; so are profiles which retain easily hydrolysable primary minerals such as feldspars or amphiboles. While each type reflects a quite precise climatic zonality, lateritic profiles are generally very old and polyphasic, that is, resulting of an imbroglio of different successive paleoclimates.

Nomenclature

Lateritic materials can be classified into several categories, distinct by nature of their mineralogical compositions, petrographical structures and degree of maturity. Three major groups are distinguished, characterized mostly by the qualities of their upper horizons, the most sensitive to climatic changes. They are *Ferricretes*, *Bauxites*, *Latosols*, and, as an extension which could be considered as abusive, *Gibbsitic Podzols*.

Lithomarges are fine saprolites, roots in common to all lateritic profiles, and continuously formed below the water table. The structure is massive and the features of the parent rock are conserved. Kaolinite is the dominant secondary mineral; goethite and hematite are also present; preserved quartz is abundant on granitic rocks and sandstones. Concretions are not observed and lithomarges are so widely distributed that they have no climatic or paleoclimatic significance—except thickness—which can sometimes exceed 100 m.

Mottle horizons are intermediate between lithomarges and ferricretes, and formed in the domain of water table fluctuations. Pedological features observed are essentially mottles, resulting of iron mobilization, exhibiting two contrasted domains: (1) yellow or red colored, in which goethite and hematite accumulate intimately associated with kaolinite, within originally clay-rich and quartz-poor litho-relict features (inherited from the parent rock weathering) or pedo-relict features (originated by secondary illuviation of kaolinite); and (2) pale or white uncolored, resulting of iron leaching out from clay-poor or quartz-rich litho-relict features. With mottles, the beginning of hardening takes place (carapace).

Ferricretes are nodular horizons, in which *concretions*, made of original kaolinite, intimately associated with abundant small sized hematite, organize by progressive iron accumulation, leading a dark red color and a significant hardening, similar to bricks (cuirasse). At the bottom of the profiles, the transition between mottle horizon and ferricrete is progressive and accompanies the iron enrichment. From the bottom to the top, quartz dissolves and may even disappear, while hematite contents increase significantly and kaolinite amounts remain constant or smoothly augment. Ferricretes develop *in situ*, sometimes over considerable thickness (when old) by leaching of soluble elements and accumulation of poorly soluble materials, which reorganize into concretions.

Granular horizons. Toward the top of the profiles, ferricretes dismantle, yielding loose horizons, made of *granules*, round in shape, and constituted essentially of relicts of concretions of hematite + kaolinite, surrounded by rings of secondary goethite (often called pisolites). In peculiar cases, together with goethite enrichment, gibbsite appears in pores of large size. The round granules are not transported but also formed *in situ*, accompanying hydration and leaching of the upper horizons of ferricretes.

Ferricretes and their associated horizons are formed specifically under tropical (warm) but seasonally contrasted climate, alternatively dry and humid. Their maximum extension is observed during the Eocene, even far away from the equatorial zone. They are the parental source of many latosols and orthobauxites, formed at their expense, under subsequent wetter or cooler tropical or subtropical climates.

Latosols are thick and soft horizons, red or yellow in color, with massive or *microgranular* structure, developed above *lithomarges*, from whose they are generally separated by a *stone line*. The microgranules are ancient nodules of hematite + kaolinite earlier formed in *ferricretes*, and later dismantled under wetter climate, in which goethite and gibbsite secondarily develop in significant proportions. Stone lines are made not only of quartz gravels derived from quartz veins, but also of blocs and granules of ancient ferricretes, buried under the latosolic horizons by termite activity. *Red latosols* are located mostly on relict plateaux of the oldest surfaces, high in altitude; they generally bear high quantities of gibbsite and rather low amounts of quartz, so that could be considered as *protobauxites*. On quartz-poor parent rocks, they could lead with time to real *orthobauxites*, also red in color. *Yellow latosols* are located downslopes on younger surfaces; they generally bear higher quantities of quartz and siliceous phytoliths, in the upper part accompanying large amounts of kaolinite associated with goethite. They generally exhibit very low if any concentrations of gibbsite at the top but may develop a bauxitic layer at the bottom, leading with time under very wet tropical climates to the formation of exploitable *cryptobauxites*, also yellow in color. Both red and yellow latosols are developed *in situ* by chemical weathering and insect bioturbation. Most, not to say all of them, derive from ancient ferricretes, later submitted to warm but very wet tropical climate or cool and humid subtropical climate, such as those developed at higher altitudes.

Lateritic bauxites are economic ores of aluminum accumulated as hydroxide (gibbsite) or oxi-hydroxide (boehmite) developed on quartz-poor materials. Several types are distinguished, according to their mineralogical compositions and to their kind of organization. *Protobauxites* are red latosols (ferralsols), yielding significant amounts of gibbsite. *Orthobauxites* result of the development with time of such protobauxites, generally under cooler and wetter subtropical climates for which altitude could be the determinant factor. In this case the top of the profile could be dark-red and bearing relicts of ancient *granules or microgranules* inherited from the ancient latosol called *conakryte*. Laying below the conakryte and above the lithomarge, the bauxite *sensu stricto* is a thick horizon, pale-pink in color and made of massive gibbsite crystallizing as septaria around pores of large size and conferring to this material a kind of consolidation which has nothing to do with concretion hardening. *Cryptobauxites* are bauxitic horizons developing in between at the top a yellow latosol (kaolinite, goethite and quartz) several meters thick, and at the bottom with a very thick lithomarge. The characteristics of the bauxitic horizon are the same except that the color may be yellow-white in this case. *Metabauxites* are ancient orthobauxites evolving under warmer and dryer climates. The red superficial horizon of conakryte has disappeared and being replaced by a ferricrete horizon located not above but below the bauxite accumulation. The bauxitic horizon, white in color, very hard and dense in consistency, is essentially made of *concretions* of boehmite (from a centimeter up to several meters in diameter), in some cases dissociated and underlined by rings of hematite (called pisolites). Metabauxites begin to form when orthobauxites are displaced in the climatic zone of ferricretes. This evolution slow down but prolonges under semiarid climates.

Gibbsitic Podzols are laterites by a strict application of the mineralogical definition, developing as ultisols at the expanse of quartz-rich parent rocks and under per-humid tropical climates. The superficial horizon is totally white (albic), rose-pale or ash colored, most of the time very thick, looking like the "Dover cliffs" in the Congo. Down below the albic horizon a layer of gibbsite cementing the quartz grains is very often observed.

Laterites are excellent indicators of continental climates and paleoclimatic successions, due to uplift, subsidence, continental drift, and relative migrations of the equator and its associated climatic zones.

Continental drift and paleoclimatic successions

One has first to realize that the equatorial zone of the great tropical forest is and was always bordered by a series two zones (N and S symmetrical) of seasonally contrasted tropical climates colonized by savannah. Those two are bordered by the two steppe and desert zones of semiarid and arid climates, farther surrounded by two other zones of subtropical climates. During the Cretaceous, the Sahara was located under the equator; bauxites were formed in Saudi Arabia, while Nigeria and Guinea were in the southern climatic zone of ferricretes which today appear as relicts in some places covering the Post-Gondwana old and high surface. At that time Congo and Amazonia were deserts covered by sand dunes. During the Eocene, continents moved northward and the equator moved southward. Furthermore the global humidity and temperature were at their maxima. Ferricretes were also at their maximum of extension in southern zones of Amazonia, Congo, Central Africa and the northern domains of the Sahara, where they still persist today but more or less dismantled. However, at higher altitudes, orthobauxites form at the expanse of red latosols on the uplifted panafrican surface of Guinea, South Mali and Burkina Faso. Europe is located in the northern zone of the subtropical climates where karst bauxites are widespread. From the Miocene to the Present, climatic zones continued to move southward while the global climate became cooler. South-Eastern Africa and Andes were uplifted. Southern ferricretes transformed into latosols in central Amazonia, Guyana, Tanzania, Congo, Gabon, while new ferricretes formed in the northern belt but on the lower surfaces (Haut-Glacis) in Burkina Faso, Ivory-Coast, and Mali. However ancient orthobauxites of Southern Mali and Northern Burkina, localized on older surface were submitted to aridity accompanying the progression southward movement of the Sahara. Orthobauxites of high surfaces were transformed into metabauxites. At the present time, Amazonia and Congo are in the equatorial zone. The uplift of Andes and of the East-African Rift makes the climate very humid on the ancient quartz-sand dune deposits, so that gibbsitic Podzols develop in the Rio Negro and Zaïre basins.

Because of the continental drift and changes in the global climate, almost all the old lateritic profiles are polygenic. Laterites by their diversity remain excellent indicators of paleoclimatic successions all over Australia, South-east Asia, Africa, Brazil, and even the United States and Europe.

Hydration-dehydration, key to laterite stability and transformations

For many years, mineralogical criteria have served as a base of soil classifications and interpretations of the leaching dynamics of cations and silica, responsible for the variety of observed soils. The hydroclimatic parameters P (rainfall) and D (drainage) as well as the P/D ratio (overall concentration factor of the weathering solutions submitted to evaporation, $E = P - D$) regulate pH and silica activity, $[SiO_2]$, in waters draining the profiles. However two other parameters: T (temperature) and $[H_2O]$ (activity of water, defined by the relative humidity of the air) are essential to control hydration-dehydration reactions occurring within the superficial horizons, and having both petrographic and climatic significance.

The equilibrium reactions among laterite minerals (formulae written, to simplify, as a sum of oxides) are the following:

$$Fe_2O_3 \cdot H_2O = Fe_2O_3 + H_2O$$
(goethite = hematite + water) \hfill (Eq. 1)

$$Al_2O_3 \cdot 3H_2O = Al_2O_3 \cdot H_2O + 2H_2O$$
(gibbsite = boehmite + water) \hfill (Eq. 2)

$$Al_2O_3 \cdot 3H_2O + 2SiO_2(aq) = Al_2O_3 \cdot 2SiO_2 \cdot 2H_2O + H_2O$$
(gibbsite + aqueous silica = kaolinite + water) \hfill (Eq. 3)

$$Al_2O_3 \cdot 2SiO_2 \cdot 2H_2O + 2SiO_2(aq) = Al_2O_3 \cdot 4SiO_2 \cdot H_2O + H_2O$$
(kaolinite + aqueous silica = smectite + water) \hfill (Eq. 4)

Compared to gibbsite and goethite, which are hydrated minerals, hematite and boehmite are simply dehydrated, while kaolinite and smectites are dehydrated but silicated mineral phases. Hydration is favored in humid and cool climates in which $[H_2O]$ is elevated and $[SiO_2]$ lowered. Excretions of gibbsite and goethite are observed in bauxites and latosols where ferricrete dismantles under wetter and cooler climates. In contrast, dehydration is favored under dry and hot climates, where $[H_2O]$ is lowered and $[SiO_2]$ elevated. Concretions of hematite and kaolinite, insuring the hardening of ferricretes, and concretions of boehmite and hematite are characteristic of the transformation of orthobauxites into metabauxites under dryer and warmer climates.

Thus, beside temperature and the activity of aqueous silica in solution, which controls the reactions of silication-desilication, temperature and the activity of water also controlled the reactions of hydration-dehydration, and finally the hardening and the dismantling conditions of the lateritic mantle, through the last 150 million years of continental drift.

Yves Tardy

Bibliography

Bardossy, G., and Aleva, G.J.J., 1990. Lateritic bauxites. In *Developments in Economic Geology*. Elsevier, Volume 27, p.1.
King, L.C., 1967. *The Morphology of the Earth*. Olivier and Boyd.
Michel, P., 1973. *Les Bassins des Fleuves Sénégal et Gambie. Etude géomorphologique*. Volume 1,2,3. ORSTOM: Paris, Mémoir, 63.
Millot, G., 1964. *Géologie des Argiles*. Masson.
Nahon, D., 1976. *Cuirasses Ferrugineuses et Encroutements Calcaires au Sénégal Occidental et en Mauritanie*. Science de Géolologie, Strasbourg, Mémoir, 44, 232pp.
Nahon, D., 1991. *Introduction to the Petrology of Soils and Chemical Weathering*. Wiley.
Segalen, P., 1994. *Les Sols Ferrallitiques et leur Répartition Géographique*. Volume 1, 198pp. Coll. Etudes Thèses, ORSTOM: Paris.
Segalen, P., 1995. *Les Sols Ferrallitiques et leur Répartition Géographique*. Volume 2, 169pp., Volume 3, 201pp. Coll. Etudes Thèses, ORSTOM: Paris.
Tardy, Y., 1969. *Géochimie des Altérations. Etude des Arènes et des Eaux de quelques Massifs Cristallins d'Europe et d'Afrique*. Science Géology: Strasbourg, Mémoir, 31, 199pp.
Tardy, Y., 1993. *Pétrologie des Latérites et des Sols Tropicaux*. Masson.
Tardy, Y., 1994. PIRAT, Programme Interdisciplinaire de Recherche de Biogéodynamique Intertropicale Périatlantique. *Science Géology*: Strasbourg, Mémoir, 96, 100pp.
Tardy, Y., 1997. *Petrology of Laterites and Tropical Soils*. Translated by Sarma, V.A.K. (ed.), Balkema.
Tardy, Y., and Roquin, C., 1998. *Dérive des Continents: Paléoclimats et Altérations Tropicales*. Editions BRGM.
Thomas, M.F., 1994. *Geomorphology in the Tropics*. John Wiley.

Cross-references

Caliche–Calcrete
Hydroxides and Oxyhydroxide Minerals
Silcrete
Weathering, Soils, and Paleosols

LEPISPHERES

Lepispheres (from the Greek, meaning "spheres of blades") are 3-micron to 20-micron-diameter rosettes of authigenic silica (SiO_2) formed during low-temperature diagenesis (Figure L5). Coined by Wise and Kelts (1972) as a term to describe a secondary siliceous cement viewed by light and scanning electron microscopy in an Oligocene deep-sea chalk, lepispheres are now regarded as morphologic expressions of the incipient mineral phase in the formation of deep-sea chert (Wise *et al.*, 1992). As such, they form in a variety of deep- to shallow-marine settings where precursor materials of amorphous silica such siliceous microfossil tests and volcanic glass are present. The unstable amorphous silica (opal-A of Jones and Signet, 1971) is converted to lepispheres (opal-CT) by a zero-order dissolution–diffusion–reprecipitation reaction. The metastable lepispheres may eventually convert to the more stable mineral phase, quartz (opal-C). In rock terms, the

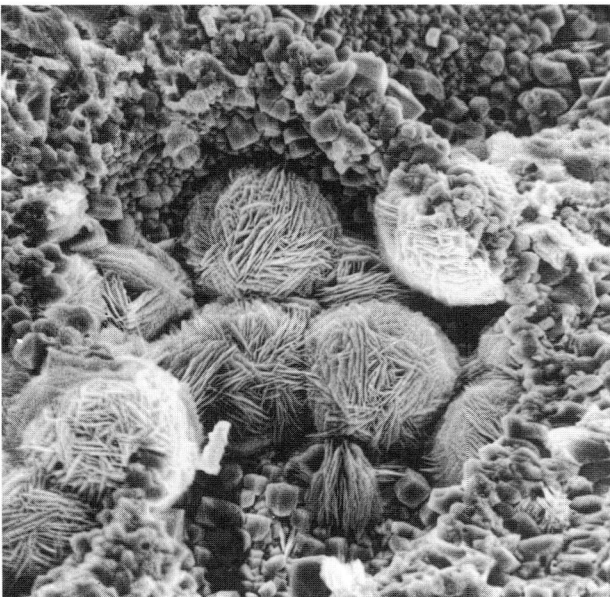

Figure L5 Cluster of opal-CT lepispheres filling a recrystallized foraminiferal test chamber in a partially silicified deep-sea chalk (scanning electron micrograph from Wise and Weaver, 1974, figure 5a, 1600X).

lepispheres may coalesce to form porcelanite, the intermediate phase in the formation of true quartz chert from a siliceous precursor material.

Opal-CT lepispheres are distinguished by a somewhat broad X-ray diffraction pattern centered about a strong cristobalite peak with a minor peak between 4.25Å and 4.30Å (Wise *et al.*, 1992, figure 1). Their mineralogy is best described as an unidimensionally disordered structure of low cristobalite with low tridymite domains (Florke *et al.*, 1976). In the light microscope, lepispheres are extinct under crossed nicols (due to the pseudoisotropic nature of the mineral), although the many fine blades diffract nonpolaraized light sufficiently to produce an apparent (but not real) high refringence.

Lepispheres have been reproduced in laboratory experiments where wet diatomaceous sediment has been subjected to moderately high temperatures for a number of months (Kastner *et al.*, 1977). Nucleation of lepispheres is favored by the presence of carbonate (such as calcareous microfossils), magnesium (as is found in ordinary seawater), and high concentrations of dissolved silica in the pore fluids. Because siliceous microfossils such as diatoms have a large surface area, their dissolution proceeds rapidly, quickly bringing the pore fluid concentration of silica above the equilibrium solubility of opal-CT. Because opal-CT precipitates at a silica concentration well below the equilibrium solubility of opal-A, however, the growth of the lepispheres drives the reaction by constantly drawing silica from the pore fluids, causing further dissolution of the parent material until the supply is virtually exhausted.

Although no gel or colloidal phase has ever been observed in the formation of lepispheres, such phases have been detected by high-resolution transmission microscopy in the formation of some zeolites. Whether or not such colloidal phases exist in the formation of lepispheres is a subject for future research.

A survey of non-marine environments in which they might form would also be in order.

Sherwood W. Wise, Jr.

Bibliography

Flörke, O.W., Hollmann, R., von Rad, U., and Rosch, H., 1976. Intergrowth and twinning in opal-CT lepispheres. *Contributions to Mineralogy and Petrology*, **58**: 235–242.

Jones, J.B., and Segnit, E.R., 1971. The nature of opal I. Nomenclature and constituent phases. *Journal of the Geological Society of Australia*, **18**: 57–68.

Kastner, M., Keene, J.B., and Gieskes, J.M., 1977. Diagenesis of siliceous oozes—I. Chemical controls on the rate of opal-A to opal-CT transformation—an experimental study. *Geochimica et Cosmochimica Acta*, **41**: 1041–1059.

Wise, S.W. Jr., Buie, B.F., and Weaver, F.M., 1972. Chemically precipitated cristobablite and the origin of chert. *Eclogae Geologicae Helvetiae*, **65**: 157–163.

Wise, S.W. Jr., and Kelts, K.R., 1972. Inferred diagenetic history of a weakly silicified deep sea chalk. *Transactions of the Gulf-Coast Association of Geological Societies*, **22**: 177–203.

Wise, S.W. Jr., and Weaver, F.M., 1974. Chertification of oceanic sediments. *Special Publications of the International Association of Sedimentologists*, **1**: 301–326.

Cross-references

Diagenesis
Diagenetic Structures
Porewaters in Sediments
Siliceous Sediments
Zeolites in Sedimentary Rocks

LIQUEFACTION AND FLUIDIZATION

Liquefaction is a process by which a liquid saturated soil loses its solid characteristics and for a short time acts like a viscous fluid. The most dramatic examples are observed during earthquakes. Buildings will then sink or fall over as their foundations are no longer supported; pipes and buried tanks rise to the surface buoyed up by the liquefied soil; buried utility lines may rupture; ground becomes displaced and slopes fail; sand-boils, the remnant of geyser-like upwelling of the liquefied material, cover the surface. Despite these dramatic effects, there are sometimes beneficial aspects to fluidization. The liquefied soils cannot transmit the earthquake's elastic waves and as result protect buildings from damage. There are many examples of buildings that, while they may have sunk slightly, survived the earthquake because the soil around them liquefied.

Two conditions are required for soil liquefaction. The first is that the soil be saturated. This means that the entire interstitial pore space between the particles is filled with liquid. Such is the condition of soils below the water table and liquefaction is a particular problem in areas with high water tables. The second condition is that the soil must be loosely packed or *under-consolidated*. When sheared, an under-consolidated soil will contract or reduce the pore space between particles. (Conversely an *over-consolidated* soil will expand, increasing the pore space.) When saturated the pore space is filled with a nearly incompressible liquid so that reducing the pore space requires forcing out the liquid. In a porous material the liquid flowrate is proportional to the pressure gradient *via* Darcy's law. The only place for the liquid to go is the surface so that forcing the fluid from the shrinking pore space requires very large fluid pressures to push the liquid toward the low pressure at the ground surface. Quickly the pressure can rise to the weight of the overburdening soil, (which is the maximum value to which the pressure can rise), at which point the entire weight is supported by the fluid pressure and the fluid dominates the behavior of the particle/liquid mixture.

The process of liquefaction can be summarized as follows. Shaking by the earthquake causes the saturated soil to compact. To do so, it builds up pore pressure to force out the interstitial liquid. The pressure builds until the liquid supports the entire weight of the overburden. This leaves no pressure on the interparticle contacts which allows the particles to slip freely. The soil then behaves as a viscous fluid. This continues until enough liquid has been driven out of the pore space to allow the particles to come back into contact and again support the applied loads elastically. Following a liquefaction event, the surface of the ground is often covered with puddles of the water driven out of the pore space.

The ground under critical structures, is often compacted to an over-consolidated state in order to prevent liquefaction. Generally over-consolidated soils are stronger than under consolidated soils, and they are less likely to shear during an earthquake. But even if the earthquake manages to shear an over-consolidated soil, the soil would expand, not contract. In this case nearly the opposite of liquefaction occurs. To expand the soil, liquid must be sucked into the additional pore space. This requires a low liquid pressure, so that during this period, more weight is supported by the particle contacts which actually strengthens the soil. But this exercise is often futile if the surrounding soil is left in an under-consolidated state. When an earthquake occurs, the surrounding soil will liquefy causing a pore pressure rise that is initially confined locally. However the pore pressure will quickly diffuse into the surrounding unliquefied soil including the material that was purposely over-consolidated. Thus, the over-consolidated material can still be put into a weakened state which may result in damage to overlying structures.

The large viscosity of a liquefied soil arises from two sources. First of all, when the bulk material is sheared, the fluid shears only within the small spaces between particles. In this way the actual shear rate experienced by the actual fluid is many times larger than that experienced by the material as a whole and results in a proportionally larger viscosity. But the largest contribution to the apparent viscosity will come from the internal momentum transport due particle collisions which results in apparent viscosities that are many orders of magnitude larger than those for the pore fluid alone. Liquefaction also causes in an increase in the apparent density. The fluid/solid mixture will behave as a single fluid with a density equal to that of the mixture density which may be several times that of the pore fluid alone. In such a situation, light rocks that would sink in the pore fluid alone, may rise to the surface of the liquefied material.

Fluidization is a similar phenomenon. A fluidized bed is a common device used in chemical processing (in fact, the catalytic crackers used in petroleum processing are fluidized beds). These blow fluid up through a bed of solid particles until the pressure drop across the bed equals the weight of the bed. Then as for liquefaction, the particles are completely supported by the fluid and, with no pressure on the particle contact points, behave much like a fluid. Heavy objects placed on fluidized beds will sink while lighter objects rise to the surface and bubbles that are strikingly similar to gas bubbles in a liquid, form within the bed, rise and break on the surface.

Similar phenomena can be observed in geological systems whenever there is an upwelling of fluid through a particle bed. Even if the fluid flow is not large enough to completely fluidize the material, it may still weaken it to a large extent. A well-known example of geological fluidization is quicksand. Quicksand is composed of a fine sand through which there is an upwelling of water that makes the sand behave like a very viscous fluid into which people and animals can become trapped in much the same as in the very viscous asphalt of a tarpit. It has been speculated that fluidization is also responsible for the emplacement of pyroclastic flows. These may be fluidized by the outgassing of recently released volcanic material or by air entrained by the motion of the flow. As the fluidized material behaves as if it has the large density of solid, lighter materials are buoyed to the surface while heavier materials sink. This explains why lighter materials such as pumice are found near the surface and heavier materials at the lower levels of the deposits.

Charles S. Campbell

Bibliography

Allen, J.R.L., 1982. *Sedimentary Structures: Their Character and Physical Basis*, Volume 2, Elsevier.

Cross-references

Compaction (Consolidation) of Sediments
Fabric, Porosity, and Permeability
Grain Flow

LOAD STRUCTURES

This name is most appropriately given to a sub-class of syndepositional to postdepositional, soft-sediment deformation structures which result from the partial to complete inversion of one or more pairs of discrete layers of granular sediment, the uppermost of which in each pair, in some cases deforming while continuing to be deposited, arguably had the greater bulk density (Allen, 1982). Load structures therefore record the action of essentially vertical forces, although the simultaneous influence of a body force related to a depositional slope is not excluded. They arise almost exclusively in aqueous environments and represent a movement within the sediment toward a state of minimum potential energy. Convolute stratification is excluded from the sub-class because these structures are restricted to single layers in which there may have been a continuous, positive gradient of bulk density, evidence for which may have survived.

Like soft-sediment deformation structures generally, the production of load structures calls for the operation of trigger mechanisms, deformation mechanisms and driving mechanisms (Allen, 1982; Owen, 1987, 1995; Harris *et al.*, 2000; Jones and Omoto, 2000). There is a predisposition or potential for their formation if one or more of a number of prior conditions are met in the sediments to be affected. Each of the layers involved should have experienced little or no compaction and be fully water-saturated. In each layer the pore-fluid pressure should be high and the grain fabric in or tending toward a metastable state, a circumstance favored by rapidity of deposition and a high water-content. The vertical arrangement should see layers of high bulk density overlying those of lower density; it is well-known that muds, with a comparatively high water content on deposition, have a significantly lower bulk density than sands, which are deposited with a closer packing. The chief trigger mechanisms are: (1) a sufficiently powerful earthquake; (2) aseismic events in the region which raise pore-fluid pressures in confined sedimentary layers; (3) the sudden appearance of an agent from which coarse sediment is very rapidly deposited on a finer substrate (e.g., density currents in lakes/ oceans, floods due to ice dam-breaks); (4) changes in hydrostatic pressure and bed shear-stress related to the passage of progressive water waves or tsunami; (5) changes in pressure and shear stress on a bed resulting from the advection of large-scale turbulence; and (6) the seasonal melting of layered alluvium in periglacial regions, remembering that water expands on freezing. These triggers act by rendering the sediments in the layers, or parts of the layers, fluid-like for a period. Such a state can persist, however, only for so long as the grains have failed to resettle. The deformation mechanisms are two-fold, as in soft-sediment deformation generally. Layers that have become effectively viscous respond to the driving forces by folding, occasionally on more than one wavelength where contrasting, minor layers are involved; those that are effectively brittle fail by faulting. Almost all load structures exhibit evidence of folding, but some show minor faulting as well. Recalling that water-saturated sediments can behave as complex non-Newtonian fluids, the occurrence of faults may record the rapid development of extreme strains during deformation. Alternatively, it may be evidence that the grains had resettled sufficiently to make the sediment again solid-like. Driving forces have already been touched on. As their action is essentially vertical, they should be expressed in load structures by geometrical patterns symmetrical about essentially vertical axes. Systematic departures from such symmetry may be expected only when a directed body force, as on a depositional slope, also came into play.

In essence, the formation of load structures reduces to the problem of the instability in the field of gravity of layered materials capable of behaving as non-Newtonian fluids for a limited period. Theoretical and experimental work on simple systems of this kind (Ramberg, 1964, 1968, 1981) suggests that the scale and geometrical character of load structures is a function largely of the scale of the layers made temporarily fluid-like and the rheological contrast between them, controlling the strength of the chief driving force and the rate of deformation. A general limit on the extent of strain is imposed by the duration of the fluid-like state. In particular, the scale of the deformation increases with the thickness of the affected layers. Thus the spatial patterns or wavelengths discernible in beds showing load structures can range in scale from millimeters and centimeters to tens or hundreds of meters, as

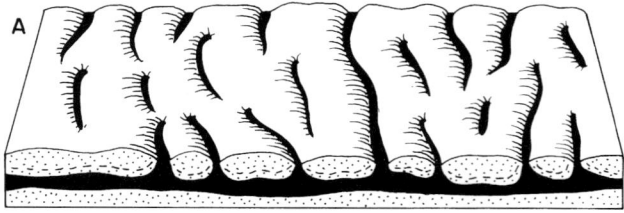
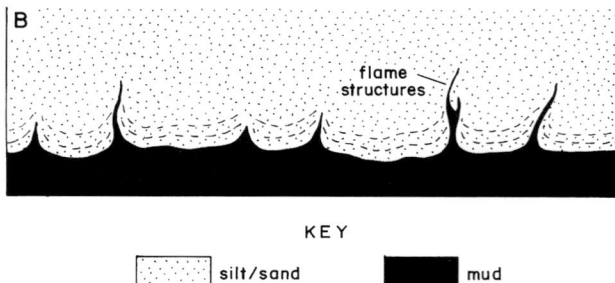

Figure L6 Schematic representations of (A) wrinkle marks, in vertical section and plan view, and (B) load casts on the underside of a sandstone bed, in vertical section.

the affected layers increase in thickness. For this reason, and because of the range of possible triggers, the occurrence of load structures has few restrictions related to depositional environment.

Amongst the smallest of load structures are the *wrinkle marks* widely found in the estuarine intertidal zone (Allen, 1984). These seem to be triggered by the action of waves as the tide retreats across a surface on which a pair of tidal laminae had just been deposited. Typically, the pair is a few millimeters thick, involving a lower, clayey-silty lamina which is overlain by one that is silty-sandy. During loading, the clayey-silty sediment penetrates upward through the coarser lamina at intervals of the order of 10 mm to define in plan a mosaic of roughly equant to elongated, silty-sandy pods (Figure L6(A)). A marked elongation parallel with the contours is observed where deposition had occurred on a significant slope and a subhorizontal body-force also came into play. Slope-parallel microfaults may in such cases slightly displace the pods relative to each other.

Load casts (Figure L6(B)) are probably the commonest and most widespread form of load structure, occurring in fluvial, lacustrine, deltaic, shallow-marine and deep-water environments (Allen, 1982). In deep water, turbidity currents suddenly emplace sand beds centimeters to meters thick upon substrates of soft mud. Either at the time of deposition or shortly afterward, the lower part of the sand sinks down some way into the mud, which rises upward for some distance between the pockets of sand as a series of what have been called flame structures. A freshet spreading across an already-drowned river floodplain on which mud has settled, or a river flood entering a muddy lake, can have a similar effect. Where deformation affects the whole thickness of the coarse layer, breaking it up into discrete portions, or where multiple coarse layers are so affected, the structures called pseudo-nodules (Macar, 1948) and ball-and-pillow (Cooper, 1943; Kuenen, 1948) can arise, as in many near-shore deltaic and shallow-marine environments. In some varieties of load cast, especially where trains of isolated sand ripples or small dunes had moved across a surface of mud, the presence internally of truncated laminae shows that deformation proceeded episodically. In this case the magnitude of the driving force varies spatially, being greatest beneath the crests of the bedforms. The largest load structures of all are those expressed most dramatically by huge mudlumps or mud diapers, as seen in the rapidly advancing Mississippi birdsfoot delta, where sandy mouth bars are advancing across shallow-marine, largely underconsolidated muds at rates of tens of meters annually (Morgan and Andersen, 1961).

Load structures are widely observed proof that sediments freshly deposited in many kinds of aqueous environment can be temporarily changed from solid-like to fluid if certain preconditions are met and if a trigger mechanism is activated. They provide no information on current directions but can provide evidence of depositional slopes.

John R.L. Allen

Bibliography

Allen, J.R.L., 1982. *Sedimentary Structures: Their Character and Physical Basis*. Volume 2, Amsterdam: Elsevier.
Allen, J.R.L., 1984. Wrinkle marks: an intertidal sedimentary structure due to aseismic soft-sediment loading. *Sedimentary Geology*, 41: 75–95.
Cooper, J.R., 1943. Flow structures in the Berea Sandstone and Bedford Shale of central Ohio. *Journal of Geology*, 51: 190–203.
Harris, C., Murton, J., and Davies, M.C.R., 2000. Soft-sediment deformation during thawing of ice-rich frozen soils: results of scaled centrifuge modelling experiments. *Sedimentology*, 47: 687–700.
Jones, A.P., and Omoto, K., 2000. Towards establishing criteria for identifying trigger mechanisms for soft-sediment deformation: a case study of Late Pleistocene lacustrine sands and clays, Onikobe and Nakayamadaira Basins, northeastern Japan. *Sedimentology*, 47: 1211–1226.
Kuenen, Ph.H., 1948. Slumping in the Carboniferous rocks of Pembrokeshire. *Quarterly Journal of the Geological Society of London*, 104: 365–380.
Macar, P., 1948. Les pseudo-nodules du Fammennien et leur origine. *Annales de la Société Géologique de Belgique*, 72: 47–74.
Morgan, J.P., and Andersen, H.V., 1961. Genesis and palaeontology of the Mississippi River mudlumps. *Louisiana Geological Survey Bulletin*, 35: 1–208.
Owen, G., 1987. Deformation processes in unconsolidated sands. In Jones, M.E., and Preston, R.M.F. (eds.), *Deformation of Sediments and Sedimentary Rocks*. Geological Society of London, Special Publication, 29, pp. 11–24.
Owen, G., 1995. Soft-sediment deformation in Upper Proterozoic sandstones (Applecross Formation) at Torridon, northwest Scotland. *Journal of Sedimentary Research*, A65: 495–504.
Ramberg, H., 1964. Selective buckling of composite layers with contrasted rheological properties, a theory for simultaneous formation of several orders of fold. *Tectonophysics*, 4: 307–341.
Ramberg, H., 1968. Instability of layered systems in the field of gravity. *Physics of the Earth and Planetary Interiors*, 1: 429–474.
Ramberg, H., 1981. *Gravity, Deformation and the Earth's Crust*, Academic Press, 2nd edn.

Cross-references

Ball-and-Pillow (Pillow) Structure
Flame Structure
Liquefaction and Fluidization
Pseudonodules

LUNAR SEDIMENTS

Introduction

The surface and surficial deposits on the Moon are radically different from Earth's. The Moon is completely dry and does not have an atmosphere; it is devoid of an internally driven magnetic field and plate tectonics is absent; large and small impacts through time have produced a thick regolith, and sediments so produced are transported ballistically. Regolith may be shock-indurated by fresh impacts, or may be buried under fresh ejecta that may be hot and may sinter unconsolidated grains together, to produce regolith breccias. All lunar samples are from the regolith; and, all remote sensing signals (*albedo*, *UV-Vis-IR reflectance* [of sunlight], *natural γ-ray*, and *X-ray fluorescence* [by solar X-rays] *spectra*) of the Moon are from its regolith. Most of the signals come from the uppermost 0.5 µm layer, that is, almost entirely from the dusty veneer of the Moon. Because the composition of the finest fractions of lunar soils is different from the bulk composition of any lunar soil, it is essential that all calibration be corrected back to the compositions of bulk soils. The remote sensing techniques and their lunar calibrations are used to explore all atmosphere-free rocky bodies in the solar system; hence the extreme importance of precise and accurate characterization of lunar soils in planetary exploration.

In the initial days of lunar sample return by Apollo missions, the sub-centimeter fraction of the samples was informally termed "fines" and then "lunar soil" came in vogue. In lunar science literature and in the community, "soil" is now entrenched with an informal size connotation; the word has no significance *vis a vis* its use for Earth materials in agricultural, engineering, geological, and hydrological sciences. In general, lunar soils are gray in color and consist of angular and irregularly shaped grains with a mean size of about 60 µm. Grains cemented with fresh glass produced by micrometeorite impacts on lunar soils produce a type of constructional particle called *agglutinate* (a term adapted from terrestrial pyroclastic vocabulary). Except for a minor (<2 percent) meteoritic component, lunar soil composition varies between anorthositic and basaltic. McKay *et al.* (1991) summarize lunar soil data gathered up to about 1990.

Petrographic composition

Two principal terrains configure the lunar surface. One is a high mountainous terrain (*highlands*) consisting of light colored anorthositic rocks that are 4.4 billion years or older in age. Flat lowlands, essentially impact basins (*mare*) filled with dark Fe-rich basalt flows ranging in age from about 3 billion years to 4 billion years in age, comprise the other. Impacts have mechanically disintegrated these rocks and have also indurated some of the clasts together. The regolith on the surfaces of these units generally reflects the composition and the color of the rock units below although there has been some mixing of distant impact ejecta in all local soils. The majority of lunar soils consist of five major grain types: mineral fragments (10–40 percent), igneous and metamorphic rock fragments (5–20 percent), breccias with various degrees of coherence (10–40 percent), and agglutinates (20–50 percent). Glass spherules of either impact or volcanic fire fountain origin make up a minor component (usually <5 percent) except in a few unusual soils and in some restricted size ranges. Genetically, lunar soil components may also be classified into two groups: bedrock derived and regolith derived. Agglutinates, regolith breccias, and glass spherules are regolith-derived components.

Orange soil

An orange layer in the regolith near the Shorty Crater in Mare Tranqullitatis consists of more than 90 percent high-titanium (up to 16 percent TiO_2) glass beads ranging in color from orange and red to black in hand specimen. At the bottom of a nearby drill core the soils consist of nearly 100 percent black beads in which skeletal crystals of olivine and ilmenite abound. These beads are products of volcanic fire fountaining with remnants of condensed volcanic gases on their surface. Beads of high magnesium iron-poor green volcanic glass are also found in significant amounts a few soils from the Apennine Mountains but not in layers.

Chemical composition

Chemical compositions of lunar soils also range from basaltic to anorthositic. Mixing of mare and highland components in different proportions give rise to diverse soil compositions. Differential comminution and differential agglutination of lunar soils have rendered different size fractions to be compositionally different from each other. The extent of mare ejecta mixing with highland soils and vice versa can be estimated better from the chemical compositions of soils than from modal compositions, partly because a few minor and trace element concentrations are characteristic of certain pristine igneous rocks. Average compositions (wt percent) of a few typical mare and highland soils are given below.

	Mare	Highland
SiO_2	40.6	45.0
TiO_2	8.4	0.54
Al_2O_3	12.0	27.3
Cr_2O_3	0.45	0.33
FeO	16.7	5.1
MnO	0.23	0.30
MgO	9.9	5.7
CaO	10.9	15.7
Na_2O	0.16	0.46
K_2O	0.14	0.17

Crustal composition of the moon

With no bedrock sampled so far, direct evidence of rock types in the crust of the Moon comes from regolith samples. Rock fragments from boulder sizes to sand sizes have been analyzed for chemical and mineralogic compositions, crystal chemistry, isotopic systematics, radiometric age dating, experimental petrology, and nuclear and ferromagnetic resonance measurements, to name a few. It is a tale of incredible provenance sleuthing. The crust of the Moon is inferred to be made mostly of ferroan and gabbroic anorthosites, norites enriched in K, REE, P, U, and Th, various kinds of Fe-rich (~20 percent FeO) basalts that have large ranges of K_2O (0.06–0.36 percent) and TiO_2 (0.4–13.0 percent), and a low-Fe basalt (~10 percent

FeO) with higher concentrations of K_2O (~0.5 percent or more), P (~2500 µg/g), and REE.

Grain size

Lunar regolith samples are not plentiful. NASA allocated approximately 0.5 g or 0.25 g of the submillimeter fraction of lunar soils for determining their grain size distribution. Sieving with minimal loss required more than 20 hours spanned in about 4 to 5 workdays per sample. The results were then combined with those of the subcentimeter fraction of the rather hastily sieved lunar regolith at NASA's curatorial facilities (Graf, 1993). The mean grain size (M_Z) of lunar soils range between 3ϕ and 4ϕ (125–62.5 µm) and their inclusive graphic sorting (σ_{IG}) generally vary between 2ϕ and 3.0ϕ, that is, they are poorly sorted by terrestrial standards. Many soils are coarsely skewed; much of the fine dust is locked up in agglutinates.

Vertical profile

Several trenches, drive tubes, and drill cores have penetrated and sampled the lunar regolith up to a depth of about 3 m. Understandably none has reached bedrock. Each section of the regolith shows evidence of layering, by way of color (shades of gray), grain size, chemical and mineralogical composition, and maturity. However, layers in closely spaced drill cores and in adjacent trenches (within a few meters) do not correlate indicating that impacts have moved the regolith considerably. Nearly all vertical profiles of soils show that the topmost layer is more mature (see below for the concept) than those below.

Origin and evolution

Large meteorite impacts up to about 1 b.y. ago have produced most of the regolith of the Moon. Impacts by micrometeorites (<1 mm) have continuously comminuted the regolith and are largely responsible for the production of lunar soils. Micrometeoritic impacts on lunar soils melt a miniscule amount, mostly the finest dust, at the point of impact. The melt incorporates soil grains as clasts and quenches quickly to produce agglutinates, that is, glass bonded aggregates. Micrometeoritic impacts affect only the topmost millimeters of the lunar regolith. With continued exposure, that is, increasing maturity, the soil in this layer becomes finer and finer simultaneously producing agglutinates that are larger. Over time the rate of comminution and agglutination becomes equal. In this steady state, continued exposure of lunar soils at the surface increases the abundance of agglutinates but does not reduce the mean grain size any more. However, slightly larger impacts excavate material from below, which may not be as mature, and mix the two together. Such events disturb the steady state. Most lunar soils presumably are not in a steady state but may be close to one.

Because the Moon does not have any magnetosphere or any atmosphere, energetic particles of the solar wind freely bombard the lunar surface and implant solar wind elements (SWE), especially hydrogen ions, in the outer rinds (~150 nm) of exposed soil grains. Abundance of SWE in lunar soils increases with maturity. Apart from melting a small fraction of lunar soil grains, micrometeoritic impacts also vaporize a part of the target and much of itself. Elemental dissociation takes place in the high temperature vapor from which nanophase Fe^0 metal condenses and is deposited on surfaces of soil grains. Additionally, the melt itself may undergo reduction reactions with implanted hydrogen and produce nanophase Fe^0 globules that are strewn in agglutinitic glass. The two processes not only darken lunar soils but also increase the abundance of nanophase Fe^0 with increasing maturity (Hapke, 2001; McKay et al., 1991).

A robust correlation between the abundance of agglutinates, SWE, nanophase Fe^0 (measured by ferromagnetic resonance), inverse grain size, and to some extent with soil color strongly supports the inferred processes of lunar soil evolution.

Abhijit Basu

Bibliography

Graf, J.C., 1993. Lunar soils grain size catalog. *NASA Reference Publication 1265*. Houston: NASA, 406pp.
Hapke, B., 2001. Space weathering from Mercury to the asteroid belt. *Journal of Geophysical Research-Planets*, **106**: 10039–10073.
McKay, D.S., Heiken, G.H., Basu, A., Blanford, G., Simon, S., Reedy, R., French, B.M., and Papike, J.J., 1991. The lunar regolith. In Heiken, G.H., Vaniman, D., and French, B.M. (eds.), *Lunar Sourcebook*. Cambridge University Press, pp. 285–356.

Cross-reference

Sediments Produced by Impact

M

MAGADIITE

Magadiite is a hydrous sodium silicate mineral that was discovered by Hans Eugster (1967) in late Pleistocene lake sediments at Lake Magadi in the southern Kenya Rift Valley. Although rare, magadiite is a well-known precursor of nonmarine chert. Quartzose "Magadi-type cherts", which have formed by the diagenetic alteration of magadiite, have been reported from rocks of Precambrian to Holocene age. Magadi-type chert is generally considered diagnostic of saline, alkaline lacustrine environments.

Magadiite [$NaSi_7O_{13}(OH)_3 \cdot 4H_2O$] and the related sodium silicates, kenyaite $Na_2Si_{22}O_{41}(OH)_8 \cdot 6H_2O$], makatite [$Na_2Si_4O_8(OH)_2 \cdot 4H_2O$], and kanemite [$NaHSi_2O_4(OH)_2 \cdot 2H_2O$], form mainly in hypersaline, sodium carbonate-rich brines that have a pH > 9 and consequently have high concentrations of dissolved silica (up to 2,700 ppm). Sodium silicate minerals are present at Lakes Magadi and Bogoria in Kenya, Lake Chad, and in some alkaline lakes in California and Oregon. When fresh, magadiite is white, soft, puttylike, and easily deformable, but it dehydrates rapidly on exposure to air, and hardens irreversibly to form a fine chalky aggregate. At high magnification, magadiite typically displays minute aggregates of thin platy monoclinic crystals (<10 μm) that may be equant or have a spherulitic habit (Icole and Perinet, 1984; Sebag et al., 2001).

At Lake Magadi, magadiite form beds a few decimeters thick, lenses, and concretions with irregular protrusions, and veins that transect bedding in zeolitic lacustrine silts (Figure M1). Elsewhere, magadiite also fills cracks in large (meter-scale) polygons (e.g., Alkali Lake, Oregon), or replaces plants (Sebag et al., 2001).

Formation of magadiite

Two general pathways have been proposed to explain the formation of magadiite in silica-rich $NaCO_3$ brines: a decrease in pH and evaporative concentration. At Magadi, subhorizontal beds have been explained by seasonal precipitation in a stratified saline, alkaline lake. Magadiite has been inferred to precipitate when dilute inflow waters flow across a dense, sodium carbonate brine rich in dissolved silica and lower the pH at the fluid interface (chemocline). At Lake Chad and in some American examples, magadiite may have precipitated by evaporative concentration or by capillary evaporation of saline, alkaline brines at a shallow subsurface water table. Other inferred mechanisms include subsurface mixing of dilute and saline, alkaline groundwaters, a reduction in pH of an alkaline brine resulting from an influx of biogenic or geothermally sourced CO_2, and precipitation from interstitial brines. Different sodium silicate minerals may form according to the concentrations of Na and SiO_2 in the brine.

Conversion to chert

At Magadi, beds of magadiite pass laterally into quartzose cherts confirming their genetic relationship. Magadi-type cherts form beds, plates, and lobate nodules, commonly with features indicating soft-sediment deformation upon compaction. Reticulate fine cracks, salt crystal pseudomorphs or molds, and relict spherulites are also common.

The conversion of magadiite to quartz is rapid and involves a loss of sodium, water, and volume. Leaching by percolating meteoric water, dehydration, and spontaneous recrystallization due to sorption of silica by clays, have been proposed. Intermediate diagenetic products, including kenyaite, amorphous silica and moganite, may form during the transformation (Icole and Perinet, 1984; Sheppard and Gude, 1986). Both magadiite and the associated cherts have a distinctive trace element signature unlike other cherts.

Although the genetic models are well established and magadiite is easily synthesized in the laboratory, few natural examples of modern precipitation have been documented. Recently Behr (2002) has shown that many of the cherts at Lake Magadi may have been precipitated as amorphous silica by bacteria and may not have had a sodium silicate precursor. Similarly, many ancient continental cherts resemble Magadi-type cherts in morphology and fabric, but are unlikely to have had a magadiite precursor. The discovery of magadiite in

Figure M1 *In situ* magadiite in the High Magadi Beds at the southern end of Lake Magadi, Kenya. Soft white masses and thin streaky lenses of magadiite are present in darker lacustrine silts that were deposited when Lake Magadi was a deeper, fresher, and probably stratified lake.

cavities in volcanic rocks in Namibia and Quebec, and in hot springs deposits at Lake Bogoria in Kenya, shows that it is not restricted to lacustrine environments. Much remains to be learned about natural sodium silicates and their diagenesis.

Robin W. Renaut

Bibliography

Behr, H.J., 2002. Magadiite and Magadi-chert: a critical analysis of the silica sediments in the Lake Magadi basin, Kenya. In Renaut, R.W., and Ashley, G.M. (eds.), *Sedimentation in Continental Rifts*. SEPM, Society for Sedimentary Geology, Special Publication, 73, (In Press).
Eugster, H.P., 1967. Hydrous sodium silicates from Lake Magadi, Kenya: precursors of bedded chert. *Science*, **157**: 1177–1180.
Icole, M., and Perinet, G., 1984. Les silicates sodiques et les milieux évaporitiques carbonatés bicarbonates sodiques: une revue. *Revue de Géologie Dynamique et de Géographie Physique*, **25**: 167–176.
Sebag, D., Verrecchia, E.P., Seong-Joo, L., and Durand, A., 2001. The natural hydrous sodium silicates from the northern bank of Lake Chad: occurrence, petrology and genesis. *Sedimentary Geology*, **139**: 15–31.
Sheppard, R.A., and Gude, A.J., 1986. Magadi-type chert—a distinctive diagenetic variety from lacustrine deposits. In Mumpton, F.A. (ed.), *Studies in Diagenesis*. US Geological Survey Bulletin, 1578, pp. 335–345.

Cross-references

Lacustrine Sedimentation
Siliceous Sediments

MAGNETIC PROPERTIES OF SEDIMENTS

Introduction—classes of magnetism

Sediments are aggregates of minerals, including silicates, carbonates and trace amounts of various oxides and sulfides. The magnetic properties of sediments can therefore be traced back to the magnetic properties of their mineralogical constituents which obviously may vary with sediment type, provenance area, and depositional setting. Magnetic properties of any material are directly related to its electronic structure, notably the presence or absence of uncompensated electron spins, and the presence or absence of collective spin behavior. Any moving electric charge creates a magnetic field; thus also electrons that orbit atomic nuclei and spin around their own axes, are microscopic magnets. The magnetic moment of an individual electron spin is termed the Bohr magneton.

The basic classes of magnetism are diamagnetism, paramagnetism and ferromagnetism. If there are no uncompensated electron spins, a material is *diamagnetic*. This situation occurs for example in calcite ($CaCO_3$), dolomite ($CaMg(CO_3)_2$), quartz (SiO_2), or forsterite ($MgSiO_3$). The *susceptibility* is the magnetic moment of a substance measured in an applied magnetic field divided by the applied field. Diamagnetic minerals have a small and negative susceptibility (Table M1), that is, the direction of the sample's magnetic moments opposes that of the applied field. Diamagnetism is independent of temperature.

If there are uncompensated electron spins in a crystal structure that behave independently of each other, the material is *paramagnetic*. This occurs for example in silicates and carbonates that contain iron or manganese (or some other transition elements). Biotite and clay minerals like illite often contain at least some Fe, and are therefore paramagnetic (Table M1). Also ankerite $Ca(Mg, Fe)(CO_3)_2$, is paramagnetic because the magnetic moment of uncompensated electron spins is much larger than the diamagnetic moment due to the orbit of the electrons around their respective atomic nuclei. Paramagnetic minerals have a small and positive susceptibility because magnetic moments tend to align to an applied field. The susceptibility of paramagnets is small because thermal jostling randomizes the electron spins. Paramagnetism is inversely proportional to (absolute) temperature.

Ferromagnetism "sensu lato" and permanent or remanent magnetization are restricted to minerals that have collective spin behavior. In certain crystalline solids that contain transitional elements in large amounts, the atomic nuclei are sufficiently close to each other to have the orbits of the outermost electrons overlapping with each other. In this situation, the uncompensated spins may line up and their individual magnetic moments can be added to get macroscopic magnetism. This situation can occur in metals, oxides and sulfides. The corresponding minerals are referred to as magnetic minerals. In nature, the transition element of most interest is iron. Various types of collective spin behavior are distinguished (Figure M2). Ferro- and ferrimagnetic material have a very high susceptibility while antiferromagnets have a susceptibility comparable to paramagnets (*cf.* Table M1). The field and temperature dependence of the susceptibility is complex. At a certain temperature—named the *Curie temperature*, characteristic of the material—thermal jostling becomes stronger than the collective spin interaction changing the ferromagnetic structure into a paramagnetic structure. On cooling through the Curie temperature, the ferromagnetic structure is restored.

Induced and remanent magnetization

Diamagnetic and paramagnetic moments are "induced" magnetic moments: they only exist in an applied magnetic field. The magnetic moment is proportional to the applied

Table M1 Typical magnetic properties of some natural minerals

	T_C	M_s	χ	ARM	IRM
Fe_3O_4 (soft)	578	92	560	18	9
Fe_3O_4 (hard)	578	92	400	110	22
$Fe_{2.4}Ti_{0.6}O_4$ (soft)	200	24	170	80	7
$Fe_{2.4}Ti_{0.6}O_4$ (hard)	200	24	200	480	12
αFe_2O_3	675	0.2–0.4	0.06–0.6	0.002	0.1–0.24
Fe_3S_4	350*	28	120	110	11
Fe_7S_8	325	18	0.1–50	80	4.5
$\alpha FeOOH$	120	0.001–0.2	0.7	0.0005	0.0005–0.005
αFe	765	220	2000	800	80
paramagnets	1/T	0	~1	0	0
biotite			~0.8		
illite			~0.15		
ilmenite			~1–1.15		
diamagnets	const	0	~−0.006	0	0
quartz			−0.0062		
$CaCO_3$			−0.0048		

soft = magnetically soft, i.e., multidomain; hard = magnetically hard, i.e., single domain. T_C = Curie temperature (°C). Values for M_s, χ, ARM and IRM are referring to values at room temperature. M_s = saturation magnetization (Am^2kg^{-1}); χ = low-field susceptibility ($10^{-6} m^3kg^{-1}$); ARM = anhysteretic remanent magnetization grown in a 100 mT peak alternating field with a 100 μT direct current bias field ($10^{-3} Am^2kg^{-1}$); IRM = isothermal remanent magnetization acquired in a 1T direct current field (Am^2kg^{-1}). * : greigite decomposes before its Curie temperature is reached. For paramagnets and diamagnets a few examples are given.

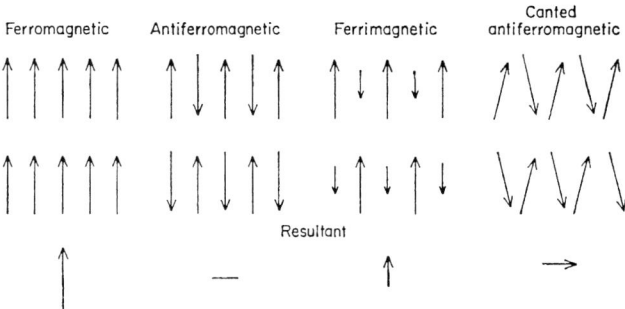

Figure M2 Schematized collective spin structures. Resultant indicates the resulting macroscopic magnetism.

field. The proportonality constant is the susceptibility (χ, also χ_{in} or κ), a material property. In low applied fields it is named the low-field or initial susceptibility. The susceptibility of individual diamagnetic and paramagnetic minerals add linearly to yield an overall susceptibility. For sediments which contain only trace amounts of magnetic minerals, the same holds for the low-field susceptibility due to ferromagnetism. The measured susceptibility is therefore a composite signal. It may be equated to the concentration of ferrimagnetic spinel, when its concentration is >~ 1 ‰ χ is frequently used as a proxy for the magnetic mineral concentration, because χ varitions due to grain size are comparatively small. In some sedimentary settings the *anisotropy of magnetic susceptibility* yields information for example concerning bottom currents, whereas in lithified sediments it may yield information on deformation patterns (e.g., Tarling and Hrouda, 1993). As a rule, comparatively slowly accumulating marine sediments are bioturbated and the sedimentary anisotropy information is lost.

Remanent magnetic moments remain in a zero field. Depending on the size of the magnetic particles, remanent magnetizations—once acquired—may last for geological times. This forms the foundation for the paleomagnetic signal recorded in rocks (e.g., Butler, 1992). When the magnetization of magnetic minerals is measured as function of the applied field, a so-called hysteresis loop (Figure M3) results. It requires energy to change the magnetic state of an ensemble of magnetic particles. Therefore, the response of the sample lags the field change, it shows hysteresis. The area contained in the loop represents the energy of the magnetic particles in the sample. The shape of the loop depends on the grain size of the magnetic particles, while the values for the coercive force (Table M1, M2) can be related to magnetic mineralogy. The concentration can be inferred from the saturation magnetization and saturation remanence. Thus, these loops contain information concerning the grain size, magnetic mineralogy, and the concentration of magnetic particles in a sample. Hysteresis parameters and other mineral-magnetic parameters yield information specifically concerning the magnetic minerals present in a sample. They are useful proxy parameters for (paleo)climate and (paleo)environmental processes (e.g., Maher and Thompson, 1999).

Natural magnetic minerals and mineral-magnetic parameters

By far the most important terrestrial magnetic mineral is magnetite (Fe_3O_4), one endmember of the titanomagnetite solid solution series. It may occur in trace amounts in virtually

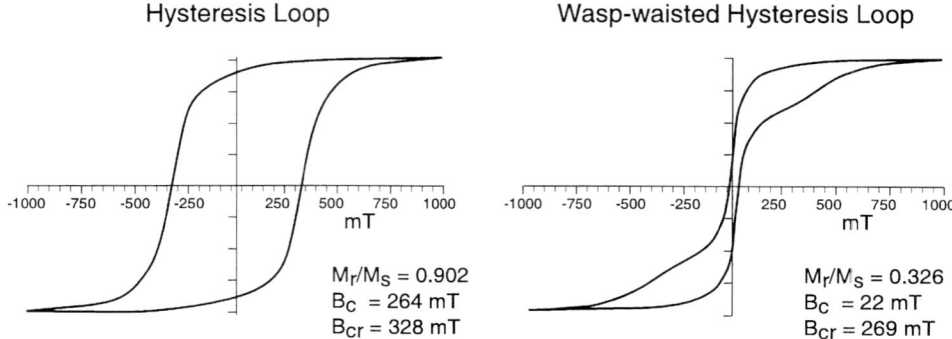

Figure M3 Examples of hysteresis loops. Left panel: typical single domain loop characterized by a high M_r/M_s ratio. Right panel: so-called wasp-waisted loop (narrow central part and wider loops in higher fields before closure in highest fields) generated by two distinct coercivity fractions, due to a soft magnetite-like magnetic mineral and a much harder antiferromagnetic mineral.

Table M2 Acronyms and definitions of some mineral-magnetic parameters

M_s	saturation magnetization expressed per unit volume (Am^{-1}); σ_s when expressed per unit mass ($Am^2\,kg^{-1}$).
M_r	remanent saturation magnetization per unit volume (Am^{-1}); σ_r when expressed per unit mass ($Am^2\,kg^{-1}$).
B_c	coercive force (mT). The oppositely directed magnetic field strength required to reduce the magnetic moment of a sample (assemblage of grains) to zero after saturation, measured in that applied field.
B_{cr}	remanent coercive force (mT). The oppositely directed magnetic field strength required to reduce the remanent magnetic moment of a sample to zero after saturation. By its virtue of being a remanence, its measurement takes place in zero field.
χ_{in}	low-field or initial susceptibility, the magnetic moment of a sample measured in a low applied magnetic field divided by the field value (mass-specific units $m^3\,kg^{-1}$; volume-specific units are dimensionless). Also χ or κ.
χ_{hf}	high-field susceptibility, the magnetic moment of a sample measured between two field-values higher than the fields needed to saturate the ferrimagnetic contribution to the magnetization. Also referred to as the paramagnetic susceptibility (units are the same as for χ_{in}).
ARM	Anhysteretic Remanent Magnetization (Am^2), acquired by the decay of an asymmetric alternating field (AF) from a peak value to zero (usually). In practice, AF demagnetization equipment is used, a small direct current (DC) bias field is imposed. ARM intensity is linear with the DC bias field for low concentrations of magnetic minerals to field values of $\sim 50\,\mu$ to $80\,\mu T$ (i.e., to values comparable to the intensity of the geomagnetic field). σ_{ARM}: ARM intensity per unit mass ($Am^2\,kg^{-1}$). Also κ_{ARM}.
χ_{ARM}	ARM intensity per unit volume divided by the value of the DC bias field (dimensionless SI units).
(S)IRM	(Saturation) Isothermal Remanent Magnetization. IRM is acquired by imposing a certain DC field to a sample at a constant temperature (usually room temperature), the maximum possible remanence value (after magnetic saturation of the sample) is termed SIRM, M_r, or M_{rs}.

any sediment or soil. Other natural magnetic minerals include hematite (αFe_2O_3), maghemite (γFe_2O_3), goethite ($\alpha FeOOH$), and the sulfides greigite (Fe_3S_4) and pyrrhotite (Fe_7S_8). Pyrite (FeS_2), ubiquitous in anoxic sediments, is not magnetic.

Mineral-magnetic measurements involve determination of the field- and temperature-dependence of various types of induced and remanent magnetizations. The techniques are sensitive, require little sample preparation, are rapid, often grain-size indicative, and usually non-destructive. The magnetic properties describe bulk properties of the sample, so they are complementary to micro-analytical techniques that usually analyze only a fraction of a total sample. The field-dependent parameters offer most grain-size dependent information, essential for a paleoclimatic and paleoenvironmental reconstruction. In addition they are comparatively easy acquired, thus suited for the processing of large numbers of samples.

Information concerning particle size can be gathered from magnetic measurements because the size of a particle is related to its magnetic domain structure. In very small grains (<~30 nm, all ranges refer to magnetite at room temperature) no stable domain structure can exist, these particles are termed superparamagnetic. Slightly larger equant particles (~30–~100 nm) are single domain, whereas particles larger than ~10 μm are multidomain, they contain many magnetic domains separated by domain walls. The grain-size range between single domain and multidomain is termed pseudo-single-domain. Single domain grains respond to a changing magnetic field by rotation of their magnetic moments, while multidomain grains basically respond by movement of domain walls, two processes that are characterized by a different field dependence. This very fact allows deduction of grain size from field-dependent measurements. The more commonly used mineral-magnetic parameters are summarized in Table M2.

For low concentrations (i.e., in the absence of magnetic interactions) mass-dependent parameters are linear with concentration at least to a first-order approximation which is applicable to the situation in sediments. By dividing two mass-dependent parameters, a concentration-independent variable is obtained which allows interpretation of grain-size trends. Field values, such as the coercivity parameters, are concentration-independent to a first-order approximation.

Large multidomain particles acquire comparatively low ARMs and IRMs. IRM, by its nature, is a much stronger remanence than ARM (Table M1). Single domain magnetite particles are particularly prone to acquire ARM, so the ARM/IRM ratio may be used to trace grain size if magnetite is magnetically the dominant mineral. Different coercivity fractions in a single sample may be quantified utilizing the IRM component analysis procedure that fits an optimal number of coercivity components to a measured IRM acquisition curve (Kruiver et al., 2001).

Sediment magnetism

In sediments, the magnetic mineral suite is a mixture of particles with two main origins: detrital particles and particles that were formed *in situ*. The detrital particles are eroded and subsequently transported from a source area by eolian, riverine, lacustrine or marine processes (only in very slowly accumulating abyssal plain sediments cosmogenic particles may constitute a significant source). Iron oxides are chemically sufficiently stable during transport to survive, although the particles abrase and surficially oxidize. Iron sulfides do not survive transport conditions, they are formed in the sediment column at the depositional site.

In situ formed authigenic particles are a consequence of diagenetic processes. These processes may also change the detrital input to some extent. Biological factors often are important. Biomineralization is a widespread phenomenon; for example a major part of the global calcium carbonate mass is of biogenic origin. In virtually all sedimentary environments and water-logged soils, bacteria which use iron in their metabolism are known to occur. The bacteria may use iron as a terminal electron acceptor: magnetite is formed outside the cells (extracellular magnetite) as a consequence of its low solubility product. The crystals are very fine-grained and usually are poorly crystalline. Alternatively, other bacteria may produce chains of highly crystalline SD magnetite grains (magnetosomes) inside their cells; these are referred to as magnetotactic bacteria (e.g., Hesse and Stolz, 1999).

The distinction between authigenesis and biogenesis is arbitrary. Authigenesis refers to inorganic chemical, rather than biological processes. However, usually at least one of the reactants in an "inorganic" chemical reaction process is bacterially produced or mediated. The sulfide in the pyrite formation process, which usually has a profound impact on the assemblage of magnetic minerals in sediments, is produced by sulfate-reducing bacteria. The sulfate reacts with available ferrous iron to form intermediate monosulfides, which usually react completely to form pyrite. Greigite and pyrrhotite have been reported as intermediate products in these reactions. Especially greigite is increasingly documented in a wide variety of (lake and marine) sediments (e.g., Snowball, 1997; Roberts et al., 1996). The reaction pathway which leads to the formation of pyrrhotite, is not yet clearly established. Both sulfides may form authigenically in sediments. Greigite in particular is documented in a variety of lake sediments that are characterized by much lower sulfate contents than marine conditions. Sulfate reduction under anoxic conditions usually does not lead to the formation of pyrite, the final product in marine sediments. Instead greigite or pyrrhotite, intermediate products, often remain. The hysteresis properties are very similar to those of magnetite and thermal methods are often required to unambiguously diagnoze greigite (and pyrrhotite) with magnetic means. Because greigite may occur in magnetotactic bacteria as well its occurrence can be used as a biomarker (Pósfai et al., 1998).

From the many examples that show the usefulness of magnetic properties to document paleoclimate and paleoenvironmental processes two are shown here: (1) the influence of diagenetic processes due to cyclic redox conditions in sediments from the Mediterranean Sea; and (2) the observation of climate variability in the North Atlantic including Heinrich events. Noteworthy as well are the well-known loess-paleosol records from the Chinese Loess Plateau that have been shown to be a continuous recorder of paleoclimate throughout the Quaternary (e.g., Maher and Thompson, 1999) with the paleosols having higher magnetic susceptibility than the intercalating loess. Although it has been known for a long time that the paleosols are correlated to warmer and moister climatic periods, only recently a robust orbital chronology could be formulated for the loess/paleosol sequence (Heslop et al., 2000).

The Mediterranean Sea

Apart from sulfide formation, the influence of diagenesis on the mineral-magnetic record can be substantial because iron through its two redox states is an important element in many diagenetic reactions, including biologically mediated ones. From the Miocene into the Quaternary many Mediterranean sediments are characterized by astronomically controlled sequences of anoxic and suboxic sediments (e.g., Hilgen, 1991). The dominant source of the original magnetic minerals is thought be dust from the Sahara with minor riverine input depending on the location in the Mediterranean Sea. The role of biogenic magnetite in oxic sediment seems to be minor. Under anoxic conditions, layers enriched in organic material may be deposited, so-called sapropels. During sapropel deposition, below each sapropel there is a zone where the formerly existing magnetic minerals have been by and large reductively dissolved; the iron is precipitated as pyrite (Figure M4). When oxic conditions are restored in the water column these sapropels are (partially) reoxidized starting from their top parts. During this oxidative diagenesis, pyrite and other sulfides are oxidized; the iron precipitates amongst others as magnetic oxides in the zone of the sapropel that is slowly oxidizing away. Compared to "average" sediment, the zones straddling sapropels are thus characterized by low magnetic intensities below the visible sapropel and higher magnetic intensities in the zone directly above the sapropel. In

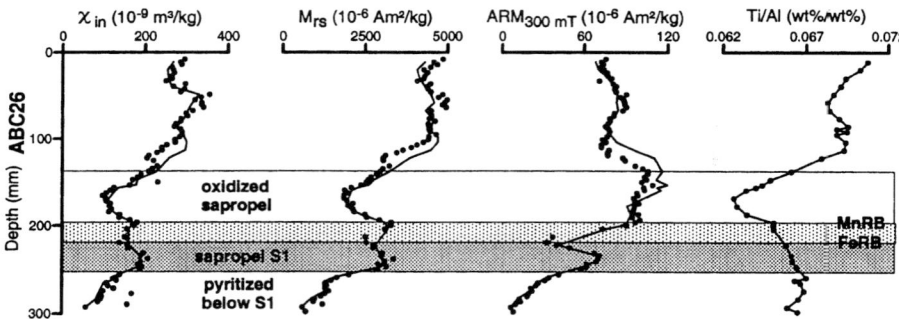

Figure M4 Susceptibility (χ_{in}), saturation remanence (M_{rs}), ARM acquired in 300 mT peak alternating field (ARM_{300mT}) and Ti/Al as a function of depth in box core ABC26 (lat 33°21'; lon 24°55'E) from the eastern Mediterranean. χ_{in} and M_{rs} follow the Ti/Al ratio that is a proxy for dust content. ARM is indicative of the precipitated magnetite in the oxidized sapropel. MnRB = manganese redox boundary; FeRB = iron redox boundary (after Passier et al., 2001).

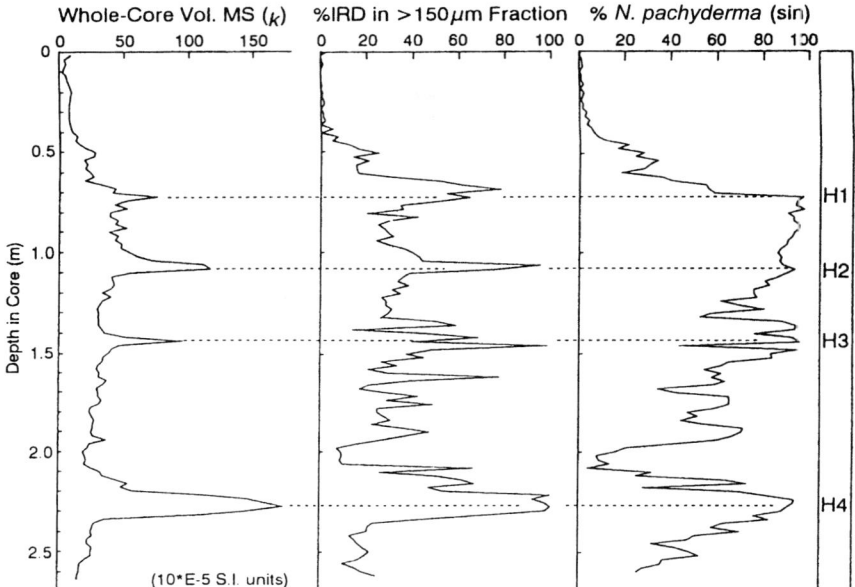

Figure M5 Downcore profiles of magnetic susceptibility (κ), the percentage of ice-rafted debris (IRD) in the >150 μm fraction, and the percentage of N. pachyderma in relation to the Heinrich events in core BOFS-5K (lat 50°41'N, lon 21°52'W). Susceptibility is most easily acquired and offers best discrimination in this oceanographic setting (after Robinson et al., 1995).

particular the ARM record traces the diagenetic enrichment. Part of the precipitated magnetite has properties that makes a biogenic origin likely.

Most research has focused on the youngest sapropel (e.g., Passier et al., 2001) but older sapropels have a similar signature. Undoubtly, sapropels represent extremely varying redox conditions; similar, albeit less extreme magnetic changes have been reported in other sedimentary systems as well (e.g., Tarduno and Wilkinson, 1996).

The north Atlantic

The circum-Atlantic bedrock is often composed of highly magnetic crystalline basement rocks and basaltic volcanics. Weathering of these rocks provides comparatively magnetic detritus which makes the detritus-rich North Atlantic sediments eminently suited for paleoclimate studies with magnetic methodology. The contrast with biogenic carbonate and opal that are both diamagnetic is very strong. The orbital periods of c.100, 41 and 21 k year are reflected in many sediment records. The glacial sediments are characterized by higher concentrations and the presence of coarser-grained pseudo-single-domain to multidomain magnetite (Robinson, 1986). Superposed on this periodicity much more rapid, millenial-scale, climate changes may be recorded as well, if the sedimentation rate is sufficiently high (> ~10 cm/k year) and the influence of diagenesis on the magnetic properties can be shown to be insignificant (Stoner and Andrews, 1999). Heinrich layers are

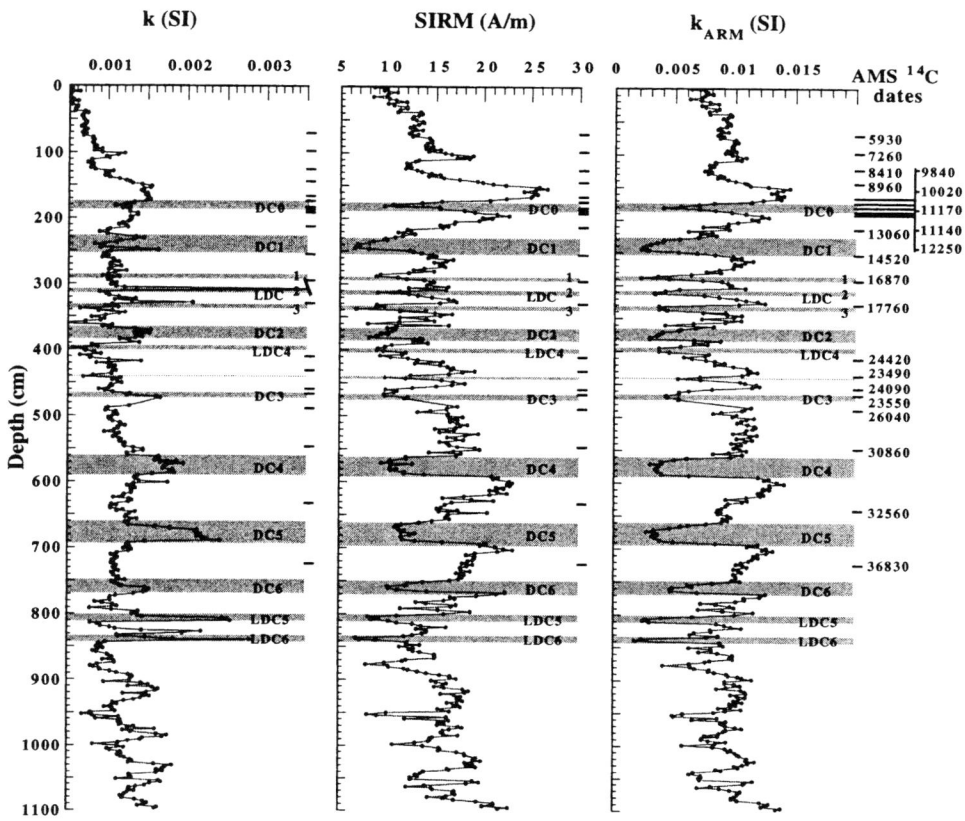

Figure M6 Downcore profiles of magnetic susceptibility (k), saturation remanence (SIRM), and anhysteretic susceptibility (k_{ARM}) in piston core P-094 (lat 50°12'N; lon 45°41'W) from the Labrador Sea. DC = detrital carbonate content between 20 wgt percentage to 60 wgt percentage; LDC = low detrital carbonate (<10 wgt percentage). All layers are associated with increased magnetic grain size whereas only part of them are identified by increased susceptibility, showing the merit of proxy parameters based on remanent magnetization properties (after Stoner et al., 1996).

characterized by a high content of iceberg-rafted debris (IRD) and can therefore be traced easily by magnetic susceptibility measurements (Figure M5).

Comparatively far out of the glacier margins IRD and Heinrich layers are closely associated. However, approaching those margins, this relationship becomes less clear: many more sediment layers contain at least some IRD. The IRD in the Heinrich layers and the "background" IRD appear to be different. When normalized for carbonate, the concentration and size of the magnetic grains in Heinrich layer IRD is larger than in "background" IRD (Robinson et al., 1995) which makes mineral-magnetic parameters attractive proxies. A piston core taken at Labrador rise (Figure M6) shows at least 13 distinct rapidly deposited layers. Note that expression of the low-field susceptibility of the layers differs through the core whereas the profiles of the remanent magnetization parameters SIRM and k_{ARM} offer much more structure than would be gathered from visible lithological changes only (Stoner et al., 1996).

Mark J. Dekkers

Bibliography

Butler, R.F., 1992. Paleomagnetism—from magnetic domains to geologic terranes, website: www.geo.arizona.edu/Paleomag/book.

Heslop, D., Langereis, C.G., and Dekkers, M.J., 2000. A new astronomical time scale for the loess deposits of Northern China. *Earth and Planetary Science Letters*, **184**: 125–139.

Hesse, P., and Stolz, J.F., 1999. Bacterial magnetite and the quaternary climate record. In Maher, B.A., and Thompson, R. (eds.), *Quaternary Climates, Environments and Magnetism*. Cambridge: Cambridge University Press, pp. 163–198.

Hilgen, F.J., 1991. Astronomical calibration of Gauss to Matuyama sapropels in the Mediterranean and implication for the Geomagnetic Polarity Time Scale. *Earth and Planetary Science Letters*, **104**: 226–244.

Kruiver, P.P., Dekkers, M.J., and Heslop, D., 2001. Quantification of magnetic coercivity components by the analysis of acquisition curves of isothermal remanent magnetisation. *Earth and Planetary Science Letters*, **189**: 269–276.

Maher, B.A., and Thompson, R. (eds.), 1999. *Quaternary Climates, Environments and Magnetism*. Cambridge: Cambridge University Press.

Maher, B.A., and Thompson, R. 1999. Palaeomonsoons I: the magnetic record of palaeoclimate in terrestrial loess and palaeosol sequences. In Maher, B.A., and Thompson, R. (eds.) *Quaternary Climates, Environments and Magnetism*. Cambridge: Cambridge University Press, pp. 81–125.

Passier, H.F., de Lange, G.J., and Dekkers, M.J., 2001. Rock-magnetic properties and geochemistry of the active oxidation front and the youngest sapropel in the Mediterranean. *Geophysical Journal International*, **145**: 604–614.

Pósfai, M., Buseck, P.R., Bazylinski, D.A., and Frankel, R.B., 1998. Reaction sequences of iron sulfide minerals in bacteria and their use as biomarkers. *Science*, **280**: 880–883.

Roberts, A.P., Reynolds, R.L., Verosub, K.L., and Adam, D.P., 1996. Environmental magnetic implications of greigite (Fe_3S_4) formation in a 3 million years lake sediment record from Butte Valley, northern California. *Geophysical Research Letters*, **23**: 2859–2862.

Robinson, S.G., 1986. The late Pleistocene palaeoclimate record of North Atlantic deep-sea sediments revealed by mineral-magnetic measurements. *Physics of the Earth and Planetary Interiors*, **42**: 22–47.

Robinson, S.G., Maslin, M.A., and McCave, N., 1995. Magnetic susceptibility variations in upper Pleistocene deep-sea sediments of the N.E. Atlantic: implications for ice rafting and palaeocirculation at the Last Glacial Maximum. *Paleoceanography*, **10**: 221–250.

Snowball, I.F., 1997. Gyroremanent magnetization and the magnetic properties of greigite-bearing clays in Southern Sweden. *Geophysical Journal International*, **129**: 624–636.

Stoner, J.S., and Andrews, J.T., 1999. The North Atlantic as a Quaternary magnetic archive. In Maher, B.A., and Thompson, R. (eds.) *Quaternary Climates, Environments and Magnetism*. Cambridge: Cambridge University Press, pp. 49–80.

Stoner, J.S., Channell, J.E.T., and Hillaire-Marcel, C., 1996. The magnetic signature of rapidly deposited detrital layers from the deep Labrador Sea: relationship to North Atlantic Heinrich Layers. *Paleoceanography*, **11**: 309–325.

Tarduno, J.A., and Wilkinson, S.L., 1996. Non-steady state magnetic mineral reduction, chemical lock-in, and delayed remanence acquisition in pelagic sediments. *Earth and Planetary Science Letters*, **144**: 315–326.

Tarling, D.H., and Hrouda, F., 1993. *Magnetic Anisotropy*. Cambridge: Cambridge University Press.

Cross-references

Authigenesis
Bacteria in Sediments
Bedding and Internal Structures
Biogenic Sedimentary Structures
Carbonate Mineralogy and Geochemistry
Cyclic Sedimentation
Diagenesis
Diagenetic Structures
Eolian Transport and Deposition
Fabric, Porosity, and Permeability
Geophysical Properties of Sediments
Hydroxides and Oxyhydroxide Minerals
Iron–Manganese Nodules
Milankovitch Cycles
Neritic Carbonate Depositional Environments
Paleocurrent Analysis
Red Beds
Sapropel
Sulfide Minerals in Sediments
Weathering, Soils, and Paleosols

MASS MOVEMENT

Introduction

Mass movement refers to processes by which soil and rock materials move downslope under their own weight (Selby, 1993; Turner and Schuster, 1996). The six principal movement mechanisms are creep, sliding, flow, toppling, free-fall, and rolling. Given the spectrum of possible processes, mass movement deposits exhibit a wide range of sedimentological features and surface morphology. Most deposits are typically poorly sorted, unstratified clastic aggregates (i.e., diamictons) with predominantly angular clasts of local provenance. Deposits associated with rapidly flowing events, such as debris flows and rock avalanches, often exhibit crude stratification and inverse grading. The typical morphologies of mass movement deposits are colluvial cones and fans, planar colluvial blankets, linear ridges and lobes, and complex hummocky forms. The latter have often been confused with glacial drift deposits. Given the shear strength of many landslide materials, some are emplaced at relatively steep angles (5°–15°) and therefore violate the Principle of Original Horizontality. Such deposits require cautious structural interpretation when encountered in ancient lithified assemblages.

Strength of hillslope materials

Detachment of material from a slope as a slide or fall requires that shear stress, due to the downslope component of the bulk weight of the material, equals shear strength along a plane of detachment, as described by the Mohr–Coulomb equation:

$$s = c' + (\sigma - u)\tan \phi' \qquad \text{(Eq. 1)}$$

where s is shear strength, c' is cohesion, σ is total normal stress, u is pore water pressure, and ϕ' the angle of shearing resistance. Term $(\sigma - u) = \sigma'$ is effective normal stress (Craig, 1997). Cohesion is associated with either clay or cementing material. Clay cohesion is a function of the load history of a material, being relatively low ($<10\,\text{kPa}$) in recently deposited muddy deposits, and relatively high ($>100\,\text{kPa}$) in overconsolidated clay-shale formations (Selby, 1993; Craig, 1997). Angle of shearing resistance, ϕ', controls the steepest stable slope for a dry frictional material, and depends on particle angularity, mineralogy, and packing density. It is highest in dense deposits of angular quartz-feldspathic grains (35°–45°), and lowest in low-density deposits of clay (5°–15°) (Kenney, 1984). Clay mineral species also controls frictional strength, being highest in kaolinite, intermediate in illite, and lowest in montmorillonite. Shear strength also depends on material displacement history. First-time failures in soil and rock mobilize peak strength, comprising *in situ* cohesion and friction due to maximum grain or rock-joint interlocking (Craig, 1997). Small shear displacements, of the order 0.1 m, cause substantial loss of cohesion and a reduction of frictional resistance due to shear dilatancy and the development of preferred grain alignments. The resultant ultimate or residual strength is appreciably lower than the original peak value (Craig, 1997). This means that surfaces which have experienced past movements are predisposed to continued movement. Similar reduced strength commonly occurs along tectonically sheared rock surfaces.

On slopes where soil is significantly reinforced by tree roots, a third strength component termed root reinforcement, c'_r, exists. It is significant on slopes with relatively thin ($<2\,\text{m}$) mantles of colluvial material or weathered glacial drift. Since most mountain soils are relatively thin, landslide events, and hence regional sediment flux to depositional areas, are strongly regulated by plant cover and its degree of anthropogenic disturbance (Sidle *et al.*, 1985).

Causes of mass movement events

Slope failure as a landslide may be induced by an increase in shear stress, a decrease in shear strength or both. Shear stress increase, coupled with shear strength decrease, typically occur when materials are water saturated by rainfall or snowmelt.

Wetting increases bulk weight and hence shear stress, while changing effective stress as the groundwater table rises above the level of the potential slip plane. Such concomitant changes in stresses may occur within a few hours, explaining why many slope failures occur during major storms (Selby, 1993).

On a longer timescale, slope undercutting by rivers and glaciers causes slope steepening and hence shear stress increase. This induces slow creep movements within a slope, causing a progressive reduction in both cohesion and frictional resistance over time and hence long-term material fatigue. Seismic shock often provides the final trigger for many catastrophic slope failures in a wide range of materials, ranging from hard, jointed rock to unconsolidated clays.

Types of mass movement

A common failure type is the *translational slide*, involving a slab-shaped detachment along a well-defined rupture surface. In rock slopes, *rock slides* typically follow discontinuities such as bedding, joints, foliation and faults. Sliding is feasible only where the dip of the discontinuity is less than the slope angle, allowing the critical plane to crop out on a slope (Hoek and Bray, 1981). Large rock slides ($20\,Mm^3$ to $200\,Mm^3$) are often seismically triggered and produce *rock avalanches*, since total disintegration of the mass allows flow to occur. Entrainment of saturated valley-floor material provides a liquefied base for the avalanche, allowing velocities of 10 m per second to 30 m per second. Exceptionally large liquefied failures, termed *lahars*, result from collapse of stratovolcanoes (Vallance and Scott, 1997). The largest lahar deposits approach $10^9\,m^3$, cover tens of km^2, and rank as the largest landslides on Earth. High mobility of lahars is caused by saturation by melting of ice and snow, drainage of volcanic crater lakes, and entrainment of saturated alluvial material.

On soil and debris covered slopes, translational *debris slides* occur along the soil–bedrock interface since soil shear strength is lower than that of rock. Debris slides are typically $10^2\,m^3$ to $10^3\,m^3$ and are triggered mainly by rainstorms. Scour of saturated material along a stream channel transforms many debris slides into *debris flows* (Iverson *et al.*, 1997). Debris flows occur in every climatic zone, attaining velocities of 5 m per second to 10 m per second and typical volumes of $10^3\,m^3$ to $10^5\,m^3$.

On slopes cut into thick clay or shale, deep-seated *slumps* are common, involving a roughly cylindrical-arc slip surface oblique to the stratification with backward rotation of the landslide mass. Slumping creates a steep, debuttressed headscarp prone to further slumping, a process termed retrogression. Prolonged retrogression may produce an elongated *earth flow*, up to several kilometers in length, moving at $1\,m\,a^{-1}$ to $5\,m\,a^{-1}$, with a total volume exceeding $10^8\,m^3$. In low density saturated silts and clays, liquefaction may occur at depth as the deposit collapses under shear loads. Frictional strength declines notably as pore pressure rises, allowing rapid acceleration as a *quick clay flow-slide* (Turner and Schuster, 1996). Flow-slide retrogression may occur at meters per minute, with flow velocity attaining 1 m per second to 5 m per second.

In permafrost areas, seasonal thaw of ice-rich soil produces a 0.5 m to 1.0 m thick water-saturated active layer above permafrost material. *Solifluction* involves movement of the active layer at $10^{-2}\,m\,a^{-1}$ to $10^{-1}\,m\,a^{-1}$ on slopes as low as $2°$. Movement on slopes well below the theoretical lowest angle predicted by Mohr-Coulomb theory occurs by a combination of *frost creep* and *gelifluction* (Harris, 1987). Frost creep is soil movement by alternating cycles of frost heave and thaw-settlement in soils subjected to ice segregation. Gelifluction is a type of soil flow, occurring during seasonal thaw, in which high water pressures develop during melting of low density, ice-rich soil. Overburden pressure is largely borne by the porewater, allowing low effective stress to develop. A transient phase of negligible frictional strength occurs, allowing soil to move on slopes substantially flatter than possible in saturated, consolidated soils. On steeper slopes in permafrost areas, *skin flows* involve rapid downslope detachment of the entire active layer. Retrogressive *thaw slumps* are triggered by river undercutting of slopes and exposure of ice-rich sediments. Thaw slumps are the largest slope failures found in permafrost areas.

Michael J. Bovis

Bibliography

Craig, R.F., 1997. *Soil Mechanics*. 6th edn., London: Spon Press.
Harris, C., 1987. Mechanisms of mass movement in periglacial areas. In Anderson, M.G., and **Richards** (eds.), *Slope Stability*. J. Wiley & Sons, pp. 531–559.
Hoek, E., and Bray, J.W., 1981. *Rock Slope Engineering*. 3rd edn., London: Institution of Mining and Metallurgy.
Iverson, R.M., Reid, M.E., and LaHusen, R.G., 1997. Debris flow mobilization from landslides. *Annual Review of Earth and Planetary Sciences*, **25**: 85–138.
Kenney, T.C., 1984. Properties and behaviours of soils relevant to slope stability. In Brunsden, D., and Prior, D.B. (eds.), *Slope Instability*. John Wiley & Sons, pp. 27–66.
Selby, M.J., 1993. *Hillslope Materials and Processes*. 2nd edn, Oxford University Press.
Sidle, R.C., Pearce, A.J., and O'Loughlin, C.L., 1985. *Hillslope Stability and Land Use*. American Geophysical Union.
Turner, A.K., and Schuster, R.L. (eds.), 1996. *Landslides: Investigation and Mitigation*. Washington, DC: National Academy Press, pp. 585–606.
Vallance, J.W., and Scott, K.M., 1997. The Osceola Mudflow from Mount Rainier; sedimentology and hazard implications of a huge clay-rich debris flow. *Geological Society of America Bulletin*, **109**: 143–163.

Cross-references

Atterberg Limits
Avalanche and Rock Fall
Bentonites and Tonsteins
Debris Flow
Earth Flows
Grain Flow
Gravity-Driven Mass Flows
Liquefaction and Fluidization
Mudrocks
Slide and Slump Structures
Slurry

MATURATION, ORGANIC

Organic maturation refers to the progressive and mainly irreversible transformation of organic matter (OM) in response initially to biological and later to thermal energy. Organic maturation is also referred to as organic diagenesis, organic metamorphism, and as coalification when referring to coal.

Organic matter is a minor component in most sedimentary rocks, however its importance by far outweighs its abundance. Organic matter comprises the fossil fuel resources petroleum and coal. The products of OM diagenesis govern many mineral reactions and are important in the transport and deposition of many ore minerals. Studies of organic maturation have mainly been pursued form two separate avenues: organic geochemists have investigated the chemical reactions and products of organic maturation particularly from the perspective of petroleum whereas coal petrologists have studied organic maturation of coal mainly microscopically.

Introduction

Diagenesis of OM has been considered in terms of three or four stages by various authors (e.g., Tissot *et al.*, 1974; Bustin, 1989; Diessel, 1992). *Eogenesis* has been used in reference to biological, physical and chemical changes in OM that occur at low temperatures where reactions are mainly biologically mediated. *Diagenesis*, at higher temperatures normally associated with greater depths of burial, is successively referred to as *catagenesis* and *metagenesis*. Diagenetic reactions of OM due to contact with free oxygen or meteoric water has been refereed to as *telogenesis*. The boundary between different diagenetic stages is gradational as a result of the heterogeneity of the OM and biological processes. The different coal ranks from peat to lignite, subbituminous, high-, medium-, and low-volatile bituminous coal, semi-anthracite and anthracite, are names sequentially assigned to successively more carbon-rich and volatile-poor coals that form in response to progressive diagenesis.

Substantial research exists on the kinetics of organic diagenesis because of the importance of predicting the timing and volume of hydrocarbon generation from the diagenesis of kerogens. As well, sophisticated numerical models have been proposed that attempt to predict organic maturation based on the thermal history of the strata and the activation energy of various kerogen types. Because many of the products of organic diagenesis are metastable and rocks are open to migration of products (hydrocarbons), thermodynamic equilibrium is rarely ever established.

Eogenesis

Eogenesis refers to low temperature diagenesis of OM, where reactions are at least in part biochemical resulting from metabolic processes of organisms, and the effects of temperature and pressure are subordinate (Figure M7). Due to the diversity of microbial processes and organic substrates, reactions during eogenesis are complicated and poorly understood. Different depositional settings give rise to different diagenetic processes and sources of OM, which in turn influence the type and rate of degradation of the OM mainly by bacteria, actinomyces and fungi. For OM to be preserved at all to further undergo diagenesis requires rather specific depositional conditions. It is clear that in subaerial environments chemical oxidation and aerobic microbial decomposition invariable leads to complete decomposition of the OM. Anaerobic microbial processes are generally slower and hence eogenetic reactions are commonly retarded such that there is greater potential for a portion of the OM to be preserved. For

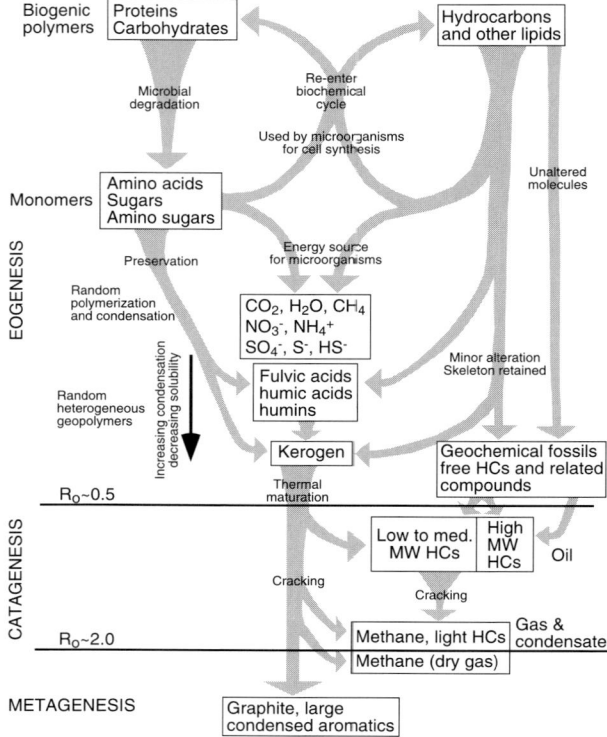

Figure M7 Schematic diagenetic pathway of organic matter in sediments. Modified after Barnes et al. (1990).

OM to survive eogenesis and to move on to catagenesis, it must be rapidly removed from oxic conditions. Proteins, carbohydrates (sugars, cellulose, and chitin), lipids (fats, waxes, and steryl esters) and lignins are the main biopolymers contributed by organisms to the sediment. Biopolymers are variously depolymerized during eogenesis forming what is colloquially referred to geopolymers, which are essentially fulvic and humic acids, humins and kerogens (Bustin, 1999). Geopolymers are defined on their solubility, their molecular weight and, because they form by random recombination, their structures. Geopolymers vary depending on available monomers and diagenetic conditions. With increasing cross-linkage and loss of acidic functional groups (carboxyl and phenolic hydroxyl groups), the humic and fulvic acids lose their base solubility and form an insoluble humin/kerogen residue. The term humin is used for base insoluble fraction of the OM in soils, whereas kerogen refers to the organic fraction in rocks, which is insoluble in organic solvents. The details of diagenetic reactions leading to kerogen formation remain poorly understood. Some kerogen forms through loss of functional groups and cross-linkage formation early in diagenesis, but others later during catagenesis and still other kerogen appears to arise directly from monomers without an intervening humic stage particularly under anoxic conditions. The lipid-rich components are particularly resistant to biochemical processes and unless degradation is particularly intense, these components survive eogenesis. An important product of eogenesis is the formation of mainly methane (up to 99 percent), carbon dioxide (0 percent to 8 percent) and minor heavier gases by fermentation reactions. This biogenic gas may source significant economic accumulations both in the coal (coalbed methane) and conventional reservoirs. Carbon dioxide generated together with organic degradation products influences the Eh, pH, and ionic composition of pore waters and thus plays a role in diagenesis of minerals.

The end of eogenesis generally occurs when the microscopic reflectance of coal maceral vitrinite is about 0.5 percent, which also corresponds to the boundary between subbituminous coal and high-volatile bituminous coal. At this stage base-soluble compounds are usually negligible and thermal energy is required for further diagenesis.

Catagenesis

Diagenesis beyond the eogenetic stage is in response to thermal exposure (temperature and time), generally accompanying burial in a sedimentary basin or high heat flow adjacent intrusive or extrusive bodies (Figure M7). The rate of organic diagenesis is exponential with temperature and linear with time. Confining pressure has little effect during organic diagenesis apart from compaction during the peat and lignite stages and some experiments suggest high confining pressure retard catagenetic and metagenetic reactions. Catagenetic reactions take place at temperatures of around 40°C to 150°C and moderate pressures (30–150 MPa), often producing significant amounts of petroleum, methane and carbon dioxide. The amount and type of liquid hydrocarbons generated varies with the type of kerogens present. Generally most coals are considered to have little liquid hydrocarbon potential whereas algal-rich, fine-grained rocks are excellent liquid hydrocarbon source rocks. During catagenesis coal rank progressively increases from subbituminous to anthracite and kerogen increases in carbon content, decreases in volatiles, aliphatic chains are cleaved and the size of aromatic clusters grows through condensation and development of ordering. Coincident with the generation of gas, the surface area of the kerogen increases such that significant quantities of the generated gas are adsorbed in the matrix (up to about 30 m^3/tonne for coal) giving rise to coalbed methane. Coalbed methane (and gas shale), is an important energy resource, however, it is also a potential explosive hazard during coal mining. The exact level of catagenesis for liquid hydrocarbon generation (oil birth line) varies with the chemical composition of the kerogen. The catagenetic level at which liquid hydrocarbons begin to crack is referred to as the oil death line. The zone of catagenesis between the birth and death "lines" is referred to as the *oil window* or oil kitchen (Figure M7).

Organic matter undergoes noticeable transformations that are observable microscopically even at the beginning of the eogenetic stage. The fraction of the OM derived from higher plants, mainly cellulose and lignin, give rise to the microscopically defined maceral huminite (at low maturation levels) and later vitrinite. The reflectivity of huminite/vitrinite progressively increases, as observed microscopically in reflected light through diagenesis. This degree of reflectance can be measured with a specially designed microscope and is referred to as vitrinite reflectance. Vitrinite reflectance is the most widely used method for determining the degree of diagenesis of coals and OM in rocks. During catagenesis vitrinite reflectance progressively increases from about 0.5 percent to 2.0 percent (Figure M7). Liptinite components change little until the oil birth line which corresponds to a vitrinite reflectance of about 0.5 percent and then it become progressively brighter and merge in brightness with vitrinite at a reflectance of about 1.4 percent (the death line of oil). The fraction of the OM rich in lipids is microscopically darker than vitrinite and is referred to as liptinite. The liptinite macerals change little with diagenesis until the onset of the oil window at which time there is a "jump" in diagenesis as HCs are generated (Figure M7). That microscopically defined fraction of the OM that is semi charred or charred at the site of deposition (charcoal) is referred to respectively as semifusinite and inertinite. These materials being variously precoalified undergo little change during diagenesis.

Metagenesis

Metagenesis refers to the level of diagenesis of the OM at which crystalline ordering of the OM begins (Figure M7). Aromatic clusters increase in size and C–C bonds are broken generating additional methane. Any remnant aliphatic molecules and any previously formed heavier hydrocarbons are cracked to methane. The remaining kerogen becomes further enriched in carbon with loss of volatiles. Microscopically the reflectance of vitrinite progressively increases from about 2.0 percent to 10 percent or more and the liptinite component and lower reflecting inertinite components become visually indistinct from vitrinite. Early studies argued that graphite is the end point of coal diagenesis however field and experimental evidence now suggest that the activation energy for formation of graphite is so high that it is unlikely that it will form in nature by thermal exposure alone (Figure M7).

Telogenesis

Organic matter diagenesis is progressive and for the most part irreversible. Organic matter at all stages of diagenesis is reduced and subsurface environments are typically anaerobic. Where, however, OM is uplifted to the surface or comes in contact with oxygenated meteoric water, it is subjected to oxidation. Oxidative processes are collectively refereed to as teleogenesis and result in marked changes in the OM. During oxidation nonaromatic groups are selectively removed and acidic functional groups such as carboxyl, carbonyl and phenolic hydroxyl are formed leading to production of humic acids. The oxidative diagenetic processes have a deleterious effect on coal quality: coking coals loose their ability to form coke, thermal coals decline in calorific value and coal particles loose their hydrophobicity and thus are difficult to separate from mineral matter during coal processing (Copard *et al.*, 2002; Teichmüller, 1982). If oxidation is intense it may be microscopically evident by distinctive darkish halo along boundaries of organic particles. With pooled hydrocarbons two distinct telogenesis processes occur: (1) water washing which results in selective removal of lighter hydrocarbons by groundwater; and (2) bacterial degradation which can lead to loss of normal alkanes and isoprenoids leaving a heavier oil residue enriched in cycloalkanes and aromatics. These two processes are considered responsible for formation of the heavy oil deposits.

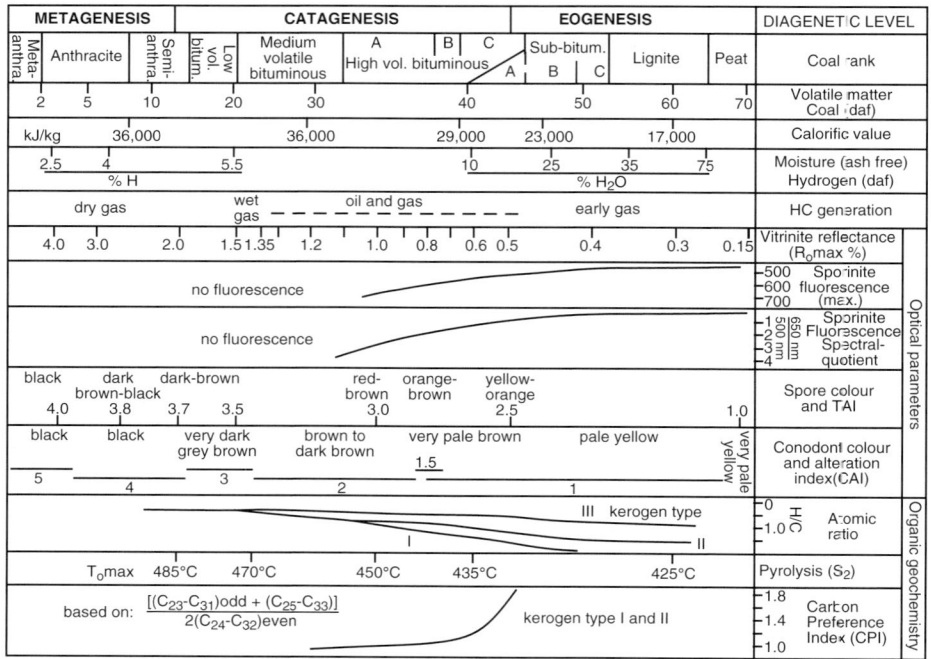

Figure M8 Correlation of major organic maturation indices. The correlation shown is based on virtinite reflectance. Modified after Bustin et al. (1990).

Quantifying organic diagenesis

Both the kerogens and bitumen fraction of the OM have been used to quantify diagenesis by chemical, physical and microscopic methods (Bustin et al., 1990; Tissot and Welte, 1978). In Figure M8, the correlation between the major different diagenesis indices is provided. A few of the more common techniques are summarized below. The reflectance of vitrinite, and its precursor huminite is the most widely accepted method of quantifying organic diagenesis and the method to which all others are compared. Vitrinite reflectance is measured microscopically using a highly polished specific and a specialized microscope equipped with a photometer. With increasing levels of diagenesis the reflectivity of vitrinite increases in a predictable fashion. Characteristics of coal have often been used as diagenetic indicators. The volatile matter, moisture, and hydrogen content of coal decrease with increasing diagenesis and the carbon content and calorific value increase in a predictable manner (Figure M8). The chemistry of kerogens has also been found to be useful indicator. For example the hydrogen to carbon ratio and pyrolysis yield of kerogens decrease with increasing diagenesis. The carbon preference index (ratio of odd to even n-alkanes) also progressively decreases with diagenesis from values greater than 1 (due to initial organic selectively) to values approaching one as diagenesis progresses.

The fluorescence of liptinite macerals has been utilized in some studies. With increasing levels of diagenesis the intensity of fluorescence decreases and the wavelength of maximum fluorescence progressively increase. Such measurements require a microscope equipped with a scanning monochromatic filter and sensitive photometer.

Kerogen coloration in transmitted light including pollen, spores and conodonts has been successively used to quantify maturation. As with toast, the OM darkens with increases with degree of thermal exposure and specialists attribute the colors to various numerical schemes based on visual examination. The color of kerogens has been referred to as the *Thermal Alteration Index*, whereas the color of conodonts is referred to as the *Conodont Alteration Index* (CAI) (Figure M8).

R. Marc Bustin and Raphael A.J. Wüst

Bibliography

Barnes, M.A., Barnes, W.C., and Bustin, R.M., 1990. Chemistry and diagenesis of organic matter in sediments and fossil fuels. In McIlreath, I.A., and Morrow, D.W. (eds.), *Diagenesis*. Geological Association of Canada, pp. 189–204.

Bustin, R.M., 1989. Diagenesis of kerogen. In Hutcheon, I.E. (ed.), *Burial Diagenesis*. GAC/MAC Short Course Handbook. Montreal, Quebec: Mineralogical Association of Canada, pp. 1–38.

Bustin, R.M., 1999. Coal: origin and diagenesis. In Marshall, C.P., and Fairbridge, R.W. (ed.), *Encyclopedia of Geochemistry*. Kluwer Academic, pp. 90–92.

Bustin, R.M., Barnes, M.A., and Barnes, W.C., 1990. Determining levels of organic diagenesis in sediments and fossil fuels. In McIlreath, I.A., and Morrow, D.W. (eds.), *Diagenesis*. Geological Association of Canada, pp. 205–226.

Copard, Y., Disnar, J.-R., and Becq-Giraudon, J.F., 2002. Erroneous maturity assessment given by Tmax and HI Rock-Eval parameters on highly mature weathered coals. *International Journal of Coal Geology*, **49** (1): 57–65.

Diessel, C.F.K., 1992. *Coal-bearing Depositional Systems*. Springer-Verlag.

Teichmüller, M., 1982. Origin of the petrographic constituents of coal. In Stach, E. *et al.* (eds.), *Textbook of Coal Petrology*. Gebrüder Bornträger, pp. 219–294.

Tissot, B., Durand, B., Espitalié, J., and Combaz, A., 1974. Influence of nature and diagenesis of organic matter in formation of petroleum. *American Association of Petroleum Geologists Bulletin*, **58** (3): 499–506.

Tissot, B.P., and Welte, D.H., 1978. *Petroleum Formation and Occurrence. A New Approach to Oil and Gas Exploration.* Springer-Verlag.

Cross-references

Coal Balls
Diagenesis
Kerogen

MATURITY: TEXTURAL AND COMPOSITIONAL

Maturity refers to the degree to which clastic sediment has been modified by physical and chemical processes at Earth's surface. *Textural maturity* refers to the degree to which physical characteristics of grains and populations of grains approach the "ultimate end product" (Pettijohn, 1975, p.491). *Compositional maturity* (stability) refers to the degree to which chemical characteristics approach the "ultimate end product," so that grains are more in equilibrium with Earth's surface conditions.

Textural maturity

As rocks approach Earth's surface during removal of overlying rocks, individual fragments of rock are defined by diverse physical, chemical and biologic weathering processes. Joints and other structural discontinuities tend to define gravel clasts, whereas preexisting grain size tends to define sand and silt clasts. Intensity of chemical weathering largely controls clay mineralogy (*q.v.*). Chemical weathering causes the least stable components in go into solution and/or to combine into more stable minerals during weathering. The result of these varied processes is the creation of sedimentary clasts of diverse grain size, shape and composition. As the clasts are eroded, transported and deposited (possibly through several cycles), they are modified by diverse physical and chemical processes. Most of these processes produce increased textural maturity through time.

Various measures of textural maturity have been developed. *Roundness* is a measure of the degree to which a grain has attained a continuously curving surface. *Sphericity* is a measure of the degree to which a grain has attained a spherical shape (equidimensional). These characteristics can be determined by painstaking measurements of curvature and dimensions of individual grains, or more commonly, by visual comparison with standard charts. A more sophisticated method for analyzing large populations of grain shapes utilizes Fourier analysis (e.g., Ehrlich and Weinberg, 1970).

Grain-size distributions reflect diverse processes at work during erosion, transportation and deposition. Some depositional environments tend to concentrate certain grain sizes; in the case of high-energy beaches, coarser clasts are left at river mouths and finer clasts are transported offshore, resulting in a concentration of fine to medium sand along beaches. Such an environment produces a narrow range of grain sizes, and is, therefore, texturally mature. In contrast, mass wasting and flash floods produce texturally immature deposits.

Standard petrographic methods of studying sandstone have shown a strong correlation between grain size and composition (e.g., Boggs, 1968). As rock fragments break into their constituent minerals (especially quartz and feldspar), QFR compositions change from dominantly R to dominantly Q and F. Thus, finer-grained sandstone (texturally more mature; higher QFR percent Q and F) is produced by the breakage of coarser grains (texturally less mature; higher QFR percent R). Interpretation of compositional maturity of sandstone is complicated by this artifact of method because of the dependence of composition on grain size, but interpretation of textural maturity is straightforward. Thus, the use of QFR petrographic methods is more useful for paleoclimate and other studies, where textural maturity is diagnostic (e.g., Suttner *et al.*, 1981; Suttner and Basu, 1985). Where reconstruction of original source rocks is the goal, the Gazzi-Dickinson method (QFL) is recommended (see below).

Compositional maturity (stability)

Most minerals and rocks are unstable at Earth's surface because they originated at higher pressures and temperatures, and in different geochemical environments. As a result, surface conditions are continuously acting on these minerals and rocks in ways that create more stable assemblages. Metastability is common at Earth's surface due to slow reaction rates at low temperatures; nonetheless, given the immensity of geologic time, even metastable minerals are gradually altered.

In general, minerals crystallized at higher pressures (P) and temperatures (T) are less stable at Earth's surface than are lower-PT minerals. Thus, the inverse of Bowen's reaction series is Goldich's weathering series, which indicates that, for example, albite is more stable at Earth's surface than is anorthite, and biotite is more stable than olivine. Thus, one measure of compositional maturity is the stability of mineral assemblages relative to their parent assemblages in their provenance areas.

The ZTR index (Hubert, 1960) is a commonly used mineral-stability measure. It is determined by dividing the quantity of zircon, tourmaline and rutile (super-stable accessory minerals) by the quantity of total accessory minerals. Thus, super-stable mineral assemblages have ZTR values close to unity.

Petrographically determined measures of stability include the proportion of total quartz in the QFL population (using the Gazzi-Dickinson method, which lessens the dependence of composition on grain size; Ingersoll *et al.*, 1984). Sub-populations also indicate relative stability; for example, Qm/(Qm + Qp), Qp/(Qp + Lm + Lv + Ls), Ls/(Lm + Lv + Ls), Qm/(Qm + Fk + Fp) values all tend toward unity in super-stable sandstone. Polycrystallinity and undulosity tend to weaken quartz grains, so that ultra-quartzose sandstone tends to have relatively low proportions of both (Basu, 1985). Super-stable conglomerate consists predominantly of quartzite clasts, analogous to super-stable quartz arenite (including orthoquartzite).

Stability of mudrock is highly dependent on climate because stability of clay minerals is determined primarily by chemical weathering regime. Thus, light leaching produces montmorillonite and illite, and related clays, depending on starting material. With increased leaching, kaolinite is favored. With extreme leaching, aluminum and iron oxides are the primary residuum (laterites, *q.v.*), thus representing extreme compositional maturity (super stable).

Ratios of major oxides also indicate compositional maturity because the oxides behave differently at Earth's surface. MgO, CaO, and Na_2O tend to decrease, whereas K_2O, SiO_2, TiO_2, and Fe_2O_3 tend to increase during weathering (Pettijohn, 1975). Various ratios of the latter oxides to total oxides approach unity with increasing stability.

Discussion

Textural maturity and compositional maturity are distinctly different measures of the degree to which sediment has been modified during the diverse processes of weathering, erosion, transportation and deposition. Distinction between these two attributes, however, has not always been clear. The "graywacke problem" is an informative example of possible confusion.

Geosynclinal theory equated "graywacke" (poorly sorted sandstone, with abundant "fine-grained matrix") with "eugeosynclinal sedimentation" (e.g., Kay, 1951). When most "graywackes" were recognized as turbidites, it was said that turbidites were poorly sorted mixtures of sand and mud. The presence of abundant "matrix" in "graywackes" was cited as evidence. Detailed petrographic study of "graywackes" has shown that, in almost all cases, most of the "matrix" is pseudomatrix, that is, physically deformed and chemically altered detrital grains (Dickinson, 1970). Recognition of pseudomatrix allows for the correct interpretation of provenance (based on assignment of pseudomatrix to detrital-grain categories) and the correct interpretation of textural maturity (little original detrital matrix, which means that sand was well sorted prior to diagenetic alteration). Recognition of pseudomatrix as detrital grains is consistent with the observation that modern turbidites are well sorted, that is, have less than 5 percent matrix (e.g., Hollister and Heezen, 1964). Thus, the "graywacke problem" resulted from the confusion of compositional maturity (unstable lithic grains and micas readily become pseudomatrix during diagenesis) with textural maturity (pseudomatrix was interpreted as detrital matrix, which led to the inference that turbidites were poorly sorted). Careful petrographic study is needed to recognize the original detrital texture and composition of sandstone, and for the correct interpretation of both textural and compositional maturity.

Raymond V. Ingersoll

Bibliography

Basu, A., 1985. Reading provenance from detrital quartz. In Zuffa, G.G. (ed.), *Provenance of Arenites*. D. Reidel, pp. 231–247.
Boggs, S. Jr., 1968. Experimental study of rock fragments. *Journal of Sedimentary Petrology*, **38**: 1326–1339.
Dickinson, W.R., 1970. Interpreting detrital modes of graywacke and arkose. *Journal of Sedimentary Petrology*, **40**: 695–707.
Ehrlich, R., and Weinberg, 1970. An exact method for characterization of grain shape. *Journal of Sedimentary Petrology*, **40**: 205–212.
Hollister, C.D., and Heezen, B.C., 1964. Modern graywacke-type sands. *Science*, **146**: 1573–1574.
Hubert, J.F., 1960. Petrology of the Fountain and Lyons Formations, Front Range, Colorado. *Colorado School of Mines Quarterly*, **55**: 242pp.

Ingersoll, R.V., Bullard, T.F., Ford, R.L., Grimm, J.P., Pickle, J.D., and Sares, S.W., 1984. The effect of grain size on detrital modes: a test of the Gazzi-Dickinson point-counting method. *Journal of Sedimentary Petrology*, **54**: 103–116.
Kay, M., 1951. *North American Geosynclines*. Geological Society of America Memoir, 48, 143pp.
Pettijohn, F.J., 1975. *Sedimentary Rocks*. Harper and Row, 3rd edn.
Suttner, L.J., and Basu, A., 1985. The effect of grain size on detrital modes: a test of the Gazzi-Dickinson point-counting method—discussion. *Journal of Sedimentary Petrology*, **55**: 616–617.
Suttner, L.J., Basu, A., and Mack, G.H., 1981. Climate and the origin of quartz arenites. *Journal of Sedimentary Petrology*, **51**: 1235–1246.

Cross-references

Climatic Control of Sedimentation
Feldspars in Sedimentary Rocks
Grain Size and Shape
Heavy Minerals
Provenance
Quartz, Detrital
Sands and Sandstones
Tectonic Controls of Sedimentation
Weathering, Soils, and Paleosols

MEANDERING CHANNELS

The planform of rivers—the channel pattern as seen from an overflying aircraft—is quite varied and has invited many attempts at classification by river scientists. An important early planform classification by Leopold and Wolman (1957) divided rivers into three categories: braided, meandering and straight. Braided channels are multiple-thread patterns and contrast with the single-thread meandering and straight channels; meandering channels are distinguished from their straight counterparts by their sinuously winding course. The term "meander" has become part of our everyday language and derives from the apparently aimless wandering habit of the Maiandros in Phrygia, a winding river named by the ancient Greeks in what is now modern Turkey.

More recently river scientists have recognized the limitations of this simple tripartite river-planform division and have stressed instead the continuum of river planforms. Nevertheless, the meandering channel remains an important end-member in all modern classifications of river planform and is one of the most common patterns found in nature. It may be defined as a single-thread river channel consisting of a continuous series of channel bends of alternating curvature together displaying elements of both periodic and random geometry.

The planform and channel geometry of river meanders

One of the most intriguing properties of meandering channels is the sometimes remarkably regular periodic waveform they display (see Figure M9). In its purest form this regular sinuous waveform has a characteristic shape (sine-generated) and a geometry that scales with channel width so that meanders tend to be geometrically self-similar regardless of the size of the river. For example, on average, meander wavelength typically

Figure M9 (A) Aerial photograph of the freely meandering Pembina River, Alberta, Canada; (B) aerial photograph of confined meandering Beaver River, Alberta, Canada.

is 8–12 times the bankfull channel width and meander-bend curvature is about 2–3 times the channel width. Because channel width (W) is related to discharge (Q), meander wavelength (L) is also related to river discharge

$$L = k\sqrt{Q}$$

(where k ranges from 30 to 60, depending on the return period of Q), a relationship which has been exploited in paleo-hydrologic studies of ancient rivers (see Dury, 1964).

Initially river scientists were preoccupied with the periodic character of river-meander traces and only more recently have acknowledged their non-periodic, intermittent, or random behavior. In most meandering channels the orderliness of highly periodic sinuosity is blended with, and disturbed by, a variety of planform disorder that relates to channel migration processes. For example, regular meander-bend development can be distorted by heterogeneities in the alluvium through which bends are migrating, meander bends can migrate laterally to form complex meander loops which are much larger than the associated meander bends, and high sinuosity meander bends can erode back on themselves to form meander cutoffs. Typical variations in the relative importance of these two planform elements led Kellerhals *et al*. (1976) to distinguish three categories of meandering: irregular, strongly repeating, and tortuous.

Although the shape of channel bends in a freely migrating meander train can be symmetrical and sinusoidal, even slight river confinement commonly leads to bend asymmetry as the train migrates downvalley. In cases of pronounced confinement (e.g., meanders formed in the confines of abandoned glacial meltwater channels of the Canadian Prairies), meander asymmetry is extreme and is an indicator of flow and migration direction (Figure M9(B)).

Channel cross-sections at or somewhat downstream of the axis of meander bends are distinctly asymmetrical with deeper flow occupying the outer bank zone and the deposition of a point bar causing shoaling at the opposite inner bank. At points of meander inflection flow typically is shallower over a channel-wide bar called a *riffle*. This "pool and riffle" sequence is a distinctive morphological characteristic of all meandering channels.

Flow in meandering channels

Although much debated, there is no generally accepted universal explanation for why meanders form in rivers (see reviews by Callander, 1978 and Knighton, 1998). Whatever the cause of meandering, however, the general pattern of flow through the alternating channel bends of meanders is well-known and understood. In the absence of secondary flow, bend flow seeks to conserve angular momentum so that it tends to conform to that of a free vortex with high velocity at the smaller radius of the inner bank and lower velocity at the outer bank where radial acceleration is lower. But secondary flow redistributes momentum in the bend to achieve a reversal of this relationship in the zone of fully developed bend flow. Near the water surface where velocity is highest, secondary flow is dominated by centrifugal "forces" and flow moves toward the outer bank. Near the bed, where velocity and thus the centrifugal effects are lowest, the balance of forces is dominated by the inward hydraulic gradient of the super-elevated water surface and secondary flow moves toward the inner bank. The three-dimensional expression of these forces is helical flow which has an alternating sense through successive meander bends. One of the important consequences of helical flow in meanders is that sediment eroded from the outside of a meander bend tends to be moved to the inner bank or point bar of the next downstream bend.

The process of meander formation and maintenance

Meander planform geometry is controlled by the process of lateral migration. Meandering channels erode the outer banks of channel bends and maintain a constant channel width by achieving a matching rate of deposition on the point bar forming the inner bank. Rates of migration may be exceedingly slow (a small fraction of channel width/year) or very rapid (several channel widths/year) depending on the ratio stream power/bank strength, and on the degree of bend curvature (Hickin, 1974; Hickin and Nanson, 1975, 1984). Depending on the particular pattern of bend erosion and deposition, meanders can exhibit a variety of evolution. Simple bends can increase in amplitude by extension until decreasing bend curvature deflects or constrains the direction and rate of motion. Phase differences in the flow and bend alignment can lead to bend rotation, lobing and compound meander loops. Where confinement constrains meander development, bend motion is mainly restricted to downvalley translation. These motions are further modified by the formation of gooseneck and chute cutoffs which short-circuit the meandering channel path, sometimes forming ox-bow lakes.

The deposits of meanders: meander floodplains

The nature of sediments deposited to form the floodplains of most meandering rivers is intimately related to the sediment available for point bar deposition and to the process of lateral migration. The classical point bar model of deposition in sand- and gravel-bed rivers, originating primarily from the work of Allen (1963), is illustrated in Figure M10. Within-channel lateral accretion deposits of gravel overlain by sand are capped by overbank fines deposited during flood discharges. Lateral accretion deposits consist of a basal gravel platform of bed material on which are deposited upward-fining sands and silts representing the declining energy of the depositional environment as the point bar aggrades to less frequently inundated elevations. Internal sedimentary structures vary from dune to ripple related cross strata and trough cross-lamination in the lateral accretion deposits to horizontal strata in the overbank deposits. This upward-fining sediment column with its associated upward downscaling of sedimentary structures, is

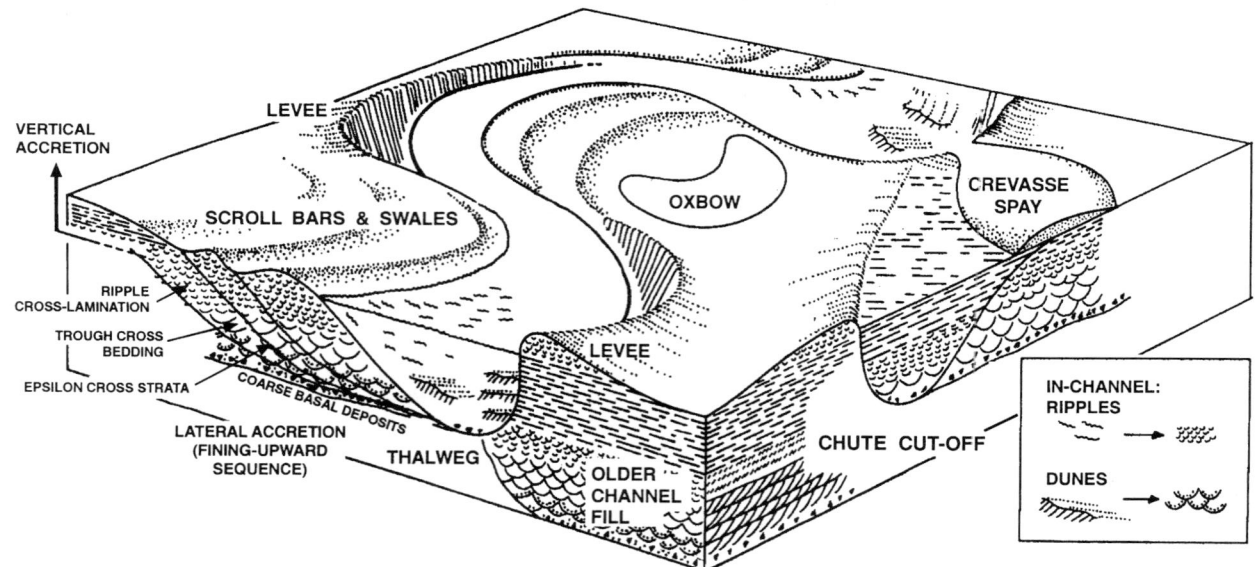

Figure M10 The geomorphic setting and sedimentological elements of the upward-fining sequence, the basis of the classical meandering-river facies model (modified from Walker and Cant, 1984).

packaged in units bound by sigmoidal epsilon cross strata (ECS), corresponding to distinct phases of channel migration and point bar deposition. These packages of sediment often form arcuate ridges or scroll bars, separated one from the other by troughs in the surface of the floodplain.

More recent investigations, however, have recognized many variations on this classical model of meander deposition and still others describe floodplain sedimentology which seems to be entirely unconventional. For example, in Miall's (1985) models of meandering river floodplain (gravel meandering, gravel-sand meandering, sand meandering and fine-grained meandering) he recognizes the important control of sediment type on the details of floodplain sedimentology. These river channel styles reflect a progressive increase in the abundance of fines although coarser basal lateral accretion deposits are common to all models. Floodplains with lateral accretion deposits consisting of sandy silt with negligible upward fining have been described by Nanson (1980). Others have also noted that meandering river floodplains formed in environments where silt is abundant and gravel and sand are much less available, particularly in low-gradient tropical regions, also do not appear to be adequately described by the classical Allen model (Jackson, 1981; Smith, 1987). Similarly, the negligible role played by lateral accretion in some steep upland valleys formed in non-cohesive sediments, as well as in low-gradient channels in cohesive sediments, is formalized in the genetic classification of floodplains by Nanson and Croke (1992).

The floodplains of meanders may also include a significant proportion of counterpoint (or eddy-accretion) sediments of horizontally bedded silt and fine sand deposited in flow separation zones at points of abrupt direction change in the channel. In confined meanders the abrupt angle bends that occur where the channel impinges on the valley wall (Figure M9(B)) form systematically alternating zones of flow separation downstream at the valley walls. These alternating gyres trap eddy accretion sediments which form concave bank benches with a curvature opposed to that of the convexly curved scroll bars on the adjoining point bar (Hickin, 1979, 1986; Nanson and Page, 1983; Burge and Smith, 1999).

River planform facies and architectural element models of meander deposits

Attempts to characterize meandering river sediments in terms of bedform-scale sedimentary structures and assemblages (planform facies models) that would distinguish them with some certainty from other river types (especially braided) have not fulfilled early promise. Although general sedimentological properties for sand/gravel-bed rivers are well represented by these facies models, overlap among planform groups is so significant that their diagnostic utility is weak and often misleading (Jackson, 1978; Brierley and Hickin, 1991; Hickin, 1993; Miall, 1996). In recent years some fluvial sedimentologists have advocated the development of bar-scale or architectural element-scale models to characterize fluvial sediments (Miall, 1985). It has been argued that such planform element models may be far more effective diagnostic tools than are bedform-scale facies models. This development represents a return to earlier geomorphic concepts such as those developed by Happ et al. (1940) and Fisk (1944, 1947) and espoused by Leopold, Wolman and Miller (1964) who recognized that floodplains of rivers such as the Mississippi consist of an assemblage of distinct morphological and sedimentological elements (such as channel fills, point bars, meander scrolls, sloughs, levees, backswamp deposits, and crevasse splays). But such an element approach to the depositional record requires three-dimensional mapping of such features and the typically poor access to exposures in the field is rarely adequate for this purpose. There is some prospect, however, that remote sensing of the subsurface architecture by means such as ground-penetrating radar (GPR), shallow seismic and electrical resistivity surveys, may yield the necessary data to allow this approach to be developed further (Alexander et al., 1994; Leclerc and Hickin, 1997).

Summary

Meandering channels, single-thread river-channels consisting of a continuous series of channel bends of alternating curvature, are an important end-member in all modern classifications of river planform and is one of the most common river patterns found in nature.

River meanders tend to have a sinuous waveform of characteristic shape (sine-generated) and a geometry that scales with channel width so that meanders tend to be geometrically self-similar regardless of the size of the river. The regularity of this sinuous pattern, also reflected in a corresponding "pool and riffle" bed geometry, can be disturbed by such effects as floodplain heterogeneities, meander cutoffs, complex loop development, and channel confinement.

In the classical point bar model of deposition in sand-bed and gravel-bed rivers within-channel lateral accretion deposits of gravel overlain by sand are capped by overbank fines deposited during flood discharges. This upward-fining sediment column with its upward downscaling of sedimentary structures, is packaged in units bound by sigmoidal epsilon cross strata (ECS), corresponding to distinct phases of channel migration and point bar deposition.

More recent investigations have recognized many variations on this classical model of meander deposition, encompassing the broader range of sediments actually found in nature, including counterpoint deposits common in confined meanders. Mud-dominated floodplains remain poorly studied and the subject of current research.

Planform facies models at the bedform scale have enjoyed limited diagnostic success. Bar-scale or architectural element-scale models of floodplain sedimentology hold more promise but require three-dimensional data dependent on remote sensing of the subsurface by techniques such as ground-penetrating radar.

Edward J. Hickin

Bibliography

Alexander, J., Bridge, J.S., Leeder, M.R., Collier, E.L., and Gawthorpe, R.L., 1994. Holocene meander belt evolution in an active extensional basin, southwestern Montana. *Journal of Sedimentary Research*, **B64**: 542–559.

Allen, J.R.L., 1963. The classification of cross-stratified units, with notes on their origin. *Sedimentology*, **2**: 93–114.

Brierley, G.J., and Hickin, E.J., 1991. Channel planform as a non-controlling factor in fluvial sedimentology: the case of the Squamish River floodplain, British Columbia. *Sedimentary Geology*, **75**: 67–83.

Burge, L.M., and Smith, D.G., 1999. Confined meandering river eddy accretions: sedimentology, channel geometry, and depositional processes. In Smith, N.D., and Rogers, J. (eds.), *Fluvial Sedimentology VI*. SEPM, Special Publication, 28, International Association of Sedimentologists.

Callander, R.A., 1978. River meandering. *Annual Review of Fluid Mechanics*, **10**: 129–158.

Dury, G.H., 1964. Principles of underfit streams. *United States Geological Survey Professional Paper* 452A.

Fisk, H.N., 1944. *Geological Investigation of the Alluvial Valley of the Lower Mississippi River*. Vicksburg, Mississippi: Mississippi River Commission.

Fisk, H.N., 1947. *Fine-grained Alluvial Deposits and their Effect on Mississippi River Activity*. Vicksburg, Mississippi: Mississippi River Commission.

Happ, S.C., Rittenhouse, G., and Dobson, D., 1940. Some Principles of Accelerated Stream Valley Sedimentation. *US Department of Agriculture Technical Bulletin*, 695.

Hickin, E.J., 1974. The development of meanders in natural river channels. *American Journal of Science*, **274**: 414–442.

Hickin, E.J., 1979. Concave-bank benches on the Squamish River, British Columbia, Canada. *Canadian Journal of Earth Sciences*, **16** (1): 200–203.

Hickin, E.J., 1986. Concave-bank benches in the floodplains of Muskwa and Fort Nelson Rivers, British Columbia. *The Canadian Geographer*, **30** (2): 111–122.

Hickin, E.J., 1993. Fluvial facies models: a review of Canadian research. *Progress in Physical Geography*, **17** (2): 205–222.

Hickin, E.J., and Nanson, G.C., 1975. The character of channel migration on the Beatton River, Northeast British Columbia, Canada. *Geological Society of America Bulletin*, **86**: 487–494.

Hickin, E.J., and Nanson, G.C., 1984. Lateral migration rates of river bends. *Journal of Hydraulic Engineering*, American Society of Civil Engineers, **110** (11): 1557–1567.

Jackson, II, R.G., 1978. Preliminary evaluation of lithofacies models for meandering alluvial streams. In Miall, A.D. (ed.), *Fluvial Sedimentology*. Canadian Society of Petroleum Geologists Memoir, 5, pp. 543–576.

Jackson, II, R.G., 1981. Sedimentology of muddy fine-grained channel deposits in meandering streams of the American middle west. *Journal of Sedimentary Petrology*, **51**: 1169–1192.

Kellerhals, R., Church, M., and Bray, D.I., 1976. Classification of river processes. *Journal of the Hydraulics Division of the American Society of Civil Engineers*, **93**: 63–84.

Knighton, D., 1998. *Fluvial Forms and Processes: A New Perspective*. London: Arnold, pp. 205–230.

Leclerc, R., and Hickin, E.J., 1997. Ground penetrating radar stratigraphy of a meandering river floodplain, South Thompson River, British Columbia. *Geomorphology*, **21**: 17–38.

Leopold, L.B., and Wolman, M.G., 1957. River channel patterns-braided, meandering and straight. *United States Geological Survey Professional Paper*, 282B: 39–85.

Leopold, L.B., Wolman, M.G., and Miller, J., 1964. *Fluvial Process in Geomorphology*. Freeman.

Miall, A.D., 1985. Architectural-element analysis: a new method of facies analysis applied to fluvial deposits. *Earth Science Reviews*, **22**: 261–308.

Miall, A.D., 1996. *The Geology of Fluvial Deposits*. Springer-Verlag.

Nanson, G.C., 1980. Point bar and floodplain formation on the Beatton River, northeastern British Columbia, Canada. *Sedimentology*, **27**: 3–30.

Nanson, G.C., and Hickin, E.J., 1986. A statistical analysis of bank erosion and channel migration in western Canada. *Geological Society of America Bulletin*, **97**: 497–504.

Nanson, G.C., and Croke, J.C., 1992. A genetic classification of floodplains. *Geomorphology*, **4**: 3–30.

Nanson, G.C., and Page, K.J., 1983. Lateral accretion of fine-grained concave benches on meandering streams. In Collinson, J.D., and Lewin, J. (eds.), *Modern and Ancient Fluvial Systems*. International Association of Sedimentologists, Special Publication, 6, pp. 133–144.

Smith, D.G., 1987. Meandering river point bar lithofacies models: modern and ancient examples compared. In Etheridge, F.G., Flores, R.M., and Harvey, M.D. (eds.), *Recent Developments in Fluvial Sedimentology*. SEPM, Special Publication, 39, pp. 83–91.

Walker, R.G., and Cant, D.J., 1984. Sandy fluvial systems. In Walker, R.G. (ed.), *Facies Models*. Geoscience Canada, Reprint Series 1, 2nd edn, 72.

Cross-references

Erosion and Sediment Yield
Floodplain Sediments
Placers, Fluvial
Rivers and Alluvial Fans

MÉLANGE; MELANGE

Mélange (spelled melange in much English-language literature) describes heterogeneous rock units that contain blocks, typically of hard or structurally competent lithologies, surrounded and supported by a fine-grained matrix of shale, slate, or serpentinite. Considerable controversy has surrounded the term mélange since its first use by Greenly (1919), in the description of a unit from Anglesey in the British Isles. Interest in mélanges was revived during the development of plate tectonics, when mélanges were identified as tectonic products of subduction (e.g., Hsu, 1968). Controversy has surrounded both the definition of the term mélange, and the interpretation and origin of units so described.

Definitions

Almost all definitions of mélange include as critical components the presence of blocks of varied sizes and lithologies, not in contact with each other, and surrounded by a finer-grained matrix. Most geologists have also agreed that a mélange has to be a unit that is mappable, typically at a scale such as 1 : 25,000 (Silver and Beutner, 1980) or 1 : 24,000 (Raymond, 1984). Authors have differed as to whether the presence of a tectonic fabric in the matrix is essential to the definition of mélange. Most described examples do have a fabric of some sort; rock bodies in which blocks are present in a clearly untectonized matrix are typically labeled using more interpretive terms such as "olistostrome" or "debris flow" (*q.v.*), which imply specifically sedimentary processes of origin. Some authors have suggested that such sedimentary units be excluded from the definition of mélange (Hsu, 1974). Nonetheless, the term has been applied to many units entirely lacking tectonic fabric, and the type example described by Greenly (1919) locally lacks fabric. A purely descriptive definition is therefore preferable; units of known tectonic origin are commonly distinguished as "tectonic mélanges."

Also in dispute is whether the blocks have to include "exotic" lithologies, such as the blueschist and ultramafic units in the well-known Franciscan mélanges of California. Although exotic blocks were regarded as essential in the definition adopted by Raymond (1984), it is often difficult to decide what lithologies are truly "exotic", and close textural and fabric similarities exist between chaotic units with and without exotics (Brandon, 1989). Thus a definition consistent with most modern usage is: "an internally disorganized unit that is mappable at 1 : 25,000 scale, that is characterized by the presence of blocks of varied lithologies and sizes supported by a fine-grained matrix."

Interpretations

Proposed mélange origins fall into three large categories. In the first group are interpretations that appeal to gravitationally driven movement on slopes to disrupt bedding in sedimentary components and to mix the blocks and matrix. Such processes include slides, slumps, and debris-flows. The term "olistostrome" (Flores, 1955, 1956), literally "slide layer," describes a mélange, with or without fabric, interpreted to be formed by slope processes. In the second category are processes, such as fluid overpressuring and diapirism, that occur within a pile of sedimentary rocks, and result in disruption and mixing without necessarily penetrating to the Earth's surface. Thirdly, there are processes related to faults that are rooted at depth in the lithosphere, including subduction zones and transform faults.

In attempts to distinguish between these potential origins, investigators have typically appealed to a variety of lines of evidence, studying both the internal fabric and structure of mélanges, and the external contacts and geometry of mapped mélange bodies.

Features and origin of blocks

Observations of the internal fabric of mélanges can provide information on the rheology of the blocks and matrix and may therefore help to constrain the mode of formation.

Many mélanges preserve evidence for the process of fragmentation that has produced blocks. In some cases it is possible to observe partially fragmented beds cut by brittle surfaces. These surfaces may be either extensional or shear fractures, that produce irregular shapes and trains of fragments derived from an original block or bed (e.g., Cowan, 1982). If fractures cross-cut grains, cements or veins in the blocks, this may indicate a post-lithification origin for the structures, and that fragmentation occurred at depth. Where extensional fractures occur, matrix may penetrate into the blocks producing veins that are partially or completely filled by shale. This has been interpreted as an indication of the presence of a fluid matrix, but in other cases it has been shown that matrix was already in a compacted state, with strong fissility, before it penetrated into veins (e.g., Waldron et al., 1988). Whatever the state of lithification, it is clear that elevated fluid pressures can play a role in promoting brittle fracturing by reducing the effective normal stress on potential fracture planes.

Not all blocks are produced by fracturing. In some cases, disruption of bedding in sedimentary protoliths leads to the dispersion of individual grains or highly attenuated beds into the mélange matrix. Fragments produced in such circumstances have less angular outlines, typically appearing as lens shapes or "phacoids." The presence of phacoids is an indication of deformation that was ductile at mesoscopic scale; however, phacoids in unmetamorphosed mélanges do not typically show evidence for intracrystalline plastic deformation at microscopic scale. In some mélanges, blocks are connected by highly thinned "wisps" representing attenuated beds (e.g., Byrne, 1984); microscopic examination indicates that such wisps are typically formed by disaggregation, and not by grain breakage or plastic flow (Brandon, 1989).

Some mélanges contain blocks from diverse sources, including intrusive igneous and metamorphic lithologies (e.g., ophiolites and blueschists) that cannot have formed in proximity to a lower grade or purely sedimentary matrix. Cloos (1982) suggests that such "exotic" blocks are incorporated by large-scale flow within an accretionary wedge. However, incorporation as large sedimentary clasts derived from fault scarps or mass flows that incorporate older rocks at the Earth's surface is a possibility that is difficult to eliminate, especially in tectonically active environments such as subduction complexes and collision zones.

Fabric and origin of matrix

Matrix fabrics in mélanges are extremely variable. Some mélanges lack matrix fabric; in these cases it is generally clear that the matrix formed by flow of wet sedimentary material. In most cases some fabric is present, and typically this takes the form of a scaly foliation (Moore et al., 1986), characterized by anastomosing planes surrounding lenticular bodies of fine-grained, fissile material. Foliation surfaces are typically polished and may show slickenside striations that indicate sense of shear. Some such surfaces may be traceable into faults or shear zones that dismember blocks. Larger scale fabrics may be present: in many mélanges the matrix is heterogeneous, and can be divided into domains of varying color, texture, etc., reflecting different protoliths. The domains are typically highly elongated in map outline, and may contribute an overall foliation to the map pattern, and an overall stratification to the mélange. Such stratification has been used as an indicator of sedimentary origin (Steen and Andresen, 1997), but it is quite possible to envisage the development of large-scale foliation during accretion and deformation in a subduction zone, so such inferences are probably unwarranted. Other mélanges display more continuous fabrics such as slaty cleavage or metamorphic foliation; in many cases it can be shown that these fabrics post-date the mixing and fragmentation processes and may be unrelated to the formation of the mélange.

Overall geometry of mélange units

Given the difficulty of using arguments based on internal fabric to distinguish between possible origins, arguments based on contacts and the overall geometry of mélanges often provide a better basis for distinguishing between possible modes of origin.

Most helpful are clear stratigraphic contacts either at the base or top of a mélange body. Brandon (1989) emphasizes evidence from depositional basal stratigraphic contacts in attributing an olistostromal origin to mélanges of the Pacific Rim Complex of Vancouver Island. Swarbrick and Naylor (1980) describe interbedded sediments between mélange units in Cyprus, that indicate that the mélange is clearly stratified. Pini (1999) describes tabular bodies with fabric that form part of a stratigraphic succession and are therefore identified as of sedimentary origin, despite the presence of scaly fabrics.

Mélanges of diapiric origin may also be identified based on large-scale geometry. Orange (1990) describes the variations in fabric within the Duck Creek mélange of the Olympic Peninsula, Washington, tracing a transition from margins with pervasive scaly foliation and inequant phacoidal blocks, to a core with poorly developed foliation, and equant, angular blocks. Using fold asymmetries, Orange (1990) is able to demonstrate opposing senses of shear on opposite sides of the mélange, further supporting an origin by diapirism.

Mélange formation by tectonic processes in fault or shear zones is most easily demonstrated by regional scale cross-cutting relationships along fault zones. Pini (1999) was able to show such relationships for units in the Apennines, termed tectonosomes. Orange (1990) described the Hogsback mélange of the Olympic Peninsula, concluding that shear-zone origin is indicated by a combination of: consistent foliation in map-pattern, consistent fold vergence, increase in foliation intensity toward the center of a mélange, and tight clustering of clast long-axis directions throughout the unit.

Conclusion

Units described as mélange undoubtedly have a variety of origins. Field and thin-section evidence may provide information about the lithification state of the material at the time of fragmentation and/or mixing, but only in cases where there is clear evidence for prior burial and lithification can a deep-seated, tectonic origin be inferred. Many mélanges involve some process of deformation of wet or incompletely lithified material, either within a gravity-driven slump or debris flow, within a diapir, or within a tectonic accretionary terrain in which effective stresses are reduced by high fluid pressures. The products of these processes (boudinage, brittle fracturing, scaly fabrics) appear similar in all three settings. Under such circumstances, determining the origin of a mélange depends on careful recording of variations across the mélange body and documentation of contact relationships.

John W.F. Waldron

Bibliography

Brandon, M.T., 1989. Deformational styles in a sequence of olistostromal mélanges, Pacific Rim Complex, western Vancouver Island, Canada. *Geological Society of America Bulletin*, **101**: 1520–1542.
Byrne, T., 1984. Structural geology of mélange terranes in the Ghost Rocks Formation, Kodiak Islands, Alaska. In Raymond, L.A. (ed.), *Melanges, Their Nature, Origin and Significance*. Geological Society of America, Special Paper, 198, pp. 21–52.
Cloos, M., 1982. Flow melanges: numerical modeling and geological constraints on their origin in the Franciscan subduction complex, California. *Geological Society of America Bulletin*, **93**: 330–345.
Cowan, D.S., 1982. Deformation of partly dewatered and consolidated Franciscan sediments near Piedras Blancas Point, California. In Leggett, J.K. (ed.), *Trench and Forearc Geology*. Geological Society of London, Special Publication, 10, pp. 439–457.
Flores, G., 1955. Les résultats des études pour la recherche pétrolifère en Sicilie: discussion. In *Proceedings, Fourth World Petroleum Congress*. Section 1/A/2. Rome: Casa Editrice Carlo Columbo, pp. 121–122.
Flores, G., 1956. The results of the studies on petroleum exploration in Sicily: discussion. *Bollettino Del Servizio Geologico d'Italia*, **78**: 46–47.
Greenly, E., 1919. *The Geology of Anglesey*. Geological Survey of Great Britain, Memoir, 1.
Hsu, K.J., 1968. The principles of melanges and their bearing on the Franciscan-Knoxville paradox. *Geological Society of America Bulletin*, **79**: 1063–1074.
Hsu, K.J., 1974. Melanges and their distinction from olistostromes. In Dott, R.H., and Shaver, R.H. (eds.), *Modern and Ancient Geosynclinal Sedimentation*. Society of Economic Paleontologists and Mineralogists, Special Publication, 19, pp. 321–333.
Moore, J.C., Roeske, S., Lundberg, N., Schoonmaker, J., Cowan, D.S., Gonzales, E., and Lucas, S.E., 1986. Scaly fabrics from Deep Sea Drilling Project cores from forearcs. In Moore, J.C. (ed.), *Structural Fabric in Deep Sea Drilling Project Cores from Forearcs*. Geological Society of America, Memoir, 166, pp. 55–68.
Orange, D.L., 1990. Criteria helpful in recognizing shear-zone and diapiric melanges; examples from the Hoh accretionary complex, Olympic Peninsula, Washington. *Geological Society of America Bulletin*, **102**: 935–951.
Pini, G.A., 1999. *Tectonosomes and Olistostromes in the Argile Scagliose of the Northern Apennines*. Geological Society of America, Special Paper, 335.
Raymond, L.A., 1984. Classification of melanges. In Raymond, L.A. (ed.), *Melanges: Their Nature, Origin and Significance*. Geological Society of America, Special Paper, 198, pp. 7–20.
Silver, E., and Beutner, E., 1980. Melanges—Penrose Conference Report. *Geology*, **8**: 32–34.
Steen, Ø., and Andresen, A., 1997. Deformational structures associated with gravitational block gliding: examples from sedimentary olistoliths in the Kalvag Melange, western Norway. *American Journal of Science*, **297**: 56–97.
Swarbrick, R.E., and Naylor, M.A., 1980. The Kathikas melange, southwest Cyprus: Late Cretaceous submarine debris flows. *Sedimentology*, **27**: 63–78.
Waldron, J.W.F., Turner, D., and Stevens, K.M., 1988. Stratal disruption and development of mélange, Western Newfoundland: effect of high fluid pressure in an accretionary terrain during ophiolite emplacement. *Journal of Structural Geology*, **10**: 861–873.

Cross-references

Debris Flow
Deformation of Sediments
Gravity-Driven Mass Flows
Mass Movement

MICRITIZATION

Bathurst (1966) coined the term "micritization" in reference to a process of alteration of original skeletal grain fabric to a cryptocrystalline texture by repeated algal microborings and subsequent filling of the microborings by micritic precipitates. Bathurst was simply applying Folk's (1959) term for cryptocrystalline carbonate (<4 μm) micrite—a contraction of the words microcrystalline calcite—to describe a process of textural diminution in carbonate grains. In fact, Wolf (1965) used the term "grain-diminution" to describe the early alteration of crustose coralline algal cellular skeletons to micrite. Bathurst, however, did not accept recrystallizated micrite in his definition of "micritization", which he wanted to restrict to referring to the process of microboring and subsequent precipitation of micrite infill. Bathurst's idea became very popular (e.g., Lloyd, 1971; Gunatilaka, 1976; Kobluk and Risk, 1977). Subsequently, Friedman (1985), pointed out that Folk (1959, 1962) introduced the term micrite to describe microcrystalline ooze matrix, not textural alteration. He felt that Bathurst and others were misusing Folk's terminology.

Despite the concerns about the original intentions of authors when they introduced the terms micrite and micritization, Alexandersson (1972) indicated that there was a need for the broadest use of both terms. He proposed that the term micrite should no longer have "any mineralogical implications" (p.206) and it should be used simply as a textural term. Alexandersson further suggested that micritization should refer to an alteration "of a pre-existing fabric to micrite" with "no genetic implications" (p.206). To him, micritization included all processes involving the destruction of an ordered crystallographic fabric to a disordered micritic texture, with either a decrease or increase of crystal size. Most recent studies of the alteration of carbonate grains have accepted Alexandersson's broad definitions (e.g., Land and Moore, 1980; Reid *et al.*, 1992; Reid and Macintyre, 1998).

As mentioned earlier, Bathurst (1966) suggested that the process responsible for micritization of carbonate grains was algal microboring and subsequent precipitation of micrite in open borings. Earlier studies based on light microscopy had concluded that the cryptocrystalline textures in many shallow-water carbonate sediments were the result of recrystallization (e.g., Illing, 1954; Purdy, 1963, 1968; Pusey, 1964; Winland, 1969). However, for almost two decades after Bathurst's initial report, shallow-water marine carbonates were generally thought to be stable, with micritic textural alterations formed by boring and infilling.

More recently, pore-water chemical studies have indicated that extensive alteration of carbonate sediments is taking place in shallow tropical waters (e.g., Morse *et al.*, 1985; Walter and Burton, 1990; Rude and Aller, 1991; Walter *et al.*, 1993; Patterson and Walter, 1994). In addition a series of detailed petrographic and mineralogical studies have illustrated that micritization by recrystallization is widespread in carbonate grains found on the seafloor of tropical shallow waters (Reid *et al.*, 1992; Reid and Macintyre, 1998). Macinytre and Reid also discovered that the original skeletal needles of the green alga *Halimeda incrassata* (Macintyre and Reid, 1995) and the porcelaneous foraminifera *Archaias angulatis* (Macintyre and Reid, 1998) are recrystallizing to *minimicrite* (crystals <1 μm

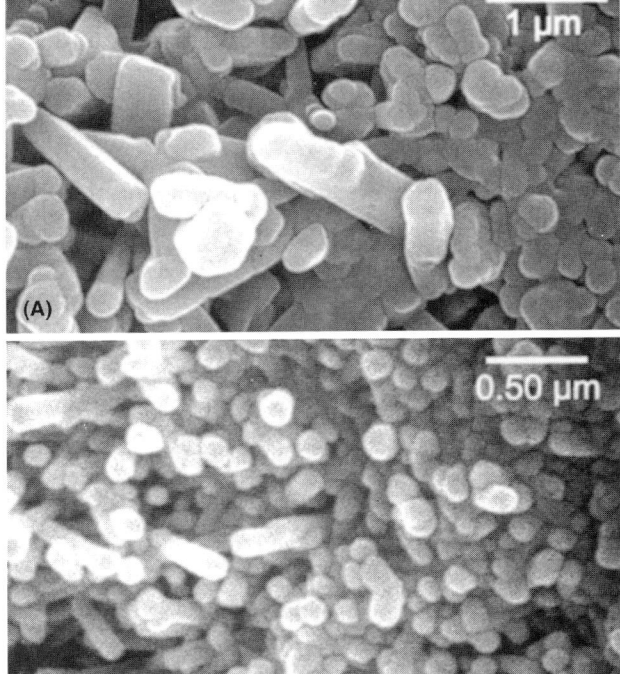

Figure M11 Original skeletal needles being micritized to minimicrite in living organisms. (A) The green alga *Halimeda incrassata*. (B) The porcellaneous foraminifera *Archaias angulatis*.

after Folk, 1974) while these organisms are still alive (Figure M11(A),(B)). *In vivo* micritization of carbonate skeletons is a textural alteration only and does not involve changes in mineralogy. In both of these studies the micritization process was thought to be related to the breakdown of the skeletal needles into their original polycrystalline aggregates or domains, which were deposited in organic envelopes that controlled the needle shape. Degradation of these organic envelopes and alternating conditions of precipitation and dissolution caused by reversals on CO_2 partial pressures associated with changes in photosynthesis and respiration in *Halimeda* (Macintyre and Reid, 1995) and algal symbionts in the foraminifera (Macintyre and Reid, 1998), may have resulted in the micritization of the skeletons of these living organisms.

Further studies of *Halimeda* and porcelaneous foraminifera grains deposited on the shallow tropical seafloor indicate that the micritization of skeletal needles continues after death, with extensive formation of minimicrite (Reid and Macintyre, 1998). As the micritization process continues, the anhedral equant minimicrite tends to increase in size to form subhedral blocky crystals of micrite (Figure M12). This continuing micritization of carbonate sediment grains can involve mineralogical changes and "together with concomitant cavity filling results in the gradual obliteration of skeletal elements and the formation of micritized grains with little evidence of original skeletal structure" (Reid and Macintyre, 1998, p.945). The processes responsible for this micritization of carbonate sediment grains may include organic alteration of pore-water chemistry, or reduction of high free surface energies associated with the increase in crystal size.

The role of microborers in the micritization process of sediment grains was re-introduced when lightly etched polished sections of micritized grains examined using scanning electron microscope (SEM) revealed the presence of multi-cyclic boring and concurrent filling of bore holes (Macintyre *et al.*, 2000; Reid and Macintyre, 2000). The endolithic cyanobacterium responsible for this boring and infilling activity is *Solentia* sp. This microborer does not limit its activity to the edge of grains but eventually reworks the entire grain with minimal damage the grain's outer walls (Figure M13(A)). The infilling cement consists of fibrous minimicrite crystals, which form banded patterns extending across the 5–10 μm diameter boreholes (Figure M13(B)) (Macintyre *et al.*, 2000). With the lack of open boreholes and the well preserved grain boundaries, it is very difficult in SEM and light petrographic microscope studies to distinguish between concurrent infillings of microborings and recrystallization

Figure M12 Subhedral micrite in an area of minimicrite in an *Archaias* sediment grain, Great Bahama Bank, Bahamas.

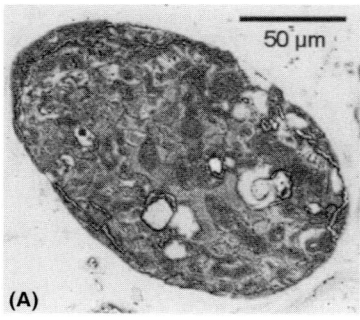
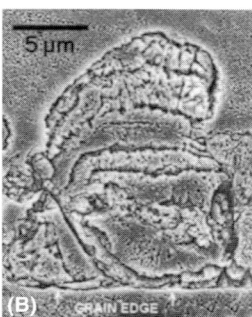

Figure M13 Scanning electron microscope photomicrographs of etched thin sections showing grains micritized by the endolithic cyanobacterium *Solentia* sp. (A) Grain almost completely micritized by boring and infilling action of *Solentia* sp. Note that there is very little damage to the original grain boundaries. (B) Detail of cyanobacterial microborings showing banded minimicrite infillings.

in the formation of micritized carbonate grains (Reid and Macintyre, 2000).

Summary

(1) Micritization of carbonate sediment grains in shallow-water tropical seas is widespread forming micritic grains, which commonly show no evidence of original structure.
(2) Micritization of carbonate skeletons can be initiated when the organism is still alive.
(3) The breakdown of original skeletal needles to form minimicrite in living organisms likely involves both the destruction of the organic envelopes, in which the needles are originally formed and changing internal chemical conditions. No mineralogical changes of the skeleton occur during this *in vivo* micritization.
(4) The micritization of skeletal grains continues after skeletons are deposited on the seafloor. Postmortem micritization involves the formation of anhedral minimicrite and subhedral micrite and may include changes in mineralogy.
(5) Driving forces for the micritization of sediment grains may include biological alteration of pore-water chemistry and inorganic reduction in free energy of crystal surfaces.
(6) Microboring activity can also be an important factor in the micritization of carbonate grains. This can involve (a) formation of open boreholes and later precipitation of infilling cement, as described by Bathurst, or (b) concurrent infilling with little evidence of open boreholes. In the latter case, micritization by microboring is difficult to distinguish from micritization by recrystallization.

Ian G. MacIntyre and R. Pamela Reid

Bibliography

Alexandersson, E.T., 1972. Micritization of carbonate particles: process of precipitation and dissolution in modern shallow-marine sediments. *Universitet Uppsala, Geologiska Institut Bulletin*, **7**: 201–236.
Bathurst, R.G.C., 1966. Boring algae, micrite envelopes, and lithification of molluscan biosparites. *Geological Journal*, **5**: 15–32.
Folk, R.L., 1959. Practical petrographic classification of limestones. *American Association of Petroleum Geologists Bulletin*, **43**: 1–38.
Folk, R.L., 1962. Spectral subdivision of limestone types. In Ham, W.E. (ed.), *Classification of Carbonate Rocks*. American Association of Petroleum Geologists, pp. 62–84.
Folk, R.L., 1974. The natural history of crystalline calcium carbonate: effect of magnesium content and salinity. *Journal of Sedimentary Petrology*, **44**: 40–53.
Friedman, G.M., 1985. The problem of submarine cement in classifying reefrock: an experience in frustration, Schneidermann, N. and Harris, P.M. (eds.), *Society of Economic Paleontologists and Mineralogists*, pp. 117–121.
Gunatilaka, A., 1976. Thallophyte boring and micritization within skeletal sands from Connemara, western Ireland. *Journal of Sedimentary Petrology*, **46**: 548–554.
Illing, L.V., 1954. Bahamian calcareous sands. *American Association of Petroleum Geologists Bulletin*, **38**: 1–95.
Kobluk, D.R., and Risk, M.J., 1977. Micritization and carbonate-grain binding by endolithic algae. *American Association of Petroleum Geologists Bulletin*, **61**: 1069–1082.
Land, L.S., and Moore, C.H., 1980. Lithification, micritization and syndepositional diagenesis of biolithites on the Jamaican island slope. *Journal of Sedimentary Petrology*, **50**: 357–370.
Lloyd, R.M., 1971. Some observations on recent sediment alteration ("micritization") and the possible role of algae in submarine cementation. In Bricker, O.P. (ed) *Carbonate Cements*. The Johns Hopkins Press, pp. 72–79.
Macintyre, I.G., and Reid, R.P., 1995. Crystal alteration in a living calcareous alga (*Halimeda*): implications for studies in skeletal diagenesis. *Journal of Sedimentary Research*, **A65**: 143–153.
Macintyre, I.G., and Reid, R.P., 1998. Recrystallization in a living porcelaneous foraminifera (*Archaias angulatis*): textural changes without mineralogic alteration. *Journal of Sedimentary Research*, **68**: 11–19.
Macintyre, I.G., Prufert-Bebout, L., and Reid, R.P., 2000. The role of endolithic cyanobacterium in the formation of lithified laminae in Bahamian stromatolites. *Sedimentology*, **47**: 915–921.
Morse, J.W., Zullig, J.J., Lawrence, D.B., Millero, F.J., Milne, P., Mucci, A., and Choppin, G.R., 1985. Chemistry of calcium carbonate-rich shallow water sediments in the Bahamas. *American Journal of Science*, **285**: 147–185.
Patterson, W.P., and Walter, L.M., 1994. Syndepositional diagenesis of modern platform carbonates: evidence from isotopic and minor element data. *Geology*, **22**: 127–130.
Purdy, E.G., 1963. Recent calcium carbonate facies of the Great Bahama Bank. *Journal of Geology*, **71**: 334–355, 472–497.
Purdy, E.G., 1968. Carbonate diagenesis: an environmental survey. *Geologica Romana*, **7**: 183–228.
Pusey, W.C., 1964. Recent Calcium Carbonate Sedimentation in Northern British Honduras (Ph.D Thesis). Houston: Rice University, 247pp.
Reid, R.P., Macintyre, I.G., and Post, J.E., 1992. Micritized skeletal grains in northern Belize lagoon: a major source of Mg-calcite mud. *Journal of Sedimentary Petrology*, **62**: 145–156.
Reid, R.P., and Macintyre, I.G., 1998. Carbonate recrystallization in shallow marine environments: a widespread diagenetic process forming micritized grains. *Journal of Sedimentary Research*, **68**: 928–946.
Reid, R.P., and Macintyre, I.G., 2000. Microboring verses recrystallization: further insight into the micritization process. *Journal of Sedimentary Research*, **70**: 24–28.
Rude, P.D., and Aller, R.C., 1991. Fluorine mobility during early diagenesis of carbonate sediment: an indicator of mineral transformation. *Geochimica et Cosmochimica Acta*, **55**: 2491–2509.
Walter, L.M., and Burton, E.A., 1990. Dissolution of recent platform carbonate sediments in marine pore fluids. *American Journal of Science*, **290**: 601–643.
Walter, L.M., Bischof, S.A., Patterson, W.P., and Lyons, T.W., 1993. Dissolution and recrystallization in modern shelf carbonates: evidence from pore water and solid phase chemistry. *Philosophical Transactions of the Royal Society (London)*, **344**: 27–36.
Whinland, H.D., 1969. Stability of carbonate polymorphs in warm, shallow seawater. *Journal of Sedimentary Petrology*, **39**: 1579–1587.
Wolf, K.H., 1965. "Grain-diminution" of algal colonies to micrite. *Journal of Sedimentary Petrology*, **35**: 420–427.

MICROBIALLY INDUCED SEDIMENTARY STRUCTURES

Coastal sedimentary systems of moderate climate zones are governed not only by physical dynamics like erosive tidal flushing or reworking by wave action, but also by biotic factors like sediment stabilizing or accumulating by epibenthic bacterial communities. The shallow-marine environments are colonized by benthic microorganisms, amongst those photo-autotroph cyanobacteria are an abundant group (ecology: Whitton and Potts, 2000). With the aid of their adhesive and slimy "extracellular polymeric substances" (EPS, see for introduction Decho, 1990), epibenthic species attach to mineral grains, and they can even form thick, tissue-like, organic layers covering extensive areas of the sedimentary surface. Such bacterial "carpets" are termed "microbial mats" (definition of term see Krumbein, 1983; overview also in Gerdes and Krumbein, 1987; Cohen and Rosenberg, 1989; Stal and Caumette, 1994; Stolz, 2000).

Colonizing at the interface between sediment and water, cyanobacteria are able to react to sediment-affecting physical agents in different ways.

Responsive behavior of epibenthic cyanobacteria to physical sedimentary dynamics

During periods of no or low rates of sediment deposition, thick microbial mat layers develop. By growth, they smoothen out or level the original tidal surface morphology, and form flat bedding surfaces (compare Noffke and Krumbein, 1999). In close up on vertical sections through thick mat layers, the growing biomass drags upward mineral grains from the sedimentary substrate. The grains are separated from each other, and become oriented with their long-axes perpendicularly to the loading pressure (microbial grain separation; Noffke et al., 1997).

Slow moving, suspension-rich bottom currents induce a vertical orientation of cyanobacterial filaments that trigger fall-out of sediment by "baffling, trapping, and binding" (Black, 1933). Over time, the sediment agglutinating microbial community grows upward (Gerdes et al., 1991). In vertical section, this produces laminated intrasedimentary patterns.

In depositional areas of higher hydrodynamic energies, only thin microbial mats can grow. They cover surface structures, like ripple marks, without altering the original shape of the relief. Buried depositional surfaces are "imprinted" by the thin, former mat layers (Gerdes et al., 2000, and references therein).

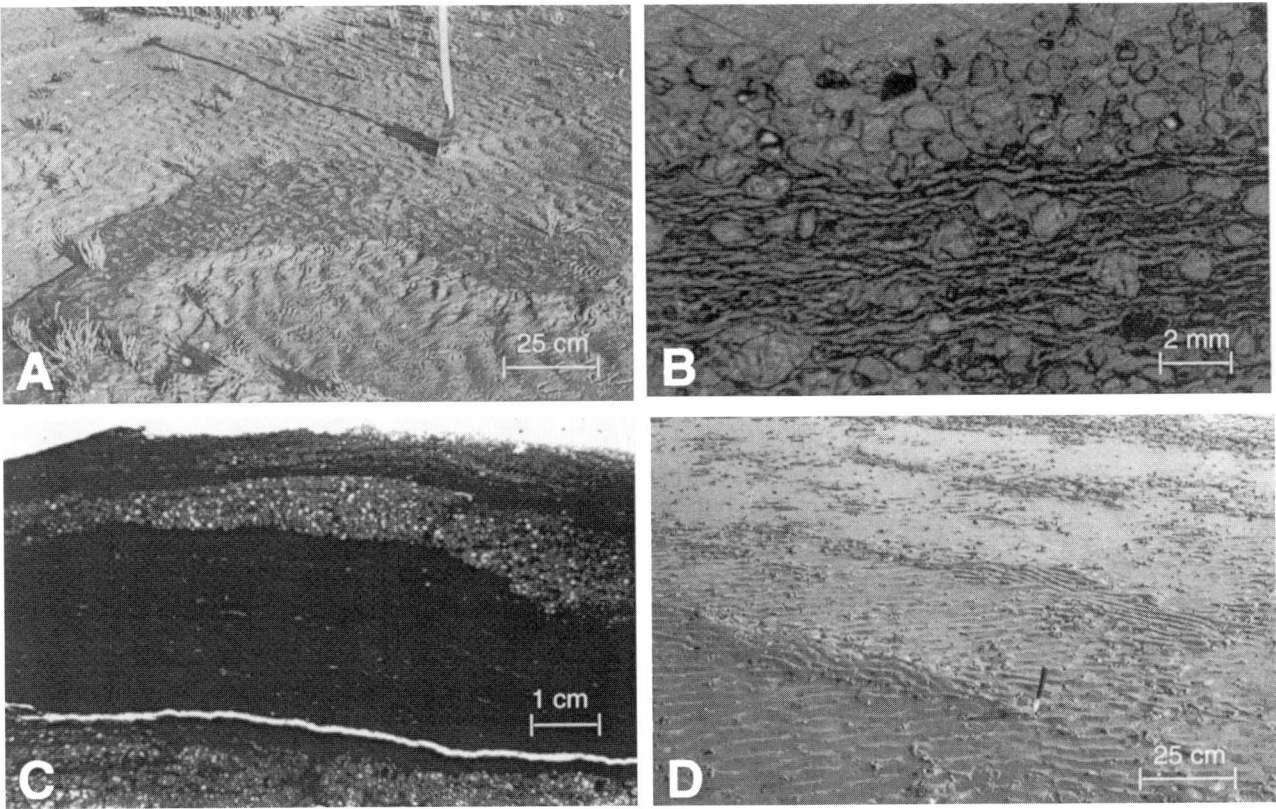

Figure M14 Microbially induced sedimentary structures in physical depositional areas of moderate climate zones—examples. (A) Sedimentary surface smoothened by microbial mat cover; (B) "Oriented grains" within microbial mat layer; thin section; (C) "Biolaminite"; thin section; (D) "Multidirected ripple marks".

Only when subjected to strong bottom currents, mat-secured sedimentary surfaces are eroded, because the dense and coherent microbial mats seal effectively their substrate and increase resistance against erosion. Additionally, cyanobacterial layers prohibit exchange of gas between sediment and water. These effects are known as "biostabilization" (Paterson, 1994; Krumbein et al., 1994; Paterson, 1997; Noffke et al., 2001).

Microbially induced sedimentary structures

The responsive behavior of the mat-constructing micro-epibenthos generates a great variety of characteristic sedimentary structures. The structures differ significantly from biogene stromatolites we are familiar from chemical lithologies, and mineral precipitation plays no role in their formation.

Flat bedding planes originated by immense mat growth are termed "leveled surfaces" (Noffke, 2000; Noffke et al., 2001), Figure M14(A). In the fossil record, they can be recognized as "wrinkle structures" (Hagadorn and Bottjer, 1997), where they document pauses in sedimentation (Noffke, in press). Characteristically, thin-sections through leveled surfaces reveal "mat layer-bound oriented grains", that is mineral particles floating independently from each other in the organic matrix like a chain of pearls, Figure M14(B). They document former microbial grain separation, and pressure-related orientation (Noffke et al., 1997).

Laminated intrasedimentary pattern caused by baffling, trapping, and binding are well-known as "biolaminites", or, after consolidation, as "planar stromatolites" (discussion of terms by Krumbein, 1983), Figure M14(C).

Typical surface morphologies that are shaped by biostabilization interfering with erosion are "erosional remnants and pockets" (Noffke, 1999; fossil example: MacKenzie, 1968), or "multidirected ripple marks" (Noffke, 1998; fossil counterparts: Pflueger, 1999), Figure M14(D). Internal features are "fenestrae fabrics" (Gerdes et al., 2000). "Sinoidal structures", visible in vertical sections through the sediment, imprint former ripple mark valleys (Gerdes et al., 2000; Noffke et al., 2001; fossil example: Noffke, 2000).

Because the structures are formed syndepositionally by biotic-physical interference, they were grouped as an own category "microbially induced sedimentary structures" into the Classification of Primary Sedimentary Structures (Noffke et al., 2001).

Nora Noffke

Bibliography

Black, M., 1933. The algal sediments of Andros Islands, Bahamas. *Royal Society of London Philosophical Transactions*, Series. B, **222**: 165–192.
Cohen, Y., and Rosenberg, E., 1989. *Microbial Mats. Physiological Ecology of Benthic Microbial Communities*. Washington, DC: American Society of Microbiologists, 494pp.
Decho, A.W., 1990. Microbial exopolymer secretions in ocean environments: their role(s) in food webs and marine processes. *Oceanographic Marine Biology Annual Review*, **28**: 73–153.
Gerdes, G., and Krumbein, W.E., 1987. *Biolaminated Deposits*. Springer, 183 p.
Gerdes, G., Krumbein, W.E., and Reineck, H.E., 1991. Biolaminations—ecological versus depositional dynamics. In Einsele, G., Ricken, W., and Seilacher, A. (eds.), *Cycles and Events in Stratigraphy*. Springer-Verlag, pp. 592–607.
Gerdes, G., Klenke, Th., and Noffke, N., 2000. Microbial signatures in peritidal siliciclastic sediments: a catalogue. *Sedimentology*, **47**: 279–308.
Hagadorn, J.W., and Bottjer, D., 1997. Wrinkle structures: microbially mediated sedimentary structures common in subtidal siliciclastic settings at the Proterozoic–Phanerozoic transition. *Geology*, **25**: 1047–1050.
Krumbein, W.E., 1983. Stromatolites—The challenge of a term in space and time. *Precambrian Research*, **20**: 493–531.
Krumbein, W.E., Paterson, D.M., and Stal, L.J., 1994. *Biostabilization of Sediments*. BIS, University of Oldenburg, 526pp.
MacKenzie, D.B., 1968. Sedimentary features of Alameda Avenue cut, Denver, Colorado. *Mountain Geologist*, **5**: 3–13.
Noffke, N., 1998. Multidirected ripple marks rising from biological and sedimentological processes in modern lower supratidal deposits (Mellum Island, southern North Sea). *Geology*, **26**: 879–882.
Noffke, N., 1999. Erosional remnants and pockets evolving from biotic–physical interactions in a Recent lower supratidal environment. *Sedimentary Geology*, **123**: 175–181.
Noffke, N., 2000. Extensive microbial mats and their influences on the erosional and depositional dynamics of a siliciclastic cold water environment (Lower Arenigian, Montagne Noire, France). *Sedimentary Geology*, **136**: 207–215.
Noffke, N., 2003. Epibenthic cyanobacterial communities in the context of sedimentary processes within siliciclastic depositional systems (present and past). In Patterson, D., Zavarzin, G., and Krumbein, W.E. (eds.), *Biofilms Through Space and Time*. Kluwer Academic Publishers (in press).
Noffke, N., and Krumbein, W.E., 1999. A quantitative approach to sedimentary surface structures contoured by the interplay of microbial colonization and physical dynamics. *Sedimentology*, **46**: 417–426.
Noffke, N., Gerdes, G., Klenke, Th., and Krumbein, W.E., 1997. A microscopic sedimentary succession indicating the presence of microbial mats in siliciclastic tidal flats. *Sedimentary Geology*, **110**: 1–6.
Noffke, N., Gerdes, G., Klenke, Th., and Krumbein, W.E., 2001. Microbially induced sedimentary structures—a new category within the classification of primary sedimentary structures. *Journal of Sedimentary Research*, **71**: 649–656.
Paterson, D.M., 1994. Microbiological mediation of sediment structure and behaviour. In Stal, L.J., and Caumette, P. (eds.), *Microbial Mats*. Springer-Verlag, pp. 97–109.
Paterson, D.M., 1997. Biological mediation of sediment erodiability: ecological and physical dynamics. In Burt, N., Parker, R., and Watts, J. (eds.), *Cohesive Sediments*. Wiley, pp. 215–229.
Pettijohn, F.J., and Potter, P.E., 1964 *Atlas and Glossary of Primary Sedimentary Structures*. Springer-Verlag.
Pflueger, F., 1999. Matground structures and redox facies. *Palaios*, **14**: 25–39.
Stal, L.J., and Caumette, P., 1994 Microbial mats. Structure, development and environmental significance. *NATO ASI Series Ecological Sciences*, 35, Berlin, Heidelberg, New York: Springer Verlag, 462pp.
Stolz, J.F., 2000. Structure of microbial mats and biofilms. In Riding, R., and Awramik, S.M. (eds.), *Microbial Sediments*. Springer Verlag, pp. 1–8.
Whitton, B.A., and Potts, M., 2000. *The Ecology of Cyanobacteria. Their Diversity in Time and Space*. Kluwer Academic Publishers, 669p.

Cross-references

Algal and Bacterial Carbonate Sediments
Bacteria in Sediments
Bedding and Internal Structures
Biogenic Sedimentary Structures
Surface Forms

MILANKOVITCH CYCLES

Variations in the Earth's orbital parameters cause quasi-periodic, 10^4–10^6 year scale changes to occur in the incoming solar radiation, or insolation. These insolation changes are commonly known as *Milankovitch cycles*, after the Yugoslav mathematician who first described the cycles (Milankovitch, 1941). Milankovitch cycles have been linked to large-scale climate cycles and, in turn, to climate-mediated sedimentary cycles (e.g., Berger *et al.*, 1984; Einsele *et al.*, 1991; Shackleton *et al.*, 1999).

Earth's orbital parameters

The Earth's rotating axis of figure is tilted with respect to the Sun, Moon and the other planets. Consequently, the axis precesses mainly in response to the luni-solar gravitational attraction on the Earth's equatorial bulge, at a rate governed primarily by the Earth's shape, rotation rate, and Earth–Moon distance. In addition, the axial precession experiences gravitational perturbations from the motions of the other planets acting on the Earth's orbit (e.g., Laskar 1988, Berger and Loutre, 1990). Some perturbations modulate the distance and timing of the Earth's closest annual approach to the Sun (orbital eccentricity and longitude of perihelion), which together affect the year-to-year duration and intensity of seasonal insolation, or *climatic precession*, also called precession parameter, precession index, or precession–eccentricity syndrome. Variations in the Earth's *orbital eccentricity*, generated by interactions among these perturbations, give rise to small changes in the Earth's total annual insolation. Other perturbations cause the Earth's orbital plane to tip and precess (orbital inclination and longitude of the ascending node), thereby modulating the Earth's axial tilt angle, or *obliquity*, relative to the Sun, and consequently, the Earth's latitudinal distribution of insolation.

The evolution of the Earth's orbital parameters over the past 10 myr is shown in Figure M15. The parameters exhibit quasi-periodic behavior with many closely spaced frequencies that are traceable to modulating effects from the planetary perturbations discussed above (see also Laskar, 1988, 1990). The relatively short time spans (<10 million years) covered by most stratigraphic sequences will usually involve analysis that averages over some of the more closely spaced modes. Current astrodynamical theory provides an accurate ephemeris back to *ca.* 16 Ma; for times prior to this, the theory can be used only in limited ways and then with extreme caution (Laskar, 1999).

Geophysical phenomena related to the Earth's tidal dissipation may have played a significant role in the long-term evolution of the orbital parameters. Specifically, the faster rotation rates, closer Earth–Moon distances, and higher ellipticities predicted for remote geologic ages would have involved faster precession rates, hence shorter periodicities for the climatic precession and obliquity (e.g., Berger *et al.*, 1992). On the other hand, some of the orbital eccentricity modes, for example, the 404 kyr cycle, are thought to have remained stable over the past 200 Ma, raising the possibility that they can be used as high-precision geochronometers (Shackleton *et al.*, 1999).

Earth's orbitally forced insolation

The insolation received at a specific terrestrial horizontal surface is determined by the Earth–Sun distance and elevation of the Sun in the sky, which together depend upon geographical latitude, time of day and position of the Earth along its annual orbit. From year to year, the orbital parameters alter the Earth–Sun distance and solar elevation angle on any given calendar day. The relative contributions of the parameters that force the insolation depend upon latitude and effective time(s) of forcing. This was first recognized by Milankovitch (1941), who showed that summer-time "caloric insolation" in the tropics is driven mainly by precession, but

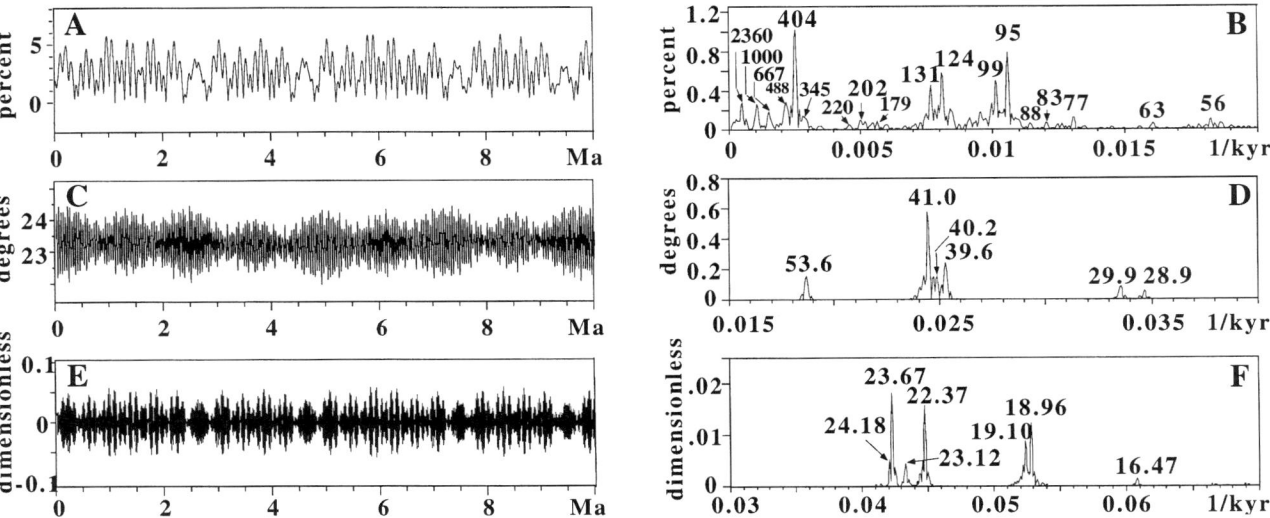

Figure M15 Earth's orbital parameters, 0 Ma to 10 Ma, according to the LA90 model (Paillard *et al.* 1996). (A) eccentricity series; (B) eccentricity spectrum; (C) obliquity series; (D) obliquity spectrum; (E) climatic precession series; (F) climatic precession spectrum. Labels indicate periodicity in kyr of significant harmonic components.

poleward mainly by obliquity. In fact, there are an infinite number of ways to compute orbitally forced insolation. For example, in the illuminated latitudes of Northern summer solstice, noon-time instantaneous insolation is characterized interannually by dominant precession forcing (Figure M16(A)). In contrast, insolation integrated over all daylight hours has increased obliquity forcing toward the North Pole (Figure M16(B)), and insolation integrated over midday solar elevations only results in dominant obliquity forcing everywhere (Figure M16(C)). These and other insolation calculations for equinoctial times and over other time intervals are discussed in Berger et al. (1993). How the orbital parameters are ultimately captured in the sedimentary record depends on what form of orbitally forced insolation drives the climatic variables that control sedimentation.

Milankovitch cycles in stratigraphy

Evidence for orbital forcing in the stratigraphic record is found in the geochemistry and lithology of sediment in both marine and continental deposits. In Milankovitch studies, the most intensely researched sedimentary parameter is the oxygen isotope ratio ($\delta^{18}O$) in foraminifera, which is thought to track global ice volume and, to a lesser extent, ocean temperature and salinity. The Plio–Pleistocene benthic $\delta^{18}O$ record in sediments cored from the tropical eastern Atlantic seafloor shows predominant obliquity forcing that would be expected of high-latitude insolation-forced glaciation cycles (Figure M17(A)). Eolian dust variations in the same sediment are dominated instead by climate precession (Figure M17(B)), which is interpreted as evidence for low-latitude insolation forcing of the monsoon over Africa. At same time, continental loess sequences of north central China developed with an oscillating concentration of iron-rich minerals linked to pedogenesis, detected as a magnetic susceptibility signal. At the Xifeng-1 locale, the signal exhibits a strong obliquity mode up to ca. 600 ka, when it abruptly gains an extremely high amplitude ~ 100 kyr mode—possibly a response to the rapid expansion of the Northern continental glaciations—causing the power spectrum to be heavily weighted in the eccentricity band (Figure M17(C)). These records represent only three of many scores of Plio–Pleistocene sedimentary sequences collected over the past 20 years that have yielded signals consistent with Milankovitch forcing.

While knowledge about the orbital parameters prior to ~ 16 Ma is uncertain, all indications are that the parameters were operating in a recognizable form at least as long ago as the Triassic. For example, the Ca/Fe ratio of Late Eocene

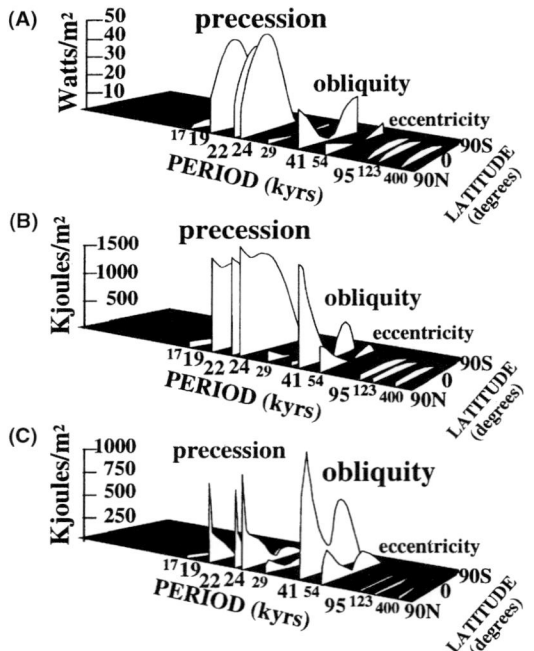

Figure M16 Global variability of the interannual insolation on 21 June, 0 Ka to 150 ka, in spectral amplitude, as a function of latitude ϕ. (A) instantaneous insolation at noon (zenith angle $z = 0$); (B) daily integrated insolation; (C) noon zenith-class ($z = [\phi - \delta, 90° - \phi]$, where δ is solar declination on 21 June) integrated insolation. Locations $\phi > 90°S - \delta$ experience polar winter night and receive no insolation. Redrawn from Berger et al. (1993, Figure 14).

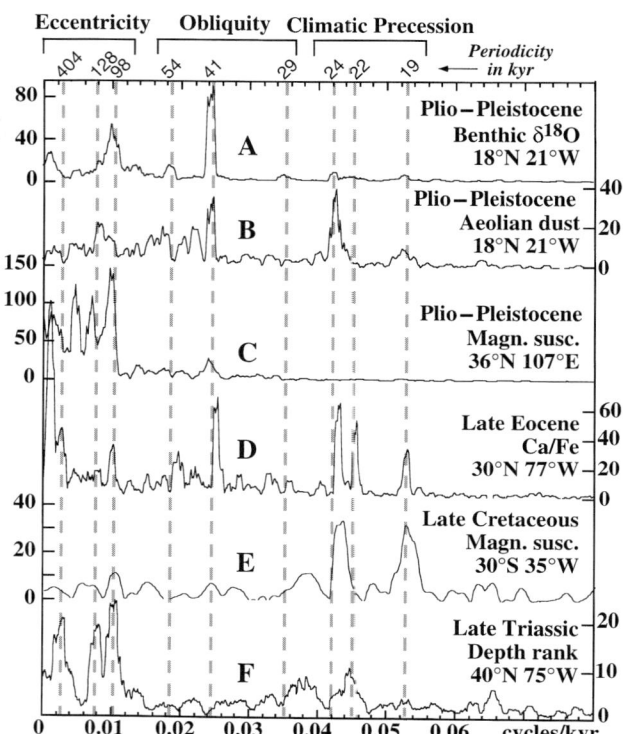

Figure M17 Power spectra of sedimentary parameters from marine and continental stratigraphy. Vertical axes indicate spectral density. (A) Plio–Pleistocene benthic $\delta^{18}O$, ODP site 659, orbitally tuned, 0 Ma to 2.5 Ma (Tiedemann et al., 1994). (B) Plio–Pleistocene aeolian dust concentration, ODP site 659, orbitally tuned, 0 Ma to 2.5 Ma (Tiedemann et al., 1994). (C) Magnetic susceptibility of loess at Xifeng-1 (Kukla et al., 1990), untuned, 0 Ma to 2.6 Ma, referred to the GPTS of Berggren et al. (1995). (D) Late Eocene Ca/Fe series, ODP site 1052, orbitally tuned, 35 Ma to 38 Ma, (Pälike et al., 2001). (E) Late Cretaceous magnetic susceptibility series, DSDP site 516F, untuned, ~ 81 Ma to 82 Ma, referred to the GPTS of Cande and Kent (1992) (Park et al., 1993). (F) Late Triassic depth rank series, Lockatong formation, Newark Basin, ca. 5 myr long, untuned (Olsen and Kent, 1996).

sediments from offshore Florida tracks carbonate versus terrigenous content, and shows a mixed obliquity–precession response consistent with middle latitude insolation forcing (Figure M17(D)). Late Cretaceous carbonate-rich sequences from the Brazil Basin were deposited with a strong climatic precession signature (Figure M17(E)). Other basinal sequences extending back into the Early Cretaceous have also been found to contain depositional patterns with orbital modes (e.g., Herbert et al., 1995). Likewise, evidence for orbital forcing occurs in Jurassic stratigraphy (Shackleton et al., 1999). Finally, one of the most striking examples of Milankovitch cycles occur in the Triassic continental deposits of eastern North America. Here, lacustrine cycles occur as facies alternating between deep perennial lake to playa, and have been interpreted as flooding-drying responses to climatic precession. The spectrum of the depth rank series constructed from the facies logs indicates that the cycles were strongly modulated by the orbital eccentricity (Figure M17(F)).

Summary

The Earth's orbital parameters dictate the amount of insolation received at the Earth's surface. The orbitally forced insolation that comes to affect climate change can take on many different forms that depend on the latitude and time of forcing of the responding climate; these are the Milankovitch cycles. Geochemical and lithological variations of many ancient cyclic sedimentary sequences have signals that are consistent with Milankovitch cycles.

Linda Hinnov

Bibliography

Berger, A., Imbrie, J., Hays, J., Kukla, G., and Saltzman, B., 1984. *Milankovitch and Climate, Parts 1 and 2*. D. Reidel.
Berger, A., and Loutre, M.-F., 1990. Origine des fréquences des éléments astronomiques intervenant dans le calcul de l'insolation. *Bulletin de la Classe des Sciences, 6e Série, Académie Royale de Belgique*, **1–3**: 45–106.
Berger, A., Loutre, M.-F., and Laskar, J., 1992. Stability of the astronomical frequencies over the earth's history for paleoclimate studies. *Science*, **255**: 560–566.
Berger, A., Loutre, M.-F., and Tricot, C., 1993. Insolation and earth's orbital periods. *Journal of Geophysical Reseseurch*, **98**: 10341–10362.
Berggren, W.A., Kent, D.V., Swisher, C.C., and Aubry, M.-P., 1995. A revised cenozoic geochronology and chronostratigraphy. In Berggren, W.A., Kent, D.V., Aubry, M.-P., and Hardenbol, J. (eds.), *Geochronology, Time Scales and Global Stratigraphic Correlation*. SEPM Special Publication No. 54, pp. 129–212.
Cande, S.C., and Kent, D.V., 1992. A new geomagnetic polarity time scale for the Late Cretaceous and Cenozoic. *Journal of Geophysical Research*, **97**: 13917–13951.
Einsele, G., Ricken, W., and Seilacher, A., 1991. *Cycles and Events in Stratigraphy*. Springer-Verlag.
Herbert, T.D., Premoli Silva, I., Erba, E., and Fischer, A., 1995. Orbital chronology of Cretaceous–Paleocene marine sediments. In Berggren, W.A., Kent, D.V., Aubry, M.-P., and Hardenbol, J. (eds.), *Geochronology, Time Scales and Global Stratigraphic Correlation*. SEPM, Special Publication, 54, pp. 81–94.
Kukla, G., An, Z.S., Mélice, J.-L., Gavin, J., and Ziao, J.L., 1990. Magnetic susceptibility record of chinese loess. *Transactions of the Royal Society of Edinburgh: Earth Sciences*, **81**: 263–288.
Laskar, J., 1988. Secular evolution of the solar system over 10 million years. *Astronomy and Astrophysics*, **198**: 341–362.
Laskar, J., 1990. The chaotic motion of the solar system: a numerical estimate of the size of the chaotic zones. *Icarus*, **88**: 266–291.
Laskar, J., 1999. The limits of earth orbital calculations for geological time-scale use. *Philosophical Transactions of the Royal Society of London*, **A357**: 1785–1759.
Milankovitch, M., 1941. *Kanon der Erdbestrahlung und seine Anwendung auf das Eiszeitenproblem*. Belgrade, YU: Royal Serbian Academy, Special Publication, 132, Section of Mathematical and Natural Sciences, 32, 633pp.
Olsen, P., and Kent, D.V., 1996. Milankovitch climate forcing in the tropics of Pangea during the Late Triassic. *Palaeogeography, Palaeoclimatology, Palaeoecology*, **122**: 1–26.
Pälike, H., Shackleton, N.J., and Röhl, U., 2001. Astronomical forcing in Late Eocene marine sediments. *Earth and Planetary Science Letters*, **193**: 589–602.
Paillard, D., Labeyrie, L., and Yiou, P., 1996. Macintosh program performs time-series analysis. *Eos Transactions*, **77**: 379.
Park, J., D'Hondt, S.L., King, J.W., and Gibson, C., 1993. Late Cretaceous precessional cycles in double time: a warm-earth Milankovitch response. *Science*, **261**: 1431–1434.
Shackleton, N.J., McCave, I.N., and Weedon, G.P. (eds.) 1999. Astronomical (Milankovitch) calibration of the geological timescale: a discussion. *Philosophical Transactions of the Royal Society of London, Series A*, **357**: 1731–2007.
Tiedemann, R., Sarnthein, M., and Shackleton, N.J., 1994. Astronomical timescale for the Pliocene Atlantic $\delta^{18}O$ and dust flux records of ODP Site 659. *Paleoceanography*, **9**: 619–638.

Cross-references

Climatic Control of Sedimentation
Cyclic Sedimentation
Isotopic Methods in Sedimentology
Neritic Carbonate Depositional Environments

MIXED SILICICLASTIC AND CARBONATE SEDIMENTATION

Introduction

Every student of the sedimentology and stratigraphy of carbonates early on appreciates the unique aspects of carbonate deposition and the fundamental differences between carbonate and siliciclastic (or clastic) deposition (e.g., Wilson, 1975; Walker and James, 1992). The majority of carbonate deposits forming today in tropical and sub-tropical settings and their ancient counterparts (e.g., rimmed shelves, ramps, etc.) require warm, shallow (within the photic zone), clear water (low turbidity) of near normal salinity which is well circulated by currents, waves and tidal processes to insure proper oxygenation and nutrient levels. The requirement of low turbidity, clear water generally means that where clastic sediment is supplied coevally to a carbonate environment, deposition of the two sediment types is generally mutually exclusive, such that the "carbonate factory" favors or is restricted to areas free of "polluting" clastic materials. However, the mixing of clastic and carbonates occurs today in several Holocene settings, including the Belize shelf, the northeastern shelf of Queensland, Australia (landward of the Great Barrier Reef), and along coastal environments within the Persian Gulf (Wantland and Pusey, 1975; Maxwell and Swinchatt, 1970; Purser, 1973). In addition, the mixing and interbedding of carbonate and clastic sediments is extremely common throughout the stratigraphic record (Mount, 1984; Budd and Harris, 1990; Doyle and Roberts, 1988; Lomando and Harris, 1991; Dolan, 1989), where mixing occurs in a variety of environments: coastal and inner shelf; middle and outer shelf (reef); slope to basin environments. Mixed deposits occur in temperate as well as tropical shelf environments.

Late Paleozoic mixed clastic–carbonate successions have probably received the most study of all mixed deposits. For example Carboniferous "cyclothems" of the Midcontinent and southwestern USA and Europe (e.g., Weller, 1930; Duff *et al.*, 1967; Wilson, 1967; Rankey, 1999) provide the underpinnings of models for cyclic stratigraphy at all scales. From these and additional studies of mixed clastic–carbonate strata of the Permian Basin in the New Mexico and Texas, the concept of "cyclic and reciprocal sedimentation" (Van Siclen, 1958; Wilson, 1967; Silver and Todd, 1969; Meissner, 1972) was introduced to explain the origin of mixed clastic–carbonate shelf-to-basin cyclic deposits. This critical concept has evolved to provide the foundation upon which present-day "sequence stratigraphy" is predicated (e.g., Sarg, 1988). From an economic viewpoint, mixed deposits are significant reservoirs of oil and gas (e.g., the Permian Basin, USA), and studies of the high-resolution stratigraphic architecture of mixed clastic–carbonate sequences has lead to the discipline of reservoir characterization (e.g., Kerans and Tinker, 1997).

An insightful approach to the problem of mixed deposits has been set forth by Budd and Harris (1990) who recognize basically two types of mixed deposition: (1) mixtures due to spatial variability, where mixing occurs principally by lateral facies mixing of coeval sedimentary environments (see also Mount, 1984); and (2) mixtures due to temporal evolution in sedimentation, induced by sea-level changes and/or variations in sediment supply, causing a vertical variation in the stratigraphic succession. Mixing can occur through a wide range of scales, from millimeters to kilometers, respond to a wide range of processes, and can be influenced by all orders of cyclicity. Trends of mixing can vary along both depositional strike and dip, ranging from coastal environments to deep basinal settings.

Important factors that influence mixing include: (i) climate: humid (fluvial-deltaic clastic input onto a carbonate shelf; e.g., the Belize shelf) versus arid (eolian influx; e.g., the Persian Gulf); (ii) sediment supply of clastic material and source terrain (proximal clastic marine shelf with a linear source versus distal tectonically active, uplifted source terrain with a fluvial, point-sourced input); (iii) transport mode (episodic storm mixing versus lowstand-induced sediment by-pass of a carbonate shelf); (iv) shelf width and paleobathymetry (carbonate ramp, distally-steepened carbonate ramp, reefal rimmed high-relief platform, etc.) (v) oceanographic factors such as prevailing wind- and storm-driven currents, tidal range, etc.; (vi) relative changes in sea-level which will act to either trap clastic sediments updip on the shelf during relative highstands or promote clastic influx onto and across a carbonate shelf during relative lowstands (Van Siclen, 1958). The rates and magnitudes and of such relative sea-level changes will determine the amount and style of mixing, such that mixed clastic-carbonate cycles and sequences will vary markedly from "ice-house" periods in geologic history to "greenhouse" periods.

Styles of clastic and carbonate mixing

Depositional mixing and stratigraphic partitioning occurs at all scales and to all degrees. For instance, such mixing may occur: (1) within a single bed whereby minor amounts of clastic material are dispersed within a carbonate bed with a low degree of "lithologic separation" (e.g., Cambro–Ordovician peritidal facies of North America, Read and Goldhammer, 1988; Goldhammer *et al.*, 1993); (2a) within a single heterolithic bed ("ribbon rocks") whereby carbonate material (ooids, peloids, etc.) comprises coarser-grained planar and ripple lamination which alternates with fine-grained clastic material (shale; e.g., Cowan and James, 1993, 1996); (2b) within a single heterolithic bed ("ribbon rocks") whereby clastic material (mature quartz sand) comprises coarser-grained planar laminations and ripple lamination which alternates with thin intercalations of finer-grained carbonate mud (e.g., Demicco and Hardie, 1994); (3a) between vertically juxtaposed beds within high-frequency cycles (meter-scale; "fifth"- and "fourth"-order) with a high-degree of lithologic separation (e.g., Pennsylvanian "cyclothems"; Wilson, 1975; Driese and Dott, 1984; Atchley and Loope, 1993; Yang, 1996; Rankey *et al.*, 1999); (3b) between vertically juxtaposed bedsets within low-frequency sequences (decameter-scale, "third-order") with a high-degree of lithologic separation (e.g., basinal portions of "composite sequences" of the Permian of west Texas; Silver and Todd, 1969; Meissner, 1972; Tinker, 1998); (4a) between vertically juxtaposed beds within high-frequency cycles (meter-scale; "fifth"- and "fourth"-order) with a low degree of lithologic separation, such that vertical contacts between clastics (typically) shale and carbonates are gradational within a cycle (e.g., Grotzinger, 1986; Osleger, 1991; Cowan and James, 1993); (4b) between vertically juxtaposed beds within low-frequency sequences (decameter-scale, "third-order") with a low degree of lithologic separation, such that vertical, as well as lateral contacts between clastics (typically shale) and carbonates are gradational within a cycle (e.g., Grotzinger, 1986; Chow and James, 1987; Sonnenfeld and Cross, 1993); (5) laterally within a single bed as time-equivalent facies changes (spatial mixing; Houlik, 1973; Shinn, 1973; Mount, 1984; Brett and Baird, 1985; Grotzinger, 1986; Calvet and Tucker, 1988; Adams and Grotzinger, 1996; Cowan and James, 1996); and (6) as regionally extensive km-scale depositional systems that can replace one another in time and space (Aitken, 1978; Byers and Dott, 1981).

Models of clastic and carbonate mixing

Spatial variations in coeval environments

In his summary of mixed clastic-carbonate systems, Mount (1984) highlighted four types of mixing of spatially separated carbonate and clastic environments for shallow Holocene shelf settings: (i) punctuated mixing—where sporadic storms and other extreme, high-intensity periodic events transfer sediment across highly contrasting environmental boundaries separating clastics from carbonates; (ii) facies mixing—where sediments are mixed along diffuse borders between contrasting facies, for example nearshore clastic belts and offshore carbonate reefs or ooid shoals; (iii) *in situ* mixing—where the carbonate fraction consists of the authochthonous death assemblages of calcareous organisms that accumulated on or within clastic substrates, for example foram-mollusc assemblages within a subtidal clastic shelf; and (iv) source mixing—where admixtures of carbonates into a clastic-dominated setting are generated in response to uplift and erosion of proximal carbonate terrains.

Punctuated and facies mixing are the most prevalent, and both will lead to the lateral mixing of coeval juxtaposed environments, producing lateral interfingering of clastics and carbonates. Spatial mixing invokes local autocyclic processes, and thus occurs along a timeline in an ancient deposit. Spatial mixing requires no changes in relative sea-level.

Holocene examples that are affected by spatial mixing include: (a) humid fluvio-deltaic systems proximal to a carbonate shelf, such as the nearshore clastic belt and offshore barrier reef of the Belize shelf (Wantland and Pusey, 1975), and the northeastern Australian shelf and offshore Great Barrier Reef (Maxwell and Swinchatt, 1970; Flood and Orme, 1988); (b) Arid eolian dunes proximal to carbonate ramp system, such as along the leeward southeast coast of the Qatar Peninsula in the Persian Gulf (Shinn, 1973); (c) the deep axial trough of Persian Gulf, where fine-grained clastics (clays and silts) from the Tigris-Euphrates delta accumulate adjacent to fine-grained carbonates of the Persian Gulf carbonate ramp system (Purser, 1973). Ancient examples include Cambrian ramp-to-deeper shale shelf transitions in the Upper Cambrian intra-shelf basin of the Virginia Appalachians (Markello and Read, 1981), and the Upper Muschelkalk (Triassic) outer ramp deposits of the Catalan Basin (Spain; Calvet and Tucker, 1988), among others (Houlik, 1973; Mount, 1984; Brett and Baird, 1985; Grotzinger, 1986; Sonnenfeld and Cross, 1993; Adams and Grotzinger, 1996; Cowan and James, 1996).

Temporal variations in environments or evolution in sedimentation

Over geologic time, mixtures may occur as a result of relative sea-level change and/or variations in sediment supply, which causes vertical variations in the stratigraphic succession of facies. The concept of "cyclic and reciprocal sedimentation" (Van Siclen, 1958; Wilson, 1967; Silver and Todd, 1969; Meissner, 1972) which invokes changes in relative sea level was developed to explain the origin of mixed clastic-carbonate shelf-to-basin cyclic deposits. This model suggests that carbonate sedimentation dominates during relative sea-level highstands and rises in sea-level, when clastics are trapped updip in flooded fluvial valleys and narrow clastic shelves while the carbonate factory is fully operational. Additionally, during such highstands, rates of carbonate sedimentation (between 20 cm to 30 cm/1000 years in lagoons to greater than 1 m/1000 years in reefs and shoals) are sufficient to keep pace with typical background subsidence rates such that the subtidal factory outcompetes any clastic influx. Shelfal carbonate deposits are typically thick and the coeval carbonate basin is starved and only thin carbonate deposits are preserved.

Clastic sedimentation dominates during lowstands when carbonate shelves are subaerially exposed and thus largely shut down. During a relative fall in sea-level, widespread braided fluvial systems, shallow deltas or nearshore shallow marine clastics may migrate across the shelf completely and ultimately source deep-water clastic deposits (Van Siclen, 1958; Wilson, 1969; Meissner, 1972; Rankey et al., 1999), or clastics may migrate laterally by longshore prcesses. During true lowstands, fluvial-deltaic systems may incise the antecedent shelf (e.g., the Pleistocene of Belize, Choi and Ginsburg, 1982) and funnel clastic sediments to point-sourced submarine canyons into the basin in humid climates, or in arid climates, eolian dunes may migrate across subaerially deflated carbonate shelves, often truncating older carbonate features (Goldhammer et al., 1991; Borer and Harris, 1991; Atchley and Loope, 1993). During such lowstands, the subaerially exposed carbonate shelf may be karstified and often completely by-passed such that clastic material is deposited on adjacent slopes and basinal areas as onlapping sheets, wedges and fans that terminate against the defunct carbonate slope, for example the Upper Permian Capitan depositional system of west Texas and New Mexico (Bebout and Kerans, 1993; Tinker, 1998). Such basinal deposits may be then subject to reworking by contour-hugging bottom currents. The cyclic and reciprocal model further suggests that lowstand shelfal clastics will be thin and coeval basinal clastics (submarine fans, turbidites) will be much thicker. However, the height and steepness of the underlying carbonate slope is an important consideration in determining the character of any lowstand clastic sand body. Alternatively, if an updip clastic source is limited, basinal settings may receive little or no clastic material and only record the shutdown of the carbonate platform in the form of submarine diastems (e.g., mineralized, cemented hardgrounds).

The character of the lowstand clastic sediments reflects both proximity to the source area and depth of deposition. For example in the late Devonian of the Canning Basin (western Australia), Southgate et al. (1993) record the evolution of several Frasnian–Fammenian "third-order" depositional sequences whereby the shelfal carbonates are completely bypassed. Within the basin, however, there exist several sequences of lowstand clastics each composed of basin-floor fans, slope fans, and prograding complexes, which overall form a fining-upward trend from basal coarse clastics to fine-grained mudstone and shales vertically within a particular lowstand. This grain-size trend is interpreted to depict initial stream rejuvenation upon initial subaerial shutdown of the carbonate shelf, followed by subsequent waning of coarse proximal sediment supply as renewed transgression of the shelf occurred.

An unequivocal and detailed example of mixed clastic-carbonate sedimentation driven by sea-level changes is provided by the Quaternary reefs in the southernmost Belize lagoon (Choi and Ginsburg, 1982). During the last glacial maxima, eustatic sea-level was about 120 m beneath its present day level, exposing and karstifying much of the Belize carbonate shelf. As coastal rivers advanced seaward onto the shelf, they created morphological features including fluvial and submarine channels, fluvial channel bars, levess and deltaic lobes. Subsequently, when eustatic sea-level rose, such clastic features with positive morphologic relief served as nucleation sites and nodes for the initial development of Holocene reefs and other biogenic carbonates. Another example, provided by Davies et al. (1989) illustrates the cyclic and reciprocal nature of Cenozoic clastic-carbonate sequences develped in response to eustasy along the northeastern shelf of Australia (Figure M18).

Conclusions

Mixed siliciclastic and carbonate sedimentation is a common theme throughout the stratigraphic record. Holocene systems enable us to examine mixtures due to spatial variability, where mixing occurs principally by lateral facies mixing of coeval sedimentary environments. Studies of spatial mixing emphasize process and provide clues to examing ancient mixed successions. Where such examples are cyclically interbedded, they are often attributed to temporal evolution in

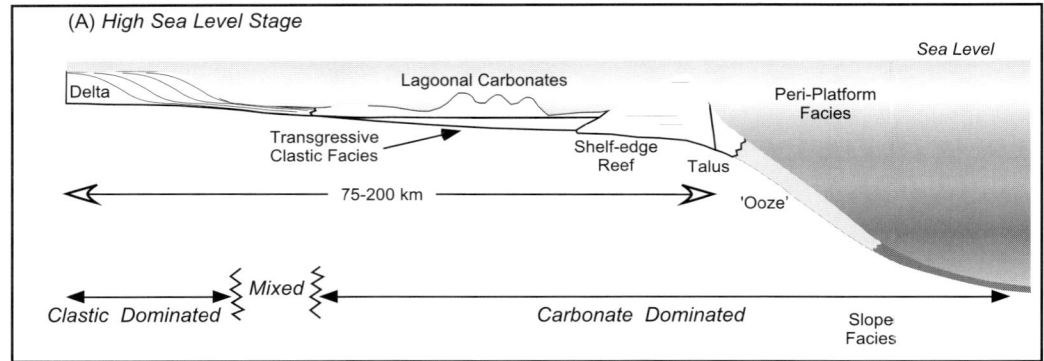

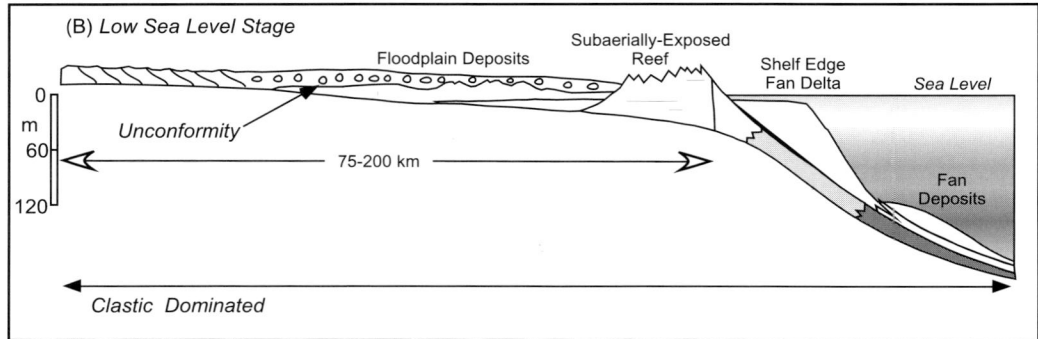

Figure M18 Schematic sections showing high (A) and low (B) sea-level control on the stratigraphic evolution and sedimentary geometry of shelf facies in the central Great Barrier Reef province. Note the predominance of siliciclastic facies in the lowstand. Modified from Davies *et al.* (1989).

sedimentation, induced by sea-level changes and/or variations in sediment supply, causing a vertical variation in the stratigraphic succession. Many, if not most, ancient mixed successions can be interpreted as examples of cyclic and reciprocal sedimentation, controlled by cyclic variations in relative sea-level.

Robert K. Goldhammer

Bibliography

Adams, R.D., and Grotzinger, J.P., 1996. Lateral continuity of facies and parasequences in Middle Cambrian platform carbonates, Carrara formation, Southeastern California, USA. *Journal of Sedimentary Research*, **66**: 1079–1090.

Aitken, J.D., 1978. Revised models for depositional grand cycles, Cambrian of the southern Rocky Mountains, Canada. *Bulletin of Canadian Petroleum Geology*, **26**: 515–542.

Atchley, S.C., and Loope, D.B., 1993. Low-stand aeolian influence on stratigraphic completeness: upper member of the Hermosa Formation (latest Carboniferous), southeastern Utah, USA. In Pye, K., and Lancaster, N. (eds.), *Aeolian Sediments, Ancient and Modern*. International Association of Sedimentologists, Special Publication, 16, pp. 127–149.

Bebout, D.G., and Kerans, C., 1993. *Guide to the Permian Reef Geology Trail, McKittrick Canyon, Guadalupe Mountains National Park, West Texas*. Texas Bureau of Economic Geology, Guidebook 26.

Borer, J.M., and Harris, P.M., 1991. Lithofacies and cyclicity of the Yates Formation, Permian Basin: implications for reservoir heterogeneity. *American Association of Petroleum Geologists Bulletin*, **75**: 726–779.

Brett, C.E., and Baird, G.C., 1985. Carbonate-shale cycles in the Middle Devonian of New York: an evaluation of models for the origin of limestones in terrigenous shelf sequences. *Geology*, **13**, 324–327.

Budd, D.A., and Harris, P.M. (eds.), 1990. *Carbonate-Siliciclastic Mixtures*. SEPM (Society for Sedimentary Geology), Reprint Series Number 14.

Byers, C.W., and Dott, R.H. Jr., 1981. SEPM research conference on modern shelf and ancient cratonic sedimentation—the orthoquartzite-carbonate suite revisited. *Journal of Sedimentary Petrology*, **51**: 329–346.

Calvet, F., and Tucker, M.E., 1988. Outer ramp cycles in the Upper Muschelkalk of the Catalan Basin, northeast Spain. *Sedimentary Geology*, **57**: 185–198.

Choi, D.R., and Ginsburg, R.N., 1982. Siliciclastic foundations of quaternary reefs in the southernmost Belize lagoon, British Honduras. *Geological Society of America Bulletin*, **93**: 116–126.

Chow, N., and James, N.P., 1987. Cambrian grand cycles: a northern Appalchian perspective. *Geological Society of America Bulletin*, **98**: 418–429.

Cowan, C.A., and James, N.P., 1993. The interactions of sea-level change, terrigenous-sediment influx, and carbonate productivity as controls on the Upper Cambrian grand cycles of western Newfoundland, Canada. *Geological Society of America Bulletin*, **105**: 1576–1590.

Cowan, C.A., and James, N.P., 1996. Autogenic dynamics in carbonate sedimentation: meter-scale, shallowing-upwards cycles, Upper Cambrian, western Newfoundland, Canada. *American Journal of Science*, **296**: 1175–1207.

Davies, P.J., Symonds, D.A., Feary, D.A., and Pigram, C.J., 1989. The evolution of the carbonate platforms of northeast Australia. In Crevello, P.D., Wilson, J.L., Sarg, J.F., and Read, J.F. (eds.), *Controls on Carbonate Platform and Basin Development*. SEPM (Society for Sedimentary Geology), Special Publication, 44, pp. 233–258.

Demicco, R.V., and Hardie, L.H., 1994. Sedimentary structures and early diagenetic features of shallow marine carbonate deposits, *Controls on Carbonate Platform and Basin Development*. SEPM (Society for Sedimentary Geology), Atlas Series No. 1.

Dolan, J.F., 1989. Eustatic and tectonic controls on deposition of hybrid siliclastic/carbonate basinal cycles: discussion with examples. *American Association of Petroleum Geologists Bulletin*. **73**: 1233–1246.

Doyle, L.J., and Roberts, H.H. (eds.), 1988. *Carbonate–Clastic, Transitions*. Developments in Sedimentology 42. Elsevier.

Driese, S.G., and Dott, R.H., 1984. Model for sandstone-carbonate "cyclothems" based on upper member of Morgan Formation (Middle Pennsylvanian) of northern Utah and Colorado. *American Association of Petroleum Geologists Bulletin*, **68**: 574–597.

Duff, P.Mcl.D., Hallam, A., and Walton, E.K., 1967. *Cyclic Sedimentation*. Developments in Sedimentology 10. Elsevier.

Flood, P.G., and Orme, G.R., 1988. Mixed siliciclastic/carbonate sediments of the Northern Great Barrier Reef Province, In Doyle, L.J., and Roberts, H. H. (eds.), 1988, *Carbonate–Clastic Transitions*. Developments in Sedimentology 42. Elsevier. pp. 175–206.

Goldhammer, R.K., Oswald, E.J., and Dunn, P.A., 1991. The hierarchy of stratigraphic forcing: an example from Middle Pennsylvanian shelf carbonates of the Paradox Basin. In Franseen, E.K., Watney, W.L., Kendall, St, G.C.C., Ross, W. (eds.), *Sedimentary Modeling: Computer Simulations and Methods for Improved Parameter Definition*. Kansas Geological Survey, Special Publication, 233, pp. 361–413.

Goldhammer, R.K., Lehmann, P.J., and Dunn, P.A., 1993. The origin of high frequency platform carbonate cycles and third-order sequences (Lower Ordovician El Paso Group, West Texas); constraints from outcrop data and stratigraphic modeling. *Journal of Sedimentary Research*, **63**: 318–359.

Grotzinger, J.P., 1986. Cyclicity and paleoenvironmental dynamics, Rocknest platform, northwest Canada. *Geological Society of America Bulletin*, **97**: 1208–1231.

Houlik, C.W., 1973. Interpretation of carbonate-detrital silicate transitions in the Carboniferous of western Wyoming. *American Association of Petroleum Geologists Bulletin*, **57**: 498–509.

Kerans, C., and Tinker, S.W., 1997. *Sequence stratigraphy and characterization of carbonate reservoirs*. SEPM (Society for Sedimentary Geology), Short Course Notes No. 35, 130pp.

Lomando, A.J., and Harris, P.M. (eds.), 1991. *Mixed Carbonate—Siliciclastic Sequences*. SEPM (Society for Sedimentary Geology), Core Workshop 15, 568pp.

Markello, J.R., and Read, J.F., 1981. Carbonate ramp-to-deeper shale shelf transitions of an Upper Cambrian intrashelf basin, Nolichucky Formation, southwest Virginai Appalachians. *Sedimentology*, **28**: 573–598.

Maxwell, W.G.H., and Swinchatt, J.P., 1970. Great Barrier Reef: regional variation in a terrigenous-carbonate province. *Geological Society of America Bulletin*, **81**: 691–724.

Meissner, F.F., 1972. Cyclic sedimentation in Middle Permian strata of the Permian Basin. In Elam, S.G., and Chuber, S. (eds.), *Cyclic Sedimentation in the Permian Basin*. 2nd edn. West Texas Geological Society, Publication, 72-60, pp. 203–232.

Mount, J.M., 1984. Mixing of siliciclastic and carbonate sediments in shallow shelf environments. *Geology*, **12**: 432–435.

Osleger, D.A., 1991. Subtidal carbonate cycles: implications for allocyclic versus autocyclic controls. *Geology*, **19**: 917–920.

Purser, B.H. (ed.), 1973. *The Persian Gulf*. Springer Verlag.

Rankey, E.C., Bachtel, S.L., and Kaufman, J., 1999. Controls on stratigraphic architecture of icehouse mixed carbonate-siliciclastic: a case study from the Holder Formation (Pennsylvanian, Virgilian), Sacramento Mountains, New Mexico. In Harris, P.M., Saller, A.H., and Simo, J.A.T. (eds.), *Advances in Carbonate Sequence Stratigraphy: Application to Reservoirs, Outcrops, and Models*. SEPM (Society for Sedimentary Geology), Special Publication, 63, pp. 127–150.

Read, J.F., and Goldhammer, R.K., 1988. Use of Fischer plots to define third-order sea-level curves in Ordovician peritidal cyclic carbonates, Appalachians. *Geology*, **16**: 895–899.

Sarg, J.F., 1988. Carbonate sequence stratigraphy. In Wilgus, S., Hastings B.S., Kendall St, G.C., Posamentier, H.W., Ross, C.A., and Van Wagoner, J.C. (eds.), *Sea Level Changes—An Integrated Approach*. SEPM (Society for Sedimentary Geology), Special Publication, 43, pp. 155–181.

Shinn, E.A., 1973. Sedimentary accretion along the leeward, SE coast of Qatar Peninsula, Persian Gulf. In Purser, B.H., (ed.), *The Persian Gulf*. Springer Verlag, pp. 199–210.

Silver, B.A., and Todd, R.G., 1969. Permian cyclic strata, northern Midland and Delaware basins, west Texas and southeastern New Mexico. *American Association of Petroleum Geologists Bulletin*, **53**: 2223–2251.

Sonnenfeld, M.D., and Cross, T.A., 1993. Volumetric partitioning and facies differentiation within the Permian Upper San Andres Formation of Last Chance Canyon, Guadalupe Mountains, New Mexico. In Loucks, R.G., and Sarg, J.F. (eds.), *Carbonate Sequence Stratigraphy: Recent Developments and Applications*. American Association of Petroleum Geologists Memoir, 57, pp. 435–474.

Southgate, P.N., Kennard, J.M., Jackson, M.J., O'Brien, P.E., and Sexton, M.J., 1993. Reciprocal lowstand clastic and highstand carbonate sedimentation, subsurface Devonian Reef Complex, Canning Basin, Western Australia. In Loucks, R.G., and Sarg, J.F. (eds.), *Carbonate Sequence Stratigraphy: Recent Developments and Applications*. American Association of Petroleum Geologists Memoir, 57, pp. 157–179.

Tinker, S.W., 1998. Shelf-to-basin facies distribution and sequence stratigraphy of a steep-rimmed carbonate margin: Capitan depositional system. McKittrick Canyon, New Mexico and Texas. *Journal of Sedimentary Research*, **68**: 1146–1175.

Van Siclen, D.C., 1958. Depositional topography—examples and theory. *American Association of Petroleum Geologists Bulletin*, **42**: 1897–1913.

Walker, R.G., and James, N.P. (eds.), 1992. *Facies Models: Response to Sea Level Change*. Geological Association of Canada.

Wantland, K.F., and Pusey, W.C. III (eds.), 1975. *Belize shelf—Carbonate Sediments, Clastic Sediments, and Ecology*. American Association of Petroleum Geologists Studies in Geology 2, 312pp.

Weller, J.M., 1930. Cyclical sedimentation of the Pennsylvanian period and its significance. *Journal of Geology*. **38**: 97–135.

Wilson, J.L., 1967. Cyclic and reciprocal sedimentation in Virgilian strata of southern New Mexico. *Geologic Society of America Bulletin*, **78**: 805–818.

Wilson, J.L., 1975. *Carbonate Facies in Geologic History*. Springer-Verlag.

Yang, W., 1996. Cycle symmetry and its causes, Cisco Group (Virgilian and Wolfcampian), Texas. *Journal of Sedimentary Research*, **66**: 1102–1121.

Cross-references

Facies Models
Neritic Carbonate Depositional Environments
Sands and Sandstones

MIXED-LAYER CLAYS

Clay mineral crystals contain from a few to few hundred silicate layers (1:1 or 2:1, dioctahedral or trioctahedral, see figure 1 in *Illite Group Minerals*), spread over tens to thousands of nm in the a* and b* crystallographic directions and stacked parallel in the c* direction. Layers are either electrically neutral or they have a negative charge, resulting from isomorphous substitutions within the tetrahedral and/or octahedral sheet. The charged layers are bounded by interlayer cations or sheets of interlayer hydroxide. If the interlayer cations are hydrated, the interlayer (smectite or vermiculite type) has swelling properties (i.e., it may change thickness and accept organic molecules). Clay crystals may contain non-identical layers and/or interlayers, that is, be mixed-layered (interstratified). Mixed-layering is defined as random and characterized by the ordering parameter (Reichweite) R0 if there is no preferred sequence in stacking of layers (or interlayers) A and B. If some sequences are privileged, such mixed-layering is called ordered (R1 if ABAB..., R2 if AABAAB... etc.; details in Moore and Reynolds, 1997). If only one type of sequence is present, the mineral is called regular. Such minerals (only of R1 type) exist in nature; they are given separate names, and their origin is explained by different mechanisms. Other names of mixed-layer clays are combinations of the components' names, for example, illite–smectite.

Mixed-layering is identified and quantified by means of X-ray diffraction (XRD), supplemented by chemical analysis and electron microscopy. 00l XRD reflections of mixed-layer minerals appear between positions of corresponding reflections of pure components. The nature of layers and swelling interlayers is revealed by XRD of specimens heated or treated with different exchange cations or organic solvents. Computer modeling of such diffraction effects is used to measure layer ratios and ordering (Moore and Reynolds, 1997). If the imaging conditions are properly selected, and a distinctive and stable under vacuum spacing of the swelling interlayer achieved, the nature of layers can be identified in transmission electron microscope by measuring the layer spacings and obtaining AEM chemical analyzes of the bulk crystal.

Mixed-layer clays with smectitic interlayers can be separated in dilute suspensions, by infinite osmotic swelling along these interlayers, into individual layers or blocks of layers bounded by stable interlayers (fundamental particles, see figure 1 in *Illite Group Minerals*). When such separation is stabilized in the solid state, the mixed-layering diffraction effect is destroyed and such blocks of layers (individual layers do not diffract coherently) can be studied by XRD, as if they were pure end member clays (Eberl *et al.*, 1998).

Mixed-layer clay minerals are most often intermediate products of reactions altering pure end-member clays. Mixed-layer minerals occur in natural environments ranging from surface to low-grade metamorphic and hydrothermal conditions (see review by Środoń, 1999). The majority of mixed-layer clays have been synthesized in the laboratory. Most often, the mixed-layering is 2-component, but more complicated interstratifications have also been documented. Mixed-layer clays are either di or trioctahedral; di/trioctahedral interstratifications are very rare. Most mixed-layer clays contain swelling interlayers of smectite type (typical of diagenetic alteration) or vermiculite type (typical of weathering alteration). In general, the weathering reactions producing mixed layering are reversals of the corresponding high temperature reactions, but the reaction paths are quite different. Solid-state transformation and dissolution/crystallization are the two mechanisms responsible for the formation of different mixed-layer clays.

Dioctahedral mixed-layer clays

Kaolinite–Smectite (K–S)

K–S is known from Recent soils developed on acidic to alkaline parent rocks and from the corresponding paleosols (Hughes *et al.*, 1993). If present in sedimentary rocks, K–S is most often a detrital component, which survives burial temperatures of at least 100°C. Its further evolution has not been traced. Most probably, the abundance of K–S is underestimated, because it is difficult to detect and also can be misidentified by XRD as halloysite.

K–S is typically very fine-grained, randomly interstratified, and it always evolves from smectite (origin from kaolinite has not been reported). The reaction has been reproduced in laboratory and is promoted by a high supply of Al.

Illite–Smectite (I–S)

I–S is the most abundant, most ubiquitous and the most studied mixed-layer clay. I–S can be detected by XRD in the <0.2 μm fraction of almost every sample of sedimentary rock. Smectite or recycled I–S is stable in sedimentary environments and starts to react toward more illitic composition at about 70°C. This reaction has been originally interpreted as fixation of potassium in smectitic (swelling) interlayers plus solid-state transformation of layers, but currently it is explained as dissolution of smectite layers and simultaneous nucleation of 2 nm illite particles in the swelling interlayers and subsequent growth of these particles (see *Illite Group Minerals*).

Evolution of I–S is continuous, always from R0 (decreasing to ca. 45 percent smectite), through R1 (down to 20 percent S), to R>1, and finally to illite. Regular 1:1 interstratification (K-rectorite) is rare, known mostly from hydrothermal environments. Chemical evolution involves enrichment of I–S in K (and in some cases NH_4) and Al, a decrease in cation exchange capacity, and liberation of Ca, Mg, Fe, and Si, which contribute to quartz and carbonate cementation. I–S crystals have a lognormal distribution of thickness with the mean of 5–6 layers/crystal, which increases below 20 percent S.

The percent S in I–S is an indicator of diagenetic grade, analogous to organic indicators (see *Maturation, Organic*), and can be used for evaluation of maximum paleotemperatures and their K–Ar dating (see *Illite Group Minerals*). Because of its abundance and swelling layers, I–S controls mechanical, electrical and exchange properties of most shales.

Weathering of illite or I–S is not a simple reversal of diagenetic reaction and is not well understood. Smectite can be produced directly by dissolution-precipitation (Šucha *et al.*, 2001) or by transformation via ordered illite–vermiculite and vermiculite (Wilson and Nadeau, 1985).

Glauconite–Smectite (G–S)

Minerals in monomineral glauconite pellets are iron-rich analogues of I–S. Glauconitization (see *Glaucony and Verdine*) proceeds at the sediment/water interface, through the same steps of random and ordered interstratifications as illitization. Regular interstratification has not been observed. Anomalous chemical characteristics of some samples may result from interlayering with berthierine (iron serpentine). Data are lacking on the evolution of G–S during burial diagenesis.

Na-illite–smectite (rectorite)

This mineral is reported only as a regular variety, called rectorite (allevardite in older literature), and its geological occurrence is restricted to hydrothermal alteration zones and low-grade metamorphic rocks. In both cases, it is associated with pyrophyllite and occasionally with cookeite, which indicates crystallization tempearatures of ca. 280°C. Sometimes, Ca may prevail over Na, which is expressed by the presence of margarite-like layers. Rectorite crystals are always very thick, as is evident from their very sharp diffraction peaks and electron microscope images.

Chlorite–smectite (tosudite)

This clay is known only as a regular interstratification (named tosudite) of beidellite (Al-smectite) and di-dioctahedral chlorite (donbassite), or di-trioctahedral chlorite (sudoite, Mg-rich; or cookeite, Li-rich). Likewise, only regular interstratification was produced in hydrothermal experiments. Tosudite is known from hydrothermal, pneumatolytic, or low-temperature metamorphic alteration zones, in association with dioctahedral chlorite, serpentine, rectorite, and pyrophyllite. This association suggests that tosudite forms during acid hydrothermal alteration, at temperatures well below 350°C.

Mica (Illite)–Vermiculite (I–V)

I–V is a product of potassium removal during weathering of micas and illite. It occurs as thick crystals, with tri-dimensional order, frequently with almost regular interstratification. The vermiculitic layer may evolve toward smectite, giving rectorite-like clay (Wilson and Nadeau, 1985). Partial chloritization of vermiculitic interlayer is, however, more common. Clays of this type are called soil vermiculites, soil chlorites, chlorite intergrades, or swelling chlorites (Barnhisel and Bertsch, 1989). Formation of soil vermiculites is a massive phenomenon in temperate climates and this material has been identified in fresh sediments. Its further evolution during diagenesis remains obscure.

Trioctahedral mixed-layer clays

Serpentine–Chlorite (Sp–C)

Sp–C is exclusively Fe-rich clay (berthierine-chamosite). It is known to form in Recent "verdine facies" sediments in very shallow tropical seas (see *Glaucony and Verdine*), often as sandstone grain coatings. Either both layers crystallize simultaneously, or chloritization starts at the sea bottom. During burial, the berthierine content decreases and disappears completely at the transition zone between shales and slates. Sp:C ratio can be now quantified precisely by XRD (Moore and Reynolds, 1997). Regularly interstratified Sp–C is called dozyite. Sp–C interstratification is accompanied by interstratification of polytypes.

Chlorite–Smectite (vermiculite) (C–S)

C–S is the second most common mixed-layer mineral in sedimentary rocks, often occurring together with I-S, and is characteristic of hypersaline facies, volcanoclastic rocks and graywackes (Reynolds, 1988). Unknown from Recent sedimentary environments, C–S is a product of smectite chloritization at elevated temperatures (commonly during burial diagenesis). The process is not fully analogous to illitization of smectite and is not fully understood. Instead of a continuous process, overlapping succession of three phases is observed: C–S with <20 percent C, corrensite (regular interstratification), and C–S > 85 percent C. Diagenetic corrensite appears in sedimentary rocks at 60°C to 160°C. A decrease of Si and interlayer exchangeable cations, and an increase of tetrahedral Al accompanies chloritization.

Weathering of chlorite commonly produces corrensite as an intermediate phase. Distinct from diagenetic corrensite, this mineral is characterized as chlorite-vermiculite, that is, high-charge corrensite. The final product is vermiculite (*q.v.*), which may evolve latter into smectite. The process involves loss of Fe and Mg to such extent that the end member vermiculite is strictly dioctahedral (Proust *et al.*, 1986).

Mica–Chlorite (M–C)

M–C is typically close to regular interlayering and the mica component is almost exclusively biotite. All M–C minerals occur as big crystals producing excellent quality electron microscope images. The majority of described occurrences are from hydrothermally altered rocks or the products of low to medium-grade regional metamorphism of granite, gneiss, basalt, or pelites. M–C may form both by alteration of biotite and chlorite.

Talc–Smectite (T–S)

T–S clays are known from saline lake and hot-spring deposits, weathered ophiolites and dolomites. They are frequently associated with sepiolite. Full range of interstratification was reported (Wiewióra *et al.*, 1982), including regular aliettite. The temperature of formation (hot springs) and solution chemistry are probably the factors responsible for the variation in expandability. It remains unclear whether the reaction proceeds from smectite to talc, or *vice versa*, or in both directions.

Mica–Vermiculite (M–V)

M–V is a common product of K-removal during weathering of biotite. The reaction is easily reproducible in laboratory. The complete compositional range is known (Pozzuoli *et al.*, 1992), including a regular variety (hydrobiotite). Little is known about the behavior of M–V at the subsequent stages of the rock cycle.

Jan Środoń

Bibliography

Barnhisel, R.I., and Bertsch, P.M., 1989. Chlorites and hydroxy-interlayered vermiculite and smectite. In Dixon, J.B., and Weed, S.B. (eds.), *Minerals in Soil Environments*. Soil Science Society of America, pp. 729–788.

Eberl, D.D., Nuesch, R., Šucha, V., and Tsipursky, S., 1998. Measurement of fundamental particle thicknesses by X-ray diffraction using PVP-10 intercalation. *Clays and Clay Minerals*, **46**: 89–97.

Hughes, R.E., Moore, D.M., and Reynolds, R.C. Jr., 1993. The nature, detection, occurrence, and origin of kaolinite/smectite. In *Kaolin Genesis and Utilization*. The Clay Minerals Society, Special Publication, 1, pp. 291–323.

Moore, D., and Reynolds, R.C., 1997. *X-Ray Diffraction and the Identification and Analysis of Clay Minerals*. Oxford University Press.
Pozzuoli, A., Vila, E., Franco, E., Ruiz-Amil, A., and de la Calle, C., 1992. Weathering of biotite to vermiculite in Quaternary lahars from Monti Ernici, central Italy. *Clay Minerals*, **27**: 175–184.
Proust, D., Eymery, J.P., and Beaufort, D., 1986. Supergene vermiculitization of a magnesian chlorite: iron and magnesium removal process. *Clays and Clay Minerals*, **34**: 572–580.
Reynolds, R.C. Jr., 1988. Mixed layer chlorite minerals. In *Hydrous Phyllosilicates (exclusive of micas)*. Mineralogical Society of America, Reviews in Mineralogy, 19, pp. 601–629.
Środoń, J., 1999. Nature of mixed-layer clays and mechanisms of their formation and alteration. *Annual Review of Earth and Planetary Sciences*, **27**: 19–53.
Šucha, V., Środoń, J., Clauer, N., Elsass, F., Eberl, D.D., Kraus, I., and Madejova, J., 2001. Weathering of smectite and illite-smectite in Central-European temperate climatic conditions. *Clay Minerals*, **36**: 403–419.
Wiewióra, A., Dubińska, E., and Iwasińska, I., 1982. Mixed-layering in Ni-containing talc-like minerals from Szklary, Lower Silesia, Poland. In *Proceedings of the International Clay Conference*. Italy: Bologna-Pavia, pp. 111–125.
Wilson, M.J., and Nadeau, P.H., 1985. Interstratified clay minerals and weathering processes. In Drever, J.I. (ed.), *The Chemistry of Weathering*. D. Reidel, pp. 97–118.

Cross-references

Bentonites and Tonsteins
Berthierine
Chlorite in Sediments
Clay Mineralogy
Diagenesis
Glaucony and Verdine
Illite Group Clay Minerals
Kaolin Group Minerals
Mudrocks
Smectite Group
Vermiculite
Weathering, Soils, and Paleosols

MIXING MODELS

In general, sediments are not faithful recordings of the history of sedimentary input at a given location because they are subject to a variety of postdepositional processes that can alter their composition, structure, and therefore the apparent record. Foremost amongst these processes is widespead mixing of sediments by biological or physical agents. Geologists, geochemists, biologists, and engineers often find it necessary to quantify the effects of mixing on the distribution of *solid* sedimentary components. For example, the fate of a buried contaminant can be controlled by its unanticipated release from the sediments by mixing. In addition, isotopic distributions can be significantly affected by mixing, such that estimates of sediment accumulation rates become highly skewed to unreasonably large values if the influence of mixing is not taken into account.

Correct prediction of the effects of mixing can be obtained through mathematical models; these models are based on the principle of mass/chemical species conservation, and their mathematical forms depend on the particular assumptions made about the nature of the mixing. For example, mixing due to the random movement of small infauna, the feeding of head-down (conveyor belt) species, the infilling of vacated burrows, and the resuspension of sediment by clams or waves all generate different types of sediment motions and, in principle, these differences are reflected in different mathematical forms for the appropriate mixing model (Boudreau, 1997). However, in practice, workers in this field tend to use a simple array of models, that is, *well-mixed* models, *diffusive* models, and *nonlocal* models. In addition, these models can be distinguished further on the basis of the nature of the inputs they require and outputs they produce. Thus, models can be either of *forward type*, where the user inputs all parameter values and expects a prediction of the distribution of a sediment component with space and/or time, or they can be of *backward type*, where parameters or the input history are calculated from the observed depth profile of a sediment property.

Well-Mixed Models. Mixing can be so intense that the sediment component has uniform distribution (concentration, C) in the mixed layer. For example, the distribution of ^{14}C often exhibits a surficial zone some 4 cm to 10 cm thick of constant "age", which reflects relatively fast mixing compared to the half-life of this isotope. This type of mixing will cause homogenization of input signals to the sediment and add long tails of the signals when they enter the sediment record. An example is schematically illustrated in Figure M19.

When mixing is so intense, the concentration of the homogenized component, C (mass or numbers per unit volume), in the mixed layer is given by

$$\frac{dC}{dt} = \frac{F_{in}(t) - wC}{L} - R \qquad \text{(Eq. 1)}$$

where t is time, $F_{in}(t)$ is the input (flux) at the sediment–water interface (mass or numbers per unit area and time), w is the burial velocity relative to the sediment–water interface (length per unit time), L is the depth of the surficial layer that is mixed, and R represents any sources or sinks for the component within the mixed layer, for example, radioactive decay of an isotopic tracer (mass or numbers per unit volume and time). Equation 1 balances the temporal change in concentration in the mixed layer (left-hand side) with the sum of the input at the

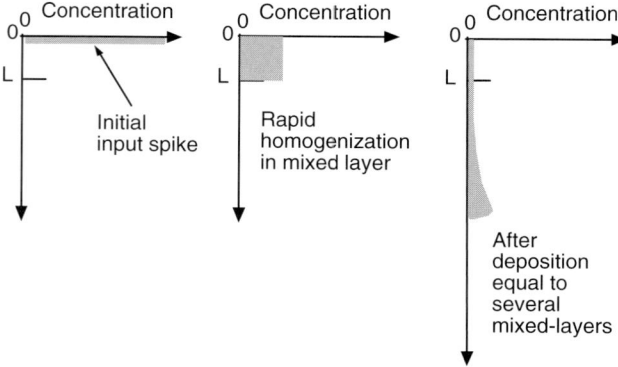

Figure M19 Schematic evolution of a concentration spike deposited at the interface of a sediment with a well-mixed layer of thickness.

interface, the output (burial) at the base of the mixed layer, and the net production (consumption) by sources and sinks on the right-hand side.

Diffusion Models. Mixing is often not strong enough to homogenize the active layer, and in this case the concentration of the sediment component will also be a function of space, as well as time. The most popular model in this case likens the effects of mixing to those of diffusion (see *Chemical Diffusion*), whereby mixing imparts small random motions to the sediment particles. The mathematical form of this model is, at constant porosity,

$$\frac{\partial C}{\partial t} = \frac{\partial}{\partial x}\left(D_B \frac{\partial C}{\partial x}\right) - w\frac{\partial C}{\partial x} - R \quad \text{(Eq. 2)}$$

where, x is depth in the sediment, D_B is the mixing coefficient, analogous to the molecular/ionic diffusion coefficient (length squared per unit time), and the other terms are as in equation (1). Here the temporal change at a given depth (left) is the result of a balance (right) between the effects of mixing, burial and reaction.

Equation 2 is typically applied to isotopic data in the mixed zone to obtain D_B, as the accumulation rate w generally cannot be extracted unambiguously from such data. Assuming steady state ($\partial C/\partial t = 0$), constant parameters, $R = \lambda C$ (λ = decay constant), and $w^2 \ll 4\lambda D_B$, the solution to equation (2) is

$$\ln[C(x)] = -(\lambda/D_B)^{1/2}x + \ln[C(0)] \quad \text{(Eq. 3)}$$

where $C(0)$ is the concentration/activity at the sediment–water interface. An example fit of equation 3 to $^{210}Pb_{ex}$ data to get D_B is illustrated in Figure M20.

A large marine data base of D_B values, calculated as above, is available; this data indicates that D_B is correlated to the sediment accumulation velocity w, i.e.,

$$D_B = 15.7w^{0.6} \quad \text{(Eq. 4)}$$

so that this equation may be used in situations where D_B cannot be calculated from data.

Nonlocal Models. Biological organisms commonly displace sediment in ways that can hardly be thought of as diffusive, i.e., small scale and/or random. In such cases another type of model must be employed, wherein the biological mixing acts like related sources and sinks of material (see Boudreau, 1997, for mathematical details). This allows sedimentary components to be transported large distances through the alimentary canal of large infauna and by directed (non-random) motions. For example, with these models it is possible to describe situations dominated by head-down deposit feeder, a.k.a. conveyor-belt feeders, where the organisms ingest sediment at depth and defecate it at the sediment–water interface; the resulting feeding voids are filled-in by compaction. A similar model can account for effects of burrowing fiddler crabs in marshes, and a nonlocal model can explain persistent subsurface maxima in radioactive isotopes in some sediments as a result of subsurface defecation.

Nonlocal models differ fundamentally from the other two models in how they operate on abrupt temporal change in the composition or structure (e.g., laminations) of sediments. In both the well-mixed and diffusion models, such discontinuities are very rapidly lost, which begs the question how any such changes can be preserved in biotic sediments. Nonlocal models, in contract, are not as effective in dissipating these structures, and their chances of survival are much higher. As the geologic record abounds with many such changes, this fact argues that real mixing is probably nonlocal in character. In addition, nonlocal mixing can actually act to segregate sediment components.

Forward and Backward Models. The two groups do not constitute alternatives to the above models, but reclassifies them in terms of what the model produces. In forward models, we solve the equations given above for C, and to do so we must input values of the parameters, that is, w, D_B, etc., and the flux at the interface, $F(t)$. This type of model describes the sensitivity studies common in the geochemical literature. With inverse models, we start with observed values of C, and then run the equations "backwards" to solve for either the value of a chosen parameter or the input function. Least-squares regression analysis of isotope data to obtain values of w or D_B in sediments is an example of a simple inverse model as in Figure M20. Other more impressive applications have included objective proof of gasoline-based lead reduction in sediments and the need for the use of nonlocal models (see the examples reviewed in Boudreau, 2000).

Bernard P. Boudreau

Bibliography

Berner, R.A., 1980. *Early Diagenesis: A Theoretical Approach.* Princeton University Press.
Boudreau, B.P., 1997. *Diagenetic Models and their Implementation.* Springer-Verlag.
Boudreau, B.P., 2000. The mathematics of early diagenesis: from worms to waves. *Reviews in Geophysics,* 38: 389–416.

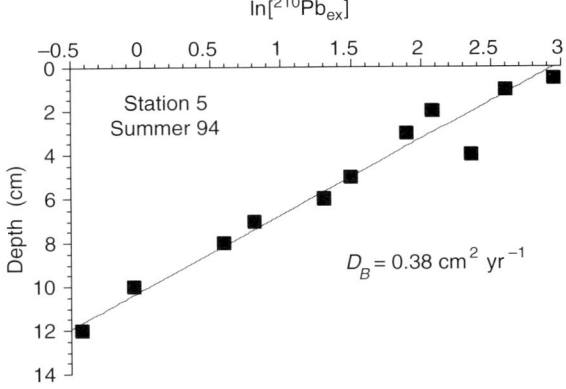

Figure M20 Change of excess ^{210}Pb with depth in marine sediments from the Canadian continental shelf. When plotted in a semi-logarithmic manner as in this figure, the data closely follow a straight line, that is, equation (3), with the stated D_B value.

MUDROCKS

Introduction

Mudrocks are one of the several names for fine-grained, argillaceous sedimentary rocks, also broadly called shales and mudstones, all of which consist mostly of terrigenous clay and

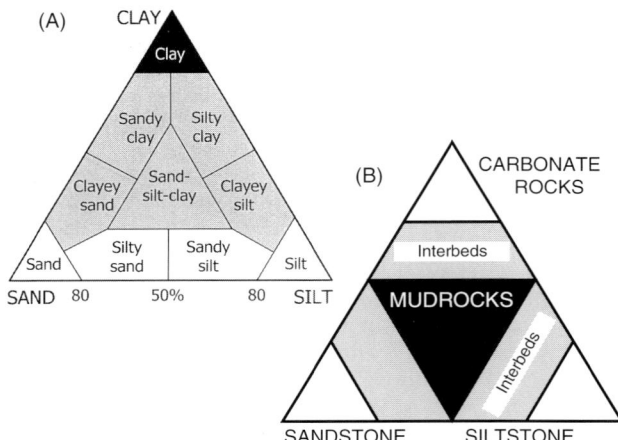

Figure M21 (A) Terminology for muddy unconsolidated mixtures (Shepard, 1954, figure 7); and (B) the three most common lithologic transitions of mudrocks.

silt along with some carbonate and organic matter. Mudrocks occur either as thin beds or in thick sections, but some siltstones, sandstone, or carbonate beds are nearly always present and rare is the section of mudrocks that does not pass laterally into interbedded carbonates or sandstones somewhere along its limits (Figure M21). Mudrocks thus may form but a small part of a deposit or almost all of it with names such as London Clay Formation, Rhinestreet Shale, Quamby Mudstone Group, Cambridge Argillite, etc.

Mudrocks deserve attention for many reasons. They are the most abundant of all the sedimentary rocks, they occur throughout the geologic column in all the major sedimentary environments, and are less studied and understood than either carbonates or sandstones. Thus there is much to learn about and from them. Unlike carbonates and sandstones, whose modern deposits have received much attention and play an important role in interpreting ancient analogs, much more of our understanding of mudrocks is based on studies of the ancient.

From a scientific viewpoint, understanding the origin of mudrocks has the potential to tell us much about ancient environments of deposition and how they may have changed throughout geologic time. Thought of in this manner, mudrocks, and sandstones nicely complement one another in the study of earth history—sandstones provide information about source rocks and their plate tectonic setting and the paleo-hydraulics of sand transport into a basin, whereas mudrocks are keys to the paleo oxygen levels of the basin, and through their microfauna and body fossils, insights to its paleo-circulation system, environment of deposition and possible paleoclimates. Exceptionally, they are also rich *Lagerstätten* for fossils such as the famous Cambrian Burgess Shale Formation of the Canadian Rockies, because in a muddy basins fossils may be quickly buried by a slurry, may die and fall into toxic bottom waters free of scavengers and, because the low permeability of mud minimizes dissolution by migrating acid solutions, escape much diagenesis. Add to this information from the study of interbedded carbonates and sandstones and a truly comprehensive and integrated view of earth history can be obtained.

Practical uses abound as well. Mudrocks serve as the source of nearly all oil and gas, have many uses ranging from barriers to fluid flow (confinement of toxic waste, caps for aquifers, and seals for petroleum traps) to bricks, china, tooth paste, cosmetics, catalysts to decoloring agents and hundreds more. It is fair to say that without mud and mudrocks, modern life as we know it could not exist—all this having started from sun dried bricks some 6,000 years ago. And, mudrocks, because of their propensity to compact, expand, or fail, also demand special attention in modern construction.

The literature of mudrocks is vast and widely scattered, because so many researchers have different interests in them—basin analysis (petroleum and stratiform ore deposits), biology and paleontology, construction, their role in environmental protection, organic, and inorganic geochemistry, industrial minerals and products, sedimentology, and many more. Four recent texts include those by Chamley (1989), Weaver (1989), O'Brian and Slatt (1990) and the two volumes of Schieber *et al.* (1998), which contain 30 research and review articles and many references. Much of the earlier literature is found in Potter *et al.* (1980).

Texture and nomenclature

By convention, the fine-grained sediments include both silts and clays and it is, in fact, their relative proportions that are used to classify them texturally (Table M3). There are several competing systems of nomenclature for mudrocks, the board general genetic term for fine-grained argilleceous rocks, but all depend on the relative proportions of clay and silt. Once this is understood, it becomes easy to use the different systems. When the clay fraction (less than 4 microns) forms more than 67 percent of the rock, the terms *claystone* or *clayshale* are used; when the clay fraction forms 33–66 percent, the terms *mudstone* or *mudshale* are used; but when silt (>4 microns) forms 67 or more percent, the term *siltstone* is used. This important "purity" distinction is easily carried out in the field as a first approximation by putting a small piece between the teeth and determining its grittiness. Fine carbonate is also common in many mudrocks associated with limestones or dolomites and locally mudrocks may also be siliceous or phosphatic. Color is exceptionally variable ranging from black to brownish black through many shades of gray (the most common) to red and purple and even whitish gray and thus far more variable than other sedimentary rocks. And, unlike sandstones and carbonates, the colors of mudrocks have much genetic significance.

Processes

Sources of clay and silt

The immediate source of the clay and silt fractions of mudrocks are soils and preexisting mudrocks. Weathering in soils is favored by high rainfall and good drainage, which together remove cations from complex silicates formed at high temperatures to make progressively simpler clay mineral structures stable at the earth's surface. Where weathering is incomplete because of either insufficient rainfall or high steep slopes, clay and silt compositions closely reflect

Table M3 Describing mudrocks[a] after Stow (1981)

Basic Terms			
Unlithified	*Lithified/non-fissile*	*Lithified/fissile*	*Proportions & grainsize*
Silt	Siltstone	Siltshale	$>\frac{2}{3}$ silt-sized[b]
Mud	Mudstone	Mudshale	silt & clay mixture
Clay	Claystone	Clay-shale	$>\frac{2}{3}$ clay-sized[c]
Texture	*Composition*		
Silty	Calcareous, siliceous, carbonaceous (all >10%), pyritiferous (1 to 5%)		
Sandy >10%	and ferruginous, micaceous, and fossiliferous		
Pebbly			
Stratification (bedding, if >1 cm & lamination, if less)			
May be parallel, wavy (rippled), cross-laminated, graded, continuous-discontinuous, bioturbated, or deformed (use appropriate combinations)			
Metamorphic equivalents			
Argillite	Slightly	metamorphosed/non-fissile	silt and
Slate	Metamorphosed/fissile		clay
Pelite	Either argilite or slate		mixture

[a] More than 50% terrigenous of which more than 50% must be smaller than 63 microns; [b] 4–63 µm; [c] <4 µm

source rocks, but where rainfall is abundant, soil drainage good, and slopes low, the resulting fine debris mostly consists of kaolinite and quartz (silica, aluminum, and iron oxides) and only poorly reflects source rock composition. Thus, *residence time* in the source area is all important for mudrock composition and is ultimately linked to tectonic stability. The volume of fines depends on relief and rainfall—a combination of high relief and rainfall (monsoons against a high Himalayas) produces large volumes of debris whereas low rainfall and low relief (Sahara craton), high rainfall and low relief (Amazon craton) or wide carbonate landscapes produce but little.

But source rocks also need some consideration too when fines are considered. Advanced weathering of granite and gneiss yields a kaolinite-rich clay fraction plus coarse and fine quartz silt, whereas volcanic rocks yield fines rich in unstable glass, feldspar, and amphiboles that transform rapidly into zeolites and smectites. Low grade metamorphic rocks, on the other hand, especially the greenschist facies, tend to yield micas, illites, chlorites, and silt and fine sand-sized quartz and feldspar. It also appears that most fines travel in jumps as streams rework their alluvium; that is, most of their time is spent in storage on the floodplain. This implies that headwater mineralogy may alter down stream, especially on warm, wet stable cratons. Another major source of the silt fraction of mudrocks is wind erosion of deserts (Sahara, central Asia, and north China) to deposit loess, which may have been far more important in pre-Devonian times before extensive plant cover. And finally, some clay minerals in mudrocks are authigenic such as the illites and chlorites formed in deep burial whereas palygorskite (saline lakes and seas) and glauconite (shallow shelves) are formed on sea bottoms. In addition, some quartz silt may be born in place at temperatures well below 90°C from the diagenic alteration of opaline quartz of biogenic origin. Although deciphering the proportions of detrital to authigenic components of a mudrock can at times be difficult, it seems safe to say that most of the clay and silt fraction of mudrocks is detrital and that much comes from the erosion of preexisting mudrocks and siltstones. Ultimately, however, most is derived from the weathering of granites, gneisses, volcanics, and low-rank metamorphic rocks.

Transport

Differential transport is responsible for the segregation of the large volumes of clay and silt that form the world's mudrocks and, on a much smaller scale, it is differential transport that also segregates by particle size and shape the clay minerals of one clay bed from another—an all important distinction in the mining and utilization of clays for their many diverse industrial uses.

In all of these, virtually all the clay fraction is transported in suspension by either wind or water as is much of the fine silt fraction, whereas coarser silt and sand, are transported intermittently by saltation. Fifteen microns seems to be an important size for the hydraulic behavior of silt—experimentally it has been found that below 15 microns most silt is deposited as planar lamination whereas above 15 microns small scale bedforms develop.

In both wind and water it is the upward component of turbulence, v_u that prevents a suspended particle from settling to the bottom. The magnitude of this component is roughly proportional to the average velocity of the flow so the stronger the flow, the greater v_u and the larger the size of the particle that can be suspended. In detail, the criteria for suspension is $v_u > w_s$ where w_s is settling or fall velocity which is given to a good approximation by Stokes Law for small particles as

$$w_s = \frac{1}{18}\frac{d^2 g}{\mu}\Delta\rho \qquad (\text{Eq. 1})$$

where d is particle diameter, $\Delta\rho$ is the density difference between particle and fluid, μ is dynamic viscosity, and g is gravity.

As particle size decreases, so too does the downward driving force acting on it. Consequently, particles a few microns in diameter have exceedingly small fall velocities, remain suspended by the weak upward components of turbulence in gentle currents for weeks and are thus carried great distances. It is this small fall velocity of clay and fine silt that is responsible for the great lateral continuity of many mudrocks, be they thick or thin. Stokes law is also the key control on the concentration of organic matter that occurs commonly in mudrocks, bottom oxygen levels permitting. The low density of most organic matter permits it to preferentially settle to the bottom in quiet water along with the clay fraction to produce organic rich shales—the source beds for petroleum. Because clays and organic material settle together in quiet water, the mudrocks of a topographic low—those deposited in a protected paleo sinkhole, in a gentle sag on the seafloor, or abandoned channel or canyon—are nearly always thicker, darker colored and more clay- and organic rich than lateral equivalents outside the low.

Several observations are needed about Stokes law. First, it only applies to single spherical particles smaller than about 180 micrometers settling in quiet, isothermal, nonturbulent water—hardly the conditions of clay flakes in nearly always turbulent natural flows. Thus predicted fall velocities are conservative ones. Still another complicating factor is that many, perhaps most, clays, and fine silts are transported not as single particles, but as loose, large, aggregates called *floccules*, *flocs* for short, which settle differently than single particles. Floccules are composed of networks of clay flakes, silt, fine skeletal debris and organic material all held together by organic slimes or, for the clay minerals, by ionic bonding. Floccules occur in streams, but are most common where fresh water mixes with saline. Saline water favors flocculation, because its positively charged sodium and negatively charged chlorine ions promote the attraction of clay particles. The large planar face surfaces of clay particles are negatively charged and thus brought together by sodium ions. Edge to edge bonding also occurs but, because the edges have both positive and negative charges, the process differs. Depending on the charge, there will be either an edge to face or face to face contact. In low pH environments, say the acid waters of a swamp, hydrogen ions are bonding agents for clay flakes. Finally and exceptionally, in semiarid climates expandable clays can aggregate to form sand-sized particles (peds) which locally form clay dunes and a few, rare alluvial sands. Biological aggregation also occurs in seas and lakes, is far more important, and is called *pelletization*—the product of infauna feeding on delicious organic–rich mud.

In sum, while both weak and strong currents can transport clay in suspension long distances, weak ones are necessary for it to finally settle to the bottom flocculated or not. This means quiet water, either shallow or deep is needed for deposition—a small lagoon fed by a muddy river, a temporary lake on a flood plain during spring floods, a deep lake in a rift, a deep shelf whose bottom is rarely stirred by storm waves, or an oceanic trench only reached by distal, slow moving turbidity currents that deposit thin graded muddy laminations.

It appears that much, perhaps most, transport of clay and silt is episodic. On land this equates to floods—maximum sheet wash erosion of slopes, gullies, and stream banks provides high suspended concentrations and volumes in streams while in lakes and seas, episodic storms and turbidity currents play a comparable role.

Sedimentary structures

Mudrocks and sandstones share the same types of sedimentary structures—hydraulic, chemical, and biologic. These differ only in abundance and scale, all the hydraulic ones in mudrocks being much smaller, because of weaker currents. Thus mudrocks mostly have lamination rather than beds (By convention all stratification thinner than 10.0 mm is called *lamination* and that thicker, *bedding*).

Lamination in mudrocks is mostly seen by variations in proportions of silt and clay, but is also defined by thin carbonates (pelagic rain or algal mats?), by minor textural variations within muddy layers and exceptionally by variations of organic matter (kerogen). Lamination takes many different forms from rare, silty planar streaks only a few grains thick in almost completely muddy sections to almost all siltstone, where each siltstone is separated by a thin clay parting—with all possibilities in between. Individual laminations may be continuous or discontinuous, wavy, or planar, massive, or graded, inclined (cross laminated) or convoluted, random, or rhythmic. Rhythmic lamination is typically defined by alterations of clay and silt and less commonly by carbonate. The presence or absence of lamination in mudrocks is used to divide them into two broad classes—those that have lamination (classically called *shales*) and those that lack lamination and are massive (often called *mudstones*). In several black shale basins individual laminations as well as packets of them have been convincingly correlated more than 100 km after cores were X-radiographed, thus demonstrating a virtual total lack of bottom currents.

Several different processes are responsible for absence of lamination in mudrocks. The principal control is the amount of bottom oxygen that inhibits benthonic organisms from scavenging on or in muddy bottoms and thus destroying primary stratification. In ancient basins paleo oxygen levels are approximated by noting the degree of bioturbation and how it varies with the color, pyrite, and organic content of the mudrock. Lighter mudrock colors, less pyrite and less organic material typically are found as bioturbation increases and lamination decreases. A few massive mudrocks also are the product of muddy slurries that flowed along the bottom, and exceptionally, some are formed by deposition from denser than normal suspensions.

The term *blocky* is used to describe the weathering of a massive, poorly, or non-laminated mudrock. Two other terms are often used in the description of mudrocks, *fissility and parting*. Fissility refers to the fanlike separation of one lamination from another during weathering. Thus, while fissility is not a primary sedimentary structure it reflects one. For example, well, and uniformly laminated mudrocks may become paper like *(Blätternton)* upon weathering as their individual laminations separate one another by swelling of expandable clays. An argillite, on the other hand, is usually so well indurated that its laminations do no separate one from the other upon weathering. The term *parting* is used for a single lamination separating more massive lithologies above and below. X-radiography is needed to see the full details of the physical, chemical, and biological structures of many muds, although careful observation of weathered outcrops captures most details.

Present on the bottoms of some thin silty laminations are small flutes and load casts plus borings, traces, and impressions made by bottom dwellers. The orientation of flutes and cross-lamination provides useful insights to the paleocurrent pattern of ancient shaly basins, which can also be estimated from oriented wood, graptolites, ostracodes, and some elongate foraminifers. It is always good to remember that bottom currents exist in both oxygenated and poorly oxygenated basins, although in mudrocks, thin sections, and polished sections may be needed to see micro cross-lamination and flutes best. Current systems in ancient muddy basins have also, in a few instances, been inferred from their pattern of silt deposition (Jones and Blatt, 1984).

In sum, little is more important in the study of mudrocks than their sedimentary structures, especially when combined with examination of body and trace fossils. Sedimentary structures help us identify types of bottom currents such as distal turbidity currents, contour currents or hemipelagic rains of fine terrigenous debris. Relating abrupt vertical changes in the sequence of structures in a mudrock with the presence of weak discontinuities, changes in fossil assemblages, pyrite, or phosphate, clay content and color is the basis for recognizing the basin-center equivalents of major stratigraphic breaks that are common along basin margins.

Oxygen

The presence or absence of dissolved oxygen in bottom waters and in the first few decimeters of muddy bottoms is all important for the color, degree of bioturbation and abundance of oxygen sensitive minerals such as pyrite and siderite and organic matter (kerogen). The concept of stratification of the water column is needed here. The water of a lake or sea is described as *stratified*, when not uniformly mixed. This may be the result of temperature, density, or oxygen differences. Mudrocks are the best indicators of stratification of the water column in ancient basins.

Three states of oxygen are recognized: *oxic* (fully oxygenated) bottom waters (more than 1.0 ml O_2/l, Eh $\gg$ 0), *dysoxic* water (1.0–0.1 ml O_2/l, Eh $\cong$ 0) and *anoxic* water (less than 0.1 ml O_2/l, Eh $\subseteq$ 0). The corresponding sediments are described as *aerobic, dysaerobic, and anaerobic*. The contact between these oxygen levels is called a *redox* front and is identified in sediments by changes in color or mineralogy (Figure M22).

Three possibilities exist at the sediment–water interface: the base of oxygenated water can be well above the bottom so the bottom is either dysaerobic or anaerobic, the contact may be at the bottom (dysaerobic) or oxygenated pore water can be below the interface. In addition, there may be small, separate, isolated redox fronts around isolated masses of buried organic matter such as a dead fish, in, and near burrows where animals introduced oxygenated water (as do fractures both early and late) and around roots. These locally discolor the sediment as iron is either oxidized or reduced.

The limits of 1.0, 0.1, and 0.0 ml O_2/l were established by observation of organisms, color, bioturbation, and lamination in modern sediments and serve well as *approximations* to the paleo oxygen levels of ancient muddy basins. As oxygen supply decreases—diversity of benthic fauna decreases, fossils becomes smaller and may be encrusted by pyrite or phosphate, bioturbation is less abundant and burrows become shorter and have smaller diameters (so lamination is better preserved), colors become darker (more preserved organic matter) and pyrite increases in abundance—all variations that can be easily seen and mapped either laterally or vertically in mud-rich

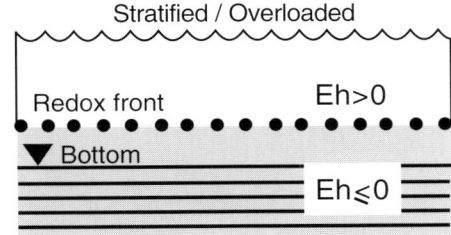

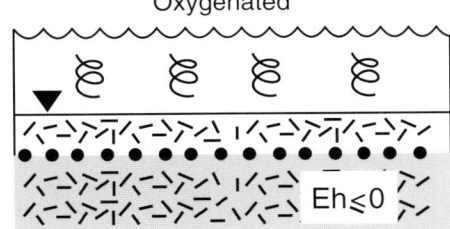

Figure M22 Two end members of a continuous bottom oxygen spectrum—*anoxic* (above) versus *oxic* bottom waters (below). Anoxic bottom waters may result from either a stratified water column or from organic overloading whereas oxic waters require ample turbulence and mixing.

sections. Collectively, these observations are fundamental to the paleo environments of mudrocks and their now vanished paleo-oxygen levels.

The above three principal outcomes depend on the current system of the basin and the local supply of organic matter. Where tidal currents and wind generated waves or an impinging oceanic current reach the bottom, there is turbulence and dissolved oxygen *unless* organic matter generated by dead pelagic organisms near the surface or woody debris transported into the basin overwhelms oxygen supply, a condition called *overloading*. Overloading is perhaps most familiar to us when we read about a fish kill in a stream with excess sewage. Exceptionally, major storms on a shelf oxygenate its water to deeper depths and intruding turbidity currents carry dissolved oxygen from the shelf into bottom waters far beyond its edge and temporally supply enough to support benthic organisms termed "doomed pioneers". In addition to these dynamic factors, basin geometry, paleogeography, and climate also need consideration. Steep-sided, deep, narrow basins, a deep canyon cutting a shelf, or even a shallow sinkhole flooded by a rise in sea level are all protected from currents and thus poorly ventilated with oxygen-deficient waters. Paleogeography plays a role too (Figure M23). For example, faulting, or volcanism may isolate a gulf from an ocean or a coastal barrier may grow across a bay and create a stratified lagoon. All of the above possibilities for creating anoxia are favored by warm tropical, climates—warm water enhances metabolic rates so productivity will be greater than in cold water, given equal nutrients, which explains why most of the world's major petroleum source beds were deposited within $\pm 45°$ of the equator. Dense saline cold water at the bottom also inhibits mixing with surface waters.

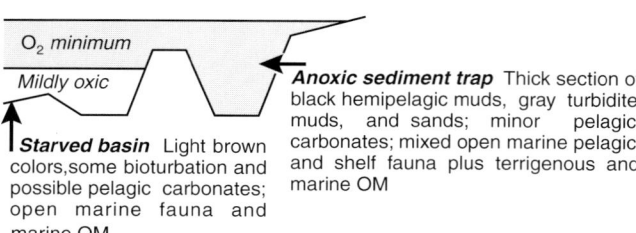

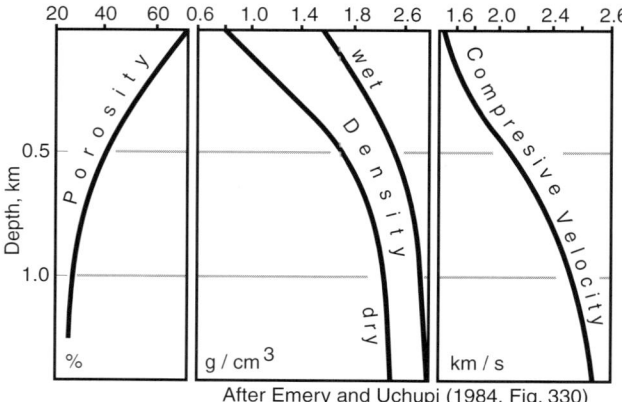

Figure M23 Some common conditions for anoxia (Adopted from Garrison, 1990, figure 3). If more subdued, the bottom topography of the faulted profile also applies to shelves.

Figure M24 Interdependence of changes in porosity, density, and sonic velocity in the first 1,500 m of deep ocean, fine-grained sediments (After Emery and Uchupi, 1984, figure 330).

Burial

There are both physical and mineralogical aspects to burial and understanding the physical compaction processes and its consequences is necessary before we can properly examine mineralogical changes during burial. What makes the compaction burial process of mudrocks so different from that of sandstones and carbonates is that its starting point is mostly water, clay minerals, fine silt and nearly always some organic debris—a very reactive mixture subject to much change as it is buried, compacted, and heated.

Mud compacts more than any other sediment, because it has the largest initial porosity, as much as 80–90 percent for soupy mud. This porosity initially decreases rapidly and exponentially with depth of burial after which decline is approximately linear. Simultaneously, pore water is expelled, and where siltstone and sandstone carrier beds are interbedded, they and early microfractures serve as conduits for its escape. Consequently, throughout such a subsiding section total pressure equals lithostatic pressure plus hydrostatic pressure. But where pore water cannot readily escape (lack of conduits, rapid sedimentation, or extra water released *in situ* from the transformation of smectite to illite), pore water pressure exceeds hydrostatic *and overpressure* results. When hydrostatic pressure exeeds lithostatic, a mud or mudrock will defom, flow, and perhaps even rise to the surface as a *mud lump* or *shale diaphir*. Water expelled by compaction is a medium for petroleum migration and, because it is also rich in ions and dissolved organic acids, the framework grains of distant sandstone reservoirs may be altered by it and their cements dissolved to create secondary porosity or new cements may be precipitated. In this changing system, water is dominant in the first few meters, but as it is expelled with depth and porosity becomes less, both density and sonic velocity increase (except where mudrocks are overpressured or unusually organic rich) as does temperature (Figure M24).

In the first few tens of centimeters of burial (beyond the reach of reworking by waves, currents and burrowing organisms), the pore water of mud rapidly becomes anoxic and becomes the principal driving force in early diagenesis. Important early diagenetic minerals are pyrite, siderite, and calcite.

Pyrite commonly develops very early in the burial of mud by the reaction of iron minerals with H_2S. This is produced by anaerobic bacteria using organic material as a feedstock or energy source to reduce dissolved sulfate in anoxic pore waters. In marine waters dissolved sulfate is widely available and, where organic matter is also, pyrite will form abundantly. On the other hand, if pore waters are initially fresh, dissolved sulfate has low abundance and is rapidly consumed by anaerobic bacteria so little pyrite forms—even when organic matter is present. It is this initial difference in availability of dissolved sulfate that explains why pyrite is much more abundant in marine rather than freshwater mudstones, why the C/S ratio helps distinguish them, and why low sulfur coals are common in alluvial deposits, whereas high sulfur coals are closely associated with marine mudrocks and limestones. When pyrite forms early, it is thought to be small and disseminated, but when formed later, say around a dead organism buried under anoxic or dysoxic bottom waters, it is larger and less concentrated.

Decaying organic matter also generates HCO_3, which may precipitate siderite or calcite at or a short distance below the interface to form early concretions (commonly with a core fossil). Careful study of the carbonate chemistry and morphology of such concretions combined with mudrock color, fossils, types, and abundances of organics and pyrite seems essential for understanding the paleoenvinments of muds. Phosphatic nodules and ferrugious encrustments also occur early and are indications of slow sedimentation.

At depth, the driving force is temperature which causes a random mixture of detrital minerals and kerogen to move to a new mineralogical equilibrium in the presence of now concentrated pore water. Some key mineral transformations include smectite and mixed lattice clay minerals into illite and chlorite, volcanic glass into smectite clays, zeolites, and others, amphorous (opaline) silica into first opal CT followed by low quartz, diverse feldspars into albite, and the progressive evolution of kerogen to yield first gas, then oil and finally dry gas. Crystallinity of the clay minerals also improves as temperatures rise. In a broad way, many of these transformations can serve as guides to burial history keeping in mind, however, that important variations will result from different geothermal gradients and how long the mixture remained at a given temperature range (its subsidence history).

The final result of the above processes is the transformation of a mud into a low permeability, low porosity, compact, tightly interconnected and inter grown, felt-like fabric of both neoformed and detrital clay minerals and quartz, commonly with some pyrite and even some supermature kerogen into a well indurated, lower Paleozoic or Precambrian mudrock.

Mudrock stratigraphy, environments, and basin fill

Although mudrocks occur in all the major depositional environments, they are most abundant by far in the marine and lacustrine relms, because they provide the best accommodation, the best traps, for fines—their wide areas of quiet water below wave base are the end of the line, so to speak, for the suspension transport of clay and fine silt from far distant headwaters. In seas and lakes, mud accumulates readily in bordering low-energy deltas and on shelves, slopes, and basin floors (notable is how closely the facies of deep rift lakes mimic deep marine facies). And, although far less abundant, mud also occurs in shallow lakes, alluvial deposits, on protected coasts, and in estuaries.

Mudrocks play an important role in understanding sedimentary basins and developing their resources. First of all, in marine and lacustrine basins they are likely to be the most widespread and continuous lithologies and thus are of great value for correlation; secondly, they are the very best lithology to study to infer paleo-oxygen levels (which are likely to be more regional than local even though many anoxic events are short lived) and thus are useful for correlation; thirdly, their silt content and gamma ray intensity provide rough proxies for distance from the shoreline (silty mudrock is more proximal, whereas more clay-rich is more distant); fourthly, mudrocks nearly always are the repository for rich assemblages of pelagic microfauna and spores; and finally, the study of their vitrinite reflectance gives a good indication of their thermal history after deposition. So to ignore mudrocks is to miss much in the study of sedimentary basins.

Shifting shorelines and changes in relative water depth in a lacustrine or marine basin best explain the origin, distribution, and kinds of their mudrocks. For example, pro delta mud, possibly even including some deposited by distal turbidity currents, occurs far inshore at highstands, but as shorelines retreat, perhaps even beyond the shelf edge, a low stand delta will develop far downdip intercalating near shore mudrocks with deepwater ones (Figure M24). Or, as relative sea level uses, thin trangressive organic-rich mudrocks (black shales are likely to occur above a low-relief scour surface as mud is temporarily trapped inshore and in estuaries.

Today, the methodology of *sequence stratigraphy* is standard to identify the migration of relative water depths in lacustrine or marine basins (Bohacs, 1998). Recognition of two surfaces is essential to apply it—*flooding surfaces and sequence boundaries*. Flooding surfaces represent an increase in *accommodation* (space for deposition), whereas sequence boundaries represent a decrease. Flooding surfaces are identified by abrupt increases in water depths (more clay, less silt, more open marine fauna and marine organic matter, etc.) as a result of either greater distance from the shoreline or an abrupt cessation of terrigenous supply. Consequently, deeper water deposits overlie shallower ones. Successive flooding surfaces define *parasequences*, which range from a few too many meters in thickness and it is always their mudrocks that record the *maximum flooding surface*. During such a highstand, far down dip, there is little or no sedimentation and a *time-rich condensed section* develops; where the bottom is well oxygenated, a hardground forms, but where it is not, organic-rich muds are deposited (toe of a cyclothem). Sequence boundaries, on the other hand, represent the converse—abrupt shallowing as evidenced by a disconformity (coarser silts and sand with cross lamination and scours and lags underlying above finer-grained deposits plus regional truncation below and onlap above). Both surfaces typically cover 100s to 1,000s of square kilometers and provide time lines for correlation. In proximal coastal deposits paleosoils and peats and coals help identify changes in relative sea level. Paleosoils, like their modern equivalents, need good drainage (low water tables, low stands), whereas coastal mires and peats require rising sea levels and water tables.

Gamma ray logging and organic content help identify flooding surfaces and sequence boundaries. More clay produces high gamma ray readings and less clay lower ones. In addition, both the abundance and kind of kerogen systematically vary as well; because a flooding surface marks deeper water and less detritus, the fines above it will have more total organic carbon and more marine kerogen (types I and II) above the boundary whereas the reverse is true for a sequence boundary (Creaney and Passey, 1993). The sedimentology of particulate organic matter (its size, sorting, and abrasion) also helps identify both surfaces (Tyson, 1995, table 25-2).

Combined use of sequence stratigraphy along with the identification and mapping of mudrock facies is needed to efficiently identify the depositional environment of mudrocks, the one nicely complementing the other. Sequence stratigraphy relates the local section to the big picture—where the section is in relation to basin center or edge and at the same time incorporates information about transitions from prior to subsequent environments (with or without stratigraphic breaks). Mudrock facies, on the other hand, are the key to the local interpretation of the section, because they provide information about current regimes, oxygenation, and water depths based on sedimentary structures plus the kinds, abundance, distribution (patchy or uniform), and preservation of body and trace fossils (Brett and Allison, 1998; Wetzel and Uchmann, 1998). Combine all of the above with the possible presence of early authigenic minerals—pyrite, siderite, calcite, glauconite, and phosphate plus total organic carbon and kerogen types—and rare is the mudrock section that cannot be interpreted (Figure M25).

Provenance

There are five mineralogical and chemical ways to determine the provenance of mudrocks—study their inter bedded sandstones and siltstones with thin sections, use heavy minerals from the sandstones and siltstones (and even possibly the mudrocks), identify, and systematically map their clay mineral composition, and study both their inorganic and organic chemistry. As with sandstones, the broad objective here is insight to their kinds of source rocks and their location,

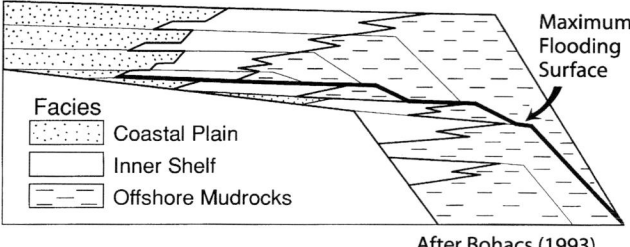

Figure M25 Idealized downdip cross section of mudrocks in a basin with a well defined shelf break; note maximum flooding surface. Lightest lines indicate stacking patterns in mudrocks (After Bohacs, 1993).

relief, and weathering regimen. Commonly, this reduces to magmatic arcs, continent-to-continent collisions, or stable cratons, although, more specific targets are possible (Morton et al., 1991). Where available, paleocurrent information should always be combined with such studies.

The most direct, simplest, and nearly always the most informative, is to apply the standard methods of thin section petrography to inter bedded sandstones and siltstones, which are commonly present. But thin section study of mudrocks has also been made with profit for provenance, especially to identify fine-grained volcanic debris (Zimmerle, 1998). Heavy minerals in mudrocks and siltstones, although rarely studied today, deserve more attention, especially to constrain geochemical investigations.

Identification and regional or basin wide mapping of clay mineral composition works well in modern muddy basins, where the relative abundance of two end members is the most telling—the kaolinite-rich facies (coarse-grained, and indicative of good drainage, ample rainfall and tectonic stabililty) and the smectite-rich facies (fine grained, volcanism, or frequent wetting and drying in the source region). The mapping of clay mineral distributions across a muddy basin largely depends on their differential settling by size (kaolinite is coarser and proximal, smectite finer and distal). Many applications to ancient basins show similar relations, although with deep burial and time, both kaolinite and smectite are replaced by illite and chlorite; hence burial diagenesis can blur original compsitional variations across a basin.

Chemical study of mudrocks to determine their provenance is truly vast and includes bulk composition and selected ratios based on it, trace elements, the rare earths and the types of organic matter. Chemical study seems most appropriate where interbedded terrigenous siltstones and sandstones are absent. Below only several of the most common techniques, mostly based on the 10 standard oxides, are reported (but see Provenance for more techniques).

The CAI index, $100 [Al_2O_3/(Al_2O_3 + CaO^* + Na_2O + K_2O]$, where CaO^* is only the calcium in the silicates, was introduced by Nesbitt and Young (1982) to assess the mineralogical maturity of a mudrock or sandstone—the more CaO, Na_2O and K_2O, the less weathered. Low values of 45–55 equate to little weathering, whereas high values of 80 or more indicate advanced weathering so the higher the index the more weathered the detritus. There have also been several attempts to use the entire bulk composition to assign a tectonic setting (Roser and Korsch, 1988; Zhang et al., 1998). The Ti/Al ratio has received much attention, because both elements vary widely in igneous rocks, but are relatively stable in weathering (Young and Nesbitt, 1998). For example, it has been used to distinguish cratonic muds (high Al) from those derived from oceanic basalts (high Ti). Another useful pair of ratios are SiO_2/Al_2O_3 and K_2O/Na_2O, which have been used to discriminat accretionary from passive margins (Roser and Korsch, 1988). The study of REE elements is very popular and one of the most informative elements is europium, Eu; when high relative to the chrondite standard, derivation from mafic rocks is inferred, but when low, acid felsic rocks are inferred. In part related to bulk composition is gamma ray logging of outcrops and cores (Bohacs, 1998), which measures the concentration of potassium (illite and feldspar), uranium (phosphate and organic matter), and thorium (volcanic ash and heavy minerals). This is rapid and easy and, while not totally a provenance tool, has the advantage of direct correlation to subsurface logs.

The three types of organic matter based on bulk hydrogen, oxygen, and carbon ratios (Tyson,1995, pp.367–382) are useful in separating woody plant organic matter of continental, Type III, from marine or lacustrine, Types I and II. When this separation can be made and mapped, proximity to land can be identified keeping in mind that land-derived organic matter can be transported far seaward. On the other hand, the direct study of pollen (polymorphs), which hydrodynamically are deposited with silt and mud, directly identifies shorelines and, to some degree, paleoclimates (Tyson, 1995, pp. 1–6, 261–284).

Paul E. Potter

Bibliography

Bohacs, K., 1993. Source quality variations tied to sequence development in the Monterey and associated formations. In Katz, B.J., and Patt, L.M. (eds.), *Source Rocks in a Sequence Framework*. American Association Petroleum Geologists Bulletin, Volume 48, pp. 166–190.

Bohacs, K., 1998. Contrasting expressions of depositional sequences in mudrocks from marine and non marine environments. In Schieber, J., Zimmerle, W., and Sethi, P.S. (eds.), *Shales and Mudstones*. Stuttgert, E. Schweizerbart'sche Verlagsbuchhandlung, I, pp. 33–78.

Brett, C.A., and Allison, P.A., 1998. Paleontological approaches to the environmental interpretation of marine mudrocks. In Schieber, J., Zimmerle, W., and Sethi, P.A. (eds.), *Shales and Mudstones*. Stuttgart: E. Schweizerbart'sche Verlagsbuchhandlung, Volume 1, pp. 301–349.

Chamley, H., 1989. *Clay Sedimentology*. Springer Verlag.

Creaney, S., and Passey, Q.R., 1993. Recuning patterns of total organic carbon and source rock quality within a sequence stratigraphy framework. *American Association Petroleum Geologists Bulletin*, 77: 386–401.

Emery, K.O., and Uchupi, E., 1984. *The Geology of the Atlantic Ocean*. Springer Verlag.

Garrison, R.E., 1990. Pelagic and hemipelagic sedimentary rock as source and reservoir rocks. In Brown, G.C.D.S., and Schweller, W.V. (eds.), *Deep Marine Sedimentation; Depositional Models and Case Histories in Hydrocarbor. Exploration and Development*. Bakersfield: Society Economic Paleontologists and Mineralogists, Pacific Section, pp. 123–149.

Jones, R.L., and Blatt, H., 1984. Mineral dispersal patterns in the Pierre Shale. *Journal Sedimentary Petrology*, 54: 17–28.

Morton, A.C., Todd, S.P., and Haughton, P.D.W. (eds.), 1991. *Developments in Sedimentary Provenance Studies*. Geological Society, Special Publication, 57.

Nesbitt, H.W., and Young, G.M., 1982. Early Proterozoic climates and plate motions inferred from major element chemistry of lutites. *Nature*, **299**: 715–717.

O'Brien, N.R., and Slatt, Roger, M., 1990. *Argillaceous Rock Atlas*. Springer Verlag.

Potter, P.E., Maynard, Barry, J., and Pryor, W.A., 1980. *Sedimentology of Shale*. Springer Verlag.

Roser, B.P., and Korsch, R.J., 1988. Provenance signatures of sandstone–mudstone suites determined using discriminant function analysis of major elements. *Chemical Geology*, **67**: 119–139.

Schieber, J., Zimmerle, W., and Stehi, S. (eds.), 1998. *Shales and Mudstones*. I and II: Stuggart: E. Schweizerbart'sche Verlagsbuchhandlung, 384 and 296pp.

Stow, D.A.V., 1981. Fine-grained sediments: terminology. *Quaternary Journal Engineering Geology*, **14**: 243–244.

Shepard, F.P., 1954. Nomenclature based on sand-silt-clay ratios. *Journal Sedimentary Petrology*, **24**: 151–158.

Tyson, R.V., 1995. *Sedimentary Organic Matter*. London: Chapman and Hall.

Weaver, C.E., 1989. *Clays, Muds, and Shales*. Elsevier.

Wetzel, A., and Uchmann, A., 1998. Biogenic structures in mudstones. In Schieber, J., Zimmerle, W., and Sethi, P.A. (eds.), *Shales and Mudstones*. I. Stuggart: Schweizerbart'sche Verlagsbuchhandlung, pp. 351–369.

Young, G., and Nesbitt, H.W., 1998. Processes controlling the distribution of Ti and Al in weathering profiles, siliclastic sediments and sedimentary rocks. *Journal Sedimentary Research*, **68**: 448–455.

Zhang, J.L., Sun, M., Wang, S., and Yu, X., 1998. The composition of shales from the Ordos Basin, China: effects of source weathering and diagenesis. *Sedimentary Geology*, **116**: 129–141.

Zimmerle, W., 1998. Petrography of the Boom Clay from the Rupelian type locality, northern Belgium. In Schieber, J., Zimmerle, W., and Sethi, P.S. (eds.), *Shales and Mudstones*. I: Stuggart: Schweizerbart'sche Verlagsbuchhandlung, pp. 13–33.

Cross-references

Black Shales
Clay Mineralogy
Colloidal Properties of Sediments
Colors of Sedimentary Rocks
Compaction (Consolidation) of Sediments
Depositional Fabric of Mudrocks
Flocculation
Oceanic Sediments

N

NEOMORPHISM AND RECRYSTALLIZATION

Recrystallization and neomorphism are processes that transform minerals *in situ* into different forms of themselves, or a polymorph. Confusion exists with the term recrystallization because it is sometimes restricted to strain induced transformation (Bathurst, 1958) but at other times it is used in a general way for any change in form without a change of mineral species. The term replacement is sometimes used for any mineral transformation, including recrystallization and neomorphism. Replacement, however, is often restricted to transformations that involve changes in bulk composition (synonymous with metasomatism) such as dolomitization or silicification of limestone. Neomorphism and recrystallization are thought to occur by simultaneous transformation across a thin fluid film and are to be distinguished from cementation (mineral crystallization in fluid filled cavities). Neomorphism, recrystallization and cementation all involve dissolution and precipitation, but cement crystallizes in visible (scale of light microscope) cavities and does not involve simultaneous dissolution and precipitation at sites separated by a few microns or less.

Folk introduced the term neomorphism in 1965 to include all transformations of older crystals that are gradually consumed, and their place simultaneously occupied by new crystals of the same mineral or a polymorph. Folk pointed out that recrystallization, used loosely, is synonymous with neomorphism but that recrystallization *sensu stricto* excludes inversion. As such, considerable overlap exists between neomorphism and recrystallization, and to avoid confusion it must be clear whether recrystallization is being used in a strict or loose sense. Folk added textural riders to the term neomorphism to indicate an increase (aggrading) or decrease (degrading) in crystal size and whether an increase in size occurred by growth from a few isolated crystals (porphyroid) or every adjacent crystal (coalescive).

The transformation of aragonite to calcite, with or without interstitial liquid films (Folk, 1965, p.21) has been called inversion but Bathurst (1975, p.234) specifically excluded water from the process and defined inversion as a dry, solid-state reaction. Folk (1965) considered dry inversion of aragonite to calcite as a special type of neomorphism. The dry inversion of aragonite to calcite, however, does occur naturally in blueschist metamorphic facies (Carlson and Rosenfeld, 1981). The reaction path and textures produced in this reaction are different from those found in water-saturated sediments. Consequently inversion and neomorphism are both polymorphic transformation process, the former is dry and the latter is wet and their reaction products are readily distinguishable.

Since 1965 neomorphism has been applied most commonly to the aragonite to calcite transformation. Investigation has concentrated on mollusc shells (Sandberg and Hudson, 1983; Hendry *et al.*, 1995; and Maliva, 1998), scleractinian corals (Pingitore, 1976; Martin *et al.*, 1986), ooids (Wilkinson, *et al.*, 1984) and botryoidal cements (Sandberg, 1985).

The overlap between recrystallization and neomorphism and the variable number of transformation processes encompassed by these terms has caused confusion. This led Machel (1997) to recommend the abandonment of the term neomorphism. The term can, however, be usefully applied to the wet aragonite to calcite transformation for which it has been used overwhelmingly since 1965. Neomorphism would then become restricted to a common and important type of polymorphic transformation. Recrystallization, on the other hand, applies to transformation without change in mineral species. Folk (1965) used neomorphism for transformations where the mineralogy of the precursor mineral is unknown but recrystallization would be just as apt and the general term transformation is recommended here to avoid confusion.

Calcitization is the transformation of any mineral to calcite and as such can be a type of recrystallization (calcite to calcite), replacement (e.g., anhydrite to calcite) or neomorphism (aragonite to calcite).

Neomorphism

Neomorphic transformation of a mass of aragonite crystals usually starts from many points within and at the margin of that mass. The fluid that exists at each initiation point moves centrifugally away from that point as a thin film until it stops against a seam of organic matter; meets another fluid film, or the transformation stops (Wardlaw et al., 1978; Martin et al., 1986). The number of initiation points and the manner in which these films move controls the resulting mosaic of comparatively few neomorphic (sometimes a single) calcite crystals. These are optically and texturally largely unrelated to the precursor aragonite (Figure N1). Mineralogical transformation across such a film is often incomplete, with aragonite relics being found encased in the neomorphic calcite. These relics retain their original orientation and indicate the structure of the transformed grain. Palimpsest structures can also be preserved in neomorphic calcite by trapped original organic matter that can give rise to a pale to dark brown pseudopleochroism that is distinctive of some neomorphic calcites (Hudson, 1962; Sandberg and Hudson, 1983). In some cases neomorphism occurs in two stages separated by a long (10^6 year) time interval (Hendry et al., 1995).

Transformation within the confines of a thin film (although one has never been directly observed; Wardlaw et al., 1978) allows chemical information to be transferred from aragonite to neomorphic calcite. Many neomorphic calcites contain ~1,000 ppm Sr (tabulated in Maliva, 1998, p.181) a concentration above that recorded for most calcite cements (~300 ppm Sr) and often used to diagnose transformed aragonite (Sandberg and Hudson, 1983). Some neomorphic calcites, however, have Sr concentrations that are similar to adjacent cements (Maliva and Dickson, 1992) and the Sr content of neomorphic calcite is often overestimated due to analytical problems (see discussion Pingitore, 1994 and reply Moshier and Kirkland, 1994).

Work in the 1960s and 1970s concentrated on neomorphic calcite that formed in meteoric waters; some neomorphic calcites described from vadose and phreatic environments have different properties (Pingitore, 1976). Neomorphic calcite can also form in modified marine waters (Maliva, 1995) and from evolved isolated connate waters (Kendall, 1975; Bathurst, 1983). In all these cases the neomorphic products are similar and are undiagnostic of the transformation fluid.

Recrystallization

The case for recrystallization is often made for the matrix of ancient carbonate rocks that are composed of calcite microspar. Modern marine, tropical lime mud—the likely precursor to calcite microspar—is composed predominantly of metastable aragonite and Mg calcite. Lasemi and Sandberg (1984) have differentiated calcitic microspar into that originally dominated by aragonite (that has neomorphosed) and that originally dominated by Mg calcite (that has recrystallized). Some workers (Steinen, 1982; Munnecke et al., 1997), however, doubt that the aragonite mud neomorphosed but some aragonite was trapped in calcite cement that filled the mud's pore system. Microspar commonly has an irregular crystal size distribution, is composed of turbid crystals and contains impurities concentrated along crystal boundaries (Figure N2). Such calcite microspar is thought to arise by recrystallization (Folk, 1965) caused by the metastability of original Mg calcite but physical processes such as Ostwald Ripening may also provide the driving force for finely crystalline calcite (Kile et al., 2000).

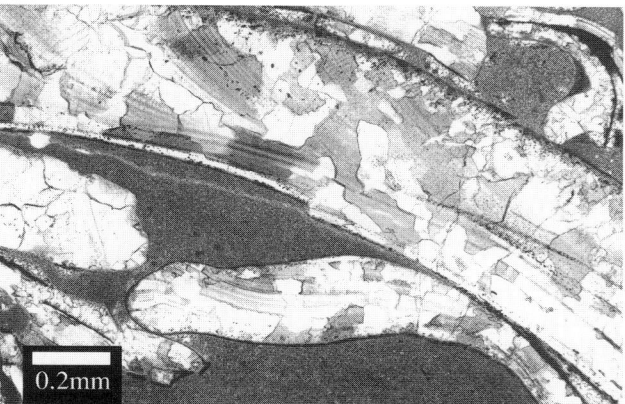

Figure N1 Photomicrograph (plane polarized light) of thin section of Unio Bed, Purbeck Beds, Lulworth Cove, Dorset, England. *Unio* valves are set in a micritic matrix. The original shell structure is preserved in coarsely crystalline calcite that neomorphosed the original aragonite shell. The mottled appearance of the neomorphic calcite is caused by pseudopleochroism, an effect that causes a color change from colorless to brown as the microscope stage is rotated.

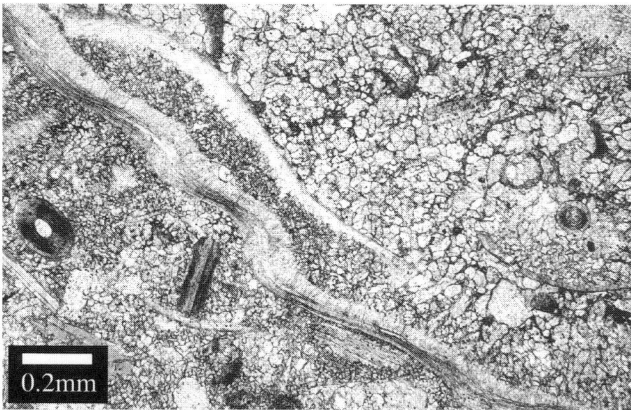

Figure N2 Photomicrograph (plane polarized light) of thin section of the Purbeck Beds, Durlston Bay, Dorset, England. Skeletal grains are separated by recrystallized lime mud. This sample shows an erratic crystal size variation and impurities along the intercrystalline boundaries of the calcite crystals. Original mineralogy unknown.

Conclusions

Recrystallization is a change in mineral form (crystal size and/or shape) without a change in mineral species or bulk composition. The term neomorphism as coined in 1965 is confusing due to overlap in meaning with recrystallization, but if restricted to wet polymorphic transformation serves a useful purpose. Inversion is a dry, solid-state polymorphic transformation that is unlikely to occur in water-saturated sediments. No specific term is available for a mineral that is interpreted to have changed form from an unidentified precursor, as is thought to occur when some lime muds are transformed to calcite microspar; here the general term transformation can be used.

J.A.D. Dickson

Bibliography

Bathurst, R.G.C., 1958. Diagenetic fabrics in some British Dinantian limestones. *The Liverpool and Manchester Geological Journal*, **2**: 11–36.
Bathurst, R.G.C., 1975. *Carbonate Sediments and Their Diagenesis*. Developments in Sedimentology, Volume 12, Elsevier.
Bathurst, R.G.C., 1983. Neomorphic spar versus cement in some Jurassic grainstones: significance for evaluation of porosity evolution and compaction. *Journal of the Geological Society of London*, **140**: 229–137.
Carlson, W.D., and Rosenfeld, J.L., 1981. Optical determination of topotactic aragonite-calcite growth kinetics: metamorphic implications. *Journal of Geology*, **89**: 615–638.
Folk, R.L., 1965. Some aspects of recrystallization in ancient limestones. In *Dolomitization and Limestone Diagenesis*. SEPM (Society for Sedimentary Geology), Special Publication, 13, pp. 14–48.
Hendry, J.P., Ditchfield, P.W., and Marshall, J.D., 1995. Two-stage neomorphism of Jurassic aragonitic bivalves: implications for early diagenesis. *Journal of Sedimentary Research*, **A65**: 214–224.
Hudson, J.D., 1962. Pseudo-pleochroic calcite in recrystallized shell limestones. *Geological Magazine*, **99**: 492–500.
Kendall, A.C., 1975. Post-compactional calcitization of molluscan aragonite in a Jurassic limestone from Saskatchewan, Canada. *Journal of Sedimentary Petrology*, **45**: 399–404.
Kile, D.E., Eberl, D.D., Hoch, A.R., and Reddy, M.M., 2000. An assessment of calcite crystal growth mechanisms based on crystal size distributions. *Geochimica et Cosmochimica Acta*, **64**: 2937–2950.
Lasemi, Z., and Sandberg, P.A., 1984. Transformation of aragonite-dominated lime muds to microcrystalline limestones. *Geology*, **12**: 420–423.
Machel, H.G., 1997. Recrystallization versus neomorphism, and the concept of "significant recrystallization" in dolomite research. *Sedimentary Geology*, **113**: 161–168.
Maliva, R.G., 1995. Recurrent neomorphic and cement microtextures from different diagenetic environments, Quaternary to Late Neogene carbonates, Great Bahama Bank. *Sedimentary Geology*, **97**: 1–7.
Maliva, R.G., 1998. Skeletal aragonite neomorphism—quantitative modelling of a two-water diagenetic system. *Sedimentary Geology*, **121**: 179–190.
Maliva, R.G., and Dickson, J.A.D., 1992. The mechanism of skeletal aragonite neomorphism: evidence from neomorphosed molluscs from the upper Purbeck Formation (Late Jurassic–Early Cretaceous), southern England. *Sedimentary Geology*, **76**: 221–232.
Martin, G.D., Wilkinson, B.H., and Lohmann, K.C., 1986. The role of skeletal porosity in aragonite neomorphism—*Strombus* and *Montastrea* from the Pleistocene key Largo Limestone, Florida. *Journal of Sedimentary Petrology*, **56**: 194–203.
Moshier, S.O., and Kirkland, B.L., 1994. Identification and diagenesis of a phylloid alga: *Archaeolithophyllum* from the Pennsylvanian Providence Limestone, Western Kentucky—reply. *Journal of Sedimentary Petrology*, **64**: 925–928.
Munnecke, A., Westphal, H., Reijmer, J.J.G., and Samtleben, C., 1997. Microspar development during early marine burial diagenesis: a comparison of Pliocene carbonates from the Bahamas with Silurian limestones from Gotland (Sweden). *Sedimentology*, **44**: 977–990.
Pingitore, N.E., 1976. Vadose and phreatic diagenesis: processes, products and their recognition in corals. *Journal of Sedimentary Petrology*, **46**: 985–1006.
Pingitore, N.E., 1994. Identification and diagenesis of a phylloid alga: *Archaeolithophyllum* from the Pennsylvanian Providence Limestone, Western Kentucky—discussion. *Journal of Sedimentary Research*, **64**: 923–924.
Sandberg, P.A., 1985. Aragonite cements and their occurrence in ancient limestones. In *Carbonate Cements*. SEPM (Society for Sedimentary Geology), Special Publication, 36, pp. 33–57.
Sandberg, P.A., and Hudson, J.D., 1983. Aragonite relic preservation in Jurassic calcite-replaced bivalves. *Sedimentology*, **30**: 879–892.
Steinen, R.P., 1982. SEM observations on the replacement of Bahaman aragonitic mud by calcite. *Geology*, **10**: 471–475.
Wardlaw, N., Oldershaw, A., and Stout, M., 1978. Transformation of aragonite to calcite in a marine gastropod. *Canadian Journal of Earth Sciences*, **15**: 1861–1866.
Wilkinson, B.H., Buczynski, C., and Owen, R.M., 1984. Chemical control of carbonate phases: implications from Upper Pennsylvanian calcite-aragonite ooids of southeastern Kansas. *Journal of Sedimentary Petrology*, **54**: 932–947.

Cross-references

Cements and Cementation
Diagenesis
Micritization
Sedimentologists

NEPHELOID LAYER, SEDIMENT

Introduction

The lower water column in most parts of the ocean, both shelf waters and the deep sea, shows a large increase in light scattering and attenuation conferred by the presence of increased amounts of fine-grained particulate material. This has been confirmed by size measurements and filtration of seawater with gravimetric analysis. This part of the water column is termed the bottom nepheloid layer (BNL) (Figure N3). Optical work shows that the BNL is up to

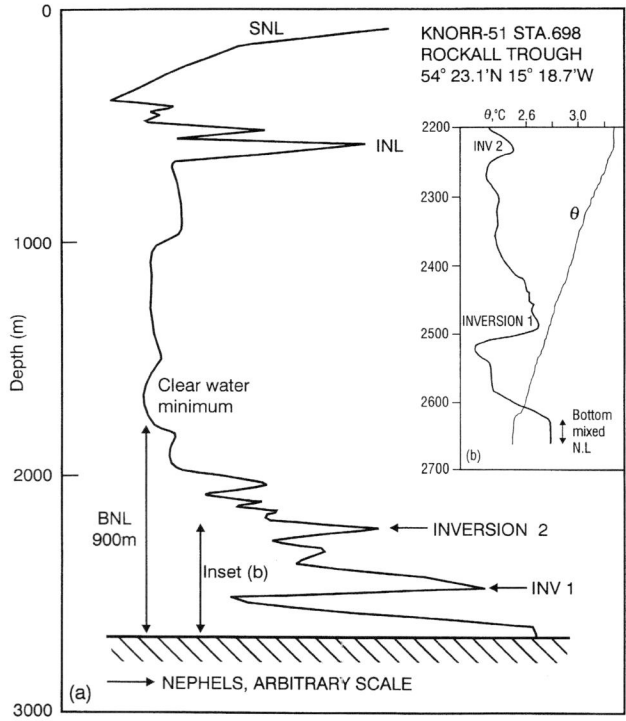

Figure N3 Full depth nephelometer profile in the Rockall Trough. Bottom, intermediate and surface nepheloid layers and the clear water minimum are apparent as well as two large inversions (after McCave, 1986). The instrument is uncalibrated but in this area the maximum concentration is probably of order $500\,\mathrm{mg\,m^{-3}}$

2000 m thick and generally has a basal uniform region, the bottom mixed nepheloid layer, corresponding quite closely to the bottom mixed layer defined by constant potential temperature. Above this light scattering falls off more or less exponentially to the clear-water minimum marking the top of the BNL.

Another class of nepheloid layer found especially at continental margins is the intermediate nepheloid layer (INL) occurring frequently at high levels off the upper continental slope and at the depth of the shelf-edge (Figure N3). These INLs are similar to the inversions observed in the BNL on some profiles (Figure N3).

Nepheloid layers of both types are principally produced by resuspension of the bottom sediments. Their distribution thus indicates the dispersal of resuspended sediment in the oceans and coastal seas. Most concentrated nepheloid layers occur on the continental shelf, upper slope or deep continental margin. They indicate the locus of active suspension and deposition by strong bottom currents. Nepheloid layers off the shelf almost always have concentrations $<1 \text{ g m}^{-3}$.

A surface nepheloid layer (SNL) occurs almost everywhere but varies in strength seasonally. In the open ocean it is mainly due to phytoplankton, whereas in shelf seas it is often the seaward extension of river outflow plumes.

Optics: what instruments "see"

Most detection of deep ocean nepheloid layers has been through measurement of light scattering or nephelometry. Because most nephelometers do not measure a defined optical parameter their outputs cannot be quantitatively compared, save via calibration with some standard such as Formazin. Nowadays transmissometry is more commonly used for precise work. Most of the light scattering recorded by nephelometers is caused by fine particles. Although larger particles are present, they are rare and are not important contributors to the signal. For a transmissometer, the transmission T is related to beam attenuation coefficient c over pathlength l as $T = e^{-cl}$. Attenuation is due to absorption a and scattering b, thus $c = a + b$. The value of c for pure seawater is $0.360 \pm 0.005 \text{ m}^{-1}$ for an instrument operating at $\lambda = 660$ nm, and any excess is due to particulate effects. Because large particles scatter more at angles very close to the beam axis, the transmissometer is more sensitive to larger particles.

Bottom nepheloid layer features and origin

The principal features of BNLs are that the concentration is generally highest close to the bed and decreases upward, but with important increases or inversions. The thickness of the BNL is generally in the region of 500 m to 1500 m, and exceptionally up to 2000 m. This is clearly greater than the thickness of the bottom mixed layer, a fact which precludes the possibility of simple mixing by boundary turbulence being a sufficient mechanism for BNL generation. Inversions, upward increases of concentration, become less common in the upper part of the bottom nepheloid layer. Pronounced inversions (INLs) are seen over slopes and on margins with canyons. These are produced by internal wave and tide resuspension (Dickson and McCave, 1986). Thus some canyons act as point sources for supply of turbid layers which mix out into the ocean interior (Gardner, 1989). These features are accounted for in the "separated mixed-layer model" in which vertical transport by turbulent mixing in bottom boundary layers of 10 m to 100 m thickness, is followed by their detachment and lateral advection along density (isopycnal) surfaces. The detachment occurs where isopycnals intersect the bottom and this is likely to occur both in areas of steep topography with relief of hundreds of meters, and also in areas of lower gradient at benthic fronts where sloping isopycnals intersect the bottom (Armi and D'Asaro, 1980; McCave, 1983). With time these layers thin by mixing at their boundaries and lateral spreading to yield, eventually, a uniform stratification.

Particles comprising the nepheloid layer are modified by aggregation with similar sized particles, scavenging by larger rapidly settling ones, and by biological activity. Particles tend to settle and be deposited from the bottom mixed layer. The larger particles should be deposited in a few weeks to months, 10 μm to 20 μm particles taking 50 days to 20 days to settle from a 60 m thick layer. This affects the transmissometer signal but has little effect on the nephelometer record because the timescale of fine particle removal involves Brownian aggregation with a "half-life" of several months to years giving fine material in dilute nepheloid layers a residence time of years (McCave, 1984). The dilute nepheloid layers in tranquil parts of the oceans could thus contain material originally resuspended very far away. The contribution of this material to the net sedimentation rate of these tranquil regions may not be negligible. The rate of deposition in the Pacific of only 0.5 mm ka^{-1} to 2 mm ka^{-1} could include up to 1 mm ka^{-1} of fine material from the nepheloid layer.

The turbidity minimum

Over the depths separating the BNL from the SNL (and from shelf edge and upper slope INLs) stratification is weak, primary production of organic matter is absent and the ocean is often under-saturated with respect to calcite, aragonite and opal. Supply of particles from the side is thus scarce and decreasing downward transport due to dissolution and bacterial consumption of particles leads to a nepheloid minimum where (in the Atlantic) concentrations are 5 mg to 20 mg m^{-3}. Concentrations given by nephelometers at the minimum differ by a factor of about three between areas under high surface productivity (high) and mid gyre regions (low). Thus, the flux of large particles from the surface in these areas appears to provide a seasonal supply of smaller particles to mid-depth rather than their removal by scavenging. Temporal variability also occurs seasonally with higher mid-water turbidity under higher summer productivity.

Regional distribution

In general, deep western boundary currents and regions of recirculation carry high particulate loads. However, the highest turbidity is also found beneath regions of high surface eddy kinetic energy (variance of current speed). The broad picture of high speed inflows on the western sides of ocean basins with a distributed return flow, demonstrated by hydrographic and current measurements, is reflected in the distribution of BNL concentration for the Atlantic and Indian Oceans. Nevertheless several areas of very high concentration or load are not obviously related to likely increases in mean bottom flow velocity, for example the center of the Argentine

Basin (Biscaye and Eittreim, 1977). These turbidity highs may be caused by intermittently high velocities.

The fact that there is more suspended material at ocean depths greater than about 4000 m partly reflects the fact that waters at those depths are in contact with a much greater potential source area of seabed in proportion to their volume than the shallower parts of the oceans. For 4 km to 5 km depth the value is 0.83 km^2/km^3, whereas for 2 km to 3 km it is 0.11 km^2/km^3. Deeper waters feel more bed.

I. Nicholas McCave

Bibliography

Armi, L., and D'Asaro, E., 1980. Flow structures of the benthic ocean. *Journal of Geophysical Research*, **85**: 469–484.
Biscaye, P.E., and Eittreim, S.L., 1977. Suspended particulate loads and transports in the nepheloid layer of the abyssal Atlantic Ocean. *Marine Geology*, **23**: 155–172.
Dickson, R.R., and McCave, I.N., 1986. Nepheloid layers on the continental slope west of Porcupine Bank. *Deep Sea Research*, **33**: 791–818.
Gardner, W.D., 1989. Periodic resuspension in Baltimore Canyon by focusing of internal waves. *Journal of Geophysical Research*, **94**: 18185–18194.
McCave, I.N., 1983. Particulate size spectra, behaviour and origin of nepheloid layers over the Nova Scotian Continental Rise. *Journal of Geophysical Research*, **88**: 7647–7666.
McCave, I.N., 1984. Size-spectra and aggregation of suspended particles in the deep ocean. *Deep-Sea Research*, **31**: 329–352.
McCave, I.N., 1986. Local and global aspects of the bottom nepheloid layers in the world ocean. *Netherlands Journal of Sea Research*, **20**: 167–181.

Cross-references

Continental Rise and Slope
Flocculation

NERITIC CARBONATE DEPOSITIONAL ENVIRONMENTS

Introduction

Marine carbonate sedimentation (Bathurst, 1975; Scholle *et al.*, 1983; Tucker and Wright, 1990; James and Kendall, 1992; Wright and Burchette, 1996) takes place in two environments, the benthic, neritic realm and the pelagic deep-sea realm (see *Oceanic Sediments*). Neritic carbonate sediments are "born" as precipitates or skeletons within the depositional environment. This attribute has profound consequences: (1) large structures such as platforms are produced entirely by sediments formed in place, they are self-generating and self-sustaining; (2) the temporal and spatial style of accumulation depends upon the nature of the sediments themselves; (3) sediment production can fill accommodation space and thus create shallowing-upward, autostratigraphic patterns; (4) grain size variations need not signal changes in hydraulic regime; and (5) sediment composition is fundamental in characterizing the depositional environment.

A *carbonate platform* (Figure N4) is a large edifice formed by the accumulation of sediment in an area of subsidence. Such structures can be several kilometers thick, extend over many hundreds of square kilometers, and be constructed by stacked carbonate ramps and/or flat-topped platforms. A *carbonate shelf* is a platform tied to an adjacent continental landmass. This hinterland is a potential source of terrigenous clastic sediment, fresh water, terrigenous organic matter and nutrients. A craton many hundreds to thousands of kilometers across is called an *epeiric platform* when covered by shallow seawater. A *carbonate bank* is an isolated platform surrounded by deep ocean water and cut off from terrigenous clastic sediment. A *carbonate atoll* is a specific type of bank developed on a subsiding volcano. Similar banks, unrelated to volcanoes, are termed *faros*. Atolls and banks can be so dominated by reefs that their geological expressions are termed *reef complexes* or *carbonate buildups*.

These morphological structures are generally either a *flat-topped platform* with narrow rim and relatively steep basin-facing margin, or a *ramp*, a structure that slopes gently basinward from the shoreline at angles of less than one degree.

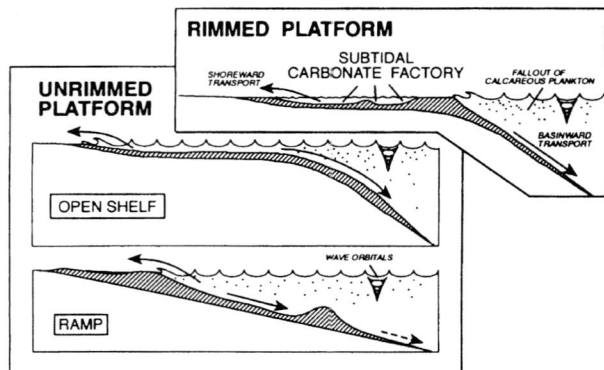

Figure N4 A sketch illustrating the different types of carbonate platforms (after James and Kendall, 1992).

Ramps

A *ramp* (Figures N4, N5) is an unrimmed edifice in which nearshore, wave-agitated facies grade into deeper water, low-energy deposits and there is no discernable break in slope (Read, 1985; Burchette and Wright, 1992; Wright and Burchette, 1998). Modern examples are found in areas where coral reefs are absent because of elevated salinity (Persian Gulf, Shark Bay), cool seawater (southern Australia—New Zealand, western Florida) or high nearshore terrigenous sedimentation (Papua New Guinea). They are "distally steepened" (with a subtle shelf break), or "homoclinal" (uniform low slope gradient with no slope break). Most are storm-dominated.

Ramps are hydrodynamically partitioned into depth-related, shore-parallel facies belts. The *Inner Ramp* lies within fairweather wave base (FWWB), is subject to high-energy agitated conditions, onshore currents push coarse material shoreward and muds into lagoons, offshore-directed storm-surge currents transport finer-grained material basinward. Facies are dominated by carbonate sand shoals, shoreface deposits, tidal deltas, seagrass banks and barrier islands backed by lagoonal and peritidal flats. The *Mid Ramp* lies

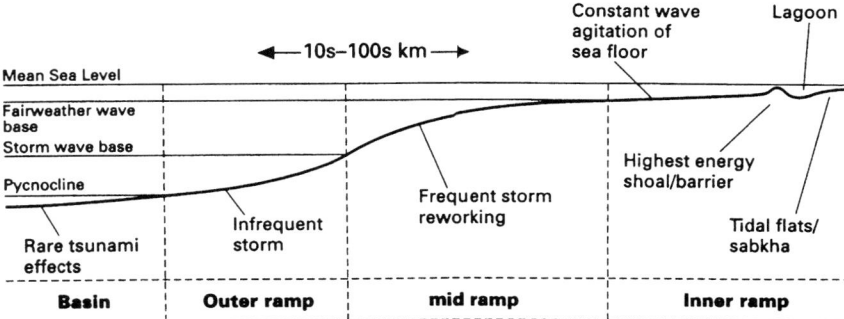

Figure N5 The main environmental subdivisions on a homoclinal carbonate ramp (from Burchette and Wright, 1992).

between FFWB and storm wavebase (SWB), is a generally quiet environment disturbed by episodic high-energy storm events and is characterized by decreased organic productivity offshore. Fair-weather muds are deposited from suspension while storm events transport coarser-grained material from shallow-water areas (storm surge currents) producing deposits that change progressively offshore from swaley-cross-stratification to amalgamated HCS to HCS to graded "tempestite" couplets. The *Deep Ramp* lies below SWB, is an area of quiet water suspension deposition of carbonate and terrigenous muds and rare storm events. Sediments there have few sedimentary structures, are typically laminated with some silty rippled beds floored by skeletal lags, and contain a sparse *in situ* macro-biota. The wackestones and packstones are locally interbedded with anoxic shales, if there is a strong pycnocline, and can have rhythmic alternations resulting from climatically, and/or oceanographically driven fluctuations.

Flat-topped platforms

A flat-topped platform, generally with a steep basinward margin, can be rimmed or unrimmed. An *unrimmed* or *open flat-topped platform* (Figure N4), is one in which there is no basin-facing barrier. Unrimmed platforms occur today on the leeward side of large tropical banks and are the norm in all cool water settings. Because oceanic waves and swells can sweep directly onto and across the shallow seafloor of open shelves and ramps, (1) the energy level of most shallow water environments is high; (2) nearshore facies are complex; (3) sediment can easily be transported into deep water; (4) only the nearshore zone can keep pace with sea level rise; and (5) subtidal accumulation space will be controlled as much by the depth of wave abrasion as by sea level. Since the same physical processes as those on terrigenous clastic shelves affect unrimmed platforms and ramps, they have similar facies. What sets them apart is the continual in place production of carbonate sediment and early cementation, which constantly alters the nature of the seafloor.

A *rimmed flat-topped platform* (Figure N4) has a segmented to continuous rampart of reefs and/or lime sand shoals along the seaward margin that absorbs ocean waves. Modern warm water platforms, typified by the Great Barrier Reef, the Belize Reef Complex, the Bahama Banks and oceanic atolls and fringing reefs worldwide, are generally rimmed because corals are prolific today and construct large reefs along the edges of shelves and banks. By absorbing waves and swells and dissipating storm energy, the rim protects a variety of lower energy environments. It also confines the movement of coarse-grained sediment to the lagoon/shallow platform and can potentially restrict water circulation, increasing the possibility of lagoonal evaporites. Accumulation space is limited by sea level and carbonate facies are strongly differentiated.

The margin is typified by extensive growth of *barrier reefs*, with associated islands and passes, resulting in rapidly accumulating carbonate reef limestones and coarse skeletal grainstones and rudstones that are periodically swept lagoonward and basinward. Some platform margins are occupied by ooid–peloid *sand shoal complexes* that are either storm- or tide-dominated. The leeward *lagoon*, 10s to many 100s of kilometers wide, ranging from less than 10 m to many 10s of meters in depth, is the site of widespread, muddy, biotically diverse, extensively bioturbated seafloor carbonate sediment in environments that range from shallow grass banks to deep, poorly illuminated burrowed muds. The area may be dotted with patch reefs and/or small islands resulting from hurricane (cyclone) resedimentation. The strandline can be the site of beaches but more commonly, is an area of muddy *tidal flat* sedimentation. The muds on the tidal flats are driven onshore during cyclonic storms.

The rapid accretion rates of reefs and ooid shoals generally leads to steep basin-facing *carbonate slopes*, typified by a mixture of pelagic and resedimented carbonate deposits. While many have the attributes of outer ramp facies, they also contain numerous sediment gravity flows and debrites. These materials pass rapidly into basinal sediments.

The sediment factory

The whole carbonate deposition system is depended upon the productivity of the carbonate factory (Figure N4), that region of the seafloor where sediment is generated by biotic growth and inorganic/microbial particle production (see *Algal and Bacterial Carbonate Sediments*). Particles of all grain sizes are formed here, either crystallizing as skeletons or precipitating directly, to generate muddy "subtidal" deposits, ooid sands or reefs and mounds. Much of the abundant fine fraction is resuspended periodically and piles up as muddy peritidal flats against highs on the platform and along the shoreline. Fine sediment is also swept seaward where, together with sediment gravity flows originating at the margin, it accumulates on the slope and on adjacent basin margin. Regardless of facies, it is the factory that is at the core of carbonate deposition. All carbonate facies and carbonate stratigraphy depend on the "health" of this production unit.

For carbonate production and sediment accumulation to be at a maximum, the environment must be just right; not too deep, not too shallow; not too warm, nor too cold; not too fresh, but not too saline either; not too much terrigenous clastic sediment; not too many nutrients but neither too few (the *Goldilocks Window* of Goldhammer et al., 1990). The warm-water, tropical factory is dominated by the *photozoan assemblage* (James and Clarke, 1997) which comprises calcareous phototrophic/mixotrophic organisms with photosymbionts, calcareous algae, ooids and mud-producing algae and microbes. These thrive in the euphotic zone (generally <120 m deep in the modern ocean) and produce most sediment in waters less than 20 m deep (Schlager, 1981). This factory is highly sensitive to oceanic perturbation and is best developed in clear oligotrophic, open ocean settings, with elevated nutrients, terrigenous clays, saline and fresh water all being deleterious to optimum accumulation. The slower-growing *heterozoan assemblage* is there in the background of tropical environments but dominates cool-water settings. It can extend to depths of ~300 m, beyond which it becomes insignificant, but maximum production is generally <150 m.

Accumulation rates

The Holocene warm-water carbonate factory (excluding reefs) produces sediment that accumulates at rates of ~10 cm/1,000 years to 60 cm/1,000 years. Cenozoic cool-water platform sediment accumulates at ~1 cm/1,000 years to 10 cm/1,000 years. It is difficult to assess rates of sediment accumulation in ancient platforms because of post-depositional diagenesis, that is, physical and chemical compaction, but most are in the range of 2 cm/1,000 years to 10 cm/1,000 years (Schlager, 1981; James and Bone, 1991).

Subtidal environments

Shallow inner to mid ramp and lagoonal seafloor settings are euphemistically called "subtidal environments", as separate from reefs, sand shoals, strandline and slope settings, and constitute the bulk of the carbonate factory. They comprise that part of the seafloor that is relatively uniform in composition and biota. Conditions are normal open marine, the biota is diverse and sediment is typically muddy.

Sediment forms through biogenic or biogenically mediated processes and if not buried quickly is altered by early, seafloor diagenesis (see *Diagenesis*). Little material is generated by physical abrasion. Mud is produced by water-column precipitation, by disintegration of calcareous algae and delicate invertebrate skeletons or by bioerosion. Sands are biofragmental, oolitic, peloidal and intraclastic. Granule and boulder-size material is formed by invertebrate skeletons and sediment lithified in the environment of deposition and eroded by physical processes. Because so much sediment is biogenic, grain-size is as much a function of organism skeletal architecture as it is hydrodynamic energy. The nature of the sediment is principally controlled by water temperature although elevated salinity dramatically reduces benthic and thus particle diversity.

Sediments, once produced, are typically burrowed and/or ingested by a host of infaunal organisms. Dwelling burrows are particularly conspicuous, creating both a labyrinth of tunnels and reworking vast amounts of sediment. Ingestion of sediment in turn passes sediment through the guts of invertebrates many times before final deposition, and these organisms produce fecal pellets, which, if lithified early, are preserved.

Storms are particularly effective in carbonate environments because they are so shallow. Warm-water, tropical environments are especially susceptible to hurricanes and cyclones. Cool-water environments are affected by winter storms where, in open ocean situations, wave-base may reach 200 m or more. On rimmed platforms storms erode and transport sediment on the shelf, sweep sediment into islands and transport fines onto tidal flats or across the shelf edge as mud plumes. On open shelves and ramps there is transport from shallow to deeper water. Thus, storm beds are a common feature of "subtidal" environments, usually typified by shell lags in Phanerozoic carbonates and storm lags of rounded intraclasts in Precambrian and early Phanerozoic rocks.

Modern, shallow, euphotic environments in all realms are characterized by prolific growth of plants, especially seagrasses. Plant leaves act as renewable substrates for delicate calcareous encrusters that eventually produce mud-size particles. The roots bind sediment together and thus impede erosion. The leaves also act as a baffle slowing water movement and as a protective canopy for the growth of a variety of other algae and invertebrates. Modern grasses are angiosperms and so this environment is Cretaceous and younger. While microbes doubtless played a similar role in the older rock record, the nature of this niche is uncertain. In cold-water, high-energy, neritic environments soft green and brown algae (kelp) are prolific, and act in the same way as renewable substrates for carbonate epibionts.

In general, sediment in warm-water, inner ramp or lagoonal environments is typically biotically diverse floatstone, wackestone and packstone that is burrowed and punctuated by storm deposits in the form of biofragmental tempestites and/or shell-intraclast lags. On cool-water shelves and ramps, muddy sediments are rare and the sediment is typically high-energy biofragmental rudstone and grainstone. Mud only accumulates below storm wave base.

Sand shoal complexes

Sand bodies occur at the strandline, on the inner ramp or along the flat-topped platform margin. Ooids are common throughout, pelmatozoans are ubiquitous in the Paleozoic and benthic foraminifers are typical of the Mesozoic and Cenozoic. Inner platform and inner ramp carbonate sand bodies (Figure N6) have the same textural attributes as siliciclastic sediments in similar settings except for common early cementation which produces beachrock that can be fragmented into clasts forming rudstones. A striking terrestrial facies adjacent to carbonate beaches is thick accumulations of *eolianites* (carbonate eolian dunes), particularly in the Cenozoic, in both warm- and cool-water realms.

Ooid and biofragmental sand complexes are common at platform margins of all ages and the best studied in the modern ocean are from the Bahamas. Their location and style is controlled by seafloor relief (presence of islands or reefs), whether they are located on the windward or leeward side of the platform and the local tidal range. Islands at the platform edge may have flood-tidal deltas between them. Alternatively, the rimmed margin may develop marine or tidal bar belts (Harris, 1984) (Figure N6). A *Marine Sand Belt* is a strike-oriented shoal on either the windward (typically ooid-rich with

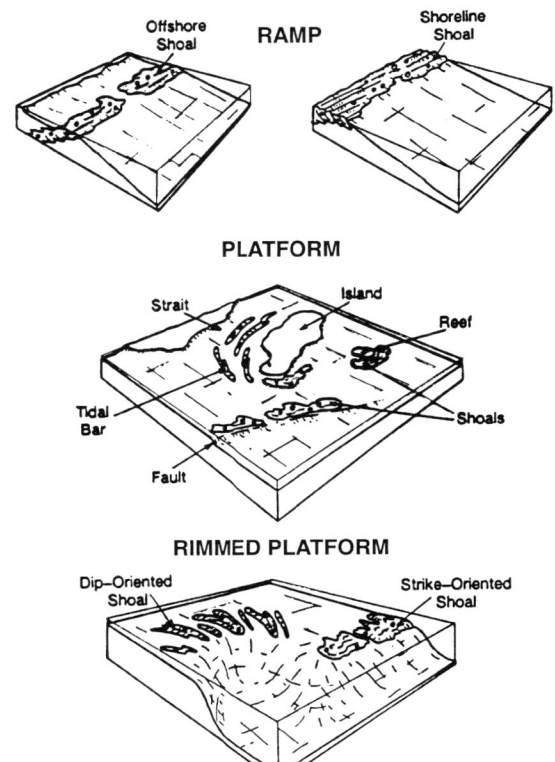

Figure N6 The style and location of carbonate sand shoals.

on-shelf directed transport) or the leeward (typically ooid-peloid-biofragmental and offshelf directed transport) margin. Such shoals are a few km wide and up to 100 km in length and comprise a "shield" of subaqueous dunes covered in asymmetrical, m-scale, flood-directed smaller dunes and broken periodically by spillover lobes. Sands grade outboard into more biofragmental deposits and inboard into muddy subtidal facies. The shoals are capped locally by islands and eolianites. In contrast, a *Tidal Bar Belt* comprises dip-oriented bars of elongate subaqueous dunes normal to the platform margin. They develop where strong tidal flow exceeds 1 m/second, with flood or ebb flow dominating different sides of the bar.

Work on high-energy cool-water platforms indicates that much of the shelf is covered with widespread subaqueous dunes composed of biofragments.

Peritidal environment

Calcareous sediments deposited in shallow seawater on and adjacent to muddy tidal flats (*peritidal* sediments and facies) are some of the most conspicuous in the rock record (Ginsburg, 1975). Such facies are typically arranged vertically in a m-scale, a *shallowing-upward sequence*, or *shallowing-upward succession*, (Pratt et al., 1992) in which marine sediments are overlain successively by muddy carbonates deposited in paleoenvironments subject to varying periods of exposure (Figure N8).

Three bathymetric zones are recognized: subtidal, intertidal and supratidal (Figure N7). The *subtidal zone* is permanently submerged and ranges from low-energy, lagoonal environments to higher energy shoals. Semimonthly neap tides may briefly expose the shallowest portions. The *intertidal*, or *littoral*, *zone* lies between normal low- and high-tide levels and is therefore submerged on a diurnal or semidiurnal basis. It is generally dissected by subtidal creeks and can be dotted with brackish or saline ponds. The *supratidal zone* is above normal

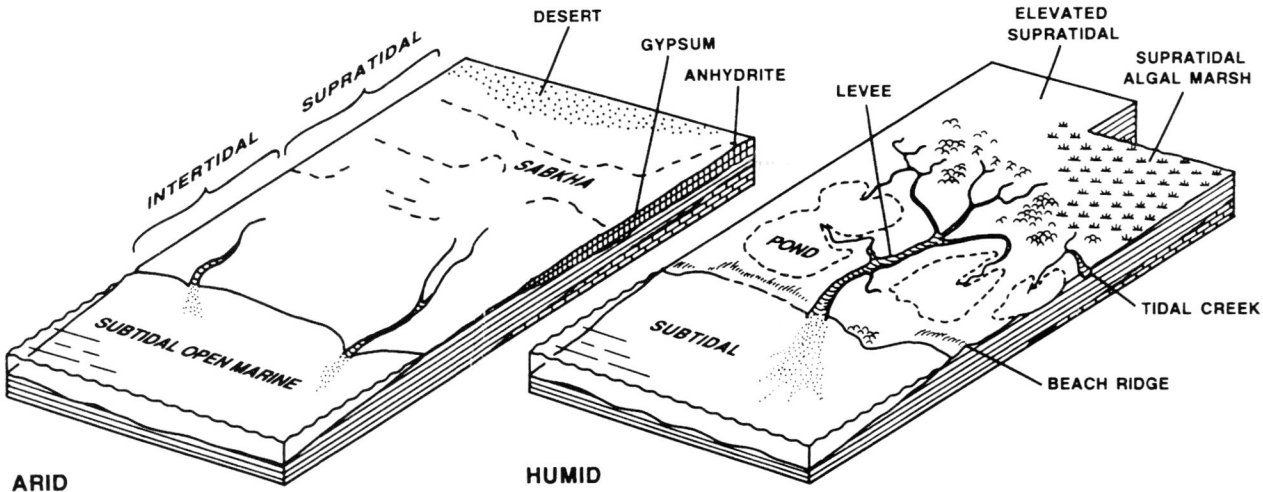

Figure N7 Block diagrams showing the main morphological elements of a muddy tidal flat under humid and arid climates (from Pratt et al., 1992).

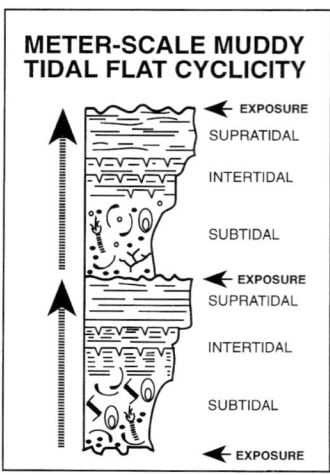

Figure N8 A diagram illustrating the nature of stacked m-scale peritidal cycles (from Pratt et al., 1992).

high tide, and is flooded only during storms and semimonthly spring tides. It can become evaporitic in semiarid and arid climates (*sabkha*).

Tidal flats are geographically complex and tidal range is modified by wind. Deposition is dominated by storms, which stir up the adjacent offshore sediment and drive mud-laden waters up the tidal creeks and onto the flats. Depositional textures and fabrics are generated by daily lunar tides, and contain the distinctive sedimentary features that allow precise location of paleoenvironments. Tidal flats are thus repositories of allochthonous sediment, and prograde as wedges seaward shorelines (e.g., Qatar, Shark Bay, Spencer Gulf), in the lee of rocky islands (e.g., northern Bahamas, Caicos, Belize), spits (e.g., Florida) and reefs and shoals (e.g., Trucial Coast), and as discrete islands and banks in shallow seas (Purser, 1973; Hardie, 1977; Hardie and Shinn, 1986).

Shallow subtidal and lower intertidal environments

In tranquil settings of normal salinity, the lowermost intertidal zone is a thoroughly bioturbated mixture of lime mud, pellets and bioclasts. Sediment is usually covered during low tide with an ephemeral microbial ("algal") slick that is the source of food for grazing organisms such as gastropods and worms. Crabs, shrimps, worms and fish bioturbate underlying sediment. Many low-energy flats are fronted by beaches of bioclastic sand winnowed from creeks and ponds or the adjacent seafloor during storms. Beach sands can be partially lithified forming gently seaward-dipping layers of beachrock.

Intertidal flat environments

Microbial mat cover here is more permanent, and forms thick, leathery carpets that can be locally shrunken, torn and folded. They exhibit various surface features such as pustules, blisters, wrinkles, crenulations or small, domical stromatolites. Such mats are a variety of filamentous and coccoid cyanobacteria ("blue-green algae") and bacteria, and are responsible for the millimeter-scale lamination exhibited by most of the sediment beneath them.

Wavy-, lenticular-, and flaser-bedded peloidal lime mudstone or grainstone (calcisiltite) and dolomitized argillaceous lime mudstone (sometimes interbedded with small hemispheroidal stromatolites), arising from the alternation between slack water and sediment transport by both unidirectional currents and waves under lower flow regime, are particularly distinctive of Precambrian and lower Paleozoic facies. These carbonates typically contain intraclastic horizons which are absent in siliciclastic counterparts. Phanerozoic intertidal facies frequently have bioclastic layers. Well-sorted coquinas were likely washed in from the subtidal zone by storms, whereas poorly sorted shelly deposits containing a low-diversity assemblage of gastropods or bivalves probably represent the *in situ* intertidal fauna. Laminae that appear laterally continuous, undulating and uniform in thickness, are typically intercalated in this facies, and record periods of stabilization or binding of the substrate by a microbial mat. Post-Paleozoic intertidal sediments are generally thoroughly churned, bioturbated, poorly fossiliferous lime mudstone units that can be interpreted as intertidal if other criteria, such as vertically juxtaposed beds with desiccation cracks, are not present.

Ponds and creeks

Ponds containing brackish or hypersaline water are common features of the intertidal zone (Figure N7), especially in more humid climates. These contain a restricted biota of microbial mats, foraminifers, gastropods, small bivalves, shrimps, ostracodes, and nematode and polychaete worms that is adapted to fluctuating salinity. This assemblage, living in a stressed environment, is typically one of high numbers of individuals but low species diversity, different from the immediate offshore biota.

Also characteristic of the intertidal zone are permanently submerged tidal creeks or channels that are the conduits for tidal flooding and draining. Such tidal creeks (least common in arid settings) are up to several meters deep and tens of meters wide with a basal lag of semilithified intraclasts eroded from the surrounding flats and flanking levees, and bars of bioclastic sand winnowed from the ponds. Supratidal "levees" protrude above high-tide level and are microbially laminated.

Supratidal flats

Most of the sediment surface here is covered by microbial mats that are typically shrunken into desiccation polygons and commonly dislodged as chips or intraclasts. Laminated sediment beneath these mats is fine grained with occasional coarser intercalations that reflect deposition by exceptional storms and, in some regions, by winds blowing off the adjacent land surface. Beds and nodules of anhydrite precipitate in these sediments in arid settings. In many areas, the supratidal zone is the locus of widespread synsedimentary cementation by microcrystalline aragonite, high-Mg calcite, or dolomite. This forms lithified pavements a few centimeters thick that are usually broken into intraclasts by forces exerted during crystal growth, groundwater pore pressure, or the roots of halophytic plants such as grasses and mangroves.

Landward parts of the supratidal zone grade into eolian deposits, soils or freshwater marshes and lakes, or onlap bedrock surfaces, depending on the region's geography and climate. Marshes and lakes, which exist in the more humid areas and have fluctuating water chemistry, are characterized by microbial mats and stromatolites; these are partially lithified by high-Mg calcite cement and calcification of organic substrates, and are interbedded with thin-bedded, locally bioturbated lime mud and bioclastic and peloidal carbonate sand deposited during storms. Much of the microbially laminated sediment shows fenestral fabric, millimeter- to centimeter-sized subhorizontal, sheet-like pores formed as voids bridged by microbial mats or as molds of degraded mats. Decimeter-scale teepee structures, consisting of disrupted and overthrust crusts of lithified, tufa-like fenestral sediment giving an inverted V-shaped cross section, form in areas of groundwater discharge (Kendall and Warren, 1987). These also contain complex generations of internal sediment and aragonite and high-Mg calcite cements.

Ancient supratidal facies

Microbially laminated limestone or dolostone, usually with desiccation cracks and coarser rippled layers, is a common peritidal rock type. Intercalated within many microbially laminated rocks are intraclastic horizons, which are analogous to pavements of microbial mat chips or fragments of cemented crusts in modern supratidal environments. Fenestral lime mudstone and peloidal grainstone are common and, by analogy with tidal flats of Florida and the Bahamas, were probably deposited in moist supratidal "algal marshes" or around ponds. This facies sometimes exhibits features, such as pendent fibrous cement, brecciated crusts and tepees, pisolites and pores with geopetal sediment floors, suggestive of flushing by downward-percolating seawater and rainwater and upward-flushing by groundwaters in a subaerial environment.

Reefs

There has for decades been an ongoing debate as to what constitutes an ancient "reef" (Heckel, 1974; James, 1983; Geldsetzer *et al.*, 1988; Wood, 2000). A useful compromise is that *reefs* are *biologically constructed reliefs* which were, like modern reefs, constructed by large, usually clonal elements (on average >5 cm in size), and capable of thriving in energetic environments; *mounds* are those structures which were built by smaller, commonly delicate and/or solitary elements in tranquil settings (see *Carbonate Mud Mounds*). Two terms, bioherms and biostromes, are commonly used to designate biogenically constructed geological structures. A *bioherm* is a lens-shaped reef or mound; a *biostrome* is a tabular rock body, usually a single bed of similar composition. Another commonly used generic epithet with no compositional, size or shape connotation is *carbonate buildup*.

Depositional facies

Reefs generally comprise three facies (Figure N9): (1) *Core facies*—massive, unbedded carbonate with or without skeletons; (2) *Flank or fore reef facies*—bedded carbonate sand and conglomerate of in place and/or core-derived material, dipping and thinning away from the core; and (3) *Interreef or open platform facies*—subtidal limestone or terrigenous clastic sediment, unrelated to reef growth.

Any living reef or mound is a delicate balance between: (1) upward growth of in place calcareous elements (metazoan and microbial); (2) continuing destruction by a host of raspers, borers and grazers; (3) prolific sediment production by rapidly growing, short-lived, attached calcareous benthos; and (4) concurrent inorganic or organically induced cementation. The

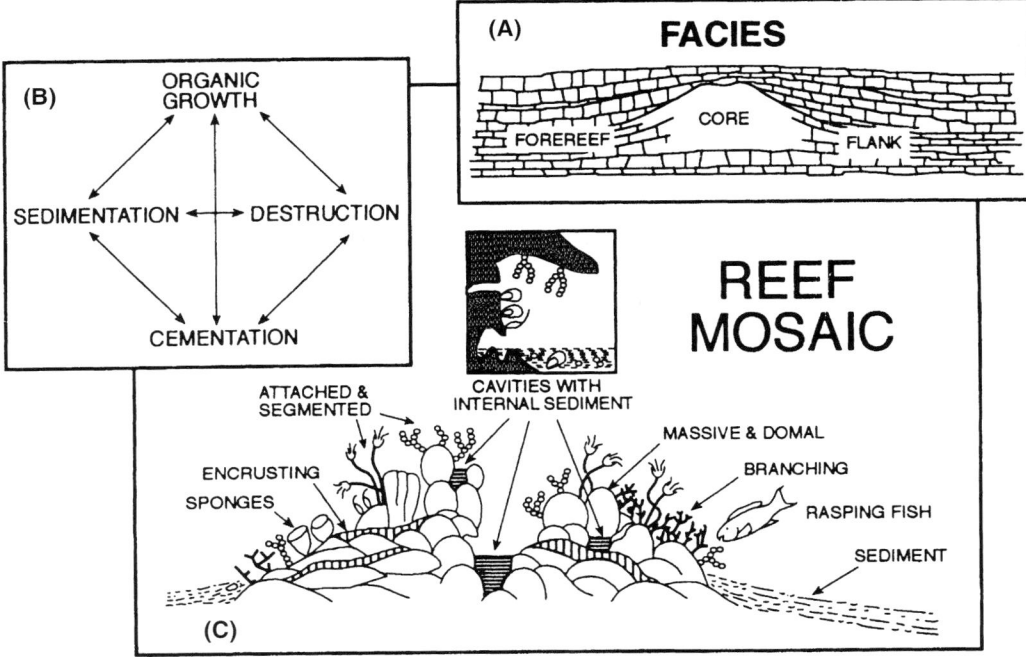

Figure N9 A sketch illustrating the facies (A), processes (B) and general attributes (C) of reefs (after James and Bourque, 1992).

large skeletal metazoans (e.g., corals) generally remain in place after death, except when so weakened by bioeroders that they are toppled by storms. Their irregular shape and growth habit result in roofed-over cavities that can be inhabited by smaller, attached calcareous benthos. Encrusting organisms growing over dead surfaces aid in stabilization. Branching reef-builders can also be preserved in place, but are just as commonly fragmented by storms into sticks and rods, which form skeletal conglomerates.

Most sediment is produced by the postmortem disintegration of segmented (e.g., calcareous green algae) or nonsegmented (e.g., bivalves, brachiopods, foraminifers) organisms. Additional sediment is generated by bioeroders: boring organisms (worms, sponges, bivalves) produce lime mud; rasping organisms, generally herbivores (echinoids, fish), graze the surface creating copious quantities of carbonate sand and silt. The sediment accumulates around the buildups as an apron, lodges between skeletons as a matrix and filters into growth cavities as geopetal internal sediment. Rigidity of many reefs is accomplished by invertebrates and microbes encrusting or growing on top of one another. Many reefs are also preferential sites for synsedimentary cement precipitation (see *Diagenesis*), and are hard limestone just below the growing surface.

Results of these processes can be seen in many Phanerozoic reefs and mounds, and especially in Mesozoic and Cenozoic structures, which are decidedly "modern" in composition. Paleozoic buildups do not appear to have been affected much by boring organisms but were typically sites of intense synsedimentary cement precipitation; some reefs contain so much cement that they have been called "cementation reefs". Precambrian buildups, with their limited biological components, were mainly constructive, clearly fractal in character, and contained few growth cavities.

The reef growth window

The modern growth window is determined by the combination of factors that control growth of the major organism, in this case corals. Hermatypic (reef-building) corals grow in waters between 18° and 36°C, but are best adapted to form reefs in waters between 25°C and 29°C. Periodic exposure is not necessarily lethal and some intertidal corals are out of water for many hours daily. The salinity window ranges from 22‰ to 40‰, but most corals grow best in waters between 25‰ and 35‰.

Hermatypic, reef-building corals are cnidarians which contain light-dependent, photosynthetic symbiotic microorganisms (zooxanthellae). Light decreases exponentially with depth and the lower limits of hermatypic coral and calcareous green algal growth are 80 m to 100 m. Because they are mixotrophs corals flourish in nutrient-impoverished oceanic regions. Increased nutrient levels, from upwelling on the outer platform and runoff on the inner platform, lead to dramatic changes in reef structure; at intermediate levels the animal-plant symbionts are replaced by more heterotrophic forms and "fouling organisms" such as filamentous algae, fleshy algae and small suspension-feeding animals (barnacles and bivalves) of the heterozoan assemblage. Reefs still grow in such regions only because herbivores graze back the algae. Fine-grained suspended sediment limits coral growth because it decreases light penetration and covers or clogs the polyps. It is difficult to decouple the effect of fine terrigenous sediment and nutrients on coral reef growth because they usually occur together.

Scleractinian coral reefs

Modern reefs exhibit depth zonation (Figure N10) because of decreasing wave energy, light intensity and to a lesser extent temperature, with increasing bathymetry.

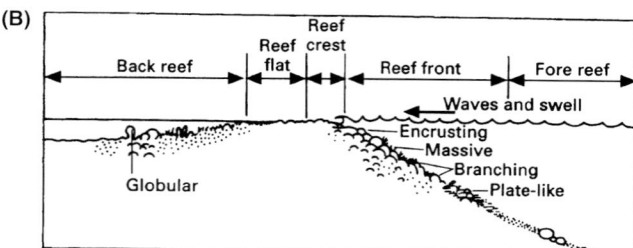

(A) Growth form		Environment	
		Wave energy	Sedimentation
	Delicate, branching	Low	High
	Thin, delicate, plate-like	Low	Low
	Globular, bulbous, columnar	Moderate	High
	Robust, dendroid branching	Mod-high	Moderate
	Hemispherical, domal, irregular, massive	Mod-high	Low
	Encrusting	Intense	Low
	Tabular	Moderate	Low

Figure N10 (A) growth forms of reef-building organisms and their general environments, (B) the zonation of a modern, windward coral reef (after James and Bourque, 1992).

Reefs on the leeward side of banks can be strikingly different, either because of reduced wave activity or because of "bad water". The bad water is formed on the platform by fine-grained sediment production, oxygen depletion, heating or evaporation. It is usually driven off the bank downwind, across the leeward margin, inhibiting reef growth to a depth of several 10s of meters. Such shallow leeward margins can, therefore, be bare rock floors covered by soft fleshy algae or hard coralline algae with active reef growth taking place only in deeper water, below this interface.

There is commonly a cross-shelf zonation in reef composition that partly reflects the outboard decrease in fine sediment and nutrients. *Inner shelf reefs* are characterized by (1) quickly growing corals with a high tolerance for fine sediment but unable to withstand turbulent waters; (2) large and abundant heterotrophic sponges; (3) low epifaunal diversity; (4) few soft corals; and (5) abundant soft algae and few calcareous algae. *Outer shelf reefs* (in areas of little upwelling) are distinguished by (1) slow-growing autotrophic corals which cannot withstand suspended sediment but are adapted to high-energy seas; (2) reduced numbers of sponges, most of which contain photosynthetic symbionts; (3) common tridacnid bivalves in the Pacific; (4) high epifaunal diversity; and (5) prolific calcareous algae.

Ancient reefs

The history of Phanerozoic reefs and mounds (Figure N11) is complex (James, 1983; Wood, 2000) Descriptions of Precambrian buildups can be found in Geldsetzer *et al.* (1988) and Grotzinger (1988).

During geological periods when large calcareous skeletal and microbial structures were common, reefs grew as fringing reefs inboard, as patch reefs across platforms, and as semicontinuous to continuous barrier reefs along platform margins. Coeval mounds developed in quiet water settings across the platform, on ramps or on the slope. During geological periods when there were no large reef builders, mounds were the early buildups and grew leeward of sand shoals on deep water ramps and slopes.

Although we cannot be certain about the physiological tolerances of all fossil reef-builders, some generalities are possible. Scleractinian corals, the modern reef-building corals, evolved in the Triassic and were probably zooxanthallate and reef-building by late Triassic, certainly by Middle Jurassic, time. Tabulate corals are a diverse group of several organisms and it is not clear whether they contained similar symbionts. Rudist bivalves, mound builders of Cretaceous seas, are related to modern, zooxanthallate tridacnid bivalves and their ability to form large skeletons is thought by many to be the result of symbiosis. Fossil sponges, both spiculate and soft bodied, and coralline sponges (hard skeletoned such as stromatoporoids, archaeocyathans) are poorly understood, but modern sponges on outer shelf reefs commonly contain cyanobacterial symbionts. Regardless, for those organisms that did contain phototrophic symbionts, light and nutrients would have been important limiting factors.

Fossil reef facies distribution is generally the same as that of modern coral reefs with the caveat that is not clear whether all stromatoporoid and tabulate coral reefs built structures up into the surf zone. Modern reef facies characteristics are directly applicable to Mesozoic and Cenozoic coral-algal reefs, and, with the above caveat, useful for the interpretation of Paleozoic and Proterozoic structures as well.

Autostratigraphy

Ecological succession as exemplified in reefs by an increase in species diversity, biomass, structural complexity, and stability through time is manifest as autostratigraphy. The stages of succession have been called *stabilization, colonization, diversification and domination*. In more general terms the first two stages are really pioneer or developmental phases whereas the last two are climax or mature phases. Such *pioneer communities* are composed of generalists which are small, solitary, generally erect, hardy and fast growing, have high recruitment and reproduction rates, and are of wide geographic distribution. They can readily adapt to new substrates and tolerate abundant suspended matter. Assemblages are of generally low diversity and dominated by one or two elements, poorly zoned and characterized by clumps of fossils with spaces between. A *climax community* is composed of clonal, long-lived, commonly endemic specialists who occupy narrow environmental niches, are slow growing and of large to gigantic size and have a wide variety of shapes. The community is one of high biomass that completely covers the seafloor and in which nutrients are conserved and efficiently recycled, but it is vulnerable to catastrophe.

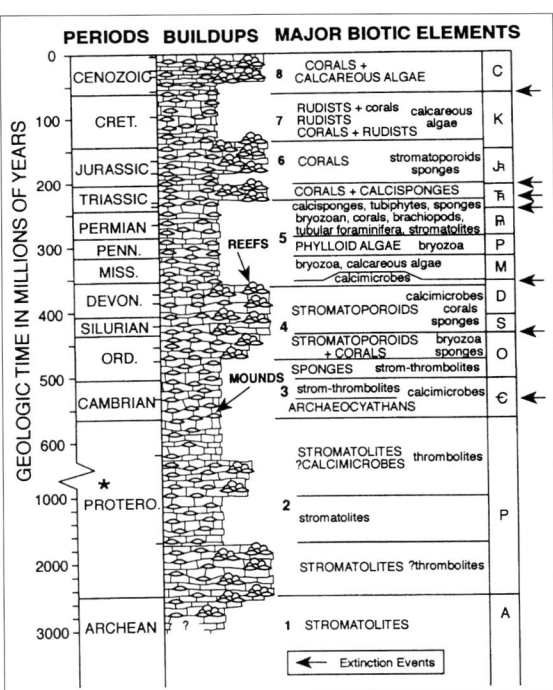

Figure N11 An idealized stratigraphic column illustrating periods when there were only biogenic mounds (see *Carbonate Mud Moulds*) and periods when there were both mounds and reefs. The main reef-building organisms are noted beside each stage (from James and Bourque, 1992).

Allostratigraphy

The allostratigraphy of reefs is a function of: (1) varying conditions in the growth window; (2) ecological succession; (3)

antecedent topography; and (4) the rates of sea level rise and organism growth (Kendall and Schlager, 1981).

Reef *Start-up* can take place: (1) during sea level rise following exposure; (2) during sea level fall when the seafloor comes into the growth window; (3) when the seafloor is raised by mound development into the growth window; (4) when factors inimical to reef growth are turned off.

Keep-up Reefs maintained their crests at or near sea level, tracking sea level rise. *Catch-up Reefs* began as shallow reefs which became deeper as rise rate exceeded accretion rate, only to grow upward and catch up with sea level. Alternatively, they started on a deep substrate and then grew up to sea level. Once reefs of either type reached sea level they maintained a near-surface crest or developed a capping facies of subtidal pavements to storm-ridge deposits. At this point growth was limited to the seaward margin and reefs prograded laterally over their own reef front and fore reef sediments. *Give-up Reefs* initially grew as did the others but then succumbed to changing environmental conditions and died. All studies note that the responses of reef growth to sea level rise varied greatly, even within the same reef.

Reefs can turn off and die in a variety of situations. The most obvious reason, that sea level rise was too rapid for the reefs to keep up, is not the answer. This is because reefs can have extraordinarily high growth rates that potentially exceed any rise in sea level except that related to tectonics. Changes in the growth window are more likely to be the cause of reefs giving up. *Drowning* (Schlager, 1981) is submergence of the platform top below the photic zone by a rise in sea level or shallowing of the photic limit. Alternatively reefs can be killed by some combination of increased turbidity, nutrient excess or hot saline waters which swept off the shelves when sea level flooded the wide shallow platforms (the "bad water" syndrome).

Basin-facing slopes

Introduction

Carbonate slopes (Cook and Enos, 1977; MacIlreath and James, 1984; Mullins and Cook, 1986) encompass a suite of environments which pass seaward from shallow water, sunlit, platform margins and inner ramp environments to the deep, dark, basin settings. They are variably affected with depth by changing seawater chemistry, temperature, pressure and biota. The best studied carbonate slopes in today's oceans are those in the northern Bahamas-Florida region, augmented by more local studies in Belize, Jamaica, Grand Cayman, and several atolls in the Pacific and Indian oceans.

Slope sediment is a variable mixture of autochthonous benthic biofragments, resedimented, allochthonous material from upslope and pelagic fallout. The pelagic component is mostly terrigenous in pre-Jurassic rocks and mostly planktic carbonate in younger strata. *Periplatform ooze* is intermixed platform-derived and pelagic carbonate muds while *hemipelagic sediment* is a mixture of carbonate and terrigenous clastic material.

Autochthonous sediment

This material includes fecal pellets derived from epifauna and infauna, seafloor Mg–calcite cement, peloidal Mg–calcite cement precipitated within foraminifer tests, and skeletal debris associated with biota indigenous to the slope environment. Siliceous sponge spicules form a locally important supply of noncarbonate sediment.

Allochthonous sediment

Such sediment is transported to and accumulates on the slope by suspension settling, gravity (resedimented) flow, rock fall, and submarine creeping and sliding. Bottom currents, particularly contour currents, can winnow and rework these deposits, or prevent accumulation entirely, resulting sediment starvation, submarine dissolution or lithification.

Fine-grained carbonates composed shallow water-derived particles are transported basinward from shallow water by tides, storms, and currents forming concentrated to dilute sediment-seawater mixtures. These plumes can dissipate, in which case material rains down to the seafloor, or progress downslope as unconfined and confined dilute turbidity currents. Such flows can also move along mid-water density interfaces (e.g., pycnoclines) from which sediment rains down on the basin floor.

This sediment is varying proportions of mud-size algal and inorganically precipitated aragonite needles, blades of Mg–calcite and aragonite, mud- to sand-sized skeletal and nonskeletal debris, lithoclasts, and bioeroded particles (e.g., sponge chips). Jurassic and younger muds are rich in calcareous plankton while older slope carbonates are mostly platform-derived sediment. Locally, coarser periplatform sediments contain gravels and boulder-sized lithoclasts derived from shallow water facies or carbonate bedrock.

A common lithology in many ancient slope sequences is interbedded, finely crystalline, dark grey to black, lime mudstone and shale, forming evenly and continuously rhythmic sequences. In places where the carbonate is finely laminated, suspension deposition may have occurred within an oxygen minimum zone that limited bioturbation. Oxygenated slopes are commonly intensively bioturbated, obliterating primary sedimentary laminations and possibly encouraging the formation of nodular bedding.

Sediment gravity flow deposits are manifest as carbonate turbidites, carbonate grain flow deposits and debrites. Most particles are polygenetic, from the platform, inner to mid ramp and the slope.

A hallmark of many carbonate slopes are massive, thick limestone conglomerates called *debrites, debris sheets or avalanches, olistostromes, and mass breccia flows*. A few decimeters to several tens of meters in thickness, such deposits are generally poorly sorted, lack stratification, and have a random or chaotic clast arrangement. Some have normal and reverse grading, clast alignment near the base, sides or top of the deposits, local clast imbrication, and swirled domains in which clasts display a gradual change in flat clast orientation. Clasts are rounded to angular, and sizes vary from gravel- to house-sized boulders with maximum clast size and bed thickness commonly being correlated.

Paleomargin proximity is commonly indicated by the most abundant and thickest conglomerates containing the largest sizes and varieties of clasts. *Polymict conglomerates* result from the mixing of platform-derived exotic clasts (which may be variable themselves) with slope-derived clasts and matrix. *Oligomict conglomerates* are almost always due to erosion and redeposition of slope limestone in the form of tabular clasts (limestone chips) and interparticle matrix. Rafts of coherently bedded, slope-derived limestone several tens of meters in size are identical to *in situ* slope limestones. Such rafts become fragmented during progressive downslope transport to yield matrix and tabular clasts. Soft-sediment folding, faulting, fracture cleavage formation and brecciation characterize such rafts.

Intraformational truncation surfaces, shear zones, and detached slide masses are all products of slope failure and form in any of the preceding facies. The first two provide evidence for an ancient slope; the third, like gravity flows, can be preserved either downslope or within an adjacent basin.

Slope styles

Depositional or accretionary slopes, particularly characteristic of ramps, demonstrate net accumulation with the locus of sedimentation along the upper slope or mid ramp and sediment thickness decreasing seaward. A *bypass slope* is one in which there is little sediment accumulation because sediment has largely moved across it, to accumulate at the base of slope as a sediment wedge. *Erosional slopes* illustrate net sediment loss and are characterized by truncated bedding and relatively steep gradients.

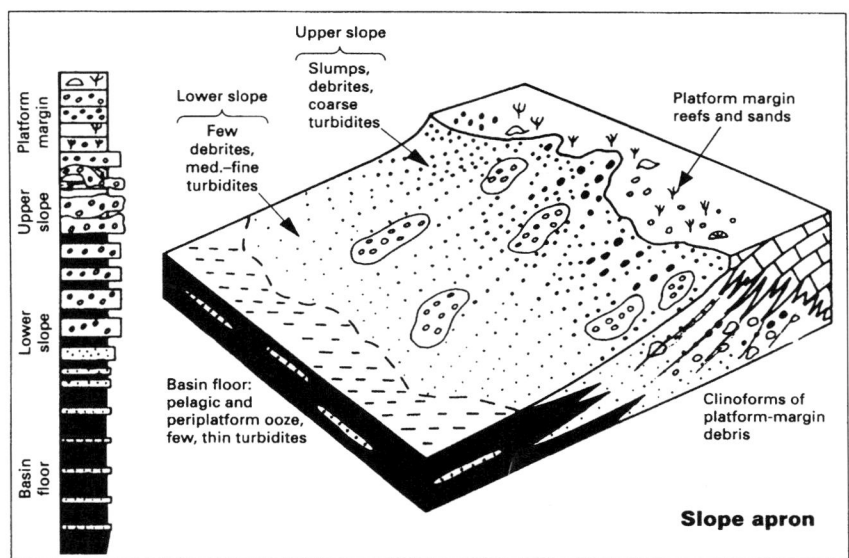

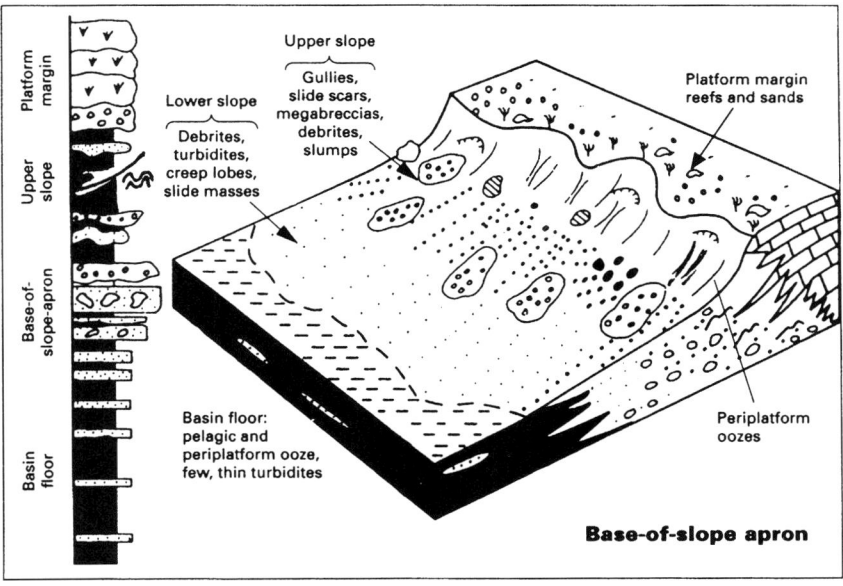

Figure N12 Deep water carbonate slope apron facies (from Mullins and Cook, 1986).

Modern carbonate slopes are generally steeper than their terrigenous counterparts and tend toward concavity with the angle of the upper slope increasing with slope height. Steepening of the carbonate slope, results from: (1) reef-building; (2) submarine cementation; and (3) shallow subsurface lithification. Early lithification may act to delay large-scale failure. Gradients are steeper (30°–40°) for sediments with cohesionless, grain-supported fabrics (with or without mud), than those of paleoslopes dominated by mud-supported, cohesive fabrics (<15°). Pure muds have slope angles of less than 5°.

Many modern carbonate slopes are typified by numerous parallel incised gullies, oriented perpendicular to the margin. Such gullies form a *line source* of sediment supply from the platform and upper slope to the lower slope.

Slope sediment aprons

The ultimate sediment accumulations are wedge-shaped to lenticular bodies that thicken toward the platform margin. Facies belts closely parallel the platform margin.

Carbonate slope aprons (depositional slopes) (Figure N12) extend without break from the basin along gentle (<4°) gradients up to a shallow water platform margin. A line source of platform-derived sediment originates from numerous channels dissecting shoals, reefs and islands along the platform margin. Downslope sediment transport along such carbonate aprons is predominantly via unchannelized sheet flows. These are like the mid- and outer-ramp facies and the two systems seem to merge.

Rimmed platform base-of-slope aprons form downslope from the platform-slope break along a relatively steep (>4°) margin. Sediment is supplied from a line source which typically takes the form of a series of closely spaced gullies transporting shallow water and slope-derived sediment through some portion of the upper to middle (bypass) slope. Sediment emerging from a gullied slope initially follows a dispersal path outlining a series of small, interleaved fan-shaped deposits while broadly defining platform margin-parallel facies belts. As with carbonate slope aprons, systematic vertical sequences are not expected.

Open (unrimmed) platform aprons occur in both warm and cool-water realms. In warm-water settings, because the platform margin is below depths at which the carbonate factory operates, such sediment aprons are likely to be dominantly pelagic in origin. In cool-water settings, however, where the sediment factory is typically deeper, the slopes are in part depositional. Mass-wasting deposits include pelagic turbidites and debrites, the latter of which can contain clasts derived from the platform margin or fragmented beds of early lithified pelagite. The generally finer-grained nature of the deeper platform margin sediment implies that submarine cementation (and therefore production of clasts) may not be as important as along shallower rimmed platform margins. Pre-Mesozoic aprons adjacent to open platform margins are argillaceous.

Carbonate submarine fans appear to be rare in the rock record, but several have been described from the Paleozoic. No modern-day examples have been documented.

Stratigraphy

Since so much of the sediment comes from the platform the stratigraphic style of slope deposits closely reflects conditions on the platform. Abundant allochthonous carbonate generally signals a healthy, generally prograding platform while starved carbonate or just shale usually means exposure, rapid accretion or subsidence of the platform below the zone of optimum production.

Noel P. James

Bibliography

Bathurst, R.G.C., 1975. *Carbonate Sediments and Their Diagenesis*. Elsevier.

Burchette, T.P., and Wright, V.P., 1992. Carbonate ramp depositional systems. *Sedimentary Geology*, **79**: 3–57.

Cook, H.E., and Enos, P. (eds.), 1977. *Deep-water Carbonate Environments*. Society of Economic Paleontologists and Mineralogists, Special Publication, 25, 336pp.

Geldsetzer, H.H.J., James, N.P., and Tebbutt, G. (eds.), 1988. *Reefs; Canada and Adjacent Areas*. Canadian Society of Petroleum Geologists, Memoir 13, 775pp.

Ginsburg, R.N. (ed.), 1975. *Tidal Deposits: A Casebook of Recent Examples and Fossil Counterparts*. Springer-Verlag.

Goldhammer, R.K., Dunn, P.A., and Hardie, L.A., 1990. Depositional cycles, composite sea-level changes, cycle stacking patterns, and the hierarchy of stratigraphic forcing: examples from alpine Triassic platform carbonates. *Geological Society of America Bulletin*, **102**: 535–562.

Grotzinger, J.P., 1988. Precambrian reefs. In *Reefs, Canada and Adjacent Areas*. Canadian Society of Petroleum Geologists, Memoir 13, pp. 9–12.

Hallock, P., and Schlager, W., 1986. Nutrient excess and the demise of coral reefs and carbonate platforms. *Palaios*, **1**: 389–398.

Hardie, L.A., and Shinn, E.A., 1986. Carbonate depositional environments, modern and ancient. Part 3: tidal flats. *Colorado School of Mines, Quarterly*. **81**: 74pp.

Hardie, L.A. (ed.), 1977. *Sedimentation on the Modern Carbonate Tidal Flats of Northwest Andros Island, Bahamas*. Johns Hopkins University, Studies in Geology 22, 202pp.

Heckel, P.H., 1974. Carbonate buildups in the geologic record: a review. In Laporte, L.F. (ed.), *Reefs in Time and Space*. Society of Economic Paleontologists and Mineralogists, Special Publication, 18, pp. 90–155.

James, N.P., 1983. Reef environment. In Scholle, P.A., Bebout, D.G., and Moore, C.H. (eds.), *Carbonate Depositional Environments*. American Association of Petroleum Geologists, Memoir 33, pp. 345–440.

James, N.P., and Bone, Y., 1991. Origin of a cool-water, Oligo-Miocene deep shelf limestone. Eucla Platform, southern Australia. *Sedimentology*, **38**: 323–341.

James, N.P., and Kendall, A.C., 1992. Introduction to carbonate and evaporite facies models. In Walker, R.G., and James, N.P. (eds.), *Facies Models—Response to Sea Level Change*. Geological Association of Canada, pp. 265–276.

James, N.P., and Bourque, P.-A., 1992. Reefs and mounds. In Walker, R.G., and James, N.P. (eds.), *Facies Models—Response to Sea Level Change*. Geological Association of Canada, pp. 323–348.

James, N.P., and Clarke, J.D.A. (eds.), 1997. *Cool-Water Carbonates*. SEPM, Special Publication, 56, 440pp.

Kendall, C.G.St.C., and Warren, J., 1987. A review of the origin and setting of tepees and their associated fabrics. *Sedimentology*, **34**: 1007–1028.

Kendall, C.G.St.C., and Schlager, W., 1981. Carbonates and relative changes in sea level. *Marine Geology*, **44**: 181–212.

Laporte, L.F. (ed.), 1974. *Reefs in Time and Space*. Society of Economic Paleontologists and Mineralogists, Special Publication, 18, 256pp.

McIlreath, I.A., and James, N.P., 1984. Carbonate slopes. In Walker, R.G. (ed.), *Facies Models*. Geoscience Canada Reprint Series 1, pp. 245–257.

Mullins, H.T., and Cook, H.E., 1986. Carbonate apron models: alternatives to the submarine fan model for paleoenvironmental analysis and hydrocarbon exploration. *Sedimentary Geology*, **48**: 37–79.

Nelson, C.S. (ed.), 1988. Cool water carbonate sediments. *Sedimentary Geology*, **60**: 367pp.
Pratt, B.R., James, N.P., and Cowan, C.A., 1992. Peritidal carbonates. In Walker, R.G., and James, N.P. (eds.), *Facies Models—Response to Sea Level Change*. Geological Association of Canada, pp. 303–322.
Purser, B.H. (ed.), 1973. *The Persian Gulf: Holocene Carbonate Sedimentation and Diagenesis in a Shallow Epicontinental Sea*. Springer-Verlag.
Read, J.F., 1985. Carbonate platform facies models. *American Association of Petroleum Geologists, Bulletin*, **69**: 1–21.
Schlager, W., 1981. The paradox of drowned reefs and carbonate platforms. *Geological Society of America Bulletin*, **92**: 197–211.
Scholle, P.A., Bebout, D.G., and Moore, C.H. (eds.), 1983. *Carbonate Depositional Environments*. American Association of Petroleum Geologists, Memoir 33, 708pp.
Stanley, G.D. Jr., and Fagerstrom, J.A. (eds.), 1988. Ancient reef ecosystems issue. *Palaios*, 3: 110–250.
Toomey, D.F. (ed.), 1981. *European Fossil Reef Models*. Society of Economic Paleontologists and Mineralogists, Special Publication 30, 546pp.
Tucker, M.E., and Wright, V.P., 1990. *Carbonate Sedimentology*. Blackwell Scientific Publications.
Wilson, J.L., 1975. *Carbonate Facies in Geologic History*. Springer-Verlag.
Wood, R., 2000. *Reef Evolution*, Oxford University Press.
Wright, P.V., and Burchette, T.P., 1996. Shallow-water carbonate environments. In Reading, H. (ed.), *Sedimentary Environments*. Blackwell Science, pp. 325–394.
Wright, V.P., and Burchette, T.P., 1998. *Carbonate Ramps*. Geological Society Special Publication, 149, 465pp.

Cross-references

Algal and Bacterial Carbonate Sediments
Beachrock
Bioclasts
Bioerosion
Calcite Compensation Depth
Carbonate Diagenesis and Microfacies
Carbonate Mineralogy and Geochemistry
Carbonate Mud Mounds
Cements and Cementation
Classification of Sediments and Sedimentary Rocks
Diagenesis
Dolomites and Dolomitization
Encrinites
Micritization
Mixed Siliciclastic and Carbonate Sedimentation
Seawater: Temporal Changes in the Major Solutes
Sedimentologists
Stromatolites

NUMERICAL MODELS AND SIMULATION OF SEDIMENT TRANSPORT AND DEPOSITION

A numerical model simulating sediment transport and deposition is an equation set describing the spatial and temporal evolution of a fluid flow carrying sediment over, and interacting with, an adjacent mobile particulate bed. Although literally scores of such models exist, each starting from quite different assumptions and incorporating quite different sedimentary processes, the most common and the focus of this entry are process-based models constructed with the mutual goals of prediction and understanding. Given water and sediment mass fluxes delivered to a geomorphic system through time, these numerical models are expected to accurately predict the total mass flux and character of sediment passing each point in the system at time and space scales appropriate for the application. To do so, each model must capture how changes in state variables of the fluid flow, the sediment transport, and the deforming bed influence one another through feedback loops. Each also must embody the physical processes thought to be relevant and consequently each is a conjecture about the dynamical behavior of the geomorphic system. Thus, for example, if the geomorphic system is a river channel reach of a given initial hydraulic geometry and sediment characteristics with a given sediment and water feed at the upstream end, the model in question should predict accurately the temporally evolving flux of water and sediment along the reach and the amount of bed erosion or deposition through time using conservation and rate laws that describe all the processes and their interactions. This entry describes the character, use, and limitations of this general class of sediment transport model. Channelized flows are emphasized, although much of the discussion also applies to sediment transport models of lakes and coastal oceans. Also, the emphasis here is on non-cohesive sediment transport because most models, if they address cohesive sediment at all, are still quite primitive in its treatment. Table N1 list some popular examples.

Model components

There are five components or computation modules in a sediment transport model: (1) a hydrodynamic module to compute the flow field; (2) a shear stress module to calculate the stress effective in transporting sediment as the bed roughness evolves and in some models, to calculate the shear stress distribution arising from turbulent fluctuations; (3) a bed-load transport module; (4) a suspended load transport module; and (5) a bed module that keeps track of bed erosion, deposition, and the evolving bed texture. The computation modules are organized into a solution algorithm, the structure of which depends upon whether the equation set is to be solved simultaneously or in series. Figure N13 gives a typical organization of the computation modules for a series solution of a 1-D, unsteady, nonuniform application. Because continuous solutions to the equation set are not known, the set is always solved by either finite-difference or finite element techniques at discrete nodes in the model space, $x = 1\Delta x, 2\Delta x, \ldots, n\Delta x$, $y = 1\Delta y \ldots$, $t = 1\Delta t \ldots$, and so on.

Each of the five components is described below. For the purposes of illustration, a right-handed Cartesian coordinate system is assumed in which x defines the primary horizontal flow direction, y the transverse direction, and z is positive upward.

Hydrodynamic module

An accurate description of the fluid flow field is a necessary condition for predicting sediment transport and deposition. Without significant loss of generality most geomorphic hydraulic models assume that the fluid is incompressible, the fluid density is everywhere equal, the pressure distribution in the fluid is hydrostatic, and the eddy viscosity approach can be used to describe the role of turbulence in momentum transfer (see Lane, 1998 for a review). Whether further simplifications are justified depends upon the application. In the most demanding applications the flow field is unsteady and

Table N1 Some popular numerical sediment transport models

Model	Authors	Comments
HEC-6 4.1	US Army Corps of Engineers	1D, movable bed, open channel flow model simulating scour and deposition from steady flows; FEMA[a] approved
TABS-SED2D 4.5	US Army Corps of Engineers	SED2D computes sediment loadings and bed elevation changes when supplied with a flow field from TABS or RMA2; treats clay beds; FEMA approved
MIKE 11	DHI Water and Environment, Inc.	1D, commercial movable bed open channel flow model simulating cohesive and non-cohesive sediment transport and deposition; FEMA approved
CH3D-SED	Gessler et al., 1999	3D model of unsteady flows in estuaries and rivers, including vertical mixing and surface heat exchange; suspended sediment transport modeled by 3-D advection–diffusion equation
MIDAS	van Niekerk et al., 1992	freely-available, 1D, open channel, uncoupled unsteady, gradually-varied flow and sediment model
FFDC	Tetratech	freely-available, curvilinear orthogonal coordinate, coupled hydrodynamic and sediment model for coastal oceans
ECOM-SED	HydroQual, Inc. & Delft Hydraulics	3D, commerical hydrodynamic and sediment model

[a] US Federal Emergency Management Agency

nonuniform and significant secondary circulation exists. Consequently, the hydraulic module must compute the instantaneous, turbulence-average flow velocities u, v, and w at all nodes in the model space, as well as the water surface elevation, h at all surface nodes. Four dependent variables require four equations for their solution. The equations are the x- and y-directed general laws of motion, the hydrostatic pressure distribution in the vertical, and conservation of mass equation:

$$\frac{\partial u}{\partial t} + \frac{\partial u^2}{\partial x} + \frac{\partial}{\partial y}uv + \frac{\partial}{\partial z}uw = fv - \frac{1}{\rho}\frac{\partial p}{\partial x} + \frac{\partial}{\partial x}\left(A_H \frac{\partial u}{\partial x}\right) + \frac{\partial}{\partial y}\left(A_H \frac{\partial u}{\partial y}\right) + \frac{\partial}{\partial z}\left(A_V \frac{\partial u}{\partial z}\right) \quad \text{(Eq. 1)}$$

$$\underbrace{\frac{\partial v}{\partial t}}_{(1)} + \underbrace{\frac{\partial}{\partial x}uv + \frac{\partial v^2}{\partial y} + \frac{\partial}{\partial z}vw}_{(2)} = \underbrace{-fu}_{(3)} - \underbrace{\frac{1}{\rho}\frac{\partial p}{\partial y}}_{(4)} + \underbrace{\frac{\partial}{\partial x}\left(A_H \frac{\partial v}{\partial x}\right) + \frac{\partial}{\partial y}\left(A_H \frac{\partial v}{\partial y}\right)}_{(5)} + \underbrace{\frac{\partial}{\partial z}\left(A_V \frac{\partial v}{\partial z}\right)}_{(6)} \quad \text{(Eq. 2)}$$

$$\frac{\partial p}{\partial z} = -\rho g \quad \text{(Eq. 3)}$$

$$\frac{\partial u}{\partial x} + \frac{\partial v}{\partial y} + \frac{\partial w}{\partial z} = 0 \quad \text{(Eq. 4)}$$

where u, v, w = velocity components in the x, y, z-directions; t = time; f = Coriolis parameter; ρ = local fluid density, accounting for temperature, salinity, and sediment concentration; p = pressure; A_H = horizontal turbulent diffusion coefficient; A_V = vertical turbulent diffusion coefficient; and g = gravitational acceleration. Term 1 above accounts for unsteadiness of flow; Term 2 accounts for nonuniformity of flow; Term 3 accounts for Coriolis accelerations; Term 4 accounts for fluid pressure gradients arising from gradients in the water surface elevation; Term 5 accounts for shearing forces per unit mass due to velocity gradients in the horizontal and Term 6 accounts for shearing forces per unit mass due to velocity gradients in the vertical.

Before the equation set can be solved for the dependent variables, the turbulent diffusion coefficients must be specified. Numerous approaches exist (cf. Nezu and Nakagawa, 1993) ranging from "zero-equation" turbulence models that specify constant coefficients to "two-equation" models that equate A_V to the square of the turbulence energy per unit mass and to the inverse of its rate of dissipation. Turbulence production and dissipation in turn are computed at all nodes using transport equations for the turbulence.

In selected applications the above equation set can be considerably simplified. For example, if only a cross-sectional average description of the flow in a rectangular channel is

Shear stress module

The suspended load is determined by the sediment concentration and velocity distributions over the vertical, both of which depend upon the total turbulence-averaged bed shear stress τ_0, exerted by a flow on its bed and banks. The bed-load on the other hand, should be computed using only the skin friction component of the bed shear stress, τ_0', and should not include the bedform shear stress, τ_0'', that is, that portion of the fluid drag expended exciting roller vortices on the lee sides of dunes. The skin friction component is called the *effective bed shear stress* and various schemes are available to compute it (e.g., Einstein and Barbarossa, 1952; Kazemipour and Apelt, 1983). For example, in the case of equation 5 where the mean properties of the flow are of interest, the effective bed shear stress law may be computed from the quadratic shear stress law as:

$$\tau_0' = \frac{f}{8}\rho V^2 \qquad \text{(Eq. 7)}$$

where f, the Darcy–Weisbach friction factor for turbulent flows, is calculated using the Colebrook–White formula. Alternatively, if the flow is steady and uniform, the total turbulence-averaged bed shear stress, given by:

$$\tau_0 = \rho g R S \qquad \text{(Eq. 8)}$$

where R = hydraulic radius and S = bed slope, may be reduced by subtracting the bedform shear stress, the latter equal to:

$$\tau_0'' = \frac{1}{8}\rho V^2 \frac{s^2}{lh} \qquad \text{(Eq. 9)}$$

where s = dune height and l = dune length.

Although most sediment transport models use the temporal mean fluid shear stress to represent the fluid forces on grains, in reality the local instantaneous bed shear stress fluctuates dramatically due to flow turbulence. Some sediment transport models incorporate this variation (e.g., van Niekerk *et al.*, 1992; Bridge and Bennett, 1992) by computing a Gaussian distribution of instantaneous effective shear stresses.

Bedload transport module

The bed material load consists of all those grains in transport that are directly supplied by, and interchange with, the alluvial bed, whether traveling as bed-load or suspended load. Its opposite is wash load. For a given flow strength and bed material it is assumed that each size fraction will be transported at a fixed rate, the sum of which is called the sediment transport capacity of the flow. Prediction of the bed material load at capacity usually proceeds by computing the bed-load and suspended load independently.

As discussed elsewhere (see entries in this volume on *Sediment Transport*), bed-load refers to all those sliding, rolling, and saltating grains supported at least in part by collisions with other grains or contact with the bed. In water, the grains travel within a few grain diameters of the bed as a low-concentration, dispersed, grain flow.

Grains are considered to be in motion when:

$$\Theta' > \Theta_{ci} \qquad \text{(Eq. 10)}$$

in which Θ' = the dimensionless effective bed shear stress of the local flow and Θ_{ci} = the critical shields parameter for the ith grain size and density in question. Accurate prediction of

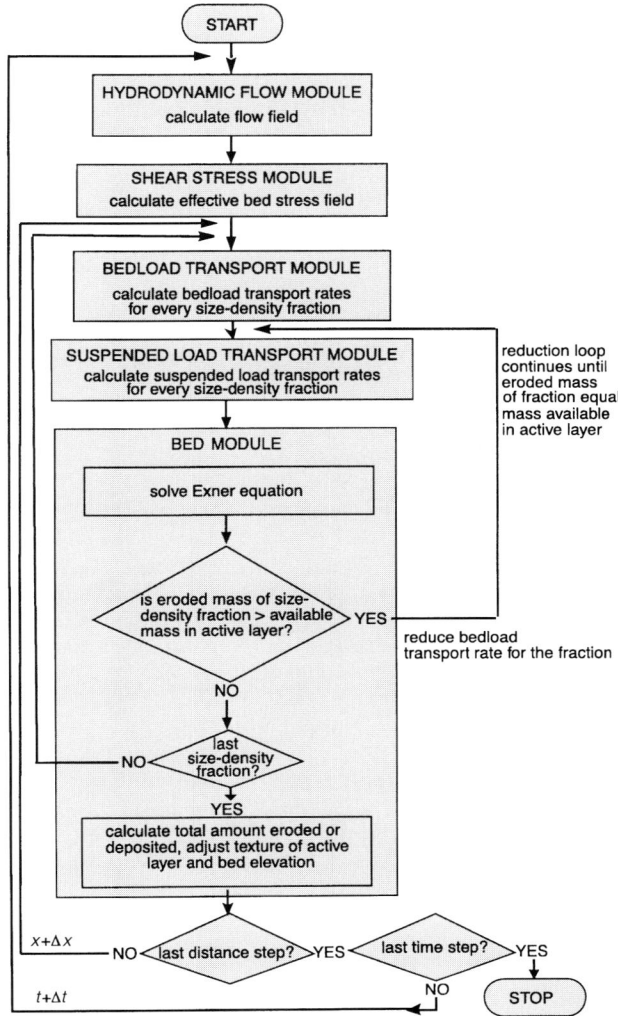

Figure N13 Flow Chart of a typical sediment transport model.

needed, the Coriolis term may be dropped and equations 1–4 integrated to yield:

$$\frac{\partial Q}{\partial t} + \frac{\partial}{\partial x}QV + gA\frac{\partial h}{\partial x} + \frac{fWV^2}{8} + gA\frac{\partial b}{\partial x} = 0 \qquad \text{(Eq. 5)}$$

$$\frac{\partial A}{\partial t} + \frac{\partial Q}{\partial x} - q_l = 0 \qquad \text{(Eq. 6)}$$

where Q = water discharge; V = cross-section average velocity; A = cross-section area; h = water depth; f = friction factor; W = channel width; b = bed elevation; and q_l = lateral inflow per unit length of channel. Equations 5 and 6 can be simplified further by assuming the flow is steady and uniform provided that the ratios V/gST and h/SL are much less than 1 (Paola, 1996), where S is the bed slope, T is a timescale over which the unsteadiness occurs, and L is a length scale over which the channel nonuniformity occurs. It should be noted however, that if the application requires dynamical adjustments in channel width, a 2D formulation is necessary at minimum.

bed-load fluxes depends strongly upon knowing these critical shear stresses.

Research over the last two decades (*cf.* Komar, 1996) shows that:

$$\Theta_{ci} = \Theta_{c50}\left(\frac{D_i}{D_{50}}\right)^{-m} \quad \text{(Eq. 11)}$$

where Θ_{c50} = the critical shields parameter of D_{50}, the median size in the bed size distribution, D_i = the grain size in question, and $m \approx 0.65$ for beds coarser than sand.

Once grains are entrained, they may pass directly into the suspended load. Grains are moving as bed-load when:

$$w \geq Bu_*' \quad \text{(Eq. 12)}$$

where w = the grain fall velocity (see entry in this volume on grain fall velocity), $B = 0.8$, and u_*' = the local effective shear velocity.

The weight or volume transport rate of bed fractions meeting the criteria of equations 11 and 12 may be calculated using one of many formulas (see reviews by Gomez and Church, 1989; Yang and Wan, 1991). Most can be shown to reduce to a function of the form:

$$i_{bi} = aP_i(\Theta' - \Theta_{ci})^b \quad \text{(Eq. 13)}$$

where i_{bi} = bed-load transport rate per unit width of the ith fraction at capacity; a is roughly a constant, P_i = volumetric proportion of the ith fraction in the bed, and $1 \leq b \leq 2$. Note that as the various size fractions are differentially entrained, the bed size distribution evolves, thereby modifying the bed-load transport rates through the coefficient P_i.

Suspended load transport module

The suspended load consists of all those grains borne aloft in the flow by an upward-directed turbulence momentum flux. Operationally, the suspended load consists of all moving grains for which equation 12 is untrue. Here too, the researcher can choose among numerous formula (see entries in this volume on sediment transport). The most basic conception assumes that if the vertical profiles of both sediment concentration $C(z)$, and forward velocity $u(z)$, are known, the volumetric discharge of suspended grains passing through a cross section of unit area at height z is given by $C(z)u(z)$. Integration of this quantity over the depth yields the suspended load transport rate:

$$i_{si} = \int_a^h C_i(z)u(z)\,dz \quad \text{(Eq. 14)}$$

where i_{si} = volumetric suspended load transport rate per unit width of the ith fraction moving in the x-direction, and a = the height off the bed at which a reference concentration is known.

The concentration function is defined by the *Rouse equation*:

$$\frac{C_i(z)}{C_i(a)} = \left[\frac{(a)(h-z)}{(z)(h-a)}\right]^R \quad \text{(Eq. 15)}$$

or more recently, by the van Rijn equation (van Rijn, 1984), where R = the Rouse Number. In either case, a reference concentration $C_i(a)$, is needed for the integration. Some researchers (e.g., van Niekerk *et al.*, 1992) take the reference height as the top of the moving bed layer and the reference concentration as the concentration of the ith fraction in the subjacent bed-load as defined by a function of the form given by equation 13. Others, such as van Rijn, point out that in the presence of bedforms another approach is needed. Van Rijn takes the reference height as one-half the dune height or the equivalent roughness height if bedform dimensions are not known and computes the concentration at that height as a function of excess effective shear stress.

The velocity profile traditionally is defined by the law of the wall for hydraulic rough flow conditions:

$$\frac{u(z)}{u_*} = \frac{1}{\kappa}\ln\left(\frac{z}{z_0}\right) \quad \text{(Eq. 16)}$$

where: u_* = flow shear velocity; κ = von Karman's constant; and $z_0 = 33$ percent of the equivalent roughness height of Nikuradse.

Predictions of suspended load flux by equation 14 are suitably accurate provided the concentrations do not increase to values that dampen turbulence.

Bed module

The core of a sediment transport model is its bed module. The bed is both a source and sink of the various size fractions in transport and it dynamically responds to and modifies the overlying flow field by changing its elevation and roughness. These roles can be described by an equation accounting for the mass fluxes of grains to and from the bed. It is a simple statement of conservation of mass of the bed, often called the *Exner equation*:

$$\frac{\partial}{\partial t}Tb_i = -\frac{1}{(1-p)\gamma'}\left(\frac{\partial}{\partial x}T(i_{bi} + i_{si})\right) + q_b \quad \text{(Eq. 17)}$$

where: T = width of the active bed, usually assumed to be channel width; b_i = bed elevation attributable to the ith size-density fraction; p = bed porosity; γ' = immersed specific gravity of grains; $(i_{bi} + i_{si})$ = the immersed weight transport rates per unit width of the bed-load and suspended loads; q_b = volumetric lateral sediment inflow per unit along-stream distance; and it is assumed for simplicity that the application is 1D. Equation 17 expresses how spatial gradients in transport rates of the various size fractions give rise to temporal changes in bed elevation. If the changes in bed elevation are nonuniform, bed slopes and cross sectional areas evolve, thereby changing the flow field. Note that per unit area, the relative proportions of the b_i represent the proportions of the various size fractions in the static bed, thereby allowing computation of a new bed grain size distribution and hydraulic roughness.

In practice, equation 17 is applied to an upper layer of the alluvium called the *active layer*, in recognition of the fact that over timescales of minutes to hours there is a finite thickness of alluvium exposed to the flow. The active layer may be conceived as a layer of mixing between the traction carpet and the static bed. The thickness of the active layer has been taken variously as the height of dunes present on the bed, the thickness of the armor layer (and hence some multiple of a characteristic grain size of the bed), or a function of excess shear stress. As grains are removed from the active layer on its upper surface, grains are added from below in proportion to their concentrations in the subjacent layer. During deposition, grains pass out of the bottom of the active layer in proportion their concentrations in the layer.

Solution of equation set

The computational modules described above require initial and boundary conditions to form a closed set of equations. In addition, if the application involves channelized flow, either the alongstream channel widths must be specified or an additional function added to relate width to the state variables. Initial conditions include the geometry and bed textures of the geomorphic domain of interest and flow velocities and sediment transport rates everywhere in the domain (both usually taken to be zero for lack of better information). To avoid errors arising from bogus initial conditions, it is common practice to "spin-up" a model before interpreting the results. Boundary conditions include temporally evolving water and sediment hydrographs along upstream open boundaries and (typically) water surface elevations along downstream open boundaries. Lateral inflows of sediment and water along closed boundaries also must be specified.

Because the equation set is analytically unsolvable, the time-space domain is subdivided into a finite number of nodes or elements, and solutions are obtained at discrete points in space and discrete times. The flow chart of MIDAS (van Niekerk et al., 1992), provides a typical example of computational flow for a 1D case (Figure N13). After the user has specified the initial and boundary information, the flow field at time $t + \delta t$ is computed across the whole domain by some combination of equations 1 through 6. The effective shear stresses are calculated next from equations 7 through 9. Starting at the upstream end of the domain, the bed-load transport rates are computed from equations 10 through 13, thereby providing the reference concentrations for the suspended load computation using equations 14 through 16. After the bed material load at node 1 has been computed, equation 17 is solved for changes in bed elevation at that node arising from erosion or deposition of each size fraction. If the mass of any size fraction to be eroded is greater than is available in the active layer, then that fraction's bed-load transport rate (and consequently its suspended load transport rate) are incrementally reduced until mass is conserved. Computation proceeds through the domain, after which the time step is incremented, the flow field is recomputed taking into account the updated water depths and hydraulic roughnesses, and the sediment transports are recomputed in light of the new flow field.

This algorithm is not the only possible computation method. Although this uncoupled approach is the most common, a few models such as CH3D-SED (Gessler et al., 1999) simultaneously solve for all dependent variables in a fully coupled solution. The advantage of a coupled solution is that uncoupled models are restricted to short time steps so that the hydrodynamic solution scheme adjusts to small changes in the bed.

An example

To gain an appreciation of model capabilities, consider a comparison of predictions from the sediment-routing model MIDAS (Table N1) with flume data collected by Little and Mayer (1972). Little and Mayer investigated the effects of sediment gradation on channel armoring. A nonuniform bed of sand and gravel was placed in a flume 12.2 m long, 0.6 m

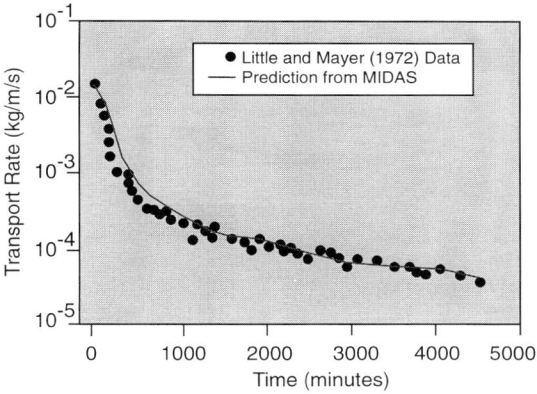

Figure N14 Comparison of observed sediment transport rates from a flume study of bed armoring by Little and Mayer (1972) with predictions from the sediment transport model MIDAS (van Niekerk et al., 1992).

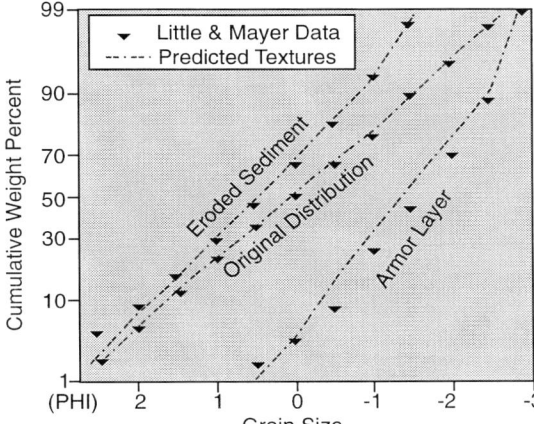

Figure N15 Comparison of observed textures from a flume study of bed armoring by Little and Mayer (1972) with predictions from the sediment transport model MIDAS (van Niekerk et al., 1992).

wide, and 0.1 m high. Clear water was passed over the bed to produce bed degradation and armoring. The eroded sediment was caught by screen separators at the downstream end of the flume, and at regular intervals was dried, weighed, and stored for later sieving. When the total transport rate was 1 percent of the initial transport rate and the armoring was thought complete, the flume was drained and the armor layer was sampled using molten beeswax as described by the authors. For numerical modeling, the flume was divided into eight sections and solutions of the MIDAS equation set were computed for every section every minute.

The computed and the measured total transport rates (Figure N14) show good agreement, the computed values being within a factor of 2 of the measured values at all times. Temporary divergence of the two curves arises because turbulence produces random and intermittent movement of the bed particles. After 75.5 hours of flow, the size distributions of the original sediment, the bed-armor sediment, and the total eroded sediment (Figure N15) show that the numerically simulated armor layer is slightly finer than observed, although the difference in mean grain sizes is only 2.7 mm versus 3 mm. The numerically simulated and physically observed grain-size distributions of the transported sediment almost coincide.

Unresolved problems

Although much progress has been made in sediment transport models, the accuracy of predictions in many applications is still regrettable. Bedload functions are still prone to order of magnitude errors (*cf.* Gomez and Church, 1989; Yang and Wan, 1991), and present formulations need to be tested in extreme events when much of the sediment transport occurs in many geomorphic systems. These errors are compounded when computing sediment flux divergences in equation 17. Also, suspended load formulations break down at hyper-concentrations such as seen in many rivers draining loess provinces and do not yet account for the wash load. There is also a pressing need to treat channel pattern changes and dynamic adjustments in channel hydraulic geometries. Secondary flows are commonly assumed to be negligible, yet in natural channels, and particularly during overbank flows, lateral gradients of flow depth and friction factor induce significant lateral velocity gradients. Although not much as been said about coastal ocean sediment transport models, a better understanding of the basic physics of combined oscillatory and unidirectional flows is needed and the processes of sediment aggregation, flocculation, and disaggregation in marine waters must be better understood.

A final thought

Given the work yet to be done, it is sobering to realize that the most severe test yet of numerical sediment transport modeling is being carried out right now on the Yangtze River in China. Construction of the Three Gorges Dam began in 1994 and is scheduled to take 20 years. It will be the largest hydroelectric dam in the world, stretching nearly a mile across and towering 575 feet above the world's third longest river. Its reservoir will stretch over 350 miles upstream and force the displacement of 1.2 million people. The role that numerical sediment transport modeling has played in the dam's conception, feasibility studies, and design is unprecedented and the predictions are controversial. Physical and in particular, mathematical modeling of sedimentation was conducted by the Yangtze Valley Planning Office (China) and reviewed by the Yangtze Joint Venture (Canadian International Development Agency). Dr Luna Leopold, a respected elder statesman of fluvial engineering in the United States has written, "The sedimentation conditions at various times during the first 100 years of operation have been forecast by use of mathematical models and physical analogues that involve many assumptions of unverified reliability". Let us hope our faith in numerical models of sediment transport and deposition is justified.

Rudy L. Slingerland

Bibliography

Bridge, J.S., and Bennett, S.J., 1992. A model of the entrainment and transport of sediment grains of mixed sizes, shapes, and densities. *Water Resources Research*, **28** (2): 337–363.
Einstein, H.A., and Barbarossa, N.L., 1952. River channel roughness. *Transactions American Society of Civil Engineers*, **117**: 1121–1146.
Gessler, D., Hall, D., Spasojevic, M., Holly, F., Pourtaheri, H., and Raphelt, N., 1999. Application of 3D mobile bed, hydrodynamic model. *Journal of Hydraulic Engineering*, **125**: 737–749.
Gomez, B., and Church, M., 1989. An assessment of bed load sediment transport formulae for gravel bed rivers. *Water Resources Research*, **25**: 1161–1186.
Kazemipour, A.K., and Apelt, C.J., 1983. Effects of irregularity of form on energy losses in open channel flow. *Australian Civil Engineering Transactions*, **CE25**: 294–299.
Komar, P.D., 1996. Entrainment of sediments from deposits of mixed grain sizes and densities. In Carling, P.A., and Dawson, M.R. (eds.), *Advances in Fluvial Dynamics and Stratigraphy*. New York: John Wiley & Sons, pp. 127–181.
Lane, S.N., 1998. Hydraulic modelling in hydrology and geomorphology: a review of high resolution approaches. *Hydrological Processes*, **12**: 1131–1150.
Little, W.C., and Mayer, P.G., 1972. The role of sediment gradation on channel armoring. *Publication No. ERC-0672*, School of Civil Engineering in Cooperation with Environmental Research Center, Georgia Institute of Technology, Atlanta, GA. pp. 1–104.
Nezu, I., and Nakagawa, H., 1993. *Turbulence in Open-channel Flows*. IAHR Monograph, Series A. A. Balkema, Rotterdam. 281pp.
van Niekerk, A., Vogel, K.R., Slingerland, R.L., and Bridge, J.S., 1992. Routing of heterogeneous size-density sediments over a movable stream bed: model development. *Journal of Hydraulic Engineering*, **118** (2): 246–262.
Paola, C., 1996. Incoherent structure: turbulence as a metaphor for stream braiding. In Ashworth, P.J., Bennett, S.J., Best, J.L., and McLelland, S.J. (eds.), *Coherent Flow Structures in Open Channels*. John Wiley & Sons, pp. 705–723.
van Rijn, L.C., 1984. Sediment transport, Part II: Suspended load transport. *Journal of Hydraulic Engineering*, **110** (11): 1613–1641.
Vogel, K.R., van Niekerk, A. Slingerland, R.L., and Bridge, J.S., 1992. Routing of heterogeneous size-density sediments over a movable stream bed: model verification and testing. *Journal of Hydraulic Engineering*, **118** (2): 263–279.
Yang, C.T., and Wan, S., 1991. Comparison of selected bed-material formulas. *Journal of Hydraulic Engineering*, **117**: 973–989.

Cross-references

Diffusion, Turbulent
Rivers and Alluvial Fans
Sediment Fluxes and Rates of Sedimentation
Sediment Transport by Unidirectional Water Flows

O

OCEANIC SEDIMENTS

Introduction

Oceanic sediments are found on the deep ocean floor (generally below 3 km depth) hundreds of kilometers seaward of the continental margins and cover more of the earth's surface (55 percent) than the continents (29 percent). Within this vast realm there is an array of distinctive sediments, *Globigerina* ooze, red clay, hemipelagic muds, manganese nodules that occur nowhere else on earth. These sediments are largely generated within the ocean although detritus from the continents can also be important, especially at the margins of the ocean basins. Ocean drilling reveals that the oldest deep-sea sediments are Jurassic (*ca.* 140 Ma) and that the type and distribution of oceanic sediments has changed with time. Ocean waters filling the ocean basins today are cold and well oxygenated but 50 million years ago were warmer and less well oxygenated. Pelagic clays are accumulating where once calcareous-rich sediments were deposited. As climate and the ocean environment have evolved, the suites of sediments that cover the modern seafloor have changed.

Oceanic sediments and deep-sea sediments are used herein interchangeably although some authors use "deep-sea sediments" as encompassing the sediments of both the deep ocean and the continental margin (shelf, slope, rise). The "deep ocean", "deep ocean basins", "oceanic floor" and "deep seafloor" refer to portions of the ocean basins that generally lie below 3 km, although some texts define the deep ocean as greater than 4 km. The emphasis in this article is on the type, distribution and sedimentation of modern oceanic sediments that occur beyond the continental rise with the exception of abyssal fans (cones). The latter are really features that originate on the slope-rise and extend into deep ocean basins. For readers interested in a more comprehensive discussion of deep-sea sediments, I suggest reviews by Berger (1976), Davies and Gorsline (1976), Barron and Whitman (1982) and texts by Kennett (1982), Lisitzin (1996), the Open University, Volume 5, *Ocean chemistry and Deep-sea sediments* (1989 and later editions) and Seibold and Berger (1996).

Historical development

Marine sediments are mentioned in writings as far back as Greek and Roman times (e.g., Strabo and Posidonius in the first century BC) but it was not until the 19th century with the advancement of microscopes and the introduction of cable and sampling devices that the systematic collection and description of deep ocean sediments really began.

The *Challenger* Expedition (1872–76) revolutionized knowledge about the deep ocean and, among many firsts, made the first systematic sampling of oceanic sediments (see *Sedimentologists*). At the urging of the Royal Society of London, the British government commissioned the corvette, *H.M.S. Challenger*, under the direction of Charles Wyville Thomson, professor of natural history at the University of Edinburgh to determine the "conditions of the deep-sea through all the great ocean basins". It was the first major scientific exploration of the ocean. The introduction of steam winches and steel cable made deep sampling practical and in four years, the expedition carried out nearly 500 deep soundings with a mechanical grab and 133 dredging. Sediment samples were collected in all of the major ocean basins and near Antarctica, even in the Mariana trench at a depth of 8,185 m. Over the next two decades, Sir John Murray, in collaboration with A. F. Renard, (Murray and Renard, 1891), described and classified the samples, introduced many of the terms still in use today and laid the foundation for the study of deep-sea sediments. Murray, the naturalist on board the *Challenger*, also collected samples of shell-bearing plankton and correctly related their distribution in the surface ocean to the calcareous and siliceous sediments on the seafloor (Murray, 1897).

The *Challenger* Expedition generated a major interest in the scientific investigation of the oceans and was followed by an era of national oceanographic expeditions spanning nearly 70 years although interrupted by two world wars. Expeditions were conducted by many nations, including England, the United States, Germany, Japan, Russia, Holland, Sweden, and France and lead to the establishment of the first oceanographic institutions, Woods Hole Oceanographic Institiution (United States) and Plymouth Marine Laboratory (England). While most expeditions were devoted to the biology, chemistry, and physics of the oceans, some important advances occurred in the study of ocean sediments. It was during the German Southern Polar Expedition (1901–03) that the gravity corer was invented and the first sediment cores were recovered.

By the 1930s, electronic depth sounders replaced wire soundings to measure ocean depths and reveal the complex bathymetry of the deep ocean. This technology was first used during the German *Meteor* Expeditions in the Atlantic. In 1935, Schott (1935) in a pioneering study based on gravity cores collected by the *Meteor*, devised a means for correlating between cores, calculating reliable rates of sedimentation and studied the effects of the Ice Ages in the Atlantic Ocean.

One of the last major national expeditions was the Swedish Deep Sea Expedition (1947–49) to the equatorial Pacific Ocean. It employed the newly developed Kullenberg piston corer that could recover sediment cores tens of meters in length and made possible the study of Pleistocene ocean history. In these cores, Arrenhius (1952) discovered distinct Pleistocene cycles in calcium carbonate content which he related to glacial to interglacial variations in productivity and preservation. He proposed that the higher carbonate content was due to increased surface productivity during glacial times, initiating a debate about the processes which control biogenic sedimentation, a debate that continues today (see Berger, 1992).

In 1968, the Deep Sea Drilling Project (DSDP) was launched to test the theory of continental drift by determining the age and evolution of the ocean basins. Prior to ocean drilling, information about deep-sea sediments was limited to the surface layer penetrated by a piston corer, a few tens of meters. The entire inventory of pre-Quaternary deep-sea sediments was fewer than 100 cores. In the past 34 years, DSDP (and its successors, the Ocean Drilling Project (ODP) and International Decade of Ocean Drilling (IDOP)) have collected a vast array of information from all of the ocean basins except the Arctic Ocean, including long cores of sediments extending back to the early Jurassic. These data have significantly expanded our understanding of the ocean basins, their origins, the sediments that fill them, life that existed in the past and the dynamics of the interplay between tectonics, climate and the oceans.

Classification and major types

Sediments of the deep ocean basins are termed *pelagic* (*pelagios = of the sea*) if they originate in the ocean or *hemipelagic* when mixed with terrigenous material derived from the continent. Pelagic sedimentation involves two processes: the packaging of biologically produced debris in the water column and dust that falls into the ocean and its transfer to the seafloor. Pelagic particles "rain" down from the surface. These processes occur everywhere within the ocean.

Sedimentation of the terrigenous component in hemipelagic sediments involves resuspension and turbidite and suspension flows. These processes move mud (silt and clay) that was originally deposited on the continental slope into the deep ocean where it becomes mixed with pelagic debris. While the distinction between pelagic and hemipelagic seems obvious, in practice it can be difficult to determine the origins of sediment particles because the terrigenous component in both types of deposits is derived from the continents and the mineralogy and grain size may be very similar. Determining the relative proportion of particles that are pelagic in hemipelagic sediments can be difficult.

There is no single widely agreed upon classification of oceanic sediments and different authors even use some of the common terms differently. Two broad approaches have been used to classify marine sediments: descriptive, based on grain size and composition or genetic, based on origins, or a combination of the two. Murray and Renard (1891), for example, used a descriptive-genetic approach when they coined the terms "red clay" and "*Globigerina* ooze" for common deep-sea deposits and this approach is followed here in a scheme modified from Berger (1974) (Table O1). A useful discussion on the philosophy of sediment classifications is given in Davies and Gorsline (1976).

There are three kinds of sediments in the deep-sea: *Terrigenous* (or Lithogenous): detrital or clastic particles produced by the weathering, erosion and transport of pre-existing rocks, primarily from the continents and volcanic eruptions; *biogeneous*: biologically produced shells or skeletons of calcium carbonate, siliceous or calcium phosphate, either in the water column or on the seafloor; and, *chemogenic* (or hydrogenous): primary (authigenetic) or diagenetic chemical precipitates from seawater or interstital waters, sometimes mediated by bacteria.

Aerially and volumetrically two types of sediments dominate in the deep ocean: biogenic oozes, especially calcareous oozes and pelagic clay (abyssal "red" clay). Ooze is the loose, unconsolidated accumulation of biogenic debris (shells of foraminifera, diatoms, radiolaria or coccoliths) usually mixed with some clay. In the modern ocean, roughly 50 percent of the seafloor is covered by foraminiferal-rich calcareous ooze, 37 percent by pelagic clay and 12 percent by siliceous-rich sediments (Berger, 1976). All others sediment types equal < 1 percent (Figure O1).

Volcanogenic materials are common in the deep sea but the coarser-grained materials are limited to the source area. Volcanic ash can be transported hundreds of kilometers from its source and is widely distributed in the ocean. Within the sediments it is an important source of silica in the formation of zeolite and clay minerals. There are several rare but distinct deposits that occur in modern and ancient sediments. One of them is cosmic dust (microtektites) particles that survive the trip through the atmosphere and fall into the ocean. They are found in pelagic red clay, mostly in the southern hemisphere. Another is laminated, diatom-rich mud that occurs in Pliocene and Miocene deposits in the eastern equatorial Pacific (Pearce *et al.*, 1996). Typically such deposits are associated with organic carbon-rich hemipelagic sediments beneath upwelling areas in continental margins and not the deep ocean.

Data from the ocean drilling indicate that the relative abundance and distribution of major sediment types has changed with time under the influence of different climate and oceanic environments and that biogenic sediments were even more widely distributed in the early Tertiary and Cretaceous. The record of deep-sea sediments provides one of the most important sources of information about the history of global wind patterns, ocean fertility, ocean geochemistry and biotic evolution.

The distribution of major sediment types in the deep ocean can be related to three factors: (1) distance from the continents; (2) water depth, which influences the preservation of biogenic sediments; and (3) ocean fertility, which controls surface productivity. There are two major sediment boundaries in the deep ocean: (1) The Calcium Carbonate Compensation Depth (calcite compensation depth) that separates carbonate-rich

Table O1 Classification of Oceanic Sediments

1. **Pelagic deposits (oozes and clays)**
   < 25 percent of fraction > 5 μm fraction is terrigenic or volcanigenic
   medium grain-size > 5 μm (except for authigenic minerals and pelagic fossils)
   A. **Pelagic clay**
      1. Calcareous clay: 1–30 percent calcareous fossils (foraminifera, coccoliths)
      2. Siliceous clay: 1–30 percent siliceous fossils (diatoms, radiolaria, sponge spicules)
      3. Zeolitic clay: > 30 percent zeolitic minerals
   B. **Oozes (loose, unconsolidated accumulation of pelagic skeletal debris)**
      1. Calcareous ooze: < 30 percent < 2/3 calcareous fossils (foraminifera, pteropods, coccoliths)
      2. Siliceous ooze: > 30 percent opal silica fossils (diatoms, radiolaria)

II. **Hemipelagic deposits (muds, mostly silt and clay)**
   > 25 percent of fraction > 5 m is terrigenic or volcanogenic
   median grain-size > 5 m (except for authigenic minerals and pelagic fossils)
   A. **Calcareous muds:** > $30CaCO_3$
   B. **Terrigenous muds:** < $30\ CaCO_3$ > 50 percent quartz, feldspar, micas, lithic fragments
   C. **Volcanogenic muds:** < 30 percent $CaCO_3$ > 50 percent ash, palagonite, etc.

III. **Chemogenic deposits (authigenetic and diagenetic precipitates from seawater)**
   A. silicate minerals-smetitic clay, zeolites (phillipsite, palagonite, palygoskite, sepiolite
   B. Ferromanganese minerals-oxyhydroxides of Fe and Mn; amorphous, fine-grained crystals, coating, nodules
   C. Sulfate minerals-barite

(after Berger, 1974, with modifications).

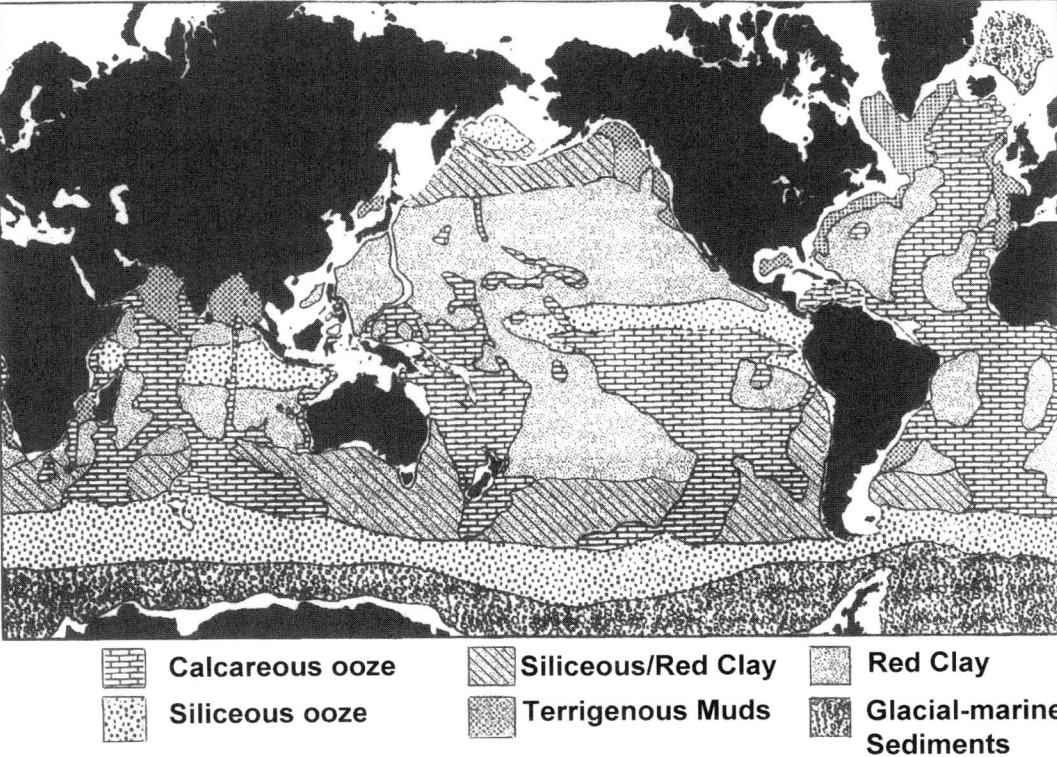

Figure O1 Distribution of the major types of oceanic sediments (after Davies and Gorsline, 1976, with modifications based on many sources).

oozes and carbonate-poor pelagic clay (see *Calcite Compensation Depth*). The depth and distribution of this boundary varies within and between ocean basins; and (2) the transition from hemipelagic to pelagic sediments at the margins of the deep ocean, which is gradual and often difficult to, delineate, exactly.

Sedimentation rates and thickness of oceanic sediments

While John Murray and other 19th century workers suspected that sediments in the deep ocean accumulate slowly, they had no means by which to access rates of sedimentation and it remained for Schott (1935) to provide the first quantitative estimates by using the occurrence of a distinctive tropical foraminifer, *Globorotalia menardii*. During glacial periods, this species is absent in cores in the central Atlantic Ocean and its first appearance coincides with the base of the Holocene. Using the age of the base of the Holocene (determined on land) as a datum, he calculated that the calcareous oozes in the Atlantic had accumulated at rates of several centimeters per thousand years. Today a variety of dating techniques (radiocarbon, Uranium series, etc.) are available to determine the rates of sedimentation. While rates vary considerably, in general, they fall in the range of millimeters to tens of centimeters per thousand years with the lowest rates occurring in abyssal pelagic clays (1 mm/kyear) and highest rates in hemipelagic muds (meters/kyear) (Table O2).

Prior to deep ocean drilling, the ocean basins were believed to be permanent, unchanging features of the earth and therefore to contain a thick sediment record extending back to earliest Earth history. In the 1950s, marine geophysical surveys revealed that the sediment cover in the ocean basin was in fact remarkably thin (Ewing *et al.*, 1964; Raitt, 1956) but it remained for the Deep Sea Drilling Project to confirm the age and thickness of deep-sea sediments. In the central Pacific Ocean the sedimentary layer resting on basaltic crust varies from 100 m to 300 m thick, except along the equator and a few plateaus, where it exceeds 500 m (Winterer, 1989; see Figure O2). Near the continental margins it may exceed one kilometer. The patttern is similar in the deep Atlantic although the average sediment thickness is thicker, closer to 500 m thick.

Terrigenous sediments

Terrigenous sediments are composed of particles that are characterized by grain size (gravel, sand, silt, clay), composition and origin.

Terrigenous material in the ocean comes from three sources, rivers, glaciers in polar latitudes and eolian dust. Between 14 billion tons/year to 19 billion tons/year of particulate matter and another 3.9 billion tons of dissolved substances are delivered annually to the ocean, largely by rivers (Garrels and Mackenzie, 1971; Lisitzin, 1996). Most of this is trapped in estuaries, deltas or offshore basins and except for the muds transferred by turbidity and suspension flows, little of the total travels beyond the continental margin and into the deep ocean. The bulk of terrigenous sediments (nearly 90 percent) are contained in the continental margin (shelf, slope and rise), and the fine-grained hemipelagic muds of the continental slope and rise are the most abundant by volume of all marine sediments (about 70 percent).

Today glaciers in polar regions erode and transport 1.5–2 billion tons/year of sediment to the oceans (Garrels and Mackenzie, 1971; Lisitzin, 1996), and the input must have been greater during periods of glaciation. While much of the material is dumped on the shelf and slope near the ice margin, ice-rafted debris and the fine grained suspended fraction reach the deep seafloor. When icebergs melt in the open sea they dump their load so that ice-rafted lithic fragments in deep-sea sediments trace the drift and extent of icebergs. During glacial times, drop-stones occurred in all high latitude sediments and ice-rafted debris was transported south to about 40°N in the North Atlantic and north to about 40°S around Antarctica. The suspended load of glaciers transports large volumes of sand, silt and clay, including the clay minerals chlorite and illite. Lisitzin (1996) believes that most of this material reaches the pelagic realm and is a more important source of pelagic clay than river-derived suspended sediments.

Dust storms generated in arid regions transport 1.6 billion tons/year of fine-grained materials over the ocean where it settles-out and eventually reaches the deep-seafloor. Most of this is silt and clay and contributes over half of the mineral component in pelagic clay (Windom, 1975).

Three important sedimentary deposits in the deep ocean are primarily terrigenous, the hemipelagic muds of abyssal cones and abyssal plains and the pelagic red clay of the deep basins.

Abyssal cones and plains

Off major river deltas, where the supply rate of terrigenous material is high, large submarine fans (*q.v.*) or abyssal cones form at the base of the continental slope (Bouma *et al.*, 1985). These features may be very large, extending to depths > 4,000 m and covering 1,000s km^2 of the seafloor. The largest of them is the Bengal debris cone of the Ganges and Brahmaputra Rivers, occupying all of the Bay of Bengal and extending south to Sri Lanka. This feature covers more than 2 million km^2 and contains 5 million km^3 of sediments (Curray and Moore, 1971). Similar large abyssal cones are found off the Amazon, Congo and Mississippi Rivers. Abyssal cones are formed by turbidite flows, fan-shaped in plan view and have channels, levees and interchannel areas typical of all submarine fans (Normark and Piper, 1991). At their distal end, cones merge gradually into abyssal deep-sea plains.

Abyssal plains are large features (1,000s km^3) formed by the accumulation of distal turbidites and suspension flows in the deep, peripheral parts of the ocean basins between the margins and the oceanic ridges. Sediment trap data show the they are the result of hemipelagic processes involving fine-grained muds (silt and clay), winnowed off the shelf and slope and transported offshore at depths of 2–3 km as a plume (nepheloid flow: see *Nepheloid Layer*) that extends 100s of km seaward. Abyssal plains are the flattest surfaces on earth, with less than a few meters of relief over thousands of kilometers. While they are present in all the basins, they are most common in the Atlantic Ocean. In the Pacific Ocean, abyssal plains are relatively rare because trenches intercept the flow of sediments from the continents. An interesting case is the Alaskan abyssal plain found in the Gulf of Alaska. It is essentially a "fossil" feature that began in the Eocene off the

Table O2 Typical Rates of sedimentation in oceanic sediments

Sediment Type	Rate (meters/million years = mm/thousand years)
Hemipelagic muds	10s–100s
Abyssal plains and cones	10s–100s
Ice-rafted material	> 10
Pelagic red clay	1–5
Chemogenic sediments	< 0.2
Manganese nodules	0.2–5 mm/myr
Calcareous ooze	3–50
Siliceous ooze	2–10

(based on many sources).

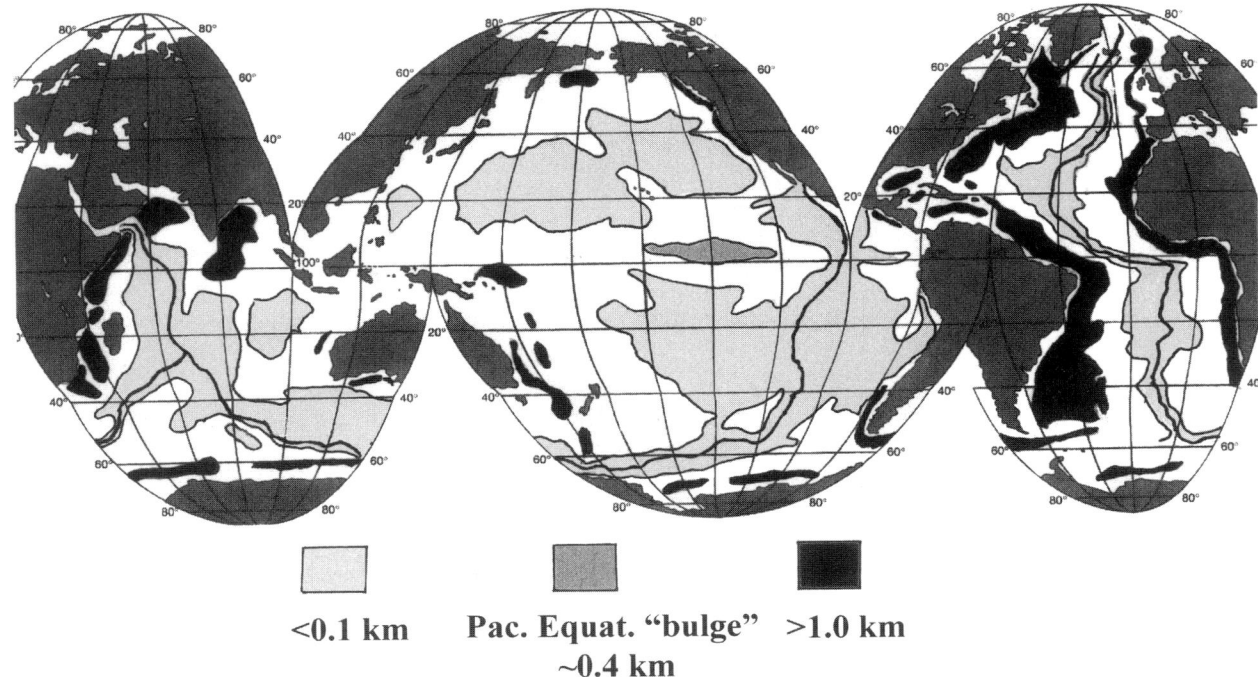

Figure O2 Thickness of the sedimentary cover overlying oceanic basalt. Note that except for the continental margin and the narrow belt along the equator, the sediment cover is less than 200 m thick on crust upto 60 million years old (after Berger, 1974 with modifications from several sources, including Heezen and Hollister, 1971; Winterer, 1989 and results of the Ocean Drilling Project).

coast of Washington but with the northward motion of the Pacific plate, is now isolated from a major sediment source.

Pelagic red clay

Pelagic clay, also commonly known as deep-sea clay or abyssal red clay, is a fine-grained lithogenous deposit, the bulk of which is clay and silt with minor amounts of sand that is unique to the deep ocean (generally below 4,000 m). The deposit is widespread, especially in the Pacific, South Atlantic and southern Indian Oceans (Figure O1). Murray and Renard (1891) named it red clay and concluded that it is the *in situ* alteration of volcanic ash or clayey material from land. The actual color is dark reddish brown or chocolate caused by amorphous or poorly crystalline iron-oxide coatings and minerals and is indicative of an oxidizing environment.

Pelagic red clay is limited to areas of the seafloor shielded from more rapidly accumulating turbidites and biogenic sediments. In the North Pacific, for example, deep trenches along the western and northern sides of the basin and oceanic ridges (Juan de Fuca and Gorda) along the eastern side prevent terrigenous materials from the continents from reaching the central region. Also, because of the great depth, generally below 4,000 m, calcareous biogenic materials are dissolved and never reach the bottom except for insoluble debris like shark teeth. The situation is similar in the South Pacific. In the Atlantic, where terrigenous input is higher and trenches are rare, areas of pelagic red clay are more limited and lie at the distal margins of abyssal plains.

One of the basic features of pelagic red clay is that it accumulates very slowly. Typical sedimentation rates are < 1–2 mm/1,000 years, which explains why the thickness of post-Cretaceous sediments in the central North Pacific is less than 200 m (Winterer, 1989). Leinen (1989) and others have shown that the rates of accumulation have varied with time, increasing significantly in the past 20 million years and since the Quaternary, apparently in response to increase atmospheric transport of dust (Rea, 1994).

While the bulk of "red clay" is very fine-grained clay and silt, it also contains coarse silt and sand size particles of authigenetic chemogenic minerals, volcanic debris and fish teeth. In high latitudes, it may contain ice-rafted debris and dropstones. The chemogenic minerals are mostly zeolites (phillipsite), Fe-Mn oxides and amorphous hydroxides (Griffin *et al.*, 1968; Piper and Heath, 1989; Leinen and Pisias, 1984). Manganese and iron oxides occur as grain coatings, nodules and encrustations. Much of the sediment (up to 50 percent) is X-ray amorphous which chemical analysis indicates is rich in amorphous iron and manganese oxyhydroxides (Piper and Heath, 1989). Typically pelagic clay deposits are 75 percent to 90 percent clay minerals less than 3 microns in size but it was not until the introduction of x-ray diffraction techniques in the 1930s that it was possible to determine their composition. In the last three decades much has been learned about the source of the clays in the ocean, where they originate and how they are transported.

Oceanic clays contain many minerals but are dominated by four: smectite (montmorillonite), illite, chlorite and kaolinite (see articles on each of these clay groups). Illite, the most pervasive clay in the ocean, is a general term for clays belonging to the mica group. It is largely land derived and transported as suspended particulates by rivers and glaciers and as dust (Kennett, 1982). Smectite is also widespread, especially in the Pacific and mostly originates in the ocean from the low temperature alteration of volcanic ash and hydrothermal activity although it is also formed by weathering on land and transported to the oceans. Kaolinite, a product of intense weathering in humid, tropical climates, and chlorite, a mineral from low-grade metamorphic rocks common in the glaciated shield areas of North America, also are land-derived.

The distribution and relative abundance of the clay minerals in pelagic clay deposits indicate that except for the South Pacific and parts of the Indian Ocean, the primary source of the clays in the deep ocean is the continents. Illite dominates in the central North Pacific, the North Atlantic and is abundant off major rivers in the South Atlantic. Koalinite is concentrated near the continents in tropical areas in the Altantic and Indian ocean and abundant chlorite is limited to the margin near Antarctica and in the Alaskan Bight (Windom, 1976; Berger, 1974; Lisitzin, 1996). Only the smectite clays of the South Pacific and Indian Ocean that reflect volcanism and hydrothermal activity are mainly of oceanic origin. Murray's conclusions of a century ago regarding the origin of abyssal red clay are basically correct.

While rivers are important suppliers of fine-grained materials to the ocean, it is now recognized that most of the clay and fine silt fraction in pelagic clay is delivered by wind transport (Rex and Goldberg, 1968; Windom, 1976; Duce *et al.*, 1991; Rea, 1994) (Figure O3). Major sources of eolian (windblown) sediments are seasonal dust storms in arid and semiarid regions that are moved by zonal winds (Trades, Westerlies) over the ocean. Rea (1994) notes that essentially all of the dust transport in the North Pacific and North Atlantic occurs in the spring, rather than winter when the trade winds and westerlies are much stronger and reflects the frequency of major storms. Dust from eastern Asia dominates the entire Pacific north of the Intertropical Convergence Zone (ITCZ). In the subtropical Atlantic north of the equator dust is dominately from the sub-Saharan Africa. Accumulation rates of eolian sediments in the deep ocean match the pattern of global flux of dust into the ocean, with the highest rates (up to $1,000\,mg/cm^2/ka$) occurring in the northwestern Pacific downwind of east central Asia, in the subtropical North Atlantic off the Sahara and Sahel and in the Arabian gulf southeast of the Arabian peninsula (Rea, 1994). In the southern hemisphere, moderate amounts enter the ocean from the deserts of the Horn of Africa and southwest of Australia.

The composition of windblown dust is related to the source area and in general is similar to the mineralogy of red clay. Comparison of the mineral aerosol collected at sea demonstrates that the clays, quartz and feldspar are virtually identical with those in surface sediments (Leinen, 1989; Rea, 1994). The silt-size quartz common in pelagic clay has long been recognized as eolian in origin (Murray and Renard, 1891) and used as an indicator of the wind-blown dust contribution in deep sea sediments. In the North Pacific, a broad band of high quartz content between $20°N$ and $40°N$, which closely matches the distribution of illite, has been used to infer the extent of quartz-rich Asian dust by the westerlies (Rex and Goldberg, 1968; Moore and Heath, 1978: see Figure O3). Rare earth geochemistry and Sr and Nd isotopes can now be used to pinpoint the provenance of the eolian materials in pelagic clays.

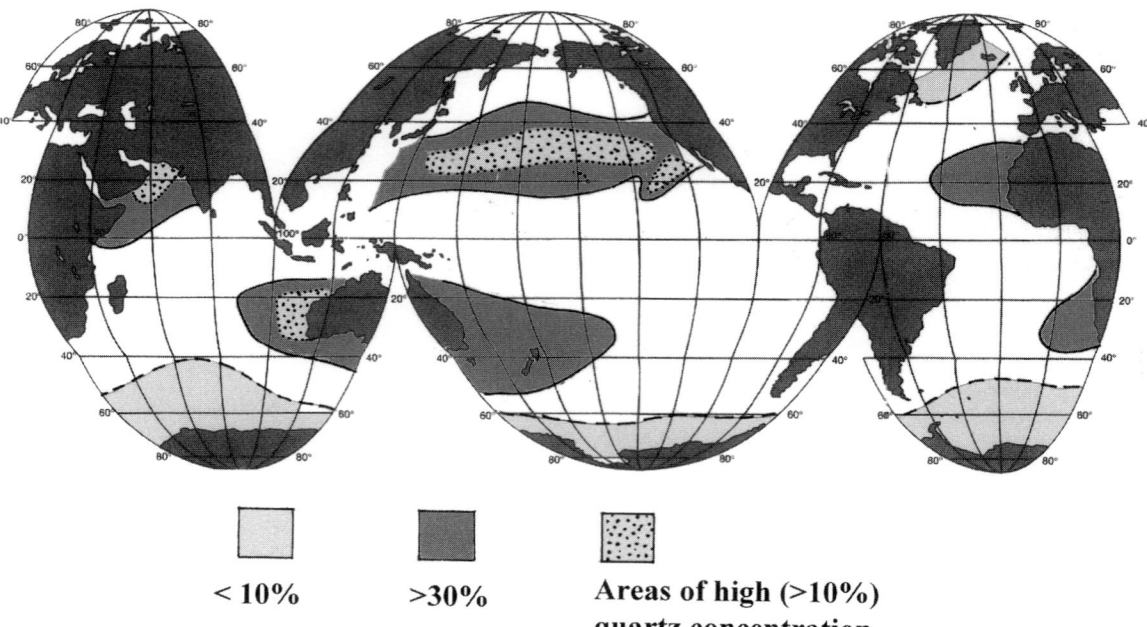

Figure O3 Contribution of eolian (windblown dust) debris to terrigenous sediments in the deep ocean. Areas of high, silt-size quartz content indicated in the central Pacific, NW Indian Ocean and SW of Australia (after Lisitzin, 1996 with modifications based on Moore and Heath, 1978; Rea, 1994).

Table O3 Common Minerals in Deep Ocean Sediments

Biogenous	Terrigenous	Chemogenous	Other
Aragonite	Quartz	Ferromanganese	cosmic spherules
Calcite	Feldspar	Minerals	
Opal	Clay minerals	Barite	
Apatite	Mica	Phillipsite	
Organic matter	Volcanic glass	Clinoptiolite	
		Smectite	
		Palygorskite	
		Sepiolite	

(modified afer Goldberg, 1963).

As noted by Rea (1994) the eolian component of pelagic clay can be confused with hemipelagic sediments because both are similar in composition and grain size and it is unclear how far offshore the influence of hemipelagic processes extend. He restricts studies of the eolian component of pelagic sediments to deposits more than 1,000 km offshore.

Chemogenic sediments

In addition to the materials introduced into the ocean, a number of minerals are precipitated from seawater, either as primary authigenic minerals or during burial and diagenesis, as diagenetic alterations of precursor minerals. They are referred to as chemogenic, authigenic or hydrogenous sediments.

Common minerals formed in the pelagic realm include metal-rich deposits, oxides and oxyhydroxides of iron and manganese, silicates, mostly zeolites and clay minerals and barite (Table O3). Normally chemogenic minerals are a minor fraction of the total sediment and masked by the terrigenic or biogenic component. The exceptions are the metal-rich sediments formed by hydrothermal activity on the flanks and at the crest of oceanic ridges and in the slowly accumulating pelagic clays of the abyssal seafloor. An excellent review of seafloor mineralization associated with mid-ocean ridges is given in Humphris *et al.* (1995).

Within pelagic clays, the chemogenic component can represent a significant or even major fraction of the bulk sediment. Zeolites, hydrous aluminosilicate minerals, especially the mineral Phillipsite, can account for up to 50 percent of the bulk sediment in pelagic clay. It is believed to be an alteration product of basaltic glass or possibly from the reaction of biogenic silica and aluminum dissolved in seawater (Piper and Heath, 1989). Its decreasing abundance below the surface suggests that the mineral is metastable. Smectite, which is widespread in the ocean, commonly occurs with zeolites in surface sediments and the co-occurence suggest it is also an alteration product of volcanic glass.

The best-known chemogenic deposits in the deep ocean are iron-manganese minerals that occur as coating, micronodules, macronodules and, when the nodules join, pavement (see *Iron–Manganse Nodules*). Manganese nodules are found in all ocean basins but are most extensive in the Pacific Ocean, covering vast areas (Cronan, 1977: see Figure O4). Similar but much smaller deposits are present in the Atlantic and Indian Oceans because the higher accumulation rates of terrigenous sediments tend to bury the nodules. Manganese nodules have been extensively studied because of the economic potential of the trace metals (Ni, Co, Cu, and Ti) contained within the principal nodule minerals, birnessite and todorokite.

Nodules usually begin as coatings around some object like a shark tooth or rock fragment, and grow outward as thin layers of manganese oxide precipitated from seawater and from within the sediments. The coatings vary from a few millimeters to tens of centimeters thick and the nodules can be from millimeters to tens of centimeters in diameters. Their formation is a complex chemical interaction between seawater, sediments, pore waters and decomposition of the tiny amount of organic carbon that survives from biogenic rain. The composition of nodules varies, apparently in relation to distance from geochemical processes at the oceanic ridges and nodules in the eastern-most Pacific are rich in Ni, Cu, and Co poor while those in the western Pacific are rich in Co and Pb and poor in Ni and Cu (Cronan, 1977). Nodules grow very slowly, at the rates of milimeters in thousands of years.

Biogeneous sediments

Biogenic sediments in the deep ocean are the accumulations of organic debris, shells and skeletal parts produced mostly by pelagic organisms, single celled phytoplankton (diatoms, silicoflagellates and coccolithophores) and zooplankton (foraminifera, radiolaria) (Table O4). The main types in the deep ocean are calcareous and siliceous oozes (Table O1) produced by plankton while biogenous sediments in shallow-water (shelf, banks, islands) are produced by benthic organisms (algae and molluscs, echinoderms and corals). Following Murray, who introduced the term, oozes are named for their dominant component, such as foraminiferal ooze or *Globigerina* ooze (a foraminifer), diatom ooze or various combinations, e.g., foraminferal–radiolarian ooze.

Biogenic deposits are the most widespread and abundant deposits in the modern deep ocean (Berger, 1976) and ocean drilling results indicate they were even more so in the geologic past. The type and distribution of biogenous materials reflects ocean–climate interactions that control primary productivity at the surface and ocean geochemistry at depth. Unlike terrigenous sediments, whose particles are the product of weathering, erosion and transport, biogenic particles are grown "in place" and generally come to the seafloor as a pelagic "rain". As a result, the distribution of biogenic sediments tends to closely resemble the biotic distributions of the organisms that produced them (Murray and Renard, 1891).

Processes that govern the delivery and accumulation of biogenic sediments in the deep ocean are understood in general but not in detail. They include primary productivity and recycling in the euphotic zone, export of material from this zone and its regeneration in the water column and seafloor. The net accumulation on the seafloor is a balance between production, the supply of organic matter and shells and preservation, the loss from chemical (dissolution) and biological destruction in the water column and seafloor. Influencing the net outcome is the rate, type and quantity of material produced, its transport to the seafloor, temperature, pressure and ocean geochemistry at depth and the rate of burial in the sediments.

Rates of primary production are controlled by nutrient supply and mixing in the ocean and regions of highest productivity are divergence zones along the equator and upwelling at the ocean's margins. Recent estimates of primary

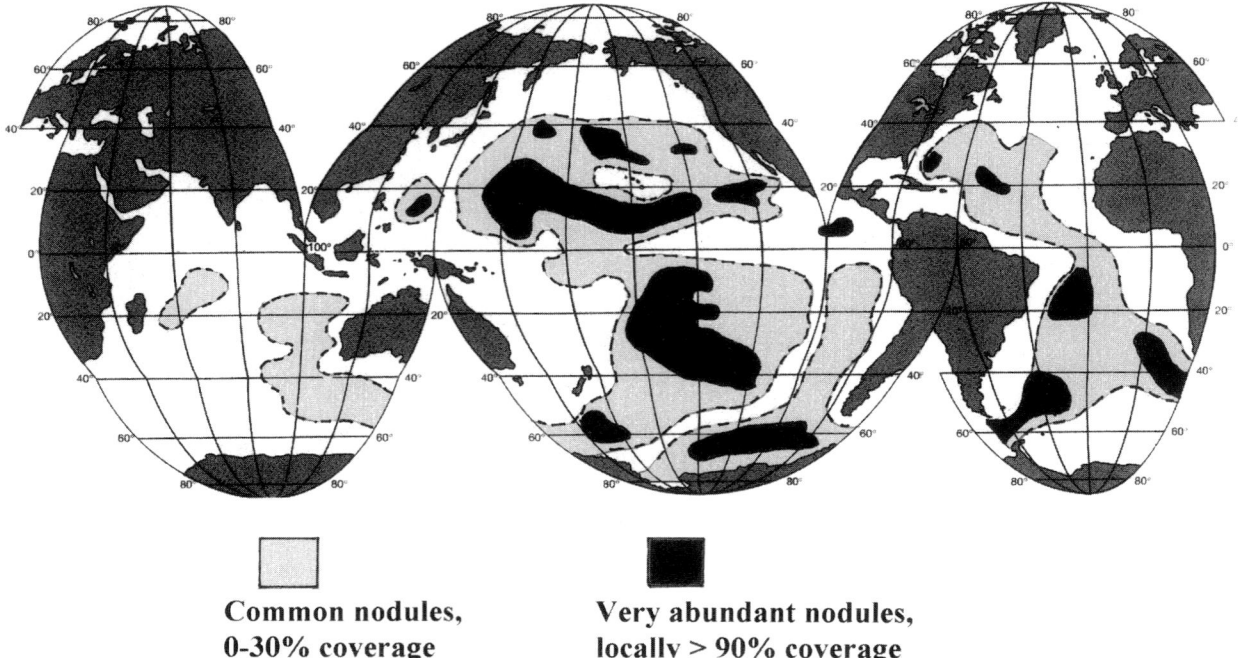

Figure O4 Distribution of Manganese nodules (after Berger, 1976, based on data from Cronan, 1977, with modifications from Piper and Heath, 1989, Lisitzin, 1996 and others.).

Table O4 Common minerals, type of plankton and potential preservation of common biogenic sediments

Type	Plankton	Shell Mineral	Preservation Potential
Calcareous oozes	Foraminifera	CaCO$_3$	good/fair
	Coccololithophores	Calcite (low Mg)	very good
	Pteropods	Calcite (Very low Mg) Aragonite	poor
Siliceous oozes	Diatoms	Opal (hydrated)	moderate/poor
	Radiolaria	Opal (hydrated)	good/excellent
	Sillicoflagellates	Opal	fair

productivity make use of satellite observations and global databases but still have difficulty in assessing the contribution of cyanobacteria (picoplankton), the largest group of phytosynthesizers in the ocean, and nannoplankton (Azam, 1998; Falkowski et al., 1998). Useful "best-guess" maps of global productivity are given in Berger et al. (1989).

Most of the organic matter and skeletal debris produced in the euphotic zone is consumed and a fraction is exported as particulate organic matter (POM). The particles range in size from less than one to several hundreds of microns in size. The smaller material is bacteria, algal cells and fine organic detritus while the larger particles are clumps of organic matter, diatoms, plankton shells and inorganic particles such as clay. Transport to the bottom occurs in a complex process of packaging and repackaging of particulate matter into fecal pellets and clumps or aggregates of detritus that because of greater size and density than individual shells provide a fast track to the bottom (Honjo et al., 1995). The larger pellets and aggregates fall as a pelagic "rain" called "marine snow".

The preservation of biogenic materials follows two pathways: organic carbon-rich debris is nearly all recycled (remineralized) by microbial consumption and only a tiny trace (<0.5 percent) reaches the deep ocean floor (Gardner, 1997). Mineralized skeletal debris is dissolved in the water column or, once buried, in the sediments. Surface waters are saturated with respect to carbonate [CO_3^{2-}] but undersaturated at depth. Silica is the inverse; surface waters are greatly undersaturated while deep waters are more silica-rich. The ocean is generally undersaturated for silica at all depths and especially in the euphotic zone because of high biological demand (Figure O5). An excellent discussion of the geochemical and physical factors which control the preservation of calcium carbonate and opal silica in the ocean is given in Broecker and Peng (1982).

Calcareous oozes

Calcareous oozes (>30 percent calcareous materials) are the remains of pteropods, foraminifera, and coccolithophorids. Pteropods are pelagic gastropods common in warm water latitudes that grow small (200–1,000 microns) aragonitic shells. As ocean waters become undersaturated for carbonate [CO_3^{2-}], aragonite is unstable and dissolves in deep water. Pteropod-rich deposits are limited to depths of less than 2500 m in the Atlantic and 1,500 m in the Pacific Ocean. Foraminifera, rhizopodean protists, live both on the bottom

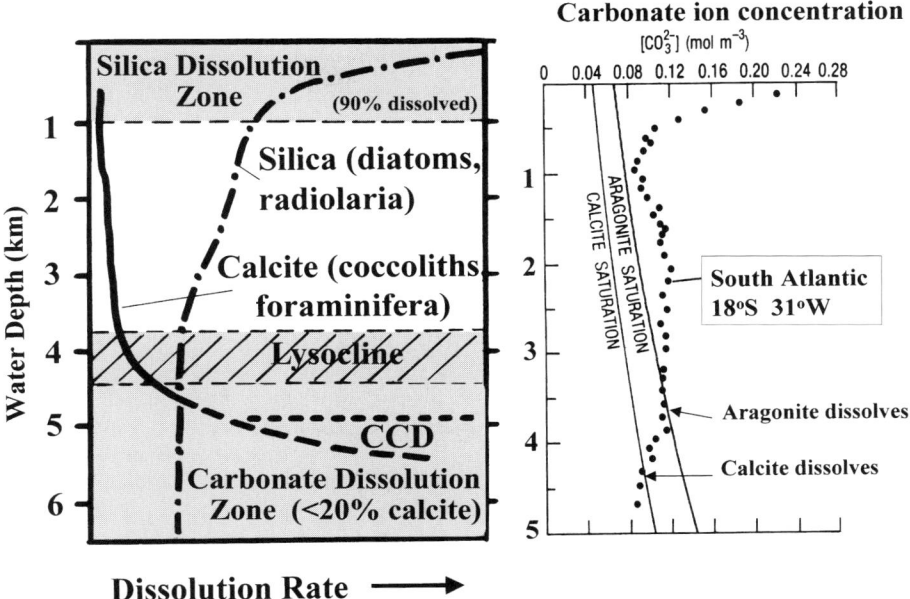

Figure O5 Generalized dissolution profiles of calcite and silica. The depth of the Lysocline, taken as the level of rapid increase in calcite dissolution and Calcium Carbonate Compensation Depth (CD), where the rate of supply equals the rate of dissolution, varies within and between ocean basins. Typical carbonate ion profile for the Atlantic Ocean with the saturation profiles for calcite and aragonite indicated (after many sources, including, Berger, 1974, Broecker and Peng, 1982, Lisitzin, 1996).

and in the plankton and secrete tiny (50–500 microns) shells of $CaCO_3$ that incorporate minor amounts of Mg and Sr. Modern species show distinct latitudinal patterns related to water temperature. *Globigerina* ooze is one of the most common deposits in the ocean. Coccoliths are very tiny (1–50 micron) calcareous plates secreted by coccolithophodidae, a member of the gloden-brown algae. Because of their size, they are referred to as nannoplankton. Coccoliths are single calcite crystals and more resistant to dissolution than the shells of foraminifera or pteropods.

Murray (1897) recognized that the distribution of calcareous sediments in the deep ocean changes with water depth and reasoned that it was related to ocean geochemistry. Calcium carbonate dissolution increases with depth (Peterson, 1966) as a result of decreasing temperature, increasing pressure and, most importantly, increasing undersaturation of calcium carbonate in seawater (Broecker and Peng, 1982). Carbonate profiles reveal that there are two critical depths, the first is a level distinct increase in dissolution, termed the *Lysocline* by Berger (1974) and below it, a point where the rate of supply of calcareous debris from the surface is equaled by the rate of carbonate dissolution. This level defines the Calcium Carbonate Compensation Depth (calcite compensation depth) and below this depth little or no carbonate accumulates (Figure O5). In practice, this depth is defined by the relative abundance and preservation of calcareous fossils (foraminifera, coccoliths) in the sediments and there is some calcite present (typically < 10 percent) even in deep pelagic clays. The calcite compensation depth describes a surface that lies between about 3 km and 5 km but differs in each ocean basin, being deepest in the Atlantic and Indian Oceans (> 5 km) and shallowest in the North Pacific and along the margins of the oceans. It indicates that the deep Atlantic is nearly saturated for calcite to 4.5 km depth while the North Pacific is undersaturated below about 1 km. The calcite compensation depth marks a major boundary between pelagic red clay and carbonate sediments and divides the ocean into carbonate-rich and carbonate-poor regions.

Siliceous oozes

Siliceous oozes (> 30 percent biogenic siliceous materials) are primarily the accumulation of diatoms and radiolarians with minor amounts of silicoflagellates. In some high latitude abyssal environments and near speading ridges, siliceous sponge spicules are important. Diatoms, a major part of the phytoplankton, secrete a variety of different types of shells (frustules), commonly pennate- or centric-shaped (5–50 microns) composed of opaline silica. Under ideal conditions, diatom concentrations exceed millions of cells/m^3 but most of the shell material is redissolved in the water column. Radiolaria are a large, complex group of protozoans that inhabit all depths from the surface to bathypelagic depths. Most secrete shells of hydrated silica; one modern group (acantharians) secretes shells of strontium sulfate that readily dissolves after the death of the organism. The siliceous shells of radiolarians are 50–500 microns in size and many are resistant to dissolution. Silicoflagellates, a minor group of phytoplankton, are golden-brown algae that secrete an internal skeleton of silica. Tiny in size (1–20 microns) they are commonly found in diatom and radiolarian oozes.

Silica secreting phytoplankton controls the silica budget in the ocean. Diatoms alone account for about 35 percent of the primary productivity in oligotropic regions of the ocean and

75 percent in nutrient-rich coastal and high latitude waters (Nelson *et al.*, 1995; Ragueneau *et al.*, 2000). Maximum production occurs in two major latitudinal belts, along the equator in the Pacific and in high latitudes in the Southern Ocean surrounding Antarctica and North Pacific, and in coastal upwelling regions (Figure O6). These account for about 12 percent of the surface of the ocean (Calvert, 1983; Lisitzin, 1996). In the central ocean gyres, silica production is low. Nelson *et al.* (1995) estimate that biogenic silica production in the ocean is 200 to 280×10^{12} mol Si/yr, about 30 percent lower than previous estimates (Calvert, 1983). Most of it (about 60 percent) is redissolved in the mixed layer (upper 100 meter) and continues to dissolve, more slowly, in transit to the bottom and after deposition. It is generally agreed that less than 4 percent of the silica incorporated in opaline shells in the photic zone is buried. A comparison of surface production and where silica accumulates on the seafloor reveals a strong bimodal pattern. About 15 percent to 25 percent of the silica produced in the waters over diatomaceous-rich sediments is preserved while very little of the silica produced elsewhere in the ocean is preserved on the seafloor (Nelson *et al.*, 1995). Siliceous-rich sediments in the deep ocean are confined to narrow belts directly beneath oceanic divergence zones associated with high diatom production (Figure O6).

Siliceous shells or skeletal parts are all composed of opal, a hydrated form of amorphous silicon dioxide. It is easily dissolved in the water column and when buried in the sediments undergoes diagenetic alteration to more stable forms of silica (Kastner, 1981; McManus *et al.*, 1995; Dixit *et al.*, 2001).

Models of biogenic sedimentation

Biogenic sediments exhibit considerable variation in space and in time. Surface sediments in the modern Atlantic Ocean are predominately calcareous, even at great depths while siliceous oozes are virtually absent in the North Atlantic. In contrast, calcareous sediments in the Pacific Ocean are limited to shallow ($<3,500$ m) oceanic ridges and plateaus and along the equator. Siliceous rich sediments, however, are widespread in the North Pacific, in a broad belt adjacent to Antactica and along the equator. The Indian Ocean contains both carbonate- and siliceous-rich sediments. What accounts for these major differences within and between ocean basins? The short answer is a combination of ocean fertility and deep ocean circulation.

Berger (1974, 1976) proposed a "fertility-depth" model for eastern central Pacific sedimentation that links the concentration of nutrients in the euphotic zone to two different biotic responses, which in turn determines the composition and fate of biogenic particles fluxed to the seafloor. These conditions can be simplified as "nutrient poor" and "nutrient rich" and correspond closely to the "carbonate ocean" and "silica ocean" modes of Honjo (1997).

The carbonate ocean mode occurs when sinking particles deliver more carbonate than silica (low Si/Ca ratio) to the seafloor and more inorganic carbon. It prevails when nutrient supply is low and surface waters become nutrient poor, especially in dissolved silica, such as in oceanic gyres and convergence zones. Low nutrient supply favors coccolithophorids and cynobacteria as the primary producers. Because of their small size (pico- and nannoplankton), first level consumers are small zooplankton (planktonic foraminifera) and long food chains develop. Export production is rich in

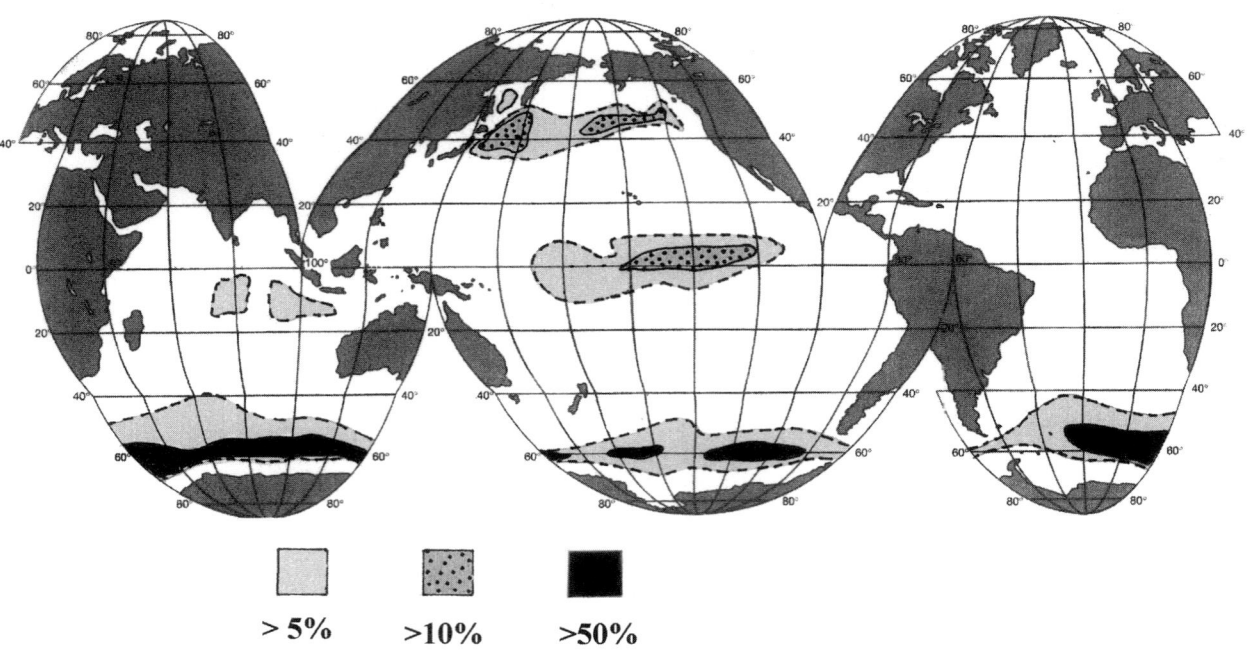

Figure O6 Distribution of Opal siliceous-rich sediments (after many sources, including Calvert, 1983, Kennett, 1982, Lisitzin, 1996 and Nelson *et al.*, 1995).

foraminifera and coccoliths and calcareous sediments accumulate on the seafloor (above the calcite compensation depth).

The silica ocean prevails in oceanic divergence zones and upwelling areas, where high nutrient concentrations in surface waters favor diatoms as the primary producers. Diatoms metabolize nutrients faster than other phytoplankton and reproduce rapidly to create blooms with cell densities of 10^7 per m^3. Short food chains develop because large diatoms can be eaten by higher consumers (large zooplankton and fish). Export output is high in organic carbon. Sinking particles are silica-rich (diatoms, silicoflagellates) and even though a high proportion of the silica is regeneration in the water column, the high flux rates lead to siliceous-rich sediments. As Berger notes, under these conditions, the bacterial decomposition of organic carbon leads to carbon dioxide, carbonic acid and dissolution of carbonate shells. Despite high production rates, dissolution rapidly removes carbonate from the sediments.

The second factor is differences in deep ocean circulation that controls ocean geochemistry, and therefore dissolution patterns and the general pattern of ocean fertility (nutrient distribution). Berger (1970, 1974) referred to this as Basin-Basin fractionation where deep ocean circulation leads to the fractionation of silica and carbonate between ocean basins. In ocean basins that exchange deep water outflow for surface water inflow (lagoonal-type circulation), like the North Atlantic, bottom waters are young, nutrient poor, well oxygenated and tend to be saturated for calcium carbonate. Such basins accumulate calcareous sediments but dissolve silica. The contrasting system is an ocean basin that exchange surface waters for deep water inflow (estuarine-type circulation), like the North Pacific. Here, deep waters pass through the south Atlantic and Indian Oceans before entering the Pacific and are old, poorly oxygenated but nutrient rich. In the long passage, the microbial breakdown of organic matter uses dissolved oxygen, produces carbon dioxide and regenerates nutrients. These waters become under saturated for carbonate but enriched in nutrients and dissolved silica. As these deep waters are upwelled to the surface they generate high surface productivity and diatom-rich debris. Sediments in this type of system accumulate siliceous-rich sediments with little calcium carbonate.

Today the Pacific is exporting carbonate and depositing silica while the North Atlantic is the reverse. However, it follows from the models that a change in climate or the connections between ocean basins that effect vertical circulation will bring about changes in ocean fertility and chemistry. The result will be changes in biogenic sedimentation like those observed in the Pleistocene and Tertiary record.

Summary

As John Murray observed over a century ago, oozes and clays, with the accumulation of *manganese* nodules and zeolites where sedimentation rates are very low, cover the deep ocean floor. At the margins of the basins, close to the continents, hemipelagic sediments encroach on the deep ocean and mix with pelagic debris. Because hemipelagic sedimentation rates are 10–1,000 times higher than pelagic sedimentation rates, hemipelagics usually mask the contribution from the overlying water. Beyond the reach of turbidite and suspension flows, the terrigenous component of pelagic clays comes from windblown dust, traveling thousands of kilometers across the ocean before finally settling out. This material reaches the seafloor as part of the pelagic rain as it is ingested by grazers along with other food particles and transported by fecal pellets and phytoplankton aggregates. The most widespread deposits in the deep ocean are biogenic in origin. Foraminiferal ooze covers nearly 50 percent of the seafloor at depths above the calcite compensation depth while radiolarian-foraminiferal-diatom rich sediments dominate beneath the equator in the Pacific and Indian Oceans. In high latitude regions of the North Pacific and in a belt surrounding Antarctica, diatom oozes cover the seafloor. The abundance and distribution of these biogenic sediments indicate much about the fertility of the oceans and the geochemistry of deep waters as it circulates between the major ocean basins.

Robert G. Douglas

Bibliography

Arrenhius, G., 1952. Sediment cores from the East Pacific. In *Swedish Deep-Sea Expedition Reports*, no. 5, pp. 1–227.

Azam, F., 1998. Microbial control of oceanic carbon flux: the plot thickens. *Science*, **280**: 694–696.

Barron, E., and Whitman, J., 1982. Oceanic sediments in time and space. In Emiliani, C. (ed.), *The Oceanic Lithosphere, The Sea*. John Wiley and Sons, pp. 689–731.

Berger, W., 1970. Biogenous deep-sea sediments: fractionation by deep-sea circulation. *Geological Society of America Bulletin*, **81**: 1385–1402.

Berger, W., 1974. Deep-sea sedimentation. In Burk, C.A., and Drake, C.D. (eds.), *The Geology of Continental Margins*. Springer-Verlag, pp. 213–241.

Berger, W.H., 1976. Biogenous deep-sea sediments: production, preservation and interpretation. In Riley, J., and Chester, K. (eds.), *Chemical Oceanography*. Volume 5, Academic Press, pp. 265–387.

Berger, W., 1992. Pacific carbonate cycles revisted: arguments for and against productivity control. In Ishizaki, K., and Saito, T. (eds.), *Centenary Japanese Micropaleontology*. Tokyo: Terra Scientific Publishing company, pp. 15–25.

Berger, W., Smetacek, V., and Wefer, G. (eds.), 1989. *Productivity of the Ocean: Present and Past*. Dahlem Konferenzen LS 44. John Wiley and Sons.

Bouma, A.H., Normark, W.R., and Barnes, N.F., 1985. *Submarine Fans and Related Turbidite Systems*. Springer-Verlag.

Broecker, W.S., and Peng, T.-H., 1982. *Tracers in the Ocean*. Columbia University Press.

Calvert, S., 1983. Sedimentary geochemistry of silicon. In Aston, S. (ed.), *Silicon Geochemistry and Biochemistry*. Academic Press, pp. 143–186.

Cronan, D.S., 1977. Deep-sea nodules: distribution and geochemistry. In Glasby, G.P. (ed.), *Marine Manganese Deposits*. Elsevier, pp. 11–25.

Curray, J.R., and Moore, D.G., 1971. Growth of the Bengal Deep-sea fan and denudation of the Himalayas. *Geological Society of America, Bulletin*, **82**: 565–572.

Davies, T., and Gorsline, D., 1976. Oceanic sediments and sedimentary processes. In Riley, J., and Chester, K. (eds.), *Chemical Oceanography*, Volume 5, Academic Press, pp. 1–80.

Dixit, S., Van Cappellen, P., and van Bennekom, A.J., 2001. Processes controlling solubility of biogenic silica and pore water build-up of silicic acid in marine sediments. *Marine Chemistry*, Volume 73, 333–352.

Duce, R.A., Unmi, C.K., and Ray, B.J., 1991. The atmospheric input of trace species to the world ocean. *Global Biogeochemical Cycles*, **5**: 193–259.

Ewing, M., Ludwig, W., and Ewing, J., 1964. Sediment distribution in the oceans: the Argentine Basin. *Journal of Geophysical Research*, **71**: 1611–1636.

Falkowski, P., Barber, R., and Smetacek, V., 1998. Biogeochemical controls and feedbacks on ocean primary production. *Science*, **281**: 200–205.

Gardner, W., 1997. The flux of particles to the deep-sea: methods, measurement and mechanisms. *Oceanography*, 3: 116–121.

Garrels, R.M., and Mackenzie, F.T., 1971. *Evolution of sedimentary rocks*. New York: W. W. Norton.

Goldberg, E.D., 1963. Mineralogy and chemistry of marine sedimentation. In Shepard, F.P., *Submarine Geology*. New York: Harper and Row, pp. 436–466.

Griffin, J.J., Windom, H., and Goldberg, E.D., 1968. The distribution of clay minerals in the world ocean. *Deep Sea Research*, 15: 433–459.

Honjo, S., 1997. The rain of ocean particles and Earth's carbon cycle. *Oceanus*, 40: 4–7.

Honjo, S., Dymond, R., Collier, R., and Manganini, W.J., 1995. Export production of particles to the interior of the equatorial Pacific Ocean during the 1992 EqPac experiment. *Deep-Sea Research II*, 44: 831–870.

Humphris, S., Zierenberg, R., Mullineaux, L., Thomson, R. (eds.), 1995. Seafloor Hydrothermal Systems, Physical, Chemical, Biological and Geological Interactions. *Geophysical Monograph*, Volume 91, American Geophysical Union.

Kastner, M., 1981. Authigenic silicates in deep-sea sediments: formation and diagenesis. In Emiliani, C., (ed.), *The Oceanic Lithosphere, The Sea*. John Wiley and Sons, pp. 915–981.

Kennett, J., 1982. *Marine.Geology*. Englewood Cliffs: Prentice-Hall.

Leinen, M., 1989. The pelagic clay province of the North Pacific Ocean. In Winterer, E.L., Houssong, E.M., and Decker, R.W. (eds.), *The Eastern Pacific Ocean and Hawaii. The Geology of North America, vol. N*. Geological Society of America, pp. 323–335.

Leinin, M., and Pisias, N., 1984. An objective technique for determing end-member compositions and for partitioning sediments according to their sources. *Geochemica Et Cosmochimica Acta*, 48: 47–62.

Lisitzin, A., 1996. *Oceanic Sedimentation, Lithology and Geochemistry*. The American Geophysical Union.

McManus, J., Hammond, D., Berelson, W., Kilgore, DeMaster, T.D., Ragueneau, O., and Collier, R., 1995. Early diagensis of biogenic opal: dissolution rates, kinetics and paleoceanographic implications. *Deep-Sea Research II*, 42: 871–903

Moore, T.C. Jr., and Heath, R.G., 1978. Sea-floor sampling techniques. In Riley, J., and Chester, K. (eds.), *Chemical Oceanography*. Volume 7, Academic Press, pp. 75–126.

Murray, J., 1897. On the distribution of the Pelagic Foraminifera at the surface and on the floor of the Ocean. *Natural Science*, 11: 17–27.

Murray, J., and Renard, A., 1891. *Report on Deep-Sea Sediments Based on the Specimens Collected during the Voyage of H.M.S. Challenger in the Years 1872–1876*. Challenger Reports. London: Government Printer.

Nelson, D.M., Treguer, P., Brzezinski, M., Leynaert, A., and Quguiner, B., 1995. Production and dissolution of biogenic silica in the ocean: revised global estimates, comparison with reginoal data and relationship to biogenic sedimentation. *Global Biogeochemical Cycle*, 9: 359–372.

Normark, W.R., and Piper, D.J., 1991. Initiation process and flow evolution of turbidity currents: implication for the depositional record. In Osborne, R.H. (ed.), *From Shoreline to Abyss*. Society of Sedimentary Geology Special Paper, 46, pp. 207–230.

Pearce, R.B., Kemp, A.E.S., Baldauf, J.G., and King, S.C., 1996. High-resolution sedimentology and micropaleontology of laminated diatomaceous sediments from the eastern equatorial Pacific Ocean (Leg 138). In Kemp, A.E.S. (ed.), *Palaeoclimatology and Palaeoceanography from Laminated Sediments*. Geological Society of London, Special Publication, 116, pp. 221–242.

Peterson, M., 1966. Calcite: rates of dissolution in a Verical Profile in the Central Pacific. *Science*, 154, 11542–11544.

Piper, D.Z., and Heath, G.R., 1989. Hydrogenous sediment. In Winterer, E.L., Houssong, E.M., and Decker, R.W. (eds.), *The Eastern Pacific Ocean and Hawaii. The Geology of North America*. Volume N, Geological Society of America, pp. 337–345.

Ragueneau, O., Treguer, P., Leynaert, A., Anderson, R.F., Brzezinski, M., DeMaster, D.J., Dugdate, R., Dymond, J., Fischer, G., Francois, R., Heize, C., Maier-Reimer, E., Martin-Jezequel, V., Nelson, D., and Queguiner, B., 2000. A review of the Si cycle in the modern ocean: recent progress and missing gaps in the application of biogenic opal as a paleoproductivity proxy. *Global and Planetary Change*, 26: 317–365.

Raitt, R., 1956. Seismic refraction studies of the Pacific Ocean Basin, Part I: crustal thickness of the Central Equatorial Pacific. *Geological Society of America Bulletin*, 67: 1623–1640.

Rea, D.K., 1994. The paleoclimatic record provided by eolian deposition in the deep sea: the geologic history of wind. *Reviews of Geophysics*, 32: 159–195.

Rex, R.W., and Goldberg, E.D., 1968. Quartz content of pelagic sediments of the Pacific Ocean. *Tellus*, 10: 153–159.

Seibold, E., and Berger, W.H., 1996. *The Sea Floor, An Introduction to Marine Geology*. Springer-Verlag.

Schott, W., 1935. Die Foraminiferen in dem Aquatorialen Teil des Atlantischen Ozeans. *Wisschafen. Ergeb., Deutschen Atlantischen Expedition.Vermuss. Forschungsschiff Meteor*. 1925–27, Volume 111, pp. 43–134.

Windom, H., 1975. Eolian contribution to marine sediments. *Journal of Sedimentary Petrology*, 45: 520–529.

Windom, H., 1976. Lithogenous material in marine sediments. In Riley, R., and Skirrow, G. (eds.), *Chemical Oceanography*. Volume 5, pp. 103–135.

Winterer, E., 1989. Sediment thickness map of the Northeast Pacific. In Winterer, E.L., Hussong, D.M., and Decker, R. (eds.), *The Eastern Pacific Ocean and Hawaii. The Geology of North America*. Volume N, Geological Society of America, pp. 307–310.

Cross-references

Calcite Compensation Depth
Carbonate Mineralogy and Geochemistry
Classification of Sediments and Sedimentary Rocks
Clathrates
Eolian Transport and Deposition
Extraterrestrial Material in Sediments
Glacial Sediments: Processes, Environments, and Facies
Gravity-Driven Mass Flows
Iron–Manganese Nodules
Mudrocks
Nepheloid Layer, Sediment
Neritic Carbonate Depositional Environments
Seawater: Temporal Changes to the Major Solutes
Sediment Fluxes and Rates of Sedimentation
Sediment Transport by Unidirectional Water Flows
Sedimentology: History in Japan
Sedimentologists
Siliceous Sediments
Slope Sediments
Submarine Fan and Channels
Turbidites

OFFSHORE SANDS

Many middle and outer continental shelves around the world are covered, at depths from about 50 m to more than 200 m, with sand deposits, a few decimeters to more than 50 m thick. These sands are often remolded by tidal or unidirectional currents and/or storms and form longitudinal sand ridges (or sand banks of some authors) or transverse dunes. They may also appear as large sand sheets (with or without superimposed bed forms) forming a "sand belt" roughly parallel to the shelf edge. Understanding these deposits requires taking into account both present day sediment dynamics and Late Pleistocene glacio-eustatic and sediment flux changes.

Main types of shelf sand bodies and related processes

From a sediment dynamic point of view, offshore sand bodies have been classified on the basis of their orientation with respect to the main current at their origin. Their main characteristics are summarized in Table O5.

Large sand dunes

Large sand dunes (or sand waves of several authors) are observed at any water depths where current velocity near the bed is more than about 0.6 m/second, and sand is available. Even though the relationship between their height (H) and water depth is a matter of discussion, the largest are observed in the deepest environments (i.e., H about 20 m, at water depths around or more than 100 m). In tidal seas or on margins where geostrophic currents are particularly high, such as SE African continental margin, (Flemming, 1980), they are in equilibrium with peak current velocities (for instance spring tidal currents in tidal environments), and they move in the direction of their steeper side. Their migration rate is inversely proportional to their height (Rubin and Hunter, 1982) and reach values of 70 m/year for dunes 8 m high in the Dover Strait (Berné et al., 1989). In places where there is a seasonal change of physical regime, they can seasonally reverse their asymmetry (Harris, 1991). On the outer shelf, sand dunes are often relict and the result of conditions that existed during the last deglacial sea-level rise, but active sand dunes are found along the shelf edge, because of amplification of the energy of internal waves in such areas (Heathershaw and Codd, 1985; Karl et al., 1986). The internal structure of some large sand dunes consists of large foreset beds dipping at the angle of repose, but most display a hierarchy of bounding surfaces that represent episodic changes of the current and wave regime (Berné et al., 1988; Bristow, 1995) (Figure O7). Offshore sand dunes contain very few (if any) silts, and tidal couplets with mud drapes, typical for estuarine dunes (Visser, 1980) have never been described in modern offshore dunes.

Sand ridges

Sand ridges (or sand banks of some authors) are present on storm and tide-dominated shelves. They are inclined at an angle to the main axis of the tidal flow (or the main direction of propagation of storm waves). The model that best predicts the formation and evolution of offshore sand ridges is that of Huthnance (Huthnance, 1982a,b; Hulscher, 1996), where interaction of friction and Coriolis effect generates vorticity favoring sand accumulation around an initial sea-bed irregularity. A detailed description of this model, and a review of hydrodynamic processes at the origin of sand ridges, is given by Dyer and Huntley (1999). The best documented tidal sand ridges are in the North Sea, the Celtic Sea and the East China Sea (Stride, 1982; Yang, 1989), whereas storm ridges are mainly known from the Mid-Atlantic Bight (Swift and Field, 1981). Some authors claim that ridges formed on the middle continental shelf at or near present day water depths during the last ca. 7,000 years (Rine et al., 1991), however, many more believe that they developed during the last deglacial sea-level rise, in relatively shallow waters. Different scenarios have proposed that, during sea-level rise, sediments are reworked into ridges at the position of ebb-tidal deltas that represent point-sources of sediment (McBride and Moslow, 1991). As sea-level rise proceeds, ridges may be disconnected from the shore (if sediment is no more supplied), and their orientation will depend upon the rate of coastal retreat and lateral inlet migration. In the tide-dominated Celtic

Table O5 Main characteristics of major offshore sand bodies

Sand bodies	Length or lateral extent (km) E (av.–max.)	Width or spacing L (av.–max.)	Height (m) H (av.–max.)	Main relationships between different parameters (with h, water depth)	Orientation with respect to dominant current (α) And associated near-bed current velocity (V_b)
Dunes	0.5–3[a]	100–500 m[a]	2–25[a]	$H = 0.0677 L^{0.8098}$[f] $H = 0.086 h^{1.19}$[g]	$\alpha = 70°$–$90°$ $V_b = 50$–100 cm/s
Sand ribbons	1–15[b]	10–200 m[b]	1–4 m	$L = 4 h$[h]	$\alpha = 0°$ $V_b = 60$–100 cm/s
Offshore sand banks				$H/L = 0.003$[i] $E/L > 40$[d]	
Tide-dominated				$L = 250 h$[j]	$\alpha = 7°$–$15°$[c]
active	5–70	2–30 km	13–43		$\alpha = 28°$[j]
moribund	5–200	2–30 km	55[c,d]		$V_b = 50$–120 cm/s
Storm-dominated	1–20	0.5–7 km	7–12[c,d]		$\alpha = 35°$–$40°$[c]
Sand patches	1–20		2–5		$V_b < 50$ cm/s
Sand Sheets[e]	5–100	10–400 km	5–12		$V_b = 30$–55 cm/s (tide-dominated)

[a] Berné (1991); [b] Belderson et al. (1982); [c] Belderson (1986); [d] Amos and King (1984); [e] M'hammdi (1994); [f] Flemming (1988); [g] Allen (1984); [h] McLean (1981); [i] Off (1963); [j] Huthnance (1982b); [k] Mosher and Thomson (2000).

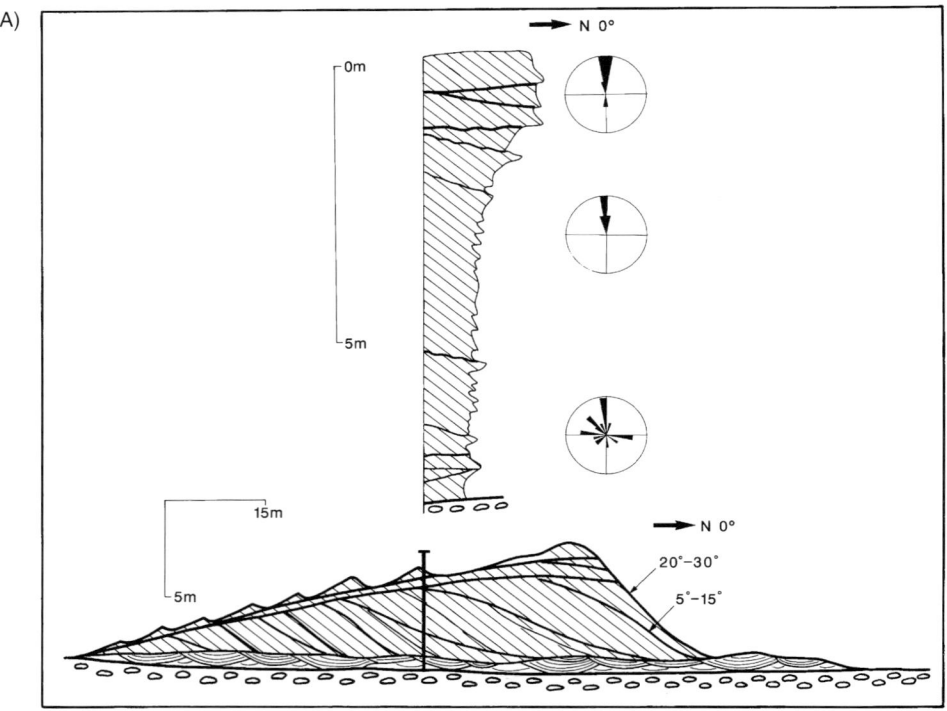

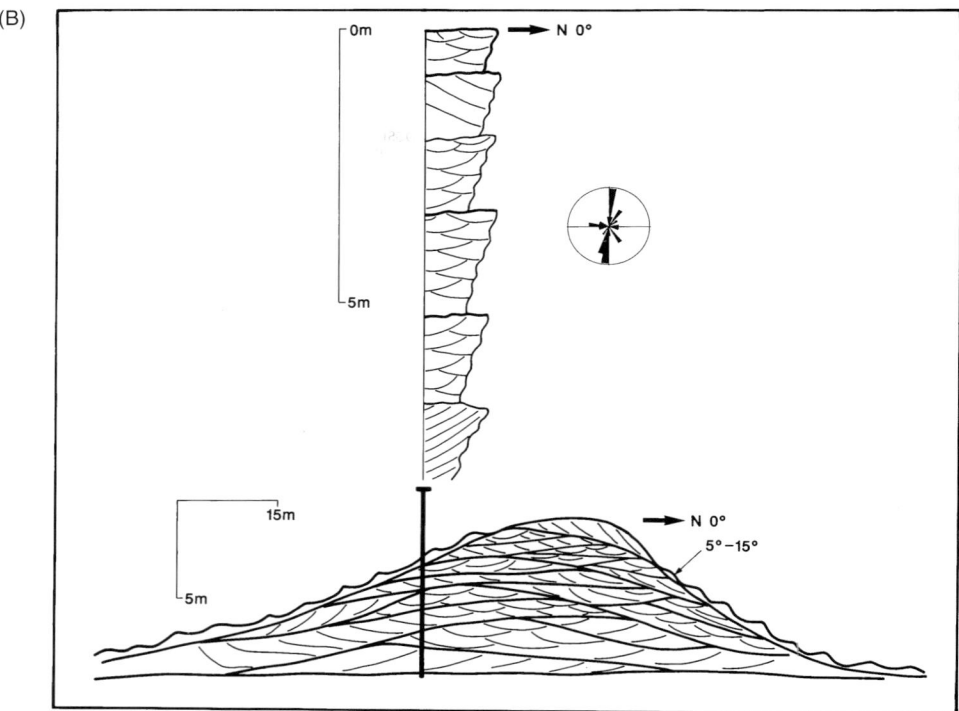

Figure O7 (*Continued*)

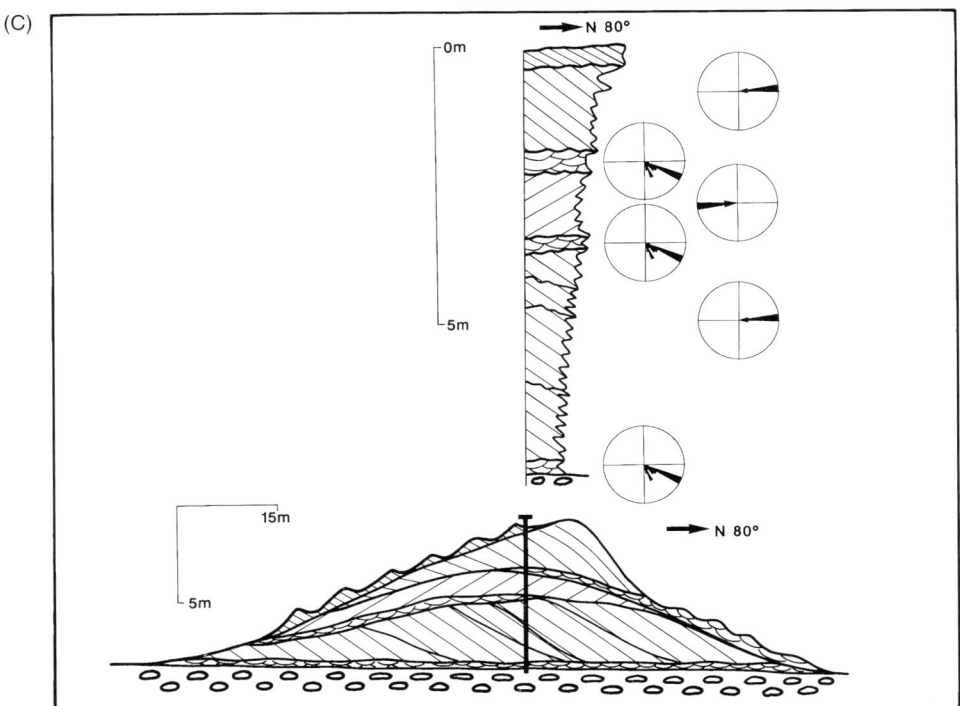

Figure O7 Scenarios for internal structure of large (>3 m) sand dunes (sand waves), based on seismic and core data (modified from Berné, 1991). The circles represent dipping angles and orientations of megaripple (small dune) cross-beds, for a crest of the large dune with an arbitrary E–W direction. (A) Asymmetrical dune with cosets several meter-thick. The master bedding results from erosion surface formed during storms. The cross-beds within each coset may have an angle of dip as high as 34°. Superimposed bedforms create an erosion surface at the top of the sand wave, and have dip angles in the same range as that of the large sand wave. Medium-scale cross-beds at the bottom of the sand wave have dip angles almost perpendicular to that of the sand wave, due to the secondary circulation along the lee-side. (B) Symmetrical dune where flood and ebb net sediment transport during spring tides are in opposite directions and of similar magnitude. Medium-scale cross-beds with divergent directions constitutes the core of the sand wave. Major bounding surfaces are related to migration of superimposed small dunes (megaripples), as in estuarine dunes (Dalrymple and Rhodes, 1995). Such dunes are encountered along the great axis of several sand ridges, or in zones of bedload parting. (C) Reversing dunes. This structure corresponds to cases where seasonal changes create inversion of net bedload transport. Large-scale cross beds correspond to periods when asymmetry of the dune is well pronounced. Major bounding surfaces form during periods of reversal of net bedload transport, when the bed form becomes symmetrical. Because no avalanche occurs at that time, megaripple cross-beds (usually dipping parallel or slightly oblique to the great axis of the dune) are preserved. Such structures are observed in estuaries where seasonal changes of stratification occur, but also on open shelves where wind-driven currents reverse seasonally (Le Bot et al., 2000).

Sea, it is believed that ridges formed between the last sea-level lowstand and the early stage of the deglacial sea-level rise (i.e., between ca. 20 ka and 14 ka) (Bouysse et al., 1976; Pantin and Evans, 1984; Stride et al., 1982). This interpretation is supported by numerical modeling of the M2 tidal constituent for a sea level lowered by 100 m, which indicates that the shelf currents would have been about twice as strong as at present, with an offset of 12° to 23° between the regional peak tidal flow and the axis of the ridges (Belderson, 1986). A similar numerical model has shown that the orientation of tidal ellipse for a sea-level lowered by 80 m matches the orientation of tidal ridges in the East China Sea (Oh and Lee, 1998). Some sand ridges display on seismic records an internal structure characterized by clinoforms dipping at about 5° in the same direction as the steeper side (presumably the direction of sand ridge migration) (Houbolt, 1968). Others present a core of coastal deposits suggesting that the ridges cannibalized pre-existing deposits (Berné et al., 1994; Laban and Schüttenhelm, 1981). This observation led Snedden and Dalrymple (1999) to classify sand ridges as a function of the amount of their migration: juvenile ridges retain their initial nucleus that represent the initial irregularity necessary for ridge development in the Huthnance model, whereas "fully evolved" ridges have migrated sufficiently that they contain no trace of their origin. Alternatively, it has been proposed that "cored" ridges represent situations where substrate is erodible and sand supply limited (Berné et al., 1998). In such situations where the spatial derivative of the transport rate is negative, ridges (or any other type of bedforms) are migrating with a negative angle of climb and incorporate previous deposits of different origins (Rubin, 1987). In any event, an important observation is that such sand ridges may in fact contain a relatively large fraction of non-sandy material.

Most of the cores from modern sand ridges show massive sand with an overall coarsening upward trend. In the East China Sea, the bottomsets of the clinoforms from sand ridges lying at about 90 m display alternating sand and silt bedding with burrowed tidal rhythmites, whereas the topsets consist of homogeneous fine to medium sand (Berné et al., 2002). Cores from the Middelkerke bank in the North Sea show, in the upper part of the ridge, ripple to medium dune bedding, with some herringbone cross-beds (Trentesaux et al., 1999). The lower part of the bank consists of estuarine sediments incorporated into the ridge.

Lowstand and transgressive shorefaces and delta fronts

In areas where erosion due to wave and tidal ravinement is limited, shoreface deposits and sandy delta fronts that formed at period of lower sea-level may be preserved and represent another type of offshore sand bodies, where the initial geometry is only partly preserved.

Forced-regressive and lowstand deposits are preserved at depths close to the position of shorelines during the Last Glacial Maximum (i.e., about 90 m to 120 m water depth). In the moderate energy, storm-dominated Western Mediterranean Sea, they form large sand bodies, up to 35 m thick and 15 km wide, that extend parallel to bathymetric contour lines along distances more than 100 km (Rabineau et al., 1998). On seismic profiles, they display internal clinoforms dipping seaward at an angle of 2° to 5° (Figure O8). They formed as prograding sandy deposits migrating seaward in response to sea-level fall, following a mechanism initially described from outcrop studies in the stratigraphic record (Plint, 1988). The clinoforms consist of glauconitic sand with shells and shell debris, passing progressively to silty bottomsets corresponding to a lower shoreface/prodeltaic environment (Berné et al., 1998). The "sand ridges" that lie parallel to the shelf edge at water depths between 75 m and 100 m in the Indian Ocean off Bombay (Wagle and Veerayya, 1996) could have the same origin, but in this case, higher wave energy has subsequently reworked the deposits. Deltas that formed during lowstand of the sea have similar shapes in two dimensions, but more lobate geometries. They are described from examples in the Eastern Mediterranean Sea (Aksu et al., 1987; Skene et al., 1998), the Gulf of Mexico (Suter and Berryhill, 1985), the South China Sea (Hiscott, 2001).

Such stranded sand bodies do not only form during the falling and low stand stages of glacio-eustatic cycles. During the last deglacial period, slowing of sea-level rise and/or increase in sediment supply have also favored formation of transgressive shoreface deposits that are now isolated on the middle or inner shelf. For instance, sand banks of the inner shelf Trinity/Sabine incised valley system in the Gulf of Mexico are interpreted as stranded shoreface deposits that formed during a phase of relatively slow sea-level rise, and where stranded on the shelf during a rapid sea-level rise (Anderson and Thomas, 1991; Rodriguez et al., 1999). An alternative interpretation for shore-parallel sand shoals lying at about 50 m water depth in the NE Gulf of Mexico is that they developed at the position of an escarpment formed during the transgression, but that they consist solely of post-transgressive deposits (McBride et al., 1999).

Sand sheets and sand patches

Sheet-like bodies of sand largely covered by sand waves (large dunes) are described in "low energy" tidal environments (where current velocity near the bed is around 55 cm/second) such as the Southern Bight of the North Sea (Belderson, 1986; Stride et al., 1982). They have no preferential orientation and their internal structure is that of (tidal) dunes (large and medium scale cross beds). In areas dominated by unidirectional currents, such as the southeast African shelf, sand sheets with very asymmetric bed forms are observed (Flemming, 1980). Where the sand cover is discontinuous, sand patches resting on a coarse lag of fluvial or transgressive deposits are observed. These patches may or may not have a preferential orientation, depending on the local hydrodynamic regime. On the Californian shelf dominated by swells of the Pacific Ocean, they form transverse sand patches or very flat dunes migrating seaward in response to large bottom stresses generated by combined waves and currents (Cacchione et al., 1987). Similar sand patches are also described in the Bay of Biscaye at depths ranging from 30 m to 90 m. Cores display stacked fining upward elementary sequences 1 cm to 10 cm thick, with erosional base, typical of storm deposits (Cirac et al., 2000).

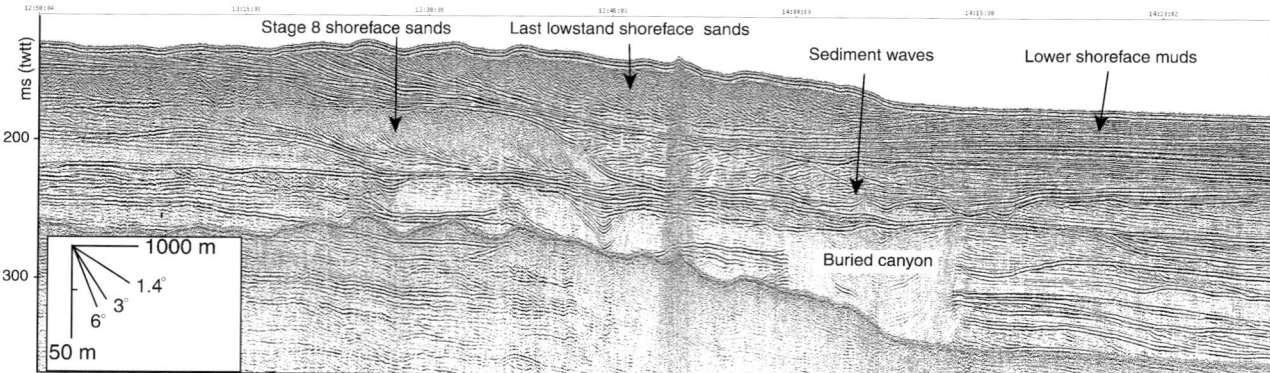

Figure O8 Recent and buried shoreface sands in the Western Mediterranean Sea. Offshore sands in the Gulf of Lions cover the outer shelf at water depths from 90 m to about 120 m. Note the abrupt seaward transition between sands and mud, marked by a step in the sea-floor morphology. Similar sand bodies formed during other glacial periods (corresponding to isotopic stage 8) are now buried and encased into prodeltaic muds. Vertical scale is in milliseconds two way travel time (mstwtt).

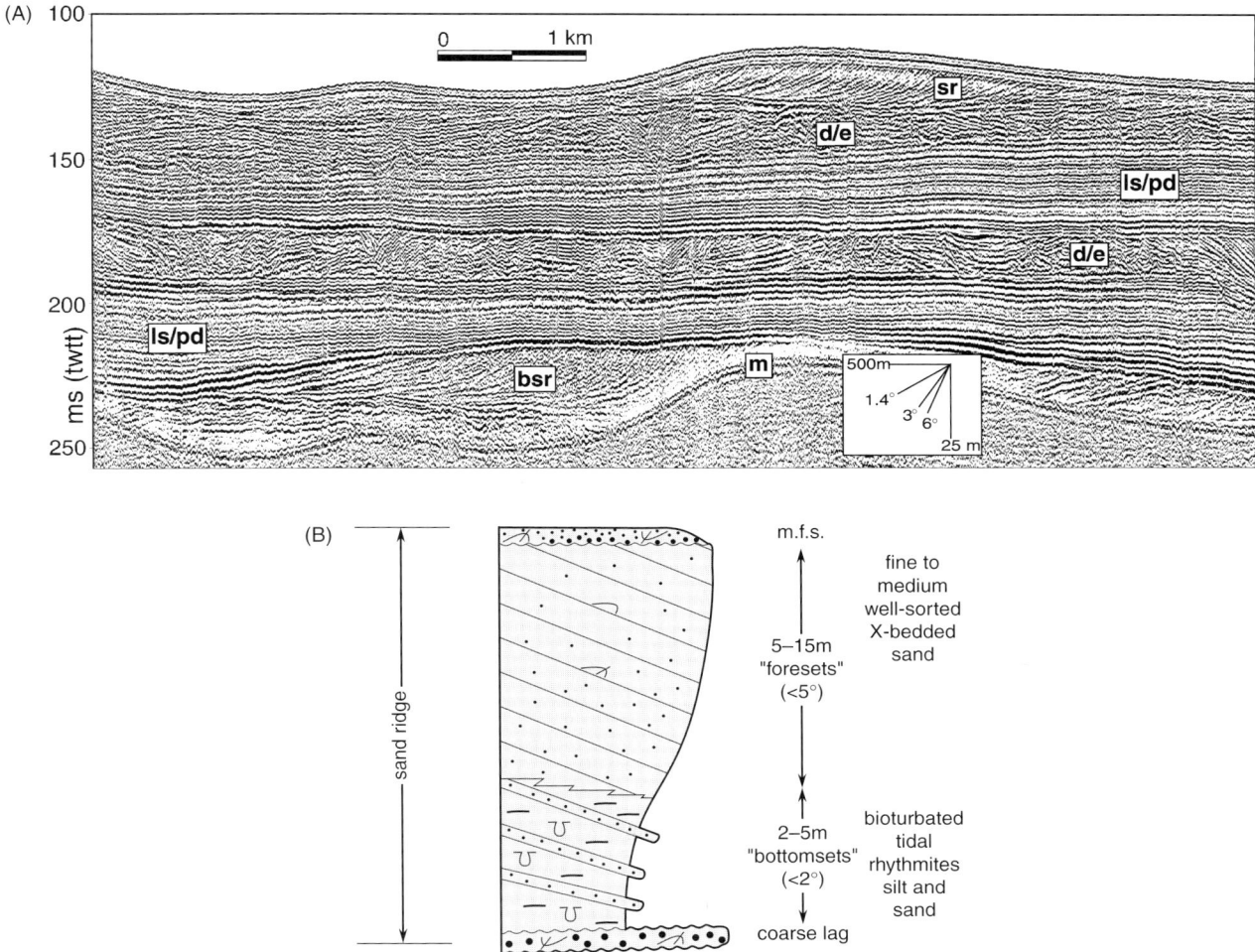

Figure O9 Seismic (A) and synthetic log (B) for a sand ridge on the outer shelf of the East China Sea (water depth about 80 m). The seismic profile, perpendicular to the great axis of the sand ridge, display internal clinoforms dipping at about 2–3° in the same direction as the "steep" side of the ridge. A buried Pleistocene sand ridge (bsr) is also observed, overlain by lower shoreface or prodeltaic silts (ls/pd) and deltaic/estuarine heterogeneous sediments with numerous channel structures (d/e). The external morphology of this buried has been preserved. "m" is the seismic multiple. Vertical scale is in milliseconds two way travel time (mstwtt). The synthesis of several cores across the ridge shows (Figure O9(B)) that the upper part of the clinoforms ("foresets") consists of fine to medium sands, whereas the "bottomsets" display tidal rhythmites. The top of the ridge is blanketed by a carpet of sediment in equilibrium with modern hydrodynamic regime, but the ridge is believed to have form during the last deglacial sea-level rise.

Preservation of Pleistocene offshore sands

There are contradictory explanation about the origin of many sand bodies encased in mudstones, for instance the Shannon Sandstone in Wyoming, that are interpreted either as fossil sand ridges (Tillman, 1999) or as regressive shoreface deposits (Bergman and Walker, 1999). Examples from modern (Pleistocene) continental shelves demonstrate that both types of sand body may be preserved. This discovery is rather recent, because most of the early studies on continental shelves focused on NW Europe and North American margins where low subsidence, limited sediment supply and strong bottom currents favor reworking of deposits during each glacial cycle. In contrast, seismic surveys and geotechnical boreholes on (deltaic) margins with high sedimentation rate and relatively high subsidence rates (around 250 m/million years at the shelf edge) demonstrate the high potential of preservation of shelf sand bodies in such areas, where they are generally encased in prodeltaic/lower shoreface regressive silty deposits. Examples of Pleistocene buried sand bodies include regressive, lowstand and transgressive shorefaces (Rabineau et al., 1998), lowstand (shelf edge) deltas (Hiscott, 2001; Skene et al., 1998; Suter and Berryhill, 1985; Sydow and Roberts, 1994; Winn et al., 1998) and transgressive sand ridges (Berné et al., 2002; Yang, 1989). These sand bodies are typically 10 m to 30 m thick, with a lateral extent of several km to more than 100 km, and their initial shape is sometimes preserved (Figure O9). Most authors agree in attributing a major role to 4th order glacio-eustatic cycles (i.e., about 100 kyr) in the construction of these sand bodies, even though shorter time-periods are the dominant factors controlling the up-building of transgressive parasequences.

Serge Berné

Bibliography

Aksu, A.E., Piper, D.J.W., and Konuk, T., 1987. Late Quaternary tectonic and sedimentary history of outer Izmir and Candarli Bays, Western Turkey. *Marine Geology*, **76**: 89–104.

Allen, J.R.L., 1984. *Principles of Physical Sedimentology*. George Allen & Unwin.

Amos, C.L., and King, E.L., 1984. Bedforms of the Canadian eastern seaboard: a comparison with global occurences. *Marine Geology*, **57**: 167–208.

Anderson, J.B., and Thomas, M.A., 1991. Marine ice-sheet decoupling as a mechanism for rapid, episodic sea-level change: the record of such events and their influence on sedimentation. *Sedimentary Geology*, **70**: 87–104.

Belderson, R.H., 1986. Offshore tidal and non-tidal sand ridges and sheets: differences in morphology and hydrodynamic setting. In Knight, R.J., and McLean, J.R. (eds.), *Shelf Sands and Sandstones*. Canadian Society of Petroleum Geologists, Calgary, pp. 293–301.

Belderson, R.H., Johnson, M.A., and Kenyon, N.H., 1982. Bedforms. In Stride, A.H. (ed.), *Offshore Tidal Sands*. London: Chapman and Hall, pp. 27–57.

Belderson, R.H., Pingree, R.D., and Griffiths, D.K., 1986. Low sea-level tidal origin of Celtic Sea sand banks- Evidence from numerical modelling of M2 tidal streams. *Marine Geology*, **73**: 99–108.

Bergman, K.M., and Walker, R.G., 1999. Campanian Shannon sandstone: an example of a falling stage systems tract deposit. In Bergman, K.M., and Snedden, J.D. (eds.), *Isolated Shallow Marine Sand Bodies: Sequence Stratigraphic Analysis and Sedimentologic Interpretation*. SEPM (Society for Sedimentary Geology), pp. 85–93.

Berné, S., 1991. *Architecture Et Dynamique Des Dunes Tidales. Exemples De La Marge Atlantique Française*. unpublished PhD Thesis, Lille, 295p.

Berné, S., Allen, G., Auffret, J.P., Chamley, H., Durand, J., and Weber, O., 1989. Essai de synthèse sur les dunes hydrauliques géantes tidales actuelles. *Bulletin de la Société Géologique de France*, **6**: 1145–1160.

Berné, S., Auffret, J.P., and Walker, P., 1988. Internal structure of subtidal sand waves revealed by high-resolution seismic reflection. *Sedimentology*, **35**: 5–20.

Berné, S., Lericolais, G., Marsset, T., Bourillet, J.F., and de Batist, M., 1998. Erosional shelf sand ridges and lowstand shorefaces: examples from tide and wave dominated environments of France. *Journal of Sedimentary Research*, **68** (4): 540–555.

Berné, S., Trentesaux, A., Stolk, A., Missiaen, T., and De Batist, M., 1994. Architecture and long-term evolution of a tidal sandbank: the Middelkerke Bank, Southern North Sea. *Marine Geology*, **121**: 57–72.

Berné, S., Vagner, P., Guichard, F., Lericolais, G., Liu, Z., Trentesaux, A., Yin, P., and Yi, H.I., 2002. Pleistocene forced regressions and tidal sand ridges in the East China Sea. *Marine Geology*, **188**(3–4): 293–315.

Bouysse, P., Horn, R., Lapierre, F., and Le Lann, F., 1976. Etude des grands bancs de sable du Sud-Est de la Mer Celtique. *Marine Geology*, **20**: 251–275.

Bristow, C.S., 1995. Internal geometry of ancient tidal bedforms revealed using ground penetrating radar. In Flemming, B.W. (ed.), *3rd International Research Symposium on Modern and Ancient Clastic Tidal Deposits; Tidal clastics 92*. Wilhelmshaven: International Association of Sedimentologists, pp. 313–328.

Cacchione, D.A., Field, M.E., Drake, D.E., and Tate, G.B., 1987. Crescentic dunes on the inner continental shelf off northern California. *Geology*, **15**: 1134–1137.

Cirac, P., Berné, S., Castaing, P., and Weber, O., 2000. Processus de mise en place et d'évolution de la couverture sédimentaire superficielle de la plate-forme nord-aquitaine. *Oceanologica Acta*, **23**: 663–686.

Dalrymple, R.W., and Rhodes, R.N., 1995. Estuarine dunes and bars. In Perillo, G.M.E. (ed.), *Geomorphology and Sedimentology of Estuaries*. Elsevier, pp. 359–422.

Dyer, K.R., and Huntley, D.A., 1999. The origin, classification and modelling of sand banks and sand ridges. *Continental Shelf Research*, **19** (10): 1285–1330.

Flemming, B.W., 1980. Sand transport and bedform patterns on the continental shelf between Durban and Port Elizabeth (southeast African continental margin). *Sedimentary Geology*, **26**: 179–205.

Flemming, B.W., 1988. Zur klassifikation subaquatischer, stromungstransversaler Transportkorper. *Bochumer Geologische Und Geotechnische Arbeiten*, **29**: 44–47.

Harris, P.T., 1991. Reversal of subtidal dune asymmetries caused by seasonally reversing wind-driven currents in Torres Strait, northeastern Australia. *Continental Shelf Research*, **11**(7): 655–662.

Heathershaw, A.D., and Codd, J.M., 1985. Sandwaves, internal waves and sediment mobility at the shelf-edge in the Celtic Sea. *Oceanologica Acta*, **8**: 391–402.

Hiscott, R.N., 2001. Depositional sequences controlled by high rates of sediment supply, sea-level variations, and growth faulting: the Quaternary Baram Delta of northwestern Borneo. *Marine Geology*, **175** (1–4): 67–102.

Houbolt, J.J.H.C., 1968. Recent sediments in the Southern Bight of the North Sea. *Geologie en Mijnbouw*, **47** (4): 245–273.

Hulscher, S.J.M.H., 1996. Tide-induced large-scale regular bedform patterns in a three-dimensional shallow water model. *Journal of Geophysical Research*, **101**: 20727–20744.

Huthnance, J.M., 1982a. On one mechanism forming linear sand banks. *Estuarine, Coastal and Shelf Science*, **14**: 279–299.

Huthnance, J.M., 1982b. On the formation of sand banks of finite extent. *Estuarine, Coastal and Shelf Science*, **15**: 277–299.

Karl, H.A., Cacchione, D.A., and Carlson, P.R., 1986. Internal-wave currents as mechanism to account for large sand waves in Navarinsky Canyon head, Bering Sea. *Journal of Sedimentary Petrology*, **56** (5): 706–714.

Laban, C., and Schüttenhelm, R.T.E., 1981. Some new evidence on the origin of the Zealand ridges. In Nio, E.D., Schüttenhelm, R.T.E., and Van Veering, T.C.E. (eds.), *Holocene Marine Sedimentation in the North Sea Basin*. Special Publication 5, International Association of Sedimentologists, pp. 239–245.

Le Bot, S., Herman, J.P., Trentesaux, A., Garlan, T., Berné, S., and Chamley, H., 2000. Influence des tempêtes sur la mobilité des dunes tidales dans le détroit du Pas-de-Calais. *Oceanologica Acta*, **23** (2): 129–141.

M'hammdi, N., 1994. *Architecture du Banc Sableux Tidal De Sercq (Îles anglo-normandes)*. PhD Thesis, Université de Lille 1, Laboratoire de Sédimentologie et Tectonique, Lille, 203pp.

McBride, R.A., Anderson, L.C., Tudoran, A., and Roberts, H.H., 1999. Holocene stratigraphic architecture of a sand-rich shelf and the origin of linear shoals: northeastern Gulf of Mexico. In Bergman, K.M., and Snedden, J.W. (eds.), *Isolated Shallow Marine Sand Bodies: Sequence Stratigraphic Analysis and Sedimentologic Interpretation*. SEPM Special Publication, Tulsa: Society for Sedimentary Geoplogy (SEPM), pp. 95–126.

McBride, R.A., and Moslow, T.F., 1991. Origin, evolution, and distribution of shoreface sand ridges, Atlantic inner shelf, USA. *Marine Geology*, **97**: 57–85.

McLean, S.R., 1981. The role of non uniform roughness in the formation of sand ribbons. *Marine Geology*, **42**: 49–74.

Mosher, D.C., and Thomson, R.E., 2000. Massive submarine sand dunes in the eastern Juan de Fuca Strait, British Columbia. In Tentesaux, A., and Garlan, T. (eds.), *Marine Sandwave Dynamics*. France: University of Lille, pp. 131–142.

Off, T., 1963. Rhythmic linear sand bodies caused by tidal currents. *American Association of Petroleum Geologists Bulletin*, **47**: 324–341.

Oh, S.I., and Lee, D.E., 1998. Tides and tidal currents of the Yellow and East China Seas during the last 13,000 years. *Journal of the Korean Society of Oceanography*, **33** (4): 137–145.

Pantin, H.M., and Evans, C.D.R., 1984. The Quaternary history of the Central and Southwestern Celtic Sea. *Marine Geology*, **57**: 259–293.

Plint, A.G., 1988. Sharp-based shoreface sequences and "offshore bars" in the Cardium Formation of Alberta: their relationship to relative changes in sea level. In Wilgus, C.K. *et al*. (eds.), *Sea-level*

Changes: An Integrated Approach. SEPM Special Publication, 42, Tulsa, pp. 357–370.

Rabineau, M., Berné, S., Le Drezen, E., Lericolais, G., Marsset, T., and Rotunno, M., 1998. 3D architecture of lowstand and transgressive Quaternary sand bodies on the outer shelf of the Gulf of Lion, France. *Marine and Petroleum Geology*, **15** (5): 439–452.

Rine, J.M., Tillman, R.W., Culver, S.J., and Swift, D.J.P., 1991. Generation of late Holocene sand ridges on the middle continental shelf of New Jersey, USA-evidence for formation in a mid-shelf setting based on comparison with a nearshore ridge. In Swift, D.J.P., Oertel, G.F., Tillman, R.W., and Thorne, J.A. (eds.), *Shelf Sand and Sandstone Bodies: Geometry, Facies and Sequence Stratigraphy*. Blackwell, pp. 395–423.

Rodriguez, A.B., Anderson, J.B., Siringan, F.P., and Taviani, M., 1999. Sedimentary Facies and Genesis of Holocene Sand Banks on the East Texas Inner Continental Shelf: Isolated Shallow Marine Sand Bodies. In Bergman, K.M., and Snedden, J.W. (eds.), *Isolated Shallow Marine Sand Bodies: Sequence Stratigraphic Analysis and Sedimentologic Interpretation*. SEPM (Society for Sedimentary Geology), Special Publication, pp. 165–178.

Rubin, D.M., 1987. *Cross-Bedding, Bedforms, and Paleocurrents*. SEPM Concepts in Sedimentology and Paleontology Series 3.

Rubin, D.M., and Hunter, R.E., 1982. Bedform climbing in theory and nature. *Sedimentology*, **29**: 121–138.

Skene, K.I., Piper, D.J.W., Aksu, A.E., and Syvitski, J.P.M., 1998. Evaluation of the global oxygen isotope curve as a proxy for Quaternary sea level by modeling of delta progradation. *Journal of Sedimentary Research*, **68** (6): 1077–1092.

Snedden, J.W., and Dalrymple, R.W., 1999. Modern shelf sand ridges: from historical perspective to a unified hydrodynamic and evolutionary model. In Bergman, K.M., and Snedden, J.W. (eds.), *Isolated Shallow Marine Sand Bodies: Sequence Stratigraphic Analysis and Sedimentologic Interpretation*. SEPM (Society for Sedimentary Geology), Special Publication, pp. 13–28.

Stride, A.H., 1982. *Offshore Tidal Sands: Processes and Deposits*. Chapman and Hall.

Stride, A.H., Belderson, R.H., Kenyon, N.H., and Johnson, M.A., 1982. Offshore tidal deposits: sand sheet and sand bank facies. In Stride, A.H. (ed.), *Offshore Tidal Sands: Processes and Deposits*. Chapman and Hall, pp. 95–125.

Suter, J.R., and Berryhill, H.L.J., 1985. Late Quaternary shelf-margin deltas, Northwest Gulf of Mexico. *Bulletin of the American Association of Petroleum Geologists*, **69** (1): 77–91.

Swift, D.J.P., and Field, M.E., 1981. Evolution of a classic sand ridge field: Maryland sector, North American inner shelf. *Sedimentology*, **28**: 461–482.

Sydow, J., and Roberts, H.H., 1994. Stratigraphic framework of a late Pleistocene shelf-edge delta, northeast Gulf of Mexico. *American Association of Petroleum Geologists Bulletin*, **78**: 1276–1312.

Tillman, R.W., 1999. The Shannon Sandstone: a review of the Sand-Ridge and Other Models. In Bergman, K.M., and Snedden, J.D. (eds.), *Isolated Shallow Marine Sand Bodies: Sequence Stratigraphic Analysis and Sedimentologic Interpretation*. SEPM (Society for Sedimentary Geology), pp. 29–53.

Trentesaux, A., Berné, S., and Stolk, A., 1999. Sedimentology and stratigraphy of a tidal sand bank in the southern North Sea. *Marine Geology*, **159**: 253–272.

Visser, M.J., 1980. Neap-spring cycles reflected in Holocene subtidal large-scale bedform deposits: a preliminary note. *Geology*, **8**: 543–546.

Wagle, B.G., and Veerayya, M., 1996. Submerged sand ridges on the western continental shelf off Bombay, India: evidence for Late Pleistocene–Holocene sea-level changes. *Marine Geology*, **136**: 79–95.

Winn, R.D., Roberts, H.H., Kohl, B., Fillo, R.H., Crux, J.A., Bouma, A.H., and Spero, H.W., 1998. Upper Quaternary strata of the upper continental slope, northeast Gulf of Mexico: sequence stratigraphic model for a terrigeneous shelf edge. *Journal of Sedimentary Research*, **68** (4): 579–595.

Yang, C.S., 1989. Active, moribund and buried tidal sand ridges in the East China Sea and the Southern Yellow Sea. *Marine Geology*, **88**: 97–116.

Cross-references

Coastal Sedimentary Facies
Cross-Stratification
Glaucony and Verdine
Milankovitch Cycles
Ripple, Ripple Mark, Ripple Structure
Sediment Transport by Tides
Sediment Transport by Waves
Surface Forms
Tides and Tidal Rhythmites

OIL SANDS

Oil sands constitute an immense petroleum resource on a worldwide scale (Figure O10) and Canada, at 1,700 billion barrels (AEUB, 1996), dominates the current estimates (Meyer, 1995, 1996). The petroleum resource in the Orinoco Oil Belt (OOB) of Venezuela is thought to be similar in size at 1,200 billion barrels (Fiorillo, 1987) and dominates the world heavy oil resources (Meyer, 1998).

Oil sands (also known as tar sands) are defined as petroleum deposits where the viscosity of the oil (bitumen) is greater than 10,000 centipoises (cP) at reservoir conditions (Meyer, 1995). The general definition applied to oil sands is a petroleum deposit where oil viscosity is so high that there is no commercial inflow to a well at reservoir conditions (AEUB, 1996; Butler, 1997).

Bitumen is deficient in the lighter hydrocarbon molecules (Meyer, 1998) and is generally in the 6°API (1,029 kg/m^3) to 12°API (986 kg/m^3) range. However, oil viscosity is greatly affected by temperature, and heavier bitumens experience large reductions in viscosity with an increase in temperature. Bitumen in the Clearwater Deposit, Alberta, Canada is lighter (10–12°API) than the oil in many of the heavy oil deposits in the OOB (8–14°API; Starr et al., 1981). However, higher reservoir temperatures (40–70°C) in the OOB give the oil mobility.

All of the significant oil sands deposits in Canada occur in the northeastern part of Alberta (Figure O11). The bitumen is thought to be a conventional oil that was degraded, mainly through biodegradation, as it migrated updip into the shallower portions of the basin (Creaney and Allan, 1992). The Wabiskaw–McMurray and Grand Rapids deposits in the

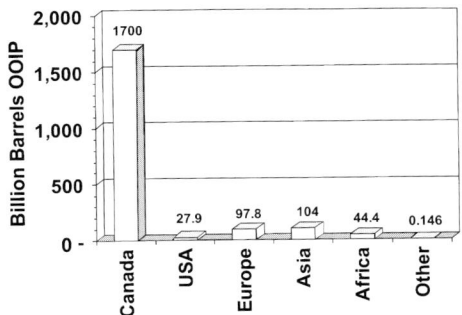

Figure O10 World oil sands resources, Original Oil In Place (OOIP).

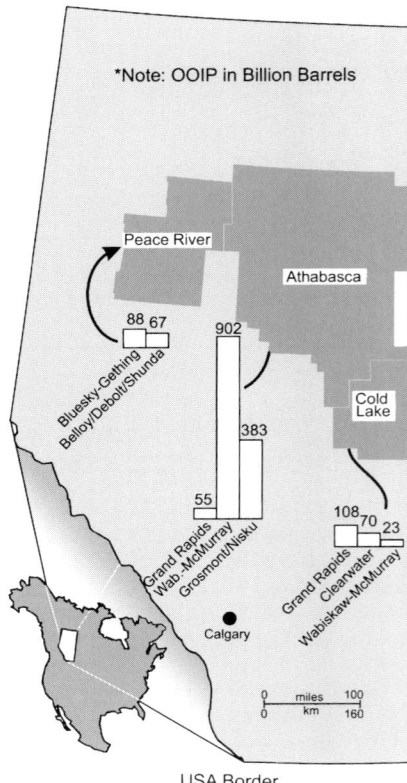

Figure O11 Oil sands areas and OOIP estimates for deposits in Alberta, Canada.

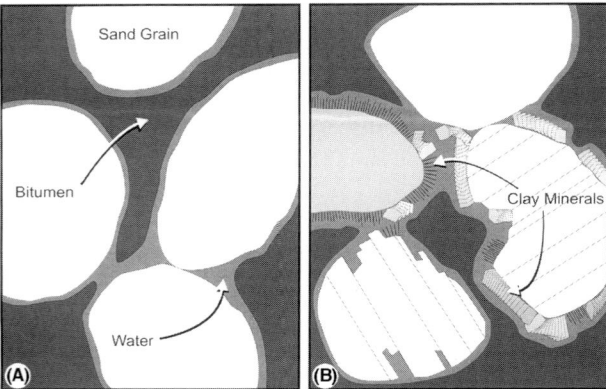

Figure O12 Schematic diagram of thin sections of water wet oil sands. (A) Clean quartz sand with low water saturation. (B) Arkose or litharenite with authigenic clay minerals and higher water saturation; porosities can be higher than in (A).

Athabasca Oil Sands Area have no discernible structural or stratigraphic closure; apparently the oil was biodegraded into immobility and trapped itself (Wightman et al., 1995).

Data are sparse for many other oil sands deposits but most occur in the shallow (less than 1,000 m), updip or marginal portions of basins (Phizackerley and Scott, 1978; Roadifer, 1987). These deposits also represent conventional oils that have been biodegraded into bitumen through contact with meteoric water (Meyer, 1996).

Most of Alberta's oil sands are comprised of consolidated (but mostly uncemented) sand, bitumen and water. The exceptions occur in Mississippian and Devonian carbonates but these are classified as oil sands because of the bitumen (AEUB, 1996). With the sand and carbonate, water coats the mineral matter and bitumen fills the remainder of the pore space (Figure O12).

Bitumen distribution and oil sands quality (porosity, permeability and oil saturation) are directly controlled by depositional facies in most of the clastic oil sands and heavy oil deposits around the world (Phizackerley and Scott, 1978). This is the result of shallow depths of burial (limited diagenesis) in basin margin settings. Clean, well sorted sands form the best oil sands whereas sands with abundant matrix material, mudstones and shales contain little or no oil. Depositional environments are generally continental to marine shoreline (Phizackerley and Scott, 1978).

In Alberta, the Lower Cretaceous Mannville Group contains most of the oil sands resource and within it, the Wabiskaw–McMurray Deposit is the largest with 902 billion barrels (AEUB, 1996). The rich oil sands generally occur in channel deposits, so lateral continuity is problematic. Trough cross bedded sand, with little or no interbedded mudstone, in the lower parts of channel deposits represents the best oil sands (33–38 percent porosity, 80–95 percent pore oil saturation and 5–10 D of permeability, up to 20 D in gravel; Wightman et al., 1995). The overlying lateral accretion beds are comprised of interbedded sands and mudstones. The mudstones reduce ore grade for surface mining operations and the lower permeabilities (generally 10–300 mD) have a significant, adverse effect on in situ operations (Strobl et al., 1997).

The sedimentological characteristics of a deposit influence the type of in situ recovery process that is most applicable. Fining upward channel deposits are particularly suited to Steam Assisted Gravity Drainage (SAGD) as the best oil sands are located in the lower part of the reservoir (Figure O13). Cyclic Steam Stimulation (CSS) is being utilized in the coarsening upward Clearwater Deposit of Alberta (Butler, 1997); pressure parting induces vertical permeability, especially in the lower portions of the reservoir (Figure O14).

Oil sands have several types of associated water sands (Figure O15) which occur within and adjacent to the bitumen (Wightman et al., 1995; Barson et al., 2001). Water sands reduce overall reservoir quality; in particular, top water can be very detrimental to in situ operations (Barson et al., 2001).

Oil sands currently (September, 2001) contribute about 650,000 barrels of oil/day in Canada, with 340,000 barrels of synthetic crude from mining and 310,000 barrels of in situ bitumen. Production is increasing, with predictions of up to 1,650,000 barrels/day by 2015 (NEB, 2000). Worldwide production from oil sands is also expected to increase as conventional oil supplies decline in the future.

Daryl M. Wightman

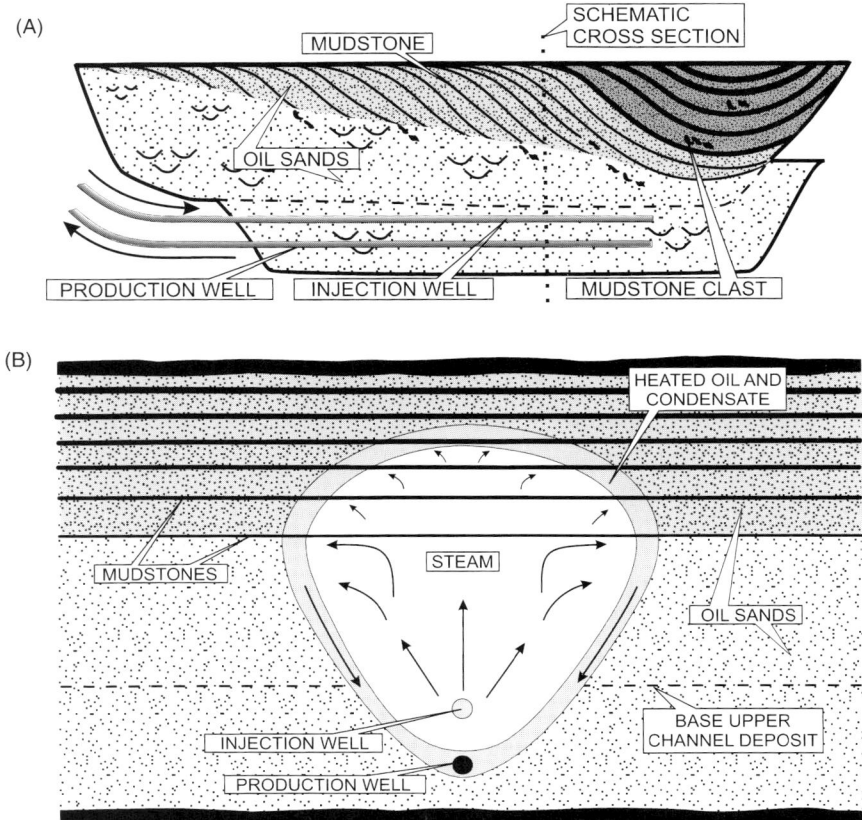

Figure O13 Schematic diagram of Steam Assisted Gravity Drainage (SAGD) in 2 stacked channel deposits. (A) Longitudinal profile of horizontal well pair. (B) Cross section of (A) showing development of steam chamber retarded by lateral accretion mudstones. Sandy to muddy channel "life cycle" from Wightman and Pemberton (1997).

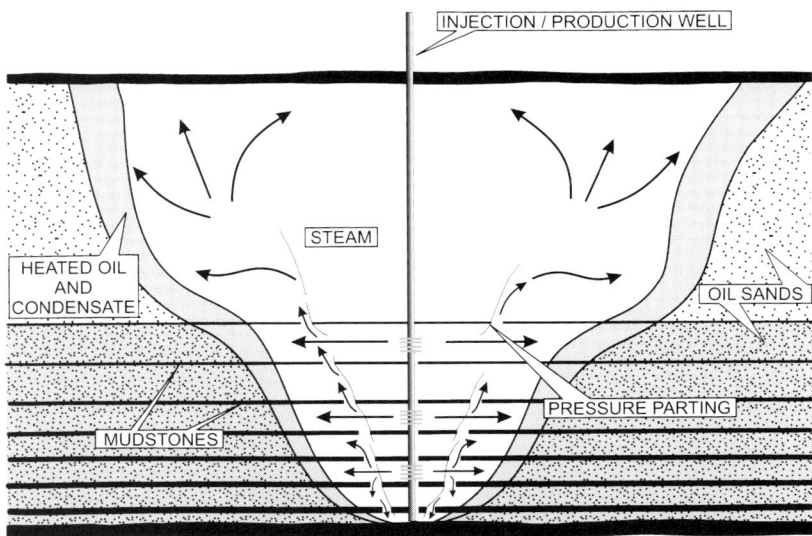

Figure O14 Schematic diagram of Cyclic Steam Stimulation (CSS), during steam injection, in a coarsening upward deposit.

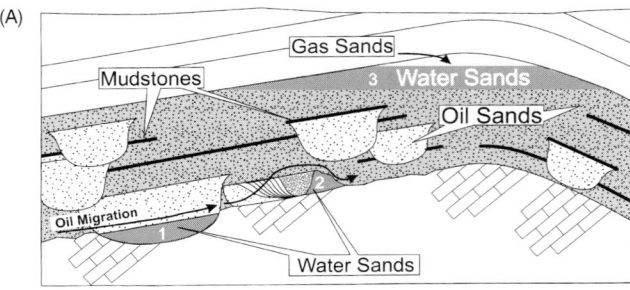

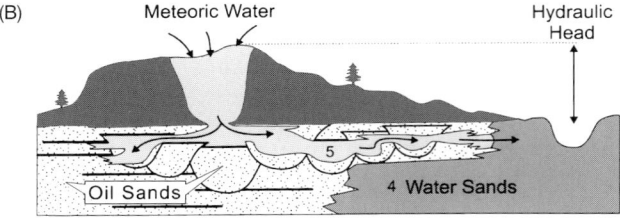

Figure O15 Schematic diagram of types of water sands. (A) In paleotopographic lows (1), behind low permeability strata (2) and in "leaked off" gas zones (3). (B) Regional water sands (4) and flushed zones (5) with low or remnant oil saturations; flushing greatest in high permeability sands/gravels.

Bibliography

AEUB (Alberta Energy and Utilities Board), 1996. *Crude Bitumen Reserves Atlas*. Statistical Series, 96-38.
Barson, D., Bachu, S., and Esslinger, P., 2001. Flow systems in the Mannville Group in the east-central Athabasca area and implications for steam-assisted gravity drainage (SAGD) operations for *in situ* bitumen production. *Bulletin of Canadian Petroleum Geology*, **49**: 376–392.
Butler, R.M., 1997. *Thermal Recovery of Oil and Bitumen*. Calgary: GravDrain Inc.
Creaney, S., and Allan, J., 1992. Petroleum systems in the Foreland Basin of Western Canada. In Macqueen, R.W., and Leckie, D.A. (eds.), *Foreland Basins and Fold Belts*. American Association of Petroleum Geologists Memoir, 55, pp. 279–308.
Fiorillo, G., 1987. Exploration and evaluation of the Orinoco Oil Belt. In Meyer, R.F. (ed.), *Exploration for Heavy Crude Oil and Natural Bitumen*. American Association of Petroleum Geologists Studies in Geology, 25, pp. 103–114.
Meyer, R., 1995. Bitumen. In *Encyclopedia of Energy Technology and the Environment*, 4 Volume Set, New York: John Wiley & Sons, pp. 433–454.
Meyer, R.F., 1996. Tars, tar sands and extra heavy oils. In Dasch, E.J. (ed.), *Encyclopedia of Earth Sciences*. Simon & Schuster Macmillan, pp. 1072–1075.
Meyer, R.F., 1998. World heavy crude oil resources. In *World Petroleum Congress Proceedings*. 15th, Beijing, pp. 459–471.
NEB (National Energy Board), 2000. *Canada's Oil Sands: A Supply and Market Outlook to 2015*.
Phizackerley, P.H., and Scott, L.O., 1978. Major tar-sand deposits of the world. In Chilingarian, G.V., and Yen, T.F. (eds.), *Bitumens, Asphalts and Tar Sands*. Developments in Petroleum Science, Volume 7, Elsevier, pp. 57–92.
Roadifer, R.E., 1987. Size distributions of the world's largest known oil and tar accumulations. In Meyer, R.F. (ed.), *Exploration for Heavy Crude Oil and Natural Bitumen*. American Association of Petroleum Geologists Studies in Geology, 25, pp. 3–24.
Starr, J., Prats, J.M., and Messulam, S.A., 1981. Chemical properties and reservoir characteristics of bitumen and heavy oil from Canada and Venezuela. In Meyer, R.F., and Steele, C.T. (eds.), *The Future of Heavy Crude Oils and Tar Sands*. First UNITAR International Conference, McGraw-Hill, pp. 168–173.
Strobl, R.S., Muwais, W.K., Wightman, D.M., Cotterill, D.K., and Yuan, L., 1997. Application of outcrop analogues and detailed reservoir characterization to the AOSTRA Underground Test Facility, McMurray Formation, Northeastern Alberta. In Pemberton, S.G., and James, D.P. (eds.), *Petroleum Geology of the Cretaceous Mannville Group, Western Canada*. Canadian Society of Petroleum Geologists Memoir, 18, pp. 375–391.
Wightman, D.M., Attalla, M.N., Wynne, D.A., Strobl, R.S., Berhane, H., Cotterill, D.K., and Berezniuk, T., 1995. *Resource Characterization of the McMurray/Wabiskaw Deposit in the Athabasca Oil Sands Area: A Synthesis*. Alberta Oil Sands Technology and Research Authority Technical Publication Series, 10.
Wightman, D.M., and Pemberton, S.G., 1997. The Lower Cretaceous (Aptian) McMurray Formation: an overview of the Fort McMurray area, northeastern Alberta. In Pemberton, S.G., and James, D.P. (eds.), *Petroleum Geology of the Cretaceous Mannville Group, Western Canada*. Canadian Society of Petroleum Geologists Memoir, 18, pp. 312–344.

Cross-references

Coastal Sedimentary Facies
Deltas and Estuaries
Rivers and Alluvial Fans
Sands and Sandstones

OOLITE AND COATED GRAINS

Oolite is a sedimentary rock, most often limestone, composed of smoothly laminated, spherical to ellipsoidal coated grains that are cemented together (Figure O16). The grain components of oolite are called *ooids* (or *ööids* or *ooliths*), terms derived from the Greek word for fish roe ("oon"), which ooids resemble. The terms "oolite" and "ooid" are considered distinct with oolite pertaining to rock and ooid referring to an individual grain (Teichert, 1970). Not all workers adhere to this technical distinction. In geologic descriptions, the term "oolitic" refers either to ooid-rich sediment or to ooid-bearing rocks. The term "pisoid", by convention, refers to those ooids larger than 2 mm. Historically, the term oolite was first used in 1727 by F.E. Bruckman (Cayeux, 1935, p.211). "Oolitic grains" were described by Lyell (1854, p.12) and Sorby (1879, p.74). The term ooid was first introduced by Kalkowsky (1908, p.72).

Ooids have rounded grain shapes and generally spherical to ellipsoidal grain forms (Figure O17). Internally, ooids consist of a *nucleus* surrounded by an evenly laminated *cortex* which increases in sphericity with distance from the nucleus. The nucleus commonly consists of a shell or fossil fragment, a peloid or a detrital sand grain whereas the cortex consists of concentric layers of calcium carbonate or other minerals.

Ooids are distinguished from other coated grains by descriptive and/or genetic criteria depending upon classification scheme. Most workers readily distinguish ooids from "oncoids" by the even lamination of the ooid cortex and the general absence of biogenic features such as algal filaments. Of these, evenness of cortex lamination is probably the most

OOLITE AND COATED GRAINS

Figure O16 Mississippian oolite, Ste. Genevieve Limestone, Kentucky. Note the laminated cortices surrounding the nuclei of some of the ooids (see arrowheads). Scale = 1 mm.

Figure O17 Modern ooids, Joulters Cay, Bahamas. Note the spherical to ellipsoid grain forms and the smooth, polished surfaces of the ooids. Scale = 1mm.

reliable criterion to distinguish ooids from oncoids, as biogenic features such as mucus films are evident in many ooids (e.g., Newell *et al.*, 1960). Although the arbitrary size distinction separating ooids from pisoids is clear, the term pisoid is additionally used by many workers to refer to nonmarine oolitic grains (e.g., Flügel, 1982). Because of this, some recent workers (Sumner and Grotzinger, 1993) have begun to use the term "giant ooid" when describing large (>2 mm) oolitic grains that appear to have formed in marine environments. Comprehensive reviews and information about ooids and other coated grains can be found in the works of Bathurst (1975), Simone (1981), Flügel (1982), Peryt (1983) and Tucker and Wright (1990). Reviews of ooids composed of iron-bearing minerals can be found in the work of Young (1989). Reviewed here are calcareous ooids and oolitic rocks as these form the majority of modern and ancient ooid-bearing deposits.

Modern environments of formation

Ooids form in a wide variety of modern depositional settings. Most form in shallow, tropical marine settings of moderate to high wave activity, at depths < 10 m (often ≤2 m), where the water is supersaturated with respect to calcium carbonate. Well-studied areas of modern marine ooid formation include the Great Bahama Bank and the Persian Gulf. Oolitic sands in these settings typically occur along the edges of platforms, in lagoons, in tidal deltas and mid-channel bars, and in beaches and tidal flats. Most marine ooids occur in or near the environment in which they are forming, although pelagic and eolian occurrences clearly indicate significant transportation. Ooids and pisoids also form in hypersaline lakes and lagoons, such as the Great Salt Lake, Utah and Baffin Bay, Texas as well as in brackish water lakes such as Pyramid Lake, Nevada (Sandberg, 1975; Land *et al.*, 1979; Popp and Wilkinson, 1983). Other environments where oolitic grains form include freshwater lakes, calcareous soils (calcrete), rivers, caves, hot springs and human-constructed features such as drainage pipes and water treatment plants.

Mineralogy and microfabric of modern ooids

Calcareous ooids are composed of aragonite, high Mg-calcite, low Mg-calcite or, in the case of bimineralic ooids, some combination of aragonite and calcite. Modern marine ooids have primarily an aragonite cortex; those with a high magnesian calcite cortex occur but are much less common. The cortex of ooids from nonmarine settings has a variable composition depending on the Mg/Ca ratio of the precipitating water and the salinity. Ooids from hypersaline settings with a high Mg/Ca ratio are dominantly aragonite, and to a lesser degree high Mg-calcite, or they are bimineralic with a cortex composed of a combination of aragonite and high Mg-calcite (e.g., Sandberg, 1975; Land *et al.*, 1979). Ooids from Pyramid Lake, Nevada, a brackish water ensialic lake with a fluctuating Mg/Ca ratio, are bimineralic with cortices composed of aragonite and low Mg-calcite (Popp and Wilkinson, 1983). Ooids from more dilute lakes as well as streams, caves and calcareous soils are commonly composed of low Mg-calcite.

The ooid cortex exhibits a range of crystal fabrics depending upon original mineralogy, the energy level of the depositional setting, organic influences and subsequent diagenesis. Most modern aragonitic ooids from the Bahamas have a *tangential microfabric* consisting of small (1–3 μm) aragonite rods and submicron nannograins oriented parallel to the cortex laminae. In thin section under cross-polarized light, modern tangential ooids commonly display a distinctive pseudouniaxial cross (Figure O18). The tangential microfabric appears limited to those ooids composed of aragonite and is considered by most workers to reflect relatively high energy conditions

Figure O18 Thin section view of modern tangential ooids under cross-polarized light, Joulters Cay, Bahamas. Note the pseudouniaxial crosses exhibited by several of the ooids. Scale = 0.1 mm.

characterized by numerous grain collisions. Some workers assert that tangential microfabrics are influenced by biological processes (e.g., Fabricius, 1977).

Modern ooids may also have a *radial microfabric* consisting of fibrous or bladed crystals of aragonite oriented normal to the ooid nucleus. Aragonitic ooids with radial microfabrics are well described from the Persian Gulf, the Great Barrier Reef and from the Great Salt Lake (Loreau and Purser, 1973; Davies and Martin, 1976; Sandberg, 1975). High Mg-calcite ooids with radial microfabrics are uncommon but have been described from the Great Barrier Reef (Marshall and Davies, 1975). Isolated radial laminae of low Mg-calcite occur in some ooids from Pyramid Lake (Popp and Wilkinson, 1983; Wilkinson et al., 1980). Although exceptions are known, radial ooids are more common to low-energy settings, settings characterized by fewer grain collisions.

Ooids lacking an orderly tangential or radial microfabric are described as having a *random microfabric* and are commonly referred to as micritic ooids. Modern examples generally consist of a random arrangement of aragonite rods or crystals, an arrangement that is more often a product of micritization rather than an original depositional fabric. Micritic ooids have been described from a variety of low and high-energy settings. Numerous additional ooid microfabrics have been described from some modern and ancient oolitic deposits (e.g., cerebroid ooids; half moon ooids). These special types are summarized in Richter (1983).

Origin of ooids

Ooids may form by a variety of physical, chemical and biological processes; however the exact mechanism(s) by which ooids form is still unresolved. The strictly physical origin proposed by Sorby (1879) involving vertical accretion (snowballing) of calcareous grains is often discounted, because grain interactions during ooid growth tend to polish the ooid surface rather than contribute to physical accumulation of calcium carbonate. The only site where substantive physical accumulation occurs is in borings within the ooid cortex (Gaffey, 1983). Formation by chemical (inorganic) processes is favored by many workers, particularly for radial ooids (Tucker and Wright, 1990). In the marine environment, inorganic formation of ooids requires supersaturated water, abundant nuclei, normal to elevated salinity, and warm temperatures ($>20°C$). Significant agitation may or may not be required for inorganic carbonate precipitation. Laboratory experiments mimicking such conditions have produced tangential ooids similar to modern Bahamian ooids (Davies et al., 1978).

Biological processes leading to the formation of ooids have been suspected since the late 19th century (Rothpletz, 1892). However, the degree to which organic compounds and organisms are involved in ooid formation is still a matter of debate. Several experimental studies have shown that the presence of organic compounds, particularly humic acids, can facilitate the formation of high Mg-calcite radial ooids, at least under laboratory conditions (Suess and Fütterer, 1972; Davies et al, 1978; Ferguson et al., 1978). Interestingly, in those solutions lacking any humic compounds, only aragonite needles or aragonite ooids with tangential fabrics formed. The role of algae and cyanobacteria in the formation of ooids remains largely unresolved, despite significant study by some workers (e.g., Fabricius, 1977). While algae and cyanobacteria certainly have a role in the formation of oncoids, their importance in the formation of ooids is less clear, especially those forming in caves or other settings with no light. Recent investigations of freshwater ooids from Lake Geneva (Davaud and Girardclos, 2001) demonstrate that nucleation of calcitic oolitic coatings can occur in close association with mucus films produced by the activities of diatoms and cyanobacteria. Those films appear to act as a catalyst or at least a substrate for carbonate precipitation and cortex growth. Similar associations between mucus films and crystal growth were described by Folk and Lynch (2001). They found minute crystals of aragonite with peculiar "baton" and composite needle morphologies imbedded in the mucus-rich laminae of some Bahamian aragonitic ooids. The formation of these crystals is attributed to the activities of minute "nannobacteria" in a manner similar to carbonate precipitation induced by larger bacteria (e.g., Chafetz and Buczynski, 1992). If nannobacteria truly exist—a matter still under debate—their presence and their micro-environmental influence strongly suggest that biological processes greatly influence the formation of ooids. Further work is needed to firmly establish the role of organic compounds and biological processes on the formation of ooids. This work will likely be the most active area of ooid research in the coming years.

Ancient ooids and oolite deposits

Oolite has a long geologic history extending at least to the late Archean. The environment of deposition of ancient oolitic deposits has been studied extensively, as has the texture and mineralogy of the ooids they contain. Ancient ooids commonly have radial textures, with or without clear concentric laminations; other ancient ooids consist of large neomorphic calcite crystals or are filled with clear calcite cement. Although radial-concentric ooids were previously thought to represent a diagenetic replacement product of aragonitic ooids, research

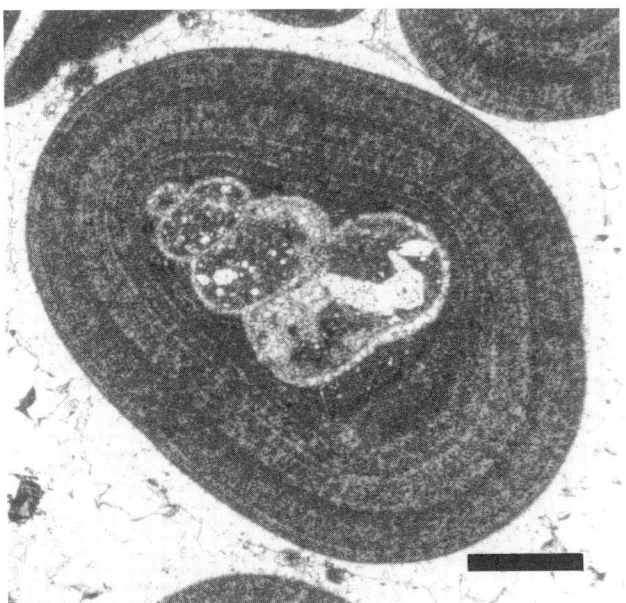

Figure O19 Thin section view of a Mississippian radial ooid, Ste. Genevieve Limestone, Illinois. The once aragonitic gastropod in the nucleus of the ooid has been replaced by clear calcite cement. The fact that no such replacement is seen in the cortex suggests strongly that the cortex was originally calcite. Scale = 0.1 mm.

has shown that these ooids were originally calcitic (Sandberg, 1975) (Figure O19). Originally aragonitic ooids have textures similar to that resulting from replacement of known aragonitic skeletal grains (Sandberg, 1985). These ooids contain coarse neomorphic spar, commonly with relic inclusions of aragonite. Aragonitic ooids are susceptible to dissolution in meteoric waters and also tend to develop oomoldic porosity that may later be filled with clear calcite cement.

Detailed petrographic studies have shown that mineralogy of ooids has oscillated throughout the Phanerozoic (Sandberg, 1983). Ooids that were originally composed of aragonite and possibly Mg-calcite occur primarily in rocks deposited during "icehouse" times (Late Precambrian–Early Cambrian, Middle Carboniferous–Triassic, and Tertiary–Recent). Ooids that were originally composed of calcite occur primarily in rocks deposited during "greenhouse" times (Ordovician-Mississippian and Jurassic–Cretaceous) (Sandberg, 1983). A few anomalies have been reported (e.g., Upper Jurassic Smackover Formation of the Gulf Coast) where aragonitic ooids occur during a calcite-facilitating episode. This may be due to local conditions such as evaporite deposition, leading to an increase in Mg/Ca ratio (Heydari and Moore, 1994). The oscillatory trend in the mineralogy of ooids has been linked to fluctuations in ocean-atmosphere pCO_2 level (Sandberg, 1983), although more recent investigations suggest that fluctuations in marine Mg/Ca ratios driven by plate tectonic processes are the governing factor (Stanley and Hardie, 1998). The general secular trend in the mineralogy of ooids correlates with the first-order, global sea-level, which is also tied to plate tectonic processes. High sea-level stands or submergent/greenhouse episodes are associated with high rates of sea-floor spreading and low Mg/Ca ratio. Such conditions favor the precipitation of low Mg-calcite ooids. Low sea-level stands or emergent/icehouse episodes are associated with low rates the sea-floor spreading rate and a relatively high Mg/Ca ratio. These conditions favor the precipitation of aragonite and high Mg-calcite ooids (Wilkinson et al., 1985; Stanley and Hardie, 1998). There were apparently four peak periods of ooid deposition during the Phanerozoic, including the Late Cambrian, Late Mississippian, Late Jurassic, and Holocene (Wilkinson et al., 1985). Peak ooid production occurred during times when overall sea-level was either rising or falling, rather than at maximum (highstands) or minimum (lowstands) sea-levels.

In addition to variations in the composition of ooids with time, recent studies have shown that ooid size has varied over geologic time, particularly during the Precambrian. Most Phanerozoic ooids are less than 2 mm, whereas those from the Precambrian have sizes up to 18 mm (Sumner and Grotzinger, 1993). Neoproterozoic ooids are particularly large, a phenomenon attributed to low nuclei supply, high growth rates and high average water velocity during that time (Sumner and Grotzinger, 1993).

Oolitic limestones form prolific petroleum reservoirs (see Keith and Zuppann, 1993). Intergranular porosity is preserved during greenhouse periods because of the calcitic mineralogy of the ooids. Calcitic ooids are resistant to dissolution compared to their aragonitic counterparts. This resistance limits the formation of oomoldic porosity and the occlusion of primary pores by secondary cements. Examples of petroleum reservoirs in oolitic rocks include the Jurassic Arab sequence in the Middle East, the Jurassic reservoirs of the Paris Basin, and the Jurassic Smackover reservoirs of the Gulf of Mexico (Moore, 2001, p.271). Significant reservoirs occur in Mississippian oolitic rocks as well. These include the Ste. Genevieve Limestone of the Illinois Basin, several Middle and Late Mississippian formations in the Anadarko Basin and Hugoton Embayment, the Greenbrier and Monteagle limestones of the Appalachian Basin, and the Madison Group of the Williston Basin (Keith and Zuppann, 1993, p.5).

Fredrick D. Siewers

Bibliography

Bathurst, R.C.G., 1975. *Carbonate Sediments and Their Diagenesis.* Developments in Sedimentology, Volume 12, Elsevier.
Cayeux, L., 1935. *Les Roches Sédimentaires De France, Roches Carbonatées.* Paris: Masson.
Chafetz, H.S., and Buczynski, C., 1992. Bacterially induced lithification of microbial mats. *Palaios*, **7**: 227–293.
Davaud, E., and Girardclos, S., 2001. Recent freshwater ooids and oncoids from western Lake Geneva (Switzerland): indications of a common organically mediated origin. *Journal of Sedimentary Research*, **71**: 423–429.
Davies, P.J., Bubela, B., and Ferguson, J., 1978. The formation of ooids. *Sedimentology*, **25**: 703–730.
Davies, P.J., and Martin, K., 1976. Radial aragonite ooids, Lizard Island, Great Barrier Reef. *Geology*, **4**: 120–122.
Fabricius, F.H., 1977. *Origin of Marine Ooids and Grapestone.* Contributions to Sedimentology, Volume 7, Stuttgart: E. Schweizerbart'sche Verlagsbuchhandlung.
Flügel, E., 1982. *Microfacies Analysis of Limestones.* Springer-Verlag.
Ferguson, J., Bubela, B., and Davies, P.J., 1978. Synthesis and possible mechanisms of formation of radial carbonate ooids. *Chemical Geology*, **22**: 285–308.
Folk, R.L., and Lynch, F.L., 2001. Organic matter, putative nannobacteria and the formation of ooids and hardgrounds. *Sedimentology*, **48**: 215–229.

Gaffey, S.J., 1983. Formation and infilling of pits of marine ooid surfaces. *Journal of Sedimentary Petrology*, **53**: 193–208.
Heydari, E., and Moore, C.H., 1994. Paleogeographic and paleoclimatic controls on ooid mineralogy of the Smackover Formation, Mississippi salt basin: implications for Late Jurassic seawater composition. *Journal of Sedimentary Research*, **64**: 101–114.
Keith, B.D., and Zuppann, C.W., 1993. *Mississippian Oolites and Modern Analogs*. AAPG Studies in Geology, Volume 35, Tulsa: American Association of Petroleum Geologists.
Kalkowsky, E., 1908. Oolith und Stromatolith im norddeutschen Buntsandstein. *Zeitschrift Der Deutschen Geologische Gesellschaft*, **60**: 68–125.
Land, L.S., Behrens, E.W., and Frishman, S.A., 1979. The ooids of Baffin Bay, Texas. *Journal of Sedimentary Petrology*, **49**: 1269–1278.
Loreau, J.-P., and Purser, B.H., 1973. Distribution and ultrastructure of Holocene ooids from the Persian Gulf. In Purser, B.H. (ed.), *The Persian Gulf*. Spring-Verlag, pp. 279–328.
Lyell, C.H., 1854. *A Manual of Elementary Geology*, 4th edn, New York: D. Appleton and Company.
Marshall, J.F., and Davies, P.J., 1975. High magnesium calcite ooids from the Great Barrier Reef. *Journal of Sedimentary Petrology*, **45**: 285–291.
Moore, C.H., 2001. *Carbonate Reservoirs, Porosity Evolution and Diagenesis in a Sequence Stratigraphic Framework*. Developments in Sedimentology, Volume 55, Elsevier.
Newell, N.D., Purdy, E.G., and Imbrie, J., 1960. Bahamian oölitic sand. *Journal of Geology*, **68**: 481–497.
Peryt, T.M., 1983. *Coated Grains*. Springer-Verlag.
Popp, B.N., and Wilkinson, B.H., 1983. Holocene lacusterine ooids from Pyramid Lake, Nevada. In Peryt, T.M. (ed.), *Coated Grains*. Springer-Verlag, pp. 142–153.
Richter, D.K., 1983. Calcareous ooids: a synopsis. In Peryt, T.M. (ed.), *Coated Grains*. Springer-Verlag: pp. 71–99.
Rothpletz, A., 1892. On the formation of oolite. *American Geologist*, **10**: 279–282.
Sandberg, P.A., 1975. New interpretations of Great Salt Lake ooids and of ancient non-skeletal carbonate mineralogy. *Sedimentology*, **22**: 497–537.
Sandberg, P.A., 1983. An oscillating trend in phanerozoic non-skeletal carbonate mineralogy. *Nature*, **305**: 19–22.
Sandberg, P.A., 1985. Recognition criteria for calcitized skeletal and non-skeletal aragonites. *Palaeontographica Americana*, **54**: 272–281.
Simone, L., 1981. Ooids: a review. *Earth-Science Reviews*, **16**: 319–355.
Sorby, HC., 1879. On the structure and origin of limestones. *Quarterly Journal of the Geological Society of London*, **35** (Proceedings), 56–95.
Stanley, S.M., and Hardie, L.A., 1998. Secular oscillations in the carbonate mineralogy of reef-building and sediment-producing organisms driven by tectonically forced shifts in seawater chemistry. *Paleogeography, Paleoclimatology, Paleoecology*, **144**: 3–19.
Suess, E., and Fütterer, D., 1972. Aragonite oöids: experimental precipitation from seawater in the presence of humic acid. *Sedimentology*, **19**: 129–139.
Sumner, D.A., and Grotzinger, J.P., 1993. Numerical modeling of ooid size and the problem of Neoproterozoic giant ooids. *Journal of Sedimentary Petrology*, **63**: 974–982.
Teichert, C., 1970. Oolite, oolith, ooid: discussion. *American Association of Petroleum Geologists Bulletin*. **54**: 1748–1749.
Tucker, M.E., and Wright, V.P., 1990. *Carbonate Sedimentology*. Blackwell.
Wilkinson, B.H., Popp. B.N., and Owen, R.M., 1980. Nearshore oolite formation in a modern temperate-region marl lake. *Journal of Geology*, **88**: 697–704.
Wilkinson, B.H., Owen, R.M., and Carroll, A.R., 1985. Submarine hydrothermal weathering, global eustasy, and carbonate polymorphism in phanerozoic marine oolites. *Journal of Sedimentary Research*, **55**: 171–183.
Young, T.P., 1989. Phanerozoic ironstones: an introduction and review. In Young, T.P., and Taylor, W.E.G. (eds.), *Phanerozoic Ironstones*. Geological Society of London, Special Publication, **46**, ix–xxv.

Cross-references

Algal and Bacterial Carbonate Sediments
Bacteria in Sediments
Bioerosion
Caliche–Calcrete
Carbonate Mineralogy and Geochemistry
Cave Sediments
Classification of Sediments and Sedimentary Rocks
Diagenesis
Neomorphism and Recrystallization
Neritic Carbonate Depositional Environments
Seawater: Temporal Changes in the Major Solutes
Tidal Flats
Tidal Inlets and Deltas
Tufas and Travertines

OPHICALCITES

Introduction

Serpentinites and ophiolitic mélanges (*q.v.*) locally contain carbonatized ophiolitic breccias known as ophicalcites. The best-documented cases are from the Jurassic of the Swiss Alps (Weissert and Bernoulli, 1984; Bernoulli and Weissert, 1985; Früh-Green *et al.*, 1990) and of the Italian Apennines (Folk and McBride, 1976; Barbieri *et al.*, 1979), and from the Cretaceous of the USA. Rockies (Carlson, 1984). Paleozoic ophicalcite is only known from the Ordovician of the Canadian Appalachians (Lavoie and Cousineau, 1995).

Hypothesis of origin

The origin of ophicalcites is controversial (Folk, 1986; Bernoulli and Weissert, 1986). In the second edition of the AGI *Glossary of Geology* (1980, p.437), ophicalcites (or ophicarbonates) are described as "a recrystallized metamorphic rock composed of calcite and serpentine, commonly formed by dedolomitization of a siliceous dolostone." Before ophiolites were interpreted to represent oceanic crust, ophicalcites were assumed to result from contact metamorphism between intrusive ultramafics and older sedimentary hosts (Peters, 1963) or from regional metamorphism (Trommsdorff *et al.*, 1980). Since plate tectonics, near-to-seafloor high temperature processes have been suggested: hydrothermal weathering (Barbieri *et al.*, 1979) and combined tectonosedimentary and hydrothermal events (Lemoine, 1980; Bernoulli and Weissert, 1985; Früh-Green *et al.*, 1990; Lavoie and Cousineau, 1995). Cold processes have also been suggested: carbonate precipitation from methane seepage on serpentine seamounts (Haggerty, 1991; Fryer, 1992) and pedogenic alteration of ophiolite (Folk and McBride, 1976).

Modern occurrences of ophicalcite

Carbonatized serpentine fragments have been dredged from hydrothermally active transform-fault settings along the mid-Atlantic ridge (Bonatti *et al.*, 1974) and were also recovered from serpentine seamounts in a forearc setting (Haggerty, 1991). Void-filling pelagic and coarser-grained carbonates are locally interbedded with lithic (ophiolite-derived) laminae. In voids, carbonate cements consisting of magnesium calcite or aragonite are present. A near-to-seafloor origin was proposed: water reached the ultramafic part of an oceanic crust and transformed it to serpentine resulting in density inversion and diapiric rise of the serpentinite. Brecciation and carbonatization follow as the serpentine diapir reaches the ocean floor (Bonatti et al., 1977).

Rock record of ophicalcite

Paleozoic ophicalcite

In the Canadian Appalachians, ophicalcites have been reported in Ordovician ophiolitic mélanges located along major terrane-bounding faults (Cousineau, 1991). The mélanges consist of serpentinized ultramafics locally brecciated and carbonatized (ophicalcites) as well as metasedimentary and igneous rocks fragments. An oceanic transform-fault setting has been proposed for the formation of the Ordovician ophiolitic mélange (Cousineau, 1991).

The ophicalcite is a breccia; it is composed of angular, millimeter- to meter-size, clast- to matrix/cement-supported carbonatized ultramafic blocks. The ultramafics were first intensely serpentinized and variably deformed and carbonatization of ultramafic material followed. The degree of carbonatization ranges from faintly fractured serpentinite with carbonatized phenocrysts and some micrite infills, up to intensely brecciated material showing various carbonates with few preserved indicators of the protolith including chromite grains and serpentinite minerals. The relative chronology between the various void fillings is obscured by complex crosscutting patterns of fractures. In many cases, fractures with the complete spectrum of sediments and cements (see below) cut through older ones totally occluded by a similar succession. Meter-sized fragments of well-cemented ophicalcites are found in the micrite. All these suggest a complex history of sedimentation—cementation—fracturing.

Carbonate sediments in the ophicalcite

Besides the pervasive initial metasomatic carbonatization of brecciated ultramafics, discrete cavity-filling carbonates are recognized. The most important consists of laminated micrite interlayered with graded coarser-grained lithic (serpentinite, chromite, mafic minerals) calcarenite. In the most carbonatized cases, the sediments can constitute up to 50 percent of total carbonates. In the least altered ultramafics, it commonly is the only carbonate phase.

Carbonate cements in the ophicalcite

Isopachous layers of calcite cement coat micrite and were precipitated either over undisturbed infills or on clasts of indurated sediment, suggesting synchroneity with some fracturing. Remaining void spaces are healed, as seen under cathodoluminescence, by multiple generations of blocky carbonate cements (calcite and minor dolomite). These cements are volumetrically important in large fractures and voids that were not totally occluded by the previous carbonates.

Geochemistry of carbonate elements

For the Appalachian ophicalcite, the oxygen and carbon stable isotope ratios of the micrite fall in the range of values for Ordovician marine carbonates. Based on oxygen stable isotope and fluid inclusions microthermometry, following cements were precipitated by hot saline fluids. The carbon isotopic values for cements indicate marine waters with input of hydrothermal-derived fluids as also suggested from low radiogenic strontium isotopic ratios of cements (Lavoie and Cousineau, 1995; Lavoie, 1997; Chi and Lavoie, 2000).

Proposed origin for Paleozoic ophicalcite

Crosscutting relationships between sediments, cements, and clasts of cemented ophicalcite together with carbonate geochemistry provide evidence for complex early seafloor fracturing, sedimentation/cementation, and fluid circulation. For the Paleozoic ophicalcite, a tectonically active seafloor hydrothermal vent system was proposed to explain the synchroneity of micrite sedimentation, fracturing, and cementation (Lavoie and Cousineau, 1995).

The Mesozoic ophicalcites

Detailed studies of Mesozoic ophicalcites have documented a similar complex pattern of crosscutting relationships between micrite, carbonate cements, and ophicalcite/ophiolite clasts altogether with multiple fracturing events. Repetitive micrite and serpentinite arenite infills coeval with diverse carbonate cements is documented. Based on these observations and stable isotope geochemistry, a similar scenario of early tectonosedimentary origin on the deep seafloor was also proposed for Jurassic ophicalcites (Früh-Green et al., 1990).

Life on the deep seafloor

Methane-based bacterial chemosynthesis associated with ophicalcites is reported from the Mesozoic of North America (Carlson, 1984). Limited biological activity occurred around the hydrothermal vents responsible for the formation of the Paleozoic ophicalcite (Lavoie, 1997). Peloidal mats coated clasts and are followed by bacterially mediated botryoidal calcite precipitation. Carbon stable isotope ratios of peloidal mats and botryoidal cements indicate that the exhalatives were hydrocarbon-free. The lack of corrosion of carbonates suggests minimal rate of H_2S venting and indicates that the venting system was similar to H_2S-poor white smokers. Thermophilic bacterial sulfate reduction was active around these Ordovician vents.

Denis Lavoie

Bibliography

Barbieri, M., Masi, U., and Tolomeo, L., 1979. Stable isotope evidence for a marine origin of ophicalcites from the north-central Apennines (Italy). *Marine Geology*, **30**: 193–204.

Bernoulli, D., and Weissert, H., 1985. Sedimentary fabrics in Alpine ophicalcites, south Pennine Arosa zone, Switzerland. *Geology*, **13**: 755–758.

Bernoulli, D., and Weissert, H., 1986. Sedimentary fabrics in Alpine ophicalcites, south Pennine Arosa zone, Switzerland—reply. *Geology*, **14**: 637–638.

Bonatti, E., Emiliani, C., Ferrara, G., Honnorez, J., and Rydell, H., 1974. Ultramafic carbonate breccias from the equatorial Mid-Atlantic Ridge. *Marine Geology*, **16**: 83–102.

Bonatti, E., Sarnthein, M., Boersma, A., Gorini, M., and Honnorez, J., 1977. Neogene crustal emersion and subsidence at the Romanche fracture, equatorial Atlantic. *Earth and Planetary Science Letters*, **35**: 369–383.

Carlson, C., 1984. Stratigraphic and structural significance of foliate serpentinite breccias, Wilbur Springs. In Carlson, C. (ed.), *Depositional Facies of Sedimentary Serpentinite: Selected Examples from the Coast Ranges, California*. SEPM, Pacific Section, Field Trip Guidebook No. 3, pp. 108–112.

Chi, G., and Lavoie, D., 2000. A combined fluid-inclusion and stable isotope study of Ordovician ophicalcite units from southern Quebec Appalachians, Quebec. In Geological Survey of Canada, Current Research, 2000-D5, 9p.

Cousineau, P.A., 1991. The Rivière des Plantes ophiolitic melange; Tectonic setting and melange formation in the Québec Appalachians. *Journal of Geology*, **99**: 81–96.

Folk, R.L., 1986. Sedimentary fabrics in Alpine ophicalcites, south Pennine Arosa zone, Switzerland—discussion. *Geology*, **14**: 636.

Folk, R.L., and McBride, E.F., 1976. Possible pedogenic origin of Ligurian ophicalcite: a mesozoic calichified serpentinite. *Geology*, **4**: 327–332.

Früh-Green, G.L., Weissert, H., and Bernoulli, D., 1990. A multiple fluid history in Alpine ophiolites. *Geological Society of London Journal*, **147**: 959–970.

Fryer, P., 1992. A synthesis of Leg 125 drilling of serpentine seamounts on the Mariana and Izu-Bonin forearcs. In Fryer, P., Pearce, J.A., Stokking, L.B. *et al.* (eds.), *Proceedings of the Ocean Drilling Program, Scientific Results*, Volume 125, pp. 593–614.

Haggerty, J.A., 1991. Evidence from fluid seeps atop serpentine seamounts in the Mariana Forearc: clues for emplacement of the seamounts and their relationships to forearc tectonics. *Marine Geology*, **102**: 293–309.

Lavoie, D., and Cousineau, P.A., 1995. Ordovician ophicalcites of southern Quebec Appalachians: a proposed early seafloor tectonosedimentary and hydrothermal origin. *Journal of Sedimentary Research*, **A65**: 337–347.

Lavoie, D., 1997. Hydrothermal vent bacterial community in Ordovician ophicalcite, southern Quebec Appalachians. *Journal of Sedimentary Research*, **67**: 47–53.

Lemoine, M., 1980. Serpentinites, gabbros and ophicalcites in the Piedmont-Ligurian domain of the western Alps: possible indicators of oceanic fracture zones and of associated serpentinite protrusions in the Jurassic-Cretaceous Tethys. *Archives des Sciences Genève*, **33**: 103–115.

Peters, T., 1963. Mineralogie und petrographie des totalpserpentins bei davos. *Schweizerische Mineralogische Und Petrographische Mitteilungen*, **43**: 529–686.

Trommsdorff, V., Evans, B.W., and Pfeifer, H.R., 1980. Ophicarbonate rocks: metamorphic reactions and possible origin. *Archives des Sciences Genève*, **33**: 361–364.

Weissert, H., and Bernoulli, D., 1984. Oxygen isotope composition of calcite in Alpine ophicarbonates: a hydrothermal or Alpine metamorphic signal? *Eclogae Geologicae Helvetiae*, **77**: 29–43.

Cross-references

Bacteria in Sediments
Cathodoluminescence
Cements and Cementation
Diagenesis
Fluid Inclusions
Isotopic Methods in Sedimentology
Mélange; Melange

P

PALEOCURRENT ANALYSIS

Introduction

The configuration of sedimentary bodies, from the smallest patch of sand or gravel to the deposits of entire depositional systems, is determined in part by the pattern of air or water currents prevailing during deposition. Evidence of the orientation and strength of these currents can commonly be deduced from evidence preserved within the rocks.

There are two types of paleocurrent evidence:

(1) Sedimentary structures and fabrics that preserve a record of the direction and orientation of the current that formed them;
(2) Rock properties that change in the direction of flow as a result of the transport process;

Paleocurrent analysis may provide information on one or more of the following:

(1) the direction of local or regional paleoslope, which reflects tectonic subsidence patterns;
(2) the direction of sediment supply (in unimodal systems, only);
(3) the geometry and trend of lithologic units; and
(4) the depositional environment.

Paleocurrent indicators

Sedimentary structures that may contain useful paleocurrent information include:

(1) *Ripple marks* (*q.v.*) *and crossbedding* (see *Cross-lamination*). Dune and ripple crests are typically oriented transverse to flow, while the foresets of crossbed structures normally dip in a downstream direction. Because of local turbulence of flow, the direction indicated by foreset dip may not always correspond exactly to local flow direction, and so it is usually advisable to use a large number of orientation measurements to ensure reliable results. This is the only type of sedimentary structure from which paleocurrent data may be obtained from the subsurface using petrophysical techniques such as the dipmeter or microresistivity scanning tools.
(2) *Channels and scours* occur in many environments and may indicate the orientation of major erosive currents, such as those generating river or tidal channels and delta or submarine fan distributaries. The larger channels, which are those most likely to be of regional significance, are usually too large to be preserved in outcrops, but may be spectacularly displayed by reflection-seismic data, especially in horizontal (time-slice) sections.
(3) *Parting lineation* (*q.v.*) *or primary current lineation*, the product of plane-bed flow conditions, is only visible on bedding-plane exposures. It usually yields directional readings of low variance because it forms during high-energy flow in river, delta, or tidal channels, when bars or other obstructions to flow are under water and flow-sinuosity is low. The structure indicates orientation but not direction of flow, because of the ambiguity between two equally possible readings at 180° to each other. Usually, this can be resolved with reference to other structures nearby.
(4) Clast transport by traction or in sediment-gravity flows commonly produces a measurable *gravel fabric*. Imbrication (*q.v.*) in traction current deposits occurs where platy clasts are stacked up in a shingled pattern, with their flattest surface dipping upstream and resting on the next clast downstream. Because gravel is only moved under high-energy flow conditions, it tends to show low directional variance, like parting lineation.
(5) *Sole markings* are typically associated with the deposits of turbidity currents and fluidized or liquified flow, where vortex erosion may occur at the base of a flow (see *Scour, Scour Marks*), and "tools," such as pebbles, twigs or bone fragments, can be swept down a bedding surface (see *Tool Marks*). They also occur less commonly in other clastic environments. Their greatest use, however, is in the investigation of submarine-canyon and fan deposits. They are best seen on the undersides of bedding surfaces, where sandstone has formed a cast of the erosional feature in the

underlying bed. Tool markings yield information on orientation but not direction, like parting lineation; flute marks are longitudinally asymmetric, with their deepest end lying upstream.

(6) *Oriented plants, bones, shells*, etc. do not respond systematically to the aligning effects of currents unless they are elongated. There may be ambiguity as to whether they are oriented transverse or perpendicular to current patterns, and there are other difficulties, for example, the tendency of fossils such as high-spired gastropods to roll in an arc. Fossils are usually only useful for local, specialized paleocurrent studies.

(7) Oriented sandstone samples may be studied in the laboratory to explore such parameters as sand-grain orientation and magnetic anisotropy.

Paleocurrent patterns may also be deduced from regional changes in rock properties such as clastic grain size and clast roundness, the downstream decrease in thickness of sediment gravity flows, or the change in proportions of detrital components.

Data collection and processing

Paleocurrent data should be carefully documented in the field, including the following information:

(1) location and (if relevant) precise position in a stratigraphic section;
(2) structure type;
(3) indicated current direction;
(4) scale of structure (thickness of crossbed, depth and width of channel, mean or maximum clast size in imbricate gravels); and
(5) local structural dip.

Current directions should be measured to the nearest 5° with a magnetic or sun-compass and corrected to true north wherever necessary. In the case of parting lineation and tool markings, the correct orientation of two possibilities at 180° can usually be identified by referring to other types of current structures nearby. Measurement accuracy greater than $\pm 5°$ is difficult to achieve and is, in any case, rarely important. Difficulties in field measurement commonly arise because of incomplete exposure of crossbed sets. For example, two-dimensional exposures in flat outcrop faces cannot provide precise orientation information.

Indicated current directions may need to be corrected for structural dip and fold plunge, otherwise significant directional distortions may result. For linear structures such as parting lineation or sole markings, structural dips as high as 30° can be safely ignored, as they result in errors of less than 4°. However, the foreset dip orientation of crossbedding is significantly affected by structural dip, and should be corrected wherever the structural dip exceeds 10°. This subject was discussed further by Potter and Pettijohn (1977, chapter 10), who illustrated a correction technique using a stereonet. Ramsay and Huber (1983) provided graphical solutions. Middleton (2000) referred to computer routines for the necessary trigonometric calculations. Wells (1988) and Miall (1999) provided practical guides to field procedures, and recommended the use of a pocket calculator to perform structural corrections and data synthesis in the field.

Sooner or later the question will arise as to how many readings should be made? There is no single or simple answer to this problem because it depends on how many measurable current indicators are available for observation and what the objectives of the study are. Olson and Potter (1954) discussed the use of grid-sampling procedures and random selection of structures to measure, followed by calculation of reliability estimators to determine how many readings were necessary in order to be sure of determining correct directional trends. In this study, they were concerned only with determining regional paleoslope. We now understand a great deal about air and water flow patterns in different depositional environments, and a case could be made for measuring and recording every visible sedimentary structure. Such detailed data can be immensely useful in amplifying environmental interpretations and clarifying local problems. Some selection may have to be made in areas of particularly good exposure. A practical compromise is to record every available structure along measured stratigraphic sections and to fill in the gaps between sections with spot (gridded or random) samples. This procedure permits the elaboration of both local and regional paleocurrent trends. When constructing lateral profiles of large outcrops, it is essential to document these with abundant paleocurrent determinations. Such data provide the third dimension to a two-dimensional outcrop and yield essential information regarding the shape and orientation of architectural elements.

If the trend itself is important, 25 readings per sample station is commonly regarded as the minimum necessary for statistically significant small samples. However, the same or fewer readings plotted in map or section form can yield a great deal of environmental detail, whether or not their mean direction turns out to be statistically significant. Several hundred or a few thousand readings may be necessary for a thorough analysis of a complete basin.

A variety of statistical data-reduction and data-display techniques is available for paleocurrent work. The commonest approach is to group data into subsets according to stratigraphic or areal distribution criteria, display them visually in current-rose diagrams, and calculate their mean and standard deviation (or variance). The method of grouping the data into subsets has an important bearing on the interpretations to be made from them, as discussed in the next section. A current rose diagram is simply a histogram converted to a circular distribution. The compass is divided into 20, 30, 40, or 45° segments, and the rose is drawn with the segment radius proportional to number of readings or percent of total readings, but this exaggerates the importance of modes. A visually more correct procedure is to draw the radius proportional to the square root of the percent number of readings, so that the segment area is proportional to percent. Examples are given in Figure P1. Wells (2000) warned that the choice of origin for rose diagrams and the method of subdivision of the segments can affect the visual appearance of the resulting diagram and the orientation of the modes. The calculation of means, which is done from raw data, is not affected, but Wells' warning indicates the need for caution in the visual interpretation of rose diagrams.

Arithmetic and vector methods may be used for calculating mean, variance, vector strength, and statistical tests for randomness (Curray, 1956; Potter and Pettijohn, 1977; Wells, 1988).

Statistical procedures, such as moving averages and trend analysis, are available for smoothing local detail and determining regional trends. A moving average map is constructed from gridded data. An arbitrary grid is drawn

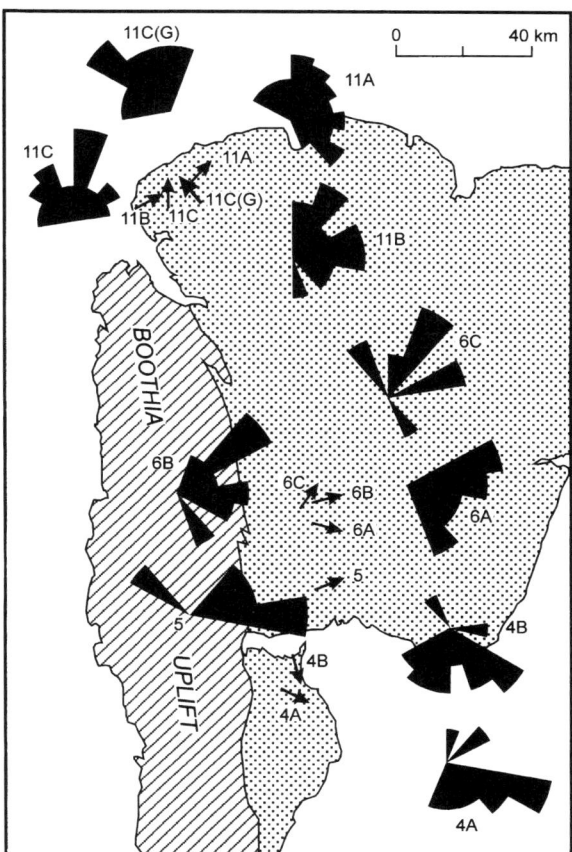

Figure P1 The use of paleocurrent rose diagrams to illustrate the spread of readings at a given location. In this case (a Devonian fluvial system in the Canadian Arctic; from Miall and Gibling, 1978), each segment of each rose is an arc of 25°, and mean flow directions for each location are shown by arrows. The arrow corresponding to each rose diagram is indicated by the location number.

on the map; the mean current direction for the data in each group of four squares is then calculated and shown by an arrow at the center of this area. Each data point is thus used four times (except at the edges of the map).

The basin analyst should be wary of becoming too deeply enmeshed in the refinements of statistical methods. The use of probability tests is a useful curb on one's wilder flights of interpretive fancy, but there is a vast literature on the statistics of the circular distribution that seems to detract attention from very simple questions: do the data make geological sense? Can they be correlated with trends derived from other methods, such as lithofacies mapping or petrographic data? Moving average and trend analyzes are useful techniques for reducing masses of data to visually appealing maps, but they inevitably result in a loss of much interesting detail.

Environment and paleoslope interpretations

A recommended first level of analysis is to plot current-rose diagrams and mean vectors for each outcrop or local outcrop group (Figure P1), separating the readings according to major facies variations. If contemporaneous basin shape and orientation can be deduced from these data or from other mapping techniques, the paleocurrent data can be used interactively with lithofacies and biofacies criteria to interpret the depositional environment and to outline major depositional systems, such as deltas or submarine fan complexes. The geologist should examine the relationships between mean current directions in different lithofacies and in different outcrops. The number of modes in the rose diagrams (modality) and their orientation with respect to assumed shoreline or lithofacies contours are also important information. Another useful approach is to plot individual readings or small groups of readings collected in measured sections at the correct position in a graphic section.

Paleocurrent distributions are commonly categorized as unimodal, bimodal, trimodal and polymodal. Each reflects a particular style of current dispersion. For example, the rose diagrams in Figure P1 are unimodal to weakly bimodal, and vector mean directors are all oriented in easterly directions, roughly normal to the basin margin, with the exception of station 11C(G). The data were derived mainly from trough and planar crossbedding and parting lineation, and the environment of deposition is interpreted as braided fluvial (Miall and Gibling, 1978). In some areas, for example, stations 6A, B, and C, there is a suggestion of a fanning out of the current systems, indicating possible deposition on large, sandy alluvial fans. The anomalous data set of station 11C(G) was derived from giant crossbed sets ranging from 1 m to 6 m in thickness. They may be eolian dunes or large sand waves in a trunk river draining the basin east of Boothia Uplift. Field evidence is inconclusive, but the paleocurrent data can be reconciled with either interpretation.

Coastal regions where rivers debouch into an area affected by waves and tides can give rise to very complex paleocurrent patterns (Figure P2). Bimodal, trimodal, or polymodal distributions may result. However, time-velocity asymmetry of tidal currents can result in local segregation of currents, so that they are locally unimodally ebb- or flood-dominated. This could cause confusion at the outcrop level, because the tidal deltas showing such paleocurrent patterns may consist of lithofacies that are very similar to some fluvial deposits. In wave-dominated environments, current reversals generate distinctive internal ripple-lamination patterns and herringbone crossbedding. Paleocurrent analyzes of these bimodal crossbeds may yield much useful information on the direction of wave attack and hence on shoreline orientation. The term paleoslope means little in this environment because there are a diversity of local slopes and because waves and tides are only marginally influenced by the presence and orientation of bottom slopes. Paleocurrent data cannot be used to infer provenance in such situations.

Paleocurrent data formed part of the basis for a long-standing debate regarding the formation of hummocky cross-stratification in shallow-marine environments. Some workers attributed this structure in part to geostrophic currents flowing obliquely across the continental shelf; others argued that shore-normal paleocurrents indicate an origin as a result of wave-generated oscillatory and return flow across the shelf floor (the debate is summarized in Walker and Plint, 1992).

Submarine-fan and other deep-marine deposits show three main paleocurrent patterns: (1) Individual submarine fans prograde out from the continental slope and therefore show radial paleocurrent patterns with vector mean directions

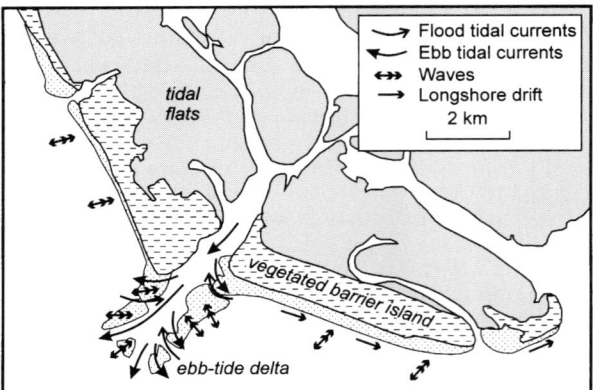

Figure P2 The complex flow patterns around a tidal inlet. Waves typically approach a shoreline obliquely, but are deflected to arrive at the shoreline with a nearly perpendicular orientation. Tidal currents are locally ebb- or flood-dominated. In such cases recorded paleoflow may be nearly unimodal. However, partial preservation of all wave and tide influences is typical, and can generate very complex paleocurrent patterns. Longshore currents are generated by wave and tide pressures at the shore, and may be particularly active during storms.

oriented perpendicular to the regional basin strike. This pattern is typical of many continental margins, particularly divergent margins. (2) In narrow oceans and many arc-related basins, deep-water sedimentation takes place in a trough oriented parallel to tectonic strike. Sediment-gravity flows, particularly low-viscosity turbidity currents, emerge from submarine fans and turn 90°, to flow longitudinally downslope, possibly for hundreds of kilometers. Current directions may be reversed by tilting of the basin. (3) Contour currents or boundary undercurrents flow parallel to continental margins and generate paleocurrent patterns oriented parallel to basin margins.

On a basinal scale paleocurrent trends may reveal large-scale paleogeographic features. For example, the east to northeastward-directed flow indicated by all the rose diagrams in Figure P1 (except 11C(G)) indicates a regional paleoslope dipping in that direction, a fact confirmed by other stratigraphic and sedimentologic data. In fluvial environments, flow directions transverse and parallel to basin axis may indicate the presence of tributary and trunk river systems. In coastal regions paleocurrents may indicate the orientation of shorelines, and deep marine paleocurrents may be used to determine the orientation of the continental slope and submarine canyons.

Andrew D. Miall

Bibliography

Curray, J.R., 1956. The analysis of two-dimensional orientation data. *Journal of Geology*, **64**: 117–131.
Miall, A.D., 1999. *Principles of Sedimentary Basin Analysis*. 3rd edn, Springer-Verlag.
Miall, A.D., and Gibling, , M.R., 1978. The Siluro-Devonian clastic wedge of Somerset Island, Arctic Canada, and some regional paleogeographic implications. *Sedimentary Geology*, **21**: 85–127.
Middleton, G.V., 2000. *Data Analysis in the Earth Sciences*. Prentice Hall.
Olson, J.S., and Potter, , P.E., 1954. Variance components of crossbedding direction in some basal Pennsylvanian sandstones of the eastern Interior Basin: statistical methods. *Journal of Geology*, **62**: 26–49.
Parks, J.M., 1970. Computerized trigonometric solution for rotation of structurally tilted sedimentary directional features. *Geological Society of America Bulletin*, **81**: 537–540.
Potter, P.E., and Pettijohn, F.J.. 1977. *Paleocurrents and Basin Analysis*. Academic Press.
Ramsay, J.G., and Huber, M.I., 1983. *The Techniques of Modern Structural Geology*. Academic Press.
Walker, R.G., and Plint, A.G., 1992. Wave- and storm-dominated shallow marine systems. In Walker, R.G., and James, N.P. (eds.), *Facies Models: Response to Sea-level Change*. Geological Association of Canada, Geotext 1, pp. 219–238.
Wells, N.A., 1988. Working with paleocurrents. *Journal of Geological Education*, **35**: 39–43.
Wells, N.A., 2000. Are there better alternatives to standard rose diagrams? *Journal of Sedimentary Research*, **70**: 37–46.

Cross-references

Bedding and Internal Structures
Cross-stratification
Fabric, Porosity, and Permeability
Imbrication and Flow-Oriented Clasts
Parting Lineations and Current Crescents
Ripple, Ripple Mark, Ripple Structure
Scour, Scour Marks
Surface Forms
Tool Marks

PARTING LINEATIONS AND CURRENT CRESCENTS

Introduction

Parting lineations and current crescents are structures that are commonly preserved on the bedding surfaces of generally planar bedded sandstones and siltstones. Such structures are useful in determining paleocurrent directions and the nature of the current that transported the sediment. The structures can form in response to unidirectional or oscillating (long and short period) currents and they may be present in the deposits of a range of environmental settings, including fluvial, beach, shallow marine, and deep marine (turbidites) deposits.

Parting lineations

Parting lineations include two related structures that may be commonly seen when a planar laminated (horizontal or inclined) rock is split, or parted, along internal bedding planes. This general term refers to any low relief, linear structure that is present on such surfaces (Crowell, 1955). Two common forms of parting lineation are termed "parting-step lineation" and "current lineation" and both may be present on the same bedding surface.

Parting-step lineations (McBride and Yeakel, 1963), are structures that reflect the manner in which a rock breaks, parallel to original bedding. Indurated stratified sandstones will commonly split parallel to bedding but not perfectly so. When such rocks are split, steps form on the top bedding surface which are remnants of immediately overlying bed that have not separated from its basal surface. These remnants form lamina-thick (a millimeter or less) steps on the surface that are longer than they are wide; these elongated steps make up parting-step lineation (Figure P3(A)). McBride and Yeakle (1963) demonstrated that the orientation of the trend of the length of these steps is approximately parallel to the orientation of the long axes of sand grains within the rock. Due to the relatively strong preferred alignment of the sand grains, when laminae break as a rock is parted, the broken laminae form elongate ridges, or steps, that are longest in the direction parallel to the grain long axes. The orientation of sand grains in such rocks is well-known to be determined by the direction of the depositing flow; the grains are aligned parallel to the current. Therefore, the orientation of parting step lineation similarly trends parallel to the depositing current. It is important to note that parting-step lineation is not a primary structure but is a reflection of the primary internal fabric which, in turn, reflects the sense (but not the absolute direction) of the depositing current.

Current lineations are primary sedimentary structures that are made up of more-or-less parallel, linear mounds (a few grain diameters high) extending parallel to parting-step lineations, internal fabric and, therefore, the depositing current (Figures P3(A) and P4(B)). The lineations may extend from a few centimeters to several tens of centimeters along their length and their transverse spacing ranges from a few millimeters to over a centimeter. Allen (1964) explained the formation of current lineation by the action of flow-parallel rotating vortices near the boundary of the flow, which sweep grains laterally into the linear mounds that characterize this structure. More recently, Allen (1982) suggested that current lineations form as sand grains are swept into low speed streaks within the viscous sublayer of a current. Weedman and Slingerland (1985) found that the spacing of "sand streaks" (active current lineations) was comparable to that of viscous sublayer streaks when fine and medium sand is in transport and that spacing increased as the flow strength and sediment transport also increased. Current lineations are most commonly reported in the deposits of unidirectional flows (e.g., fluvial deposits, B-division of turbidites) but have also been reported on bedding surfaces in hummocky cross-stratified sandstones.

Both current and parting-step lineation are useful indicators of the sense of the paleocurrent direction. They can be used in conjunction with other directional indicators to infer the absolute paleocurrent direction. Thin sections, cut perpendicular to bedding and parallel to the lineation, will show the imbrication direction of grains which, providing the absolute paleocurrent direction. In the field, heavy mineral shadows and current crescents can indicate the absolute paleocurrent direction along the lineations.

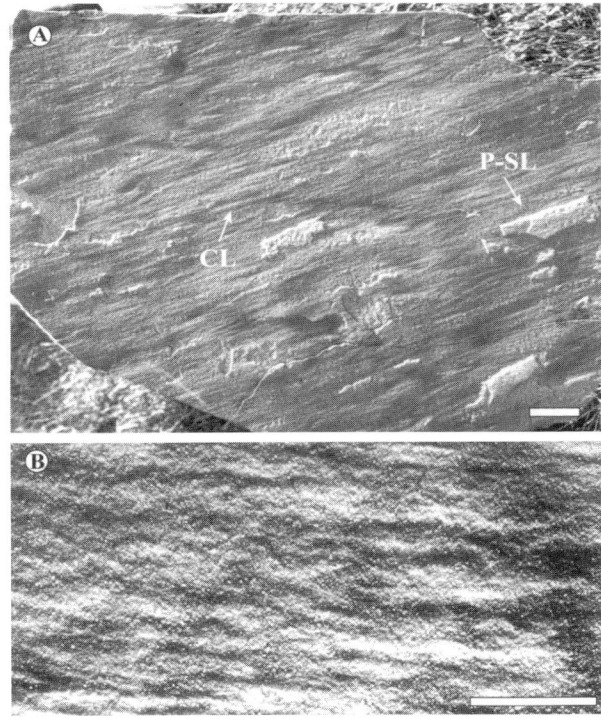

Figure P3 (A) Parting-step (P-SL) and current lineations (CL) on a bedding surface from the Silurian Whirlpool Sandstone of southern Ontario. Note that one of each structure is labeled whereas there are several examples of each on in the photograph. The white scale bar is approximately 5 cm long. (B) Current lineation on a fine sand bedding surface that was deposited under upper flow regime plane bed conditions. The white scale bar is approximately 8 cm long.

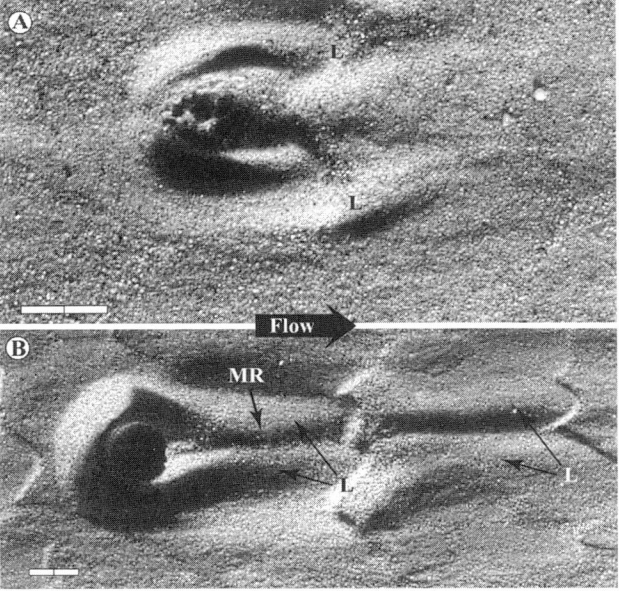

Figure P4 (A) A current crescent formed in a flume on a bed of fine sand under flow strengths just fast enough to move sand over the bed; the ripples are the stable bedform for the flow conditions; this crescent formed after one minute of flow. The object was a metal sphere but it has been removed for the photograph. Note that only the scour along the leading edge and sides of the crescent are well developed whereas the depositional lobes (labeled "L") are poorly developed. (B) A current crescent on the same sand bed after approximately 2 min of flow. Note the pair of well-developed lobes separated by a medial ridge (labeled "MR"). The lobes extending from the scour (black "L") have generated a second set of ripple-like lobes (white "L").

Current crescents

Current crescents are arcuate scours that may be preserved on bedding surfaces (Figure P4). The scours form as a current flows around an obstacle on the bed (e.g., a single particle, shells, debris). The form of the crescent reflects the pattern of flow around the obstacle. At the leading edge of the obstacle, a rolling vortex develops which produces a curved scour with "arms" extending downstream from the sides of the obstacle (Figure P4). Further downstream sediment that has been eroded from the bed is deposited in two lobes, on either side of the object (but downstream from it) forming areas of higher relief than the ambient bed. In well-developed crescents a medial ridge separates the two lobes (Figure P4(B)). The combination of erosional and depositional features collectively make up the current crescent.

The form of a current crescent appears to depend on the size and shape of the obstacle, the flow conditions and the duration of time over which the structure forms. In general, when flows are of short duration or if the obstacle is moved away from the scour shortly after the current becomes active the structure may resemble a small, crescent-shaped scour immediately around the upstream edge and sides of the object; there may be little deposition such that the depositional lobes are difficult to recognize (Figure P4(A)). In contrast, when the current is of significant duration and the obstacle remains stationary the depositonal lobes may generate two trains of low ripples that can extend for decimeters to meters downstream (Figure P4B), depending on the size of the obstacle.

Unlike lineations, current crescents may be used to infer the absolute paleoflow direction and their orientation can be measured with a compass in the field. Current crescents are most commonly unidirectional (pointing in one direction, the paleoflow direction), particularly when found in the deposits of unidirectional currents (e.g., rivers, turbidity currents). However, it is a common structure on beach faces (foreshore) where they can form in response to both the swash and backwash. In this setting the structures may be bidirectional. Furthermore, swash- and backwash-oriented crescents may not be at 180° to each other, depending on the angular relationship between incoming waves and the orientation of the slope of the beach face.

Rick Cheel

Bibliography

Allen, J.R.L., 1964. Primary current lineations in the Lower Old Red Sandstone (Devonian), Anglo-Welsh Basin. *Sedimentology*, 3: 89–108.
Allen, J.R.L., 1982. *Sedimentary Structures: Their Character and Physical Basis*. Volume II: Elsevier.
Crowell, J.C., 1955. Directional-current structures from the Prealpine Flysh, Switzerland. *Geological Society of America Bulletin*, 66: 1351–1384.
McBride, E.F., and Yeakel, L.S., 1963. Relationship between parting lineation and rock fabric. *Journal of Sedimentary Petrology*, 33: 779–782.
Weedman, S.D., and Slingerland, R., 1985. Experimental study of sand streaks formed in turbulent boundary layers. *Sedimentology*, 32: 133–145.

Cross-references

Fabric, Porosity, and Permeability
Heavy Mineral Shadows
Paleocurrent Analysis
Planar and Parallel Lamination
Scour, Scour Marks
Surface Forms

PEAT

Peat is a highly variable substance. It is highly variable in space because there are a diverse number of plant taxa characteristic of the different geographic biomes that contribute to peat formation. Peat, ranging in age from less than a year to hundreds and thousands of years, is also highly variable in time because it changes character as it ages.

It is difficult to consider peat without considering how it originates and the landform in which it occurs. Functional criteria must be considered, especially hydrological conditions and associated biological features that give rise to peat and peatland formation (Ingram, 1978). Not doing so has led to much confusion and misunderstanding of the nature of peat and peat lands. Von Bülow (1929) and Overbeck (1975) are good examples of attempts to characterize peat and peatlands.

Peat has different meanings to different people in many disciplines, which has also confounded the problems. Peat, as a fresh material or as a source of various chemical and biotic compounds extracted from it, has a number of industrial applications in horticulture, energy generation, medicine, mining, and specialized product industries (Fuchsman, 1980). Pristine and managed peatlands are of interest in ecology, soil science, biogeochemistry, hydrology, archaeology, cultural histories, agriculture, horticulture, forestry, and engineering. All of these have developed, to varying degrees, a discipline-specific and sometimes conflicting vocabulary to describe peat.

Functional criteria

After plants and animals living in waterlogged soils complete their life cycle, their dead remains become peat. Given that plants are the dominant component of the living community and its litter, peat is considered to be largely plant matter. The newly dead plant matter undergoes a series of changes, the onset of which is variously referred to as decay, decomposition, humification or breakdown processes. The processes involve: (a) weakening of plant structure and loss of organic matter, as gas or in solution, as a result of leaching and of attack by living aerobic soil microfauna and bacteria; (b) loss of physical structure of broken plant parts; and (c) change of chemical state (Clymo, 1983). This occurs in a near-surface layer, or *acrotelm*, which is characterized by oscillating water table, variable water content, high hydraulic conductivity, periodic oxygen entry, and de-watering. Not all of the plant matter decays before new plant matter from the following year is added to that of the previous year. The addition of new peat to the older leads to accumulation. The addition of peat year after year gives rise to thick accumulations and to the buildup of peat landforms or peatlands (Clymo, 1984). Eventually the deeper and older peat will become completely engulfed by the water table and will no longer be influenced by oscillations in the water table. It will have a more or less constant water content and negligible oxygen entry, small hydraulic conductivity, and be devoid of aerobic microfauna and bacteria. A lower layer or *catotelm* will develop with characteristics quite distinct from the acrotelm layer on the surface (Ingram, 1978). Decay processes are generally fast immediately following death of the plant and slow down considerably as the peat enters and continues to reside in the catotelm. The acrotelm is the primary peat-forming layer where decay processes predominate compared to the catotelm. Decay continues in the catotelm but at rates several orders of magnitude less than in the acrotelm. Peat that makes it into the catotelm is preserved for several hundreds and thousands of years following death of the plants.

Waterlogging, therefore, is the key ingredient leading to peat accumulation. Because the diffusion of oxygen is many orders of magnitude slower in water than in air, the low oxygen regime brought about by waterlogging, leads to accumulation. If waterlogging is interrupted in any one year, no peat will make it into the catotelm and no net accumulation will occur. If dry conditions persist over several years, it is possible that catotelm peat may become re-exposed to oxygen, and the peatland will degrade.

Peat landforms with acrotelm and catotelm layers, and hence, actively forming and accumulating peat are referred to as *mires*. Such two-layered mires are *diplotelmic* (Ingram, 1978). The term *peatland* originates from the early writings where *peatland* was used to describe any land area covered by peat (Gorham, 1953). It is also probably a shortened version for "peat landform" in geomorphology that refers to the biotic landforms formed of peat. In North America *mire* is generally not recognized, and instead, *peatland* is used in the broadest sense to denote any peat landform irrespective of whether it is diplotelmic or not and whether it is actively forming and accumulating peat or not. In Europe and Russia, there is a tendency to differentiate actively peat-accumulating mires from the more generic peatland term, which applies to all peatlands regardless of actively peat-accumulating or not.

The term *haplotelmic* is used when only the catotelm remains (Ingram, 1978). Erosion or human alteration are the primary forces responsible for eliminating the acrotelm. Whether natural aging of the peat-accumulating system can also contribute to disappearance of the acrotelm is less clear, such as polygon peatlands in Arctic Canada (Vardy and Warner, 2002).

A maximum depth for peatlands is unbounded. Average depth globally has been estimated to be 1.3–1.4 m (Lappalainen, 1996) and a maximum of 15 m is rarely exceeded (Clymo, 1983). At the other end of the spectrum, it has been customary to take an arbitrary minimum depth of 30, 40, or 50 cm. In diplotelmic systems, the catotelm is the thicker being several meters compared to the thinner acrotelm, which usually is not much more than 0.5–0.75 m.

Attributes of peat

It is possible to define *peat* in two ways. The first, and most common, is to define peat as a substance derived from peatlands and/or mires that is composed of the partially decomposed remains of plants with over 65 percent organic matter (dry weight basis) and less than 20–35 percent inorganic content (Clymo, 1983; Heathwaite *et al.* 1993). It is also possible to define peat in its intact or natural state (i.e., in the mire or peat land) as 88–97 percent water, 2–10 percent dry matter and 1 percent to 7 percent gas (Heathwaite *et al.*, 1993). Botanical composition and state of decomposition are generally agreed upon as the two most important characteristics.

Botanical composition

It is the plant communities that will dictate the litter source for the formation of the peat under them. This has led to first-order classifications that differentiate peat into four broad categories: *moss, herbaceous, wood* and *detrital* or *humified peat*. Moss is often subdivided into *Sphagnum* (i.e., Sphagnaceae—the peat mosses) and brown mosses (i.e., Amblystegiaceae) because both groups are common peat formers. *Sphagnum* is the single most important moss genus. Few other moss groups, except for members of the Polytrichaceae are important peat-formers, for example, in the southern hemisphere (Clymo, 1983). Identification of the moss remains to family and genus is often possible in the field with a hand lens, and where preservation is good, it is possible to identify to species level with the aid of a microscope (e.g., Janssens, 1983). The same is true for flower and seed parts, leaves, stems and roots of sedge (i.e., Cyperaceae), grass (i.e., Poaceae) and rush (i.e., Juncaceae) remains, which are probably, the most important groups comprising herbaceous peats. It is easy to differentiate, *Carex, Eriophorum, Phragmites, Cladium, Juncus* peats, and in the case of the Andes in South America, for example, *Oxychloe* peats (Earle *et al.*, 2003).

Woody peats are common in forested and shrub-dominated peatlands. Wood, if there were trees or tall and dwarf shrubs, is commonly encountered in peat. Much of it is identifiable depending upon the state of preservation. Stumps, logs or large branches can be sectioned and identified using standard wood identification techniques. Seeds, nuts, samaras, strobili, cones, leaves, needles, buds and bud scales, and bark can also be identified, almost certainly to family, often to genus and even species.

The last major category refers to peat where the plant remains are so disarticulated (i.e., detrital peat) or decomposed (i.e., humified peat) that the bulk of the plant is no longer recognizable and identifiable. Where macroscopic remains are no longer recognizable, it is possible to use paleoecological techniques such as pollen and spores analyzes to infer plant origins of the peat. Other plant groups including lichens, such as *Siphula* or *Cladina* communities in western Canada, and algae may leave distinctive thin horizons in moss or herbaceous-rich peats, but not usually in accumulations of any significance.

State of decomposition

This character, usually called the state of humification, is partly assessed by the extent to which plant structure is visible, and partly by color. Field assessment is usually made using the H scale (Von Post and Granlund, 1926). The scale is a subjective description by a numerical scale of 1–10 based on the color of the water squeezed in the hand. Numerous other more sophisticated techniques to assess density of color of water or chemical extractants also exists (Tolonen and Saarenmaa, 1979), but comparisons of techniques show reasonable agreement, so the simplest H scale technique, is the one most often used.

Other characteristics

Other important characteristics that can be measured to further characterize peats are inorganic matter content, inorganic solutes (i.e., calcium, potassium), cation exchange capacity, microorganism activity, bulk density (dry matter mass/volume), water content, gas content, water-retaining capacity, proportion of fibre (i.e., as size class), structure, heat of combustion, color, and age (Clymo, 1983).

Occurrence and distribution of peatlands

Peatlands develop where geological stability and physiography allow water to collect on the land surface long enough in a growing season to support peat-forming plant communities. A positive or balanced hydroperiod is, therefore, required in order to maintain waterlogging and to sustain the peat-forming plants and peatland.

Peatlands are cosmopolitan in distribution. About 95 percent of the worlds peatlands occur in North America, Asia and Europe (Lappalainen, 1996). Canada has the largest area of peatlands in the world with $1.1 \times 10^6 \, km^2$ (Tarnocai et al., 2000). Most of the peat in the world lies in the Boreal and Temperate Biomes of the northern hemisphere where geological history, cool and moist climate and productive plant communities have given rise to widespread peatland formation. These are regions where water is especially abundant, waterlogging is widespread, thus favoring peat-forming plants and peatlands.

Barry G. Warner

Bibliography

Clymo, R.S., 1983. Peat. In Gore, A.J.P. (ed.), *Mires: Swamp, Bog, Fen and Moor. General Studies.* Elsevier Science, pp. 159–224.
Clymo, R.S., 1984. The limits to peat bog growth. *Philosophical Transactions of the Royal Society of London,* Series B, 303: 605–654.
Earle, L.R., Warner, B.G., and Aravena, R., 2003. Rapid development of an unusual peat-accumulating ecosystem in the Chilean Altiplano. *Quaternary Research* (In press).
Fuchsman, C.H., 1980. Peat: Industrial Chemistry and Technology. Academic Press.
Gorham, E., 1953. Some early ideas concerning the nature, origin and development of peat lands. *Journal of Ecology,* 41: 257–274.
Heathwaite, A.L., Göttlich, K., Burmeister, E.G., Kaule, G., and Grospietsch, Th., 1993. Mires: definitions and form. In A.L. Heathwaite (ed.), *Mires: Process, Exploitation and Conservation.* J. Wiley and Sons, pp. 1–75.
Ingram, H.A.P., 1978. Soil layers in mires: function and terminology. *Journal of Soil Science,* 29: 224–227.
Janssens, J.A., 1983. A quantitative method for stratigraphical analysis of bryophytes in Holocene peat. *Journal of Ecology,* 71: 189–196.
Lappalainen, E., 1996. General review on world peatland and peat resources. In Lappalainen, E. (ed.), *Global Peat Resources.* Geological Survey of Finland, pp. 53–56.
Overbeck, F., 1975. *Botanisch-Geologische Moorkunde.* Neumunster: Wachholtz.
Tarnocai, C., Kettles, I.M., and Lacelle, B., 2000. *Peatlands of Canada.* Geological Survey of Canada, Open File Report 3834.
Tolonen, K., and Saarenmaa, L., 1979. The relationship of bulk density to three different measures of the degree of peat humification. *Proceedings, International Symposium On Classification of Peat and Peatlands.* Hyytiälä, Finland: International Peat Society, pp. 227–238.
Vardy, S.R., and Warner, B.G., 2002. Holocene ecosystem variability of polygon fens in central Nunavut, Canada. *Canadian Journal of Earth Sciences.*
Von Bülow, K., 1929. *Allgemeine Moorgeologie.* Berlin: Verlag von Gebrüder Borntraeger.
Von Post, L., and Granlund, E., 1926. Södre Sveriges torvtillgångar I. *Sveriges Geologische Undersølgelse,* **C335**: 1–127.

Cross-references

Humic Substances in Sediments
Kerogen
Maturation, Organic

PETROPHYSICS OF SAND AND SANDSTONE

Introduction

Porosity in sand is the only component that forms during the sedimentation process. During most of its existence porosity does not represent a simple void but the presence of a fluid or gas. Many of the physical properties of sand and sandstone (hereafter referred to as "sandstone") are functions of the details of the porous part of sandstone. Porosity differs from other constituents of sandstone in that it forms a physically continuous three-dimensional web-like complex. The intimate properties of this complex strongly affect such properties as permeability (Ehrlich et al., 1991b) and fracture toughness (Ferm et al., 1990).

Models for porous microstructure prior to image analysis

For most of the 20th Century sandstone porosity was the domain of physicists and engineers. They deduced structure consistent with physical measurements, especially permeability. Darcy's permeability carries with it no intrinsic information concerning the microstructure. Darcy's Law relates flow (caused by a differential between input and outlet pressure) in a cylinder containing a porous medium to the geometry of the cylinder. Taking the geometry into account, the rate of flow through each cylinder is proportional to the difference between input and output pressures. However the constant of proportionality (commonly called the "permeability") varies from sample to sample. Permeability thus involves the nature of the porous medium. However an infinite number of significantly different porous microstructures can yield the same value of permeability. Permeability would reach a maximum if the cylinder were simply an empty tube. Hence one way to form a geometric model of the microstructure is to assume a homogeneous (or "homogeneously heterogeneous") medium where the micro flow paths are longer than the length of the sample cylinder; that the flow "sees" a greater length than the length of the sample cylinder. This added length was postulated to be due to the tortuous flow paths through the medium. Although the concept of "tortuosity" was very effectively demolished by Scheidegger, (1974) on theoretical grounds, and by Ehrlich et al. (1991b) on empirical (quantitative petrographic) grounds, the concept of tortuosity is very popular with probabilistic modelers (Dullien, 1979). As discussed below, the permeability (holding grain size constant) is due to the presence of hyper efficient flow paths along packing flaws.

Pores and pore throats

The porosity complex consists of two major parts: pores and pore throats (Berg, 1975; Dullien, 1979). Pores account for all of the volume of porosity and tend to occupy the interstices between grains thus providing a cuspate shape bearing concave facets each representing the boundary between a pore and an adjacent sand grain. Adjacent pores communicate at the cusps, the cross sectional area of minimum volume of a plane cutting through a cusp is termed a "pore throat" (Scheidegger, 1974). Pore throats, then, are the bottlenecks that affect transmissivity of fluids, pressures, diffusion, and other transport properties. Pore throats are important mediators of the rate of diagenesis (Prince and Ehrlich, 2000) and are extremely important in hydrocarbon reservoirs. Pores, on the other hand, represent the capacity of the sandstone but are seldom important in fluid transport.

Analysis of the porous microstructure

A functional description of sandstone porosity should produce output that consistent with subjective porosity classification but also quantitative and precise in order to describe physical properties that are continuous variables. This calls for the application of digital data acquisition and an image processing capability. Such capabilities only became commonly available after 1975, so the preexisting subjective classification (inter-granular pore, intra-granular pores and moldic pores) was really most valuable in tracking diagenetic processes than relating thin section data to physical response.

Imaging

Commonly sandstones are studied by imaging a planar section *via* a light microscope, a scanning electron microscope or an electron luminescence apparatus. The plane of section interacts with the three-dimensional sample by (given simple assumptions) such that the ratio of the total area of each constituent to the others approximates the volume of the ratios of the components.

Pore types and throat sizes

Information concerning the size and shape of grains or pores is hazy in that the plane of section can penetrate a grain anywhere between the "equator" and the "pole" such that a wide size range of two-dimensional representatives represents each grain size class. Pore throats are seldom encountered by the plane of section as shown by the abundance of "isolated" intergranular pores in most sandstones, (the pore throat lying above or below the plane of section). Even when throats are encountered, their true size is impossible for estimate from section. Accordingly additional ways to observe sandstone must be used to supplement two-dimensional images. These include permeameters, mercury injection apparatus, and measurement of electrical properties of saturated media (Etris *et al.*, 1989; Ehrlich *et al.*, 1991a,b). By building a model consistent with all of these, we can deduce the values of the critical parameters that control the porous complex.

Petrographic image analysis

"Petrographic Image Analysis" ("PIA") (Ehrlich *et al.*, 1984, 1991a,b) is a generic term associated with the move from subjective totally visual microscopic analysis to acquisition and processing of digital imagery. Areas containing the entire plane of section can be imaged and the collection of image may form a seamless mosaic before the porous complex is analyzed. Having stated that only throats direct influence on transport properties, one might infer that analysis of pores in section would be fruitless. For instance the relationship between porosity and permeability is commonly very weak. However it can be demonstrated that there is a strong correlation between pore type and throat size (McCreesh *et al.*, 1991b). For this to occur and be relevant to transport properties, some types of pores must form continuous networks that are relatively efficient with respect to transport. Quantitative determination of pore types can only be accomplished by deducing the three dimensional sizes and shapes of pores from their two dimensional sections. We define the porosity objects exposed in section as *porels (Porosity Elements)* (Ehrlich *et al.*, 1991a). A porel is defined as a patch of porosity that is internally continuous with all of its parts. In a loose packed sand for instance, the porosity may represent a lacelike filigree in section and so the image might contain only two or three really extensive porels. As porosity declines as a consequence of diagenesis, porels systematically evolve into small isolated patches of porosity when viewed in section. One form of PIA involves the use of a variant of a common image analysis algorithm termed "erosion/dilation differencing" (Ehrlich *et al.*, 1991a) that produces porosity size and shape spectra that can be analyzed by a spectral unmixing algorithm, Polytopic Vector Analysis, an offshoot of factor analysis, (Johnson *et al.*, 2001) into an objective classification of pore types, including their sizes, shape, and volumetric proportions. This provides a link to earlier subjective pore type classifications as well as to physics. The output of this procedure can be related to the physical measurements on order to construct a plausible three-dimensional model of the porosity consistent with all methods of measurement.

Mercury injection—determination of throat size

Mercury injection porosimetry is the prime physical test, followed by permeability (Ehrlich *et al.*, 1991b) and electrical conductivity / resistivity (Etris *et al.*, 1989). Mercury injection involves placing a sample (commonly a core plug about 2 cm in diameter) into a sealed pressure vessel connected to a supply of mercury. Mercury is a strongly nonwetting phase and will not imbibe into the sample spontaneously. Pressure is progressively increased until mercury is forced into the sample. At the lowest pressure of injection only a fraction of the porosity is invade. The relationship between injected volume and pressure is recorded. The volume of porosity invaded in a pressure increment represents the amount of porosity that can be accessed by pore throats of a given size range through the Laplace equation (Scheidegger, 1974; Dullien, 1979). Thus mercury porosimetry carries strong throat size information and poor pore type information, the inverse of imaging in that regard. At a high enough pressure, almost all porosity is filled except for isolated inclusions within grains. The injection data can be precisely related to the pore type information obtained by image analysis *via* multiple regression with care taken that the imagery is taken from a section at the end of a sample cylinder before injection. The results consistently show a strong relationship between pore type and throat size (McCreesh *et al.*, 1991). For this to occur pores of like type must be mutually connected into circuits characterized by a common size range.

Relationship between microstrucure and physics

The results of this combination of image analysis and mercury injection data are sufficient to successfully estimate permeability and electrical properties such as conductivity of the sample when saturated with fluids.

Microcircuits and packing flaws

The concept of hyper efficient microcircuits (those associated with the largest throats) dominating transport was directly verified by another image analytical procedure involving analysis of the Fourier transform of the binary image (porosity 1.0, matrix 0.0) (Prince et al., 1995). Filtering the transform produces images where the circuits are prominently displayed. The Fourier transform is exactly equivalent to an X-ray diffraction pattern and so the numerical tools used in the analysis are much the same. Both the optical transform of sandstones and the X-ray transforms of crystals, share common characteristics in that in neither are the particle/atoms perfectly arrayed. Instead packing flaws separate closely packed domains. Graton and Fraser (1935) demonstrated *via* a series of experiments with monolayers of equal size spheres the nature of the flaws and their inevitability just about the same time as crystal chemists were using bead packs to illustrate packing flaws in ionic crystals. Unfortunately the work of Graton and Fraser has largely been ignored due to its prolixity and length.

Origin of packing flaws: the Graton and Fraser model

According to Graton and Fraser (1935) the packing flaws are a consequence of sedimentation. A depositional episode commonly occurs as the sedimentation of a collection of suspended grains that accumulate in a layer that is many grains thick under the influence of gravity. Closest packing is the arrangement of lowest potential energy and so nearby grains tend to cluster in this fashion. Such clusters form spontaneously throughout the sedimenting mass but there is no process that will force the clusters into a common orientation. As the accreting closely packed clusters grow, the eventually interfere with one another producing bounding compromise zone (packing flaws) characterized by large pores connected by large throats—the permeability circuits. In theory and observation, once generated, packing flaws propagate without limit, ending only when intersecting another set of flaws. This provides the continuity that ensures agreement between two-dimensional images and the volumetrically based physical measurements such as mercury injection porosimetry.

Graton and Fraser (1935) also came up with the basis of a natural history of packing flaws. As mentioned above, sand is commonly deposited episodically producing a sand accumulation consisting of a large number of sand on sand beds. The trivial unconformities defined by bed boundaries represents a sand water interface of relatively long duration resulting in a sort of annealing (densification) of the sand exposed to the moving water. The impulse of the next episode of sedimentation results in a major oriented mismatch between that interface and the newly deposited sand. Thus packing flaws are more densely spaced just above bedding planes. Outcrop observation verifies that many outcrops "weep" from a small interval (few cm thick) from the base of an overlying bed. As discussed below, this effect is also observed in partially oil-saturated core, where the dark oil decorates the bases of beds and may be absent in the middle and upper parts of a bed.

Multiphase flow

The porous microstructure is critically important in the case of multiphase flow such as oil and water. In this case two mutually immiscible fluids are competing for the pore space. Commonly but not always water is a "wetting" phase and so can spontaneously imbibe into the sandstone *via* capillary forces (Berg, 1975). Oil is commonly a "nonwetting" phase and must be forced into the sandstone by pressure just as mercury (another extremely nonwetting phase) is forced into a sandstone sample. Thus water will tend to segregate and be held in parts of microporosity containing small throats and the nonwetting phase will initially be forced into the porous circuits connected by the largest throats. The wetting phase will also coat all mineral surfaces and so be somewhat of a thin barrier between the oil and the pore walls. "Relative permeability" tests measure the degree of mobility of either phase as a function of saturation. Such are in agreement with the microstructural model described above (Prince and Ehrlich, 2000). The nonwetting phase is loosely held in pores with large throats (packing flaws) and the wetting phase is tightly held in pore complexes connected by small throats. Presumably water completely fills the pores of sandstone before migration of petroleum. The buoyant force of the oil column provides the pressure needed for to displace this initial water with oil. The longer the oil column, the greater the pressure gradient from the oil water content to the seal (Berg, 1975). Thus oil near the top of a reservoir displaces more water than rock near the oil water contact. However changes in grain size (which also affects throat sizes) as well as disordered zones near bed boundaries will ensure various values of water saturation for a given height above the oil water contact.

Summary and conclusions

The petrography of sandstones and other porous media can best be determined by a combination of image analysis and physical measurement. The two dimensional information taken from plane of thin section can be turned into a three dimensional microstructural complex by convolving that data with physical properties that arise from the three dimensional volume. The common availability of high quality relatively inexpensive digital image capability along with a wealth of image processing algorithms allow such analysis to be undertaken at almost any level. The nuances gained have shown their worth in the petroleum industry and also as a way to quantify the nature and extent of diagenesis. This field is still very young and many opportunities exist for new findings. A caveat is that the practitioner must have intermediate-level competence in multivariate analysis.

Robert Ehrlich

Bibliography

Berg, R.R., 1975. Capillary pressures in stratigraphic traps. *American Association of Petroleum Geologists Bulletin*, **59** (6): 939–956.

Dullien, F.A.L., 1979. *Porous Media: Fluid Transport and Pore Structure*, 2nd edn, Academic Press.
Ehrlich, R., Crabtree, S.J., Kennedy, S.K., and Cannon, R.L., 1984. Petrographic image analysis I—analysis of reservoir pore complexes. *Journal of Sedimentary Petrology*, **54** (4): 1365–1376.
Ehrlich, R., Horkowitz, K.O., Horkowitz, J.P., and Crabtree, S.J., 1991a. Petrography and reservoir physics I: objective classification of reservoir porosity. *American Association of Petroleum Geologists Bulletin*, **75** (10): 1547–1562.
Ehrlich, R., Etris, E.L., Brumfield, D., and Yuan, L.P., 1991b. Petrography and reservoir physics III: physical models for permeability and formation factor. *American Association of Petroleum Geologists Bulletin*, **75** (10): 1579–1592.
Etris, E.L., Brumfield, D.S., and Egrlich, R., 1989. Petrographic insights into the relevance of Archie's Equation: formation factor without "M" and "A". *SPWLA Journal Thirtieth Annual Logging Symposium*, June, 1989, I (F): pp. 1–18.
Ferm, J.B., Ehrlich, R., Kranz, R.L., and Park, W.C., 1990. The relationship between petrographic image analysis data and fracture toughness. *Association of Engineering Geologists Bulletin*, **27** (3): 327–339.
Graton, L.C., and Fraser, H.J., 1935. Systematic packing of spheres with particular relation to porosity and permeability. *Journal of Geology*, **43**: 785–909.
Johnson, G.W., Ehrlich, R., and Full, W., 2001. Principal component and receptor models. In **Murphy, and Morrison** (eds.), *Introduction to Environmental Forensics*. Academic Press, Chapter 12, pp. 461–515.
McCreesh, C.A., Ehrlich, R., and Crabtree, S.J., 1991. Petrography and reservoir physics II: relating thin section porosity to capillary pressure: the association between pore types and throat size. *American Association of Petroleum Geologists Bulletin*, **75** (10): 1563–1578.
Prince, C.R., Ehrlich, R., and Anguy, Y., 1995. Analysis of spatial order in sandstones II: grain clusters, packing flaws, and the small-scale structure of sandstones. *Journal of Sedimentary Research*, **A65**: 13–28.
Prince, C.M., and Ehrlich, R., 2000. A test of hypotheses regarding quartz cementation in sandstones: a quantitative image analysis approach. In Worden, R. (ed.), *Quartz Cementation in Sandstones*, IAS Special Publication, 29.
Scheidegger, A.E., 1974. *The Physics of Flow Through Porous Media*, 3rd edn. University of Toronto Press, 353pp.

Cross-references

Fabric, Porosity, and Permeability
Sands, Gravels and their Lithified Equivalents

PHOSPHORITES

Introduction

Phosphorites are rocks enriched in phosphorus relative to average crustal abundances, an enhancement usually expressed in terms of P_2O_5 concentrations. Whereas the average P_2O_5 content of continental crustal rocks is estimated as 0.23 percent (Ronov and Yaroshevsky, 1969) and sedimentary rocks average 0.03–0.16 percent (McKelvey, 1973), rocks typically designated as phosphorites have 15–37 percent P_2O_5 (Bentor, 1980). Phosphorites thus have phosphate contents that are 60 to 160 times greater than the crustal average and on the order of 100 to over 1,000 times greater than the averages for common sedimentary rocks. The most contentious issues concerning phosphorites center on the mechanism or mechanisms by which this enrichment has taken place in the geologic past and whether formation of phosphorites represents a significant perturbation of the biogeochemical cycle for phosphorus, questions discussed below.

The element phosphorus averages about 70 ppb ($\sim 2.3\,\mu\mathrm{mol/L}$) in seawater, is a limiting nutrient to biological productivity on geological timescales, and regulates the global carbon cycle and climate. Because of its low abundance and because it is closely tied to short-lived biological–chemical cycles of growth and decay, phosphorus has a relatively brief residence time in the ocean, estimated for the surface ocean by Mackenzie *et al.* (1993) as 0.07 years based on phytoplankton uptake. The overall oceanic residence time for phosphorus is estimated to range between *ca.* 10,000 years and 40,000 years (Delaney, 1998; Colman and Holland, 2000; Guidry *et al.*, 2000).

Initial interest in phosphorites stemmed from their importance as a raw material for the production of fertilizer phosphate. Along with potassium and nitrogen, phosphorus is critical for plant growth; but, whereas K and N are readily available from several sources (seawater, evaporite deposits, the atmosphere), phosphorus can only be obtained in large quantities from phosphorite deposits. Following the discovery in the mid-19th Century of the role of mineral nutrition in plant metabolism by the German chemist Justus von Liebig, phosphorite deposits began to be exploited after more readily available P sources such as guano, manure, and crushed bones became inadequate to support expanding agricultural systems. Phosphorites are now the main source of fertilizer P, and mining of phosphorites is a world-wide enterprise, with major centers of production in the USA, Morocco, and several countries in the Middle East. Excluding China, global production in 2000 was close to 92,200 thousand metric tons (IFA, 2001).

The main mineral in phosphorites is carbonate fluorapatite (CFA) or francolite, which according to Slansky (1986) has the simplified general formula $Ca_{10}[(PO_4)_{6-x}(CO_3)_x]F_{2+x}$, with numerous substitutions of both cations and anions (Nathan, 1984; Jarvis *et al.*, 1994). The most important ancient deposits are marine, granular phosphorites that formed in continental margin or epeiric sea settings (Figure P5). Two of the key questions about this kind of phosphorite concern: (1) *phosphogenesis*: how do CFA particles initially form in marine environments? (2) *concentration*: how do such particles become dominant in granular phosphorites?

Phosphogenesis

Studies in modern environments in which CFA forms have demonstrated that this mineral commonly precipitates during early diagenesis in the upper few tens of centimeters of sediment. The most notable of these environments are the continental margins of Perú, Baja California [México], southwest Africa, and eastern Australia. Whereas the first three of these are regions of pronounced eastern boundary currents, strong coastal upwelling, prominent oxygen minimum zones, and organic-rich sedimentation (1–20 percent organic carbon), the eastern Australia margin is an area of low productivity, oxygen-rich bottom waters, and sediments low in organic matter (<0.5 percent organic carbon). These differences suggest that CFA may form in different ways, depending upon the environmental conditions. Studies of Perú margin sediments, for example, have linked phosphogenesis with high organic carbon burial rates and anoxic diagenesis (Burnett, 1977). In this setting, microbial degradation of organic matter

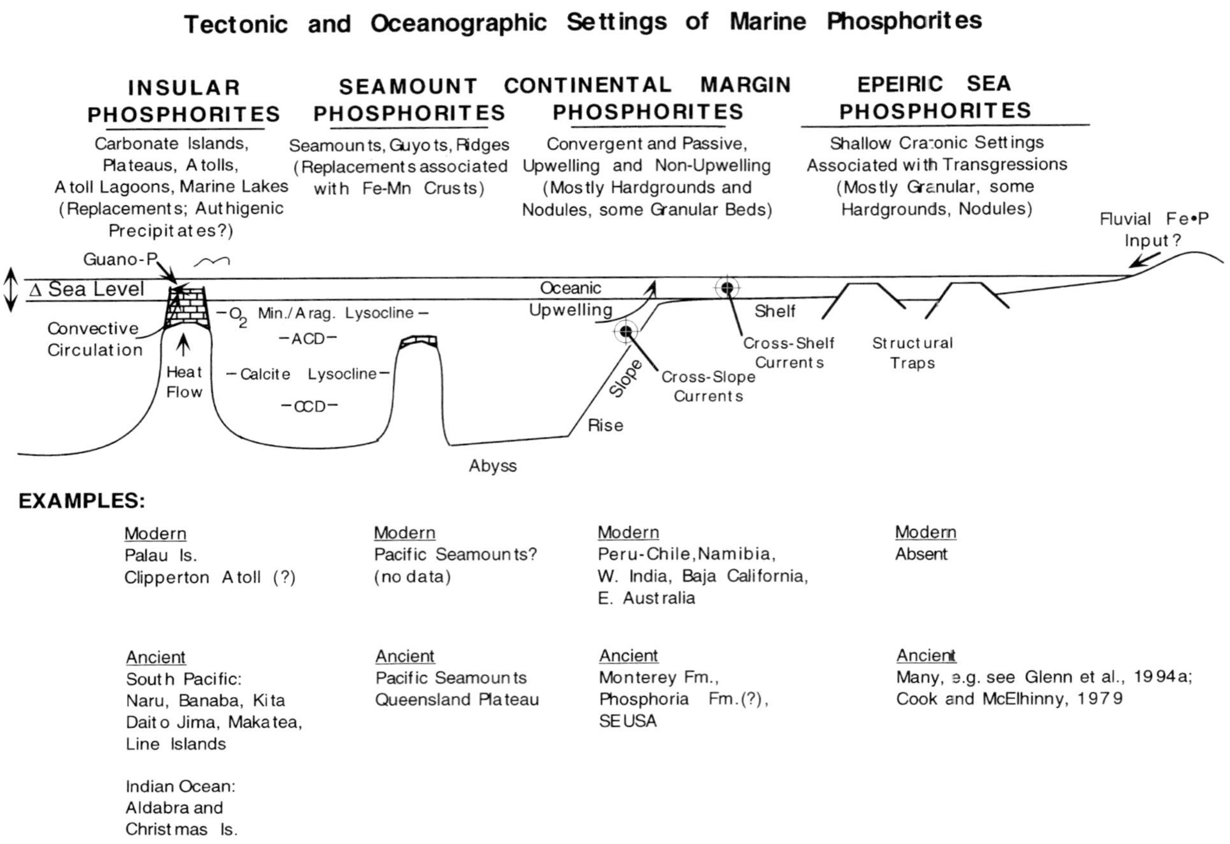
Tectonic and Oceanographic Settings of Marine Phosphorites
INSULAR PHOSPHORITES
SEAMOUNT PHOSPHORITES
CONTINENTAL MARGIN PHOSPHORITES
EPEIRIC SEA PHOSPHORITES
Carbonate Islands, Plateaus, Atolls, Atoll Lagoons, Marine Lakes (Replacements; Authigenic Precipitates?)
Seamounts, Guyots, Ridges (Replacements associated with Fe-Mn Crusts)
Convergent and Passive, Upwelling and Non-Upwelling (Mostly Hardgrounds and Nodules, some Granular Beds)
Shallow Cratonic Settings Associated with Transgressions (Mostly Granular, some Hardgrounds, Nodules)
Fluvial Fe•P Input?
Guano-P
Δ Sea Level
Convective Circulation
Heat Flow
$-O_2$ Min./Arag. Lysocline
–ACD–
– Calcite Lysocline –
–CCD–
Oceanic Upwelling
Shelf
Cross-Shelf Currents
Structural Traps
Slope
Cross-Slope Currents
Rise
Abyss
EXAMPLES:
Modern
Palau Is.
Clipperton Atoll (?)
Modern
Pacific Seamounts?
(no data)
Modern
Peru-Chile, Namibia,
W. India, Baja California,
E. Australia
Modern
Absent
Ancient
South Pacific:
Naru, Banaba, Kita
Daito Jima, Makatea,
Line Islands
Indian Ocean:
Aldabra and
Christmas Is.
Ancient
Pacific Seamounts
Queensland Plateau
Ancient
Monterey Fm.,
Phosphoria Fm.(?),
SE USA
Ancient
Many, e.g. see Glenn et al., 1994a;
Cook and McElhinny, 1979

Figure P5 Tectonic and oceanographic settings of marine phosphorites as derived from studies of the modern and ancient record (after Glenn *et al.*, 1994a, reprinted with permission of Birkhäuser Verlag AG).

increases reactive dissolved phosphate (PO_4^{3-}) in pore waters to supersaturation levels with respect to CFA, probably via transformation from a metastable amorphous precursor phase (e.g., see Jarvis *et al.*, 1994). Froelich *et al.* (1983, 1988) measured pronounced increases in dissolved phosphate in pore waters in the uppermost few centimeters of Perú margin sediments (an interfacial "P-spike"), directly below which dissolved F^- decreased rapidly in concert with declining phosphate concentrations. They interpreted this pattern as reflecting CFA precipitation, with diffusion of seawater-derived F^- acting to limit the zone of phosphogenesis to very shallow burial depths. In addition, Glenn *et al.* (1988) postulated that CFA precipitation was also confined to the sediment-water interface due to mineral lattice poisoning by excessive carbonate ion concentrations with sediment depth. Although continued increases in dissolved phosphate are provided to Perú margin pore waters deeper in the sediment column (due to progressive bacterial organic matter degradation), the locus of CFA precipitation thus appears to be most strongly associated with the interfacial phosphate maxima that occurs closely adjacent to the sediment-water interface. Similar interfacial phosphate spikes are also found in association with the formation of Recent phosphorites along the Mexican contential margin (Jahnke *et al.*, 1983; Schuffert *et al.*, 1994) and off the eastern coast of Australia (Cook and O'Brien, 1990; Heggie *et al.*, 1990).

To account for the formation of CFA in organic-poor sediments on the eastern Australian margin, O'Brien *et al.* (1990) outlined a mechanism whereby iron oxyhydroxides scavenge F^- from seawater and PO_4^{3-} derived both from seawater and from organic matter subjected to oxic and suboxic microbial degradation within the top few centimeters of sediment (Figure P6). When these particles are buried into the suboxic zone, they dissolve, releasing both F^- and PO_4^{3-} to pore waters, thus promoting CFA precipitation. Ferrous iron along with any remaining F^- and PO_4^{3-} then diffuses upward into the oxic zone to be refixed by ferric oxyhydroxides and become available for further recycling. This iron redox-P cycle may also promote the formation of glauconite which is commonly associated with modern and ancient phosphorites. Froelich *et al.* (1988) invoked a similar process to help explain phosphogenesis in Perú margin sediments.

Other potential sources for the buildup of PO_4^{3-} in pore waters include dissolution of fish debris, metabolic activities of sulfide-oxidizing or other bacteria, and P release to solution by bacteria in response to redox changes. Microbial microstructures are present in many CFA grains, and microbes may affect elevation of phosphorus concentrations in pore waters, but it has not been convincingly demonstrated that they are directly involved in the formation of CFA (see discussion in Krajewski *et al.*, 1994).

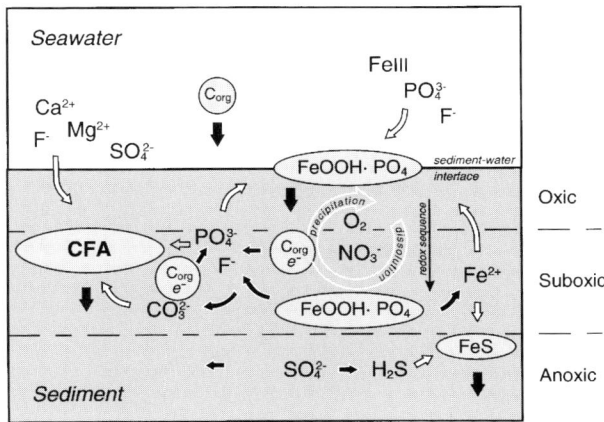

Figure P6 Schematic of CFA precipitation in surficial marine sediments illustrating phosphorus derived from either the Fe-redox cycle or directly from the microbial breakdown of organic matter. Light stippled areas represent solid phases, black arrows are solid-phase fluxes. White-outlined black arrows indicate reactions, white arrows are diffusion pathways. Phosphorus for CFA precipitation may be derived from either or both of direct microbial decomposition of organic matter (C_{org}) or from a redox-coupled iron oxyhydroxide-phosphorus pumping mechanism (right side of illustration). During burial and mixing, microbial decomposition of organic matter utilizes a sequence of electron acceptors in order of decreasing thermodynamic advantage. Oxygen is used first, followed by nitrate (and nitrite), manganese- and iron-oxyhydroxides and sulfate, in that order (Froelich et al., 1979). Degradation of organic matter liberates PO_4^{3-} to solution, and dissolution of FeOOH liberates Fe^{2+}, PO_4^{3-}, and F^- causing elevated concentrations of these ions in porewaters. If sufficient levels are attained, PO_4^{3-} reacts with Ca^{2+}, Mg^{2+}, SO_4^{2-} and F^- ions diffusing into the sediment from seawater, and CO_3^{2-} derived from the oxidation of organic matter, to precipitate a precursor phosphate mineral which subsequently recrystallizes to CFA (francolite; see text). Excess PO_4^{3-} may also diffuse upward toward the sediment-water interface where it is resorbed by ferric iron oxyhydroxides. Iron may diffuse both downward to be fixed as glauconite (e.g., Glenn, 1990) under suboxic conditions, or as FeS under anoxic conditions, and upwards to the oxic layer where it is reprecipitated as FeOOH. In this later case, the Fe-redox cycle provides an effective means of trapping PO_4^{3-} in the sediment and promotes precipitation of CFA, especially in more organic-lean sediments (*cf.* O'Brien et al., 1990). Modified after Jarvis et al., 1994, reprinted with permission of Birkhäuser Verlag.

Types of phosphate particles and concentration processes

The precipitation of CFA may produce a variety of particle types, including peloids, coated grains, laminae, and small nodules dispersed in muds. In this type of sediment, termed "pristine" phosphate by Föllmi et al. (1991), the primary phosphate particles appear to be *in situ*, with no reworking by mechanical or biological processes. For the most part, however, these dispersed forms have whole rock P_2O_5 concentrations well below those of economic phosphorites. Baturin (1971) accounted for this disparity by proposing a model of sea level "highstand" phosphogenesis alternating with "lowstand" reworking by currents to produce winnowed phosphorite layers in Quaternary deposits from the continental shelf of southwest Africa.

Marine phosphorites in the stratigraphic record commonly have two attributes: (1) they are grain-supported, granular concentrations of silt-, sand-, and pebble-size phosphatic grains; and (2) they occur as stratified sediment bodies that display evidence for transport and redeposition by bottom currents. The most common grains are structureless peloids, which are probably mostly authigenic rather than phosphatized biogenic fecal pellets. Other grain types may include concentrically coated grains, phosphatic intraclasts and nodules, primary phosphatic bioclasts (inarticulate brachiopods, vertebrate bones, teeth, fish scales), and phosphatized carbonate skeletal grains. The richest phosphorites have CFA cements, but granular phosphorites may also be cemented by carbonate or silica minerals.

Some stratified phosphorites have traction current structures such as cross-bedding (Glenn and Arthur, 1990), others have characteristic of turbidites and indicate deposition from gravity flows (Grimm and Föllmi, 1994), still others appear to be tempestite beds (Trappe, 1992). Many phosphorites occur in condensed intervals that contain evidence for multiple episodes of primary phosphogenesis, shallow burial, exhumation, and hydraulic reworking (Föllmi, 1989, 1990; Glenn and Arthur, 1990). Phosphatic hardgrounds are common in such condensed sections. Depositional and biological amalgamation may lead to the formation of thick (meter scale) phosphorite beds, and it appears that post-depositional burrowing may commonly destroy current induced structures in many massive phosphorite beds.

The characteristics above lead most workers to accept Baturin's (1971) notion of a two-stage mechanism (i.e., primary phosphogenesis followed by reworking) for the formation of large, granular phosphorite deposits. However, nontransported phosphorites such as hardgrounds are associated with some granular phosphorites, and only in a few cases are the primary phosphatic mudrocks, the supposed parental sediment for the redeposited phosphatic grains, clearly identifiable in the accompanying sediment layers. Moreover, the dominant types of Proterozoic phosphorites are primary phosphorite mudrocks and stromatolitic phosphorites which were clearly not transported (see papers in Cook and Shergold, 1986a).

Classification of sedimentary phosphorite rocks

Most past attempts to classify phosphorites were based on petrologic characteristics such as grain types and/or textures (e.g., Riggs, 1979; Cook and Shergold, 1986b; Slansky, 1986). More recent schemes have focused on field-based characteristics such as sedimentary structures and bedding properties. Among the most useful of these is the classification of Föllmi et al. (1991), who recognized three interpretive genetic categories based on relative rates of sediment accumulation and erosion (Figure P7):

(1) *Pristine*: phosphates which lack any signs of reworking; this includes phosphatic rocks as well as more concentrated phosphorites such as phosphatized stromatolites and phosphatic hardgrounds.
(2) *Condensed*: phosphatic particles, laminae, and beds which have been concentrated by winnowing and reworking processes or bioturbation.
(3) *Allochthonous*: phosphatic particles that were entrained by and redeposited from turbulent or gravity-driven flows.

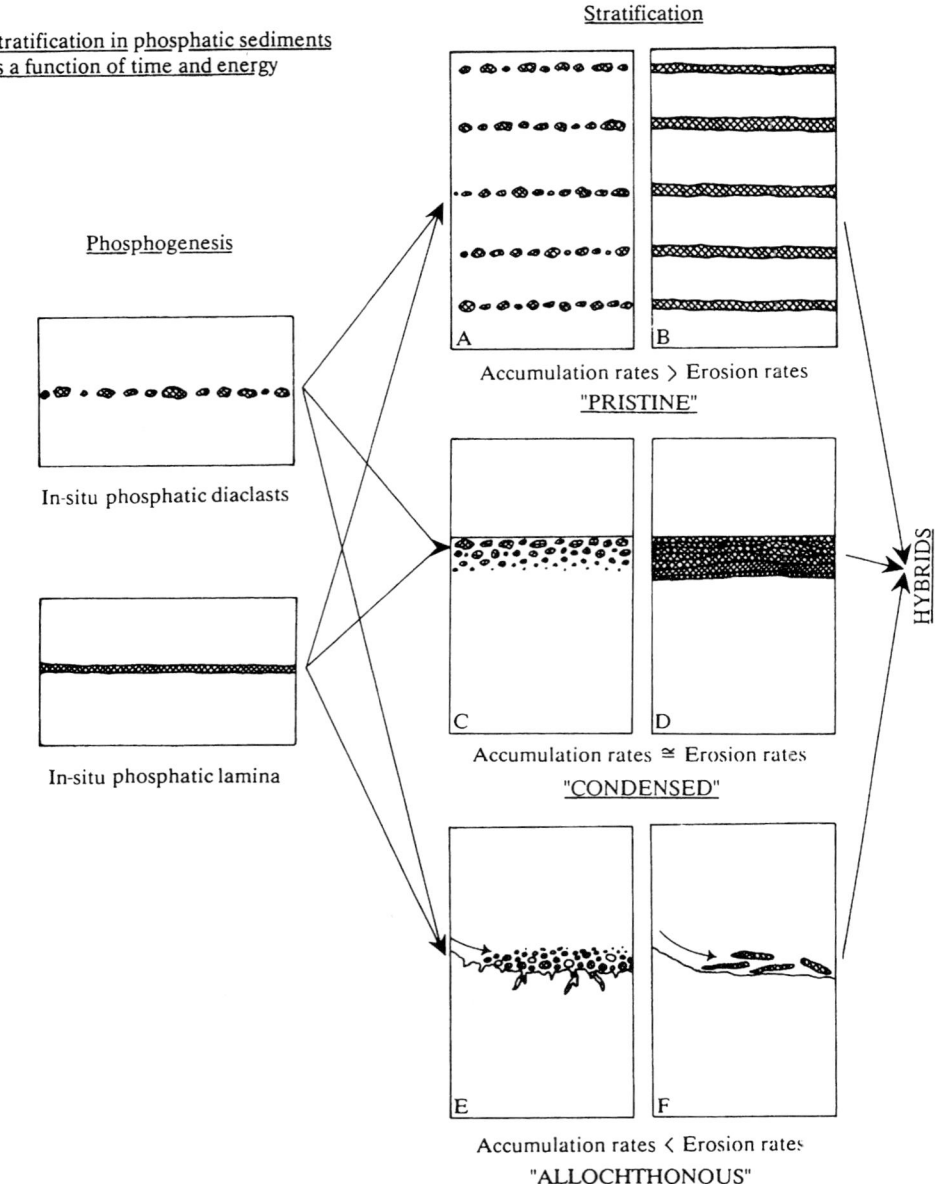

Figure P7 Genetic classification of stratification types (from Follmi et al., 1991 and Glenn et al., 1994a). Reprinted with the permission of Springer Verlag.

Hybrids of these categories are common. One example is that pristine phosphatic grains may experience redeposition followed by multiple episodes of burial, renewed phosphogenesis and phosphatic cementation, and subsequent erosional exhumation, resulting in a complex and laterally variable phosphorite condensed bed.

The formation of "phosphorite giants"

Very large phosphorites deposits (typically with reserves greater than 10^9 metric tons of P_2O_5) have been termed "phosphorite giants" by Glenn and Arthur (1990) and account for a substantial proportion of world phosphate production. Examples include the Cretaceous-Eocene Tethyan deposits and the Permian Phosphoria Formation of western North America.

Kazakov (1937) made one of the first attempts to explain such deposits by invoking upwelling and inorganic precipitation of apatite from seawater in coastal regions, an idea subsequently modified to include the role of organic degradation in shallowly buried sediments, as outlined above. Most phosphorite giants were deposited beneath relatively shallow waters of marginal seas and epeiric platforms (Figure P5), and most seem to correlate in a general way with elevated sea levels and, more specifically, with marine transgressions. The interrelated factors involved may include (Glenn et al., 1994a): (1) highstands of sea level increase the potential surface area for phosphorite accumulation on shelves and in continental interiors; (2) highstands may also increase the potential for upwelling into shelf seas; (3) sediment starvation and phosphogenesis may be favored during transgressions by the trapping of diluting siliciclastic sediment in proximal parts of depositional systems; (4) wave-induced and other cross-shelf currents may develop along flooded margins and platforms, thus aiding in winnowing, reworking, and concentration of pristine CFA particles. In addition to transgressive phases, relative falls in sea level would lower wave base and may aid in reworking and concentrating such particles, in the manner proposed by Baturin (1971). The majority of phosphorite giants are of the condensed or allochthonous variety (Föllmi et al., 1991).

The abundance of modern phosphorites in upwelling regions has led many workers to postulate similar settings for ancient phosphorites. Sheldon (1980), for example, subdivided large phosphorites into those associated with equatorial upwelling (e.g., the Cretaceous Tethyan deposits) and those associated with mid-latitude boundary currents and trade-wind upwelling (e.g., the Phosphoria Formation). However, the efficiency of sustained upwelling within large, shallow epeiric seas far from the open ocean, the setting for many phosphorite giants including the Tethyan deposits, has been questioned. Alternative models include the delivery of fluvial-borne P derived from intensively weathered continental regions to epeiric platforms (Glenn and Arthur, 1990).

Episodicity of phosphorite giants and the global phosphorus cycle

The distribution of phosphorite giants in the Phanerozoic is episodic, with peaks in abundance in the Early Cambrian, Ordovician, Permian, Late Cretaceous–Eocene, and Miocene (Cook and McElhinny, 1979). This has been variously explained as the consequence of favorable positions of continental margins vis a vis upwelling zones, equable climates, global cooling, widespread anoxia in the oceans, paleolatitude, and other conditions. However, as noted by Glenn et al. (1994a), none of these mechanisms provides comprehensive explanations for all phosphorite giants, each of which has its own particularities.

A related question is whether phosphorite giants record accelerated P withdrawal from the ocean into continental margin and epicontinental sediments, or whether they represent a combination of unusual and localized geological circumstances unrelated to global controls. Sheldon (1980), for example, postulated that the episodicity of marine phosphorite deposition resulted from variations in global deep-ocean circulation. A number of recent studies, however, suggest that several phosphorite giants required a net P withdrawal rate comparable to that observed in modern Perú margin sediments (cf. Filippelli and Delaney, 1992; Glenn et al., 1994b; Compton et al., 2000). Thus, if phosphorite giants of the past do indeed record episodic accelerated P withdrawal, the present day must be included among these episodes.

The global phosphorus cycle is complex and poorly understood. Phosphorus input to the ocean is mainly from rivers, secondarily from eolian phosphorus particle transport. Major oceanic sinks include organic matter, fine-grained disseminated authigenic CFA in detrital-rich continental margin sediments, and adsorption on iron oxides formed at mid-ocean ridges and in non-upwelling continental margin sediments (Delaney, 1998). Due to variations in sedimentation rates, phosphorus accumulation rates are several orders of magnitude higher in upwelling-phosphogenic and detrital continental margin settings than in open ocean sediments (Filippelli, 1997). Present consensus holds that past variations in the phosphorus cycle were responses to changes in rates of chemical weathering, sea-level fluctuations, spreading rates of mid-ocean ridges, glaciation, and oceanic circulation. Compton et al. (2000) postulated that phosphorite giants most likely formed during episodes of marine transgression that coincided with increased chemical weathering rates and decreases in production of iron oxides at mid-ocean ridges and, for the last 25 million years, convincingly showed that major phosphorite accumulations occurred in near-synchronicity with proposed periods of episodic uplift of the Himalayan-Tibetan Plateau. Elevated sea levels shift primary biological productivity to enlarged shallow water settings where organic phosphorus is released in sediments to form CFA. Subsequent sea level fluctuations make such sediments susceptible to reworking, generating phosphorites. Viewed in this manner, phosphorites may be a proxy for times in which phosphorus input to the oceans was greater than the sinks provided by organic carbon and iron-oxide burial.

Phosphorites, condensed sections and sequence stratigraphy

Condensed sections or sequences represent extremely slow net sedimentation over long periods of time. Such intervals have been long recognized as characterized by the following features: (1) enrichment of well preserved fossils and fossil fragments; (2) faunal mixing, fossils from different paleontological zones within a single bed; and (3) widespread distribution of sediments with negligible thicknesses. In addition, it has become increasingly recognized that intervals of stratigraphic condensation are also frequently marked by the occurrence of authigenic minerals, including phosphorites and glauconites. With the advent of "sequence stratigraphy," condensed sections have taken on new significance as an integral component in the stratigraphic architecture of sequences controlled by fluctuations in relative sea level. Loutit et al. (1988) and their colleagues define condensed sections in light of sequence stratigraphy as thin marine units of pelagic to hemipelagic sediments that: (1) are characterized by very low sedimentation rates that are areally most extensive at the time of maximum transgression and coincident with the surface of maximum flooding; (2) are associated with apparent marine hiatuses; and (3) occur as omission surfaces or marine hardgrounds. They result from sediment starvation during times of maximum rates of sea level rise. These occur along the break separating transgressive and highstand systems tracts (Figure P8). Assuming that seismic reflectors reflect time-lines, sequence condensation may, hypothetically, also occur anywhere that reflectors converge such as, for example, along surfaces of onlap, backstepping, downlap and toplap (Kidwell, 1991; Figure P8). In addition, while phosphorites may be generally associated with transgressive phases of second and third orders changes of sea level (5–50 and 0.5–5 Ma, respectively), much reworking of these may occur during intervening higher orders of sea level change (forth, fifth, and sixth order; 0.1–0.5, 0.01–0.1, and <0.01 Ma, respectively).

However, as noted above, most major episodes of phosphorite genesis do appear to be related to transgressive phases, although the occurrence of reworked deposits within the upper portions of highstand systems tracts may also reflect episodes of condensation and bypassing that occur in association with the convergence of toplapping seismic reflectors (Figure P8). In sum, the actual distribution of condensed deposits within seismic sequences is probably more complex than depicted by recent models.

Other types of phosphorite deposits

In addition to phosphorites formed in continental margin or epeiric sea environments (Figure P5), lesser quantities of phosphorites occur in three other settings:

(1) *Island or insular phosphorites* are commonly composed of guano and the replacement products of reactions between bird droppings and the underlying host rock; the latter is commonly a carbonate, and these types of phosphorites are important economic resources on Quaternary carbonate islands in the Pacific and Indian Oceans. The largest deposits of this kind occur on the island of Nauru in the western equatorial Pacific (Piper *et al.*, 1990). As noted by Glenn *et al.* (1994a), not all island phosphorites are associated with guano, and some may have precipitated in insular marine lakes or other settings.

(2) *Seamount and Guyot phosphorites* and phosphatized limestones, many associated with iron-manganese encrustations, are common on elevated portions of the seafloor in all the world's oceans. Although a few are submerged insular deposits, most appear to be of marine origin and formed in areas of slow sedimentation and a pronounced oxygen minimum zone. Burnett *et al.* (1987) suggested they formed preferentially at low latitudes, possibly in equatorial upwelling regions of high productivity. They linked the association of iron-manganese crusts to the enhanced concentration of phosphate and metals within an oxygen minimum zone that bathes the upper parts of seamounts.

(3) *Igneous rocks* such as carbonatites and other alkaline igneous intrusive complexes which are enriched in fluorapatite, $Ca_{10}(PO_4)_6F_2$ (Slansky, 1986). Important economic deposits of this kind occur in northern Russia, Brazil, and eastern and southern Africa (Cook, 1984).

Environmental issues connected to phosphorites

Along with rises in world populations and living standards, exploitation of phosphorites as a source of fertilizers will inevitably also increase. This may be accompanied by several

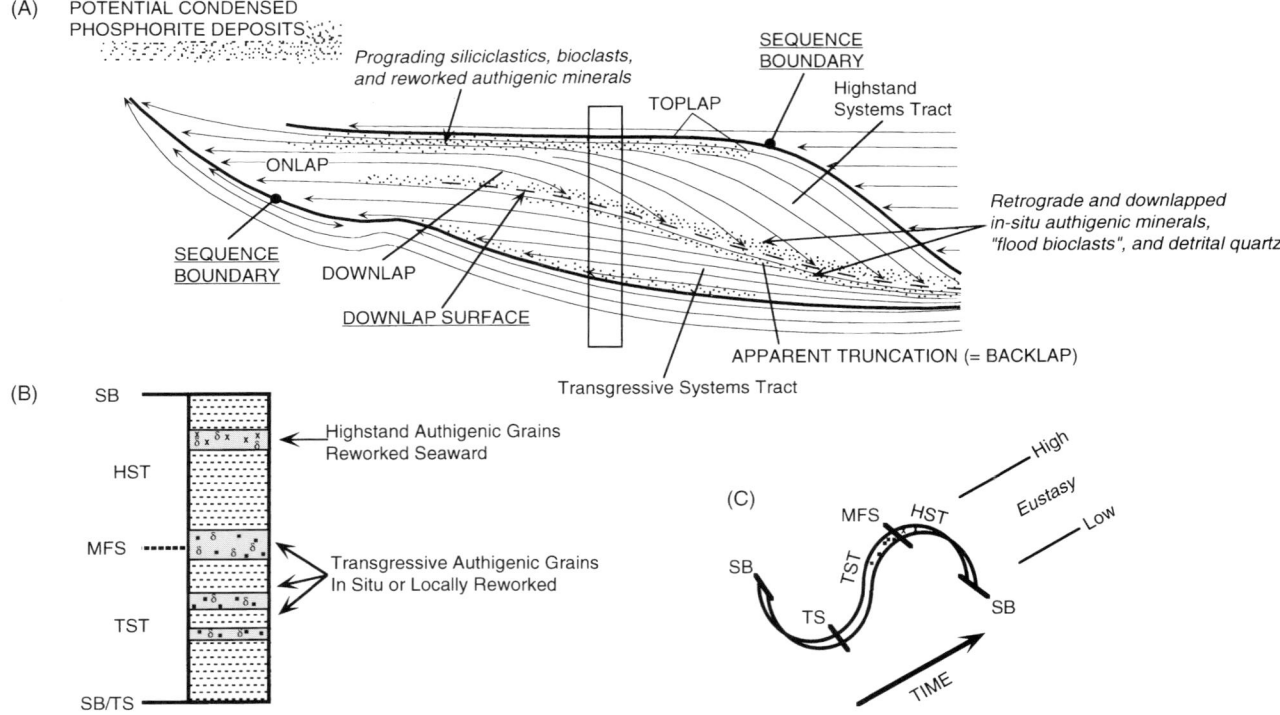

Figure P8 Potential relationships between sequence stratigraphy, condensed sections and phosphorites. (A) Possible positions of sequence condensation within an idealized depositional sequence (stipples), (after Kidwell, 1991). (B) Schematic illustration of the placement of non-reworked pristine authigenic phosphorite (and/or glauconite) and reworked phosphorite (and/or glauconite). Pristine phosphates may form within transgressive systems tracts and at the maximum flooding surface, whereas later phases are reworked seaward within highstand systems tracts. (C) The timing of systems tracts development with respect to one cycle of sea level change. LST lowstand systems tract, TST transgressive systems tract, HST highstand systems tract, TS transgressive surface, MFS maximum flooding surface, SB sequence boundary. Dots on the sea level curve represent locations of maximum phosphorite emplacement. (After Glenn *et al.*, 1994a, reprinted with permission of Birkhäuser Verlag AG).

environmental problems (see Jarvis *et al.*, 1994), among which three are especially significant:

(1) *Mining and processing*: most phosphorites are extracted from open-pit mines with their attendant problems of landscape disruption and disposal of mine tailings. In addition, beneficiation processes commonly produce phosphatic "waste clays" which are extremely fine-grained and have poor settling properties, hence may need decades to settle and thus require numerous settling ponds covering large areas.

(2) *Phosphogypsum*: phosphogypsum is a by-product of the conversion, using sulfuric acid, of phosphorite ore to phosphoric acid, an intermediate compound in the manufacture of phosphatic fertilizers. The amount of phosphogypsum produced is substantial. Jarvis *et al.* (1994) estimate worldwide annual production on the order of 100 metric Mt. Although this form of gypsum has commercial applications in many parts of the world, governmental regulations in the USA prevent its usage, in part due to an association with radionuclides (apparently in a separate phase) in the ^{238}U decay chain, especially radon. Stockpiling of this material is thus necessary, leading in turn to potential problems of air and water contamination in the vicinity of the stockpiles.

(3) *Trace elements*: many phosphorites contain small but significant amounts of potentially toxic elements (e.g., cadmium, selenium, arsenic) or radionuclides, particularly uranium. The possibility thus exists that harmful amounts of these elements could be released into the environment during the processing of phosphorite ore or the application of fertilizers produced from phosphorites.

Summary

Phosphorites are unique sedimentary deposits that have piqued the interest of academicians and economic geologists for decades. Due to their high phosphorus content, they are the world's most important economic resource of elemental phosphorus and thus are an essential additive for fertilizers and phosphate-based chemicals. The most common phosphate-bearing mineral in the majority of phosphorites punctuating the geologic record is carbonate fluorapatite (CFA), which has an abbreviated chemical formula of formula $Ca_{10}[(PO_4)_{6-x}(CO_3)_x]F_{2+x}$, but actually also contains numerous substitutions of both cations and anions. Field observations plus isotopic and pore water studies of modern marine phosphorite occurrences demonstrate that most form initially as "pristine," nonreworked deposits during early diagenesis within the upper few tens of centimeters beneath the seafloor. Most occur beneath strong upwelling systems which provide a high delivery rate of phosphorus-bearing sedimentary organic matter to the seafloor; this material is subsequently broken down by microbial degradation processes in pore waters. Many are then in turn "upgraded" to what are termed "condensed" or "allochthonous" phosphorites by processes such as sediment winnowing and reworking, largely by bottom currents. Most ancient examples of these deposits thus usually occur as stratified granular phosphorites containing admixtures of silt to very large sand-sized phosphatic grains, the most common cements of which are silica or carbonate minerals. Concretionary and hardground phosphorites also occur in both modern and ancient deposits. Additionally, there is evidence that phosphorus may also be delivered to the sites of phosphogenesis by a redox cycle whereby phosphorus sorbed to iron oxyhydroxides is released as soluble reactive P in suboxic pore waters and thus becomes available for solid phase fixation into the CFA compound. Very large phosphorite deposits, termed "phosphorite giants," occur episodically throughout the Phanerozoic during discrete time periods. Most were deposited on oceanic slopes and shelves and in epeiric seaways. In addition, phosphatic sediments are also found on oceanic islands and atop seamounts, guyots and submarine plateaus. Present consensus holds that past variations in the phosphorus cycle and phosphorite output of the oceans were responses to a variety of factors including sea-level fluctuations, changes in rates of continental chemical weathering, shifts in the paleolatitudes of continental margins and in positions of coastal upwelling zones, and in patterns of oceanic circulation.

Craig R. Glenn and Robert E. Garrison

Bibliography

Baturin, G.N., 1971. Stages of phosphorite formation on the ocean floor, *Nature Physical Sciences*, **232**: 61–62.

Bentor, Y.K., 1980. Phosphorites—The unsolved problems. In Bentor, Y.K. (ed.), *Marine Phosphorites—Geochemistry, Occurrence, Genesis*. Tulsa: Society Economic Paleontologists and Mineralogists Special Publication, 29, pp. 3–18.

Burnett, W.C., 1977. Geochemistry and origin of phosphorite deposits from off Peru and Chile. *Geological Society of America Bulletin*, **88**: 813–823.

Burnett, W.C., Cullen, D.J., and McMurtry, G.M., 1987. Open-ocean phosphorites—in a class by themselves? In Teleki, P.G., Dobson, M.R., Moorelan, J.R., and von Stackelberg, U. (eds.), *Marine Minerals*. D. Reidel (Kluwer Academic Press), pp. 119–134.

Colman, A.S., and Holland, H.D., 2000. The global diagenetic flux of phosphorus from marine sediments to the oceans: redox sensitivity and the control of atmospheric oxygen levels. In Glenn, C.R., Prévôt-Lucas, L., Lucas, J. (eds.), *Marine Authigenesis: From Microbial to Global*. SEPM (Society for Sedimentary Geology), Special Publication, 66, pp. 53–76.

Compton, J., Mallinson, D., Glenn, C.R., Filippelli, G., Föllmi, K., Shields, G., and Zanin, Y., 2000. Variations in the global phosphorus cycle. In Glenn, C.R., Prévôt-Lucas, L., and Lucas, J. (eds.), *Marine Authigenesis: From Global to Microbial*. SEPM (Society for Sedimentary Geology) Special Publication, 66, pp. 21–33.

Cook, P.J., 1984. Spatial and temporal controls on the formation of phosphate deposits-a review. In Nriagu, J.O., and Moore, P.B., (eds.), *Phosphate Minerals*. Springer-Verlag, pp. 242–274.

Cook, P.J., and McElhinny, M.W., 1979. A reevaluation of the spatial and temporal distribution of phosphate deposits in the light of plate tectonics. *Economic Geology*, **74**: 315–330.

Cook, P.J., and Shergold, J.H., (eds.). 1986a. *Phosphate Deposits of the World: Volume 1–Proterozoic and Cambrian Phosphorites*. Cambridge University Press.

Cook, P.J., and Shergold, J.H., 1986b. Proterozoic and Cambrian phosphorites—an introduction. In Cook, P.J., and Shergold, J.H., (eds.), *Phosphate Deposits of the World: Volume 1–Proterozoic and Cambrian Phosphorites*. Cambridge University Press, pp. 1–8.

Cook, P.J., and O'Brien, G.W., 1990. Neogene to holocene phosphorites of Australia. In Burnett, W.C., and Riggs, S.R., (eds.), *Phosphate Deposits of the World: Volume 3–Neogene to Modern Phosphorites*. Cambridge University Press, pp. 98–121.

Delaney, M.L., 1998. Phosphorus accumulation in marine sediments and the oceanic phosphorus cycle. *Global Biogeochemical Cycles*, **12**: 563–572.

Filippelli, G.M., 1997. Controls on phosphorus concentration and accumulation in oceanic sediments. *Marine Geology*, **139**: 231–240.

Filippelli, G.M., and Delaney, M.L., 1992. Similar phosphorus fluxes in ancient phosphorite deposits and a modern phosphogenic environment. *Geology*, **20**: 709–712.

Föllmi, K.B., 1989. *Evolution of the Mid-Cretaceous Triad.* Lecture Notes in Earth Sciences, 23. Berlin: Springer-Verlag.

Föllmi, K.B., 1990. Condensation and phosphogenesis: example of the Helvetic Mid-Cretaceous (Northern Tethyan Margin). In Notholt, A.J.G., and Jarvis, I. (eds.), *Phosphorite Research and Development*. Geological Society of London, Special Publication, 52, pp. 237–252.

Föllmi, K.B., Garrison, R.E., and Grimm, K.A., 1991. Stratification in phosphatic sediments: illustrations from the Neogene of California. In Einsele, G., Ricken, W., and Seilacher, A. (eds.), *Cycles and Events in Stratigraphy*. Springer-Verlag, pp. 492–507.

Froelich, P.N., Klinkhamer, G.P., Bender, M.L., Luetke, N.A., Heath, G.R., Cullen, D., Dauphin, P., Hammond, D., Hartman, N., and Mynard, V., 1979. Early oxidation of organic matter in pelagic sediments of the Eastern Equatorial Atlantic: suboxic diagenesis. *Geochemica et Cosmochimica Acta*, **43**: 1075–1090.

Froelich, P.N., Kim, K.H., Jahnke, R., Burnett, W.C., Soutar, A., and Deakin, M., 1983. Pore water fluoride in Peru Continental Margin sediments: uptake from seawater. *Geochimica et Cosmochimica Acta*, **47**: 1605–1612.

Froelich, P.N., Arthur, M.A., Burnett, W.C., Deakin, M., Hensley, V., Kaul, L., Kim, K.-H., Roe, K., Soutar, A., and Vathakanon, C., 1988. Early diagenesis of organic matter in Peru Continental Margin sediments: phosphorite precipitation. *Marine Geology*, **80**: 309–343.

Glenn, C.R., 1990. Pore water, petrologic and stable carbon isotopic data bearing on the origin of modern peru margin phosphorites and associated authigenic phases. In Burnett, W.C., and Riggs, S.R. (eds.), *Phosphate Deposits of the World: Volume 3—Neogene to Modern Phosphorites*. Cambridge University Press, pp. 46–61.

Glenn, C.R., Arthur, M.A., Yeh, H.-W., and Burnett, W.C., 1988. Carbon isotopic composition and lattice-bound carbonate of Peru–Chile margin phosphorites. *Marine Geology*, **80**: 287–307.

Glenn, C.R., and Arthur, M.A., 1990. Anatomy and origin of a Cretaceous phosphorite-greensand giant, Egypt. *Sedimentology*, **37**: 123–154.

Glenn, C.R., Föllmi, K.B., Riggs, S.R., Baturin, G.N., Grimm, K.A., Trappe, J., Abed, A.M., Galli-Olivier, C., Garrison, R.E., Ilyin, A., Jehl, C., Rohrlich, V., Sadaqah, R.M., Schidlowski, M., Shelton, R.E., and Siegmund, H., 1994a. Phosphorus and phosphorites: sedimentology and environments of formation. *Eclogae Geologicae Helvetiae*, **87**: 747–788.

Glenn, C.R., Arthur, M.A., Resig, J.M., Burnett, W.C., Dean, W.E., and Jahnke, R.A., 1994b. Are modern and ancient phosphorites really so different? In Iijima, A., Abed, A.M., and Garrison, R.E., (eds.), *Siliceous, Phosphatic and Glauconitic Sediments of the Tertiary and Mesozoic*. Proceedings of the 29th International Geological Congress, Part C. Utrecht: VSP BV, pp. 159–188.

Grimm, K.A., and Föllmi, K.B., 1994. Doomed pioneers: allochthonous crustacean tracemakers in anaerobic basinal strata, Oligo-Miocene San Gregorio Formation, Baja California Sur, Mexico. *Palaios*, **9**: 313–334.

Guidry, M.W., Mackenzie, F.T., and Arvidson, R.S., 2000. Role of tectonics in phosphorus distribution and cycling. In Glenn, C.R., Prévôt-Lucas, L., and Lucas, J. (eds.), *Marine Authigenesis: From Global to Microbial*. SEPM (Society for Sedimentary Geology), Special Publication, 66, pp. 35–51.

Heggie, D.T., Skyring, G.W., O'Brien, D.W., Reimers, C., Herczeg, A., Moriarty, D.J.W., Burnett, W.C., and Milnes, A.R., 1990. Organic carbon cycling and modern phosphorite formation on the East Australian continental margin: an overview. In Notholt, A.J.G., and Jarvis, I. (eds.), *Phosphorite Research and Development*. Geological Society Special Publication, 52, pp. 87–117.

IFA (International Fertilizer Industry Association), 2001. *Quarterly Phosphate Rock Statistics, January–December 2000*. A/01/15r, pp. 1–14.

Jahnke, R.A., Emerson, S.R., Roe, K.K., and Burnett, W.C., 1983. The present day formation of apatite in Mexican continental margin sediments. *Geochimica et Cosmochimica Acta*, **47**: 259–266.

Jarvis, I., Burnett, W.C., Nathan, Y., Almbaydin, F., Attia, K.M., Castro, L.N., Flicoteaux, R., Hilmy, M.E., Hussain, V., Qutawna, A.A., Serjani, A., and Zanin, Y.N., 1994. Phosphorite geochemistry: state of the art and environmental concerns. *Eclogae Geologicae Helvetiae*, **87**: 643–700.

Kazakov, A.V., 1937. The phosphorite facies and the genesis of phosphorites. In *Geological Investigations of Agricultural Ores. Transactions, Scientific Institute of Fertilizers and Insecto-Fungicides*. Volume 142, pp. 95–113.

Kidwell, S.M., 1991. Condensed deposits in siliciclastic sequences: expected and observed features. In Einsele, G., Ricken, W., and Seilacher, A. (eds.), *Cycles and Events in Stratigraphy*. Springer-Verlag, pp. 682–695.

Krajewski, K.P., Van Cappellen, P., Trichet, J., Kuhn, O., Lucas, J., Martin-Algarra, A., Prévôt, L., Tewari, V.C., Gaspar, L., Knight, R.I., and Lamboy, M., 1994. Biological processes and apatite formation in sedimentary environments. *Eclogae Geologicae Helvetiae*, **87**: 701–745.

Loutit, T.S., Hardenbol, J., Vail, P.R., and Baum, G.R., 1988. Condensed sections: the key to age dating and correlation of continental margin sequences. In Wilgus, C.K., Hastings, B.S., Kendall, C.G.S.C., Posamentier, H.W., Ross, C.A., and Van Wagoner, J.C., (eds.), *Sea-Level Changes: An Integrated Approach*. Society of Economic Paleontologists and Mineralogists, Special Publication, 42, pp. 183–213.

Mackenzie, F.T., Ver, L.M., Sabine, C., Lane, M., and Lerman, A., 1993. C, N, P, S global biochemical cycles and modeling of global change. In Wollast, R., Mackenzie, F.T., and Chou, L. (eds.), *Interactions of C, N, P and S Biochemical Cycles and Global Change*. NATO Asi Series, Series 1: Global Environmental Change, 4. Springer-Verlag, pp. 1–61.

McKelvey, V.E., 1973. Abundance and distribution of phosphorus in the lithosphere. In *Environmental Phosphorus Handbook*. Wiley, pp. 13–31.

Nathan, Y., 1984. The mineralogy and geochemistry of phosphorites. In Nriagu, J.O., and Moore, P.B. (eds.), *Phosphate Minerals*. Springer-Verlag, pp. 275–295.

O'Brien, G.W., Milnes, A.R., Veeh, H.H., Heggie, D.T., Riggs, S.R., Cullen, D.J., Marshall, J.F., and Cook, P.J., 1990. Sedimentation dynamics and redox iron-cycling: controlling factors for the apatite–glauconite association on the East-Australian margin. In Notholt, A.J.G., and Jarvis, I. (eds.), *Phosphorite Research and Development*. Geological Society, Special Publication, 52, pp. 61–86.

Piper, D.Z., Loebner, B., and Aharon, P., 1990. Physical and chemical properties of the phosphate deposit on Nauru, Western Equatorial Pacific Ocean. In Burnett, W.C., and Riggs, S.R., (eds.), *Phosphate Deposits of the World: Volume 3—Neogene to Modern Phosphorites*. Cambridge University Press, pp. 177–194.

Riggs, S.R., 1979. Petrology of the tertiary phosphorite system of Florida. *Economic Geology*, **74**: 195–220.

Ronov, A.B., and Yaroshevsky, A.A., 1969. Chemical composition of the Earth's crust. In Hart, P.J. (ed.), *The Earth's Crust and Upper Mantle*. American Geophysical Union Geophysical Monograph 13, pp. 37–57.

Schuffert, J.D., Jahnke, R.A., Kastner, M., Leather, J., Sturz, A., and Wing, M.R., 1994. Rates of formation of modern phosphorite off Western Mexico. *Geochemica et Cosmochemica Acta*, **58**: 5001–5010.

Sheldon, R.P., 1980. Episodocity of phosphate deposition and deep ocean circulation—a hypothesis. In Bentor, Y.K. (ed.), *Marine Phosphorites—Geochemistry, Occurrence, Genesis*. Society of Economic Paleontologists and Mineralogists, Special Publication, 29, pp. 239–247.

Slansky, M., 1986. *Geology of Sedimentary Phosphates*. Tiptree, Essex: North Oxford Academic.

Trappe, J., 1992. Microfacies zonation and spatial evolution of a carbonate ramp: marginal Moroccan phosphate sea during the paleogene. *Geologisches Rundschau*, **81**: 105–126.

Cross-references

Authigenesis
Diagenesis
Oceanic Sediments
Upwelling

PHYSICS OF SEDIMENT TRANSPORT: THE CONTRIBUTIONS OF R. A. BAGNOLD

During the twentieth century a huge amount of time, money and research effort was expended in attempting to develop accurate and reliable models of sediment transport by wind and water. Amongst the numerous significant contributions that resulted, that of R. A. Bagnold is characterized particularly by his appreciation of the pivotal importance and great usefulness of fundamental physics in describing and defining sediment transport mechanisms and processes.

It was as an officer in the British Army, stationed in Egypt and India between 1926 and 1932, that Bagnold's fascination with sediment movement began. Bagnold and a small group of like-minded officers began exploring the Libyan Sand Sea using Model T Ford motor cars. Early forays quickly evolved into serious geographical exploration and the group were responsible for mapping an area of desert the size of India, work that earned Bagnold the Gold Medal of the Royal Geographical Society in 1934.

During these expeditions Bagnold became fascinated by the movement of sand by wind and he wished to understand the physics of the processes responsible for the formation of desert dunes. Bagnold was astonished and appalled to find that no theory of dune formation existed and that no serious scientific investigation of the movement of sand by wind had at that time been performed. He decided that, despite having no formal training in research and no higher degree, he would retire from the army and devote all of his time and energy to the study of wind blown sands.

After making preliminary measurements of the physical properties of desert sands and the near-boundary wind speeds, Bagnold built himself a wind tunnel with glass sides, for easy observation and photography, within which to perform controlled experiments. He used high-speed cinematic photography (then in its infancy) to observe and trace the entrainment of individual sand grains. His results showed that detaching grains left their pockets at very steep angles, proving that lift forces were of equal importance to drag forces in detaching grains. His movie-films demonstrated that entrained sand grains move along ballistic trajectories, unaffected by turbulence—a motion he described as "a series of bounds", and that moving grains dislodge further grains when they bounce—a process later described as saltation. Bagnold compared the hopping distance of saltating grains to the crest to crest spacing of ripples formed in the sand surface, concluding that,

"The length of the characteristic path from impact to impact was found in all experiments to be the same as the measured wavelengths of the ripples formed on the surface" (Bagnold, 1936, p.23).

These early experiments further established that, not only does the rate of sand transport vary with the cube of the wind speed, but also the presence of moving sand itself generates drag that affects the velocity profile in a transport-limiting mechanism (Bagnold, 1936).

Having obtained controlled data elucidating the transport process, Bagnold returned to the field to check his results against reality. After spending eight days alone in the desert waiting for a sandstorm, he obtained reliable measurements that confirmed the wind-tunnel results. He was also able to apply his understanding of the saltation process to explain why sand moving over a rocky plain becomes size sorted and tends to collect into patches. Patches occur because grains moving over a bare rock surface bounce very efficiently, maintaining motion, while the grain–grain collisions of a grain moving over a sand surface sap energy from the moving grain, tending to damp transport so that sand accumulates in a patch.

The complete results of his research on sand transport by wind during the 1930s were published in *The Physics of Blown Sand and Desert Dunes* (Bagnold, 1941), a book that has been reprinted many times and which remains a standard text.

In 1937, Bagnold was persuaded to perform experiments on the shock pressures generated by waves breaking against a sea wall (Bagnold, 1939, 1940), but characteristically, he took the opportunity to extend this work to the inter-action of waves in shallow water with a mobile sand bed. This line of research was interrupted by World War II, but resumed and published immediately afterward (Bagnold, 1946, 1947). From his wave-tank experiments, Bagnold was able to explain the formation of rolling-grain ripples under oscillatory flow, their evolution into vortex ripples and the occurrence of "brick pattern" ripples under certain conditions. He further related ripple formation to increased drag on the moving water and produced a simple expression for the first disturbance of a planar sand bed by oscillatory water motion.

Recalled to the army at the outbreak of World War II, Bagnold returned to Cairo and used the results of his desert exploration and mapping expeditions to persuade General Wavell that a real threat existed of an Italian raid on the Aswan dam from a base deep within the Libyan Desert. Commissioned to establish the Long Range Desert Group (LRDG), identify any threat and nullify it, Bagnold then performed what may be the quintessential example of application of sedimentology to warfare. He used his knowledge of the physics of wind blown sands and the morphology of desert dunes to take LRDG vehicles deep behind enemy lines, scouting and reporting troop movements, mapping enemy positions, collecting intelligence and creating havoc through lightening raids on forts, convoys and airfields. When enemy forces attempted to follow and track down the LRDG they quickly became stuck in the soft sand of dune troughs (Bagnold knew that you must drive over the "accretion deposits" of the dune crest) or got hopelessly lost in the maze of sand dunes. Without Bagnold's grasp of sedimentology and geomorphology, pursuit was impossible. The formation and leadership of the LRDG uniquely combined Bagnold's talents as a scientist and soldier and must rank as one of his most noteworthy achievements. A contemporary account of the exploits of the LRDG may be found in Bagnold (1945) while for a longer term perspective readers should consult his autobiography (Bagnold, 1990).

Bagnold again retired from the army in 1944 and, after a period as Director of Research for Shell, during which he was instrumental in the establishment of the Hydraulics Research Station at Wallingford, England, he again decided to pursue fundamental research full time.

Bagnold's post-war work focused on the movement of sediment by water. His approach rested on the application of theory to identify key variables and formulate relationships between them, supported by carefully conceived and executed experiments to quantify physical constants and material properties. The conception and construction of experimental apparatus and instrumentation continued to be incidental to his work, but stemmed from a Bagnold family tradition that, "a properly equipped workbench was as important to study as was a well stocked library".

He began by conceiving a series of fundamental experiments on the physics of sediment movement conducted in a rotating-drum test rig he himself designed, to quantify the physics of the gravity-free dispersion of large, solid spheres in a Newtonian fluid (Bagnold, 1954). He demonstrated that collisions between suspended solids generate a "dispersive pressure" on the boundary resisting further dispersion. This experiment established for the first time that the coefficient of solid-to-solid friction for sheared dispersion of granular solids is virtually the same as for the same solids in continuous contact—a piece of knowledge that lies at the heart of practically every sediment transport equation in common usage today. A further discovery of some portent was summed up eloquently by Bagnold himself in his autobiography,

"The rate of mass transport of an immersed weight of solids is easily measured, in the laboratory at least. And the fluid shear stress times the velocity of its action is the rate at which the fluid does transporting work. So I now saw clearly that the transport rate of solids must depend primarily on a rate of fluid energy supply, or, in other words on *stream power* and not as previously thought, on either stream velocity or bed shear stress, but the product of the two. This, I felt, began to put the subject of solids transport by water flows on a sound footing." (Bagnold, 1990, pp. 163–164). In 1956 Bagnold was approached by Luna B. Leopold, then Head of the Water Resources Division of the United States Geological Survey to "help stir the pool of complacent tradition with the stick of reality" concerning sediment transport. The first area of cooperative research concerned river meandering (Leopold, *et al.*, 1960; Bagnold, 1960). Bagnold and his co-workers challenged the "square-law" then generally accepted as governing flow resistance in alluvial rivers. The "square law" argued that flow resistance should increase with the square of the mean velocity because this was known to be the case for simple boundary layer flow past a flat plate, provided that boundary conditions remain unchanged. Their results showed that, while the square law holds for straight channels, deformation of the water surface due to transverse deflections of all or part of the flow in meandering rivers could lead to a doubling of the rate at which resistance increases with the square of the velocity (Leopold *et al.*, 1960). Further theoretical analysis, coupled with carefully designed experiments, demonstrated that energy losses at bends were minimized for ratios of bend radius to width between 2 and 3. A great deal of subsequent empirical research has demonstrated that natural meanders do, in fact, tend to evolve to a bend ratio to width ratios in this range (Hickin and Nanson, 1984).

Another Anglo-American research initiative, with Douglas Inman of Scripps Oceanographic Institute, led to Bagnold's main contribution to the understanding of beach and near-shore sedimentation. Application of knowledge gained through earlier studies of wave mechanics, the transport of grains by fluids, the movement of sand by waves and the concept of stream power was employed to construct a comprehensive explanation of on-shore and long-shore sediment dynamics in the littoral zone (Inman and Bagnold, 1963). Embedded within this treatise were a power-based sediment transport equation for waves, a theoretical explanation of the mechanics of auto-suspension in off-shore turbidity currents, a theoretically-based explanation of the equilibrium beach-profile, and a power-based model for the transport of sand by long-shore currents.

It was in the realm of sediment transport by rivers that Bagnold's final and possibly most lasting contribution was made. Adopting an approach to the sediment transport problem based in general physics, he was able to attribute the suspension of granular material in a fluid shear flow to the effect of anisotropic turbulence creating an upward supporting stress (Bagnold, 1966). He went on to differentiate definitively between suspended and bed-load components of the total load, recognizing that the immersed weight of the suspended load is supported by turbulence, while the immersed weight of bed-load is supported by (intermittent) solid–solid contact with the bed. This explained why bed-load can therefore occur even under laminar flow, in the complete absence of turbulence while suspension cannot (Bagnold, 1973).

Discovery of the difference between the physics of suspended and bed loads allowed Bagnold to derive sediment transport equations for each process from a fundamentally sound basis (Bagnold, 1973; 1980). It is perhaps his bed-load equation which has proven the most useful in practical applications and this equation remains in widespread use, especially in the UK. His use of stream power has also been replicated in several of the most popular sediment transport equations used in North America (see, e.g., Yang, 1976 and Chang, 1988).

Advances in fluid mechanics, turbulence theory and the sophistication of instrumentation in the late twentieth have supported new findings that, to varying degrees replaced some of Bagnold's interpretations, which were necessary based on incomplete knowledge of the nature of turbulent flow structures. However, his contribution to sedimentology lies not only in the original knowledge and useful sediment transport equations that he crafted—both certain to be superseded with the passage of time, but also in his exemplary approach to conducting basic research in this difficult and often frustrating field of enquiry. In sum, his contributions to science and engineering must place him in the first rank of twentieth century researchers in sedimentology and sediment transport.

Colin R. Thorne

Bibliography

Bagnold, R.A., 1936. The movement of desert sand. *Geographical Journal*, **85**: 342–369.

Bagnold, R.A., 1939. Committee on wave pressures: interim report on wave-pressure research. *Journal of the Institution Civil Engineers*, **12**: 201–226.

Bagnold, R.A., 1940. Beach formation by waves: some model experiments in a wave tank. *Journal of the Institution Civil Engineers*, paper number 5237, 27–53.

Bagnold, R.A., 1941. *The Physics of Blown Sands and Desert Dunes*. William Morrow and Company in 1971, Republished by Chapman and Hall, Methuen Halsted Press, ISBN X-76–050873–8.

Bagnold, R.A., 1945. Early days of the Long Range Desert Group. *Geographical Journal*, **105**: 30–46.

Bagnold, R.A., 1946. Motion of waves in shallow water. Interaction between waves and sand bottoms. *Proceedings of the Royal Society of London*, Series A, **187**: 1–18.

Bagnold, R.A., 1947. Sand movement by waves: some small-scale experiments with sand of very low density. *Journal of the Institution Civil Engineers*, paper number 5554, **27**: 447–469.

Bagnold, R.A., 1954. Experiments on the gravity-free dispersion of large solid spheres in a Newtonian fluid under shear. *Proceedings of the Royal Society of London*, Series A, A225: 49–63.
Bagnold, R.A., 1960. Some aspects of the shape of river meanders. *US Geological Survey Professional Paper 282-E*, 135–144.
Bagnold, R.A., 1966. An approach to the sediment transport problem from general physics. *US Geological Survey Professional Paper 422-I*, 1–37.
Bagnold, R.A., 1973. The nature of saltation and of bed-load transport in water. *Proceedings of the Royal Society of London*, Series A, **332**: number 1591, 473–504.
Bagnold, R.A., 1980. An empirical correlation of bedload transport rates in flumes and natural rivers. *Proceedings of the Royal Society of London*, Series A, **372** paper number 1751, 453–473.
Bagnold, R.A., 1990. *Sand, Wind and War: Memoirs of a Desert Explorer.* University of Arizona Press.
Chang, H.H., 1988. *Fluvial Processes in River Engineering.* Wiley Interscience.
Hickin, E.J., and Nanson, G., 1984. Lateral migration rates of river bends. *Journal Hydraulic Engineering*, **110**: 1557–1567.
Inman, D.L., and Bagnold, R.A., 1963. Beach and nearshore processes—Part 2, Littoral processes. In Hill, M.N. (ed.), *The Sea—Ideas and Observations on Progress in the Study of the Seas, Volume 3—The Earth Beneath the Sea.* Interscience Wiley: pp. 529–553.
Leopold, L.B., Bagnold, R.A., Brush, L.M. Jr, and Woman, M.G., 1960. Flow resistance in sinuous or irregular channels. *U.S. Geological Survey Professional Paper 282-D*, 111–134.
Kenn, M.J., 1988. Biography. In Thorne, C.R., MacArthur, R.C., and Bradley, J.B. (eds.), *The Physics of Sediment Transport: A Collection of Hallmark Papers by R.A. Bagnold.* American Society of Civil Engineers, Book Number 665, xiii–xvii.
Yang, C.T., 1976. Minimum unit stream power and fluvial hydraulics. *Proceedings of the American Society of Civil Engineers, Journal of the Hydraulics Division*, **102**: 919–934.

Cross-references

Eolian Transport and Deposition
Sediment Transport by Unidirectional Water Flows
Sediment Transport by Waves
Sedimentology, History
Sedimentologists

Figure P9 Small subvertical pillar structures (light-colored streaks) in crudely layered muddy sandstone deposited by a sediment gravity flow. Pillars mark sheet-like zones from which darker, argillaceous and organic grains have been removed by water escape. Lower Cretaceous Britannia Formation, North Sea. Scale on left in tenths of a foot (3 cm units).

PILLAR STRUCTURE

Vertical water-escape channels, variously termed pillar structures, elutriation pipes, elutriation columns, fluidization pipes, fluidization channels, tube dewatering structures, and water-escape sheets, are elongated vertical to subvertical zones of sediment that are usually lighter-colored and contain less clay, mica, organic detritus, and other hydraulically mobile grains than surrounding sediments (Figure P9). They form through localized fluidization of sediment along discrete subvertical zones and accompanying elutriation of fine-grained components during either post- or syndepositional water escape. They are frequently associated with dish structures (see *Dish Structure*, figure D26).

The term "pillar structure" was introduced by Wentworth (1966) from studies of turbidites in western California. He interpreted the "pillars" as vertical fluid-escape channels. Similar structures had been previously described in fluvial deposits of the Old Red Sandstone by Allen (1961). Vertical water-escape channels have also been described from volcaniclastic mud flows (Best, 1989). Lowe and LoPiccolo (1974) discussed pillar structures associated with dish structures, and Lowe (1975) provided a broader classification and discussion of vertical fluid-escape channels. The formation of these structures has been studied experimentally by Nichols *et al.* (1994) and Druitt (1995).

Vertical water-escape channels occur in sediments ranging from coarse silt to pebbly sand or, more rarely, fine gravel, usually in discrete beds at least a few decimeters thick. Although they were initially thought to have pillar- or column-like geometries, based largely on two-dimensional exposures, most vertical water-escape channels are isolated to anastomosing subvertical sheets. The widths of individual sheets or pillars range from about a millimeter to a few centimeters with heights ranging from a few millimeters to more than a meter. When sediment is moved over significant distances from one bed to others, the terms "sediment intrusion" or "dike" should be used. Subvertical sediment intrusions can be up to several tens of meters high, and their formation is related both to water escape and density instabilities. In most cases, however, water-escape through vertical channels or conduits involves only local fluidization and restricted sediment movement.

Sediment in water-escape channels (pillar structures) is usually homogenized and the margins of the channels may cut sharply across primary sedimentary structures in surrounding sediments.

Localized fluidization can take place during either rapid sedimentation or post-depositional dewatering. Druitt (1995) investigated experimentally the formation of "elutriation pipes" during sedimentation of poorly sorted aqueous dispersions. He observed that at concentrations of 25–55 percent solids vertical water-escape channels propagated upward at the same rate as the depositional surface. A strong upflow of escaping fluid above each escape channel depleted the settling dispersion in fines and resulted in the upward propagation of the stable water-escape channel. At higher concentrations the water-escape channels nucleated throughout the dense dispersion. This behavior resembles the aggregative behavior and self-organized segregation of particles in sedimenting bidisperse systems that lead to the formation of streaming columns and blobs (Weiland et al., 1984; Batchelor and Van Rensburg, 1986).

Mud-rich deep-water sediment gravity flow deposits called slurry beds (Lowe and Guy, 2000) often contain a variety of vertical water-escape channels that appear to develop during the dewatering of sediments possessing strength. Some channels are a few centimeters long and develop in sets with individual pillars spaced 1–2 cm apart ("stress pillars" of Lowe, 1975). They might have formed through a process similar to convective streaming (Weiland et al., 1984). Others occur as sets of small columns, 1–2 mm wide and 1–25 cm long (Figure P9). The wide range of rheological properties, particle densities, and hydraulic sizes in such sediments favors formation of a wide variety of water-escape structures.

Zoltán Sylvester and Donald R. Lowe

Bibliography

Allen, J.R.L., 1961. Sandstone-plugged pipes in the Lower Old Red sandstone of Shophire, England. *Journal of Sedimentary Petrology*, **31**: 325–335.
Batchelor, G.K., and Janse Van Rensburg, R.W., 1986. Structure formation in bidisperse sedimentation. *Journal of Fluid Mechanics*, **166**: 379–407.
Best, J.L., 1989. Fluidization pipes in volcaniclastic mass flows, Volcan hudson, Southern Chile. *Terra Nova*, **1**: 203–208.
Druitt, T.H., 1995. Settling behavior of concentrated dispersions and some volcanological applications. *Journal of Volcanology and Geothermal Research*, **65**: 27–39.
Lowe, D.R., and LoPiccolo, R.D., 1974. The characteristics and origins of dish and pillar structures. *Journal of Sedimentary Petrology*, **44**: 484–501.
Lowe, D.R., 1975. Water-escape structures in coarse-grained sediments. *Sedimentology*, **22**: 157–204.
Lowe, D.R., and Guy, M., 2000. Slurry-flow deposits in the Britannia Formation (Lower Cretaceous), North Sea: a new perspective on the turbidity current and the debris flow problem. *Sedimentology*, **47**: 31–70.
Nichols, R.J., Sparks, R.S.J., and Wilson, C.J.N., 1994. Experimental studies of the fluidization of layered sediments and the formation of fluid escape structures. *Sedimentology*, **41**: 233–253.
Wentworth, C.M. Jr., 1966. *The Upper Cretaceous and Lower Tertiary Rocks of the Gualala Area, Northern Coast Ranges, California.* Unpublished Ph.D. Thesis, Stanford University.
Weiland, R.H., Fessas, Y.P., and Ramarao, B.V., 1984. On instabilities arising during sedimentation of two-component mixtures of solids. *Journal of Fluid Mechanics*, **142**: 383–389.

Cross-references

Dish Structure
Fluid Escape Structures
Gravity-Driven Mass Flows
Liquefaction and Fluidization
Turbidites

PLACERS, FLUVIAL

Fluvial placer deposits form when mechanical sorting causes concentration of heavy minerals of economic interest. The heavy minerals must be chemically stable to survive so placers usually consist of native metals such as gold (SG 15–19) and the platinum group (14–21); resistate minerals, for example, cassiterite (~ 7), chromite (4–5), ilmenite (4.5–5), rutile (4.2), magnetite (5.2), zircon (4.7); and, diamonds (3.5) and other gemstones. Placer deposits can be of considerable economic importance—for example, the tin (cassiterite) deposits of South East Asia (Taylor, 1986) and the gold paleoplacers of the Witwatersrand (Smith and Minter, 1980).

Small accumulations of heavy minerals can form as planar laminae (Ljunggren and Sundborg, 1968); thin ripples on the upstream slopes of dunes (Brady and Jobson, 1973); or develop on topsets and foresets of dunes (Brady and Jobson, 1973). At intermediate scales heavy minerals accumulate at bar heads (Hanson, 1979; Day and Fletcher, 1991); in zones of flow separation in the wake of obstacles and at channel confluences (Best and Brayshaw, 1985); or where a channel narrows and flow is constricted (Smith and Beukes, 1983; Paopongsawan and Fletcher, 1993).

Cumulative concentration of heavy minerals at small to intermediate scales can upgrade average concentrations in fluvial sediments so they become greater than in their source materials and continue to increase as drainage basin size increases (Fletcher and Muda, 1999; Halfpenny and Mazzucchelli, 1999). Weathering in the humid tropics is probably especially favorable to this insofar as large volumes of bedrock are converted to fine grained weathering products that are easily removed in suspension to leave a coarse bedload enriched in resistate heavy minerals (Sirinawin et al., 1987). Provided sediment continues to be preferentially transported, changes in stream gradient (Toh, 1978; Day and Fletcher, 1991; Hou and Fletcher, 1996) are often sites of accumulation of heavy minerals within fluvial systems. Exits of rivers from highlands onto plains can be particularly favorable for placers if the upstream basin contains an extensive source of heavy minerals (Toh, 1978).

Placers form because the density of heavy minerals causes them to lag behind similarly sized grains of light minerals during entrainment, suspension and transport of grains, and to settle, by dispersive shearing, within a moving bed (Slingerland, 1984; Reid and Frostick, 1985; Slingerland and Smith, 1986; Fletcher and Loh, 1997). However, grain size of the heavy mineral, and the roughness and nature of the streambed voids are also important factors in their accumulation. For example, cobble-gravels at the heads of point bars have greater accumulations of coarse heavies than sands at bar tails but the difference in concentration between these sites decreases with decreasing grain size of the heavy mineral (Saxby and Fletcher, 1986; Fletcher and Day, 1989). At larger scales coarse heavy minerals form placers closest to the source whereas finer heavy minerals tend to accumulate further downstream (Toh, 1978).

The association of fine heavies with coarser sediments leads to the concepts of *hydraulic (or settling) equivalence, entrainment equivalence*, and *transport equivalence*. *Hydraulic equivalence*, introduced by Rubey (1933), describes particles of different densities and size that settle through a column of fluid at the same velocity, whereas *entrainment equivalence* considers grains of light and heavy minerals that are entrained together at the same critical shear stress Slingerland (1977). *Transport equivalence*, as defined by Fletcher *et al.* (1992), refers to those grains that, despite differences in their physical properties, have the same net average transport rate.

Sorting, transport and grain size equivalence of mineral grains can only be described quantitatively by theories of grain settling, pivoting and entrainment under idealized conditions (see Slingerland, 1984; Slingerland and Smith, 1986; Fletcher and Loh, 1997). Transport of grains of different sizes and densities by unsteady, nonuniform turbulent flows acting on non-uniform streambeds cannot, as yet, be predicted quantitatively. Field observations show that fluvial processes concentrate heavy mineral grains that are small enough to hide in the bed ($d_i/D_m < 0.2$, where d_i and D_m are grain diameter and the median diameter of the bed, respectively). For grain sizes finer than about 100 µm concentration of heavy minerals in bedload is increased by transfer of fine sand, silt and clay to suspended load at the early stages of sediment transport (Fletcher and Loh, 1997).

Preferential entrainment of light minerals grains becomes less effective as a concentration process as the size of the heavies approaches the upper limit of hiding ($d_i/D_m \sim 0.2$) and their exposure to the flow and the probability of their entrainment increases. However, as flows increase scouring of the bed releases buried heavy minerals and a wide range of grain sizes and densities are mobilized. As discharge falls, and the bed is re-established, coarse heavy minerals are among the first to be trapped at favorable sites such as the voids at bar heads. At the same time fine heavies travelling higher in the flow partly overpass trap sites and develop less spatial variation in concentration on the stream bed (Fletcher and Wolcott, 1991).

Improved understanding of fluvial placer deposits, and hence development of improved exploration guidelines for economic placers, requires more detailed descriptions of their characteristics, at grain to system scales, and the use of these observations to test and refine theories of sediment transport and grain sorting in fluvial channels.

W.K. Fletcher

Bibliography

Best, J.L., and Brayshaw, A.C., 1985. Flow separation—a physical process for the concentration of heavy minerals within fluvial channels. *Quarterly Journal of Geological Society of London*, **142**: 747–755.

Brady, L.L., and Jobson, H.E., 1973. An experimental study of heavy mineral segregation under alluvial-flow conditions. *US Geological Survey Professional Paper 562-K*, 38pp.

Day, S.J., and Fletcher, W.K., 1991. Concentrations of magnetite and gold at bar and reach scales in a gravel-bed stream, British Columbia, Canada. *Journal of Sedimentary Petrology*, **61**: 871–882.

Fletcher, W.K., and Muda, J., 1999. Influence of selective logging and sedimentological processes on geochemistry of tropical rainforest streams. *Journal of Geochemical Exploration*, **67**: 211–222.

Fletcher, W.K., and Loh, C.H., 1997. Transport and deposition of cassiterite by a Malaysian stream. *Journal of Sedimentary Research*, **67**: 763–765.

Fletcher, W.K., Church, M., and Wolcott, J., 1992. Fluvial transport equivalence of heavy minerals in the sand size range. *Canadian Journal of Earth Sciences*, **29**: 2017–2021.

Fletcher, W.K., and Wolcott, J., 1991. Transport of magnetite and gold in Harris Creek, British Columbia; and implications for exploration. *Journal of Geochemical Exploration*, **41**: 253–274.

Fletcher, W.K., and Day, S.J., 1989. Behaviour of gold and other heavy minerals in drainage sediments: some implications for exploration geochemical surveys. *Institution of Mining and Metallurgy Transactions*, **98**: B130–B136.

Halfpenny, R., and Mazzucchelli, R.H., 1999. Regional multi-element drainage geochemistry in the Himalayan Mountains, northern Pakistan. *Journal of Geochemical Exploration*, **57**: 223–234.

Hanson, C.G., 1979. *Geochemical and Mineralogical Investigations of Methods for Detecting Kimberlites in the Area of Stockdale Kimberlite, Riley County, Kansas*. Unpub. M.S. Thesis, Pennsylvania State University, 119pp.

Hou, Z., and Fletcher, W.K., 1996. The relations between false gold anomalies, sedimentological processes and landslides in Harris Creek, British Columbia, Canada. *Journal of Geochemical Exploration*, **57**: 21–30.

Ljunggren, P., and Sundborg, A., 1968. Some aspects of fluvial sediments and fluvial morphology, II. A study of some heavy mineral deposits in the valley of the river Lule. *Geografiska Annaler*, **50A**: 121–135.

Paopongsawan, P., and Fletcher, W.K., 1993. Distribution and dispersion of gold in point bar and pavement sediments in the Huai Hin Laep, northeastern Thailand. *Journal of Geochemical Exploration*, **47**: 251–268.

Reid, I., and Frostick, L.E., 1985. Role of settling, entrainment and dispersive equivalence and of interstice trapping in placer formation: *Journal of the Geological Society of London*, **142**: 739–746.

Rubey, W.W., 1933. The size distribution of heavy minerals within a water-laid sandstone: *Journal of Sedimentary Petrology*, **3**: 3–29.

Saxby, D., and Fletcher, W.K., 1986. The geometric mean concentration ratio (GMCR) as an estimator of hydraulic effects in geochemical data for elements dispersed as heavy minerals. *Journal of Geochemical Exploration*, **26**: 223–230.

Sirinawin, T., Fletcher, W.K., and Dousset, P.E., 1987. Evaluation of geochemical methods in exploration for primary tin deposits: Batu-Gajah-Tonjong Tualang area, Perak, Malaysia. *Journal of Geochemical Exploration*, **29**: 165–181.

Slingerland, R.L., 1977. The effects of entrainment on the hydraulic equivalence relationships of light and heavy minerals in sands. *Journal of Sedimentary Petrology*, **47**: 753–770.

Slingerland, R.L., 1984. Role of hydraulic sorting in the origin of fluvial placers. *Journal of Sedimentary Petrology*, **54**: 137–150.

Slingerland, R.L., and Smith, N.D., 1986. Occurrence and formation of water-laid placers. *Annual Review of Earth and Planetary Sciences*, **14**: 113–147.

Smith, N.D., and Beukes, N.J., 1983. Bar to bank flow convergence zones: a contribution to the origin of fluvial placers. *Economic Geology*, **75**: 1–14.

Smith, N.D., and Minter, W.E.L., 1980. Sedimentological controls of gold and uranium in two Witwatersrand paleoplacers. *Economic Geology*, **75**: 1–14.

Taylor, D., 1986. Some thoughts on the development of the alluvial tinfields of the Malay-Thai Peninsula. In *GEOSEA V Proceedings* 1. *Geological Society of Malaysia Bulletin* 19, pp. 375–392.

Toh, E.S.C., 1978. Comparison of exploration for alluvial tin and gold. In Jones, M.J. (ed.), *Proceeding of the 11th Commonwealth Mining and Metallurgy Congress*. Hong Kong, Transactions of the Institution of Mining and Metallurgy, pp. 269–278.

Cross-references

Grain Threshold
Heavy Minerals
Sediment Transport by Unidirectional Water Flows

PLACERS, MARINE

Introduction

A placer can be defined as a surficial mineral deposit formed by the mechanical concentration of mineral particles from weathered debris (Bates and Jackson, 1984). Marine placers are mostly metallic minerals or gems which have been transported to the seafloor in a solid form. They are mainly mechanically and chemically resistant minerals which have been liberated on breakdown of their parent rocks. Such rocks are usually igneous in origin but metamorphic and sedimentary rock can also liberate placer minerals on breakdown. Marine placers occur almost entirely in shallow continental shelf environments.

Nature of marine placer minerals

Marine placer minerals can be classified by composition (Table P1) or in terms of their specific gravity. Emery and Noakes (1968) defined heavy heavy minerals as those with specific gravities of 6.8–21, light heavy minerals as those with specific gravities of 4.2–5.3, and gems with specific gravities from 2.9–4.1. The heavy heavy minerals include native gold and platinum and the tin bearing placer cassiterite, light heavy minerals generally comprise the resistant accessory minerals of igneous rock and include monazite, zircon, ilmenite and rutile as important varieties together with magnetite. The most important gem placer mineral is diamond.

Formation of marine placer deposits

The formation of marine placer mineral deposits requires several separate processes. There must be a large primary source of minerals. For this reason major marine placer deposits are much more common off continental margins than off islands because the latter are rarely large enough to contain sufficient source rock. Resistant accessory minerals are liberated on source rock weathering and are released for transport. Chemical weathering is more important than mechanical weathering in this regard, which is one of the reasons why beach and marine placer mineral deposits are more common off mid and low latitude coastal areas than they are off high latitude areas. Mechanical weathering predominates in the latter. Transport of the placer minerals is largely by water, during which the less resistant minerals are separated from the heavier more resistant ones. On reaching the coast, placer minerals are usually concentrated in sinks or traps because of their high specific gravity, or concentrated on the upper parts of beaches by wave action, sometimes leading to a concentration of the deposits to economic grades.

The depositional environments of marine placer minerals have been reviewed in Cronan (1980, 2000) and need not be considered in detail here. In the offshore area, placer minerals can occur in several situations (Figure P10). It should be remembered however that some of the placer mineral deposits currently offshore were formed in river valleys which subsequently became inundated during the post glacial rise in sea level. Certain deposits of cassiterite, scheelite, gold and platinum fall into this category (Cronan, 1992). Further, many placer minerals have been deposited on beaches near the mouths of the rivers down which the minerals were transported and

Table P1 Principal placer minerals and their composition (from Cronan, 1992)

	Specific gravity	Composition and principal element
Gems		
Diamond	3.5	C
Garnet	3.5–4.27	$(CaMgFeMn)_3 (FeAlCrTi)_2 (SiO_4)_3$
Ruby	3.9–4.1	Al_2O_3
Emerald	3.9–4.1	Al_2O_3
Topaz	3.4–3.6	$Al_2F_2SiO_4$
Heavy noble metals		
Gold	20	Au
Platinum	21.5	Pt
Light heavy minerals		
Beryl	2.75–2.8	$Be_3Al_2Si_6O_{18}(Be)$
Corundum	3.9–4.1	Al_2O_3 (Al)
Rutile	4.2	TiO_2 (Ti)
Zircon	4.7	$ZrSiO_4$ (Zr)
Chromite	4.5–4.8	$FeCr_2O_4$ (Cr)
Ilmenite	4.5–5.0	$FeOTiO_2$ (Ti)
Magnetite	5.18	Fe_3O_4 (Fe)
Monazite	5.27	$(CeLaYt) PO_4 ThO_2 SiO_2$ (Th REE)
Scheelite	5.9–6.1	$CaWO_4$ (W)
Heavy heavy minerals		
Cassiterite	6.8–7.1	SnO_2 (Sn)
Columbite, Tantalite	5.2–7.9	$(FeMn) (NbTa)_2O_6$ (Nb, Ta)
Cinnabar	8–10	HgS (Hg)

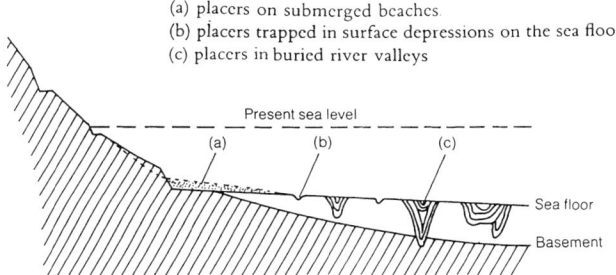

Figure P10 Environments of marine placer deposition (from Cronan, 1992).

have subsequently been concentrated in coastal dunes by wind action. Placer minerals transported from the land into the sea can be further transported by wave and current action, but when the energy level of the environment decreases to below that at which the minerals can be transported, they are deposited on the seafloor. Once deposited, a significantly higher energy level is required to pick them up again. For example, Hosking and Ong (1963) noted that off North Cornwall in the UK free cassiterite grains tend to be concentrated in elongate zones below low water and parallel to the coast. Further, Moore and Welkie (1976) have suggested that very fine grained gold and platinum placer deposits can occur in quiet bays and protected estuaries off Alaska. Initially supplied to the offshore area under higher energy conditions, their distribution would be related to circulation gyres of slackening currents within protected embayments, coupled with agglomeration of particles on contact with electrolytes in seawater.

Marine placer mineral deposit types

Light heavy minerals

Certain light heavy minerals, particularly rutile, ilmenite, magnetite, monazite, zircon, sillimanite, and garnet can be concentrated by the huge panning system of sandy beaches (Kudrass, 2000). In the surf zone, various processes tend to separate minerals according to their density, size, and shape. Heavy minerals are concentrated by the combination of these processes in the upper part of the beach where wind may erode them and form heavy mineral rich coastal dune deposits. According to Kudrass (2000), with rising sea level, the amount of sand stored on the beach and shelf, the intensity of long shore currents, the wind direction and wave energy, and the morphology and width of the shelf are all parameters determining the dispersal of heavy mineral bearing sand during marine transgression. In the case of large sand volumes, and a wide gently sloping shelf, the transgressing sea cannot move the total amount of sand stored in the coastal system. The transgression then results in an extensive mixture of the available mobile beach sand and dune placer deposits, and the latter become widely distributed and disseminated. The ilmenite–rutile–zircon placer deposit on the middle shelf off northern Mozambique, which is related to the Zambezi River, is one example of this type of disseminated placer deposit.

Where there are moderate to low volumes of heavy mineral bearing sand stored in the coastal system, and a narrow steep shelf, the transgressing sea moves the deposits by storm washover processes (Kudrass, 2000). Due to the vertical differentiation of heavy mineral assemblages, some heavy minerals are preferentially concentrated in the landward migrating sand while others are predominantly left behind in the trailing sand sheet covering the transgressed shelf. The sand arriving at the coastline undergoes further differentiation in the surf zone, which leads to the formation of important beach and dune placer deposits, such as those on the southeast coast of Australia.

Tin

As mentioned, drowned river valleys can contain alluvial placer mineral deposits, as for example in the tin deposits off south east Asia. Often in these situations the placer minerals are at the bottom of the sediment column overlying bedrock. In the formation of giant tin deposits on continental shelves, the drowning of old river systems is recognized to be the most important genetic process by far (Yim, 2000). Thus the bulk of the economic tin deposits occurring on continental shelves were formed under subaerial conditions prior to submergence below sea level. Consequently, in terms of understanding the geology of tin placer deposits on continental shelves, it is essential to learn from their comparatively well studied onshore counterparts (Yim, 2000).

Gold

According to Garnett (2000a), the simplest form of offshore gold concentration is, like tin, in a buried fluvial channel which is usually covered by recent transgressive marine sediments. The highest gold grades are usually at the base of the alluvial sequence resting on the bedrock. Examples exist in the Philippines and other areas where tropical and subtropical weathering has eroded coastal primary gold deposits.

In Arctic regions, gold bearing source rocks have been subjected to massive weathering and grain reduction by glaciation. The resulting sediments containing liberated gold have been transported by glacial and fluvial processes to be distributed over relatively large regions. Particulate gold was then further transported by marine processes (Garnett, 2000a). Marine gold deposits of such origin occur off Alaska, southern Chile and Argentina, New Zealand, and elsewhere. Because of its high density, gold is less mobile than other heavy minerals of equal particle size. Only very fine particles travel far under marine conditions. Most economic concentrations of placer gold are thus rarely situated more than a few kilometers from the site of the original accumulation.

The most important marine placer gold deposit is that off Nome, Alaska. According to Garnett (2000a) all writers attribute the spatial distribution of gold in the sediments covering the coastal plain there to glacial events combined with marine transgression and regression. During the Pleistocene glaciation of the region, changing sea levels caused the Bering Sea to intermittently become a coastal plain. Primary gold was released from hard rock sources by extreme frost action to be deposited in down slope alluvial sediments and in fluvial channels which were incised into the coastal plain. These deposits were later swept up together with eroded bedrock by advancing glaciers. Later, as sea level rose, the topography was molded, levelled and partially eroded. The high storm energy, especially in the surf zone and shallow waters, reworked the sediments. Marine invasion alternating with retreat of the sea acted on the exposed glacial deposits, leaving relict thin gold-bearing layers on the seafloor. Sediment-hosted gold exposed on the seabed was then upgraded by the winnowing action of marine currents to create a lag deposit.

Diamonds

The principal gem placer minerals occurring in the offshore environment are diamonds. These were discovered in southwest Africa in raised beach sands in 1908 (Murray, 1969), the diamond bearing beds extending down to the coast and offshore. The diamonds were derived from kimberlites in the interior and had been transported to the coast by fluvial processes. According to Garnett (2000b), longshore currents combined with high energy wind and wave action during periods of considerable sea-level change concentrated the diamonds in trap sites on paleo coastlines and other marine geological features. The host gravels generally exist as a thin veneer on an irregular bedrock. Discontinuous deposits lie on the inner and middle sections of the continental shelf along the South African and Namibian coastlines.

Diamondiferous deposits on the inner shelf off southern Africa are similar to those onshore. With a density of $3.5\,g/cm^3$ diamond is the lightest of the economic placer minerals. Thus it travels farther than gold and heavy placer minerals. According to Garnett (2000b), the diamonds were redistributed from paleo river mouths along the coast and on ancient beaches that developed in response to sea-level changes. Powerful south-westerly winds, combined with high energy swells, created mostly northward moving longshore drift currents. These transported the diamonds in waters less than about 35 m deep. The resulting diamond-bearing deposits have been continually upgraded by marine action. The winnowing of the host sediments by waves has been the single most important process that contributed to the enrichment (De Decker and Woodborne, 1996). Present day diamond concentrations have resulted from the continual reworking of older deposits and the subsequent redeposition of diamonds in preferred trap sites throughout periods of sea-level change. Deposits in the surf zone were continually sorted, which concentrated the diamonds close to the bedrock. They were trapped in gravels which collected in gulleys, potholes and channels eroded into the wave cut terraces. Their final site of entrapment was usually in a medium energy environment (Garnett, 2000b).

On the middle shelf off southern Africa, resistant ridges up to 8 m high trend generally north-northwest and provide a rough bedrock on which diamonds have been trapped. Cemented diamond bearing gravels on the ridges were submerged during marine transgressive events. Their consequent reworking produced a diamondiferous lag gravel. According to Garnett (2000b), the diamonds are dispersed neither uniformly nor totally at random. Irregularly shaped well mineralized areas exist in places as widely separated "islands" within extensive tracts which are either low grade or barren. Inside such areas, the diamonds exhibit very localized distribution characteristics.

David S. Cronan

Bibliography

Aleva, G.J.J., 1973. Aspects of the historical and physical geology of the Sunda Shelf essential for the exploration of submarine tin placers. *Geologie en Mijnbouw*, **52**: 79–91.

Bates, R.L., and Jackson, J.A. (eds.), 1984. *Dictionary of Geological Terms*, 3rd edn, American Geological Institute, Anchor Press.
Cronan, D.S., 1980. *Underwater Minerals*. Academic Press.
Cronan, D.S., 1992. *Marine Minerals in Exclusive Economic Zones*. Chapman and Hall.
Cronan, D.S. (ed.), 2000. *Handbook of Marine Mineral Deposits*. Bocca Raton: CRC Press.
De Decker, R.H., and Woodborne, M.W., 1996. Geological and technical aspects of marine diamond exploration in Southern Africa. *Offshore Technology Conference Proceedings*. OTC 8018, 561.
Emery, K.O., and Noakes, L.C., 1968. *Economic Placer Deposits of the Continental Shelf*. Bangkok: ECAFE CCOP Tech. Bull I, CCOP.
Garnett, R.H., 2000a. Marine placer gold with particular reference to Nome, Alaska. In Cronan, D.S. (ed.), *Handbook of Marine Mineral Deposits*. Bocca Raton: CRC Press, pp. 67–101.
Garnett, R.H., 2000b. Marine placer diamonds with particular reference to Southern Africa. In Cronan, D.S. (ed.), *Handbook of Marine Mineral Deposits*. Bocca Raton: CRC Press, pp. 103–141.
Hosking, K.F.G., and Ong, P.M., 1963. The distribution of tin and certain other heavy minerals in the superficial portions of the Gwithian/Hayle beach of West Cornwall. *Transactions of the Royal Geological Society of Cornwall*, **19**: 351–390.
Kudrass, H.R., 2000. Marine placer deposits and sea level change. In Cronan, D.S. (ed.), *Handbook of Marine Mineral Deposits*. Bocca Raton: CRC Press, pp. 3–12.
Moore, J.R., and Welkie, C.J., 1976. Metal bearing sediments of economic interest, coastal Bering Sea. In *Proceedings of the Symposium on Sedimentation, Geological Society of Alaska*. K1–K17.
Murray, L.G., 1969. Exploration and sampling methods employed in the offshore diamond industry. *Proceedings of the Ninth Commonwealth Mining and Metallurgy Congress*, **2**: pp. 71–94.
Yim, W.W.S., 2000. Tin placer deposits on continental shelves. In Cronan, D.S. (ed.), *Handbook of Marine Mineral Deposits*. Bocca Raton: CRC Press, pp. 27–66.

Cross-references

Grain Threshold
Heavy Minerals
Placers, Fluvial
Sediment Transport by Waves
Swash and Backwash, Swash Marks

PLANAR AND PARALLEL LAMINATION

Laminae are defined as sedimentary strata that are less than 10 mm thick, according to McKee and Weir (1953). Laminae are recognizable because of variation in sediment texture and/or composition within and between them (Figure P11). For example, the grain size within a lamina may increase or decrease upward, or be different in adjacent laminae. Other laminae may have distinctive concentrations of dark (heavy) minerals. *Planar laminae* have planar bounding surfaces. The term planar lamination is commonly taken to indicate planar laminae that are more-or-less horizontal (within a few degrees) when originally deposited, and that have more-or-less parallel bounding surfaces (but laminae do vary in thickness laterally). Hence, planar lamination has also been called horizontal lamination, even lamination and parallel lamination. The term *parallel lamination* has also been used to describe laminae with parallel but nonplanar boundaries (Figure P11). Planar laminated sandstones (popularly known as flagstones) are valuable building materials. Commonly, these sandstones contain near-parallel ridges of sand grains along lamina

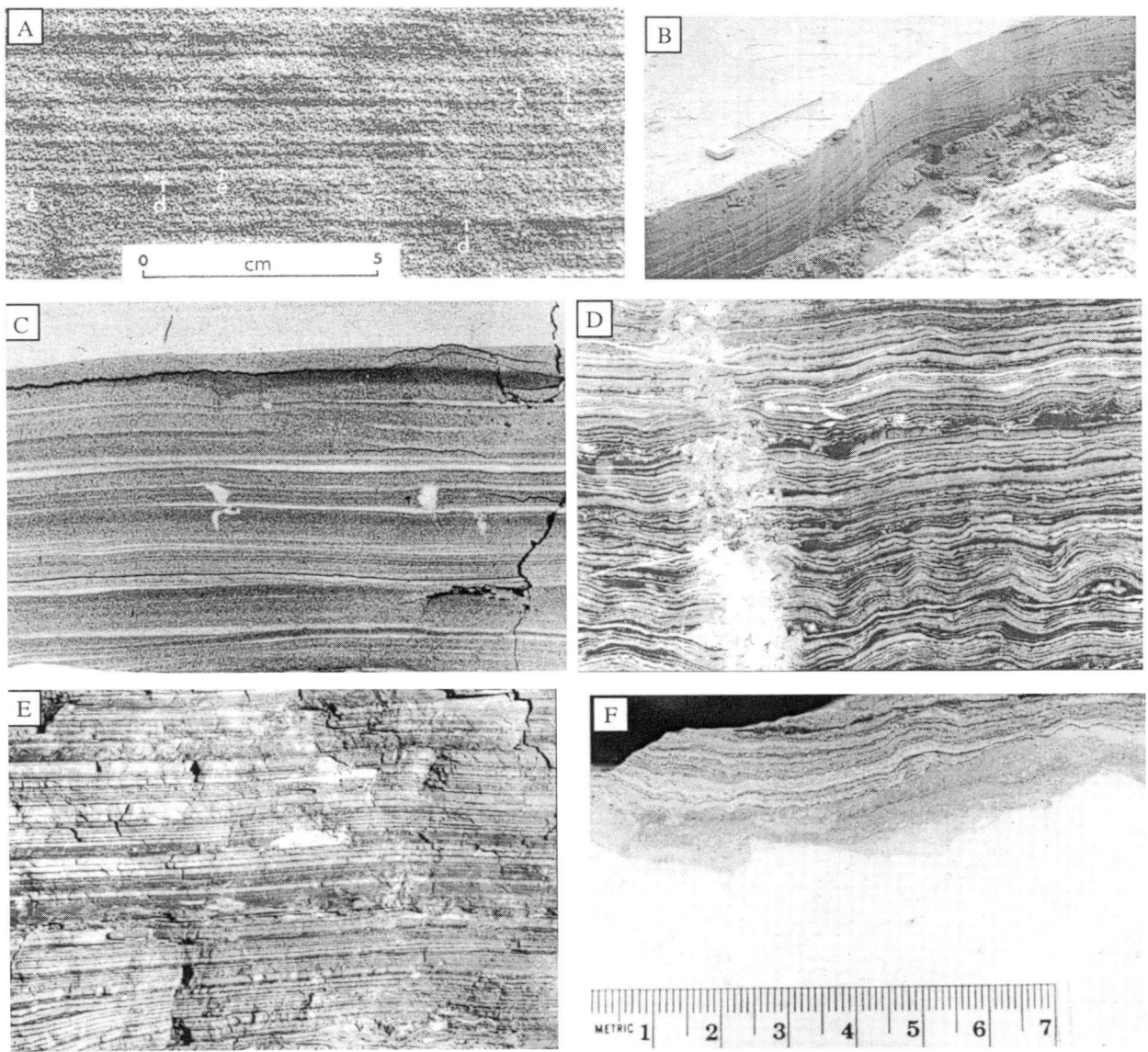

Figure P11 (A) Planar laminae thought to have formed on an upper stage plane bed, Permo-Triassic St Bees Sandstone, UK. Low-angle laminae labeled c. Sharp bases to relatively thick laminae labeled d. Discontinuous (truncated) laminae labeled e (from Best and Bridge, 1992). (B) Modern beach laminae from Virginia, USA. (C) Planar laminae composed of very fine grainstone (light) and mudstone (dark), Middle Ordovician St. Paul Group, Maryland, USA (photo courtesy of Robert Demicco). (D) Wavy and crinkled parallel laminae (stromatolite) composed of grainstone (dark) and dolomitic limestone (light), Upper Cambrian Conocoheague Limestone, Maryland, USA (photo courtesy of Robert Demicco). (E) Planar, parallel laminae in dolomitic mudstone, Lower Proterozoic Wittenoom Dolomite, Western Australia (photo courtesy of Robert Demicco). (F) Parallel laminated calcium carbonate crusts on Pleistocene limestone, Florida Keys, USA (photo courtesy of Robert Demicco).

boundaries, referred to as *primary current lineation* or *parting lineation*. The ridges are a few grains high and have a spacing of about 1 cm.

Laminae commonly occur in sets (*laminasets*) that are recognizable because of changes in texture, composition, or orientation of laminae (Figure P11). Sets of planar and parallel laminae normally have bounding surfaces that are more-or-less parallel to the laminae within them. If the angle between the laminae and the laminaset boundary exceeds a few degrees, the sedimentary structure is referred to as *cross lamination* or *inclined lamination*. However, some sets of planar lamination do contain low-angle inclined laminae in places.

The term lamina has been used to denote a thin sediment layer produced by momentary fluctuations in current velocity and sediment transport rate (Otto, 1938, and many others). Such genetic implications for the term lamina should be avoided, as it cannot be known conclusively how all laminae form. Indeed, planar and parallel laminae may form in different ways in different depositional environments. Their origin is also different depending on sediment size. Some examples are given below, but the list is not exhaustive.

Origin of planar laminated sand under unidirectional currents

Upper-stage plane beds occur beneath fast, shallow unidirectional water flows in which there is appreciable transport of bed load and suspended load. Deposition on upper-stage plane beds gives rise to planar lamination (Figure P11(A)). Most early explanations for the formation of planar laminae beneath upper-stage plane beds involved the influence of turbulent variations in bed shear stress and sediment transport (reviewed by Bridge, 1978; Allen, 1982, 1984). Turbulence, and specifically the turbulent bursting process linked to large coherent structures in the flow (Bridge, 1978; Allen, 1984; Cheel and Middleton, 1986), cannot explain the thickness, lateral extent and size grading within planar laminae. However, the migration of low-relief bed waves can (Jopling, 1964, 1967; Bridge and Best, 1988, 1997; Paola et al., 1989; Best and Bridge, 1992). The fine-grained bases of laminae are due to deposition from suspension on the distal lee sides of the bed waves. Most low-relief bed waves do not have avalanche faces, and the grain size of bed load tends to decrease from the crest to the toe of the lee side, resulting in the coarsening upward parts of laminae. However, the largest low-relief bed waves may have a short lee-side avalanche face at the base of which coarse sediment accumulates. This relatively coarse sediment lies with sharp contact on the underlying laminae. A predominance of fining-upward laminae may result from preferential preservation of the lower parts of the largest bed waves that moved over the bed. Low-angle, inclined laminae within the thicker planar laminae are due to turbulent fluctuations in sediment transport on to the depositional lee side of the bed waves (Figure P11(A)). Parting lineation results from the imprint of high-speed and low-speed streaks within the viscous sublayer of hydraulically smooth and transitional turbulent boundary layers (Allen, 1982).

Transcurrent lamination is the name given to sets of more-or-less planar lamination that are produced by the migration of low-relief bed waves and/or those with a very small angle of climb (Allen, 1982). Transcurrent laminae have also been associated with migration of low-relief ripples (in water and air) and dunes, bed-load sheets, and antidunes (Jopling, 1967; Smith, 1971; Reineck and Singh, 1973; McBride et al., 1975; Bennett and Bridge, 1995).

Origin of planar laminated sand under reversing currents

Upper-stage plane sand beds also occur under high-velocity water currents that reverse in direction periodically, particularly associated with gravity waves in shallow water. Deposition under these plane beds also gives rise to planar lamination. However, under these conditions, the deposition of laminae is probably due to the unsteady sediment transport associated with the passing of individual waves. In the surf zone of sandy beaches, the swash and backwash associated with wave breaking gives rise to planar laminae with characteristic concentrations of heavy minerals (Figure P11(B)). This is the well-known *beach lamination* (Clifton, 1969; Reineck and Singh, 1973; Allen, 1982). The character of sets of these laminae can be influenced by tides, storms or storm seasons. It is uncertain whether low-relief bed waves play any role in the formation of laminae by wave currents.

Origin of planar and parallel laminated mud and sand

Lamination of mud and very fine sand involves episodic deposition from suspension, provided that the mud was not transported as sand-sized pellets (Figure P11(C)). The laminae will be planar if they are deposited on a planar surface, and if deposition rate does not vary appreciably in space. The laminae arise because of temporal changes in the sediment supply or the ability of the flow to suspend sediment of a particular type. For example, changes in the turbulence intensity of water currents may be associated with the passage of individual floods over floodplains and lakes, seasonal variations in sediment supply and transport rate in lakes (e.g., *glacial varves*), passing of gravity waves, periodic storms at sea, and tidal currents that vary over a number of different timescales (reviews in Reineck and Singh, 1973; Smith et al., 1991; Demicco and Hardie, 1994). The general term for these types of laminae is *rhythmite*. Although turbulent fluctuations in suspended sediment transport (time scale of seconds or less) have been invoked to explain planar laminae in muds (review in Bridge, 1978), it is unlikely that a recognizable mud layer could be deposited in such a short period of time.

Parallel, nonplanar lamination in muds and sands occurs where the sediment is periodically draped over a surface with preexisting topography. Carbonate mud laminae deposited on cyanobacterial mats during storms or high tides give rise to parallel, nonplanar laminated *stromatolites* (Figure P11(D)).

Chemically and biologically formed planar laminae

Planar laminae may be formed by seasonal variations in the type and abundance of the tests of microorganisms (e.g., foraminifera, diatoms) that accumulate on the bed of the sea and lakes. Variations in the nature of chemical precipitation of minerals in some depositional environments may also give rise to planar lamination (e.g., laminated calcium carbonate crusts). Some lamination in muds is manifested only by color changes, associated with changes in chemical composition that could be primary or diagenetic (Figure P11(E), (F)).

John S. Bridge

Bibliography

Allen, J.R.L., 1982. *Sedimentary Structures: Their Character and Physical Basis.* Developments in Sedimentology 30, Amsterdam: Elsevier Science Publishers.

Allen, J.R.L., 1984. Parallel lamination developed from upper-stage plane beds: a model based on the larger coherent structures of the turbulent boundary layer. *Sedimentary Geology*, **39**: 227–242.

Bennett, S.J., and Bridge, J.S., 1995. The geometry and dynamics of low-relief bed forms in heterogeneous sediment in a laboratory channel, and their relationship to water flow and sediment transport. *Journal of Sedimentary Research*, Series A, **65**: 29–39.

Best, J.L., and Bridge, J.S., 1992. The morphology and dynamics of low amplitude bedwaves upon upper stage plane beds and the preservation of planar laminae. *Sedimentology*, **39**: 737–752.

Bridge, J.S., 1978. Origin of horizontal lamination under turbulent boundary layers. *Sedimentary Geology*, **20**: 1–16.
Bridge, J.S., and Best, J.L., 1988. Flow, sediment transport and bedform dynamics over the transition from upper-stage plane beds: implications for the formation of planar laminae. *Sedimentology*, **35**: 753–763.
Bridge, J.S., and Best, J.L., 1997. Preservation of planar laminae arising from low-relief bed waves migrating over aggrading plane beds: comparison of experimental data with theory. *Sedimentology*, **44**: 253–262.
Cheel, R.J., and Middleton, G.V., 1986. Horizontal lamination formed under upper flow regime plane bed conditions. *Journal of Geology*, **94**: 489–504.
Clifton, H.E., 1969. Beach lamination: nature and origin. *Marine Geology*, **7**: 553–559.
Demicco, R.V., and Hardie, L.A., 1994. *Sedimentary Structures and Early Diagenetic Features of Shallow Marine Carbonate Deposits*. SEPM Atlas Series No.1, 265pp.
Jopling, A.V., 1964. Interpreting the concept of the sedimentation unit. *Journal of Sedimentary Petrology*, **34**: 165–172.
Jopling, A.V., 1967. Origin of laminae deposited by the movement of ripples along a streambed: a laboratory study. *Journal of Geology*, **75**: 287–305.
McBride, E.F., Shepherd, R.G., and Crawley, R.A., 1975. Origin of parallel near-horizontal laminae by migration of bed forms in a small flume. *Journal of Sedimentary Petrology*, **45**: 132–139.
McKee, E.D., and Weir, G.W., 1953. Terminology for stratification and cross-stratification in sedimentary rocks. *Bulletin of the Geological Society of America*, **64**: 381–390.
Otto, G.H., 1938. The sedimentation unit and its use in field sampling. *Journal of Geology*, **46**: 569–582.
Paola, C., Wiele, S.M., and Reinhart, M.A., 1989. Upper-regime parallel lamination as the result of turbulent sediment transport and low-amplitude bedforms. *Sedimentology*, **36**: 47–60.
Reineck, H.E., and Singh, I.B., 1973. *Depositional Sedimentary Environments*. Springer.
Smith, D.G., Reinson, G.E., Zaitlin, B.A., and Rahmani, R.A. (eds.), 1991. *Clastic Tidal Sedimentology*. Canadian Society of Petroleum Geologists Memoir, 16: pp. 29–39.
Smith, N.D., 1971. Pseudo-planar stratification produced by very low amplitude sand waves. *Journal of Sedimentary Petrology*, **41**: 624–634.

Cross-references

Bedding and Internal Structures
Bedset and Laminaset
Biogenic Sedimentary Structures
Microbially Induced Sedimentary Structures
Parting Lineations and Current Crescents
Stromatolites
Surface Forms
Swash and Backwash, Swash Marks
Tidal Flats
Tides and Tidal Rhythmites
Varves

POREWATERS IN SEDIMENTS

Approximately 20 percent by volume of the sedimentary portion of the earth's crust consists of water that occupies most of the matrix and fracture porosity of sediments and sedimentary rocks. These porewaters are variously called *groundwater* in freshwater aquifer systems, *interstitial water* in marine sediments, and *formation water* or *basinal brines* in deep sedimentary basins. Porewaters as a group range widely in chemical and isotopic composition and in their physical properties. There has been considerable interest in the origin and migration of porewaters because of the role these fluids play in the geological, geochemical, and economic resource evolution of sediments and sedimentary basins. The presence of porewaters under pressure facilitates or may even induce tectonic deformation of sedimentary basins and continental margins. Porewaters serve as transport agents for heat and reactive solutes in sediments, and variations in the composition of porewaters in space and time yield important information on mechanisms and rates of solute transport and diagenetic reaction. The study of porewaters thus complements the study of diagenesis obtained by the sedimentological, mineralogical, and geochemical study of the solid phases present in a sediment.

Historical development and outline

The origin of waters in the earth's crust has been speculated upon since ancient times (Hanor, 1988). Prior to the 1600s, there was a general belief that fresh groundwaters were formed by the subsurface distillation of seawater which had infiltrated deep into the earth. These freshwaters get discharged at the surface to form rivers and streams. Seawater salt was left behind in the form of subsurface brines. Beginning with the publication of Perrault's mass balance studies in 1674, it was eventually recognized that rivers, streams, and fresh groundwaters are meteoric in origin. Debate on the origin of subsurface brines in sediments, however, has continued up until the present time. With the advent of modern geochemical techniques in the last half of the past century, much has been learned about the chemical and isotopic composition of porewaters in sediments and their origin. Chief sources of information have been the analysis of produced waters from oil and gas fields, the analysis of potable groundwaters, and the extensive body of analyzes of waters in deep sea sediments generated as part of the JOIDES and ODP drilling projects.

This article reviews the types of solutes found in sedimentary porewaters and the general types of processes which control or influence porewater compositions. This background information is then applied to three major types of sedimentary pore waters: (1) shallow continental waters; (2) marine interstitial waters in continental shelf and deep sea sediments; and (3) waters in deep sedimentary basins. The focus is primarily on the geochemistry of porefluids, but it should be recognized that many important physical properties of porewaters and bulk sediments, such as density, viscosity, thermal and electrical conductivity, and sonic transit time, are directly dependent on fluid temperature, pressure, and chemical composition.

General composition of porewaters

Concentration units

A wide variety of units have been used to describe the concentration of solutes in sedimentary porewaters. Concentrations are typically expressed in terms of mass, moles, or equivalents of solute per liter of solution, kilogram of solution, or kilogram of H_2O. Many groundwater analyzes are presented in mg/L or g/L, units which are convenient because of the volumetric methods used in most analytical techniques. Many in the marine geochemistry community, however, use mass or moles per kg of solution, which are pressure and temperature independent concentration units. The concentration unit molality, mol/kg H_2O, is used in thermodynamic calculations of saturation state. One final unit is the equivalent (eq), where the number of equivalents of a solute is equal to the number of moles of solute times absolute charge of the solute. The units eq/L or meq/L are widely used in making charge balance calculations from water analyzes and in defining hydrochemical facies, as described below.

Salinity and major solutes

The polar nature of the H_2O molecule makes water an excellent solvent for charged ions and other polar molecules. The salinity or total dissolved solute (TDS) concentrations of porewaters varies greatly, even within individual sedimentary units. Salinities of a few hundred mg/L or less are typical of sediments flushed by meteoric waters. The average salinity of seawater is approximately 35,000 mg/L, and the porewaters in many marine sediments have similar salinities. The salinities of waters in the deeper parts of sedimentary basins, however, can reach levels of 350,000 mg/L or more.

Although a wide range of dissolved species exist in natural waters, the greatest proportion of TDS is contributed by a relatively few species, typically Na^+, K^+, Mg^{2+}, Ca^{2+}, Cl^-, HCO_3^-, and SO_4^{2-}. Several factors determine the major solute concentration of a porewater. These include relative abundance of the potential solute, the ease of accommodating the solute into liquid water, and reaction rates and solubility constraints imposed by coexisting mineral phases and reactive mineral and organic surfaces. Cation accommodation in water is determined in part by ionic potential, the ratio of cation charge to ionic radius in Ångstroms. Cations of small ionic potential, roughly 3 or less, are readily accommodated in aqueous solution as hydrated cations. These include most of the group IA and IIA cations, including Li^+, Na^+, and K^+, and Mg^{2+}, Ca^{2+}, Sr^{2+}, and Ba^{2+}. Small, highly charged cations having an ionic radius of 12 or greater form soluble charged anionic complexes with oxygen. Common examples include the anionic complexes of the oxidized forms of C, N, P, and S, which include HCO_3^- and CO_3^{2-}, NO_3^-, PO_4^{3-}, and SO_4^{2-}. Cations of intermediate ionic potential, however, are accommodated in aqueous solution with much greater difficulty. These include the abundant rock-forming cations Al^{3+}, Fe^{3+}, and Mn^{4+}, which tend to react with water to form insoluble hydroxides or to remain bound in silicate or oxide mineral lattices. Silica forms the sparingly soluble complex, silicic acid, H_4SiO_4. Valence state plays a key role in the relative solubility and mobility of a cation. The reduced forms of iron and manganese, Fe^{2+} and Mn^{2+}, are far more soluble than the oxidized forms Fe^{3+} and Mn^{4+}. Halogen elements form soluble, hydrated anions: F^-, Cl^-, Br^-, I^-.

Minor inorganic species

Sedimentary porewaters as a group contain measurable concentrations of virtually every naturally occurring element in the periodic chart. Some elements, such as Pb, Zn, and Cu, sometimes occur in sufficient concentration to make the porewater in question, generally a basinal brine of high salinity, a potential ore-forming fluid. Others, such as Br, are useful in tracing the origin of high salinity in porewaters. Some minor and trace elements, such as Ba, Ra, and As, pose potential health problems when they are present in drinking-water supplies.

Gases and dissolved organics

Most gases and nonpolar organic molecules are only sparingly soluble in aqueous solution, but their presence can be of importance in driving diagenetic reactions and in deducing the origin of porewater. The microbial consumption of dissolved oxygen and production of dissolved CO_2 are two important subsurface reactions which influence the dissolved gas composition of natural porewaters (Chapelle, 1993). Groundwaters derived from rainwater typically contain atmospherically derived noble gases, whose concentration can be used to estimate the temperature of the water at the time of recharge. Waters in the deeper parts of sedimentary basins can contain primordial helium, ^3He, thought to be derived from the mantle or deep crust. Very shallow groundwaters typically contain minor concentrations of fulvic and humic acids derived from the breakdown of organic matter (Drever, 1997), and basinal waters often contain a variety of low molecular weight organic compounds produced by the thermal maturation of organic matter.

pH and redox state

The pH is defined as minus the log of the activity of the hydrogen ion, $-\log a(H^+)$ and is routinely determined on porewater samples as a measure of acid-base conditions. The pH of natural porewaters varies from over 9 for some shallow groundwaters to less than 4 for some subsurface brines. Several conventions are in use to describe the redox state of natural waters: Eh, or electrode potential; pe, minus the log of the activity of the electron; and foe, the fugacity of oxygen. Sediments and fluids in close proximity to the atmosphere or to oxygenated seawater are typically oxidized because of the availability of dissolved O_2. In organic-rich sediments, however, conditions generally become progressively reducing with depth or distance from recharge area.

Isotopic composition

Several major isotopic systems have been utilized to deduce the origin of solutes and of H_2O in different types of porewater. Hydrogen-oxygen systematics have been used to distinguish between H_2O derived from meteoric precipitation and H_2O left behind in residual brines during the evaporation of seawater (Hanor, 1988). Chlorine and I isotopic systematics have been used to help constrain the apparent age of groundwaters. Carbon and Sr, and more recently Li and B, isotopic systematics have been used to deduce the sources of these solutes and type and extent of diagenetic reaction between fluids and mineral phases.

Graphical representation of porewater compositions and hydrochemical facies

The composition of shallow fresh groundwaters varies considerably in terms of the relative proportion of major cations and anions, more so than either marine-derived interstitial waters or waters in deep sedimentary basins. It has been a challenge to develop simple graphic approaches to portray the spatial and temporal variations in typical groundwater composition (Domenico and Schwartz, 1990). A useful convention is the concept of hydrochemical facies (Back, 1961), where the composition of porewater is described in terms of those few cations and anions which contribute the bulk of the ionic charge in equivalents (eq/L) or milliequivalents (meq/L). A Ca-HCO$_3$ porewater, for example, is one dominated in terms of equivalents of charge by Ca^{2+} and HCO$_3^-$, although lesser amounts of many other solutes may be present. Prefixes such a F$^-$, B$^-$, S$^-$, and HS$^-$, representing fresh, brackish, saline, and hypersaline, may be added so that some qualitative information on TDS is conveyed as well.

General physical and chemical controls on the composition of porewaters

The general types of chemical reactions and physical processes which control porewater compositions are general to each of the three end-member types of pore waters discussed here. The physical processes include advection, which is the bulk transport of solutes through the pore spaces of a sediment by the fluid flow, and fluid mixing which occurs as a result of molecular diffusion and hydrodynamic dispersion around mineral grains. Advection can be induced by sediment compaction, by differences in topographic elevation, by differences in fluid density, and by fluid pressures in excess of hydrostatic (Domenico and Schwartz, 1990).

The types of chemical reactions which influence porewater composition include: hydrolysis or dissolution–precipitation reactions involving porewater and bulk mineral phases, many of which involve hydrogen ion and are thus acid–base reactions; redox reactions in which there is a transfer of electrons, usually between C, N, Fe, Mn, and S-bearing solutes and solids; and mineral and organic surface reactions involving adsorption and ion exchange. Many reactions involving porewaters and ambient mineral phases occur nearly simultaneously. For example, the production of H$^+$ by the oxidation of a sulfide mineral may induce the hydrolysis or dissolution of a silicate mineral. Some of the cations released in the process may be adsorbed onto the surface of newly formed clay minerals produced by the destruction of the silicate precursor.

The combined effects of advection, dispersion, and chemical reaction on the changes in spatial and temporal concentration of a solute with time (dC_i/dt) along a one-dimensional flow path can be described by the following mass balance equation:

$$\left(\frac{\partial \phi C_i}{\partial t}\right) = -v\left(\frac{\partial \phi C_i}{\partial x}\right) + D_i\left(\frac{\partial^2 \phi C_i}{\partial x^2}\right) + \phi \Sigma(R_{ij}) \quad \text{(Eq. 1)}$$

where t is time, v is advective fluid velocity, ϕ is porosity, and x is distance. D_i is the coefficient of hydrodynamic dispersion for solute i, and R_{ij} is the net source-sink term for solute i for chemical reaction j (Lichtner et al., 1996).

The first term on the right accounts for the net increase or decrease in C_i as the result of the advection or bulk flow of fluid of varying C_i through the sediment. The second term describes the mass balance resulting from mixing due to molecular or Fickian diffusion and dispersion. The final term is the net sum of the rate of the chemical reactions involving other phases and solute i. Variants on this mass balance equation in one, two, and three dimensions are at the heart of mathematical computations designed to couple the physical and chemical processes which control porewater compositions. These calculations are also used to quantify the mass transfer which occurs between mineral and fluid phases during diagenesis. In one-dimensional systems, it is sometimes possible to solve the equation analytically (Berner, 1980), but in complex heterogeneous systems numerical techniques are required (Lichtner et al., 1996).

Continental groundwaters

Topographically-driven meteoric groundwater systems

The sources of porewaters found in shallow freshwater aquifers and interbedded aquicludes are meteoric precipitation and recharge by lakes and streams during high surface water levels. That portion of rain or snow melt which does not form runoff infiltrates downward into the unsaturated zone, where it may be returned to the atmosphere via evapotransporation and/or migrate further downward to recharge an underlying aquifer. The groundwater within an aquifer will then typically migrate both downward and laterally along preferred flow paths and eventually discharge into either another continental surface environment, such as a spring, lake, or stream, or into the offshore submarine environment. Most meteoric groundwater flow is driven by differences in topographic elevation between areas of recharge and discharge (Domenico and Schwartz, 1990).

Rainwater typically contains several mg/L or more of dissolved salts. Diagenetic alteration of meteoric water compositions, however, occurs immediately within the soil environment (Drever, 1997). The carbonic acid and organic acids produced by soil bacteria are important weathering agents, and the consequent dissolution of carbonate and silicate minerals results in the leaching of the more soluble cations, such as Na$^+$, K$^+$, Mg^{2+}, and Ca^{2+}, from the soil. Although there is some complexing of Al and Fe by organics, the less soluble cations, such as Al^{3+}, Si^{4+}, and Fe^{3+} are preferentially left behind in residual mineral phases. Oxidation of sulfide minerals produces hydrogen ions, which can induce further weathering, and dissolved sulfate. Cation exchange on newly formed clay minerals favors retention of the divalent cations Mg^{2+} and Ca^{2+} in freshwater environments. High rates of evapotranspiration relative to rainfall in arid climates can serve to increase the TDS of soil waters.

Mineral-fluid and organic reactions continue as groundwater migrates along flow paths within the saturated zone, flow paths which are determined by spatial variations in permeability and hydraulic head. The local presence of evaporite minerals such as halite, anhydrite, and gypsum can add Na, Cl, Ca, and SO$_4$ to the groundwater. As the groundwater flows down gradient from recharge areas it may encounter progressively reducing conditions as dissolved oxygen is depleted by aerobic respiration. Nitrate, manganese, amino acids, iron, and organic carbon are the typical succession of electron acceptors microbes then use to oxidize organic carbon as subsurface conditions pass progressively from oxidizing to reducing (Chapelle, 1993). When conditions become sufficiently reducing, the reduction and dissolution of Fe and Mn minerals commonly gives rise to significant concentrations of these metals in shallow groundwaters. With a further decrease in redox state, however, sulfate is reduced to sulfide, and Fe may be precipitated out as an Fe-sulfide.

In carbonate aquifers, it is not surprising to find groundwaters dominated by Ca-HCO$_3$ or Ca-Mg-HCO$_3$ as the principal solutes. A well-studied example is the Floridian

Aquifer in central Florida (Sprinkle, 1989). In siliciclastic aquifers, however, there is typically a progression from Ca-Na-HCO_3 waters in shallow recharge areas to Na-Ca-HCO_3 and then Na-HCO_3 waters further down dip. Examples include the US. Atlantic and Gulf coastal plain aquifers, where bicarbonate concentrations increase down flow paths from initial values of less than 50 mg/L to over 500 mg/L (Chapelle, 1993). The probable source of CO_2 is bacterially oxidized organic material present in the aquifers and/or the interbedded aquicludes. The origin of high Na-HCO_3 groundwaters in siliciclastic aquifers has been a matter of some debate. One school of thought holds that the removal of Ca is predominantly by ion exchange, the other that Ca is removed as $CaCO_3$ by the introduction of HCO_3.

Continental brines

In arid continental climates where potential evapotranspiration far exceeds rainfall, circumstances can exist which give rise to the formation of brine lakes. Well-known examples include Lake Magadi, Kenya; Lake Tyrrell, Murray Basin, Australia; and Great Salt Lake, Utah. The composition of the brines produced depends critically on the relative proportions of Ca, HCO_3, SO_4, and Mg in the fluid being evaporated. Documented brine types include Na-Ca-Mg-Cl, Na-Mg-SO_4-Cl, and Na-CO_3-SO_4-Cl saline waters (Eugster and Hardie, 1978). These brines recharge into underlying sediments as a result of density-driven flow, where dissolved Mg is removed by the formation of Mg-rich smectite, sepiolite, Mg-calcite, and/or dolomite, and sulfate can be reduced by reduction to sulfide. In contrast to freshwater conditions where ion exchange favors the adsorption of divalent cations, monovalent cations, particularly K, are preferentially adsorbed on clays in saline conditions and removed from solution.

Porewaters in marine sediments

Fluid flow and solute transport in many marine sediments, particularly in deep sea pelagic and hemipelagic sediments are dominated by compaction-driven advection and molecular diffusion (Berner, 1980). More dynamic fluid environments are found on continental shelves, where topographically driven meteoric fluids can mix with marine porewaters; in accretionary prisms, where there is well-documented seaward expulsion of fluids; and along mid-ocean rises and ridges and other submarine volcanic settings, where thermal convection operates (Hesse, 1990).

Seawater is a Na-Cl dominated fluid with SO_4, Mg, Ca, and K as the other major species. The relative proportion of these solutes is nearly constant throughout the world's oceans, with the exception of SO_4-depleted, anoxic bottom waters in closed submarine basins. Much of the water trapped in pelagic and hemipelagic sediments at the time of deposition is seawater. This water, however, undergoes continuous chemical diagenesis and a change in composition with time and depth of burial (Hesse, 1990; Schulz and Zabel, 2000). There is typically a systematic increase in Ca and systematic decreases in Mg and K with depth. The variations in Ca and Mg are generally thought to be the result of hydrolysis of silicate minerals in underlying oceanic basement basalts. The downward decrease in dissolved K may be related to the formation of the potassium-bearing zeolite, phillipsite, and adsorption by ion exchange on clays. There is often a distinct downward decrease in $\delta^{18}O$ values of marine interstitial waters, commonly interpreted as resulting from preferential removal of heavy oxygen by the formation of phyllosilicates in altered igneous rocks and volcaniclastic sediments. Increases in $\delta^{18}O$ increases, however, have been noted in gas-hydrate bearing sections of DSDP cores. Elevated temperatures and enhanced reaction rates associated with submarine hydrothermal activity, such as in the Guyamas Basin, methase release of Ca and uptake of Mg by hydrothermally altered basalts and volcaniclastic sediments. There are also documented increases in dissolved Li, K, and Rb in these areas.

Silica concentrations are typically highly variable with depth and are dependent on lithology. Elevated silica concentrations, for example, are found in siliceous oozes. Recrystallization of biogenic pelagic carbonates releases substantial amounts of Sr but does not seem to have a significant effect on Ca concentration gradients.

With the exception of dissolution of evaporite minerals, chloride does not participate in mineral-fluid reactions in the subsurface. Its behavior is thus said to be conservative. As a conservative species chloride often does not display vertical concentration gradients in marine sediments. The changes in chloride which have been observed in evaporite-free sections result from mixing with freshwater released from gas hydrates, loss of water in hydration reactions, and possibly slight changes in seawater chlorinity during glacial–interglacial cycles. There is typically a reduction in porewater chloride and salinity in decollments and faults in accretionary wedges as a result of dehydration reactions and large-scale fluid expulsion.

In more organic-rich hemipelagic sediments redox conditions become progressively reducing with depth. As a result of a progression of redox reactions similar to those which characterize continental groundwaters, dissolved ammonia, phosphate, and alkalinity typically increase in the upper part of the sediment column, and dissolved sulfate is reduced (Berner, 1980).

Saline waters in deep sedimentary basins

The hydrogeology of deep sedimentary basins is less well-known quantitatively than that of shallow groundwater systems, but deep basin circulation plays a key role in controlling the salinity and composition of basinal fluids through advection and dispersive mixing of fluids of varying salinity. The deep pore fluids in many sedimentary basins are overpressured, i.e., they exhibit fluid pressures in excess of hydrostatic pressures. There has been considerable debate on the origin of overpressuring (Hanor, 1988), but rapid deposition of fine-grained sediment of low permeability is one likely mechanism. There is evidence from variations in salinity and fluid composition that overpressured fluids can be episodically expelled upward into overlying hydropressured sediments (Roberts and Nunn, 1995). The presence of bedded and diapiric evaporites provide other mechanisms for inducing fluid flow. The dissolution of halite produces dense brines which can convect downward. Salt, however, is also an excellent conductor of heat, and there have been documented examples of upward fluid flow along the flanks of warm salt domes (Hanor, 1994).

Sources of chloride and controls on salinity

Pore waters in the deeper parts of sedimentary basins, including basins now part of the continental crust, and basins containing tectonically deformed sedimentary sequences, are typically saline. The salinities are commonly higher than that of seawater (35,000 mg/L) and can range from 100,000 mg/L to 150,000 mg/L in basins with salt domes to over 350,000 mg/L in basins with bedded basal evaporites. In some sedimentary basins, such as the Illinois, Alberta, and Michigan basins, there is a progressive increase in salinity with depth. In other sedimentary basins, such as the Gulf of Mexico basins there can be reversals in salinity with depth (Hanor, 1988). In the mid-20th century, much discussion was given to the origin of high salinity by the process of membrane filtration or reverse osmosis. In this process overpressuring forces uncharged water molecules through clay membranes, but the negative charge on the clay particles repels charged ions and produces high concentrations of solutes on the influent side of the membranes. Most workers today, however, consider the chloride in basinal brines to have been introduced by some combination of the subsurface dissolution of halite (Land, 1997) and the infiltration of subaerially evaporated marine waters (Carpenter, 1978). During the evaporation of seawater, Br preferentially remains in solution while Cl is precipitated out, first as halite and then as K and Mg salts (McCaffrey et al., 1987). Brines formed by dissolution of halite thus typically have low Br/Cl ratios, which reflect the low Br/Cl ratio of the parent halite. Brines formed by the subaerial evaporation of seawater, in contrast, typically have elevated Br/Cl ratios.

Major solutes

Most basinal brines of moderate salinity are dominated by dissolved Na and Cl. As salinity increases, however, there is a progression from Na-Cl to Na-Ca-Cl to Ca-Cl brines. The controls on the composition of basinal brines have been discussed and debated for a century and a half (Hanor, 1988). In the mid-1800s it was proposed that the high Ca concentrations were connate or syngenetic in origin and reflected high Ca/Na ratios of ancient seawater, a hypothesis which is no longer accepted. Although most of the chloride in basinal brines is derived from halite dissolution or from marine and evaporated marine waters, basinal brines have neither the composition of a NaCl solution or of evaporated marine waters. Instead, there are systematic increases in Na, K, Mg, Ca, and Sr with increasing salinity. The 1:1 slope of the monovalent cations and 2:1 slope of the divalent cations on log–log plots is consistent with the buffering of brine compositions by multi-phase silicate and carbonate mineral assemblages (Hanor, 1994). The phases involved most likely include minerals such as quartz, K-feldspar, albite, illite, smectite, chlorite, calcite, and dolomite. The general decrease in pH and carbonate alkalinity with increasing salinity is also consistent with rock-buffering of fluid compositions. Sulfate concentrations, in contrast to the other major solutes, do not show any systematic variations with salinity. Sulfate concentrations instead are controlled by anhydrite dissolution and the reduction of sulfate to sulfide. Basinal brines and gases in iron-poor sediments can contain significant concentrations of H_2S.

Organic and heavy metals

Volatile fatty acids (VFAs), such as acetate, propionate, and butyrate, are found in concentrations exceeding 100 or even 1000 mg/L in some basinal waters. These waters tend to be of low salinity and may reflect water released from hydrocarbon source rocks during maturation of organic matter. Under conditions of high fluid pressure, concentrations of dissolved methane exceeding several thousand mg/L can be forced into solution.

Basinal brines have long been considered the fluids involved in the formation of some sediment-hosted ore deposits, including Mississippi Valley Type (MVT) and sedimentary-exhalative (SEDEX) deposits. Although much consideration has been given to the role of organic complexing agents, such as acetate, in solubilizing heavy metals such as Pb and Zn, more recent work has shown that chloride is the key complexing agent (Hanor, 1997). At chloride levels in excess of 100 g/L, the $PbCl_4^{2-}$ and $ZnCl_4^{2-}$ complexes become effective solubilizing agents for Pb and Zn. Barite ($BaSO_4$) is also a common mineral phase in MVT and SEDEX deposits. High concentrations of Ba are found in those basinal brines that have low sulfate concentrations.

Isotopic composition

The isotopic composition of basinal brines is significantly influenced by diagenetic reactions. Exchange of oxygen with silicate minerals can increase the $\delta^{18}O$ values of basinal waters and the introduction of H from maturation of organic matter can lower δ^2H values. Other changes include the introduction of radiogenic ^{87}Sr and changes in the Li and B isotopic compositions from the diagenesis of siliciclastic sediments.

Summary

Even though the salinity and composition of porewaters in sediments and sedimentary rocks varies widely, many of these variations are systematic and can be explained in terms of basic principles of solute transport and chemical reaction. In terms of major solutes, redox reactions involving organic matter are important in influencing the concentrations of SO_4 and HCO_3 in all porewaters. Reactions involving silicates, carbonates, and evaporites, however, are the ultimate controls on the addition and removal of Na, K, Mg, Ca, and Cl.

As a group, meteoric groundwaters have the highest variation in proportions of major solutes, which reflect highly dynamic flow systems and local lithologic control on water compositions. Many interstitial marine waters still retain a strong seawater signature despite on-going organic and mineral diagenesis. Diffusional exchange with overlying seawater and within the sediment column is one factor contributing to the buffering of fluid compositions in these sediments. Saline waters in deep sedimentary basins show systematic changes in the concentrations and relative proportions of most major solutes, reflecting rock buffering involving multiple sediment lithologies.

Much work remains to be done in quantifying our understanding of the controls on the properties of pore fluids, including establishing the driving forces and rates of solute transport and the kinetics of the diagenetic reactions which cause mass transfer to occur between porefluids and their sedimentary matrix.

Jeffrey S. Hanor

Bibliography

Appelo, C.A.J., and Postma, D., 1993. *Geochemistry, Groundwater, and Pollution*. Rotterdam: A.A. Balkema.
Back, W., 1961. Techniques for mapping hydrochemical facies. *US Geological Survey Professional Paper*, **424** (D): 380–382.
Berner, R.A., 1980. *Early Diagenesis: A Theoretical Approach*. Princeton University Press.
Carpenter, A.B., 1978. Origin and chemical evolution of sedimentary brines in sedimentary basins. *Oklahoma Geological Survey Circular*, **79**: 60–77.
Chapelle, F.H., 1993. *Ground-Water Microbiology and Geochemistry*. Wiley.
Domenico, P.A., and Schwartz, F.W., 1990. *Physical and Chemical Hydrogeology*. New York: Wiley.
Drever, J.I., 1997. *The Geochemistry of Natural Waters, Surface and Groundwater Environments*, 3rd edn. Prentice-Hall.
Eugster, H.P., and Hardie, L.A., 1978. Saline Lakes. In Lerman, A. (ed.), *Lakes—Geochemistry, Geology, Physics*. New York: Springer-Verlag, pp. 237–293.
Hanor, J.S., 1988. *Origin and Migration of Subsurface Sedimentary Brines*, SEPM Short Course No. 21. Tulsa: Society of Economic Paleontologist and Mineralogists.
Hanor, J.S., 1994. Physical and chemical controls on the composition of waters in sedimentary basins. *Marine and Petroleum Geology*, **11**: 31–45.
Hanor, J.S., 1997. Controls on the solubilization of dissolved lead and zinc in basinal brines. In Sangster, D.F. (ed.), *Carbonate-Hosted Lead–Zinc Deposits*, Littleton: Society of Economic Geologists, Special Publication, 4, pp. 483–500.
Hesse, R., 1990. Early diagenetic porewater/sediment interaction: modern offshore basins. In McIlreath, I.A., and Morrow, D.W., (eds.), *Diagenesis, Geoscience Canada Reprint Series 4*, pp. 277–316.
Land, L.S., 1997. Mass transfer during burial diagenesis in the Gulf of Mexico Sedimentary Basin: an overview. In Montanez, I., Gregg, J.M., and Shelton, K.S. (eds.), *Basinwide Fluid Flow and Associated Diagenetic Patterns*, Tulsa: (SEPM) Society for Sedimentary Geology, Special Publication, 56, pp. 29–40.
Lichtner, P.C., Steefel, C.I., and Oelkers, E.H. (eds.), 1996. *Reactive Transport in Porous Media, Reviews in Mineralogy 34*. Mineralogical Society of America.
McCaffrey, M.A., Lazar, B., and Holland, H.D., 1987. The evaporation path of seawater and the coprecipitation of Br^- and K^+ with halite. *Journal of Sedimentary Petrology*, **57**: 928–937.
Roberts, S.T., and Nunn, J.A., 1995. Episodic fluid expulsion from geopressured sediments. *Marine and Petroleum Geology*, **12**: 195–204.
Schulz, H.D., and Zabel, M. (eds.), 2000. *Marine Geochemistry*. Springer-Verlag.
Sprinkle, C.L., 1989. Geochemistry of the Floridian aquifer system in Florida and in parts of Georgia, South Carolina, and Alabama. *U.S. Geological Survey Professional Paper*, 1403-I.

Cross-references

Diagenesis
Diffusion, Chemical
Evaporites
Fabric, Porosity, and Permeability
Hydrocarbons in Sediments
Isotopic Methods in Sedimentology
Oceanic Sediments
Weathering, Soils, and Paleosols

PRESSURE SOLUTION

A grain-scale mechanism of ductile and water-assisted deformation

The transition from loose sediments to hard rocks arises through physico-chemical processes at the grain scale. Pressure solution, in addition to cataclasis, grain sliding, and plastic deformation, is one of these processes. Pressure solution takes place when some aqueous fluid coating is present around the grains. It is a water-assisted diffusional mass transfer normally occurring in the top few kilometers of sedimentary basins and in other geological environments such as fault gouges and low-grade metamorphic rocks. This slow mechanism of deformation can induce large amounts of ductile strain over geological times when stress and temperature are not high enough to promote brittle or plastic deformation.

The classical pressure solution structures have been described since the last century, for example by Sorby in 1863. They include grain and pebble indentations, stylolites, partly dissolved voids, dissolution seams, and crenulation cleavage. Since, extensive field evidence has accumulated which confirms the prevalence of this mechanism in nature.

Following Sorby's pioneering observations, Weyl (1959) performed the first quantitative interpretation of pressure solution and proposed that the driving force for the deformation be related to stress concentration at grain or pebble contacts. From natural observations such as those of Figure P12, he deduced that deformation by pressure solution involves several successive steps: dissolution at the grain contacts, transport of the dissolved solutes toward the open pore, precipitation in the pore or transport to other pores. This results in an indentation of the grains into each other and a decrease of rock porosity resulting in a ductile compaction.

Mechano-chemical processes

A crucial parameter for pressure solution is the presence of water in the pores, which acts as a medium of reaction and transport with the minerals. If a porous medium is not saturated with respect to water, pressure solution will be localized in the pores where water is present. This effect can create compacted regions in the sediment whereas the porosity of other regions will not be modified.

At the grain scale, pressure solution deformation occurs through three serial steps, the whole process being driven by stress gradients along the grain surface. First, minerals dissolve at grain contacts because of a concentration of stress. The stress variation between the grain contact and the pore surface can be related to the chemical potential of the dissolving mineral (Paterson, 1973). The result is that the concentration in aqueous species is greater in the contact film and induces a gradient of concentration between the contact film and the pore. The second step involves solutes in the contact film diffusing along the grain boundaries. The nature of the interface between two grains is crucial because it is a medium of dissolution and diffusion of matter. It is assumed that some water is trapped inside this interface. And the third and last step is precipitation on the free face of the grains. Note that solutes can be transported by diffusion in the pore fluid and precipitate at some distance from the dissolution area. This provides an explanation of mass transfer at centimeter scale in sedimentary environments under local gradients of stress.

If one of these three steps is slower than the two others, it will control the kinetics of the overall deformation. For example, diffusion within grain contact films controls the compaction of salt aggregates at room temperature. In sedimentary basins, compaction of quartz-rich sediments is limited by the step of quartz precipitation between 3 km and 5 km and by the diffusion at the grain contact below 5 km.

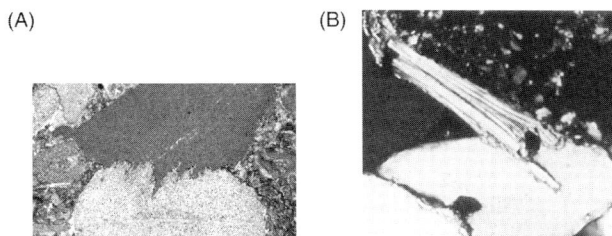
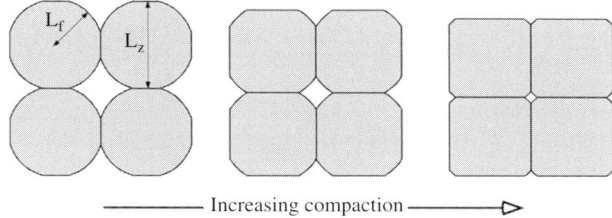

Figure P12 Pressure solution at grain scale: (A) Indentation of two carbonate fossils in a limestone, Mons, Belgium; (B) Indentation of a grain of quartz by a mica flake in a sandstone from the Norwegian shelf.

Figure P13 A cross-section view of a cubic packed network of truncated spheres used to model pressure solution. The grain shapes evolve due to pressure solution: the grain radius L_f increases while the grain flattens (L_z decreases) resulting in the porosity and the pore surface decrease. Such models can be applied in sedimentary basin conditions to model compaction.

Above 3 km, pressure solution remains so slow that its rate can be neglected.

Experimental insights

Attempts to reproduce the mechanism and to obtain quantitative information in the laboratory have proved difficult however, owing to incomplete analysis of samples' geometry used in experimental study and to the confounding influence of other deformation mechanisms that can operate concurrently in the experiments. The experiments can be divided in two categories: single contact (Hickman and Evans, 1991) and aggregate compaction (Rutter, 1976; Gratier and Guiguet, 1986) experiments. The former experiments give insights on the mechanism of deformation itself on a single grain whereas the results of the latter are average creep laws. Recent experiments on rock salt, gypsum and quartz aggregates have shown that pressure solution rate increases when: (1) temperature increases (with an activation energy between 15 kJ/mole and 80 kJ/mole); (2) effective stress increases; (3) grain size decreases.

Such experiments indicate that geometrical parameters control the pressure solution rate such as grain size and the geometry of the grain–grain interface. Knowledge of the geometry permits the estimation of the path length for diffusion of the solutes from the contacts to the pore space. Physico-chemical variables of importance include temperature, stress, chemistry of the pore water, mineralogy of the rock, and transport properties of the grain contacts.

Ductility of the upper crust

Pressure solution, as a time-dependent process, is responsible for a slow ductile creep of the rocks. Modeling pressure solution requires taking into account the evolution of the rock texture and grain geometry (Figure P13) during deformation as the contact area and the pore surface vary during deformation. Such models allow investigating the sensitivity to the main input parameters: transport properties at grain contacts, grain size, temperature, and stress. Typical results of these models concern the viscosity of compaction of sedimentary rocks and the time-scale relevant to compaction processes. For example, preliminary results indicate that the viscosity of compaction of silico-clastic sedimentary rocks ranges between 10^{19} and 10^{23} Pa·s depending on temperature, grain size, stress, and mineralogy; values which are similar to what can be deduced from field-based observations.

Still unresolved questions associated with the basic understanding of the process of pressure solution remain. For example, the dominating driving force and mechanism of dissolution, the geometry of the grain boundaries, the structure and transport properties of the aqueous fluid in that grain boundary, and the influence of pore fluid compositions represent challenging issues for actual research.

In the upper crust pressure solution results in a slow process of dissipation of stress energy stored during sedimentation and burial or by tectonic forces. Field observations indicate that pressure solution is intimately associated with brittle processes of deformation, for example on active faults where earthquakes are characteristic of a brittle behavior whereas pressure solution remains the main mechanism of deformation during the interseismic period.

François Renard and Dag Dysthe

Bibliography

Gratier, J.P., and Guiguet, R., 1986. Experimental pressure solution-deposition on quartz grains: the crucial effect of the nature of the fluid. *Journal of Structural Geology*, **8**: 845–856.

Hickman, S.H., and Evans, B., 1991. Experimental pressure solution in halite: the effect of grain/interphase boundary structure. *Journal of the Geological Society of London*, **148**: 549–560.

Paterson, M.S., 1973. Nonhydrostatic thermodynamics and its geologic applications. *Reviews of Geophysics and Space Physics*, **11**: 355–389.

Rutter, E.H., 1976. The kinetics of rock deformation by pressure solution. *Philosophical Transactions of the Royal Society of London*, **283**: 203–219.

Sorby, H.C., 1863. On the direct correlation of mechanical and chemical forces. *Proceedings of the Royal Society of London*, **12**: 538–550.

Weyl, P.K., 1959. Pressure-solution and the force of crystallization—a phenomenological theory. *Journal of Geophysical Research*, **64**: 2001–2025.

Cross-references

Cements and Cementation
Compaction (Consolidation) of Sediments
Deformation of Sediments
Diagenesis
Diagenetic Structures
Grain Size and Shape
Stylolites

PROVENANCE

Introduction

On Earth, sedimentation is driven principally by tectonic forces, subordinately by climatic and eustatic forces, and to a minor extent in response to impact processes. Detrital sediments retain a record of eroded crustal material. Knowledge of crusts of the past is gleaned from provenance studies. Tracking the immediate and ultimate sources and origins of sediments, principally of siliciclastic sediments, is the domain of provenance studies. In addition, estimating amounts, proportions, and rates of supply of sediments from multiple sources is also within the purview of provenance studies. Contemporary investigations, therefore, focus not only on types, identification, and tectonic settings of source rocks, or the climatic conditions in source areas, or paleogeographic localities of source areas, but also on quantitative distributions of source material and the kinematics of source area tectonics. Approaches, i.e., methodologies of investigations, depend generally on the purpose of the investigators and the scientific questions address by them, but are largely multidisciplinary and diverse (Zuffa, 1985; Morton *et al.*, 1991; Johnsson and Basu, 1993; Bahlbourg and Floyd, 1999; DeCelles *et al.*, 1998). No parent rock or parent rock-assemblage is ever fully preserved in any sediment. Because post-derivation weathering, transport, and depositional, and especially diagenetic processes determine the preservation potential of original detrital signatures and lead to disproportionate representation of parent material in sediment, methodologies have to be appropriately adapted in specific cases. We discuss some of the methods before describing the results and the kinds of inferences made in a few recent studies.

The detective work to infer provenance is thus equivalent to restoring a jigsaw puzzle with many missing and damaged pieces.

Historical

Provenance studies have prompted and progressed with newer instrumentation to determine sediment properties. More than 50 years ago color and shapes of heavy minerals were guides to parent rocks; some 25 years ago detrital zircons were separated on the basis of color and shape to obtain U–Pb age-dates to identify sources in the igneous basement of the Potsdam Sandstone; currently, age-dating of and oxygen isotope systematics in single crystals of detrital zircon provides information on the genesis and recycling of crustal rocks. Similarly, optical microscopic observations are now supplemented by backscattered electron image analyzes; chemical compositions of sediments extend beyond major element to trace and rare earth element distribution; and, chemical and isotopic compositions of single detrital grains are determined by using electron and ion microprobes.

Approaches

Methodologies of provenance determination are broadly divisible into three types: petrographic, chemical, and isotopic. All three types are increasingly used in provenance interpretation. Except for the uncommon occurrence of some index mineral (e.g., glaucophane to indicate a blueschist-type source rock) or an index fossil (e.g., coeval taxa from a disintegrating bank into a deep sea turbidite), most contemporary data are quantitative in nature. Statistical principles of point-counting methods are applied in most measurements.

Petrographic

Mineralogic compositions and classification of sandstones have been linked to source rocks for a long time. For example, arkosic sandstones and conglomerates, many with granitic rock-fragments, have been linked to continental shields, and, lithic graywacke sandstones, many with volcanic and schistose rock fragments, have been linked to orogenic sources. Relative concentrations of detrital quartz (Q), feldspar (F), and rock fragments (R) define sandstone classification. Because the proportion of rock fragments increases with the increasing grain size of sediments, and, because quartz and feldspars also occur in rock fragments, estimating the proportions of QFR is difficult. Results of QFR determination differ depending on the methodology used especially in treating the grain-size effect. Some investigators restrict the grain size range of samples, for example, to medium sand size (0.25–0.50 mm) for sandstones. Others point-count quartz and feldspars in coarse-grained rock fragments as mineral fragments, and restrict the counts of rock fragments, designated as lithics (L), to very fine-grained sedimentary, schistose, and volcanic grains. Composition fields in QFL space (triangular diagrams) are sensitive to detritus from different tectonic regimes (Dickinson, 1985). This is especially true if very fine-grained aggregates of quartz (polycrystalline quartz; chert) are assigned to the L-pole leaving the Q-pole reserved for monocrystalline quartz only. This converts the QFL diagram to a Q_mFL_t diagram. Although the fields in QFL/Q_mFL_t diagrams are intended to be guides, many sandstone petrologists tend to use a classificatory approach to infer tectonic provenance. The diagram was modified later to erase definitive compositional fields emphasizing the essence of the relationship between sandstone composition and tectonic provenance (Dickinson, 1988: Figure P14).

The QFL approach has two limitations. First, errors associated with the boundaries of compositional fields are large. Second, it ignores the information contained in rock fragments; especially that carbonate grains may be intrabasinal and coeval with the host sediment or extrabasinal and older detritus. Both are useful as provenance indicators. Nevertheless, point counting by the Gazzi–Dickinson method, using QFL/Q_mFL_t, is the preferred mode for inferring tectonic provenance from petrographic analysis of bulk sediments.

Climate profoundly affects mineralogical and chemical compositions of sediments. Under warm and wet conditions relative to cold and dry conditions, rock-fragments break down to their constituent minerals more rapidly; they become depleted in a sediment while quartz, being the most durable mineral at Earth's surface, is enriched relative to other minerals. Duration of weathering, be it at rock-outcrops or while a body of sediments is parked temporarily (e.g., river valleys, deltas, coastal plains, beaches), contribute to the degree of quartz enrichment. Common upper crustal metamorphic rocks are generally enriched in quartz and are finer grained than common igneous rocks (feldspar-rich and coarser grained). Therefore, it is possible to differentiate between sediments derived principally from metamorphic and igneous rocks under similar climatic conditions; and, to assess relative

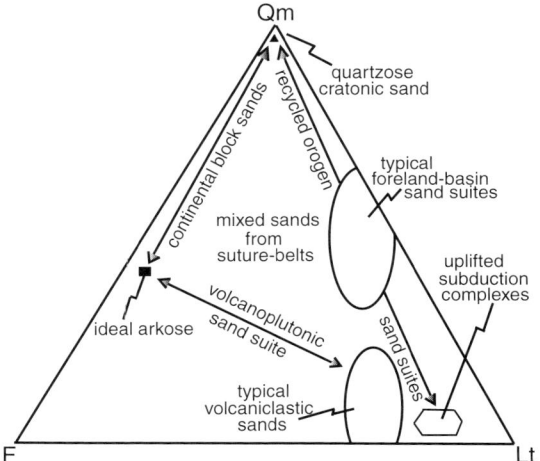

Figure P14 A Q_mFL_t plot of detrital modes (modified after Dickinson, 1988) serves better as a guide a provenance rather than serving as a classification (e.g., as in Dickinson, 1985).

climatic influence on bodies of sediments derived principally from similar source rocks. Point-counting methods in these studies define rock fragments as multi-mineral sand-sized grains in which at least two sub-grains are larger than 0.0625 mm or no single phase constitutes >90 percent of the whole grain.

Properties of rock fragments. Rock fragments, if identifiable despite alteration suffered through weathering and diagenesis, are infallible guides to source rocks. Carbonate rock-fragments may contain identifiable fossils, which may specifically identify and date a source rock. Particularly informative are fossils in coeval carbonate fragments derived from shelfal banks in off-shelf sediments. Overall, the rarity of good preservation and differential preservation of different rock-types lead to incomplete and possibly misleading inferences. For example, quartzite rock fragments are far better preserved than schistose rock fragments. Inference based solely on grain counts of rock fragments derived from a sequence of metamorphosed sedimentary rocks would underestimate the contribution from schistose parent rocks. Based on such arguments, Palomares and Arribas (1993) developed the concept of Sand Generation Index (SGI) quantifying the proportions of schistose, gneissic, and granitoid rock fragments expected to be delivered to daughter sands from identical climatic and geomorphic units. A follow-up study on sedimentary source rocks confirms earlier conclusions.

Properties of minerals. All minerals suffer different degrees of alteration and obliteration through processes of weathering, transportation, deposition, and especially diagenesis. Feldspar, the most abundant mineral in crystalline basement rocks, is subordinate to quartz in most sandstones. If preserved, types of feldspars, their chemical compositions, twinning habits and zoning, and their structural states are useful guides to parent rocks. However, feldspar grains do not survive abrasion and transportation well. Fragmentation along cleavage planes and composition planes (of twinned feldspars) may preserve a biased population in detrital sands. For example, if twinned feldspars from igneous source rocks break along composition planes during transportation, the detrital population may look like one from metamorphic source rocks that contain, in general, fewer twinned feldspars. Moreover, partial to complete dissolution and albitization render provenance determination from feldspars prone to substantially inadequate interpretation much like that from rock-fragments alone. Hence, degrees of alteration of feldspars cannot be assigned unequivocally to weathering of parent rocks to estimate climatic conditions prevalent in the provenance (e.g., Folk, 1980; after Krynine and the "Dogma of the Immaculate Feldspar").

Many in search of provenance-related properties have studied quartz because it is the most abundant and the most durable detrital mineral. Additionally, some believe that "quartz is an ideal mineral to study because many of them could be assigned to a definite type of source rock" (Folk, 1980). Properties of quartz are related to its formation and subsequent overprinting by geological events. The genetic classification of quartz types based on optical petrographic properties as taught by Krynine (in 1940s) and later encoded by Folk (p.69 in Folk, 1980) is still valid (but see *Quartz, Detrital,* and caveats below). Principal criteria are extinction types (straight or undulose), inclusions (vacuoles; microlites), idiomorphism and embayment, and nature of sub-grain boundaries (straight; crenulated; wavy) of polycrystalline grains. Based on these criteria six genetic types of quartz are defined: (a) common plutonic quartz, (b) volcanic quartz, (c) vein quartz, and metamorphic quartz of three principal types, i.e., (d) recrystallized, (e) schistose, and (f) stretched. The last three are polycrystalline quartz grains and the first three are presumably all monocrystalline. However, local environment of quartz generation does not necessarily correlate with specific source rock assemblages. For example, vein quartz may come from quartz-bearing veins in several kinds of rocks from several tectonic environments. Additionally, not every detrital quartz grain in a sedimentary rock can be unequivocally assigned to a particular genetic type. Folk devised a pragmatic empirical classification of quartz types for detrital grain generally in the size range between 0.15 mm and 0.35 mm based on the nature and degree of undulose extinction and the nature and abundance of inclusions. An empirical distribution-diagram of detrital quartz-types from different source rocks is constructed to infer the provenance of sand-sized quartz grains in a sedimentary rock (p.67 and p.73 in Folk, 1980). Although neither Krynine nor Folk published data-tables in their summary reports, their observations (published in various journal articles) are based on examining numerous thin sections from numerous rock samples.

Two commentaries on the above remain important. Detailed quantitative optical petrographic studies of quartz in modern sand from known source rocks to test the empiricism of Krynine and Folk showed that: (1) apparent undulosity of any grain depends on the crystallographic orientation of the plane of the thin section and flat-stage estimates may be random, i.e., not significant; and (2) polycrystalline and strained quartz grains (hence undulose) are selectively destroyed in transportation relative to those without strain, i.e., exhibiting straight extinction. This invalidates much of the empirical approach of Krynine and Folk. However, because strain in quartz under natural stress is finite and bending of the c-axis is possibly limited to not more than 30°, one would expect a statistically higher proportion of undulose quartz grains in detritus from metamorphic source rocks. Based on a combination of universal-stage and flat-stage measurements of undulosity and polycrystallinity in medium sand-sized (0.25–0.50 mm) quartz grains from igneous granitic rocks and various grades of regionally metamorphosed rocks, separate provenance fields can be delineated in a "diamond diagram" that show some overlap between granitic and gneissic source rocks (Basu et al., 1975). In summary, principles espoused by Krynine and Folk are validated although some modification of detail has become necessary. Additionally, post-lithification deformation of a sedimentary unit, for example, such as those involved in tight folding, imparts new strain on detrital quartz grains modifying their original characteristics.

Quartz, feldspar, carbonate, and clay minerals are light constituents ($\rho < 2.89$ g/cc) of detrital sediments and usually add up to about 99 percent of all sedimentary rocks. Heavy minerals are rare because common mafic minerals are weathered easily and because durable heavy minerals (e.g., zircon, tourmaline, rutile) are uncommon in parent rocks. However, minerals characteristic of certain rock types, if preserved, are very useful for provenance determination. For example, the simple presence of detrital glaucophane in a sandstone indicates a high-pressure low-temperature blueschist source that, hardly any property of detrital quartz or feldspar could reveal. Some minerals, such as sillimanite, kyanite, rutile, and most detrital garnet are exclusively metamorphic in origin. Most olivine and pyroxene, especially those rich in Fe, are igneous in origin. Magnetite and ilmenite can be either igneous or metamorphic. However, intergrowth and oxidation-exsolution textures controlled by crystallographic properties and mode of origin (observed with reflected light microscopy or in backscattered electron images) in detrital magnetite–hematite–ilmenite may distinguish igneous and metamorphic source rocks.

A recent and still rudimentary development in provenance studies is to trace the origin of seemingly inorganic material and structures to its biological parent. Unlike permineralized body fossils or impressions, some detrital material may be biogenic but may not have preserved the morphology of its biologic precursor. If and when the biological origin of such grains is proven and accepted by the community, their study would move from the realm of provenance to that of paleontology. For example, Schieber (1996) has discovered remnants of fossilized algal cysts associated with micron-scale detrital oval quartz grains that have generally been considered to be clastic quartz in mudstones, but are possibly silica-replacement of the cysts. A debate rages over a string of nanometer-scale euhedral magnetite crystals of hexaoctahedral prismatic morphology, which if exclusively biogenic, would support the existence of fossils in a Martian meteorite (Thomas-Keprta et al., 2000).

In summary, many properties of minerals are characteristic of their pedigree. In practice, it is not always easy to isolate such properties but may be used in conjunction with other variables for very specific identification of source rocks.

Chemical

Weathering and diagenesis may modify the bulk chemical composition of detrital sediments rather severely even if petrographic identities of constituent grains are preserved. For example, feldspar grains weathered or altered (into clay) may still be recognized petrographically as feldspars, but their chemical composition is close to that of clays. A quartz-cemented feldspathic sandstone in which detrital feldspars might have been silicified shows an anomalously high SiO_2 content. Albitization of calcic plagioclases changes the CaO/Na_2O ratio of the rock; calcite cementation moves the bulk composition in the opposite direction. K or Ba metasomatism during diagenesis may foster a false sense that the original detritus might have come from a K or Ba-rich source rock. However, not all K, even in shales with known K-metasomatism, results from diagenesis, and it is possible to identify the proportion of detrital and diagenetic K in some shales. There may also be cases where the bulk chemical composition may actually identify the family of a source rock, even though it can not be resolved petrographically. This is more likely in sediment that has lost its porosity very early in diagenesis and in which subsequent chemical diagenesis has been minimal. For example, if a siliciclastic rock has a large proportion of diagenetic matrix (cf. pseudomatrix, epimatrix), its chemical composition may tell if the precursor parent grains were volcanic (generally Fe–Mg-rich) or metamorphic (generally Al-rich) rock fragments. Completely argillized rock-fragments may not be readily identifiable in a thin section yet plug porosity and reduce the ease with which diagenetic fluids can permeate a rock. Thus, although weathering, transport, and depositional processes fractionate elemental abundances and ratios between source rocks and detritus in a basin, further fractionation during diagenesis is minimized in material of low porosity.

Bulk rocks. No element is absolutely immobile in the sedimentary system. Some are less mobile than others (e.g., Ta, Th, Zr, Sc, Co, Ti, V, Nb, Y, Th, some REE) and many pairs of these elements (e.g., Th/U, Zr/Y, La/Sm, Ti/Zr) maintain the same ratio relative to other pairs, as parent rocks are converted to sediments and sedimentary rocks. Most of these elements occur in very small quantities in the bulk of sedimentary rocks; most reside in minor and rare minerals, especially in heavy minerals (*q.v.*) and in or on clay minerals (*q.v.*). Because the distribution of heavy minerals, especially those rich in trace elements and rare earth elements, between samples of coarser detrital sediments may not be uniform, most informative data come from siltstones and shales. Provenance interpretation of TE and REE distribution in shales and siltstones is now common.

Mud and mudstones or graywackes comprised of significant argillaceous or argillized rock fragments derived from a large hinterland, transported by large rivers, and deposited on platforms or in large turbidite units, have large-scale provenances, such as the continental shield and the mountain chains beyond. Mudstones and graywackes with such large provenances, representing different time-slices of Earth-history, and collected from around the world, have been analyzed to monitor the temporal changes in the distribution of trace and rare earth elements in Earth's crust (Taylor and McLennan, 1995) and to model the growth of Earth's crust. The ability to infer the "big picture" in space and time through studies of muddy sediments is a consequence of using trace and rare earth elements (TE and REE) as the principal indicators. A pronounced increase in REE content takes place in sediments across the Archean–Proterozoic boundary, indicating a rapid growth of continental crust at that time.

Distribution and partitioning of TE and REE in igneous and metamorphic processes are far better known than in sedimentary processes. Inverse modeling to back-calculate the source composition of sediments is not possible yet.

Additionally, minerals rich in TE and REE (e.g., apatite, monazite, garnet, zircon) may not be homogeneously distributed through different grain size fractions of sediments; some may be susceptible to diagenetic alteration (e.g., apatite) and others may be so durable (e.g., zircon) that they can recycle not only through many sedimentary cycles but also through one or more plutonic cycles. Therefore, application of TE and REE studies to provenance analysis on a regional or a smaller scale is limited. Interpreting of the ultimate origin of the crystalline rocks (e.g., depleted or enriched mantle; crustal melting) from which sediments of a basin have been derived is more readily done than figuring out the nature and the identity of the immediate source rock.

The grain size effect can be offset by analyzing only the <2 μm fraction of sediments, which however excludes contribution of minerals that do not normally alter or disintegrate into the <2 μm fraction. Diagenetic effects in the distribution of TE and REE in shales, and by extrapolation in the <2 μm fraction, are well documented. Nevertheless, studies at smaller scales and in conjunction with petrographic data are gradually defining and constraining the use of TE and REE as provenance indicators (Cullers, 2000). Because TE and REE analyzes can be made on small samples of <2 μm fractions, provenance and sediment dispersal patterns of carbonate sediments may also be traced using the small amounts of detrital clayey material dispersed in such rocks. At present, however, TE and REE characteristics are mostly useful for identifying arc-derived sediments including eolian dust.

Single minerals. Detrital sediments are but a mechanical collection of many single mineral grains each of which may retain a chemical memory of its origin. The more complex the composition of a mineral the higher is the probability of TE and REE substitutions in its crystal structure, which are sensitive to modes of formation of the mineral. Thus low quartz (SiO_2) does not allow much substitution in its lattice, but feldspar ($(Ca, Na, K, Ba)[Al_{2-x}Si_{2+x}]O_8$) provides ample opportunities for substitution. Of particular interest is the easy ability of Eu^{2+} to enter the plagioclase structure relative to other REE. The extent of Eu anomaly in a collection of consanguineous detrital plagioclase determined by instrumental neutron activation analysis (INAA), or determined on individual detrital grains by ion microprobes (SIMS; secondary ion mass spectrometer) indicates if their parental crystalline rock was derived from a source enriched or depleted in REE. Much less abundant detrital minerals, i.e., the heavy minerals, may have definitive chemical signatures of their origin, which can be determined by electron probe microanalysis or using analytical SEMs (scanning electron microscopes).

Luminescence, including *cathodoluminescence* (CL; *q.v.*) and *thermoluminescence* (TL) are physical properties of minerals. Both depend on a combination of the distribution of minor (and trace) elements and crystallographic properties and it is rarely possible to separate the physical from the chemical causes of luminescence. CL depends on the distribution of exciter and inhibiter ions that usually reside in minor to trace quantities in minerals, and on crystal defects such as vacancy in the oxygen site. CL of detrital silicate grains may be diagnostic with some unusual spectrum or fabric and may identify a specific source rock (Seyedolali *et al.*, 1997). More importantly, CL is commonly used to identify populations of zircons for ion microprobe age-dating of single grains. It is common to find igneous cores (CL-bright) of zircon with metamorphic overgrowths (CL-dull), and to date the core and overgrowths separately. In addition, specific identification of CL spectra of particular source rock, for example, marble with a characteristic stress-history may identify the source-quarry of an ancient statue (e.g., Lapuente *et al.*, 2000); or, identical CL spectra of two diamonds may identify the jewels to be from a single raw stone (Sunagawa *et al.*, 1998). *Thermoluminescence* (TL) can be used, under favorable circumstances to age-date bodies of mostly Holocene sediments. Relative degrees of weathering and exposure to the atmospheres affect TL properties of sediments. Thus, TL is a good guide in identifying the origin and in tracing movements of coastal sands, sand dunes, and loess.

Isotopic

Stable isotopes. Principles and constraints that apply to the use of TE and REE in provenance analysis also apply to ratios of stable isotopes and radioisotopes. Robust stable isotopic signatures of source rocks are commonly preserved in selected mineral and rock fragments. Large sand grains seem to be less affected by diagenetic processes than smaller clay-sized grains and are the preferred material for analyzes. Ratios of $^{18}O/^{16}O$ and $^{13}C/^{12}C$ in carbonate detritus, such as foraminiferal tests or nodules in siliciclastic sediments, may uniquely identify the source. Isotopic identification of provenance of iceberg-rafted detritus, e.g., for Heinrich events or to track recycling of carbonate dusts is now routine. Isotopic ratios of carbon in organic matter are routinely used to trace the source of petroleum, as are certain organic molecules. Combined molecular-isotopic tracing of organic particles is indeed an extremely discriminating parameter. Oxygen isotopic ratios of detrital quartz and fluid inclusions therein may indicate their temperature of formation. For example, $\delta^{18}O/^{16}O$ of quartz from very high temperature rhyolitic tephra and low-temperature authigenic products are different although both types of grains may reside in a single body of sediment. Preservation of high temperature isotopic signature, however, may depend on the degree and selectivity of diagenetic reactions, which in the extreme may homogenize all $\delta^{18}O/^{16}O$ values of detrital quartz. The uniformity of oxygen isotopic ratios may also indicate recycling of sedimentary material such as in Vendian and Cambrian shales in Spain.

Radioisotopes. Precise radiometric age-dates perhaps connect detrital grains to source rocks most definitively. For example, from a U–Pb concordia, a 422 million year old granite clast in a Devonian conglomerate is uniquely tied to a specific Silurian granite the original source-terrain of which is identified to be 2445 million years old. Because weathering and diagenetic alterations may disturb the original isotope systematics, the precision in dating is often compromised. If source rocks had suffered retrograde events then some of the isotope systems might have been reset complicating the direct tie-in with daughter sediments. Sedimentary differentiation may preferentially concentrate certain minerals in different size fractions such that analyzes of size separates produce pseudo-isochrons, for example of the Rb–Sr system. Understanding of radioisotope systematics in sedimentary systems has increased substantially in the last 25 years and many of these methodological and natural artifacts have become known.

Determination of radioisotope systematics (mostly Sm–Nd, Rb–Sr, ^{40}Ar–^{39}Ar, Pb–Pb, and U–Th disequilibria) and dating Quaternary dust have immensely improved the understanding of its origin, movement, and its relation to geomorphologic changes and climatic conditions through time. For example, Innocent et al. (2000) trace the silt fraction in deep-sea sediments to riverine contribution and the movement of the <2 μm fraction to ocean bottom currents through Nd-isotopic measurements; Grousset et al. (1998) and Walter et al. (2000) recognize specific dust plumes produced during glacial and interglacial periods and deposited as discrete units from Sr and Nd isotopic studies; and, Aleinikoff et al. (1999) trace the sources of loess in Colorado using the Pb isotopic compositions of K-feldspars.

Ancient lithified sediments are commonly far too removed from their sources and their distribution systems (e.g., rivers, wind pattern) are mostly obliterated. Therefore, the origin and modes of movements of ancient sediments are much more difficult to trace than those of their Recent counterparts. If relatively unaltered large polymineralic rock fragments such as pebbles and boulders are present, it is possible to obtain mineral-separates for determining an isochron and estimating the age of the clast for provenance. However, polymineralic pebbles and boulders are not abundant in the sedimentary record nor are they necessarily unaltered. Disaggregation, mineral-separation, and age-dating many pebbles in one conglomerate may be much too time-consuming and perhaps not cost-effective. Populations of a detrital mineral if free from substantial weathering and diagenetic modification of the radioisotope system of interest may produce reliable isochrons and ages (Faure, 1986). Such populations could be separated by grain size (e.g., for feldspars) or color and shape (e.g., for zircon) to obtain appropriate isochrons and concordia for detrital minerals. Improvements in measurement techniques, such as secondary ion mass spectrometry (SIMS), laser ablation microprobe-inductively coupled plasma-mass spectrometry (LAM-ICP-MS), and thermal ionization mass spectrometry (TIMS) now allow age-dating individual detrital grains (especially zircon). The potential of such analyzes in provenance studies is immense. As SIMS becomes more readily available, one may expect to obtain age-dates of several individual grains from all samples of interest and truly collect representative and comprehensive range of ages of source rocks represented in a Formation. Much of current isotopic investigations of bulk sediments and single grains only provide an average provenance age of a sedimentary unit. This is because many detrital grains in clastic sediments are recycled and several source rocks commonly contribute to a single sedimentary unit. However, very ancient age of some detrital zircons indicates the existence of an early crust of the earth and is an important contribution by itself.

Fission tracks in apatite and zircon are robust indicators of ages of thermal events, which are commonly related to unroofing and cooling histories. Thus, apart from identifying source rocks under limited conditions, they provide good evidence for tectonic events especially about the exhumation history of an orogen. Fission track data contribute, albeit somewhat indirectly, to a necessary correction of U–Pb dating of detrital zircons that have endured post-deposition heating and cooling (Carter and Moss, 1999).

Purpose and future

The purpose of any provenance study dictates the choice of methods and the scope of the investigation. For example, diverse approaches with varied samples are necessary to address a segment of the kinematics of Himalayan orogeny (e.g., DeCelles et al., 1998) whereas a single confirmatory test can identify two fragments of a single diamond (Sunagawa et al., 1998). For most geological problems, application of more than one method decreases the degree of uncertainty in the results. Part of the uncertainty is attributable to not only to the variability inherent in nature but also to the uncertainties associated with the methods themselves. It is necessary to be cognizant of both.

Very few provenance studies have attempted to quantitatively estimate the relative contribution of different rock units or tectonic units to a body of detrital sediments. For a more complete paleogeologic reconstruction of the earth through time, and estimating rates of relevant geologic processes, such quantitative analysis may be the future of provenance studies (Valloni and Basu, 2003).

Abhijit Basu

Bibliography

Aleinikoff, J.N., Muhs, D.R., Sauer, R.R., and Fanning, C.M., 1999. Late quaternary loess in northeastern Colorado; Part II, Pb isotopic evidence for the variability of loess sources. *Geological Society of America Bulletin*, **111**: 1876–1883.

Arribas, J., Marfil, R., and De La Pena, J.A., 1985. Provenance of Triassic feldspathic sandstones in the Iberian Range (Spain): significance of quartz types. *Journal of Sedimentary Petrology*, **55**: 864–868.

Bahlburg, H., and Floyd, P.A. (eds.), 1999. Advanced techniques in provenance analysis of sedimentary rocks. *Special Issue Sedimentary Geology*, **124**: 224pp.

Basu, A., Young, S.W., Suttner, L.J., James, W.C., and Mack, G.H., 1975. Re-evaluation of the use of undulatory extinction and polycrystallinity in detrital quartz for provenance interpretation. *Journal of Sedimentary Petrology*, **45**: 873–882.

Basu, A., 1985. Reading provenance from detrital quartz, In Zuffa, G.G., (ed.), *Provenance of Arenites*. Reidel: NATO-ASI, Series C, 148, pp. 231–248.

Carter, A., and Moss, S.J., 1999. Combined detrital-zircon fission-track and U-Pb dating: a new approach to understanding hinterland evolution. *Geology*, **27**: 235–238.

Cullers, R.L., 2000. The geochemistry of shales, siltstones and sandstones of Pennsylvanian-Permian age, Colorado, USA; implications for provenance and metamorphic studies. *Lithos*, **51**: 181–203.

DeCelles, P.G., Gehrels, G.E., Quade, J., Ojha, T.P., Kapp, P.A., and Upreti, B.N., 1998. Neogene foreland basin deposits, erosional unroofing, and the kinematic history of Himalayan fold-thrust belt, western Nepal. *Geological Society of America Bulletin*, **110**: 2–21.

Dickinson, W.R., 1985. Interpreting provenance relations from detrital modes of sandstones. In Zuffa, G.G. (ed.), *Provenance of Arenites*. D. Reidel Publishing Company, NATO-ASI Series C148, pp.333–361.

Dickinson, W.R., 1988. Provenance and sediment dispersal in relation to paleotectonics and paleogeography of sedimentary basins. In Kleinspehn, K.L., and Paola, C. (eds.), *New Perspectives in Basin Analysis*. Springer, pp. 3–25.

Faure, G., 1986. *Principles of Isotope Geology*. Wiley.

Folk, R.L., 1980. *Petrology of Sedimentary Rocks*. Austin, TX: Hemphill's.

Grousset, F.E., Parra, M., Bory, A., Martinez, P., Bertrand, P., Shimmield, G., and Ellam, R.M., 1998. Saharan wind regimes traced by the Sr–Nd isotopic composition of subtropical Atlantic sediments; last glacial maximum versus today. *Quaternary Science Reviews*, **17**: 395–409.

Innocent, C., Fagel, N., and Hillaire, M.C., 2000. Sm-Nd isotope systematics in deep-sea sediments; clay-size versus coarser fractions. *Marine Geology*, **168**: 79–87.

Johnsson, M.J., and Basu, A. (eds.), 1993. *Processes Controlling the Composition of Clastic Sediments*. Geological Society of America Special Paper, Volume 284: 342pp.

Lapuente, M.P., Turi, B., and Blanc, P., 2000. Marbles from Roman Hispania; stable isotope and cathodoluminescence characterization. *Applied Geochemistry*, 15: 1469–1493.

Morton, A.C., Todd, S.P., and Haughton, P.D.W. (eds.), 1991. *Developments in Sedimentary Provenance Studies*. Geological Society Special Publication, 57, 370p.

Palomares, M., and Arribas, J., 1993. Modern stream sands from compound crystalline sources; composition and sand generation index. In Johnsson, M.J., and Basu, A. (eds.), *Processes Controlling the Composition of Clastic Sediments*. Geological Society of America Special Paper, 284, pp.313–322.

Pearce, T.J., Besly, B.M., Wray, D.S., and Wright, D.K., 1999. Chemostratigraphy: a method to improve interwell correlation in barren sequences; a case study using onshore Duckmantian/Stephanian sequences (West Midlands, UK.). *Sedimentary Geology*, 124: 197–220.

Schieber, J., 1996. Early diagenetic silica deposition in algal cysts and spores: a source of sand in black shales? *Journal of Sedimentary Research*, 66: 175–183.

Seyedolali, A., Krinsley, D.H., Boggs, S. Jr, O'Hara, P.F., Dypvik, H., and Goles, G.G., 1997. Provenance interpretation of quartz by scanning electron microscope cathodoluminescence fabric analysis. *Geology*, 25: 787–790.

Sunagawa, I., Yasuda, T., and Fukushima, H., 1998. Fingerprinting of two diamonds cut from the same rough. *Gems and Gemology*, 34: 270–280.

Taylor, S.R., and McLennan, S.M., 1995. The geochemical evolution of the continental crust. *Reviews of Geophysics*, 33: 241–265.

Thomas-Keprta, K.L., Bazylinski, D.A., Kirschvink, J.L., Clemett, S.J., McKay, D.S., Wentworth, S.J., Vali, H., Gibson, E.K. Jr, and Romanek, C.S., 2000. Elongated prismatic magnetite crystals in ALH84001 carbonate globules: potential Martian magnetofossils. *Geochimica et Cosochimica Acta*, 64: 4049–4081.

Valloni, R., and Basu, A. (eds.), 2003. *Quantitative Provenance Studies in Italy*. Roma: Memorie Descrittve della Carta Geologica d'Italia (in press).

Walter, H.J., Hegner, E., Diekmann, B., Kuhn, G., and Rutgers, L.M.M., 2000. Provenance and transport of terrigenous sediment in the South Atlantic Ocean and their relations to glacial and interglacial cycles; Nd and Sr isotopic evidence. *Geochimica et Cosmochimica Acta*, 64: 3813–3827.

Zuffa, G.G. (ed.), 1985. *Provenance of Arenites*. Dordrecht, Reidel: NATO-ASI, Series C, 148: 408pp.

Cross-references

Cathodoluminescence
Climatic Control of Sedimentation
Extraterrestrial Material in Sediments
Features Indicating Impact and Shock Metamorphism
Feldspars in Sedimentary Rocks
Fluid Inclusions
Heavy Minerals
Isotopic Methods in Sedimentology
Quartz, Detrital
Sand and Sandstone
Tectonic Controls of Sedimentation

PSEUDONODULES

Definition and description

Pseudonodules are a soft-sediment deformation structure comprising rounded masses of clastic sediment set in similar or finer-grained matrix. The terminology of pseudonodules and ball-and-pillow structure (*q.v.*) is highly confusing. It is proposed here that pseudonodules can refer both to individual rounded masses, and to horizons comprising a single row of masses. Pseudonodule horizons can be subdivided into attached pseudonodules, in which the upper surfaces of the pseudonodules are level with the upper surface of the matrix (Figure P15(A),(B)), and detached pseudonodules that are overlain, as well as surrounded by matrix (Figure P15(F),(G)).

Pseudonodules are usually composed of sand or sandstone, or occasionally gravel or clastic limestone. The matrix is usually mud, mudstone or shale, although it may be sandstone, and may differ little in lithology from the pseudonodules (Figure P15(B)), in which case the structure may resemble convolute lamination (*q.v.*). Pseudonodules are usually semicircular, oval or kidney-shaped, although some are contorted into zigzags or spirals. They range in length from a few centimeters to several meters, and in thickness from a few centimeters to a meter or so. Their outer surfaces may preserve structures such as ripple marks. Lamination within pseudonodules typically conforms to their outer margins, but may become more contorted toward the center. Lamination may be contorted in the matrix. There may be a lateral passage from pseudonodules to "normal" load casts affecting only the base of a bed. Pseudonodules are particularly common in shallow marine and deltaic deposits, but occur in most depositional environments, and similar structures are known from volcaniclastic and intrusive igneous rocks. Some periglacial involutions (cryoturbation structures) resemble pseudonodules.

Although pseudonodules resemble chemically formed diagenetic concretions or the products of spheroidal weathering, the presence of deformed laminae shows that they form by the physical disturbance of unconsolidated sediment. They are a type of load structure (*q.v.*), formed by the foundering of sediment into its substrate, but whereas "normal" load casts affect the base of a laterally continuous bed, pseudonodules seem to represent more extreme loading that led either to the detachment of sinking masses from the base of a bed or to the puncturing of a bed by rising plumes of buoyant sediment.

Historical development

The term pseudonodule (as "pseudo-nodule") was introduced by Macar (1948), who recognized that contorted laminae within rounded sandstone masses encased in mudstone demonstrated that the structures had formed through deformation prior to lithification. There are few significant differences between the structures described by Macar (1948) and those described by Smith (1916) and Potter and Pettijohn (1963) as ball-and-pillow structure, and most authors have subsequently considered the terms ball structure (Smith, 1916), pillow-form structure (Smith, 1916), pseudonodule or pseudonodule (Macar, 1948) and ball-and-pillow structure (Potter and Pettijohn, 1963) to be synonymous. Other synonyms that have been introduced include: balled-up structure, crumpled ball, droplets, flow-fold, flow layer, flow rolls, flow structure, force-aparts, hassock structure, intra-stratal flowage, kneaded sandstone, mammillary structure, sand rolls, sandstone balls, sandstone flow, slump balls, snowball structure and storm rollers.

Various criteria have been used by other authors to differentiate between pseudonodules and ball-and-pillow structure, including: the recognition of a source layer for ball-and-pillow structure; the occurrence of pseudonodules in

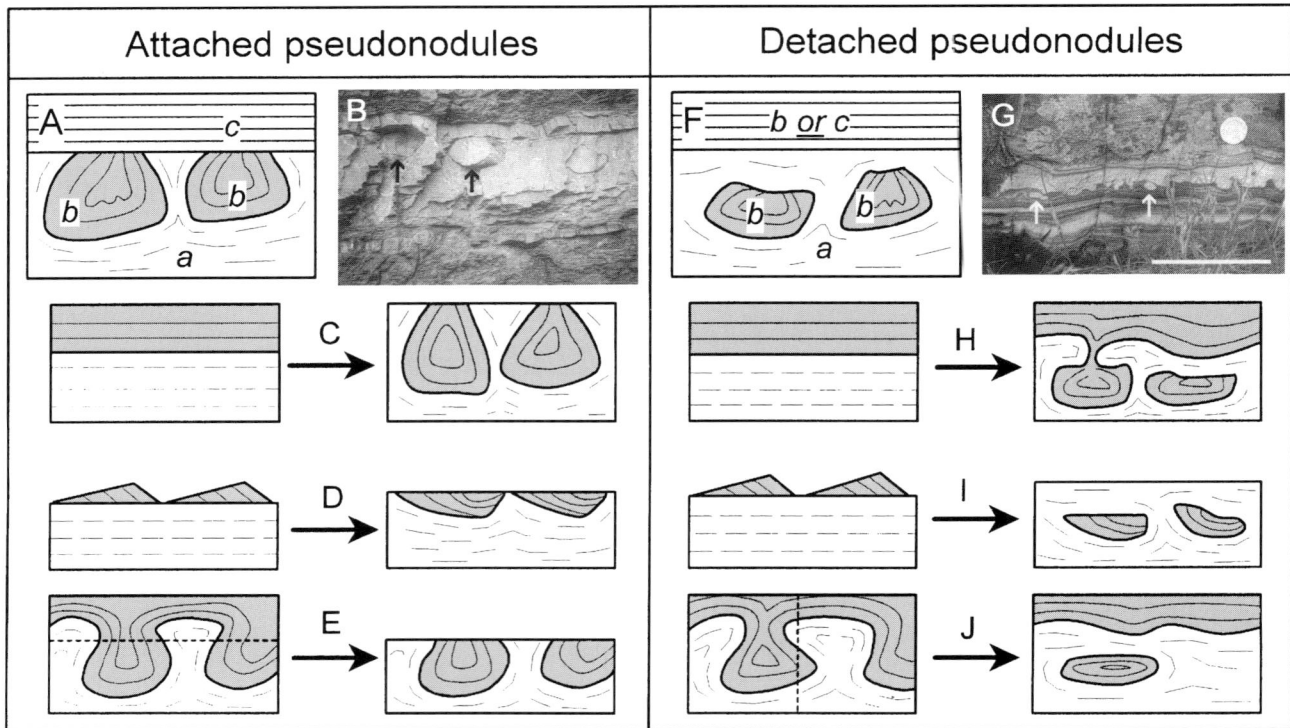

Figure P15 Definition and interpretation of pseudonodules. See text for explanation. B shows attached pseudonodules (arrowed) from Upper Carboniferous delta mouth bar sandstones at Amroth, west Wales, UK. The pseudonodules and their matrix are of similar lithology. The illustrated section is 1.5 m thick. G shows detached pseudonodules (arrowed) and load casts on the base of a tuff bed from the Ordovician of Ramsey Island, west Wales, UK. Scale bar (bottom right) is 10 cm.

a single row, as opposed to vertically stacked masses in ball-and-pillow structure; the use of pseudonodule to describe a single mass and ball-and-pillow structure for a horizon of pseudonodules; or a difference in the inferred origin. What seems clear is that soft-sediment deformation structures comprising rounded masses isolated in matrix are sufficiently distinct from load casts to merit a separate name, and that sufficient variation exists to define several structures in a continuum from load casts to pseudonodules and ball-and-pillow structure. The definitions presented here combine several of the above criteria in an easily applied, objective scheme.

Interpretation

Some early workers interpreted pseudonodules as the products of lateral movements on slopes, but experiments by Kuenen (1958) demonstrated that most pseudonodules are a type of load structure (q.v.), in which vertical displacements are driven by variations in gravitational potential energy. Most load structures develop where denser sediment overlies less dense, but similar deformation results from the foundering of an unevenly distributed load (e.g., bedform relief). Deformation can occur in a system in which potential energy is not at a minimum if a lower layer becomes liquidized (i.e., undergoes a temporary reduction in strength). This allows the upper layer to founder and the lower to rise buoyantly. Several mechanisms of liquidization are associated with the expulsion of excess pore fluid, and some load structures represent a type of fluid escape structure (q.v.).

Examples of deformation pathways that can lead to pseudonodules are shown in Figure P15. These can operate in combination, and the end products may be further complicated through the action of lateral forces related to slopes or currents, giving an asymmetry to the pseudonodules. The morphology of pseudonodules is influenced by the relative viscosity of the sediment layers (Anketell et al., 1970), and sinking pseudonodules can deform if they encounter an obstruction, such as non-liquidized sediment.

Methods by which attached pseudonodules can form include: the puncturing of an upper layer by rising buoyant plumes (Figure P15(C); cf. experiments described by Owen, 1996); the foundering of an unevenly distributed load (Figure P15(D)); or erosion of load casts (Figure P15(E)). Detached pseudonodules can form by: detachment of masses sinking from the base of an upper layer (Figure P15(H)); complete disruption and foundering of a denser layer (cf experiments by Kuenen, 1958); continued sinking of an unevenly distributed load (particularly if the load is denser than its substrate: e.g., loading of sand ripples into mud as in Figure P15(I)); certain cross-sections of load casts (Figure P15(J)); or isolated slump folds.

Significance of pseudonodules

Pseudonodules record the post-depositional disturbance of sediment by loading processes driven by gravitationally unstable density gradients or by unevenly distributed sediment load. Their presence has implications for the physical state of sediment shortly after deposition, which may relate to depositional processes (Rijsdijk, 2001).

Several factors may favor the formation of pseudonodules over load casts confined to the base of a continuous layer. These include: (a) a driving force of high magnitude (e.g., an unusually large density contrast); (b) a combination of driving forces (e.g., sand bedforms on a mud substrate); (c) a lower layer of very low strength (e.g., deposition of sand onto freshly-deposited mud); or (d) prolonged deformation (e.g., a very thick or fine-grained lower layer, since for some liquidization mechanisms the time needed for strength to be restored is proportional to layer thickness and inversely proportional to grain-size).

Several liquidization mechanisms are initiated by the action of a triggering agent, and pseudonodule horizons have been used as evidence of paleoseismicity, although load structures can readily be triggered by other agents such as the rapid deposition of sediment. The common association of pseudonodules with shallow marine deposits may point to storm activity as a common trigger.

Geraint Owen

Bibliography

Allen, J.R.L., 1982. *Sedimentary Structures: Their Character and Physical Basis*. Elsevier.
Anketell, J.M., Cegla, J., and Dzulynski, S., 1970. On the deformational structures in systems with reversed density gradients. *Annales de la Société Géologique de Pologne*, **40**: 3–30.
Kuenen, Ph.H., 1958. Experiments in geology. *Transactions of the Geological Society of Glasgow*, **23**: 1–28.
Macar, P., 1948. Les pseudo-nodules du Famennien et leur origine. *Annales de la Société Géologique de Belgique*, **72B**: 47–74.
Owen, G., 1996. Experimental soft-sediment deformation: structures formed by the liquefaction of unconsolidated sands and some ancient examples. *Sedimentology*, **43**: 279–293.
Potter, P.E., and Pettijohn, F.J., 1963. *Paleocurrents and Basin Analysis*. Springer-Verlag.
Rijsdijk, K.F., 2001. Density-driven deformation structures in glacigenic consolidated diamicts: examples from Traeth y Mwnt, Cardiganshire, Wales, UK. *Journal of Sedimentary Research*, **71**: 122–135.
Smith, B., 1916. Ball or pillow-form structures in sandstones. *Geological Magazine*, **53**: 146–156.

Cross-references

Ball-and-Pillow (Pillow) Structure
Bedding and Internal Structures
Convolute Lamination
Deformation of Sediments
Fluid Escape Structures
Liquefaction and Fluidization
Load Structures

Q

QUARTZ, DETRITAL

Detrital quartz forms about 35 percent of the mineral grains in sedimentary rocks, an amount subequally divided between sandstones and mudrocks. Because of its abundance its internal features, grain size, and shape have been the topic of thousands of journal articles since the 1850s. However, despite this extraordinary effort the amount of information that can be obtained from the mineral is not large.

Internal structure

Many investigations of detrital quartz have concerned its internal structure, particularly the degree of crystal deformation and the character of the crystals in polycrystalline grains (pure quartz rock fragments). Results reveal that nearly all quartz crystals are visibly plastically deformed, a property described in thin section as undulatory extinction. Polycrystalline quartz grains contain crystals that may be elongated or equant, crystallographically aligned or not, unimodal or bimodal in size, and the crystals may have smooth contacts with their neighbors or be interpenetrating (sutured). Elongated and/or aligned crystals and bimodal crystal size distribution indicate recrystallization and, therefore, derivation from a metamorphic rock. Undulatory extinction in crystals and variable types of crystal contacts are common in both metamorphic and igneous rocks.

As the total amount of kinetic energy applied to quartz grains during their sedimentary life increases, polycrystalline grains and grains with undulatory extinction are preferentially destroyed (Figure Q1). Polycrystalline grains fracture along crystal boundaries to produce smaller single crystals of quartz. Crystals with undulatory extinction are destroyed more easily than crystals without undulatory extinction so that pure quartz sandstones are preferentially enriched in nonundulatory grains. The average pure quartz sandstone contains about 40 percent nonundulatory grains, compared to 15 percent to 20 percent in the igneous and metamorphic rocks from which the grains were derived.

Quartz crystals in porphyritic volcanic rocks are generally not plastically deformed and, therefore, are abundant in volcaniclastic sandstones. But in most sandstones quartz grains from porphyritic volcanic rocks are trivial in abundance compared to quartz grains from granites, gneisses, and schists.

Mineral inclusions

Most detrital quartz grains lack mineral inclusions but a wide variety of these inclusions have been described from detrital grains. Commonly the inclusions are too small to be identified with certainty in thin section using only a polarizing microscope. Any mineral found in igneous and metamorphic rocks can be found in detrital quartz. Insofar as the mineral is diagnostic of a specific type of rock it can be useful in provenance determinations. Although there have been numerous studies of mineral inclusions in crystalline rocks, only one investigator (Tyler, 1936) has used inclusions in sandstones for provenance determination. Unfortunately, the sandstone was the St. Peter (Cambrian, Wisconsin), now known to be composed largely of recycled grains from a wide variety of crystalline rocks so that there can be no check on Tyler's conclusions regarding the accuracy of mineral inclusions as an indicator of provenance.

Grain size

Detrital quartz grains vary in size from cobbles to clay. Cobble and pebble size grains are derived largely from either metaquartzites or hydrothermal veins. Those from hydrothermal veins commonly are milky or white in color, an effect produced by inclusions of water bubbles. As grain size decreases into the granule and coarse sand sizes, the proportion of quartz grains derived from granitoid rocks and gneisses increases. In finer sizes the proportion of quartz grains from schists increases. In granule and coarse sand sizes the grains are mostly polycrystalline but in fine sand sizes single crystals dominate. In silt sizes polycrystalline grains are rare.

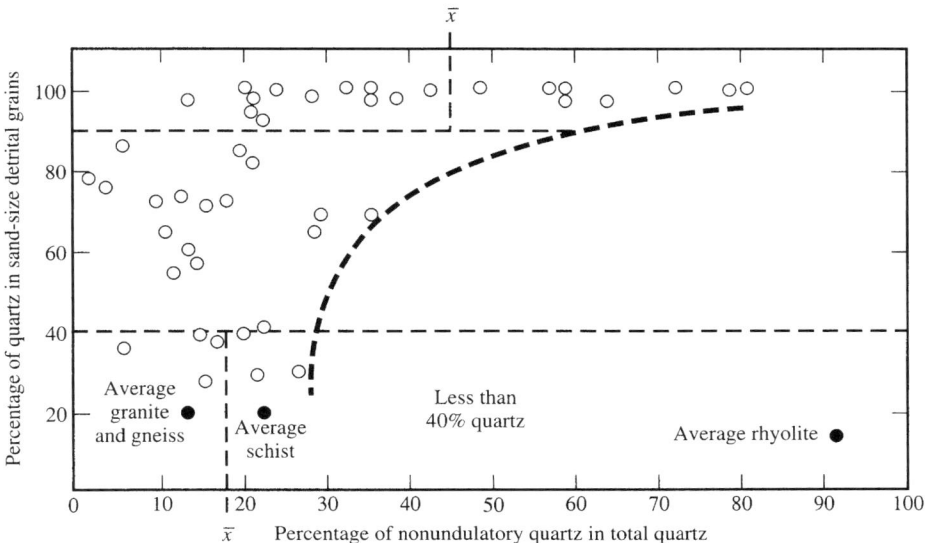

Figure Q1 Relationship between percentage of quartz in crystalline rocks and sandstones and percentage of that quartz that has undulatory extinction (Blatt and Christie, 1963).

This indicates that chert is not an important source of quartz silt.

Part of the change in internal character of quartz with decreasing grain size results from the disintegration of polycrystalline grains during soil formation, impacts during stream transport, and high velocity impacts in environments of deposition such as beaches, marine sand bars, and desert dunes. More important as an explanation, however, is the source rock of most silt size quartz grains. Isotopic evidence indicates that the bulk of detrital silt size quartz grains are generated by disintegration of low-grade metamorphic rocks such as slates, phyllites, and low-grade schists (Figure Q2). These source rocks contain quartz crystals mostly of silt size, which are released into the sedimentary environment as the rocks decay and the grains are eroded. Most silt size detrital quartz grains are not produced by abrasion of coarser quartz particles released from granites and gneisses and coarse grained schists. This fact reflects the hardness of quartz, lack of cleavage, chemical stability, and the fact that quartz sand is transported in streams by saltation, in which impacts with other quartz grains are limited. Apparently, disintegration of polycrystalline quartz grains produces fine quartz sand but little silt.

Detrital quartz is rare in sizes smaller than about two micrometers.

Depositional environment

Detrital quartz tends to be more abundant in some depositional environments than in others. This reflects the fact that feldspars and polymineralic rock fragments are more easily disintegrated than quartz in environments of high kinetic energy. Hence, beaches, marine sand bars, and desert dunes tend to be richer in quartz than are rivers and alluvial fans. But exceptions are common. The amount of quartz in a sedimentary rock is not a reliable environmental indicator.

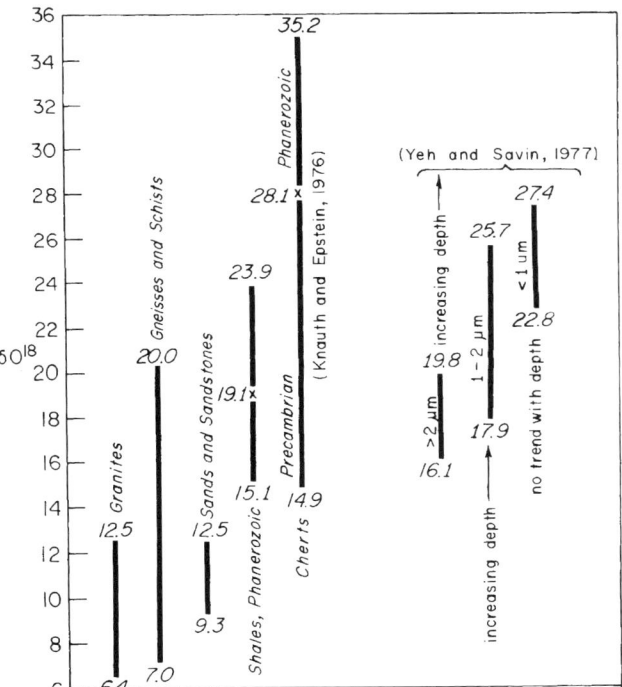

Figure Q2 Range in $\delta^{18}O$ values in quartz from sedimentary rocks and various types of coarse grained source rocks. Quartz in the few slates and phyllites that have been analyzed (not shown) have $\delta^{18}O$ values that average 19 (Blatt, 1987).

Quartz cobbles and pebbles travel in streams by rolling and hence are nearly always rounded. The degree of roundness of quartz sand has been used as an indicator of depositional environment but this also is not reliable. Quartz sand grains can be rounded only in environments of high kinetic energy (not in rivers) but such rounded grains can be subsequently incorporated into environments of low kinetic energy following burial, uplift, and erosion. Recycling is a common phenomenon but the percentage of recycled quartz in sedimentary rocks is unknown. Probably it is quite high.

The percentage and grain size of detrital quartz have been used as an indicator of distance from shore in shallow marine mudrocks. As distance increases, quartz grain size decreases, as does the quartz/clay mineral ratio. Alignment of elongate quartz grains can reveal dominant current directions but directions are more easily deciphered using cross-bedding or other sedimentary structures.

Harvey Blatt

Bibliography

Blatt, H., 1963. Original characteristics of clastic quartz grains. *Journal of Sedimentary Petrology*, **37**: 401–424.

Blatt, H., 1987. Oxygen isotopes and the origin of quartz. *Journal of Sedimentary Petrology*, **57**: 373–377.

Blatt, H., and Christie, J.M., 1963. Undulatory extinction in quartz of igneous and metamorphic rocks and its significance in provenance studies of sedimentary rocks. *Journal of Sedimentary Petrology*, **33**, 559–579.

Blatt, H., and Schultz, D.J., 1976. Size distribution of quartz in mudrocks. *Sedimentology*, **23**: pp. 857–866.

Blatt, H., and Totten, M.W., 1981. Detrital quartz as an indicator of distance from shore in marine mudrocks. *Journal of Sedimentary Petrology*, **51**: 1259–1266.

Chandler, F.W., 1988. Quartz arenites: review and interpretation. *Sedimentary Geology*, **58**: 105–126.

Gotze, J., and Zimmerle, W., 2000. *Quartz and Silica as Guides to Provenance in Sediments and Sedimentary Rocks*. Stuttgart: E. Schweizerbart'sche Verlagsbuchhandlung, 91pp.

Kennedy, S.K., and Arikan, F., 1990. Spalled quartz overgrowths as a potential source of silt. *Journal of Sedimentary Petrology*, **60**: 38–44.

Lechak, P., and Ferrell, R.E. Jr., 1988. Morphologies of suspended clay-sized quartz particles in the Mississippi River and their relation to sedimentary sources. *Geology*, **16**: 334–336.

Suttner, L.J., Basu, A., and Mack, G.H., 1981. Climate and the origin of quartz arenites. *Journal of Sedimentary Petrology*, **51**: 1235–1246.

Suttner, L.J., and Leininger, R.K., 1972. Comparison of the trace element content of plutonic, volcanic, and metamorphic quartz from southwestern Montana. *Geological Society of America Bulletin*, **83**: 1855–1862.

Tyler, S.A., 1936. Heavy minerals of the St. Peter Sandstone in Wisconsin. *Journal of Sedimentary Petrology*, **6**: 55–84.

Cross-references

Classification of Sediments and Sedimentary Rocks
Diagenesis
Isotopic Methods in Sedimentology
Mudrocks
Provenance
Sands, Gravels, and their Lithified Equivalents
Weathering, Soils, and Paleosols

R

RED BEDS

"Red Beds" is the term applied to sedimentary sequences which are predominantly red in color but usually associated with variable proportions of interdigitated drab strata, typically grey, grey-green, brown, or black. They comprise a wide range of facies representing the whole spectrum of non-marine depositional environments from alluvial fans, river floodplains, deserts, lakes, and deltas, and range in age from Early Proterozoic to Cenozoic. Red beds are economically significant; they host a large number of oil and gas reservoirs and are associated with tabular copper and uranium–vanadium deposits (Rose, 1976; Hansley and Spirakis, 1992). Oxidation–reduction reactions play a key role in the formation of red beds in both the depositional and post-depositional settings. The nature and abundance of organic material is thus a key factor in determining the particular characteristics of a red bed sequence.

Red beds are colored by finely disseminated ferric oxides, usually in the form of hematite (α-Fe_2O_3) and generally have a color between 5R and 10R of the rock color chart (Goddard, 1951). Oxidation-reduction (Eh) and pH control hematite formation and the observed minor color variations seen in red beds. In general, the color variations relate to changes in pore water chemistry resulting from fluctuations in depositional or early diagenetic environment. In particular, intercalated marine units are normally non-red because of the reducing conditions generated below the sediment–water interface and such color changes can be a sensitive indicator of base level shifts in marginal marine sequences. The occurrence of ferric oxides in red beds indicates that they formed under oxidizing conditions and that the Earth's atmosphere was oxidizing. In this context they hold great significance for the evolution of life and much of the key evidence for the evolution of terrestrial faunas and floras including amphibians, reptiles (including dinosaurs) and mammals is found in red bed sequences. However, although it is clear, no particular paleoclimatic significance can be attached to them, since they are known to form in both arid and moist tropical climates.

In the past the actual mechanisms of red bed formation have been widely debated and we now know that red beds are polygenetic in origin. The conditions for red bed formation necessarily coincide with those that are needed for hematite formation since this mineral causes the red coloration. These include positive Eh and neutral to alkaline pH, conditions which, at the surface at least, are met in hot, continental low latitude areas. Marine red beds are rare because of the abundance of organic material in the marine environment. They do, however, constitute a distinct group of red beds and are included in the classification here. Three principle mechanisms have been proposed to explain the origin of the red color in continental red beds, each relating to a different stage in the depositional and diagenetic history. Four groups of red beds are recognized as follows:

(1) *Primary red beds*. Deposition and early diagenesis from the erosion of lateritic soils in wet uplands and the aging of ferric hydroxides, especially in the pedogenic environment;
(2) *Diagenetic red beds*. Burial diagenesis and the intrastratal solution of ferromagnesian silicates;
(3) *Secondary red beds*. Weathering and ferruginization of non-red strata after uplift;
(4) *Marine red beds*. Very low organic productivity and slow sedimentation rates resulting in oxic conditions below the sediment–water interface.

Primary red beds

Krynine (1950) suggested that red beds were formed primarily by the erosion and redeposition of red soils or older red beds. A fundamental problem with this hypothesis is the relative scarcity of recent red colored alluvium, and Van Houten (1961) developed the idea to include the *in situ* (early diagenetic) reddening of the sediment by the dehydration of brown or drab colored ferric hydroxides. These ferric hydroxides commonly include goethite (α-FeOOH) and socalled "amorphous ferric hydroxides" or limonite. In fact, much of this material may be the mineral ferrihydrite which has a composition near $2.5\ Fe_2O_3 \cdot 4.5\ H_2O$. This dehydration or "aging" process is now known to be intimately associated with pedogenesis in alluvial floodplains and desert environments.

Berner (1969) showed that ferric hydroxide (goethite) is normally unstable relative to hematite and in the absence of water or at elevated temperature will readily dehydrate according to the reaction:

$$2\alpha\text{-FeOOH(goethite)} \to \alpha\text{-Fe}_2\text{O}_3\text{(hematite)} + \text{H}_2\text{O} \quad \text{(Eq. 1)}$$

The Gibbs free energy (ΔG_r) for this reaction at 25°C is -2.76 kJ/mol and Langmuir (1971) showed that ΔG_r becomes increasingly negative with smaller particle size. Thus detrital ferric hydroxides including goethite and ferrihydrite will spontaneously transform into red colored hematite pigment with time. This process not only accounts for the progressive reddening of alluvium but also the fact that older desert dune sands are more intensely reddened than their younger equivalents. Folk (1976) showed that the red color of dune sands in the Simpson Desert of Australia was formed during an older period of lateritization and subsequent dehydration of the ferric hydroxides. This paper formed a good model for the reddening process in ancient eolian dune sequences.

Diagenetic red beds

The formation of red beds during burial diagenesis was clearly described by Walker (1967) and Walker *et al.* (1978). The key to this mechanism is the intrastratal alteration of ferromagnesian silicates by oxygenated groundwaters during burial. Walker's studies show that the hydrolysis of hornblende and other iron-bearing detritus follows Goldich's stability series. This is controlled by the Gibbs free energy (ΔG_r) of the particular reaction. For example, the most easily altered material would be olivine:

$$\text{FeSiO}_4\text{(fayalite)} + 0.5\text{O}_2 \to \alpha\text{-Fe}_2\text{O}_3\text{(hematite)} + \text{SiO}_2\text{(quartz)} \quad \text{(Eq. 2)}$$

with $\Delta G_r = -27.53$ kJ/mol

A key feature of this process, and exemplified by the reaction, is the production of a suite of by-products which are precipitated as authigenic phases. These include mixed layer clays (illite–montmorillonite), quartz, potash feldspars, and carbonates as well as the pigmentary ferric oxides. Reddening progresses as the diagenetic alteration becomes more advanced and is thus a time-dependent mechanism. The other implication is that reddening of this type is not specific to a particular depositional environment. However, the favorable conditions for diagenetic red bed formation, i.e., positive Eh and neutral–alkaline pH are most commonly found in hot, semi-arid areas, and this is why red beds are traditionally associated with such climates.

Secondary red beds

Secondary red beds are characterized by irregular color zonation, often related to sub-unconformity weathering profiles. The color boundaries may cross-cut lithological contacts and show more intense reddening adjacent to unconformities. Johnson *et al.* (1997) have also showed how secondary reddening phases might be superimposed on earlier formed primary red beds in the Carboniferous of the southern North Sea. The general conditions leading to post-diagenetic alteration have been described by Mücke (1994). Important reactions include pyrite oxidation:

$$3\text{O}_2 + 4\text{FeS}_2 \to \text{Fe}_2\text{O}_3\text{(hematite)} + 8S° \quad \Delta G_r = -789 \text{kJmol}^{-1} \quad \text{(Eq. 3)}$$

and siderite oxidation:

$$\text{O}_2 + 4\text{FeCO}_3 \to 2\text{Fe}_2\text{O}_3\text{(hematite)} + 4\text{CO}_2 \quad \Delta G_r = -346 \text{kJmol}^{-1} \quad \text{(Eq. 4)}$$

Secondary red beds formed in this way are an excellent example of telodiagenesis. They are linked to the uplift, erosion, and surface weathering of previously deposited sediments and require conditions similar to Primary and Diagenetic red beds for their formation.

Marine red beds

The scarcity of marine red beds is due to the fact that in the marine environment there is abundant organic material and reducing conditions prevail below the sediment–water interface and the stable iron-bearing phase is pyrite (FeS_2). This is a consequence of bacterial activity which results in marine sulfate reduction and the destruction of detrital iron oxides. The process may be summarized using the following equations in which organic material is given the general formula CH_2O. Initially sulfate reduction results in the formation of bicarbonate and hydrogen sulfide:

$$2\text{CH}_2\text{O} + \text{SO}_4^{2-} \to 2\text{HCO}_3^- - \text{HS}^- + \text{H}^+ \quad \text{(Eq. 5)}$$

This followed by the reduction of iron ($\text{Fe}^{3+} \to \text{Fe}^{2+}$) and the combination of ferrous ion, initially as an intermediate phase such as mackinawite ($\text{FeS}_{0.9}$) or greigite (Fe_3S_4) and subsequently as pyrite as shown in the reaction:

$$\text{FeOOH} + 2\text{HS}^- \to \text{FeS}_2 + \text{H}_2\text{O} + \text{H}^+ \quad \text{(Eq. 6)}$$

When organic productivity is very low or the rate of sedimentation is very slow the organic material may be completely destroyed and oxidizing conditions (the Eh fence) migrates downward through the sediment column. Under these conditions detrital iron oxides remain stable and further hematite may be produced by equation 1. This is the mechanism by which deep marine red clays and other pelagic red beds form (Froelich *et al.*, 1979; see *Oceanic Sediments*).

Paleomagnetism of red beds

Red beds have long been a source of paleomagnetic data since they have a very stable natural remanent magnetization (NRM). The magnetization is carried by hematite due to its intrinsic spin-canted antiferromagnetism. In some red beds the ferrimagnetic magnetite Fe_3O_4 is also present but typically this is only found in younger (Mesozoic and Tertiary) sections. There has been much debate regarding the validity of paleomagnetic data from red beds; clearly for detailed paleomagnetic analysis the remanence must have been acquired at, or shortly after, deposition. This mechanism is called depositional (DRM) or post-depositional remanent magnetization (PDRM) and appears to be important in fine grained sediments where magnetic oxides are small enough to rotate into the ambient magnetic field prior to compaction. Magnetizations of this type enable red beds to accurately record the geomagnetic field and there have been many magnetostratigraphic applications, particularly in dating, and correlation of paleontologically-barren sequences (e.g., Baag and Helsley, 1974; Johnson *et al.*, 1982)

The growth of authigenic hematite in red beds results in the formation of chemical remanent magnetization (CRM) and frequently because of the protracted nature of mesodiagenesis such magnetizations are multicomponent and complex (see Turner, 1980 for a detailed discussion). In the future, paleomagnetism will prove to be a valuable technique in dating diagenetic events in red beds and will help resolve the difficulties of distinguishing between different types of red bed sequences.

Peter Turner

Bibliography

Baag, C., and Helsley, C.E., 1974. Evidence for penecontemporeneous magnetization of the Moenkopi formation. *Journal of Geophysical Research*, **79**: 3308–3320.
Berner, R.A., 1969. Goethite stability and origin of red beds. *Geochimica et Cosmochimica Acta*, **35**: 267–273.
Folk, R.L., 1976. Reddening of desert sands: Simpson Desert, N.T., Australia. *Journal of Sedimentary Petrology*, **46**: 604–615.
Froelich, P.N. *et al.*, 1979. Early oxidation of organic matter in pelagic sediments of the eastern equatorial Atlantic: suboxic diagenesis. *Geochimica et Cosmochimica Acta*, **43**: 1075–1090.
Goddard, E.N., 1951. *Rock Color Chart*. Geological Society of America.
Johnson, S.A., Glover, B.W., and Turner, P., 1997. Multiphase reddening and weathering events in Upper Carboniferous red beds from the English West Midlands. *Journal of the Geological Society of London*, **154**: 735–745.
Johnson, N., Opdyke, N.D., Johnson, G.D., Lindsay, E.H., and Tahirkheli, R.A.K., 1982. The paleomagnetism of the Middle Siwalik formations of Northern Pakistan. *Paleogeography, Paleoclimatology, Paleoecology*, **37**: 43–62.
Hansley, P.L., and Spirakis, C.S., 1992. Organic-matter diagenesis as the key to a unifying theory for the genesis of tabular Uranium–Vanadium deposits in the Morrison formation—Colorado Plateau. *Bulletin of the Society of Economic Geologists*, **87**: 352–365.
Krynine, P.D., 1950. Petrology, stratigraphy, and origin of the Triassic sedimentary rocks of Connecticut. *Bulletin of the Connecticut Geology and Natural History Survey*, **73**: 239p.
Langmuir, D., 1971. Particle size effect on the reaction Goethite = Hematite + Water. *American Journal of Science*, **271**: 147–156.
Mücke, A., 1994. Part 1. Postdiagenetic ferruginization of sedimentary rocks (sandstones, oolitic ironstones, kaolins and bauxites)—including a comparative study of the reddening of red beds. pp. 361–395, In Wolf, K.H., and Chilingarian, G.V. (eds.), *Diagenesis, IV*. Developments in Sedimentology, 51, Elsevier.
Rose, A.W., 1976. The effect of cuprous chloride complexes in the origin of red-bed copper and related deposits. *Economic Geology*, **71**: 1036–1048.
Schwertmann, U., and Taylor, R.M., 1989. Iron oxides. pp. 379–438. In Dixon, J.B., and Weed, S.B. (eds.), *Minerals in Soil Environments*. Soil Science Society of America.
Turner, P., 1980. *Continental Red Beds*. Elsevier.
Van Houten, F.B., 1961. Climatic significance of red beds. In Nairn, A.E. M. (ed.) *Descriptive Paleoclimatology*. Interscience Publishers pp. 89–139.
Van Houten, F.B., 1973. Origin of red beds: a review—1961–1972. *Annual Review Earth and Planetary Sciences*, **1**: 39–61.
Walker, T.R., 1967. Formation of red beds in modern and ancient deserts. *Bulletin of the Geological Society of America*, **78**: 353–368.
Walker, T.R., Waugh, B., and Crone, A.J., 1978. Diagenesis in first cycle desert alluvium of Cenozoic age, southwestern United States and northwestern Mexico. *Bulletin of the Geological Society of America*, **89**: 19–32.

Cross-references

Colors of Sedimentary Rocks
Desert Sedimentary Environments
Diagenesis
Magnetic Properties of Sediments
Maturation, Organic
Oceanic Sediments
Rivers and Alluvial Fans
Sands, Gravels, and their Lithified Equivalents
Weathering, Soils, and Paleosols

REEFS

Definitions and classifications

While originally defined by mariners as any shallow and hard submarine structure potentially hazardous to vessels, the term reef is restricted in earth sciences to mean *laterally confined carbonate bodies built by sessile benthic aquatic organisms*. This definition is wide enough to include reef structures of all ages, yet precise enough to identify reefs as a recurring biological phenomenon.

Another common definition restricts reefs to rigid, wave-resistant biogenic structures with a significant syndepositional relief and composed mainly of *in-situ* framework builders. These attributes are often difficult to demonstrate in the fossil record and should not be included in a geologic reef definition. Moreover, even Holocene coral reefs often are internally composed of reworked framework and bioclastic debris (Hubbard *et al.*, 1998) rather than consisting of an *in-situ* framework.

Reefs are unique sedimentary systems because they significantly modify the environment and structure their own habitat. Shallow water reefs are protectors of tropical coasts and stabilize platform margins. Reefs and their sand- to boulder-sized debris form important hydrocarbon reservoirs.

The most widely used reef classifications focus on environment, geometry, and composition. For modern coral reefs, Charles Darwin (1842) recognized fringing reefs, barrier reefs, and atolls. He identified these environmental reef types as belonging to a genetic sequence linked to subsidence history. A fringing reef grows along the shore. With subsidence and continued reef growth, lagoonal sediments fill the space between the former shore and the reef becomes a barrier reef. If all preexisting land is submerged and only the reef and lagoonal sediments remain, the reef will eventually form an atoll. Barrier reefs and atolls exhibit a pronounced facies zonation including (from lagoon to open ocean): back reef, reef flat, reef crest, reef front, and fore reef. Darwin's model, although still valid today and often applicable to the fossil record, must be broadened to include other environmental reef types with or without modern counterparts. Patch reefs (equivalent to platform reefs) were common throughout the geological record. They are isolated reef structures mostly in lagoonal settings, on carbonate platforms, or in epeiric seaways. Reefs growing on submarine ramps and slopes are rare today but were common in the past.

Geologists traditionally emphasize geometrical reef attributes subdividing bioherms and biostromes. A *bioherm* is a mound-shaped structure, while a biostrome is a roughly tabular structure without a significant syndepositional relief. The term bioherm is roughly equivalent to ecologic reef or buildup, and *biostrome* roughly agrees with a stratigraphic reef (Dunham, 1970). Other common geometrical terms in reef classification include banks (tabular structures similar to biostromes but mostly micritic in composition) and pinnacle reefs (nearly cylindrical isolated structures often surrounded by basinal sediments). Terms such as ball, dish, haystack, kalyptra, mamelon, spread-eagle, and spruce describe the length/width ratio and overall shape of reef bodies in more detail.

A compositional classification of reefs is usually based on three principal components: skeletons, carbonate matrix, and carbonate cement. Framework reefs, reef mounds, and mud mounds are identified. A *framework reef* consists of skeletal organisms in mutual contact. Skeletal organisms range from calcified microbes to complex metazoans and framework. *Reef mounds* consist of approximately equal proportions of skeletal organisms and a matrix of carbonate mud and marine cement; they appear to be matrix-supported. A *mud mound* is dominated by carbonate mud and marine cement, whereas skeletal framework is almost absent. The origin of micrite and marine cement in reefs is continuously disputed and should therefore not be included in the reef definition. If microbes are identified in micrite-rich mounds the term microbial mound can be applied. Evidence for the both active and passive microbial participation in micrite production is increasing, even in mounds without obvious stromatolitic or thrombolitic structures (Riding, 2000). Common features of many ancient reefs are large volumes of marine cement often in the form of Stromatactis (*q.v.*). Additional compositional attributes used to classify reefs include the dominant biota (e.g., coral reef or sponge reef), the growth forms and constructional role of reef builders, and the density of framework.

Complex reef classifications combining constructional, compositional, and genetic attributes have been published (Insalaco, 1998), but their utilization is limited because genetic interpretations, often changing with scientific views, are intrinsic to the classification. Descriptive geometrical and compositional classifications should be preferred in the study of ancient reefs.

Reef builders

Reef builders are organisms that contribute to reef growth either directly by the accumulation of their skeletal material or by their metabolic activity. Cyanobacteria and other bacteria, microalgae, and fungi represent the microbial consortium that played a substantial role in reef-building throughout the Proterozoic and Phanerozoic. The functional role of microbes in reef-building is to trap and bind sediment, and to induce carbonate precipitation (Reitner and Neuweiler, 1995). Microbial crusts are often important to provide physical strength to a reef and if possessing external calcified walls microbes could play a substantial role in framework construction. Calcareous algae have been notable reef builders throughout the Phanerozoic and became especially important when global climate was deteriorating (Kiessling, 2001).

The most important sessile invertebrate animals acting as reef builders are sponges and corals. Although coralline sponges almost exclusively live in reef caves today, where they are reef dwellers rather than builders, various sponge groups such as archaeocyaths, stromatoporoids, pharetronids, chaetetids, and lithistids were very important reef builders in particular Phanerozoic periods (Figure R1). The great dominance of coral reefs in modern tropical seas has led to an almost synonymous treatment of coral reefs and reefs in general. Yet although corals—Tabulata and Rugosa in the Paleozoic, Scleractinia in the Mesozoic and Cenozoic—were important reef builders from the Upper Ordovician until today, it is only in the Mesozoic that coral reefs became predominant (Figure R1). Even in the Mesozoic, bivalves as reef builders, most notably the rudists in the Cretaceous, occasionally outnumbered corals. The great success of scleractinian corals in reef-building is related to the efficient symbiosis with unicellular algae (zooxanthellae), which permit corals to grow rapidly in nutrient-depleted waters. Other Phanerozoic reef builders were bryozoans and occasionally brachiopods, pelmatozoans, and worms.

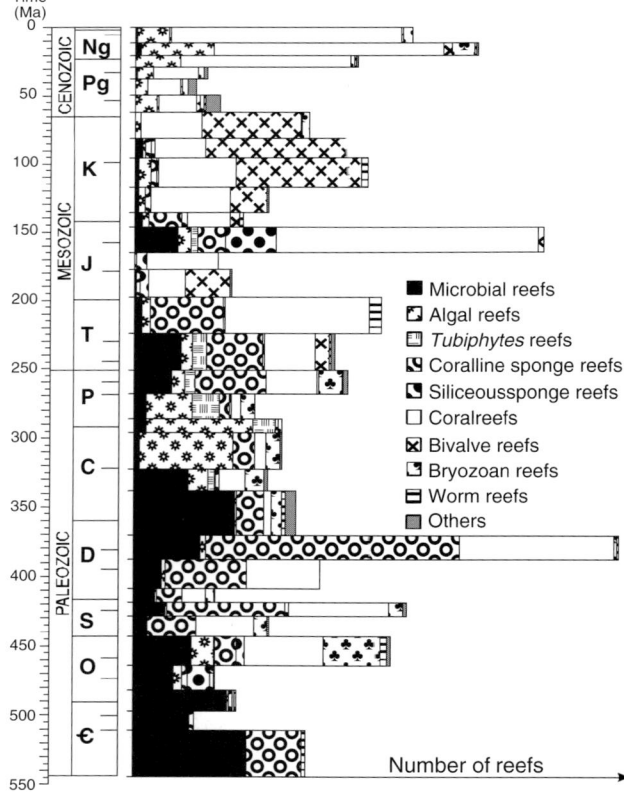

Figure R1 Relative and absolute abundance of Phanerozoic biotic reef types. Time slices represent supersequences as defined in Kiessling *et al.* (1999).

Reef distribution

The reef definition presented in the first paragraph is wide enough to include "tropical" reefs and nontropical or cool-water reefs. Tropical coral reefs, mostly composed of zooxanthellate scleractinian corals and coralline algae are limited today between 35°S and 34°N at western oceanic margins, where warm nutrient-depleted currents reach fairly high latitudes. The tropical reef zone is confined toward the eastern side of oceans where cool nutrient-rich currents reach low latitudes (Figure R2(A)). However, in modern high latitudes (up to more than 70° latitude) reefs are widespread as well. In shallow water, coralline algae may form reefs up to 500 m across (Freiwald and Henrich, 1994), whereas deep water slopes fed by nutrient-rich water sometimes bear impressive reef mounds constructed by azooxanthellate corals (Freiwald et al., 1999). While modern tropical reefs are largely controlled by temperature and light availability, cool-water reefs require high nutrient concentrations and low terrigenous input. Cool-water reefs differ in biological composition from tropical reefs and are geographically isolated from the tropical reef zone. Applying these criteria to detect cool-water reefs in the fossil record, Kiessling (2001) noted that cool-water reefs have a much less continuous record than tropical reefs. Cool-water reefs apparently only proliferated during icehouse climatic intervals, possibly because only then were high latitudes significantly enriched in nutrients.

A changing ecology of reefs is suggested by variable global reef distribution patterns. For much of the Paleozoic, reefs grew inside vast epeiric seaways and were more common along the eastern margin of oceans than today (Figure R2(B)). Since eastern oceanic margins and shelf seas are usually enriched in nutrients, Paleozoic reefs were apparently more

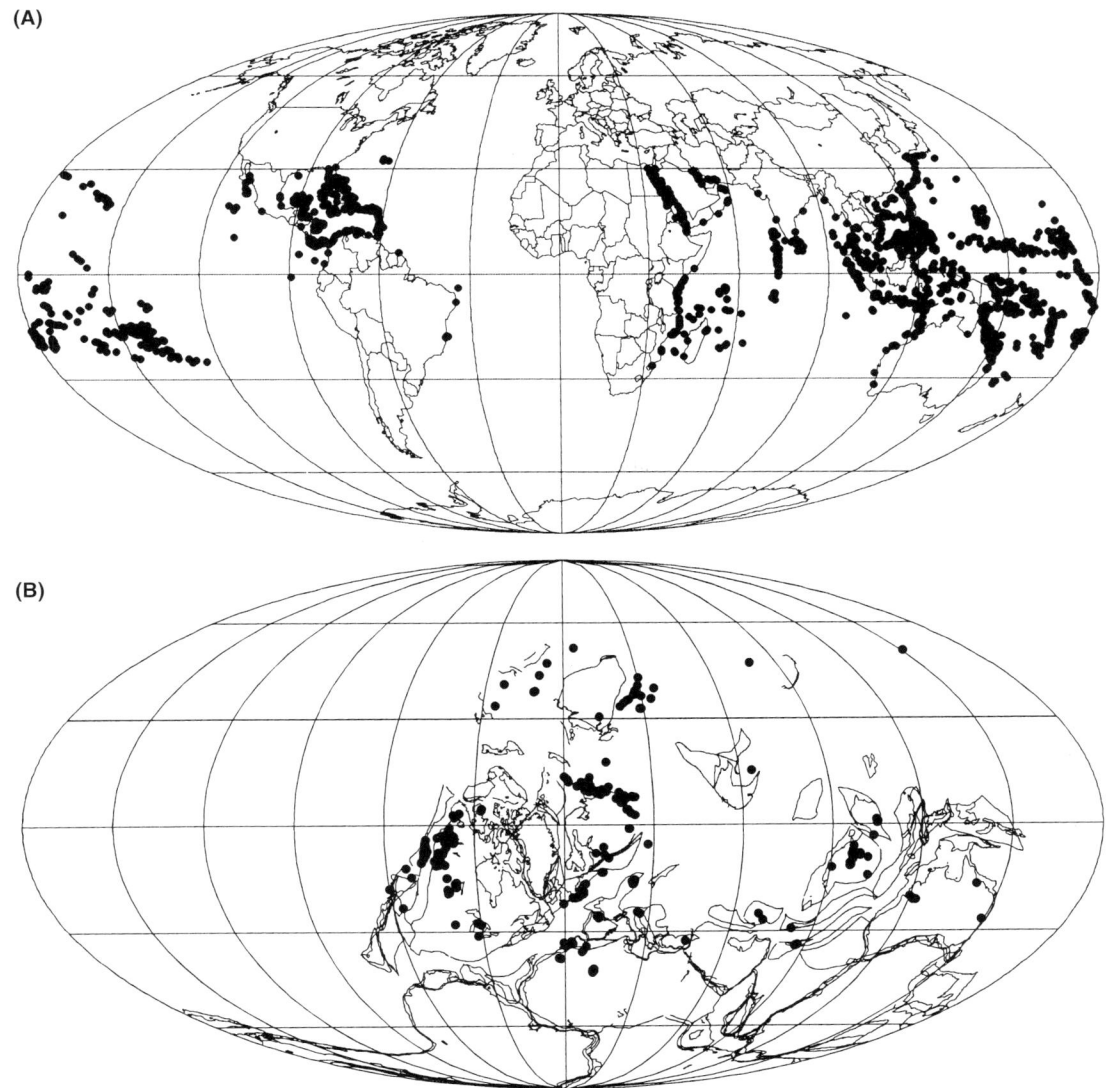

Figure R2 Geographic "tropical" reef distribution pattern today (A) and in the Middle Devonian (B).

nutrient-opportunistic than Mesozoic and Cenozoic reefs, possibly because Paleozoic reef builders did not host photoautotrophic symbionts or the host-symbiont interactions were not as efficient as today.

Temporal trends and fluctuations in reefs

Whether reefs represent a global evolving ecosystem, or a chance association of species with similar ecological requirements (Wood, 1999, p.9) is still disputed. However, within the wide range of reef structures found in the fossil record, several significant temporal trends are evident in a large database incorporating all Phanerozoic reef types (Kiessling et al., 1999). The most significant trends are related to the declining participation of microbes in reef-building through the Phanerozoic (Figure R1). The number of mud mounds and the relative abundance of both micrite and spar declined, whereas the number of framework reefs increased through time. The number of barrier reefs, the amount of reefal debris, and the intensity of bioerosion apparently also increased during the Phanerozoic. The skeletal mineralogy of reef builders showed a systematic trend from predominantly calcitic reef builders in the Paleozoic to aragonitic reef builders in the Cenozoic. Reef diversity, reef abundance and preserved reef dimensions all vary substantially through time without an underlying trend.

The controls behind these trends and fluctuations are largely unknown. Attempts to correlate long-term changes in reefs with geological changes have so far produced equivocal results. This has three main reasons: (1) different reef types and organisms respond differently to the same climatic or chemical changes (Webb, 1996); (2) intrinsic evolutionary changes in reefs may be more important than extrinsic forcing, and (3) geological as well as reef ecosystem changes are currently insufficiently resolved both temporally and spatially to test rigorously whether they are causally linked.

Wolfgang Kiessling

Bibliography

Darwin, C., 1842. *The structure and distribution of coral reefs*. London, Smith-Elder.
Dunham, R.J., 1970. Stratigraphic reefs versus ecologic reefs. *AAPG Bulletin*, **54**: 1931–1932.
Freiwald, A., and Henrich, R., 1994. Reefal coralline algal build-ups within the Arctic Circle: morphology and sedimentary dynamics under extreme environmental seasonality. *Sedimentology*, **41**: 963–984.
Freiwald, A., Wilson, J.B., and Henrich, R., 1999. Grounding Pleistocene icebergs shape recent deep-water coral reefs. *Sedimentary Geology*, **125**: 1–8.
Hubbard, D.K., Burke, R.B., and Gill, I.P., 1998. Where's the reef: the role of framework in the Holocene. *Carbonates and Evaporites*, **13**: 3–9.
Insalaco, E., 1998. The descriptive nomenclature and classification of growth fabrics in fossil scleractinian reefs. *Sedimentary Geology*, **118**: 159–186.
Kiessling, W., 2001. Paleoclimatic significance of Phanerozoic reefs. *Geology*, **29**: 751–754.
Kiessling, W., Flügel, E., and Golonka, J., 1999. Paleoreef maps: evaluation of a comprehensive database on Phanerozoic reefs. *AAPG Bulletin*, **83**: 1552–1587.
Reitner, J., and Neuweiler, F., 1995. Mud mounds: a polygenetic spectrum of fine-grained carbonate buildups. *Facies*, **32**: 1–70.
Riding, R., 2000. Microbial carbonates: the geological record of calcified bacterial-algal mats and biofilms. *Sedimentology*, **47** (1): 179–214.
Webb, G.E., 1996. Was Phanerozoic reef history controlled by the distribution of nonenzymatically secreted reef carbonates (microbial carbonate and biologically induced cement)? *Sedimentology*, **43**: 947–971.
Wood, R., 1999. *Reef Evolution*. Oxford University Press.

Cross-references

Algal and Bacterial Carbonate Sediments
Carbonate Mud-Mounds
Neritic Carbonate Depositional Environments
Stromatactis
Stromatolites

RELICT AND PALIMPSEST SEDIMENTS

In 1952, K.O., Emery defined relict sediments as "remnant from a earlier and different environment," and in 1968, expanded the definition to "...sediments that were deposited long ago in equilibrium with their environments; afterward, the environments changed so that the sediments are no longer in equilibrium even though they remain unburied by later sediments." Although the qualifier does not appear in this extracted definition, both papers applied the term to continental shelf sediments. The term is a codification of earlier thought about the continental shelf, in which the "coastal profile," a concave-upward exponential curve, was seen as an equilibrium response to the coastal regime of waves and tides (Gulliver, 1899). The response was thought to involve not only the shape of the curve, but the distribution of grain sizes along it. "the subaqueous profile is steepest near the land where the debris is coarsest and most abundant, and progressively more gentle further seaward where the debris has been ground finer and reduced in volume by the removal of part in suspension. At every point, the slope is precisely of the steepness required to enable the amount of wave energy there developed to dispose of the volume and size of debris there is in transit" (Johnson, 1919). Such comments amounted to "thought experiments" on the part of Edwardian geomorphologists such as Douglas Johnson, who had no access to research vessels. Francis Shepard (1932), who did, and who was willing to compile notations from maps in areas which he could not sample, argued to the contrary, "the most outstanding feature on continental shelves is the general scarcity of outward decreasing gradation in texture." This condition he attributed to lowered sea level during Pleistocene glacial episodes, which exposed the shelf, and subjected it to deposition of till at high latitudes and fluvial sand and gravel at low latitudes. The explanation was given formal expression by Emery (1952, 1968), Francis Shepard's student. Emery contrasted *relict* sediments with such presently forming (*modern*) classes as *detrital* (water, ice, and wind), *biogenic, residual, authigenic*, and *volcanic* sediments. The main contrast was that between modern detrital sediments on shorefaces and shelves near river mouths (usually fine-grained, dark, and muddy) and the relict sediment found further seaward usually coarser-grained and iron-stained. Many modern detrital sediments showed seaward-fining grain size gradients, but grain size in the relict sediment, at least at the 10 km or greater sample spacing characteristic of the early surveys, seemed to vary in a random manner.

During the period of rapid geological exploration of continental shelves in the 1970s and 1980, the concepts of modern, detrital, and relict sediment filled a need, and relict sediments were described from many of the world's shelves (Slatt and Lew, 1973; Herzer, 1981; Barrie *et al.*, 1984; Shen, 1985). More thought was given to the issue. Swift *et al.* (1971) described a *palimpsest* sediment as "one which exhibits petrographic attributes of an earlier depositional environment and in addition of a later [modern] environment." McManus (1975) distinguished between *in situ* relict and palimpsest deposits on one hand, and freshly sedimented materials which might be completely derived from reworking of the bottom (*proteric* sediments), partially so derived (*amphoteric* sediments), or might be entirely new to the environment (*neoteric* sediments). At the same time, advances in automating textural analysis (Ehrlich and Weinberg, 1970) lead to quantitative criteria for applying these terms (Brown *et al.*, 1980; Mazzullo *et al.*, 1983).

A persistent problem in the application of "relict" and related terms to the interpretation of shelf sediment has been the lack of quantitative and analytical understanding of the equilibrium from which the relict sediment is supposed to be a departure. While Ehrlich and Weinberg (1970) and their coworkers achieved considerable success in distinguishing among sources and agents of sediment input on continental shelves by means of grain shape analysis, the progression of the sediment toward textural maturity defined by it is not an equilibrium concept. The equilibrium parameter implicit in the early discussions (Emery, 1952, 1958) is grain size. Grain size frequency distributions have long been seen as equilibrium responses to the hydraulic regime (e.g., Johnson, 1919) however, the problem is a complex one, encompassing aggregate behavior at a range of spatial scales. Not only must grain size frequency distributions of individual bottom samples be understood in terms of boundary layer fluid dynamics (e.g., Sengupta *et al.*, 1991), but regional patterns of grain size must be understood in terms of advective-diffusive grain transport (Clarke *et al.*, 1983; Swift *et al.*, 1986).

Donald J.P. Swift

Bibliography

Barrie, J.V., Lewis, C.F.M., Fader, G.B., and King, L.H., 1984. Seabed processes on the northeastern Grand Banks of Newfoundland: modern reworking of relict sediments. *Marine Geology*, 57: 209–227.
Brown, P.J., Ehrlich, R., and Colquehoun, D., 1980. Origin of pattern of quartz sand types on the southeastern United States continental shelf and implications on contemporary shelf sedimentation: Fourier grain shape analysis. *Journal of Sedimentary Petrology*, 50: 1095–1100.
Clarke, T.L., Swift, D.J.P., and Young, R.A., 1983. A stochastic modeling approach to the fine sediment budget of the New York Bight. *Journal Geophysical Research*, 88: 9653–9660.
Ehrlich, R., and Weinberg, B., 1970. An exact method for the characterization of grain shape. *Journal of Sedimentary Petrology*, 40: 205–212.
Emery, K.O., 1952. Continental shelf sediments off southern California. *Geological Society of America*, 63: 1105–11108.
Emery, K.O., 1968. Relict sediments on continental shelves of the world. *American Association of Petroleum Geologists Bulletin*, 52: 445–464.
Gulliver, F., 1899. Shoreline topography. *American Academy of Arts and Sciences Proceedings*. 34: 151–258.
Herzer, R.H., 1981. *Late Quaternary Stratigraphy and Sedimentation of the Canterbury Continental Shelf, New Zealand*. New Zealand Oceanographic Institute Memoir, 899.
Johnson, D.W., 1919. *Shore Processes and Shoreline Development*. Columbia University Press (Hafner Facsimile edition, 1952).
Mazullo, J., Ehrlich, R., and Hemming, M.A., 1983. Provenance and areal distribution of Late Pleistocene and Holocene quartz sand on the southern New England continental shelf. *Journal of Sedimentary Petrology*, 54: 1135–1348.
McManus, D.A., 1975. Modern versus relict sediment on the continental shelf. *Geological Society of America Bulletin*, 86: 1154–1160.
Sengupta, S., Ghosh, J.K., and Mazumder, B.S., 1991. Experimental-theoretical approach to the interpretation of grain size frequency distributions. In Syvitski, P.M. (ed.), *Principles, Methods, and Application of Particle Size Analysis*. Cambridge University Press, pp. 264–280.
Shepard, F.P., 1932. Sediments of the continental shelves. *Geological Society of America Bulletin*, 43: 1017–1040.
Shen Huati, 1985. Age and genetic model of relict sediments on the East China Sea shelf. *Haiyang Xuebao* (Acta Oceanologica Sinica), 7: 67–77.
Slatt, R., and Lew, A.B., 1973. Provenance of Quaternary sediments on the Labrador continental shelf and slope. *Journal of Sedimentary Petrology*, 43: 1054–1060.
Swift, D.J.P., Stanley, D.J, and Curray, J.R., 1971. Relict sediments of continental shelves: a reconsideration. *Journal of Geology*, 16: 221–250.
Swift, D.J.P., Thorne, J.A., and Oertel, G.F., 1986. Fluid process and sea floor response on a storm dominated shelf: Middle Atlantic shelf of North America. Part iI: response of the shelf floor. In Knight, R.J., and Mc Lean, J.R. (eds.), *Shelf Sands And Sandstone Reservoirs*. Canadian Society of Petroleum Geologists Memoir, Volume 11, pp. 191–211.

Cross-references

Diffusion, Turbulent
Grain Size and Shape
Offshore Sands

RELIEF PEELS

Relief peels are produced by impregnating a thin, surficial layer of unconsolidated, granular sediment. Because of spatial differences in the porosity and permeability, the binding agent penetrates to different depths, producing relief on the peel's surface (Figure R3). The differences in porosity and permeability are themselves determined by subtle differences in grain size, sorting, and packing that are created during deposition of the sediment and/or during any post-depositional bioturbation, deformation, or incipient diagenesis. Such peels have been used extensively since the early 1950s to study the physical and biogenic structures present in modern deposits and unconsolidated older sediments. Commonly, the impregnation process yields a more detailed record of those structures than can be seen in the original core or exposure. The permanence of the peel allows for more thorough study than would be possible otherwise, and photography can be done under controlled lighting conditions, providing better documentation of the structures than might be possible in the field. Relief peels obtained from two or more, mutually perpendicular surfaces offer the possibility of preparing three-dimensional reconstructions of the structures (*cf*. Bouma, 1969), while multiple contiguous peels can be used to represent large sections. Small samples can be cut from peels, in order to study lamina-specific grain-size distributions, particle orientations, and fabrics. Relief peels make excellent teaching aids in introductory sedimentology courses, as well as in petroleum-geology courses to demonstrate microscale permeability heterogeneity. They also make attractive displays.

Relief peels have been made from both siliciclastic and carbonate sediments with grain sizes ranging from gravel to mud. Many different binding agents and techniques have been used over the years; readers are referred to Bouma (1969) and Klein (1971) for comprehensive reviews. Of the many different binding agents used, epoxy resins, polyester resins, and lacquers are perhaps the most commonly used, although readily available materials such as spray-on photomount adhesive have also been employed (Whyte, 1992). The selection is largely a matter of personal preference and availability, although the binding agent should not react with water if the sediment is wet, and must have a viscosity that is appropriate for the situation. For very coarse sediments (very coarse sands and gravels), high-viscosity agents must be used to avoid excessive penetration, which requires large volumes of the binding agent and creates a heavy peel. On the other hand, muddy sediments are difficult to peel because of their very low permeability (Figure R3(C)); only very low-viscosity binding agents will work. This commonly requires the addition of a thinner and/or heating of the binding agent to lower its viscosity. (Note that all work with such chemicals should be performed in a well ventilated place.) Experimentation is needed to obtain optimal results, especially for heterolithic deposits.

The surface to be peeled must be cleaned to remove any disturbed material and smoothed to remove surface irregularities that would otherwise be visible on the peel. In the case of field exposures, care must be taken to ensure that the face is stable, to avoid collapse and potential injury. In very loose sediments, this may require the introduction of a temporary binding agent. In dry sand, wetting the sediment before excavation may be sufficient; however, in gravels, Hattingh *et al.* (1990) have found it necessary to add wallpaper glue to the water and allow it to set before beginning excavation. Water-saturated sediments, including cores, must be allowed to dry somewhat, in order to permit excavation and/or entry of the binding agent

The method of application of the binding agent is dependant on the nature of the agent and the situation where the peel is being made. For cores in the laboratory, it is common to pour the agent directly and evenly onto the level surface, taking care to avoid disturbing the surface. A single application is generally sufficient. Strips of plastic film can be inserted around the margins of the core, to prevent the binding agent from adhering to the core box or tube, and to stop the excess material from flowing off the surface. In the case of vertical faces in outcrops or trenches, the binding agent is most commonly brushed or sprayed on gently, usually in several coats to ensure saturation of the surface layer. Spraying is least likely to disturb the surface, but requires more equipment than other approaches (cf. Getzen and Levey, 1982). Barr *et al.* (1970) have described a single-application technique that may work in some circumstances.

In order to strengthen peels, some type of backing is typically applied. Loosely woven fabric (e.g., cheesecloth) or wire mesh is sufficient for many situations, and work especially well in cases where the sediment surface cannot be smoothed

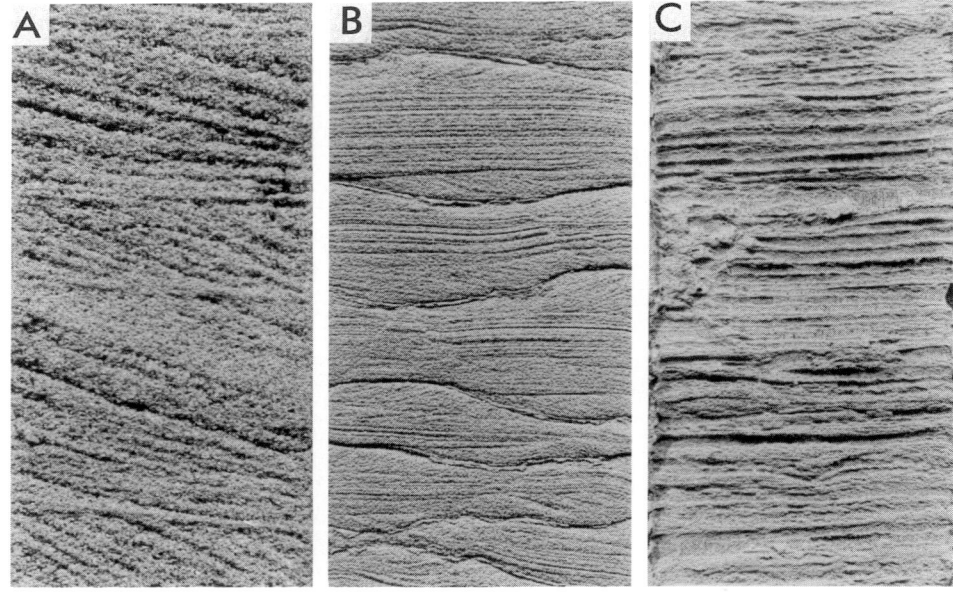

Figure R3 Three 8-cm wide relief peels from modern tidal deposits, Bay of Fundy, prepared from oriented cores using epoxy resin: (A) medium to coarse grained, cross-bedded sand (epoxy penetration 1 cm); (B) interbedded, mud-draped current ripples, and upper-flow-regime parallel lamination in fine-very fine sand (epoxy penetration 2 mm to 2.5 mm); and (C) muddy tidal rhythmites in which very thin sand layers stand out (epoxy penetration 0–1.5 mm). The mud present in B and C remains on the peel because of cohesion, not because of epoxy penetration.

completely (e.g., in the case of gravels). More rigid backings such as masonite, plexiglass, or stiff card are also used. They may be applied during the peel-making process, or after the peel has been removed. Any backing should be carefully labeled with all relevant sample information, including any orientation data.

The peel must be left in place until the binding agent has set sufficiently that the peel and the surface relief are capable of retaining their integrity. This may take several tens of minutes to many hours depending on the binding agent used and the temperature; heat lamps can be used, with caution, to speed drying. Complete curing may take several days. During this time, it may be necessary to weigh the peel down to prevent warping. Once the binding agent has set completely, loose sediment can be removed by brushing and/or washing (unless the binding agent is soluble in water). A thin, transparent coating may then be applied to protect the surface, if desired. The peel is then ready for study; additional mounting may be required for display purposes.

Robert W. Dalrymple

Bibliography

Barr, J.L., Dinkelman, M.G., and Sandusky, C.L., 1970. Large epoxy peels. *Journal of Sedimentary Petrology*, **40**: 445–449.
Bouma, A.H., 1969. *Methods for the Study of Sedimentary Structures*. New York: Wiley-Interscience.
Getzen, R.T., and Levey, R.A., 1982. Rigid-peel technique of preserving structures in coarse-grained sediments. *Journal of Sedimentary Petrology*, **52**: 652–654.
Hattingh, J., Rust, I.C., and Reddering, J.S.V., 1990. A technique for preserving structures in unconsolidated gravels in relief peels. *Journal of Sedimentary Petrology*, **60**: 626–627.
Klein, G.deV., 1971. Peels and impressions. In Carver, R.E. (ed.), *Procedures in Sedimentary Petrology*. Wiley-Interscience, pp. 217–250.
Whyte, M.A., 1992. The use of "photomount" adhesive as a medium for peels of unconsolidated sediments. *Journal of Sedimentary Petrology*, **62**: 741–742.

Cross-references

Bedding and Internal Structures
Biogenic Sedimentary Structures
Fabric, Porosity, and Permeability
Grain Size and Shape
Mudrocks
Weathering, Soils, and Paleosols

RESIN AND AMBER IN SEDIMENTS

Many plants produce resinous exudates in response to injury or stress. These exudates, which serve to seal injuries and discourage predators, are organic materials, often composed of mixtures of semi-volatile and volatile organic compounds know as terpenes. After exudation, these mixtures typically harden into solid masses, either by evaporation of volatile components and/or by polymerization. These masses are often highly resistant to biological attack, and hence, tend to resist degradation processes which break down other plant tissues.

For this reason, these materials often accumulate in developing sediments and become part of the sedimentary record.

Resins are fairly common constituents of sediments derived from terrestrial systems. They do not usually occur as massive deposits, although in some cases significant concentrations are known. The most famous of these is the very well-known Eocene "Blue Earth" found in and around the Baltic region, which is the source strata for Baltic ambers, which have been known and collected for millennia. So-called "gem" quality ambers, those that are clear and are most suitable for us in ornamentations or artistic applications are usually less common than opaque samples, but in many instances opaque samples are overlooked or discarded by prospectors and collectors and hence these are not as common in collections as they probably are in sediments. Not surprisingly, resinous materials are commonly found in association with coals and coally shales, although in many of these settings resins may exist only as finely dispersed particles which are identified by microscopy on the basis of their optical properties.

Distribution

Resin-derived materials are both geographically and chronologically widely distributed in the sedimentary record. The oldest confirmed terpenoid ambers have been reported in Triassic sediments, but optically-defined resinites are commonly reported in Carboniferous and Permian coals. The vast majority of well-known ambers are found in sediments of Cretaceous age or younger.

Ambers are known from every continent except Antarctica (and probably exist there but have not been observed). The best known deposits are located in Europe and western Asia, but extensive deposits are also known from the Caribbean Basin (especially the Dominican Republic), South East Asia and North America (Figure R4). Ambers are also known in Africa and South America, but fewer deposits have been described for these locales. This probably does not reflect a greater absence of ambers in these regions, but rather the political realities in many of these regions which tend to discourage prospecting.

Terminology and nomenclature

The nomenclature of ambers is a mess. Historically, because the appearance and to some extent the physical properties of ambers vary significantly, many have been given informal names based on the geographic origins or on the name of the individual who first reported the existence of a particular deposit. Hence, the literature is full of references to burmite, rumanite, beckerite, glessite, etc. In reality, most of these names are not particularly useful since they are not based on any useful objective characteristic of the amber which would allow a user to determine the relationship of samples from different deposits. An exception to this is "succinite," the term generally applied to ambers from the Baltic region. This term derives from the characteristic presence of succinic acid (also sometimes known as amber acid) in ambers from this region.

Adding to this confusion, use of the term "amber" can itself be controversial when the age of the sample is in question or is known to be modest. Because resins are continuously being produced and deposited, recent, and modern deposits are well-known. In fact in some places, these recent deposits, (which often contain materials on the order of a few thousand years

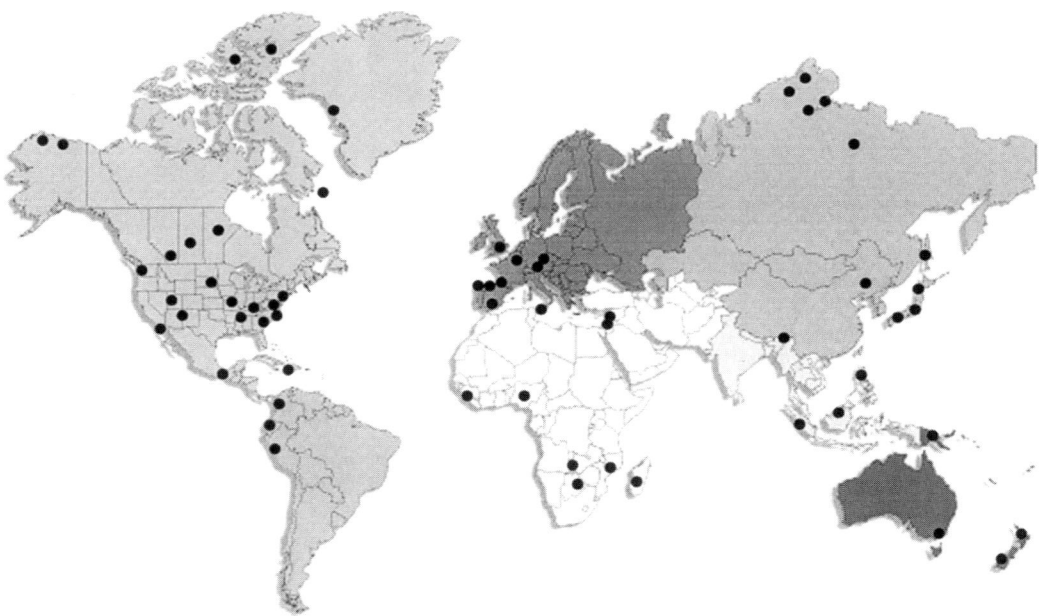

Figure R4 Global distribution of known significant amber deposits. (Many smaller, less well-characterized deposits are also known.)

Table R1 Radiocarbon (^{14}C) scale for definition and description of sedimentary resins. (Modified after Anderson, 1996)

^{14}C Age (years BP)	
0–250	Modern Resin or Recent Resin or Copal (in specific circumstances[a])
250–5,000	Ancient Resin
5000–40,000	Sub-fossil Resin
>40,000	Amber, Fossil Resin, or Resinite ("resinite" reserved for samples identified by (optical) microscopic analysis)

[a] See Langenheim, 1995 and references cited therein.

old or less), were actively "mined" until quite recent times to provide raw materials for varnishes and polishes. Although often structurally related to much older samples, use of the term "amber" to describe these recent materials is usually considered inappropriate. Loosely (or un-) defined terms such as "copal" or "young amber" are sometimes applied to these materials (See Langenheim, 1995, 2002 for a discussion). However, this raises the (controversial) question of when a sedimentary resin becomes amber. This is often hotly debated since application of the term "amber" increases the value of the material considerably. An objective basis for clarification of the nomenclature of ambers has been proposed (see Table R1 below). This uses objective analyzes (radiocarbon dating) of the amber itself to determine the correct nomenclature, but this has yet to be widely adopted and remains controversial.

Structural characteristics and classification

Unlike many inorganic minerals, ambers do not have a continuous, or repeating structure. In fact, ambers are typically amorphous materials. Most ambers are, however, based on polymeric structures. This is probably due to the fact that non-polymeric resins are simply softer, more soluble, and volatile and generally less resilient than their polymeric analogues, and hence, are less likely to survive burial and diagenesis in a recognizable form.

Modern analyzes of amber have shown that many ambers share common structural characteristics, and that ambers can be grouped, or classified on this basis. For many purposes, these groupings are more useful than conventional nomenclature (for the reasons set out above) and classification systems based on structural characteristics have been proposed by several authors. The most widely used and accepted of these systems classifies ambers on the basis of the specific structural characteristics of the resin. A summary of this system is illustrated below (Figure R5).

Class I ambers, which are based on polymers of labdanoid diterpenes, are clearly the most common form of amber. Possibly accounting for 80 percent or more of known amber reserves. This class of ambers is further subdivided on the basis of the presence or absence of succinic acid, (butane-1, 4-dioic), and the specific stereochemistry of the labdanoids on which the polymer backbone of the amber is based. *Class II* ambers, which are based on a not yet fully described (Anderson and Muntean, 2000) polysesquiterpenoid structure, are less

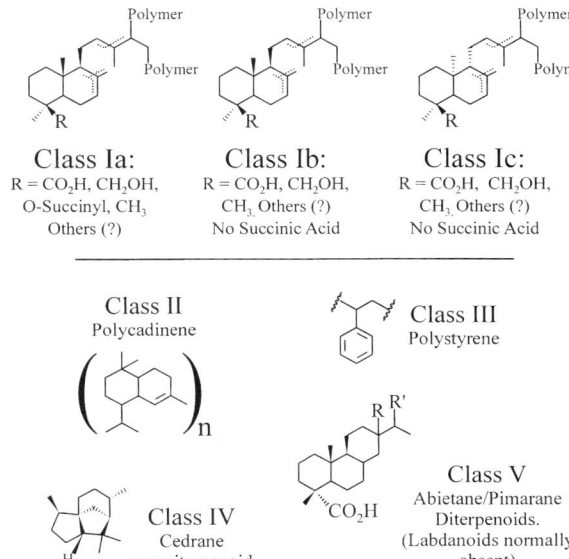

Figure R5 Classification system for sedimentary resins based on structural characteristics of major organic constituents. Modified after Anderson and Crelling, 1995.

common overall than Class I ambers but are nevertheless very abundant in some regions, especially South East Asia and parts of western and southern North America. Class III, Class IV, and Class V ambers are generally rare and represent resins of specific families or non-polymeric resins which are preserved only in unusual sedimentary circumstances.

Class I ambers are by far the best characterized ambers at this time, but even for these materials complete structural characterization remains elusive. It is known, however, that the structures of these materials undergo regular systematic changes over time in response to (primarily) temperature. This raises the possibility that these materials be useful as thermal maturity indicators, and if so these materials would not suffer from complicating factors such as multiple sources, migration, or mixing, which can be problematic in other organic maturity markers. Also, in most cases ambers occlude a variety of terpenes in addition to the common structural components which make up the polymeric backbone. In at least some cases these occluded compounds may be characteristic of the original botanical source, allowing investigators to determine the presence of particular genera or species at a particular location at a particular point in geologic time. Such information is likely to be highly valuable in paleoenvironmental reconstructions.

Inclusions

Finally, it is not possible to conclude this article without some comment about inclusions, which are often found in ambers. Due to their "sticky" nature when first exuded, it is not uncommon for extraneous materials to become embedded within resins. These inclusions often include plant parts, small insects, and (rarely) small vertebrates and other materials such as feathers. Once enveloped, the quality of preservation of these included materials is unequaled by any other medium. As a result, these types of fossils are highly prized by collectors and paleontologists and are sometimes of considerable value. Insects, for example, are preserved in exquisite detail, often allowing paleo-entomologists to study detailed characteristics. This extraordinary degree of preservation extends to some extent to soft tissues, and even cellular structure (Grimaldi, 1996). Claims of preservation of DNA, and even viable bacterial spores in amber have received considerable press, but for the most part remain controversial at this point.

Ken B. Anderson

Bibliography

Anderson, K.B., and Crelling, J.C., 1995. Introduction. In Anderson, K.B., and Crelling, J.C. (eds.), *Amber, Resinite, and Fossil Resins*. ACS Symposium Series Volume 617. pp. ix–xvii.

Anderson, K.B., 1996. The nature and fate of natural resins in the geosphere. VII. A radiocarbon (^{14}C) age scale for description of immature natural resins. An invitation to scientific debate, *Organic Geochemistry*, **25** (3/4): 251–253.

Anderson, K.B., and Muntean, J.V., 2000. The nature and fate of natural resins in the geosphere. X. structural characteristics of the macromolecular constituents of modern Dammar resin and Class II ambers. *Geochemical Translation*, Article No. 1. http://www.rsc.org/ej/gt/2000/b000495m/index.html

Grimaldi, D.A., 1996. *Amber: Window to the Past*. American Museum of Natural History, and Harry N. Abrams.

Langenheim, J.H., 1995. Biology of Amber Producing Trees: Focus on Case Studies of Hymenaea and Agathis. In Anderson, K.B., and Crelling, J.C. (eds.), *Amber, Resinite, and Fossil Resins*. ACS Symposium Series, Volume 617, pp. 1–31.

Langenheim, J.H., 2002. Plant Resins: their Value to Plants and Humans. Timber Press. In Press.

RIPPLE, RIPPLE MARK, RIPPLE STRUCTURE

Definition

Ripples, *ripple marks*, or *ripple structures* can be defined as small-scale, flow-transverse ridges of silt or sand produced by fluid shear at the boundary between moving water or air and an erodible sediment bed. Principal ripple types are *current ripples*, formed by unidirectional water flows, *wave ripples*, generated by oscillatory wave action, and *wind ripples*, formed by eolian currents (see *Desert Sedimentary Environments*). Large-scale equivalents of ripples are *dunes* (see *Surface Forms*).

Basic concepts

Ripple marks are quasi-triangular in vertical cross-section parallel to flow direction (Figures R6(A),(C–E)) or wave propagation (Figure R6b). Current ripples are asymmetric, with gentle upstream face (*stoss side*) and steep downstream face (*lee side*) approaching or at angle-of-repose. Individual current ripples can be up to 60 cm long and 6 cm high, but the mean length and height of a field of current ripples are usually <20 cm and <2 cm, respectively. Wave ripples are symmetric or slightly asymmetric in cross-section (Figure R6(B)). They can be 200 cm long and 23 cm high, but typical dimensions are an order of magnitude less.

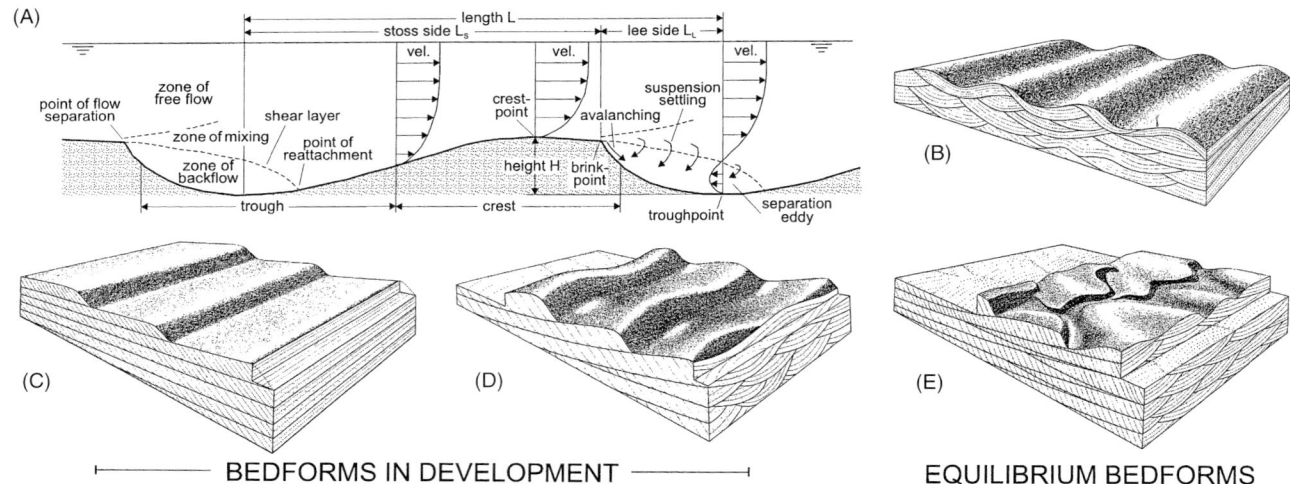

Figure R6 (A) Ripple marks in vertical profile parallel to flow. Terminology is based on Allen (1968) and Reineck and Singh (1980). Note that avalanching and suspension settling generate cross-lamination depicted in (C)–(E). vel. = flow velocity. Water depth note to scale. (B) Plan form and internal structure of symmetrical wave ripples with rounded crests. Wave ripple crests can also be pointed. Wave propagation direction is perpendicular to crest lines. Note bundle-wise arrangement of cross-laminae. (C) Straight-crested, (D) Sinuous-crested, and (E) Linguoid current ripples, with characteristic cross-lamination. Flow is from left to right. Note that current ripples evolve from straight-crested via sinuous-crested to linguoid, independent of flow velocity. Ripples in (C) and (D) are therefore developing, and (E) shows equilibrium forms. Figures (B)–(E) after Reineck and Singh (1980). Copyright © 1980 by Springer-Verlag. All rights reserved.

The geometry of ripples is expressed by the *vertical form index* (or *ripple index*), L/H, the *symmetry index*, L_S/L_L, and the *plan form* (Figure R6). Most common vertical form indices are 8–9 for current ripples and 3–8 for wave ripples, although lower indices and indices up to 20 are no exception for both types. The mean symmetry index for a field of current ripples is mostly >2.5, and dissimilar from typical indices of <2.5 for wave ripples. However, co-existence of oscillatory and unidirectional currents may produce *combined-flow ripples*, whose symmetry index overlaps with current ripples, and thus complicates genetic process reconstruction. Weak cross-sectional asymmetry can also be produced by intrinsic resultant mass transport of water in the direction of wave propagation, mostly in shallow water. Current ripples comprise four common styles of plan form: *incipient*, characterized by mm-high solitary ripples or ripple patches on a flat bed; *straight-crested*, with straight, continuous crest lines (Figure R6(C)); *sinuous-crested*, with sinuous, continuous crestlines (Figure R6(D)); and *linguoid*, with discontinuous, tongue-shaped crestlines (Figure R6(E)). Although wave ripples can have three-dimensional plan form, particularly in the presence of unidirectional currents, wave interference, or under high waves, straight crests with diagnostic bifurcation and abrupt termination of crest lines are most common.

Flow over current ripples separates at the brinkpoint in an upper *zone of free flow* and a *zone of backflow* with *separation eddy* in the lee of the ripple (Figure R6(A)). These zones are divided by a *zone of mixing*, a shear layer along which Kelvin-Helmholtz vortices generate high instantaneous shear. In the ripple trough, the shear layer gradually trends downward until the main flow becomes *reattached* to the bed. Current ripples migrate in the flow direction by erosion of the stoss side and deposition on the lee side. The migration rate is inversely proportional to ripple size and proportional to local sediment transport rate, which varies with the degree of sheltering by adjacent ripples. Depositional processes associated with ripple migration comprise suspension settling through the zone of mixing and recurrent avalanching of bed load material from the brinkpoint onto the lee slope (Figure R6(A)). Each avalanche-settling cycle produces a *cross lamina*. Sets of cross-laminae are diagnostic of migrating current ripples. Preservation of cross-lamination is enhanced by net deposition in flows that are oversaturated with sediment and produce *climbing ripple cross-lamination* (see *Cross-Stratification*).

Flow over wave ripples comprises three stages: (1) During passage of a water wave crest, flow separation occurs over the ripple crest and a separation eddy and avalanche deposit are produced; (2) During subsequent decelerating flow in the wave cycle, the eddy rises, and may take suspended sediment with it; (3) During passage of the water wave trough (with reversed flow direction), the suspended sediment moves over the ripple crest, a new separation eddy develops and avalanching occurs on the opposite ripple face. Wave ripple cross-lamination produced by net deposition from successive waves is characterized by bundles of oppositely inclined laminae (Figure R6(B)). For combined-flow ripples, the dominant inclination of cross-laminae is parallel to the unidirectional flow component.

Historical development

Ripple marks have been described extensively since the beginning of modern observational geology. Sorby (1859), Darwin (1884), Hunt (1904), Kindle (1917) and Bucher (1919) established the foundations for modern ripple analysis. Allen (1968), Potter and Pettijohn (1977), Reineck and Singh (1980) and Allen (1984) summarized further benchmark papers, which include studies on ripples in recent and ancient sediments, ripples in experimental flumes, and the theory of ripple formation and stability. In the last two decades, ripple marks have been used primarily in facies analysis. Yet, innovative work continued in experimental flumes and through mathematical modeling (e.g., Miller and Komar, 1980; Diem, 1985; Arnott and Southard, 1990; Gyr and Schmid, 1989; Nelson and Smith, 1989; Best, 1992; Baas, 1994, 1999; Werner and Kocurek, 1999; Baas et al., 2000). Best (1996) and Raudkivi (1997) presented excellent reviews of recent current-ripple literature.

Applications

Ripples and cross-lamination have been used as indicators of stratigraphic younging and to reconstruct (paleo)flow direction or (paleo)wave crest orientation. Moreover, ripples can be used as estimators of local flow properties, notably flow type, flow strength and sediment accumulation rate. Flow strength can be constrained from *bedform stability diagrams* (see *Surface Forms*). Current ripples form at relatively low flow velocities between the threshold for sediment movement and *upper-stage plane bed* (for particles $< \sim 0.12$ mm) or *dunes* (for particles ~ 0.12–0.7 mm). They do not form in cohesive clay and in sand coarser than ~ 0.7 mm. Wave ripples exist in all silt- and sand-grades, and between maximum oscillatory velocities for the threshold of sediment movement and upper-stage plane bed.

Unidirectional flows occur in many depositional settings, which renders current ripples of limited use in paleoenvironmental reconstruction. Nevertheless, hydraulic conditions are most favorable in sandy rivers, on intertidal flats, and in turbidites. Wave ripples require relatively weak near-bed oscillatory flow, such as on lake beaches, intertidal flats and the marine shoreface. They are therefore better suited for paleoenvironmental analysis than current ripples. Combined-flow ripples have been found on intertidal flats and in other shallow marine and coastal environments.

Ripple marks provide the sediment bed with form roughness. Form roughness is essential for calculating flow resistance, bed shear stress and velocity distribution in sediment transport analysis.

Current investigations and gaps in knowledge

Refinement of existing models and development of new ideas continually improve our knowledge of ripples. Present-day work on current ripples concentrates on the mechanisms controlling their formation (Best, 1992; Raudkivi, 1963; Williams and Kemp, 1971), their rate of development toward equilibrium morphology (Baas, 1999) and the feedback relationships between ripple geometry and flow conditions (e.g., based on "wave instability" theory by Richards, 1980). A field of current ripples can be generated from a single, artificial, or flow-induced, bed defect if the height of the defect is large enough to induce flow separation at its lee side.

Baas (1994, 1999) demonstrated that current ripples developing on fine- and very fine sand-grade flat beds evolve from incipient, via straight-crested and sinuous-crested, to linguoid equilibrium plan form with constant mean height and length. Contrary to earlier work (e.g., Allen, 1968), the equilibrium dimensions were found to be independent of flow velocity. Velocity governs merely the time needed to reach equilibrium geometry, which may range from several minutes to hundreds of hours. Baas (1993) proposed relationships between sediment size, D, and equilibrium current ripple height, H_e, and length, L_e (cf. Raudkivi, 1997):

$$H_e = 3.4 \log D + 18$$
$$L_e = 75.4 \log D + 197$$
(in mm) (Eq. 1)

Although the above model has been applied successfully to current ripple dynamics in tidal environments (Oost and Baas, 1994) and used to reconstruct sedimentation rates in turbidites (Baas et al., 2000), several gaps in our knowledge remain. First, the role of turbulence in the growth rate of current ripples has never been assessed in detail. Second, aggradation promotes current ripple preservation, but no comprehensive study of its influence on ripple dynamics exists, particularly under conditions of turbulence modulation. This knowledge would strengthen the use of current ripples in (paleo)hydraulic reconstructions, and in the determination of drag coefficients in sediment transport calculations.

The latest studies on wave-induced bedforms concentrate on the sediment entrainment under oscillatory and combined flows, and the dynamics of such flows over wave and combined-flow ripples (e.g., Traykovski et al., 1999; Paphitis et al., 2001). This work will remain important, considering the complexity of the feedback relationships between flows and bedforms. A better process-based distinction between combined-flow and current ripples is needed, as the use of indices is unsatisfactory. This could be achieved by exploring potential differences in grain shape fabric of current, wave, and combined-flow ripples.

Jaco H. Baas

Bibliography

Allen, J.R.L., 1968. *Current ripples: their relation to patterns of water and sediment motion*. Amsterdam: North-Holland Publishing Company.

Allen, J.R.L., 1984. *Sedimentary structures: their character and physical basis*. Elsevier.

Arnott, R.W., and Southard, J.B., 1990. Exploratory flow-duct experiments on combined-flow bed configurations, and some implications for interpreting storm-event stratification. *Journal of Sedimentary Petrology*, **60**: 211–219.

Baas, J.H., 1993. Dimensional analysis of current ripples in recent and ancient depositional environments. *Geologica Ultraiectina*, **106**: 199pp.

Baas, J.H., 1994. A flume study on the development and equilibrium morphology of small-scale bedforms in very fine sand. *Sedimentology*, **41**: 185–209.

Baas, J.H., 1999. An empirical model for the development and equilibrium morphology of current ripples in fine sand. *Sedimentology*, **46**: 123–138.

Baas, J.H., Van Dam, R.L., and Storms, J.E.A., 2000. Duration of deposition from decelerating high-density turbidity currents. *Sedimentary Geology*, **136**: 71–88.

Best, J.L., 1992. On the entrainment of sediment and initiation of bed defects: insights from recent developments within turbulent boundary layer research. *Sedimentology*, **39**: 797–811.

Best, J.L., 1996. The fluid dynamics of small-scale alluvial bedforms. In Carling, P.A., and Dawson, M.R. (eds.), *Advances in Fluvial Dynamics and Stratigraphy*. John Wiley & Sons, pp. 67–125.

Bucher, W.H., 1919. On ripples and related sedimentary surface forms and their paleogeographic interpretations. *American Journal of Science*, **47**: 149–210, 241–269.

Darwin, G.H., 1884. On the formation of ripple-mark. *Proceedings of the Royal Society of London*, **36**, 18–43.
Diem, B., 1985. Analytical method for estimating palaeowave climate and water depth from wave ripple marks. *Sedimentology*, **32**: 705–720.
Gyr, A., and Schmid, 1989. The different ripple formation mechanism. *Journal of Hydraulic Research*, **27**: 61–74.
Hunt, A.R., 1904. The descriptive nomenclature of ripple-mark. *Geological Magazine*, **1**: 410–418.
Kindle, E.M., 1917. *Recent and Fossil Ripple-Marks*. Museum Bulletin of the Geological Survey of Canada, Volume 25, pp.121.
Miller, M.C., and Komar, P.D., 1980. Oscillation sand ripples generated by laboratory apparatus. *Journal of Sedimentary Petrology*, **50**: 173–182.
Nelson, J.M., and Smith, J.D., 1989. Mechanics of flow over ripples and dunes. *Journal of Geophysical Research, C, Oceans*, **94**: 8146–8162.
Oost, A.P., and Baas, J.H., 1994. The development of small-scale bedforms in tidal environments: an empirical model and its applications. *Sedimentology*, **41**: 883–903.
Paphitis, D., Velegrakis, A.F., Collins, M.B., and Muirhead, A., 2001. Laboratory investigations into the threshold of movement of natural sand-sized sediments under unidirectional, oscillatory and combined flows. *Sedimentology*, **48**: 645–659.
Potter, P.E., and Pettijohn, F.J., 1977. *Paleocurrents and Basin Analysis*. Second, corrected, and updated edition. Springer-Verlag.
Raudkivi, A.J., 1963. Study of sediment ripple formation. *Journal of the Hydraulics Division, Proceedings of ASCE*, **89**: 15–33.
Raudkivi, A.J., 1997. Ripples on stream bed. *Journal of Hydraulic Engineering*, **123**: 58–64.
Reineck, H.E., and Singh, I.B., 1980. *Depositional Sedimentary Environments, with Reference to Terrigenous Clastics*. Second, revised and updated edition. Springer-Verlag.
Richards, K.J., 1980. The formation of ripples and dunes on an erodible bed. *Journal of Fluid Mechanics*, **99**: 597–618.
Sorby, H.C., 1859. On the structures produced by the currents present during the deposition of stratified rock. *The Geologist*, **2**: 137–147.
Traykovski, P., Hay, A.E., Irish, J.D., and Lynch, J.F., 1999. Geometry, migration, and evolution of wave orbital ripples at LEO-15. *Journal of Geophysical Research, C, Oceans*, **104**: 1505–1524.
Werner, B.T., and Kocurek, G., 1999. Bedform spacing from defect dynamics. *Geology*, **27**: 727–730.
Williams, P.B., and Kemp, P.H., 1971. Initiation of ripples on flat sediment beds. *Journal of the Hydraulics Division, Proceedings of ASCE*, **97**: 505–522.

Cross-references

Angle of Repose
Bedding and Internal Structures
Bedset and Laminaset
Cross-Stratification
Desert Sedimentary Environments
Flow Resistance
Flume
Paleocurrent Analysis
Sediment Transport by Tides
Sediment Transport by Unidirectional Water Flows
Sediment Transport by Waves
Sedimentologists
Surface Forms
Turbidites

RIVERS AND ALLUVIAL FANS

Introduction

A river is a natural stream of water that under the influence of gravity flows regularly or intermittently in a channel or channels toward a receiving basin, commonly an ocean or lake. Rivers are open systems in which energy and matter are exchanged with the external environment (Knighton, 1998). They are supplied with water almost entirely from precipitation (meteoric water) routed to the channel via overland flow, groundwater, swamps, lakes, snowfields, and glaciers, and they acquire most of their sediment load by dissecting uplands or by reworking unconsolidated debris from previous erosional events. Sediment deposited by rivers in subaerial settings is called *alluvium*. Consequently, rivers are divisible into *bedrock reaches* with rigid boundaries, and *alluvial reaches* that have a mobile boundary of relatively unconsolidated material, often with vegetation assisting channel stabilization. Rivers drain catchments (drainage basins or watersheds), are organized into complex patterns of trunk, tributary, and distributary channels, and commonly support adjacent floodplains that are inundated when the channels exceed bankfull capacity. Their headwaters are principally in upland areas and they may flow throughout the year (perennial) or cease for part of each year (intermittent if flow is seasonal and ephemeral if flow is irregular). Due largely to tributary contributions, river systems generally increase in discharge and channel dimensions downstream, however, in areas of permeable sediment or aridity, percolation, and evaporation may cause channels to reduce or disappear.

While only about 0.005 percent of continental water is stored in rivers at any one time (Knighton, 1998), rivers transport to depositional sinks the great majority of subaerially weathered and eroded sediment as well as dissolved material. River drainage networks occupy about 69 percent of the land area, transporting an estimated 19 billion tonnes of material annually, about 20 percent of which is in solution. Denudation rates and river load reflect primarily the effects of climate, relief, and tectonic uplift, as well as rock, and soil type, catchment size, vegetation cover, and human influence. Upland rivers are commonly bedrock systems that form a sediment *source zone* (Schumm, 1981). They are typically erosional with very limited sediment storage. Intermediate reaches usually form a *transfer zone* with disjunct areas of bedrock erosion and temporary deposition. Lowland rivers are dominantly alluvial. They lose competence to form a *sink zone*, depositing their load in floodplains, alluvial fans, lakes, desert basins (internal sinks), or in marine deltas, estuaries, lagoons, continental shelves, and ocean basins (external sinks). Many rivers have responded to sea-level fluctuations by cutting and filling their lowermost valleys.

Fluvial strata constitute about 10–20 percent of the Phanerozoic rock record, with many predominantly alluvial deposits in excess of 10 km thick. The oldest known fluvial deposits are in the Archean of southern Africa, where they date back to more than 3.5 Ga and reflect the emergence of continental crust and establishment of subaerial weathering. Alluvial detritus is formed principally of silicate gravel, sand, and mud, with volcanic, carbonate, and evaporite grains, and particles of gold, tin-bearing minerals, and other economic materials (see *Placers, Fluvial*). Some deposits contain huge boulders derived from rock falls and catastrophic floods in steeplands. Rivers that traverse peatlands and densely vegetated regions transport large amounts of plant material and dissolved and colloidal organic matter, and these blackwater streams contribute significantly to the carbon content of sediments in sink zones. Alluvial strata include a wide range of sedimentary structures mostly generated by unidirectional flow. Floodplain deposits are usually associated with soil formation and consequently, in combination, alluvial stratigraphy can reveal a detailed story of variable flow conditions and changing terrestrial environments over space and time.

Historical developments

The importance of river erosion and landscape formation was clearly recognized in classical times. Whereas Thales of Miletus (b. ~624 BC) thought that deltas building into the sea were evidence that water could change into sediment, Herodotus (b. 484 BC) identified sedimentation and described the alluvial plains of Egypt as the "gift" of the Nile. Plato (b. 428 BC) clearly recognized fluvial erosion and transportation. Aristotle (b. 384 BC) was the first scholar known to have conceptualized the modern hydrological cycle and he also identified river erosional and depositional processes, particularly those on the Nile and the silting up of river channels entering the Black Sea. Strabo (b. 64/63 BC) described floodplains as having been formed of sediment from their adjacent rivers. However, between Roman times and the Renaissance a long scientific interregnum prevailed in Europe, with a widespread belief that the Earth's surface had been shaped by catastrophic processes associated with the biblical Great Flood (the *Diluvialist* theory).

After the geological importance of hydraulics and river processes was rediscovered by Leonardo da Vinci, Agricola, Bernard Palissy, George Bauer and others in the 15th and 16th Centuries, the importance of fluvial processes through geological time was questioned again in the 18th Century due to a group lead by a German scholar, Abraham Werner. Known as the *Neptunists*, they proposed that the rock record represented deposition in deep oceans where powerful currents had carved valleys and ridges, restricting alluvial influence to modern rivers and their sediments. James Hutton in the 18th Century was the first post-classical scholar to clearly recognize and articulate the importance of modern-day processes to explain features visible in ancient rocks, the basis of his principle of Uniformitarianism. He saw that long-term fluvial erosion was the erosive mechanism active in the repetitive cycles of landscape denudation: *No trace of a beginning, no prospect of an end*. He argued that fluvially eroded debris from the land created most of the detritus appearing in ancient sedimentary strata, and hence he and his followers became known as the *Fluvialists*. Hutton's concepts were clearly espoused by Charles Lyell in the many editions of his seminal book *The Principles of Geology* (1830–1833).

It was not until the late 19th and early 20th Centuries that modern and ancient river deposits and fluvially eroded landscapes received widespread and detailed investigation, especially from North Americans such as J.W. Powell during exploration and settlement of that continent. G.K. Gilbert (1877) was the first to apply the concept of *dynamic equilibrium* to rivers, and W.M. Davis in the late 1890s proposed that the normal or fluvial cycle of erosion and landscape change that, although now not wholly accepted, fostered a widespread interest in modern geology and geomorphology. Davis was probably the first person to recognize braiding as a river pattern distinctly different from meandering. In 1925, J. Barrell demonstrated that sedimentary strata, hitherto widely thought to be deposited in marine environments, were in many cases terrestrial including fluvial. The outstanding study of the Mississippi River by H. Fisk (1944) described the fluvial geomorphology, sedimentology, and stratigraphy of a very large river system, Miall (1996) regarding this work as the first major advance in modern fluvial sedimentology.

Post-war fluvial facies analysis, strongly linked to drilling and exploitation of hydrocarbon reservoirs, led in the 1960s to J.R.L. Allen's comprehensive model for meandering-river deposits, the first for any environmental setting. Undoubtedly the most significant modern advances in understanding river form and process originated from research by L.B. Leopold, M.G. Wolman, S.A. Schumm, J. M. Miller and their associates in the 1950s and 1960s. R.A. Bagnold (see *Physics of Sediment Transport: The Contributions of R.A. Bagnold*) greatly advanced sediment transport theory, and A. Sundborg described in detail meandering river processes and floodplain sedimentology, providing important European contributions. Sediment dynamic experiments, following studies in the 19th Century by H.C. Sorby, allowed detailed facies interpretation in hydrodynamic terms (see *Flumes; Sediment Transport by Unidirectional Flows*). With the development of radiocarbon dating in the late 1940s and luminescence dating in the 1970s, the means of dating Quaternary alluvium became available, allowing fluvial sediments and paleoflow regimes to be related more precisely to wider controls, principally tectonic activity, eustasy, and climate change.

Theoretical underpinnings

It has been recognized since the 18th Century that quantitative analysis in fluvial studies has been of necessity a compromise between theory (based on "ideal nonviscous fluids") and empiricism (based on real viscous fluids), a problem known as the d'Alembert paradox. Despite major advances, the development of unifying theories in fluid mechanics and sediment transport remain elusive. Empirical research into stochastic relationships has shown that, as flows vary, rivers construct highly predictable channel forms and sedimentary structures. However, a truly rational or deterministic explanation for such consistency has not been possible due to a lack of mathematical closure, for while there are four flow variables (width, depth, velocity, and slope) there are only three determining equations (continuity, resistance, and sediment transport). Solutions have been sought by adopting *extremal hypotheses*. If fluvial and thermodynamic systems are compared, rivers may be visualized as adopting a *most probable* state by minimizing both the variance of the system's flow components and the total work done. More strongly physically-based extremal hypotheses have been proposed by imposing assumed conditions such as maximum sediment transport rate or minimum stream power. In a recent reassessment of some of these approaches, Huang and Nanson (2000) have demonstrated mathematically that straight reaches of alluvial rivers appear to operate at maximum efficiency and illustrate the basic physical *Principle of Least Action*. However, such theoretical analyzes remain contentious.

Drainage networks

Drainage basins are usually characterized by interconnected lengths of river channel that converge from the elevated basin margins toward a trunk stream that exits at the lowest point. The resulting combination of channels forms a *drainage network* (Figure R7). Drainage networks refer to the pattern of the network of stream channels within a basin or sub-basin,

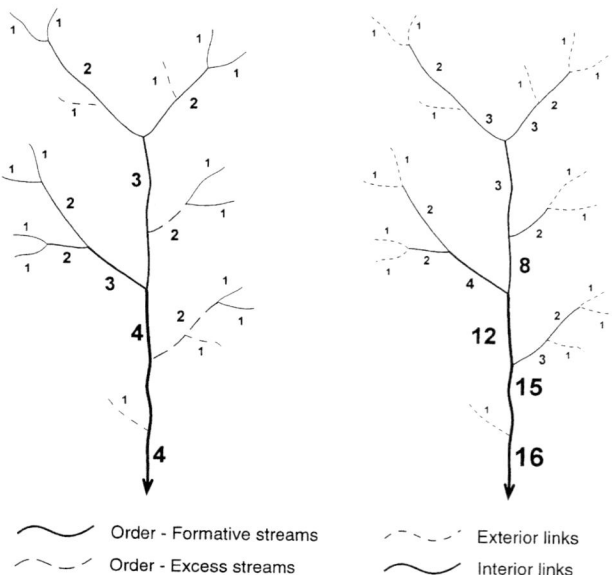

Figure R7 Drainage-basin stream networks with stream ordering: (A) as by Horton (1945) (modified by Strahler); (B) channel links ordered by magnitude (after Knighton, 1998).

and should not be confused with *stream patterns* that refer to the planform appearance of individual reaches of river (e.g., meandering and braiding, see below). They are usually erosional features and have been mapped in the stratigraphic record where unconformable surfaces separate bedrock from overlying alluvium. Headwater streams are normally relatively small, steep, often bedrock controlled with uneven gradients, and the bases of the hillslopes are connected directly to the stream channels such that rates of channel incision and sediment removal relate closely to slope processes. Downstream, the basin channels are of lower and more uniform gradient, and are commonly *self-formed*, being shaped within their own alluvium. Although in places they may connect with the hillslope, it is more common that alluvial terraces and floodplains act as a temporal and spatial buffer between channel and slope processes. Channel links are a length of channel from the source to the first tributary (*external link*) or between two successive tributaries (*internal link*). In 1945, Horton developed a system of ordering channel links within drainage networks in a downstream direction (Figure R7(A)). This provides a useful classification of the position of streams within a network, although link order is not a reliable indicator of channel size or discharge. To determine *link magnitude*, each link is assigned the value of the sum of all the external links supplying it, providing a useful surrogate to link discharge (Figure R7(B)).

In terrain of homogeneous or alternatively highly complex geological structure, drainage networks tend to take on a strongly irregular *dendritic* pattern. However, where well-defined and relatively simple geological structures are present (faults, folds, domes, volcanic cones, rectangular joints etc.), visually distinctive and somewhat regular drainage patterns can form. Drainage networks are highly amenable to numerical analysis and in consequence they have been studied in great detail in attempts to interpret their controlling variables, internal relationships, sediment yields and their relationship to both catchment and overall landscape evolution. However, because most networks are dendritic, their essentially irregular pattern means that few physically definitive relationships have been identified, except perhaps those from topological studies that have pointed out the spatially nonrandom elements present.

The most widely used and probably useful statistic derived from drainage network analysis is that of *drainage density* (the total length of channels per unit area of the basin). It expresses the degree of basin dissection and sediment production and transport by surface streams, and is influenced by such variables as climate, rock, and soil type, vegetation density, and landuse. Highly impermeable, soft, and poorly vegetated rocks can produce high drainage densities and sediment yields (e.g., badlands). Drainage density is also broadly correlated with precipitation. It is low in very arid areas, greatest in semi-arid areas where there is sufficient rainfall to cause erosion but insufficient vegetation to impede it, and lower again in more humid areas with denser vegetation. It appears to partially rise again in very wet areas where the stabilizing effects of even dense vegetation are overcome.

Stream gradients

Stream gradients are the result of two broad causes. An independent gradient is imposed on the stream by the gross valley form, the product of geological and drainage history. However, an adjustable and therefore dependent component of gradient develops from the interaction of stream discharge, width, depth, velocity, sediment size, sediment load, boundary roughness, and path sinuosity. Mackin (1948) stated that a *graded or equilibrium stream* (see below) is one in which, over a period of years, slope is delicately adjusted to provide, with available discharge and prevailing channel characteristics, just the velocity required for the transportation of the sediment supplied from upstream. Confined upland streams are generally not equilibrium alluvial systems because their gradients are largely imposed by a bedrock valley that is the product of ancient stream erosion or forces such as glaciation and tectonism. It is in the middle or lower reaches where the valley is wider that the stream can more readily adjust its gradient by altering its sinuosity and hence its path length in response to contemporary conditions. Change in channel slope accounts for only part of the adjustment of alluvial rivers to their controlling factors and, more realistically, the channel's various morphological and hydraulic parameters tend to show mutual adjustment.

Over their full distance, the longitudinal profiles of natural rivers show a strong tendency to become concave upward. This concavity is associated with at least three possible causes, although the directions of cause and effect are difficult to resolve. Firstly, there is a tendency in most rivers for discharge to increase downstream but for velocity to remain fairly constant. Discharge (a volume) increases as a cubic function whereas the resisting boundary of the channel (the area of the bed and banks) increases only as a squared function. If the gradient did not decline substantially an imbalance between impelling and resisting forces would result and the flow would accelerate rapidly downstream. Profiles tend to be most concave where discharges increase downstream most rapidly. Secondly, grain size commonly declines downstream due to clast abrasion (see *Attrition, Fluvial*) and size sorting, so the gradients required for sediment entrainment and transport will decline downstream. Stream profiles are more concave where clast sizes decline in size downstream most rapidly and show little concavity where grain size is constant or increases. Slope decreases most rapidly with a decline in gravel size and least with a decline in sand size. Because of the production of finer clasts, streams over shale tend to have gentler gradients than those over sandstone or limestone, although concavity depends on the combined effects of grain size and discharge. A third possible reason for profile concavity is that antecedent relief and potential energy conditions along a river from headwaters to mouth cause random-walk models to develop concavity as the *most probable* profile. In relation to this proposal it has been found that shorter streams tend to have less concave profiles and streams with greater relief exhibit greater concavity.

Marked changes in river gradient are termed *knickpoints*, and may reflect the presence of more resistant bedrock, changes in sediment load (for example, from tributary streams), tectonic activity, or base-level changes in the past. Eustatic or other base level changes tend not to be propagated over long distances upstream unless the stream is confined and can not readily adjust other parameters such as sinuosity.

Biota and soils

Prior to the mid Paleozoic, subaerial erosion was dominantly physical and produced abundant coarse material, with river forms probably reminiscent of today's active glaciofluvial or volcanic landscapes. In the late Silurian and Devonian the evolution of terrestrial plant communities and associated soils greatly enhanced chemical weathering and the production of clays. The development of cohesive stream banks and stabilizing root systems must have changed rivers dramatically in the late Paleozoic, promoting the accumulation of substantial argillaceous and organic deposits in terrestrial environments and strongly affecting global patterns of carbon production and storage. The development of terrestrial vegetation also reduced runoff and allowed the diversification of a complex array of floodplains and associated soil types in environments ranging from rainforests to grasslands (after the mid Cenozoic) and deserts. The first peats (precursors to coal) were deposited in the Devonian and peat has subsequently been an important component of fluvial strata. The rapid succession of plant communities in response to climate change, termed *floral overturn*, should have exerted an important influence on sediment flux to depositional basins.

There is now appreciation of the importance and complexity of river-vegetation interactions. Particular attention has been given to the effect of vegetation growing within channels on flow resistance, of bankline vegetation and debris dams on bank strength and channel morphology, and of riparian vegetation generally on floodplain and channel sedimentation and erosion. Wildfires destroy vegetation and promote landscape degradation, greatly increasing sediment yield and adding charcoal to the fossil record.

Floodplain soils can be diagnostic of particular environments, containing evidence in the form of pollen and spores, invertebrate remains, burrows, rhizomorphs, evaporites, and desiccation cracks (see *Weathering, Soils, and Paleosols*). For this reason and because soils develop through close interaction with the atmosphere and the groundwater prism, they are important repositories of information about past climates. Large animals, including hippopotamus, beaver, and perhaps dinosaurs, have affected rivers by trampling sediments, damming channels, and creating paths across banks that can lead to avulsion.

River hydraulics and sediment entrainment

Flow in open channels is subject to two roughly opposing principal forces. Acceleration due to gravity (g) acts to move water and sediment downslope with its effectiveness modified by stream gradient (S), while flow resistance (channel roughness) (see *Flow Resistance*) opposes this downslope motion. The interaction of these two forces ultimately determines the ability of flowing water to erode and transport sediment, and thereby generate landforms. Flow in rivers is almost always turbulent. The flow boundaries including the water surface are deformable, thereby changing the shape, roughness, and energy conditions in the channel as the flow changes. Frictional losses are the product of boundary roughness, internal distortion, and changing flow phases (super- to subcritical). All these factors make fluid mechanics much more complicated than solid mechanics, however, relatively simple empirical formulas have been developed to obtain approximate relationships suitable for most straightforward analytical procedures.

River discharge (Q):

$$Q = wdv \quad \text{(Eq. 1)}$$

where w = channel width; d = mean depth; v = mean velocity. This is known as the *flow continuity equation*.

Mean boundary shear stress (τ):

$$\tau = \gamma R S \quad \text{(Eq. 2)}$$

where γ = specific weight of water; R = hydraulic radius $\sim d$, ($\gamma = \rho g$, where ρ is water density). This is known as the *Du Boys equation*.

Channel roughness or flow resistance is commonly obtained from one of two equations:

Manning equation:

$$n = 1/v\, R^{2/3}\, S^{1/2} \quad \text{(Eq. 3)}$$

Darcy–Weisbach equation:

$$f = 8 g\, RS/v^2 \quad \text{(Eq. 4)}$$

The work done by a river can be expressed in terms of stream power in two forms:
Power per unit length of the channel:

$$\Omega = \gamma Q S \quad \text{(Eq. 5)}$$

Power per area of the bed:

$$\omega = \gamma Q S / w \quad \text{(Eq. 6)}$$

While suspended and dissolved sediment load can be relatively easily measured the more complex problem of determining bed load is usually attempted using one of many available transport equations, none of which are reliable over a wide range of conditions (see *Numerical Models and Simulation of Sediment Transport and Deposition*; *Sediment Fluxes and Rates of Sedimentation*). These equations can be classified into three

basic types (Knighton, 1998):
Excess shear stress ($r_o - r_{cr}$):

$$q_{sb} = X' r_o (r_o - r_{cr}) \quad \text{Du Boys type} \quad \text{(Eq. 7)}$$

Excess discharge ($q - q_{cr}$):

$$q_{sb} = X'' s^k (q - q_{cr}) \quad \text{Schoklitsch type} \quad \text{(Eq. 8)}$$

Excess stream power ($\omega - \omega_{cr}$):

$$q_{sb} \approx (\omega - \omega_{cr})^{3/2} d^{2/3} D^{-1/2} \quad \text{Bagnold type} \quad \text{(Eq. 9)}$$

In the above, q_{sb} is bed load per unit channel width, X' and X'' are sediment coefficients, D is grain size, q is flow discharge per unit width of channel and the subscript "cr" denotes the critical condition for sediment motion.

In a straight river cross section, the flow velocity is usually fastest at or just below the surface near the center of the channel and declines toward the bed and banks, the flow field deforming through river bends (Figure R8). A narrow deep channel usually directs a relatively gentle velocity gradient to a fine-grained bed that requires low shear stresses for transport, and directs relatively steep gradients to banks that are often cohesive, well vegetated, and erosion resistant. Wide shallow channels tend to exhibit erodible banks and coarse and/or abundant bed load that requires high shear stress for transport, braided rivers being a classic type.

The erosion of cohesive material such as bedrock or partially indurated alluvium occurs in three ways. *Corrosion* is the chemical dissolution of rock in water and can be important where rocks are readily soluble (e.g., limestone) or where water is chemically active. *Corrasion* occurs when rock is mechanically braided by water armed with particles either in transport or stationary but vibrating on the rock surface. *Cavitation* results from the production of shock waves due to the formation and implosion of vapor bubbles on rock surfaces at very high velocities. This is a rare process in most subaerial streams but can be common in subglacial or underground streams where flow is confined.

The entrainment of noncohesive sediment from the boundary is a complex process but largely a function of flow velocity and grain size. In the case of gravels, the sorting, and packing arrangements of the clasts on the bed are also important, for an armored or imbricated bed can be difficult to entrain (see *Armor*). As flow velocity increases two dominant forces act to entrain individual clasts. A *drag force* caused by friction between the flow and the surface of the particle causes it to move downstream and to slide or roll (or pivot) over any obstructing particle. However, as velocity increases a pressure gradient develops between the top and bottom of the grain setting up a *lift force* that acts to lift the particle out of its recess on the bed. The lift force decreases rapidly away from the bed but once the particle is elevated it is far more susceptible to the drag force and to turbulent eddying within the body of the flow that acts to keep particles in motion. The entrainment of sediment on the boundary can be expressed in terms of mean flow velocity and mean grain size, but the problem is to define the threshold velocity for motion on an imbricated or armored bed, or where grain size is highly variable.

The competence of historic or prehistoric flows is commonly judged from the size of the largest particle in the resulting deposit, although such estimates may suffer from the limited availability of coarse sediment. Fine-grained, cohesive deposits require much higher flow velocity for entrainment and transport than the size of individual grains would suggest. Such materials include bed and bank material, sand-sized mud aggregates reworked from soils, and flocs of aggregated clay, organisms, and detrital organic material in rivers and estuaries (e.g., Gibling *et al.*, 1998). Because aggregates and flocs are commonly destroyed by compaction during burial, flow competence may be difficult to assess for some fine-grained material, which may have originated as sand-sized bed-load rather than from settling of suspended flakes.

Equilibrium and threshold theories

Because alluvial rivers are open systems with mobile and deformable boundaries, they have the ability to self regulate. If perturbed, they will often return to something like their original condition (*homeostasis*), or they will adopt a new form but one that minimizes the effects of the original change. This reflects *dynamic equilibrium*, a condition first described for rivers by Gilbert (1877) and applied widely in science, especially chemistry (Le Chatelier's Principle). Because rivers are relatively slow to adjust and because their morphological parameters adjust at different rates, it can be difficult to determine if a particular reach is in an equilibrium or balanced condition. Nevertheless, the recognition that alluvial rivers adjust toward equilibrium conditions has become a basic tenet, almost a paradigm of modern fluvial research. Richards (1982) has suggested that equilibrium conditions in rivers can be identified on the basis of four criteria: (1) essentially stable relationships between form and process; (2) continuity of sediment transport in a reach over time and space; (3) strong correlations between system variables; (4) an adjustment to maximum operational efficiency. In many studies roughly balanced conditions have been shown to exist over short periods between erosion and deposition in a single reach, or between width, depth, and velocity. If one variable is altered, the others adjust in a way that minimizes the overall change and maintains a balanced condition. Because flow conditions and sedimentary bedforms are usually in dynamic equilibrium, ancient flow structures permit an accurate interpretation of paleoflow conditions.

Most changes in rivers are the result of external factors such as changes in base level, climate, vegetation, or sediment supply, and occur gradually. Rivers in dynamic equilibrium generally resist change, however, Schumm (1973) has shown that an *extrinsic threshold* can be reached when a progressive

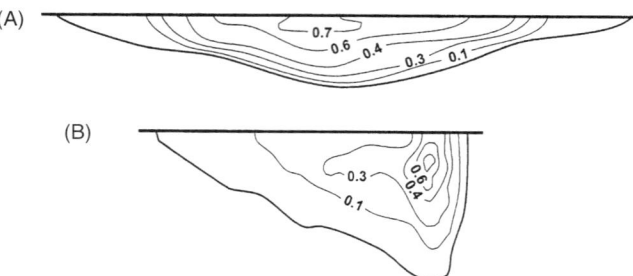

Figure R8 Velocity fields in cross sections of: (A) a wide shallow channel (note the steep velocity gradient to the bed); (B) a narrow deep channel bend viewed downstream and curving to the left (note the steep velocity gradient against the outer cutbank).

change in an external variable triggers a sudden change in the system's response. At an imposed critical change of slope or sediment load, a meandering channel can change abruptly to a braided channel. Similarly, a gradual and progressive increase in the flow velocity will suddenly achieve the threshold for sediment entrainment, after which the whole streambed becomes mobile. There are numerous thresholds above critical entrainment when bedforms change from a motionless bed, to ripples, to dunes, to transitional, and to super-critical flow bedforms (see *Surface Forms*). Schumm (1973) also showed that changes may be initiated intrinsically when, with no external change, one of the variables reaches a critical condition (an *intrinsic threshold*). While operating within its normal range of water discharge and sediment load variation, a river may reach a point of incipient instability when a threshold is crossed and a sudden internal change is initiated. A meander cutoff is an example where gradual, ongoing adjustments to equilibrium conditions prevail until a threshold is reached, with breaching of the meander neck and a dramatic change in local sinuosity and gradient (Knighton, 1998).

Crossing an intrinsic threshold will normally produce only a localized change recognizable in the stratigraphic record as, for example, an avulsion channel or crevasse splay.

Hydraulic geometry, regime theory and dominant discharge

Channel geometry is the cross-sectional form of a stream channel (width, mean, or maximum depth, cross-sectional area) fashioned over a period of time in response to formative discharges (approximately bankfull) and sediment characteristics (Figure R9). Because the above three geometric parameters and the additional four flow parameters (velocity, water-surface slope, flow resistance, and sediment concentration) vary with discharge, the term *hydraulic geometry* is used to describe the relationships of all seven parameters to discharge as the independent variable. Discharge changes can be measured at-a-station as the channel fills during a flood, or at bankfull in the downstream direction. There are significantly different relationships for *at-a-station* and *downstream*

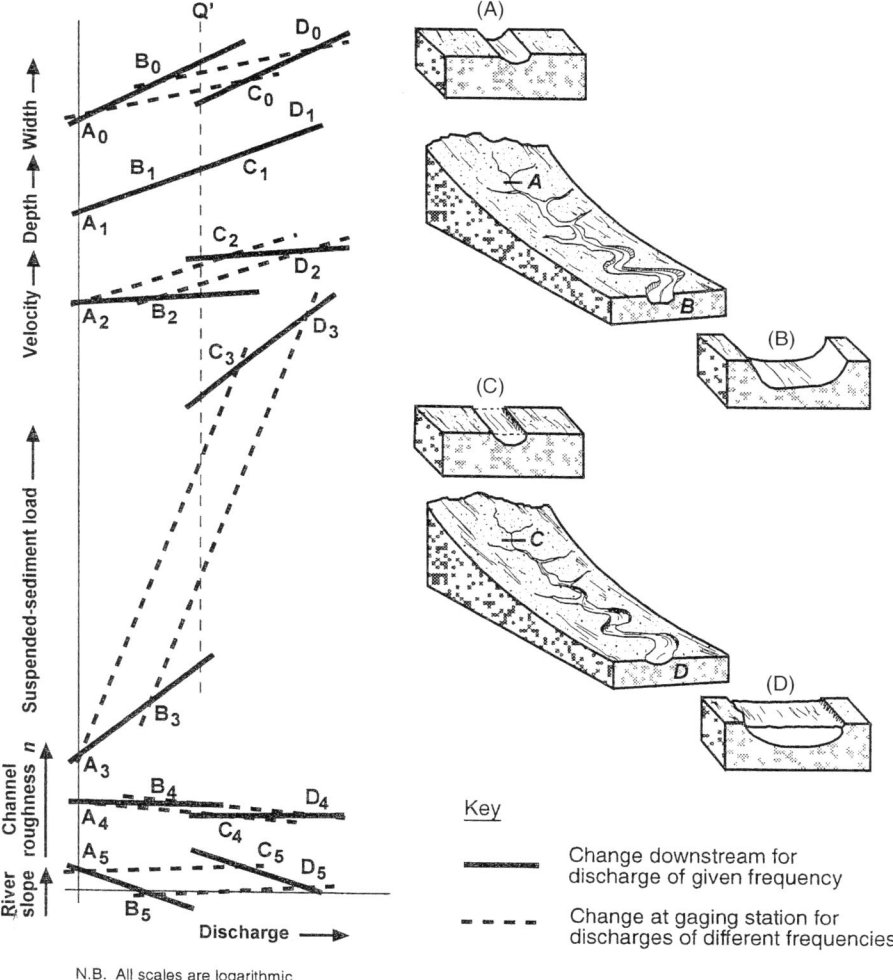

Figure R9 Hydraulic geometry relationships of river channels, comparing variations in width, depth, velocity, suspended load, toughness, and slope to variations in discharge, both at-a-station and downstream (after Leopold and Maddock, 1953).

hydraulic geometry (Figure R9). Holding discharge constant, at-a-station hydraulic geometry is largely controlled by variations in bank strength and available sediment load. Streams with low sediment loads and cohesive or well-vegetated banks tend to be relatively narrow and deep whereas those with abundant loads and weak banks tend to be wide and shallow. However, because bank strength has only a moderate range but river discharges vary by many orders of magnitude, hydraulic geometry is remarkably consistent across the full range of river discharges (Figure R9). Because channel depth is greatly restricted by limited bank strengths, rivers increase in width relative to their depth as their size and discharge increases—a prominent downstream tendency (Church, 1992).

Research during the construction and maintenance of irrigation canals with near-constant discharge in India and Pakistan in the late 19th and early 20th Centuries independently determined very similar hydraulic geometry relationships to those in natural rivers, engineering findings that have been termed *regime theory*. Consequently, stable alluvial rivers exhibiting consistent and predictable hydraulic geometries are said to be *in equilibrium* or *in regime*.

Hydraulic geometry shows that river channel dimensions are closely adjusted to water discharge. However, discharge varies from perhaps no flow in droughts through to catastrophic flood events: so which discharge(s) define the channel's characteristics? Wolman and Leopold (1957) showed that in the USA, bankfull flows occur with the surprising regularity of about once every 1–2 years across a diverse range of rivers, something that would be an extraordinary coincidence if bankfull flows did not in themselves play a large part in determining channel dimensions. It has also been shown that with increasing at-a-station discharge, flow velocity tends to increase until near bankfull flow conditions and then stabilizes at higher discharges because of a marked increase in roughness near the bank crests and over the floodplains. In other words, most flows beyond bankfull are not notably more effective in altering the channel and transporting sediment than is bankfull flow. Furthermore, while in some cases exceptional floods may undertake significant work in the form of sediment transport and channel reconstruction, they are sufficiently rare that on an average annual basis, they usually achieve far less than do smaller but more frequent events of about bankfull (Figure R10). However, there is evidence that in a limited number of environments (e.g., pro-glacial streams, confined gorges, and some dryland rivers), catastrophic events can achieve substantial and long-lasting changes. In other words, here these become the dominant discharge. Extreme, high-velocity events cause considerable channel enlargement, followed by a long period of "recovery" from smaller flows. Thus, channel dimensions at a given time in such an environment may reflect the period that has elapsed since the last extreme event, which also terminated "memory" of the pre-existing channel geometry.

Sediment transport

As discharge increases, detrital sediment transport in alluvial channels mostly occurs when the flow conditions (largely velocity) exceed the critical condition for entrainment of sediment at rest on the bed and banks (see *Numerical Models and Simulation of Sediment Transport and Deposition*). The flow conditions required to entrain stationary sediment are greater

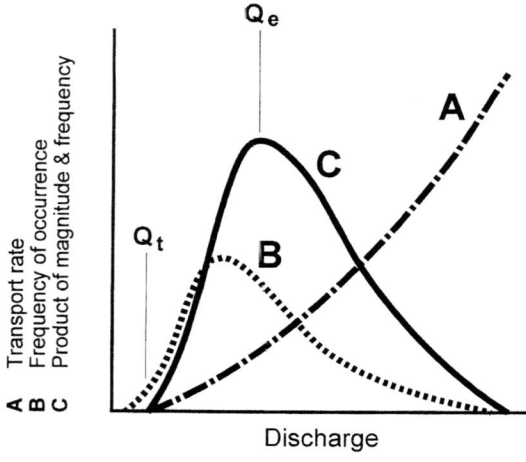

Q_e Most effective transporting discharge
Q_t Threshold discharge

Figure R10 Dominant discharge. Curve A is the transport rate of supended load rising with discharge. Curve B is the frequency of the full range of possible discharges. Curve C is the product of curves A and B and shows that the most effective discharges for transporting a river's load are moderate floods, generally occurring about once every 1 to 5 years.

than those required to keep sediment moving. Rivers transport their sediment load in essentially four ways. *Bed load* or *traction load* is almost constantly in contact with the bed, *saltating load* bounces or skips over the bed (effectively intermediate between bed load and suspended load), *suspended load* is held in the water column by turbulence and *dissolved load* is transported in solution. The first three are composed of detrital sediment, and the concentrations and relative proportions of the first two are highly dependent on the energy of the flow, which increases with water discharge. Suspended load is energy dependent but can also be strongly influenced by the rate of sediment supply to the river. The concentration of dissolved material depends largely on water temperature, catchment geology, groundwater chemistry, and vegetation. In contrast to detrital load, dissolved load commonly decreases in concentration due to dilution effects with increasing flood discharge. The three detrital types are the primary constituents of alluvial strata, with bed, and saltating load commonly forming sedimentary flow structures diagnostic of variable flow conditions. Soluble matter, once in solution, is mostly flushed from the catchment, being precipitated only under unusual geochemical conditions. Because of the difficulty of separating different sediment types on the basis of transport, another commonly used classification defines the detrital sediment load in terms of the bed composition prior to entrainment. *Bed-material load* is that which is contained largely within the bed material at rest, whereas *wash load* is the muddy suspended component not found in significant quantities in the bed (it is derived largely from fines in the stream banks, floodplains, and valley side-slopes).

Yields of suspended sediment and dissolved material are usually closely related to water discharge and are therefore relatively easily measured. In contrast, bed load (especially coarse bed load) is transported in a highly variably fashion, both spatially, and temporally, and is therefore extremely difficult to measure accurately, especially in large rivers. Bed-load is mostly (but certainly not entirely) a function of the highly variable shear stresses acting on the bed, and is often estimated from average flow and sediment characteristics. Because bed load is the coarsest fraction and moves mostly short distances during relatively infrequent, high magnitude flows, it is commonly the smallest proportion of transported sediment (often <10 percent of the total load). It is therefore sometimes ignored, however, this overlooks the importance of bed-load in determining river form, as well as its contribution to the stratigraphic record—in laterally migrating and accreting systems sediment deposited within channels has an especially high probability of being preserved. The *capacity* of a river to transport bed-load is usually limited, but few flows reach their limit for transport of suspended load, the concentration of which is largely controlled by the rate of supply.

Reliable relationships have been determined between sedimentary structures, sediment size, and flow velocity (or shear stress) (see *Surface Forms*; *Sediment Fluxes and Rates of Sedimentation*). Bed roughness changes in a complex but predictable fashion with increasing velocity and sediment transport. It reaches a maximum with dune formation under lower regime flow and antidune formation under upper regime flow.

Sediment character has a profound influence on river style. As a consequence, Schumm (1960) developed a classification of rivers based on bed load, mixed load and suspended load systems, with width:depth ratios of >40, 10–40, and <10, respectively. His approach has influenced later classifications of rivers, floodplains, and stratigraphic sequences.

Sediment deposition

Alluvium results from fluvial sedimentation. This takes place for suspended or saltating sediment when the flow velocity drops below that of the settling velocity (the velocity of a particle falling through a column of still water), and for bed load when flow velocity drops below that maintaining sliding or rolling of particles over the bed. When velocity wanes or decreases locally within the channel or on the floodplain surface, the coarsest fractions are deposited first. As a result, sediment sizes are sorted vertically, and laterally within the total alluvial system. On laterally migrating meandering rivers, upward fining successions within point bar and floodplain deposits result from flow velocities that decline from near the deepest part of channel (the thalweg) and adjacent point bar (depositing gravel or coarse sand), to the upper point bar, convex bank, and floodplain surface (depositing fine sand and mud). In braided rivers, coarse braid bars characterize the lowermost deposits, and with braid-channel, and braid-bar migration or abandonment, overbank fines, and channel fills characterize the uppermost deposits. Adjacent to laterally stable channels, or on floodplains away from the zone of active channels, floodplain strata broadly fine upward as each successive stratum makes the surface higher and less accessible to overbank flooding. However, in detail such upward fining successions are usually complex, exhibiting numerous size variations, and reversals resulting from temporal and spatial variations in flow conditions within individual floods and between successive floods. Secondary currents are known to play a major role in producing the broad spatial variations in sedimentation in bends, bars, and floodplain scrolls, as well as in innumerable smaller flow structures.

In a genetic classification of different floodplain types, Nanson and Croke (1992) identified six floodplain-forming processes: lateral point-bar accretion, overbank vertical accretion, braid-channel accretion, oblique accretion, counterpoint accretion, and abandoned channel accretion. Both geomorphologically and sedimentologically, the major distinction is between coarser bar deposits laid down by the lateral shifting of channels, and fine-grained vertical accretion deposits resulting from overbank sedimentation, but there are many varied combinations of all six. Based largely on the interrelated properties of stream power and sediment character, Nanson, and Croke identified three floodplain classes: (1) high-energy noncohesive; (2) medium-energy noncohesive; and (3) low-energy cohesive (Figure R11). To these they added 13 derivative orders and suborders that are based largely on the floodplain forming processes. Miall (1996) incorporated this essentially geomorphic model with his 1985 architectural element analysis to develop a detailed system of fluvial sedimentary facies and styles (see *Floodplain Sediments*). Although generally viewed as depositional elements, floodplains can also be subject to periods of intense erosion during flood episodes, resulting in *catastrophic stripping*.

River deposits include a wide range of sediment types, but a distinctive suite of facies is common to many channel and floodplain settings. The reader is referred to Miall (1996) for a detailed description of fluvial facies.

Channel patterns: their geomorphology and sedimentary facies

Rivers respond to imposed discharges and sediment conditions by adjusting their channel pattern or planform in conjunction with their hydraulic geometry. Because river patterns are so easily recognized on air photos and maps they have become a primary basis for river classification (Figure R12). Generalizations from planform can be made in terms of lateral stability, sediment load, sediment size, bed/suspended load ratio, width/depth ratio, floodplain type, and stratigraphy.

Leopold and Wolman (1957) proposed the first widely adopted geomorphological classification with their concept of a continuum of river patterns between end-members of *straight*, *meandering*, and *braided*. A significant problem is that these are not mutually exclusive; for example, many braided channels have a sinuous (meandering) planform. Nevertheless, in an effort to retain the continuum approach, more recent classifications have essentially adopted the original tripartite system but with additional types added, the most distinctive being low energy, fine-grained multiple-channel rivers termed *anastomosing*. The existence of a broad continuum of river patterns is evidenced in slope-discharge plots that, in order of declining stream power and grain size, separate rivers into braided, meandering, straight, and anastomosing (Figure R13). However, to date there is insufficient evidence to determine whether straight and anastomosing channels can be separated from each other on this basis alone. The concept that stream power and sediment-caliber control the continuum has lead some to argue that the relative ease of eroding and transporting bank material determines channel patterns (Brotherton, 1979). Both meandering and braiding patterns appear to reflect a need to consume excess energy were the valley slope is greater than

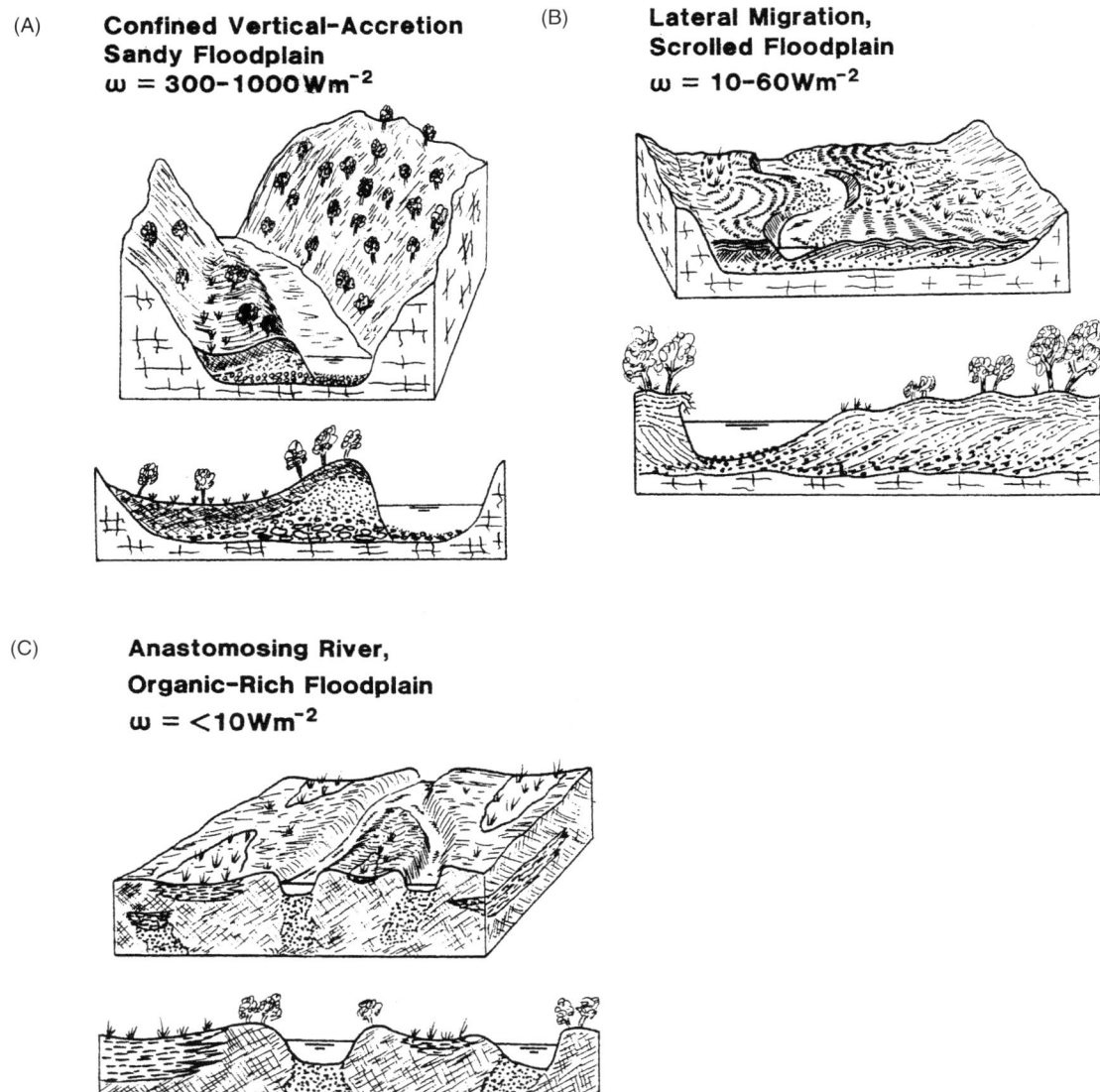

Figure R11 Examples of Nanson and Croke's (1992) three main floodplain classes: (A) A High-energy noncohesive floodplain; (B) A medium-energy noncohesive floodplain; (C) A low-energy cohesive floodplain.

that required for an equilibrium channel slope (Bettess and White, 1983), but the issue remains unresolved.

The term *anabranching* describes rivers that flow in multiple channels separated by stable, vegetated, alluvial islands that divide flows up to bankfull regardless of their energy or sediment size, with anastomosing rivers simply a low-energy fine-grained type (Nanson and Knighton, 1996) (see *Anabranching Rivers and Anastomosing Rivers*). Importantly, neither term is now used as a synonym for braided rivers in which the flow is divided by braid bars overtopped below bankfull. Individual anabranches can be straight, meandering, or braided. The term *wandering* river has become popular to describe anabranching rivers intermediate between meandering and braided. It is important to recognize that the association between river planform and sedimentology is not always clear, especially in transitional zones (Brierley and Hickin, 1991).

These classifications apply mainly to self-formed, alluvial channels. Church (1992) noted the problem of including rivers from mountains to basins within one classification scheme. Taking this problem into account, he divided channels into small, intermediate, and large categories based not on channel dimensions but on the relationship between grain diameter (D) and depth (d). In *small channels* (D/d>1.0), individual boulders are significant form elements, leading to irregular steps and pools. In *intermediate channels* (D/d 0.1–1.0), flows are often wake-dominated, with a variety of

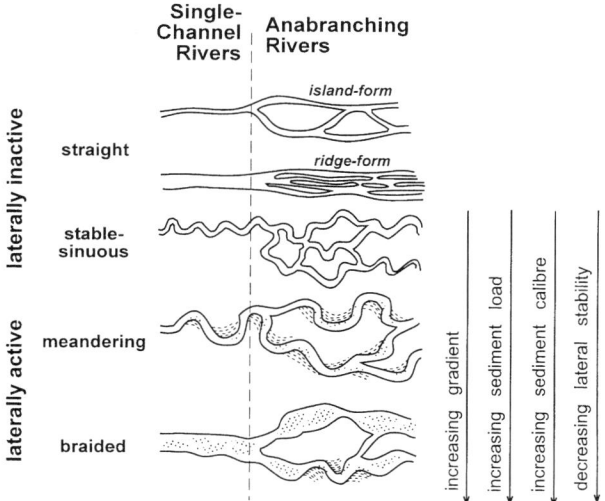

Figure R12 The common range of river patterns in single-channel and anabranching rivers. There is a trend of increasing gradient and sediment calibre and load from laterally inactive, to meandering, to braided (modified from Nanson and Knighton, 1996).

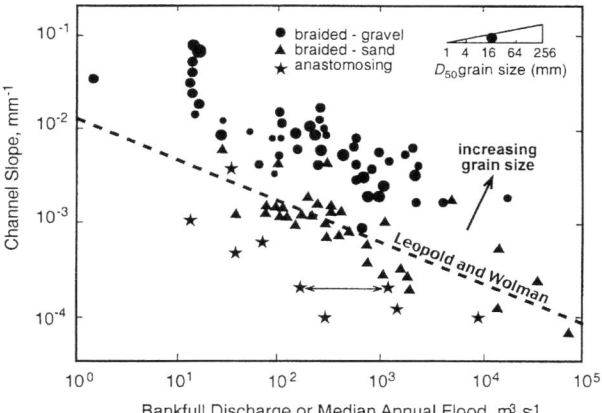

Figure R13 Gravel braided, sand braided, and anastomosing channels differentiated on slope-discharge, with median grain size (D_{50}) (after Knighton, 1998).

pools, riffles, rapids, and bars, as well as steps, and jams caused by logs and branches. *Large channels* (D/d<0.1) are dominated by the water-flow regime, and exhibit deep shear flow, and a well-defined velocity profile, with bedforms that dissipate energy. The channel types described in the previous paragraphs are typically large channels following this scheme, and their fills predominate in the fluvial record. Small and intermediate channel fills are probably important in the rock record where unconformities include bedrock valleys with very coarse sediment.

Braided Rivers are relatively high-energy systems with large width-depth ratios and at low stage have multiple channels that divide and rejoin around alluvial bars (see *Braided Channels*). They tend to occur in settings with steep gradients, weakly cohesive banks, abundant coarse sediment, and variable discharge (Leopold and Wolman, 1957; Knighton, 1998). Sediments are typically dominated by gravel and sand, but silt-dominated systems are also known. Braided rivers are prominent near mountain and glacial fronts, range from humid to arid regions, and commonly change downvalley into finer grained, single-channel systems, although some feed directly into oceans, and lakes as *braid deltas*. The Brahmaputra River, one of the world's largest braided systems, has a channel-belt width of up to 20 km, with individual channels several kilometers wide and maximum scour depths of up to 50 m, however, braiding can also occur in small streams crossing a sandy beach.

While processes and relationships between variables can be described, a widely accepted *rational* explanation for braiding has been elusive. Leopold and Wolman (1957) argued that braiding is an equilibrium adjustment to erodible banks and a debris load excessive for a single channel. Some workers have differentiated between braiding caused by the above, and that designed to consume energy in less-efficient multiple channels as a result of excessively steep valley gradients, following observations that degree of braiding may increase as slope steepens (Bettess and White, 1983).

In a detailed flume study, Ashmore (1991) found that braiding can develop in four ways: (1) growth of a central bar without avalanche faces, commonly linked to stalling of coarse bed-load in the channel center (Leopold and Wolman, 1957); (2) scour at pools leading to deposition of a downchannel lobe, with stalling of bed-load sheets on the lobe top and subsequent development of a mid-channel bar with avalanche faces; (3) chute cut-off of mid-channel and point (bank-attached) bars, with the formation of a new, discrete channel; and (4) dissection of lobate bars by multiple channels. Others cite avulsion within the channel belt as a bar-forming process (Kraus and Wells, 1999). Repetition of these dynamic processes along the river produces a complex array of interacting channels and bars. In different environments, channels can be aggradational, vertically stable, and incisional but all situations are characterized by frequent shifts in channel position. Many changes in braided river morphology occur episodically, even at constant discharge, as a result of short-term fluctuations in the transport rate related in particular to bed-load pulses.

Friend and Sinha (1993) defined the intensity of braiding using a *braid-channel ratio*, B, which measures the tendency of a channel reach to develop multiple channels. $B = L_{ctot}/L_{cmax}$, where L_{ctot} is the sum of the mid-channel lengths of all primary channel segments in a reach, and L_{cmax} is the mid-channel length of the widest channel in the reach.

Braided river deposits in the rock record are typically composed of suites of channel fills or stories that are stacked to form complex (multistory) bodies dominated by sandstone and conglomerate. Bar deposits are identified as large-scale sets of inclined strata that tend to dip downstream and oblique to flow, although lateral accretion is common. The channel bodies range from a few meters to hundreds of meters thick, with some especially thick occurrences exceeding 1 km of coarse-grained strata, and they vary from small bodies embedded in floodplain mudstones to extensive, basin-scale sheets.

Meandering Rivers are usually moderate-energy, single-chaneled, and sinuous rivers with moderate width-depth ratios and gradients (see *Meandering Channels*). Their planforms can be described precisely in terms of wavelength, amplitude, and radius of curvature, although in reality their planform can be irregular. Included within the group are sinuous channels, with point bars developed at convex bends, and relatively straight channels within which the thalweg winds between alternate bars. Channel sinuosity is measured as the ratio of channel length to valley length with the term "meandering" applied when this ratio is greater than about 1.3. Such rivers contain relatively few bars that are not bank attached, thus having braid-bar ratios close to 1.0. Meanders are self-similar over a wide range of scales and their planform (especially width and wavelength) can be related to channel discharge (see Knighton, 1998, table 5.9), and these relationships have been used to determine paleodischarges from ancient meander traces (e.g., Smith, 1987).

"There is no general agreement as to how or why streams meander." (Knighton, 1998, p. 225). The regularity and widespread occurrence of meanders shows they are not simply the product of random disturbance. The "bar theory" proposes that periodic deformation of the bed is the fundamental cause of meandering, especially through the development of alternate bars that deflect flow into the banks. However, this does not explain the initiation of the alternate bars. Furthermore, meanders occur without sediment deposition in supraglacial streams cut into ice and in ocean currents. Using "probability theory" Langbein and Leopold (1966) proposed that meanders reduce stream gradients to an equilibrium slope for the transport of an imposed sediment load, producing a longer path length with minimum variance and minimum total work. This approach suggests that alternate bars and point bars are a response to controlling factors, not themselves the prime cause of meandering.

Meandering rivers tend to occur in settings with moderate gradients, cohesive and/or well-vegetated banks, mixed loads of sand (sometimes gravel) and mud, and commonly perennial flow. The concave banks in bends are typically erosional and adjacent to pools. The locus of the lowest point in the channel (the thalweg) changes to the opposite side of the channel at the riffle (sometimes called the crossover), a shallow zone in the long profile between each bend (pool). The convex side of each bend usually has a point bar with surfaces that slope gently into the adjacent pool. Point-bar surfaces generally dip at low angles ($\sim 4°$) but locally exceed $12°$, and their width is about two-thirds that of the channel. Meandering rivers are especially common in downvalley, lower-gradient reaches of rivers, and in estuaries they support bi-directional, tidal flow, and more complex bars.

As the outer bank of each bend erodes, the point bar migrates systematically laterally and downstream, usually balanced by erosional retreat of the concave bank. Lateral migration rates are most rapid where bends have radius of curvature to channel width ratios of about 2 to 3 (Hickin and Nanson, 1975). Where a meander tightens through time, the point bar may be partially truncated by erosion along chute channels (chute cutoff) or completely truncated with full isolation of the bend (neck cutoff), resulting in an oxbow lake. Repeated cutoffs and local channel-migration result in a meander belt composed of many laterally juxtaposed channel segments (Fisk, 1944).

Meandering-river facies and grain size distribution primarily reflect the pattern of helicoidal flow that develops around each bend (see *Meandering Channels*). As the channel migrates, sediment eroded from the concave bank is transported obliquely across the channel and up the opposing point bar, with flow waning progressively toward the bar top. Sediments grade and bedform scale decrease systematically up the bar surface (Allen, 1970). These lateral-accretion deposits are termed *inclined stratification*, or in mixed load streams, *inclined heterolithic stratification*. Point-bar accretion is commonly linked to scroll bars which can develop as a within-channel bar on the point bar platform, and through vertical and up- and downstream growth become effectively a levee bank around the convex bank of a meander bend. With lateral accretion, repeated scroll bar, and associated levee bank growth generate a distinctive ridge-and-swale or scroll topography on the floodplain surface.

Meandering rivers on the scale of the lower Mississippi may be expected to generate moderately wide (e.g., 10 km) and relatively thick (e.g., 10–30 m) meander-belt deposits (Fisk, 1944). These broad bodies (an order of magnitude wider than the parent channel) form by a combination of progressive lateral accretion, meander cutoff, and channel avulsion, and they consist of juxtaposed channel segments that can be identified by abrupt changes in accretion direction and paleoflow. Such bodies with a set of discrete segments or stories can be termed multilateral; however, stories within many channel bodies are stacked both vertically and laterally, and the bodies are best described as "multistory". Exhumed meander belts with mappable segments and cutoffs have been documented from well-exposed terrains (Smith, 1987). The majority of meandering-fluvial bodies documented in the geological record are less than 30 m thick and 6 km wide, with width : thickness ratios that commonly exceed 100 and rarely approach 1,000. In some instances, channel fills with point-bar deposits form components of much thicker, complex fluvial bodies, and valley fills.

Straight rivers rarely persist in an alluvial setting for a distance of more than 7–10 channel widths. While many rivers contain straight reaches, for the purpose of definition straight rivers are classed as those without significant bends for more than this distance (sinuosity of <1.1). Where compared to other patterns they are argued to be at the low end of the flow-strength to bank-strength ratio, so are usually fine grained. Flume experiments suggest that straight channels develop at very low gradients, an inference supported by the relatively straight channels (even prior to canalization) in the lower Mississippi River and Delta. Where naturally sinuous channels in readily erodible material have been artificially straightened, alternate bars usually form rapidly and subsequent bank erosion leads to the development of a meandering pattern. Some straight reaches reflect incision along tectonic lineaments which controlled the channel orientation.

Surprisingly, straight channels have never been specifically studied as a pattern type so very little is known about their geomorphology or stratigraphy, and there is no proposed facies model for straight rivers. Because of the relative absence of lateral activity, the predominant mechanism for alluvial sedimentation along straight reaches of river is likely to be vertical aggradation of both channel and overbank deposits, and a fixed-channel model (Friend, 1983) may be appropriate (see below).

Anabranching rivers are a system of multiple channels characterized by vegetated or otherwise stable alluvial islands which are either excised by avulsion (see *Avulsion*) from a previously continuous floodplain, or formed by the accretion of sediment in a previously wide channel. The islands divide flow up to bankfull (Nanson and Knighton, 1996). Anastomosing channels are low gradient, laterally stable, straight (most common) to highly sinuous, with low width-depth ratio and well-vegetated or highly cohesive banks.

Anabranching rivers include a wide range of settings and facies which greatly complicate attempts to explain the form and origin of these unusual rivers (see *Anabranching Rivers*). Because of their propensity in certain environments to accumulate sandy channels and fine-grained overbank muds and organics, anastomosing rivers can be important locations for the preservation of coal and hydrocarbons. An anastomosing reach of the valley-bound Columbia River of western Canada comprises a suite of mixed-load channels that avulse frequently through rapidly accreting organo-detrital wetlands, with prominent levees through which crevasse splays feed into floodbasins (Smith and Smith, 1980). Anastomosing systems of the Okavango Delta in Botswana are sandy bed-load channels with very subdued levees that transport low suspended loads through nonaccreting to very slowly accreting reed-swamps and lakes (e.g., McCarthy *et al.*, 1988). In stark contrast, rivers of the semi-arid Australian Channel Country are entirely detrital, mud-dominated systems, with high suspended load, subdued levees and insignificant floodbasins. They traverse broad alluvial plains that are accreting very slowly in a cratonic setting, and exhibit infrequent avulsions (Gibling *et al.*, 1998). On the Ganges plains in the Himalayan Foreland Basin, a rapidly accreting anastomosing reach of the Baghmati River experiences avulsion on a decadal scale, with frequent crevassing through levees onto monsoonally flooded plains. Anastomosing rivers were also prominent during the early Holocene history of the Rhine-Meuse and Mississippi deltas. However, in a detailed review, Makaske (2001) found no standard sedimentary succession for anastomosing rivers.

As a group, anastomosing rivers are not diagnostic of climatic conditions, ranging as they do through sub-arctic, alpine, temperate, wet tropical, and semi-arid settings (Makaske, 2001). However, individual types may well be diagnostic of certain climatic conditions. Because most rivers are single thread, the question arises as to why multi-thread anabranching rivers form at all. They appear to develop under conditions where there is insufficient gradient to move water and particularly bed-load through a single channel system (deltas are classically anabranched). However, as with other river patterns, a truly rational explanation is still sought.

Alluvial fans

Alluvial fans are fluvial depositional zones that are essentially fan-shaped in planview, convex upward in transverse section, and occur at the exit of a confining basin. Their location reflects a sudden decline in flow competence due to valley widening, a reduction in gradient, and a loss of water due to infiltration. Fans vary greatly in size but exhibit well defined relationships between their area and slope and the size and lithology of their catchments. They may exist as single features or coalesce to form a piedmont or *bajada* many kilometers in lateral extent where closely spaced rivers exit a mountain front. They typically extend for a few kilometers to tens of kilometers basinward, but some volcanoclastic fans and aprons that border active volcanic arcs extend for more than 100 km, on account of their huge yields of unconsolidated material, enormous discharge and long runout distances for debris and stream flows. Fans that debouch into an ocean or lake are termed *fandeltas*, with good examples in the Red Sea, Gulf of Suez, and East African rifts. In relatively arid environments, flow volume declines basinward, and the size, and proportion of channels decreases, with a concomitant increase in sheetfloods (sheets of unconfined water moving rapidly downslope).

Talus or *scree cones* are steep accumulations of debris at an abrupt break of gradient such as at the base of a cliff or at the exit of a small hanging gully. Although also fan-shaped, they are not alluvial, formed instead of loose, highly angular clasts transported subaerially by gravity, sometimes aided by snow avalanches. They are distinguishable from fans by being much steeper ($26°-36°$) and with sediment that *increases* in size down slope.

Large rivers such as the Gandak and Kosi that enter the Gangetic Basin from the Himalayan Front in India and the Okavango River in Botswana deposit sediment cones that extend hundreds of kilometers from the mountains. However, such low-gradient features are not considered by some to be alluvial fans. Blair and McPherson (1994) argued that there is a hiatus in the range of longitudinal slopes of alluvial distributary systems, with few fluvial depositional systems having slopes in the range 0.007–0.026. Most meandering and braided systems have gradients of <0.007, and they recommended restricting the term alluvial fan to systems with gradients of 0.026–0.466, as such fans are characterized by viscous debris flows and sheet floods. Stanistreet and McCarthy (1993) proposed a tripartite classification with end-members of *debris-flow dominated fans* with gradients of 0.1 to 0.001, *braided fluvial fans* such as the Kosi system with gradients of 0.001 to 0.0003 and *low sinuosity/meandering fluvial fans* such as the Okavango system with gradients of <0.0003. In an earlier classification, Schumm (1977) called debris-flow dominated fans *dry fans* and did not differentiate between the two types of fluvial fan that were called collectively *wet fans*. Low gradient coalescing alluvial surfaces have been termed *riverine plains*, as for example the Canterbury Plains of New Zealand and the Riverine Plain in the Murray basin of Australia.

The main advantage of the more prescriptive approach of Blair and McPherson (1994) is that fan deposits are potentially recognizable in the rock record through the predominance of debris-flow deposits with matrix-supported fabrics (dry fans). However, the Stanistreet and McCarthy (1993) approach more readily incorporates what have been widely regarded to date geomorphically as alluvial fans, while still retaining the debris-flow category as their steepest type. An additional problem with the Blair and McPherson approach of linking fans with debris flows is the strong overlap of mechanisms between end-members of debris and stream flows and the difficulty of distinguishing true debris-flow deposits in some cases. In some debris flows, larger clasts are not suspended but remain in contact during motion, generating deposits with clast-supported textures and low clay content. There has also been widespread recognition of hyperconcentrated flood flows, intermediate between debris and stream flows in hydraulic behavior and sediment properties (commonly clast-supported) and representing extraordinary floods with unusually large sediment volumes. As Miall (1996) noted, it remains to be seen which approach proves to be the most useful and widely adopted.

Dry fan development is known to be prevalent today, especially in arid upland regions with sparsely vegetated slopes subject to infrequent flash flooding. Wet fans were predominant in the geologic past, especially in glacial periods with abundant seasonal snowmelt, and are still active today in some humid upland regions with frequent runoff (Schumm, 1977). Because of the limited travel-distance and viscosity of debris flows, dry fans tend to be small, and relatively steep. Debris flow deposits in the near-source part of the fan are poorly sorted and often matrix-supported with angular clasts and outsize boulders. Running water reworks them through sheet and channel flow into a downslope progression of gravel and sand, with mud predominant near the fan toe. The lower fan can be characterized by braided stream deposition. Debris flows aggrade the upper fan, but fluvial action can periodically cause entrenchment, producing a complex topography. "Sieve deposits" of well-sorted gravel without matrix, identified as a distinctive sediment type on some fans, may be debris-flow deposits from which fines have largely been winnowed at the surface.

Wet fans can be small or large and show pronounced sediment-size sorting, clast rounding and traction-current deposition. Alternating aggradation and incision near the fan head produces coarse overbank sheetflow deposits and splays as well as trenched channels with longitudinal gravel bars. The mid fan consists of typically braided channels that switch back and forth depositing longitudinal gravel and transverse sand bars. The lower fan is characterized by transverse bars or dunes with overbank mud and the channels meander in places, as in the Kosi Fan of northern India. Experimental work by Schumm (1977) and Schumm et al. (1987) identify two main autogenic processes in the upper to middle part of braided wet fans: the avulsive shifting of distributary channels and a cyclic trenching and backfilling of channels. These alternations can be controlled entirely by intrinsic thresholds relating to sediment supply from the basin and deposition on the fan. During the trenching phases, coarse sediment is moved more effectively down the fan whereas with backfilling the fan aggrades and sediment is distributed across the surface by sheet flow. With fan building, segmentation can produce individual depositional zones or lobes, each associated with an individual distributary channel (Schumm et al., 1987). In combination they build to form the fan as a whole. Climate undoubtedly affects fan development. During wet periods wet fans are usually strongly aggradational whereas extensive trenching can occur in dry periods.

Coarse-grained, permeable fan sediment promotes groundwater recharge, resulting in strong transmission losses and the demise of perennial rivers over the fan surface. In arid regions the groundwater may discharge beyond the fan margin, resulting in evaporitic playa lakes in arid regions such as Death Valley, California, as well as promoting enhanced vegetation growth in some areas. Large groundwater resources are associated with many fans. Some interfan areas have little fluvial activity, and arid-zone fans may interfinger with eolian deposits. However, very active rivers can be present between large cones, for example the Baghmati River that lies between the Gandak and Kosi cones in northern India. Channel switching on fans results in inactive areas where soil formation and chemical effects, such as formation of desert varnish, are enhanced.

Alluvial fan deposits in the rock record may be hundreds of meters thick and commonly occupy rapidly subsiding extensional and foreland basins where faults controlled basin-margin relief. Most are a few tens of square kilometers in area, but some large, wedge-shaped volcanoclastic aprons cover areas of $10^3 \, km^2$ and are up to 1 km thick. Fan deposits are typically poorly channelized and comprise amalgamated narrow channel (ribbon) bodies and broad (sheet) bodies of conglomerate and sandstone. Downflow, ribbons, and narrow sheets of conglomerate and sandstone are interbedded with mudstones, and mark the positions of major distributaries and fan lobes in more basinward settings, sometimes passing distally into sheetflood deposits.

DeCelles et al. (1991) documented wet fan deposits in the Cenozoic of the western US with cycles of aggradation and degradation. Periods of aggradation and flow expansion near the fan apex generated broad sheets which were subsequently incised, allowing sediment to bypass the upper fan through deep channels. The cycle was completed with channel back-filling and renewed dispersal over the fan surface. The resulting lithosomes are 10 m to 50 m thick and up to several kilometers wide, and are composed of smaller channel bodies in the form of scour-based sheets up to 10 m thick and hundreds of meters wide. Similar episodes of sheet aggradation under conditions of high supply, followed by incision as supply waned are also prominent in volcanoclastic fan and apron deposits. Other fan deposits consist of overlapping lobes of gravel separated by thin paleosols. These studies underline the importance of sediment supply in controlling fan architecture, commonly linked to climatic change or volcanism. Trenching may also have extrinsic causes, such as tilting or base-level rise in fan deltas, and extensional basins have complex fault systems where rapid shifts of activity and sediment supply along fault segments can result in complex sedimentation patterns in fans and bajadas.

Many fan deposits consist of cycles or megasequences, tens to hundreds to meters thick, that coarsen or fine upward. Coarsening upward cycles represent the basinward advance of fan lobes or entire fans, bringing coarse sediment over basinal fines. In contrast, fining upward cycles reflect declining sediment supply due to scarp retreat, fault-block evolution, denudation of topography, or gradual abandonment of fan segments. In strike-slip basins, entire fan bodies may represent the lateral migration of the depocenter past the fan entry point, so that the fan is progressively detached from its feeder channel. The presence of fan wedges in basinal fills may imply tectonic activity and uplift of a nearby source area, or may imply periods of tectonic quiescence followed by trenching and dispersal of fan sediment farther out into the basin. Fan and other fluvial deposits may also build basinward as thrusts advance, with deformation and cannibalization of fan sediments below the thrust mass. The location of fans close to tectonically active areas can cause early tectonic deformation, including the development of unconformities.

Deltas and estuaries

Rivers of all types are influenced by coastal processes. Most large rivers with high sediment loads terminate in delta distributary systems on open marine shelves (shelf-phase deltas) or on the shelf margin at lowstand or where the shelf is unusually narrow (shelf-margin deltas) (see Deltas). They may also terminate as bayhead deltas in restricted embayments, in back-barrier lagoons or in freshwater lakes. Under the low gradients typical of rivers approaching the coast, distributaries tend to have low sinuosities and many are essentially straight, commonly forming anastomosing and distributary networks.

Other rivers enter the sea through estuaries which are the drowned seaward parts of river valleys (see *Estuaries*). Estuarine facies commonly have a tripartite nature, passing landward from sandy wave-influenced facies at the estuary mouth to muddy central-basin facies, to sandy river-dominated facies (Dalrymple *et al.*, 1992). Although flocs of clay and organic material are widespread in inland rivers, fluvial deposition is especially enhanced by flocculation of clay where the suspended load encounters saline water, with preferential settling of mud during low-velocity turnaround points in the tidal cycle. Where sea-level rise inundated low-gradient plains, such as on the North American craton during the Late Paleozoic and Cretaceous, tidally influenced channel and valley fills are prominent. Important depositional zones on many modern river systems, deltas, and estuaries are described in detail elsewhere in this volume.

Major controls on accumulation of alluvial deposits

Tectonism, eustasy, and climate are key factors that govern the nature and accumulation of alluvial sediments. The effects of tectonism and eustasy are mediated principally through the creation and loss of accommodation, the space available for sediment to accumulate. Because insufficient sediment may be available to fill all the space generated in a basin, it is important in stratigraphic analysis to distinguish the thickness of strata preserved at a given place and time (*accommodation*) from the maximum possible space available which corresponds closely to the height of a water column at a given place and time (*potential accommodation*).

Tectonism operates through basin-scale subsidence and local fault activity, over timescales of years to millions of years in orogenic belts, and locally over just a few decades within sedimentary basins. Gravelly alluvial fan wedges accumulate where faults create strong basin-margin relief, and unusually thick alluvium, including peat, accumulates in rapidly subsiding parts of foreland and extensional basins. Such deposits commonly contain cycles and megasequences that may reflect tectonic events. Asymmetric subsidence in half grabens leads to drawdown of axially supplied sediment into the most rapidly subsiding zones, strongly influencing the proportion and connectedness of channel bodies in the succession.

Eustasy operates principally through glacioeustatic change in sea-level, related to systematic variation in the amount of insolation received as the earth orbits the sun (the Milankovitch effect). Sea level may vary by >100 m on timescales of tens of thousands to a few hundred thousand years. Such high-frequency and high-amplitude changes generate prominent sequences in coastal alluvium. Sea-level fall and lowstand are marked by widespread erosion, valley formation, and paleosol development on interfluves, whereas transgression and highstand are marked by flooding surfaces, backfilling of valleys with alluvium, and peat accumulation in coastal wetlands (Posamentier and Allen, 1999). On geological timescales, sea-level change typically affects fluvial facies and accumulation rates for only a few hundred kilometers inland. During periods of earth history when continental ice-sheets were lacking, sea-level changes of low amplitude may have been mediated by other factors, including buildup of valley glaciers, methane hydrates, and groundwater storage. Lake-level change affects fluviolacustrine systems and, due to the confined nature of lakes, can produce unusually large changes in level over short periods, with profound effects on fluvial architecture, as in the Volga River where it enters the Caspian Sea. Groundwater level acts as base level for some fluvio-eolian systems.

Climate change involves many parameters, including total precipitation, temperature, and nature and degree of seasonality (Blum and Tornqvist, 2000). Such change operates on timescales of years to millions of years, and is mediated to alluvial systems through changes in discharge, sediment flux, sediment type and soil type. Basinwide changes in fluvial style and sediment grade in the Channel Country of Australia have been linked to climatic and discharge changes that correlate with glacial cycles (Gibling *et al.*, 1998). Discharge variation may be especially important in monsoonal areas, such as southern Asia, where the intensity of monsoon circulation and precipitation is closely linked to the glacial-interglacial cycle. In such areas, precipitation may increase at least twofold over periods of a few thousand years, causing basinwide changes in fluvial style and vegetation and strong incision of drainage networks. The type and density of vegetation in riparian and upland regions is especially sensitive to climate and strongly influences sediment flux because of the crucial role that vegetation plays in stabilizing soils. Thus, floral overturn in response to climatic variation may have had a strong influence on sediment supply to basins, as may vegetation destruction by wildfire.

A key effect of the three major controls is the creation of river valleys, as outlined in earlier paragraphs. Valleys are linear troughs with dimensions much larger than those of the channels they contain (Schumm and Ethridge, 1994). Valley fills are most commonly recognized in the bedrock record where prominent unconformities represent long gaps in time, whereas valleys incised into alluvium and marine sediments may represent short-term effects, and are more difficult to recognize. Although many coastal valleys reflect sea-level lowering, valley formation may also reflect discharge variation, uplift, and localization of incision along faults.

River terraces are paleo-floodplains, abandoned usually as a result of climate change, eustasy, or tectonism. They are normally elevated above the floodplain along the valley margins but can dip below the floodplain surface, and below sea level off shore as a result of sea-level rise. Their sedimentology is diagnostic of the environmental and associated flow conditions that prevailed at the time of their formation.

Geometry of fluvial bodies in the ancient record

The three-dimensional geometry, permeability framework, and degree of connection (connectedness) of fluvial deposits have considerable practical importance for aquifers, hydrocarbon reservoirs, and mining of coal and economic minerals that are hosted by fluvial strata. Important studies of channel-body geometry were published by Krynine (1948) and Friend (1983). Channel and floodplain sediment bodies in preserved strata can be divided on the basis of width : thickness ratio (measured normal to the flow direction) into ribbons (<15 : 1), narrow sheets (<100 : 1) and broad sheets (>100 : 1). As noted earlier, a link between channel proportions and sediment load was noted by Schumm (1960), and these arbitrary divisions have counterparts in modern channel dimensions, for many self-formed single channels have width:mean depth ratios of less than 20, with progressively fewer with ratios up to 100 and above. Large braided rivers and braidplains may have width:depth ratios of several hundred and upward. Krynine also suggested divisions in terms of stratal area. No upper limit has been placed on sheets which, in the case of widespread flood layers and braided-fluvial sheets, can have very high aspect ratios, as noted earlier. This geometric classification is independent of scale, and can apply equally to scour fills a few decimeters and valley fills that are several kilometers thick. Dimensional divisions become more difficult to apply where fluvial bodies are stacked to great thicknesses without intervening extra-channel strata, as in some alluvial fan deposits localized at basin margins and basinwide braided-fluvial formations. Sediment accumulation as levees and crevasse splays on channel margins can result in channel fills with *central bodies* and marginal *wings*.

As noted by Friend (1983), braided and meandering rivers tend to have *mobile channel belts* that migrate laterally, and generate widespread sheets of gravel and sand. The sheets may contain discrete channel fills of ribbon form that represent the fills of confluences, anabranches, and localized scours that were cut and filled rapidly. In contrast, channels in cohesive fine-grained material can be categorized as *fixed channels* that are laterally stable between episodes of abrupt switching and generate bodies in the form of ribbons and narrow sheets. These represent an initial channel form, perhaps widened slightly through time by bank migration. A third category is *sheetflood deposits* with very extensive sediment sheets that represent poorly channelised settings.

Where floodplains aggrade rapidly, typically due to rapid subsidence or sea-level rise, channel bodies tend to form ribbons or narrow sheets that are encased in floodplain fines and poorly connected. In contrast, where floodplains aggrade slowly, rivers tend to migrate laterally and rework their alluvium, resulting in sheets of well connected channel bodies. Growth of peat domes can restrict channel migration and promote thick ribbon bodies. Thus, channel-body geometry commonly represents an interplay between initial channel form, aggradation rates of channel and floodplain sediments, and the rate of lateral channel expansion through sediment of varied consistency. Avulsion patterns have an important influence on channel body form, as avulsing channels can create new courses or can reoccupy pre-existing courses, thus promoting the generation of stacked, multistory bodies (Jones and Schumm, 1999; Kraus and Wells, 1999) . There is no unique relationship between channel planform and 3D channel-body form, because the final form usually represents the evolution of one or more channels over a prolonged period, during which the channels migrate and avulse laterally and aggrade vertically. Commonly, however, fluvial processes and alluvial facies are related to the four dominant river patterns of straight, meandering, braided, and anabranching, and to the two dominant terminal depositional systems, fans, and deltas.

A descriptive framework for studying the geometry of alluvial bodies was outlined by Miall (1996). This framework involves the use of facies codes for sediment types, recognition of bounding surfaces with a hierarchical order (from bed boundaries to major valley bases), and the identification of architectural elements that are characteristic of many fluvial systems. The latter include sediment gravity sheets, sandy bedform sheets, downstream, and laterally accreting elements, laminated sand sheets, and overbank fines, as well as bodies with distinctive channel form. The combination of these elements and bounding surfaces can be used to present an objective view of the channel body.

Gerald C. Nanson and Martin R. Gibling

Bibliography

Allen, J.R.L., 1970. Studies in fluviatile sedimentation: a comparison of fining-upward cyclothems, with special reference to coarse-member composition and interpretation. *Journal of Sedimentary Petrology*, **40**: 298–323.

Ashmore, P.E., 1991. How do gravel-bed rivers braid? *Canadian Journal of Earth Sciences*, **28**: 326–341.

Bettess, R., and White, W.R., 1983. Meandering and braiding of alluvial channels. *Proceedings of the Institution of Civil Engineers*, **75**: 525–538.

Blair, T.C., and McPherson, J.G., 1994. Alluvial fans and their natural distinction from rivers based on morphology, hydraulic processes, sedimentary processes, and facies assemblages. *Journal of Sedimentary Research*, **A64**: 450–489.

Blum, M.D., and Tornqvist, T.E., 2000. Fluvial responses to climate and sea-level change: a review and look forward. *Sedimentology*, **47**: 2–48.

Brierley, G.J., and Hickin, E.J., 1991. Channel planform as a non-controlling factor in fluvial sedimentology: the case of the Squamish River floodplain, British Columbia. *Sedimentary Geology*, **75**: 67–83.

Brotherton, D.I., 1979. On the origin and characteristics of river channel patterns. *Journal of Hydrology*, **44**: 211–230.

Church, M., 1992. Channel morphology and typology. In Calow, P., and Petts, G.E. (eds.), *The River's Handbook: Hydrological and Ecological Principles*. Blackwell, pp. 126–143.

Dalrymple, R.W., Zaitlin, B.A., and Boyd, R., 1992. Estuarine facies models: conceptual basis and stratigraphic implications. *Journal of Sedimentary Petrology*, **62**: 1130–1146.

DeCelles, P.G., Gray, M.B., Ridgway, K.D., Cole, R.B., Pivnik, D.A., Pequera, N., and Srivastava, P., 1991. Controls on synorogenic alluvial-fan architecture, Beartooth Conglomerate (Palaeocene), Wyoming, and Montana. *Sedimentology*, **38**: 567–590.

Fisk, H.N., 1944. Geological investigation of the alluvial valley of the lower Mississippi River Vicksburg. Mississippi River Commission, **78** pp.

Friend, P.F., 1983. Towards the field classification of alluvial architecture or sequence. In Collinson, J.D., and Lewin, J. (eds.), *Modern and Ancient Fluvial Systems*. International Association of Sedimentologists, Special Publication, 6, pp. 345–354.

Friend, P.F., and Sinha, R., 1993. Braiding and meandering parameters. In Best, J.L., and Bristow, C.D. (eds.), *Braided Rivers*. Geological Society of London, Special Publication, 75, pp. 105–111.

Gibling, M.R., Nanson, G.C., and Maroulis, J.C., 1998. Anastomosing river sedimentation in the Channel Country of Central Australia. *Sedimentology*, **45**: 595–619.

Gilbert, G.K., 1877. Report on the geology of the Henry Mountains. United States Geological Survey, Rocky Mountain Region, Washington, DC.

Hickin, E.J., and Nanson, G.C., 1975. The character of channel migration on the Beatton River, north-east British Columbia, Canada. *Geological Society of America Bulletin*, **86**: 487–494.

Horton, R.E., 1945. Erosional development of streams and their drainage basins: hydrophysical approach to quantitative morphology. *Geological Society of America Bulletin*, **56**: 275–370.

Huang, H.Q., and Nanson, G.C., 2000. Hydraulic geometry and maximum flow efficiency as products of the principle of least action. *Earth Surface Processes and Landforms*, **25**: 1–16.

Jones, L.S., and Schumm, S.A., 1999. Causes of avulsion: an overview. In Smith, N.D., and Rogers, J. (eds.), *Fluvial Sedimentology VI*. International Association of Sedimentologists, Special Publication, 28, pp. 171–178.

Knighton, D., 1998. *Fluvial Form and Processes: A New Perspective*. Arnold, London.
Kraus, M.J., and Wells, T.M., 1999. Recognizing avulsion deposits in the ancient stratigraphic record. In Smith, N.D., and Rogers, J. (eds.), *Fluvial Sedimentology VI*. International Association of Sedimentologists, Special Publication, 28, pp. 251–268.
Krynine, P.D., 1948. The megascopic study and field classification of sedimentary rocks. *Journal of Geology*, **56**: 130–165.
Langbein, W.B., and Leopold, L.B., 1966. River meanders- theory of minimum variance. *United States Geological Survey Professional Paper*, 422H.
Leopold, L.B., and Maddock, T., 1953. The hydraulic geometry of stream channels and some physiographic implications. *United States Geological Survey Professional Paper*, 252.
Leopold, L.B., and Wolman, M.G., 1957. River channel patterns- braided, meandering, and straight. *United States Geological Survey Professional Paper*, Series B, 282, 39–85.
Mackin, J.H., 1948. Concept of the graded river. *Geological Society of America Bulletin*, **59**: 463–512.
Makaske, B., 2001. Anastomosing rivers: a review of their classification, origin, and sedimentary products. *Earth-Science Reviews*, **53**: 149–196.
McCarthy, T.S., Stanistreet, I.G., Caincross, B., Ellery, W.N., and Ellery, K., 1988. Incremental aggradation on the Okavango Delta-fan, Botswana. *Geomorphology*, **1**: 267–278.
Miall, A.D., 1996. *The Geology of Fluvial Deposits*. Springer-Verlag.
Nanson, G.C., and Croke, J.C., 1992. A genetic classification of floodplains. *Geomorphology*, **4**: 459–486.
Nanson, G.C., and Knighton, A.D., 1996. Anabranching rivers: their cause, character, and classification. *Earth Surface Processes and Landforms*, **21**: 217–239.
Posamentier, H.W., and Allen, G.P., 1999. Siliciclastic sequence stratigraphy—concepts and applications. SEPM, Society for Sedimentary Geology, *Concepts in Sedimentology and Paleontology*, 7.
Richards, K.S., 1982. *Rivers: Form and Process in Alluvial Channels*. Methuen, London.
Schumm, S.A., 1960. The shape of alluvial channels in relation to sediment type. *United States Geological Survey Professional Paper*, Series B, 352, 17–30.
Schumm, S.A., 1973. Geomorphic thresholds and complex response of drainage systems. In Morisawa, M. (ed.), *Fluvial Geomorphology*. Binghamton, NY: New York State University Publications in Geomorphology. pp. 299–309.
Schumm, S.A., 1977. *The Fluvial System*. Wiley.
Schumm, S.A., 1981. Evolution and response to the fluvial system, sedimentologic implications. In Ethridge, F.G., and Flores, R.M. (eds.), *Recent and Ancient Nonmarine Environments: Models for Exploration*. Society of Economic Paleontologists and Mineralogists, Special Publication, 31, pp. 19–29.
Schumm, S.A., and Ethridge, F.G., 1994. Origin, evolution, and morphology of fluvial valleys. In Dalrymple, R.W., Boyd, R., and Zaitlin, B.A. (eds.), *Incised-Valley Systems: Origin and Sedimentary Sequences*. Society of Economic Paleontologists and Mineralogists, Special Publication, 51, pp. 11–27.
Schumm, S.A., Mosley, M.P., and Weaver, W.E., 1987. *Experimental Fluvial Geomorphology*. Wiley.
Smith, D.G., and Smith, N.D., 1980. Sedimentation in anastomosed river systems: examples from alluvial valleys near Banff, Alberta. *Journal of Sedimentary Petrology*, **50**: 157–164.
Smith, R.M.H., 1987. Morphology and depositional history of exhumed Permian point bars in the southwestern Karoo, South Africa. *Journal of Sedimentary Petrology*, **57**: 19–29.
Stanistreet, I.G., and McCarthy, T.S., 1993. The Okavango fan and the classification of subaerial systems. *Sedimentary Geology*, **85**: 115–133.
Wolman, M.G., and Leopold, L.B., 1957. River flood plains: some observations on their formation. *United States Geological Survey Professional Paper*, Series C, 282, pp. 87–109.

Cross-references

Alluvial Fans
Anabranching Rivers
Armor
Attrition (Abrasion), Fluvial
Avulsion
Braided Channels
Debris Flow
Deltas and Estuaries
Erosion and Sediment Yield
Floodplain Sediments
Flow Resistance
Flume
Grain Size and Shape
Mass Movement
Meandering Channels
Numerical Models and Simulation of Sediment Transport and Deposition
Physics of Sediment Transport: The Contributions of R.A. Bagnold
Fluvial Placers
Sediment Fluxes and Rates of Sedimentation
Sediment Transport by Unidirectional Water Flows
Surface Forms
Weathering, Soils, and Paleosols

S

SABKHA, SALT FLAT, SALINA

The term *sabkha* (Arabian for *salt flat*) is used by sedimentologists to designate an arid, subaerially exposed environment (typically a salt-encrusted mudflat). The term came into wide use among sedimentologists following the pioneering studies of the modern carbonate tidal flat deposits of the arid Trucial Coast, Persian Gulf (now the Arabian Gulf) by D.J. Shearman's Imperial College group (Curtis *et al.*, 1963; Kinsman, 1969; Butler, 1970). The term *continental sabkha* has been used for the arid mudflat of a non-marine closed basin, better known to students of the closed basins of the western US as a saline *playa* (Spanish for "shore" or "beach"). In South America the term *salina* is used for a salt-encrusted mudflat. The factors common to both marine and non-marine sabkhas are (1) an arid climate, (2) the dominance of subaerial exposure of the mudflats, punctuated by occasional flooding (seawater on marine sabkhas and meteoric waters on playas), and (3) deposition of evaporite minerals (such as gypsum and halite), subaqueously in ephemeral "lakes" but in the vadose zone and as efflorescent surface crusts during periods of subaerial exposure.

The existing studies of the Holocene Arabian Gulf marine sabkhas have focused on vadose zone gypsum and anhydrite in the sabkha environment. Warren (1989, p.38) defines sabkhas "as marine and continental mudflats where displacive and replacive evaporite minerals are forming in the capillary zone above a saline water table". All the spectacularly deformed layers of nodular anhydrite (enterolithic folds, overthrust and underthrust ridges, diapiric structures, etc.; see Butler, 1970; Demicco and Hardie, 1994, figures 160(A–C)) in the Arabian Gulf are attributed to intrasediment growth of gypsum and anhydrite (e.g., Warren, 1989, figure 2.5). While saline mineralization in the vadose zone is a most significant and characteristic process in any ephemeral evaporite environment, the Arabian Gulf sabkha "model", based on vadose processes, does not include a most important stage in the marginal marine sabkha depositional cycle, the "gypsum pan" stage. This pivotal step in evaporite mineral formation in the "sabkha cycle" is outlined below.

The following description of the formation of *bedded gypsum* in a modern marine sabkha environment is based on observations made on the large (80–100 km^2) "ephemeral gypsum pan" of the extensive arid supratidal flat at the southwest margin of the Colorado River delta, Baja California, Mexico (Castens-Seidell, 1984; Hardie and Shinn, 1986, pp. 46–49 and table 1; Demicco and Hardie, 1994, p.197). This active gypsum pan cycles through the following sequence of stages: a seawater flood stage → an evaporative concentration stage (ephemeral shallow saline lake) → a desiccation stage (dry gypsum pan) → a halite pan stage, and back to a flood stage. The length of each cycle is a function of the interval between storms (months to a few years). After a major seawater flooding event the supratidal flat is covered with a sheet of mud-laden seawater (<1–2 m deep). As the flood finally subsides, most of the seawater drains back to the sea leaving an isolated sheet of seawater ponded in the shallow depression (the "evaporating pan") on the supratidal flat. Deposition begins with a cm-scale layer of siliciclastic mud that settles out over the gypsum bed formed during the last gypsum pan cycle. This mud layer is soon covered by a cyanobacterial mat that develops low-amplitude polygonal undulations as the mat grows. Evaporative concentration eventually leads to shrinking of the "lake" and the nucleation of very small aragonite needles and gypsum euhedra at the brine-air interface, held there by surface tension. As these surface-nucleated crystals grow, their increasing weight eventually leads to their settling to the bottom of the shallow "lake" where they act as nuclei for syntaxial bottom growth. Competitive growth and space constraints favor the vertically-oriented nuclei, resulting in a subaqueous gypsum layer composed of an upward radiating array of blade-shaped and "swallow-tail twin" crystals (e.g., Demicco and Hardie, 1994, figure 154A). With progressive evaporative concentration, the saline lake progressively shrinks, exposing the newly-formed subaqueous gypsum layer to the hot, dry air. The gypsum of this subaerially-exposed layer then undergoes diagenetic modification as evaporative pumping draws brine from the shallow subsurface brine table (<1 m) up through the vadose zone. This brine pumping process promotes the precipitation of gypsum in the intercrystalline cavities in the gypsum bed as vadose cement, which in turn causes the bed to expand laterally. This generates small-scale buckles in the gypsum crust and leads to the formation of larger antiform polygonal ridges (several cms. in height) spaced approximately 2 m apart ("tepees", see Demicco and Hardie, 1994, figure 161). Ultimately, the continued lateral expansion of the gypsum layer gives rise to fracturing at the polygon boundaries and the development of overthrusting and underthrusting at the ridges. In turn, these thrust fracture zones act as localized high permeability conduits for enhanced evaporative pumping of brine up through the vadose zone. When these brines emerge at the surface, they evaporate rapidly to dryness and deposit finely crystalline gypsum beneath the fractured ridges, adding to their volume and height, a process that produces diapir-like masses of gypsum beneath the ridges. This tepee-forming process is only halted by the next flooding event when a new gypsum pan cycle begins. The cycle starts again as seawater, undersaturated with respect to gypsum, floods over the dry gypsum pan partially dissolving the buckled gypsum surface layer, enlarging existing intercrystalline cavities and rounding the margins of exposed euhedral gypsum crystals. This dissolution phase soon results in the floodwaters becoming almost saturated with respect to gypsum. This allows erosion, transportation and redeposition on the pan of loose gypsum crystals and cleavage fragments as sand and gravel sheets. Isolated patches of gypsum sands may be sculptured into classic unidirectional bedforms and wave ripples. Ultimately, as the flood waters subside, seawater left ponded on the gypsum pan evaporates to the point of gypsum saturation and a new gypsum bed is formed. During this stage of this new gypsum precipitation cycle, the dissolution features of the underlying gypsum bed are "repaired" by syntaxial overgrowth (Demicco and Hardie, 1994, figure 154C) and by cementation on the walls of intercrystalline cavities.

Most significantly, the layered gypsum of the Baja California supratidal flats *has not been converted to nodular anhydrite*, yet it has all the basic morphological features found in the Persian Gulf layered nodular anhydrite, that is, cm-scale layers of detrital sediment alternating with cm-scale layers of gypsum displaying enterolithic folding, diapiric structures, and polygonal tepees with overthust and underthrust "faults" located at the buckled tepee ridges. At the landward margin of the sabkha, gypsum crystals are in the early stages of dehydration to anhydrite (Castens-Seidell, 1984). It is easy to imagine that with time, as the Baja California tidal flats continue to prograde seaward, complete conversion of these gypsum layers to anhydrite will mimic the spectacular evaporite features observed in trenches cut into the landward regions of the Persian Gulf sabkha. By analogy, then, *the Persian Gulf layered nodular anhydrite would have been formed by dehydration of layered gypsum that crystallized subaqueously in shallow ephemeral gypsum pans*, inheriting the principal deformation structures from the parent gypsum bed. The Persian Gulf "gypsum pan" deposits, converted completely to anhydrite, are now defunct. They have been eroded by deflation and covered by windblown sands as the tidal flats have prograded seaward over the past 5000 years. This same model could apply to other, more ancient, layered nodular anhydrite deposits in the geological record.

For each flood event, as evaporative concentration proceeds and the shallow saline "lake" shrinks, the surface brine body may reach saturation with respect to halite, forming an *ephemeral halite pan* at the center (the "bullseye") of the gypsum pan. All the essential characteristics outlined above for ephemeral gypsum pan deposition have been documented and described by Lowenstein and Hardie (1985) for bedded halite in the ephemeral salt pan that lies in the center of the Baja California gypsum pan (see also pioneering study of Shearman, 1970). Significantly, however, the halite of the salt pan on this arid tidal flat is short-lived because, typically, the subaerially-exposed, highly soluble halite crusts of the salt pan are completely dissolved by large inundations of seawater driven onshore by strong storms. When the onshore winds subside, most of the floodwaters drain off the tidal flats carrying dissolved NaCl back to the sea, leaving the relatively insoluble gypsum crusts essentially intact. This observation suggests that we should not expect to find thick accumulations of salt pan halite capping arid tidal flat cycles. Instead, such depositional cycles would be capped by ephemeral gypsum pan deposits (see *Anhydrite and Gypsum*, figure A8).

Analogous processes and products can occur in nonmarine closed basins on ephemeral playas. The main difference would be that the floodwaters would be relatively dilute meteoric waters with a much greater potential for widespread dissolution during the initial flooding stage. In addition, for gypsum to be a significant precipitate, the inflow waters must be of the "neutral" Na-K-Ca-Mg-SO$_4$-Cl chemical group (see *Evaporites*). In modern ephemeral playa settings in arid closed basins, halite pans are far more abundant than gypsum pans because in such basins there are no outlets that would allow flushing away of the halite formed in the "bullseye" of the pan, as occurs on marine supratidal flats as described above for the Baja California sabkha.

Lawrence A. Hardie

Bibliography

Butler, G.P., 1970. Holocene gypsum and anhydrite of the Abu Dhabi sabkha: an alternative explanation of origin. In Rau, J.L., and Dellwig, L.F. (eds.), *Proceedings of the Third Salt Symposium*. Cleveland: Northern Ohio Geological Society, pp. 120–152.

Castens-Seidell, B., 1984. *The Anatomy of a Modern Marine Siliciclastic Sabkha in a Rift Valley Setting: Northwest Gulf of California, Baja California*. Unpublished PhD dissertation, Baltimore: Johns Hopkins University, 386pp.

Curtis, R., Evans, G., Kinsman, D.J.J., and Shearman, D.J., 1963. Association of dolomite and anhydrite in the recent sediments of the Persian Gulf. *Nature*, **197**: 679–680.

Demicco, R.V., and Hardie, L.A., 1994. *Sedimentary Structures and Early Diagenetic Features of Shallow Marine Carbonate Deposits*. SEPM Atlas Series Number 1, 265pp.

Hardie, L.A., and Shinn, E.A., 1986. Carbonate depositional environments modern and ancient, part 3: tidal flats. *Colorado School of Mines Quarterly*, **81**: 1–74.

Kinsman, D.J.J., 1969. Modes of formation, sedimentary associations and diagnostic features of shallow water and supratidal evaporites. *American Association of Petroleum Geologists Bulletin*, **53**: 830–840.

Lowenstein, T.K., and Hardie, L.A., 1985. Criteria for the recognition of salt pan evaporites. *Sedimentology*, **32**: 627–644.

Shearman, D.J., 1970. Recent halite rock, Baja California, Mexico. *Institute of Mining and Metallurgy, Transactions*, **B79**: 155–162.

Warren, J.K., 1989. *Evaporite Sedimentology*. Prentice Hall.

Cross-references
Anhydrite and Gypsum
Desert Sedimentary Environments
Evaporites

SALT MARSHES

Marshes are wetland environments dominated by herbaceous plants. Sea-coast tidal salt marshes are a specific type found in sheltered back-barrier, estuarine and deltaic settings where they form important environments and ecosystems. They contain halophytic (salt tolerant) grasses, rushes, sedges and forbes that are able to survive in intertidal and fringing environments hostile to upland vegetation. Mature marshes are dominated by living plants and are usually constructed on peats, the partially preserved remnants of former marsh plants. Marshes usually stand higher than flats, from mid-tide to highest spring tides, and are often best developed as a broad flat within a few decimeters of mean high water (MHW). Marsh sediments have been used as tools for stratigraphic reconstruction of coastal evolution (Kraft, 1971) and bases for interpretation of Holocene sea-level change (Redfield and Rubin, 1962; Kraft and Belknap, 1977; van de Plaasche, 1986; Gehrels et al., 1996) using radiocarbon dating of peats built from plants that grow near mean high sea level.

Morphology

The morphology of salt marshes depends on tidal range, slope of substrate, and geometry of the enclosing embayment (Kelley et al., 1988). The simplest salt marsh, often an early or ephemeral phase, is created through the colonization of tidal flats with halophytes. Later successional phases create more complex morphology, to a large degree controlled by the ecology of the vegetation. Fringing marshes occur adjacent to uplands. They may occupy relatively steep margins in tidal rivers and estuaries, particularly in rocky or paraglacial regions such as New England. More widespread are those on the landward edge of lagoons, on low-angle coasts. Estuarine and Fluvial salt marshes are elongated in a marginal fringe parallel to the channel and the upland. They may vary from fringing to broad types. Back-barrier marshes are broad and may fill the majority of backbarrier systems (South Carolina tide-dominated mesotidal barrier systems, many New England marshes), or may occur as a relatively narrow band on the lagoon side of wave-dominated microtidal barriers (Long Island, New Jersey, Delaware). Broad, well-developed marshes are generally flat, graded to near mean high water (± 30 cm or less), with the majority of their area in the high marsh zone. Fringing marshes are steeper, because they compress all the marsh zones, controlled by frequency of flooding, into a narrower band.

Ecosystems

Within a marsh there are commonly: (1) a low marsh, colonizing zone; (2) a broad high marsh; (3) tidal creeks; (4) creek levees or margins; (5) salt pannes; (6) higher-high marsh; and (7) upland fringe. Marshes are generally zoned by the ecology of the plants (Niering and Warren, 1980). Primarily tolerance of the halophytes is to frequency of tidal inundation. Some plants have specialized structures in subsurface rhizomes and roots for gas transport, and glands for excreting excess salt. In some fringing marshes with high tidal ranges, the zonation can be very sharp. In broad marshes or in lower tidal ranges, the zonation tends to be more mosaic in nature, and controlled by proximity to tidal creeks.

Low salt marsh has a very restricted diversity, dominated by salt marsh cord grass *Spartina alterniflora* in much of North America and other areas. Low salt marsh occurs from Mean Sea Level (MSL) to MHW, and is flooded by most tides. High salt marsh is more diverse, generally from MHW to SHW, flooded every fortnight. High salt marsh is more diverse, and varies by climate zone. In New England high marsh is dominated by *Spartina patens* grass, in the mid-Atlantic by *S. patens* and *Distichlis spicata*, in Georgia and Florida by the rush *Juncus roemarianus*. On the Pacific coast of the US, it is dominated by *Salicornia pacifica*, while *Pucinellia maritima* is common in the British Isles.

Higher high marsh is flooded a few times per month, occurring above MHW to Spring High Water. The ecology also varies by latitude. In Maine the dominant plant is generally *Juncus gerardii*, while the Mid-Atlantic is less distinct, with *Distichlis spicata* and shrubs *Iva fructescens* and *Baccharus halimifolia* important.

Upland Fringe is primarily a freshwater system, flooded only during storms or astronomical highest tides, several times per year. Cattails (*Typha angustifolia*) predominate in the Upland Fringe in Maine, while in the mid-Atlantic *Phragmites australis* is dominant. *P. australis* is an aggressive colonizer of disturbed lands, and in the last few decades has expanded greatly in mid-Atlantic marshes, on dredge spoil islands, and is being carried to marshes in northern New England, particularly near major roads.

Marshes may also grade from primarily saline systems, as described above, to more brackish and freshwater systems along an estuary. Many species of sedges such as *Scirpus* and *Carex* characterize these transitional brackish marshes. Salt pannes are flooded low areas in high marsh, with variable salinities that vary from fresh to hypersaline. Their origins are controversial, but usually ascribed to "rotten spots" or ice plucking. They are dominated by cyanobacterial mats, and fringed by *Salicornia* spp., and are a critical ecosystem for larval and juvenile fish.

Processes

Marshes grow through *in situ* primary production, from roots, rhizomes, and stems of living plants, but also build up through accumulation of organic detritus. Most salt marshes are composed not of pure organic peat, but have a majority of inorganic sediment that is carried into the marsh in suspension (or by ice rafting). "Peat" is a field term for fibrous organic rich sediments found in marshes, but its true definition (>50 dry weight percent organic carbon) is found mainly in freshwater bogs. Highly fibrous New England marsh peats often contain less than 10 percent organics by dry weight (although 90 percent water and organics by volume). New England marshes are very firm and fibrous, Georgia and Florida marshes are muddy or sandy, while mid-Atlantic marshes are intermediate in composition.

Tidal currents predominate in channels. Flow is progressively channelized during late ebb through early flood tides, whereas sluggish sheet flow predominates at high water over broad marshes. Blocks constantly slump from margins of creeks, particularly in New England type marshes (due to high tidal range ice effects). However, rapid growth and sediment accumulation in the low marsh environment builds up and heals the slump, resulting in only slow channel migration.

Clastic sediments in fringing or narrow marshes often fine seaward, with lesser influence of runoff from upland sources. This inorganic sediment is extracted from suspension by: (1) direct settling from suspension—enhanced by a quiet water baffling effect, increased boundary layer thickness caused by the plants; (2) pelletization by organisms (mollusks, decapods, worms) and trapping in organic films; (3) adhering to grass stems, washed off by rain or accumulated as the plant dies; and (4) ice rafting in colder climates.

Broad marshes receive sediment from tidal creeks and may have coarser-grained levees, fining toward their interiors. Sedimentation rates in high marshes are an order of magnitude higher at the creek margin than the interior, due to supply. Sedimentation rates on low marsh are another order of magnitude higher (e.g., Wood *et al.* (1989) found sediment accumulation rates up to 8 cm per year in low marsh-tidal flat colonization zone, 0.5 cm per year in fringing marshes, and 2 mm/yr or less in broad marshes). Thus, broad marsh accumulates in equilibrium with sea-level rise and local compaction of marsh, while low marsh can colonize and accumulate much more rapidly.

Autocompaction occurs as a marsh compresses under the weight of accumulating sediment and through decay. Organic structures rot to some degree, but cellulose is preserved in saturated, anoxic, and lowered pH conditions. Kaye and Barghoorn (1964) demonstrated autocompaction to 40 percent of original cross-sections in predominantly freshwater marsh sediments, while Belknap and Kraft (1977) showed compaction-displacement of stratigraphic surfaces related to overall thickness of peats in paleovalleys.

Subfossil plant remains (roots, rhizomes, seeds) are commonly preserved well enough to identify species (Kraft *et al.*, 1979). Insect parts, diatoms, and pollen are also preserved and identifiable. Calcareous shells are not commonly preserved, because of the acidic conditions. Agglutinated foraminifera fossils are often well preserved, and have become critical tools in identification of marsh paleoenvironments and as sea-level indicators (Scott and Medioli, 1978; Gehrels *et al.*, 1996), providing more precise levels than the plants can.

Environmental evolution and stratigraphy

Frey and Basan (1985) suggest that marshes pass through stages. They identify: (1) a Youthful stage with mostly *S. alterniflora*, extensive lagoons, and a high accumulation rate; (2) a Mature stage with approximately equal proportions of high marsh and low marsh, less lagoon area, and the overall accumulation rate decreasing; and (3) Old Age, which consists of almost all high marsh, with poor drainage, and essentially no lagoon.

Salt marsh stratigraphic evolution depends on creation of new accommodation space for peat accumulation through autocompaction and sea-level rise. Rising sea level also allows transgression of the marsh's leading edge over upland surfaces. In a classic model of development of New England-type marshes Mudge (1858) considered upbuilding and encroachment on the land (transgressive stratigraphy) in an environment of subsidence. Conversely, Shaler (1885) emphasized the progradation of marshes through colonization of flats and lagoons (regressive stratigraphy). Redfield (1972) investigated Barnstable marsh, behind the Sandy Neck spit on Cape Cod, Massachusetts, and found stratigraphy supporting both the Shaler and Mudge models, in a setting of relative sea-level rise, with expansion over mudflat and encroachment onto the upland. Recent work supports this conclusion, but it is clear that there are more complex relationships to rates of sea-level change and sediment introduction in New England salt marshes (Gehrels *et al.*, 1996).

Conclusions

Present knowledge of marsh environments provides a solid basis for facies models in estuarine, deltaic, and barrier settings. Sea-level change analysis is firmly established for general trends. Current research is focused on extracting the most details and establishing the greatest accuracy possible for addressing climate change on the millennial to decadal scale. Debates continue over limits of accuracy and precision of plant, foraminiferal and other indicators (Kelley *et al.*, 2001; van de Plassche, 2001), but it is clear that salt marshes contain a rich record of climate proxies.

Daniel F. Belknap

Bibliography

Belknap, D.F., and Kraft, J.C., 1977. Holocene relative sea-level changes and coastal stratigraphic units on the northwest flank of the Baltimore Canyon Trough geosyncline. *Journal of Sedimentary Petrology*, **47**: 610–629.

Gehrels, W.R., Belknap, D.F., and Kelley, J.T., 1996. Integrated high-precision analyses of holocene relative sea-level changes: lessons from the coast of Maine. *Geological Society of America Bulletin*, **108**: 1073–1088.

Kelley, J.T., Belknap, D.F., and Daly, J.F., 2001. Comment on "North Atlantic Climate–Ocean variations and sea level in Long Island Sound, Connecticut, since 500 cal yr A.D.". *Quaternary Research*, **55**: 105–107.

Kelley, J.T., Belknap, D.F., Jacobson, G.L. Jr., and Jacobson, H.A., 1988. The morphology and origin of salt marshes along the glaciated coastline of Maine, USA. *Journal of Coastal Research*, **4**: 649–665.

Kraft, J.C., 1971. Sedimentary facies patterns and geologic history of a Holocene marine transgression. *Geological Society of America Bulletin*, **82**: 2131–2158.

Kraft, J.C., Allen, E.A., Belknap, D.F., John, C.J., and Maurmeyer, E.M., 1979. Processes and morphologic evolution of an estuarine and coastal barrier system. In Leatherman, S.P. (ed.), *Barrier Islands*. New York: Academic Press, pp. 149–183.

Mudge, B.F., 1858. The salt marsh formations of Lynn. *Proceedings of Essex Institute*, **2**: 117–119.

Niering, W.A., and Warren, R.S., 1980. Vegetation patterns and processes in New England salt marshes. *Bioscience*, **30**: 301–307.

van de Plassche, O., 1986. *Sea-level Research: a Manual for the Collection and Evaluation of Data*. Norwich, England: Geo Books, 618pp.

van de Plassche, O., 1991. Late Holocene sea-level fluctuations on the shore of Connecticut inferred from transgressive and regressive overlap boundaries in salt-marsh deposits: origin of the paleovalley system underlying Hammock River Marsh, Clinton, Connecticut. *Journal of Coastal Research*, Special Issue, **11**: 159–179.

van de Plassche, O., 2001. Reply (to Kelley *et al.*). *Quaternary Research*, **55**: 108–111.

Redfield, A.C., 1972. Development of a New England salt marsh. *Ecological Monographs*, **42**: 201–237.
Redfield, A.C., and Rubin, M., 1962. The age of salt marsh peat and its relation to recent changes in sea level at Barnstable, Massachusetts. *Proceedings of the National Acadamy of Sciences*, **48**: 1728–1735.
Scott, D.B., and Medioli, F.S., 1978. Vertical zonations of marsh foraminifera as accurate indicators of former sea levels. *Nature*, **272**: 528–531.
Shaler, N.S., 1885. Preliminary report on sea-coast swamps of the Eastern United States. *US Geological Survey 6th Annual Report, 1886.* pp. 353–398.
Wood, M.E., Kelley, J.T., and Belknap, D.F., 1989. Pattern of sediment accumulation in the tidal marshes of Maine. *Estuaries*, **12**: 237–246.

Cross-references

Barrier Islands
Deltas and Estuaries
Sediment Transport by Tides
Tidal Flats
Tidal Inlets and Deltas

SANDS, GRAVELS, AND THEIR LITHIFIED EQUIVALENTS

Introduction

Sand is defined petrographically as detrital particles between 1/16th and 2 mm in diameter. Geologically, sand and its lithified equivalent, sandstone, are defined as deposits consisting of 75 percent or more of sand particles, the remainder typically being silt or mud. Gravel is the general term applied to detrital particles larger than 2 mm, with such terms as pebbles, cobbles and boulders defined by specific size limits (see below). A lithified gravel is termed a breccia if the gravel fragments are angular, and conglomerate if they have been rounded by attrition. Very few sandstones and conglomerates consist entirely of sand- or gravel-sized particles, respectively. Most conglomerates include a matrix of sand and finer particles, and many sandstones are, to a greater or lesser degree, muddy, silty, or conglomeratic. The third component in many sandstones and conglomerates comprises one or more cements of silica, calcium carbonate or other, less common minerals. In most sandstones and conglomerates there are remaining pore spaces (up to 40 percent but typically <25 percent by volume) that are filled with pore fluids.

Sandstones and conglomerates typically preserve considerable internal evidence of their origins. The mineralogical composition of the detrital particles reflects their lithologic source, and may be studied to yield evidence of sedimentary provenance and unroofing history of uplifted areas surrounding a sedimentary basin. Textures and structures of a deposit provide evidence of the sedimentary processes involved in its transport and deposition.

Sands and gravels constitute about 25 percent of sedimentary rocks, by volume. They have economic importance as hosts for about one third of the world's conventional oil and gas and most of the heavy oil, plus several important economic minerals mined for such commodities as copper and uranium.

Grain size

Grain-size limits for sand and gravel, in mm and in ϕ (phi) units ($-\log_2$ of the grain diameter) are given below. The subdivision is logarithmic. That is, from very-fine sand to pebble grade, each grade size limit is double its predecessor. Similar grade-size subdivisions are used for gravel-size particles but are not shown.

	boulder	
256 mm		-8ϕ
	cobble	
64		-6ϕ
	pebble	
2		-1ϕ
	very coarse sand	
0.5		1ϕ
	medium sand	
0.25		2ϕ
	fine sand	
0.125		3ϕ
	very fine sand	
0.0625		4ϕ

Modifiers may be used to indicate minor components, for example, silty, very fine-grained sandstone, pebbly sandstone, muddy pebble-conglomerate.

For most practical purposes, for example for regional basin analysis or petroleum exploration, grain size may be estimated from hand specimens using a grain-size comparison chart (Miall, 1999, p.26). Vertical stratigraphic changes in grain size of several classes, for example, from coarse-grained sandstone upward to fine-grained sandstone, are significant in terms of changes in transport processes and depositional environments, and are obvious enough to not require confirmation by extensive laboratory analysis. For example, graded bedding is a common result of transport by sediment gravity flow processes, including turbidity currents and debris flows, while a cruder upward decrease in grain size in successive beds is a common result of the gradual filling of channels in fluvial, deltaic, and tidal settings.

Regional variations in sand or gravel grain size may yield information on proximity to sediment sources and on transport directions. Measurements of the largest particles in a gravel (or conglomerate) are particularly useful for such purposes, because gravel clast size decreases rapidly with distance of transport. For example, the downstream diminution of clast size on an alluvial fan at the margins of a basin (as a result of attrition and sorting), and the transition from a fan environment into a larger river may be readily seen from a map of clast-size variations. Mapping such variations may be particularly useful in lithified and deformed rocks where the paleogeography may not be readily apparent from basic stratigraphic data. Grain size changes resulting from attrition or sorting during transport are much less marked in sands, and may require detailed study if regional changes are to be examined.

For more detailed work, laboratory procedures and statistical data analysis methods are available. Use of these methods has declined in recent years because they do not yield additional information that is significant in most basin analysis applications. However, studies of sediment transport and bedform generation may require more precise and detailed information.

The grain size of loose sand may be measured by the use of sieves, settling tubes or electronic grain counting methods. Lithified sandstones may be analyzed in thin section, using a moveable stage to count a set number of grains (typically several hundred). When logarithmic grain-size classes are used, as in the table above, size distributions approximate that of the normal (Gaussian) curve. Cumulative distributions, plotted on a probability ordinate, approximate a straight line. The steepness of the line (the narrowness of the peak about the mean in a histogram) indicates the degree of "sorting" of the sand. Wind transport and wave sorting tend to generate well-sorted sands, that is, sand with the greater bulk of the sand concentrated over a limited size range, while fluvial transport generates more poorly sorted sands. Many attempts have been made to use various statistical measures of the grain-size distribution to identify sediment transport processes, in order to facilitate sedimentological work on very small samples, such as drill cuttings. However, none of these methods is very reliable, especially in lithified sediments, for a variety of reasons. For example, grain-size distribution may be affected by the size distribution of grains in the source rock, and may be modified by diagenesis. Facies analysis methods are quicker to apply and more reliable.

Grain shape

Shape may be defined in terms of three different parameters, roundness, sphericity, and form. Roundness may be measured following a detailed laboratory procedure, but is more commonly estimated by comparison with sets of images of known roundness. These show variations from very angular, through angular, sub-angular, sub-rounded, rounded to well rounded. Fragments are typically angular immediately following breakage from a parent rock, and acquire roundness through attrition during transportation. High degrees of roundness usually are displayed only by particles that have undergone intense wind abrasion, or are multicycle grains.

Sphericity may be defined as the ratio between the diameter of the sphere with the same volume as the particle, and the diameter of the sphere circumscribed around the grain. A more significant measure in terms of its hydraulic significance is the "maximum projection sphericity", which depends on the maximum projection area, and is a better indicator of the fluid resistance of the particle.

Detrital particles may be characterized as three-dimensional ellipsoids with three measurable axes, a short axis, an intermediate axis, and a long axis. Form is defined by the use of two shape indices, the ratio of the intermediate to long axis and the ratio of the short to intermediate axis. The relationships between these two indices define four form classes: oblate (flattened sphere), bladed, prolate (rod-shaped) and equant (round disk).

Studies of these three parameters have some limited usefulness in investigations of sediment transport. Roundness and sphericity increase with distance of transport, while form affects efficiency of transport. Rod-shaped particles are more easily transported in rivers; disks are more common on wave-formed beaches.

Grain orientation

Traction-current transport tends to align sand and gravel grains such that elongate (rod-shaped) particles commonly align themselves either parallel or perpendicular to the direction of movement. Flat, disk-shaped clasts tend to assume a position on the bed with their maximum projection plane dipping toward the direction of flow—upstream in fluvial environments, facing the sea in wave-formed beaches. This is referred to as an imbricated fabric. Particles transported by sediment gravity flow may also assume preferred orientations. In debris flows, clasts may be rotated by internal shear immediately prior to the cessation of flow, with imbricated, flat (bedding-plane-parallel) or vertical orientations common. Observation and measurement of these fabrics may provide useful information on paleoflow directions (Martini, 1971; Schaefer and Teysen, 1987).

Composition

Sand and gravel represent the detrital products of weathering. Most detrital particles are generated by mechanical weathering, with subsequent fragmentation and rounding of grains taking place by attrition during downslope movement and during movement by wind and water. Chemical weathering typically removes much of the source materials in solution, but may leave sand- and gravel-sized particles as a residue.

The composition of sand and gravel deposits reflects that of the source materials, as modified by weathering and transportation (Blatt, 1992). Some types of detrital particle are much more resistant to weathering and erosion than others. Quartz, an abundant mineral in igneous and metamorphic sources, is highly resistant to mechanical and chemical weathering processes, and is by far the most common detrital sand particle. Minerals such as zircon, rutile, tourmaline and garnet, which occur as accessories in igneous and metamorphic sources, are also resistant and constitute minor components of many sandstones. Grains of this resistant group may be recycled through more than one episode of sandstone formation and erosion, preserving evidence of their lengthy history as detrital particles in the form of well-rounded shapes, and remnants of earlier grain outlines preserved beneath fragments of cement.

After quartz, the next most common detrital particles are feldspars, although these tend to weather readily. After this come a wide range of rock fragments. Most types of igneous, metamorphic and sedimentary rocks occur as fragments in sand and gravel deposits. The smaller, sand-sized particles may consist of single mineral crystals of the source rock, such as rounded fragments of quartz or feldspar, or more or less homogeneous pieces of fine-grained rocks such as shale or slate. Larger fragments, including many gravel particles, may consist of recognizable rock types, such as rounded pieces of an earlier sandstone, or a cobble of a coarse-grained plutonic rock.

Sands and gravels derived by erosion of older rocks, as described in the previous paragraphs, are classified, together with mudrocks and silts, as siliciclastic sediments, indicating the predominance of silica-rich detrital particles in their composition.

By contrast, in temperate to tropical marine settings, sands composed entirely of calcium carbonate may accumulate in shelf and shoreline environments from the fragments of carbonate-secreting organisms such as invertebrate shells and corals. The seawater in such places is commonly saturated with the calcium and carbonate ions, and lime precipitation is common. A particularly distinctive carbonate sand is one composed of ooliths, which are particles consisting of concentric laminae of lime crystallized around a small nucleus, such as a tiny shell fragment or a wind-blown quartz particle. In contrast to siliciclastic sediments, carbonate sands are rarely transported far, typically being confined to the shoreline or shelf environment in which they are formed.

In rare instances sands may be composed of evaporite particles. Evaporite deposits, such as halite and gypsum, may form by evaporation of water bodies in hot, enclosed basins. Erosion and resedimentation of this material may take place during flash flood events, during which evaporite bodies may be fragmented, and the clasts redistributed as a form of sand.

Siliciclastic sandstones may be classified according to their detrital composition. Many different classification schemes have been devised. That shown in Figure S1 is one of the more popular. In most sandstones, each detrital particle is composed of a single mineral species, such as a crystal or crystal cluster of quartz, feldspar, calcium carbonate, etc. Arenites are distinguished from wackes by the proportion of detrital matrix (grains <30 microns in diameter). Arenites are defined as sandstones containing less than 15 percent matrix, wackes as those with 15–75 percent matrix, while rocks consisting of more than 75 percent fine-grained particles are termed mudrocks. Particles larger than 30 microns are grouped into three categories that define the corners of the triangular classification. Sandstones are classified according to the proportions of the various detrital particles, summed to 100 percent. Determinations of these proportions are typically carried out by thin-section analysis using a movable stage. Except in cases of unusually pure (monomineralic) sands, counts of 300–400 grains are typically required to obtain reliable (stable) compositional data.

Sand-sized quartz grains, consisting of a few intergrown crystals of silica, are derived from coarse-grained granitic igneous or metamorphic rocks. Sand-sized quartz grains consisting of numerous small intergrown crystals may have been derived from finer grained igneous or metamorphic sources. A metamorphic origin is commonly revealed by a "strained" crystal fabric—individual crystals go into extinction under cross polarizers under slightly different positions as the microscope stage is rotated, indicating slight distortions of the crystals lattice. Silica grains of sedimentary origin may also be present in the form of chert, which occurs as mosaics of tiny interlocking grains, commonly preserving replaced fossil structures. Chert is normally counted as a rock fragment for the purpose of sandstone classification. Feldspathic sandstones typically are derived from granites or granite gneisses. Lithic sandstones may come from a variety of sources, and are

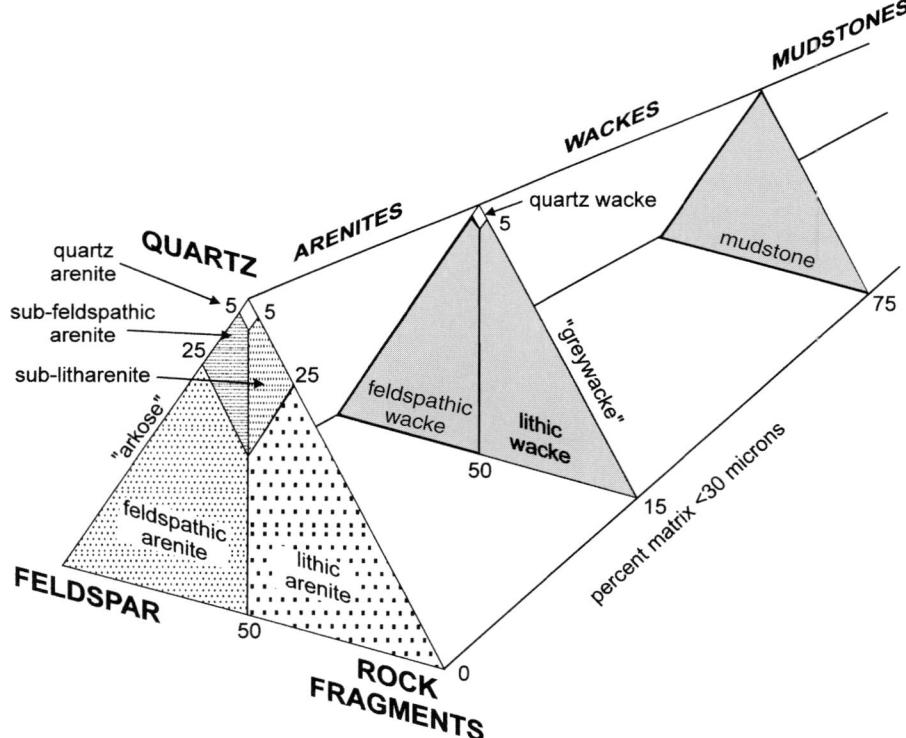

Figure S1 Classification of sand and sandstone according to composition. Sand is classified according to the relative proportion of three major classes of detrital composition. Separation of the three triangles is on the basis of the proportion of detrital matrix (modified from Dott, 1964).

characterized by the predominance of recognizable rock fragments. For example, modern "black-sand beaches" in Hawaii are composed of basalt fragments derived from wave erosion of fresh lava flows. Green beaches are composed largely of olivine grains, accumulated by selective wave sorting from the same volcanic sources. Sands with such unusual compositions are rare in the rock record because of the chemical and mechanical instability of basalt grains under normal weathering conditions. Quartz arenites typically represent recycled deposits from which other minerals have been removed by mechanical or chemical destruction. It is likely that many pure quartz arenites contain grains that were originally derived from a primary igneous or metamorphic source hundreds of millions of years before ending up in the rock in which they are found today. Some rock fragments may oxidize or otherwise decompose in place, and may be represented by clots of clay, that can be mistaken for detrital matrix. This has been called "pseudomatrix."

Figure S1 includes two old petrographic terms that are not recommended, but are still in use. "Arkose" is an old term for feldspathic arenites, and "greywacke" is an old term for lithic wackes.

Variations in detrital composition areally across a basin or vertically, through a stratigraphic section, can yield much useful information about the paleogeography of a basin and the uplift history of surrounding source terranes, especially if such studies are carried out in concert with analysis of grain-size variations, as noted in the previous section. Given the varying resistance of detrital species to destruction by the processes of transportation, areal variations in composition may yield information about the location of specific sediment sources, and distances of sediment transport away from them. Certain components, particularly distinctive rock fragments, may be capable of being traced back to specific source rocks exposed in a source terrane. Direction and distance of transport may be indicated qualitatively by a decrease in the proportion and grain size of such distinctive components, with a concomitant increase in the proportion of stable grains, particularly quartz. For example, a localized carbonate reef deposit or a small igneous intrusion may yield floods of cobbles or boulders of a distinctive composition to an alluvial fan deposit banked against the adjacent basin margin, with the size and abundance of such grains decreasing toward the basin center as the particles are destroyed by attrition and

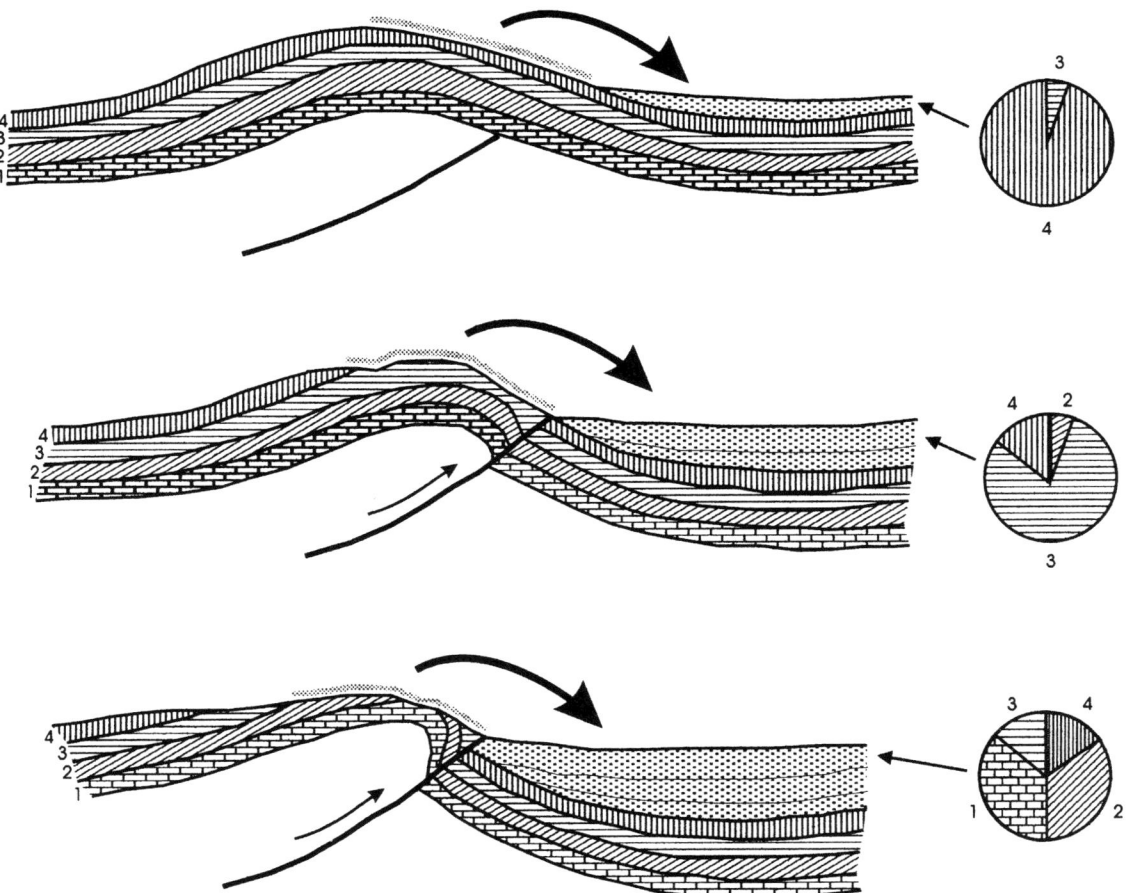

Figure S2 Three successive cross-sections illustrating the erosional unroofing of an uplifted fault block. Detritus from successive stratigraphic units 1, 2, 3, and 4 appear in inverse order in the sand detritus transported to the basin at right, as shown in the pie diagrams (after Graham et al., 1986).

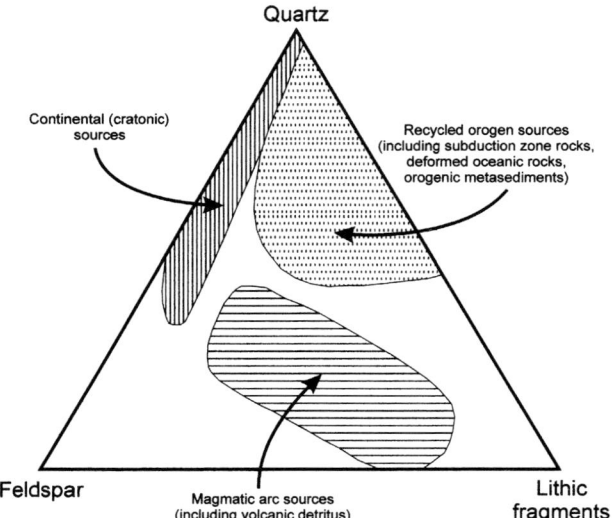

Figure S3 An example of a detrital sand composition plot showing the fields representing derivation from three broad classes of plate-tectonic setting (after Dickinson and Suczek, 1979).

weathering and are diluted by intermingling with other species from other sediment sources.

Vertical variations in detrital composition through a stratigraphic succession may yield invaluable information about the tectonic evolution of a source terrane. For example, proximal sand and gravel deposits in fault-bounded basins commonly display an "inverted stratigraphy" effect, in which younger strata are composed of fragments of successively older source rocks, that are exposed to erosion by gradual unroofing and exposure of deeper levels to surface weathering. Careful analysis of such detrital sequences may provide much information about the tectonic events involved in the building of a source terrane, for example, the emplacement of particular thrust plates or the unroofing of certain igneous bodies (e.g., Graham et al., 1986; see Figure S2). Suturing and uplift of terranes on convergent continental margins may yield "overlap assemblages" of detrital particles derived from one terrane but transported onto and deposited on the adjacent terrane. Such occurrences can help in the dating of the amalgamation events.

Siliciclastic rocks may be characterized by their petrofacies, which is the term used to encompass stratigraphic variations in their detrital composition. Petrofacies characteristics may comprise a useful mapping tool, and may convey valuable information about regional controls on sediment sources (Dickinson and Rich, 1972).

In broad terms, detrital composition reflects the plate-tectonic setting of the source terranes (Potter, 1978, 1984; Dickinson and Suczek, 1979), a relationship that may be useful in the analysis of ancient and deformed rocks, where the plate-tectonic evolution of the area may be one of the problems to be investigated. For example, crystalline cratonic rocks tend to yield quartz- and feldspar-rich detritus; sands derived from magmatic arcs are commonly rich in lithic fragments, particularly volcanic detritus, and so on (Figure S3). Interpretations need to be made with caution, however, because of the many local factors that affect petrographic composition, including weathering, transport history, depositional environment, and diagenesis (Miall, 1999, p.564).

Andrew D. Miall

Bibliography

Blatt, H., 1992. *Sedimentary Petrology*, 2nd edn. W. H. Freeman and Company.
Dickinson, W.R., and Rich, E.I., 1972. Petrologic intervals and petrofacies in the Great Valley sequence, Sacramento Valley, California. *Geological Society of America Bulletin*, **83**: 3007–3024.
Dickinson, W.R., and Suczek, C.A., 1979. Plate tectonics and sandstone compositions. *American Association of Petroleum Geologists Bulletin*, **63**: 2164–2182.
Dott, R.H. Jr., 1964. Wacke, greywacke and matrix—what approach to immature sandstone classification? *Journal of Sedimentary Petrology*, **34**: 625–632.
Graham, S.A., Tolson, R.B., DeCelles, P.G., Ingersoll, R.V., Bargar, E., Caldwell, M., Cavazza, W., Edwards, D.P., Follo, M.F., Handschy, J.F., Lemke, L., Moxon, I., Rice, R., Smith, G.A., and White, J., 1986. Provenance modeling as a technique for analyzing source terrane evolution and controls on foreland sedimentation. In Allen, P.A., and Homewood, P. (eds.), *Foreland Basins*. International Association of Sedimentologists, Special Publication, 8, pp. 425–436.
Martini, I.P., 1971. A test of validity of quartz grain orientation as a paleocurrent and paleoenvironmental indicator. *Journal of Sedimentary Petrology*, **41**: 60–68.
Miall, A.D., 1999. *Principles of Sedimentary Basin Analysis*, Third edition. Springer-Verlag.
Potter, P.E., 1978. Petrology and composition of modern big-river sands. *Journal of Geology*, **86**: 423–449.
Potter, P.E., 1984. South American modern beach sand and plate tectonics. Nature, **311**: 645–648.
Schaefer, A., and Teysen, T., 1987. Size, shape and orientation of grains in sands and sandstones: image analysis applied to rock thin-sections. *Sedimentary Geology*, **52**: 251–271.

Cross-references

Classification of Sediments and Sedimentary Rocks
Climatic Control of Sedimentation
Diagenesis
Feldspars in Sedimentary Rocks
Grain Size and Shape
Provenance
Quartz, Detrital
Rivers and Alluvial Fans
Tectonic Control of Sedimentation

SAPROPEL

An unlithified, fine-grained, often laminated, black organic-rich deposit containing mainly macrophytic (benthic) plant remains that have been decomposed anaerobically. The original definition (Wasmund, 1930) refers to lacustrine environments where the bottom water contains dissolved H_2S and a benthic fauna is consequently absent. The term has subsequently been extended to include marine deposits that are olive-green, gray or brown in color and contain both planktonic as well as macrophytic organic material. Examples of such sediments occur in the Holocene–Pleistocene sediments of the Black Sea (Strakhov, 1971) and the Holocene–Pliocene section of the Mediterranean Sea (Kullenberg, 1952;

Kidd et al., 1978), including poorly-lithified rocks outcropping in the northern borderlands of the Mediterranean (van Os et al., 1994).

Composition

According to the definition of Kidd et al. (1978), sapropels contain more than 2 wt. percent organic carbon and are at least 1 cm thick; sapropelic sediments are similarly defined but contain 0.5–2 wt. percent carbon. Organic carbon contents in Black Sea and Mediterranean sapropels can range up to 16 wt. percent and 30 wt. percent, respectively. The sapropels of the Black Sea contain both terrestrial and marine organic material (Wakeham et al., 1991), whereas in the Mediterranean examples the organic fraction is overwhelmingly marine (ten Haven et al., 1987). Sapropels also contain variable amounts of biogenous carbonate and silica together with fine-grained aluminosilicates. In common with other organic-rich marine sediments, sapropels have high concentrations of pyrite and a suite of minor and trace metals that often include Ba, Cu, Mo, Ni, U, V, and Zn (Calvert, 1983).

Black Sea sapropels

A black, very fine-grained sapropel has accumulated in the deep basin of the Black Sea over the last 7,500 years (Degens and Ross, 1972; Jones and Gagnon, 1994). The lower part of this deposit is especially organic rich (up to 20 wt. percent organic C), whereas the modern section contains 5 wt. percent organic C. A benthic fauna is absent in this permanently anoxic basin. The black coloration is caused by the presence of iron monosulfide, which readily oxidizes when it is exposed to the atmosphere, causing the color of the sediment to fade to dark grey (Berner, 1974). Sapropel formation began when saline Mediterranean waters entered the Black Sea basin via the Bosphorus Strait as a result of post-glacial sea level rise. This changed the Pleistocene Black Lake into the modern brackish (salinity 11^a), anoxic basin (Middelburg et al., 1991). The accumulation rate of organic material increased markedly at this transition, and the accumulation rate of coccolith calcite increased about 2,700 years ago (Jones and Gagnon, 1994), thereby diluting the organic flux and initiating the formation of the modern laminated calcareous marls (Calvert et al., 1991). Sapropels are also known to occur in the Pleistocene section of the deep Black Sea (Calvert, 1983).

Mediterranean sapropels

Multiple horizons of discrete centimeter- to decimeter-thick olive green to brown sapropels are found intercalated in the normal nanofossil marls of the Mediterranean Sea. They were discovered in the eastern basin during the Swedish Deep-Sea Expedition in 1947 (Kullenberg, 1952) confirming a prediction of Bradley (1938) that laminated, organic-rich glacial sediments should have formed when lowered sea level caused stagnation of the partially restricted eastern basin. Drilling in the Mediterranean over the last 25 years has extended the occurrence of these deposits throughout the eastern and western basins back to the Early Pliocene (Murat, 1999; Emeis et al., 2000). The sapropels (>2 percent $C_{organic}$) are up to 70 cm thick and contain up to 30 wt. percent organic carbon. They may be faintly laminated, and either contain an impoverished benthic microfauna or are barren (Jorissen, 1999). They contain relatively high concentrations of pyrite but only trace amounts of iron monosulfide (Passier et al., 1997); consequently their color does not change on exposure to the atmosphere. They are restricted to periods of climatic amelioration in the Mediterranean when the monsoon-driven freshwater balance changed drastically, and bottom water oxygen levels decreased leading to dysoxic or anoxic conditions on the seafloor.

Formation of sapropels

Sapropels are considered to form by one or a combination of the following processes: (1) increased preservation of organic matter in the bottom sediments; (2) high rates of supply of organic matter to the seafloor; (3) decreased dilution of the organic fraction by other sediment components. The preservation hypothesis focuses on the lower rate of decomposition of deposited organic matter under anoxic or dysoxic conditions because of the lower energy yield of microbially-driven degradation of organic matter using oxidants other than oxygen (nitrate, Mn and Fe oxides and sulfate). Thus, the factors that promote the establishment of bottom water anoxia (or dysoxia) regardless of organic matter supply ultimately drive sapropel formation (Ryan and Cita, 1977). An alternative view centers on situations where the supply of organic matter relative to that of other sediment components increases, either by higher planktonic production in the overlying waters or by higher rates of supply of terrestrial organic matter (Calvert, 1987). Under these circumstances, lower bottom water oxygen levels are due to the consumption of dissolved oxygen by the degradation of settling organic matter and not necessarily by basin restriction, sea level change, etc.; anoxia is a consequence of higher production. Decreased dilution of the organic fraction of a sediment would amplify the effects of anoxia or organic supply emphasized in the other two mechanisms of sapropel formation.

The balance of evidence suggests that the Black Sea sapropel began to form when anoxic conditions were established in the basin following the influx of saline waters from the Mediterranean when sea level reached the Bosphorus sill (Degens and Ross, 1974). Whether the anoxic conditions promoted the preferential preservation of organic matter or production in the basin increased markedly following the transformation of the basin from the lake to the modern marine phase are still matters of dispute. Nevertheless, the bulk accumulation rate of organic matter in the modern facies is similar to rates determined in oxygenated settings at similar water depths and sedimentation rates in the open ocean (Calvert et al., 1991). In the Mediterranean, detailed studies of the geochemistry and micropalaeontology of the sapropels and associated sediments (Wehausen and Brumsack, 1999; Calvert and Fontugne, 2001; Mercone et al., 2001) show that bottom waters ranged between full anoxia and dysoxia, reflected in the high redox-sensitive trace metal (Cu, Mo, Ni, U, V) contents, and that planktonic production was significantly higher (high Ba) during their formation. They formed during periods of higher fresh water input to the eastern basin, driven by more intense monsoonal rainfall over northern Africa and the northern Mediterranean borderlands at the precessional frequency (Rossignol-Strick et al., 1982; Rohling, 1994), which probably led to increased stratification and more restricted vertical circulation. The evidence for higher production requires that nutrient supply was higher, and this has been ascribed to a reversal in circulation (Berger, 1976) such that an estuarine circulation regime (surface water outflow and subsurface water inflow) replaced the modern anti-estuarine circulation (deep water outflow and surface water inflow).

Summary

Sapropel is an organic-rich sediment or sedimentary rock that may be finely laminated; it lacks, or has a depauperate, benthic fauna and high concentrations of redox-sensitive trace metals. Holocene and Pleistocene examples of this facies evidently formed by high organic matter supply and/or preferential preservation of deposited organic matter under anoxic/dysoxic bottom water or sediment conditions in partially restricted environments.

Stephen E. Calvert

Bibliography

Berger, W.H., 1976. Biogenous deep-sea sediments: production, peservation and interpretation. In Riley, J.P., and Chester, R. (eds.), *Chemical Oceanography*. Academic Press, pp. 265–388.
Berner, R.A., 1974. Iron sulfides in Pleistocene deep Black Sea sediments and their paleo-oceanographic significance. In Degens, E.T., and Ross, D.A. (eds.), *The Black Sea—Geology, Chemistry and Biology*. American Association of Petroleum Geologists Memoir, pp. 524–531.
Bradley, W.H., 1938. Mediterranean sediments and Pleistocene sea levels. *Science*, **88**: 376–379.
Calvert, S.E., 1983. Geochemistry of Pleistocene sapropels and associated sediments from the eastern Mediterranean. *Oceanologica Acta*, **6**: 255–267.
Calvert, S.E., 1987. Oceanographic controls on the accumulation of organic matter in marine sediments. In Brooks, J., and Fleet, A.J. (eds.), *Marine Petroleum Source Rocks*. Geological Society of London, pp. 137–151.
Calvert, S.E., and Fontugne, M.R., 2001. On the Late Pleistocene–Holocene sapropel record of climatic and oceanographic variability in the eastern Mediterranean. *Paleoceanography*, **16**: 78–94.
Calvert, S.E., Karlin, R.E., Toolin, L.J., Donahue, D.J., Southon, J.R., and Vogel, J.S., 1991. Low organic carbon accumulation rates in Black Sea sediments. *Nature*, **350**: 692–694.
Degens, E.T., and Ross, D.A., 1972. Chronology of the Black Sea over the last 25,000 years. *Chemical Geology*, **10**: 1–16.
Degens, E.T., and Ross, D.A., 1974. Recent sediments of the Black Sea. In Degens, E.T., and Ross, D.A. (eds.), *The Black Sea—Geology, Chemistry and Biology*. American Association of Petroleum Geologists Memoir 20, pp. 183–199.
Emeis, K.-C., Sakamoto, T., Wehausen, R., and Brumsack, H.-J., 2000. The sapropel record of the eastern Mediterranean Sea—results of Ocean Drilling Program Leg 160. *Palaeogeography Palaeoclimatology Palaeoecology*, **158**: 371–395.
Jones, G.A., and Gagnon, A.R., 1994. Radiocarbon chronology of Black Sea sediments. *Deep-Sea Research*, **41**: 531–557.
Jorissen, F.J., 1999. Benthic foraminiferal successions across late Quaternary Mediterranean sapropels. *Marine Geology*, **153**: 91–101.
Jorissen, F.J., Barmawidjaja, D.M., Puskaric, S., and van der Zwaan, G.J., 1992. Vertical distribution of benthic foraminifera in the northern Adriatic Sea: the relation with the organic flux. *Marine Micropaleontology*, **19**: 131–146.
Kidd, R.B., Cita, M.B., and Ryan, W.B.F., 1978. Stratigraphy of eastern Mediterranean sapropel sequences recovered during Leg 42A and their paleoenvironmental significance. In Hsu, K.J., and Montadert, L. (eds.), *Initial Reports of the Deep-Sea Drilling Project*. U.S. Government Printing Office, pp. 421–443.
Kullenberg, B., 1952. On the salinity of the water contained in marine sediments. *Goteborgs Kungl. Vetenskaps. Vitt-Samhal. Handlingar*, **6B**: 3–37.
Mercone, D., Thomson, J., and Troelstra, S.R., 2001. High-resolution geochemical and micropalaeontological profiling of the most recent eastern Mediterranean sapropel. *Marine Geology*, **177**: 25–44.
Middelburg, J.J., Calvert, S.E., and Karlin, R.E., 1991. Organic-rich transitional facies in silled basins: response to sea-level change. *Geology*, **19**: 679–682.
Murat, A., 1999. Pliocene–Pleistocene occurrence of sapropels in the western Mediterranean Sea and their relation to eastern Mediterranean sapropels. In Zahn, R., Comas, M.C., and Klaus, A. (eds.), *Proceedings of the Ocean Drilling Program Scientific Results*. Volume 161, pp. 519–528.
Passier, H.E., Middleburg, J.J., De Lange, J.J., and Böttcher, M.E., 1997. Pyrite contents, microtextures, and sulfur isotopes in relation to the formation of the youngest eastern Mediterranean sapropel. *Geology*, **25**: 519–522.
Rohling, E.L., 1994. Review and new aspects concerning the formation of eastern Mediterranean sapropels. *Marine Geology*, **122**: 1–28.
Rossignol-Strick, M., Nesteroff, W., Olive, P., and Vergnaud-Grazzini, C., 1982. After the deluge: Mediterranean stagnation and sapropel formation. *Nature*, **295**: 105–110.
Ryan, W.B.F., and Cita, M.B., 1977. Ignorance concerning episodes of ocean-wide stagnation. *Marine Geology*, **23**: 193–215.
Strakhov, N.M., 1971. Geochemical evolution of the Black Sea in the Holocene. *Lithology and Mineral Resources*, **3**: 263–274.
ten Haven, H.L., Baas, M., de Leeuw, J.W., Schenck, P.A., and Brinkhuis, H., 1987. Late Quaternary Mediterranean sapropels, II, Organic geochemistry and palynology of S_1 sapropels and associated sediments. *Chemical Geology*, **64**: 149–167.
van Os, B.J.H., Lourens, L.J., Hilgen, F.J., de Lange, G.J., and Beaufort, L., 1994. The formation of Pliocene sapropels and carbonate cycles in the Mediterranean: diagenesis, dilution and productivity. *Paleoceanography*, **9**: 601–617.
Wakeham, S.G., Beier, J.A., and Clifford, C.H., 1991. Organic matter sources in the Black Sea as inferred from hydrocarbon distributions. In Izdar, E., and Murray, J.W. (eds.), *Black Sea Oceanography*. Kluwer Academic Publishers, pp. 319–341.
Wasmund, E., 1930. Bitumen, Sapropel, und Gyttja. *Geologiska Forenigens Forhandlingar*, **52**: 315–350.
Wehausen, R., and Brumsack, H.-J., 1999. Cyclic variations in the chemical composition of eastern Mediterranean Pliocene sediments: a key for understanding sapropel formation. *Marine Geology*, **153**: 161–176.

Cross-references

Black Shales
Oceanic Sediments

SCOUR, SCOUR MARKS

Scour is a general sedimentary process which brings about the sustained lowering of a surface by the direct or indirect action upon it of a current of water or air. The process normally acts differentially, resulting in a range of distinctive forms, called scour or erosional marks. Differential solution, as of limestone surfaces or beds of halite or gypsum, gives rise to structures resembling some kinds of scour mark, but the concept of scour excludes this particular mechanism. Scour has been much studied by engineers in connection with the stability of bridge piers and the control of drifting snow along highways.

The mechanisms of scour (Allen, 1982a) are: (1) entrainment; (2) stripping; and (3) corrasion. *Entrainment* affects surfaces composed of sand or gravel, and sees particles removed one by one as the result of the direct action of the shear and pressure forces exerted by the moving fluid, to which may be added impacts due to particles already in transport as they return toward and strike the bed. Particles become entrained when friction and their immersed weight are insufficient to retain them on the bed. *Stripping* is restricted to surfaces of soft mud in aqueous environments. It involves the tearing of clumps of mud from the bed by the direct action of the current. Depending on the consistency of the bed and the magnitude of the stress, these clumps range in size from tenuous wisps a few millimeters across to plumose-marked plastic lumps measured in decimeters. *Corrasion* sees the removal of material from the surface by the action of fluid-driven debris, which either pounds the bed or chisels fragments from it, depending on its size and the strength of the bed. Generally speaking, corrasion affects materials—stiff or dried muds and rocks—which are too strong to be removed by stripping. The tools available to currents are various: sand grains, gravel, shells and even pieces of ice.

From a genetic standpoint, scour marks may for convenience be grouped into two largely distinct classes, termed current-excited and defect-excited. *Current-excited scour marks* are those which result when a bed is affected by an imposed, stationary pattern of currents, for example, a secondary flow, which create over the bed a corresponding variation in the value of the shear stress or the concentration plus effectiveness of any transported tools. There is no requirement for any significant inhomogeneity in the consistency of the bed. *Defect-excited scour marks* are those which arise when a current flows over a bed containing some gross, localized iregularity or defect. Some defects stand bluff to the current, for example, an upright plant stem, a pebble or shell, a block of stranded ice, or a half-emergent concretion. Others are negative features that reach down from the surface, for example, an open invertebrate burrow on the sea bed, or shrinkage cracks on a surface of dried mud awaiting the next river flood. In all of these cases, the defect creates around itself a pattern of currents which lead to differential scour. Of course, no real surface acted on by a current can be perfectly homogeneous, and it is therefore possible for an unstable interaction between the surface and the current to arise which, in the long term, creates a pattern of scour and a corresponding set of apparently organized forms. Strictly, such scour marks are defect-excited, but the defects in such cases were never gross and remain unseen.

Scour marks are a diverse group and have been reported and discussed from many different environments in a very considerable literature (Allen, 1982b). The main types are as follows:

Current-excited scour marks

    gutters and gutter casts
    ridges-and-furrows (wind)
    yardangs
    furrows-and-ridges (wave-related)

Defect-excited scour marks

    current crescents
    flute marks
    potholes and pothole casts

Gutters, fossilized as gutter casts, are sets of long, occasionally branching, flow-parallel furrows, and ridges found where eroding river and tidal currents drive coarse debris over surfaces of mud and occasionally gravel or rock. Typically, in fluvial, intertidal and shallow subtidal depths, their transverse spacing ranges between a few decimeters and a few meters, and the sides may overhang. There is evidence that they are localized by secondary currents (paired corkscrew vortices) which concentrate the bed-material in transport into flow-parallel bands, but single gutters could be defect-related. Depositional events, mainly on the ridges, commonly make a modest contribution to the relief of the intertidal forms. In subtidal environments such as the English Channel, where depths measure tens of meters, the grooves occur in sandy-shelly gravels. They have a characteristic transverse spacing of the order of 100 m, and may be many hundreds of meters long. Entrainment is probably the chief mechanism involved in their formation.

Sets of flow-parallel ridges-and-furrows occur in deserts, chiefly the Sahara, where vigorous, sand-laden winds of roughly constant direction are able to corrade comparatively level and uniform rock surfaces. They are probably related to the sets of corkscrew vortices that form in the thick atmospheric boundary-layer when thermal instability occurs. Ridges-and-furrows are huge structures found over vast areas. In depth they range from a few to many meters, lie hundreds of meters apart parallel with the wind, and take extreme lengths of tens of kilometers. As their length shortens, ridges-and-furrows grade into the desert landforms known as yardangs. These are comparatively short ridges of lemniscate plan, the bluff end facing the wind, which arise where comparatively soft materials, such as mudstone, diatomite or friable sandstone, are subject to corrasion. Erosional forms similar to yardangs have been shaped by extreme floods, on both the Earth when ice-dams have broken and, apparently, in the past on other planetary bodies.

Wave-related furrows-and-ridges are found on muddy shores in erosional retreat, such as some saltmarsh edges, and on intertidal rock platforms at the feet of cliffs. In the first of these contexts, where the term mud mound is often applied, they appear as sets of furrows and commonly overhanging ridges arranged at right-angles to the shore. Typically, the sides of the ridges are striated, and the tools responsible for the corrasion, chiefly pebbles and shells, are visible on the floors of the furrows. The transverse spacing of the structures, ranging from a decimeter or so to a few meters, appears to increase with the degree of exposure of the shore to wave activity, supporting the idea that furrows-and-ridges reflect the complex current patterns due to edge waves. Also oriented at right-angles to shore are the furrows-and-ridges that can occur at the feet of cliffs. Less is known about them, but they are similar to the structures developed in mud and may have the same origin.

Current crescents are defect-related scour marks that form because of stripping and/or corrasion around bluff bodies in fluvial and tidal environments and, occasionally, in deep water where turbidity currents operate. The body has a number of effects (Paola *et al.*, 1986). Firstly, it compresses the current around the sides, enhancing its erosive powers. Secondly, the flow decelerates ahead of the body, resulting in the separation of the flow from the bed and the creation of a vigorous, corkscrewing vortex which curves around the front and sides of the body. Thirdly, on the downstream side a sluggish wake is formed in which the rate of erosion is much reduced. The overall result is a U-shaped groove which encircles the front and sides of the body, in the lee of which a dagger-shaped ridge may be found.

Frequently formed by turbidity currents as they drive over mud beds, and occasionally fashioned by river floods, flute marks are probably the commonest and most widespread of scour structures. They are horseshoe-shaped depressions centimeters to decimeters long with arms that open out downcurrent. Typically, they occur in dense groups and are preserved as casts on the undersides of sandstone beds. Experimentally, flute marks are generated most readily by stripping and/or corrasion at negative defects, such as the tops of burrows or shrinkage cracks. Flow separation occurs along the upstream edge of the defect, with the result that enhanced turbulence is advected to the downstream side, where the flow reattaches. The erosion rate is greatest here. Consequently, the defect gradually evolves through differential erosion into a horseshoe-shaped hollow which is symmetrical about the line of flow. Each mature flute mark contains a well-established separated flow, similar to that found in the lee of many sand ripples and dunes. If a surface is scattered with defects, a field of intersecting flute marks can quickly become established. However, a single, isolated flute mark when mature can trigger the formation of secondary marks.

Potholes have most often been reported from powerful rock-bound rivers, but are also known from river floodplains and even intertidal and shallow-marine environments where mud beds can be affected. Wave-action alone may generate them in the marine realm. Fossilized as pothole-casts, they range from shallow hollows to deep, steep-sided shafts drilled into the substrate by the action of swirling masses of sand and gravel which corrade the surface. Potholes range in size from a decimeter or so wide and deep to giants with depths in the neighborhood of 20 m and widths of about 10 m. They seem almost invariably to be initiated at defects in the substrate, such as minor faults, joints or shrinkage cracks. Potholes in fluvial environments are economically important, for when being filled with sediment, they act as traps for placer minerals.

Scour marks record the fact that surfaces affected by aqueous currents or the wind can experience differential erosion reflecting either patterns of flow and fluid force existing in the current or patterns excited in the current by the presence of some gross irregularity of the bed. Most have directional properties which can be used to infer ancient current directions.

John R.L. Allen

Bibliography

Allen, J.R.L., 1982a. *Sedimentary Structures: Their Character and Physical Basis*. Volume 1, Elsevier.
Allen, J.R.L., 1982b. *Sedimentary Structures: Their Character and Physical Basis*, Volume 2, Elsevier.
Paola, C., Gust, G., and Southard, J.B., 1986. Skin friction behind isolated hemispheres and the formation of obstacle marks. *Sedimentology*, 33: 279–293.

Cross-references

Bedding and Internal Structures
Paleocurrent Analysis
Parting Lineations and Current Crescents
Surface Forms

SEAWATER: TEMPORAL CHANGES IN THE MAJOR SOLUTES

Introduction

The exceptional paper by Rubey (1951), entitled "Geologic history of seawater: an attempt to state the problem", crystallized thought concerning the factors affecting seawater composition. After review of the available information, Rubey concluded that the composition of seawater had varied within narrow limits over much of geological time. The systematic nature of his approach and the strength of his arguments resulted in a 50-year consensus that the conclusion was correct. His arguments did not preclude temporal changes to minor and trace solute contents and there are numerous recent papers which indicate that minor and trace element compositions of seawater varied over time (deRonde *et al.*, 1997; Cicero and Lohmann, 2001).

The arguments of Rubey were sufficiently persuasive that only the "more profitable" lines of evidence were pursued and documented in detail until recently. Canfield *et al.* (2000) challenged Rubey's conclusion by presenting evidence that sulfate was low in Archean seawater and had increased in abundance since about 2.5 Ga. The conclusion, however, was most strongly, and convincingly challenged by Lowenstein *et al.* (2001) who demonstrated that Mg/Ca ratio varied between latest Precambrian time and the present, and that the variation was sufficient for aragonite rather than calcite to have precipitated from seawater. These Mg/Ca variations result primarily from the rates of generation of oceanic crust through which seawater circulates *via* thermally driven convection cells (near oceanic ridges). The greater the rate of seafloor generation, the greater the amount of interaction with oceanic crust and the higher Mg/Ca in seawater.

Generation of oceanic crust and implications concerning seawater composition were not considered by Rubey, primarily because such concepts awaited elucidation some 30 years later (Edmond *et al.*, 1979; Sleep *et al.*, 1983; Thompson, 1983). These ideas coupled with the study of Lowenstein *et al.* (2001) and Canfield *et al.* (2000) demonstrate that the concentrations of the major solutes of seawater varied through the Proterozoic and Phanerozoic Eons. There is also evidence that its composition varied appreciably through the Archean Eon. Following Rubey, "lines of evidence" for continual change to major solute contents of seawater are developed here. Unfortunately, the dearth of reliable information (and detailed study) prevents firm conclusions from being drawn about seawater compositions during early periods.

Major solute fluxes to the seawater reservoir

Rubey was unaware of the important role that oceanic crust-seawater interactions had on seawater composition. Neither was he aware of recent findings concerning the compositional evolution of the crust because fundamentally important aspects of its evolution, now apparent, were unknown 50 years ago. The arguments of Rubey (1951) concerning seawater composition are reconsidered in light of recent findings regarding geochemical cycles.

A simple geochemical cycle applicable to all major elements of seawater (Figure S4) emphasizes that seawater constituents

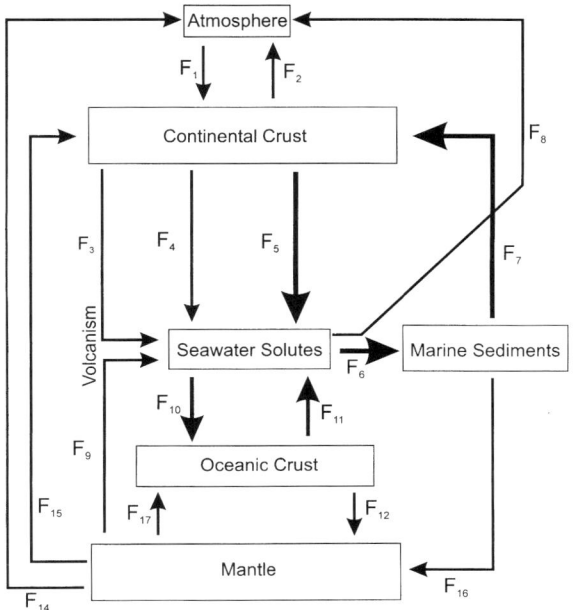
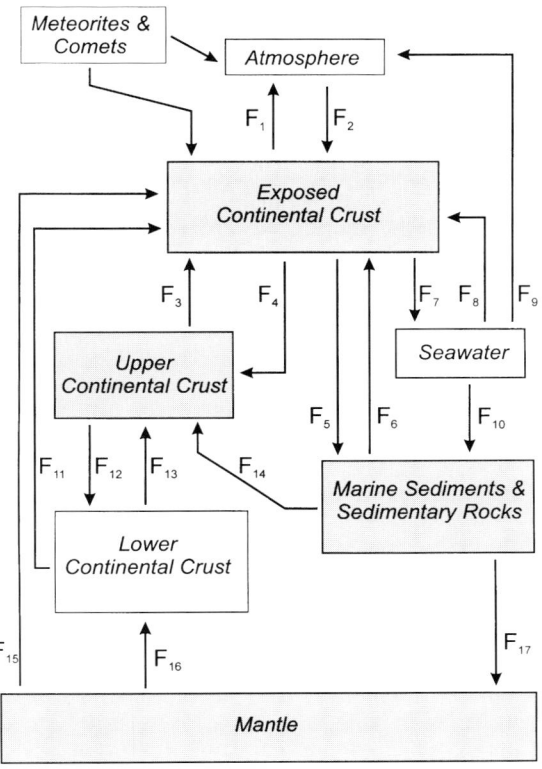

Figure S4 Geochemical cycle applicable to the major solutes of seawater. Each box represents a separate reservoir and each arrow a flux (labelled "F"). The contribution from each reservoir and the magnitude of each flux differs for each major solute. The thicker arrows indicate fluxes important to major solutes, although this is a generalization.

Figure S5 Geochemical cycle applicable to the composition of the Exposed Continental Crust (ECC). Boxes and arrows has the same meaning as in Figure S4. The shaded boxes represent the reservoirs representing the major compositional input to the ECC.

are derived from the continental crust, oceanic crust, and the mantle. Details of the mantle- and oceanic crust-seawater solute fluxes have been addressed recently (Edmond *et al.*, 1979; Sleep *et al.*, 1983; Thompson, 1983; Wolery and Sleep, 1988; Lowenstein *et al.*, 2001). The continental crust is the single largest source for solutes today and may have been in the past (Berner and Berner, 1996), and of these fluxes (Figure S4, F_3, F_4, F_5) river water fluxes (F_5) are the greatest (Berner and Berner, 1996).

Exposed continental crust and seawater composition

The major solutes or seawater are derived in greatest part from chemical, and biological weathering of the Exposed Continental Crust (ECC) (Berner and Berner, 1996) so that the major solute content of seawater is expected to evolve as the composition of the ECC evolves. Aspects controlling the compositional evolution of the ECC may be addressed *via* another geochemical cycle. Figure S5 sub-divides the continental crust into different reservoirs. Most importantly, the ECC is separated from other crustal components. Thus Figure S5 accentuates the various contributions affecting the composition of the ECC.

The composition of the ECC has changed between the Hadean and (at least) late Precambrian time, and much of its compositional evolution can be assessed from geological observations and considerations summarized below. Blatt and Jones (1975) provide a sound estimate of the major geologic units contributing to the ECC. Of the rocks exposed at the surface of the continents; 66 percent are sedimentary and 34 percent are crystalline (defined to include plutonic, extrusive and metamorphic rocks). Of the crystalline rocks, the extrusives (volcanic rocks) constitute about a quarter (8–10 percent of the exposed crust), the remainder (about 25 percent) being plutonic and metamorphic.

Rejuvenation of the ECC proceeds continuously in response to tectonic processes, thus replenishing materials removed by chemical weathering and erosion. The sources of materials for rejuvenation are apparent from Figure S5. They are: marine sedimentary rocks; the upper crust from which volcanic, plutonic and metamorphic rocks are provided; the mantle from which volcanic rocks primarily are provided. The lower crust, seawater, the atmosphere and the extraterrestrial reservoir contribute to the exposed crust, but their contributions are smaller than those from the previously-mentioned reservoirs (Figure S5). The lower crust, for example, may contribute extrusive materials directly to the exposed crust but few other avenues are available for transport directly to the ECC. Lower crustal material more likely enters the upper crustal reservoir before being exposed at the surface. Seawater provides H_2O and salts directly to the continents by subsurface infiltration and indirectly *via* the sedimentary rock (pore waters) and atmospheric (aerosols and solutes of rainwater) reservoirs.

The extraterrestrial contribution (Figure S5, Meteorites and comets reservoir) supplies small amounts of debris to the continents *via* micrometeorites to large bolides. This flux to the

ECC is now negligible but must have been important during the first few hundred Ma of Earth history.

Temporal compositional changes to the ECC

The sedimentary rock reservoir includes carbonates, sandstones and shales. Their proportions exposed at the surface likely is similar to their abundance in the rock record, measurements and estimates of which lead to evaporite:sandstone:carbonate:shale proportions of 3:15:20:62 (Garrels and Mackenzie, 1971, p.206, p.247 and p.249). Because they represent 66 percent of the ECC, and because the ECC is the major contributor to seawater composition, a change in sedimentary rock proportions likely would be reflected in seawater composition. There are well recognized temporal changes to the sedimentary rock reservoir (Vinogradov and Ronov, 1956; Garrels and Mackenzie, 1971; Veizer, 1979, 1985; Godderis and Veizer, 2000) and sympathetic changes to seawater composition almost certainly have occurred.

Sedimentary rock–ECC fluxes

The development of large sedimentary platforms and deposition of compositionally mature platformal sediments commenced at about 3 Ga, the Witswatersrand, Pongola and Buhwa platforms being examples (Fedo et al., 1996). Platform development was diachronous and extended beyond the end of the Archean (2.5 Ga). The first extensive platformal sequence associated with the Canadian Shield is the Huronian sequence (Young, 1995) with an age no older than 2.3 Ga.

Addition of a mature sedimentary rock platformal "blanket" to the exposed continents near the end of the Archean should have changed appreciably the rate of supply of solutes to streams and to seawater. These changes to rates of supply likely resulted in changes to the major alkali and alkaline earth solutes of seawater. Mature siliciclastic sediments are depleted of Na, K, Mg, and Ca relative to the crystalline or volcanic rocks exposed at the surface. Seawater Na, K, and Mg likely declined in abundance as these weathered materials covered at least half of the previously exposed crystalline rocks (preventing their weathering).

Mantle–ECC fluxes

The mantle supplies materials to the ECC *via* volcanism primarily, producing large volcanic fields and "basaltic plateaus". Extrusive rocks today represent about 8–10 percent of the exposed crust but as Blatt and Jones (1975) emphasize, their contribution to marine sediments probably is double this value (contributes 15–20 percent to the sedimentary reservoir). Chemical weathering of volcanic materials is rapid and they should yield solutes to stream and river waters at a rate akin to the rate of production of detritus from the ECC (at least 15–20 percent of seawater solutes supplied by weathering and erosion of mantle-derived volcanics).

The nature of mantle volcanic materials supplied to the ECC has changed over time. Komatiites are found in Archean volcanic sequences but are absent from volcanic sequences of younger age. With the change in mantle input to the ECC, seawater composition likely changed in response. The high Mg content and the labile nature of the constituents of komatiites and high Mg-basalts volcanic rocks probably affected Ca, Mg, and Si contents of seawater, and its salinity may have been greater during the Archean.

Veizer (1979) noted that the sedimentary rocks are considerably more mafic (on average) than average upper crust. He argues that this was due to an early mafic–ultramafic crustal component being added to, and recycled within the sedimentary mass. Veizer estimated that about 30 percent of oceanic basalt would have to be added to 70 percent upper continental crust to generate compositions consistent with the sedimentary rock compositions. The finding attests to the temporal changes to the ECC resulting from mantle–ECC interaction.

Continental crust–ECC fluxes

Theories for genesis of continental crust can be grouped into three categories: (1) homogeneous accretion; (2) inhomogeneous accretion; (3) terrestrial models (Condie, 1989). Brown and Mussett (1981) summarize arguments related to the first two groups and both imply development of a crust during accretion (before the first 300 Ma of Earth history). Much study has been focused on terrestrial models. They consist of two groups, those based on the assumption that the majority of the crust evolved very early without its volume having change greatly over geological time (Armstrong, 1968; Fyfe, 1978; Rudnick, 1995; Sylvester et al., 1997) and those based on the assumption that the early crust was initially of small volume and that its volume increased over time, perhaps dramatically at approximately 2.5 Ga (Godderis and Veiser, 2000). Models of all types have associated ambiguities and inconsistencies and little can be concluded with certainty. The moon, however, has an anorthositic crust which formed prior to, or early during, the late heavy bombardment (3.8 Ga). If the moon and Earth are genetically related as commonly proposed (Taylor, 1997), the Earth almost certainly generated a crust (perhaps anorthositic?) at the same time as (or before) the moon. If the initial crust were primarily anorthositic its weathering would have produced solute contents much different from present-day seawater.

Taylor and McLennan (1985) conclude that the Archean crust was compositionally different from the post-Archean crust, the dramatic change occurring near 2.5 Ga, although recognized to be diachronous. The change was from a dominantly plagioclase, quartz and mafic minerals-dominated Archean crust to a post-Archean crust which included K-rich granites.

As previously noted, the upper continental crust constitutes about 25 percent of the ECC but must have been as much as 50 percent of the ECC prior to 2.5 Ga. Development of sedimentary platforms bordering continents and their subsequent exposure (incorporation into the ECC reservoir), coupled with the simultaneous decrease in the amount of exposed crystalline crust, must have had significant effect on the composition of seawater during the period since about 3 Ga. Canfield et al. (2000) corroborate the conclusion; they demonstrate that sulfate content was low during the Archean but increased to present levels beginning at about the Archean–Proterozoic transition (about 2.5 Ga).

Extraterrestrial–ECC fluxes

The history of the Hadean Eon (4.5 Ga to 3.8 Ga), although highly speculative, has been deduced in some detail (Taylor,

1997). Noble gas ratios and abundances indicate that the present atmosphere and hydrosphere are secondary (Cogley and Henderson-Sellers, 1984; Taylor, 1997). The implication is that the hydrosphere developed either during accretion (after separation of the Moon) or between about 4.2 Ga and 3.8 Ga, the period of "late heavy bombardment". If the crust formed during accretion its composition would have been modified by incorporation of material captured during the late heavy bombardment. Taylor (1997) summarizes the effects on the composition of the mantle of such capture. By analogy, there must have been similar or greater compositional effect on the upper crust and on the ECC. These materials probably were derived from meteorites of carbonaceous chondritic composition or from comets (Chyba, 1990; Taylor, 1997), both of which contain appreciable amounts of water (generally greater than 10 wt. percent, Brown et al., 2000). If ECC composition were affected by the "late heavy bombardment" then seawater composition undoubtedly would have been affected by the extraterrestrial input. The effects of such contributions await study.

Summary

There is compelling evidence that seawater composition has changed continuously during the entire history of the Earth. Compositional change probably was rapid during the transition from the Hadean Eon through the late heavy bombardment and into the Precambrian Era. Changes to the composition of the upper continental crust during the late Archean likely caused change to the major solute concentrations of seawater. The development of stable platforms during the late Archean and Proterozoic, and the incorporation of chemically mature platformal sedimentary rocks into the Exposed Continental Crust (ECC) must have resulted in change to seawater composition. Canfield et al. (2000) demonstrate that sulfate content has changed during the Proterozoic Eon and Lowenstein et al. (2001) demonstrate major solute changes during the Phanerozoic Eon.

Chloride has the longest residence time of any major element in the oceans (about 100,000 years; Berner and Berner, 1996). Based on this residence time, major constituents of seawater may achieve steady state within 100,000–300,000 years. Some trace constituents may take longer to achieve steady state (e.g., Br) in part because their crustal abundances may vary much more. Although seawater has evolved continuously over geological time, an essentially steady state composition may have been achieved many times, the steady state composition likely being different during each steady state stage.

H. Wayne Nesbitt

Bibliography

Armstrong, R.L., 1968. A model for Sr and Pb isotope evolution in a dynamic Earth. *Reviews in Geophysics*, **6**: 175–199.
Berner, K.E., and Berner, R.A., 1996. *Global Environment*. Prentice Hall.
Blatt, H., and Jones, R.L., 1975. Proportions of exposed igneous, metamorphic, and sedimentary rocks. *Ecological Society of America Bulletin*, **81**: 255–262.
Brown, G.C., and Mussett, A.E., 1981. *The Inaccessible Earth*. Unwin Hyman Ltd.
Brown, P.G., Hildebrand, A.R., and Zolensky, M.E. et al., 2000. The fall, recovery, orbit and composition of the Tagish Lake meteorite: a new type of carbonaceous chondrite. *Science*, **290**: 320–324.
Canfield, D.E., Habicht, K.S., and Thamdrup, B., 2000. The Archean sulfur cycle and the early history of atmospheric oxygen. *Science*, **288**: 658–661.
Chyba, C.F., 1990. Impact delivery and erosion of planetary oceans in the early inner solar system. *Nature*, **343**: 129–133.
Cicero, A.D., and Lohmann, K.C., 2001. Sr/Mg variation during rock-water interaction: implications for secular changes in the elemental chemistry of ancient seawater. *Geochimica et Cosmochimica Acta*, **65**: 741–761.
Cogley, J.G., and Henderson-Sellers, A., 1984. The origin and earliest state of the Earth's hydrosphere. *Reviews of Geophysics and Space Physics*, **22**: 131–175.
Condie, K.C., 1989. *Plate Tectonics and Crustal Evolution*. Oxford: Pergamon Press.
deRonde, C.E.J., Channer, D.M.DeR., Faure, K., Bray, C.J., and Spooner, E.T.C., 1997. Fluid chemistry of Archean seafloor hydrothermal vents: implications for the composition of circa 3.2 Ga seawater. *Geochimica et Cosmochimica Acta*, **61**: 4025–4042.
Edmond, J.M., Measures, C., McDuff, R.E., Chan, L.H., Collier, R., Grant, B., Gordon, L.J., and Corliss, J.B., 1979. Ridge crest hydrothermal activity and the balances of the major and minor elements in the ocean: the Galapagos data. *Earth and Planetary Science Letters*, **46**: 1–18.
Fedo, C.M., Eriksson, K.A., and Krogstad, E.J., 1996. Geochemistry of shales from the Archean (3.0 Ga) Buhwa greenstone belt, Zimbabwe: implications for provenance and source-area weathering. *Geochimica et Cosmochimica Acta*, **60**: 1751–1763.
Fyfe, W.S., 1978. The evolution of the Earth's crust: modern plate tectonics to ancient hot spot tectonics. *Chemical Geology*, **23**: 89–114.
Garrels, R.M., and Mackenzie, F.T., 1971. *Evolution of Sedimentary Rocks*. New York: Norton & Co.
Godderis, Y., and Veizer, J., 2000. Tectonic control of chemical and isotopic composition of ancient oceans: the impact of continental growth. *American Journal of Science*, **300**: 434–461.
Lowenstein, T.K., Timofeett, M.N., Brennan, S.T., Hardie, L.A., and Demicco, R.V., 2001. Oscillations in Phanerozoic seawater chemistry: evidence from fluid inclusions. *Science*, **294**: 1086–1088.
McLennan, S.M., and Taylor, S.R., 1991. Sedimentary rocks and crustal evolution: tectonic setting and secular trends. *Journal of Geology*, **99**: 1–21.
Rubey, W.W., 1951. Geologic history of seawater: an attempt to state the problem. *Bulletin of the Geological Society of America*, **62**: 1111–1147.
Rudnick, R.L., 1995. Making continental crust. *Nature*, **378**: 571–578.
Sleep, N.H., Morton, J.L., Burns, L.E., and Wolery, T.J., 1983. Geophysical constraints on the volume of hydrothermal flow at ridge axes. In *Hydrothermal Processes at Seafloor Spreading Centers*. Plenum Press, NATO Conference Series 4, Volume 12: 53–70.
Sylvester, P.J., Campbell, I.H., and Bowyer, D.A., 1997. Niobium/Uranium evidence for early formation of the continental crust. *Science*, **275**: 521–523.
Taylor, S.R., 1997. The origin of the Earth. *Journal of Australian Geology and Geophysics*, **17**: 27–31.
Taylor, S.R., and McLennan, S.M., 1985. *The Continental Crust: Its Composition and Evolution*. Blackwell Scientific Publishers.
Thompson, G., 1983. Basalt-seawater reaction. In *Hydrothermal Processes at Seafloor Spreading Centers*. Plenum Press, NATO Conference, Series 4, Volume 12, 225–278.
Veizer, J., 1979. Secular variations in chemical composition of sediments: a review. *Physics and Chemistry of the Earth*, **11**: 269.
Veizer, J., 1985. Basement and sedimentary recycling, 2, time dimension to global tectonics. *Journal of Geology*, **93**: 624–643.
Vinogradov, A.P., and Ronov, A.B., 1956. Composition of the sedimentary rocks of the Russian Platform in relation to the history of its tectonic movements. *Geochemistry*, **6**: 533–559.
Wolery, T.J., and Sleep, N.H., 1988. Interactions of geochemical cycles with the mantle. In *Chemical Cycles in the Evolution of the Earth*. John Wiley, pp. 77–103.
Young, G.M., 1995. The Huronian Supergroup in the context of a Paleoproterozoic Wilson cycle in the Great Lakes region. *Canadian Mineralogist*, **33**: 921–922.

Cross-references

Carbonate Mineralogy and Geochemistry
Diagenesis
Dolomites and Dolomitization
Evaporites
Ironstones and Iron Formations
Oceanic Sediments
Oolite and Coated Grains
Phosphorites
Siliceous Sediments

Table S1 Relationships (see text for terminology details)

$C_s = M/V$	[M/L^3]
$Q_s = C_s Q$	[M/T]
$Q_T = Q_s + Q_b$	[M/T]
$M/T = Q_D + Q_T$	[M/T]
$Y = M/A_{db}/T$	[M/(L^{2}T)]
$D = Y/\rho$	[L/T]
$Q_T = Q_{T0} + \rho(E_R - D_R)$	[M/(L^{2}T)]
$\eta = (E_R - D_R)/\rho$	[L/T]
$\eta = Q_T/\rho/A_{dep}$	[L/T]

SEDIMENT FLUXES AND RATES OF SEDIMENTATION

An understanding of sediment fluxes, both at the local and global scale, involves identifying sites where sediment is produced, the processes involved in liberating this sediment, and processes and rates involved in the transport of the sediment to depositional environments. Sediment production occurs at sites where sedimentary particles form, across sites of temporary storage of these sediment particles, and even at sites of long-term storage, given tectonic movement within geological systems. While terminology remains an important part of compartmentalizing and understanding earth science, consistency in use of terms is not universal. Here we describe a few of the important terms related to sediment fluxes and sedimentation rates.

Definitions and concepts

Weathering is the rate-limiting step in the entire sequence of events. Weathering involves both chemical and biogeochemical processes that work to dissolve minerals within rock faces and the cements holding together the detrital grains within sedimentary rocks. Using a complex of natural inorganic and organic compounds found universally within aerosols and surface water, grains are liberated, fractures are weakened and material is generated for subsequent transport. Active weathering also involves mechanical processes that apply physics to the destruction of rock surfaces. Mechanical processes include the freeze thaw action of water, wind or water abrasion involving grain on grain impact, the action of glaciers (ice on rock abrasion, rock carried by ice on rock; quarrying), and the generation and impact of rockslides. Younger mountains shed more sediment (up to a factor of seven) than indurated and older mountains (Pinet and Souriau, 1988).

Sediment undergoes remobilization through sheet, rill and gully erosion of hillslopes, through landslides, and through channel erosion. The mass of sediment that passes by a given cross-section per unit time is known as the *transport load* (M), given in units of kg per second or MT per year. The terms *transport rate* and *sediment discharge* are used interchangeably with transport load. The transport load consists of the *dissolved* load (Q_D) and the *detrital load* (Q_T), or just one if specified. *Yield* (Y) is the transport load normalized per unit drainage basin area (A_{db}), often given in units of T/km^2 per year, kg/m^2 per year, or g/cm^2 per year, and can include both detrital and dissolved load (Table S1). The *rate of erosion* (E_R) has the same meaning as yield except that it tends to be used in more local environments and to reflect shorter events. Over time the exposed land surface is weathered and lowered with respect to a datum, at a rate known as the *denudation rate* (D), in units of length per time (mm/kyr). The denudation rate is a measure of earth surface removal, and takes into account both dissolved and detrital surface lowering.

Most dissolved load is transported within surface and subsurface waters flowing to the ocean. A much smaller portion is transported in the form of aerosols, within the atmospheric circulation. While most of the dissolved load initially ends up in the ocean as dissolved salt, uptake by plankton and other primary producers converts much of the dissolved load back into a solid form, including tests of silica and carbonate, larger shells and coral reefs. Subsequent chemical and mechanical degradation of reefs, and transport of these biogenic grains along with other shell fragments and tests, adds to the detrital load being transported across the depositional environments.

Detrital grains are transported primarily by water and wind, as *bed load* (Q_b) and *suspended load* (Q_s), but regionally may also be transported by flowing ice (ice-sheets, glaciers, icebergs, sea ice). Bed load, sometimes called *contact load*, consists of coarser particles that travel along the bed by traction and or saltation, essentially in continuous contact with the bed. Bed load particles move through the influence of shear stress or drag imposed by the moving fluid (water or air), along with other grain-to-grain interactions. If the flowing fluid is unable exceed some threshold in stream power or critical shear stress, bed load transport ceases. The suspended load consists of particles suspended within the fluid by turbulence or by colloidal suspension. The suspended load consists of both a *bed-material load* (i.e., of the same range of grain size as found on the bed), and a *wash load* (finer material than bed particles).

Often the term *sediment load* is used interchangeably with suspended load, as bed load is typically a minor component of the total load. The local sediment load is then the upstream load (Q_{T0}) plus the density-corrected difference between the erosion rate and the depositional rate, a relationship known as the *Exner equation*. Sediment concentration (C_s) is the dry weight of suspended sediment per unit volume of water or air (V), at any given time and place. Thus the suspended sediment load is simply the sediment concentration times the *discharge rate*, (Q), i.e., the fluid volume moved per unit time.

The *sedimentation rate*, also known as the *deposition rate* (D_R), is the vertical flux of sediment to the depositional surface, such as the seafloor in the case of a marine environment (e.g., kg/m^2 per year). Sedimentation rates vary across time, because periods of rapid sedimentation alternate with periods of slower deposition, quiescence, or erosion. Thus sedimentation rates in various depositional environments refer to average values of net sedimentation.

The rate of sedimentation within a basin strongly depends on the proximity to the sediment source, such as a river mouth or ice terminus. Typically, the sedimentation rate decreases exponentially away from the source. If the sediment source location is fixed in time, fluctuations in sedimentation rate are a function of sediment supply. However, the location of a sediment source is commonly variable in geological time, determined by variations in sea level, the position of an ice terminus or river mouth. During a period of rising sea level, the river mouth moves landward and thus the rate of offshore sedimentation will decrease. Sedimentation rate therefore depends on both the yield of sediment from land and the distance that the sediment load has to travel to a particular location. This leads to two axioms (Syvitski, 1993):

- Constant or increasing accumulation rates indicates either: (i) increasing sediment input; or (ii) decreasing distance from the source.
- Decreasing accumulation rates indicate either: (i) constant or decreasing sediment input; or (ii) increasing distance from source.

The shoreline is therefore an arbitrary boundary within the detrital sediment transport system.

Accumulation rate (η) is the vertical change in the depositional surface due to the composite sedimentation history (e.g., mm per year), and equals the deposition rate minus the erosion rate times the sediment bulk density. *Sediment bulk density* (ρ) is a function of distance from the sediment supply (grain size and excess pore pressure as conditioned by the rate of sedimentation), burial depth and time (compaction). Erosion events (such as sediment failures or wave resuspension) work to decrease the accumulation rate in the shallower portions of a basin and increase the rates in the deeper water sections. Accumulation rates are influenced by biological, chemical, geological, and geographical factors, and are highly variable within depositional environments, between depositional environments and across geological time.

Accumulation rate is often reported as a site-specific measurement, such as determined from the thickness of sediment between two dated horizons within a sediment core. However, basin shape can greatly affect the accumulation rate. Given a constant depositional rate, the local accumulation rate will decrease as a power function with time, due to the ever-increasing width of a basin floor concomitant with sedimentary fill of the basin (Syvitski, 1993). For example, the initially high (up to 9 m per year , Powell, 1983) accumulation rates within fjords seem extremely high because the basin floors are very narrow. After a few thousand years, when fjords are filled with hundreds of meters of sediment, accumulation rates will appear to have decreased by orders-of-magnitude, even if the depositional rate is held constant. Therefore when comparing different basins, it is useful to calculate accumulation rates at the regional level (Syvitski, 1993): i.e., the total mass of sediment between dated unit boundaries, normalized to the depositional area (A_{dep}).

Influences and rates

The erosion of bedrock takes place almost entirely in the headwaters of the catchment. This newly eroded sediment is then transported toward the coastal zone. How long it takes for this transport and how sediment makes the journey are important questions. However, denudation rate and/or regional erosion rates are similarly independent of the magnitude of the terrestrial flux of sediment to the ocean (Kirchner et al., 2001). Very little (≈ 10 percent) of the sediment that is eroded from the mountains and hilltops actually makes it to the ocean. Most of the sediment is either in flux, or being sequestered in interior drainage basins, lakes, subsiding alluvial fans, flood plains, coastal plains and deltaic plains (Meade, 1996). Holeman (1980) notes that 20th century erosion across the conterminous US (5.3 GT per year) is an order-of-magnitude higher than the fluvial discharge of sediment (0.445 GT per year: Curtis et al., 1973). And while eolian transport to the global ocean is small (Table S2), wind erosion dominates water erosion at the continental scale (Smith et al., 2001).

At the seasonal scale, sediment is stored in riverbeds and along their banks at low or falling discharge, only to be resuspended at high or rising discharges. Often temporarily stored sediment is washed out of the system before peak discharge is reached. At the decade to century timescale, sediment moves down the drainage system as a series of kinematic waves, wherein it might take decades for the sediment to reach the lower reaches of the flood basin. At even longer timescales (millennia), tectonics plays an important role and there are large variations between subsiding basins and basins being exhumed (uplifted). Understanding these scales plays an important role in deciphering the difference between soil erosion and sediment discharge by rivers (Meade, 1996).

Most of the sediment delivered to the ocean is by rivers, which accounts for approximately 95 percent of the global flux from land to sea (Table S1). Most of that flux is as suspended sediment load. Dissolved load and bed load are nearly an order-of-magnitude less than the suspended load. Detrital denudation rates range from 1 mm/kyr for the St. Lawrence River basin, to 670 mm/kyr for the Brahmaputra River system. Small mountainous rivers have detrital denudation rates that can exceed 40,000 mm/kyr (Milliman and Syvitski,

Table S2 Global estimates of the flux of sediment from land to the ocean

Transport mechanism	Global flux GT per year	Estimate grade	Reference
Rivers: suspended load	18	B$^+$	Milliman and Syvitski, 1992
bed load	2	B$^-$	Milliman and Syvitski, 1992
dissolved load	5	B$^+$	Summerfield and Hulton, 1994
Glaciers, sea ice, icebergs	2	C	Hay, 1994
Wind	0.7	C	Garrels and Mackenzie, 1971
Coastal Erosion	0.4	D	Hay, 1994

1992). Chemical denudation rates are typically smaller and range from 1 mm/kyr for the Kolyma, Niger, and Nile basins, to 27 mm/kyr for the Chiang Jiang basin (Summerfield and Hulton, 1994). For anomalously low sediment-discharge rivers, like the Dnepr or St. Lawrence, chemical denudation can account for between 86 percent and 94 percent of the basin denudation (Summerfield and Hulton, 1994).

The "natural" sediment load is influenced by a variety of nonanthropogenic factors, including:

(1) Size of drainage basin (Wilson, 1973; Milliman and Syvitski, 1992);
(2) Large scale relief (Hay and Southam, 1977; Pinet and Souriau, 1988; Milliman and Syvitski, 1992; Harrison, 1994);
(3) Local relief (Schumm 1965; Ahnert, 1970; Jansen and Painter, 1974);
(4) Geology of the drainage basin (Pinet and Souriau, 1988; Milliman and Syvitski, 1992; Inman and Jenkins, 1999);
(5) Climate (Fournier, 1960; Schumm, 1965; Jansen and Painter, 1974; Hay et al., 1987);
(6) Runoff (Walling, 1987; Milliman and Syvitski, 1992);
(7) Vegetation (Douglas, 1967); and
(8) Lakes (Milliman, 1980);

Relief and area-dependent runoff are the two most important factors (Table S3). Competence of rivers to transport sediment is governed by the volume flow, gradient, and the sediment load itself. The runoff on large rivers is influenced by snowmelt, in some cases ice-melt, rainfall, and groundwater efflux (which accounts for 30 percent of runoff at the global scale).

For large rivers, interannual differences in sediment discharge are significantly smaller than the interseasonal differences (Meade, 1996). On decadal timescales, large floods can cause marked increases in the sediment discharge of rivers of moderate size (e.g., Eel where 3 days of peak flooding in 1964, transported more sediment than during the previous 8 years of river flow combined: Syvitski and Morehead, 1999). On multi-decadal timescale, river loads relate more to fluctuations in climate. One of the largest impacts of climate change is through changes in the overall water balance with subsequent impacts on land cover density and thus erosion rates. Knox (1993) emphasizes that relatively modest shifts in average climate conditions (i.e., 1–2 °C; <20 percent precipitation) have large impacts on the behavior of a river's flood response and thus sediment yield. Bed load transport is particularly susceptible to the frequency-magnitude distribution of floods. The ratio of magnitude of 100-yr floods and annual floods is greater in arid than in humid environments (Molnar, 2001). Thus a shift toward more arid conditions, despite a decrease in precipitation and discharge, could double the bed load transport and thus the incision rate (Molnar, 2001).

Human impact on the flux of sediment to the global ocean is well recognized (Milliman et al., 1987; Hu et al., 2001). Land use is probably the dominant control on particulate fluxes in areas of low relief and large-scale urbanization, in difference to mountainous regions where natural processes are likely to still dominate. Land clearing in low relief areas increases sediment yields by more than an order of magnitude (Douglas, 1993). The effect increases with decreasing drainage area. Changes due to man affect small river basins more dramatically than larger river basins, due to the modulating ability of large rivers. Some investigators of large rivers therefore argue that modern values of sediment flux remain relatively constant across geological time (Summerfield and Hulton, 1994). While there is no accepted value for the "pre-anthropogenic" influenced flux of sediment to the coastal oceans, Milliman and Syvitski (1992) argue that the modern 20 GT per year global detrital flux estimate might be 50 percent smaller circa 2000 years ago before human impact was great.

The Mediterranean landscape is arguably the most human impacted terrain on earth. Dedkov and Mozzherin (1992) note that 75 percent of the average sediment yield (1,100 T/km^2 per year) of Mediterranean basins may be attributed to human activity. The terrain is naturally vulnerable to processes of erosion with its steep slopes, high relative relief, fissile sedimentary rocks, thin erodible soil covers, and active tectonic settings (Woodward, 1995). Severe land degradation has taken place including badland formation as representative of acute land deterioration. Elsewhere, deforestation in tropical areas may account for the extraordinary loads from Oceania (Hu et al., 2001; Milliman and Syvitski, 1992). Countering this increased erosion due to human activity is the rapid development of reservoirs worldwide. About 2,500 years ago the sediment discharge of the Yellow River was about 1/10th what it was 30 years ago when it peaked with rapid cultivation of the loess plateau (Milliman et al., 1987; Saito et al., 1994). A combination of soil preservation practices of the 1980s and dam construction (3,380 reservoirs and another 30,000 diversion works of various scales) have sharply reduced both water and sediment discharge. The decrease in sediment load carried by the Yellow River has in turn rapidly increased coastal erosion (Saito et al. 1994).

Table S3 The global flux of suspended sediment (Q_s) as influenced by maximum basin relief and contributing land surface area (Milliman and Syvitski, 1992)

Basins with Relief (m)	Land surface		Load	Global load	
	Mkm^2	%*	MT per year	MT per year	%**
>3,000	16.3	12.5	4,500	9804	52.5
1,000–3,000	26.4	20.4	3,783	8,248	44.1
500–1,000	4.9	3.8	200	436	2.3
100–500	11.9	9.2	96	209	1.1
<100	0.2	0.2	1	2	—
Total	59.7	46.1	8,580	18,692	100

* Given the total land area of earth is $\approx 130 \times 10^6$ km^2
** Proportioned load assumes that the distribution of measured rivers is representative of the earth's topography and land area.

Global sediment fluxes are known to change across broad geologic time. Hay (1994) notes that Holocene fluxes to marginal seas are typically 1.5 to 4 times higher than that for similar periods during the Pleistocene. Quaternary sediment was deposited at globally averaged accumulation rates of 4 g/cm^2 per year, which is 54 percent higher than the Pliocene and 3 times higher than the Miocene rates (Hay, 1994). Carbonate sediment accounts for 18.4 percent of the total Quaternary sediments in the ocean basins, whereas it accounts for 40 to 50 percent of Miocene and earlier sediments. Peizhen *et al.* (2001) show that global sediment accumulation rates increased by 2–10 times beginning 2–4 Myr ago, when climate switched from a long period of climate stability, to a time of frequent abrupt changes in temperature, precipitation, and vegetation. The result was that during the Quaternary, the landscaped moved away from equilibrium configurations to periods of high sediment production from landscapes out of equilibrium.

Ice sheets and glaciers create sediment through abrasion, quarrying, and channelized and nonchannelized fluid flow at their base. Estimates of the global flux of glacial detritus range from 1.5 GT per year to 50 GT per year with the lower value probable (Table S2). The higher estimates do not effectively take into account the recycling of glacimarine sediment back into the ice mass (R. Powell, Pers. comm., 2001). The sediment delivery rate from ice sheets and glaciers to the ocean depends on a variety of processes. Their marine termini melt at a rate dependent on the temperature of the ocean water and the strength of the currents flowing across the termini (Syvitski, 1989). The sediment delivery rate will thus depend on the melt rate and the concentration of sediment (supraglacial, englacial and subglacial) within the glacier. Sediment is also transported at the base of a glacier as basal till, through a variety of subglacial processes (Andrews and Syvitski, 1994). Fast flowing ice streams are more able to transport till compared with cold-based or slow flowing ice sheets (Syvitski, 1993). Meltwater flowing within and beneath the ice mass is able to carry large volumes of sediment, and this is particularly true for Alaskan glaciers. Sediment delivery from the melt of drifting icebergs is perhaps the greatest sediment-dispersal mechanisms as icebergs are able to transit hundreds to thousands of kilometers (Syvitski *et al.*, 1996a,b).

The wind borne flux of sediment to the ocean is in the order of 0.5 GT per year to 0.9 GT per year (Table S2), but greatly varies with time (Garrels and Mackenzie, 1971). While river discharge is the largest global contributor of sediment (Table S2), eolian transport is of equal magnitude to that of river flux for the continent of Africa (*cf.* Prospero, 1981; and Milliman and Syvitski, 1992). While eolian transport is important today, fluvial transport dominated the supply of sediment to the Arabian Sea during the last glacial period.

Coastal erosion is not well-known but global estimates vary from 0.2 GT per year to 0.9 GT per year (Hay, 1994), with anthropogenic effects of significance at the regional level. Sea ice, where ubiquitous, is also able to carry large volumes of coastal sediment into the deeper ocean, including sediment that was initially transported to the coast by wind or streams (Gilbert, 1983).

Rates of sedimentation

There are a variety of methods used in the determination of sedimentation rates (Table S4). In brief these are:

Direct Observations: systematic comparison between old and new soundings (best suited to high sedimentation rate environments such as deltas, lagoons, bays and estuaries), sediment trap studies, and repeated coring (Nittrouer, 1999).

Theoretical Calculations: detrital sedimentation rates may be obtained by subtracting the amount of sediment leaving from the amount of sediment entering a depositional setting. The method is particularly good for semi-closed basins such as deep silled fjords (Syvitski *et al.*, 1988). As an example, Piper (1991) quantified both sources and sinks of Holocene sediment on the Scotian and Labrador shelves. On the Scotian shelf, sediment supply from rivers is roughly 5.5×10^6 T per year and supply from current and wave erosion of older Quaternary shelf deposits is 10.5×10^6 T per year. Together these sources balance the annual sediment sink of 16×10^6 T per year, measured from the volume of Holocene sediment on the shelf

Table S4 Modern Accumulation Rates (from Olsen, 1978; Syvitski *et al.*, 1987; Hay, 1994; and references therein)

Environment	Accumulation Rates
Flood Plains	mm/yr to cm/yr
Lakes	mm/yr to cm/yr*
Eolian loess	0.1 to 1 mm/yr
Deltas	cm/yr to m/yr
Prodeltas	mm/yr to cm/yr
Estuaries and bays	mm/yr to cm/yr
Fjords	mm/yr to m/yr
Tidal marshes and bogs	mm/yr to cm/yr
Carbonate Reefs/Platforms	mm/yr to cm/yr
Continental Shelves	0 (bypassing) to 10 cm/yr**
Submarine Canyons	0.1–1 mm/yr
Inland Seas	0.01 to 2 mm/yr
Slope and Rise	0.01 to 4 mm/yr**
Abyssal Plain	0.01 to 2 mm/yr

*larger in glacial and proglacial lake environments, or after catastrophic events like debris flow emplacements.
** highest values from the California STRATAFORM site: Sommerfield and Nittrouer (1999); Alexander and Simoneau (1999).

and slope. On the Labrador shelf, the Holocene sediment sources included rivers (50 km^3), submarine erosion by iceberg turbation (20 km^3), and ice rafting (2.4 km^3). The sinks include coastal basins (35 km^3), shelf troughs (24 km^3) and the continental slope (13.4 km^3).

Radiometric Dating: measures that use the decay constant of radionuclides and determining their decrease in concentration to secular equilibrium values, with depth in the sediment column. Some of these decaying isotopes include U^{234}, Th^{230}, Pa^{231}, C^{14}, Pb^{210} (with half lives of 250,000, 75,000, 35,000, 5,730 and 22.3 years, respectively). Other radionuclides used in the determination of sedimentation rates include Be^{10}, Al^{26}, Si^{32} and the K–Ar method (see Olsen, 1978, and references therein).

Amino Acid Racemization: proteins of organisms buried in sediment undergo slow racemization or molecular change in their structure. The rate of change is affected by temperature and thus the method works best for environments of relative constant temperatures (Miller *et al.*, 1997).

Varves: in certain lakes and fjords and other deep coastal basins, the annual sedimentation is visually recognized by the color of winter layers often in contrast to the color of summer sedimentation layer(s). Assuming that sedimentation occurs every year, then the layers are counted to produce a time versus sediment depth history of the sedimentation rates (Hughen *et al.*, 1996).

Pollutants and other Tracers: the distribution of pollutants in sediments can be used to date through historical forensic analysis the onset of layers within sediments associated with anthropogenic events. For example Fe^{55} and Cs^{137} are radionuclides produced by nuclear testing that peaked world-wide in the late 1950s and early 1960s (Alexander and Simoneau, 1999). Other tracers include DDT (dichlorodiphenyltrichloroethane), products of synthetic detergents, paint lead, and PCPs.

Depositional environments

River courses are divided between a degradational section where the stream gradient is relatively steep and little sediment storage takes place, and an aggradational section where gradient is sharply reduced through meandering and *flood plain* storage (Table S4). Not all rivers flow to the ocean, as large areas of the earth's landmass drain internally. Along stream pathways, the sediment load may also become trapped *in lakes*. Mid to high-latitude rivers are particularly affected by the filtering effect of glacial lakes (Table S4). Man-made reservoirs have reduced the sediment load of some rivers by two-orders of magnitude (Nile, US Colorado), while others such as the Indus or Mississippi have had their load cut by a factor of between 5 and 2, respectively (Milliman and Syvitski, 1992). Elsewhere on continents, wind-borne loess deposits accumulate episodically at relatively low rates (Table S4).

While most of the sediment discharge is reported as delivery to the ocean, a more realistic wording would be delivery to the *delta plain* which could trap much of the sediment. The ultimate sink for river sediments, on a millennial to multi-millennial timescale, is the coastal zone. The role between sediment supply and sea level rise distinguishes between coastlines with *coastal embayments* (undersupply), and those with long coastlines (balanced supply), from those with coastal protrusions, i.e., *deltas* (over-supply). All of the modern river deltas appear to be <10,000 years old (Stanley and Warne, 1994). In the case of the Amazon, 20 percent of the annually delivered load (1 BT per year) is retained by its delta; the remaining 80 percent is deposited on the continental shelf and coast with none reaching the deep sea. In the case of the Ganges and Brahmaputra Rivers, 55 percent of the annually delivered load (1.1 BT per year) is retained by its delta; the remaining 45 percent is deposited on the continental shelf and coast (36 percent) with (9 percent) reaching the deep sea. In the case of the Yellow River, 82 percent of the annually delivered load (1.1 BT per year) is retained by its delta; the remaining 18 percent is deposited on the *continental shelf* and coast (36 percent) with none reaching the deep sea (see Meade, 1996).

Deltaic sedimentation (Table S4) is dominated by four processes (Syvitski *et al.*, 1988): (1) deposition of bed material load at the river mouth the main point for hydraulic transition; (2) sedimentation under the seaward-flowing river plume that carries the wash load across the *prodelta* region; (3) sediment bypassing processes, such as turbidity currents that may result from delta-front failure, or hyperpycnal flows that result from the super-elevated sediment concentrations carried by some rivers (Mulder and Syvitski, 1995); and (4) diffusive processes that work to smear sediment down-slope from tidal and wave action.

Sediment accumulation in *estuaries* (Table S4) depends on the sediment supply from both river and offshore sources, and the mixing energy within the estuary (Wang *et al.*, 1998) i.e., a balance of buoyancy forces set up by river discharge and processes such as those associated with tidal action that work to mix the fresh water with the denser salt-water layers. A *fjord* is a deep high-latitude estuary which has been, or is presently being, excavated or modified by land-based ice (Syvitski *et al.*, 1987) and can range from low to high sedimentation systems. Fjords are sites of net sediment accumulation, and globally may have retained 25 percent of the sediment removed from the land over the last 100,000 years (Syvitski and Shaw, 1995). Typically sediments are spread over a great distance from the river mouth due to the fjord margins confining the fluvial plume emanating from the river, and the propensity of fjords to experience long-traveling sediment gravity flows.

In *continental shelf* settings, tide, storm, and wave-driven currents operate to produce an equilibrium shelf gradient and to change the configuration of the coastline (Wang *et al.*, 1998). In the *upper slope* region, sediments are delivered from the shelf *via* boundary layer currents including fluid muds, and nepheloid—hemipelagic sedimentation. These upper slope sediments are then reworked down slope by a variety of processes characterized as diffusive (i.e., submarine landslides, mudflows, and creep). Sediment gravity flows comprise the major processes operating in the mid- to *lower slope* region of the profile. Upper-slope diffusive processes operate in concert with sediment gravity flow processes in the lower slope and *rise* environments to produce a wide range in accumulation rates (Table S4: Syvitski *et al.*, 1996c; Hay, 1994).

About 40 percent of the world's continental shelves are carbonate shelves (Hay, 1994). Biogenic production (Table S4) was highest in glacial times, possibly as a result of the changes in the rates of ocean mixing and nutrient supply to the photic zone (Hay, 1994). Subaerial exposure and subsequent flooding of the continental shelves affected the sites of carbonate deposition and the global carbonate budget (Milliman, 1993).

Summary

(1) Weathering is the rate-limiting step in the entire sequence of events determining global sediment fluxes and rates of sedimentation.

(2) Sedimentation rates vary across time, because periods of rapid sedimentation alternate with periods of slower deposition, quiescence, or erosion. Sedimentation rates in various depositional environments are therefore average values of net sedimentation.

(3) The rate of sedimentation within a basin strongly depends on the proximity to the sediment source, such as a river mouth or ice terminus.

(4) Very little (≈ 10 percent) of the sediment eroded from mountains and hilltops actually makes it to the ocean. Most of the sediment is either in flux, or being sequestered in interior drainage basins, lakes, subsiding alluvial fans, flood plains, coastal plains and deltaic plains.

(5) Sediment delivery *via* rivers accounts for approximately 95 percent of the global flux from land to sea. Most of that flux is as suspended sediment load. Dissolved load and bed load are almost an order-of-magnitude less than the suspended load, with important regional exceptions.

(6) Drainage basin area and relief are the most important "geological" factors in controlling the sediment load of rivers. However, modern rivers are greatly impacted by human activity, either through increases in soil erosion, or through sediment load reduction from reservoir filtration.

(7) While contributing less to the modern global flux of sediment, sediment transport by ice, wind and waves, remains important in regions dominated by these processes, and achieve more global significance at earlier times in the Quaternary.

James P.M. Syvitski

Bibliography

Ahnert, F., 1970. Functional relationships between denudation, relief, and uplift in large mid-latitude drainage basins. *American Journal of Science*, **268**: 243–263.

Alexander, C.R., and Simoneau, A.M., 1999. Spatial variability in sedimentary processes on the Eel continental slope. *Marine Geology*, **154**: 243–254.

Andrews, J.T., and Syvitski, J.P.M., 1994. Sediment fluxes along high latitude glaciated continental margins: Northeast Canada and Eastern Greenland. In Hay, W. (ed.), *Global Sedimentary Geofluxes*. Washington: National Academy of Sciences Press, pp. 99–115.

Clague, J.J., and Evans, S.G., 2000. A review of catastrophic drainage of moraine-dammed lakes in British Columbia. *Quaternary Science Reviews*, **19**: 1763–1783.

Curtis, W.F., Culbertson, K., and Chase, E.B., 1973. Fluvial-sediment discharge to the oceans from the conterminous United States. *US Geological Survey Circular*, **670**: 1–17.

Dedkov, A.P., and Mozzherin, V.I., 1992. Erosion and sediment yield in mountain regions of the world. In Walling, D.E., Davies, T.R., and Hasholt, B. (eds.), *Erosion, Debris Flows and Environment in Mountain Regions*. Proceedings of the Chengdu Symposium, July 1992). IAHS Publication, 209, pp. 29–36.

Douglas, I., 1993. Sediment transfer and siltation. In Turner, B.L., Clark, W.C., Kates, R.W., Richards, J.F., Mathews, J.T., and Meyer, W.B. (eds.), *The Earth as Transformed by Human Action*. Cambridge University Press, pp. 215–234.

Douglas, J., 1967. Man, vegetation and the sediment yield of rivers. *Nature*, **215**: 925–928.

Fournier, F., 1960. *Climat et Erosion*. Paris: Presses Universitaires de France.

Garrels, R.M., and Mackenzie, F.T., 1971. *Evolution of Sedimentary Rocks*. Norton.

Gilbert, R., 1983. Sedimentary processes of Canadian arctic fjords. *Sedimentary Geology*, **36**: 147–175.

Harrison, C.G., 1994. Rates of continental erosion and mountain building. *Geologische Rundschau*, **83**: 431–447.

Hay, W.H., 1994. Pleistocene-Holocene fluxes are not the earth's norm. In Hay, W. (ed.), *Global Sedimentary Geofluxes*. National Academy of Sciences Press, pp. 15–27.

Hay, W.W., and Southam, J.R., 1977. Modulation of marine sedimentation by the continental shelves. In Andersen, N.R., and Malahoff, A. (eds.), *The Fate of Fossil Fuel CO_2 in the Ocean*. Plenum Press, pp. 569–604.

Hay, W.W., Rosol, M.J., Jory, D.E., and Sloan, J.L. II, 1987. Tectonic control of global patterns and detrital and carbonate sedimentation. In Doyle, L.J., and Roberts, H.H. (eds.), *Carbonate Clastic Transitions: Developments in Sedimentology*. Elsevier, pp. 1–34.

Holeman, J.N., 1980. Erosion rates in the US estimated from the soil conservation services inventory. *EOS*, **61**: 954.

Hu, D., Saito, Y., and Kempe, S., 2001. Sediment and nutrient transport to the coastal zone. In Galloway, J.N., and Melillo, J.M. (eds.), *Asian Change in the Context of Global Climate Change: Impact of Natural and Anthropogenic Changes in Asia on Global Biogeochemical Cycles*. Cambridge University Press, IGBP Publication, Series 3, pp. 245–270.

Hughen, K.A., Overpeck, J.T., Anderson, R.F., and Williams, K.M., 1996. The potential for paleoclimate records from varved Arctic lake sediments: Baffin Island, eastern Canadian Arctic. In Kemp, A.E.S. (ed.), *Paleoclimatology and Paleoceanography from Laminated Sediments*. Geological Society Special Publication, 116, pp. 57–71.

Inman, D.L., and Jenkins, S.A., 1999. Climate change and the episodicity of sediment flux of small California rivers. *Journal of Geology*, **107**: 251–270.

Jansen, J.M., and Painter, R.B., 1974. Predicting sediment yield from climate and topography. *Journal of Hydrology*, **21**: 371–380.

Kirchner, J.W., Finkel, R.C., Riebe, C.S., Granger, D.E., Clayton, J.L., King, J.G., and Megahan, W.F., 2001. Mountain erosion over 10 yr, 10 k.yr., and 10 m.yr. time scales. *Geology*, **29**: 591–594.

Knox, J.C., 1993. Large increase in flood magnitude in response to modest changes in climate. *Nature*, **361**: 430–432.

Meade, R.H., 1996. River-sediment inputs to major deltas. In Milliman, J.D., and Haq, B.U. (eds.), *Sea-Level Rise and Coastal Subsidence*. Kluwer Academic Publishers, pp. 63–85.

Miller, G.H., Magee, J.W., and Jull, A.J.T., 1997. Low-latitude glacial cooling in the Southern Hemisphere, from amino-acid racemization in emu eggshells. *Nature*, **385**: 241–244.

Milliman, J.D., and Syvitski, J.P.M., 1992. Geomorphic/tectonic control of sediment discharge to the ocean: the importance of small mountainous rivers. *Journal of Geology*, **100**: 525–544.

Milliman, J.D., 1980. Transfer of river-borne particulate material to the oceans. In Martin, J., Burton, J.D., and Eisma, D. (eds.), *River Inputs to Ocean Systems*. Rome, SCOR/UNEP/UNESCO Review and Workshop. FAO, pp. 5–12.

Milliman, J.D., 1993. Production and accumulation of calcium carbonate in the ocean: budget of a nonsteady state. *Global Biochemical Cycles*, **7**: 927–957.

Milliman, J.D., Qin, Y.S., and Ren, M.E.Y., 1987. Man's influence on the erosion and transport of sediment by Asian rivers: the Yellow River (Huanghe) example. *Journal of Geology*, **95**: 751–762.

Molnar, P., 2001. Climate change, flooding in arid environments, and erosion rates. *Geology*, **29**: 1071–1074.

Mulder, T., and Syvitski, J.P.M., 1995. Turbidity currents generated at river mouths during exceptional discharge to the world oceans. *Journal of Geology*, **103**: 285–298.

Nittrouer, C.A., 1999. STRATAFORM: overview of its design and synthesis of its results. *Marine Geology*, **154**: 3–12.

Nordin, C.F., 1978. Fluvial sediment transport. In Fairbridge, R.W., and Bourgeois, J. (eds.), *Encyclopedia of Sedimentology*. Dowden, Hutchison & Ross, pp. 339–343.

Olsen, C.R., 1978. Sedimentation rates. In Fairbridge, R.W., and Bourgeois, J. (eds.), *Encyclopedia of Sedimentology*. Dowden, Hutchison & Ross, pp. 687–692.

Peizhen, Z., Molnar, P., and Downs, W.R., 2001. Increased sedimentation rates and grain sizes 2 m.yr. to 4 m.yr. ago due to the influence of climate change on erosion rates. *Nature*, **410**: 891–897.
Pinet, P., and Souriau, M., 1988. Continental erosion and large-scale relief. *Tectonics*, **7**: 563–582.
Piper, D.J.W., 1991. Seabed geology of the Canadian eastern continental shelf. *Continental Shelf Research*, **11**: 1013–1035.
Powell, R.D., 1983. Glacial-marine sedimentation processes and lithofacies of temperate glaciers, Glacier Bay, Alaska. In Molnia, B.F. (ed.), *Glacial-marine Sedimentation*. Plenum Press, pp. 185–232
Prospero, J.M., 1981. Eolian transport to the world ocean. In Emiliani, C. (ed.), *The Sea, Volume 7, The Oceanic Lithosphere*. John Wiley & Sons, 801–874.
Saito, Y., Ikehara, K., Katayama, H., Matsumoto, E., and Yang, Z., 1994. Course shift and sediment discharge changes of the Huanghe recorded in sediments of the East China Sea. *Chishitsu News*, **476**: 8–16.
Schumm, S.A., 1965. Quaternary paleohydrology. In Wright, H.E. Jr., and Frey, D.G. (eds.), *The Quaternary of the United States*. Princeton University Press, pp. 783–794.
Smith, S.V., Renwick, W.H., Buddemeier, R.W., and Crossland, C.J., 2001. Budgets of soil erosion and deposition for sediment and sedimentary organic carbon across the conterminous United States. *Global Biogeochemical Cycles*, **15**: 697–707.
Sommerfield, C.K., and Nittrouer, C.A., 1999. Modern accumulation rates and a sediment budget for the Eel shelf: a flood-dominated depositional environment. *Marine Geology*, **154**: 227–241.
Stanley, D.J., and Warne, A.G., 1994. Worldwide initiation of Holocene marine deltas by deceleration of sea-level rise. *Science*, **265**: 228–231.
Summerfield, M.A., and Hulton, N.J., 1994. Natural controls of fluvial denudation rates in major world drainage basins. *Journal of Geophysical Research*, **99**: 13871–13883.
Syvitski, J., Field, M., Alexander, C., Orange, D., and Gardner, J., 1996c. Continental-slope sedimentation. *Oceanography*, **9**: 163–167.
Syvitski, J.P., and Morehead, M.D., 1999. Estimating river-sediment discharge to the ocean: application to the Eel margin, northern California. *Marine Geology*, **154**: 13–28.
Syvitski, J.P.M., 1989. On the deposition of sediment within glacier-influenced fjords: oceanographic controls. *Marine Geology*, **85**: 301–329.
Syvitski, J.P.M., 1993. Glacimarine environments in Canada: an overview. *Canadian Journal of Earth Sciences*, **30**: 354–371.
Syvitski, J.P.M., and Shaw, J., 1995. Sedimentology and geomorphology of fjords. In Perillo, G.M.E. (ed.), *Geomorphology and Sedimentology of Estuaries*. Elsevier, pp. 113–178.
Syvitski, J.P.M., Lewis, C.F.M., and Piper, D.J.W., 1996a. Paleoceanographic information derived from acoustic surveys of glaciated continental margins: examples from eastern Canada. In Andrews, J.T., Austin, W.E.N., Bergsten, H., and Jennings, A.E. (eds.), *Late Quaternary Palaeoceanography of the North Atlantic Margins*. Geological Society, Special Publication, 111, pp. 51–76.
Syvitski, J.P.M., Andrews, J.T., and Dowdeswell, J.A., 1996b. Sediment deposition in an iceberg-dominated glacimarine environment, East Greenland: basin fill implications. *Global and Planetary Change*, **12**: 251–270.
Syvitski, J.P.M., Burrell, D.C., and Skei, J.M., 1987. *Fjords: Processes and Products*. Springer-Verlag.
Syvitski, J.P.M., Smith, J.N., Boudreau, B., and Calabrese, E.A., 1988. Basin sedimentation and the growth of prograding deltas. *Journal of Geophysical Research*, **93**: 6895–6908.
Syvitski, J.P.M., in press. Supply and flux of sediment along hydrological pathways: research for the 21st century. *Global and Planetary Change*.
Walling, D.E., 1987. Rainfall, runoff, and erosion of the land: a global view. In Gregory, K.J. (ed.), *Energetics of Physical Environment*. Wiley, pp. 89–117.
Wang, Y., Ren, M.-E., and Syvitski, J.P.M., 1998. Sediment transport and terrigenous fluxes. In Brink, K.H., and Robinson, A.R. (eds.), *The Sea: Volume 10—The Global Coastal Ocean: Processes and Methods*. John Wiley & Sons, 253–292.
Wilson, L., 1973. Variations in mean annual sediment yield as a function of mean annual precipitation. *American Journal of Science*, **273**: 335–349.
Woodward, J.C., 1995. Patterns of erosion and suspended sediment yield in Mediterranean river basins. In Foster, I.D.L., Gurnell, A.M., and Webb, B.W. (eds.), *Sediment and Water Quality in River Catchments*. John Wiley & Sons., pp. 365–389.

Cross-references

Climatic Control of Sedimentation
Compaction (Consolidation) of Sediments
Erosion and Sediment Yield
Numerical Models and Simulation of Sediment Transport and Deposition
Sediment Transport by Unidirectional Water Flows
Tectonic Controls of Sedimentation
Weathering, Soils and Paleosols

SEDIMENT TRANSPORT BY TIDES

Tidal currents

Tidal currents are unique among the processes responsible for sediment transport and deposition because of their regularity, with the speed and direction varying with the frequency of the governing astronomical period (see *Tides and Tidal Rhythmites*). In coastal settings where the shorelines constrain the flow, the landward- (flood) and seaward-directed (ebb) currents typically have directions 180° apart, in a pattern that is termed rectilinear. A period of little or no current (i.e., *slackwater*) varying in length from a few to several tens of minutes generally accompanies each flow reversal. As a result, sediment transport is intermittent, with episodes of sand transport (if the currents are sufficiently strong) alternating with periods of mud deposition during the slack-water intervals (Figure S6). In open-shelf settings removed from the confining influence of coastlines, tidal currents are rotary because of the influence of the Coriolis force (Allen, 1997), with the direction changing progressively through 360° over one complete tidal period. In such situations, there is no slack-water period, although the currents transverse to the primary flow directions generally are slower than the main ebb and flood currents. As a result, mud deposition is inhibited. Typically, the highest maximum speeds are found in constricted, inshore areas, where they may be as high as 1–2 m/s. As a result, the transport and deposition of sediment by tides is most important in coastal zones, with tidal currents dominating sedimentation inside the sheltered confines of estuaries, lagoons, and deltaic distributary channels where wave action is minimal (see *Neritic Carbonate Depositional; Environments Deltas and Estuaries*).

In the ideal case, tidal-current speeds vary sinusoidally over an individual ebb or flood period, with equal durations and speeds in the opposing directions. In this case, there is no net (residual) sediment transport (Figure S6(A)). In reality, however, various factors cause deviations from the ideal, producing a *time–velocity asymmetry* that causes an inequality in the amount of sediment transported in the two directions (Figure S6(B)).

Tidal-wave deformation: in the shallow water of the coastal zone, the crest of the tidal wave travels faster than the trough because the latter experiences greater frictional retardation.

Convergence of the shoreline, creating a landward decrease in the cross-sectional area, compounds this distortion. As a result, the flood tide typically has a shorter duration and higher maximum speed than the ebb tide (Dyer, 1995).

Hypsometric influences: because the area-elevation distribution (the *hypsometry*) of coastal areas is generally not linear, the volume of water passing through any channel will vary in response to temporal changes in size of the area being flooded or drained (Boon and Byrne, 1981). Thus, the maximum current speeds will tend to occur when the largest area is being flooded or drained rather than at mid-tide. Such hypsometric influences can also cause inequalities between the maximum ebb and flood currents (Friedrichs and Aubrey, 1988).

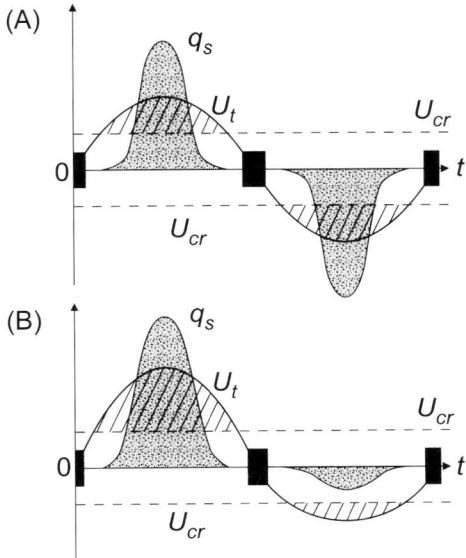

Figure S6 Transport of bedload sediment by: (A) an idealized, symmetrical tidal current; and (B) a tidal current with a pronounced time–velocity asymmetry. The bedload discharge (stippled area) is approximated as: $q_s \propto [U_t - U_{cr}]^3$, for $U_t > U_{cr}$ (the cross-hatched areas), where U_t is the depth-averaged tidal-current speed and U_{cr} is the threshold of bedload movement. In (A) the bedload discharges in the opposite directions are equal, such that there is no net transport. In (B) the short-duration high speeds on one half of the tide generate more transport than the slower, longer-duration flow in the opposite direction, yielding a residual transport in the direction of the faster currents. The black rectangles at the times of near-zero current speeds represent the periods when mud can be deposited.

Residual, estuarine circulation: at tidally influenced river mouths, fresh water rides above denser seawater (Figure S7). This generates a net seaward flow in the upper layer and a residual landward flow in the near-bed salt wedge (Nichols and Biggs, 1985; Dyer, 1995). This so-called *estuarine circulation*, which occurs in estuaries and delta distributary channels, becomes progressively less important as the tidal range increases, because strong tidal currents and intense turbulence destroy the salinity stratification.

Superimposed currents: wind forcing and oceanic circulation are responsible for slow water movement on timescales ranging from hours to years. Such currents may not be sufficient by themselves to initiate sediment movement, but may lead to residual transport when they are superimposed on the stronger tidal flows.

Bathymetric influence: local bathymetric irregularities, including bedrock promontories and current-generated sea-bed topography (e.g., tidal bars), cause spatial variability in the bed shear stress, with areas facing into the current experiencing stronger currents than areas sheltered from the main flow. Because of tidal flow reversals, these areas of exposure and sheltering alternate positions on each ebb and flood tide. This in turn produces inequalities between the ebb and flood currents at most locations.

Bedload transport

The transport of sand-sized bed-load sediment by tidal currents is generally describable using the concepts developed for unidirectional flow (see *Sediment Transport by Unidirectional Flows*). The unsteadiness associated with flow acceleration and deceleration over each tidal cycle has relatively little affect, because sand-sized sediment responses rapidly to changes in bed shear stress (Van Rijn, 1993). Consequently, bed-load-transport equations developed for unidirectional flows can be integrated over a complete tidal cycle, using the instantaneously measured mean current speeds and water depths, to obtain the residual bed-load discharge, provided salinity stratification is not pronounced. Because the discharge of bed-load is a power function of the current speed, the direction of residual sand transport can commonly be determined by comparing only the maximum speeds in the ebb and flood directions, with lesser regard for the durations of weaker currents (Figure S6(B)). Thus, the factors that determine the residual transport of sand are those that most strongly influence the inequality of the peak speeds.

Bathymetric and hypsometric influences, coupled with any fluvial discharge, create systems of *mutually evasive channels*: channels that open in a seaward direction tend to be flood

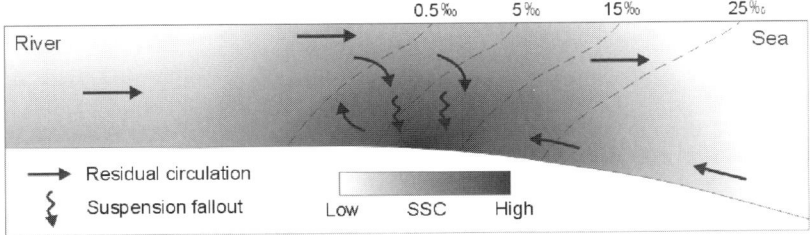

Figure S7 Formation of a salt wedge, as shown by the inclined salinity contours (dashed lines), in the zone of mixing between fresh water and seawater. The resulting, residual estuarine circulation (outward flow in the surface layer; landward flow near the bed) traps fine-grained material and develops a zone of elevated suspended-sediment concentrations (SSCs) called the *turbidity maximum*.

dominant, whereas channels that connect directly to a river tend to be ebb dominant. Elongate tidal bars separate such channels and have opposite directions of net sand transport on either side of their crest (Dalrymple and Rhodes, 1995). Erosional *bed-load partings* are produced at locations where the directions of residual transport diverge (i.e., there are opposing directions of net transport1 on either side of the parting; Harris et al., 1995). Conversely, net deposition occurs in areas where the residual transport converges. Such a *bed-load convergence* appears to be present within all estuaries and in the mouth-bar area of tidally influenced or dominated deltas: sand can be moved landward from the shelf because of the general flood dominance caused by the shallow-water distortion of the tidal wave, whereas the seaward movement of river-supplied sediment is slowed by the periodic current reversals. On tide-dominated continental shelves, bed-load partings and convergences are separated by areas with a consistent direction of net sand transport. These *sediment-transport pathways* may be tens to hundreds of kilometers long and display a predictable succession of bedforms (Belderson et al., 1982; see also *Surface Forms*).

Suspended-load transport

Compared to bed-load, the transport of suspended material is more strongly influenced by flow unsteadiness and weak residual currents, because fine-grained material continues to move for some time after currents have dropped below the threshold for bed-load movement. Thus, the transport of fine-grained sediment may be strongly influenced by weak, residual currents, and the net transport of bed-load and suspended-load sediment may be in different, even opposing, directions.

In detail, the movement of suspended sediment lags behind the water movement, both during deposition from a decelerating current and during erosion by an accelerating current. Van Straaten and Kuenen (1958; *cf.* Nichols and Biggs, 1985; Dyer, 1995) developed a simplified, yet elegant, model to account for the more important affects (Figure S8). In this model, the tide is assumed to be symmetrical and to show a linear, landward decrease in the maximum current speed. Two lags are considered: *settling lag*—the distance (and time) taken for suspended sediment to settle to the bed as the current speed decreases toward slack water (Figure S8(A)); and *scour lag*—a delay in resuspension from the bed because the current speed needed to erode fine-grained sediment is greater than the current speed at which deposition occurs (Figure S8(B)). These two lags operate at both the landward and seaward ends of a sediment-particle's excursion over a half tidal cycle. However, because the distance–velocity distributions are asymmetrical (even though speeds at any point are sinusoidal), the lags are longer at the landward end. Furthermore, the water and particle excursion distances decrease toward the land (i.e., E–E' is shorter than F–F'; Figure S8). Thus, there is a net landward migration of the particle and a resulting tendency for fine-grained sediment to accumulate near the landward margin of tidal flats and at the head of estuaries.

The foregoing illustrates that the transport and deposition of fine-grained sediment is strongly influenced by the factors governing the erosion threshold. Unlike noncohesive sand for which the threshold-of-motion is essentially constant for a give grain size, the erosion threshold for mud is not constant and increases as the degree of consolidation increases (Parchure and Mehta, 1986). Thus, the longer slack-water period that

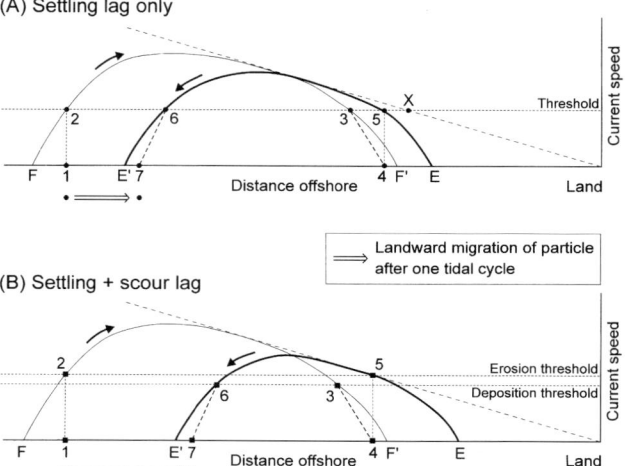

Figure S8 Transport history of suspended sediment in a tidal environment, based on Van Straaten and Kuenen (1958), in a situation where the maximum current speeds decrease toward the land (dashed diagonal line in (A) and (B)). The presence of a single upper bound for both the flood and ebb current speeds implies that the water movement is symmetrical, with no net water discharge. In (A) only the effect of settling lag is examined. Water originating at point F on the flood tide entrains a particle at location 1 when the speed exceeds the threshold velocity (2). The particle then travels landward with the current, eventually being deposited at 4, which is located some distance landward of point 3 where the current speed falls below the threshold for continued movement. The distance (and corresponding time interval) between 3 and 4 is termed the *settling lag*. On the ensuing ebb tide, water originating at F' retraces the F–F' trajectory, but is not able to re-entrain the particle because the speed has not reached the threshold by the time the water reaches location 4. Instead, water originating at E re-entrains the particle, carries it seaward along the trajectory E–E', and deposits it at 7. The distance of travel between 6 and 7 is also a result of the settling lag. However, because of the distance-velocity asymmetry and the shorter excursion on the ebb tide, the particle experiences a net landward displacement (1→7). Note that a particle coming to rest landward of point X in (A) cannot be re-entrained by the tidal currents shown because the maximum speed is always below the threshold. (B) shows a similar transport history but introduces separate erosion and deposition thresholds, which is more realistic than the situation shown in (A). As a result of the higher erosion threshold, point E is located further landward than in (A). This causes an increased delay in the re-entrainment of the particle which Van Straaten and Kuenen (1958) termed the *scour lag*. (Note that the definition of scour lag given by Nichols and Biggs (1985) and Dyer (1995) is not precisely the same as the original definition). Because of the scour lag, the particle does not travel as far seaward and experiences a greater, net landward displacement than in (A). (A) and (B) after Figures S7 and 4, respectively, of Van Straaten and Kuenen (1958).

characterizes more landward locations promotes higher erosion thresholds, longer scour lags, and preferential mud deposition. Exposure to the atmosphere, especially at times of high evaporation potential, and the growth of microbial coatings are particularly effective at producing cohesion (see *Tidal Flats*).

The residual estuarine circulation discussed above is also significant with regard to the trapping of suspended sediment in coastal settings. The suspended sediment introduced by the river moves seaward in the surficial layer, but then settles into the landward-moving lower layer (Figure S7), a process that is aided by the *flocculation* of fine-grained sediment that occurs in this region (Nichols and Biggs, 1985; Burban *et al.*, 1989; Dyer, 1995). This circulation leads to the increase of suspended-sediment concentrations (SSCs) near the limit of salt-water intrusion (Nichols and Biggs, 1985; Dyer, 1995). The higher transport capacity that characterizes the flood tide in many cases, because of the shallow-water deformation of the tidal wave (see above), also contributes to the trapping of fine-grained sediment near the landward limit of tidal influence. The combined influence of these factors produces a *turbidity maximum* in which SSCs are elevated relative to areas further landward or seaward (Figure S7). SSCs in the turbidity maximum tend to increase as the tidal energy increases, and thus tend to be highest in macrotidal areas and at spring tides, because of greater resuspension of mud. Near-bed SSCs can exceed $10\,\mathrm{gl}^{-1}$ producing mobile sediment bodies that are called *fluid mud* (Faas, 1991). These dense suspensions occupy channel-bottom locations where they can form anomalously thick mud deposits. The high SSCs promote this deposition by decreasing the bed shear stress (Li and Gust, 2000) during the ensuing tidal current. As a result, higher mean current speeds are needed to resuspend the sediment than would be the case in the absence of the high SSCs.

Summary

The transport of sediment by tidal currents is complex because of the pronounced temporal and spatial variability of flow strength and direction. Sediment particles of different sizes respond to the flow unsteadiness with different amounts of lag, such that bed-load and suspended load can display different directions of residual transport: bed-load tends to move in the direction of the current with the higher peak speed, whereas the suspended load commonly moves in the direction of the residual water movement. Bathymetric and hypsometric influences produce spatial inequalities in the strengths of the flood and ebb currents, creating complex transport pathways characterized in coastal areas by mutually evasive channels. The common development of a flood dominance because of the distortion of the tidal wave in shallow water, coupled with the presence of a density-driven estuarine circulation, leads to the trapping of both bed-load and suspended-load material within estuaries and/or at the mouth of deltaic distributaries. Fluid mud is commonly formed beneath the turbidity maximum that forms as a result of this trapping.

Robert W. Dalrymple and Kyungsik Choi

Bibliography

Allen, P.A., 1997. *Earth Surface Processes*. Blackwell Science.
Belderson, R.H., Johnson, M.A., and Kenyon, N.H., 1982. Bedforms. In Stride, A.H. (ed.), *Offshore Tidal Sands: Processes and Deposits*. New York: Chapman and Hall, pp. 27–57.
Boon, J.D., and Byrne, R.J., 1981. On basin hypsometry and the morphodynamic response of coastal inlet systems. *Marine Geology*, **40**: 27–48.
Burban, P.-Y., Lick, W., and Lick, J., 1989. The flocculation of fine-grained sediments in estuarine waters. *Journal of Geophysical Research*, **94**: 8323–8330.
Dalrymple, R.W., and Rhodes, R.N., 1995. Estuarine dunes and barforms. In Perillo, G.M.E. (ed.), *Geomorphology and Sedimentology of Estuaries*. Elsevier, Developments in Sedimentology 53, pp. 359–422.
Dyer, K.R., 1995. Sediment transport processes in estuaries. In Perillo, G.M.E. (ed.), *Geomorphology and Sedimentology of Estuaries*. Amsterdam: Elsevier, Developments in Sedimentology 53, pp. 423–449.
Faas, R.W., 1991. Rheological boundaries of mud: where are the limits? *Geo-Marine Letters*, **11**: 143–146.
Friedrichs, G.T., and Aubrey, D.G., 1988. Non-linear tidal distortion in shallow well mixed estuaries: a synthesis. *Estuarine, Coastal and Shelf Science*, **27**: 521–546.
Harris, P.T., Pattiaratchi, C.B., Collins, M.B., and Dalrymple, R.W., 1995. What is a bedload parting? In Flemming, B.W., and Bartoloma, A. (eds.), *Tidal Signatures in Modern and Ancient Sediments*. International Association of Sedimentologists, Special Publication, 24, pp. 2–10.
Li, M.Z., and Gust, G., 2000. Boundary layer dynamics and drag reduction in flows of high cohesive sediment suspensions. *Sedimentology*, **47**: 71–86.
Nichols, M.M., and Biggs, R.B., 1985. Estuaries. In Davis, R.A. Jr. (ed.), *Coastal Sedimentary Environments*. Springer-Verlag, pp. 77–186.
Parchure, T.M., and Mehta, A.J., 1986. Erosion of soft cohesive sediment deposits. *Journal of Hydraulic Engineering*, **111**: 1308–1326.
Van Rijn, L.C., 1993. *Principles of Sediment Transport in Rivers, Estuaries, and Coastal Seas*. Amsterdam: Aqua Publications.
Van Straaten, L.M.J.U., and Kuenen, Ph.H., 1958. Tidal action as a cause of clay accumulation. *Journal of Sedimentary Petrology*, **28**: 406–413.

Cross-references

Deltas and Estuaries
Flocculation
Neritic Carbonate Depositional Environments
Surface Forms
Tidal Flats
Tides and Tidal Rhythmites

SEDIMENT TRANSPORT BY UNIDIRECTIONAL WATER FLOWS

Sediment properties

Description of sediment movement involves terms such as grain size, shape and density, settling velocity, and sediment transport rate. For the range of grain size, shape and density of sediment available for transport see *Grain Size and Shape*. Methods of measuring and analyzing these grain properties are described also in Carver (1971) and McManus (1988). A sediment grain immersed in water experiences a hydrostatic pressure that is greater on its base than on its top: an upward directed buoyancy force equal to the weight of water displaced by the grain. If the weight of the grain exceeds the buoyancy force, the grain will sink. The *immersed weight* of the grain is the gravity force minus the buoyancy force, given by $Vg\,(\sigma-\rho)$, where V is grain volume, g is gravitational acceleration, σ is sediment density, and ρ is fluid density.

Drag on sediment grains

A fluid moving relative to a solid boundary, such as water flowing over a bed of sediment or a sediment grain settling in water, experiences resistance to motion due to the action of various types of frictional forces (e.g., viscous shear stress, dynamic pressure). This resistance to motion is called *drag*. Drag due to viscous shear is called surface drag or skin-friction drag, whereas drag due to fluid pressure is called form drag. In turbulent flows, form drag is normally much more important than surface drag.

A general equation for the drag force is:

$$F_D = C_D\, a(\rho V_r^2/2) \qquad \text{(Eq. 1)}$$

where C_D is a drag coefficient, a is the cross-sectional area exposed to the drag, and V_r is the relative velocity of the solid and fluid. The term in brackets is actually the dynamic pressure on the area of the solid that faces the fluid flow, but this term also accounts for the pressures on the lee-side of the solid associated with flow separation, and for viscous forces. Flow separation produces a pressure against the flow (form drag). The drag coefficient takes into account whether or not flow separation occurs behind a solid body, and the nature of such flow separation, including the size of the separation eddies. The drag coefficient therefore depends on the geometry of the solid body, and specifically how streamlined it is. The drag coefficient is also dependent on an appropriate Reynolds number, as this determines the relative importance of viscous drag and form drag. The drag equation is used in two important cases: (1) the settling of grains in fluids (See *Grain Threshold, Grain Settling*); and, (2) fluid moving over beds of sediment.

Drag and lift on bed grains

The mean drag force acting on bed grains is commonly resolved into components parallel to the bed (drag) and normal to the bed (lift). The mean drag on a sediment grain can be expressed as:

$$F_D = C_D\, a(\rho u^2/2) \qquad \text{(Eq. 2)}$$

where u is the mean velocity of the fluid at the level of effective mean drag. This velocity is proportional to the square root of the bed shear stress, τ_o. The level of effective drag depends on the shape of the grain and its position in the bed. The drag coefficient for single grains resting on the bed is similar to that for settling grains. The area exposed to drag depends on the shape of the grain, and how it is packed in the sediment bed. This area will be proportional to the square of grain size. Thus, it can be shown that

$$F_D \propto \tau_o D^2 \qquad \text{(Eq. 3)}$$

where the constant of proportionality depends on the grain shape and orientation, its relative protrusion above the mean bed level, and the grain Reynolds number.

Both turbulent and nonturbulent fluid lift forces can act on bed grains (summary in Bridge and Dominic, 1984). Nonturbulent lift forces are pressures caused by asymmetrical flow around near-bed grains. As turbulence originates very close to the bed of a stream, bed grains are undoubtedly affected by turbulent lift, and this is most likely to be dominant over nonturbulent lift (outside the viscous sublayer). Fluid lift is commonly expressed in a similar way to drag, but using a lift coefficient instead of a drag coefficient.

The threshold of transport of cohesionless sediment (sand and gravel)

If fluid drag and lift forces exceed the forces keeping sediment grains in place on the bed (gravity and cohesive forces associated with vegetation, microbial coatings, cementation, and clay minerals), the grains will be entrained by the flow. In the case of *cohesionless* grains, the threshold of entrainment depends on the balance between the drag and gravity forces, F_D and F_G respectively, as long as fluid lift is assumed to be a linear function of drag. As

$$F_D \propto \tau_o D^2 \qquad \text{(Eq. 4)}$$

and

$$F_G \propto (\sigma - \rho) g D^3 \qquad \text{(Eq. 5)}$$

the threshold of entrainment should depend on

$$F_D/F_G \propto \tau_o/(\sigma - \rho)g\, D = \Theta \qquad \text{(Eq. 6)}$$

where Θ is referred to as the *dimensionless bed shear stress* (a measure of sediment mobility) and D is a grain size representative of all grains in the bed. Values of dimensionless bed shear stress at the threshold of motion are typically between 0.03 and 0.06 (Figure S9: based on numerous theoretical and experimental studies, reviewed in Miller *et al.*, 1977; Yalin and Karahan, 1979; Wiberg and Smith, 1987; Bridge and Bennett, 1992; Komar, 1996: Buffington and Montgomery, 1997). The exact value of Θ depends on the degree to which the grains are immersed in the viscous sublayer, the relative magnitude of the lift component of drag, the bed slope, and the relative size, shape and arrangement of the grains in the bed.

The boundary grain Reynolds number (Figure S9) is defined as

$$Re_b = U*D/\nu \approx 10 D/\delta \qquad \text{(Eq. 7)}$$

where $U* = \sqrt{(\tau_o/\rho)}$ is the shear velocity, ν is kinematic viscosity, and δ is the thickness of the viscous sublayer. Thus,

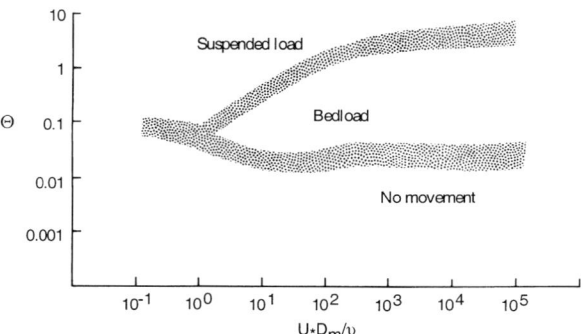

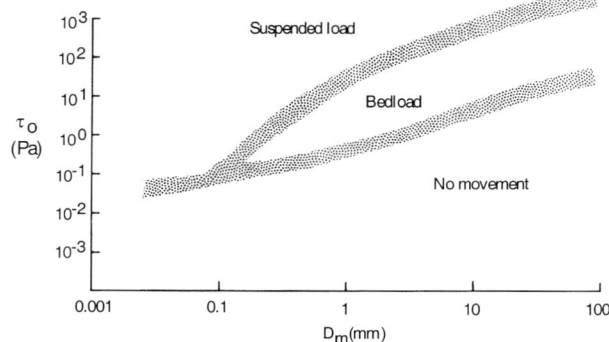

Figure S9 Threshold of motion (shaded zones) for bed load and suspended load for the case of cohesionless sediment. Based on numerous experimental and theoretical studies discussed in text.

the boundary grain Reynolds number is proportional to the ratio of grain diameter/thickness of the viscous sublayer. For Re_b less than 5 to 10 (hydraulically smooth flow), the grains are immersed in the viscous sublayer, are subjected mainly to viscous forces, and are sheltered from the much greater drag forces of the turbulent flow above. This is the reason why the dimensionless bed shear stress must be relatively large to entrain these grains (Figure S9). However, this is also partly to do with increasing effects of cohesion in very small grains. If bed grains project high above the viscous sublayer (Re_b>70 to 100; hydraulically rough flow), turbulent shear stresses and form drag are dominant, viscous forces are negligible, and the threshold of motion becomes independent of Reynolds number (Figure S9). For fixed values of fluid and sediment density, the threshold of entrainment can be expressed in terms of bed shear stress and grain size (Figure S9).

The threshold of motion shown in Figure S9 is composed of a range of experimentally determined values. This range exists partly because not all of the controlling parameters (such as bed slope and relative size, shape and arrangement of grains on the bed) are represented on plots of dimensionless bed shear stress and boundary grain Reynolds number. However, perhaps the main reason for the range is the difficulty in defining the threshold of motion. Turbulent flow is associated with fluid drag and lift forces that vary greatly in time and space. These forces are acting upon grains that vary in size, shape, and arrangement on the bed. Thus, a locally high instantaneous fluid force may dislodge a few of the most easily moved grains, but may not produce general sustained movement. Such intermittent movement may be difficult to observe (see *Grain Threshold*).

The threshold of motion defined in Figure S9 only applies to *planar beds* with relatively small slopes. Some theories take into account the effects of bed slope (e.g., Wiberg and Smith, 1987; Bridge and Bennett, 1992). If there are bed waves such as ripples or dunes present on the bed, the threshold of motion is complicated by non-negligible bed slope and the fact that the spatially averaged bed shear stress includes components of bed-form drag that do not effectively act on bed grains. Thus, in order to define the threshold of motion on bed waves, it is necessary to define local fluid drag and lift on their upstream sides, and to account for the upstream sloping bed surface.

The threshold of motion defined in Figure S9 only applies to the *mean* grain size of grains that do not vary much in size, shape, and density. Prediction of the threshold of entrainment of individual grains in a mixture of grains of different size, shape, and density requires much more detailed analysis of the forces involved, including turbulent variations in fluid drag and lift. There are a number of different theoretical and empirical approaches to this problem (reviews by Bridge and Bennett, 1992; Komar, 1996; Buffington and Montgomery, 1997; Bridge, 2003). All of the approaches predict that the threshold of motion for a particular grain size in the mixture depends on its size relative to the median size of the mixture. Larger-than-average grains are relatively easier to entrain than if all grains were the same size, whereas the opposite is true for the smaller-than-average grains. This is because large grains protrude relatively higher into the flow where average fluid velocity is greater, and because friction angles are relatively low. Such grains may have dimensionless bed shear stress as low as 0.01 at the threshold of motion. In contrast, small grains sit low in the flow and have large friction angles, or are hidden by the larger grains. In rare cases, the threshold of motion of the large and small grains may be so similar that they are almost equally mobile. However, these predictions are based on consideration of time-averaged fluid drag and lift. It is important to consider turbulent variations in fluid drag and lift, as these cause different fractions to move at different times. Small grains may not actually be hidden from turbulent lift associated with flow separation to the lee of larger grains. The largest grains may only move under the influence of the largest instantaneous fluid drag or lift, and not by the mean values. Lack of consideration of turbulence may lead to underestimation of the threshold shear stresses for the larger grains, and vice versa for small grains.

In view of the fact that fluid turbulence has a dominating influence on sediment entrainment, and that turbulent motions vary in time and space, it is not surprising that when a particular size fraction starts to move on the bed, not all of the grains of that size fraction on the bed will be moving. This will be particularly true for the coarsest fractions that require the largest instantaneous turbulent stresses for movement. The condition where some of the grains of a given size on the bed move within a given time interval, but not all of them move, has been called partial transport by Wilcock and McArdell (1997). Partial transport is thought to be common in many gravel-bed rivers. As would be expected, the proportion of grains present in the bed that are mobile is greatest for the finest sizes and least for the coarsest sizes.

The threshold of transport of cohesive sediment (mud)

Cohesion in muds is the dominant control on resistance to entrainment (review by Black et al., 2002). Cohesion is partly due to electro-chemical forces, the magnitude of which depends on mineralogy (e.g., exchangeable cations), the size, shape, and spatial arrangement of the clay flakes, and the ionic properties of the pore water and eroding water. However, cohesion can also be due to microbial organisms (bacteria, microphytobenthos). Resistance properties also depend on the state and history of consolidation. For example, there may be a flocculated structure, a pelleted structure due to desiccation and bioturbation, or fissility due to compaction. Certainly, compacted and dry clays are more difficult to erode than soft, wet clays. Many experimental studies have been conducted to establish relationships between critical bed shear stress for entrainment and some gross property of the clay such as cohesive shear strength, liquid and plastic limits, percent clay, density, permeability and porosity. These parameters have usually been linked to critical shear stress one or two at a time using only one type of clay-water complex. There are no universal correlations for all types of clays in all aqueous environments. The nature of entrainment also depends on the state and history of consolidation, which will determine whether the clay is eroded as individual flakes, floccules, or fragments of consolidated mud. Also, direct fluid stress may be augmented by impacts on the bed of grains already in motion.

Bed load

Sediment grains with diameters greater than approximately 0.1 mm (sand and gravel) generally move close to the boundary (within ten grain diameters) by rolling, sliding and saltating (jumping). These grains move more slowly than the surrounding fluid because of their intermittent collisions with the bed and each other. These grains form the *bedload*. The continuing movement of bed-load grains requires the existence of an upward dispersive force (Bagnold, 1966, 1973) that must be exactly balanced by the immersed weight of the moving grains for steady bed-load transport. The two ways of dispersing grains normal to the flow boundary are by collisions between the grains and bed, and by fluid lift.

Saltation is the dominant mode of bed-load transport, with rolling and sliding only occurring near the threshold of entrainment and between individual jumps (summary in Bridge and Dominic, 1984). As the downstream velocity of rolling and sliding grains increases over an irregular bed, the grains are frequently launched upward as they are forced over protruding immobile grains, thus initiating saltations. At low transport rates, saltation trajectories are unlikely to be interrupted by collisions with other moving grains. However, at higher sediment transport rates (greater concentration of bed-load grains), saltations will be more interrupted by collisions between moving grains, and these collisions will contribute to the upward dispersive stress.

Both turbulent and non-turbulent fluid lift forces can act on bed-load grains (summary in Bridge and Dominic, 1984). Non-turbulent lift forces are pressures arising from a net relative velocity between grains and surrounding fluid. They are caused by asymmetrical flow around near-bed grains, the reduced grain velocity relative to the velocity gradient in the surrounding fluid (shear drift), and the effects of grain rotation (Magnus lift). It is very difficult to specify their individual contribution to net lift, particularly when complex grain collisions occur within the bed load, and when turbulent lift forces are dominant.

As turbulence originates very close to the bed of a stream, bed-load grains are undoubtedly affected by turbulent lift as well as by drag. As near-bed turbulent fluctuations increase with bed shear stress (or shear velocity), grain saltations are modified more and more by turbulence as bed shear stress and sediment transport rate increase. Turbulence affects different sized grains in different ways, depending on immersed weight of grains and the direction and magnitude of turbulent fluid motions (review by Bridge, 2003).

The mean height of the bed-load (saltation) zone controls the effective roughness of the bed, and must be known to calculate bed-load transport rate. Many approaches to determining the mean height of the bed-load zone are theoretical-empirical, predicting that saltation height increases with sediment size and some measure of sediment transport rate (reviews by Bridge and Dominic, 1984; Bridge and Bennett, 1992; Bridge, 2003). In general, saltation height is proportional to grain diameter. Theoretical models of saltation trajectories of individual grains have also been used to determine saltation height. However, it is very difficult to model the influence on saltating grains of turbulent fluid motions, and the effects on grain trajectories of impacts among grains are not considered.

Suspended load

Grains of clay, silt and very fine sand generally travel within the flow, supported by upward-directed fluid turbulence. These grains, the *suspended load*, travel at approximately the same speed as the surrounding fluid because they are not decelerated by intermittent collisions with the bed. Maintenance of a suspended-sediment load requires the existence of a net upward-directed fluid stress to balance the immersed weight of the suspended grains. This can only arise if the vertical component of turbulence is anisotropic (Bagnold, 1966). This anisotropy can be associated with the turbulent bursting phenomenon, whereby sediment is suspended mainly in vigorous upward ejections of fluid, and then returns to the bed under the influence of gravity and downward-directed turbulent motions. This idea is supported by experimental observations of discrete clouds of suspended sediment associated with turbulent ejections, and the wavy trajectories of suspended grains (review in Bridge, 2003). However, sediment is also suspended in vortices generated by eddy shedding from the separated boundary layer in the trough regions of ripples and dunes. With dunes, these vortices can reach the water surface as boils or kolks. Provided that a grain experiences more upward directed turbulent motions than the combination of downward directed turbulent motions and gravitational settling, it will remain in suspension.

Two criteria for the suspension of sediment are:

$$rms \; v'_+ > V_g \qquad (Eq.\ 8)$$

$$rms \; v'_+ - rms v'_- > V_g \qquad (Eq.\ 9)$$

where $rms \; v'$ is the root mean square of the vertical turbulent fluctuations of the flow, the subscripts refer to upward motions (+) or downward motions (−), and V_g is the mean terminal settling velocity of the sediment. The first criterion indicates that, as long as time-averaged upward-directed turbulent velocities are in excess of V_g, sediment can be moved upward within the flow, even if the sediment eventually returns to the bed. The second criterion indicates that for a grain to *remain* in suspension the time-averaged upward-directed turbulent velocity experienced by the grain must exceed the downward-directed velocity plus the settling velocity of the grain. There is an empirical relationship between the time-averaged upward and downward directed velocities and the overall $rms \; v'$ near the bed, such that the two suspension criteria above become

$$a \, rms v' > V_g \quad \text{where } a \text{ is } 1.2 \text{ to } 1.5 \qquad (Eq.\ 10)$$

$$b \, rms v' > V_g \quad \text{where } b \text{ is } 0.4 \text{ to } 0.9 \qquad (Eq.\ 11)$$

Furthermore, $rms v'/U* \approx 1$ near the bed, so that the suspension criteria for near-bed conditions are

$$aU* > V_g \quad \text{and} \quad bU* > V_g \qquad (Eq.\ 12)$$

Also, as the shear velocity and the settling velocity can be related to the dimensionless bed shear stress and the grain Reynolds number, the threshold of suspension can be represented on the threshold of entrainment diagrams (Figure S9). As the bed shear stress varies in time and space, sediment with a particular settling velocity (or grain size) may travel as both bed load and sediment load. This is particularly true of the sand sizes coarser than about 0.1 mm. This is called *intermittently suspended load*.

The *washload* is commonly distinguished from *suspended bed-material load*. Wash load is very fine grained and suspended even at low flow velocity. Furthermore, the volume concentration of wash load does not vary much with distance from the bed, whereas the volume concentration of suspended bed-material load decreases markedly with distance from the bed. Whereas the suspended bed-material load originates from the bed, the wash load can come from bank erosion and overland flow. Perhaps, the two different suspension criteria effectively distinguish wash load from suspended bed-material load.

Effect of sediment transport on flow characteristics

In uniform, turbulent boundary layers over planar beds with sediment transport, it is generally recognized that the moving sediment modifies flow characteristics. Many scientists in the 1940s to 1960s (reviewed in ASCE, 1975) suggested that the presence of sediment in the flow caused a reduction in turbulent eddy sizes and in the degree of turbulent mixing. The main effect of sediment on turbulence was expected to be near the bed where sediment concentration is highest. These conclusions have been substantially confirmed by recent studies (review by Bridge, 2003). The nature of the interaction between sediment grains and fluid turbulence actually depends on the relative velocity of the grains and the turbulent fluid (which depends on grain weight), and the relative size of the grains and the turbulent eddies. If the relative velocity of the grains and fluid is large (i.e., large, heavy grains), turbulence intensity (particularly of the smallest eddies) is increased as a result of vortex shedding or inertial effects. If the relative velocity is small (i.e., small grains), turbulence intensity is decreased as the turbulent energy is transferred to the grains. More recently it has been suggested that turbulent mixing is suppressed because of the *density stratification* brought about by an upward decrease in suspended sediment concentration, as represented by a *flux Richardson number*. However, this idea is not well supported by natural data (Bennett *et al.*, 1998). In natural flows, bed forms have an overwhelming effect on turbulence intensities and eddy sizes.

Sediment transport rate (bed load)

The *sediment transport rate* is the amount (weight, mass, or volume) of sediment that can be moved past a given width of flow in a given time. The sediment transport rate controls the development of bed forms such as ripples and dunes, the nature of *erosion* and *deposition*, and the dispersal of physically transported pollutants. There is a vast and complicated literature on the subject (reviews in Graf, 1971; ASCE, 1975; Yalin, 1977; Van Rijn, 1984a,b,c; Chang, 1988; and many others) and only the basics will be presented here.

Bed-load transport rate can be expressed as

$$i_b = W u_b \qquad \text{(Eq. 13)}$$

where i_b is the bed-load transport rate as immersed weight passed per unit width, W is the immersed weight of bed-load grains over unit bed area, and u_b is the mean bed-load grain velocity. The contacts between bed-load grains and the bed result in resistance to forward motion. Therefore, the fluid must exert a mean downstream force on the grains to maintain their steady motion. For bed-load transport on a bed of slope angle β, the force balance parallel to the bed can be expressed as

$$\tau_o + W \sin \beta = T + \tau_r \qquad \text{(Eq. 14)}$$

where τ_o is the bed shear stress applied to the moving grains and immobile bed (but excluding form drag due to bed forms and banks), $W \sin \beta$ is the down-slope weight component of the moving grains per unit bed area (with $\sin \beta$ positive if the bed slopes down flow), T is the shear resistance due to the moving bed load, and τ_r is the residual shear stress carried by the immobile bed. Bed-load shear resistance can be expressed as

$$T = (W \cos \beta - F_L) \tan \alpha \qquad \text{(Eq. 15)}$$

where F_L is the net fluid lift on bed-load grains per unit bed area, taken as resulting dominantly from anisotropic turbulence, and given by

$$F_L = W(bU*/V_g)^2 \qquad \text{(Eq. 16)}$$

and $\tan \alpha$ is the dynamic friction coefficient. From experiments on shearing of grains in dry grain flows and coaxial drums, where volumetric grain concentrations are approximately 0.5, $\tan \alpha$ commonly lies between 0.4 and 0.75 (review in Bridge and Dominic, 1984; Bridge and Bennett, 1992). The residual fluid stress τ_r equals the bed shear stress at the threshold of motion τ_c according to Bagnold (1956, 1966). Thus, once the applied fluid stress exceeds the value necessary for entrainment, grains will be entrained until the bed-load resistance caused by their motion reduces the applied fluid stress to the threshold value of τ_c. Therefore, the immersed weight of grains moving per unit bed area is given, from equations 14 to 16, as

$$W = \frac{\tau_o - \tau_c}{\tan \alpha \left[\cos \beta - (bU*/V_g)^2 \right] - \sin \beta} \qquad \text{(Eq. 17)}$$

As the threshold of suspension is reached, the term in square brackets approaches zero, indicating that the weight of bed-load grains approaches infinity. In reality, this situation could not arise because high grain concentrations in the bed load layer would occlude the bed from fluid drag and lift, thus limiting the weight of bed-load grains (e.g., Bagnold, 1966). Furthermore, high concentrations of near-bed grains modify turbulence characteristics above the bed. In many natural flows, the bed-slope is small enough such that $\sin \beta$ tends to zero, and the term in square brackets tends to 1, so that

$$W = \frac{\tau_o - \tau_c}{\tan \alpha} \qquad \text{(Eq. 18)}$$

The mean velocity of the average bed-load grain is given by

$$u_b = u_D - u_R \qquad \text{(Eq. 19)}$$

where u_D is the time-averaged flow velocity at the level of effective drag on the grain, and u_R is the relative velocity of the fluid and the grain. The relative velocity is that at which the fluid drag plus down-slope weight component on the grain is in equilibrium with the bed-load shear resistance:

$$F_D + F_G \sin \beta = F_G \tan \alpha (\cos \beta - (bU*/V_g)^2) \qquad \text{(Eq. 20)}$$

Using the general drag equation for F_D, and the fact that F_G is equal to the fluid drag at a grain's terminal settling velocity, results in

$$u_R = V_g (\tan \alpha (\cos \beta - (bU*/V_g)^2) - \sin \beta)^{0.5} \qquad \text{(Eq. 21)}$$

Finally, as $u_D/U* = a$, the mean velocity of bed-load grains is given by

$$u_b = aU* - V_g (\tan \alpha (\cos \beta - (bU*/V_g)^2) - \sin \beta)^{0.5} \qquad \text{(Eq. 22)}$$

or approximately as

$$u_b = aU* - V_g (\tan \alpha)^{0.5} \qquad \text{(Eq. 23)}$$

Furthermore, as the grain velocity must be zero at the threshold of grain motion, this equation may also be written as

$$u_b = a(U* - U*_c (\tan \alpha / \tan \alpha_c)^{0.5}) \qquad \text{(Eq. 24)}$$

Determination of the value of a requires knowledge of the height above the bed level of the effective drag on the grains (that varies with the saltation height) and details of the near-bed velocity profile. There have been numerous attempts to relate saltation height to some measure of flow stage (review in Bridge and Bennett, 1992). The flow velocity at the height of effective drag depends on whether the flow is hydraulically smooth, transitional or rough. For transitional and rough flows, a is about 8–12 (Bridge and Dominic, 1984). Also, at the threshold of suspension, the grain velocity approximately equals the fluid velocity and $u_b = aU_*$, which is another suspension criterion.

Thus, the final approximate bed-load transport equation can be written as

$$i_b = \frac{a}{\tan \alpha}(U_* - U_{*c})(\tau_o - \tau_c) \quad \text{(Eq. 25)}$$

Equations of this form have been developed, using similar principles, by a number of workers (details in Bridge and Dominic, 1984). This equation agrees very well with experimental data for flows over plane beds. For low sediment transport rates (lower-stage plane beds), $a/\tan \alpha$ is approximately 10, but is approximately 17 for high sediment transport rates (upper-stage plane beds). For beds covered with bed forms such as ripples and dunes, the bed-load transport equation can be applied by determining the average bed shear stress over a number of bed forms, and to remove the part of the total average stress that is due to form drag in the lee of the bed forms, because it is ineffective in moving bed-load grains (e.g., Engelund and Hansen, 1972; Engelund and Fredsoe, 1982; Nelson and Smith, 1989). However, bed-load transport rates associated with ripples and dunes can also be calculated from knowledge of their mean height and downstream migration rate.

The bed-load transport theory developed above is for mean flow and sediment characteristics, but it can be adapted for use with heterogeneous sediment and with temporal variations in bed shear stress (e.g., Bridge and Bennett, 1992). In this case, it is necessary to specify the proportion of the different grain fractions available in the bed for transport, and the thresholds of entrainment and suspension for each of the grain fractions. It is also necessary to specify the proportion of time that a particular bed shear stress acts upon the bed-load, requiring a frequency distribution of bed shear stress. Bridge and Bennett's (1992) bed-load transport model for heterogeneous sediment agrees well with natural data. Models such as this (e.g., Wiberg and Smith, 1989; Parker, 1990) are essential for quantitative understanding of sediment sorting during erosion, transport and deposition.

Sediment transport rate (suspended load)

Suspended-sediment transport rate at a point in the flow is commonly expressed as

$$i_s = u_s C \quad \text{(Eq. 26)}$$

where u_s is the average speed of the sediment (approximately the same as the fluid velocity), and C is the volume concentration of suspended sediment. The units of i_s are therefore volume of sediment transported per unit cross-sectional area normal to the flow direction per unit time. The vertical variation of the velocity of suspended sediment and fluid can be calculated using an appropriate velocity profile law (e.g., the law of the wall). Calculation of the vertical variation of C in steady, uniform water flows is traditionally based on the balance of downward settling of grains and their upward diffusion in turbulent eddies, i.e.,

$$V_g C + \varepsilon_s \, dC/dy = 0 \quad \text{(Eq. 27)}$$

where the first term is the rate of settling of a particular volume of grains per unit volume of fluid, the second term is rate of turbulent diffusion of sediment per unit volume, and ε_s is the diffusivity of suspended sediment (equivalent to a kinematic eddy viscosity, ε). This balance can also be written for individual grain fractions. In order to determine C at any height above the bed, y, it is necessary to calculate the vertical variation in ε_s. Assuming that $\varepsilon_s = \beta \varepsilon$, where β is close to 1, and that the law of the wall extends throughout the flow depth, results in the well-known *Rouse equation*

$$\frac{C}{C_a} = \left[\frac{d-y}{y} \cdot \frac{a}{d-a}\right]^z \quad : z = \frac{V_g}{\beta \kappa U_*} \quad \text{(Eq. 28)}$$

where C_a is the value of C at $y = a$. This equation predicts that C decreases continuously and smoothly with distance from the bed, as would be expected in view of the fact that turbulence intensity also decreases with distance from the bed. The distribution of suspended sediment becomes more uniform throughout the flow depth as exponent z decreases, that is as settling velocity (grain size) decreases and/or as shear velocity (near-bed turbulence intensity) increases (Figure S10). For example, a decrease in water temperature causes an increase in fluid viscosity, which may in turn cause a decrease in settling velocity and an increase in suspended sediment concentration. The Rouse equation has been shown to agree with measured suspended-sediment concentrations in the lower part of stream flows but concentrations tend to be overestimated in the upper parts (figure S2). This discrepancy is addressed below.

Some of the assumptions used in developing the Rouse equation have been criticized: mainly the choice of velocity profile (law of the wall), and the fact that the interaction between suspended sediment, fluid properties and turbulence is not considered. Perhaps the most serious concern with the Rouse equation is that the sediment diffusivity has rarely been calculated directly using quantitative observations of the motion of sediment in turbulent eddies. Thus, there is doubt about the vertical variation of ε_s and β (e.g., ASCE, 1975; Coleman, 1970; Van Rijn, 1984b). Bennett *et al.* (1998) have measured turbulent motions of suspended sediment and concluded that β is indeed close to 1. However, suspended sediment concentration in the upper half of the flow depth in natural flows is commonly larger than predicted by the Rouse equation. One reason for this discrepancy could be that sediment grains that are suspended to the higher levels in the flow are not associated with mean turbulence characteristics as specified in the Rouse equation, but with turbulent eddies with the greatest turbulence intensities and mixing lengths (Bennett *et al.*, 1998). Thus, the Rouse equation, based on average turbulence characteristics, may only be strictly applicable for the lower parts of the flow.

Application of the Rouse equation in practice requires calculation of C_a and a, and there are various methods for doing this (review by Bridge, 2003). Normally, C_a is calculated at a position within or at the top of the bed-load layer. If bed waves are present, a can be taken as half bed-wave height or equivalent roughness height (Van Rijn, 1984c). The mean volume concentration of grains in the bed-load layer, C_b, is the

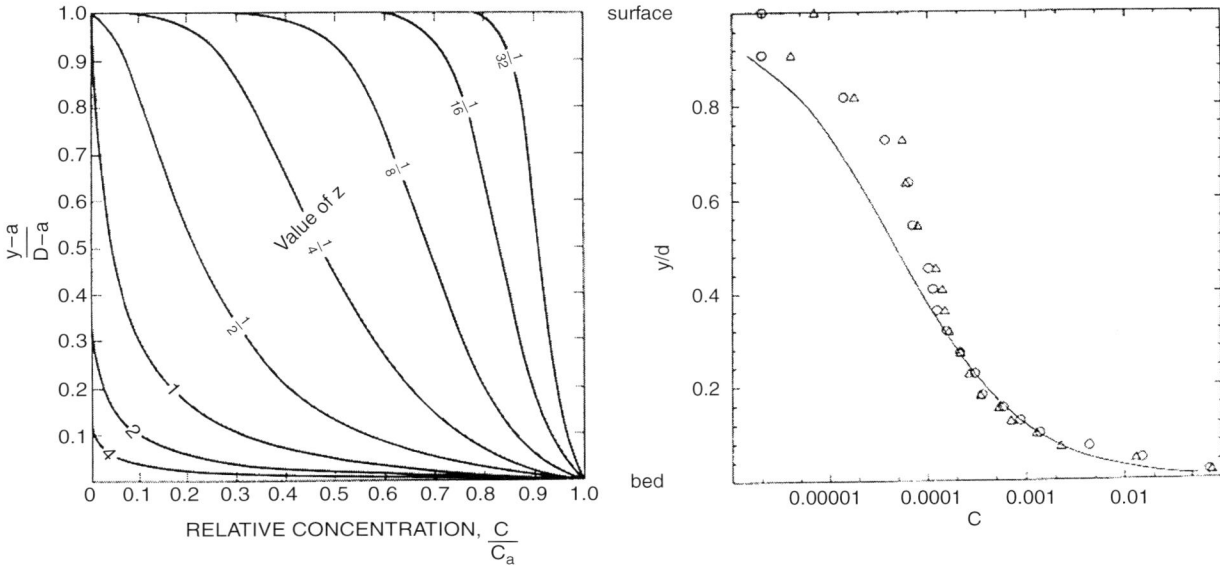

Figure S10 A) Variation of relative concentration of suspended sediment with distance from the bed as a function of z according to the Rouse equation (after ASCE, 1975). The distribution of suspended sediment becomes more uniform throughout the flow depth as exponent z decreases, that is as settling velocity (grain size) decreases and/or as shear velocity (near-bed turbulence intensity) increases. B) Comparison between the Rouse equation and data from an upper stage plane bed (from Bennett et al., 1998). Predicted suspended sediment concentration is less than observed in the upper part of the flow.

volume of grains per unit bed area divided by the mean thickness of the bed-load layer, y_b, given using Bridge and Dominic's (1984) model as

$$C_b = \frac{\tau_o - \tau_c}{(\sigma - \rho)gy_b \tan \alpha} \quad \text{(Eq. 29)}$$

The mean grain concentration can be assumed to occur at a distance from the bed of half the thickness of the bed-load layer, or half the equivalent roughness height.

An alternative to the convection-diffusion approach to calculating suspended sediment concentration in water flow is that based on the mechanics of two-phase flows (e.g., McTigue, 1981; Cao et al., 1995; Greimann et al., 1999; Greimann and Holly, 2001). The two-phase flow approach is useful for simulating interactions between grains, and the effects of grain inertia. Interactions between particles lead to an increase in the ability of the flow to suspend particles. Particle interactions are only important near the bed where local grain concentrations exceed about 0.1. Also, particle inertia is only important for large grains near the bed. Unfortunately, this approach is not fully developed, and cannot yet be applied easily to real rivers.

Evaluation of sediment transport rate models

There are many different models for sediment transport rate in steady, uniform flows. Some are for bed load only, whereas others predict bed load and suspended load either separately or together (total load). Important differences in these models depend on whether or not they consider individual sediment size-density fractions, the sediment available in the bed, and the effects of turbulent variations in fluid drag and lift on bed load; the way in which flow over bed forms such as ripples and dunes is treated; the description of the vertical variation in flow velocity, turbulence intensity, and diffusivities of fluid momentum and suspended sediment, and; how moving sediment interacts with and modifies the water flow. Most sediment transport models are both theoretical and empirical, but it is desirable to minimize the use of empirical factors that limit the general applicability of a model. There have been many comparisons of the performances of the various sediment transport models (e.g., Graf, 1971; ASCE, 1975; Yalin, 1977; Chang, 1988; Gomez and Church, 1989).

Variation of sediment transport rate in time and space

Sediment transport rate varies over a large range of time and space scales, associated for example with fluid turbulence, the migration of bed waves (such as ripples, dunes, and bars), the break-up of armored or partially cemented beds, bank slumps, seasonal floods, flow diversions, and earthquake-induced changes in sediment supply. A characteristic feature of unsteady, nonuniform flows is that sediment transport rate commonly varies incongruently with temporal and spatial changes in water discharge: i.e., there is a lag or hysteresis. This may be due to variations in available sediment supply, or due to the lag of sediment and water flow behind a faster moving flood wave. Wash load is never well correlated with water discharge, and commonly reaches a peak during rising discharge because of an overland flow source. The lag of bed-load transport rate behind a spatial change in flow characteristics is very small (order of grain diameters); however, for suspended sediment this spatial lag is on the order of flow depth.

Sorting and abrasion of grains during transport

The fact that bed load and suspended load travel at different speeds and in different parts of a turbulent flow leads to the possibility of *sorting*, or physical separation of grains with different settling velocities (determined partly by grain size, shape and density). Figure S11 indicates *approximately* the range of grain sizes (for a given grain shape and density) that can be transported as bed load for a particular bed shear stress. The actual grain size range in the bed load depends on what is available for transport, the threshold of entrainment or suspension for individual grain fractions, and turbulent fluctuations in bed shear stress (Bridge and Bennett, 1992). Nevertheless, Figure S11 explains in a general way why intermittently suspended load exists, and why it requires temporal variations in bed shear stress. It is now possible to calculate theoretically the range of sizes of grains with a specific shape and density that will form the bed load and intermittently suspended load, given a distribution of available sediment (Bridge and Bennett, 1992; Bennett and Bridge, 1995). In general, the mean and standard deviation of grain size in the bed load increases as the bed shear stress increases (Figure S11). The size distribution of suspended sediment is more difficult to specify. Although the coarse end of the distribution is controlled by bed shear stress, the overall distribution is controlled by sediment availability.

Collisions between near-bed grains in transport results in *attrition* (*abrasion*) of the grains (see *Attrition, Fluvial*). Experimental studies of abrasion of gravel during river transport indicate that grain size is reduced and grain shape is modified, depending on the hardness of the grains and distance of transport. Abrasion of sand in water is negligible (in contrast to that in air flows).

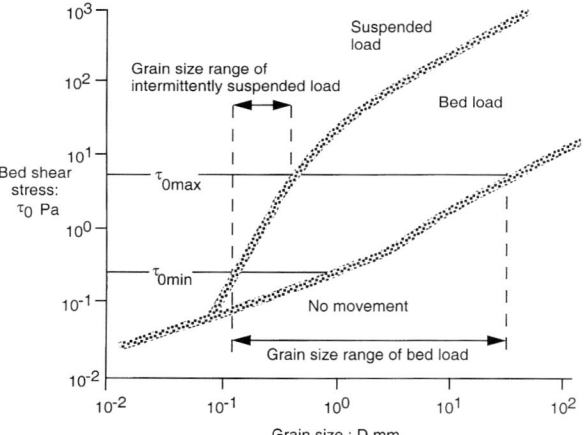

Figure S11 Approximate prediction of grain-size range of bed load and intermittently-suspended load where bed shear stress is allowed to vary between limits τ_{omin} and τ_{omax}. The mean grain size and range of sizes increases as instantaneous bed shear stress increases. The prediction is only approximate because the thresholds of motion and suspension of individual grains in a mixture of different-sized grains are not necessarily exactly as shown. Threshold curves are based on sediment of constant grain size and density, and time-averaged bed shear stress.

Erosion and deposition

Erosion and deposition are due to changes in sediment transport rate in space (mainly) and in time (minor), as expressed in the equation of conservation of sediment mass (sediment continuity equation). A one-dimensional version of this equation is

$$-C_b dh/dt = di_x/dx + di_x/u_{sx} dt \qquad \text{(Eq. 30)}$$

where C_b is the volume concentration of sediment in the bed (1−porosity), h is bed height, t is time, i_x is the downstream sediment transport rate by volume, x is downstream distance, and u_{sx} is downstream velocity of the sediment. In uniform flow, $di_x/dx = 0$, but an increase in sediment transport rate in time ($di_x/u_{sx} dt > 0$) would result in erosion. Such erosion would involve entrainment of a layer of bed grains amounting to an average thickness not exceeding a few grain diameters. In other words, erosion and deposition in uniform, unsteady flows is normally negligible. In contrast, rates of erosion and deposition in nonuniform flows can be substantial, depending on the magnitude of di_x/dx. Erosion occurs in spatially accelerating flows where di_x/dx is positive, and deposition occurs in spatially decelerating flows where di_x/dx is negative.

It is very common in nature for relatively long periods of little net erosion or deposition to be punctuated by relatively short periods (e.g., during seasonal floods) when there is marked erosion followed by marked deposition or *vice versa* (i.e., the terms in equation 30 vary over time). This explains why sedimentary strata are normally bounded by planes that record erosion or nondeposition. If di_x/dx is negative and does not vary very much in space, a layer of sediment will be deposited that varies little in thickness laterally. If di_x/dx is negative but changes markedly in space, sediment will be deposited as mounds or ridges (e.g., ripples, dunes, and bars). Similarly, a surface of erosion will tend to be planar if di_x/dx is positive and does not vary very much in space, whereas it will tend to be strongly curved (e.g., trough-shaped or spoon-shaped) if di_x/dx is positive and changes locally. In reality, di_x/dx varies in space on various different scales, such that there are different scales of variation in the geometry of strata and erosion surfaces superimposed.

Deposition rates and erosion rates depend on the time scale and space scale of measurement. Vertical deposition rates averaged over a few years and over bed waves like ripples, dunes and bars, can range from fractions of a millimeter per year to tens of meters per year (Bridge and Leeder, 1979; Sadler, 1981; Enos, 1991). However, as many deposits will be completely or partially eroded before final burial, deposition rates over progressively longer time periods tend to decrease. Deposition rates and erosion rates averaged over millions of years are typically on the order of 0.1 mm per year. The deposits with the best *preservation potential* were deposited in the topographically lowest areas (e.g., deepest parts of channels) and those that subsided rapidly.

Sediment sorting and orientation during deposition

Application of equation 30 to individual grain fractions leads to prediction of *sorting* in deposited sediment. Rational prediction of the sorting of deposits requires knowledge of: (1) the sediment types available for transport; (2) the time and space variations in flow and sediment transport rates of individual grain fractions, and; (3) the flow and sediment transport over bed forms such as ripples, dunes and bars. Such prediction has been attempted in only very limited cases (e.g., Bennett and Bridge, 1995; Cui et al., 1996), using sediment routing models as discussed below. In general, when erosion occurs, the coarsest and/or densest grains are most likely to remain in the bed. The large, immobile grains that are left on an eroded bed surface are commonly referred to as a *lag deposit*. The term *armoring* refers to the situation where there are sufficient large immobile (or only intermittently mobile) grains on an eroding bed to protect the potentially mobile grains underneath from entrainment (see *Armor*). Armored beds commonly occur in gravel-bed rivers during falling flow stages after floods. The development of armoring results in an increase in bed roughness and a decrease in sediment transport rate (Gomez, 1993; Hassan and Church, 2000). The increase in bed roughness is partly to do with increasing bed sediment size, and partly due to the development of bed structures such as pebble clusters. When deposition occurs, the coarsest and/or densest grain fractions in the bed-load are preferentially deposited, but the texture of the deposited sediment is closely related to that of the bed-load from which it was derived. It is common for the deposits associated with flood conditions (high bed shear stress) to be coarser grained than those associated with falling and low flow stages (low bed shear stress), because mean grain size of sediment in transport and being deposited is directly related to bed shear stress. Thus, layers of sediment deposited during floods tend to have fine-grained tops, unless bed armoring occurred.

Mean grain size of sediment in the bed surface and in transport commonly decreases exponentially in the direction of transport. This is mainly due to downstream decrease in the bed shear stress and turbulence intensity of the transporting and depositing flow, such that the coarsest bed load grains are progressively lost in the downstream direction (review in Bridge, 2003). Downstream reduction in size due to progressive abrasion associated with grain collisions is of minor importance, as only the larger and softer gravel grains suffer appreciable abrasion. Intermittently suspended grains may be deposited with bed load grains, either as a relatively fine-grained surficial layer, or by infiltrating the pore spaces between the larger grains in the bed (thus reducing the degree of sorting). Most of the suspended load is not deposited with bed load, but accumulates in places where turbulence intensity is low (e.g., flood basins, lakes, and ponds). Suspended sediment can be sorted by selective deposition according to settling velocity.

Heavy minerals (e.g., iron compounds, gold) in the bed load and suspended load are commonly finer grained than associated light minerals (see *Heavy Minerals*). Heavy minerals may be concentrated in the bed where a flow that is powerful enough to entrain and transport all grain fractions decelerates and deposits some of them. The relatively small heavy minerals become protected from re-entrainment by the larger light minerals, and some may infiltrate into the pore spaces between the larger grains in the bed. Subsequent erosion may remove the large light minerals but not the small heavies. Such conditions are found in association with various types of bed forms.

Grains tend to be oriented during transport on the bed immediately prior to deposition. On plane beds, ellipsoidal or rod-shaped grains tend to be oriented with their long axes approximately parallel to flow, and disc- and tabular-shaped pebbles tend to be oriented with their maximum projection planes dipping upstream (*imbrication*, *q.v.*). However, these characteristic orientations are modified where bed forms are present.

Grain size, shape, grain orientation and packing (fabric) in sediments influence *porosity and permeability*, with the most important controls being mean grain size, size sorting, and fabric (Brayshaw et al., 1996; see *Fabric, Porosity, Permeability*). The spatial variation of porosity and permeability varies greatly with spatial variation in texture and fabric (i.e., with stratification).

Nature of erosion

The amount of erosion at a point is limited by two main factors. The first is bed armoring, such that the progressive accumulation of the heaviest grains on the bed protects the underlying sediment from further erosion. The other factor is the generation during vertical erosion of a progressively larger depression in the sediment bed. Such a depression leads to expansion and deceleration of the flow, thereby leading to cessation of vertical erosion, or at least to relocation of the zone of erosion.

Surfaces of erosion are recognizable by truncation of underlying strata and by a single overlying layer of the coarsest and/or densest grains from the underlying material that could not be entrained (i.e., a *lag deposit*). Features indicative of local erosion include *current crescents*, horseshoe-shaped scours around the upstream margins of immobile obstacles to the flow (e.g., large grains, tree trunks). The eroded material is commonly deposited on the lee side as a ridge of sediment or *sand shadow*. Other erosional features include elongate, flow-parallel channels called *gutter casts*, and large-scale features such as channels.

Erosion of cohesive mud depends on the degree and nature of consolidation of the mud. Soft mud that is entrained as mud-sized particles travels in suspension. However, various bed features may be eroded into the mud by the flow according to the bed shear stress (Allen, 1982; see *Scour, Scour Marks*): *straight or meandering longitudinal grooves and ridges*; *flute marks*; *transverse ridge marks*. Soft mud is also eroded and molded by various hard objects (tools) carried by the flow. *Tool marks* are commonly long grooves of varying width and depth due to objects (e.g., shells, bits of wood, shale chips) sliding and rolling along the bed and eroding into the mud. There are also marks due to intermittent contacts of objects with the bed (e.g., bounce, skip and prod marks). These and the current-formed marks are normally preserved as coarser sediment covers the mud more-or-less immediately following mark formation (e.g., *sole marks* on the base of sandstone strata). Consolidated (compacted) mud is commonly eroded as pellets or flat chips of sand or gravel that then travel as bed load. Such relatively large grains may occur as lag deposits. Erosion marks in consolidated mud include steep-walled gutters and potholes.

Sediment routing models

The only rational way of analyzing and predicting bed-elevation changes and sediment sorting during erosion, transport, and deposition is with *sediment routing models* (reviewed by Bridge, 2003). Sediment routing models are used to deal with problems such as bed erosion and armoring downstream of dams, and reservoir sedimentation. Sediment routing models normally require treatment of unsteady, gradually varied flow acting upon movable, heterogeneous sediment beds. A realistic sediment routing model requires specification of: (1) sediment types available for transport; (2) the mean and turbulent fluctuating values of fluid force acting upon sediment grains, and how these vary in time and space; (3) the interaction between fluid forces and individual sediment fractions, resulting in their entrainment and transport as bedload or suspended load; (4) erosion and deposition as determined by a sediment continuity equation applied to different grain fractions; (5) the concept of an active bed layer to which sediment can be added or be removed; and (6) an accounting scheme to record the nature of deposited sediment or armored, eroded surfaces. Many sediment routing models do not have all of these desirable features, and contain many arbitrary correction factors. Examples of the use of sediment routing models to predict the nature of sediment transport, erosion and deposition, the development of armor layers, and the sorting of sediment by size and density are given by Vogel et al. (1992), Bennett and Bridge (1995), and see *Numerical Models and Simulation*.

Sediment transport and bed forms

One of the most striking and important characteristics of sediment transport over a bed of cohesionless sediment is the development of bed forms such as ripples, dunes and bars. These are important to sedimentary geologists because they are associated with certain kinds of sedimentary structures, and to engineers because they influence flow resistance and sediment transport rate. The origin, geometry, flow and sedimentary processes of bed forms are reviewed by Allen (1982), Bridge (2003) and see *Surface Forms*.

John S. Bridge

Bibliography

Allen, J.R.L., 1982. *Sedimentary Structures: Their Character and Physical Basis*. Developments in Sedimentology 30, Elsevier Science.
American Society of Civil Engineers, 1975. *Sedimentation Engineering*, 745pp.
Bagnold, R.A., 1956. The flow of cohesionless grains in fluids. *Philosophical Transactions of the Royal Society of London*, **A249**: 235–297.
Bagnold, R.A., 1966. An approach to the sediment transport problem from general physics. *US Geological Survey Professional Paper* **442-I**.
Bagnold, R.A., 1973. The nature of saltation and of "bed-load" transport in water. *Proceedings of the Royal Society of London*, **A332**: 473–504.
Bennett, S.J., and Bridge, J.S., 1995. An experimental study of flow, bedload transport and bed topography under conditions of erosion and deposition and comparison with theoretical models. *Sedimentology*, **42**: 117–146.
Bennett, S.J., Bridge, J.S., and Best, J.L., 1998. The fluid and sediment dynamics of upper-stage plane beds. *Journal of Geophysics Research*, **103**: 1239–1274.
Black, K.S., Tolhurst, T.J., Paterson, D.M., and Hagerthey, S.E., 2002. Working with natural cohesive sediments. *Journal of Hydraulic Engineering, ASCE*, **128**: 2–8.
Brayshaw, A.C., Davies, G.W., and Corbett, P.W.M., 1996. Depositional controls on primary permeability and porosity at the bedform scale in fluvial reservoir sandstones. In Carling, P.A., and Dawson, M.R. (eds.), *Advances in Fluvial Dynamics and Stratigraphy*. Wiley, pp. 374–394.
Bridge, J.S., 2003. *Alluvial Rivers and Floodplains*. Blackwell.
Bridge, J.S., and Bennett, S.J., 1992. A model for the entrainment and transport of sediment grains of mixed sizes, shapes and densities. *Water Resources Research*, **28**: 337–363.
Bridge, J.S., and Dominic, D.F., 1984. Bed load grain velocities and sediment transport rates, *Water Resources Research*, **20**: 476–490.
Bridge, J.S., and Leeder, M.R., 1979. A simulation model of alluvial stratigraphy. *Sedimentology*, **26**: 617–644.
Buffington, J.M., and Montgomery, D.R., 1997. A systematic analysis of eight decades of incipient motion studies, with special reference to gravel-bedded rivers. *Water Resources Research*, **33**: 1993–2029.
Cao, Z., Wei, L., and Jainheng, X., 1995. Sediment-laden flow in open channels from two-phase flow viewpoint. *Journal of Hydraulic Engineering, ASCE*, **121**: 725–735.
Carver, R.E. (ed.), 1971. *Procedures in Sedimentary Petrology*. Wiley.
Chang, H.H., 1988. *Fluvial Processes in River Engineering*. Wiley.
Coleman, N.L., 1970. Flume studies of the sediment transfer coefficient. *Water Resources Research*, **6**: 801–809.
Cui, Y., Parker, G., and Paola, C., 1996. Numerical simulation of aggradation and downstream fining. *Journal of Hydraulic Research*, **34**: 185–204.
Engelund, F., and Fredsoe, J., 1982. Sediment ripples and dunes. *Annual Review of Fluid Mechanics*, **14**: 13–37.
Engelund, F., and Hansen, E., 1972. *A Monograph on Sediment Transport in Alluvial Streams*. Copenhagen: Teknisk Forlag.
Enos, P., 1991. Sedimentary parameters for computer modeling. In Franseen, E.K., Watney, W.L., Kendall, C.G.St.C., and Ross, W. (eds.), *Sedimentary Modeling: Computer Simulations and Methods for Improved Parameter Definition*. Kansas Geological Survey Bulletin 233, pp. 63–99.
Gomez, B., 1993. Roughness of stable, armored gravel-bed rivers. *Water Resources Research*, **29**: 3631–3642.
Gomez, B., and Church, M., 1989. An assessment of bed load sediment transport formulae for gravel bed rivers. *Water Resources Research*, **25**: 1161–1186.
Graf, W.H., 1971. *Hydraulics of Sediment Transport*. McGraw-Hill.
Greimann, B.P., and Holly, F.M., 2001. Two-phase flow analysis of concentration profiles. *Journal of Hydraulic Engineering, ASCE*, **127**: 753–762.
Greimann, B.P., Muste, M., and Holly, F.M., 1999. Two-phase formulation for suspended sediment transport. *Journal of Hydraulic Research*, **37**: 479–500.
Hassan, M.A., and Church, M., 2000. Experiments on surface structure and partial sediment transport on a gravel bed. *Water Resources Research*, **36**: 1885–1895.
Komar, P.D., 1996. Entrainment of sediments from deposits of mixed grain sizes and densities. In Carling, P.A., and Dawson, M.R. (eds.), *Advances in Fluvial Dynamics and Stratigraphy*. Wiley: pp. 127–181.
McManus, J., 1988. Grain size determination and interpretation. In Tucker, M. (ed.), *Techniques in Sedimentology*. Blackwell: pp. 63–85.
McTigue, D.F., 1981. Mixture theory for suspended sediment transport. *Journal of the Hydraulic Division, ASCE*, **107**: 659–673.
Miller, M.C., McCave, I.N., and Komar, P.D., 1977. Threshold of sediment motion in unidirectional currents. *Sedimentology*, **24**: 507–528.
Nelson, J.M., and Smith, J.D., 1989. Mechanics of flow over ripples and dunes. *Journal of Geophysical Research*, **94**: 8146–8162.
Parker, G., 1990. Surface-based bedload transport relation for gravel rivers. *Journal of Hydraulic Research*, **28**: 417–436.
Sadler, P.M., 1981. Sediment accumulation rates and the completeness of stratigraphic sections. *Journal of Geology*, **89**: 569–584.
Van Rijn, L.C., 1984a. Sediment transport, part I: bed load transport. *Journal of Hydraulic Engineering, ASCE*, **110**: 1431–1456.
Van Rijn, L.C., 1984b. Sediment transport, part II: suspended load transport. *Journal of Hydraulic Engineering, ASCE*, **110**: 1613–1641.
Van Rijn, L.C., 1984c. Sediment transport, part III: bed forms and alluvial roughness. *Journal of Hydraulic Engineering, ASCE*, **110**: 1733–1754.
Vogel, K., Van Niekerk, A., Slingerland, R., and Bridge, J.S., 1992. Routing of heterogeneous size-density sediments over a moveable bed: model verification and testing. *Journal of Hydraulic Engineering, ASCE*, **118**: 263–279.
Wiberg, P.L., and Smith, J.D., 1987. Calculations of critical shear stress for motion of uniform and heterogeneous sediments. *Water Resources Research*, **23**: 1471–1480.

Wiberg, P.L., and Smith, J.D., 1989. Model for calculating bed load transport of sediment. *Journal of Hydraulic Engineering, ASCE*, **115**: 101–123.

Wilcock, P.R., and McArdell, B.W., 1997. Partial transport of a sand-gravel sediment. *Water Resources Research*, **33**: 233–245.

Yalin, M.S., 1977. *Mechanics of Sediment Transport*, 2nd edn, Pergamon Press.

Yalin, M.S., and Karahan, E., 1979. Inception of sediment transport. *Journal of the Hydraulics Division, ASCE*, **105**: 1433–1443.

Cross-references

Attrition (Abrasion), Fluvial
Armor
Cross-stratification
Erosion and Sediment Yield
Fabric, Porosity and Permeability
Flow Resistance
Grain Settling
Grain Size and Shape
Grain Threshold
Gutters and Gutter Casts
Heavy Minerals
Imbrication and Flow-Oriented Clasts
Parting Lineations and Current Crescents
Physics of Sediment Transport: The Contributions of R.A. Bagnold
Planar and Parallel Lamination
Ripple, Ripple Mark, Ripple Structure
Scour, Scour Marks
Surface Forms
Tool Marks
Tracers for Sediment Movement

SEDIMENT TRANSPORT BY WAVES

Water waves

Water waves are periodic undulations of the air–water interface (Figure S12) defined by their *height* (H), *wavelength* (L), *period of oscillation* (T) and *speed of propagation* (C). The water moves in quasi-circular or elliptical *orbits*; the *orbital diameter* (d_o) equals the wave height at the air–water interface and decays with increasing water depth. At depths $>L/2$ the oscillation is essentially dissipated; at depths $<L/2$, the oscillation interacts with the bed to form an *oscillatory boundary layer*. Waves are found wherever wind blows over a significant body of water (e.g., the deep ocean, the continental shelf, the shoreface, estuaries, lakes, etc.). They can be *progressive* (e.g., *wind waves*) or *standing* (e.g., a lake *seiche*) and either actively *forced* (*sea*) or *free* (propagate freely; e.g., *swell*). Swift (1976) defined *hydraulic provinces* stretching from the deep ocean to the coastline. In the innermost zone (*shoreface*), both the hydrodynamics and sediment transport are controlled by waves, while further offshore (*continental shelf*) waves play a major role in the *entrainment* of sediment (Cacchione and Drake, 1990), although the transport may be dominated by other flows. Modern wave theory dates back to the classic work of Stokes (1847) and a recent theoretical treatment of the hydrodynamics of water waves is given by Le Mehauté (1976; especially Part 3). Smith (1977), Sleath (1984), Grant and Madsen (1979; 1982), Nielsen (1992), Fredsøe and Deigaard (1992) and Van Rijn (1993) review the theory of sediment transport under waves.

Water waves are classified by their period (frequency) of oscillation and the force generating them as well as the force restoring equilibrium of the water surface (Kinsman, 1975). The most important waves outside the zone of wave breaking (*surf zone*) are *gravity waves* (forced by wind and restored by gravity), with periods ranging from 1–25 s. Their magnitude depends upon the wind speed, duration of time the wind blows and the *fetch* length (length of open water over which the wind blows). A modulation of wave height is common and *gravity waves* propagate as *groups* of large waves separated by several smaller waves (Figure S13). This modulation forces a secondary wave, the *group-bound long wave*, which propagates at the *group* speed and has a period equal to the *group* period. Waves with periods of ∼25 s to several minutes or longer (frequencies of 0.004–0.04 Hz) are *infragravity waves* (called *surf beat* by Munk, 1949). They are generally small (<1 cm high) in the deep ocean, but increase in size shoreward and are important to water circulation and sediment transport within the *surf zone* (Bowen and Huntley, 1984). In intermediate water depths, infragravity energy results from wave–wave interactions and consists of a mixture of *forced* and *free* wave motions (Herbers *et al.*, 1995). Several mechanisms associated with wave breaking in the *surf zone* have been proposed for their generation including: (i) release of the *group bound long*

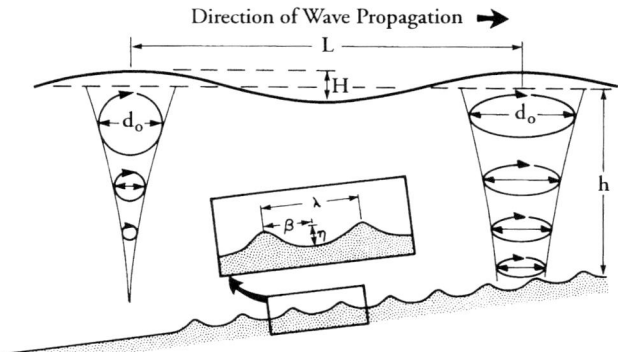

Figure S12 The surface form of deep water and shoaling surface gravity waves, the orbital motion beneath the surface, the interaction of the wave with a rippled bed and the characteristic wave and ripple properties (modified after Clifton, 1984). L = wave length; H = wave height; d_o = orbital diameter; h = water depth; λ = ripple wavelength; η = ripple height; β = ripple crest-to-trough distance.

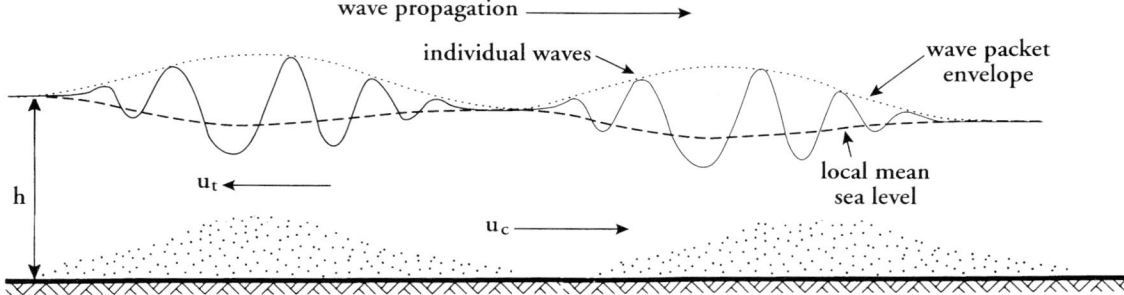

Figure S13 The surface profiles of high frequency surface gravity waves (solid line), the wave envelope (dotted line) and the forced group bound long wave (dashed line). The horizontal near-bed velocities under the long wave crest and trough (u_c and u_t) and the relative sediment concentrations (stippled) are also shown, as is the water depth (h). Note the coupling of large sediment concentrations induced by the large gravity waves with the upwave (offshore) velocity under the trough of the long wave; there is often a small phase lag in this relationship.

wave (List, 1992); (ii) periodic shift in the position of wave breaking (Symonds *et al.*, 1982); (iii) persistence of *groupiness* into the *surf zone* causing a rise and fall of water at the shoreline (Watson and Peregrine, 1992). Nearshore *infragravity energy* can take several forms including standing and progressive *edge waves* (trapped to the shoreline by refraction) and long *leaky waves* (reflected seaward without trapping; Huntley, 1976). An important characteristic of *infragravity waves* in the *surf zone* is that they increase in height as the offshore wave height increases. In contrast, *gravity waves* are *saturated* in shallow water and thus decrease in importance proportionally during storms. *Shear waves* are a subset of the *infragravity* field (periods range from several tens of seconds to several tens of minutes) and they form within the *surf zone* through *instabilities* in the *longshore current*. They appear as large fluctuations in the current (Bowen and Holman, 1989; Oltman-Shay *et al.*, 1989) and can contribute significantly to both the cross-shore and alongshore transport of sediment (e.g., Aagaard and Greenwood, 1995).

Wave propagation

Propagating *gravity waves* transfer energy and momentum, and as they move into shallow water they *shoal*. With increased frictional drag, the orbital motions are deformed, wave speed and length decreases and height increases. The *specific energy density* (according to linear theory, $E = 1/8\rho_f g H^2$) increases until the wave breaks. Energy may also be re-distributed laterally (parallel to the wave crests) through the processes of *refraction* and *diffraction*. Further, the simple *near-sinusoidal* shape, characteristic of deep water, becomes increasing non-symmetrical about the horizontal and vertical axes (*wave skewness* and *asymmetry*). These non-linearities (Figure S14) are critical to the *net* transport by waves as they introduce asymmetries in the oscillatory velocities. In very shallow water, waves become unstable and break when $H/h \approx 0.4$–1.0. The wave form may be destroyed in a *plunging breaker*, producing a large *roller vortex*, or the wave may propagate further as a *spilling breaker* or *surf bore* with reduced height. The *surf zone* is a complex hydrodynamic environment, where interactions with *secondary waves* and *quasi-steady currents* of various origins occur. Ultimately *wave energy* is dissipated in the reversing currents of the *swash* on the *beach face*; the *uprush* and *backwash* induce high frequency, reversing sediment transport similar to that of waves. Transport is complicated

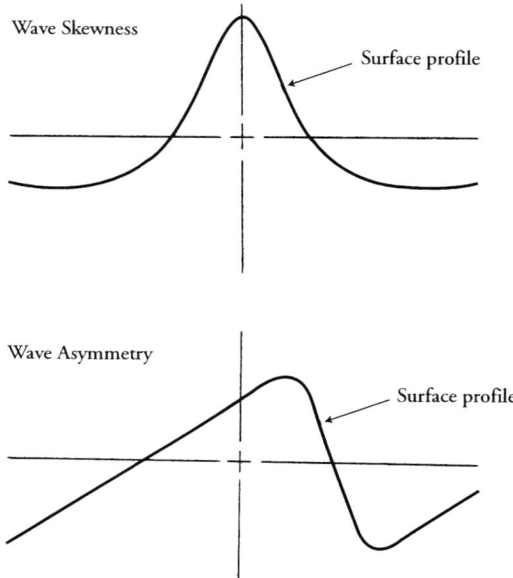

Figure S14 Surface gravity wave profiles for nonsymmetric waves. The *skewed* surface profile denotes a lack of symmetry with respect to the horizontal axis; oscillatory velocities will increase under the crest relative to those under the trough, although they will last for a shorter period of time. The *asymmetrical* surface profile denotes a lack of symmetry with respect to the vertical axis, resulting in marked differences in accelerations between the leading and trailing faces of the wave.

by extremely small water depths, large Froude Numbers, large pressure gradients near *bore* fronts and the infiltration or exfiltration of water from the *beach face water table* (for a review see Butt and Russell, 2000). With propagation over a sedimentary bed, a distinct *wave boundary layer* develops and decays each half-wave cycle, passing from laminar ($Re < 10^5$) to turbulent ($Re > 10^6$) and back to laminar flow (Jensen *et al.*, 1989). This layer may be small (~ 10 cm) but the velocity gradients and thus the instantaneous stresses may be large. Theoretically, gravity waves propagating over a permeable bed induce an oscillating flow even within the bed, as a result of the spatial variation in pressures over a single wavelength. Flows normal to the bed surface are also predicted; flow is out of the

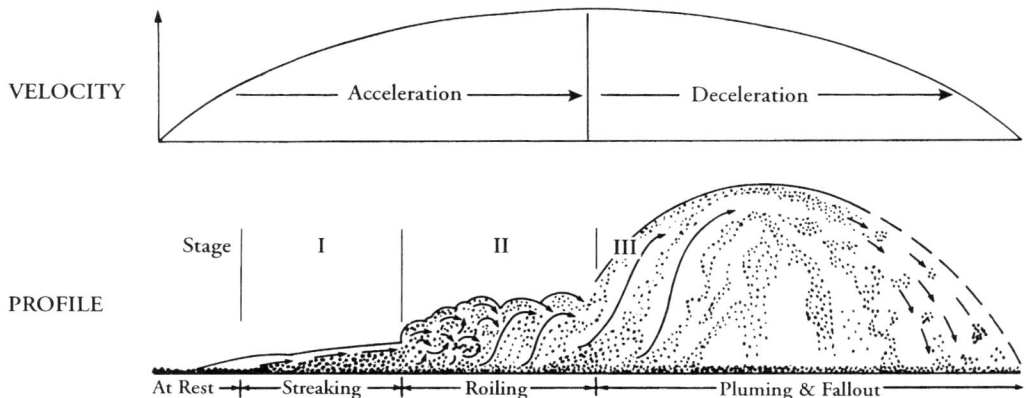

Figure S15 Sequence of development of the fluid-granular boundary layer under stresses associated with flow acceleration and deceleration under waves (modified after Conley and Inman, 1992). Note that the *streaking-roiling* transition marks the change from laminar to turbulent flow and that the boundary layer expands upward dramatically during the deceleration phase.

bed and into the fluid (*injection*) under the wave trough, and out of the fluid and into the bed (*suction*) under the crest, resulting in a "*ventilation*" of the boundary layer. For detailed reviews of wave and combined-flow boundary layers see Grant and Madsen (1986) and Sleath (1990).

Bed stress and modes of transport

The total bed stress under waves can be divided into two components: (i) *form drag*; (ii) *skin friction*. The latter is the stress primarily responsible for transport (the *effective stress*) and is defined using the quadratic stress law as:

$$\tau_w^{sf} = \tfrac{1}{2} \rho_f f \, u_w^2 \qquad \text{(Eq. 1)}$$

where τ_w^{sf} = skin friction (*sf*) stress under waves (*w*), ρ_f = fluid density, f = a wave friction factor, and u_w = maximum horizontal oscillatory velocity at the bed. The skin friction *Shields parameter (or number)* under waves is defined as:

$$\theta_w = \frac{\tau_w^{sf}}{(\rho_s - \rho_f)\,g\,D} \qquad \text{(Eq. 2)}$$

where ρ_s = particle density and D = particle diameter. Theoretically, the *effective stresses* imposed on the bed are periodic and decay to zero at the flow reversal. Once the *threshold* for the *initiation of motion* (see *Grain Threshold*) is exceeded, particles will roll back-and-forth in place. As flow accelerates, a *fluid-granular* boundary layer may develop, with three distinct phases (Figure S15): (i) *streaking*: as flow starts, grains roll and may separate by density into *streaks* or *lambda-shaped* patterns under laminar conditions; (ii) *roiling*: continuing acceleration causes *sheetflow*, with sediment ejected upward into the fluid sporadically. In this phase also, stream-wise periodicities in the sediment may form *transition ripples*; (iii) *pluming*: as flow decelerates, the *sheetflow* layer may lift off the bed, forcing sediment up through the boundary layer and into the free stream, where it ultimately diffuses and settles. The change from *streaking* to *roiling* marks the transition from laminar to turbulent flow and also the transition from *bed-load* to combined *bed* and *suspended* load.

Sediment transport modes are directly related to *bed roughness* (Figure S12) and distinct regimes of flow occur similar to

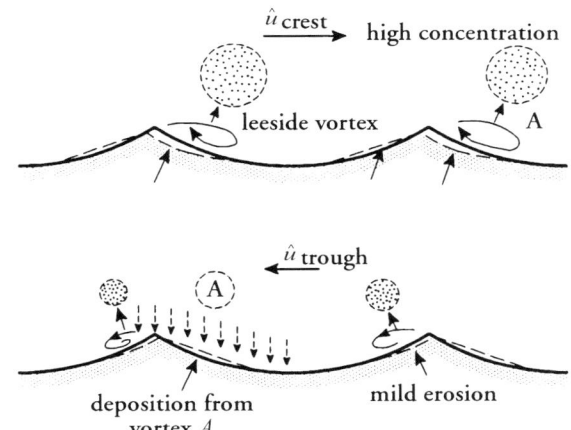

Figure S16 Lee vortex generation and ejection over a steep rippled bed during passage of a successive wave crest and trough ($\hat{u}$ = maximum oscillatory velocity; A = ejected vortex). Note that the larger vortex produced during the passage of the *gravity wave* crest may be advected in a direction opposite to that of the wave advance. The opposite would be true for the smaller vortex formed under the wave trough, resulting in a net transport in a direction opposite to that of the wave propagation direction (modified after Van Rijn, 1993).

that of quasi-steady flow (Clifton *et al.*, 1976; Davidson-Arnott and Greenwood, 1976; see *Bars, Littoral; Surface Forms*). Two distinct regimes occur under near-symmetric waves: (i) a *ripple regime* (*Shields numbers* < 0.8); and (ii) a *sheetflow regime* (Ribberink and Al-Salem, 1995), where bedforms are erased (*Shields numbers* > 0.8–1.0). Theoretically, *sheetflow* is supported by inter-granular forces and sediment-flow interaction. Mass concentrations exceed $\sim 200\,\mathrm{kg\,m^{-3}}$ and the *sheetflow* layer increases in thickness with decreasing grain size. In contrast, *suspended load* is supported by Reynolds stresses, and several mechanisms are responsible for the entrainment and vertical mixing of sediment: (i) *flow separation* and *vortex generation* in the lee of bedform crests, and the release of these vortices from the bed as flows reverse (Figure S16). The vortex

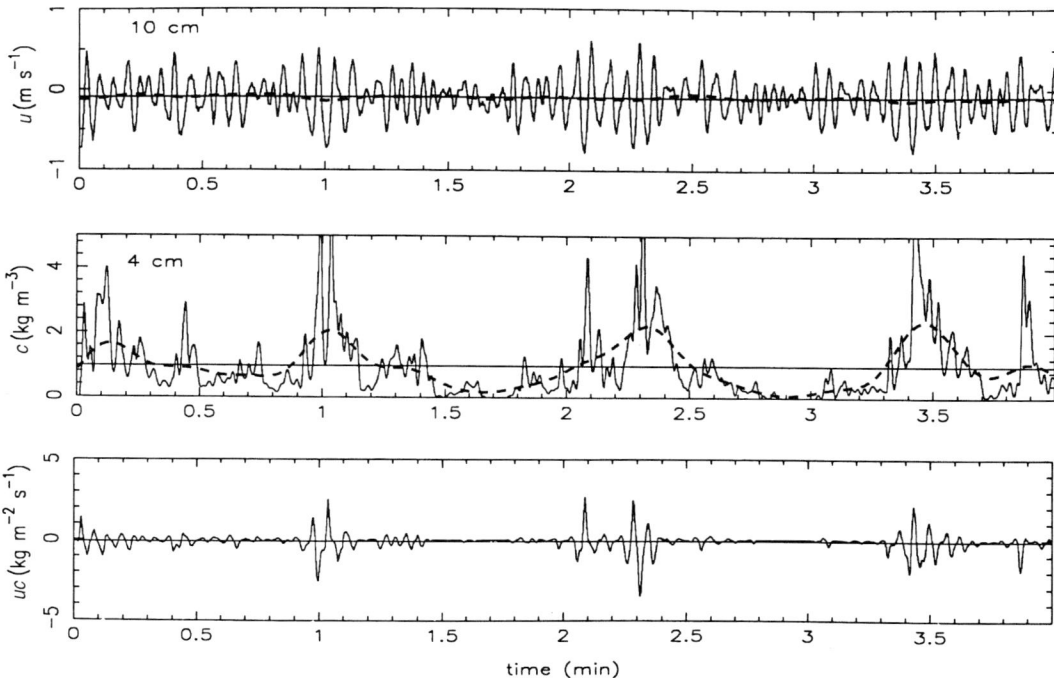

Figure S17 Time series of near-bed horizontal velocity (u; $z = 0.01$ m), suspended sediment concentration (c; $z = 0.04$ m) and net sediment transport (uc) measured in the field under shoaling "*groupy*" waves over a bed with a mean grain diameter of 130 µm. The dashed line in the upper panel is the filtered long wave motion; a similar filter is applied to the concentrations in the center panel. Note the "*episodic*" nature of sediment suspension, the response to individual waves and wave groups, the good correlation between suspension "*episodes*" and the long wave motion, and the small overall down-wave (onshore) net sediment transport.

carries sediment away from the bed and then, rapidly unwinds, diffusing sediment outward, resulting in sediment settling, creating a strongly *episodic* pattern of suspension transport (Figure S17). A distinct *phase lag* between the horizontal currents and sediment concentration at any elevation is created. These *phase lags* determine both the rate and direction of the net suspension transport, and may result in a net transport directed opposite to that of the oscillatory current initiating the sediment motion. Bedform asymmetry will cause significant differences in the size and intensities of the vortices associated with the two directions of motion; this inequality may also lead to complex net transport patterns; (ii) turbulence generated through interaction between the fluid and the bed, and between the vortices and the surrounding fluid induces a vertical diffusion of sediment; (iii) *ventilation* of the wave boundary layer may also contribute to the suspended load.

The relative importance of *bed-load* and *suspended load* is an open question (*cf.* Komar, 1978; Sternberg *et al.*, 1989) and almost certainly depends on the level of applied stress. In deep water, sediment will move as *bed-load* just above the threshold and *bedforms* will develop; their migration may play a major role in the *net sediment transport* (e.g., Traykovski *et al.*, 1999) even though suspension is enhanced by bedforms. As depth decreases and waves become asymmetric and break, suspension load usually dominates, at least for the *net transport* (e.g., Aagaard *et al.*, 2002). Time-averaged sediment concentrations do increase continuously from the top of the suspended load layer to the immobile particles at the *bed*, and follow a logarithmic law similar to that in quasi-steady flows (Osborne and Greenwood, 1993). However, as the *granular layer* accelerates and decelerates during each half-wave cycle, sediment may move from *bed-load* into *suspended load* and back again. In practice, the two modes cannot be separated even with fast response optical and acoustic sensors (White, 1998).

Transport rates

Net sediment transport rates are a product of the sediment mass in motion and the velocity with which the mass moves, integrated over space and time:

$$\vec{i} = (\rho_s - \rho_f)\, g \int_0^T \int_0^{z_o} N_z \vec{U}_z\, dz \qquad \text{(Eq. 3)}$$

where i = immersed weight transport rate across a unit width, ρ_s = mass density of the solid, ρ_f = mass density of the water, g = acceleration due to gravity, N = sediment volume concentration, U = horizontal sediment velocity, z = elevation, z_o = elevation at the top of the mobile sediment, and T = some measure of time (usually the wave period). Theory indicates that the *local net sediment transport* is a function of the balance between *perturbations* of the symmetric oscillatory velocity field and gravity acting on a slope (e.g., Bagnold, 1963;

Bowen, 1980; Bailard, 1981). It is rare that oscillatory motion is perfectly symmetrical in nature, and rare to find a perfectly horizontal, flat bed where gravity has no effect. Thus, there is always a secondary stress superimposed on the symmetrical oscillatory stress. This may be induced by *streaming* in the wave boundary layer (mass transport), by *skewness* and/or *asymmetry* of the waveform (Figure S13), or by some other process in shallow water (e.g., *undertows, longshore currents, drift velocities* associated with *infragravity waves*, etc.). Differences exist between the onshore and offshore velocities as shoaling occurs; generally, larger velocities under the crest will entrain larger volumes of sediment and larger sizes than the smaller velocities under the trough. Since transport rates have been related to the velocity raised to some exponent (values from 3 to 7 have been suggested; Dyer and Soulsby, 1988; Sleath, 1995), small variations in the on-offshore velocities can be very important to sediment transport.

A simple concept of waves *"stirring"* and currents *"transporting"* sediment, has often been used to *model* transport under waves (Bagnold, 1963). The shoreface is viewed as a balance between the cross-shore component of *mass transport* associated with the incident waves, which is landward and a seaward directed mean current (e.g., *undertow*; Roelvink and Stive, 1988). The oscillation itself may be responsible for a significant part of the transport, at least for the suspended load owing to the nature of the *flux coupling* between velocity and sediment concentration (Jaffe et al., 1985), especially over bedforms. Outside the surf zone, oscillatory transport associated with the incident wave band is directed onshore, while an offshore transport is associated with the group-forced bound long wave. Sediment transport at low frequencies results from the increased sediment concentrations due to the passage of groups of large waves and transport directed offshore by the currents associated with the trough of the long wave (see Figure S13). Outside the surf zone, the *net sediment transport rate* is generally positive (i.e., sediment moves in the direction of wave propagation), except on very steep beaches sloping upward in the direction of wave propagation, where gravity may reverse the transport (Fredsøe and Deigaard, 1992). On plane, horizontal beds the important mechanisms contributing to this positive flux are: wave *skewness* and *asymmetry*, *wave drift* and *streaming* in the wave boundary layer. In contrast, transport by second order waves, such as the *group bound long wave*, is directed up-wave or offshore. Greater sediment concentrations in the water column are associated with the larger waves in a group which coincide with the trough (upwave or offshore flow) of the long wave. Shi and Larsen (1984) demonstrated this theoretically, but it has also been well documented in the field (e.g., Osborne and Greenwood, 1992). *Infragravity waves* influence transport in the surf zone, especially in the case of deposition in periodic bar forms (e.g., Bauer and Greenwood, 1990); it is assumed that the transport is induced either by the second order drift velocities associated with standing infragravity waves or by the primary orbital motions.

However, in all instances the direction and rate of transport is strongly influenced by the *phase coupling* between the horizontal velocities in the wave-current boundary layer and the concentrations of sediment produced by the ejection of separation vortices from bottom roughness. At times the phase lag may cause the suspended load to be transported up wave or offshore (see Figure S17).

Cross-shore and longshore sediment transport

The *wave orbital motion* and *cross-shore mean currents* within and immediately above the *wave boundary layer* dominate the cross-shore transport. Complex nonlinear interactions occur between the *oscillatory* currents, the *mean* currents and the small-scale *bed roughness* (e.g., vortex ripples; see *Surface Forms; Bars, Littoral*). Assuming a no-slip condition between the fluid and sediment grains, the time-averaged local net sediment transport rate $<q_s>_n$ is given by:

$$<q_s>_n = \frac{1}{T} \int_0^T \int_0^h u_{(z,t)}' c_{(z,t)}' \, dz \, dt \qquad (Eq.\ 4)$$

where T = wave period, t = time, z = elevation, h = water depth, u' = instantaneous velocity and c' = instantaneous mass concentration. In terms of the steady, periodic and random components, this may be written as:

$$<q_s>_n = \int_0^h <u_{(z,t)}'> \cdot <c_{(z,t)}'> + <\tilde{u}_{(z,t)} \tilde{c}_{(z,t)}> + <u_{(z,t)}^r c_{(z,t)}^r> \, dz$$

where $<>$ = time average, $<u'_{(z,t)}><c'_{(z,t)}>$ = transport by time-averaged, quasi-steady, mean currents, and $\sim$ and r are the periodic and random components respectively. The latter is generally considered to be small where the periodic component is large. According to Jaffe et al. (1985) the term $u'_{(z,t)} c'_{(z,t)}$ is a measure of the correlation between fluctuations in velocity and sediment concentration (typically nonzero) and represents the time-varying sediment flux due to waves. A convenient way of computing this quantity is through cross-spectral analysis (Huntley and Hanes, 1987), as the cospectrum (the real part of the cross-spectrum) yields the cross product of velocity and concentration as a function of frequency. The local net oscillatory suspended sediment transport rate may thus be computed from:

$$<q_s>_{osc} = \frac{\Delta f}{F} \int_0^\infty C_{uc}(f) \qquad (Eq.\ 6)$$

where $C_{uc}(f)$ = the cospectral variance at a given frequency (f), Δf = the cospectral resolution and F = the cospectral frequency range. The cospectrum reveals the relative contribution of different frequency bands to both the direction and rate of suspended sediment transport and can be used to identify transport due to wind waves, infragravity waves (such as the group-bound long wave, edge waves, etc.) and far infragravity waves such as shear waves.

Osborne and Greenwood (1992) found good agreement between the quantities $<q_s>_n$ and $(<q_s>_m + <q_s>_{osc})$. While this procedure incorporates some of the nonlinear effects in a shoaling, irregular wave field, others are omitted. For example, part of the transport associated with *wave skewness* is included in the cospectral estimates through the coupling between larger onshore (or offshore) current velocities and larger sediment concentrations. However, other non-linear effects such as *wave asymmetry* or the higher order moments of the velocity field (e.g., Guza and Thornton, 1985) are not accounted for.

Time-averaged oscillatory currents

Stokes second order theory suggests a *Lagrangian* water motion (*wave drift*) occurs under shoaling waves, being largest at the surface and decreasing exponentially with depth. Longuet-Higgins (1953) defined a *Eulerian mass transport* velocity, with a vertical structure which varied with the wave number ($\kappa = 2\pi/L$). This mass transport was directed down-wave near the bed and at the surface, and directed up-wave at intermediate depths. At the bed it was:

$$\bar{u}_b = \frac{5}{4}\left(\frac{\pi H}{L}\right)^2 C \frac{1}{(\sinh(\kappa h))^2} \quad \text{(Eq. 7)}$$

where u_b = velocity at the bed, overbar = time-averaged mean, H = wave height, C = wave celerity, h = water depth and κ = wave number ($2\pi/L$). It has been suggested that it is this current that maintains the onshore sediment flux to create the upwardly sloping shoreface profile in balance with gravity.

As waves break, *secondary waves* and *quasi-steady currents* are common. Transport becomes complex and the net transport vector will depend upon the interaction of several mechanisms, resulting in both *shore-parallel* as well as *cross-shore* motion. Generally, onshore transport still results from the *skewed* and *asymmetric* breaking waves and *bores*, although turbulence induced by the breaking wave at the surface becomes important. Offshore transport may be induced by: (i) *undertow*; (ii) *rip* currents; (iii) oscillatory and drift velocities associated with secondary waves, such as *edge waves* or *leaky waves* or *shear waves*. Within the *surf zone*, pressure gradient driven, time-averaged mean flows directed seaward (*undertows*) are caused by water level *setup* at the shoreline. *Wave setup* results from the onshore flux of momentum due to waves (*radiation stress*) and is combined with a *mass flux* of water directed onshore by the waves plus any applied *wind stress*.

When waves break at an oblique angle to the shoreline, the alongshore component of the *radiation stress* creates a shore-parallel current or *longshore current*. Pressure gradients due to setup differentials along shore, as well as the shore-parallel component of the *wind stress* can enhance this forcing. The volume of sediment transported alongshore by these quasi-steady *longshore currents* is termed *littoral drift* and essentially confined to the *surf zone*. Transport reaches its maximum at a cross-shore position between that of the maximum current velocity and the breaker line and decreases rapidly offshore; there is a subsidiary peak of transport in the *swash zone*. *Rip currents* are discrete, narrow, high velocity *jets* of offshore-directed flow across the *surf zone*. They are often associated with bathymetric forms such as cross-shore oriented depressions (*rip channels*) or breaks in a quasi-shore parallel bar, but are found also on uniformly sloping beaches. Rip currents are not generally *steady state* phenomena, but are capable of transporting large volumes of both coarse bed-load, especially in migrating megaripples (e.g., Gruszczynski *et al.*, 1993), and suspended load.

Size fractionation

An important characteristic of waves shoaling across the shoreface are bed-sediments of similar density which increase in size from deep to shallow water, in association with increasing local slope (e.g., Guillen and Hoekstra, 1996). On the upper shoreface grain-size distributions are more variable, reflecting local variations in bathymetry and the rapid changes in hydrodynamics associated with wave breaking and complex nearshore circulation, especially in the presence of *bars* (see *Bar, Littoral*). Suggested mechanisms for size-sorting include: (i) onshore coarse bed-load transport by mass transport within the wave boundary layer, and an offshore movement of fines by the return flow at mid-water depths; (ii) a near-bed velocity asymmetry (amplitude and duration) induced by wave shoaling, enabling all sediment sizes to move landward under the wave crests, but only finer sediments move offshore under the trough; (iii) setup-induced undertow and/or rip currents, which continue offshore from the breaker zone; and (iv) a net transport induced by group-forced long waves. Greenwood and Xu (2001) review size-fractionation processes and show that under shoaling waves over a near-horizontal, rippled bed, a distinct horizontal fractionation of sediment by size occurs by suspension transport alone.

Transport models

Models of sediment transport under waves depend upon the spatial and temporal scales of integration used in equation 3: (i) at the individual wave cycle-storm cycle scale (10^{-2}–10^0 years); these models incorporate the fundamental physical processes as far as possible; (ii) at the decadal time scale (10^1 to 10^2 years); these may be time-integrated, physical-based models, but more usually are highly parameterized models or morphological models (see List and Terwindt, 1995 for examples); and (iii) at the geologic time scale ($>10^3$ years); these models are based either on simple geometric rules (e.g., the "*Bruun Rule*") and sediment conservation (e.g., Cowell *et al.*, 1999) or diffusion rules (Kauffman *et al.*, 1992); a more detailed process–response approach has been attempted recently by Storms *et al.* (2002). Physically based models are developed on the basis of the physics of individual wave cycles and then integrated to the scale of storms. However, it is clear that integrating small-scale processes to larger spatial and temporal scales is problematic, as it involves interactions between the wave climate, sea level, availability of sediment and the shoreface morphology (slope). In this complex geomorphic system, non-linear processes are often driven by stochastic boundary conditions.

Two approaches to wave-scale modeling have been adopted: (i) the energetics approach of Bagnold (1963; 1966; Bailard, 1981); (ii) the applied stress approach (Grant and Madsen, 1982). Soulsby (1997) gives an extensive review of a large range of models for suspended and bed-load transport in both waves and combined flows and Schoonees and Theron (1995) evaluated ten different time-dependent sediment transport models and concluded that the *energetics* approach was generally the best. However, this model does not consider the *threshold of motion*, assumes an instantaneous response of the bed and does not consider the increased turbulence associated with wave breaking. One major difficulty encountered in the modeling of suspended sediment transport is the existence of phase lags between the near-bed horizontal velocity and sediment concentration noted earlier.

Longshore transport under quasi-steady, wave-forced currents in the *surf zone* has been relatively well understood and modeled for the last three decades (for a review see Komar, 1998). The basic energy flux approach of R.A. Bagnold is again prominent; the flux rate is:

$$I_L = (\rho_s - \rho_f) \, g \, a' \, S_L = K \, P_L \qquad \text{(Eq. 8)}$$

where $K =$ a proportionality constant, $I_L =$ immersed weight transport rate; $\rho_s, \rho_f =$ mass densities of solid and fluid respectively; $g =$ gravitational constant; $a' =$ porosity; $S_L =$ longshore transport volume, and:

$$P_L = E \, C \, n \cos \alpha \sin \alpha \qquad \text{(Eq. 9)}$$

where $n =$ ratio of group speed to wave speed, $E =$ specific energy density; $C =$ wave speed; $n =$ ratio of wave speed to group speed; and $\alpha =$ the angle of wave breaking at the shoreline. For a recent detailed evaluation of some well-known models, see Bayram et al. (2001).

Brian Greenwood

Bibliography

Aagaard, T., and Greenwood, B., 1995. Longshore and cross-shore suspended sediment transport at far infragravity frequencies in a barred environment. *Continental Shelf Research*, **15**: 1235–1249.

Aagaard, T., Black, K.P., and Greenwood, B., 2002. Cross-shore suspended sediment transport in the surf zone: a field-based parameterization. *Marine Geology*, **185**: 283–302.

Bagnold, R.A., 1963. Mechanics of marine sedimentation. In Hill, M.N. (ed.), *The Sea*. Volume 3, Wiley-Interscience, pp. 507–528.

Bagnold, R.A., 1966. An approach to the sediment transport problem from general physics. *US Geological Survey Professional Paper* 422–1, 37pp.

Bailard, J.A., 1981. An energetics model for a plane sloping beach. *Journal Geophysical Research*, **86**: 10938–10954.

Bauer, B.O., and Greenwood, B., 1990. Modification of a linear bar-trough system by a standing edge wave. *Marine Geology*, **92**: 177–204.

Bayram, A., Larson, M., Miller, H.C., and Kraus, N.C., 2001. Cross-shore distribution of longshore sediment transport: comparison between predictive formulas and field measurements. *Coastal Engineering*, **44**: 79–99.

Bowen, A.J., 1980. Simple models of nearshore sedimentation: beach profiles and longshore bars. In McCann, S.B. (ed.), *The Coastline of Canada*. Geological Survey of Canada, Paper 80–10, pp. 1–11.

Bowen, A.J., and Huntley, D.A., 1984. Waves, long waves and nearshore morphology. *Marine Geology*, **60**: 1–13.

Bowen, A.J., and Holman, R.A., 1989. Shear instabilities of the mean longshore current, 1, Theory. *Journal of Geophysical Research*, **94**: 18,023–18,030.

Butt, T., and Russell, P., 2000. Hydrodynamics and cross-shore sediment transport in the swash zone of natural beaches: a review. *Journal of Coastal Research*, **16**: 255–268.

Cacchione, D.A., and Drake, D.E., 1990. Shelf sediment transport: an overview with applications to the northern California continental shelf. In Le Mehauté, B., and Hanes, D.M. (eds.), *The Sea*. Volume 9, Part B, Chapter 21, Wiley-Interscience, pp. 729–773.

Clifton, H.E., 1976. Wave-formed sedimentary structures—a conceptual model. In Davis, R.A., and Ethington, R.L. (eds.), *Beach and Nearshore Sedimentation*. Society Economic Paleontologists and Mineralogists, Special Publication, 24, pp. 126–148.

Clifton, H.E., and Dingler, J.R., 1984. Wave-formed structures and paleoenvironmental reconstruction. *Marine Geology*, **60**: 165–198.

Conley, D.C., and Inman, D.L., 1992. Field observations of the fluid granular boundary layer under near-breaking waves. *Journal of Geophysical Research*, **97**: 9631–9643.

Cowell, P.J., Roy, P.S., Cleveringa, J., and de Boer, P.L., 1999. Simulating coastal systemstracts using the shoreface translation model. In Harbaugh, J.W., Watney, W.L., Rankey, E.C., Slingerland, R., Goldstein, R.H., and Franseen, E.K. (eds.), *Numerical Experiments in Stratigraphy: Recent Advances in Stratigraphic and Sedimentologic Computer Simulations*. SEPM, Special Publication 62, pp. 165–175.

Davidson-Arnott, R.G.D., and Greenwood, B., 1976. Facies relationships on a barred coast, Kouchibouguac, New Brunswick, Canada. In Davis, R.A., and Ethington, R.L. (eds.), *Beach and Nearshore Sedimentation*. Society Economic Paleontologists and Mineralogists, Special Publication, 24, pp. 149–168.

Dyer, K., and Soulsby, R.L., 1988. Sand transport on the continental shelf. *Annual Review of Fluid Mechanics*, **20**: 295–324.

Fredsøe, J., and Deigaard, R., 1992. *Mechanics of Coastal Sediment Transport*. Singapore: World Scientific Publishing Co., Advanced Series on Ocean Engineering, Volume 3, 369pp.

Grant, W.D., and Madsen, O.S., 1979. Combined wave and current interaction with a rough bottom. *Journal of Geophysical Research*, **84**: 1797–1808.

Grant, W.D., and Madsen, O.S., 1982. Movable bed roughness in unsteady oscillatory flow. *Journal of Geophysical Research*, **87**: 469–481.

Grant, W.D., and Madsen, O.S., 1986. The continental-shelf bottom boundary layer. *Annual Review of Fluid Mechanics*, **18**: 265–305.

Greenwood, B., and Zhiming Xu, 2001. Size fractionation by suspension transport: a large-scale flume experiment with shoaling waves. *Marine Geology*, **176**: 157–174.

Gruszczynski, M., Rudowski, S., Semil, J., Slominski, J., and Zrobek, J., 1993. Rip currents as a geological tool. *Sedimentology*, **40**: 217–236.

Guillen, J., and Hoekstra, P., 1996. The "equilibrium" distribution of grain size fractions and its implications for cross-shore sediment transport: a conceptual model. *Marine Geology*, **135**: 15–33.

Guza, R.T., and Thornton, E.B., 1985. Velocity moments in nearshore. *Journal Waterways, Ports, Coastal, Ocean Engineering*, **111**: 235–256.

Herbers, T.H.C., Elgar, S., and Guza, R.T., 1995. Generation and propagation of infragravity waves. *Journal of Geophysical Research*, **100**: 24683–24872.

Huntley, D.A., 1976. Long period waves on a natural beach. *Journal of Geophysical Research*, **81**: 6441–6449.

Huntley, D.A., and Hanes, D.M., 1987. Direct measurement of suspended sediment transport. *Proceedings, Coastal Sediments '87, New Orleans*. American Society of Civil Engineers, pp. 723–737.

Jaffe, B.E., Sternsberg, R.W., and Sallenger, A.H., 1985. The role of suspended sediment in shore-normal beach profile changes. *Proceedings, 19th International Conference on Coastal Engineering, Houston*. American Society of Civil Engineers, pp. 1983–1996.

Jensen, B.L., Sumer, B.M., and Fredsøe, J., 1989. Turbulent oscillatory boundary layers at high Reynolds numbers. *Journal of Fluid Mechanics*, **206**: 265–297.

Kaufman, P., Grotzinger, J.P., and McCormick, D.S., 1992. Depth-dependent diffusion algorithms for simulation of sedimentation in shallow marine depositional systems. In Franseen, E.K., Watney, W.L., Kendall, C.G.St.C., and Ross, W. (eds.), *Sedimentary Modeling; Computer Simulations and Methods for Improved Parameter Definition*. Kansas Geological Survey, Special Publication, 233, pp. 489–508.

Kinsman, B., 1965. *Wind waves*. Prentice-Hall.

Komar, P.D., 1978. Relative quantities of suspension versus bedload transport on beaches. *Journal Sedimentary Petrology*, **48**: 921–932.

Komar, P.D., 1998. *Beach Processes and Sedimentation*, 2nd edn, Prentice-Hall.

Le Mehauté, B., 1976. *An Introduction to Hydrodynamics and Water Waves*. Part 3: Water Wave Theories. New York: Springer-Verlag, pp. 195–272.

List, J.H., 1992. A model for the generation of two-dimensional surf beat. *Journal of Geophysical Research*, **97**: 5263–5635.

List, J.H., and Terwindt, J.H.J. (eds.), 1995. Large-Scale Coastal Behaviour. *Special Issue, Marine Geology*. 126, 331pp.

Longuet-Higgins, M.S., 1953. Mass transport in water waves. *Philosophical Transactions Royal Society London*, **A245**: 535–581.

Munk, W.H., 1949. Surf beat. *EOS, Transactions American Geophysical Union*, **30**: 849–854.

Nielsen, P., 1992. *Coastal Boundary Layers and Sediment Transport*. Singapore: World Scientific Publishing Company, Advanced Series on Ocean Engineering, Volume 4.

Oltman-Shay, J., Howd, P., and Birkemeir, W., 1989. Shear instabilities of the mean longshore current, 2, Field observations. *Journal of Geophysical Research*, **88**: 2579–2591.

Osborne, P.D., and Greenwood, B., 1992. Frequency dependent cross-shore suspended sediment transport. 1. a non-barred shoreface. *Marine Geology*, **106**: 1–24.

Osborne, P.D., and Greenwood, B., 1993. Sediment suspension under waves and currents: time scales and vertical structure. *Sedimentology*, **40**: 599–622.

Ribberink, J.S., and Al-Salem, A.A., 1995. Sheet flow and suspension of sand in oscillatory boundary layers. *Coastal Engineering*, **25**: 205–225.

Roelvink, J.A., and Stive, M.J.F., 1989. Bar generating cross-shore flow mechanisms on a beach. *Journal of Geophysical Research*, **94**: 4785–4800.

Schoonees, J.S., and Theron, A.K., 1995. Evaluation of 10 cross-shore transport/morphological models. *Coastal Engineering*, **25**: 1–41.

Shi, N.C., and Larsen, L.H., 1984. Reverse sediment transport induced by amplitude modulated waves. *Marine Geology*, **54**: 181–200.

Sleath, J.F.A., 1984. *Sea Bed Mechanics*. Wiley Interscience.

Sleath, J.F.A., 1990. Seabed boundary layers. In Le Mehaute, B., and Hanes, D.M. (eds.), *The Sea*. Volume 9, Part B, Chapter 20, p. 693–727. Wiley-Interscience.

Sleath, J.F.A., 1995. Sediment transport by waves and currents. *Journal Geophysical Research*, **100**: 10977–10986.

Soulsby, R., 1997. *Dynamics of Marine Sands*. London: Thomas Telford Services.

Sternberg, R.W., Shi, N.C., and Downing, J.P., 1989. Continuous measurements of suspended sediment. In Seymour, R.J. (ed.), *Nearshore Sediment Transport*. Plenum Press, pp. 231–257.

Storms, J.E.A., Weltje, G.J., van Dijke, J.J., and Kroonenberg, S.B., 2002. Process-response modelling of wave-dominated coastal systems: simulating evolution and stratigraphy on geological timescales. *Journal of Sedimentary Petrology*, **72**: 226–239.

Smith, J.D., 1977. Modeling of sediment transport on continental shelves. In Goldberg, E.D., McCave, I.N., O'Brien, J.J., and Steele, J.H. (eds.), *The Sea*, Volume 6, Chapter 13. Wiley-Interscience, pp. 539–578.

Stokes, G.G., 1847. On the theory of oscillatory waves. *Transactions Cambridge Philosophical Society*. 8 and Supplementary Scientific Paper 1.

Symonds, G., Huntley, D.A., and Bowen, A.J., 1982. Two-dimensional surf beat. Long wave generation by a varying break point. *Journal of Geophysical Research*, **80**: 492–498.

Swift, D.J.P., 1976. In Stanley, D.J., and Swift, D.J.P. (eds.), *Marine Sediment Transport and Environmental Management*. Wiley Interscience, pp. 311–350.

Traykovski, P., Hay, A.E., Irish, J.D., and Lynch, J.F., 1999. Geometry, migration and evolution of wave orbital ripples at LEO-15. *Journal of Geophysical Research*, **104**: 1505–1524.

Van Rijn, L.C., 1993. *Principles of Sediment Transport in Rivers, Estuaries and Coastal Seas*. Amsterdam: Aqua Publications.

Watson, G., and Peregrine, D.H., 1992. Low frequency waves in the surf zone. *Proceedings 23rd International Conference on Coastal Engineering*. American Society of Civil Engineers, pp. 818–831.

White, T.E., 1998. Status of measurement techniques for coastal sediment transport. *Coastal Engineering*, **35**: 17–45.

Cross-references

Bar, Littoral
Bedding and Internal Structures
Coastal Sedimentary Facies
Physics of Sediment Transport: the Contributions of R.A. Bagnold
Sediment Fluxes and Rates of Sedimentation
Sediment Transport by Tides
Sediment Transport by Unidirectional Water Flows
Surface Forms
Swash and Backwash, Swash Marks

SEDIMENTARY GEOLOGY

The discipline devoted to the scientific study of sediments and sedimentary rocks has been described by a number of names. Petrology is the science that studies rocks, particularly their mineralogy, textures, and origins. From the beginnings of the discipline, some petrologists devoted themselves to sedimentary rocks, and texts on petrology included sections or volumes devoted to Sedimentary Petrology: early examples include Hatch and Rastall's *Petrology of the Sedimentary Rocks* (1913) and Milner's *Introduction to Sedimentary Petrography* (1922)—the latter being largely devoted to the study of heavy minerals and size analysis. When the Society for Economic Paleontologists and Mineralogists was founded in 1927, it derived its name from two groups of specialized researchers, then largely working in petroleum exploration: those who studied microfossils (the "economic paleontologists") and those who studied heavy minerals ("the economic mineralogists"): it was thought at that time that heavy minerals would be useful to correlate oil sands that contained no microfossils. In 1931, SEPM began publication of the first journal devoted exclusively to sediments and sedimentary rocks, under the name *Journal of Sedimentary Petrology*.

Before 1920 most authors did not differentiate the study of sedimentary rocks from *Stratigraphy*, which is the study of stratified rocks, concerned "not only with the original succession and age relations of rock strata, but also with their form, distribution, lithologic composition, fossil content, geophysical and geochemical properties—indeed, with all characters and attributes of rocks as *strata*; and their interpretation in terms of environment or mode of origin, and geologic history." (AGI *Glossary of Geology*, 2nd edn, 1980). In the 1920s, however, most research in stratigraphy consisted in classifying, dating, and correlating rock units, and it became apparent that more attention should be devoted to understanding how sedimentary rocks had been deposited. Accordingly, the National Research Council of the United States established a Committee on Sedimentation, which began gathering material for a *Treatise on Sedimentation*, eventually published in 1926 (2nd edn, 1932). The term *Sedimentology* was originally proposed by A.C. Trowbridge in 1925, and published by Waddell in 1933: it was essentially synonymous with "*sedimentation*" (as used by Twenhofel) and may be defined as the study of natural sediments, both lithified and unlithified, and of the processes by which they were formed.

Sedimentation and Sedimentology are distinguished from Stratigraphy by the fact they they confine attention to sediments, and do not concern themselves directly with dating or correlating sedimentary strata, or with fossils, except in so far as they contribute to the formation of sediment (e.g., as grains, or as organisms binding or burrowing into sediment; see *Bioerosion, Biogenic Structures, Taphonomy*). The term *Sedimentology* was not widely used until the International Association of Sedimentologists was founded in Europe and its first Congress held in Belgium in 1946.

In the 1980s there was a renaissance of Stratigraphy, largely under the impetus of first *Seismic Stratigraphy*, and then *Sequence Stratigraphy*. When the Geological Society of America decided to set up a section devoted to sedimentary rocks, therefore, they wished to emphasize that it would be devoted not only to sedimentary petrology, or to sedimentology (in its restricted definition) but also to the "new" stratigraphy and the broader aspects of sedimentary basin analysis: the name approved in 1985 was the Division of Sedimentary Geology. In the next year SEPM formally changed its name, retaining only the initials, and adding "The Society for Sedimentary

Geology." In 1994, the *Journal of Sedimentary Petrology* became the *Journal of Sedimentary Research*.

Thus, *Sedimentary Geology* includes geological sedimentology, stratigraphy, and geological (rather than geophysical) aspects of basin analysis. Sedimentary Petrography and Petrology include descriptive and theoretical aspects of natural sediments and sedimentary rocks. Sedimentology is somewhat broader, in that it includes studies of all kinds of sediments, and of the fundamental chemical, biological and physical processes that lead to their formation, but not all practitioners would distinguish it from Sedimentary Petrology. A few authors have interpreted sedimentology in terms as broad as sedimentary geology, but most exclude those aspects of stratigraphy that deal with the dating, correlation, and classification of large bodies of stratified rocks.

Gerard V. Middleton

Cross-references

Sedimentologists
Sedimentology, History
Sedimentology—Organizations, Meetings, Publications

SEDIMENTARY STRUCTURES AS WAY-UP INDICATORS

Criteria for determining the original stratigraphic facing direction or way-up in vertical or overturned strata are important both for determining correct stratigraphic succession and for correctly interpreting complex tectonic structures. Any original feature of sedimentary or igneous rocks that reflects the way-up during their formation qualifies. Examples include features known to indicate unambiguously either the bottom or top of sedimentary beds or of lava flows. Cross bedding, symmetric ripples, and graded bedding were among the first sedimentary features to be invoked for way-up determination with the many kinds of sole marks coming into play much later.

The value of cross-bedding was recognized in southwestern Ireland in the mid-nineteenth century. In 1856 Patrick Ganly published clear diagrams showing that the truncated tops and asymptotic bases of cross laminae in folded Silurian sandstones on the Dingle Peninsula provide an unambiguous criterion of the original up direction. He also emphasized that individual cross laminae tend to thin toward their bases. He even showed sketches of ripple cross lamination in modern fluvial sands for analogy. Ganly's shrewd insight was generally ignored and soon forgotten, so had to be re-discovered.

While mapping Precambrian strata in the Florence iron district of Wisconsin in 1910, William O. Hotchkiss wrote in a letter to C.R. Van Hise (June 26, 1910) that "an inspiration came to me that the cross-bedding gives the basal direction for a formation as well as any other criterion." Way-up criteria prior to that time included only such things as ripple marks, drag folds, and cleavage-bedding relationships. Hotchkiss' independent discovery (Figure S18) might well have suffered the same fate as Ganly's, for he himself never published it. Instead, it was reported in publications by other Wisconsin workers (Cox and Dake, 1916; Shrock, 1948; Dott, 2001).

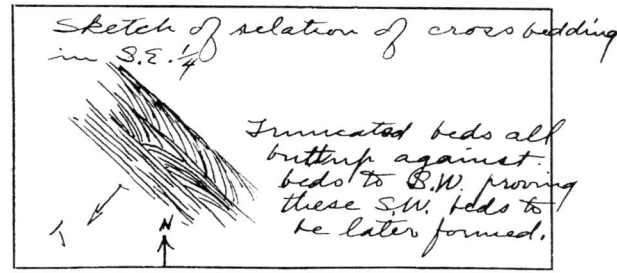

Figure S18 Field notebook sketch of cross bedding in overturned quartzite by W.O. Hotchkiss (1910, Hotchkiss Notebook No. 1, Wisconsin Geological and Natural History Survey).

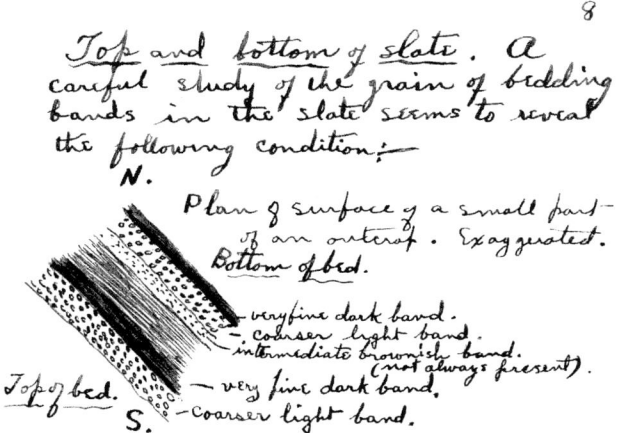

Figure S19 Field notebook sketch of graded bedding in overturned strata by C.L. Dake (1910, Dake Notebook No. 1, Wisconsin Geological and Natural History Survey).

Hotchkiss also recognized the potential of graded bedding for way-up determination as reported by his field assistant, Charles L. Dake, who was assigned to test the value of graded bedding with observations in the field (Figure S19) and in oriented thin sections (Cox and Dake, 1916, 50–52).

Ripple marks were cited as a way-up criterion almost as early as cross-bedding by Irish geologist J. Beetes Jukes in *The Student's Manual of Geology* (1862, 192). It was but a passing observation, however, with no distinction made between symmetric and asymmetric ripples for this determination; asymmetric ones are ambiguous because bottom and top surfaces appear much the same. In 1896, Charles R. Van Hise, one of Hotchkiss' mentors, cited Jukes and discussed how ripples could distinguish way-up. Although Van Hise's illustration was of an unusual example of superimposed symmetric ripples of two scales, he gave no overt indication that asymmetric ripples are of no value. His protégé, Charles K. Leith, did make this distinction clearly, however, when he reproduced Van Hise' diagram in his 1913 structural geology textbook. Leith's book was the first widely available reference to way-up criteria, although the discussion was brief. Included were cross-bedding, symmetric ripple marks, and graded bedding.

Dissemination of the re-discovered value of sedimentary way-up indicators had a complex history. In 1924 two young University of Wisconsin graduates, Olaf Rove and Sherwood Buckstaff, showed an older Norwegian geologist, Thorolf Vogt, that a complex succession at Ballachulish in the Scottish Highlands must be inverted because of the orientation of truncated cross-bedding (Vogt, 1930; Dott, 2001). Vogt conveyed this revelation to the man who had mapped the area, Edward B. Bailey, who received the news with skepticism; he did not know of Leith's 1913 textbook, although several of his countrymen had cited it. Bailey's edification did not begin until 1927, when he was invited to accompany a Princeton University field trip across Canada. In the Precambrian shield north of Lake Superior, Geological Survey of Canada geologist Thomas L. Tanton, who had received his PhD. from Wisconsin (1915), showed the Princeton group how both cross-bedding and graded bedding indicate way-up. In 1929, Bailey hosted the Princeton field trip in the Highlands. Tanton came, too, and finally convinced Bailey that, indeed, the entire Ballachulish succession was inverted. Once converted, Bailey quickly arranged for a joint publication in the 1930 *Geological Magazine* of companion papers by Vogt (reinterpretation of the Ballachulish structure), by Tanton (summary of way-up criteria), and one by himself (cross-bedding and graded bedding as hallmarks of two fundamentally different facies groups). Bailey's paper became a classic.

General recognition of the value of sedimentary structures for way-up determination during the 1920s is evidenced by brief notes in the *Treatise on Sedimentation* (Twenhofel, 1926) about cross-bedding (p. 619) and symmetric ripples (pp. 665–666). Tanton's 1930 list of indicators included cross-bedding, channelling, mud cracks, texture gradation (graded bedding), pebble dints (depression of underlying laminae), symmetric ripple marks, vesiculated tops of lava flows, pillow lavas, curved fracture cleavage, drag folds, and cleavage-bedding relations. In later years, such additional criteria as sole marks, rain drop impressions, convexity of shells and of stromatolites and the like were recognized and catalogued in detail by Shrock (1948), who from 1928 to 1937 had been on the University of Wisconsin faculty. The value of sedimentary features for structural interpetation has never again been forgotten.

Robert H. Dott, Jr.

Bibliography

Bailey, E.B., 1930. New light on sedimentation and tectonics. *Geological Magazine*, **67**: 77–92.
Cox, G.H., and Dake, C.L., 1916. Geological criteria for determining the structural position of sedimentary beds. *Bulletin of the University of Missouri School of Mines and Metallurgy*, **2**: 1–59.
Dott, R.H. Jr., 2001. Wisconsin roots of the modern revolution in structural geology. *Bulletin of the Geological Society of America*, **113**: 996–1009.
Ganly, P., 1856. Observations on the structure of strata. *Journal of the Geological Society of Dublin*, **7**: 164–166.
Jukes, J.B., 1862. *Student's Manual of Geology*. Edinburgh: Adam and Charles Black, 2nd edn.
Leith, C.K., 1913. *Structural Geology*. New York: Holt.
Shrock, R.R., 1948. *Sequence in Layered Rocks*. McGraw-Hill.
Tanton, T.L., 1930. Determination of age-relations in folded rocks. *Geological Magazine*, **67**: 73–76.
Twenhofel, W.H., 1926. *Treatise on Sedimentation*. Baltimore: Williams and Wilkins.
Van Hise, C.R., 1896. Principles of North American Pre-Cambrian Geology. *Sixteenth Annual Report of the United States Geological Survey (1894–95)*. Part 1, pp. 571–843.
Vogt, T., 1930. On the chronological order of deposition of the Highland schists. *Geological Magazine*, **67**: 68–73.

SEDIMENTOLOGY, HISTORY

Sedimentology may be defined as the study of natural sediments, both lithified (sedimentary rocks) and unlithified, and of the processes by which they were formed. The term *sedimentology* was first used by A.C. Trowbridge in 1925 (Waddell, 1933), but did not come into common usage until the 1950s (see *Sedimentary Geology*). An interest in sedimentation and the origin of sedimentary rocks dates back to the very beginnings of geology. For an annotated list of earlier reviews of its history, see Potter (1978). In the following discussion, the history of the subject has been divided into six periods, divided by the dates: 1830, 1894, 1931, 1950, and 1977.

The major developments marking the end of each period were: (i) general acceptance of actualism as a basis for geology (Lyell's *Principles*, 1830); (ii) the theoretical elaboration of actualism as the basis of a specific science of sedimentary rocks (the "comparative lithology" of Walther's *Einleitung*, 1893–1894) and the practical demonstration of actualistic methods by such major investigations as the *Challenger* expedition, and the Funafuti boring; (iii) growth in professionalization culminating in the founding of SEPM and the publication of the first sedimentological journal (*Journal of Sedimentary Petrology*, 1931); (iv) the beginning of large-scale studies of modern sediments and introduction of many new techniques, particularly in marine geology. These developments coincided with major conceptual and instrumental advances in geochemistry, and in the interpretation of carbonate rocks, sandstones, and sedimentary structures; and (v) the revival of interest in stratigraphic interpretation and the large-scale geometry of sedimentary bodies brought about by seismic stratigraphy.

Such a division is inevitably arbitrary, and the criteria used to select the dates are not based on uniform criteria. The first two dates are defined conceptually, the third date is defined mainly in terms of a social phenomenon (the definition of "sedimentation," "sedimentary petrology," or "sedimentology" as a subdivision of geology, with its own professional specialists), and the fourth and final dates by advances in both concepts and techniques. Before 1894, sedimentology scarcely existed, except as a part of stratigraphy. Only after 1950 was it common to find specialists who studied sedimentary rocks, but declined to be called stratigraphers, and since 1977 an increasing number of specialists refuse to make a hard distinction between sedimentology and at least some aspects of paleontology and stratigraphy, which they include together as "sedimentary geology"(*q.v.*)

First period (–1830)

The end of the first period is marked by the date when uniformitarianism, or the actualistic method, became generally accepted as the basis for geology. As with most changes in science, this was a gradual process. Actualism ("the present is the key to the past") had been proposed before James Hutton and John Playfair, and its general acceptance was not achieved by them. It was not even achieved at first by Charles Lyell (Cannon, 1960; Hallam, 1989, Chapter 2) but the date of appearance of the first volume of Lyell's *Principles* (1830) is a convenient one with which to limit a history of sedimentology. At some time not long after 1830, there was widespread acceptance of the general actualistic paradigm: "that, during the ages contemplated by geology, there has never been any interruption to the agency of the same, uniform laws of change. The same assemblage of general causes...may have been sufficient to produce by their various manifestations, the endless diversity of effects of which the shell of the earth has preserved the memorials..." Since the beginning of the age of solar system exploration (Sputnik I, 1957) it became possible to recognize the importance of some processes on earth that had not previously been given much attention, such as impact cratering, and to begin to speculate scientifically about the very early history of the earth, when conditions were certainly very different from the present. Since then, an exaggerated contrast between "uniformitarianism" (the doctrine that the *intensity* of natural causes remained nearly constant in the past) and "catastrophism" (that much of geological history was characterized by much more intense causes than at present) is no longer an issue.

After 1830, geology rapidly became professionalized, largely by the establishment of state and national surveys, and the use of geology as an aid in the search for and exploitation of coal and mineral deposits. The central paradigm for professionalization, however, was not actualism as much as it was the concept of "strata identified by fossils," propounded slightly earlier by William Smith, and by Georges Cuvier and Alexandre Brongniart. Many accomplished stratigraphers continued to hold "nonuniformitarian" views on the origin of sedimentary rocks.

Although actualism is a central paradigm for sedimentology, the science itself can scarcely be said to have begun with Lyell. Actualism only made the science possible, but Lyell and his contemporaries did not develop its principles. They were more concerned with establishing the basic working method of historical geology as a whole, than they were with detailed investigations of the origin of sedimentary rocks. Lyell's *Principles* contained many observations that we would now regard as within the scope of sedimentology, for example, his discussion (borrowed in part from C.E.A. Von Hoff; see Hamm, 1993) of rivers, deltas, and coastal processes, but these phenomena were presented more to prove uniformitarianism than to form the basis for an interpretation of sedimentary rocks. Lyell's own systematic discussion of sedimentary rocks (volume 3 of the *Principles*, which later became the *Elements*) tended to follow Smith's stratigraphic paradigm, rather than his own actualistic one. Most of the *Elements* could equally well have been written by a catastrophist, such as William Smith himself, Georges Cuvier, or Adam Sedgwick.

For many years, the great success of stratigraphy diverted the attention of geologists away from investigating sedimentary rocks themselves, to the study of their correlation and enclosed fossils. It is, nevertheless, worth mentioning a few exceptions. The Comte de Buffon (in his *Natural History*, 1778) recognized the vegetable origin of coal, and in the same year von Beroldingen gave a more detailed account, proposing the *in situ* (authochontous) origin and describing how peat was formed and transformed into coal (see Moore, 1940). The vegetable origin of coal was also held by James Hutton, and later confirmed by microscopic investigation (by W. Hutton, no relation to James, and by Whitham, both in 1833, see Moore, 1940). Abraham Werner (1786) gave precise descriptions of many sedimentary rock types, including "graywacke". Antoine Lavoisier (1789) proposed a model of how sediments vary with distance from the shore, distinguished *littoral* from *pelagic* deposits, and discussed the principles of transgressive and regressive overlap (see Carozzi, 1965: unfortunately only the littoral/pelagic distinction was known to most later investigators). Dolomieu described dolomite (1791; see Carozzi and Zenger, 1981; and McKenzie, in Müller *et al.*, 1991 for the long and complex history of this mineral). William Buckland (1823; see Cadée, 1991) investigated cave sediments in a thoroughly actualistic manner, despite his reputation as a catastrophist. Thomas Webster (1826; see Cadée, 1991) interpreted enigmatic "dirt beds" as fossil soils. Hutton's idea of the "geological cycle" was neglected, then revived many times, perhaps most fruitfully in the form of geochemical cycles, after the development of plate tectonics (Gregor, 1992; Oldroyd, 1996, Chapter 12).

Second period (1830–1894)

Almost as soon as it was recognized that fossils characterized sedimentary rocks of a given age, it was also realized that different types of sediments, deposited at the same time, do not all contain the same types of fossils, because of the different environments in which they were deposited. The general aspect of sediments deposited in different environments was called *facies*, and the concept was developed particularly by the Swiss geologist Amanz Gressly (1838; see *Facies Models*; Wegmann, 1963; Nelson, 1985; Cross and Homewood, 1997).

Perhaps the first major success of the actualistic method was, however, the glacial theory developed by Louis Agassiz (1840; after earlier work by other Swiss geologists, see Hallam, 1989, Chapter 3; Tinkler, 1985, Chapter 7) from observations of modern Swiss glaciers and their sediments. It was early supported by Buckland and, at first, by Lyell: but Lyell's uniformitarianism later persuaded him that the role of continental ice sheets had been exaggerated, and that most glacial phenomena were better explain by the action of floating ice. After Agassiz moved to the United States he soon persuaded most American geologists to adopt the theory of continental glaciation, though there were some nonbelievers, such as J.W. Dawson (Sheets-Pyenson, 1996, p.138–140).

Early investigations of modern sediments were limited to accessible areas (e.g., the Mediterranean coast, Gressly, 1861; see Cadée, 1991) but were extended by investigation of coral reefs (Darwin, 1837, 1842; see Stoddart, 1976; see also *Sedimentologists: Vaughan, Murray, Walther*), deserts (Walther, 1890; see *Sedimentologists: Walther*), evaporites (Ochsenius, 1877, see Kirkland and Evans, 1973), and deep-sea sediments (Murray and Renard, 1891; see *Sedimentologists: Murray*). The origin of coal was a subject of intense investigation (the allochthonous versus autochthonous debate, essentially resolved by the recognition of the significance of "underclays" by William Logan, 1840, confirmed by H.D. Rogers and J.W. Dawson in America ; see Moore, 1940; Torrens, 1999).

The earliest microscopic observations, begun with the investigation of coal, were extended to other types of sedimentary rocks with spectacular results. Ehrenburg (1854; see Jahn, in Gillispie, 1971, Volume 4) showed the organic origins of diatomite, radiolarian chert, and chalk. In 1850 H.C. Sorby (see *Sedimentologists: Sorby*) began the use of the petrographic microscope, and by the end of the period he had, almost single-handedly, established the science of sedimentary petrology, using a technique originally developed by Nicol to study fossil wood.

In this period, sedimentary structures were first clearly described. Sedgwick (1847) distinguished between bedding and cleavage, and Thomson (1861: see Durney, 1978) developed a theory of pressure-solution. Sorby (1852; see *Sedimentologists*) described how ripples were formed and their migration produced ("ripple drift") cross-lamination, and his explanation was independently rediscovered in America. George Darwin (1883) studied current ripples experimentally, and recognized the importance of flow separation. De la Beche and others had earlier described cross-bedding, but early attempts to explain its origin (by micro-delta growth and truncation) by Sorby (1852) and W.M. Davis (1890) were less successful. Eolian cross-bedding was well-known, but the connection between subaqueous cross-bedding and dune migration was not made until much later. The earliest observations on the orientation of grains and fossils were made by Hall, Miller, Gressly, and Ruedemann (see Potter and Pettijohn, 1973).

Observations made in several different disciplines advanced our understanding of hydraulics and sediment transport (see Miall, 1978; Dooge, 1987, for references in this paragraph). Brahms (1753) had already developed the "sixth-power law" (relating the weight of a grain to the sixth power of the velocity needed to move it in water) and it was rediscovered independently by Airy (1834). Humphreys and Abbott (1861) produced a major study of the hydraulics of the Mississippi River (used by Lyell in later editions of the *Principles*). Sternberg (1875) described the exponential decline downstream of the weight of pebbles in some rivers, and Daubrée (1879) studied attrition (abrasion) of particles experimentally. Du Boys (1879) developed the first theory of bed-load movement, and Kennedy (1895) proposed a regime theory for Indian canals.

The first work to draw together the many scattered observations on modern sediments, to mark out clearly the stages of the sedimentary cycle (weathering, transportation and abrasion, deposition, and diagenesis: see Dunoyer de Segonzac, 1968), and to document the actualistic method as the basis for a specific science of sedimentary rocks ("comparative lithology") was Johannes Walther's *Introduction to Geology as Historical Science* (1893–1894; see *Sedimentologists: Walther*). The method was demonstrated by major investigations, including Walther's own studies of deserts, studies of coral reefs including the report of the Royal Society of London's expedition to Funafuti (Bonney, 1904), and, most important of all, publication by Murray and Renard of the sediment studies made by the *Challenger* expedition to the deep seas (1872–1876; see *Sedimentologists: Murray*).

This period was also the "heroic age" of geomorphology, which established not only the significance of glacial action (thus solving a long-standing problem posed by the peculiar properties of "drift"), but also the effectiveness of fluvial, rather than marine, processes for erosion and sediment transport (Tinkler, 1985, chapters 4–8). Notable contributions were by G.K. Gilbert (on the Henry Mountains and the concept of *grade*, 1877, and on Lake Bonneville, and its deltas and beaches, 1890; see *Sedimentologists: Gilbert*), and by W.M. Davis (1884, 1889; see Chorley *et al.*, 1973) on geomorphic models and the "cycle of erosion." These studies established a process, and/or model-based approach that changed the nature of sedimentological studies and provided a firm foundation for future work (Loewinson-Lessing, 1954; for contemporary views see Cross, 1902; Teall, 1902). Thus, 1894, the date of publication of the third volume of Walther's work, a volume prophetically titled *Modern Lithogenesis*, seems an appropriate one to choose as the beginning of a systematic science of sedimentology.

Third period (1894–1931)

This period was marked by the further development of lines of investigation established during the previous period. Investigations of modern sediments continued, and were summarized in the *Treatise on Sedimentation* (edited by Twenhofel, 1926; revised 1932), and in *Recent Marine Sediments* (edited by Trask, 1939). Both volumes were produced under the sponsorship of the US National Research Council, which set up a Committee on Sedimentation in 1920. Important works applying actualistic methods to the sedimentary record were published by Sorby, C. Schuchert (*Paleogeography of North America*, 1910), J. Barrell (see *Sedimentologists: Barrell*), and A.W. Grabau (*Principles of Stratigraphy*, 1913—dedicated to Johannes Walther). Other manifestations of the power of the actualistic method were the development of soil science and from it the study of paleosols (see Retallack, 1990), and the development of taphonomy and therefore of the processes of fossilization and bioturbation (see Cadée, 1991, and *Sedimentologists: Richter*).

The major new techniques developed were size analysis (by Udden, Wentworth, and Atterberg (see *Sedimentologists: Udden*; Krumbein and Pettijohn, 1938); and shape analysis (by Wentworth, Waddell, and Zingg). The petrographic approach developed in the previous period by Sorby, was continued mainly by Lucien Cayeux in France (see *Sedimentologists: Cayeux*). It was extended to include the study of accessory minerals by separation using heavy liquids (a technique pioneered by Thoulet, 1881), and applied by many investigators early in the nineteenth century (see Milner, 1962; Luepke, *Heavy Minerals*, in Good, 1998). The value of heavy minerals to the study of provenance was early demonstrated by Thomas (1902), Brammall and Harwood (1923) then by many others (see Milner, 1962). Only in the 1920s did such studies become popular in the petroleum industry, in the (misguided) hope that heavy minerals would be useful for correlating oil sands.

The foundations of coal petrology (coal types, the distinction between macerals and phyterals) were established using thin and polished section techniques by White and Thiessen (1913) in America, Winter (1913) in Germany, and Stopes (1919) in England (see Van Krevelen, 1961, appendix) and an international agreement on petrographic nomenclature was reached at the Heerlen conference in 1935.

A German tradition of chemical studies, already established toward the end of the previous period by K.G. Bischopf, J. Roth and others, was continued by the monumental studies of J.H. Van't Hoff on the evaporation of seawater (see Eugster, in Ginsburg, 1973; and Kirkland and Evans, 1973).

Physical experimentation was carried to new levels of sophistication in the use of flumes by a German school including Engels (from 1898; see Rouse and Ince, 1957), Kramer (1932) and Shields (1936; see Dooge, 1987; Buffington,

1999). Gilbert's work in 1914 (see *Sedimentologists: Gilbert*) was stimulated by his interest in the physiographic activity of rivers and in the environmental problems produced by hydraulic gold-mining in California, and was carried out in a large, open-air flume on the Berkeley campus—these studies together produced his distinctions between saltation and suspension, and between the competence and capacity of flows. They also led to his formulation of flow regimes, with "dunes," formed in a lower regime, separated from plane bed and antidunes, in an upper regime. Unfortunately, the flume was not quite large enough to make clear the distinction, eventually discovered in the 1960s, between ripples and true subaqueous dunes.

Perhaps the main characteristic of the period, however, is not to be found in the investigations themselves, so much as in a tendency to define sedimentology as a distinct discipline within geology. This process of professionalization can be traced through the appearance of textbooks, journals, societies, and research institutes; and it took place in America, Russia, and to varying degrees, in England and Germany at about the same time, during the 1920s and 1930s. A convenient date is 1931, making the appearance of the first journal devoted exclusively to sedimentology, the *Journal of Sedimentary Petrology*. This journal was published by the Society of Economic Paleontologists and Mineralogists, founded in 1927 as an offshoot of the American Association of Petroleum Geologists (see Russell, 1970; for more recent history, see Clinton, 2001). The first American textbooks were by Krumbein and Pettijohn (1938) and Twenhofel (1939).

In Russia, the influence of the *Challenger* reports and Walther's work was strong. Strakhov (1970) dates the "independence of lithology as a scientific discipline" from 1923, the date when Y.V. Semoilov clearly expressed the "endeavour to systematically develop lithology as a science..." The main period of professionalization, however, was during the 1920s and 1930s (Anonymous, 1967). The first textbooks on sedimentology in Russian were produced by Nalivkin in 1932, and by Shvetsov in 1934. National lithological conferences were not held until 1952, and it was not until 1963 that there appeared *Lithology and Mineral Resources*, a journal devoted exclusively to sediments and sedimentary rocks.

In Germany, there were many early sedimentological studies (reviewed in Brinkmann, 1948). Marine institutes existed in Naples (Italy), Helgoland, and Kiel, but most of their work was biological. In 1929 Rudoph Richter (see *Sedimentologists: Richter*) established the Senckenberg am Meer as an institute for the study of modern shallow-marine sediments, to aid in the study of sedimentary rocks. In 1937, Erich Wasmund established a marine geological station at Kiel. In 1918, Richter had begun publication of a journal *Senkenbergiana*, in which most of the sedimentological work carried out at the marine institute was published. But there is still no separate national sedimentological journal or society of sedimentologists in Germany.

The European countries were active in founding the International Association of Sedimentologists, and the first Congress was held in Belgium in 1946. The IAS journal *Sedimentology*, appeared for the first time in 1962. The first major French textbooks were by Cayeux (1916: see *Sedimentologists: Cayeux*), Carozzi (1953), and Vatan (1954). The first German textbook, after Walther, was by Correns (1939: see *Sedimentologists: Correns*), and this was followed, in 1964–1973, by a major three-volume work by Von Engelhardt, Müller and Füchtbauer.

An active group of sedimentologists existed in Great Britain at the beginning of the twentieth century, and the first textbooks were by Hatch and Rastall (1913), Milner (1922; see Milner, 1962), Marr (1929), and Boswell (1933). In the 1920s and 1930s, however, heavy mineral studies predominated, the Sorby tradition was neglected in favor of classificatory biostratigraphy, and the professionalization of sedimentology had to wait upon the founding of IAS.

For developments in Japan, see *Sedimentology: History in Japan*.

In summary, the professionalization of sedimentology took place first in the USA in the 1920s and 1930s. The first results, particularly the new journal and the two NRC-sponsored symposia (which included many contributions from outside the United States), exerted a profound influence on the development of sedimentology elsewhere.

Fourth period (1931–1950)

The early professional period was characterized by the revival of studies pioneered earlier (petrography, diagenesis) and the development of new techniques (x-ray methods, particularly for clay mineralogy, statistical methods, coring techniques). Techniques were standardized for routine sedimentological analysis, particularly of modern sediments. These techniques and the dissemination of sedimentological observations and ideas through the publication of the many books that appeared at the close of the period (particularly Pettijohn, 1949) set the stage for the rapid progress in and growth of sedimentology in the next period.

An example is the book by Shrock (1948), devoted to the use of sedimentary structures for "way-up" determination. There were numerous scattered early references to such techniques, as far back as the work of Ganly (1856) on the use of cross-bedding (see *Sedimentary Structures as Way-up Indicators*) but the technique itself was not widely used until the 1930s and the literature was not fully known until Shrock's book was published. The book had a much wider impact: it was the first to give a systematic description and discussion of a wide range of structures, and it did this by making extensive use of block diagrams, which had rarely been used earlier for this purpose. Their first application to sedimentary structures was perhaps by Knight in 1929, though the technique itself has a much longer history in the earth sciences (see Howarth, 1999).

Though Sorby was aware of the potential of measuring the preferred orientation of structures to determine paleocurrents, the first systematic basin studies using this technique were begun by R. Brinkmann and E. Cloos in Germany in the 1930s (see Pettijohn, 1962). Statistical techniques to handle this type of data were described by Reiche, Shotton, Brinkmann and others, before being fully developed by R.A. Fisher (see Potter and Pettijohn, 1963; Howarth, 1999). In England and America, attention had earlier been focused mainly on paleogeography (by Jukes-Browne and Schuchert; see Schuchert, 1910), but shifted under Twenhofel's influence to the recognition of sedimentary environments. The key concept of facies (*q.v.*) long neglected by English-speaking geologists, was rediscovered in America, and popularized at a meeting of the Geological Society of America held in 1948.

The concept of geosynclines, though originated by Hall (1859) and Dana in America (Dott, 1974, 1978) was developed after that mainly in Europe (see Hsu, in Ginsburg, 1973), along with concepts of "tecto-facies" such as flysch and molasse, until interest in geosynclines was revived in America by Kay (1951). Further interest in the tectonic control of sedimentation was stimulated by the revival of sedimentary petrology in the broad sense, as opposed to heavy mineral studies. The broader studies had been continued in the previous period mainly by Cayeux (see Taylor, 1950). Shvetsov and others in Russia developed theories of the control of sedimentary mineralogy by tectonics, and these theories were further developed and popularized in America by P.D. Krynine (see Sedimentologists: *Krynine*), E.C. Dapples, W.C. Krumbein (see *Sedimentologists: Krumbein*) and L.L. Sloss, who later became better known as one of the founders of sequence stratigraphy (Miall and Miall, 2001).

X-ray techniques established the crystalline nature of even the finest clay fractions beginning with the work of L. Pauling in 1930 on kaolin, chlorite and talc and a complete summary of such studies was first published by G.W. Brindley in 1951 (for other contributors, see Grim, 1942; and *Sedimentologists: Correns*). Clay mineralogists rapidly formed a separate discipline, with their own societies and journals (*Clay Mineral Bulletin*—later, *Clay Minerals*, 1947; *Clays and Clay Minerals*, 1952; also see Volume 34, number 1 of *Clay Minerals* for historical papers celebrating the fiftieth anniversary of the Clay Minerals Group). The sedimentological significance of this work was elucidated by Millot (1964, summarizing work from the 1940s and 1950s; see Carozzi, 1975, 162–165 for a brief history of this, largely French, series of environmental studies).

Statistical techniques were developed for the processing of petrographic and particularly grain-size data (see *Sedimentologists: Krumbein: Statistics Applied to Sediments*). Heavy mineral and insoluble residue studies continued but began to wane in popularity, as their use in the petroleum industry was replaced by geophysical logging techniques.

Studies of physical processes of sedimentation were carried out mainly by hydraulic engineers, using the boundary layer concepts developed at the beginning of the twentieth century by Prandtl and Von Karman (e.g., Shields, 1936; Rouse, 1937; Einstein, 1937; Vanoni, 1946: see Middleton and Southard, 1984 for references), but most of these studies were not well-known to geologists. Exceptions were the work of Rubey (e.g., 1933, 1938; see Ernst, 1978), Hjulström (1935, summarized by him in Trask, 1939), and particularly the work of Bagnold on wind-blown sand (see *Sedimentologists: Bagnold*).

Fifth period (1950–1977)

A number of books appeared at the close of the fourth period (Shrock, 1948; Shepard, 1948; Pettijohn, 1949; Twenhofel, 1950; Trask 1950; Kuenen, 1950; Krumbein and Sloss, 1953), and a re-reading of these books is a simple way of establishing the "state of the art" just before the beginning of the major expansion of sedimentological studies in the 1950s and 1960s. During this period, courses entitled "sedimentology" became widespread in university curricula (many of them using Pettijohn's book as a text), and many new staff were hired specifically to teach sedimentology, rather than stratigraphy. By the late 1970s, probably "sedimentologists" outnumbered "stratigraphers" in university geology departments. Many sedimentologists were also hired by oil company research laboratories. For some "personal perspectives" of this period, see Pettijohn (1984), Van Houten (1989), Picard (1989) and Friedman (1998). The influence of Pettijohn's book, strongly oriented toward sedimentary rocks, rather than processes, continued until the appearance of a new, more process- and facies-oriented text by Blatt *et al.* (1972) and the classic text by Bathurst (1975) on carbonate rocks (see *Sedimentologists: Bathurst*).

The first major change in this period was the great expansion in the number and scale of investigations of modern sediments, due partly to the interests of petroleum companies, partly to the spectacular growth in oceanography (stimulated by submarine warfare in World War II), and partly to the renewed interest in facies. For example, the American Petroleum Institute project on the Gulf Coast, and most of H.N. Fisk's work on the Mississippi alluvial plain and delta postdate 1950.

L.V. Illing's work on the Bahama Banks in 1954 was the first of many investigations of modern carbonate sediments in the Caribbean and Persian Gulf. These investigations and a renewed interest in carbonate petrography had a revolutionary effect on the study of ancient limestones, dolomites, and evaporites. Notable advances included the first modern classifications of limestones by Folk and Dunham (see Ham, 1962; Folk, in Ginsburg, 1973; and Morse and Mackenzie, 1990, p.599–602, for the historical background to modern carbonate studies), the discovery of modern, early-diagenetic dolomites (see Pray and Murray, 1965), and the discovery of sabkha gypsum and anhydrite (see Kirkland and Evans, 1973). Concepts derived from these studies strongly influenced the interpretation of ancient carbonates and evaporites.

The second major change was in both the techniques and theory of sedimentary geochemistry. Analytical advances included electromagnetic and centrifugal techniques for mineral separation, flame photometry, colorimetric analysis, optical emission spectroscopy and atomic absorption spectroscopy, for both trace and major element analysis, all of which became widespread in geology in the 1950s (see summaries in Smales and Wager, 1960; and for later developments, particularly ICPMS developed in the 1980s, see Gill, 1997). The electron microscope was another powerful tool, first applied in sedimentology to clays and cherts (see *Sedimentologists: Folk*), and, after the development of scanning electron microscopy, to microfossils and grain surface textures (see *Surface Textures of Grains*). X-ray fluorescence spectroscopy developed rapidly and led in the 1960s to the electron microprobe. Radioactivation analysis allowed analysis of trace elements in the ppb range. Carbon-14 dating, developed by Libby (beginning in 1949; see Taylor, 2000) allowed accurate dating of sediments up to about 50,000 years BP, and in the 1980s, use of accelerator mass spectrometry allowed the sample size to be reduced. Mass-spectrometric methods were applied to radioactive elements (e.g., K–Ar dating, and later Ar–Ar dating) present in sedimentary minerals such as illite and glauconite, and in the 1980s, the Uranium series method has been developed to fill the gap in dating between K–Ar and Carbon-14. The most important applications of mass spectroscopy in sedimentology, however, were on the natural distributions of the stable isotopes of hydrogen, carbon, oxygen and sulfur (see Epstein, 1997; *Isotopic Methods in Sedimentology*). The first modern review of the analytical geochemistry of sediment was published by Degens in 1965.

Mason's paper on oxidation and reduction, in 1949, and Krumbein and Garrel's paper of 1952 on the use of Eh and pH set the stage for many subsequent theoretical, experimental and field studies of weathering, chemical sedimentation and diagenesis under low-temperature, aqueous conditions (see Garrels and Christ, 1965; Eugster, in Ginsburg, 1973; Berner, 1971; Morse and Mackenzie, 1990).

The third major change was a conceptual one: the experimental development of the hypothesis of high-density turbidity currents by Kuenen in 1948 (see Walker, in Ginsburg, 1973). This was the first entirely new theory of sediment transport and deposition developed since the glacial theory, and it revolutionized the study of flysch sediments, and made possible a deep-water interpretation of sands and gravels previously thought to be necessarily deposited in shallow water, because they were too coarse to be carried by wave-action into deep water. It also made possible an interpretation of submarine canyons, which had been a long-standing enigma since their discovery in the nineteenth century. Since the hypothesis could not be tested by direct observation, it also stimulated the observation, explanation, classification, and mapping of sedimentary structures, and the further development of basin analysis using paleocurrent techniques (Bouma, 1962; Potter and Pettijohn, 1963; Pettijohn and Potter, 1964; Conybeare and Crook, 1968; Ricci Lucchi, 1970).

The fourth major change was the revival in tectonic control of sediment types, and the use of sedimentary mineralogy to determine provenance, brought about by the development of plate tectonics in 1963–1967 (see *Provenance*). Plate tectonics also led to the Deep Sea Drilling Project, begun in 1968, which led to greatly increased understanding of the deposition and diagenesis of deep-sea calcareous and siliceous sediments (see Warme *et al.*, 1981) and threw new light on the old controversy about the depth of ancient evaporite basins (see Cita, in Müller *et al.*, 1991).

The fifth major change was that experimental work on sediment transport and bed forms, carried out largely by engineers since the 1930s, was reintroduced to geologists at an SEPM symposium on sedimentary structures (Middleton, 1965; and Middleton, 1977 for later developments). The revolution in understanding of cross-bedding, in particular, began earlier in Germany with work that finally recognized the link between dune migration and cross-bedding (based in part on observations of intertidal dunes; see Hülsemann, 1955), but in the English speaking world it was largely due to work carried out by US Geological Survey hydraulicians in a newly constructed large flume in Colorado (see Simons, in Middleton, 1965). After this, there was a rapid increase in the number of experimental studies carried out by geological sedimentologists, and more recently by geomorphologists. New theoretical and experiment studies of sediment transport were published in the 1950s by Bagnold (see *Physics of Sediment Transport: the Contributions of R.A. Bagnold*). Re-education of sedimentologists in fluid mechanics was promoted by Allen (1968, 1985) and Middleton and Southard (1978, 1984), and the application to sedimentary structures was exhaustively reviewed by Allen (1982).

Finally, toward the end of this period there was a major change in interest from textural or mineralogical studies of sediments ("sedimentary petrology") toward the recognition and explanation of sedimentary facies in terms of their environment of deposition (Reineck and Singh, 1973); these studies were further organized into a number of models (see *Facies Models*), based largely on concepts dating back to Walther's "comparative sedimentology." Crucial works included Van Straaten (1955) and others on tidal sediments (see Middleton, 1991), Bernard, LeBlanc and Major (1962, on the Brazos River and Gulf Coast barriers); De Raaf, Reading and Walker (1965, on Pennsylvanian cycles), and the work by Shearman (1965, on modern evaporites). This work was systematized by Walker (1976; see 1979) and Reading (1978, and later editions: which may be consulted for detailed references). By the time that Walker and coauthors came to revise their volume in 1989, the influence of the new "sequence stratigraphy" was strongly felt, to the point that an entirely new volume was thought necessary (Walker and James, 1992; and see Moore, 2001).

Final period (1977–present)

This period is too recent to be reviewed as history, but its characteristics will become apparent, I hope, from the articles contained in this encyclopedia. For the history of sequence stratigraphy see Miall and Miall (2001).

Gerard V. Middleton

Bibliography

Agassiz, L., 1840. *Etudes sur les Glaciers*. Neuchatel, Petitpierre (Reprinted 1966, London, Dawson of Pall Mall; translated and edited by Carozzi, A.V., 1967, Hafner).
Allen, J.R.L., 1968. *Current Ripples*. North Holland.
Allen, J.R.L., 1982. *Sedimentary Structures*. Elsevier, 2 volumes.
Allen, J.R.L., 1985. *Principles of Physical Sedimentology*. George Allen and Unwin.
Anonymous, 1967. The development of lithology in the USSR and its immediate objectives. *Lithology and Mineral Resources*, **1967** (5): 527–544.
Bathurst, R.G.C., 1975. *Carbonate Sediments and their Diagenesis*. Elsevier.
Berner, R.A., 1971. *Principles of Chemical Sedimentology*. McGraw-Hill.
Blatt, H., Middleton, G., and Murray, R., 1972. *Origin of Sedimentary Rocks*, 2nd edn, Prentice-Hall, 1980.
Bonney, T.G. (ed.), 1904. *The Atoll of Funafuti*. London: Royal Society.
Boswell, T.G.H., 1933. *On the Mineralogy of the Sedimentary Rocks*. Thomas Murby.
Bouma, A.H., 1962. *Sedimentology of Some Flysch Deposits*. Elsevier.
Brindley, G.W. (ed.), 1951. *X-ray Identification and Crystal Structures of Clay Minerals*. Mineralogical Society [of London]. Second edition, edited by Brown, G., 1961.
Brinkmann, R., 1948. Die allgemeine Geologie (inbes. die exogene dynamik) in letzten Jahrhundert. *Zeitschrift der Deutchen Geologischen Gesellschaft*, **100**: 25–49.
Buffington, J.M., 1999. The legend of A.F. Shields. *Journal of Hydraulic Engineering*, **125**: 376–387.
Cadée, G.C., 1991. The History of Taphonomy. In Donovan, S.K. (ed.), *The Process of Fossilization*. Chapter 1. Columbia University Press.
Cannon, W.F., 1960. The uniformitarian-catastrophist debate. *Isis*, **51**: 38–55.
Carozzi, A.V., 1953. *Pétrographie des Roches Sedimentaire*. Neuchatel: Griffon.
Carozzi, A.V., 1965. Lavoisier's fundamental contribution to stratigraphy. *Ohio Journal of Science*, **65**: 71–85.
Carozzi, A.V. (ed.), 1975. *Sedimentary Rocks*. Stroudsberg PA, Dowden, Hutchinson & Ross.
Carozzi, A.V., and Zenger, D.H., 1981. Sur un genre de pierres calcaires très-peu effervescentes avec les acides, & phosphorescentes par la collision, by Déodat de Dolomieu, 1791. Translation with notes of Dolomieu's paper reporting his discovery of dolomite. *Journal of Geological Education*, **29**: 4–10.
Chorley, R.J., Beckinsale, R.P., and Dunn, A.J., 1973. *The History of the Study of Landforms*. 2 *The Life and Work of W.M. Davis*. London: Methuen.

Clinton, H.E., 2001. A history of the society from 1976–2001. *Journal of Sedimentary Research*, **71** (6): 1040–1055.

Conybeare, C.E., and Crook, K.A.W., 1968. *Manual of Sedimentary Structures*. Bulletin 102, Australian Bureau of Mineral Resources, Geology and Geophysics.

Cross, T.A., and Homewood, P.W., 1997. Amanz Gressly's role in founding modern stratigraphy. *Geological Society of America Bulletin*, **109**: 1617–1630.

Cross, W., 1902. The development of systematic petrography in the nineteenth century. *Journal of Geology*, **10**: 331–376, 451–499.

Darwin, G.H., 1883. On the formation of ripple-mark in sand. *Proceedings of the Royal Society* [of London], 18–43.

Davis, W.M., 1890. Structure and origin of glacial sand plains. *Bulletin of the Geological Society of America*, **1**: 195–202.

Degens, E.T., 1965. *Geochemistry of Sediments*. Prentice-Hall.

Dooge, J.C.I., 1987. Historical development of concepts in open channel flow. In Garbrecht, G. (ed.), *Hydraulics and Hydraulic Research: A Historical Review*. A.A. Balkema, pp. 205–230.

Dott, R.H. Jr., 1974. The geosynclinal concept. In *Modern and Ancient Geosynclinal Sedimentation*. Society of Economic Paleontologists and Mineralogists, Special Publication, 19, pp. 1–13.

Dott, R.H. Jr., 1978. Tectonics and sedimentation a century later. *Earth Science Reviews*, **14**: 1–34.

Dunoyer de Segonzac, G., 1968. The birth and development of the concept of diagenesis (1886–1966). *Earth Science Reviews*, **4**: 153–201.

Durney, D.W., 1978. Early theories and hypotheses on pressure-solution-redeposition. *Geology*, **6**: 369–372.

Epstein, S., 1997. The role of stable isotopes in geochemistries of all kinds. *Annual Review of Earth and Planetary Sciences*, **25**: 1–21.

Ernst, W.G., 1978. William Walden Rubey. *Biographical Memoirs of the National Academy of Sciences*, **49**: 204–223.

Friedman, G.M., 1998. Sedimentology and stratigraphy in the 1950s to mid-1980s: the story of a personal perspective. *Episodes*, **21**: 172–177.

Garrels, R.M., and Christ, C.L., 1965. *Solutions, Minerals and Equilibria*. Harper and Row.

Gill, R. (ed.), 1997. *Modern Analytical Geochemistry: An Introduction to Quantitative Chemical Analysis Techniques for Earth, Environmental and Material Scientists*. Longman.

Gillispie, C.C. (ed.), 1970–1981. *Dictionary of Scientific Biography*. Charles Scribner's Sons, 18 volumes.

Ginsburg, R.N. (ed.), 1973. *Evolving Concepts in Sedimentology*. Johns Hopkins University Press.

Good, G.A. (ed.), 1998. *Sciences of the Earth*. Garland Publishing, 2 volumes.

Gregor, B., 1992. Some ideas on the rock cycle: 1788–1988. *Geochimica et Cosmochimica Acta*, **56**: 2993–3000.

Grim, R., 1942. Modern concepts of clay minerals. *Journal of Geology*, **50**: 225–275.

Hallam, A., 1989. *Great Geological Controversies*, 2nd edn, Oxford University Press.

Ham, W.E. (ed.), 1962. *Classification of Carbonate Rocks*. American Assciation of Petroleum Geologists, Memoir 1.

Hamm, E.P., 1993. Bureaucratic *statistik* or actualism? K.E.A. Von Hoff's *History* and the history of geology. *History of Science*, **31**: 151–176.

Hatch, F.H., and Rastall, R.H., 1913. *Petrology of the Sedimentary Rocks*. Thomas Murby.

Howarth, R.J., 1999. Measurement, portrayal and analysis of orientation data and the origins of early modern structural geology (1670–1967). *Proceedings of the Geological Association*, **110**: 273–309.

Hülsemann, J., 1955. Grossrippeln und Schragschichtens-Gefüge im Nordsee-Watt und in der Molasse. *Senckenbergiana*, **36**: 359–388.

Kay, G.M., 1951. *North American Geosynclines*. Geological Society of America, Memoir 48.

Kirkland, D.W., and Evans, R. (eds.), 1973. *Marine Evaporites: Origin, Diagenesis, and Geochemistry*. Dowden, Hutchinson, & Ross.

Krumbein, W.C., and Pettijohn, F.J., 1938. *Manual of Sedimentary Petrology*. Appleton-Century.

Krumbein, W.C., and Sloss, L.L., 1953. *Stratigraphy and Sedimentation*. W.H. Freeman.

Kuenen, Ph.H., 1950. *Marine Geology*. Wiley.

Loewinson-Lessing, F., 1954. *A Historical Survey of Petrology*. Edinburgh: Oliver and Boyd. Translated from the Russian by S.I. Tomkieff.

Marr, J.E., 1929. *Deposition of the Sedimentary Rocks*. Cambridge University Press.

Miall, A.D., 1978. Fluvial sedimentology: a historical review. In Miall, A.D. (ed.), *Fluvial Sedimentology*. Canadian Society of Petroleum Geologists, Memoir 5, pp. 1–47.

Miall, A.D., and Miall, C.E., 2001. Sequence stratigraphy as a scientific enterprise: the evolution and persistence of conflicting paradigms. *Earth-Science Reviews*, **54**: 321–348.

Middleton, G.V. (ed.), 1965. *Primary Sedimentary Structures and their Hydrodynamic Interpretation*. Society of Economic Paleontologists and Mineralogists, Special Publication, 12.

Middleton, G.V., 1977. Introduction—progress in hydraulic interpretation of sedimentary structures. *SEPM Reprint Series*, **3**: 1–15.

Middleton, G.V., 1991. A short historical review of clastic tidal sedimentology. In Smith *et al.* (eds.) *Clastic Tidal Sedimentology*. Canadian Society of Petroleum Geologists, Memoir 16, pp. ix–xv.

Middleton, G.V., and Southard, J.B., 1984. *Mechanics of Sediment Movement*. SEPM Short Course 3, second edition.

Millot, G., 1964. *Géologie des Argilles*. Paris: Masson. Translated into English as The Geology of Clays. Springer-Verlag, 1970.

Milner, H.B., 1962. *Sedimentary Petrography*. George Allen and Unwin, 4th edn, 2 volumes.

Moore, C.H., 2001. *Carbonate Reservoirs: Porosity Evolution and Diagenesis in a Sequence Stratigraphic Framework*. Elsevier.

Moore, E.S., 1940. *Coal*, 2nd edn. John Wiley and Sons.

Morse, J.W., and Mackenzie, F.T., 1990. *Geochemistry of Sedimentary Carbonates*. Elsevier.

Müller, D.W., McKenzie, J.A., and Weissert, H. (eds.), 1991. *Controversies in Modern Geology*. Academic Press.

Nelson, C.M., 1985. Facies in stratigraphy: from "terrains" to "terranes". *Journal of Geological Education*, **33**: 175–187.

Oldroyd, D.R., 1996. *Thinking About the Earth*. Harvard University Press.

Pettijohn, F.J., 1949. *Sedimentary Rocks*. Harper and Row. 2nd edn, 1957; 3rd edn, 1975.

Pettijohn, F.J., 1962. Paleocurrents and paleogeography. *American Association of Petroleum Geologists Bulletin*, **46**: 1468–1493.

Pettijohn, F.J., 1984. *Memoirs of an Unrepentant Field Geologist*. University of Chicago Press.

Pettijohn, F.J., and Potter, P.E., 1964. *Atlas and Glossary of Primary Sedimentary Structures*. Springer-Verlag.

Picard, M.D., 1989. The education of a sedimentary petrologist. *Journal of Geological Education*, **37**: 232–234.

Potter, P.E., 1978. Sedimentology—yesterday, today, and tomorrow. In *Encyclopedia of Sedimentology*. Stroudsburg PA, Dowden: Hutchinson & Ross, pp. 718–724.

Potter, P.E., and Pettijohn, F.J., 1963. *Paleocurrents and Basin Analysis*, 2nd edn, Springer-Verlag, 1977.

Pray, L.C., and Murray, R. (eds.), 1965. *Dolomitization and Limestone Diagenesis*. Society of Economic Paleontologists and Mineralogists, Special Publication, 13.

Reading, H.G. (ed.), 1978. *Sedimentary Environments and Facies*, 2nd edn, 1986; 3rd edn, 1996, Blackwell Scientific.

Reineck, H.-E., and Singh, I.B., 1973. *Depositional Sedimentary Environments*, 2nd edn 1980, Springer-Verlag, .

Retallack, G.J., 1990. *Soils of the Past*. Boston: Unwin Hyman.

Ricci Lucchi, F., 1970. *Sedimentografia*. Bologna, Zanichelli, 2nd English edn, 1995, Columbia University Press.

Rouse, H., and Ince, S., 1957. *History of Hydraulics*. Dover edition, 1963.

Russell, D., 1970. History of the SEPM *Journal of Sedimentary Petrology*, **40**: 7–28.

Schuchert, C., 1910. Paleogeography of North America. *Geological Society of America Bulletin*, **20**: 427–606.

Sheets-Pyenson, S., 1996. *John William Dawson*. McGill-Queen's University Press.

Shepard, F.P., 1948. *Submarine Geology*. Harper and Row.

Shrock, R.R., 1948. *Sequence in Layered Rocks*. McGraw-Hill.

Smales, A.A., and Wager, L.R. (eds.), 1960. *Methods in Geochemistry*. Interscience.

Stoddart, D., 1976. Darwin, Lyell, and the geological significance of coral reefs. *British Journal for the History of Science*, **9**: 199–218.

Strakhov, N.M., 1970. Evolution of concepts of lithogenesis in Russian geology (1870–1970). *Lithology and Mineral Resources*, **1970** (2): 157–177.

Taylor, J.H., 1950. The contribution of petrology to the study of sedimentation. *Science Progress*, **38**: 652–667.

Taylor, R.E., 2000. Fifty years of radiocarbon dating. *American Scientist*, **88**: 60–67.

Teall, J.J.H., 1902. The evolution of petrological ideas. *Proceedings of the Geological Society [of London]*, **58**: lxiii–lxxviii.

Tinkler, K.J., 1985. *A Short History of Geomorphology*. London: Croom Helm.

Torrens, H.S., 1999. William Edmond Logan's geological apprenticeship in Britain 1831–1842. *Geoscience Canada*, **26**: 97–110.

Trask, P.D. (ed.), 1939. *Recent Marine Sediments*. Second printing 1955, with new preface and bibliography. American Association of Petroleum Geologists.

Trask, P.D. (ed.), 1950. *Applied Sedimentation*. Wiley.

Twenhofel, W.H., 1939. *Principles of Sedimentation*, 2nd edn, McGraw-Hill, second edition, 1950.

Van Houten, F.B., 1989. Krynine, Pettijohn, and sedimentary petrology. *Journal of Geological Education*, **84**: 1973–1976.

Van Krevelen, D.W., 1961. *Coal: Typology–Chemistry–Physics–Constitution*. Elsevier.

Vatan, A., 1954. *Pétrographie Sédimentaire*. Paris: Editions Technip.

Von Engelhardt, W., Füchtbauer, H., and Müller, G., 1964–1973. *Sediment Petrologie*. E. Schweizerbart'sche Verlagsbuchhandlung, 3 volumes. Translated into English in 1967–1977.

Waddell, H., 1933. Sedimentation and sedimentology. *Science*, **77**: 536–537.

Walker, R.G. (ed.), 1979. *Facies Models*. Geological Association of Canada, Reprint Series 1, 2nd edn, 1984.

Walker, R.G., and James, N.P. (eds.), 1992. *Facies Models: Response to Sea Level Change*. Geological Association of Canada.

Warme, J.E., Douglas, R.G., and Winterer, E.L., 1981. *The Deep Sea Drilling Project: A Decade of Progress*. Society of Economic Paleontologists and Mineralogists, Special Publication, 32.

Wegmann, E., 1963. L'exposé original de la notion de faciès par A. Gressly (1814–1865). *Sciences de la Terre*, **9**: 83–119.

Werner, A.G., 1971. *Short Classification and Description of the Various Rocks*. Translated from the German, and edited by Ospovat, A.M. Hafner.

Cross-references

Carbonate Diagenesis and Microfacies
Facies Models
Physics of Sediment Transport: The Contributions of R.A. Bagnold
Sedimentary Geology
Sedimentology: History in Japan
Sedimentologists:
 Ralph Alger Bagnold
 Joseph, Barrell
 Robin G.C. Bathurst,
 Lucien Cayeux
 Carl Wilhelm Correns
 Robert Louis Folk
 Grove Karl Gilbert
 Amadeus William Grabau
 Amanz Gressly
 William Christian Krumbein
 Paul Dmitri Krynine
 Philip Henry Kuenen
 John Murray and the Challenger Expedition
 Francis J. Pettijohn
 Rudolf Richter and the Senckenberg Laboratory
 Henry Clifton Sorby
 William H. Twenhofel
 Johan August Udden
 T.Wayland Vaughan
 Johannes Walther
 Sedimentology: History in Japan

SEDIMENTOLOGY: HISTORY IN JAPAN

The study of sediments and sedimentary rocks in Japan started with the introduction of Geology in the 1870s shortly after the Meiji Restoration in 1867 which modernized Japanese society. Since then, the study of sediments and sedimentary rocks developed rapidly. The study of sedimentology in Japan was reviewed by Shoji (1966), Shoji et al. (1968) and Okada (1998, 2002).

The history of the development of sedimentology in Japan can be divided into six stages.

Stage 1 (before 1930)

Immediately after the Restoration, the Japanese government invited European and American scientists to establish modern science and technology in Japan. Among them, an American geologist Benjamin S. Lyman (1835–1920) and a German specialist Edmund Naumann (1854–1927) became the founders of geology in Japan. In 1873 Lyman conducted a geological survey of Hokkaido mainly for developing coal resources and published the first geologic map of Hokkaido in 1876. Naumann was appointed as the first professor of geology (1877–1879) at the University of Tokyo, founded the Geological Survey of Japan in 1878, and determined the tectonic and stratigraphic framework of the Japanese Islands. The Geological Survey of Japan, by systematic mapping, quickly developed the study of regional stratigraphy and sedimentary rocks over the whole country. The geology department of the University of Tokyo, prepared a number of able stratigraphers, and contributed to the establishment of many other geology departments and institutions as places for the study of stratigraphy and sedimentary rocks. In 1893 the Geological Society of Japan was founded to support and enhance the quality of geological science in Japan.

Stage 2 (1930–1950)

In the young stage of Japanese geology, the study of sedimentary rocks was called *chiso-gaku* (lithology in English), under the influence of classic text-books with the same title by Hanjiro Imamura (1931) and Masao Minato (1953), meaning the science of all aspects of sediments and sedimentary rocks. In this period, Jun-ichi Takahashi and Tsugio Yagi made significant contributions to petroleum geology and the study of authigenic minerals such as glauconite (Takahashi, 1935; Takahashi and Yagi, 1929). It is worthy of special mention that Yagi (1933) introduced a new term *"taiseki-gaku"* (sedimentology in English) in his study of Miocene oil-bearing sediments in Northeast Japan as early as H. Wadell's (1932) proposal of sedimentology. The word "sedimentology", however, had not been commonly used before 1950.

Stage 3 (1950–1960)

The new discipline of sedimentology quickly became popular and grew into an important field of geology in Japan. This trend in Japan was much earlier than in the international community, in which sedimentology seems to have become established in the early 1960s, and the study of sedimentary facies, sedimentation and sedimentary environments became prominent among stratigraphers and paleontologists. In 1951, the Sedimentology Subcommittee was established under the Science Council of Japan in order to promote sedimentological studies in Japan. It continues to the present and liaises with the International Association of Sedimentologists (IAS). With the support of the Sedimentology Subcommittee, systematic collaborative studies of the recent sediments of the Matsukawa-ura Estuary near Sendai in northeastern Japan and of the bottom sediments of Tokyo Bay were carried out in 1953 and 1955, respectively. On the other hand, the Sedimentology Research Group founded the Association of Japanese Sedimentologists in 1957 and published a journal called "*Sedimentology Research*" in mimeograph. Sedimentological studies, mainly on the petrology of sandstones and heavy mineral analysis, were active at Tokyo, Tohoku, Nagoya, Kyoto, and Kyushu Universities.

Stage 4 (1960–1980)

In 1961, the Science Council of Japan requested the Sedimentology Subcommittee to discuss the creation of a National Institute of Experimental Sedimentology in order to promote long term basic science in Japan. A proposal for a new institute comprising seven research sections (dynamics, physical, mineralogical, chemical, biochemical, lithological, and mathematical studies), was submitted to the government in 1967 (Omori, 1968), but unfortunately it was not adopted. Instead, however, the Ocean Research Institute was established in 1963 at the University of Tokyo, and the Marine Geology Department of the Geological Survey of Japan started in 1974, both of which have contributed much to create a collaborative system of research on submarine sediments and have greatly encouraged sedimentological studies in Japan. In particular, the official participation of Japan in the International Phase of Ocean Drilling (IPOD) in 1971 provided a large boost to the further development of Japanese sedimentology.

Additionally, a very successful symposium on "Some Problems in Sedimentology" held in Nagoya University in 1967 made a deep impact on many geologists and sedimentologists (Shoji *et al.*, 1967). In 1971 the Sedimentology Research Group reorganized the Association of Japanese Sedimentologists into the Sedimentological Society of Japan, but its journal remained unimproved.

Many sedimentologists in this period studied the flysch and turbidites of the so-called geosynclines of various ages from Paleozoic to Tertiary, particularly of the Shimanto Belt in southwestern Japan. Though the new concept of plate tectonics or new global tectonics was being applied to Japanese geology in the 1970s, it then had not been widely accepted.

Stage 5 (1980–1990)

A remarkable feature of Japanese sedimentology in this stage was the diversity of interests of young sedimentologists many of whom came to occupy important positions at universities, institutions and the councils of the Geological Society of Japan. Their activities included facies analysis, sedimentary petrology, mathematical sedimentology, experimental sedimentology, sequence stratigraphy, and chemical sedimentology; many of them attaining international levels in quality.

Sedimentological studies of accretionary complexes were improved remarkably not only because the new paradigm of plate tectonics became firmly established in Japanese geology,
but also because stratigraphical and chronological studies of these highly deformed strata were developed rapidly through the successful application of radiolarian biostratigraphy.

Stage 6 (1990–)

The Sedimentological Society of Japan was reformed in 1994 with a stimulating and high-quality biannual journal entitled "*Journal of the Sedimentological Society of Japan*". Papers from India, Korea, Russia and USA were also included in the journal. The Society has about 600 members and holds scientific meetings twice a year. Another important activity of the Society was the bench-mark publication of "*The Encyclopedia of Sedimentology*" in 1998.

Special mentions should be made of the recent progress in the study of sequence stratigraphy in Japan (Saito *et al.*, 1995) and of the provenance study of siliciclastic rocks mainly from their chemical aspects (Kumon *et al.*, 2000), in which an effective provenance indicator, Basicity Index, was proposed.

In the 21st century, it is suggested that Japanese sedimentology should contribute more to environmental sedimentology, and achieve greater internationalization through links with the International Association of Sedimentologists.

Hakuyu Okada

Bibliography

Imamura, H., 1931. *Lithology*. Tokyo: Kokonshoin (in Japanese).
Kumon, F., Kiminami, K., Hoyanagi, K., Takeuchi, M., Musashino, M., and Miyamoto, T. (eds.), 2000. Compositions of siliciclastic rocks in relation to provenance, tectonics and sedimentary environments. *Memoirs of the Geological Society of Japan*, **57**: 1–240 (in Japanese with English abstracts, some papers in English).
Minato, M., 1953. *Lithology*. Tokyo: Iwanami (in Japanese).
Okada, H., 1998. A short history of sedimentology in Japan. *Journal of Sedimentological Society of Japan*, **48**: 5–12 (in Japanese with English abstracts).
Okada, H., 2002. *Sedimentology: A way to the new discipline of earth sciences.* Tokyo: Kokonshoin (in Japanese).
Omori, M., 1968. On the long term research plan in geological sciences in Japan. In Geological Society of Japan (ed.), *Geology of Japan—Present and Future—*. Tokyo: Geological Society of Japan, pp. 592–605 (in Japanese).
Saito, Y., Hoyanagi, K., and Ito, M. (eds.), 1995. Sequence stratigraphy—Toward a new dynamic stratigraphy. *Memoirs of the Geological Society of Japan*, **45**, 1–249 (in Japanese with English abstracts, some papers in English).
Sedimentological Society of Japan, 1998. *The Encyclopedia of Sedimentology*. Tokyo: Asakura Shoten (in Japanese with an index in English).
Shoji, R., 1966. Recent progress in sedimentology. *Journal of Geography*, **75**: 11–35, 70–82 (in Japanese).
Shoji, R., Mizuno, A., Okada, H., and Mizutani, S. (eds.), 1967. *Some Problems in Sedimentology*. Abstracts Book for Symposium of Joint Meeting of 5 Geoscience Societies in Japan. Nagoya University, pp. 1–302 (in Japanese).
Shoji, R., Taguchi, K., and Iijima, A., 1968. Present and future of sedimentology. In Geological Society of Japan (ed.), *Geology of Japan—Present and Future—*Tokyo: Geological Society of Japan, pp. 99–122 (in Japanese).
Takahashi, J., 1935. The marine kerogen shales from the oil fields of Japan. A contribution to the origin of petroleum. *Science Reports, Tohoku Imperial University*, Series III, 1, pp. 63–156.
Takahashi, J., and Yagi, T., 1929. The peculiar mud-grains in the recent littoral and estuarine deposits, with special reference to the origin of glauconite. *Economic Geology*, **24**: 838–852.
Wadell, H., 1932. Sedimentation and sedimentology. *Science*, **75**: 20.

Yagi, T., 1933. Sedimentological study of oil-bearing lower formations of the Tsugaru- Matsumae regions. Part 1. *Journal of the Japanese Association of Mineralogists, Petrologists and Economic Geologists*, **10** (3): 82–92 (in Japanese).

Cross-reference

Sedimentology, History

SEDIMENTOLOGY—ORGANIZATIONS, MEETINGS, PUBLICATIONS

Organizations

Scientific organizations, "societies", play a key role in the intellectual vitality and growth of the discipline by providing the means for scientists to communicate with each other, learn new ideas, methods and skills, as well as attract new scientists to the profession, particularly students. Societies act as locus of activity in the form of published journals, special topic volumes and memoirs; field trips and field conferences; meetings featuring talks, posters, short courses, and workshops; awards, i.e., recognition for outstanding scientific accomplishments; research grants; and general social and scientific networking through newsletters and the internet. In addition, societies contribute to educational programs at all levels (K-12, college, and continuing education) and provide the important connection between the scientific community and general public via outreach (e.g., public lectures). Societies are also a valuable resource of scientific information for local and national government personnel developing public policy.

Originally sedimentology and paleontology were included within large broad-based geological societies, along with the closely associated discipline of stratigraphy. But, during the early part of the 20th century the scientific needs of the fossil fuel industry raised the profiles of sedimentology and paleontology as separate entities and spawned the formation of the first scientific organization for them in 1926, the Society of Economic Paleontologists and Mineralogists (SEPM), a division of the American Association of Petroleum Geologists (AAPG). SEPM became an independent society in 1986 and has since changed its name to SEPM Society for Sedimentary Geology. It is now the largest society in the world (4500 members) focused on sedimentary geology.

At the beginning of the 21st century the nature and practice of the discipline of sedimentology is clearly different from what it was a few decades ago. It is more broad based and diverse with less focus strictly on sediments themselves, i.e., physical and chemical processes and deposits. A sedimentologist now may work with seismic data, study aquifer characteristics, interpret paleoclimate from stable isotopes, or consider the impact of biological processes on weathering or diagenesis. Sedimentologists like other disciplines are becoming more interdisciplinary; however, the need for organizations, meetings, and publications remains strong.

There are easily dozens of regional societies dedicated to sedimentology and/or sedimentary rocks in various countries around the world. No single listing of them exists, however, and there is little communication among them. Activity in the English-speaking realm is focused in a few large organizations: SEPM Society for Sedimentary Geology (4,500 members); International Association of Sedimentologists (IAS), (~2,000 members); Sedimentary Geology Division (~1,500 members) of the Geological Society of America (GSA); the British Sedimentological Research Group, BSRG, (~500 members); and Indian Association of Sedimentologists. The sedimentological communities in Canada, South Africa, Hong Kong, New Zealand and Australia meet as specialty groups or Divisions within the larger geological societies. The Sedimentological Society of Japan (550 members) publishes in Japanese (with English abstracts); and societies in National Republic of China communicate only in Chinese.

Meetings

Meetings bring scientists together, usually annually, to provide a forum for exchanging ideas, presenting research results, and educating in the form of field trips, short courses, and workshops. Meetings also provide a means for helping members with professional development, such as retooling for a career change, or offering courses for students on writing resumes, grant proposals, or applying for employment. Social activities and networking contribute to the health of the profession by establishing long term bonds and making new connections.

Meetings also fulfill the growing need for utilizing interdisciplinary and multidisciplinary research to address scientific questions that fall between the natural boundaries of existing disciplines. Small societies find a logistical and economical advantage to hold meetings with other societies. For examples, the Mineralogical Society, Paleontological Society, Society of Economic Geologists, and Geochemical Society all have potential for interacting with sedimentologists and sparking new ideas and new approaches to answering questions that cannot be addressed by any group alone.

Publications

There are three major journals devoted to sedimentology. Two are sponsored by scientific societies and are based on volunteer efforts of the community, *Journal of Sedimentary Research* (SEPM) and *Sedimentology* (IAS), whereas *Sedimentary Geology* is published commercially by Elsevier Publishers. All three have sedimentologists as Editor-in-Chiefs and Associate Editors and they draw from the sedimentological community for reviews of the manuscripts. Papers based substantially on sedimentology may also be published in broad-based journals such as *Nature, Science, and Geology* and in specialty journals such as *Marine Geology*, *Journal of Coastal Research*, *American Association of Petroleum Geologists (AAPG) Bulletin*, *Geoarchaeology*, *Quaternary Research*, and *Earth Surface Processes and Landforms*.

Specialty volumes are a published collection of papers on a related topic often spawned from a symposium or conference. SEPM and IAS regularly produce 2–5 topical volumes a year. GSA and The Geological Society (of London) frequently create "special publications" that focus on sedimentology and are sold to geological community. As a group of papers focused on a relatively narrow topic and placed between two covers, these publications are valued resources and preferred alternatives to widely decimated papers appearing in a number of journals over several years.

At the beginning of the new millennium, the publishing of sedimentological journals, as all scientific journals, is in a stage of gradual transition between paper and electronic format. Problems of permanent archival have yet to be worked out, but the continuous flow of paperbound journals into and filling of libraries is definitely a thing of the past. Visions of electronic publishing involving papers electronically linked to maps, references cited, and even animation will likely become reality.

Gail M. Ashley

Cross-references

Sedimentary Geology
Sedimentology, History
Sedimentology: History in Japan

RALPH ALGER BAGNOLD (1896–1990)

Ralph Alger Bagnold was born on April 3rd, 1896 in Davenport, England. He was commissioned as an officer in the Royal Engineers in 1915 and served in France and Flanders during World War I, subsequently taking leave from the Army to earn an honors degree in engineering at the University of Cambridge. Rejoining the army, official service took Bagnold to Egypt, from where he began exploring the Libyan Sand Sea with fellow bachelor officers, using Model T Ford motor cars. Weekend trips evolved into serious geographical exploration and mapping and Bagnold developed specialized techniques for taking these simple automobiles deep into the desert without getting completely stuck or lost.

Bagnold became fascinated by the movement of sand by wind and wished to understand the physics of the processes responsible for the formation of desert dunes. He retired from the army in 1935 to devote all his time and energy to the study of sand movement, but still found time to write *Libyan Sands*, an exciting chronicle of his desert expeditions (Bagnold, 1935). The results of his pre-war research were published in *The Physics of Blown Sand and Desert Dunes* (Bagnold, 1941).

Returning to Cairo in 1939 at the outbreak of World War II, he established and led the Long Range Desert Group (LRDG). Operating deep behind enemy lines, the LRDG scouted and reported troop movements, mapped enemy positions, collected intelligence and created havoc through lightening raids on forts, convoys and airfields. So important were the achievements of the LRDG that Bagnold was promoted to the rank of Brigadier and awarded the Order of the British Empire (military). A good account of the exploits of the LRDG may be found in Bagnold's autobiography (Bagnold, 1990).

Bagnold again retired from the army in 1944 and, after a short period as Director of Research for Shell, he decided to pursue fundamental research full time. His primary goal remained the elimination of the empiricism that had long dominated study of the movement of granular materials by flowing fluids. His approach rested on the application of theory to identify key variables and formulate relationships between them, supported by carefully conceived and executed experiments to elucidate and quantify physical constants and material properties. The conception and construction of innovative experimental apparatus and instrumentation was incidental to his work.

Bagnold's post-war work produced hallmark papers on the morphology and sedimentology of near-shore zones, beaches, submarine density currents, and meandering rivers. He also introduced two of the most significant concepts underpinning our understanding of the sediment transport process: "dispersive stress" and "stream power". A collection of his finest papers was published by the American Society of Civil Engineers in 1988 (Thorne *et al.*, 1988).

Bagnold's achievements as an explorer won him the 1934 Gold Medal of the Royal Geographical Society and his contribution to the engineering profession earned him the Telford Premium of the Institution of Civil Engineers in 1940 (repeated in 1946). His calibre as a scientist was marked by his election as a Fellow of the Royal Society in 1944. His post-war work brought numerous awards, medals and honorary degrees.

Recent advances in fluid mechanics, boundary layer theory and the sophistication of instrumentation have supported new findings that, to some extent, supersede Bagnold's work, which were necessary based on incomplete knowledge of the nature of turbulent flow structures. However, his contributions to science and engineering form essential components of progress in sedimentology and the study of sediment transport during the twentieth century.

Brigadier Bagnold was married to Dorothy Alice Plank (who died 1989) in 1946. They had two children, Stephen (born in 1947) and Jane (born in 1948). He died peacefully following a short illness on May 28th, 1990.

Colin R. Thorne

Bibliography

Bagnold, R.A., 1935. *Libyan Sands: Travel in a Dead World*. London: Hodder and Stoughton, 288p. Republished with epilogue by Michael Haag Ltd., P.O. Box 369, London, NW3 4ER, UK (1987), ISBN 0-902743600 (UK), 0-870523848 (US), 288pp.
Bagnold, R.A., 1941. *The Physics of Blown Sands and Desert Dunes*. New York: William Morrow and Company, 265p. Republished by Chapman and Hall, Methuen Inc, Halsted Press (1971), ISBN X-76-050873-8, 265pp.
Bagnold, R.A., 1990. *Sand, Wind and War: Memories of a Desert Explorer*. Tucson, Arizona, USA: University of Arizona Press, ISBN 0-8165-121-6, 202pp.
Thorne, C.R., MacArthur, R.C., and Bradley, J.B., 1988. *The Physics of Sediment Transport: A Collection of Hallmark Papers by R.A. Bagnold*. New York: American Society of Civil Engineers, Book Number 665, 350pp.

JOSEPH BARRELL (1869–1919)

American geologist Joseph Barrell was a pioneer in integrating tectonics, denudation, and climate for the understanding of sedimentation. Barrell graduated from Lehigh University in

1892 with high honors and promptly earned two advanced degrees (EM, 1893; MS, 1897). Joseph then accepted an instructorship at Lehigh for four more years to teach mining, metallurgy, mechanical drawing, and mine surveying. Recognizing Barrell's exceptional potential, one of his professors urged him to take more advanced work at Yale. While studying at Yale, he worked during the summers for the US Geological Survey in Montana mining districts. In 1900, he received the PhD and returned to Lehigh as a professor of geology and biology. In 1903, Barrell was called to Yale to be the professor of dynamic and structural geology.

Joseph Barrell's early publications dealt with mining and regional geology. Next they emphasized erosion, sedimentation, ancient climate, and paleogeography. His later ones treated the strength of the earth's crust, isostasy, the origin and genesis of the earth, and radioactivity and geologic time. He also wrote and spoke on epistemology and the transition from a qualitative to a more quantitative geology. Barrell perfected a talent for generalization from the work of others and in constructing his syntheses, he was an ardent exponent of multiple working hypotheses. In sedimentary geology, Barrell early concerned himself with criteria for distinguishing ancient non-marine from marine sediments by giving close attention to sedimentary structures, and in doing so he refined criteria of paleogeography and paleoclimate and destroyed the ruling dogma that practically all of the ancient stratigraphic record was marine. Barrell was a devotee of William Morris Davis's idealized evolution of landscapes through cycles of erosion, to which he added the idea of complementary cycles of deposition related to changing base level. He explicated his views in two papers on ancient deltas (1912 and 1913–1914). As a river passes through its cycle from youth to maturity to old age, its sediments must change, too, with one part of the depositional cycle being a delta stage. In 1916 he combined his knowledge of the Devonian Catskill delta complex of the Appalachian region with published work on the contemporaneous fluvial Old Red Sandstone of Britain. From these Devonian researches, Barrell later suggested that environmental factors, especially climate, had influenced the evolution of the first air-breathing animals; drying of climate stimulated the evolution of amphibians from river fishes. Similarly, he suggested that the vicissitudes of Pleistocene climate had contributed to the appearance of the human species.

Barrell's most important contribution to sedimentary geology was a long article, "Rhythms and the measurements of geologic time," published in 1917 two years before he died unexpectedly from pneumonia and meningitis. His thesis was that "Nature vibrates with rhythms, climatic and diastrophic." He first discussed rhythms of denudation, arguing that erosion is pulsatory; the Davis erosion cycle represents a single rhythm, but small, partial cycles can be superimposed on larger ones. Next he argued that sedimentation is also discontinuous, resulting in a stratigraphic record riddled with breaks of varying durations. Barrel coined the term diastem for the lesser but numerous breaks in contrast with the longer but rarer breaks known as disconformities. His figure 5 (p. 796) so effectively illustrated this concept that it has been reproduced many times; figure 6 (p. 800) and the accompanying discussion illustrates that he already understood the implications of climbing dune forms for time equivalences of the cross laminae and diastems produced by migrating dunes, thus anticipating important revelations in eolian sedimentology during the 1970s–1980s. Finally, by invoking estimates of rates of erosion and sedimentation along with the loss of primeval earth heat, Barrell estimated geologic time spans and then compared these with the measurement of geologic time by radioactivity, which was still in its infancy. He concluded that the two approaches converged upon the conclusion that between 550 million years and 700 million years had elapsed since the beginning of Cambrian time, a remarkably modern figure!

Robert H. Dott, Jr.

Bibliography

Barrell, J., 1906. Relative geological importance of continental, littoral, and marine sedimentation. *Journal of Geology*, **14**: 316–356, 430–457, 524–568.
Barrell, J., 1908. Relations between climate and terrestrial deposits. *Journal of Geology*, **16**: 159–190.
Barrell, J., 1912. Criteria for the recognition of ancient delta deposits. *Bulletin of the Geological Society of America*, **23**: 377–446.
Barrell, J., 1913–1914. The Upper Devonian Delta of the Appalachian geosyncline. *American Journal of Science*, Series 4, **36**, 429–472; **37**, 87–109, 225–253.
Barrell, J., 1917. Rhythms and the measurements of geologic time. *Bulletin of the Geological Society of America*, **28**: 745–904.
Kendall, M.B., 1981. Joseph Barrell. In *Dictionary of Scientific Biography*. New York: Scribner, pp. 469–471.
Schuchert, C., 1919. Joseph Barrell (1869–1919). *American Journal of Science*, Series 4, **48**: 251–280.

ROBIN G.C. BATHURST (1920–)

Bathurst was born in Chelsea, London, on March 21, 1920. His undergraduate studies in geology, begun at the Chelsea Polytechnic in 1939, were interrupted by World War II and continued at Imperial College, London where he earned his B.Sc. in 1948. After this he spent three years at Cambridge studying Wealden sands with Percy Allen. He was appointed to teach sedimentology at University of Liverpool, retired from teaching in 1982, but remained there as a Senior Research Fellow until 1987.

His work at Liverpool shifted from siliciclastic sediments to carbonate rocks, and particularly their diagenesis. In this field, he was largely self-taught, though influenced by the work of H.C. Sorby (*q.v.*), Maurice Black, and especially Bruno Sander. Sander's work on Triassic carbonates, originally published in German in 1936, was at that time largely unknown in the English-speaking world (it was translated into English in 1951). In 1958 and 1959 Bathurst published three papers dealing with limestone diagenesis, all of which are now classics that have had an immense influence on the study of carbonate diagenesis. The earliest paper, published in the *Liverpool and Manchester Geological Journal*, was the first (in English) to develop criteria distinguishing carbonate cavity-filling cements from calcite mosaics formed by recrystallization. It was not only important scientifically but also had significant applications to understanding the diagenetic evolution of limestones and their pore systems, which at that time had become very important in the scientific community and the petroleum industry. His interest in these topics persisted throughout his career (Bathurst, 1986, 1993). Later contributions include documenting the importance of interpenetration of sediment grains ("fitted fabric"), dissolution-seams, and stylolites in the development of rhythmic limestone bedding (Bathurst, 1991).

In 1971, Bathurst published the first edition of his book *Carbonate Sediments and their Diagenesis* (revised in 1975) which remained the standard reference work for many years. Bathurst travelled widely, working with many of the leading carbonate petrologists in many other countries. In 1959 he began a series of research seminars on carbonates at Liverpool, which inspired a generation of sedimentologists in the United Kingdom and professional colleagues in the international community: out of this developed the meeting, held every four years, now known as the Bathurst Meeting of Carbonate Sedimentologists. He received the Lyell Medal (1978) from the Geological Society of London, the Twenhofel Medal (1983) from SEPM, and the Sorby Medal (1986) from the International Association of Sedimentologists.

Gerard V. Middleton

Bibliography

Bathurst, R.G.C., 1958. Diagenetic fabrics in some British Dinantian limestones. *Liverpool and Manchester Geological Journal*, **2**: 11–23.

Bathurst, R.G.C., 1971. *Carbonate Sediments and Their Diagenesis*. Amsterdam: Elsevier, 2nd edn, 1975.

Bathurst, R.G.C., 1986. Carbonate diagenesis and reservoir development: conservation, destruction and creation of pores. *Colorado School of Mines Quarterly*, **81** (4): 1–24.

Bathurst, R.G.C., 1991. Pressure-dissolution and limestone bedding: the influence of stratified cementation. In Einsele, G., Ricken, W., and Seilacher, A. (eds.), *Cycles and Events in Stratigraphy*. Springer-Verlag, pp. 450–463.

Bathurst, R.G.C., 1993. Microfabrics in carbonate diagenesis: a critical look at forty years in research. In Rezak, R., and Lavoie, D.L. (eds.), *Carbonate Microfabrics*. Springer-Verlag, pp. 3–14.

Pray, L.C., 1984. Robin G.C. Bathurst, William H. Twenhofel Medalist. *Journal of Sedimentary Petrology*, **54**: 1379–1380.

Reading, H.G., 1987. Citation for Sorby Medallist, Robin Gilbert Charles Bathurst. *Sedimentology*, **34**: 177–179.

Sander, B., 1951. *Contributions to the Study of Depositional Fabrics: Rhythmically Deposited Triassic Limestones and Dolomites*. Tulsa: American Association of Petroleum Geologists (translated from the German edition of 1936).

LUCIEN CAYEUX (1864–1944)

Cayeux's long academic career culminated with his appointment in 1912 to the chair of geology at the Collège de France in Paris. He spent the next thirty years developing the investigation and the teaching of sedimentary petrography, a field in which he is considered one of its founders. In his doctoral dissertation (1897) on the chalk and certain Mesozoic and Cenozoic siliceous rocks of the Paris basin, Cayeux gave the first outline of his outstanding analytical method of investigation. For a given sedimentary rock, he first described its detrital and authigenic components, the microfossils, and finally the interstitial matrix or cement. The textural relationships between all the constituents combined with chemical analysis led to a synthesis attempting to establish the original characters of the sediment, the provenance of the components, and the post-depositional diagenetic modifications. Cayeux's avant garde techniques of 1897 became increasingly sophisticated with time and for him, the science of sedimentary rocks was essentially the complete and comparative natural history of ancient and recent sediments.

Following his investigation of the chalk in which he demonstrated that, in spite of apparent similarities with the recent deep-sea Globigerina ooze, it was a shallow-water deposit, Cayeux undertook a series of comprehensive and well-illustrated monographs on the sedimentary rocks of France and its colonies. He successively investigated the Tertiary sandstones of the Paris basin (1906); Paleozoic and Mesozoic oolitic iron ores (1909, 1922); banded and nodular siliceous rocks (1929); carbonate rocks: limestones and dolomites (1935), and phosphates (1939, 1941a, 1950 [posthumous]). Although Cayeux's interpretation of the oolitic iron ores remained controversial, his synthetic views on the deposition and diagenesis of siliceous rocks, phosphates and carbonates show an unusually modern character. Particularly in carbonate rocks, he anticipated many modern trends concerning early diagenetic processes such as cementation, recrystallization, and dolomitization.

During his years of teaching at the Collège de France, Cayeux realized the need for an introductory book to the field of sedimentary petrography through the use of high-quality extensive illustrations of mineral constituents, microstructures of skeletal remains, and depositional-diagenetic textures. The result was his *Introduction à l'étude pétrographique des roches sédimentaires*, in 2 volumes, a precursor of the numerous atlases of microfacies published half-a-century later to meet the needs of an expanding oil exploration. It was published in 1916, out of print in 1927, and reprinted in 1931.

Cayeux's philosophical approach to geology developed from his studies on iron ores, siliceous rocks, phosphates, and mostly carbonates. He concluded that many fundamental features of these deposits had been generated contemporaneously or penecontemporaneously with deposition by extensive submarine processes of reworking, cementation, recrystallization, and replacement. He felt that most of these processes, so clearly displayed in ancient sediments, were inactive or incipiently active in present-day marine environments. This led to his last book (1941b) *Causes Anciennes et Causes Actuelles en Géologie* in which he concluded that serious consideration should be given to the dual concept of past and present causes in geology. This controversial challenge to uniformitarianism has not yet been fully answered.

Albert V. Carozzi

Bibliography

Cayeux, L., 1897. *Contribution à l'étude micrographique des terrains sédimentaires. I. Étude de quelques dépôts siliceux, secondaires et tertiaires du bassin de Paris. II. Craie du bassin de Paris*. Mémoires Société Géologique du Nord, **4**: No. 2.

Cayeux, L., 1906. *Structure et Origine des Grès du Tertiaire Parisien*. In the series *Études des gîtes minéraux de la France*. Paris: Imprimerie Nationale.

Cayeux, L., 1909. *Les Minerais de Fer Oolithique de France. Fascicule I: Minerais de Fer Primaires*. In the series *Études des gîtes minéraux de la France*. Paris: Imprimerie Nationale.

Cayeux, L., 1916. *Introduction à l'Étude Pétrographique des Roches Sédimentaires*. In the series *Mémoires explicatifs de la carte géologique détaillée de la France*. Paris: Imprimerie Nationale, 2 vols, (reprinted 1931).

Cayeux, L., 1922. *Les Minerais de Fer Oolithique de France. Fascicule II: Minerais de Fer Secondaires*. In the series *Études des gîtes minéraux de la France, Service carte géologique de la France*. Paris: Imprimerie Nationale.

Cayeux, L., 1929. *Les Roches Sédimentaires de France. Roches Siliceuses*. In the series *Mémoires explicatifs de la carte géologique détaillée de la France*. Paris: Imprimerie Nationale.

Cayeux, L., 1935. *Les Roches Sédimentaires de France. Roches Carbonatées (Calcaires et Dolomies)*. Paris: Masson. Annotated English translation by Albert V. Carozzi, 1970, Darien: Hafner Publishing Company.

Cayeux, L., 1939. *Les Phosphates de Chaux Sédimentaires de France (France Métropolitaine et d'Outre-Mer) I*. In the series *Études des gîtes minéraux de la France, Service carte géologique de la France*. Paris: Imprimerie Nationale.

Cayeux, L., 1941a. *Les Phosphates de Chaux Sédimentaires de France (France Métropolitaine et d'Outre Mer) II. Egypte, Tunisie, Algérie*. In the series *Études des gîtes minéraux de la France, Service carte géologique de la France*. Paris: Imprimerie Nationale.

Cayeux, L., 1941b. *Causes Anciennes et Causes Actuelles en Géologie*. Paris: Masson. Annotated English translation by Albert V. Carozzi, 1971, New York: Hafner Publishing Company.

Cayeux, L., 1950. [posthumous]. *Les Phosphates de Chaux Sédimentaires de France (France Métropolitaine et d'Outre Mer) III. Maroc et Conclusions Générales*. In the series *Études des gîtes minéraux de la France, Service carte géologique de la France*. Paris: Imprimerie Nationale.

CARL WILHELM CORRENS (1893–1980)

C.W. Correns was born on May 19, 1893 at Tuebingen (Germany), the son of the botanist, Carl Correns, who rediscovered Mendel's laws of heredity. After World War I, Correns started as a field geologist mapping several areas of the Rheno-Hercynian belt in the Variscan orogen in Germany. Shales are a major constituent of this belt and were the only abundant rock type of which the mineral composition was unknown in those days. This fact was not accepted by Correns' supervisors in the Prussian Geological Survey. They claimed that clay minerals are the fraction of fine grained sediments, soluble in sulfuric and hydrochloric acid. Correns' statement about the unknown nature of clay minerals was confirmed by chemists investigating colloidal systems at the Kaiser Wilhelm Institute of Physical Chemistry in Berlin. They offered him a part-time position to study the adsorption of certain metals on kaolinite. He became increasingly interested in the composition of Recent sediments and was invited to join the scientific party in one of the first extended oceanographic expeditions. From July 1926 to June 1927 he participated in two traverses through the equatorial Atlantic Ocean by the German research vessel "Meteor." On this cruise a primitive piston corer was used to sample the upper few decimeters of sediments at more than 100 stations between South America and Africa.

On returning from this introduction to practical oceanography, Correns was appointed to the chair of geology at the University of Rostock, close to the shore of the Baltic Sea. He arrived with his valuable collection of "red clay," "blue mud" and "globigerina ooze" from the bottom of the equatorial Atlantic. Correns anticipated that X-ray investigation of the clays was required to recognize their distinctive minerals. Simple X-ray cameras and high-voltage equipment were available 15 years after M. von Laue's discovery of X-ray diffraction by crystal lattices. Correns and his coworkers then systematically investigated a selection of Atlantic pelagic clays, as well as shales of Pleistocene to Jurassic age from northeast Germany, using the microscope, X-ray diffraction and chemical analysis. By studying separate grain-size fractions, they discovered that even the smallest consists of crystalline rather than amorphous substances. The latter result was earlier claimed by many scientists. Correns and coworkers observed that the grain size distributions were comparable in both Recent pelagic clays and ancient shales. Mica was the most abundant mineral of these pelites.

Crystallographic studies revealed that mica as well as the other clay minerals are sheet silicates. In the 1930s Correns (1937) compiled the results of the Rostock laboratory on the grain size, and mineral and (partial) chemical composition of the pelagic clays and foraminiferal oozes from the "Meteor" expedition in the Atlantic. At about the same time he revised the old-fashioned textbooks on sediment formation and composition. Correns (1939) presented the state of the art on sedimentary petrology ("Die Sedimentgesteine") condensed to 146 pages in a unique international book on the origin of rocks. His coauthors were Tom Barth (magmatic rocks) and Pentti Escola (metamorphic rocks).

At the beginning of World War II, most of the copies of this book were lost during the bombing of Leipzig. This event, and the fact that the book was written in German, precluded a wide international readership, so it is worth quoting the main headings of the part on sedimentary rocks: mechanical weathering, chemical weathering, transport and deposition of clastic sediments, fabrics and constituents of clastic sediments, chemical and biogenic sediments, and diagenesis.

Because of gaps in knowledge revealed by his second book, Correns turned his attention to the experimental investigation of the solubility, as a function of pH, of simple compounds such as quartz and aluminum oxide. Next came experimental weathering of feldspar, mica and other minerals, as well as glass, in open systems. In 1939 Correns was appointed to the chair of sedimentary petrology, and later to that of mineralogy, at the University of Goettingen. By the time the book "Die Enstehung der Gesteine" was reprinted (1960), Correns had already turned his interest toward geochemistry, investigating the crustal distribution of the halogens and titanium. He became cofounder of *Geochimica et Cosmochimica Acta* and of *Contributions to Mineralogy and Petrology*.

Correns died on August 29, 1980, and is known as one of the pioneers of sedimentary petrology, based on generalizations from carefully observed details.

Karl Hans Wedepohl

Bibliography

Correns, C.W., 1937. *Die Sedimente des äquatorialen Atlantischen Ozeans. Wissenschaftliche Ergebnisse der Deutschen Atlantischen Expedition auf dem Forschungs- und Vermessungs-schiff "Meteor" 1925—1927*, Volume 3, Part 3. Berlin: Leipzig: de Gruyter.

Correns, C.W., 1939. *Die Sedimentgesteine*. In Barth, T.F.W., Correns, C.W., and Eskola, P. (eds.), *Die Entstehung der Gesteine*. Berlin: Springer-Verlag, pp. 116–262.

Cross-references
Illite Group Clay Minerals
Mudrocks
Oceanic Sediments
Sedimentology, History

ROBERT LOUIS FOLK (1925–)

Robert Louis ("Luigi") Folk has an unusual and exceptional record as a geologic educator, researcher, and author of articles on sedimentary rocks. His teaching and published contributions on sandstone, limestone, dolomite, shale, and chert during a 50-year career have earned him numerous accolades, including the two highest honors for research in sedimentary geology: the W. H. Twenhofel Gold Medal from the SEPM (Society for Sedimentary Geology) in 1979 and the H.C. Sorby Medal from the International Association of Sedimentologists in 1990. He also received the Penrose Medal, the premier medal for research, from the Geological Society of America in 2000.

Bob was born in September 1925 in Shaker Heights, Ohio. He earned all his degree—BS, MS, and PhD, the PhD in 1952—from Pennsylvania State College. His mentor at Penn State was Paul D. Krynine, legendary at the time as a leader in the emerging field of sedimentary rocks. Bob did not own a car when he was a graduate student, so topics for his Master's thesis and PhD dissertation were determined by the rocks that he could reach by bus from his apartment in State College, prefacing later expeditions by train and bus to localities in Italy. His dissertation was on Cambro-Ordovician carbonate rocks in the vicinity of Penn State. Unraveling the primary and diagenetic textures of these beds led to the first comprehensive carbonate classification scheme in English. Publication of the classification in the Bulletin of the American Association of Petroleum Geologists led to the best paper of the year award; it was a preview of Bob's subsequent outstanding research.

Bob worked two years for the Gulf Oil Company before joining the University of Texas at Austin in 1952, retiring in 1988. At the University of Texas-Austin he supervised 26 Master's and 14 PhD students. His teaching style would now be considered unusual—use of an Aboriginal wadi stick as a pointer, lab exercises dominated by rocks, extensive reviews of the literature, random-walk field trips, and so on. He repeatedly earned local and national awards, including the Neil Miner Award of the National Association of Geology Teachers (1989) and the Distinguished Educator Award of the AAPG (1998). He also has served as a visiting professor at the Australian National University, University of Milano, and Tongji University (PRC). At the University of Texas at Austin he served as the J. Nalle Gregory Professor in Sedimentary Geology from 1977–1982; and from 1982 until retirement in 1988 as the Dave P. Carleton Professor. He presently holds the title of Carleton Professor Emeritus.

Most of Bob's major contributions are based on observations at the hand-specimen or thin-section scale, although in the last 10 years the scanning electron microscope has become a research instrument of choice. Repeatedly he has seen textures in rocks that other people overlooked or paid no attention to, in the process arriving at new and novel interpretations of features. Bob has made fundamental contributions on the classification of sandstone, limestone, and mudrock and codified such common parameters as textural maturity, grain-size statistics, rock color, grain roundness, grain shape, identifying and naming architectural components of sandstone, limestone, dolomite, mudrock, and chert. His *Petrology of Sedimentary Rocks*, a soft-bound locally published semi-text, which first appeared in 1957, and was revised periodically until 1980; sold for little more than production costs. The 1980 version is available "on line"; it remains a fundamental resource of sedimentary petrologists. His curiosity-generated projects outside mainstream sedimentary geology have led to publications on the petrology of avian urine, the petrography of roofing tiles, enhanced stereo vision using two hands, black phytokarst from Hell, shape development on Tahiti, a unit of scuffle abrasion on stone steps, vitrified rat feces of aragonite, and a challenge to the concept that the pyramids of Egypt are made of epoxy-cemented crushed stone. His citation index is off the chart when compared with most other academicians; this reflects the fundamental importance of his contributions over the years. His research has been accomplished with almost no external funding.

Not all of Bob's contributions have been widely accepted. His ideas on sediment transport, as he says, sank like a stone. His suggestion that ultra-small bacteria (<0.1 µm), called nannobacteria, are directly or indirectly responsible for the precipitation of many carbonate, silicate, oxide, sulfide, and possibly other minerals has been rejected by most microbiologists and questioned by many geologists. However, there is also strong support for his hypothesis. The verdict on this issue has not been rendered, but the research illustrates once again how Bob recognized textural elements in SEM images of rocks that had been ignored or overlooked by other workers.

In 1972, when Bob was a visiting professor at the University of Milano, he fell in love with Italy. This led to his admiring most things Italian and Italian rocks became a major research focus. His nickname now is Luigi rather than Bob.

Earle F. McBride

GROVE KARL GILBERT (1843–1918)

One of the greatest American geologists of the nineteenth century—a man who should be ranked with Lyell, Agassiz, and Smith—is Grove Karl Gilbert, born May 6, 1843 in Rochester, NY, and who died May 1, 1918 while preparing for another field season in Utah. His intellectual contributions were theoretical and institutional, ranging from the concept of dynamic equilibrium of landscapes to founding member and Chief Geologist of the United States Geological Survey. He was one of the premier scientific explorers of the American West and it is not coincidental that his lifetime is called the heroic age of American geology.

Born the son of a self-taught portrait painter, Gilbert was the youngest of three children. He received a classical education studying at home and from the University of Rochester where he emphasized Greek and Latin; he took but one geology course. After a failed attempt at teaching school on the Michigan frontier, he landed a job at Cosmos Hall, the famous distributor of natural history artifacts run by Henry A. Ward. This in turn led to a position with the Ohio Geological Survey as a volunteer, where he came under the tutelege of John Strong Newberry. Two years later Newberry recommended him for the position of geologist on the US Army's Geographical Survey West of the 100th Meridian. He remained an employee of the US Government for the rest of his life, joining the Powell Survey in 1875, and the fledging United State Geological Survey in 1879.

Gilbert's principal contributions to sedimentology are contained in his four great monographs: *Report on the Geology of the Henry Mountains* in 1877; *Lake Bonneville*, 1890; *The Transportation of Debris by Running Water*, 1914; and *Hydraulic-Mining Debris in the Sierra Nevada*, 1917. In the third and final part of the Henry Mountains report he erected the foundations of modern geomorphology. Conceiving of a stream as an engine which performs work and applying the laws of conservation of energy and least action, he codified three laws of land sculpture and invented the fruitful concept of dynamic equilibrium of landscapes and its derivitive, the graded stream. He also presented the concept of bed-load as a corrasion tool in bedrock streams and articulated for the first time the conditions under which streams will form lateral planation surfaces.

In his Lake Bonneville studies the objective was twofold: "the discovery of the local Pleistocene history and the discovery of the processes by which the changes constituting that history were wrought." Here is the first description of the origin of coastal features such as spits, bars, and wave-cut terraces as a product of the balance between wave energy and sediment supply as it affects littoral drift and the first sedimentologic description of the delta type which still bears his name. Its everlasting contribution however, was primarily methodological. In this one quarto volume Gilbert showed how sedimentary processes could be deduced from geomorphic forms using the proper application of mechanical laws.

Gilbert's monograph on bed-load transport (with E. C. Murphy) is a seminal document in experimental sedimentology. It is the first systematic attempt to formulate the functional relationships between bed-load flux and flow energetics, the only exception possibly being the work of C. Lechalas in 1871. The experiments were conducted in four flumes of varying sizes. In each, the capacity of the flow was measured as a function of seven variables: discharge, slope, fineness of debris, depth, width, and to a lesser extent, mean velocity and channel curvature. Although Gilbert was never able to reduce the data to a rational theory of bed-load transport, he succeeded in formulating the basic questions and concepts. Besides providing practical formulae relating capacity to discharge, slope, and width/depth ratio, he described the particle dynamics of the moving bedlayer, the threshold of motion as a function of grain diameter, and the role played by fines in increasing the capacity of the coarse fraction in bimodal mixtures. While others, especially French engineers, had conducted flume experiments before Gilbert, his methodology, mathematical formulations, and physical explanations were unrivaled. The experimental design was so well conceived and the observations so meticulously taken, the data are still used today.

In his last monograph Gilbert applied these flume results to address the natural and man-made problems arising from excess sediment supply to the Sacarmento River system. Using his idea of a graded stream, he predicted the effectiveness of various engineering structures to control flooding along its course. And in a final rhetorical flourish he predicted that the tidal bar at the mouth of the Golden Gate would move inland as a consequence of reduction in tidal prism as the bay was filled by mining debris and agricultural sediment runoff. The power of this work was not in the social change it effected, but in the template it provided for subsequent conservation literature.

In separate contributions he published on the origin of bedforms, in particular ripples marks, ripple drift lamination, and "giant wave" ripples, from which he attempted to deduce ancient water wave heights.

Rudy Slingerland

Bibliography

Pyne, Stephen, J., 1980. *Grove Karl Gilbert, A Great Engine of Research*. University of Texas Press.
Chamberlin, T.C., 1918. Grove Karl Gilbert. *Journal of Geology*, **26** (4): 375–376.

AMADEUS WILLIAM GRABAU (1870–1946)

Born of German heritage in Cedarburgh, Wisconsin, January 9, 1870, the young Grabau was raised in a bilingual family. His fluency in German proved later to be an asset in his drive to bring advanced concepts in sedimentology and stratigraphy to American classrooms. In particular, Grabau was strongly influenced by the German geologist Johannes Walther (*q.v.*), to whom in 1913 he dedicated his textbook, *Principles of Stratigraphy*. A correspondence course in mineralogy from the Massachusetts Institute of Technology started Grabau's formal education in geology. He graduated from that institution with a BS degree in 1896. Grabau earned his MS and DSc degrees from Harvard University in 1896 and 1900, respectively. His subsequent career may be divided into an American phase from 1900–1919, when he was associated principally with Columbia University, and a Chinese phase from 1920–1946, when he held joint appointments at the National University of Peking (now Beijing University) and the Geological Survey of China. Grabau remained in China during the Second World War and died in Peking, March 20, 1946. His gravesite and memorial is located on the campus of Beijing University.

Regarded overall as a stratigrapher and paleontologist, Grabau's impact on sedimentology came during his early career through various attempts to devise a comprehensive classification scheme for sedimentary rocks. The most detailed treatment occurs in *Principles of Stratigraphy* in a single 30-page chapter out of a total text of 1,185 pages in 32 chapters. Grabau divided all rocks into two groups: exogenetic as opposed to endogenetic. He defined exogenetic rocks as those formed by agents acting from the outside to break down preexisting materials. Endogenetic rocks were classified as those formed by agents acting from within, for example precipitation of minerals from solutions. Essentially, his dichotomy corresponds to clastic and nonclastic rocks. In an earlier 1911 paper, Grabau stressed that after the question of origin, texture was the next most important consideration. He applied the terms rudaceous, arenaceous, and lutaceous to clastic particles that range in size from boulders to clay. Other terms coined by Grabau, such as calcirudytes, calcarenytes, and calcilutytes still are used for various grades of bioliths, or limestones. Derivation of new terms from classical roots was a proclivity of Grabau's. In a state of mild exasperation, Joseph Barrell (*q.v.*) offered a written response to an abstract by Grabau on the classification of marine deposits, suggesting that English terms "perfectly clear and quite as sharp" might be used in place of those derived from Greek. As an example, Barrell preferred "shallow-sea deposits" for the "Thalassigenous deposits" proclaimed by Grabau (1913).

Grabau's lasting influence was as a compiler with an encyclopedic mastery of materials and concepts. To regard his massive *Principles of Stratigraphy* as a textbook devoted strictly to the study of layered rocks would be a gross under estimation. His command of the topic was expansive, as indicated by the array of contents in separate sections on the atmosphere, hydrosphere, lithosphere, pyrosphere (on volcanism), and centrosphere (on internal earth dynamics). Grabau's outlook was entirely holistic and careful examination of the contents of *Principles of Stratigraphy* reveals that well over half the book is devoted to topics closely allied with sedimentology. Individual chapters, for example, deal with evaporites, eolian deposits, fluvial deposits, swamp deposits, reef and algal deposits, and deep-sea deposits. His handling of stratigraphy in the broadest sense, was thus prescient of contemporary texts that combine the studies of sedimentary environments and stratigraphy. Grabau (1917) may have shown his ultimate passion, however, with the prediction that "In future, the study of lithogenesis must go hand in hand with the study of paleogeography."

Markes E. Johnson

Bibliography

Grabau, A.W., 1911. On the classification of sand grains. *Science*, 33: 1005–1007.
Grabau, A.W., 1913. *Principles of Stratigraphy*. New York: A.G. Seiler.
Grabau, A.W., 1913. A classification of marine deposits. *Geological Society of America Bulletin*, 24: 711–714.
Grabau, A.W., 1917. Problems of the interpretation of sedimentary rocks. *Geological Society of America Bulletin*, 28: 735–744.
Johnson, M.E., 1985. A.W. Grabau and the fruition of a new life in China. *Journal of Geological Education*, 33: 106–111.
Shimer, H.W., 1947. Memorial to Amadeus William Grabau. *Proceedings Volume of the Geological Society of America Annual Report for 1946*, pp. 155–166.

Cross-reference

Sedimentology, History

AMANZ GRESSLY (1814–1865)

Gressly was born in Switzerland, and became interested in geology while a medical student in Strasbourg. In 1838, he published in French "Geological observations on the Soleurois Jura." This work was carried out under the direction of Jules Thurmann, and contains the analysis of facies that is generally considered to be Gressly's main contribution to geology. The paper, and his extensive collections of fossils, led to his appointment as an assistant to Louis Agassiz: when Agassiz left for America in 1846 he took many of Gressly's fossils with him, but left Gressly and his other assistants (Eduard Desor and Carl Vogt) behind. Gressly was then employed in engineering studies for railroads in Switzerland, but found time to explore ecological zones on the Mediterranean coast (Schneer, 1972).

The parts of Gressly's 1838 paper related to facies have been reprinted, with a critical commentary by Wegmann (1963) and discussed by many other authors, most recently by Cross and Homewood (1997). Gressly explained how he was led to develop the concept of facies:

"In the regions that I have studied, perhaps more than anywhere else, very variable modifications, both petrographic and paleontologic, everywhere interrupt the universal uniformity that, up until now, has been maintained for different stratigraphic units ("terrains" in the original French) in different countries. They even reappear successively in many stratigraphic units and astonish the geologist who wishes to study the nature of our Jurassic ranges..."

"There are two principal points that always characterize the group of modifications that I call *facies* or *aspects of stratigraphic units*: one is that *a similar petrographic aspect of any unit necessarily implies, wherever it is found, the same paleontological assemblage; the other, that a similar paleontological assemblage rigorously excludes the genera and species of fossils frequent in other facies.*" (Gressly, 1838, pp. 10–11, translated from the French: the emphasis is Gressly's own.)

Gerard V. Middleton

Bibliography

Cross, T.A., and Homewood, P.W., 1997. Amanz Gressly's role in founding modern stratigraphy. *Geological Society of America Bulletin*, 109: 1617–1630.
Schneer, C.J., 1972. Gressly, Amanz. In Gillespie, C.C. (ed.), *Dictionary of Scientific Biography*. Volume 5, pp. 533–534.
Wegmann, E., 1963. L'exposé original de la notion de facies par A. Gressly (1814–1865). *Science de la Terre*, IX: 83–119.

WILLIAM CHRISTIAN KRUMBEIN (1902–1979)

William Christian Krumbein, father of computer geology, pioneer in application of statistical techniques to sediments and sedimentary rocks, and innovator of sediment analyzes, was born in Beaver Falls, Pennsylvania on January 28, 1902 and died of a heart attack in Los Angeles, California on August 18, 1979. His specialties were the study of physical properties of sediments, application of statistics to sedimentology, dynamics of sedimentary processes, and regional sedimentary analysis. His degrees included a PhB in business administration (1926), and a MS in geology (1930) and PhD (1932) both from the University of Chicago, and a DSc (honora causa) from Syracuse University in 1979. He advanced through the academic ranks from assistant to associate professor at the University of Chicago and after working for the Beach Erosion Board of the US Corp of Engineers during WWII and a short stint with Gulf Research and Development Company (1945–1946), joined Northwestern University as a full professor in 1946, later made William Deering Professor of Geological Sciences, retiring in 1970.

His first papers in 1932 were on the mechanical analysis of fine-grained sediments and his last was on probabilistic modeling in 1978. (He was senior author posthumously on, *CORSURF: a Covariance-Matrix Trend Analysis Program*, published in *Computers & Geosciences* in 1995.) He explored sampling, textures, size distribution, diagenesis, transport, properties, and classification of modern and ancient sediments. He was one of the early workers to apply quantitative techniques, including descriptive statistics, Latin square experiments, regression analysis, Markov chains, and probabilistic modeling, extensively to geological problems.

With introduction of the computer in the mid 1950s, he transferred his statistical analyzes from the calculator to the computer; in 1958, he published (with L.L. Sloss) the first geologically oriented computer program (in SOAP). His last publications were philosophical in nature as he explored the directions and influences of quantification on sedimentology. He had an uncanny analytical mind and could cut through the extraneous material directly to the problem.

He carefully formulated and presented his ideas in his 130+ publications including 5 books. His manual with Francis Pettijohn, *Manual of Sedimentary Petrography* (1938), was a standard text for several decades as was his book with Larry Sloss on *Stratigraphy and Sedimentation* (1951, 1963). However, it was his book in collaboration with Frank Graybill in 1965, *An Introduction to Statistical Models*, that put him in the forefront of quantitative sedimentology. Krumbein coauthored several papers with leading statisticians of the day including John Tukey, Geoff Watson, Frank Graybill, Ron Shreve, and Mike Dacey.

He was a founding member of the International Association for Mathematical Geology (IAMG), was elected the first Past-President of the Association, and their premier award was named the William Christian Krumbein Medal in his honor; these tributes indicate the esteem held for him by his peers. He was a fellow of GSA, AAAS, and ASA; member of SEPM (President in 1950), SEG, AAPG, and AGU. He received a Guggenheim Fellowship, was a Fulbright Lecturer, and a President's Fellow at Northwestern University.

Daniel F. Merriam

Bibliography

Krumbein, W.C., 1932. A history of the principles and methods of mechanical analysis. *Journal of Sedimentary Petrology*, 2 (2): 89–124.

Krumbein, W.C., 1932. The mechanical analysis of fine-grained sediments. *Journal of Sedimentary Petrology*, 2 (3): 140–149.

Krumbein, W.C., 1934. Size frequency distributions of sediments. *Journal of Sedimentary Petrology*, 4 (2): 65–77.

Krumbein, W.C., 1937. (and Aberdeen, E.J.). The sediments of Barataria Bay (La.). *Journal of Sedimentary Petrology*, 7 (1): 3–17.

Krumbein, W.C., 1953. (and Miller, R.L.). Design of experiments for statistical analysis of geological data. *Journal of Geology*, 61 (6): 510–532.

Krumbein, W.C., 1956. (with Slack, H.A.). Statistical analysis of low-level radioactivity of Pennsylvanian black fissile shale in Illinois. *Geological Society of America Bulletin*, 67 (6): 739–762.

Krumbein, W.C., 1958. (and Sloss, L.L.). High-speed digital computers in stratigraphic and facies analysis, *American Association of Petroleum Geologists Bulletin*, 42 (11): 2650–2669.

Krumbein, W.C., 1960. The "geological population" as a framework for analysing numerical data in geology, *Liverpool and Manchester Geology Journal*, 2 (3): 341–368.

Krumbein, W.C., 1962. Open and closed number systems in stratigraphic mapping. *American Association of Petroleum Geologists Bulletin*, 46 (12): 2229–2245.

Krumbein, W.C., 1968. Statistical models in sedimentology. *Sedimentology*, 10 (1): 7–23.

Krumbein, W.C., 1969. (and Dacey, M.F.). Markov chains and embedded markov chains in geology. *Journal of Mathematical Geology*, 1 (1): 79–96.

Krumbein, W.C., 1974. The pattern of quantification in geology. In Merriam, D.F. (ed.), *The Impact of Quantification on Geology, Syracuse University Geology Contributions*. Volume 2, 51–66.

Krumbein, W.C., 1975. Probabilistic modeling in geology. In Merriam, D.F. (ed.), *Random Processes in Geology*. Springer-Verlag, pp. 39–54.

Cross-reference

Statistical Analysis of Sediments and Sedimentary Rocks

PAUL DIMITRI KRYNINE (1902–1964)

Paul D. Krynine was born in Siberia, the son of Dimitri P. Krynine, an engineer who travelled with his family to Argentina (1909–1917), then back to Moscow, where Paul graduated with a BSc in geology in 1924. Paul then emigrated to the United States, obtained a second bachelor's degree in geology at Berkeley (1927), and worked for Standard Oil of California for three years in Mexico. At this time his father also left Russia and became professor of soils engineering at Yale: Paul moved to Yale in 1931, and completed his doctorate there in 1936. He then accepted a teaching position at Penn State University where he remained until his death in 1964.

Krynine's doctoral thesis was a detailed study of the Triassic of the Connecticut Basin, a study which set new standards for sedimentary petrographic work in America (publication was delayed by the Second World War until 1950). Krynine made use of his experience in Mexico to argue that arkoses were not necessarily, or even most probably, deposited in arid climates. Humid climates, combined with active tectonic uplift of suitably feldspar-rich source rocks, and burial in a rapidly subsiding basin, was the most probable explanation. He also noted that red soils, which he considered important in the deposition of red arkosic beds, were confined to humid climates (his study of red beds in Mexico was, however, flawed by inadequate field work). These views started a long fruitful debate about the relative importance of tectonics, climate, and diagenesis in controlling the major mineralogy of sandstones (with most sedimentologists now inclined to the opinion that Krynine underestimated the importance of diagenesis).

In Pennsylvania, Krynine produced petrographic studies of the Third Bradford Sand, a major Devonian oil reservoir, and a sandstone with many "low rank metamorphic" rock fragments, and the Oriskany Sandstone, a pure quartz sandstone. Out of these, and other studies came Krynine's famous "megascopic" classification of sandstones into four main types: arkoses (with few fine-grained rock fragments), low and high rank greywackes (rich in fine grained rock fragments, and either poor—low rank, or rich—high rank, in feldspar), and the Orthoquartzites that were derived from them by much sedimentary recycling. Krynine presented these ideas in 1948 with the aid of the first triangular diagram used in sedimentary petrology: the classification was designed to be used for *field* classification using a hand-lens, so exact numerical boundaries between petrographic fields were never specified. This work, and Krynine's earlier ideas about tectonic control, were presented at meetings but published, if at all, only in obscure conference proceedings (Folk and Ferm, 1966, give a complete bibliography, and an exposition of some of his obscure publications: see also Bates and Griffiths, 1971). Nevertheless, Krynine's ideas were very influential, focused attention on refined petrographic interpretation of the major (light) constituents of sandstones (not just the heavy minerals, which were the focus of much attention in earlier decades), and set off many other attempts at sandstone classification. Krynine did not neglect heavy minerals, either and was one of the first since Sorby to expose the wealth of information that can be obtained by studying the compositional and textural *varieties* of key minerals, particularly stable minerals such as quartz and tourmaline. His basic concepts were extended by Robert L. Folk (one of his doctoral students) to produce the first generally accepted classification of limestones.

Krynine was an eccentric personality, no field geologist (he was almost blind), but a highly perceptive petrologist and an extraordinarily effective teacher. His teaching style and influence have been amply documented by his students (Folk, Ferm, Williams) and contemporaries (Griffiths, van Houten): together he and Griffiths made Penn State one of the three major schools for sedimentology in America in the 1940s and 1950s (the others were Chicago, with Pettijohn and Krumbein, and Northwestern with Dapples, Sloss, and Krumbein after he left Chicago). These were the decades when sedimentology came to maturity in the United States.

Gerard V. Middleton

Bibliography

Bates, T.F., and Griffiths, J.C., 1971. Memorial of Paul Dimitri Krynine. *American Mineralogist*, **56**: 690–698.
Folk, R.L., and Ferm, J.C., 1966. A protrait of Paul D. Krynine. *Journal of Sedimentary Petrology*, **36**: 851–863.
Van Houten, F.B., 1989. Krynine, Pettijohn, and sedimentary petrology. *Journal of Geological Education*, **37**: 241–242.
Williams, E.G., 1965. Memorial to Paul D. Krynine (1901–1964). *Proceedings of the Geological Society of America*, 1965, P63–P67.

Cross-references

Sedimentology, History
Sedimentologists: Robert L. Folk, Francis J. Pettijohn

PHILIP HENRY KUENEN (1902–1976)

Philip Kuenen was born in Scotland. His father was a Dutch physicist and his mother was English: they moved to Leiden in 1907, where Kuenen received all his education (including his doctorate in economic geology in 1925). He served as geologist on the *Snellius* expedition in 1929–1930. In 1934 he was appointed lecturer in Groningen, and remained there until his retirement in 1972.

Kuenen made major contributions in three main areas: experimental geology, marine geology, and sedimentology. In experimental geology, his main studies of sedimentological interest were: (1) a long series of experiments on the attrition of grains by sediment transport both in air (using a wind-tunnel) and in water, using a horizontal "racetrack" flume with the flow driven by a paddle wheel; (2) experiments on sedimentary structures, including ripples, plane lamination, and load structures ("pseudonodules"); and (3) studies of turbidity currents and their deposits (see below).

His paper of 1965 defends the experimental approach, and gives a summary of many of the results. He was himself an excellent experimentalist, with a fine physical intuition, but with almost no knowledge of mathematics or theoretical mechanics. His studies on attrition established that aqueous transport, even for long distances, was incapable of producing major changes in the size and shape of resistant mineral sand grains, such as quartz: thus well-rounded quartz grains must have undergone at least one cycle of wind transport. Earlier studies, using tumbling mills, did not duplicate natural processes, and so gave an exaggerated estimate of the attrition resulting from transport in water.

Kuenen's interest in marine geology dated from his experience as geologist on the *Snellius* expedition and included studies of coral reefs, eustatic changes of sea level, and calculations of the total mass of sediments in the oceans based on geological–geochemical cycles. His views were summarized in his influential textbook *Marine Geology*, published in 1950. Later he collaborated with his colleague L.M.J.U. van Straaten in series of studies of sediment transport and deposition on Dutch tidal flats (see *Sediment transport by Tides*).

Kuenen is undoubtedly best known for his development of the turbidity current theory for the origin of graded sand beds. The studies began as an experimental attempt (1937) to test Daly's hypothesis that muddy density flows were responsible for the erosion of submarine canyons. Experimental work continued during the war, while Kuenen was confined to the laboratory. The emphasis soon changed to the ability of turbidity currents (a term suggested by Johnson) to transport and deposit sand in deep water. The results were first presented at the International Geological Congress in London in 1948, and led immediately to collaborative papers with Migliorini, who had been studying graded beds in the field in the Apennines, and with Natland, a micropaleontologist who had faunal evidence that many of the sands and conglomerates in the Cenozoic Ventura Basin in California had been deposited in deep water. Natland organized a symposium at the SEPM meeting in Chicago in 1950, and he and Kuenen carried out field work together in California. The symposium volume was published in 1951. A detailed history of the period has been given by Walker (1973). Kuenen's work, together with that of coworkers and that of Heezen and other marine geologists at Lamont undoubtedly changed the way that geologists interpreted thick series of graded beds, previously thought to have been deposited somehow in shallow water, even though they did not display many of the structures (e.g., cross-bedding) thought to be typical for sands deposited in shallow environments.

Following the war, Kuenen was a strong advocate of turbidity currents and conducted field studies in many different regions. He recognized, described and explained many of the structures typical of such deposits, and helped to develop the growing field of paleocurrent studies in the 1960s and 1970s. One result from these studies was the discovery that in many sedimentary basins, sediment derived from the sides or end of the basin was transported mainly longitudinally, along the basin floor.

Gerard V. Middleton

Bibliography

Bourgeois, J., 1981. Kuenen, Philip Henry. In Gillispie, C.C. (ed.), Dictionary of Scientific Biography. **Supplement II Vol. 17**: pp. 509–514.

Walker, R.G., 1973. Mopping up the turbidite mess. In Ginsburg, R.N. (ed.), *Evolving Concepts in Sedimentology*. Baltimore: Johns Hopkins University Press, pp. 1–37.

Cross-references

Attrition (Abrasion), Fluvial
Attrition (Abrasion), Marine
Ball-and-Pillow (Pillow) Structure
Grading, Graded Bedding
Load Structures
Sedimentology, History
Turbidites

JOHN MURRAY (1841–1914) AND THE CHALLENGER EXPEDITION

The voyage of HMS Challenger (1872–1876) was a major event in the history of oceanography and in the knowledge of the deep sea. It was initiated and led by C. Wyville Thomson, but after his stroke he was forced to resign the Directorship, and turn over the publication of the results to John Murray. Murray oversaw the publication of 50 volumes in the years from 1880 to 1895: five of the volumes he coauthored himself, including one on the *Deep-sea Deposits* (1891).

John Murray was born in Cobourg, Ontario and attended school there until 1868, when he moved to Scotland, living with his grandfather, and attending Stirling High School, and then the University of Edinburgh. Though it was originally intended that he should become a medical doctor, he followed an irregular course of study, earning himself a reputation as a "chronic student," and never graduated. Besides some medical courses, he studied physics with Tait, and geology with Geikie. He also participated in a voyage to Spitsbergen, when he became interested in marine biology. By chance, one of the assistant naturalists appointed to the *Challenger* expedition was forced to resign, and, at Tait's recommendation, Thomson appointed Murray to replace him.

Murray's interests remained very broad, but on the *Challenger* he was assigned the bottom sediments as his main responsibility. A preliminary report, sent to the Royal Society from Valparaiso, and published in the *Proceedings* in 1875, already recognized "Shore deposits," "*Globigerina* ooze," "Radiolarian ooze," "Diatomaceous ooze," and "Red and Green clays." The *Challenger* travelled 110,000 km in the Atlantic and Pacific Oceans, and took soundings, dredgings, and bottom samples at 362 stations, thus achieving a major advance in knowledge of the biology and sediments of the deep sea. The final report, written with the Abbé Alphonse Renard, of the Brussels Museum, was not restricted to the *Challenger* samples, but also used comparative material from many other sources. It provided, for the first time, a classification and detailed description of deep-sea sediments (including also "Pteropod ooze" and iron-manganese nodules), which could be compared with sediments in the geological record.

The shallow-water nature of most sedimentary rocks was rarely questioned in America, but many European authors believed that some geosynclinal sediments, notably radiolarian cherts were deposited in the deep sea (e.g., Neumayer in 1876; Haug in 1900; and Steinmann in 1905). Though this interpretation remains controversial, plate tectonics suggests that some radiolarian sediments, associated with ophiolites, may in fact have been deposited at great depths. At first the Cretaceous chalks of Europe were compared with *Globigerina* ooze, but more detailed comparison with the *Challenger* samples soon showed significant differences. Undoubted deep-sea deposits were, however, described by Gregory in 1889 from Barbados and by Guppy in 1892 from Trinidad.

Besides his work on deep-sea sediments, Murray became a leading authority on coral reefs, opposing Darwin's subsidence theory of atolls and barrier reefs. The controversy was in part responsible for the drillings (1896–1898) on Funafuti commissioned by the Royal Society. The results, reported in 1904, remained inconclusive. After discovering a sample of phosphate rock from Christmas Island (in the Indian Ocean) in collections made for comparison with *Challenger* material, Murray first persuaded the British Government to annex this uninhabited island, and then organized a company to explore and ultimately exploit this resource. The population reached 1,500 inhabitants, and Murray claimed that the government received much more in royalties and taxes than the entire cost of the *Challenger* expedition! He used his own profits to finance further oceanographic research.

Late in life, Murray participated in several other expeditions, including investigations of Scottish fjords and lochs, and a four-month cruise (1905) of the Norwegian ship *Michael Sars*, together with Johan Hjort. The results were later included in their book *The Depths of the Ocean* (1912). Three years later Murray published a popular book on oceanography. He remained active until he died in 1914 as a result of an automobile accident.

Gerard V. Middleton

Bibliography

Burstyn, H.L., 1981. Murray, John. In Gillispie, C.C. (ed.), *Dictionary of Scientific Biography*. New York: Charles Scribner's Sons, Volume **9**, pp. 588–590.

Herdman, Sir, W.A., 1923. *Founders of Oceanography and Their Work*. New York: Longman, Green.

Linklater, E., 1972. *The Voyage of the Challenger*. London: John Murray.

Cross-references

Oceanic Sediments
Sedimentology, History

FRANCIS J. PETTIJOHN (1904–1999)

His life, 1904–1999, spanned most of the 20th century and he influenced many. By the time he finished high school, Francis Pettijohn knew he wanted to be a geologist. All his academic training was at the University of Minnesota (BA, 1924; MA, 1925; and PhD, 1930), where Professor Grout was his advisor and the Precambrian of Minnesota was the great center of geologic interest, which was studied through fieldwork, thin sections, and a few wet chemical analyzes. He taught briefly as a graduate student at nearby Macalester College, but spent 1925–1927 at Oberlin College in Ohio. His article "Geology at Oberlin a Half Century Ago," published in the Oberlin alumni magazine in 1980, is a fine insight to a small Midwestern geology department in the late 1920's. Francis developed his teaching skills and studied the beaches of Lake Erie, but he spent some of his summers in the North Woods. In 1927–1928 he was a Fellow at the University of California, where he became acquainted with Andrew E. Lawson from whom he learned much.

In 1929, Francis moved to the University of Chicago and started a program in sedimentary rocks with two sedimentology courses. He and W. C. Krumbein published *A Manual of Sedimentary Petrology* in 1938, a landmark book that was recently reprinted, and *Sedimentary Rocks* (1949), which established his reputation worldwide. His most famous paper at that time was "Archean Sedimentation" published in the GSA Bulletin in 1943. During these Chicago years he spent nearly every summer in field studies in the Lake Superior region for the Michigan and US Geological Surveys and the Geological Survey of Canada. During these prewar Chicago years he was active with the National Research Council in Washington, the forerunner of today's National Science Foundation, and often commented on how stimulating these small meetings were. Someone would bring along a puzzling rock and a wild discussion would break out. Later he edited the *Journal of Geology* for six years and traveled abroad, mostly to international meetings.

When he was 48 years old, Francis moved to The John Hopkins University in Baltimore in 1952, where he remained for the rest of his life, becoming emeritus in 1973. He left Chicago because fieldwork was considered old fashioned. At Johns Hopkins, Francis discovered the Appalachians, had new colleagues and had the most productive part of his career. He published *Paleocurrents and Basin Analysis* (1963) and *Sand and Sandstone* (1972), later translated into Russian and Chinese, and he revised *Sedimentary Rocks* (1976) for the third time. He also published *Atlas and Glossary of Sedimentary Structures* (1964) and *Memoirs of an Unrepentant Field Geologist* (1988), and helped edit *Studies of Appalachian Geology, Central and Southern* (1970), and wrote *A Century of Geology, 1884–7984*, at the Johns Hopkins University (1988). While at Johns Hopkins, he received many foreign visiting geologists who came to the United States at the invitation of the US Geological Survey. With all of these activities, Francis had time to be a Councilor of the Geological Society of America and to serve on several advisory boards.

The following special qualities and viewpoints were typical of Francis and served him well throughout a spectacularly successful, long life. He had sufficient self-confidence that he could fully appreciate and easily work with talented people and learn from them. He had the ability to read an article critically (thanks to Andrew Lawson). He persuaded with good observations, rationality, and tact. Conversely, even though Prof. Pettijohn had some strong opinions, he was conservative in his expression of them. Francis followed the rule, "Clarity of expression is best achieved by simplicity of expression." His favorite short and simple rule was, "Only outcrops keep geologists honest," which goes far to explain the common-sense attitudes seen in his writing.

Francis's influence on the profession was great and extended worldwide. Even as late as 1998, 25 years after he retired, his publications were cited 57 times; from 1990 to 1998, they were cited 497 times. His influence was transmitted through his books, his key articles, and his many students and visitors. Many of his students became prominent in the US geological and Canadian Geological Surveys, the petroleum industry, and university teaching. Many people had the opportunity to work with him and watch how he approached and resolved geologic problems—by asking questions that could be answered by the direct logical study of rocks or sediments rather than questions based on broad physical, chemical, or geological principles. Right or wrong, his analysis of a problem always employed a strong internal logical consistence and a full review of the prior literature.

Paul E. Potter

RUDOLF RICHTER (1881–1957), AND THE SENCKENBERG LABORATORY

Rudolf Richter (Figure S20(A)) was born on November 7, 1881 at Glatz in Silesia and studied at the Universities of Munich and Marburg. Richter worked under the tutelage of Emanuel Kayser and completed his doctorate at Marburg in 1909, his doctoral dissertation was published under the title "Beitraege zur Kenntnis devonischer Trilobiten aus dem Rheinischen Schiefergebirge" (Contributions to the knowledge of Devonian trilobites from the Rheinische Schiefergebirge). Richter is generally regarded as having been one of the most influential geologists of his time, his influence reaching far beyond the boundaries of Germany. Biographical works include Stubblefield (1957), Kegel (1957), Schmidt (1982), and Burghard Flemming (personal communication) provided insights on both Richter and the Senckenberg Laboratory.

After leaving Marburg Richter earned his living as a high school teacher in Frankfurt, his scientific work being carried out in his spare time. In 1908 he joined the Senckenberische

Figure S20 The Richters; (A) Rudolf Richter 1881–1957; and (B) Emma Richter 1888–1956. (Photographs courtesy of the University of Frankfurt.)

Naturforschende Gesellschaft and in 1919 he launched the journal *Senckenbergiana* to mark the centenary of the society. Richter maintained the editorship of that journal until his death in 1957. He also founded and edited its popular magazine *Natur und Museum*. In 1920 Richter was appointed Privat Dozent in Geology and Paleontology at the Johann Wolfgang Goethe University in Frankfurt; in 1925 he became Extraordinary professor, and in 1934 Ordinary Professor and Director of the Geological and Paleontological Institute a position he held until his retirement in 1949. In 1932 Richter took control of the Society and in 1933 was named "Leiter" and also took over the directorship of the Natur-Museum at Frankfurt.

Richter married Emma Hüther (Figure S20(B)) in 1913 and they enjoyed a remarkable collaboration. Emma was a skilled artist and was invaluable in their joint work on trilobites and the series of ichnological papers. The papers on animal-sediment relationships were fundamental in establishing basic principles on how organisms affect the substrate. Their ichnological work also established the zoological affinities of many trace fossils and became the corner stone for further work in actuopaleontology and actuoichnology. Emma died on November 15, 1956.

In 1929 with the support of the Senckenberg Society and the German Navy Richter established the Senckenberg Vorschungstelle für Meeresgeologie und Meerespaläontologie at Wilhelmshaven which was subsequently known as "Senckenberg am Meer" (Senckenberg by the Sea). The institute is presently housed in a building named after him, the "Rudolf Richter Haus". Richter's interest in the modern marine environment was documented as early as 1920 when he gave an inaugural lecture at Frankfurt University entitled "Die Erscheinungen des Wattenmeers in ihrer Bedeutung für die Geologie" (The manifestations of the Wadden Sea and their significance for geology). The station was founded with the specific aim of studying animal-sediment relationships in the intertidal environment in the Wadden Sea. It was the first institution founded with the specific aim of actively applying the actualistic concepts of Charles Lyell, following the principle "the present is the key to the past". The Wadden Sea has been the enormously successful hunting grounds of well-known Senckenbergian researchers such as Walther Häntzschel, Wilhelm Schäfer, and Hans-Erich Reineck. It was instrumental in establishing Germany as the home of ichnological research and shallow water marine sedimentology. The Senckenberg Lab has been used as the model for other facilities around the world including the Sapelo Island Marine Institute in Georgia. In fact these two institutes had very close ties in the 1970s and did joint work on the sedimentology and animal-sediment relationships of modern shoreface and estuarine environments (see Howard *et al.*, 1972 and Howard and Frey, 1975). Today the station houses the Marine Science Division of the Senckenberg Institute and the legacy of excellence left by Richter, Häntzschel, Schäfer and Reineck continues.

Richter received many honors including honorary membership in the Société belge de Géologie, Paléontologie et d'Hydrogeologie, the Paläontologische Gesellschaft and the Paleontology Society of America. He was also the recipient of the Gold Medal of the Paläontologische Gesellschaft and the Hans Stille Medal given by the Deutsche Geologische Gesellschaft. To celebrate his seventieth birthday former students and friends put together a special Festschrift volume published by the Senckenbergische Naturforschende Gesellschaft. Rudolf Richter passed away at Frankfurt on January 5, 1957.

S. George Pemberton

Bibliography

Howard, J.D., Frey, R.W., and Reineck, H.-E., 1972. Georgia coastal region, Sapelo Island, USA: sedimentology and biology. *Senckenbergiana Maritima*, **4**, 3–222.

Howard, J.D., and Frey, R.W., 1975. Estuaries of the Georgia coast, USA: sedimentology and biology. *Senckenbergiana Maritima*, **7**, 1–305.

Kegel, W., 1957. *Nachruf für Rudolf Richter—Emma Richter. Zeitschrift Deutsche Geologische Gesellschaft*. Volume **110**: 637–642.

Schmidt, H., 1982. Rudolf Richter (1881–1957) in seinen Worten. *Aufsätze Reden Senckenbergische Naturforschende Gesellschaft*, **32**, 1–95.

Stubblefield, C.J., 1957. Rudolf Richter. *Proceedings Geological Society of London*, **1554**, 137–138.

HENRY CLIFTON SORBY (1826–1908)

Henry Clifton Sorby lived all his life in Sheffield, England. He never married, and inherited enough money from his father to pursue a life of independent scholarship and scientific research. He made major contributions to microscopy applying the newly discovered polarized light technique to the study of rocks, meteorites and metals, earning the titles "Father of Petrography" and "Founder of Metallography". Later he worked on spectrum analysis, and coastal marine biology, using his own 35-ton vessel.

Sorby never attended university: his main forum for the exchange of ideas was the local Literary and Philosophical Society, though from the beginning of his active research career (in 1851) he presented papers to the Geological Society of London, the British Association for the Advancement of

Science, and other well-known societies. Later in life he helped found the Royal Microscopical Society (1874) and the Mineralogical Society (1876: he was the first president), and was active in the Geological Society, becoming president in 1878–1880.

Sorby's first petrographic study was of a "Calcareous Grit," and he went on to make major discoveries about the nature of calcareous rocks and their diagenesis: many of these were summarized in his Presidential Address to the Geological Society (1879). He distinguished the different types of grains making up limestones, and described the mineralogy and textures of bioclasts, and their influence on resistance to mechanical attrition and chemical alteration. He demonstrated the original calcitic nature of Paleozoic corals, as opposed to the aragonitic nature of most later corals. He described oolites, and proposed a mechanical "snowball" origin, followed by recrystallization, and recognized that some ancient oolites were originally aragonitic, while others were calcitic. His petrographic studies were soon extended to minerals found in other rocks, particularly sandstones (1877, 1880), and he paid particular attention to the mineral and fluid inclusions found within mineral grains (1858), showing experimentally how fluid inclusions could be used to deduce the temperature and pressure at which the minerals had originally formed. His combination of microscopic and experimental studies also provided the first clear evidence that slaty cleavage was formed by deformation under anisotropic stress (1853).

Sorby was also interested in the hydraulics of sediment transport and structure formation. He observed in a small stream how ripples formed soon after sediment began to move, but were washed away leaving a plane bed at higher flow intensities. He observed ripple cross-lamination and attributed it to "ripple drift" and described climbing sets (1859). He proposed that cross-bedding ("drift bedding") formed by migration of microdeltas. He also systematically recorded the orientation of ripples and cross-bedding and proposed that they be used to study paleocurrents, long before this became standard practice.

In retrospect it seems strange that Sorby's work did not inspire similar studies by his contemporaries. The work on limestones lay fallow (with rare exceptions, such as Cullis' study of the Funafuti cores, and Cayeux's work in France) until the revival of carbonate petrology in the 1950s. The textural and mineralogical study of sands fared better, but the main emphasis was on the "heavy minerals" and on mineral inclusions. Sorby's work on ripples and cross-bedding remained largely unknown, even after the revival of interest in America, due to the work of Gilbert, Kindle, and Bucher, until rediscovered in the 1960s. Paleocurrent techniques were developed in Germany (Brinkmann and others) and America, before being taken up again in Sorby's homeland.

Gerard V. Middleton

Bibliography

Allen, J.R.L., 1993. Sedimentary structures: Sorby and the last decade. *Journal of the Geological Society of London*, **150**: 417–425.
Folk, R.L., 1965. Henry Clifton Sorby (1826–1908), the founder of petrography. *Journal of Geological Education*, **13**: 43–47.
Sellwood, B.W., 1993. Structure and origin of limestones. *Journal of the Geological Society of London*, **150**: 801–809.

Smith, C.S., 1975. Sorby, Henry Clifton. In Gillispie, C.C. (ed.), *Dictionary of Scientific Biography*. New York: Charles Scribner's Sons, Volume 12, pp. 542–546.
Summerson, C.H., (ed.), 1976. *Sorby on Sedimentology: A Collection of Papers from 1851 to 1908 by Henry Clifton Sorby*. Miami: Comparative Sedimentology Laboratory, Geological Milestones I.

Cross-references

Carbonate Diagenesis and Microfacies
Sedimentology, History

WILLIAM H. TWENHOFEL (1875–1957)

William H. Twenhofel was prominent in the emergence during the 1920s–1930s of a new specialty called sedimentation, which became known as sedimentology during the 1950s. He was born on a northern Kentucky farm to German immigrant parents. Twenhofel's formal education began in public country schools and a private secondary school. For eight years, he taught grade school and studied at a normal (teachers) college until he qualified for a position at East Texas Normal College. After three years at Commerce, Texas, he entered Yale University at the age of 32, where he earned AB, MA, and PhD degrees within four years. He had intended to pursue mathematics, which he had been teaching, but an emergency assignment to teach geology showed him his true calling. At Yale, Professor Charles Schuchert suggested a dissertation on fossils of the Ordovician–Silurian boundary in Maritime Canada, which became the foundation for his first career, paleontology and stratigraphy. But also at Yale, Joseph Barrell had planted interests in weathering, erosion and sedimentation, which would blossom into Twenhofel's later and more famous sedimentation career. In his presidential address to the Paleontological Society in 1931, Twenhofel blended the two with a talk about the importance of sedimentary environments to paleoecology.

After six years at the University of Kansas, Twenhofel joined the University of Wisconsin in 1916. His sedimentological career began in 1919 with appointment to the National Research Council's Committee on Sedimentation. As chair of the Committee (1923–1931), he was the principal author of A *Treatise on Sedimentation*, a 661 page synthesis published in 1926, which defined the new specialty. In that same year, he was also a co-founder of the Society of Economic Paleontologists and Mineralogists (SEPM), the first professional society for sedimentology. From 1933 to 1946, he was editor of the first journal in the discipline. *The Journal of Sedimentary Petrology*, which was published by SEPM. After the *Treatise* appeared, the National Research Council capitalized upon his unusual talents by enlisting him to organize a Committee on Stratigraphy and another on Paleoecology.

Like many sedimentary geologists of his day, Twenhofel's early publications were chiefly on stratigraphy and paleontology. Later he contributed papers on a very wide range of sedimentary topics, which included the origin of black shales, Cambrian glauconitic greensands and their potential as a source of potash fertilizer, marine conglomerates and unconformities, deep-sea sediments, reefs, and black beach sands.

In 1939 Twenhofel published the important text book, *Principles of Sedimentation*. He is also remembered for emphasizing the environments of deposition and paleoecology of sediments, for pioneering the study of temperate lake sediments, and for stressing the geology of soils and the importance of soil conservation.

As an educator, Twenhofel inspired countless introductory geology students as well as many post-graduate students; he was also active in public outreach. Twenhofel's importance to sedimentology was recognized by the creation in 1973 of the Twenhofel Medal as SEPM's highest honor. Other recognitions included an honorary doctorate from Belgium's University of Louvain awarded to him in 1947.

Robert H. Dott, Jr.

Bibliography

Dunbar, C.O., 1960. Memorial to William Henry Twenhofel. In *Proceedings Volume of the Geological Society of America Annual Report for 1960*, pp. 151–156.
Shrock, R.R., 1981. William Henry Twenhofel. In *Dictionary of Scientific Biography*, Volume 6, pp. 518–519.
Twenhofel, W.H., 1926. *A Treatise on Sedimentation*. Baltimore: Williams and Wilkins.
Twenhofel, W.H., 1931. Environments in sedimentation and stratigraphy. *Bulletin of the Geological Society of America*, **42**: 407–424.
Twenhofel, W.H., 1939. *Principles of Sedimentation*. McGraw-Hill.
Twenhofel, W.H., and Tyler, S.A., 1941. *Methods of Study of Sediments*. McGraw-Hill.

JOHAN AUGUST UDDEN (1859–1932)

Johan August Udden was the "father of sedimentology" in North America, although during his 49-year career he made contributions in stratigraphy, areal geology, eolian processes, till in the upper Mississippi valley, and other topics. He was born at Lekasa, Sweden, March 19, 1859, and in 1861 moved with his parents to Carver, Minnesota. He attended Augustana College in Rock Island, Illinois, and received a bachelor's degree in 1881, a master's degree in 1889, an honorary PhD degree in 1900, and an honorary Doctor of Laws in 1929. He taught natural science and other courses at Bethany College, Lindsborg, Kansas, from 1881 to 1888 as the first teacher of an institution established to enhance the education of Swedish emigrants to Kansas. Udden returned to Augustana College as the Oscar II Professor of Geology and Natural History and taught geology, botany, zoology, meteorology, astronomy, physiology between 1888 and 1911. Even with this crushing teaching load he published 46 papers. In 1911, Udden joined the Bureau of Economic Geology and Technology of the University of Texas and served as its director from 1915 until his death. Prior to moving to Texas he worked short periods for the Iowa Geological Survey, University of Texas Mineral Survey, and US Geological Survey.

Udden is best remembered as the originator of the Udden grain-size scale for sediments published in 1898 and 1914. In order to document the eolian origin of loess, which influential geologists, including J.D. Dana disputed, he collected till, glacial outwash, and wind-blown sediment around the Mississippi River valley and analyzed their grain size distributions. He rejected a partially geometric scale used by soil scientists in favor of a totally geometric scale. He devised sieves to determine grain size distributions for gravel and sand to 0.06 mm and laboriously measured the diameter of finer particles, which he called dust, down to 4 μm under the microscope. He used histograms to help visualize his data. Chester Wentworth later changed some of the grain-size terms to establish the Udden-Wentworth scale that we use today. Udden was inventive in collecting eolian sediment. In addition to cylindrical sediment traps, he used glycerin-coated cloth attached to flagpoles, collected dust from melted snow, and swept sediment from the clothes of train passengers and from tree and corn leaves. After years of collecting sediment from various environments, in 1914 he published the first sediment grain-size database of 337 samples.

Udden's contributions extend far beyond grain-size analyzes. In 1905, he flew a model helicopter of his design. Using a horse, he studied the geology within 30 miles of Rock Island in great detail. He compiled a detailed picture of the artesian groundwater system in the area and used well cuttings and driller's logs to establish the subsurface stratigraphy. He recognized the potential value of well cuttings in a variety of geological endeavors and demonstrated this to the oil industry by establishing the first subsurface geology laboratory when he assumed directorship of the Texas Bureau of Economic Geology. In 1914, from a study of cuttings from a deep well in West Texas, he demonstrated the occurrence of extensive deposits of potash salts in the Permian basin. In 1916, on the basis of theoretical reasons and fieldwork done years previously, Udden advised the regents of the University of Texas of the likely occurrence of oil and gas on university lands in Reagan County, Texas. This led to the discovery of the first deep giant oil field in Texas and led to the development of the West Texas and eastern New Mexico oil province. Oil and gas revenues from university lands generated the permanent endowment of the Texas state university system. Udden's early work on the Permian in the Glass Mountains led to the establishment of that area as the type marine Permian of the world.

According to Charles Laurence Baker, a colleague of Udden in Texas, Udden's success can be attributed to his mastery of the basic natural sciences chiefly through his teaching, his ingenuity, and his ability to rely on his own powers of observation and reasoning. His ingenuity is exemplified by his invention of one of the earliest successful adding machines, and his ability to accomplish important results with cheap and simple devices. Udden early came to the conclusion that the best contributions he could make, given the limited opportunities available to him, was in fields neglected by others. In 1911, Udden was awarded the Swedish Order of the North Star by the king of Sweden.

Earle F. McBride

Bibliography

Baker, C.L., 1933. Memorial of Johan August Udden. *Geological Society of America Bulletin*, **44**: 402–413.
Heiman, Monica, 1963. *A Pioneer Geologist: Johan August Udden—a Biography*. Texas: Privately published, Kerrville.
Sellards, E.H., 1932. Memorial to Dr. Johan August Udden. *University of Texas, Bureau of Economic Geology*, Bulletin 3201, pp. 8–12.

Udden, J.A., 1898. *The Mechanical Composition of Wind Deposits.* Augustana Library Publications No. 1, Rock Island, Lutheran Augustana Book Concern.

Udden, J.A., 1914. Mechanical composition of clastic sediments. *Geological Society of America Bulletin*, **25**: 655–744.

T. WAYLAND VAUGHAN (1870–1952)

Thomas Wayland Vaughan, (born Jonesville, Texas, September 20, 1870—died Bethesda, Maryland, January 16, 1952) entered Tulane University in 1885, receiving a BS in Physical Sciences in 1889. From 1889 to 1892 he taught physics and chemistry at Mount Lebanon, Louisiana, where he encountered Eocene fossils, and became especially interested in corals. In 1892, Vaughan published his first two scientific papers (his bibliography lists about 350 titles until his last paper in 1946). The same year he enrolled in Harvard University and received a BA (1893) and a MA (1894). His PhD (1903) thesis was on Eocene and Oligocene corals.

Vaughan was a US Geological Survey, Assistant Geologist, 1894–1903; 1907–1923 he was geologist in charge of the Coastal Plain Subsection. Attending the 1897 International Geological Congress allowed him to examine European fossil coral collections and meeting Sir John Murray sparked his interest in oceanography. Two major Federal projects, resulting in more than 100 papers, occupied him between 1901 and 1923. First was Carribean geologic and paleontologic studies. Vaughan assisted in Cuba in 1901, and in 1911 worked in the Panama Canal Zone. In 1914 he led an expedition to islands north of Guadalupe and west of Puerto Rico. In 1919 he examined strata in the Virgin Islands and eastern Puerto Rico. From 1919 to 1921 he directed reconnaissance geological studies in Dominica and Haiti. His second major effort concerned Atlantic and Gulf Coast coastal plain investigations, ranging from Mexico north to Cape Cod. In addition to these duties, with support from the Carnegie Institution of Washington, Vaughan studied recent and fossil corals and coral reefs in the Florida Keys and in the Bahamas from 1908 to 1915. He is reputed to have published more papers on corals than any of his predecessors or contemporaries. In 1922 he also began the study of larger Foraminifera, which subsequently became particularly important for stratigraphic dating in the Pacific Islands.

During his early days, stratigraphy and paleontology for correlation were mainstays of the geologic community. The first world war and the need for petroleum began to focus attention on rocks as such and sedimentation started to develop as a discrete discipline. The formation of the National Research Council in 1915 and its subsequent war efforts, served as a national focus for comprehensive geological investigations. Vaughan was member of the NRC, Division of Geology and Geography, 1919–1926 and chairman of the Committee on Sedimentation, 1919–1923. This committee laid the foundation for the classic "Treatise on Sedimentation" complied by W. H. Twenholfel (1926). Vaughan wrote few papers directly on sedimentation, but it was his vision which led this work. He then went on to organize the Committee on Submarine Configuration and was its chairman, 1992–1932. Prior to the advent of the second world war, this committee stimulated what little American investigation there was of the oceanic sedimentation.

February 1, 1924, Vaughan moved to California, though he continued as a nominal member of the USGS, with the title of senior geologist (1924–1928) and principal scientist (1928–1939). His new position was Director of the Scripps Institution for Biological Research which shortly became the Institution of Oceanography. Vaughan made the place into a world leader in the growing field of ocean investigations, and remained its head until 1936. Biological work continued, but Vaughan instituted programs in dynamic and physical oceanography, as well as chemical and geological oceanography. As a member of the National Academy of Sciences, in 1927 Vaughan was appointed to the Committee on Oceanography. In large measure this committee was responsible for the founding of Woods Hole Oceanographic Institution and the Oceanographic Laboratories of the University of Washington. The 1926 Pan-Pacific Congress formed an International Committee on Oceanography of the Pacific. Vaughan served as its chairman for a decade. Thus, a significant portion of our present understanding of the oceans can be traced to his influence; his 1924 paper on oceanography is noteworthy. Simultaneously with his directorship, Vaughan was Professor of Oceanography; upon his retirement in 1936, he was appointed professor emeritus.

In retirement, Vaughan returned to the Natural History Museum in Washington. He continued his studies of Cenozoic fossil corals and foraminifers until 1947 when pneumonia led to failing eyesight which curtailed his scientific efforts. "His interest remained unabated to the end, and his circle of close friends kept him in close contact with current developments in his many fields of interest to the day of his death on January 16, 1952" (Wells, 1952, p.1495).

Just as Vaughan fused his combined interests in stratigraphy and paleontology into formalizing sedimentation at the NRC, he also sparked studies of the allied discipline of paleoecology. Indeed drawing a sharp line between these two may be difficult. The efforts of the Committee on Ecology and Paleoecology took longer to crystalize than did that on Sedimentation. It is sufficient to note that both volumes of "Treatise on Ecology and Paleoecology" (Geological Society of America Memoir, 67) are dedicated to him.

Vaughan's outstanding efforts were recognized by his peers and in 1915 he was president of the Geological Society of America, among other organizations. He was awarded the Agassiz Medal (1935) and the Mary Clark Thompson Medal (1945) of the National Academy of Sciences, and the Penrose Medal (1946) of the Geological Society of America, along with many honorary memberships, degrees, and other honors. In the last year of his life, Scripps Institution dedicated a new building to him in appreciation of his manifold activities.

Ellis L. Yochelson

Bibliography

Thompson, T.G., 1958. Thomas Wayland Vaughan: September 20, 1870–January 16, 1952. *Memorial Volumes of the National Academy of Sciences*, **32**: 398–437 [with photograph and bibliography].

Vaughan, T.W., 1947. Response [to receipt of Penrose medal]. *Proceedings Volume of the Geological Society of America, Annual Report for 1946*, 70–75 [with photograph].

Wells, J.W., 1952. Memorial—Thomas Wayland Vaughan (1870–1952). *Bulletin of the American Association of Petroleum Geologists*, **36**: 1495–1497 [photograph].

JOHANNES WALTHER (1860–1937)

Biographical data

Johannes Walther was one of the pioneers in sedimentology combining lithological and ecological aspects. He was born July 20, 1860 in Neustadt/Orla, Germany and died May 4, 1937 in Hofgastein. After his studies of zoology in Jena (PhD 1882) he studied geology more intensely in Leipzig and Munich. In 1886, he became lecturer in Jena; in 1894 associate "Haeckel-Professor" and from 1906 until his retirement 1928 he held the post of director of the Geological Institute in Halle. He received honorary degrees in Perth and Melbourne (1914!) and in 1928 he was guest professor at the John Hopkins University; from 1924 to 1931 he served as president of the Academy "Leopoldina" in Halle. (for biography and bibliography, see I. Seibold, 1992).

At the time when Walther wrote his principal works, paleontology, stratigraphy and tectonics played a leading role in geology. Walther took little interest in them but followed his line of "dynamic geology". His scientific thinking was deeply influenced by the famous zoologist Ernst Haeckel, the German apostle for Darwin whose thoughts thus shaped Walther's view on geology. He looked at the *processes* causing the formation of rocks and not only at their characteristics. He wanted to revive Lyell's uniformitarianism, but he showed critically that some phenomena of the past are not occurring today. On his travels he tried to study as many different geological conditions as possible. His topics were:

Shallow marine sediments

In 1884, 1885 and 1910, Walther went to the Zoological Station in Naples to carry out investigations of the recent seabottom. The 1910 study on the sediments of the "Taubenbank", a shoal in the Gulf of Naples, became classic for its quantitative and experimental approaches to sedimentology (e.g., ashes of Mt. Vesuvius) and biogeology (e.g., bioturbation). For instance, he kept crawfish together with mussels in an aquarium to find out how long the former would need to work the shells into detritus.

Reefs

In the Sinai Peninsula (1886), Walther studied recent and fossil coral reefs, their growth, subsidence, tectonic uplift and diagenetic processes. He completed his investigations 1888 in India (see R.N. Ginsburg *et al.*, 1994, for commented translations).

Deserts

His main desert voyages in the Sinai and Egypt (1886), USA (1891) and Central Asia (1897) revealed to him the importance of the eolian erosion, which had hitherto found less recognition. After a first publication in 1892, his desert book of 1900 went through four editions (English edition by E. Gischler and K.W. Glennie, 1997) and made him widely known as the "Desert-Walther". Besides a vast material of observations about erosion and sedimentation he dealt also with fossil deserts and the shift of climatic zones. His theory of the origin of salt deposits in desert basins stands against Ochsenius' bar theory.

Lithogenesis

Finally, his name became worldwide linked with the "Law of the Facies" originally expressed by A. Gressly (1839) but explained in detail in volume III ("Lithogenesis of the present") of Walther's principal work "Einleitung in die Geologie als historische Wissenschaft" (1893/94). Historical geology had previously been treated almost exclusively under its paleontological aspects. He pointed to the priority of lithological research in connection, of course, with paleontological findings. One can regret that he published only few case studies to illustrate his ideas. In his time, this inspiring work received only limited attention in Anglo-American circles, though Amadeus Grabau welcomed it with enthusiasm (1913). Yet Walther's impact on Russian geology (including translations of his main works) was remarkable.

Since the 1970s his work found renewed attention by the facies research in the oil industry. Middleton (1973), dealt with the Law of facies. The recent English translations of his works will make him better known as a forerunner of modern biogeology.

Eugen and Ilse Seibold

Bibliography

Ginsburg, R.N., Gischler, E., and Schlager, W. (eds.), 1994. *Johannes Walther on Reefs*. University of Miami, Geological Milestones II, 140pp.
Gischler, E., and Glennie, K.W. (eds.), 1992. *Johannes Walther: The Law of Desert Formation—Present and Past*. University of Miami, Geological Milestones IV, 273pp.
Grabau, A.W., 1913. *Principles of Stratigraphy*, **2**, Dover.
Middleton, G.V., 1973. Johannes Walther's law of the correlation of facies. *Bulletin of the Geological Society of America*, **84**: 979–988.
Seibold, I., 1992. *Der Weg zur Biogeologie—Johannes Walther*. Springer.
Walther, J., 1893/94. *Einleitung in die Geologie als historische Wissenschaft*, Volume 3, Jena: Fischer.

SEDIMENTS PRODUCED BY IMPACT

Impact cratering is now recognized to be a very important (if not *the* most important) surface-modifying process in the planetary system. The reason for this becomes clear if one takes a look at the Moon, or peruses spacecraft images of the surfaces of Mercury, Venus, Mars, the asteroids, and the moons of the outer gas planets. On Earth, the situation seems to be different. Lunar-like craters are not an obvious or common landform, and impact cratering is not described as an important process in textbooks on terrestrial geology. There are a number of reasons for this deficiency, as impacts happen on Earth just as they do on other planets, but endogenic forces cover or obliterate their traces. This would explain why impact cratering, on first glance, does not seem to be a particularly important geological process on Earth and why it has been ignored by generations of earth scientists.

One of the most important driving forces for impact research in the past decades was the study of rocks from the *Cretaceous–Tertiary (K–T) boundary*. At the beginning of this important development was a serendipitous discovery. Alvarez and colleagues were trying to use the concentration of the extraterrestrial tracer element iridium (Ir) as a proxy for the sedimentation rate of Cretaceous and Tertiary age limestones in Italy. It turned out that Ir was not useful as such a tracer because the abundances were too low, but they found that the concentrations of the rare platinum group elements (PGEs; Ru, Rh, Pd, Os, Ir, and Pt) and of other siderophile elements (e.g., Co, Ni) in the thin clay layer that marks the K–T boundary are considerably enriched compared to those found in adjacent Cretaceous and Tertiary rocks. These significant enrichments (up to four orders of magnitude) and the characteristic interelement ratios were interpreted by Alvarez *et al.* (1980) to be the result of a large asteroid or comet impact, which also caused extreme environmental stress. There are several excellent overviews of this discovery and its implications for the earth sciences (e.g., Glen, 1994; Alvarez, 1997; Powell, 1998). Subsequently, shock-metamorphosed minerals (quartz, feldspar, etc.) were found in K–T boundary layer samples from all over the world (e.g., Bohor *et al.*, 1984, 1987), which confirmed the impact origin of the K–T boundary layer and, thus, that a large-scale impact event was responsible for the end-Cretaceous mass extinctions. Details on shock metamorphism can be found in Features Indicating Impact and Shock Metamorphism, and other reviews (e.g., Melosh, 1989; Stöffler and Langenhorst, 1994; Grieve *et al.*, 1996; French, 1998; Montanari and Koeberl, 2000; Koeberl, 2001).

An important result of K–T boundary studies is the demonstration how the various types of distal ejecta can be correlated with proximal ejecta and, eventually, an impact structure. The determination of the enrichment in the concentration of the element iridium, a specific marker of extraterrestrial material, in the sediments from the K–T boundary provided the first direct evidence for a large impact event at the end of the Cretaceous. This led, in turn, to the theory that the mass extinction of the end of the Cretaceous was caused by a huge asteroid or comet impact. Studies of the K–T boundary event provided data that helped with the understanding of the origin and significance of other, possibly impact-related, boundaries and exotic layers in the stratigraphic record (*cf.* Montanari and Koeberl, 2000; Peucker-Ehrenbrink and Schmitz, 2001). As such, K–T boundary research played several important roles: to increase the visibility and acceptance of impact research, and to provide applications for decades of impact-related research.

An important basic concept is the definition of *proximal and distal ejecta*. Proximal ejecta are those that are found in the immediate vicinity of an impact crater, within <5 crater radii from the rim. In contrast, distal ejecta are those ejecta that occur at considerable distances from the source crater (>5 crater radii from the crater rim; see, e.g., Melosh, 1989). Distal ejecta commonly consist of (usually fine-grained) rock and mineral fragments or of glassy spherules or fragments. Whereas it is often not possible to right away recognize their connection to a specific impact structure, distal ejecta can act as a guide to major impact events and even lead to the discovery of large impact structures. During crater formation, at the end of the excavation stage, about 90 percent of all material ejected (excavated) from the crater is deposited as proximal ejecta. This conclusion has been obtained from impact experiments, simulation calculations, and comparison with nuclear and chemical explosions. The mass of the ejecta decreases outward: about half of the ejected material can be found within about one crater radius from the rim. At greater distances the ejecta blankets become increasingly thinner and discontinuous. The presence of an atmosphere significantly influences the ejection and distribution mechanisms. On an airless planet, ejecta will be emplaced ballistically. Calculations and experiments show that not only the total ejecta mass decreases with increasing distance from the crater, but also the average particle size. Ejection of material from a crater is a complicated, multistage process. Some material is ejected early on during the contact and compression stage at very high velocities, but most material is ejected during the excavation phase. The material seems to be ejected at relatively low angles, with very little material being thrown out at angles of more than 45° (Oberbeck, 1975). On Earth (or another planet with an atmosphere), ejecta near the crater can also be deposited in a base surge, which is a gravity-driven density current (composed of air and dust entrained by the fireball) that flows down (and outward) from the expanding mushroom-shaped cloud.

Material launched early in the crater formation will still be in flight after the crater is completely excavated and after the more massive local ejecta have been deposited to form the continuous deposits around the crater (Oberbeck, 1975). Late during the crater formation (when the shock wave has already decayed somewhat), material is ejected at lower velocities and will be deposited close to the crater rim, whereas material that was ejected early and at higher velocities is deposited at greater distances from the crater. The low velocity ejecta deposited at the crater rim often preserve the initial stratigraphic relationship, which leads to the inverted stratigraphy (the overturned flap) at the crater rim (because material from greater depth at the target is deposited last on the rim). This can well be seen in the field at, for example, the famous Meteor Crater in Arizona. In general, material from greater stratigraphic depth at a crater location is, thus, deposited closer to the crater compared to distal ejecta, which more commonly consist of the uppermost stratigraphic layers of the target. The higher energies available early in the crater formation sequence make it also more likely that the earliest ejecta are molten or at least strongly shocked, while the later ejecta may only be slightly shocked or not at all. Examples of molten distal ejecta include tektites (see below). On Earth the interaction with the atmosphere complicates matters somewhat. Early ejecta may also be entrained in the rapidly expanding fireball and, if the event is large enough, they will be ejected outside of the atmosphere, leading to a global distribution. Size sorting will result in larger particles settling out first and finer dust being deposited later.

There are several empirically derived equations that describe the thickness of ejecta blankets with distance from the crater (*cf.* Melosh, 1989, p. 90). One of the most commonly used ones gives a power law for the ejecta blanket thickness t as a function of distance r from the crater center:

$$t = 0.14 R^{0.74} (r/R)^{-3.0} \text{ for } r \geq R \qquad \text{(Eq. 1)}$$

where R is the radius of the transient cavity and dimensions are in meters. The exact values for the function of R are debated, and Melosh (1989) proposed a more general version in which the term $0.14 R^{0.74}$ is replaced by $f(R)$, a poorly defined function that depends on a variety of parameters.

From the study of the distribution of microtektites (see below) in the Australasian strewn field, Glass and Pizzuto (1994) derived the following equation:

$$t = 0.02\ R(r/R)^{-4.4 \pm 0.3} \quad \text{(Eq. 2)}$$

which can be rewritten as a function of the distance alone:

$$t = 10^{24.0 \pm 0.4} r^{-4.4 \pm 0.3} \quad \text{(Eq. 3)}$$

This relation seems to be valid not only for proximal, but also for distal ejecta, such as tektites and microtektites.

Impact ejecta are found in Archean and Proterozoic rocks sequences in Australia and South Africa. Investigations of these layers are still in progress (see chapter by Koeberl, this volume). However, the study of the K–T boundary ejecta provided the most influence for the discussion about the importance of impact events. It is easy to detect the K–T boundary layer in the field, where is appears as a distinct break in lithology, with a thin (usually up to 1 or 2 cm thick) layer commonly composed of clay or claystone. A detailed description of the field occurrences of various K–T boundary layers relations was recently given by Smit (1999), and references therein. Information on a variety of aspects of the study of the K–T boundary, the K–T impact event, and its implications can be found, for example, in the "Snowbird" series of conference proceedings (Silver and Schultz, 1982; Sharpton and Ward, 1990; Ryder et al., 1996; Koeberl and MacLeod, 2002). From the sedimentological perspective, K-T boundary layers in marine and terrestrial sections have a somewhat different appearance. Terrestrial K-T boundaries (e.g., in the Western Interior of North America) are characterized by an up to 3-cm-thick dual layer, whereas in most marine sections it does not display a dual-layered nature. At the Western Interior locations, the lower layer is mainly kaolinitic with some smectite; it contain hollow spheres up to 1 mm in diameter, that are most often also composed of (and filled by) kaolinite, but at some sites (e.g., in Wyoming) they are composed of goyazite (a hydrous alumino-phosphate; an alteration product). Bohor and Glass (1995) give a detailed description of the formation and alteration of these spherules. The upper (thinner) layer is commonly laminated and consists of smectite and kaolinite clay with variable amounts of organic matter, giving it a dark gray to black appearance (Bohor, 1990). Most of the shocked minerals and the siderophile element anomaly (see below) are restricted to the upper layer; it has been called the "fireball" layer (e.g., Bohor, 1990). However, the composition and internal structure of the K/T boundary layer varies with distance from the source and depositional environment (e.g., Smit, 1999). In marine sites in Italy, for example, a cm-thick clay layer occurs within well-bedded limestones (Figure S21). Besides high contents of the PGEs and the presence of shocked minerals, evidence for impact in the various K-T boundary layers includes also the presence of osmium and chromium with extraterrestrial isotope ratios, fragments of the original bolide, impact-related diamonds, high-pressure polymorphs of quartz (coesite, stishovite), and impact glass. In the early 1990s, the 200-km-diameter Chicxulub structure in Mexico (Yucatan Peninsula) was confirmed as an impact structure of K–T boundary age. Thus, the source crater of the K–T boundary impact was found.

There are several other cases where impact craters were identified after careful studies of distal ejecta layers. For example, Gostin et al. (1986) reported on the discovery of an

Figure S21 The Cretaceous-Tertiary boundary layer at Frontale, Umbria–Marche sequence, central Italy. The grayish clay layer is about 1.5 cm thick and situated within limestones. End-Cretaceous rocks are marked with a K, and earliest Tertiary rocks are marked with a T.

impactoclastic layer within late Precambrian shales of the 590 Ma Bunyeroo Formation in the Adelaide geosyncline, South Australia. This layer was found to contain abundant shocked quartz grains, zircons, shattered mineral grains, and small shatter cones (e.g., Wallace et al., 1996). The ejecta were found in outcrops and drill cores over several hundred kilometers in south-central Australia. Around the same time, Williams (1986) identified the Acraman structure in South Australia as an impact structure, which was rapidly confirmed as the source crater of the Bunyeroo impact ejecta layer. Gostin et al. (1989) and Wallace et al. (1990) detected enrichments of the PGEs in the ejecta layer; however, post-formational redistribution had altered the PGE patterns. The diameter of the deeply eroded Acraman structure is probably about 90 km. Impact ejecta have been found at distances of up to 450 km from the Acraman structure (i.e., about 10 crater radii), making this a true distal ejecta layer.

Another important group of distal impact ejecta are *tektites and microtektites*. They are a group of natural glasses that have been known for several centuries by now, which earlier had been interpreted to represent glassy meteorites or ejecta from terrestrial or lunar volcanoes. Mainly due to chemical studies, it is now commonly accepted that tektites are the product of melting and quenching of terrestrial rocks during hypervelocity impact on the Earth. The chemistry of tektites is in many respects identical to the composition of upper crustal material. Tektites are currently known to occur in four strewn fields of Cenozoic age on the surface of the Earth. Strewn fields can be defined as geographically extended areas over which tektite material is found. The four strewn fields are: the North American, Central European (moldavite), Ivory Coast, and Australasian strewn fields. Tektites found within each strewn field have the same age and similar petrological, physical, and chemical properties.

In addition to the (macroscopic) tektites found on land, microtektites were found in three of the four strewn fields. Microtektites occur in deep-sea cores, within layers that correspond in stratigraphic age to the radiogenic age of tektites on land, are generally less than 1 mm in diameter, and show a somewhat wider variation in chemical composition than tektites on land, but with an average composition that is very close to that of "normal" tektites. Microtektites have been very important for defining the extent of the strewn fields, as well as for constraining the stratigraphic age of tektites, and to provide evidence regarding the location of possible source craters. For details see Koeberl (1994) and Montanari and Koeberl (2000).

Tektites are chemically homogeneous, often spherically symmetric natural glasses, with most being a few centimeters in size. Those found on land have commonly been classified into three groups (see, e.g., O'Keefe, 1963): (1) normal or splash-form tektites; (2) aerodynamically shaped tektites; and (3) Muong Nong-type tektites (sometimes alsocalled layered tektites). The aerodynamic ablation results from partial remelting of glass during atmospheric re-entry after it was ejected outside the terrestrial atmosphere and solidified through quenching. The shapes of splash-form tektites include spheres, droplets, teardrops, dumbbells, etc., or fragments of such shapes. These are not to be confused with aerodynamical forms. Muong Nong-type tektites were named after the type-locality in Laos. They are usually considerably larger than normal tektites and are of chunky, blocky appearance. Muong Nong-type tektites show a layered structure with abundant vesicles. These tektites are less depleted in volatile elements than the splash-form tektites.

Microtektites are, by definition, less than 1 mm in size, and show a variety of shapes, ranging from spherical to dumbbell, disc, oval, and teardrop. In color they range from colorless and transparent to yellowish and pale brown. They often contain bubbles and lechatelierite inclusions. Microtektites occur in the stratigraphic layers of the deep-sea sediments that correspond in age to the radiometrically determined ages of the tektites found on land. Thus, they are distal ejecta and represent an impact marker. For example, in deep sea sediments in the Indian Ocean and Pacific Ocean in the general vicinity of Australia and Indochina they have a stratigraphic occurrence near the Brunhes–Matuyama boundary; this and their composition indicate that they are indeed part of the Australasian strewn field.

The geographical distribution of microtektite-bearing cores defines the extent of the respective strewn fields, as tektite occurrences on land are much more restricted. Furthermore, microtektites have been found together with melt fragments, high-pressure phases, and shocked minerals (e.g., Glass, 1989; Glass and Wu, 1993) and, therefore, provide confirming evidence for the association of tektites with an impact event. The variation of the microtektite concentrations in deep-sea sediments with location increases toward the assumed or known impact location. Glass and Pizzuto (1994) determined the geographic variation in abundance of Australasian microtektites (ranging from <1 to 3,255 microtektites per cm^2 in the >125 µm size range), and found that the abundances increased toward Indochina; and deduced a possible source region in Cambodia.

Within the past few decades the impact origin of tektites was established, raising the question regarding their source craters. Since then, numerous suggestions and educated guesses have been made regarding the location of the possible source craters for the tektite strewn fields. Relatively reliable links between craters and tektite strewn fields have been established between the Bosumtwi (Ghana) and the Ries (Germany) craters, and the Ivory Coast and the Central European fields, respectively. Only recently, a plausible link was established between the newly discovered Chesapeake Bay crater and the North American strewn field. No large crater with a compatible age has yet been identified for the Australasian strewn field. The exact process of tektite formation is still not quite clear, although from chemical and isotopic studies it is fairly well established that tektites have formed from the immediate surface layers of the target rocks (*cf.* Koeberl, 1994).

Distal ejecta ("impactoclastic layers") can be used as markers for impact events in the stratigraphic record. "Impact markers" can be described as all chemical, isotopic, and mineralogical species derived from the encounter of cosmic bodies (such as cometary nuclei or asteroids) with the Earth, as discussed in more detail by Montanari and Koeberl (2000). Such markers are quite important in the detection and study of accretionary events in the sedimentary record, to identify their origin, and to evaluate their possible role in global change and on the Earth's biotic and climatic evolution throughout geological time. Distal ejecta layers can be used to study a possible relationship between biotic changes and impact events, because it is possible to study such a relationship in the same outcrops, whereas correlation with radiometric ages of a distant impact structure is always associated with larger errors. Impactoclastic layers are composed of distal ejecta. For example, microtektites occur in well-defined and thin layers in deep-sea sediments and provide an excellent time marker. The sedimentology of impactoclastic layers can provide important information not only on source craters and impact processes, but also on the effects (local, regional, or global) of the related impact events.

Christian Koeberl

Bibliography

Alvarez, L.W., Alvarez, W., Asaro, F., and Michel, H.V., 1980. Extraterrestrial cause for the Cretaceous-Tertiary extinction. *Science*, **208**: 1095–1108.

Alvarez, W., 1997. *T. Rex and the Crater of Doom*. Princeton University Press.

Bohor, B.F., 1990. Shocked quartz and more: impact signatures in Cretaceous/Tertiary boundary clays. In Sharpton, V.L., and Ward, P.D. (eds.), *Global Catastrophes in Earth History*. Geological Society of America, Special Paper, 247, pp. 335–347.

Bohor, B.F., and Glass, B.P., 1995. Origin and diagenesis of the K/T impact spherules—From Haiti to Wyoming and beyond. *Meteoritics*, **30**: 182–198.

Bohor, B.F., Foord, E.E., Modreski, P.J., and Triplehorn, D.M., 1984. Mineralogical evidence for an impact event at the Cretaceous/Tertiary boundary. *Science*, **224**: 867–869.

Bohor, B.F., Modreski, P.J., and Foord, E.E., 1987. Shocked quartz in the Cretaceous/Tertiary boundary clays: evidence for global distribution. *Science*, **236**: 705–708.

French, B.M., 1998. *Traces of Catastrophe: A Handbook of Shock-Metamorphic Effects in Terrestrial Meteorite Impact Structures*. LPI Contribution 954, Houston: Lunar and Planetary Institute, 120pp.

Glass, B.P., 1989. North American tektite debris and impact ejecta from DSDP Site 612. *Meteoritics*, **24**: 209–218.

Glass, B.P., and Pizzuto, J.E., 1994. Geographic variation in Australasian microtektite concentrations: implications concerning the location and size of the source crater. *Journal of Geophysical Research*, **99**: 19075–19081.

Glass, B.P., and Wu, J., 1993. Coesite and shocked quartz discovered in the Australasian and North American microtektite layers. *Geology*, **21**: 435–438.

Glen, W., 1994. *The Mass Extinction Debates*. Stanford University Press.

Gostin, V.A., Haines, P.W., Jenkins, R.J.E., Compston, W., and Williams, I.S., 1986. Impact ejecta horizon within late Precambrian shales, Adelaide Geosyncline, south Australia. *Science*, **233**: 198–200.

Gostin, V.A., Keays, R.R., and Wallace, M.W., 1989. Iridium anomaly from the Acraman impact ejecta horizon: impacts can produce sedimentary iridium peaks. *Nature*, **340**: 542–544.

Grieve, R.A.F., Langenhorst, F., and Stöffler, D., 1996. Shock metamorphism in nature and experiment: II. Significance in geoscience. *Meteoritics and Planetary Science*, **31**: 6–35.

Koeberl, C., 1994. Tektite origin by hypervelocity asteroidal or cometary impact: target rocks, source craters, and mechanisms. In Dressler, B.O., Grieve, R.A.F., and Sharpton, V.L. (eds.), *Large Meteorite Impacts and Planetary Evolution*. Geological Society of America, Special Paper 293, pp 133–152.

Koeberl, C., 2001. The sedimentary record of impact events. In Peucker-Ehrenbrink, B., and Schmitz, B. (eds.), *Accretion of Extraterrestrial Matter Throughout Earth's History*. New York: Kluwer Academic/Plenum Publishers, pp. 333–378.

Koeberl, C., and MacLeod, K. (eds.), 2002. *Catastrophic Events and Mass Extinctions: Impacts and Beyond*. Geological Society of America, Special Paper, 356, 746 pp.

Melosh, H.J., 1989. *Impact Cratering: A Geologic Process*. New York: Oxford University Press.

Montanari, A., and Koeberl, C., 2000. *Impact Stratigraphy: The Italian Record*. Lecture Notes in Earth Sciences, Volume 93, Heidelberg: Springer Verlag, 364pp.

Oberbeck, V.R., 1975. The role of ballistic erosion and sedimentation in lunar stratigraphy. *Reviews of Geophysics and Space Physics*, **13**: 337–362.

O'Keefe, J.A. (ed.). 1963. *Tektites*. Chicago: University of Chicago Press.

Peucker-Ehrenbrink, B., and Schmitz, B. (eds.), 2001. *Accretion of Extraterrestrial Matter Throughout Earth's History*. New York: Kluwer Academic/Plenum Publishers.

Powell, J.L., 1998. *Night Comes to the Cretaceous*. New York: W.H. Freeman.

Ryder, G., Fastovsky, D., and Gartner, S. (eds.), 1996. *The Cretaceous-Tertiary Event and Other Catastrophes in Earth History*. Geological Society of America, Special Paper, 307, 576pp.

Sharpton, V.L., and Ward, P.D. (eds.), 1990. *Global Catastrophes in Earth History*. Geological Society of America, Special Paper 247, 631pp.

Silver, L.T., and Schultz, P.H. (eds.), 1982. *Geological Implications of Impacts of Large Asteroids and Comets on the Earth*. Geological Society of America, Special Paper 190, 528pp.

Smit, J., 1999. The global stratigraphy of the Cretaceous-Tertiary boundary impact ejecta. *Annual Reviews of Earth and Planetary Science*, **27**: 75–113.

Stöffler, D., and Langenhorst, F., 1994. Shock metamorphism of quartz in nature and experiment: I. Basic observations and theory. *Meteoritics*, **29**: 155–181.

Wallace, M.W., Gostin, V.A., and Keays, R.R., 1990. Acraman impact ejecta and host shales: evidence for low-temperature mobilization of iridium and other platinoids. *Geology*, **18**: 132–135.

Wallace, M.W., Gostin, V.A., and Keays, R.R., 1996. Sedimentology of the Neoproterozoic Acraman impact ejecta horizon, South Australia. *AGSO: Journal Australian Geology and Geophyics*, **16**: 443–451.

Williams, G.E., 1986. The Acraman impact structure: source of ejecta in late Precambrian shales, South Australia. *Science*, **233**: 200–203.

Cross-references

Extraterrestrial Material in Sediments
Features Indicating Impact and Shock Metamorphism

SEPTARIAN CONCRETIONS

Septarian structures are former cracks, often filled with cement and are most commonly found in concretions hosted in mudrocks, although rare occurrences are known from siltstones and sandstones. Septarian structures occur in concretions or concretionary sheets, which may be chemically and mineralogically the same as non-septarian concretions in the same mudrock. The host concretions are most commonly calcite, dolomite or siderite dominated, although rare occurrences have been interpreted from silica concretions. Septarian concretions are predominantly a pre-Pleistocene phenomenon, although a potential "proto" version has been described from the late Pleistocene (Duck, 1995).

Crack morphology

Septarian structures were initially formed as open fractures, and are most often concentrated in the central regions of concretions and reduce in width and frequency toward the outer parts of the concretion, which may or may not be cracked. The fractures take a variety of forms, from lenticular cracks, which are widest toward the center of the concretion, to straight-sided cracks, to dense networks of hairline cracks. The orientation of the cracks can be diverse, but always include cracks which are predominantly sub-vertical to the bedding plane. Some sub-vertical cracks in sections perpendicular to the bedding plane may appear inclined, which is often due to the low angle intersection of the crack and section. Cracks can also be sub-horizontal or curved, sometimes reflecting the shape of the concretion (Figure S22A). Within horizontal sections the cracks often demarcate approximately equidimensional blocks of the concretion (Figure S22B). Multiple generations of cracks are apparent in some septarian concretions, cross cutting earlier formed cracks. Cracks are mostly displacive and may cut body fossils, or occasionally cracks may show a component of shear displacement.

Crack filling cements

Cracks may range from largely unfilled to fully cement filled, often with a variety of distinctively colored spar cements. The fills may also include geopetal sediment, or sediment injected from the surrounding mudrocks, through cracks that reached the concretion surface. The cements grew on the fracture walls within a pore-fluid filled crack. This is evident by the preservation of partially unfilled cracks displaying crystal terminations in the voids. The cracks are usually filled with calcite, but sometimes include other carbonates, sulfide, sulfates or clay minerals, sometimes quite different to the mineralogy of the concretion body. In most occurrences the volume of the cracks make up less than 20 percent of the volume of the concretion body, although in rare occurrences the cement-filled cracks may constitute up to ~70 percent of the concretion body.

Timing of crack formation

The timing of the crack formation relative to the formation of the concretion body is indicated by the geochemistry of the body and crack cements. Detailed carbon and oxygen isotopic studies (see *Isotopic Methods in Sedimentology*) indicate the central region of the concretion body generally formed first. The outside of the concretion body may have a similar cement chemistry to the earlier formed crack cements, hence for some concretions the outer part of the concretion formed at the same time as the inner cracks fills. The fracturing of some of

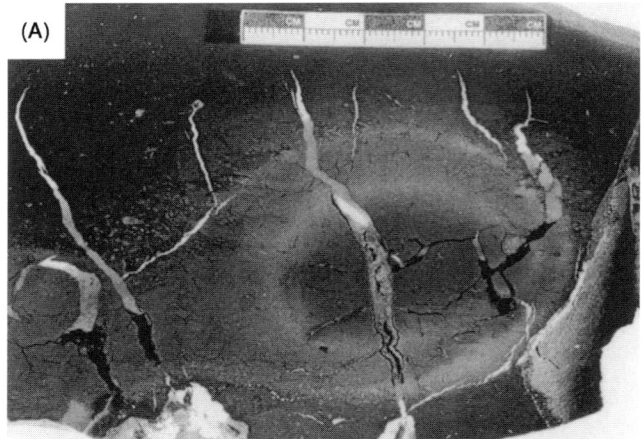

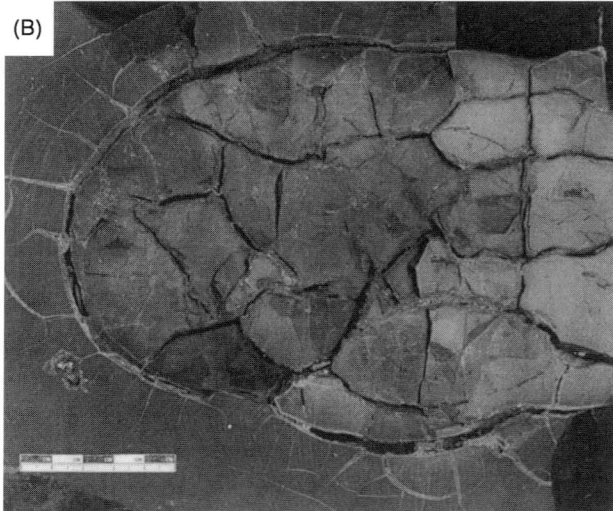

Figure S22 (A) vertical section through a compound calcite (pale central part) and siderite (dark outer) septarian concretion. The lenticular and parallel-sided cement-filled cracks are mostly vertical to bedding. Three generations of cement fill occur a clear calcite spar (dark) in the calcite concretion (large and small cracks), a pale brown spar (pale gray) and a barite (white), which mostly fills the later tapering terminations to cracks. Some crack voids are preserved in the inner calcite concretion (Ironstone Shales, Lower Jurassic, UK). (B) horizontal section through dolomitic septarian concretion with septarian cracks partially filled by dolomite (Lower Carboniferous of Berwickshire, Scotland).

the early cement fills by later cracking, and subsequent filling by later cements has indicated that some concretions have a repeated history of cracking and successive crack fills. The linkage of cement chemistry to likely changes in pore-fluid chemistry with sediment burial (see *Diagenesis*), has also indicated that in some septarian concretions the crack-fills formed later than the concretion body.

Depth of formation

Estimates of the depth of formation of septarian cracks are closely linked to the origin of the body of the concretion. There are three main clues to the depth of formation. Firstly, the variation in the amount of cement in the concretion body has been used as a relative indicator of burial depth, through its assumed passive filling of the early sediment pore-space. This is not without its problems, due to the possible displacive or replacive growth of some concretion body cements. Secondly, the isotope geochemistry of the crack and body cements, linked to models of organic matter stimulated diagenesis (see *Diagenesis*) has suggested estimates of likely burial depth, based on typical present-day depths of pore-water chemical boundaries. Thirdly, occurrences of septarian concretions exhumed at disconformities and subsequently re-buried, in which estimates of the amount of sediment removed can be ascertained (Hesselbo and Palmer, 1992). These three estimates suggest that some septarian cracking was initiated at a minimum of a few meters to a few 10s of meters depth. Maximum cracking depths are often not well constrained, but may extend to several kms of burial, or occur during uplift.

Origin of cracks

Septarian structures have attracted debate since the late 19th Century, when the accepted hypothesis was that they formed by shrinkage of the interior parts of the concretion. This idea has attracted many proponents since that time involving increasingly more complicated growth models (Pratt, 2001). However, a plausible model that explains this shrinkage process in firm physical terms has largely eluded its proponents. The physical process that may provide a comparable analogue for the shrinkage hypothesis is the production of syneresis cracks (see *Syneresis*). The shrinkage model is also deficient in explaining the physical and chemical nature of the cementing medium which shrinks in the early-formed parts of the concretion, and deficiencies pertaining to the textural and geochemistry data about the relative timing of textural features (Astin, 1986). More recent hypotheses suggest the cracks are generated by tensional failure, either through stresses produced by excess pore pressure (see *Compaction*; Astin, 1986; Hounslow, 1997) or shaking motions produced by earthquakes (Pratt, 2001). The essential differences between these models are that the Astin and Pratt models implicate formation-wide stresses, which are through unknown processes, localized to the concretion body. Without a process of tensional failure localization, both the Astin and Pratt models would also produce septarian cracks in rocks other than the concretion. The Hounslow model implicates stresses localized to the cementing concretion, whose cementation is the cause of the tensional stress. The driving motivation for tensional fracture models is the realization that large instantaneous tensional forces are needed to fracture rigid body fossils and the concretion body in its consolidated state. Tensional stresses applied over a long period of time (sub-critical crack growth) appear to provide a solution in lowering the threshold for fracture growth, and partially cemented concretions, as implicated in some growth models, may lower the tensional strength of the concretion body. A deficiency in the tensional fracture models are that in some concretions the tensional strains implied by the large volumes of crack-filling cement are far in excess of what could be expected. Hence, a hybrid shrinkage- tensional fracture model may be appropriate for some septarian concretions (Pratt, 2001).

Ultimately the origin of septarian cracks is tied to the debate about the origin of concretion bodies. One of the problems is the lower number of case studies linking the timing of concretion body and septarian cement formation, which utilize integrated and detailed geochemical, textural and mineralogical data. Study of hiatus septarian concretions could be a fruitful avenue for further study, since exhumation provides a relative timing and depth indicator (Hesselbo and Palmer, 1992). A second gap in knowledge is the physical properties of concretions when they first form, which is fundamental in dictating the likely processes of septarian crack formation. The absence of a present day analogue for septarian concretions either suggests the formation processes are absent or very rare at the present day, or that they take longer to form than current understanding suggests.

Mark W. Hounslow

Bibliography

Astin, T.R., 1986. Septarian crack formation in carbonate concretions from shales and mudstones. *Clays Clay Minerals*, **21**: 617–631.

Duck, R.W., 1995. Subaqueous shrinkage cracks and early sediment fabrics preserved in Pleistocene calcareous concretions. *Journal Geological Society of London*, **152**: 151–156.

Hesselbo, S.P., and Palmer, T.J., 1992. Reworked early diagenetic concretions and the bioerosional origin of a regional discontinuity within British Jurassic marine mudstones. *Sedimentology*, **39**: 1045–1065.

Hounslow, M.W., 1997. Significance of localized pore pressures to the genesis of septarian concretions. *Sedimentology*, **44**: 1133–1147.

Pratt, B.R., 2001. Septarian concretions: internal cracking caused by syn-sedimentary earthquakes. *Sedimentology*, **48**: 189–213.

Cross-references

Compaction (Consolidation) of Sediments
Diagenesis
Diagenetic Structures
Isotopic Methods in Sedimentology
Syneresis

SILCRETE

Silcretes or siliceous duricrusts are defined by Summerfield (1983) as the indurate products of surficial and penesurficial (near-surface) silica accumulation. They are formed by cementation and/or low-temperature replacement of bedrocks, weathering deposits, unconsolidated sediments, soil or other materials. Summerfield proposes silcretes should contain at least 85 percent silica, however some silcretes may not contain this percentage as they retain parts of the parent material. Silcretes form nodular, columnar and/or massive layers in outcrops. In hand specimens they show many different features inherited from the parent material or acquired during silica accumulation and afterward in silica diagenesis. Silcrete horizons usually vary from several cm to 3 m, but can occasionally exceed this thickness. Although silcretes exist globally, the principal studies on them have been carried out in Australia and Africa.

Once the petrology of a silcrete is known, any further discussion of its formation needs to explain: (1) the source of silica; (2) the silica transfer from the source to the site of accumulation; (3) the conditions and mechanisms that give rise to silica precipitation; and (4) their relation to the climate and paleosurfaces.

Mineralogical composition and petrology

The main SiO_2 minerals deposited during silcrete formation are: quartz, moganite (silica polymorph with a crystal structure closely related to that of quartz), opal CT (stacked sequences of cristobalite/tridymite and occasionally Opal A (amorphous silica). These silica phases display many fabrics and textures, depending on the parent material, chemical composition of the pore fluids, and whether the mineral formation process is replacement or cementation. Consequently the petrography of silcretes is highly complex.

Common crystalline quartz of different sizes is frequently associated with several varieties of chalcedony. Chalcedony is fibrous microcrystalline quartz and different varieties (with distinct names in French and English) can be observed using a polarizing microscope. These are: calcedonite (length-fast chalcedony, in which the elongation of the fibers is perpendicular to the crystallographic c-axis), quartzine (length-slow chalcedony, in which the elongation is parallel), lutecite (another type of length slow, in which the fiber axis is inclined by approximately 30°) and helicoidal calcedonite or zebraic chalcedony (systematic helical twisting of the fiber axes around the crystallographic c-axis). The varieties of chalcedony allow the definition of formation environments as acid or nonsulfate (length-fast) or basic or sulfate/magnesium rich-environments (length-slow) (see review in Hesse, 1991). Unfortunately, there are exceptions to these rules and rigid application of these criteria can lead to error in interpretation.

The microscopic characteristics of moganite have not yet been studied in depth, but it is generally described as uniformly flaky or fibrous (length-slow). Moganite, in amounts of over 20 percent, is considered indicative of special evaporitic environments which have still not been clearly determined (Heaney, 1995).

The opal in silcretes is generally opal CT and appears throughout the rock as massive or glaebular. Glaebules are concentric rings of silica and matrix components and are usually less than one centimeter in diameter, although they can occasionally reach several centimeters. The opaline cements are fibrous and named lussatite (length-slow) or lussatine (length-fast). Opal A is not common, but of great interest, because in some cases its existence propitiates the formation of opaline gemstones. The opaline phases in silcretes appear where the parent material is mudstone or the silica concentration of the pore fluids is very high (120 ppm of silica for amorphous silica and 89 ppm for opal CT). These opaline phases can recrystallize to quartz (aging). In the case of silcretes, temperature and time, the two main factors favoring aging, are not always necessary.

The major silcrete types and genesis

Parent materials determine the majority of the mineralogical and micromorphological characteristics of silcretes and as there are many parent materials, there are many varieties of silcretes. It is therefore difficult to find a universally acceptable petrographic classification of silcrete. Summerfield's classification (1983) has been commonly used, although this classification is not appropriate for silcretes formed from nondetrital parent materials. Thiry and Milnes (1991) developed a general genetic classification of silcretes, which subdivided them into pedogenic and groundwater types. Pedogenic silcretes are the result of silica precipitation within the soil profile. The structures and fabric of the parent materials are disturbed or destroyed during silcrete formation being substituted by others more typical of paleosols. Intermittent and repetitive leaching, infiltration and illuviation alternate with evaporation, which causes the precipitation of silica minerals. This occurs in areas with alternating wet and dry periods.

Groundwater silcretes are massive, with sharp upper and lower boundaries, and lack pedogenic features such as nodules and columns. The structures and fabrics of the parent material are retained in the silcrete. They probably represent silicification on or close to the water table and are related to the position of the paleowater table. The silica is imported either by infiltration through the upper part of the profile or by lateral groundwater flow. The mixing zone between meteoric ground water and saline waters from a lake or the sea facilitates silification processes and the silica rocks formed there are considered to be silcretes by some authors. Characteristics of the two types of silcretes, vaclose and groundwater, sometimes appear together because silicification might have commenced with groundwater and continued into the unsaturated zone (Bustillo and Bustillo, 2000). Generally the silica precipitation is dependent upon geochemical parameters at the microscale and the pore water movement at the mesoscale. The movement of water is largely conditioned by geomorphological and hydrological controls, and to climatically influenced water-table fluctuations.

Silcretes and climate

Varied studies of silcretes in Australia (Langford-Smith, 1978) revealed these occur in two geological contexts: (1) associated with deep-weathering profiles where their formation is attributed to wetter conditions; and (2) arid/semiarid or alkaline conditions. In the first context silcretes can coexist with ferricretes and laterites and in the second with calcretes, dolocretes and gypcretes. Within deep weathering profiles, chemical alteration breaks the silicate minerals down and releases the silica. The mechanism of silica precipitation is unclear but sufficiently sluggish drainage could lead to silica precipitation. In arid/semiarid conditions, the source of silica is sometimes a consequence of the high amount of silica liberated when other duricrusts (calcretes, dolocretes, and gypcretes) are formed on a silicate parent material. The dissolution of opaline particles (phytolites, diatoms...) and eolic quartz sands and silica-rich groundwaters are other sources. Silica precipitation mainly occurs in response to concentration through evaporation or by a localized decrease in pH.

The existence of two distinct climatic formation environments, does not mean all silcretes should be assigned to them. Moreover, efficient geochemical and petrological tools capable of definitively assigning a silcrete to one of these two contexts have not yet been developed. After many years use, criteria such as the amount of TiO_2 or some petrographic characteristics, are not now considered acceptable (Nash et al., 1994).

Oxygen isotopic studies of the quartz that forms silcretes may yield environmental information, including variations in temperature (Knauth, 1992). The opaline phases and to a lesser extent, the chalcedonic quartz, are not appropriate for the isotopic studies due mainly to bound/included water. Once we know the oxygen isotopic composition of the quartz, the oxygen isotopic composition of fluids from which it precipitated can be calculated if the temperature range can be established. For this calculation, a fractionation formula can be used. Webb and Golding (1998) consider the Knauth and Epstein formula gives more realistic results under near surface conditions, which is where silcretes are formed. Detailed geochemical and oxygen isotopic studies of a silcrete and its parent material can shed light upon the mechanism of silcrete formation and if evaporation is a major factor in silcrete formation (Webb and Golding, 1998). Silcrete formation depends on a complex interaction between climate and silica supply and needs tectonic stability.

Mª Angeles Bustillo

Bibliography

Bustillo, M^a.A., and Bustillo, M., 2000. Miocene silcretes in argillaceous playa deposits, Madrid Basin, Spain: petrological and geochemical features. *Sedimentology*, **47**: 1023–1039.

Heaney, P., 1995. Moganite as an indicator for vanished evaporites: a testament reborn? *Journal of Sedimentary Research*, **A65**: 633–638.

Hesse, R., 1990. Silica diagenesis: origin of inorganic and replacement cherts. In McIlreath, A., and Morrow, D.W. (eds.), *Diagenesis.*, Canada: Geoscience Reprint Series, Volume 4, pp. 253–275.

Knauth, L.P., 1992. Origin and diagenesis of cherts: an isotopic perspective. In Clauer, N., and Chaudhuri, S. (eds.), *Isotopic Signatures and Sedimentary Records*. Berlin: Springer-Verlag, pp. 123–152.

Langford-Smith, T., 1978. *Silcrete in Australia*. Australia: Department of Geography, University of New England.

Nash, D.J., Thomas, D.S.G., and Shaw, P.A., 1994. Siliceous duricrusts as palaeoclimatic indicators: evidence from the Kalahari Desert of Botswana. *Palaeogeography, Palaeoclimatology, Palaeoecology*, **112**: 279–295.

Thiry, M., and Milnes, A.R., 1991. Pedogenic and groundwater silcretes at Stuart Creek opal field, south Australia. *Journal of Sedimentary Petrology*, **61**: 111–127.

Summerfield, M.A., 1983. Silcrete. In Goudie, A.S., and Pye, K. (eds.), *Chemical Sediments and Geomorphology*. London: Academic Press, pp. 59–61.

Webb, J.A., and Golding, S.D., 1998. Geochemical mass-balance and oxygen-isotope constraints on silcrete formation and its paleoclimatic implications in southern Australia. *Journal of Sedimentary Research*, **68A**: 981–993.

Cross-references

Laterites
Siliceous Sediments

SILICEOUS SEDIMENTS

Introduction

Siliceous sediments are composed of silica that has actually precipitated at or near the site of deposition or has replaced pre-existing sediments. They are distinguished from clastic or terrigeneous sediments which are made of grains derived from rocks elsewhere and physically transported to the site of deposition. In today's oceans, the dominant examples of siliceous sediments are oozes composed of microscopic silica particles precipitated biologically by diatoms and, to a lesser extent, by radiolarians, sponges, and silicoflagellates. Diatom silica is precipitated on a colossal scale in surface waters and slowly settles to the ocean floor to produce diatomaceous oozes. Although most of the opaline tests dissolve in transit, a small percentage survive and accumulate in thicknesses up to hundreds of meters. The purity of the ooze is greatest for depths below the carbonate compensation depth and in areas far removed from landmasses that can supply abundant clays. Siliceous oozes are regularly encountered on the deep ocean floor and in deep-sea drill cores. Locally, radiolarians and silica sponges are so abundant that oozes of radiolarian skeletons and opaline spicules can also accumulate.

Diatomites, porcelanites, and bedded cherts

Along the coast of California, the Miocene Monterey Formation has been uplifted to produce superb exposures of diatomaceous layers displaying every gradation from pure, unconsolidated diatomite to porcelain-like beds and even brittle, glass-like beds of chert. The diatomite is extensively mined for insulation, paint filler, filtration material, and a wide variety of other industrial uses (Figure S23). The classic study by Bramlette (1946) first demonstrated that the gradations in lithology represented various steps in the burial transformation of opaline diatoms into microcrystalline quartz. Uplifted exposures along coastal California allowed detailed examination of this process where it had been arrested in various states of progress. Subsequent mineralogic and isotopic studies revealed that the solution-resistant particles that had accumulated continued to resist dissolution during burial until elevated temperatures provoked an *in situ* solution-reprecipitation conversion into hard, porcelain-like beds (Figure S24) of hydrous opal with weak X-ray diffraction peaks of cristobalite and trydimite (opal-CT). Upon further burial, this opal-CT converted to α-quartz. Pore spaces and fractures that developed at various times during burial became filled directly with microcrystalline quartz or with fibrous quartz that grades outward into microcrystalline aggregates of single, well-formed crystals. The most recent studies (Behl and Garrison, 1994) suggest that much of the quartz in the Monterey Formation formed throughout the burial history rather than only at maximum burial as originally thought. At burial depths such that temperatures reach 80–100°C, all opal and opal-CT becomes transformed into microcrystalline quartz. Most of the great oil fields in California have developed where abundant organic matter in diatomite converted to oil and accumulated in fractured masses of brittle porcelanite and chert.

The stepwise transformation so clearly displayed in the Monterey Formation apparently also operates during burial of all deep sea oozes and readily explains the lithologic/mineralogic variations of siliceous oozes encountered during deep-sea drilling. Diatoms have become particularly abundant in the last 30 million years and now dominate silica deposition in the oceans. Tertiary examples of Monterey-type diatomites, porcelanites, and cherts are particulary common in the circum-Pacific area and have received much attention. Examples of Eocene radiolarian oozes that transformed into porcelanite and bedded chert are also regularly encountered during deep drilling of the ocean floor.

Early observers noted peculiar regional beds of Paleozoic and Mesozoic strata composed of interlocking grains of microquartz that were clearly not clastic deposits. Examples include bedded cherts in the Alps (Gruneau, 1965) and the Devonian Arkansas/Caballos novaculite in Texas/Arkansas (Miser and Purdue, 1929; Folk and McBride, 1976). In many fold belts, the cherts display a distinctive "pinch and swell" texture and are known as "ribbon cherts". Chemical etching of the microquartz now reveals radiolarian forms confirming prior inferences that most of these bedded cherts originated as radiolarian opal. The etched-out forms are distinct enough for assigning biostratigraphic ages useful in tectonic studies. Apparently, at times and in various places, radiolarians were abundant enough to accumulate as opaline oozes similar to the Tertiary diatomaceous examples. It is now inferred that these transformed into bedded cherts through the stepwise opal to opal-CT to quartz sequence so beautifully documented for the Miocene Monterey Formation.

Novaculite is a white radiolarian chert made almost exclusively of intergrown grains of microquartz and is famous for its widespread use as a whetstone. Other variations in color and texture of radiolarian cherts probably relate to the clay mineral content. The nature of the depositional basins combined with dilution effects of shale and calcareous plankton and possible diagenetic segregation of silica from shale (Murray *et al.*, 1992) may all play a role in determining

Figure S23 Quarry in Monterey Diatomite. Lompoc, California.

Figure S24 Transition of laminated organic-rich Monterey diatomite to opal-CT. Hammer head is on opal-CT horizons. Beach cliff at Pt. Pedernales, California.

the final lithologic character of the chert. Inasmuch as the mobile seafloor carries all sediments into accretionary wedges and subduction zones within about 100 million years, most of these older bedded cherts are today found in highly deformed and altered mountain belts.

Silicification

Silica is soluble in aqueous solution as H_4SiO_4. At pH>9, these units polymerize and solubility of all solid forms of silica increases dramatically. At lower pH and room temperature, solubility of quartz is about 6 ppm and opal is about 120 ppm. The solubility increases greatly with temperature, but because of the long times available, the low-temperature values are high enough to allow large-scale transfer of silica in the pore fluids of sediments prior to compaction. Supersaturation with respect to quartz is achieved if ground water moves through uncompacted sediment containing opal or grains of more soluble silica such as volcanic glass and/or clay minerals. Silicification can then occur if some other phase such as metastable aragonite/calcite shell fragment (or a tree trunk buried in volcanic ash) slowly dissolves or disintegrates and is transported away by the through-going, quartz-saturated ground water. Quartz can grow in the microscopic pore spaces so vacated and replacement can occur on such an incremental basis that extraordinary preservation of the precursor material is achieved. Even forms as small as bacteria can be preserved in this manner. Fibrous quartz and small quartz crystals can also grow in the larger void spaces from such saturated solutions. Quartz crystals up to 10 μm have been grown in seawater at room temperature in 2 years in solutions with low silica concentration (MacKenzie and Gees, 1971). If dissolved silica concentrations are >120 ppm, silicification by opal can occur. The long time-scales required to achieve such incremental silica replacement preclude realistic laboratory simulations; the silicification process must be inferred from studies of the rock record.

Mineralogy

Silica occurs in sedimentary rocks as several varieties of opal and quartz. Opal is the most common form in Cenozoic deposits, but older strata are always composed of quartz. Amorphous, hydrous opal (opal-A) is today precipitated by diatoms, radiolarians, sponges, and silicoflagellates, and also forms inorganically in unusual depositional environments and in pore spaces during shallow burial. This material transforms to quartz with increased time, temperature, and/or depth of burial. Both nodular and bedded cherts are composed of microcrystalline quartz with individual grains ranging in size from about 8–20 μ (Figure S25) . The grains are thoroughly intergrown with each other and this makes chert one of the hardest of the sedimentary rocks. Each individual grain is crystallographically complex, with many lattice faults and up to 1 wt. percent OH and H_2O. Fibrous quartz is typically associated with the micro quartz (Figure S25) and has an even more complex internal structure. Coarser crystals of quartz with well-defined grain boundaries and more normal internal crystallographic ordering often occur together with fibrous and microcrystalline quartz and also form as recrystallization or annealing products of microquartz.

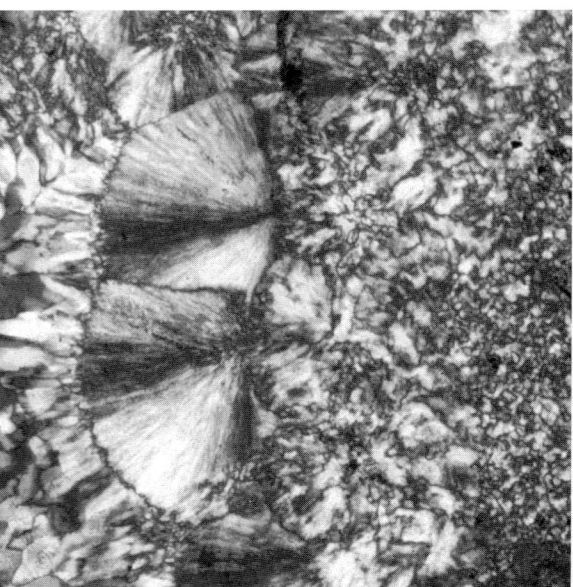

Figure S25 Thin-section of chert under crossed nicols. Microquartz on right half, 2 bundles of fibrous quartz grade to left to coarser individual crystals of quartz. From 2.7 Ga silicified stromatolite in Fortesque Group, W. Australia. Scale bar is 10 μm.

Nodular replacement cherts

The earliest examples of silica in sediments to be studied were the nodular cherts in limestones (Figure S26). Famous as the source rock for anthropological artifacts and widely used as flints for starting fires and igniting munitions (even used as cannonballs in one battle of the American Civil War), nodular cherts were originally interpreted as balls of opaline gel that rolled around on the seafloor and hardened into quartz during burial (Tarr, 1917). A long-standing debate started when Van Tuyl (1918) argued that the nodules were actually replacement features where zones of limestone had been incrementally replaced by silica precipitated from later diagenetic fluids moving through the sediment. Among others, Biggs (1957) provided thorough documentation for early replacement and the basic idea has not been challenged since.

Replacement of carbonate to produce nodular chert occurs during early sediment burial when shell fragments and other primary grains of carbonate undergo dissolution and re-precipitation into interlocking calcite and/or dolomite crystals to form limestone and/or dolostone (Knauth, 1979, 1994). In Phanerozoic rocks, the most common silica source is opaline sponge spicules which dissolve and re-precipitate as microquartz during this transformation of sediment into sedimentary rock. The amount of chert that forms depends on the abundance of the silica secreting sponges in the original depositional environment and the vagaries of hydrologic movement of pore fluids during stabilization. The nodules are often among the first stable phases to form, and cross-bedding, oolitic texture, and even organic-walled microfossils are often exquisitely preserved. Primary textures in the surrounding carbonate are typically degraded or obliterated during this early diagenetic "stabilization" event making cherts important "windows" to the original depositonal

Figure S26 Nodular replacement chert in 1.2 Ga Bass Limestone, Grand Canyon, Arizona.

Figure S27 Nodular cherts displaying primary pisolitic fabric which was largely obliterated in host dolostone by later neomorphism. Chuar Formation, Grand Canyon, Arizona.

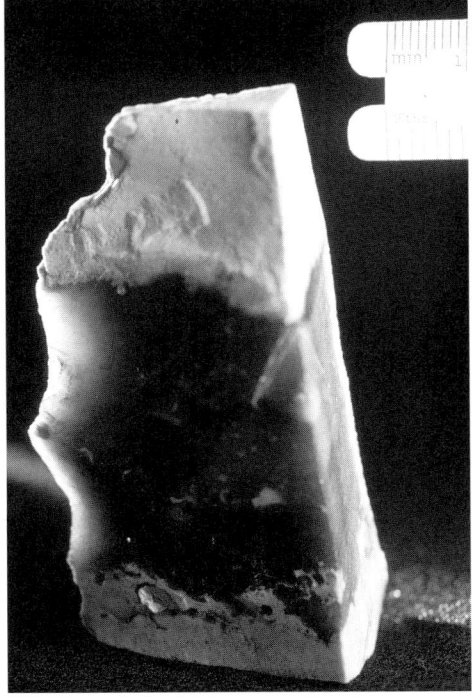

Figure S28 Dark nodular chert displaying translucence and conchoidal fracture typical of near 100% replacement of limestone with microquartz. Cretaceous Edwards Formation, Texas.

structures and fabrics that have been otherwise lost (Figure S27). This aspect of cherts has been generally under-utilized and offers much potential in the investigation of carbonate depositional environments.

In addition to the prominent nodules, replacement chert also occurs as small blebs, thin stringers, silicified fossils, and disseminated patches throughout the parent carbonate mass. The color of the chert is black if unreplaced impurities are organic matter, gray if remnant carbonate, and red if iron oxide. The purer cherts are translucent and display conchoidal fracture (Figure S28). If pieces of such chert are reassembled, they often do not fit back together perfectly. Apparently, there is a slight flexing of the sample after breakage. Such pieces may also luminesce when rubbed together in the dark. Other cherts have splinterey fracture and this usually indicates low-grade metamorphic recrystallization. Metamorphism can transform the purer cherts into interlocking grains of coarse quartz that are among the hardest known rocks. In thin section, cherts display a large range of petrographic textures that relate to the manner in which they formed and their subsequent burial and metamorphic history (Hesse, 1990; Carozzi, 1993)

While samples of chert from the geologic column may appear somewhat similar with regard to lithology and petrographic texture, the oxygen isotope variation in them is larger than for any other group of terrestrial rocks (Knauth, 1992). The $^{18}O/^{16}O$ ratio is determined by the temperature of quartz precipitation and by the $^{18}O/^{16}O$ ratio of the pore fluid. Because there are 2 variables, there has been much controversy over the origin of the enormous isotopic variation encountered in these rocks. One implication is that nodular cherts formed most abundantly in coastal environments where the transformation of primary metastable grains to stable crystals occurred in meteoric ground waters mixed with marine pore fluids. Archean examples are all enormously depleted in ^{18}O relative to Phanerozoic examples, and it has been argued that this indicates that the temperature of stabilization of initially deposited precipitates (and thus the climatic temperature) was up to 40°C warmer in the Archean (Knauth and Lowe, 2003).

Replacement cherts are found in numerous Precambrian carbonates where they can display well-preserved microfossils. Fine laminations in many Precambrian limestones and dolostones are commonly defined by thin siliceous laminae composed of microquartz, clay, and detrital quartz. These are often classified as "stromatolitic" layers formed by trapping on biologic mats or precipitation of metastable carbonate grains which are subsequently stabilized as limestone, dolostone, or chert. The chert can form classic nodules, but more typically is arrayed as thin discontinuous bands, blebs, and coarse laminae (Figure S26). Black, dull-luster cherts in Precambrian stromatolites have provided much, if not most, of the Precambrian organic-walled microfossils that record the paleontologic record of early life. The silica source for these cherts remains problematic. Ocean water itself almost certainly had much higher dissolved silica in the absence of abundant silica secreting organisms to remove it and also due, possibly, to higher saturation values from much warmer climatic temperatures. It is also possible that some varieties of Precambrian bacteria precipitated silica (and carbonate) directly. In any case, pre-compaction silica diagenesis was extremely active, and distinct zones and laminae were often replaced with microquartz before bacteria could degrade or decompose.

Although nodular replacement cherts in carbonates are common in rocks older than mid-Tertiary, they have never been found forming anywhere today. The "disappearance" of the process of nodular chert formation is roughly coincident with the explosive radiation of diatoms in the mid-Tertiary and with the apparent reduction of silica sponge populations in shallow waters. Silica precipitation by diatoms is so prolific today that dissolved silica in surface and shallow waters is usually <5 ppm. Opaline sponge spicule debris is still produced in shallow waters in relatively minor amounts, but the spicules appear to quickly re-dissolve in the undersaturated, shallow seawater and do not contribute significant dissolved silica for diagenetic pore fluids. It seems that the advent of diatoms so reduced dissolved silica concentrations in shallow waters that opaline sponge spicules can no longer accumulate in shallow water carbonate depositional environments. The inability to observe cherts forming today means that all knowledge of replacement cherts must presently be based on inferences deduced from investigations of the rock record.

In addition to the nodular replacement cherts and the bedded cherts, there are numerous other examples of silica accumulations and replacements that occur during early diagenesis in sediments of all types. These included silicified fossils of everything from bacteria to trees, interbedded layers of chert and iron oxide (iron formations), silicified evaporites, silicified volcaniclastic materials, and vein/cavity in-fills in all types of sedimentary and volcanic rocks. The processes and timing of silica diagenesis in many of these examples remain poorly understood.

Silicified evaporites

Soluble evaporite minerals such as gypsum, anhydrite, and halite often occur in the supratidal facies of limestones and dolostones. These phases are the first to dissolve during early diagenesis when coastal ground waters containing a meteoric water component percolate through the original marine precipitates. Vugs and molds are thus encountered by silica-saturated percolating groundwaters and can become filled with quartz, fibrous quartz, and/or microquartz. Small, cm-sized cauliflower-shaped nodules of chert that mimic the original morphology of gypsum/anhydrite masses are thus common in supratidal carbonates (Chowns and Elkins, 1974). The surrounding carbonate groundmass is subsequently dolomitized, but the distinctively shaped chert nodules provide instantaneous recognition of a probable supratidal depositional environment. Even in cases where chert nodules also replace the areas immediately surrounding evaporite crystals, coarse quartz in-fills of dove-tailed twins of gypsum or even halite cubes can be observed in petrographic thin-section. These examples of silicification provide powerful indicators of the supratidal environment in carbonate rocks so old that all traces of the primary evaporitic minerals have been removed by later dissolution.

Magadi cherts

The youngest cherts are probably those that formed in and around Pleistocene lakes in the rift valleys of East Africa. There, hydrous Na-silicates such as magadiite are precipitated in varve-like layers in alkaline lakes which undergo seasonal evaporation. Grid-work arrays of submicroscopic quartz crystals apparently replace the Na-silicate phases prior to compaction (Schubel and Simonson, 1990). Oxygen isotope studies suggest that the diagenetic fluids had become evaporatively enriched in dissolved silica. Silicification of trona and other lacustrine evaporite minerals are also evident. Several Tertiary examples of cherts that formed in this manner have been identified (Sheppard and Gude, 1986).

Precambrian cherts

Bedded cherts and silicified strata of all kinds are widespread in Archean Greenstone Belts. Silica was apparently vastly more mobile in these depositional environments and early silicification of sediments appears to have occurred on a scale not observed since. A possible explanation is that dissolved silica was fed into the deep oceans by hydrothermal vents on a vast scale and that silicification in the vent areas was intense. However, massive silicification does not appear to occur around modern examples and modern deep sea hydrothermal silica deposits are limited to local precipitation of opaline particles ("white smokers") induced by the sudden encounter of silica-saturated hydrothermal water with cold ocean water. It is possible that the Archean climates were dramatically warmer and that the Archean oceans therefore had dramatically higher silica solubility. Silicification in hydrothermal cells is thereby facilitated as more silica-rich ocean water is circulated through the adjacent sediment pile. More extensive deposits of opaline flocs and particles around hydrothermal vents may have been widespread, and these would have converted to quartz chert during early diagenesis.

Also abundant in the Archean as well as in the early Paleoproterozoic are the famous and enigmatic "Banded Iron Formations" of interbedded chert and iron oxide. Peaking in abundance at prior to 2.0 Ga, these also formed on stable platforms in water shallow enough for stromatolites produced by photosynthetic cyanobacteria to occasionally accumulate. The iron in these deposits was apparently injected as Fe^{+2} into deep anoxic ocean water from far-away hydrothermal vents. The dissolved Fe^{+2} was oxidized in surface waters and/or in shallow waters oxygenated by agitation or by cyanobacteria. It then accumulated either in pulses of deposition or else steadily amidst silica particles that were accumulating in such pulses because the deposits are beautifully banded with alternating layers of iron oxide (or iron carbonate) and chert. The nature of the original iron and silica precipitates is unknown, but both were likely hydrous, amorphous phases which underwent early diagenetic transformation to crystalline phases. The demise of iron formation deposition has been widely attributed to gradual oxygenation of the ocean, but this does not explain the disappearance of the associated silica deposition. One possibility is that climatic temperatures also declined with time which greatly reduced the amount of dissolved silica transported in the oceans. Numerous other types of silica-replaced sediments occur in the Archean. The silicification is often so extensive that the nature of the replaced sediments can be inferred only by study of petrographic fabrics and preserved sedimentary structures.

Evolution of siliceous sediments with time

For at least the past 540 million years, siliceous sediments have formed primarily by the deposition of biologically secreted opal particles that dissolve and reprecipitate as quartz beds and nodules during the compaction history of the sediments. Prior to this, there is no current fossil evidence that the sponges, radiolaria, and diatoms were present to supply biogenic silica to sediments. Cherts and silicified sediments are nevertheless widespread in Precambrian strata. In the Archean, it is possible that much higher ocean temperatures allowed greatly enhanced transport of dissolved silica and that large areas of the ocean floor were blanketed with opaline flocs derived from intense weathering and from deep-ocean hydrothermal vents. It is also possible that bacterial precipitation of opal was widespread and produced equivalent amounts of silica as that produced by the Phanerozoic organisms. However introduced into sediments, silica diagenesis in the Archean must have been particulary vigorous to produce such extensive occurrences of bedded cherts and silicified sediments.

In the late Archean and early Proterozoic, interbedded cherts and iron oxides became common. Inasmuch as the iron is thought to have been exhaled from submarine hydrothermal vents, it is likely that much of the silica was also. However, the role of organisms and the mechanisms that produced rhymthic deposition of iron oxide and silica are still unknown. Large-scale siliceous sedimentation declined during the Proterozoic, but replacement cherts remained common in stromatolitic and other laminated carbonates throughout the Precambrian. Carbonates and associated evaporites were apparently replaced with microcrystalline quartz during early diagenetic stabilization via the same processes that produced Phanerozoic nodular cherts, but the silica source remains problematic. The amount of silica introduced into the Precambrian carbonates appears no different from the younger cases.

The evolution of sponges and radiolarians in the earliest Paleozoic was coincident with the evolution of prodigeous numbers of carbonate-secreting organisms. Limestones and dolostones throughout the Paleozoic and Mesozoic are filled with nodular and disseminated replacement chert that resulted when opaline sponge spicules dissolved and reprecipitated during early diagenetic stabilization. Thick radiolarian oozes commonly accumulated in the deep sea and these were transformed into great bedded chert sequences now displayed in deformed and uplifted accretionary wedges.

Silica secreting diatoms evolved within the last 100 million years and completely transformed the nature of siliceous sediments. These prolific photosynthesizers now dominate silica geochemistry in the ocean, lakes, rivers, streams, and soils. Silica is so efficiently extracted from water by these organisms that dissolved silica concentrations in surface seawater and virtually all fresh water environments are kept well below saturation for quartz and all forms of opal. Silica mobilization and re-precipitation in pore fluids has thus been arrested in sediments except for deep-sea diatomaceous oozes. Any biologic opal introduced into shallow marine carbonate sediments simply dissolves back into the sea; nodular replacement cherts have essentially stopped forming. Siliceous sedimentation is now mainly a case of diatom accumulation. The geologic history of siliceous sediments has thus undergone significant changes that provide another example of the profound influence of biologic evolution on the evolution of sedimentary rocks.

L. Paul Knauth

Bibliography

Behl, R.J., and Garrison, R.E., 1994. The origin of chert in the Monterey Formation of California (USA). In Iijima, A., Abed, A.M., and Garrison, R.E. (eds.), *Siliceous, Phosphatic and Glauconitic Sediments of the Tertiary and Mesozoic. Proceedings of the 29th International Geological Congress.* Part C. pp. 101–132, Utrecht: VSP.

Biggs, D.L., 1957. Petrography and origin of Illinois nodular cherts. *Illinois State Geological Survey Circular*. 245.

Bramlett, M.N., 1946. The Monterey Formation of California and the origin of its siliceous rocks. United States Geological Survey Professional Paper 212.

Carozzi, A.V., 1993. *Sedimentary Petrography*. Prentice Hall.

Chowns, T.M., and Elkins, J.E., 1974. The origin of quartz geodes and cauliflower cherts through silicification of anhydrite nodules. *Journal of Sedimentary Petrology*, **44**: 885–903.

Folk, R.L., and McBride, E.F., 1976. The Caballos Novaculite revisited Part I: Origin of novaculite members. *Journal of Sedimentary Petrology*, **46**: 659–669.

Grunan, H.R., 1965. Radiolarian rocks and associated rocks in space and time. *Eclogae Geological Helvetiae*, **58**: 157–208.

Hesse, R., 1990. Origin of chert and silica diagenesis. In McIlreath, I.A., Morrow, D.W. (eds.), *Diagenesis*. Geological Association of Canada, Geoscience Canada Reprint Series 4, pp. 227–275.

Knauth, L.P., 1979. A model for the origin of chert in limestone. *Geology*, **7**: 274–277.

Knauth, L.P., 1992. Origin and diagenesis of cherts: an isotopic perspective. In *Isotopic Signatures and Sedimentary Records*. Lecture Notes in Earth Sciences, 43, New York: Springer-Verlag, pp. 123–152.

Knauth, L.P., 1994. Petrogenesis of chert. *Reviews of Mineralogy*, **29**: 233–258.

Knauth, L.P., and Lowe, D.R., 2003. High Archean climatic temperature inferred from oxygen isotope geochemistry of cherts in the 3.5 Ga Swaziland Supergroup, South Africa.. *Geological Society of America Bulletin*, **115** no. 2 (in press).

Mackenzie, F.T., and Gees, R., 1971. Quartz: synthesis at earth-surface conditions. *Science*, **173**: 533–535.

Miser, H.D., and Purdue, A.H., 1929. Geology of the Dequeen and Caddo Gap Quadrangles, Arkansas. *US Geological Survey Bulletin*, 808.

Murray, R.W., Jones, D.L., and Bucholtz ten Brink, M.R., 1992. Diagenetic formation of bedded chert: evidence from chemistry of the chert-shale couplet. *Geology*, **20**: 271–274.

Schubel, K.A., and Simonson, B.M., 1990. Petrography and diagenesis of cherts from Lake Magadi, Kenya. *Journal of Sedimentary Petrology*, **60**: 761–776.

Sheppard, R.A., and Gude, A.J., 1986. Magadi-type chert—a distinctive diagenetic variety from lacustrine deposits. *United States Geological Survey Bulletin*, **1578**: 335–345.

Tarr, W.A., 1917. Origin of the chert in the Burlington Limestone. *American Journal of Science*, **44**: 409–452.

Van Tuyl, F.M., 1918. The origin of chert. *American Journal of Science*, **45**: 449–456.

Cross-references

Diagenesis
Diagenetic Structures
Geodes
Ironstones and Iron Formations
Isotopic Methods in Sedimentology
Magadiite
Neomorphism and Recrystallization
Silcrete

SLIDE AND SLUMP STRUCTURES

Slide and slump structures are common geological phenomena on subaerial and subaqueous slopes (Maltman, 1994). They range in size and volume from a few tens of centimenters to several thousand cubic kilometers (Jansen *et al.*, 1987; Martinsen, 1994), and encompass gliding unconsolidated sediments as well as lithified blocks (olisotliths). Slides and slumps are significant geological processes which have an important impact on human life, known from recent catastrophies (Maltman, 1994, and references therein). In addition, sliding and slumping can seriously affect the stability of offshore installations (Prior and Coleman, 1982), and generate tsunamis (Driscoll *et al.*, 2000). Products of slides and slumps provide important information on the depositional setting of a sedimentary rock unit.

Basic characteristics and theory

Slumps and slides are downslope movements of sediments above a basal shear surface where there is, respectively, significant and insignificant internal distortion of the bedding (Stow, 1986). In slumps, the bedding should be recognizable, otherwise they classify as debris flows. There is a continuous transition between slides, slumps and debris flows, and some units may show characteristics of all three modes of transport depending on the internal deformation state (Martinsen, 1994).

Slumping and sliding are common especially where there are significant fine-grained sediments. The structures form above a basal shear surface (Figures S29 and S30), the depth to which is decided mainly by the pressure gradient in the sediment. Where the pore pressure approaches or balances the normal stress induced by the weight of the overburden, the shear strength is sufficiently reduced to allow slippage along the basal shear surface, given a sufficiently high shear stress. These relationships are given by the equations:

Shear strength $(\tau) = C + (\sigma - p)\tan\varphi$ (Eq. 1; Hampton, 1979)

and

Shear stress $(S) = \rho g s Y \tan\alpha$ (Eq. 2; Middleton and Southard, 1978)

where C is sediment cohesion, σ is normal stress (or weight of overburden), p is pore pressure, φ is the angle of internal friction, ρ is sediment density, g is acceleration due to gravity, s is solidity (or the complementary of porosity), Y is sediment thickness and α is slope angle.

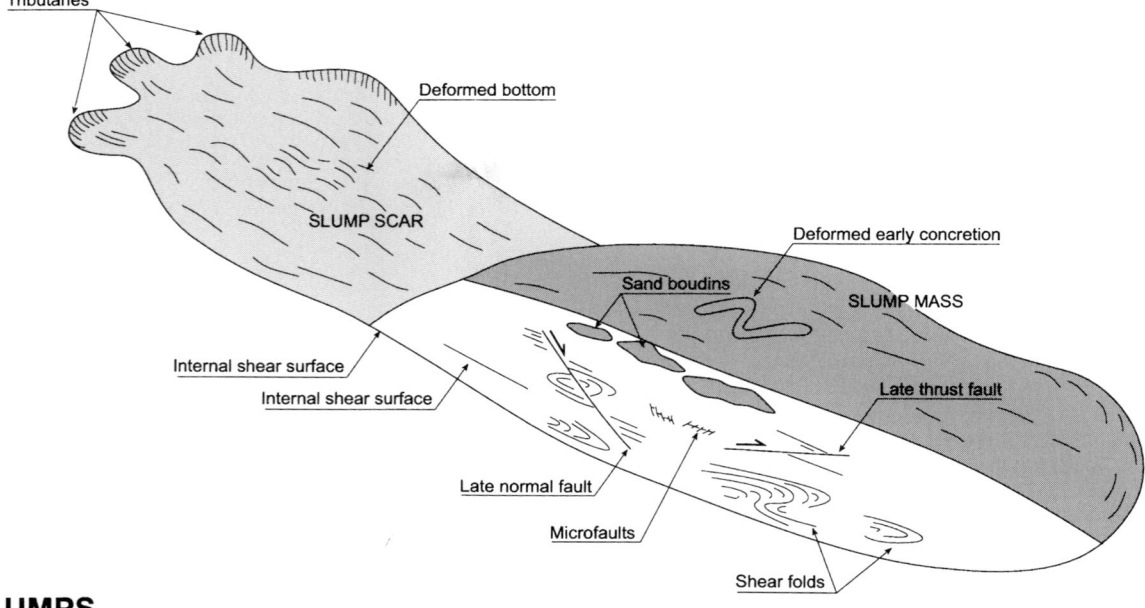

Figure S29 Idealized slump model. Modified from Martinsen (1994).

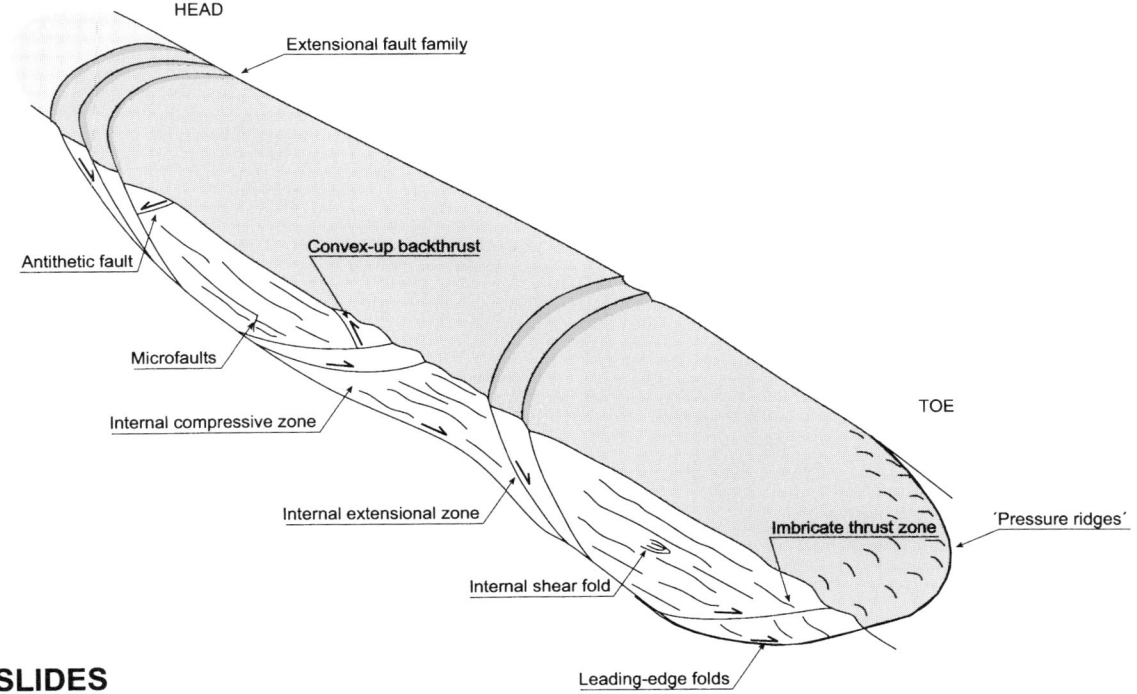

SLIDES

Figure S30 Idealized slide model. Modified from Martinsen (1994).

Once initiated, the slide or slump basal shear surface will propagate upslope in a radial fashion from its nucleation point (Williams and Chapman, 1983), leading to the formation of a scoop-shaped, concave-downslope depression or scar, often with an irregular outline with tributaries (Martinsen, 1994). The shear surface is probably initiated as a slope-parallel feature, but at some point steepens to intersect the sediment surface. The transition may occur at lithofacies boundaries, or at sites of pore pressure jumps where material strength contrasts are present (Crans *et al.*, 1980).

Slumps and slides are triggered by a variety of processes. Seismic triggering, cyclic wave-loading, high sedimentation rates and methane generation causing overpressuring, slope oversteepening, and tectonic movements are all possible initiators (Martinsen, 1994).

Slumps and slides form on low-gradient slopes, as little as 0.1° (Prior and Coleman, 1978). They range in thickness from 0.5 m (Martinsen, 1994), to several hundreds of meters on continental slopes (Jansen *et al.*, 1987).

Slump structures

Moving slumps deform intensely internally and produce a wide variety of deformational structures. Folds, boudins, microfaults, internal shear surfaces and faults are common structures in slumps (Figure S29). These structures suggest that the slumps go through a main phase of plastic/ductile deformation whereby folds and boudins are formed. The ductile phase is followed by a late brittle phase where the faults form. It is common to see strain overprinting where early-formed folds are truncated by late faults (Figure S31).

Figure S31 Slump unit from the Upper Carboniferous Gull Island Formation, County Clare, Ireland. Note the internal distorted bedding of the unit in the near cliff, which is approximately 15 m high.

The slump folds are mainly sheath folds formed by simple shear (Figure S31; Martinsen, 1994), although buckle folds also occur (Woodcock, 1976). Faults can be extensional and contractional (Figure S29), related to local obstacles or shear surface irregularities. Models of slumps and slides (e.g., Allen, 1985) show a well-defined upper extensional zone and a downslope contractional zone. However, it is likely that a *lateral* compaction will occur when fine-grained slumps come to a halt, preventing development of toe contractional zones, leaving *open-ended zones* (Garfunkel, 1984).

Slump scars often show evidence of retrogressive failure, from footwall unloading (Martinsen, 1994). The slump scars are generally filled by fine-grained sediment, or expanded by current erosion and filled by coarser-grained turbidity current deposits. In this case, it may be impossible to differentiate the slump scar from an erosional channel.

Slumps form hummocky sediment masses in plan view, particularly at their downslope ends if toe-zones are developed. The topography may be organized into lobate, convex-downslope ridges (Prior et al., 1984) which are the surface manifestation of toe-zone thrust faults (Martinsen and Bakken, 1990).

Slide structures

Slides include bank collapse in river channels, subaerial mudslides, sediments displaced by delta-front growth faults, shelf-edge faults and submarine glide-blocks and olistoliths. Slides are either *rotational* or *translational* (Allen, 1985), depending on the thickness to movement distance ratio.

In their most simple form, slides are spoon-shaped features with a three-partite morphology of an extensional head region, a middle, translational, "rigid" zone and downslope, contractional toe zone (Figure S30; Brunsden, 1984). In rigid zones, there may internal slip between beds in the form of sheath folds or microfaults (Figure S31). The downslope toe region can be "open-ended" with no contractional structures developed because of lateral compaction (Garfunkel, 1984; see also above).

The head region usually develops a "family" of listric faults (Crans et al., 1980) which soles out at a basal decollement. Antithetic extensional faults are common. Fault families in slides is an essential difference from slumps, where most faults are single faults. The margins of the slides are dominated by strike-slip deformation, but the width variability of the slide scar can cause both transpressional and transtensional movement if the slide scar narrows or widens.

Contractional deformation is most commonly expressed as thrust faults. The thrusts form classic duplex and imbricate zone geometries (Figure S31; Dingle, 1977). Three different types of contractional zones occur in slides (Martinsen and Bakken, 1990): (1) The thick package zones; similar to mountain belt thrust zones, but several orders of magnitude smaller; (2) Thin package zones developed at the frontal margin of heterogeneous slides; (3) Basal contractional zones related to obstacles such as ramps.

Toe zones of slides may be confused with contractional zones formed by *gravity spreading*. Gravity spreading forms in progradational settings (e.g., deltas) due to progressive sediment loading (Schack Pedersen, 1987).

Ole J. Martinsen

Bibliography

Allen, J.R.L., 1985. *Principles of Physical Sedimentology*. London: George Allen and Unwin.
Brunsden, D., 1984. Mudslides. In Brunsden, D., and Prior, D.B. (eds.), *Slope Instability*. Chichester: John Wiley and Sons, pp. 363–418.
Crans, W., Mandl, G., and Haremboure, J., 1980. On the theory of growth-faulting—a geomechanical delta model based on gravity sliding. *Journal of Petroleum Geology*, **2**: 265–307.
Dingle, R.V., 1977. The anatomy of a large submarine slump on a sheared continental margin (SE Africa). *Journal of the Geological Society, London*, **134**: 293–310.
Driscoll, N.W., Weissel, J.K., and Goff, J.A., 2000. Potential for large-scale submarine slope failure and tsunami generation along the US mid-Atlantic coast. *Geology*, **28**: 407–410.
Garfunkel, Z., 1984. Large-scale submarine rotational slumps and growth faults in the eastern mediterranean. *Marine Geology*, **55**: 305–324.
Hampton, M.A., 1979. Buoyancy in debris flows. *Journal of Sedimentary Petrology*, **49**: 753–758.
Jansen, E., Befring, S., Bugge, T., Eidvin, T., Holtedahl, H., and Sejrup, H.P., 1987. Large submarine slides on the Norwegian continental margin: sediments, transport and timing. *Marine Geology*, **78**: 77–107.
Maltman, A., 1994. *The Geological Deformation of Sediments*. London: Chapman and Hall.
Martinsen, O.J., 1994. Mass movements. In Maltman, A. (ed.), *The Geological Deformation of Sediments*. London: Chapman and Hall, pp. 127–165.
Martinsen, O.J., and Bakken, B., 1990. Extensional and compressional zones in slumps and slides in the Namurian of County Clare, Ireland. *Journal of the Geological Society, London*, **147**: 153–164.
Middleton, G.V., and Southard, J.B., 1978. *Mechanics of Sediment Movement*. Tulsa: SEPM Short Course, 3.
Prior, D.B., Bornhold, B.D., and Johns, M.W., 1984. Depositional characteristics of a submarine debris flow. *Journal of Geology*, **92**: 707–727.
Prior, D.B., and Coleman, J.M., 1978. Disintegrating retrogressive landslides on very low-angle subaqueous slopes, Mississippi delta. *Marine Geotechnology*, **3**: 37–60.
Prior, D.B., and Coleman, J.M., 1982. Active slides and flows in underconsolidated marine sediments on the slopes of the Mississippi delta. In Saxov, S., and Nieuwenhuis, J.K. (eds.), *Marine Slides and Other Mass Movements*. NATO Workshop, Plenum Press, pp. 21–49.
Schack Pedersen, S.A., 1987. Comparative studies of gravity tectonics in quaternary sediments and sedimentary rocks related to fold belts. In Jones, M.E., and Preston, R.M.F. (eds.), *Deformation of Sediments and Sedimentary Rocks*. Geological Society of London, Special Publication, Volume 29, pp. 165–180.
Stow, D.A.V., 1986. Deep clastic seas. In Reading, H.G. (ed.), *Sedimentary Environments and Facies*, 2nd edn, Blackwell, pp. 399–444.
Williams, G.D., and Chapman, P., 1983. Strains developed in the hangingwalls of thrusts due to their slip/propagation rate: a dislocation model. *Journal of Structural Geology*, **5**: 563–572.
Woodcock, N.H., 1976. Structural style in slump sheets: Ludlow series, Powys, Wales. *Journal of the Geological Society, London*, **132**: 399–415.

Cross-references

Bedding and Internal Structures
Debris Flow
Gravity-Driven Mass Flows
Rivers and Alluvial Fans
Slope Sediments
Storm Deposits
Submarine Fans and Channels
Synsedimentary Structures and Growth Faults
Tectonic Controls of Sedimentation
Tsunami Deposits

SLOPE SEDIMENTS

The *slope* is a relatively steep transition zone between a low-gradient marine shelf (<1° gradient) and the deep seafloor. Mostly, slopes occupy the *bathyal zone* of the ocean basins (200–2000 m). Benthic macrofauna are rare, including thin-shelved bivalves, brachiopods, and echinoderms. Benthic foraminifera, where present, are useful in providing water-depth estimates along ancient slopes. Many slopes are located at the edges of continents; others fringe volcanic islands or intracratonic basins on continental crust. Around the passive margins of the Atlantic Ocean, the *continental slope* extends from the *shelf–slope break* (~120 m average water depth) to the top of the *continental rise* (~2,400 m average depth; Emery and Uchupi, 1984). Along subduction margins, the base of the slope is much deeper at the landward edge of the trench. The depth of the *shelf–slope break* varies widely between 20–500 m (Fairbridge and Bourgeois, 1978), and may be within the photic zone along carbonate platform margins where reefs grow into the surf zone immediately landward of a precipitous upper-slope escarpment.

Basin slopes are distinguished from adjacent shelves by a steeper gradient and by a lack of significant wave and tidal-current action. Instead, *sediment gravity flows* predominate, sometimes associated with *thermohaline currents* which sweep the distal parts of some slopes. Gradients range from 3–6° or less on mud-dominated passive-margin slopes (Heezen, 1959), to 10–30° on erosional slopes flanking strike-slip basins like those along the North Anatolian Transform Fault of northwestern Turkey (Aksu *et al.*, 2000), to local near-vertical escarpments seaward of variably cemented carbonate barrier reefs. The prodelta slopes of marine coarse-grained deltas commonly include sands and gravels transported by gravity currents and slope failures (Nemec, 1990).

Cementation, organic binding and reef development at carbonate platform margins may lead to the development of a very steep upper-slope segment (Coniglio and Dix, 1992). Carbonate slopes of this type are strongly concave in shape because the steep upper slope segment gives way in deeper water to a much gentler gradient. Some carbonate slopes can be steepened by erosion, which may cause the upper slope to retreat tens of kilometers (Jansa, 1981; Mullins *et al.*, 1986). Sandy carbonate sediments form steeper slopes than carbonate muds (30–40° versus <5°; Kenter, 1990).

A fraction of the sediment that moves across the shelf–slope break accumulates on the slope itself, leading to aggradation and progradation, or partial filling of slope basins along convergent continental margins. A greater portion, however, bypasses the slope to deeper water areas like the continental rise and abyssal plain (passive margins) or oceanic trenches (convergent margins). Seismic data provide considerable information on continental slopes. Various patterns of reflections, designated as *seismic facies*, have been recognized (e.g., Sangree *et al.*, 1978). These include: (1) mounded chaotic facies, the result of slumps, slides and other mass movements; (2) onlapping fill facies, due to infilling of slope irregularities by turbidity currents; (3) divergent and parallel-layered facies, a result of deposition from low-energy turbidity currents and hemipelagic sedimentation; and (4) sheet drape facies, resulting from blanketing of slope irregularities by hemipelagic sediments. Hemipelagic sediments accumulate sufficiently slowly that they are usually thoroughly bioturbated by representatives of the *Zoophycos* trace-fossil assemblage. Gravity-flow deposits and sediment slides are emplaced rapidly, so preserve physical structures (e.g., lamination formed by traction transport; deformation structures formed by shearing).

The gradients and morphologies of siliciclastic slopes are strongly controlled by tectonics, sediment supply, and salt or mud diapirism. On passive margins, gradients are low and the seabed may be quite featureless. Exceptions are erosional gullies and submarine canyons in the upper slope that funnel shelf-derived sediment seaward, and slide scars and mound-shaped slide and slump deposits that characterize some slopes. In high latitudes, where continental ice sheets recently delivered sediment directly to the upper slope, gullies are replaced by widely separated, glacially cut transverse troughs, and the slope tends to be smooth, with or without down-to-the-basin listric faults (Aksu and Hiscott, 1989, 1992). During interglacial times, these glacially fed slopes may become extensively dissected by headward erosion of gullies and canyons (Hesse and Klaucke, 1995).

The slopes along convergent plate boundaries lead from a narrow shelf, across an accretionary prism, to the floor of a deep-sea trench. Thrust faults and local mud diapirs in the accretionary prism produce a complex slope morphology (Figure S32) with intraslope basins that are elongated parallel to the plate boundary (Pickering *et al.*, 1995; Moore *et al.*, 2001). Turbidity currents travelling down the slope are diverted into these intraslope basins where they can become ponded.

On basin slopes seaward of major river deltas (e.g., Gulf of Mexico, Niger Delta area), loading of water-rich muds or deeply buried salt by a thick delta and slope succession leads to complex diapiric intrusions, thrust faulting, and down-to-the-basin growth faulting which can create a rugged seafloor characterized by intraslope basins and hills (e.g., Sigsbee Knolls on the Texas–Louisiana slope; Satterfield and Behrens, 1990).

Sedimentary processes and typical deposits

Particulate sediments are transported beyond the shelf–slope break by three main mechanisms: (1) bottom-hugging sediment gravity flows; (2) thermohaline bottom currents; and (3) surface wind-driven currents or river plumes that carry suspended sediment off the shelf. Tidal currents and surface waves are only locally important as transport agents on the upper parts of slopes and in the heads of some submarine canyons (Shepard *et al.*, 1979). Sediment gravity flows that develop on basin-margin slopes include *turbidity currents*, *liquefied flows*, and *debris flows*.

Where shelves are narrow (e.g., offshore California), plumes of suspended sediment from river deltas can extend beyond the

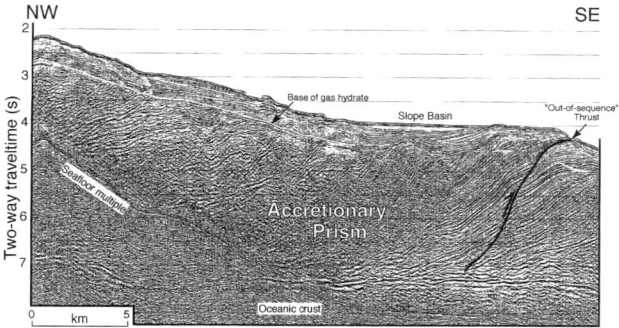

Figure S32 Multichannel seismic dip profile midway down the slope of the Nankai accretionary prism near longitude 134.5E, south of Japan (water depth ~3,000 m). A turbidite-filled *slope basin* has been created atop the stacked thrust sheets of the prism. The white line marks a bottom-simulating reflector, attributed to the presence of gas hydrates. Seismic profile discussed in Moore *et al.* (2001). Reproduced by permission of G.F. Moore.

shelf–slope break (Emery and Milliman, 1978; Thornton, 1981, 1984), directly contributing fine-grained sediments to the slope. In polar areas, sediment-laden spring meltwater may actually flow from the land across the surface of floating sea ice, to deposit its load directly onto the continental slope (Reimnitz and Bruder, 1972).

Mud that leaves the shelf either by high-level escape in river plumes, by dilute turbidity-current flow, or by movement along density interfaces in the water column over the slope eventually settles to the seafloor to form *hemipelagic deposits*. Deposition rates are on the order of 10–60 cm/1000 years (Krissek, 1984; Nelson and Stanley, 1984; Coniglio and Dix, 1992). Hemipelagic sediments contain a pelagic component (biogenic skeletons, volcanic and windblown dust, ice-rafted detritus), generally 5–50 percent but locally as much as 75 percent, with the remainder consisting of terrigenous mud supplied from the adjacent land mass. The finest particles are probably carried to the bottom as aggregates in the form of faecal pellets (Calvert, 1966; Schrader, 1971; Dunbar and Berger, 1981). Unlike turbidites, which fill low areas on the slope (Figure S32), hemipelagic deposits form even-thickness drapes (Figure S33).

Dilute low-concentration turbidity currents tend to travel at relatively low velocities, are potentially many tens of meters thick, and may be turned by the Coriolis Effect to flow oblique or parallel with the bottom contours before reaching the base of the slope (Hill, 1984). On carbonate slopes, the fine-grained material exported from the platform is referred to as *periplatform sediment* (Schlager and James, 1978; Coniglio and Dix, 1992). Suspended sediment concentrations in shelf areas may be quite high due to input of mud-laden river water, or stirring of the bottom by strong waves, tidal currents, or internal waves at density interfaces (Cacchione and Southard, 1974). This suspended sediment may be advected off the shelf by ambient currents, possibly wind-driven, or by transport in cascading cold water that may flow off the shelf in the winter months (McCave, 1972; McGrail and Carnes, 1983; Wilson and Roberts, 1995). Suspensions of fine-grained sediment may also leave the shelf as dilute turbidity currents (lutite flows),

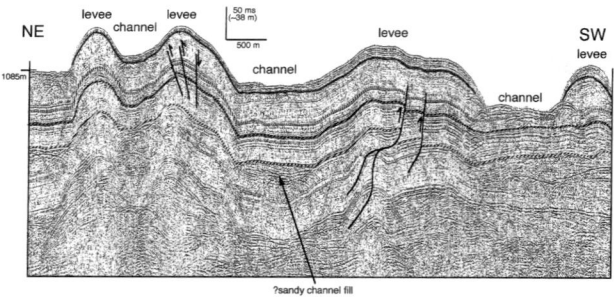

Figure S33 Single-channel seismic strike profile across the middle slope offshore Brunei, Borneo (Hiscott, in press). Water depth is ~1,000 m. The slope gradient, toward the reader, is ~2°. A number of leveed channels have grown vertically during the Quaternary, principally through fallout of suspended load from the Baram Delta to form an even-thickness hemipelagic drape. The channel to the SW has been most recently active—levees have enhanced sediment thicknesses and the floor of the channel appears to have a central thalweg channel flanked by low, possibly sandy inner levees (cf. Piper et al., 1999). The patterned lines mark reflections that can be widely traced across the Brunei shelf and upper slope.

moving along the bottom onto the lower slope and rise, or along density interfaces in the ocean water. These dilute suspensions may move down a smooth upper slope as unconfined sheet flows, or may be confined by gullies or canyons in which suspension is assisted by weak tidally-forced flows (Shepard et al., 1979; Gorsline et al., 1984). Most mud transported off the shelf by dilute turbidity currents (lutite flows) bypasses the slope, leading to maximum rates of accumulation on the continental rise (Nelson and Stanley, 1984). In carbonate settings, platform-derived sediment can be traced >100 km from the shelf edge (Heath and Mullins, 1984).

High-concentration turbidity currents carrying significant sand load are confined to gullies and larger canyons on the slope, permitting the sediment to *bypass* this area to accumulate in deeper water (base-of-slope fans, or on the abyssal plain). Komar (1971) argued that turbidity currents are likely to be *supercritical* (Froude number >1.0) on slopes greater than ~0.5°. Such turbidity currents can effectively carry sand-sized detritus across basin-margin slopes without leaving a deposit: there may even be net erosion in these areas. For example, Weaver and Thomson (1993) demonstrated slope bypassing by large mud-rich turbidity currents entering the Madeira Abyssal Plain by examining their reworked coccolith assemblages. Such bypassing requires that the turbidity current be at the least self-sustaining on the slope, or while flowing through slope channels. This condition of *self maintenance* (Southard and Mackintosh, 1981) was named *autosuspension* by Bagnold (1962).

Gully margins and adjacent areas of the upper slope may be sites of sediment failures as coherent slides and internally folded and deformed slumps. The downslope component of gravity can cause sediment masses previously deposited on a slope to move by increments or during a single episode into deeper water. Slow downslope movement without slip along a single detachment surface, i.e., without failure, is referred to as *creep*. No structures in ancient successions have been unambiguously attributed to creep, although features in seismic profiles have been interpreted to have formed by this mechanism (Hill et al., 1982; Mulder and Cochonat, 1996). More rapid downslope movements generate sediment slides and debris flows. Sediment slides can result in little-deformed to intensely folded, faulted and brecciated masses (Barnes and Lewis, 1991) that have translated downslope from the original site of deposition, leaving a headwall scarp and slide scar where the failure occurred. If the primary bedding is entirely destroyed by internal mixing, with soft muds and/or water mixed into the slide, then it may transform into a debris flow. The base of the slide mass is marked by an *intraformational truncation surface* and shear zones (Coniglio, 1986).

Initiation of slides in muddy terrigenous sediments depends on: (1) bottom slope (Moore, 1961); (2) sedimentation rates (Hein and Gorsline, 1981); and (3) the response of the sediment to cyclic stress produced by earthquake shaking (Morgenstern, 1967). Sedimentation rates on basin-margin slopes vary widely, but Hein and Gorsline (1981) conclude that a rate of 30 mg cm^{-2} per year must be attained before slope failures become common. For example, in the Santa Barbara Basin, California Borderland, sedimentation rates exceed 50 mg cm^{-2} per year, and debris flows are widespread on slopes of <1° (Hein and Gorsline, 1981). In many areas, sediment slides occur on very low slopes, suggesting the involvement of water-rich, rapidly deposited sediments.

The sedimentation rate effectively determines the water content and shear strength of the sediment, although shear strength is also a function of other variables such as content of organic matter (Keller, 1982) and generation of gas in the sediment by decay of organics or by decomposition of gas hydrates (Carpenter, 1981). According to Keller (1982, p.197), "cohesive sediments with greater than about 4–5 percent organic carbon [have]... (1) unusually high water content; (2) very high liquid and plastic limits; (3) unusually low wet bulk density; (4) high undisturbed shear strength; (5) high sensitivity; (6) high degrees of apparent overconsolidation; and (7) high potential for failure [by liquefaction] in situations of excess pore pressure".

Slides and debris-flow deposits (*debrites*) are common slope sediments. In carbonate settings, the individual blocks in debrites may be larger in diameter than 100 m (Hiscott and James, 1985; Hine *et al.*, 1991). Single flow deposits can be many tens of meters thick (Figure S34), and taper to convex-upward *snouts* at their downslope and lateral margins (Hiscott and James, 1985; Hiscott and Aksu, 1994). Shingled, down-slope-elongate debrites are common in high-latitude slopes (Aksu and Hiscott, 1992), where glaciers delivered large quantities of poorly sorted sediment directly to the upper slope during Quaternary glacial advances.

The deeper segments of slopes on the western sides of ocean basins are affected by geostrophic, thermohaline currents moving at mean speeds of 10–30 cm per second (McCave *et al.*, 1980; Hollister and McCave, 1984), with short "gusts" reaching about 70 cm per second (Richardson *et al.*, 1981). These currents follow bathymetric contours, and are known as *contour currents*. They carry a dilute suspended load, generally < 0.1–$0.2 \, \mathrm{g \, m^{-3}}$, that forms the thick bottom *nepheloid layer* (Ewing and Thorndike, 1965; Biscaye and Eittreim, 1977). Concentrations may briefly exceed $12 \, \mathrm{g \, m^{-3}}$ (Biscaye *et al.*, 1980, Gardner *et al.*, 1985). Most of the fine-grained suspended material is derived by winnowing of the seafloor; the rest is probably added to the current by cascades of cold shelf water or lutite flows originating at the shelf–slope break. The water depths to which thermohaline currents influence the seabed have changed from glacial to interglacial times in high-latitude oceans (Hillaire-Marcel *et al.*, 1994), in some cases shifting from the continental rise to the lower continental slope. It is reasonable to assume that similar excursions of the axes of contour currents occurred along ancient slopes and rises.

Contour current deposits, or *contourites* (Hollister and Heezen, 1972), can be treated as two end members: (1) muddy contourites, and (2) sandy contourites. Muddy contourites are fine grained, mainly homogeneous and structureless, thoroughly bioturbated, and only rarely show irregular layering, lamination and lensing. They are poorly sorted silt- and clay-size sediments with up to 15 percent sand. They range from finer grained homogeneous mud to coarser grained mottled silt and mud, and their composition is most commonly mixed biogenic and terrigenous grains. According to Hollister and McCave (1984), short-term depositional rates can be extremely high, about 17 cm per year, followed by rapid biogenic reworking by burrowers. It may be difficult to distinguish between mud turbidites and muddy contourites, although some criteria have been suggested (Bouma, 1972; Stow, 1979).

Sandy contourites comprise irregular layers <5 cm thick that are either structureless and thoroughly bioturbated, or that possess some primary parallel or cross lamination which may be accentuated by heavy-mineral or foraminiferal concentrations (Bouma and Hollister, 1973). Grading may be normal or inverse, and bed contacts may be sharp or gradational. Grain size ranges from coarse silt to, rarely, medium sand, with poor to moderate sorting. The sandy facies, which may be rich in biogenic sand grains, is produced

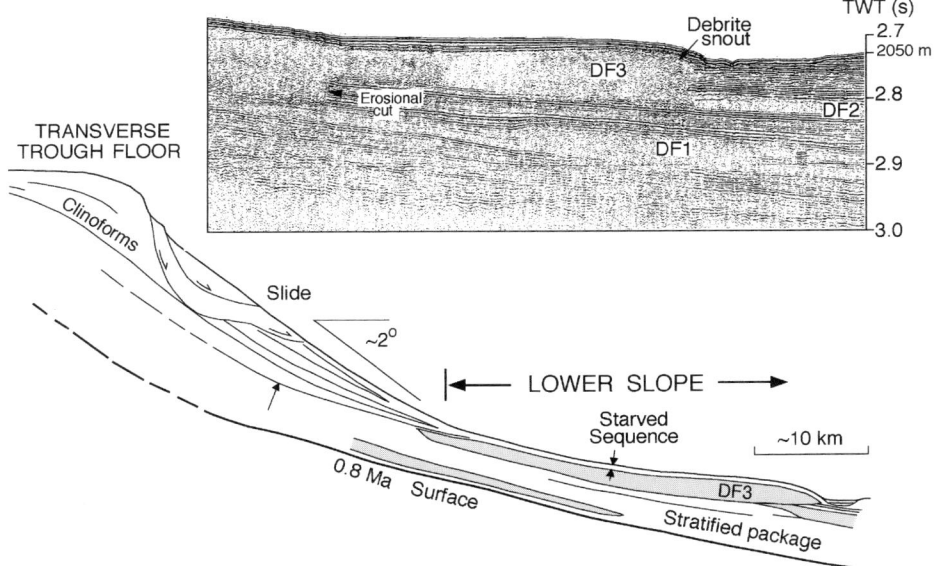

Figure S34 Single-channel seismic dip profile from the lower slope offshore Baffin Island, Canadian Arctic (modified from Hiscott and Aksu, 1994). The line drawing shows the relationship of the lower slope to the entire margin. Stratified hemipelagic deposits (including ice-rafted sediments) are punctuated by large-volume, mounded debris-flow tongues (DF1–DF3) derived from the upper slope when glaciers were depositing till and outwash at the shelf edge. The debrites have distal snouts that testify to the intrinsic strength of the viscous muddy debris.

by winnowing of fines by stronger flows (Driscoll *et al.*, 1985). Where currents are focused by seabed topography, as in the Florida Strait, coarse grained contourites may include lag gravels and cemented *hardgrounds* (Coniglio and Dix, 1992). Reworking of sandy turbidites by contour currents can result in bottom-current-modified turbidite sands, believed to be common on some continental slopes.

Muddy and sandy contourites commonly occur together in vertical upward coarsening to upward fining sequences (Faugères *et al.*, 1984; Stow and Piper, 1984). A complete sequence shows inverse grading from a fine homogeneous mud, through a mottled silt and mud, to a fine-grained sandy contourite facies, and then back to a muddy contourite. The changes in grain size, sedimentary structures and composition probably are related to long-term (1 kyr to 30 kyr) fluctuations in the mean current velocity (Stow and Piper, 1984).

Oxygenation and organic-matter deposition

In some slopes, particularly along the eastern sides of ocean basins (e.g., offshore west Africa, California, and Peru), the wind-driven surface circulation causes upwelling of nutrient-rich deep water, and enhanced biological productivity above the slope. The high biological productivity leads to organic carbon and phosphate deposition and, locally, accumulation of diatomaceous deposits. Oxygen is consumed to oxidize this material, and eventually sediment pore water and possibly even the marine bottom water become anoxic, creating an *oxygen minimum* along the slope. For example, along the eastern Pacific continental margin, an oxygen minimum layer occurs in intermediate water masses, commonly at depths of 250–1500 m (Ingle, 1981).

Two levels of oxygen depletion of bottom waters are recognized by Byers (1977): *disaerobic* conditions (0.1–1.0 ml O_2/L H_2O) under which only resistant infauna survive to bioturbate sediment; and *anaerobic* conditions (<0.1 ml O_2/L H_2O) where no benthic organisms live and the sediment remains undisturbed by burrowers, commonly being finely laminated. Where the sediment pore water is *anaerobic* but the overlying bottom water is oxygenated, some filter-feeding organisms may populate the seabed, but there will be no infauna and therefore little bioturbation.

Facies successions and architecture of slope deposits

There is consensus that slope successions do not consist of predictable, depositional cycles (Pickering *et al.*, 1989; Coniglio and Dix, 1992). Instead, the facies architecture is disordered. Stow (1986) identified differences in facies associations between muddy facies of the upper versus lower slope. The upper slope is marked by more abundant coarse grained gravity-flow deposits, slide scars, erosional gullies and channels; in contrast, the lower slope is characterized by more fine-grained turbidites, relatively rare submarine channel fills, debrites, slide and slump scars, and local interfingering with contourite drifts and/or submarine-fan deposits.

Summary

Basin slopes extend from the shelf–slope break at ~100 m depth, to the top of the continental rise on passive margins (~2500 m depth), or to the landward edge of the trench along subduction margins. This includes the entire bathyal zone. Siliciclastic slopes typically have gradients of 3–6°, whereas the upper segments of carbonate slopes can be much steeper because of biological reef development and early cementation. The upper slope is generally an area of bypass, cut by gullies and slide scars. Any draping sediments are hemipelagic or periplatform deposits (the latter for carbonate settings). Farther downslope, turbidites and debrites punctuate the slope succession, particularly in bathymetric lows like those atop accretionary prisms at convergent margins. The distal slope may be influenced by thermohaline bottom currents producing muddy and sandy contourites. In areas where oceanic upwelling enhances biological productivity, organic-rich deposits, phosphates and diatomites may accumulate under dysaerobic to anaerobic conditions.

Richard N. Hiscott

Bibliography

Aksu, A.E., and Hiscott, R.N., 1989. Slides and debris flows on the high-latitude continental slopes of Baffin Bay. *Geology*, **17**: –885–888.

Aksu, A.E., and Hiscott, R.N., 1992. Shingled Upper Quaternary debris flow lenses on the NE Newfoundland slope. *Sedimentology*, **39**: 193–206.

Aksu, A.E., Calon, T.J., Hiscott, R.N., and Yaşar, D., 2000. Anatomy of the North Anatolian Fault Zone in the Marmara Sea, western Turkey: extensional basins above a continental transform. *GSA Today*, **10** (6): 3–7.

Bagnold, R.A., 1962. Auto-suspension of transported sediment: turbidity currents. *Proceedings of the Royal Society of London*, Series A, **265**: 315–319.

Barnes, P.M., and Lewis, K.B., 1991. Sheet slides and rotational failures on a convergent margin: the Kidnappers Slide, New Zealand. *Sedimentology*, **38**: 205–222.

Biscaye, P.E., and Eittreim, S.L., 1977. Suspended particulate loads and transports in the nepheloid layer of the abyssal Atlantic Ocean. *Marine Geology*, **23**: 155–172.

Biscaye, P.E., Gardner, W.D., Zaneveld, J.R.V., Pak, H., and Tucholke, B., 1980. Nephels! Have we got nephels! *EOS, Transactions of the American Geophysical Union*, **61**: 1014.

Bouma, A.H., 1972. Recent and ancient turbidites and contourites. *Transactions of the Gulf Coast Association of Geological Societies*, **22**: 205–221.

Bouma, A.H., and Hollister, C.D., 1973. Deep ocean basin sedimentation. In Middleton, G.V., and Bouma, A.H. (eds.), *Turbidites and Deep Water Sedimentation*. Anaheim: Pacific Section, Society of Economic Paleontologists and Mineralogists, Short Course Notes, pp. 79–118.

Byers, C.W., 1977. Biofacies patterns in euxinic basins: a general model. In Cook, H.E., and Enos, P. (eds.), *Deep-Water Carbonate Environments*. Society of Economic Paleontologists and Mineralogists, Special Publication, 25, pp. 5–17.

Cacchione, D.A., and Southard, J.B., 1974. Incipient sediment movement by shoaling internal gravity waves. *Journal of Geophysical Research*, **70**: 2237–2242.

Calvert, S.E., 1966. Accumulation of diatomaceous silica in the sediments of the Gulf of California. *Geological Society of America, Bulletin*, **77**: 569–596.

Carpenter, G., 1981. Coincident sediment slump/clathrate complexes on the US Atlantic continental slope. *Geo-Marine Letters*, **1**: 29–32.

Coniglio, M., 1986. Synsedimentary submarine slope failure and tectonic deformation in deep-water carbonates, Cow Head group, western Newfoundland. *Canadian Journal of Earth Sciences*, **23**: 476–490.

Coniglio, M., and Dix, G.R., 1992. Carbonate slopes. In Walker, R.G., and James, N.P. (eds.), *Facies Models: Response to Sea Level Change*. Geological Association of Canada, pp. 349–374.

Driscoll, M.L., Tucholke, B.E., and McCave, I.N., 1985. Seafloor zonation in sediment texture on the Nova Scotian lower continental rise. *Marine Geology*, **66**: 25–41.

Dunbar, R.B., and Berger, W.H., 1981. Fecal pellet flux to modern bottom sediment of Santa Barbara Basin (California) based on sediment trapping. *Geological Society of America, Bulletin* **92**: 212–218.

Emery, K.O., and Milliman, J.D., 1978. Suspended matter in surface waters: influence of river discharge and of upwelling. *Sedimentology*, **25**: 125–140.

Emery, K.O., and Uchupi, E., 1984. *The Geology of the Atlantic Ocean*. New York: Springer-Verlag.

Ewing, M., and Thorndike, E.M., 1965. Suspended matter in deep ocean water. *Science*, **147**: 1291–1294.

Fairbridge, R.W., and Bourgeois, J., 1978. Submarine (bathyal) slope sedimentation. In Fairbridge, R.W., and Bourgeois, J. (eds.), *Encyclopedia of Sedimentology*. Stroudsburg: Dowden, Hutchinson and Ross, pp. 774–777.

Faugères, J.-C., Stow, D.A.V., and Gonthier, E., 1984. Contourite drift moulded by deep Mediterranean outflow. *Geology*, **12**: 296–300.

Gardner, W.D., Biscaye, P.E., Zaneveld, J.R.V., and Richardson, M.J., 1985. Calibration and comparison of the LDGO nephelometer and the OSU transmissometer on the Nova Scotian rise. *Marine Geology*, **66**: 323–344.

Gorsline, D.S., Kolpack, R.L., Karl, H.A., Drake, D.E., Fleischer, P., Thornton, S.E., Schwalbach, J.R., and Svarda, C.E., 1984. Studies of fine-grained sediment transport processes and products in the California Continental Borderland. In Stow, D.A.V., and Piper, D.J.W. (eds.), *Fine-Grained Sediments: Deep-Water Processes and Facies*. Oxford: Geological Society of London, Special Publication 15, pp. 395–415.

Heath, K.C., and Mullins, H.T., 1984. Open-ocean, off-bank transport of fine-grained carbonate sediment in the northern Bahamas. In Stow, D.A.V., and Piper, D.J.W. (eds.), *Fine-Grained Sediments: Deep-Water Processes and Facies*. Geological Society of London, Special Publication, 15, pp. 199–208.

Heezen, B.C., 1959. Dynamic processes of abyssal sedimentation: erosion, transportation, and redeposition on the deep-sea floor. *Geophysical Journal*, **2**: 142–163.

Hein, F.J., and Gorsline, D.S., 1981. Geotechnical aspects of fine grained mass flow deposits: California Continental Borderland. *Geo-Marine Letters*, **1**: 1–5.

Hesse, R., and Klaucke, I., 1995. A continuous along-slope seismic profile from the upper Labrador slope. In Pickering, K.T., Hiscott, R.N., Kenyon, N.H., Ricci Lucchi, F., and Smith, R. (eds.), *Atlas of Architectural Styles in Turbidite Systems*. London: Chapman and Hall, pp. 18–22.

Hill, P.R., 1984. Facies and sequence analysis of nova scotian slope muds: turbidite versus "hemipelagic" deposition. In Stow, D.A.V., and Piper, D.J.W. (eds.), *Fine-Grained Sediments: Deep-Water Processes and Facies*. Geological Society of London, Special Publication, 15, pp. 311–318.

Hill, P.R., Moran, K.M., and Blasco, S.M., 1982. Creep deformation of slope sediments in the Canadian Beaufort Sea. *Geo-Marine Letters*, **2**: 163–170.

Hillaire-Marcel, C., de Vernal, A., Lucotte, M., Mucci, A., Bilodeau, G., Rochon, A, Vallières, S., and Wu, G., 1994. Productivité et flux de carbone dans la mer du Labrador au cours des derniers 40 000 ans. *Canadian Journal of Earth Sciences*, **31**: 139–158.

Hine, A.C., Locker, S.D., Hallock, P., and Mullins, H.T., 1991. Strongly contrasting modes of slope failure and erosion along the carbonate margins of three open seaways, northwest Nicaraguan rise. *American Association of Petroleum Geologists Bulletin*, **75**: 595–596.

Hiscott, R.N., in press. Latest Quaternary Baram prodelta, northwestern Borneo. In Sidi, F.H., Nummedal, D., Imbert, P., Darman, H., and Posamentier, H.W. (eds.), *Tropical Deltas of Southeast Asia – Sedimentology, Stratigraphy, and Petroleum Geology*. Tulsa: Society of Economic Paleontologists and Mineralogists, Special Publication 76.

Hiscott, R.N., and Aksu, A.E., 1994. Submarine debris flows and continental slope evolution in front of Quaternary Ice Sheets, Baffin Bay, Canadian Arctic. *American Association of Petroleum Geologists Bulletin*, **78**: 445–460.

Hiscott, R.N., and James, N.P., 1985. Carbonate debris flows, Cow Head Group, western Newfoundland. *Journal of Sedimentary Petrology*, **55**: 735–745.

Hollister, C.D., and Heezen, B.C., 1972. Geologic effects of ocean bottom currents: western North Atlantic. In Gordon, A.L. (ed.), *Studies in Physical Oceanography*, 2. New York: Gordon and Breach, pp. 37–66.

Hollister, C.D., and McCave, I.N., 1984. Sedimentation under deep-sea storms. *Nature*, **309**: 220–225.

Ingle, J.C. Jr., 1981. Origin of Neogene diatomites around the North Pacific rim. In Garrison, R.E., and Douglas, R.G. (eds.), *The Monterey Formation and Related Siliceous Rocks of California*. Anaheim: Pacific Section Society of Economic Paleontologists and Mineralogists, Special Publication, pp. 159–179.

Jansa, L.F., 1981. Mesozoic carbonate platforms and banks of the eastern North American margin. *Marine Geology*, **44**: 97–117.

Keller, G.H., 1982. Organic matter and the geotechnical properties of submarine sediments. *Geo-Marine Letters*, **2**: 191–198.

Kenter, J.A., 1990. Carbonate platform flanks: slope angle and sediment fabric. *Sedimentology*, **37**: 777–794.

Komar, P.D., 1971. Hydraulic jumps in turbidity currents. *Geological Society of America Bulletin*, **82**: 1477–1488.

Krissek, L.A., 1984. Continental source area contributions to fine-grained sediments on the Oregon and Washington continental slope. In Stow, D.A.V., and Piper, D.J.W. (eds.), *Fine-Grained Sediments: Deep-Water Processes and Facies*. Geological Society of London Special Publication 15, pp. 363–375.

McCave, I.N., 1972. Transport and escape of fine-grained sediment from shelf areas. In Swift, D.J.P., Duane, D.B., and Pilkey, O.H. (eds.), *Shelf Sediment Transport: Process and Pattern*. Stroudsburg: Hutchinson and Ross, pp. 225–248.

McCave, I.N., Lonsdale, P.F., Hollister, C.D., and Gardner, W.D., 1980. Sediment transport over the Hatton and Gardar contourite drifts. *Journal of Sedimentary Petrology*, **50**: 1049–1062.

McGrail, D.W., and Carnes, M., 1983. Shelfedge dynamics and the nepheloid layer in the northwestern Gulf of Mexico. In Stanley, D.J., and Moore, G.T. (eds.), *The Shelfbreak: Critical Interface on Continental Margins*. Society of Economic Paleontologists and Mineralogists, Special Publication, 33, pp. 251–264.

Moore, D.G., 1961. Submarine slumps. *Journal of Sedimentary Petrology*, **31**: 343–357.

Moore, G.F., Taira, A., Bangs, N.L., Kuramoto, S., Shipley, T.H., Alex, C.M., Gulick, S.S., Hills, D.J., Ike, T., Ito, S., Leslie, S.C., McCutcheon, A.J., Mochizuki, K., Morita, S., Nakamura, Y., Park, J.O., Taylor, B.L., Toyama, G., Yagi, H., and Zhao, Z., 2001. Data report: structural setting of the Leg 190 Muroto Transect. In Moore, G.F., Taira, A., and Klaus, A. (eds.), *Proceedings of the Ocean Drilling Program*. Initial Reports, Leg 190. College Station, Texas: Ocean Drilling Program.

Morgenstern, N.R., 1967. Submarine slumping and the initiation of turbidity currents. In Richards, A.F. (ed.), *Marine Geotechnique*. Urbana: Illinois University Press, pp. 189–220.

Mulder, T., and Cochonat, P., 1996. Classification of offshore mass movements. *Journal of Sedimentary Research*, **A66**: 43–57.

Mullins, H.T., Gardulski, A.F., and Hine, A.C., 1986. Catastrophic collapse of the west Florida carbonate platform margin. *Geology*, **14**: 167–170.

Nelson, T.A., and Stanley, D.J., 1984. Variable depositional rates on the slope and rise off the Mid-Atlantic states. *Geo-Marine Letters*, **3**: 37–42.

Nemec, W., 1990. Aspects of sediment movement on steep delta slopes. In International Association of Sedimentologists, Special Publication, 10, pp. 29–73.

Pickering, K.T., Hiscott, R.N., and Hein, F.J., 1989. *Deep-Marine Environments: Clastic Sedimentation and Tectonics*. London: Unwin Hyman.

Pickering, K.T., Underwood, M.B., Taira, A., and Ashi, J., 1995. IZANAGI sidescan sonar and high-resolution multichannel seismic reflection line interpretation of Nankai accretionary prism and trench, offshore Japan. In Pickering, K.T., Hiscott, R.N., Kenyon, N.H., Ricci Lucchi, F., and Smith, R. (eds.), *Atlas of Architectural Styles in Turbidite Systems*. London: Chapman and Hall, pp. 34–49.

Piper, D.J.W., Hiscott, R.N., and Normark, W.R., 1999. Outcrop-scale acoustic facies analysis and latest Quaternary development of Hueneme and Dume fans, offshore California. *Sedimentology*, **46**: 47–78.

Reimnitz, E., and Bruder, K.F., 1972. River discharge into an ice covered ocean and related sediment dispersal, Beaufort Sea, coast of Alaska. *Geological Society of America, Bulletin*, **83**: 861–866.

Richardson, M.J., Wimbush, M., and Mayer, L., 1981. Exceptionally strong near-bottom flows on the continental rise of Nova Scotia. *Science*, **213**: 887–888.

Sangree, J.B., Waylett, D.C., Frazier, D.E., Amery, G.B., and Fennessy, W.J., 1978. Recognition of continental-slope seismic facies, offshore Texas–Louisiana. In Bouma, A.H., Moore, G.T., and Coleman, J.M. (eds.), *Framework, Facies and Oil-Trapping Characteristics of the Upper Continental Margin*. American Association of Petroleum, Geologists Studies in Geology, 7, pp. 87–116.

Satterfield, W.M., and Behrens, E.W., 1990. A Late Quaternary canyon/channel system, northwest Gulf of Mexico continental slope. *Marine Geology*, **92**: 51–67.

Schlager, W., and James, N.P., 1978. Low-magnesian calcite limestones forming at the deep-sea floor, Tongue of the Ocean, Bahamas. *Sedimentology*, **25**: 675–702.

Schrader, H.J., 1971. Fecal pellets: role in sedimentation of pelagic diatoms. *Science*, **174**: 55–57.

Shepard, F.P., Marshall, N.F., McLoughlin, P.A., and Sullivan, G.G., 1979. *Currents in Submarine Canyons and Other Sea Valleys*. American Association of Petroleum Geologists, Studies in Geology, 8.

Southard, J.B., and Mackintosh, M.E., 1981. Experimental test of autosuspension. *Earth Surface Processes and Landforms*, **6**: 103–111.

Stow, D.A.V., 1979. Distinguishing between fine-grained turbidites and contourites on the Nova Scotian deep water margin. *Sedimentology*, **26**: 371–387.

Stow, D.A.V., 1986. Deep clastic seas. In Reading, H.G. (ed.), *Sedimentary Environments and Facies*, 2nd edn, Blackwell Scientific, pp. 399–444.

Stow, D.A.V., and Piper, D.J.W., 1984. Deep-water fine-grained sediments: facies models. In Stow, D.A.V., and Piper, D.J.W. (eds.), *Fine-Grained Sediments: Deep-Water Processes and Facies*. Geological Society of London, Special Publication, 15, pp. 611–646.

Thornton, S.E., 1981. Suspended sediment transport in surface waters of the California Current off southern California: 1977–78 floods. *Geo-Marine Letters*, **1**: 23–28.

Thornton, S.E., 1984. Basin model for hemipelagic sedimentation in a tectonically active continental margin: Santa Barbara basin, California Continental Borderland. In Stow, D.A.V., and Piper, D.J.W. (eds.), *Fine-Grained Sediments: Deep-Water Processes and Facies*. Geological Society of London, Special Publication, 15, pp. 377–394.

Weaver, P.P.E., and Thomson, J., 1993. Calculating erosion by deep-sea turbidity currents during initiation and flow. *Nature*, **364**: 136–138.

Wilson, P.A., and Roberts, H.H., 1995. Density cascading: off-shelf sediment transport, evidence and implications, Bahama Banks. *Journal of Sedimentary Research*, **A65**: 45–56.

Cross-references

Debris Flow
Gravity-Driven Mass Flows
Nepheloid Layer, Sediment
Oceanic Sediments
Slide and Slump Structures
Substrate-Controlled Ichnofacies
Synsedimentary Structures and Growth Faults
Turbidites
Upwelling

SLURRY

Introduction

A slurry is a mixture of solids and fluid that is highly mobile under gravitational and low magnitude stresses, and remains coherent (no bulk phase separation) during transport. In geophysical settings, a slurry is part of a spectrum of mass-flow phenomena known as sediment gravity flows. In agricultural, industrial, and engineering settings, a slurry is a coherent two-phase mixture that commonly is pumped into place rather than driven by gravity. The criterion of high-mobility is included here to distinguish a slurry from concentrated, low-mobility industrial suspensions of solids and fluid such as gels, greases, and pastes. The fluid medium of both geophysical and industrial slurries commonly consists of water but in some circumstances may consist of a gas or oil. The solids component of a slurry can have widely varying composition, size distribution, and concentration. Common geophysical and industrial slurries can consist of cohesive fine sediment, inorganic granular debris, rock fragments, coal, animal waste, sewage sludge, snow, or ice. Some familiar examples of slurries include freshly mixed concrete, drilling mud, and debris flows. The volume concentration of solids in a slurry can vary widely. In general, the finer grained the solids, the lower the concentration needed to obtain a coherent solids-fluid mixture. For example, a slurry composed of smectite and water can become coherent at a solids concentration of as low as 3 percent by volume (Hampton, 1972), whereas a granular debris flow typically does not become coherent until the solids concentration reaches about 50 percent or more (e.g., Iverson and Vallance, 2001).

Distinguishing characteristics of a slurry

A slurry differs from a dilute suspension or solids-laden fluid in terms of momentum transfer, the pressure in the suspending fluid, and phase separation of the solid and fluid constituents. In dilute suspensions or solids-laden fluids, momentum is transferred by fluid forces, and the solids are a passive component transported by the fluid (see *Gravity-Driven Mass Flows*). In a slurry, however, significant momentum transfer occurs through solid-phase interaction, owing to the greater concentration of particles. Solids are no longer transported passively, but instead move the suspending fluid in response to a pressure gradient or gravitational body force.

Nonequilibrium fluid pressure, the total fluid pressure minus the local hydrostatic fluid pressure, further distinguishes a slurry from a solids-laden fluid. In a concentrated slurry, nonequilibrium fluid pressure arises from gravitational settling of the particles, which transfers stress to the fluid phase. If the fluid can migrate easily through pore spaces around the solid particles, the nonequilibrium fluid pressure dissipates readily, otherwise it remains elevated and helps balance the weight of the overlying mixture. As long as pore-fluid pressure remains elevated, particle settling is hindered, and separation of the solid and fluid phases is limited.

A lack of phase separation over timescales relevant to the transport process can be used to further distinguish a slurry from a solids-laden fluid. In a solids-laden fluid, the phases are not intimately bound, and gravitational settling of particles is controlled by the size, shape, and density of the particles, by the density and viscosity of the fluid, and by fluid dynamics (such as turbulence). Gravitational phase separation in slurries, however, is related to dissipation of nonequilibrium fluid pressure (e.g., Major, 2000; also see *Gravity-Driven Mass Flows*). An approximate estimate of the time necessary for phase separation to occur in most geophysical slurries is obtained by examining the ratio h^2/D, where h is the thickness of the slurry and D is the hydraulic diffusivity, a property that depends on the compressibility and permeability of the mixture and the viscosity of the fluid. Values of D are influenced by the concentration and size distribution of solids in a slurry, and can range over several orders of magnitude. Concentrated suspensions of sand that contain at least a few percent silt and clay have diffusivities on the order of $10^{-7} \mathrm{m}^2$ per second or smaller, whereas dilute suspensions of sand have diffusivities on the order of $0.1 \mathrm{m}^2$ per second. In the context of mixture transport, the wide range in diffusivity values illustrates why the mixture dimension (h), as well as its solids and fluid properties, influences whether it remains coherent over relevant transport timescales.

Applications

Although naturally occurring slurries, such as debris flows, subaqueous slumps, and snow-rich slushflows, commonly pose significant environmental hazards, artificial slurries are used for a number of agricultural, industrial, and geotechnical purposes. For example, fluvial and estuarine bed sediments have been dredged, mixed into slurries, and pumped to facilities for bioremediation treatment to remove contaminants (e.g., Glaser, 1997); by-products of coal treatment and other industrial wastes commonly are mixed into slurries and pumped into impoundment ponds for storage or disposal; agricultural and animal wastes are converted into slurries and spread across the land surface or injected underground for disposal; and clay-slurry barrier walls have been developed and used for industrial waste containment and other geotechnical applications (e.g., Guy, 1995).

Jon J. Major

Bibliography

Glaser, J.A., 1997. Utilization of slurry bioreactors for contaminated solids treatment: an overview. In *Proceedings of the 4th International In Situ and On-Site Bioremediation Symposium*. New Orleans, April 28–May 1, 1997, 5, pp. 123–130.

Guy, D.G., 1995. A slurry wall groundwater barrier used to isolate a major trunk road from an exposed aquifer. *Soil and Environment*, 5: 1235–1236.

Hampton, M.A., 1972. The role of subaqueous debris flow in generating turbidity currents. *Journal of Sedimentary Petrology*, 42: 775–793.

Iverson, R.M., and Vallance, J.W., 2001. New views of granular mass flows. *Geology*, 29: 115–118.

Major, J.J., 2000. Gravity-driven consolidation of granular slurries: Implications for debris-flow deposition and deposit characteristics. *Journal of Sedimentary Research*, 70: 64–83.

Cross-references

Atterberg Limits
Autosuspension
Coal Balls
Colloidal Properties of Sediments
Debris Flow
Earth Flows
Grain Settling
Gravity-Driven Mass Flows
Hindered Settling
Liquefaction and Fluidization
Smectite Group

SMECTITE GROUP

Smectite, also known as "swelling clay", is a diverse group of clay minerals with a 2:1 layer silicate structure that can expand and contract upon wetting and drying. All 2:1 layer silicates (which also include illite/mica, chlorite, and vermiculite groups) have 2 tetrahedral sheets surrounding a central octahedral sheet. Smectite 2:1 layers have a negative charge which is balanced by weakly bonded, exchangeable cations, located in the interlayer region (between 2:1 layers). Smectite interlayers also contain water, which results in little to no stacking order between 2:1 layers, a property known as turbostratic stacking.

Chemical composition, morphology, and identification

The smectite group (derived from the Greek, smectos, which means soap) shows large chemical variability, resulting in many specific mineral names (Güven, 1988). The most common smectite mineral is montmorillonite, which has the following representative chemical formula: $X_{.3} \cdot nH_2O [(Al_{1.5}Fe^{3+}_{.2}Mg_{.3})Si_4O_{10}(OH)_2]^{-.3}$, where $X =$ exchangeable cations such as Na^+ and Ca^{2+}; $n =$ number of interlayer water molecules, which varies but commonly is ~ 4–5; Al, Fe, and Mg = octahedral cations; and Si = tetrahedral cation. Tetrahedral substitutions of Al^{3+} for Si^{4+} are also common and develop negative 2:1 layer charge, which can vary from ~ -0.25 to -0.6. Smectite has a crystal structure similar to vermiculite, except that vermiculite has a larger negative charge on 2:1 layers. There are two subgroups of smectite minerals: dioctahedral smectite, which has 2 octahedral cations per $O_{10}(OH)_2$ and trioctahedral smectite, which has 3 octahedral cations per $O_{10}(OH)_2$. Fe-rich dioctahedral smectite is nontronite and Mg-rich trioctahedral smectite has several specific mineral names including saponite (with a tetrahedral charge) and stevensite (with an octahedral charge). Much rarer examples include Co-rich, Cr-rich, Cu-rich, Mn-rich, Ni-rich, V-rich, and Zn-rich smectites. Many of the above elements, and others such as Li and F, occur in small amounts in common smectite minerals.

Smectite morphology, which depends on conditions of formation, includes subhedral flakes, euhedral elongate laths and ribbons, and less commonly, euhedral rhombohedra and hexagons (Güven, 1988). Because natural smectite is nearly always very fine-grained (typically $<2 \mu m$), identification is normally by X-ray diffraction. Upon exposure to ethylene glycol, smectite has a diagnostic 1.7 nm (17Å) basal spacing, which represents the thickness of the 2:1 layer plus interlayer.

Sedimentary petrology

Smectite is relatively common in soil, sediment, and Cenozoic to Mesozoic sedimentary rock and paleosol. It forms in a variety of surficial environments associated with waters that have relatively high cation (Na, Ca, +K)/H activity ratios and $SiO_{2(aq)}$ activities. It is thermodynamically metastable compared to other more ordered silicates; it forms readily, however, because of fast reaction kinetics (Essene and Peacor, 1995).

Smectite forms commonly in soils, particularly in areas with moderate to relatively low rainfall, poor drainage, and neutral to basic soil waters (Chamley, 1989, pp. 21–50). The relatively weak leaching capacity of such soil waters allows retention of soluble elements (Na, Ca, Mg) and silica necessary for smectite formation. Smectite forms by alteration of volcanic glass, feldspars, mafic silicates, and other aluminosilicate minerals as well as by direct precipitation from solution. Soil smectite also forms from clay mineral precursors such as illite, chlorite, and vermiculite as well as the disordered phases imogolite and allophane. Most commonly dioctahedral Al-rich smectite forms in soils; however, dioctahedral Fe-rich smectite (nontronite) forms from parent rocks rich in mafic silicates. Smectite formed in a soil environment is notoriously difficult to characterize chemically because of its heterogeneity and the presence of other minerals in the clay size-fraction, making chemical characterization difficult. Soil smectite commonly occurs with an incomplete Al- or Fe-octahedral sheet in the interlayer, making it intermediate between smectite and chlorite. Soil smectite can also have a relatively high negative charge on the 2:1 layers, making it transitional to vermiculite. Such minerals are called hydroxy-interlayered smectite, hydroxy-interlayered vermiculite, or simply hydroxy-interlayered expandable. Finally, soil smectite occurs commonly as a mixed-layer mineral (i.e., 2 different structures present within a single crystallite) with chlorite, illite, or vermiculite. The presence of smectite in a paleosol can suggest an arid paleoclimate.

Smectite has a detrital origin in many other surficial depositional environments such as glacial, fluvial, desert, freshwater lacustrine, and deltaic (Chamley, 1989, pp. 53–116). Study of minerals in such settings can give insight into provenance and paleoenvironmental conditions in the sediment source area. Evaporative (saline and alkaline) lakes can produce trioctahedral Mg-rich smectite, which forms by either alteration of a precursor dioctahedral smectite or direct precipitation. Smectite tends to stay in suspension and pass directly into the marine environment rather than settle in an estuary because it is not affected as much as other clay minerals by differential settling resulting from flocculation. Flocculation (q.v.) involves aggregation of fine-grained minerals into larger masses commonly because of their interaction with saline water. As smectite is transported from continental to marine environments there is little change other than interlayer cation exchange, which commonly involves a change from Ca-rich smectite to Na-rich smectite. Authigenic Fe-rich smectite forms as a precursor to Fe-rich illite (glauconite) in shallow, low-latitude marine environments near rivers that supply abundant dissolved iron (Chamley, 1989, 213–234). Smectite in a deep marine environment commonly has a mixed origin; much of it is detrital and Al-rich but some of it can be authigenic and Fe-rich (Chamley, 1989, 259–289).

Al-rich smectite forms during low-temperature ($<\sim 50°C$) burial diagenesis or hydrothermal alteration of silicic volcanic glass and feldspar (Chamley, 1989, 333–358). Bentonites (q.v.) are soft, plastic, light-colored clay layers comprised predominantly of smectite that typically form by alteration of volcanic ash. With increasing temperature of burial or hydrothermal alteration as well as geologic age, smectite reacts to mixed-layer illite-smectite and eventually to pure illite (Chamley, 1989, pp. 359–389). Interlayer water lost during this reaction is considered important to geopressure development and petroleum migration out of source rocks (Eslinger and Pevear, 1988, Section 5, pp. 33–42). Low-temperature burial diagenesis or hydrothermal alteration of mafic volcanic glass and mafic silicates results in formation of *palagonite*, which consists of a variety of authigenic minerals including abundant smectite that ranges from Fe-rich to Mg-rich (Chamley, 1989, pp. 291–329). High-temperature hydrothermal alteration of basalt produces mixed-layer chlorite-smectite.

Physical properties and environmental applications

The very fine grain size of smectite and presence of a negative 2:1 layer charge as well as exchangeable cations and water in the interlayer result in environmentally important physical properties including high cation exchange capacity (ability to exchange cations between the interlayer and pore water), catalytic action (ability to increase the rate of a chemical reaction), swelling (ability to incorporate large amounts of interlayer water causing expansion of 2:1 layers from each other), and low permeability. The high cation exchange capacity of smectite can influence the chemistry of natural waters and mobility of pollutants in groundwater. Smectite can catalyze chemical reactions because of its extremely large and chemically reactive surface area, including the interlayer region. The catalytic activity of smectite along with its ability to adsorb polar organic molecules, has lead some to propose that smectite was important to the origin of life and that smectite minerals can even be considered crystal genes because of their ability to grow (replicate) and introduce defects (evolve) (Cairns-Smith and Hartman, 1986, pp. 130–158). In dilute water, smectite with interlayer Na can expand to up to 10–15 times the volume of the dry clay, exerting a great upward pressure. As it dries, it shrinks back to its original volume. This shrink-swell action can damage walls, foundations, and roads. Average annual damage from swelling clays in USA is several billion dollars, making it one of the most costly natural hazards (Keller, 2000, pp. 68–69). Problematic areas in USA are in Montana, Wyoming, South and North Dakota, Colorado, Texas, and Louisiana. The swelling of Na-smectite along with its relatively low engineering (shear) strength can also greatly increase damage due to liquefaction (transformation from a solid to liquid state) during an earthquake and landsliding. Swelling of Na-smectite in an oil reservoir rock can greatly reduce production potential (Eslinger and Pevear, 1988, Section 8, pp. 6–10). The extremely low permeability of smectite-rich earth materials make them effective aquitards and very useful as liners in waste disposal systems. Conversely, smectite-rich soils (known as vertisols) are not particularly productive because of their impermeable nature and their tendency to be very soft and plastic when wet and very hard when dry, which makes plowing difficult (Grim and Güven, 1978, pp. 245–246).

Commercial deposits and uses

Commercial deposits of smectite are mainly from bentonite (defined above) and fuller's earth (any natural material that has a high absorptive capacity and can decolorize oil; smectite is commonly abundant in fuller's earth). Large deposits of bentonite and fuller's earth are located in the USA (Wyoming, Mississippi, Texas, Arizona, and California), England, Germany, Russia, Japan, and many other countries (Grim and Güven, 1978, pp. 13–126). The major commercial uses of smectite are for foundry molding sands and drilling mud.

Many other minor commercial uses of smectite are listed by Grim and Güven, (1978, pp. 217–248). Molding sands, which consist of sand and most commonly smectite clay, are used by foundries to receive molten metal during the casting process. The function of smectite is to bind the sand and maintain the mold shape after hot metal is poured into it. Smectite is a major component in mud used in drilling of oil and water wells. The function of smectite is to create a viscous fluid with water that cools and lubricates the drill bit and removes rock chips by circulating drilling mud down through the hollow drill stem and up through the area between drill stem and hole. Smectite in drilling mud gives it the desirable property of *thixotropy*, which is the ability to go from a liquid state when stirred to a semi-solid state upon standing. Smectite thixotropy, which is related to its ability to undergo swelling, allows the drilling mud to stiffen and prevent rock chips from settling back into the hole when pumping of drilling mud is stopped temporarily. There are many other uses of smectite including decolorization of oil, catalysis in petroleum refining, ceramics, animal bedding, adhesives, cement, clarification of wine and water, animal feed, grease, ink, pharmaceutical products, paint, paper, pelletizing iron ore, pesticides, rubber, and soap.

Summary

Smectite represents a chemically diverse group of clay minerals with a 2:1 layer silicate structure that swells and shrinks with wetting and drying. It occurs commonly in soil, sediment, and post-Paleozoic sedimentary rock associated with waters that have relatively high cation (Na, Ca, +K)/H activity ratios and $SiO_{2\ (aq)}$ activities. It forms most commonly in soils from subhumid to arid climates, however, it also forms in saline and alkaline lacustrine, deep marine, and low-temperature burial diagenetic and hydrothermal environments. Aluminosilicates and silicic volcanic glass alter to Al-smectite whereas mafic silicates and mafic volcanic glass alter to Fe- or Mg-smectite. Environmentally important physical properties include high cation exchange capacity, catalytic action, swelling, and low permeability. Commercial uses are mainly as drilling mud and foundry molding sands.

Stephen P. Altaner

Bibliography

Cairns-Smith, A.G., and Hartman, H. (eds.), 1986. *Clay Minerals and the Origin of Life*. Cambridge University.
Chamley, H., 1989. *Clay Sedimentology*. Springer-Verlag.
Eslinger, E., and Pevear, D., 1988. *Clay Minerals for Petroleum Geologists and Engineers*. Society Economic Paleontologists and Mineralogists Short Course 22.
Essene, E.J., and Peacor, D.R., 1995. Clay mineral thermometry: a critical perspective. *Clays and Clay Minerals*, **43**: 540–553.
Grim, R.E., and Güven, N., 1978. *Bentonites: Geology, Mineralogy, Properties, and Uses*. Developments in Sedimentology 24, Elsevier.
Güven, N., 1988. Smectites. In Bailey, S.W. (ed.), *Hydrous Phyllosilicates (Exclusive of Micas)*. Reviews in Mineralogy 19. Washington: Mineralogical Society of America, pp. 497–559.
Keller, E.A., 2000. *Environmental Geology*. Prentice-Hall.

Cross-references

Bentonites and Tonsteins
Cation Exchange
Chlorite in Sediments
Diagenesis
Illite Group Clay Minerals
Kaolin Group Minerals
Mixed-Layer Clays
Mudrocks
Vermiculite
Weathering, Soils and Paleosols

SOLUTION BRECCIAS

Solution breccias are common and widespread in the geological record. Brecciation may occur entirely within one soluble rock (usually limestone or dolostone) or where more soluble evaporites are removed in mixed evaporite–carbonate and evaporite–redbed sequences. It may be extended upward by mechanical failure (stoping) into overlying insoluble strata, e.g., dissolution of ~180 m of salt has propagated through 400 m of carbonates overlain by ~650 m of clays, mudstones, and sandstones at a site in Saskatchewan (Christiansen and Sauer, 2001).

Brecciation can occur during early diagenesis when evaporites are dissolved in supratidal carbonate sequences (Choquette and James, 1988) or where caves in case-hardened carbonate sand dunes collapse. It is common at inter- and intra-stratal sites during burial and later diagenesis, caused by expulsion of basinal fluids, or by hydrothermal or meteoric waters: it may extend to pressure solution (stylolite) depths. It is also common in superficial karst terrains, where meteoric groundwater is known to brecciate carbonate rocks at depths as great as 2 km.

Breccia fabrics and matrix

Three principal fabrics are recognized (Stanton, 1966): (1) crackle breccia, where beds sag apart and crack upon dissolutional removal of support but there is little displacement; (2) mosaic pack breccia, where there is clast support. The clasts have usually dropped and vary from partly rotated to completely disorganized in orientation. Globally, clast sizes range from small pebbles to "cyclopean breccias" with blocks of hundreds of cubic meters but are normally more limited within any given breccia; and (3) rubble float breccia, which is matrix-supported. A genetic association of these fabrics is often found, consisting of crackle breccias at the top and around the perimeter of a body, pack breccia within it where beds are thicker, and float breccias where beds are thin or at the base. In some Mississippi Valley Type (MVT) deposits, these may be underlain by a basal "trash" zone of insoluble residua (Sangster, 1988).

Clast-supported breccias may be comprised of clasts and voids only, with little or no matrix or cement. This is most common in breccias in sinkholes and caves that are actively developing today. Matrix is usually present in early diagenetic, inter- and intrastratal breccias, consisting of carbonate fines and insoluble residua that have filtered down or (more rarely) allochthonous sands and clays. Older breccias of all types will normally display partial or complete cementation. Calcite is the principal cement but aragonite, dolomite and gypsum are also common. In MVT the chief cements are dolomite, FeS_2, PbS, and ZnS.

Figure S35 A dolomite and dedolomite solution breccia created by preferential dissolution of gypsum in a sabkha sequence, the Bear Rock Formation (Lower–Middle Devonian) at Bear Rock near Fort Norman, Northwest Territories, Canada. The section is ~15 m in height.

Form and location of the principal types of breccia bodies

As noted, solution breccias are common on the surface and underground in modern karst terrains, chiefly where sinkholes, river cliffs, or shafts and chambers in caves, have collapsed. Individual collapses may be millions of cubic meters in volume. Where surficial collapses are preserved in buried paleokarsts, Choquette and James (1988) term them "mantle breccias".

Extensive brecciation is usually present where the proximal margins of buried salt bodies on the continents are subject to dissolution by groundwaters penetrating the cover rocks, creating a receding "salt slope" (Ford and Williams, 1989, p.460). In Manitoba and Saskatchewan, Canada, the salt slope of a Devonian deposit is ~800 km in length and now at a mean depth of 200 m below the surface. It has been subroded westward an average of 130 km since burial in the Devonian.

Breccia pipes ("geological organs", "breakthroughs", "prismatic bodies" of different authors) are the most widely reported breccia bodies, thousands of examples being described in the world literature (Bosak *et al.*, 1989). Diameters range from <10 m to >10 km, with a majority being 20–250 m. Most are plumb vertical, with reported heights ranging from 20 m to more than 1000 m. Higher examples have usually stoped upward through one or more cover formations, which may include siliciclastics, coal measures, extrusive volcanics, etc. The upward termination may be in undisturbed rock, in downfaulted but not brecciated strata, in a closed depression at the surface, or even upstanding as a firmly cemented and resistant "castille" above an erosional plain (e.g., Hill, 1996, in New Mexico). Most breccia pipes originate in point dissolution of salt (occasionally, of gypsum or anhydrite) above a fracture junction, anticlinal crest or buried reef that channels groundwater. They may be targets for the later precipitation of economic ore, such as the well-known uranium pipes of the Grand Canyon, Arizona (Huntoon, 1996).

Other forms are described in most detail from MVT deposits (e.g., Sangster, 1988). Breccia domes typically 30–100 m in diameter are similar to breakdown rooms in modern caves or transitional to breccia pipes. Tabular or sinuous forms are wide but shallow breccias extending for one km or more, often along paleoreef margins. Straight linear features ("runs") and reticulate mazes include individual breccia-filled corridors up to 30 m high, 150 m or more wide, and one or more km in length. Most of these features occur at relatively shallow depths beneath long-exposure paleokarst surfaces and are believed to be genetically associated with them. Some are attributed to brecciation and subsequent sulfide ore (cement) deposition from invading hydrothermal solutions (Dzulynski and Sass-Gutkiewicz, 1989).

Derek Ford

Bibliography

Bosak, P., Ford, D.C., Glazek, J., and Horacek, I. (eds.), 1989. *Paleokarst; a Systematic and Regional Review*. Prague: Academia Praha/Elsevier.
Choquette, P.W., and James, N.P., 1988. Introduction. In James, N.P., and Choquette, P.W. (eds.), *Paleokarst*. New York: Springer Verlag. pp. 1–21.
Christiansen, E.A., and Sauer, E.K., 2001. Stratigraphy and structure of a Late Wisconsin salt collapse in the Saskatoon Low, south of Saskatoon, Saskatchewan, Canada: an update. *Canadian Journal of Earth Sciences*, **38**: 1601–1613.
Dzulynski, S., and Sass-Gutkiewicz, M., 1989. Pb–Zn ores. In Bosak *et al.* (op. cit), pp. 377–398.
Ford, D.C., and Williams, P.W., 1989. *Karst Geomorphology and Hydrology*. London: Chapman and Hall.
Hill, C.A., 1996. *Geology of the Delaware Basin, Guadalupe, Apache and Glass Mountains, New Mexico and Texas*. Society of Economic Paleontologists and Mineralogists, Publication, 96–39, 480pp.
Huntoon, P.W., 1996. Large-basin ground water circulation and paleo-reconstruction of circulation leading to uranium mineralization in Grand Canyon breccia pipes, Arizona. *The Mountain Geologist*, **33** (3): 71–84.
Sangster, D.F., 1988. Breccia-hosted Lead-Zinc deposits in carbonate rocks. In James, N.P. and Choquette, P.W. (eds.), *Paleokarst*. New York: Springer Verlag. pp. 102–116.
Stanton, R.J., 1966. The solution brecciation process. *Geological Society of America, Bulletin*, **77**: 843–848.

Cross-references

Anhydrite and Gypsum
Cave Sediments
Dedolomitization
Micritization
Pressure Solution
Red Beds
Stylolites
Sulfide Minerals in Sediments

SPELEOTHEMS

Introduction

"Speleothem" (Greek; "spele"—cave, "them"—make)—a general term for secondary minerals deposited in natural caves (Moore, 1952). More than 250 different minerals are recorded

(Hill and Forti, 1997). At the global scale calcite precipitates in solutional caves in carbonate rocks are overwhelmingly predominant, both in frequency of individual deposits and in aggregate volume. They accumulate in air-filled (vadose) and water-filled (phreatic) caves. Aragonite is second in abundance but rare as a subaqueous deposit. "Travertine" and "sinter" are alternative general terms for these two precipitates, or "tufa" when they are precipitated partially or entirely on organic frameworks, usually vegetal and in the illuminated entrance zones of caves. Gypsum is third in abundance, in gypsum caves but also in limestone caves with gypsum interbeds or inclusions, or where H_2SO_4 can be created by oxidation and react with the rock. Hydrated carbonates and sulfates (e.g., hydromagnesite, epsomite) are also quite common, usually as pastes or powders of crystallites that are small in amount. Other secondary minerals are more localized, associated with particular source conditions in the bedrock or clastic sedimentary fillings, and quite rare at the global scale. They include native sulfur, many oxides and hydroxides, halides, nitrates, phosphates, silicates, vanadates, and a few organic minerals (Hill and Forti, 1986, 1997). Perennial ice is found in many cold caves.

Calcite and aragonite speleothems

Crystallography and diagenesis

Calcite ($CaCO_3$) is trigonal, with rhombohedral or scalenohedral habit. In most speleothems it is present in microcrystalline form with the c axis oriented roughly across the direction of growth ("length slow" of Folk and Assereto, 1976) or as coarser palisade, columnar or equant ("coconut meat") crystal aggregates with the c axes oriented to growth ("length fast"; see Railsback, 2000). Large monocrystalline speleothems occur but are rare. Density and size of fluid inclusions are very variable. Many vadose speleothems display alternations of coarser and finer crystalline growth, often with temporary cessations due to drying or dissolution, zones with dust, flood mud or organic grains: existing crystals may grow through these or new crystals form upon them. Subaqueous deposits are more homogeneous. Pure calcite is translucent, or opaque white due to fluid inclusions. Strong color banding (yellow, brown, red-brown) is common, however, due chiefly to incorporation of fulvic acid chromophores from soil waters above the cave (Van Beynen et al., 2001). Metals such as iron (red), copper (blue), and nickel (bright green) also provide color but rarely in visible concentrations (see Cabrol and Mangin, 2000).

Aragonite ($CaCO_3$) is orthorhombic. In caves it almost always occurs in acicular or tabular crystal habit, as clusters of small needles radiating from the rock or calcite speleothems, or as massive botryoidal aggregates forming larger deposits such as stalagmites. Twinned tabular habit is occasionally found. The deposits are more homogeneous in crystal size and more uniformly white than calcite, lacking strong color banding from organic chromophores. There are terminations due to drying, dissolution, or flood mud deposition.

Deposition of aragonite rather than calcite is usually attributed to enrichment of Mg^{2+} ions in the feedwater. This may be due to presence of dolostone or be part of an evaporative depositional sequence beginning with a core of calcite, proceeding to a surround of aragonite when the Ca^{2+} ions are sufficiently depleted and concluding with an aureole of hydromagnesite ($4MgCO_3 \cdot Mg(OH_2) \cdot 4H_2O$) when they are exhausted. Because evaporation is important aragonite speleothems are more common in warmer climates. Railsback et al. (1994) report wet-dry seasonal alternations of calcite and aragonite in a Botswana stalagmite. Thicker, nonrhythmic aragonite layers are also found in some calcite speleothems.

Diagenesis is common in aragonite speleothems. Smaller botryoids are replaced by larger ones or by equant or columnar calcite. It is unusual to find aragonite older than ~100 kyr that does not display at least some calcite replacement. In purely calcite deposits there may be partial or complete replacement of microcrystalline layers by columnar calcite. Vugs become infilled with zoned cements.

Vadose morphology and size

Shapes of speleothems in vadose caves are created by gravity, or by crystal and capillary forces. Gravity forms are overwhelmingly predominant in volume. Principal types are stalactites, stalagmites, and sheets of "flowstone" on floors and/or walls (Figure S36). A "column" is a stalactite-stalagmite pair grown together.

The fundamental form is the soda straw stalactite, a thin sheath of calcite enclosing the feedwater canal. The c-axis is oriented downward. New growth occurs only at the tip. Diameter is determined by droplet size and ranges 2–9 mm. Calcite straws can extend for several meters before they fall under their own weight. Aragonite straws are rare, taking a clumped form with narrower necks that may represent temporary halts in deposition. Leakage or obstruction of the canal overplates the sheath, creating tapered (carrot-like) calcite stalactites up to one meter in diameter and several in length. There may be accelerated deposition on protuberances, creating a myriad of subsidiary forms such as crenulations, corbels, drapes and lesser stalactites. "Draperies" develop where feedwater trickles down an inclined roof and grow downward with c-axes perpendicular to growth edges. If initiated on rough surfaces they may be quite sinuous, like a drawn household drape.

Shapes of calcite stalagmites are similarly varied. The most simple is the "candlestick", which adds all new growth at the top under a nearly constant drip feed. Minimum diameters are ~3 cm. Varying drip or greater fall height causes terraced or corbelled thickening; at the extreme the form is like a pile of soup plates. More common are conical or tapered forms, broadening into domes with flowstone sheets around them. Some corbelled stalagmites are >30 m high. Domes may be 50 m or more in basal diameter. Aragonite stalagmites are much smaller, with narrow taper.

Calcite flowstones are deposited from film flow and accrete roughly parallel to the host surface, even where this is vertical. They may extend tens to hundreds of meters downstream of a single source and accumulate to thicknesses of several meters. Aragonites are again smaller and thinner. Flowstone layers interbedded with sands and clays are common in river caves. "Gours" or "rimstones" are calcite dams building upward from irregularities in small channels or on flowstone and stalagmite surfaces. The greatest impound water to depths of several meters. Rims are often strikingly crenulated.

Calcite "helictites" or "excentrics" grow where crystal or capillary forces predominate, skewing c-axis orientation to create narrow, curvilinear tubes extending out, up and down from rock or parent stalactites, etc. Most are short, <10–20 cm

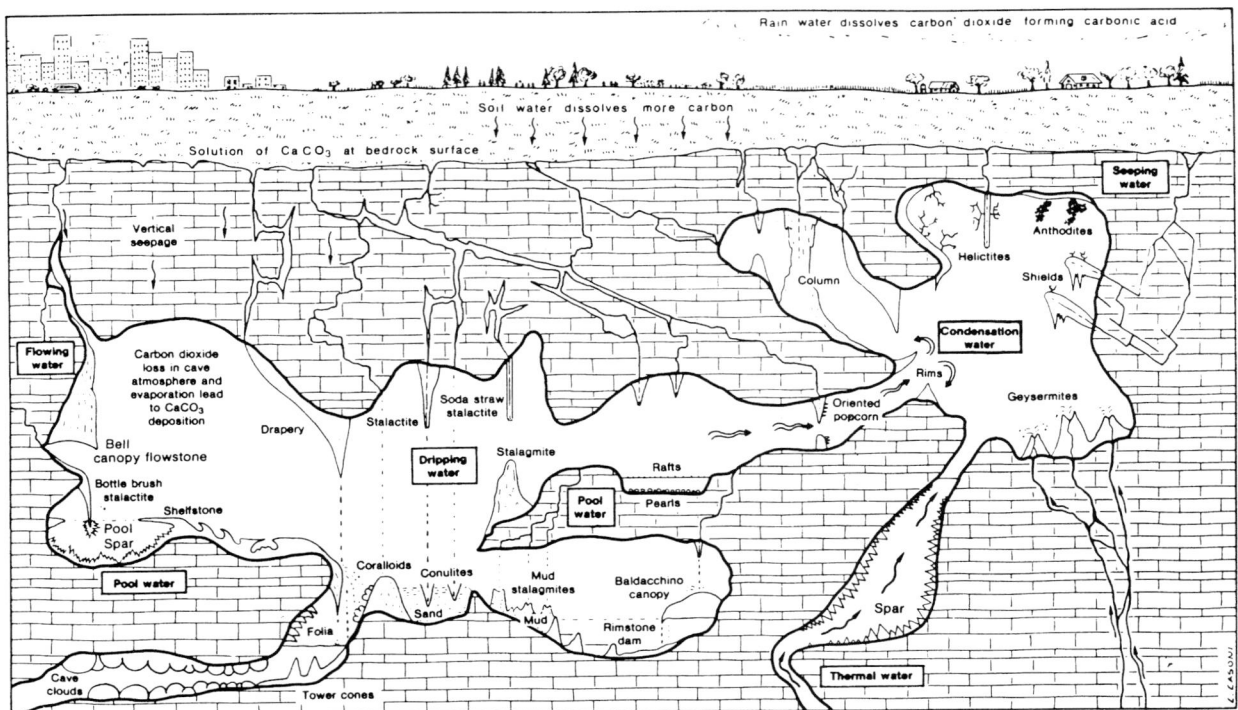

Figure S36 Model to show vadose and subaqueous speleothem precipitates; from Hill and Forti (1986) with permission.

in length. Dense clustering may form tangled masses like the Medusa's hair. "Anemolites" grow upwind into prevailing drafts. Clusters of needles fanning outward ("frostwork") are the most common type of aragonite deposit encountered in caves.

Globular forms are also common. "Cave pearls" are spheroidal accretions about a nucleus such as grit agitated by water dripping into a pool. "Popcorn" or "cave coral" are semi spherical accretions on host flowstone or other surfaces, often in dense, multi-layered clusters.

Subaqueous speleothems

Calcite may precipitate from thermal and meteoric waters. The principal thermal form is a spar lining on all surfaces, e.g., most of the 150+ km of passages in Jewel Cave, South Dakota. Deposition extends from the watertable to a limiting depth determined by pressure and Ca^{2+} saturation state. Aggregate thicknesses of one meter more are known. Spar linings are also common in meteoric caves. A wealth of more complex, branching crystal structures and rounded microcrystalline "clouds" can form in static pools. Water surfaces are marked by growth of shelfstone around the edges and floating rafts of microcrystalline calcite accreting around dust particles.

Distribution and abundance

Stalactites, stalagmites and flowstones may occur as isolated individuals in a cave, in clusters or aligned along feedwater fractures, or broadcast. Density may increase until all surfaces are covered.

Vadose speleothems grow most readily at shallow depths beneath soils rich in CO_2 in humid tropical and temperate conditions. These conditions permit year-round deposition, which may be accelerated by evaporative effects during dry periods. The largest individuals and greatest densities are found in these settings. Accumulations generally diminish in drier and cooler climates. Large stalagmites found in some desert or arctic caves tend to be relics from earlier, more favorable conditions.

Many caves have several levels. Speleothems are often fewer and smaller in the lower levels, which are usually younger. In any setting speleothem deposition is largely prohibited where there are impermeable beds, e.g., shales, above a cave.

Growth rates, age and environmental studies

Under optimal conditions (large thermodynamic excess of Ca^{2+} ions, drip rate and evaporation) soda straws may extend several cm in a year. Normal accretion rates (increase of thickness of a growth layer) in other speleothems probably range between ~ 1.0 mm/10^3 years in cold climate flowstones to >1.0 m/10^3 year in warm cave entrances. The lowest rate reliably measured (by U series dating) is ~ 0.001 mm/10^3 year in a thermal water spar close to the pressure limit for deposition. Some speleothems appear to have accreted at constant rates but others vary by factors of ten or more. Many contain hiatuses caused by drought, cold or change of groundwater routing.

Many speleothems can be dated accurately (± 1 percent error) by the ^{230}Th/^{234}U method if they are less than ~ 500 kyr in age. Variations of ^{18}O/^{16}O isotope ratios along the dated growth trend may indicate paleotemperature changes at the site and ^{13}C/^{12}C ratios suggest correlative changes of vegetation amount or type. Where present, annual or event banding revealed by UV fluorescence and other techniques now permits very high resolution reconstructions of past conditions above a cave (see Hill and Forti, 1997, pp. 271–284).

Gypsum speleothems

Crystallography and morphology

Gypsum occurs as acicular, curved, equant, prismatic or tabular crystals as contact or penetration twins and as long fibers. Individual prismatics >1.0 m in length are known. Gypsum is deposited in three principal modes in caves: (1) as evaporitic inclusions of coarse, fibrous crystals growing as lenses up to several cm in thickness in bedrocks or cave sediments, which they rupture—"evapoturbation". (2) as scattered tiny patches, thicker encrustations or excentric extrusions, on rock, sediment or calcite speleothems. Most frequent are "flowers", extruded, twisting bundles of fibrous crystals up to 50 cm in length. Epsomite and mirabolite, the most common Mg, and Na sulfates, also display this form. Clusters of gypsum needles grow from sulfate-rich sediments and fibers of "hair" from roofs. Much larger, bifurcating stalactites are reported from a few caves, e.g., "The Chandeliers", Lechuguilla Cave, New Mexico. (3) as regularly bedded floor deposits or wall encrustations in evaporating lakes or pools: thicknesses of several meters occur in Carlsbad Caverns, New Mexico, where much is reprecipitated from alteration crusts formed by H_2SO_4 reacting with the limestone walls.

Derek Ford

Bibliography

Cabrol, P., and Mangin, A., 2000. *Fleurs de Pierre*. Lausanne: Delachaux et Niestle.
Folk, R.L., and Assereto, R., 1976. Comparative fabrics of length-slow and length-fast calcite and calcitized aragonite in a Holocene speleothem, Carlsbad Caverns, New Mexico. *Journal of Sedimentary Petrology*, **46** (3): 486–496.
Hill, C., and Forti, P., 1986. *Cave Minerals of the World*. Huntsville, AL: National Speleological Society of America.
Hill, C., and Forti, P. (eds.), 1997. *Cave Minerals of the World*, 2nd edn. Huntsville, AL: National Speleological Society of America.
Moore, G.W., 1952. Speleothem—a new cave term. *National Speleological Society of America, News*, **10** (6): 2.
Railsback, L.B., 2000. *An Atlas of Speleothem Microfabrics*. Athens:University of Georgia, GA. http://www.gly.uga.edu/railsback/speleoatlas/SAindex1.html
Railsback, L.B., Brook, G.A., Chen, J., Kalin, R., and Fleisher, C.J., 1994. Environmental controls on the petrology of a late Holocene speleothem from Botswana with annual layers of aragonite and calcite. *Journal of Sedimentary Research*, **64**: 147–155.
Van Beynen, P.E., Bourbonniere, R., Ford, D.C., and Schwarcz, H.P., 2001. Causes of color and fluorescence in speleothems. *Chemical Geology*, **175** (3–4): 319–341.

Cross-references

Anhydrite and Gypsum
Caliche–Calcrete
Cave Sediments
Geodes
Hydroxides and Oxyhydroxide Minerals
Neomorphism and Recrystallization
Siliceous Sediments
Sulfide Minerals in Sediments
Travertine and Other Spring Deposits

SPICULITES AND SPONGOLITES

Introduction

Spongolites and spiculites are common in the rock record, and represent environments that were dominated by sponges, a phylum of organisms that inhabit practically all modern aquatic environments (Reitner and Keupp, 1991). Sponges use silica and/or carbonate and/or organic tissue to build skeletons composed of discrete spicules or amalgamated spicules that form a rigid skeleton. A sediment predominantly composed of discrete spicules is termed a spiculite, whilst one predominantly composed of rigid bodied sponges is termed a spongolite.

Taxonomy

Phylum Porifera contains an estimated 15,000 living sponges making it one of the most diverse and successful groups of aquatic organisms. Fossil and extant sponges are divided into two subphyla: Cellularia (classes Demospongia, Calcarea, Archaeocyatha); and Symplasia (class Hexactinellida; Hooper, 2000). Most demosponges and hexactinellids form their skeletal material from siliceous spicules and/or spongin (organic tissue), with some calcareous "corraline" demosponges. Calcarea and Archeocyatha have calcareous skeletal components. Most sponges contain discrete spicules within a mass of cells ("soft sponges"), with relatively few taxa containing rigid skeletons formed by articulation, secondary silicification/calcification, or fusion of spicules. Spicule morphologies are extremely diverse, with taxonomic identification determined from the complete spicule complement, and arrangement. Relatively few spicules are diagnostic, making identification from isolated sponge spicules generally only possible to a high taxonomic level. This has resulted in paleontological concentration on rigid skeleton sponges. Accurate identification from isolated spicules is one of the most pressing problems in sponge paleontology. Rigid skeleton sponges traditionally include: lithistids (siliceous), stromatoporoids (calcareous), sphinctozoans (calcareous), chaetitids (calcareous), and sclerosponges (calcareous basal skeleton, siliceous spicules). However, it is now generally accepted that these "fossil groups" are primarily optimal sponge body plans rather than monophyletic lineages, and that these body plans have evolved convergently from various demosponge and calcarea lineages during the Phanerozoic (e.g., Debrenne, 1999; Pisera, 1999). The current paleontological revision will hopefully clarify the many problems with fossil sponge identification. Hexactinellid (siliceous) and archaeocyathid (calcareous) are apparently the sole monophyletic groups amongst rigid bodied sponge fossils (see figure 6 of Debrenne, 1999, Hooper and van Soest, in preparation).

Paleoecology

Sponges are sessile benthic invertebrates that gather food by filtering particulate, colloidal and dissolved nutrients from water actively pumped through internal canals and chambers (Hooper, 2000; Rigby and Stearn, 1983). In general they are slow growing, contain bacterial symbionts, and have a cell-colony metabolic organization. Sponges reproduce both sexually and asexually, with sponge larvae motile for only a short period before encrusting a firm substrate. They generally have low growth rates and long lifespans (Reitner and Keupp, 1991), and are some of the most evolutionarily conservative animals known (e.g., Pisera, 1999), Freshwater sponges are restricted to a few lineages, and fossil spiculites and spongolites are generally interpreted as indicators of marine environments (Hooper, 2000). Sponges are most common on temperate carbonate shelves, within reef and rocky substrate environments, and within muddy, low-energy environments (van Soest et al., 1994). Their general preference is for oligotrophic, low energy, fine-grained, low sedimentation-rate environments. However, sponges exploit many different habitats and such a diverse and evolutionarily conservative group of animals will have developed species-dependent environmental controls (e.g., Maldonado et al., 1999). Paleoecological issues deserve considerable further investigation.

Sponge-dominated deposits have previously been interpreted as indicating deep and/or cold-water environments (Beauchamp and Desrochers, 1997) due to modern hexactinellid spiculite and spongolite accumulations in arctic and antarctic regions (Conway et al., 1991). However, fossil spiculites and spongolites, including hexactinellid communities, have been interpreted as shallow and warm (Gammon et al., 2000), and the presence of hexactinellid and demosponge faunas throughout modern oceans, including tropical assemblages, suggests modern sponge habitats may not be representative of the past because of the inability of modern sponges to compete with more highly evolved organisms such as photosymbiotic corals (e.g., Lévi, 1991; Tabachnick, 1991). The presence of shallow-marine deposits composed primarily of deep-marine sponges also suggests that water-depth is not a major control factor for sponges (Gammon et al., 2000).

The modern ocean is undersaturated with respect to silica, leading to rapid sea-floor dissolution of sponge skeletal remains. Additionally, carbonates and silica buffer solutions to substantially differing pH's. Opaline biogenic silica is therefore generally unstable within predominantly calcareous sediments, and vice versa (Johnson, 1976). Siliceous sponge assemblages are probably significantly impoverished by diagenesis and assemblage reconstruction is particularly challenging.

Spiculite and spongolite

Modern Arctic and Antarctic spiculite and spongolite deposits contain matted, interlocked spicule-mat textures developed on glacial moraine substrates in deep shelf localities (Conway et al., 1991). The very high length:width ratio of interlocked spicules produces a cohesive sediment structure that resists erosive currents and provides greater substrate aeration which promotes microbial populations. The resulting muddy, microbially-influenced matrix may be a good analogue for ancient spicular mudmounds. Ancient spiculites are occasionally cross-bedded, but more commonly massive, possibly due to such spicule networks. Detailed textural work on spiculites is a promising avenue for further sponge community research.

Modern lithistid localities are muddy, deep shelf localities, which is consistent with muddy spongolite deposits in the rock record (e.g., Wiedenmayer, 1994). Such fossil sponge faunas represent in situ assemblages, but are often poorly preserved due to diagenesis. There are no modern analogues for ancient calcareous or siliceous sponge reef/bioherm spongolites (e.g., Reitner and Keupp, 1991). These have coarse-grained whole and fragmented sponges with or without copious synsedimentary cements. Interpretations exist for both deep and shallow waters reefs (Vacelet, 1988), consistent with the generally large depth ranges and nonphotosymbiotic metabolisms of modern sponges.

Spiculites and spongolites have been formed repeatedly throughout the Phanerozoic. The earliest known sponge fossils are Late Neoproterozoic, but sponge-dominated environments started with early Cambrian archaeocyathids, the world's first metazoan reef builders (Debrenne, 1999). Sphinctozoans and stromatoporoids were common reef builders during the Paleozoic (mainly Ordovician—Devonian), whilst siliceous sponges tended to form spongolite bioherms and spiculitic mudmounds.

Siliceous sponges particularly enjoyed Mesozoic tropical Tethyan environments, where they arose to dominance repeatedly through the Jurassic and Cretaceous (mainly lithistids with some hexactinellids in relatively deep shelf environments; Pisera, 1999). Calcareous stromatoporoids and sphinctozoans made brief appearances as reef-builders, but were not as prolific as Paleozoic forms (Debrenne, 1999).

The general lack of Tertiary spiculites and spongolites has been correlated to the diatom radiation that is inferred to have decreased oceanic dissolved silica, a key nutrient for siliceous sponges (Maldonado et al., 1999). However, sponge-dominated sediments occur through the Eocene (Europe and Australia; Gammon and James, 2001; Pisera, 1999). These deposits are primarily shallow-marine deposits in unusual basinal settings, emphasizing the Phanerozoic trend of sponge displacement to more marginal habitats by more evolved taxa.

Paul R. Gammon

Bibliography

Beauchamp, B., and Desrochers, A., 1997. Permian warm-to very cold-water carbonates and cherts in northwest Pangea. In James, N.P., and Clarke, J.D.A. (eds.), *Cool-Water Carbonates*. SEPM (Society for Sedimentary Geology), Special Volume 56, pp. 327–347.

Conway, K.W., Barrie, J.V., Austin, W.C., and Luternauer, J.L., 1991. Holocene sponge bioherms on the western Canadian continental shelf. *Continental Shelf Research*, **11**: 771–790.

Debrenne, F., 1999. The past of sponges—sponges of the past. In Hooper, J.N.A. (ed.), *Proceedings of the 5th International Sponge Symposium: Origin and Outlook*. Brisbane: Memoirs of the Queensland Museum, 707pp.

Gammon, P.R., and James, N.P., 2001. Palaeogeographical influence on Late Eocene biosiliceous sponge-rich sedimentation, southern Western Australia. *Sedimentology*, **48**: 559–584.

Gammon, P.R., James, N.P., and Pisera, A., 2000. Eocene spiculites and spongolites in southwestern Australia: not deep, not polar, but shallow and warm. *Geology*, **28** (9): 855–858.

Hooper, J.N.A., 2000. *Sponge Guide: Guide to Sponge Collection and Identification*. Brisbane: Queensland Museum, 141pp.

Hooper, J.N.A., and van Soest, R. (eds.), 2002. *Systema Porifera* (in preparation—a complete revision of fossil and extant sponge taxa).

Johnson, T.C., 1976. Controls on the preservation of biogenic opal in sediments of the eastern tropical Pacific. *Science*, **192**: 887–890.

Lévi, C., 1991. Lithistid sponges from the Norfolk Rise–Recent and Mesozoic genera. In Reitner, J., and Keupp, H. (eds.), *Fossil and Recent Sponges*. Berlin: Springer-Verlag, pp. 72–82.

Maldonado, M., Carmona, M.C., Uriz, M., and Cruzado, A., 1999. Decline in Mesozoic reef-building sponges explained by silicon limitation. *Nature*, **401**: 785–788.

Pisera, A., 1999. Postpaleozoic history of the siliceous sponges with rigid skeleton. In Hooper, J.N.A. (ed.), *Sponge Guide: Guide to Sponge Collection and Identification*. Brisbane: Queensland Museum, pp. 463–472.

Reitner, J., and Keupp, H. (eds.), 1991. *Fossil and Recent Sponges*. Springer-Verlag.

Rigby, J.K., and Stearn, C.W., 1983. *Sponges and Spongiomorphs*. Short course of The Paleontological Society. Indianapolis: University of Tennessee, 220pp.

Tabachnick, K.R., 1991. Adaptation of the hexactinellid sponges to deep-sea life. In Reitner, J., and Keupp, H. (eds.), *Fossil and Recent Sponges*. Springer-Verlag, pp. 378–386.

Vacelet, J., 1988. Indications de profondeur données par les Spongiaires dans les milieux benthiques actuels. *Géologie Méditerraneenne*, 15: 13–26.

van Soest, R.W.M., van Kempen, T.M.G., and Braekman, J.-C., 1994. *Sponges in Time and Space: Biology, Chemistry, Paleontology*. Rotterdam: A.A. Balkema.

Wiedenmayer, F., 1994. *Contributions to the Knowledge of Post-Palaeozoic Neritic and Archibenthal Sponges (Porifera)*. Schweizerische Paläontologische Abhandlungen 116. Basel: Kommission der Schweizerischen Paläontologischen Abhandlungen.

STAINS FOR CARBONATE MINERALS

Introduction

Stains have been devised for the rapid identification of many common minerals; Reid (1969) provides an excellent compendium. The identification of minerals, however, can now be achieved with much greater certainty using modern analytical techniques (SEM, microprobe and so on). Stains have been devised that display intracrystalline variations in chemical composition although techniques such as cathodoluminescence, fluorescence and backscatter electron imaging can display this type of information more comprehensively and reliably than staining. Some stains, however, are very versatile, they can be used in the field, on rock slabs or thin sections; they are inexpensive, easy to use and rapid to apply.

Staining carbonate minerals

The staining of carbonate minerals has a long history (Lemberg, 1887) and involves a plethora of different methods (Freidman, 1959, 1971; Reid, 1969).

General carbonate stain

The most widely used stain for carbonates employs a mixture of Alizarin red S and potassium ferricyanide dissolved in a dilute hydrochloric acid solution. Set procedures are published for this dual stain (Dickson, 1965, 1966; Evamy, 1963, 1969) but the stain can, within certain limits, be modified to suit the needs of the observer and the type of material being stained. The organic dye Alizarin red S (ARS) will produce a pink to red stain on any carbonate that will react with dilute acid. The reaction between carbonates and acid is usually controlled (1–2 minutes at 25°C for thin sections) so that the more reactive minerals, such as calcite and aragonite, stain red but the less reactive ones, such as dolomite and siderite, remain unstained. Calcite is however an anisotropic mineral that is more soluble in a direction parallel to its c-axis than normal to the c-axis. The ARS stain (at HCl concentrations between 0.2 percent and 1.0 percent, vol./vol.) can differentiate calcite sections that are normal to the c-axis (stained pink) from those that are parallel to the c-axis (stained red). This appears counterintuitive—the c-axis normal section reacts more vigorously with the acid yet stains less intensely, and vice versa. This is due to greater evolution of CO_2 bubbles from the c-axis normal surface preventing the stain from settling. The c-axis parallel sections produce fewer bubbles and stain precipitation is unimpeded. This differentiation is only effective at acid concentrations between 0.2 percent and 1.0 percent (vol./vol.) given stain times of between 1 minute and 2 minutes.

The intensity of the ARS stain is affected by HCl concentration. At 1.5 percent the stain is faint, vigorous evolution of CO_2 bubbles occurs, and the section is etched thinner (from 30 μm to ~15 μm) after being treated for 1 minute. At 0.1 percent HCl concentration and 1 minute staining time the calcite c-axis orientation is weakly differentiated, and the intensity of the stain masks the underlying fabric. Staining for more than 2 minutes at 0.1 percent HCl causes the layer of stain to crack and peel off the stained surface. Reaction between calcite and acid in the 1–2 minutes recommended for staining is very sensitive to acid concentration, which can be adjusted to develop the desired stain intensity. Dolomite does not stain using these acid concentrations but as iron is substituted into the dolomite lattice it becomes more reactive. Ferroan dolomite and ankerite react with dilute HCl, effervescing and staining with ARS. The ARS stain color of ankerite (mauve) can be distinguished from calcite (red) but in the dual stain the mauve color is suppressed due to intense staining by potassium ferricyanide.

Potassium ferricyanide (PF) produces a precipitate of Turnbull's blue when ferrous iron is released to the staining solution. It might be expected that siderite (ferrous carbonate) would react with this stain but siderite does not react with dilute cold HCl, iron is not released to the staining solution and consequently no stain is precipitated. The rate of reaction of the PF stain, as with ARS, is controlled by reaction rate between carbonate and HCl. Calcites and dolomites containing ferrous iron do react with the stain but dolomite reacts less vigorously than calcite so the intensity of PF stain is not proportional to the iron concentration of the two minerals. The PF stain is very sensitive to ferrous iron concentrations in calcite and will distinguish differences in concentration in the order of a few 100 ppm. Lindholm and Finkelman (1972) correlated iron concentration of calcite with stain color, but to match their color range their staining procedure must be reproduced exactly. The relationship between iron concentration and stain color is non-linear as the reactivity of carbonate minerals increases with rising iron concentration (Reeder, 1983). Ferroan calcite stains blue and ferroan dolomite stains green to turquoise. As iron concentration increases in these two minerals the colors converge. The difficulty in distinguishing such iron-rich carbonates can be overcome by staining with ARS alone.

When iron concentrations are low the dual staining procedure can be modified to increase iron differentiation. Extending the staining time in the dual stain beyond 90 seconds is inadvisable as the stain cracks and thick deposits of ARS obscure the section. Material with low iron concentrations can be stained first in a PF/HCl solution for 60 seconds before staining with the dual (ARS + PF) stain. The distribution of ferrous iron (resolvable to 1 μm) in thin section or rock slab can be differentiated in a few minutes using this stain (Figure S37*).

*Figure S37 appears on Plate II, facing page 115.

Distinguishing aragonite from calcite

Feigl's stain (Feigl, 1937), a solution of silver and manganese sulfates to which sodium hydroxide is added (for preparation see Friedman, 1959), stains aragonite black. Other orthorhombic carbonates also react positively with Feigl's stain. The black color is due to a precipitate of manganese oxide and metallic silver. Calcite remains unstained during limited exposure to Feigl's solution. The stain differentiates aragonite from calcite due to their different solubilities and also causes differential etching between the two minerals. Schneidermann and Sandberg (1971) have used this selective etching (imaged by SEM) as an additional feature of the stain to help distinguish between the two minerals. The latter authors also recommend using $MnSO_4 \cdot H_2O$, rather than $MnSO_4 \cdot 7H_2O$, to prepare the stain, diluting the original recipe with water, and warming the solution. Variations in crystal size and orientation have led to differing results using Feigl's stain so that many regard the stain as unreliable.

Distinguishing Mg calcite from calcite

The Titan-yellow (Clayton yellow) stain for assessing the Mg content of calcite was described by Friedman (1959) modified by Winland (1971) and again improved by Choquette and Trusell (1978). Choquette and Trusell give the staining and fixing procedure for this stain which barely stains calcite with up to 3 percent $MgCO_3$ stains Mg calcite with 5–8 percent $MgCO_3$ pink to pale red, and Mg calcite with >8 percent $MgCO_3$ a deep red color. This indication of Mg content is qualitative because crystal orientation and size affect the stain in addition to Mg content. The stain is cheap and simple to use but has not gained wide acceptance.

Conclusions

Stains have been used to determine mineralogy and intracrystalline chemical variations although both these properties can now be more reliably and quantitatively determined using analytical techniques. Stains can however be applied to outcrop, rock slabs or thin sections in a few minutes at negligible expense. Their application requires little training other than some basic safety training in the use of chemical substances and an understanding of how the stain operates.

J.A.D. Dickson

Bibliography

Choquette, P.W., and Trusell, F.C., 1978. A procedure for making the Titan-yellow stain for Mg-calcite permanent. *Journal of Sedimentary Petrology*, **48**: 639–641.
Dickson, J.A.D., 1965. A modified staining technique for carbonates in thin section. *Nature*, **105**: 587.
Dickson, J.A.D., 1966. Carbonate identification and genesis as revealed by staining. *Journal of Sedimentary Petrology*, **36**: 491–505.
Evamy, B.D., 1963. The application of a chemical staining technique to a study of dedolomitization. *Sedimentology*, **2**: 164–170.
Evamy, B.D., 1969. The precipitational environment and correlation of some calcite cements deduced from artificial staining. *Journal of Sedimentary Petrology*, **39**: 787–821.
Feigl, F., 1937. *Qualitative Analysis by Spot Test*. New York: Nordemann Publications Company.
Friedman, G.M., 1959. Identification of carbonate minerals by staining methods. *Journal of Sedimentary Petrology*, **29**: 87–97.
Friedman, G.M., 1971. Staining. In *Procedures in Sedimentary Petrology*. New York:Wiley-Interscience, pp. 511–530.
Lemberg. J., 1887. Zur microchemischen Untersuchung von Calcit, Dolomit und Predazzit. *Zeitschrift der Deutschen Geologischen Gesellschaft*, **40**: 357–359.
Lindholm, R.C., and Finkelman, R.B., 1972. Calcite staining: semiquantitative determination of ferrous iron. *Journal of Sedimentary Petrology*, **42**: 239–242.
Reeder, R.J., 1983. Crystal chemistry of the rhombohedral carbonates. In *Carbonates: Mineralogy and Chemistry*. Reviews in mineralogy 11, Mineralogical Society of America, pp. 1–47.
Reid, W.P., 1969. Mineral staining tests. *Colorado School of Mines, Mineral Industries Bulletin*, **12**: 1–20.
Schneidermann, N., and Sandberg, P.A., 1971. Calcite–aragonite differentiation by selective staining and electron microscopy. *Gulf Coast Association of Geological Societies, Transactions*, **21**: 349–352.
Winland, H.D., 1971. Non-skeletal deposition of high-Mg calcite in the marine environment and its role in the retention of textures. In *Carbonate Cements*. John Hopkins University Studies in Geology, 19, 278–284.

Cross-references

Carbonate Mineralogy and Geochemistry
Diagenesis
Dolomite Textures
Dolomites and Dolomitization

STATISTICAL ANALYSIS OF SEDIMENTS AND SEDIMENTARY ROCKS

Statistical analysis is a standard tool used by sedimentologists, who are concerned with sampling and describing properties of populations of particles. They use summary statistics such as the range, mean, mode, median, standard deviation, variance, skewness, and kurtosis to transmit information about sediments. Univariate, bivariate, and multivariate statistical tests are among their basic statistical tools. Descriptive statistics in geology date back to the 19th Century when Charles Lyell used the ratio of living to extinct invertebrate fossils to subdivide the Tertiary into epochs as reported in his *Principles of Geology* (1833) (see Merriam, 1981).

Because much of the historical data in the earth sciences are qualitative, nonparametric statistics are used for their analysis. Nonparametric tests do not require assumptions about the populations from which the samples were taken as do parametric tests. There also is an inherent amount of uncertainty in geological properties because of irregularly spaced or missing data. Methods for determining the spatial relations and display of sediment properties, therefore, are especially important. Dieter Marsal's (1987) book is an excellent introduction to statistics for the geoscientist; John Davis' (1986) book on statistics and data analysis is an introductory to immediate level text with many examples and a computer program, STAT, on disk; and Nick Rock's (1988) numerical geology is an intermediate text for university courses on the subject with an extensive reference list and glossary of terms.

With introduction of quantitative methods, the subject of sedimentation gradually transformed into sedimentology. One of the first papers in this transition was the classic *On the*

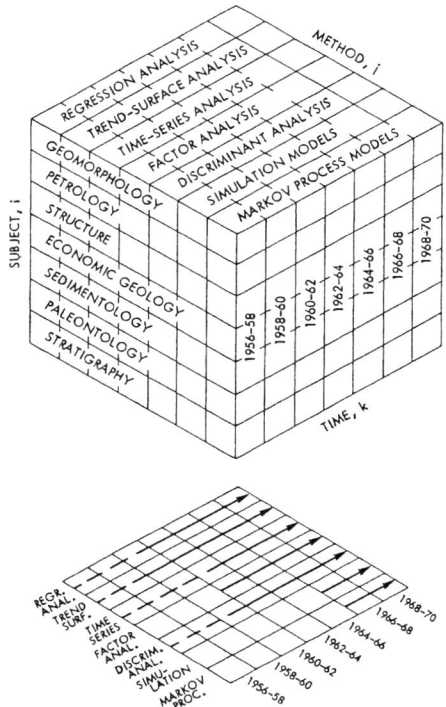

Figure S38 Subset of geological subjects, methods of data analysis, and time in two-year periods (upper); horizontal slice to show time of entry of computer applications in sedimentology (Krumbein, 1969, p. 254). see section on: William Christian Krumbein.

Application of Quantitative Methods to the Study of the Structure and History of Rocks by Henry C. Sorby in 1908, which presented results based on a mass of data accumulated from stream experiments; W. Mackie described rounding of particles in a 1897 paper, *On the Laws that Govern the Rounding of Particles of Sand*; and Johann A. Udden published his paper, *Mechanical Composition of Clastic Sediments*, in 1914 (see Merriam, 1981, for references). This early work on clastic sediments was followed by the contributions of C.K. Wentworth, William C. Krumbein (arguably the most influential of the numerical sedimentologists; see article in this volume), A.B. Vistelius, and J.C. Griffiths, who applied regression analysis, time-series analysis, factor analysis, discriminant analysis, and Markov chain analysis to sedimentological problems (Figure S38).

Bill Krumbein and Francis Pettijohn's 1938 textbook, *Manual of Sedimentary Petrography*, was a primer on the analysis of sediments and included two chapters on statistical methods, a forerunner of Krumbein's later work. Robert Folk's manual, *Petrology of Sedimentary Rocks* (1961) contains basic statistical techniques for the analysis of clastic rocks; Robert Miller and James Kahn published a book on a general survey of statistical methods in 1962; the Krumbein and Franklin Graybill book on statistical models in 1965 was an introduction to the subject with emphasis on sediments; but John Cedric Griffiths' *Scientific Method in Analysis of Sediments* published in 1967 was at the time the ultimate publication on statistics applied to sediments and sedimentary rocks. Griffiths' book was followed by Frits Agterberg's tome on geomathematics (1974) and Walther Schwarzacher's book on sedimentation models (1975), both for the advanced student and professional.

Carbonate sediments have been treated differently than clastic sediments because they are not amenable to disaggregation and particle analysis. John Imbrie and Edward Purdy were among the first to apply factor analysis to carbonates in their 1962 paper, *Classification of Modern Bahamian Carbonate Sediments*, followed by Imbrie's *Factor and Vector Analysis Programs for Analyzing Geological Data* in 1963 and Purdy's *Recent Calcium Carbonate Facies of the Great Bahamas Bank* in the same year (see Merriam, 1981, for references).

Statisticians, too, have applied their tools to problems in the earth sciences since the turn of the 20th Century. Karl Pearson studied the variation in homogeneous material; A.N. Kolmogorov used the lognormal law of distribution of particle sizes to analyze crushed rock; Churchill Eisenhart was interested in significance tests on lithological variation; C.J.C. Ewing discussed operator variation in heavy-mineral separation; S.S. Wilks explored statistical inference; K.V. Mardia worked with directional data; G.S. Watson contributed to statistics of orientation data; D.M. Hawkins wrote on robust statistics; J.C. Gower was concerned with sediment classification; M.F. Dacey applied Markov chain analysis; and J. Tukey suggested that methods of confirmatory data analysis, such as the jackknife approach, were logical and useful for geological problems.

In the early 1960s a French engineer, Georges Matheron, introduced to the earth sciences a new branch of statistics known as *geostatistiques* (geostatistics), in his papers *Traité de geostatistique appliquée*, parts 1 (1962) and 2 (1963). For an introduction in English, see Matheron (1963). The new school is based on the idea of regionalized variables or partially dependent observations located in two- or three-dimensional space. Kriging is the name applied for methods of interpolating spatial geological data that makes use of geostatistical concepts.

Daniel F. Merriam

Bibliography

Agterberg, F.P., 1974. *Geomathematics*: Elsevier Scientific.
Cubitt, J.M., and Henley, S. (eds.), 1978. *Statistical Analysis in Geology*. Benchmark Papers in Geology 37, Stroudsburg, PA: Dowden, Hutchinson, and Ross.
Davis, J.C., 1986. *Statistics and Data Analysis in Geology*, 2nd edn. John Wiley & Sons.
Folk, R.L., 1961. *Petrology of Sedimentary Rocks*. Austin, TX: Hemphill's.
Gill, D., and Merriam, D.F. (eds.), 1979. *Geomathematical and Petrophysical Studies in Sedimentology*. Pergamon Press.
Griffiths, J.C., 1967. *Scientific Method in Analysis of Sediments*. McGraw-Hill.
Krumbein, W.C., 1969. The computer in geological perspective. In Merriam, D.F. (ed.), *Computer Applications in the Earth Sciences*. Plenum Press, pp. 251–275.
Krumbein, W.C., and Pettijohn, F.J., 1938. *Manual of Sedimentary Petrography*. Appleton-Century.
Krumbein, W.C., and Graybill, F.A., 1965. *An Introduction to Statistical Models in Geology*. McGraw-Hill.
Marsal, D., 1987. *Statistics for Geoscientists*. Pergamon Press.
Matheron, G., 1963. Principles of geostatistics. *Economic Geology*, **58**: 1246–1266.
Merriam, D.F. (ed.), 1970. *Geostatistics, a Colloquium*. Plenum Press.
Merriam, D.F. (ed.), 1972. *Mathematical Models of Sedimentary Processes*. Plenum Press.

Merriam, D.F. (ed.), 1976. *Quantitative Techniques for the Analysis of Sediments*. Pergamon Press.

Merriam, D.F. (ed.), 1981. Down-to-earth statistics: solutions looking for geological problems. *Syracuse University Geological Contributions*, **8**: 97pp.

Miller, R.L., and Kahn, J.S., 1962. *Statistical Analysis in the Geological Sciences*. John Wiley & Sons.

Rock, N.M.S., 1988. *Numerical Geology*. Lecture Notes in Earth Sciences 18, Springer-Verlag.

Sorby, H.C., 1908. On the application of quantitative methods to the study of the structure and history of rocks. *Quarterly Journal of the Geological Society of London*, **44** (2): 171–233.

Schwarzacher, W., 1975. *Sedimentation Models and Quantitative Stratigraphy*. Elsevier Scientific.

Cross-reference

Sedimentologists: William Christian Krumbein

STORM DEPOSITS

Large storms have wide-ranging effects in a spectrum of environments from outer shelves to coastal plains. The most common deposits of storms in the rock record are sandstone beds deposited under powerful near-bottom water motions produced by waves and currents. Early studies of the effects of storms on modern shelves (e.g., Hayes, 1967) sparked interest in finding analogues in the rock record. This resulted in the development of storm facies models, with particular emphasis on hummocky cross-stratification (HCS) (Walker, 1984). The origin and dynamics of the depositing flows were considered controversial (see Swift *et al.*, 1983; Walker, 1984) and initial models relied on cross-shelf transport by turbidity currents (Hamblin and Walker, 1979). Actualistic models of storm dynamics subsequently constructed by oceanographers and marine geologists largely focused on nearly shore-parallel geostrophic currents. Duke (1990) suggested that ancient shallow-marine storm deposits (tempestites) were deposited by geostrophic combined flows and subsequent work indicates that combined flows are in fact recorded in ancient sandy tempestites (e.g., Midtgaard, 1996). However, many ancient storm-deposited sandstone beds show greater thickness and cross-shelf distribution than their modern counterparts (Leckie and Krystinik, 1989; Myrow and Southard, 1996). In addition, the time between successive beds in many deposits in the rock record may be many orders of magnitude too large to explain as a simple series of storm events.

The stratification preserved in tempestites, particularly HCS, has been studied experimentally (Arnott and Southard, 1990) and through the analysis of grain fabric (Cheel, 1991). Such studies, and those of modern shelves (e.g., Amos *et al.*, 1996), confirm that HCS forms as a result of complex oscillatory flow or combined flow (waves and currents). A wide range of vertical stratification sequences is possible within individual storm-deposited beds (Myrow and Southard, 1991). Three processes operate during storms to produce the wide range of tempestites encountered in the sedimentary record: geostrophic currents, wave oscillations, and excess-weight forces (density-induced flow) (Myrow and Southard, 1996). The relative importance of these three processes may change during individual events and between events both within and between basins. Thus considerable variability is possible in the character of individual beds, although most beds display graded bedding and evidence for flow deceleration.

Geostrophic flows are near-shore-parallel flows that result from Coriolis deflection of water masses that are driven offshore in response to pressure gradients associated with coastal setup. Combined flows result from the superposition of storm-generated waves with these flows. Because these geostrophic flows are oriented slightly obliquely offshore, instantaneous bed shear stresses in the combined-flow boundary layer show asymmetry in the offshore direction (Duke, 1990). Sediment transport results from the nonlinear addition of the two components of flow in the boundary layer, with the net result that wave oscillations are stronger in the offshore direction and thus preferentially move sediment in that direction. Whether this asymmetry in transport can account for significant cross-shelf transport is unclear. Regardless of transport mechanisms, most tempestite successions show patterns of decreased bed thickness and evidence for less powerful currents from onshore to offshore. An exception to this rule occurs in areas of nearshore sediment bypass (Myrow, 1992).

In many cases, ancient deposits contain evidence for powerful offshore-directed flow, which is at odds with deposition under combined flows. Such flow is possibly driven, at least in part, by excess-weight forces. These forces result from high suspended sediment concentrations (SSCs) near the bed. High SSCs can result from resuspension of shoreline sand by storm processes. A number of studies indicate that very high SSCs (up to 4 g/L) are produced on shorefaces and inner shelves and that this leads to powerful seaward-directed sediment-laden flows (e.g., Wright *et al.*, 1994). Plumes associated with fluvial discharge events may also produce hyperpycnal (bottom-hugging) flows, and such flows are termed oceanic floods (Wheatcroft, 2000). These could be pre-ignited with regard to autosuspension, but this is unnecessary for significant cross-shelf transport because storm waves can provide the extra turbulence needed to maintain flow. Results of both oceanographic (e.g., Cacchione *et al.*, 1999) and modeling experiments of the northern California coast indicate that gravity driven underflows associated with ocean floods are an important factor for cross-shelf transport of fine-grained sediment. Such flows have high SSCs (up to 2.5 g/L) even at mid-shelf depths (50 m to 70 m).

The sequence-stratigraphic context of tempestites may have a strong control on their abundance, nature, and stratigraphic distribution. Transgressive to early highstand conditions may have higher accommodation space and thus greater potential for Coriolis effect to produce geostrophic flow, whereas late highstand conditions promote greater bottom friction and more offshore-directed flow (McKie, 1994). Late highstand to early lowstand systems tracts of third- or higher order sequences tend to have abundant tempestites due to mobilization of coastal sand during relative sea level fall (Einsele, 1996). Sediment supply by rivers, or *in situ* carbonate production, has a first-order control on tempestite thickness and abundance. More work is needed to better constrain the recurrence intervals of ancient deposits. Comparison of these data to those from the Modern will help to establish the degree to which actualistic models apply to the ancient.

Paul Myrow

Bibliography

Amos, C.L., Li, M.Z., and Choung, K.-S., 1996. Storm-generated, hummocky stratification on the outer-Scotian shelf. *Geo-Marine Letters*, **16**: 85–94.
Arnott, R.W., and Southard, J.B., 1990. Exploratory flow-duct experiments on combined-flow bed configurations, and some implications for interpreting storm-event stratification. *Journal of Sedimentary Petrology*, **60**: 211–219.
Cacchione, D.A., Wiberg, P.L., Lynch, J., Irish, J., and Traykovski, P., 1999. Estimates of suspended-sediment flux and bedform activity on the inner portion of the Eel continental shelf. *Marine Geology*, **154**: 83–97.
Cheel, R.J., 1991. Grain fabric in hummocky cross-stratified storm beds: genetic implications. *Journal of Sedimentary Petrology*, **61**: 102–110.
Duke, W.L., 1990. Geostrophic circulation or shallow marine turbidity currents? the dilemma of paleoflow patterns in storm-influenced prograding shoreline systems. *Journal of Sedimentary Petrology*, **60**: 870–883.
Einsele, G., 1996. Event deposits: the role of sediment supply and relative sea-level changes—overview. *Sedimentary Geology*, **104**: 11–37.
Hamblin, A.P., and Walker, R.G., 1979. Storm-dominated shallow marine deposits: the Fernie-Kootenay (Jurassic) transition, southern rocky mountains. *Canadian Journal of Earth Sciences*, **16**: 1673–1690.
Leckie, D.A., and Krystinik, L.F., 1989. Is there evidence for geostrophic currents preserved in the sedimentary record of inner to middle-shelf deposits? *Journal of Sedimentary Petrology*, **59**: 862–870.
McKie, T., 1994. Geostrophic versus friction-dominated storm flow: paleocurrent evidence from the Late Permian Brotherton formation, England. *Sedimentary Geology*, **93**: 73–84.
Midtgaard, H., 1996. Inner-shelf to lower shoreface Hummocky sandstone bodies with evidence for geostrophic-influenced combined flow, lower Cretaceous, west Greenland. *Journal of Sedimentary Research*, **66**: 343–353.
Myrow, P.M., 1992. Bypass-zone tempestite facies model and proximality trends for an ancient muddy shoreline and shelf. *Journal of Sedimentary Petrology*, **62**: 99–115.
Myrow, P.M., and Southard, J.B., 1991. Combined-flow model for vertical stratification sequences in shallow marine storm-deposited beds. *Journal of Sedimentary Petrology*, **61**: 202–210.
Myrow, P.M., and Southard, J.B., 1996. Tempestite deposition. *Journal of Sedimentary Research*, **66**: 875–887.
Swift, D.J.P., Figueiredo, A.G. Jr, Freeland, G.L., and Oertel, G.F., 1983. Hummocky cross-stratification and megaripples: a geological double standard? *Journal of Sedimentary Petrology*, **47**: 1242–1260.
Walker, R.G., 1984. Shelf and shallow marine sands. In Walker, R.G. (ed.), *Facies Models*, 2nd edn. Toronto: Geoscience Canada, Reprint Series 1, pp. 141–170.
Wheatcroft, R.A., 2000. Oceanic flood sedimentation: a new perspective. *Continental Shelf Research*, **20**: 2059–2066.
Wright, L.D., Xu, J.P., and Madsen, O.S., 1994. Across-shelf benthic transports on the inner shelf of the Middle Atlantic Bight during the "Halloween storm" of 1991. *Marine Geology*, **118**: 61–77.

Cross-references

Bedding and Internal Structures
Coastal Sedimentary Facies
Facies Models
Gutter and Gutter Casts
Hummocky and Swaley Cross-Stratification
Paleocurrent Analysis
Sands, Gravels, and their Lithified Equivalents

STROMATACTIS

Stromatactis is a spar body common in Paleozoic carbonate mud-mounds but much less common in Mesozoic ones. Although it has been regarded originally as a fossil (Dupont, 1881), and later as a permineralized unfossilizable organism or as recrystallized patches of the host carbonate, most carbonate workers now agree that it is the result of centripetal cementation in a cavity system, regardless of the origin of the cavity (e.g., Bathurst, 1982; Lees and Miller, 1985; Bourque and Boulvain, 1993; several papers in Monty *et al.*, 1995). Because it is a diagenetic structure, and not a fossil, stromatactis should not be capitalized, nor italicized. There is no consensus on the definition of stromatactis and the origin of the cavities into which it developed.

Some workers define stromatactis in an all-inclusive manner and consider that stromatactis is all spar bodies having a flat base and digitate top resulting from the cement filling of cavities, no matter if the spar bodies, and hence the cavities, are connected or isolated in the host rock (e.g., Monty, *in*

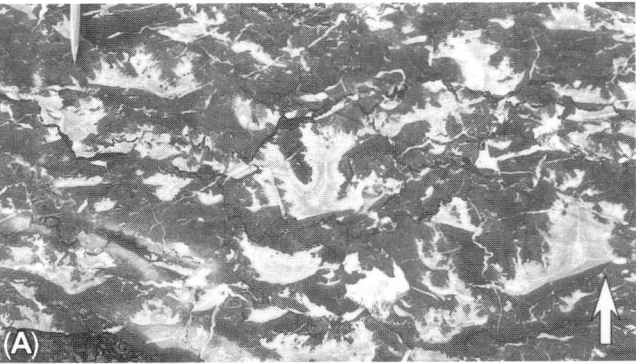

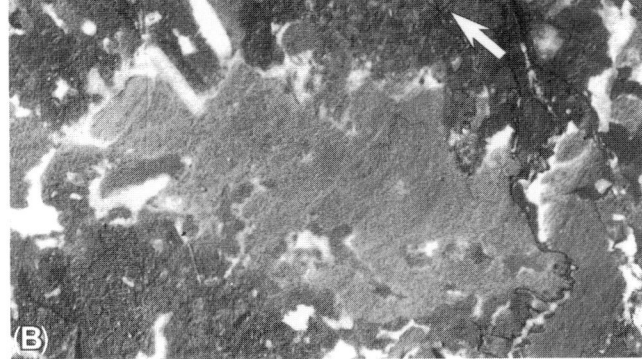

Figure S39 Stromatactis and related structure from Devonian (Frasnian) mud-mounds in the Belgian Ardennes. (A) Typical stromatactis limestone showing cross-section in a stromatactis network. Although the spar bodies look as individuals, they are interconnected in the third dimension. Les Maquettes Quarry. Width of picture is 35 cm. (B) Cross-section of filling of layered sediment in the original cavity network that has inhibited stromatactis development; Croisette Quarry. This kind of structure has been misidentified occasionally as stromatolite. Width of picture is 9 cm. Arrows indicate sedimentary polarity.

Monty et al., 1995). Others consider that too wide-open a definition of stromatactis has led to a misuse of the term, and that stromatactis is a specific fabric that should be more restrictively defined (e.g., Bathurst, 1982, 1998). Based on original material described by Dupont (1881) in the Belgian Ardennes, and following Bathurst (1982) who drew attention to the occurrence of stromatactis in reticulate swarms, Bourque and Boulvain (1993, p.608) proposed to restrict the term stromatactis to "a spar network, whose elements have flat to undulose smooth lower surfaces and digitate upper surfaces, made up principally of isopachous crusts of centripetal cement and embedded in finely crystalline limestone".

Stressing the network nature of stromatactis brings into relief that stromatactis is one of a two end-members continuum of a specific cavity network filling controlled by two antagonistic processes, internal sedimentation versus centripetal marine cementation. When the rate of internal sedimentation is lower than the rate of marine cementation, centripetal cementation of the cavity network forms stromatactis (Figure S39(A)); when the rate of internal sedimentation is higher than the rate of marine cementation, layered sediment inhibits stromatactis development and fills entirely the cavity network (Figure S39(B)) (see Neuweiler et al., 2001, for discussion on the processes involved and their causes).

Several scenarios were proposed to account for the formation of the cavity system, among which: (1) winnowing of noncohesive mud sandwiched between layers of early cemented sediment crusts (Bathurst, 1982) or lithified microbial mats (Pratt, 1982) on the sea-floor; (2) collapse of muddy sediment bridged by cohesive microbial mats on the sea-floor (Lees and Miller, in Monty et al., 1995); (3) fluid escape after the collapse of decayed microbially bound sediment in the shallow burial environment (Monty, in Monty et al., 1995); and (4) collapse of uncemented material in an early cementing sponge/mud mixture in the shallow subsurface environment (Bourque and Boulvain, 1993).

Pierre-André Bourque

Bibliography

Bathurst, R.C.G., 1998. The world's most spectacular carbonate mud mounds (Middle Devonian, Algerian Sahara)—Discussion. *Journal of Sedimentary Research*, **68**: 1051.
Bathurst, R.G.C., 1982. Genesis of stromatactis cavities between submarine crusts in Palaeozoic carbonate mud buildups. *Journal of the Geological Society of London*, **139**: 165–181.
Bourque, P.-A., and Boulvain, F., 1993. A model for the origin and petrogenesis of the red stromatactis limestone of Paleozoic carbonate mounds. *Journal of Sedimentary Petrology*, **63**: 607–619.
Dupont, E., 1881. Sur l'origine des calcaires dévoniens de la Belgique. *Bulletin de l'Académie Royale des Sciences de Belgique*, Third Series, **2**: 264–280.
Lees, A., and Miller, J., 1985. Facies variation in Waulsortian buildups, Part 2; Mid-Dinantian buildups from Europe and North America. *Geological Journal*, **20**: 159–180.
Monty, C.L.V., Bosence, D.W.J., Bridges, P.H., and Pratt, B.R. (eds.), 1995. *Carbonate Mud-Mounds, Their Origin and Evolution*. International Association of Sedimentologists, Special Publication, 23.
Neuweiler, F., Bourque, P.-A., and Boulvain, F., 2001. Why is stromatactis so rare in Mesozoic carbonate mud mounds? *TerraNova*, **13**: 333–337.
Pratt, B.R., 1982. Stromatolitic framework of carbonate mud-mounds. *Journal of Sedimentary Petrology*, **52**: 1203–1227.

Cross-references

Bacteria in Sediments
Carbonate Mud-Mounds
Cements and Cementation
Diagenesis
Diagenetic Structures

STROMATOLITES

Stromatolites are critical geological features as they are the only biogenic sedimentary structures that have an age range extending fully from the early Archean to the Recent, thereby demonstrating a fundamentally uniformitarian phenomenon. First described from the Cambrian of New York State by J. H. Steel in 1825, and already widely recognized by the time the term "stromatolite" was coined by E. Kalkowsky in 1908 for Triassic examples, it was C. D. Walcott in 1914 who noted their predominance in Precambrian strata and first deduced a microbial origin (Riding, 1999). M. Black was the first to document, in 1933, the stromatolite-forming activities of living microbial mats, on tropical tidal flats of Andros Island, Bahamas. The discoveries of modern analogs of columnar forms in Shark Bay, Western Australia in 1961 (Figure S40(A), (B)) and at Lee Stocking Island, Bahamas in 1986 are milestones in the earth sciences (see Golubic, 1991). 3.5 Ga stromatolites from Western Australia represent the oldest physical evidence of biological activities, most certainly photosynthetic, on Earth. In turn, stromatolites signpost the search for evidence of life on other planets.

Definition

Even though the evidence for prior microbial activity is usually circumstantial, the definition of stromatolite has evolved as the composition and metabolism of benthic microbial mats and biofilms in various settings and the wide variety of resulting structures and their evolutionary context have become better understood. Mucilagenous mats and biofilms are complex communities dominated by bacteria and, in the photic zone, by cyanobacteria, although other microorganisms may contribute. While shallow-marine and lacustrine calcareous stromatolites are the most commonly recognized (Riding, 2000), other deposits that can be embraced by the term include many tufas or travertines, phosphate crusts, silica sinters, and manganese nodules, for which a microbial control or influence can be determined or inferred. Thus, Riding (1999; see also Krumbein, 1983) advocated a broad approach: a stromatolite is simply a *laminated benthic microbial deposit*. Although there is no constraint of scale, stromatolites are typically centimeter- to meter-sized, and laminae micrometers to millimeters in thickness. Spheroidal counterparts that rolled around during formation are oncoids (or oncolites). Non-laminated structures similar to stromatolites are thrombolites (Pratt, 1995).

Morphology and microstructure

Stromatolites include beds and domes with planar, wavy, or convex lamination, and reefal masses composed of conical or

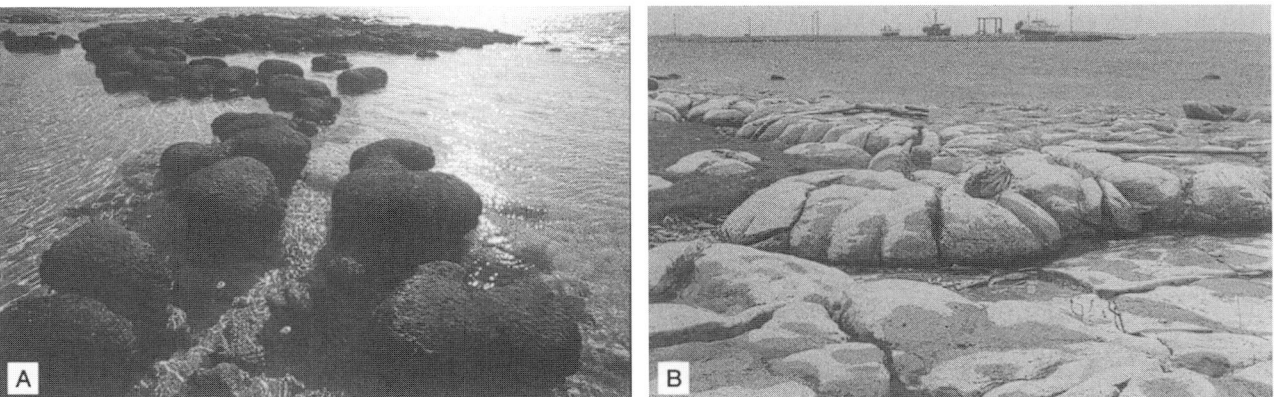

Figure S40 Meter-sized stromatolites. (A) Intertidal flat, Holocene, Shark Bay, Western Australia. (Photograph courtesy of R. A. Wood). (B) Coastal exposure, Petit Jardin Formation (Upper Cambrian), northwestern Newfoundland.

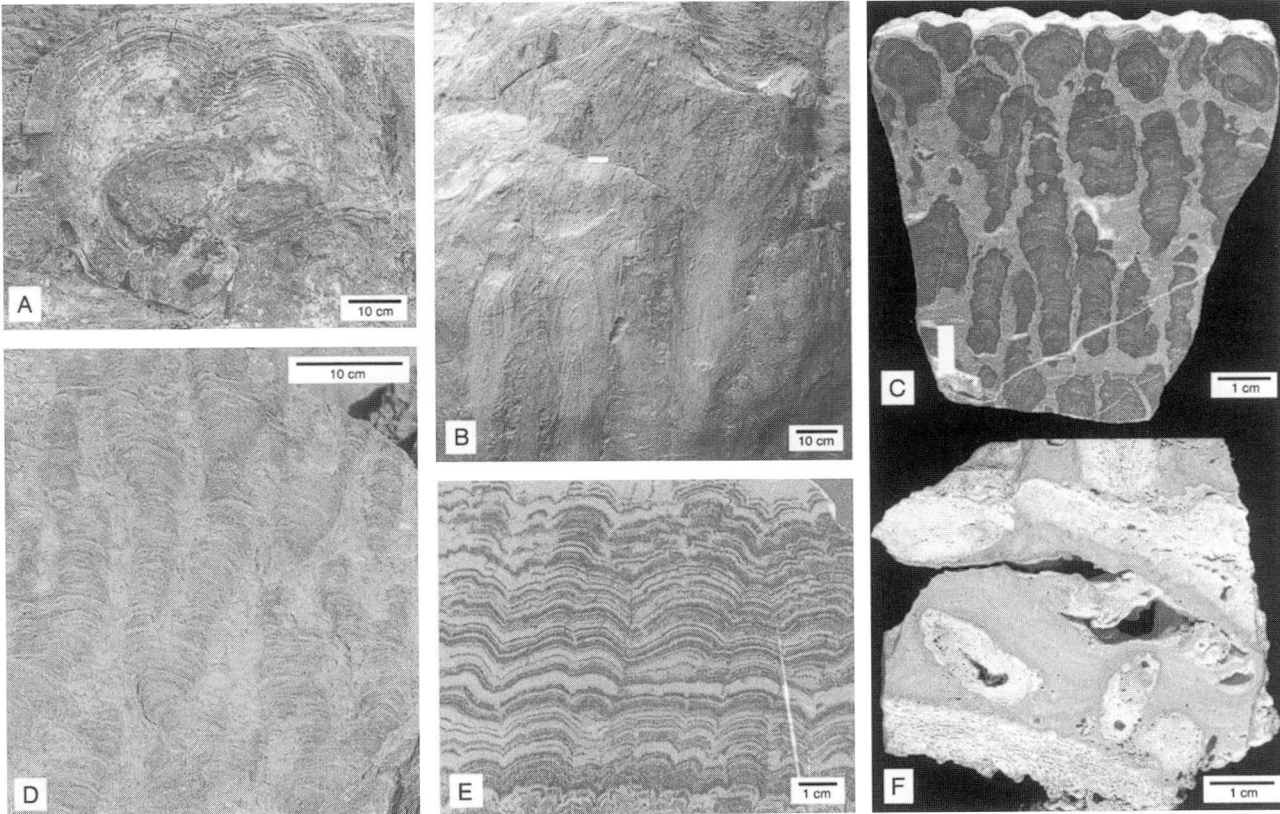

Figure S41 Stromatolite morphology, outcrops and polished surfaces. (A) Hemispheroidal mound, *Collenia* form. Helena Formation (Mesoproterozoic), southwestern Alberta. (B) Conical columns, *Conophyton* form. I-5 member, Atar Group (Neoproterozoic), Mauritania. (C) Branching columns. Cass Fjord Formation (Upper Cambrian), Devon Island, Canadian Arctic. (D) Branching columns, *Baicalia* form. I-5 member, Atar Group. (E) Wavy lamination. Lockport Formation (Lower Silurian), northern Ohio. (F) Laminae coating *Porites* coral rubble. Forereef (Holocene), northern Jamaica.

convexly laminated columnar and branching forms (see examples in Walter, 1976) (Figure S41(A)–(E)). In addition, Phanerozoic marine stromatolites are commonly intergrown with sessile invertebrate skeletons (Figure S41(F)). Beginning in the 19th century, many fossil stromatolites were given linnéan binomial names. This practice has declined, although a number of recurring "form-genera" are still commonly used as descriptors.

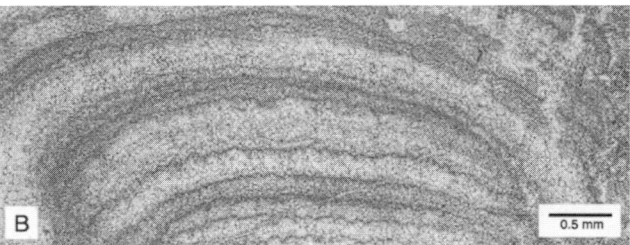

Figure S42 Stromatolite microstructure, photomicrographs of thin sections. (A) Laminae of micrite and clotted micrite, separated by finely crystalline calcite cement. I-5 member, Atar Group (Neoproterozoic), Mauritania. (B) Laminae of micrite and recrystallized micrite. Peg Formation (Paleoproterozoic), Northwest Territiories.

Lamination consists of rhythmic combinations of microbially triggered and sometimes inorganic precipitates, permineralized microbes, physically trapped sedimentary particles, and primary pore space that is subsequently cemented (see examples in Riding and Awramik, 2000) (Figure S42(A), (B)). Calcareous stromatolites are especially vulnerable to diagenetic modification, such as recrystallization, dolomitization, and silicification. The last process may fortuitously preserve microbial remains before substantial degradation (e.g., Knoll et al., 1991). In general, some degree of synsedimentary lithification is a prerequisite to preservation.

Evolution

Stromatolites are conspicuous in younger Archean and Proterozoic shallow-marine carbonate successions (e.g., Walter et al., 1992; Hofmann, 2000). A Precambrian biostratigraphy based on morphology and microstructure has been attempted but has not proved convincing. Still common in the Cambro–Ordovician, they also occur sporadically in younger strata. The apparent decline in abundance of marine stromatolites has been attributed to competition for substrate space by encrusting invertebrates and calcareous algae, the effects of grazing and burrowing animals, chemical changes in seawater, and evolving sedimentological conditions (Pratt, 1982; Grotzinger and Knoll, 1999). Evolutionary aspects of stromatolites from other settings are uncertain at present.

Frontiers

Microbial biofilms and mats are complex systems, constantly in a state of flux between growth and decay, and much remains to be learned about microbially induced diagenesis and the process and temporal significance of lamina formation and overall growth rates in both ancient and modern examples of all types. A thorough assessment of stromatolitic structures of non-carbonate mineralogy and from non-marine settings, such as hot springs, is still in progress. Lastly, if present, the remains of past microbial activities on other planets—astrobiology—may come from stromatolitic rocks, the cue for continued research on Earth's own record.

Brian R. Pratt

Bibliography

Golubic, S., 1991. Modern stromatolites: a review. In Riding, R., and Awramik, S.M. (eds.), *Calcareous Algae and Stromatolites*. Springer-Verlag, pp. 541–561.
Grotzinger, J.P., and Knoll, A.H., 1999. Stromatolites in Precambrian carbonates: evolutionary mileposts or environmental dipsticks? *Annual Reviews of Earth and Planetary Sciences*, **27**: 313–358.
Hofmann, H.J., 2000. Archean stromatolites as microbial archives. In Riding, R., and Awramik, S.M. (eds.), *Microbial Sediments*. Springer-Verlag, pp. 315–327.
Knoll, A.H., Swett, K., and Mark, J., 1991. Paleobiology of a Neoproterozoic tidal flat/lagoon complex: the Draken Conglomerate formation. *Journal of Paleontology*, **65**: 531–570.
Krumbein, W.E., 1983. Stromatolites—the challenge of a term in space and time. *Precambrian Research*, **20**: 493–531.
Pratt, B.R., 1982. Stromatolite decline—a reconsideration. *Geology*, **10**: 512–515.
Pratt, B.R., 1995. The origin, biota and evolution of deep-water mud-mounds. In Monty, C.L.V., Bosence, D., Bridges, P.H., and Pratt, B.R. (eds.), *Carbonate Mud-mounds: Their Origin and Evolution*. International Association of Sedimentologists, Special Publication, 23, pp. 49–123.
Riding, R., 1999. The term stromatolite: towards an essential definition. *Lethaia*, **32**: 321–330.
Riding, R., 2000. Microbial carbonates: the geological record of calcified bacterial-algal mats and biofilms. *Sedimentology*, **47** (1): 179–214.
Riding, R., and Awramik, S.M. (eds.), 2000. *Microbial Sediments*. Springer-Verlag.
Walter, M.R. (ed.), 1976. *Stromatolites*. Developments in Sedimentology 20. Elsevier.
Walter, M.R., Grotzinger, J.P., and Schopf, J.W., 1992. Proterozoic stromatolites. In Schopf, J.W., and Klein, C. (eds.), *The Proterozoic Biosphere: A Multidisciplinary Study*. Cambridge University Press, pp. 253–260.

Cross-references

Algal and Bacterial Carbonate Sediments
Carbonate Mud-Mounds
Diagenesis
Microbially Induced Sedimentary Structures
Neritic Carbonate Depositional Environments
Phosphorites
Reefs
Stromatactis
Tidal Flats
Tufas and Travertines

STYLOLITES

Introduction

Stylolites are macroscopic serrate subplanar surfaces at which mineral material has been removed by pressure dissolution, resulting in a decrease in total volume of rock. Stylolites are

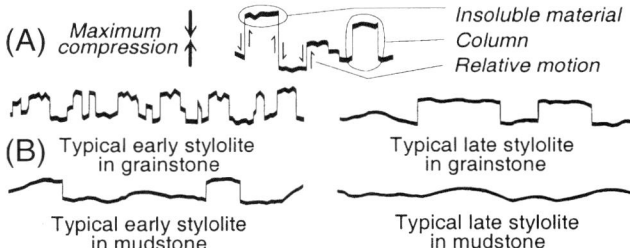

Figure S43 (A) Components of a stylolite. (B) Typical profiles of stylolites in different carbonate lithologies (after Andrews and Railsback, 1997, figure 15).

made readily visible by the accumulation of residues of relatively insoluble minerals, such as pyrite, oxides, clays, and other silicates. If no such residue has accumulated, a stylolite may be recognizable only by its truncation of elements of the host rock's preexisting fabric.

Stylolites commonly consist of interfitting striated columns capped by subplanar laminae of insoluble material (Figure S43(A)). The columns form parallel to maximum compressive stress, as parcels of rock on each side of the stylolite are pushed past each other. The columns are commonly perpendicular to the general plane of the stylolite, but they can be inclined relative to it. The columns are typically parallel to the direction of maximum compression during their formation.

Stylolites take their name from these columns, in that "stylolite" is derived from the Greek word "stylos" (for column or pillar). In older works and some recent Eurasian literature, individual columns are sometimes called "stylolites", but more recent publications generally use that term for the entire surface defined by the columns as a group. Thus a horizontal stylolite in modern American usage consists of many vertical stylolites as defined in some older or Eurasian usage.

Stylolites are common in many limestones and dolostones, can be common in evaporites and deeply buried sandstones, and have been reported in igneous rocks (Golding and Conolly, 1962). Stylolites parallel to bedding of sedimentary rocks are commonly thought to result from compression by overburden, whereas stylolites perpendicular to bedding (transverse stylolites) are usually attributed to tectonic compression. Transverse stylolites can predate or postdate bedding-parallel stylolites in the same rock (Andrews and Railsback, 1997).

Development

Development of a stylolite requires a solution into which minerals can dissolve and a pore network through which dissolved solids can diffuse or advect from the developing stylolite. Porosity also enhances stylolite development by localizing stress on nonpore areas and thus increasing stress there. Some models of stylolite development have therefore suggested that bedding-parallel stylolites form in zones of prior high porosity (Merino et al., 1983), and that most if not all transverse stylolites form along preexisting fractures.

Because its formation involves movement of two rock volumes toward each other, development of a stylolite requires a previously lithified host rock. Compression of unlithified

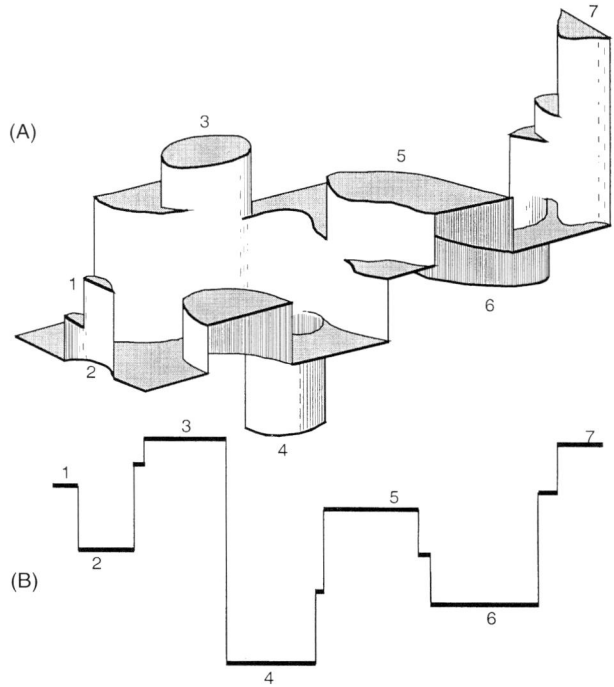

Figure S44 (A) Three-dimensional image of a stylolite (after Smith, 2000, figure 7e). (B) One two-dimensional profile of stylolite shown in A. Numbers identify like positions in the two images.

grainy sediments commonly results in intergranular, rather than whole-rock, compaction, and serrate intergranular contacts can form. Such contacts are microscopic stylolites, the lateral extent of which is limited to the area of contact between two grains. Either intergranular compaction and suturing or (more commonly) cementation can lithify a sediment sufficiently for a true macroscopic stylolite to form.

Stylolites vary from serrate to undulatory, but even most undulatory surfaces ("dissolution seams") are serrate at microscopic scales. Development of a serrate surface requires dissolution that is alternately localized on opposite sides of the stylolite. In limestones, stylolites in grain-rich rocks (grainstones and packstones) are generally more serrate than those in wackestones and micstones, presumably because juxtaposition of various grain types and of intergranular pores localizes destruction of rock volume (Railsback, 1993) (Figure S43(B)). Stylolites that formed early in a rock's history are commonly more serrate than those formed later, presumably because more extensive lithification reduces variability of rock fabrics and of solubility on opposite sides of a stylolite (Andrews and Railsback, 1997) (Figure S43(B)).

Problems and progress

Many studies of stylolites have examined their host rocks as a means to determine lithologies or rock characteristics favoring formation of stylolites. Such studies are potentially flawed, however, because the lithology in which the stylolite formed has inevitably been destroyed during that stylolite's growth. The lithology (or lithologies) in which a stylolite is presently observed may in fact have characteristics that halted, rather than promoted, stylolite development. Otherwise, the present host lithology would have been destroyed by further development of the stylolite.

Traditional studies of stylolites have examined planar sections cut parallel to the columns. Such studies have used the resulting two-dimensional images to classify stylolites according to their general appearance (Park and Schot, 1968; Buxton and Sibley, 1981), quantify their morphology using as many as five parameters (Railsback, 1993; Andrews and Railsback, 1997), or measure their fractal dimension (Drummond and Sexton, 1998). More recently, sequential sectioning has been used to generate three-dimensional images of stylolites (Smith, 2000), which demonstrate that the sides of columns form an anastomosing interconnected surface that can serve as a pathway for fluids (Figure S44).

Significance

Stylolites are significant in several fields. Stylolites are important petrologically because they modify rock fabrics and because they generate dissolved solids that precipitate elsewhere as cement. Stylolites are significant to stratigraphy because weathering of stylolites generates apparent bedding in many stratigraphic sections and because loss of material along stylolites can cause significant stratigraphic shortening unrelated to erosion. Hydrologically, stylolites have been cited as barriers to fluid flow and, in other settings, as conduits for fluid flow. Stylolites are also commonly used as records of compressive stress in tectonic studies, and development of transverse stylolites can contribute to crustal shortening parallel to the direction of their columns.

L. Bruce Railsback

Bibliography

Andrews, L.M., and Railsback, L.B., 1997. Controls on stylolite development: morphologic, lithologic, and temporal evidence from bedding-parallel and transverse stylolites from the US Appalachians. *Journal of Geology*, **105**: 59–73.
Buxton, T.M., and Sibley, D.F., 1981. Pressure solution features in a shallow buried limestone. *Journal of Sedimentary Petrolology*, **51**: 19–26.
Drummond, C., and Sexton, D., 1998. Fractal structure of stylolites. *Journal of Sedimentary Research*, **68**: 8–10.
Golding, H.G., and Conolly, J.R., 1962. Stylolites in volcanic rocks. *Journal of Sedimentary Petrology*, **32**: 534–538.
Merino, E., Ortoleva, P., and Strickholm, P., 1983. Generation of evenly-spaced pressure-solution seams during (late) diagenesis: a kinetic theory. *Contributions to Mineralogy and Petrology*, **82**: 360–370.
Park, W.C., and Schot, E.K., 1968. Stylolites: their nature and origin. *Journal of Sedimentary Petrology*, **38**: 175–191.
Railsback, L.B., 1993. Lithologic controls on morphology of pressure-dissolution surfaces (stylolites and dissolution seams) in Paleozoic rocks from the Mideastern United States. *Journal of Sedimentary Petrology*, **63**: 513–522.
Smith, J.V., 2000. Three-dimensional morphology and connectivity of stylolites hyperactivated during veining. *Journal of Structural Geology*, **22**: 59–64.

Cross-references

Compaction (Consolidation) of Sediments
Diagenesis
Diagenetic Structures
Pressure Solution

SUBMARINE FANS AND CHANNELS

Like alluvial fans (*q.v.*), submarine fans are cone-like accumulations of sediment, developed at a change of slope, generally below a single major feeder channel (though a few fans may have more than one channel, generally they are not all active at the same time). Submarine fans and channels, however, are much larger and more common than subaerial fans, and have been constructed mainly by turbidity currents and other sediment gravity flows (see *Gravity-Driven Mass Flows*). The coarser sediment is largely confined to the distributary channels, and much of the fan may be composed of silt or mud deposited by overflows on levees (Figure S45). The need for a single major feeder channel to produce a fan means that, although carbonate turbidites are common (see *Turbidites*), carbonate fans are rare. The large size of many submarine fans often produces a geometry that is constrained by preexisting topography (which may, in turn be related to active tectonics) so that a typical fan shape is not well developed. In addition, examples of thick siliciclastic turbidites constructed at the base of submarine slopes but lacking fan geometry have also been described (as submarine ramp deposits: Heller and Dickinson, 1985; Reading and Richards, 1994).

Modern submarine fans-channel systems were first described from offshore California in the late 1950s and recognition of small ancient submarine fans in the stratigraphic record followed shortly after (see historical review in Pickering *et al.*, 1989, pp. 1–4). The facies model of Mutti and Ricci Lucchi (1972: see figure S47) established a pattern for interpretation that persisted, with variations, for the next 20 years. Subsequent mapping of the ocean basins revealed that submarine fans include some of the largest geomorphic features on the earth's surface, reaching lengths of 1,500 kms and widths of almost 1,000 kms (Indus fan: Kolla and Coumes, 1987; McHargue and Webb, 1986; and see McHargue in Weimer and Link, 1991). Such fans are much too large to be recognizable in exposed ancient rocks, and the deposits would be difficult to recognize even in subsurface basin studies. Large submarine channels have been described from the deep sea, with channel depths of more than a km, and widths of more than 10 kms, flanked by levees that rise tens of meters above the surrounding surface and extend many kms away from the channel. Such channels are also difficult to recognize in ancient rock outcrops, though a few of comparable size have been identified in subsurface studies (Walker, 1992).

Large fans are generally composed of finer sediment than the "classical" small fans described by Mutti and Ricci Lucchi, and attempts have recently been made to describe alternative fan models characteristic of finer grained turbidite systems (Reading and Richards, 1994; Bouma and Stone, 2000: see Figure S51). It is claimed that large volumes of sediment and the presence of mud increases the "efficiency" of sand transport, thereby allowing more sand to be carried through the midfan channels to the distal fan and basin plain. Bouma (in Bouma and Stone, 2000) suggests that fine grained fans are typical of passive margin tectonic settings, with long fluvial transport leading to deltas, and from them to the head of the main distributary channel. The sand in them is concentrated at the base of the slope, and/or is transported efficiently through

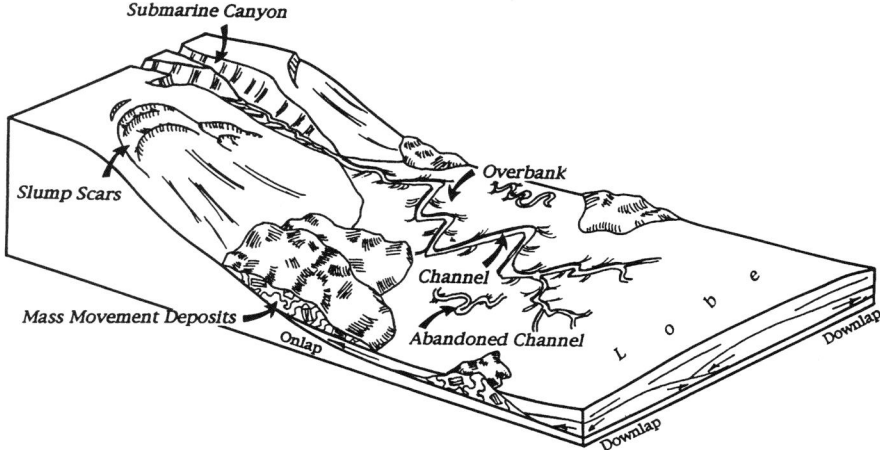

Figure S45 Surface morphology of a submarine fan. From Normark *et al.* (1993, their Figure 6 on p.100).

channels to distal sandy lobes. Coarse grained fans, in contrast, are typical of more tectonically active regions, with only short fluvial feeder systems, and with no prominent delta: much sand may be supplied into the feeder channel by longshore drift. Much of the coarse sediment is deposited in the midfan region. Though large fans may be dominated overall by finer sediment, many do contain substantial amounts of sand, for example the Amazon fan (Piper and Normark, 2001: see figure S48).

Not all submarine channels are directly related to submarine fans: the best known exception being the Northwest Atlantic Mid-ocean channel, described by Hesse and Rakofsky (1992). For an ancient large-scale canyon/channel complex, not obviously related to a fan, see Bruhn and Walker (1997).

Ancient submarine fans include many large petroleum reservoirs: they have become increasingly important as exploration has moved into deeper water, offshore. For a review of the special problems posed by exploration of such fans see Weimer *et al.* (2000).

Canyons, channels, and depositional lobes

Canyons

The upper parts of submarine channels, particularly those that incise the edge of the continental shelf are generally called canyons. They have been known since the end of the nineteenth century. In most cases, they are clearly erosional features, with some even eroded through hard igneous or metamorphic rocks. The eroding agent remained controversial until the 1950s. Though the main agent of erosion is now thought to be turbidity currents, there is still debate about the role played by other agents, including rivers (during lower stages of sea level), mass-wasting processes (including slides, slumps, and debris flows), sand flows, earthquakes, and structural or tectonic features. At lower sea levels (especially during the last ice age) rivers flowed across the continental plains (the shelves at higher sea-level stands), and entered what are, at higher stands, the heads of submarine canyons. Then, because of the lower water temperatures and higher suspended sediment loads in the rives, the river flows plunged below the surface of the sea to form "hyperpicnal flows:" that is, turbidity currents formed directly from river flows, without the deposition and resuspension that is generally necessary today (Mulder and Syvitski, 1995).

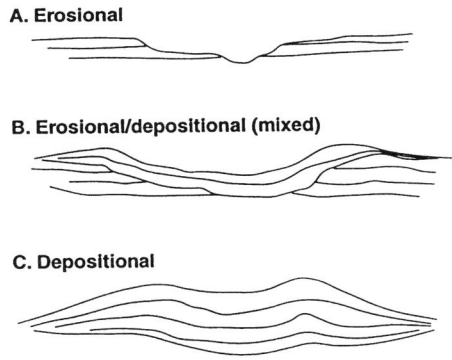

Figure S46 Erosional and depositional channels, as indicated by seismic profiles. From Clark and Pickering (1996, their Figure 1 on p.195, based on the earlier work of Normark).

During the greater part of the existence of a canyon, there is only temporary storage of sediment, but filling of the canyon, and therefore its preservation in the geological record, can certainly take place. Many, but not all, modern submarine canyons are related, if not indeed directly connected with major rivers. A good compilation of the early literature is given in Whitaker (1976) and the more recent literature is reviewed by Pickering *et al.* (1989).

Channels

The term channel is reserved for features underlain by nearly contemporaneous sediment, on fan surfaces, or in deep-sea basins. Channels are of all sizes, and may be either erosional or depositional in character (Figure S46), as clearly indicated by seismic profiling, for both modern and ancient channels (Clark and Pickering, 1996). Most active channels are directly connected with the main feeder canyon or channel, but some in the mid to lower fan have local "headwaters" on submarine slopes or on the flanks of large levees. In the midfan region, the position of the active channel changes frequently due to avulsion. The scale of submarine channels is generally larger than that of subaerial river channels, because they are adjusted to transmit intermittent turbidity current flows, many of which are large events, and most of which travel more slowly, on a given submarine slope, than a river of comparable size would on a comparable subaerial slope.

Generally, many of the flows exceed the capacity of the channel, so most submarine channels have very large constructional levees, compared with rivers. In large fans, the thickness of levee deposits may reach hundreds of meters (Figure S49). Coriolis effects are proportionately larger for turbidity currents than for rivers, causing more frequent overbank flow on one side of submarine channels, so that one levee is generally higher than the other. In other respects, submarine channels show many features in common with subaerial channels, including internal terracing, braiding, and meandering.

Though many flows overflow channels, particularly in the middle and lower parts of fans, most of the coarser sediment (sand, and especially gravel) is carried in the lower part of the flow, so is confined to the channels, and only spread out a sheet-like deposits where the flows emerge from the distal end of the channel (Figure S50).

On submarine fans, well-developed channels are characteristic of the middle region of the fan. While the channels are active, there are generally sites of sediment by-passing, or accumulation only of the coarsest lag deposits. Aggradation and catastrophic events cause channels to change position frequently on the fan surfaces, and abandoned channels are filled by smaller flows carrying finer sediment, so in the geological record they can frequently be recognized by their basal erosion surface and lag, by amalgamation surfaces, and by fining and thinning upward sequences in the sandstone filling, even when the channel geometry cannot be clearly seen in the outcrop. For further details see Clark and Pickering (1996).

Depositional lobes

Depositional lobes, with only minor channels, and with a generally convex-up profile can be mapped below (distal to) distributary channels on some modern fans. Their presence is inferred in ancient fans, by the existence of "packets" of relatively coarse turbidite beds (some meters or tens of meters thick), separated from other packets by finer sediments. Though some authors have claimed that these "packets" show thickening and coarsening upward sequences (supposedly produced by progradation of the lobes) statistical studies have failed to support this as a general trend. Attempts to relate the statistical trends actually exhibited by turbidite sequences to fan environments have been reviewed by Carlson and Grotzinger (2001).

Facies models for small, coarse fans

Many authors have proposed models (almost all of them variations of the original model of Mutti and Ricci Lucchi, 1972) to aid in the identification of ancient fans and in the description and interpretation of the observed facies. For succinct accounts, see Walker (1992) and Normark et al. (1993; and in Weimer and Link, 1991), and for a more extended account see Pickering et al. (1989). Besides the use of trends in bed thickness and grain size, these schemes use the frequency

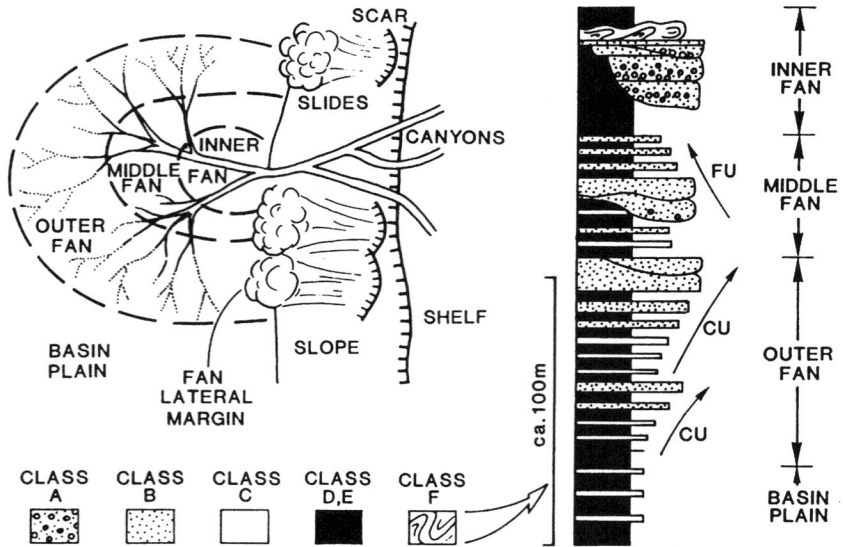

Figure S47 The fan model of Mutti and Ricci Lucchi (1972) as redrawn by Pickering et al. (1989, their Figure 7.2 on p.162). "Class A", etc. refer to facies: A are conglomerates, B are coarse sandstones, both generally massive and amalgamated. C are "classical turbidites" of the proximal type, that can be described by the "Bouma sequence" of internal structures (found mainly on the outer fans and basin floor). D and E are thin alternations of mud and sand. F are chaotic facies. FU refers to fining and thinning upward, CU to coarsening and thickening upward.

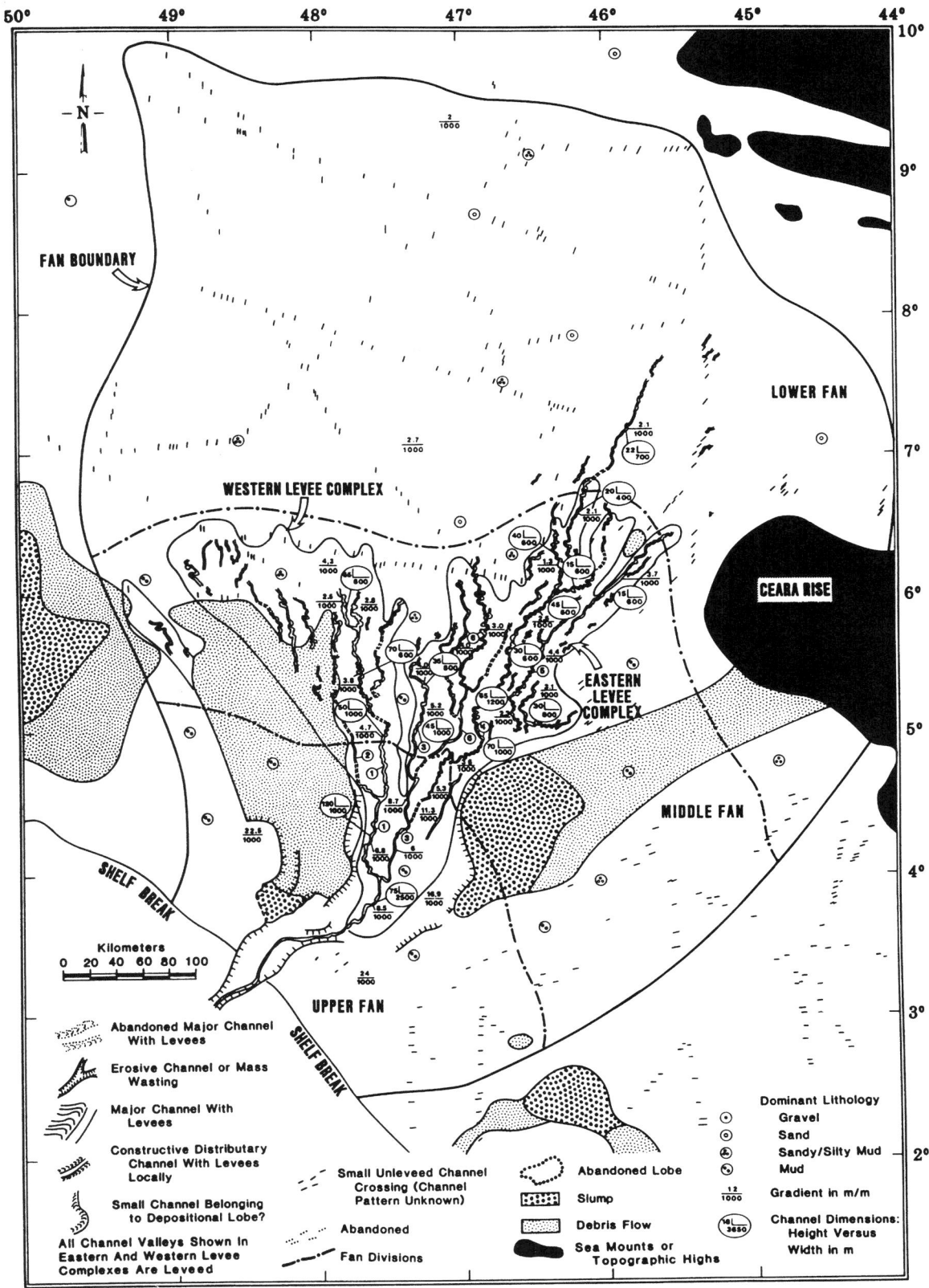

Figure S48 Map of the Amazon fan, from Pickering *et al.* (1989, their Figure 7.21 on p.179).

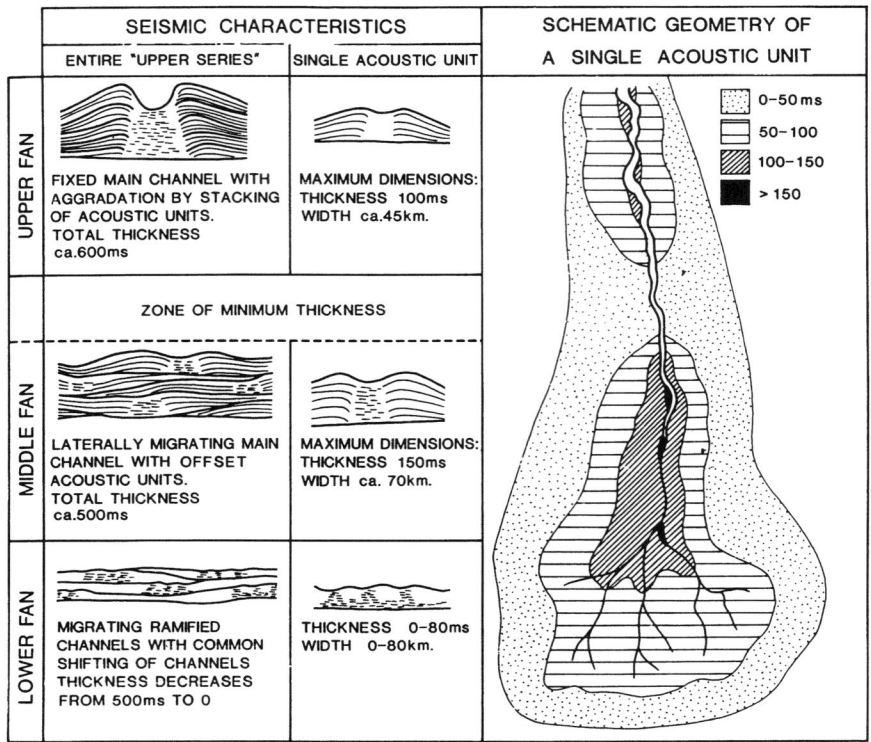

Figure S49 Schematic diagram showing channel–levee systems on the Rhone Fan. Thiknesses are in two-way seismic travel times, where 1 ms is about 0.75 m. From Pickering et al. (1989, their Figure 7.19 on p.177).

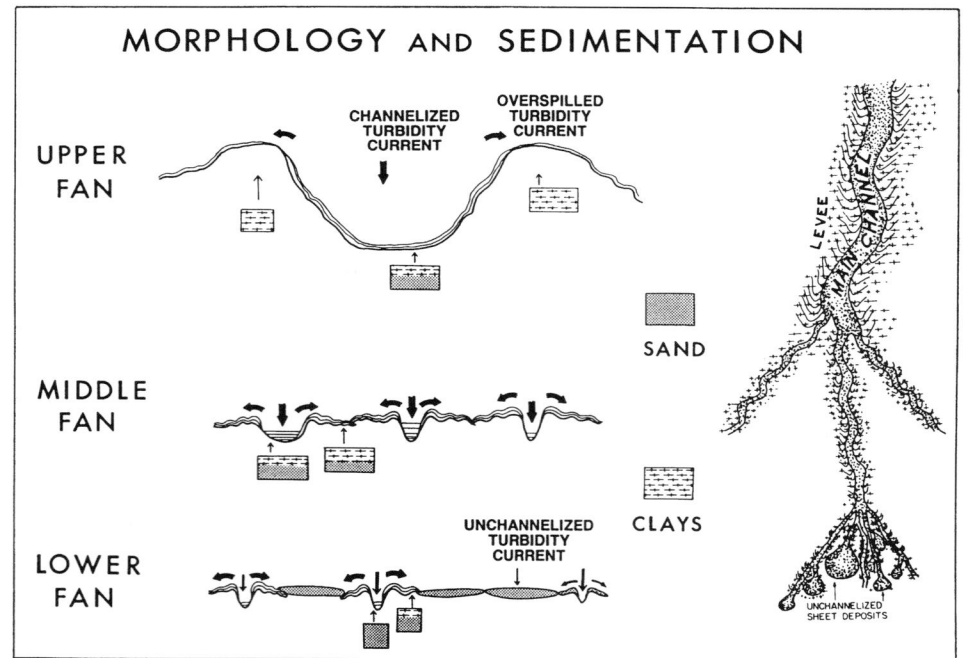

Figure S50 Model of sedimentation for the Indus fan channels, proposed by Kolla and Coumes (1986, their Figure 18 on p.674). The decreasing influence of channelized turbidity current down channel is shown in the sections on the left, along with the increasing importance of overbank flows, and apparent branching of channels shown on the plan on the right.

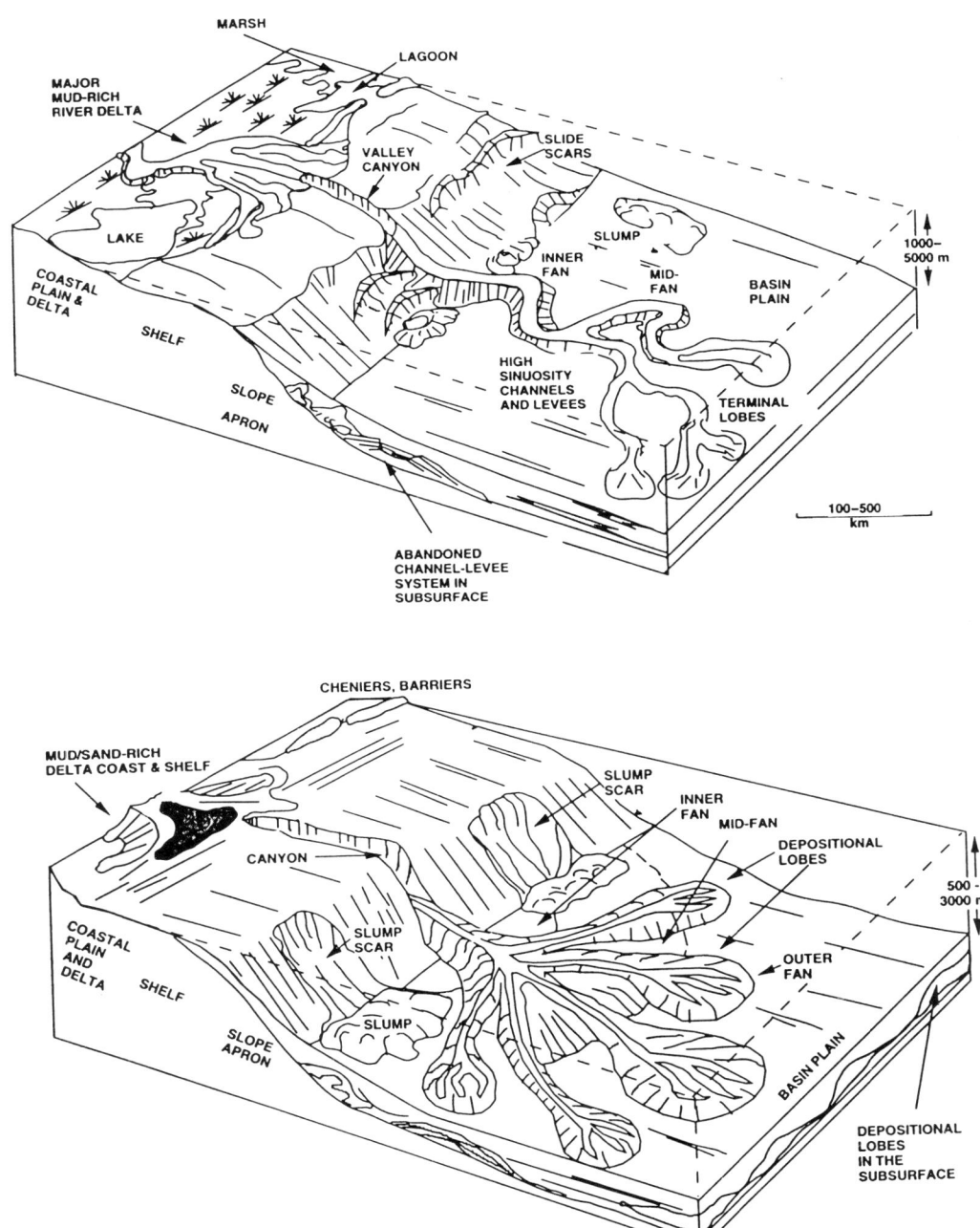

Figure S51 Two "end member" models for large fans (from Reading and Richards, 1994: top, their Figure 3, on p.801; bottom, their Figure 5 on p.803). The upper figure shows a model for a "point source mud-rich submarine fan" and the lower shows a model for a "point-source mud/sand-rich submarine fan." Note the difference in scale, and in the abundance of levees, channels, and lobes.

of amalgamation of beds, and the types of internal structures displayed by the sand beds, to identify not only channels and lobes, but also ancient levee deposits, and to distinguish fan deposits from the sediment gravity flow deposits of other environments, such as basin plains. Though the models are still useful, sequence stratigraphic concepts, developed since the 1980s, show that they need to be modified to include the effects on submarine systems of tectonics and changes in relative sea level, both of which can affect the nature and frequency of the events supplying sediment into the feeder channel, and therefore the facies and stratigraphic sequences that are produced by the fan.

Models for large fans

Thus far, few attempts have been made to develop general models for very large fans, such as the Mississippi and Amazon fans. Individual modern fans, however, are discussed and analyzed in Kolla and Coumes (1987), Weimer and Link (1991), Droz *et al.* (1996), Bouma and Stone (2000), and Piper and Normark (2000). Figure S48 shows a map of the Amazon fan, Figure S49 shows a "model" based on the Rhone fan, and Figure S50 shows a "model" for the deposits of an active channel on the Indus fan. Reading and Richards (1994) have presented a spectrum of models, two of which re shown in Figure S51. Larger physical scale implies that large scale controls (climate, tectonic, and sea-level change over long periods of time) become more important. So sequence stratigraphic concepts are just as crucial as sedimentological processes, and are more useful when outcrop is not available and only large features can be observed, using seismic methods in the subsurface.

Some authors prefer to abandon the idea of "facies models," and instead prefer to use the concept of "architectural analysis." In this approach, each fan, or larger "turbidite system" is unique, but is constructed from recognizable "architectural elements" of several different scale (e.g., in order of increasing scale: beds, facies associations, stages, systems: see Mutti and Normark, in Weimer and Link, 1991). Thus there are not just a few "models" but an almost limitless number of possible "complexes" that may be constructed using the various "elements." In assessing the recent literature, the reader should keep in mind the difficulty in observing the smaller-scale "elements" in deep water, and the present limited understanding of the physical processes involved (turbidity currents, debris flows, etc.) at very large scales.

Gerard V. Middleton

Bibliography

Bouma, A.H., and Stone, C.G. (eds.), 2000. *Fine-grained Turbidite Systems*. American Association of Petroleum Geologists Memoir, 72.
Bruhn, C.H.L., and Walker, R.G., 1997. Internal architecture and sedimentary evolution of coarse-grained, turbidite channel-levee complexes, Early Eocene Regência Canyon, Espíritu Santo Basin, Brazil. *Sedimentology*, **44** (1): 17–46.
Carlson, J., and Grotzinger, J.P., 2001. Submarine fan environment inferred from turbidite thickness distributions. *Sedimentology*, **48**: 1331–1351.
Clark, J.D., and Pickering, K.T., 1996. Architectural elements and growth patterns of submarine channels: application to hydrocarbon exploration. *American Association of Petroleum Geologists Bulletin*, **80**: 194–221.
Droz, L., Rigaut, F., Cochonat, P., and Tofani, R., 1996. Morphology and recent evolution of the Zaire turbidite system (Gulf of Guinea). *Geological Society of America Bulletin*, **108**: 253–269.
Hartley, A.J., and Prosser, D.J. (eds.), 1995. *Characterisation of Deep Marine Clastic Systems*. Geological Society [of London], Special Publication No. 94
Heller, P.L., and Dickinson, W.R., 1985. Submarine ramp facies model for delta-fed sand-rich turbidite systems. *American Association of Petroleum Geologists Bulletin*, **69**: 960–976.
Hesse, R., and Rakofsky, A., 1992. Deep-sea channel/submarine yazoo system of the Labrador Sea: a new deep-water facies model. *American Association of Petroleum Geologists Bulletin*, **76**: 680–707.
Kolla, V., and Coumes, F., 1987. Morphology, internal structure, seismic stratigraphy, and sedimentation of Indus Fan. *American Association of Petroleum Geologists Bulletin*, **71**: 650–677.
McHargue, T.R., and Webb, J.E., 1986. Internal geometry, seismic facies, and petroleum potential of canyons and inner fan channels of the Indus Submarine Fan. *American Association of Petroleum Geologists Bulletin*, **70**: 161–180.
Mulder, T., and Syvitski, J.P.M., 1995. Turbidity currents generated at mouths of rivers during exceptional discharges to the world oceans. *Journal of Geology*, **103**: 285–299.
Mutti, E., and Ricci Lucchi, F., 1972. Le torbiditi dell'Apennino settentrionale: introduzione all'analisi di facies. *Memorie della Societa Geologica Italiana*, **11**: 161–199. English translation by T.H. Nilsen, *International Geology Review*, 20, 125–160, 1978).
Normark, W.R., Posamentier, H., and Mutti, E., 1993. Turbidite systems: state of the art and future directions. *Reviews of Geophysics*, **32**: 85–113.
Pickering, K.T., Hiscott, R.N., and Hein, F.J., 1989. *Deep Marine Environments: Clastic Sedimentation and Tectonics*. London: Unwin Hyman.
Piper, D.J.W., and Normark, W.R., 2001. Sandy fans—from Amazon to Hueneme and beyond. *American Association of Petroleum Geologists Bulletin*, **85**: 1407–1438.
Reading, H.G., and Richards, M., 1994. Turbidite systems in deepwater basin margins classified by grain size and feeder system. *American Association of Petroleum Geologists Bulletin*, **78**: 792–822.
Walker, R.G., 1992. Turbidites and submarine fans. In Walker, R.G., and James, N.P. (eds.), *Facies Models: Response to Sea Level Change*. Geological Association of Canada, pp. 239–263.
Weimer, P., and Link, M.H. (eds.), 1991. *Seismic Facies and Sedimentary Processes of Submarine Fans and Turbidite Systems*. Springer-Verlag
Weimer, P., Slatt, R.M., Dromgoole, P., Bowman, M., and Leonard, A., 2000. Developing and managing turbidite reservoirs: case histories and experiences: results of the 1998 EAGE/AAPG Research Conference. *American Assocation of Petroleum Geologists Bulletin*, **84**: 453–465.
Whitaker, J.H.McD. (ed.), 1976. *Submarine Canyons and Deep-Sea Fans*. Dowden: Hutchinson & Ross.

Cross-references

Alluvial Fan
Climatic Control of Sedimentation
Cyclic Sedimentation
Debris Flow
Facies Models
Gravity-Driven Mass Flows
Liquefaction and Fluidization
Mass Movement
Slide and Slump Structures
Slope Sediments
Slurry
Tectonic Controls of Sedimentation
Turbidites

SUBSTRATE-CONTROLLED ICHNOFACIES

Three substrate-controlled ichnofacies have been established (Ekdale *et al.*, 1984): *Glossifungites* (firmground), *Trypanites* (hardground), and *Teredolites* (woodground).

The *Glossifungites* ichnofacies

The *Glossifungites* ichnofacies is environmentally wide ranging, but only develops in firm, unlithified substrates such as dewatered muds. In clastic settings dewatering results from burial and the substrates are made available to tracemakers if

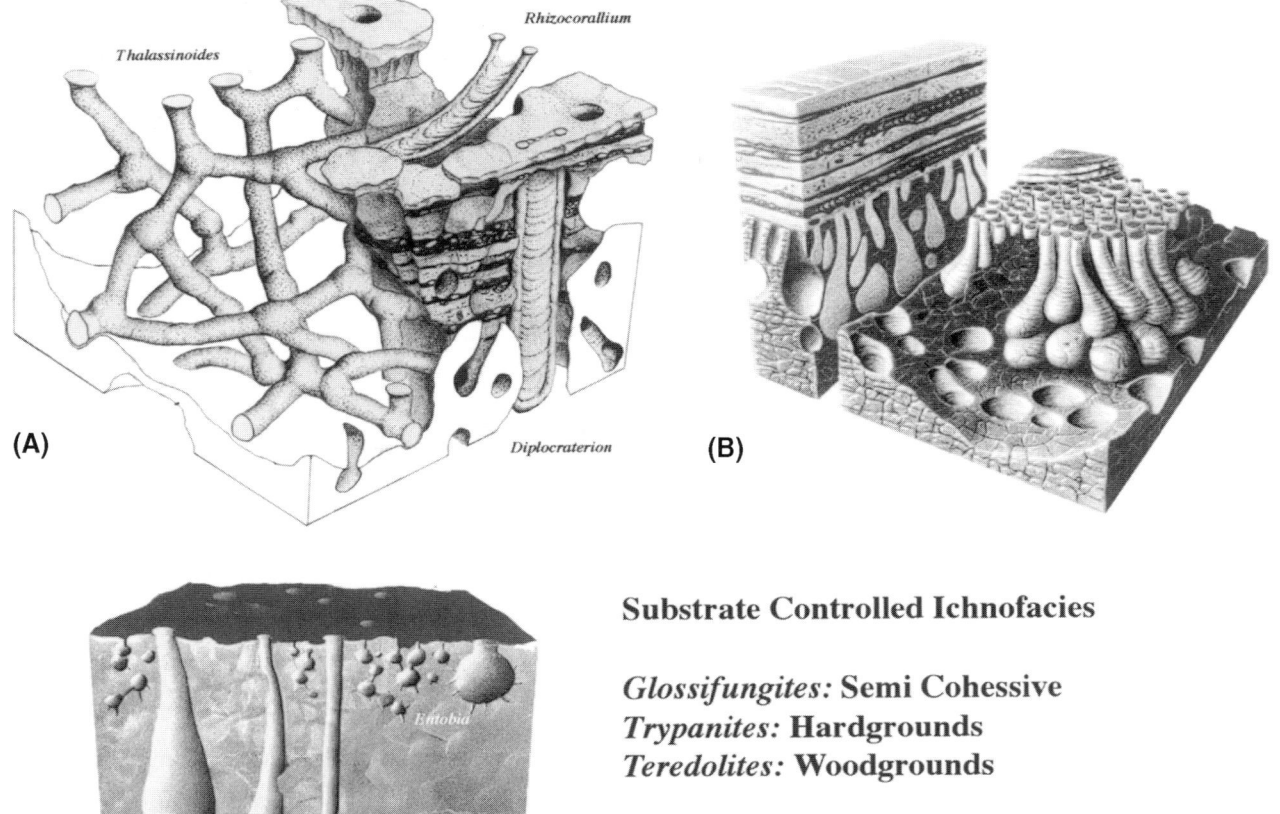

Figure S52 Substrate controlled ichnofacies; (A) The *Glossifungites* ichnofacies is indicative of semi cohesive substrates like dewatered muds or compacted sands; (B) The *Teredolites* ichnofacies is indicative of a woodground like a peat deposit; and (C) The *Trypanites* ichnofacies is indicative of fully cemented substrates like hardgrounds, disconformities, and beachrock.

exhumed by later erosion (Pemberton and Frey, 1985). Exhumation can occur in terrestrial environments, as a result of channel meandering or valley incision, in shallow-water environments as a result of meandering tidal channels, coastal erosion, erosive shoreface retreat, or as a result of submarine channels cutting through previously deposited sediments. Such horizons commonly form at bounding discontinuities and may be critical in the evolving concept of sequence stratigraphy (Pemberton *et al.*, 1992; Pemberton and MacEachern, 1995).

The *Glossifungites* ichnofacies (Figure S52(A)) is characterized by: (1) vertical, cylindrical U or tear shaped pseudo borings, sparsely to densely branching dwelling burrows, and/or mixtures of burrows and pseudo borings; (2) protrusive spreiten in some burrows that develop mostly through animal growth (funnel shaped *Rhizocorallium* and *Diplocraterion* [formerly *Glossifungites*]); (3) animals that leave the burrow to feed (e.g., crabs) as well as suspension feeders; and (4) low diversity, but commonly abundant individual structures.

The *Teredolites* ichnofacies

The *Teredolites* ichnofacies (Figure S52(B)) consists of a characteristic assemblage of borings or burrows in woody or highly carbonaceous substrates. Woodgrounds differ from this substrates in three main ways: (1) they may be flexible instead of rigid; (2) they are composed of carbonaceous material instead of mineral matter; and (3) they are readily biodegradable (Bromley *et al.*, 1984). Such differences dictate that the means by which, as well as the reasons for which these two types of substrates are penetrated are different. Because currents can raft woody substrates, it is important to determine whether the borings are autochthonous or allochthonous. Only the autochthonous forms are true members of the *Teredolites* ichnofacies. These assemblages may also be important in defining sequence and parasequence boundaries (Savrda, 1991). The use of bored logs to define bounding surfaces as in the concept of log-grounds (i.e., Savrda *et al.*,

1993) should be avoided, as it is very difficult to ascertain when the logs were bored. Such logs are clasts and should not be treated in the same manner as the *Teredolites* ichnofacies.

The *Teredolites* ichnofacies is characterized by: (1) sparse to profuse club shaped borings; (2) boring walls that are generally ornamented with the texture of the host substrate (i.e., tree ring impressions); (3) stumpy to elongate subcylindrical excavations in marine or marginal marine settings; and (4) shallower, sparse to profuse nonclavate etchings (isopod borings) in freshwater settings.

The *Trypanites* ichnofacies

The *Trypanites* ichnofacies (Figure S52(C)) develops in fully lithified substrates such as hardgrounds, reefs, rocky coasts, beachrock and other omission surfaces. As such, development of this ichnofacies also corresponds to discontinuities that have major sequence stratigraphic significance. Bromley and Asgaard (1993) subdivided the *Trypanites* ichnofacies into two ichnofacies; the *Entobia* ichnofacies for rocky shorelines (also see De Gilbert *et al*., 1998) and the *Gnathichnus* ichnofacies for bored shells and boulders found further offshore. Bored shells and boulders, however, are not substrates that can be correlated because it is difficult to ascertain when the boring activity was initiated.

The traces in the *Trypanites* ichnofacies are characterized by: (1) cylindrical to vase, tear or U shaped to irregular domiciles of suspension feeders or passive carnivores; (2) raspings and gnawings of algal grazers and similar organisms (mainly chitons, limpets, and echinoids); (3) moderately low diversity, although the borings and scrapings of individual ichnogenera may be abundant; and (4) borings oriented perpendicular to the substrate which may include large numbers of overhangs. In contrast to the *Glossifungites* ichnofacies, the walls of the borings cut through hard parts of the substrate rather than skirting around them.

Significance of substrate controlled ichnofacies

In clastic settings, most of these trace assemblages are associated with erosionally exhumed (dewatered and compacted or cemented) substrates, and hence, correspond to erosional discontinuities. Woodgrounds consist of xylic clasts or interwoven mats of vegetation, forming resilient substrates that are not necessarily erosionally exhumed. In carbonate settings, firmground or hardground surfaces may occur at the sediment water interface, due either to erosional exhumation or non-depositional breaks with associated submarine cementation (Bromley, 1975). Depositional breaks, in particular condensed sections, may also be semilithified or lithified presumably at the upper contact (or downlap surface) and may be colonized without associated erosion. In general, however, the recognition of substrate-controlled ichnofacies may be regarded as equivalent to the recognition of discontinuities in the stratigraphic record.

Although certain insect and animal burrows in the terrestrial realm may be properly regarded as firmgound (e.g., Fürsich and Mayr, 1981) or more rarely, hardground suites, they have a low preservational potential and constitute a relatively minor component in the geologic record of such associations. The overwhelming majority of these assemblages originate in marine or margin marine settings. A discontinuity may be generated in either subaerial or submarine settings; but colonization of the surface may be regarded to be marine influence, particularly in pre Tertiary intervals. This circumstance has important implications, so far as interpretation of the origin of the discontinuity is concerned.

The substrate controlled ichnocoenose typically cross cuts a preexisting softground suite and, hence, reflects conditions post dating both initial deposition of the unit and erosion of that unit. The suite also corresponds to a hiatus between the erosion event (which exhumes the substrate) and deposition of the overlying unit. During this time gap, organisms colonize the substrate. By observing (1) the softground ichnofacies (contemporaneous with deposition of the unit); (2) the ichnofacies of the exhumed substrate; and (3) the ichnofacies of the overlying unit, it is possible to make some interpretation regarding the origin of the surface and the allocyclic or autocyclic mechanisms responsible.

S. George Pemberton

Bibliography

Bromley, R.G., 1975. Trace fossils at omission surfaces. In Frey, R.W. (ed.), *The Study of Trace Fossils*. Springer Verlag, pp. 399–428.

Bromley, R.G., and Asgaard, U., 1993. Two bioerosion ichnofacies produced by early and late burial associated with sea level change. *Geologische Rundschau*, **82**: 276–280.

Bromley, R.G., Pemberton, S.G., and Rahmani, R.A., 1984. A Cretaceous woodground: the Teredolites ichnofacies. *Journal of Paleontology*, **58**: 488–498.

De Gilbert, J.M., Martinell, J., and Domenech, R., 1998. Entobia ichnofacies in fossil rocky shores, lower Pliocene, northwestern mediterranean. *Palaios*, **13**: 476–487.

Ekdale, A.A., Bromley, R.E., and Pemberkon, S.E., 1984. *Ichnology: The Use of Trace Fossil in Sedimentology and Stratigraphy*. Society of Economic Paleontologists and Mineralogists, Short Course Notes 15.

Fürsich, F.R., and Mayr, H., 1981. Non-marine Rhizocorallium (trace fossils) from the Upper Freshwater molasse (Upper Miocene) of southern Germany. *Neues Jahrbuch für Geologie und Palaontologie, Monatshefte*, **6**: 321–333.

Pemberton, S.G., and Frey, R.W., 1985. The Glossifungites ichnofacies: modern examples from the Georgia coast, USA. In Curran, H.A. (ed.), *Biogenic Structures: Their Use in Interpreting Depositional Environments*. Society of Economic Paleontologists and Mineralogists, Special Publication, 35, pp. 237–259.

Pemberton, S.G., and MacEachern, J.A., 1995. The sequence stratigraphic significance of trace fossils: examples from the cretaceous foreland basin of Alberta, Canada. In Van Wagoner, J.C., and Bertram, G. (eds.), *Sequence Stratigraphy of Foreland Basin Deposits- Outcrop and Subsurface Examples from the Cretaceous of North America*. Tulsa: American Association Petroleum Geologists, Memoir 64, pp. 429–475.

Pemberton, S.G., MacEachern, J.A., and Frey, R.W., 1992. Trace fossil facies models: environmental and allostratigraphic significance. In Walker, R.G., and James, N. (eds.), *Facies Models: Response to Sea Level Change*. Geological Association of Canada, pp. 47–72.

Savrda, C.E., 1991. Teredolites, wood substrates, and sea-level dynamics. *Geology*, **19**: 905–908.

Savrda, C.E., Ozalas, K., Demko, T.H., Hutchinson, R.A., and Scheiwe, T.D., 1993. Log grounds and the ichnofossil Teredolites in transgressive deposits of the clayton formation (Lower Paleocene), western Alabama. *Palaios*, **8**: 311–324.

Cross-references

Biogenic Sedimentary Structures
Taphonomy: Sedimentological Implications of Fossil Preservation

SULFIDE MINERALS IN SEDIMENTS

The formation processes of metal sulfides in sediments, especially iron sulfides, have been the subjects of intense scientific research because of linkages to the global biogeochemical cycles of iron, sulfur, carbon, and oxygen. Metal sulfide precipitation can occur during several stages of diagenesis and burial of continental-margin, pelagic, and non-marine sediments. During early stages of diagenesis, anaerobic sulfate-reducing bacteria consume organic carbon and sulfate and produce hydrogen sulfide and bicarbonate. Pyrite (FeS_2) is formed by the reaction of this biogenic sulfide with reactive, detrital iron-bearing minerals near the sediment–water interface (Kaplan et al., 1963; Berner, 1970, 1984). The initial reaction products are iron monosulfide minerals that may subsequently transform to pyrite by reaction with polysulfides or other reactive oxidants (e.g., Morse et al., 1987). Pyrite formation in sediments is thought to be limited either by the microbial production of sulfide (organic carbon or sulfate limitation) or by the availability of reactive iron minerals (iron limitation). During later stages of diagenesis and deep sediment burial, thermochemical sulfate reduction may become important and more refractory iron oxides, typically unreactive during early diagenesis, are sources of iron in sulfidation reactions. Studies of metal sulfides in sediments and sedimentary rocks employ analytical techniques such as microscopy, sulfur isotope ratio measurements, and trace element measurements. Relationships between the concentrations and partitioning of sedimentary Fe, S, and C provide important tools for reconstructing paleoenvironments over a range of temporal and spatial scales (e.g., Goldhaber and Kaplan, 1974).

Transition metal sulfides (e.g., NiS, CuS, ZnS, CdS, HgS) have exceedingly low solubility products and might be expected to form during early diagenesis. However, with the exception of iron, d-transition metals are typically present in trace amounts in sediments, which does not allow for any significant accumulation of sulfide minerals other than those of iron. Consequently, iron sulfides are the only metal sulfides commonly recognized in marine and non-marine sediments. In contaminated sediments or systems with high metal loadings, sulfides of Zn, Cd, and Cu have been reported (e.g., Luther et al., 1980). Trace metals are frequently found associated with pyrite and include Co, Ni, Cu, Zn, As, Mo, Cd, Sb, Hg, and sometimes Mn. In particular, As, Hg, and Mo have a high affinity for pyrite and are generally enriched in pyrite relative to bulk sediments (Raiswell and Plant, 1980; Huerta-Diaz and Morse, 1992). Questions remain as to whether the trace metals found associated with the sulfide fraction are as discrete phases or as coprecipitates with the abundant iron sulfides.

Several iron sulfide phases have been synthesized in the laboratory, either as transient intermediates or as stable end products, and are therefore likely to form in sedimentary environments. These phases are: disordered mackinawite, $Fe_{1+x}S$; mackinawite, $Fe_{1+x}S$; cubic iron sulfide, FeS; hexagonal pyrrhotite, $Fe_{1-x}S$; greigite, Fe_3S_4; smythite, Fe_9S_{11}; marcasite, orthorhombic FeS_2; and pyrite, cubic FeS_2. Pyrrhotite and pyrite represent the thermodynamically stable phases at the temperatures and pressures of early diagenesis (Berner, 1967). Disordered mackinawite, mackinawite, and

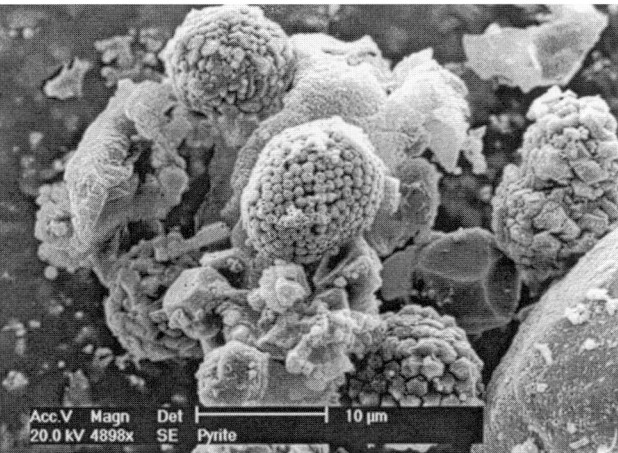

Figure S53 Group of pyrite framboids (modern slope sediments, Black Sea).

Figure S54 Framboid with pyritohedral microcrystals (Cretaceous sediments).

greigite are metastable with respect to pyrite and/or stoichiometric pyrrhotite but are considered to be the principal precursor phases to pyrite (e.g., Schoonen and Barnes, 1991). Smythite and cubic FeS are exceedingly rare (Krupp, 1994). Marcasite has not been identified in modern sedimentary environments but is present in some ancient sedimentary rocks. Marcasite formation is thought to require a low pH and high concentration of polysulfide species. Greigite was originally reported from Tertiary lacustrine deposits but has been more recently documented in sediments as old as Upper Cretaceous siliciclastics (Reynolds et al., 1994). Despite its predicted stability, hexagonal pyrrhotite is apparently very rare as an authigenic phase in sediments and sedimentary rocks. In marine and non-marine sediments, pyrite is the normal end product of sulfate reduction and iron mineral sulfidation. The kinetics of transformation from precursor iron sulfides to pyrite may be slow (years) or fast (days) depending on factors such as pH, redox state, and temperature.

Pyrite is the most common authigenic mineral in recent marine and lacustrine sediments, as well as in sedimentary rocks, especially gray- to black-colored shales (Arthur and Sageman, 1994). It occurs typically as single grains 1 μm to 10 μm in size, as framboids mainly <20 μm in diameter, as groups of framboids termed polyframboids, and as replacements of organic matter (Figures S51–S53). Microcrystals and framboids are the basic textures that form during sediment deposition and diagenesis (Raiswell, 1982). Microcrystals are usually subhedral to euhedral; octahedrons and pyritohedrons are common. The term *framboid* refers to a unique raspberry-like microtexture, which has been the subject of numerous studies. The ordered, cell-like arrangement of pyrite microcrystals in framboids has resulted in a long-lived debate on whether they are organic or inorganic in origin (see Wilkin and Barnes, 1997 and references therein). These textures can survive to low-greenschist facies metamorphism. Pressure solution and overgrowth may occur during low-grade metamorphism and deformation. With progressive deformation, pressure solution produces flattened grains perpendicular to the maximum compressive stress that results in the development of foliations and lineations (McClay and Ellis, 1983). Mass transfer during pressure solution has important implications with respect to the distribution of trace elements and the sulfur isotopic systematics of sulfide ores.

There are other examples of localized, macroscopic to microscopic varieties of pyrite found in modern and ancient sediments. Some examples are pyritized organic matter, pyritized burrow tubes, shell infillings of pyrite, carbonate shell replacement by pyrite, pyrite concretions and nodules, and perhaps most spectacularly, the pyritization of soft-bodied materials (e.g., Briggs *et al.*, 1991; Canfield and Raiswell, 1991).

A second setting where metal sulfides accumulate in sedimentary systems is in areas such as mid-ocean ridges where hydrothermal fluids discharge into seawater. Rapid cooling of the hydrothermal fluids as they interact with cold ocean water causes the precipitation of dissolved solutes, often as very-fine grained sulfide minerals ("black smokers"). Metal sulfides may also precipitate on the walls of the vents or in vertical tubular structures ("chimneys"). Pyrite again dominates the sulfide mineralogy of these submarine volcanogenic deposits although sulfide precipitates can also include those of Cu and Zn. Precious metals such as Ag and Au may also be enriched in the products of these hydrothermal processes. Volcanogenic massive sulfide deposits have been important sources of base metals; notable are the deposits in Japan (Kuroko-type) and the Iberian Peninsula (Pyrite Belt). Such volcano-sedimentary deposits grade into bedded metal sulfide ores that often show no clear associations with volcanism or hydrothermal activity. The origin of these sediment-hosted sulfide ores historically has been controversial; however, this important class of mineral deposits includes some of the most famous ore districts (e.g., Broken Hill, Australia; Sullivan, North America; Kupferschiefer, Europe; Zambian Copper Belt, Africa). In these deposits, metal sulfides are thought to have formed syngenetically, precipitated from hydrogen sulfide of hydrothermal or microbiological origin.

Geologic evidence suggests the persistence for over 3 Ga of deep-ocean hydrothermal systems. The discovery of life and diverse ecosystems that have adapted to the warm, anoxic environments of seafloor hydrothermal systems, along with the discovery of ancient sulfur-metabolizing bacteria has spawned hypotheses that such environments could be the sites for the origin of life on earth. Metal sulfides in sediments may have played a role in catalyzing reactions and/or providing energy sources through redox reactions that ultimately led to the buildup of complex organic molecules, thereby guiding the transition from a prebiotic earth to a biotic earth (e.g., Wächterhäuser, 1988).

Richard T. Wilkin

Bibliography

Arthur, M.A., and Sageman, B.B., 1994. Marine black shales: depositional mechanisms and environments of ancient deposits. *Annual Review of Earth and Planetary Sciences*, **22**: 499–551.

Berner, R.A., 1967. Thermodynamic stability of sedimentary iron sulfides. *American Journal of Science*, **265**: 773–785.

Berner, R.A., 1970. Sedimentary pyrite formation. *American Journal of Science*, **268**: 1–23.

Berner, R.A., 1984. Sedimentary pyrite formation: an update. *Geochimica et Cosmochimica Acta*, **48**: 605–615.

Briggs, D.E.G., Botrell, S.H., and Raiswell, R., 1991. Pyritization of soft-bodied fossils: Beecher's Trilobite Bed, Upper Ordovician, New York State. *Geology*, **19**: 1221–1224.

Canfield, D.E., and Raiswell, R., 1991. Pyrite formation and fossil preservation. In Allison, P., and Briggs, D. (eds.), *Taphonomy: Releasing the Data Locked in the Fossil Record. Volume 9 of Topics in Geobiology*. Plenum Press, pp. 337–387.

Goldhaber, M.B., and Kaplan, I.R., 1974. The Sulfur Cycle. In Goldberg, E.D., (ed.), The Sea, Volume 5, Marine Chemistry. Wiley, pp. 569–655.

Huerta-Diaz, M.A., and Morse, J.W., 1992. Pyritization of trace metals in anoxic marine sediments. *Geochimica et Cosmochimica Acta*, **56**: 2681–2702.

Kaplan, I.R., Emery, K.O., and Rittenberg, S.C., 1963. The distribution and isotopic abundance of sulfur in recent sediment off Southern California. *Geochimica et Cosmochimica Acta*, **27**: 297–331.

Krupp, R.E., 1994. Phase relations and phase transformations between the low-temperature iron sulphides mackinawite, greigite, and smythite. *European Journal of Mineralogy*, **6**: 265–278.

Luther, G.W. III, Meyerson, A.L., Krajewski, J.J., and Hires, R., 1980. Metal sulphides in estuarine sediments. *Journal of Sedimentary Petrology*, **50**: 1117–1120.

McClay, K.R., and Ellis, P.G., 1983. Deformation and recrystallization of pyrite. *Mineralogical Magazine*, **47**: 527–538.

Figure S55 Pyritohedral grain of pyrite (Cretaceous sediments).

Morse, J.W., Millero, F.J., Cornwell, J.C., and Rickard, D., 1987. The chemistry of the hydrogen sulphide and iron sulphide systems in natural waters. *Earth-Science Reviews*, 24: 1–42.

Raiswell, R., 1982. Pyrite texture, isotopic composition and the availability of iron. *American Journal of Science*, 282: 1244–1263.

Raiswell, R., and Plant, J., 1980. The incorporation of trace metals into pyrite during diagenesis of black shales, Yorkshire, England. *Economic Geology*, 75: 684–689.

Reynolds, R.L., Tuttle, M.L., Rice, C.A., Fishman, N.S., Karachewski, J.A., and Sherman, D.M., 1994. Magnetization and geochemistry of greigite-bearing Cretaceous strata, North Slope Basin, Alaska. *American Journal of Science*, 294: 485–528.

Schoonen, M.A.A., and Barnes, H.L., 1991. Reactions forming pyrite and marcasite from solution. II Via FeS precursors below 100°C. *Geochimica et Cosmochimica Acta*, 55: 1505–1514.

Wächterhäuser, G., 1988. Pyrite formation, the first energy source for life: a hypothesis. *Systematic and Applied Microbiology*, 10: 207–210.

Wilkin, R.T., and Barnes, H.L., 1997. Formation processes of framboidal pyrite. *Geochimica et Cosmochimica Acta*, 61: 323–339.

Cross-references

Bacteria in Sediments
Diagenesis
Heavy Minerals
Isotopic Methods in Sedimentology
Toxicity of Sediments

SURFACE FORMS

The nature of surface forms

Surface forms are geometrical features that develop on a surface of cohesive or noncohesive sediment by the action of a flow of fluid over that surface. Surface forms are commonly produced by flows of water in streams and rivers, by tidal currents in shallow coastal areas, by continental-shelf bottom flows, and by various kinds of flows in the deep ocean. The generating flows on continental shelves may be purely unidirectional, or purely oscillatory as a result of the passage of surface waves, or combinations of unidirectional and oscillatory flows. They are also common in environments on land where winds are strong enough to move sand along the surface. Some surface forms are generated solely by localized erosion of the sediment bed; others are generated by transportation of sediment along and over the bed in a pattern of local erosion and deposition without overall net erosion of the surface, and perhaps with overall net deposition.

The range in both the scale and the geometry of surface forms is extremely wide. The scale of common surface forms ranges from centimeters to kilometers horizontally and from centimeters to at least tens of meters vertically. Most surface forms show more or less elongation in their geometry, either transverse to the flow or parallel to the flow, although some are oblique to the flow direction and others are approximately equidimensional rather than elongate. In part this wide range of diversity is owing to a wide range in conditions over which a generating process acts, but in part also because the processes that generate surface forms are diverse. It is clear that surface forms are a polygenetic phenomenon: a number of disparate effects of fluid flow give rise to them.

Surface forms may or may not exist in a state of equilibrium with the generating flow. Equilibrium of surface forms is common when the flow persists in a nearly unchanging state for a time long enough that the surface forms can adjust their characteristics fully in response to the flow. In other cases, the development of the surface forms lags behind changes in the flow. Such disequilibrium makes for even greater variety in surface forms, and it is a complicating factor when attempts are made to use the characteristics of preserved surface forms for making interpretations of the former flow that generated the forms.

The size of sediment in the bed of sediment on which surface forms can exist ranges widely, from cohesive or semicohesive clays and muds to gravels. Distinctive erosional forms, like flutes and flow-parallel furrows, are characteristic of fine, cohesive sediment surfaces. Rugged flow-transverse forms, like ripples and dunes, are characteristic of coarse silts, sands, and (with lesser frequency) gravels.

The significance of surface forms

Surface forms are of interest to a wide range of specialists in the natural sciences, including geologists, geomorphologists, geographers, and oceanographers, because they are so varied in their scale, geometry, and origin and are so characteristics of natural flow environments, both subaqueous (under water) and subaerial (under air). Moreover, even apart from their intrinsic interest, surface forms are of interest for practical reasons as well. Large subaqueous surface forms many meters high can be obstacles to navigation, and their movement can be a threat to submarine structures. The rugged topography of surface forms in rivers and tidal channels causes flow separation at the crests and therefore large values of form drag. Surface forms are thus the most important determinant of resistance to channel flow, and hydraulic engineers have expended much effort on the development of depth–discharge predictors based on the hydraulic relationships of the surface forms (Vanoni, 1975, p. 114–152). The nature of the surface forms is also closely bound up with the sediment transport rate in unidirectional flows, in that the downcurrent movement of rugged flow-transverse surface forms largely involves recycling of bed load within the forms. Sedimentologists have given much attention to certain kinds of surface forms because of their role in generating stratification, which is one of the most useful tools available for interpreting ancient sedimentary environments (see *Bedding and Internal Structures*).

Terminology and classification of surface forms

Terminology for surface forms is diverse and can be confusing. Nonspecialists are confronted with a wide variety of terms, among the most common being ripples, dunes, antidunes, sand waves, sediment waves, furrows, and flutes. The general term surface form is nearly synonymous with the common terms bed form or bedform. Some workers use the term bed form for an individual, distinctive geometrical element that is formed on a sediment bed by a flow of overlying fluid. In that context, the individual bed forms constitute a bed configuration, which is the aggregate of all of the bed forms that are present in a given area of the sediment bed at a given time. Others use the term bed form (more commonly, bedform) for both the individual element and the entire configuration, with the intended sense derived from the context of the description or discussion.

Even under an unchanging flow, and with unchanging sediment characteristics, the particular geometry of the bed configuration that is actively coupled with the flow at a given time differs in detail from that of the configuration at any other time. It is sometimes useful to think of the average characteristics of such a collection of equivalent configurations as the bed state. Bed states with closely related geometry and kinematics, formed under a particular range of flow and sediment characteristics and readily distinguishable from those formed under different ranges of conditions, are sometimes called bed phases, in loose analogy with thermodynamic phases. Some examples of such bed phases are current ripples, subaqueous dunes, subaerial dunes, flutes, or erosional furrows (see classification below).

Surface forms can be classified in a variety of ways. An ideal classification would distinguish among genetically different kinds of forms, and within those kinds, characteristic subclasses based on differences in geometry and kinematics as a function of differences in flow and sediment characteristics. There is no universally accepted classification scheme for surface forms. Widely recognized categories of surface forms are shown in Table 1. Owing to the complexity and diversity of processes that generate surface forms, such a classification is bound to be inadequate to express the full range of the phenomenon, but it can serve as a guide to the principal kinds of surface forms discussed in later sections of this article.

Flows that generate surface forms

Flows of water that generate surface forms range from unidirectional to oscillatory. In unidirectional flows, water velocity is always in the same direction, although speeds may vary on timescales ranging from seconds to tidal periods and longer. In purely oscillatory flows, water movements range from back-and-forth movement in a single orientation to more complex oscillations in which water motions change direction seemingly at random owing to the superposition of more than one oscillatory component, as for example at the bed of an ocean or a lake where the water surface is characterized by a complex multidirectional wave spectrum. In most subaerial environments, the wind is notoriously changeable in both speed and direction, although in some areas of the world, sand-moving winds blow very consistently from a narrow range of directions.

Subaqueous surface forms generated by unidirectional flows

Flow-transverse surface forms

Flows of water over noncohesive beds of silt, sand, or gravel generate a variety of rugged surface forms. These forms are most common in sands. It has been customary to think in terms of a progression of regimes of surface forms with increasing flow strength: current ripples, dunes, plane-bed transport, and antidunes. The picture is actually more complicated, however, because the succession of different kinds of surface forms varies not only with flow strength but also with sediment size.

Flow-transverse surface forms have been studied in water channels and flumes, by both hydraulic engineers and earth scientists, since the beginning of the twentieth century. Particularly notable studies are those by Gilbert (1914) and Simons and coworkers (Simons and Richardson, 1963; Guy et al., 1966). For a summary of laboratory data, see Southard (1991). Theoretical studies of the existence of rugged flow-transverse surface forms have had a long and varied history. A long-standing approach has been through stability analyzes, in which the equation of motion, supplemented with a relationship between flow strength and sediment transport, is appropriately linearized and then a small perturbation is introduced. If the perturbation is damped, then a planar transport surface is stable; if the perturbation grows, then the bed develops into rugged surface forms. The most successful of such approaches has been that of Richards (1980).

Graphical representation

There have been many attempts to represent the range of conditions that produce each kind of surface form in graphs. Such graphs are loosely analogous to the thermodynamic phase diagrams. Such graphs are useful ways of portraying the relationships among the various kinds of surface forms, to the extent that the existence fields in the graphs do not overlap. Because of the effect of water temperature on water viscosity, these variables must be in nondimensionalized form in order to represent the existence of the various kinds of surface forms in a rational way. In the graphs presented below, these nondimensionalized variables have been converted to their 10°C equivalents, to make comparison with actual flow conditions more readily apparent.

There have been two main approaches to the construction of graphs of the existence fields of flow-transverse subaqueous surface forms. In one approach, the shear stress the flow exerts on the sediment bed is used to represent the strength of the flow, from the standpoint that it is this shear stress that actually moves the sediment. The problem with this approach is that when the bed is roughened with rugged surface forms, most of the bed shear stress is from drag rather than skin friction, which is the part of the bed shear stress that actually moves the sediment. Such graphs are successful only to the extent that the skin friction can somehow be separated out from the total bed shear stress. The most recent example of such a graph is by Van den Berg and Van Gelder (1993), shown here as Figure S56. The graph in Figure S56 uses a nondimensionalized form of the skin friction, computed according to a rational method, and nondimensional median sediment size.

The other approach to graphs of the existence fields is to use the mean velocity of the flow as a surrogate for the flow strength. The problem introduced by using the mean flow velocity is that the flow depth must be introduced as a leading variable as well, because various combinations of flow depth and flow velocity correspond to the same bed shear stress. The advantage of such a representation, however, is that it sidesteps the problem of partitioning the bed shear stress and leads to unambiguous characterization of the existence fields. The most recent example of such a graph is by Southard and Boguchwal (1990), shown here as Figure S57. The graph in Figure S57, which shows surface forms in terms of nondimensional mean flow velocity and nondimensional median sediment size, is drawn for just one value of nondimensional mean flow depth. For greater (lesser) flow depths, the existence fields shift their position up (down), and the shapes of the fields change somewhat as well.

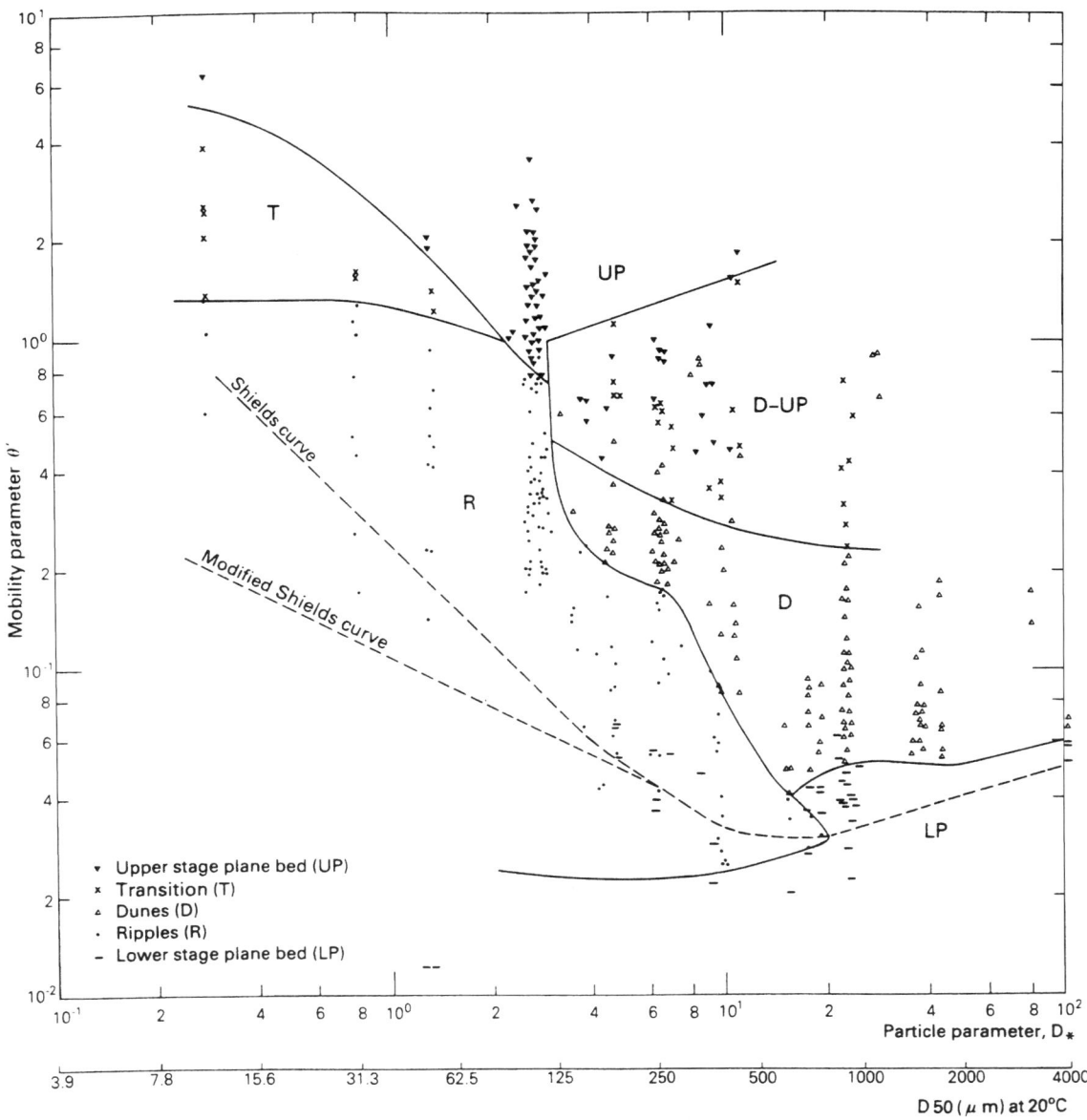

Figure S56 Graph showing existence fields of surface forms generated by unidirectional water flows in shallow channels. Data are from various sources. The mobility number θ' is a nondimensional measure that expresses the shear stress exerted on the particles by the flowing water, and $D*$ is a nondimensional measure of the median particle diameter. The lower vertical axis gives the equivalent value of actual median particle diameter at a reference water temperature of 20°C. (From Van den Berg and Van Gelder, 1993.)

Current ripples

Over a wide range of sediment sizes, from the finest noncohesive silts up to about 0.7 mm, and for flow strengths ranging upward from the threshold for sediment movement, the equilibrium surface form is current ripples. For sediment sizes less than about 0.2 mm, ripples give way to a planar transport surface with increasing flow strength; for coarser sizes, ripples pass abruptly into dunes with increasing flow strength.

Current ripples are rugged small-scale bed forms with crests and troughs oriented dominantly transverse to flow. Their downcurrent surfaces have slopes usually at or near the angle of repose of the sediment (of the order of 30°), although tending to be less at high flow strengths, and their upcurrent slopes are usually much gentler. Spacings are usually 0.1–0.2 m, and heights are usually no more than a few centimeters. Height and spacing are nearly independent of flow strength, flow depth, and sediment size.

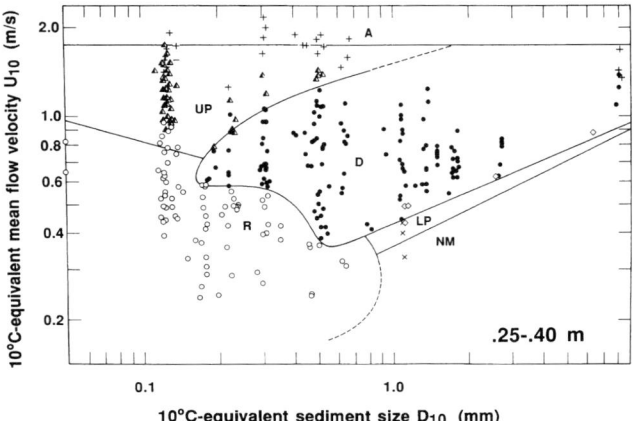

Figure S57 Graph showing the existence fields of surface forms generated by unidirectional water flows in shallow channels, with flow depths ranging from 0.25 m to 0.40 m. Data are from various sources. On the vertical axis, U_{10} is a nondimensional measure of the mean flow velocity, expressed as the equivalent value of the actual mean flow velocity at a reference temperature of 20°C. On the horizontal axis, D_{10} is the corresponding measure of the median particle diameter. A, antidunes; UP, upper-stage plane bed; D, dunes; R, ripples; LP, lower-stage plane bed; NM, no movement. (From Southard and Boguchwal, 1990.)

When current ripples first make their appearance on a planar sediment surface, they are strongly regular and straight-crested, but as they grow toward equilibrium they develop substantial irregularity in geometry: crests and troughs tend to be sinuous and variable in height, and when traced laterally individual ripples crests end by merging with another ripple crest of disappearing into a trough.

Ripples move in the downcurrent direction at speeds orders of magnitude less than the flow velocity by means of erosion of sediment from the upstream surface and deposition on the downstream surface. Individual ripples exist for time periods corresponding to downcurrent movement typically equal to less than ten ripple spacings before merging with another ripple. Meanwhile, new ripples are formed by division of a preexisting single ripple.

Subaqueous dunes

Over a range of sediment sizes from about 0.2 mm upward into gravels, and over a wide range of moderate to high flow strengths (Figures S56, S57) subaqueous dunes are the stable surface form under unidirectional water flows. In sediments finer than about 0.7 mm, dunes succeed ripples with increasing flow strength; in coarser sediments, dunes make their appearance at flow strengths not much greater than the threshold for sediment movement. With increasing flow strength, dunes are washed out to a planar transport surface.

In the past, surface forms here called dunes have variously been termed megaripples and sand waves as well as dunes. In recent years (see Ashley, 1990) it has become widely accepted to describe all such large-scale rugged flow-transverse forms as dunes. Some investigators, however, consider large-scale sand waves in tidal environments to be fundamentally different from dunes in unidirectional flows (Allen, 1980).

Dunes are much like ripples in geometry and movement. At relatively low flow strengths, dunes tend to be regular and straight crested (often called "two-dimensional dunes"), whereas at relatively high flow strengths they tend to be sinuous and discontinuous (often called "three-dimensional dunes"), with strong variability of crest and trough heights. For three-dimensional dunes, the existence of deep scour pits along troughs is especially characteristic.

The spacing of dunes ranges upward from somewhat less than half a meter to thousands of meters, and the height of dunes ranges upward from a several centimeters to many tens of meters. The controls on dune size are complex, but it seems clear that there is no fundamental limit to dune size, if the combination of flow strength and sediment size are appropriate and flow depth is sufficiently great. There is a well-defined break in size between current ripples and dunes.

Antidunes

Over a wide range of noncohesive sediment sizes, and for high flow strengths and relatively shallow depths, the equilibrium surface form is antidunes. Antidunes are trains of undulatory surface forms that develop on a planar transport surface. For a given flow depth, antidunes always develop if the flow strength is increased sufficiently, although the common limitations to possible flow strengths in nature mean that antidunes are characteristic only of rather shallow flows, except perhaps under conditions of catastrophic flood flows. Antidunes tend to move slowly upstream, growing in height and therefore in steepness as they move. Eventually the entire train of antidunes becomes unstable; the antidunes then break, causing much suspension of sediment, and the bed surface reverts to being nearly planar, whereupon the antidunes slowly reform.

Mud waves

In many regions in the abyssal ocean, at the foot of the continental slope, the surfaces of large bodies of fine sediment known as sediment drifts are covered with large-scale, low-amplitude wavelike features; for a well-documented recent example, see Flood and Shor (1988). These features have variously been called sediment waves, mud waves, or abyssal antidunes (the last because they seem to have shifted upcurrent; see below). Similar but smaller waves have been observed on the flanks of channels on submarine fans (e.g., Normark et al., 1980; McHugh and Ryan, 2000).

The sediment of which the waves consist seems invariably to be fine mud. The spacing of these waves ranges from one kilometer to as much as twenty kilometers, and heights are in the range of five meters to one hundred meters. Slopes of the flanks of these waves are seldom greater than one degree. Despite their gentle slopes, the waves are strikingly revealed by subbottom echo-sounding techniques. In some cases these waves are symmetrical, but typically they have shifted slowly in the direction that is known or assumed to be upcurrent. The subbottom record shows clearly that in most cases there has been deposition on both flanks of the wave, but at a lower rate on the flank toward which the wave has shifted. The internal stratification of waves shows clearly that they existed, and were the site of active continuous or semicontinuous deposition, for geologically long times. In some cases they are seen to be draped with different sediment and therefore inactive; in other cases they seem likely to be active still.

The dynamics of the mud waves are not fully understood. The waves seem to have been formed by slow deposition of fine sediment from bottom currents with moderate velocities and presumably low sediment concentrations. In the case of the waves on sediment drifts, the depositing currents seem to be bottom-hugging contour currents carrying low concentrations of suspended sediment, which however are substantially greater than the much cleaner water higher in the water column. In the case of the waves on submarine fans, the evidence points to mild turbidity flows that result from overbank spillage of channelized episodic turbidity currents from channels onto broad levees. The similarity in features between these two different environments suggests a common origin. The most widely accepted theory (Normark *et al.*, 1980; Flood, 1988) is that these waves are a manifestation of a pattern of nearly stationary internal waves generated in the near-bottom interval of vertical density gradient that is occasioned by the upward decrease in suspended-sediment concentration.

Flow-parallel surface forms

Furrows are a kind of flow-parallel surface form that is produced by localized erosion of cohesive sediments subjected to moderately strong unidirectional or bidirectional currents in a generally depositional regime. They have been reported from the abyssal ocean (e.g., Hollister *et al.*, 1974), estuaries (e.g., Flood, 1981), and lakes (e.g., Viekman *et al.*, 1992).

Furrows range in depth from a meter to more than ten meters, and in width from a meter to as much as one hundred meters. In cross section they often appear to be roughly trapezoidal, with steep and often rippled side slopes and a nearly flat floor. Spacing of the furrows ranges from more than ten times width to even less than width. The sediment surface between furrows is generally flat. Individual furrows can extend for distances greater than one kilometer, and perhaps much farther, given the difficulties of continuous observation over broad areas in inaccessible environments. Furrows commonly bifurcate in tuning-fork-like junctions, which open toward the oncoming current. Measured current strengths in furrowed areas range from five centimeters per second in abyssal examples to more than half a meter per second in shallow continental shelves and large lakes. In at least some cases there is good evidence of continuing deposition in the areas between the furrows, and this has to be reconciled with the obviously erosional nature of the furrows themselves.

It is generally believed that the furrows are a manifestation of the existence of patterns of helicoidal streamwise secondary circulation superimposed upon the mean flow. The furrows become localized along streamwise zones of flow convergence, where low-density aggregates collect in windrows and abrade the underlying cohesive sediment as they are moved downcurrent. The sediment put in suspension then tends to be carried upward in the ascending branch of the circulation cell. In this view, the furrows can be maintained by occasional strong erosive flows on a background of overall fine-sediment deposition during long periods of weaker flow. See Flood (1983) for an exposition of this theory.

Erosional forms: flutes

When an erosive current passes over a bed of cohesive sediment, the sediment surface tends to take on a characteristic fluted geometry. The individual flutes vary in their shape, but their distinctive characteristic is an approximately bilateral symmetry about a vertical plane parallel to the flow but a strong asymmetry when viewed normal to that plane, whereby the depth of the scour and the steepness of the surface are much greater at the upstream end than at the downstream end. The adjective "heel-shaped" (Allen, 1984) is an apt descriptor. The typical flute, when viewed from above, has an upcurrent end with approximately parabolic shape, although in many cases the upstream end of the flute is more pointed or has an asymmetrical corkscrew shape. The geometrical pattern of the flutes shows considerable variety, as does the density of flutes on the sediment surface. Most flutes are on streamwise scales of the order of ten centimeters, but they range upward in size to at least a few meters, and there have been reports of giant flute-shaped scours that seem to be the result of especially powerful turbidity currents (e.g., Shor *et al.*, 1990). Much of our understanding of the nature and origin of flutes has come about through the experimental work of J.R.L. Allen (for a summary, see Allen, 1984).

Flutes tend to be formed by powerful currents, like turbidity currents or river floods, which form the flutes during an initial erosive phase and then bury them during a later depositional phase. Much of what is known about flutes is derived from observations of the bases of sandstone beds less resistant overlying shales in the ancient sedimentary record; the sandstones record the casts (i.e., the negative imprints) of the flutes. There has been much success in producing realistic flutes experimentally; see Allen (1984) for a summary of such work (and see *Scour, Scour Marks*). It is clear that flutes are the result of net erosive removal of material from the sediment surface without further interaction with the eroded material with the sediment bed downstream; the eroded material goes directly into suspension rather than being transported as bed load. The close similarity between flutes formed on cohesive mud beds and those formed by dissolution on soluble surfaces of rock or ice by flowing water supports that notion.

Subaqueous surface forms generated by oscillatory and combined flow

Oscillation ripples

Noncohesive sediments over a wide range of sediment sizes assume surface forms in a wide range of sizes and geometries in bidirectional purely oscillatory flow. Such forms have been the subject of systematic study in both natural environments and laboratory tanks since early in the twentieth century. Among the earliest of modern laboratory studies is that of Bagnold (1946); the classic field study was by Inman (1957).

Field studies of oscillatory-flow surface forms suffer from the difficulties of observation under conditions of heavy seas, and laboratory studies have concentrated mainly on the fairly short oscillation periods to which small apparatus are limited. More recent laboratory studies at much longer oscillation periods in a large oscillatory-flow tunnel are reported by Southard *et al.* (1990).

A useful way to describe the conditions of existence of oscillatory-flow surface forms is to consider them as a function of three leading variables: median sediment size, maximum oscillatory velocity, and oscillation period. As with unidirectional-flow surface forms in the preceding section, the ranges of existence of the various oscillatory-flow surface forms can

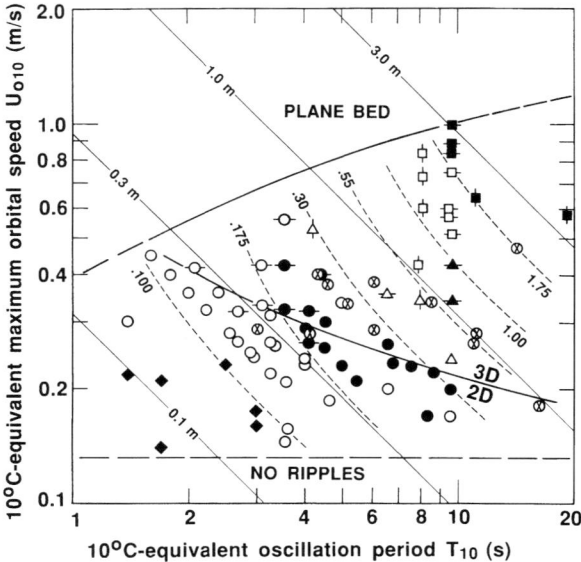

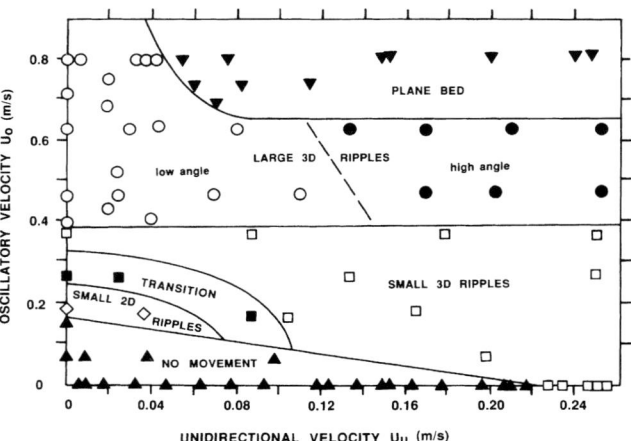

Figure S58 Relationships of oscillatory-flow surface forms in a graph of nondimensional maximum oscillatory flow speed U_{o10} versus nondimensional oscillation period T_{10}, expressed as equivalent values at a reference temperature of $10°C$, for particle sizes of about 0.1 mm. Symbols for spacing of surface forms: solid diamonds, <0.100 mm; open circles, 0.10–0.175 mm; solid circles, 0.175–0.30 m; open triangles, 0.30–0.55 m; solid triangles, 0.55–1.0 m; open squares, 1.0–1.75 m; solid squares, >1.75 m. The dashed curves are approximate contours of surface-form spacing. Horizontal and vertical tick marks represent three-dimensional configurations; symbols without tick marks represent two-dimensional configurations. (From Southard, 1991.)

Figure S59 Relationships among combined-flow surface forms in a plot of nondimensional maximum oscillatory flow speed U_{o10} versus nondimensional maximum unidirectional flow speed U_{u10}, expressed as equivalent values at a reference temperature of $10°C$, for particle sizes of around 0.1 mm and oscillation periods of around 9.5 second. Solid circles, combined-flow ripples; open circles, plane-bed sediment movement (for large U_{o10}), no sediment movement (for small U_{o10}). Data are from various sources. (From Arnott and Southard, 1990.)

be viewed in a three-dimensional graph with nondimensionalized versions of these three variables along the axes.

Figure S58 is a velocity–period for a sediment size of about 0.1 mm, with axes expressed as $10°C$-equivalent maximum oscillatory velocity and oscillation period. For a wide range of oscillatory velocities above the threshold for sediment movement, ripples are the equilibrium surface form. At even higher oscillatory velocities, the ripples are washed out to a planar transport surface. Over most of the ripple field, ripple spacing increases upward to the right, with increasing velocity and period. The spacing of the ripples scales closely with the orbital diameter (the distance traversed by water particles in the course of a single oscillation). For relatively small velocities and periods the ripples are geometrically regular and straight-crested, but with increasing velocity and period there is a well-defined change to less regular geometry. The graph for coarser sand sizes is similar, but the transition to plane-bed transport is at a higher velocity, and the ripples are regular and straight-crested over the entire existence field.

The nature of oscillation ripples formed by more complex multidirectional oscillatory flows is poorly understood. In the case of what are called ladderback ripples, for which two different sets of oscillation ripples are superimposed, it is generally assumed that a second set of ripples become superimposed on an preexisting set without, however, completely obliterating the earlier set. Geometrically complex small-scale ripples showing mixtures of hexagonal and even pentagonal shapes might be generated by multidirectional oscillatory flow, inasmuch as normal-to-crest profiles are close to being symmetrical, but observational data are not available to confirm such a view. Hummocky forms with rounded crests interspersed with swales, in a seemingly random pattern on lateral scales of up to a meter, are known from bedding planes in ancient shallow-marine deposits. It is not clear whether such forms are generated by strong bidirectional or strong multi-directional oscillatory flows, because of the difficulties of observation in modern field and laboratory flow environments under such conditions.

Combined-flow forms

Combined flows are common in natural flow environments, especially shallow marine and lacustrine settings, where strong winds associated with storms cause both unidirectional and oscillatory water motions. The range of such combined flows is enormous, and it should be expected that the surface forms generated by those flows should accordingly show a wide range of scales and geometries.

Much less is known about the nature of combined-flow surface forms than about those generated by unidirectional and oscillatory flows. There have been few recent field studies; that by Amos *et al.* (1988) is representative of the problems and possibilities of making such observations in shallow marine settings. Moreover, there have been few laboratory studies, and even an ambitious laboratory program can hope to cover only a narrow subset of combined-flow surface forms there have been few laboratory studies. Among the few recent studies, Arnott and Southard (1990) made a systematic survey of combined-flow surface features in a large combined-flow duct but for only a single oscillation period, a single sediment size, and a wide range of oscillatory flow velocities but only a narrow range of unidirectional flow velocities (Figure S59).

It is clear from studies of combined-flow surface forms up to now that only a small unidirectional flow component is needed to make the forms strongly asymmetrical. The consequence is that combined-flow forms, at least in fine sands, look superficially like current ripples. A broad range of larger-scale combined-flow forms generated by stronger oscillatory flows with a subordinate unidirectional component look superficially like small unidirectional-flow dunes, although the sediment size is much finer than can support unidirectional-flow dunes. This may be the main reason why reports of combined-flow surface forms preserved in the sedimentary record have been so much scarcer than their importance would seem to imply. Much more observation is needed before even a first-order picture of the range of combined-flow forms is available.

Subaerial surface forms

When the wind blows over a surface of loose sediment at a speed sufficient to move the particles, the surface is molded into a variety of configurations with characteristic scale and geometry. The smallest-scale features are called ripples, and larger-scale features are called dunes, although with much variation in terminology. Modern studies of subaerial surface forms date from the classic field and laboratory work by Bagnold (1941). For recent summaries of the nature, classification, and dynamics of subaerial surface forms, see Greeley and Iversen (1985), Pye and Tsoar (1990), Nickling (1994), and Lancaster (1994).

It is generally agreed that there is a fundamental dynamical distinction between subaerial ripples and dunes. This is reflected in a well-defined break or gap in the range of sizes of the two kinds of features. It is less clear whether there are natural breaks in the range of dune sizes. Wilson (1972) reported a clear break between two classes of large-scale subaerial surface forms, which he called dunes and draa, but later workers (e.g., Ellwood *et al.*, 1975) have been more inclined to believe that there is a smooth variation in large-scale eolian forms, without any natural breaks.

A complicating factor in classifying subaerial surface forms and understanding their dynamics is that in most settings wind speed and direction are at least to some degree variable. This is less of a problem for ripples than for dunes, however, because typical small-scale ripples respond to sand-moving winds on timescales of minutes to hours, whereas dunes, with their far greater bulk relative to rates of sand transport, require much longer times to become equilibrated to a steady wind.

Another reason behind controversy over the dynamics and classification of subaerial surface forms is that nature tends to give us biased realizations of the phenomena. Fractionation of particle size along transport paths causes particle-size distributions to evolve in particular ways, and this has a tendency to obscure the picture of dynamics and classification by making it difficult obtain an even view of the process and features involved.

Comprehensive data on equilibrium subaerial forms generated by steady winds, of the kind that have been obtained for subaqueous forms in laboratory water channels and tanks, would constitute a natural basis for evaluating the fundamental nature of subaerial forms. Only for small-scale subaerial ripples, however, have studies been carried out in large wind tunnels in which not only wind strength but also particle size and sorting are varied over a wide and unbiased range. Such studies are virtually impossible for large-scale subaerial features.

The dynamics of subaerial surface forms and subaqueous surface forms should be expected to be at least in part fundamentally different, because of the great difference in the ratio of sediment density to fluid density in the two cases. In transport of mineral-density particles by water, the density ratio is less than about three, except for local and uncommon concentrations of higher-density mineral particles, whereas the density ratio in most eolian settings is of the order of 10^3. In consequence, particle transport by wind is much less affected by fluid turbulence; the sediment particles, once set into motion, tend to follow trajectories, usually described as ballistic, that are governed much more by particle inertia than by fluid turbulence. It should also be expected, however, that this difference is much more significant for subaerial ripples than for subaerial dunes. The scale of ripples is of the same order as the typical excursion lengths of the transported particles, whereas dunes are far larger. The dynamics of wind ripples are in some way governed by instabilities associated with the nature and scale of particle excursions, whereas the dynamics of dunes is in some way related to instabilities associated with larger-scale streamwise variations in sediment transport rate, in ways that are likely to be broadly similar to subaqueous dunes.

Mineral-density particles in water and in air, the two cases of density ratio that are important on Earth, represent only two points along the spectrum of density ratios. It seems likely to many if not most astronomers that there are planets in other solar systems with surface conditions represented by a far larger range of sediment-to-fluid density ratios. Very little is known about such cases; in particular, the question arises: do the surface forms we know as current ripples and wind ripples grade continuously from one to the other, or is there a break in existence in some intermediate range? For some comments on ripples on other planets, see Greeley and Iversen (1985).

Subaerial ripples

Subaerial ripples are small-scale features that are oriented transverse to the wind. They range in spacing from less than a centimeter to as much as a few tens of meters, and in height from a few millimeters to as much as a meter. The most common ripples range in spacing from a few centimeters to between ten and fifteen centimeters. These are variously called impact ripples or ballistic ripples, which are genetic terms that reflect commonly held views on their dynamics (see below). Some investigators refer to these simply as "normal" ripples. They are strikingly regular in their geometry, with long, straight crests and troughs that commonly extend for some meters before either ending or bifurcating. Crest height and trough depth vary little in the transverse direction. In vertical profile the ripples are noticeably asymmetric, with upwind (stoss) slopes of $8-10°$ and downwind (lee) slopes of $20-30°$. For an especially illuminating field study of wind ripples, see Sharp (1963).

From a number of field and laboratory studies, summarized by Pye and Tsoar (1990) and Lancaster (1994), it is known that ripple size increases with both wind velocity and particle size. There is some evidence as well that ripple size increases as the sorting of the sediment decreases (i.e., the spread of sizes around the mean increases).

Many theories have been proposed for the dynamics of wind ripples, and there is as yet no general agreement. The classical viewpoint, dating from the pioneering work of Bagnold (1941), is that the ripple spacing is governed by the saltation distance of the moving particles. This view has been challenged by Sharp (1963) and more recently by Anderson (1987).

Larger subaerial ripples in coarser sediment, well into the pebble size range (>2 mm), although less common that the smaller ripples in sands, are well-known. These have variously been called granule ripples or megaripples. Sediment size distributions are usually bimodal, and the ripples tend to be less regular in the crest pattern. There is disagreement about whether these are dynamically distinct from the more common smaller ripples. Also reported in the literature are small-scale but much less regular ripples, variously called fluid drag ripples or aerodynamic ripples, which are formed by strong winds in very fine sand or coarse silt.

When saltating sand particles move across a damp sand surface, as on a beach at low tide during a strong offshore wind that transports sand from coastal dunes onto the beach, some of the particles adhere to the surface and remain there, held by surface tension. They are wetted by capillarity, and in turn capture more saltating particles. In some cases the surface remains approximately planar as it aggrades; in other cases, a pattern of irregularly transverse ridges, called adhesion ripples, develops. Adhesion ripples commonly have spacings of the order of a centimeter and heights of a fraction of a millimeter to a few millimeters. Less regular patterns without any appreciable transverse orientation have been called adhesion warts.

Subaerial dunes

Subaerial dunes are mounds or ridges of sand that are shaped by the wind. In a narrower definition, dunes are independent of surrounding topography; in that respect, dunes are features that develop in broad areas that are underlain by loose sand and are subjected to sand-moving winds. In addition, however, certain features that are commonly termed dunes are accumulations of sand that form to the windward or leeward sides of cliffs, hills, or other topographic features, large and small. One of the most extensive studies of dunes in the broad regions of dunes in subtropical deserts, known as sand seas, is by McKee (1979).

An often bewildering variety of terms have been used to describe or classify dunes. Many of these terms are partly or wholly synonymous, and many have been used only in local areas. This diversity in terminology is in great part a reflection of the great variability in conditions under which dunes form, as noted below. Breed and Grow (1979) provide an extensive correlation among the various classifications schemes.

Dunes vary extremely widely in scale and geometry. This should not come as a surprise, inasmuch as several factors exert important controls: the size, sorting, and thickness of loose sediment available to be formed into dunes; wind speed and its variability; wind direction, and its variability; vegetation; and downwind change in time-mean sediment transport rate, which determines whether the dunes are experiencing net aggradation, net degradation, or a neutral condition. Leaving aside the stabilizing effect of vegetation, it is generally considered that the most important of these are the nature of the wind regime and the thickness of available sediment.

Many classifications of dunes have been proposed. These classification are of two kinds: morphologic or morphodynamic. A morphologic classification is based solely on observable dune geometry, whereas a morphodynamic classification takes account of assumed characteristics of wind and transport regimes. One of the most common morphodynamic classifications divides dunes into longitudinal, transverse, and oblique with respect to the direction of net sand transport. This classification is not without problems, however, because it is commonly difficult or even impossible to establish the direction of net sand transport, owing to the variability of the wind and the long timescales over which the wind and transport must be averaged.

In some regions with dunes, the sand-moving winds are very consistently from a narrow range of directions, so that there is a simple relationship between the dominant wind and the net sediment transport direction. In other cases, however, the sediment-transporting wind is complexly multidirectional, and the relationship between the wind and the net transport direction is less obvious. In the case of star dunes, the dunes are formed in areas with little or no net transport.

Much of the variability in dune characteristics is a consequence of the effective average thickness of transportable sand in a given area in relation to the potential size of the dunes. For a given dune size, some minimum thickness of sand is necessary to assure full coverage of movable sand even in dune troughs; conversely, for a given thickness of movable sand, there is an upper limit to dune size without "starvation" (i.e., exposure of rigid substrate) in dune troughs.

The effect is especially relevant to variations in geometry of transverse dunes. When only a minimum of movable sand is available, barchans are the typical dune form. Barchans are isolated, crescent-shaped dunes with sharp horns pointing downwind. They have a gentle upwind slope and generally an angle-of-repose downwind slope. With increasing availability of movable sand, the barchans begin to join at the horns, and they eventually form continuous but sinuous and irregular transverse ridges. Eventually even troughs are occupied by movable sand.

Linear dunes, often strikingly regular in their geometry and commonly oriented nearly parallel to what is thought to be the direction of net sand transport, are common in regions of abundant sand supply. Such dunes are especially characteristic of the world's great subtropical deserts. In most such regions, the troughs between the linear dunes are nonetheless starved of sand. In such areas individual linear dunes can be continuous for distances of as great as two hundred kilometers. The origin of these linear dunes continues to be controversial. In one view, they are formed under the influence of helical secondary circulations superimposed on the mean wind in the atmospheric boundary layer. Such circulations are known to exist, but no convincing direct evidence has yet to be adduced that they are instrumental in the generation and maintenance of linear dunes.

Smaller-scale dunes are commonly seen to be superimposed on larger-scale dunes. Dunes with superimposed smaller dunes are termed compound dunes. There is often uncertainty about whether the superimposition represents an equilibrium condition or a manifestation of adjustment of the dunes to changing conditions. In at least some cases it seems clear that the former is the case.

The fundamental question as to why dunes form is an enduring one. Two approaches, which are not fundamentally different, are possible. In one approach, which is relevant to conditions of minimum sand supply, one can think in terms of the conditions needed to localize a patch of sand moving over a rigid substrate and then cause the patch to grow into one or a series of dunes. This approach was first developed systematically by Bagnold (1941). In the other approach, one can assume a thick continuous bed of sand and think in terms of how the sand-moving wind generates one or another dune type. In the latter approach, which is the one usually taken in the case of subaqueous surface forms, the problem of the dynamics and classification of dunes might be considered from the perspective of a "thought experiment" in which a full bed of sand is subjected to a wide range of wind conditions but also a gradually shrinking thickness of the sand layer, to reveal changes as sand starvation makes its appearance and increases in importance. For subaerial dunes, of course, the feasibility of an actual experiment of that kind is extremely limited, owing to the large scales involved.

John B. Southard

Bibliography

Allen, J.R.L., 1980. Sand waves: a model of origin and internal structure. *Sedimentary Geology*, **26**: 281–328.
Allen, J.R.L., 1984. *Sedimentary Structures; Their Character and Physical Basis, Volume I, Volume II*. Amsterdam: Elsevier.
Amos, C.L., Bowen, A.J., Huntley, D.A., and Lewis, C.F.M., 1988. Ripple generation under the combined influence of waves and currents. *Continental Shelf Research*, **8**: 1129–1153.
Anderson, R.S., 1987. A theoretical model for aeolian impact ripples. *Sedimentology*, **34**: 943–956.
Arnott, R.W., and Southard, J.B., 1990. Experimental study of combined-flow bed configurations in fine sands, and some implications for stratification. *Journal of Sedimentary Petrology*, **60**: 211–219.
Ashley, G.M., 1990. Classification of large-scale subaqueous bed forms: a new look at an old problem. SEPM Bedforms and Bedding Structures Research Symposium. *Journal of Sedimentary Petrology*, **60**: 160–172.
Bagnold, R.A., 1941. *The Physics of Blown Sand and Desert Dunes*. London: Methuen.
Bagnold, R.A., 1946. Motions of waves in shallow water. Interaction of waves and sand bottoms. *Royal Society [London], Proceedings*, **A187**: 1–15.
Breed, C.S., and Grow, T., 1979. Morphology and distribution of dunes in sand seas observed by remote sensing. In McKee, E.D., (ed.), A Study of Global Sand Seas. *US Geological Survey, Professional Paper 1052*, pp. 305–397.
Ellwood, J.M., Evans, P.D., and Wilson, I.G., 1975. Small-scale aeolian bedforms. *Journal of Sedimentary Petrology*, **45**: 554–561.
Flood, R.D., 1981. Distribution, morphology, and origin of sedimentary furrows in cohesive sediments, Southampton Water. *Sedimentology*, **28**: 511–529.
Flood, R.D., 1983. Classification of sedimentary furrows and a model for furrow initiation and evolution. *Geological Society of America, Bulletin*, **94**: 630–639.
Flood, R.D., 1988. A lee wave model for deep-sea mudwave activity. *Deep-Sea Research*, **35**: 973–983.
Flood, R.D., and Shor, A.N., 1988. Mud waves in the Argentine Basin and their relationship to regional bottom circulation patterns. *Deep-Sea Research*, **35**: 943–971.
Gilbert, G.K., 1914. The Transportation of Debris by Running Water. *US Geological Survey, Professional Paper 86*.
Greeley, R., and Iversen, J.D., 1985. *Wind as a Geological Process on Earth, Mars, Venus and Titan*. Cambridge University Press.
Guy, H.P., Simons, D.B., and Richardson, E.V., 1966. Summary of Alluvial Channel Data from Flume Experiments, 1956–1961. *US Geological Survey, Professional Paper 462-I*.
Hollister, C.D., Flood, R.D., Johnson, D.A., Lonsdale, P., and Southard, J.B., 1974. Abyssal furrows and hyperbolic echo traces on the Bahama Outer Ridge. *Geology*, **2**: 395–400.
Inman, D.L., 1957. *Wave-Generated Ripples in Nearshore Sands*. U.S. Army Corps of Engineers, Beach Erosion Board, Technical Memorandum 100.
Lancaster, N., 1994. Dune morphology and dynamics. In Abrahams, A.D., and Parsons, A.J., (eds.), *Geomorphology of Desert Environments*. London: Chapman and Hall.
McHugh, C.M.G., and Ryan, W.B.F., 2000. Sedimentary features associated with channel overbank flow: examples from the Monterey Fan. *Marine Geology*, **163**: 199–215.
McKee, E.D., 1979. A Study of Global Sand Seas. *US Geological Survey, Professional Paper 1052*.
Nickling, W.G., 1994. Aeolian sediment transport and deposition. In Pye, K. (ed.), *Sediment Transport and Depositional Processes*. Blackwell, pp. 293–350.
Normark, W.R., Hess, G.R., Stow, D.A.V., and Bowen, A.J., 1980. Sediment waves on the Monterey Fan Levee: a preliminary physical interpretation. *Marine Geology*, **37**: 1–18.
Pye, K., and Tsoar, H., 1990. *Aeolian Sand and Sand Dunes*. London: Unwin Hyman.
Richards, K.J., 1980. The formation of ripples and dunes on an erodible bed. *Journal of Fluid Mechanics*, **99**: 597–618.
Sharp, R.P., 1963. Wind ripples. *Journal of Geology*, **71**: 617–636.
Shor, A.N., Piper, D.J.W., Hughes Clarke, J.E., and Mayer, L.A., 1990. Giant flute-like scour and other erosional features formed by the 1929 Grand Banks turbidity current. *Sedimentology*, **37**: 631–645.
Simons, D.B., and Richardson, E.V., 1963. Forms of bed roughness in alluvial channels. *American Society of Civil Engineers, Transactions*, **128**: 284–302.
Southard, J.B., 1991. Experimental determination of bed-form stability. *Annual Review of Earth and Planetary Sciences*, **19**: 423–455.
Southard, J.B., and Boguchwal, L.A., 1990. Bed configurations in steady unidirectional water flows. Part 2: Synthesis of flume data. *Journal of Sedimentary Petrology*, **60**: 658–679.
Southard, J.B., Lambie, J.M., Federico, D.C., Pyle, H.T., and Weidman, C.R., 1990. Experiments on bed configurations in fine sands under bidirectional purely oscillatory flow, and the origin of hummocky cross-stratification. *Journal of Sedimentary Petrology*, **60**: 1–17.
Van den Berg, J.H., and Van Gelder, A., 1993. A new bedform stability diagram, with emphasis on the transition of ripples to plane bed over fine sand and silt. In Marzo, M., and Puigdefabregas, C. (eds.), *Alluvial Sedimentation*. International Association of Sedimentologists, Special Publication, Volume 17, pp. 11–21.
Vanoni, V.A., 1975. *Sedimentation Engineering*. New York: American Society of Civil Engineers.
Viekman, B.E., Flood, R.D., Wimbush, M., Faghri, M., Asako, Y., and Van Leer, J.C., 1992. Sedimentary furrows and organized flow structure: a study in Lake Superior. *Limnology and Oceanography*, **37**: 797–812.
Wilson, I.G., 1972. Universal discontinuities in bedforms produced by the wind. *Journal of Sedimentary Petrology*, **42**: 667–669.

Cross-references

Bedding and Internal Structures
Bedset and Laminaset
Cross-Stratification
Dunes, Eolian
Eolian Transport and Deposition
Flaser
Flow Resistance
Flume
Gutters and Gutter Casts
Hummocky and Swaley Cross-Stratification
Microbially Induced Sedimentary Structures
Offshore Sands
Paleocurrent Analysis
Parting Lineations and Current Crescents

Physics of Sediment Transport: The Contributions of R.A. Bagnold
Ripple, Ripple Mark, Ripple Structure
Scour, Scour Marks
Sedimentary Structures as Way-up Indicators
Tool Marks

SURFACE TEXTURES

Introduction

The advent of the scanning electron microscope (SEM) brought a new tool to the sedimentologists armory. In particular, it allows images to be obtained at a variety of magnifications (10–10 k times) together with a depth of field unobtainable with a light microscope, especially at high magnifications. Despite the high cost and considerable complexity of the instruments introduced in the late 1960s a number of workers exploited the relative ease of geological sample preparation for the SEM. It was, for example, no longer necessary to produce surface replicas, necessary with transmission electron microscope techniques, as the surface itself was imaged. Subsequent advances in ease of use of the SEM has, in effect, meant that bench top instruments can now be used as "super microscopes" for a wide range of sedimentological purposes and that even casual users can obtain a high degree of proficiency at a basic level.

One early geological use of the SEM, by Krinsley and co-workers, was to view surface textures of individual sand-size grains as an extension of optical microscopic examination of detrital grains. An outline description of the SEM itself and a recent review of surface textures together with a general overview of the use of the SEM in glacial geology is given in Whalley (1996) and a wide-ranging review of the technique and associated methods can be found in various papers in Marshall (1987) and the more general books on the use of SEM in sediments by Smart and Tovey (1981, 1982).

Principles of the technique

Large energy inputs into transportive sedimentary systems usually produce attrition of particles as well as sorting. It is the small scale, inter-particle, attrition, which produces changes in shape and rounding, as well as impacts of surfaces, that are revealed by SEM examination. Thus, the fundamental concept behind the examination of surface textures is that detrital grains, undergoing transport, will abrade and that the core (or original) grain will reflect the process of sediment movement by the development of a distinctive surface texture. This texture is in addition to, but associated with, any changes produced in the grain rounding or shape modification that may occur. Chemical action on grain surfaces may be found under suitable environments and surface textures are also produced in as a result of such modifications. For many years it was recognized that "frosting" was a typical texture seen on detrital grains from both present-day deserts as well as grains from the geological record. Krinsley's exploitation of the SEM was to show that a range of distinctive microscale features could be identified from various sedimentological environments (e.g., Krinsley and Marshall (1987) for an overview). The technique thus associates specific surface textures (often a range of features) on the surfaces of grains with corresponding sedimentological events, thereby giving clues as to the sediment's history. Quartz, most usually sand-sized, grains are generally used for identification of surface textures because of their ubiquity, although garnets or heavy minerals (e.g., Stieglitz, 1969) have also been used. There are however dangers in using surface textures as a unique method of sediment fingerprinting.

Sampling and preparation of material

Cost, both "beam time" and subsequent viewing time, is a problem with SEM examination. As SEM surface texture examination is an individual rather than a bulk technique, the representativeness of a sample is, or may be, problematical. Even a minimum of 30 grains can be expensive if multiple samples are to be examined. As the technique is largely dependent upon visual characterization (see below) a bias may be introduced so blind sampling and sediment preparation is recommended. Indeed, to be as objective as possible, the person viewing grains and identifying the textures should not be the person who needs the data or interprets the findings. Such precautions may not be necessary if the technique is used as check or as an adjunct to other methods of characterization.

Grains for mounting should be cleaned by gentle washing, adhering material being removed with hydrochloric acid and stannous chloride. Ultrasonic treatment has been discouraged by some workers although it may be useful unless grains are very delicate. Sieving should be avoided at any stage of the sampling/selection process as this may introduce new surface textures. It is usual to examine grains in the 250–1000 μm range. At the simplest level, preparation involves sticking grains to an SEM specimen stub and grains of this size can be picked up by micro-forceps under a binocular microscope and arranged on the specimen stub. This allows easy identification of quartz (or other mineral type) but may introduce a sampling bias. Under the SEM it may not be possible to identify the mineralogy without the use of an elemental detector (usually an energy-dispersive spectrometer, EDS/EDX) which, although useful, is still an expensive ancillary item for desk-top SEMs. The specimen stub surface is best coated with double-stick tape as grains placed carefully on this usually adhere well. Coating by gold (sputtering) or carbon (evaporation) provides good electrical continuity over the surface, poor contact gives charging under the electron beam which degrades the image. Grains may be placed on stubs in rows or in a spiral to allow subsequent "revisiting". Typically, rather more than 30 grains should be placed to allow for poor photography, etc at the observation stage. Sprinkling is an alternative way of providing many grains for examination but experience shows this frequently gives poor electrical connection for larger grains and less easy identification due to overlapping. However, sprinkling may be the only way to mount small grains (<250 μm). An electronic spatula is one way to distribute grains carefully and a gentle jet of clean dry air both removes surplus grains and improves adhesion.

Observations under the SEM

The normal (secondary/emitted) mode of SEM observation is generally employed for visual characterization. There may be advantages in cathodo-luminescence (Smart and Tovey, 1981) and back-scattered images for certain specialized needs but are rarely necessary. The traditional way of observing has been to identify a row of grains and move progressively along, identifying grains and recording on a lab sheet, and to take a low power photograph of each with one or more higher power (200–1000×) photographs of each grain and its surface textures. Because of the expense and/or pressure on beam time, photographs are best examined for subsequent feature recognition rather than when observing by SEM. A certain amount of experience is necessary to identify useful features and this tends to slow the direct examination process. Conventional photography may take a minute or more to obtain an image. With the advent of quality video photography it has become much easier to take many images, and at a variety of scales, so it is most convenient to carry out all the detailed examination "off-line". This reduces the cost of both beam time and the time needed to perform conventional photography. Identification of grains on lab sheets is important as it may be necessary to revisit grains to obtain quality photography for publication purposes.

Identification of textures

That it should be feasible to differentiate between the basic environments of eolian, beach, fluvial attrition, etc. is perhaps remarkable. However, the early experiments did show that grains from an undisputed, single component provenance examined by SEM did provide unique and different textures. A "surface texture" is often identified by a suite of microfeatures and early papers listed a few of these components and showed representative micrographs. To identify provenance therefore, grain textures are identified from photographs of sample grains which are then compared with catalogued images. The "SEM atlas" of Krinsley and Doornkamp (1973) is still widely used for this purpose. A few example illustrations are given in this article. There has been discussion about the degree to which interpretation can be made. Care must be taken to avoid over-interpretation, especially where grains have undergone multiple transport cycles. Further discussion on this can be found in Whalley (1996).

An apparently more quantitative method than texture comparison was subsequently evolved whereby the individual features or texture components were recorded and then compared with lists of features categorized from known environments. These lists were built up from both original and previously published papers. The number of texture components may vary from paper to paper but percentages/histograms or presence/absence have been used to provide assessment of a suite of textures (Bull and Culver, 1979). The paper by Higgs (1979) is useful as it provides both a semi-quantitative assessment and a selection of micrographs. It should be noted that the range of environments which can be identified uniquely is somewhat limited, although various authors have tried to extend the possibilities. Some of these possibilities are mentioned below (see also Whalley, 1996).

Sedimentological information about environments is imprinted upon the *tabula rasa* of an original grain. This is easily appreciated when grains are freshly weathered but it is less easy to identify environments unambiguously if grains are part of a multi-stage cycle. Further, it should be noted that many of the features or texture components are not independent. For example, a grain released from a rock mass by weathering or by impact, such as a block falling on talus, both produce sharp edges and corners, stepped and rippled surfaces ("Wallner lines") as a direct product of brittle fracture (see Gallagher, 1987, for a detailed discussion). These texture suites are typical of mechanically-weathered sediments but also of some glacial sediments. Thus, although there has been considerable energy involved in the creation of new surfaces, the surface texture is no different. This has produced some problems in the recognition of a truly "glacial" surface texture. Although modified grain edges, gouges and chattermarks may be found and which may be the result of true subglacial action, their significance has been debated. Orr and Folk (1983) suggested that chattermarks are not necessarily environmentally controlled, rather that the visibility of chattermarks trails is related to mineralogy and weathering. It is probable that there be several ways to produce such features.

Modifications to an original grain surfaces are of two basic kinds; mechanical and chemical. As well as the complete fracture through a grain, the former also relates to modification of surfaces or edges by impacting, usually spinning, grains in a fluid medium. Edges and corners are abraded first because spinning grains catch each other (Figure S60). This action also

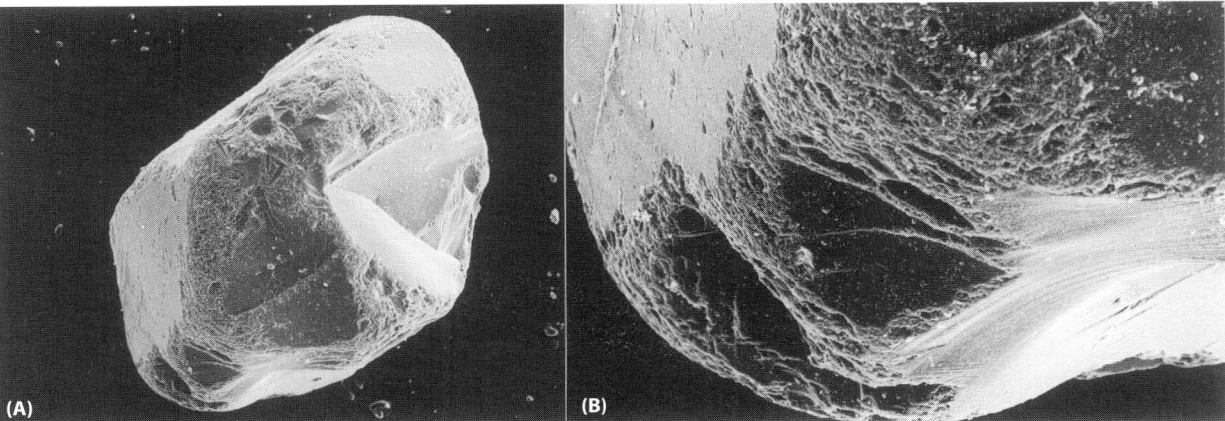

Figure S60 (A) A grain of crushed Brazilian quartz from a s sample which has been abraded in a closed air jet system for 4 hours. Picture width = 500 μm. (B) Detail of (A) showing the abrasion of the edge and corners although the main faces are almost devoid of any impact textures. Picture width = 230 μm.

produces a gradual modification in grain angularity to produce more rounded grains. Thus, the basic shape may also be modified. Surfaces, being less exposed to spinning grain protuberances, are less prone to modification than edges. In experiments of rounding of sand grains in eolian environments it has been shown (Whalley *et al.*, 1982) that textures are of mechanical formation even though they appear to be similar to some textures resulting from chemical action.

Chemical action on grains is usually found in the form of surface etching, or as silica re-precipitation; some crystallographic orientations are particularly susceptible to etching. The typical forms are varieties of etching and diagenetic features (Krinsley and Doornkamp, 1973). Crystallographic orientation with respect to etched surfaces may also play a part in determining the size and prominence of etching and silica re-deposition. Weathering on mineral grains has been used as a means of dating deposits. For example, Black and Dudas (1987) used a method of estimating surface weathering of soils of glacial origin in Alberta, although others have found no clear association between post-depositional diagenesis and time. It would appear that there needs to be further research in elucidating the actual formative mechanisms for chemical as well as mechanical surface texture formation.

It should be noted that identification of these surface textures, both as individual components and suites, is ultimately the result of a pattern matching processes in the brain. This is a surprisingly effective and discriminating process but is very difficult to make quantitative. Only the feature counting method allows for some quantification but it too is applied after the pattern recognition process has been accomplished and is still subject to vagaries of feature identification. Categories may overlap or result from differences in crystallographic orientation or of applied stress field. Only one published experiment (Culver *et al.*, 1983), using the feature counting method, shows a measure of the operator variance of the surface texture method. It may be significant however that (from unpublished data) five experienced users of SEM texture analysis, when given a set of micrographs "blind", correctly and uniformly identified the provenance/transport history of six samples in less than five minutes. The conclusion suggests that using the brain as a classifier may work at least as well as lengthy and tedious feature counting methods.

Particle form: surface texture, shape, angularity, and roundness

Traditionally, "size" is measured and interpreted as a bulk attribute whereas components of a particle's "form" are analyzed individually and then aggregated. The components of form F (effectively in 2 dimensions, the observation plane, if the sphericity is ignored) can, following Whalley (1972), be related by:

$$F = f(sh, sp, a, r, t)$$

Where sh = shape, sp = sphericity, a = angularity, r = roundness, and t = surface texture. These inter-related concepts are associated with the attrition of grains undergoing transport processes. Shape/sphericity, angularity and roundness are numerically ascribed but the chart comparisons for angularity-roundness are of only ordinal measure. True quantitative measures, such as Fourier and fractal methods, have been employed by perimeter quantification (see Orford and Whalley, 1991, for review and papers in Marshall, 1987). However, there are still very few published studies which incorporate quantitative assessment of form components together with surface texture analysis. Automatic image analysis methods exist for some simple shapes but these have not, to date, been used with the rather complex surface textures found on the three dimensions of quartz grains. However, it is possible that in the future neural networks and high powered computing may produce some degree of feature extraction which can aid quantification and improve objectivity as well as speed analysis.

Uses and limitations of the technique

After the first flush of papers using SEM to examine surface textures there were some sceptics about the viability of the method. This was less a failure of the technique but rather that it was sometimes used to "over-interpret" samples. Subsequently, despite recognition of the technique's utility, it has not developed into a *necessary* form of sediment analysis. However, it is a useful adjunct, in association with techniques such as size analysis, to assist identification of sedimentary environments. It can be especially useful where there are only a few grains available such as from an ice core or wind-blown collector or in forensic sedimentology.

The following sections summarize the main "texture-environments" which have been used with SEM surface texture analysis and add to comments made above. Photomicrographs are provided to show some of the ideal textures. However, published books and papers must be used to show the range of textures for any analyzes. The main works which show good examples of textures and the use of the technique include: Krinsley and Doornkamp (1973), Marshall (1987), Whalley (1978, 1996), a bibliography of older references is given in Bull *et al.* (1986).

Weathered and glacial textures

As noted above, "weathered" (i.e., untransported) grains have a similar, if not identical, surface texture to certain grains of glacial provenance. In particular, grains transported on a glacier surface as well as perhaps some englacially-transported grains have not been subjected to any form of modification. Thus, it is not possible to use surface textures to differentiate these environments without considerable care. It is not possible to say that a texture is definitively glacial without other, supporting, evidence. However, it has been possible to show that certain modifications to the angular characteristics can be given by subglacial grinding or fluvio-glacial action. Mahaney (1995) has suggested that glacial grains can be used as indicators of age and ice sheet characteristics. The transport mode by glaciers must be clearly differentiated: supraglacial transport is unlikely to produce any textures which are different from weathered grains or grus, subglacial action may produce distinctive abrasion of some edges and faces.

Glacio-marine deposits

It has been suggested that glacio-marine environments can be identified by SEM texture evaluation. However, this has usually necessitated the identification of "glacial" textures.

Examination of micrographs in these papers does not equivocally support a glacial transportive environment.

Periglacial environments and possible relative dating of deposits

SEM surface textures have been used to help unravel the complexities of periglacial environments but the difficulty of distinguishing between "weathered" and "glacial" environments applies here too. If left long enough, usually in an alkaline environment, even quartz grains weather. This can be related to the formation of diagenetic features on grain surfaces. It is possible that grains in a near-surface environment (e.g., moraine ridges) can produce differences of weathering textures which allow relative ages to be ascribed to the sampled features.

Eolian processes

It is generally believed that eolian grains are distinctive and have characteristic surface textures (Figure S61). Various stages in the rounding process have been identified and it is probable that silt-sized particles can be produced as rounding took place. This latter characteristic has also been used to assist in the interpretation and characterization of loess deposits. Silt, especially as loess (Figure S62), could be useful in the identification of terrigenous sediments in, for example, marine deposits. Early investigators suggested that features known as "upturned plates" were not only characteristic but indicative of the velocity of the transporting wind. This has not however, been followed up as the rotational velocity of spinning grains appears to be the significant factor involved (Whalley et al., 1982).

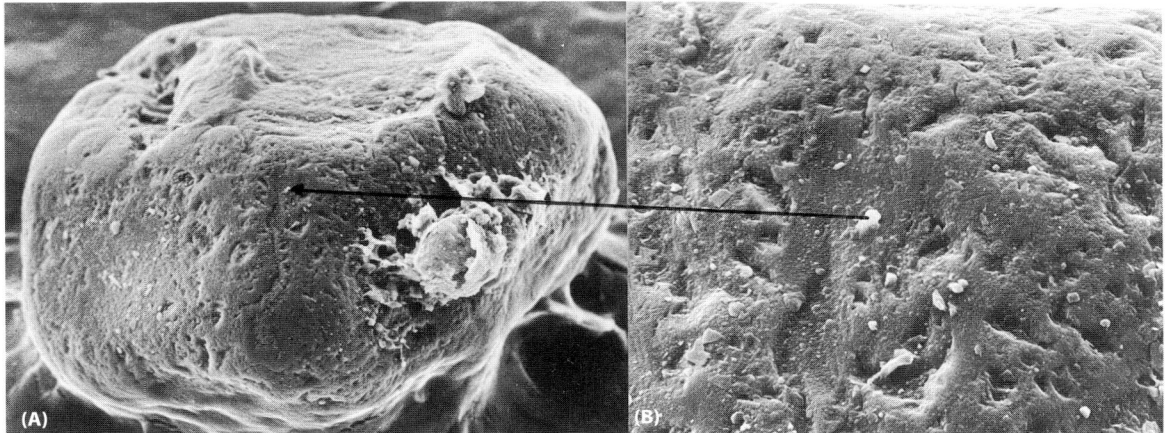

Figure S61 (A) Present day eolian grain from Bahrain dune showing the distinctive rounded outline and smoothed edges and corners while (cf. Figure S60) the faces also show impact features. Picture width = 250 μm. (B) Detail showing a typical eolian texture with micro-scale impact features. Picture width = 45 μm.

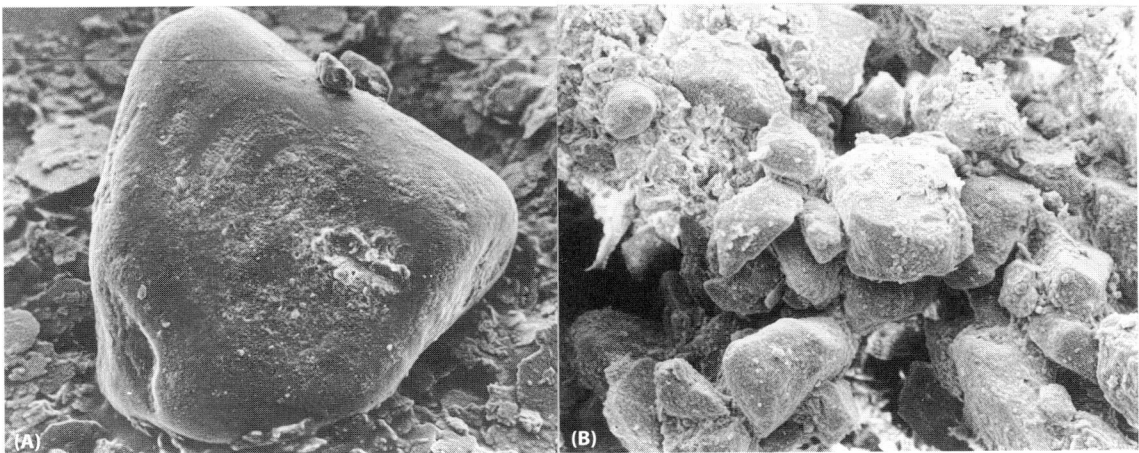

Figure S62 (A) A sand-sized grain from a loessic deposit (East Kent, England) with rounded outline but some surface precipitation which smoothes out some of the surface detail. Picture width = 320 μm. (B) A more general view of some of the same loess deposit showing a variety of angular grain outlines in the fine sand fraction as well as silt-sized particles. This is a fracture surface of a small piece of the loess. Picture width = 200 μm.

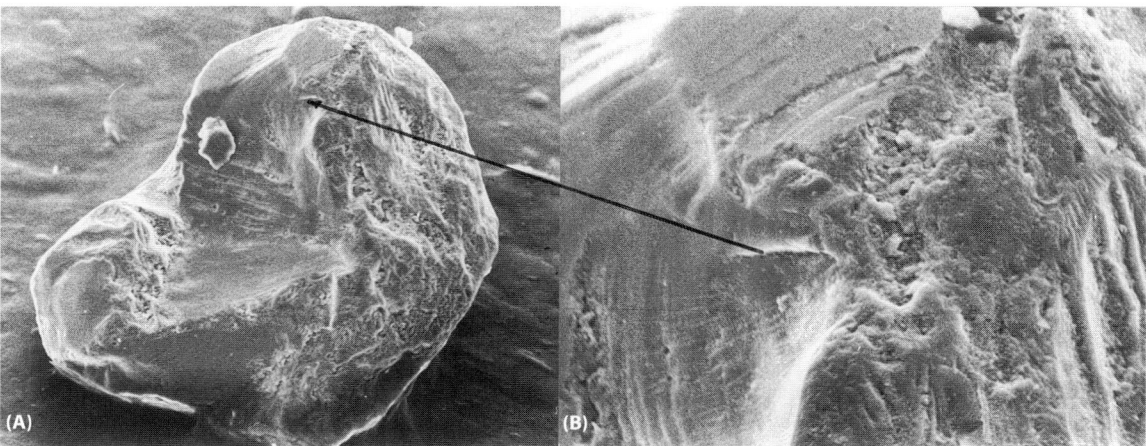

Figure S63 (A) Present day grus material (Ben Ledi, Scotland, derived from gneiss) showing both fracture characteristics as well as some modification of the edges and corners which might be due to eolian activity. This is a good illustration of how difficult it can be to discern the mode of origin of a grain/deposit unless something is known about the environmental history before interpretation. Picture width = 220 μm. (B) A detail of Figure S63(A) showing the conchoidal fracture and parallel arcuate lines which are typical of brittle fracture. The edges and corner however do show that there has been some modification of the textures and that they are akin to eolian textures (Figure S60). Picture width = 50 μm.

Beach processes

There are distinctive surface textures found on beach sediments with small impact versus resulting from the highly turbulent and energetic processes of breaking waves on beaches. Fractured grains, with the fractures cutting across the rounded grain profile and texture, are common and distinctive.

Volcano-clastic material

Tephra particles identified by SEM are distinctive (Marshall, 1987) and have been found frequently in glacial deposits and peats and have been used, together with chemical composition, to mark horizons.

Summary

With relatively inexpensive SEMs and simplicity of use, taking micrographs of quartz (and heavy minerals, etc.) for surface texture analysis is a useful adjunct to conventional sedimentological techniques. How and how intensively it is used, depends upon the ultimate objective. Only rarely will the technique provide unequivocal evidence of a transport environment but it is certainly useful in identifying problem cases and assisting with interpretations. It seems clear that there is still room for experimentation to define mechanisms more closely and perhaps to elucidate the energetics of sedimentological transport processes and temporal factors affecting diagenesis.

W. Brian Whalley

Bibliography

Black, J.M.W., and Dudas, M.J., 1987. The scanning electron microscopic morphology of quartz in selected soils from Alberta. *Canadian Journal of Soil Science*, **67**: 965–971.

Bull, P.A., Whalley, W.B., and Magee, A.W., 1986. An annotated bibliography of environmental reconstruction by SEM. *British Geomorphological Research Group, Technical Bulletin*, 35. Norwich: Geo Books.

Bull, P.A., and Culver, S.J., 1979. An application of scanning electron microscopy to the study of ancient sedimentary rocks from the Saionia Scarp, Sierra Leone. *Palaeogeography, Palaeoclimatology and Palaeoecology*, **26**: 159–172.

Culver, J.J., Bull, P.A., Campbell, S., Shakesby, R.A., and Whalley, W.B., 1983. Environmental discrimination based on quartz grain surface textures: a statistical investigation. *Sedimentology*, **30**: 129–136.

Gallagher, J.J. Jr, 1987. Fractography of sand grains broken by uniaxial compression. In Marshall, J.R. (ed.), *Clastic Particles*. Van Nostrand Reinhold, pp. 189–228.

Higgs, R., 1979. Quartz-grain surface features of Mesozoic-Cenozoic sands from the Labrador and Western Greenland continental margins. *Journal of Sedimentary Petrology*, **49**: 599–610.

Krinsley, D.H., and Doornkamp, J.C., 1973. *Atlas of Quartz Sand Surface Textures*. Cambridge University Press.

Krinsley, D.H., and Marshall, J.R., 1987. Sand grain textural analysis: an assessment. In Marshall, J.R. (ed.), *Clastic Particles*. Van Nostrand Reinhold, pp. 2–15.

Mahaney, W.C., 1995. Glacial crushing, weathering and diagenetic histories of quartz grains inferred from scanning electron microscopy. In Menzies, J. (ed.), *Modern Glacial Environments*. Butterworth-Heinemann, pp. 487–506.

Marshall, J.R. (ed.), 1987. *Clastic Particles*. Van Nostrand Reinhold.

Orford, J.D., and Whalley, W.B., 1991. Quantitative grain form analysis. In Syvitski, J.P.M. (ed.), *Principles, Methods and Applications of Particle Size Analysis*. Cambridge University Press, pp. 88–108.

Orr, E.D., and Folk, R.L., 1983. New scents on the chattermark trail: weathering enhances obscure microfractures. *Journal of Sedimentary Petrology*, **53**: 121–129.

Smart, P., and Tovey, N.K., 1981. *Electron Microscopy of Soils and Sediments: Examples*. Oxford: Clarendon Press.

Smart, P., and Tovey, N.K., 1982. *Electron Microscopy of Soils and Sediments: Techniques*. Oxford: Clarendon Press.

Stieglitz, R.D., 1969. Surface textures of quartz and heavy mineral grains from fresh-water environments: an application of scanning electron microscopy. *Geological Society of America Bulletin*, **80**: 2091–2094.

Whalley, W.B., 1972. The description of sedimentary particles and the concept of form. *Journal of Sedimentary Petrology*, **42**: 961–965.

Whalley, W.B. (ed.), 1978. *Scanning Electron Microscopy in the Study of Sediments*. Norwich: Geo Abstracts.

Whalley, W.B., 1996. Scanning electron microscopy. In Menzies, J. (ed.), *Past Glacial Environments*. Butterworth-Heinemann, pp. 357–375.

Whalley, W.B., Smith, B.J., and Marshall, J.R., 1982. Origin of desert loess from some experimental observations. *Nature*, **300**: 433–435.

Cross-references

Attrition (Abrasion), Fluvial
Attrition (Abrasion), Marine
Cathodoluminescence
Desert Sedimentary Environments
Diagenesis
Eolian Transport and Deposition
Features Indicating Impact and Shock Metamorphism
Glacial Sediments: Processes, Environments and Facies
Grain Size and Shape
Peat
Provenance
Sediment Transport by Unidirectional Water Flows
Swash and Backwash, Swash Marks
Tills and Tillites
Weathering, Soils, and Paleosols

SWASH AND BACKWASH, SWASH MARKS

The terms swash and backwash collectively refer to the oscillatory motion of the shoreline due to the continuous arrival of waves. They also describe the associated thin lens of water behind the moving shoreline that periodically covers and uncovers the beach face. Some researchers use the term swash to describe the complete cycle of shoreline oscillation (i.e., both landward and seaward motion), whereas others use the term swash to describe the landward motion of the shoreline and use the term backwash to describe its seaward motion. Other terms synonymous with swash and backwash are wave-runup and -rundown, respectively.

Most of the early work relating to swash was carried out by applied mathematicians with an interest in nearshore oceanography. Benchmark papers from this discipline are those by Carrier and Greenspan (1958) for the case of non-breaking waves, and Shen and Meyer (1963) for the case of breaking waves. Coastal engineers principally interested in wave overtopping of structures have also contributed extensively to our knowledge, with the paper by Battjes (1971) providing the foundation for much of the engineering literature. Prior to the mid 1970s comprehensive field measurements of swash and backwash are absent from the literature. Since that time the number of publications has increased rapidly. Currently swash-backwash hydrodynamics and swash-zone morphodynamics are major research topics in the fields of coastal-oceanography and coastal-geomorphology/sedimentology (see Komar, 1998 and Hughes and Turner, 1999 for recent reviews).

Process description

The nature of swash and backwash varies between two end-member beach types: steeply-sloped *reflective beaches* and gently-sloped *dissipative beaches*. The nomenclature relates to the predominance of reflection or dissipation of energy in the wind- and swell-wave bands of the nearshore wave spectrum.

On steeply-sloped beaches the swash-energy spectrum (calculated from measured time series of shoreline position or elevation) typically displays a narrow peak somewhere between 0.05 Hz and 0.2 Hz. This frequency of shoreline motion is related to the continuous arrival of wind- and swell-waves at the beach face, either as surging or plunging breakers. There is often an additional peak in the swash spectrum at a frequency that is half that of the incident wind- or swell-wave frequency, and arises due to the reflective nature of steeply-sloped beaches. The interference between incident and reflected waves produces a standing wave field in the nearshore zone, which can influence the swash-backwash process. Possible standing wave types are leaky modes whose crests are parallel to the shoreline and edge wave modes whose crests are normal to the shoreline. The latter results in a rhythmic variation of the maximum swash limit along the beach.

On gently-sloped beaches a significant amount of the incident wave energy is dissipated via bed friction and turbulent wave-breaking across a broad surf zone. Consequently the shoreline motion associated with the arrival of each wind- or swell-wave is often negligible, and the swash is said to be saturated in this frequency band. In contrast with steep beaches, the swash-energy spectrum typically displays a broad-banded peak somewhere between 0.05 Hz and 0.005 Hz (termed the infragravity frequency band) and a logarithmic decrease in energy toward the wind- and swell-wave frequency band. The infragravity shoreline motion dominant on dissipative beaches is associated with long waves that are initially bound to incident wave groups, but subsequently released at the seaward margin of the surf zone. These long waves have such a low steepness that they are reflected by even the gentlest of beach slopes. So despite the strong dissipation of high frequency waves on gently-sloped beaches, standing waves still develop in the nearshore due to the interaction of incident and reflected long waves. Again the standing wave types can be either leaky modes or edge wave modes.

Several aspects of swash-backwash behavior have been quantified using the *Irribaren number*, ξ_o,

$$\xi_o = \frac{\beta}{(H_o/L_o)^{0.5}}$$

where β is the beach slope, H and L are the wave height and length, and the subscript o denotes deepwater measurements. ξ_o is effectively the ratio of the beach steepness to wave steepness. Swash-backwash driven by the arrival of wind- and swell-waves on reflective beaches is associated with values of $\xi_o > 1$, and swash-backwash driven by long waves on dissipative beaches is associated with values of $\xi_o \ll 1$. In general both swash frequency and swash height increase with increasing values of ξ_o. The statistical distribution of swash heights on both reflective and dissipative beaches typically follows a Rayleigh distribution.

The dynamic geometry and internal flow field of the swash-backwash lens is reasonably well established for wind- and swell-waves, but less well established for long waves. When a wind- or swell-wave arrives at the beach face it causes a nearly instantaneous acceleration of the shoreline to its maximum velocity, termed the initial swash velocity. As a result of this impulse the shoreline climbs the beach with a velocity that decreases toward zero at a roughly constant rate. This deceleration is due largely to gravity opposing the motion. Throughout the swash the lens of water travelling behind the shoreline progressively shallows with both time and distance up the beach. When the shoreline velocity reaches zero the shoreline is at its maximum landward displacement, termed the swash limit, and the swash part of the cycle is complete. The shoreline now recedes back down the beach with a velocity that increases at a roughly constant rate due to gravity assisting the motion. The maximum shoreline velocity during the backwash occurs near the completion of the cycle. Both shoreline and internal flow velocities are therefore largest on the lower beach face. Toward the end of the backwash a hydraulic jump may occur on the lower beach face.

The internal flow field within the swash–backwash lens differs from the shoreline motion. At any position on the beach the landward-directed flow during the swash decreases to zero and then reverses direction before the swash limit is reached. As a consequence the duration of landward-directed flow is shorter than seaward-directed flow. This asymmetry in flow duration is most pronounced on the lower beach face and decreases toward the swash limit. In order to sustain a seaward-sloping beach face with this type of flow asymmetry, the relationship between the sediment transport flux and the flow field cannot be straightforward.

Swash marks and swash-related internal structures

The most widely recognized surface form in the swash zone is the upper-stage plane bed, which is consistent with the near-critical to super-critical flow regime that is characteristic of both swash and backwash. Other bed features observed include parting lineations, rolling-grain ripples, rhomboid ripples, antidunes, and swash marks.

Swash marks are typically low (<5 mm high), narrow (<20 mm wide) ridges of sand that mark the landward swash limit. In plan-form they are convex toward the landward direction, widely spaced on gentle slopes and more closely spaced on steep slopes. Two explanations have been proposed. The first is favored on fine-sand beaches and involves sediment flotation. When the bed sediment is not completely wetted (e.g., landward of the groundwater effluent zone) low density or platy grains (e.g., mica flakes and shell detritus) can be supported by surface tension on the narrow wedge of water at the leading edge of the swash lens. At the swash limit, this thin wedge of water seeps into the beach and deposits the floating grains in the form of a swash mark (Emery and Gale, 1951). The second explanation is favored for swash marks on steep, medium-coarse-sand beaches. The reverse grading observed within these swash marks (sediment fines with depth and also in the seaward direction) suggests that they are the result of a narrow, highly concentrated grain flow driven by the leading edge of the swash (Sallenger, 1981).

Because of the subtle relief of bed features in the swash zone the preservation potential of related internal stratification is low. The most likely internal structures to be preserved are those relating to beach-slope changes resulting from: (1) storm erosion and post-storm recovery; (2) profile cut-and-fill due to tidal translation of the swash zone; or (3) migration of beach cusps. Beach deposits are characterized by seaward dipping, continuous planar to curved parallel laminae, often with alternating fine and coarse grained laminae. Coarsening upward sequences of beach laminae occur in relation to beach cusps. Reverse grading within individual beach laminae is ubiquitous, and results from dispersive pressure within the highly-concentrated bed-load transport layer that is characteristic of swash and backwash flows (see *Grain flow; Physics of Sediment Transport: the contributions of R.A. Bagnold*, for dispersive pressure). There is currently no proposed methodology for quantifying paleohydraulic conditions from ancient swash-backwash deposits.

Michael G. Hughes

Bibliography

Battjes, J.A., 1971. Runup distributions of waves breaking on slopes. *Journal of Waterways, Harbours, and Coastal Engineering Division, American Society of Civil Engineers*, **97**: 91–114.
Carrier, G.F., and Greenspan, H.P., 1958. Water waves of finite amplitude on a sloping beach. *Journal of Fluid Mechanics*, **4**: 97–109.
Emery, K.O., and Gale, J.F., 1951. Swash and swash mark. *Transactions, American Geophysical Union*, **32**: 31–36.
Hughes, M.G., and Turner, I.L., 1999. The beach face. In Short, A.D. (ed.), *Handbook of Beach and Shoreface Morphodynamics*. John Wiley and Sons, **pp. 119–144.**
Komar, P.D., 1998. *Beach Processes and Sedimentation*. Prentice-Hall.
Sallenger, A.H.Jr, 1981. Swash mark and grain flow. *Journal of Sedimentary Petrology*, **51**: 261–264.
Shen, M.C., and Meyer, R.E., 1963. Climb of a bore on a beach—Part 3. Run-Up. *Journal of Fluid Mechanics*, **16**: 113–125.

Cross-references

Coastal Sedimentary Facies
Grading, Graded Bedding
Grain Flow
Planar and Parallel Lamination
Physics of Sediment Transport: The Contributions of R.A. Bagnold
Surface Forms

SYNERESIS

Syneresis [*aka* synaeresis] is the process first identified in colloidal solutions whereby spontaneous contraction of a gel results in the expulsion of liquid. It is an intensively researched and industrially important transformation that takes place during the maturation of substances as diverse as polymers, cement paste and foams. An everyday example is provided by the gel formed by the acidification of milk that, during subsequent syneresis, expels fluid (whey) as the curd contracts to <30 percent of its original volume. In earth science it has been proposed that an analogous process takes place when recently-deposited clay-rich sediment comes into contact with a saline solution. As a result of syneresis, water is expelled from the sediment, and this causes the formation of shrinkage cracks on the sediment surface. In some instances these cracks may later be filled with sand or silt, and preserved on the base of the overlying bed, as in the case of desiccation cracks.

There are many published descriptions from the geological record of sand-filled cracks which have been interpreted as syneresis structures. These structures, which are also referred to as "subaqueous shrinkage cracks", are inferred to have formed at, or near, the sediment–water interface. However, in all of these cases there is a lack of unambiguous evidence to confirm their mode of origin, and in recent years a number of workers have questioned the applicability of the syneresis mechanism in such situations. Likewise, no examples have been reported of syneresis cracks forming at the present day on the sediment–water interface in marine or brackish water environments, although these structures have been reported from sediments in more saline situations such as salt pans.

In recent years it has been proposed that, in suitable mudrocks having a high content of swelling clays (smectites), syneresis may take place at various depths *below* the sediment–water interface, during burial of these sediments. On a small scale this may result in the formation of interstratal dewatering structures, seen as complexes of sand-filled veins which commonly form polygonal patterns in plan view (see *Fluid Escape Structures*), or on a larger scale they may be seen as km-scale stratabound sets of randomly oriented normal faults.

History

Jüngst (1934) was responsible for introducing the syneresis concept into the realm of geological processes. White (1961) later carried out experimental work on clays and proposed that syneresis cracks may be found in sediments, in particular those containing swelling clays (smectites). Application of the syneresis mechanism to sedimentation gained further credence following the experiments of Burst (1965) on the action of strong electrolytes on clays sedimented in fresh water, and syneresis cracks were reported from many different geological settings around the world during the 1965–1985 period (e.g., Picard, 1966). The problem of correctly identifying syneresis cracks (Plummer and Gostin, 1981), which have proved difficult to distinguish from trace fossils, especially in Neoproterozoic rocks, was exemplified by the dispute over the origin of quartz-rich sand- and silt-filled cracks from Middle Devonian lake sediments of NE Scotland. Originally interpreted as syneresis cracks by Donovan and Foster (1967), they were reinterpreted by Astin and Rogers (1991) as being formed during subaerial desiccation, with many of the cracks having nucleated upon gypsum crystals. This disagreement heralded an increasing uncertainty over the reality of the traditional syneresis model as applied to mudrocks at the sediment surface (Pratt, 1998; Tanner, 1998), with the implication that it should be abandoned. At the same time, the importance of seismic shock as a catalyst for promoting a syneresis-like loss of fluid from mudrocks after they had been buried, to form either interconnected sets of sand-filled, interstratal dewatering structures (Pratt, 1998; Tanner, 1998), or basin-scale sets of normal faults (Cartwright and Dewhurst, 1998; Dewhurst *et al.*, 1999), was recognized.

Mode of formation

Syneresis occurs when a subaqueous layer of sediment rich in flocculated clays, especially smectites (swelling clays with a high water content), is brought into contact with a saline solution and spontaneously contracts in volume due to interparticle attraction. This causes the expulsion of water which had previously occupied the spaces between the randomly organized clay particles, and the resulting isotropic contraction of mud-rich layers in the plane of the bedding at, or close to, the water–sediment interface leads to the formation of shrinkage cracks. These cracks have the potential to be filled with grains of sand and silt, and preserved in the geological record, in much the same way as desiccation or mud cracks (see *Desiccation Structures*).

Following the experiments reported by Burst (1965), the common mode of occurrence of syneresis cracks was thought to be as isolated, lenticular, filled-cracks. The cracks are small in size and restricted to a single mud layer (*cf.* interstratal dewatering cracks). The problem which has inhibited further work on the syneresis mechanism in sediments, is that it has proved impossible to erect criteria to distinguish between fossil sand-filled cracks formed by the passive contraction of a muddy layer due to sudden loss of part of its pore fluid (syneresis), and those resulting from the forced evaporation of the pore fluid (desiccation). The two processes have many features in common: as the tensile stress fields in the muddy layers in each case will be similar, both have the potential to form linked patterns of polygonal cracks. The precise morphologies of the resultant cracks, and of the patterns they make, will be controlled by features such as bed thickness, rheology, and anisotropy and, with one exception, will not be diagnostic of the geological process or setting. The exception is that, as desiccation results in a greater water loss than syneresis, special features such as mud curls may develop and confirm that sub-aerial exposure of the wet mudrock was responsible for their formation. No such positive feature for the recognition of syneresis cracks has been identified, and the first geological examples of such structures, figured by White (1961, his plates 3 and 4), could just as readily be interpreted as desiccation or interstratal shrinkage structures.

The interstratal mode of "syneresis" is thought to be caused by earthquake-induced ground motion which triggers off a combination of syneresis (*s.s.*), liquifaction, and water-escape processes relatively close to the sediment surface (Pratt, 1998). It differs from that thought by Cartwright and Dewhurst (1998) and Dewhurst *et al.* (1999) to be responsible for the development of a km-scale system of polygonal normal faults, in that the latter is inferred to occur by syneresis during compaction, without the mediation of earthquake shaking. The driving force in this case is the three-dimensional contraction of colloidal smectitic gels through interparticle forces in a saline medium, on a major scale.

Summary

It has been proposed that syneresis cracks form in smectite-bearing muds at two different levels in a sedimentary basin: at the sediment-water interface, and in rocks that have already been buried. The processes that occur at each level are also different. That taking place at the sediment surface in a marine environment is true syneresis, due to the spontaneous loss of fluid from a gel in contact with a saline solution; that taking place at depth apparently requires the intervention of seismic activity to act as a triggering mechanism, could also involve compaction-driven expulsion of fluid, and results in sets of interstratal, filled fractures. The nomenclature used to describe the small-scale interstratal cracks formed in this way is a problem. They are not true syneresis cracks and should be referred to by a descriptive, non-genetic, name such as "interstratal dewatering structures". To name km-scale fractures as syneresis structures is also misleading: syneresis may be involved in the production of these structures but it is probably not the only process involved.

"True" syneresis cracks formed at the sediment–water interface, if they exist in sedimentary systems apart from highly saline lakes, are impossible to distinguish from other shrinkage cracks, especially desiccation cracks, and there is a strong case for ceasing to use the term in such situations. As its use to describe the other types of fracture or crack is debatable, the term is best abandoned.

P.W. Geoff Tanner

Bibliography

Astin, T.M., and Rogers, D.A., 1991. "Subaqueous shrinkage cracks" in the Devonian of Scotland reinterpreted. *Journal of Sedimentary Petrology*, **61**: 850–859.

Burst, J.F., 1965. Subaqueously formed shrinkage cracks in clay. *Journal of Sedimentary Petrology*, **35**: 348–353.

Cartwright, J.A., and Dewhurst, D.N., 1998. Layer-bound compaction faults in fine-grained sediments. *Geological Society of America Bulletin*, **110**: 1242–1257.

Dewhurst, D.N., Cartwright, J.A., and Lonergan, L., 1999. The development of polygonal fault systems by syneresis of colloidal sediments. *Marine and Petroleum Geology*, **16**: 793–810.

Donovan, R.N., and Foster, R.J., 1972. Subaqueous shrinkage cracks from the Caithness Flagstone Series (Middle Devonian) of northeast Scotland. *Journal of Sedimentary Petrology*, **42**: 309–317.

Jüngst, H., 1934. Zur geologischen bedeutung der synärese. (Geological significance of syneresis). *Geologische Rundschau*, **25**: 321–325. [in German]

Picard, M.D., 1966. Oriented, linear-shrinkage cracks in Green River Formation (Eocene), Raven Ridge area, Uinta Basin, Utah. *Journal of Sedimentary Petrology*, **36**: 1050–1057.

Plummer, P.S., and Gostin, V.A., 1981. Shrinkage cracks: desiccation or synaeresis? *Journal of Sedimentary Petrology*, **51**: 1147–1156.

Pratt, R.B., 1998. Syneresis cracks: subaqueous shrinkage in argillaceous sediments caused by earthquake-induced dewatering. *Sedimentary Geology*, **117**: 1–10.

Tanner, P.W.G., 1998. Interstratal dewatering origin for polygonal patterns of sand-filled cracks: a case study from Late Proterozoic metasediments of Islay, Scotland. *Sedimentology*, **45**: 71–89.

White, W.A., 1961. Colloid phenomena in the sedimentation of argillaceous rocks. *Journal of Sedimentary Petrology*, **31**: 560–570.

Cross-references

Colloidal Properties of Sediments
Desiccation Structures
Flocculation
Fluid Escape Structures

T

TALC

Talc is a 2:1 phyllosilicate with an ideal composition of $Mg_3Si_4O_{10}(OH)_2$. The 2:1 layer structure comprises two tetrahedral sheets of silica tetrahedra bound to one octahedral sheet of edge-linked octahedra. The octahedral sheet in talc is magnesium-rich, trioctahedral, and referred to as brucite-like. The 2:1 talc layer is electrostatically neutral and is bound to other layers by van der Waals forces. A small amount of ionic attraction between the layers may be present and result, in part, from cation substitutions (Giese, 1975). Although most talcs have compositions close to the ideal, substantial amounts of Fe may replace Mg in octahedral sites, minor amounts of Al may replace both Si and Mg in tetrahedral and octahedral sites, respectively, and small amounts of Ni, Mn, Cr, and Ti may substitute into the structure. Traces of Ca, Na, and K may occupy the interlayer site to satisfy any layer charge imbalance. The dioctahedral analog of talc is pyrophyllite, an aluminum-rich, dioctahedral 2:1 phyllosilicate.

Talc is triclinic (1-layer), although many earlier references report that it takes a 2-layer monoclinic form (Bailey, 1980). The mineral is green to white to silvery white, has a specific gravity of 2.58–2.83, and a hardness of 1, making it the softest mineral on the Moh's scale. It has a greasy feel, a pearly luster and is sectile. It often occurs in massive or foliated aggregates, but rare crystals are tabular and show perfect cleavage {001}. Massive talc is referred to as soapstone or steatite. The mineral forms in high-magnesian rocks, such as ultramafics, basic igneous rocks and siliceous dolomites that have been subjected to low grade metamorphism, contact metamorphism, hydrothermal alteration, or weathering (Evans and Guggenheim, 1988). Talc is associated with serpentinization and the transformation of magnesian-rich silicates such as olivine, enstatite, tremolite, and chlorite. Talc is used in the production of ceramics, paint, rubber insecticides, cosmetics, paper, and plastics. Soapstone blocks are used for ornamental materials and carvings.

Talc is not a major component of sediments and sedimentary rocks, but its occurrence is more common and widespread than previously thought. In modern ocean hydrothermal environments talc forms as a direct precipitate when hot solutions emanating from fissures and vents mix with cold, Mg-rich bottom waters. The main hydrothermal vent field in the Tjornes Fracture Zone (depth of 400 m) north of Iceland consists of about 20 large mounds, chimneys, and spires of anhydrite and talc (Hannington et al., 2001). Sediment coring through talc- and anhydrite-rich sedimentary layers that occur up to 7 m below the mounds revealed pore fluids with temperatures as high as 250°C. Talc also forms as an alteration product of basalts through which these hot fluids flow. Shau and Peacor (1992) studied trioctahedral minerals in DSDP drillhole 504B and found a sequence of hydrothermal alteration products in the basalt as follows: saponite (to a depth of 624 m); mixed-layer corrensite-chlorite + chlorite + corrensite (from 624 m to 965 m); chlorite + talc + mixed-layer talc–saponite (from 725 m to 1076 m). Mineral occurrences were not continuous as they were related to rock permeability and temperature variations. Seafloor hydrothermal activity has also resulted in the formation of talc from the alteration of epiclastic sediments and pumaceous tuffs in the Okinawa Trough (Marumo and Hattori, 1999) and from transformation of detrital clays in the Dead Sea Rift (Sandler et al., 2001). In both these examples mineral assemblages changed with depth because mixing proportions of hydrothermal fluids and seawater resulted in different reaction temperatures. Talc has been observed in sediments and in metalliferous deposits accumulating in the Red Sea, on the Mid-Atlantic Ridge, in the Gulf of California and along the East Pacific Rise. Temperatures of the hot, brine solutions that give rise to the formation of talc on the seafloor are estimated to vary between 200°C and 400°C. Minerals associated with talc formation in these hydrothermal environments include sulfides of copper, iron, and zinc, magnesium-rich clays (e.g., smectite, chlorite, corrensite), anhydrite, sepiolite, and zeolites.

In some older sedimentary rocks talc has been reported to occur as a replacement product. Nesbitt and Prochasaka (1998) suggested that dolomite, magnesite, and talc mineralization in the Middle Cambrian carbonate rocks of the southern Canadian Rocky Mountains occurred as warm,

Mg-enriched brines (originating from seawater) circulated through the rock units in the Late Devonian or Early Mississippian. Noack et al. (1989) reported that oolitic talc in the Upper Proterozoic rocks of the Congo formed from the transformation of original stevensite (or sepiolite) during diagenesis. Contact metamorphism resulted in the reaction of an assemblage of saponite + quartz + dolomite to produce grain-coating flakes of talc at temperatures estimated to be between 130 and 180 °C in the arkosic rocks of the Triassic Sherwood Sandstone Group in Northern Ireland (McKinley et al., 2001). Talc along with chlorite formed in and near fractures in Cambrian dolostones in the vicinity of Winterboro, Alabama from hydrothermal solutions and metasomatic processes (Blount and Vassiliou, 1980).

Talc forms authigenically in evaporitic sedimentary environments and during burial diagenesis in evaporitic sediments when fluids and pore waters become highly alkaline (pH ≥ 9) and concentrated with magnesium. Delicate rosettes and cornflake aggregates of talc in salt beds of the Paradox Formation, Paradox Basin, Utah, indicate an authigenic origin for the mineral (Weaver, 1989, p.413). Droste (1963) studied phyllosilicates in a variety of North American salt deposits and found talc, corrensite, serpentine, and other clays commonly present. Talc also is present in the Permian Zechstein salt deposits of Germany (Braitsch, 1971) and the Silurian evaporite deposits of New York (Bodine and Standaert, 1977), among others. Scrivenor and Sanderson (1982) reported that isolated flakes of talc in Triassic halite deposits in Somerset, England, likely formed from the reaction of dolomite + silica + water during burial diagenesis. In some evaporite deposits, talc may exist within a mixed-layer clay structure, such as talc–saponite or talc–chlorite. Talc in these sedimentary rocks is often associated with other syngenetic and epigenetic Mg-rich phyllosilicates such as saponite, stevensite, corrensite, trioctahedral chlorites, sepiolite, and serpentine, to name but a few.

Talc is found in soils derived from the weathering of ultramafic rocks and rocks containing Mg-rich minerals such as enstatite and hornblende (Nahon and Colin, 1982; Proust, 1982; Zelazny and White, 1989). The mineral may form both by transformation of preexisting silicates and by neoformation involving congruent dissolution of the parent mineral and subsequent precipitation of talc from solution (Noack et al., 1986). Talc occasionally is present in soils as a minor inherited component.

Finally, talc is present in oceanic surface waters, but it is rarely found as a significant component of modern-day marine sediments (Hathaway, 1979). The mineral has been identified in the suspended load of the Atlantic Ocean, the Caribbean and Mediterranean Seas, and in the Amazon River (Gibbs, 1967; Jacobs and Ewing, 1969; Poppe et al., 1983). It has also been found in small quantities in Upper Pliocene-Holocene marine sediments from the Alboran Basin (Skillbeck and Tribble, 1999), and in beach sands along the coast of the Gulf of Mexico and Florida (Griffin, 1963). Because talc is widely used commercially, especially as a carrier of pesticides in crop dusting, it is suspected that some talc in the suspended load of surface waters derives from anthropogenic dust and industrial effluents. A second important source is likely long-range transport by atmospheric winds and dust storms.

Richard H. April

Bibliography

Bailey, S.W., 1980. Structures of layer silicates. In Brindley, G.W., and Brown, G. (eds.), *Crystal Structures of Clay Minerals and Their X-ray Identification*. Mineralogical Society Monograph 5, pp. 1–123.

Blount, A.M., and Vassiliou, A.H., 1980. The mineralogy and origin of the talc deposits near Winterboro, Alabama. *Economic Geology*, **75**: 107–116.

Bodine, M.W., and Standaert, R.R., 1977. Chlorite and illite compositions from Upper Silurian rock salts, Retsof, New York. *Clays and Clay Minerals*, **25**: 57–71.

Braitsch, O., 1971. *Salt Deposits, Their Origin and Composition*. Springer-Verlag.

Droste, J.B., 1963. Clay mineral composition of evaporite sequences. *Northern Ohio Geological Society Monograph*, **1**: 47–54.

Evans, B.W., and Guggenheim, S., 1988. Talc, pyrophyllite, and related minerals. In Bailey, S.W. (ed.), *Hydrous Phyllosilicates (exclusive of mica)*. Reviews in Mineralogy 19. Mineralogical Society of America, pp. 225–294.

Gibbs, R.J., 1967. The geochemistry of the Amazon River system; part I, the factors that control the salinity and the composition and concentration of the suspended solids. *Geological Society of America Bulletin*, **78**: 1203–1232.

Giese, R.F., 1975. Interlayer bonding in talc and pyrophyllite. *Clays and Clay Minerals*, **23**: 165–166.

Griffin, G.M., 1963. Occurrence of talc in clay fractions from beach sands of the Gulf of Mexico. *Journal of Sedimentary Petrology*, **33**: 231–233.

Hannington, M., Scholten, J., Botz, R., Garbe-Schonberg, D., Jonasson, I.R., Roest, W., Herzig, P., and Stoffers, P., 2001. First observations of high-temperature submarine hydrothermal vents and massive anhydrite deposits off the north coast of Iceland. *Marine Geology*, **177**: 199–220.

Hathaway, J.C., 1979. Clay minerals. In Burns, R.G. (ed.), *Marine Minerals*. Reviews in Mineralogy, 6. Mineralogical Society of America, pp. 123–150.

Jacobs, M.B., and Ewing, M., 1969. Mineral source and transport in waters of the Gulf of Mexico and Caribbean Sea. *Science*, **163**: 805–809.

Marumo, K., and Hattori, K.H., 1999. Seafloor hydrothermal clay alteration at Jade in the back-arc Okinawa Trough: mineralogy, geochemistry, and isotope characteristics. *Geochimica et Cosmochimica Acta*, **63**: 2785–2804.

McKinley, J.M., Worden, R.H., and Ruffell, A.H., 2001. Contact diagenesis: the effect of an intrusion on reservoir quality in the Triassic Sherwood Sandstone Group, Northern Ireland. *Journal of Sedimentary Research*, **A71**: 484–495.

Nahon, D.B., and Colin, F., 1982. Chemical weathering of orthopyroxenes under lateritic conditions. *American Journal of Science*, **282**: 1232–1243.

Nesbitt, B.E., and Prochaska, W., 1998. Solute chemistry of inclusion fluids from sparry dolomites and magnesites in Middle Cambrian carbonate rocks of the southern Canadian Rocky Mountains. *Canadian Journal of Earth Sciences*, **35**: 546–555.

Noack, Y., Decarreau, A., and Manceau, A., 1986. Spectroscopic and oxygen isotope evidence for low and high temperature origin of talc. *Bulletin de Minéralogie*, **109**: 253–263.

Noack, Y., Decarreau, A., Boudzoumou, F., and Trompette, R., 1989. Low-temperature oolitic talc in Upper Proterozoic rocks, Congo. *Journal of Sedimentary Petrology*, **59**: 717–723.

Poppe, L.J., Hathaway, J.C., and Parmenter, C.M., 1983. Talc in the suspended matter of the northwestern Atlantic. *Clays and Clay Minerals*, **31**: 60–64.

Proust, D., 1982. Supergene alteration of hornblende in an amphibolite from Massif Central, France. In Van Olphen, H., and Veniale, F. (eds.), *Proceedings of the International Clay Conference*. 1981. Elsevier, pp. 367–364.

Sandler, A., Nathan, Y., Eshet, Y., and Raab, M., 2001. Diagenesis of trioctahedral clays in a Miocene to Pleistocene sedimentary-magmatic sequence in the Dead Sea Rift, Israel. *Clay Minerals*, **36**: 29–47.

Scrivenor, R.C., and Sanderson, R.W., 1982. Talc and aragonite from the Triassic halite deposits of the Burton Row borehole, Brent

Knoll, Somerset. *Report, UK Institute of Geological Sciences*, **82** (1): 58–60.
Shau, Y-H., and Peacor, D.R., 1992. Phyllosilicates in hydrothermally altered basalts from DSDP Hole 504B, Leg 83; a TEM and AEM study. *Contributions to Mineralogy and Petrology*, **112**: 119–133.
Skillbeck, C.G., and Tribble, J.S., 1999. Description, classification, and the origin of Upper Pliocene-Holocene marine sediments in the Alboran Basin. *Proceedings of the Ocean Drilling Program: Scientific Results*, **161**: 83–97.
Weaver, C.E., 1989. *Clays, Muds, and Shales*. Developments in sedimentology 44. Elsevier.
Zelazny, L.W., and White, G.N., 1989. The pyrophyllite-talc group. In Dixon, J.B., and Weed, S.B. (eds.), *Minerals in Soil Environments*, 2nd edn, Madison: soil science society of America, pp. 527–550.

Cross-references

Anhydrite and Gypsum
Authigenesis
Berthierine
Chlorite in Sediments
Desert Sedimentary Environments
Diagenesis
Dolomites and Dolomitization
Evaporites
Mixed-Layer Clays
Mudrocks
Porewaters in Sediments
Smectite Group
Weathering, Soils, and Paleosols

TAPHONOMY: SEDIMENTOLOGICAL IMPLICATIONS OF FOSSIL PRESERVATION

Taphonomy is the study of processes that influence preservation of potential fossils; for general reviews, see Schäfer (1972), Müller (1979), Seilacher *et al.* (1985), Allison and Briggs (1991), Donovan (1991), Kidwell and Behrensmeyer (1993), Martin (1999). This field encompasses two major aspects: *biostratinomy*, the study of processes affecting organism remains or their traces, prior to their final burial, and *fossil diagenesis*, investigation of phenomena affecting potential fossils after burial (Figures T1–T3). Recently, taphonomy has developed, both as a means of assessing bias in the fossil record, and more positively, as a critical tool for paleoenvironmental analysis. Taphonomic analyzes compliment sedimentological study and yield critical information on the physico-chemical parameters of ancient environments. Uniformitarianistic assumptions are applicable in taphonomy because the physical and chemical properties of organism skeletons have probably been relatively constant throughout geologic time, despite the nonuniformity imparted by evolution.

Biostratinomy: fossils as sedimentary particles

Disarticulation and rates of burial

Certain fossils, including bivalved shells and, especially, multi-element skeletons, are sensitive indicators of rapid, episodic sediment accumulation. The rate at which organism skeletons disintegrate after death is a function of their delicacy, environmental energy, temperature, oxygen levels, and residence time on the seafloor.

Articulated, closed bivalve shells indicate burial that was rapid enough to prevent the shells from gaping open at the hinge (Schäfer, 1972). Conversely, abundant "butterflied" (splayed) bivalve shells indicate moderate rates of sedimentation. Hinge ligaments may remain intact for periods of up to months, but will ultimately decay and allow the valves to become disassociated. Brachiopod shells tend to remain closed after death. Closed shells filled with calcite spar, pyrite, or other minerals, as opposed to sediment, indicate pulses of burial that covered the shells while interiors were still occupied by soft parts, preventing infiltration of sediment. Tissues subsequently decayed, leaving a void space within the shells.

Ratios of articulated to disarticulated shells of bivalves or brachiopods may be an important indicator of the extent of time-averaging in a deposit (Kidwell and Bosence, 1991; Brett and Baird, 1993). Deposits containing very high proportions of articulated specimens, will be minimally time-averaged.

Organisms with multielement skeletons, composed of multiple articulated ossicles, e.g., trilobites, crabs, echinoderms, and vertebrates, are only rarely preserved intact (Figure T1). Field and experimental studies indicate that degradation of muscle tissues occurs within hours to days after death, while destruction of collagenous ligaments ensues very rapidly, such that most skeletons are completely disarticulated within days to months (Schäfer, 1972; Donovan, 1991; Plotnick, 1986; Brett *et al.*, 1997). Not all multielement skeletons are equally subject to disarticulation. Starfish and crinoid crown skeletons disintegrate within days of death (Figure T1). Likewise, most arthropods disintegrate very rapidly following death (Schäfer, 1972; Plotnick, 1986), but some barnacles, remain intact for much longer periods. Echinoid tests have interlocking sutures that may remain intact for periods of several years (Figure T1), depending on energy levels, oxygenation, and water temperature (Allison, 1990; Kidwell and Baumiller, 1990). Crinoid columns may disarticulate into more durable increments (pluricolumnals) of approximately equal length because of slight differences in the ligamental articulation at certain joints within the column (Baumiller and Ausich, 1992).

Some beds of intact pelagic fossils, such as fish, squids, or marine reptiles may represent carcasses that settled onto anoxic seafloors (Martill, 1991). Furthermore, Allison's (1988) calculations demonstrate that larger organism bodies become anoxic microenvironments internally during early phases of decay. Inhibition of scavenging in these environments may prolong the association of skeletal elements and even preserve traces of soft parts (Seilacher *et al.*, 1985). However, even under conditions of anaerobiosis, bacterial decay of ligaments is rapid and the slightest currents will serve to disarray pieces (Allison, 1988), implying that burial is a critical prerequisite to articulated preservation.

Groupings of well-preserved, articulated, multi-element skeletons occur on certain planes that would not otherwise be recognizable as event-beds. Because such skeletons can not be reworked without dissociation, even a single intact specimen of a trilobite, crinoid crown or vertebrate skeleton, provides unambiguous evidence that the enclosing sediment accumulated rapidly and was not subsequently disturbed.

Under extraordinary conditions entire bodies with soft parts may be preserved (see Seilacher *et al.*, 1985). Such conservation Lagerstätten deposits reflect combinations of rapid

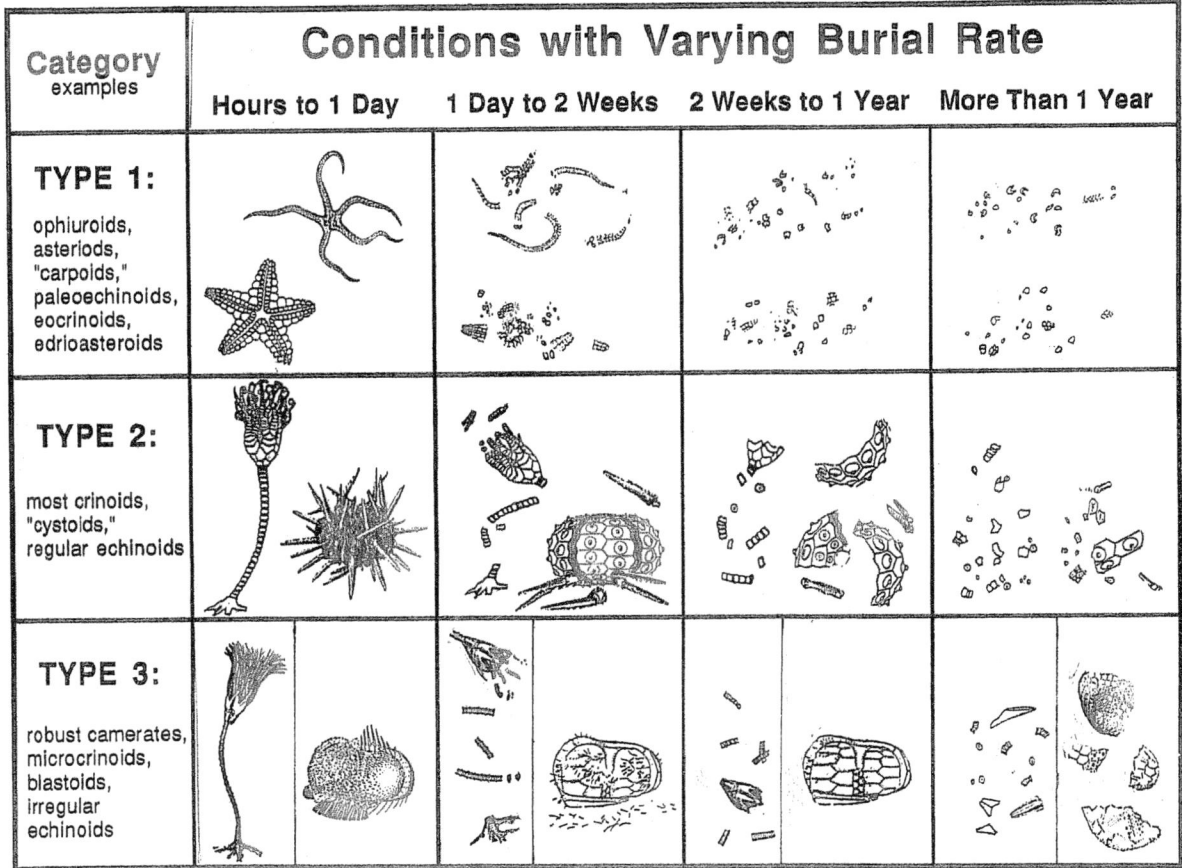

Figure T1 Preservational condition (grades) of three taphonomic categories of echinoderm fossils for varying times of post-mortem exposure. From Brett et al. (1997).

burial, anoxic sediments and early, bacterially-mediated mineralization.

Life orientations and *in situ* burial

Examples of *in situ* preservation, including trace fossils, are common among burrowing and hard substrate encrusting/endolithic organisms. Some assemblages of well preserved fossils, such as cemented oysters, tube worms, bryozoans, and rooted crinoids are demonstrably *in situ* and mark colonized discontinuity surfaces, such as hardgrounds.

Organisms, preserved in unstable orientations, such as ceratoid corals, or pedically-attached brachiopods preserved in upright position are very unlikely to have rolled into this position accidentally. Such life orientations record episodes of sediment input within low-energy settings.

Brachiopods and clams, preserved in life positions, may provide indicators of sedimentation events. Burrowing bivalves may die within their burrows; however, experimental and observational studies have shown that, in most cases, when bivalves are stressed, they rise from their burrows (Schäfer, 1972). Thus, large numbers of articulated infaunal shellfish on bedding planes may be associated with depositional events.

Another sensitive indicator is provided by associated, though disarticulated arthropod (e.g., trilobite) molt parts. It is virtually impossible for different portions of the skeleton to be transported any distance and remain associated. Hence, their occurrence proves *in situ* biological activity in low energy environments.

Transport and reorientation

Fossils also provide evidence for reworking and/or transport of skeletons that would otherwise remain unsuspected (Allen, 1992). Articulated fragile fossils, e.g., crinoids or trilobites, are commonly thought to have been buried *in situ* in low energy environments. However, Allison (1986) demonstrated in tumbling barrel experiments that even soft-bodied organisms may be potentially moved considerable distances without disarticulating, providing that this happens very shortly (hours) after their death. Such may apply to articulated aligned skeletons found at the bases of turbidites or storm beds.

Many skeletal elements, including valves of most brachiopods, ostracodes, and clams, and trilobite cephala and pygidia, have approximately concavo-convex or dish-shapes. Random orientations, including some lateral or edgewise shells, occur primarily in heavily bioturbated sediments or low energy traps.

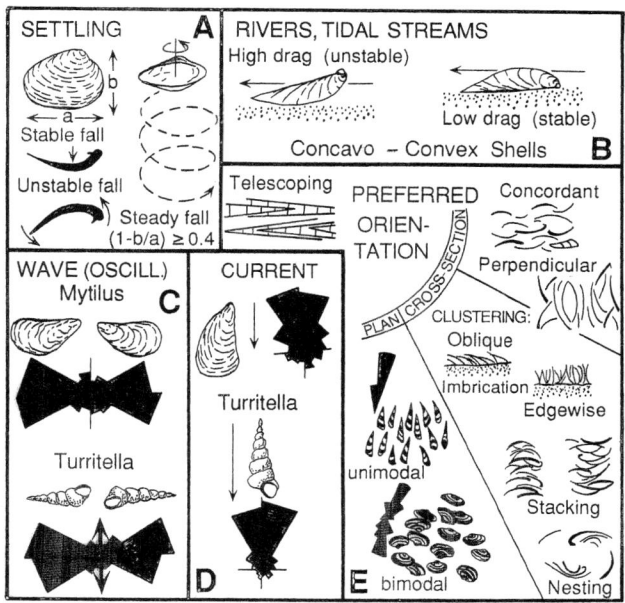

Figure T2 Orientations and fabrics of shells under varying conditions. (A) settling of concavo-convex shells under free fall conditions; (B) re-orientation of concavo-convex shells by currents; (C) bimodal orientations of shells in oscillating currents. (D) alignment of shells in unidirectional currents; rose diagrams show directions of apices of shells; arrows indicate current directions; (E) various types of shell fabrics in plan and cross sectional view. Modified from Allen (1992) and Kidwell and Holland (1991).

Concentrations of shells in localized pockets commonly display convex-downward or lateral-oblique positions along linear features that may be interpreted as subtle scours or gutter casts (Futterer, 1978). Their presence is a significant indicator of storm-generated erosion and redeposition.

Concavo-convex shells most commonly display preferred convex-upward orientations (Figure T2(B)). Even gentle currents will cause shells to flip into a hydrodynamically stable convex upward orientation (Middleton, 1967; Brenchley and Newall, 1970). Thus, the occurrence of abundant shells in convex-upward positions on bedding planes provides evidence for reworking of shells by one or more storm-generated gradient currents.

Beds of preferentially concave-upward shells are not as common. When the shells are allowed to resettle from suspension they will generally settle concave-upward (Figure T2(A)). Such reorientation might occur in areas close to storm wave-base, where gentle storm-generated waves would lift shells off the bottom and allow free-fall back to the substrate. Concave-upward accumulations in rippled, cross bedded sands also may arise from free-fall of shells down the lee sides of migrating bedforms (Brenchley and Newall, 1970).

Preferred orientation of elongate skeletons has been the subject of numerous experimental studies (see Nagle, 1967; Brenchley and Newall, 1970; Reyment, 1980; Kidwell and Bosence, 1991). Elongate particles typically orient themselves selectively in a current. Cylindrical objects may roll perpendicular to the current direction. Elongate objects that do not roll, e.g., rhabdosomes of graptolites, will become oriented parallel to the current. Similarly, conical shells, including many nautiloid cephalopods, tentaculitids, and gastropods, will become aligned parallel to the current with the apex directed up-current (Figure T2(D)). Hence, depending upon the type of fossil, it may be possible to determine not only the line of current action, but also the direction of propagation.

Bimodal orientations are common and may result from more than one cause. In some instances, the bimodality is size dependent; for example, objects, such as fragments of large nautiloids, may approach a cylindrical shape and roll with the current, whereas smaller cone-shaped shells will orient in a preferred up-current direction. Conversely, conical shells of similar size and shape may show nearly perfect 180° bimodality, aligned parallel to subtle wave ripples (Figure T2(C)). Such fossil evidence indicates oscillatory currents or waves.

Stacks of edgewise or shingled shells (Figure T2(E)) apparently occur where densely packed shells were affected by oscillatory, storm-generated waves and indicate deposition within storm wave-base (Seilacher and Meischner, 1965; Grinnell, 1974). A similar mode of preservation in shell accumulations involves cup and saucer stacking in which concavo-convex shells are bundled together in nested groupings. These, too, seem to reflect the effects of settling and concentration of shells during turbulence events and they are commonly associated with minor sediment grading.

Another, perhaps related, phenomenon involves the cone-in-cone interpenetration of nautiloids observed in some densely packed cephalopod limestones (Figure T2(E)). These appear to represent turbulence effects that first align and then telescope cephalopods; perhaps septa had been weakened somewhat by dissolution or corrosional effects.

Upright ammonoids and nautiloids have been observed in a number of sections in the geologic record. Cephalopod phragmocones will initially settle in a vertical attitude during the flooding of the chambers. There is considerable variation based on shell shape and type, but most shells will settle vertically only up to about 10 m (Raup, 1973); beds with abundant upright ammonoids may thus provide a useful depth indicator. However, in orthoconic nautiloids, ballasting of the apical end of the phragmocone by cameral deposits may also cause shells to sink vertically and embed with apex downward following decay and removal of the soft body parts. In either case, preservation of this attitude demonstrates that the sediments were relatively soft to allow shells to penetrate some distance into the seafloor.

Sorting and fragmentation

Skeletal parts may be acted upon by various hydrodynamic processes, which serve to sort and fragment them. Sorting of skeletal elements by size, density, and even shape, is common in environments above wave base. Differential transport of left versus right valves of equivalved pelecypods (e.g., Figure T2(C)) may occur in environments where waves approach a coast obliquely (Müller, 1979). Hence, left-right sorting may be an excellent indicator of wave swash environments.

Biased ratios of skeletal parts may provide an indicator of time-averaging in shell-rich deposits. For example, brachiopod assemblages commonly show several times as many whole pedicle as brachial valves (Holland, 1988; Brett and Bordeaux, 1990). This indicates that the pedicle valves, which possess heavy hinge dentition and a thickened interarea, were better able to withstand destructive processes. Offshore, low energy facies that show strongly biased valve ratios represent time-averaged accumulations. During prolonged intervals of exposure, occasional storm disturbances fragmented shells, preferentially removing the more fragile skeletal parts.

Fragmentation of shells may occur *via* biotic or abiotic means. Much fragmentation is thought to reflect physical impact of shells with one another. Generally, high proportions of fragmented skeletal material indicate repeated processing by high energy events.

Shattering of septa in nautiloid and ammonoid cephalopod shells is common in some offshore facies and has been used as a paleobathymetric indicator (Westermann, 1985). The shattered phragmocones may reflect hydrostatic pressures on an empty body chamber that shattered septa. Given this assumption, and septal strength indices for internally shattered cephalopods, approximate depths of implosion can be inferred. However, more recent work by Hewitt and Westermann, (1996) suggests that some of the shattering may actually take place due to explosion; i.e., septa blown forward owing to back pressures. Punctures in the shells may allow water to penetrate into back chambers of the shell to cause shattering of weaker early septa at much shallower depths than those predicted by the implosion model. If these two aspects of shell internal fragmentation can be sorted out, then shattered phragmocones may indeed provide a useful paleobathymetric indicator.

Corrasion and encrustation

It is often difficult to distinguish between shells that have been physically abraded and those corroded by biogeochemical processes. Brett and Baird (1986) suggested the term corrasion for cases of unspecified corrosion and abrasion. Even such unspecified shell damage may indicate exposure in the depositional environment.

Physically abraded fossils are common in many sandstones and siltstones, but rare in mudrocks, as clay sized particles are not affective abrasive agents. Abraded fossils in shales might indicate a complex burial history in which shells had been, at some point and time, affected by sediments coarser than the matrix in which they reside.

Corrosion and/or bioerosion of shells predominates over physical abrasion in offshore, low-energy environments (Kidwell and Bosence, 1991). Many apparently abraded shells in mudrocks have probably been chemically etched and/or acted upon by microboring organisms.

Microborings of endolithic organisms, identified by SEM, may be useful indicators of exposure and relative water depth. Endolithic algae of various sorts, for example, are confined to various portions of the euphotic to upper dysphotic zone (Vogel *et al.*, 1987). Absence of these sorts of endoliths on shells from sediments, in which independent evidence indicates shallow water deposition, might be used as an indicator of turbidity or elevated sedimentation rates.

Even in life, skeletons may become encrusted with epibiontic organisms, such as bryozoans. After death of their occupants, shells may be encrusted both externally and internally. Internal encrustation provides an excellent indication that shells remained unburied after death. Skeletons that have been corroded, or abraded may be subsequently encrusted and provide evidence for more prolonged exposure. Fossil corals that show one side with the epitheca perfectly intact and the other side corroded and subsequently encrusted (Brett and Baird, 1986) must have lain partially buried in the mud; the upper surface was exposed to destructive and encrusting agents.

Fossil diagenesis: geochemical processing

Fossil diagenetic features also prove valuable in the interpretation of sedimentary environments. Early diagenetic phenomena may provide information regarding the geochemistry of bottom waters and the upper, "taphonomically active zone" (TAZ) of the sediment column (Martin, 1999; Figure T3). Diagenetic features of note include evidence for early dissolution, compaction, and mineralization of fossils.

Skeletal minerals and dissolution

Organisms secrete a variety of minerals, including calcite, aragonite, silica, and phosphate (apatite). Different skeletal mineralogies have varying stabilities.

Carbonate skeletons are commonly preserved in normal marine environments, although those composed of aragonite are more prone to dissolution. The large biocrystals of echinoderm plates are exceptionally resistant to dissolution and may be the only skeletal carbonates remaining after prolonged exposure.

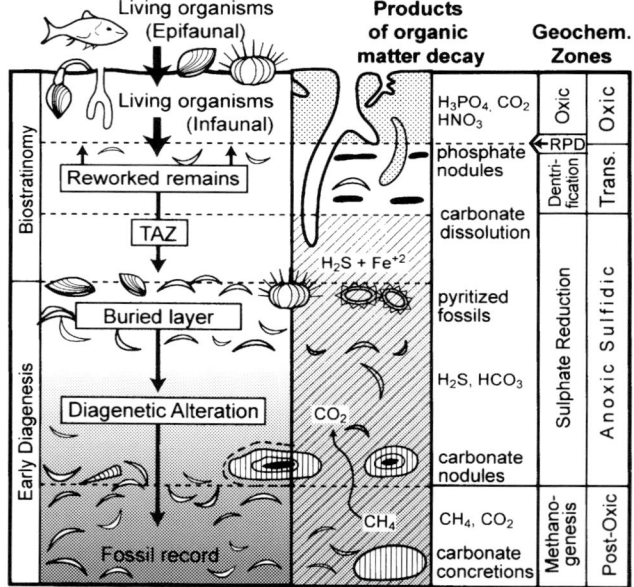

Figure T3 The realms of biostratinomy and early diagenesis in marine sediments overlain by oxic water. Left columns show transition of organism remains from epifaunal and infaunal community into the physically/biologically mixed or taphonomically active zone (TAZ) and its transition to the permanent fossil record. Note some upward reworking of skeletal remains in the TAZ; also note intact skeletons in abruptly buried obrution deposit, which is also the site of pyrite nucleation. Middle columns highlight bacterially-mediated diagenetic processes and products; right columns identify geochemical zones. Abbreviations: RPD: redox potential discontinuity (oxic-anoxic boundary); Trans: transitional redox conditions. Adapted from diagrams in Martin (1999) and Wignall (1994).

Carbonate skeletons that dissolved prior to compaction may be preserved as plastically deformed molds. Such preservation would indicate undersaturation with respect to aragonite and/or calcite in the upper sediments. Conversely, many fossil molds display mosaic fracture patterns on their surfaces (Brett and Baird, 1986). Such shells survived without dissolution until after early phases of compaction and brittle fracture of the shells.

Even more stable than echinoderm plates, are skeletal elements composed of dense calcium fluophosphate (apatite). In certain environments, particularly dysoxic, organic-rich sediments, bone apatite (organic fluophosphate) may be replaced by the far more stable form calcium carbonate fluorapatite. Fluorapatite is stable over a wider pH/Eh range than calcite, and is one of the most stable biogenically associated components (Lucas and Prévôt, 1991). Once bones, teeth, conodonts, or fossil molds have been replaced by calcium fluorapatite, they are resistant to further dissolution. Such materials may form bone beds, which remain stably on the sea bottom for periods of hundreds of thousands to perhaps millions of years (Kidwell and Bosence, 1991; Martill, 1991).

Early diagenetic mineralization

Early diagenetic minerals in fossils, form as a result of the action of anaerobic bacteria (Allison and Briggs, 1991; Briggs, 1995). They may provide valuable information on sediment geochemistry and rates of burial.

Pyrite is common in marine mudrocks because of the availability of iron in terrigenous muds and the ubiquitous abundance of dissolved sulfate in marine (but not in fresh) water. In anoxic sediments generally low in organic matter early diagenetic pyritization tends to be localized in the vicinity of anaerobically decaying organisms (Hudson, 1982; Brett and Baird, 1986). As a result of local sulfate reduction, H_2S is liberated around decomposing organic material (Figure T3). Dissolved iron will react at these sites of sulfide production

Well preserved, pyritized fossils tend to occur in bioturbated mudstones, indicating environments with limited bottom water oxygenation, and anoxic, sulfidic microenvironments within the sediment (Hudson, 1982; Brett and Baird, 1986; Canfield and Raiswell, 1991a).

Fossils enclosed within carbonate concretions are typically three-dimensional; hence, the concretions formed before shells could be crushed by overburden pressure.

Most anaerobic decay processes generate bicarbonate (HCO_3) which may initiate the precipitation of calcite or siderite concretions around decaying organic matter in the zones of nitrate or sulfate reduction (Berner, 1981; Canfield and Raiswell, 1991b; figure 3).

However, mass balance calculations have shown that the bicarbonate generated by decaying carcasses is insufficient to form a concretion. Carbonate concretions, including those which enclose well preserved fossils, were largely generated by the decay of disseminated organic matter within the sediment.

The development of large concretions requires that the sediment surrounding a local nucleation center remains within the zone of sulfate reduction or methanogenic zone (Figure T3) in the upper few decimeters of the sediment column for prolonged periods, probably at least a few thousand years (Raiswell, 1976; Canfield and Raiswell, 1991b; Wignall, 1994).

Hence, horizons of fossil-bearing concretions may reflect episodes of sedimentation that were then followed by sediment starved intervals. Not surprisingly, many concretion beds underlie recognized diastems and concretions may also become reworked into erosion lag beds (Brett and Baird, 1986).

The decay of organic matter and reduction of phosphatic iron hydroxides liberates phosphate-bearing compounds (Lucas and Prévôt, 1991; Wignall, 1994; Figure T3). Dissolved phosphate may be released to the water column, if anoxia persists to the sediment-water interface. However, if a thin oxic zone exists in the upper sediment then the phosphates may be reprecipitated (Berner, 1981; Allison, 1988), especially around phosphatic skeletal nuclei, such as bones or lingulid brachiopods.

If the sedimentation rate is very low and geochemical conditions remain relatively constant, then phosphorus compounds will be sufficiently concentrated that phosphate minerals can precipitate. These minerals may replace organic remains or can form concretions. Thus, the occurrence of phosphatic fossil molds or concretions is nearly always an indicator of sediment starvation, commonly associated with deepening episodes.

Under less strongly reducing conditions the precursors of ironstones, such as berthierine or chamosite may precipitate in the upper part of the sediment, commonly as oolitic coatings or impregnations of skeletal grains. Later diagenesis may further chemically alter these precursors to hematite or goethite. "Fossil ores" commonly occur near the interfaces of reduced iron rich muds and carbonate shoals, suggesting critical redox boundary conditions that existed in relatively low sedimentation areas of mixed siliciclastic-carbonate deposition. Widespread beds of "fossil ore", i.e., hematite or chamosite coated fossils are also excellent indicators of sediment starved discontinuities; they commonly intergrade with phosphatic nodule beds (Brett, 1995).

Prefossilized material may be identified by breakage along formerly continuous structures, rotated geopetal structures, and shell infillings that do not match surrounding matrix sediment. Perhaps the most diagnostic feature is abrasion, bioerosion, or encrustation of diagenetically altered surfaces (e.g., abraded or bored phosphatic steinkerns) and preferential alignment (Kidwell and Bosence, 1991). In rare cases, even trace fossils may be diagenetically cemented and then reworked.

Summary discussion: preservational processes and taphofacies

Various aspects of fossil preservation may be combined to recognize *taphofacies* characteristic of various sedimentary environments (Speyer and Brett, 1986, 1988). Together with litho-, bio-, and ichnofacies, taphofacies tend to vary predictably with sedimentary environments as shown by studies in modern marine settings (e.g., Parsons and Brett, 1991). Fossils and their modes of preservation may provide important insights into a number of features of sediment deposition. These include: (i) sedimentary environment (depth, temperature, salinity, oxygen level, substrate consistency); (ii) dynamics of sediment accumulation, average rates, as well as evidence for episodicity of sedimentation and erosion; (iii) temporal scope of individual mudrock units; and (iv) sediment geochemistry and early diagenetic environments.

Taphonomic features, (e.g., articulation of delicate skeletal elements) may provide unambiguous evidence for episodic sedimentation; conversely, highly corroded fossil material provides a distinctive signature of time-averaging. The overall condition of skeletons may be assessed qualitatively or quantitatively, but certainly should be noted in the field. Semi-quantitative indices (e.g., of corrosion or fragmentation) may include the proportion of skeletal parts in different, arbitrarily defined preservational states. In such cases, it is commonly useful to develop a set of standards with which particular shells may be compared and assigned to a category in much the same way that grain shapes and roundness indices have long been assessed on the basis of standardized profiles by sedimentologists. Skeletal condition is particularly valuable for recognizing qualitatively the differing relative extents of sedimentary time-averaging that may be related to sequence stratigraphy (Brett, 1995).

Overall, condensed, highly time averaged, shelly accumulations will display high frequencies of disarticulation, fragmentation, valve ratio biasing, corrosion, and/or abrasion (Fürsich, 1978; Kidwell, 1991; Brett, 1995). Condensed intervals, associated with marine transgressions, thus may be readily recognized.

Conversely, rapidly accumulated sediments, typical of later highstands, will display higher proportions of articulated material with little or no shell breakage, including intact, delicate forms, such as branching bryozoans, little or no biasing of valve ratios, and few, if any, fossils that are heavily corroded, abraded, or otherwise altered in their surfaces (Kidwell, 1991; Brett, 1995).

Taphonomic observations should accompany any interpretation of fossiliferous sediments. By making even a qualitative assessment of the preservation states of fossils, a sedimentologist may be able to determine a great deal about the dynamics of sediment accumulation and many other paleoenvironmental parameters.

Carlton E. Brett

Bibliography

Allen, J.R.L., 1992. Transport hydrodynamics. In Briggs, D.E.G., and Crowther, P.R. (eds.) *Palaeobiology: A Synthesis*. Blackwell Scientific Publications, pp. 227–230.

Allison, P.A., 1986. Soft-bodied animals in the fossil record: the role of decay in fragmentation during transport. *Geology*, **14**: 979–981.

Allison, P.A., 1988. The role of anoxia in decay and mineralization of proteinaceous organic macrofossils. *Paleobiology*, **14**: 139–154.

Allison, P.A., 1990. Variation in rates of decay and disarticulation of Echinodermata: implications for the application of actualistic data. *Palaios*, **5**: 432–440.

Allison, P.A., and Briggs, D.E.G., 1991. *Taphonomy: Releasing the Data Stored in the Fossil Record*. Plenum Press.

Baumiller, T.K., and Ausich, W.I., 1992. The broken-stick model as a null hypothesis for crinoid stalk taphonomy and as a guide to the distribution of connective tissue in fossils. *Paleobiology*, **18**: 288–298.

Behrensmeyer, K., and Kidwell, S.M., 1985. Taphonomy's contribution to paleobiology. *Paleobiology*, **11**: 105–119.

Berner, R.A., 1981. Authigenic mineral formation resulting from organic matter decomposition. *Fortschrift Mineralogie*, **59**: 117–135.

Brenchley, P.J., and Newall, G., 1970. Flume experiments on the orientation and transport of models and shell valves. *Palaeogeography, Palaeoclimatology, Palaeoecology*, **7**: 185–220.

Brett, C.E., 1995. Sequence stratigraphy, biostratigraphy, and taphonomy in shallow marine environments. *Palaios*, **10**: 597–616.

Brett, C.E., and Baird, G.C., 1986. Comparative taphonomy: a key to paleo-environmental interpretation based on fossil preservation. *Palaios*, **1**: 207–227.

Brett, C.E., and Baird, G.C., 1993. Taphonomic approaches to temporal resolution in stratigraphy. In Kidwell, S.M., and Behrensmeyer, A.K. (eds.), *Taphonomic Approaches to Time Resolution in Fossil Assemblages*. Paleontological Society Short Course 6, pp. 250–274.

Brett, C.E., Baird, G.C., and Speyer, S.E., 1997. Fossil Lagerstätten: stratigraphic record of paleontological and taphonomic events. In Brett, C.E., and Baird, G.C. (eds.), *Paleontological Events: Stratigraphic Paleoecological and Evolutionary Implications*. Columbia University Press, pp. 3–40.

Brett, C.E., and Bordeaux, Y.L., 1990. Taphonomy of brachiopods from a Middle Devonian shell bed: implications for the genesis of skeletal accumulations. *Proceedings of the Second International Brachiopod Congress*. Dunedin, New Zealand: Balkema Press, pp. 219–226.

Briggs, D.E.G., 1995. Experimental taphonomy. *Palaios*, **10**: 539–550.

Canfield, D.E., and Raiswell, R., 1991a. Pyrite formation and loss of preservation. In Allison, P.A., and Briggs, D.E.G. (eds.), *Taphonomy: Releasing the Data Locked in the Fossil Record*. Plenum Press, pp. 337–387.

Canfield, D.E., and Raiswell, R., 1991b. Carbonate precipitation and dissolution: its relevance to fossil preservation. In Allison, P.A., and Briggs, D.E.G. (eds.), *Taphonomy: Releasing the Data Locked in the Fossil Record*. Plenum Press, pp. 411–463

Donovan, S.K., 1991. *The Processes of Fossilization*. London, Belhaven Press.

Fürsich, F.T., 1978. The influence of faunal condensation and mixing on the preservation of fossil benthic communities. *Lethaia*, **11**: 151–172.

Futterer, E., 1978. Hydrodynamic behavior of biogenic particles. *Neues Jahrbuch fur Geologie und Paläontologie Abhandlungen*, **157**: 37–42.

Grinnell, R.S., 1974. Vertical orientation of some shells on Florida oyster reefs. *Journal of Sedimentary Petrology*, **41**: 116–122.

Hewitt, R., and Westermann, G.E.G., 1996. Post-mortem behaviour of Early Palaeozoic nautiloids and paleobathymetry. *Paläontologisches Zeitschrift*, **70**: 405–424.

Holland, S.M., 1988. Taphonomic effects of sea-floor exposure on an Ordovician brachiopod assemblage. *Palaios*, **3**: 588–597.

Hudson, J.D., 1982. Pyrite in ammonite-bearing shales from the Jurassic of England and Germany. *Sedimentology*, **25**: 639–667.

Kidwell, S.M., 1991. The stratigraphy of shell concentrations. In Allison, P.A., and Briggs, D.E.G. (eds.), *Taphonomy: Releasing the Data Locked in the Fossil Record*. Plenum Press, pp. 211–290.

Kidwell, S.M., and Baumiller, T., 1990. Experimental disintegration of regular echinoids: roles of temperature, oxygen, and decay thresholds. *Paleobiology*, **16**: 247–271.

Kidwell, S.M., and Behrensmeyer, A.K., (eds.), 1993. *Taphonomic Approaches to Time Resolution in Fossil Assemblages*. Knoxville, Tennessee: The Paleontological Society, Short Courses in Paleontology G, 302pp.

Kidwell, S.M., and Bosence, D.W., 1991. Taphonomy and time-averaging of marine shelly faunas. In Allison, P.A., and Briggs, D.E.G. (eds.), *Taphonomy: Releasing the Data Locked in the Fossil Record*. Plenum Press, pp. 115–209.

Kidwell, S.M., and Holland, S.M., 1991. Field description of coarse bioclastic fabrics. *Palaios*, **6**: 426–434.

Lucas, J., and Prévôt, L.E., 1991. Phosphates and fossil preservation. In Allison, P.A., and Briggs, D.E.G. (eds.), *Taphonomy: Releasing the Data Locked in the Fossil Record*. Plenum Press, pp. 389–409.

Martin, A.E., 1999. *Taphonomy: A Process Approach*. Cambridge University Press.

Martill, D.M., 1991. Bones as stones: the contribution of vertebrate remains to the geologic record. In Donovan, S.K. (ed.), *The Processes of Fossilization*. London: Bellhaven Press, pp. 270–292.

Middleton, G.V., 1967. The orientation of concave-convex particles deposited from experimental turbidity currents. *Journal of Sedimentary Petrology*, **37**: 229–232.

Müller, A.H., 1979. Fossilization (taphonomy). In Robison, R.A., and Teichert, C. (eds.), *Treatese on Invertebrate Paleontology, Part A*,

Introduction. Lawrence, Kansas: Geological Society of America and University of Kansas Press, pp. 2–78.

Nagle, J.S., 1967. Wave and current orientation of shells. *Journal of Sedimentary Petrology*, **37**: 1124–1138.

Parsons, K.M., and Brett, C.E., 1991. Taphonomic processes and biases in modern marine environments: an actualistic perspective on fossil assemblage preservation. In Donovan, S.K. (ed.), *The Processes of Fossilization*. London, Belhaven Press, 22–65.

Plotnick, R.E., 1986. Taphonomy of a modern shrimp: implications for the arthropod fossil record. *Palaios*, Volume 1, p.286–293.

Raiswell, R., 1976. The microbiological formation of carbonate concretions in the Upper Lias of NE England. *Chemical Geology*, **18**: 227–244.

Raup, D.M., 1973. Depth inference from vertically embedded cephalopods. *Lethaia*, **6**: 217–226.

Reyment, R.A., 1980. Floating orientations of cephalopod shell models. *Palaeontology*, **24**: 931–936.

Schäfer, W., 1972. *Ecology and Palaeoecology of Marine Environments*. University of Chicago Press.

Seilacher, A., Reif, W.-E., and Westphal, F., 1985. Sedimentological, ecological, and temporal patterns of fossil Lagerstätten. *Philosophical Transactions of the Royal Society*, London, B 311: 5–23.

Seilacher, A., and Meischner, D., 1965. Fazies-analyze im Paläozoikum des Oslo-Gebiets. *Geologisches Rundschau*, **54**: 596–619.

Speyer, S.E., and Brett, C.E., 1986. Trilobite taphonomy and Middle Devonian taphofacies. *Palaios*, **1**: 312–327.

Speyer, S.E., and Brett, C.E., 1988. Taphofacies models for epeiric sea environments: middle Paleozoic examples. *Palaeogeography, Palaeoclimatology, Palaeogeography*, **63**: 225–262.

Vogel, K., Golubic, S., and Brett, C.E., 1987. Endolith associations and their relation to facies distribution in the middle devonian of New York. *Lethaia*, **20**: 263–290.

Westermann, G.E.G., 1985. Post-mortem descent and septal implosion in Silurian nautiloids. *Paläontologisches Zeitschrift*, **59**: 79–87.

Wignall, P.W., 1994. *Black Shales: Oxford Monographs on Geology and Geophysics*, Volume 30, 127pp.

Cross-references

Algal and Bacterial Carbonate Sediments
Bacteria in Sediments
Bioclasts
Bioerosion
Biogenic Sedimentary Structures
Black Shales
Coal Balls
Diagenetic Structures
Micritization
Mudrocks
Paleocurrent Analysis
Phosphorites
Reefs
Sediment Fluxes and Rates of Sedimentation
Sedimentologists
Storm Deposits
Substrate-Controlled Ichnofacies
Sulfide Minerals in Sediments
Upwelling

TECTONIC CONTROL OF SEDIMENTATION

Introduction

Sedimentation is influenced by three extrinsic variables: tectonics, climate, and sea level, with the latter two potentially dependent upon each other and upon tectonics. In addition, sedimentation is affected by processes inherent to the depositional system, which act independently of the extrinsic variables. A primary goal in the study of sediment and sedimentary rocks is to decipher the relative effects of the extrinsic and intrinsic processes on the depositional history of a sedimentary basin.

Tectonics affects sedimentation in two different, but often related ways: subsidence of the crust and uplift of the land. Although sediment may be deposited temporarily in tectonically inactive, topographically low areas of the crust, it is subsidence over geologically significant intervals of time that continually provides the accommodation space for the accumulation and burial of sediment in sedimentary basins. Subsidence may be driven by vertical offset of faults, lateral flow of the mantle, cooling and concomitant increase in density of the crust and/or mantle, and isostatic response to loading of the crust, including the load provided by sediment within the basin (Ingersoll and Busby, 1995). Often working in concert with subsidence, uplift of the land provides detrital sediment to depositional basins through weathering and erosion. Vertical tectonic movement of the crust is required to maintain high regions or they will ultimately be reduced in elevation to near sea level.

Early attempts to classify sedimentary basins were hampered by the lack of a comprehensive model to explain crustal deformation (Schuchert, 1923; Kay, 1951). It is now widely accepted that the dominant tectonic process affecting the crust and upper mantle of the earth during most of its history is plate tectonics, with mantle plumes of secondary importance. Consequently, the following discussion of sedimentary basins is organized according to their plate-tectonic settings. Other references on this topic include Allen and Allen (1990) and Busby and Ingersoll (1995).

Sedimentary basins associated with diverging plates

Initial breakup of continental crust produces a long, linear extensional terrane referred to as a continental rift. In some cases rifting initially takes place simultaneously along three arms (triple junction), one of which eventually fails while the other two evolve into the rifted continental margin. The failed arm (aulacogen) represents a region of thick sediment that extends far into the continental interior (Hoffman *et al.*, 1974).

Continental rift basins are either bordered on both sides (full graben) or one side (half graben) by normal faults that extend at least into the middle crust. It is movement on the normal faults that provides the primary mechanism of basin subsidence and adjacent uplift. The margins of a continental rift basin commonly contain coarse detritus deposited on alluvial fans. In a half graben, small catchments on the footwall scarp result in small fans, while larger catchments in the hanging-wall mountains produce broad fans that extend far into the basin (Figure T4; Leeder and Gawthorpe, 1987). Basin-center facies in rift basins may include lakes, an axial river, and/or eolian sand field. Large lakes are particularly common in the East African rift (Soreghan and Cohen, 1996), while the Rio Grande rift of the southwestern United States has axial rivers (Mack *et al.*, 1997). Continental rift basins also may be inundated by shallow seas, particularly if the rift is near sea level at the onset of extension (Gawthorpe *et al.*, 1997).

Normal faults in continental rifts have finite lengths usually measured in a few tens to a hundred kilometers. Strain is transferred from the tip of one active fault to another via an

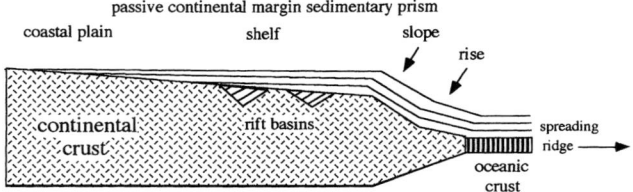

Figure T5 Cross-section of a passive continental margin showing the major depositional settings.

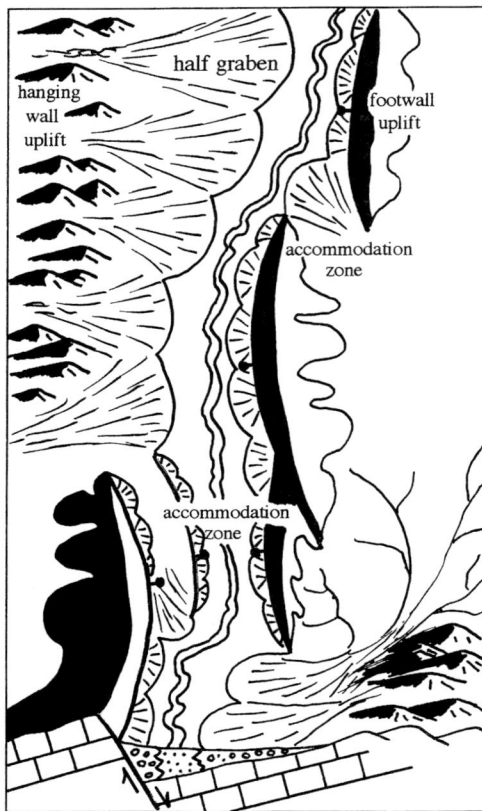

Figure T4 Schematic map and cross-section of a continental rift, showing basin-bounding normal faults (ball and stick on down side of fault), an accommodation zone in which the border faults overlap and dip in the same direction (upper part of diagram), an accommodation zone in which the border faults overlap and dip toward each other, and the asymmetrical distribution of alluvial-fan and axial-fluvial environments in half grabens.

accommodation zone, whose structural style depends on the degree of overlap and dip direction of the faults (Morley et al., 1990). Because of large catchments, accommodation zones may be the locus of sedimentation by large alluvial fans or deltas (Figure T4).

Continental rifting may eventually split the continental land masses apart, creating between them a narrow, incipient ocean basin floored by nascent oceanic crust, such as exists today in the Red Sea (Purser and Hotzl, 1988). The narrow seaway may be initially flanked by high escarpments, narrow coastal plains, and marine shelves, whose depositional environments include alluvial fan, fluvial, siliciclastic shoreline, including small deltas, and marine carbonates. With time the marginal rift blocks are reduced by erosion and onlapped by sediment. Because of a narrow connection to the open ocean and poor circulation, the incipient ocean basin may accumulate fine, organic-rich sediment and/or evaporites.

With increasing distance from the oceanic spreading ridge, the previously rifted continental margin subsides slowly and asymmetrically about a hinge on the landward side. Subsidence, which decreases exponentially with time, is primarily related to the amount of crustal thinning, cooling of the lithosphere, and sediment loading (Bond et al., 1995). The resultant passive continental-margin prism of sediment is deposited in coastal plain, shallow-marine shelf, and deep-water rise environments, and consists predominantly of fine siliciclastic and/or carbonate sediment (Figure T5). Sedimentation may be locally influenced by and strata disrupted by contemporaneous rise of diapirs of salt deposited in the incipient ocean basin phase. Along-strike variation in passive margin sedimentation is related to continental transform faults that produce an orthogonal continental margin (Thomas, 1991).

The thickness of sediment deposited at active oceanic spreading ridges is initially influenced by half grabens in the oceanic crust. Although the earliest sediment may be umber, the iron- and manganese-rich fallout from black smokers (Robertson and Hudson, 1973), sedimentation on oceanic crust is dominated by pelagic sediment composed of carbonate and siliceous shells and clay that settle from suspension. The global distribution of pelagic sediment is primarily related to proximity to continental land masses and to the productivity of planktonic organisms (Davies and Gorsline, 1976). Tectonism comes into play in pelagic sedimentation through subsidence of the oceanic crust, a time-dependent process driven by cooling of the oceanic plate as it moves away from the spreading ridge. The subsiding plate may eventually fall below the carbonate and silica compensation depths, inhibiting the deposition of carbonate and siliceous oozes and promoting deposition of clay (Heezen et al., 1973). Voluminous outpouring of basaltic lava on the ocean floor, such as occurs above mantle plumes, may build intraoceanic islands. Once the volcanism has slowed or ceased, however, the oceanic plateau will subside to abyssal depths, producing a stratigraphy that may include carbonate reef and shoal deposits overlain by pelagic sediment (Clague, 1981).

Sedimentary basins associated with subduction

The process of subduction of an oceanic plate beneath another oceanic plate or a plate with continental crust on its leading edge results in a series of depositional basins that are bordered by tectonic elements related to the subduction process (Figure T6). The topographic low at the junction of the two plates is the trench, and oceanic sediment and basaltic crust scraped off of the subducting slab onto the overriding plate creates the topographically high subduction complex. Partial melting of the asthenosphere above the subducting slab generates magma that rises to create a volcanic arc and its underlying plutons.

Mostly pelagic sediment, locally affected by thermohaline currents, is deposited on the outer trench slope, where flexure of the subducting slab may result in extensional basins

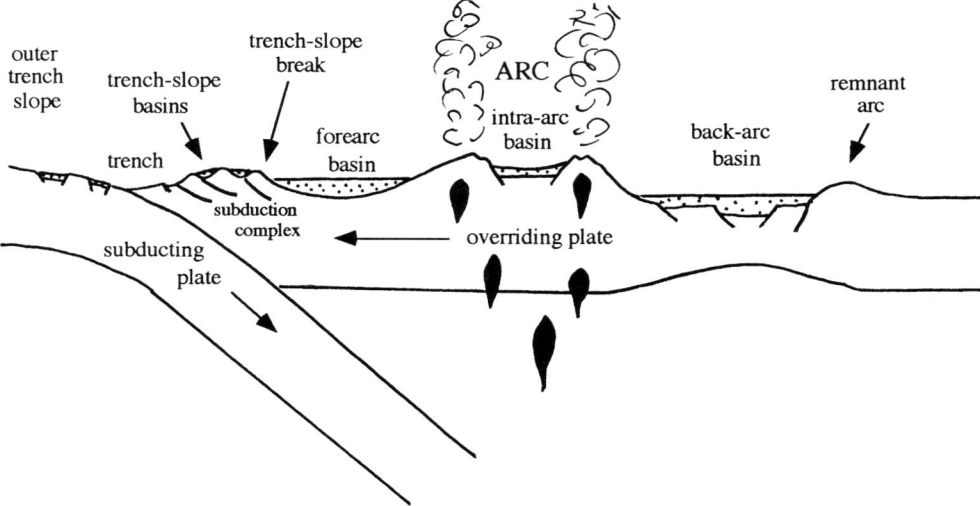

Figure T6 Cross-section of converging plates showing the subduction complex and arc, and their associated sedimentary basins.

(Figure T6; Underwood and Moore, 1995). In addition to pelagic sediment, the trench and trench-slope basins may receive transverse- or longitudinally dispersed gravity flows and submarine slides. The amount of sediment in the trench is highly variable and depends on the ratio of rate of sediment influx to the rate of subduction. Developing on the subduction complex, trench-slope basins may consist of grabens, linear valleys separated by fault- and fold-generated ridges, and/or a thin apron of sediment.

The forearc basin, which is situated between the magmatic arc and the subduction complex, is generally the largest of the subduction-related basins (Figure T6; Dickinson, 1995). Usually less than 150 km wide, forearc basins may extend laterally for thousands of kilometers or be segmented into a series of smaller basins. Sediment thickness in forearc basins may be up to 10 km, and the sediment commonly progressively onlaps both the arc and the subduction complex. Subsidence mechanisms of forearc basins are poorly understood, but may include downward pull of the subducting plate, tectonic loading by the subduction complex, cooling of a relict arc, and sediment loading. The principle source of siliciclastic sediment is the volcanic arc, although the plutonic core of the arc may be periodically exposed. Less commonly, the subduction complex may provide sediment to the forearc basin via submarine slides or by erosion of a complex that builds above sea level. Many forearc basins are primarily deep marine, in which hemipelagic muds and turbidite sands are deposited. It is also possible for forearc basins to infill with shallow-marine and nonmarine siliciclastic sediment, and marine forearc basins bordered by only a few volcanic islands may experience significant carbonate deposition.

Many active volcanic arcs contain intra-arc basins, which may be fault-bound or simply occupy topographically low regions between the volcanoes (Figure T6; Smith and Landis, 1995). Accumulation rates of sediment and volcanic rocks are high, because of active eruptions and erosion of high-relief volcanoes. Proximal to the volcanoes, lava flows and pyroclastic flows are interbedded with coarse gravity flow deposits, while in more distal settings volcanic ash is interbedded with fine-grained alluvial-fan, fluvial, and lacustrine sediment. In oceanic areas, it is common for part of the intra-arc basin to be marine and contain fan-delta, delta, deep-sea fan, hemipelagic, and carbonate deposits. Intra-arc basins provide important information about the evolution of the arc, but may be difficult to distinguish from other arc-related basins in ancient rocks.

In some cases, lithospheric extension occurs behind the arc, creating a back-arc basin (Figure T6; Marsaglia, 1995). Back-arc basins may develop in intraoceanic arc (e.g., Mariana arc) or continental-margin arc (e.g., Sea of Japan) settings and are especially common in the western Pacific associated with rapid subduction of old, steeply dipping oceanic lithosphere. Extension is driven by mantle flow above the subducting plate and/or by relative movement of the trench toward the subducting plate (roll back) related to steepening of the angle of subduction. Initial breakup commonly takes place in the thermally weakened arc, splitting a remnant arc from the active arc. Tectonics and sedimentation of back-arc basins resemble that in rifts and acting spreading ridges, with the exception that the active arc is an important source of detrital sediment and volcanic rocks.

Sedimentary basins associated with compressional mountain building

Contraction of continental crust results in long (1,000s km), narrow (few 100 km) mountains characterized by thrust faults and associated folds. These fold-thrust belts may develop by continental collision, in which the edge of a continental plate collides with another continent, microcontinent, or oceanic terrane, such as an arc, subduction complex, or aseismic ridge. Fold-thrust belts comparable in style and scale to those formed by collision may also develop in a back-arc setting, such as the subandes of South America. The sedimentary basins that are complementary to fold-thrust belts are termed foreland basins and can be separated into peripheral (associated with collisional fold-thrust belts) and retroarc (associated with back-arc fold-thrust belts) (Dickinson, 1974). Tectonic subsidence of foreland basins is primarily driven by loading of the crust by thrust sheets (Jordan, 1981), but also may be influenced by downward pull of the crust by the subducting slab (Royden, 1993) or mantle flow above the subducting slab (Gurnis, 1992).

DeCelles and Giles (1996) divided foreland basins into four potential depozones that may migrate laterally through time (Figure T7). The wedge-top depozone represents locally derived sediment deposited on top of the active, frontal part of the fold-thrust belt, including piggyback basins (Ori and Friend, 1984). A common feature of wedge-top strata are progressive unconformities, which develop by rotation of syntectonic sedimentary beds on fault-bend folds (Riba, 1976; Burbank et al., 1996). The greatest volume of sediment in foreland basins is deposited in the foredeep depozone, which is typically 100–300 km wide and receives the bulk of its sediment from the fold-thrust belt. Foredeeps of peripheral foreland basins often begin with deep-water sedimentation (*flysch*), followed by shallow-marine, shoreline, and nonmarine sedimentation (*molasse*). Fluvial systems within the foredeep may be transverse or axial, and large drainages may emerge onto the foredeep at the tips of major thrust faults. Flexural bending of the floor of the foreland basin may result in a forebulge located on the craton side of the foredeep. Generally hundreds of kilometers wide and a maximum of a few hundred meters high, the forebulge is the site of erosion or deposition of thin sediment derived from the fold-thrust belt or the craton. The slowly subsiding back-bulge depozone lies cratonward of the forebulge and contains relatively thin, fine siliciclastic sediment derived from the fold-thrust belt and/or the craton, or is a site of marine carbonate deposition.

Another type of back-arc contractional deformation of continental crust occurs today in the Sierras Pampeanas of Argentina and occurred in latest Cretaceous–Eocene time in the Western Interior of the United States (Jordan and Allmendinger, 1986). This type of deformation is located over 500–1000 km from the active trench and is associated with subhorizontal subduction. Deformation produces discrete, basement-cored uplifts tens of kilometers wide and hundreds of kilometers long bordered by thrust and reverse faults. The uplifts are separated by intermontane basins the same scale or larger than the uplifts. In the modern and ancient examples cited above, the basins are entirely continental, include alluvial-fan, fluvial, and lacustrine environments, and may display subsidence histories and sediment distribution similar to that of the wedge-top or foredeep depozones of foreland basins (Seager et al., 1997).

Several other sedimentary basins may be associated with continental collisions. Deep-sea sediment derived from collisional mountains may be deposited in remnant ocean basins, which exist between and may eventually be deformed by the colliding continents (Graham et al., 1975). The best modern example of a remnant ocean basin is the Bay of Bengal and associated Bengal deep-sea fan. Continental collisions may also generate deformational structures hundreds to thousands of kilometers inboard of the orogen. Such structures may include extensional basins (impactogens) and strike-slip faults (Sengor et al., 1978; Tapponnier et al., 1982).

Sedimentary basins associated with strike-slip faults

Strike-slip faults exist in a variety of different tectonic settings and can be divided into transform and transcurrent types (Sylvester, 1988). Transform faults define plate boundaries and penetrate the lithosphere. Examples of transform strike-slip faults include those at oceanic spreading ridges and the San Andreas fault, which forms the boundary between the North American and Pacific plates. In contrast, a transcurrent strike-slip fault, such as the North Anatolian fault in Turkey, is confined to the crust and is located within a plate. Most strike-slip basins are related to bends in the trace of strike-slip faults or are associated with en echelon strike-slip faults (Figure T8). Extensional fault-bound basins form at releasing bends in strike-slip faults and tend to be elongate or lens-shaped (Crowell, 1974). Restraining bends tend to produce thrust faults and associated transpressional basins, which subside in response to thrust loading (Nilsen and Sylvester, 1995). In addition, parallel but offset strike-slip faults may form extensional stepover basins, such as in the Dead Sea rift (Aydin and Nur, 1985), and transrotational basins form among crustal blocks rotated about a vertical axis between strike-slip faults (Ingersoll, 1988).

Strike-slip basins may contain a variety of marine and nonmarine depositional systems arranged similar to that in rift or foreland basins. Most diagnostic of strike-slip basins are high sediment accumulation rates, abrupt lateral and vertical facies changes, lateral migration of the depocenter producing stratal shingling, and stratigraphic variations in provenance related to lateral displacement of source terranes (Sylvester, 1988; Nilsen and Sylvester, 1995). Although generally not

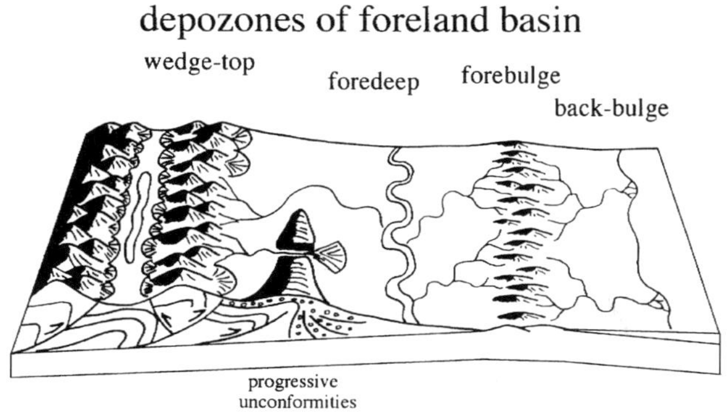

Figure T7 Schematic map and cross-section, not to scale, of the depozones of a foreland basin, based on DeCelles and Giles (1996).

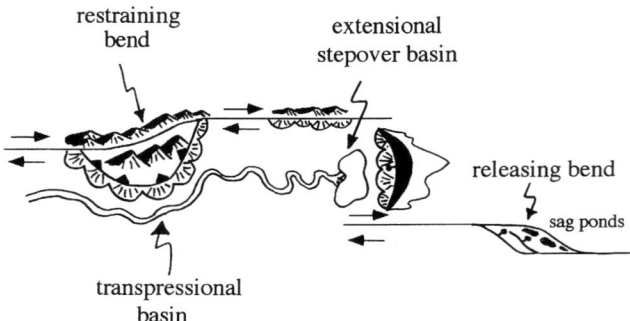

Figure T8 Schematic map of three types of sedimentary basins associated with en echelon, right-lateral strike-slip faults. Ball and stick on down side of normal fault; bold triangles on up side of thrust fault.

associated with distinct sedimentary basins, transform faults at oceanic spreading ridges may develop vertical scarps that supply gravity flows to the adjacent ocean floor (Simonian and Gass, 1978).

Cratonic sedimentary basins

Cratonic basins are large ($>10^5$ km^2), generally elliptical basins that develop within the interior of continents far from active plate boundaries. Despite relatively slow and erratic subsidence, they may accumulate thousands of meters of sediment because of histories measured in hundreds of millions of years. The basin fill consists predominantly of nonmarine and shallow-marine fine-grained siliciclastics, carbonates, and evaporites, and unconformities are common. Sloss (1963) defined six Phanerozoic depositional sequences separated by continental-scale unconformities in the cratonic basins of North America, some of which could be correlated with cratonic basins in Asia (Sloss, 1972), South America (Soares et al., 1978), and Africa (Peters, 1979). Stratigraphic similarities of cratonic basins in Europe and India were also noted by Klein and Hsui (1987). Throughout their histories, cratonic basins may be affected by a variety of tectonic processes, including various types of thermal and isostatic subsidence, lower crustal phase changes or melting, and stress related to plate interactions (Klein, 1995). A comprehensive model to explain the origin and apparent synchroneity of cratonic basins remains elusive, although Klein (1995) proposed that they evolve in response to breakup of supercontinents.

Summary

(1) Plate tectonics is the primary driving force for crustal uplift and basin subsidence through the processes of faulting, isostatic response to loading and unloading, lateral flow of the mantle, and change in density of the crust and mantle due to heating and cooling.
(2) Sedimentary basins associated with diverging plates include continental rifts (e.g., East African rift), incipient ocean basins (e.g., Red Sea), passive continental margins (e.g., Atlantic coast of North America), and oceanic basins created at active spreading ridges (e.g., Mid Atlantic Ridge).
(3) Subduction-related basins may develop on the outer-trench slope of the subducting slab, in the trench, and on the overriding plate, including trench-slope basins, forearc basins, and intra-arc basins. Extensional back-arc basins may develop in areas of trench roll back or asthenospheric upwelling.
(4) Continental collision and back-arc compression produce fold-thrust belts and their complementary foreland basins, which can be divided into wedge-top, foredeep, forebulge, and back-bulge depozones. An usual type of back-arc compression associated with subhorizontal subduction results in discrete uplifts separated by intermontane basins. Continental collision may produce extensional basins and strike-slip faults far inboard of the collisional orogen.
(5) Strike-slip faults may define plate boundaries (transform faults) or reside within a plate (transcurrent fault). Basins are created at releasing and restraining bends in fault traces, between en echelon faults, and by rotation of fault-bound blocks about a near vertical axis.
(6) Cratonic basins develop far from active plate margins and may have long histories of intermittent subsidence and erosion. Cratonic basins on different continents may have similar histories, but their origin remains poorly understood.

Greg H. Mack

Bibliography

Allen, P.A., and Allen, J.R., 1990. *Basin Analysis, Principles, and Applications*. London: Blackwell Science.
Aydin, A., and Nur, A., 1985. The types and roles of stepovers in strike-slip tectonics. In *Strike-Slip Deformation, Basin Formation, and Sedimentation*, SEPM (Society for Sedimentary Geology) Special Publication, 37, pp. 35–44.
Bond, G.C., Kominz, M.A., and Sheridan, R.E., 1995. Continental terraces and rises. In Busby, C.J., and Ingersoll, R.V. (eds.), *Tectonics of Sedimentary Basins*. Blackwell Science, pp. 149–178.
Burbank, D., Meigs, A., and Brozovic, N., 1996. Interactions of growing folds and coeval depositional systems. *Basin Research*, **14**: 199–223.
Busby, C.J., and Ingersoll, R.V. (eds.), 1995. *Tectonics of Sedimentary Basins*. Blackwell Science.
Clague, D.A., 1981. Linear island and seamount chains, aseismic ridges, and intraplate volcanism: results from DSDP. In *The Deep Sea Drilling Project: A Decade of Progress*. SEPM (Society for Sedimentary Geology), Special Publication, 32, pp. 7–22.
Crowell, J.C., 1974. Origins of Late Cenozoic basins in southern California. In *Tectonics and Sedimentation*. SEPM (Society for Sedimentary Geology), Special Publication, 22, pp. 190–204.
Davies, T.A., and Gorsline, D.S., 1976. Oceanic sediments and sedimentary processes. In Riley, J.P., and Chester, R. (eds.), *Chemical Oceanography*. Volume 5: pp. 1–80.
DeCelles, P.G., and Giles, K.A., 1996. Foreland basin systems. *Basin Research*, **8**: 105–123.
Dickinson, W.R., 1974. Plate tectonics and sedimentation. In *Tectonics and Sedimentation*. SEPM (Society for Sedimentary Geology), Special Publication, 22, pp. 1–27.
Dickinson, W.R., 1995. Forearc basins. In Busby, C.J., and Ingersoll, R.V. (eds.), *Tectonics of Sedimentary Basins*. Blackwell Science, pp. 221–261.
Gawthorpe, R.L., Sharp, I., Underhill, J.R., and Gupta, S., 1997. Linked sequence stratigraphic and structural evolution of propagating normal faults. *Geology*, **25**: 795–798.
Graham, S.A., Dickinson, W.R., and Ingersoll, R.V., 1975. Himalayan-Bengal model for flysch dispersal in Appalachian-Ouachita system. *Geological Society of America Bulletin*, **86**: 273–286.
Gurnis, M., 1992. Rapid continental subsidence following the initiation and evolution of subduction. *Science*, **255**: 1556–1558.
Heezen, B.C., and eleven co-authors, 1973. Diachronous deposits, a kinematic interpretation of the post-Jurassic sedimentary sequence on the Pacific Plate. *Nature*, **241**: 25–32.
Hoffman, P.F., Dewey, J.F., and Burke, K., 1974. Aulacogens and their genetic relation to geosynclines with a Proterozoic example from Great Slave Lake, Canada. In *Modern and Ancient Geosynclinal Sedimentation*, SEPM (Society for Sedimentary Geology), Special Publication, 19, pp. 38–55.
Ingersoll, R.V., 1988. Tectonics of sedimentary basins. *Geological Society of America Bulletin*, **100**: 1704–1719.
Ingersoll, R.V., and Busby, C.J., 1995. Tectonics of sedimentary basins. In Busby, C.J., and Ingersoll, R.V. (eds.), *Tectonics of Sedimentary Basins*. Blackwell Science, pp. 1–51.
Jordan, T.E., 1981. Thrust loads and foreland basin evolution, Cretaceous, western United States. *American Association of Petroleum Geologists Bulletin*, **65**: 2506–2520.
Jordan, T.E., and Allmendinger, R.W., 1986. The Sierras Pampeanas of Argentina: a modern analogue of Rocky Mountain foreland deformation. *American Journal of Science*, **286**: 737–764.
Kay, M., 1951. North American geosynclines. *Geological Society of America Memoir*, **48**: 1–143.

Klein, G.D., 1995. Intracratonic basins. In Busby, C.J., and Ingersoll, R.V. (eds.), *Tectonics of Sedimentary Basins*. Blackwell Science, pp. 459–478.
Klein, G.D., and Hsui, A.T., 1987. Origin of intracratonic basins. *Geology*, **15**: 1094–1098.
Leeder, M.R., and Gawthorpe, R.L., 1987. Sedimentary models for extensional tilt-block/half-graben basins. In *Continental Extensional Tectonics*. Geological Society of London, Special Publication, 28, pp. 139–152.
Mack, G.H., Love, D.W., and Seager, W.R., 1997. Spillover models for axial rivers in regions of continental extension: the Rio Mimbres and Rio Grande in the southern Rio Grande rift, USA. *Sedimentology*, **44**: 637–652.
Marsaglia, K.M., 1995. Interarc and backarc basins. In Busby, C.J., and Ingersoll, R.V. (eds.), *Tectonics of Sedimentary Basins*. Blackwell Science, pp. 299–329.
Morley, C.K., Nelson, R.A., Patton, T.L., and Munn, S.G., 1990. Transfer zones in the East African rift system. *American Association of Petroleum Geologists Bulletin*, **74**: 1234–1253.
Nilsen, T.H., and Sylvester, A.G., 1995. Strike-slip basins. In Busby, C.J., and Ingersoll, R.V. (eds.), *Tectonics of Sedimentary Basins*. Blackwell Science, pp. 425–457.
Ori, G.G., and Friend, P.G., 1984. Sedimentary basins, formed and carried piggyback on active thrust sheets. *Geology*, **12**: 475–478.
Peters, S.W., 1979. West African intracratonic stratigraphic sequences. *Geology*, **7**: 528–531.
Purser, B.H., and Hotzl, H., 1988. The sedimentary evolution of the Red Sea rift: a comparison of the northwest (Egyptian) and northeast (Saudi Arabian) margins. *Tectonophysics*, **153**: 193–208.
Riba, O., 1976. Syntectonic unconformities of the Alto Cardener, Spanish Pyrenees: a genetic interpretation. *Sedimentary Geology*, **15**: 213–233.
Robertson, A.H.F., and Hudson, J.D., 1973. Cyprus umbers: chemical precipitates on a Tethyan ocean ridge. *Earth and Planetary Science Letters*, **18**: 93–101.
Royden, L.H., 1993. Tectonic expression of slab pull at continental convergent boundaries. *Tectonics*, **12**: 303–325.
Schuchert, C., 1923. Sites and nature of the North American geosynclines. *Geological Society of America Bulletin*, **34**: 151–230.
Seager, W.R., Mack, G.H., and Lawton, T.F., 1997. Structural kinematics and depositional history of a Laramide uplift-basin pair in southern New Mexico: implications for development of intraforeland basins. *Geological Society of America Bulletin*, **109**: 1389–1401.
Sengor, A.M.C., Burke, K., and Dewey, J.F., 1978. Rifts at high angles to orogenic belts: tests for their origin and the upper Rhine graben as an example. *American Journal of Science*, **278**: 24–40.
Simonian, K.O., and Gass, I.G., 1978. Arakapas fault belt, Cyprus: a fossil transform belt. *Geological Society of America Bulletin*, **89**: 1220–1230.
Sloss, L.L., 1963. Sequences in the intracratonic interior of North America. *Geological Society of America Bulletin*, **74**: 93–114.
Sloss, L.L., 1972. Synchrony of Phanerozoic sedimentary-tectonic events in the North American craton and the Russian platform. *XXIV International Geological Congress Proceedings, Montreal*, Section 6, pp. 24–32.
Smith, G.A., and Landis, C.A., 1995. Intra-arc basins. In Busby, C.J., and Ingersoll, R.V. (eds.), *Tectonics of Sedimentary Basins*. Blackwell Science, pp. 263–298.
Soares, P.C., Landim, P.M.B., and Fulfaro, V.J., 1978. Tectonic cycles and sedimentary sequences in the Brazilian intracratonic basins. *Geological Society of America Bulletin*, **89**: 181–191.
Soreghan, M.J., and Cohen, A.S., 1996. Textural and compositional variability across littoral segments of Lake Tanganyika: the effects of asymmetric basin structure on sedimentation in large rift lakes. *American Association of Petroleum Geologists Bulletin*, **80**: 382–409.
Sylvester, A.G., 1988. Strike-slip faults. *Geological Society of America Bulletin*, **100**: 1666–1703.
Tapponnier, P., Peltzer, G., LeDain, A.Y., and Armijo, R., 1982. Propagating extrusion tectonics in Asia: new insights from simple experiments with plasticine. *Geology*, **10**: 611–616.
Thomas, W.A., 1991. The Appalachian-Ouachita rifted margin of southeastern North America. *Geological Society of America Bulletin*, **103**: 415–431.
Underwood, M.B., and Moore, G.F., 1995. Trenches and trench-slope basins. In Busby, C.J., and Ingersoll, R.V. (eds.), *Tectonics of Sedimentary Basins*. Cambridge, MA: Blackwell Science, pp. 179–219.

Cross-references

Calcite Compensation Depth
Coastal Sedimentary Facies
Deltas and Estuaries
Fan Delta
Gravity-Driven Mass Flows
Lacustrine Sedimentation
Oceanic Sediments
Rivers and Alluvial Fans
Slope Sediments
Submarine Fans and Channels
Turbidite

TIDAL FLATS

Definition

Tidal flats are sandy-muddy depositional systems along marine and estuarine shores periodically submerged and exposed in the course of the rise and fall of the tide (*cf.* Bates and Jackson, 1987). They differ from tidal beaches in spatial extent, topographic complexity, and facies differentiation. Perhaps the oldest reference to tidal flats dates back to the 1st Century AD when the Roman historian Pliny the Elder, after visiting the Wadden Sea coast in the year 47 AD, described it in his epochal work "Naturalis historia" as an area which is inundated by the sea with forceful currents twice a day and of which it was doubtful whether it formed part of the land or the sea.

Physical setting, dimensions, and hydrologic classification

On a global scale, tidal flat environments are exposed to two dominant tidal types, a semi-diurnal and a diurnal one which are separated by two mixed types consisting of a predominantly semi-diurnal and a predominantly diurnal one. The local tidal type controls the daily frequency and rate of inundation of a tidal flat (Allen, 1997). The most prominent tidal period recorded in rhythmic tidal deposits is the fortnightly spring-neap cycle (e.g., Visser, 1980). Another short-term astronomical feature influencing depositional processes on tidal flats is the daily inequality of the tide which results from the inclination of the earth's axis relative to the plane of the ecliptic, and hence the tidal bulge (e.g., De Boer *et al.*, 1989). The daily inequality increases from the equator toward higher latitudes. The largest astronomical tides in the course of a year, i.e., the highest spring and lowest neap tides, occur at the vernal and autumnal equinoxes. These are the periods of maximum tidal energy flux and hence of greatest sediment transport potential. In addition to these short-period cycles, the tide is also modulated over longer astronomical periods such as the 18.6 year nodal cycle (Oost *et al.*, 1993). Good examples of rhythmic tidal deposits have been described from ancient tidal deposits (e.g., Archer *et al.*, 1991; Williams, 1991).

A second important factor affecting depositional processes on tidal flats is the tidal range. In combination with the size of the tidal basin it determines the volume of the tidal prism and hence the tidal discharge. Being an easily determined parameter, the tidal range has been used to classify tidal coasts. The most widely applied scheme to date is that of Davies (1964) in which microtidal shores (0–2 m) are distinguished from mesotidal (2–4 m) and macrotidal (>4 m) ones. An alternative scheme comprising five subdivisions, and thus offering greater differentiation, is that of Hayes (1979) in which microtidal shores (0–1 m) are distinguished from low-mesotidal (1–2 m), high-mesotidal (2–3.5 m), low-macrotidal (3.5–5.0 m), and high-macrotidal (>5 m) ones. This latter scheme takes distinct, process-related geomorphic features into consideration.

Modern tidal-flat depositional systems are commonly found in estuaries and along low-lying coastal-plain shores between the equator and the polar seas at tidal ranges above about 0.5 m. They can attain shore-normal widths of several kilometers, alongshore extensions of many 10s to 100s of kilometers, and vertical accretion heights of several 10s to 100s of meters (Davis, 1994; Ehlers, 1998). They occur along barred and non-barred sea shores, and within estuaries. Along non-tidal or microtidal shores with less than 0.5 m tidal range, "pseudo-tidal" flats may be found which have much in common with true tidal flats, but lack the regular tidal rhythm. Here the inundation interval is controlled by wind set-up, the resulting depositional environment being known as "wind flats". Back-barrier tidal flats are limited to tidal ranges between about 0.5 m and 3.5 m, the transition to nonbarred tidal flats being controlled by the interaction between the tidal range and wave action (e.g., Hayes, 1979). At tidal ranges above 3.5 m tidal flats are always nonbarred, irrespective of the wave climate. Extreme macrotidal ranges occur where the length of a shallow, funnel-shaped bay approximates a quarter wavelength of the tidal wave. In such cases, the natural oscillation of the water body resonates with the tidal period and the tidal wave is hence superimposed on a standing wave. This results in strong amplification of tidal water levels at the head of such bays as, for example, in the Bay of Fundy, Canada (Dalrymple and Zaitlin, 1989), in the Severn Estuary, Great Britain (Allen, 1990), or in the Bay of Mont St Michel, France (Larsonneur, 1994). French (1997) distinguishes no less than seven intertidal coastal types based mainly on morphology and facies successions.

Sedimentology, stratigraphy, and facies associations

Tidal flats occur in both siliciclastic and carbonate depositional systems (e.g., Ginsburg, 1975) and have been documented in the rock record since Archaean times over 3,300 Ma ago (e.g., Hobday and Eriksson, 1977). Important benchmark papers in tidal flat research published before 1970 can be found in Klein (1976). Following the settling lag/scour lag mechanism along the tidal energy gradient (Van Straaten and Kuenen, 1957), grain sizes progressively fine toward the shore. The result is a sedimentary facies succession that has recently been classified on the basis of mud content (cf. Flemming, 2000). It commences with sand flats (<5 percent mud) at the high-energy end near the low-tide level, and proceeds through slightly muddy sand flats (5–25 percent mud), muddy sand flats (25–50 percent mud), sandy mud flats (50–75 percent mud), slightly sandy mud flats (75–95 percent mud), to mud flats (>95 percent mud) at the low-energy end (high-tide level). Sand flats can usually be further subdivided into a variety of size classes, the total range depending on overall sediment supply and energy flux. Numerous aspects of mudflat properties and processes, including a typological classification of mud flats and a discussion of mass physical properties, can be found in Dyer (2000).

Along the energy gradient, tidal flats display numerous physical and biological sedimentary structures, both on the surface and in cross-section (e.g., Klein, 1977; Reineck and Singh, 1980), only a few (such as tidal rhythmites) being unequivocally diagnostic (e.g., Visser, 1980). Exposed sandy tidal flats are dominated by physical structures generated by waves and currents (complex wave and current ripple structures, dune cross-bedding, upper plane bed lamination, fluid escape structures, etc.). Strong bioturbation is rare in sand flats. As energy levels decrease, burrowing organisms such as polychaete worms and molluscs in temperate climates or callianassid shrimps in tropical and subtropical climates begin to bioturbate the sediment, the degree of bioturbation generally increasing toward the finer-grained sediment facies. Mud enters the sedimentary system in the form of either sortable silt particles or flocs and aggregates (including faecal pellets) composed of finer-grained components (generally <10 µm). Deposition of the low-density mud flocs and other aggregates follows the principle of hydraulic equivalence as reflected in the progressive increase in mud content along the energy gradient (Flemming, 2000). As a result, muddy intertidal sediments are actually well sorted. At the same time, primary sedimentary structures such as mud-draped ripple cross-beds evolve into flaser, wavy, and lenticular bedding units before merging into horizontally laminated muds at the landward margin. Contorted and convoluted mud beds with mud-cracked surfaces appear at the transition to the lower supratidal environment as desiccation at neap tide promotes compaction, while storm-generated silt layers may be interspersed along more exposed parts of the system. Above the mean high-tide level, root-structures of salt-marsh plants begin to intensely bioturbate the sediment.

Tidal flat depositional systems evolve in response to the longer-term sediment budget which, in turn, is a function of the rate of sediment supply from external sources and the rate of sea-level rise (or fall). Thus, if the sediment supply exceeds the deficit created by sea-level rise, the system will prograde seawards while aggrading. If both are balanced, then the system will simply aggrade in place. On the other hand, if the deficit exceeds the supply, then reworking along the seaward margin results in transgressive landward displacement of the individual facies belts (Figure T9).

Biological aspects

Besides the influence of physical parameters such as currents, wave action and grain size, the nature of intertidal deposits is strongly affected by the presence or absence of particular animal and plant associations. In recent years, the importance of microbial communities has received increasing attention (e.g., Krumbein et al., 1994). The nature of biological influence on tidal-flat depositional systems depends on species composition and community structure which both change with geographic locality and climatic setting (e.g., Chapman, 1974; Reise, 2001). As a result, tidal flats in different parts of the world can often be distinguished on the basis of characteristic biological surface structures and bioturbation patterns. A systematic study of biologically generated

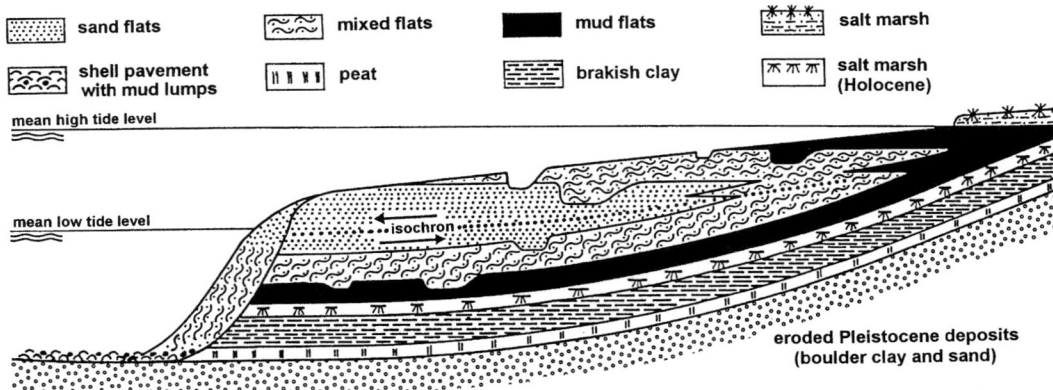

Figure T9 Typical facies associations and stratigraphy of a nonbarred tidal flat depositional system of the southern North Sea (modified after Reineck and Singh, 1980). Below the line marked "isochron" the sequence is transgressive (retrogradational), above the line it is regressive (progradational).

sedimentary structures in the Wadden Sea can be found in Schäfer (1972). Furthermore, the transition from upper intertidal to lower supratidal environments is characterized by well-developed microbial mats and distinct floral zones. In tropical regions these are associated with mangrove, in temperate regions with salt marsh plant communities (e.g., Allen and Pye, 1992; Augustinus, 1995).

Outlook

Tidal deposits, both modern and ancient, have remained a focus of ongoing research to this day the world over. Up to date summaries of this work can be found in the proceedings and field trip guides of the 4-yearly international tidalite conferences held in different parts of the world since 1985 (De Boer, Van Gelder and Nio, 1988; Flemming and Hertweck, 1994; Flemming and Bartholomä, 1995; Alexander, Davis, and Henry, 1998; Park and Davis, 2001). An aspect that requires more attention in the future is the integration of different research fields, especially sedimentology, hydrology, geochemistry, biogeochemistry, microbiology, paleontology, and ecology.

Burghard W. Flemming

Bibliography

Alexander, C.R., Davis, R.A., and Henry, V.J. (eds.) 1998. *Tidalites: Processes and Products*. SEPM, Special Publication, 61. Tulsa: Society for Sedimentary Geology.

Allen, J.R.L., 1990. The Severn Estuary in Southwest Britain: its retreat under marine transgression, and fine sediment regime. *Sedimentary Geology*, 66: 13–28.

Allen, J.R.L., and Pye, K. (eds.), 1992. *Saltmarshes: Morphodynamics, Conservation, and Engineering Significance*. Cambridge University Press.

Allen, P.A., 1997. *Earth Surface Processes*. Blackwell Science.

Archer, A.W., Kvale, E.P., and Johnson, H.R., 1991. Analysis of modern equatorial tidal periodicities as a test of information encoded in ancient tidal rhythmites. In Smith, D.G., Reinson, G.E., Zaitlin, B.A., and Rahmani, R.A., (eds.), *Clastic Tidal Sedimentology*. Canadian Society of Petroleum Geologist, Memoir 16. pp. 189–196.

Augustinus, P.G.E.F., 1995. Geomorphology and sedimentology of mangroves. In Perillo, G.M.E. (ed.), *Geomorphology and Sedimentology of Estuaries*. Elsevier Science, pp. 333–357.

Bates, R.L., and Jackson, J.A. (eds.), 1987. *Glossary of Geology*, 3rd edn. American Geological Institute.

Chapman, V.J., 1974. *Salt Marshes and Salt Deserts of the World*, 2nd edn. Lehre: Cramer.

Dalrymple, R.W., and Zaitlin, B.A., 1989. *Tidal deposits in the macrotidal Cobequid Bay–Salmon River estuary, Bay of Fundy*. Canadian Society of Petroleum Geologists, Field Guide, Second International Research Symposium on Clastic Tidal Deposits, August 22–25, Calgary, Alberta, 84p.

Davies, J.L., 1964. A morphogentic approach to world shorelines. *Zeitschrift für Geomorphologie*, 8: 127–142.

Davis, R.A. Jr, (ed.) 1994. *Geology of Holocene Barrier Island Systems*. Berlin: Springer.

De Boer, P.L., Van Gelder, A., and Nio, S.D. (eds.), 1988. *Tide-influenced Sedimentary Environments and Facies*. Dordrecht: D. Reidel.

De Boer, P.L., Oost, A.P., and Visser, M.J., 1989. The diurnal inequality of the tide as a parameter for recognizing tidal influences. *Journal of Sedimentary Petrology*, 9: 912–921.

Dyer, K.R. (ed.), 2000. *Intertidal Mudflats: Properties and Processes. Part I: Mudflat Properties. Part II: Mudflat Processes*. Special Issue, Continental Shelf Research 20, (10/11/12/13), pp. 1037–1788.

Ehlers, J., 1988. *The Morphodynamics of the Wadden Sea*. Rotterdam: A. A. Balkema.

Flemming, B.W., 2000. A revised textural classification of gravel-free muddy sediments on the basis of ternary diagrams. *Continental Shelf Research*, 20: 1125–1137.

Flemming, B.W., and Bartholomä, A. (eds.) 1995. *Tidal Signatures in Modern and Ancient Sediments*. International Association of Sedimentologists, Special Publication, 24. Blackwell Science.

Flemming, B.W., and Hertweck, G. (eds.), 1994. Tidal flats and barrier systems of continental Europe: a selective overview. *Senckenbergiana maritima*, 24: 1–209.

French, P.W., 1997. *Coastal and Estuarine Management*. London: Routledge.

Ginsburg, R.N. (ed.), 1975. *Tidal Deposits—A Casebook of Recent Examples and Fossil Counterparts*. Springer.

Hayes, M.O., 1979. Barrier island morphology as a function of tidal and wave regime. In Leatherman, S.P., (ed.), *Barrier Islands*. Academic Press, pp. 1–27.

Hobday, D.K., and Eriksson, K.A. (eds.), 1977. Tidal sedimentation with special reference to South African examples. *Sedimentary Geology*, 18: 1–356.

Klein, G. de Vries, 1976. *Holocene Tidal Sedimentation*. Stroudsburg: Dowden, Hutchinson & Ross, Benchmark Papers in Geology, 30.

Klein, G. de Vries, 1977. *Clastic Tidal Facies*. Champaign: CEPCO.

Krumbein, W.E., Paterson, D.M., and Stal, L.J. (eds.), 1994. *Biostabilization of Sediments*. Oldenburg: BIS-Verlag.
Larsonneur, C., 1994. The bay of Mont-Saint-Michel: a sedimentation model in a temperate macrotidal environment. *Senckenbergiana maritima*, **24**: 3–63.
Oost, A.P., de Haas, H., Ijnsen, F., van den Boogert, J.M., and de Boer, P.L., 1993. The 18.6 year nodal cycle and its impact on tidal sedimentation. *Sedimentary Geology*, **87**: 1–11.
Park, Y.A., and Davis, R.A. Jr (eds.), 2001. *Proceedings of Tidalites 2000*. Seoul: Korean Society of Oceanography.
Reineck, H.-E., and Singh, I.B., 1980. *Depositional Sedimentary Environments*, 2nd edn. Springer-Verlag.
Reise, K., 2001. *Ecological Comparisons of Sedimentary Shores*. Berlin: Springer-Verlag.
Schäfer, W., 1972. *Ecology and Palaeoecology of Marine Environments*. Edinburgh: Oliver & Boyd.
Smith, D.G., Reinson, G.E., Zaitlin, B.A., and Rahmani, R.A. (eds.), 1991. *Clastic Tidal Sedimentology*. Canadian Society of Petroleum Geologists, Memoir 16.
Van Straaten, L.M.J.U., and Kuenen, P.H., 1957. Accumulation of fine grained sediments in the Dutch Wadden Sea. *Geologie en Mijnbouw*, **19**: 329–354.
Visser, R., 1980. Neap-spring cycles reflected in Holocene subtidal large-scale bedform deposits: a preliminary note. *Geology*, **8**: 543–546.
Williams, G.E., 1991. Upper Proterozoic tidal rhythmites, South Australia: sedimentary features, deposition, and implications for the earth's paleorotation. In Smith, D.G., Reinson, G.E., Zaitlin, B.A., and Rahmani, R.A. (eds.), *Clastic Tidal Sedimentology*. Canadian Society of Petroleum Geologists, Memoir 16, pp. 161–178.

Cross-references

Barrier Islands
Bedding and Internal Structures
Biogenic Sedimentary Structures
Coastal Sedimentary Facies
Flaser
Grain Settling
Grain Size and Shape
Ripple, Ripple Mark, and Ripple Structure
Salt Marshes
Sediment Transport by Tides
Surface Forms
Tides and Tidal Rhythmites

TIDAL INLETS AND DELTAS

A *tidal inlet* is an opening in the shoreline through which water penetrates the land thereby providing a connection between the ocean and bays, lagoons, marsh, and tidal creek systems. Tidal currents maintain the main channel of a tidal inlet. The second half of this definition distinguishes tidal inlets from large, open embayments or passageways along rocky coasts. Tidal currents at inlets are responsible for the continual removal of sediment dumped into the main channel by wave action. Thus, according to this definition tidal inlets occur along sandy or sand and gravel barrier coastlines, although one side may abut a bedrock headland. Some tidal inlets coincide with the mouths of rivers (estuaries) but in these cases inlet dimensions and sediment transport trends are still governed, to a large extent, by the volume of water exchanged at the inlet mouth (tidal prism) and the reversing tidal currents, respectively.

Tidal inlets are found along barrier coastlines throughout the world. They provide a pathway for ships and small boats to travel between the open ocean to sheltered waters. Diversity in the morphology, hydraulic signature, and sediment transport patterns of tidal inlets attests to the complexity of their processes. The variability in oceanographic, meteorological, and geologic parameters, such as tidal range, wave energy, sediment supply, storm magnitude and frequency, fresh water influx, and geologic controls, and the interactions of these factors, are responsible for this wide range in tidal inlet settings. At most inlets over the long-term, the volume of water entering the inlet during the flooding tide equals the volume of water leaving the inlet during the ebbing cycle. This volume is referred to as the *tidal prism*. The tidal prism is a function of the open water area and tidal range in the backbarrier as well frictional factors, which govern the ease of flow through the inlet.

The formation of a tidal inlet requires the presence of an embayment and the development of barriers. In coastal plain settings, the embayment or backbarrier was often created through the construction of the barriers themselves, like much of the East Coast of the United States or the Friesian Island coast along the North Sea (FitzGerald and Penland, 1987; FitzGerald, 1996). Breaching of a barrier and spit accretion across a bay are the common mechanisms of inlet formation. In other instances, flooding of former river valleys has also produced embayments associated with tidal inlet development.

Tidal deltas

A tidal inlet is specifically the area between two barriers or between the barrier and the adjacent bedrock or glacial headland. The deepest part of an inlet, the *inlet throat*, is located normally where spit accretion of one or both of the bordering barriers constricts the inlet channel to a minimum width and minimum cross sectional area. Here tidal currents normally reach their maximum velocity. Commonly, the strength of the currents at the throat causes sand to be removed from the channel floor leaving behind a lag deposit consisting of gravel or shells or in some locations exposed bedrock or indurated sediments. Closely associated with tidal inlets are sand shoals and tidal channels located on the landward and seaward sides of the inlets. Flood tidal currents deposit sand landward of the inlet forming a *flood-tidal delta* and ebb-tidal currents deposit sand on the seaward side forming an *ebb-tidal delta*.

Flood-tidal deltas

Their presence or absence, size, and development are related to a region's tidal range, wave energy, sediment supply and backbarrier setting. Tidal inlets that are backed by a system of tidal channels and salt marsh (mixed-energy coast) usually contain a single horseshoe-shaped flood-tidal delta (e.g., Essex River Inlet, Massachusetts; Figure T10). Contrastingly, inlets that are backed by large shallow bays may contain multiple flood-tidal deltas. Along some microtidal coast, such as Rhode Island, flood deltas form at the end of narrow inlet channels cut through the barrier. Changes in the locus of deposition at these deltas produce a multi-lobate morphology resembling a lobate river delta (Boothroyd *et al*., 1985). Flood delta size commonly increases as the amount of open water area in the backbarrier increases.

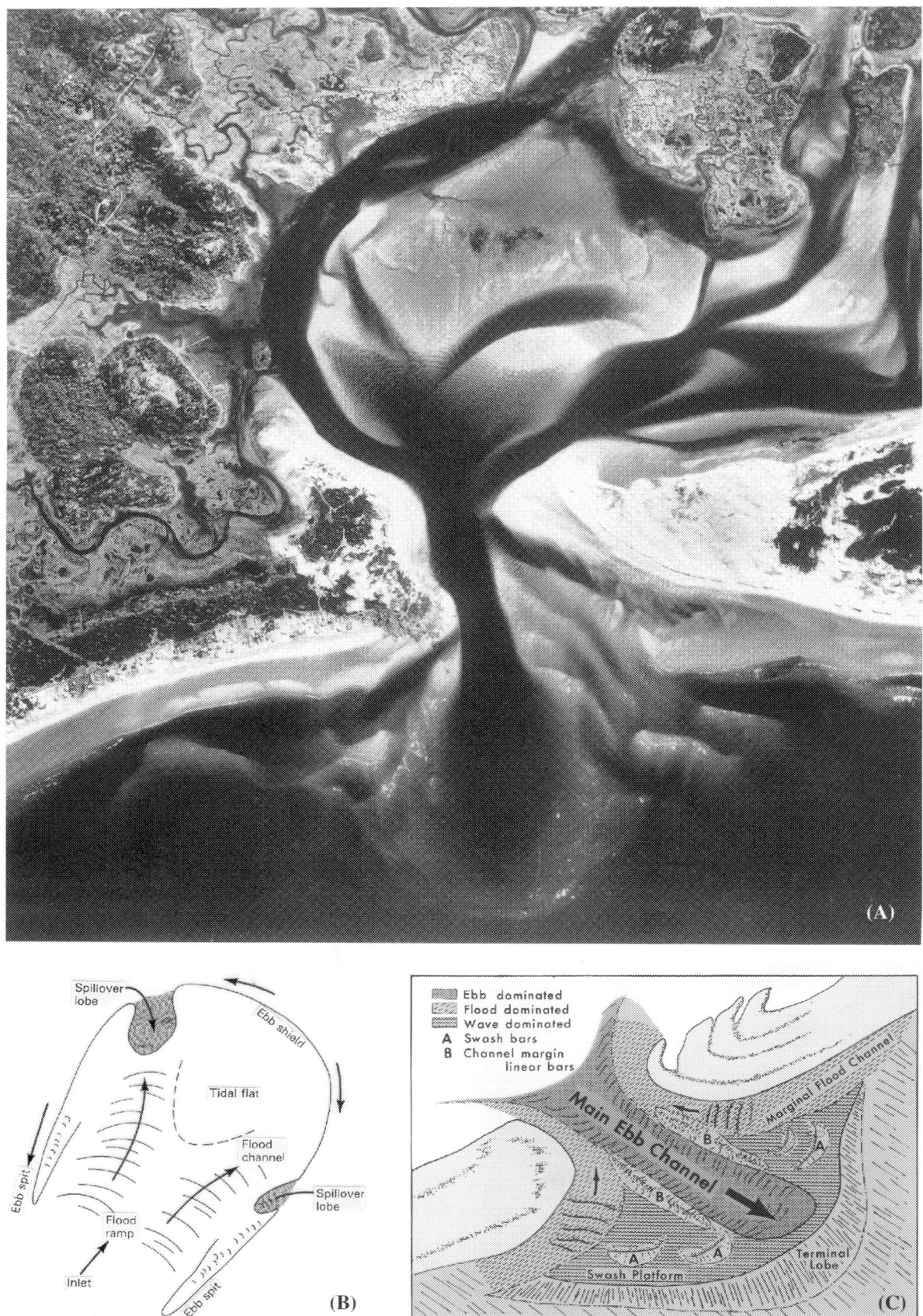

Figure T10 (A) Vertical aerial photograph of Essex River Inlet, Massachusetts with well developed flood and ebb-tidal deltas. (B) Flood-tidal delta model (from Hayes, 1975). (C) Ebb-tidal delta model (from FitzGerald *et al.*, 1976, after Hayes, 1975).

Flood-tidal deltas are best revealed in areas with moderate to large tidal ranges (1.5–3.0 m) because in these regions they are well exposed at low tide. As tidal range decreases, flood deltas become largely sub-tidal shoals. Most flood-tidal deltas have similar morphologies consisting of the following components (Hayes, 1979). *Flood ramp* is a landward shallowing channel that slopes upward toward the intertidal portion of the delta. Strong flood-tidal currents and landward sand transport in the form of landward-oriented sandwaves dominate the ramp. The flood ramp splits into two shallow *flood channels*. Like the flood ramp, these channels are dominated by flood-tidal currents and flood-oriented sand waves. Sand is delivered through these channels onto the flood delta. *Ebb shield* defines the highest and landwardmost part of the flood delta and may be partly covered by marsh vegetation. It shields the rest of the delta from the effects of the ebb-tidal currents. *Ebb spits* extend from the ebb shield toward the inlet. They form from sand that is eroded from the ebb shield and transported back toward the inlet by ebb-tidal currents. *Spillover lobes* are deposits of sand that form where the ebb currents have breached through the ebb spits or ebb shield depositing sand in the interior of the delta.

Through time, some flood-tidal deltas accrete vertically and/or grow in size. This is evidenced by an increase in areal extent of marsh grasses, which require a certain elevation above mean low water to exist. At migrating inlets new flood-tidal deltas are formed as the inlet moves along the coast and encounters new open water areas in the backbarrier. At most stable inlets, however, sand comprising the flood delta is simply recirculated. The transport of sand on flood deltas is controlled by the elevation of the tide and the strength and direction of the tidal currents. During the rising tide, flood currents reach their strongest velocities near high tide when the entire flood-tidal delta is covered by water. Hence, there is a net transport of sand up the flood ramp, through the flood channels and onto the ebb shield. Some of the sand is moved across the ebb shield and into the surrounding tidal channel. During the falling tide, the strongest ebb currents occur near mid to low water. At this time, the ebb shield is out of the water and diverts the currents around the delta. The ebb currents erode sand from the landward face of the ebb shield and transport it along the ebb spits and eventually into the inlet channel where once again it will be moved onto the flood ramp thus completing the sand gyre (Boothroyd, 1985).

Ebb-tidal deltas

Ebb-tidal deltas are accumulations of sand that has been deposited by ebb-tidal currents and which have been modified subsequently by waves and tidal currents. Ebb deltas exhibit a variety of forms dependent on the relative magnitude of wave and tidal energy of the region as well as geological controls. Along mixed energy coasts, most ebb-tidal deltas contain a set of general features. *Main ebb channel* is a seaward shallowing channel that is scoured in the ebb-tidal delta sands. It is dominated by ebb-tidal currents. Sediment transported out the main ebb channel is deposited in a lobe of sand forming the *terminal lobe*. The deposit slopes relatively steeply on its

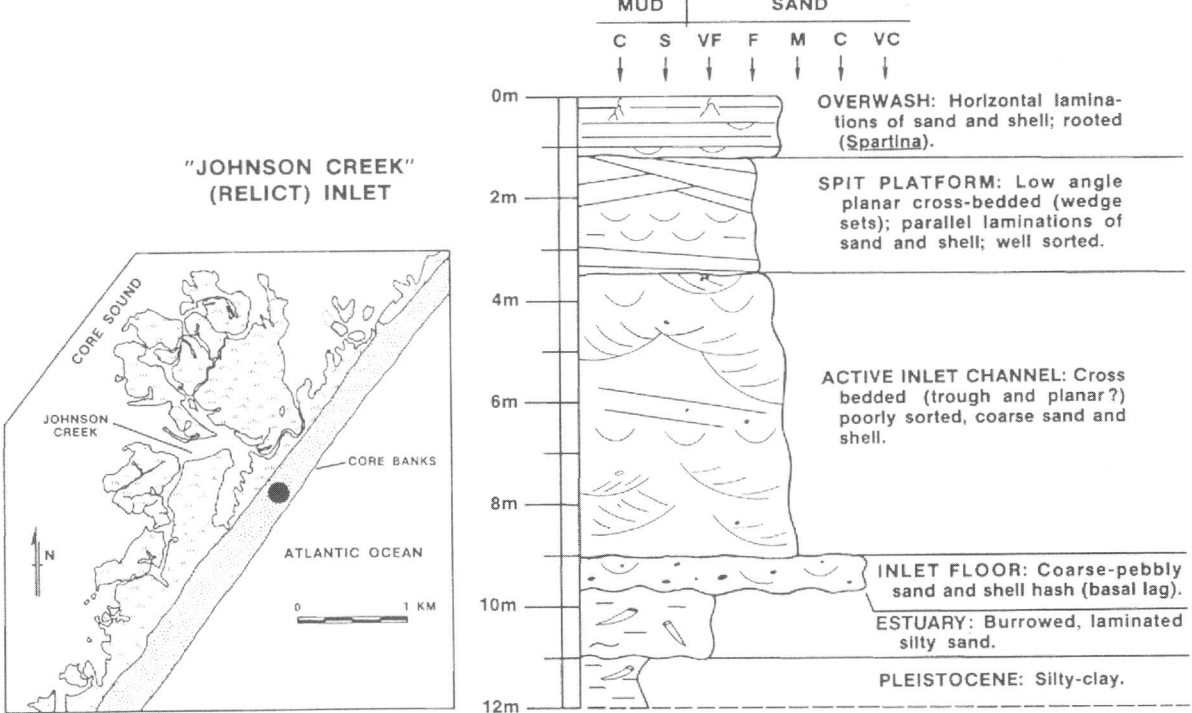

Figure T11 Sedimentary sequence through a former tidal inlet along Core Bank, North Carolina as determined from a sediment core (from Moslow and Tye, 1985). This model does not depict the large-scale accretionary beds that have been observed in ground-penetrating radar profiles across tidal inlet fills (FitzGerald et al., 2001).

seaward side. The outline of the terminal lobe is well defined by breaking waves during storms or periods of large wave swell at low tide. *Swash platform* is a broad shallow sand platform located on both sides of the main ebb channel, defining the general extent of the ebb delta. *Channel margin linear bars* border the main ebb channel and sit atop the swash platform. These bars tend to confine the ebb flow and are exposed at low tide. Waves breaking over the terminal lobe and across the swash platform form arcuate-shaped *swash bars* that migrate onshore. The bars are usually 50–150 m long, 50 m wide, and 1–2 m in height. *Marginal-flood channels*, the shallow channels (0–2 m deep at mean low water) located between the channel margin linear bars and the onshore beaches, are dominated by flood-tidal currents.

The general shape of an ebb-tidal delta and the distribution of its sand bodies tell us about the relative magnitude of different sand transport processes operating at a tidal inlet (FitzGerald, 1988). Ebb-tidal deltas that have elongated main ebb channel and channel margin linear bars that extend far offshore are tide- dominated inlets. Wave-generated sand transport plays a secondary role in modifying delta shape at these inlets. Because most sand movement is in onshore-offshore direction, the ebb-tidal delta overlaps a relatively small length of inlet shoreline. This has important implications concerning the extent to which the inlet shoreline undergoes erosional and depositional changes.

Wave-dominated inlets tend to be small relative to tide-dominated inlets. Their ebb-tidal deltas are driven onshore, close to the inlet mouth by the dominant wave processes. Commonly, the terminal lobe and/or swash bars form a small arc outlining the periphery of the delta. In many cases the ebb-tidal delta of these inlets is entirely subtidal. In other instances, sand bodies clog the entrance to the inlet leading to the formation of several major and minor tidal channels. At mixed energy tidal inlets the shape of the delta is the result of tidal and wave processes. These deltas have a well-formed main ebb channel, which is a product of ebb-tidal currents. Their swash platform and sand bodies substantially overlap the inlet shoreline many times the width of the inlet throat due to wave processes and flood-tidal currents. Ebb-tidal deltas may also be highly asymmetric such that the main ebb channel and its associated sand bodies are positioned primarily along one of the inlet shorelines. This configuration normally occurs when the major backbarrier channel approaches the inlet at an oblique angle or when a preferential accumulation of sand on the updrift side of the ebb delta causes a deflection of the main ebb channel along the downdrift barrier shoreline.

Sand transport patterns

The movement of sand at a tidal inlet is complex due to reversing tidal currents, effects of storms, and interaction with the longshore transport system. The inlet contains short-term and long-term reservoirs of sand varying from the relatively small sandwaves flooring the inlet channel that migrate meters each tidal cycle to the large flood-tidal delta shoals where some sand is recirculated but the entire deposit may remain stable for hundreds or even thousands of years. Sand dispersal at tidal inlets is complicated because in addition to the onshore-offshore movement of sand produced by tidal and wave-generated currents, there is constant delivery of sand to the inlet and transport of sand away from the inlet produced by the longshore transport system. In the discussion below the patterns of sand movement at inlets are described including how sand is moved past a tidal inlet.

The ebb-tidal delta has segregated areas of landward versus seaward sediment transport that are controlled primarily by the way water enters and discharges from the inlet as well as the effects of wave-generated currents. During the ebbing cycle the tidal flow leaving the backbarrier is constricted at the inlet throat causing the currents to accelerate in a seaward direction. Once out of the confines of the inlet, the ebb flow expands laterally and the velocity slows. Sediment in the main ebb channel is transported in a net seaward direction and eventually deposited on the terminal lobe due to this decrease in current velocity. One response to this seaward movement of sand is the formation of ebb-oriented sandwaves (dunes) having heights of 1–2 m.

In the beginning of the flood cycle, the ocean tide rises while water in main ebb channel continues to flow seaward as a result of momentum. Due to this phenomenon, water initially enters the inlet through the marginal flood channels, which are the pathways of least resistance. The flood channels are dominated by landward sediment transport and are floored by flood-oriented bedforms. On both sides of the main ebb channel, the swash platform is most affected by landward flow produced by the flood-tidal currents and breaking waves. As waves shoal and break, they generate landward flow, which augments the flood tidal currents but retards the ebb tidal currents. The interaction of these forces acts to transport sediment in a net landward direction across the swash platform. In summary, at many inlets there is a general trend of seaward sand transport in the main ebb channel, which is countered by landward sand transport in the marginal flood channels and across the swash platform.

Preservation of tidal inlets sequences

A typical sandy barrier-inlet model of sedimentary facies succession grades upward from coarse channel sand, and in some cases fine gravel, into medium to fine beach and dune sand (Reinson, 1984). The fining upward sequence characterizing tidal inlets contrasts with coarsening upward facies of most barriers. Tidal inlet and tidal delta facies have been described in the rock record by Barwis and Makurath (1978). Channel-fill facies and large-scale accretionary bedding resulting from inlet migration characterize tidal inlet sequences. Small scale bi-directional stratification is also typical of inlet facies. Tidal inlet deposits are distinguished from riverine and tidal channel facies by their discontinuous nature in dip section. Because tidal inlets commonly scour to depths of 8 m and deeper, their fill sequences have a high preservation potential. Ebb-tidal delta facies sit atop the shoreface and have a low preservation potential. The exception to this trend are fill sequences associated with migrations of the main ebb channel, which may be partially preserved due to the depth of channel scour. Flood-tidal deltas may be preserved as a series of stacked sand lobes (1–3 m thick) dominated by landward dipping stratification (Boothroyd, 1985).

Migrating inlets may account for a net loss of sand from the littoral system. If the inlet channel scours below the base of the barrier sands, then beach sand, which fills the channel, will not be entirely compensated by the deposits excavated on the eroding side of the channel. Because up to 40 percent of the length of barriers is underlain by tidal inlet fill deposits ranging in thickness from 2 m to 10 m (Moslow and Heron, 1978;

Moslow and Tye, 1985) this volume represents a large, long-term loss of sand from the coastal sediment budget.

Duncan M. FitzGerald and Ilya V. Buynevich

Bibliography

Barwis, J., and Makurath, J.H., 1978. Recognition of ancient tidal inlet sequences: an example from the Upper Silurian Keyer Limestone in Virginia. *Sedimentology*, **25**: 61–82.
Boothroyd, J.C., 1985. Tidal inlets and tidal deltas. In Davis, R.A. Jr (ed.), *Coastal Sedimentary Environments*. Spinger-Verlag, p. 445–532.
Boothroyd, J.C., Friedrich, N.E., and McGinn, S.R., 1985. Geology of microtidal coastal lagoons, RI. In Oertel, G.F., and Leatherman, S.P. (eds.), Barrier Islands. *Marine Geology*. Volume 63, pp. 35–76.
FitzGerald, D.M., 1988. Shoreline erosional-depositional processes associated with tidal inlets. In Aubrey, D.G., and Weishar, L. (eds.), *Hydrodynamics and Sediment Dynamics of Tidal Inlets*. Springer, pp. 186–225.
FitzGerald, D.M., 1996. Geomorphic variability and morphologic and sedimentological controls on tidal inlets. In Mehta, A.J. (ed.) *Understanding Physical Processes at Tidal Inlets. Journal of Coastal Research Special Issue*, 23, pp. 47–71.
FitzGerald, D.M., Nummedal, D., and Kana, T., 1976. *15th Coastal Engineering Conference, Honolulu*. American Society of Civil Engineers, pp. 1868–1880.
FitzGerald, D.M., Buynevich, I.V., and Rosen, P.S., 2001. Geological evidence of former tidal inlets along a retrograding barrier: Duxbury Beach, Massachusetts, USA. *Journal of Coastal Research*, SI 34, (Healy, T.R, (ed.), 437–448).
FitzGerald, D.M., and Penland, S., 1987. Backbarrier dynamics of the East Friesian Island. *Journal of Sedimentary Petrology*, **57**: 746–754.
Hayes, M.O., 1975. Morphology of sand accumulations in estuaries. In Cronin, L.E. (ed.), *Estuarine Research*, Volume 2, Academic Press, pp. 3–22.
Hayes, M.O., 1979. Barrier island morphology as a function of tidal and wave regime. In Leatherman, S.P. (ed.), *Barrier Islands: From the Gulf of St. Lawrence to the Gulf of Mexico*. Academic Press, pp. 1–28.
Moslow, T.F., and Heron, S.D., 1978. Relict inlets: preservation and occurrence in the Holocene stratigraphy of southern Core Banks, North Carolina. *Journal of Sedimentary Petrology*, **48**: 1275–1286.
Moslow, T.F., and Tye, R.S., 1985. Recognition and characteristic of Holocene tidal inlet sequences. *Marine Geology*, **63**: 129–151.
Reinson, G.E., 1984. Barrier island and associated strand-plain systems. In Walker, R.G. (ed.), *Facies Models*. Geoscience Canada Reprint Series 1, pp. 119–140.

Cross-references

Sediment Transport by Tides
Tidal Flats

TIDES AND TIDAL RHYTHMITES

Introduction

Oceanic tides are the regular (periodic) daily rise and fall of the oceans; a direct consequence of the gravitational attraction of the Moon and Sun on the Earth. Tidal rhythmites, as used herein, include small-scale primary sedimentary structures consisting of: (1) stacked sets of clay-draped ripples that as cosets can be classified as lenticular, wavy, and flaser bedding; and (2) flat-laminated siltstone and silty claystone with thin intercalated claystone layers. In the rock record the tidal influence on the origin of these features is indicated by the very regular progressive vertical thickening and thinning of individually accreted sets in response to changing current velocities associated with lunar or lunar-solar tidal cycles (Kvale *et al.*, 1999). Recognition of the changing set thickness as a function of these cycles is important because a repetitive sequence of similar-appearing bedding can be generated via processes other than tidal (e.g., lacustrine varves).

What generates tides?

To understand oceanic tides and tidal rhythmites it is sometimes useful to think in terms of purely astronomical tides and equilibrium tidal theory. By definition, equilibrium tides are ideal and defined by the tractive gravitational forces of the Moon, and to a lesser extent the Sun, on an idealized Earth completely covered by deep water of uniform depth that is capable of instantly responding to changes in tractive forces (MacMillan, 1966). In the equilibrium model, the tidal forces from the Moon and Sun in combination with centrifugal forces associated with the spin of the Earth produce oceanic bulges on opposite sides of the Earth (Figure T12(A)). The bulges tend to more closely follow the motion of the Moon as the Moon accounts for approximately 70 percent of the tide-raising forces because of its closer proximity to the Earth than the Sun. The rotation of the Earth through each of these bulges once a day produces two tides a day (the semidiurnal tide). As a point on the Earth's ocean enters the bulge, tides rise (flood tide), and as that point moves out of the bulge, tides fall (ebb tide). The relative difference in elevation between the heights of the flood and ebb tides is referred to as the "tidal range."

It is important to note that in the real world there are not two tidal bulges through which the Earth spins but rather water within the ocean basins is displaced and forced to rotate as discrete waves or bulges about a series of fixed (amphidromic) points by a combined force generated by the spin of the Earth and the gravitational forces of the Moon and Sun (Figure T13). Because of the Coriolis effect, these discrete tidal bulges or waves typically rotate about the amphidromic point in a counterclockwise direction in the northern hemisphere and clockwise in the southern hemisphere. In an open-ocean semidiurnal tidal system, a tidal bulge completes one rotation about an amphidromic point every 12.42 hours.

As the open-ocean tides approach continental shelf areas, tidal energy is concentrated and tidal ranges generally increase independent of latitude. In some cases, the geometry of an embayed area along a coastline is such that the period of time it takes for a tide to move in and out of the basin is close to one of the astronomical tidal periods (discussed below) and significant tidal amplification takes place. An extreme example of this is the Bay of Fundy in eastern Canada where there is a significant resonant amplification of the open-ocean tides (generally 1 m or less) with tides rising and falling up to 16 m/day.

Because of the inadequacy of the equilibrium tidal theory to explain the complexities of real world tidal systems, the concept of harmonic analysis of tidal constituents was developed and should be considered when attempting to understand tides (see discussions in Pugh, 1987; and Kvale *et al.*, 1999).

Tidal cycles

The intensity or height of the daily or twice daily tides can vary in a number of ways. The semimonthly variations in daily or semidiurnal tidal heights are known as neap-spring tidal cycles. In a semidiurnal tidal system, neap-spring tidal cycles are the result of the phase changes of the Moon. Spring tides occur when the Earth, Moon, and Sun are nearly aligned at new and full Moon every 14.76 days (Figure T12(B)). Neap tides occur when the Sun and Moon are aligned at right angles from the Earth at first and third quarter phases of the Moon. The result is that spring tides are higher than neap tides (Figure T12(E)). The time it takes for the Moon to go from new Moon to new Moon is called the synodic month, which has a modern period of 29.53 days.

The tidal force also depends on the declination of the Moon, the tide-raising force at a given location being greater when the Moon is closer to the zenith. The period of the variation in lunar declination is called the tropical month and is the interval of time it takes the Moon to complete one orbit moving from its maximum northerly declination to its maximum southerly declination and then returning (Figure T12(C)). The current length of the tropical month is 27.32 days. The effect of the tropical month in most semidiurnal systems is to cause the diurnal inequality of the tides in which the morning high tide is greater or lesser than the evening high tide. Ideally, the diurnal inequality is reduced to zero when the Moon is over the equator (Figure T12(E)). However, in all dominantly diurnal systems having primarily one tide per day, tropical monthly tides are responsible for generating neap-spring cycles every 13.66 days (Figure T12(F)). In such cases, the dominant tidal force depends on the declination of the Moon, with the force being greatest (thereby generating spring tides) when the Moon is at zenith.

Because the lunar orbit is slightly elliptical, the Moon's orbital travel carries it between perigee (closest approach to the Earth) and apogee (farthest distance from the Earth) (Figure T12(D)). The period of time for the Moon to move from perigee to perigee is called the anomalistic month. The anomalistic month currently is 27.55 days. During a lunar month there are two spring tides that are often of unequal

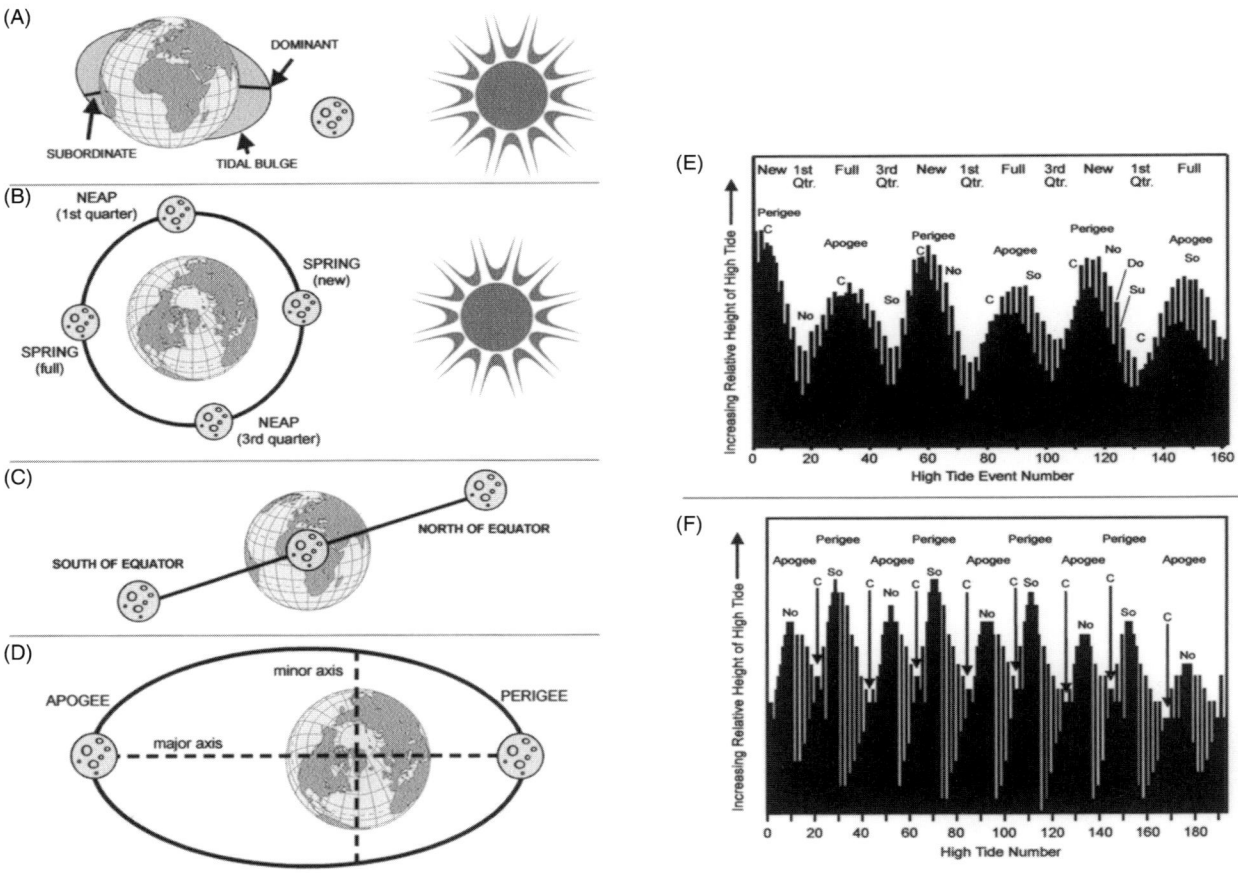

Figure T12 Idealized equilibrium tidal models that illustrate semidiurnal tides (A), the synodic month (B), the tropical month (C), and the anomalistic month (D). Figure T12E depicts a segment of the 1991 predicted high tides from Kwajalein Atoll, Pacific Ocean. Figure F depicts a segment of the 1994 predicted high tides from the Barito River, Borneo. "Su"—Subordinate semidaily tide; "Do"—Dominant semidaily tide; "C"—Equatorial passage of the Moon (Crossover); "No"—Moon at maximum northern declination; "So"—Moon at maximum southern declination. Figures modified from Kvale et al. (1998).

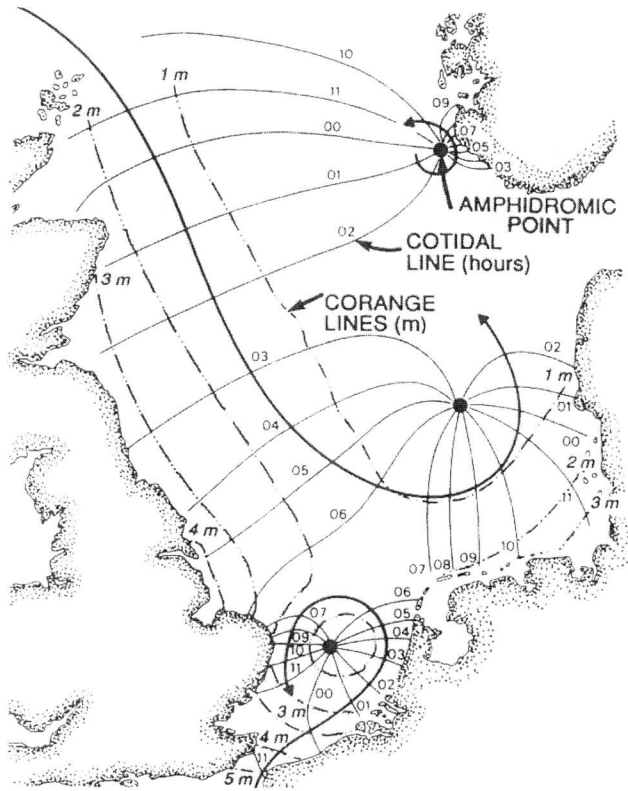

Figure T13 North Sea amphidromic tidal system. Lines of equal tidal range are labeled as "corange lines." The cotidal lines show times of high water. Note the counterclockwise rotation of the tidal wave about each amphidromic point. Modified from Dalrymple (1992).

magnitude (Figures T12(E) and (F)). These are most noticeable when spring tides occur during or near a time of lunar perigee and apogee.

Beyond the semidaily, daily, semimonthly, and monthly tidal cycles mentioned above, there are also longer period tidal cycles that can impact sedimentary processes. These include the semiannual tidal cycle and the 18.6 year lunar nodal cycle. These are discussed in detail in Kvale et al. (1999) and Miller and Eriksson (1997). Even longer multiyearly tidal periods exist but any impact on sedimentation by these longer tidal periods is undocumented.

Tidal rhythmites

Tidal rhythmites occur in mud- and tide-dominated systems and have been identified through a significant part of Earth's history. In modern settings they have been identified in subtidal delta front settings (Cowen et al., 1998) and subtidal to intertidal tidal flats associated with deltaic distributary channels (Staub et al., 2000) and estuaries (e.g., Dalrymple and Makino, 1989; Tessier, 1993). Ancient analogs to these modern rhythmite depositional systems are also known (e.g., Feldman et al., 1995; Miller and Eriksson, 1997; Brenner et al., 2000; Choi and Park, 2000). Typical of modern tidal rhythmites are tidal ranges generally greater than or equal to 3 m, although a sub-Recent example that formed in a 2 m tidal range is known (Roep, 1991); significant accommodation space; high sedimentation rates and significant suspended load; a general absence of bioturbators; and brackish to fresh water chemistries at the depositional site (typically within fluvial-tidal transition zones). Normal marine salinities, however, have been suggested for at least one heavily bioturbated ancient example (Martino and Sanderson, 1993).

A basic understanding of tidal theory allows for the interpretation of precise paleoastronomical information from tidal rhythmites (Figure T14). Because tidal rhythmites are small-scale sedimentary features, relatively small amounts of accommodation space (tens of centimeters to a few meters) allow for the build up of relatively long records of ancient semidaily to even multiyearly tidal cycles. These long tidal rhythmites tidal records have been used to interpret ancient paleoclimates (Kvale et al., 1994), paleo-ocean seiches (Archer, 1996), and even ancient Earth–Moon distances (e.g., Kvale et al., 1999).

Erik P. Kvale

Bibliography

Archer, A.W., 1996. Panthalassa: paleotidal resonance and a global paleocean seiche. *Paleoceanography*, **11**: 625–632.
Brenner, R.L., Ludvigson, G.A., Witzke, B.J., Zawistoski, A.N., Kvale, E.P., and Joeckel, R.M., 2000. Late Albian Kiowa-Skull Creek marine transgression, Dakota Formation, eastern margin of Western Interior Seaway. *Journal Sedimentary Research*, **70**: 868–878.
Choi, K.S., and Park, Y.A., 2000. Late Pleistocene silty tidal rhythmites in the macrotidal flat between Youngjong and Yongyou Islands, west coast of Korea. *Marine Geology*, **167**: 231–241.
Cowen, E.A., Cai, J., Powell, R.D., Seramur, K.C., and Spurgeon, V.L., 1998. Modern tidal rhythmites deposited in a deep-water estuary. *Geo-Marine Letter*, **18**: 40–48.
Dalrymple, R.W., 1992. Tidal Depositional Systems. In Walker, R.G., and James, N.P. (eds.), *Facies Models: Response to Sea Level Changes*, Geological Association of Canada, p.196.
Dalrymple, R.W., and Makino, Y., 1989. Description and genesis of tidal bedding in Cobequid Bay-Salmon River estuary, Bay of Fundy, Canada. In Taira, A., and Masuda, F. (eds.), *Sedimentary Facies of the Active Plate Margin*. Tokyo: Terra Publishing Co., pp. 151–177.
Feldman, H.R., Gibling, M.R., Archer, A.W., Wightman, W.G., and Lanier, W.P., 1995. Stratigraphic architecture of the Tonganoxie paleovalley fill (Lower Virgilian) in northeastern Kansas. *AAPG Bulletin*, **79**: 1019–1043.
Kvale, E.P., Fraser, G.S., Archer, A.W., Zawistoski, A., Kemp, N., and McGough, P., 1994. Evidence of seasonal precipitation in Pennsylvanian sediments of the Illinois Basin. *Geology*, **22**: 331–334.
Kvale, E.P., Johnson, H.W., Sonett, C.P., Archer, A.W., and Zawistoski, A., 1999. Calculating lunar retreat rates using tidal rhythmites. *Journal of Sedimentary Research*, **69**: 1154–1168.
Kvale, E.P., Sowder, K.H., and Hill, B.T., 1998. *Modern and ancient tides: Poster and explanatory notes*. SEPM (Society for Sedimentary Geology), and Indiana Geological Survey, Bloomington, IN.
MacMillan, D.H., 1966. *Tides*. New York: American Elsevier.
Martino, R.L., and Sanderson, D.D., 1993. Fourier and autocorrelation analysis of estuarine tidal rhythmites, lower Breathitt Formation (Pennsylvanian), eastern Kentucky, USA. *Journal of Sedimentary Petrology*, **63**: 105–119.
Miller, D.J., and Eriksson, K.A., 1997. Late Mississippian prodeltaic rhythmites in the Appalachian Basin: a hierarchical record of tidal and climatic periodicities. *Journal of Sedimentary Research*, **67**: 653–660.
Pugh, D.T., 1987. *Tides, Surges, and Mean Sea Level*. John Wiley and Sons.
Roep, Th.B., 1991. Neap-spring cycles in a subrecent tidal channel (3665 BP) at Schoorldam, NW Netherlands. *Sedimentary Geology*, **71**: 213–230.

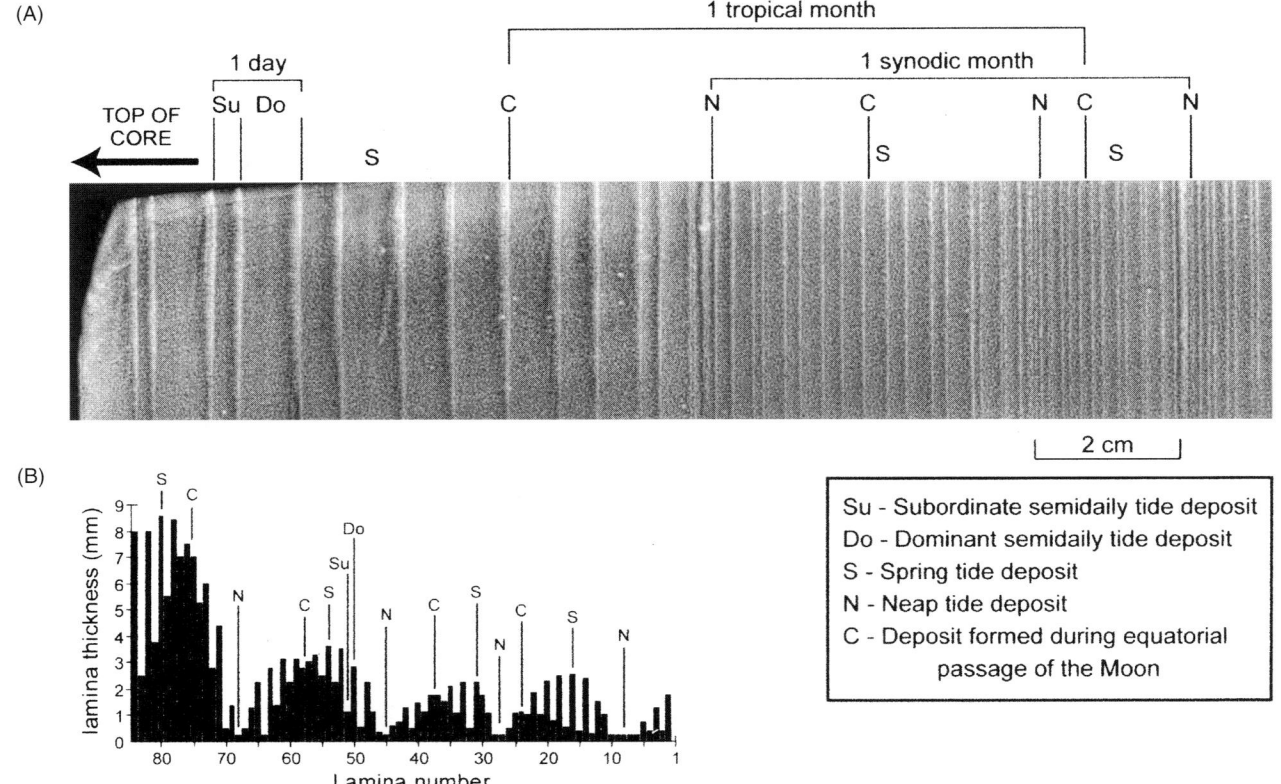

Figure T14 Photograph of a core of tidal rhythmites from the Mansfield Fm. of Indiana, USA. (A) Rock has been stained with a red vegetable dye to enhance laminae contacts. Figure T14(B) is a bar chart showing thickness variability of laminae, as well as the paleoastronomic interpretation.

Staub, J.R., Among, H.L., and Gastaldo, R.A., 2000. Seasonal sediment transport and deposition in the Rajang River delta, Sarawak, East Malaysia. *Sedimentary Geology*, **133**: 249–264.

Tessier, B., 1993. Upper intertidal rhythmites in the Mont-Saint-Michel Bay (NW France): perspectives for paleoreconstruction. *Marine Geology*, **110**: 355–367.

Cross-references

Bedding and Internal Structures
Cyclic Sedimentation
Deltas and Estuaries
Flaser
Sediment Fluxes and Rates of Sedimentation
Sediment Transport by Tides
Tidal Flats
Varves

TILLS AND TILLITES

Till (often termed *diamict* or *diamicton* today, to avoid the genetic connotation that the term till carries) has been defined by Dreimanis and Lundqvist (1984, p.9) as "*a sediment that has been transported and is subsequently deposited by or from glacier ice, with little or no sorting* (disaggregation) *by water*". However, controversy surrounds the usage and definition of the term (Hambrey and Harland, 1981; Dreimanis, 1988; Menzies and Shilts, 2002). Tills exhibit a wide range of grain size, from very poorly to well-sorted. Tills contain a variable percentage of clasts (usually sub-angular and often striated) and mineral grains that have been transported considerable distances from up-ice sources. In general, the majority of the clasts and finer particles are of local origin (<15 km up-ice). Many tills contain intraclasts of exotic sediments often stratified or laminated. Some tills are crudely bedded and have been termed stratified tills. In the past the terms deformation and comminution tills indicative of secondary mobilization were used but, today, all tills are acknowledged as having suffered some form of deformation (Ruszczyńska-Szenajch, 2001). Boulton and Deynoux (1981) introduced the concept of primary and secondary deposition applied to tills but with ancient tills such terminology is difficult to apply. Tillites are lithified tills. The AGI (1997, p.666) defines tillite as "*a consolidated or indurated sedimentary rock formed by the lithification of glacial till*". Both till and tillite are genetic terms that presuppose in their application a knowledge of the nature of the processes and mechanics of deposition.

The term till, a dialect term from Scotland, appears to have been first applied by Geikie (1863, p.185) to a "*a stiff hard clay full of stones varying in size up to boulders produced by abrasion carried on by the ice sheet as it moved over the land*". Synonymous with till are "boulder clay" and "ground moraine". Tillite seems to have been first used by Penck (1906) when describing the Dwyka Formation of South Africa.

Table T1 Classification of Tills (modified from Dreimanis, 1988: with pers comm, van der Meer, 2002)

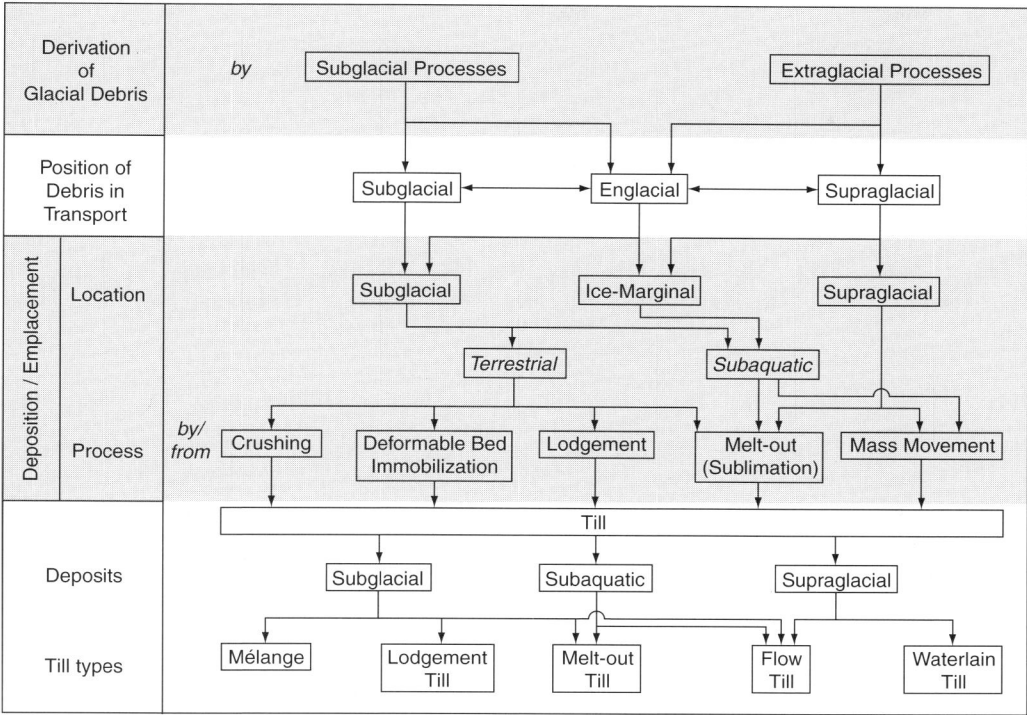

Tills may form within several distinct glacial sub-environments: subglacial, englacial, proglacial (terrestrial/subaquatic) and supraglacial; and may be deposited from active temperate wet-based and polar cold-based ice masses, under stagnant ice conditions, and subaquatic environments of both grounded and floating ice masses (Figure T15(A),(B)). Table T1 illustrates the complexity and interrelationships that exist within till-forming glacial environments.

Under wet-based, active temperate conditions considerable input from meltwater can be expected, as well as thermal fluctuations, and transient fluctuations in ice velocity and stress conditions. Subglacial tills formed under active temperate conditions are recognized as Lodgement, Flow, and Mèlange, with a variety of local till sub-types. Sub-types of tills are often identified owing to some peculiar characteristic of grain size, provenance and/or structural attribute. Where ephemeral passive ice conditions occur within active ice masses Melt-out tills can be deposited. Although comparatively rare, under dry polar climates sublimation or meltout tills can form. Both lodgement and meltout tills have been extensively described in the literature (Dreimanis, 1988; Brodzikowski and Van Loon, 1991), whilst subglacial mèlanges have only recently been recognized as separate, distinct subglacial sediments (Curry *et al.*, 1994; Hoffman and Piotrowski, 2001; Menzies and Shilts, 2002). These subglacial tills are formed as a consequence of: (1) high shear stress resulting in debris being released from basal ice in a "smearing on" process where high pervasive shear stress results in ice pressure melting and debris release (*Lodgement till*); or (2) by slow undermelt releasing debris gradually (*Melt-out till*) (Shaw, 1987); or (3) as a result of portions of a soft deforming subglacial bed either reaching shear strengths higher than the mobilizing basal ice stresses due to onset of freezing temperatures, or reduced porewater content, or porewater pressure reduction due to reduced overburden stress (Menzies, 1989; Fuller and Murray, 2000) producing *Mèlange*; and finally; (4) by the glacial crushing of basal debris and bedrock (H-bed conditions, Menzies, 1989) resulting in a form of pulverized sediment being produced (*pers.comm.* van der Meer, 2002).

Within the Supraglacial environment (Figure T15(A)) tills have been recognized as Flow tills where sediment approaching an effective stress level of zero slip, slide, and flow off the ice (Boulton, 1968); and, where temperatures and debris cover permit, sediments are directly released (melt-out till) from the ice by melting (Bouton, 1970) or, in rare arid conditions, by sublimation (Shaw, 1977). The degree to which meltwater is associated with and influences the mechanics and subsequent deposition of a melt-out till remains somewhat controversial (Rappol, 1987; Paul and Eyles, 1990).

A complex set of conditions can be found within Proglacial environments whether terrestrial or subaquatic. In terrestrial environments (Figure T15(A)) the dominant process of till formation tends to be by mass movement thus flow tills are formed, and where debris-laden ice, in some instances buried and re-exhumed in the proglacial environment, melt-out thus producing melt-out tills. Where lakes exist deposition may

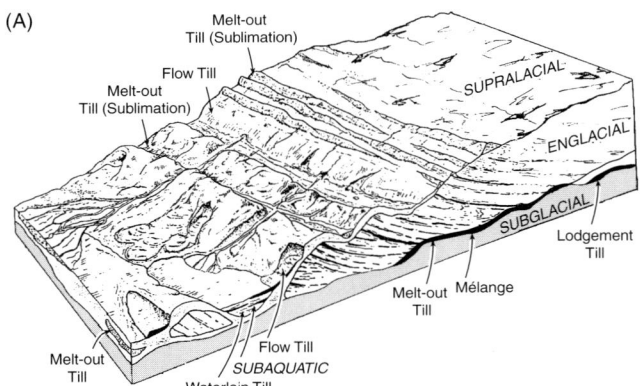

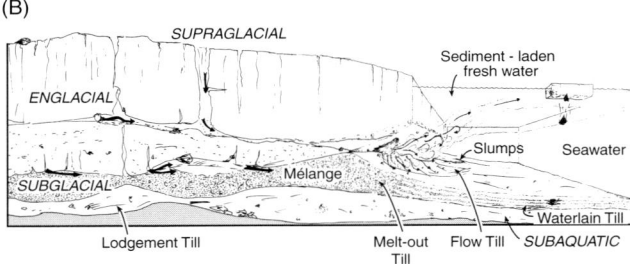

Figure T15 Locations of till types within terrestrial and equatic ice margins (adapted and revised from Menzies and Shilts, 2002).

occur by rain out through the water column from debris-laden meltwater streams or floating ice rafted debris producing *Waterlain Tills*.

As Figure T15(B) illustrates a complex set of sub-environments exists where ice masses either float or are grounded within large bodies of fresh or brackish water. Where a floating ice mass exists large amounts of debris are likely to come off the surface of the ice, from ice cliff melting, from calving ice bergs, and from englacial and subglacial meltwater tunnels. The result of these sediment release processes is to cause an enormous rain-out of sediment through the water column resulting in a Waterlain Till being deposited on the lake or ocean floor. Beneath grounded ice masses sediment will be squeezed, pushed or otherwise extruded from the subglacial contact at the ice/bed grounding line. Where debris is mobile and extruded from the grounding line a Mèlange, perhaps in the form of a "till delta or tongue", may form (Alley *et al.*, 1986; King and Fader, 1986; Anderson *et al.*, 2001).

There are several mistaken ideas concerning tills and tillites that persist, *viz.*, that tills typically exhibit a bi-modal grain size distribution, are overconsolidated, have a very low permeability, are unstratified, homogenous, massive, and unstructured, and, finally, that tills of differing colors are of different depositional phases and therefore indicative of more than one glaciation. As in all myths there are occasional truths but, as a general rule, all are essentially incorrect. Where tills can be seen forming at present the usage of the term till seems appropriate. Where, as is so often the case, either the actual means of deposition is obscured or the sediments are ancient, then the till-forming mechanisms and environments within which they were originally deposited are unknown; then the nongenetic terms diamict or diamicton for unlithified sediments and diamictite for lithified sediments may be preferred. A persistent, and perhaps unsolvable problem, remains in identifying as tills those sediments that are marine as opposed to glaciomarine. With the advent of glacial micromorphology a clearer understanding of the internal *in situ* stratigraphy and microstructures within till has great potential in furthering our understanding of this complex, and often enigmatic, glacial sediment (Van der Meer, 1996; Menzies, 2000).

John Menzies

Bibliography

Alley, R.B., Blankenship, D.D., Bentley, C.R., and Rooney, S.T., 1986. Deformation of till beneath ice stream B, West Antarctica. *Nature*, **322**: 57–59.

America Geological Institute (AGI), 1997. *Glossary of Geology*, 4th edn, Alexandria, Virginia, 769pp.

Anderson, J.B., Smith Wellner, J., Lowe, A.L., Mosola, A.B., and Shipp, S.S., 2001. Footprint of the expanded WestAntartic Ice Sheet: Ice stream history and behavior. *GSA Today*, **11** (10): 4–9.

Brodzikowski, K., and Van Loon, A.J., 1991. *Glacigenic Sediments*. Developments in Sedimentology, 49. Elsevier Science Publications, 674pp.

Boulton, G.S., 1968. Flow tills and some related deposits on some Vestspitsbergen glaciers. *Journal of Glaciology*, **7**: 391–412.

Boulton, G.S., 1970. On the deposition of subglacial and melt-out tills at the margins of certain Svalbard glaciers. *Journal of Glaciology*, **9**: 231–245.

Boulton, G.S., and Deynoux, M., 1981. Sedimentation in glacial environments and the identification of tills and tillites in ancient sedimentary sequences. *Precambrian Research*, **15**: 397–422.

Curry, B.B., Troost, K.C., and Berg, R.C., 1994. Quaternary Geology of the Martinsville Alternative Site, Clark County, Illinois. *Illinois State Geological Survey*, Circular **556**: 85pp.

Dreimanis, A., 1988. Tills: their genetic terminology and classification. In Goldthwait, R.P., & Matsch, C.L. (eds.), *Genetic Classification of Glacigenic Deposits*. A.A. Balkema, pp. 17–83.

Dreimanis, A., and Lundqvist, J., 1984. What should be called till? *Striae*, **20**: 5–10.

Fuller, S., and Murray, T., 2000. Evidence against pervasive bed deformation during the surge of an Icelandic glacier. In Maltman, A.J., Hubbard, B., and Hambrey, M.J. (eds.), *Deformation of Glacial Materials*. Geological Society of London, Special Publication, 176, pp. 203–216.

Geikie, A., 1863. On the phenomena of the Glacial Drift of Scotland. *Geological Society of Glasgow, Transactions*, **1**(2): 190pp.

Hambrey, M.J., and Harland, W.B., (eds.), 1981. *Earth's Pre-Pleistocene Glacial Record*. Cambridge University Press.

Hoffman, K., and Piotrowski, J.A., 2001. Till mèlange at Amsdorf, central Germany: sediment erosion, transport, and deposition in a complex, soft-bedded subglacial system. *Sedimentary Geology*, **140**: 215–234.

King, L.H., and Fader, G.B.J., 1986. Wisconsinan glaciation of the Atlantic continental shelf of southeast Canada. *Geological Survey of Canada Bulletin*. **363**: 72pp.

Menzies, J., 1989. Subglacial hydraulic conditions and their possible impact upon subglacial bed formation. *Sedimentary Geology*, **62**: 125–150.

Menzies, J., 2000. Micromorphological analyses of microfabrics and microstructures, indicative of deformation processes in glacial sediments. In Maltman, A.J., Hubbard, B., and Hambrey, M.J. (eds.), *Deformation of Glacial Materials*. Geological Society of London, Special Publication, 176, pp. 245–257.

Menzies, J., and Shilts, W.W., 2002. Subglacial Environments. In Menzies, J., (ed.), *Modern and Past Glacial Environments*. Butterworth-Heineman Publishers. Chapter 8, 183–278.

Paul, M.A., and Eyles, N., 1990. Constraints on the preservation of diamict facies (Melt-out tills) at the margins of stagnant glaciers. *Quaternary Science Reviews*, **9**: 51–69.
Penck, A., 1906. Süd-Afrika und Sambesifalle. *Geographiche Zeischrift*, **12**: 600–611.
Rappol, M., 1987. Saalian till in The Netherlands: a review. In van der Meer, J.J.M., (ed.), *Tills and Glaciotectonics*. A.A. Balkema Publishers, pp. 3–21.
Ruszczyńska-Szenajch, H., 2001. "Lodgement till" and "deformation till". *Quaternary Science Reviews*, **20**: 579–581.
Shaw, J., 1977. Tills deposited in arid polar environments. *Canadian Journal of Earth Sciences*, **14**: 1239–1245.
Shaw, J., 1987. Glacial sedimentary processes and environmental reconstruction based on lithofacies. *Sedimentology*, **34**: 103–116.
Van der Meer, J.J.M., 1996. Micromorphology. In Menzies, J. (ed.) *Past Glacial Environments, Sediments, Forms, and Techniques, Glacial Environments II*. Oxford: Butterworth-Heineman, pp. 335–355.

Cross-references

Classification of Sediments and Sedimentary Rocks
Deformation of Sediments
Glacial Sediments: Processes, Environments, and Facies
Mélange; Melange

TOOL MARKS

A tool mark is a mark produced by the impact against a muddy bottom of a solid object driven by a current moving over the bed. It is generally preserved as a cast, seen on the base of a sand or silt bed deposited on the muddy bottom soon after the marks have been formed. The term was originally proposed by Dzulynski and Sanders (1962, p.72). A large number of different types were named during the 1960s by geologists, mostly from observations on the soles of turbidite sandstones (see *Turbidites*; Dzulynski and Walton, 1965; Allen 1982, volume 2, chapter 13). Tool marks are also found in other environments, however, notably in fluvial deposits and marine storm deposits (see *Storm Deposits, Gutter and Gutter Cast*). They are valuable indicators of paleocurrent direction (see *Paleocurrents*) and also give some information about the nature of the clay bottom and the sediment transport mechanism operating in the flow that deposited the overlying sand.

Allen (1982, pp. 509–520) classified tool marks into: (i) *drag marks*, formed by protracted contact of a tool moved by traction; (ii) *roll marks*, formed by protracted contact of a tool rolling over the bed; (iii) *prod marks*, formed by a brief contact of a tool with the bed; (iv) *tumble marks* (also called skip marks) made by a saltating tool that produced a series of marks; (v) *skim marks*, produced by grazing contact of a tool with the bed; and (vi) marks related to near approaches, a category which includes chevron marks (see Figure T16; and see Allen, 1982, for references to original sources for these terms). Some authors (e.g., Ricci Lucchi, 1995, pp.134–139) distinguish *bounce casts*, with an almost symmetrical shape, from prod marks which are asymmetrical (deeper on the down-current side) and therefore permit determination of the current direction. *Brush marks* are similar to prod marks but have wrinkles produced by drag of mud on the up-current side. For good photographs see Allen (1982) or Ricci Lucchi (1995).

Gerard V. Middleton

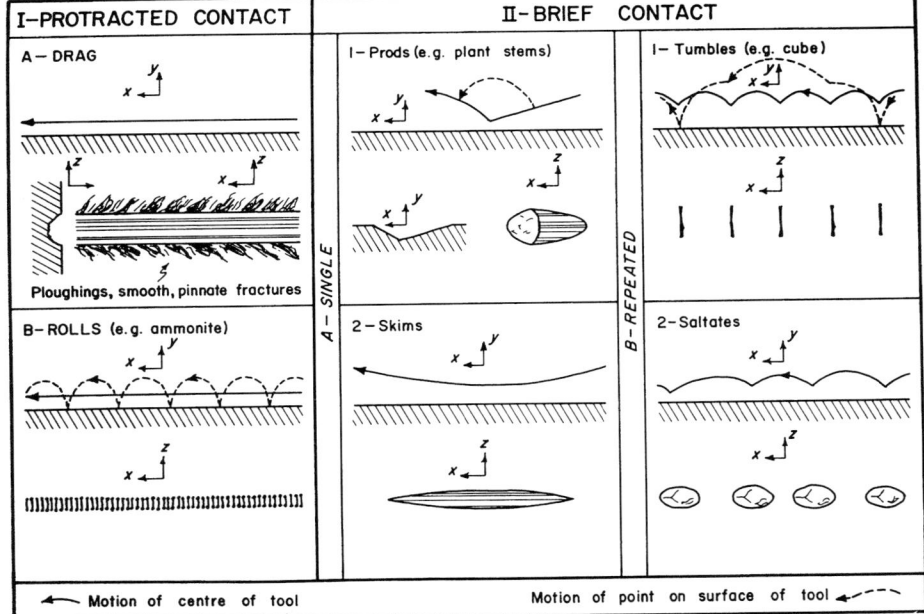

Figure T16 Classification of tool marks, based on tool kinematics (from Allen, 1982, p.510).

Bibliography

Allen, J.R.L., 1982. *Sedimentary Structures: Their Character and Physical Basis*, Volume 2, Elsevier.
Dzulynski, S., and Sanders, J.E., 1962. Current marks on firm mud bottoms. *Transactions of the Connecticut Academy of Arts and Sciences*, **43**: 57–96.
Dzulynski, S., and Walton, E.K., 1965. *Sedimentary Features of Flysch and Greywackes*. Elsevier.
Ricci Lucchi, F., 1995. *Sedimentographica: A Photographic Atlas of Sedimentary Structures*, 2nd edn. Columbia University Press.

Cross-references

Gutters and Gutter Casts
Storm Deposits
Turbidites

TOXICITY OF SEDIMENTS

Overview

As water quality has improved over the past three decades in North America, diffuse sources of pollution such as stormwater runoff and sediments are now recognized as long-term, widespread pollutant sources to aquatic systems. Substantial impacts on the ecosystem from sediment-associated contaminates range from direct effects on benthic communities to substantial contributions to contaminant loads and effects on upper trophic levels through food chain contamination (e.g., McCarty and Secord, 1999).

Sediment contamination primarily occurs because many chemicals bind to organic or inorganic particles that eventually settle to the bottom of our streams, rivers, reservoirs, lakes, estuaries, or marine waters. Once contaminants bind to a particle surface or sorb into its interior matrix, they become less likely to be biotransformed and desorption is usually very slow; therefore, sorbed contaminants will reside for long periods in the sediment. Sediment-associated contaminants tend to accumulate onto small, fine-grained particles and settle in depositional areas. This is promoted largely by the very high surface area and the tendency for higher concentrations of organic matter in the fine particles.

From a contamination perspective, we understand that sediments are extremely important to the food web and serve as a habitat for the benthic community as well as contaminant reservoir for bioaccumulation and trophic transfer. Once chemical contamination reaches a concentration whereby it causes adverse effects to biota, then it is considered polluted. The greater the evidence of sediment contamination within a watershed, the more likely a fish consumption advisory exists (US EPA, 1997).

Extent of the problem

There are over 9 billion cubic meters of surface sediments (upper 5 cm) in the United States of which approximately 1.2 billion can currently be considered contaminated to levels that pose potential risks to fish and to humans and wildlife who eat fish (US EPA, 1997). This estimate is derived from a literature search on sediment contamination and represents 65 percent of the watersheds in the continental US. However, only 11 percent of the nation's rivers had data with any toxicity information available and of those 77 percent had at least one station that was sufficiently contaminated such that adverse effects are probable or possible. This leaves 89 percent of the data sets with no available toxicity data (US EPA, 1997). Therefore, the true extent of sediment contamination remains unknown.

Sediment associated contaminants are found in every type of aquatic environment within the US from mountain streams to large rivers, from small lakes to the Great Lakes and in estuaries and bays (US EPA, 1997). Further, recent investigations have found sediment pollution in areas previously not thought to be contaminated (e.g., estuarine areas of North Carolina, Pelley, 1999). While the true extent of sediment contamination is not known, it is apparent that huge quantities of sediments in industrialized countries are contaminated with metal and organic chemicals at levels that pose risks to both aquatic life and to humans consuming aquatic species.

Effects on ecosystems and components

Despite a concentrated research effort over the past two decades, our ability to assess the impact of contaminated sediments on ecosystems remains a challenge that is inherent in the very nature of sediments. The most obvious effect in most assessments of sediment contamination is that of a direct adverse impact on the benthic (or bottom-dwelling) community, (e.g., Peterson et al., 1996). Because alteration of the benthic community structure can result from the mixture of contaminants present as well as from potential for confounding factors, such as poor habitat, high flows, or low dissolved oxygen in the water column, it is often difficult to establish cause and effect for sediment associated contaminants. Studies of field sites where direct biological effects are obviously contaminant-induced, are often located in watersheds containing high levels of industrial and shipping activities. Here, sediments are likely to be visibly contaminated with petroleum products and may contain a myriad of metals, pesticides, and other synthetic nonpolar organic chemicals, e.g., Grand Calumet River, Indiana. Thus, assigning cause-effect between single, specific chemicals and benthic communities becomes nearly impossible. Most of the detailed biological assessments dealing with contaminated sediments have dealt with freshwater systems, although several sediment risk assessments are currently underway in coastal harbor areas. Unfortunately, most of these risk assessments are only published in the "grey" literature and difficult to obtain.

It is not just infaunal (burrowing) benthic invertebrate organisms that demonstrate effects from sediment-associated contaminants. Many bottom-feeding fish have hepatic and epidermal neoplasms (tumors). There is a strong link between contaminated sediment and fish neoplasms in all the major water bodies of the US (Harshbarger and Clark, 1990). The sediment link for these neoplastic growths was clearly demonstrated in the Black River in Lorain, OH. The high incidence of neoplasms in brown bullhead showed a high correlation with high PAH sediment concentrations. These neoplasms resulted in age specific mortality in the bullheads. Once most of the PAH contaminated sediments were removed by dredging, the incidence of neoplastic growths declined sharply (Baumann and Harshbarger, 1995).

Sediment-associated contaminants have also become linked via food web transfer to impacts on upper trophic levels. Such transfer occurs with mercury and some organochlorines, such as PCBs and DDT that are poorly biotransformed and are hydrophobic; however with other chemicals these connections are more difficult to establish. From modeling exercises, food web transfer of persistent contaminants is important for maintaining the chemical concentrations observed in upper tropic levels and the benthic component is essential to account for the observed concentrations (Thomann et al., 1992; Morrison et al., 1996). Trophic transfer of sediment-associated contaminants has been documented in both freshwater systems (e.g., Lester and McIntosh, 1994) and marine systems (e.g., Maruya and Lee, 1998). In Saginaw Bay, Lake Huron, tree swallows were found to accumulate PCB from sediments. In some areas of the Great Lakes and in the Hudson River, NY, system reproductive damage has been observed for this species directly linked to PCB (Bishop et al., 1999; McCarty and Secord, 1999). Further, impacts to birds were observed in the Saginaw Bay system with reproductive effects on the Caspian terns following a 100 year flood event suggesting impacts through food web transfer from a latent sediment source (Ludwig et al., 1993). Thus, fresh pulses of contaminated sediments from the watershed can recharge the system resulting in impacts on the receiving system.

Developing cause–effect links

The complexity of the distributions of contaminants, nutrients, other sediment/habitat characteristics and resident biota make accurate determinations of exposure and effects difficult, but not impossible. This reality suggests that accurate assessments of contaminant effects use a weight-of-evidence approach (i.e., gathering supporting data from several lines of investigation) to determine the extent or magnitude and cause of the contamination problem. The initial ideas on the weight-of-evidence approach took data from sediment chemistry, benthic community structure and sediment laboratory bioassays to establish a triad of information that could be used to determine hazard (Chapman, 1986). This approach is continually improving with additional development on sediment bioassays (toxicity tests), interpretability of sediment chemistry, and the addition of new types of information such as in situ toxicity tests and residue (tissue chemical concentration)-effects information.

Sediment toxicity testing on whole field collected sediments and laboratory dosed sediments are an essential line-of-evidence that the contaminants in sediment can produce the effects that are observed in benthic communities. The earliest sediment bioassay was performed in the early 1970s and demonstrated avoidance behavior by amphipods for contaminated sediments. However, much of the work demonstrating the effect of sediment-associate contaminants came in the 1980s. By the mid 90's, standardized methods for whole sediment toxicity testing occurred within the US EPA, American Society for Testing and Materials (ASTM), and Environment Canada. These tests measured acute (short-term ≤ 10 days) toxicity in benthic macroinvertebrates such as the amphipods *Hyalella azteca, Rhepoxynius abronius, Ampelisca abdita, Eohaustorius estuarius* and *Leptocheirus plumulosus* and the midges *Chironomus tentans* and *Chironomus riparius* (US EPA, 1994a,b; ASTM, 2000a). The primary measured response was mortality, but in the case of the midge, growth was included and reburial was an additional endpoint for the *Rhepoynmius abronius*. Unfortunately, most sediment toxicity test methods are focused on acute (short-term exposure) and not chronic (long-term exposure) toxicity. The measures of acute toxicity are not often adequate to detect the impacts on benthic communities. For instance, the 10-d test with *Rhepoxynius abronius* was not sensitive enough to describe the loss of amphipods from Lauritizen Channel in San Francisco Bay (Swartz et al., 1994). In reality, chronic toxicity is the more pervasive problem and it is the chronic responses, such as changes in reproduction that lead to population level responses. While recently developed chronic methods (US EPA, 2000, 2001) will greatly aid our ability to determine if sediments are toxic, their long duration and increased costs may impede their widespread adoption.

A large number of other species have been used for determining the toxicity of sediments, ranging from bacteria to fish and amphibians (Burton, 1991). The commonly used test species encompass trophic levels ranging from decomposers and producers to predators and representing multiple levels of biological organization. Comparisons of their sensitivities have shown a wide range of responses to different types of sediment contamination, with an equally wide range of discriminatory power (ability to detect differences between samples) (Burton et al., 1996). This reality suggests that more than one or two species may be necessary to determine with certainty whether or not sediment contamination is ecologically significant.

The freshwater species that have been used successfully and most often in assessments of sediment toxicity include both water-column and benthic species: *Selenastrum capricornutum, Daphnia magna, Ceriodaphnia dubia, Pimephales promelas, H. azteca, C. tentans* and *C. riparius*, and *Hexagenia limbata*. Standard guides for whole sediment toxicity testing have also been published by the American Society for Testing and Materials for invertebrates, including *H. azteca, C. tentans, C. riparius,* the cladocerans *Daphnia magna* and *Ceriodaphnia dubia*, the mayfly *Hexagenia*, the Great Lakes amphipod *Diporeia*, and the oligochaete *Tubifex tubifex* (ASTM, 2000a–e). Dredged material evaluations in the US follows the protocols published by the US EPA and US Army Corps of Engineers; in 1991 for dredged material proposed for ocean disposal, and protocols published in 1998 for dredged material proposed for disposal in "inland" waters of the US (USEPA–USACE, 1998). Canada's Ontario Ministry of the Environment has also published solid-phase test methods (Bedard et al., 1992).

Toxicity tests for marine and estuarine species also range from fish to bacteria. The marine and estuarine species frequently and successfully used in studies of sediment contamination include: the fish *Cyprinodum variegatus* and *Menidia beryllina*, and amphipods *Rhepoxynius abronius, Ampelisca abdita, Eohaustorius estuarius, Grandidierella japonica* and *Leptocheirus plumulosus*, the bivalves *Haliotis rufescens, Crassostrea gigas* and *Mytilus* sp. the echinoderms *Strongylocentrotus purpuratus, Dendraster excentricus,* and *Arbacia punctulata*, the polychaetes *Dinophilus gyrociliatus* and *Neanthea arenaceodentata*, the mysid *Americamysis bahia* (formerly *Mysidopsis bahia)*, the alga *Champia parvula*, and the bacterial assay Microtox using *Vibrio fischeri*. As with freshwater tests, several "standard" methods for marine and estuarine organisms exist. The US EPA has standardized protocols for toxicity testing with the amphipods *Rhepoxynius, Ampelisca, Eohaustorius, and Leptocheirus* using a 10-d survival assay, a chronic 28-d survival, growth reproduction assay with *Leptocheirus*, and 28-d bioaccumulation assays with *Macoma* and *Nereis*

(US EPA 1994b, 2001). The European OECD (Organization of Economic Cooperation and Development) is discussing the possibility of developing standardized sediment tests. Environment Canada has developed methods for sediments, elutriates, and leachate samples using *Vibrio fisheri* luminescence, sea urchins and sand dollars (echinoids), and seven amphipod species. Methods were recently published by the Chesapeake Bay Program for benthic testing using *Lepocheirus plumulosus, Ampelisca abdita, Lepidactylus dytiscus,* and *Monoculodes edwardsi*. Dredged materials in the US are tested using US EPA and US Army Corps of Engineers protocols. The American Society for Testing and Materials has standard procedures for amphipods, fish, mysids, echinoderm, and oyster early life stages, polychaetes, and algae (ASTM, 2000b–e).

Another component that is useful in establishing causality is linking sediment contaminant exposure and tissue-residue effects in organisms. Because many factors appear to alter the bioavailability of contaminants in sediments (Hamelink *et al.*, 1994), approaches to establish links between the body-residue concentrations and effects in aquatic organisms provide the insight to better link the toxic response directly to contaminants. Data has been amassing over the course of the past several years that allow the direct comparison of some residue levels with acute and chronic effects (www.wes.army.mil/el/ered; Jarvinen and Ankley, 1999). However, the data base is very limited at this time, thus there is still need to establish a weight-of-evidence approach for developing the link between the observed response and the presence of contaminants in sediments.

The second recent development to assist in the formation of the link between sediment contaminants, responses in the laboratory and observed responses in the field is that of *in situ* tests. These are tests that are conducted in the field. One approach uses confined organisms, such as traditional surrogates (e.g., *Daphnia magna, Ceriodaphnia dubia, H. azteca, C. tentans, Lumbriculus variegatus, Pimephales promelas*) or indigenous species in chambers, cages, bags, or corrals. Organisms are placed in test chambers, which optimize exposures to sediments, pore waters, and/or overlying waters, and are exposed in the field from time periods ranging from a day to months. This approach has a number of advantages, in that it can better simulate real-world exposures, which may fluctuate dramatically over a period of hours, reduces sampling/experimental related artifacts, and allows for more natural interactions of potentially critical physical and chemical constituents than do laboratory conditions (e.g., Chappie and Burton, 2000).

Traditionally sediment contamination was determined by assessing the bulk chemical concentrations of individual compounds often comparing to some background or reference value. The desire to have chemical parameters to evaluate the hazard of sediments has focused primarily on two approaches: (1) a statistical approach to establish the relationship between sediment contamination and toxic response; and (2) a theoretically based approach that attempted to account for differences in bioavailability through equilibrium partitioning.

The original sediment quality guidelines (SQGs) that were compared to a reference or to background provided little insight into the impact of sediment contaminants. Therefore, SQGs for individual chemicals were developed that relied on paired field sediment chemistry with field or laboratory biological effects data. They are often based on frequency distributions and account for the impact of all chemicals present, but do not establish cause and effect. These approaches have also been shown to be useful and predictive of biological effects in many (but not all) marine and freshwater systems (e.g., Long *et al.*, 1998).

However, the issue of bioavailability remains (that portion of the chemical that is available for biological uptake and subsequent adverse effects). The equilibrium partitioning approach attempts to address this issue specifically. This approach suggested that interstitial water concentrations represented the equivalent chemical activity for exposure of aquatic organisms, and chemical concentrations could be determined from an equilibrium partitioning calculation (Di Toro *et al.*, 1991). The equilibrium partitioning calculation was particularly important for attempting to account for the changes in bioavailability that occur with changes in amounts of organic matter in sediments.

This method has also been applied to metals by accounting for the interactions with acid volatile sulfide. Five metals, Cd, Ni, Pb, Zn, and Cu, form insoluble sulfides and so their toxicity is limited by the amount of sulfide in the sediment. Toxicity is observed when the amount of metal stoichiometrically exceeds the amount of sulfide that can bind it. A clear demonstration of the utility of this approach was shown for Cd toxicity to amphipods, *Ampelisca abdita* and *Rhepoxynius hudsoni* in marine sediments. Like the case for organic contaminants, the toxicity predicted from this approach has been sometimes over predicted usually because of the presence of other ligands that bind the metals (Ankley *et al.*, 1993).

When comparisons have been made between the equilibrium-partitioning (EQP) method and other measures of sediment quality to estimate toxic response, the sediment quality guideline values and their predictive ability were comparable, i.e., in greater than 70 percent of samples toxicity or absence of toxicity were correctly predicted. However, when comparing the predictive ability for field sites, EQP consistently gave more type II errors, false negatives, than other sediment quality guidelines (Ingersoll *et al.*, 1997). This likely results from failure to account for the wide variety compounds that exist in sediment and their interactions.

Risk-assessment for sediment-associated contaminants is a relatively new field and is developing rapidly (Ingersoll *et al.*, 1997). The strengths and limitations of the various assessment methods dictates that a weight-of-evidence approach define the significance and role of sediment-associated contaminants on the aquatic environment and also through food web transfer to terrestrial species. This approach should include an integrated approach, which identifies and ranks stressors using habitat, physical, and chemical measures, biological community structure, toxicity, and bioaccumulation descriptors. Use of these tools can provide essential characterizations of key watershed sources, sensitive receptors, natural variation, and both natural and anthropogenic stressors. Only with the use of multiple tools and an understanding of their interactions can reliable determinations of sediment pollution and long-term consequences be made. Then cost-effective, environmentally protective management decisions can be made about the type, extent, and need for sediment remediation.

G. Allen Burton, Jr. and Peter F. Landrum

Bibliography

ASTM, 2000a. Test method for measuring the toxicity of sediment-associated contaminants with freshwater invertebrates. E1706–00. In *Annual Book of ASTM Standards*. Volume 11.05, American Society for Testing and Materials, West Conshohocken, PA.
ASTM, 2000b. *Annual Book of ASTM Standards*. Section 11 Water and Environmental Technology E1367. Biological Effects and Environmental Fate; Biotechnology; Pesticides: Standard guide for conducting 10-day static sediment toxicity tests with marine and estuarine amphipods. Section 11.05 American Society for Testing and Materials. West Conshohocken, PA.
ASTM, 2000c. *Annual Book of ASTM Standards*. Section 11 Water and Environmental Technology E1191. Biological Effects and Environmental Fate; Biotechnology; Pesticides: Standard guide for conducting life-cycle toxicity tests with saltwater mysids. Section 11.05 American Society for Testing and Materials. West Conshohocken, PA.
ASTM, 2000d. *Annual Book of ASTM Standards*. Section 11 Water and Environmental Technology E729. Biological Effects and Environmental Fate; Biotechnology; Pesticides: Standard guide for conducting acute toxicity tests on test materials with fishes, macroinvertebrates, and amphipods. Section 11.05 American Society for Testing and Materials. West Conshohocken, PA.
ASTM, 2000e. *Annual Book of ASTM Standards*. Section 11 Water and Environmental Technology E1498. Biological Effects and Environmental Fate; Biotechnology; Pesticides: Standard guide for conducting sexual reproduction tests with seaweeds. 11.05 American Society for Testing Materials. West Conshohocken, PA.
Ankley, G.T., Mattson, V.R., Leonard, E.N., West, C.W., and Bennett, J.L., 1993. Predicting the acute toxicity of copper in freshwater sediments: evaluation of the role of acid volatile sulfide. *Environmental Toxicology and Chemistry*, **12**: 315–320.
Baumann, P.C., and Harshbarger, J.C., 1995. Decline in liver neoplasms in wild brown bullhead catfish after coking plant closes and environmental PAHs plummet. *Environmental Health Perspectives*, **103**: 168–170.
Bedard, D.C., Hayton, A., and Persaud, D., 1992. Ontario Ministry of the Environment Laboratory Sediment Biological Testing Protocol, Ontario Ministry of the Environment, Toronto, Ontario, Canada, draft report.
Bishop, C.A., Mahony, N.A., Trudeau, S., and Pettit, K.E., 1999. Reproductive success and biochemical effects of tree swallows (Tachycineta bicolor) exposed to chlorinated hydrocarbon contaminants in wetlands of the Great lakes and St Lawrence River Basin, USA, and Canada. *Environmental Toxicology and Chemistry*, **18**: 263–271.
Burton, G.A. Jr, 1991. Assessing freshwater sediment toxicity. *Environmental Toxicology and Chemistry*, **10**: 1585–1627.
Burton, G.A. Jr, Ingersoll, C., Burnett, L., Henry, M., Hinman, M., Klaine, S., Landrum, P., Ross, P., and Tuchman, M., 1996. A comparison of sediment toxicity test methods at three Great Lake Areas of concern. *Journal of Great Lakes Research*, **22**: 495–511.
Chapman, P.M., 1986. Sediment quality criteria from the sediment quality triad: an example. *Environmental Toxicology and Chemistry*, **5**: 957–964
Chappie, D.J., and Burton, G.A. Jr, 2000. Applications of aquatic and sediment toxicity testing in situ. *Journal of Soil and Sediment Contamination*, **9**: 219–246.
Di Toro, D.M., Zarba, C.S., Hansen, D.J., Berry, W.J., Swartz, R.C., Cowen, C.E., Pavlou, S.P., Allen, H.E., Thomas, N.A., and Paquin, P.R., 1991. Technical basis for establishing sediment quality criteria for nonionic organic chemicals by using equilibrium partitioning. *Environmental Toxicology and Chemistry*, **12**: 1541–1583.
Hamelink, J.L., Landrum, P.F., Bergman, H.L., and Benson, W.H., 1994. *Bioavailability: Physical, Chemical, and Biological Interactions*. Boca Raton, FL: Lewis Publishers.
Harshbarger, J.C., and Clark, J.B., 1990. Epizootiology of neoplasms in bony fish of North America. *Science of the Total Environment*, **94**: 1–32.
Ingersoll, C.G., Dillon, T., and Biddinger, G.R., 1997. *Ecological Risk Assessment of Contaminated Sediments*. Pensacola, FL: Society of Environmental Toxicology and Chemistry Press.
Jarvinen, A.W., and Ankley, G.T., 1999. *Linkage of Effects to Tissue residues: Development of a Comprehensive Database for Aquatic Organisms Exposed to Inorganic and Organic Chemicals*. Penscola, FL: Society of Environmental Toxicology and Chemistry Press.
Lester, D.C., and McIntosh, A., 1994. Accumulation of polychlorinated biphenyl congeners from Lake Champlain sediments by *Mysis relicta*. *Environmental Toxicology and Chemistry*, **13**: 1825–1841.
Long, E.R., Field, L.J., and MacDonald, D.D., 1998. Predicting toxicity in marine sediments with numerical sediment quality guidelines. *Environmental Toxicology and Chemistry*, **17**: 714–727.
Ludwig, J.P., Auman, H.J., Kurita, H., Ludwig, M.E., Campbell, L.M., Giesy, J.P., Tillitt, D., Jones, P., Yamashita, N., Tanabe, S., and Tatsukawa, R., 1993. Caspian tern reproduction in the saginaw bay ecosystem following a 100-year flood event. *Journal of Great Lakes Research*, 96–108.
Maruya, K.A., and Lee, R.F., 1998. Biota-sediment accumulation and trophic transfer factors for extremely hydrophobic polychlorinated biphenyls. *Environmental Toxicology and Chemistry*, **17**: 2463–2469.
McCarty, J.P., and Secord, A.L., 1999. Reproductive ecology of tree swallows (*Tachycineta bicolor*) with high levels of polychlorinated biphenyl contamination. *Environmental Toxicology and Chemistry*, **18**: 1433–1439.
Morrison, H.A., Gobas, F.A.P.C., Lazar, R., and Haffner, G.D., 1996. Development and verification of a bioaccumulation model for organic contaminants in benthic invertebrates. *Environmental Toxicology and Technology*, **30**: 3377–3384.
Pelley, J., 1999. North Carolina considers controls to protect contaminated waters. *Environmental Toxicology and Technology*, **33**: 10A.
Peterson, C.H., Kennicutt II, M.C., Green, R.H., Montagna, P., Harper, D.E. Jr., Powell, E.N., and Roscigno, P.F., 1996. Ecological consequences of environmental perturbations associated with offshore hydrocarbon production: a perspective on long-term exposures in the Gulf of Mexico. *Canadian Journal of Fisheries and Aquatic Sciences*, **53**: 2637–2654.
Swartz, R.C., Cole, F.A., Lamberson, J.O., Ferraro, S.P., Schults, D.W., DeBen, W.A., Lee II, H., and Ozretich, R.J., 1994. Sediment toxicity, contamination, and amphipod abundance at a DDT- and Dieldrin-contaminated site in San Francisco Bay. *Environmental Toxicology and Chemistry*, **13**: 949–962.
Thomann, R.V., Connolly, J.P., and Parkerton, T.F., 1992. An equilibrium model of organic chemical accumulation in aquatic food webs with sediment interaction. *Environmental Toxicology and Chemistry*, **11**: 615–629.
US Environmental Protection Agency, 1994a. Methods for measuring the toxicity and Bioaccumulation of Sediment-associated Contaminants with Freshwater Invertebrates. Office of Research and Development. EPA/600/R-94/024. Washington, DC.
US Environmental Protection Agency, 1994b. Methods for assessing the Toxicity of sediment-associated contaminants with estuarine and marine amphipods. *EPA/600/R-94/025*, Office of Research and Development, Washington, DC.
US Environmental Protection Agency, 1997. *The Incidence and Severity of Sediment Contamination in Surface Waters of the United States*. Volumes 1–3. EPA 823-R-97-006. Science and Technology Office. Washington, DC.
US Environmental Protection Agency, 2000. Methods for assessing the toxicity and bioaccumulation of sediment-associated contaminants with freshwater invertebrates. Second edition, EPA/600/R-99/064, Duluth, MN.
US Environmental Protection Agency, 2001. Method for assessing the chronic toxicity of marine and estuarine sediment-associated contaminants with the amphipod *Leptocheirus plumulosus*, 1st edn, EPA/600/R-01/020. Washington, DC.
USEPA-USACE, 1998. Evaluation of dredged material for discharge in waters of the US testing manual. EPA-823-B-98-005. Washington, DC.

TRACERS FOR SEDIMENT MOVEMENT

For the purposes of studying sediment dynamics, tracers are defined here as marked particles introduced into a river, estuary, or coastal system, to obtain general information on the characteristics of sediment movement within such environments. Tracers provide a relatively simple means of overcoming technical and sampling problems without the need for a detailed kinematic study of the sedimentary regime (Crickmore et al., 1990). The type of tracer to be used will largely depend on the objectives, environment characteristics, and intended observation period of the experiments (Hassan and Ergenzinger, in press). In addition, one must consider each method and materials, the sample size and expected recovery rate, detection, data collection and analyzes.

Both artificial and natural tracers are available for tracking coarse and fine material. In selecting an artificial tracer, the geochemical and physical properties, and hydrodynamic conditions must be considered. Some tracing techniques can be used for fine and coarse sediment, though tagging silt and clay sizes is complex. With fine material, physical characteristics make tracing and detection difficult for several reasons: (i) the fine sediment is usually carried in suspension, and its movement dispersion over large areas is rapid; (ii) coating or attaching tags onto the fine sediment may change the physical characteristics of the particle; and (iii) the cohesive nature of clay and silt causes flocculation, and labeling may lead to changes in the chemical properties of the sediment.

A variety of detection methods have been tried, ranging from simple visual identification of the tracers, to magnetic or metal detection, radioactive detection, neutron activation, and radio wave detection. Some detection methods, using paint, magnets, and radio transmitters, permit individual identification of tracer particles whilst others such as radioactive material do not. The approach will clearly influence how much information can be gathered, and hence subsequent data analyzes and interpretations. Detection strategies include: individual location of particles; fixed automatic detection of particle passage; periodic dynamic detection and automatic dynamic detection (Hassan and Ergenzinger, in press).

A brief background and explanation is provided below on the most commonly-used tracers.

Painted particles are probably the simplest method of tracing gravels. Paint is easy to apply and can be detected visually, but field detection is limited to the surface. In the fluorescent method a special dye is applied to a set amount of sediment. Upon light stimulation, the treated sediment emits light of a wavelength characteristic of the dye (Crickmore et al., 1990). Samples can then be taken within the study area to assess the dispersion of sediment. The fluorescent method is suitable for sand tracing, and even larger material. Suitable fluorescent dyes include rhodamine, auramine, eosine, primulin, fluorescein, and anthracine.

The *radioactive technique* is very versatile and can be used in both fine and coarse sediment (Crickmore et al., 1990). It allows *in situ* continuous detection of exposed and buried particles and, hence, permits immediate data processing and rapid evaluation of the tracking strategy. The presence of the radioactively labeled sediment is detected from the radiation emitted by the tag (e.g., Crickmore et al., 1990). Wide ranges of radioisotopes provide a choice of tracers with a half-life to match most objectives. According to the method of labeling, the radioactive tracers can be grouped into three categories: manufacture of radioactive grains of the same density and shape as the bed material, application of the isotope to the outside or within the natural grains, and introduction of natural radioactive minerals. The most commonly used isotopes are ^{140}La (half-life of 40.2 hours), ^{51}Cr (half-life of 27.8 days), ^{46}Sc (half-life of 83.9 days), and ^{110}Ag (half-life of 253 days). The method is toxic and dangerous to the environment and therefore it is very difficult to obtain permission to use.

In the *neutron activation method*, a chemical element is fixed to the sediment and its concentration is measured after activation in a nuclear reactor. The sediment is labeled by an element with a high thermal neutron which, after irradiation, is detectable above the natural radioactive background. Studies have used, among other elements, cobalt, tantalum, iridium, rhenium, and gold (e.g., Crickmore et al., 1990).

The *fingerprinting technique* can be used to define sediment sources and transfer within a drainage basin, and sedimentation rates in lakes and reservoirs. This method is based on (i) the selection of physical and/or chemical properties that distinguish a particular source material, and (ii) the comparison of measurements of the same properties of the sediment with the equivalent values obtained for the source material (e.g., Oldfield et al., 1979; Walling and Woodward, 1992).

Fallout radionuclides have been used to pursue a wide range of studies within drainage basins (e.g., Walling et al., 1999). On reaching the earth's surface by either wet or dry fallout, radionuclides are adsorbed onto the soil and become useful for tracing the movement of fine sediment. Applications include the study of suspended sediment sources, sedimentation rates and sediment yield, gully erosion, floodplain sedimentation, sediment budget, residence time of fine sediment, redistribution of soil, and lake and reservoir sedimentation. Three fallout radionuclides have primarily been used: ^{137}Cs, unsupported ^{210}Pb, and ^{7}Be, of which ^{137}Cs is the most favored. Caesium-137 is an artificial radionuclide produced as a result of global fallout from the testing of nuclear weapons. With a half-life of 30.1 years, ^{137}Cs provides an average value of surface erosion for a period of about 45 years. It has been found that eroded soils lose ^{137}Cs in proportion to the severity of erosion. At erosion sites ^{137}Cs will be present in quantities smaller than the reference value, whereas at deposition sites the opposite is observed. Both ^{210}Pb and ^{7}Be are natural radionuclides that reach the earth's surface *via* atmospheric fallout at a constant rate through time. Because of a relatively short half-life, the ^{7}Be technique is useful for short-term measurements that can be related to specific events.

Ferruginous tracers include metal strips, acid coating, artificial magnets, and mineral magnetite (Hassan and Ergenzinger, in press). The underlying principle is that when a tracer passes over an iron-cored coil of wire, a measurable electrical pulse is generated. Magnet or metal locators are used to find tagged particles, and automatic detection systems are also available. The latter type of system can provide *in situ* continuous measurements of sediment movement to record an unlimited number of clasts. Due to the high recovery rate, magnetic methods are the most widely used technique. The method permits the location of buried particles, as well as those at the surface. Both natural and artificial magnetic tracers have been used. If no naturally magnetic material is present and suitable for use where the research is to be carried out, then it is necessary to use artificially magnetized tracers. For tracing gravel larger than 16 mm, a magnet can be inserted into pre-drilled holes. Manufactured synthetic magnetic clasts, made from resin or concrete with a magnetic core in the center, have been used for tagging small and large particles (>8 mm). This approach allows the manufacture of clasts of different shapes and densities. Magnetic enhancement is another method, which is based on the enhancement of natural magnetism by strong heating of naturally iron-rich pebbles. Through the heat treatment the particle mineralogy is altered and the magnetism is enhanced up to 300 times its original strength to a level that can be detected and distinguished from the bed material. Magnetic properties of the soil can be used to trace sediment sources and fine sediment transport in alluvial streams (e.g., Oldfield *et al.*, 1979). The upper part of the soil profile has a higher concentration of magnetic content than the lower part or the parent material due to pedogenic processes and burning. The enhanced magnetism (see artificial magnetic enhancement, above) helps to differentiate between the surface and subsurface soil and hence the suspended sediment sources.

The *Radio transmission* method permits the active detection of tracers in cobble size material (Schmidt and Ergenzinger, 1992). A transmitter capsule, about 65 mm long and 20 mm in diameter and with a battery life expectancy of about three months, is inserted into the clast. Within a set of tracers, the stones will each emit slightly different frequencies so that they can be individually identified. The detection system consists of a transmitter, an antenna, receiver, and a data logger.

Marwan A. Hassan

Bibliography

Crickmore, M.J., Tazioli, G.S., Appleby, P.G., and Oldfield, F., 1990. *The Use of Nuclear Techniques in Sediment Transport and Sedimentation Problems*. Technical Documents in Hydrology, IHP-III Project 5.2, Paris: UNESCO, 170pp.

Hassan, M.A., and Ergenzinger, P., in press, Use of tracers in fluvial geomorphology. In Kondolf, G.M., and Piegay, H. (eds.), *Methods in Fluvial Morphology*. Wiley and Sons. Expected to be published in April 2003.

Ingle, J.C. Jr, 1966. *The Movement of Beach Sand*. Elsevier.

Oldfield, F., Rummery, T.A., Thompson, R., and Walling, D.E., 1979. Identification of suspended sediment sources by means of magnetic measurements: some preliminary results. *Water Resources Research*, **15**: 211–217.

Schmidt, K.-H., and Ergenzinger, P., 1992. Bedload entrainment, travel lengths, step lengths, rest periods—studies with passive (iron, magnetic) and active (radio) tracer techniques. *Earth Surface Processes and Landforms*, **17**: 147–165.

Walling, D.E., He, Q., and Blake, W., 1999. Use of ^{7}Be and ^{137}Cs measurements to document short and medium term rates of water induced soil erosion on agricultural lands. *Water Resources Research*, **35**: 3865–3874.

Walling, D.E., and Woodward, J.C., 1992. Use of radiometric fingerprints to derive information on suspended sediment sources. In Bogen, J., and Walling, D.E. (eds.), *Erosion and Sediment Monitoring Programmes in River Basins*. IAHS Publication Number 210, pp. 153–164.

Cross-references

Attrition (Abrasion), Fluvial
Attrition (Abrasion), Marine
Coastal Sedimentary Facies
Deltas and Estuaries
Grain Size and Shape
Rivers and Alluvial Fans
Sediment Transport by Unidirectional Water Flows
Sediment Transport by Tides
Sediment Transport by Waves

TUFAS AND TRAVERTINES

Tufas and travertines are rare in Paleozoic and Mesozoic successions, but become frequent in the Cenozoic. Nevertheless, the majority of recorded examples are Quaternary and Holocene (Waring, 1965; Pentecost, 1995; Ford and Pedley, 1996). The term "tufa" is recommended for the description of all ambient temperature freshwater carbonate deposits with "travertine", an alternative name, being reserved for all hydrothermal freshwater deposits (Pedley, 1990). In contrast, many employ the term "travertine" for all types with distinction into high temperature "thermogene" and ambient temperature "metogene" deposits (Pentecost, and Viles, 1994). Tufas commonly contain abundant evidence of mosses and higher plants, whereas thermal sites encourage physico-chemical precipitates and colonization is restricted to microbes. Travertine (hydrothermal) deposits frequently contain unusual geochemical signatures (e.g., high sulfur and carbon dioxide components). Many deposits have been dated using radiocarbon methods (Srdoc *et al.*, 1986), and uranium series isotopes (Hennig *et al.*, 1983). Stable isotopes of oxygen and carbon have proved valuable as proxy-environmental indicators (Andrews *et al.*, 2000).

Speleothems (*q.v.*) are similar to travertines but are physico-chemical cave and cavity precipitates.

Tufas

Tufas typically are rich in plants, but may also contain insects, pollen, molluscs, and ostracods (Preece, 1991; Vadour, 1994; Taylor *et al.*, 1998). Mosses, liverworts, and semi-aquatics commonly provide a framework upon which fringes of low-magnesian calcite cements precipitate. Precipitation is partly controlled physico-chemically by CO_2 degassing which renders the dissolved calcium carbonate less soluble (Lorah and Herman, 1988). However, there is also an important biomediated contribution triggered as a by-product of photosynthesis and associated metabolic processes connected with diatoms, cyanobacteria, and heterotropic bacteria. These microorganisms are ever present within surficial (procaryote-microphyte) biofilms and many tufas and travertines are commonly entrapped within the associated fringe cements. Lithification is virtually instantaneous in many fluvial, waterfall, and lake margin settings and forms laminated, freshwater plant-dominated framestone reefs or "phytoherms" (Buccino *et al.*, 1978; Pedley, 1990). Oncoids are common and are often associated with allochthonous tufa debris (phytoclast tufa) in channels and point bars. Fine lime mud-grade tufa formed in lakes is derived either from marginal stromatolite microherms and phytoherms or is a by-product of microbial ("picoplankton") metabolism within the water column (Thompson, 2000).

Tufa classifications are diverse (e.g., Chafetz and Folk, 1984; Ordonez and Garcia del Cura, 1983; Pedley, 1990). Confusion is reduced if studies are based on facies associations (Buccino et al., 1978; Pentecost and Viles, 1994; Ford and Pedley, 1996). These demonstrate that lithofacies complexity is a function of slope angle, irregularities in substrate, and velocity and flow-frequency of carbonate-enriched water supply. Although there is a facies continuum, probably four common models can be derived from the spectrum.

Fluvial tufa model

This may be flow dominated (braided fluvial tufas) and dominated by oncoid and phytoclast channel fills. Small phytoherms and microherms (especially stromatolites) may occur on the stream bed or along the channel margins. Alternatively, gorges may be dominated by several cycles of fluvial barrage tufas (Golubic, 1969). Barrages are narrow, downstream arcuate dams of phytoherm framestone that create lakes. The pool depocenters commonly contain thick lime mud and organic-rich deposits (Pedley et al., 2000). The contained pollen assemblages are useful proxy-climatic indicators.

Perched springline tufa model

This consists of a multilobate mound that develops in close proximity of valley side resurgences. Typically, such mounds have a virtually flat top and steep distal face, the latter often associated with cascades and trailing mosses (Violante et al., 1994). Detrital (phytoclastic) deposits, developed distal to the lobe, typically contain land snails and paleosols.

Lacustrine tufa model

This model develops in association with large standing water bodies. Their margins are sites of "bacterioherms" which develop characteristic flat-tops controlled by the position of the air/water interface. In the shallow marginal waters oncoids occur and *Chara* stands are common. Lime silts and muds dominate the deepest waters.

Paludal tufa model

Poorly drained slopes colonized by hydrophytic vegetation (especially moss hummocks and grass cushions) are sites of lime mud precipitation (e.g., Belgian "Crons"). More extensive tufa sheets are developed in valley bottom situations and may be associated with shallow ephemeral pools and ill-defined watercourses. Small tufa mounds may form where lime-rich springs issue into otherwise organic dominated ponds. Frequent desiccation encourages paleosol development throughout many deposits.

Saline tufas

Tufa variants which do not fit into the four models are the "mound tufas" and "towers" which develop within saline lakes (e.g., Pyramid, USA, Benton, 1995; also in China Arp et al., 1998). They consist of bacterial microherms. They develop at freshwater resurgence points on the lake bed where microbial colonization is encouraged by local salinity dilution. Marginal microbial phytoherms can also occur.

Travertines

Waring (1965) lists several hundred thermal (travertine-forming) springs. Many deposits are physico-chemical calcium carbonate precipitates from simple degassing and cooling of lime-rich waters close to thermal springs. Cyanobacteria live in the near boiling water and trigger further calcium carbonate precipitation. However, toxic sulfides can inhibit biofilm colonization. Temperature and toxicity generally decline down-flow. Consequently, higher plants are able to colonize these cooler water (tufa) domains. The two commonest associations are:

Fissure ridges

In tectonically active areas calcium-rich thermal waters can rise along fault-lines and crystalline travertine are deposited to form elongate mounds with cleft-like median valleys. Pinacles and mounds can develop from the same process where the water is point sourced (e.g., Mammoth Hot Springs, Wyoming). The deposits are often composed of thin laminae, which dip steeply away from the fissure.

Terraces

Alternatively, lobate, planar sheets can develop into an intricate series of microterraces and gutters associated descending flights of small rimstone pools (e.g., Guo and Riding, 1998).

Diagenesis

The vast majority of tufas and travertines are composed of low magnesian calcite. They are prone to early "spar-micritization" by microborers, however, many retain sharp internal fabric boundaries. The addition of physico-chemical cements further reinforces older deposits. Metastable carbonates (e.g., aragonite in some travertines) soon neomorphose to calcite spar. Selective dissolution within tufas and travertines appears common. However, many voids are primary cavities lined by later cements.

Martyn H. Pedley

Bibliography

Andrews, J.E., Pedley, H.M., and Dennis, P., 2000. Palaeoenvironmental records in Holocene Spanish tufas: a stable isotope approach in search of reliable climatic archives. *Sedimentology*, **47**: 961–978.

Arp, G., Hofmann, J., and Reitner, J., 1998. Microbial fabric formation in spring mounds ("Microbialites") of alkaline salt lakes in the Japan Sand Sea, PR China. *Palaios*, **13(6)**: 581–592.

Benson, L., 1995. Carbonate deposition, Pyramid Lake subbasin, Nevada:2. Lake levels and polar jetstream positions reconstructed from radiocarbon ages and elevations of carbonates (tufas) deposited in the Lahontal basin. *Palaeogeography Palaeoclimatology Palaeoecology*, **117**: 1–30.

Buccino, G., D'Argenio, B., Ferreri, V., Brancaccio, L., Panichi, C., and Stanzione, D., 1978. I travertini della bassa Valle del Tanagrio (Campania): studio geomorphologico, sedimentologico e geochimico. *Bollitino Societa Geologica Italia*, **97**: 617–646.

Chafetz, H.S., and Folk, R.L., 1984. Travertines: depositional morphology and the bacterially-constructed constituents. *Journal of Sedimentary Petrology*, **54**: 289–316.

Ford, T.D., and Pedley, H.M., 1996. A review of tufa and travertine deposits of the World. *Earth Science Reviews*, **41**: 117–175.
Freytet, P., and Plet, A., 1996. Modern freshwater microbial carbonates: the *Phormidium* stromatolites (tufa-travertine) of southeastern Burgundy (Paris Basin, France). *Facies*, **34**: 219–238.
Golubic, S., 1969. Cyclic and noncyclic mechanisms in the formation of travertine. *Verhandlungen der Internationalen Vereinigung fuer Theoretische und Angewandte Limnologie*, **17**: 956–961.
Guo, L., and Riding, R., 1998. Hot-spring travertine facies and sequences, Late Pleistocene. Rapolano Terme, Italy. *Sedimentology*, **45**: 163–180.
Henning, G.J., Grun, R., and Brunnacker, K., 1983. Speleothems Travertines and Paleoclimates. *Quaternary Research*, **20**: 1–29.
Lorah, M.M., and Herman, J.S., 1988. The chemical evolution of a travertine depositing stream: geochemical processes and mass transfer reactions. *Water Resources Research*, **24**: 1541–1552.
Ordonez, S., Gonzalez-Martin, J.-A., and Garcia del Cura, M.A., 1983. Recent and tertiary fluvial carbonates in central Spain. In Collinson, J.D., and Lewin, J. (eds.), *Ancient and Modern Fluvial Systems*. International Association of Sedimentologists, spec. publ. Special Publication, 6, pp. 485–497.
Pedley, H.M., 1990. Classification and environmental models of cool freshwater tufas. *Sedimentary Geology*, **68**: 143–154.
Pedley, H.M., 1994. Prokaryote-microphyte biofilms and tufas: a sedimentological perspective. *Kaupia, Darmstadter Beitrage zur Naturgeschichte*, **4**: 45–60.
Pedley, H.M., Hill, I., and Denton, P., 2000. Three dimensional modelling of a Holocene tufa system in the Lathkill valley, north Derbyshire, using ground penetrating radar. *Sedimentology*, **47**: 721–735.
Pentecost, A., 1995. The Quaternary travertine deposits of Europe and Asia Minor. *Quaternary Science Reviews*, **14**: 1005–1028.
Pentecost, A., and Viles, H.A., 1994. A review and reassessment of travertine classification. *Géographie Physique et Quaternaire*, **48**: 305–314.
Preece, R.C., 1991. Radiocarbon-dated molluscan successions from the Holocene of central Spain. *Journal of Biogeography*, **18**: 409–426.
Srdoc, D., Horvatincic, N., Obelic, B., Krajcar-Bronic, I., and O'Malley, P., 1986. The effects of contamination of calcareous sediments on their radiocarbon ages. *Radiocarbon*, **25**: 510–514.
Taylor, D.M., Pedley, H.M., Davies, P., and Wright, M.W., 1998. Pollen and mollusc records for environmental change in central Spain during the mid- and late Holocene. *The Holocene*, 8 (5): 605–612.
Thompson, J.B., 2000. Microbial whitings. In Riding, R.E., and Awramik, S.M. (eds.), *Microbial Sediments*. Springer-Verlag, pp. 250–260.
Vadour, J., 1994. Evolution Holocene des vallées dans le Midi Mediteranée Franċais. *Geographie physique et Quaternaire*, **48**(3): 315–326.
Violante, C., Ferreri, V., D'Argenio, B., and Golubic, S., 1994. Quaternary travertines at Rochetta a Volturno (Isernia, Central Italy): facies analysis and sedimentary model of an organogenic carbonate system. In *Fieldtrip Guidebook A1*. International Association of Sedimentolgists Ischia 94, 15th Regional Meeting, Italy, pp. 3–23.
Waring, G.A., 1965. Thermal springs of the United States and other countries of the World. *U.S. Geological Survey Professional Paper*, Volume 492, 383pp.

Cross-references
Algal and Bacterial Carbonate Sediments
Bacteria in Sediments
Micritization
Speleothems
Stromatolites

TSUNAMI DEPOSITS

A tsunami deposit is a sedimentary layer deposited in low-lying coast regions by a tsunami generated by some form of seafloor deformation, such as earthquakes and submarine landslides, or by a subaerial event that significantly disturbs the water surface, such as a landslide or glacier calving event. Tsunami deposits most commonly form sheets of well-sorted sand that extend inland as much as several kilometers across coastal lowlands. Sheets of coral gravel or pebbles have also been reported. The sand sheets typically contain marine fossils, lack bedform sedimentary structures, and fine both upward and landward. A tsunami may produce a single upward-fining bed, or its waves may produce a vertical series of such beds.

Tsunamis may also deposit boulders. Modern tsunamis are known to have moved boulders as much as 30 m in diameter (Simpkin and Fiske, 1983). Diamicts and imbricated boulder fields have also been ascribed to tsunami, but they so far lack modern analogs.

Although early scientific papers about tsunamis mention their deposits (Platania, 1909; Billings, 1915), tsunami sedimentology probably begins with a paper describing onshore erosion and deposition by the 1960 Chile tsunami on the northeastern coast of Japan (Konno, 1961). The 1960 locally deposited a sheet of sand up to 20 cm thick blanketing coastal lowlands. The sand is described as structureless except for vertical grading, which formed one or more cycles within the bed (Figure T17). The sand filled low places, such as furrows in plowed fields, but it also thinned landward in some places. Konno interpreted the sand as depositing from suspended load.

Tsunamis since 1960 have also produced sand "sheets" on coastal lowlands (Dawson et al., 1996; Minoura et al., 1997; Goldsmith et al., 1999; Caminade et al., 2000). In general, "modern" tsunami deposits have consisted of well-sorted sand sheets up to about 40 cm thick extending inland up to about 300 m inland. The sand source is generally inferred to have been either at the shoreline or just offshore, and the presence of erosional features along the shoreline tends to support the idea that sand is mined from the coastline and moved inland. These later surveys have shown that individual fining upward "pulses" of sedimentation can be traced in the deposit, possibly correlated to individual waves in the tsunami train, and have established the concept of overall landward fining in the direction of wave motion. One result of these recent surveys is a set of criteria for identifying ancient tsunami deposits (Dawson and Shi, 2000).

Criteria for identifying tsunami deposits include both sediment composition and physical properties. In addition to deposit fining landward and upward, researchers have noted that modern tsunami deposits appear to be well-sorted to moderately well sorted and positively skewed (Dawson et al., 1996). Local overwash fans caused by tsunami typically show foreset bedding, but sand sheets have not yielded sedimentary structures other than normal grading. Tsunami deposits contain marine microfossils (Hemphill-Haley, 1996) and often retain evidence of landward movement in the form of bent-over plant debris at the base of the deposit (Atwater and Hemphill-Haley, 1997). Like their modern counterparts, ancient tsunami deposits show some subset of these criteria,

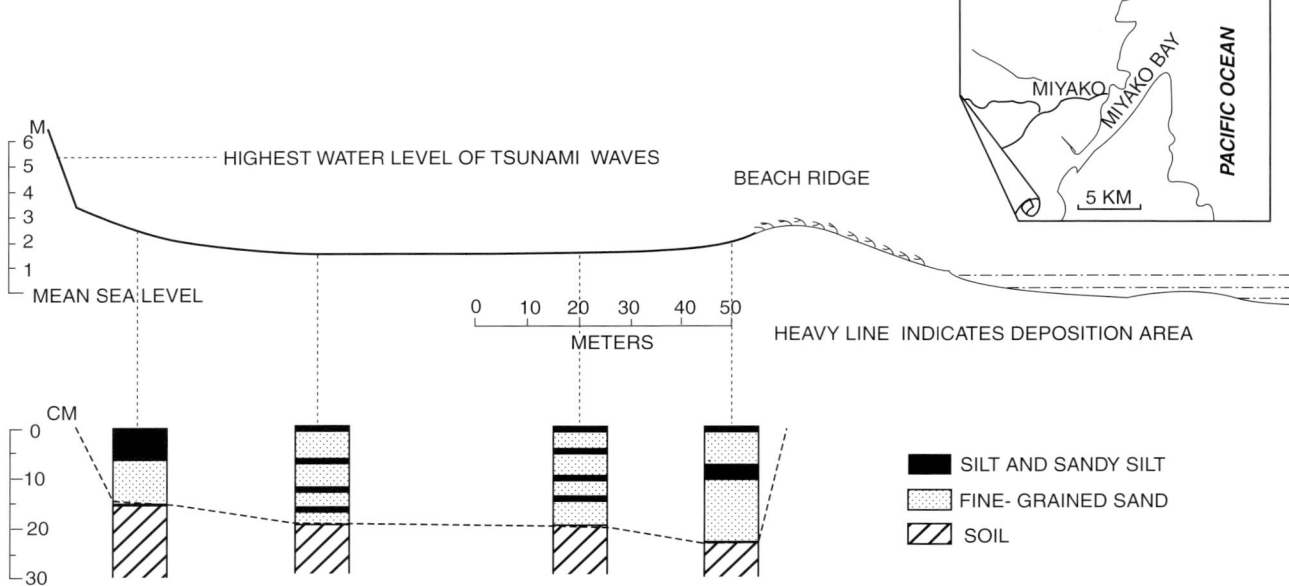

Figure T17 Deposit from the 1960 Chile tsunami near Miyako, Japan, showing four graded units (Konno et al., 1961).

and are generally closely associated with a geologically inferred causal mechanism (e.g., earthquake evidence such as coastal subsidence for subduction zone related tsunamis) (Atwater, 1987).

As early as 1984, onshore deposits ascribed to tsunami have been described in sequences of coastal sediments, and have often served as the first warning of the tsunami hazard faced by coastal communities. Sandy paleotsunami deposits have been described from Scotland (Dawson et al., 1988), Norway (Bondevik et al., 1997), Kamchatka, Japan (Minoura and Nakaya, 1991), New Zealand (Goff et al., 2000), the Mediterranean (Dominey-Howes et al., 1999), and the Pacific coast of North America (Clague et al., 2000). Most are sand sheets in sequences of peaty or muddy deposits of coastal lowlands.

As with sandy deposits, onshore movement of large boulders has been noted in many early descriptions of tsunamis (Makino, 1980; Simpkin and Fiske, 1983; Yamashita, 1997). Modern tsunamis have also moved boulders (McSaveney et al., 2000), and (Yeh et al., 1993) describe boulders from the 1992 Flores tsunami within sandy deposits. These deposits differ, however, from poorly sorted diamictons ascribed to tsunami (Bryant, 2001) in that the Flores boulders lie as individual large blocks protruding from an extensive, thin sand sheet rather than incorporated into the deposit. Ancient tsunamis have also been inferred based solely on the presence of onshore boulder deposits (Bryant et al., 1992; Moore et al., 1994; Hearty, 1997), but these ideas remain controversial because boulders can be transported inland by coastal processes other than tsunamis, such as storms.

Andrew Moore

Bibliography

Atwater, B.F., 1987. Evidence for great Holocene earthquakes along the Outer coast of Washington State. *Science*, **236**: 942–944.

Atwater, B.F., and Hemphill-Haley, E., 1997. Recurrence intervals for great earthquakes of the past 3,500 years at Northeastern Willapa Bay, Washington. *U.S. Geological Survey Professional Paper 1576*.

Billings, L.G., 1915. Some personal experiences with earthquakes. *National Geographic*, **27**: 57–71.

Bondevik, S., Svendsen, J.I., and Mangerud, J., 1997. Tsunami sedimentary facies deposited by the Storegga tsunami in shallow marine basins and coastal lakes, Western Norway. *Sedimentology*, **44**: 1115–1131.

Bryant, E., 2001. *Tsunami: The Underrated Hazard*. Cambridge University Press.

Bryant, E.A., Young, R.W., and Price, D.M., 1992. Evidence of tsunami sedimentation on the southeastern coast of Australia. *Journal of Geology*, **100**: 753–765.

Caminade, J.-P., Charlie, D., Kanoglu, U., Koshimura, S.-I., Matsutomi, H., Moore, A., Ruscher, C., Synolakis, C., and Takahashi, T., 2000. Vanuatu earthquake and tsunami cause much damage, few casualties. *Eos, Transactions, American Geophyiscal Union*, 81, **641**: 646–647.

Clague, J.J., Bobrowsky, P.T., and Hutchinson, I., 2000. A review of geological records of large tsunamis at Vancouver Island, British Columbia, and implications for hazard. *Quaternary Science Reviews*, **19**: 849–863.

Dawson, A.G., Long, D., and Smith, D.E., 1988. The Storegga slides: evidence from eastern Scotland for a possible tsunami. *Marine Geology*, **82**: 271–276.

Dawson, A.G., and Shi, S., 2000. Tsunami deposits. *Pure and Applied Geophysics*, **157**: 875–897.

Dawson, A.G., Shi, S., Dawson, S., Takahashi, T., and Shuto, N., 1996. Coastal sedimentation associated with the June 2nd and 3rd, 1994 tsunami in Rajegwesi, Java. *Quaternary Science Reviews*, **15**: 901–912.

Dominey-Howes, D., Cundy, A., and Croudace, I., 1999. High energy marine flood deposits on Astypalaea Island, Greece: possible

evidence for the Ad 1956 southern Aegean Tsunami. *Marine Geology*, 163: 303–315.
Goff, J.R., Rouse, H.L., Jones, S.L., Hayward, B.W., Cochran, U., McLea, W., Dickinson, W.W., and Morley, M.S., 2000. Evidence for an earthquake and tsunami about 3100–3400 years ago, and other catastrophic saltwater inundations recorded in a coastal lagoon, New Zealand. *Marine Geology*, 170: 231–249.
Goldsmith, P., Barnett, A., Goff, J.R., McSaveney, M., Elliott, S., and Nongkas, M., 1999. Report of the New Zealand reconnaissance team to the area of the 17 July 1998 tsunami at Sissano Lagoon, Papua New Guinea. *Bulletin of the New Zealand Society for Earthquake Engineering*, 33: 102–117.
Hearty, P.J., 1997. Boulder deposits from large waves during the last Interglaciation on North Eleuthera, Bahamas. *Quaternary Research*, 48: 326–338.
Hemphill-Haley, E., 1996. Diatoms as an aid in identifying Late-Holocene tsunami deposits. *The Holocene*, 6: 439–448.
Konno, E. (ed.), 1961. Geological observations of the Sanriku Coastal Region damaged by Tsunami due to the Chile Earthquake in 1960. *Contributions to the Institute of Geology and Paleontology, Tohoku University*, 52, pp. 1–40.
Makino, K., 1980. *Yaeyama No Meiwa Otsunami [the Great Meiwa Tsunami in the Yaeyama Islands]*. Kumamoto, Japan: Jyono Press.
McSaveney, M., Goff, J.R., Darby, D., Goldsmith, P., Barnett, A., Elliott, S., and Nongkas, M., 2000. The 17th July 1998 tsunami, Sissano Lagoon, Papua New Guinea—evidence and initial interpretation. *Marine Geology*, 170: 81–92.
Minoura, K., Imamura, F., Takahashi, T., and Shuto, N., 1997. Sequence of sedimentation processes caused by the 1992 Flores tsunami. *Geology*, 25: 523–526.
Minoura, K., and Nakaya, S., 1991. Traces of tsunami preserved in inter-tidal lacustrine and marsh deposits: some examples from northeast Japan. *Journal of Geology*, 99: 265–287.
Moore, J.G., Bryan, W.B., and Ludwig, K.R., 1994. Chaotic deposition by a giant wave, Molokai, Hawaii. *Geological Society of America Bulletin*, 106: 962–967.
Platania, G., 1909. Il maremoto dello Stretto di Messina del 28 Dicembre 1908. *Bollettino della Società Sismologica Italiana*, 13: 369–458.
Simpkin, T., and Fiske, R.S., 1983. *Krakatau, 1883–the Volcanic Eruption and Its Effects*. Washington, DC: Smithsonian Institution Press.
Yamashita, F., 1997. *Tsunami [in Japanese]*. Tokyo: Ayumi Shuppan.
Yeh, H.H., Imamura, F., Synolakis, C., Tsuji, Y., Liu, P., and Shi, S., 1993. The Flores Island tsunamis. *Eos, Transactions, American Geophyiscal Union*, 74: 371–373.

Cross-reference

Storm Deposits

TURBIDITES

Turbidites are the deposits of turbidity currents, which are gravity-driven turbid suspensions of fluid (usually water) and sediment. They form a class of subaqueous sediment gravity flow (see *Gravity-Driven Mass Flows*) in which the suspended sediment is supported during transport largely or wholly by fluid turbulence. Turbidites range in grain-size from mud to gravel, and may be of any composition—siliciclastic, carbonate, or even sulfide near deep-sea hydrothermal vents. Turbidite beds range in thickness from millimeters to tens of meters, and individual events can, in extreme cases, involve the resedimentation of hundreds of cubic kilometers of sediment. They are amongst the commonest of sedimentary deposits, and turbidite depositional systems such as submarine fans and basin plains form the largest individual sedimentary accumulations on the Earth. Thick sequences of synorogenic turbidites (*flysch*) are common in the ancient record.

Turbidity currents were first so named in 1938, although the existence of density underflows in lakes had been recognized earlier (see review by Walker, 1973). Their potential importance as agents of sediment transport into deep water was not fully recognized till Kuenen and Migliorini (1950) (see *Sedimentologists: Kuenen*) combined experimental work with field observations to suggest that turbidity currents were the cause of graded sandstone beds deposited in deep water. The name *turbidite* for the deposits of turbidity currents came into use about 1960, and Bouma's (1962) analysis of sandstone turbidites shortly thereafter generated the first and, for some time, the sole depositional model. The last three decades of the 20th Century saw an exponential increase in turbidite research. This was driven in part by new techniques of investigation in the deep sea, including the development of side-scan sonar, and drilling by the Deep Sea Drilling Project and its successor, the Ocean Drilling Program. However this increase was also due to the rising importance of turbidites as hydrocarbon reservoirs. Turbidites contain some of the worlds largest and most prolific reservoirs outside of the Middle East, and Cenozoic turbidite systems of the continental margins are a major target for hydrocarbon exploration.

Turbidity currents move by virtue of gravity acting upon the density excess due to the presence of suspended sediment, and flow down-slope, or spread by collapse over low-gradient basin floors. Concentrations of suspended sediment range from a few parts per thousand up to 10 percent or more, though the upper limit for what can truly be regarded as a turbidity current is a matter of debate. Deposition of sediment from the current occurs when fluid turbulence decays, generally as a result of deceleration of the current. Although turbidity currents tend to flow toward gravity base (i.e., the lowest point on the transport path), deposition—especially of coarser grained sediment—may also occur on the slope. Turbidite deposits thus occur on submarine slopes (see *Slope Sediments*), in canyons, slope channels, submarine fans (see *Submarine Fans and Channels*), slope aprons and ramps, and basin plains. They also occur as fans and basin plain deposits in lakes and artificial reservoirs.

Initiation

Turbidity currents are commonly initiated by the failure of unstable material on subaqueous slopes, the failure being triggered by earthquakes, pore fluid pressure imbalance, decomposition of clathrate (*q.v.*), or simply by over-steepening of the slope by deposition (Normark *et al.*, 1991; Rothwell *et al.*, 1998). In these circumstances turbidity currents may arise by the entrainment of water into flows that were initially more viscous and concentrated (such as slumps and debris flows, *q.v.*) that were generated high on the slope or within a submarine canyon. Such currents may deposit sediment over the course of hours. However, some turbidity currents may be generated by highly sediment-charged river outflow where the concentration of suspended material is sufficiently high to render the river effluent denser than the water into which it flows, and where the river discharges directly onto a slope (Mulder and Syvitski, 1995). Such hyperpycnal flows are

GRAIN SIZE	BOUMA (1962) DIVISIONS	INTERPRETATION
Mud	E Laminated to homogeneous mud	Deposition from low-density tail of turbidity current ± settling of pelagic or hemipelagic particles
Silt	D Upper mud/silt laminae	Shear sorting of grains & flocs
Sand	C Ripples, climbing ripples, wavy or convolute laminae	Lower part of lower flow regime of Simons *et al.* (1965)
	B Plane laminae	Upper flow regime plane bed
Coarse Sand	A Structureless or graded sand to granule	Rapid deposition with no traction transport, possible quick (liquefied) bed

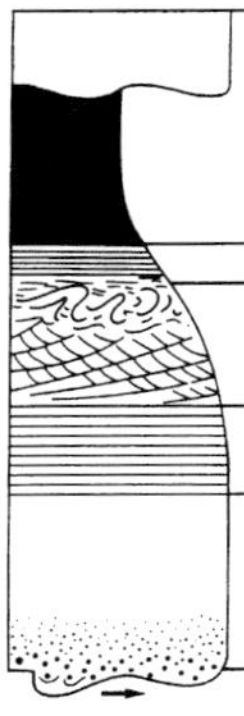

Figure T18 The Bouma sequence of sedimentary structures (and its interpretation) formed by turbidity currents of low to moderate density (after Pickering *et al.*, 1989, reproduced with permission).

more commonly achieved in fresh water, but may occur in the marine realm as frequently as annually in association with "dirty" rivers, or less frequently in association with catastrophic events in the subaerial realm such as major floods, landslides or glacial lake outbreaks (e.g., Zuffa *et al.*, 2000). On their passage down submarine canyons and channels, turbidity currents may erode substantial amounts of sediment from the seafloor, increasing their momentum and erosive power in a positive feedback loop known as autosuspension (*q.v.*) or ignition (Bagnold, 1962; Fukushima *et al.*, 1985). In this way, coarse sediment introduced into canyons between turbidity current events is flushed out, significantly increasing the volume and altering the grain-size mix of the resulting current (e.g., Hughes Clarke *et al.*, 1990).

Depositional models

Turbidites commonly show graded bedding (*q.v.*) and a sequence of sedimentary structures indicative of waning flow during passage of the turbidity current, and progressively declining bed shear stress during deposition of the bed. The rather common occurrence of these structures in a particular order led Bouma (1962) to propose an idealized vertical sequence for graded sand-to-mud turbidites (Figure T18) that remains the standard for comparison in normally-graded sandy turbidites. These divisions have been interpreted in terms of the hydraulic regime that produced them (Figure T18; Walker, 1973) and reflect the longitudinal structure of the current (Allen, 1991). The lowest part of the bed (often overlying a basal erosion surface) commonly consists of structureless graded sandstone—Bouma's T_a division. This is interpreted as the result of deposition from suspension so rapidly as to inhibit movement of material in traction; this may occur via a transient liquefied bed (Middleton, 1967) or zone of hindered settling (*q.v.*). This precludes the formation of traction-generated structures (Allen, 1991, Kneller and Branney, 1995). The T_b division consists of a parallel-laminated interval (see Plane and Parallel Lamination) indicative of an upper stage plane bed, indicating slower deposition rates and an extended period of traction. The ripple-laminated T_c division (see Ripples) commonly shows upward-increasing rates of climb, and by inference increasing rates of deposition. Convolute lamination is common but by

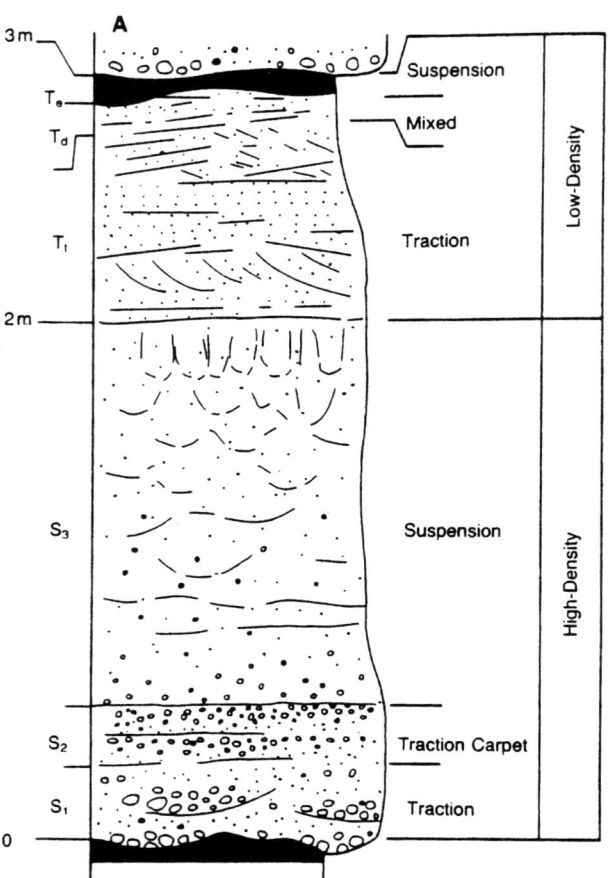

Figure T19 The idealized sequence of sedimentary structures formed by turbidity currents of high density (after Lowe, 1982, reproduced with permission).

no means universal within the T_c division. The T_d division consists of parallel laminated silt or hydraulically equivalent flocculated clay formed under the weakest of currents, while the T_e division represents sedimentation of mud from suspension in the absence (or virtual absence) of a current.

Complete Bouma sequences are rather uncommon. Partial sequences in which lower divisions may be missing ("base-absent") are widespread, especially in the distal parts of turbidite systems, or in overbank settings where the finer-grained upper parts of turbidity currents have spilled out of channels (see Submarine Fans and Channels). Divisions within the sequence may also be absent or poorly developed (e.g., T_{abce}), sometimes with a sharp grain-size break at the level of the missing division.

The Bouma sequence describes some of the features associated with deposits of currents of low to moderate density. An analogous model for the deposits of high density turbidity currents was proposed by (Lowe, 1982) (Figure T19). A lowermost S_1 division is dominated by traction of coarse-grained material. This is overlain by an S_2 division dominated by inverse-graded layers, interpreted by Lowe as the deposits of a succession of "frozen" traction carpets; subtle inverse-to-normally graded layers corresponding to this division may represent deposition during the passage of large eddies (Hiscott, 1994; Kneller et al., 1999). Corresponding R1 and R2 intervals are differentiated in gravelly turbidites. The overlying S_3 division (more or less corresponding to an expanded T_a division) consists of sand that is structureless or dominated by fluid escape structures (q.v.). The ensuing T_t division is dominated by traction and in part this corresponds to Bouma's T_b and/or T_c divisions, but may also include cross-stratified sands, sometimes well-sorted, that reflect partial bypassing of the flow (Mutti, 1992). As with the Bouma sequence, the model for high-density turbidity current deposits is idealized. Basal divisions may be absent, and the deposit may include grain-size breaks, especially common above or below intervals with cross-stratification (q.v.).

Comparable models for fine grained turbidites have been developed by, amongst others, Piper (1978) and Stow and Shanmugam (1980). Facies classifications of turbidites attest to the wide diversity of depositional facies (Ghibaudo, 1992; Mutti, 1992; Pickering et al., 1989), most of which do not conform to the idealized models, and reflect the range of variables that affect deposition (Kneller, 1995).

Other depositional and post-depositional features

The base of coarse-grained turbidites commonly reveals evidence of erosion. This may take the form of casts of erosional structures such as flutes (see *Scour, Scour Marks*) or tool marks (q.v.) formed in fine grained substrate (commonly the muddy top of the underlying turbidite) during passage of the frontal, nondepositional parts of the flow (Dzulynski and Sanders, 1959). These form good paleocurrent indicators (see *Paleocurrent Analysis*). Such structures often contain grains that are significantly coarser than those in the remainder of the bed, suggesting that the head of the current carries the coarsest grains. Trace fossils preserved on the sole of the bed also bear witness to this erosion. Where erosion has completely removed the muddy top of an underlying graded sand-to-mud bed, the sandy portions of successive beds become amalgamated. Where repeated this can produce composite "beds" of sand many meters thick in which the amalgamation surfaces may be hard to discern. Such amalgamated beds are common in thick-bedded sandy and gravelly turbidite sequences such as those found in submarine canyons and channels.

Bases of turbidite sandstones commonly show load structures (q.v.), often associated with liquefaction structures (see *Liquefaction and Fluidization*) in the underlying and/or overlying bed. The load casts often amplify pre-existing perturbations of the bed, such as flute casts. Dewatering structures (commonly developed in S_3 divisions) may be arranged in a vertical sequence from consolidation laminae (wavy and/or anastomosing) near the base, through dish structures (q.v.) to pipes or sheets, which in some instances may be inclined down-current due to shear within the dewatering sediment mass. Liquefaction of the deposit can result in the creation of sand volcanoes on the tops of beds (where liquefaction of the bed occurred at the surface) or injection of clastic dykes and sills (q.v.) into surrounding sediments (Nicholls, 1995), upward, downward or sideways from the parent bed. Another common feature of the deposits of presumed moderate to high concentration turbidity currents is the presence of shale clasts, some of which may have experienced considerable transport after incorporation into the flow during initiation or by erosion of the bed, while others may be incorporated during deposition via a highly concentrated (perhaps liquefied) depositional layer that may behave as a transient debris flow (Stow and Johansson, 2000).

Some turbidity currents experience reflection from basin-bounding surfaces and intrabasinal topography, resulting in distinctive current changes within the bed, commonly associated with small reversals in grading (Pickering and Hiscott, 1985; McCaffrey and Kneller, 2001).

Current controversies

Considerable debate exists about what constitutes a turbidity current, and thus what may or may not correctly be interpreted as a turbidite (Shanmugam, 2000). Some authors (e.g., Middleton and Hampton, 1973; Mulder and Alexander, 2001) restrict their definition of turbidity currents to flows in which the grains are supported almost exclusively by fluid turbulence, whereas others (e.g., Lowe, 1982) take a broader view. It is useful to recall that the root of the term turbidity current is the word *turbid* (i.e., opaque with sediment) as distinct from *turbulent* (i.e., disturbed by eddies). Much uncertainty arises from the lack of direct observations from large natural currents, and from the difficulty in producing in the laboratory nondepositional sand-bearing currents of even moderate density. This renders the debate largely speculative.

Ben C. Kneller

Bibliography

Allen, J.R.L., 1991. The Bouma division A and the possible duration of turbidity currents. *Journal of Sedimentary Petrology*, **61**: 291–295.
Bagnold, R.A., 1962. Auto-suspension of transported sediment; turbidity currents. *Proceedings of the Royal Society of London, Series*, **A265**: 315–319.
Bouma, A.H., 1962. *Sedimentology of Some Flysch Deposits; a Graphic Approach to Facies Interpretation*. Elsevier.
Dzulynski, S., and Sanders, J.E., 1959. Bottom marks on firm lutite substratum underlying turbidite beds. *Geological Society of America Bulletin*, **70**: 1594.
Fukushima, Y., Parker, G., and Pantin, H.M., 1985. Prediction of ignitive turbidity currents in Scripps submarine canyon. *Marine Geology*, **67**: 55–81.
Ghibaudo, G., 1992. Subaqueous sediment gravity flow deposits; practical criteria for their description and classification. *Sedimentology*, **39**: 423–454.
Hampton, M.A., 1972. The role of subaqueous debris flow in generating turbidity currents. *Journal of Sedimentary Petrology*, **42**: 775–793.
Hiscott, R.N., 1994. Traction-carpet stratification in turbidites; fact or fiction? *Journal of Sedimentary Research*, **A64**: 204–208.
Hughes Clarke, J.E., Shor, A.N., Piper, D.J.W., and Mayer, L.A., 1990. Large-scale current-induced erosion and deposition in the path of the 1929 Grand Banks turbidity current. *Sedimentology*, **37**: 613–629.
Kneller, B., 1995. Beyond the turbidite paradigm; physical models for deposition of turbidites and their implications for reservoir prediction. In Hartley, A., and Prosser, D.J. (eds.), Characterization of deep marine clastic systems. *Geological Society Special Publications*, Volume 94, pp. 31–49.
Kneller, B.C., Bennett, S.J., and McCaffrey, W.D., 1999. Velocity structure, turbulence and fluid stresses in experimental gravity currents. *Journal of Geophysical Research, Oceans*, **104**: 5281–5291.
Kneller, B.C., and Branney, M.J., 1995. Sustained high-density turbidity currents and the deposition of thick massive sands. *Sedimentology*, **42**: 607–616.

Kuenen, P.H., and Migliorini, C.I., 1950. Turbidity currents as a cause of graded bedding. *Journal of Geology*, **58**: 91–127.

Lowe, D.R., 1982. Sediment gravity flows: II, depositional models with special reference to the deposits of high-density turbidity currents. *Journal of Sedimentary Petrology*, **52**: 279–297.

McCaffrey, W., and Kneller, B., 2001. Process controls on the development of stratigraphic trap potential on the margins of confined turbidite systems and aids to reservoir evaluation. *AAPG Bulletin*, **85**: 971–988.

Middleton, G.V., 1967. Experiments on density and turbidity currents; [Part] 3, deposition of sediment. *Canadian Journal of Earth Sciences*, **4**: 475–505.

Middleton, G.V., and Hampton, M.A., 1973. Sediment gravity flows; mechanics of flow and deposition. In Middleton, G.V., and Bouma, A.H., (eds.), *Turbidites and Deep-Water Sedimentation*. Society of Economic Paleontologists and Mineralogists, Pacific Section pp. 1–38.

Mulder, T., and Alexander, J., 2001. The physical character of subaqueous sedimentary density flows and their deposits. *Sedimentology*, **48**: 269–299.

Mulder, T., and Syvitski, J.P.M., 1995. Turbidity currents generated at river mouths during exceptional discharges to the world oceans. *Journal of Geology*, **103**: 285–299.

Mutti, E., 1992. *Turbidite Sandstones*. Agip: San Donato Milanese, 275pp.

Nicholls, R.J., 1995. The liquification and remobilization of sandy sediments. In Hartley, A., and Prosser, D.J., (eds.), *Characterization of Deep Marine Clastic Systems*. Geological Society Special Publications, Volume 94, pp. 63–76.

Normark, W.R., Piper, D.J.W., and Osborne, R.H., 1991. Initiation processes and flow evolution of turbidity currents; implications for the depositional record. In Osborne, R.H., (ed.), *From Shoreline to Abyss; Contributions in Marine Geology in Honor of Francis Parker Shepard*. Society of Economic Paleontologists and Mineralogists, Volume **46**, pp. 207–230.

Pickering, K.T., and Hiscott, R.N., 1985. Contained (reflected) turbidity currents from the Middle Ordovician Cloridorme Formation, Quebec, Canada; an alternative to the antidune hypothesis. *Sedimentology*, **32**: 373–394.

Pickering, K.T., Stow, D.A.V., Watson, M.P., and Hiscott, R.N., 1986. Deep-water facies, processes and models: a review and classification scheme for modern and ancient sediments. *Earth Science Reviews*, **23**: 75–174.

Piper, D.J.W., 1978. Turbidite muds and silts on deepsea fans and abyssal plains. In Stanley, D.J., and Kelling, G. (eds.), *Sedimentation in Submarine Canyons, Fans, and Trenches*. Dowden: Hutchinson & Ross, pp. 163–175.

Rothwell, R.G., Thomas, J., and Kaehler, G., 1998. Low-sea-level emplacement of very large late Pleistocene "megaturbidite" in the Western Mediterranean Sea. *Nature*, **392**: 377–380.

Shanmugam, G., 2000. 50 years of the turbidite paradigm (1950s–1990s); deep-water processes and facies models; a critical perspective. *Marine and Petroleum Geology*, **17**: 285–342.

Stow, D.A.V., and Johansson, M., 2000. Deep-water massive sands; nature, origin and hydrocarbon implications. *Marine and Petroleum Geology*, **17**: 145–174.

Stow, D.A.V., and Shanmugam, G., 1980. Sequence of structures in fine-grained turbidites; comparison of Recent deep-sea and ancient flysch sediments. *Sedimentary Geology*, **25**: 23–42.

Walker, R.G., 1973. Mopping up the turbidite mess. In Ginsburg, R.N. (ed.), *Evolving Concepts in Sedimentology*. Johns Hopkins University, Studies in Geology, Volume 21, pp. 1–37.

Zuffa, G.G., Normark, W.R., Serra, F., and Brunner, C.A., 2000. Turbidite megabeds in an oceanic rift valley recording jokulhlaups of late Pleistocene glacial lakes of the Western United States. *Journal of Geology*, **108**: 253–274.

Cross-references

Autosuspension
Bedding and Internal Structures
Clastic (Neptunian) Dykes and Sills
Clathrates
Convolute Lamination
Cross-Stratification
Debris Flow
Dish Structure
Fluid Escape Structures
Grading, Graded Bedding
Grain Flow
Gravity-Driven Mass Flows
Hindered Settling
Liquefaction and Fluidization
Load Structures
Paleocurrent Analysis
Planar and Parallel Lamination
Ripple, Ripple Mark, Ripple Structures
Sedimentologists
Slide and Slump Structures
Slope Sediments
Submarine Fans and Channels
Tool Marks

U

UPWELLING

Introduction
One principle characteristic of upwelling systems is that they occur predominantly in the same location due to the combination of physiographic factors (coastline, shelf bathymetry) and seasonal features such as wind strength and direction. Upwelling may occur in the open ocean as a consequence of divergence between major current and frontal systems, or along continental margins (often referred to as coastal upwelling). Although both open ocean and coastal upwelling influence the underlying sedimentary record, the predominant diagnostic sedimentary facies relation to upwelling are developed along continental margins. The hydrographic features of an upwelling system commonly drive undepleted nutrients (especially nitrate and phosphate) into the euphotic zone resulting in intense, and often seasonal, primary production of phytoplankton and related trophic grazing by zooplankton, fish, and higher predators. This high primary and secondary production produces large quantities of biomass which subsequently sinks and decays during transit through the water column resulting in major decreases in dissolved oxygen levels caused by aerobic microbial respiration of the decomposing organic matter. Where the "oxygen minimum zone" impinges on the seafloor the accumulating sediments are commonly anaerobic giving rise to a particular set of facies relationships and diagenetic processes (Figure U1). Frequently, at continental margins where wind-driven upwelling occurs, there are high levels of lithogenic detritus derived from the adjacent land mass added to the organic-rich anaerobic sediment. At the present day, such coastal upwelling systems are characteristic of continental margins off California, Peru, Mauritania, Namibia, Oman, Pakistan, and Australia with open ocean upwelling systems characterized by the Eastern Equatorial Pacific and the Antarctic Polar Front.

Historical development
The study of upwelling systems and their resulting imprint on seafloor sediments has long attracted significant scientific study (e.g., see the following edited volumes; Theide and Suess, 1983; Summerhayes *et al.*, 1992; Summerhayes *et al.*, 1995). In part, this has been due to the long-known association between organic matter, black metalliferous shales and siliceous deposits characteristic of petroleum source rocks (e.g., the Miocene Monterey Formation of the western USA), but more latterly, the emerging significance of upwelling systems as part of the oceanic carbon cycle and resulting implication for the climate system. Through the study of such systems, there have grown up important geological paradigms, such as the enhanced preservation of organic matter under anaerobic depositional conditions (Demaison and Moore, 1980), and subsequent challenges to the paradigm based on detailed modern oceanographic studies (Pedersen and Calvert, 1990). Other major areas of study have centered on the issues of allocthonous versus autocthonous carbon burial at shelf depocenters (Walsh, 1991; Jahnke and Shimmield, 1995), and increased levels of carbon burial at upwelling centers during past glacial episodes (Muller *et al.*, 1983). Current research aims to use the complex composition of biogenic microfossils, sediment organic and inorganic geochemistry, and microfabric lithology (laminations) to reconstruct the history of upwelling systems over a variety of timescales from decades (e.g., reconstructing El Nino variations), to tens of thousands of years relevant to the study of the waxing and waning of the ice sheets and associated climatic change.

Current understanding and applications to paleoceanography
Coastal upwelling systems, although small in areal extent compared to whole ocean, account for over 50 percent of oceanic primary production and more than 95 percent of the sedimentary accumulation of organic carbon (Romankevitch, 1984). The accumulating sediment, derived from organic matter, biogenic skeletal material (calcite and opal), fish scales and bones, and lithogenic minerals carried by wind and fluvial

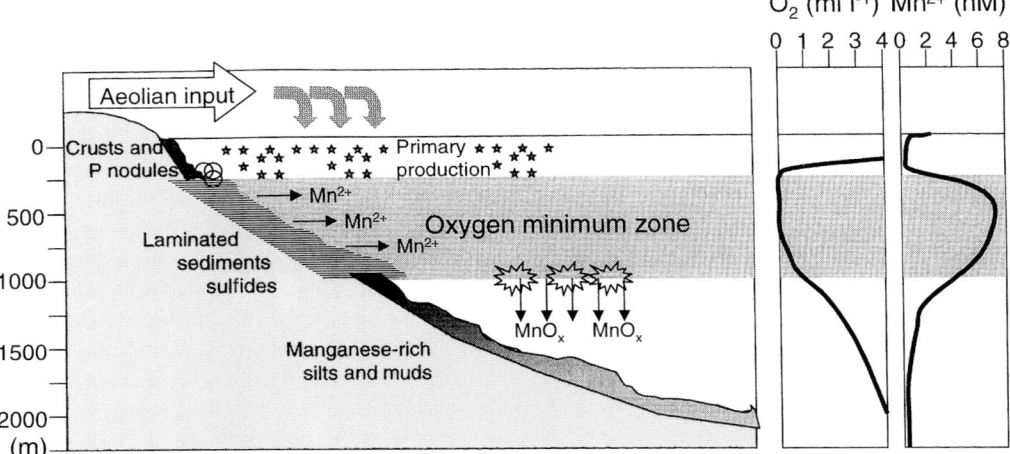

Figure U1 Schematic diagram illustrating the effect of an intense oxygen minimum zone, produced as a consequence of upwelling water fuelling primary production and subsequent microbial decay of the organic matter, on the sediment lithology and geochemistry. Such OMZ formation can result in substantial leakage of dissolved Mn into the water column and resulting recycling and reprecipitation in underlying sediments. The profiles of dissolved oxygen and manganese are typical of a modern upwelling system, such as in the northwest Arabian Sea.

discharge, become augmented by diagenetic processes leading to the concentration of phosphorites and sulfide minerals under the suboxic and anoxic sedimentary conditions prevailing. Linking the processes sediment supply to the locus of sedimentary deposition is notoriously difficult at continental margins (Jahnke and Shimmield, 1995; Bertrand *et al.* 1996). Frequently, the organic matter actually accumulating may be older than the surrounding sediment based on ^{14}C age dating (Anderson *et al.*, 1994), or the quantity of carbon in the sediment does not appear to be related to the bottom water oxygen (Pedersen *et al.*, 1992), but instead to the intensity of sediment resuspension and reworking. Such physical reworking may be relatively common at coastal upwelling centers—the Oman margin is characterized by refractory carbon diagenesis (Pedersen *et al.*, 1992) and deep sulfate reduction in the pore waters, and the Peru margin has large areas of hard pavement and phosphorite pellet beds (Jahnke and Shimmield, 1995). In contrast, the absence of oxygen in bottom waters at continental margin upwelling sites can lead to an absence of benthic infaunal and minimal sediment disturbance through bioturbation. Under these conditions quite labile organic matter is found and the sediment retains a remarkable microfabric with annual laminations related to the changing surface ocean productivity (Brodie and Kemp, 1994). Even the changing percentages of diatoms, radiolaria, and foraminifera in the sedimentary laminations may be used to interpret interannual variablity in upwelling associated with events such as El Nino (Lange *et al.*, 1987). Even under open ocean upwelling conditions, such as those associated with the eastern equatorial Pacific divergence, Kemp and Baldauf (1993) have found historical evidence for very rapid deposition of giant diatom mats and resulting excellent preservation in the sedimentary record.

The geochemistry of sediments underlying upwelling centers is varied and fascinating. A wide range of redox environments are commonly found, often with redox gradients in the sediments that can be measured on a scale of millimeters, rather than meters commonly found in open ocean sedimentary settings. Rapid consumption of oxygen, nitrate, and sulfate is commonly observed with attendant dissolution of carbonate and degradation of the organic matter. Sulfide mineral precipitation occurs during sulfate reduction and the release of iron III from the oxidized coatings on mineral grains transported by winds from the adjacent, often arid, landmass. One key factor influencing organic matter preservation under upwelling systems is burial rate. Often the sediments are accumulating relatively rapidly (say $>5\,\text{cm}\,\text{ky}^{-1}$) giving rise to burial efficiencies of greater than 10 percent. The elemental and isotopic composition of the organic matter (*Kerogen, q.v.*) has long been used as a proxy for the changes in upwelling, circulation, and bottom water oxygen (Reimers and Suess, 1983). Over the past 15 years the discovery that changes in the C_{37} alkenone unsaturation index (U_{37}^k) could be related to the water temperature in which the phytoplankton lived, as enabled reconstructions of past upwelling history such as for the Peru margin (McCaffrey *et al.*, 1990). Upwelling centers also provide inorganic proxy indicators of changes in oceanic productivity, such as the sedimentary concentration of the element barium in the mineral phase, barite. Bishop (1988) showed the precipitation of barite in the decaying organic matter within frustules of sinking diatoms. Von Breymann *et al.* (1992) summarized the known information on high barium in hemipelagic sediments but also identified the potential Achilles' heel for the use of Ba as a paleoproductivity proxy in organic-rich sediments; namely that strong sulfate reduction diagenesis caused sedimentary barite to dissolve.

The changes in the faunal composition of the upwelling sediments can give important clues as to the temperature of the upwelled waters, the intensity of the primary and secondary production, and the species succession through a seasonal upwelling event. Typically, the calcareous tests of the foraminifera are most often studied, both for species diversity and abundance, and the oxygen and carbon isotopic record they contain. Particular species have now come to be associated typically with upwelling conditions, such as *Globigerina bulloides* (Prell and Curry, 1981). However, close examination of both the species succession, and the isotopic make-up (particularly the $\delta^{13}C$ signature) reveals complexity in the response of production of the foraminifera, the depth of the thermocline, and the origin of the upwelled water (Kroon and Ganssen, 1989).

Summary

The sediments underlying upwelling systems, and particularly those at continental margins, are a very important repository of organic matter in the world's biogeochemical cycle of carbon. The upwelling process fuels high levels of marine primary productivity which induces low levels of dissolved oxygen in the overlying water column during decomposition of the sinking organic matter. The resulting sediment, rich in organic carbon and with accelerated diagenetic reactions, ultimately forms some of the most valuable hydrocarbon source rock. In certain cases, phosphorite formation augments the mineralogical value. Modern upwelling systems have become the type areas for testing paradigms on organic matter preservation and for proving inorganic and organic proxy paleo-indicators of productivity and ocean temperature. Intense episodes of oceanic productivity have been recorded in massive diatom mat accumulations, and under certain conditions, well preserved laminated sediments deposited under anaerobic conditions are helping to provide a detailed historic record of interannual climatic variability, such as that associated with the El Nino episodes in the Pacific.

Graham Shimmield

Bibliography

Anderson, R.F., Rowe, G.T., Kemp, P., Trumbore, S., and Biscaye, P.E., 1994. Carbon budget for the mid-slope depocenter of the Mod-Atlantic Bight. *Deep-Sea Research*, **41**: 669–703.

Bertrand, P., Shimmield, G., Martinez, P., Grousset, F., Jorissen, F., Paterne, M., Pujol, C., Bouloubassi, I., Buat-Menard, P., Peypouquet, J.-P., Beaufort, L., Sicre, M.-A., Lallier-Verges, E., Foster, J.M., and Ternois, Y., 1996. The glacial ocean productivity hypothesis: the importance of regional temporal and spatial studies. *Marine Geology*, **130**: 1–9.

Bishop, J.K.B., 1988. The barite-opal-organic carbon association in oceanic particulate matter. *Nature*, **332**: 341–343.

Brodie, I., and Kemp, A.E.S., 1994. Variation in biogenic and detrital fluxes and formation of laminae in late Quaternary sediments from the Peruvian coastal upwelling zone. *Marine Geology*, **116**: 385–398.

Demaison, G.J., and Moore, G.T., 1980. Anoxic environments and oils source bed genesis. *Organic Geochemistry*, **2**: 9–31.

Jahnke, R.A., and Shimmield, G.B., 1995. Particle flux and its conversion to the sediment record: coastal ocean upwelling systems. In Summerhayes, C.P., Emeis, K.-C., Angel, M.V., Smith, R.L., and Zeitzschel, (eds.), *Upwelling in the Ocean: Modern Processes and Ancient Records*. John Wiley and Sons, pp. 83–100.

Kemp, A.E.S., and Baldauf, J., 1993. Vast Neogene laminated diatom mat deposits from the eastern equatorial Pacific Ocean. *Nature*, **362**: 141–144.

Kroon, D., and Ganssen, G., 1989. Northern Indian Ocean upwelling cells and stable isotope composition of living planktonic foraminifers. *Deep-Sea Research*, **36**: 1219–1236.

Lange, C., Berger, W.H., Burke, S.K., Casey, R.E., Schimmelmann, A., Soutar, A., and Weinheimer, A.L., 1987. El Nino in the Santa Barbara basin: diatom, radiolarian, and foraminiferal responses to the 1983 El Nino event. *Marine Geology*, **78**: 153–160.

McCaffrey, M.A., Farrington, J.W., and Repeta, D.J., 1990. The organic geochemistry of Peru margin surface sediments–1. A comparison of the C_{37} alkenone and historical El Nino records. *Geochimica Cosmochimica Acta*, **54**: 1671–1682.

Muller, P.J., Erlenkeuser, H., and van Grafenstein, R., 1983. Glacial-interglacial cycles in oceanic productivity inferred from organic carbon contents in eastern North Atlantic sediment cores. In Suess, E., and Theide, J. (eds.), *Coastal Upwelling: Its Sediment Record, Part B*. Plenum, pp. 365–398.

Pedersen, T.F., and Calvert, S.E., 1990. Anoxia versus productivity: what controls the formation of organic carbon-rich sediments and sedimentary rocks. *American Association of Petroleum Geologists Bulletin*, **74**: 454–466.

Pedersen, T.F., Shimmield, G.B., and Price, N.B., 1992. Lack of enhanced preservation of organic matter in sediments under the oxygen minimum on the Oman Margin. *Geochimica et Cosmochimica Acta*, **56**: 545–551.

Prell, W., and Curry, W.B., 1981. Faunal and isotopic indices of monsoonal upwelling: western Arabian Sea. *Oceanologica Acta*, **4**: 91–98.

Reimers, C.E., and Suess, E., 1983. Late Quaternary fluctuations in the cycling of organic matter off central Peru: A proto-kerogen record. In Suess, E., and Theide, J. (eds.), *Coastal Upwelling and its Sedimentary Record, Part A*. Plenum, pp. 497–526.

Romankevitch, E.A., 1984. *Geochemistry of Organic Matter in the Ocean*. Springer.

Summerhayes, C.P., Emeis, K-C., Angel, M.V., Smith, R.L., and Zeitzschel, B., 1995. *Dahlem Workshop on Upwelling in the Ocean: Modern Processes and Ancient Records*. John Wiley and Sons.

Summerhayes, C.P., Prell, W.P., and Emeis, K.C. (eds.), 1992. *Upwelling Systems: Evolution Since the Early Miocene*. Geological Society of London Special Publication, 64.

Theide, J., and Suess, E., (eds.), 1983. *Coastal Upwelling, Its Sedimentary Record, Part A and Part B*. Plenum.

Von Breymann, M.T.K., Emeis, K.-C., and Suess, E., 1992. Water depth and diagenetic constraints on the use of barium as a paleoproductivity indicator. In *Upwelling Systems: Evolution Since the Early Miocene*. Geological Society of London, Special Publication, 64, pp. 273–284.

Walsh, J.J., 1991. The importance of continental margins in the marine biogeochemical cycling of carbon and nitrogen. *Nature*, **350**: 53–55.

Cross-references

Black Shales
Kerogen
Mudrocks
Phosphorites
Sapropel
Slope Sediments

VARVES

A varve is a demonstrably annual sedimentary deposit. Implied is rhythmicity in one or more observable characteristics occurring as a response to seasonal fluctuations of physical, chemical, or biological processes. The term originated with de Geer (1912) from the Swedish word *varv* meaning a cyclic event or form regardless of periodicity. However, de Geer and others since have emphasized that a varve represents a yearly cycle, and this remains fundamental to the value of the concept.

The depositional environment is normally aquatic, although varves may occur subaerially, for example, as a result of seasonally varying eolian processes (Stokes, 1964) or snowfall as it transforms to glacial ice. Aquatic varves are best known from lacustrine sediments, although they have also been reported extensively in marine deposits (Fisher, 1990). *Clastic varves* normally consist of fine-grained sediment deposited in low-energy conditions when inflow of water and sediment is small, and coarser sediment deposited in response to large inflows and vigorous circulation. This produces the respective dark and light couplets that distinguish classical glacilacustrine varves (Figure V1) but clastic varves are also produced in nonglacial lakes where there is a large seasonal variation in sediment input, for example in the monsoon climate of southeast Asia (Ross and Gilbert, 1998). Very low settling rates of clay-sized particles (about 1 mm/d to 1 m/d) according to Stokes Law, necessitate flocculation to permit deposition in one winter, especially in deep lakes. In summer, turbidity currents commonly transport coarser sediment throughout the lake to create numerous graded laminae.

Biogenic varves form by deposition of allochthonous and autochthonous organic matter produced in seasonal cycles, including pollen in lakes and diatoms in the sea. In lakes having water charged with dissolved sediments, *chemogenic varves* may form by precipitation of salts, especially calcium carbonate, from supersaturated solution, for example during summer when water temperature is elevated (Anderson, 1988). In some situations, higher concentration of carbonate in summer deposits is due to rapid settling and quick burial of larger particles, whereas in winter, fine particles settle slowly and carbonate is dissolved in the cold lake water. Changes in redox conditions in glacimarine environments have been shown to produce varves in glacimarine sediments (Stevens, 1986). In both lakes and the ocean, fine-scale varves are best preserved in the absence of bioturbation, commonly in permanently or intermittently anoxic basins.

The value of varves depends on demonstrating that the observed cyclicity is annual. This may be done from independent evidence, including radiometric dating, correlation to other time series proxies such as tree-ring records, relation to dated time-stratigraphic markers such as tephra, comparison to measured or calculated rates of sediment accumulation (e.g., in sediment traps), or by comparison of characteristics of the deposits with hydrologic or climatic controls (Figure V1). However, many accounts of varves depend only on the inference that strong, regular cyclicity must be annual. Where the annual character cannot be established, cyclic deposits should be referred to as *rhythmites*.

Counting varves provides an important dating tool, although errors may occur if more than one rhythmite occurs in a given year, or if seasonal variation is sufficiently muted in one or more years that a varve is not recognized. The former is common where extreme events, such as major storms (Figure V1), interspaced with low-energy conditions occur irregularly. The latter may occur, for example, when a mild winter maintains water and sediment input to the lake or sea, or during a cool dry summer when inflow is reduced or precipitation of salts fails to occur. Over- or under-counting of varve time series can be reduced by comparison of the thickness time series of varve records both from the same depositional environment (e.g., a single lake) and from different, widely separated environments.

More valuable, however, is the application of varve records as proxies in paleoenvironmental assessment. Thickness of the whole varve or of its seasonal components provides a measure of the inter-annual variations of the climatic, hydrologic, or aquatic processes that produced it (Kemp, 1996). Thickness variation also provides a measure of the fluctuations in the

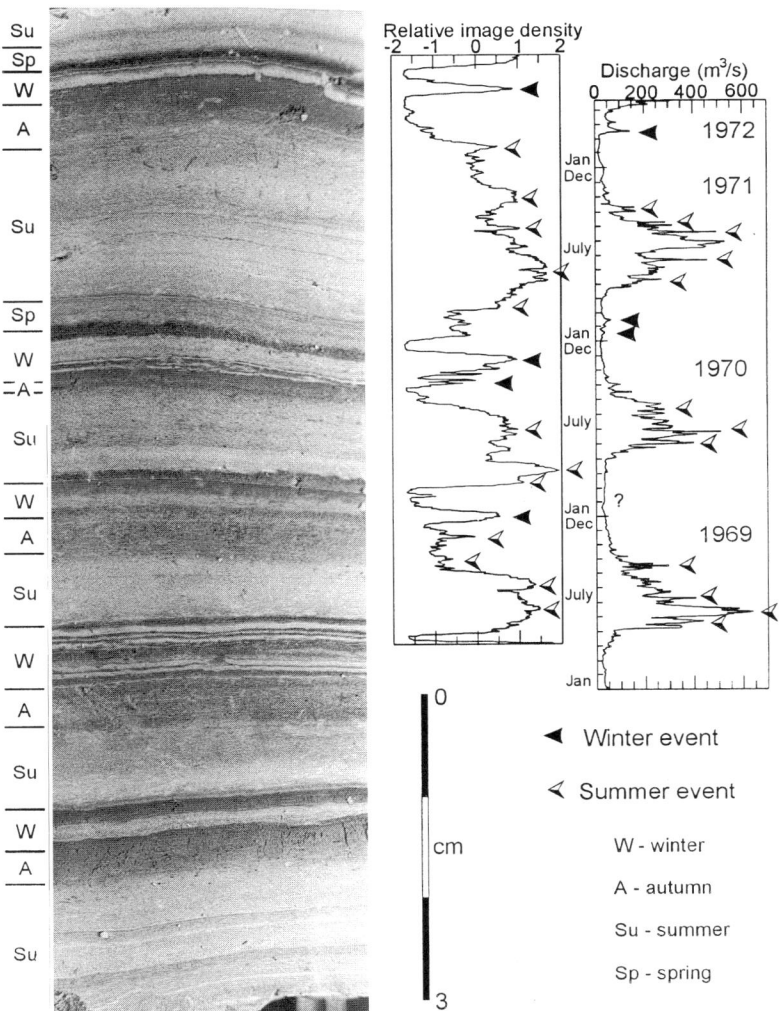

Figure V1 Glacilacustrine sediment from Lillooet Lake, British Columbia, Canada (50.7°N 123°W) showing varves deposited between 1967 and 1972. Winter deposits dominated by deposition of clay-sized sediment are dark (except where winter storm events raise inflow enough to flush sediment stored in the inflowing stream bed); summer silt and fine sand are light. Image density is presented as a relative scale in standard deviations from the mean density of the image. Symbols mark inferred corresponding peaks of light tone (coarser sediment) and higher mean daily inflow to the lake, distinguished by winter and summer events. Evidence of "spring" deposits in glacial varves is commonly missing because of the rapid increase in discharge associated with nival melt. The correspondence is not exact because (1) discharge of other streams representing 47 percent of the drainage basin does not correspond exactly with the discharge of the principal stream shown, (2) sediment load and discharge do not correspond exactly (e.g., sediment concentration during nival (spring) melt is lower than during glacial (summer) melt), (3) wind- and inflow-induced lacustrine circulation varies, and (4) sediment of varying grain size settles from the water column at different rates.

rates of sediment accumulation from year to year. Such time series have been successfully constructed from single lakes over periods of thousands of years (Zolitschka and Negendank, 1996), or by teleconnection between lacustrine or marine records having overlapping portions. Intra-annual variations based on the sedimentary characteristics of laminae within varves may be examined microscopically, for example in thin section, to assess climatic events, including temperature and precipitation, hydrologic events, especially the inflow of water and sediment (Figure V1), and the magnitude and frequency of extreme events affecting sedimentation.

Robert Gilbert

Bibliography

Anderson, R.Y., 1988. Lacustrine varve formation through time. *Paleogeography, Paleoclimatology, Paleoecology*. **62**: 215–235.

De Geer, G., 1912. A geochronology of the last 12,000 years. *Proceedings of the International Geological Congress*, Stockholm (1910), **1**: 241–253.
Fisher, C.G., 1990. Bibliography and inventory of holocene varved and laminated marine sediments. *National Oceanic and Atmospheric Administration (NOAA) Paleoclimate Publications Series Report No. 1*, 107p.
Kemp, A.E.S. (ed.), 1996. *Palaeoclimatology and Palaeooceanography from Laminated Sediments*. Geological Society, Special Publication, 116, London, 258p.
Ross, J., and Gilbert, R., 1998. Lacustrine sedimentation in a monsoon environment, the record from Phewa Tal, Middle Mountain region of Nepal. *Geomorphology*, **27**: 307–323.
Stevens, R., 1986. Glaciomarine varves and the character of deglaciation, Savean valley, southwestern Sweden. *Boreas*, **15**: 289–299.
Stokes, W.L., 1964. Eolian varving in the Colorado Plateau. *Journal of Sedimentary Petrology*. **34**: 429–432.
Zolitschka, B., and Negendank, J.F.W., 1996. Sedimentology, dating, and palaeoclimatic interpretation of a 76.2 ka record from Lago Grande de Monticchio, southern Italy. *Quaternary Science Reviews*, **15**: 101–112.

Cross-references

Climatic Control of Sedimentation
Colors of Sedimentary Rocks
Cyclic Sedimentation
Glacial Sediments: Processes, Environments, and Facies
Grain Size and Shape
Impregnation
Lacustrine Sedimentation
Upwelling

VERMICULITE

Introduction

"Vermiculite" is derived from the Latin word *vermiculus* (= small worm) which refers to the wormlike projections this mineral forms when subjected to high temperatures. Vermiculite is a hydrous 2 : 1 phyllosilicate similar in structure to mica, but has a lower layer charge, and contains hydrated exchangeable cations in the interlayers versus "non-exchangeable" K^+ in the interlayers of mica. Vermiculites occur in macroscopic and clay-sized forms. The macroscopic vermiculite is invariably trioctahedral in composition, whereas the clay-sized vermiculite may be either dioctahedral or trioctahedral and is frequently found in soil environments. The structure, crystal chemistry, mineralogy, and surface chemistry of vermiculite are responsible for the important roles they play in natural environments as well as their useful environmental and industrial applications.

Structure and composition

The basic structure of vermiculite is similar to that of mica. As in mica, vermiculite consists of one octahedral sheet sandwiched between two tetrahedral sheets T_2O_5 or T_4O_{10} (T = tetrahedral cation). Normally, the tetrahedral sheet is composed of tetrahedral cations, such as, Si^{4+} and Al^{3+} coordinated with four O^{2-}. The three O^{2-} anions of each tetrahedron lie in the same plane (basal oxygens) and are shared by three nearest neighbor tetrahedra. The fourth O^{2-} (apical oxygen) points away from the plane of basal oxygens and is shared with the octahedral sheet.

The octahedral sheet is composed of octahedral cations, such as, Mg^{2+}, Fe^{2+}, Fe^{3+}, Al^{3+} and two planes of apical oxygen and hydroxyl anions. The octahedral sheet could be trioctahedral, dioctahedral, or a composite of di- and trioctahedral composition. In trioctahedral sheet, all the possible octahedral cation sites (that is, three out of three) are filled. This demands that the octahedral cations be divalent, for example, $Mg_3(OH)_6$ or $Mg(OH)_2$. When the octahedral cation sites are occupied by a trivalent cation, Al^{3+}, only two out of three octahedral sites will be occupied to preserve an electrical neutrality. The octahedral sheet can be represented by $Al_2(OH)_6$ or $Al(OH)_3$. Some vermiculites have octahedral composition which is intermediate between ideal dioctahedral and trioctahedral composition. A structural scheme of dioctahedral vermiculite based on octahedral and tetrahedral sheets in shown in Figure V2.

Trioctahedral vermiculites resemble trioctahedral micas such as phlogopite and dioctahedral vermiculites resemble dioctahedral micas such as muscovite. Trioctahedral vermiculites exist both as well crystallized macroscopic particles and clay-sized form while dioctahedral vermiculites exists only as small particles. About one-seventh to slightly less than one-fourth of the Si^{4+} tetrahedral sites are substituted by Al^{3+} giving rise to a net negative charge (layer charge) to the structure. The negative layer charge is balanced by hydrated exchangeable cations (Mg^{2+}, Ca^{2+}, etc.) occupying the interlayer spaces. In contrast, K^+ ions are the main charge

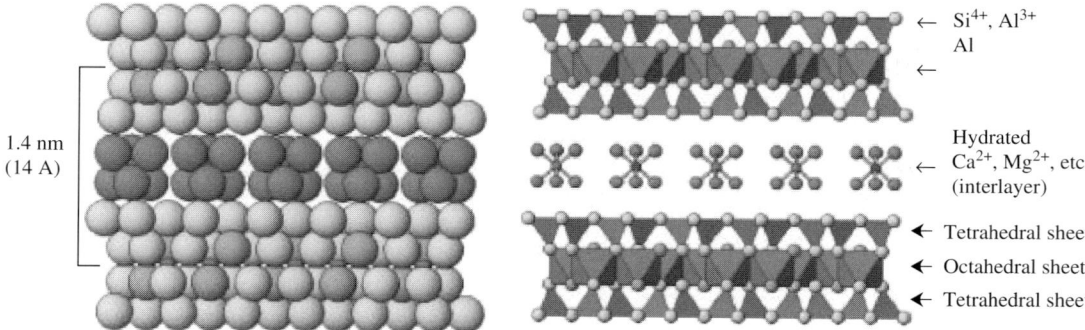

Figure V2 Structural scheme of dioctahedral vermiculite based on octahedral and tetrahedral sheets (Schulze, 2002).

Table V1 Chemical composition and structural formulas of ideal and typical trioctahedral and dioctahedral vermiculites (Malta, 2002)

Vermiculite type	Interlayer	Octahedral cations	Tetrahedral cations	Coordinating anions
Ideal Trioctahedral	$(nH_2O, Mg_{z/})$ $z = 0.6–0.9$	$(Mg)_3$	(Al_xSi_{4-x}) $x = 0.6–1.5$	$O_{10}(OH)_2$
Examples of Tri-octahedral Vermiculites				
Kenya	$(Mg_{0.32})$	$(Mg_{2.36}Fe^{3+}_{0.48}Al_{0.16})$	$(Al_{1.27}Si_{2.72})$	$O_{10}(OH)_2$
Santa Olalla	$(Mg_{0.39}Ca_{0.0})$	$(Mg_{2.59}Fe^{3+}_{0.24}Al_{0.06}Fe^{2+}_{0.03}Ti_{0.08})$	$(Al_{1.28}Si_{2.82})$	$O_{10}(OH)_2$
Llano	$(Mg_{0.47})$	$(Mg_{2.81}Fe^{3+}_{0.07}Al_{0.11})$	$(Al_{1.11}Si_{2.90})$	$O_{10}(OH)_2$
Ideal Dioctahedral	$(nH_2O, Mg_{z/})$ $z = 0.6–0.9$	$(Al)_2$	(Al_xSi_{4-x}) $x = 0.6–0.9$	$O_{10}(OH)_2$
Examples of Soil Vermiculites				
Sample A	$(Na_{0.61})$	$(Al_{1.44}Ti_{0.14}Fe^{3+}_{0.16}Mn_{0.05}Mg_{0.27}H_{0.03})$	$(Al_{1.10}Si_{2.90})$	$O_{10}(OH)_2$
Sample B	$(Na_{0.64})$	$(Al_{1.24}Ti_{0.18}Fe^{3+}_{0.20}Mn_{0.02}Mg_{0.35}H_{0.25})$	$(Al_{0.76}Si_{3.24})$	$O_{10}(OH)_2$
Sample C	$(Na_{0.67})$	$(Al_{1.12}Ti_{0.16}Fe^{3+}_{0.29}Mn_{0.03}Mg_{0.75}H_{0.19})$	$(Al_{1.31}Si_{2.69})$	$O_{10}(OH)_2$

balancing cations in mica which are not exchangeable. On the basis of layer charge, vermiculites are intermediate between smectites and micas, and resemble smectites in exhibiting reversible swelling in water and organic liquids. Vermiculites also have similarities with chlorites except that the interlayer cations in vermiculites are about 1/9 of those in chlorites.

In Mg-vermiculite, H_2O molecules occupy two sheets of sites arranged in a distorted hexagonal pattern around Mg ions. Each pair of water sheets is held together by Mg ions which lie equidistant from the adjacent tetrahedral sheets. In turn, the water molecules in the same sheet are bonded to each other and to the oxygen in the nearest 2 : 1 layer surface by hydrogen bonding. In a fully hydrated 1.48 nm (basal spacing) phase, each Mg ion contains 16 H_2O molecules. Dehydration of the interlayer spaces (at lower humidity or higher temperature) occurs at definite steps yielding several less hydrated phases with d(001) spacings of 1.436, 1.382, 1.159, 1.159/0.990, and 0.902 nm.

The chemical composition of vermiculites varies greatly depending on the parent material from which they are derived and the environment in which they are formed. Almost all of the macrocrystalline vermiculites have trioctahedral compositions, whereas the soil vermiculites are intermediate between di- and trioctahedral compositions. The chemical compositions of ideal and typical vermiculites are presented in Table V1.

Formation, distribution, and abundance

Vermiculites are primarily formed by the alteration of micaceous minerals as a result of weathering, hydrothermal action, percolating ground water, or a combination of these three factors. The major commercial deposits of vermiculite are found in ultrabasic and basic rocks. However, there are a few reports which indicate that vermiculite may have been derived from feldspars or formed from the precipitation of gels. The preponderance of evidence shows that the trioctahedral vermiculites are formed from the action of supergene solutions (solutions descending from the surface) on biotite and phlogopite present in the host rocks (basic and ultrabasic rocks, gneisses, and schists, carbonates, granites, etc). This implies that the interlayer K^+ ion from mica is being replaced by a hydrated cation, generally Mg^{2+}. This process is reversible, that is, expanded vermiculite layers (1.4 nm) can contract to 1.0 nm mica type layers in the presence of K^+ ions. Vermiculites are not stable even at mildly hydrothermal conditions ($>200–300°C$). In these conditions, chlorite is formed instead of vermiculite, although the chlorite in turn could alter to vermiculite. In soils, trioctahedral vermiculites are formed from the alteration of biotite, phlogopite, or chlorite, while dioctahedral vermiculites are formed from muscovite. However, there are several reports that the trioctahedral mica such as biotite could also be converted to dioctahedral vermiculite. The major steps involved in the transformation of mica to vermiculite are: (1) removal of potassium by hydrated exchangeable cations; and (2) reduction of layer charge.

Vermiculites are widely distributed in soils and rocks. They can be found from polar to tropical climates including areas of low and high rainfalls. They are more common in soils from temperate and subtropical climates than in tropical climates. Vermiculites formed from mica are not common in marine environment, since K^+ ions present in seawater readily causes layer collapse. It is believed that the vermiculites formed in marine sediments are derived from non-micaceous sources, for example volcanic material, chlorite, and hornblende.

In acidic environments, vermiculites are commonly interlayered with hydroxy-aluminum species, with subsequent loss of expansion and collapse and cation exchange capacity properties. The hydroxy-Al interlayering is an essential conditon for the persistence of vermiculites, especially the trioctahedral variety, in soils under intense weathering environments. The normal sequence of weathering under intense weathering conditions may be represented as mica → vermiculite → hydroxy-Al interlayered vermiculite. In the absence of hydroxy aluminum interlayering, vermiculite would be a short-lived intermediate phase leading to kaolinite or smectite.

Chemical and physical properties

Layer charge and cation exchange capacity

Vemiculites have relatively high layer charge and cation exchange capacity. Layer charge and cation exchange capacity are closely related. Layer charge refers to the net negative charge of the 2 : 1 layer arising from the isomorphous substitution of a higher valence cation by a lower valence cation. The cation exchange capacity (CEC) refers to the quantity of readily exchangeable cations neutralizing negative charge. Both the permanent negative charge from isomorphous substitution and the negative charge (pH dependent) from the broken bonds at the edges contribute to CEC. The cation exchange capacity of vermiculite ranges from 130 to 210 $cmol_c/kg^1$. The layer charge expressed per half unit cell ($O_{10}OH_2$) range from 0.6 to 0.9. The high CEC and layer charge of vermiculites thus dominates the exchange properties of sediments that contain it in significant amounts. Because of its high CEC, vermiculite is suitable material for removing large quantity of several types of smaller heavy metal cations such as Cu^{2+}, Pb^{2+}, Cd^{2+}, Zn^{2+}, Ni^{2+}, etc. from industrial wastes. However, in most cases the sorption appears to be non-selective.

Selective sorption and fixation of cations

The unique structure of vermiculite is also favorable for selective sorption and fixation of low hydration energy cations such as NH_4^+, K^+ and Cs^+. In selective sorption, certain cations are favored for sorption compared to others. These selectively sorbed cations may or may not be held tightly. If the cations are held tightly and resist replacement by other cations, they are considered to have been fixed. Selective sorption and fixation of NH_4^+ and K^+ are important for devising effective soil management strategies, while that of Cs^+ is important for radioactive waste disposal. Selective sorption and fixation of cations in vermiculite are influenced by several factors such as cation size, cation valence, cation hydration energy, vermiculite structural and crystal chemical parameters (layer charge, charge location, composition of octahedral sheet) hydroxy interlayering, particle size, frayed edges or wedge zones, etc.

Organo-vermiculite

Vermiculites may also be modified using organic cations to adsorb and trap varieties of nonionic and anionic organic compounds that are detrimental to our environment. The modified clays are usually referred to as organo-clays (organo-vermiculite). Potentially the organo-vermiculite can be used for an *in situ* treatment of contaminated sediments.

Osmotic swelling

Two types of swelling occur in vermiculites: normal swelling as a result of uptake of as much as 2 molecular layers of water (type 1), and osmotic swelling (type 2) which involves the uptake of much larger volumes of water. The type 1 swelling, also referred to as interlayer or crystalline swelling, is limited to ~1.5 nm with an inorganic cation, and 1.94 nm with an n-butylammonium cation. The osmotic swelling in vermiculite is characterized by a reversible large volume expansion leading to the formation of a coherent gel. The layer expansion could be as much as 91.0 nm with butylammonium ion. Although Li^+ is the only inorganic cation that causes vermiculite to swell osmotically several organic molecules, such as propyl and butylammonium and certain amino acid cations are known to exhibit osmotic swelling with vermiculite. The degree of swelling increases as the concentration of the electrolyte decreases. Osmotically swollen vermiculite can be delaminated by mechanical agitation to produce high aspect ratio vermiculite particles. The high aspect ratio vermiculites can be used to produce non-burning paper, or fire proofing or resistant films and coatings on combustible and noncombustible substrates.

Exfoliation

Macrocrystalline vermiculite expands (exfoliates) to as much as 8–12 times its original volume upon heating at high temperature, >300°C and typically 870–1090°C, as the interlayer and structural water is converted to steam. The expanded vermiculite forms light weight granules that have the appearance of a large worm. The thermal expandability and light weight characteristics of exfoliated vermiculite have been exploited for various industrial and agricultural applications, including gaskets for high temperature sealing, such as in catalytic converters; insulation and fire retardants; various construction products; potting soils; soil conditioners; carrier for fertilizers; insecticides and herbicides; various livestock applications; and ammonia filtering in aquaculture.

Identification

Vermiculite is routinely identified by X-ray diffraction based on the 1.4 nm peak produced by a Mg-saturated sample. The Mg-saturated sample is further treated with glycerol or ethylene glycol to differentiate it from smectite, which expands with later treatments. The vermiculite peak collapses to 1.0 nm upon K-saturation. Chlorite also has a peak at 1.4 nm but it neither expands with ethylene glycol/glycerol nor collapses with K treatment. Smectites peaks tend to be broader than vermiculite peaks. Vermiculite is more likely to occur in the coarse fraction and smectite in the fine clay fraction (<0.2 µm). The hydroxyAl-interlayered vermiculite will resist both collapse and expansion. K-saturated hydroxy-Al interlayered vermiculite will not collapse to ~1.0 nm, but to a broad 1.1–1.3 nm peak. It may be necessary to heat the sample at 300°C or higher temperature to collapse the layers to 1.0 nm depending upon the polymeric nature of the hydroxyAl-interlayer species.

Vermiculite can also be identified by determining its layer charge using the alkylammonium ion exchange technique. This method essentially involves the treatment of each sample with n-alkylammonium cations, with n ranging from 6 to 18 cabon atoms followed by XRD measurements. In addition to the total charge, the alkylammonium method also provides the information on charge heterogeneity or distribution. This method when combined with the Green-Kelly test (Hofmann and Klemen effect) also allows one to estimate the magnitude of both octahedral and tetrahedral charges of dioctahedral vermiculite.

Prakash B. Malla

Bibliography

Calle, C. De La, and Suquet, H., 1988. Vermiculite. In Bailey, S.W. (ed.), *Hydrous Phyllosilicates (Exclusive of Mica)*. Reviews in Mineralogy 19. Mineralogical Society of America, pp. 455–496.

Douglas, L.A., 1989. Vermiculites. In Dixon, J.B., and Weed, S.B. (eds.), *Minerals in Soil Environments*, 2nd edn, Soil Science Society of America, pp. 635–674.
Deer, W.A., Howie, R.A., and Zussman, J., 1992. *An Introduction to the Rock-Forming Minerals*. Longman.
Malla, P.B., 2002. Vermiculites—chemistry, mineralogy, and applications. In Dixon, J.B., and Schulze, D.G. (eds.), *Soil Mineralogy with Environmental Applications, SSSA Series No. 7*. Soil Science Society of America, pp. 501–529.
Schulze, D.G., 2002. An introduction to soil mineralogy. In Dixon, J.B., and Schulze, D.G. (eds.), *Soil Mineralogy with Environmental Applications, SSSA Series No. 7*. Soil Science Society of America, pp. 1–35.

Cross-references

Clay Mineralogy
Mixed-Layer Clays
Weathering, Soils and Paleosols

W

WEATHERING, SOILS, AND PALEOSOLS

Soils are fundamental to life on this planet. The mineral nutrients provided by weathering within soil and the degree of drainage of soil control what kind of life can thrive in a particular place. On the other hand, living creatures with their roots, jaws, and other means of acquiring nutrients do much to determine the nature of soil. Soils include billions of bacteria, millions of nematodes and a few plants in just about every square centimeter. Soil's diverse microbes and internal absorptive surfaces of clay neutralize poisons and purify water. By fueling photosynthesis, soil regulates the composition of the atmosphere. Through the engine of soil, life has far-reaching effects on land, water, and air. The intimate interrelationship between soil, life, and surface environments also has a long fossil record in the form of fossil soils, or paleosols. These remains of soils of the past are now known as old as 3,500 million years on Earth. Even more ancient soils and paleosols are now known on the Moon and Mars, and certain kinds of meteorites may be fragments of paleosols as old as 4,600 million years.

Weathering

Weathering is the relentless alteration of sediments and rocks by a variety of chemical, biological, and physical agents at the surface of planetary bodies. In humid forested environments of good soil fertility, as in the oak forests of northern Europe and eastern North America, the general weathering regime is a soil-forming process called *lessivage* (Figure W1). Chemically, lessivage is destruction of primary minerals, such as feldspars, by weak acids, such as carbonic acid, which liberate into soil solution cationic nutrients such as calcium (Ca^{2+}), magnesium (Mg^{2+}), sodium (Na^+) and potassium (K^+). This chemical reaction, known as *hydrolysis*, is not simple dissolution of minerals, but incongruent dissolution which leaves a residue of silica (Si) and sesquioxides (Al^{3+}, Fe^{3+}) in clays, such as smectite (q.v.), and oxide and hydroxide minerals (q.v.), such as goethite. The process of lessivage thus creates soil with fewer feldspar grains and more reddish clay. Biologically, hydrolysis is important because base cations are fundamental plant nutrients: calcium for cell walls, magnesium as a critical component of chlorophyll and potassium and sodium as cell electrolytes. Plants promote hydrolysis in many ways. Their root respiration of carbon dioxide (CO_2) enriches soil water in carbonic acid (H_2CO_3). Other acids and chelates produced by plants and their associated microbes accelerate release of nutrients by weathering. Physically, plants hold the soil together with their roots, and regulate water loss by means of ground cover. Earthworms, squirrels, and gophers physically churn the soil.

A variety of weathering regimes is found in different parts of the world (Figure W1). In humid regions of high water table, such as swamps, the principal soil building process is accumulation of plant debris in peat, within which decay is suppressed by lack of oxygen and acidic leachates from plants. Chemically reducing soil below the peat can be leached of iron in its ferrous (Fe^{2+}) ionic form which is green to gray in color and more soluble than red, oxidized (Fe^{3+}) iron. The process by which reddish brown soils are turned gray by leaching of reduced iron is called *gleization*. In humid regions of conifer forests on sediments and rocks rich in quartz, plants produce unusual amounts of acid, with the result that even clay is destroyed in the soil, which builds a subsurface horizon reddened by sesquioxides and humus in a process called *podzolization*. Humid tropical rain forests grow in soils of low fertility because of long and deep weathering of nutrient cations promoted by abundant moisture and warmth. They have thick soils rich in oxide minerals, such as hematite, and clay minerals, such as kaolinite (q.v.). Their chemical enrichment in iron (Fe^{3+}) and aluminum (Al^{3+}) is a soil forming process called *ferallitization*. The deep and thorough weathering of tropical soils can enrich sesquioxides (Fe^{3+} and Al^{3+}) and silicon (Si) elements to the grade of economically valuable ores of iron (laterite, q.v.), aluminum (bauxite, q.v.) and silica (silcrete, q.v.). In deserts (q.v.) there is limited water for hydrolysis and mobilization of nutrient cations. Calcium, and also some magnesium, hydrolytically released from feldspars and other minerals, is not washed from the profile or taken up

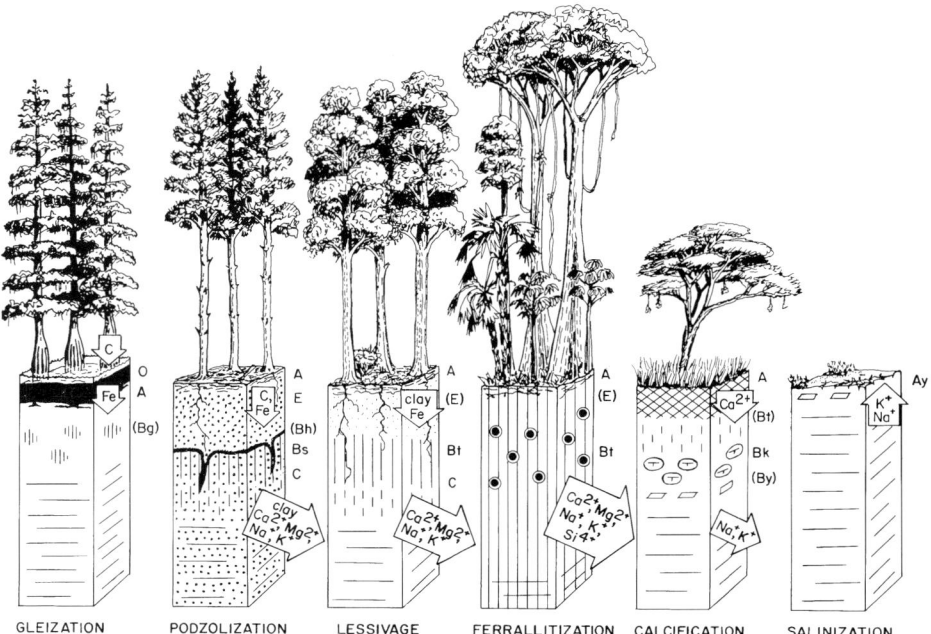

Figure W1 Common weathering regimes.

by plants. It accumulates by calcification in a subsurface horizon as hard, white, nodules, or bands, called caliche (q.v.) or calcrete, mainly fine-grained, low-magnesium calcite ($CaCO_3$), but rarely dolomite [$CaMg(CO_3)_2$]. Over time, large nodules grow and coalesce into thick layers. In very dry climates hydrolytic weathering is very limited, and there is insufficient moisture to leach away the cations, which accumulate in the soil as crystals of rock salt (NaCl) or gypsum ($CaSO_4$): a process called *salinization*.

The result of these various weathering processes is to produce a weathering profile of alteration that diminishes in intensity down from the surface to the parent material below. Weathering profiles are tens of meters thick in tropical humid regions but only centimeters thick in desert playas. The upper part of a weathering profile is commonly called soil or solum. It represents that part of the profile most altered by roots, burrows, and other processes that completely obscure structures of the parent material. In deeply weathered profiles there is additional alteration well below the soil, in a thick horizon called *saprolite*. Typically this is clayey, reddish, or yellowish, soft material in which folding, crystal outlines, schistosity, or bedding from the parent material remain visible. Although saprolitic weathering may appear to be largely chemical because it is beyond the reach of most roots and burrows, there is a biological component to this alteration from termites, fungi, and other microbes. *Regolith* is another term used for deep weathering profiles, but regolith includes sediments as well as soil and saprolite. The terms weathering profile and soil refer to alteration in place, without transport. Soil is commonly eroded and the material redeposited. Sediment of recognizable soil clasts is called a *pedolith* or soil sediment. A redeposited laterite is the original and remains the best example, because clasts so rich in kaolinite and hematite are distinctive. Although most sediments are derived from soils, it is best not to extend the term pedolith to widespread and far-travelled alluvium, but restrict it to distinctive soil materials. The distinction between sedimentation and soil formation is fundamental because soils develop in profiles from the top downward, but sediments accumulate in sequences from the bottom upward. The theory and practice of soil science (pedology) and sedimentary geology (sedimentology) are very different, and have further diverged because of the traditional association of pedology with agricultural studies and sedimentology with geological studies.

Soils

Soil profiles are the tangible and diverse products of weathering. We classify them in order to understand the processes that form them, and to manage their use. The US Soil Conservation Service recognizes a dozen soil orders (Figure W2), which are the largest units in a comprehensive, hierarchical soil classification that is widely used throughout the world. The basis for soil classification is the sequence and development of *horizons*. The process of lessivage, described above, produces a soil with a leached, sandy, quartz-rich surface horizon (E or eluvial) over a red, clayey subsurface (Bt or argillic) horizon. The classification has strict definitions of the degree of alteration needed for specific horizons, for example at least 8 percent additional clay is needed for an argillic horizon, with some exceptions. If the soil has an argillic horizon that is relatively fertile (with abundant available base cations) it is an Alfisol, but a soil with an argillic horizon impoverished in base cations is an Ultisol. Soils can be identified using a dichotomous key (Figure W3), but the following paragraphs convey the gist of each soil order and some of the critical terminology in their definition.

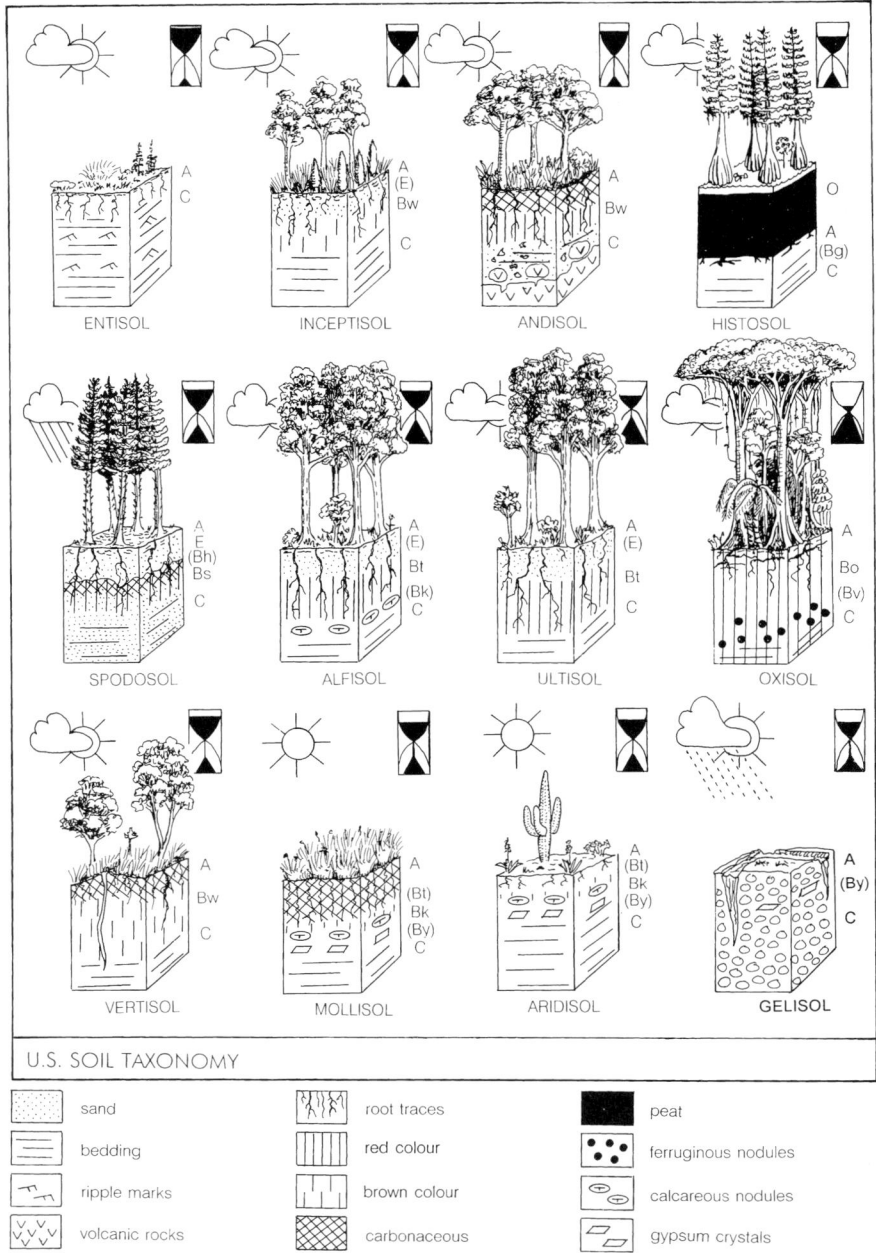

Figure W2 Cartoons of climate, vegetation, and profile form of various orders of soils defined by the US soil taxonomy.

Entisols (incipient soils) are immature, showing little weathering beyond the initial growth of pioneering plants. They develop on flood deposits, landslides, and other geologically young surfaces. These are important soils for market gardening.

Inceptisols (young soils) have recognizable horizons, but none of these are developed to the extent found in other soil orders because of a short time of development or conditions hostile to soil formation. Because they are an early stage in the development of other soil orders, Inceptisols are extremely diverse. Typically their subsurface shows only minor accumulation of clay, carbonate, or iron stain in what is termed a *cambic horizon* (Bw in soil science shorthand). Inceptisols are important soils for crops, orchards, and pasture.

Andisols (volcanic ash soils) are formed on volcanic ash rich in glassy shards that weather to noncrystalline colloids with low bulk density and very high fertility. In profile form, they are generally similar to Inceptisols. These soils are very important for cropping in tropical regions, where other soils are deeply weathered of plant nutrients.

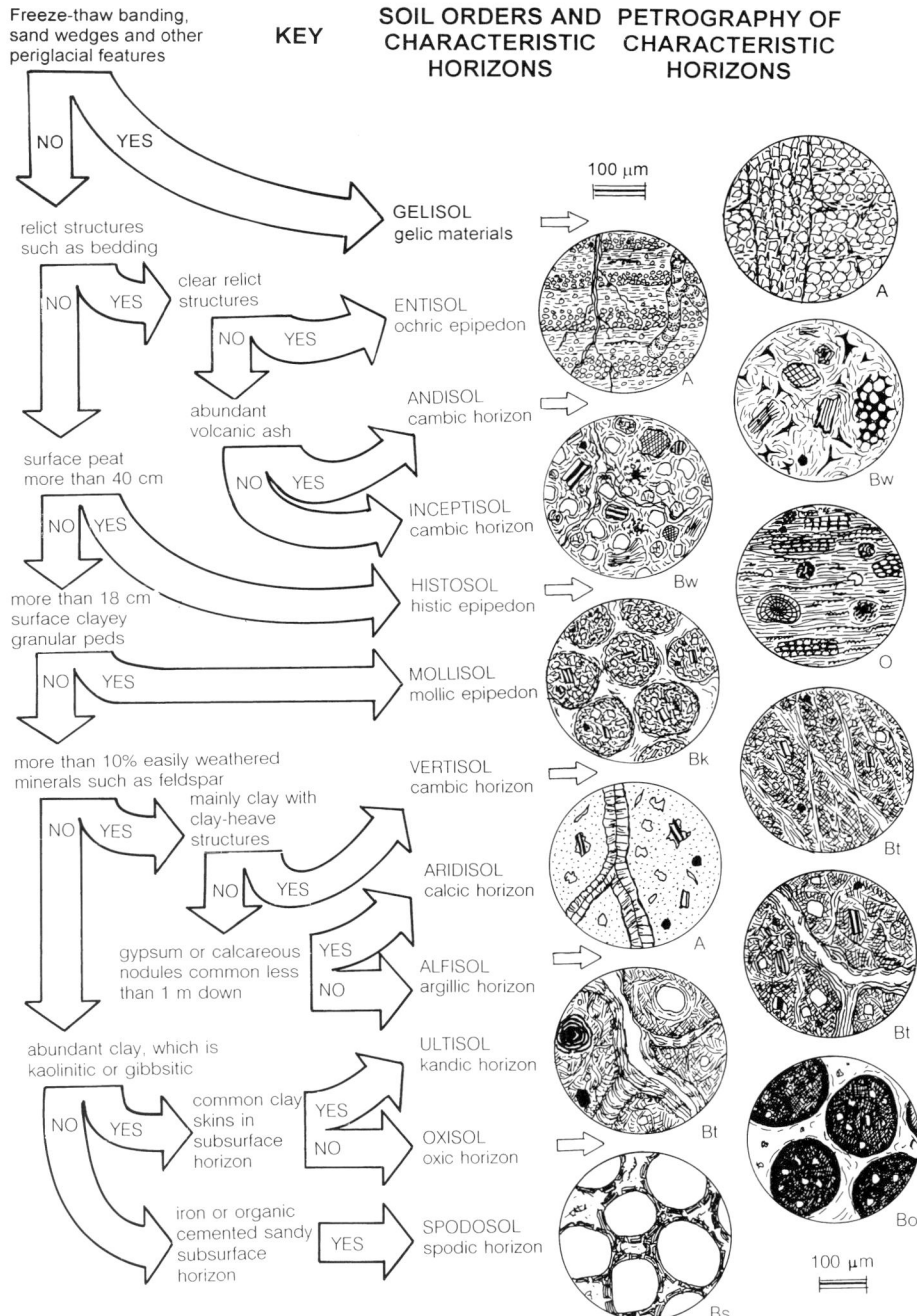

Figure W3 A key for recognition of the soil orders of the US soil taxonomy, with emphasis on features that can be recognized in outcrop and in petrographic thin sections of paleosols.

Histosols (peaty soils) are soils in the sense that they support bald cypress swamps, papyrus marshes and other wetland vegetation. Their surface layer (histic epipedon or O horizon), is what a sedimentary geologist would call peat (*q.v.*). It forms by successive increments of plant material, with decay suppressed by stagnant water conditions, in a way analogous to the accumulation of sediments. They are best left alone to preserve water quality, but some Histosols are logged for specialty lumber and dried for domestic fuel.

Spodosols (sandy forest soils) have attractive profiles with white (*eluvial*) surface horizons contrasting with reddish brown (*spodic* or Bs) subsurface horizons enriched in iron, aluminum, and organic matter. They are quartz-rich, clay-poor, acidic, and infertile. They support conifer forests and heath that can tolerate such low fertility, and are used mainly for softwood lumber production and water quality preservation.

Alfisols (fertile forest soils) have a subsurface horizon enriched in clay (*argillic* or Bt horizon) that is rich in nutrient cations. Such clays are typically smectite and illite. These soils support broadleaf forest vegetation, but are widely cleared for crops and grazing.

Ultisols (base-poor forest soils) have a clayey subsurface horizon that is poor in nutrient cations and usually dominated by kaolinitic clay (*kandic* and Bt horizon). Other than this mineral and chemical difference, they appear generally similar to Alfisols. Mixed conifer-broadleaf forest is typical. Once cleared these soils are fertile enough for cattle grazing, orchards, and vineyards, but are best left forested for lumber and water quality preservation.

Oxisols (deeply weathered tropical soils) are thick, nutrient poor, and highly aluminous and ferruginous (oxic or Bo horizon). A common micromorphology is sand-sized clods of hematite and quartz. These soils support tropical rain forest, but are used for sugar cane, as well as tropical tree crops such as cocoa, mango, and papaya.

Vertisols (swelling clay soils) are rich in smectite clays, which have the physical property of swelling when wet and then cracking as they shrink and dry. They form in climates with pronounced seasonality of rainfall. The most distinctive physical feature of these soils is a pattern of troughs or pits a few to several decimeters deep (*gilgai* microrelief) between the pressure ridges around the deep cracks. At depth, the criss-crossing slickensided planes and deformed pressure ridges create a characteristic thinning and thickening of soil horizons (*mukkara structure*). Vertisols support mainly grassland and wooded grassland, and are used primarily for grazing. Their physical instability is very destructive of roads, fences, and buildings.

Mollisols (grassland soils) have a thick, dark, clayey surface of high fertility (*mollic epipedon* or A horizon). This consists of small rounded clayey crumbs enriched and bound with finely decayed organic matter. Mollisols are widely used for grazing, as well as herbaceous crops such as corn and wheat.

Aridisols (desert soils) are little weathered, clay-poor, and have nodules or layers of calcite or dolomite (calcic or Bk horizon) within a meter of the surface. Salts such as gypsum may also be found at depth. Some Aridisols are irrigated for crops, but water flushes salts to the surface with disastrous results. Others are used for grazing at low stocking densities, but most Aridisols remain unused.

Gelisols (permafrost soils) have ground ice or other permafrost features (gelic materials) within a meter of the surface. Ice wedges, stone stripes and other deformation features are characteristic. Some Gelisols support taiga forest at high latitudes and krummholz at high altitudes. Other Gelisols support tundra and alpine fellfield. A short growing season limits agricultural use of Gelisols, although forestry, reindeer herding and caribou hunting does support some human activity.

Another way of looking at soils useful in understanding how they form such diversity is analyzing the factors that create them: climate, organisms, topographic relief, parent material and time for formation. The diversity of soils can be considered a vast natural laboratory of concurrent experiments in soil formation. If we wish to study one of the factors in soil formation such as time, then we should find a group of soils of comparable climates, vegetation, geomorphic setting and parent materials but varied time for formation. Such a suite of soils is called a *chronosequence*, and mathematical relationships between soil properties and time derived from such a group of soils is called a *chronofunction*. Examples include soils of a flight of alluvial terraces excavated by a river cutting lower into its valley during uplift, or soils of moraines left behind by retreating glaciers. Chronofunctions may quantify the accumulation of clay in argillic horizons, of carbonate in calcic horizons and of peat in histic epipedons with the progress of time. Soils of the chronosequence may have been dated by radiocarbon, human artifacts or fossils, but the chronofunctions derived from them can be used to assess the age of soils nearby that lack dateable materials. Landscape histories reconstructed in this way are important for siting permanent facilities such as bridges, dams, and nuclear power plants. In addition to chonofunctions, comparable approaches can be used to quantify the role of climate (*climofunctions*), organisms (*biofunctions*), topographic relief (*topofunctions*) and parent material (*lithofunctions*). In this way it is possible to investigate the process of soil formation with rigor.

Paleosols

Soils have a fossil record as paleosols. Most of these are fossilized by burial in flood deposits or volcanics (Figure W4), but some are still at the surface, either by exhumation or by

Figure W4 A modern grassland soil, and two comparable paleosols, all Mollisols, buried in volcanic sandstones at the middle Miocene (14 Ma) site of Fort Ternan, Kenya. Hammer for scale has a handle 25 cm long. These are the earliest known well-drained grassland soils in Africa.

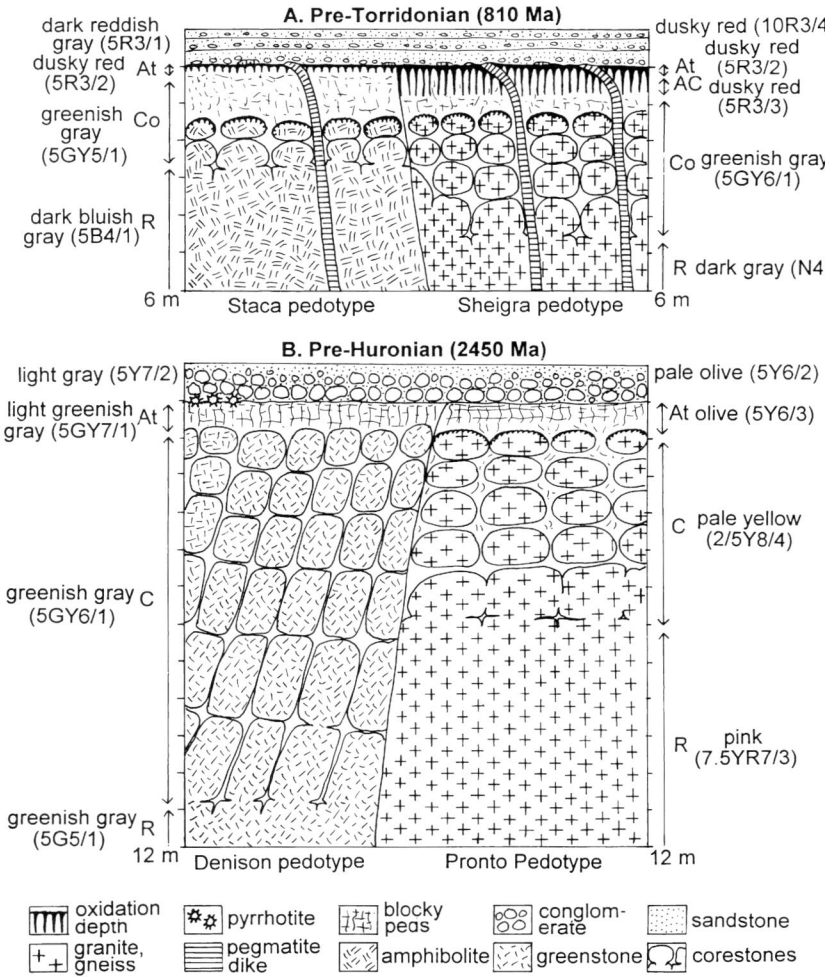

Figure W5 Diagrammatic sketch of Precambrian paleosols at major unconformities, including pre-Torridonian (810 Ma) profiles from near Sheigra, Scotland (above), and pre-Huronian profiles from near Elliot Lake, Ontario, Canada (below). All the paleosols were well drained, as indicated by corestones and clay formation, but the Scottish profiles were more oxidized and indicate higher levels of atmospheric oxygen than the Canadian profiles.

outlasting the conditions that formed them. Lateritic Ultisols and Oxisols from the middle Miocene (16 Ma) thermal maximum are widespread surface paleosols at high latitudes, and are easily recognized because they are so much more deeply weathered, kaolinitic, and ferruginous than associated soils. Paleosols also are commonly preserved at major geological unconformities (Figure W5).

The fossilization process of soils is a hindrance to interpretation of paleosols because it alters many of their features, including those important to their classification and reconstruction of past soil-forming processes. Three common alterations after burial of soils in sedimentary successions are: (1) decomposition of organic matter such as leaf litter and roots at the surface of the soil; (2) mobilization of reduced iron (Fe^{2+}) during organic matter decomposition under low oxygen conditions; and (3) dehydration of yellow iron-hydroxides, such as goethite, to red iron oxides, such as hematite. These three processes alone can convert a dark brown to yellowish brown soil to a carbon-lean claystone, brick-red in color, and riddled with green-gray alteration haloes that coalesce toward the top of the profile. Such substantial changes in appearance have hindered the recognition of many paleosols. With deep burial, paleosols are compacted from the weight of overburden and cemented by deep groundwater. Metamorphism can further obscure soil features and minerals by the development of cleavage and metamorphic minerals.

Despite problems of burial alteration, many features of soils remain in paleosols and can be used to recognize them within sedimentary or volcanic sequences. Features most diagnostic of paleosols are: (1) tubular structures that branch and thin irregularly downward or show anatomy of fossilized root traces; (2) gradational alteration down from a sharp lithological contact like that of a land surface and soil horizons; and (3) the complex patterns of cracks and mineral replacement like those of soil clods (*peds*) and planar features (*cutans*). It is also possible to classify paleosols within systems devised for

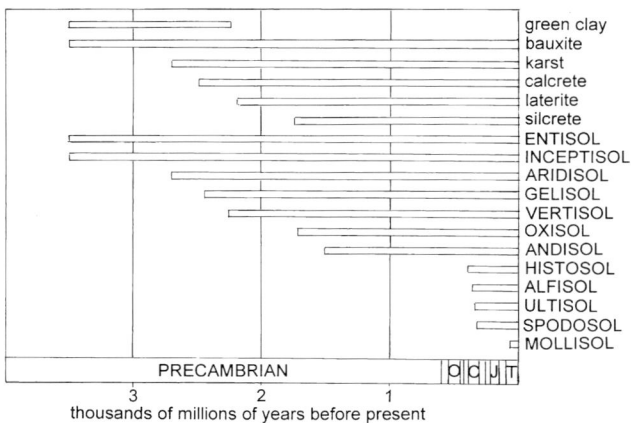

Figure W6 Geological range of different weathering products and soil orders of the US taxonomy. The diversification of soils and weathering products through time reflects evolutionary changes in the atmosphere and terrestrial ecosystems.

soils, provided allowance is made for burial alteration. Geochemical analysis, X-ray diffraction, and thin section petrography are particularly helpful in the classification of lithified paleosols. Tests of soil fertility used to distinguish Alfisols from Ultisols do not work with paleosols because the originally reactive soil surfaces have been compacted, cemented, and recrystallized during burial. However Alfisols have molar ratios of alumina/bases less than 2, smectite, and illite as dominant clays, and thin section fabrics with scattered highly birefringent clay streaks when viewed under crossed nicols. Ultisols by comparison have alumina/bases more than 2, consist mainly kaolinite and retain a fabric dominated by highly oriented and highly birefringent clay. Each of the US soil orders has a characteristic petrographic appearance (Figure W3) and a long fossil record (Figure W6).

Soils and paleosols of the Moon and planets provide new challenges for Earth-bound classifications of soil, and the geological history of soil. Conditions on our water-less and atmosphere-less Moon are most unlike Earth. The most important soil forming process on the Moon is continual bombardment of the surface by sand-sized micrometeoroids. These pulverize and locally melt the soil, darkening the surface with added metal. Dark metal-rich horizons found in lunar cores represent fossil soils buried by the debris of exceptionally powerful impacts. Each of these thin paleosols, like the surface soil, took many millions of years to form (see *Lunar Sediments*).

On Mars there is a thin atmosphere of carbon dioxide and evidence from shorelines and paleochannels of water in the distant geological past. Soils analyzed there by robotic landers are rich in iron and swelling-clay (smectite), with subsurface hardpans of salts. These soils are similar to salty Aridisols now forming in the Dry Valleys of Antarctica. Such soils require warmer temperatures and more water than currently available on Mars, and they may be relict paleosols dating back to the time of fossil channels on Mars more than 2,000 million years ago.

Some scientists have also suggested that certain kinds of meteorites may be parts of paleosols. Like Martian paleosols, carbonaceous chondrites are rich in iron and smectite, cracked, and stained, and veined with salts. By this interpretation, some carbonaceous chondrites may be pieces of the oldest paleosols in the solar system, some 4,600 million years old.

The most ancient known paleosols on Earth, some 3500 years old, are thick, deeply weathered and green gray in color, unlike ancient soils and paleosols of the Moon, Mars, and meteorites. They may represent an extinct soil order, for the moment informally called "Green Clays". Soils on Earth were alive with microbes at least back 2,700 million years, judging from the life-like carbon isotopic composition of paleosols. Microbial crusts also would explain landscape stabilization, deep weathering, and microtextures of paleosols back to 3,500 million years. Nevertheless, Precambrian paleosols do not fit easily within modern soil classifications and may reflect an early atmosphere rich in carbon dioxide but with very low amounts of oxygen (Figure W5). Red and oxidized paleosols including Oxisols and weathering products such as laterites that indicate the rise of oxygen in the atmosphere do not appear until about 2,200 million years ago. With the appearance of the first large continents by amalgamation of smaller island arcs at this time, came the first paleosols recognizable as Vertisols and Aridisols of dry climate. Gelisol paleosols also have been found among the tillites and striated pavements remaining from late Precambrian ice ages.

Land plants and animals have left traces in Late Ordovician Entisols, Inceptisols, and Andisols. Histosols do not appear until the advent of substantial land vegetation during the Early Devonian. Forested soils such as Alfisols are not known earlier than Late Devonian. Spodosols and Ultisols may be as old, but are currently not known among paleosols older than Carboniferous. Mollisols of grasslands appeared relatively late in geological history with the Tertiary rise of grasses and grazers. The role of grassland ecosystems in the evolution of humans and the emergence of agriculture and civilizations also has been investigated using fossil soils associated with human fossils and artifacts.

The long fossil record of soils is complimentary to evidence of fossils and sediments for the history of life and environments on Earth in the geological past. Soils and life have both diversified though geological time, and paleosols record this fundamental part of terrestrial ecosystems.

Gregory J. Retallack

Bibliography

Jenny, H., 1941. *Factors of Soil Formation*. McGraw-Hill.
Retallack, G.J., 1997. *A Color Guide to Paleosols*. Wiley.
Retallack, G.J., 2001. *Soils of the Past*, 2nd edn. Blackwell.
Soil Survey Staff, 1999. *Keys to Soil Taxonomy*. Blacksburg, Virginia: Pocahontas Press.

Cross-references

Bauxite
Bentonites and Tonsteins
Biogenic Sedimentary Structures
Caliche–Calcrete
Clay Mineralogy
Climatic Control of Sedimentation
Floodplain Sediments
Humic Substances in Sediments
Laterites
Peat
Silcrete

X

X-RAY RADIOGRAPHY

X-ray radiography (commonly shortened to radiography) is a non-destructive technique based on differential passage of X-rays through a sample onto a specific film. The differences in density in the sample, caused by variations in composition, result in differences in attenuation.

This method was introduced into sedimentology by Hamblin (1962). Basically it does not differ from X-rays used in medicine, biology, and paleontology, except that an industrial quality film is used. X-ray radiography can significantly enhance large and small variations in lithology and often brings out sedimentary structures that were not detected by the naked eye. In addition to those observations, radiography reveals microfaults and folds, presence of and disturbances by fauna and flora, and the presence of nodules and other postdepositional chemical inclusions (Krinitzski, 1970).

The technique is not limited to the shape or thickness of a sample. It is advised, however, to work with a thin plane-parallel slice. A core is too thick in the center to distinguish phenomena because those are projected on top of each other. The sides are too thin and thus will be overexposed. Imbedding the core, or irregular sample, in a plexiglas box with fine sand can reduce those thickness variations. Because the pores of the loose sand are not filled, the core has a higher density than the sand, and the thickness differences are only partly eliminated. If one deals with a lot of indurated or non-indurated cores, it is advised to make a mold of fine sand impregnated with a plastic (Bouma, 1969).

Indurated samples should be cut to obtain a plane-parallel slice about 1 cm thick. Soft sediments should also be sliced. For those it is advised to construct a plexiglas tray a little narrower and shorter than the core width and film length, respectively. Glue or tape 1-cm high sides onto the base of the tray. Cut the sampling tube lengthwise and remove the upper half of the tube. Cut sufficient of the upper part off the core to fit the tray. Place the tray over the upper part. Next remove the core segment and the tray from the lower half of the sampling tube and remove all sediment that protrudes the tray (for details see Bouma, 1969). A nice, thin slice of sediment results.

Plexiglas and most plastics are transparent to X-rays. Saran wrap is transparent also but the wrinkles show up on the film. Grains of coarse sediment are visible on the film (Bouma, 1969).

There are several X-ray units on the market. Commonly used are small units with a small focal spot. The smaller the focal spot the sharper the picture. A focal distance of 90–100 cm is common. The X-ray beam has to be perpendicular

Figure X1 Print of a radiograph of clayey marsh deposits. Distance focal point to film 92 cm, 40 kV, 5 mA, 45 seconds, industrial film. Slice 5 mm thick, 15 cm wide. Note the layering and lenses as well as the low density of Spartina roots (double white) and small grass roots near the top of the film.

to the film. A maximum tube current of 100–140 kV is sufficient for samples up to 9 cm thick. An increase in tube current results in a change of intensity of the emitted radiation. The intensity is approximately proportional to the milliamperage. An increase in voltage adds shorter wavelengths to the beam which results in a decrease of the picture contrast. A lowering of the kV increases the scattering.

One has to get used to interpreting radiographs (Figure X1). The developed film shows dark for low-density material, and light for dense material (sand). Making a print of the radiograph reverses the dark and light presentations. Bouma (1969) shows a large number of photographs and radiographs of different rock types, making it clear that trial exposures are needed to reach the best results. In general the muds and mudstones give the best radiographs because present sandy laminae and plant fragments stand out. Pure sands and many carbonates lack sufficient density contrast.

Arnold H. Bouma

Bibliography

Bouma, A.H., 1969. *Methods for the Study of Sedimentary Structures.* New York: Wiley-Interscience. (Can be obtained from Krieger Publishing Co., Huntington, New York, 1979.)

Hamblin, W.K., 1962. X-Ray radiography in the study of structures in homogeneous sediments. *Journal of Sedimentary Petrology*, **32**: 201–210.

Krinitzsky, E.L., 1970. *Radiography in the Earth Sciences and Soil Mechanics.* New York: Plenum Press.

Cross-references

Bedding and Internal Structures
Coring Methods, Cores
Impregnation
Relief Peels

Z

ZEOLITES IN SEDIMENTARY ROCKS

Introduction

Zeolites are among the most common authigenic silicates in sedimentary rocks. They occur in a wide variety of rocks types and are especially common in altered vitric tuffs. Zeolites are tectosilicates and have a 3-dimensional anion framework with an atomic ratio of $(Al+Si):O = 2.0$. The charge of the framework is balanced by cations, principally Ca, Na, and K. Zeolites can be viewed as hydrated equivalents of the feldspars in which the Si:Al ratio varies from about 1.5–5. The structural framework is relatively porous and contains interconnected cavities in which cations and water molecules reside. The cations and H_2O molecules are bound loosely, which gives properties of cation exchange and reversible dehydration. More than 35 different zeolite minerals have been identified and the seven most common in sedimentary rocks are analcime, phillipsite, natrolite, laumontite, heulandite, clinoptilolite, erionite, and mordenite (Table Z1).

The report of Murray and Renard in 1891 was the first to document a zeolite as widespread in sedimentary deposits. This zeolite was phillipsite, a significant constituent of deep-sea sediments sampled by the Challenger expedition. Ross and Bradley in 1928 reported the occurrence of analcime as the major alteration product in tuffs deposited in saline, alkaline lakes. The common silicic zeolite clinoptilolite was first reported in 1933 as an alteration product in marine silicic tuffs by Bramlette and Posnjak. Research on the zeolites of sedimentary rocks greatly increased in the 1950s, due to the use of X-ray diffraction for mineral identification and to interest in natural zeolites for adsorption and as molecular sieves. Coombs (1954) recognized a vertical zoning of zeolite types in a 10,000 m sequence of clastic sediments, which led to considering zeolite mineral assemblages in terms of burial depth and pressure-temperature stability fields. Studies of saline-lake deposits (Hay, 1966) showed that variations in pH, salinity, and cation content of lacustrine pore fluids resulted in mineralogic effects comparable to those caused by different burial depths (see Hay, 1966, for these early references).

Essential concepts

Zeolites occur in a wide variety of sedimentary rocks. They are postdepositional minerals and form from varied types of materials including volcanic glass, feldspar, feldspathoids, smectite, and kaolinite. Zeolites and some clay minerals can form from the same aluminosilicate materials, and the most critical requirement for zeolites is a high activity ratio of $Na^+ + K^+ + Ca^{2+}/H^+$. Thus zeolites are generally formed in alkaline environments. The nature of cations is important in determining which cationic type is formed. Zeolites are subdivided into three groups based on the silica activity in pore fluid at the time of origin. Silica-poor zeolites (e.g., natrolite) are favored by environments undersaturated with respect to quartz; some other zeolites co-exist with quartz (e.g., heulandite); and silicic zeolites may only be stable in solutions supersaturated with respect to quartz (e.g., clinoptilolite). Zeolites may alter to other zeolites under changed physiochemical conditions. Temperature is an important control on both the rate of reaction in forming zeolites and the species of zeolite, with higher temperatures and pressures favoring the less hydrous zeolites.

Types of geologic occurrence

Most zeolite occurrences in sedimentary rocks can be grouped into several types of geologic environment or hydrogeologic system. These are: (1) saline and highly alkaline lakes; (2) deep-sea sediments; (3) low-temperature open to closed tephra systems; (4) burial diagenesis; and (5) hydrothermal alteration. These occurrences generally exhibit characteristic patterns of zeolite zoning in tephra deposits (Hay, 1977).

Zeolites are common in deposits of saline, highly alkaline lakes (pH = 9.5–10). Vitric tephra is readily altered to zeolites, and the largest relatively pure concentrations of zeolites are found here. Early-formed zeolites may alter to analcime and or authigenic K-feldspar. Small amounts of zeolite, most commonly analcime, are commonly formed by reaction of smectite in lacustrine clays. Zoning in ash layers is chiefly lateral and reflects salinity gradients in the original lake water (Sheppard and Gude, 1968). Vitric ash may alter to zeolites in less than 1,000 years (Taylor and Surdam, 1981).

Table Z1 Main Compositional Features of Zeolites in Sedimentary Rocks[a]

Zeolite	Si/(Al + Fe^{3+})	H$_2$O[b]	Dominant cations
Clinoptilolite	4.0–5.1	3.5–4.0	K > Na
Mordenite	4.3–5.3	3.0–3.5	Na > K
Heulandite	2.9–4.0	2.5–3.1	Ca, Na
Erionite	3.0–3.6	3.0–3.5	Na, K
Chabazite	1.7–3.8	2.7–4.1	Ca, Na
Phillipsite	1.3–3.4	1.7–3.3	K, Na, Ca
Analcime	1.7–2.9	1.0–1.3	Na
Laumontite	2.0	2.0	Ca
Wairakite	2.0	1.0	Ca
Natrolite	1.5	1.0	Na

[a] Compositional data are modified after Iijima, 1988.
[b] Number of water molecules per aluminum atom.

Zeolites have formed widely in deep-sea sediments at temperatures of ≤20°C. Phillipsite and clinoptilolite are the principal zeolites, and their average total amount has been estimated as 3.5 percent of deep-sea sediments. Phillipsite is found at or near the sediment-water interface and is most common a depths of 150 m, whereas clinoptilolite usually occurs at greater depths. This feature may reflect a dissolution–precipitation relation between phillipsite and clinoptilolite.

Zeolites have been widely formed at low temperatures in vitric tephra deposits as a result of hydrolysis in which glass reacts first to form a clay mineral, generally smectite, that raises the pH and activities of Na$^+$, K$^+$, and SiO$_2$ into a zeolite stability field, where zeolite is formed by interaction of glass and pore fluid. In closed-system alteration, reactions proceed to completion without substantial ionic diffusion or interchange of pore fluid from outside the reacting system. In open-system alteration, fluids moving through the tephra deposits are changed progressively by the same water-rock reactions as in closed systems. Large, relatively pure concentrations of zeolite have been formed by open-system alteration of silicic tephra deposits.

Burial diagenesis comprises the zeolites and associated minerals formed on a regional scale in thick accumulations of sedimentary rock. The vertical zonation of zeolites primarily reflects the increase in temperature with depth. Most reported examples are in the Circum-Pacific area, and the most instructive are volcaniclastic strata in the Green Tuff region of Japan, where zeolites are being formed at known temperatures. Four principal zones have been recognized: I, fresh glass; II, clinoptilolite–mordenite; III, analcime; and IV, albite (Iijima, 1988). The upper limit of zone II is 41–55°C, for zone III, 84–91°C, and zone IV, 120–124°C. Temperature limits are lowered somewhat by saline and alkaline pore fluids.

Zeolites are common in active thermal areas with steep geothermal gradients. Zeolites and associated minerals are zoned according to temperature, but temperature limits differ from those of the same zeolites in burial diagenesis. Mordenite in the Wairakei geothermal area, for example, forms at 150–230°C compared to 41–55°C for burial diagenesis in Japan. Mordenite and wairakite are among the zeolites that are more common as hydrothermal minerals than in burial diagenesis.

Current controversy

Probably the greatest controversy regarding zeolites in sedimentary rocks is the temperature at which zeolites were formed in subaerial tephra deposits known or inferred to have been deposited at elevated temperatures. A few examples are: (1) tephra deposits of Monte Nuovo, near Naples, Italy, which were erupted in 1538 AD; (2) tephra deposits of Vesuvius that buried Ercolano (Herculaneum) in 79 AD; and (3) the Neapolitan Yellow Tuff (NYT) of Italy, which is a voluminous deposit erupted 12,000 yr BP (de' Gennaro et al., 1995). Some studies have concluded that the zeolites were formed at elevated temperatures during cooling of the deposits. Other studies of the same deposits consider the zeolites to be the product of open-system alteration at low temperature. Hydrothermal alteration has also been proposed for some of the larger zeolitic tephra deposits including the NYT (Hall, 2000). Hydrothermal fluid of the NYT is attributed to a shallow magma body beneath the caldera formed during eruption.

Richard L. Hay

Bibliography

Coombs, D.S., 1954. The nature and alteration of some Triassic sediments from Southland, New Zealand. *Transactions of the Royal Society of New Zealand*, **82**: 65–103.

de' Gennaro, M., Adabbo, M., and Langella, A., 1995. Hypothesis on the genesis of zeolites in some European volcaniclastic deposits. In Ming, D.W., and Mumpton, F.A. (eds.), *Natural Zeolites '93: Occurrence, Properties, Use*. New York: International Committee on Natural Zeolites, Brockport, pp. 51–67.

Hall, A., 2000. Large eruptions and large zeolite deposits. In Colella, C., and Mumpton, F.A. (eds.), *Natural Zeolites for the Third Millennium*. Naples, Italy, De Frede Editore, pp. 161–175.

Hay, R.L., 1966. *Zeolites and Zeolitic Reactions in Sedimentary Rocks*. Geological Society of America Special Paper 85.

Iijima, A., 1988. Diagenetic transformation of minerals as exemplified by zeolites and silica minerals. In Chilingarian, G.V., and Wolf, K.H. (eds.), *Diagenesis II, Developments in Sedimentology*. Amsterdam: Elsevier Science Publishers, pp. 147–211.

Sheppard, R.A., and Gude, A.J., 3rd., 1968. Distribution and genesis of authigenic silicate minerals in tuffs of Pleistocene Lake Tecopa, Inyo County, California. *US Geological Survey Professional Paper 597*.

Taylor, M.W., and Surdam, R.C., 1981. Zeolite reactions in the tuffaceous sediments at Teels Marsh, Nevada. *Clays and Clay Minerals*, **29**: 145–174.

Cross-references

Authigenesis
Desert Sedimentary Environments
Diagenesis
Oceanic Sediments
Sabkha, Salt Flat, Salina
Sedimentologists

Index of Authors Cited

Aagaard, P. 29, 66
Aagaard, T. 625
Abbott, P.L. 190
Abdel-Wahab, A.A. 224
Abed, A.M. 526
Abel, W.A. 27
Abrajano, T. 292
Abt, S.R. 294
Acrivos, A. 360
Adabbo, M. 780
Adam, D.P. 424
Adams, A.E. 69
Adams, J. 25
Adams, R.D. 446
AEUB 502
Agassiz, L. 633
Aggarwal, P.K. 394
Agrawal, Y.C. 285
Agricola, G. 249
Agterberg, F.P. 685
Aguilar, C. 384
Aharon, P. 526
Ahlbrandt, T.S. 253
Ahn, J.H. 65, 126
Ahnert, F. 605
Ahr, W.M. 397
Aigner, T. 354
Aiken, G.R. 362
Ainsworth, P. 371
Ainsworth, R.B. 202
Aissaoui, D.M. 395
Aitken, J.D. 446
Aksu, A.E. 292, 498, 499, 672, 673
Al Farraj, A. 7
Al-Salem, A.A. 626
Albritton, C.C. 292
Aleinikoff, J.N. 548
Aleva, G.J.J. 145, 411, 534
Alex, C.M. 673
Alexander, C.R. 605, 606, 736
Alexander, H.S. 354
Alexander, J. 433, 760
Alexandersson, E.T. 438
Algeo, T.J. 183
Ali, Y.A. 18

Allan, G.L. 397
Allan, J. 502
Allen, E.A. 587
Allen, G.P. 498, 583
Allen, H.E. 751
Allen, J.R.L. 16, 40, 42, 59, 61, 137, 168,
    172, 183, 195, 213, 214, 231, 267, 271,
    287, 296, 335, 356, 413, 414, 433, 498,
    514, 530, 536, 551, 567, 582, 596, 618,
    633, 650, 668, 711, 728, 733, 736, 748
Allen, P.A. 183, 354, 358, 364, 609, 733, 736
Aller, R.C. 438
Alley, R.B. 331, 746
Allison, K.R. 287
Allison, P.A. 458, 728
Allmendinger, R.W. 733
Allouc, J. 71
Almbaydin, F. 526
Almon, W.R. 126
Alonso-Zarza, A.M. 7, 9
Alvarez, L.W. 656
Alvarez, W. 656
Amårk, M. 137
American Geological Institute (AGI) 746
American Society of Civil Engineers 618
Amery, G.B. 674
Amit, R. 8
Amodio, S. 183
Anderson, E.J. 184
Among, H.L. 744
Amorosi, A. 333
Amos, C.L. 498, 687, 711
Amthor, J.E. 241, 242
An, Z.S. 443
Anbar, A.D. 384
Andersen, H.V. 414
Andersen, M.A. 120
Andersen, T. 299
Anderson, J.B. 330, 498, 499, 746
Anderson, K.B. 565
Anderson, L.C. 498
Anderson, R.F. 396, 492, 605, 763
Anderson, R.S. 253, 711
Anderson, R.Y. 765
Anderson, S.P. 395

Anderson, T.F. 394
Anderton, R. 271
Andresen, A. 436
Andrews, J.E. 754
Andrews, J.T. 424, 605, 606
Andrews, L.M. 223, 692
Angel, M.V. 763
Anguy, Y. 519
Anketell, J.M 551
Ankley, G.T. 751
Anonymous 223, 633
Anselmetti, F.S. 314
Antia, E.E. 157
Aomine, S. 4
Apelt, C.J. 480
Aplin, A.C. 167
Appelo, C.A.J. 542
Appleby, P.G. 753
Appleman, D.E. 139
Arakel, A.V. 91
Aravena, R. 516
Arbey, F. 63
Archer, A.W. 736, 743
Archer, J. 137
Archie, G.E. 267, 314
Arikan, F. 554
Armi, L. 464
Armijo, R. 734
Armstrong, R.L. 599
Arnott, R.W. 364, 567, 687, 711
Aronson, J.L. 29
Arora, V.K. 344
Arp, G. 3
Arrhenius, G. 379, 341
Arribas, J. 548, 549
Arthur, M.A. 85, 384, 394, 526, 702
Arvidson, R.S. 99, 100, 241, 526
Asako, Y. 711
Asaro, F. 656
Asgaard, U. 82, 700
Ashi, J. 673
Ashley, G.M. 711
Ashmore, P.E. 87, 582
Aslan, A. 35, 287
Asmerom, Y. 394

INDEX OF AUTHORS CITED

Asserto, R. 234, 681
Astin, T.M. 720
Astin, T.R. 183, 223, 659
ASTM (American Society for Testing and Materials) 344, 751
Atchley, S.C. 446
Athy, L.F. 167
Attalla, M.N. 502
Attia, K.M. 526
Atwater, B.F. 183, 756
Aubrey, D.G. 609
Aubry, M.-P. 443
Auffret, J.P. 498
Augustinus, P.G.E.F. 157, 736
Auman, H.J. 751
Ausich, W.I. 249, 728
Austen, I. 285
Austin, W.C. 682
Autin, W.J. 287
Auton, T.R. 229
Awramik, S.M. 3, 690
Awwiller, D.N. 394
Ayalon, A. 394, 396
Aydin, A. 733
Azam, F. 491
Azevedo, T.M. 91

Baag, C. 557
Baas, J.H. 567, 568
Baas, M. 594
Baba, J. 338
Bachtel, S.L. 447
Bachu, S. 502
Back, W. 190, 241, 542
Badiozamani, K. 241
Bagnold, R.A. 31, 172, 229, 253, 335, 352, 528, 529, 618, 625, 638, 672, 711, 759
Bahlburg, H. 548
Bailard, J.A. 625
Bailey, E.B. 335, 628
Bailey, E.M. 403
Bailey, S.D. 246
Bailey, S.W. 65, 126, 400, 722
Bain, D.C. 126, 167
Baird, G.C. 446, 728
Baker, A.J. 394
Baker, C.L. 651
Baker, J.C. 53
Baker, P.A. 134, 241, 242
Baker, V.R. 292, 292–293
Bakken, B. 668
Baldauf, J. 763
Baldauf, J.G. 492
Bale, A.J. 285
Ballantyne, C.K. 7
Ballard, R.R.B. 224
Balling, R.C. 123
Balme, B.E. 148
Baltzer, F. 241
Banfield, J.F. 219
Bangs, N.L. 673
Banks, N.L. 195
Banner, J.L. 394
Bao, H. 368, 395
Baozhen, Z 299
Bar-Matthews, M. 396
Barbarossa, N.L. 294, 480
Barber, D.J. 100

Barber, R 491
Barbieri, M. 507
Bardaji, T. 274
Bardossy, G. 145, 411
Bargar, E. 592
Barkan, E. 396
Barmawidjaja, D.M. 594
Barnaby, R.J. 105
Barnes, H.L. 703
Barnes, M.A. 403, 428
Barnes, N.F. 491
Barnes, P.M. 672
Barnes, P.W. 157
Barnes, S.S. 378
Barnes, W.C. 403, 428
Barnett, A. 757
Barnhisel, R.I. 126, 449
Barr, J.L. 563
Barrell, J. 172, 639
Barrett, P.J. 330
Barrett, T.J. 384
Barrie, J.V. 561, 682
Barron, E.J. 146, 491
Barson, D. 502
Bartholomä, A. 27, 736
Barwis, J. 741
Baryosef, B. 4
Bascom, W.N. 22, 27
Bassler, R.S. 308
Basu, A. 416, 430, 548, 549, 554
Basu, A.R. 394
Batchelor, G.K. 530
Bateman, R.M. 148, 314
Bates, R.L. 224, 283, 534, 736
Bates, T.F. 646
Bathurst, R.G.C. 69, 93, 134, 219, 224, 438, 462, 474, 505, 633, 640, 688
Battjes, J.A. 718
Baturin, G.N. 525, 526
Bauer, A. 371
Bauer, B.O. 246, 625
Bauer, R.A. 35
Bauld, J. 39
Baum, G.R. 333, 526
Baum, R.L. 248
Baumann, P.C. 751
Baumiller, T. 728
Baumiller, T.K. 728
Bayliss, P. 126
Bayram, A. 625
Bazykin, D.A. 183
Bazylinski, D.A. 424, 549
Beard, B.L. 384
Beauchamp, B. 682
Beaufort, D. 450
Beaufort, L. 594, 763
Bebout, D.G. 19, 242, 446, 475
Beck, J.W. 394, 395
Beck, V.H. 91
Beckinsale, R.P. 633
Becq-Giraudon, J.F. 403, 428
Bedard, D.C. 751
Befring, S. 668
Behl, R.J. 665
Behling, P.J. 146
Behr, H.J. 418
Behrens, E.W. 506, 674
Behrensmeyer, A.K. 287, 728
Behrensmeyer, K. 728
Behringer, R.P. 353

Beier, J.A. 594
Bein, A. 234
Belderson, R.H. 498, 499, 609
Belknap, D.F. 587, 588
Bender, M.L. 526
Benitez-Latorre, J. 516
Benito, M.I. 394
Benn, D.I. 292, 331
Bennett, J.G. 344
Bennett, J.L. 751
Bennett, M.R. 331
Bennett, R.H. 206
Bennett, S.J. 301, 480, 536, 618, 759
Benson, L. 754
Benson, W.H. 751
Bentley, C.R. 746
Bentor, Y.K. 134, 525
Berelson, W. 492
Berendsen, H.J.A. 36
Berezniuk, T. 502
Berg, R.C. 746
Berg, R.R. 518
Bergan, M. 281
Berger, A. 443
Berger, A.L. 183
Berger, W.H. 89, 120, 491, 492, 594, 673, 763
Berggren, W.A. 443
Bergman, H.L. 751
Bergman, K.M. 498
Berhane, H. 502
Berné, S. 498, 499
Bernard, H.A. 47, 271
Berner, K.E. 599
Berner, R.A. 85, 145, 160, 219, 226, 408, 451, 542, 557, 594, 599, 633, 702, 728
Bernoulli, D. 507, 508
Berry, W.J. 751
Berryhill, H.L.J. 499
Berthois, L. 338
Bertling, M. 71
Bertram, M.A. 99
Bertrand, P. 548, 763
Bertsch, P.M. 126, 449
Besly, B.M. 549
Best, J.L. 301, 530, 531, 536, 537, 567, 618
Bettess, R. 582
Beukes, N.J. 384, 531
Beutner, E. 436
Beveridge, T.J. 38, 39
Bevins, R. 137
Bhattacharya, J.B. 203
Bhattacharya, J.P. 202, 203
Bhattacharyya, D.P. 65, 385
Bice, K.L. 396
Biddinger, G.R. 751
Bideau, D. 267
Bierkens, M.F.P. 287
Bierman, P.R. 394
Biggs, D.L. 665
Biggs, R.B. 202, 609
Bildgen, P. 394
Billings, L.G. 756
Bilodeau, G. 673
Biot, M.A. 314
Bird, D.K. 29
Bird, M.I. 123, 394, 395, 397
Birkemeir, W. 625
Biscaye, P.E. 464, 672, 673, 763
Bischof, S.A. 438

Bischoff, A. 278
Bischoff, W.D. 99, 100
Bish, D.L. 400
Bishnoi, P.R. 138
Bishop, C.A. 751
Bishop, F.C. 99, 100
Bishop, J.K.B. 763
Bissell, H.J. 249
Bittencourt, ACSP 202
Bjørkum, P.-A. 118, 167, 168
Bjørlykke, K. 29, 118, 167
Blachère, J.R. 241
Black, J.M.W. 716
Black, K.P. 625
Black, K.S. 618
Black, M. 440
Blackwell, D.D. 316
Blair, T.C. 7, 374, 582
Blais, J.M. 408
Blake, W. 753
Blanc, P. 105, 549
Blanford, G. 416
Blankenship, D.D. 746
Blasco, S.M. 673
Blatt, H. 60, 118, 458, 554, 592, 599, 633
Blissenbach, E. 7
Blockley, J.G. 384
Bloom, A.L. 183, 395
Bloos, G. 354
Blount, A.M. 400, 722
Bluck, B.J. 22, 27
Blum, J.D. 395
Blum, M.D. 582
Boardman, M.R. 242
Bobrowsky, P.T. 756
Bock, W.D. 135
Bodine, M.W. 722
Boersma, A. 507–508
Boersma, J.R. 59
Bøggild, O.B. 69
Bögli, J. 15
Boggs, S.Jr. 184, 430, 549
Boguchwal, L.A. 301, 711
Bohacs, K.M. 85, 458
Bohn, H.L. 108
Bohor, B.F. 63, 264, 656
Boichard, R. 241
Boiron, M.C. 105
Boles, J.R. 20, 29
Bolt, G.H. 108
Bonatti, E. 507, 507–508
Bond, G.C. 733
Bondevik, S. 756
Bone, Y. 474
Bonnecaze, R.T. 352
Bonney, T.G. 633
Bookin, A.S. 400
Boon, J.D. 609
Boothroyd, J.C. 741
Bora, J.A. 185
Borchert, H. 263
Bordeaux, Y.L. 728
Borer, J.M. 446
Bornhold, B.D. 31, 668
Borre, M 120
Bory, A. 548
Bosak, P. 678
Bosellini, A. 183
Bosence, D.W. 102, 688, 728
Boswell, T.G.H. 633

Botrell, S.H. 702
Böttcher, M.E. 594
Bottjer, D. 440
Botz, R. 722
Boudreau, B.P. 226, 451, 606
Boudzoumou, F. 722
Bouloubassi, I. 403, 763
Boulton, G.S. 193, 331, 746
Boulvain, F. 688
Bouma, A.H. 59, 168, 170, 271, 375, 491, 499, 563, 633, 672, 698, 759, 778
Bourbonniere, R. 681
Bourgeois, J. 59, 364, 647, 673
Bourillet, J.F. 498
Bouroz, A. 63
Bourque, P.-A. 101, 474, 688
Boutakoff, N. 308
Bouysse, P. 498
Bowen, A.J. 31, 625, 626, 711
Bowen, R. 395
Bowman, D. 7
Bowman, J.R. 224, 395
Bowman, M. 698
Bowyer, D.A. 599
Boyce, F.M. 408
Boyd, R. 47, 202, 582
Brørs, B. 31
Bradley, J.B. 638
Bradley, W.C. 25
Bradley, W.H. 594
Brady, L.L. 356, 531
Brady, P.V. 225, 242
Braekman, J.-C. 683
Braitsch, O. 263, 722
Brakenridge, G.R. 287
Bramlett, M.N. 89, 665
Brancaccio, L. 754
Brandon, M.T. 436
Brandtstädter, H. 224, 242
Branney, M.J. 759
Brassell, S.C. 85
Bray, C.J. 599
Bray, D.I. 294, 344, 434
Bray, J.W. 425
Brayshaw, A.C. 531, 618
Brazier, V. 7
Breed, C.S. 711
Brenchley, P.J. 364, 728
Brennan, S.T. 263, 599
Brenner, R.L. 364, 743
Brett, C.A. 458
Brett, C.E. 206, 446, 728, 729
Bretz, J..H. 292
Bricker, O.P. 53, 118
Bridge, J.S. 35, 60, 87, 202, 287, 433, 480, 536, 537, 618
Bridges, E.M. 4
Bridges, P.H. 102, 688
Brien, D.L. 34
Brierley, G.J. 433, 582
Briggs, D.E.G. 702, 728
Brindley, G.W. 65, 142, 400, 633
Brinkhuis, H. 594
Brinkmann, R. 633
Bristow, C.S. 246, 498
Brodie, I. 763
Brodzikowski, K. 746
Broecker, W.S. 183, 396, 491
Bromley, R.E. 120, 700
Bromley, R.G. 71, 82, 120, 700

Brook, G.A. 681
Brookfield, M.E. 60, 253
Brooks, J.R.V. 206
Brooks, N.H. 301
Brørs, B. 31
Brosse, E. 167
Brotherton, D.I. 582
Broussard, M.L. 202
Brown, G.C. 142, 599
Brown, L.F., Jr. 183
Brown, P.G. 599
Brown, P.J. 561
Brown, R.G. 234
Brown, R.J.E. 8
Brown, S.M. 109
Brozovic, N. 733
Bruder, K.F. 674
Bruhn, C.H.L. 698
Brumer, C.A. 293
Brumfield, D.S. 519
Brumsack, H.-J. 594
Brune, G.M. 408
Brunn, P. 157
Brunnacker, K. 755
Brunner, C.A. 760
Brunsden, D. 668
Brunsdon, C.F. 344
Brush, L.M. Jr. 529
Bryan, W.B. 757
Bryant, E.A. 292, 756
Bryant, I.D. 202
Bryant, W.R. 170, 206, 274
Brzezinski, M.A. 395, 492
BSI (British Standards Institution) 344
Buat-Menard, P. 763
Buatois, L.A. 82
Bubela, B. 505
Buccino, G. 754
Bucher, W.H. 567
Bucholtz ten Brink, M.R. 665
Buck, S.G. 385
Buczynski, C. 462, 505
Budai, J.M. 190
Budd, D.A. 241, 446
Buddemeier, R.W. 606
Buffington, E.C. 31
Buffington, J.M. 294, 618, 633
Buffler, R. 139
Bugge, T. 668
Buie, B.F. 412
Bull, P.A. 716
Bull, W.B. 7, 8, 145, 7–8
Bullard, T.F. 430
Bullen, S.B. 234
Bullen, T.D. 395
Burban, P.-Y. 609
Burbank, D. 733
Burchette, T.P. 474, 475
Burden, J. 362
Burge, L.M. 433
Burke, E.A.J. 299
Burke, K. 733, 734
Burke, R.B. 560
Burke, S.K. 763
Burke, W.H. 118, 395
Burland, J.B. 121
Burley, S.D. 20, 105, 106
Burmeister, E.G. 516
Burnett, L. 751
Burnett, W.C. 525, 526

Burnham, A.K. 306
Burns, L.E. 599
Burns, R.G. 378
Burns, S.J. 21, 118, 134, 241, 396
Burns, V.M. 378
Burr, G.S. 395
Burrell, D.C. 606
Burruss, R.C. 299
Burst, J.F. 333, 720
Burstyn, H.L. 647
Burt, D.M. 100
Burt, N.T. 285
Burt, R. 4
Burton, E.A. 106, 118, 438
Burton, G.A. Jr. 751
Burton, J.H. 65, 126
Busby, C.J. 733
Busby, J.F. 190
Busch, D.A. 202
Buschkuehle, B.E. 242
Buseck, P.R. 127, 424
Busenberg, E. 99
Bustillo, M. 660
Bustin, R.M. 85, 403, 428
Butler, G.P. 19, 585
Butler, J.B. 344
Butler, R.F. 423
Butler, R.M. 502
Butt, T. 625
Butterfield, G.R. 253
Butterfield, N.J. 394
Buxton, T.M. 692
Buynevich, I.V. 47, 741
Byerly, G.R. 264
Byers, C.W. 446, 672
Byrne, J.V. 19
Byrne, R.J. 609
Byrne, T. 436

Cabrerra, J.G. 248
Cabrol, P. 681
Cacchione, D.A. 498, 625, 672, 687
Cadée, G.C. 633
Cai, J. 743
Caincross, B. 583
Cairns-Smith, A.G. 677
Calabrese, E.A. 606
Calder, E.S. 31
Caldwell, M. 592
Caldwell, W.B.E. 397
Callander, R.A. 434
Calle, C. 768
Calon, T.J. 672
Calvache, M. 8
Calvert, S. 491
Calvert, S.E. 66, 378, 379, 384, 594, 672, 763
Calvet, F. 446
Cameron, A.R. 403
Caminade, J. 756
Campbell, C.S. 33, 336
Campbell, C.V. 59, 60, 364
Campbell, I.H. 599
Campbell, L.M. 751
Campbell, S. 716
Campion, K.M. 185, 203
Cande, S.C. 443
Canfield, D.E. 395, 599, 702, 728
Cannon, J.R. 18

Cannon, R.L. 519
Cannon, W.F. 633
Cant, D.J. 434
Cao, Z. 618
Capo, R.C. 241
Carey, S. 292
Carling, P.A. 292
Carlson, C. 508
Carlson, C.P. 42
Carlson, P.R. 498
Carlson, W.D. 462
Carmona, M.C. 682
Carnes, M. 673
Carozzi, A.V. 633, 665
Carpenter, A.B. 241, 542
Carpenter, G. 672
Carpenter, S.J. 395
Carpenter, J.R. 185
Carr, S.J. 375
Carr, T.R. 183
Carrier, G.F. 718
Carrigy, M.A. 16
Carroll, A.R. 85, 506
Carson, M.A. 33, 256
Carstens, H. 224
Carter, A. 548
Cartwright, J.A. 720
Carver, R.E. 358, 618
Cas, R.A.F. 296, 335
Casas, E. 299
Casey, M. 167
Casey, R.E. 763
Cassells, J.B.C. 294
Castaing, P. 498
Castens-Seidell, B. 585
Castillo, M.M. 138
Castro, L.N. 526
Cathelineau, M. 105
Caumette, P. 39, 440
Cavazza, W. 592
Cavell, P.A. 242
Cawood, P.A. 358
Cayeux, L. 505, 640, 640–641
Cegla, J. 551
Cembranos, M.L. 4
Centineo, M.C. 333
Cerling, T.E. 91, 395
Čermák, V. 315
Chadwick, O.A. 91, 241
Chafetz, H.S. 53, 395, 505, 754
Chaloner, W.G. 123
Chamberlain, C.P. 397
Chamberlin, T.C. 643
Chamley, H. 127, 134, 142, 458, 498, 677
Chan, L.H. 599
Chan, M.A. 224
Chandler, F.W. 554
Chandler, J.H. 344
Chandler, R.J. 248
Chang, H.H. 529, 618
Channell, J.E.T. 424
Channer, D.M.DeR. 599
Chapelle, F.H. 542
Chapman, C.R. 278
Chapman, D.L. 109
Chapman, D.M. 668
Chapman, P.M. 751
Chapman, R.E. 267
Chapman, S. 229

Chapman, V.J. 736
Chappell, J.M.A. 395
Chappie, D.J. 751
Charaklis, W.G. 39
Charlie, D. 756
Chase, E.B. 605
Chaudhuri, S. 395
Chazot, G. 395
Cheel, R.J. 356, 364, 537, 687
Chen, C.L. 352
Chen, J. 681
Chen, W. 224
Chen, Y. 4
Cheng, D.C.-H. 360
Cherry, J.A. 268
Chesnut, D.R. 148
Chi, G. 508
Childs, C.W. 395
Chilingar, G.V. 249
Chillingarian, G.V. 168
Chivas, A.R. 358, 395, 397
Choi, K.S. 743
Choi, D.R. 446
Choppin, G.R. 438
Choquette, P.W. 15, 118, 267, 678, 684
Chorley, R.J. 34, 633
Chough, S.K. 274
Choung, K.-S. 687
Chow, N. 446
Chowns, T.M. 224, 308, 665
Christ, C.L. 63, 634
Christiansen, E.A. 678
Christie, J.M. 554
Christie, P.A.F. 167
Christman, R.F. 362
Chrzastowski, M.J. 35
Chuber, S. 183
Chuhan, F.A. 167
Church, M. 8, 11, 22, 25, 345, 434, 480, 531, 582, 618
Church, M.A. 344
Chyba, C.F. 599
Cialone, M.A. 43
Cicero, A.D. 599
Cirac, P. 498
Cisne, J.L. 183
Cita, M.B. 594
Clague, D.A. 733
Clague, J.J. 33, 605, 756
Clark, I.D. 395
Clark, J.B. 751
Clark, J.D. 698
Clark, J.S. 123
Clarke, J.D.A. 474
Clarke, T.L. 561
Clauer, N. 333, 371, 395, 450
Claxton, B.L. 300
Claypool, G.E. 306, 395
Clayton, C.J. 306
Clayton, J.L. 605
Clayton, R.H. 385
Clayton, R.N. 395, 396
Clayton, W. 137
Clements, R.G. 83
Clemett, S.J. 549
Clemmensen, L. 137
Clenell, M.B. 167
Cleveringa, J. 625
Clifford, C.H. 594
Clift, R. 229

INDEX OF AUTHORS CITED

Clifton, H.E. 42, 157, 537, 625
Clinton, H.E. 634
Cloos, M. 436
Cloud, P. 384
Clough, J.G. 183
Clough, S.R. 35, 287
Clymo, R.S. 516
Cobb, J.C. 157
Cochonat, P. 673, 698
Cochran, J.B. 170
Cochran, U. 757
Codd, J.M. 498
Cogley, J.G. 599
Cohen, A.S. 734
Cohen, Y. 39, 440
Cole, F.A. 751
Cole, R.B. 582
Cole, R.D. 206
Colella, A. 202, 274
Coleman, J.M. 36, 202, 283, 668
Coleman, M.L. 29, 306, 395, 396
Coleman, N.L. 229, 618
Coleman, S.E. 301
Colin, F. 722
Collier, E.L. 433
Collier, R. 274, 492, 599
Collins, M.B. 568, 609
Collinson, J. 137
Collinson, J.D. 195, 203, 271
Collinson, M.E. 123, 148
Colman, A.S. 525
Colquehoun, D. 561
Colquhoun, D.J. 47
Combaz, A. 429
Compston, W. 656
Compton, J. 525
Condie, K.C. 599
Congdon, R.D. 63
Coniglio, M. 395, 672
Conley, D.C. 625
Connolly, J.P. 751
Conolly, J.R. 692
Conway, K.W. 682
Conybeare, C.E.B. 283, 634
Cook, H.E. 474
Cook, P.J. 525, 526
Cooke, R.U. 22, 211, 253
Coombs, D.S. 29, 780
Cooper, J.R. 414
Copard, Y. 403, 428
Corbett, P.W.M. 618
Corliss, J.B. 599
Cornelisse, J. 285
Cornell, R.M. 368
Cornwell, J.C. 703
Correns, C.W. 118, 641
Corte, A.E. 214
Corvi, P.J. 306
Costa, J.E. 8, 33, 292
Cotterill, D.K. 502
Coumes, F. 698
Cousineau, P.A. 508
Cowan, C.A. 446, 475
Cowan, D.S. 436
Cowell, P.J. 47, 625
Cowen, C.E. 751
Cowen, E.A. 743
Cowling, T.G. 229
Cox, G.H. 628
Cox, L. 384

Crabtree, S.J. 519
Craig, H. 366, 395
Craig, R.F. 24, 425
Crans, W. 668
Crawford, M.L. 299
Crawley, R.A. 537
Creaney, S. 85, 458, 502
Creaser, R.A. 397
Crelling, J.C. 565
Crerar, D.A. 118, 136
Cresswell, R.G. 397
Crickmore, M.J. 753
Cripps, J. 123
Croke, J.C. 287, 434, 583
Cronan, D.S. 379, 491, 534
Crone, A.J. 557
Crook, K.A.W. 283, 634
Cross, B.L. 308
Cross, H. 159
Cross, T.A. 35, 185, 287, 634, 644
Cross, W. 634
Cross, T.A. 447
Crossland, C.J. 606
Croudace, I. 757
Crowder, D.W. 344
Crowell, J.C. 184, 331, 514, 733
Crowley, T.J. 145
Cruden, D.M. 33, 248
Crunan, H.R. 665
Crux, J.A. 499
Cruzado, A. 682
Csanady, G.T. 229
Cubitt, J.M. 685
Cui, B. 338, 344
Cui, Y. 618
Culbertson, K. 605
Cullen, D.J. 525, 526
Cullers, R.L. 548
Culver, J.J. 716
Culver, S.J. 499, 716
Cundy, A. 757
Cuomo, M.C. 206
Curiale, J.A. 366
Curray, J.R. 491, 512, 561
Curry, B.B. 746
Curry, W.B. 763
Curtis, C. 306, 396
Curtis, C.D. 29, 395
Curtis, R. 585
Curtis, W.F. 605
Cys, J.M. 118

Dabrio, C.J. 42, 274
Dacey, M.F. 184
Dainyak, L.G. 371
Dake, C.L. 628
Daley, B. 354
D'Almeida, G.A. 253
Dalrymple, R.W. 202, 498, 499, 582, 609, 736, 743
Daly, J.F. 587
Daly, R.A. 31
Dan, J. 8
Dana, J.D. 308
Dapples, E.C. 224
Darby, D. 757
Darby, S.E. 294
D'Argenio, B.D. 183, 754, 755

Darwin, C. 560
Darwin, G.H. 568, 634
D'Asaro, E. 464
Dasch, E.J. 397
Datta, P. 400
Daughtry, K.L. 397
Dauphin, P. 526
Davaud, E. 53, 505
David, P.P. 246
Davidson, D.F. 135
Davidson, M.A. 43
Davidson-Arnott, R.G.D. 42, 625
Davies, A.M. 224
Davies, D.K. 126
Davies, G.R. 241, 263
Davies, G.W. 618
Davies, J.L. 736
Davies, M.C.R. 414
Davies, P. 755
Davies, P.J. 53, 447, 505, 506
Davies, R. 360
Davies, R.K. 202
Davies, T. 491
Davies, T.A. 733
Davies, T.R.H. 301
Davis, A. 85, 403
Davis, B.S. 39
Davis, E.E. 138
Davis, J.C. 183, 685
Davis, J.M. 224, 267
Davis, R.A. 736
Davis, R.A., Jr 47, 736–737
Davis, R.H. 360
Davis, W.M. 634
Davoli, G. 59, 292
Dawans, J.M. 242
Dawson, A.G. 756
Dawson, S. 756
Day, S.J. 531
De Batist, M. 498
de Boer, P.L. 183, 625, 736, 737
De Decker, R.H. 534
De Geer, G. 766
de' Gennaro, M. 780
De Gilbert, J.M. 700
de Haas, H. 737
De Josselin de Jong, G. 336
De La 768
de Lange, G.J. 423, 594
De La Pena, J.A. 548
De La Rocha, C.L. 395
De Lange, J.J. 594
de Leeuw, J.W. 148, 594
de Ligny, D. 400
de Raaf, J.F.M. 59
de Vernal, A. 673
de la Calle, C. 450
de Ronde, C.E.J. 599
DeBen, W.A. 751
DeCaritat, P. 126
DeCelles, P.G. 224, 548, 582, 592, 733
DeMaris,P.J. 148
DeMaster, D.J. 492
DeMaster, T.D. 492
DeNiro, M.J. 395
DePaolo, D.J. 395
DePasquale, F. 224
Deakin, M. 526
Dean, W.E. 85, 526
Dearnaley, M.J. 285

Debrenne, F. 682
Decarreau, A. 722
Decho, A.W. 39, 440
Dedkov, A.P. 605
Deer, W.A. 769
Défarge, C. 101
Degens, E.T. 19, 594, 634
Deigaard, R. 625
Dekkers, M.J. 423
Delaney, M.L. 120, 525, 526
Deloule, E. 105
Delour, J. 353
Demaison, G.J. 763
Demars, C. 105–106
Demicco, R.V. 19, 157, 263, 447, 537, 585, 599
Demko, T.H. 700
Demoulin, A. 137
Denaix, L. 4
Denison, R.E. 118, 395
Denlinger, R.P. 188, 352–353
Dennis, P. 754
Denny, C.S. 8
Denton, P. 755
Derenne, S. 403
Derry, L.A. 395
Desrochers, A. 15, 682
Deutsch, A. 278
Dewers, T.A. 190
Dewey, J.F. 733, 734
Dewhurst, D.N. 720
Deynoux, M. 331, 746
D'Hondt, S.L. 443
Di Toro, D.M. 751
Diamond, L.W. 29
Dickens, G.R. 138
Dickinson, W.R. 430, 548, 592, 698, 733
Dickinson, W.W. 757
Dickson, J.A.D. 118, 121, 234, 462, 684
Dickson, R.R. 464
Diedrihc, N.W. 185
Diekmann, B. 549
Diem, B. 568
Diepenbroek, M. 27
Diessel, C.F.K. 63, 85, 403, 428
Dietrich, W.E. 22, 338, 344, 408
Difficy, J.P. 287
Dijkstra, T.A. 268
Dill, R.F. 31
Diller, G. 137
Dillon, T. 751
Dillon, W. 139
DiMichele, W.A. 145, 148
Dimitrov, P. 292
Dimroth, E. 384
DIN (Deutsche Ingenieure Normung) 344
Ding, T. 395
Dingle, R.V. 668
Dingler, J.R. 42, 625
Dinkelman, M.G. 563
Diplas, P. 344
Disnar, J.-R. 403, 428
Ditchfield, P.W. 224, 462
Dix, G.R. 672
Dixit, S. 491
Dobkins, J.E. 27
Dobson, D. 434
Dockal, J.A. 190
Dodds, J. 267

Dohrenwend, J.C. 9
Dolan, J.F. 447
Domack, E.W. 331
Domenech, R. 700
Domenico, P.A. 542
Domenico, S.N. 314
Dominey-Howes, D. 757
Dominguez, J.M.L. 202
Dominic, D.F. 618
Donahue, D.J. 395, 594
Donaldson, J.A. 384
Donovan, D.T. 183
Donovan, R.N. 374, 720
Donovan, S.K. 728
Dooge, J.C.I. 634
Doornkamp, J.C. 716
Dorale, J.A. 395
Dore, W.H. 109
Dorn, R.I. 8
Dorobek, S. 241
Dott, R.H. 364, 447
Dott, R.H. Jr. 59, 446, 592, 628, 634
Douglas, I. 605
Douglas, J. 605
Douglas, L.A. 769
Douglas, R.G. 635
Douglas, S. 39
Doukhan, J.-C. 278
Dousset, P.E. 531
Doveton, J.H. 314
Dowdeswell, J.A. 331, 606
Dowling, C.B. 394
Downing, J.P. 626
Downing, R.A. 120
Downs, W.R. 606
Doyle, L.J. 447
Doyle, R.J. 38
Drake, D.E. 498, 625, 673
Drake, T.G. 301
Dreimanis, A. 331, 746
Drescher, A. 336
Dressler, B.O. 278
Drever, J.I. 262, 542
Dria, K. 362
Driese, S.G. 447
Driscoll, M.L. 672
Driscoll, N.W. 668
Drits, V. 371
Drits, V.A. 234, 371, 400
Drivet, E. 241
Dromgoole, P. 698
Droppo, I.G. 408
Droste, J.B. 722
Droz, L. 698
Druffel, E.R.M. 395
Druitt, T.H. 360, 530
Drummond, C. 692
Drummond, C.N. 183, 185
DuMontelle, P.B. 35
Dubińska, E. 450
Duce, R.A. 491
Duck, R.W. 659
Dudas, M.J. 716
Dudley, S.J. 294
Duff, P.McL.D. 183, 447
Dufficy, J.P. 35
Dugdate, R. 492
Duggan, J.P. 241
Duivenvoorden, C. 202
Duke, W.L. 354, 364, 687

Dullien, F.A.L. 519
Dumont, J.L. 91
Dunbar, C.O. 135, 651
Dunbar, R.B. 673
Dunham, J.B. 243
Dunham, R.J. 110, 135, 249, 272, 560
Dunlop, D.J. 368
Dunn, A.J. 633
Dunn, P.A. 183, 447, 474
Dunne, K.C. 287
Dunoyer de Segonzac, G. 634
Dupont, E. 688
Durand, A. 418
Durand, B. 362, 403, 429
Durand, J. 498
Durney, D.W. 634
Durrance, E.M. 224
Dury, G.H. 434
Duschatko, R.W. 268
Dutton, S.P. 20, 219
Dvorkin, J. 314
Dyer, K. 625
Dyer, K.R. 229, 285, 498, 609, 736
Dymond, J. 492
Dymond, J.R. 378
Dymond, R. 492
Dypvik, H. 549
Dysthe, D. 167
Dzulynski, S. 59, 282, 551, 678, 748, 759

Earle, L.R. 516
Eastoe, C. 396
Eberhardt, P. 106
Eberl, D.D. 142, 234, 371, 449, 450, 462
Eberli, G.P. 183, 314
Eble, C.F. 148
Ebneth, S. 395
Edelvang, K. 285
Edie, B.J. 135
Edmond, J.M. 395, 599
Edwards, D.P. 592
Edwards, K.J. 123
Edwards, M.B. 195
Edwards, R.L. 395
Egeberg, P.K. 167
Eggenkamp, H.G.M. 395
Egrlich, R. 519
Ehlers, J. 736
Ehrenberg, S.N. 126
Ehrlich, H.L. 39
Ehrlich, R. 430, 519, 561
Eidsvik, K.J. 31
Eidvin, T. 668
Einsele, G. 93, 183, 443, 687
Einstein, A. 229, 352, 360
Einstein, H.A. 294, 480
Eisma, D. 285
Eittreim, S.L. 464, 672
Ekdale, A.A. 82, 700
El-Younsy, A.R.M. 224
Elam, J.C. 183
Elders, W.A. 371
Elgar, S. 625
Elimelech, M. 159
Elkins, J.E. 224, 308, 665
Ellam, R.M. 548

INDEX OF AUTHORS CITED

Ellery, K. 583
Ellery, W.N. 583
Elliot, T. 183
Elliott, S. 757
Ellis, C. 188, 353
Ellis, P.G. 702
Ellwood, J.M. 711
Elmore, D. 396
Elorza, J.J. 308
Elsass, F. 371, 450
Elverhøi, A. 188
Embry, A.F. 134
Emeis, K.C. 594, 763
Emerson, S.R. 526
Emery, D. 184
Emery, K.O. 19, 272, 458, 534, 561, 673, 702, 718
Emiliani, C. 395, 507
Emms, P.W. 31
Engel, M.H. 85, 362
Engel, R.J. 4
Engelund, F. 618
Englezos, P. 138
Engström, F. 121
Enos, P. 267, 474, 618
Environmental Protection Agency (US) 751
Enzel, Y. 8
Epstein, S.M. 397, 634
Erba, E. 443
Erdmann, A.L. 35
Erdmer, P. 397
Erel, Y. 395
Ergenzinger, P. 753
Eriksson, K.A. 183, 384, 599, 736, 743
Erlenkeuser, H. 763
Ernst, W.G. 308, 634
Erskine, W.D. 35
Eshet, Y. 722
Eslinger, E.V. 29, 127, 397, 677
Espitalié, J. 429
Essene, E.J. 100, 105, 677
Esslinger, P. 502
Esteban, M. 15
Ethington, R.L. 243
Ethridge, F.G. 354, 583
Etris, E.L. 519
Eugster, H.P. 19, 145, 184, 211, 262, 263, 418, 542
Evamy, B.D. 190, 684
Evans, B. 543
Evans, B.W. 508, 722
Evans, C.D.R. 498
Evans, D.J.A. 292, 331
Evans, G. 585
Evans, H.T. Jr. 63
Evans, J.M. 397
Evans, M.J. 395
Evans, P.D. 711
Evans, R. 634
Evans, S.G. 33, 605
Ewen, D.F. 358
Ewers, W.E. 384
Ewing, J. 491
Ewing, M. 491, 673, 722
Eyles, C.H. 145, 331
Eyles, N. 145, 331, 747
Eymery, J.P. 450
Faas, R.W. 609
Fabricius, F.H. 505
Fabricius, I.L. 120

Fader, G.B. 561
Fader, G.B.J. 746
Fagel, N. 548
Fagerstrom, J.A. 82, 475
Faghri, M. 711
Fahnestock, R.H. 1
Fahnestock, R.K. 25, 272
Fairbridge, R.W. 183, 190, 219, 673
Fairchild, I.J. 243, 331
Faleide, J.I. 168
Falkowski, P. 491
Fallick, A.E. 20, 21, 127, 234, 394
Fanning, C.M. 548
Farley, K.A. 264
Farmer, V.C. 4, 400
Farquhar, J. 395
Farrell, K.M. 287
Farrington, J.W. 763
Fastovsky, D. 657
Faugères, J.-C. 673
Faure, G. 395, 400, 548
Faure, K. 599
Faust, G.T. 65
Faust, L.Y. 314
Fawcett, P. 146
Feary, D.A. 447
Feder, J. 168
Federico, D.C. 711
Fedo, C.M. 599
Feigl, F. 684
Feldman, H.R. 743
Fendorf, S.E. 109
Fennessy, M.J. 285
Fennessy, W.J. 674
Ferguson, J. 505
Ferguson, R. 344
Ferguson, R.I. 87
Ferland, M.A. 47
Ferm, J.B. 519
Ferm, J.C. 646
Fernandes, A. 4
Fernandez, J. 8
Ferrara, G. 507
Ferraro, S.P. 751
Ferrell, R.E. Jr. 554
Ferreri, V. 183, 754, 755
Ferris, F.G. 39
Fessas, Y.P. 360, 530
Field, L.J. 751
Field, M. 606
Field, M.E. 498, 499
Fieller, N.R.J. 344
Fifield, L.K. 397
Figueiredo, A.G. Jr 687
Filippelli, G. 525
Filippelli, G.M. 525, 526
Fillo, R.H. 499
Finkel, R.C. 605
Finkelman, R.B. 684
Finkleman, R. 308
Fiorillo, G. 502
Fischer, A. 443
Fischer, A.G. 183, 185
Fischer, G. 492
Fischer, J. 106
Fishbaugh, D.A. 42
Fishenich, J.C. 294
Fisher, C.G. 766
Fisher, I.S. 308
Fisher, M.J. 396

Fisher, Q.J. 167, 224
Fisher, R.S. 20–21
Fisher, R.V. 63, 134
Fishman, N.S. 703
Fisk, H.N. 272, 287, 434, 582
Fiske, R.S. 757
Fitchen, W.M. 183
FitzGerald, D.M. 47, 741
Flörke, O.W. 412
Flügel, E. 69, 505, 560
Flanningan, D.T. 408
Fleet, A.J. 167
Fleischer, P. 673
Fleisher, C.J. 681
Fleming, R.W. 248
Flemming, B.W. 157, 498, 736
Flenley, E.C. 344
Fletcher, W.K. 531
Flick, R.E. 31
Flicoteaux, R. 526
Flinsch, M.R. 396
Flint, R.F. 135, 331
Flint, S.S. 202
Flood, P.G. 447
Flood, R.D. 711
Flores, G. 436
Flores, R.M. 287
Floyd, P.A. 63, 548
Foerster, R. 308
Folk, R.L. 27, 118, 135, 219, 224, 234, 249, 272, 344, 345, 438, 462, 505, 508, 548, 557, 646, 650, 665, 681, 685, 716, 754
Föllmi, K.B. 525, 526
Follo, M.F. 592
Fontes, J.Ch. 395
Fontugne, M.R. 594
Foord, E.E. 264, 656
Ford, D. 110
Ford, D.C. 678, 681
Ford, R.L. 430
Ford, T.D. 755
Forman, S.A. 65
Forti, P. 681
Fortin, D. 39
Foscolos, A.E. 66
Foster, J.M. 763
Foster, R.J. 720
Fournier, F. 605
Fowler, B. 379
Früh-Green, G.L. 508
Frakes, L.A. 145
Fralick, P.W. 384
France-Lanord, C. 395
Francis, J.E. 145
Francis, J.R.D. 301
Franco, E. 450
Francois, R. 492
Francu, J. 395
Frank, A. 246, 253
Frankel, R.B. 424
Franks, P.C. 224
Franks, S.G. 29
Franseen, E.K. 183
Frape, S.K. 366, 396
Fraser, G.S. 42, 743
Fraser, H.J. 519
Frazier, D.E. 674
Frazier, S. 362
Fredsøe, J. 618, 625
Freeland, G.L. 687

Freeze, R.A. 268
Freiwald, A. 560
French, B.M. 278, 416, 656
French, P.W. 736
Frexxotti, M.-L. 299
Frey, M. 126, 371
Frey, R.W. 82, 83, 649, 700
Freyssinet, P. 395
Freytet, P. 91, 755
Friedman, G.M. 19, 135, 438, 634, 684
Friedman, I. 395
Friedrich, G. 379
Friedrich, N.E. 741
Friedrichs, G.T. 609
Friend, P.F. 11, 582
Friend, P.G. 734
Fripp, J.B. 344
Frishman, S.A. 506
Frisia, S. 100
Fritz, P. 366, 395
Fritz, S.J. 65, 66
Froelich, P.N. 526, 557
Frolov, E.B. 366
Frost, R.L. 51
Frostick, L.C. 22
Frostick, L.E. 8, 531
Fry, W.H. 109
Fryberger, S.G. 211, 246, 253
Fryer, P. 508
Frykman, P. 121
Fu, L. 224
Füchtbauer, H. 118, 135, 368, 635
Fuchsman, C.H. 516
Fukushima, H. 549
Fukushima, Y. 31, 353, 759
Fulfaro, V.J. 734
Full, W. 519
Fullagar, P.D. 333
Fuller, S. 746
Fullerton, L.B. 126
Fung, J.C.H. 229
Futterer, E. 354, 728
Fürsich, F.R. 700
Fürsich, F.T. 728
Fütterer, D. 506
Fyfe, W.S. 394, 395, 396, 599

Gabel, S.L. 203
Gaffey, S.J. 506
Gage, K.L. 397
Gagliano, S.M. 283
Gagnon, A.R. 594
Gaines, A.M. 99, 100
Gaines, G.L. 109
Gale, A.S. 120
Gale, J.F. 718
Gallagher, J.J. Jr 716
Galli-Olivier, C. 526
Galloway, W.E. 157, 183, 202
Galtier, J. 149
Galvin, C. 374
Galy, A. 395
Gammon, P.R. 682
Ganly, P. 628
Ganssen, G. 763
Gapon, E.N. 109
Garbe-Schonberg, D. 722
Garber, R.A. 19

Garcette-Lepecq, A. 403
Garcia del Cura, M.A. 755
García, M. 31
Gardeweg, M.C. 31
Gardner, J. 606
Gardner, L.W. 314
Gardner, W. 492
Gardner, W.D. 464, 672, 673
Gardulski, A.F. 673
Garfunkel, Z. 668
Garlan, T. 498
Garlick, G.D. 224
Garnett, R.H. 534
Garrels, R.M. 63, 256, 263, 492, 599, 605, 634
Garrison, R.E. 118, 121, 458, 526, 665
Gartner, S. 657
Garvie, L.A.J. 127
Gaspar, L. 526
Gass, I.G. 734
Gassmann, F. 314
Gastaldo, R.A. 145, 744
Gat, J. 408
Gat, J.R. 397
Gaudette, H.E. 242
Gaupp, R. 371
Gautret, P. 102
Gavin, J. 443
Gawthorpe, R.L. 433, 733, 734
Gecol, H. 360
Gees, R. 665
Gehrels, G.E. 548
Gehrels, W.R. 587
Geikie, A. 746
Geiss, J. 106
Gektidis, M. 71
Geldsetzer, H.H.J. 474
Gerdes, G. 39, 440
Germann, K. 135
Gersib, G.A. 287
Gerson, R. 8
Gessler, D. 480
Gessler, J. 22
Getzen, R.T. 563
Ghibaudo, G. 759
Ghosh, J.K. 561
Gibbons, G.S. 374
Gibbs, M.T. 146
Gibbs, R.J. 722
Gibling, D.A. 364
Gibling, M.R. 11, 512, 582, 743
Gibson, C. 443
Gibson, E.K. Jr 549
Gidaspow, D. 352
Gidman, J. 234
Giese, R.F. 400, 722
Gieskes, J.M. 412
Giesy, J.P. 751
Gilbert, G.K. 301, 582, 711
Gilbert, R. 605, 766
Gile, L.H. 91
Giles, J.R.A. 91
Giles, K.A. 733
Giles, M.R. 167, 219, 371
Gilg, H.A. 397
Gill, D. 685
Gill, I.P. 560
Gill, R. 634
Gillette, D.A. 253

Gillispie, C.C. 634
Gillman, R.A. 224
Gingras, M.K. 82, 202
Ginsburg, R.N. 53, 118, 157, 183, 272, 446, 474, 634, 653, 736
Giosan, L. 202
Giovanoli, R. 379
Girard, J.-P. 395
Girardclos, S. 505
Gischler, E. 53, 653
Given, R.K. 241
Gjessing, E.T. 362
Glaser, J.A. 675
Glass, B.P. 656
Glassby, G.P. 379
Glasser, N.F. 331
Glazek, J. 678
Gleason, J.D. 396
Glen, W. 656
Glenn, C.R. 333, 525, 526
Glennie, K.W. 211, 272, 653
Glover, B.W. 557
Gobas, F.A.P.C. 751
Goddard, E.N. 557
Godderis, Y. 599
Goff, J.A. 668
Goff, J.R. 757
Golabics, S. 755
Gold, T. 366
Goldberg, E.D. 492
Goldberg, S. 109
Goldhaber, M.B. 702
Goldhammer, R.K. 183, 185, 447, 474
Golding, H.G. 692
Golding, S.D. 660
Goldsmith, J.R. 99, 100, 99–100
Goldsmith, P. 757
Goldstein, A. 224, 308
Goldstein, R.H. 118, 299, 299
Gole, M.J. 384
Goles, G.G. 549
Golonka, J. 560
Golubic, S. 71, 690, 729, 755
Gomez, B. 480, 618
Gonthier, E. 673
Gonzales, E. 436
Gonzalez, L.A. 395
Gonzalez-Martin, J-A. 755
Good, G.A. 634
Goodbody, Q.H. 135
Goodwin, L. 224
Goodwin, P.W. 184
Gordon, L.J. 599
Gorham, E. 408, 516
Gorini, M. 507–508
Gorsline, D.S. 272, 491, 673, 733
Gorton, N.T. 105
Gorur, N. 292
Göttlich, K. 516
Gostin, V.A. 656, 657, 720
Gotze, J. 554
Goudie, A.S. 91, 211, 253
Gouy, G. 109
Goy, J.L. 7, 8, 9, 274
Grabau, A.W. 184, 644, 653
Grace, J.R. 229
Graf, D.L. 99, 100, 395, 99–100
Graf, J.C. 416
Graf, W.H. 618
Graham, J. 354

INDEX OF AUTHORS CITED

Graham, S.A. 592, 733
Grahame, D.C. 109
Grammer, G.M. 53, 184
Grams, J.C. 105
Granger, D.E. 605
Granlund, E. 516
Grant, B. 599
Grant, W.D. 625
Grass, A.J. 229
Gratier, J.P. 167, 543
Graton, L.C. 519
Graves, W.J. 303
Gray, J.M.N.T. 352
Gray, M.B. 582
Graybill, F.A. 685
Greb, S.F. 148
Greeley, R. 253, 711
Green, R.H. 751
Greenbaum, N. 8
Greenkorn, R.A. 268
Greenly, E. 436
Greenspan, H.P. 718
Greenwood, B. 42, 43, 625, 626
Gregg, J.M. 234, 241, 242
Gregor, B. 634
Gregory, A.R. 314
Greimann, B.P. 618
Gressly, A. 272
Grieve, D.A. 403
Grieve, R.A.F. 278, 657
Griffin, G.M. 722
Griffin, J.J. 492
Griffiths, D.K. 498
Griffiths, J.C. 344, 646, 685
Grim, R. 634
Grim, R.E. 63, 142, 400, 677
Grimaldi, D.A. 565
Grimm, J.P. 430
Grimm, K.A. 526
Grinnell, R.S. 728
Grospietsch, Th. 516
Gross, G.A. 384
Grossman, R.B. 91
Grossman, S. 8
Grosswald, M. 292
Grotzinger, J.P. 118, 184–185, 384, 446, 447, 474, 506, 625, 690, 698
Grousset, F. 763
Grousset, F.E. 548
Grow, T. 711
Grun, R. 755
Gruner, J.W. 400
Grunnan, H.R. 665
Gruszczynski, M. 625
Gruver, B.L. 42
Guéguen, Y. 314
Gude, A.J. 418, 666
Gude, A.J.3rd. 780
Guggenheim, S. 142, 722
Guichard, F. 498
Guidotti, C.V. 358
Guidry, M.W. 526
Guiguet, R. 543
Guilbert, J.M. 135
Guillen, J. 625
Gulick, S.S. 673
Gulliver, F. 561
Gunatilaka, A. 234, 438
Guo, L. 755
Guo, Y. 397

Gupta, S. 733
Gurnis, M. 733
Gust, G. 596, 609
Gustavson, T.C. 145
Gutjahr, A.L. 268
Gutschick, R.C. 224
Güven, N. 63, 677
Guy, D.G. 675
Guy, H.P. 301, 711
Guy, M. 530
Guza, R.T. 625
Gyr, A. 568

Håkanson, L. 408
Håkansson, E. 120
Häntzschel, W. 283
Habicht, K.S. 395, 599
Haenel, R. 316
Haff, P.K. 253
Haffner, G.D. 751
Hagadorn, J.W. 440
Hagdorn, H. 249
Hagerthey, S.E. 618
Haggerty, J.A. 508
Haibo, F. 299
Haines, P.W. 656
Halbach, P. 379
Halfpenny, R. 531
Halicz, L. 396
Hall, A. 780
Hall, C.A. Jr. 308
Hall, D. 480
Hall, I.R. 396
Hallam, A. 183, 447, 634
Hallermeier, R.J. 344
Hallet, B. 253
Halley, R.B. 118, 242, 243
Halliday, A.N. 395, 396
Hallock, P. 474, 673
Hallsworth, C.R. 358
Halverson, G.P. 384, 395
Ham, W.E. 634
Hamblin, A.P. 687
Hamblin, W.K. 778
Hambrey, M.J. 331, 746
Hamelink, J.L. 751
Hamilton, P.J. 371
Hamm, E.P. 634
Hammond, D. 492, 526
Hampton, B.P. 224
Hampton, M.A. 408, 668, 675, 759, 760
Hamza, V. 316
Hancock, J.M. 184
Handford, C.R. 263
Hands, E.H. 42
Handschy, J.F. 592
Hanes, D.M. 625
Hannington, M. 722
Hanor, J.S. 53, 542
Hanselmann, K. 39
Hansen, D.J. 751
Hansen, E. 618
Hanshaw, B.B. 190
Hanshaw, D.B. 241
Hansley, P.L. 557
Hanson, C.G. 531
Hanson, G.N. 394
Hapke, B. 416

Happ, S.C. 434
Haq, B.U. 184
Harbaugh, J.W. 203
Harbour, J. 8
Hardenbol, J. 184–185, 333, 526
Hardie, L.A. 19, 100, 135, 145, 183–184, 211, 242, 262, 263, 474, 506, 537, 542, 585, 599
Hardie, L.H. 447
Hardman, R. 121
Haremboure, J. 668
Harker, R.I. 100
Harland, W.B. 331, 746
Harms, J.C. 59, 172, 272, 364
Harper, D.A. 395
Harper, D.E. Jr. 751
Harris, C. 414, 425
Harris, M.T. 183
Harris, P.M. 446
Harris, P.T. 498, 609
Harris, T. 145
Harrison, A.S. 301
Harrison, C.G. 605
Harriss, R.C. 379
Harshbarger, J.C. 751
Hart, B.S. 202
Hartley, A.J. 698
Hartman, H. 677
Hartman, N. 526
Harvey, A.M. 7, 8, 9
Harvie, C.E. 263
Hassan, M.A. 22, 618, 753
Hassler, S.W. 384
Haszeldine, R.S. 20, 21
Hatch, F.H. 634
Hatcher, P. 362
Hathaway, J.C. 722
Hattingh, J. 563
Hattori, I. 184
Hattori, K.H. 722
Hauff, P.L. 127
Haughton, P.D.W. 458, 549
Havholm, K. 253
Hawkesworth, C. 396
Hawksley, P.G. 360
Hay, A.E. 568, 626
Hay, R.L. 780
Hay, W.H. 605
Hay, W.W. 89, 605
Hayes, J.B. 224, 308
Hayes, J.M. 396
Hayes, M. 362
Hayes, M.J. 29
Hayes, M.O. 47, 736, 741
Hays, J. 443
Hays, J.D. 145, 184
Hayton, A. 751
Hayward, B.W. 757
Hazzard, J. 33
He, Q. 753
Heald, M.T. 29
Heaney, P. 660
Heard, H.C. 99, 99–100
Hearty, P.J. 757
Heath, G.R. 89, 492, 526
Heath, K.C. 673
Heath, R.G. 492
Heathershaw, A.D. 498
Heathwaite, A.L. 516
Hebisch, V. 379

Heckel, P.H. 184, 474
Heezen, B.C. 272, 430, 673, 733
Heggie, D.T. 526
Hegner, E. 549
Heijen, W. 106
Heiken, G.H. 416
Heiman, Monica 651
Hein, F.J. 673, 698
Heize, C. 492
Héquette, A. 157
Heling, D. 135
Heller, P.L. 35, 698
Helmold, K.P. 281
Helsley, C.E. 557
Hem, J.D. 263
Hemming, M.A. 561
Hemming, N.G. 105
Hemphill-Haley, E. 756, 757
Henderson-Sellers, A. 599
Hendricks, S.B. 109, 400
Hendry, J.P. 20, 106, 462
Henk, B. 82
Henley, S. 685
Henning, G.J. 755
Henrich, R. 560
Henry, D.J. 358
Henry, M. 751
Henry, V.J. 736
Hensley, V. 526
Herbers, T.H.C. 625
Herbert, T.D. 443
Herbertson, J.G. 301
Herczeg, A. 526
Herdman Sir, W.A. 647
Heritage, G.L. 344
Herman, J.P. 498
Herman, J.S. 241, 755
Hermanns, K. 71
Heron, S.D. 47, 741
Hertweck, G. 736
Hervig, R.L. 395, 397
Herzer, R.H. 561
Herzig, P. 722
Herzog, B.L. 35
Heslop, D. 423
Hess, G.R. 711
Hesse, P. 423
Hesse, R. 542, 660, 665, 673, 698
Hesselbo, S.P. 659
Hetherington, E.A. 118, 395
Heward, A.P. 202
Hewitt, R. 728
Hey, R.D. 344
Heydari, E. 506
Heyward, A.P. 8
Hickin, E.J. 11, 433, 434, 529, 582
Hickman, S.H. 543
Higashi, A. 214
Higgins, J.D. 24
Higgs, R. 716
Hildebrand, A.R. 599
Hilgen, F.J. 423, 594
Hill, B.T. 743
Hill, C. 681
Hill, C.A. 678
Hill, G.W. 157
Hill, I. 755
Hill, P.R. 157, 673
Hillaire, M.C. 548
Hillaire-Marcel, C. 424, 673

Hillier, S. 127, 65–66
Hills, D.J. 673
Hilmy, M.E. 526
Hilton, D.R. 366
Hinckley, D.N. 400
Hindmarsh, R.C.A. 193
Hine, A.C. 157, 184, 673
Hines, M.E. 242
Hinman, M. 751
Hinnov, L.A. 183–184
Hinton, R.W. 385
Hinze, J.O. 229
Hipple, K. 4
Hires, R. 702
Hirst, J.P.P. 8
Hiscott, R.N. 292, 335, 498, 672, 673, 698, 759, 760
Hjulstrøm, F. 347
Hobday, D.K. 736
Hoch, A.R. 462
Hoefs, J. 385, 395
Høeg, K. 167
Hoek, E. 425
Hoekstra, P. 625
Hoffman, J. 371
Hoffman, K. 746
Hoffman, P.F. 384, 395, 733
Höfler, K. 360
Hofmann, H.J. 396, 690
Hofstra, A.H. 242
Holbrook, J. 42
Holden, P. 395
Holeman, J.N. 605
Holland, H.D. 100, 263, 384, 525, 542
Holland, S.M. 728
Hollister, C.D. 272, 430, 672, 673, 711
Hollister, L.S. 299
Hollmann, R. 412
Holly, F. 480
Holly, F.M. 618
Holman, R.A. 43, 625
Holser, W.T. 118, 395, 397
Holtedahl, H. 668
Homewood, P.W. 184, 634, 644
Honda, H. 224
Hondzo, M. 188, 353
Honjo, S. 89, 492
Honnorez, J. 507, 507–508
Hooke, R. le B. 8
Hooper, J.N.A. 682
Hopkins, T.S. 157
Hopley, D. 53
Horacek, I. 678
Horbury, A.D. 219
Horkowitz, J.P. 519
Horkowitz, K.O. 519
Horn, R. 498
Hornemann, U. 278
Hornibrook, E.R.C. 66, 396
Horowitz, A.S. 69
Horton, R.E. 582
Horvatincic, N. 755
Hosking, K.F.G. 534
Hosterman, J.W. 63
Hotzl, H. 734
Hou, Z. 531
Houbolt, J.J.H.C. 498
Houghton, B.F. 335
Houlik, C.W. 447

Hounslow, M.W. 659
House, M.R. 184
House, P.K. 292
Houseknecht, D.W. 21, 106
Houston, M. 139
Hovland, M. 306
Howard, J.D. 42, 157, 364, 649
Howard, R. 149
Howarth, R.J. 634
Howd, P. 625
Hower, J. 29, 127, 333, 371
Hower, J.C. 148
Hower, M.E. 29, 127
Howie, R.A. 769
Hoyanagi, K. 636
Hoyt, J.H. 47
Hrouda, F. 424
Hsü, K. 33
Hsieh, J.C.C. 397
Hsieh, M. 366
Hsu, K.J. 436
Hsui, A.T. 734
Hu, D. 605
Huang, H.Q. 11, 582
Hubbard, B. 331
Hubbard, D.K. 560
Hubbert, M.K. 167
Huber, B.T. 145
Huber, M.I. 512
Hubert, J.F. 161, 430
Huddart, D. 331
Hudelson, P.M. 364
Hudson, J.D. 83, 462, 728, 734
Huerta-Diaz, M.A. 702
Huff, W.D. 63
Huffman, A.R. 278
Hughen, K.A. 605
Hughes, C.R. 127
Hughes, M.G. 718
Hughes, R.E. 66, 449
Hughes Clarke, J.E. 31, 711, 759
Hulbert, M.H. 206
Hulscher, S.J.M.H. 498
Hülsemann, J. 634
Hulton, N.J. 606
Humphrey, J.D. 118, 234, 242
Humphrey, N.F. 25
Humphris, S. 492
Hungr, O. 33
Hunt, A.R. 568
Hunt, J.C.R. 229
Hunt, J.M. 135, 366, 403
Hunter, I.G. 15
Hunter, R.E. 42, 157, 172, 246, 253, 384, 499
Huntley, D.A. 285, 498, 625, 626, 711
Huntoon, P.W. 678
Huppert, H.E. 352
Hurtig, E. 316
Hussain, V. 526
Huston, T.J. 395
Hutcheon, I. 126, 397
Hutchings, P.A. 71
Hutchinson, I. 756
Hutchinson, R.A. 700
Huthnance, J.M. 498
Hutter, K. 352, 353
Hyde, R. 246
Hyndman, R.D. 138

Ibbeken, H. 27, 344
Icole, M. 418
IFA 526
Iijima, A. 21, 66, 135, 636, 780
Ijnsen, F. 737
Ike, T. 673
Ikeda, H. 22
Ikehara, K. 606
Illing, L.V. 242, 272, 438
Ilyin, A. 526
Imamura, F. 757, 757
Imamura, H. 636
Imboden, D. 408
Imbrie, J. 145, 184, 443, 506
Imlay, R.W. 19
Imran, J. 352
Ince, S. 634
Ingall, E.D. 85
Ingersoll, C.G. 751
Ingersoll, R.V. 430, 592, 733
Ingle, J.C. Jr. 673, 753
Ingram, H.A.P. 516
Inman, D.L. 31, 47, 344, 529, 605, 625, 711
Innocent, C. 548
Insalaco, E. 560
Irish, J. 687
Irish, J.D. 568, 626
Irwin, H. 20, 306, 396
Iseya, F. 22
Isla, F.I. 22
Isley, A.E. 384
ISO (International Standards Organization) 344
Ito, E. 395
Ito, M. 636
Ito, S. 673
Iversen, J.D. 253, 711
Iversen, N. 306
Iverson, R.M. 188, 248, 352–353, 425, 675
Iwasińska, I. 450

Jackson, J.A. 224, 283, 534, 736, 746
Jackson, M.J. 447
Jackson, R.G. II 434
Jacob, K.H. 224, 242
Jacobs, M.B. 722
Jacobsen, S.B. 394, 395
Jacobson, G.L. Jr. 587
Jacobson, H.A. 587
Jaeger, H.M. 353
Jaffe, B.E. 625
Jago, C. 285
Jahnke, R. 526
Jahnke, R.A. 526, 763
Jahren, J. 167
Jahrens, J.S. 66
Jainheng, X. 618
James, H.L. 135, 384
James, J.P. 15
James, N.P. 15, 101, 118, 157, 184, 272, 395, 446–447, 474, 475, 635, 673, 674, 678, 682
James, W.C. 146, 548
Jamieson, T.F. 374
Jamtveit, B. 167
Janecek, T.R. 120
Jansa, L.F. 673
Janse Van Rensburg, R.W. 530

Jansen, E. 668
Jansen, J.M. 605
Janssen, H.A. 336
Janssens, J.A. 516
Jansson, M. 408
Jarvinen, A.W. 751
Jarvis, I. 135, 526
Jaunet, A.M. 101
Jeans, C.V. 93, 396
Jedoui, Y. 53
Jefferson, G.T. 110
Jefferson, T.H. 331
Jeffery, D.L. 397
Jehl, C. 526
Jell, J.S. 53
Jenden, P.D. 366
Jenette, D.C. 202
Jenkins, D.A. 4
Jenkins, R.J.E. 656
Jenkins, S.A. 605
Jennings, J.N. 15
Jenny, H. 776
Jensen, B.L. 625
Jercinovich, M.J. 8, 22
Jessop, A.M. 316
Jessup, R.W. 22
Jiang, S 395
Jiang, W.T. 66, 127
Jimenez De Cisneros, C. 184
Jobson, H.E. 356, 531
Jodry, R.L. 242
Joeckel, R.M. 743
Johansson, M. 760
John, C.J. 47, 587
Johns, M.W. 668
Johnson, A.M. 188, 248
Johnson, C.C. 146
Johnson, C.M. 384
Johnson, D.A. 711
Johnson, D.W. 157, 561
Johnson, G.D. 397, 557
Johnson, G.W. 519
Johnson, H.R. 736
Johnson, H.W. 743
Johnson, M.A. 498, 499, 609
Johnson, M.E. 644
Johnson, N. 557
Johnson, S.A. 557
Johnson, S.W. 109
Johnson, T.C. 682
Johnson, W.J. 299
Johnsson, M.J. 256, 358, 549
Johnston, G.H. 8
Jol, H.M. 47
Jonasson, I.R. 722
Jones, A.P. 414
Jones, B. 15, 85, 110, 135
Jones, C.M. 172
Jones, C.R. 202
Jones, D.L. 665
Jones, E. 183
Jones, F.T. 224
Jones, G. 242
Jones, G.A. 292, 594
Jones, G.D. 242
Jones, G.P. 121
Jones, J.B. 135, 412
Jones, L.S. 25, 35, 582
Jones, M.E. 193, 195
Jones, P. 751

Jones, R.L. 458, 599
Jones, S.L. 757
Jones, T.P. 123
Jongmans, A.G. 4
Jopling, A.V. 59, 172, 537
Jordan, T.E. 733
Jordt, H. 168
Jørgensen, B.B. 306
Jorissen, F.J. 594, 763
Jory, D.E. 605
Joswig, W. 400
Judd, A.G. 306
Jukes, J.B. 628
Jull, A.J.T. 53, 123, 256, 605
Jump, C.J. 202
Jung, W. 19
Jüngst, H. 720
Junger, E.P. 302

Kaehler, G. 760
Kahle, C.F. 110
Kahn, J.S. 268, 686
Kairo, S. 42
Kajihara, M. 338
Kaldi, J. 234
Kalff, J. 408
Kalin, R. 681
Kalkowsky, E. 506
Kalkreuth, W.D. 403
Kamb, B. 138
Kaminski, M.A. 292
Kana, T. 741
Kanji, M.A. 24
Kanoglu, U. 756
Kantorowicz, J.D. 20
Kaplan, I.R. 366, 394, 395, 702
Kapp, P.A. 548
Kappenburg, J. 285
Karachewski, J.A. 703
Karahan, E. 347, 619
Karl, H.A. 498, 673
Karlin, R.E. 594
Karpoff, A.M. 100
Kastner, M. 138, 281, 412, 492, 526
Katayama, H. 606
Katopodes, N. 352
Katz, A. 100
Kaufman, A.J. 384, 395, 396, 397
Kaufman, J. 447
Kaufman, P. 625
Kaufmann, R.S. 396
Kaul, L. 526
Kaule, G. 516
Kawamura, R. 253
Kay, G.M. 634
Kay, M. 430, 733
Kazakov, A.V. 526
Kazemipour, A.K. 480
Keays, R.R. 657
Keefer, D.K. 248
Keene, J.B. 412
Kegel, W. 649
Keith, B.D. 506
Keller, B. 287
Keller, E.A. 677
Keller, G.H. 31, 673
Keller, P.C. 308
Keller, W.D. 224, 400

Kellerhals, R. 344, 434
Kelley, J.T. 587, 588
Kelley, W.P. 109
Kelts, K.R. 412
Kemp, A.E.S. 268, 492, 763, 766
Kemp, N. 743
Kemp, P. 763
Kemp, P.H. 568
Kempe, S. 605
Kendall, A.C. 462, 474
Kendall, C. 396
Kendall, G.St.C. 183, 184–185, 474
Kendall, M.B. 639
Kenig, F. 241
Kenn, M.J. 529
Kennard, J.M. 447
Kennedy, S.K. 519, 554
Kennedy, W.J. 118, 120
Kennett, J. 492
Kenney, C. 24
Kenney, T.C. 425
Kennicutt II, M.C. 751
Kenny, R. 190
Kent, D.V. 145, 443
Kenter, J.A. 673
Kenyon, N.H. 498, 499, 609
Keppens, E. 234, 333
Keranen, R. 42
Kerans, C. 184, 446, 447
Kerr, H.W. 109
Kerr, P.F. 4
Kesel, R.H. 8, 287
Kettles, I.M. 516
Keulegan, G.H. 294
Keupp, H. 683
Khan, S.U. 362
Kidd, R.B. 594
Kidder, D.L. 224
Kidwell, S.M. 526, 728
Kiene, W.E. 71
Kiessling, W. 560
Kile, D.E. 462
Kilgore 492
Killey, M.M 35
Killops, S.D. 366
Killops, V.J. 366
Kilps, J.R. 344
Kim, K.H. 526
Kim, S. 362
Kiminami, K. 636
Kimura, H. 396
Kindle, E.M. 568
King, E.L. 498
King, J.G. 605
King, J.W. 443
King, L.C. 411
King, L.H. 561, 746
King, S.C. 492
Kingston, K.S. 43
Kinsey, D.W. 53
Kinsman, B. 625
Kinsman, D.J.J. 118, 242, 585
Kirchner, J.F. 22
Kirchner, J.W. 605
Kirkby, M.J. 145, 256
Kirkland, B.L. 371, 450
Kirkland, D.W. 634
Kirkland, I.K. 395
Kirschvink, J.L. 549
Kittleman, L.R.Jr. 344

Kjeldstad, A. 167
Klaine, S. 751
Klappa, C.F. 15
Klaucke, I. 673
Klein, C. 384
Klein, G. 135, 563, 736
Klein, G.D. 734
Kleinberg, R.L. 314
Klemme, H.D. 85
Klenke, Th. 39, 440
Klimentidis, R.E. 190
Kline, S.J. 229
Klingeman, P.C. 22
Klinkhamer, G.P. 526
Kloprogge, J.T. 51
Klovan, J.E. 134
Knapp, R.T. 31
Knarud, R. 281
Knauth, L.P. 660, 665
Kneller, B.C. 335, 759, 760
Knight, P.G. 331
Knight, R.I. 526
Knight, W.C. 63
Knighton, A.D. 11, 434, 583
Knipe, R.J. 167
Knock, D.G. 203
Knoll, A.H. 39, 118, 384, 394, 396, 690
Knox, J.C. 605
Kobluk, D.R. 438
Koch, P.L. 368
Kochel, R.C. 8, 292
Kocurek, G. 172, 246, 253, 568
Kodama, H. 66
Koeberl, C. 264, 278, 657
Koederitz, L.F. 234
Koehnken, L. 256
Koepnick, R.B. 118, 395
Koerschner III, W.F. 184
Kohl, B. 499
Kohler, C.S. 42
Kohnke, K. 51
Kolb, C.R. 272
Kolla, V. 698
Kolodny, Y. 135, 397
Kolpack, R.L. 673
Komar, P.D. 31, 157, 292, 338, 344, 347, 480, 568, 618, 625, 673, 718
Kominz, M.A. 733
Konikov, L.F. 242
Konno, E. 757
Konuk, T. 498
Kopaliani, Z.D. 87
Kopaska-Merkel, D.C. 135
Korsch, R.J. 85, 459
Koshimura, S.-i. 756
Kostaschuk, R.A. 8
Koster van Groos, A.F. 395
Kosters, E.H. 274
Kraft, J.C. 47, 587
Krajcar-Bronic, I. 755
Krajewski, J.J. 702
Krajewski, K.P. 526
Kramer, H. 301
Kranenburg, C. 285
Kranz, R.L. 519
Kraus, I. 371, 450
Kraus, M.J. 35, 287, 583
Kraus, N.C. 625
Krauskopf, K.B. 29
Kreisa, R.D. 354

Krenvolden, K.A. 139
Kreulen, R. 395
Krinitzsky, E.L. 778
Krinsley, D.H. 65, 126, 549, 716
Krishnamoorty, C. 109
Krissek, L.A. 673
Kroenke, L.W. 120
Krogstad, E.J. 599
Krone, R.B. 285
Kronen, J.D.Jr 333
Kroon, D. 763
Kroonenberg, S.B. 626
Krouse, H.R. 299
Krug, H.-J. 224, 242
Kruiver, P.P. 423
Krumbein, W.C. 27, 184, 344, 634, 645, 685
Krumbein, W.E. 39, 53, 440, 690, 737
Krupp, R.E. 702
Krynine, P.D. 135, 557, 583
Krystinik, L.F. 211, 687
Ku, T.L. 379, 397
Kudrass, H.R. 534
Kuenen, P.H. 737, 760
Kuenen, Ph.H. 27, 31, 40, 168, 272, 282, 414, 551, 609, 634
Kuhn, G. 549
Kuhn, O. 526
Kukla, G. 443
Kulke, H. 127
Kullenberg, B. 594
Kumon, F. 636
Kunk, M.J. 397
Kuramoto, S. 673
Kurita, H. 751
Kusakabe, M. 397
Kutzbach, J.E. 146
Kvale, E.P. 42, 736, 743
Kvenvolden, K.A. 138
Kyser, T.K. 396
Kyte, F.T. 264

Laban, C. 498
Labaume, P. 335
Labendeira, C. 148
LaBerge, G.L. 384
Labeyrie, L. 443
Lacasse, C. 292
Lacelle, B. 516
Lachenbruch, A.H. 214
LaHusen, R.G. 425
Lajoie, J. 135
Lallier-Verges, E. 763
Lamarck, J.B.P.A.de M.de 249
Lambe, T.W. 24, 167
Lamberson, J.O. 751
Lambert, I.B. 396
Lambie, J.M. 711
Lamboy, M. 333, 526
Lamoureux, S. 408
Lancaster, N. 211, 246, 253, 253–254, 711
Land, L.S. 20, 21, 29, 118, 135, 20–21, 219, 242, 394, 396, 438, 506, 542
Landim, P.M.B. 734
Landing, E. 161
Landis, C.A. 734
Landis, G.P. 242

INDEX OF AUTHORS CITED

Landrum, P.F. 751
Lane, M. 100, 526
Lane, N.G. 249
Lane, S.N. 344, 480
Langbein, W.B. 145, 583
Lange, C. 763
Lange, H. 118
Langella, A. 780
Langenheim, J.H. 565
Langenhorst, F. 278, 657
Langereis, C.G. 423
Langevin, P. 229
Langford-Smith, T. 660
Langmuir, D. 557
Langnes, O. 168
Lanier, W.P. 743
Lao, Y. 396
Lapierre, F. 498
Laporte, L.F. 474
Lappalainen, E. 516
Lapuente, M.P. 549
Larese, R.E. 29, 167
Large, A.R.G. 344
Largeau, C. 403
Laronne, J.B. 22
Larsen, L.H. 626
Larson, M. 625
Larson, R.R. 29
Larsonneur, C. 737
Laschet, C. 135
Lasemi, Z. 242, 462
Laskar, J. 443
Last, W.M. 408
Lattman, L.H. 8
Laudon, P.R. 234
LaValle, P.D. 27
Lavoie, D. 508
Lawless, M. 294
Lawrence, D.B. 438
Lawrence, J.R. 396
Lawton, T.F. 734
Lazar, B. 542
Lazar, R. 751
le-Campion-Alsumard, T. 71
Le Bot, S. 498
Le Drezen, E. 499
Le Lann, F. 498
Le Mehauté, B. 625
LeBlanc, R.J. 47, 202
LeDain, A.Y. 734
LeFebvre, Guy 248
Leach, D.L. 242
Leadbeater, A.D. 248
Leather, J. 526
Leatherman, S.P. 47
Lechak, P. 554
Leckie, D.A. 59, 687
Leckie, D.L. 364
Leclerc, R. 434
Lee II, H. 751
Lee, A.C. 27
Lee, D.E. 498
Lee, J.A. 375
Lee, M. 29, 397
Lee, R.F. 751
Lee, R.W. 190
Lee, Y.I. 214
Leeder, M. 202
Leeder, M.R. 8, 90, 145, 274, 433, 618, 734
Lees, A. 102, 688

Lefebvre, G. 24
Leggett, R.F. 8
Leggewie, R. 135
Lehmann, P.J. 183, 447
Lehrmann, D.J. 184
Leinen, M. 492
Leinin, M. 492
Leininger, R.K. 554
Leith, C.K. 384, 628
Lemberg, J. 684
Lemke, L. 592
Lemoine, M. 508
Leonard, A. 698
Leonard, E.N. 751
Leopold, L.B. 287, 434, 529, 583
Leppard, G.G. 408
Lericolais, G. 498, 499
Lerman, A. 226, 408, 526
Leroux, H. 278
Leslie, S.C. 673
Lespinasse, M. 105
Lester, D.C. 751
Leszcynski, S. 161
Létolle, R. 333
Lettau, H.H. 253
Lettau, K. 253
Leung Tack, D. 71
Levey, R.A. 563
Lévi, C. 682
Levinson, R.A. 203
Levish, D.R. 292
Levy, Y. 19
Lew, A.B. 561
Lewis, C.F.M. 561, 606, 711
Lewis, K.B. 672
Leynaert, A. 492
Li, J. 395
Li, M.Z. 609, 687
Li, Y. 395
Lichtner, P.C. 542
Lick, J. 609
Lick, W. 609
Lidz, B.H. 135
Lieberkind, K. 121
Liesegang, R.E. 224
Lin, T.C. 31
Lin, Z.H. 31
Lind, I.L. 120, 121
Lindgreen, H. 371
Lindholm, R.C. 684
Lindsay, E.H. 557
Link, M.H. 698
Linklater, E. 648
Lippmann, F. 224
Lippmann, T.L. 43
Lisitzin, A. 492
Liss, S.N. 408
List, J.H. 625
Lister, J.R. 352
Littke, R. 85, 403
Little, W.C. 480
Liu, P. 757
Liu, Z. 395, 498
Livingstone, I. 253, 254
Ljunggren, P. 531
Lloyd, R.M. 438
Lloyd, R.V. 100
LoPiccolo, R.D. 195, 231, 296, 530
Locker, S.D. 673
Lodding E. 51

Loebner, B. 526
Loewinson-Lessing, F. 634
Logan, B.E. 344
Loh, C.H. 531
Lohmann, K.C. 190, 394, 395, 397, 462, 599
Loijens, M. 100
Lomando, A.J. 53, 447
Lonergan, L. 720
Long, B.F. 202
Long, D. 756
Long, E.R. 751
Longstaffe, F.J. 66, 394, 395, 396, 397
Longuet-Higgins, M.S. 625
Lonne, J. 242
Lonsdale, P. 711
Lonsdale, P.F. 673
Loope, D.B. 446
Lopez, F. 294
Lorah, M.M. 755
Loreau, J-P. 506
Lorsong, J.A. 203
Loucks, R.G. 263
Lourens, L.J. 594
Loutit, T.S. 333, 526
Loutre, M.-F. 443
Love, D.W. 734
Love, L.G. 29, 395
Lowe, A.L. 746
Lowe, D. 137
Lowe, D.J. 4
Lowe, D.R. 8, 31, 59, 193, 195, 231, 264, 296, 353, 360, 530, 665, 760
Lowenstein, T.K. 184, 212, 262, 263, 299, 585, 599
Lucas, J. 526, 728
Lucas, S.E. 436
Lucia, F.J. 234, 242, 268
Luckman, B.H. 33
Lucotte, M. 673
Ludvigson, G.A. 743
Ludwig, J.P. 751
Ludwig, K.R. 757
Ludwig, M.E. 751
Ludwig, W. 491
Luetke, N.A. 526
Lugmair, G.W. 264
Lui, S. 397
Lumsden, D.N. 100, 242
Lundberg, N. 358, 436
Lundegard, P.D. 219
Lundqvist, J. 746
Lunine, J.I. 139
Luternauer, J.L. 682
Luther, G.W. III 702
Luz, B. 396
Ly, C.K. 43
Lyell, C.H. 506
Lynch, F.L. 396, 505
Lynch, J. 687
Lynch, J.F. 568, 626
Lyness, J.F. 294
Lyons, P.C. 63
Lyons, T.W. 161, 438
Lyons, W.B. 242
Lyons, W.J. 35
MacArthur, R.C. 638
MacCarthy, P. 362
MacDonald, D.D. 751

MacDonald, G.M. 8
MacEachern, J.A. 83, 202, 700
MacKenzie, D.B. 440
MacLeod, K. 264, 657
MacMillan, D.H. 743
Macar, P. 414, 551
Macaulay, C.I. 21
Machel, H.G. 100, 106, 241, 242, 462
Machette, M.N. 8, 91
Macintyre, I.G. 438
Mack, G.H. 8, 146, 430, 548, 554, 734
Mack, L.E. 394
Mackenzie, F.T. 99, 100, 118, 224, 241, 256, 263, 492, 526, 599, 605, 634, 665
Mackenzie, W.S. 69
Mackey, S.D. 35, 202
Mackin, J.H. 583
Mackintosh, M.E. 31, 674
Macko, S.A. 85, 362, 366
Macquaker, J-H-S 168
Macumber, P.G. 211
Maddock, T. 583
Madejova, J. 371, 450
Madsen, O.S. 625, 687
Magee, A.W. 716
Magee, J.W. 605
Magno, E.A. 396
Maguire, D.J. 123
Maguregui, J. 202
Mahaney, W.C. 716
Maher, B.A. 423
Mahony, N.A. 751
Maier-Reimer, E. 492
Maiklem, W.R. 19
Majewske, O.P. 69
Major, C.F. 47
Major, C.F. Jr. 271
Major, C.O. 292
Major, J.J. 188, 353, 360, 675
Major, R.P. 242
Makaske, B. 11, 583
Makino, K. 757
Makino, Y. 743
Makurath, J.H. 741
Malcolm, R. 362
Maldiney, M.A. 285
Maldonado, M. 682
Maliva, R.G. 121, 219, 224, 308, 462
Maljaars Scott, M. 253
Malla, P.B. 769
Mallinson, D. 525
Malone, A.E. 42
Maltman, A. 137, 193, 668
Maltman, A.J. 331
Mamay, S.H. 148
Manasseh, R. 360
Manceau, A. 722
Mandelbrot, B.B. 206
Mandl, G. 668
Manganini, W.J. 492
Mángano, M.G. 82
Mange, M.A. 358
Mange-Rajetzky, M.A. 358
Mangerud, J. 756
Mangin, A. 681
Mankiewicz, P.J. 85
Manning, A.J. 285
Manning, D.A.C. 85
Manning, P.G. 224
Mansour-Tehrani, M. 229

Manville, V. 335
Marchig, V. 379
Marfil, R. 548
Mark, J. 690
Markello, J.R. 447
Maroulis, J.C. 11, 582
Marr, J.E. 634
Marriott, S.B. 91
Marsaglia, K.M. 734
Marsal, D. 685
Marshall, D.J. 106
Marshall, J.D. 224, 462
Marshall, J.F. 506, 526
Marshall, J.R. 716, 717
Marshall, K. 39
Marshall, N.F. 674
Marsset, T. 498, 499
Martill, D.M. 728
Martin, A.E. 728
Martin, D.A. 256
Martin, G.D. 462
Martin, K. 505
Martin, L. 202
Martin, R.T. 142
Martin-Algarra, A. 526
Martin-Jezequel, V. 492
Martinell, J. 700
Martinez, P. 548, 763
Martini, I.P. 268, 592
Martino, R.L. 743
Martinsen, O.J. 202, 668
Marumo, K. 722
Maruya, K.A. 751
Marwick, P.J. 145
Marzo, M. 274
Mas, R. 394
Masi, U. 507
Maskalenko, V. 292
Maslin, M.A. 424
Mason, T.R. 83, 385
Masters, J.C. 121
Masters, J.M. 35
Mather, A.E. 8
Matheron, G. 685
Mathis, R.L. 118, 234, 242
Matsumoto, E. 606
Matsumoto, R. 21, 66
Matsutomi, H. 756
Matter, A. 20, 105, 106, 127, 333
Mattey, D. 148
Mattey, D.P. 396
Matthews, A. 100
Matthews, E.R. 27
Matthews, R.K. 183
Matthews, S.J. 31
Matti, J.C. 185
Mattiat, B. 206
Mattson, V.R. 751
Matzko, J.J. 308
Maurer, H.F.W. 358
Maurmeyer, E.M. 587
Mavko, G. 314
Maxey, M.R. 229
Maxwell, C. 253
Maxwell, W.G.H. 447
Mayeda, T.K. 395
Mayer, L. 674
Mayer, L.A. 31, 711, 759
Mayer, P.G. 480
Maynard, Barry, J. 459

Maynard, J.B. 66, 85, 135, 161, 206
Mayr, H. 700
Mazullo, J. 561
Mazumder, B.S. 561
Mazzucchelli, R.H. 531
Mazzullo, S.J. 118, 234, 242
McGivery, Th.A. 27
McArdell, B.W. 301, 619
McBride, E.F. 27, 161, 224, 356, 508, 514, 537, 665
McBride, R.A. 157, 498
McCabe, P.J. 172, 287
McCaffrey, M.A. 542, 763
McCaffrey, W. 760
McCaffrey, W.D. 759
McCarthy, T.S. 583
McCarty, J.P. 751
McCave, I.N. 285, 347, 443, 464, 618, 672, 673
McCave, N. 424
McClay, K.R. 702
McCormick, D.S. 625
McCreesh, C.A. 519
McCulloch, M.T. 397
McCutcheon, A.J. 673
McDermott, F. 396
McDonald, I. 264
McDonald, R.C. 287
McDonald, R.R. 253
McDonnell, J.J. 396
McDowell, S.D. 371
McDuff, R.E. 599
McElhinny, M.W. 525
McEwan, I.K. 254
McFadden, L.D. 8, 9, 22
McFarlan, E. Jr. 272
McGinn, S.R. 741
McGivery, Th.A. 27
McGough, P. 743
McGrail, D.W. 673
McHardy, W.J. 371
McHargue, T.R. 698
McHugh, C.M.G. 711
McIlreath, I.A. 474
McIntosh, A. 751
McKay, D.S. 416, 549
McKay, J.L. 396
McKee, E.D. 60, 172, 184, 246, 537, 711
McKelvey, V.E. 526
McKenna Neuman, C. 246, 253
McKenzie, B. 224, 308
McKenzie, J. 397
McKenzie, J.A. 100, 118, 241, 243, 634
McKie, T. 354, 687
McKinlay, P.A. 47
McKinley, J.M. 722
McKnight, D.M. 362
McLain, A.A. 371
McLea, W. 757
McLean, D.G. 344
McLean, S.R. 301, 498
McLennan, S.M. 549, 599
McLeod, K.G. 278
McLoughlin, P.A. 674
McManus, D.A. 561
McManus, J. 492, 618
McMurtry, G.M. 525
McNabb, D.H. 256
McNeal, B.L. 108
McNeill, D.F. 53

INDEX OF AUTHORS CITED

McNutt, R. 396
McNutt, R.H. 395
McPherson, J.G. 7, 161, 582
McSaveney, M. 757
McTigue, D.F. 618
McVicar, M.J. 303
Mcbride, E.F. 135
Meade, R.H. 256, 358, 605
Meads, R.E. 224
Measures, C. 599
Medioli, F.S. 588
Meents, W.F. 395
Megahan, W.F. 605
Megson, J. 121
Mehta, A.J. 609
Meigs, A. 733
Meike, A. 224
Meischner, D. 729
Meisheng, H. 100
Meissner, F.F. 447
Mélice, J.-L. 443
Melim, L.A. 118, 242
Melosh, H.J. 278, 657
Melosh, J.H. 33
Melville, B.W. 301
Menard, H.W. 282
Mendoza, C. 82
Mengel, J.T. 384
Menzies, J. 331, 746
Mercone, D. 594
Merino, E. 224, 692
Merriam, D.F. 184, 685, 686
Mesolella, K.J. 183–184
Mesolella, K.L. 184
Messulam, S.A. 502
Metcalfe, G. 360
Metzger, W.J. 224
Meyer, G.A. 123, 256
Meyer, R. 502
Meyer, R.E. 718
Meyer, R.F. 502
Meyers, J.H. 53
Meyers, R.A. 47
Meyers, W.J. 105, 394
Meyerson, A.L. 702
M'hammdi, N. 498
Miall, A.D. 8, 59, 60, 145, 184, 202, 272, 287, 331, 434, 512, 583, 592, 634
Miall, A.E. 202
Miall, C.E. 634
Michael, K. 242
Michel, H.V. 656
Michel, P. 411
Middelburg, J.J. 594
Middleton, G.V. 33, 59, 60, 118, 254, 268, 272, 335, 344, 353, 408, 512, 537, 633, 634, 653, 668, 728, 760
Midtgaard, H. 687
Migliorini, C.I. 272, 760
Mikkelsen, O.A. 285
Mikos, M. 25
Milan, D.J. 344
Milankovitch, M. 443
Millar, R.G. 294
Miller, D.J. 743
Miller, D.N. 224
Miller, G.H. 605
Miller, H.C. 625
Miller, J. 102, 118, 434, 688

Miller, J.F. 395
Miller, J.M.G. 331
Miller, J.R. 8
Miller, M.C. 347, 568, 618
Miller, R.L. 686
Millero, F.J. 438, 703
Milligan, T. 285
Milliken, K.L. 29, 224, 308, 358
Milliman, J.D. 69, 256, 605, 673
Millot, G. 127, 142, 411, 634
Milne, P. 438
Milner, H.B. 634
Milnes, A.R. 91, 526, 660
Mimura, K. 366
Minato, M. 636
Minoura, K. 757
Minter, W.E.L. 531
Miser, H.D. 665
Missiaen, T. 498
Mitchell, J.G. 396
Mitchum, R.M. 184–185, 203
Mitchum, R.M. Jr. 185, 272
Mittler, P.R. 42
Miyamoto, T. 636
Miyata, Y. 231, 296
Mizota, C. 396
Mizuno, A. 636
Mizutani, S. 636
Mochizuki, K. 673
Modeer, V.A. 24
Modreski, P.J. 264, 656
Moehle, S. 9
Mohrig, D. 35, 188, 353
Mojzsis, S.J. 264
Moldowan, J.M. 366
Moller, N. 263
Molnár, Á. 366
Molnar, P. 605, 606
Moncreif, A.C.M. 331
Monicard, R.P. 268
Montañez, I.P. 184–185, 242
Montagna, P. 751
Montanari, A. 264, 278, 657
Montgomery, D.R. 294, 618
Monty, C.L.V. 102, 688
Moody, J.A. 256
Moore, A. 756
Moore, C.H. 42, 118, 438, 475, 506, 634
Moore, C.H. Jr. 53
Moore, D. 450
Moore, D.G. 491, 673
Moore, D.M. 66, 142, 371, 449
Moore, E.S. 634
Moore, G.F. 673, 734
Moore, G.T. 763
Moore, G.W. 681
Moore, J.C. 436
Moore, J.G. 757
Moore, J.R. 534
Moore, P.D. 149
Moore, T.C. 89
Moore, T.C. Jr. 492
Moore, T.E. 8
Morad, S. 281
Moran, K.M. 673
Morehead, M.D. 606
Moresby, R. 53
Morgan, D.J. 63
Morgan, J.P. 414

Morgenstern, N.R. 673
Moriarty, D.J.W. 526
Morita, S. 673
Morley, C.K. 734
Morley, M.S. 757
Moro, M.C. 4
Morozova, G.S. 36
Morris, A.W. 285
Morris, R.C. 384
Morrison, D. 278
Morrison, H.A. 751
Morrow, D.W. 242
Morse, A.P. 338
Morse, J.W. 100, 224, 438, 634, 702, 703
Mortensen, J. 121
Mortimer, R.J.G. 396
Morton, A.C. 358, 458, 549
Morton, J.L. 599
Morton, J.P. 396
Moshe, R. 8
Mosher, D.C. 498
Moshier, S.O. 462
Mosley, M.P. 345, 583
Moslow, T.F. 47, 157, 498, 741
Mosola, A.B. 746
Moss, S.J. 548
Mossop, G.D. 184
Mothersill, J.S. 43
Mount, J.M. 447
Mountjoy, E.W. 100, 241, 242
Moxon, I. 592
Mozley, P.S. 118, 224, 396
Mozzherin, V.I. 605
Mucci, A. 438, 673
Muda, J. 531
Mudge, B.F. 587
Mudie, P.J. 292
Muehlenbachs, K. 396
Muhs, D.R. 548
Muir, R.O. 263
Muirhead, A. 568
Mukerji, T. 314
Mulder, T. 202, 605, 673, 698, 760
Muller, P.J. 763
Mücke, A. 557
Müller, A.H. 728
Müller, D.W. 634
Müller, G. 214, 635
Müller, M. 360
Mullineaux, L. 492
Mullins, H.T. 474, 673
Mullis, J. 20, 105, 106, 242
Mulrennan, M.E. 43
Multhauf, R.P. 263
Mumpton, F.A. 29
Munk, W.H. 625
Munn, S.G. 734
Munnecke, A. 462
Munoz, M. 316
Muntean, J.V. 565
Murat, A. 594
Murata, K.J. 29
Murphy, C.P. 375
Murphy, W.F. 314
Murphy, W.M. 118, 167
Murray, H.H. 185, 400
Murray, J. 492
Murray, L.G. 534
Murray, R. 60, 633, 634

Murray, R.C. 19, 118, 303
Murray, R.W. 665
Murray, T. 746
Murton, J. 414
Musashino, M. 636
Mussett, A.E. 599
Muste, M. 618
Mutti, E. 59, 184, 272, 283, 292, 335, 698, 760
Mutti, M. 242
Muwais, W.K. 502
Myers, K.J. 183
Myers, W.R.C. 294
Mykura, H. 224
Mynard, V. 526
Myrow, P.M. 161, 354, 687
Mysak, L.A. 146

Nadeau, P.H. 118, 167, 371, 450
Nagel, S.R. 353
Nagle, J.S. 729
Nahon, D. 411
Nahon, D.B. 722
Naiman, E.R. 234
Nakagawa, H. 480
Nakamura, Y. 673
Nakaya, S. 757
Nalpanis, P. 254
Nanson, G. 529
Nanson, G.C. 11, 206, 287, 434, 582, 583
Naqvi, J.M. 308
Nash, D.J. 660
Nash, W.P. 281
Nathan, Y. 526, 722
Naumann, R. 51
Navrotsky, A. 400
Naylor, M.A. 436
Neal, J.E. 85
Nealson, K. 39
Nealson, K.H. 219, 384
NEB 502
Neev, D. 19
Negendank, J.F.W. 766
Neill, C.R. 347
Nelson, C.M. 634
Nelson, C.S. 475
Nelson, D.M. 492
Nelson, H.F. 118, 395
Nelson, J.M. 301, 568, 618
Nelson, R.A. 734
Nelson, T.A. 673
Nemec, W. 8, 274, 673
Neretnieks, I. 268
Nesbitt, B.E. 722
Nesbitt, H.W. 459
Nesteroff, W. 594
Netterberg, F. 90
Nettleton, W.D. 90
Neuman, C.M. 22
Neuman, S.P. 268
Neumann, A.C. 71
Neuweiler, F. 102, 560, 688
Newall, G. 364, 728
Newell, N.D. 506
Newnham, R.E. 400
Nezu, I. 480
Nicholls, R.J. 760
Nichols, G. 123, 335

Nichols, G.J. 8, 123
Nichols, M.M. 202, 609
Nichols, R.J. 231, 296, 530
Nicholson, H. 243
Nicholson, W.L. 345
Nickling, W.G. 22, 246, 253–254, 711
Nicolaysen, L.O. 278
Nicoll, R.S. 308
Niedoroda, A.W. 157, 157–158
Nielsen, P. 234, 625
Nielson, R.L. 397
Niering, W.A. 587
Nieuwenhuyse, A. 4
Nikuradse, J. 294
Nilsen, H. 9
Nilsen, T.H. 8, 734
Nio, S.D. 59, 736
Nisbet, E.G. 39, 139
Nishiizumi, K. 396
Nissim, S. 8
Nittrouer, C.A. 605, 606
Noack, Y. 722
Noakes, L.C. 534
Nobles, L.H. 188
Noffke, N. 39, 440
Nongkas, M. 757
Nordin, C.F. 605
Nordstrom, C.E. 31, 47
Normack, W.R. 293
Normark, W.R. 272, 491, 492, 673, 698, 711, 760
Norris, R.D. 396
North, G.R. 145
Notholt, A.J. 135
Notholt, A.J.G. 135
Nuesch, R. 371, 449
Nummedal, D. 741
Nunn, J.A. 542
Nur, A. 314, 733
Nygaard, R. 121
Nystuen, J.P. 281

Obelic, B. 755
Oberbeck, V.R. 657
Oberhänsli, R. 358
Obermeier, S. 137
O'Brien, D.W. 526
O'Brien, G.W. 525, 526
O'Brien, N.R. 206, 459
O'Brien, P.E. 447
O'Connor, G.A. 108, 109
O'Connor, J.E. 292, 292–293
Odin, G.S. 66, 127, 333, 384
Oelkers, E.H. 118, 167, 542
Oertel, G.F. 561, 687
Off, T. 498
Ogren, D.E. 374
Oh, S.I. 498
O'Hara, P.F. 549
Ohr, M. 396
Ojakangas, R.W. 384
Ojha, T.P. 548
Okada, H. 636
O'Keefe, J.A. 657
Okuda, S. 9
Olah, G.A. 366
Olbricht, W. 344
Oldershaw, A. 225, 462

Oldfield, F. 753
Oldroyd, D.R. 634
Olive, P. 594
Ollerhead, J.W. 43
O'Loughlin, C.L. 425
Olsen, C.R. 605
Olsen, P. 443
Olsen, P.E. 145, 184
Olson, J.S. 512
Oltman-Shay, J. 625
O'Malley, P. 755
Omori, M. 636
Omoto, K. 414
O'Neil, J.R. 395
Ong, P.M. 534
O'Nions, K. 397
Onstott, T.C. 395
Oomkens, E. 272
Oost, A.P. 568, 736, 737
Opdyke, B.N. 146
Opdyke, N.D. 557
Orange, D. 606
Orange, D.L. 436
Ord, D.M. 274
Ordonez, S. 755
Orford, J.D. 716
Ori, G.G. 734
Orme, G.R. 447
Orr, E.D. 716
Ort, M. 335
Ortoleva, P. 224, 692
Orton, G.J. 202, 272, 274
Osborne, P.D. 43, 625, 626
Osborne, R.H. 27, 760
Osleger, D. 185
Osleger, D.A. 184–185, 447
Osmond, J.K. 185
O'Sullivan, J.J. 294
Oswald, E.J. 447
Otto, G.H. 537
Otto, J.B. 118, 395
Outerbridge, W.F. 63
Overbeck, F. 516
Overpeck, J.T. 605
Overstreet, R. 109
Owen, G. 137, 195, 296, 414, 551
Owen, M.W. 285
Owen, P.R. 254
Owen, R.M. 190, 462, 506
Ozalas, K. 700
Özdemir, Ö. 368
Ozretich, R.J. 751

Page, K.J. 434
Pagel, M. 105
Paik, I.S. 214
Paillard, D. 443
Painter, R.B. 605
Pak, H. 672
Palciauskas, V. 314
Palenik, S.J. 303
Pälike, H. 443
Palmer, A.R. 263
Palmer, R.J. 15, 110
Palmer, T.J. 659
Palomares, M. 549
Panichi, C. 754
Pantin, H.M. 31, 353, 498, 759

INDEX OF AUTHORS CITED

Paola, C. 35, 87, 173, 301, 344, 480, 537, 596, 618
Paopongsawan, P. 531
Paphitis, D. 568
Papike, J.J. 416
Paquette, J. 100
Paquin, P.R. 751
Parchure, T.M. 609
Parea, G.C. 19
Parfitt, R.L. 4
Park, A. 224
Park, C.F. Jr. 135
Park, J. 184, 443
Park, J.O. 673
Park, W.C. 519, 692
Park, Y.A. 737, 743
Parker, G. 22, 25, 31, 188, 301, 352, 353, 618, 759
Parkerton, T.F. 751
Parks, G.A. 118
Parks, J.M. 512
Parmenter, C.M. 722
Parra, M. 548
Parrish, J.T. 145
Parrott, B.S. 47
Parry, W.T. 224
Parsons, K.M. 729
Partridge, T.C. 278
Passey, Q.R. 85, 458
Passier, H.E. 594
Passier, H.F. 423
Patchett, P.J. 396
Paterne, M. 763
Paterson, D.M. 440, 618, 737
Paterson, M.S. 543
Paterson, W.S.B. 331
Patterson III, W.A. 123
Patterson, R.J. 242
Patterson, W.P. 438
Pattiaratchi, C.B. 609
Patton, T.L. 734
Paul, M.A. 331, 747
Paulik, F. 51
Paulik, J. 51
Pauling, L. 400
Paull, C.K. 139
Pavlou, S.P. 751
Payton, C.E. 272
Peacor, D.R. 65, 66, 100, 126, 127, 396, 677, 723
Pearce, A.J. 425
Pearce, R.B. 492
Pearce, T.J. 549
Pearson, T.H. 85
Pedersen, T.F. 763
Pedley, H.M. 754, 755
Peizhen, Z. 606
Pejrup, M. 285
Pell, S.D. 358
Pelley, J. 751
Pelosi, N. 183
Peltzer, G. 734
Pemberkon, S.E. 700
Pemberton, S.G. 82, 83, 157, 202, 502, 700
Penck, A. 747
Peng, T.-H. 491
Pengxi, Z. 299
Penland, S. 47, 202, 741
Pentecost, A. 755
Pepper, A.S. 306

Pequera, N. 582
Perch-Nielsen, K. 120
Peregrine, D.H. 626
Pérez-Arlucea, M. 35, 36, 287
Perinet, G. 418
Perkins, R.D. 185
Perkins, R.J. 229
Perry, E.A. 29, 127
Persaud, D. 751
Peryt, T.M. 506
Peters, K.E. 366, 403
Peters, S.W. 734
Peters, T. 508
Peterson, C.H. 751
Peterson, F. 146
Peterson, F.F. 90
Peterson, G. 137
Peterson, M. 492
Petit, J.R. 396
Pettijohn, F.J. 40, 63, 135, 161, 168, 173, 206, 219, 430, 440, 512, 551, 568, 634, 685
Pettijohn, F.P. 219
Pettit, K.E. 751
Petzold, D. 51
Peucker-Ehrenbrink, B. 264, 657
Pevear, D.R. 396, 677
Peypouquet, J-P. 763
Pfab, G. 368
Pfefferkorn, H.W. 145
Pfeifer, H.R. 508
Pfeiffer, A. 285
Pflueger, F. 440
Phillips, C.R. 360
Phillips, F.M. 267, 396
Phillips, R.L. 42
Phillips, T.L. 148
Philp, R.P. 366
Phizackerley, P.H. 502
Picard, M.D. 206, 224, 634, 720
Piccolo, A. 362
Pichler, T. 234
Pickering, K.T. 203, 673, 698, 760
Pickle, J.D. 430
Pickup, G. 292
Piepgras, D.J. 397
Pierson, T.C. 188
Pigott, J.D. 53
Pigram, C.J. 447
Pimentel, N.L. 90
Pinet, P. 606
Pingitore, N.E. 462
Pingree, R.D. 498
Pini, G.A. 436
Piotrowski, J.A. 746
Piper, D.J.W. 31, 492, 498, 499, 606, 673, 674, 698, 711, 759, 760
Piper, D.Z. 379, 492, 526
Pirrie, D. 224
Pisciotto, K.A. 29
Pisera, A. 682
Pisias, N. 492
Pitman, E.D. 268
Pitman, W.C. 185
Pitman, J.K. 234
Pittman, E.D. 167
Pittman, III, W.C. 292
Pivnik, D.A. 582
Pizzuto, J.E. 656
Plafker, G. 185

Plancon, A. 400
Plant, J. 703
Platania, G. 757
Platt, N.H. 91, 287
Plaziat, J.C. 90, 241
Plet, A. 755
Plint, A.G. 396, 498, 512
Plotnick, R.E. 729
Plumlee, G.S. 242
Plumley, W.J. 25
Plummer, L.N. 100, 190
Plummer, N.J. 99
Plummer, P.S. 720
Poag, C.W. 135
Pollack, J.B. 253
Polo, M.D. 42
Popp, B.N. 506
Poppe, L.J. 722
Porcelli, D. 396
Poreda, R.J. 394
Porfir'ev, V.B. 366
Porrenga, D.H. 66, 127
Posamentier, H.W. 185, 583, 698
Pósfai, M. 424
Post, J.E. 438
Postma, D. 542
Postma, G. 8, 274
Potter, P.E. 40, 69, 85, 135, 161, 168, 173, 206, 219, 440, 459, 512, 551, 568, 592, 634
Potts, M. 440
Pottsmith, H.C. 285
Poty, B. 105
Pouliquen, O. 353
Poulsen, C.J. 146
Pourtaheri, H. 480
Powell, D.M. 22
Powell, E.N. 751
Powell, J.L. 657
Powell, R.D. 331, 606, 743
Powers, D.W. 300
Pozzuoli, A. 450
Prashnowsky, A.A. 308
Prats, J.M. 502
Pratt, B.R. 3, 102, 224, 475, 659, 688, 690
Pratt, R.B. 720
Prave, A.R. 364
Pray, L.C. 267, 634, 640
Preece, R.C. 755
Prell, W.P. 763
Premoli Silva, I. 443
Prensky, S.E. 268
Presley, B.J. 161
Preston, R.M.F. 193, 195
Prévôt, L.E. 526, 728
Prezbindowski, D.R. 53, 190
Price, D.M. 756
Price, M. 120
Price, N.B. 66, 378, 379, 384, 763
Prince, C.M. 519
Prince, C.R. 519
Prior, D.B. 31, 202, 274, 668
Prochaska, W. 722
Prospero, J.M. 606
Prosser, D.J. 698
Protz, R. 397
Proust, D. 450, 722
Prud'homme, M. 229
Prufert-Bebout, L. 438
Pryor, W.A. 85, 135, 161, 206, 459

Pugh, D.T. 743
Pujol, C. 763
Pulham, A.J. 83
Puls, D.D. 203
Puls, W. 285
Purdue, A.H. 665
Purdy, E.G. 438, 506
Purser, B. 242
Purser, B.H. 241, 447, 475, 506, 734
Purucker, M.E. 66, 385
Pusey, W.C. 438
Pusey, W.C. III 447
Puskaric, S. 594
Putnam, P.E. 8
Pye, K. 65, 126, 246, 254, 345, 711, 736
Pyle, H.T. 711
Pyne, Stephen, J. 643

Qian Ning 35
Qin, Y.S. 605
Qing, H. 242
Quade, J. 395, 548
Quguiner, B. 492
Quinn, T.M. 242
Qutawna, A.A. 526

Raab, M. 722
Raab, W. 379
Rabineau, M. 499
Rachmanto, B. 42
Rachocki, A.H. 8
Radke, B.M. 118, 234, 242, 308
Radtke, G. 71
Ragueneau, O. 492
Rahmani, R.A. 537, 700, 737
Rahmanian, V.D. 185, 203
Rahn, P.H. 190
Railsback, L.B. 223, 681, 692
Raines, M.A. 190
Raiswell, R. 224, 702, 703, 728, 729
Raitt, R. 492
Rakofsky, A. 698
Ramarao, B.V. 360, 530
Ramberg, H. 414
Ramm, M. 167
Ramsay, A.T.S. 89
Ramsay, J.G. 512
Ramsbottom, W.H.C. 185
Ramseyer, K. 29, 106, 397
Rankey, E.C. 447
Rao, V.P. 333
Raphelt, N. 480
Rappol, M. 747
Rasmussen, D.L. 202
Rastall, R.H. 634
Raudkivi, A.J. 294, 568
Raup, D.M. 729
Raupach, M.R. 254
Rawson, P.F. 93
Ray, B.J. 491
Raymond, L.A. 436
Rea, D.K. 492
Read, J.F. 184–185, 447, 475
Reading, H.G. 157, 185, 202, 203, 272, 634, 640, 698
Reddering, J.S.V. 563
Reddy, M.M. 462

Redfield, A.C. 588
Reed, A.A. 161
Reed, S.J.B. 118
Reeder, R.J. 100, 106, 118, 684
Reedy, R. 416
Rees, P.McA. 146
Reiche, P. 173
Reid, I. 8, 22, 531
Reid, M.E. 34, 248, 425
Reid, R.P. 438
Reid, W.P. 684
Reif, W.-E. 729
Reijmer, J.J.G. 462
Reimer, A. 3
Reimers, C. 526
Reimers, C.E. 338, 763
Reimnitz, E. 674
Reimold, W.U. 264, 278
Reinech, H.E. 8
Reineck, H.E. 43, 59, 60, 157, 170, 206, 283, 440, 537, 568, 634, 649, 737
Reinhart, M.A. 537
Reinick, H.-E. 42
Reinson, G.E. 47, 83, 537, 737, 741
Reise, K. 737
Reitner, J. 3, 102, 560, 683
Ren, M.-e. 606
Ren, M.E.Y. 605
Renard, A. 492
Renard, F. 167
Renwick, W.H. 8, 606
Repeta, D.J. 763
Resig, J.M. 121, 526
Retallack, G.J. 91, 146, 634, 776
Rex, G. 148
Rex, R.W. 492
Reyment, R.A. 729
Reynolds, B.C. 397
Reynolds, R.C. 127, 371, 397, 450
Reynolds, R.C. Jr. 66, 142, 449, 450
Reynolds, R.C.J. 127
Reynolds, R.L. 424, 703
Reynolds, T.J. 118, 299
Reynolds, W.C. 229
Rhoades, L. 42
Rhoads, D.C. 206
Rhodes, R.N. 498, 609
Riba, O. 734
Ribberink, J.S. 626
Ricci Lucchi, F. 19, 59, 168, 272, 283, 634, 698, 748
Rice, C.A. 703
Rice, D.D. 306
Rice, M. 254
Rice, M.A. 254
Rice, R. 592
Rice, S.P. 25, 345
Rich, C.I. 65
Rich, E.I. 592
Rich, J.L. 168
Richards, K.J. 568, 711
Richards, K.S. 583
Richards, M. 272, 698
Richardson, E.V. 301, 711
Richardson, J.F. 360
Richardson, M.J. 673, 674
Richardson, W.A. 224
Richter, D.K. 506
Richter, R. 283
Rickard, D. 703

Ricken, W. 93, 183, 443
Ridgway, K.D. 582
Riding, J.B. 83
Riding, R. 3, 560, 690, 755
Riebe, C.S. 605
Rieke, H.H. 167
Rigaut, F. 698
Rigby, J.K. 683
Riggs, N. 335
Riggs, S.R. 135, 526
Rightmire, C.T. 190
Rijsdijk, K.F. 551
Riley, C.M. 19
Riley, J.J. 229
Rimstidt, J.D. 105
Rine, J.M. 157, 499
Risk, M.J. 438
Rittenberg, S.C. 702
Rittenhouse, G. 434
Ritter, D.F. 8
Ritter, J.B. 8
Ritz, K. 375
Roadifer, R.E. 502
Robbins, D. 83
Robert, A. 294
Robert, M. 101
Robert, P. 85, 403
Roberts, A.P. 424
Roberts, H.H. 202, 447, 498, 499, 674
Roberts, N. 8–9
Roberts, S.M. 299
Roberts, S.T. 542
Robertson, A.H.F. 734
Robertson, P. 308
Robinson, A.G. 219
Robinson, D. 126
Robinson, K. 400
Robinson, R. 366
Robinson, S.G. 424
Rochon, A. 292, 673
Rock, N.M.S. 686
Rodgers, J. 134, 331
Rodriguez, A.B. 499
Rodriguez-Lazaro, J. 308
Roe, K.K. 526
Roe, S.L. 9
Roedder, E. 299, 308
Roelvink, J.A. 626
Roen, J.B. 63
Roep, Th.B. 743
Roeske, S. 436
Roest, W. 722
Rogers, C.D.F. 268
Rogers, D.A. 720
Röhl, U. 443
Rohling, E.L. 594
Rohrlich, V. 66, 384, 526
Romanek, C.S. 549
Romankevitch, E.A. 763
Ronen, D. 397
Ronov, A.B. 526, 599
Rooney, S.T. 746
Roquin, C. 411
Rosch, H. 412
Roscigno, P.F. 751
Rose, A.W. 557
Rosen, M.R. 211
Rosen, P.S. 47, 741
Rosenberg, E. 39, 440
Rosenberg, P.E. 100

INDEX OF AUTHORS CITED

Rosenfeld, J.L. 462
Rosenquist, I.Th. 167
Rosenzweig, W.D. 300
Roser, B.P. 85, 459
Rosler, H.J. 118
Rosol, M.J. 605
Ross, C.A. 185
Ross, C.S. 4, 63
Ross, D.A. 19, 594
Ross, G.M. 396
Ross, J. 766
Ross, J.R.P. 185
Ross, P. 751
Rossignol-Strick, M. 594
Rostron, B.J. 242
Rothpletz, A. 506
Rothman, E.D. 185
Rothwell, R.G. 335, 760
Rotunno, M. 499
Rouse, H. 634
Rouse, H.L. 757
Rowan, E.L. 242
Rowe, G.T. 763
Rowekamp, E.T. 25
Roy, P.S. 27, 47, 625
Royden, L.H. 734
Ruan, H.D. 51
Rubey, W.W. 167, 263, 358, 531, 599
Rubin, D.M. 172, 499
Rubin, M. 190, 588
Ruddiman, W.F. 272
Rude, P.D. 438
Rudnick, R.L. 599
Rudolf, D.L. 4
Rudowski, S. 625
Rudoy, A.N. 292
Ruessink, B.G. 43
Ruffell, A.H. 722
Ruiz, J. 242
Ruiz-Amil, A. 450
Rummery, T.A. 753
Runstadler, P.W. 229
Ruscher, C. 756
Rush, P.F. 395
Russell, D. 634
Russell, J.D. 4
Russell, P. 625
Russell, S.C. 51
Rust, B.R. 206, 364
Rust, I.C. 563
Ruszczyńska-Szenajch, H. 747
Rutgers, L.M.M. 549
Rutter, E.H. 543
Ruz, M.-H. 157
Ryan, J.N. 159
Ryan, P.C. 127
Ryan, W.B.F. 292, 594, 711
Rybach, L. 316
Rydell, H. 507
Ryder, G. 264, 657
Ryder, J.N. 9

Saarenmaa, L. 516
Sabine, C. 526
Sackett, W.M. 135
Sadaqah, R.M. 526
Sadler, P.M. 185, 618
Sageman, B.B. 702
Saito, Y. 605, 606, 636

Sakai, H. 395
Sakamoto, T. 594
Sakharov, B.A. 371
Sakinc, M. 292
Saliot, A. 403
Sallenger, A.H. 625
Sallenger, A.H.Jr. 718
Saltzman, B. 443
Salyn, A.L. 371
Samtleben, C. 462
Sandberg, P.A. 118, 242, 263, 462, 506, 684
Sander, B. 640
Sanders, J.E. 135, 331, 748, 759
Sanderson, D.D. 743
Sanderson, D.J. 374
Sanderson, R.W. 722
Sandler, A. 722
Sandusky, C.L. 563
Sanford, W.E. 242
Sangree, J.B. 272, 674
Sangster, D.F. 15, 678
Sanlung, M. 202
Sanson, F.J. 101
Sanz, E.M. 91
Sares, S.W. 430
Sarg, J.F. 447
Sarnthein, M. 443, 507–508
Sarre, R.D. 254
Sass-Gutkiewicz, M. 678
Satterfield, W.M. 674
Satterley, A.K. 185
Sauer, E.K. 678
Sauer, R.R. 548
Saunders, T. 83
Saunders, T.D.A. 82
Savage, S.B. 353
Savin, S.M. 29, 146, 397
Savrda, C.E. 700
Sawatsky, L.H. 267
Saxby, D. 531
Schack Pedersen, S.A. 688
Schaefer, A. 592
Schäfer, W. 729, 737
Scheidegger, A.E. 519
Scheiwe, T.D. 700
Schenck, P.A. 594
Schenk, C.J. 211, 253
Scherhag, C. 379
Schertmann, U. 51
Schidlowski, M. 526
Schieber, J. 85, 206, 549
Schimmelmann, A. 763
Schlager, W. 184, 474, 475, 653, 674
Schleyer, R. 27
Schlichting, H. 294
Schmid 568
Schmidt, G.A. 146
Schmidt, H. 649
Schmidt, K-H. 753
Schmincke, H.U. 63, 135
Schmitz, B. 264, 657
Schmoker, J.W. 118, 242
Schneer, C.J. 644
Schneidermann, N. 684
Schnitzer, M. 362
Scholle, P.A. 69, 118, 121, 225, 308, 475
Scholten, J. 722
Schön, J.H. 314
Schüttenhelm, R.T.E. 498
Schoonees, J.S. 626

Schoonen, M.A.A. 703
Schoonmaker, J. 100, 436
Schopf, J.W. 690
Schot, E.K. 692
Schott, W. 492
Schrader, H.J. 674
Schrag, D.P. 384, 395
Schraub, F.A. 229
Schreiber, B.C. 135, 263
Schroeder, P.A. 371
Schroter, T. 135
Schubel, K.A. 665
Schubert, C. 27
Schuchert, C. 634, 639, 734
Schuffert, J.D. 526
Schults, D.W. 751
Schultz, D.J. 554
Schultz, J.L. 29
Schultz, L.G. 127
Schultz, P.H. 657
Schultz-Guttler, R. 100
Schulz, H.D. 542
Schulz, M.S. 395
Schulze, D.G. 769
Schulze-Lam, S. 39
Schumm, S.A. 34, 35, 145, 272, 582, 583, 606
Schuster, R.L. 33, 425
Schwalbach, J.R. 673
Schwarcz, H.P. 396, 397, 681
Schwartz, F.W. 542
Schwartz, M.L. 47
Schwarz, A. 283
Schwarzacher, W. 185, 686
Schwarzer, S. 360
Schwertmann, U. 368, 557
Sclater, J.G. 167
Scoffin, T.P. 53, 118, 135
Scott, A.C. 123, 148
Scott, D.B. 588
Scott, G.A.J. 256
Scott, H. 234
Scott, K.M. 188, 425
Scott, L.O. 502
Scourse, J.D. 331
Scrivenor, R.C. 722
Scruton, P.C. 203
Seager, W.R. 734
Seal, R. 301
Searl, A. 21
Sebag, D. 418
Secord, A.L. 751
Sedimentological Society of Japan 636
Seeber, L. 137
Segalen, P. 411
Segnit, E.R. 135, 412
Seguret, M. 335
Seibold, E. 492
Seibold, I. 653
Seilacher, A. 82, 83, 93, 183, 443, 729
Sejrup, H.P. 668
Selby, M.J. 24, 425
Sellards, E.H. 651
Selles-Martinez, J. 224
Sellier, M. 105
Sellwood, B.W. 650
Semil, J. 625
Sen Gupta, B.K. 333
Sengor, A.M.C. 734
Sengupta, S. 561

Seong-Joo, L. 418
Seramur, K.C. 743
Serjani, A. 526
Serra, F. 293, 760
Sethi, P. 85, 206
Seuss, E. 506
Sexton, D. 692
Sexton, M.J. 447
Seyedolali, A. 549
Shackleton, N.J. 145, 146, 395, 397, 443
Shakesby, R.A. 716
Shaler, N.S. 308, 588
Shanmugam, G. 272, 760
Shannon, E.V. 63
Sharma, P. 396
Sharma, S.K. 99
Sharp, I. 733
Sharp, J.M.Jr. 224
Sharp, R.P. 9, 188, 711
Sharpe, C.F.S. 248
Sharpton, V.L. 278, 657
Shau, Y-H. 723
Shaw, H.F. 397
Shaw, J. 292, 344, 606, 747
Shaw, P.A. 660
Shawe, D.R. 224
Shearman, D.J. 224, 272, 585
Sheets-Pyenson, S. 634
Sheldon, R.P. 135, 526
Shelton, R.E. 526
Shemesh, A. 395
Shen Huati 561
Shen, M.C. 718
Shepard, C.U. 308
Shepard, F.P. 43, 459, 561, 634, 674
Shepherd, R.G. 537
Sheppard, R.A. 418, 666, 780
Sheppard, S.M.F. 397
Shergold, J.H. 395, 525
Sheridan, R.E. 733
Sherman, D.J. 42, 246
Sherman, D.M. 703
Sherwood Lollar, B. 366
Sheu, D.-D. 161
Shi, N.C. 626
Shi, S. 756, 757
Shields, A. 301
Shields, G. 525
Shields, G.A. 395
Shields, M.J. 242
Shilts, W.W. 746
Shimer, H.W. 644
Shimkus, K. 292
Shimmield, G.B. 548, 763
Shimp, N.F. 395
Shinn, E.A. 19, 135, 184, 447, 474, 585
Shipley, T.H. 139, 673
Shipp, R.C. 43
Shipp, S.S. 746
Shoji, R. 636
Shor, A.N. 31, 711, 759
Short, A.D. 43
Shotton, F.W. 173
Shreve, R.L. 34, 301
Shrock, R.R. 214, 628, 634, 651
Shukla, V. 242
Shukolyukov, A. 264
Shuto, N. 756, 757
Shvidchenko, A.B. 87

Sibley, D.F. 21, 100, 118, 234, 242, 692
Sicre, M-A. 763
Sidle, R.C. 425
Siedlecka, A. 308
Siegmund, H. 526
Siehl, A. 66
Sieser, W.G. 121
Siever, R. 219, 281
Sigurdsson, H. 292
Silber, A. 4
Sillen, L.G. 263
Silva, A.J. 170
Silva, P.G. 7, 8, 9
Silver, B.A. 447
Silver, E. 436
Silver, L.T. 657
Simo, J.A. 242
Simon, D.B. 301
Simon, S. 416
Simone, L. 506
Simoneau, A.M. 605
Simonian, K.O. 734
Simons, D.B. 711
Simonson, B.M. 264, 384, 665
Simpkin, T. 757
Simpson, E.L. 183
Simpson, J.E. 353
Sinclair, I. 83
Sinclair, I.R. 408
Sindelar, S.T. 203
Singer, A. 4
Singh, I.B. 8, 43, 60, 157, 206, 283, 537, 568, 634, 737
Sinha, R. 582
Sinha, S.K. 301
Sinotte, S.R. 308
Sippel, R.F. 118
Sir, W.A. 647
Sirinawin, T. 531
Siringan, F.P. 499
Sisson, T.W. 34
Skei, J.M. 606
Skempton, A.W. 248
Skene, K.I. 499
Skillbeck, C.G. 723
Skinner, B.J. 139
Skyring, G.W. 526
Slack, J.F. 66
Slansky, M. 526
Slatt, R.M. 206, 459, 561, 698
Sleath, J.F.A. 626
Sleep, N.H. 39, 599
Slezak, L.A. 408
Slingerland, R.L. 35, 36, 358, 480, 514, 531, 618
Sloan, D.D., Jr. 139
Sloan, J.L. II 605
Slominski, J. 625
Sloss, L.L. 185, 634, 734
Sly, P.G. 408
Smales, A.A. 634
Small, L.F. 338
Smalley, I.J. 248, 268
Smart, P.L. 15, 110, 242, 243, 716
Smetacek, V. 491
Smit, D.E. 190
Smit, J. 657
Smith Wellner, J. 746
Smith, B. 40, 551
Smith, B.J. 717

Smith, C.S. 650
Smith, D.E. 756
Smith, D.G. 11, 47, 183, 287, 433–434, 537, 583, 737
Smith, D.S. 15
Smith, G.A. 292, 592, 734
Smith, I.J. 408
Smith, J.D. 568, 618, 619, 626
Smith, J.N. 606
Smith, J.V. 692
Smith, L.R. 35
Smith, N.D. 11, 35, 36, 287, 531, 537, 583
Smith, R.L. 763
Smith, R.M.H. 83, 583
Smith, S.V. 606
Smith, T.N. 360
Smits, L.J.M. 314
Smol, J.P. 408
Smoot, J.P. 145, 211, 212, 262, 263
Snedden, J.W. 499
Sneed, E.D. 27, 345
Snipe, L.G. 100
Snowball, I.F. 424
Snyder, R. 400
Soares, P.C. 734
Sobecki, T.M. 4
Sommerfield, C.K. 606
Sonett, C.P. 743
Song, H. 395
Sonnenfeld, M.D. 185, 447
Sonnenfeld, P. 263
Sorby, H.C. 173, 219, 506, 543, 568, 686
Soreghan, M.J. 734
Soter, S. 366
Soulsby, R.L. 229, 625, 626
Souriau, M. 606
Soutar, A. 526, 763
Southam, J.R. 605
Southard, J.B. 31, 59, 172, 254, 301, 364, 567, 596, 634, 668, 672, 674, 687, 711
Southgate, P.N. 447
Southon, J.R. 397, 594
Sowder, K.H. 743
Spötl, C. 21, 234, 397
Sparks, D.L. 109
Sparks, R.S.J. 31, 231, 296, 530
Spasojevic, M. 480
Spearing, D.R. 59, 172, 364
Spears, D.A. 63
Spencer, R.J. 262, 263, 299
Spencer-Cervato, C. 242
Spero, H.W. 499
Speyer, S.E. 728, 729
Spicer, B.E. 8, 287
Spicer, P.J. 358
Spila, M.V. 83
Spinnangr, A. 9
Spirakis, C.S. 557
Spiro, B. 331
Spitznas, R.L. 308
Spooner, E.T.C. 599
Sprenger, A. 185
Sprinkle, C.L. 542
Spurgeon, V.L. 743
Srdoc, D. 755
Srivastava, P. 582
Środoń, J. 142, 371, 395, 450
Srodon, L. 234
Stöffler, D. 278, 657
Stacey, M.W. 31

INDEX OF AUTHORS CITED

Stal, L.J. 39, 440, 737
Stalder, P.J. 127
Stallard, M.L. 29
Stallard, R.F. 256, 358
Standaert, R.R. 722
Stanistreet, I.G. 364, 583
Stanley, D.J. 561, 606, 673
Stanley, G.D.Jr. 475
Stanley, S.M. 242, 263, 506
Stanton, R.J. 397, 678
Stanzione, D. 754
Staples, L.W. 224
Starinsky, A. 397
Starr, J. 502
Stasiuk, L.D. 241
Staub, J.R. 744
Stauble, D.K. 43
Stauffer, P.H. 231
Stearn, C.W. 683
Stearns, H.T. 242
Stearns, N.D. 242
Steefel, C.I. 542
Steel, R.J. 9, 274
Steele, J.L. 316
Steele, R.J. 9
Steelink, C. 362
Steen, Ø. 436
Stegena, L. 316
Stehi, S. 459
Steinen, R.P. 135, 462
Steinmetz, J.C. 184
Stern, L.A. 397
Stern, O. 109
Sternberg, H. 25
Sternberg, R.W. 626
Sternsberg, R.W. 625
Stevens, K.M. 436
Stevens, R. 766
Stevenson, D.J. 139
Stevenson, F.J. 362
Stevenson, G.M. 184
Stewart, F.H. 263
Stieglitz, R.D. 716
Stille, P. 333
Stive, M.J.F. 626
Stockton, P.H. 253
Stoddart, D. 635
Stoddart, D.R. 53, 118
Stoermer, E.F. 408
Stoessell, R.K. 190
Stoffers, P. 722
Stokes, G.G. 626
Stokes, M. 8
Stokes, S. 253
Stokes, W.L. 766
Stolk, A. 498, 499
Stolz, J.F. 39, 423, 440
Stone, C.G. 698
Stone, J.O.H. 397
Stoner, J.S. 424
Stopes, M.C. 149
Storms, J.E.A. 567, 626
Stout, M. 462
Stout, W.E. 185
Stouthamer, E. 36
Stow, D.A.V. 272, 459, 668, 673, 674, 711, 760
Strahler, A.N. 408
Strakhov, N.M. 594, 635
Strasser, A. 53, 185

Strauch, F. 137
Strickholm, P. 692
Stride, A.H. 499
Strobl, R.S. 502
Strong, G.E. 91
Stuart, R.J. 229
Stubblefield, C.J. 649
Stubblefield, W.L. 157
Stumm, W. 159
Sturz, A. 526
Suarez, D.L. 109
Šucha, V. 371, 395, 449, 450
Suczek, C.A. 592
Suess, E. 506, 763
Sugisaki, R. 366
Sullivan, G.G. 674
Sultan, R. 224
Sumer, B.M. 625
Summerfield, M.A. 606, 660
Summerhayes, C.P. 763
Summerson, C.H. 650
Sumner, D.A. 506
Sun, H. 384
Sun, M. 459
Sunagawa, I. 234, 549
Sundborg, A. 531
Suquet, H. 768
Surdam, R.C. 780
Suter, J.R. 202, 499
Sutherland, A.J. 22
Sutherland, J.K. 224
Suttner, L.J. 430, 548, 554
Suwa, H. 9
Svarda, C.E. 673
Svendsen, J.I. 756
Swanson, F.J. 256
Swarbrick, R.E. 436
Swart, P.K. 53, 135, 242, 243, 397
Swartz, R.C. 751
Sweeney, J.J. 306
Sweet, M.L. 246, 253
Sweeting, M.M. 15
Swennen, R. 234
Swett, K. 190, 394, 396, 690
Swift, D.J.P. 157–158, 364, 499, 561, 626, 687
Swinchatt, J.P. 447
Swisher, C.C. 443
Sydow, J. 499
Syktus, J.I. 145
Sylvester, A.G. 734
Sylvester, P.J. 599
Symonds, D.A. 447
Symonds, G. 626
Synolakis, C. 756, 757
Syvitski, J.P.M. 202, 499, 605, 606, 698, 760
Szetu, J. 51

Tabachnick, K.R. 683
Tada, R. 135
Taguchi, K. 636
Tahirkheli, R.A.K. 557
Taira, A. 673
Tait, J. 371
Takahashi, J. 636
Takahashi, T. 188, 756, 757
Takeuchi, M. 636

Tanabe, S. 751
Tanner, P.W.G. 214, 720
Tanton, T.L. 628
Tapponnier, P. 734
Tarduno, J.A. 424
Tardy, Y. 411
Tarling, D.H. 424
Tarnocai, C. 516
Tarr, W.A. 224, 666
Tartaglia, P. 224
Tate, G.B. 498
Tate, R. 362
Tatsukawa, R. 751
Taviani, M. 396, 499
Taylor, B.L. 673
Taylor, D. 531
Taylor, D.M. 755
Taylor, F.W. 395
Taylor, G.H. 85, 403
Taylor, H.P. Jr. 396, 397
Taylor, J.H. 635
Taylor, J.M. 268
Taylor, K.G. 66
Taylor, M.W. 780
Taylor, R.E. 635
Taylor, R.M. 557
Taylor, S.R. 549, 599
Taylor, T.N. 149
Taylor, T.R. 21
Taylor, W.E.G. 135
Taylor, W.L. 206
Tazaki, K. 395
Tazioli, G.S. 753
Tchoubar, C. 400
Teall, J.J.H. 635
Tebbutt, G. 474
Tedrow, J.C.F. 303
Teichert, C. 506
Teichmüller, M. 85, 403, 429
ten Have, T. 106
ten Haven, H.L. 594
Ten Brink, N.W. 8
Ten Kate, W.G. 185
Ternois, Y. 763
Terwindt, J.H.J. 272, 625
Terzaghi, K. 167
Tessier, B. 59, 744
Tetzlaff, D.M. 203
Tewari, V.C. 526
Teysen, T. 592
Thamban, M. 333
Thamban, N. 333
Thamdrup, B. 599
Theide, J. 763
Thein, J. 66
Theo, S. 202
Theron, A.K. 626
Thiel, G.A. 27
Thiemens, M. 395
Thiry, M. 660
Thom, B.G. 47
Thomann, R.V. 751
Thomas, D.S.G. 212, 254, 660
Thomas, H.C. 109
Thomas, J. 760
Thomas, M.A. 498
Thomas, M.F. 411
Thomas, N.A. 751
Thomas, R.L. 379
Thomas, W.A. 734

Thomas-Keprta, K.L. 549
Thompson, A.M. 161
Thompson, C.H. 4
Thompson, G. 599
Thompson, H.S. 109
Thompson, J.B. 755
Thompson, P. 364
Thompson, R. 423, 753
Thompson, R.I. 397
Thompson, S., III. 185
Thompson, T.A. 42
Thompson, T.G. 652
Thomson, J. 594, 674
Thomson, R. 492
Thomson, R.E. 498
Thorndike, E.M. 673
Thorne, C.R. 344, 638
Thorne, J.A. 561
Thornton, E.B. 625
Thornton, S.E. 673, 674
Thorson, R. 137
Threadgold, I.M. 400
Thurber, D.L. 183–184
Thurman, E.M. 362
Thyberg, B.I. 168
Thyne, G. 29
Tian Baolin 149
Tidwell, V.C. 268
Tiedemann, R. 443
Tilleard, J.W. 35
Tilley, B.J. 397
Tillitt, D. 751
Tillman, R.W. 157, 499
Tilton, G.R. 29
Timofeeff, M.N. 263
Timofeett, M.N. 599
Tindale, D.S. 100
Tinker, S.W. 184–185, 447
Tinkler, K.J. 635
Tinterri, R. 59, 292
Tippkötter, R. 375
Tissot, B. 429
Tissot, B.P. 366, 403, 429
Tobin, R.C. 300
Todd, J.E. 225
Todd, R.G. 185, 447
Todd, S.P. 458, 549
Tofani, R. 698
Toh, E.S.C. 531
Tolhurst, T.J. 618
Tolomeo, L. 507
Tolonen, K. 516
Tolson, R.B. 592
Tomlinson, C.W. 161
Toolin, L.J. 594
Toomey, D.F. 475
Tornqvist, T.E. 582
Torrens, H.S. 635
Toth, T.A. 65, 66
Totten, M.W. 554
Tovey, N.K. 716
Townsley, M. 253
Toyama, G. 673
Trappe, J. 526
Trask, P.D. 635
Traykovski, P. 568, 626, 687
Treguer, P. 492
Trendall, A.F. 384
Trenhaile, A.S. 27
Trentesaux, A. 498, 499

Trevena, A.S. 281
Trewin, N. 206
Tribble, J.S. 100, 101, 723
Trichet, J. 101, 526
Tricot, C. 443
Triplehorn, D.M. 63, 656
Troelstra, S.R. 594
Trommsdorff, V. 508
Trompette, R. 722
Troost, K.C. 746
Troup, A.G. 379
Trudeau, S. 751
Trudgill, S.T. 15, 71
Truesdell, P.E. 345
Trumbly, N.I. 53
Trumbore, S. 763
Trusell, F.C. 684
Truswell, J. 137
Tsipursky, S. 371, 449
Tsoar, H. 246, 254, 711
Tsui, T.F. 300
Tsuji, T. 231, 296
Tsuji, Y. 757
Tuchman, M. 751
Tucholke, B.E. 672
Tucker, M. 242
Tucker, M.E. 90, 135, 446, 475, 506
Tudoran, A. 498
Turcotte, D.L. 185, 345
Turi, B. 549
Turner, A.K. 425
Turner, D. 436
Turner, I.L. 718
Turner, J. 394
Turner, P. 557
Tuttle, M.L. 703
Tuttle, O.F. 100
Twenhofel, W.H. 628, 635, 651
Tye, R.S. 36, 203, 741
Tyler, N. 202
Tyler, S.A. 554, 651
Tyson, R.V. 85, 459

Uchmann, A. 459
Uchupi, E. 458, 673
Udden, J.A. 652
Ulmer, D.S. 118
Ulmer-Scholle, D.S. 225, 308
Ulmishek, G.F. 85
Underhill, J.R. 733
Underwood, M.B. 673, 734
UNEP 212
USEPA–USACE 751
Unmi, C.K. 491
Unterschutz, J.L.E. 397
Upreti, B.N. 548
Uriz, M. 682
Usdowski, E. 242
Ussler, W. 139

Vacelet, J. 683
Vadour, J. 755
Vagner, P. 498
Vahrenkamp, V.C. 242, 243
Vail, P.R. 184–185, 272, 333, 526
Vali, H. 549

Vallance, J.W. 188, 353, 425, 675
Vallières, S. 673
Valloni, R. 549
Van Andel, T.H. 89
Van Bemmelen, J.M. 109
van Bennekom, A.J. 491
van Bergen, P.F. 148
Van Beynen, P.E. 681
Van Buchem, F.S. 185
Van Burkalow, A. 16
Van Capellen, P. 85
Van Cappellen, P. 491, 526
Van Dam, R.L. 567
van Dijke, J.J. 626
van Driel, J.N. 241
van Enckevort, I.M.J. 43
Van Gelder, A. 59, 711, 736
van Grafenstein, R. 763
van Heerden, I.L. 202
van Heteren, S. 47
Van Hise, C.R. 384, 628
Van Houten, F.B. 66, 557, 635, 646
Van Houton, F.B. 135, 384, 385
van Kempen, T.M.G. 683
Van Krevelen, D.W. 635
Van Leer, J.C. 711
van Leussen, W. 285
van Lith, Y. 100
Van Loon, A.J. 746
Van Niekerk, A. 480, 618
van Olphen, H. 285
van Os, B.J.H. 594
Van Rijn, L.C. 43, 480, 609, 618, 626
Van Siclen, D.C. 447
van Soest, R. 682
van Soest, R.W.M. 683
Van Straaten, L.M.J.U. 283, 609, 737
Van Tuyl, F.M. 225, 308, 666
Van Wagoner, J.C. 157, 184, 185, 203
van de Plassche, O. 587
van de Poll, H.W. 224
van den Boogert, J.M. 737
van der Nol, L.V. 27
van der Zwaan, G.J. 594
Van den Berg, J.H. 43, 87, 711
Van der Heide, S. 185
Van der Meer, J.J.M. 747
Vanden Bygaart, A.J. 397
Vaneslow, A.P. 109
Vanoni, V.A. 229, 301, 711
Vanoort, F. 4
Vardy, S.R. 516
Varnes, D.J. 248, 345
Vasconcelos 243
Vasconcelos, C. 100, 241
Vassiliou, A.H. 722
Vatan, A. 635
Vathakanon, C. 526
Vaughan, T.W. 652
Veeh, H.H. 526
Veerayya, M. 499
Veizer, J. 385, 394, 395, 397, 599
Velde, B. 127, 371, 65–66
Velegrakis, A.F. 568
Vella, P. 185
Vengosh, A. 397
Ver, L.M. 526

INDEX OF AUTHORS CITED

Vera, J.A. 184
Verburg, P. 4
Vergnaud-Grazzini, C. 594
Vernik, L. 314
Verosub, K.L. 424
Verrecchia, E.P. 91, 418
Verrecchia, K.E. 91
Viekman, B.E. 711
Viets, J.G. 242
Vila, E. 450
Viles, H.A. 755
Vincent, C.E. 157
Vinogradov, A.P. 599
Violante, C. 755
Viseras, C. 8
Visser, M.J. 59, 499, 736
Visser, R. 737
Vitali, F. 397
Vivit, D.V. 395
Vogel, J.S. 594
Vogel, K. 71, 618, 729
Vogel, K.R. 480
Vogt, T. 628
Voight, B. 24
Von Breymann, M.T.K. 763
von Brunn, V. 278, 385
Von Bülow, K. 516
von Dreele, R. 400
Von Engelhardt, W. 635
von Morlot, A. 190
Von Post, L. 516
von Rad, U. 412
von Stackelberg, U. 379
Vreeland, R.H. 300
Vreugdenhil, C.B. 353

Waag, C.J. 374
Wach, G.D. 157
Wächterhäuser, G. 703
Waddell, H. 635
Wadell, H. 345, 636
Wadleigh, M.A. 395, 397
Wager, L.R. 634
Wagle, B.G. 499
Wakefield, M.I. 83
Wakeham, S.G. 594
Walden, J. 9
Walderhaug, O. 29, 118, 168
Waldron, J.W.F. 436
Walker, G. 105, 138
Walker, J.R. 127
Walker, P. 498
Walker, R.G. 59, 157, 172, 185, 202, 272, 335, 364, 434, 447, 498, 512, 635, 647, 687, 698, 760
Walker, T.R. 118, 557
Wallace, K.A. 202
Wallace, M.W. 657
Walling, D.E. 606, 753
Walsh, J.J. 763
Walsh, J.N. 224
Walshe, J.L. 126
Walter, H.J. 549
Walter, L.M. 118, 395, 438
Walter, M.R. 690
Walter, W. 83
Walther, J. 653
Walton, E.K. 183, 282, 447, 748

Wan, D. 395
Wan, S. 480
Wang, L. 397
Wang, S. 459
Wang, Y. 395, 606
Wanless, H.R. 185, 243
Wantland, K.F. 447
Ward, L.F. 83
Ward, P.D. 657
Ward, W.C. 243, 344
Wardlaw, N. 462
Waring, G.A. 242, 263, 755
Warme, J.E. 635
Warne, A.G. 606
Warner, B.G. 516
Warren, A. 211, 253, 254
Warren, C.J. 4
Warren, J. 219, 263, 474
Warren, J.K. 19, 135, 212, 263, 585
Warren, R.S. 587
Warthmann, R. 100
Wasmund, E. 594
Wasserburg, G.J. 395, 396, 397
Wasson, R.J. 9, 246
Watanabe, Y. 396
Watney, W.L. 183
Watson, A. 146
Watson, D.M. 149
Watson, G. 626
Watson, M.P. 760
Waugh, B. 557
Waxman, M.H. 314
Way, J.T. 109
Waylett, D.C. 674
Weare, J.H. 263
Weaver, C.E. 63, 127, 135, 142, 459, 723
Weaver, F.M. 412
Weaver, P.P.E. 335, 674
Weaver, W.E. 583
Webb, G.E. 53, 560
Webb, J.A. 660
Webb, J.E. 698
Webb, R.H. 292
Weber, M.E. 229
Weber, O. 498
Weedman, S.D. 514
Weedon, G.P. 185, 443
Weerts, H.J.T. 287
Wefer, G. 491
Wegmann, E. 635, 644
Wehausen, R. 594
Wei, L. 618
Weidman, C.R. 711
Weiland, R.H. 360, 530
Weimer, P. 698
Weinberg 430
Weinberg, B. 561
Weinberger, R. 214
Weinheimer, A.L. 763
Weir, G.W. 60, 172, 537
Weise, S.M. 366
Weissel, J.K. 668
Weissert, H. 507, 508, 634
Welhan, J.A. 366
Welkie, C.J. 534
Weller, J.M. 185, 447
Wells A.J. 185
Wells, J.T. 285
Wells, J.W. 652

Wells, N.A. 512
Wells, S.G. 8, 9, 22, 123, 256
Wells, T.M. 35, 583
Welte, D.H. 366, 403, 429
Weltje, G.J. 626
Wenbo, Y. 299
Wenk, H.R. 100, 224
Wentworth, C.K. 345
Wentworth, C.M.Jr. 231, 530
Wentworth, S.J. 549
Werner, A. 137
Werner, A.G. 635
Werner, B.T. 246, 568
Werner, F. 283
Wershaw, R.L. 362
West, C.W. 751
West, I.M. 18
Westermann, G.E.G. 728, 729
Westphal, F. 729
Westphal, H. 462
Wetzel, A. 459
Weyl, P.K. 219, 225, 543
Whalley, W.B. 716, 717
Whateley, M.K.G. 203
Wheatcroft, R.A. 687
Wheeler, H.E. 185
Whinland, H.D. 438
Whipkey, C.E. 241
Whipple, K.X. 188, 353
Whitaker, F. 15, 110, 242
Whitaker, F.A. 242
Whitaker, F.F. 242, 243
Whitaker, J.H. McD., 354, 698
White, A.F. 395
White, B.R. 253, 254
White, C.D. 203
White, D.E. 243, 263
White, G.N. 109, 723
White, J. 592
White, J.D.L. 335
White, K. 9
White, T.E. 626
White, W.A. 720
White, W.R. 582
Whitehead, S. 308
Whitman, J. 491
Whitman, R.V. 24, 167
Whitten, D.G.A. 206
Whittington, G. 7
Whitton, B.A. 440
Whyte, M.A. 563
Wiberg, P.L. 618, 619, 687
Wiedenmayer, F. 683
Wieland, M. 352
Wiele, S.M. 537
Wiewióra, A. 450
Wigand, P.E. 8
Wiggs, G.F.S. 254
Wightman, D.M. 83, 157, 502
Wightman, W.G. 743
Wignall, P.B. 85
Wignall, P.W. 729
Wilbert, L.J. Jr. 272
Wilcock, P.R. 33, 268, 301, 619
Wilcoxon, J.A. 121
Wilderer, P.A. 39
Wilgus, C.K. 397
Wilke, B.M. 51
Wilkin, R.T. 85, 703
Wilkinson, B.H. 183, 185, 241, 462, 506

Wilkinson, M. 20
Wilkinson, S.L. 424
Willard, D.A. 148
Willemann, J.H. 185
Willetts, B.B. 253, 254
Williams, E.G. 646
Williams, G.D. 668
Williams, G.E. 657, 737
Williams, I.S. 358, 656
Williams, K.M. 605
Williams, L.A. 118, 135
Williams, L.B. 395, 397
Williams, P.B. 568
Williams, P.W. 678
Williams, W. 137
Williamson, M.F. 379
Willis, B.J. 202, 203, 287
Wilson, C.J.N. 231, 296, 335, 530
Wilson, I.G. 711
Wilson, J.B. 560
Wilson, J.L. 102, 185, 267, 268, 272, 447, 475
Wilson, L. 606
Wilson, M.A. 4
Wilson, M.D. 168
Wilson, M.J. 142, 371, 450
Wilson, P.A. 146, 396, 674
Wimbush, M. 674, 711
Winchester, J.A. 63
Windom, H. 492
Windom, H.L. 127, 185
Winfrey, M.R. 306
Wing, M.R. 526
Winker, C.D. 195
Winkler, K.W. 314
Winland, H.D. 684
Winn, R.D. 499
Winston, R.B. 149
Winter, A. 121
Winterer, E.L. 492, 635
Wise, S.W. Jr. 412
Wiseman, W.J. Jr., 31, 274
Witzke, B.J. 743
Wogelius, R.A. 243
Wolcott, J. 345, 531
Wolcott, J.F. 22, 344
Wolery, T.J. 599
Wolf, K.H. 438
Wolfe, J.A. 146
Wolfe, S.A. 246
Wolfi, W. 396
Wolfstein, K. 285
Wollast, R. 100, 118
Wolman, M.G. 287, 345, 434, 529, 583
Wong, P.K. 225
Wood, D.M. 24
Wood, M.E. 588
Wood, R. 475, 560

Woodborne, M.W. 534
Woodcock, N.H. 668
Woodsworth, G.J. 33
Woodward, J.C. 606, 753
Woody, R.E. 234
Worden, R.H. 722
World Glacier Monitoring Service 331
Woronick, R.E. 21
Wrang, P. 371
Wray, D.S. 396, 549
Wright, D.K. 549
Wright, J.V. 296, 335
Wright, L.D. 31, 43, 687
Wright, M.W. 755
Wright, P.V. 475
Wright, V.P. 9, 15, 91, 110, 135, 474, 475, 506
Wroth, C.P. 24
Wu, G. 673
Wu, J. 360, 656
Wunderlich, F. 59, 283
Wyllie, M.R.J. 314
Wynne, D.A. 502

Xu, J.P. 687
Xun, Z. 243

Yaalon, D.H. 8
Yagi, H. 673
Yagi, T. 636, 637
Yalin, M.S. 347, 619
Yamashita, F. 757
Yamashita, N. 751
Yang, C.S. 59, 499
Yang, C.T. 480, 529
Yang, Z. 606
Yang, Z.-S. 31
Yang, W. 447
Yao, X. 395
Yapp, C.J. 397
Yariv, S. 159
Yaroshevsky, A.A. 526
Yasar, D. 292, 672
Yasuda, T. 549
Yeakel, L.S. 356, 514
Yechielli, Y. 397
Yeh, E. 253
Yeh, H.-W. 397, 526
Yeh, H.H. 757
Yi, H.I. 498
Yim, W.W.S. 534
Yin, P. 498
Yiou, P. 443
Yitshak, Y. 22
Yochelson, S.H. 148
Yoshinaga, N. 4
Young, G.M. 331, 385, 459, 599

Young, G 459
Young, R.A. 561
Young, R.W. 756
Young, S.W. 548
Young, T. 385
Young, T.P. 135, 385, 506
Young, W.J. 301
Yu, W. 31
Yu, X. 459
Yuan, L. 502
Yuan, L.P. 519
Yuce, H. 292
Yurtsever, Y. 397

Zabel, M. 542
Zaitlin, B.A. 202, 537, 582, 736, 737
Zak, I. 395
Zaki, W.N. 360
Zaneveld, J.R.V. 672, 673
Zanin, Y. 525
Zanin, Y.N. 526
Zarba, C.S. 751
Zavala, C. 59, 292
Zawistoski, A.N. 743
Zazo, C. 7, 8, 9, 274
Zeikus, J.G. 306
Zeitzschel, B. 763
Zelazny, L.W. 109, 723
Zeng, J. 360
Zenger, D.H. 190, 242, 243, 633
Zenkovich, V.P. 43
Zhang, J.L. 459
Zhao, Z. 673
Zharkov, M.A. 263
Zheng, H. 400
Zheng, Y.-F. 21
Zhiming Xu 625
Zhu, X.-K. 397
Ziao, C.J.L. 443
Ziegler, A.M. 146
Ziegler, K. 397
Zierenberg, R. 492
Zimmerle, W. 85, 206, 459, 554
Zimmermann, H. 100
Zingg, T. 345
Zinkernagel, U. 106
Zolensky, M.E. 599
Zolitschka, B. 766
Zreda, M.G. 396
Zrobek, J. 625
Zuffa, G.G. 293, 549, 760
Zullig, J.J. 438
Zuppann, C.W. 506
Zussman, J. 769
Zwolinski, Z. 287

Subject Index

abiogenic hydrocarbons 365–6
abrasion
 fluvial **24–5**
 glaciers 318
 grains 616
 heavy minerals 356
 marine **25–7**
abyssal cones and plains 484–5
accommodation/sediment supply ratios 176–7
accretion 252, 598
accumulation rate 514–15, 601, 603
acoustical properties 308–12
 pore fluids 310–11
 pressure and temperature effects 311
 sediment types 309–10
 wave velocities 309–12
Activity Index, soils 23
actualism theory 628–30, 649
acute toxicity 749
Agassiz, Louis 629
aggradation 46–7, 580
Alberta, Canada, oil sands 499–500
albitization of feldspars 28, 280
alfisols 774
algae 66–7
algal carbonate sediments **1–3**
Algoma-type iron formations 381, 382
Alizarin red S (ARS) 683, Plate I
alkali feldspars 278, 279
alkanes 365
Allen, J.R.L. 747
allochthonous sediments
 classification 132–4
 glaucony 332
 lakes 405
 neritic carbonates 472–3
allocycles 173, 274
allophane **3–4**
alluvial deposits 568–9, 575
 facies models 270
 geometry of ancient deposits 581–2
 ridges 34
alluvial fans **5–9, 579–80**
 see also fan deltas

controls 6–7
 deserts 209
 processes and sediments 5–6
aluminum, bauxite **48–51**
Amazon submarine fan 695
amber **563–5**
ambient fluid 348, 350
American southwest, alluvial fans 5
amorphous structure, amber 564
anabranching rivers **9–11**, 576, 578–9
analyses see techniques
anastomosing rivers 9–10, 575, 577, 579
ancient karst **11–15**
 features 12–13
 karst breccia 13–14
ancient oolites 504–5
ancient potash, evaporites 261
ancient reefs 471
angle of repose **15–16**, 336
anhydrite **16–19**, 584–5
animals 255, 649
ankerite **19–21**
annelids 69
antidunes 706
aprons, carbonate sedimentation 474
aqueous inclusions 297–8
aquifers 539–40
aragonite
 calcitization 93
 geochemistry 93–4
 neomorphism 460–1
 ooids 129, 503–5
 sediments 127
 shells 66, 68
 speleothems 679–81
 stains 684
archetypal ichnofacies 74
architecture
 bedding 54
 fan deltas 273–4
arenites 590–1
arenites see sandstones
arkoses 278, 591, 645
armor **21–2**, 617
aromatic hydrocarbons 365

ARS see Alizarin red S
arthropods 69
ash flow 348
asteroid/comet impact 275, 654
atmospheric oxygen, iron formations 382
Atterberg Limits **22–4**
attractive energy, colloids 158–9
attrition
 fluvial **24–5**
 marine **25–7**
 quartz 646
authigenesis **27–9**, 421
authigenic minerals 216
 feldspars 280
 geodes 306–7
 zeolites **779–80**
authigenic rocks 131–2
autochthonous sediments
 classification 127–32
 glaucony 332
 lakes 405
 neritic carbonates 472
 siliceous sediments **660–6**
autocompaction, salt marshes 587
autocycles, sedimentation 173, 181–2
autosuspension **30–1**, 670
avalanches **31–3**, 252
avulsion **34–6**, 578–9, 582

back-arc basins 731
back-barrier marshes 586
backwash **717–18**
bacteria **37–9**
 carbonate sediments **1–3**
 epibenthic cyanobacteria 439–40
Bagnold, Ralph Alger, physics of sediment transport 249, 250, **527–9, 638**
Bahama platform, dolomitization 239–40
Bailey, Edward B. 628
Baja California tidal flats 585
bajadas 579
ball-and-pillow structures **39–40**, 58, 549–50
banded iron formations (BIFs) 367, 379, 380, 381

bands, liesegang banding 221, 223
barchans 710
Barrell, Joseph **638-9**
barrier islands **43-7**, 155
　environments 43-4
　morphosedimentary types 45-6
　thoeries of origin 44-5
barrier reefs 465
bars
　see also point bars
　barrier islands **43-7**
　eddy bars 290
　ephemeral braid 85
　littoral **40-3**
　tidal 740
basalt 591, 598
base-level conditions, alluvial fans 7
basinal facies
　carbonate mud-mounds **100-2**
　fan deltas 273-4
　mudrocks 452
basins
　brines 540-1
　rift basins 258, 729-30
　slopes 472-4, 669, 672
　stresses 161-2
　tectonic control 730-3
Bathurst, Robin G.C. **639-40**
bathyal zone, slopes 668-9
bathymetric forms
　glacial deposition 324-5
　glacial erosion 324-5
　tidal influence 607
bauxites **48-51**, 409
beaches
　armor 22
　shoreface facies 152-3
　swash and backwash 717-18
beachrock **51-3**, 112, Plate I
bed load 574
　braided channels 86-7
　G.K. Gilbert 643
　tidal currents 607-8
　transport model 477-8
　transport rates 613-14
　unidirectional flow 611-12
　wave action 622
bed model 478
bed slope 611
bed stress 621-2
bedded minerals
　anhydrite 17
　chert 661-2
　gypsum 16-17, 584-5
bedding **53-4**
　see also cross-stratification; sedimentary structures
　bedsets **59-61**
　composite sets 60, 61
　internal structures **54-9**
　iron formations 381
　laminated 53
bedforms 618, 633, 703-4
　cross-stratification 170-3
　eolian 252-3
　floods 290-1
　flume studies 300-1
　littoral bars 41-2
　ripples 566, 567
bedsets **59-61**

behavior, trace fossils 72, 73
benthic invertebrates, toxicity 748
bentonites **61-3**
berthierine **64-6**, 125
BIFs see banded iron formations
binding agents 561-3
bioclasts **66-70**
biodepositional structures 71
bioerosion **70-1**
biofilms 37, 688, 690
biogenic processes
　calcite 95-7
　coastal facies 150
　hydrocarbons 365-6
　sediment magnetism 421
biogenic sediments 482, 487-91
　accumulation 487-8
　carbonate ocean mode 490-1
　oozes 488-90
　silica ocean mode 491
　varves 764
biogenic structures **71-83**
　ichnofacies 74-9
　paleoenvironmental significance 79-82
　stromatolites **688-90**
　trace fossils 72-3
bioherms 558
biological processes, ooids 504
biological sediments, caves 109-10
biomarkers (hydrocarbons) 365
biopolymers 426
bioreactors, epibenthic bacteria 37
biostratification structures 71
biostratinomy 723-6
biostrome 558
biota 571, 753
bioturbation 407
　mudstones 204
　structures 71
　tidal flats 735
birds, contaminants 749
bitumen 499-500
Black Sea
　black shales 83
　sapropels 593
black shales **83-5**
Blue Earth amber 563
boehmite 48
borings 70-1, 699-700
bottom nepheloid layer 462-3
boudins 667
boulders, imbrication 372, 373
Bouma sequences, turbidites 56-7, 757, 758
boundary layers 620-1
boundstones 129-30
brachiopods, bioclasts 67, 68
brackish water ichnofacies 75, 76, 81
braided channels **85-7**, 323, 575, 577, 580, 582
breccias
　ancient karst 13-14
　impact breccia 275
　ophicalcites 506, 507
　pipes 678
　sedimentary 134
　solution **677-8**
brines
　see also saline environments; seawater
　continental 540
　deep basins 540-1

evaporites deposition 257, 260-1
　magadiite formation 417
broad marshes 586-7
Brownian motion 226-7
bryozoans 67
bulk density, compaction 162
burial
　authigenic minerals 28-9
　cementation 114-15, 116-17
　deformation 192
　diagenetic reactions 65
　dolomitization 240-1
　fossils 723-4
　mudrocks 456-7
　paleosols 775-6
burrows 699-700

calcareous sediments
　H.C. Sorby 650
　oozes 488-9
　pelagic grains 119-21
calcification, algae and bacteria 1-3
calcite
　see also aragonite
　burial-stage cements 114, 117
　cathodoluminescence 104
　compensation depth **88-9**
　cracks 657-8
　crystal structure and geochemistry 95-7
　dedolomitization 188-90
　dissolution 489
　neomorphism 460-1
　ooids 505
　sediments 127
　shells 66, 68
　speleothems 679-81
　stains 683-4
calcitization, aragonite 93
calcium carbonate
　see also aragonite; calcareous sediments; calcite; carbonate sediments and rocks
　cements 51-3
　polymorphs 93-4
calcrete/caliche **89-91**
canyons 693
carbon, stable isotopes 392
carbon cycle, upwelling 761
carbon dioxide 305
carbonaceous sediments and rocks
　charcoal **121-3**
　classification 132
　coal balls **146-9**
　peat **514-16**, 586-7
carbonate cement fabrics 112
carbonate depositional environments
　mixed sedimentation **443-7**
　neritic environments **464-75**
　platforms 464, 465
　ramps 464-5
　sediment gravity flows 472-3
　slopes 669-71
carbonate fluorapatite 519-21
carbonate minerals, stains **683-4**
carbonate sediments and rocks 632
　see also dolomites; limestones
　acoustic velocities 309-10
　algal and bacterial **1-3**
　allochthonous 472-3

autochthonous 472
R.G. Bathurst 639
bioerosion 70–1
caliche/calcrete **89–91**
cathodoluminescence 103, 104–5
L. Cayeux 640
cements 92–3, 111, 114–15
classification 127–9, 642
compaction 166–7
concretions 220
diagenesis **91–3**
facies models 270–1
factor analysis 685
fine-grained 472
geochemistry **93–100**
microfabrics **91–3**
mineralogy **93–100**
mud-mounds **100–2**
ophicalcite 507
properties 94
sands 590
carbonate skeletons 726–7
Carboniferous coal seams 146–7
casts
  gutter casts **353–4**
  load casts 58, 192, 193, 414
  pot casts 353
  tool marks **747–8**
catagenesis 366, 402, 427
catalytic properties, smectite 676
catastrophic deposits 56–7
catastrophic events **287–93**, 318
  see also floods
catastrophism theory 288, 569, 629
cathodoluminescence (CL) **102–6**
cation exchange **106–9**
  clay minerals 106, 140
  vermiculites 767–8
cauliflower nodules 307
causality, toxicity 749–50
caves
  ancient karst 13
  sediments **109–10**
  speleothems **678–81**
cavitation 572
cavities, stromatactis 687–8
Cayeux, Lucien **640**
cements and cementation **110–19**, 215
  beachrock 51–2
  burial stage 114–15
  carbonates 92–3, 111, 114–15
  cracks 657–8
  dolomites 234
  early diagenetic (non-marine) 113–14
  silica 115
  siliciclastic sediments 115–16
  syndepositional (marine) 111–13
Central Asia, paleofloods 291–2
chalcedony 117, 659–60
chalk **119–21**
Challenger Expedition (1872–76) 481, 647–8
chamosite 65, 125
channels
  see also flumes
  anabranching **9–11**, 576, 578–9
  anastomosing 9–10, 575, 577, 579
  avulsion **34–6**
  braided **85–7**
  current analysis 509

distributary (deltas) 200
floodplain sediments 286
geometry 430–2
meandering **430–4**, 575, 577–8, 582
rivers 571–9, 581–2
submarine **692–8**
tidal 738–40
charcoal **121–3**
Chattanooga Shale 83, 84
chemical compaction 163, 165–7
chemical constraints 235
chemical diffusion **225–6**
chemical divides, brines 260
chemical weathering 429
chemistry
  ankerite 19
  berthierine 64
  calcite 95–6
  dolomitization 235
  kaolin group minerals 399
  phanerozoic seawater 261–2
  porewaters 537–9
chemogenic sediments 482, 483, 487, 764
cherts
  bedded 661–2
  cements 117
  conversion from magadiite 417–18
  magadi 664
  nodular replacement 662–4
  origin 131
  Precambrian 664–5
chiso-gaku, Japan 635
chlorite **123–7**, 449
  detrital 124–5
  evaporite/dolomite association 125
  red beds association 125–6
  shallow marine 125
chronic toxicity 749
CL see cathodoluminescence
Cladophora, microbial carbonates 3
classification of sediments and sedimentary rocks **127–36**
  allochthonous deposits 132–4
  authigenic rocks 131–2
  autochthonous deposits 127–32
  bentonites 62–3
  caliche, calcrete 89–90
  carbonates 127–9, 642
  clay minerals 139–41, 140–1
  cross-stratification 171–2
  dolomites 236–7
  dolostones 131
  evaporites 258–60
  limestones 129–30, 268–9
  oceanic sediments 482–3
  phosphorites 521–2, 524
  sands 590
  sandstones 590, 646
  silcretes 659–60
  tills and tillites 745
classifications
  anabranching rivers 9
  dunes 243–4
  eolian dunes 243–4
  facies 268–9
  fluid escape structures 295–6
  glaciers 318
  reefs 557–8
  soils 771–4
  stylolites 692

subaerial dunes 710–11
surface forms 703–4
tool marks (Allen) 747
trace fossils 72–4
tuffs 62–3
clastic dykes and sills 59, **136–7**, 194
clastic facies, open coasts **149–57**
clastic rocks see mudrocks; sandstones; siliciclastic sediments
clastic varves 764
clasts, flow-oriented **371–4**
clathrates **137–9**
clay, definition 139
clay minerals **139–42**
  authigenic 117
  bentonites 62–3
  cation exchange 106
  C.W. Correns 641
  definition 139
  identification 141
  illite group **369–71**
  kaolin group **398–400**
  K–Ar dating 386
  mixed-layer **447–50**
  mudrocks 458
  oceanic sediments 485–6
  origin and distribution 141–2
  Rb–Sr isochron dating 386–7
  smectite group **675–7**
  structure and classification 139–41
  vermiculite **766–9**
  X-ray techniques 632
clay sediments
  mixed-layer **447–50**
  pelagic red clay 485–7
  syneresis 719
Clayton yellow stain see Titan-yellow stain
climate
  evaporites formation 257
  past conditions 315
  sediment provenance 544–5
  silcretes 660
climate change 145
  alluvial deposits 581
  fan delta indicator 272–3, 274
climatic control
  alluvial fans 7
  sedimentation **142–6**
climatic precession 441
climatic rhythms, Milankovitch cycles 180
cnidarians 67
coal 132, 146–7, 629–30
  see also charcoal; peat
  acoustic properties 310
  ankerite 20
  balls **146–9**
  diagenesis 426–8
  resin 563
  resistivity 313
coastal facies **149–57**
  inlet areas 153
  preservation 153–7
  processes 149–50
  shorefaces 150–3, 496
coated grains **502–6**
coccoliths 489
cohesionless sediments 86, 610–11
  see also gravels; sands
cohesive sediments see muds
colloidal properties **157–9**

colluvial processes 406
colored impregnation media 218
colors of sedimentary rocks **159–61**
columns, stylolites 690
combined flows 55–6, 708–9
comets 275, 597–8, 654
compaction **161–8**
   chemical 163, 165–7
   deformation 192
   diagenesis 215
   dolomitization model 240
   mechanical 162–5
   mudrocks 456–7
   mudstones 164–5
   porosity 161, 162, 165–6, 215
   sands 164, 165, 166–7
   stresses 161–2
compensation depth, calcite **88–9**
compositional maturity **429–30**
compound cross-stratification 171–2
compression, stylolites 691–2
computer simulation models 182–3
concave-upward shells 725
concentrated flows 347, 350
concentration fields, turbulent diffusion 227–8
concentrations
   heavy minerals 530
   porewaters 537–8
   suspensions 359–60
concretions 219–20, 223
   fossils 727
   septarian **657–9**
condensed sections, phosphorites 523–4
conductivity, heat flow 315
cone-in-cone structures 220–1, 223
conglomerates 134, 472–3, 588
consolidation *see* compaction
consolidation laminations 295
contamination 748–51
continental crust 597–9
continental evaporites 258–60
continental goundwaters 539–40
continental rifts 729–30
continental shelves 560–1, 604
continental slopes 669
contourites 269, 671–2
convection, dolomitization 240
convolute laminations 58, **168**, 192, 193
coral reefs 470, **557–60**
   J. Murray 647
   T.W. Vaughan 652
   J. Walther 653
coralline algae, reefs 2
cores and coring methods **168–70**
Coriolis effect 143, 686, 694
corrasion 572, 595, 726
correlation, mudrocks 457
Correns, Carl Wilhelm **641–2**
corrosion 572, 726
counter-flow ripples 55
cracks
   concretions 220, 657–9
   desiccation 194, 204, 209, 210, **212–14**, 719
   syneresis 718–19
craters, impact 275, 653, 654, 656
cratonic basins, deposition 733
creep 250, 424
crescentic eolian dunes 243–4, 245–6

Cretaceous-Tertiary (K-T) boundary 654–5
crevasse-channel and crevasse-splay sediments 286
crinoids, encrinites **248–9**
cross lamination 535
cross-bedding *see* cross-stratification
cross-sections *see* geometry
cross-shore sediment transport 623
cross-stratification **170–3**, 633
   see also hummocky cross-stratification; swaley cross-stratification
   classification 171–2
   current analysis 509
   cyclic 172
   deformation 59
   dunes 207
   hummocky 55, **362–4**, 686
   large offshore sand dunes 494–5
   origin 170–1
   overturned 193
   swaley 55, **363–4**
   unidirectional flows 54–5
   way-up indicator 627–8
crust
   see also tectonic control of sedimentation
   composition, Moon 415–16
   continental 597–9
   dutile deformation 543
crusting, eolian transport 251
Cruziana, ichnofacies 76, 79, 80
cryptocrystalline texture 436–8
crystal structure
   calcite 95–6
   chlorite 124
   diagenesis 216
crystallography
   dolomites 231–3
   speleothems 679
crystals
   geodes 306–8
   sand 222, 223
CSS *see* Cyclic Steam Stimulation
current crescents **513–14**, 595
current lineations 513
current ripples 565, 566, 705–6
currents
   see also flow (process)
   autosuspension 30
   gutter casts 353
   orientation 509–14
   planar laminated sands 536
   scour marks 595
   superimposed currents 607
   thermohaline 671
   tidal 55–6, 606–7, 737
   time-averaged oscillatory 624
cyanobacteria, calcification 2, 504
cyclic cross-stratification 172
cyclic sedimentation **173–83**, 444, 445, 639
   allocycles v autocycles 173–4
   Carboniferous deposits 444
   fan deposits 580
   hierarchies 174–8
   orders of cyclicity 174–5, 178–82
   quantitative analyses 182–3
   terminology 173–4
   varves **764–6**
Cyclic Steam Stimulation (CSS) 500–1
cyclicity, glaucony 332

cyclothems *see* cyclic sedimentation; rhythmites

Dake, Charles L. 627
Darcy equation 293–4
Darcy (unit of permeability) 266
Darcy's Law 266, 516
data collection and processing 510–11
dating methods
   see also isotopic methods
   radiometric 385–97, 547
   varves 764
debris avalanches 31–2
debris entrainment, glaciers 319–20
debris flows 56, **186–8**, 348, 425, 434
   deposits 579, 580
   processes, alluvial fans 5–6
debris slides 425
debrites 671
decomposition, organic matter 426, 514–15
dedolomitization **188–90**
deep basins, brines 540–1
Deep Sea Drilling Project 88–9, 169
deep water deposition
   see also oceanic sediments
   facies models 271
   paleocurrents 511–12
   zeolites 780
deepwater nepheloid layer **462–4**
deformation
   see also deformation of soft sediments
   burial 92
   cross-stratification 172
   features 277
   glaciotectonic 321
   grain-scale (pressure solution) 542
   laminae 168
   slides 668
   subglacial deposits 319, 320
   tidal-waves 606–7
deformation pathways 39, 550
deformation of soft sediments 57–9, **190–3**
   see also load structures
   dish structures 193, **230–1**, 295, 529
   forces 191–2
   processes 192
   slump structures 194, **667–8**
   structures **193–5**
   styles 194
dehydration 375, 440
deltas 153, **195–203**, 580–1, 639
   see also submarine fans
   braid deltas 577
   definition 195
   delta fronts 200, 496
   environments 200
   facies 200–1, 270
   fan deltas **272–5**
   ice-contact 324–5
   lacustrine 406
   morphology 196
   prodelta 200
   research history 197–8
   river mouth processes 198–200
   scale and importance 197
   sedimentation 604
   tidal **737–41**
density, river outflows 198, 199
density inversions, soft sediments 192

denudation rate 600, 601
deposition
    see also sediment deposition
    anhydrite and gypsum 16–18
    bathymetric forms 324–5
    charcoal 121
    eolian **249, 252–4**
    floods 289–91
    glaciers 320–4
    meander floodplains 432–3
    numerical models **475–80**
depth, calcite compensation **88–9**
deserts **207–12**
    alluvial fans 5
    armor 22
    eolian environments 207–8
    fluvial environments 208–9
    lacustrine environments 209–10
    landforms, ridges-and-furrows 595
    marine coastal environments 211
    playa environments 210–11
    sediment transport 527
    J. Walther 653
desiccation structures (cracks) 194, 204, 209, 210, **212–14**, 719
detection methods
    see also forensic sedimentology
    extraterrestrial materials 264
    tracers 752
detrital grains
    chlorite 124–5
    diagenesis 214, 216
    feldspars 278–9
    quartz **552–4**
    sand 591–2
    smectite 676
dewatering 231, 698–9
diagenesis **214–19**
    see also authigenesis; burial; compaction
    berthierine 65
    carbonates **91–3**
    cements 113–14, 116
    chemical processes 215–17
    chlorites 126
    conditions 214
    feldspars 279–80
    fossils 726–8
    free gas formation 304
    humic substances 362
    hydrocarbons 366
    organic matter 402, 426–8
    physical processes 215, 216–18
    red beds 556
    rock components 214–15
    sapropels 593–4
diagenetic structures **219–25**
    concretions 219–20, 223
    cone-in-cone 220–1, 223
    geodes 221, 223
    liesegang banding 221, 223
    sand crystals 222, 223
    stylolites 222–3
diamagnetism 418
diamictites 134, 325, 327
diamictons 32, 134, 325, 326, 330
    see also tills and tillites
diamonds, marine placers 534
diapirs, mud 194–5
diaspore 48
diatomites 661

diatoms 489–90, 491
dickite 399
differential thermal analysis (DTA) 48–50
diffuse double layer models 108
diffusion
    chemical **225–6**
    turbulent **226–30**
dilatation 335
dilute flows 347, 349, 350–1, 352
dip direction, cross-bedding 172, 200, 201, 510
disarticulation, fossils 723–4
discharge, river mouth 198
dish structures 57–8, 193, **230–1**, 295, 529
    see also pillow structures
dispersive stress 638
dissolution
    calcite compensation depth **88–9**, 172
    compaction 165
    diagenesis 216
    dolomitization 235
    fossils 726–7
    stylolites 223, 690–1
dissolved gases 304
dissolved load 574
distributary channels, deltas 200
distributary mouth bars 198
dolomites 20, **234–7**, 683
    burial-stage cements 114
    chemistry 99
    chlorite 125
    classification 236–7
    crystallography 231–3
    dedolomitization **188–90**
    diagenesis 92
    geochemistry 94, 237
    mineralogy 97
    natural environments 237–41
    occurrence 98–9
    penecontemporaneous 237–8
    textures **231–4**, 236–7
dolomitization **234–41**
    chemical constraints 235
    environments 237–41
    fluid flow 240–1
    hypersaline environments 239
    hyposaline environments 238–9
    mass balance constraints 235–6
    models 237–41
    porosity and permeability 236
dolostones 131, 234, 241
domes, breccias 678
downslope movements, slides and slumps 191, **666–8**
drag 229, 609–10
drainage networks 568, 569–70
drilling rock cores **168–70**
dry fans 579, 580
DTA see differential thermal analysis
dunes
    antidunes 706
    barchans 710
    barrier islands 44
    classification 243–4
    eolian 207, **243–6**, 252–3
    large offshore 493, 494–5
    processes and dynamics 244–5, 252–3
    sediments 245–6
    subaqueous 706
    surface forms 709–11

Dunham classification, limestones 129–30
duricrusts, silcretes **659–60**
dust, extraterrestrial 263
dutile deformation 543–4
dykes, clastic (neptunian) 59, **136–7**, 194
dynamic equlibrium, rivers 569, 572
dynamics
    debris flows 187
    eolian dunes 244–5
    eolian processes 252–3
    glaciers 317
    mass movement 347–9
    turbulent flow 226–9

Earth
    orbital parameters 441, 442–3
    orbitally forced insolation 441–2
earth flows **247–8**, 425
earthquakes
    syneresis 719
    turbidity currents trigger 757
ebb-tidal deltas 153, 738, 739–40
ECC see Exposed Continental Crust
echinoderms 69, 724
ecology
    coral reefs 471–2
    ichnofacies 77
economic uses
    calcrete 90
    evaporites 257
    smectite 676–7
ecosystems
    contamination 748–9
    salt marshes 586
eddy bars 290
eddy diffusivity 228, 229
effective porosity 266
effective stress, compaction 162
ejecta, impacts 263–4, 654–6
elastic wave velocities 311–12
electrical properties 312–13
elutriation 215, 295
embedding media 375, 562
emissions, cathodoluminescence 102–3
encrinites **248–9**
encrustation, fossils 726
endogenetic sediments, caves 109
energy, colloidal particles 158–9
entrainment 250, 319–20, 571–2, 594–5, 610
entrapment, fluid inclusions 297
environmental applications 676
environmental issues 524–5
environments
    see also eolian environments; peritidal environments; saline environments; subtidal environments
    barrier islands 43–4
    deltas 200
    deserts **207–12**, 234–6
    detrital quartz 553–4
    dolomitization 237–41
    evaporites formation 257–8
    facies models 270–1
    glacial and periglacial 327, 328, 330, 715
    low energy 332, 496
    mudrocks 457
    mudstone fabrics 206
    neritic carbonate deposition **464–75**
    oolites 503

environments (*Continued*)
   paleocurrent analysis 511–12
   phosphorites 519–20
eogenesis 402, 426–7
eogenetic cements
   marine carbonates 111–13
   non-marine carbonates 113–14
   siliciclastic deposits 116
eolian environments 207–8
   see also wind transport
   cross-stratification 170
   deposition **249**, **252–4**
   dunes **243–6**
   textures 486, 715–16
   transport **249–52**
ephemeral braid bars 85
epibenthic bacteria 37, 439–40
episodic deposits, structures 56–7
epoxy resins 375, 562
equations
   cation exchange 108
   Darcy 293–4
   isotopic methods 385–6
   motion 227
   Rouse 614–15
   shear strength 191
   Stokes settling 337
equilibrium partitioning, toxicity 750
erosion
   see also abrasion; attrition; bioerosion; corrasion
   J. Barrell 639
   bathymetric forms 324
   episodic 255
   fluvial 569, 571–2, 578
   glacial 256, 318–19, 603
   grain threshold **345–7**, 608, 610–11
   meandering channels 578
   organisms **70–1**, 255
   sediment transport rate 616–17
   submarine channels 693
   susceptibility 254
   unroofing 591–2
erosion and sediment yield **254–6**
   chlorite 124–5
   climate effects 143–4, 145
   glacial erosion 256
   hillslope factors 254–5
   human activities 255–6
   organisms 255
eskers 324
estuaries **195–203**, 581
   circulation 607–9
   definition 195–7
   processes 197
   sedimentation 604
   tidal flats 735
etching 714
ethological classification, trace fossils 73
eustasy
   alluvial deposits 581
   cyclic sedimentation 179–81
evaporites **257–63**, 269
   acoustic properties 310
   anhydrite and gypsum **16–19**
   authigenic talc 722
   brine evolution 260–1
   burial-stage cements 114–15
   chlorite occurrence 125
   dedolomitization 189

   environmental conditions 257–8
   marine and non-marine 258–60
   mineral sequences 132, 260–1
   nodules 307
   phanerozoic seawater chemistry 261–2
   resistivity 313
   sabkha 584–5
   sands and gravels 590
   silicified 664
   solution breccias 677
evidential value, sediments 301–2
exfoliation, vermiculites 768
exogenetic sediments, caves 109
exotic lithologies 434–5
experimentation 633
   see also techniques
   angle of repose 15
   flumes 300–1
   pressure solution 543
exploration, oceanic sediments 481–2
Exposed Continental Crust (ECC) 597–9
extrabasinal deposits 132–4
extraterrestrial materials **263–4**, 597–9, 653–7
   see also Mars; Moon
extreme floods 287–8
extruded sheets, soft sediments 194

fabrics **265–6**
   see also microfabrics
   breccias 677
   carbonate cements 112
   environment 206
   mélange 434, 435
   mudstones **203–6**
facies
   associations 269–70
   carbonate mud-mounds 100–1
   classifications 268–9
   coastal **149–57**
   deep water carbonate slope apron 473
   deltas 200–1
   glacigenic sediments 328, 329
   A. Gressly 629, 644
   ichnofacies concept 74–9
   iron formations 379
   ironstones 383
   littoral bars 42
   meandering channels 433
   models **268–72**
   reefs 469–70
   river channels 575–9
   sequences 54, 57
   shoreface 150–3
   slopes 672
   submarine fans 694–8
   tidal flats 735
   tidal inlets 740–1
   J. Walther 653
factor analysis, carbonates 685
fallout radionuclides 752
fans
   alluvial fans **5–9**, 209, **579–80**
   deltas **272–5**
   outwash fans 323, 324
   submarine 271, 289–90, **692–8**
faults 667–8
   growth faults **193–5**
fauna

   see also animals; organisms
   sediment mixing 450–1
   toxicity 749–50
Feigl's stain 684
feldspars **278–81**
   detrital 589–90, 592
   overgrowths 279–80
   provenance 279
ferallitization 770, 771
ferric oxides *see* iron
ferricretes 409
ferroan dolomites 19, 20, 188–90
ferromagnetism 418
ferromanganese nodules **376–9**, 487, 488
ferruginous tracers 752–3
fine-grained sediments *see* mudrocks
fingerprinting technique 752
fining-upwards sequences 57, 755
fire erosion 255
Fischer Plots, cyclic sedimentation 182
fish, neoplasms 748
fissility, mudrocks 454
fissures, filled 136
fjords 322, 601, 604
flame structures **281–2**
   see also load structures
flaser bedding 56, **282–3**
flat-topped platforms, carbonates 464
flatness, grains 339, 341
flocculation **284–5**, 454, 609
flocs 284, 340, 454
flood ramp 739
flood-basin sediments 286
flood-tidal deltas 153, 737–9
floodplains
   avulsion 35
   classes 576
   deserts 208–9
   meander deposition 432–3
   sediments **285–7**, 575, 582
floods **287–93**
   catastrophic 287–8, 289–90, 318
   depositional forms and structures 289–91
   dynamics and processes 288–9
   Mars 288, 291
   subglacial 291
flow (process)
   see also currents; thresholds; velocities
   acceleration 244
   combined 708–9
   competence 346
   direction 172
   flood dynamics 289
   fluid 217, 228–9, 240–1
   geostrophic 686
   grain threshold stress 346
   granular **335–6**
   hydrodynamic model 475–7
   littoral bars 41–2
   meandering channels 432
   multiphase 226–7, 518, 615
   oscillatory 55–6, 704, 707–8
   rapid-flow regime 336
   resistance **293–4**
   ripples 566, 567
   river channels 571–2
   sediment transport 613
   shallow flow theory 347–8
   unidirectional 536, **609–18**, 704–7
flow-oriented clasts **371–4**

flows (forms)
see also debris flows; fluidization; gravity-driven mass flows
earth flows **247–8**, 425
gravitational 424, 425
internal structures 54–6
mud flows 348
pyroclastic flows 348
surface forms 703–9
flowstones, calcite 679–80
fluid dynamics, rivers 569
fluid escape structures **294–7**
fluid flow 217, 228–9, 240–1
fluid inclusions **297–300**
composition 298
pressure 299
temperature 298–9
fluid pressure 674
fluid stress 251
fluidization 136, 295, **412–13**, 530
flumes **300–1**, 643
braided channels 577
first use 630–3
grain threshold 345
straight rivers 578
fluorapatite 727
flute marks 595–6, 707
fluvial processes and deposits
alluvial fans 5–6, 273
armor 21–2
attrition **24–5**
deserts 208–9
placers **530–2**
tufa model 754
Fluvialists 569
fluxes
global sediment 601–3
mantle 598
sediments **600–5**
solutes 596–7
flysch 732, 757
folds, soft-sediment 296, 667
foliation, mélange 435
Folk, Robert Louis **642**
limestone classification 130
foraminifera 67, 68
force chains 335–6
forces, deformation 191–2
forearc basins 731
foreland basins 731–2
forensic sedimentology **301–3**
formation water 391
fossils
bioclasts **66–70**
calcareous nannofossils 199–21
trace fossils **71–83**
fossils, preservation **723–9**
biostratinomy 723–6
diagenesis 726–8
silicification 307
trace fossils 72
fractionation, stable isotopes 392–4
fractures see cracks; faults
fragmentation
fossils 725–6
mélange blocks 435
framboids 702
framework reefs 558
francolite 519–21
free gases 304

freeze drying 375
freshwater carbonates **753–4**
freshwater fauna, toxicity 749
fringing marshes 586–7
frost creep 425
fuller's earth 676
furrows (waves) 595, 707
fusain, charcoal 122

Ganges plains 579
garnet 357
gas hydrates, clathrates 137–8
gases **304–6**, 538
gelifluction 425
gelisols 774
gem minerals, marine placers 532, 534
genetic packages 57
geochemical cycles, seawater 596–7
geochemistry
burial-stage cements 115
carbonates **93–100**
dolomites 237
evaporites 260–1
techniques 632
geochronometers 385, 388–9
see also isotopic methods
geodes 221, 223, **306–8**
geology, sedimentary **626–7**
geomechanical properties 162–3
geometry
ancient alluvial deposits 581–2
gutter casts 353
littoral bars 40–1
meandering channels 576–8
mélange units 435–6
pore spaces 267
ripple marks 565–6
river channels 430–2, 573–4
surface forms 703–4
geomorphology
see also surface forms
channels 575–9
heroic age of 630
karst 12
geophysical properties **308–14**
acoustical 308–12
electrical 312–13
radioactive 313–14
geopolymers 426
geostatistics 685
geostrophic flows 686
geosynclines 631–2
geothermic characteristics **314–16**
geothermometry, stable isotopes 392–4
Germany 631
giant current ripples, floods 290–1
giant wave deposits 291
gibbsite 48, 410
GIFs see granular iron formations
Gilbert, Grove Karl 631, **642–3**
Gilbert-type fan delta 273
glacial processes 318–21
bathymetric forms 324–5
debris entrainment and transport 319–20
deposition 320–1
erosion 256, 318–19, 603
glaciotectonism 321
glacial sediments **316–31**
depositional landforms 321–4

diamictons 32, 134, 325, 326, 330
distribution and significance 316
environment criteria 330
facies analysis 328
geological record 328–30
glacigenic 325–8
glaciolacustrine 328
glaciomarine 326–8, 328
ice-marginal landforms 321–4
landform associations 328
reworking 321
surface textures 713–15
terminology and classification 325–8
tills and tillites 325–6, **744–51**, 745
yield 256
glaciers
dynamics and structure 316–17
fluctuations 318
glacial theory 629–30
hydrology 317–18
types 317, 318, 322, 325
glacio-eustatic cycles 179–81
glaciotectonic deformation 321
glauconite see glaucony
glaucony **331–3**, 386, 448
see also verdine
gleization 770, 771
global climate **142–6**
global sediment fluxes 601–3
Globigerina ooze 489
Glossifungites ichnofacies 698–9
gold, marine placers 532, 533–4
Gouy–Chapman theory 108, 158
Grabau, Amadeus William **643–4**
graded bedding 56–7, **333–5**
distribution grading 334
normal or reverse 334
sands 646–7
tsunami deposits 756
turbidites 758
turbidity currents 334
way-up indicators 627
graded streams, G.K.Gilbert 643
gradients, streams 570–1
grading see graded bedding
grain flow **335–6**
grain orientation 265–6
see also imbrication
sands and gravels 589, 616–17
grain settling **336–8**
see also settling behaviour
grain shapes 714
distribution 344
early analysis 630
elongation and flatness 339, 341
sands and gravels 589
smoothness/roundness 341
grain sizes **338–45**
b-axis 339
determination **339–41**
detrital quartz 552–3
distribution 343–4, 429
early analysis 630
erosion threshold 611
fractionation 624
frequency 342
heavy minerals 530–1
lunar sediments 416
sampling 342–3
sands and gravels 588–9

grain sizes (*Continued*)
    settling 338
    tills 744
    J.A.Udden 651
grain threshold **345–7**
grains
    abrasion 616
    bed load 611–12
    drag/lift 610–11
    eolian 715–16
    pressure solution **542–3**
    sands and gravels 588–9
    sixth-power law 630
    sorting 616–17
    spinning 713–14
    surface textures 712–13
    transport 530–2
    turbulence 612–13
granular flow 335
granular iron formations (GIFs) 379, 380, 381
graphical representation 704–6, 708
grasses, salt marsh 586
gravels **588–92**
    anabranching rivers 9–10
    armor 21
    current analysis 509
    lithification 51–3
gravitation, turbulent diffusion 229
gravity corers 169
gravity flows *see* autosuspension
gravity waves (water) 619–20
gravity-driven mass flows 56, **347–53**
    macrodynamics 347–9
    micromechanics 349–52
    names and attributes 348
    slurry 530, **674–5**
    sorting effects 352
    turbidites 757
gray coloration 159
graywackes 133, 430
green coloration 159, 331, 332
greigite 701
Gressly, Amanz **644**
groundwaters 537
    see also porewaters
    continental 539–40
growth fabrics, cathodoluminescence 105
growth faults **193–5**
guano 109, 524
Gulf of St. Lawrence, littoral bars 41
gutters and gutter casts **353–4**, 595
gypsum **16–19**, 584–5, 681

Halimeda segment reefs 3
halite 585
hardground, ancient karst 13
HCS *see* hummocky cross-stratification
heat flow, geothermal 314–15
heavy metals 367, 541
heavy mineral shadows **355–6**
heavy minerals **356–8**, 617, 630, 646
    placers 530–1, 532–3
    provenance 356–7, 546
hematite 159, 367, 368, 555–7
hemipelagic deposits 482, 483, 670
herringbone cross-stratification 55–6
heterogeneity 217, 222
heterolithic beds 444

hierarchies, cyclic sedimentation 174–8
high-frequency stochastic cycles 181
high-stability minerals 357
hillslope factors, erosion 254–5
hindered settling **358–60**
histosols 773
Hjulström curve 345–6
Hjulström-type fan delta 273
Hotchkiss, William O. 627
HS *see* humic substances
human impact, sediment fluxes 255–6, 602
humic substances (HS) **361–2**, 426
humification, peat 514–15
huminite 427, 428
hummocky cross-stratification (HCS) 55, **362–4**, 686
hydrates, clathrates 137–8
hydration-dehydration, laterites 410–11
hydraulics, rivers 266, 571–2
hydrocarbons **364–6**
    see also kerogen
    diagenesis 426–8
    occurrences 365
    oil sands **499–502**
    origin 365–6
hydrodynamic model, fluid flow 475–7
hydrogen sulfide 304–5
hydrology
    evaporite basins 257–8
    glaciers 317–18
    paleofloods 291
hydrolysis, weathering 770
hydrothermal deposits
    dolomites 240
    iron formations 382
    talc 721
    travertine 753, 754
    zeolites 780
hydroxides and oxyhydroxides **366–8**
    formation 367–8
    properties 367
    transformation 368
hypersaline environments 239
hyposaline environments 238–9
hypsometric influences, tides 607
hysteresis loops 419, 420

ice sheets 603
    Milankovitch cycles 180
    soft-sediment deformation 58–9, 192
ice shelves 318, 322
ichnofacies 74–9, **698–700**
ichnology **71–83**, 649
Illing, L.V. 632
illite group clay minerals **369–71**, 448, 449
    crystal size and shape 370
    oceanic sediments 485–6
    origin and evolution 369
    structure 369–70
illitization 126, 369
imbrication 57, **371–4**
    classification 373
    downstream dip 372–4
    upstream dip 372
immature sediments, alluvial fans 6
imogolite **3–4**
impact sediments 263–4, **653–7**
    see also extraterrestrial materials
impact and shock metamorphism

    features **275–8**
    craters 263
    metamorphic features 275–7
    structures and products 275
impregnation 218, **374–5**, 561–3
inclusions
    amber 565
    cement 115
    detrital quartz 552
    fluid **297–300**
indicators
    see also way-up indicators
    climate change 272–3, 274
    paleoclimate 144
    paleocurrents 509–10
    provenance 546–8
individualization of evidence 302
induced and remanent magnetization 418–19
Indus submarine fan 696
industry, evaporites 257
infrared emission spectroscopy 50–1
injection bodies, dykes and sills **136–7**, 194, 296
inlets
    coastal facies 153, 155–6
    tidal **737–41**
insolation, orbitally forced 441–2
insoluble organic matter 400–3
interfacial forces, colloids 158–9
intergranular fluid 348, 350
intergranular porosity 266
internal structures (bedding) **54–9**
    see also crystal structure; fabrics; sedimentary structures; textures
    deformation-related 57–9
    large sand dunes 494–5
    microstructures 68, 516–19
    tranportation-related 54–7
International Association of Sedimentology 631, 636
intertidal flats 468
intra-arc basins 731
intrabasinal deposits, classification 127–32
intraclasts 128
intraformational mud clasts 57
intrinsic permeability 266
intrusions, soft-sediment **136–7**, 194, 296
inversion 460
    see also neomorphism
iridium 654
iron
    classification of ores 131–2
    hydroxides and oxyhydroxides 367–8
    occurrence 367
    oxidation states 159–60
    oxides 555–7
    sulphides 701
iron formations 367, **379–82**, 389
    banded and granular 367, 369, 379–81
    origins and implications 382
iron-rich chlorites 125
ironstones 379, **382–5**
    oolitic 380
    origins 383–4
iron–manganese nodules **376–9**, 487, 488
    growth and composition 376–7
    origin 378
islands, barrier islands **43–7**
isotopic methods **385–97**

basic equations 385–6
K–Ar methods 386
model ages and epsilon values 387
$S^{143}Nd/S^{144}Nd$ ratios 388
provenance indicators 547–8
radionuclides 389
Rb–Sr and Sm–Nd isochron methods 386–7
$S^{87}Sr/S^{86}Sr$ ratios 387–8
stable isotopes 389–94
tracers 385
U–Th–Pb 388–9

Japan, sedimentology **635–7**
journals 637

K-feldspars 278, 279–80
K–T boundary *see* Cretaceous-Tertiary boundary
kame terraces 324
kaolin group minerals **398–400**
   chemistry 399
   kaolinite 398–9, 448
   oceanic sediments 486
   origin and occurrence 399–400
   structure 398–9
kaolinite *see* kaolin group minerals
karst, ancient **11–15**, 109–10
kerogen 366, **400–3**, 426, 427
kinetic sieving 352
kinetics
   cation exchange reactions 108
   dolomitization 235
Krumbein, William Christian **644–5**, 685
Krynine, Paul Dimitri 133, **645–6**
Kuenen, Philip Henry 633, **646–7**
K–Ar dating 386
   illite 371
   iron–manganese nodules 376

laboratory analysis, forensic 302
lacustrine sedimentation **404–8**
   charcoal 121, 122
   classification 405–6
   deserts 209–10
   limnologic processes 404–5
   processes 405–7
   sediment record 408
   tufa model 754
   varves 764, 765
lagoonal deposits 18, 155–6
lags 608
lahars 348, 425
lakes 209–10, 404–8
   see also lacustrine sedimentation
laminae 53–5, 534–6
   contorted 549
   deformation 168
   mudstones 203–5
   planar 534–5
laminasets **59–61**
laminations 592–4
   bedding 53
   consolidation 295
   convolute 58, **168**, 192, 193
   cross 535
   deformation 192, 193
   mudrocks 454
   muds and sands 54–5

planar and parallel **534–7**
stromatolites 688–9
landforms
   see also surface forms
   bathymetric 324–5
   glacial deposition 321–4
   ice-marginal 321
   subglacial 324
landsystems approach 328
lateral accretion 57, 432–3
lateral facies, deltas 201
laterites 65, **408–11**
   continental drift influence 410
   nomenclature 409–10
latosols 409
layered structures *see* laminations; sheet structures (minerals); stratification
Leith, Charles K. 627–8
lenticular bedding *see* flaser bedding
lepispheres **411–12**
lessivage 770, 771
levees 692, 694, 696
liesegang banding 221, 223
lift, grains 610–11
light heavy minerals 532, 533
lime muds 127, 128
limestones
   see also carbonate sediments and rocks; chalk; dolomites
   ankerite 20
   cherts 662
   classification 268–9
   crinoidal 248–9
   dedolomitization 188–90
   Dunham classification 129–30
   Folk classification 130
   micrite 436
   oolites **502–6**
limnologic processes 404–5
linear dunes 243–4, 246, 710
lineations
   see also orientation
   current 513
   parting **512–14**, 535
   parting-step 512–13
liptinite 427, 428
liquefaction 57, **412–13**
   soft sediments 136, 191, 192
   structures 295, 296, 759
Liquid Limit, soils 23
Liquidity Index, soils 23
lithification
   beachrock 51–3
   sands and gravels 588–92
   stylolites 691
   Trypanites ichnofacies 700
lithofacies, glacial environments 327, 328
lithogenesis, J.Walther 653
lithomarges 409
littoral bars **40–3**
load
   see also bed load; suspended load
   fluxes 600–2
   rivers 574–5
   tides 607–9
load structures **413–14**
   load casts 58, 192, 193, 414
   pseudonodules 549, 550–1
   trigger mechanisms 413
   turbidites 759

wrinkle marks 414
lobes, depositional 694
loess 715
log-normal distribution 343
Long Range Desert Group 638
longshore sediment transport 27, 623–4
low angle lamination 54, 55
low energy environments 332, 496
lower shoreface facies 152
lowstands 445, 446, 496
luminescence, provenance indicator 547
lunar sediments **415–16**
lunettes 243–4
Lyell, Charles 628–30
Lyman, Benjamin S. 635
lyophilic colloids 158
lyophobic colloids 158
Lysocline 489

Maastrictian chalk, North Sea 120
mackinawite 701
macroborers 70–1
magadi cherts 664
magadiite **417–18**
magnesian calcite 99
magnesium carbonates *see* dolomites; dolomitization
magnesium-rich chlorites 125
magnetic properties **418–24**
   classes of magnetism 418
   induced and remanent 418–19
   natural magnetic minerals 419
   sediment magnetism 421–3
   susceptibility 422, 423
manganese nodules 487, 488
mantle, fluxes 598
marcasite 701
marine attrition **25–7**
marine environments
   deserts 211
   ichnofacies 76–9
marine fauna, toxicity 749–50
marine sediments 653
   carbonate cements 111–13
   diagenesis 92
   evaporites 258, 259
   glaucony 332
   humic substances 362
   $S^{143}Nd/S^{144}Nd$ ratios 388
   paleoclimate evidence 144
   phosphorites 520–3
   placers **532–4**
   porewaters 540–1
   red beds 556
   $S^{87}Sr/S^{86}Sr$ ratios 387–8
   sand belt 466
Mars 288, 291, 776
marshes, salt **586–8**
mass movement **424–5**
   see also flows (forms)
   causes 424–5
   earth flows **247–8**
   gravity-driven mass flows **347–53**
   rock avalanches **31–3**
   rock falls 33
   slides and slumps 191, 425, **666–8**
mass spectrometry 632
matrix
   breccias 677
   mélange 435

maturation of organic matter 402, **425–9**
  catagenesis 402, 427
  eogenesis 402, 426–7
  metagenesis 402, 427
  telogenesis 427
maturity, textural and compositional **429–30**
meandering channels **430–4**, 575, 577–8, 582
  flow 432
  formation and deposits 432–3
  planform and channel geometry 430–2
  river planform facies 433
mechanical compaction 163–5
mechanics
  avalanches and rock falls 32–3
  debris flows 186–7
  unsteady flows 351–2
Mediterranean Sea
  mineral magnetism 421–2
  sapropels 593
  sediment fluxes 602
  shoreface sands 496
meetings, sedimentology 637
megaturbidites 334
mélange (melange) **434–6**
meltwater, glacial 318, 323–4
mercury injection test 267, 517–18
mesogenetic cements 114–15, 116–17
Mesozoic bentonites 61
metagenesis 366, 402, 427
metals
  heavy 367, 541
  redox-sensitive 593–4
metamorphism
  cherts 633
  diagenesis relationship 217–18
  impact and shock features **275–8**, 654
  ophicalcites 506
metastability, diagenesis 216
meteoric carbonate cement 112
meteoric water 240–1, 391, 539–40
meteorites 597–8
  shock metamorphism 275–8
methane 305
methane hydrate 137–8
Mg/Ca ratio, seawater 596, 598
micas 449, 767
  see also phyllosilicates
micrites 128, 436
micritization **436–8**
microbial action
  diagenesis 217
  dolomitization 237–8
  kinetic isotope fractionation 394
  reef builders 558, 560
  sedimentary structures **439–40**
microbial deposits 688–90
  carbonates 1–3
  epibenthic bacteria 37
  mud-mounds 101
microborers 71, 437
microcrystalline calcite see micrites
microcrystals, sulphides 702
microfabrics
  carbonates **91–3**
  mudstones 205–6
  ooids 503–4
microscopy, carbonates 91–2

microspar, calcite 461
microstructures
  bioclasts 68
  sands and sandstones 516–19
microtektites 655–6
middle shoreface facies 152
migrated gases 305
Milankovitch cycles 145, 179–81, **441–3**
mimetic replacement, dolomites 232–4
mineralization, fossils 727
mineralogy
  ankerite 19
  bentonites 62
  carbonates **93–100**
  clay **139–42**
  dolomites 97, 98
  mudrocks 453
  ooids 503–4, 503–5, 505
  siliceous sediments 662
minerals
  see also clay minerals; feldspars; quartz
  authigenesis **27–9**
  carbonates **683–4**
  cavity precipitation 306, 307
  evaporites 257, 260–1
  fluid inclusions **297–300**
  fossil skeletons 726–7
  heavy minerals **356–8**, 617, 630, 646
  magnetic properties **418–24**
  marine placers 532, 533–4
  oceanic sediments 487
  precipitation 37–8
  provenance indicators 546–7
  shock metamorphism 276–7
  stability **429–30**
  sulfides **701–3**
  weathering properties 545–6
mining
  evaporites 257
  iron formations 379
  phosphorites 525
mires, peat 515
mixed siliciclastic and carbonate sedimentation **443–7**
mixed-layer clays 369–70, **447–50**
mixing models 238–9, 262, 444–5, **450–1**
mobile armor 21
models
  biogenic sedimentation 490–1
  computer simulation 182–3
  cyclic sedimentation 182–3
  diffuse double layer 108
  dolomitization 237–41
  facies **268–72**
  flumes 300
  grain entrainment 250
  mass transport rate 251
  meandering river floodplain 433
  mixed sedimentation 444–5
  mixing 444–5, **450–1** 238–9, 262
  numerical **475–80**
  packing flaws 518
  paleoclimate 144
  porosity 516–17
  sediment routing 617–18
  sediment transport 477–8, 615, 624–5
  submarine fans 694–8
  terrestrial 598
  tufa 754

turbidites 269
moganite 659
molasse 732
mollusks 68, 69
montmorillonite 675
Moon **415–16**, 776
moraines 321–4
motion
  equation 227
  initiation 249–50
  particles 226–7
  threshold 610–11
mottle horizons 409
mountain building, sedimentary basins 731–2
mountains, alluvial fans 5
mud cracks 204, 209, 210, **212–14**
mud flows 348
mud waves 706–7
mud-mounds 558, 560
  carbonates **100–2**
mud-silt 3
mudflats 735
  deserts 210–11
  shorefaces 153
mudrocks **451–9**, 590
  see also clays; shales; siltstones
  ankerite 20
  burial and compaction 456–7
  clasts 57
  clay and silt sources 452–3
  intraformational clasts 57
  oxygen levels 455–6, 457
  provenance 457–8, 546
  sedimentary structures 454–5
  stratigraphy and environment 457
  texture and nomenclature 452–3
  transport 453–4
muds 692
  anabranching rivers 10
  dewatered 698–9
  diapirs 194–5
  fluid 609
  laminated 536
  shrinkage cracks **212–14**
  tidal flat environments 467–8
  transport threshold 611
mudstones
  see also mudrocks
  compaction 164–5
  depositional fabric **203–6**
  microfabrics 205–6
multiphase flow 266–7, 518
Murray, John **647–8**

nacrite 399
nannobacteria, ooids 504
nannofossils, calcareous 119–21
natural environments, dolomitization 237–41
natural-levee sediments 286
Naumann, Edmund 635
Na–Ca, cation exchange 107
Na–K, cation exchange 107
S143sNd/S144sNd ratios, seawater and sediments 388
near infrared spectroscopy (NIR) 50
neomorphism 189, **460–1**
neptunian dykes and sills **136–7**

Neptunists 569
Nereites 76
neritic carbonate depositional
    environments **464–75**
  basin-facing slopes 472–4
  flat-topped platforms 464, 465
  peritidal environments 467–9
  ramps 464–5
  reefs 469–72
  sand shoal complexes 466–7
  sediment factory 465–6
  subtidal environments 466
neutron activation method 752
NIR *see* near infrared spectroscopy
nitrogen 305
Noah's flood 292
nodules
  see also geodes; pseudonodules
  anhydrite and gypsum 17–18, 585
  cherts 662–4
  iron-manganese **376–9**, 487, 488
  microbial carbonates 3
nomenclature
  amber 563–4
  bentonites 61–2
  laterites 409–10
  mudrocks 452, 453
non-calcareous rocks, bioerosion 70
non-cohesive sediments 86, 610–11
non-marine deposits
  berthierine 64–5
  carbonate cements 113–14
  evaporites 258–60
  ichnofacies 74–5
  red beds **555–7**
non-mimetic replacement, dolomites 232–4
nonparametric statistics 684
nonplanar dolomites, textures 231–2, 233
normal grading, episodic deposits 56
North Atlantic, mineral magnetism 422
North Sea, Maastrictian chalk 120
novaculite 661
numerical models of sediment transport and
    deposition **475–80**
  components 475–9
  equation set 479
  example 479–80

ocean circulation
  basin-basin fractionation 491
  upwelling systems **761–3**
Ocean Drilling Program, chalk 119
oceanic sediments **481–92**
  see also pelagic sediments
  biogenic sediments 482, 487–91
  chalk 119
  chemogenic sediments 482, 483, 487
  deep water deposition 271, 280
  distribution 483
  exploration 481–2
  sedimentation rates and thickness 483–4, 485
  terrigenous deposits 482, 484–7
  volcanogenic deposits 482, 483
oceanic waters, calcite compensation depth 88–9
oceanography, T.W.Vaughan 652
odinite 64, 65, 125

offshore bar theory 44
offshore sands **492–9**
  classification 493–6
  Pleistocene deposits 497–8
oil 364–6
  see also hydrocarbons
oil sands **499–502**
Old Red Sandstone facies 270
olistostrome 435
OMZ *see* oxygen minimum zone
oncoids 3, 502, 753
oncolites 129
ooids **502–5**
  carbonate 129
  environments 503
  ironstones 380, 381, 383
  mineralogy and microfabric 503–5
  origin 504
oolites **502–6**
oozes 482, 483
  calcareous 488–9
  siliceous 489–90, 660–1
opal 490, 659–60, 661–2
open coasts, clastic facies **149–57**
ophicalcites **506–8**
orange soil 415
organic compounds
  humic substances **361–2**
  hydrocarbons **364–6**
organic matter (OM)
  black shales 83–5
  colors 159–60
  decomposition 37–8
  kerogen **400–3**
  maturation **425–9**
  peat **514–16**
  sapropels 592–4
  slope deposition 672
  upwelling systems 761–2
organisms, erosion 255
organizations, sedimentology **637**
organogenic/microbial model, dolomitization 237–8
orientation
  see also dip direction; imbrication
  clasts 371–4
  currents 509–14
  fossils 724–5
  grains 265–6, 589, 616–17
  gutter casts 353–4
oscillations
  currents 362, 364, 624
  flows 55–6, 704, 707–8
  ripples 708
osmotic swelling, vermiculites 768
outwash fans and plains 323, 324
overbank deposits 286
overgrowths 216, Plate I
overturned cross-bedding 193
overwash, barrier islands 45–6
oxidation 159–60
  see also telogenesis
oxides, iron 366–8
oxisols 774
oxygen, mudrocks deposition 455–6
oxygen isotopes
  cherts 663–4
  feldspars 281
  ratio 442

  stable 389, 390–1
oxygen minimum zone (OMZ), upwelling 762
oxygenation, slopes 672
oxyhydroxides 48, **366–8**, 376–7

packages, shales 205
packing, loose 191
packing flaws 518
painted particles 752
paleoclimate
  indicators 144
  interpretation 144–5
  laterites 410
paleocurrent analysis **509–14**
  current indicators 509–10
  environment and paleoslope 511–12
  heavy mineral shadows 355–6
  parting lineations **512–14**
  statistical methods 510–11
paleoecology 652, 681–2
paleoenvironments
  see also environments
  lacustrine sedimentation 407–8
  trace fossils 79–82
  varves 764–5
paleoflood hydrology 291
paleogeographic reconstructions 356, 357
paleogeothermometer, illite 370–1
paleohydraulic calculations 288–9
paleokarst *see* ancient karst
paleomagnetism 421, 556–7
paleontology 650–1
paleoslope interpretations 511–12
paleosols **774–6**
Paleozoic, carbonate mud-mounds 101
palimpsest sediments **560–1**
pans, deserts 211
paragenesis, authigenic minerals 27
parallel lamination **534–7**
paramagnetism 418
parasequence, cyclic sedimentation 174, 175–6
particle settling *see* settling behaviour
particles
  see also boulders; grains; pebbles
  gravity flow 30–1
  orientation 265–6
  packing 265
particulate organic matter (POM) 488
parting, mudrocks 454
parting lineations 509, **512–14**
patterns, streams 570, 575–7
pavements, fluvial armor 21–2
PDFs *see* planar deformation features
peat **514–16**, 586–7
  characteristics 515–16
  coal balls **146–9**
  humification 514–15
pebbles
  see also gravel
  imbrication 373
  marine attrition 25–7
  sphericity 26–7
Peclet number, diffusion 226, 228
pedogenic carbonates 90
pedoliths 771
peels 374–5
  see also relief peels

PEG *see* polyethylene glycol
pelagic sediments   269, 482, 483, 730–1
   calcareous   **119–21**
   clays   641
   'rain'   482, 487, 488
   red clay   482, 483, 485–7
pellets, calcium carbonate   128
peloids, calcium carbonate   128
penecontemporaneous dolomites   237–8
periglacial environments   715
peritidal environments, carbonates   467–9
permeability   236, **266**, 516
petrographic image analysis (PIA)   517
petrography
   cathodoluminescence   104–5
   L.Cayeux   640
   definitions   627
   diagenesis   218
   lunar sediments   415
   mudstone fabrics   203
   provenance determination   544–6
petroleum   364–6
   oil sands   **499–502**
   reservoirs   257
petrology
   C.W. Correns   641
   definitions   626–7
   historical development   630
petrophysics   266–7
   acoustic velocities   311–12
   electrical properties   313
   sands and sandstones   **516–19**
Pettijohn, Francis J.   648
phacoids, mélange   435
phase separation, slurry   674
phosphates   131, 727
phosphogenesis   519–21
phosphorites   **519–26**
   classification   521–2, 524
   concentration   521
   environments   519–20
   'giant' deposits   522–3
   stratigraphy   523–4
phosphorus   519, 523
photography   340–1, 713
photomicrographs Plates I-II
phylloid algae   2
phyllosilicates   398
   see also clay minerals; sheet structures (minerals)
   chlorite   **123–7**
   micas   449, 767
   structure and classification   139–41
   talc   **721–3**
phylogenetic classification, trace fossils   72, 74
physical properties   265–8
physics of sediment transport (R.A.Bagnold)   249, 250, **527–9**
PIA *see* petrographic image analysis
pillar structures   193, **529–30**
pillows *see* ball-and-pillow structures
pipes, breccia   678
pisoids   503
pisolites   129
piston corers   169
placers
   fluvial   **530–2**
   heavy minerals   358
   marine   **532–4**

plains, deltas   200
planar deformation features (PDFs)   277
planar dolomites   231–2, 233
planar lamination   **534–7**
plane-parallel lamination   54
planform
   see also geometry
   meandering channels   578
plankton   487, 761
plants
   charcoal   **121–3**
   coal balls   146–8
   erosion   255
   peat   515
   resin   **563–5**
   subtidal environments   466
   tufas   753
Plastic Limit, soils   23
Plasticity Index, soils   23–4
plate tectonics
   development of concept   633
   diverging plates   729–30
   sands and gravels   592
   subduction zones   730–1
platforms, carbonates   239–40, 464, 465, 598
platinum group elements   654
playas   210–11, 584
plucking or quarrying (glacial)   318
podzolization   410, 770, 771
point bars   57, 432, 578
polyethylene glycol (PEG)   375
polymorphs
   see also neomorphism
   calcium carbonate   93–4
polytype, illite   370
POM *see* particulate organic matter
ponds and creeks, carbonates   468
porcelanites   661
pore fluids *see* porewaters
pore spaces   517
   diagenesis   214–15
   geometry   267
   pore throats   517
   porels   517
porewaters   310–11, 315, **537–42**
   composition   537–9
   debris flows   187–8
   marine sediments   391, 540–1
porosity   **266**
   compaction   161, 162, 165–6
   dolomitization   236
   models   516–17
   reduction   215
   stylolites   691
pot casts   353
potassium   313, 331
potassium ferricyanide   683
Precambrian
   cherts   664–5
   paleosols   775
precipitation of minerals
   cements   111
   concretions   220
   diagenesis   216
   geodes   306, 307
   lime muds   128
preservation
   coastal facies   153–7
   fluid inclusions   297
   glacigenic sediments   328

trace fossils   72
pressure
   acoustic effects   311
   authigenesis   28
   fluid inclusions   299
   nonequilibrium fluid   674
   porewater   187–8
   resistivity effects   313
   shock metamorphism   275–6
pressure solution   223, **542–3**, Plate I
pressure–temperature behaviour, fluid inclusions   298
primary lacustrine sediments   405
primary red beds   555–6
Principles of Geology (C.Lyell)   628–30
Principles of Stratigraphy (A.W.Grabau)   643–4
process facies models   269
prodelta   200
professionalization sedimentology   629, 631–3
proglacial tills   745–6
progradation
   barrier islands   45–6
   coastal facies   153–5, 157
   deltas   200–1
properties of sediments
   see also grain shapes; grain sizes; permeability; porosity
   carbonates   94
   colloids   **157–9**
   geomechanical   162–3
   geophysical   **308–14**
   iron oxides   367
   magnetic   **418–24**
   physical   265–8
   weathering of minerals and rocks   545–6
provenance   **544–9**
   chemical determination   546–7
   feldspars   279
   heavy minerals   356–7, 546
   isotope determination   547–8
   mineral and rock properties   545–6
   mudrocks   457–8, 546
   petrographic determination   544–6
pseudomorphs, calcite   99
pseudonodules   39, 58, **549–51**
   see also ball-and-pillow structures
   soft sediments   193
publications, sedimentology   **637–8**
pyrite   701–2
   marine mudrocks   727
   mudrocks   456
pyroclastic flows   348
pyroclastic rocks, classification   134

QFL determination, provenance   544
QFR determination, provenance   544
quantitative analyses   182–3, 569, 645, 684–5
quartz
   attrition   646
   authigenic   28
   burial-stage cements   116–17
   cathodoluminescence   103–4
   conversion from magadiite   417–18
   detrital   **552–4**
   durability   545–6
   geodes   306, 307

lepispheres **411–12**
  sands and gravels 589–92
  silcretes 659–60
  siliceous sediments 661–2
  surface textures 712
quasistatic flows 336

radio transmission, tracers 753
radioactive properties 313–14, 315
radioactive tracers 752
radiocarbon ages, resins 564
radiogenic isotopes *see* isotopic methods
radiography *see* X-ray radiography
radiolaria 67, 489
radiometric dating 385–97, 547
  see also isotopic methods
Raman spectra, bauxite 49, 50
ramps, carbonates 464–5
rapid earth flows 247
rapid-flow regime 336
rare earth elements, provenance indicators 546–7
rasping organisms, bioerosion 70–1
rate of...
  accumulation 514–15, 601, 603
  burial of fossils 723–4
  denudation 600, 601
  sediment transport 251, 614–17, 621–3
  sedimentation **600–5**
Rb–Sr isochron dating 386–7
reaction kinetics, cation exchange 108
recognition key, soils 773
recovery, oil sands 500–1
recrystallization **460–2**
  see also micritization
red beds 125–6, **555–7**
red clay, pelagic 483, 485–7
red coloration 159
redox conditions 159–60, 762
redox-sensitive trace metals 593–4
reduction features 222, 223
reefs 469–72, **557–60**
  see also stromatolites
  ancient reefs 471
  barriers reefs 465
  classification 557–8
  depositional facies 469–70
  distribution 559–60
  growth window 470
  microbial carbonates 2–3
  reef builders 558
  scleractinian corals 470–1
  stratigraphy 471–2
reflux model, dolomitization 239
regime, glaciers 317
regolith 771
regressive coastal deposition 196
regressive stratigraphy 47
relict sediments **560–1**
relief peels **561–3**
remobilization, erosion products 256
replacement
  mimetic and dolomitic dolomites 232–3
  minerals 215, 216, 721–2
repose, angle of **15–16**
reptation, eolian transport 250
repulsive energy, colloids 159
research
  barrier islands 44

caliche/calcrete 90–1
cathodoluminescence 105
deltas 197–8
diagenesis 218
resin **563–5**
resistance, flow **293–4**
resistivity
  chemical diffusion 225
  sediments 312–13
  temperature and pressure effects 313
retrogradation, barrier islands 45–6
reversing currents, lamination 536
reversing eolian dunes 243–4
reworked sediments 321, 406–7
Reynolds flux 228, 229
Reynolds numbers 337, 351, 358, 610–12
Reynolds stresses 228
rhodoliths 3
Rhone submarine fan 696
rhythmic sedimentation *see* cyclic sedimentation
rhythmites
  tidal **741–4**
  varves **764–6**
Richter, Emma 649
Richter, Rudolph **648–9**
ridges
  see also bars
  alluvial 34
  sand 493–6
  shoreface-attached 152
  subglacial 324
ridges-and-furrows 595
rift basins 258, 729–30
ripple-drift cross-lamination 55
ripples, ripple marks, ripple structures **565–8**, 627, 630
  see also dunes
  currents 509, 705–6
  definition 565
  formation 527
  oscillatory flows 708
  H.C.Sorby 650
  subaerial 709–10
risk-assessment, toxicity 750
rivers **568–83**
  see also alluvial deposits; channels; fluvial processes and deposits; streams
  anabranching **9–11**, 576, 578–9
  anastomosing 9–10, 575, 577, 579
  braided **85–7**, 323, 575, 577, 580, 582
  channel geomorphology 575–9
  deserts 208
  equilibrium and threshold theories 569, 572–3, 574
  erosion 569, 571–2, 578
  hydraulics and geometry 571–2, 573–4
  longitudinal profiles 570–1
  meandering **430–4**, 575, 577–8, 582
  river mouth processes 198–200
  sediment transport and deposition 571, 574–5, 576, 581, 601–2, 604
  straight 575, 577, 578
rock avalanches **31–3**, 348, 425
rock components, diagenesis 214–15
rock falls 33
rock slides 425
rock strength, chemical compaction 167
rocks
  see also carbonaceous sediments and rocks;

carbonate sediments and rocks
  chemical determination of provenance 546–7
  classification **127–36**
  weathering properties 545
rose diagrams, paleocurrent analysis 510, 511
rosettes, lepispheres 411–12
Rosin distribution 343
roundness 341, 429, 589
  pebbles 27
  quartz sand 553–4
Rouse equation 614–15
Rouse parameter 226
Russia, sedimentology 631

sabkhas 211, **584–6**
  dolomitization 239
  gypsum 18
SAGD *see* Steam Assisted Gravity Drainage
salina **584–6**
saline environments
  dolomitization 238–9
  minerals 210–11, 257
  salt wedges 607
  tufas 754
  zeolites 779
salinity, porewaters 538, 540–1
salinization 771
salt bodies 678
salt flats **584–6**
salt flocculation 284
salt marshes **586–8**
saltation 250–1, 574, 611–12, 710
sampling
  coring methods **168–70**
  diagenesis 218
  grain size 342–3
  grain surfaces 712
  paleocurrent directions 510
  X-ray radiography 777
sand bodies, deformations 58–9
sand crystals 222, 223
sand dunes *see* dunes
sand flats 735
sand injection structures 194
sand shoal complexes 466–7
sand volcanoes 57–8, 194
sands **588–92**
  avulsion deposits 35
  R.A.Bagnold 638
  classification 590
  compaction 164, 165–7
  composition 589–92
  deformation 192
  eolian dunes 243–4
  graded beds 646–7
  grain flow 335–6
  grain size and shape 588–9
  laminated 536
  lithification (beachrock) **51–3**
  marine attrition 25
  petrophysics **516–19**
  submarine fans 692–4
  tidal deltas 739–40
  tsunami deposits 755
sands, offshore **492–9**
  banks 493
  dunes 493, 494–5

818 SUBJECT INDEX

sands, offshore (*Continued*)
   ridges 493–6, 497
   sheets and patches 496
   shorefaces 496
sands, oil **499–502**
sandstones 588–90, Plate I
   see also arkoses; graywackes; turbidites
   acoustic velocities 309–10
   ankerite 19–20
   cathodoluminescence 104–5
   cements 115–16
   chemical compaction 165
   classification 132–3, 590, 646
   dish structures 230–1
   laminated (flagstones) 534–5
   petrophysics **516–19**
   reservoirs, feldspars 280–1
   wood charcoal 121, 122
saprolite 771
sapropels **592–4**
scablands debate 288
scanning electron microscope (SEM) 712–16
Science Council of Japan 636
scleractinian coral reefs 470–1
scour and scour marks **594–6**
   current analysis 509
   current crescents **513–14**
   turbidites 759
scree *see* talus cones
SCS *see* swaley cross-stratification
sea-floor ferromanganese nodules 376, 377–8
sea-level control, shelf facies 445, 446
seawater
   calcite compensation depth **88–9**
   chemistry 261–2, 540
   dolomitization model 239–40
   evaporation concentration 258, 259
   gypsum 16
   Mg/Ca ratio 596, 598
   $^{143}Nd/^{144}Nd$ ratios 388
   nepheloid layer **462–4**
   $^{87}Sr/^{86}Sr$ ratios 387–8
   solute changes **596–600**
secondary lacustrine sediments 405
secondary red beds 556
sediment deposition
   debris flows 187–8
   eolian dunes 245
   glacial 320–4
   meandering channels 575, 578
   river mouths (deltas) 198–200
   rivers 571, 574–5, 576, 581, 601–2, 604
   slopes 669–72
   submarine channels 693–4
   transport rate 616–17
   turbidity currents 757–9
sediment entrainment 250, 319–20, 571–2, 594–5, 610
sediment factory, carbonates 465–6
sediment fluxes **600–5**
   depositional environments 604
   influences 601–3
   rates of sedimentation 603–4
sediment transport
   see also currents; load; rivers
   bed stress 621–2
   braided channels 86–7
   cross-shore 623

   drag 609–11
   eolian **249–52**
   erosion 616–17
   experimental work 633
   floods 289
   R.L.Folk 642
   glaciers 319–20
   longshore 623–4
   models 615, 617–18, 624–5
   numerical models **475–80**
   oscillatory currents 624
   physics (R.A.Bagnold) **527–9**
   rates of transport 622–3
      bed load 613–14
      suspended load 614–15
   rivers 571–2, 574–5
   sand and gravel 610–11
   sorting 616–17
   tidal inlet sands 740
   tides **606–9**
   tracers **752–3**
   turbulent diffusion **226–30**
   unidirectional water flows **609–18**
   waves **619–26**
sediment yield 124–5, 143–4, 145, **254–6**
sedimentary basins
   see also basin facies
   mountain building 731–2
   strike-slip faults 732–3
   subduction zones 730–1
sedimentary dykes 59, **136–7**, 194
sedimentary geology **626–7**
sedimentary iron ores 131–2
sedimentary structures
   see also cross-stratification; laminations; scour and scour marks
   dunes 246
   early recognition 630
   microbially induced **439–40**
   mudrocks 454–5
   mudstones 203–5
   paleocurrent indicators 509–10
   tidal flats 735
   tidal rhythmites **741–4**
   turbidites 758–9
   way-up indicators **627–8**
sedimentation
   climatic control **142–6**
   definition 626
   nepheloid layer **462–4**
   rates of 483–4, 485, **600–5**
   slopes 669–72
   tectonic control **729–34**
sedimentologists 631, **638–53**, 685
sedimentology 626–7
   bedding definition 53
   history **628–37**
      first period (pre-1830) 628–9
      second period (1830–94) 629–30
      third period (1894–1931) 630–1
      fourth period (1931–1950) 631–2
      fifth period (1950–77) 632–3
   Japan **635–7**
   meetings **637**
   organizations **637**
   publications **637–8**
Sedimentology Society of Japan 636
seepage 295
seismic profiling, facies models 271
seismites 57

selective grain entrainment 346
selective sorption, vermiculites 768
SEM *see* scanning electron microscope
SEMP *see* Society of Economic Paleontologists and Mineralogists
Senckenberg Laboratory (R.Richter) **648**
septarian concretions 220, **657–9**
sequence stratigraphy 636
   cyclic sedimentation 176
   mudrocks 457
   phosphorites 523–4
serpentine group 64–6, 449
serration, stylolites 691
settling behavior
   curves 337–8
   hindered settling **358–60**
   scour lag 608
   velocities 284–5, 337, 358, 359
shadows, heavy minerals **355–6**
shales 132, 641
   acoustic properties 310
   depositional fabric 203–6
   diagenesis 126
   packages 205
   resistivity 313
shallow depths
   authigenic minerals 28
   marine evaporites 258
   marine sediments 653
shallow flow theory 347–8
shape analysis, grains 589, 630
shear strength, equation 191
shear stress 249–50, 477
shear surfaces 666–7
shearing resistance, soils 24
sheet flow, imbrication 373
sheet structures (minerals)
   chlorite 123
   clay minerals 139–41, 447–50
   illite 370
   vermiculite 766–7
shelf sediments, diamond placers 534
shells 723–5
   see also bioclasts
shelves *see* continental shelves; slope sediments
shock metamorphism 263, **275–7**, 654
shoreface deposits 496
shoreface facies 79–80, 150–3
shoreline deposits, deserts 210
shrinkage cracks 204, 209
Shrinkage Limit, soils 23
silcretes **659–60**
silica
   see also chalcedony; opal; quartz
   burial-stage cements 115
   clathrate structure 138
   geodes 221
   lepispheres **411–12**
   spongolites 682
silicates
   see also clay minerals; phyllosilicates
   iron 383
   magadiite **417–18**
siliceous sediments 489–90, **660–6**
siliciclastic sediments
   cements 115–17
   climate effects 143–4, 145
   geothermometry 393
   mixed **443–7**

sandstones 590, 592
slopes 669, 672
silicification 307, 662, 664
sills, clastic (neptunian) **136–7**
silt, classification 128
siltstones, fern charcoal 121, 122
simulation of sediment transport and deposition 182–3, **475–80**
sinuosity, channels 578
sixth-power law 630
size analysis, grains 588–9, 611, 624, 630, 651
skeletal fossils
  bioclasts 69
  disarticulation 723
  micritization 437
  minerals 726–7
  orientation 725
skeletal mud-mounds 101
Skolithos ichnofacies 72, 76, 78, 80, 152
slack-water deposits 291
slides 194, 248, 424, 425, **666–8**
slopes
  avalanches and rock falls **31–4**
  failure 424–5
  neritic carbonate slopes 473–4
  sediments 474, **668–74**, 717–18
slow earth flows 247–8
slump structures 58, 194, 425, **666–8**
slurry 530, **674–5**
smectite group 655, **675–7**, 719
  bentonite 61
  illitization 126
  mixed layer structure 448, 449
  pelagic deposits 486
smoothness 341
Sm–Nd isochron dating 386–7
snout, mass flows 349
soapstone 721
Society of Economic Paleontologists and Mineralogists (SEMP) 650–1
soft-sediment deformation 57–9, **190–3**, **193–5**
  see also dish structures
  ball-and-pillow structures **39–40**, 58
  folds 296
  intrusions 194, 296
  load structures 193, **413–14**
  pseudonodules 193, **549–51**
soils **770–6**
  Atterberg Limits 22–4
  avalanches 255
  classification 771–4
  deserts 207
  liquefaction **412–13**
  river valleys 571
  smectite 676
  weathering processes 770–1
solar radiation 143, 404
sole markings 353, 509–10
  see also gutter casts
solenoporaceae 3
solid solution behavior 96–7
solifluction 425
solubility, calcite analyses 96–7
solutes
  evaporative concentration 207
  porewaters 538, 539–40, 541
  seawater 596–600
solution breccias **677–8**

Sorby, Henry Clifton 542, 630, **649–50**
sorting
  fossils 725–6
  grains 616–17
  gravity-driven mass flows 352
source rocks, mudrocks 452–3
source-area factors, alluvial fans 6
spar bodies, stromatactis 687
spatial variations in sediments 444–5
specific gravities, heavy minerals 356
spectrometry 632
spectroscopy, bauxite 49–51
speleothems 13, **678–81**
sphericity 26–7, 341, 429, 689
spiculites **681–3**
spinning, grains 713–14
spit accretion theory, offshore bars 44
sponges 67, 68, 558
spongolites **681–3**
S87sSr/S86sSr ratios, seawater sediments 387–8
stability of composition **429–30**
stability fields, anhydrite and gypsum 16, 17
stable isotopes 389–94
  applications 385
  fractionation and geothermometry 392–4
  isotopic exchange 393
  provenance indicators 547
  sediment variations 391–2
  variations in water 390–1
stains, carbonate minerals **683–4**, Plates I–II
stalactites 679–80
stalagmites 679–80
star dunes 243–4
statistical analyses 645, **684–6**
  cyclic sedimentation 182–3
  paleocurrents 510–11
  rivers 569
steady uniform shear flow 349–51
Steam Assisted Gravity Drainage (SAGD) 500–1
steatite 722
Stokes' Law 339, 359, 406, 453–4
Stokes' settling equation 337
storage, erosion products 256
storm deposits 466, **686–7**
straight rivers 575, 577, 578
strain, compaction 162
strata *see* bedding
stratification
  see also bedding; lamination
  cross-stratification **170–3**, **362–4**, 633
  glacial sediments 320
  littoral bars 41–2
  phosphorites 522
  terminology 60
stratigraphic younging, ripples 567
stratigraphy
  ancient karst 12
  barrier islands 46–7
  coal balls 148
  cyclic sedments 174–5, 178–82
  definition 626
  A.W.Grabau 643
  iron formations 381–2
  Japan 636
  Milankovitch cycles 442–3
  reefs 471–2

salt marshes 587, 591–2
W.H.Twenhofel 650
streams
  see also rivers
  R.A.Bagnold 638
  deserts 208
  G.K.Gilbert 643
  gradients 570–1
  patterns 570, 575–7
  power 528
strength, loss of 191, 194
stresses
  bed stress 621–2
  compaction 161–2
  dispersive 638
  flow stress 346
  fluid stress 251
  Reynolds 228
  septarian cracks 658
  shear 249–50, 477
strike, deltaic sediments 201
strike-slip faults, sedimentary basins 732–3
stripping 595
stromatactis **687–8**
stromatolites 664, **688–90**
structure, crystalline 95–6, 124, 216
  see also sheet structures (minerals)
structure, internal
  see also fabrics; petrophysics of sands and sandstones; textures
  bedding **54–9**
  clay minerals 139–41, 370, 447–50
  humic substances 361–2
structures
  see also deformation structures; sedimentary structures
  biogenic **71–83**
  deformation **193–5**
  desiccation **212–14**
  diagenetic **219–25**
  dish **230–1**
stylolites 222–3, **690–2**
subaerial surface forms 709–11
subaqueous speleothems 680
subaqueous surface forms 704–9
subduction zones, sedimentary basins 730–1
subglacial sediments and processes 291, 319–20, 324, 745
submarine fans
  channel systems **692–8**
  facies models 271
  floods 289–90
submergence theory, offshore bars 44–5
subplanar surfaces, stylolites 690–1
subsidence 729
substrates
  bioerosion 70–1
  ichnofacies 76, **698–700**
subsurface features, ancient karst 12–13
subtidal environments 466, 468
successions, shoreface 154–6
sulfide minerals **701–3**, 750
sulfur, stable isotopes 389, 390–1
sulphates *see* anhydrite; gypsum
superimposed currents, tides 607
superimposed strata 60, 61
Superior-type iron formations 381, 382
supply limited transport, eolian 251–2
supraglacial tills 745
supratidal flats 468–9, 585

surface accumulation, calcrete/caliche 89–91
surface forms **703–12**
  see also bedforms
  ancient karst 12–13, 14
  subaerial 709–11
  subaqueous 704–9
surface shearing stress 249–50
surface textures **712–17**
susceptibility, erosion 254
suspended load 574
  concentrations 68
  sediment transport 612
  tidal currents 608–9
  transport model 477–8
  transport rate 614–15
  wave action 622
suspension 30
  see also autosuspension
  clay fraction 453–4
  eolian transport 250–1
  particle settling 358–60
swales 44, 363
swaley cross-stratification (SCS) 55, **363–4**
swash and swash marks **717–18**
swelling, smectite 676
syndepositional cements 111–13, 116
syneresis 658, **718–20**
synthesis, dolomites 97–8

table salt, evaporites 257
talc 449, **721–3**
talus cones 33, 579
taphofacies 727
taphonomy 71, **723–9**
tar sands 499–502
taxonomy, soils 772, 776
teaching 642, 646
techniques
  cathodoluminescence **102–6**
  cores and coring methods **168–70**
  detrital grain determination 548
  geochemistry 632
  historical introduction 629–33
  impregnation **374–5**
  isotopic methods **685–97**
  mass spectrometry 632
  microstructure determination 516–18
  relief peels **561–3**
  scanning electron microscope 712–16
  soils recognition key 773
  spectroscopy 49–51
  statistical analyses 182–3, 510–11, 569, 645, **684–6**
  tracers for sediment movement **752–3**
  X-ray radiography **777–8**
tectonic control of sedimentation 632, **729–34**
  see also plate tectonics
  alluvial deposits 581
  alluvial fans 6–7
  compressional mountain building 731–2
  cratonic basins 733
  cyclic sedimentation 182
  diverging plates 729–30
  dolomitization 241
  fan deltas 274
  marine phosphorites 520, 522–3
  sands and gravels 592

strike-slip faults 732–3
  subduction 730–1
tectonic mélanges 434
tectosilicates, zeolites **779–80**
tektites 654–6
telogenesis 115, 427
temperature
  acoustic effects 311
  authigenesis 28
  fluid inclusions 298–9
  resistivity effects 313
tempestites 686
temporal climate change 145
temporal variations in sediments 445
tephra deposits 716
  allophane and imogolite **3–4**
  classification 134
  zeolites 780
Teredolites ichnofacies 699–700
terminal velocity, settling 337
terminology
  bedsets 60, 61
  cyclic sedimentation 173–4
  dedolomitization 190
  A.W. Grabau 644
  stratification 60
  surface forms 703–4
terrain, ancient karst 14
terrestrial models, crustal evolution 598
terrestrial sediments 362, 482, 484–7
textural maturity **429**
textures 265–8
  see also fabrics; grain shapes; grain sizes
  beachrock 52
  bentonites 62
  dolomites **231–4**, 236–7
  glacigenic sediments 326
  limestone classification 130
  littoral bars 41
  mudrocks 452
  surface **712–17**
theories
  barrier island formation 44–5
  glacial 629–30
  Gouy–Chapman 108, 158
  historical development 628–30
  shallow flow 347–8
thermal transformations, bauxite 48–50
thermodynamics, dolomitization 235
thermohaline currents 671
thixotrophy 677
thorium, abundance 313
thresholds
  eolian transport 250
  erosion **345–7**, 608, 610–11
  fluvial erosion 573
  transport 610–11
thrust faults 668
tidal currents 55–6, 587
tidal flats 467–9, **734–7**
  biological aspects 735–6
  muddy carbonates 467–8
  sedimentology and facies 735
tidal inlets and deltas **737–41**
  coastal facies 153
  flow patterns 512
tidal sedimentation
  bundles 56
  rhythmites 269, **741–4**
  sand ridges 393–6

tides 734, **741–4**
  coastal facies 149
  cycles 742–3
  ranges 734–5
  sediment transport **606–9**
tills and tillites 325–6, **744–51**, 745
time-averaged oscillatory currents 624
tin, marine placers 533
tissue-residue levels, toxicity 750
Titan-yellow stain 684
tonsteins **61–3**
tool marks **747–8**
topography
  see also geometry; landforms; surface forms
  heat flow 315
  karst **11–15**
total overburden stress 161–2
tourmaline 357
toxicity **748–51**
trace elements 546–7
trace fossils **71–83**
  classification 72–4
  paleoenvironmental significance 79–82
tracers for sediment movement 385, **752–3**
traction load 56, 252, 574
tractive currents 54–5
transformation 188, 460–1
  see also neomorphism; recrystallization
transgressions, glaucony 332
transgressive coastal deposition 196
transgressive erosion 155
transgressive stratigraphy 47
transport
  see also sediment transport
  charcoal 121
  eolian **249–52**
  erosion products 254–5
  floods 289
  flumes 300–1
  fossils 724–5
  glaciers 319–20
  grain thresholds **345–7**, 611
  mudrocks 453–4
travertines 754
tropical coral reefs 559
tropical month 742
trough cross-stratification 171
Trypanites ichnofacies 699–700
tsunami deposits **755–7**
tufas **753–4**
tuffs 62–3
turbidites **757–60**
  see also graded bedding
  Bouma sequences 757, 758
  convolute lamination 192
  depositional features 759
  high density deposits 758–9
  model 269
  tool marks **747–8**
turbidity, nepheloid layer 463
turbidity currents 348, 757–8
  autosuspension 30–1
  experiments 633
  graded bedding 334
  P.H.Kuenen 646
  slope sediments 670
turbidity maximum 607, 609
turbulence, grains 612–13
turbulent diffusion **226–30**

Brownian motion 226–7
   random fluid motions 228–9
Twenhofel, William H. **650–1**
two-phase flow 266–7, 518, 615

Udden, Johan August **651–2**
Udden-Wentworth grain-size scale 339, 340, 651
ultisols 774
unconsolidated sediments 310
unidirectional water flow **609–18**, 704–7
   internal structures 54–5
   planar lamination 536
   ripples 567
Uniformitarianism 288, 569, 628–9
   see also actualism
unroofing, erosion 591–2
unsorted sediments (glacial) 325
unsteady flows 351–2
uplift 729
uplift-stage cements 115
upper shoreface facies 151–2
upward-fining sequence 432
upwelling systems **761–3**
uranium, abundance 313
U–Th–Pb dating 388–9

vadose zone
   sabkha 584–5
   speleothems 679–80
varves **764–6**
vaterite, geochemistry 94
Vaughan, Thomas Wayland **652**
vegetation
   coal ball evidence 148
   deserts 207
   eolian transport 251
   erosion 255
velocities
   channel flow 572
   eolian transport 250
   erosion threshold 345
   grain entrainment 250
   settling 284–5, 337, 358, 359
   sixth-power law 630
   waves 309–12
   wind 244–5
verdine **332–3**

vermiculite 449, **766–9**
vertical facies, deltas 200–1
vertical pillar structure 529–30
vertisols 774
viscosity, soil 412
viscous liquid 412
vitrinite 427, 428
volcanic materials 61–3, 598
volcanic rocks, charcoal 122
volcaniclastic sediments
   see also tephra deposits
   chlorites 126
   classification 134
   oceanic sediments 482, 483
volcanoes, sand 57–8, 194
vugs *see* geodes

wackes 590
Walther, Johannes **653**
Walther's Law of Facies 270
wandering rivers 576
wash load 612, 615
water
   see also seawater
   groundwater 537
   mass exchange (lakes) 404–5
   meteoric 240–1, 391, 539–40
   porewaters **537–42**
   stable isotopes variations 390–1
   unidirectional flows **609–18**, 704–7
water transport 528
   see also rivers; sediment transport; wave transport
water-escape channels 529–30
waterlain tills 746
waterlogging, peat 515
wave ripples 55, 565, 566
wave transport **619–26**
   R.A.Bagnold 527
   modes 621–2
   rates of transport 621–3
   swash and backwash 717–18
   water waves 619–20
   wave propagation 620–1
wave-dominated inlets 740
wave-formed bars **40–3**
waves
   coastal facies 149

   lakes 405
   mud 706–7
   velocities 309–12
wavy bedding *see* flaser bedding
way-up indicators **627–8**
   cross-stratification 627–8
   fining-upwards sequences 57, 755
   first use 631
   graded bedding 334, 627
weathering 600, 713–15, **770–1**
   bauxite 48
   chemical 429
   clay minerals 142
   detrital chlorite 124–5
   diagenesis relationship 217–18
   mineral and rock properties 545–6
   mudrocks 452–3
   provenance 544–5
   sediment yield 254
weight loss, attrition 24, 25
well cuttings 651
Wentworth, Chester 651
wet fans 580
wind transport **249–52**
   R.A. Bagnold 527–9
   dunes 244–5, 710–11
   fluxes 603
   pelagic clay 486
   ripples 565, 709–10
   surface forms 709–11
   wave structures 55
wood
   bioerosion 70
   charcoal 121, 122, 123
wrinkle marks 414

X-ray radiography 631–2, 641, **777–8**

yield, sediment **254–6**
   see also erosion and sediment yield

zeolites 487, **779–80**
zircon 357
Zoophycos, ichnofacies 76, 80
ZTR index 429

Kluwer Academic Encyclopedia of Earth Sciences Series

Previous Volumes in the Series (currently in print)

C.W. Finkl: *Encyclopedia of Applied Geology*   ISBN 0-442-22537-7

J.E. Oliver & R.W. Fairbridge: *Encyclopedia of Climatology*   ISBN 0-87933-009-0

C.W. Finkl: *Encyclopedia of Field and General Geology*   ISBN 0-442-22499-0

D.R. Bowes: *Encyclopedia of Igneous and Metamorphic Petrology*   ISBN 0-442-20623-2

D.E. James: *Encyclopedia of Solid Earth Geophysics*   ISBN 0-442-24366-9

J.H. Shirley & R.W. Fairbridge: *Encyclopedia of Planetary Sciences*   ISBN 0-412-06951-2

E.M. Moores & R.W. Fairbridge: *Encyclopedia of European & Asian Regional Geology*   ISBN 0-412-74040-0

R.W. Herschy & R.W. Fairbridge: *Encyclopedia of Hydrology and Water Resources*   ISBN 0-412-74060-5

D.E. Alexander & R.W. Fairbridge: *Encyclopedia of Environmental Science*   ISBN 0-412-74050-8

C.P. Marshall & R.W. Fairbridge: *Encyclopedia of Geochemistry*   ISBN 0-412-75500-9

New and Forthcoming Volumes

M. Schwartz: *Encyclopedia of Coastal Science*

Chesworth: *Encyclopedia of Soil Science*

Online Edition

The online version of this Encyclopedia book series is the *Earth Sciences Encyclopedia Online* and can be accessed from <www.eseo.com>.